D0937236

OCEANOGRAPHIC INDEX

Subject Cumulation 1946-1971

Compiled by

Dr. Mary Sears

of the

Woods Hole Oceanographic Institution

Volume 1

A – Ch

G. K. HALL & CO., 70 LINCOLN STREET, BOSTON, MASSACHUSETTS

1972

This publication is printed on permanent/durable acid-free paper.

ISBN 0-8161-0932-X

PREFACE

The convenience of a cumulative oceanographic card index was demonstrated to the author during World War II. Consequently, on returning to the Woods Hole Oceanographic Institution in 1946, the present index was initiated to meet personal needs. Its arrangement has had to be extremely informal as it had to be put together and maintained as time allowed. For expediency, although it attempts to cover "all" oceanography, many pertinent related topics have been for the most part excluded: limnology, terrestrial geology, basic chemistry, fisheries biology, malacology, algology, etc. There are, however, good indices available for these general areas. Indeed, the limitations are perhaps more of an asset than a handicap.

In twenty-five years, many people at odd moments have spent time on the index, the chief of whom are Miss Constance W. Chadwick, Mrs. Joan B. Hulburt, Mrs. Esther N. Wilson, Mrs. Juanita A. Mogardo and Mrs. Marcia H. Schweidenback and other individuals too numerous to list from the Job Shop in Woods Hole. Handwritings have changed with time too. There are many opportunities for errors to have crept in as the index has been put together intermittently without checking. Despite these inaccuracies which persist now that it is published, it has proven invaluable in all sorts of ways as a source of "information retrieval" when requests are received from serious scientists, from school children or as an editorial tool in connection with the preparation of the bibliography-abstract section of *Deep-Sea Research.* Indeed, the author, geographical and subject indexes of my personal card catalogue also serves as a cumulative index to that section of *Deep-Sea Research* through Vol. 17, No. 6 and to the earlier mimeographed bibliography distributed by Mr. A. R. Miller of the Woods Hole Oceanographic Institution. The subject cumulation contains those items in the *Deep-Sea Research* abstract-bibliography section through Vol. 19, No. 4, except for titles dated 1972. (These will appear in supplemental volumes.) All are, however, much more inclusive and contain references extending back to the turn of the century and even earlier, some of them almost classics of their kind. It should be understood, on the other hand, that the cumulations do not pretend to be complete, but rather that they have assisted me in recalling earlier work without being forced to a time-consuming library search. As a biologist, I early learned how much time it takes to go back year by year, volume by volume through such compendia as the *Zoological Record, Biological Abstracts,* and the *Bibliographia Oceanographia.* Incomplete though it is one can go back over the years almost instantaneously.

I have drawn on the nearly 4000 journals (approximately 2200 being received currently) held in the Library of the Marine Biological Laboratory, which is also the repository of the library of the Woods Hole Oceanographic Institution. My index is, however, quite different from theirs (also published by the G. K. Hall & Co.) in that my titles have been selected from the periodicals week by week as they arrived on the shelves. In recent years, since the discontinuance of their reprint collection, my index includes a few titles which will not be found in that library for they do not have the journals and they have not retained the separates. The M.B.L. catalogue, on the contrary, indexes their vast reprint collection, collected since the inception of that library, which places more emphasis on all phases of biology and on separates of those who have worked at the laboratory from its earliest days and before that at the former U. S. Fish Commission laboratory. Hence, the two series more or less supplement each other.

DIRECTIONS FOR USE OF THE SUBJECT CUMULATION

The subject cumulation is arranged alphabetically by author under each subject heading. In transliteration from the Russian, Japanese, etc. different systems have been used over the years, sometimes with variations in the same journal. Or again an individual may sometimes use his given name without an initial, but in his next paper he may use one or more initials (often up to four). It may then be difficult to identify several reports as belonging to one individual. When it is possible, these then are placed in chronological sequence. In the few short weeks preparatory to reproduction, some attempt was made to put all entries by one author in one place no matter how it was spelled and some guides as to variant spellings were inserted. Yet, the changes made were not always logical and may not be correct. Thus, many Indians have the last name Rao, but at times the name immediately preceding it, as Subba Rao, was treated as if the two together were the actual last name. Then some names such as Krishnamurty may have been written as one word or two, etc. In gathering all variants of a Russian or Japanese name in one place, it may on first glance, appear as if they were out of alphabetical order! Some Spanish names were listed under what appeared to be the last name which is in reality the mother's original surname, not the actual surname of the individual, which immediately precedes it. In short, if you do not find the author you seek, look for a variant spelling.

As filing under subject headings has been done sporadically, alternative words for the same subject have frequently been used. Some effort was made to rectify this and cards have been inserted where appropriate to call attention to other entries.

Attention is called to the fact that in places, there are subindices within the index: *bottom sediments, chemistry, instrumentation* and *methods.* Certain subjects during earlier years were filed without subheadings, but when say 500-1000 references were accumulated, it became necessary to subdivide them: circulations, currents, salinity, temperature, waves, etc., etc. The subject index is not entirely devoid of biological subjects. There may be biological terms having to do with anatomy, physiology, etc. and for groups of organisms which have received little emphasis and which are not for the most part included in the organismal cumulation to appear in 1974 or 1975. (Topics on phytoplankton, zooplankton, etc. are in this latter index.) "Data only" was a subject heading in the early days, but was discontinued after a number of years and such citations were entered as a subheading under the subject for which the data were available. Because of the various inconsistencies, most (but not all) subject headings are listed beginning on page v.

MARY SEARS

Woods Hole
3 August 1972

LIST OF SUBJECT HEADINGS

abnormalities
abrasion
absorption, coefficient
absorption, light
absorption, sound
absorption, solar radiation
abyssal benthos, see benthos
abyssal fan
abyssal fauna
abyssal floor
abyssal gaps
abyssal hills
abyssal plain
abyssal water
abyssal zonation
acceleration potential
acclimatization
accretion
achondrites
acoustic bottom loss
acoustic energy
acoustic field
acoustic properties
acoustic signal processing
acritarchs
adaptations
adiabatic cooling
adsorption
adsorption, sound
advection
advection, heat
aerial photography, see: photography
aerial scouting
aeromagnetic survey
aerosols
aftershock activity
age determination (biological)
age determination (geological)
age of water masses
age (antiquity) of the oceans
aggregates, organic (aggregations)
aids to navigation
aircraft performance
airplanes
air-sea boundary (interface, interaction)
albedo
aleurite
algae, see organismal file for taxonomy, physiology and ecology
algae, boring
algae as fertilizer
algae, as food
algae, chemistry of
algae, drifting of
algae, effect of
algae, harvesting of
algae, radioactivity of
algae, resources
algae, uses of
allergies, marine
allovs
amber
ambergris
ambiguity function analysis
amphibious warfare
<u>Amphioxus</u>
anaerobic basins
anaerobiosis
analysis harmonic
analysis, power spectrum
analysis, statistical
anaphylaxis
anchor drift
anchor ice
anchoring
anchor stations
andesite
andesite line
anelasticity
animal attacks
animal distribution
animal diversity
anisotrophy
anorthosite
anoxia
anoxic water
antibiotics
antibiotics, effect of
anticlines
antiquity of the oceans (see also: age of the oceans)
antiyeast activity
anticorrosives
antifouling compounds
antifouling protection
antimetabolites
antisubmarine warfare
apatite
APEX
aphotic zone
aquaculture
aqualung and use
aquanauts
aquaria
aquifer
archaeology
archeopyle
arcs (island), see: island arcs
archiannelids
artificial media
Asdic
assimilation
Atlantis (lost)
atlases
atmospheric carbon dioxide
atmospheric dust
atmospheric heating
atmospheric pressure, effect of
atolls
atomic energy
atomic wastes
attapulgite
attenuation, light
attenuation, seismic
attenuation, sound
attenuation, waves
augite
austausch
AUTEC
autocorrelation
autoinhibitors
autolysis
automatic stations (see also: buoys)
auxospores
aviation
avoidance
azoic zone
back-passing
back radiation
back scattering
bacteria
bacteria, antiyeast
bacteria, bioassay
bacteria, biomass
bacteria, chemistry of
bacteria, chemosynthesis by
bacteria, chitinoclastic
bacteria, chitinolytic
bacteria, chitin decomposing
bacteria, coliform
bacteria, data only
bacterial decomposition, effect of
bacteria, deep
bacteria, distribution of
bacteria, effect of
bacteria, habitats of
bacteria, haemolytic
bacteria, heterstrophic
bacteria, lists of species
bacteria, littoral
bacteria, nitrifying
bacteria, nitrogen-fixing
bacteria, pathogenic
bacteria, psychrophilic
bacteria, salt requirement
bacteria, sterility
bacteria, structure of
bacteria, sulphur
bacteria, species of
bacterial viability

bactericidal agent
bactericidal effects
bacterio-neuston
bacteriophage
balance (equations of)
baleen plates
banks
banks, effect of
baroclinic flow
baroclinic instability
baroclinic layer
baroclinic ocean
baroclinic sea
baroclinic waves
barometric pressure
barotropic fluid
barotropic models
barotropic ocean
barrier beaches
barriers, effect of
barrier islands
bars
baryte
basalt
basaltic magma
basaltic rock
basement
basins
basins, anoxic
batholith
bathymetric chart
bathymetric distribution
bathymetry
bathymetry (data only)
bathymetry, data processing
bathymetry, effect of
bathymetry, seamounts
bathynauts
bathyphotometer
bathyscaphe
bathysphere
bathysphere, observations from
bathythermograph
bathythermograph (data only)
bauxite
bays
beaches
beach cusps
beach erosion
beach gradients
beach penetration
beach sand
beach surf zone
beach trafficability
bearing, ships
Beaufort scale
bedding
bed rock
behavior
benches
"bends"
benthic parameters
benthos
benthos, abyssal
benthos, biomass
benthos, colonization
benthos, deep
benthos, dispersal of
benthos, effect of
benthos, infauna
benthos, intertidal
benthos, lists of spp.
benthos, mass occurrence
benthos, productivity of
benthos, zonation of
Bernard cells
bibliographies, acoustics
bibliographies, air-sea interface
bibliographies, areal
bibliographies, Africa
bibliographies, Antarctic
bibliographies, Arctic
bibliographies, Atlantic
bibliographies, Australia
bibliographies, Australasia
bibliographies, Baltic
bibliographies, Black Sea
bibliographies, Bay of Biscay
bibliographies, Canada
bibliographies, Caribbean
bibliographies, Gulf of Mexico
bibliographies, India
bibliographies, Indian Ocean
bibliographies, Japan
bibliographies, Jugoslavia
bibliographies, Mediterranean
bibliographies, New Zealand
bibliographies, Norway
bibliographies, Polynesia
bibliographies, South America
bibliographies, USSR
bibliographies, biology, algae
bibliographies, biology, annelids
bibliographies, biology, antibiotics
bibliographies, biology, bacteria
bibliographies, biology, benthos
bibliographies, biology, bioluminescence
bibliographies, biology, birds
bibliographies, biology, cephalopods
bibliographies, biology, chaetognaths
bibliographies, biology, cirripeds
bibliographies, biology, coelenterates
bibliographies, biology, crustaceans
bibliographies, biology, crustaceans (amphipods)
bibliographies, biology, crustaceans (cladocerans)
bibliographies, biology, crustaceans (copepods)
bibliographies, biology, crustaceans (cumaceans)
bibliographies, biology, crustaceans (decapods)
bibliographies, biology, crustaceans (isopods)
bibliographies, biology, crustaceans (ostracods)
bibliographies, biology, crustaceans (pycnogonids)
bibliographies, biology, crustaceans (schizopods) (euphausiids, mysids)
bibliographies, biology, crustaceans (stomatopods)
bibliographies, biology, crustaceans (tanaids)
bibliographies, biology, diatoms
bibliographies, biology, dinoflagellates
bibliographies, biology, echinoderms
bibliographies, biology, ecology
bibliographies, biology, faecal pellets
bibliographies, biology, fishes
bibliographies, biology, fisheries
bibliographies, biology, fish anatomy
bibliographies, biology, fish behavior
bibliographies, biology, fish culture
bibliographies, biology, fish detection
bibliographies, biology, fish diseases
bibliographies, biology, fish flour
bibliographies, biology, fish parasites
bibliographies, biology, fish tagging
bibliographies, biology, foraminifera
bibliographies, biology, fouling
bibliographies, biology, fungi
bibliographies, biology, insects (marine)
bibliographies, biology, larvae
bibliographies, biology, mammals
bibliographies, biology, marine borers
bibliographies, biology, micro-organisms
bibliographies, biology, molluscs (includes pteropods)
bibliographies, biology, nannoplankton
bibliographies, biology, neuston
bibliographies, biology, parasites
bibliographies, biology, peridinians
bibliographies, biology, plankton
bibliographies, biology, Pogonophora
bibliographies, biology, poisonous animals
bibliographies, biology, priapulids
bibliographies, biology, primary productivity
bibliographies, biology, protozoa
bibliographies, biology, red water, red tides
bibliographies, biology, rotifers
bibliographies, biology, sea farming
bibliographies, biology, self-purification
bibliographies, biology, silicoflagellates

charting
checklists
chemistry
chemistry, acid-iron waste
chemistry, adenine
chemistry, adenosine
chemistry, activity coefficients
chemistry, aequorin
chemistry, aggregates
chemistry, albumen
chemistry, algin
chemistry, alginates
chemistry, alkali
chemistry, alkaline earths
chemistry, alkalinity
chemistry, alkalinity (data only)
chemistry, alkalinity reserve
chemistry, alkanes
chemistry, alloys
chemistry, aluminosilicates
chemistry, aluminum
chemistry, aluminum 26
chemistry, amino acids
chemistry, ammonia
chemistry, ammonia (data only)
chemistry, anions
chemistry, anoxic waters
chemistry, antimony
chemistry, apparent oxygen utilization
chemistry, arabinose
chemistry, argon
chemistry, arsenates
chemistry, arsenic
chemistry, ascorbic acid
chemistry, astacin (includes non-astacin)
chemistry, auxins
chemistry, barium
chemistry, barium sulphate
chemistry, base
chemistry, benzo 3.4 pyrene
chemistry, bicarbonate
chemistry, biotin
chemistry, bitumen
chemistry, blood pigments
chemistry, blood proteins
chemistry, borate
chemistry, boric acid
chemistry, boron
chemistry, boron/salinity
chemistry, bromide
chemistry, bromide/chlorinity
chemistry, bromine
chemistry, buffers
chemistry, cadmium
chemistry, caesium
chemistry, caesium 134
chemistry, caesium 137
chemistry, calcareous deposits (see also under bottom sediments)
chemistry, calcification
chemistry, calcium
chemistry, calcium anomalies
chemistry, calcium - 45
chemistry, calcium carbonate
chemistry, calcium, data only
chemistry, calcium/chlorinity
chemistry, calcium uptake
chemistry, carbohydrates
chemistry, carbon
chemistry, carbon assimilation
chemistry, carbonate, etc.
chemistry, carbon/chlorophyll
chemistry, carbon cycle
chemistry, carbon dioxide
chemistry, carbon fixation
chemistry, carbon, total
chemistry, carbon inorganic
chemistry, carbon particulate
chemistry, carbon reservoir
chemistry, carbon, surface and bottom
chemistry, carbonic acid
chemistry, carcinogens
chemistry, carotene
chemistry, carotenoids
chemistry, carotenoproteins
chemistry, carotin
chemistry, catalytic activity
chemistry, cations
chemistry, cellulose
chemistry, cerium
chemistry, cerium isotope
chemistry, chelators
chemistry, constituents
chemistry, concentration (accumulation) within organisms
chemistry, chitin
chemistry, chloride
chemistry, chlorine
chemistry, chlorinity
chemistry, chlorellin
chemistry, chlorophyll
chemistry, chlorophyll a
chemistry, chlorophyll b
chemistry, chlorophyll c
chemistry, chlorophyll budget
chemistry, chlorophyll data only
chemistry, cholesterol
chemistry, chromatophorotropins
chemistry, chromium
chemistry, chromium-50
chemistry, chromolipoids
chemistry, cobalamin
chemistry, cobalt
chemistry, cobalt-60
chemistry, co-enzymes
chemistry, conchidin
chemistry, copper
chemistry, copper, effect of
chemistry, cytosine toxicity
chemistry, DDT
chemistry, decalcification
chemistry, denitrification
chemistry, deuterium
chemistry, deuterium - hydrogen ratio
chemistry, deuterium oxide
chemistry, diadinoxanthin
chemistry, Eh
chemistry, electrolytic conductance
chemistry, electrons, hydrated
chemistry, elements
chemistry, enzymes
chemistry, equivalent volumes
chemistry, esters
chemistry, EDTA
chemistry, europium
chemistry, fats
chemistry, fatty acids
chemistry, ferric hydroxide
chemistry, ferritin
chemistry, ferro-manganese nodules
chemistry, ferrous sulphate
chemistry, ferrous sulfide nodules
chemistry, fluoride
chemistry, fluoride/chlorinity
chemistry, fluorine
chemistry, fucoxanthin
chemistry, galacturonic acid
chemistry, gallium
chemistry, gamma ray emitters
chemistry, gases
chemistry, gases atmospheric
chemistry, gases dissolved
chemistry, gas, effect of
chemistry, gas exchange, ocean-atmosphere
chemistry, germanium
chemistry, giberellic acid
chemistry, giberellins
chemistry, glucose
chemistry, glucuronic acid
chemistry, glycollic acid
chemistry, gold
chemistry, guanine
chemistry, haemocyanin
chemistry, haemoglobin
chemistry, halides
chemistry, halogens
chemistry, heavy metals
chemistry, helium
chemistry, hexachlorocycohexane
chemistry, humic acid
chemistry, hydrocarbons
chemistry, hydrogen isotopes
chemistry, hydrogen peroxide
chemistry, hydrogen sulphate
chemistry, hydrogen sulphide
chemistry, hydrogen sulphide, effect of
chemistry, hydroxylamine
chemistry, hydroxypatite
chemistry, hydroxyl ions
chemistry, indium
chemistry, iodate
chemistry, iodide
chemistry, iodine

earths crust
earth currents
earths mantle
earthquakes
earths rotation
earths sphericity
ecdysis
echinoderms
echiuroids
echoes
echograms
echolocation
echolocation, ranging
echolocation, sounding
eclipse, effect of
ecology
economics
ecosystem
ectocrines
eddies
eddy conductivity
eddy correlation
eddy diffusion
eddy pressure
eddy viscosity
edge waves
education
eel grass
effectors
effluents
eggwhite
eighteen degree water
Ekman layer
Ekman spiral
Ekman transport
electric charge
electric currents
electric field
electric organs
electric potentials
electric conductivity
electric power
electricity
electrocardiogram
electrochemistry
electrolytic conductance
electromagnetic field
electromagnetic induction
electronics
electron microscope
electro-perception
embryology
encapsulants
encyclopaedias
endemism
endocrines
enemies
energy
energy budget
energy considerations
energy conversion
energy dissipation
energy exchange
energy flow
energy, kinetic
energy loss
energy transfer
energy transport
engineers
engineering oceanography
enrichment
enteropneust
environments
epibiota
epineuston
equations
equator
equatorial flow
equivalent conductance
errosion
errors
eruptions
escarpments
estuaries
euphotic layer
euryhalinity
eutrophication
evaporation
evaporites
evolution
excavation, nuclear
exchange, chemical
exchange, data
exchange rate
exchange, water
excretion
excretion, nutrients
excretion, amino acids
excretion, organic matter
excretion, salt
exhalations, hydrothermal
expanding earth
expatriation
exploration
exploitation
explosions
explosives, effect of
exposure
extension of range
extinction coefficient
extracellular products
extraterrestrial material
extracts, effect of
eyes
faecal pellets
false echoes
fans
fan valleys
farming the sea
farming, crustaceans
farming, fish
farming, pearls
farming, shellfish
fat
fat content
fat formation
faults
faults, block
faults, transcurrent
faults, transform
faulting
faunas
faunal diversity
faunal extinction
faunal provinces
feeding
feldspar
fertility
fertilization
fetch
films
filter feeders
filtrates, effect of
fish, anatomy - physiology
fish, behavior
fish, "fishbites"
fish, blood
fish, chemistry
fish, culture
fish, ecology
fish, effect of
fish, feeding mechanisms
fish, flour
fish, growth
fish, hearing in
fish, lateral lines
fish, migration
fish, mortality
fish, "nursery"
fish, oil
fish, otoliths
fish, physiology
fish, propulsion
fish, rare
fish, scales
fish, schooling
fish, sense of smell
fish, stranding
fish, swimming
fish, tagging
fish, taste bud
fish, vertebrae
fish, warm-blooded
fish, by geographical areas
fish, by habitat
fish, taxonomic
fish, taxonomic, by class or order
fish, taxonomic, by family
fish, taxonomic, by genera
fish and/or fisheries by common names
"fisheries" of invertebrates
"fisheries" of molluses
"fisheries" of crustaceans
Fourier analysis
fracture pattern
fracture zone
frazil ice
freezing

methods, chemistry, carbon dioxide
methods, chemistry, carbon and its compounds
methods, chemistry, carotenoids
methods, chemistry, cesium
methods, chemistry, chloride
methods, chemistry, chlorinity
methods, chemistry, chromatography
methods, chemistry, chlorophyll
methods, chemistry, chromium
methods, chemistry, cobalt
methods, chemistry, copper
methods, chemistry, deoxyribonucleic acid
methods, chemistry, enzymes
methods, chemistry, fluorescence
methods, chemistry, fluorides
methods, chemistry, gases
methods, chemistry, geochemistry
methods, chemistry, glucose
methods, chemistry, halogens
methods, chemistry, hydrocarbons
methods, chemistry, hydrogen sulphide
methods, chemistry, iodate
methods, chemistry, iodine
methods, chemistry, iron
methods, chemistry, lactic acid
methods, chemistry, lead
methods, chemistry, lithium
methods, chemistry, magnesium
methods, chemistry, manganese
methods, chemistry, mercury
methods, chemistry, micronutrients
methods, chemistry, molybdenum
methods, chemistry, nickel
methods, chemistry, niobium
methods, chemistry, nitrate
methods, chemistry, nitrite
methods, chemistry, nitrogen
methods, chemistry, nutrients
methods, chemistry, organic acids
methods, chemistry, organic matter
methods, chemistry, osmium
methods, chemistry, oxygen
methods, chemistry, particulate matter
methods, chemistry, pH
methods, chemistry, phaeophytin (pheophytin)
methods, chemistry, phosphorus and its compounds
methods, chemistry, phosphoric acid
methods, chemistry, pigments
methods, chemistry, potassium
methods, chemistry, productivity
methods, chemistry, proteins
methods, chemistry, radioactivity
methods, chemistry, radionuclides
methods, chemistry, rare earths
methods, chemistry, rH
methods, chemistry, rubidium
methods, chemistry, salinity
methods, chemistry, scandium
methods, chemistry, selenium
methods, chemistry, sodium hydroxide
methods, chemistry, spectrometry
methods, chemistry, strontium
methods, chemistry, sugars
methods, chemistry, sulphate
methods, chemistry, titanium
methods, chemistry, titration
methods, chemistry, trace elements
methods, chemistry, tracers
methods, chemistry, tungsten
methods, chemistry, uranium
methods, chemistry, urea
methods, chemistry, vitamins
methods, chemistry, yttrium
methods, chemistry, zinc
methods, chemistry, zirconium
methods, circulation
methods, coasts
methods, computers
methods, "core"
methods, corrosion
methods, currents
methods, data analysis
methods, data collection
methods, data processing
methods, density
methods, density, currents
methods, desalination
methods, diffusion
methods, diving
methods, drafting
methods, dynamic heights
methods, echo-ranging
methods, echo-sounding
methods, evaporation
methods, fisheries (far from complete)
methods, geodesy
methods, geology
methods, geology, bathmetry
methods, geology, beaches and coasts
methods, geology, beaches and tracers
methods, geology, bottom sediments
methods, geology, coring
methods, geology, drilling
methods, geology, interstitial water
methods, geology, minerals
methods, geology, topography
methods, geophysics
methods, hydrodynamics
methods, hydrographic surveys
methods, ice
methods, internal waves
methods, light
methods, meteorology
methods, microseisms
methods, navigation
methods, oil
methods, optical properties
methods, particles
methods, photogrammetry
methods, pollution
methods, sampling
methods, scattering layer
methods, sea level
methods, seiches
methods, storm surges
methods, suspended matter
methods, swell
methods, temperature
methods, tides
methods, transparency
methods, transport
methods, ultra-violet
methods, upwelling
methods, vertical motion
methods, vertical transport
methods, water masses
methods, water samples
methods, waves
methods, wind
mica
microbenthos
microbial oxidation
microbiology
microfiltration
microfossils
micronekton
micronutrients
microorganisms
micropaleontology
microplankton
micropulsations
microrelief
microscopy
microseisms
microtektites
microtrees
microwaves
microzooplankton
midnight sun
mid-ocean ridges
mid-ocean rises
migrants
migrations
military oceanography
<u>milieu interieur</u>
minerals (individual minerals arranged alphabetically under this heading)
mines
mining
minutes
miogeosynclines
miscellaneous (i.e. general)
mixed layer
mixing
models

slopes
slope water
slumps
smell
smog
smoke (see also: sea smoke)
snakes (see also: sea snakes)
"snow" haze
snowfall
solar eclipse, effect of
solar elevation
solar radiation
sonic booms
sorption
sorting (geol.)
sound
sound channels
sound, effect of
sound transmission (propagation)
sound velocity
soundings
space charge
space oceanography (see chiefly: oceanography from space)
Spartina
spawning
speciation
species diversity (see also: diversity)
specific gravity
specific heat
specific volume
spherules
spits
sponges
spores
spray
squid beaks
stability
stagnation
standard sea water
standing crop
station lists
St. Elmo's fire
stomach contents
storms
storms, effect of
storm floods
storm forecasting
storm fury
storm surges
submarine canyons
submarine geology
submarine geophysics
submarine topography (see: bottom topography)
submarine valleys
submergence
submersibles
subsidence
substrate, effect of
sun, effect of
sun, glitter
sun, rays
sunshine
supercooled water
surf
surf beats
surf break
surf, effect of
surf zone
surface, effect of
surface films
surface flow
surface potentials
surface tension
surface water
surges (see also: storm surges)
surveys
survival
suspended matter
suspension
swales
swarming
swash
swell
swell, data only
swell, effect of
swell prediction
swimbladders
swimmers itch
swimming
symbiosis
symposia
synonymies
systematics
tablemount
tables
tagging
taxonomy
tectonics
tektites
temperature
temperature, air
temperature, annual variation
temperature anomalies
temperature, average
temperature, bottom
temperature chlorinity
temperature, data only
temperature, effect of
temperature fluctuations
temperature forecasting (see also: forecasting, temperature)
temperature gradients
temperature increase
temperature, infrared
temperature inversions
temperature, maximum
temperature, monthly means
temperature microstructure
temperature-nitrite
temperature-oxygen
temperature, paleo-
temperature, pollution (see chiefly: pollution)
temperature potential
temperature range
temperature-salinity (T-S)
temperature-sigma-*t*
temperature, surface
temperature, surface, anomalies
temperature, surface, data only
temperature, surface, effect of
temperature, surface, mean
temperature, surface, -bottom
temperature variation
tephra
terminology
terraces
territorial behavior
Thalassia
theoretical hydrodynamics
thermal balance
thermal conductivity
thermal energy
thermal gradient
thermal history
thermal structure
thermocline
thermocline, depth of
thermocline, effect of
thermodynamics
thermograms
thermostat
thermosteric anomaly
theta (potential temperature)-salinity
tholeiites
tidal anomalies
tidal basins
tidal constants
tidal currents
tidal currents, effect of
tidal cusps
tidal cycle
tidal energy
tidal flats
tidal flow
tidal mixing
tidal waves
tides
tides, data only
tides, effect of
tides, high sea
tides, internal
tides, oceanic
tide rips
tides, solar
tide tables
tides, wind
time scale
TIROS
tornadoes
towers
towing
toxicity
toxins, see: chemistry, toxins
tracers
tracks

abnormalities
Desmuth, V. M., and Kuber Vidyasagar 1968.
A prawn with double rostrum.
J. mar. biol. Ass. India 10(2): 391.

abnormalities
Marichamy, R. 1968.
On certain abnormalities in the short-jaw anchovy Thrissina baelama (Forsk.) (Family: Engraulidae).
J. mar. biol. Ass. India, 10(2): 395-397.

abnormalities

abnormalities
Sivalingam, D., and P. Vedavyasa Rao 1968
A case of abnormal petasma in the penaeid prawn, Metapenaeus affinis (H. Milne Edwards).
J. mar. biol. Ass. India 10(2): 386-387.

abnormalities
Suseelan, C., 1967.
On an abnormality in the penaeid prawn
Metapenaeus affinis (H.Milne Edwards).
J.mar.biol.Ass.,India, 9(2):438-440.

abrasion of pebbles

abrasion
Edgerton,H., E. Seibolf,K. Vollbrecht und F. Werner,1966.
Morphologische Untersuchungen am Mittelgrund (Eckernförder Bucht, west liche Ostsee).
Meyniana, 16:37-50.

abrasion of pebbles
Kuenen, Ph. H., 1956.
Experimental abrasion of pebbles. 2. Rolling by current. J. Geol., 64(4):336-368.

abrasion
*Werner,Friedrich,1967.
Sedimentation und Abrasion am Mittelgrund (Eckernförder Bucht, westliche Ostsee).
Meyniana,17:101-110.

abrasion
Zhiliaev, A.P., and N.V. Esin, 1965.
Methods for quantitative estimation of abrasion based on flysch coastal studies. (In Russian).
Okeanologiia, Akad. Nauk, SSSR, 5(6):1107-1109.

methods, geological
abrasion
coasts

absorption coefficient
Zege, E.P., A.P. Ivanov, B.A. Kagin, and I.L. Katzev 1971.
Determination of the extinction and absorption coefficients in water and atmosphere by distribution in time of the reflected pulse signal. (In Russian; English abstract).
Fizika Atmosfer. Okean. Izv. Akad. Nauk SSSR 7(7): 750-757

absorption light

absorption, light
Bauer, D., J.C. Brun-Cottan et A. Saliot 1971.
Principe d'une mesure directe dans l'eau de mer du coefficient d'absorption de la lumière.
Cah. océanogr. 23(9): 841-858.

absorption, light
Lecompte, Pierre, 1968.
Étude théorique de l'absorption du rayonnement solaire dans la mer.
Cah. océanogr., 20(3):195-201.

absorption, light
Pelevin, V. N., 1965.
Some experimental results of determination of absorption coefficient of radiation in the sea. (In Russian; English abstract).
Trudy Inst. Okeanol, Akad. Nauk SSSR 77: 67-72.

absorption,light
Semenchenko,I.B., 1966.
Determination of the optical characteristics of sea water on the basis of measurements of the spectral brightness coefficients from an aircraft. (In Russian;English abstract).
Okeanologiia,Akad. Nauk,SSSR,6(6):1073-1080.

absorption (sound)

absorption, sound
Albers, Vernon M., 1965.
Underwater acoustics handbook II.
The Pennsylvania State University Press, 356 pp. $12.50.

absorption, sound
Busby, J., and E.G. Richardson, 1957.
The absorption of sound in sediments.
Geophysics, 22(4):821-828, 6 figs.

absorption (sound)
Mulders, C. E., 1949.
Ultrasonic absopption inwater in the region of 1 mc/s. Nature 164:347-348.

absorption, sound
Schulkin, M., 1963.
Eddy viscosity as a possible acoustic obsorptio mechanism in the ocean.
J. Acoust. Soc., Amer., 35(2):253-254.

absorption, sound
Schulkin, M., and H.W. Marsh, 1963.
Errata: Sound absorption in sea water. J. Acoust. Soc., Amer., 1962, 34:864.
J. Acoust. Soc., Amer., 35(5):739.

absorption, sound
Shumway, George, 1960.
Sound speed and absorption studies of marine sedi- ments by a resonance method. II.
Geophysics, 25(3):659-682.

absorption, sound
Watson, John D., and Robert Meister, 1963.
Ultrasonic absorption in water containing plankton in suspension.
J. Acoust. Soc., Amer., 35(10):1584-1589.

absorption, solar radiation

absorption, solar radiation
Darby, Hugh H., E.R.Fenimore Johnson and George W. Barnes, 1937.
Studies on the absorption and scattering of solar radiation by the sea. Spectrographic and photoelectric measurements.
Papers from the Tortugas Lab., 31:

Originally printed in:
Carnegie Inst., Washington, Publ., No. 475:191-205.

absorption, surface
Ivanov, V. N. 1970.
Absorbing surface of suspension in sea water as studied by means of radioisotopes.
Dokl. Akad. Nauk SSSR 195(6): 1441-1442.
(In Russian)

abstracts, fish
India,Central Marine Fisheries Research Institute 1968.
Advance abstracts of contribution on fisheries and aquatic sciences in India.
Adv.Abstr.Contr.Fish.Aquat.Sci.India, 11(4):120pp (mimeographed).

abstacts, Station Marine de Tulear
Thomassin, Bernard A. 1971.
Revue bibliographique des travaux de la station marine de Tuléar (République Malgache) 1961-1970.
Tethys Suppl. 1: 3-50

abyssal benthos, see: benthos

abyssal benthos: SEE benthos

abyssal fan
Moore, George T. 1969.
Interaction of rivers and oceans - Pleistocene petroleum potential.
Bull. Am. Ass. Petrol. Geol., 53(2): 2421-2430

Pleistocene
submarine canyons
rivers, effect of
abyssal fan.
California
petroleum potential

ABYSSAL FAUNA

abyssal faunas
Bruun , A. Fr., 1961.
New light on the abyssal and hadal faunas of the Pacific (Abstract).
Proc. Ninth Pacific Sci. Congr., Pacific Sci. Assoc., 1957, (Fish.), 10:14.

abyssal fauna
Menzies, Robert J., John Imbrie and Bruce C. Heezen, 1961.
Further considerations regarding the antiquity of the abyssal fauna with evidence for a changing abyssal environment.
Deep-Sea Res., 8(2):79-94.

abyssal floor

abyssal floor
Heezen, Bruce C., and H.W. Menard, 1963.
12. Topography of the deep-sea floor.
In: The Sea, M.N. Hill, Editor, Interscience Publishers, 3:233-280.

abyssal sea floor
Menzies, R.J., 1965.
Condition for the existence of life on the abyssal sea floor.
In: Oceanography and Marine Biology, H. Barnes, editor, George Allen and Unwin, 3:195-210.

abyssal gaps

abyssal gaps
Heezen, Bruce C., and H.W. Menard, 1963.
12. Topography of the deep-sea floor.
In: The Sea, M.N. Hill, Editor, Interscience Publishers, 3:233-280.

abyssal hills

abyssal hills
Andrews James E., 1971.
Abyssal hills as evidence of transcurrent faulting on North Pacific fracture zones.
Bull. geol. Soc. Am, 82(2): 463-470.

Pacific, north

abyssal hills
Belderson,R.H., and A.S. Laughton,1966.
Correlation of some Atlantic turbidites.
Sedimentotogy 7(2):103-116.

abyssal hills
*Conolly,John R., and C.C. von der Borch,1967.
Sedimentation and physiography of the sea-floor south of Australia.
Sedimentary Geol., 1(2):181-220.

abyssal hills
Heezen, Bruce C., and H.W. Menard, 1963.
12. Topography of the deep-sea floor.
In: The Sea, M.N. Hill, Editor, Interscience Publishers, 3:233-280.

abyssal hills
Ilyin, A.V., 1971.
Main features of geomorphology of Atlantic bottom. (In Russian; English abstract).
Okeanol. Issled. Rezult. Issled. Mezhd. Geofiz. Proekt. 21: 107-246.

abyssal hills

Krause, Dale C., 1961
Geology of the sea floor east of Guadalupe Island.
Deep-Sea Res., 8(1): 28-38.

Abyssal hills

Luyendyk, Bruce P., 1970.
Origin and history of abyssal hills in the northeast Pacific Ocean
Bull. geol. Soc. Am. 81 (8): 2237-2260

abyssal hills
Pacific, northeast

abyssal hills
Luyendyk, Bruce P., and Celeste G. Engel, 1969.
Petrological magnetic and chemical properties of basalt dredged from an abyssal hill in the north-east Pacific.
Nature, Lond., 223(5210):1049-1050.

abyssal hills
*Menard, H.W., and Jacqueline Mammerickx, 1967.
Abyssal hills, magnetic anomalies and the East Pacific Rise.
Earth Planet Sci. Letters, 2(5):465-472.

abyssal hills
*Moore, Theodore C., Jr., and G. Ross Heath, 1967.
Abyssal hills in the central equatorial Pacific: detailed structure of the sea floor and sub-bottom reflectors.
Mar. Geol., 5(3):161-179.

abyssal hills
Naugler, Frederic P., and David E. Rea 1970
Abyssal hills and sea-floor spreading in the central North Pacific.
Bull. geol. Soc. Am. 81 (10): 3123-3128

abyssal hills (not salt dominated)
* Watson, Jerry A., and G. Leonard Johnson, 1969.
The marine geophysical survey in the Mediterranean.
Int. hydrogr. Rev., 46(1): 81-107.

Abyssal plain

abyssal plains
Bellaiche, Gilbert, 1970.
Géologie sous-marine de la marge continentale au large du massif des Maures (Var, France), et de la plaine abyssale Ligure.
Revue Géogr. phys. Géol. dyn. (2) 12 (5): 403-440.

abyssal plain
Bellaiche Gilbert 1968
Précisions apportées à la connaissance de la pente continentale et de la plane abyssale à la suite de trois plongées en bathyscaphe Archimède.
Rev. Géogr. phys. Géol. dynam. (2) 10 (2): 137-145

Abyssal plains
Bullard, E.C., 1962.
The deeper structure of the ocean floor.
Proc. R. Soc., London, (A) 265(1322):386-395.

abyssal plain
Bunce, Elizabeth T., 1966.
The Puerto Rico Trench.
In: Continental margins and island arcs, W.H. Poole, editor, Geol. Surv. Pap., Can., 66-15:165-175.

abyssal plains
*Bye, John A.T., 1971.
The slope of abyssal plains. J. geophys. Res., 78(18): 4188-4194.

abyssal plains

abyssal plains
Collette, B.J., and K.W. Rutten 1970.
Differential compaction vs. diapirism in abyssal plains.
Mar. geophys. Res. 1(1): 104-107.

abyssal plains
*Conolly, John R., 1969.
Western Tasman sea floor.
N.Z. Jl Geol. Geophys. 12:310-343.

abyssal plains
*Conolly, John R. and Maurice Ewing, 1969.
Redeposition of pelagic sediment by turbidity currents; a common process for building abyssal plains. Trans. Gulf Coast Ass. geol. Socs, 19: 506.

abyssal plain
*Conolly, John R., and C.C. von der Borch, 1967.
Sedimentation and physiography of the sea-floor south of Australia.
Sedimentary Geol., 1 (2):181-220.

abyssal plains
Emery, K.O., Bruce C. Heezen and T.D. Allan, 1966.
Bathymetry of the eastern Mediterranean Sea.
Deep-Sea Res., 13(2):173-192.

abyssal plain
Ewing, John, J.L. Worzel, Maurice Ewing and Charles Windisch, 1966.
Ages of Horizon A and the oldest Atlantic sediments: coring at an outcrop of Horizon A establishes it as a buried abyssal plain of Upper Cretaceous age.
Science, 154(3753):1125-1132.

Abyssal Plains
Fairbridge, Rhodes W., 1965.
The Indian Ocean and the status of Gondwanaland.
Progress in Oceanography, 3:83-136.

abstract

abyssal plains
Hamilton, Edwin L., 1967.
Marine geology of abyssal plains in the Gulf of Alaska.
J. geophys. Res., 72(16):4189-4213.

Gulf of Alaska
abyssal plains
bottom sediments (thickness)
seamounts
reflection, seismic
trenches
Aleutian Trench
turbidites
bottom sediments, pelagic

abyssal plain
Heezen, B.C., M. Ewing and D.B. Ericson, 1955.
Reconnaissance survey of the abyssal plain south of Newfoundland. Deep-Sea Res. 2(2):122-133, 6 textfigs.

abyssal plain
Heezen, Bruce C., Charles D. Hollister and William F. Ruddiman, 1966.
Shaping of the continental rise by deep geostrophic contour currents.
Science, 152(3721):502-508.

abyssal plains
Heezen, Bruce C., and A.S. Laughton, 1963.
14. Abyssal plains.
In: The Sea, M.N. Hill, Editor, Interscience Publishers, 3:312-364.

abyssal plains
Heezen, Bruce C., and H.W. Menard, 1963.
12. Topography of the deep-sea floor.
In: The Sea, M.N. Hill, Editor, Interscience Publishers, 3:233-280.

abyssal plains
Ilyin, A.V., 1971.
Main features of geomorphology of Atlantic bottom. (In Russian; English abstract).
Okeanol. Issled. Rezult. Issled. Mezhd. Geofiz. Proekt. 21: 107-246.

abyssal plains
Laughton, A.S., 1959.
The sea floor.
Science Progress 47(186):230-249.

abyssal plains
*Mammerickx, Jacqueline (Mrs. Edward L. Winterer) 1970.
Moroph† of the Aleutian Abyssal Plain.
Bull. geol. Soc. Am., 81(11):3457-3464.

abyssal plains
Menard, H.W., S.M. Smith and R.M. Pratt, 1965.
The Rhône Deep-Sea Fan.
In: Submarine geology and geophysics, Colston Papers, W.F. Whittard and R. Bradshaw, editors, Butterworth's, London, 17:271-284.

fans, deep-sea
Mediterranean, west
abyssal plains
bathymetry
1594

abyssal plains
Pratt, R.M., 1965.
Ocean-bottom topography: the divide between the Sohm and Hatteras Abyssal plains.
Science, 148(3677):1598-1599.

abstract

abyssal plain
Rona, Peter A., and C.S. Clay, 1967.
Stratigraphy and structure along a continuous seismic reflection profile from Cape Hatteras, North Carolina to the Bermuda Rise.
J. geophys. Res., 72(8):2107-2130.

continental slope
continental rise
abyssal plain
unconformities

abyssal water
Craig, H., 1971.
The deep metabolism: oxygen consumption in abyssal ocean water. J. geophys. Res. 76(21): 5078-5086.

abyssal water
Lynn, Ronald J., and Joseph L. Reid, 1968.
Characteristics and circulation of deep and abyssal waters.
Deep-Sea Res., 15(5):577-598.

abyssal zonation

abyssal zonation
Bruun, A. Fr., 1957.
The ecological zonation of the deep-sea.
Proc. UNESCO Symp., Phys. Ocean., Tokyo, 1955: 160-168.

acceleration potential

Acceleration potential
Reid, Joseph L., Jr., 1965.
Intermediate waters of The Pacific Ocean.
Johns Hopkins Oceanogr. Studies, 85 pp.

acclimatization

acclimatization
Kinne, Otto, 1964.
Non-genetic adaptation to temperature and salinity.
Helgol. Wiss. Meeresuntersuch., 9(1-4):433-458.

Accretion

accretion, continental
Byrne, John V., Gerald A. Fowler and Neil J. Maloney, 1966.
Uplift of the continental margin and possible continental accretion off Oregon.
Science, 154(3757):1654-1656.

accretion
Kestner, F.J.T., and C.C. Ingles, 1956?
A study of erosion and accretion during cyclic changes in an estuary and their effect on reclamation of marginal land. J. Agric. Eng. Res. 1(1):1-8.

accretion
*Silvester, Richard, 1967.
Sediment transport and accretion around the coastline of Japan.
Proc. 10th Conf. Coast. Engng. Tokyo, 1966, 1:469-488.

achondrites

achondrites
Deuser, W.G. 1968
Iron-magnesium-aluminum relationships in achondrites.
Chem. Geol. 3(2): 81-87.

acoustic bottom loss
Hamilton, Edwin L. 1970.
Reflection coefficients and bottom losses at normal incidence computed from Pacific sediment properties.
Geophysics 35(16): 995-1004

acoustical energy

acoustical energy
Daniels, F.B., 1952.
Acoustical energy generated by the ocean waves.
J. Acoust. Soc., Amer., 24:83.

acoustic energy
Neidell, Norman S., 1968.
Data processing for controlled energy acoustic sources.
Under Sea Techn., 9(10):28-29, 44-46.

acoustic field
*Kuo, Edward Y.T., 1968.
Acoustic field generated by a vibrating boundary. I. General formulation and sonar-dome noise loading.
J. acoust. Soc., Am., 43(1):25-31.

acoustic properties
Stoll, Robert D., John Ewing and George M. Bryan 1971
Anomalous wave velocities in sediments containing gas hydrates.
J. geophys. Res. 76(8): 2090-2094

acoustic signal processing
Angelari, Richard D. 1970.
The ambiguity function applied to underwater acoustic signal processing: a review.
Ocean Engng 2(1): 13-26

acritarchs

acritarchs
Downie, Charles, and William Antony S. Sarjent, 1964.
Bibliography and index of fossil dinoflagellates and acritarchs.
Geol. Soc., Amer., Mem., 94:180 pp.

actinians
*Sassaman, C. and C.P. Mangum, 1970.
Patterns of temperature adaptation in North American Atlantic actinians.
Marine Biol., 7(2):123-130.

Adaptations

adaptations, abyssal
Allen, J.A., and H.L. Sanders, 1966.
Adaptations to abyssal life as shown by the bivalve Abra profundorum (Smith).
Deep-Sea Res., 13(6):1175-1184.

adaptations
Dunbar, M.J., 1968.
Ecological development in polar regions: a study in evolution.
Prentice-Hall, Inc. 119 pp. $4.95

adaptations
*Kähler, H.H., 1970.
Über den Einfluss der Adaptationstemperatur und des Salzgehaltes auf die Hitze- und Gefrierresistenz von Enchytraeus albidus (Oligochaeta).
Marine Biol., 5(4): 315-324.

adaptation
Kinne, Otto, 1967.
Non-genetic adaptation in Crustacea.
Proc. Symp. Crustacea, Ernakulam, Jan. 12-15, 1965, 3: 999-1007.

adaptation
Kohn, Alan J., 1971.
Diversity, utilization of resources, and adaptive radiation in shallow-water marine invertebrates of tropical oceanic islands.
Limnol. Oceanogr. 16(2): 332-348.

adaptations
Schlieper, Carl, 1960
Genotypische und phaenotypische Temperatur- und Salzgehalts Adaptationen bei marinen Bodenevertebraten der Nord-und Ostsee.
Kieler Meeresf., 16(2): 180-184.

adaptation
Squires, H.J., 1967.
Some aspects of adaptation in decapod Crustacea in the north-west Atlantic.
Proc. Symp. Crustacea, Ernakulam, Jan. 12-15, 1965, 3: 987-995

adiabatic cooling

adiabatic cooling
Crease, J., D. Catton, & R.A. Cox, 1962
Tables of adiabatic cooling of sea water.
Deep-Sea Res., 9(1): 76

adsorption

adsorption
Baylor, E.R., W.H. Sutcliffe and D.S. Hirschfeld, 1962
Adsorption of phosphates onto bubbles.
Deep-Sea Res., 9(2):120-124.
WHOI Contrib. No. 1259

adsorption
Jarvis, N.L., 1967.
Adsorption of surface-active material at the sea-air interface.
Limnol. Oceanogr., 12(2):213-221.

air-sea interface
surface tension
surface potential

adsorption
Schwarcz, H.P., E.K. Agyei and C.C. McMullen, 1969.
Boron isotopic fractionation during clay adsorption from sea-water.
Earth Planet. Sci. Lett. 6(1): 1-5.

clay minerals
adsorption
boron

adsorption, sound
Vigoureux, P., and J.B. Hersey, 1962
Sound in the sea. Ch. 12, Sect. IV. Transmission of energy within the sea. In: The Sea Vol. 1, Physical Oceanography, Interscience Publishers, 476-497.
(Mss received July 1960)
WHOI Contrib. No. 1163.

Advection

advection
Adem, Julian 1970.
Incorporation of advection of heat by mean winds and by ocean currents in a thermodynamic model for long range weather prediction
Mon. Weath. Rev. U.S. Dept. Comm. 98(10): 776-786

advection
Alexander R.C. 1971.
On the advective and diffusive heat balance in the interior of a subtropical ocean.
Tellus. 23(4/5): 393-403

advection
Arcenieva, N. Ia., 1965.
A method for calculating water temperature in the active layer of the White Sea. (In Russian).
Trudy, Gosudarst. Okeanogr. Inst., No. 86:75-94.

advection
Arons, Arnold B., and Henry Stommel, 1967.
On the abyssal circulation of the World Oceans. III. An advection-lateral mixing model of the distribution of a tracer property in an ocean basin.
Deep-Sea Research, 14(4):441-457.

circulation, abyssal
advection
lateral mixing

advection
Baer, Ledolph, Louis C. Adamo and S.I. Adelfang, 1968.
Experiments in oceanic forecasting for the advective region, by numerical modeling. 2. Gulf of Mexico.
J. geophys. Res., 73(16):5091-5104.

forecasting
advection
Gulf of Mexico
models

advection
Bathen, Karl H., 1971.
Heat storage and advection in the North Pacific Ocean. J. geophys. Res., 76(3): 676-687.

Pacific, north

advection
Bulatov, R.P., 1971.
On the structure and circulation of the bottom layer in the Atlantic Ocean. (In Russian; English abstract). Okeanol. Issled. Rezult. Issled. Mezhd. Geofiz. Proekt. 21: 43-59.

advection
Eber, L.E., 1961
Effects of wind-induced advection on sea surface temperature.
J. Geophys. Res., 66(3): 839-844.

advection
Gruzinov, V.M., 1965.
Hydrological front as a natural boundary of inherent zones in the ocean. (In Russian).
Trudy, Gosudarst. Okeanogr. Inst., No. 84:252-262.

advection
Hamm, D.P., and R.M. Lesser, 1968.
Experiments in oceanic forecasting for the advective region by numerical modeling. 1. The model.
J. geophys. Res., 73(16):5081-5089.

MODELS
FORECASTING
advection

advection
Ichie, T., 1950.
[On the mutual effect of the coastal and offshore waters.] J. Ocean. Soc., Japan, (Nippon Kaiyo Gakkaisi) 6(2):1-7 (In Japanese; English summary)

advection
Ivanov, Y.A., 1965.
The role of boundary conditions and advection in the formation and distribution of extreme values of oceanological characteristics according to depth. (In Russian).
Okeanologiia, Akad. Nauk, SSSR, 5(1):40-44.

advection, density
Kozlov, V.F., 1966.
Some exact solutions of the nonlinear equation of density advecyion in the ocean. (In Russian).
Fizika Atmosferi i Okeana, Izv., Akad. Nauk, SSSR 2(11):1205-1207.

advection, heat
Kort, V.G., 1968.
The role of heat advection by sea currents in large-scale interaction between the ocean and the atmosphere. (In Russian).
Dokl. Akad. Nauk, SSSR, 182(5):1059-1062.

advection
Laevastu, T., et P.M. Wolff 1965
Quelques principes d'analyse et de prévision océanographique synoptique.
Météorologie 1965: 305-319

advection
Lee, Arthur, and John Ramster, 1968.
The hydrography of the North Sea. A review of our knowledge in relation to pollution problems.
Helgoländer wiss. Meeresunters. 17: 44-63

advection
Currents, residual (bottom)

advection
McEwen, G. F., 1950.
A statistical model of instantaneous point and disk sources with applications to oceanographic observations. Trans. Am. Geophys. Union 31(1): 35-46, 3 textfigs.

SIO Contrib., n.s., No. 437.

advection
O'Connor, Donald J., and Dominic M. DiToro, 1965.
The solution of the continuity equation in cylindrical coordiantes with dispersion and advection for an instantaneous release.
Symposium, Diffusion in oceans and fresh waters, Lamont Geol. Obs., 31 Aug.-2 Sept., 1964, 80-85.

abstract

advection
Okubo, Akira, 1971.
4. Horizontal and vertical mixing in the sea.
In: Impingement of man on the oceans, D.W. Hood, editor, Wiley Interscience: 89-168.

advection
Olson, Boyd E., 1962
Variations in radiant energy and related ocean temperatures.
J. Geophys. Res., 67(12):4705-4712.

advection
Penin, V.V., 1966.
Hydrological conditions in the Norwegian Sea in the period of summer feeding of herring from 1958 to 1963. (In Russian).
Trudy, Poliarn. Nauchno-Issled. Proektn. Inst. Morsk. Ribn. Khoz. Okeanogr., N.M. Knipovich, PINRO, 17:55-68.

currents, dynamic
Norwegian Sea

advection, temperature
Potaichuk, M.S., 1962.
Tests on the calculation of the amount of heat advection in North Atlantic currents based on the data of the I.G.Y. (In Russian).
Trudy Gosud. Okeanogr. Inst., 67:86-103.

advection
Pritchard, D.W., Akira Okubo and Emanuel Mehr, 1962
A study of the movement and diffusion of an introduced contaminant in New York Harbor waters.
Chesapeake Bay Inst., Techn. Rept. (Ref. 62-21) 31: 89 pp. (multilithed)

advection, heat
Rodgers, G.K., 1968.
Heat advection within Lake Ontario in spring and surface water transparency associated with the thermal bar.
Proc. 11th Conf. Great Lakes Res. 1968: 480-486

advection
Seckel, Gunter R., 1960.
Advection - a climatic character in the mid-Pacific.
California Coop. Oceanogr. Fish. Invest. Rept., 7: 60-65.

advection
*Seckel, Gunter R., 1969.
The Hawaiian oceanographic climate, July 1963-June 1965. Bull. Jap. Soc. fish. Oceanogr. Spec. No. (Prof. Uda Commem. Pap.): 105-114.

heat budget
advection
Hawaii
Pacific, central

advection
Soliankin, E.V., 1962
The advective component in the annual heat turnover of the active layer of the Black Sea. (In Russian).
Okeanologiia, Akad. Nauk, SSSR, 2(6):988-998.

advection
Spencer, Derek W. and Peter G. Brewer, 1971.
Vertical advection, diffusion and redox potentials as controls on the distribution of manganese and other trace metals dissolved in waters of the Black Sea. J. geophys. Res, 76(24): 5877-5892.

advection
Tsujita, Tokim, 1966
Cold water conditions observed in the water around Japan in 196 3 and there effects on the Coastal water fisheries. (In Japanese; english abstract. Bull. Tokai Reg. fish. res. Lab., 26:1-8.

advection, effect of a
*Sapozhnikov, V.V., 1971.
The role of advection and the influence of biochemical processes on vertical phosphorus distribution in the tropical ocean. (In Russian; English abstract). Okeanologiia 11(2): 223-230.

advection, effect of
Pacific, tropical
chemistry, phosphorus (vertical)

advection, horizontal
*Roden, Gunnar I., 1971.
Aspects of the transition zone in the northeastern Pacific. J. geophys. Res. 76(15): 3462-3475.

divergence
heat flux

advection, lateral
Watanabe, N., 1955.
Hydrographic conditions of the north-western Pacific. 1. On the temperature change in the upper layer in summer. J. Ocean. Soc., Japan, 11(3):111-122.

Also: Bull. Tokai Reg. Fish. Res. Lab., 11(B-224)

advection, effect of
Wickett, W. Percy 1967.
Ekman transport and zooplankton concentration in the North Pacific Ocean.
J. Fish. Res. Bd, Can. 24(3): 581-594.

advection
Zhukov, L.A., 1961
[On the advection of heat by the currents in the upper water layer of the Atlantic Ocean.]
Issled. Severnoi Chasti Atlanticheskogo Okeana, Mezhd. Geofiz. God, Leningradskii Gidrometeorol. Inst., 1: 38-42.

aerial exposure, effect of
Coleman, N. and E.R. Trueman, 1971.
The effect of aerial exposure on the activity of the mussels Mytilus edulis L. and Modiolus modiolus (L). J. exp. mar. Biol. Ecol, 7(3): 295-304.

Aerial photography
Sea photography

aerial photography, see: photography

aerial scouting

aerial scouting
Cushing, D.H., Finn Devold, J.C. Marr, H. Kristjonsson, 1952.
Some modern methods of fish detection: echo sounding, echo ranging and aerial scouting.
F.A.O. Fish. Bull. 5(3/4): 95-119, 9 text-figs.

aeromagnetic survey
Girdler, R.W. 1970.
An aeromagnetic survey of the junction of the Red Sea, Gulf of Aden and Ethiopian rifts - a preliminary report.
Phil. Trans. R. Soc. (A) 267 (1181): 359-368.

Aerosols

aerosols
Assaf, G., and J.R. Gat 1970.
Sea surface as a sink of emanation products.
Earth Planet. Sci. Letts 7(5): 385-388.

aerosols
Belaev, L.I., 1951.
[Determination of the quantity of marine salts suspended in air.]
Doklady Akad. Nauk, SSSR, 81(6):1039-

aerosols
Belyaev, V.I., 1961(1962).
[Equations of system of water drops suspended in air.]
Izv., Akad. Nauk, SSSR, Ser. Geofiz. (9):1408-1417.

English Edit. (9): 915-923

aerosols
Blanchard, Duncan C., 1963.
The electrification of the atmosphere by particles from bubbles in the sea.
Progress in Oceanography, 1:71-202.

aerosols
Bruevich, S.V., and E. Z. Kulik, 1967.
The chemical interaction between the ocean and the atmosphere (salt exchange). (In Russian; English abstract)
Okeanologiia, Akad. Nauk, SSSR, 7(3):363-379.

aerosols
salt exchange

aerosols

Büchen, Mathias, und Hans Walter Georgii 1971.
Ein Beitrag zum atmosphärischen Schwefelhaushalt über dem Atlantik.
Meteor-Forsch. Engebnisse (B) 7: 71-77.

aerosols (lead)

*Chow, Tsaihwa J. and John L. Earl, 1970.
Lead aerosols in the atmosphere: increasing concentrations. Science 169(3945): 577-580.

lead
aerosols (lead)

aerpsols

Duce, Robert A., Theodore van Nahl and John W. Winchester, 1964.
Iodine, bromine and chlorine in the aerosol and gaseous phase of marine air from Hawaii. (Abstract).
Trans. Amer. Geophys. Union, 45(1):67.

aerosols

Jaenicke, Ruprecht, Christian Junge, und Hans Joachim Kanter 1971.
Messungen der Aerosolgrössenverteilung über dem Atlantik.
Meteor-Forsch.- Engebnisse (B) 7: 1-54.

aerosols

Junge, C.E., C.W. Chagnon and J.E. Manson, 1961
Stratospheric aerosols. J. Meteorology, 18(1): 81-108.

aerosols

Junge, C.E., Elmer Robinson and F.L. Ludwig, 1969.
A study of aerosols in Pacific air masses.
J. appl. Met., 8(3): 346-347

aerosols

Kachurin, L.G., L.E. Atlant'eva and Hsia-Yu-Ken, 1961.
[Vapor concentration and rate of growth of condensate drops in water aerosols.]
Izv., Akad. Nauk, SSSR, Ser. Geofiz., (9):1418-1425.
English Edit., (9):924-928.

aerosols

Kano, M., 1960.
A new optical method for the determination of atmospheric aerosol size distribution.
Pap. Meteorol. & Geophys., Tokyo, 11(2-4):356-360.

aerosols

Kopcewitz, Teodor, Roman Tomaszenko, and Jan Tomczak, 1959
La radioactivité des aerosols, des aerosols, des precipitations, de l'eau de la Vistule et de la mer observée par les stations de l'Institut Hydrologique et Météorologique d'Etat en 1958.
Acta Geophysica Polonica, 7(2):103-129.

aerosol particles — Yu. S. Sedunov

Krasnogorskaya, N.V., ., 1961.
[An induction method for measuring charges on individual particles.]
Izv. Akad. Nauk, SSSR, Ser. Geofiz., (5):775-785. Eng. Ed. (5):507-513
Eng. Abstr. An induction method for measuring charges on aerosol particles.
OTS-61-11147-17 JPRS:8710:2-4.

aerosols

Laktionov, A.G., and L.M. Levin, 1960.
[Comparative measurements of concentration and function of the distribution of the water aerosol particles.]
Izv. Akad. Nauk, SSSR, Ser. Geofiz., (7):1056-1058.
Translation:
Eng. Ed., (7):704-705.
Pergamon Press

aerosols

*Loucks, Ronald H. and John W. Winchester, 1970.
Some features of chlorine and bromine in aerosols.
J. geophys. Res., 75(12): 2311-2315.

aerosols

aerosols

Mikirov, A.E., 1962.
[Aerosol layer in the upper atmosphere.]
Doklady Akad. Nauk, SSSR, 142(3):587-588.

aerosols, sea salt

Peterson, M.P., and K.T. Spillane, 1969.
Surface films and the production of sea-salt aerosol.
Q. Jl R. met. Soc. 95(405): 526-534.

aerosols

Petrov, G.D., 1959.
[Optical methods for determining the spectrum of the aqueous aerosol particles.]
Izv., Akad. Nauk, SSSR, Geofiz. Ser., (5):796-797
Engl. Transl. by V.A. Salkind:566-567.

aerosols

Rasool, S.I. and S.H. Schneider, 1971.
Atmospheric carbon dioxide and aerosols: effects of large increases on global climate.
Science, 173(3992): 138-141.

aerosols

Rich, T.A., L.W. Pollak, and A.L. Metnieks, 1962
On the time required for aerosols to reach electrical equilibrium.
Geofisica Pura e Applicata, 51(1):217-224.

aerosols

Rooth, C., 1963.
The effect of weak mesoscale circulations on the distribution of marine salt aerosols.
Berichte Deutsch. Wetterdienstes, 12(91):89-91.

aerosols

Ruppersberg, Gerhard H., 1971.
Die Änderung des maritimen Dunst-Streukoeffizienten mit dem relativen Feuchte.
Meteor Forsch.-Engebn. (B) 6: 37-60

aerosols

Vernidub, I.I., A.S. Zhikarev, Kh. Kh. Medaliev, N.S. Pravdum, G.K. Sulakvelidze and G.G. Chumakova, 1962
Investigation of the ice-forming properties of lead iodide aerosols.
Izv. Akad. Nauk, SSSR, Ser. Geofiz., 9: 1286-1293
Eng. Ed. 9:801-806.

Aerosols

Volz, F.E., 1970.
Spectral skylight and solar radiance measurements in the Caribbean: maritime aerosols and Sahara dust.
J. atmos. Sci., 27 (7): 1041-1047.

dust (Sahara)
Caribbean

aerosols, vertical distribution

Werner, Ch., 1970.
Ein Lidar-System zur Messung der vertikalen Dunstverteilung in der Atmosphäre des Atlantischen Passat Experiments (APEX) auf W.F.S. Planet.
II Messergebnisse.
Meteor Forsch.-Engebnisse (B) 5: 62-84.

Aftershocks

aftershocks

Berg, Eduard, 1966.
Triggering of the Alaskan earthquake of March 28, 1964, and major aftershocks by low ocean tide loads.
Nature, 210 (5039): 893-896.

earthquakes
aftershocks
Pacific, northeast
Gulf of Alaska

aftershock activity

Nagumo, Shozaburo, Shuji Hasegawa, Sadayuki Koresawa and Heihachiro Kobayashi, 1970.
Ocean-bottom seismographic observation off Sanriku - aftershock activity of the 1968 Tokachi-Oki Earthquake and its relation to the ocean-continent boundary fault.
Bull. Earthq. Res. Inst. 48 (5): 793-809.

age determination (biological)

age determination

Miller, D.J., 1955.
Studies relating to the validity of the scale method for age determination of the northern anchovy (Engraulis Mordax).
Calif. Fish and Game, Mar. Fish. Br., Fish. Bull. No. 101: 6-34.
Dept.

AGE DETERMINATION, whales

van Utrecht, W.L., and C.N. van Utrecht-Cock, 1968.
Some comments on the significance of the cortical layer of the baleen plat in age determination of baleen whales.
Norsk Hvalfangsttid. 57 (2): 39-40.

baleen plates
age determination, whales

age determination (geological)

age determination, geological

See also: chronology
geological ages

age determination

Amin, B.S., D.P. Kharkar and D.Lel., 1966.
Dating of sediments using cosmic ray produced isotopes. (Abstract only)
Second, Int. Oceanogr. Congr., 30 May-9 June 1966, Abstracts, Moscow: 7.

age determination (geological)

Amirkhanov, K. I., K.S. Magataev and E.B. Brandt, 1957.
[Isotope dating of sedimentary minerals.]
Doklady Akad. Nauk, SSSR, 117(4):675-677.

age determination

Anderson, E.C., and H. Levi, 1952.
Some problems in radiocarbon dating.
K. Danske Vidensk. Selskab., Mat.-fysiske Medd. 27(6):22 pp.

age determination (geological)

Arrhenius, Gustaf, 1967.
Deep-sea sedimentation: a critical review of U. S. work.
Trans. Am. geophys. Un., 48(2):604-631.

age determination

Arrhenius, G., G. Kjellberg, and W.F. Libby, 1951.
Age determination of Pacific chalk ooze by radiocarbon and titanium content. Tellus 3: 222-229.

age determination

Audley-Charles, M.G., 1966.
The age of the Timor Trough.
Deep-Sea Res., 13(4):761-763.

age determination

Ault, Wayne U., 1959.
Oxygen isotope measurements on Arctic cores.
Sci. Studies, Fletcher's Ice Island, T-3, Vol. 1, Air Force Cambridge Res. Center, Geophys. Res. Pap. No. 63:159-168.

age determination

Aumento, F., 1969.
The Mid-Atlantic Ridge near 45°N. V
Fission track and ferro-manganese chronology.
Can. J. Earth Sci. 6(6): 1431-1440.

age determination

*Aumento, F., R.K. Wanless and R.D. Stevens, 1968.
Potassium-argon ages and spreading rates on the Mid-Atlantic Ridge at 45° North.
Science, 161(3848):1338-1339.

age determination

Bada, Jeffrey L., Bruce P. Luyendyk and J. Barry Maynard, 1970.
Marine sediments: dating by the racemization of amino acids.
Science, 170(3959):730-731.

age determination

Baranov, V.J., and L.R. Kuzmina, 1954.
[Ionium method of the age determination of sea sediments.] Dokl. Akad. Nauk, SSSR, 97:483-

age determinations

Baranov, V.I., and Kuz'mina, L.A., 1954.
[An ionium method for determining the age of sea cores. The direct determination of ionium.]
Dokl. Akad. Nauk, SSSR, 92(3):483.

age determination
geological

Bé, Allan W.H. and Jan Van Donk, 1971.
Oxygen-18 studies of Recent planktonic Foraminifera. Science, 173(3992): 167-168.

age determination

Berger, Rainer, R. E. Taylor and W. F. Libby, 1966.
Radiocarbon content of marine shells from the California and Mexican west coast.
Science, 153 (3738):864-866.

California
Mexico, west
carbon-14
age determination

age determination

Berggren, W.A. 1971.
Multiple phylogenetic zonations of the Cenozoic based on planktonic Foraminifera.
Proc. II Plankton. Conf. Roma 1970, A. Farinacci, editor, 41-56

age determination

Bezrukov, P.L., A. Ya. Krylov and V.I. Chernysheva, 1966.
Petrography and the absolute age of the Indian Ocean floor basalts. (In Russian; English abstract).
Okeanologiia, Akad. Nauk, SSSR, 6(2):261-266.

Indian Ocean
basalt
age determination

age determination

Bien, George S., Norris W. Rakestraw and Hans E. Suess, 1965.
Radiocarbon dating of deep water of the Pacific and Indian Oceans.
Bull. Inst. Oceanogr., Monaco, 61(1278):1-16.

age determination

Birkenmajer, Krzysztof, and Behngrell W. Brown, 1970.
Zn-enriched whale bones on raised marine terraces at Hornsund, Spitzbergen.
Anbok nrrox Polarnist. 1969: 44-54.

age determination

Birkenmajer, Krzysztof, and Ingrid U. Olsson, 1970.
Radiocarbon dating of (raised) marine terraces at Hornsund, Spitzbergen, and the problem of land uplift.
Anbok nrrox Polarnist. 1969: 17-43

age dterminaticn

Blackman, Abner, and B.L.K. Somayajulu, 1966.
Pacific Pleistocene cores: faunal analyses and geochronology.
Science, 154(3751):886-889.

age determination

*Blanchard, R.L., M.H. Cheng and H.A. Potratz, 1967.
Uranium and thorium series disequilibria in Recent and fossil marine molluscan shells.
J. geophys. Res., 72(18):4745-4757.

age determination

Blanchard, Richard L., 1965
U 234/U 238 ratios in coastal marine waters and calcium carbonates.
J. Geophys. Res., 70(16):4055-4061.

age determination a

Blow, W.H., 1970.
Deep sea drilling project, leg 2 Foraminifera from selected samples. Initial Reports of the Deep Sea Drilling Project, Glomar Challenger 2: 357-365.

age determination

Boillot, Gilbert, Pierre-Alain Dupeuble, Michel Durand-Delga et Laurent d'Ozouville 1971
Age minimal de l'Atlantique Nord d'après la découverte de calcaire tithonique à Calpionelles dans le Golfe de Gascogne.
C.r. hebd. Séanc. Acad. Sci. Paris (D) 271 (7): 671-674

age determination

*Bonatti, E., K. Bostrum, B. Eyl and E. Rona, 1969.
Geochemistry, mineralogy and absolute ages of a Caribbean sediment core. Trans. Gulf Coast Ass. Geol. Socs, 19 : 506.

age determination

Broecker, Wallace S., 1963
A preliminary evaluation of uranium series inequilibrium as a tool for absolute age measurement on marine carbonates.
J. Geophys. Res., 68(9):2817-2834.

Age determination

Broecker, Wallace, 1962.
Ra226 to U238 ratios in marine shells of known radiocarbon age. (Abstract).
J. Geophys. Res., 67(4):1629-1630.

age determination

Broecker, Wallace S., and David L. Thurber, 1965
Uranium-series dating of corals and oolites from Bahaman and Florida ey li estones.
Science, 149(3679):58-60.

age determinations

Broecker, Wallace L., and David L. Thurber, 1964
Dating of marine terraces by the Th 230/U234 method. (Abstract).
Trans. Amer. Geophys. Union, 45(1):1209

age determination

*Broecker, Wallace S., David L. Thurber, John Goddard, Teh-Lung Ku, R.K. Matthews and Kenneth J. Mesolella, 1968.
Milankovitivh hypothesis supported by precise dating of coral reefs and deep-sea sediments.
Science, 159(3812)297-300.

age determination a

Bukry, David, 1970.
Coccolith age determinations leg 4, deep sea drilling project. Initial Reports of the Deep Sea Drilling Project, Glomar Challenger, 4: 375-381.

age determination a

Bukry, David, 1970.
Coccolith age determinations leg 2, deep sea drilling project. Initial Reports of the Deep Sea Drilling Project, Glomar Challenger, 2: 349-355.

age determinations a

Bukry, David and M.N. Bramlette, 1970.
Coccolith age determinations leg 5, deep sea drilling project. Initial Repts. Deep Sea Drilling Project, Glomar Challenger 5: 487-494.

age determinations a

Bukry, David and M.N. Bramlette, 1970.
Coccolith age determinations leg 3, deep sea drilling project. Initial Reports of the Deep Sea Drilling Project, Glomar Challenger 3: 589-611.

Atlantic, equatorial
Atlantic, South

age determination a

Bukry, David and M.N. Bramlette, 1969.
Coccolith age determinations leg 1, deep sea drilling project. Initial Reports of the Deep Sea Drilling Project, Glomar Challenger 1: 369-387.

age determination

Burling, R. W., and D. M. Garner, 1959.
A section of 14C activities of sea water between 9°S and 66°S in the south-west Pacific Ocean.
N.Z. J. Geol. Geophys., 2(4):799-824.

age determination

Cann, J.R., and F. J. Vine, 1966.
An area on the crest of the Carlsberg Ridge: petrology and magnetic survey.
Phil. Trans. R. Soc., London, A, 259 (1099): 198-217.

bathymetry
basalt
breccias
abstract
magnetism
magnetic anomalies

age determination

Carr, D.R., and L.J. Kulp, 1954.
Dating with natural radioactive carbon.
Trans. N.Y. Acad. Sci., Ser. 2, 16(4):175-181.

age determination

Carr, D.R., and J.L. Kulp, 1953.
Age of a Mid-Atlantic Ridge basalt boulder.
Bull. G.S.A. 64:253-254.

age determination

Chalov, P.I., T.V. Tuzova and Ia. A. Musin, 1964.
The isotopic ratio of U^{234} / U^{238} in natural waters and their use for nuclear geochronology.
Izv., Akad. Nauk, SSSR, Ser. Geofiz., (10):1552-1561.

(In Russian)

age determination

Clarke, A.H., Jr., D.J. Stanley, J.C. Medcof, and R.E. Drinnan, 1965.
Ancient oyster and bay scallop shells from Sable Island.
Nature, Lond., 215(5106): 1146-1145

Age Determination

Cooper, L.H.N., 1965.
Radiolarians as posssible chronometers of continental drift.
Progress in Oceanography, 3:71-82.

abstract

age determination

*Dalrymple, G. Brent and Allan Cox, 1968.
Palaeomagnitism, potassium-argon ages and petrology of some volcanic rocks.
Nature, 217(5126):323-326.

age determination

Degens, Egon T., 1965.
Geochemistry of sediments: a brief survey.
Prentice-Hall, Inc., 342 pp.

age determination

Devries, H.L., and G.W. Barensen, 1954.
Measurement of age by the carbon-14 technique.
Nature 174(4442):1138-1141, 3 textfigs.

age determination

Douglas, Robert G. and Samuel M. Savin, 1971.
Isotopic analyses of planktonic Foraminifera from the Cenozoic of the northwest Pacific, leg 6. Initial Repts. Deep Sea Drilling Project, Glomar Challenger 6: 1123-1127.

age determination

Dymond, Jack R., 1966.
Potassium-argon geochronology of deep-sea sediments.
Science, 152(3726):1239-1241.

age determination
Pacific, northeast
bottom sediments, deep-sea

age determination

Dymond, Jack, and Herbert L. Windom 1968
Cretaceous K-Ar ages from Pacific Ocean seamounts.
Earth Planet. Sci. Letts 4(1): 47-52

age determination

Emery, K.O., and E.E. Bray, 1962.
Radiocarbon dating of California Basin sediments.
Bull. Amer. Assoc. Petr. Geol., 46(10):1839-1856.

age determination

Emiliani, Cesare, 1967.
The Pleistocene record of the Atlantic and Pacific oceanic sediments; correlations with the Alaskan stages by absolute dating; and the age of the last reversal of the geomagnetic field.
Progress in Oceanography, 4:219-224.

age determination

Emiliani, Cesare, 1966.
Paleotemperature analysis of Caribbean cores P6304-8 and P6304-9 and a generalized temperature curve for the past 425,000 years.
J. Geol., 74(2):109-124.

age determination

Emiliani, C., 1957.
Temperature and age analysis of deep-sea cores.
Science 125(3244):383-387.

age determination

Emiliani, Cesare, and Toshiko Mayeda, 1964.
Oxygen isotopic analysis of some molluscan shells from fossil littoral deposits of Pleistocene age.
Amer. J. Sci., 262:107-113.

age determination (oxygen)

Epstein, S., and R.P. Sharp, 1959
Oxygen isotope studies. Trans. Amer. Geo. U., 40(1): 81-84. I.G.Y. Bull., No. 21: 81-84

age determination

Ericson, D.B., 1963.
31. Cross-correlation of deep-sea sediment cores and determination of relative rates of sedimentation by micropaleontological techniques.
In: The Sea, M.N. Hill, Editor, Interscience Publishers, 3:832-842.

age determination

Ericson, David B., Maurice Ewing and Gosta Wollin. 1964
The Pleistocene Epoch in deep-sea sediments. A complete time scale dates the beginning of the first ice age at about 1 1/2 million years ago.
Science, 146(3645):723-732.

age determination

Eriksson, K. Gösta, and Ingrid U. Olson, 1963.
Some problems in connection with C14 dating of tests of Foraminifera.
Bull. Geol. Inst., Univ. Uppsala, 42:1-13.

Also:
Publ., Inst. Quarternary Geol., Univ. Uppsala, Octavo Ser., No. 7.

age determination

Fanale, Fraser P., Oliver A. Schaeffer, Aaron Kaufman, David L. Thurber and Wallace S. Broecker, 1964.
Comparison between He:U and Th 230:U234 ages of fossil shells and corals. (Abstract).
Trans. Amer. Geophys. Union, 45(1):119-120.

age determination

Feral, Alain, Claude Caratini, André Klingebiel et Paul-Charles Levèque 1971
Concordance entre les résultats obtenus à l'aide de la sédimentologie, de la palynologie et de la datation au ^{14}C dans l'étude d'un faciès flandrien de l'estuaire de la Gironde.
C.r. hebd. Séanc. Acad. Sci., Paris (D)272 (9): 1201-1203.

age determination

Feyling-Hanssen, R.W., & I. Olsson, 1959
Five radiocarbon datings of post glacial shorelines in Central Spitzbergen. Norsk Geogr. Tids., 17(1): 122-131.

age determination

Fleischer, R.L., J.R.M. Viertl, P.B. Price and F. Aumento, 1968.
Mid-Atlantic Ridge: age and spreading rates.
Science, 161(3848):1339-1342.

age determination

Folsom, T.R., R. Grismore and D.R. Young 1970.
Long-lived γ-ray emitting nuclide silver-108m found in Pacific marine organisms and used for dating.
Nature, Lond. 227(5261): 941-943

age determination

Funnell, B.M., and A. Gilbert Smith, 1968.
Opening of the Atlantic Ocean.
Nature, Lond., 219(5161):1328-1333.

age determination

Gartner, Stefan, Jr., 1971.
Calcareous nannofossil age determinations Leg 7, Deep Sea Drilling Project, Initial Repts Deep Sea Drill. Proj, 7(2): 1509-1511.

age determinations

Gartner, Stefan, Jr., 1970.
Coccolith age determinations leg 5, deep sea drilling project. Initial Repts. Deep Sea Drilling Project, Glomar Challenger 5: 495-500.

age determinations

Gartner, Stefan, 1970.
Coccolith age determinations leg 3, deep sea drilling project. Initial Reports of the Deep Sea Drilling Project, Glomar Challenger 3: 613-627.

Atlantic, equatorial
Atlantic, south

age determination

*Gennesseaux, M., et Y. Thommeret, 1968.
Datation par le radiocarbone de quelques sédiments sous-marins de la région niçoise.
Revue Géogr. phys. Géol. dynam. (2)10(4):375-382.

age determination

Godwin, H., R.P. Suggate and E.H. Willis, 1958.
Radiocarbon dating of the eustatic rise in sea level. Nature 181(4622):1518-1519.

age determination

Goldberg, Edward D. 1968.
Ionium/thorium geochronologies
Earth Planet. Sci. Letts 4(1): 17-21.

age determination

Goldberg, Edward D., 1965.
An observation on marine sedimentation rates during the Pleistocene.
Limnol. and Oceanogr., Redfield Vol., Suppl. to 10:R125-R128.

age determination

Goldberg, Edward D., and Minoru Koide, 1962.
Geochronological studies of deep-sea sediments by the ionium/thorium method.
Geochimica et Cosmochimica Acta, 26:417-450.

Age determination

Goldberg, Edward D., and Minoru Koide, 1962.
Ionium/thorium goechronology and the depth distribution of members of the U238 series in sediments of the Indian Ocean. (Abstract.)
J. Geophys. Res., 67(4):1638

chronology, ionium-thorium

Goldberg, E.D. and Koide, M., 1958
Ionium-thorium chronology in deep sea sediments of the Pacific. Science, Vol. 128, No. 3330, p. 1003.

age determination

Goldberg, Edward D., Minoru Korde and John J. Griffin, 1964.
Ionium/thorium geochronology on Miami Core A254-BR-C.
In: Recent researchers in the fields of hydrosphere, atmosphere and nuclear geochemistry, Ken Sugawara festival volume, Y. Miyake and T. Koyama, editors, Maruzen Co., Ltd., Tokyo, 117-126.

age determination

Grazzini, Colette Vergnaud, et Yvonne Herman Rosenberg, 1969.
Etude paléoclimatique d'une carotte de Méditerranée orientale.
Revue Géogr. phys. Géol. dynam. (2)11(3): 279-292

Mediterranean, east
age determination
cores

age determination

Harrington, H.J. and I.C. McKellar, 1958
A radiocarbon date for penguin colonization of Cape Hallett, Antarctica. N.Z.J. Geol. Geophys. 1(3): 571-576.

age determination

Hay, W.W., 1969.
Preliminary dating by fossil calcareous nanno-plankton deep sea drilling project, leg 1.
Initial Reports of the Deep Sea Drilling Project, Glomar Challenger 1: 388-391.

Gulf of Mexico

age determination geological

Hecht, Alan D. and Samuel M. Savin, 1971.
Oxygen-18 studies of Recent planktonic Foraminifera. Science, 173(3992): 168-169.

age determination (geol.)

Hoshino, Michihei 1967.
Absolute dating of marine sediments. (In Japanese; English abstract).
Quaternary Res. 6(4): 379-384.
Also in: Coll. Repr. Coll. mar. techn. Tokai Univ. 1967-68, 3.

age determination

Hoshino, Michihei, Masatake Nishihara and Hitoshi Aoki, 1967.
The absolute age of molluscan remains collected from the sea bottom at the mouth of Enoura Bay. (In Japanese)
Chikyukagaku 21(6): 7-8.
Also in: Coll. Repr. Coll. mar. Sci. Techn. Tokai Univ. 1967-68, 3.

age determination

Hubbs, C.L., G.S. Bien, & H.E. Suess, 1962
La Jolla natural radiocarbon measurements II.
Amer. J. Sci., Suppl.; Radiocarbon, 4: 239-249.

age determination

*Hurley, P.M., F.F.M. de Almeida, G.C. Melcher, U.G. Cordani, J.R. Rand, K. Kawashita, P. Vandoros, W.H. Pinson, Jr., and H.W. Fairbairn, 1967.
Test of continental drift by comparison of radiometric ages: a pre-drift reconstruction shows matching geologic age provinces in West Africa and northern Brazil.
Science, 157(3788):495-500.

age determination

Hurley, P.M., B.C. Heezen, W.H. Pinson and H.W. Fairbairn, 1963.
K-Ar age values in pelagic sediments of the North Atlantic.
Geochimica et Cosmochimica Acta, 27(4):393-399.

age determination,

Hurley, Patrick M. and John R. Rand, 1970.
1. General observations. 15. Continental radiometric ages. In: The sea: ideas and observations on progress in the study of the seas, Arthur E. Maxwell, editor, Wiley-Interscience 4(1): 575-596.

age determination

Jones, E.J.W. and J.I. Ewing 1969.
Age of the Bay of Biscay: evidence from seismic profiles and bottom samples.
Science 166 (3901): 102-105

age determination

Kaufman, A., W.S. Broecker, T.-L. Ku, and D.L. Thurber 1971.
The status of U-series methods of mollusk dating.
Geochim. Cosmochim. Acta 35(11):1155-1183.

age determination

Keith, M.L., and G.M. Anderson, 1963
Radiocarbon dating: fictitious results with mollusk shells.
Science, 141(3581):634-637.

age determination, foraminiferal

Kennett, James P., and Richard E. Casey, 1969.
Foraminiferal evidence for a pre-middle Eocene age of the Chatham Rise, New Zealand.
N.Z. Jl mar. Freshwat. Res. 3: 20-28 (6)

Chatham Rise
age determination, foraminiferal
New Zealand

age determination

*Kennett, J.P., N.D. Watkins and P. Vella, 1970.
Paleomagnetic chronology of Pliocene- early Pleistocene climates and the Plio-Pleistocene boundary in New Zealand. Science, 171(3968): 276-279.

age determination
New Zealand
Pleistocene
Plio-Pleistocene boundary

age determination

Kharkar, D.P., D. Lal, N. Narsappaya, & B. Peters, 1958.
Dating of ocean sediments.
C.R. Congr. Assoc. Sci. Pays Ocean Indien, 3, Tananarive, 1957, Sec. C: 39-42.
Abs. in Nucl. Sci. Abs., 15(13): #17134.

age determiniation

Kim, Wan Soo, 1970.
Studies on the Spanish mackerel populations 1. Age determination. (In Korean; English abstract).
J. oceanogr. Soc., Korea, 5(1): 37-40.

Age Determination

Koczy, F.F., 1965.
Remarks on age determination in deep-sea sediments.
Progress in Oceanography, 3:155-171.

abstract

age determination

Koczy, F.F., 1963.
30. Age determination in sediments by natural radioactivity.
In: The Sea, M.N. Hill, Editor, Interscience Publishers, 3:816-831.

age determination

Konishi, Kenji, Seymour O. Schlanger and Akio Omura, 1970.
Neotectonic rates in the central Ryukyu islands derived from ^{230}Th coral ages.
Marine Geol. 9(4): 225-240

age determination
vertical displacement

age determination (geological)

Krishnamoorthy, T.M., V.N. Sastry and T.P. Sarma 1971.
U^{234}/U^{238} as a tool for dating marine sediments.
Current Sci. 40 (11): 279-281

age determination

Kroll, V.S., 1954.
On the age-determination in deep-sea sediments by radium measurements. Deep-Sea Res., 1(4): 211-215.

age determination

Krueger, Harold W., 1964.
K-Ar age of basalt cored in the Mohole Project (Guadalupe site).
J. Geophys. Res., 69(6):1155-1156.

age determination

*Ku, Teh-Lung, 1969.
Uranium series isotopes in sediments from the Red Sea hot-brine area.
In: Hot brines and Recent heavy metal deposits in the Red Sea, E.T. Degens and D.A. Ross, editors, Springer-Verlag, New York, Inc., 512-524.

age determination

*Ku, Teh-Lung, 1968.
Protactinium 231 method of dating coral from Barbados Island.
J. geophys. Res., 73(6):2271-2276.

Age Determination

Ku, Teh-Lung, 1965.
An evaluation of the U234/U235 method as a tool for dating pelagic sediments.
J. Geophys. Res., 70(14):3457-3474.

age determination

Ku, Teh-lung, 1964.
Ratios A_{U234}/A_{U238} of deep-sea sediments. (Abstract).
Trans., Amer. Geophys. Union, 45(1):120.

age determination

Ku, Teh-Lung, and Wallace S. Broecker, 1966.
Atlantic deep-sea stratigraphy: extension of absolute chronology to 320,000 years.
Science, 151(3709):448-450.

age determination

Ku, Teh-lung, Wallace S. Broecker and Neil Opdyke 1968.
Comparison of sedimentation rates measured by paleomagnetic and the ionium methods of age determination.
Earth Planet. Sci. Letts 4(1): 1-16

age determination

*Ku, Teh-Lung, David L. Thurber and Guy G. Mathieu, 1969.
Radiocarbon chronology of Red Sea sediments.
In: Hot brines and Recent heavy metal deposits in the Red Sea, E.T. Degens and D.A. Ross, editors, Springer-Verlag, New York, Inc., 348-359.

age determination

Kulp, J.L., G.I. Bate and B.J. Giletti 1954.
New age determinations by the lead method.
Ann. N.Y. Acad. Sci. 60(3):509-520.

age determination

Kulp, J.L., H.L. Volchok and H.D. Holland, 1952.
Age from metamict minerals. Am. Mineral. 37: 709-718.

age determination

Lal, Devandra, E.D. Goldberg, and Minoru 1960
Cosmic-ray produced silicon-32 in nature.
Silicon-32, discovered in marine sponges, shows promise as a means for dating oceanographic phenomena.
Science, 131(3397):332-337.

age determinations

Lalou, Claude, Jacques Labeyrie et Georgette Delibrias, 1966.
Datation des calcaires coralliens de l'atoll de Mururoa (archipel des Tuamotu) de l'époque actuelle jusqu' à - 500,000 ans.
C.r.hebd.Seanc., Acad.Sci., Paris, (D(263(25): 1946-1949.

age determination

Libby, W.F., 1961
Radiocarbon dating.
Angew. Chem., 73(Suppl. to 7): 225-231.

age determination

Libby, W.F., 1961.
Radiocarbon dating. The method is of increasing use to the archeologist, the geologist, the meteorologist and the oceanographer.
Science, 133(3454):621-629.

age determination
* Maxwell, W.G.H., 1969.
Radiocarbon ages of sediment: Great Barrier Reef.
Sediment. Geol., 3 (4):331-333.

Great Barrier Reef
Age determination

age determination
McDougall, Ian, and F.H. Chamalaun
Isotopic dating and geomagnetic polarity studies on volcanic rocks from Mauritius, Indian Ocean.
Bull. geol. Soc. Am., 80(8):1419-1442

Mauritius
volcanic rock

age determination
Miller, A.J., A.E. Daugherty, F.E. Felin and J. MacGregor, 1955.
Age and length composition of the northern anchovy catch off the coast of California in 1952-53 and 1953-54.
Calif. Dept. Fish and Game, Mar. Fish. Br., Fish. Bull. No. 101:36-66.

age determination
Miller, John A., 1960
Potassium-argon ages of some rocks from the South Atlantic.
Nature, 187(4742): 1019-1020.

*Miyake, Y., and K. Saruhashi, 1967. age of water mass
On the radio-carbon age of the ocean waters.
Pap. Met. Geophys., Tokyo, 17(3):218-223.

age determination
Miyake, Y. and K. Saruhashi 1966.
On the radio-carbon age of the ocean waters.
Pap. Met. Geophys. Tokyo, 17(3): 218-223

age determination
Miyake, Y., and Y. Sugimura, 1961.
Ionium-thorium chronology of deep-sea sediments of the western North Pacific Ocean.
Science, 133(3467):1823-1824.

age determination
Miyake, Yasuo, and Yukio Sugimura, 1961.
Ionium-thorium chronology of deep-sea sediments of the western North Pacific Ocean. (Abstract).
Tenth Pacific Sci. Congr., Honolulu, 21 Aug.-6 Sept., 1961, Abstracts of Symposium Papers, 379

age determination
*Miyake, Yasuo, Yukio Sugimura and Eiji Matsumoto, 1968.
Ionium-thorium chronology of the Japan Sea cores.
Rec. oceanogr. Wks., Japan, n.s., 9(2):189-195.

age determination
*Naidu, A.S., 1969.
Radiocarbon date of an oolitic sand collected from the shelf off the East Coast of India.
Bull. natn. Inst. Sci., India, 38(1): 467-471.

India
bottom sediments
age determination

age determination
Naughton, John J., and I. Lynus Barnes, 1965.
Geochemical studies of Hawaiian rocks related to the study of the upper mantle.
Pacific Science, 19(3):287-290

age determination
*Neev, David, and K.O. Emery, 1967.
The Dead Sea: depositional processes and environments of evaporites.
Bull. Geol. Surv., Israel, 41: 146 pp.

age determination
Neprochnov, Iu. P., 1961
[Date on the thickness of sediments in the depression of the Arabian Sea.]
Doklady Akad. Nauk, 139(1):177-179.

age determination
Nesteroff, Wladimir D., 1959
Age des derniers mouvements du graben de la mer Rouge déterminé par la méthode du C14 appliquée aux récifs fossiles.
Bull. Soc. Geol., France (7)1:415-418.

age determination
*Noble, C.S., and J.J. Naughton, 1968.
Deep-ocean basalts: inert gas content and uncertainties in age dating.
Science, 162(3850):265-267.

Age Determination
Olausson, Eric, 1965.
Evidence of climatic changes in North Atlantic deep-sea cores with remarks on isotopic paleotemperature analysis.
Progress in Oceanography, 3:221-252.

bo abstract

age determination
Olson, Boyd E., 1967.
Chronology of sea water.
U.S. Naval Oceanogr. Off., Informal Rep. IR 67-56:44 pp. (multilithed).

Age Determination
Olsson, Ingrid U., and K. Gosta Eriksson, 1965.
Remarks on C14 dating of shell material in sea sediments.
Progress in Oceanography, 3:253-266.

abstract

age determination
Osmond, J.K., J.R. Carpenter and H.L. Windom, 1965.
Th^{230}/U^{234} Age of the Pleistocene corals and oolites of Florida.
J. Geophys. Res., 70(8):1843-1847.

age determinations
Osmond, J.K., J.R Carpenter and H.L. Windom, 1964
Th230/U234 age of Pleistocene Key Largo corals (Abstract).
Trans. Amer. Geophys. Union, 45(1):120.

age determination
Oversby, V.M., and P.W. Gast, 1968.
Oceanic basalt leads and the age of the earth.
Science, 162(3856):925-927.

age determination
*Ozima, Minoru, Mituko Ozima and Ichiro Kanoeka, 1968.
Potassium-argon ages and magnetic properties of some dredged basalts and their geophysical implications.
J. geophys. Res., 73(2):711-723.

age determination
Pearn, W.C., E.E. Angino, D. Stewart, 1963.
New isotopic age measurements from the McMurdo Sound area, Antarctica.
Nature, 199(4894):685.

age determinations
Piggot, C.S., and W.D. Urry, 1942.
Time relations in ocean sediments. Bull. G.S.A. 53:1187-1210.

age determination
Rafter, T.A., 1963.
Recent developments in the interpretation and the reporting of carbon-14 activity measurements from New Zealand. Radioactive tracers in oceanography. Symposium of the 10th Pacific Science Congres, 1961.
Union Géodes. Géophys. Int. Monogr., 20:33-41.

age determination
Richards, J.R., and W.F. Willmott 1970.
K-Ar age of biotites from Torres Strait
Austr. J. Sci. 32(9):369-370

age determination (geological)
Romankevich, E.A., P.L. Beznukov, V.I. Baranov and L.A. Khristianova 1966.
Stratigraphy and absolute age of deep-sea sediments in the western Pacific. (In Russian, English abstract).
Rez. Issled. Mezhd. Geofiz. Proekt. Okeanol. Mezhd. Geofiz. Komitet, Presid. Nauk, SSSR 14:165 pp.

age determination
*Rona, E., and C. Emiliani, 1969.
Absolute dating of Caribbean cores P6304-8 and P6304-9.
Science, 163(3862):66-68.

Age Determination
Rona, Elizabeth, L.K. Akers, John E. Noakes and Irwin Supernaw, 1965.
Geochronology in the Gulf of Mexico, Part 1.
Progress in Oceanography, 3:289-295.

abstract

age determination
Rosholt, J.N., C. Emiliani, J. Geiss, F.F. Koczy and P.J. Wangersky, 1961.
Absolute dating of deep-sea cores by the Pa 231/Th 230 method.
J. Geology, 69(2):162-185.

age determination
*Ross, David A., Egon T. Degens and Joseph MacIlvaine, 1970.
Black Sea: recent sedimentary history.
Science, 170(3954):163-165.

age determination
Rusnak, Gene A., 1963.
"Absolute dating of deep-sea cores by the Pa 231/Th 230 method" and accumulation rates: a discussion.
J. Geology, 71(6):809-810.

Rosholt, J.N., C. Emiliani, J. Geiss, F.F. Koczy, and P.J. Wangersky, 1963.
"Absolute dating of deep-sea cores by the P231/Th 230 method" and accumulation rates: a reply.
J. Geology, 71(6):810.

age determination
Rusnak, Gene A., and W.D. Nesteroff, 1964.
Modern turbidites: terrigenous abyssal plain versus bioclastic basin.
In: Papers in Marine Geology, R.L. Miller, Editor, Macmillan Co., N. Y., 488-507.

age determination
Schott, W., 1962.
Zweck und Ziel der geologischen Arbeiten auf der Karibischen Tiefsee-Expedition "Carib" 1960.
Zeits., Deutsch. Geol. Gesells., 114(2):427-429.

age determination
*Stearns, Charles E., and David L. Thurber, 1967.
Th^{230}/U^{234} dates of Late Pleistocene marine fossils from the Mediterranean and Moroccan littorals.
Progress in Oceanography, 4:293-305.

age determination
Sugimura, Y., 1969.
Geochronological studies of deep sea sediment by means of natural radioactive nuclides. (In Japanese)
J. oceanogr. Soc. Japan 25(5): 261-268.

age determination
Thommeret, Jean et Yolande Thommeret
Validité de la datation des sédiments du proche Quaternaire par le dosage du carbone 14 dans les coquilles marines
Rapp. P.-v. Réun., Comm. int. Explor. scient. Mer Méditerr., 18(3):837-844.

age determination
Thurber, David L., 1965.
The dating of molluscs from raised marine terraces.
Narragansett Mar. Lab., Univ. Rhode Island, Occ. Publ., No. 3:1-27.

age determination
Thurber, David L., Wallace S. Broecker, Richard L. Blanchard and Herbert A. Potratz, 1965
Uranium-series ages of Pacific atoll coral.
Science, 149(3679):55-58.

age determination
Thurber, David, Wallace S. Broecker and Aaron Kaufman, 1964.
U 234-Th230 method of age determination and its application to marine carbonates. (Abstract).
Geol. Soc., Amer., Special Paper, No. 76:166.

age determination
Turekian, Karl K., 1965.
Some aspects of the geochemistry of marine sediments.
In: Chemical oceanography, J.P. Riley and G. Skirrow, editors, Academic Press, 2:81-126.

age determination
Turekian, Karl. K., and Minze Stuiver, 1964.
Clay- and carbonate-accumulation rates in three South Atlantic, deep-sea cores.
Science, 146(3640):55-56.

age determination
Ulrych, T.J., 1968.
Oceanic basalt leads and the age of the earth.
Science, 162(3856):928.

age determination
Valentine, James W., and H. Herbert Veeh, 1969.
Radiometric ages of Pleistocene terraces from San Nicolas Island, California.
Bull. geol. Soc. Am., 80(7): 1415-1418.

age determination, paleomagnetic
Van Montfrans, Hendrik Martin, 1971
Palaeomagnetic dating in the North Sea Basin.
Prurico N.V. | Rotterdam, 113 pp.

age determination
VanStraaten, L.M.J.U., 1954.
Radiocarbon datings and changes of sea level at Velzan (Netherlands). Symposium: Quaternary changes in level, especially in the Netherlands.
Geol. en Mijnbouw, n.s., 16 Jaargang:247-253.

age determination
Veeh, H. Herbert, 1966.
Th 230/U238 and U234/U238 ages of Pleistocene high sea level stand.
J. geophys. Res., 71(14):3379-3386.

age determination
Vinogradov, A.P., A.L. Devirts, E.I. Dobkina, and N.G. Markova, 1963.
Age determination by the radiocarbon method. IV. (In Russian).
Geokhimia, (9):795-812.

translations:
Geochemistry, (9):823-840.

age determination
Volchok, H.L., and J.L. Kulp, 1957.
The ionium method of age determination.
Geochimica et Cosmochimica Acta, 11(4):219-246.

age determinations
Wiseman, J.D.H., 1954.
The determination and significance of past temperature changes in the upper layers of the equatorial Atlantic Ocean. Proc. R. Soc., London 222(1150):296-323, 3 textfigs.

age determination
Wolf, Karl H., Göte Östland, 1967.
14C dates of calcareous samples, Heron Island, Great Barrier Reef.
Sedimentology 8(3):249-251.

age determination
Yalkovsky, Ralph, 1964
Time series analysis of Caribbean core A172-6. (Abstract).
Trans. Amer. Geophys. Union, 45(1): 71.

age of water masses

age of water masses
Brodie, J.W., and R.W. Burling, 1958.
Age determinations of Southern Ocean waters.
Nature 181(4602):107-108.

age (antiquity) of oceans

aggregates, organic

aggregates, organic
Barber, Richard T., 1966.
Organic aggregates formed from dissolved organic material by bubbling. (Abstract only).
Second Int. Oceanogr. Congr., 30 May-9 June, 1966, Abstracts, Moscow; 19.

aggregates, flocculent
Kajihara, Masahiro 1971.
Settling velocity and porosity of large suspended particle.
J. oceanogr. Soc. Japan, 27(4): 158-162.

aggregates, organic
Riley, Gordon A., Denise Van Hamert and Peter J. Wangersky, 1965.
Organic aggregates in surface and deep waters of the Sargasso Sea.
Limnol. Oceanogr., 10(3):354-363.

aggregations
Brun, Einar, 1969.
Aggregation of Ophiothrix fragilis (Abildgaard) (Echinodermata: Ophiuroidea).
Nytt Mag. Zool. 17: 153-160

aggregates
Johannes, R.E., 1967.
Ecology of organic aggregates in the vicinity of a coral reef.
Limnol. Oceanogr., 12(2):189-195.

aggregates (organic)
Kane, Jasmine E., 1967.
Organic aggregates in surface waters of the Ligurian Sea.
Limnol. Oceanogr., 12(2):287-294.

aggregates
*Marshall, N., 1968.
Observations on organic aggregates in the vicinity of coral reefs.
Marine Biol., 2(1):50-53.

aids to navigation

aids to navigation
Aslakson, C.I., 1947
Shoran investigations for triangulation.
Int. Hydr. Rev. 24:178-180, 1 text fig.

aids to navigation
Burmister, C.A., 1947
Shoran in hydrographic surveying. Int. Hydr. Rev. 24:181-185, 3 figs.

aids to navigation
Hogben, H.E. and R.F. Hansford, 1947
Ship-borne radar; factors in the design of Navigational radar for marine transport.
Int. Hydr. Rev. 24:137-155, 20 figs.

air masses
*Kool, L.V., 1971.
Thermodynamic properties of the air masses in the tropical Atlantic. (In Russian; English abstract). Okeanologiia 11(4): 606-614.

airplanes

aircraft performance
Amstutz, D.E., and S. Neshyba, 1964.
Location of an aircraft impact from gravity waves.
Science, 145(3635):921-923.

airplanes
Itskevich, L.I., 1963.
A flying oceanographic laboratory. (In Russian).
Okeanologiia, Akad. Nauk, SSSR, 3(5):944.

U.S.A. from U.S. Naval Inst. Proc., No. 4, 1963.

air-sea boundary

A

air-sea interface
Abramov, R.V., 1966.
On possible climatological frequencies of an oscillating system ocean-atmosphere. (In Russian).
Fisika Atmosferi i Okeana, Izv., Akad. Nauk, SSSR 2(6):664-667.

air-sea interface
Akiyama, Tsutomu 1969.
Carbon dioxide in the atmosphere and in sea water in the adjacent seas of Japan.
Oceanogr. Mag., 21(1): 53-59

air-sea interface
*Aldaz, Luis 1969.
Flux measurements of atmospheric ozone over land and water. J. geophys. Res., 74(28): 6943-6946.

air-sea interface
Aliverti, G., 1952.
L'atmosfera e gli oceani, analogi e interdipendenze. An. Ist. Univ. Nav., Napoli, 16-19:135-150.

air-sea interface
Allkofer, O.C., R.D. Andresen und W.D. Dau, 1970.
Der Einfluss des Erdmagnetfeldes auf die kosmische Strahlung. II Untersuchungen der Myonenkomponente der kosmischen Strahlung während der Atlantischen Expedition IQSY 1965 auf dem Forschungsschiff Meteor.
Meteor Forsch.-Ergebnisse (B) 5:1-22
Atlantischen Exped. IQSY 1965

air-sea interface
Anderson, Robert, Sveingjorn Bjornsson, Duncan C. Blanchard, Stuart Gathman, James Hughes Sigurgeir Jónasson, Charles B. Moore, Henry J. Survilas and Bernard Vonnegut, 1965.
Electricity in volcanic clouds: investigations show that lightening can result from charge-separation processes in a volcanic crater.
Science, 148(3674):1179-1189.

air-sea interface
*Andreev, E.G., V.S. Lavorko, A.A. Pivovarov, and G.G. Khundzhua, 1969.
On vertical temperature profile at the sea-atmosphere interface. (In Russian; English abstract). Okeanologiia, 9(2): 348-352.

air-sea interface
Angot, Michel, 1965.
Cycle annuel de l'hydrologie dans la région proche de Nosy-Be.
Cahiers, O.R.S.T.R.O.M. - Océanogr., 3(1):55-66.

generalities, p. 66.

air-sea interface
Anisimova, E.P., A.S. Orlov, and A.A. Speranskaya, 1971.
A Kinematics of the near-water air layer as generated by waves on the surface of a liquid. (In Russian). Fisika Atmos. Okean. Izv. Akad. Nauk SSSR 7(11): 1221-1223.

air-sea interface
Anon., 1965.
Interaction between sea and air.
Travelers Research Center, Inc., 1965, 13-14.

air-sea interface

Ashburn, Edward V., 1963
The radiative heat budget at the ocean-atmosphere interface.
Deep-Sea Res., 10(5):597-606.

air-sea interface

*Azernikova,O.A.,1967.
On the interrelation between the changes of some behaviour elements of the White and Barents seas. (In Russian; English abstract)
Okeanologiia,Akad. SSSR,7(4):607-616.

B

air-sea interface

Babenkov, E.F., 1965.
On the role of advection in heat interaction of the sea and atmosphere. (In Russian).
Trudy, Gosudarst. Oceanogr. Inst., No. 84:132-170.

air-sea interface

Baczyk,J.,1966.
Liasons relatives entre la circulation atmosphérique, la différentiation des masses d'eaux et leurs conditions dynamiques dans la partie méridionale de la Baltique.(Abstract only).
Second Int. Oceanogr. Congr.,30 May-9 June 1966. Abstracts,Moscow:15-17.

air-sea interface

Bagge, E., and S. Skorka, 1958.
Der Übergangseffekt der Ultrastrahlungsneutronen an der Grenzfläche Luft-Wasser.
Zeits. f. Physik, 152(1):34-48.

air-sea interface

*Bang, N.D., 1971.
The southern Benguela Current region in February 1966: Part II. Bathythermography and air-sea interactions. Deep-Sea Res., 209-224.

Benguela Current
air-sea interface
divergence
upwelling
Walvis Bay
Africa, southwest

air-sea interface

Banke, Erik G. and Stuart D. Smith, 1971.
Wind stress over ice and over water in the Beaufort Sea. J. geophys. Res. 76(30): 7368-7374.

air-sea interface

Basharinov, A.E., and L.M. Mitnik, 1970.
Characteristics of the moisture field over oceans from data of radiometric microwave measurements made from the Cosmos-243 meteorological satellite. (In Russian; English abstract).
Met. Gidrol. (12): 13-18.

air-sea interface

Bashkirov, G.S., 1966.
The limiting wind current velocity in the surface layer and phenomena in the contact layer. (In Russian: English abstract).
Okeanologiia, Akad. Nauk, SSSR, 6(2):217-223.

Air-sea interface

Beliaev, L. I., 1959.
[Molecular migration of sodium chloride from sea water.]
Trudy, Morskoi Gidrofiz. Inst., Akad. Nauk, SSSR. 15:106-

air-sea interface

*Belyaev,V.I. and A.I. Zhilina,1970.
On relations between statistical characteristics of sea temperature and components of heat balance on its surface. (In Russian;English abstract).
Fisika Atmosfer. Okean., Izv.Akad Nauk,SSSR, 6(8):804-809.

air-sea interface

*Benilov,A. Yu. and B.N. Filjushkin,1970.
An application of the linear filtration technique for the analysis of fluctuations in the sea surface layer. (In Russian;English abstract).
Fisika Atmosfer. Okean., Izv.Akad Nauk,SSSR, 6(8):810-819.

air-sea-interface

Berenbeim, D. Ia., 1957
Influence of atmospheric processes on the thermal regime of the active layer of the Black Sea. Meteorol. i Gidrol. (8): 45.

air-sea interface

Bjerknes, J., 1969.
Atmospheric teleconnections from the equatorial Pacific.
Mon. weath. Rev., 97(3): 163-172

Pacific, equatorial
air-sea interface
temperature (surface) effect of

Air-Sea Interface

Bjerknes, J., 1962.
Synoptic survey of the interaction of sea and atmosphere in the North Atlantic.
Geofys. Publik., Geophysica Norvegica, 24:115-147.

air-sea interface

Bjerknes, Jacob, Lewis J. Allison, Earl R. Kreins, Frederic A. Godshall and Günter Warnecke, 1969.
Satellite mapping of the Pacific tropical cloudiness.
Bull. Am. met. Soc., 50(5): 313-322.

air-sea interface

Blanchard, Duncan C., 1967.
From raindrops to volcanoes: adventures with sea surface meteorology.
Doubleday, 180 pp. ($1.25 paper back $4.95 hard cover)

air-sea interface

Bogorodsky, M.M., 1964.
A study of tangential friction, vertical turbulent heat exchange and evaporation in the open ocean. (In Russian).
Okeanologiia, Akad. Nauk, SSSR, 4(1):19-26.

air-sea exchange

Bolin, B., 1960
On the exchange of carbon dioxide between the atmosphere and the sea. Tellus, 12(3): 274-281.

air-sea interface

Bortkovsky, R.S., 1971.
Calculation of turbulent fluxes of heat, moisture and momentum over the sea from data of ship measurements. (In Russian)
Met. Gidrol. 1971 (3): 93-98.

air-sea interface

Bortkovsky, R.S., and E.K. Buetner, 1970.
A theoretical model of the air-sea turbulent heat exchange: comparison with experiment. (In Russian; English abstract).
Fizika Atmosfer. Okean. Izv. Akad. Nauk, SSSR, 6(1): 37-44.

air-sea interface

Bortkovsky, R.S., and E.K. Buetner 1969.
Calculation of the heat-exchange coefficient above the sea. (In Russian; English abstract).
Fizika Atmosfer. Okean., Izv. Akad. Nauk, SSSR, 5(5): 494-503

air-sea interface

Bossolasco, Mario, Ignazio Dagnino e Giuseppe Flocchini, 1969.
Il contrasto mare-terra nella genesi dei temporali.
Geofisica Met. 18(3/4): 45-50

air-sea interface

Bøyum, Gunnvald, 1966.
The energy exchange between sea and atmosphere at Ocean Weather Stations M, I and A.
Geofys. Publr, 26(7):19 pp.

heat (sensible) exchange
evaporation
air-sea interface
energy exchange
Atlantic, north

air-sea interface

Bøyum, Gunnvald, 1962
A study of evaporation and heat exchange between the sea surface and the atmosphere.
Geofys. Publikasjoner, Oslo, 22(7):15 pp.

air-sea interface

Brocks, K., 1963.
Probleme der maritimen Grenzschicht der Atmosphäre.
Berichte Deutsch. Wetterdienstes, 12(91):34-46.

air-sea interface

Broecker, Wallace S., 1965.
An application of natural radon to problems in ocean circulation.
Symposium, Diffusion in oceans and fresh waters, Lamong Geol. Obs., 31 Aug.-2 Sept., 1964, 116-145.

air-sea interface

Broecker, W., 1963
4. Radioisotopes and large scale oceanic mixing. In: The Sea, M.N. Hill, Edit., Vol.2. The composition of sea water, Interscience Publishers, New York and London, 88-108.

air-sea interface

Brujewicz, S.W. and V.D. Korzh, 1971.
On boron interchange between the sea and the atmosphere. (In Russian; English abstract).
Okeanologiia 11(3): 414-422.

air-sea interface

Brujewicz, S.W., and V.D. Korzh, 1969.
Salt exchange between the ocean and the atmosphere. Okeanologiia, 9(4): 571-583.
(In Russian; English abstract)

air-sea interface

Bruevich, S.V., and E. Z. Kulik, 1967.
The chemical interaction between the ocean and the atmosphere (salt exchange) (In Russian; English abstract)
Okeanologiia, Akad. Nauk, SSSR, 7(3):363-379.

aerosols
salt exchange

air-sea interface (chemistry)

Bruevich, S.V., and E.Z. Kulik, 1967.
The change in the basic salt composition of oceanic water accompanying its transition into the atmosphere. (In Russian).
Dokl. Akad. Nauk, SSSR, 175(3): 697-699.

air-sea interface (chemistry)

air-sea interface

Büchen, Matthias 1971.
Ergebnisse der CO_2-Konzentrationsmessung in der ozeannahen Luftschicht und im Oberflächenwasser während der Atlantischen Expedition 1969.
Meteor-Forschungsergebnisse (B) 7: 55-70

air-sea interface

Budyko, M.I., and L.S. Gandin,1966.
About the determination of the turbulent ocean-atmosphere heat exchange. (In Russian).
Meteorologiya Gidrol., (11):15-25.

air-sea interface
*Byshev, V.I., 1967.
On the time variations of some parameters of the near-water layer of the atmosphere. (In Russian; English abstract).
Okeanologiia, Akad. Nauk, SSSR, 7(6):957-972.

air-sea interface
Byshev, V.I. and Yu. A. Ivanov, 1969.
The time spectra of some characteristics of the near-water layer of the atmosphere. (In Russian; English abstract).
Fizika Atmosfer. Okean. Izv. Akad. Nauk SSSR 5(1): 17-28.

air-sea interface
Byshev, V.I, and O. A. Kuznetsov, 1969.
The structure characteristics of atmospherical turbulence in near-water layer over the open ocean. (In Russian; English abstract).
Fizika Atmosfer. Okean., Izv. Akad. Nauk SSSR 5(6):575-585

C

air-sea interface
Chang, P.C., E.J. Plate and G.M. Hidy, 1971.
Turbulent air flow over the dominant component of wind-generated water waves.
J. fluid Mech. 47(1):183-208.

air-sea interface
Charnell, Robert L., 1967.
Long-wave radiation near the Hawaiian Islands.
J. geophys Res., 72(2):489-495.

air-sea interface
Charnock, H., 1964.
Energy transfer by the atmosphere and the Southern Ocean.
Proc. R. Soc., London, (A), 281(1384):6-14.

air-sea interface
Charnock, H., 1962.
Atmosphere and ocean.
Proc. R. Soc., London, (A) 265 (1322):331-334.

air-sea interface
Clauss, E., H. Hinzpeter und J. Müller-Glewe, 1970.
Messungen zur Temperaturstruktur im Wasser an der Grenzfläche Ozean-Atmosphäre.
Meteor Forsch-Ergebnisse (B) 5: 90-94.

air-sea interface
Clauss, E., H. Hinzpeter und J. Müller-Glewe, 1970.
Ergebnisse von Messungen des Temperaturfeldes der Atmosphäre nahe der Grenzfläche Ozean-Atmosphäre.
Meteor Forsch.-Ergebnisse (B) 5:85-89

air-sea interface
Coantic, Michel, 1969.
Les interactions atmosphère-océans: les processus physiques et les équations qui les governent. 3.
Cah. océanogr. 21(3):223-249.

air-sea interface

air-sea interface
Coantic, Michel, 1969.
Les interactions atmosphère-océans: les processus physiques et les équations qui les governent. 2.
Cah. oceanogr., 21(2): 105-143.

air-sea-interface
*Coantic, Michel, 1969.
Les interactions atmosphère -océans: les processus physiques et les équations qui les governent.
Cah. océanogr., 21(1):17-46.

air-sea interface
Colacino, Michele, et Enrico Rossi, 1971.
Exemples de calcul du bilan thermique dans la mer Tyrrhénienne (mai 1969)
Cah. océanogr. 23(3):267-281

air-sea interface
heat balance
Mediterranean, west

air-sea interface
Colón, José A., 1963
Seasonal variation in heat flux from the sea surface to the atmosphere over the Caribbean Sea.
J. Geophys. Res., 68(5):1421-1430.

air-sea interface
Copeland, B.J., and W.R. Duffer, 1964
Use of a clear plastic dome to measure gaseous diffusion rates in natural waters.
Limnology and Oceanography, 9(4):494-499.

air sea interface
Craig, H., 1957.
The natural distribution of radiocarbon and the exchange time of carbon dioxide between atmosphere and sea. Tellus 9(1):1-17.

air-sea interface
Craig, H., and L.I. Gordon, 1965.
Isotopic oceanography: deuterium and oxygen 18 variations in the ocean and the marine atmosphere.
Narragansett Mar. Lab., Univ. Rhode Island, Occ. Publ., No. 3:277-374.

D

air-sea interface
Deacon, E.L., P.A. Sheppard and E.K. Webb, 1957.
Wind profiles over the sea and the drag at the sea surface. Australian J. Phys., 9:511-541.
Phys. Abstr., 60:532.

air-sea interface
Deacon, E.L., and E.K. Webb, 1962
Small-scale interactions. Ch. 3, Sect. II. Interchange of properties between sea and air
In: The Sea, Interscience Publishers, Vol. 1, Physical Oceanography, pp. 43-87.
(Mss received July 1960)

air-sea interface
Deardorff, James W., 1968.
Dependence of air-sea transfer coefficients on bulk stability.
J. geophys. Res., 73(8):2549-2557.

air-sea interface
Demin, V.V. 1971.
On the influence of the ocean surface radiation variations on the accuracy of determination of the total water vapour content in the atmosphere. (In Russian).
Fizika Atmosfer. Okean. Izv. Akad. Nauk SSSR 7(7):811-813.

air-sea interface
Desai, B.N., 1968.
Interaction of the summer monsoon current with water surface over the Arabian Sea.
Indian J. Met. Geophys., 19(2):159-166

air-ocean interface
Donn, William L., and William T. McGuinness, 1959.
Some results of the IGY Island Observatory in the Atlantic.
Trans. A.G.U., 40(2):167-170.

air-sea interface
Duvanin, V.P., 1968.
A model for the interaction between macroprocesses in the ocean and in the atmosphere. (In Russian; English abstract).
Okeanologiia, Akad. Nauk, SSSR, 8(4):571-580.

E

air-sea interface
Efimov, V.V., 1970.
On the structure of the wind velocity field in the atmospheric near-water layer and the transfer of wind energy to sea waves. (In Russian; English abstract).
Fizika Atmosfer. Okean. Izv. Akad. Nauk SSSR 6(10):1043-1058.

air-sea interface.
Efimov, V.V., and V.L. Pososhkov, 1970.
Wave disturbance dynamics of the lower atmosphere by swell waves. (In Russian; English abstract).
Fizika Atmosfer. Okean., Izv. Akad. Nauk SSSR, 6(6):617-626.

air-sea exchange
Eriksson, Erik, 1960.
The yearly circulation of chloride and sulfur in nature; meteorological, geochemical and pedological implications. II.
Tellus, 12(1):63-109.

Air-Sea boundary
Ewing, G., and E.D. McAlister, 1960
On the thermal boundary layer of the ocean.
Science, 131(3410): 1374-1376.

F

air-sea interface
Fairhall, A.W., R.W. Buddemeier, I.C. Yang and A.W. Young, 1969.
Radiocarbon from nuclear testing and air-sea exchange of CO_2.
Antarctic J., U.S., 4(5): 184-185

air-sea interface
Fleagle, R.G., J.W. Deardorff and F.I. Badgley, 1958
Vertical distribution of wind speed, temperature and humidity above a water surface. J. Mar. Res., 17:131-157.

air-sea interface
Francis, J.R.D., 1954.
Laboratory models of sea surface phenomena.
Weather 9(6):163-168, 2 textfigs.

air-sea interface
Fukuoka, Jiro, 1965 (1967).
Condiciones meteorologicas e hidrograficas de los mares adyacentes a Venezuela 1962-1963.
Memoria Soc. Cienc. nat. La Salle, 25(70/71/72):11-38.

air-sea interface
Fukuoka, Jiro, 1966.
Analisis de las condiciones hidrograficas del Mar Caribe (VII) De la relacion con las condiciones hidrograficas cerca de des embocadura del Orinoco.
Memoria, Soc. Ciencias Nat., La Salle, Venezuela, 24 (69):277-307.

G

air-sea interface
Gall, Robert L., and Donald R. Johnson 1971.
The generation of available potential energy by sensible heating: a case study.
Tellus 23(6):465-482

air-sea interface
Gasanov, M.V., 1960
On the reduction of sulphates in mixtures of sea and layer water in the atmosphere of molecular hydrogen with the participation of sulphate reducing bacteria. Microbiologie, Akad. Nauk, SSSR 29(3): 419-421. (In Russian with English abstract).

air-sea interface
Gat, Joel R., 1968.
The isotopic composition of atmospheric waters in the Mediterranean Sea area and their interpretation in terms of air-sea interactions.
Rapp. P.-v. Reun. Commn Explor. scient. Mer Médit. 19(5): 923-926

air-sea interface
Gates, W.L., 1966.
On the dynamical formulation of the large-scale momentum exchange between atmosphere and ocean.
J. Mar. Res., 24(2):105-112.

air-sea interface
*Gathman, Stuart, and Eva Mae Trent, 1968.
Space charge over the open ocean.
J. atmos. Sci., 25(6):1075-1079.

air-sea interface
Gavrilin, B.L., and A.S. Monin, 1967.
A model of long term interactions between the ocean and atmosphere. (In Russian).
Dokl. Akad. Nauk. SSSR, 176(4):822-825.

air-sea interface
Godbole, Ramesh V., 1963.
A preliminary investigation of radiative heat exchange between the ocean and the atmosphere.
J. Appl. Meteorol., 2(5):674-681.

air-sea interface
Gonella, J., 1968.
Le courant de dérive, exemple des échanges atmosphère-océan. Sci. Prog La Nature, (3397): 178-184.

air-sea interface
Gonella, J., 1965.
Contribution à l'étude des échanges entre l'air et la mer en Méditerranée occidentale (Bassin Provençal) (abstract only)
Rapp. P.-v. Réun. Comm. int. Explor. scient. Mer Médit. 18(3): 795.

air-sea interface
Gonella, Joseph, 1964.
Contribution à l'étude des échanges entre l'air et la mer en Méditerranée occidentale (Bassin Provençal).
Cahiers Océanogr., C.C.O.E.C., 16(7):547-556.

air-sea interface
Goreleva, M.V., 1965.
Symposium on air-sea interaction. (In Russian).
Fisika Atmosferi i Okeana, 1(4):465-466.

air-sea interface
Gusev, A.M., 1971.
The state of studies in the interaction between the ocean and the atmosphere. (In Russian; English abstract). Okeanologiia 11(5):802-810.

H

air-sea interface
Hachey, H.B., 1961.
Oceanography and Canadian Atlantic waters.
Fish. Res. Bd., Canada, Bull., No. 134:120 pp.

air-sea inter face
Hankimo, Juhani, 1964.
Some computations of the energy exchange between the sea and the atmosphere in the Baltic area. Ilmatieteellisen Keskuslaitoksen Toimituksia (Finnish Meteorol. Off. Contrib.), No. 57:26 pp.

air-sea interface
Hankimo, Juhani, 1964
Some computations of the energy exchange between the sea and the atmosphere in the Baltic area.
Ilmatiet. Keskuslait. Toimituk. (Finnish Meteorol. Off., Contrib.), No. 57: 26 pp.

air-ocean interface
Hanzawa, M., 1958.
Studies on the inter-relationship between the sea and the atmosphere (Pt. 4)
Ocean. Mag., 10(2):215-226.

air-ocean interface
Hanzawa, M., 1958.
Studies on the inter-relationship between the sea and the atmosphere (part 3)
The Ocean. Magazine, Vol. 10: No. 1 pps. 91-96
Japan Meteor. Agency
Tokyo

air-sea interface
Hanzawa, M., 1957
Studies on the inter-relationship between the sea and the atmosphere. (Part 2).
Ocean. Mag., Tokyo, 9(1):87-93.

air-sea interface
Hanzawa, M., 1956.
Preliminary report on a close relationship between surface water temperatures at the two separated ocean stations, Papa and Tango. Studies on the interrelationship between sea and atmosphere. Oceanogr. Mag., Tokyo, 8(2):157-160.

air sea interface
Hanya, Takashisha and Ryoshi Ishiwatari, 1961
Carbon dioxide exchange between air and sea water. 1.
J. Oceanogr. Soc., Japan, 17(1):28-32.

air-sea interface
Hatzikakidis, Athan. D., 1963.
Relations between air and sea temperatures. (In Greek).
Praktika, Hellenic Hydrobiol. Inst., Athens, 8(2): 109 pp.

air-sea interface
Hicks, B.L., N. Knable, J.J. Kovaly, G.S. Newell, J.P. Ruina and C.W. Sherwin, 1960
The spectrum of X-band radiation backscattered from the sea surface. J. Geophys. Res. 65(3) 825-838.

air-sea interface
Hildreth, William W., 1962
Interaction of the ocean and atmosphere.
First Interindustrial Oceanogr. Symposium, Lockheed Aircraft Corp., 6-12.

Air-Sea Interface
Hildreth, W. W., Jr., and R. M. Lesser, 1962.
Microscale ocean-atmosphere interaction as related to the mesoscale environment of the Santa Barbara Channel area. (Abstract).
J. Geophys. Res., 67(4):1640.

air-sea interface
Hinzpeter, Hans, 1968.
Tagesperiodische Änderungen des Oberflächennahen Temperaturfeldes über dem Meer als Folge von Strahlungsquellen und -senken.
Kieler Meeresforsch. 24(1): 1-13

air-sea interface
Hoeber, Heinrich 1970.
Measurements in the atmospheric surface layer above the tropical Atlantic.
Weather 25(9): 419-424.

air-sea interface
*Hoeber, Heinrich, 1968.
Einige Ergebnisse von Windprofilmessungen über See bei stabiler Schichtung.
Tellus, 20(3):495-500.

age determination
Holmes, Charles W., J.K. Osmond and H.G. Goodell, 1968.
The geochronology of foraminiferal ooze deposits in the Southern Ocean.
Earth Planet. Sci. Letters, 5(4): 368-374

air-sea interface
*Hoover, Thomas E., and David C. Berkshire, 1969.
Effects of hydration on carbon dioxide exchange across an air-water interface.
J. geophys. Res., 74(2):456-464.

I

air-sea interface
Ichiye, Takashi, and Edward J. Zipser, 1967.
An example of heat transfer at the air-sea boundary over the Gulf Stream during a cold air outbreak.
J. Met. Soc., Japan, 45(3):261-270.

air-sea interface (resolution)
India, Indian National Committee on Oceanic Research, Council of Scientific and Industrial Research, New Delhi, 1964.
International Indian Ocean Expedition, Newsletter India, 2(4):48 pp.

air-sea interface
Iselin, C.O'D., 1942.
Interaction between the hydrosphere and the atmosphere. Trans. N.Y. Acad. Sci., Ser. II, 4(3):99-106.

air-sea interface (surface)
Ito, N., 1960.
On the energy exchange between sea surface and the air. J. Meteorol. Soc., Japan, (2) 38(2):73-79.
Also in: Coll. Reprints. Oceanogr. Lab., Meteorol. Res. Inst., Tokyo, 6.

air-sea interface
Ivanenkov, V.N., 1964.
Distribution of PCO2, pH, alk, cl in the northern part of the Indian Ocean. Investigations in the Indian Ocean (33rd voyage of E/S "Vitiaz").
(In Russian).
Trudy Inst. Okeanol., Akad. Nauk, SSSR, 64:128-143.

J

air-sea interface
Jacobs, L., 1962.
A report on a discussion "Interaction between the atmosphere and the oceans".
Meteorol. Mag., 91(1081):209-224.

air-sea interface
Japan, Hakodate Marine Observatory, 1970.
Report of the marine meteorological observations on sea fog and heat exchange at the ocean-air interface, in June, 1969. Supplement: on the drift-ice east of South Saghalien in the Okhotsk Sea in June 1969. (In Japanese)
Mar. met. Rept., Hakodate mar. Obs. 30: 67 pp.

air-sea interface
Japan, Hakodate Marine Observatory, 1969.
Report of the marine meteorological observations on sea fog and heat exchange at the ocean-air interface in June, 1968. (In Japanese)
Mar. Met. Rept. Hakodate mar. Obs., 28: 1-40

fog
heat exchange
air-sea interface

air-sea interface
Japan, Hakodate Marine Observatory, 1968.
Report of the marine meteorological observations in the North Japan Sea (1966): On the interaction between lower atmosphere and sea surface. (In Japanese)
Mar. Met. Rept. 25: 20-31.

air-sea interface
Japan, Kobe Marine Observatory, 1971.
On some weather features in the Seto Inland Sea on board the liner. (In Japanese; English abstract.)
Bull. Kobe mar. Obs. 185: 22-32

Seto Inland Sea
air-sea interface

air-sea interface
*Jarvis, N.L., 1967.
Adsorption of surface-active material at the sea-air interface.
Limnol. Oceanogr., 12(2):213-221.

adsorption
air-sea interface
surface tension
surface potential

air-sea interface
Jarvis, N.L., W.D. Garrett, M.A. Scheiman and C.O. Timmons, 1967.
Surface chemical characterization of surface-active material in seawater.
Limnol. Oceanogr., 12(1):88-96.

air-sea interface
Jayaratne, O.W., and B.J. Mason, 1964.
The coalescence and bouncing of water drops at an air/water interface.
Proc. R. Soc., London, (A), 280(1383):545-565.

air-sea interface.
Johansen, H., 1958.
On continental and oceanic influences on the atmosphere. Meteorol. Annaler., 4(8):143-158.

air-sea interface
Junge, C, B. Bockholt, K. Schütz and R. Beck 1971.
N_2O measurements in air and seawater over the Atlantic.
Meteor Forsch.-Ergebn. (B) 6: 1-11.

K

air-sea interface
Kagan, B.A., 1968.
On the relationship of drift currents and turbulence in the upper layer of the sea with a changing wind stress. (In Russian).
Fisika Atmosfer. Okean., Izv. Akad. Nauk, SSSR. 4(9):1004-1007.

air-sea interface
Kagan, B.A., 1965.
The use of the quasistationary model of the interaction between atmospheric and oceanic boundary layers for the computation of the water temperature and current in the North Atlantic. (In Russian; English abstract).
Fizika Atmosferi i Okeana, Izv., Akad. Nauk, SSSR 1(8):845-852.

Air-Sea Interface
Kagan, B. A., 1961.
(The turbulent thermal exchange between the sea surface and the atmosphere and the loss of heat by evaporation in the Arctic seas.)
Probl. Arkt. i Antarkt., (8):78-84.

air-sea interface
Kagan, B.A., and Z.M. Utina, 1963
On the theory of the thermodynamic interaction between the sea and the atmosphere. (In Russian).
Okeanologiia, Akad. Nauk, SSSR, 3(2):250-259.

air-sea interface
Kanwisher, John W., 1963
On the exchange of gases between the atmosphere and the sea.
Deep-Sea Research, 10(3):195-207.

Air-sea interface
Katsaros, Kristina, and Konrad J.K. Buettner, 1969.
Influence of rainfall on temperature and salinity of the ocean surface.
J. appl. Met. 8(1): 15-18.

air-sea interface
Khanaichenko, N.K., 1964
Concerning the approach to the study of hydrometeorological processes. (In Russian).
Materiali Vtoroi Konferentsii, Vzaimod. Atmosfer. i Gidrosfer. v Severn. Atlant. Okean. Mezhd. Geofiz. God, Leningrad. Gidrometeorol. Inst., 260-263.

air-sea interface
Kitaigorodsky, S.A., 1969.
Small scale interactions between ocean and atmosphere. (In Russian; English abstract).
Fizika Atmosfer. Okean. Izv. Akad. Nauk SSSR, 5(11):1114-1131.

air-sea interface
*Kitaigorodsky, S.A., 1967.
On the analysis of the data of fluctuation measurements in the atmosphere surface layer over the sea. (In Russian; English abstract).
Fisika Atmosfer. Okean., Izv. Akad. Nauk, SSSR, 3(12): 1312-1325.

air-sea interface
*Kitaigorodsky, S.A., 1967.
Small-scale interactions between the ocean and the atmosphere. (In Russian; English abstract).
Okeanologiia, Akad. Nauk, SSSR, 7(5):774-785.

air-sea interface
Kitaigorodsky, S.A., V.I. Makova, V.A. Gorbanev, and V.M. Tarasenko 1971.
On the conditions of hydrostatic stability in the atmospheric surface layer above the ocean. (In Russian; English abstract).
Okeanologiia 11(1): 3-15

air-sea interface
Kitaigorodsky, S.A. and Yu. Z. Miropolsky 1970.
On the theory of the open ocean active layer. (In Russian; English abstract)
Fizika Atmosfer. Okean. Izv. Akad. Nauk SSSR 6(2): 177-188

air-sea interface
Kitaigorodsky, S.A., and Yu. S. Miropolsky 1968.
The dissipation of turbulent energy in the surface layer of the sea. (In Russian; English abstract).
Fizika Atmosfer. Okean., Izv. Akad. Nauk SSSR 4(6): 647-659.

air-sea interface
Kitaigorodsky, S.A., and Yu.A. Volkov, 1965.
On the calculation of turbulent heat and humidity fluxes in the near-water layer of the atmosphere. (In Russian; English abstract).
Fizika Atmosferi i Okeana, Izv., Akad. Nauk, SSSR, 1(12):1319-1336.

air-sea interface
Kitaigorodsky, S.A., and Yu. A. Volkov, 1965.
On the roughness parameter of sea surface and the calculation of momentum flux in the near water layer of the atmosphere. (In Russian; English abstract).
Fizika Atmosferi i Okeana, Isv., Akad. Nauk, SSSR, 1(9):973-988.

air-sea interface
Konaga, Shunji 1967.
Water temperature at the sea surface VII. A calculation of energy budget near the sea surface. (In Japanese; English abstract).
Bull. Kobe Mar. Obs. 178: 148-154.

air-sea interface
*Kort, V.G., 1970.
On the large-scale interaction between the ocean and the atmosphere. Okeanologiia, 10(2): 222-239. (In Russian; English abstract)

air sea interface
Kort, V.G., 1968.
The role of heat advection by sea currents in large-scale interaction between the ocean and the atmosphere. (In Russian).
Dokl. Akad. Nauk, SSSR, 182(5):1059-1062.

air-sea interface
Korzh, V.D. 1972.
Correlation of salt concentrations at the sea surface and in the sea atmospheric moisture.
Dokl. Akad. Nauk SSSR 202(2): 322-324

air-sea interface
*Kouznetsov, O.A., 1970.
Experimental study of the air flow above the ocean. (In Russian; English abstract).
Fisika Atmosfer. Okean., Izv. Akad. Nauk, SSSR, 6(8): 798-803.

air-sea interface
Krasjuk, V.S., and O.I. Sheremetevskaya, 1963.
Inflow of the sun heat to the ocean surface. (In Russian).
Meteorol. i Gidrol., (7):18-24.

air-sea interface
Kraus, E.B. 1972.
Atmospheric ocean interactions.
Oxford Univ. Press 265pp.

air-sea interface
Kraus, E.B., 1967.
Sea-air interaction.
Trans. Am. geophys. Un., 48(2):581-584.

air-sea interface
Kraus, E.B., and R.E. Morrison, 1968.
Local interactions between the sea and the air at monthly and annual time scales.
Q. Jl R. met. Soc., 94(401):409-411.

air-sea interface
Kraus, E.B., and R.E. Morrison, 1966.
Local interactions between the sea and the air at monthly and annual time scales.
Q.J.R. Meteorol. Soc., 92(391):114-127.

air-water interface
Kurooka, H., 1957.
Modification of Siberian air mass caused by flowing out over the open sea surface of northern Japan. J. Met. Soc., Japan, (2)35(1):53-59.

air-sea interface
Kuznetsov, O.A., 1963.
On the questions of aerodynamic roughness on the surface of the sea. Marine Meteorological Investigations. (In Russian).
Trudy Inst. Okeanol., Akad. Nauk, SSSR, 72:114-138.

L

air-sea exchange
Laevastu, T., 1965.
Daily heat exchange in the North Pacific; its effect on the ocean and its relations to weather.
ICNAF, Spec. Publ., No. 6:885-889.

Pacific, north
heat exchange
air-sea interface
weather, effect of

air-sea interface
Laevastu, Taive, 1965.
Daily heat exchange in the North Pacific, its relation to weather and its oceanographic consequences.
Commentat. physico-math., 31(2):5-53.

Air-sea interface
Laevastu, T., et P.M. Wolff 1965.
Quelques principes d'analyse et de prévision océanographique synoptique.
Météorol. 1965: 305-319

air-sea interface
*Lamontagne, R.A., J.W. Swinnerton and V.J. Linnenbom, 1971.
Nonequilibrium of carbon monoxide and methane at the air-sea interface. J. geophys. Res. 76(21): 5117-5121.

air-sea interface
Lavorko, V.S., 1970.
On the turbulent exchange and heat fluxes in the water near the sea surface. (In Russian).
Fizika Atmosfer. Okean., Izv. Akad. Nauk SSSR, 6(9): 970-972.

air-sea interface
Laykhtman D.L., and Yu. P. Doronin, 1959.
The coefficient of turbeulent exchange in the sea and an estimate of the heat flow from ocean waters. (In Russian).
Arktich. i Antarktich. Nauchno-Issled. Inst., Trudy, SSSR, 226:61-65.

Translation:
Lc or SLA, mi $1.80, ph $1. 80.

air-sea interface

Laykgtman, D.L., and G.P. Orlenko, 1956
Concerning the intensity of turbulent exchange over a water surface.
Glavnaya Geofiz. Observ., Leningrad, Trudy, (60):51-52.

OTS-SLA $1.10

air-sea interface

Lecompte Pierre et Jacqueline Lenoble, 1966.
Étude théorique des échanges radiatifs mer-atmosphere en grandes longueurs d'onde.
Cahiers Océanogr., C.C.O.E.C., 18(6):497-506.

air-sea boundary

LePichon, Xavier, and Jean-Paul Troadec, 1963.
La couche superficielle de la Méditerranée au large des côtes provencales durant les mois d'été. (3ème partie).
Cahiers Océan., C.C.O.E.C., 15(8):527-539.

air-sea interface

Lyakhin, Yu.I., 1971.
Carbon dioxide exchange between the ocean and the atmosphere in the southeastern Atlantic.
Okeanologiia 11(1): 48-52. (In Russian; English abstract).

Atlantic, southeast
air-sea interface
chemistry, carbon dioxide

M

air-sea interface

Makove, V.I., 1965.
The relation of the turbulence spectra in the near sea surface layer of the atmosphere with the spectrum of the surface waves. (In Russian).
Okeanologiia, Akad. Nauk, SSSR, 5(4):592-605.

air-sea interface

Malevoky-Malevich S.P. 1969.
Formation of negative temperature gradients near water surface. (In Russian; English abstract)
Meteorologiya Gidrol. (5): 53-59

air-sea interface

Malinski, J., and W. Robakiewicz, 1965.
Interrelation between the water and air temperature in autumn of Polish coastal waters of the Baltic Sea. (In Polish; English and Russian summaries).
Wiadomosci Sluzby Hydrol. i Meteorol., Panst. Inst. Hydrol.-Meteorol., 1(2):41-48.

air-sea

Malkus, Joanne S., 1962
Large-scale interactions. Ch. 4, Sect. II. Interchange of properties between sea and air. In: The Sea, Interscience Publishers, Vol. 1, Physical Oceanography, 88-294.

air-sea interface

Manabe, Syukuro, 1969.
Climate and the ocean circulation. I. The atmospheric circulation and the hydrology of the earth's surface. II. The atmospheric circulation and the effect of heat transfer by ocean currents.
Mon. Wea. Rev. 97(11): 739-774; 775-805.

air-sea interface

Manabe, Syukuro, and Kirk Bryan, 1969.
Climate calculations with a combined ocean-atmosphere model.
J. atmosph. Sci., 26(4): 786-789

air-sea interface
models

air-sea boundary processes

Mandelbaum, H., 1956.
Evidence for a critical wind velocity for air-sea boundary processes. Trans. Amer. Geophys. Union, 37(2):685-690.

air-sea interface

Manier, G., 1962
Zur Berechning der latenten und Fühlbaren Wärmeströme von der Meeresoberfläche an die Luft.
Geofisica Pura e Applicata, 52(2):189-213.

English and German summaries

air-sea interface

Marchuk, G.I., 1965.
On the role of research on physics of the atmosphere and ocean for weather forecast. (In Russian).
Fizika Atmosfer. i Okeana, Izv., Akad. Nauk, SSSR 1(1):5-7.

air-sea interface

Matsumoto, Seiichi 1967.
Some remarks on the convective transfer under the north-westerly winter monsoon situation.
Pap. Met. Geophys., Tokyo 18(4): 183-191.

air-sea interface

Matsumoto, S., and K. Ninomiya, 1966.
Some aspects of the cloud formation and its relation to the heat and moisture supply from the Japan Sea surface under a weak winter monsoon situation.
J. Meteorol. Soc., Japan, (2), 44(1):60-75.

Sea of Japan
air-sea interface
monsoon, effect of cloud formation

air-sea interface

Matveyev, D.T., 1971.
On the spectrum of the microwave radiation of the wavy sea surface. (In Russian; English abstract). Fiz. Atmosfer. Okean. Izv. Akad. Nauk SSSR 7(10): 1070-1083.

Mc

air-sea interface

McAlister, E.D., 1967.
Using infrared to measure heat flow from the sea.
Ocean Industry, 2(6):35-39

air-sea interface
heat exchange

air-sea interface

McAlister, E.D., and William McLeish 1969.
Heat transfer in the top millimeter of the ocean.
J. geophys. Res. 74(13): 3408-3414.

air-sea interface

McEwen, G.F., 1938.
Some energy relations between the sea surface and the atmosphere.
J. Mar. Res., 1(3):217-238.

Reviewed by A. Defant in:
J. du Conseil, 14(3):409-410.

air-sea interface

*McFadden, James D., and John W. Wilkerson, 1967.
Compatibility of aircraft and shipborne instruments used in air-sea interaction research.
Mon. Weath. Rev., U.S. Dep. Comm. 95(12):936-941.

air-sea interface

MacIntyre, Ferren 1970.
Geochemical fractionation during mass transfer from sea to air by breaking bubbles.
Tellus 22(4): 451-462.

air-sea interface

Mehta, A.J., 1971.
A note on wind-stress in open sea. J. geophys. Res., 76(9): 2202-2204.

anticyclonic vortex

Meincke, Jens, 1971.
Observation of an anticyclonic vortex trapped above a seamount. J. geophys. Res, 76(30): 7432-7440.

air-sea interface

Menshov, Yu. A., and G.M. Degtjarjov, 1963.
About ocean surface effective emission. (In Russian).
Meteorol. i Gidrol., (7):44-46.

air-sea interface

Miller, Banner I., 1966.
Energy exchanges between the atmosphere and the ocean.
Hurricane Symposium, Oct. 10-11, 1966, Houston, Publ. Am. Soc. Oceanogr., 1:134-157.

air-sea interface

Miropolsky, Yu. Z., 1970.
Unstationary model of the wind-convection mixing layer in the ocean. (In Russian; English abstract).
Fizika Atmosfer. Okean., Izv. Akad. Nauk SSSR 6(12): 1284-1294

air-sea interface

Mitnik, L.M., and A.M. Shutko, 1970.
On the effect of the sea surface state on the accuracy of determining the moisture content of the atmosphere and water content of clouds when making radiometric measurements from artificial earth satellites.
Met. Gidrol. (10): 72-75.

air-sea interface

Miyake, Y., and S. Matsuo, 1962
A note on the deuterium content in the atmosphere and the hydrosphere.
Papers Meteor. Geophys., 13(3-4):245-259.

air-sea interface

Miyake, Yasuo, and Yoshio Sugiura, 1962
Vertical distribution of CO_2 in air and ocean near sea surface.
Rec. Oceanogr. Wks., Japan, Spec. No. 6: 219-221.

air-sea interaction

Monin, A.S., 1969.
Fundamental effects of the air-sea interaction. (In Russian; English abstract).
Fizika Atmosfer. Okean., Izv. Akad. Nauk SSSR, 5(11): 1102-1113.

air-sea interface

Monin, A.S., 1967.
Turbulent exchange in the air over oceans. (In Russian).
Dokl. Akad. Nauk, SSSR, 175(4):819-822.

air-sea interface

Morita, Yasuhiro 1971.
The diurnal and latitudinal variation of electric field and electric conductivity in the atmosphere over the Pacific Ocean.
J. met. Soc. Japan 49 (1): 56-58.

air-sea interface

Motohashi, Keinosuke, and Michitaka Ode, 1967.
On the energy exchange between the atmosphere and the sea east of Japan.
J. oceanogr. Soc. Japan, 23(6): 267-277.

Air-sea interface

Murray, R., and R.A.S. Ratcliffe, 1969.
The summer weather of 1968: related atmospheric circulation and sea temperature patterns
Met. Mag (Met. O. 816) 98 (1164): 201-219

N

air-sea interface

Nakajima, Chotaro, and Hiroshi Yoshioka, 1971.
On air-sea interaction in the Kii Channel.
Bull. Disas. Prev. Res. Inst. Kyoto Univ. 20 (3) (179): 217-226.

air-sea interface
Japan

air-sea interface

Namias, Jerome, 1971.
The 1968-69 winter as an outgrowth of sea and air coupling during antecedent seasons.
J. phys. Oceanogr., 1(2): 65-81.

air-sea interface

*Namias, Jerome, 1969.
Autumnal variations in the North Pacific and North Atlantic anticyclones as manifestations of air-sea interactions.
Deep-Sea Res., Suppl. 16: 153-164.

air-sea interface

* Namias, Jerome, 1969.
Seasonal interactions between the North Pacific ocean and the atmosphere during the 1960's.
Mon. Wea. Rev., 97(3): 173-192.

air-sea interaction

Namias Jerome 1967.
Further studies of drought over northeastern United States
Mon. Weath. Rev. U.S. Dep. Comm. 95 (8): 497-508.

air-sea interface

Namias, Jerome, 1965.
Macroscopic association between mean monthly sea-surface temperature and the overlying winds.
Jour. Geophys. Res., 70(10):2307-2318.

air-sea interface

Namias, Jerome, 1965.
Short-period climatic fluctuations: the nature and cause of climatic abnormalities lasting from a month to a few years are discussed.
Science, 147(3659):696-706.

air-sea interface

Namias Jerome, 1959.
Recent seasonal interactions between North Pacific waters and the overlying atmospheric circulation.
J. Geophys. Res., 64(6):631-646.

air-sea interface

Nan-Niti, Tosio, 1969.
Effect of sea and ground surface g metry to the wind profile. Bull. Jap. Soc. rish. Oceanog Spec. No. (Prof Uda Commem. Pap.): 73-75.

Air-sea interface

Nan-niti, Tosio, Akimitsu Fujiki and Hideo Akamatsu, 1968.
Micro-meteorological observations over the sea (1).
J. oceanogr. Soc., Japan, 24 (6): 281-294

air-sea interface

Naumov, A.P. and V.M. Plechkov, 1971.
On the determination of integral water content in the atmosphere over ocean by radiometer method. (In Russian). Fisika Atmosfer. Okean., Izv. Akad. Nauk SSSR, 7(3): 352-354.

methods, marine meteorology
air-sea interface

air-sea interface

Navrotsky, V.V., 1964.
Some results of research studies on the interaction of the ocean and atmosphere in the Gulf Stream during the IGY. (In Russian).
Okeanologiia, Akad. Nauk, SSSR, 4(4):603-611.

air-sea interface

Navrotsky, V.V., 1964.
The study of the interaction between oceanic currents and the atmospheric processes in the northern Atlantic. (In Russian).
Okeanologiia, Akad. Nauk, SSSR, 4(3):396-407.

AIR SEA INTERFLOW

Navrotsky, V.V., and B.N. Filyushkin, 1970.
Statistic analysis of wind velocity measurements in the near-water layer of the atmosphere. (In Russian; English abstract).
Fisika Atmosfer. Okean. Izv. Akad. Nauk SSSR, 6(3): 292-298.

air-sea interface

Neumann, G., 1960.
Some meteorologically important relationships between the ocean and atmosphere.
Trans. N.Y. Acad. Sci., (2), 22(8):615-633.

air-sea surface

Neumann, G., 1954.
Notes on the relationship of sea dn the atmosphere. Air Weather Service Bull., Sept., 45-47.

air-sea interface

Neumann, Gerhard, and Willard J. Pierson, Jr., 1963.
Known and unknown properties of the frequency spectrum of a wind-generated sea.
In: Ocean wave spectra, Proc. Conf., Easton, Md., May 1-4, 1961, Prentice-Hall, Inc., 9-21.

O

air-sea interface

*O'Brien, Edward E. and Thore Omholt, 1969.
Heat flux and temperature variation at a wavy water-air interface. J. geophys. Res., 74(13): 3384-3385.

air-sea interface

Ogata, Tetsu, 1966.

On the interactions between lower atmosphere and sea surface. (In Japanese; English abstract.)

Bull. Hakodate Mar. Obs. 16(4).

Reprinted pp. 1-47.

air-sea interface

Opevinskia, S.K., 1961.
On the theory of temperature interaction between the sea, atmosphere and land.
Izv. Akad. Nauk, SSSR, ser. Geofiz., (6):926-932.

In Russian

air-sea interface

Osborne, M.F.M., 1964.
The interpretation of infrared radiation from the sea in terms of its boundary layer.
Deutsche Hydrogr. Zeits., 17(3):115-136.

air-sea interface

*Östlund, H. Göte, 1968.
Hurricane tritium II: air sea exchange of water in Betsy 1965.
Tellus, 20(4):577-594.

P

air-sea interface

Pagava, S.T., 1968
On some peculiarities of relation between processes in ocean and atmosphere. (In Russian; English abstract.
Met. Gidrol. (1): 35-39.

air-sea interface

Pagava, S.T., 1965.
On the joint analysis of the ocean heat state and processes in the atmosphere. (In Russian).
Meteorol. i gidrol., (7):11-19.

air-sea interface

*Pandolfo, J.P., 1969.
Motions with inertial and diurnal period in a numerical model of the navifacial boundary layer. J. mar. Res., 27(3): 301-317.

air-sea interface

Paquin, J.E., and S. Pond, 1971
The determination of the Kolmogoroff constants for velocity, temperature and humidity fluctuations from second- and third-order structure functions.
J. fluid Mech. 50 (2): 257-269.

air-sea interface

Park, Kilho, and Donald W. Hood, 1963
Radiometric field measurements of CO_2 exchange from the atmosphere to the sea.
Limnol. and Oceanogr., 8(2):287-289.

air-sea interface

Paulson, Clayton A., 1969.
Comments on a paper by J.W. Deardorff, 'Dependence of air-sea transfer coefficients on bulk stability.'
J. geophys. Res., 74(8): 2141-2142.

air-sea interchange

Peixoto, J.P., 1959-1960.
O campo da divergencia do transporte do vapor de água na atmosfera.
U. Lisboa, Rewist. Fac. Ciên., 2A ser., B Ciencias fisico-quimicas, 7:25-56.

air-sea interface

Pereslegin, S.V., 1967.
On the relationship between thermal and radio emissivity contrasts of the sea surface. (In Russian; English abstract).
Fisika Atmosfer. Okean., Izv. Akad. Nauk, SSSR, 3(1):47-57.

air-sea interface

*Perry, A., 1968.
Turbulent heat flux patterns over the North Atlantic during some recent winter months.
Met. Mag., 97(1153):247-254.

air-sea interface

Peterson, Clifford L., 1960.
The physical oceanography of the Gulf of Nicoya, Costa Rica, a tropical estuary.
Inter-American Tropical Tuna Comm., Bull., 4(4): 139-216.

air-sea, interface

Perán Willard J., Jr., 1968?
Understanding and forecasting phenomena at the air-sea interface

Selected Papers from the Governor's Conference on Oceanography, October 11 and 12, 1967, at the Rockefeller University, New York, N.Y.: 162-171.

air-sea interface

Pike, Arthur C. 1971.
Intertropical convergence zone studied with an interacting atmosphere and ocean model.
Mon. Weath. Rev. U.S. Dept. Comm. 99 (6): 469-477.

air-sea interface

Pivovarov, A.A., 1968.
On the daily temperature variations in the surface layer of the ocean and in the near water surface layer of the atmosphere. (In Russian).
Fisika Atmosfer, Okean., Izv. Akad. Nauk, SSSR, 4(1): 102-107.

air-sea interface
*Plate, E.J., and G.M. Hidy, 1967.
Laboratory study of air flowing over a smooth surface onto small water waves.
K. geophys. Res., 72(18):4627-4641.

air-sea interface
Plechkov, V.M., A.S. Gurvich and V.G. Snopkov 1970.
Experimental investigations of integral steam content over the ocean by means of radiometric measurements of atmospheric thermal radiation from a ship. (In Russian)
Dokl. Akad. Nauk SSSR 193(5): 1041-1043

air-sea interface
Pond, Stephen, 1970.
Air-sea interaction. EOS. Trans. Am. geophys. Un. 52(6): 389-394.

air-sea interface
Pond, S., G.T. Phelps, J.E. Paquin, G. McBean and R.W. Stewart 1971.
Measurements of the turbulent fluxes of momentum, moisture and sensible heat over the ocean.
J. atmos. Sci. 28(6): 901-917

Air - sea interface
Pond, S., S. D. Smith, P. F. Hamblin and R. W. Burling 1966.
Spectra of velocity and temperature fluctuations in the atmospheric boundary layer over the see.
J. Atmospheric Sci., 23(4):376-386.

air-sea interface
Portman, Donald J., Floyd C. Elder and Edward Ryznar, 1961.
Research on energy exchange processes. (Summer 1960). Proceedings Fourth Conf., Great Lakes Res. Div., Inst. Sci. & Techn., Univ. Michigan Publ. (7):96-109.

air-sea interface
Polushkin, V.A., 1964.
On the practical use of some results of the investigation of the interaction between atmosphere and hydrosphere in the North Atlantic. (In Russian).
Materiali Vtoroi Konferentsii, Vzaimod. Atmosfer i Gidrosfer. v Severn. Atlant. Okean. Mezhd. Geofiz God, Leningrad. Gidromereorol. Inst., 27-31.

air-sea interface
Popova, N.M., and G.V. Svinukhov, 1970
Radioactivity of the atmosphere over the oceans in January-March of 1969, and its connection with aerosynoptic conditions. (In Russian; English abstract).
Met. Gidrol. 1970(9):62-67

air-sea interface
*Pond, S., 1968.
Air-sea interaction studies by the Department of Meteorology, Imperial College, London. Bull. Am. Met. Soc., 49(8): 832-834. Also in: Coll. Reprints, Dept. Oceanogr. Univ. Oregon, 7(1968).

air-sea interface
Postma, H., 1964.
The exchange of oxygen and carbon dioxide between the ocean and the atmosphere.
Netherlands J. Sea Res., 2(2):258-283.

air-sea interface
Preobrazhensky, L. Yu., 1969.
A calculation of the exchange coefficient in the surface layer of air above the sea. (In Russian, English abstract).
Fizika Atmosfer. Okean, Izv. Akad. Nauk SSSR 5(6): 601-607.

air-sea interface
Preobrazhensky, L. Yu., 1968.
Some characteristics of the air flow in the lower layer of the marine atmosphere. (In Russian).
Fisika Atmosfer. Okean., Izv. Akad. Nauk, SSSR, 4(9):994-999.

air-sea interface
Privalova, I.V. 1971
Meridional and vertical circulation of waters in the northern part of the Atlantic Ocean. (In Russian)
Okeanol. Issled. Rezult. Issled. Meghd.-Geofiz. Proekt. 22:154-219.

air-sea energy exchange
Privett, D. W., 1960.
The exchange of energy between the atmosphere and the oceans of the southern hemisphere.
Meteorol. Off., London, Geophys. Mem., 104 (M.O. 631d) 61pp

air-sea interaction
Pyke, Charles B., 1965.
On the role of air-sea interaction in the development of cyclones.
Bull. Amer. Meteorol. Soc., 46(1):4-15.

air-sea interface
Pond, S., 1968.
Microscale phenomena in the ocean and the atmospheric boundary layer. Proc. Region SIX IEEE Confr. 1-C-4. 10 pp. Also in: Coll. Reprints, Dept. Oceanogr. Univ. Oregon, 7(1968)

air-sea interface

Q

air-sea interface
*Quinn, J.A. and N.C. Otto, 1971.
Carbon dioxide exchange at the air-sea interface: flux augmentation by chemical reaction.
J. geophys. Res., 76(6): 1539-1549.

R

air-sea interface
*Radikevich, V.M., 1971.
Effects of stratification of the atmospheric boundary layer and the sea friction layer on the vertical profiles of turbulence parameters, current velocities and wind. Okeanologiia 11(2) 187-194. (In Russian; English abstract)

air-sea interface
*Radikevich, V.M., 1970.
On computing heat, moisture and momentum fluxes. Oceanologiia, 10(5): 878-882. (In Russian; English abstract)

air-sea interface
heat flux
moisture flux
momentum flux

air-sea interface
*Radikevich, V.M., 1969.
The effects of humidity stratification on the processes in the atmospheric near-water layer. (In Russian; English abstract).
Okeanologiia, 9(5): 773-780.

air-sea interactions
Raghavan, K., 1969.
Satellite evidence of sea-air interactions during the Indian monsoon.
Mon. Wea. Rev. 97(12):905-908

monsoons
air-sea interactions
Indian Ocean
satellites

air-sea interface
Ramage, C.S., 1968.
Problems of a monsoon ocean.
Weather, 23(1):28-37.

air-sea interface
Richards, F. A., 1965.
Dissolved gases other than carbon dioxide.
In: Chemical oceanography, J.P. Riley and G. Skirrow, editors, Academic Press, 1:197-225.

air-sea interface
Robinson, G.D., 1966.
Another look at some problems of the air-sea interface.
Q.J.R. Meteorol. Soc., 92(394):451-465.

Air-sea Interface
Roden, Gunnar I., and Joseph L. Reid, Jr., 1961.
Sea surface temperature, radiation and wind anomalies in the North Pacific Ocean.
Rec. Oceanogr. Wks., Japan, 6(1):36-52.

air-sea boundary
Roll, H.U., 1951.
Gibt es eine "kritische Windgeschwindigkeit" für Prozesse an der Grenzfläche Wasser-Luft?
Atti Convegno Int. Meteorol. Marit., Genova, 20-22/IV/1950:110-126, 11 textfigs.

air-sea interface
Romanov, Yu.A., 1971.
On amplitudes of the diurnal variations of air temperature above the sea with altitude.
(In Russian; English abstract). Okeanologiia 11(5):900-906.

air-sea interface
Ryzhkov, Yu. G., 1963.
Advective heat transfer and formation of a zone of intensified exchange in wind "tide" circulation in the deep sea. (In Russian).
Izv., Akad. Nauk, SSSR, Ser. Geofiz., (5):825-827
Translation: Amer. Geophys. Union, (5):505-507.

S

air-sea interface
Saunders, Peter M. 1967.
The temperature at the ocean-air interface.
J. atmos. Sci. 24(3): 269-273

Air-sea interface
Schell, I.I., 1965.
On the nature of the water movements in the surface layer of the ocean at OWS Stations "I" and "J", south of Iceland.
Deutsche Hydrogr. Zeits. 18(2):70-75.

air-sea interface
Scherhag, Richard, 1963.
Die wechselseitigen Einflüsse zwischen den Kontinenten, den Ozeanen und der Atmosphäre.
Ber. Deutsch. Wetterdienstes, 12(91):58-72.

air-sea interface
Schooley, Allen H., 1967.
Temperature differences near the sea-air interface.
J. mar. Res., 25(1):60-68.

air-sea interface
Sellers, William D., 1970.
The effect of changes in the earth's obliquity on the distribution of mean annual sea-level temperatures.
J. appl. Met. 9(6): 960-961.

air-sea interface
earth's obliquity, effect of

air-sea interface
Semenov, V.G., 1962
Interaction between atmosphere and hydrosphere.
Meteorol. i Gidrol. (5):22-28.

air sea interface

Semenov, V.G., 1960

[The role of ocean surface in the formation of blocking anticyclones] Meteorol. i Gidrol., 6: 17-19.

air-sea interface

Shonting, David H., 1964
Some observations of short-term heat transfer through the surface layers of the ocean.
Limnology and Oceanography, 9(4):576-588.

air-sea interface

Shuleikin, V.V., 1952.
[The temperature field above the sea and the land in winter when the coefficient of exchange is varying.]
Doklady Akad. Nauk, SSSR, 83(2):389-

air-sea interface

Shuleikin, V.V., and P.F. Shakurov, 1956.
[Lines of sodium in the air absorption spectrum above the sea.] Doklady Akad. Nauk, SSSR, 106(6): 991-994.

air-sea interface

*Siever, Raymond, 1968.

Sedimentological consequences of a steady state-ocean-atmosphere.

Sedimentology, 11 (1/2): 5-29.

sedimentation
carbonate rocks
silicates

air-sea interface

Silkin, B.I., 1968.
Aerophotographic survey of a tropical hurricane. Navitational air to oceanographic work. Studies of the ocean-atmosphere interaction. (In Russian).
Okeanologiia, Akad. Nauk, SSSR, 8(5):933-934.

air-sea interface

*Simpson, Joanne, 1969.
On some aspects of sea-air interaction in middle latitudes.
Deep-Sea Res., Suppl. 16: 233-261.

air-sea interface

Simpson, Robert H. 1968.
Understanding ocean weather.
Oceanol. intl 3(7): 42-45

Skirrow, Geoffrey, 1965. air-sea interface
The dissolved gases - carbon dioxide.
In: Chemical oceanography, J.P. Riley and G. Skirrow, editors, Academic Press, 1:227-322.

alkalinity
carbonic acid
carbon isotopes
air-sea interface
radio-carbon

air-sea interface

Smagorinsky, Joseph, 1969.
Problems and promises of deterministic extended range forecasting.
Bull. Am. met. Soc., 50(5): 286-311.

air-sea interface
forecasting, extended range

air-sea interface

Smith, F.I.P., and Alex D.D. Craik, 1971.
Wind-generated waves in thin liquid films with soluble contaminant.
J. fluid Mech. 45(3): 527-544.

air-sea interface

*Smith, Stuart D., 1970.
Thrust-anemometer measurements of wind turbulence, Reynolds stress, and drag coefficient over the sea. J. geophys. Res., 75(33): 6758-6770.

air-sea interface

Solantie, Reijo 1968.
The influence of the Baltic Sea and the Gulf of Bothnia on the weather and climate of northern Europe, especially Finland, in autumn and in winter.
Ilmatiet. Laitoks. Toim. 70: 28pp.

air-sea inter face

Solyankin, E.V., 1966.
On the relation of hydrometeorologic processes in the northern and southern hemispheres. (In Russian).
Trudy vses. Nauchno-issled. Inst. morsk. ryb. Khoz. Okeanogr. (VNIRO), 60:49-58.

air-sea interface

Soliankin, E.V., 1963.
Radiational balance in the surface layers of the Black Sea. (In Russian).
Trudy Inst. Okeanol., Akad. Nauk, SSSR, 72:155-166.

air-sea interface

Sorkina, A.I., I.V.P. Gratchev, 1957.
[Définition des caractéristiques du regime thermique et de la turbulence dans les échanges atmosphère-mer.] Trudy Gosud. Inst. Okeanol., 38:

air-sea interface

Sorkina, A.I., 1961.
Survey of research on typicafication of atmoslheric processes above the ocean. (In Russian).
Trudy Gosud. Okean. Inst., Leningrad, 61:159-168

air-sea interface

*Stewart, R.W., 1967.
Mechanics of the air-sea interface.
Physics Fluids, 10(9-2):S47-S 55.

air-sea interface

Strokina, L.A., 1963
Heat exchange between the ocean surface and lower water layer. (In Russian).
Meteorol. i Gidrol. (1):25-30.

air-sea interface

*Strokina, L.A. and A.I. Smirnova, 1969.
Heat content of the active layer of the North Atlantic and the change of the surface temperature. Okeanologiia 9(3): 395-403.

(In Russian, English abstract)

air-sea interface

Suess H.E., 1970.
The transfer of carbon 14 and tritium from the atmosphere to the ocean. J. geophys. Res., 75(12): 2363-2364.

carbon 14
tritium
air-sea interface

air-sea interface (chemistry)
Sugawara, K., 1965.
Exchange of chemical substance between air and sea.
In: Oceanography and Marine Biology, H. Barnes, editor, George Allen and Unwin, 3:59-77.

air-sea interface

Sugiura, Yoshio, 1969.
Distribution of AOU in the surface waters of Kuroshio and Oyashio and its oceanographical significance. The ability of gas exchange between air and sea.
Oceanogrl Mag. 21(1):31-35
apparent oxygen utilization (AOU)

air sea interface

Sugiura, Yoshio, E.R. Ibert and D.W. Hood, 1963
Mass transfer of carbon dioxide across sea surfaces.
J. Mar. Res., 21(1):11-24.

T

air-sea interface

*Takahashi, Tadas, 1969.
A note on the annual heat exchange across the air-sea boundary.
Rec. oceanogr. Wks. Japan, 10(1):13-22.

air-sea interface

Takenouti, Y., K. Hata and M. Tori, 1959
On the forecast of surface water temperature for the frontal zone of western North Pacific.
Bull. Hakodate Mar. Obs., (6):96-100.
Eng. with French Resumes.

air-sea interface

Takenouti, Yoshitada, and Masao Hanzawa, 1961.
Relation between anomalies of sea-surface temperatures and meteorological conditions over the North Pacific Ocean. (Abstracts).
Tenth Pacific Sci. Congr., Honolulu, 21 Aug.-6 Sept., 1961, Abstracts of Symposium Papers, 327.

air-sea interface

Takenouti, Yositada, Masaaki Hyuga, and Yurin Sakamoto, 1970.
Measurements of vertical air temperature gradients with dipmeter.
J. oceanogr. Soc. Japan 26(4): 226-232.

air-sea interface

Timeofeyev, Yu. V., 1966.
Thermal radiation sounding of water surface layer. (In Russian).
Fisika Atmosferi i Okeana, Izv., Akad. Nauk, SSSR, 2(7):772-774.

air-sea interface

Timofeev, N.A., 1965.
On the determination of the water supply in the atmosphere over the ice-free surface of the oceans. (In Russian).
Meteorol. i Gidrol., (3):24-28.

air-sea boundary

Timonov, V.V., and A.A. Gure, 1960.
[On experimental of the investigation of the change in state of the ocean-atmosphere system.]
Trudy Okeanograf. Komissii, Akad. Nauk, SSSR, 10(1):47-49.

air-sea interface

Timonov, V.V., A.I. Smirnova and K.I. Nepop, 1970.
On the centers of air-sea interaction in the north Atlantic. Okeanologiia, 10(5): 745-749.
(In Russian; English abstract)

air-sea interface

Toba, Yoshiaki, 1966.
Sea-salt particles: a factor in the air-sea interaction. (In Japanese; English abstract).
Umi to Sora, 41(3/4):71-118.

air-sea interface

Toba, Yoshiaki, 1962
The air-sea coupling and the sea-salt nuclei.
J. Oceanogr. Soc., Japan, 20th Ann. Vol., 421-431.

air-sea interface

Toba, Yoshiaki, and Sanshiro Kawai, 1970.
A method of simultaneous analysis of wind and temperature profile data over the water surface by use of the method of least squares.
J. oceanogr. Soc., Japan. 26(2): 87-94

air-water interface

Tribble, D.T., 1967.
The measurement of turbulence parameters in the near-surface layer of the atmosphere.
In: The collection and processing of field data, E.F. Bradley and O.T. Denmead, editors, Interscience Publishers, 41-46.

U

air-sea interface
U.S., Joint Panel on Air-Sea Interaction, 1963. (G.S. Benton, R.G. Fleagle, D.F. Leipper, R.B. Montgomery, N.W. Rakestraw, W.S. Richardson, H. Riehl, J. Snodgrass).
Interaction between the atmosphere and the oceans.
Bull. Amer. Meteorol. Soc., 44(1):4-17.

Complete report in NAC-NRC, Publ. No. 968.

air-sea interface
U.S. National Academy of Sciences-National Research Council, 1962.
Interaction between the atmosphere and the oceans. Report of the Joint Panel on the Air-Sea Interaction to the Committee on Atmospheric Sciences and the Committee on Oceanography.
NAS-NRC, Publ., No. 983:43 pp.

air-sea interface
U.S. National Academy of Sciences-National Research Council, Committee on Atmospheric Sciences, 1960.
Meteorology on the move.
NAS-NRC, Publ., No. 794:30 pp.

V

air-sea interface
VALERIANOVA, M.A. and E.I. SERYAKOV 1970.
On the many-year variations in the ocean-atmosphere system. Okeanologiia, 10(5): 750-756.
(In Russian; English abstract)

air-sea interface
Van Atta, C.W. and W.Y. Chen 1970
Structure functions of turbulence in the atmospheric boundary layer over the ocean.
J. fluid Mech. 44(1): 145-149.

air-sea interface
Van Loon, Henry 1971.
A half-yearly variation of the circumpolar surface drift in the Southern Hemisphere.
Tectonophysics 12(9): 511-

air-sea interface
Viebrock, Herbert, 1962
The transfer of energy between the ocean and the atmosphere in the Antarctic region.
J. Geophys. Res., 67(11):4293-4302.

air-sea interface
Vladimirov, O.A., 1960.
[Computation of increase in the area of the ocean under the influence of wave action.]
Meteorol. i Gidrol., (10):36-37.

air-sea interface
Voitova, K.V., D.L. Laikhtman, N.A. Romanova and V.G. Snopkov, 1971.
Quantitative parameters of the interaction between the ocean and the atmosphere in the Atlantic. (In Russian; English abstract).
Okeanologiia 11(4): 598-605.

air-sea interface
Volkov, P.A., 1960.
[On the transfer of liquid in the presence of surface waves.]
Meteorol. i Gidrol., (4):38-41.

air-sea interface
Volkov, Yu. A., 1970.
Turbulent fluxes of momentum and heat in the atmospheric surface layer over the wavy sea surface. (In Russian; English abstract).
Fizika Atmosfer. Okean., Izv. Akad. Nauk SSSR, 6(12): 1295-1302.

air-sea interface
*Volkov, Yu.A., 1969.
Spectra of velocity and temperature fluctuations in the air flow over the wavy sea surface.
Fizika Atmosfer. Okean., Izv. Akad. Nauk. SSSR, 5(12): 1251-1265.
(In Russian; English Abstract).

air-sea interface
*Volkov, Yu.A., V.P. Kukharets and L.R.Tsvang, 1968.
Turbulence in the boundary layer of the atmosphere above the steppe and the sea. (In Russian;English abstract).
Fisika Atmosfer.Okean.,Izv.Akad.Nauk,SSSR, 4(10): 1026-1041.

air-sea interface
*Voskanyan, A.G., A.A. Pivovarov, G.G. Khundzhua, 1970.
Experimental studies of temperature and turbulent heat exchange in the surface layer of the sea. Okeanologiia, 10(4): 588-595.
(In Russian; English abstract)

air-sea interface
*Voskanyan,A.G., A.A. Pivovarov and G.G. Khundzhua,1967.
Diurnal temperatures variation and turbulent exchange at the heating of the sea surface layer. (In Russian;English abstract).
Fisika Atmosfer.Okean.,Izv.Akad.Nauk,SSSR,3(11): 1210-1216.

W

air-sea interface
*Watson,A.G.D.,1968.
Air-sea interaction.
Sci.Prog.,Oxf.56(223):303-323.

air-sea interface
*Weber, Jon N.,1967.
Possible changes in the isotopic composition of the oceanic and atmospheric carbon reservoir over geologic time.
Geochim. cosmochim. Acta, 31(12):2343-2352.

air-sea interface
*Weiler, H.S., and R.W. Burling,1967.
Direct measurements of stress and spectra of turbulence in the boundary layer over the sea.
J. atmos. Sci., 24(6):653-664.

air-sea interface
Wilkniss, P.E. and D.J. Bressan, 1971.
Chemical processes at the air-sea interface: the behavior of fl rine. J. geophys. Res., 76(3): 736-741.

air-sea interface
Williams, Jerome, John J. Higginson and John D. Rohrbough, 1968
Sea & Air: the naval environment
U.S. Naval Inst. 338 pp.

air-sea interface
Winchester, John W., and Robert A. Duce, 1965.
Geochemistry of iodine, bromine, and chlorine in the air-sea-sediment system.
Narragansett Mar. Lab., Univ. Rhode Island, Occ. Publ., No. 3:185-201.

air-sea interface
*Wu.Jin,1969.
Wind stress and surface roughness at air-sea interface.
J. geophys. Res., 74(2):444-455.

air-sea interface
Wüst, Georg, 1959.
Sulle componenti del bilancio idrico fra atmosfera, oceano e Mediterraneo.
Annali Ist. Universitario Navale, Napoli, 28: 1-18.

air-sea interface(energy exchange)
Wyrtki, Klaus, 1958
Precipitation, evaporation and energy exchange at the surface of the southeast Asian waters. (Abstract).
Proc. Ninth Pacific Sci. Congr., Pacific Sci. Assoc., 1957, 16(Oceanogr.):179.
Published in:
Penjelidikan Laut Di Indonesia (Mar. Res. Indonesia), No. 3:7-40, Djakarta, 1957.

XYZ

air-sea interface
Yakovlev, G.N., 1957.
The turbulent heat exchange between the ice cover and the air in the central Arctic. (In Russian).
Problemy Arktiki, SSSR, (2):193-204.

Translation:
LC or SLA mi $2.70, ph $4.80.

air-sea interface
Yeo, S.-A., 1957.
The problem of wind drift near a free water surface (Abstract). Ann. Rept. Challenger Soc., 3(9):35.

air-sea interface
*Yoshihara, Hideo,1968.
Sea-air interaction: a simplified model.
J. atmos. Sci., 25(5):729-735.

air-sea interface
Zhukov, L.A., 1965 (Also: Joukov)
Calculation of the transport and changing of the water temperature of the upper layer of the North Atlantic. (In Russian; English abstract).
Okeanolog. Issled., Rezult. Issled. Programme Mezhd. Geofiz. Goda, Mezhd. Geofiz. Komitet Presidiume, Akad. Nauk, SSSR, No. 13:82-89.

air-sea interface
Zilitinkevich S.S., 1969
On the calculation of global phenomena of air-sea interaction. (In Russian; English abstract)
Fisika Atmosfer. Ocean. Izv. Akad. Nauk SSSR, 5(11): 1143-1159.

air-sea interface
* Zilitinkevich, S.S., 1969.
On the computation of the basic parameters of the interaction between the atmosphere and the ocean, Tellus 21(1): 17-24.

air-sea interface
*Zilitinkevich,S.S.,1967.
Dynamic and thermal interaction between the atmosphere and the ocean. (In Russian;English abstract).
Fisika Atmosfer. Okean., Izv.Akad.Nauk,SSSR, 3(10):1069-1077.

air-sea interface
*Zillman,J.W., and J.A. Bell,1968.
Sea-to-air heat fluxes in the southwest Indian Ocean in summer.
J.geophys.Res.,73(22):7057-7064.

air-sea interface
Zimmerman, Leon I., 1969
Atmospheric wake phenomena near the Canary islands.
J. appl. Met., 8(6): 896-907

air-sea interface
Zubov, N.N., 1955.
[Index of atmospheric turbulence, the interaction of sea and atmosphere, some features of evaporation in the sea.] Izbrannye Trudy po Okeanologii Moscow, Ch. 8, 12, 28, 171-176, 204-214, 443-456

air-sea interface
Zverev, N.I., and G.I. Morskoy, 1961.
[On the problem of atmosphere and hydrosphere interaction.]
Meteorol. i Gidrol., (5):37-41.

air-sea interface (chemical exchange)
Korzh, V.D., 1971.
Computation of relationships between chemical components of the sea water transported from the ocean to the atmosphere through evaporation. (In Russian; English abstract). Okeanologiia 11(5): 881-888.

Air-sea water cycle

air-sea water cycle
Wüst, Georg, 1959.
Sulle componenti del bilancio idrico fra atmosfera, oceano e Mediterraneo.
Annali Ist. Universitario Navale, Napoli, 28: 1-18.

air stability; see STABILITY—

air supply, undersea

air supply, undersea

Vind, Harold P., 1966.
Undersea air supply.
Science, 153(3738):873-875.

abstract

air supply, undersea

albedo

albedo

Bloch, M.R., 1965.
A hypothesis for the change of ocean levels depending on the albedo of the polar ice caps.
Palaeogeogr., Palaeoclimatol., Palaeoecol., 1:(2):127-142.

albedo

Bryazgin, N.N., 1959
[To the problem of the albedo of the drifting ice surface] Probl. Artiki i Antarktiki (1): 33-40.

albedo

Burt, W.V., 1954.
Albedo over wind-roughened water. J. Met. 11: 283-290.

albedos (on land)

Chia, Lin-Sien, 1967.
Albedos of natural surfaces in Barbados.
Q.Jl.R.Met.Soc., 93(395):116-120.

albedo

Donn, William L., and David M. Shaw, 1967.
The maintenance of an ice-free Arctic Ocean.
Progress in Oceanography, 4:105-113.

albedo

Hanson, K.J., 1961.
The albedo of sea-ice and ice islands in the Arctic Ocean basin.
Arctic, 14(3):188-196.

albedo

Hishida, Kozo, 1963.
On the albedo of radiation of the sea surface. (1). On the upward light in the sea.
J. Oceanogr. Soc., Japan, 19(1):37-40.

albedo

Hishida, Kozo, and Motoaki Kishino, 1965.
On the albedo of radiation of the sea surface.
J. Oceanogr. Soc., Japan, 21(4):148-153.

albedo

Komissarov, L.G., 1965.
Shipboard study of short wave radiation and albedo of water in the Baltic Sea in the summer of 1957. (In Russian).
Trudy, Gosudarst. Okeanogr., Inst., No. 87:77-88.

albedo

Langleben, M.P., 1969.
Albedo and degree of puddling of a melting cover of sea ice.
J. Glaciol. 8(54): 407-419.

albedo

Langleben, M.P., 1968.
Albedo measurements of an Arctic ice cover from high towers.
J. Glaciol., 7(50): 289-297

albedo

Larsson, Peter, 1963.
The distribution of albedo over Arctic surfaces.
Geogr. Rev., 53(4):572-579.

albedo

Langleben, M.P., 1971.
Albedo of melting sea ice in the southern Beaufort Sea.
J. Glaciol., 10(58): 101-104.

Arctic, Beaufort Sea
ice

albedo

Neumann, G., and R. Hollman, 1961
On the albedo of the sea surface. Symposium on the radiant energy of the sea surface, 4-5 Aug. 1960.
U.G.G.I., Monogr., No. 10:72-83.

albedo

Olson, Boyd E., 1962
Variations in radiant energy and related ocean temperatures.
J. Geophys. Res., 67(12):4705-4712.

albedo

Padmanabhamurty, B., and V. P. Subrahmanyam, 1964
Some studies on radiation at Waltair.
Indian J. Pure and Applied Physics, 2(9):293-295.

albedo

Pivivarov, A.A., V.S. Lavorko, 1961.
On the diurnal course of the sun radiation components and sea albedo. (In Russian).
Meteorol. i Gidrol., (1):43-46.

albedo

Predoehl, Martin C., 1964.
Aerial measurements of albedo of sea ice in the Antarctic. (Abstract).
Trans. Amer. Geophys. Union, 45(1):57.

albedo

Predoehl, Martin C., and Angelo F. Spano, 1965.
Airborne albedo measurements over the Ross Sea, October-November, 1962.
Mon. Weather Rev., 93(11):687-696.

albedo

Rutkovskaya, V.A., 1962.
Influence of physico-geographical factors on the albedo of land basins and some seas.
Izv., Akad. Nauk, SSSR, Ser. Geograf., (1):72-80

albedo

Sivkov, S.I., 1952.
[Geographical distribution of the effective amounts of albedo of water surfaces.]
Izv. Vses. Geograf. Obshch., 84(2):200-

albinism

McKenzie, Michael D., 1970.
First record of albinism in the hammerhead shark, Sphyrna lewini.
J. Elisha Mitchell scient. Soc. 86(1): 35-37.

aleurite

aleurite

Baranova, D.D. and G.G. Polikarpov 1965
Sorption of strontium-90 and cesium-137 in the aleurite silts of the Black Sea. (In Russian)
Okeanologiia 5(4): 646-648

Algae for classification & morphology see taxonomy file.

ALGAE - here limited to economic aspects. See ORGANISMAL INDEX for taxonomy, physiology and ecology - a section which is far from complete.

References here were accumulated from various sources to answer questions. Many (or most) are not in the library of the Marine Biological Laboratory, Woods Hole

algae

Doty, Maxwell S., 1971.
Physical factors in the production of tropical benthic marine algae. (Abstracts in English and Portuguese). In: Fertility of the Sea, John D. Costlow, editor, Gordon Breach, 1: 99-121.

algae, biomass

Mann, K. H. 1972.
Ecological energetics of the seaweed zone in a marine bay on the Atlantic Coast of Canada. I. Zonation and biomass of seaweeds. Mar. Biol. 12(1): 1-10.

ALGAE, boring

algae, boring

Swinchatt, J.P., 1969.
Algal boring: a possible depth indicator in carbonate rocks and sediments.
Bull. geol. Soc. Am., 80(7): 1391-1396.

bottom sediments, carbonate
algal boring
USA, east

ALGAE as fertilizer

algae, as fertilizer

Aitkin, J.B., and T.L. Senn, 1965.
Seaweed products as a fertilizer and soil conditioner for horticultural crops.
Botanica Marina, 8(1):144-148.

algae, as fertilizer

Boney, A. D., 1965.
Aspects of the biology of the seaweeds of economic importance.
In: Advances in Marine Biology, Sir Frederick S. Russell, editor, Academic Press, 3:105-253.

algae, as fertilizer

Booth, E., 1965.
The manurial value of seaweed.
Botanica Marina, 8(1):138-143.

algae, as fertilizer

Booth, C.O., 1964.
Seaweed has possibilities apart from its fertilizer use.
Grower, 62:442- (not seen)

The fecundity of black bean aphids was reduced on plants treated with Maxicrop (a liquid seaweed preparation) and the aphids preferred untreated plants. The incidence of mildew was reduced on Maxicrop-treated turnips and tomatoes.

algae, as fertilizer

Booth, E., 1953.
Seaweed as a fertilizer. Org. Gdng. 21(6):14.

algae, as fertilizer

Francki, R.I.B., 1960.
Studies in manurial values of seaweeds. 1. Effects of Pachymenia himantophora and Durvillea antarctica meals on plant growth. 2. Effects of Pachymenia himantophora and Durvillea antarctica on the immobilisation of nitrogen in soil.
Plant and Soil, The Hague, 12(4):297-310;311-323.

algae, as fertilizer

Little, E.C.S., 1953.
The decomposition rate and manurial value of some common brown seaweeds. Proc. Seventh Pacific Science Congress 5:1-9, 2 textfigs.

algae, commercial

Rybakov, O.S., 1971.
Commerce seaweeds of Shantar coastal waters. (In Russian). Izv. Tikhookean. nauchno-issled. Inst. ribn. Khoz. Okeanogr. 75: 160-164.

ALGAE, AS FOOD + USES OF

ALGAE as FOOD or FOOD ADDITIVE
Also includes edible forms
Algae, chemistry of, should also b checked.

Adrian, M., 1918. algae as food
Note on the use of certain seaweeds for feeding horses. C.R. Acad. Sci., Paris, 166:54-56.

Ash, A.S.F., 1954. algae, as food
Seaweeds as a food. Food Preservation Quarterly (Homebush, N.S.W., Australia), 14(4):71-73.

"Seaweeds compare favorably with land plants as sources of dietary constituents. They are, howeve less digestible and less palatable than are the common vegetables ----- seaweeds are probably of greater value as a stock food. --- Pigs, sheep and horses can benefit ---. 10-20%seaweed m meal. --- Poultry trials encouraging, but more than 10% seaweed meal upsets mineral metabolism of birds. Cows find meal unpalatable, but when accepted a positive response in fat content of milk.

Berry, M.H., and K.L. Turk, 1943. algae as food
The value of kelp meal in rations for dairy cattle. J. Dairy Sci., 26:754-755.

algae as food
Bersamin, S.V., S.V. Laron, F.R. Gonzales and R.B. Banania, 1962
Some seaweeds consumed fresh in the Philippines.
Indo-Pacific Fish. Council, FAO, Proc., 9th Sess. (2/3):115-119.

Black, W.A.P., 1955 algae as food
Seaweed in animal foodstuffs. 1. Availability and composition. Agriculture 62:12-15, 57-62.

2. Feeding and digestibility trials.

algae, as food
Black W.A.P., 1955.
Seaweeds and their constituents in foods for man and animal. Chemistry and Industry: 51 1640-1645.
"Brown seaweeds - unique in having exceptionally high mineral matter content and are rich in iodi both in the organic and inorganic forms. In addition to inorganic sulphates, they contain mannitol and fucoidin, all of which have been shown to have laxative effect. A somewhat similar effect may result from th e alginates as they have the power of absorbing considerable quantit -ies of water and therefore providing bulk.

Black, W.A.P., 1953. algae, as food
Seaweed as a stockfood. Agriculture, Min. Agric. 60:126-130.

reprint in MBL.

ALGAE, as food
Black, W.A.P., 1953.
Seaweeds and their value in foodstuffs. Proc. Nutr. Soc., 12:32.

algae, as food
Boney, A. D., 1965.
Aspects of the biology of the seaweeds of economic importance.
In: Advances in Marine Biology, Sir Frederick S. Russell, editor, Academic Press, 3:105-253.

algae, as food
Burlew, J.S., Ed., 1953.
Algal culture: from laboratory to pilot plant. Carnegie Inst., Washington, Publ., 600:357 pp.

algae, as food
Burt, A.W.A., S. Bartlett, and S.J. Rowland, 1954
The use of seaweed meals in concentrate mixtures for dairy cows. J. Dairy Research (Cambridge Univ. Press. 200 Euston Rd., London n.w. 1) 21(3):299-305.

"In view of the relatively high cost, low palatability and low nutritive value of seaweed meals, their possible uses in feeding dairy cows appear very limited"

Cameron, C.D.T., 1954. algae, as food
Seaweed meal in the ration for bacon hogs. Canadian J. Agric. Sci., 34:181-186.

algae, as food
Canada, Roy Wrigley Publications Ltd., 1956
German scientist makes bread with seaweed flour. Western Fish. 52(3):40.

Chritensen, O., 1964. algae as food
Carrageenan - a useful food additive. Food Manuf., 39(3):44- (not seen).

algae, as food
Dam,R., S. Lee, P.C. Fry and H. Fox,1965.
Utilization of algae as a protein source for humans.
J. Nutrition,86(4):376-382.

algae as food
Droop, M.R., 1966.
The role of algae in the nutrition of Heteramoeba clara Droop, with notes on Oxyrrhis marina Dujardin and Philodina roseola Ehrenberg.
In: Some contemporary studies in marine biology H. Barnes, editor, George Allen & Unwin, Ltd., 269-282.

Dunlop, G., 1953. algae as food
Feeding of seaweed to lactating cows. Nature 171 :439-440.

Fangauf, R., and G. von Barlöwen, 1953. algae, as food
Seaweed meal for chicken feeding.
Deutsche Wirtschaftsgeflügelzucht 5:648.

algae, as food
Galutira, Ernesta C., and Gregorio T. Velasquez, 1964.
Taxonomy, distribution and seasonal occurrence of edible marine algae in Ilocos Norte, Philippines.
Philippine J. Sci., 92(4):483-522.

algae, as food
Gibsen, K.F., and L.B. Rothe, 1955.
Algin-versatile food improver. Food Eng. 27(10):87-90.

algae, as food
Glabe, E.F., 1964. Carrageenan and hydroxylated lecithin, a stabilizer for process bread. Baker's Digest, 38(3):42- (not seen).

algae, as food
Glabe, E.F., P.F. Goldman and P.W. Anderson, 1957
How Irish moss extractive improves protein content of foods. Food Engin., McGraw-Hill, 29(1):65-68.

algae, as food
Hand, C.J.E., and C. Tyler, 1955.
The effect of feeding different seaweed meals on the mineral and nitrogen metabolism of laying hens. J. Sci. Food and Agric., 6:743-754.

10-20%. When seaweed low in energy birds lose weight and lowered egg production. No beneficial effects and 20% in some ways detrâmental.

algae as food
Hoppe,Heinz A.,1966.
Nahrungsmittel aus Meeresalgen.
Botanica Marina,9 (Suppl.) 19-40.

Lapique, L., 1918. algae as food
The use of marine algae for the feeding of horses. C.R. Acad. Sci., 167:1082-1085.
Paris.

Lunde, G., 1940. algae as food
Feedstuffs from seaweed. Papirjournalen 28:147-152.

Loose,Gunther,1966. algae as food
Die Bedeutung mariner Algen für die Welternährung.
Botanica Marina 9 (Suppl.): 7-18.

MacIntyre, T.A., and M.H. Jenkins, 1952. algae as food
Kelp meal in the ration of growing chickens and laying hens. Sci. Agric. 32:559-567.

algae, as food
McNaught, M.L., J.A.B. Smith and W.A.P. Black, 1954.
Seaweed ingested byb rumen microflora in vitro. J. Sci. Food & Agric. 5(7):350-352.

"only laminarin is readily and rapidly utilized under same conditions that starch and maltose are used"

Milner, H.W., 1953. algae as food
Algae as food; the one-celled plant Chlorella may become an important part of the world's food supply. An account of the first primary attempts to grow it economically on a mass scale. Sci. Amer., 189(4):

Algae, as food
Moor, B.J., 1964.
Seaweeds, their uses as food and in industry.
Commercial Fishing 3:26, 29, 30

Algae

algae, as food
Oishi, K., A. Okumura, K. Murata and H. Murayama, 1963.
Quality of kamba, edible seaweed belonging to the Laminariaceae. VI. Effect of the first processing. (In Japanese; English abstract). Bull. Jap. Soc., Sci. Fish., 29(4):354-358.

algae, as food
Okumura, A., Y. Tamura, K. Oishi and K. Murata, 1962.
Quality of kombu, edible seaweeds belonging to the Laminariaceae. V. Difference in the thickness and in the content of nitrogen in kombu blade, Lamina. (In Japanese; English abstract). Bull. Jap. Soc. Sci. Fish., 28(11):1123-1127.

algae, as food
Ringen, J., 1939
The nutritive value of seaweed.
Meld. Norg. Landbr. Høisk., 19: 451 - 541.

Algae as food
Schmid, Otto J., 1966.
Für die menschliche Ernährung wichtige Inhaltsstoffe mariner Algen.
Botanica Marina, 9 (Suppl.) 41-46.

algae as food
Senior, B.J., P. Collins and M. Kelly 1946
The feeding value of seaweeds
Econ. Proc. R. Dublin Soc. 3:273-291

algae as food
Sheehy, E.J., J. Brophy, T. Dillon and P. O'Muireachtain, 1942.
Seaweed (Laminaria) as stock food.
Econ. Proc. R. Dublin Soc. 3:150-161.

Shirreff, J., 1912. algae as food
Account of seaweed being exhibited to cows by way of condiment. Sinclair's Husbandry Pt. 2 No. 9:53-54.

algae, as food
Sperling, L., 1953.
Seaweed meal as fodder. Futter und Fütterung No. 33:250-253.

algae as food
Sumita, E., A. Kawabata and Y. Fujioka, 1936.
Influence of kelp meal feed on iodine content of hens eggs. Wiss. Ber., 6, Weltgeflügelkongr. Berlin, Leipzig, 1:343-346.

algae, as food
U.S. Fish and Wildlife Service, 1956.
Seaweed flour. Comm. Fish. Rev., 18(1):48.

algae, as food
Woodward, F.H., 1952.
Seaweeds - a new source of chemicals and food. World Crops 4(12):403.

Abstr.:
Tropical Abstracts 8(3):95.

ALGAE, chemistry of

ALGAE, chemistry of

includes use as drugs, antibiotic properties, vitamins, etc.

See also: Algae, uses of, for agar

algae, chemistry of
Almodovar, L.R., 1964.
Ecological aspects of some antibiotic algae from Puerto Rico.
Botanica Marina, 6; 143.

In the brown algae, but not in the red and green algae, antibiotic activity is only found in the winter months.

algae, chemistry of
Anon., 1952.
Wonder drugs from seaweeds. Chem. Eng. News 30 (36):3690.

algae, chemistry of
Black, W.A.P., 1954.
Constituents of the marine algae. Rep. Progr. Chem.(1953) 53:322-335.

172 references.

algae, chemistry of
Black, W.A.P., 1953.
Constituents of the marine algae.
Ann. Rept. Chem. Soc., 50:322.

algae, chemistry of
Chuecas, L., and J.P. Riley, 1966.
The component fatty acids of some sea-weed fats. J. mar. biol. Ass., U.K., 46(1):153-159.

algae, chemistry of
Cowey, C.B., and E.D.S. Corner, 1966.
The amino-acids composition certain unicellular algae, and of the faecal pellets produced by Calanus finmarchicus when feeding on them.
In: Some contemporary studies in marine biology, H. Barnes, editor, George Allen & Unwin, Ltd., 225-231.

algae, red (chemistry)
Eriksson, Caj E.A., and Per Halldal, 1965.
Purification of phycobilins from red algae and their fluorescence excitation spectra in visible and ultraviolet.
Physiol. Plantarum, 18:146-152.

algae, chemistry of
Etcheverry, H., 1958.
Algas marinas chilenas productoras de ficocoloides.
Rev. Biol. Mar., Valparaiso, 8(1-3):153-174.

algae, chemistry of
Hayashi, Koichiro, 1963.
Studies on Vitamin B_6 in marine products. IV.

The distribution of Vitamin B_6 in seaweeds.

Rept. Fac. Fish., Pref. Univ. of Mie, 4(3):429-435.

algae, chemistry of
*Hoyt, J.W., 1970.
High molecular weight algal substances in the sea.
Marine Biol., 7(2):93-99.

algae, chemistry of
Idson, B., 1956.
Seaweed colloids. $10 million now - and growing fast. Chem. Week (McGraw-Hill) 79(29):57-80.

used as (1) thickeners (2) humectants (3) coagulants (4) bulking agents (5) flocculation agents (6) antibiotic carriers) - only from red and brow

algae, chemistry of
Ishibashi, Masayoshi, Toshio Yamamoto, and Tetsuo Fujita, 1964
Chemical studies on the ocean (Part 93).
Chemical studies on the seaweeds (18). Nickel content in seaweeds.
Rec. Oceanogr. Wks., Japan, 7(2):25-32.

algae, chemistry of
Ishibashi, Masayoshi, Toshio Yamamoto and Tetsuo Fukita, 1964
Chemical studies on the ocean (Part 92).
Chemical studies on the seaweeds (17). Cobalt content in seaweeds.
Rec. Oceanogr. Wks., Japan, 7(2):17-24.

algae, chemistry of
Ishibashi, Masayoshi, Taitiro Funinaga, Fuji Morii, Yoshihiko Kanchiku, and Fumio Kamiyama, 1964
Chemical studies on the ocean (Part 94).
Chemical studies on the seaweeds (19). Determination of zinc, copper, lead, cadmium and nickel in seaweeds using dithizone extraction and polarigraphica method.
Rec. Oceanogr. Wks., Japan, 7(2):33-63.

algae, chemistry of
Ishibashi, Masayoshi, Toshio Yamamoto and Fuji Morii, 1962
Chemical studies on the ocean (Pt. 85).
Chemical studies on the seaweeds (LL).
Copper content in seaweeds.
Rec. Oceanogr. Wks., Japan, n.s., 6(2):157-168

algae chemistry of
Ishibashi, Masayoshi, Taitito Fujinaga, Tooru Kuwamoto, 1960

Chemical studies on the ocean (82). Chemical studies on the seaweeds. Quantitative determination of beryllium in seaweeds. Rec. Oceanogr. Wks., Japan, Spec. No. 4: 87-90.

algae chemistry
Ishibashi, Masayoshi, Taitiro Fujinaga, Tooru Kuwamota and Takahiro Kumamaru, 1960

Chemical studies on the ocean (83). Chemical studies on seaweeds. Quantitative determination of strontium in seaweeds. Rec. Oceanogr Wks., Japan, Spec. No. 4: 91-94.

algae, chemistry of
Ishbashi, M., and T. Yamamoto, 1960.
Inorganic constituents of seaweeds.
Rec. Oceanogr. Wks., Japan, 5(2):55-62.

algae chemistry of
Ishibashi, Masayoshi, and Toshio Yamamoto, 196[illegible]

Chemical studies on the ocean (Pts. 78-79).
Chemical studies on the seaweeds (VI & VII).
The content of calcium, magnesium and phosphorus in seaweeds. Iron content in seaweeds.
Rec. Oceanogr. Wks., Japan, Spec. No. 4: 73-78 79-86.

algae, chemistry of
Kappanna, A.N., and V. Sitakara Rao, 1962
Iodine content of marine algae from Gumarat Coast.
J. Sci. Industr. Res., (B), 21(11):559-560.

algae, chemistry of
Katayama, T., 1959.
Chemical studies on the volatile constituents of seaweed. 14. On the volatile constituents of Laminaria sp. (2).
Bull. Jap. Soc. Sci. Fish., 24(11):925-932.

algae, chemistry of
Krishna Pillai, V., 1957.
Chemical studies on Indian sea weeds. II. Partition of nitrogen.
Proc. Indian Acad. Sci., (B), 45(2):43-63.

algae, chemistry of
Krishna Pillai, V., 1957.
Chemical studies on Indian Seaweeds. III. Partition of sulphur and its relation to the carbohydrate content.
Proc. Indian Acad. Sci., B, 45(3):101-121.

algae, chemistry of
Kuenzler, Edward J., and James P. Perras 1965
Phosphatases of marine algae.
Biol. Bull. mar. biol. Lab. Woods Hole, 128(2): 271-284

algae Chemistry of
Landau, E., 1964.
Pond-producing protein. New Scientist, No. 417: 445- (not seen).

an account of large-scale production of Chlorella in the Micro-algae Research Institute of Tokyo.

algae, chemistry of
Lewis, E.J., 1962.
Studies on the proteins, peptides and free amino acid contents in some species of brown algae from southeastern coast of India.
Rev. Algologique, n.s., 6(3):209-216.

algae, chemistry of
Lopez-Benito, Manuel, 1963
Estudio de la composicion quimica del Lithothamnium clacareum (Aresch) y su aplicacion como corrector de terrenos de cultivo.
Inv. Pesq., Barcelona, 23: 53-70.

algae, chemistry of
Lyman, C.M., K.A. Kuiken, and F. Hale, 1956.
Essential amino acid content of farm feeds.
J. Agricult. and Food Chem., 4(12):1008-1014.

algae, chemistry of
Munda, Ivka, 1964.
The quantity of alginic acid in Adriatic brown algae, (In Jugoslavian; English resume).
Acta Adriatica, 11(1):205-213.

algae, chemistry of
Munda, Ivka, 1962
Relation between the amount of phydodes and reducing compounds in some Adriatic brown algae. (In Jugoslavian: English summary).
Acta Adriatica, Inst. Oceanogr. i Ribarst., Split, 10(1):3-11.

algae, chemistry of
Percival, Elizabeth, and J.K. Wold, 1963.
The acid polysaccharide from the green seaweed Ulva lactuca. II. The site of the ester sulfate. J. Chem. Soc., London, November (1040):5459-5468.

algae, chemistry of
Sieburth, John McN., 1964.
Antibacterial substances produced by marine algae.
Developments in Industrial Microbiology, 5:124-134.

algae, chemistry of
Skoenya, S.C., T.M. Paul and D. Waldron-Edward, 1964.
Studies on inhibition of intestinal absorption of radioactive strontium. 1. Prevention of absorption from ligated intestinal segments. II. Effects of administration of sodium alginate by orogastric intubination feeding. III. The effect of administration of sodium alginate in food and drinking water. Canadian Med. Assoc. J., 91: 235- , 553- , 1006- (not seen
A series of papers which show that sodium alginate effectively reduces the absorption of radioactive strontium without interfering with the absorption of calcium!

algae, biochemistry of
Strusi, Angelo, 1962.
Sulla composizione della proteine e ammino acidi liberi di alcune specie di alghe nel Mar Piccolo di Taranto.
Boll. Pesca, Piscicolt. Idrobiol., n.s., 17(1): 107-118.

algae, chemistry of
*Takagi, Mitsuzo, Keiichi Oishi, and Ayako Okumura, 1967.
Free amino acid composition of some species of marine algae. (In Japanese; English abstract).
Bull. Jap. Soc. scient. Fish., 33(7):669-673.

algae, chemistry of
Suzuki, N., 1955.
Studies on the manufacture of algin from Brown algae. Mem. Fac. Fish., Hokkaido Univ., 3(2): 93-258.

algae, decomposition of
Thorstenson, Donald C., and Fred T. Mackenzie 1971
Experimental decomposition of algae in seawater and early diagenesis.
Nature, Lond. 234 (5331): 543-545

algae, chemistry of
Woodward, F.H., 1952.
Raw materials from seaweeds. Chem. & Ind. No. 11: 243.

algae, chemistry of
Woodward, N., 1951.
Sea weeds as a source of chemicals and stockfeed. J. Sci. of Food and Agric. 2(11):477-487.

algae, chemistry of
*Yamamoto, Toshio, Tetsuo Fujita, Tsunenobu Shigematsu and Masayoshi Ishibashi, 1968.
Chemical studies on the seaweeds. (23). Molybdenum content in seaweeds.
Rec. oceanogr. Wks., Japan, n.s., 9(2):209-217.

algae, chemistry of
Youngblood, W.W., M. Blumer, R.L. Guillard and F. Fiore, 1971.
Saturated and unsaturated hydrocarbons in marine benthic algae. Marine Biol. 8(3): 190-201.

algae, chemistry of hydrocarbons 2582

algae, drifting of

algae, drifting
Milton, Daniel J., 1966.
Drifting organisms in the Precambrian sea.
Science, 153(3733):293-294.

algal drift
Segawa, Sokichi, Takeo Sawada, Masahiro Higaki, Tadao Yoshida, Hajime Ohshiro and Fumio Hayashida, 1962
Some comments on the movement of the floating seaweeds.
Rec. Oceanogr. Wks., Japan, Spec. No. 6, 153-159.
Also in: Contrib. Dept. Fish. and Fish. Res. Lab., Kyushu Univ., No. 8.

algae, ecology of
Mann, K. H. 1972.
Ecological energetics of the seaweed zone in a marine bay on the Atlantic Coast of Canada. I. Zonation and biomass of seaweeds. Mar. Biol. 12(1): 1-10.

ALGAE, effect of

algae, effect of
Anraku, Masateru, and Masanori Azeta 1969.
The feeding activities of yellowtail larvae Seriola quinqueradiata Temmink et Schlegel, associated with floating seaweeds. (In Japanese; English abstract)
Bull. Seikai reg. Fish. Res. Lab. 35: 41-50.

algae, effect of
Hommeril, P., and M. Rioult, 1965.
Étude de la fixation des sédiments meubles par deux algues marines: Rhodothamniella floridula (Dellwyn) J. Feldm. et Microcoleus chtonoplastes Thur.
Marine geology, Elseirer Publ. Co., 3(1/2):131-155.

Algae, effect of
Hommeril, P., and M. Rioult, 1962.
Phenomènes d'erosion et de sedimentation marines entre Sainte-Honorine-des-Pertes et Port-en-Bessin (Calvados). Role de Rhodothamniella floridula dans la retenue des sediments fins.
Cahiers Océanogr., C.C.O.E.C., 14(1):25-45.

algae, effect of
Mitani, Fumio, 1965.
An attempt to estimate the population size of the "majoko" the juvenile of the yellowtail, Seriola quinqueradiata Temminck et Schlegel, from the amount of the floating seaweeds based on the observations made by the means of aeroplane and vessels. 1. State of distribution of the floating seaweeds. 2. Estimate of the amount of floating seaweeds. (In Japanese; English abstract).
Bull. Jap. Soc. Sci. Fish., 31(6):423-428; 429-434.

algae, effect of.
Stockman, K.W., R.N. Ginsburg and E.A. Shinn 1967.
The production of lime mud in South Florida.
J. sedim. Petrol. 37 (2): 633-648.

algae, effect of
Sturdy, Gladys, and William H. Fischer 1966.
Surface tension of slick patches near kelp beds.
Nature, Lond. 211 (5052): 951-952.

ALGAE; harvesting

algae, harvesting
Anon., 1969.
Compressor 'vacuum' harvests Irish sea moss.
Undersea Techn., 10(2):50-51.

algae, harvesting
Boney, A. D., 1965.
Aspects of the biology of the seaweeds of economic importance.
In: Advances in Marine Biology, Sir Frederick S. Russell, editor, Academic Press, 3:105-253.

algae, harvest
Jackson, P., and R. Wolff, 1955.
Sublittoral seaweed harvester. Research 8(11):435-444.

algae, harvesting of
Mathieson, Arthur C., 1969.
The promise of seaweed: a kelp harvester- the Kelmar - is shown mowing Macrocystis pyrifera along Southern California coast.
Oceanol. Int., 4(1):37-39. (popular).

algae, harvesting of
Scholes, William A., 1969.
Harvesting seaweed off Australia.
Ocean Industry, 4(3):69-70.

algae, harvest
Woodward, F.N., 1955.
The importance of the algae. 1. World resources and composition. Times Review of Industry, Oct. 4 pp.
2. Harvesting and Technical Uses.
"it would seem that the use of red and brown seaweeds as raw materials for the chemical industry and of brown seaweeds as feeding stuff additives is likely to expand significantly in the years to come. It is unlikely that marine algae will be used in human food any more extensively in the future than at the present time. If, however, the techniques for the mass culture of microscopic algae can be so improved

algae, radioactivity
Bernhard, M., preparator, 1965.
Studies on the radioactive contamination of the sea, annual report 1964.
Com. Naz. Energ. Nucleare, La Spezia, Rept., No. RT/BIO (65) 18:35 pp.

algae, radioactivity
Folsom, Theodore R., 1958.
Approximate doses close to submerged radioactive layers of biological interest.
Proc. Ninth Pacific Sci. Congr., Pacific Sci. Assoc., 1957, Oceanogr., 16:170-175.

ALGAE, resources

algae, resources
Boney, A. D., 1965.
Aspects of the biology of the seaweeds of economic importance.
In: Advances in Marine Biology, Sir Frederick S. Russell, editor, Academic Press, 3:105-253.

algae, resources
Ciszewski, P., K. Demel, Z. Ringer and M. Szatybelko, 1962.
Resources of Furcellaria fastigiata in Puck Bay estimated by the diving method.
Prace Morsk. Inst., Ryback., A. Oceanogr.-Ichtiol., Gdyni, 11(A):9-36.
In Polish with Polish, English and Russian summaries

algae, resources
Endo, T., and Y. Matsudaira, 1960.
Correlation between water temperature and the geographical distribution of some economic seaweeds.
Bull. Jap. Soc. Sci. Fish., 26(9):871-872.

seaweeds
Jerome, William C., Jr., Arthur P. Chesmore, and Charles D. Anderson, Jr. 1968.
A study of the marine resources of the Parker-River-Plum Island Sound Estuary.
Monogr. Ser. Comm. Mass. Div. Mar. Fish. 6: 79 pp.

algae, resources
Kireeva, M.S., 1965.
Algal resources of the seas of the Soviet Union. (In Russian).
Okeanologiia, Akad. Nauk, SSSR, 5(1):14-21.

algae, resources
Rao, M. Umamaheswara, 1970.
The economic seaweeds of India.
Bull. Centr. mar. Fish. Res. Inst. 20: 68 pp. (mimeographed)

algae, resources
Scagel, Robert, F., 1961.
Marine plant resources of British Columbia.
Fish. Res. Bd., Canada, Bull., No. 127:1-39.

algae, resources
Schwenke, Heinz, 1965.
Beiträge zur angewandten marinen Vegetationskunde der westlichen Ostsee (Kieler Bucht).
Kieler Meeresforsch., 21(2):144-152.
(not profitable in Baltic)

algae, resources

USN Hydrographic Office, 1957.
Oceanographic atlas of the Polar seas. 1. Antarctic. H.O. Pub., No. 705:70 pp.

algae, resources

Wohnus, J.F., 1942
The Kelp resources of southern California.
California Fish and Game 28:199-205.

ALGAE standing crop

algae, standing crop

Boney, A. D., 1965.
Aspects of the biology of the seaweeds of economic importance.
In: Advances in Marine Biology, Sir Frederick S. Russell, editor, Academic Press, 3:105-253.

ALGAE, uses

ALGAE, uses and products

algae, uses

Anon., 1950.
Expansion of seaweeds products industries planned (Norway). Comm. Fish. Rev. 12(1):48.

algae, uses of

Canada, Nova Scotia Research Foundation, 1966.
Selected bibliography on algae, No.7:105 pp.

algae, uses of

Chapman, V.J., 1951.
The economic uses of seaweeds. Methuen, London.

algae, uses of

Chase, F.M., 1942.
Useful "algae". Ann. Rept. Smithsonian Inst., 1941(Publ. 3651):401-452.

algae, uses of

De Virville, Ad. Davy, and J. Feldman, 1964
Comptes Rendus du IVe Congrès International des Algues Marines, Biarritz, Sept. 1961.
Pergamon Press, 467 pp.

algae, uses of

Jackson, Daniel F., Editor, 1964.
Algae and man.
Plenum Press, N.Y., 434 pp.

mostly freshwater

ALGAE, uses of

Krishnamurthy V., editor, 1967
Proceedings of the Seminar on Sea, Salt and Plants held at CSMCRI-Bhavnagar on December 20-23 1965, Central Salt + Marine Chemicals Research Institute, 372 pp.

algae, uses of
algae, chemistry of

algae, uses of

Levring, Tore, Heinz A. Hoppe and Otto J. Schmid 1969
Marine algae: a survey of research and utilization.
Botanica Marina Handbooks, Cram, De Gruyter & Co., Hamburg, 1:421 (140 DM)

algae, uses of

North, Wheeler J., and Carl L. Hubbs, editors, 1968.
Utilization of kelp-bed resources in Southern California.
Fish Bull., Dept.Fish Game, Calif. 139:264.

algae, uses of

Rhydwen, W., 1954.
The many uses of Irish moss. Trade News, Canada, 7(2):8-10.

$300,000 in 1953, in Maritime Provinces food industry, stabilizer in chocolate milk, ice cream, pudding powders, hand lotions, shoe polishes, shaving cream, cough medecine, also leather dressings, as a size for cloth, a thickener for colors in calico printing, in candy, and for clearing beer.

algae, uses of

Stirni, A.R., 1955.
More about kelp. J.U.S.C. & G.S., No. 6:75-79.

algae, uses

Tseng, C.K., 1947.
Seaweed resources of North America and their utilization. Economic Botany 1:

algae, uses

Tseng, C.K., 1944.
Utilization of seaweed. Sci. Mon. 59:37-46.

algae, uses

Tseng, C.K., 1944.
Agar: a valuable seaweed product. Sci. Mon. 58: 24-32.

algae, uses

U.S. Fish and Wildlife Service
118FL Agar, agaroids and the American agar industry
173FL Agar and other seaweed gums: a summary of data on chemical and physical properties
261FL Partial bibliography relating to agar
263FL Japanese agar agar industry
306FL Strength and measurement of agar gels
307 FL Syneresis of agar gels
335FL Studies of bacteriological agar.

allergies, marine

allergies, marine

Bonnevie, P., 1948.
Fishermen's "Dogger Bank Itch", an allergic contact-eczema due to the corralline Alconidium hirsutum, the "Sea Chervil". Acta allergologica 1 (1):40-46.

alloys

Anon 1967.
Aluminum alloys keyed for deep ocean applications: profile - Aluminum Company of America oceanographic program.
UnderSea Techn. 8(4): 20-21

Alloys

alloys

Boyd, W.K., 1969.
Comments on "Coping with the problem of the stress-corrosion cracking of structural alloys in sea-water" by B.F. Brown.
Ocean Engng 1(3): 297

Corrosion
alloys

alloys

* Brown, B.F., 1969.
Coping with the problem of the stress-corrosion cracking of structural alloys in sea water.
Ocean Engng. 1(3): 291-296.

~~alloys~~

Campbell Hector S., 1969.
The compromise between mechanical properties and corrosion resistance in copper and aluminum alloys for marine applications.
Ocean Engng, 1(4): 387-393

alloys

Charnock, H., and S.A. Yeo, 1959
Thermoelectric e.m.f. of tin-cadmium alloys.
J. Sci. Instr., 36: 478-479.

alloys - metals, fatigue

Christopher, P.R., and D.R. Crabbe, 1969
A discourse on factors which control fatigue behavior in high yield low alloy steel structures.
Ocean Engng 1(5): 497-520

alloys

Fleetwood, M.J., 1969.
Non-ferrous alloys for ocean engineering.
Oceanol. int. 69, 1: 6 pp., 6 tables, 4 figs.

Alloys

Gatzek, L.E., 1965.
Corrosion prevention techniques for a ship-board missle system.
Trav. Centre de Recherches et d'Études Océanogr. n.s., 6(1/4):83-92.

alloys

Geld, Isidore, and Samuel H. Davang, 1970
Stress corrosion of a coated titanium alloy plate in sea water.
Ocean. Engng 1(6): 611-616

Alloys, aluminum

Godard, Hugh P., and F.F. Booth, 1965.
Corrosion behaviour of aluminium alloys in seawater.
Trav. Centre de Recherches et d'Études Océanogr. n.s., 6(1/4):37-52.

Alloys

May. T. P., and B.A. Weldon, 1965.
Copper-nicket alloys for service in sea water.
Trav. Centre de Recherches et d'Études Océanogr. n.s., 6(1/4):141-156.

alloys

Williams W. Lee, 1969.
Development of structural titanium alloys for marine applications.
Ocean Engng, 1(4): 375-383

materials, alloys

Amber

amber

Osing, Olga, 1967.
Baltic sea-gold.
Sea Frontiers, Univ. Miami, 13(2): 88-92

Ambergris

ambergris

Anon., 1957.
Ambergris and grisambrol. Drug & Cosmetic Ind., 80(7):100.

ambergris

Anon., 1956.
Free of the whale's whim (ambergris).
Chemical Week (McGraw-Hill) 78:61.

ambergris

Clarke, R., 1954.
A great haul of ambergris. Nature 174:155-156.

Also in: Norsk Hvalfangstid 43 Årg(8):450.

ambergris
Janistyn, H., 1956.
Ambergris. Drug and Cosmetic Industry (101 W. 31st St., N.Y. 1) 79(3):322-325.

ambergris
Muniz Angues, L., 1957.
El ambar gris, substancia de las ballenas negras. Industrias Pesqueras, 31(713):9.

ambergris
U.S. Fish and Wildlife Service
280FL Ambergris
281FL Ambergris in fisheries abstracts.

ambiguity function analysis
Angelari, Richard D., 1970.
The ambiguity function applied to underwater acoustic signal processing: a review. Ocean Engng 2(1):13-26.

amphibious warfare
Sanders, I.T., 1953.
Hydrographic surveys for amphibious wartime operations. J. CGS No. 5:65-78.

amphidromes
Irish, James, Walter Munk and Frank Snodgrass 1971.
M_2 amphidrome in the northeast Pacific.
Geophys. fluid Dynam. 2(4): 355-360.

Amphioxus — lancelets
*Cory, Robert L., and E. Lowe Pierce, 1967.
Distribution and ecology of lancelets (Order Amphioxi) over the continental shelf of the southeastern United States.
Limnol. Oceanogr., 12(4):650-656.

Amphioxus
Lancelets
Pierce, E. Lowe, 1965.
The distribution of lancelets (Amphioxi) along the coasts of Florida.
Bull. Mar. Sci., 15(2):480-494.

abstract

amphipacific distribution
Uschakov, P.V. 1971.
Amphipacific distribution of polychaetes.
J. Fish. Res Bd Can. 28(10): 1403-1406

amplitudes, sound
Borisova, E.A., V.P. Kodanev and V.A. Tuzikov 1971.
Results of an experimental investigation of sound field amplitude fluctuations in the sea. (In Russian)
Akust. Zh. 17(3): 463-464
Translation: Soviet Phys. Acoustics 17(3): 387-388

anaerobic basins
Richards, F.A., and R.F. Vaccaro, 1956.
The Cariaco Trench, an anaerobic basin in the Caribbean Sea. Deep-Sea Res., 3(3):214-228.

Anaerobiosis, effect of
Berger, Wolfgang H., and Andrew Soutar, 1970.
Preservation of plankton shells in an anaerobic basin off California.
Bull. geol. Soc. Am., 81(1): 275-282

Santa Barbara Basin
anaerobiosis effect of shell preservation

anaerobiosis
Lewin, Ralph A., Editor, 1962.
Physiology and biochemistry of algae.
Academic Press, New York and London, 929 pp.

analysis, harmonic
Munkelt, Karl, 1959.
Formelin zur harmonischen Analyse von Gezeitenerscheinungen, denen ein unbekannter Gang überlagert ist.
Deutsche Hydrogr. Zeits., 12(4): 189-195.

analysis, harmonic
Ursell, F., 1950.
On the application of harmonic analysis to ocean wave research. Science 111:445-446, 1 fig.

With a reply by H.R. Seiwell.

analysis, power-spectrum
Carrozzo, M.T., e F. Mosetti, 1965.
Alcune osservazioni sullo studio dei fenomeni fluttuanti ordinati e casuali.
Boll. G ofis. Teorica ed Appl., 7(26):120-143.

English summary, pp. 121-122.

analysis, statistical
Lohse, E.A., 1955.
A theoretical curve for statistical analysis of sediments. J. Sed. Petrol. 25(4):293-296.

anaphylaxis
Shadewaldt, H., 1965.
La croisière du Prince Albert 1er de Monaco en 1901 et la découverte de l'anaphylaxie.
Collogue Internat., Hist. Biol. Mar., Banyuls-sur-Mer, 2-6 sept. 1963, Suppl., Vie et Milieu, No. 19:305-313.

anchor drift
Sisoev, N.N., 1951.
[Anchor drift graph] Trudy Inst. Okeanol., 5:46-49.

anchor drift
Sisoev, N.N., 1951.
[Free movement of a ship at anchor and the method of their calculation] Trudy Inst. Okeanol., 5:35-45.

anchor ice
Benedicks, C., and P. Sederholm, 1943.
Regarding the formation of anchor (ground) ice.
Ark. Mat. Astron. och Fysik A 29(22): 7 pp.

anchor ice
*Dayton, Paul K., Gordon A. Robilliard and Arthur L. DeVries, 1969.
Anchor ice formation in McMurdo Sound Antarctica, and its biological effects.
Science, 163(3864):273-274.

anchor ice
*DeVries, Arthur L., and Donald E. Wohlschlag, 1969.
Freezing resistance in some Antarctic fishes.
Science, 163(3871):1073-1075.

anchor ice
Nybrand, G.L., 1943.
Bildning av Bottenis och sörpa i ett rinnende Wattendrag.
Teknisk Tidskr. Väg- och Vattenbygadskonst samt Husbyggnadsteknik 1(4):12 pp., 8 figs.

anchor ice
Piotrovich, V.V., 1956.
[Formation of depth-ice] Priroda, 1956(9):94-95.

T235R

anchoring
Abe, Shigeo, 1967.
On the thrust of Nagasaki-maru in keeping anchoring in a strong wind. (In Japanese; English abstract)
Bull. Fac. Fish. Nagasaki Univ. 22:115-119

anchoring
Abe, Shigeo, 1966.
On the anchoring in strong wind at single anchor of the variable pitch propeller ship.
(In Japanese; English abstract).
Bull. Fac. Fish., Nagasaki Univ., No. 20:70-86.

anchoring
Foster, K.W., 1966.
Dynamic anchoring of floating vessels.
Exploiting the Ocean, Trans. 2nd Mar. Techn. Soc., Conf., June 27-29, 1966, 158-172.

anchorage systems
Hromadik, Joseph J., 1968.
Deep ocean installations and fixed structures.
Ch. 10 in: Ocean engineering: goals, environment, technology, J.F. Brahtz, editor, John Wiley & Sons, 310-349.

anchoring, deep-sea
Isaacs, John D., 1963
24. Deep-sea anchoring and mooring. In: The Sea, M.N. Hill, Edit., Vol. 2 (V) Oceanographic Miscellanea, Interscience Publishers, New York and London, 516-527.

anchoring
O'Brien, T.J., and B.J. Mayo, 1963.
Sea tests on a swing-moored aircraft carrier.
Naval Civil Engineering Lab., Port Hueneme, AD-417 203 Div. 31, 78 pp.
(TISTE/OHD) OTS $8.00

Abstr.
U.S. Govt. Res. Repts., 1964, 39(2):73-74.

anchoring
Seiwell, H. R., 1940.
Anchoring ships on the high seas. Proc. U. S. Naval Inst. 66(12)[Whole No. 454]:1733-1740, 3 pls.

WHOI Contrib. No. 237.

anchoring
Sergeev, I.V., 1959
[Experience of placing the ship at anchor at great depths.]
Trudy Inst. Okeanol., 35: 206-224.

anchoring
Sharpey-Schafer, J.M., 1954.
Anchor dragging. J. Inst. Navig. 7(3):290-300.

anchoring
Snyder, Robert M., 1968.
Free-fall anchoring systems.
Mar. Sci. Instrument. 4: 189-195

anchor stations
Syssoev, N.N., 1957.
[The development of the technics and the methods of work for the observation of currents in the ocean at the anchored stations] Trudy Inst. Okean 25:7-24.

Anchor stations
see also: Buoy stations
ANCHOR STATIONS, see also: BUOY STATIONS

"anchor stations"
Azbel M. Ye., 1957.
Ship remote hydrometeorological station. (In Russian).
Trudy N.-I. In-ta. Gidrometeorol. Priborostr. Works, Sci. Res. Inst. Hydrometeorol. Instr. Bldg.,), No. 4, 7:87-97.
RZHGeofiz 9/57-8377.

anchor stations
Burkov, V.A., 1957.
[About the methods of the observations for the currents at the deep-sea diurnal anchored stations.] Trudy Inst. Okeanol., 25:25-44.

anchor stations
Ekman, V.W., 1953.
Studies on ocean currents. Results of a cruise on board the "Armauer Hansen" in 1930 under the leadership of Bjørn Helland-Hansen. Geograf. Publ. 19(1):106 pp. (text), 122 pp. (92 pls.).

anchor stations
Gieskes, J.M., J. Meincke und A. Wenck 1970.
Hydrographische und chemische Beobachtungen auf einer Ankerstationen im östlichen Nordatlantischen Ozean.
Meteor Forschungsergebn. (A) 8: 1-11

anchor stations
Hansen, W., 1954.
Die atlantischen Ankerstationen des "Armauer Hansen" im Jahre 1930 in der Bearbeitung von V.W. Ekman. Deutsche Hydrogr. Zeits. 7(1/2):55-58, 2 textfigs.

anchor station
Hussels, W. 1971.
Die Vertikalgeschwindigkeit von Bewegungen mit Gezeitenperiode im Gebiet der Grossen Meteorbank.
Meteor Forsch.-Engebn. (A): 58-66.
9

anchor stations
*Joseph, J., 1967.
Current measurements during the International Iceland-Faroe Ridge Expedition, 30 May to 18 June, 1960.
Rapp. P.-V. Réun. Cons. perm. int. Explor. Mer, 157:157-172.

anchor stations
Joseph, J., 1954.
Die Sinkstoffführung von Gezeitenströmungen als Austauschproblem. Arch. Met., Geophys., Bioklim. A, 7:482-501, 12 textfigs.

anchor stations
Parr, A. E., 1937
Report on hydrographic observations at a series of anchor stations across the Straits of Florida. Bull. Bingham Ocean. Coll. VI (3):1-62, 36 text figs.
Contribution No. 155 of the Woods Hole Oceanographic Institution.

anchor stations
Saelen, Odd H., 1963
Studies in the Norwegian Atlantic Current. II. Investigations during the years 1954-1959 in an area west of Stad.
Geofysiske Publikasjoner, 23(6):1-82.

anchor stations
v. Schubert, O., 1939
II Die ozeanographischen Arbeiten auf der zweiten Teilfahrt des Deutschen Nordatlantischen Expedition, Januar bis Juli 1938. Bericht über die zweite Teilfahrt der Deutschen Nordatlantischen Expedition des Forschungs-und Vermessungsschiffes "Meteor" Januar bis Juli 1938. Ann. Hydrogr. u. Marit. Meteorol. 1939, 11-18, 6 text figs.

anchor stations
Seiwell, H. R., 1940.
Preliminary results of measurements of temperature and salinity variations in the Gulf Stream. Trans. Am. Geophys. Union 1940:349-352, 2 text-figs.
WHOI Contrib. No. 263.

anchor stations
*Siedler, Gerold, 1968.
Schichtungs- und Bewegungsverhälnisse am Südausgang des Roten Meeres.
Meteor Forschungergeb. (A)4:1-76.

anchor stations
Trites, R.W., 1956.
The oceanography of Chatham Sound, British Columbia. J. Fish. Res. Bd., Canada, 13(3):385-434.

Andesites

andesites
*Dickinson, William R., and Trevor Hatherton, 1967.
Andesitic volcanism and seismicity around the Pacific.
Science, 157(3790):801-803.

Andesite Line

Andesite Line
*Hédervári, P., 1967.
Investigations regarding the earth's seismicity. 5. On the earthquake geography of the Pacific Basin and the Seismotectonical importance of the Andesite Line.
(Gerlands) Beitr. Geophys. 76(5):393-405.

Anelasticity

anelasticity
Eckart, C., 1948
The theory of the anelastic fluid. Rev. Modern Phys. 20(1):232-235, 2 text figs. (Comm. No.4, Mar. Phys. Lab., Univ. Calif.)

anelasticity
Eckart, K., 1948
The thermodynamics of irreversible processes the theory of elasticity and anelasticity. Phys. Rev. 73(4):373-382, 3 text figs. (Comm. No.3, Mar. Phys. Lab., Univ. Calif.)

Animal attacks

Animal attacks?
Snoke, L.R., 1957.
Resistance of organic materials and cable structures to marine biological attack.
Bell. System Techn. J., 36(5):1095-1128.

Animal distribution

animal distribution
Spärck, R., 1957
The importance of metabolism in the distribution of marine animals. L'Année Biol., (3)33(5/6): 233-240.

animal diversity
Johnson Ralph Gordon 1971.
Animal-sediment relations in shallow water benthic communities.
Mar. Geol. 11 (2): 93-104.

Anistrophy, seismic

Anisotrophy
Backus, George E., 1965.
Possible forms of seismic anisotropy of the uppermost mantle under oceans.
J. Geophys. Res., 70(14):3429-3439.

anistrophy
Keen, C.E., and D. L. Barrett 1971.
A measurement of seismic anistropy in the northeast Pacific.
Can. J. Earth Sci. 8 (9): 1056-1064.

anisotropy
*Raitt, R.W., G.G. Shor, Jr., T.J.G. Francis, and G.B. Morris, 1969.
Anisotropy of the Pacific upper mantle. J. Geophys. Res., 74(12): 3095-3109.

anistropy
Raitt, R.W., G.G. Shor, Jr., G.B. Morris and H. K. Kirk, 1971.
Mantle anisotropy in the Pacific Ocean.
Tectonophysics 12 (3): 173-186.

annual cycle
Tett, P.B., 1972.
An annual cycle of flash induced luminescence in the euphausiid Thysanoessa raschii. Mar. Biol. 12(3): 207-218.

Anomalies

anomalies, biological
Dawson, C.E. 1971.
A bibliography of anomalies of fishes.
Gulf Res. Repts. 3(2): 215-239.

anomalies, climatic
*Namias, Jerome, 1970.
Climatic anomaly over the United States during the 1960's.
Science, 170(3959):741-743.

anomalies, geomagnetic
Nagata, T., T. Oguti and S. Kakinuma, 1958.
Geomagnetic anomaly between South Africa and Antarctic continent roughly along 40 E meridian line.
Geophys. Notes, 11(2):Contrib. 26.
Originally published in:
Proc. Japan. Acad., 34(7):427-431.

anomalies, thermostéric
Montgomery, R.B., and W.S. Wooster, 1954.
Thermosteric anomaly and the analysis of serial oceanographic data. Deep-Sea Res. 2(1):63-70, 3 textfigs.

anomalies, thermosteric
Yoshida, K., and M. Tsuchiye, 1958.
Note on the thermosteric anomaly.
Geophys. Notes, Tokyo, 11(1):No. 9.

anorthosite
*Engel, C.G. and R.L. Fisher, 1969.
Lherzolite, anorthosite, gabbro and basalt dredged from the Mid-Indian Ocean Ridge.
Science, 166(3909): 1136-114 .

anoxia
Fujita, Yuji, Shoji Iizuka and Buhei Zenitani, 1969.
Microbiological studies on shallow marine areas. V. On the distribution of the sulfate-reducing bacteria and sulfur bacterial groups in oxygen-deficient or anoxic bottom layer of bay in summer. (In Japanese)
Bull. Fac. Fish. Nagasaki Univ. 28: 153-160

anoxic basins
Richards, Francis A. 1965.
Chemical observations in some anoxic sulfide-bearing basins and fjords.
Proc. Int. Water Pollution Res. Conf., Tokyo, 1964: 215-243.

anoxic basins

*Richards, Francis A., James J. Anderson and Joel D. Cline, 1971.
Chemical and physical observations in Golfo Dulce, an anoxic basin on the Pacific Coast of Costa Rica. Limnol. Oceanogr. 16(1): 43-50.

anoxic fjords

Piper, David Zink 1971.
The distribution of Co, Cr, Cu, Fe, Mn, Ni and Zn in Framvaren, a Norwegian anoxic fjord.
Geochim. cosmochim. Acta 35 (6): 531-550.

anoxic water
see also basins, anoxic

anoxic water

Mori, Isamu, and Haruhiko Irie, 1966.
The hydrographic conditions and the fisheries damages by the red water occurred in Omura Bay in summer 1965 - III. The oceanographic condition in the offing region of Omura Bay during the term of the red water. (In Japanese; English abstract).
Bull. Fac. Fish. Nagasaki Univ., No. 21:103-113.

antibiotic

antibiotics

Alain-Regnault, M., 1964.
Quand l'oceanographie devient medicale. Le dernier antibiotique dont on parle.
Sciences et Avenir, No. 210:544-549. (not seen).

antibiotics

Almodôvar, Luis R., 1964.
Ecological aspects of some antibiotic algae in Puerto Rico.
Botanica Marina, 6(1/2):143-146.

Also in:
Contrib. Inst. Mar. Biol. Univ., Puerto Rico, 5(1964).

antibiotics

Almodovar, L.R., 1964.
Ecological aspects of some antibiotic algae from Puerto Rico.
Botanica Marina, 6; 143.

In the brown algae, but not in the red and green algae, antibiotic activity is only found in the winter months.

antibiotics

Aubert, J., and Cl. Jores 1971.
Action antibiotique de quelques espèces phytoplanctoniques marines vis-à-vis de différentes salmonelles.
Rev. int. Océanogr. Méd. 22-23: 143-149

antibiotics

Aubert, M., J. Aubert et M. Gauthier, 1965.
Pouvoir autoépurateur de l'eau de mer et substances antibiotiques produites par les organismes marins.
Rev. intern. Océanogr. Med., 10: 137-207.

Antibiotic

Aubert, M., J. Aubert, M. Gauthier et S. Daniel, 1966.

Origine et nature des substances antibiotiques présentes dans le milieu marin.
1. Étude bibliographique et analyse des travaux.
2. Méthodes d'étude et techniques expérimentales.
Revue Int. Océanogr. Médicale, C.E.R.B.O.M.

antibiotics

*Aubert, M., J. Aubert, M. Gauthier et D. Pesando, 1967.
Etude de phenomènes antibiotiques liés à une efflorescence de peridiniens.
Rev. intern Océanogr. Med. 6/7:43-52.

antibiotics

Aubert, M., J. Aubert, M. Gauthier, D. Pesando et S. Daniel 1966.
Origine et nature des substances antibiotiques présentes dans le milieu marin. 6. Étude biochimique des substances antibactériennes extraites d'Asterionella japonica (Cleve).
Rev. intern. Océanogr. Med. 4: 23-32.

antibiotics

Aubert, J., D. Chaussy et M.P. Vallard, 1964.
Le pouvoir antibiotique de l'eau de mer.
Cahiers du C.E.R.B.O.M., 16:80-86.

antibiotics

Aubert, M., et M. Gautier 1967.
Origine et nature des substances antibiotiques présentes dans le milieu marin. 7. Étude systématique de l'action antibactérienne d'espèces phytoplanctoniques vis-à-vis de certains germs telluriques aérobies.
Rev. int. Océanogr. Méd. 51:63-71.

antibiotics

Aubert, M. et M. Gauthier 1966.
Origine et nature des substances antibiotiques présentes dans le milieu marin. 7. Note sur l'activité antibactérienne d'une diatomée marine: Chaetoceros teres (Cleve).
Rev. intern. Océanogr. Med. 33-37.

antibiotics

Aubert, M., et M. Gauthier, 1966.
Origine et nature des substences antibiotiques présents dans la milieu marin. IV. Étude comparative de l'action antibiotique due à un organisme phytoplanctonique produit en culture axénique et en culture non axénique.
Rev. Intern. Océanogr. Méd., 2:53-61.

antibiotics

Aubert, M., et M. Gauthier, 1966.
Origine et nature des substance antibiotiques présentes dans le milieu marin. V. Mise en évidence des fractions antibiotiques d'un extrait phytoplanctonique par séparation sur Sephadex.
Rev. Intern. Océanogr. Méd., 2:63-66.

antibiotics

Aubert, M., M. Gautier et S. Daniel, 1966
Origine et nature des substances antibiotiques présentes dans le milieu merin. 3. Activité antibactèrienne d'une diatomée marine Asterionelle japonica (Cleve.)
Revue Int. Oceanogr. Médicale, C.E.R.B.O.M. 1: 35-43.

antibiotics

Aubert, M., H. Lebout et J. Aubert, 1963.
Action du plancton dans le pouvoir antibiotique marin (Note préliminaire).
Rev. Hyg. Méd. Soc., 11:533-539. (Not seen).

64-11170 in Bull. Inst. Pasteur, 62(10):2847.

antibiotics

Aubert, Maurice, Henri Lebout and Jacqueline Aubert, 1963.
Le pouvoir antibiotique du milieu marine.
Cahiers du Centre d'Études et de Recherches de Biologie et d'Océanographie Medical, 12(4):1-85.

antibiotics

Aubert, M. et D. Pesando 1971.
Télémédiateurs chimiques et équilibre biologique océanique. 2. Nature chimique de l'inhibiteur de la synthèse d'un antibiotique produit par une diatomée.
Rev. int. Océanogr. Méd. 21:17-22

antibiotics

Aubert, M., et D. Pesando, 1969.
Variations de l'action antibiotique de souches phytoplanctoniques en fonction de rythmes biologiques marins.
Rev. int. Océanogr. méd. 25-26: 29-31

antibiotics

Aubert, M., D. Pesando et J.M. Pincemin 1970.
Médiateur chimique et relations inter-espèces mise en évidence d'un inhibiteur de synthèse métabolique d'une diatomée produit par un péridinien (étude "in vitro").
Rev. intern. Océanogr. Méd. 17: 5-21

antibiotics

Aubert, J., D. Pesando et H. Thouvenot 1968
Action antibiotique d'extraits planctoniques vis-à-vis de germes anaérobes.
Rev. int. Océanogr. Med. 10: 259-265

antibiotics

Aubert, M., et R. Vaissiere, 1964.
Présentation d'antibiogrammes d'origine planctonique.
Cahiers du C.E.R.B.O.M., 16:93-99.

antibiotics

Berland, B.R. and S.Y. Maestrini, 1969.
Study of bacteria associated with marine algae in culture. II. Action of antibiotic substances.
Marine Biol., 3(4): 334-335.

antibiotics, effect of

*Berland, Brigitte R., et Serge Y. Maestrini, 1969.
Action de quelques antibiotiques sur le développement de cinq diatomées en cultures.
J. mar. Biol. Ecol., 3(1):62-75.

antibiotics

Brisou, J., 1964.
A propos du pouvoir antibiotique du milieu marin.
Cahiers du C.E.R.B.O.M., 16:49-71.

antibiotics

Brisou, J., 1964.
Que penser du pouvoir antibiotique de l'eau de mer.
La Presse Medicale, 72(48):2887-2888.

antibiotics

Brisou, J., et Y. de Rautling de la Roy, 1964.
Action de l'acide acrylique sur St. aureus et quelques coliformes.
Comptes Rendus, Soc. Biol., France, 158(3):642.

antibotics

Burkholder, Paul R., and Lillian M., 1958.
Antimicrobial activity of horny corals.
Science, 127(3307):1174-1175.

antibiotics

Burkholder, Paul R., Lillian M. Burkholder, and Luis R. Almodovar, 1960.
Antibiotic activity of some marine algae of Puerto Rico.
Botanica Marina, 2(1/2):149-156.

antibiotics, effect of

Christiansen, Marit E. 1971.
Larval development of Hyas araneus (Linnaeus) with and without antibiotics (Decapoda, Brachyura, Majidae).
Crustaceana 21 (3): 307-315.

antibiotic action

Cioglia, L., and B. Loddo, 1961.
Sui fenomeni di autodepurazione nell'ambiente marino. II. Sopravvivenza e variabilita nella flora fecale.
Nuov. Ig. Microbiol., 12:445-462.

Ref. in: Bull. Inst. Pasteur, 1963, 61(8):2438.

antibiotics
Conover, John T., and John McN. Sieburth, 1964.
Effect of Sargassum distribution on its epibiota and antibacterial activity.
Botanica Marina, 6(½):147-157.

antibiotics
Daniel S., 1969.
Étude de l'influence de Bdellovibrio bacteriovorus dans l'auto-épuration marine.
Rev. int. Océanogr. méd. 25-26: 61-102.

pollution
Antibiotics

antibiotics
Duff, D.C.B., B.L.Bruce and N.J. Antia, 1966.
The antibacterial activity of marine planktonic algae.
Canadian J. Microbiol., 12(5):877-884.

antibiotics
Fogg, G.E., 1966.
The extracellular products of algae.
In: Oceanography and marine biology, H. Barnes, editor, George Allen & Unwin, Ltd., 4:195-212.

antibiotics
Gauthier, M., 1970.
Substances antibactériennes produites par les bactéries marines. 2. Lipo-polysaccharides antibiotiques produits par certains germes marins appartenant aux genres Pseudomonas et Chromobacterium.
Rev. intern. Océanogr. méd. 17:23-45

antibiotics
Gauthier, M. 1969.
Activité antibactérienne d'une diatomée marine: Asterionella notata (Grun).
Rev. int. Océanogr. méd. 25-26: 103-171.

antibiotics
Gauthier, M., 1969.
Substances antibactériennes produites par les bactéries marines. 1. Étude systématique de l'activité antagoniste de souches bactériennes marines vis-à-vis de germes telluriques aérobies.
Rev. int. Océanogr. méd. 25-26: 41-59.

antibiotics
Glombitza, K.-W. 1969.
Antibakterielle Inhaltsstoffe in Algen.
1. Mitteilung.
Helgoländ. wiss. Meeresunters. 19(3): 376-384.

antibiotics, effect of
Glombitza, K.-W. und R. Heyser, 1971.
Antimikrobielle Inhaltsstoffe in Algen. 4. Mitteilung. Wirkung der Acrylsäure auf Atmung und Makromolekülsynthese bei Staphylococcus aureus und Escherichia coli. Helgoländer wiss. Meeresunters 22(3/4): 442-453.

antibiotics
Grein, A., and S.P. Meyers, 1958.
Growth characteristics and antibiotic production of Actinomycetes isolated from littoral sediments and materials suspended in sea water.
J. Bacteriol., 76(5):457-463.

antibiotics
Gruber, Michael, 1968.
The healing sea. (popular).
Sea Frontiers, 14(2):74-86.

antibiotics
Guélin, Antonina, et Louis Cabioch, 1970.
Sur l'utilisation du phénomène de bactériolyse spontanée pour la connaissance de l'état sanitaire des eaux douces et marines.
C.r. hebd. Séanc. Acad. Sci., Paris, 271(1): 137-140.

antibiotics
Karssilnikova, E.N., 1961
[On the antibiotic properties of microorganisms isolated from various depths of the world ocean.]
Mikrobiologiia, 30(4):651-657.

antibiotics
Koch, 1964.
Reflexions d'ordre hygienique sur le pouvoir antibioltique de la mer.
Cahiers de. C.E.R.B.O.M., 11:111-112.

antibiotics, effect of
Lagarde, E., 1967.
1. Étude de l'action des antibiotiques sur les microflores hétérotrophes marines. Utilisation des antibiotiques dans la réalisation des cultures de foraminifères sous faible volume.
Vie Milieu (A)18(1-A): 27-35.

antibiotics
Lagarde, E. et J. Castellvi 1965
Étude du pouvoir antagoniste d'organismes marins vis-à-vis de divers groupes bacteriens.
Rapp. P.-v. Réun. Comm. int. Explor. scient. Mer Méditerr. 18(3): 619-623

antibiotics
Lebedeva, M.N., and E.M. Markianovic 1971.
Antibiotic properties of heterotrophic bacteria in South Seas. (abstract only)
Rev. int. Biol. Océanogr. méd. 24: 136-137.

antibiotics
Lewin, Ralph A., Editor, 1962.
Physiology and biochemistry of algae.
Academic Press, New York and London, 929 pp.

antibiotics
Li, C.P., B. Prescott and E.C. Martino, 1964.
Antimicrobials from sea food. (Abstract).
Gulf and Caribbean, Fish. Inst., Proc. 16th Ann. Sess., 1963:77-78.

"antibiotics"
McLachlan, J., and J.S. Craigie, 1966.
Antialgal activity of some simple phenols.
J. Phycology, 2(4):133-135.

antibiotics
Miloy, L.F. 1968.
New life-saving drugs are on the way.
Ocean Industry, 3(6): 74-76.

antibiotics
Provasoli, L., 1963
8. Organic regulation of phytoplankton fertility. In: The Sea, M.N. Hill, Edit., Vol. 2. (II) Fertility of the oceans, Interscience Publishers, New York and London, 165-219.

antibiotics
Ramamurthy, V.D. 1970.
Antibacterial activity of the marine blue-green alga Trichodesmium erythraeum in the gastro-intestinal contents of the sea gull Larus brunicephalus.
Mar. Biol. 6(1): 74-76.

antibiotics
Ramamurthy, V.D. 1970.
Antibacterial activity traceable to the marine blue green alga Trichodesmium erythraeum in the gastro-intestinal contents of two pelagic fishes.
Hydrobiologia 36(1): 159-163.

antibiotics
Roos, H., 1957.
Untersuchungen über das Vorkommen antimikrobieller Substanzen in Meeresalgen.
Kieler Meeresforsch., 13(1):41-58.

antibiotics
Rosenfeld, W. D., and C.E. Zobell, 1947.
Antibiotic production by marine microorganisms.
J. Bact. 54(3):393-398.

antibiotics
Seki, Humitake, and Nobuo Taga, 1963
Microbiological studies on the decomposition of chitin in marine environment. IV. Disinfecting effect of antibacterial agents on the chitinoclastic bacteria.
J. Oceanogr. Soc., Japan, 19(3):152-157.

antibiotics
Seshadri, R., S. Krishnamurthy and V.D. Ramamurthy, 1966.
The antibiotic properties of the tissues of some bivalves.
Proc. Indian Acad. Sci. (B) 64(2): 110-113.
Also in: Collected Reprints, Portonovo mar. biol. Sta. 1966-1967.

antibiotics
Sieburth, John McN., 1964.
Substances antibiotiques produites par les algues marines.
Cahiers du C.E.R.B.O.M., 16:101-110.

antibiotics
Sieburth, John McN., 1964.
Antibacterial substances produced by marine algae.
Developments in industrial Microbiology, 5:124-134.

Antibiotics
Sieburth, John McNeill, 1963.
Bacterial habitats in the Antarctic environment
Ch. 49 in: Symposium on Marine Microbiology, C.H. Oppenheimer, Editor, C.C. Thomas, Springfield, Illinois, 533-548.

antibiotics
Sieburth, John McN., and John T. Conover, 1965.
Sargassum tannin, an antibiotic which retards fouling.
Nature, Lond., 208(5005):52-53.

antibiotics
Sieburth, John McN., and David M. Pratt, 1962.
Anticoliform activity of sea water associated with the termination of Skeletonema costatum blooms.
Tran. N.Y. Acad. Sci., (2), 24(5):498-501.

antibiotics
Tourneur, C., 1970.
Contribution à l'étude de la production d'antibiotiques et de vitamine B12 par des bactéries isolées des vases soumises à l'influence du milieu marin.
Cah. océanogr. 22(6): 613-614.

antibiotics
Tourneur, Colette, and Jacques Kauffmann, 1963.
Contribution à l'étude de la richesse du milieu marin en bactéries productrices d'antibiotiques.
C.R. Acad. Sci., Paris, 257(7):1507-1508.

antibiotics, effect of
Steemann Nielsen, 1955.
The production of antibiotics by plankton algae and its effect upon bacterial activities in the sea.
Pap. Mar. Biol. and Oceanogr., Deep-Sea Res., Suppl. to Vol. 3:281-286.

antibiotics, effect of
Steemann Nielsen, E., 1955.
An effect of antibiotics produced by plankton algae. Nature 176(4481):553.

anticlines

anticlines
Barthe, André, Gilbert Boillot et Raoul Deloffre 1967.
Anticlinaux affectant le Crétacé à l'entrée de la Manche occidentale.
C. r. hebd. Séanc. Acad. Sci. Paris (D) 264 (24): 2725-2728.

anti-corrosive compositions

anti corrosive compositions
Fancutt, F., and J.C. Hudson, 1947.
The formation of anti-corrosive compositions for ships' bottoms and under-water service on steel. Pt. II. J. Iron and Steel Inst., No. II, for 1946: 273-296, 8 figs. Pls. XLI-XLIII.

anti-corrosives
Hunter, A.D., and C.H. Horton, 1960.
Cathodic protection checks corrosion on underwater structures.
Underwater Eng., 1(3):40-41.

anticyclones
Abramov, R.V. 1971.
Semidiurnal migrations of the North-Atlantic and South-Atlantic subtropical anticyclones. (In Russian).
Dokl. Akad. Nauk SSSR 201 (1): 67-68.

anti-dunes

"anti-dunes"
Fox, P.J., B.C. Heezen and A.M. Harian, 1968.
Abyssal anti-dunes.
Nature, Lond., 220(5166):470-472.

ANTIFOULING

antifouling compounds
Arias, E., S. Feliu y M.A. Guillén, 1970.
Algunas consideraciones y resultados de la aplicación de compuestos órgano-metálicos de plomo en la formulación de pinturas "antifouling".
Inv. pesquera, Barcelona, 34 (2): 319-354.

antifouling compounds
Christie, A.O., and D.J. Crisp, 1963.
A diffusion technique for assessing antifouling activity.
Ann. appl. Biol., 51:361-366.

Also in:
Contrib. Mar. Sci., Univ. Wales, Mar. Sci. Labs., Menai Bridge, 1962-1963.

antifouling compounds
Partington, A., 1960.
Mode of action of soluble-matrix antifouling compositions.
Nature, 187(4743):1105-1106.

antifouling
Relini, Giulio, e Sebastiano Geraci 1969.
Controllo del "fouling" e comportamento di alcuni organismi marini nei confronti di sostanze tossiche.
Pubbl. Staz. Zool. Napoli 37 (2 Suppl.): 317-326

antifouling compounds
Rischbieth, J.R., & F. Marson, 1961
A coating with antifouling properties from the reaction between cuprous oxide and sodium silicate.
Nature, 192(4804): 748.

anti-fouling compounds
Wisely, B., and R.A.P. Blick 1967.
Mortality of marine invertebrate larvae in mercury, copper, and zinc solutions.
Aust. J. mar. Freshwat. Res., 18(1): 63-72.

antifouling protection
Anderton, W.A., 1965.
Problems of adhesion with cathodic protection.
Proc. Symposium on Marine Paints, New Delhi, 1964: 28-33.

antifouling measures
Suzuki, Hiroshi and Kenjiro Konno, 1970.
Basic studies on the antifouling by ultrasonic waves for ship's bottom fouling organisms.
II. Influences of ultrasonic waves on the larvae of barnacles, Balanus amphitrite hawaiiensis, and mussels Mytilus edulis used together with poisonous substance. (In Japanese; English abstract). J. Tokyo Univ. Fish. 57(1): 9-16.

antifouling protection
Suzuki, Hiroshi, and Kenjiro Konno, 1970.
Basic studies on the antifouling by ultrasonic waves for ship's bottom fouling organisms. I. Influences of ultrasonic waves on the larvae of barnacles, Balanus amphitrite hawaiiensis, and mussels, Mytilus edulis. (In Japanese; English abstract)
J. Tokyo Univ. Fish. 56 (1/2): 31-48.

anti-fouling protection
Wuerzer, D.R., R.L. Senderling and N.F. Cardarelli, 1967.
Protecting buoys with antifouling rubber coverings.
Trans. 2nd Int. Buoy Techn. Symp., 183-191

antiquity of the ocean
Zenkevitch, L. A., 1966.
On the antiquity of the ocean and the role of the marine fauna history in the solution of this problem. (In Russian; English abstract).
Okeanologiia, Akad. Nauk, SSSR, 6(2):195-207.

Antiyeast activity

antiyeast activity
Buck, John D., and Samuel P. Meyers, 1965.
Antiyeast activity in the marine environment.
1. Ecological considerations.
Limnol. Oceanogr., 10(3):385-391.

Antimetabolites

antimetabolites
Johnston, R., 1963
Antimetabolites as an aid to the study of phytoplankton nutrition.
J. Mar. Biol. Assoc., U.K., 43(2):409-425.

Antisubmarine warfare

antisubmarine warfare
Martell, Charles B. 1968.
Airborne antisubmarine warfare
Oceanology intl 3(5): 28-31

Apatite

Apatite
Arrhenius, Gustaf, and Enrico Bonatti, 1965.
Neptunism and vulcanism in the ocean.
Progress in Oceanography, 3:7-22.

APEX
*Clauss, E., H. Hinzpeter und J. Müller-Glewe, 1970.
Ergebnisse von Messungen des Temperaturfeldes der Atmosphäre nahe der Grenzfläche Ozean-Atmosphäre.
Meteor Forsch.-Ergebnisse (B) 5: 85-89

air-sea interface
temperature (air)
APEX
heat flux

APEX
*Clauss, E., H. Hinzpeter und J. Müller-Glewe, 1970.
Messungen zur Temperaturstruktur im Wasser an der Grenzfläche Ozean-Atmosphäre.
Meteor Forsch-Ergebnisse (B) 5: 90-94.

air-sea interface
temperature, surface
APEX

APEX
*Helbig, H., und Ch. Werner 1970.
Ein Lidar-System zur Messung der vertikalen Dunstverteilung in der Atmosphäre während des Atlantischen Passat Experiments (APEX) auf W.F.S. Planet. 1. Technischer Teil.
Meteor Forsch.-Ergebnisse (B) 5: 52-61
aerosols, vertical distribution
instrumentation, aerosols
"Planet"
APEX

aphotic zone
Bernard, Francis, 1967.
Research on phytoplankton an pelagic Protozoa in the Mediterranean Sea from 1953 to 1966.
Oceanogr. Mar. Biol., Ann. Rev., H. Barnes, editor, George Allen and Unwin, Ltd., 5:205-229.

aphotic zone

Applied oceanography

applied oceanography
Abel, R.B., 1967.
Ocean science and technology: a review of the past; a glimpse of the future.
1967 NEREM Record, Inst. Electr. Electron. Engrs, 9:166-167.

aquaculture

see also: sea-farming, cultures, etc.

AQUACULTURE, see also:
SEA FARMING, CULTURES, etc.

aquaculture
Fontaine, M. 1969.
Mariculture: espoirs et limites.
Connaissance de la Mer (2): 24-29.
not seen

aquaculture
*Leccese, A. and N. Clerico, 1970.
Coastal waters pollution and molluscs culture: proposal for production, commerce and purification of molluscs.
Rev. int. Océanogr. méd. 15-19: 253-260

pollution
aquaculture
p. 260

aquaculture
Rabanal, Herminio R., 1970
Aquacultural development and public health.
Rev. int. Océanogr. méd., 15-19: 155-167.

aquaculture
Ryther, J.H., and G.C. Matthiessen, 1969.
Aquaculture, its status and potential.
Oceanus, 14(4):2-14.

Aquaculture
*Uyeno, Fukuzo, 1969.
Relationship between sea water and bottom condition and mechanism of fisheries production in estuarine areas,.....with special reference to the deterioration of pearl oyster farm. (In Japanese; English abstract). Bull. Jap. Soc. fish. Oceanogr. Spec. No. (Prof. Uda Commem. Pap.): 193-196

aqualung

aqualung
Bascom, W.N., and R.R. Revelle, 1953.
Free diving: a new exploratory tool.
Am. Sci. 41(4):624-627, 2 figs.

aqualung use
Hasler, A.D., and J.R. Villemonte, 1953.
Observations on the daily movements of fishes.
Science 118(3064):321-322, 2 textfigs.

aqua lung
Ivanoff, A., 1953.
Progrès récents de la photographie sous-marine.
La Nature No. 3221:257-261, 8 figs.

aqualung, use of
Pavlov, D.S., and D.S. Nikolaev, 1962.
On under ice work with an aqualung. Methods and results of submarine investigations. (In Russian).
Trudy, Okeanogr. Komissii, Akad. Nauk, SSSR, 14: 93-94.

aquanauts

AQUANAUTS, see also:
MAN-IN-THE-SEA

aquanauts
Bussey, Larry, 1968.
The U.S. Navy Sealab III aquanaut training and equipment.
J. Ocean Techn., 2(4):27-30.

aquanauts
Miller, James W., 1966.
The measurement of human performance, SEALAB II. Man's extension into the sea, Trans. Symp., 11-12 Jan. 1966, Mar. Techn. Soc., 156 169.

aquanauts
Pothier Richard J., 1969.
Textite - two months beneath the sea.
Sea Frontiers 15(3): 130-139. (popular)

aquaria

aquaria
*Lickey, M.E., R.L. Emigh and F.R. Randle, 1970.
A recirculating seawater aquarium system for inland laboratories.
Marine Biol., 7(2):149-152.

aquaria, salt water
McDaniel, H. R., 1961
Maintenance of sea urchins in laboratory aquaria
Texas J. Sci., 13(4): 490 - 492

aquaria
Müllegger, S., 1948.
Krebstiere im Seeaquarium. Deutschen Aquarien und Terrarienk. Zeitschr. 1(3):36-38, illus.

aquaria
Simkatis, H., 1958.
Salt-water fishes for the home aquarium.
J. Lippincott Co., Phila., and New York, 264 pp. $6.00.

Reviewed by: Cargo, D.G., 1962. Chesapeake Sci., 3(1):51.

aquaria
Thomopoulos, A., 1955.
Sur un aquarium d'eau de mer en circuit fermé installé au Laboratoire d'Anatomie et d'Histologie Comparées à la Sorbonne.
Bull. Inst. Océan., Monaco, No. 1066:1-10.

aquifer

aquifer, coastal
Chermonman, Srisakdi, 1965.
A solution of the pattern of fresh-water flow in an unconfined coastal aquifer.
J. Geophys. Res., 70(12):2813-2819.

archeology
Bascom, Willard, 1971.
Deep-water archeology. Science 174(4006): 261-269.

archeology

archeology
Bass, George F., 1964.
The promise of underwater archeology.
Ann. Rept., Smithsonian Inst., 1963, Rept., 4530: 461-471.

archeology
du Plat, Joan, 1965
Marine archaeology
Hutchinson, London.

books
archaeology

archaeology
Marx, Robert F. 1969.
Port Royal
Oceans Mag. 1(5):66-77 (popular)

archaeology
Ninkovich, Dragoslav, and Bruce C. Heezen, 1965.
Santorini tephra.
In: Submarine geology and geophysics, Colston Papers, W.F. Whittard and R. Bradshaw, editors, Butterworth's, London, 17:413-452.

archaeology
Woods, J.D., and J.N. Lythgoe, editors, 1971.
Underwater science: an introduction to experiments by divers.
Oxford Univ. Press 330pp.

archeopyle

archeopyle
Evitt William R., 1967.
Dinoflagellate studies II. The archeopyle.
Publ. Stanford Univ., Geol. Sci., 10(3):1-82

dinoflagellates, cont. physiol.
archeopyle

arcs

archiannelids

archiannelids
Gray, John S., 1966.
Selection of sands by Protodrilus symbioticus Giard.
Veröff. Inst. Meeresforsch., Bremerh., Sonderband II:105-115.

arc tectonics
Plafker, George, 1972.
Alaskan earthquake of 1964 and Chilean earthquake of 1960: implications for arc tectonics.
J. geophys. Res. 77(5): 901-925.

arrugado
Krause, Dale C., 1961
Geology of the sea floor east of Guadalupe Island.
Deep-Sea Res., 8(1): 28-38.

artificial media

artificial media
Plessis, Y., 1956.
Note sur le contôle de la salinité en milieu marin artificiel. Bull. Mus. Nat. Hist. Nat., 28 :585-589.

Asdic

asdic
Anon., 1957.
Asdic's future in fishfinding.
World Fish., London, 6(4):40-41.

Asdic
Haines, R.G., 1963
Developments in ultrasonic instruments.
Int. Hydrogr. Rev., 40(1):49-57.

Asdic
Stride, A.H., 1961
Geological interpretation of Asdic records.
Int. Hydrogr. Rev., 38(1): 131-139.

Asdic
Stubbs, A.R., 1963
Identification of patterns on Asdic records.
Int. Hydrogr. Rev., 40(2):53-63.

ASDIC
Stubbs, A.R., and R.G.G. Laurie, 1962.
Asdic as an aid to spawning ground investigations.
J. du Conseil, 27(3):248-260.

ash

ash
*Nelson, C. Hans, L.D. Kulm, Paul R. Carlson and John R. Duncan, 1968.
Mazama ash in the northeastern Pacific.
Science 161(3836):47-49.

ash
Pastouret, Léo, 1970
Étude sédimentologique et paléoclimatique de carottes prélevées en Méditerranée orientale.
Téthys 2(1):227-266.

ash
van Straaten, L.M.J.U., 1967.
Turbidites, ash layers and shell beds in the bathyal zone of the southeastern Adriatic Sea.
Revue Géogr.phys.Géol.dyn., (2),9(3):219-240.

assimilation

assimilation
*Angot, Michel, 1967.
Rapports entre la concentration en chlorophylle a, le taux d'assimilation du carbone et la valeur de l'energie lumineuse en eau tropicale littorale.
Cah. ORSTOM, Sér. Océanogr., 5(1):39-45.

assimilation
Conover, Robert J., 1966.
Factors affecting the assimilation of organic matter by zooplankton and the question of superfluous feeding.
Limnol. Oceanogr., 11(3):346-354.

assimilation
Conover, Robert J., 1966.
Assimilation of organic matter by zooplankton.
Limnol. Oceanogr., 11(3):338-345.

assimilation (nitrogen)
Eppley, Richard W., Jan N. Rogers, James J. McCarthy and Alain Sournia, 1971.
Light/dark periodicity in nitrogen assimilation of the marine phytoplankters Skeletonema costatum and Coccolithus huxleyi in N-limited chemostat cultures.
J. Phycol. 7(2): 150-154

assimilation
Forster, J.R.M. and P.A. Gabbott, 1971.
The assimilation of nutrients from compounded diets by the prawns Palaemon serratus and Pandalus platyceros. J. mar. biol. Ass. U.K. 51(4): 943-961.

assimilation
Glooschenko, Walter A. and Herbert Curl, Jr. 1971.
Influence of nutrient enrichment on photosynthesis and assimilation ratios in natural North Pacific phytoplankton communities.
J. Fish. Res. Bd Can. 28 (5): 790-798.

assimilation
*Seki, Humitake, 1967.
Effect of organic nutrients on dark assimilation of carbon dioxide in the sea.
Inf.Bull.Planktol.Japan, Comm.No.Dr.Y.Matsue, 201-205.

assimilation, carbon dioxide
*Seki, Humitake, and Claude E. ZoBell, 1967.
Microbial assimilation of carbon dioxide in the Japan Trench.
J. Oceanogr.Soc., Japan, 23(4):182-188.

assimilation, iodine
*Sugawara, Ken, and Kikuo Terada, 1967.
Iodine assimilation by a marine Navicula sp. and the production of iodate accompanied by the growth of the algae.
Inf.Bull.Planktol.Japan, Comm.No.Dr.Y.Matsue, 213-218.

assimilation
Tsukidate, Jun-ichi and Shunzo Suto, 1971.
Tracer experiments on the effect of micronutrients on the growth of Porphyra plants - 1. Iron-59 assimilation in relation to environmental factors. Bull. Tokai reg. Fish. Res. Lab. 64: 89-100.

assimilation ratios
Malone, T.C., 1971.
Diurnal rhythms in netplankton and nannoplankton assimilation ratios. Mar. Biol. 10(4): 285-289.

ATLANTIS (lost)

Atlantis (lost)
*Wetzel, Walter, 1967.
Vom gegenwärtigen Stand des Atlantis-Problems.
Meyniana, 17:111-115.

Atlases

atlases
Armstrong, Terence, 1964.
Ice atlases.
Polar Record, 12(77):161-163.

atlas
Barkley, Richard A., 1968.
Oceanographic atlas of the Pacific Ocean.
University of Hawaii Press. 156 figures.

atlases
Churkin, V.G., 1961.
[National atlases and different ways of making them.]
Izv. Vses. Geograf. Obshch., 93:83-85.

atlases
Dubach, H.W., 1959
Oceanographic Atlas of the polar seas. Part II, Arctic, H.O. Publ. 705. Wash., D.C., U.S. Hydro. Off. 1958.
Rev. in Arctic, 12(2): 115-116.

Atlases
Dubach, H.W., and M. A. Slessers, 1958
An evaluation of the "Morskoi Atlas" (with translations from Volumes I and II.) H.O. Tech. Rept. TR 38: 441 pp.

atlas
Dunbar, M.J., 1963.
The purpose and significance of the Serial Atlas of the Marine Environment.
In: Marine distributions, R. Soc., Canada, Spec. Publ., No. 5:3-9.

atlases
Engelmann, Gerhard 1971.
Das preussische Seekartenwerk vor Gründung der Kriegsmarine.
Dt. hydrogr. Z. 24 (4): 173-184.

Atlas
Hvalfangernes Assuranceforening 1946
Atlas over Antarktis og Sydishavet utgitt av Hvalgangernes Assuranceforening i anledning av Foreningens 25 aars Jubileum. Red. av. --- H.E. Hansen. 1 p.l., 17 pp. incl. col. charts. 41.5x35 cm. Sandefjord, Norge

atlas
India, Indian Ocean Biological Center, Cochin, 1970
Distribution of fish eggs and larvae in the Indian Ocean.
Int. Indian Ocean Exped., Plankton Atlas 2(2): 10 charts

Atlas
Japan, Hydrographic Office, Maritime Safety Board, Undated
State of the adjacent seas of Japan, 1955-1959.
Vol. 1: numerous charts.

atlas
Japan, Japanese Oceanographic Data Center 1970.
CSK Atlas, 4. Winter 1967: 92 pp.

Atlas
Krivonosova, N.M., 1962
On the method of compiling atlases reflecting the dynamics and the morphology of marine coasts. (In Russian).
Okeanologiia, Akad. Nauk, SSSR, 2(5):912-916.

atlases
Timonov, V.V., editor, 1967.
Atlas: Izmeneniia sostoianiia sistemy okean-atmosfera v severnoi Atlantike. 2. Godovoi tsikl izmeneniia temperatury i termokhalinnogo polia deiatel'nogo sloia okeana. (Variations in the state of the ocean-atmosphere systems of the North Atlantic. 2. Variations in the annual cycle of the circulation and of the temperature and salinity in the active layer of the ocean. Gidromet. Izdatel. Leningrad, 69 pp.

atlases
United States, Weather Bureau and Hydrographic Office, 1961.
Climatological and oceanographic atlas for Mariners. Vol. II. North Pacific Ocean.
Unnumbered pages.

atlas
United States Navy Hydrographic Office, 1959.
Climatological and oceanographic atlas for mariners. Vol. 1.
North Atlantic Ocean. 182 charts.

atlas
Webster, Wilfrid, 1961
The American Geographical Society's serial atlas of the marine environment.
Geographical Review, 51(4):570-574.

atlas
Worthington, L.V. and W.R. Wright 1970.
North Atlantic Ocean atlas of potential temperature and salinity in the deep water, including temperature, salinity and oxygen profiles from the Erika Dan cruise of 1962.
Woods Hole Oceanogr. Inst. Atlas Ser. 2: 24 pp., 58 pls.

atlases, oceanographic
Yonge, Ena L., 1962
World and thematic atlases: a summary survey.
Geogr. Rev., New York, 52(4):583-596.

atmosphere, CO_2 change in

atmosphere, carbon dioxide
Callendar, G.S., 1958.
On the amount of carbon dioxide in the atmosphere. Tellus 10(2):243-248.

atmosphere, carbon dioxide
Fonselius, S., F. Koroleff and K.-E Wärme, 1956.
Carbon dioxide variations in the atmosphere.
Tellus 8(2):176-183.

atmosphere, carbon dioxide cycle
Miyake, Y., and S. Matsuo, 1963
A role of sea ice and sea water in the Antarctic on the carbon dioxide cycle in the atmosphere.
Papers in Meteorol. and Geophys., Tokyo, 14 (2):120-125.

atmosphere, chemistry (chlorine)
Duce, Robert A., 1969
On the source of gaseous chlorine in the marine atmosphere. J. geophys. Res., 74(18): 4597-4599.

Atmospheric dust

atmospheric dust
Abramov, R.V. 1971.
Atmospheric dust over the Atlantic Ocean. (In Russian; English abstract)
Usloviia Sediment. Atlant. Okean. Reg. Issled. Meghd. Geofiz. Proekt. Meghd. Geofiz. Kom. Prezid. Akad. Nauk SSSR, 17-30.

atmospheric dust
Bruevich, S.V., and M.P. Gudkov, 1954.
[Atmospheric dust over the Caspian Sea.]
Izv. Akad. Nauk, SSSR, Ser. Geogr., (4):18-28.

RT-3171 Bibl. Transl Russ. Sci. Tech. Lit. 29.

atmospheric dust
Kool, L.V., 1971.
Jet streams over the Atlantic Ocean. (In Russian; English abstract). Usloviia sediment. Atlantich. Okeana, Rez. Issled. Mezhd. Geofiz. Proekt. Mezhd. Geofiz. Kom. Prezid. Akad Nauk SSSR, : 31-42.

atmospheric dust
*Prospero, Joseph M. and Enrico Bonatti, 1969.
Continental dust in the atmosphere of the eastern equatorial Pacific. J. geophys. Res. 74(13): 3362-3371.

Pacific, equatorial (east)
atmospheric dust

atmospheric electricity
*Mühleisen, R., 1968.
Luftelektrische Messungen aux fem Meer: Ergebonisse von der Atlantik-Fahrt des Forschungschiffes Meteor 1965. 1. Feldstärke-und Ionenmessungen.
Meteor Forschungsergeb, (A)2:57-82.

atmospheric heating

atmospheric heating
Gambo, K., 1963
The role of sensible and latent heats in the baroclinic atmosphere.
J. Meteor. Soc. Japan, Ser. II, 41(4):233-246.

atmospheric pressure, effect of

atmospheric pressure, effect of
Dobson, Fred W., 1971.
Measurements of atmospheric pressure on wind-generated sea waves.
J. fluid Mech 48(1): 91-127.

atmospheric pressure, effect of
Baczyk, Józef, 1968.
Les masses d'eaux de la mer Baltique méridionale et l'influence de leurs mouvements sur la zone littorale polonaise. (In Polish; French resumé)
Prace Geograf. Warsaw 65: 1-120.

atmospheric pressure, effect of
Carritt, D.E., 1954.
Atmospheric pressure changes and gas solubility.
Deep-Sea Res. 2(1):59-62.

atmospheric pressure, effect of
Konaga Shunji 1970.
On the short period fluctuation of Kuroshio (In Japanese; English abstract)
Bull. Kobe mar. Obs. 183: 83-95.

atmospheric pressure, effect of
Lisitzin, Eugénie, 1964.
La pression atmosphérique comme cause primaire des processus dynamique dans les océans.
Cahiers Océanogr., C.C.O.E.C., 16(1):17-22.

atmospheric pressure, effect of
Mosetti, F. 1969.
Oscillazioni del livello marino a Venezia in rapporto con le oscillazioni di pressione atmosferica.
Boll. Geofis. teor. appl. 11 (43-44): 264-278

atmospheric pressure, effect of
Roden, G.I., 1958
Spectral analysis of a sea surface temperature and atmospheric pressure record off southern California.
J. Mar. Res., 16(2):90-95.

atolls

ATOLLS, see also: CORAL ATOLLS, CORAL REEFS.

atolls
Bakus, Gerald J., 1964.
The effects of fish-grazing on invertebrate evolution in shallow tropical waters.
Allan Hancock Found. Publ. Occ. Pap. No. 27:1-29

atolls (coralline)
Boyd, Donald W., Louis S. Kornicker and Richard Rezak, 1963.
Coralline algae microatolls near Cozumel Island, Mexico.
Contrib. to Geol., 2(2):105-108.

atolls
Emery, K.O. J.I. Tracy, jr., and H.S. Ladd, 1954.
Geology of Bikini and nearby atolls. Bikini and nearby atolls: Part 1, Geology.
Geol. Survey Prof. Pap. No. 260-A:1-265, 64 pls., 11 separate charts

Atoll
Fosberg, F.R., 1949
Atolls vegetation and salinity. Pacific Sci, 3(1):89-92, 6 photos.

atolls
Fosberg, F.R., and M.-H. Sachet, editors, 1953.
Handbook for atoll research. Atoll Res. Bull. 17:129 pp., (mimeographed).

second preliminary edition

atolls
Gaskell, G.F., and J.C. Swallow, 1953.
Seismic experiments on two Pacific atolls.
Occ. Papers, Challenger Soc., No. 3:8 pp., 6 figs

atolls
Kuenen, Ph.H., 1947.
Two problems of marine geology:atolls and canyons. Verhandel. Kon. Nederl. Akad. Wetersch., Afd. Natuurk., Tweede Sectie, XLIII(3):68 pp., 4 pls., 14 textfigs.

atolls
Ladd, H.S., J.I. Tracy, jr., J.W. Wells, and K.O. Emery, 1950.
Organic growth and sedimentation on an atoll.
J. Geol. 58(4):410-425, 7 pls., 2 figs.

atolls
MacNeil, F.S., 1954.
The shape of atolls: an inheritance from subaerial erosion forms. Amer. J. Sci. 252:402-428.

atolls
Menard, H.W., and H.S. Ladd, 1963.
15. Oceanic islands, seamounts, guyots and atolls.
In: The Sea, M.N. Hill, Editor, Interscience Publishers, 3:365-387.

atolls
*Milliman, John D.,1967.
The geomorphology and history of Hogsty Reef, a Bahamian atoll.
Bull. mar. Sci., Miami, 17(3):519-543.

atolls
Milliman, John D. 1969.
Four southwestern Caribbean atolls: Courtown Cays, Albuquerque Cays, Roncador Bank and Serrana Bank
Atoll Res. Bull. 129: 1-22

atolls
Milliman, John D., 1967.
Carbonate sedimentation on Hogsty Reef, a Bahamian atoll.
J. Sedim. Petrol. 37(2): 658-676

Atolls
* Newell, Norman D. and Arthur L. Bloom, 1970
The reef flat and two-meter eustatic terrace of some Pacific atolls.
Bull. geol. Soc. Am. 81 (7): 1881-1894.

reefs
atolls
Pacific

atolls
Perkins, F., jr., and G.G. Lill, 1953.
Velocity studies on Bikini Island.
Proc. Seventh Pacific Sci. Congr., 2(Geol.):558.

Abstract.

atolls
Robertson,E.I., and A.C. Kibblewhite,1966.
Bathymetry around isolated volcanic islands and atolls in the South Pacific Ocean.
New Zealand J. Geol. Geophys., 9(1/2):111-121.

atolls
Shipek, C.G., 1962
Photographic survey of sea floor on southwest slope of Eniwetok Atoll.
Geol. Soc. Amer. Bull., 73(7): 805-812.

atolls
Silvester, R., 1965.
Coral reefs, atolls and guyots.
Nature, 207(4998):681-688.

atolls
Stoddart, D.R., 1962
Three Caribbean atolls: Turneff Islands, Lighthouse Reef, and Glover's Reef, British Honduras.
Atoll Res. Bull., 87: 151 pp.

atolls
Vallaux, F., 1955.
Anomalies magnétiques dans les atolls.
Ann. Hydrogr., (4)5:131-138.

atomic energy, effect of

atomic energy, effects of
Fontaine, M., 1956.
Les océans et les dangers résultants de l'utilisation de l'energie atomique. J. du Cons., 21(3):241-249.

atomic energy, effect of
Nakano, M., 1955.
Atomic energy and the oscillation of sea level.
J. Ocean. Soc., Japan, 11(4):165-169.

atomic wastes

atomic wastes
Klocke, F.J., and H. Rahm, 1959.
No place to hide atomic wastes.
Fisheries Newsletter, 18(8):15.

attapulgite

attapulgite
Chamley, Hervé, and Georges Millot, 1970.
Séquences sédimentaires à attapulgite dans un carotte profondé prélevée en Mer Ionienne (Méditerranée orientale).
C.r. hebd. Séanc. Acad. Sci. Paris (D) 270(8): 1084-1087.

Mediterranean, west
attapulgite

attapulgite
Heezen, Bruce C. Wladimir D. Nesteroff, Agnes Oberlin et Germaine Sabatier, 1965.
Découverte d'attapulgite dans les sédiments profonds du Golfe d'Aden et de la Mer Rouge.
C.R. Acad. Sci., Paris, 260(22):5819-5821.

attapulgite
Latouche, Claude, 1971.
Découverte d'attapulgite dans les sédiments carottés sur le dôme Catabria (Golfe de Gascogne). Conséquences paléogéographiques.
C. r. hebd. Séanc. Acad. Sci. Paris (D) 272 (16): 2064-2066

attenuation

attenuation light

attenuation, light
Barteneva, O.D., and E.A. Polyakova, 1965.
A study of attenuation and scattering of light in a natural fog due to its microphysical properties. (In Russian; English abstract).
Fisika Atmosferi i Okeana, Izv. Akad. Nauk SSSR 1(2):193-207.

attenuation, light
Bulkhurgin, M.N. and V.P. Nikolaev, 1965
On the attenuation of the collimation light beam in the sea. (In Russian; English abstract)
Trudy Inst. Okeanol. Akad. Nauk SSSR 77: 98-101.

Attenuation, light
Dera, Jerzy 1971.
Irradiance in the euphotic zone.
Oceanologia, Kom. Bad. Polsk. Akad Nauk 1: 9-98

attenuation light
light

attenuation, light
*Gordon, Howard R. and Jerzy Dera, 1969.
Irradiance attenuation measurements in sea water off southeast Florida. Bull. mar. Sci., 19(2): 279-285.

USA, east
irradiance
attenuation, light

attenuation, light

Hishida, Kozo 1966.
On the scattering and the attenuation of light in the sea water.
J. Oceanogr. Soc. Japan 22 (1): 1-6.

attenuation, light

Horrer, Paul L., 1962
Oceanographic studies during Operation "Wigwam". Physical oceanography of the test area.
Limnol. and Oceanogr., Suppl. to Vol. 7: vii-xxvi.

attenuation, light

Jerlov, N.G., and K. Nygård, 1968.
Influence of solar elevation on attenuation of underwater irradiance.
Rept. Københavns Univ. Inst. fys. Oceanogr., 4: 6 pp. (multilithed).

attenuation, light

Karabashev, G.S., 1966.
Photometer for the measurements of irradiance spectral attenuation in the sea. (In Russian; English abstract).
Okeanologiia, Akad. Nauk, SSSR, 6(5):886-891.

attenuation, light

Kinder, Floyd, 1966.
Underwater light attenuation measurements.
Jour. Sci. Photo-Opt. Instr. Eng., 4(2):50-54.
(not seen).

attenuation, light

Kullenberg, G., B. Lundgren, Sv. Aa. Malmberg, K. Nygård and N. Højerslev, 1970.
Inherent optical properties of the Sargasso Sea.
Rept. Inst. fys. Oceanogr. Univ. København 11: 12 pp. (multilithed)

attenuation, light

Ochakovsky, Yu. E., 1965.
On dependence of total attenuation coefficient upon the sea suspensions. Hydro-optical investigations. (In Russian; English abstract).
Trudy Inst. Okeanol., Akad. Nauk, SSSR, 77:35-40.

attenuation, light

Otto, L., 1966.
Light attenuation in the North Sea and the Dutch Wadden Sea in relation to Secchi disc visibility and suspended matter.
Netherlands J. Sea Res., 3(1):28-51.

attenuation, light

Pivovarov, A.A., and V.S. Lavorko, 1961.
[On the vertical attenuation of solar radiation in the sea.]
Vestnik, Moskva Univ., (3- Fizik.-Astron.) 6: 59-63.

Engl. Abstr. Vertical attenuation of solar radiation in the sea.
OTS-61-11147-18 JPRS:8807:15

ATTENUATION, RADIO WAVES

Finkel'stein M.I., V.G. Glushnev, A.N. Fetisov, and V. Ya. Ivashchenko, 1970
On the anisotropy of radio wave attenuation in sea ice. (In Russian)
Fizika Atmosfer. Okean., Izv. Akad. Nauk SSSR. 6(3): 311-313.

attenuation, seismic

attenuation, seismic waves

Barazangi, Muawia and Brian Isacks, 1971.
Lateral variations of seismic-wave attenuation in the upper mantle above the inclined earthquake zone of the Tonga Island arc: deep anomaly in the upper mantle. J. geophys. Res. 76(35): 8493-8516.

attenuation, seismic

Bose, S.K., 1961
On low-period sub-oceanic Rayleigh waves and their attenuation.
Geofis. Pura e Appl., 49: 15-35.

attenuation, seismic

Frantti, G.E., 1965.
Attenuation of Pn from offshore Maine explosions.
Bull. Seismol. Soc., Amer., 55(2):417-423.

attenuation, seismic

Iida, Kumizi, and Harumi Aoki, 1960.
Attenuation of seismic waves in the vicinity of explosive seismic origin.
J. Earth Sciences, Nagoya Univ., 8(2):109-119.

attenuation seismic

McCann, C., and D. M. McCann 1969
The attenuation of compressional waves in marine sediments.
Geophysics 34 (6): 882-892

attenuation

Shumway, G., 1956.
Resonance chamber sound velocity and attenuation measurements in sediments. Geophysics 21(2):305-319.

attenuation seismic wave

Walsh, J.B., 1966.
Seismic wave attenuation in rock due to friction.
J. geophys. Res., 71(10):2591-2599.

Attenuation sound

attenuation, sound

Albers, Vernon M., 1965.
Underwater acoustics handbook II.
The Pennsylvania State University Press, 356 pp. $12.50.

attenuation (sound)

Arase, T., 1959.
Some characteristics of long-range explosive sound propagation.
J. Acoust. Soc., Amer., 31(5):588-595.

attenuation, sound

Blickley, L.F., A.U. Ayres and P.C Curzan, 1962
Verification of attenuation and ambient noise in the ocean. (Abstract).
J. Acoust. Soc., Amer., 34(12):1974-1975.

attenuation, sound

Cole, B.F., 1965.
Marine sediment attenuation and ocean-bottom-reflected sound.
J. Acoust. Soc., Amer., 38(2):291-297.

attenuation

Gruber, G.J., & R. Meister, 1961
Ultrasonic attenuation in water containing brine shrimp in suspension.
J. Acous. Soc. Amer., 33(6): 733-740.

attenuation, sound

Hansen, Palle G., and Eric G. Barham, 1962.
Resonant cavity measurements of the effects of "Red water" plankton on the attenuation of underwater sound.
Limnology & Oceanography, 7(1):8-13.

attenuation, sound

Kibblewhite, A.C., and R.N. Denham 1971
Low-frequency acoustic attenuation in the South Pacific Ocean.
J. acoust. Soc. Am. 49 (3-2): 810-815

attenuation, sound

Kibblewhite, A.C., and R.N. Denham, 1965.
Experiment on propagation in surface sound channels.
J. Acoust. Soc., Amer., 38(1):63-71

attenuation, sound

Leroy, C.C., 1967.
Sound propagation in the Mediterranean Sea.
In: Underwater acoustics, V.M. Albers, editor, Plenum Press, 2: 203-241.

attenuation, sound

Langleben, M.P., 1969.
Attenuation of sound in sea ice, 10-500 kHz.
J. Glaciol. 8 (54): 399-406

attenuation, sound

Marsh, H.W., 1963.
Attenuation of explosive sounds in sea water.
J. Acoustic Soc., Amer., 35(11):1837.

attenuation, sound

McElroy, E.G., 1963.
In situ measurements of the attenuation of high-frequency sonar pulses in sea water.
J. Acoust. Soc., Amer., 35(8):1295-1296.

attenuation, sound

McElroy, E.G., 1962
Measurements of high-frequency pulse attenuation and bottom backscattering in the sea. (Abstract).
J. Acoust. Soc., Amer., 34(12):1986-1987

attenuation, sound

McElroy, E.G., and A. DeLoach, 1968.
Sound speed and attenuation from 15 to 1500 kHZ, measured in natural sea-floor sediments.
J. acoust. Soc. Am., 44(4):1148-1150.

attenuation, sound

Medwin, Herman, 1970.
In situ acoustic measurements of bubble populations in coastal ocean waters. J. geophys. Res., 75(3): 599-611.

attenuation (sound)

Paolini, E., 1952.
L'attenuazione degli ultrasouni usati nelle applicazione navali. Riv. Mar., Suppl. Tech. Aug.-Sept.:71-91.

attenuation, sound

Sheehy, M.J., and R. Halley, 1957.
Measurement of the attenuation of low frequency underwater sound. J. Acoust. Soc., Amer., 29(4): 464-469.

attenuation, sound

Skretting, Audun, and Claude C. Leroy, 1971
Sound attenuation between 200 Hz and 10 kHz.
J. acoust. Soc. Am. 49 (1-2): 276-282.

attenuation, sound

Stoll, Robert D. and George M. Bryan 1970.
Wave attenuation in saturated sediments.
J. acoust. Soc. Am. 47 (5-2): 1440-1447

attenuation sound

Thorp, William H., 1967.
Analytic description of the low-frequency attenuation coefficient.
J. acoust. Soc. Am., 42(1): 270.

attenuation, sound
Thorp, William H., 1965.
Deep-ocean sound attenuation in the sub- and low-kilocycle-per-second region.
J. acoust. Soc., Amer., 38(4):648-654.

attenuation, sound
Tolstoy, Ivan, 1958.
Shallow water test of the theory of layered wave guides.
J. Acoust. Soc., Amer., 30(4):348-361.

attenuation, sound
*Ulonska, Arno, 1968.
Versuche zur Messung der Schallgeschuidigkeit und Schalldämfung im Sediment in situ.
Dt. hydrogr.Z., 21(2):49-58.

attenuation, sound
Urick, R.J. 1966.
Long-range deep-sea attenuation measurement.
J. acoust. Soc. Am. 39 (5-1): 904-906

attenuation, sound
Urick, R.J., 1964.
Attenuation over RSR paths in the deep sea.
J. Acoust. Soc., Amer., 36(4):786-788.

attenuation, sound
Urick, R.J., 1963.
Low-frequency sound attenuation in the deep ocean.
J. Acoust. Soc., Amer., 35(9):1413-1422.

attenuation, sound
*Weston, D.E., 1967.
Contradiction concerning shallow-water sound attenuation.
J. acoust. Soc. Am., 42(2):526-527.

attenuation, sound
Weston, D.E., 1966.
Fish as a possible cause of low-frequency acoustic attenuation in deep water.
J. acoust. Soc., Am., 40(6):1558.

attenuation, sound
Weston, D.E., A.A. Horrigan, S.J.L. Thomas and J. Revie, 1969.
Studies of sound transmission fluctuations in shallow coastal waters.
Phil. Trans. R. Soc., 265 (1169): 567-608

attenuation, sound
Wiener, F.M., 1961.
Sound propagation over ocean waters in fog.
J. Acoust. Soc., Amer., 33(9):1200-1205.

Attenuation, waves (and swell, etc.)

attenuation, waves
Draper, L. 1964
Comments on a paper by Tsyplukhin "The results of instrumental wave attenuation within the deep sea".
Okeanologiia, 4:549-550.
Translation in:
Collected Reprints, Nat. Inst. Oceanogr., Wormley, 12. 1964.

attenuation, waves
Draper, L., 1957.
L'attentuation de la houle en fonction de la profondeur.
La Houille Blanche, No. 6: 5 pp.

attenuation, wind waves
Grushevskii, M.S., 1960
[Computations of the attenuation of sea waves with weakening winds. Wind waves.]
Trudy Okeanogr. Komissii, Akad. Nauk, SSSR, 9:117-132.

attenuation, waves
Hishida, Kozo, Toshiyoshi Tada and Sozaburo Koizumi 1969
On the attenuation of surface waves with depth.
Umi to Sora 44(2/3): 39-45.
(In Japanese; English abstr.)

attenuation, waves
Hunt, J.N., and S.K.A. Massoud, 1962
On mass transport in deep-water waves.
Geofis. Pura e Applicata, 53(3):65-76.

attenuation, waves
Iwasa, Y., 1957.
Attenuation of solitary waves on smooth bed.
Proc. Amer. Soc. Civ. Eng., 83:15 pp.
Rev. in:
Appl. Mech. Rev. 11(2):#700.

attenuation, swell
Korneva, L.A., 1965.
The energy of sporadic attenuation of swell in the sea and the potential energy of waves for turbulent friction. (In Russian).
Trudy Morsk. Gidrofiz. Inst. Akad. Nauk, SSSR, 31:22-30.

attenuation
Kornhauser, E.T., and N.P. Raney, 1955.
Attenuation in shallow-water propagation due to an absorbing bottom. J. Acoust. Soc., Amer., 27(4):689-691.

attenuation, waves
Robin, G. De Q., 1963
Wave propagation through fields of pack ice.
Phil. Trans., R. Soc., London, (A), 255(1057) 313-339.

attenuation, waves
Snodgrass, R.E., G.W. Groves, K.F. Hasselmann, E. R. Miller, W. H. Munk, and W. H. Powers, 1966.
Propagation of ocean swell across the Pacific.
Phil. Trans. R. Soc., (A), 259(1103): 431-497.

attenuation, waves
Tsyplukhin, V.F., and V.A. Sergeyev, 1962
The instrumental study of the attenuation of waves with depth.
Okeanologiia, Akad. Nauk, SSSR, 2(1):134-138.
Abstracted in: Soviet Bloc Res., Geophys., Astron., and Space, 1962(35):24 (OTS61-11147-35 JPRS13739)

augite

augite
Norris, Robert M., 1964.
Sediments of Chatham Rise.
New Zealand Dept. Sci. Ind. Res., Bull., No. 159
New Zealand Oceanogr. Inst., Memoir, No. 16:39pp.

austausch

austausch
Ichie, T., 1950.
[On the mutual effect of the coastal and offshore water.] J. Ocean. Soc., Japan (Nippon Kaiyo Gakkaisi) 6(2):1-7. (In Japanese; English summary)

Austausch coefficient
Nanniti, T., 1951.
On the Austausch coefficient in the sea.
Ocean. Mag., Tokyo, 3(1):49-52, 2 textfigs.

AUTEC

AUTEC
Covey, Charles W. 1967.
AUTEC [Atlantic Undersea Test and Evaluation Center] is commissioned.
Undersea Techn. 8(4): 22-23

Autocorrelation

autocorrelation
Burling, R.W., 1950.
Analysis of gepphysical records. N.Z. Dept. Sci. Ind. Res., Geophys. Conf., 1950, Paper No. 12: 10 pp. (mimeographed).
In WHOI mss. file.

autocorrelation
Seiwell, H.R., 1950.
Ocean wave analyses. Symposium on applications of autocorrelation analysis to physical problems, Woods Hole, Massachusetts, 13-14 June 1949. NAVEXOS-P-735:74-79, 3 textfigs.
WHOI Contrib. No. 521.

Autoinhibitors

autoinhibitors
Fogg, G.E., 1966.
The extracellular products of algae.
In: Oceanography and marine biology, H. Bernes, editor, George Allen & Unwin, Ltd., 4:195-212.

Autolysis

autolysis
*Aldrich, David V., Sammy M. Ray and William B. Wilson, 1967.
Gonyaulax monilata: population growth and development of toxicity in cultures.
J. Protozool., 14(4):636-639.

Automatic stations

automatic stations
Burnard, Francis, 1967.
Réalisation d'une station automatique d'enregistrement des facteurs physico-climiques dans la zone des marées.
Helgoländer wiss. Meeresunters. 15(1/4): 353-360

Autospectra

autospectra
Shonting, David H., 1968.
Autospectra of observed particle motions in wind waves.
J. mar. Res., 26(1):43-65.

Auxospores

auxospores
Holmes, Robert W., 1966.
Short-term temperature and light conditions associated with auxospore formation in the marine centric diatom Coscinodiscus Concinnus W. Smith.
Nature, 209 (5019):217-218.

auxospores
Sato, Shigekatsu, and Hisashi Kan-no 1967
Synchrony to the auxospore phase in the cultured population of a centric diatom Chaetoceros simplex var. calcitrans Paulsen. (In Japanese; English abstract)
Bull. Tohoku reg. Fish. Res. Lab. 27: 101-109

aviation

aviation
Golenchenko, A.P., 1959
[Use of aviation in the sea fish detection and fishery investigations.]
Biull. Okeanograf. Komissii, Akad. Nauk, SSSR., (3): 91-98.

Avoidance

avoidance
*Brinton, Edward, 1967.
Vertical migration and avoidance capability of euphausiids in the Califronia Current.
Limnol. Oceanogr., 12(3):451-483.

Avoidance reactions
*Mackie, A.M., 1970.
Avoidance reactions of marine invertebrates to either steroid glycosides of starfish or synthetic surface-active agents. J. exp. mar. Biol. Ecol., 5(1): 63-69.

avoidance
McGowan, John A., and Vernie J. Fraundorf, 1966.
The relationship between size of net used and estimates of zooplankton diversity.
Limnol. Oceanogr., 11(4):456-469.

avoidance
Nambiar, K.P.P., Yoshio Hiyama and Takaya Kusaka 1970.
Behaviour of fishes in relation to moving nets. 1. Effect of differently coloured lights on the catch by model nets.
Bull. Jap. Soc. scient. Fish 36(2): 135-138.

azoic zone

azoic zone
*Merriman, Daniel, 1968.
Speculations on life at the depths: a XIXth century prelude.
Bull. Inst. océanogr., Monaco, No. special 2: 377-385.

back-passing

back-passing
Bruun, Per M. 1967.
By-passing and back-passing off Florida.
J. WatWays Harb. Div. Am. Soc. civ. Engrs 93 (WW2): 101-128.

back radiation
Matveev, D.T. 1970.
On interpretation of measurements of ocean surface outgoing radiation on the basis of satellite data. (In Russian; English abstract).
Met. Gidrol. 1970 (8): 36-43.

backscattering

backscattering
Andreeva, I.B., 1967.
Sound scattering in the surface layers of the ocean. (In Russian; English abstract).
Okeanologiia, Akad. Nauk, SSSR, 7(1):71-77.

backscattering
*Buckley, J.P., and R.J. Urick, 1968.
Backscattering from the deep sea bed at small grazing angle.
J. acoust. Soc., Am., 44(2):648-650.

backscattering, radio waves
*Crombie, D.D., and J.M. Watts, 1968.
Observations of coherent backscatter of 2-10 MHZ radio surface waves from the sea.
Deep-Sea Res., 15(1):81-87.

backscattering (sound)
Hall, M. 1971.
Volume backscattering in the South China Sea and the Indian Ocean.
J. acoust. Soc. Am. 50 (3-2): 940-945.

backscattering acoustic
*Pruitt, R.A., S.W. Marshall, and R.R. Goodman, 1969.
Acoustic backscattering from a wind-driven water surface.
J. mar. Res., 27(1):45-54.

backscattering
Reeves, Jon, Yoshiya Igarashi, Lloyd Beck and R. Stern 1969.
Azimuthal dependence of sound back-scattered from the sea surface.
J. acoust. Soc. Am. 46 (5-2): 1284-1288.

backscattering (sound)
Schmidt, P.B., 1971.
Monostatic and bistatic backscattering measurements from the deep ocean bottom.
J. acoust. Soc. Am. 50 (1-2): 326-331

bacteria

A

bacteria
Adair, Frank W., and K. Gundersen 1969.
Chemoautotrophic sulfur bacteria from the marine environment. II. Characterization of an obligately marine facultative autotroph.
Can. J. Microbiol. 15 (4): 355-359

bacteria
Adair, Frank W., and K. Gundersen, 1969.
Chemoautotrophic sulfur bacteria in the marine environment. 1. Isolation, cultivation and distribution.
Can. J. Microbiol., 15 (4): 345-353.

bacteria
Ahrens, Renate, 1971.
Untersuchung zur Verbreitung von Phagen der Gattung Agrobacterium in der Ostsee.
Kieler Meeresforsch. 27 (1): 102-112.

bacteria
Ahrens, Renate, 1970.
3.3. Über das Vorkommen von sternbildenden Bakterien. Chemische, mikrobiologische und planktologische Untersuchungen in der Schlei im Hinblick auf deren Abwasserbelastung. Kieler Meeresforsch 26(2): 159-161.

bacteria
estuaries
Baltic

bacteria
Ahrens, Renate, 1970.
Weitere sternbildende Bakterien aus Brackwasser.
Kieler Meeresforsch. 26 (1): 74-78

Baltic
bacteria

bacteria
Ahrens, Renate, 1969.
Ökologische Untersuchungen an sternbildenden Agrobacterium-Arten aus der Ostsee.
Kieler Meeresforsch. 25 (1): 190-204

Baltic
bacteria

bacteria
*Ahrens, Renate, 1968.
Taxonomische Untersuchungen an sternbildenden Agrobacterium - Arten aus der Westlichen Ostsee.
Kieler Meeresforsch., 24(2):147-173.

bacteria
Ahrens, Ranate, und Gerhard Rheinheimer, 1967.
Über einige sternbildende Bakterien aus der Ostsee.
Kieler Meeresforsch., 23(2):127-136.

bacteria
Altschuler, Susan J., and Gordon A. Riley 1967.
Microbiological studies in Long Island Sound.
Bull. Bingham oceanogr. Coll. 19(2): 81-88.

bacteria
Anishenko Elita 1967.
Distribución de bacterias heterotróficas en aguas del Golfo de México y del Mar Caribe.
Estudios Inst. Oceanol. Acad. Cienc. Cuba 2(2): 5-20.

bacteria
Appolova, T.A., 1963.
The bacteria quantity dynamics in the Kursk Gulf. (In Russian).
Atlantich. Nauchno-Issled. Inst. Ribn. Khoz. i Okeanogr. (ATLANTNIRO), Trudy, 10:54-60.

bacteria
*Aubert, M.J. Aubert, S. Daniel, J.-P. Gambarotta, L.-A. Romey, J.-Ph. Mangin, P.-F. Bulard, P. Irr et Y. Chevalier, 1967.
Dynamique d'eaux résiduaires rejetées en bassin maritime.
Rev. intern. Océanogr. Med. 8:5-40.

bacteria
Aubert, M., J. Guizerix, J. Aubert, J. Molinari et S. Daniel, with the collaboration of M. Gauthier, J. P. Gambarotta, G. Torde, B Gaillard, H. Santos-Cottin et A. Mornes, 1966
Étude de la dispersion bactèrienne d'eaux residuairès en mer au moyen de traceurs radioactifs.
Revue Int. Océanogr. Medicale, C. E. R. B. O. M 1: 56-91.

bacteria
Aubert, M., et H. Lebout, 1963.
Recherche bactériologique des pollutions de planctonts.
Rapp. Proc. Verb., Réunions, Comm. Int., Expl. Sci., Mer Méditerranée, Monaco, 17(3):719-722.

bacteria
Aubert, M., and H. Lebout, 1962.
Étude de la situation hydrologique et bacteriologique des zones marine situées au large du delta du Var et des regions adjacentes (Suite).
Cahiers Centre Étude Recherches Biol. Océan. Med. (4):50-106.

bacteria
Aubert, M., et H. Lebout, 1962.
Étude de la situation hydrelogique et bacteriologique des zones marines situées au large du delta du Var et des regions adjacentes.
Cahiers Centre Étude Recherches Biol. Océan. Med. (3):1-49.

bacteria
Aubert, Maurice, Henri Lebout et Jacqueline Aubert, 1964.
Etude de la situation hydrologique et bacteriologique de la partie ouest de la Baies des Anges (face est et sud du Cap d'Antibes).
Cahiers Centre Etudes Recherches Biol. Océanogr. Medicale, Nice, 14(2):11-143.

B

bacteria
Babudieri, B., e R. Fravento, 1969.
Ricerche di microbiologia marina nel Mediterraneo orientale.
Boll. Pesca, Piscic. Idrobiol 24(2): 151-158

bacteria
temperature (data)

BACTERIA
Bansemir, Klaus 1969.
Bakteriologische Untersuchungen von Wasser und Sedimenten aus dem Gebiet der Island-Färöer Schwelle
Ber. dt. wiss. Komm. Meeresforsch. 20(3/4): 282-287.

bacteria
*Barbaree, J.M., and W.J. Payne, 1967.
Products of denitrification by a marine bacterium as revealed by gas chromatography.
Marine Biol., 1(2):136-139.

Bacteria
Barbour, Richard, T., 1966.
Interaction of bubbles and bacteria in the formation of organic aggregates in sea-water.
Nature, 211 (5046):257-258.

bacteria

Barghorn, E.S., & R.L. Nichols, 1961

Sulfate-reducing bacteria and pyritic sediments in Antarctica.
Science, 134(3473): 190.

bacteria

Batoosingh, E., and E.H. Anthony 1971.
Direct and indirect observations of bacteria on marine pebbles.
Can. J. Microbiol. 17(5): 655-664

bacteria

Bein, S.J., 1954.
A study of certain chromogenic bacteria isolated from "red tide" water with a description of a new species. Bull. Mar. Sci., Gulf and Caribbean 4(2):110-119.

bacteria

Belser, William L., 1964.
DNA base composition as an index to evolutionary affinities in marine bacteria.
Evolution, 18(2):177-182.

bacteria

*Berland, B.R., M.G. Bianchi et S.Y. Maestrini 1969.
Etude des bactéries associées aux Algues marines en culture. 1. Détermination préliminaire des espèces. Marine Biol., 2(4): 350-355.

bacteria

bacteria

Berland, B.R., D.J. Bonin and S.Y. Maestrini, 1970.
Study of bacteria associated with marine algae in culture. III. Organic substrates supporting growth.
Mar. Biol. 5(1):68-76

bacteria

*Berland, B.R. and S.Y. Maestrini, 1969.
Study of bacteria associated with marine algae in culture. II. Action of antibiotic substances.
Marine Biol., 3(4): 334-335.

bacteria
cultures
antibiotics

bacteria

*Bianchi, A.J.M., 1968.
A perçu sur la distribution de certains groupements fonctionnels bactériens au niveau de trois stades d'évolution des feuilles de Posidonies.
Recl.Trav.Stn.mar.Endoume, 43 (59):251-257.

bacteria

Blanchard, Duncan C., and Laurence Syzdek 1970.
Mechanism for the water-to-air transfer and concentration of bacteria.
Science 170 (3958): 626-628.

bacteria

*Blanton, W. George, and Carol J. Blanton, 1968.
Polymorphism of a deep marine benthic bacterium from the Gulf of Mexico.
J. oceanogr.Soc.Korea, 3(1):8-15.

bacteria

Bogoyavlensky, A.N., 1962
On the problem of the distribution of the heterotrophic microorganisms in the Indian Ocean and in the Antarctic waters.
Okeanologiia, Akad. Nauk, SSSR, 2(2):293-297.

bacteria

Bonde, G.J. 1968.
Studies on the dispersion and disappearance phenomena of enteric bacteria in the marine environment.
Rev. int. Océanogr. Méd. 9:17-44

bacteria

Bonde, G.J., and A. Mørk Thomsen 1971.
Computer analysis of a bacteriological monitoring system for pollution control
Rev. int. Biol. Océanogr. Méd. 24: 48-49 (abstract only)

bacteria

Boyd, William L., and Josephine W. Body, 1963
Enumeration of marine bacteria of the Chukchi Sea.
Limnology and Oceanography, 8(3):343-348.

bacteria

Brisou, Jean, et Yves de Rautlin de la Roy, 1966.
Répartition verticale des bactéries hétérotrophes en haute mer (Atlantique Nord, Station Juliett 52° N.-20° W.).
Bull Inst. Océanogr., Monaco, 66 (1361). 8pp.

bacteria

Brisou, Jean et Yves Rautlin de la Roy, 1962.
Étude microfloristique du littoral de la Rochelle. Répartition de quelques grands groupes microbiens.
Comptes Rendus, Soc. Biol., France, 156:344.

bacteria

Brisou, J., et H. Vargues 1965.
Étude sur l'halophilie des bactéries isolées du milieu marin.
Rapp. P.-v. Réun. Comm. int. Explor. scient. Mer Méditerr. 18(3): 607-608.

bacteria

Brisou, J., et H. Vargues, 1964.
Contribution à l'étude de la nutrition minérale d'un Microqie "hyperhalophile".
Comptes Rendus, Soc. Biol., France, 158(1):170.

bacteria

Brisou, Jean, et H. Vargues, 1963.
Action des cations Ca et Mg sur la croissance de quelques bactéries halophiles d'origin marine.
Ann. Inst. Pasteur, 105:586-596.

bacteria

Brisou, Jean, Huguette Vargues et Jean Cadeillan, 1961.
Vie des microbes terrestres en milieu marin. Action des composes iodes.
Comptes Rendus, Soc. Biol., France, 155(4):887.

bacteria

Brock, Thomas D., 1967.
Bacterial growth rate in the sea: direct analysis by thymidine autoradiography.
Science, 155(3758):81-83.

bacteria

Brock, T.D., 1964
Knots in Leucothrix mucor.
Science, 144(3620):870-871.

bacteria

Brown, A.D., 1964.
Aspects of bacterial response to the ionic environment.
Bacteriological Rev., 28(3):296-329.

bacteria

Brown, A.D., D.G. Drummond and R.J.M. North, 1962
The periferal structure of gram-negative bacteria II. Membranes of bacilli and spheroblasts of a marine pseudomonad.
Biochim. Biophys. Acta, 58:514-531.

bacteria

Bryan, Charles A., and Arthur H. Bryan, 1959.

A new approach to the study of marine bacteria in Gulf and Atlantic waters.
Ecology 40(4):712-715.

bacteria

Buck, J.D., and R.C. Cleverdon, 1960.
Hydrogen sulphide reduction by some agarolytic marine bacteria.
Canadian J. Microbiol., 6(5):594-595.

bacteria

Buck, John D., and Robert C. Cleverdon, 1960.

The spread plate as a method for the enumeration of marine bacteria.
Limnol. & Oceanogr., 5(1):78-80.

bacteria

*Budd, J.A., 1969.
Catabolism of trimethylamine by a marine bacterium, Pseudomonas NCMB 1154. Marine Biol., 4(3): 257-266.

bacteria
catabolism

bacteria

Burchard, Robert P. 1971.
Chesapeake Bay bacteria able to cycle carbon, nitrogen, sulfur and phosphorus.
Chesapeake Sci. 12(3): 179-180

bacteria

Burkholder, Paul R., and Seymour Lewis 1968.
Some patterns of B vitamin requirements among neritic marine bacteria.
Can. J. Microbiol. 14(5): 537-543.

C

bacteria

Campanile, E., V. Ferro, E. de Simone, G. Grasso et R. de Fusco 1970.
Souches bactériolytiques dans l'eau de mer.
Rev. int. Oceanogr. Méd. 18-19:117-124.

bacteria

Campbell, Ann E., Johan A. Hellebust and Stanley W. Watson, 1966.
Reductive pentose cycle in Nitrosocystis oceanus
J. Bacteriol., 91(3):1178-1185.

WHOI Contrib. No. 1711.

bacteria

bacteria

Carey, C.L., 1938.
The occurrence and distribution of nitrifying bacteria in the sea. J. Mar. Res. 1(4):291-304.

WHOI Contrib. No. 188.

bacteria
Carey, C.L., and S.A. Waksman, 1934
The presence of nitrifying bacteria in deep seas. Science 79 (2050):349-350.

bacteria
*Carlucci, A.F. and P.M. McNally, 1969.
Nitrification by marine bacteria in low concentrations of substrate and oxygen. Limnol. Oceanogr., 14(5): 736-739.

nitrification
bacteria

bacteria
Carlucci, A.F., and D. Pramer, 1959.
Factors affecting the survival of bacteria in sea water.
Appl. Microbiol., 7(6):388-392.

bacteria
Carroll, J.J., and L.J. Greenfield and R.F. Johnson, 1965.
The mechanism of calcium and magnesium uptake from sea water by a marine bacterium.
J. Cell. Comp. Physiol., 66(1):109-118.

bacteria
Castellví, Josefina, 1967.
Bacteriología marina.
In: Ecología marina, Monogr. Fundación la Salle de Ciencias Naturales, Caracas, 14:201-229.

BACTERIA
Castellvi, Josefina, 1967.
Ciclo de las bacterias planctonicas marines en la costa catalana.
Investigación pesq., 31(3): 611-620.

bacteria
Mediterranean, west

bacteria
Chan, E.C.S, and Elizabeth A. McManus, 1969.
Distribution, characterization and nutrition of marine microorganisms from the algae Polysiphonia lanosa and Ascophyllum nodosum
Can. J. Microbiol. 15(5): 409-420.

bacteria

bacteria a
*Choe, Sang and Geon Chee Kim, 1970.
The effect of seawater concentration of the survival of fecal pollution bacteria. (In Korean; English abstract). J. oceanol. Soc. Korea, 5(2): 65-72.

seawater, effect of
bacteria
pollution

bacteria
Chu, S. P., 1946
Note on the technique of making bacteria-free cultures of marine diatoms. JMBA, XXVI, No. 3, pp. 296-302

bacteria
Cioglia, L., and B. Loddo, 1961.
Sui fenomeni di autodepurazione nell'ambiente marino. II. Sopravvivenza e variabilita nella flora decale.
Nuov. Ig. Microbiol., 12:445-462.

Ref. in: Bull. Inst. Pasteur, 1963, 61(8):2438.

bacteria
Collins, V.G., 1957.
Planktonic bacteria.
J. Gen. Microbiol., 16(1):268-272.

bacteria
Colwell, R.R., J. Liston, W. Wiebe and M. Gochnauer, 1963
Bacterial taxonomy.
Res. in Fisheries, Coll. Fish., Fish. Res. Inst. Univ. Washington, Contrib. No. 147:49-50

bacteria
Connor, R.M., 1957.
Comparison of fermentation tube and membrane filter techniques for estimating coliform densities in sea water. 1. Preliminary report.
Appl. Microbiol., 5(3):141-144.

bacteria
Cviić, Vlaho, (posthumons), 1965.
Primary organic production, distribution, and reproduction of bacteria in the middle Adriatic euphotic zone.
Acta Adriatica, 10(9):21pp.

bacteria
Cviic, Vlaho, 1964.
Contribution a la connaissance de la distribution de la biomasse bacterienne dans l'Adriatique moyenne. (In Jugoslavian; French resume).
Acta Adriatica, 11(1):59-64.

bacteria
Cviic, Vlaho, 1963
Distribution des bacteries et de la biomasse bactérienne dans l'Adriatique méridionale. (In Jugoslavian; French resumé).
Acta Adriatica, 10(7):15 pp.

bacteria
Cviic, V., 1956.
Multiplication of heterotrophic sea bacteria in various H-ion concentrations.
Acta Adriatica, 8(8):1-15.

bacteria
Cviic, V., 1955.
Distribution of bacteria in the waters of the mid Adriatic Sea. Inst. Ocean. i Ribarstvo, Split Repts., 4(1):43 pp., 20 tables.

The M.V. "Hvar" cruises in fisheries biology, 1948-1949.

bacteria
Cviic, V., 1953.
On the ecological relations of marine bacteria and plankton.
Atti VI Congreso Int. di Microbiol., Roma, 6-12 Settembre 1953 (22)7:367-368.

bacteria
Cviic, V., 1953.
Attachment of bacteria to slides submerged in sea water. Biljeska-Notes, Inst. Ocean. i Ribarstvo, Split, No. 6:7 pp., 1 textfig.

D

bacteria
DiSalvo, Louis H., 1971.
Regenerative functions and microbial ecology of coral reefs: labelled bacteria in a coral reef microcosm. J. exp. mar. Biol. Ecol. 7(2): 123-136.

bacteria
de Silva, N.N., 1964.
A comparison of the bacterial flora of surface and sub-surface water in the sea off Ceylon.
Bull. Fish. Res. Sta., Ceylon, 17(2):151-157

Indian Ocean
bacteria

bacteria
de Silva, N.N., 1963.
Marine bacteria as indicators of upwelling in the sea
Bull. Fish Res. Sta., Ceylon, 16(2):1-10)

indicator species
bacteria

bacteria
Deveze, L., 1964.
Activités bactériennes et matières organiques dans les sediments marins.
Ann. Inst. Pasteur, 107(Suppl., No. 3):123-135.

bacteria
Devèze, L., 1955.
Parallélisme d'évolution des populations planctoniques et bactériennes marines durant la période estivale 1955. C.R. Acad. Sci., Paris, 241(22):1629-1631.

bacteria
Devège, L., 1950.
Contribution à l'étude des bacteries des eaux en relation avec les variations du milieu. Bull. Océan. Inst., Monaco, No. 973:10 pp.

bacteria
Devèze, Louis, Jean Le Petit, et Robert Matheron, 1966.
Note préliminaire sur la présence dans les eaux et les sédiments marins de bactéries solubisent certains sals minéraux insolubles (carbonates, phosphates et silicates).
Bull. Inst. Océanogr., Monaco, 66(1370):8 pp.

bacteria
Droop. M.R., and K.G.R. Elson, 1966.
Are pelagic diatoms free from bacteria?
Nature, 211 (5053):1096-1097.

E

bacteria
Eagon, R.G., 1962.
Pseudomonas natriegens, a marine bacterium with a generation time of less than 10 minutes.
J. Bacteriol., 83:736-737.

bacteria
Eagon, R.G., 1962.
Pyridine nucleotide-linked reactions of Pseudomonas natriegens.
J. Bacteriol., 84:819-821.

bacteria
Eagon, R.G., and C.H. Wang, 1962.
Dissimilation of glucose and gluconic acid by Pseudomonas natriegens.
J. Bacteriol., 83:879-886.

bacteria a
Ezura, Yoshio, and Minoru Sakai, 1970.
Numerical taxonomy of micrococci isolated from sea water. (In Japanese; English abstract).
Bull. Fac. Fish. Hokkaido Univ. 21 (2): 152-159.
bacteria

F

bacteria
Floodgate, G.D., 1966.
Factors affecting the settlement of a marine bacterium.
Veröff. Inst. Meeresforsch., Bremerh., Sonderband II:265-269.

bacteria
Floodgate, G.D., 1965.
The enumeration of bacteria in coastal waters.
Jour. Mar. Biol. Assoc., U.K., 44(2):365-372.

bacteria a
Forsyth, M.P., D.B. Shindler, M.B. Gochnauer and D.J. Kushner, 1971.
Salt tolerance of intertidal marine bacteria.
Can. J. Microbiol. 17 (6): 825-828.

bacteria
Føyn, Ernst, 1971.
Municipal wastes. (16) In: Impingement of man on the oceans, D.W. Hood, editor, Wiley Interscience: 445-459.

bacteria
Fujita, Yuji, Shoji Iizuka and Buhei Zenitani, 1969.
Microbiological studies on marine areas. V. On the distribution of the sulfate-reducing bacteria and sulfur bacterial groups in oxygen-deficient or anoxic bottom layer of bay in summer. (In Japanese; English abstract).
Bull. Fac. Fish. Nagasaki Univ. 28:153-160.

G

bacteria
Gandhi, N.M., and Yvonne M. Freitas, 1964.
The production of vitamin B12-like substances by marine micro-organisms. II. Studies on nutrition.
Proc. Indian Acad. Sci., (B), 59(1):33-46.

bacteria
Gandhi, N.M., and Yvonne M. Freitas, 1964.
The production of vitamin B12-like substances by marine organisms. 1. General survey.
Proc. Indian Acad. Sci., (B), 59(1):24-32.

bacteria
Gauthier, Gambarotte, Vallard et Daniel, 1964.
Dispersion des bactéries d'un emissaire d'eaux usées au de bouche en Mer.
Cahiers du C.E.R.B.O.M., 16:29-47.

bacteria
Genovèse, Sébastiano, 1967.
Sur quelques problèmes de bactériologie marine.
Rev.intern.Océanogr.Med., 8:41-50.

bacteria
Genovèse, S., 1961.
Sur la presence d'"eau rouge" dans le lac de Faro (Messine).
Rapp. Proc. Verb., Reunions, Comm. Int. Expl. Sci. Mer Méditerranée, Monaco, 16(2):255-256.

bacteriology
Ginés, Hno., y R. Margalef, editores, 1967.
Ecología marina.
Fundación La Salle de Ciencias Naturales, Caracas Monografia 14: 711 pp.

bacteria
*Girard, Arthur E., John D. Buck and Benhamin J. Cosenza, 1968.
A nutritional study of some agarolytic marine bacteria.
Can.J. Microbiol., 14(11):1193-1198.

bacteria
Gol'din, M. I., 1960
[Microbes in the depths of the sea] Nauka i Zhizn' (9): 37-40.

bacteria
*Gorbenko, Yu.A., 1969.
A community of periphytic microorganisms as a biological system. (In Russian; English abstract)
Okeanologiia, 9(2): 318-328.

bacteria
Gordon, R.D., 1939
Estimating bacterial populations by the dilution method. Biometrika:167-180.

bacteria
Gordon, R.D., 1938
Note on estimating bacterial populations by the dilution method. Proc. Nat. Acad. Sci., 24:212-215.

bacteria
Gordon, R.D., and C.E. Zobell, 1938
Note on the successive dilution method for extimating bacterial ppopulation. Zentralb. f. Bacteriol. II Abt., 99:318-320.

bacteria
Grant, C.W., 1942
Studies in marine microbiology at the Scripps Institution of Oceanography, University of California, La Jolla.
Amer. Phil. Soc. Year Book 1941:157-158.

bacteria
Grant, C.W., and C.E. ZoBell, 1942
Oxidation of hydrocarbons by marine bacteria. Soc. Exper. Biol. Med., Proc., 51:266-267.

bacteria
Grein, A., and C. Spalla, 1962.
Ecologia ed esigenze nutritizie della flora microbia del Golfo di Napoli.
Pubbl. Sta. Zool., Napoli, 33:32-49.
Ref. in: Bull. Inst. Pasteur, 1963, 61(8):2438.

bacteria
Gundersen, Kaare, 1966.
The growth and respiration of Nitrosocystis oceanus at different partial pressures of oxygen.
J. gen. Microbiol., 42(3):387-396.

bacteria
Gunkel, Wilfried 1968.
Die Fluctuationen der Bakterien im jahreszeitlichen Verlauf in der Nordsee.
Mar. Mykologie, Veröff. Inst. Meeresforsch Bremerh. Sonderband 3:121-122

bacteria
Gunkel, Wilfried, 1966.
Bakteriologische Untersuchungen im Indischen Ozean.
Veroff. Inst. Meeresforsch., Bremerh., Sonderband II:255-263.

bacteria
Gunkel, Wilfried, 1964.
Die Verwendung des Ultra-Turrax zur Aufteilung von Bakterienaggregaten in marinen Proben.
Helgoländer Wiss. Meeresuntersuchungen, 11(3/4): 287-295.

bacteria
Gutveib, L.G., A.G. Benghitski and M.N. Lebedeva 1971.
Synthesis of biologically active substances of vitamin B_{12} group of the bacterianeuston of the tropical Atlantic. (abstract only).
Rev. int. Biol. Océanogr. méd. 24:53-55.

H

bacteria
Haight, Janet J., and Richard Y. Morita, 1966.
Some physiological differences in Vibrio marinus grown at environmental and optimal temperatures.
Limnol. Oceanogr., 11(4):470-470.

bacteria
*Hamilton, Robert D., and Osmund Holm-Hansen, 1967.
Adenosine triphosphate content of marine bacteria
Limnol. Oceanogr., 12(2):319-324.

bacteria
chemistry, adenosine triphosphate

bacteria
Harvey, H. W., 1941.
On changes taking place in sea water during storage. J.M.B.A., n.s., 25:225-233, 1 textfig

bacteria
Hasegawa, Y., M. Yokoseki, E. Fukuhara, and K. Terai, 1952.
On the environmental conditions for the culture of laver in Usu-Bay, Iburi Prov., Hokkaido.
Bull. Hokkaido Regional Fish. Res. Lab. 6:1-24, 8 textfigs., 9 tables.

bacteria
Hayes, P.R., 1963.
Studies on marine flavobacteria.
J. Gen. Microbiol., 30(1):1-20.

bacteria
Hayes, F. R., and E. H. Anthony, 1959.
Lake water and sediment. VI. The standing crop of bacteria in lake sediments and its place in the classification of lakes.
Limnol. & Oceanogr., 4(3):299-315.

Bacteria
*Hickel, W., 1969.
Sedimentbeschaffenheit und Bakteriengehalt im Sediment eines zukünftigen Verklappungsgebietes von Industrieabwässern nordwestlich Helgolands.
Helgoländer wiss. Meeresunters, 19(1): 1-20.

bacteria
Hidaka, Tomio, 1964
Studies on the marine bacteria. 1. Comparative observations on the inorganic salt requirements of marine and terrestrial bacteria. (In Japanese; English abstract).
Mem. Fac. Fish., Kagoshima Univ., 12(2):135-152.

bacteria
Hidaka, Tomio, 1963.
Studies on the marine bacteria. 1. Comparative observations on the inorganic salt requirements of marine and terrestrial bacteria. (In Japanese English abstract).
Mem. Fac. Fish., Kagoshima Univ., 12(2):135-152.

bacteria
Hidaka, T., and K. Saito, 1957.
Studies on the cellulose decomposing bacteria in the digestive organs of the ship-worm (Teredo navalis sp.). 2.
Mem. Fac. Fish., Kagoshima Univ., 5:172-177.

bacteria
*Hobbie, John E., and Claude C. Crawford, 1969.
Respiration corrections for bacterial uptake of dissolved organic compounds in natural waters.
Limnol. Oceanogr. 14(4):528-532.

bacteria
Hock, C. W., 1941.
Marine chitin-decomposing bacteria. J. Mar. Res. 4(2):99-106.

bacteria of mud bottom
Hutton, R.F., B. Eldred, K.D. Woodburn and R.M. Ingle, 1956.
The ecology of Boca Ciega Bay with special reference to dredging and filling operations.
Florida State Bd., Conserv., Mar. Lab., St. Petersburg, Tech. Ser., 17(1):87 pp.

I

bacteria
Isachenko, B.L., and A.A. Yegorova, 1939.
[On the "bacterial plate" in the Black Sea.]
USSR. Inst. Mar. Fish., Ocean., Sbornik Knipovich :47-58.

bacteria
Ischenko, N.I., and I.B. Ulanovsky, 1963.
The protective effect of aerobic bacteria upon corrosion of carbon steel in sea water. (In Russian; English abstract).
Mikrobiologia, 32(3):521-525.

J

bacteria
*Jannasch, H.W., 1967.
Enrichments of aquatic bacteria in continuous culture.
Archiv. für Mikrobiol. 59: 165-173.

bacteria
*Jannasch, Holger W., 1967.
Growth of marine bact-ria at limiting concentrations of organic carbon in sea water.
Limnol. Oceanogr., 12(2):264-272.

bacteria (data only)
Japan, Faculty of Fisheries, Hokkaido University, 1961
Data record of oceanographic observations and exploratory fishing, No. 5:391 pp.

bacteria
*Johnson, Roy M., Rose Mary Schwent and Wesley Press, 1968.
The characteristics and distribution of marine bacteria isolated from the Indian Ocean.
Limnol. Oceanogr., 13(4):656-664.

bacteria
Jones, Galen E., 1970.
Metal organic complexes formed by marine bacteri[a]
Occ. Pap. Inst. mar. Sci., Alaska, 1: 301-319.

organic complexes, metallic
bacteria

bacteria
Jones, Galen E., 1967.
Growth of Escherichia coli in heat- and copper-treated synthetic seawater.
Limnol. Oceanogr., 12(1):167-172.

bacteria
Jones, Galen E., 1964.
Effect of chelating agents on the growth of Eschericia coli in seawater.
J. Bacteriol., 87(3):483-499.

bacteria
Jones, Galen E., 1964.
Suppression of bacteial growth by sea water.
In: Symposium on Marine Microbiology, Carl H. Oppenheimer, Editor, Charles C. Thomas, Publisher, Springfield, Illinois, pp.572-579.

bacteria
Jones, G.E., 1958.
Attachment of marine bacteria to zooplankton.
U.S.F.W.S. Spec. Sci. Rept., Fish., No. 279(2): 77-78.

bacteria
Jones, Galen E., and Holger W. Jannasch, 1959.
Aggregates of bacteria in sea water as determined by treatment with surface active agents.
Limnol. & Oceanogr., 4(3):269-276.

bacteria
Jones, G.E., and W.H. Thomas, 1958.
The effects of organic and inorganic micronutrients on the assimilation of C14 by planktonic communities and on bacterial multiplication in tropical Pacific sea water.
U.S.F.W.S. Spec. Sci. Rept., Fish., No. 279(2): 87-99.

bacteria
Jones, G.E., W.H. Thomas and F.T. Haxo, 1958.
Preliminary studies of bacterial growth in relation to dark and light fixation of C1402 during productivity determinations.
U.S.F.W.S. Spec. Sci. Rept., Fish., No. 279(2): 79-86.

K

bacteria
Kadota, H., 1956?
A study on the marine aerobic cellulose-decomposing bacteria. Mun. Coll. Agric., Kyoto Univ., Fish. Ser., 6, 74:128 pp.

bacteria
Kakimoto, Daiichi and Tomio Hidaka, 1968.
Variations of the bacteria living in the marine environment.
Bull. Misaki mar. biol. Inst., Kyoto Univ. 12:151-161

bacteria
Kalinenko, V.O., 1952.
Bacteria and invertebrates inhabiting the sea-bed. (In Russian).
Mikrobiologiia, 21(4):555-

bacteria
Kalinenko, V.O., 1961.
[The atmospheric nutrition (aerotrophic features) of the marine bacteria.]
Okeanologiia, Akad. Nauk, SSSR, 1(4):734-740.

bacteria
Kalinenko, V.O., 1959.
Bacterial colonies on metal plates in sea water.
Mikobiologiia, 28(5):750-756.

bacteria
Karssilnikova, E.N., 1961
[On the antibiotic properties of microorganisms isolated from various depths of the world ocean.]
Mikrobiologiia, 30(4):651-657.

bacteria
*Kenis, Paul R., and Richard Y. Morita, 1968.
Thermally induced leakage of cellular material and viability in Vibrio Marinus, a psychrophilic Marine bacterium.
Can.J.Microbiol., 14(11):1239-1244.

bacteria
Ketchum, B.H., J.C. Ayers, and R.F. Vaccaro, 1952.
Processes contributing to the decrease of coliform bacteria in a tidal estuary. Ecology 33(2): 247-258, 4 textfigs.

WHOI Contrib. No. 561.

bacteria
Kimata, M., and Y. Hata, 1954.
Studies on urea-splitting bacteria in the sea.
Spec. Publ. Japan Sea Region, Fish. Res. Lab., 3r Ann., 135-138.

bacteria
Kimata, Masao, Akira Kawai and Yoichi Yoshida, 1963.
Studies on marine nitrifying bacteria (nitrite formers and nitrate formers). III. On the nitrite formation of the marine nitrite formers. (In Japanese; English abstract).
Bull. Jap. Soc. Sci. Fish., 29(11):1031-1036.

bacteria
Kimata, Masao, and Yoichi Yoshida, 1969.
Studies on the marine bacteria utilizing nitrogen compounds. III. On the bacterial activities in bottom muds. (In Japanese; English abstract).
Bull. Jap. Soc. Sci. Fish., 35(2): 215-217

bacteria
Kimata, Masao, Yoichi Yoshida and Michiko Taniguchi, 1969.
Studies on the marine microorganisms utilizing inorganic nitrogen compounds. II. On the marine heterotrophic bacteria assimilating inorganic nitrogen compounds. (In Japanese. English abstract).
Bull. Jap. Soc. Sci. Fish., 35(2):211-214

bacteria
*Kimata, Masao, Yoichi Yoshida and Michiko Taniguchi, 1968.
Studies on the marine microorganisms utilizing inorganic nitrogen compounds. 1. On the marine denitrifying bacteria. (In Japanese; English abstract).
Bull. Jap. Soc. scient. Fish., 34(12):1114-1117.

bacteria
Kimura, Takahisa, 1962.
A method for rapid detection of alginic acid digesting bacteria. (IN Japanese; English abstract).
Bull. Fac. Fish., Hokkaido Univ., 12(1):41-47.

bacteria
Kimura, Takahisa, 1962.
Studies on marine agar-digesting bacteria.
(In Japanese: English abstract).
Bull. Fac. Fish., Hokkaido Univ., 12(1):33-48.

bacteria
Kopp, F.I., and E.M. Markianovich, 1950.
[Chitin-decomposing bacteria in the Black Sea.]
Doklady, Akad. Nauk, SSSR, 75(6):859-

bacteria
Koske, P.H., H. Krumm, G. Rheinheimer und K.-H. Szekields, 1966.
Untersuchungen über die Einwirkung der Tide auf Salzgehalt, Schwebstoffgehalt, Sedimentation und Bakteriengehalt in der Unterelbe.
Kieler Meeresforsch., 22(1):47-63.

bacteria
Kriss, A. E., 1960.
Micro-organisms as indicators of hydrological phenomena in seas and oceans. 1.
Deep-Sea Res., 6(2):88-94.

bacteria
Kriss, A.E., 1955.
[The bacterial world of the ocean in the region of the North Pole.] Priroda 9:61.

bacteria
Kriss, A.E., 1955.
[Microbial populations of the deepwater regions of the Sea of Okhotsk and the Pacific Ocean.]
Priroda, 44(7):65-72.

bacteria
Kriss, A.E., M.N. Lebedeva and A.V. Tsiban, 1966.
Comparative estimate of a Nansen and microbiological water bottle for sterile collection of water samples from depths of seas and oceans.
Deep-Sea Res., 13(2):205-212.

bacteria
Kriss, A.E., I.E. Mishustina, I.N. Mitskevich and E.V. Zemtsova, 1964
Microbial stratification in the World Ocean. (In Russian).
Akad. Nauk, SSSR, Isdatel. "Nauka", Moskva, 296 pp.

bacteria
Kriss, A.E., I.E. Mishustina and E.V. Zemtsova, 1962.
Biochemical activity of heterotrophic microorganisms isolated from various regions of the world ocean.
Mikrobiologiia, Akad. Nauk, SSSR, 31(2):314-323.

bacteria
Kriss, A.E., I.N. Mitzkevich, I.E. Mishustina and S.S. Abyzov, 1961
Microorganisms as hydrological indicators in seas and oceans. IV. The hydrological structure of the Atlantic Ocean, including the Norwegian and Greenland Seas, based on microbiological data.
Deep-Sea Res., 7(4): 225-236.

bacteria
Kriss, A.E., and E.A. Rukina, 1953.
[Purple aerobacteria in the H"S depths of the Black Sea.] Dokl. Akad. Nauk, SSSR, 93(6):1107.

bacteria
Krumbein, W.E., 1971.
Sediment microbiology and grain-size distribution as related to tidal movement, during the first mission of the West German Underwater Laboratory Helgoland. Mar. Biol. 10(2): 101-112.

bacteria
Krumbein, W.E., 1971.
Manganese-oxidizing fungi and bacteria.
Naturwissenschaften 58(1):56-57.
fungi
bacteria
manganese, oxidation of

bacteria
Krumm, Hans, und Gerhard Rheinheimer, 1966.
Untersuchungen zur Hydrographie, Schwebstoffzusammensetzung und Bakteriologie des Nord-Ostsee-Kanals. 2. Zur Schwebstoffzusammensetzung und Bakteriologie des Nord-Ostsee-Kanals.
Kieler Meeresforsch., 22(1):121-127.

bacteria
Kühl, Heinrich, und Gerhard Rheinheimer 1968.
Veränderungen der Bakterienflora des Planktons und einiger chemischer Faktoren während einer Tide in der Elbmündung bei Cuxhaven.
Kieler Meeresforsch. 24(1): 27-37.

L

bacteria
Lackey, James B., 1967.
The microbiota of estuaries and their roles.
In: Estuaries, G.H. Lauff, editor, Publs Am. Ass. Advmt Sci., 83:291-302.

ciliates, lists of spp.
summary 301.

bacteria
Lagarde, Edmond, 1964.
Méthode d'estimation du pouvoir dénitrifiant des eaux et des sédiments marins.
Vie et Milieu, Bull. Lab. Arago, 15(1):213-218.

bacteria
Lagarde, E., and P. Forget, 1961.
Contribution à l'étude du métabolisme de l'azote mineral en milieu marin. Étude de souche pure de bacteries dénitrificatrices et réductrices des nitrates.
Rapp. Proc. Verb., Réunions, Comm. Int. Expl. Sci. Mer Méditerranée, Monaco, 16(2):245-250.

bacteria
Lalou, C., 1957.
Étude expérimentale de la production de carbonates par les bactéries des vases de la Baie de Villefranche-sur-mer.
Ann. Inst. Océan., Monaco, 23(4):201-267.

bacteria
Lascelles, J., 1962.
The chromatophores of photosynthetic bacteria.
J. Gen. Microbiology, 29(1):47-53.

bacteria
Lebedeva, M.N., E. Ya. Anichenko and Yu. A Gorbenko, 1961.
[Distribution of heterotrophic bacteria based on microbiological studies in the Mediterranean basins.]
Trudy Sevastopol Biol. Sta., (14):1-32.

bacteria
Lebedeva, M.N., E.J. Anischenko and Ju. A. Gorbenko, 1961.
[Quantitative development of bacterial life (heterotrophs) in the seas of the Mediterranean basin.]
Doklady Akad. Nauk, SSSR, 141(6):1465-1468.

bacteria
Lebedeva, M.N., and Iu. A. Gorbenko 1964.
Some patterns of the hydrological structure of the Mediterranean Sea based on microbiological data. (In Russian).
Trudy Sevastopol. Biol. Stants. Akad Nauk 15:512-533.

bacteria
Lebout, H., 1961.
Étude bactériologique du plancton marin.
Les Cahiers du C.E.R.B.O.M., (1):12-23 (mimeographed).

bacteria
Lehr, E., and H.J. Wegner, 1954.
Der Keimgehalt des Meerwassers und seine Bedeutung für die Hochseefischerei.
Die Fischwirtschaft 6(7):173-174.

bacteria
Lewin, Ralph A., Donald M. Crothers, David L. Correll, Bernhard E. Reimann, 1964.
A phage infecting Saprospira grandis.
Canadian J. Microbiol., 10(1):75-85.

bacteria
Liston, J., W.J. Wiebe and B. Lighthart, 1963
The activities of marine benthic bacteria.
Res. in Fisheries, Coll. Fish., Fish. Res. Inst. Univ. Washington, Contrib. No. 147:47-48.

M

bacteria
MacLeod, R.A., H. Hogenkamp, and E. Onofrey, 1958.
Nutrition and metabolism of marine bacteria. VII. Growth response of a marine flavobacterium to surface active agents and nucleotides.
J. Bacteriol., 75(4):460-466.

bacteria
MacLeod, R.A., and T.I. Matula, 1962.
Nutrition and metabolism of marine bacteria. XI. Some characteristics of the lytic phenomenon.
Canadian J. Microbiol., 8(6):883-896.

bacteria
MacLeod, R.A., and T. Matula, 1961.
Solute requirements for preventing lysis of some marine bacteria.
Nature 192(4808):1209-1210.

bacteria
*Majori, L. M.L. Morelli, L. Diana e G. Rausa, 1967.
L'inquinamento delle acque del mare nell'alto Adriatico. L. Ricerche microbiologiche.
Arch. Oceanogr. Limnol. 15(Suppl.):115-123.

bacteria
Margalef, D. Ramon, 1963.
El ecosistema pelagico de un area costera del Mediterraneo occidental.
Memorias, Real Acad. Ciencias y Artes de Barcelona, 35(1):3-48.

bacteria
Marshall, K.C., Ruby Stout and Ralph Mitchell 1971.
Selective sorption of bacteria from seawater.
Can. J. Microbiol. 17 (11): 1413-1416

Mc

bacteria
Melchiorri-Santolini, U., 1963
5. Microbiologia oceanografica.
Rapp. Attività Sci. e Tecn., Lab. Studio della Contaminazione Radioattiva del Mare, Fiascherino, La Spezia (maggio 1959-maggio 1962), Comit. Naz. Energia Nucleare, Roma, RT/BIO(63) 8:125-142. (multilithed) plus Appendice 5.

bacteria
Mefedova, N.A., 1955.
Characteristics of bacteria from the bed of the north-west Pacific Ocean.] Mikrobiol., 24(3):325-

bacteria
*Mitchell, Ralph, 1968.
The effect of water movement on lysis of non-marine microorganisms by marine bacteria.
Sarsia, 34:263-266.

bacteria
Mitchell, R., and C. Wirsen, 1968.
Lysis of non-marine fungi by marine micro-organisms.
J. gen. Microbiol. 52(3): 335-345.

bacteria
Mitskevich, I.N., and A.E. Kriss, 1966.
High pressure tolerance in Pseudomonas sp., Str. 8113, isolated from a bottom layer of an abyssal trough of the Black Sea. (In Russian).
Dokl., Akad. Nauk, SSSR, 171(1):209-211.

bacteria
*Moll, Georg, Renate Ahrens und Gerhard Rheinheimer, 1967.
Elektronenoptische Untersuchungen über sternbildende Bakterien aus der Ostsee.
Kieler Meeresforsch., 23(2):137-147.

bacteria
Morita, Richard Y., and Roger D. Haight, 1964.
Temperature effects on the growth of an obligate psychrophilic marine bacterium.
Limnology and Oceanography, 9(1):103-106.

bacteria
Morita, R.Y., and R.A. Howe, 1957.
Phosphatase activity by marine bacteria under hydrostatic pressure. Deep-Sea Res., 4(4):454-458

bacteria
Morita, R.Y., and C.E. ZoBell, 1955.
Occurrence of bacteria in pelagic sediments collected during the Mid-Pacific Expedition.
Deep-Sea Res. 3(1):66-73.

bacteria
Moskovits, George and Anita Flanagan, 1967.
Growth of marine bacteria on some organic acids and its application to the selective isolation of pseudomonads.
Can. J. Micrbiol., 13(11):1561-1563.

bacteria
Murakami, Masatada, and Hiroaki Fujisaua, 1968.
Studies on xylan-decomposing bacteria in the marine environment. II. Distribution of B-1,4'-xylan-decomposing bacteria in the coastal region. (In Japanese; English abstract).
Bull. Jap. Soc. scient. Fish. 24(12):1124-1130.

N

bacteria
Nikitina, N.S., 1955.
[Seasonal changes in bacterial composition of littoral zones of eastern Murman.] Mikrobiol., 24(5):580-

bacteria
*Norton, Cynthia F., anf Galen E. Jones, 1968.
A marine isolate of Pseudomonas nigrifaciens. I. Classification and nutrition.
Can. J. Microbiol., 14(12):1333-1340.

bacteria
Novelli, G.D., 1943
Bacterial oxidation of hydrocarbons in marine sediments. Proc. Soc. Exp. Biol. and Med. 52:133-134.

O

bacteria
Oppenheimer, Carl H., 1963.
Marine bacteriology.
(Special Issue on Marine Biology) AIBS Bull.,

bacteria
Oppenheimer, C.H., 1961
Note on the formation of spherical aragonitic bodies in the presence of bacteria from the Bahama Bank.
Geochim. et Cosmochim. Acta, 23(3/4): 295-296.

bacteria
Oppenheimer, C.H., 1960.
Bacterial activity in sediments of shallow marine bays.
Geochim. et Cosmochim. Acta, 19(4):244-260.

bacteria
Oppenheimer, C.H., and A.L. Kelley, 1952.
Escherichia coli in the intestine of a wild sea lion. Sci. 115(2998):527-528.

bacteria
Oppenheimer, Carl H., and Detlef A. Warnke, 1966.
Geomicrobiological parameters along the Antarctic peninsula shelf.
Antarctic J., U.S., 1(4):128.

bacteria
Oppenheimer, C.H., and C.E. ZoBell, 1952.
The growth and viability of sixty-three species of marine bacteria as influenced by hydrostatic pressure. J. Mar. Res. 11(1):10-18.

bacteria
Orlob, G.T., 1956.
Evaluating bacterial contamination in sea water samples.
Publ. Health Rept. 71:1246-1252.

Abstr. in:
Publ. Health Eng. Abstr., 37(5):24.

P

bacteria
*Palmer, Douglas S. and Lawrence J. Albright, 1970.
Salinity effects on the maximum hydrostatic pressure for growth of the marine psychrophilic bacterium, Vibrio marinus. Limnol. Oceanogr., 15(3): 343-347.

bacteria
salinity, effect of

bacteria
Peroni, C., 1971.
Heterotrophic level of microorganisms.
CNEN Rept. RT/BIO (70)-11, M. Bernhard, editor: 69-87.

Gulf of Taranto
bacteria
bottom sediments

bacteria
Persoone, Guido, 1966.
Contributions à l'étude des bactéries marines du littoral belge. III Milieux de culture et ensemencements.
Bull. Unst. r. Sci. Nat. Belg., 42(6):1-14.

bacteria
Persoone, G., et N. De Pauw, 1968.
Contributions à l'étude des bactéries marines du littoral belge. IV. Recherche quantitative de la richesse microbienne de l'eau de chasse d'Ostende par lecture directe sur membranes filtrantes.
Bull. Inst. r. Sci. nat. Belg. 44(26): 1-9

bacteria
*Peters, J., 1970.
Hygienisch-bakteriologische Untersuchungen des Küsten-Meerwassers im Bereich der Nordsee-Insel Sylt. Helgoländer wiss. Meeresunters, 21(3): 310-319.

Q

bacteria
Polikarpov, G.G., & V.P. Parchevsky, 1961
[The radioactivity in algae of the Adriatic and Black Seas.]
Okeanologia, 2: 338-339.

bacteria
Prévot, A.-R., 1958.
Les bactéries marines et les problèmes de biologie qu'elles soulèvent. La Nature 3280:325-328.

bacteria
Prevot, A.-R., 1958.
Utilité de la bacteriologie marine dans le présent et l'avenir. Bull. Inst. Océan., Monaco, No. 1114:1-22.

bacteria
Pshenin, L.N., 1964.
Azotobacter miscellum nov. sp., an inhabitant of the Black Sea. (In Russian).
Mikrobiologiia, 33(4):684-691.

Translation:
Microbiology, Consultants Bureau, 33(4):615-620.

bacteria
Psenin, L.N., 1959.
Nitrogen-fixing bacteria in the off-shore bottom deposits of the Black Sea. (In Russian).
Doklady, Akad. Nauk, SSSR, 129(4):930-933.

bacteria
Pshenin, L.N., 1961.
[Concerning the connection between azobacter and phytoplankton in the Black Sea]
Trudy Sevastopol Biol. Sta., (14):33-43.

Photobacteria
Puisségur, C., 1946.
Les Êtres vivants lumineux. Sic. et Vie, Vol. LXX (348), translated by R. Widmer, David Taylor Model Basin Translation 221, dated May 1927 (as Luminous Living Organisms), 17 pp., 9 text figs.

R

bacteria
Renn, C.E., 1937.
Bacteria and the phosphorus cycle in the sea.
Biol. Bull. 72(2):190-195.

bacteria
Reuszer, H.W., 1933
Marine bacteria and their role in the cycle of life in the sea. III. The distribution of bacteria in the Ocean waters and muds about Cape Cod. Biol. Bull. 65(3):480-497, 2 text figs.

bacteria
Rheinheimer, Gerhard 1971.
Über das Vorkommen von Brackwasser bakterien in der Ostsee.
Vie Milieu Suppl. 22(1): 281-291

bacteria
Rheinheimer, Gerhard 1970.
3.4 Einfluss verschiedener Faktoren auf die Bakterienflora. 3.5. Bakterien und Stickstoffkreislauf. Chemische, mikrobiologische und planktologische Untersuchungen in der Schlei im Hinblick auf deren Abwasserbelastung. Kieler Meeresforsch 26(2): 161-168; 168-170.

bacteria
Rheinheimer, Gerhard 1970.
3. Mikrobiologie. 3.1. Bakterienverteilung. 3.2. Colizahlen. Chemische, mikrobiologische und planktologische Untersuchungen in der Schlei im Hinblick auf deren Abwasserbelastung. Kieler Meeresforsch 26(2): 150; 150-156; 156-159.

pollution, bacterial
estuaries

bacteria
Rheinheimer, Gerhard, 1970.
Mikrobiologische und chemische Untersuchungen in der Flensburger Förde.
Ber. dt. wiss. Komm. Meeresforsch. 21(1/4): 420-429.

bacteria
*Rheinheimer, Gerhard, 1968.
Beobachtungen über den Einfluss von Salzgehalts-schwankungen aug die Bakterienflora der westlichen Ostsee.
Sarsia, 34:253-262.

bacteria
Rheinheimer, Gerhard, 1966.
Einige Beobachtungen über den Einfluss von Ostseewasser auf limnische Bakterien-populationen.
Veröff. Inst. Meeresforsch., Bremerh., Sonderband II:237-243.

bacteria
Rheinheimer, Gerhard, 1964.
Beobachtungen über den Einfluss des strengen Winters 1962/63 auf das Bakterienleben eines Flusses.
Kieler Meeresf., 20(2):218-226.

bacteria
Rittenberg, S.C., 1940
Bacteriological analysis of some long cores of marine sediments. J. Mar. Res.3:191-201.

bacteria
Rosenberg, L.A., 1963.
Depolarizing role of some sulphate reducing and saprophyte bacteria in the electrochemical corrosion of stainless and carbon steels. (In Russian).
Trudy Inst. Okeanol., Akad. Nauk, SSSR, 70:231-245.

bacteria
Rosenberg, L.A., 1963.
On the role of bacteria in the process of electro-chemical steel corrosion in sea water. (In Russian; English abstract).
Mikrobiologiia, 32(4):689-694.

bacteria
Ruschke, Rainier, 1964.
Zur Charakteristik einiger proteolytischer und nichtproteolytischer Bakterienarten des Genus Pseudomonas.
Helgoländer Wiss. Meeresuntersuchungen, 11(3/4):296-300.

S

bacteria (E. coli)
Scarpino, P.V., and D. Pramer, 1962.
Evaluation of factors affecting the survival of Escherichia coli in sea water.
Applied Microbiol., 10(5):436-440.

Abstract in:
Publ. Health Eng. Abstr., 43(4):195 #1107 (1963)

bacteria
Seki, Humitake, 1970.
Microbial biomass in the euphotic zone of the North Pacific Subarctic Water.
Pacific Sci., 24(2): 269-274

bacteria
North Pacific Subarctic Water

bacteria, violet
*Seki, Humitake, 1967.
Occurrence of marine violet pigmented bacteria in Sagami Bay.
J. oceanogr.Soc.,Japan,23(3):124-128.

bacteria
*Seki,Humitake,1967.
Ecological studies on the lipolytic activity of microorganisms in the dee of Aburatsubo Inlet.
Rec. oceanogr.Wks,Japan,9(1):75-113.

bacteria
Seki,Humitake,1966.
Studies on microbiol participation of food cycle in the sea. III. Trial cultivation of brine shrimp to adult in a chemostat.
J. Oceanogr. Soc.,Japan, 22(3):105-110.

bacteria
Seki,Humitaka,1966.
Seasonal fluctuation of heterotrophic bacteria in the sea of Abarasubo Inlet.
J. Oceanogr. Soc.,Japan,22(3):93-104.

Bacteria
Seki, Humitake, 1965.
Studies on micorbial participation to food cycle in the sea II. Carbohydrate as the only organic source in the micrococm.
Jour. Oceanogr. Soc., Japan, 20(6):278-285.

bacteria
*Seki,Humitake, and Nobuo Taga,1967.
Removal of radio nuclides at different multiplication phases of marine bacteria.
J. Oceanogr., Soc.,Japan, 23(3):136-140.

bacteria
*Seki,Humitake, and Claude E. ZoBell,1967.
Microbial assimilation of carbon dioxide in the Japan Trench.
J.Oceanogr.Soc.,Japan,23(4):182-188.

bacteria
Seshadri, R., K. Krishnamurthy and V.D. Ramamurthy 1966.
Bacteria and yeasts in marine and estuarine waters of Portonova (S. India).
Bull. Dept. Mar. Biol. Oceanogr. Univ. Kerala, 2:5-11.

bacteria
*Shieh,H.S., 1969.
Use of choline or betaine for fatty synthesis in a marine microbe.
Can.J. Microbiol.15(3):307-308.

bacteria
Shieh,H.S.,1966.
Further studies on the oxidation of betaine by a marine bacterium,(Achromobacter cholinophagum).
Canadian J. Microbiol.,12(2):299-302.

bacteria
Shineno,Harus,1965.
Agar softening bacteria isolated from sea water. (In Japanese;English abstract).
Bull. Jap. Soc., Sci. Fish,31(10: 840-847.

bacteria
Sieburth, J. McN., 1971.
An instance of bacterial inhibition in oceanic surface water. Mar. Biol. 11(1): 98-100.

bacteria
Sieburth, John McN., 1971.
Distribution and activity of oceanic bacteria.
Deep-Sea Res. 18(11):1111-1121.

bacteria
*Sieburth,John McN.,1967.
Seasonal selection of estuarine bacteria by water temperature.
J. exp. mar. Biol. Ecol., 1(1):98-121.

bacteria
Sieburth, John McN., 1964.
Polymorphism of a marine bacterium (Arthrobacter) as a function of multiple temperature optima and nutrition.
Narragansett Mar. Lab., Univ. Rhode Island, Occ. Publ., No. 2:11-16.

bacteria
Sieburth, John McN., 1960.
Soviet aquatic bacteriology: a review of the past decade.
Quart. Rev. Biol.., 35(3):179-205.

bacteria
Simidu, U., and K. Aiso, 1962.
Occurrence and distribution of heterotrophic bacteria in sea water from the Kamogawa Bay. (In Japanese; English summary).
Bull. Jap. Soc. Sci. Fish., 28(11):1133-1141.

bacteria
Simon, G., and C.H. Oppenheimer 1968.
Bacterial changes in sea water samples, due to storage and volume.
Z. allg. Mikrobiol. 8(3): 209-214.

bacteria
Sisler, F.D., and C.E.ZoBell, 1951.
Nitrogen fixation by sulfate-reducing bacteria indicated by nitrogen-argon ratio. Science 113 (2940):511-512.

SIO Contrib. No. 514.

Bacteria
Skripchenko, N.S. 1970.
"Mineralized bacteria" from oceanic oozes (In Russian).
Dokl. Akad. Nauk SSSR 192(6): 1346-1348

"bacteria"
oozes

bacteria
Sorokin, Ju. I., 1971.
On the role of bacteria in the productivity of tropical oceanic waters.
Int. Revue ges. Hydrobiol. 56(1):1-48.

bacteria a
*Sorokin, Yu.I., 1971.
On bacterial numbers and production in the water column on the central Pacific. (In Russian; English abstract). Okeanologiia, 11(1): 105-116.

bacteria
Sorokin Iu.I., 1970.
On the 'aggregated' condition of marine bacterial plankton. (In Russian).
Dokl. Akad. Nauk, SSSR, 192(4): 908-910

bacteria
Sorokin, Iu. I. 1970.
Number and productivity of bacteria in the water and bottom sediments of the central Pacific. (In Russian).
Dokl. Akad. Nauk SSSR 192(3): 655-658.

bacteria a
*Sorokin, Yu.I., 1970.
Numbers, activity and production of bacteria in bottom sediments of the central Pacific. (In Russian; English abstract). Okeanologiia, 10(6): 1055-1065.

Pacific, central
bacteria

bacteria
Sorokin, J.I., 1964
A quantitative study of the microflora in the central Pacific Ocean.
Journal du Conseil, 29(1):25-40.

Kriss' Krassilnikoviae in reality the tentacles of Pleurobrachia pileus!

bacteria
Sorokin, J.I., 1964.
On the primary production and bacterial activities in the Black Sea.
Journal du Conseil, 29(1):41-60.

bacteria
Sorokin, Yu. I., 1963
The quantitative estimation of the bottom microflora in the central area of the Pacific Ocean. (In Russian).
Okeanologiia, Akad. Nauk, SSSR, 3(3):500-503.

bacteria
Sorokin, Y.I., 1962.
[Experimental study of bacterial sulphate reduction in the Black Sea using S35.]
Mikrobiologiia, 31(3):402-410.

bacteria
Sorokin, Ju. I., 1962
On V.O. Kalinenko's paper "Atmospheric nutrition (aerotrophy) in marine bacteria".
Okeanologiia. Akad. Nauk. SSSR, 2(3):561-562.

BACTERIA
*Sorokin, Yu.I., T.S. Petipa, and E.V. Pavlova, 1970.
Quantitative estimate of marine bacterioplankton as a source of food. (In Russian; English abstract). Okeanologiia, 10(2): 332-340.

bacteria
*Stanley, Simon O. and Richard Y. Morita, 1968. Salinity effect on the maximal growth temperature of some bacteria isolated from marine environments. J. Bact., 95(1): 169-173. Also in: Coll. Reprints, Dept. Oceanogr. Univ. Oregon 7(1968).

bacteria
salinity, effect of

bacteria
Starr, T.J., and M.E. Jones, 1957.
The effect of copper on the growth of bacteria isolated from marine environments.
Limnol. Oceanogr., 2(1):33-36.

bacteria
Steinberg, P.L., 1961
The resistance of organic materials to attack by marine bacteria at low temperatures.
Bell System Techn. J., 40(5):1369-1396.

bacteria
Stewart, J.E., P. L. Hoogland, H.C. Freeman, and A.E.J. Waddell, 1963.
Amino acid composition of representatives of eight bacterial genera with reference to aquatic productivity.
J. Fish. Res. Bd., Canada, 20(3):729-734.

bacteria
Strakhov, N.M., 1948.
[The true role of bacteria in the formation of carbonate rocks.] Izvest. Akad. Nauk, USSR, Ser. Geol. 3:9-30.

T

bacteria
Taga, Nobuo 1968.
Some ecological aspects of marine bacteria in the Kuroshio Current.
Bull. Misaki mar. biol. Inst. Kyoto Univ. 12:65-76.

bacteria
*Taga, Nobuo, 1967.
Microbial coloring of sea water in tidal pool, with special reference to massive development of photosynthetic bacteria.
Inf. Bull. Planktol. Japan, Comm. No. Dr. Y. Matsue, 219-229.

Taga, Nobuo, 1965. bacteria
Marine bacteria in the Koroshio. (In Japanese).
Inf. Bull. Planktol., Japan, No. 12:6-7.

bacteria
Kuroshio
Pacific, northwest

no abstract

bacteria
Taga, Nobuo, and Fumitake Seki, 1962.
V. Preliminary report on the microbiological survey made during the fourth cruise of the Japanese Deep Sea Expedition (JEDS-4).
Oceanogr. Mag., Japan Meteorol. Agency, 13(2): 143-147.

JEDS Contrib. No. 33.

bacteria
Tajima, Kenichi, Yoshio Ezura and Minoru Sakai, 1971.
Variation in the bacterial flora in chill stored sea water. (In Japanese; English abstract).
Bull. Hokkaido Univ. Fac. Fish. 22(1): 80-90.

bacteria
Tajima, Kenichi, Yoshio Ezura and Minoru Sakai, 1971.
Procedure for the isolation of psychrophilic marine bacteria. (In Japanese; English abstract)
Bull. Hokkaido Univ. Fac. Fish. 22(1): 73-79.

bacteria
Teixeira, Clovis, and Miryam B. Kutner, 1962.
Plankton studies in a mangrove environment.
1. First assessment of standing stock and principal ecological factors.
Bol. Inst. Oceanogr., Sao Paulo, Brasil, 12(3) 101-124.

bacteria
Teixeira, C., J. Tundisi and M.B. Kutner, 1965.
Plankton studies in a mangrove environment. II. The standing stock and some ecological factors.
Bolm Inst. Oceanogr., S Paulo, 14(1):13-41.

bacteria
Tennant, A.D., and J.E. Reid, 1961.
Coliform bacteria in sea water and shellfish. 1. Lactose fermentation at 35.5° and 44° C.
Canadian J. Microbiol., 7(5):725-731.

bacteria
Tomlinson, N., and R.A. MacLeod, 1957.
Nutrition and metabolism of marine bacteria. IV. The participation of Na^+, K^+, and Mg^{++} salts in the oxidation of exogenous substrates by a marine bacterium.
Canadian J. Microbiol., 3(4):627-638.

bacteria
Tourneur, Colette, and Jacques Kauffmann, 1963.
Contribution à l'étude de la richesse du milieu marin en bactéries productrices d'antibiotiques.
C.R. Acad. Sci., Paris, 257(7):1507-1508.

bacteria
Tsyban, A.V. 1971
Sea foam as an ecological niche for bacteria. (In Russian; English abstract)
Gidrol. Zh. 7(3): 14-24

bacteria
Tsyban, A.V., 1966.
Seasonal distribution of saprophytic bacteria in the northwestern part of the Black Sea.
(In Russian)
Gidrobiol. Zhurn. 2(2): 42-46.

bacteria
Turner, M., and T.R.G. Gray, 1962.
Salt-marsh bacteria.
Nature, 194(4828):559-561.

U

bacteria (data)
Uyeno, Fukuzo, 1966.
Nutrient and energy cycles in an estuarine oyster area.
J. Fish. Res. Bd., Can., 23(11):1635-1652.

Uyeno, Fukuzo, 1964. bacteria
Relationships between production of foods and oceanographical condition of sea water in pearl farms. II. On the seasonal changes of sea water constituents and of bottom condition, and the effect of bottom cultivation. (In Japanese; English abstract)
J. Fac. Fish., Pref. Univ. Mie, 6(2):145-169.

V

bacteria
Vaccaro, R.F., M.P. Briggs, C.L. Carey and B.H. Ketchum, 1950.
Viability of Escherichia coli in sea water.
Amer. J. Pub. Health 40(10):1256-1266.

bacteria
Vacelet, Eveline 1970.
Variations au cours de l'année de la proportion de germes viables et non viables dans l'eau de mer littorale.
Tethys 2(1):19-34

bacteria, littoral
*Vacelet, Eveline, 1969
Rôle des populations phytoplanctoniques et bactériennes dans le cycle du phosphore et de l'azote en mer et dans les flaques supralittorales du Golf de Marseille.
Tethys 1(1): 5-118

Mediterranean, west
phytoplankton, littoral
nutrients

bacteria, physiology
Vargues, H., 1962.
Contribution à l'étude du caractère halophile chez les bactéries isolées du milieu marin.
Bull. Inst. Océanogr., Monaco, 59(1230):176 pp.

bacteria
Velankar, N.K., 1957.
Bacteria isolated from sea-water and marine mud off Mandapam (Gulf of Manaar and Palk Bay).
Indian J. Fish., 4(1):208-227.

bacteria
Velankar, N.K., 1955.
Bacteria in the inshore environment at Mandapam.
Indian J. Fish. 2(1):96-112.

bacteria
Venkataraman, R., and A. Sreenivasen, 1954.
Bacteriology of off-shore sea-water of the west coast. Proc. Indian Acad. Sci., B, 40(6):161-166.

bacteria
Venkatesan, V., and V.D. Ramamurthy 1971.
Marine microbiological studies of mangrove swamps of Killai backwaters.
J. oceanogr. Soc. Japan 27(2): 51-55

bacteria
Vlajnic, O., 1955.
Some new species of marine bacteria.
Acta Adriatica 7(2):1-34.

bacteria
Voitov, V.I., A.A. Egorova and N.I. Tarasov, 1960.
On the luminescence of a free Black Sea bacterium, Bacterium issatchenkoi Egorova. (In Russian)
Doklady, Akad. Nauk, SSSR, 132(6):1424-1426.

W

bacteria
Wai, Nganshou, and Chu-Ming Wu, 1961.
Isolation of non-sulfur bacteria in Taiwan sea water.
Bull. Inst. Chem. Acad., Sinica, (4):29-32.

bacteria
Waksman, S.A., 1955.
Marine bacteria: recollections and problems.
Pap. Mar. Biol. and Oceanogr., Deep-Sea Res., Suppl. to Vol. 3:15-19.

bacteria
Waksman, S.A., 1936.
Decomposition of organic matter in sea water by bacteria. III. Factors influencing the rate of decomposition. Biol. Bull. 70(3):472-483, 2 textfigs.

WHOI Contrib. No. 78.

bacteria
Waksman, S.A., 1934.
The role of bacteria in the cycle of life in the sea. Sci. Monthly 38:35-49, textfigs.

WHOI Contrib. No. 23.

bacteria
Waksman, S.A., 1934
The distribution and conditions of existence of bacteria in the sea. Ecol. Mon. 4(4): 523-529.

bacteria
Waksman, S.A., M. Hotchkiss, C.L. Carey, and Y. Hardman, 1938.
Decomposition of nitrogenous substances in sea water by bacteria. J. Bact. 35(5):477-486.

WHOI Contrib. no. 160.

bacteria
Waksman, S.A., and U. Vartiovaara, 1938.
The adsorption of bacteria by marine bottom.
Biol. Bull. 74(1):56-63.

WHOI Contrib. No. 161.

bacteria
Waksman, S.A., and M. Hotchkiss, 1937.
Viability of bacteria in sea water. J. Bact. 33(4):389-400.

WHOI Contrib. 124.

bacteria
Waksman, S.A., and M. Hotchkiss, 1938.
On the oxidation of organic matter in marine sediments by bacteria. J. Mar. Res. 1(2):101-118, textfigs. 34-35.

WHOI Contrib. No. 165.

bacteria
Waksman, S.A., J.L. Stokes, and M.R. Butler, 1937.
Relation of bacteria to diatoms in sea water.
J.M.B.A. 22:359-373.

WHOI Contrib. 157.

bacteria
Waksman, S.A., and C.L. Carey, 1935.
Decomposition of organic matter in sea water by bacteria. 1. Bacterial multiplication in stored sea water. J. Bact. 29(5):531-543, 4 textfigs.

WHOI Contrib. No. 57.

bacteria
Waksman, S.A., and C.L. Carey, 1935.
Decomposition of organic matter in sea water by bacteria. II. Influence of addition of organic substances upon bacterial activities. J. Bact. 29(5):545-561, 2 textfigs.

WHOI Contrib. No. 68.

bacteria
Waksman, S.A., C.L. Carey, M.C. Allen, 1934
Bacteria decomposing alginic acid. J. Bacteriol. 28(2):213-220.

bacteria
Waksman, S.A., H.W. Reuszer, C.L. Carey, M. Hotchkiss, and C.E. Renn, 1933.
Studies on the biology and chemistry of the Gulf of Maine. III. Bacteriological investigations of the sea water and marine bottoms. Biol. Bull. 64(2):183-205.

WHOI Contrib. No. 10.

bacteria
Waksman, S.A., C.L. Carey, H.W. Reuszer, 1933
Marine bacteria and their role in the cycle of life in the sea. I. Decomposition of marine plant and animal residues by bacteria. Biol. Bull. 65(1):57-79, 4 text figs.

bacteria
Waksman, S.A., and C.L. Carey, 1933.
Role of bacteria in decomposition of plant and animal residues in the ocean. Proc. Soc. Exp. Biol and Med. 30:526-527.

WHOI Contrib. No. 6.

bacteria
Waksman, S.A., M. Hotchkiss, and C.L. Carey, 1933
Marine bacteria and their role in the cycle of life in the sea. II. Bacteria concerned in the cycle of nitrogen in the sea. Biol. Bull. 65:137-167, 1 pl.

bacteria
Waksman, S.A., D.B. Johnstone, and C.L. Cary, 1943
The effect of copper upon the development of bacteria in sea water and the isolation of specific bacteria. J. Mar. Res., 5(2):136-152, 5 figs.

bacteria
Walls, Nancy W., 1967.
Bacteriology of Antarctic region waters and sediments.
Antarctic Jl., U.S.A., 2(5):192-193.

bacteria
Watanabe, Ko, and Miryam B. Kutner, 1965.
Plankton studies in a mangrove environment III. Bacteriological analysis of waters in Cananeia.
Bolm. Inst. Oceanogr., S Paulo, 14(1):43-51.

bacteria
*Westheide, W., 1968.
Zur quantitativen Verteilung von Bakterien und Hefen in einem Gezeitenstrand der Nordseeküste.
Marine Biol., 1(4):336-347.

beaches
bacteria
fungi
North Sea, south

bacteria
Weyland, H. 1967.
Beitrag zur quantitativen Verteilung mariner und "terrestrischer" Bakterien im Wasser und in Sedimenten der Deutschen Bucht.
Helgoländer wiss. Meeresunters. 15(1/4): 226-242.

bacteria (E.coli)
*Weyland, Horst, 1967.
Über die Verbreitung von Abwasserbakterien im Sediment des Weserästuars.
Veröff. Inst. Meeresforsch. Bremerh. 10(3):173-182.

bacteria
Weyland, Horst., 1966.
Untersuchungen über die Vermehrungsrate mariner Bakterien in Seewasser.
Veröff. Inst. Meeresforsch., Bremerh., Sonderband II:245-252.

bacteria
Wood, E.J.F., 1963
Heterotrophic micro-organisms in the oceans.
In: Oceanography and Marine Biology, H. Barnes Edit., George Allen & Unwin, 1:197-222.

bacteria
Wood, E.J.F., 1950.
Investigations on underwater fouling. Pt. 1. The role of bacteria in the early stages of fouling.
Australian J. Mar. and Freshwater Res. 1:85-91.

bacteria
*Wright, Richard T., 1970.
Glycollic acid uptake by planktonic bacteria.
Occ. Pap. Inst. mar. Sci., Alaska, 1: 521-536.

bacteria
glycollic acid

XYZ

bacteria
Yaphe, W., 1962.
Detection of marine alginolytic bacteria.
Nature, 196(4859):1120-1121.

bacteria
Yoshida, Yoichi 1967.
Microdeterminations of nitrogenous compounds in sea water, using marine bacteria. 1. Determination of ammonia using a marine nitrifying bacterium. 2. Determination of nitrate using a marine nitrate-reducing bacterium. (In Japanese; English abstracts).
Bull. Jap. Soc. scient. Fish. 33(4): 343-347; 348-350.

bacteria
Yoshida, Yoichi, and Masao Kimata, 1969.
Studies on the marine microorganisms utilizing inorganic nitrogen compounds.
IV. On the liberation rates of inorganic nitrogen compounds from bottom muds to sea water.
V. On the uptake or liberation of inorganic nitrogen compounds by the microorganisms as a whole in sea water. (In Japanese; English abstract)
Bull. Jap. Soc. scient. Fish., 35(3):303-306; 307-10.

bacteria
Zanetti, M., et L. Malavasi, 1961.
Ricerca sull' inquinamento batterico delle acque del litorale adriatico-romagnolo.
Nuov. Ann. Ig. Microbiol., 12:438-444.

Ref. in: Bull. Inst. Pasteur, 1963, 61(8):2438.

bacteria
Zarma, Mircea, 1969.
Données préliminaires sur le rapport en pourcentage entre les bactéries hétérotrophes de types respiratoires différent se trouvant sur le fond de la mer Noire, près de la côte dobroudgéenne.
Rapp. P.-v. Réun. Commn int. Explor. scient. Mer Médit. 19(5):865.

bacteria
Zarma, Mircea, 1963.
Contribution à l'étude des microorganismes marins anaérobies stricts. II. Bactéries ammonifiantes anaérobies isolées du fond de la Mer Noires au niveau de la Station Zoologique Marine "Prof. I. Borcea", d'Agigea (Dobroudja).
Rapp. Proc. Verb., Réunions, Comm. Int. Expl. Sci., Mer Méditerranée, Monaco, 17(3):675-677.

bacteria
Zarma, Mircea, 1963.
Les bactéries sulfo-oxydantes dénitrifiantes de la Mer Noire.
Rapp. Proc. Verb., Réunions, Comm. Int., Expl. Sci., Mer Méditerranée, Monaco, 17(3):679-686.

bacteria
Zemtsova, E.V., and A.E. Kriss, 1962.
The survival of sea microorganisms (heterotrophs) grown under laboratory conditions.
Doklady Akad. Nauk, SSSR, 142(3):695-698.

bacteria
Zobell, Claude E., 1964.
Hydrostatic pressure as a factor affecting the activities of marine microbes.
In: Recent researches in the fields of hydrosphere, atmosphere and nuclear geochemistry, Ken Sugawara festival volume, Y. Miyake and T. Koyama, editors, Maruzen Co., Ltd., Tokyo, 83-116.

bacteria
ZoBell, C.E., 1952;
Bacterial life at the bottom of the Philippine Trench. Science 115(2993):507-508.

bacteria
Zobell, C.E., 1950.
Annotated bibliography on the ecology of marine bacteria. Comm. N.R.C., Div. Geol., Geogr., 10: 31-54.

bacteria
Zobell, C.E., 1943
The effect of solid surfaces upon baterial activity. J. Bact. 46(1):39-56, 2 textfigs.

bacteria
ZoBell, C.E., 1943
Influence of Bacterial Activity on Source Sediment. Oil Weekly, 26 Apr. 1943:1-12.

bacteria
ZoBell, C.E., 1942
Bacteria of the marine world. Sci. Monthly, 55:320-330.

bacteria
ZoBell, C.E., 1941
Studies on marine bacteria. I. The cultural requirements of heterotrophic aerobes. J. Mar. Res.4:42-75.

bacteria
Zobell, C.E., 1940
Effect of oxygen tension on the rate of oxidation of organic matter in sea water by bacteria. J. Mar. Res.3:211-223.

bacteria
Zobell, C.E., 1940.
The effect of oxygen tension on the rate of oxidation or organic matter in sea-water by bacteria. J. Mar. Res. 3(3):211-223.

WHOI Contrib. No. 280.

bacteria
Zobell, C.E., 1939
Primary film formation by bacteria and fouling. Collecting Net, 14: 2 pp.

bacteria
ZoBell, C.E., 1939
Occurrence and activity of bacteria in marine sediments. Recent Marine Sediments, a symposium: 416-427 (A.A.P.G., Tulsa Oklahoma)

bacteria
Zobell, C.E., 1938
Studies on the bacterial flora of marine bottom sediments. J. Sed. Petrol. 8:10-18.

bacteria
ZoBell, Claude E., and Keith M. Budge, 1965.
Nitrate reduction by marine bacteria at increased hydrostatic pressures.
Limnology and Oceanography, 10(2):207-214.

abstract

pressure, effect of
nitrate reduction

bacteria
Zobell, C.E., and C.B. Feltham, 1938
Bacteria as food for certain marine invertebrates. J. Mar. Res.1:312-327.

bacteria
Zobell, C.E., and C.W. Grant, 1943
Bacterial utilization of low concentrations of organic matter. J. Bact. 45(6):555-564, 1 textfig.

bacteria
Zobell, C.E., C.W. Grant, and H.F. Haas, 1943
Marine microörganisms which oxidize petroleum hydrocarbons. Bull. Am. Assoc. Pet. Geol. 27(9):1175-1193

bacteria
ZoBell, Claude E., and Leslie L. Hittle, 1969
Deep-sea pressure effects on starch hydrolysis by marine bacteria.
J. oceanogr. Soc., Japan, 25(1): 36-47.

bacteria
Zobell, C.E., and J.H. Long, 1938
Studies on the isolation of bacteria-free cultures of marine phytoplankton. J. Mar. Res., 1:328-334.

bacteria
ZoBell, C.E., and H.D. Michener, 1938
A paradox in the adaptation of marine bacteria to hypotonic solutions. Science 87:328-329.

bacteria
Zobell, C.E., and S.C. Rittenberg, 1948.
Sulfate-reducing bacteria in marine sediments.
J. Mar. Res. 7(3):602-617.

bacteria
Zobell, C.E., and S.C. Rittenberg, 1938
The occurrence and characteristics of chitinoclastic bacteria in the sea. J. Bact 35:275-287.

bacteria
Zyban A.V., and V.A. Shnaidman 1969.
On the statistical approach to the study of the ecology of marine bacteria. (In Russian)
Dokl. Akad. Nauk, SSSR 187(5): 1159-1162

bacteria

bacteria, antiyeast

bacteria, antiyeast.
Buck, John D., and Samuel P. Meyers 1966.
Growth and phosphate requirements of Pseudomonas piscicida and related antiyeast pseudomonads.
Bull. Mar. Sci., 16(1): 93-101.

bacteria, bioassay

Belser, W., 1958.
Possible applications of a bacterial bioassay in productivity studies.
U.S.F.W.S., Sp. Sci. Rept. Fisheries No. 279. pt. 2:55-58.

bacteria, biomass

bacteria, biomass
Zarma, Mircea, 1965
Quelques données concernant le nombre des bactéries et la biomasse bactérienne dans l'eau de la plate-forme continentale de la Mer Noire au niveau de la ville de Constantza.
Rapp. P.-v. Réun., Comm. int. Explor. scient., Mer Méditerr., 18(3):635-641.

bacteria, chemistry of

bacteria, chemistry of
*Budd, J.A., and C.P. Spencer, 1968.
The utilisation of alkylated amines by marine bacteria.
Marine Biol., 2(1):92-101.

bacteria, chemosynthesis by

Bacteria, chemosynthesis by
Sorokin, Yu. I., 1965.
On bacterial chemosynthesis in the Black Sea. (In Russian; English abstract).
Izv., Akad. Nauk, SSSR, Ser. Biol., (3):413-422.

bacteria, chitin-decomposing

bacteria, chitinoclastic
Kihara, Kojo, and Nobuichi Morooka, 1962.
Studies on marine chitin-decomposing bacteria. 1. Classification and description of species.
J. Oceanogr. Soc., Japan, 18(3):147-152.

bacteria, chitinoclastic
Lear, Donald W., Jr., 1963.
Occurrence and significance of chitinoclastic bacteria in pelagic waters and zooplankton.
Ch. 56 in: Symposium on Marine Microbiology, Carl H. Oppenheimer, Editor, Charles C. Thomas, Publisher, 594-560.

bacteria, chitinoclastic
Seki, Humitake, and Nobuo Tago, 1965.
Microbiological studies on the decomposition of chitin in marine environment. VIII. Distribution of chitinoclastic bacteria in the pelagic and neretic waters.
J. Oceanogr. Soc., Japan, 21 (4):174-187.

bacteria, chitinoclastic
Seki, Humitake, and Nobuo Taga, 1965.
Microbiological studies on the decomposition of chitin in the marine environment. VII. Experimental considerations on survival conditions during the transport of chitinoclastic bacteria in the air and in the sea.
J. Oceanogr. Soc., Japan, 21(1):6-17.

Bacteria, chitinoclastic
Seki, Humitake, and Nobuo Taga, 1965.
Microbiological studies on the decomposition of chitin in marine environment. VI. Chitinoclastic bacteria in the digestive tracts of whales from the Antarctic Ocean.
Jour. Oceanogr. Soc., Japan, 20(6):272-277.

bacteria, chitinoclastic
Seki, Humitake, and Nobuo Taga, 1963
Microbiological studies on the decomposition of chitin in marine environment. V. Chitinoclastic bacteria as symbionts.
J. Oceanogr. Soc., Japan, 19(3):158-161.

bacteria, chitinoclastic
Seki, Humitake, and Nobuo Taga, 1963
Microbiological studies on the decomposition of chitin in marine environment. IV. Disinfecting effect of antibacterial agents on the chitinoclastic bacteria.
J. Oceanogr. Soc., Japan, 19(3):152-157.

bacteria, chitinoclastic
Seki, Humitake, and Nobuo Taga, 1963
Microbiological studies on the decomposition of chitin in marine environment. III. Aerobic decomposition of chitin by the isolated chitinoclastic bacteria.
J. Oceanogr. Soc., Japan, 19(3):143-151.

bacteria, chitinolytic
Brisou, J., C. Tysset, Y. de Rautlin de la Roy, R. Curcier and R. Moreau, 1964.
Etude sur la chitinolyse en milieu marin.
Ann. Inst. Pasteur, 106(3):469-478.

bacteria (chitin decomposing)
Seki, Humitake, and Nobuo Taga, 1963.
Microbiological studies on the decomposition of chitin in marine environment. II. Influence of some enviromental factors on the growth and activity of marine chitinoclastic bacteria.
J. Oceanogr. Soc., Japan, 19(2):109-111.

bacteria (chitin-decomposing)
Seki, Humitake, and Nobuo Taga, 1963.
Microbiological studies on the decomposition of chitin in marine environment. 1. Occurrence of chitinoclastic bacteria in the neritic region.
J. Oceanogr. Soc., Japan, 19(2):101-108.

bacteria, coliform
Pavletić, Zlatko, and Božidar Stilinović, 1969
Preliminary bacteriological investigations in the bay of Valdibora. (In Jugoslavian; Italian and English abstracts)
Thalassia Jugoslavica, 5: 251-255.

bacteria, coliform
Pike, E.B., A.L.H. Gameson and D.J. Gould, 1970.
Mortality of coliform bacteria in sea water samples in the dark.
Rev. int. Océanogr. méd., 15-19: 97-107

bacteria (data only)

bacteria (data only)
Japan, Hokkaido University, Faculty of Fisheries, 1967.
The Oshoro Maru cruise 18 to the east of Cape Erimo, Hokkaido, April, 1966.
Data Record Oceanogr. Obs. Explor. Fish. 11: 121-164.

bacteria (data only)
Japan, Shimonoseki University of Fisheries, 1965.
Data of oceanographic observations and exploratory fishings, International Indian Ocean Expedition 1962-63 and 1963-64. No. 1: 453 pp.

Bacteria (data only)
Japan, Shimonoseki University of Fisheries 1965.
International Indian Ocean Expedition 1962-63 and 1963-1964.
Data Oceanogr. Obs. Explor. Fish. 1: 453 pp.

Koyomaru
Indian Ocean
temperature (data only)
salinity (data only)
oxygen (data only)
pH (data only)
phosphate (data only)
total phosphorus (data only)
silicon (data only)
exploration lists of spp. ([illegible]

bacteria, decomposition by

bacterial decomposition, effect of
Berner, Robert A., 1969.
Chemical changes affecting dissolved calcium during the bacterial decomposition of fish and clams in sea water.
Marine Geol. 17(3): 253-274

bacteria, deep
ZoBell, Claude E., 1968.
Bacterial life in the deep sea.
Bull. Misaki mar. biol. Inst., Kyoto Univ. 12: 77-96

bacteria, distribution

BACTERIA, DISTRIBUTION OF

bacteria (on plankton)
Aubert, M., and H. Lebout, 1962.
Etude microbiologique du plancton de la Baie des Anges, à Nice.
Pubbl. Staz. Zool., Napoli, 32(Suppl.):483-489.

bacteria, vertical distribution
Cviic, V., 1960
Méthodes de determination, directe et indirecte, du nombre des bactéries dans l'eau de mer et certaines données sur leur distribution verticale dans la partie méridionale de l'Adriatique.
Comm. Int. Expl. Sci. Mer Medit., Monaco, Rapp. Proc. Verb., 15(3): 39-43.

bacterial distribution
Kriss, A.E., S.S. Abyzov, M.N. Lebedeva, I.E. Mishustina and I.N. Mitskeirch, 1960
[Geographic distribution patterns of the distribution of the microbe population (heterotrophs) in the world ocean.] Izv. Akad. Nauk, SSSR, Ser. Geograf. (5): 34-41.

bacteria, vertical distribution
Sorokin, Iu. I., 1962.
[Vertical distribution of saprophytic bacteria within the water layers of the central part of the Pacific Ocean.]
Doklady Akad. Nauk, SSSR, 145(1):198-194.

bacteria, effect of

bacteria, effect of
*Aizatullin, T.A. and K.M. Khailov, 1970.
Kinetics of enzymic hydrolysis of macromolecules solved in sea water in the presence of bacteria.
Gidrobiol. Zh., 6(6): 49-55. (In Russian; English abstract)

bacteria, effect of
macromolecules

bacteria, effect of
Bansemir, Klaus, und Gerhard Rheinheimer 1970.
3.6. Bakterielle Sulfatreduktion und Schwefeloxydation. Chemische, mikrobiologische und planktologische Untersuchungen in der Schlei im Hinblick auf deren Abwasserbelastung. Kieler Meeresforsch 26(2): 170-173.

bacteria, effect of
Benghitsky, A.G., L.G. Gutveib, and M.N. Lebedeva, 1970
Synthesis of cobalamines with bacteria isolated from the digestive organs of Black Sea fish. (In Russian)
Gidrobiol. Zh. 6(6): 72-75

cobalamines
bacteria, effect of.
Black Sea

bacteria, effect of
Berland, B.R., D.J. Bonin and S.Y. Maestrini, 1972.
Are some bacteria toxic for marine algae?
Mar. Biol. 12(3): 189-193.

bacteria, effect of
*Billy, Cecile, 1967.
Alginolyse et association bactérienne en milieu marin.
Vie Milieu (A)18(1-A):1-26.

Bacteria, effect of
Booth, G.H., Pamela M. Shinn and D.S. Wakerly, 1955.
The influence of various strains of actively growing sulfate-reducing bacteria on the anaerobic corrosion of mild steel.
Trav. Centre de Recherches et d'Etudes Oceanogr. n.s., 6(1/4):363-371.

Bacteria, effect of
Brisou, J., et U. de Rautlin de la Roy, 1965.
Le role des bacteries aerobies sulfhydrogenes dans la corrosion des metaux.
Trav. Centre de Recherches et d'Etudes Oceanogr. n.s., 6(1/4):373-375.

bacteria, effect of
Buck, J.D., S.P. Meyers and K.M. Kamp, 1962
Marine bacteria with antiyeast activity.
Science, 138(3547):1339-1340.

bacteria, effect of
Carroll, James J., and Leonard J. Greenfield, 1964
Mechanism of calcium and magnesium uptake from the sea by marine bacteria. (Abstract).
Geol. Soc., Amer., Special Paper, No. 76:30.

bacteria, effect of
Chandramohan, D., 1971.
Indole acetic acid synthesis in sea.
Proc. Indian Acad. Sci. (B) 73(3): 105-109.

chemistry, indole acetic acid.
bacteria, effect of

bacteria, effect of
Gorbenko, Iu. A., and Z. S. Kuchepova 1964.
Correlation between diatoms and rod-shaped bacteria in primary films. (In Russian).
Trudy Sevastopol. Biol. Stants. Akad. Nauk 15: 485-492.

bacteria, role of
Gunkel, Wilfried, 1962
Überlegungen zur Rolle der Bakterien im Stoffkreislauf des Meeres.
Kieler Meeresf., 18(3) (Sonderheft): 136-144.

bacteria, effect of
Guillard, R. R. L., 1959.
Further evidence of the destruction of bivalve larvae by bacteria.
Biol. Bull., 117(2):258-266.

bacteria, effect of
Hastings, J.W. and George Mitchell, 1971.
Endosymbiotic bioluminescent bacteria from the light organ of pony fish. Biol. Bull. mar. biol. Lab. Woods Hole 141(2): 261-268.

bacteria, effect of
Hata, Yoshihiko, 1963
Microbial activities in the development of reducing conditions in marine sediments (Preliminary report). (In Japanese; English abstract).
J. Shimonoseki Univ., Fish., 12(2/3):1-10.

bacteria, effect of
Hayes, F.R., 1955.
The effect of bacteria on the exchange of radio-phosphorus at the mud-water interface.
Proc. Int. Assoc. Theor. Appl. Limnol., 12:111-116.

bacteria, effect of
Hayes, F.R., 1955.
The effect of bacteria on the e xchange of radio-phosphorus at the mud-water interface.
Verh. Int. Ver. Limnol. 12:111-116, 4 textfigs.

bacteria, effect of
Ischenko, N.I., and I.B. Ulanovskii, 1963.
Protective effect of aerobic bacteria against corrosion of carbon steel in sea water. (In Russian).
Mikrobiologiya, 32(3):521-525.

Translation for NSF
32(3):445-448.

bacteria, effect of
*Jannasch, Holger W., Kjell Eimhjellen, Carl O. Wirsen and A. Farmanfarmaian, 1971.
Microbial degradation of organic matter in the deep sea. Science, 171(3972): 672-675.
W.H.O.I. Contrib. 2573.
organic matter
bacteria, effect of
2573

bacteria, effect of
Kobayashi, Michiharu, Atsuo Kawamura, Sinotaro Oya, Kyoji Mikami, Hiroshi Nakanishi, Kiyomi Murota, Yoshihiro Kurugose and Tadaaki Kawasugi, 1969.
Sewage purification by photosynthetic bacteria and its use as fish-feed. (In Japanese; English abstract).
Bull. Jap. Soc. scient. Fish. 35(10): 1021-1026

bacteria, effect of
Lalou, Mlle., 1955.
La précipitation expérimentale des carbonates par les bactéries.
Bull. d'Info., C.C.O.E.C., 7(8):358-359.

bacteria, effect of
Muller, William A., and John J. Lee 1969.
Apparent indispensability of bacteria in foraminiferan nutrition.
J. Protozool. 16(3): 471-478.

Bacteria, effect of
Plessis, A., et M.C. Gatelier, 1965.
Le role des associations bacteriennes dans le development et la resistance des reducteurs de sulfate.
Trav. Centre de Recherches et d'Etudes Oceanogr. n.s., 6(1/4):377-380.

bacteria, effect of
Provasoli, L., 1971.
Nutritional relationships in marine organisms.
In: Fertility of the Sea, John D. Costlow, editor, Gordon Breach, 2: 369-382.

bacteria, effect of
Rozenberg, L.A., 1963.
Role of bacteria in the electrochemical corrosion of steel in sea water. (In Russian).
Mikrobiologiia, 32(4):689-694.

Translation for NSF:
Microbiology, 32(4):586-590. 1964.

bacteria, effect of
Sieburth, J. M, 1959.
Antibacterial activity of Antarctic marine phyto-plankton.
Limnol. & Ocean., 4(4):419-424.

bacteria, effect of
Shinano, Haruo, and Minoru Sakai, 1969.
Studies of marine bacteria taking part of the precipitation of calcium carbonate. 1. Calcium carbonate deposited in peptone medium prepared with natural sea water and artificial sea water. (In Japanese; English abstract).
Bull. Jap. Soc. scient. Fish. 35(10). 1001-1005.

bacteria, effect of
Sorokin, Iu. I., 1970
On the quantitative evaluation of the role played by bacteria-plankton in the rotation of organic matter in tropical waters. (In Russian).
Dokl. Akad. Nauk, SSSR 193(4): 923-925

bacteria, effect of
Sorokin, Yu.I. and Yu.A. Bogdanov, 1971.
Transformation of iron in the process of bacterial decomposition of the plankton organic substance. (In Russian). Gidrobiol. Zh., 7(2): 106-107.

bacteria, effect of
Spencer, C.P., 1956.
The bacterial oxidation of ammonia in the sea.
J.M.B.A., 35(3):621-630.

bacteria, effect of
Taniguti, Tadataka, 1964.
Consumption of dissolved oxygen by bacterial decomposition of various organic compounds added into inshore sea water.
Bull., Fac. Fish., Nagasaki Univ., No. 17:104-109 abstract, p. 104.

Bacteria, Effect of
Tourneur, C., 1970.
Contribution à l'étude de la production d'antibiotiques et de vitamine B12 par des bactéries isolées des vases soumises à l'influence du milieu marin
Cah. océanogr. 22(6): 613-614

bacteria, effect of
Uyeno, Fukuzo, Kyoichi Kawaguchi, Nagao Terada and Tadashi Okada 1970.
Decomposition, effluent and deposition of phytoplankton in an estuarine pearl oyster area.
Rept. Fac. Fish. Prefect. Univ. Mie 7(1): 7-41

bacteria (iron), effect of
Wiseman, J.D.H., M.W. Strong and H.J.M. Bowen, 1956.
Marine organisms and biogeochemistry. 1. The rates of accumulation of nitrogen and calcium carbonate on the equatorial Atlantic floor. 2. Marine iron bacteria as rock forming organisms. 3. The biogeochemistry of strontium.
Adv. Sci., 12(49):579-588.

bacteria, fimbriation of
Ahrens, Renate, und Georg Moll 1971.
Über die Fimbriierung sternbildender Bakterien aus Brackwasser.
Kieler Meeresforsch. 27(1): 113-116.

bacteria, habitats of
Sieburth, John McNeill, 1963.
Bacterial habitats in the Antarctic environment
Ch. 49 in: Symposium on Marine Microbiology, C.H. Oppenheimer, Editor, C.C. Thomas, Springfield, Illinois, 533-548.

bacteria, haemolytic
Davies, David H., 1961.
Shark research in Natal.
Marine Studies off the Natal Coast, C.S.I.R. Symposium, No. S2:81-88. (Multilithed).

bacteria, heterotrophic
Bianchi, A.J.M., 1971.
Distribution de certaines bactéries hétérotrophes aérobies dans les sédiments profonds de Méditerranée Orientale. Mar. Biol. 11(2): 106-117.

bacteria, heterotrophic
*Fujita, Yuji, Tadataka Taniguti and Buhei Zenitani, 1967.
Microbiological studies on shallow marine areas. III. On relation of the heterotrophic bacteria to the changes in carbon dioxide, oxygen consumption, organic acid and sulfides in the mud sediment. (In Japanese: English abstract).
Bull. Fac. Fish. Nagasaki Univ., 23:187-196.

bacteria, heterotrophic
Kriss, A.E., 1962.
Distribution of heterotrophic bacteria in the waters of the Pacific. (In Russian).
Doklady, Akad. Nauk, SSSR, 146(6):1422-1425.

bacteria, heterotrophic
Kriss, A.E., and I.N. Mitzkevich, 1970
Quantitative distribution of heterotrophic bacteria in the equatorial-tropical zone of the Pacific Ocean near South America. (In Russian; English abstract)
Mikrobiologiia, 39(6): 1087-1094

bacteria, heterotrophic
Lebedeva, M.N., and E.M. Markianovic 1971.
Antibiotic properties of heterotrophic bacteria in South Seas. (abstract only)
Rev. int. Biol. Océanogr. méd. 24: 136-137.

bacteria, heterotrophic
Lebedeva, M.N., E.M. Markianovich and L.G. Gutveib 1971.
Spreading of heterotrophic bacteria auxotrophic in vitamin B_{12} in southern seas. (In Russian; English abstract).
Gidrobiol. Zh. 7(2): 20-26.

bacteria, heterotrophic
Liston, J., 1968.
Distribution, taxonomy and function of heterotrophic bacteria on the sea floor.
Bull. Misaki mar. biol. Inst., Kyoto Univ., 12: 97-104

bacteria, heterotrophic
Melchiorri-Santolini, Ulderico, 1966.
Pelagic heterotrophic bacteria in the Ligurian Sea and Lago Maggiore.
Memorie Ist. ital. Idrobiol. 20:261-287.

bacteria, heterotrophic

*Murchelano, R.A. and C. Brown, 1970.
Heterotrophic bacteria in Long Island Sound.
Marine Biol., 7(1): 1-6.

---

bacteria,heterotrophic

*Seki,Humitake,1968.
Some characteristics of marine microorganisms in the Pacific Water.
Rec.oceanogr.Wks,Japan,n.s.,9(2):247-255.

---

bacteria, heterotrophic

Shewan, J.M. 1971
Recent progress in the taxonomy and identification of some genera of marine bacteria. (abstract only)
Rev. int. Biol. Océanogr. méd. 24:47.

---

bacteria heterotrophic

Taniguchi, Tadataka, Yuji Fujita and Buhei Zenitani 1966.
Microbiological studies of shallow marine areas. II. Seasonal change in constituent groups of heterotrophic bacterial populations in respect of polysaccharide decomposition. (In Japanese; English abstract).
Bull. Fac. Fish. Nagasaki Univ. 21: 243-249.

---

bacteria, lists of species

bacteria, lists of spp.

Lackey, James B., and Elsie W. Lackey,1963
Microscopic algae and protozoa in the waters near Plymouth in August 1962.
J. Mar. Biol. Assoc., U.K., 43(3):797-805.

---

bacteria, littoral

Kufferath, Jean 1970.
Contributions à l'étude des bactéries des eaux marines du littoral Belge.
Bull. K. Belg. Inst. Nat. Wet. 46(36): 6pp.

---

bacteria, luminous

Srivastava, Vinod S., and Robert A MacLeod, 1971.
Nutritional requirements of some luminous marine bacteria.
Can. J. Microbiol. 17(5): 703-711.

---

bacteria (nitrifying)

bacteria nitrifying

Carlucci, A.F., and J.D.H. Strickland 1968.
The isolation, purification and some kinetic studies of marine nitrifying bacteria.
J. exp. mar. Biol. Ecol. 2(2): 156-166.

---

bacteria, nitrifying

Kawai, Akira, Yoichi Yoshida and Masao Kimata 1968.
Nitrifying bacteria in the coastal environments.
Bull. Misaki mar. biol. Inst. Univ. Kyoto 12: 181-194

---

bacteria (nitrifying)

Kimata, Masao, Akira Kawai and Yoichi Yoshida, 1963.
Studies on marine nitrifying bacteria (nitrite formers and nitrate formers). II. On the distribution of marine nitrifying bacteria in Maizuru Bay. (In Japanese; English abstract).
Bull. Jap. Soc. Sci. Fish., 29(11):1027-1030.

---

bacteria, nitrifying

Yoshida, Yoichi, Akira Kawai and Masao Kimata 1967.
Studies on marine nitrifying bacteria (nitrite and nitrate formers). V. Effects of environmental factors on the nitrite formation of cell-free extracts of a marine nitrifying bacterium. VI. Distribution of limnetic nitrifying bacteria in a waste region. (In Japanese; English abstracts).
Bull. Jap. Soc. scient. Fish. 33(5): 421-425; 426-429.

---

bacteria, nitrifying

Watson, Stanley W. 1968.
Marine nitrifying bacteria.
Bull. Misaki mar. biol. Inst. Kyoto Univ. 12:179-180.

---

bacteria nitrifying

Yoshida, Yoichi, and Masao Kimata 1967.
Studies on marine nitrifying bacteria (nitrite and nitrate formers). VII. Distribution of marine nitrifying bacteria and the role played by them in the offshore region. (In Japanese; English abstract).
Bull. Jap. Soc. scient. Fish. 33(6): 578-585.

---

bacteria, nitrogen-fixing

Maruyama Yoshiharu, Nobuo Taga, and Osamu Matsuda, 1970.
Distribution of nitrogen-fixing bacteria in the central Pacific Ocean.
J. oceanogr. Soc. Japan 26(6): 360-366.

---

bacteria, nitrogen-fixing

Pshenin, L.N. 1964.
On the nitrogen-fixing bacteria in the near-surface layers of the Black Sea. (In Russian).
Trudy Sevastopol. Biol. Stants. Akad. Nauk SSSR 15: 3-7.

---

bacteria, pathogenic

bacteria, pathogenic

Brisou, J., 1962.
Survie des bacteries pathogènes dans les eaux douces et le milieu marin.
Sem. Hôp., Paris, 38:3913-3915.

Ref. in: Bull. Inst. Pasteur, 1963, 61(8):2434-2435.

---

bacteria, pathogenic

*Guelin, A., I. Bychovskaja, P. Lepine et D. Lamblin, 1970.
Distribution des germes parasites des bactéries pathogènes dans les eaux mondiales.
Rev. int. Océanogr. méd. 18-19: 77-83
bactericidal effects
germs, parasitic
bacteria, pathogenic
p.82.

---

bacteria, phosphate solubilizing

Ayyakkannu, K. and D. Chandramohan, 1971.
Occurrence and distribution of phosphate solubilizing bacteria and phosphatase in marine sediments at Porto Novo. Mar. Biol. 11(3): 201-205.

---

bacteria physiology

bacteria,physiol

Hamilton,R.D., K.M. Morgen, and J.D.H. Strickland,1966.
The glucose uptake kinetics of some marine bacteria.
Canadian J. Microbiol., 12(5):995-1003.

---

bacteria/plankton

bacteria/plankton

Campello, F., J. Brisou, et Y. de Rautlin de la Roy, 1963.
Étude sur les relations existant entre le plancton et les bactéries dans les eaux portuaires de La Pallice.
Comptes Rendus, Soc. Biol., France, 157(3):618.

---

bacteria, planktonic

Sieburth, John McN. 1968.
Observations on bacteria planktonic in Narragansett Bay, Rhode Island: a résumé.
Bull. Misaki mar. biol. Inst. Kyoto Univ. 12: 49-64.

---

bacterial populations

Sorokin, Yu.I., 1971.
Bacterial populations as components of oceanic ecosystems. Mar. Biol. 11(2): 101-105.

---

bacteria, psychrophilic

bacteria, psychrophilic

Morita, Richard Y., 1968.
The basic nature of marine psychrophilic bacteria.
Bull. Misaki mar. biol. Inst., Kyoto Univ. 12: 163-177

bacteria, psychrophilic

---

bacteria psychrophilic

Morita,Richard Y,1966.
Marine psychrophilic bacteria.
In: Oceanography and marine biology, H. Barnes, editor,George Allen & Unwin,Ltd., 4:105-121.

---

bacteria, salt requirement

Hidaka, Tomio, and Minoru Sakai, 196[illegible]
Comparative observation of the inorganic salt requirements of the marine and terrestrial bacteria.
Bull. Misaki mar. biol. Inst., Kyoto Univ., 12: 125-149

---

bacteria,sterility

*Watson,Stanley W., and John B. Waterbury,1969.
The sterile hot brines of the Red Sea.
In: Hot brines and Recent heavy metal deposits in the Red Sea,E.T.Degens and D.A.Ross,editors, Springer-Verlag,New York,Inc., 272-281.

WHOI Contrib. No. 2190.

---

bacteria, streptococci

Buck, John D. 1969.
Occurrence of false-positive most probable number tests for fecal streptococci in marine waters.
Appl. Microbiol. 18(4): 562-565

---

bacteria, structure

Kennedy, S.F., R.R. Colwell and G.B. Chapman, 1970.
Ultrastructure of a marine psychrophilic Vibrio.
Can. J. Microbiol., 16(11): 1027-1031

---

bacteria, purple-sulphur

bacteria (purple-sulphur)

Kriss, A.E., 1959.
The role of microorganisms in the primary production of the Black Sea.
J. du Cons., 24(2):221-233.

---

bacteria, sulphur

Lialikova, N.N., 1957.
[A study of assimilation of free CO2 by purple bacteria in Belovod Lake.]
Mikrobiologiia, 26(1):92-98.

bacteria, sulfur
Tilton, Richard C. 1968.
The distribution and characterization of marine sulfur bacteria.
Rev. int. Océanogr. Méd. 9:237-253

sulfur bacteria
*Tilton, R.C., A.B.Cobet, and G.E.Jones,1967.
Marine thiobacilli. I. Isolation and distributior
Can. J. Microbiol., 13(11):1521-1528.

sulfur bacteria
*Tilton,R.C., G.J.Stewart, and G.E. Jones,1967.
Marine thiobacilli. II. Culture and ultra structure.
Can. J. Microbiol., 13(11):1529-1534.

bacteria, sulfur
Trüper, Hans G., and Sebastiano Genovese 1968.
Characterization of photosynthetic sulfur bacteria causing red water in Lake Faro (Messina, Sicily).
Limnol. Oceanogr. 13(2):225-232.

bacteria, sulfate-reducing

Bacteria, sulfur ~~xxxx~~ ducing
Dufour, M., J, Galliano et R. Molinier, 1961.
Sur l'activité des bactéries sulfato-réductrices dans les sols marins superficiels de la baie du Busc (Var).
Rapp. Proc. Verb., Réunions, Comm. Int. Expl. Sci. Mer Méditerranée, Monaco, 16(2):433-438.

bacteria, sulfate-reducing
Kimata, Masao, Hajime Kadota, Yoshihiko Hata and Hideo Miyoshi, 1960.
Distribution of sulfate-reducing bacteria in Maizuru Bay.
Rec. Oceanogr. Wks., Japan, Spec., No. 4:53-54.

bacteria, sulphate-reducing
Sénez, J.C., 1962.
Rôle écologique des bactéries sulfato-réductrices
Problemi ecologici delle zone litorale del Mediterraneo, 17-23 luglio 1961.
Pubbl. Staz. Zool., Napoli, 32(Suppl.):427-441.

bacteria, sulfate-reducing
Vacelet, Eveline 1971.
Écologie de la microflore sulfato-réductrice en milieu marin et en milieu à salinité variable.
Téthys 3(1):3-15

bacteria, sulphatereducing (data)
Zarma, M., 1961.
Contributions à l'étude des microorganismes marins anaérobies stricts. IV. Les bactéries sulfato-réductrices du fond de la Mer Noire au niveau de la côte dobrogéenne.
Rapp. Proc. Verb., Réunions, Comm. Int. Expl. Sci. Mer Méditerranée, Monaco, 16(2):257-262.

bacteria, species of

Nitrosocystis oceanus sp.n.
Watson, Stanley W., 1965.
Characteristics of a marine nitrifying bacterium Nitrosocystis oceanus sp. n.
Limnol. and Oceanogr., Redfield Vol., Suppl. to 10:R274-R289.

bacteria, sulfur
van Gemerden, H., and H.W. Jannasch, 1971.
Continuous culture of Thiorhodaceae. Sulfide and sulfur limited growth of Chromatium vinosum.
Arch. Mikrobiol. 79(4):345-353.

bacterial viability

bacterial viability
Orlob, G.T., 1956.
Viability of sewage bacteria in sea water.
Sew. & Ind. Wastes, 28:1147-1167.

Abstr. Publ. Health Eng. Abstr. 36(12):27.

bacteria, viability of
Suñer Pi, J., 1968.
Viabilité bactérienne dans l'eau de mer.
Rev. intern. Océanogr. Méd., 10:209-220.

bacteria, viability

bacteriocidal activity
Glombitza, K.-W. und R. Heyser, 1971.
Antimikrobielle Inhaltsstoffe in Algen. 4. Mitteilung. Wirkung der Acrylsäure auf Atmung und Makromolekülsynthese bei Staphylococcus aureus und Escherichia coli. Helgoländer wiss. Meeresunters 22(3/4): 442-453.

bactericidal agent

bactericidal agent
Johannesson, J.K., 1957.
Nature of the bactericidal agent in sea water.
Nature 180(4580):285-286.

bactericidal effects
*Guelin, A., I. Bychovskaja, P. Lepine et D. Lamblin, 1970.
Distribution des germes parasites des bactéries pathogènes dans les eaux mondiales.
Rev. int. Océanogr. Méd. 18-19:77-83
bactericidal effects
germs, parasitic
bacteria, pathogenic

bacterioneuston
Tsyban, A.V., 1971.
Marine bacterioneuston.
J. oceanogr. Soc. Japan 27(2): 56-66.

bacteriophage

bacteriophage
Chen, Peter K., Ronald V. Cirarella, Omar Salazar and Rita R. Colwell, 1966.
Properties of two marine bacteriophages.
J. Bacteriol. 91(3):1136-1139.

abstract?

bacteriophage

bacteriophage
Kott, Y., and H. Ben Ari 1968
Bacteriophage as marine pollution indicators.
Rev. int. Océanogr. Méd. 9:207-217.

bacteriophage
Monteverde, Jose Julio, and M. Adela Caria, 1958.
Investigación de Bacteriófage en agua de mar.
Bol. Mus. Argentino de Ciencias Nat. "Bernardino Rivadavia", No. 17:14 pp.

bacteriophage
Spencer, R., 1955.
A marine bacteriophage. Nature 174(4459):890-891.

bacteriophage
Valentine, Artrice F., Peter K. Chen, Rita R. Caldwell and George B. Chapman, 1966.
Structure of a marine bacteriophage as revealed by the negative staining technique.
J. Bacteriol., 91(2):819-822.

bacteriophage

abstract???

bacteriostasis
*Sieburth, J. McN., 1971.
An instance of bacterial inhibition in oceanic surface water. Mar. Biol. 11(1): 98-100.

balance

balance
Ivanoff-Frankzkevich, G.N., 1962.
Some aspects of smoothing the equations of balance. (In Russian; English summary).
Trudy Inst. Okean., Akad. Nauk, SSSR, 60:96-113.

baleen plates

baleen plates
van Utrecht, W.L., and C.N. van Utrecht,1968.
Some comments on the significance of the cortical layer of the baleen plate in age determination of baleen whales.
Norsk. Hvalfangsttid. 1968(2):39-40.

banks

banks
Damodaran, R., and C. Hridayanathan 1966.
Studies on the mud banks of the Kerala coast.
Bull. Dept. mar. Biol. Oceanogr. Univ. Kerala, 2:61-68.

banks
Erofeev, P.N., 1963.
The structure of the Herrindge bank (the Atlantic). (In Russian).
Doklady, Akad. Nauk, SSSR, 151(5):1159-1161.

banks (shell)
Greensmith,J.T., and E.V. Tucker,1968.
Imbricate structure in Essex offshore shell banks.
Nature,Lond., 220(5172):1115-1116.

banks
Hoshino, Michihei, and Hiroji Homma 1966.
Geology of submarine banks in the Japan Sea. (In Japanese; English abstract)
Chikyukagaku 82:
Reprinted in: Coll. Repr. Coll. mar. Sci. Technol. Tokai Univ. 2(1966): 43-51.

banks
*James, Noel P., and Daniel J. Stanley,1968.
Sable Island Bank off Nova Scotia:Sediment dispersal and Recent history.
Bull.Am.Ass.Petr.Geol., 52(11):2208-2230.

banks
Maloney, Neil J., 1967.
Geomorphology of the continental margin of central Oregon, U.S.A.
Bol. Inst. Oceanogr. Univ. Oriente 6(1): 116-146

banks
Okusa, Shigeyasu 1966.
Physical properties of rocks from Yamato Bank, Japan Sea.
Work. Pap. Symp. Geol. Probl. Japan Sea Reg., Kanagawa Univ. 1966.
Reprinted in: Coll. Repr. Coll. mar. Sci. Technol. Tokai Univ. 2(1966): 123-130.

banks
Shor, G.G., Jr., and D.D. Pollard, 1963.
Seismic investigations of Seychelles and Saya de Malha banks, northwest Indian Ocean.
Science, 142(3588):48-49.

banks

Ulrich, Johannes, 1964.
Zur Topographie der Rosemary-Bank.
Kieler. Meeresf., 20(2):95-100.

banks

Yong, G. A., and A. L. Kontis, 1964.
A study of aeromagnetic component data, Plantagenet Bank.
U. S. Naval Oceanogr. Off., Techn. Rept., TR-144: 18 pp. (multilithed).

banks, effect of

Cooper, L.H.N., 1960 banks, effect of
Some theorems and procedures in shallow-water oceanography applied to the Celtic Sea. JMBA 39:155-171.

baroclinic flow

Derome, Jacques, and C.L. Dolph, 1970.
Three-dimensional non-geostrophic disturbances in a baroclinic zonal flow.
Geophys. fluid Dyn., 1(1/2): 91-122.

baroclinic flow

Fofonoff, N.P., 1962
Sect. III. Dynamics of Ocean Currents.
In: The Sea, Interscience Publishers, Vol. 1
Physical Oceanography, pp. 323-395.

baroclinic instability

Tareev, B.A. 1968. BAROCLINIC INSTABILITY
A non-geostrophic two-layer model of baroclinic instability in the atmosphere and in the ocean. (In Russian).
Doкl. Akad. Nauk SSSR, 178(6): 1304-1306.

baroclinic instability

baroclinic layer

baroclinic layer
*Kozlov, V.F., 1967.
On the theory of a baroclinic layer at the equator. (In Russian; English abstract).
Okeanologiia, Akad. Nauk, SSSR, 7(4):577-585.

baroclinic layer

Lineikin, P.S., 1955.
Determination of thickness of the baroclinic layer of the sea. Doklady Akad. Nauk, SSSR, 101(3): 461-464.

baroclinic ocean

baroclinic ocean
*Bolgurtsev, B.N., and V.F. Kozlov, 1969.
An approximate method of computation of velocity and density fields in the baroclinic ocean (In Russian; English abstract).
Fisika Atmosfer. Okean., Izv. Akad. Nauk SSSR, 5(7): 704-713.

baroclinic ocean

*Gutman, L.N., 1970.
On large-scale currents in a baroclinic ocean. (In Russian; English abstract). Fizika Atmosfer. Okean., Izv. Akad. Nauk SSSR, 6(9): 908-921.

baroclinic ocean

*Kozlov, V.F., 1970.
The model of the meandering of the inertial currents in the baroclinic ocean. Fizika Atmosfer. Okean., Izv. Akad. Nauk SSSR, 6(9): 922-933.
(In Russian; English abstract)

baroclinic sea

Marchuk, G.I. 1970.
On the statement of the problem on the structure of flows in a baroclinic sea taking into account the macroturbulent exchange. (In Russian; English abstract).
Met. Gidrol. 1970(3): 12-17.

baroclinic ocean

Pedlosky, Joseph, 1965.
A necessary condition for the existence of an inertial boundary layer in a baroclinic ocean.
J. Mar. Res., 23(2):69-72.

baroclinic waves

baroclinic waves

Tareev, B.A., 1965.
Internal baroclinic waves in the course of flowing about the roughness of the ocean floor and their influence upon the process of sedimentation in the depth. (In Russian).
Okeanologiia, Akad. Nauk, SSSR, 5(1):45-51.

BAROMETRIC PRESSURE

barometric pressure

LeFloch, J., 1961.
Influence de la pression barométrique et du vent sur le niveau moyen au Havre.
Cahiers Océanogr., C.C.O.E.C., 13(8):547-551.

Barometric Pressure

Perez Sori, Jose A., 1961.
Salinidad, temperatura, presion atmospherica y pH en la playa Habana: II.
Centro Invest. Pesqueras, Cuba, Contrib. No. 13:19 pp.

barometric pressure

Yamaguti, S., 1960.
On the changes in the heights of mean sea-levels, before and after the great earthquakes.
Bull. Earth Res. Inst., Tokyo, 38(1):145-175.

barotropic fluid

barotropic fluid

Blumen, William, and Warren M. Washington, 1969
The effect of horizontal shear flow on geostrophic adjustment in a barotropic fluid.
Tellus 21(2): 167-176

barotropic models

Sarkisjan, A.S. 1969.
On the defects of the barotropic oceanic circulation. (In Russian; English abstract).
Fizika Atmosfer. Okean. Izv. Akad. Nauk SSSR 5(8): 818-835.

barotropic ocean

*Egorov, K.L., D.L. Laikhtman, and V.M. Radikevich, 1970.
Model of a barotropic ocean. (In Russian; English abstract). Okeanologiia, 10(2): 249-255.

barotropic ocean

Endoh, Masahiro and Takashi Nitta, 1971.
On Rossby waves forced by traveling atmospheric disturbances in a barotropic bounded ocean.
J. Met. Soc. Jap. (2) 49(3): 137-145.

BAROTROPIC OCEAN

*Johnson, J.A., 1970.
Oceanic boundary layers. Deep-Sea Res., 17(3): 455-465.

boundary layers
friction
inertia
barotropic ocean

barriers

barrier beaches

Dillon, William P., 1970.
Submergence effects on a Rhode Island barrier and lagoon and inferences on migration of barriers.
J. Geol., 78(1): 94-106.

barriers, effect of

Matushevsky, G.V., 1963
A study of wind wave fields close to barriers (In Russian).
Okeanologiia, Akad. Nauk, SSSR, 3(3):395-404.

barrier islands

barrier islands

Hoyt, John H., 1970. (deceased 6 Sept. 1970)
Development and migration of barrier islands, northern Gulf of Mexico: discussion.
Bull. Geol. Soc. Am. 81(12): 3779-3782.

Gulf of Mexico
barrier islands

barrier islands

*Hoyt, John H., 1967.
Occurrence of high-angle stratification in littoral and shallow neritic environments, central Georgia coast.
Sedimentology, 8(3):229-238.

barrier islands

Hoyt, John H., and John R. Hails, 1967.
Pleistocene shoreline sediments in coastal Georgia: deposition and modification.
Science, 155(3769):1541-1543.

barrier islands

Kumar, Naresh, and J.E. Sanders, 1970.
Are basal transgressive sands chiefly inlet-filling sands?
Marit. Sed. 6(1): 12-14

USA (Long Island)
barrier islands

barrier islands

Kwon, H.J. 1969.
Barrier islands of the northern Gulf of Mexico coast: sediment source and development.
Coast. Stud. Ser. Louisiana State Univ. Press 25: 51 pp. (multilited)

barrier islands

Otvos, Ervin G., Jr. 1970.
Development and migration of barrier islands, northern Gulf of Mexico: reply.
Bull. Geol. Soc. Am., 81(12): 3783-3788.

barrier islands.

Otvos, Ervin G., Jr., 1970.
Development and migration of barrier islands, northern Gulf of Mexico.
Bull. geol. Soc. Am. 81(1): 241-246.

barrier islands

Schwartz, Maurice L. 1971.
The multiple causality of barrier islands.
J. Geol., 79(1): 91-94.

barrier reefs

Coudray, Jean, 1971.
Nouvelles données sur la nature et l'origine du complexe récifal côtier de la Nouvelle-Calédonie: Étude sédimentologique et paléoécologique préliminaire d'un forage réalisé dans le récif-barrière de la côte sud-ouest.
Quatern. Res. 1(2): 236-246.

barrier reef

Jennings, J.N., and E.C.F. Bird, 1967.
Regional geomorphological characteristics of some Australian estuaries.
In: Estuaries, G.H. Lauff, editor, Publs Am. Ass. Advmt Sci., 83:121-128.

barrier reefs

Pasenau, Horst 1971.
Morphometrische Untersuchungen an Hangterrassen der Grossen Meteorbank.
Meteor-Forsch.-Engebnisse (C) 6: 69-82.

barrier reefs

*Wolf, Karl H., and Göte Östland, 1967
14C dates of calcareous samples, Heron Island, Great Barrier Reef.
Sedimentology 8(3):249-251.

barriers, tidal

Duff, G.F.D., 1970.
Tidal resonance and tidal barriers in the Bay of Fundy system.
J. Fish. Res. Bd, Can. 27 (10): 1701-1728.

Canada, east
resonance, tidal
barriers, tidal
tides

bars, effect of

Andersen, T., F. Beyer and E. Føyn 1970.
Hydrography of the Oslofjord: report on the study course in chemical oceanography arranged in 1969 by ICES with support of UNESCO.
Coop. Res. Rept (A), Int. Cons. Explor. Sea, 20: 62 pp. (multilithed)

bars

*Bajorunas, L., and D.B. Duane, 1967.
Shifting offshore bars and harbor shoaling.
J. geophys. Res., 72(24):6195-6205.

bars (crescentic)

Bowen, Anthony J. and Douglas L. Inman, 1971.
Edge waves and crescentic bars. J. geophys. Res. 76(36): 8662-8671.

bars

Bowler, J.M., 1966.
The geology and geomorphology. Port Philip survey 1957-1963.
Mem. nat. Mus. Vict., 27:19-67.

bars

Brauckhoff K., 1970
Der morphologische Strandaufbau der Flachuferküste Ostrügens und seine Gesetzmässigkeiten.
Acta Hydrophys. 15(2): 93-103

bars

Ertel, H., und G. Kobe, 1961.
Hydrodynamische Erklärung der "Seebär"-Erscheinung.
Beitr. Geophys., 75(5):409-413.

bars

Leontiev, O.K., and L.G. Nikiforov, 1965.
On the causes of planetary distribution of coastal bars. (In Russian).
Okeanologiia, Akad. Nauk, SSSR, 5(4):653-661.

bars

Mann, I.J., 1881.
River bars. London, Crosby Lockwood & Co., 77 pp., 23 figs.

BARS

Nikiforov, L.G., 1964.
On the problem of the formation of coastal bars based on the study of the Ogurchinsky Island. (In Russian).
Okeanologiia, Akad. Nauk, SSSR, 4(4):654-659.

bars

Romashin, V.V., 1962.
Formation of the estuary bar in shallow water. (In Russian).
Trudy Gosud. Okeanogr. Inst., 66:116-120.

bars

Seibold, Eugen, 1963
Geological investigation of near-shore sand-transport-examples of methods and problems from the Baltic and North seas.
Progress in Oceanography, M. Sears, Edit., Pergamon Press, London, 1:1-70.

bars

Shuliak, B.A., and V.L. Boldyrev, 1966.
On the problem of beach bar formation processes. (in Russian).
Okeanologiia, Akad. Nauk, SSSR, 6(1):109-116.

bars

Varjo, Uuno, 1964.
Über Riffbildungen und ihre Entstehung an den Küsten des Sees Oulujarvi (Finnland).
Die Küste, 17: 51-80.

bars

Williams, W.W., and C.A.M. King, 1951.
Observations made on Karentes Beach. Bull. B.E.B. 5(2):19-23, 2 figs.

From Bull. d'Info, C.O.E.C. Dec. 1950.

Baryte

Arrhenius, Gustaf, and Enrico Bonatti, 1965.
Neptunism and vulcanism in the ocean.
Progress in Oceanography, 3:7-22.

basalts

Ade-Hall, J.M., 1971.
Discussion of paper by Minoru Ozima and Mituko Ozima, 'Characteristic thermomagnetic curve in submarine basalts'. J. geophys. Res. 76(32): 8077-8079.

basalts

Balashov, Yu. A., L.V. Dmitriev and A. Ya. Sharas'kin, 1970.
Distribution of the rare earths and yttrium in the bedrock of the ocean floor. (In Russian)
Geokhimiya (6): 647-660.
Translation: Geochem. Int. 7 (3): 456-468.

basalts

Bass, Manuel N., 1971.
Variable abyssal basalt populations and their relation to sea-floor spreading rates.
Earth Planet. Sci. Letts 11 (1): 18-22

basalt

Bezrukov, P.L., A. Ya. Krylov and V.I. Chernysheva, 1966.
Petrography and the absolute age of the Indian Ocean floor basalts. (In Russian; English abstract).
Okeanologiia, Akad. Nauk, SSSR, 6(2):261-266.

Indian Ocean
basalt
age determination

basalts

Bonatti, Enrico, 1968.
Rocks and sediments from the Barracuda Fracture Zone. (Abstract only). Trans. Fifth Carib. Geol. Conf., St. Thomas, V.I., 1-5 July 1968, Peter H. Mattson, editor. Geol. Bull. Queens Coll., Flushing, 5: 45.

basalts

Bonatti, Enrico, 1969.
Fissure basalts and ocean-floor spreading on the East Pacific Rise. Science, 166(3909): 1182-1183.

basalt

Bonatti, Enrico 1968.
Fissure basalts and ocean-floor spreading on the East Pacific Rise.
Science 161 (3844): 886-888.

basalts

Bonatti, Enrico, and David E. Fisher, 1971.
Oceanic basalts: chemistry versus distance from oceanic ridges.
Earth planet. Sci Letts. 11 (4): 307-311.

basalt

Bonatti, Enrico, and Oiva Joensuu, 1966.
Deep-sea iron deposit from the South Pacific.
Science, 154(3749):643-645.

basalt

Bonnati, E., J. Honnorez and G. Ferrara 1971.
1. Ultramafic rocks. Peridotite-gabbro-basalt complex from the equatorial Mid-Atlantic Ridge.
Phil. Trans. R. Soc. (A) 268 (1192): 385-402

basin (shelf)

Bott, M.H.P. 1971.
Evolution of young continental margins and formation of shelf basins.
Tectonophysics 11 (5): 319-327.

basalt

Bowin, C.O., A.J. Nalwalk and J.B. Hersey, 1966.
Serpentinized peridotite from the north wall of the Puerto Rico Trench.
Geol. Soc. Amer., Bull., 77(3):257-270.

Basalt

Brooke J., E. Irving and J.K. Park, 1970.
The Mid-Atlantic Ridge near 45°N., XIII Magnetic properties of basalt bore-core.
Can. J. Earth Sci., 7(6): 1515-1527.

basalts

Cann, J.R. 1970.
Petrology of basalts dredged from the Gulf of Aden.
Deep-Sea Res. 17(3): 477-482

basalt

Cann, J.R., and F.J. Vine 1966.
An area on the crest of the Carlsberg Ridge: petrology and magnetic survey.
Phil. Trans. R. Soc. (A) 259(1099): 198-217.

basalt

Chase, Richard L., 1969.
Bsalt from the axial trough of the Red Sea.
In: Hot brines and Recent heavy metal deposits in the Red Sea, E.T. Degens and D.A. Ross, editors, Springer-Verlag, New York, Inc., 122-128.

basalts

*Chase, R.L., and J.B. Hersey, 1968.
Geology of the north slope of the Puerto Rico Trench.
Deep-Sea Research, 15(3):297-317.

pyroxene
olivine
nontronite (saponite)

basalts

Christensen, Nikolas I., 1970.
Compressional-wave velocities in basalts from the Juan de Fuca Ridge. J. geophys. Res., 75 2773-2775.

basalt

Cifelli, Richard, 1965.
Late Tertiary planktonic Foraminifera associated with a basaltic boulder from the Mid-Atlantic Ridge.
J. Mar. Res., 22(3): 73-87.

abstract.

Basalts

*Dalrymple, G. Brent, and James G. Moore, 1968.
Argon-40: excess in submarine pillow basalts from Kilauea Volcano, Hawaii.
Science, 161(3846):1132-1135.

chemistry, argon-40
basalts
Hawaii

basalts

*De Boer, Jelle, Jean-Guy Schilling and Dale C. Krause, 1969.
Magnetic polarity of pillow basalts from Reykjanes Ridge. Science, 166(3908): 996-998.

Reykjanes Ridge
basalts
magnetic polarity

basalts

Deneufbourg, G., 1971.
Étude géologique du port de Papeete Tahiti (Polynésie Française). Cah. pacifiques 15: 75-82.

basalts

Engel, A.E.J. 1971.
Characteristics, occurrence and origins of basalts of the oceans. (Abstract only).
Phil. Trans. R. Soc. (A) 268 (1192): 493.

basalts

Engel, C.G., and A.E.J. Engel, 1963
Basalts dredged from the northeastern Pacific Ocean.
Science, 140(3573):1321-1324.

basalts

Engel, A.E.J., Celeste Engel and R.G. Havens, 1965.
Chemical characteristics of oceanic basalts and the upper mantle.
Geol. Soc., Amer., Bull., 76(7):719-734.

basalt

Engel, C.G. and R.L. Fisher, 1969.
Lherzolite, anorthosite, gabbro and basalt dredged from the Mid-Indian Ocean Ridge.
Science, 166(3909): 1136-114 .

basalt, tholeiitic

Engel, Celeste G., Robert L. Fisher and A.E.J. Engel, 1965.
Igneous rocks of the Indian Ocean floor.
Science, 150(3696):605-610.

basalts

Fisher, David E., 1971.
Incorporation of Ar in East Pacific basalts.
Earth Planet. Sci. Letts 12(3): 321-324.

basalts

Fisher, David E. 1970.
Comments on "K-Ar ages of submarine basalts dredged from seamounts in the western Pacific area and discussion of oceanic crust" by M. Ozima, I. Kaneoka, S. Aramaki. Earth Planet. Sci. Letts. 8(1970): 237-249 (with reply)
Earth planet. Sci. Letts 9(4): 310-312.

basalts

Fisher, David E., 1969.
Fission track ages of deep sea glasses.
Nature, Lond., 221(5180)549-550.

basalts

*Fisher, David E., Enrico Bonatti, Oiva Joensuu and J. Funkhouser, 1968.
Ages of Pacific deep-sea basalts and spreading of the sea floor.
Science, 160(3832):1106-1107.

Pacific
basalts
sea-floor spreading

basalt

Fisher, Robert L., and Celeste G. Engel, 1969.
Ultramafic and basaltic rocks dredged from the nearshore flank of the Tonga Trench.
Bull. geol. Soc. Am., 80(7): 1373-1378.

Tonga Trench
basalt
ultramafic rocks

basalts

Fisher, Robert L., Celeste G. Engel and Thomas W.C. Hilde, 1968.
Basalts dredged from the Amirante Ridge, western Indian Ocean.
Deep-Sea Res., 15(5):521-534.

basalts

Frey, Fred A., and Larry Haskin, 1964.
Rare earths in oceanic basalts.
J. Geophys. Res., 69(4):775-780.

basalts a

*Gast, P.W., 1971.
II. Fresh basalts. Dispersed element chemistry of oceanic ridge basalts. (abstract only).
Phil. Trans. R. Soc. Lond. (A) 268(1192): 467.

basalt a

*Green, D.H., 1971.
V. Petrogenesis and discussion. Composition of basaltic magmas as indicators of conditions of origin: application to oceanic volcanism. Phil. Trans. R. Soc. Lond. (A) 268(1192): 707-725.

basalt

Hamilton, Warren, 1966.
Origin of the volcanic rocks of eugeosynclines and island arcs.
In: Continental margins and island arcs, W.H. Poole, editor, Geol. Surv. Pap., Can., 66-15:348-355.

basalt

Harris, P.G. 1969.
Basalt type and African rift valley tectonism.
Tectonophysics 8(4/6): 427-436.

basalts

Hekinian, Roger, 1971.
Chemical and mineralogical differences between abyssal hill basalts and ridge tholeiites in the eastern Pacific Ocean.
Mar. Geol. 11(2): 77-91.

basalt

*Irving, E., 1968.
Measurement of polarity in oceanic basalt.
Can. J. earth Sci., 5(5):1319-1321.

basalt

Irving, E., W.A. Robertson and F. Aumento 1970.
The Mid-Atlantic Ridge near 45° N. XI. Remanent intensity, susceptibility and iron content of dredged samples.
Can. J. Earth Sci. 9(2-1): 226-238.

basalt

Kovylin, V.M., 1966.
Results of seismic research in the southwestern part of the abyssal basin of the Japan Sea. (In Russian; English abstract).
Okeanologiia, Akad. Nauk, SSSR, 6(2):294-305.

Sea of Japan
reflection, seismic
refraction, seismic
bottom sediments
basalt

basalt

Krueger, Harold W., 1964.
K-Ar age of basalt cores in the Mohole Project (Guadalupe site).
J. Geophys. Res., 69(6):1155-1156.

basalt

Kuno, Hisashi, 1966.
Lateral variation of basalt magma across continental margins and island arcs.
In: Continental margins and island arcs, W.H. Poole, editor, Geol. Surv. Pap., Can., 66-15:317-335.

basalt

Krause, Dale C., and Jean-Guy Schilling, 1969.
Dredged basalt from the Reykjanes Ridge, North Atlantic.
Nature, Lond., 224(5221): 791-793.

basalt

*Ladd, Harry S., Joshua I. Tracey, Jr., and M. Grant Gross, 1967.
Drilling on Midway Atoll, Hawaii.
Science, 156(2778):1088-1094.

Hawaii
Miocene
basalt
bottom sediments, reef

basalt

Luyendyk, Bruce P., and Celeste G. Engel, 1969.
Petrological magnetic and chemical properties of basalt dredged from an abyssal hill in the northeast Pacific.
Nature, Lond., 223(5210):1049-1050.

basalt

Macdougall, D., 1971
Deep sea drilling: age and composition of an Atlantic basaltic intrusion. Science 171(3977): 1244-1245.

basalt

MacDonald, Gordon A., 1969
Petrology of the basalt cores from midway atoll Hawaiian islands: description of altered basalt lava flows from beneath the reef cap.
Prof. Pap. U.S. geol. Surv. 680-B: B1-B10.
Geology of the midway area.

basalts a

Malahoff, Alexander and G.P. Woollard, 1970.
II. Regional observations. 3. Geophysical studies of the Hawaiian Ridge and Murray Fracture Zone. In: The sea: ideas and observations on progress in the study of the seas, Arthur E. Maxwell, editor, Wiley-Interscience 4(2): 73-131.

subsidence
gravity
basalts
magnetic anomalies

basalt

*Masuda, Akimasa, 1968.
Nature of the experimental Mohole basalt - redetermination of lanthanides.
J. geophys. Res., 73(16):5425-5428.

Mohole
basalt
lanthanides
rare earths

basalts a

Matthews, D. H. 1971.
III. Weathered and metamorphosed basalts. Altered basalts from Swallow Bank, an abyssal hill in the NE Atlantic and from a nearby seamount.
Phil. Trans. R. Soc. (A) 268 (1192): 551-571.

basalt

Matthews, D.H., F.J. Vine and N.R. Cann 1965.
Geology of an area of the Carlsberg Ridge, Indian Ocean.
Bull. Geol. Soc. Am. 76(6): 675-682.

basalt

Melson, William G., Vaughn T. Bowen, Tjeerd H. van Andel and Raymond Siever, 1966.
Greenstones from the central valley of the Mid-Atlantic Ridge.
Nature, 209 (5023):604-605.

basalt, olivine (alkali)

Melson, William G., Eugene Jarosewich, Richard Cifelli and Geoffrey Thompson 1967.
Alkali olivine basalt dredged near St. Pauls rocks, Mid-Atlantic Ridge.
Nature, Lond., 215 (5099):381-382.

basalts a

Melson, W. G., and G. Thompson 1971.
Petrology of a transform fault zone and adjacent ridge sediments.
Phil. Trans. R. Soc. (A) 268 (1192): 423-441.

basalt

Melson, W. G., Geoffrey Thompson and Tjeerd H. van Andel 1968
Volcanism and metamorphism in the Mid-Atlantic Ridge, 22°N. Latitude.
J. geophys. Res. 73(18): 5925-5941.

basalts, abyssal

Melson, William G., and Tjeerd H. Van Andel, 1966.
Metamorphism in the Mid-Atlantic Ridge, 22°N. latitude.
Marine Geol., 4(3):165-186.

basalt

Mindeli, P. Sh., Yu. V. Neprochnov and Ye. I. Partaraya, 1966.
Granite-Free area in Black Sea Trough, from seismic data.
Int. Geol. Rev., 8(1):36-43. (not seen).

basalt

Moberly, Ralph, Jr., 1963.
Amorphous marine muds from tropically weathered basalt.
Amer. J. Sci., 261: 767-772.

basalt (mud from)

Moberly, Ralph, Jr., 1961.
Amorphous marine muds from tropically weathered basalt. (Abstract).
Tenth Pacific Sci. Congr., Honolulu, 21 Aug.-6 Sept., 1961, Abstracts of Symposium Papers, 379-380.

basalt

Moore, James G., 1970.
Submarine basalt from the Revillagigedo islands region, Mexico.
Marine Geol. 9(5): 331-345.

basalt

Moore, James G., 1965.
Petrology of deep-sea basalt near Hawaii.
Amer. J. Sci., 263(1):40-52.

basalt

Morgan, J. W., and J. F. Lovering, 1965.
Uranium and thorium abundances in the basalt cored in Mohole Project (Guadalupe Site).
J. geophys. Res., 70(18):4724-4725.

basalt

Muir, I.D., C.E. Tilley and J.H. Scoon 1966.
Basalts from the northern part of the Mid-Atlantic Ridge. II The Atlantis collections near 30°N.
J. Petrol. 7(2): 193-201.

Basalts

Nayudu, Y. Rammohanroy, 1965.
Petrology of submarine volcanics and sediments in the vicinity of the Mendicino Fracture Zone.
Progress in Oceanography, 3:207-220.

basalts a

Nicholls, G. D. and M. R. Islam 1971.
Geochemical investigations of basalts and associated rocks from the ocean floor and their implications.
Phil. Trans. R. Soc. (A) 268 (1192): 469-486.

basalt

*Noble, C.S., and J.J. Naughton, 1968.
Deep-ocean basalts: inert gas content and uncertainties in age dating.
Science, 162(3850):265-267.

basalts (tholeiitic)

Noe-Nygaard, Arne. 1966.
Chemical composition of tholeiitic basalts from the Wyville-Thompson Ridge belt.
Nature, 212(5059):272-273.

basalts

Noe-Nygaard, Arne, and Jóannes Rasmussen, 1968.
Petrology of a 3000 metre sequence of basaltic lavas in the Faeroe Islands.
Lithos, 1(3): 286-304.

Faeroes
Atlantic northeast

basalts

O'Hara, M.J., 1968.
Are ocean floor basalts primary magma?
Nature, Lond., 220(5168):683-687.

basalt

Oversby, V.M., and P.W. Gast, 1968.
Oceanic basalt leads and the age of the earth.
Science, 162(3856):925-927.

basalts a

Opdyke, N.D., 1970.
1. General observations. 5. Paleomagnetism (June 1968). In: The sea: ideas and observations on progress in the study of the seas, Arthur E. Maxwell, editor, Wiley-Interscience, 4(1): 157-182.

basalts

Ozima, Minoru 1971.
Magnetic processes in oceanic ridge.
Earth planet. Sci. Letts 13 (1):1-5.

basalts

Ozima, Minoru and Mituko Ozima, 1971.
Characteristic thermomagnetic curve in submarine basalts. J. geophys. Res., 76(8): 2051-2056.

basalts

*Ozima, Minoru, Mituko Ozima and Ichiro Kaneoka, 1968.
Potassium-argon ages and magentic properties of some dredged basalts and their geophysical implications.
J. geophys. Res., 73(2):711-723.

basalts

Ozima, Mituko, and Minoru Ozima, 1967.
Self-reversal of remanent magnetization in some dredged submarine basalts.
Earth Planet. Sci., Letters, 3(3):213-215.

basalt

Ozima, Minoru, Shigeo Zashu and Naoko Ueno, 1971.
K/Rb and $(^{87}Sr/^{86}Sr)_0$ ratios of dredged submarine basalts.
Earth Planet. Sci. Letts, 10 (2): 239-244.

basalt
Park, J.K. and E. Irving, 1970.
The Mid-Atlantic Ridge near 45°N. XII. Coercivity, secondary magnetization, polarity, and thermal stability of dredge samples.
Can. J. Earth Sci., 7(6):1499-1514

basalts a
Schilling, J.-G., 1971.
Sea-floor evolution: rare-earth evidence.
Phil. Trans. R. Soc. (A) 268 (1192):663-706.

basalts
Tatsumoto, M., 1966.
Genetic relations of oceanic basalts as indicated by lead isotopes.
Science, 153(3740):1094-1101.

basalt
Ulrych, T.J., 1968.
Oceanic basalt leads and the age of the earth.
Science, 162(3856):928.

basalts
*Ulrych, T.J., 1967.
Oceanic basalt leads: a new interpretation and an independent age for the earth.
Science, 158(3798):252-256.

basalt a
Vallier, Tracy L., 1970.
Volcanism. Initial Repts. Deep Sea Drilling Project, Glomar Challenger 5: 531-534.

basalt
Verschure, R.H. 1966.
Possible relationships between continental and oceanic basalt and kimberlite.
Nature, Lond., 211 (5056): 1387-1389.

basalts
Von Herzen, R.P., 1969.
Fissure basalts and ocean-floor spreading on the East Pacific Rise. Science, 166(3909): 1181-1182.

basalts, chemistry
Cann, J.R. 1971.
Major element variations in ocean-floor basalts.
Phil. Trans. R. Soc. (A) 268 (1192): 495-505.

basalts, chemistry of
Cann, J.R., 1970.
Rb, Sr, Y, Zr and Nb in some ocean floor basaltic rocks.
Earth planet. Sci. Letts. 10(1): 7-11

basalts, chemistry of a
*Hart, S.R., 1971.
K, Rb, Cs, Sr and Ba contents and Sr isotope ratios of ocean floor basalts. Phil. Trans. R. Soc. Lond. (A) 268(1192): 573-587.

basalts chemistry
Hart, Roger, 1970.
Chemical exchange between sea water and deep ocean basalts.
Earth Planet. Sci. Letts. 9(3): 269-279.

basalts, chemistry
*Hubbard, Norman J., 1969.
A chemical comparison of oceanic ridge Hawaiian tholeiitic and Hawaiian alkalic basalts.
Earth Planet. Sci. Letters, 5(5):346-352.

basalts
Peterman, Zell E., and Carl E. Hedge, 1971.
Related strontium isotopic and chemical variations in oceanic basalts.
Bull. geol. Soc. Am., 82(2):493-500.

basalts - chemistry
Tatsumoto, Mitsunobu, 1969.
Lead isotopes in volcanic rocks and possible ocean-floor thrusting beneath island arcs.
Earth Planet. Sci. Letts 6(5): 369-376

basaltic magma
basaltic magma
*Sigurdsson, Haraldur, 1968.
Petrology of acid xenoliths from Surtsey.
Geol. Mag., 105(5):440-453.

basalt, magnetic properties
Bagin, V.I., S. Yu. Brodskaya, V.A. Jilyaeva, G.N. Petrova, D.M. Pechersky and O.J. Schmidt 1969.
Magnetic characteristics of basalts and peridotites of rift zones.
Tectonophysics 8 (4/6): 437-442.

basalt, magnetic properties
Irving, E., 1970.
The Mid-Atlantic Ridge at 45°N. XIV. Oxidation and magnetic properties of basalt; review and discussion.
Can. J. Earth Sci. 7 (6): 1528-1538

Basalt magnetic
Watkins, N.D., T. Paster and J. Ade-Hall 1970.
Variation of magnetic properties in a single deep-sea pillow basalt.
Earth Planet. Sci. Letts. 8(4): 322-328

Pillow basalt
Barrett, D.L., and F. Aumento, 1970.
The Mid-Atlantic Ridge near 45° N. XI. Seismic velocity, density and layering of the crust.
Can. J. Earth Sci., 7(4): 1117-1124

basalts, pillow
Corliss, John B., 1971.
The origin of metal-bearing submarine hydrothermal solutions. J. geophys. Res. 76(33): 8128-8138.

basalt pillows
Dymond, Jack, 1970.
Excess argon in submarine basalt pillows.
Bull. geol. Soc. Am., 81(4): 1229-1232.

basalts (tholeiitic)
Schilling, Jean-Guy, 1969.
Red Sea floor origin: rare-earth evidence.
Science, 165 (3900): 1357-1360

basaltic rock
basaltic rock
Manghnani, Murli H., and George P. Woollard, 1965.
Ultrasonic velocities and related elastic properties of Hawaiian basaltic rocks.
Pacific Science, 19(3):291-295.

basalts, tholeiitic a
Aumento, F., 1971.
Uranium content of mid-oceanic basalts.
Earth planet. Sci. Letts 11 (2): 90-94.

basalt-trachyte
Chayes, F., 1963
Relative abundance of intermediate members of the oceanic basalt-trachyte association.
J. Geophys. Res., 68(5):1518-1534.

basement
basement a
Cann, J.R. 1971
Petrology of basement rocks from Palmer Ridge, NE Atlantic.
Phil. Trans. R. Soc. (A) 268 (1192): 605-617.

basement
Drake, Charles L., 1966.
Recent investigations of the continental margin of eastern United States.
In: Continental margins and island arcs, W.H. Poole, editor, Geol. Surv. Pap., Can., 66-15:33-47.

basement
*Efendiyeva, M.A., 1967.
Crystalline basement relief beneath the Baltic aquatorium from magneto metric data.
Int. Geol. Rev., 9(10):1304-1308.

(Translated from: Sovet Geol. 1967(4):88-94.

basement
Komukai, R., 1956.
On the bottom topography and sediments in the western passage of the Tugaru Strait.
Rept. Hydrogr. Office, 31(4):45 pp.

basement
Moskalenko, V.N., 1966.
New data on the structure of sedimentary strata and solid basement in the Levant Sea (seismic profiling). (In Russian; English abstract).
Okeanologiia, Akad. Nauk, SSSR, 6(6):1030-1040.

basement
Scholl, David W., Edwin C. Buffington and David M. Hopkins, 1966.
Exposure of basement rock on the continental slope of the Bering Sea.
Science, 153(3739):992-994.

basement a
Sclater, J.G., E.J.W. Jones and S.P. Miller, 1970.
The relationship of heat flow, bottom topography and basement relief in Peake and Freen deeps, northeast Atlantic.
Tectonophysics 10(1/3): 283-300

basement
Uchupi, Elazar, 1966.
Structural framework of the Gulf of Maine.
J. Geophys. Res., 71(12):3013-3028.

basement
Vogt, P.R., 1970
Magnetized basement outcrops on the southeast Greenland continental shelf.
Nature, Lond., 226 (5247): 743-744

Greenland
basement
outcrops
continental shelf

basement rocks

Van Andel, Tj. H. 1971.
Tectonic evolution and trace element composition of basement rocks of the Mid-Atlantic Ridge: 8°S. (Abstract only).
Phil. Trans. R. Soc. (A) 268 (1192): 661.

basin

basins
*Antoine, J.W., and B.R. Jones, 1967.
Geophysical studies of the continental slope, scarp, and basin, eastern Gulf of Mexico.
Trans. Gulf Coast Assoc. Geol. Soc., 17th Ann. Meet. 268-277.

Basins
Bott, M.H.P. and A.B. Watts 1970.
Deep sedimentary basins proved in the Shetland-Hebridean continental shelf and margin.
Nature, Lond. 225 (5229): 265-268.

magnetic anomalies
Hebridean-Scotland shelf

basins
Dehlinger, Peter, R.W. Couch, D.A. McManus and Michael Gemperle, 1970.
II. Regional observations. 4. Northeast Pacific Structure. In: The sea: ideas and observations on progress in the study of the seas, Arthur E. Maxwell, editor, Wiley-Interscience 4(2): 133-189.

basins
Elmendorf, C.H., and B.C. Heezen, 1957.
Oceanographic information for engineering submarine cable systems. Bell Syst. Tech. J., 36(5): 1047-1094.

basin
Emery, K.O., 1953.
A newly surveyed submarine basin off Mexico.
Am. J. Sci. 251(9):656-660, 2 textfigs.

basins
Ewing, John I., 1968?
History of the ocean basins as recorded in the sediments
Selected Papers from the Governor's Conference on Oceanography, October 11 and 12, 1967, at the Rockefeller University, New York, N.Y.: 144-155.

basins
Ewing, Maurice, Xavier Le Pichon and John Ewing, 1966.
Crustal structure of the mid-ocean ridges.
4. Sediment distribution in the South Atlantic Ocean and the Cenozoic history of the Mid-Atlantic Ridge.
J. Geophys. Res., 71(6):1611-1636.

basins
Frakes, Lawrence A., and John C. Crowell, 1968.
Late Palaeozoic glacial facies and the origin of the South Atlantic Basin.
Nature, Lond., 217(5131):837-838.

basins
Frederickson, A.F., 1967.
Ownership and exploration of the Atlantic offshore basins.
1967 NEREM Record, Inst. Electr. Electron. Engrs, 9:162-163.

basins
Karig, Daniel E., 1970
Ridges and basins of the Tonga-Kermadec Island arc system. J. geophys. Res., 75(2): 239-254

island arcs
ridges
basins
Tonga-Kermadec Island Arc

basins
*Maloney, Neil J., 1967.
Geomorphology of the continental margin of Venezuela. 3. Bonaire Basin (66oW to 70oW longitude.
Bol. Inst. Oceanogr., Univ. Oriente, 6(2):286-302.

basins
Menard, H.W., 1967.
Transitional types of crust under small ocean basins.
J. geophys. Res., 72(12):3061-3075.

BASINS
Scholl, David W., and David M. Hopkins, 1969.
Newly discovered Cenozoic basins, Bering Sea shelf, Alaska.
Bull. Am. Ass. Petrol. Geol. 53 (10): 2067-2078.

basins, marginal
Sleep, Norman, and M. Nafi Toksöz 1971.
Evolution of marginal basins.
Nature Lond. 233 (5321): 548-550.

basins
Wilson, J. Tuzo, 1966.
Patterns of growth of ocean basins and continents.
In: Continental margins and island arcs, W.H. Poole, editor, Geol. Surv. Pap., Can., 66-15:388-391.

basins
Woollard, G.P., 1967.
Recent geological and geophysical evidence on the orgin of ocean basins.
1967 NEREM Record, Inst. Electr. Electron. Engrs, 9:160.

basins, marginal
*Karig, Daniel E., 1971.
Origin and development of marginal basins in the western Pacific. J. geophys. Res., 76(11): 2542-2561.

batholith

batholith
Larsonneur, Claude, 1967.
Etude de la partie immergée du massif granitique de Barfleur (Manche)
Bull. Soc. Linn. Normandie (10) 8(2): 231-241

bathymetric chart

bathymetric charts
Ritchie, G.S., 1968.
The limits of hydrospace.
Hydrospace, 1(4):36-37.

bathymetric chart
Wiseman, J.D.H., and C.D. Ovey, 1955.
The general bathymetric chart of the oceans.
Deep-Sea Res. 2(4):269-273.

bathymetric distribution

bathymetric distribution
Enequist, P., 1949.
Studies on the soft-bottom amphipods of the Skagerak. Zool. Bidrag, Uppsala, 28:297-492, 67 textfigs. 5 charts.

bathymetry

A

bathymetry
*Aagaard, K., and L. Coachman, 1968.
The East Greenland Current north of Denmark Strait: I.
Arctic 21(3):181-200.

bathymetry
Aasen, O., 1952.
The Lusterfjord herring and its environment.
Repts. Norwegian Fish. & Mar. Invest., 10(2): 63 pp., 3 pls., 20 textfigs.

bathymetry
*Abe, Shigeo, Shigeaki Yada, Shoroku Inoue, and Yusho Akishige, 1967.
Survey of summer trawl grounds off the northwest coast of Australia. (In Japanese; English abstract).
Bull. Fac. Fish., Nagasaki Univ., 23: 205-215.

bathymetry
Abramov, Ye. P., and O. N. Sokolov, 1963.
Experience of using sounding for geological mapping of the sea bottom. (In Russian).
Geofiz. Priborostroyeniye, Gostoptechisdet., Leningrad., No. 15:116-128.

Abstracted in:
Soviet Bloc Res., Geophys. Astron. and Space, 96: 46.

bathymetry
Agapova, G.V., 1966.
Some characteristics of bottom geomorphology in the northwestern Atlantic Ocean. (In Russian; English abstract).
Okeanologiia, Akad. Nauk, SSSR, 6(4):666-671.

Atlantic, northwest
bathymetry
submarine geology

bathymetry
Agapova, G.V., 1961
[The geomorphologic map of the Caspian sea bottom (scale 1: 3 000 000).]
Okeanologia, 2: 274-277.

bathymetry
*Agapova, G.V. and V.F. Kanaev, 1969.
The experience of the statistical processing of bottom relief data on the 2nd cruise of the R/V Akademik Kurchatov. Okeanologiia, 9(4): 724-729.
(In Russian; English abstract)

bathymetry
Agassiz, A., 1902.
1. Preliminary report and list of stations. With remarks on the deep-sea deposits by Sir John Murray. Reports on the scientific results of the expedition to the Tropical Pacific, in charge of Alexander Agassiz, by the U.S. Fish Commission Steamer "Albatross", from August, 1899, to March, 1900, ----. Mem. M.C.Z. 26(1):1-114, 19 pls.

bathymetry
Ahlmann, H.W., 1955.
Forskning och händelser i Arktis efter Vega-Expeditionen. Ymer 75(2):81-97, 8 textfigs.

bathymetry
Ahrens, E., 1959.
Mesures bathymetriques en mer à l'aide des sondeurs à ultra-sons.
Navigation, Rev. Tech. de Nav. Mar. et Aer., 7(27): 238-257.

bathymetry
Akagawa, M., 1956.
On the oceanographical conditions of the North Japan Sea (west of the Tsugaru Straits) in summer (Part 2).
Bull. Hakodate Mar. Obs., No. 3:5-11. 1-7.

bathymetry
Alarcón, Elías, 1970.
Descripción oceanográfica preliminar del Golfo de Arauco.
Bol. Cient. Inst. Fomento pesq., Chile, 13: 51pp.
T-O

bathymetry

Algeria, Centre de Geologie Marine et de Sedimentologie, 1962
Utilisation des radio-traceurs à courbet-marine. Étude de l'ensablement d'un port du littoral algerois. Methodes sedimentologiques utilisation de traceurs radio-actifs.
Cahiers Océanogr., C.C.O.E.C., 14(8):526-542.

bathymetry

Alinat, Jean Jacques-Yves Cousteau, Günter Giermann, Olivier Leenhardt, Christian Perrin et Serge Puirot, 1969.
Lever de la carte bathymétrique de la mer Ligure.
Bull. Inst. océanogr., Monaco, 68(1395): 12 pp.

bathymetry
Mediterranean, west

bathymetry

Aliverti, G., M. Picotti, L. Trotti, A. DeMaio, O. Lauretta e M. Moretti 1968.
Atlante del Mar Tirreno: isoterme ed isoline dedotte dalle misure eseguite durante le crociere per l'Anno Geofisico Internazionale 1957-1958.
Consiglio Naz. Ricerche, Roma, Ist. Univ. Navale, Napoli 115 pls.

bathymetry

Allan, T.D., 1966.
A bathymetric chart of the Red Sea.
Int. Hydrogr. Rev., 43(1):33-36.

bathymetry

Allan, T.D., 1965.
Red Sea bathymetric chart. Hydrographic Department British Admiralty, Chart No. 6359.

bathymetry

Allan, T.D. and C. Morelli, 1970.
II. Regional observations. 13. The Red Sea (May 1969). In: The sea: ideas and observations on progress in the study of the seas, Arthur E. Maxwell, editor, Wiley-Interscience 4(2): 493-542.

bathymetry

Allan, T.D. and M. Pisani 1965.
Gravity, magnetic and depth measurements in the Ligurian Sea.
Rapp. P.-V. Réun. Comm. int. Explor. scient. Mer Méditerr. 18(3): 907-909.

bathymetry

Allard, M.P., 1948
Compte Rendu des Travaux. Mission Hydrographique du Maroc (Mai 1947). No. 1018:18 pp., 1 pl., 1 chart.

bathymetry

Anderson, E.R., 1964.
Single-depth charts of the World's ocean basins at depths to 3500 fathoms. Set 2. U.S. Navy Electronics Laboratory, Research and Development Rept. 1252: unnumbered charts (folded and bound).

bathymetry

Anma, Kei, 1968.
Geomagnetic and bathymetric study of the Okhotsk Sea. (2):
Oceanogrl Mag., 20(1):65-72.

bathemetry

Anon., 1970.
Report on a preliminary survey in Malacca Strait and Singapore Strait.
Int. hydrogr. Rev. 47(1): 7-22

bathymetry
Malacca Strait
Singapore Strait

bathymetry

Anon., 1969.
Riches from Government files: historical oceanography data collected by government agencies may be used to provide clues to mineral deposits.
Ocean Industry, 4(7):44-50. (secondary sources)

bathymetry

Anon, 1962
Les reliefs de l'Océan Indien précisés par le Vitiaz.
La Nature, 3327: 303-304.

bathymetry

Anon., 1955.
Tiefenkarte des Nordpolarbeckens, Kenntnisstand 1948 u. 1954. Petermanns Geogr. Mitt. 99(1):pl. 7.

bathymetry

Anon., 1953.
Recent French bathymetrical operations in the Antarctic Ocean. Int. Hydrogr. Rev. 30(1):65-71, 3 pls.

bathymetry

Anon., 1882.
Sondages de l'aviso "Le Travailleur" dans le Golfe de Gascogne. Ann. Hydrogr., Paris, (2)4: 354-398.

bathymetry

Anon., 1882.
Sondages. Ann. Hydrogr., Paris, (2)4:89-123.

bathymetry

Anon., 1950.
Exploracion oceanográfica del Africa Occidental. Campaña del "Malaspina" en enero de 1950 en aguas del Sáhara desde Punta Durnford a Cabo Barbas. Bol. Inst. Español Ocean., No. 38:12 pp. 2 figs.

bathymetry

Anon., 1950.
Rapport preliminaire sur les sondages bathymetriques executes par l'Aviso Polaire "Commandant Charcot" au cours de la campagne 1949-1950.
Bull. d'Info., C.O.E.C., 2(9):335-338, 1 fig.

bathymetry

Anonymous, 1949-1950.
[Bathymetric survey in the area east of Petrico and north of Kymi, Alieytikon Deltion. Bathymetric survey of the area of the islands of Hydra-Ag. Georgios; bathymetric survey in the area of the islands Amorgos-Denous.] Greek Hydrobiological Inst. Nos. 1-3, 24 pp., 1 chart.

bathymetry

Anon., 1949.
Campañas del "Xauen" en 1947 y 1948 en el mar de Alboran y en el estrecho de Gibraltar. Registro de operaciones. Bol. Inst. Español Ocean. 18: 1-53.

bathymetry (data)

Anon., 1949.
Exploracion oceanografica del Africa Occidental. Campañas del "Malaspina" en 1947 y 1948 en aguas del Sahara desde Cabo Juby a Punta Durnford. Registro de operaciones. Bol. Inst. Español Ocean. No. 23: 28 pp., 2 textfigs.

bathymetry

Anon., 1939 (?)
Synopsis of the Hydrographical investigations (July-December 1939). Semiannual Rept. Oceanogr. Invest. (July-Dec. 1939), Imp. Fish. Expt. Sta., Tokyo, No. 65, 147 pp.

bathymetry

Anon., 1925-1928.
Ricerce e scandagliamento del Banco Graham. Ann. Idrografico, 12(1):4 pp., 4 figs. (typed)

In MBL Library

Bathymetry

Arctowski, H., 1945
Bathymetry of the Pacific. Am. Geophys. Union Trans. of 1944, Part IV:608-609.

bathymetry

Aric, K., H.B. Hirschleber und W. Weigel, 1968.
Ergebnisse seismischer Messungen im Bereich der Grossen Meteor-Bank.
Z. Geophys. 34(5):531-534.

bathymetry

Asashina, H., 1927.
Changes of the depths of Hiuti Nada (central part of the Inland Sea) during the past 30 years. (In Japanese). Hydrogr. Bull. (Suiro Yoho) 6: 255-262.

bathymetry

Association d'Océanographie Physique, Union Géodésique et Géophysique, 1940.
Report of the Committee on the Criteria and Nomenclature of the Major Divisions of the Ocean Bottom. Publ. Sci. No. 8:124 pp., 3 fold-in charts.

bathymetry

Athearn, William D., 1968.
Upper slope bathymetry off Guayanilla and Ponce, Puerto Rico. Trans. Fifth Carib. Geol. Conf. St. Thomas, V.I., 1-5 July 1968, Peter H. Mattson, editor. Geol. Bull. Queens Coll., Flushing, 5:41-43.

bathymetry

Athearn, William D., 1963.
Bathymetry of the Straits of Florida and the Bahama Islands. Part II. Bathymetry of the Tongue of the Ocean, Bahamas.
Bull. Mar. Sci., Gulf and Caribbean, 13(3):365-377.

WHOI Contrib. No. 1399.

bathymetry

Auffret, G.A., L. Berthois et L. Cabioch 1971.
Aperçu sur la bathymétrie, la sédimentologie et les peuplements benthiques au large de Roscoff d'après la carte bathymétrique au 1/40.000e et les photographies sous-marines.
Cah. Biol. mar. 12(4): 497-504

bathymetry

Avilov, I.K., 1965.
Relief and bottom sediments of the shelf and continental slope of the northwestern Atlantic. (In Russian).
Trudy vses. nauchno-issled. Inst. morsk. ryb. Khoz. Okeanogr. (VNIRO), 57:173-234.

bathymetry

Avilov, I.K., 1965.
Some data on the bottom topography and grounds of the West African shelf. Investigations in line with the programme of the International Geophysical Year, 2. (In Russian).
Trudy, vses. Nauchno-Issled. Inst. Morsk. Ribn. Choz. i Okeanogr., (VNIRO), 57:235-259.

bathymetry

Avilov, I.K., 1965.
Relief and bottom sediments of the shelf and continental slope of the Northwestern Atlantic. Investigations in line with the programme of the International Geophysical Year, 2. (In Russian).
Trudy, Vses. Nauchno- Issled. Inst. Morsk. Ribn. Choz. i Okeanogr. (VNIRO), 57:173-234.

B

bathmetry

Bakaev, V.G., editor, 1966.
Atlas Antarktiki, Sovetskaia Antarktich-eskaia Ekspeditsiia. 1.
Glabnoe Upravlenie Geodesii i Kartografii MG SSSR, Moskva-Leningrad, 225 charts.

bathymetry

*Ball, M.M., C.G.A. Harrison, R.J. Hurley and C.E. Leist, 1969.
Bathymetry in the vicinity of the northeastern scarp of the Great Bahama Bank and Exuma Sound.
Bull. mar. Sci., 19(2): 243-252.

bathymetry
scarps
Great Bahama Bank
Exuma Sound

bathymetry

Ballard, James A., 1966.
Structure of the lower continental rise hills of the western North Atlantic.

Geophysics, 31(3):506-523.

bathymetry

Barahona Fernandes, Jose Augusto, 1962
The use of Raydist in the survey of a bank offshore Lourenco Marques.
Int. Hydrogr. Rev., Suppl., 3:7-13.

bathymetry

*Barber, F.G., 1968.
On the water of Tuktoyaktuk Harbour.
Manuscript Rept., Ser.Mar.Sci.Br.Dept.Energy, Mines,Resources,Ottawa, 9:32 pp.

bathymetry

Barber, F.G., 1967.
A contribution to the oceanography of Hudson Bay.
Manuscript Rep. Ser., Dept.Energy,Mines,Resources, Can., No. 4:69 pp. (multilithed).

BAthymetry

Barber, F.G., 1958.
Currents and water structure in Queen Charlotte Sound, British Columbia.
Proc. Ninth Pacific Sci. Contr., Pacific Sci. Assoc., 1957, Oceanogr., 16:196-199.

bathymetry

Barber, F.G., and C.J. Glennie, 1964.
On the oceanography of Hudson Bay, an atlas presentation of data obtained in 1961.
Manuscr.Rep.Ser.,mar.Sci.Br.,Dept.Mines tech Surv.Ottawa, 1: numerous pp. (Unnumbered) (multilithed).

bathymetry

Barker, P.H., 1967.
Bathymetry of the Fiordland continental margin
N.Z. Jl. Sci. 10(1): 128-137.

bathymetry
New Zealand
continental margin

bathymetry

Barker, P.H. 1966.
A reconnaissance survey of the Murray Ridge.
Phil. Trans. R. Soc. (A) 259 (1099): 187-197.

bathymetry

Barker, P.H., and A.C. Kibblewhite, 1965.
Physical oceanographic data from the Tui cruise 1962.
New Zealand J. Sci., 8(4):604-634.

bathymetry

Barnard, J. Laurens and John R. Grady, 1968.
A biological survey of Bahia de los Angeles, Gulf of California, Mexico. I. General account.
Trans. San Diego Soc. nat. Hist 15 (6): 51-66

bathymetry

Barrett, D.L., 1966.
Lancaster Sound shipborne magnetometer survey.
Canadian J. Earth Sci., 3(2):223-235.

bathymetry

Beach Erosion Board, 1948.
Proof test of water transparency method of depth determination. Eng. Notes No. 29:29pp., 34 figs. data sheets (multilith).

bathymetry

Belknap, G.E., 1874.
Deep sea soundings in the North Pacific Ocean obtained in the United States Steamer "Tuscarora". USHO 54:51 pp., 12 pls., 9 profiles, 1 chart.

bathymetry

Beloussov, I.M.B., 1970.
Certain geomorphological peculiarities of the southern part of the Gulf of Mexico and the region of Cuba. (In Russian; Spanish abstract).
Okeanol. Issled. Rez. Issled. Mezhd. Geofiz. Proekt. Mezhd. Geofiz. Kom. Prezid. Akad. Nauk SSSR 20: 11-21. Also English abstract

bathymetry

Belousov, I.M. 1965.
A bathymetric chart of the northern Indian Ocean. (In Russian) (Not seen)
Seismologiya: Sbornik Statei, SSSR 13:163-171.

(National Lending Library for Science and Technology, Boston Spa, Yorkshire, England)

Belousov, I.M., 1965. bathymetry
Bathymetric chart of the northwestern Indian Ocean. (In Russian; English abstract).
Okeanolog. Issled., Rez. Issled. po Programme Mezhd. Geofiz. Goda, Mezhd. Geofiz. Komitet, Prezidiume Akad. Nauk, SSSR, No.13:163-171.

bathymetry

Bennett, Edward B., and Milner B. Schaefer, 1960.
Studies of physical, chemical and biological oceanography in the vicinity of the Revilla Gigedo Islands during "Island Current Survey" of 1957.
Bull. Inter-American Tropical Tuna Comm. 4(5): 219-317. (Also in Spanish)

bathymetry

Beresnov, A.F., 1962
Geomorphology of the Davis Sea bottom.
Mezhd. Geofiz. Komitet, Prezidiume Akad. Nauk. SSSR, Rezult. Issled. Programme Mezhd. Geofiz. Goda, Okeanol. Issled., No. 5:43-59.

bathymetry

Berger, Jon, A.E. Cok, J.E. Blanchard and M.J. Keen, 1966.
Morphological and geophysical studies on the eastern seaboard of Canada: The Nova Scotian Shelf.
In: Continental drift, G.D. Garland, editor, Spec. Publs. R.Soc.Can., No.9:102-113.

bathymetry

Berrit, G.R., 1962.
Campagne No. 11. Campagne Jonas. 1. Resultats d'observations. II. Observations du niveau marin à la station littorale de l'A.G.I. à Gorée (Dakar).
Cahiers Oceanogr., C.C.O.E.C., 14(2):132-135.

bathymetry

Berrit, G.R., 1961
Etude des conditions hydrologiques en fin de saison chaude entre Pointe-Noire et Loanda.
Cahiers Oceanogr., C.C.O.E.C., 13(7): 456-461.

bathymetry

Berthois, L., 1969
Bathymétrie de l'Atlantique Nord entre les Féroé et les Hébrides et lithologie des blocs épars sur le fond.
Bull. Ass. fr. Etud gr. profondeurs océaniques 6:1-28 (not seen)

bathymetry

Berthois, Léopold, René Battistini et Alain Crosnier, 1964.
Recherches sur le relief et la sédimentologie du plateau continental de l'extrême sud de Madagascar.
Cahiers Océanogr., C.C.O.E.C., 16(7):511-527.

Berthois, L., and R. Brenot, 1960 bathymetry

La morphologie sous-marine du talus du plateau entre le sud de l'Irlande et le Cap Ortegal. J. du Cons., 25(2):111-114.

bathymetry

Berthois, L., and R. Brenot, 1957.
Note préliminaire sur la topographie du talus du plateau continental dans le Golfe de Gascogne.
Rev. Trav. Inst. Pêches Marit., 21(3):435-438.

bathymetry

Berthois, L., and R. Brenot, 1957.
Topographie sous-marine de l'"Outer Silver Pit" et du "Markham's Hole", Lat. N 54° Long. E. Gr. 2°.
Rev. Trav. Inst. Pêches Marit., 21(3):431-434.

bathymetry

Berthois, L., R. Brenot et P. Ailloud, 1965.
Essai d'interprétation morphologique et géologique de la pente continentale à l'ouest de la Peninsule Ibérique.
Rev. Trav. Inst. Pêches marit., 29(3):343-350.

bathymetry

Berthois, L., R. Brenot et P. Ailloud, 1965.
Essai d'interprétation morphologique et tectonique des levés bathymétriques exécutés dans la partie sud-est du Golfe de Gascogne.
Rev. Trav. Inst. Pêches marit., 29(3):321-342.

bathymetry

Berthois, L., R. Brenot, G. Auffret et Marie-Henriette De Buit, 1968.
La sédimentation dans la région de la Bassurelle (Manche orientale): Etude hydraulique, bathymétrique, dynamique et granulométrique.
Trav. Cent. Rech. Etud. océanogr. ns. 8(3): 13-29

tides
English Channel
currents
bathymetry
bottom sediments

bathymetry

Berthois, L., R. Brenot et J. Debyser, 1968.
Remarques sur la morphologie de la marge continentale entre l'Irlande et le Cap Finisterre.
Revue Inst.français Pétrole, 23(9):1046-1049.

bathymetry

Berthois L., et A. Crosnier 1966.
Etude dynamique de la sédimentation au large de l'estuaire de la Betsiboka.
Cah. ORSTOM, Sér. Océanogr. 4(2): 49-130.

bathymetry

Besnard, W., 1952.
Resultados cientificos do Cruzeiro do "Baependi" e do "Vega" na Ilha da Trinidada. Bol. Inst. Ocean., Univ. Sao Paulo, 3(1/2):49-54, 1 fold-in.

bathymetry

Besnard, W., 1951.
Resultados cientificos do cruzeiro do "Baependi" e do "Vega" à Ilha da Trinidade. Contribuição para o conhecimento da plataforma insular da Ilha da Trinidade. Bol. Inst. Paulista Ocean. 2(2):37-48, 5 pls.

bathymetry

Betz, F., and H.H. Hess, 1940.
Floor of the North Pacific Ocean. Trans. Am. Geophys. Union 21:348-349.

bathymetry
Bezrukov, P.L., 1957.
New data on the regularities of the morphology of submarine relief. (In Russian).
Doklady, Akad. Nauk, SSSR, 116(5):841-

bathymetry
*Birch, Francis S., 1970.
The Barracuda Fault Zone in the western North Atlantic: geological and geophysical studies.
Deep-Sea Res., 17(5): 847-859.

Barracuda Fault Zone
Atlantic, northwest
bathymetry

bathymetr
Blanc, Jean et Colerine Froget, 1967.
Recherches de géologie marine et sédimentologie. Campagne de la Calypso en Méditerranée orientale (quatrième mission, 1964).
Annls Inst. Oceanogr., n.s. 45(2): 257-292.

bathymetry
Blumenstock, D. I., and D. F. Rex, 1959
Microclimatic observations at Eniwetok NAS-NRC, Pac. Sc. Bd., Atoll Res. Bull., No. 71: 158.

bathymetry
Bonnefille, R., 1959.
The generalization of Vantroys' similarity laws allowing for variations in the depth of the sea.
La Houille Blanche, 14(5):547-555.
Rev. in:
Appl. Mech. Rev., 13(7):#3751.

bathymetry
48-64.
Bortnikov, V.S., and I.M. Belousov, 1968.
Preliminary results of geological investigations in the western Indian Ocean. (In Russian).
Trudy, Vses. Nauchno-Issled. Inst. Morsk. Ribn. Okeanogr (VNIRO) 64, Trudy Azovo-Chernomorsk. Nauchno-Issled. Inst. Morsk. Ribn. Khoz. Okeanogr. (AscherNIRO), 28:
48-64
Indian Ocean
bathymetry
bottom sediments

bathymetry
Bossolasco, Mario, and Ignazio Dagnino, 1961.
Ricerche di fisica marina.
Contr. Ist. Geofis. e Geodet., Univ., Genova, all'Anno Geofis. Internaz., 1957-58 & 1959, Memoria, No. 1: 62 pp.
Reprinted from:
Geofisica e Meteorologia (Boll. Soc. Ital. di Geofis. e Meteorol., Genova), Vol. 7(1959): 99-113; Vol. 8(1960):22-32; 87-96; 142-155; 9(1961):55-64.

bathymetry
Bott, M.H.P., C.W.A. Browitt and A.P. Stacey 1971.
The deep structure of the Iceland-Faroe Ridge.
Mar. geophys. Researches 1(3): 328-351

bathymetry
Bouma, A.H., 1965.
Sedimentary characteristics of samples collected from some submarine canyons.
Marine Geology Elsevier Publ. Co., 3(4):291-320.

bathymetry
Bourcart, Jacques, 1960
Carte topographique fond de la Méditerranée occidentale. Bull. Inst. Océanogr., Monaco, No. 1163: 20 pp.

bathymetry
Bourcart, Jacques, 1958.
Rapport sur une mission effectuée à bord de la Calypso en mai 1957.
Bull. d'Info., C.C.O.E.C., 10(4):188-191.
Reprinted in:
Trav. Lab. Geol. Sous-Marine, 8, 1958.

bathymetry
Bourcart, J., 1957.
Essai de carte sous-marin de l'ouest de la Corse.
Géogr. Phys. et de Géol. Dynamique (2)1(1):31-35.

bathymetry
Bourcart, J., 1958.
Rapport sur une mission effectuée à bord de la Calypso en mai 1957.
Bull. d'Info., C.C.O.E.C., 10(4):188-191.

bathymetry
Bourcart, J., 1957.
Géologie sous-marine de la Baie de Villefranche.
Ann. Inst. Océan., Monaco, 23(3):127-199.

bathymetry
Bourcart, J., 1957.
Géologie sous-marine de la Baie de Villefranche.
Ann. Inst. Océan., Monaco, 33(3):137-200.

bathymetry
Bourcart, J., 1955.
Recherches sur le plateau continental de Banyuls-sur-Mer.
Vie et Milieu, Bull. Lab. Arago, 6(4):435-524.

bathymetry
Bourcart, J., 1949.
Géographie du fond des mers. Étude du relief des Océans. Paris, Payot, 307 pp.

bathymetry
Bourcart, J., 1948.
Rapport sur une mission oceanographique sur le plateau continental marocain. Bull. Sci., C.O.E.C., Maroc, 2:22-36, 5 figs.

bathymetry
Bourcart, J., H. Horiot, and C. Lalou, 1950.
Sur la topographie sous-marine au large de Toulon. C.R. Acad. Sci., Paris, 230:561-563, 1 textfigs.

bathymetry
Bourcart, Jacques, Claude Lalou and Maurice Gennesseaux, 1958.
Le relief sous-marin du precontinent entre le Rhone et la Ciotat.
Bull. d'Info., C.C.O.E.C., 10(3):144-152.
Reprinted in:
Trav. Lab. Geol. Sous-Marine, 8, 1958.

bathymetry
Bouysse, Philippe et Jean-René Vanny, 1966.
La Baie de la Vilaine: étude sédimentologique et morphologique.
Cahiers Oceanogr., 18(4):319-341. (C.C.O.E.C.)
abstract, p. 329.
sedimentation
France, northwest
bathymetry
bottom sediments

bathymetry
Bowin, Carl O. 1968.
Geophysical study of the Cayman Trough.
J. geophys. Res. 73(16): 5159-5173.

bathymetry
Boyd, L. A., 1948.
The coast of northeast Greenland with hydrographic studies in the Greenland Sea. Am. Geogr. Soc. Spec. Pub. 30:340 pp., illus., maps.

bathymetry
Brennecke, W., 1915.
Oceanographische Arbeiten S.M.S. "Planet" im westlichen Stillen Ozean 1912-1913. Ann. Hydr., usw., 43:145-158.

bathymetry(data)
Brennecke, W., 1920.
Lotungen des Kabeldampfers "Grossherzog von Oldenburg" im Östlichen Teil des Nordatlantischen Ozeans. Arch. Deutschen Seewarte 33(3):1-20, 1 textfig., 2 pls.

bathymetry
Brenot, R., & L. Berthois, 1962
Bathymetrie du secteur atlantique du banc Porcupine (ouest de l'Islande) au cap Finisterre (Espagne).
Rev. Trav. Inst. Peches Marit., 26(2):219-246.

bathymetry
Brodie, J.W., 1964.
The bathymetry of the New Zealand region.
New Zealand Dept. Sci. Ind. Res., Bull., No. 161;
New Zealand Oceanogr. Inst., Memoir, No. 11:54 pp

bathymetry
Brodie, J.W., 1964.
The fjordland shelf and Milford Sound.
New Zealand Dept. Sci. Ind. Res., Bull., 157;
New Zealand Oceanogr. Inst., Memoir, No. 17: 15-23.

bathymetry
Brodie, J.W., 1952.
Features of the sea floor west of New Zealand.
N.Z. J. Sci. Tech., Sect. B, 33(5):373-384, 6 textfigs.

bathymetry
Brouwer, H.A., 1939.
Exploration of the Lesser Sunda Islands. Geogr. Jour. 94:1-10.

bathymetry
Bruun, A.Fr., J.W. Brodie and C.A. Fleming, 1955.
Submarine geology of Milford Sound, New Zealand.
N.Z.J. Sci. Tech., B, 36:397-410, 7 textfigs.

bathymetry
Bruun, A. Fr., and A. Kiilerich, 1957.
Bathymetrical features of the Bali-Lombok Strait.
Penjelidikan Laut, Indonesia (Mar. Res.), No. 3: 1-6.

bathymetry
Bryan, G.S., 1940.
A bathymetric chart of the North Pacific Ocean.
Assoc. d'Ocean. Phys., Pub. Sci., Spec. Art. No. 8:78-80, 1 fold-in chart.

bathymetry
Bryant, William R., John Antoine, Maurice Ewing, and Bill Jones, 1968.
Structure of Mexican continental shelf and slope, Gulf of Mexico.
Bull. Am. Ass. Petr. Geol., 52(7): 1204-1228.

bathymetry
Budanova, L.J., 1961
The bathymetric map of the northwestern part of the Pacific ocean (scale 1:5 000 000)
Okeanologiia, Akad. Nauk, SSSR, (2):278-282.

bathymetry
Budinger, Thomas F., 1967.
Cobb Seamount.
Deep-Sea Research, 14(2):191-201.
seamounts
Cobb Seamount
bathymetry

bathymetry
Bunce, Elizabeth T., 1966.
The Puerto Rico Trench.
In: Continental margins and island arcs, W.H. Poole, editor, Geol. Surv. Pap., Can., 66-15:165-175.

W.H.O.I. Contrib. No. 1718.

bathymetry
Bunce, Elizabeth T., and Davis A. Fahlquist, 1962
Geophysical investigation of the Puerto Rico Trench and outer ridge.
J. Geophys. Res., 67(10):3955-3972.

WHOI Contrib. No. 1287

bathymetry
Bunce, Elizabeth T. and J.B.Hersey,1966.
Continuous seismic profiles of the outer ridge and Nares Basin north of Puerto Rico.
Geol. Soc., Am., Bull., 77 (8):803-812.

bathymetry
Bunce, Elizabeth T., M.G. Langseth, R.L. Chase and M. Ewing, 1967.
Structure of the western Somali Basin.
J. geophys. Res., 72(10):2547-2555.

Contrib. No. 1855 from the Woods Hole Oceanographic Institution

Indian Ocean
Somali Basin
1855
reflection, seismic
bottom sediments, terrigenous
bathymetry
"Chain"
gravity anomalies
magnetic intensity

bathymetry
Bunce, Elizabeth T., Joseph D. Phillips, Richard L. Chase, and Carl O. Bowin, 1970.
II. Regional observations. 11. The Lesser Antilles Arc and the eastern margin of the Caribbean Sea. In: The sea: ideas and observations on progress in the study of the seas, Arthur E. Maxwell, editor, Wiley-Interscience 4(2): 359-385.

bathymetry
*Burckle, Lloyd H., Tsunemasa Saito and Maurice Ewing, 1967.
A Cretaceous (Turonian) core from the Naturaliste Plateau, southeast Indian Ocean.
Deep-Sea Research, 14(4):421-426.

bathymetry
Burke, Kevin, 1967.
The Yallahs Basin: a sedimentary basin southeast of Kingston Jamaica.
Marine Geol., 5(1):45-60

Yallahs Basin
Caribbean
bathymetry
insular shelf
island slope
bottom sediments
cores
cable breaks

BT
Burkov, V.A., 1963
Some results of oceanographic observation with express methods to the east and south of Japan. (In Russian; English abstract).
Okeanolog. Issled. Rezhult. Issled. po Programme Mezhd. Geofiz. Goda, Mezhd. Geofiz. Kom., pri Presidiume Akad. Nauk, SSSR, No. 9: 32-41.

C

bathymetry
Calmels, Augusto P., y Sergio P. Andreoli, 1968.
Contribución al conocimiento oceanográfico del area exterior de la Ria de Bahia Blanca.
Contrib. Inst. Oceanogr., Univ. Nac. Sur, Bahia Blanca 1: 46 pp. (mimeographed).

bathymetry
Cann, J.R., and F.J. Vine 1966.
An area on the crest of the Carlsberg Ridge: petrology and magnetic survey.
Phil. Trans. R. Soc. (A) 259 (1099): 198-217.

bathymetry
Capronniers, Capt., 1926.
Mission hydrographique de l'Indochinie (1923-1924). Ann. Hydro., Paris, ser. 3, 7:99-116.

bathymetry
Capart, A., 1955.
Quelques échosondages des fonds et des poissons aux environs de Monaco.
Bull. Inst. Océan., Monaco, No. 1068:1-10.

bathymetry
Capart, J.J., et F. Gbesset, 1964.
Expéditions Antarctiques Belges Mission IRIS - 1960, C.N.R.P.B. (1). Rapport de la section océanographique.
Bull. Inst. R. Sci. Nat. Belgique, 40(2):47 pp.

bathymetry (data)
Capurro, L.R.A., 1955.
Expedicion Argentina al Mar de Weddell (diciembre 1954 a enero de 1955). Ministerio de Marina, Argentina, Direccion Gen. de Navegacion e Hidrografia, 184 pp.

bathymetry
Carsola, A.J., 1955.
Bathymetry of the Arctic Basin. J. Geol. 63(3): 274-278, 2 textfigs.

bathymetry
Carsola, A.J., R.L. Fisher, C.J. Shipek and George Shumway, 1961.
Bathymetry of the Beaufort Sea.
In: Geology of the Arctic, Univ. Toronto Press, 678-689.

bathymetry
Caspers, H., 1951.
Quantitative Untersuchungen über die Bodentierwelt des Schwarzen Meeres im bulgarischen Küstenbereich. Arch. f. Hydrobiol. 45(1/2):192 pp., 66 textfigs.

bathymetry
*Cavalcanti, Lourinaldo B., Petronio A. Coelho, Marc Kempf, Jannes M. Mabesoone, e Olimpio C. Da Silva, 1965/1966.
Shelf off Alagoas and Sergipe (northeastern Brazil). 1. Introduction.
Trabh Inst. Oceanogr., Univ. Fed. Pernambuco, Recife, (7/8):137-150.

bathymetry
Chambers, S.W., 1929.
Vertical sections of one thousand meters and over in the northeast Pacific Ocean. Scripps Inst. Ocean. 23 pp. (mimeographed).

bathymetry
Charcot, J.B., 1934.
Rapport préliminaire sur la campagne du "Pourquoi Pas?" en 1933. Ann. Hydrogr., Paris, (3)13:1-85.

bathymetry
*Chase, R.L., and J.B. Hersey, 1968.
Geology of the north slope of the Puerto Rico Trench.
Deep-Sea Research, 15(3):297-317.

basalts
feldspar
pyroxene
olivine
nontronite (saponite)

bathymetry
Chang, Sun-duck, 1969.
The submarine topography of Chinju Bay.
Bull. Pusan Fish. Coll. 9(2): 89-94
(In Korean; English abstract)

bathymetry
Chavanier, C., 1932.
Mission hydrographique de l'Indochine (de 1er juin 1929 au 1er juin 1930). Ann. Hydro., Paris, ser. 3, 11:153-214.

bathymetry
Chemino, W., 1869.
Sondes et températures dans le Gulf Stream.
Ann. Hydrogr., Paris, (2) 32:501-515.

CHILE bathymetry
Republica de Chile, Instituto Hidrografico de la Armada, 1965.
Operación Oceanografica Marchile II: datos fisico-quimicos y batimetria, realizafa por el A.G.S. Yelcho entre el 5 de Julio y el 4 de Agosto de 1962. Unnumbered pp.

bathymetry
*Cholet, J., B. Damotte et G. Grau, J. Debyer et L. Montadert, 1968.
Recherches préliminaires sur la structure géologique de la marge continentale du Golf de Gascogne: commentaires sur quelques profiles de sismique réflexion "Flexotir".
Revue Inst. français Pétrole, 23(9):1029-1045.

bathymetry
Chosen. Fishery Experiment Station, 1938.
Research in the neighboring sea of the Korean Gulf on board the R.M.S. Misago-maru in Feb-Mar, 1933. Ann. Rept. Hydrogr. Obs., Chosen Fish. Exp. Sta., 8(IIA):7-22.

bathymetry
Chosen. Fishery Experiment Station, 1938.
Oceanographic investigations in the Japan Sea made during Aug. and Oct. 1933, on board the R.M.S. Misago-maru. (In Japanese; English headings for tables and charts). Ann. Rept. Hydrogr. Obs., Chosen Fish. Exp. Sta. 8:77-113, tables, charts.

bathymetry
Chosen. Fishery Experiment Station, 1938.
Hydrographic observations during Aug. 1933, on board the R.M.S. Misago-maru in the southwest part of the Japan Sea. (In Japanese; English abstract). Ann. Rept. Hydrogr. Obs., Chosen Fish. Exp. Sta. 8:iii-iv; 67-76.

bathymetry
Chosen, Fishery Experiment Station, 1937.
Results of the soundings in the western part of the Japan Sea. (In Japanese). Bull. Chosen Fish. Exp. Sta. 5:1-44.

bathymetry
Clements, T., and K.O. Emery, 1947
Seismic activity and topography of the sea floor off southern California. Bull. Seis. Soc. Amer., 87:307-313.

bathymetry
Clowes, A.J., 1934.
Hydrology of the Bransfield Strait. Discovery Rept. 9:1-64, 68 textfigs.

bathymetry
Codispoti, L.A., and F.A. Richards 1968.
Micronutrient distributions in the East Siberian Sea and Laptev seas during summer 1963.
Arctic 21 (2): 67-83

bathymetry
Cohen, Philip N., 1959.
Directional echo sounding on hydrographic surveys.
Int. Hydrogr. Rev., 36(1):29-42.

bathymetry
Collier, A., and J.W. Hedgpeth, 1950.
An introduction to the hydrography of tidal waters of Texas. Publ., Inst. Mar. Sci., 1(2):120-198, 32 textfigs.

bathymetry
Collin, A.E., and M.J. Dunbar, 1964
Physical oceanography in Arctic Canada.
In: Oceanography and Marine Biology, Harold Barnes, Editor, George Allen & Unwin, Ltd. 2:45-75.

bathymetry
Colman, J.S., 1933.
The Woods Hole Oceanographic Institution: Sonic soundings and an account of a winter cruise to the Somers Islands. Geogr. Jour. 82(4):326-336, 2 textfigs.

bathymetry
*Conolly, John R., and C.C. von der Borch, 1967. Sedimentation and physiography of the sea floor south of Australia. Sedimentary Geol., 1(2):181-220.

bathymetry
Conseil Permanent International pour l'Exploration de la Mer, 1909. II. Summary of the results of investigations. Rapp. Proc. Verb. 10:24-43, 11 textfigs.

bathymetry
Cook, Kenneth L., 1966. Rift system in the Basin and Range province. In: World rift system, T.N. Irvine, editor, Dept. Mines Techn. Surveys, Geol. Survey, Can., Paper, 66-14:246-277.

bathymetry
Cooke, R.J.S., 1966. Some seismological features of the North Maquarie Ridge. Nature, Lond. 211 (5052): 953-954

bathymetry
Cot, M., 1940. The western Mediterranean Sea. Assoc. d'Ocean., Spec. Art., Publ. Sci., No. 8:54-60.

bathymetry
Couillault, M., 1945. Mission hydrographique d'Indochine (Octobre 1937-Septembre 1938). Ann. Hydrogr., Paris, (3)17:115-169, Fig. 8.

not presently in MBL
not checked.

bathymetry
*Countryman, Kenneth A., and William L. Gsell, 1966. Operations Deep Freeze 63 and 64, summer oceanographic features of the Ross Sea. Tech. Rept., U.S. Naval Oceanogr. Off., Tr-190:193pp. (multilithed).

bathymetry
Crary, A.P., 1955. Bathymetric chart of the Arctic Ocean along the route of T-3 April 1952 to October 1953. Int. Hydrogr. Rev. 23(1):165-174.

Reproduced from Bull. G.S.A. 65:709-772, 2 figs. (1954).

bathymetry
Crary, A.P., 1954. Bathymetric chart of the Arctic Ocean along the route of T-3, April 1953 to October 1953. Bull. G.S.A. 65(7):709-712, 2 textfigs.

bathymetry
Crary, A.P., R.D. Cotell, and J. Oliver, 1952. Geophysical studies in the Beaufort Sea, 1951. Trans. Amer. Geophys. Union 33(2):211-216, 3 textfigs.

bathymetry
Crary, A. P., and Norman Goldstein, 1959. Geophysical studies in the Arctic Ocean. Sci. Studies at Fletcher's Ice Island, T-3, Vol., 1, Air Force Cambridge Research Center, Geophys. Res. Pap. 63:7-38.

bathymetry
Creager, Joe S., and Dean A. McManus, 1965. Pleistocene drainage patterns on the floor of the Chukchi Sea. Marine Geology, Elsevier Publ. Co., 3(4):279-290.

bathymetry
Crean, P.B., 1967. Physical oceanography of Dixon Entrance, British Columbia. Bull. Fish. Res. Bd. Can. 156:1-66.

bathymetry
Crosnier, A., 1964. Fonds de peche le long des cotes de la Republique Federale du Camaroun. Cahiers, O.R.S.T.R.O.M., No. special:133 pp.

abstract

bathymetry
Cullen, David J. 1967. The Antipodes Fracture Zone, a major structure of the south-west Pacific. N.Z. Jl mar. Freshwat. Res. 1(1): 16-25

bathymetry
*Cullen, D.J., 1967. A note on the regional structure of the south-west Pacific. N.Z. Jl Sci., 10(3):813-815.

bathymetry a
Curray, Joseph R., and David G. Moore, 1971. Growth of the Bengal Deep-Sea Fan and denudation in the Himalayas. Bull. geol. Soc. Am. 82(3): 563-572

D

bathymetry
Damiani, L., 1937. Mission hydrographique de l'Indochine (de juin 1933 à juin 1934). Ann. Hydrogr., Paris, Ser. 3; 14:123-158.

bathymetry
Damiani, L., 1928. Mission hydrographique de l'Indochine (janvier 1926, fevrier 1927). Ann. Hydrogr., Paris, Ser. 3, 8:165-186.

bathymetry
D'Anglejean, B.F. 1967. Origin of marine phosphorites off Baja California, Mexico. Marine Geol., 5(1): 15-44.

oxygen
upwelling
phosphorites
Mexico, west
bathymetry
minerals, apatite
[illegible]

bathymetry
Dannstedt, G., 1947. Bottentopografien i Södra Kalmarsund. Geografiska Annaler 29(1/2):1-19, 5 textfigs., with English summary.

bathymetry
Davidson, C., 1924. Distortion of the sea-bed in the Japanese earthquake of 1 September 1923. Geogr. Jour. 63:241-243.

bathymetry
DeBuen, F., 1937. Sur quelques resultats de la IIIème croisière du Garde-Côte "Xauen" dans les eaux de Guipuzea (Espagne) en 1934. Cons. Perm. Int. Expl. Mer, Rapp. Proc. Verb. 104:18-25, 3 textfigs.

bathymetry
Defant, A., 1940. Vorschlag des Instituts für Meereskund, Berlin, betreffs einer Fortführung des "Grundkarte der ozeanischen Lotungen des Atlantischen Ozeans" und Ausdehnung dieser Arbeit auch auf die anderen Ozeane. Assoc. Océanogr. Phys., Proc. Verb., No. 3:106-107.

bathymetry
Defant, A., 1939. Die Altair-Kuppe. Abhandl. Preuss. Akad. Wiss., 1939, Phys.-math. Kl., No. 5:40-45, 1 textfig., 1 fold-in.

bathymetry
Defant, A., 1938. Über die Aufnahme morphologischer Einzelheiten d des Meeresbodens mittels des Echolotes. Geol. Rundschau 30(1/2):121-131, 8 textfigs., 1 fold-in

bathymetry
Defant, A., 1931 Bericht über die Ozeanographischen Untersuchungen des Vermessungsschiffes "Meteor" in der Dänemarkstrasse und in der Irmingersee. Zweiter Bericht. Sitzber. Preuss. Akad. Wiss., Phys. Math. Kl.19: 17 pp., 5 text figs.

bathymetry
Defant, A., G. Böhnecke, H. Wattenberg, 1936. I. Plan und Reiseberichte die Tiefenkarte das Beobachtungsmaterial. Die Ozeanographischen Arbeiten des Vermessungsschiffes "Meteor" in der Dänemarkstrasse und Irmingersee während der Fischereischutzfahrten 1929, 1930, 1933 und 1935. Veroffentlichungen des Instituts für Meereskunde, n.f., A. Geogr.-naturwiss. Reihe, 32:1-152 pp., 7 text figs., 1 plate.

bathymetry
Defant, A., and B. Helland-Hansen, 1939. Bericht über die oceanographischen Untersuchungen im zentralen und östlichen Teil des Nordatlantischen Ozeans im Frühsommer 1938 (Internationale Golfstrom-Expedition). 2. Die Altair-Kuppe. Abhandl. Preuss. Akad. Wiss., Phys.-Math. Kl., 1939(5):40-45, 1 textfig., 1 fold-in.

bathymetry a
Dehlinger, Peter, R.W. Couch, D.A. McManus and Michael Gemperle, 1970. II. Regional observations. 4. Northeast Pacific Structure. In: The sea: ideas and observations on progress in the study of the seas, Arthur E. Maxwell, editor, Wiley-Interscience 4(2): 133-189.

bathymetry
*DeLeeuw, M.M., 1967. New Canadian bathymetric chart of the Western Arctic Ocean, north of 72°. Deep-Sea Res., 14(5):489-504.

bathymetry
DeMiro Orell, Manuel, 1968. Caracteristicas generales de los sedimentos recientes de los fondos marinos de Venezuela. Memoria Soc. Cienc. nat. La Salle, 28(79):97-137.

bathymetry
Den, N., W.J. Ludwig, S. Murauchi, M. Ewing, H. Hotta, T. Asanuma, T. Yoshii, A. Kubotera and K. Hagiwara, 1971. Sediments and structure of the Eauripik-New Guinea Rise. J. geophys. Res., 76(20): 4711-4723.

bathymetry
d'Erceville, I., 1938-1939. Mission hydrographique de l'Indo-Chine, 1936-1937. Ann. Hydrogr., Paris, Ser. 3, 16:79-118, figs.

bathymetry
*Destombes, Marcel, 1968. Les plus anciens sondages portés sur les cartes nautiques aux XVIe et XVIIe Siècles: contribution à l'histoire de l'océanographie. Bull. Inst. océanogr., Monaco, No. special 2: 199-222.

bathymetry
Deutsche Seewarte, Hamburg, 1912. Die grosste Meerestiefe. Ann. Hydrogr., usw., 40: 393.

bathymetry

Dietrich, Gunter, 1964.
Oceanic Polar Front Survey in the North Atlantic.
En: Research in Geophysics. Solid Earth and Interface Phenomena, 2:291-308.

Oceanic Polar Front Survey
Atlantic, North
bottom morphology
bathymetry
Anton Dohrn Seamount
seamounts
Rosemary Bank
temperature
salinity
dynamic topography
pot
oceanic fronts
oxygen
currents

bathymetry

Dietrich, G., 1957.
Schichtung und Zirkulation der Irminger See in Juni 1955. Ber. Deutsch. Wiss. Komm. Meeresf., n.f., 14(4):255-312.

bathymetry

Dietrich, G., 1939
IV. Einige morphologische Ergebnisse der "Meteor"-Fahrt Januar bis Mai 1938. Ann. Hydr. u. Mar. Meteorol., 1939:20-23, 2 textfigs.

bathymetry

Dietrich, Günter, und Johannes Ulrich, 1961
Zur Topographie der Anton-Dohrn-Kuppe.
Kieler Meeresf., 17(1):3-7.

bathymetry

Dietz, R.S., 1954.
Marine geology of northwestern Pacific: Description of Japanese Bathymetric Chart 6901.
Bull. G.S.A. 65(1):1199-1224, 6 textfigs.

bathymetry

Dietz, R.S., A.J. Carsola, and E.C. Lafond, 1948.
Topography and sediments of the Bering and Chukchi seas. (Abstract). Bull. G.S.A. 59:1318.

bathymetry

Dietz, R.S., and M.F. Sheehy, 1953.
TransPacific detection by underwater sound of Myojin volcanic explosions.
J. Ocean. Soc., Japan, 9(2):53-83, 6 textfigs.

Also published:
Bull. G.S.A. 65(10):941-956, 4 pls., 4 textfigs.
in 1954

bathymetry

Dieuzeide, D.R., 1951.
Résultats de quelques sondages dans la baie de Castiglione. Trav. C.O.E.C., Algerie,
Bull. Aquic. et Pêches, Castiglione, n.s., No. 3: 139-158, 6 pls.

bathymetry

Dimmler, R., 1958
Das Beaufort-Becken im Nordpolarmeer.
Deutsche Hydrogr. Zeits. 11(2): 86-88.

bathymetry

Dishon, Menachem, and Bruce C. Heezen, 1967.
Computerized library of deep-sea soundings.
Nature, 215(5109):1439-1441.

bathymetry

Dodimead, A.J., F. Favorite and T. Hirano, 1964.
Review of oceanography of the subarctic Pacific Region. Salmon of the North Pacific Ocean. II.
Collected Reprints, Tokai Reg. Fish. Res. Lab., No. 2:187 pp.
(In Japanese).

heat budget
oxygen
thermocline (interesting diagrams)
halocline

bathymetry

Duffo, Henri, 1965.
Hydrographic soundings in surf areas.
Int. Hydrogr. Rev., 42(2):47-57.

bathymetry

Düing, Walter, 1965.
Strömungsverhältnisse im Golf von Neapel.
Pubbl. Staz. Zool., Napoli, 34:256-316.

bathymetry

Dunbar, M.J., 1966.
The sea waters surrounding the Quebec-Labrador peninsula.
Cahiers Geogr., Quebec, 10(19):13-35.

bathymetry

Dunbar, M.J., 1958.
Physical oceanographic results of the "Calanus" Expeditions in Ungava Bay, Frobisher Bay, Cumberland Sound, Hudson Strait and Northern Hudson Bay, 1949-1955.
J. Fish. Res. Bd., Canada, 15(2):155-201.

bathymetry

Dunbar, M.J., 1955.
Arctic and subarctic marine ecology.
Arctic 7(3/4):213-228.

bathymetry

Dunbar, M.J., 1951.
Eastern Arctic waters. Fish. Res. Bd., Canada, Bull. No. 88:131 pp., 32 textfigs.

bathymetry

Duncan, A.R. 1970.
Petrology of rock samples from seamounts near White Island, Bay of Plenty.
N.Z. Jl Geol. Geophys. 13(3): 690-696

bathymetry

Dybern, Bernt I. 1967.
Topography and hydrography of Kvitudvikpollen and Vågsböpollen on the west coast of Norway.
Sarsia 30:1-27.

E

bathymetry

Ellison, J. J., D. E. Powell, and H. H. Hildebrand, 1950
Exploratory fishing expedition to the northern Bering Sea in June and July 1949.
Fishery Leaflet 369: 56 pp., 23 figs. (multilith).

bathymetry

Emery, K.O., 1953.
A newly surveyed submarine basin off Mexico.
Am. J. Sci. 251(9):656-660, 2 textfigs.

bathymetry

Emery, K. O., 1950.
A deep gathogram across the North Atlantic Ocean. Am. J. Sci. 248(2):100-106, 8 figs.

bathymetry

Emery, K. O., 1949.
Topography and sediments of the Arctic basin.
J. Geol. 57(5):512-521, 1 map.

bathymetry

Emery, K.O., Bruce C. Heezen and T.D. Allen, 1966.
Bathymetry of the eastern Mediterranean Sea.
Deep-Sea Res., 13(2):173-192.

Bathymetry

Emery, K. O., and J. Hülsemann, 1961 (1962).
The relationships of sediments, life and water in a marine basin.
Deep-Sea Res., 8(3/4):165-180.

bathymetry

Emery, K.O., and David A. Ross 1968.
Topography and sediments of a small area of the continental slope south of Martha's Vineyard.
Deep-Sea Res. 15(4): 415-422.

bathymetry

Emilsson, Inquar, A. Garcia Occhipinti, A.S. Kutner, I.C. Miniussi e M. Vannucci, 1963.
Levantamento oceanográfico-meteorológico da enseada do Mar Virado Ubatuba, Estado de São Paulo.
Contrib. Avulsas, Inst. Oceanogr., Univ. São Paulo, Oceanogr. Fisica, No. 5:116 pp.

S. America, east
bottom sediments (data)
salinity (data)
temperature (data)
zooplankton, lists of spp.
pollution, atmospheric

summery p 5-7, p.52

bathymetry

Ermel, Hans, 1966.
Der deutsche Beitrag zur Neuherstellung der General Bathymetric Chart of the Oceans (GEBCO).
Deutsche Hydrogr. Zeits., 19(2):49-57.

bathymetry

Evgenov, N.I., 1935.
High latitude expedition on the ice-breaker Sadko. Biull. Arktich. Inst., Leningrad, 5(10): 322-328; 355-360.

bathymetry

Ewing, J.I., N.T. Edgar and J.W. Antoine, 1970.
II. Regional observations. 10. Structure of the Gulf of Mexico and Caribbean Sea. (Dec. 1968)
In: The sea: ideas and observations on progress in the study of the seas, Arthur E. Maxwell, editor, Wiley-Interscience 4(2): 321-358.

bathymetry

Ewing, John, and Maurice Ewing, 1962
Reflection profiling in and around the Puerto Rico Trench.
J. Geophys. Res., 67(12):4729-4740.

bathymetry

Ewing, Maurice and John Ewing, 1965.
The sediments of the Argentine Basin.
Anais Acad. bras. Cienc., 37(Supl):31-61.

bathymetry

Ewing, Maurice, William J. Ludwig and John Ewing, 1965.
Oceanic structural history of the Bering Sea.
J. geophys. Res., 70(18):4593-4600.

bathymetry

Ewing, Maurice, William J. Ludwig and John I. Ewing, 1964
Sediment distribution in the oceans: The Argentine Basin.
J. Geophys. Res., 69(10):2003-2032.

bathymetry

Ewing, Maurice, Xavier Le Pichon and John Ewing, 1966.
Crustal structure of the mid-ocean ridges.
4. Sediment distribution in the South Atlantic Ocean and the Cenozoic history of the Mid-Atlantic Ridge.
J. Geophys. Res., 71(6):1611-1636.

F

bathymetry
Fail, J.P., L. Montadert, J.R. Delteil, P. Valery, Ph. Patriat et R. Schlich, 1970.
Prolongation des zones de fractures de l'Océan Atlantique dans le Golfe de Guinée.
Earth Planet. Sci. Letts 7(5): 413-419.

bathymetry
Fairbridge, R.W., 1955.
Some bathymetric and geotectonic features of the eastern part of the Indian Ocean. Deep-Sea Res. 2(3):161-171.

bathymetry
Fairbridge, R.W., & H.B. Stewart, Jr., 1960
Alexa Bank, a drowned atoll on the Melanesian Border Plateau.
Deep-Sea Res., 7(2): 100-116.

bathymetry
Fairhead, J.D. and R.W. Girdler, 1970.
The seismicity of the Red Sea, Gulf of Aden and Afar triangle.
Phil. Trans. R. Soc. (A) 267 (1181): 49-74

Gulf of Aden earthquakes
Owen fracture zone

bathymetry
Farenholz, I.V., 1963
A multicoloured map of the world. (In Russian)
Okeanologiia, Akad. Nauk, SSSR, 3(3):565.

bathymetry
Farquharson, W. I., 1936.
Topography with an appendix on magnetic observations. John Murray Exped., 1933-34, Sci. Repts. 1(2):43-61, 6 pls., 6 charts.

bathymetry
Favorite, Felix, 1967.
The Alaskan Stream.
Bull. Int. North Pac. Fish. Comm., 21:1-20.

temperature
currents
currents, dynamic
transport

bathymetry
Fierro, G. 1966.
Profils échographiques dans les bouches de Bonifacio et dans le Golfe de l'Asinara. (Abstract only).
Rapp. P.-v. Réun. Comm. int. Explor. scient. Mer Méditerr. 18 (3): 931.

bathymetry
Fierro, Giuliano, 1965.
Observations morphologiques et sédimentologiques sur les bouches de Bonifacio et le Golfe de l'Asinara.
Cahiers Océanogr., C.C.O.E.C., 17(8):565-571.

sedimentation
"Bannock"
Mediterranean, west
bathymetry
submarine canyons

conclusions, p. 570.

bathymetry
Fishel, Nathan, 1962
Hydrographic survey of Melville Bay (West Greenland) in 1959 using two range Radist.
Int. Hydrogr. Rev., Suppl., 3:15-27.

bathymetry
Fisher, Robert L., 1966.
The median ridge in the south central Indian Ocean.
In: The world rift system, T.N. Irvine, editor, Dept. Mines Techn. Surveys, Geol. Survey, Can., Paper, 66-14:135-147.

bathymetry
Fisher, Robert L., 1961.
Middle America Trench: topography and structure.
Bull. Geol. Soc., Amer., 72(5):703-720.

bathyme try
Fisher, R.L., A.J. Carsola and G. Shumway, 1958.
Deep-Sea bathymetry north of Point Barrow.
Deep-Sea Res., 5(1):1-6.

bathymetry
Fisher, R.L., G.L. Johnson and B.C. Heezen 1967.
Mascarene Plateau, western Indian Ocean.
Bull. Geol. Soc. Am. 78: 1247-1266

bathymetry
Fisher, Robert L. and Robert M. Norris, 1960
Bathymetry and geology of Sala y Gomez, Southeast Pacific. Bull. Geol. Soc. Amer., 71(4): 497-502.

bathymetry
Fisher, Robert L., and Russell W. Raitt, 1962
Topography and structure of the Peru-Chile trench.
Deep-Sea Res., 9(5):423-443.

bathymetry
Fleming, C.A., and J.J. Reed, 1951.
Mernoo Bank, east of Canterbury, New Zealand.
N.Z.J. Sci. Tech., B, 32(6):17-30.

bathymetry
Fofonoff, N. P. and S. Tabata, 1966.
Variability of oceanographic conditions between Ocean Station P and Swiftsure Bank off the Pacific coast of Canada.
J. Fish. Res. Bd., Canada, 23(6):825-868.

bathymetry
Fomin, L.M. 1969.
An assessment of the bottom relief ruggedness. (In Russian; English abstract).
Okeanologiia 9(4): 667-675

bathymetry
France, Comité Central d'Océanographie et d'Études des Côtes, 1953.
Les recents travaux bathymetriques français dans l'Océan Antarctique. Bull. d'Info. 5(2):61-66 (mimeographed), 3 pls. (multilithed).

bathythermograms
France, Service Hydrographique de la Marine, 1963
Resultats d'observations.
Cahiers Océan., 15(10):738-758.

bathymetry
Francis, T.J.G., 1962
Black Mud Canyon.
Deep-Sea Res., 9(5):957-964.

bathymetry
Franco, Paolo 1970.
Oceanography of northern Adriatic Sea. 1. Hydrologic features: cruises July-August and October-November 1965.
Archo Oceanogr. Limnol. 16 (Suppl.): 1-93.

bathymetry
Frassetto, R., and J. Northrop, 1957.
Virgin Islands bathymetric survey. Deep-Sea Res. 4(2):138-146.

G

bathymetry
Gade, Herman G., 1961.
On some oceanographic observations in the southeastern Caribbean Sea and adjacent Atlantic Ocean with special reference to the influence of the Orinoco River.
Bol. Inst. Oceanogr., Univ. de Oriente, Venezuela, 1(2):287-342.

bathymetry
Gardner, James V. 1970.
Submarine ecology of the western Coral Sea.
Bull. geol. Soc. Am. 81(9): 2599-2614.

a
bathymetry
Garner, D.M. 1970.
Hydrological studies in the New Zealand region 1966 and 1967: oceanic hydrology north-west of New Zealand, hydrology of the north-east Tasman Sea.
Bull. N.Z. Dept. scient. industr. Res. 202: 7-49.
Also: Mem. N.Z. oceanogr. Inst. 58.

bathymetry
Garner, D.M., 1962.
Analysis of hydrological observations in the New England region, 1874-1955.
New Zealand Oceanogr. Inst. Mem., No. 9:45 pp.

Also:
New Zealand, Dept. Sci. Ind. Res., Bull. 144.

bathymetry
Gass, I.G. 1967
Geochronology of the Tristan da Cunha group of islands
Geol. Mag., 104(2): 160-170

Tristan da Cunha
bathymetry

bathymetry
Geffrier, C.de, et J. Milliau, 1938-1939.
Mission hydrographique de la Guinée Française, de la Mauritanie et du Sénégal, 1936-1937.
Ann. Hydrogr., Paris, (3)16:119-183, figs.

bathymetry
Germany, Oberkommando der Kriegsmarine, 1940.
Die Naturverhaltnisse des sibirischen Seeweges.
169 pp.

bathymetry
Gershanovich, D.E., 1963
Bottom relief of the fishing grounds and some features of the Bering Sea geomorphology. (In Russian).
Sovetsk. Ribokh. Issled. B Severo-Vostokh. Chasti Tikhogo Okeana, VNIRO 48, TINRO 50(1): 13-76.

bathymetry
Gershanovich, D.E., 1962.
New data on Recent sediments of the Bering Sea.
Trudy Vses. Nauchno-Issledov. Inst. Morsk. Ribn. Khoz. i Okean., VNIRO, 46:128-164.

In Russian

bathymetry
Gershanovich, D.E., 1962.
Relief and Recent sediments of the Bering Sea shelf. (In Russian).
Trudy Vses. Nauchno-Issledov. Inst. Morsk. Ribn. Khoz. i Okean., VNIRO, 46:164-185.

bathymetry
Gibson, William M., 1960
Submarine topography in the Gulf of Alaska.
Bull. Geol. Soc. Amer., 71(7): 1087-1108.

bathymetry
Giermann, Günter, 1961.
Erläuterungen zur bathymetrischen Karte der Strasse von Gibraltar.
Bull. Inst. Océanogr., Monaco, 58(1218A,B):1-28, 5 charts.

bathymetry
Giorgetti, F., and F. Mosetti, 1969.
General morphology of the Adriatic Sea.
Boll. Geofis. teor. appl. 11 (42-43): 49-56
submarine geology
bathymetry

bathymetry
Glangeaud, L., R. Schlich, G. Pautot, G. Bellaiche, P. Patriat et M. Ronfard, 1965.
Morphologie, tectonophysique et évolution géodynamique de la bordure sous-marine des Maures et de l'Esterel. Relations avec les régions voisines.
Bull. Soc. géol. Fr., (7), 7:998-1009.
submarine canyons
magnetic anomalies

bathymetry
Godo, T., and M. Kuroiwa, 1955.
On the exploration of the Naruto Straits.
Rec. Ocean. Wks., Japan, 2(3):63-70, 8 textfigs.

bathymetry
Gödecke, E., 1940.
Die heutige Kenntnis von der Morphologie und Hydrographie der Helgoländer Tiefen Rinne im Zusammenhang mit entsprechenden Verhältnissen der südöstlichen Deutschen Bucht. Ann. Hydr. usw. 68:405-421, Pls. 30-36.

bathymetry
Goncharov, V.P., and O.V. Mikhailov, 1963
New data on the floor topography in the Mediterranean Sea. (In Russian).
Okeanologiia, Akad. Nauk, SSSR, 3(6):1056-1060.

bathymetry
Goodacre, A.K., 1964.
A shipborne gravimeter testing range near Halifax, Nova Scotia.
J. Geophys. Res., 69(24):5373-5381.

bathymetry
Gordienko, P.A., 1961
The Arctic Ocean.
Scientific American, 204(5): 88-102.

bathymetry
Gorshkova, T.I., 1965.
Carbonates in the sediments of the Norwegian-Greenland Sea basin as indicators of the distribution of water masses. (In Russian).
Trudy vses. nauchno-issled. Inst. morsk. ryb. Khoz. Okeanogr. (VNIRO), 57:297-312.

bathymetry
Gorshkova, T.I., 1965.
Sediments of the Norwegian Sea. (In Russian; English abstract).
Okeanolog. Issled., Rez. Issled. po Programme Mezhd. Geofiz. Goda, Mezhd. Geofiz. Komitet, Prezidiume, Akad. Nauk, SSSR, No.13:212-224.

bathymetry
Gorshkova, T.I., 1965.
Carbonates in the sediments of the Norwegian-Greenland Sea basin as indicators of the distribution of water masses. Investigations in line with the programme of the International Geophysical Year, 2. (In Russian).
Trudy, Vses. Nauchno-Issled. Inst. Morsk. Ribn. Choz. i Okeanogr. (VNIRO), 57:297-312.

bathymetry
Gorsline, Donn S., 1963.
Oceanography of Apalachicola Bay, Florida.
In: Essays in Marine Biology in honor of K.O. Emery, Thomas Clements, Editor, Univ. Southern California Press, 69-96.

bathymetry
Graham, H.W., 1952.
A contribution to the oceanography of the Sulu Sea. Proc. Seventh Pacific Sci. Congr., Met. Ocean., 3:225-266, 29 textfigs.

bathymetry
#Grant, A.C., 1968.
Seismic-profiler investigation of the continental margin northeast of Newfoundland.
Can.J. earth Sci., 5(5):1187-1198.

bathymetry
Grant, A.C., 1966.
A continuous seismic profile on the continental shelf off NE Labrador.
Canadian J. Earth Sci., 3(5):725-730.

bathymetry
Grant, A.C., and H.S. Manchester 1970.
Geophysical investigations in the Ungava Bay-Hudson Strait region of northern Canada.
Can. J. Earth Sci. 7(4): 1062-1076.

bathymetry
Great Britain, Hydraulics Research Station, Wallingford, 1963.
Karnafuli River, East Pakistan.
Notes, No. 5: 2 pp.

bathymetry
Greece, National Hellenic Oceanographic Society, 1957.
The great depths of the oceans and the geological pattern of their floor. Marine Pages, No. 5: 67-85.

bathymetry
Greve, Sv., 1938.
Echo soundings: an analysis of results. Dana Rept. No. 14: 25 pp., 4 textfigs., 19 pls.

bathymetry
Grim, Paul J., 1970.
Computer program for automatic plotting of bathymetric and magnetic anomaly profiles.
ESSA Techn. Memo, ERLTM-AOML8: 31 pp. (mimeographed)
computer programs
magnetic anomalies
bathymetry

bathymetry
Grousson, R., 1957.
Bathymetrie de la côte atlantique du Maroc.
Bull. d'Info., C.C.O.E.C., 9(8):411-412, 1 pl.

bathymetry (data)
Gudkovich, Z.M., 1954/55.
Depth soundings. Material. Nabluid. Nauch-Issledov. Dreifuius, 1950/51, Morskoi Transport, 1: pages not given
E.R. Hope, translator, AMS-Astia Doc. 117132.
reference room

bathymetry
Guilcher, A., 1950
Reconnaissance du Banc Lawson. Bull. d'Info. 2eme Ann., No. 7: 253-254, 1 pl. (mimeographed)

bathymetry
Guilcher, A., 1950
Leve du banc de la schôle. Bull. d'Info., 2nd Annee, No. 7: 251-252, 2 pls. (mimeographed)

H

bathymetry
Hachey, H.B., L. Lauzier and W.B. Bailey, 1956.
Oceanographic features of submarine topography.
Trans. R. Soc., Canada, (3)50:67-81.

bathymetry
Hadley, M.L., 1964
The continental margin southwest of the English Channel.
Deep-Sea Res., 11(5):767-779.

bathymetry
Hall, G.P.D., 1954.
Survey of a newly discovered feature (Genista Bank off the Arabian coast). Deep-Sea Res. 2(1): 80-83, 3 textfigs.

bathymetry
Harrison, J.C., and R.E. von Huene, 1965.
The surface-ship gravity meter as a tool for exploring the geological structure of continental shelves.
Ocean Sci. and Ocean Eng., Mar. Techn. Soc.,-Amer. Soc. Limnol. Oceanogr., 1:414-431.

bathymetry
#Harrison, W., J.J.Norcross, N.A.Pore and E.M. Stanley, 1967.
Circulation of shelf waters off the Chesapeake Bight: surface and bottom drift of continental shelf waters between Cape Henlopen, Delaware and Cape Hatteras, North Carolina, June 1963-December 1964.
ESSA Prof.Pap.U.S.Dept.Comm.3:1-82.

bathymetry
Hart, T. John, and Ronald I. Currie, 1960
The Benguela Current.
Discovery Repts., 31: 123-298.

bathymetry
Harvey, John, 1965.
The topography of the South-Western Faroe Channel.
Deep-Sea Res., 12(2):121-127.

bathymetry
Hashimoto, Tomiju, and Minoru Nishimura, 1959.
Reliability of bottom topography obtained by ultrasonic echo-sounding.
Int. Hydrogr. Rev., 36(1):43-50.

bathymetry
Hatori, Tokutaro, 1966
Vertical displacement in a tsunami source area and the topography of the sea bottom.
Bull. Earthq. Res. Inst., Univ. Tokyo, 44 (4): 449-1464.

bathymetry
Hawkins, James W., Edwin C. Allison and Doug Macdougall, 1971.
Volcanic petrology and geologic history of Northeast Bank, Southern California Borderland.
Bull. geol. Soc. Am. 82(1): 219-228.

bathymetry

Hawort, R.T., 1971.
A geophysical profile in the south-east Pacific.
Earth planet. Sci. Letts 11(2):83-89.

Hudson Deep magnetism
bathymetry

bathymetry

Hayes, Dennis E., 1966.
The Peru-Chile Trench. (Summary only).
In: Continental margins and island arcs, W.H. Poole, editor, Geol. Surv. Pap. Can., 66-15:238-242.

bathymetry

Hayes, Dennis E., 1966.
A geophysical investigations of the Peru-Chile Trench.
Marine Geol., 4(5):309-351.

bathymetry (Aransas Bay)

Hedgpeth, J.W., 1953.
An introduction to the zoogeography of the northwestern Gulf of Mexico with reference to the invertebrate fauna. Publ. Inst. Mar. Sci. 3(1): 111-224, 46 textfigs.

Bathymetry

Heezen, B.C., E.T. Bunce, J.B. Hersey and M. Tharp, 1964.
Chain and Romanche Fracture zones.
Deep-Sea Research, 11(1):11-33.

WHOI Contrib. No. 1424.

bathymetry

Heezen, Bruce C., Maurice Ewing and G. Leonard Johnson, 1966.
The Gulf of Corinth floor.
Deep-Sea Research, 13(3):381-411.

Gulf of Corinth
Mediterranean, east
bathymetry
turbidity currents

abstract

bathymetry

Heezen, Bruce C., Bill Glass and H. W. Menard, 1966.
The Manihiki Plateau.
Deep-Sea Research, 13(3):445-458.

Manihiki Plateau
Pacific, equatorial (west)
plateau
bathymetry
reflection, seismic
cores
underwater photography
bottom sediments
currents, effect of
escarpments

bathymetry

Heezen, Bruce C., and G. Leonard Johnson, 1969
Mediterranean undercurrent and microphysiography west of Gibraltar.
Bull. Inst. océanogr., Monaco, 67 (1380): 95 pp.

Baggy Pedis
Gulf
bottom topography
bottom sediments
undercurrents
Strait of Gibraltar
current measurements
p. 50

bathymetry

Heezen, Bruce C., and G. Leonard Johnson, 1965.
The South Sandwich Trench.
Deep-Sea Research, 12(2):185-197.

bathymetry

Heezen, B.C. and A.C. Pimm, 1971.
Underway observations. Initial Repts. Deep Sea Drilling Project, Glomar Challenger, 6: 691-708.

bathymetry

Helland, Palmar, 1963.
Temperature and salinity variations in the upper layers at Ocean Weather Station M (66° N, 2° E).
Acta Univ. Bergensis, Ser. Mathematica Rerumque Naturalium, Arbok, Univ. Bergen, Nat.-Naturv. Serie, 1963, No. 16:26 pp.

Also in:
Collected Papers, Weather Ship Station M, 66° N, 2° E, Geophys. Inst., Univ. Bergen, 1963.

bathymetry(data)

Henjes, F., 1910.
Ein Beitrag zur Morphographie des Meeresbodens im südwestlichen Pazifischen Ozean.
Arch. Deutschen Seewarte 32(3):1-42, 5 pls., 6 textfigs.

bathymetry

Herdman, H.F.P., 1957.
Recent bathymetric charts and maps of the Southern Ocean and water around Antarctica.
Deep-Sea Res., 4(2):130-137.

bathymetry

Herdman, H.F.P., 1948.
Soundings taken during the Discovery investigations, 1932-39. Discovery Repts., 25:41-106, Pls. 23-31.

bathymetry

Hersey, J.B., 1965.
Sedimentary basins of the Mediterranean Sea.
In: Submarine Geology and Geophysics, Colston Papers, W.F. Whittard and R. Bradshaw, editors, Butterworth's, London, 75-89. 1965

Mediterranean
bathymetry
reflection, seismic

abstract

bathymetry

Hess, H.H., 1966.
Caribbean Research Project, 1965, and bathymetric chart.
Geol. Soc. Amer. Mem., 98:1-10.

bathymetry

Hess, H.H., 1948
Major structural features of the western North Pacific: an interpretation of H.O. 5485, Bathymetric chart, Korea to New Guinea. Bull. G.S.A., 59(5):417-466, H.O. 5485, 9 text figs. fold-in.

bathymetry

Hess, H.H., 1946
Drowned ancient islands of the Pacific basin
Trans. Am. Geophys. Un. 27(6):875-

bathymetry

Hess, H. H., and M. W. Buell, jr., 1950.
The greatest depth in the oceans. Trans. Am. Geophys. Union 31(3):401-405, 8 textfigs.

bathymetry

*Hilde, Thomas W.C., John M. Wageman and Willis T. Hammond, 1969.
The structure of Tosa Terrace and Nankai Trough off southeastern Japan.
Deep-Sea Res., 16(1):67-75.

bathymetry

Hinschberger, Félix, 1964.
La répartition des fonds sousmarins dans le vestibule du Goulet de Brest.
C.R. Acad. Sci., Paris, 258(26):6497-6499.

bathymetry

*Hinschberger, Félix, et Jean-Pierre Corlay, 1967.
Reconstitution d'un réseau hydrographique immergé autour d'Ouessant et du plateau de Molène (Finistère).
Norois, 14(56):569-584.

bathymetry

Hirota, Reiichiro, 1961
Zooplankton investigations in the Bingo-Nada region of the Setonaikai (Inland Sea of Japan).
J. Sci. Hiroshima Univ., (B) Div. 1, 20: 83-145.
Also in:
Contrib. Mukaishima Mar. Biol. Sta., Hiroshima Univ., No. 67.

bathymetry

Hoang Quoc Truong, 1961
Preliminary plankton research in the bay of Nhatrang, Vietnam.
Ann. Fac. Sci., Saigon, 1961:91-100.

Also:
Contrib., Inst. Oceanogr., Nhatrang, No. 51.

bathymetry

Holtedahl, O., 1940.
The submarine relief off the Norwegian coast with bathymetrical map in seven sheets of the Norwegian coastal waters and adjoining seas.
Norske Videnskaps Akad. i Oslo:43 pp., 7 textfigs., 6 pls. (This apparently put out separately and not in any series of the academy???)

bathymetry

*Hoskin, Hartley, 1967.
Seismic reflection observations on the Atlantic continental shelf, slope, and rise southeast of New England.
J. Geol., 75(5):598-611.

WHOI Contrib. No. 1800.

bathymetry

Houbolt, J.J.H.C., 1957.
Surface sediments of the Persian Gulf near the Qatar Peninsula. Mouton & Co., Den Haag, 113 pp. 31 figs., fold-ins.

bathymetry

Houtman, Th. J., 1965.
Winter hydrological condition of coastal waters south of Kaikoura Peninsula.
N.Z.J. geol. geophys. 8(5):807-819.

bathymetry

Houtz, R.E., 1962.
The 1953 Suva earthquake and tsunami.
Bull. Seismol. Soc., Amer., 52(1):1-12.

bathymetry

*Houtz, R., J. Ewing, M. Ewing and A.G. Lonardi, 1967.
Seismic reflection profiles of the New Zealand Plateau.
J. geophys. Res., 72(18):4713-4729.

bathymetry

*Houtz, R. and R. Meijer, 1970.
Structure of the Ross Sea shelf from profiler data.
J. geophys. Res., 75(32):6592-6597.

bathymetry

Huang, Tunyow, 1971.
Foraminiferal trends in the surface sediments of Taiwan Strait.
Techn. Bull. Econ. Comm. Asia Far East, 4:23-61

bathymetry

Hughes, A. J., 1950.
Influence of echo sounding on hydrography. Hydr. Rev. 27(1):29-39.

bathymetry

Hunkins, Kenneth, 1966.
The Arctic continental shelf north of Alaska.
In: Continental margins and island arcs, W.H. Poole, editor, Geol. Surv. Pap. Can., 66-15:197-205.

bathymetry
Hunt, John M., Earl E. Hays, Egon T. Degens and David A. Ross, 1967.
Red Sea: detailed survey of hot-brine areas.
Science, 156(3774):514-516.

cores
rift valleys
hot-brines
high salinity
iron oxides
bottom sediments, chemistry
gravity anomalies

bathymetry
Hurley, Robert J., 1964.
Bathymetric data from the search for USS "Thresher".
Int. Hydrogr. Rev., 41(2):43-52.

bathymetry
Hurley, Robert J., 1964.
Bathymetry of the Straits of Florida and Bahama Islands. III. Southern Straits of Florida.
Bull. Mar. Sci., Gulf & Caribbean, 14(3):373-380.

bathymetry
Hurley, Robert J., 1964.
Analysis of bathymetric data from the search for U.S.S. Thresher. (Abstract).
Geol. Soc., Amer., Special Paper, No. 76:85-86.

bathymetry
Hurley, R.J., V.B. Siegler and L.K. Fink, Jr., 1962.
Bathymetry of the Straits of Florida and the Bahama Islands. I. Northern Straits of Florida.
Bull. Mar. Sci., Gulf and Caribbean, 12(3):313-321.

bathymetry
Husby, David M., 1969.
Oceanographic observations North Pacific Ocean Station November, 30°00'N, 140°00'W, March 1967-March 1968.
Oceanogr. Rept. U.S. Coast Guard 26 (CG 373-26): 217 pp.

I

bathymetry
Ichie, T., K. Tanioka, and H. Yoshizawa, 1949.
On the tidal currents and other hydrological conditions at Matsuyama Port. Papers and Repts., Ocean., Kobe Mar. Obs., Ocean. Dept., No. 1:15 pp 15 figs., followed by tables of data. (Odd atlas-sized pages - mimeographed).

bathymetry
Il'in, A.V., 1960.
Geomorphological investigations in the North Atlantic on the Research Vessel "Mikhail Lomonosov". (In Russian).
Trudy, Morsk. Gidrofiz. Inst., 19:116-

Translation:
Scripta Tecnica for AGU, 1963. Soviet Oceanogr. Transactions, Issue, No. 3:85-98.

bathymetry
Inderbitzen, Anton L., 1965.
An investigation of submarine slope stability.
Ocean Sci. and Ocean Eng., Mar. Techn. Soc., Amer. Soc. Limnol. Oceanogr., 2:1309-1344.

bathymetry
Inman, D.L., 1950.
Submarine topography and sedimentation in the vicinity of Mugu Submarine Canyon, California.
B.E.B. Tech. Memo, No. 19:45 pp., 10 figs.

S.I.O. Sub. Geol. Rept., No. 10:
S.I.O. Contrib. No. 473.

bathymetry
Irving, E.M., 1951.
Submarine morphology of the philippine Archipelago and its geologic significance. Philippine J. Sci. 80(1):55-88, 2 pls.

bathymetry
Isezaki, Nobuhiro, Katsumi Hata and Seiya Uyeda 1971.
Magnetic survey of the Japan Sea (I.)
Bull. earthq. Res. Inst. Tokyo Univ. 49(1/3): 77-83.

J

bathymetry
Jacoby, G., 1956.
Über die Entwicklung des Helgoländer Gebietes und die Karten von Johannes Mejer.
Deutsches Hydrogr. Zeits., 9(4):165-174.

bathymetry
Jakleln, A., 1936
Oceanographic investigations in East Greenland waters in the summers of 1930-1932. Skr. om Svalbard og Ishavet No. 67:79 pp., Pl. 2, 28 text figs.

bathymetry
Japan, Japanese Oceanographic Data Center 1970.
CSK Atlas 4. Winter 1967: 32 pp.

bathymetry
Japan, Kobe Marine Observatory, 1951
Meteorology and oceanography in the Seto Inland Sea.
Bull. Kobe Mar. Obs., No. 161: 211 pp.

bathymetry
Japan, Maritime Safety Board, 1958.
[On the depth in the mud zone]
Hydrogr. Bull. (Publ. 981), No. 55:54-72, pls.

bathymetry
Japan, Maritime Safety Board, 1958.
[On the bottom configuration and sediments in the adjacent sea of Mogami Bank, Japan Sea.]
Hydrogr. Bull., Publ. (981):55:37-53.

bathymetry
Japan, Science Council, National Committee for IIOE, 1966.
General report of the participation of Japan in the International Indian Ocean Expedition.
Rec. Oceanogr. Wks., Japan, n.s., 8(2):133 pp.

International Indian Ocean Expedition
bathymetry
Indian Ocean
geomagnetism
gravity anomaly
temperature
salinity
mixed layer
salinity minimum layer
currents
drift bottles
T-S
(over)

bathymetry
Japan, Tokai Regional Fisheries Research Laboratory, 1952.
Report oceanographical investigation (Jan.-Dec. 1950, collected), No. 74:143 pp.

bathymetry
Johnson, G. Leonard, AH. W. Closuit and James A. Pew 1969.
Geologic and geophysical observations in the northern Labrador Sea.
Arctic 22(1): 56-68.

bathymetry
Johnson, G. Leonard, and Oscar B. Eckhoff, 1966.
Bathymetry of the north Greenland Sea.
Deep-Sea Res., 13(6):1161-1173.

bathymetry
Johnson, G.L., J.S. Freitag and J.A. Pew, 1970.
Structure of the Norwegian Basin
Arbok norsk Polarinst. 1969: 7-16.

bathymetry
Johnson, G. Leonard, and Donald B. Milligan, 1967
Some geomorphological observations in the Kara Sea.
Deep-Sea Research, 14(1):19-28.

bathymetry
Jones, Bennett G., and William Shofnos, 1961.
Mapping the low-water line of the Mississippi delta.
Int. Hydrogr. Rev., 38(1):63-76.

bathymetry
Jordan, G.F., R.J. Malloy, and J.E. Kofoed, 1964.
Bathymetry and geology of Pourtales Terrace, Florida.
Marine Geology, 1(3):259-287.

bathymetry
Jordan, J.F., & H.B. Stewart, Jr., 1961
Submarine topography of the western Straits of Florida.
Bull. Geol. Soc. Amer., 72(7): 1051-1058.

bathymetry
*Joseph, J., 1967.
The bottom topgraphy of the Iceland-Faroe Ridge region.
Rapp. P.-V. Réun.Cons.perm.int.Explor. Mer, 157:16-17.

bathymetry (Baltic)
Jurva, R., 1952.
Seas. Fennia 72:136-160.

general

bathymetry
Just, Jean, 1970.
Marine biological investigations of Jørgen Brønlund Fjord, North Greenland: physiographical and bathygraphical survey, methods and lists of stations.
Medd. Grønland, 184(5): 42 pp

K

bathymetry
Kagami, Hideo, Noriyuki Nasu and Masuoki Horikoshi, 1962
Submarine topography at the entrance of Tokyo Bay (Co-operative survey at the entrance of Tokyo Bay in 1959 - Part 2). (In Japanese; English abstract).
J. Oceanogr. Soc., Japan, 20th Ann. Vol., 82-89.

bathymetry
Kanayev, V.F., 1960
[New data about the bottom relief of the western part of the Pacific Ocean.] Okean. Issle., IGY Com., SSSR, 33-44.

bathymetry
Kanaev, V.F., 1959
[Bathymetrical features of the Kronotski Bay.]
Trudy Inst. Okeanol., 36: 5-20.

bathymetry
Kanaev, V.F., 1956.
[Bottom bathymetry and its accomplishment in oceanographic investigations.] Trudy Inst. Okean. 19:164-168.

bathymetry
Kanaev, V.F., and N.I. Larina, 1959
[Bathymetry of the North Kuril area.]
Trudy Inst. Okeanol., 36: 158-168.

bathymetry
Kanaiev, V.F., and N.A. Marova, 1965.
Bathymetric chart of the North Indian Ocean. (In Russian; English abstract).
Okeanolog. Issled., Rez. Issled. po Programme Mezhd. Geofiz. Goda, Mezhd. Geofiz. Komitet, Prezidiume Akad. Nauk, SSSR, No.13:157-162.

the chart is attached to the article.

bathymetry a
Karig, Daniel E., 1971.
Structural history of the Mariana island arc system.
Bull. geol. Soc. Am., 82(2):323-344.

Marianas
bathymetry
seismic profiles

bathymetry
Kato, Kenji, 1961
Some aspects on biochemical characteristics of sea water and sediments in Mochima Bay, Venezuela.
Bol. Inst. Oceanogr., Univ. de Oriente, Venezuela 1(2):343-358.

bathymetry
Kato, K., 1955.
Regional distribution of organic matter in bottom sediments in the vicinity of Okushiri Strait, Hokkaido.
Bull. Fac. Fish., Hokkaido Univ., 6(1):19-24, 1 table, 6 textfigs.

bathymetry
Kato, K., 1955.
Accumulation of organic matter in marine bottom and its oceanographic environment in the sea to the southeast of Hokkaido.
Bull. Fac. Fish., Hokkaido Univ., 6(2):125-151, 15 textfigs.

bathymetry a
Keen, M.J., B.D. Loncarevic and G.N. Ewing, 1970.
II. Regional observations. 7. Continental Margin of Eastern Canada: Georges Bank to Kane Basin. In: The sea: ideas and observations on the progress in the study of the seas, Arthur E. Maxwell, editor, Wiley-Interscience 4(2): 251-291.

bathymetry
Keller, G.H., and G. Peter, 1968.
East-west profile from Kermadec Tremch to Valparaiso, Chile.
J.geophys.Res., 73(22):7154-7157.

bathymetry
Keller, George H., and Adrian F. Richards, 1967.
Sediments of the Malacca Strait, Southeast Asia.
J.sedim. Petrol., 37(1):102-127.

bathymetry
Kent, P.E., 1967.
North Sea exploration- a case history.
Georg. J. 133(3):289-301.

bathymetry
Kibblewhite, A.C., and R.N. Denham 1967
The bathymetry and total magnetic field of the South Kermadec Ridge seamounts.
N.Z. Jl. Sci., 10(1): 52-67.

seamounts
South Kermadec Ridge
Pacific, southwest
magnetic field
magnetic anomalies
ridges
seismic activity

bathymetry
Kiilerich, A., 1945
On the Hydrography of the Greenland Sea.
Medd. om Grønland 144(2):63 pp., 22 text figs., 3 pls.

bathymetry
King, Lewis H., and Brian MacLean, 1970.
Origin of the outer part of the Laurentian Channel.
Can. J. Earth Sci., 7(6):1470-1484

Bathymetry
Kluiev, E.V., 1965.
The role of permafrost factors in the dynamics of bottom relief of polar seas. (In Russian).
Okeanologiia, Akad. Nauk, SSSR, 5(5):863-869.

permafrost, effect of
Arctic
Bathymetry

bathymetry
Knott, S.T., E.T. Bunce and R.L. Chase, 1966.
Red Sea seismic reflection studies.
In: The world rift system, T.N. Irvine, editor, Dept. Mines Techn. Surveys, Geol. Survey, Can., Paper, 66-14:33-61.

WHOI Contrib. No. 1747.

bathymetry
Knox, G.A. 1957.
General account of the Chatham Islands 1954 Expedition.
N.Z. D.S.I.R. Bull. (Mem. N.Z. Oceanogr. Inst. 2) 122:1-17

bathymetry
Koczy, F.F., 1956.
Echo soundings. Rept. Swedish Deep Sea Exped., 4(Bottom Invest. 3):99-158, 23 fold-ins.

bathymetry
Koczy, F.F., and M. Burri, 1958.
Essai d'interpretation de quelques formes du terrain sous-marin. D'Après les échogrammes de l'expedition suédoise de l'Albatross.
Deep-Sea Res., 5(1):7-71.

bathymetry
Koldewijn, Bernard William, 1958.
Sediments of the Paria-Trinidad Shelf.
Repts. Orinoco Shelf Exped., 3:109 pp., 11 pls., fold-in maps.

Geol. Inst., Groningen, Publ.

bathymetry
Komukai, R., 1956.
On the bottom topography and sediments in the western passage of Tugaru Strait.
Rept. Hydrogr. Off., 31(4):45 pp.

bathymetry
Komukai, R., and R. Nakayama, 1958.
On the bottom topography of Japan Trench off Kasimanda.
Hydrogr. Bull., Mar. Safety Board, Tokyo, (Publ. 981), No. 57:45-52.

Bathymetry
Korea, Republic of, Fisheries Research and Development Agency, 1964.
Oceanographic Handbook of the Neighboring Seas of Korea, 214 pp.

bathymetry
Kornicker, Louis S., 1964.
Form replica of a submerged barrier chain with lagoonal basin off South Cat Cay, Bahamas.
Bull. Mar. Sci., Gulf and Caribbean, 14(1):168-171.

bathymetry
Kort, V.G., 1966.
On the hydrology of the Drake Passage. (In Russian; English abstract).
Okeanol. Issled., Mezhd. Geofiz. Kom., Presid., Akad. Nauk, SSSR, No. 15:28-44.

bathymetry
Krause, Dale C., 1968.
Bathymetry, geomagnetism, and tectonics of the Caribbean Sea north of Colombia. (Abstract only). Trans. Fifth Carib. Geol. Conf., St. Thomas, V.I., 1-5 July, 1968, Peter H. Mattson, editor. Geol. Bull. Queens Coll., Flushing 5: 7.

bathymetry
Krause, Dale C., 1967.
Sea floor relief.
Trans. Am. geophys. Un., 48(2):631-638.

topography
bathymetry

bathymetry
Krause, Dale C., 1966.
Tectonics, marine geology, and bathymetry of the Celebes Sea- Sulu Sea region.
Geol. Soc., Am., Bull., 77(8):813-832.

bathymetry
Krause, Dale C., 1966.
Geology and geomagnitism of the Bounty region east of the South Island, New Zealand.
New Zealand Dept. Sci. Ind. Res., Bull., 170:32pp.
New Zealand Oceanogr., Inst. Mem. No.30:32pp.

bathymetry
Krause, Dale C., 1965.
Submarine geology north of New Guinea.
Geol. Soc., Amer., Bull., 76(1):27-42.

bathymetry
Krause, Dale C., 1964.
Guinea fracture zone in the Equatorial Atlantic.
Science, 146(3640):57-59.

bathymetry
Krause, Dale C., 1964.
Lithology and sedimentation in the southern continental borderland.
In: Papers in Marine Geology, R.L. Miller, Editor, Macmillan Co., N.Y., 274-318.

Bathymetry
Krause, Dale C., 1962.
Interpretation of echo sounding profiles.
Int. Hydrogr. Rev., 39(1):65-123.

bathymetry
Krause, D.C., H.W. Menard and S.M. Smith 1964.
Topography and lithology of the Mendocino Ridge.
J. mar. Res. 22 (3): 236-249.

bathymetry
Kreffer, J.C., 1971.
Navigation and bathymetry. Hydrogr. Newsletter, R. Netherlands Navy, Spec. Publ. 6: 9.

bathymetry
Kubo, I., and E. Asada, 1957.
A quantitative study on crustacean bottom epifauna in Tokyo Bay. J. Tokyo Univ., Fish., 43(2): 429-289.

bathymetry
Kuenen, Ph. H., 1949.
Ambon and Haroekol. Verhand. Neder. Geol.-Mijnbouwkundig Genootschap., Geol. ser., 15(1): 44-62, 7 textfigs.

bathymetry
Kutschale, Henry 1966
Arctic Ocean geophysical studies: the southern half of the Siberia Basin.
Geophysics 31 (4): 683-710.

L

bathymetry (data)
Lacombe, Henri, 1959.
Mission hydrographique des cotes de France et d'Afrique du Nord, 1953-1954.
Ann. Hydrogr., No. 729:1-166.

bathymetry
Lacombe, H., and ? Mannevy, 1951.
Étude bathymetrique de la Mediterranée entre le cote prevencale et la Corse. Bull. d'Info., C.O.E.C., 3(4):147-148.

bathymetry
Lacombe, Henri, Paul Tchernia, Claude Richez and Lucien Gamberoni, 1964.
Deuxième contribution à l'étude du régime du détroit de Gibraltar (Travaux de 1960).
Cahiers Océanogr., C.C.O.E.C., 16(4):283-327.

Laktionov, A. F., V. A. Shamontyev and A. V. Yanes, 1960 bathymetry
Oceanographic characteristics of the North Greenland Sea.
Soviet Fish. Invest., North European Seas, VNIRO, PNIRO, Moscow, 1960:51-65.

bathymetry
Lalou, Claude, et Maurice Gennesseaux, 1959
Travaux de la Station Océanographique de Villefranche. Rev. Géogr. Phys. et Géol. Dyn. (2) 2(4): 231-252.

bathymetry
Larson, Roger L., 1971.
Near bottom geologic studies of the East Pacific Rise Crest.
Bull. geol. Soc. Am. 82 (4): 823-842.

bathymetry
Laughton, A.S., 1970.
A new bathymetric chart of the Red Sea.
Phil. Trans. R. Soc., (A) 267 (1181): 21-22
Chart

Red Sea
bathymetry

bathymetry
Laughton, A.S. 1966.
The Gulf of Aden.
Phil. Trans. R. Soc. (A) 259 (1099): 150-171.

bathymetry
Laughton, A.S., 1965.
Bathymetry of the north-eastern Atlantic Ocean and Recent geophysical studies. (abstract)
In: Submarine geology and geophysics, Colston Papers, W.F. Whittard and R. Bradshaw, editors, Butterworth's, London, 17:175.

Atlantic, northeast
bathymetry

bathymetry
Laughton, A.S., 1959.
The sea floor.
Science Progress, 47(176):230-249.

bathymetry
Lavrov, V.M. 1967.
Bottom sediments and relief of the North Atlantic. (In Russian)
Atlant. nauchno-issled. Inst. rybn. khoz. okeanogr. (AtlantNIRO). Materialy Konferentsii po Resul'tatam Okeanologicheskikh Issledovanii v Atlanticheskom Okeane, 251-257.

bathymetry
Lawson, A.C., 1950.
Sea bottom off the coast of California. A review of Special Paper 31, Geological Society of America, by Shepard and Emery, 1941. Bull. G. S.A. 61:1225-1242, 1 fig.

bathymetry
Lee, Arthur, 1963.
The hydrography of the European Arctic and subarctic seas.
In: Oceanography and Marine Biology, H. Barnes, Editor, George Allen and Unwin, 1:47-96.

bathymetry
Leenhardt, O., S. Pierrot et A. Rebuffet 1969
A propos de la cote 1980 metres au sud du Toulon.
Cah. océanogr., 21 (6): 581-553

Mediterranean, west
bathymetry

bathymetry
Le Floch, Jean, et V. Romanovsky, 1966.
L'eau intermédiaire en Mer Tyrrhénienne en régime d'été. (no abstract).
Cahiers océanogr., 18(3):185-228.

bathymetry, (inshore)
Lefur, A., 1951.
Note sur la détermination du gradient des plages par examen de la houle sur les photographies aériennes. Ann. Hydrogr., Paris, 2(723):167-176.

bathymetry
Leontyeva, V.V., 1960
Some data on the hydrology of the Tonga and Kermadec Trenches.
Trudy Inst. Okeanol., 40:72-82.

Bathymetry
Le Pichon, Xavier, 1966.
Étude géophysique de la dorsale médio-Atlantique.
Cahiers océanogr., 18(7):551-620.

bathymetry
Le Pichon, Xavier, 1960
The deep water circulation in the southwest Indian Ocean. J. Geophys. Res., 65(12): 4061-4074.

bathymetry
Le Pichon, Xavier, Roy D. Hyndman and Guy Pautot 1971.
Geophysical study of the opening of the Labrador Sea.
J. geophys. Res. 76(20): 4724-4743

bathymetry
*Lepley, Larry K., 1966.
Submarine geomorphology of eastern Ross Sea and Sulzberger Bay, Antarctica.
Tech. Rept., U.S. Naval Oceanogr. Off., TR-172:34 pp. (multilithed).

bathymetry
Lepley, L.K., 1964.
Submarine geomorphology of eastern Ross Sea and Sulzberger Bay, Antarctica.
Marine Geol., 2(4):253-261.

bathymetry
Lepley, L.K., 1964.
Submarine geomorphology of eastern Ross Sea and Sulzberger Bay, Antarctica. (Abstract).
Geol. Soc., Amer., Special Paper, No. 76:311-312.

bathymetry
Lim, Du Byung, and Sun-duck Chang 1969
On the cold water mass in the Korea Strait.
J. oceanol. Soc. Korea 4 (2): 71-82.

bathymetry
Lipps, J.H., and J.E. Warme, 1966.
Planktonic foraminiferal biofacies in the Okhotsk.
Contrib. Cushman Found., Foram. Res., 17(4):125-134

bathymetry
Lisicki, Andrzej, 1968.
Average depth and water volume of the Baltic Sea. (In Polish; English abstract).
Zesz. Geograf., WSP, Gdansk., 10:121-130.

bathymetry
Lisitzin, E., 1951.
A brief report on the scientific results of the hydrological expedition to the Archipelago and Åland Sea in the year 1927. Fennia 73(4):21 pp., 8 textfigs.

bathymetry
Litvin, V.M., 1967.
The bottom relief in the Faroe-Shetland Channel area. (In Russian).
Mater. Rybokhoz. Issled. Severn. Basseina, (PINRO), 10:157-161.

bathymetry
Litvin, V.M., 1966.
New data on the structure of the shelf and island slope in the southwest Iceland area. (In Russian).
Materiali, Sess. Uchen. Soveta PINRO Rez. Issled., 1964, Minist. Ribn. Khoz., SSSR, Murmansk, 97-107.

bathymetry
*Litvin, V.M. and M.V. Rudenko, 1971.
New data on bottom geomorphology of the south-eastern Atlantic. (In Russian; English abstract
Okeanologiia 11(2): 231-238.

Atlantic, southeast
bathymetry

bathymetry
Litvin, V.M., and V.D. Rvachev, 1962
A study of bottom relief and types of gound in the fishing areas of Labrador and Newfoundland. (In Russian; English summary).
Sovetskie Ribochoz. Issledov. Severo-Zapad. Atlant. Okeana, VNIRO PINRO, 99-111.

bathymetry
Løken, O.H., and D.A. Hodgson, 1971.
On the submarine geomorphology along the east coast of Baffin Island.
Can. J. Earth Sci., 8(2): 185-195

bathymetry
Loncarevic, B.D., 1964
Geophysical studies in the Indian Ocean.
Endeavor 23(88):43-47.

bathymetry
Loncarevic, B. D., C. S. Mason and D. H. Matthews, 1966.
Mid-Atlantic Ridge near 45° north. 1. The median valley.
Canadian J. Earth Sci., 3(3):327-349.

bathymetry
*Ludwig, William J., Dennis E. Hayes and John I. Ewing, 1967.
The Manila Trench and West Luzon Trough. I. Bathymetry and sediment distribution.
Deep-Sea Res., 14(5):533-544.

bathymetry
Luskin, B., B.C. Heezen, M. Ewing and M. Landisman, 1954.
Precision measurement of ocean depth.
Deep-Sea Res. 1(3):131-140, 7 textfigs.

M

bathymetry
Maeda Sonosuke 1968.
On the cold water belt along the northern coast of Hokkaido in the Okhotsk Sea. I. Hydrography of the Okhotsk Sea.
Umi to Sora 43(3):71-90.

bathymetry
Malahoff, Alexander, 1971.
Magnetic lineations over the Line Islands Ridge.
Bull. geol. Soc. Am. 82(7):1977-1983.

bathymetry
Malahoff, Alexander and Floyd McCoy, 1967.
The geologica structure of the Puna Submarine Ridge, Hawaii.
J. geophys. Res., 72(2):541-548.

bathymetry a
*Maley, Terry S. and G. Leonard Johnson, 1971.
Morphology and structure of the Aegean Sea.
Deep-Sea Res., 18(1): 109-122.

Aegean Sea
Mediterranean east
bathymetry
island arcs
troughs
ridges

bathymetry a
Mallory, J.K., 1970.
The bathymetry and microrelief of False Bay.
Trans. roy. Soc. S. Afr. 39(2): 109-112.

Bathymetry
Malloy, R.J., and R.J. Hurley 1970.
Geomorphology and geologic structure: Straits of Florida.
Bull. geol. Soc. Am., 81(7): 1947-1972

bathymetry
Straits of Florida

bathymetry
Maloney, Neil J. 1968.
Geomorphology of the continental margin of Venezuela. 2. Continental terrace off Carupano (64°W to 62°W longitude).
Bol. Inst. Oceanogr. Univ. Oriente 6(1): 147-155.

bathymetry
*Maloney, Neil J., 1967.
Geomorphology of the continental margin of Venezuela. 3. Bonaire Basin (66oW to 70oW. Longitude).
Bol. Inst. Oceanogr., Univ. Oriente, 6(2):286-302.

bathymetry
Maloney, Neil J. 1967.
Geomorphology of the continental margin off central Oregon, U.S.A.
Bol. Inst. Oceanogr. Univ. Oriente 6(1): 116-146.

bathymetry
*Maloney, Neil J., 1967.
Geomorphology of continental margin of Venezuela, Cariaco Basin.
Boln Inst. Oceanogr. Univ. Oriente, 5(1/2):38-53.

bathymetry
Maloney, Neil J., 1965
Geomorphology of the central coast of Venezuela
Bol. Inst. oceanogr., Univ. Oriente, 4(2): 246-265.

bathymetry
bottom sediment

bathymetry
Mal'tsev, V.N., 1959.
[Basic structural forms in the bottom relief of Davis Sea.]
Inform. Biull. Sovetskoi Antarkt. Exped., (12): 14-16.

bathymetry
Marlowe, James I. 1969
A succession of Tertiary strata off Nova Scotia as determined by dredging.
Can. J. Earth Sci., 6(5): 1077-1094

bathymetry
Canada east
submarine canyons
Tertiary
reflection, seismic

bathmetry
Marlowe, J.I., 1966.
Mineralogy as an indicator of long-term current fluctuations in Baffin Bay.
Canadian J. Earth Sci., 3(2):191-202.

Bathymetry
Marukawa, H., and T. Kamiya, 1926.
Outline of the hydrographical features of the Japan Sea.
Annot. Oceanogr. Res., 1(1):1-7, tables and charts

bathymetry
Mallory, J.K., 1961.
Bathymetric and hydrographic aspects of marine studies off the Natal coast.
Marine Studies off the Natal Coast, C.S.I.R. Symposium, No. S2:31-39 (multilithed).

bathymetry
Mann, A., 1907
Report on the diatoms of the Albatross voyages in the Pacific Ocean, 1888-1904.
Contrib. U. S. Nat. Herb. 10(5):221-419, Pls. XLIV-LIV.

bathymetry
Matthews, D.H., 1966.
The northern end of the Carlsberg Ridge.
In: The world rift system, T.N. Irvine, editor, Dept. Mines Techn. Surveys, Geol. Survey, Can., Paper, 66-14:124-134.

bathymetry
Matthews, D.H. 1965.
The Owen Fracture Zone and the northern end of the Carlsberg Ridge.
Phil. Trans. R. Soc. (A) 259 (1099): 172-186.

bathymetry
Matthews, L.H., 1934.
The marine deposits of the Patagonian continental shelf. Discovery Repts. 9:177-204, Pls. 2-14.

topography
Mattison, G.C., 1948
Bottom configuration in the Gulf of Mexico.
J. Coast and Geodetic Survey, 1:76-81, 4 figs. with comment by P.A. Smith pp.81-82.

Mc

bathymetry
McConnell, James J., Jr., 1964.
Bathymetry.
U.S.C.G. Oceanogr. Rept., No. 1(373-1):25-26.

bathymetry
McGregor, Bonnie A. 1968.
Features at the Sohm Abyssal Plain terminus.
Marine Geol., 6(5): 401-414.

Sohm Abyssal Plain
magnetic anomalies
bathymetry

bathymetry
McLellan, H.J., and W.D. Nowlin, 1963
Features of deep water in the Gulf of Mexico.
J. Mar. Res., 21(3):233-245.

bathymetry
McManus, Dean A. 1967.
Physiography of Cobb and Gorda rises, northwest Pacific Ocean.
Bull. geol. Soc. Am. 78(4): 527-546.

bathymetry
McManus, Dean A., 1965.
Blanco Fracture Zone, northeast Pacific Ocean.
Marine Geol., 3(6): 429-455

Blanco Fracture zone
fracture zones
Pacific, northeast
bathymetry
seismicity
magnetism

bathymetry a
McMaster, Robert L., John D. Milliman and Asaf Ashraf, 1971.
Continental shelf and upper slope sediments off Portuguese Guinea, Guinea, and Sierra Leone West Africa.
J. sedim. Petrol. 41(1): 150-158.

bathymetry
Melson, W.G., Geoffrey Thompson and Tjeerd H. van Andel 1968.
Volcanism and metamorphism in the Mid-Atlantic Ridge, 22°N Latitude.
J. geophys. Res. 73(18): 5925-5941.

bathymetry
Menard, H.W., T.E. Chase and S.M. Smith, 1964
Galapagos Rise in the southeastern Pacific.
Deep-Sea Res., 11(2):233-242.

bathymetry
Menard, H.W., and R.S. Dietz, 1952.
Mendicino submarine escarpment. J. Geol. 60(3): 266-278, 5 textfigs.

bathymetry
Menard, H.W., S.M. Smith and R.M. Pratt, 1965.
The Rhône Deep-Sea Fan.
In: Submarine geology and geophysics, Colston Papers, W.F. Whittard and R. Bradshaw, editors, Butterworth's, London, 17:271-284.

fans, deep-sea
Mediterranean, west
abyssal plains
bathymetry
1594

bathymetry
Menzies, R.J., O.H. Pilkey, B.W. Blackwelder, D. Dexter, P. Huling and L. McCloskey, 1966.
A submerged reef off North Carolina.
Int. Rev. ges. Hydrobiol., 51(3):393-431.

bathymetry
Midttun, L., and J. Natvig, 1957.
Pacific Antarctic waters. Scientific results of the "Brategg" Expedition, 1947-48, No. 3.
Publikasjon, Komm. Chr. Christensens Hvalfangstmus, Sandefjord, No. 20:1-130, 39 textfigs.

bathymetry
*Milliman, John D., 1967.
The geomorphology and history of Hogsty Reef, a Bahamian atoll.
Bull. mar. Sci., Miami, 17(3):519-543.

bathymetry
Milliman, John D., and Peter R. Supko, 1968.
On the geology of San Andres Island, western Caribbean.
Geologie Mijnb., 47(2): 102-105.

bathymetry
magnetism

bathymetry, coastal
Misawa, Yoshifumi, Norihiro Sugai and Koji Murakoshi, 1971.
Seabottom topography and geology at Kitami offshore in Hokkaido. (In Japanese; English abstract). J. Tokai Univ. Coll. mar. Sci. Technol. 5: 55-71.

bathymetry
Mogi, Akio, 1962
Submarine topography of the adjacent seas of Japan Islands. (In Japanese).
J. Oceanogr. Soc., Japan, 20th Ann. Vol., 52-63.

bathymetry
Mogi, A., 1959
On the depth change at the time of Kanto-Earthquake in Sagami Bay.
Publ. 981, Hydrogr. Bull. No. 60:52-60.

bathymetry
Mogi, Akio, and Takahiro Sato, 1958
[On the bottom configuration and sediments in the adjacent sea of Mogami Bank, Japan Sea.]
Hydrogr. Bull., Maritime Safety Bd., Japan, (Publ. 981): 37-53.

bathymetry
*Molloy, Arthur E., 1968.
Sea floor bathymetry near the Clarion and Clipperton fracture zones.
Int. hydrogr. Rev., 45(1):107-121.

bathymetry
Moore, David G., and George Shumway, 1959.
Sediment thickness and physical properties: Pigeon Point Shelf, California.
J. Geophys. Res., 64(3):367-374.

bathymetry
Moore, David G., and Joseph R. Curray, 1963
Structural framework of the continental terraces northwest Gulf of Mexico.
J. Geophys. Res., 68(6):1725-1747.

bathymetry
Moore, J. G., 1947.
The determination of the depths and extinction coefficients of shallow water by air photography using colour filters. Phil. Trans. Roy. Soc., London, Ser. A., Math. Phys. Sci., 240(816): 163-217.

bathymetry
Morelli, C., 1970.
Physiography, gravity and magnetism of the Tyrrhenian Sea. Boll. Geofis. teor. appl. 13(48): 275-309.

bathymetry
Morelli, C., 1966.
The geophysical situation in Italian waters (1966.0).
Int. hydrogr. Rev., 43(2):133-147.

bathymetry
Morimoto, R., R.L. Fisher and N. Nasu, 1955.
Bathymetry and petrography of the Bayonnaise Rocks, Japan.
Proc. Jap. Acad., 31(10):637-641.

bathymetry
Morton, Robert W., 1968.
Bathymetry and mineralogy of a submerged plain, south of Roosevelt Roads, Puerto Rico. (Abstract only). Trans. Fifth Carib. Geol. Conf., St. Thomas, V.I., 1-5 July 1968, Peter H. Mattson, editor. Geol. Bull. Queens Coll., Flushing, 5: 39.

bathymetry
Mosby, H., 1940.
Nomenclature of the submarine features of Southern Seas.
Spec. Art., Publ. Sci., Assoc. d'Océanogr. Phys. No. 8:95-98.

bathymetry
Mosby, H., 1938
Svalbard waters. Geophys. Publ. 12(4): 85 pp., 34 text figs.

bathymetry
Mosetti, F., 1966.
Morfologia dell'Adriatico settentrionale.
Boll. Geofis. teor. appl., 8(30):138-150.

Adriatic
bathymetry
summary

bathymetry
Mulicki, Z., 1957.
[Ecology of the more important Baltic invertebrates.] Prace Morsk. Inst. Ryback., Gydni, No. 9 :313-379.

bathymetry
Murakami, A., 1954.
[Oceanography of Kasaoka Bay in Seto Inland Sea.]
Bull. Nakai Regional Fish. Res. Lab. 6:15-57, 42 textfigs.

bathymetry
Murray, H.W., 1946
Submarine relief of the Aleutian trench.
Trans. Am. Geophys. Un. 27(6):871-875, 4 textfigs.

N

bathymetry
*Nachtigall, Karl Hans, 1968.
Über die Unterwasserhangnorphologie vor Rantum und Kampen auf Sylt.
Meyniania, 18:43-63.

bathymetry (experimental)
Nakano, M., 1947
An investigation on the effect of prevailing winds upon the depths of bays.
Geophys. Mag. 14(2/4):57-86, 7 figs.

bathymetry
*Nalbant, Teodor T., 1968.
Observatii asupra conditiilor de media din zonele Georges Bank si Nantucket in legatura cu procesul de pescuit.
Buletinul Inst. Cercetari Proiectari Pisicole, 27(4):75-93.

bathymetry
Nasu, Keiju, 1963
Oceanography and whaling ground in the Subarctic region of the Pacific Ocean.
Sci. Repts., Whales Res. Inst., No. 14:105-155.

bathymetry
Navarro, F. de P., 1947.
Exploración oceanografía del Africa occidental desde el Cabo Ghir al Cabo Judy: resultados de las campañas del "Malaspina" y del "Xauen" en Mayo 1946. Trab. Inst. Español Ocean. No. 20: 40 pp., 8 figs.

bathymetry
Nayudu, Y.R., and B.J. Enbysk, 1964.
Bio-lithology of northeast Pacific surface sediments.
Marine Geol., 2(4):310-342.

bathymetry
Nesteroff, W.D., 1955.
Les récifs coralliens du Banc Farsan Nord (Mer Rouge). Rés. Sci. Camp. Calypso. 1. Camp. 1951-1952 en Mer Rouge. Ann. Inst. Océan., Monaco, 30:8-53, 21 pls., 11 textfigs.

bathymetry
Nesteroff, Wladimir D., Solange Duplaix, Jacqueline Sauvage, Yves Lancelot, Frédéric Melières et Edith Vincent, 1968.
Les dépôts récents du canyon de Cap Breton.
Bull. Soc. géol. France (7) 10: 218-252.

submarine canyons
bottom sediments, foraminifera, etc.

bathymetry
Neumann, G., 1944.
Das Schwarze Meer. Zeitsch. Gesellschaft f. Erdkunde zu Berlin. 1944 (3/4):92-114, 25 text figs.

bathymetry
Newton, John G. and Orrin H. Pilkey, 1969.
Topography of the continental margin off the Carolinas.
SEast. Geol., Duke Univ., 10(2):87-92

Continental margin
Continental rise
Continental slope
bathymetry

bathymetry
Niino, H., and T. Kumagori, 1956.
Preliminary re port on the new bank at the southern end of Emperor Sea Mount.
J. Tokyo Univ. Fish., 42(1):101-102.

ba thymetry
Nishimura, S., 1969.
The zoogeographical aspects of the Japan Sea. Part V. Publ. Seto mar. biol. Lab., 17(2): 67-142.

salinity
water masses

bathymetry
Nizery, 1949
Etude hydrographique et oceanographique du Trou-sans-fond pres D'Abidjan (Cote d'Ivorre).
Bull. d'Info. Com. Antr. Ocean & Etudes des Cotes, No.7:13-19.

bathymetry
Norris, Robert M., 1964.
Sediments of Chatham Rise.
New Zealand Dept. Sci. Ind. Res., Bull., No. 159:
New Zealand Oceanogr. Inst., Memoir, No. 16:39pp.

bathymetry
Northrop, J., 1954.
Bathymetry of the Puerto Rico Trench.
Trans. Amer. Geophys. Union 35(2):221-225, 4 textfigs.

Bathymetry
Northrop, John, Maurice Blaik and Roberto Frassetto, 1959.
Bathymetry of the Gibbs Hill area, Bermuda.
Deep-Sea Res., 5:290-296.

bathymetry
Nowlin, W.D., Jr., and H.J.McLellan, 1967.
A characterization of the Gulf of Mexico waters in winter.
J.mar.Res., 25(1):29-59.

bathymetry
Nutt, D.C., and L.K. Coachman, 1956.
The oceanography of Hebron Fjord, Labrador.
J. Fish. Res. Bd., Canada, 13(5):709-858.

O

bathymetry
Oberkommando der Kriegsmarine, 1940
Die Naturverhältnisse des Sibirischen Seeweges. Beilage zum Handbuch des Sibirischen Seeweges. 169 pp., 64 text figs., 2 bathymetric charts.

bathymetry
*Okuda, Setsuo, 1968.
On the change in salinity distribution and bottom topography after the closing of the mouth of Kojima Bay.
Bull.Disas.Prev.Inst.Kyoto Univ., 18(1):35-48.

bathymetry
Oren, O.H., 1964.
Hydrography of Dahlak Archipelago (Red Sea).
Sea Fish.Res. Sta., Israel, Bull. No.35:3-22.

bathymetry
*Ostenso, N.A., 1968.
Geophysical studies in the Greenland Sea.
Bull.Geol.Soc.Am., (1):107-132.

bathymetry
Ostenso, N.A., 1963.
Physiography and structure of the Arctic Ocean Basin.
I.G. Bull., 70:1-12. In: Trans. Amer. Geophys. Union, 44(2):637-648.

Original report was:
Geophysical investigations of the Arctic Ocean Basin. Univ. Wisconsin (1962) Geophys. Polar Res. Cen. Res. Rept., No. 62-4.

bathymetry
Owaki, N., 1963
A note on depth when the bottom is soft mud.
Int. Hydrogr. Rev., 40(2):41-43.

bathymetry
Ozawa, Keijiro and Tsotomu Isouchi, 1966.
Continuous echo sounding and bottom confriguration in the eastern Indian Ocean.
J. Tokyo Univ.Fish., (Spec.ed.)8(2):55-69.

bathymetry
Ozawa, Keijiro and Tsutomu Isouchi, 1968.
Continuous echo-sounding and bottom configuration in the Southern Ocean. J. Tokyo Univ. Fish., 9(1): 13-32.

P

bathymetry
Paccagnini, Ruben N., and Alberto O. Casellas, 1961
Estudios y resultados preliminares sobre trabajos oceanograficos en el area del Mar de Weddell. (In Spanish; Spanish, English, French, German, and Italian resumes).
Contrib., Inst. Antartico Argentino, No. 64: 12 pp.

bathymetry
Pacheco, F.H., 1961
Origen y relieve submarino del Estrecho de Gibraltar.
Bol. Inst. Espan. Ocean., 105: 26 pp.

bathymetry
Paisley, J.T.K., 1955.
Survey of a seapeak in the Mozambique Channel.
Int. Hydrogr. Rev. 32(2):149-151, 3 fold-ins.

bathymetry
Panov, D.G., 1948.
Results of charting the floors of Soviet seas.
Izvest. Vses. Geogr. Obshch. 80:329-339.

bathymetry
Pantin, H.M., 1966.
Sedimentation in Hawke Bay.
New Zealand Dept. Sci. Ind. Res., Bull., 171:1-64.
Also: New Zealand Oceanogr.Inst. Mem.No.28:1-64.

bathymetry
Parfait, 1883.
Mission scientifique du Talisman.
Ann. Hydrogr., Paris, (2) 5:259-311, figs.

bathymetry
Parker, R.H., and J.R. Curray, 1956.
Fauna and bathymetry of banks on continental shelf, northwest Gulf of Mexico. Bull. Amer. Assoc. Petr. Geol., 40(10):2428-2439.

bathymetry
Parr, A. E., 1937.
A contribution to the hydrography of the Caribbean and Cayman Seas, based upon the observations made by the Research Ship "Atlantis" 1933-1934. Bull. Bingham Ocean. Coll., V(4):1-110, 82 textfigs.

bathymetry
Pautot, Guy, 1970.
La marge continentale au large de l'Esterel (France) et les mouvements verticaux pliocènes.
Mar. Geophys. Res. 1(1): 61-84

bathymetry
Pelczar, Maria, 1971.
Changes in the bottom bathymetry of the Indian Ocean. (In Polish; English abstract). Zeszyty Naukowe UG, Geogr. 1: 41-94.

bathymetry
Pelletier, B.R., 1968.
Submarine physiography, bottom sediments, and modern sediment transport in Hudson Bay.
In: Earth Science Symposium on Hudson Bay, Ottawa February, 1968, Peter J. Hood, editor, GSC Pap. Geol. Surv. Can., 68-53:100-136

circulation
transport, sediments

bathymetry
Pelletier, B.R., 1966.
Development of submarine physiography in the Canadian Arctic and its relation to crustal movements.
In: Continental drift, G.D.Garland, editor, Spec.Publs.R.Soc.Can., No.9:77-101.

bathymetry
Petelin, B.P., 1957.
The relief of the floor and the bottom deposits in the north-west Pacific.
Proc. UNESCO Symp., Phys. Ocean., Tokyo, 1955: 225-237.

bathymetry
Peter, George, 1966.
Preliminary results of a systematic geophysical survey south of the Alaska peninsula.
In: Continental margins and island arcs, W.H. Poole, editor, Geol. Surv.Pap., Can., 66-15:223-237.

bathymetry
Peter, George, and Omar E. DeWald 1969.
Geophysical reconnaissance in the Gulf of Tadjura.
Bull. geol. Soc. Am. 80(11):2313-2316

bathymetry
Peter, George, Barrett H. Erickson and Paul J. Grim, 1970.
II. Regional observations. 5. Magnetic structure of the Aleutian Trench and Northeast Pacific Basin. In: The sea: ideas and observations on progress in the study of the seas, Arthur E. Maxwell, editor, Wiley-Interscience 4(2): 191-222.

bathymetry
Peter, G., L.A. Weeks and R.E. Burns, 1965.
A reconnaisssance geophysical survey in the Andaman Sea and across the Andaman-Nicobar Island Arc.
In: Int. Indian Ocean Exped., USC&GS Pioneer, 1964, U.S. Dept. Commerce, Environmental Sci. Services Administration, 1:91-107.

bathymetry
Peter, G., L.A. Weeks, and R.E. Burns, 1966.
A reconnaissance geophysical survey in the Andaman Sea and across the Andaman-Nicobar Arc.
J. geophys. Res., 71(2):495-509.

bathymetry
Petersen, G. Høpner, 1964.
The hydrography, primary production, bathymetry, and "tagsåq" of Disko Bugt, West Greenland.
Medd. om Grønland, 159(10):45 pp.

bathymetry
Pettersson, H., 1955.
Résumé des résultats de la croisière de l'"Albatross" en Méditerrannée.
Trav. Lab. Géol. Sous-Marine 6:1-5.

bathymetry
Pettersson, H., 1947.
The Swedish Deep-Sea Expedition. Nature 160 (4069):559-560.

bathymetry
Pfannenstiel, Max, 1960
Erläuterungen zu den bathymetrischen Karten des östlichen Mittelmeeres.
Bull. Inst. Oceanogr., Monaco, No. 1192: 60pp. 9 charts.

Summary in French, German and English

bathymetry
Phillips, J.D., G. Thompson, R.P. Von Herzen and V.T. Bowen, 1969.
Mid-Atlantic Ridge near 43°N Latitude. J. geophys. Res., 74(12): 3069 -3081.

magnetic field
petrology
heat flow

bathymetry
Phipps, Charles V.G., 1966.
Evidence of Pleistocene warping of the New South Wales continental shelf.
In: Continental margins and island arcs, W.H. Poole, editor, Geol.Surv.Pap.,Can., 66-15:280-293.

bathymetry
Pickard, G.L., 1961.
Oceanographic features of inlets in the British Columbia mainland coast.
J. Fish. Res. Bd., Canada, 18(6):907-999.

bathymetry
Pigneres, P., 1949.
Le sondage par le son. Bull. Tech. Veritas 31(1): 2-14.

bathymetry
Pincus, H.J., 1950.
Depth-sounding data from coordinated vessels. (Abstract). Trans. Am. Geophys. Union 31(2):331.

bathmetry
Poizat, Claude, 1970.
Hydrodynamisme et sédimentation dans le Golfe de Gabès (Tunisie).
Téthys 2(1): 267-296.

bottom sediments, bioclastic fragments
bathymetry

bathymetry
Pollak, M. J., 1950.
The water structure in the Brownson Deep. Trans. Am. Geophys. Union 31(3):393-397, 3 textfigs.

bathymetry
*Pratt, Richard M., 1968.
Atlantic continental shelf and slope of the United States-physiography and sediments of the deep-sea basin.
Prof.Pap.U.S.geol.Surv., 529B:B1-B44, fold in chart.

bathymetry
Pratt, R.M., 1966.
The Gulf Stream as a graded river.
Limnol. and Oceanogr., 11(1):60-67.

bathymetry
Pratt, R.M., and Bruce C. Heezen, 1964
Topography of the Blake Plateau.
Deep-Sea Res., 11(5):721-728.

WHOI Contr. No. 1492

Bathymetry
Prebble, M.M., 1968.
Some implications arising from bottom soundings taken around Pram Point, Ross Island Antarctica.
N.Z. Jl. Geol. Geophys. 11(4):900-907.

bathymetry
Puri, Harbans S., Gioacchino Bonaduce and John Malloy, 1964.
Ecology of the Gulf of Naples.
Pubbl. Staz. Zool., Napoli, 33(Suppl.):87-199.

Q

R

bathymetry
Ramamurthy, S., 1954.
Hydrobiological studies in the Madras coastal waters. J. Madras Univ., B, 23(2):148-163, 3 textfigs.

bathymetry
Rao, T.C.S., 1969.
Bathymetric features of Bay of Bengal. Bull. natn. Inst. Sci., India, 38(1): 421-423.

bathymetry a
*Rea, David K. and Barrett H. Erickson, 1971.
Bathymetric and magnetic profile along 143°W, northeast Pacific Ocean. J. geophys. Res., 76(8): 1948-1953.

Pacific, northeast
bathymetry
magnetism

bathymetry
Reineck, Hans Erich, 1965.
Sedimentgefüge im Golf von Neapel.
Pubbl. Staz. Zool. Napoli 34:112-137

bathymetry
Reyes, F Eduardo, 1967.
Carta batilitologica de Valparaiso.
Rev.Biol.mar.Chile, 13(1):59-69.

bathymetry
Ritchie, G.S., 1958.
Sounding profiles between Fiji, Christmas and Tahiti Islands.
Deep-Sea Res., 5(3):162-168.

Bathymetry
Roa Morales, Pedro, and Francois Ottmann, 1961.
Primer estudo topografico y geologico del Golfo de Cariaco.
Bol. Inst. Oceanograf., Univ. Oriente, Cumana, Venezuela, 1(1):5-20.

bathymetry
Robertson, E.I., and A.C. Kibblewhite, 1966.
Bathymetry around isolated volcanic islands and atolls in the South Pacific Ocean.
New Zealand J. Geol. Geophys., 9(1/2):111-121.

bathymetry
Robinson, Edwin S., 1963
Geophysical investigations in McMurdo Sound, Antarctica.
J. Geophys. Res., 68(1):257-262.

Bathymetry
Rochford, D. J., 1966.
Some hydrological features of the eastern Arafura Sea and the Gulf of Carpenteria in August 1964.
Australian J. Mar. freshw. Res., 17(1):31-60.

bathymetry
Rodin, G.I., 1958.
Oceanographic and meteorological aspects of the Gulf of California. Pacific Science 12(1):21-45.

bathymetry
Rodolfo, Kelvin S., 1969.
Bathymetry and marine geology of the Andaman Basin, and tectonic implications for southeast Asia.
Bull. Geol. Soc. Am., 80(7): 1203-1230.

Andaman Basin
Indian Ocean
bathymetry

bathymetry
Rogalla, Ernst-Helmuth, 1960
Über die Aufnahme einer untermeerischen Kuppe zwischen Rockall und St. Kilda.
Deutsche Hydrogr. Zeits., 13(1):24-27.

bathymetry
Rona, Peter A., 1961
Gibbs seamount.
Deep-Sea Res., 8(1): 76-77.

bathymetry
Roos, S. E., 1937.
Some geographican results of the Second Byrd Antarctic Expedition, 1933-1935. I. The submarine topography of the Ross Sea and adjacent waters. Geogr. Rev. 27(4):574-583, 12 figs.

WHOI Contrib. No. 91.

bathymetry
Romanovsky, V., 1945
Résultats des sondages effectués dans la baie du Roi. Bull. Inst. Ocean., Monaco, No.877, 9 pp., 4 figs.

bathymetry
Romanovsky, V., 1943
Oscillations de rivage et bathymétre dans la région sud de la Baie du Roi (Spitzberg). Bull. Soc. Géol. de France, No.1-2-3, :81-90, 4 figs., 1 pl.

bathymetry a
*Rona, Peter A., 1971.
Bathymetry off central northwest Africa.
Deep-Sea Res., 321-327.

Africa, northwest
bathymetry

bathymetry
*Rona, Peter A., Eric D. Schneider and Bruce C. Heezen, 1967.
Bathymetry of the continental rise off Cape Hatteras.
Deep-Sea Res., 14(5):625-633.

bathymetry a
Ross, David A., 1971.
Sediments of the northern Middle America Trench.
Bull. geol. Soc. Am. 82(2): 303-322.

Mexico, west
bathymetry
bottom sediments, minerals

bathymetry
*Ross, David A., Earl E. Hays and Frank C. Allstrom 1969.
Bathymetry and continuous seismic profiles of the hot brine region of the Red Sea.
In: Hot brines and Recent heavy metal deposits in the Red Sea, E.T. Degens and D.A. Ross, editors, Springer-Verlag, New York, Inc., 82-97.

bathymetry
Ross, D.I. 1967.
Magnetic and bathymetric measurements across the Pacific-Antarctic Ridge south of New Zealand.
N.Z. Jl Geol. Geophys. 10(6): 1452-1465.

Bathymetry
*Rossi, S., F. Mosetti e B. Cescon, 1968.
Morfologia e natura del fondo nel Golfo di Trieste (Adriatico settentrionale fra Punta Tagliamento e Punta Salvore).
Boll. Soc. adriat. Sci., 56(2):187-206.

bathymetry
Rotschi, Henri, 1959
Resultats des observations scientifiques du "Tiare", Croisière DAUPHIN (8-14 août 1959) sous le commandement du Lieutenant de Vaisseau Cerbelaud.
C.L.O.E.C., Inst. Francais d'Océanie, Rapp. Sci., No. 14: 18 pp.

bathymetry
Rotschi, H., and L. Lemasson, 1967.
Oceanography of the Coral and Tasman seas.
Oceanogr. Mar. Biol. Ann. Rev., Harold Barnes, editor, George Allen and Unwin, Ltd., 5:49-97.

bathymetry
Rouch, J., 1943.
Traité d'Océanographie physique. Sondages.
Payet, Paris, 256 pp., 100 textfigs.

bibliography, red tides
Rounsefell, George A., and Walter R. Nelson, 1966.
Red-tide research summarized to 1964 including an annotated bibliography.
Spec. Scient. Rep. U.S. Fish Wildl. Serv., Fish., 535: 85pp. (multilithed).

bathymetry
Rusnak, Gene A., Robert L. Fisher and Francis P. Shepard, 1964.
Bathymetry and faults of Gulf of California.
In: Marine geology of the Gulf of California, a symposium, Amer. Assoc. Petr. Geol., Memoir, T. van Andel and G.G. Shor, Jr., editors, 3:59-75.

bathymetry
Rvachev, V.D., 1964
Bottom contour and sediments on the Georges and Banquereau banks. (In Russian).
Materialy Ribochozh. Issled. Severn. Basseina, Gosudarst. Kom. Ribn. Chozh. SNCH, SSSR, Poliarr Nauchno-Issled. i Proektn. Inst. Morsk. Ribn. Chozh i Okanogr., N.M. Knipovich (PINRO), 2: 78-87.

bathymetry
Rvachev, V.D., 1963
The topography and bottom sediments of the shelf area in southwestern Greenland. (In Russian).
Okeanologiia, Akad. Nauk, SSSR, 3(6):1046-1055.

S

bathymetry
Saemundsson, B., 1948.
Description of Faxa Bay.
Rapp. Proc. Verb., Cons. Perm. Int. Expl. Mer, 120:26-29, 3 textfigs.

bathymetry
Said, R., 1953.
Foraminifera of Great Pond, Falmouth, Massachusetts. Contr. Cushman Found. Foram. Res. 4(1): 7-14, 1 table, 3 textfigs.

WHOI Contrib. No. 588.

bottom sediments
Said, R., 1951.
Organic origin of some calcareous sediments from the Red Sea. Science 113(2940):517-518.

bathymetry
Samalea, J. A., 1949.
Exploración del "banco del Xauen" accidente topográfico descubierto frente al Peñón de Vélez de la Gomera. Bol. Inst. Español Ocean., No. 16, 7 pp., 2 textfigs., 1 chart.

bathymetry
Sato, Takahiro, and Akio Yogi, 1965.
Guyots found from the Marshall and East Caroline Ridges. Researches on the GEBCO, No.1.
J. Oceanogr. Soc., Japan, 21(4):139-147.

bathymetry
Sato, T., and M. Tamaki, 1962
Bottom sediments of the Aomori Bay. (In Japanese; English abstract).
Hydrogr. Bull., (Publ. No. 981), Tokyo, No. 72: 16-25.

bathymetry
Sauzay, A., 1950.
Étude bathymétrique de la côte de Provence du Cap d'Antibes à la frontière italienne. Bull. d'Info., C.O.E.C., 2(10):395-396, 1 pl.

bathymetry
Saville, T., jr., and J.M. Caldwell, 1953.
Accuracy of hydrographic surveying in and near the surf zone. B.E.B. Tech. Memo. No. 32:28 pp., 8 figs.

bathymetry
Scaccini-Cicatelli, Marta, 1959.
Caratteri idrodinamici e batimetrici del Golfo dell'Asinara.
Note Lab. Biol. mar. Fano, 1(13):109-120.

bathymetry
Scaccini-Cicatelli, 1960
Sur les caractères hydrodynamiques et bathymétriques du Golfe d'Asinara. Comm. Int. Sci., Mer Méditerranée, Rapp. Proc. Verb., Monaco, 15(3): 265-269.

bathymetry
Schmidt-Ries, H., 1949.
Kurzgefasste Hydrographie Griechenlands.
Arch. Hydrobiol. 43(1):95-141.

bathymetry
Schneider, Eric., and Bruce C. Heezen, 1966.
Sediments of the Caicos Outer Ridge, The Bahamas.
Bull. geol. Soc., Am., 77(12):1381-1398.

Bathymetry
Scholl, David W., Edwin C. Buffington and David M. Hopkins, 1968.
Geologic history of the continental margin of North America in the Bering Sea.
Marine Geol., 6(4): 297-330.

bathymetry
Schott, G., 1942
Geographie des Atlantischen Ozeans im auftrage der Deutschen Seewarte vollständy neu bearbeitete dritte auflage. 438 pp., 141 text figs., 27 plates. C. Boysen. Hamburg.

Owned by MS

bathymetry (data)
Schott, G., and P. Perlewitz, 1906.
Lotungen I.N.M.S. "Edi" und des Kabeldampfers "Stephen" im westlichen Stillen Ozeans.
Arch. Deutschen Seewarte 29(2):1-38, 4 pls.

bathymetry
Schulz, B., 1934.
Die Ergebnisse der Polarexpedition auf dem U-Boat "Nautilus". Ann. Hydrogr., usw., 62:147-152, Pl. 16 with 2 figs.

bathymetry
Schulz, B., and A. Wulff, 1929
Hydrographie und Oberflächen plankton des westlichen Barentsmeeres im Sommer 1927.
Ber. deutschen wissensch. Komm. F. Meeresforsch. n.s. 4(5):232-372, 13 tables, 25 text figs.

bathymetry
Schumacher, A., 1958.
Die Lotungen der "Schwabenland".
Deutsch. Antark. Exped., 1938/39, Wiss. Ergeb., 2(2):41-62.

bathymetry
Schuster, O., 1952.
Die Vareler Rinne im Jadebusen. Die Bestandteile und das Gefüge einer Rinne im Watt. Abhandl. Senckenbergischen Naturforsch. Gesellscheft 486: 1-38, 14 textfigs.

bathymetry
Seiwell, H.R., 1949.
Oceanographic factors in underwater depth determination by aerial photography. Photogram. Eng. 15:171-176.

bathymetry
Senin, Y.M., 1963. Preliminary results of the relief and ground survey of the continental shelf of West Africa. (In Russian).
Atlantich. Nauchno-Issled. Inst. Ribn. Khoz. i Okeanogr. (ATLANTNIRO), Trudy, 10: 5-9.

bathymetry
Seymour Sewell, R. B., 1940.
The Indian Ocean. Assoc. d'Océan. Phys., Spec. Art., Publ. Sci. No. 8:81-86, Fig. 4.

Filed under Association
ms

bathymetry
Shalowitz, A.L., 1951.
Cartography in the submerged lands oil cases.
J. C.& G.S. No. 4:68-78, 7 textfigs.

bathymetry
Shannon, L.V. 1966.
Hydrology of the south and west coasts of South Africa.
Investl Rep. Div. Fish. Un. SAfr. 22 pp. 76 figs.

bathymetry
Shannon, L.V., and M. van Rijswijk, 1969.
Physical oceanography of the Walvis Ridge region.
Investl Rep. Div. Sea Fish. SAfr. 70: 1-19.

bathymetry
Shaver, Ralph, and Kenneth Hunkins, 1964.
Arctic Ocean geophysical studies: Chukchi Cap and Chukchi Abyssal Plain.
Deep-Sea Res., 11(6):905-916.

bathymetry
Shcherbakov, D.I., 1954.
First results of the High-Latitude Arctic Expedition of 1954. Vestnik Akad. Nauk, SSSR (9):10-16.

bathymetry
Shepard, Francis P. 1965
Submarine canyons explored by Cousteau's diving saucer.
In: Submarine geology and geophysics, Colston Papers, W. F. Whittard and R. Bradshaw, editors, Butterworths, London 17: 303-309

bathymetry
Shepard, F. P., 1950.
1940 E.W. Scripps cruise to the Gulf of California. III. Submarine topography of the Gulf of California. 43:32 pp., 6 textfigs., 3 pls., 11 charts.

bathymetry
*Shepard, F.P., R.F. Dill and Ulrich von Rad, 1969.
Physiography and sedimentary processes of La Jolla Submarine Fan and Fan-Valley, California.
Bull. Am. Ass., Petrol. Geol., 53(2):390-420.

bathymetry
Shibata, Keishi, 1966.
An echo survey near the Ojin Seamount.
Bull. Fac. Fish., Nagasaki Univ., No. 20:55-57.

bathymetry
Shumway, G., 1957.
Carnegie Ridge bathymetry. Deep-Sea Res., 4(4): 250-253.

~~bathymetry~~
Simpson, E.S.W., and Erica Forder, 1968.
The Cape Submarine Canyon. 1. Bathymetry.
Fish. Bull. Misc. Contrib. Oceanogr. Fish. Biol. S. Afr., 5:35-38.

bathymetry
*Simpson, E.S. W., and A. du Plessis, 1968.
Bathymetric, magnetic, and gravity data from the continental margin of southwestern Africa.
Can. J. Earth Sci., 5(4-2):1119-1123.

bathymetry
Simpson, E.S.W., A. du Plessis and Erica Forder, 1970.
Bathymetric and magnetic traverse measurements in False Bay and west of the Cape Peninsula.
Trans. roy. Soc. S Afr. 39 (2): 113-116

bathymetry
Skogsberg, T., 1936.
Hydrography of Monterey Bay, California. Thermal conditions, 1929-1933. Trans. Am. Phil. Soc. 29(1):1-152, 45 textfigs.

bathymetry
Slaucitajs, L., 1947
Ozeanographie des Rigaischen Meerbusens. Teil 1, Statik. Contrib. Baltic Univ. No.45, Pinneberg, 110 pp., 69 text figs.

bathymetry
Somma, A., 1952.
Elementi di Meteorologia ed Oceanografia. Pt. II. Oceanografia. Casa Editrice Dott. Antonio Milano, Padova, Italia, xvii 758 pp., 322 textfigs., maps, tables. (4500 lire).

bathymetry
Smith, E.H., F.M. Soule, and O. Mosby, 1937
The Marion and General Greene Expeditions to Davis Strait and Labrador Sea under Direction of the United States Coast Guard 1928-1931-1933-1934-1935. Scientific Results. Part 2. Physical Oceanography. C.G. Bull. 19: 259 pp., 155 figs., tables.

bathymetry
Sorensen, F.N., and J. Alan Ballad., 1968.
Preglacial structure of Georges Basin and Northeast Channel, Gulf of Maine.
Bull. Am. Ass. Petrol. Geol., 52(3):494-500.

bathymetry
Spiess, Fred N., Bruce P. Luyendyk, Roger L. Larson, William R. Normark and John D. Mudie, 1969.
Detailed geophysical studies on the northern Hawaiian Arch using a deeply towed instrument package.
Marine Geol., 7 (6): 501-527.

bathymetrie
Starr, Robert B. and Robert G. Bassinger, 1968.
Marine geophysical observations of the eastern Puerto Rico - Virgin Islands region. Trans. Fifth Carib. Geol. Conf., St. Thomas, V.I., 1-5 July 1968, Peter H. Mattson, editor. Geol. Bull. Queens Coll., Flushing 5: 25-28.

bathymetry
Stearns, Franklin, 1969.
Bathymetric maps and geomorphology of the middle Atlantic continental shelf.
Fish. Bull. U.S. Dept. Comm. 68(1): 37-66.

bathymetry
Stefansson, Unnstein, 1962.
North Icelandic waters.
Rit Fiskideildar, Reykjavik, 3:269 pp.

bathymetry
Stocchino, C., ed A. Testoni, 1969.
Le correnti nel canale di Corsica e nell' Arcipelago Toscano.
Raccolta Dati Oceanogr. Comm. ital Oceanogr. Consig. naz. Ricerche (A) (19): 1-15.

bathymetry
Stocks, Th., 1961
Eine neue Tiefenkarte der Deutschen Bucht.
Berichte zur deutschen Landskunde, 27(2):280-305.

bathymetry
Stocks, T., 1960
Zur Bodengestalt des Indischen Ozeans: Bericht uber den gegenwärtigen Stand der Forschung.
Erdkunde, 14(3):161-170.
Abstr. in:
Meteorol. & Geoastrophys. Abstr., 12(5): 910.
Also in:
Deutsches Hydrogr. Inst., Ozeanogr., 1960

bathymetry
Stocks, Theodor, 1958.
Untermeerische Bodenformen und ihre Nutzung für die Schiffahrt. In: Ozeanographie, 1958, Deutsches Hydrographisches Institut, 36-64 (The original journal is not obvious from this reprint)

bathymetry
Stocks, T., 1956.
Der Boden der südlichen Nordsee. 2. Beitrag. Eine neue Tiefenkarte der südlichen Nordsee.
Deutsche Hydrogr. Zeits., 9(6):265-280.

bathymetry
Stocks, T., 1950.
Die Tiefenverhältnisse des Europäischen Nordmeeres.
Deutsche Hydro. Zeits. 3(1/2):93-100, Pl. 6.

bathymetry
Stocks, T., 1950.
Tiefenkarte des Hawaii-Sockels. Mitt. Geogr. Gesellschaft, Hamburg, 49:134-142, Pls. 3-4.

bathymetry (sources of data).
Studds, F.A., 1950.
Coast and Geodetic Survey data; an aid to the coastal engineer. Inst. Coastal Eng., Univ. Ext. Univ. Calif., Long Beach, 11-13 Oct. 1950:29 pp. pls. (multilithed).

bathymetry
Summerhayes, C.P., 1969.
Submarine geology and geomorphology off northern New Zealand.
N.Z. Jl Geol. Geophys. 12(2/3): 507-525

bathymetry
Summerhayes, C.P., 1967
Bathymetry and topographic lineation in the Cook Islands.
N.Z. Jl. Geol. Geophys. 10(6): 1382-1399.

bathymetry
Summerhayes, C.P., 1967.
The marine geology of the Auckland Islands area.
Trans. R. Soc., N.Z., Geol., 4 (13): 235-244

bathymetry
Sutton, G.H., 1971.
Scan-Challenger Surveys. Initial Repts. Deep Sea Drilling Project, 8: 951-989.

bathymetry
Sverdrup, H.U., 1950.
Physical oceanography of the North Polar Sea.
Arctic 3(3):178-186, textfigs.

bathymetry
Sverdrup, H. U., 1940.
The Arctic regions. Assoc. d'Ocean. Phys., Spec Art., Publ. Sci. No. 8:50-53, 1 fig.

filed under Assoc.

BATHYMETRY
Swift, Donald J. P., and Anil K. Lyall, 1968.
Origin of the Bay of Fundy, an interpretation from sub-bottom profiles.
Marine Geol., 6 (4): 331-343

T

Bathymetry
Tagg, A. R., and E. Uchupi, 1966.
Distribution and geologic structure of Triassic rocks in the Bay of Fundy and the northeastern part of the Gulf of Maine.
Geol. Surv. Res., 1966 (B), Geol. Surv. Prof. Paper. 550-B:B95-B98.

bathymetry
Tapager, J.R.D., 1955.
Local environmental factors affecting ice formation in Søndre Strømfjord, Greenland.
U.S.H.O. Tech. Rept., TR-22:26 pp.

bathymetry
Tarassov, A.N., 1961.
[Question of the coordinating of sounding tracks and oceanographic profiles.]
Trudy Inst. Okeanol., Akad. Nauk, SSSR, 47:3-14.

bathymetry
Tatara K., 1955.
Preliminary note on the bottom topography of the "Damo" area. Rec. Ocean. Wks., Japan, 2(3):71-78.

bathymetry
Tayama, R., 1952.
Coral reefs of the South Seas. Publ. No. 941, Bull. Hydrogr. Office 11:292 pp., (English resume) pp. 184-292) with Appendix 1 of 133 pp. of photos and Appendix 2 of 18 charts.

bathymetry
Tayama, R., 1950.
[The submarine configuation off Sikoku especial -ly the continental slope.] Hydro. Bull. Spec. Publ. 7:54-82.

bathymetry
Tchernia P., 1951.
Compte-rendu préliminaire des observations océanographiques faites par le Batiment Polaire "Commandant Charcot" pendant la campagne 1949-1950. Bull. d'Info., C.O.E.C., 3(1):13-22, 11 figs.

bathymetry
Thijsse, J. Th., 1950.
Veranderingen in Waterbewegingen en Bodemrelief in de Waddenzee. Waddensymposium, Tijdschr. Kon. Nederl. Aardrijkskundig Genootschap:74-77.

bathymetry
Thomsen, H., 1938.
Hydrographical observations made during the DANA Expedition, 1928-30. Dana Rept. 12:46 pp.

bathymetry
Tiphane, Marcel 1965
Topographie de la Baie-des-Chaleurs.
Cah. Inf. Stn. biol. mar. Grande-Rivière, 29: 3pp., fold-in chart (multilithed)

bathymetry
Tiphane, Marcel 1966.
Topographie de la région des bancs de pêche gaspésiens.
Cah. Inf. Stn. biol. mar. Grande-Rivière 33: 4pp., fold-in chart.

bathymetry
Tolstoy, I., 1951.
Submarine topography in the North Atlantic.
Bull. G.S.A. 62(5):441-450.

bathymetry
Tolstoy, I., and M. Ewing, 1949.
North Atlantic hydrography and the Mid-Atlantic Ridge. Bull. G.S.A. 60(10):1527-1540, 8 pls.

WHOI Contrib. No. 498.

No mention of hydrography!

bathymetry
Trechnikov, A.F., 1960.
[Nouvelles données sur la relief du fond et sur les eaux du bassin Arctique.] Priroda, Feb. 1960: 25-32.

Translation in: Cahiers Océanogr., C.C.O.E.C., 12(9):622-630.

Bathymetry
Treherne, P. C., 1962.
Hydrographic surveys off the Lybian coast for an oil terminal.
Int. Hydrogr. Rev., 39(1):39-50.

bathymetry
Tressler, W.L., S. Bershad and W.H. Berninghausen 1956.
Rochedos Saõ Pedro e São Paulo (St. Peter and St. Paul Rocks). H.O. Tech. Rept. Tr-31:63 pp.

bathymetry
Trites, R.W., 1960.
An oceanographical and biological reconnaissance of Kennebecasis Bay and the Saint John River estuary.
J. Fish. Res. Bd., Canada, 17(3):377-408.

bathymetry
Trotti, Leopoldo, 1969.
A bathymetric and geological survey in the middle Adriatic Sea.

Rapp. P.-v. Reun. Commn int. Explor. scient. Mer Mediterr., 19(4): 611-612.

bathymetry
Trotti, Leopoldo 1968.
A bathymetric and geological survey in the middle Adriatic Sea.
Int. hydrog. Rev. 45(2): 59-71.

bathymetry
Trotti, L., and M. Bini, 1954.
Topografia della soglia sottomarina compresa tra l'Isola di Capraia e Capo Corso.
An. Idrografici, 14 pp.

incomplete reference - cited Trans AGU

U

bathymetry
Uchupi, Elazar, 1968
Atlantic Continental shelf and slope of the United States - physiography. Description of the submarine physiography of the Continental margin between Nova Scotia and the Florida Keys.
Prof. Pap. U.S. Geol. Surv., 529 C: C1-C30

bathymetry
*Uchupi, Elazar, 1968.
Tortugas Terrace, a slip surface?
Prof. Pap. U.S. Geol. Surv., 600-D:D231-234.

WHOI Contrib. No. 1989.

bathymetry
*Uchupi, Elazar, 1967.
Bathymetry of the Gulf of Mexico.
Trans. Gulf Coast Assoc. Geol. Soc. 17th Ann. Meet.: 161-172.

bathymetry
*Uchupi, Elazar, 1967.
The continental margin south of Cape Hatteras, North Carolina.
SEast. Geol., Duke Univ., 8(4):155-177.

WHOI Contrib. No. 1918.

bathymetry
Uchupi, E. 1966
Topography and structure of Cashes Ledge, Gulf of Maine.
Maritime Sediments, Halifax 2(3): 117-120.

bathymetry
Uchupi, Elazar, 1966.
Shallow structure of the Straits of Florida.
Science, 153(3735):529-531.

bathymetry
Uchupi, Elazar, 1966.
Structural framework of the Gulf of Maine.
J. geophys. Res., 71(12):3013-3028.

bathymetry
Uchupi, Elazar, and K.O. Emery, 1968.
Structure of continental margin off Gulf coast of United States.
Bull. Am. Ass. Petr. Geol. 52 (7): 1162-1193.

bathymetry
bottom sediments
diapirs

bathymetry
Uchupi, Elazar, J.D. Milliman, Bruce P. Luyendyk, C.O. Bowin and K.O. Emery, 1971
Structure and origin of southeastern Bahamas.
Bull. Am. Ass. Petrol. Geol. 55 (5): 687-704.

bathymetry
Uda, Michitaka, 1962
Subarctic oceanography in relation to whaling and salmon fisheries.
Sci. Repts., Whales Res. Inst., No. 16: 105-119.

bathymetry
Udintsev, G.B., 1966.
Results of the upper mantle project studies in the Indian Ocean by the research vessel "Vityaz"
In: The world rift system, T.N. Irvine, editor, Dept. Mines Techn. Surveys, Geol. Survey, Can., Paper, 66-14:148-171.

bathymetry
Udintsev, G.B., 1960
[On the bottom relief of the western region of the Pacific Ocean.]
Okean. Issle., IGY Com., SSSR 5-32.

bathymetry
Udintsev, G.B., 1959.
Investigation of the relief of seas and oceans.
Akad. Nauk, SSSR, Itogi Nauki, Dostizheniia Okeanol. 1. Uspechi b Izuchenii Okean. Glubin (Biol. i Geol.):27-90.

bathymetry
Udinzev, G.B., 1957.
[Bottom relief of the Okhotsk Sea]
Trudy Inst. Okeanol., 22:3-76.

bathymetry
Udintsev, G.B., G.V. Agapova, A.F. Beresnev, L. Ya. Boudanova, L.K. Zatonskiy, N.L. Zenkevich, A.G. Ivanov, V.F. Kanaiev, I.P. Koutcherov, N.I. Larina, N.A. Marova, V. Mineiev, and E.I. Rautskiy, 1963.
The new bathymetric map of the Pacific.
Okeanol. Issled. Rezhult. Issled., Programme Mezhd. Geofiz. Goda, Mezhd. Geofiz. Komitet, Presidiume, Akad. Nauk, SSSR, No. 9:60-99.

bathymetry
Udintsev, G.B., I.G. Boichenko and V.F. Kanaev, 1959
[Relief of the Bering Sea.] Trudy Inst. Okeanol. 29: 17-64. (In Russian)

bathymetry
Ulrich, Johannes 1971.
Zur Topographie und Morphologie der Grossen Meteorbank.
Meteor-Forsch-Ergebnisse (C) 6: 48-68

bathymetry
Ulrich, Johannes 1968
Die Echolotungen des Forschungsschiffe Meteor im Arabischen Meer während der Internationalen Indischen Ozean Expedition.
Meteor Forschungsergebn (C) 1:1-12

bathymetry
Ulrich, J., 1962
Echolotprofile der Forschungsfahrten von F.F.S. Anton Dohrn und V.F.S. Gauss im Internationalen Geophysikalischen Jahr, 1957/1958.
Deutsch. Hydrogr. Zeits., Erganzungsheft, 4(6) 15 pp.

bathymetry
Ulrich, J., 1960
Zur Topographie des Reykjanes-Rückens.
Kieler Meeresf., 16(2): 155-163.

Bathymetry
United States, Department of Commerce, Environmental Sciences Services Administration, 1965
International Indian Ocean Expedition, USC&GS Ship Pioneer - 1964.
Vol. 1. Cruise Narrative and scientific results 139 pp.
Vol. 2. Data report: oceanographic stations, BT observations, and bottom samples, 183 pp.

bathymetry
United States Naval Oceanographic Office, 1965
Oceanographic Atlas of the North Atlantic Ocean, Section V.
U.S. Oceanogr. Off., Publ., No. 700:71 pp. (quarto)

Bay of Biscay
Mediterranean
Bosporus

Gulf of Persia
underwater photography
oceanography
... effect of ... action

V

bathymetry
Vallaux, C., 1927.
Expédition scientifique du "Meteor" au sud de l'Atlantique et dans l'Ocean Austral (1925-1926): premiers résultats. Ann. Inst. Océan., n.s., 4(1):1-24, 6 textfigs.

bathymetry
Van Andel, Tjeerd H., Richard P. Von Herzen and J.D. Phillips 1971.
The Vema Fracture Zone and the tectonics of transverse shear zones in oceanic crustal plates.
Mar. geophys. Researches 1 (3): 261-283.

bathymetry
van der Linden, Willem J.M., 1970
Morphology of the Tasman Sea floor
N.Z. Jl Geol. Geophys. 13(1): 282-291.

seamounts
rises

bathymetry
van Huyster, Th., 1944.
Soundings and bathymetric charts. III. Echo soundings. Snellius Rept., Vol. 2, Pt. 2, Ch. 3: 134 pp.

bathymetry
*Vanney, J.R., 1968.
La pente continentale au large des Charentes (Golfe de Gascogne): présentation géomorphologique préliminaire.
Trav. Cent. Rech. Etud. oceanogr., n.s., 7(3/4):15-23.

bathymetry
Van Riel, P. M., 1943.
The bottom water, Introductory remarks and oxygen content. Snellius Exped. Vol. 2, Pt. 5, Ch. 1:77 pp., 34 textfigs.

bathymetry
Van Riel, P.M., 1940.
The bottom configuration of the Austral-Asian Archipelago. Assoc. d'Ocean. Phys., Spec. Art., Publ. Sci., No. 8:87-92, fig. 5.

bathymetry
Van Voorhis, Gerald D., and Thomas M. Davis, 1964
Magnetic anomalies north of Puerto Rico: trend removal with orthogonal polynomials.
J. Geophys. Res., 69(24):5363-5371.

bathymetry
Van Weel, K. M., 1923.
Meteorological and hydrographical observations made in the western part of the Netherlands East Indian Archipelago. Treubia, 4:559 pp.

bathymetry
*Viglieri, Alfredo, 1968.
La carte générale bathymétrique des océans établie par S.A.S. le Prince Albert 1er.
Bull. Inst. océanogr., Monaco, No. spécial 2: 243-253.

bathymetry
Viglieri, A., 1959
Carte générale bathymétrique des océans.
Scientia, 80(5): 103-111.

bathymetry
*Villegas, L., H. Trujillo y E. Löhre, 1967.
Informe sobre las investigaciones exploratorias realizadas en la zona de Chiloe-Guitecas, durante febrero y marzo de 1967, con el B/I Carlos Darwin.
Publ. Inst. Fomento Pesquero, Chile, 32:19, pp. (mimeographed).

bathymetry
Vinogradova, P.S., 1966.
On the structure of the shelf and continental slope off West Spitzbergen. (In Russian).
Materiali Sess. UChen. Soveta PINRO Rez. Issled. 1964, Minist. Ribn. Khoz. SSSR, Murmansk, 49-58.

bathymetry
Vinogradova, P. S., 1964.
Some results of the study of the continental slope along the western border of the Barents Sea. (In Russian).
Material. Sess. Uchen. Sov. PINRO, Rez. Issled., 1962-1963, Murmansk, 20-34.

bathymetry
Vinogradova, P.S., and V.M. Litvin, 1960
Studies on the bottom relief and sediments in the Barents and Norwegian seas. Soviet Fish. Invest., North European Seas, VNIRO, PNIRO, Moscow, 1960:97-110.

ocean depths
Virgili, Carmina, 1967.
El límite de los océanos.
In: Ecología marina, Monogr. Fundación La Salle de Ciencias Naturales, Caracas, 14:1-34.

bathymetry
Vogt, P.R., G.L. Johnson, T.L. Holcombe, J.G. Gilg and O.E. Avery 1971.
Episodes of sea-floor spreading recorded by the North Atlantic basement.
Tectonophysics 12 (3): 211-234.

bathymetry
Volkov, P., 1961.
New exploration of the bottom topography in the Greenland Sea.
Morskoi Flot, 1961(3):35-37.

Translation: E.R. Hope T356R Copy: Reflom

bathymetry (data)
von Bonde, C., 1950.
Twentieth annual report for the year ended December 1949. Commerce and Industry, April 1950: 412 pp., 12 charts.

bathymetry (data)
von Bonde, C., 1949
Nineteenth Annual Report for the year ended December 1947. Division of Fisheries, Dept. Commerce and Industries. Commerce & Industry, S. Africa, Apr. 1949:415-476, 4 charts.

bathymetry
Von Huene, Roland, George G. Shor, Jr., and Erk Reimnitz, 1967.
Geological interpretation of seismic profiles in Prince William Sound, Alaska.
Bull. geol. Soc. Am., 78(2):259-268.

bathymetry
Vroman, M., 1961.
On the Red Sea rift problem.
Bull. Res. Council, Israel, (G) 10(1-4):321-338.

bathymetry
Vuletic, A., 1953.
Structure geologique du fond du Malo et du Veliko Jesero, sur l'ile de Mjet.
Acta Adriatica 6(1):3-63, 16 textfigs.

W

bathymetry
* Watson, Jerry A., and G. Leonard Johnson, 1969.
The marine geophysical survey in the Mediterranean.
Int. hydrogr. Rev., 46(1): 81-107.

bathymetry
Wellman, Peter 1967
The aeromagnetic anomalies and the bathymetry of the central part of the Hawaiian Ridge.
N.Z. Jl Geol. Geophys. 10(6): 1407-1423

bathymetry
Wennekens, M.P., 1959
Water mass properties of the Straits of Florida and related waters.
Bull. Mar. Sci., Gulf and Caribbean, 9(1): 1-52.

bathymetry
Weyl, R., 1949.
Eine neue Tiefenkarte der Caribischen See und ihre tektonische Ausdeutung. Petermanns Geogr. Mitt., 93 Jahrgang, 173-174, 1 textfig.

bathymetry
Whitcroft, H.J., 1945
The bathymetry of the Central American Region., West Coast of Mexico and Northern Central America. Am. Geophys. Union Trans. of 1944, Part IV:606-608, 1 textfig.

bathymetry a

*Whitmarsh, R.B., 1971.
Interpretation of long range sonar records obtained near the Azores. Deep-Sea Res., 18(4): 433-440.

bathymetry

Wilde, Pat, 1964
Deep-sea channels on the Cocos Ridge, East-Central Pacific. (Abstract).
Trans. Amer. Geophys. Union, 45(1):70.

bathymetry

Williams, W.W., 1947
The determination of gradients on enemy-held beaches. Geograph. Jour. CIX (1/3):76-90, 6 textfigs, 6 photos.

bathyscaphe

Willm, P., 1958
Le bathyscaphe.
Journées des 24 et 25 février 1958: 187-192.
Publ. by Centre Belge d'Océan. Recherches Sous-marines.

bathyscaphes

Willm, P., 1958.
Résultats scientifiques des campagnes du Bathyscaphe F.N.R.S. III, 1954-1957. 3. Notes techniques.
Ann. Inst. Océanogr., Monaco, n.s., 35(4):245-254.

bathymetry

*Windisch, Charles C., R.J.Leyden, J.L. Worzel, T. Saito and J. Ewing, 1968.
Investigation of Horizon Beta.
Science, 162(3861):1473-1479.

bathymetry

Winston, Edward J., and Kenneth D. Etulain, 1969.
Mapping the ocean bottom --- the sea depths on a continuing, systematic, and ecological basis has become a virtual necessity.
Oceanology intl, 4(3):40-42. (popular)

bathymetry

Wold, Richard J. and Ned A. Ostenso, 1971.
Gravity and bathymetry survey of the Arctic and its geodetic implications. J. geophys. Res. 76(26): 6253-6264.

bathymetry a

Wold, R.J., T.L. Woodzick and N.A. Ostenso 1970.
Structure of the Beaufort Sea continental margin.
Geophysics 35(5):849-861.

bathymetry

Wüst, G., 1942.
Die morphologischen und ozeanographischen Verhältnisse des Nordpolarbeckens. Veröffentlichungen des Deutschen Wissenschaftlichen Instituts zu Kopenhagen. Reihe 1: Arktis. Herausgegeben von Prof. Dr. Phil. Hans Freibold, No. 6:21 pp., 7 textfigs., 1 pl. (Gebrüde Borntrager, Berlin-Zehlendorf, 1942).

bathymetry

Wüst, G., 1941.
Relief und Bodenwasser im Nordpolarbecken.
Zeitschr. d. Ges. f. Erdkunde, 1941:163-180.
Berlin

bathymetry

Wüst, G., 1940.
Das Relief des Azorensockels und des Meeresbodens nördlich und nordwestlich der Azoren. Wiss. Ergeb. Internat. Golfstrom-Unternehmung 1938, Lief. 2, In August Beiheft, Ann. Hydr., usw., 68:19 pp., 7 textfigs., 5 pls.

bathymetry

Wüst, G., 1939.
Das submarine Relief bei den Azoren.
Abhandl. Preussischen Akad. Wissenschaften, Jahrgang 1939, Phys.-math. Kl. No.5:46-58, 7 text figs.

XYZ

bathymetry

Yamazi, Isamu, 1966.
Zooplankton communities of the Navesink and Shrewsbury rivers and Sandy Hook Bay, New Jersey.
Tech.Pap.Bur.Sport Fish.Wildl.,U.S.,(2):1-44. (multilithed).

bathymetry

Yamazi, I., 1953.
Plankton investigations in inlet waters along the coast of Japan. IX. The plankton of Onagawa Bay on the eastern coast of Tohoku District.
Publ. Seto Mar. Biol. Lab. 3(2)(Article 17):173-187, 10 textfigs.

bathymetry

Yamazi, I., 1950.
Plankton investigations in inlet waters along the coast of Japan. Publ. Seto Mar. Biol. Lab. 1(3):93-113, 14 textfigs.

bathymetry

*Yasui, Masashi, Yuichi Hashimoto and Seiya Uyeda, 1967.
Geomagnetic and bathymetric study of the Okhotsk Sea -(1).
Oceanogrl Mag., 19(1):73-85.

bathymetry

Yasui, Masashi, Jotaro Masuzawa, Tsutomu Sawara and Toshisuke Nakai, 1962(1961)
Sounding of the Ramapo Deep.
Repts. JEDS, Deep-Sea Res. Comm., Japan Soc., Promotion of Science, 3:41-49.

Originally published (1961) in:
Oceanogr. Mag., Tokyo, 13(1):41-49.

bathymetry

Yoshida, K., K. Kajiura, and K. Hidaka, 1953.
Preliminary report of the observation of ocean waves at Hachijo Island. Rec. Ocean. Wks., Japan, n.s., 1(1):81-87, 8 textfigs.

bathymetry

Yoshikawa, T., 1953.
Some consideration on the continental shelves around the Japanese Islands.
Nat. Sci. Rept., Ochanomizu Univ., 4(1):138-150, 9 textfigs.

bathymetry

Zatonsky, L.K., 1965.
New bathymetric chart of the Atlantic Ocean. (In Russian; English abstract).
Okeanolog. Issled. Rezult. Issled. Programme Mezhd. Geofiz. Goda, Mezhd. Geofiz. Komitet Presidiume, Akad.Nauk, SSSR, No. 13:172-180.

bathymetry

Zatonsky, L.K., V.F. Kanaev, V.I. Tikhonov, G.B. Udintsev, 1961
The bottom topography of the Kuril-Kamchatka arch and its vulcanism.
Okeanologiia, Akad. Nauk, SSSR., (2):265-269.

bathymetry

Zeigler, J.M., W.D. Athearn and H. Small, 1957.
Profiles across the Peru-Chile Trench.
Deep-Sea Res. 4(4):238-249.

bathymetry

Zeleginskaya, L.M. 1966.
On the quantitative indicies of mortality of plankton components in shallow places of the Black Sea. (In Russian; English abstract.
Zool. Zh. 45(8):1251-1253.

bathymetry

Zenkevich, N.L., 1959.
Recent data on the bottom relief of the Sea of Japan.
Izv., Akad. Nauk, SSSR, Ser. Geogr., (3):86-

bathymetry

Zenkevich, N.L., 1957.
Bottom relief of the northern part of the Japan Sea. Trudy Inst. Okeanol., 22:252-259.

bathymetry

Zhivago, A. V., and A. P. Lisitsin, 1958.
Bottom relief and the deposits of the Southern Ocean.
Inform. Biull. Sovetsk. Antarkt. Exped., (3):21-22.

bathymetry

Zhivago, A.V., and V.V. Patrikeev, 1949.
Method of recording the variations in the littoral relief of the sea bottom during storms. Izvest Akad. Nauk. SSSR, Ser. Geogr. i. Geofiz., 13(2): 151-153.

bathymetry

Zorell, F., 1935.
Beiträge zur Hydrographie der Deutschen Bucht auf Grund der Beobachtungen von 1920 bis 1932 -- Arch. Deutschen Seewarte 54(1):1-69, 24 pls.

bathymetry, ancient

Tanner, William F., 1971. a
Numerical estimates of ancient waves, water depth and fetch.
Sedimentology, 16(1/2): 71-88

bathymetry, data only

bathymetry (data only)

Brandhorst, Wilhelm, Manuel Mendez y Omar Royas 1966.
Observaciones oceanográfico-biologicas sobre los recursos de la anchoveta (Engraulis ringens Jenyns) en la zona norte de Chile.
Publ. Inst. Fomento Pesquero Chile, 22: 3 pp., 9 tables (mimeographed).

BT(data only)

Canada, Canadian Committee on Oceanography, 1965.
Ocean Weather Station "P", North Pacific Ocean, May 16 to August 12, 1964.
Canadian Oceanographic Data Centre, 1965 Data Record Series, No. 3:112 pp. (Unpublished manuscript).

BT(data only)

Canada, Canadian Committee on Oceanography, 1965.
Data Record, Gulf of St. Lawrence and Halifax section, August 26 to September 3, 1963.
Canadian Oceanogr. Data Centre, 1965 Data Rec. Ser. No. 2:98 pp. (Unpublished manuscript).

bathymetry, data for

Dishon, Menachem, and Bruce C. Heezen 1968.
Digital deep-sea sounding library: description and index list.
Int. hydrogr. Rev. 45(2):23-39.

bathymetry (data only)

Great Britain, Ministry of Defence, Hydrographic Department, 1966.
Bathymetric, magnetic and gravity investigations. H.M.S. Owen, 1962-1963.
Admiralty Marine Sci. Publ. (9)(H.D. 567): 19pp., numerous charts. (in 2 parts).

bathymetry (data only)

Hayes, Dennis E., James R. Heirtzler, Ellen M. Herron and Walter C. Pitman III
A. Navigation. B. Bathymetric and geomagnetic measurements.
Prelim. Rept. U.S.N.S. Eltanin Cruises 22-27, Jan. 1966 - Feb. 1967, Lamont-Doherty Survey of the World Ocean, 21 (Tech Rept. 2-CU-2-69):130 pp. (multilithed)
1969

BT (data only)
Japan, Maritime Safety Agency 1966.
Results of oceanographic observations in 1963.
Data Rept. Hydrogr. Obs., Ser. Oceanogr., Publ. 792 (3): 74 pp.

bathymetry (data only)
Japan, Shimonoseki University of Fisheries, 1970
Data of oceanographic observations and exploratory fishings. 6. Oceanographic surveys of the Kuroshio and its adjacent waters, 1967 and 1968: 117 pp.

bathymetry (data only)
Japan, Shimonoseki University of Fisheries 1968
Oceanographic surveys of the Kuroshio and its adjacent waters 1965 and 1966.
Data Oceanogr. Obs. Explor. Fish. 4: 1-178.

bathymetry (data only)
Japan, Shimonoseki University of Fisheries, 1966.
Fisheries oceanography and exploratory trawl fishings in the Bering Sea.
Data of Oceanographical Observations and Exploratory Fishings, No. 2: 109 pp.

bathymetry (data only)
Japan, Shimonoseki University of Fisheries 1965
International Indian Ocean Expedition 1962-63 and 1963-64.
Data Oceanogr. Obs. Explor. Fish. 1: 453 pp.

bathymetry (data only)
Nato Subcommittee on Oceanographic Research, 1962
Current measurements, meteorological observations and soundings of the M/S "Helland-Hansen" near the Strait of Gibraltar, May -June 1961. Tables. 29 pp. (multilithed).

bathymetry (data only)
United States U.S. Department of Commerce, Environmental Science Services Administration, 1969.
Bathymetry, geomagnetic and gravity data, USC&GS Ship Pioneer - 1964.
Int. Indian Ocean Exped. 3: 250 pp.

bathymetry, data processing

bathymetry, data processing
Smith, Stuart M., 1963
An investigation of automatic bathymetric data processing.
Int. Hydrogr. Rev., 40(1):131-147.

bathymetry, effect of

bathymetry, effect of
Allard, P., 1953.
Influence de la configuration des bassins sur le régime des marées littorales.
Ann. Inst. Océan., Monaco, 28(2):63-112, 31 text-figs.

bathymetry, effect of
*Ripper, I.D. and R. Green, 1967.
Tasmanian examples of the influence of bathymetry and crustal structure upon seismic T-wave propagation.
N.Z. Jl Geol. Geophys. 10(5):1226-1230.

bathymetry, micro-
Trabant, Peter K., 1969.
Coastal microbathymetry off Fort Lauderdale Florida.
J. mar. Techn. Soc., 3(3): 47-52

bathymetry, seamounts
Scrimger, J.A. and W. Halliday, 1971.
Bathymetry of Pacific seamounts Bowie and Union.
Deep-Sea Res., 18(1): 123-126.

Bowie Seamount
Union Seamount
bathymetry, seamounts

bathynauts

bathynauts (Deep-Research vessel)
Dietz, R.S., 1963
23. Bathyscaphs and other deep submersibles for oceanographic research. In: The Sea, M.N. Hill, Edit., Vol. 2 (V) Oceanographic Miscellanea, Interscience Publishers, New York and London, 497-515.

bathyphotometer

bathyphotometer, see also: INSTRUMENTATION, light

bathyphotometer
Clarke, G.L., and C.J. Hubbard, 1959.
Quantitative records of the luminescent flashing of oceanic animals at great depths.
Limnol. & Oceanogr., 4(2):163-180.

WHOI Contrib 1019

bathyscaphe

bathyscaphe
Anonymous, 1961.
Polish bathyscaphe
Vechernaya Moskva, 15 July '61:4.

Abstr.: Polish scientist designs bathyscaphe.
OTS-61-11147-19:5

"bathyscaphe"
Anon., 1956.
Unterwasserbeobachtungen aus einem Hydrostat.
Deutsche Fishherei Zeitung, 3(6):185-189.

translated from Rybnoe Khozjaistvo (12):

bathyscaphe
Azhazha, V., 1961.
Mastery of the oceans
Pravda, 16 July: 3.

Engl. Abstr.: Soviet oceanographic research not without impediments.
OTS-61-11147-18 JPRS:8807:13.

bathyscaphs
Barham, Eric G., 1963
The deep scattering layer as observed from the Bathyscaph Trieste.
Proc., XVI Int. Congr., Zool., Washington, D.C. Aug. 20-27, 1963:298-300.

bathyscaphe
Bernard, F., 1962.
Contribution du bathyscaphe à l'étude du plancton. 1 avantages et inconvénients. Contributions to symposium on zooplankton production, 1961.
Rapp. Proc. Verb., Cons. Perm. Int. Expl. Mer, 153:25-28.

Bathyscaphe
Bernard, Francis, 1958.
Plancton et benthos observés durant trois plongées en Bathyscaphe au large de Toulon.
Ann. Inst. Océan., 35:287-326.

bathyscaphe
Bernard, F., 1957.
Plancton observé durant trois plongées en bathyscaphe au large de Toulon. C.R. Acad. Sci., Paris 245(22):1968-1971.

Bathyscaphe
Bernard, F., 1955.
Zooplancton vu au cours d'une plongee du bathyscaphe F.N.R.S. III au large du Toulon.
C.R. Acad. Sci., Paris, 240:2565-2566.

Bathyscaphe
Bernard, F., 1955.
Densité du plancton vu au large de Toulon depuis le Bathyscaphe F.N.R.S. III.
Bull. Inst. Océan., Monaco, No. 1063:16 pp.

bathyscaphe
Botteron, Germain, 1958.
Etude de sédiments recoltés au cours de plongées avec le bathyscaphe "Trieste" au large de Capri
Bull. Lab., Géol., Minéralog. Geophys., et du Musée Geol., Univ. Lausanne, No. 124: 19 pp.

bathyscaphe
Botteron, G., 1958
Etude de sédiments récoltés au cours de plongées avec le bathyscaphe "Trieste" au large de Capri.
Bull. Soc. Vaudoise Sci. Nat., 67 fas 2 (298): 73-92.

bathyscaphe
Brouardel, J., 1958.
Appareils de prélèvement (F.N.R.S. III).
Ann. Inst. Océan., 35:255-258.

bathyscaphe
Cita, M.B., 1955.
Studio sulla microfauna contenata in un campione di fondo raccolto del batiscafo "Trieste" nel Mare di Capri.
Atti Soc. Ital. Mus. Civ., Milan, 94(2):209-221.

bathyscaphe
Delauze, Henri, 1963
Les installations scientifiques du bathyscaphe "Archimède".
Bull. Assoc. Française pour l'Étude des Grandes Profondeurs Océaniques, No. 2:6-8.
(multilithed)

bathyscaphes
Delauze, H., et J.M. Pérès, 1963.
Aperçu sur les résultats de la campagne au Japon du bathyscaphe "Archimède".
Rec. Trav. Sta. Mar., Endoume, Bull., 30(45):3-8.

bathyscaphs
Dietz, R.S., 1963
23. Bathyscaphs and other deep submersibles for oceanographic research. In: The Sea, M.N. Hill, Edit., Vol. 2 (V) Oceanographic Miscellanea, Interscience Publishers, New York and London, 497-515.

bathyscaph
Dietz, Robert S., 1959.
1100 Meter Dive in the Bathyscaph Trieste.
Limn. & Ocean. Vol. 4, No. 1, pp. 94-101.

bathyscaphs
Diomidov, M.N., 1962.
The plans for a Soviet bathyscaph. Methods and results of submarine investigations. (In Russian)
Trudy, Okeanogr. Komissii, Akad. Nauk, SSSR, 14: 127-130.

Bathyscaphe
Fage, Louis, 1958.
Introduction: Les campagnes scientifiques du Bathyscaph F.N.R.S. III, 1954-1957.
Ann. Inst. Ocean., 35:237-242.

bathyscaphe

Page, L., 1958.
Resultats scientifiques des campagnes du Bathyscaphe F.N.R.S. III 1954-1957.
1) Introduction: Les campagnes scientifiques du Bathyscaphe F.N.R.S. III 1954-1957.
Ann. Inst. Ocean. New Ser. 35(4):237-242.

bathyscaph

France, Institut Français de Navegation, 1964.
Le bathyscaphe "Archimède".
Navigation, 12(46):187-188.

bathyscaphe

Furnestin, J., 1955.
Une plongée au bathyscaphe. Rev. Trav. Inst. Pêches Marit. 19(4):435-442.

Bathyscaphe

Houot, G., 1958.
Le Bathyscaphe F.N.R.S. III et la recherche scientifique
Ann. Inst. Ocean., 35:243-244.

Bathyscaphe

Houot, G., 1958.
Resultats scientifiques des campagnes du Bathyscaphe F.N.R.S. III 1954-1957.
2) Le Bathyscaphe F.N.R.S. III et la recherche scientifique.
Ann. Inst. Ocean. New Ser. 35(4):243-244.

instruments, bathyscaphe

Houot, Capt. de Corvette, 1955.
Le bathyscaphe F.N.R.S. 3 au service de l'exploration des grandes profondeurs. Ltr. to Editors, Deep-Sea Res. 2 (3):247-249.

bathyscaphe

Jerlov, N.G., and Jacques Piccard, 1959.
Bathyscaphe measurements of daylight penetration into the Mediterranean.
Deep-Sea Res., 5(3):201-204.

bathyscaph

LaFond, Eugene C., 1962
Bathyscaph dive Eighty-four. (Abstract).
J. Geophys. Res., 67(9):3573.

bathyscaphe

Laughton, A.S., 1959.
The sea floor.
Science Progress, 47(186):230-249.

bathyscaphe

Lyman, John, 1960.
La profondeur atteinte par le Bathyscaphe "Trieste".
Cahiers Oceanogr., C.C.O.E.C., 12(6):389-391.

bathyscaphe

MacKenzie, K.V., 1962
In situ sound-speed measurements aboard the French Bathyscaph ARCHIMEDE at Japan. (Abstract).
J. Acoust. Soc., Amer., 34(12):1974.

bathyscaph

Mackenzie, K.V., 1961
Sound-speed measurements utilizing the bathyscaph "Trieste".
J. Acoustical Soc., Amer., 33(8):1113-1119.

bathyscaphe

Monod, Th., 1954.
Sur un premier essai d'utilisation scientifique du bathyscaphe F.N.R.S. 3. C.R. Acad. Sci., Paris 238:1951.

bathyscaphe

Peres, J.M., 1960.
Le bathyscaphe, instrument d'investigation biologique des mers profonds.
Rec. Trav. Sta. Mar. d'Endoume, 33(20):17-27.

bathyscaphe

Pérès, J.-M., 1959.
Deux plongées au large du Japon avec le bathyscaphe français F.N.R.S. III.
Bull. Inst. Océanogr., Monaco, 1134:28 pp.

bathyscaphe

Pérès, J.-M., 1959.
Observations en bathyscaphe de l'instabilité des vases bathyales méditerranéenes.
Rec. Trav. Sta Mar. d'Endoume, 29(17):3-4.

bathyscaphe

Pérès, J. -M., 1958.
Le benthos du système bathyal de la Méditerranée : plongées en bathyscaphe et pêches profondes.
Centre Belge d'Océanogr., Recherches Sous-Marines Journees, 24 et 25 février, 1958:175-186.

Bathyscaphe

Pérès, J. M., 1958.
Remarques générales sur un ensemble de quinze plongées efectuées avec le Bathyscaphe F.N.R.S. III.
Ann. Inst. Océan., 35:259-286.

bathyscaphe

Pérès, J.M., 1958.
Resultats scientifiques des campagnes du Bathyscaphe F.N.R.S. III, 1954-1957. 5). Remarques générales sur un ensemble de quinze plongées effectuées avec le Bathyscaphe F.N.R.S. III.
Ann. Inst. Océanogr., Monaco, n.s., 35(4):259-286.

Bathyscaphe

Peres, J.-M., 1958.
Trois plongées dans le canyon du Cap Sicié, effectuées avec le bathyscaphe F.N.R.S. III de la Marine Nationale.
Bull. Inst. Océanogr., Monaco, No. 1115:21 pp.

bathyscaphe

Pérès, J.M., and J. Picard, 1956.
Nouvelles observations biologiques effectuées avec le Bathyscaphe FNRS III et considerations sur le système aphotique de la Méditerranée.
Bull. Inst. Océan., Monaco, 1075:1-9.

bathyscaphe

Pérès, J.M., and J. Picard, 1955.
Observations biologiques effectuées au large de Toulon avec le Bathyscaphe F.N.R.S. III de la Marine Nationale. Bull. Inst. Océan., Monaco, No. 1061:1-8.

bathyscaphe

Pérès, J.M., and J. Picard, 1955.
Observations biologiques effectuees avec le bathyscaphe FNRS III. C.R. Acad. Sci., Paris, 240(23):2255-2256.

bathyscaphe

Pérès, J.M., J. Picard and M. Ruivo, 1957.
Résultats de la campagne de recherches du Bathyscaphe FNRS III. Bull. Inst. Ocean., Monaco, No. 1092:29 pp.

bathyscaphe

Picard, J., 1961.
11000 Meter unter dem Meerespiegel. Die Tauchfahrten des Bathyskaphs "Trieste.
F.A. Brockhaus, Wiesbaden.

Reviewed by Th. Stocks, Deutsche Hydrogr. Zeits. 15(1):38(1962).

bathyscaphe

Piccard, Jacques, 1958
Le bathyscaphe et les plongées du Trieste, 1953-1957. Comité pour la Recherche océanographique au moyen du Bathyscaphe Trieste (C.R.O.), Lausanne, 29 pp.

In MBL

bathyscaphe

Piccard, J., and R.S. Dietz, 1957.
Oceanographic observations by the bathyscaph Trieste (1953-1956). Deep-Sea Res., 4(4):221-229.

bathyscaphe

Suzjumov, E.M., 1962
Deep sea research with a bathyscaphe. The Xth Pacific Science Congress.
Okeanalogiia, Akad. Nauk, SSSR, 2(3):530-534.

bathyscaphe

Tregouboff, G. 1961.
Prospection biologique sous-marine dans la région de Villefranche-sur-mer en juillet-août 1960.
Bull. Inst. océanogr. Monaco 58(1220): 14 pp.

bathyscaphe

Tregouboff, G., 1961
Rapport sur les travaux intéressant la planctonologie méditerranéene publiés entre juillet 1958 et octobre 1960.
Rapp. Proc. Verb., Reunions, Comm. Int. Expl. Sci. Mer Méditerranée, Monaco, 16(2):33-68.

Bathyscaphe

Trégouboff, G., 1958.
Le Bathyscaphe au service de la planctonologie.
Ann. Inst. Océan., 35:327-341.

bathyscaphe

Trégouboff, G., 1958.
Prospection biologique sous-marine dans la région de Villefranche-sur-Mer au cours de l'année 1957.
1. Plongées en bathyscaphe.
Bull. Inst. Océan., Monaco, No. 1117:37 pp.

bathyscaphe

Tregouboff, G., 1956.
Prospection biologique sous-marine dans la region de Villefranche-sur-mer en juin 1956.
Bull. Inst. Ocean., Monaco, No. 1085: 24 pp.

bathyscaph

USSR, Akademia Nauk, SSSR, 1962
The world's largest bathyscaph.
Okeanologiia, Akad. Nauk, SSSR, 2(2):378-379.

bathyscaphe

Willm, P., 1958.
Notes techniques (F.N.R.S. III).
Ann. Inst. Océan., 35:245-254.

bathysphere

Bathysphere ("hydrostat")

Kiselev, O.N., 1960.
[On tests of a deep-water "hydrostat" PINRO.]
Nauchno-Techn. Biull., PINRO, 3(13):42-43.

bathysphere, observation from

Tsujita, T., 1953.
Studies on naturally occurring suspended organic matter in waters adjacent to Japan II.
Rec. Ocean. Wks., Japan, n.s., 1(2):94-100, 2 textfigs., 1 pl., of 4 figs.

bathythermograph
see also Temperature (BT)

BT= bathythermograph, i.e. temperature information has been collected with this instrument and is therefore shallow.

Bt (data)

Australia, Commonwealth Scientific and Industrial Organization, 1962.
Oceanographic observations in the Indian Ocean in 1961.
Div. Fish. Oceanogr., Cruise, DM 1/61:88 pp.

BT

Berrit, G.R., 1961
Etude des conditions hydrologiques en fin de saison chaude entre Pointe-Noire et Loanda.
Cahiers Oceanogr., C.C.O.E.C., 13(7): 456-461.

BT

Berrit, G.R., R. Gerard, L. Lemasson, J.P. Rebert et L. Vercesi, 1968.
Observations océanographiques exécutées en 1967 II. Bathythermogrammes. Doc. Centre Rech. océanogr. Abidjan, 027.

BT

Biays, P., 1960.
Le courant du Labrador et quelques-unes de ses conséquences géographiques.
Cahiers Géogr., Quebec, 4(8):237-302.

BT

*Boudreault, F. Robert, 1967.
Régime thermique saisonnier d'une station-pilote à l'entrée de la baie des Chaleurs.
Naturaliste Can., 94:695-698.

Also: Trav. Pêch., Quebec, 18.

BT

Bourkov, V.A., 1963
Some results of oceanographic observations with express methods to the east and south of Japan.
Okeanol. Issled., Rezhult. Issled., Programme Mezhd. Geofiz. Goda, Mezhd. Geofiz. Komitet, Presidiume Akad. Nauk, SSSR, No. 9:32-41.

BT data

Bumpus, D.F., 1958.
Oceanographic observations, 1956, East coast of the United States. USFWS Spec. Sci. Rept., Fish., 233:132 pp.

BT

Bumpus, D.F., and Joseph Chase, 1965.
Changes in the hydrography observed along the east coast of the United States.
ICNAF Spec. Publ. No. 6:847-853.

BT

Canada, Fisheries Research Board of Canada, 1959.
Bathythermograms and meteorological data record. Swiftsure Bank and Umatilla Reef Lightships 1958.
Mss. Rept. Ser. (Oceanogr. Limnol.) No. 37:121 pp.

BT

Chew, Frank, K.L. Drennan and W.J. Demoran, 1962.
On the temperature field east of the Mississippi Delta.
J. Geophys. Res., 67(1):271-279.

BT

Conseil Permanent International pour l'Exploration de la Mer, 1963.
ICES oceanographic data lists, 1957 (1):277 pp.

Bt (data)

Forest, Jacques, 1966.
Compte rendu et liste des stations. Campagne de la Calypso au large des cotes Atlantiques de l'Amerique du sud (1961-1962).
Ann. Inst. Oceanogr., Monaco, n.s. 44:329-350.

BT

Giovando, L. F., 1962.
The OCEAN system of assessment of bathythermograms.
Fish. Res. Bd., Canada, Mss. Rept. Ser., (Oceanogr. & Limnol.), No. 105:numerous pp. (Multilithed).

BT- (data)

Hanzawa, Masao, 1966.
Verification of the Fourier approximation of the space-wise interpolation of BT data in the Pacific Ocean.
Oceanogr. Mag., Jap. Met. Soc., 18(1/2): 113-135

Pacif.
BT (data)

BT

Hermann, F., 1952.
Hydrographic conditions in the southwestern part of the Norwegian Sea, 1951. Cons. Perm. Int. Expl. Mer, Ann. Biol. 8:23-26, Textfigs. 4-12.

BT (data)

Heath, R. J., 1967.
Water masses and fronts in the Southern Ocean south of New Zealand.
Bull. N.Z. Dept. Sci. Ind. Res., 174: 1-40

Also: N.Z. Oceanogr. Inst. Mem. 36.

BT

India, Naval Headquarters, Office of Scientific Research & Development, 1957.
Indian Oceanographic Stations List, No. 1/57: unnumbered pp. (mimeographed).

BT

Japan, Maizuru Marine Observatory, Oceanographical Section, 1961
[Report of the oceanographic observations in the Japan Sea from June to July, 1959.]
Bull. Maizuru Mar. Obs., No. 7:57-64.

BT

Japan, Nagasaki Marine Observatory, Oceanographical Section 1960.
Report of the oceanographic observation in the sea west of Japan from January to February, 1960. Report of the Oceanographic observation in the sea north-west of Kyushu from April to May, 1960.
Results Mar. Meteorol. & Oceanogr., J.M.A., 27:42-50; 51-67.

Also in:
Oceanogr. & Meteorol., Nagasaki Mar. Obs., (1961) 11 (202).

BT

Japan, Nagasaki Marine Observatory, Oceanographical Section, 1960
[Report of the oceanographic observation in the sea west of Japan from January to February 1959.]
Res. Mar. Meteorol. & Ocean., J.M.A., 25: 48-56.
Also in:
Oceanogr. & Meteorol., Nagasaki Mar. Obs., (1961) 11(199).

BT

Joseph, J. (compiler), 1961.
Trübungs- und Temperatur-Verteilung auf den Stationen und Schnitten von V.F.S. "Gauss" sowie Bathythermogramme von F.F.S. "Anton Dohrn" und V.F.S. "Gauss" im Internationalen Geophysikalischen Jahr, 1957/1958.
Deutschen Hydrogr. Zeits., Erganz., B(4)(5):1-131.

BT

Lacombe, Henri, and Mme C. Riches, 1961.
Contribution à l'étude du régime du Détroit de Gibraltar. II. Etude hydrologique. Année Géophysique Internationale, 1957-1958. Participation française.
Cahiers Océanogr., C.C.O.E.C., 13(5):276-291.

bathythermograms

LaFond, E.C., 1956.
Bathythermograms - an oceanographic tool.
Current Sci., 25(2):40-41.

bathythermograph

LaFond, E. C., 1949
The use of bathythermograms to determine ocean currents. Trans. Am. Geophys. Union, 30(2):231-237, 6 text figs.

BT

Lalou, C., 1963.
Travaux de la Station Océanographique de Villefranche sur Mer: hydrologie superficielle de la Baie de Villefranche sur Mer d'octobre 1958 à septembre 1960.
Rev. Géogr. Phys. et Géol. Dyn., (2), 5(2):85-100.

Also in:
Trav. Lab. Géol. Dyn. et Centres de Recherches Géodynamiques, Fac. Sci., Univ. Paris, No. 2 (1964).

BT

Lalou, Claude, et Maurice Gennesseaux, 1959

Travaux de la Station Océanographique de Villefranche. Rev. Géogr. Phys. et Géol. Dyn., (2) 2(4): 231-252.

BT

*Lee, A.J., 1967.
Temperature-depth records by the Research Vessel Ernest Holt.
Rapp. P.-V. Reun. Cons. perm. int. Explor. Mer., 157:223-237.

BT

Minas, H.J., and B. Coste, 1964.
Etude de la structure hydrologique et de quelques aspects de la productivité de la zone euphotique en fin d'été au niveau d'une station fixe (Bouée-Laboratoire du Comexo) en rade de Villefranche s/mer.
Rec. Trav. Sta. Mar., Endoume, 34(50):133-155.

bathythermograph data

Lovett, R.F. 1968.
A voyage to the Arctic by submarine.
Mar. Obs. 38(221): 131-136

BT

Megia, T.G., and A.R. Sebastian, 1953.
The thermal structure of the surface waters off western Philippines based on BT observations.
Phil. J. Fish. 1(2):131-143, 8 textfigs.

BT

Montgomery, R.B., and E.D. Stroup, 1962.
Equatorial waters and currents at 150°W in July-August 1952.
The Johns Hopkins Oceanogr. Studies, No. 1:68 pp.

temperature (bathythermograph)

Nasu, K., 1960

Oceanographic investigation in the Chukchi Sea during the summer of 1958.
Sci. Repts., Whales Res. Inst., 15: 143-157.

BT (data)
Nichols, M.M., and M.P. Lynch, 1964.
Shelf observations - hydrography: cruises of January 22-25, July 15-19, 1963.
Virginia Inst. Mar. Sci., Spec. Sci. Rept., (48): 33 pp. (multilithed).

bathythermograph
Pavlov, V.M., Kukoa, V.I., 1957.
[The experience of works with the bathythermograph TB-52 (T-52).] Trudy Inst. Okeanol., 25:88-97.

BT
Robinson, Margaret K., 1961
Comparison of NORPAC temperature sections with average BT sections along the same tracks.
J. Mar. Res., 19(1): 21-27.

BT
Rodin, G.I., 1958.
Oceanographic and meteorological aspects of the Gulf of California.
Pacific Science 12(1):21-45.

BT
Rogalla, E., (1959) 1961
Hydrographic conditions in the Dogger Bank region in October 1959.
Ann. Biol., Cons. Perm. Int. Expl. Mer, 16: 45-46.

BT
Sherman, Kenneth and Robert P. Brown, 1961.
Oceanographic and biological data, Hawaiian waters, January-October 1959.
U.S.F.W.S. Spec. Sci. Rept., Fish., No. 396:71 pp.

bathythermograph observations
Smith, O.R. and E.H. Ahlstrom, 1948
Echo-ranging for fish schools and observations on temperature and plankton in waters off Central California in the spring of 1946. U.S. Fish and Wildlife Ser., Spec. Sci. Rept. No.44, 30 pp. plus 13 figs.

bathythermograms
Tabata, S., C.D. McAllister, J.H. Meikle and H.J. Hollister, 1962
Oceanographic data record, Ocean Weather Station "P", January 17 to August 5, 1962.
Fish. Res. Bd., Canada, Mss. Rept. Ser. (Ocean and Limnol.), No. 139: 113 pp. (multilithed).

B T (data)
United States, Department of Commerce, Environmental Sciences Services Administration, 1965
International Indian Ocean Expedition, USC&GS Ship Pioneer - 1964.
Vol. 1. Cruise Narrative and scientific results 139 pp.
Vol. 2. Data report: oceanographic stations, BT observations, and bottom samples, 183 pp.

BT
USN Hydrographic Office, 1957.
Oceanographic atlas of the Polar seas. 1. Antarctic. H.O. Pub., No. 705:70 pp.

bathythermograph
Voit, S.S., D.A. Aksenov, M.M. Bogorodsky, V.V. Sinukov and V.A. Vladimirzev, 1961
[Some peculiar features of water circulation in the Black Sea and the water regime in the Pre-Bosphorus area.]
Okeanologiia, Akad. Nauk, SSSR, 1(4):613-625.

BT
Wilson, Robert C., Eugene L. Nakamura, and Howard O. Yoshida, 1958
Marquesas area fishery and environmental data, October 1957-June 1958.
USFWS Spec. Sci. Rept. Fish., No. 283: 105 pp.

BT
Wooster, W.S., and T. Cromwell, 1958.
An oceanographic description of the Eastern Tropical Pacific. Bull. S.I.O., 7(3):169-282.

BT
Zalogin, B.S., and M.S. Edel'man, 1959
[Use of the bathythermograph in an Arctic sea.]
Meteorol. i Gidrol. (4): 58-61.
LC or SLA mi $1.80 ph $1.80

BT - data only

BT (data only)
Aliverti, G., e L. Trotti, 1965.
Atlante dei batitermogram raccolti nel Mare Tirreno durante l'A.G.I., 1957-1958.
Pubbl. Ist. Talassogr., No. 415:25 pp.

BT (data)
Bernhard, Michel, 1963.
1. Introduzione. 2. Chemical oceanografica.
Rapp. Attività Sci. e Tecn., Lab. Studio della Contaminazione Radioattiva del Mare, Fiascherino, La Spezia (maggio 1959-maggio 1963), Comit. Naz. Energia Nucelare, Roma, RT/BIO, (63), 8:7-39.

BT(data only)
Berrit, G.R., 1962
Resultats d'observations. Campagne No. 11, Campagne JONAS.
Cahiers Oceanogr., C.C.O.E.C., 14(1):54-76.

BT(data only)
Berrit, G.R. and J.R. Donguy, 1964
Radiale de Pointe-Noire.
Cahiers Océanogr., C.C.O.E.C., 16(3):231-247.

bathythermograph data
Blackburn, M., R.C. Griffiths, R.W. Holmes and W.H. Thomas, 1962.
Physical, chemical and biological observations in the eastern tropical Pacific Ocean: three cruises to the Gulf of Tehuantepec, 1958-1959.
U.S.F.W.S. Spec. Sci. Rept., Fish., No. 420:170pp

BT (data only)
Berrit, G.R., R. Gerard, L. Lemasson, J.P. Rebert et L. Vercesi, 1968.
Observations océanographiques executees en 1967. II. Bathythermogrammes.
Document sci. Provis., Centre Rech. océan., Cote d'Ivoire, ORSTOM, 027: 20 pp. (mimeographed).

bathythermograph (data only)
California, State of, Department of Fish and Game, Marine Research Committee, 1966.
California Cooperative Oceanic Fisheries Investigations, Data Report No. 4:51 pp. (Multilithed).

BT (data only)
Canada, Canadian Oceanographic Data Center, 1969.
Operation Tanquary, Ellesmere Island, N.W.T., 1963-1966.
1969 Data Record Ser. 13: 152 pp.

Arctic, Canadian
temperature (data only)
salinity (data only)
density (data only)
sound velocity (data only)
BT (data only)

bathythermograph (data only)
Canada, Canadian Oceanographic Data Center 1969.
Ocean Weather Station 'P' North Pacific Ocean, May 17 to August 15, 1968.
1969. Data Record Ser. 12: 163 pp.
(mimeographed).

BT (data only)
Canada, Canadian Oceanographic Data Center, 1969.
Ocean Weather Station 'P' North Pacific Ocean, December 3, 1967, to February 25, 1968.
1969 Data Record Series 6: 116 pp.
(multilithed)

BT (data only)
Canada, Canadian Oceanographic Data Centre, 1968.
Hudson Bay, Hudson Strait and Arctic, July 21 to September 9, 1967.
1968 Data Record Ser. 4:123 pp. (multilithed)

BTs (data only)
Canada, Canadian Oceanographic Data Centre, 1968.
Baffin Bay bathythermograms, 1964: September 4, to October 24, 1964.
1968 Data Record Ser., 6:33 pp. (multilithed)

BT (data only)
Canada, Canadian Oceanographic Data Center, 1968
Ocean Weather Station 'P', North Pacific Ocean, April 7 to July 6, 1967.
1968 Data Record Ser. 5:140 pp. (multilithed)

bathythermograph (data only)
Canada, Canadian Oceanographic Data Center, 1967.
Ocean Weather Station 'P', North Pacific Ocean, October 28, 1966 to January 9, 1967.
1967 Data Record Ser., 8: 111 pp.

BT (data only)
Canada, Canadian Oceanographic Data Centre, 1967.
Ocean Weather Station "P", August 5 to October 31, 1966.
1967 Data Record Ser., 6:164 pp. (mimeographed).

BT (data only)
Canada, Canadian Oceanographic Data Center, 1967.
Ocean Weather Station 'P' North Pacific Ocean, May 27 to August 10, 1966.
1967 Data Record Ser., 5: 185 pp.

BT (data only)
Canadian Oceanographic Data Center 1966
Ocean Weather Station 'P' North Pacific Ocean, December 11, 1965, to March 9, 1966.
1966 Data Record Ser. 8: 144 pp.
(multilithed)

BT (data only)
Canada, Canadian Oceanographic Data Centre, 1966.
Arctic, Hudson Bay and Hudson Strait, August 5 to October 4, 1962.
1966 Data Record Series, No. 4:247 pp.

BT (data only)
Canada, Canadian Oceanographic Data Centre, 1966.
Ocean Weather Station "P", January 23, 1965 to April 19, 1965.
1966 Data Record Ser., No.1:122 pp. (multilithed).

BT (data only)
Canadian Oceanographic Data Centre, 1965.
Data record, Saguenay and Gulf of St. Lawrence, August 19 to August 30, 1963.
1965 Data Record Series, No. 12:123 pp. (multilithed).

BT(data only)
Canada, Canadian Oceanographic Data Center, 1964
Ocean weather station "P", North Pacific Ocean. 1964 Data Record Series, No. 15:95 pp. (multilithed).

bathythermograph (data only)
Canada, Department of Mines and Technical Surveys, 1960.
Tidal and oceanographic survey of Hudson Strait, August and September, 1959. Data Record, 55 pp. (multilithed.)

BT(data only)
Canada, Fisheries Research Board, 1959.
Physical and chemical data record, coastal seaways project, November 12 to December 5, 1958 MSS Rept. Ser. (Oceanogr. & Limnol.), No. 36: 120 pp.

BT (data only)
Canada, Pacific Oceanographic Group, Nanaimo, British Columbia, 1963
Data record of bathythermograms observed in the Northeast Pacific Ocean during the Gulf of Alaska Salmon Tagging Program, April to July, 1962.
Fisheries Res. Bd., Canada, Mss Rept. Ser. (Oceanogr. and Limnol.), No. 143: 74 pp. (multilithed).

BT(data only)
Canada, Pacific Oceanographic Group, Nanaimo, 1961
Data record of oceanographic observations and bathythermograms observed in the northeast Pacific Ocean by Fisheries Research Board chartered fishing vessels, May to July, 1960.
Fish. Res. Bd., Canada, Manuscript Rept. Ser. (Oceanogr. and Limnol.), No. 87: 30 pp.

BT (data only)
Champagnat, C., F. Conand, J.L. Cremoux et J.P. Rebert 1969
Campagne océanographique du Jean Charcot (Dakar - Cap Blanc - Iles du Cap Vert) du 29-7 au 5-8-68.
ORSTOM, Sénégal, DSP 17:87pp (mimeographed)

BT
Congo, Office de la Recherche Scientifique et Technique Outre-Mer, 1967.
Bathythermogrammes, 1966.
Centre de Pointe-Noire, Oceanographie, Doc. No. 359 S.R.:27 pp. (multilithed).

BT(data only)
Crean, P.B., R.B. Tripp, and H.J. Hollister, 1962
Oceanographic data record, Monitor Project, January 16 to February 5, 1962.
Fish. Res. Bd., Canada, Mss. Rept. Ser. (Oceanogr. & Limnol.), No. 113:169 pp. (multilithed)

BT (data only)
Cremoux, J.L. 1970.
Observations océanographiques effectuées en 1969. II. Observations de surface et de fond — bathythermogrammes.
Centre Rech. océanogr. Dakar-Thiaroye ORSTOM, DSP 25: 23+ unnumbered pp.

BT (data only)
Dodimead, A.J. 1968.
Temperatures and salinities of the Northeast Pacific Ocean during 1965 and 1966 with appendix: data record of bathythermograms April-August 1965 and 1966.
Techn Rept. Fish. Res. Bd. Can. 54: unnumbered pp. (mimeographed)

BT (data only)
Favorite, Felix, and Glenn Pedersen, 1959.
North Pacific and Bering Sea oceanography, 1958.
U.S.F.W.S. Spec. Sci. Rept., Fish., No. 312:230pp

BT (data only)
Favorite, Felix, and Glenn M. Pedersen, 1959.
North Pacific and Bering Sea oceanography, 1957. USFWS Spec. Sci. Rept., Fish. No. 292:106 pp.

Bt (data only)
Fuglister, F.C., Edit., 1960.
Atlantic Ocean atlas of temperature and salinity profiles and data from the Internationa Geophysical Year, 1957-1958, Vol. 1:209 pp.

BT(data only)
Graham, Joseph J., and William L. Craig, 1961
Oceanographic observations made during a cooperative survey of albacore (Thunnus germo) off the North American west coast in 1959.
U.S.F.W.S. Spec. Sci. Rept., Fish., No. 386: 31 pp.

BT(data only)
Herlinveaux, R.H., 1970.
Japack 8/68 - oceanographic and biological observations.
Techn. Rept. Fish. Res. Bd. Can. 159: 60pp.

BT (data only)
Herlinveaux, R.H., 1963
Data record of oceanographic observations made in Pacific Naval Laboratory underwater sound studies, November 1961 to November 1962.
Fish. Res. Bd., Canada, Mss. Rept. Ser. (Ocean. and Limnol.), No. 146:101 pp. (multilithed).

BT (data only)
Herlinveaux, R.H., 1963
Oceanographic observations in the Canadian Arctic Basin, Arctic Ocean, April-May, 1962
Fish. Res. Bd., Canada, Mss. Rept. Ser. (Ocean. and Limnol.), No. 144: 25 pp. (multilithed).

BT (data only)
Herlinveaux, R.H., 1961.
Data record of oceanographic observations made in Pacific Naval Laboratory underwater sound studies.
Fish. Res. Bd., Canada, Mss. Rept. Ser. (Ocean. & Limnol.), No. 108:85 pp.

BT(data only)
Hollister, H.J., 1961
Bathythermograms and meteorological data record, Swiftsure Bank and Umatilla Reef Lightships, January 1, 1960 to June 30, 1961.
Fish. Rec. Bd., Canada, Manuscript Rept. Ser. (Oceanography & Limnology), No. 99: 89 pp. (multilithed).

BT (data only)
India, Naval Headquarters, New Delhi, 1958.
Indian oceanographic station list, Ser. No. 2: numerous pp., (Mimeographed)

BT data
Japan, Central Meteorological Observatory, 1956.
The results of marine meteorological and oceanographic observations (NORPAC Expedition Spec. No.) 17:131 pp.

BT (data only)
Japan, Japan Meteorological Agency, 1970.
The results of marine meteorological and oceanographical observations, July-December, 1968, 44:311 pp.

BT (data only)
Japan, Japan Meteorological Agency, 1970
The results of marine meteorological and oceanographical observations. (The results of the Japaneses Expedition of Deep Sea (JEDS-11); January-June 1967 41: 332 pp.

BT (data only)
Japan, Japan Meteorological Agency 1970.
The results of marine meteorological and oceanographical observations, July-December 1966, 40:336pp.

BT (data only)
Japan, Japanese Meteorological Agency, 1968.
The results of the Japanese Expedition of Deep Sea (JEDS-10).
Res.mar.met. oceanogr.Observ., Jan-June 1965, 37: 385.

BT (data only)
Japan, Japanese Meteorological Agency, 1967.
The results of marine meteorological and oceanographical observations, July-December, 1964, 36: 367 pp.

BT (data only)
Japan, Japan Meteorological Agency. 1965.
The results of marine meteorological and oceanographical observations, July-December 1963, No. 34: 360 pp.

BT (data only)
Japan, Japan Meteorological Agency, 1964.
The results of marine meteorological and oceanographical observations, January-June 1963, No. 33:289 pp.

BT (data only)
Japan, Japan Meteorological Agency, 1964.
Oceanographic observations.
Res. Mar. Meteorol. & Oceanogr. Obs., No. 31 :220 pp.

BT(data only)
Japan, Japan Meteorological Agency, 1962
The results of marine meteorological and oceanographical observations, January-June 1961, No. 29: 284 pp.

BT(data only)
Japan, Japan Meteorological Agency, 1962
The results of marine meteorological and oceanographical observations, July-December 1961, No. 30:326 pp.

BT(data only)
Japan, Japan Meteorological Agency, 1962
The results of marine meteorological and oceanographical observations, July-December, 1960, No. 28: 304 pp.

BT (data only)
Japan, Japan Meteorological Agency, 1961
The results of marine meteorological and oceanographical observations, January-June 1960. The results of the Japanese Expedition of Deep-Sea (JEDS-2, JEDS-3), No. 27: 257 pp.

BT(data only)
Japan, Japan Meteorological Agency, 1960.
The results of marine meteorological and oceanographical observations, July-December, 1959, No. 26:256 pp.

BT (data only)
Japan, Japan Meteorological Agency, 1960.
The results of marine meteorological and oceanographical observations. Jan-June 1959, No. 25: 258 pp.

BT(data only)
Japan, Japan Meteorological Agency, 1958
The results of marine meteorological and oceanographical observations, July-December 1957, No. 22: 183 pp.

BT (data only)
Japan, Japan Meteorological Agency, 1958.
The results of marine meteorological and oceanographical observations, January-June, 1957, No. 21:168 pp.

BT(data only)
Japan, Kobe Marine Observatory, 1961
Data of the oceanographic observations in the sea south of Honshu from February to March and in May, 1959.
Bull. Kobe Mar. Obs., No. 167(27):99-108; 127-130;149-152;161-164;205-218.

BT(data only)
Japan, Maizuru Marine Observatory, 1963
Data of the oceanographic observations (1960-1961) (35-36):115-272.

BT (data only)
Japan, Maritime Safety Agency 1967.
Results of oceanographic observations in 1964.
Data Rept. Hydrogr. Obs., Ser. Oceanogr. Publ. 792(4): 88pp.

BT (data only)
Japan Maritime Safety Agency 1967.
Results of oceanographic observations in 1965.
Data Rept. Hydrogr. Obs. Ser. Oceanogr. (Publ. 792) Oct. 1967, 5: 115pp.

BT (data only)
Japan, Maritime Safety Agency, 1965.
Results of oceanographic observations in 1962.
Data Rept. Hydrogr. Obs. Ser. Oceanogr. (Publ. 792) Nov. 1965, 1:65 pp.

oxygen(data only)
BT (data only)
GEK (data only)

BT (data only)
Japan, Maritime Safety Agency, Tokyo, 1964.
Tables of results from oceanographic observations in 1961.
Hydrogr. Bull., Tokyo. (Publ. No 981) No. 77: 82 pp.

BT (data only)
Japan, Maritime Safety Agency, Tokyo, 1964.
Tables of results from oceanographic observations in 1960.
Hydrogr. Bull., (Publ. No. 981), No. 75:86 pp.

BT (data only)
Japan, Maritime Safety Board, 1961
Tables of results from oceanographic observation in 1959.
Hydrogr. Bull. (Publ. No. 981), No. 68: 112 pp.

BT (data only)
Japan, Maritime Safety Board, 1961
Tables of results from oceanographic observations in 1958.
Hydrogr. Bull., Tokyo, No. 66 (Publ. No. 981): 153 pp.

BT (data only)
Japan, Nagasaki University Research Party for CSK, 1966.
The results of the CSK - NU65S.
Bull. Fac. Fish. Nagasaki Univ., No. 21:273-292.

waves (data only)
"Nagasaki-maru"
GEK (data only)

BT (data only)
Japan, Shimonoseki University of Fisheries 1970.
Data of oceanographic observations and exploratory fishings. 5. Oceanographic surveys of the Kuroshio and its adjacent waters, 1967 and 1968: 117pp.

BT (data only)
Korea, Fisheries College, Pusan, 1968.
Baek-Kyung-Ho cruise to the central Pacific Ocean, 1967.
Data Rept. Oceanogr. Obs. Expl. Fish., 1:30 pp.

BT (data only)
Korea (South), R.O.K. Navy Hydrographic Office, 1962
Technical reports.
H.O. Publ. South Korea, No. 1101: 85 pp.

BT (data only)
Lane, R.K., R.H. Herlinveaux, W.R. Harling and H.J. Hollister, 1960.
Oceanographic data record, Coastal Seaways Projects, October 3 to 26, 1960.
Fish. Res. Bd., Canada, Mss Rept. Ser. (Oceanogr. & Limnol.), No. 83:142 pp. (multilithed)

BT(data only)
(grids)
Lane, R.K., A.M. Holler, J.H. Meikle and H.J. Hollister, 1961
Oceanographic data record, Monitor and Coastal Projects, March 20 to April 14, 1961.
Fisheries Res. Bd., Canada, Manuscript Rept. Ser. (Oceanogr. & Limnol.), No. 94: 188 pp. (multilithed).

BT (data only
McGary, James W., and Joseph J. Graham, 1960.
Biological and oceanographic observations in the central north Pacific, July-September 1958.
U.S.F.W.S. Spec. Sci. Rept., Fish., No. 358: 107 pp.

BT(data)
McGary, James W., Edward D. Stroup, 1958
Oceanographic observations in the central North Pacific, September 1954-August 1955.
U.S.F.W.S. Spec. Sci. Rept., Fish., No. 252: 250 pp.

bathythermograph (data only)
*Netherlands, Hydrographer, Royal Netherlands Navy, 1967.
Navado III, bathythermograph observations, H. Neth. M.S. Snellius, 1964-1965.
Hydrogr. Newsletter, Spec. Publ. 4: 280 pp.

BT (data only)
Republique du Congo, Centre de Pointe Noire 1970
Bathythermogrammes 1969.
Doc. scient. Centre Pointe Noire ORSTOM n.s. 9: 4pp., 74pls.

BT(data only)
Rotschi, Henri, Yves Magnier, Maryse Tirelli, et Jean Garbe, 1964.
Résultats des observations scientifiques de "La Dunkerquoise" (Croisière "Guadalcanal").
Océanographie, Cahiers O.R.S.T.R.O.M., 11(1):49-154.

BT (data only)
*Saur, J.F.T., and Dorothy D. Stewart, 1967.
Expendable bathythermograph data on subsurface thermal structure in the eastern North Pacific Ocean.
Spec. scient. Rep. U.S. Fish. Wildl. Serv. Fish., 548: 70 pp.

BT(data only)
Stefansson, Unnstein, 1963
The "Aegir" Redfish Larvae Expeditions to the Irminger Sea in May 1961. Oceanographical observations.
Rit Fiskideildar, 4(1):1-35.

BT (data only)
Tabata, S., C.D. McAllister, R.L. Johnston, D.G. Robertson, J.H. Meikle, and H.J. Hollister, 1961
Data record. Ocean Weather station "P" (Latitude 50 00'N, Longitude 145 00'W), December 9, 1959 to January 19, 1961.
Fish. Res. Bd., Canada, MSS. Rept. Ser. (Ocean & Limnol.), No. 98: 296 pp. (Multilithed).

BT (data only)
Tabata, S., D.G. Robinson, W. Atkinson and H.J. Hollister, 1962.
Oceanographic data record, Ocean Weather Station "P", September 12, 1961 to January 21, 1962.
Fish. Res. Bd., Canada, Mss. Rept. Ser., (Ocean. and Limnol.), No. 125:187 pp.

BT (data only)
Trotti, L., 1966.
Atlante dei batitermogrammi, Mare Adriatico - crociere 1965-1966.
Pubbl. Ist. sper. tallass., 430:12 pp. (quarto).

BT (data only)
Trotti, L., 1966.
Atlante dei batitermogrammi, Golfo Palmas-Canale di Sardegna, Bannock 1965.
Pubbl. Ist. sper. talassogr., 421:7 pp. (quarto).

BT (data only)
Trotti, L., e A. de Maio, 1966.
Atlante dei batitermogrammi, Mare Tirreno, crociera Bannock 1963.
Pubbl. Ist. sper. talassogr., 420:10 pp. (quarto).

BT (data only)
United States, National Oceanographic Data Center, 1965.
Data report EQUALANT III.
Nat. Oceanogr. Data Cent., Gen. Ser., G-7:339pp. $5.00
Publ.

"Zvezda"
"Reine Pokou"
Salinity (data only)
density (data only)
sound velocity (data only) 1971

BT(data only)
Wilson, R.C., and M.O. Rinkel, 1957.
Marquesas area oceanographic and fishery data, January-March 1957.
USFWS Spec. Sci. Rept., Fish. No. 238:136 pp.

bauxite

bauxite
*Kaye, C.A., 1967.
Fossiliferous bauxite in glacial drift, Martha's Vineyard, Massachusetts.
Science, 157(3792):1035-1037.

bays

bays
Gorsline, D.S., 1967.
Contrasts in coastal bay sediments on the Gulf and Pacific coasts.
In: Estuaries, G.H. Lauff, editor, Publs Am. Ass. Advmt Sci., 83:219-225.

bays
Kulm, L.D., and John V. Byrne, 1967.
Sediments of Yaquina Bay, Oregon.
In: Estuaries, G.H. Lauff, editor, Publs Am. Ass. Advmt Sci., 83:226-238.

sedimentation
flats, tidal
Newport, sediment
p. 236, 238

bays
Lidz, Louis, 1965.
Sedimentary environment and foraminiferal parameters: Nantucket Bay, Massachusetts.
Limnol. Oceanogr., 10(3):392-402.

BAYS
Moore, Joseph G., Jr., 1968.
Bays and estuaries and the Texas Water Plan.
Proc. Gulf Carib. Fish. Inst. 20th Ann. Sess. 60-68

bays (topography of)
Nozawa, Koji, 1964.
Topographical consideration on the shallow water fishing ground inside bay. (In Japanese; English abstract).
Mem., Fac. Fish., Kagoshima Univ., 13:5-25.

bays
Park, Kilho, W.T. Williams, J.M. Prescott and D.W. Hood, 1963.
Amino acids in Redfish Bay, Texas.
Publ. Mar. Sci., Port Aransas, 9:59-63.

bays
Silvester, Richard, 1970.
Growth of crenulate shaped bays to equilibrium.
J. WatWays Harb. Div. Am. Soc. civ. Engrs, 96(WW2):275-287.

beaches

A

beaches
Adams, J.A.S., A. Mahdavi and J.J.W. Rogers, 1964.
Thorium, uranium and potassium in Gulf and Atlantic coast beach sands. (Abstract).
Geol. Soc., Amer., Special Paper, No. 76:2.

beaches
Aksenov, A.A., and V.P. Petelin, 1964
On the distribution of heavy minerals in the beach belt. (In Russian).
Okeanologiia, Akad. Nauk, SSSR, 4(2):295-299.

beaches
Allard, M.P., 1950.
Le regime de la cote marocaine entre Safi et Mogardor. Bull. d'Info., C.O.E.C., 2(10):369-378, 6 pls.

beaches
Allard, P., 1949.
Relations hydrodynamiques dans les mers littorales. Application au bassin oriental de la Manche. Ann. Geophys. 5:25-60.

(entry not checked)

beaches
Allen, J.R.L., 1967.
A beach structure due to wind-driven foam
J. Sed. Pet. 37(2): 691-692

beaches
Amanieu, M. 1969.
Recherches écologiques sur les faunes des plages abritées de la region d'Arcachon.
Helgoländer wiss. Meeresuntersuch. 19(4): 455-557.

beaches
*Andrews, Peter B., and G.J. van der Lingen, 1969.
Environmentally significant sediments logic characteristics of beach sands.
N.Z.Jl.Geol.Geophys.12:119-137.

beaches
*Ansell, Alan D. and Ann Trevallion, 1969.
Behavioural adaptations of intertidal molluscs from a tropical sandy beach. J. exp. mar. Biol. Ecol., 4(1): 9-35.

beaches
Arbey, F., 1961.
Études littorales sur la côte des Maures. 2. Les sinuosités de plage.
Cahiers Océanogr., C.C.O.E.C., 13(8):569-576.

beaches
Arbey, F., 1959.
Observations nouvelles sur les croissants et sinusites de plage.
Comptes Rendus Acad. Sci. Paris 248(22): 3187-3189.

beaches
Aubert, M., and H. Lebout, 1961.
Étude de la pollution des plages de Nice par les residus ou déchets amenes par les courants ou les fleuves cotiers.
Cahiers d'Études et de Recherches Biol./Océanogr. Med. (4):74 pp. (mimeographed).

beaches
Auzel, M., 1953.
Evolution récente des plages de Normandie.
C.R.S. Soc. Géol., France, 15:302-303.

beaches
Auzel, M., 1951.
Étude de la plage de Deauville. Bull. d'Info., C.O.E.C., 3(5):175-179, 1 fig.

beaches
Auzel, M., and J. Bourcart, 1950.
Sur l'erosion de la plage de Deauville pendent le printemps de 1950. Bull. d'Info., C.O.E.C., 2(10):379-386, 8 textfigs.

beaches
Azmon, E., 1961
Marine to nonmarine transition of sediments across Little Sycamore Beach, California.
Bull. Geol. Soc. Amer., 72(5): 763-766.

B

beaches
Bagnold, R.A., 1963.
21. Beach and nearshore processes. 1. Mechanics of marine sedimentation.
In: The Sea, M.N. Hill, Editor, Interscience Publishers, 3:507-528.

beaches
Bajard, Jacques et Marc Gautier, 1969.
Dynamiques des plages du Nord-Cotentin.
Cah. océanogr. 21(7): 635-651

beaches
France, northwest

beaches
Bascom, Willard, 1964
Waves and beaches: the dynamics of the ocean surface.
Anchor Books, Science Study Series, Doubleday & Co., Inc., New York, 267 pp. $1.45.
(Popular paperback)

beaches
Bascom, W., 1960.
Beaches.
Scientific American, 203(2):80-97.

beaches
Bascom, W.N., 1951.
The relationship between sand size and beach-face slope. Trans. Amer. Geophys. Union, 32(6): 866-874, 7 textfigs.

beaches
Batham, E.J., 1958
Ecology of Southern New Zealand exposed rocky shore at Little Papanui, Otago Peninsula.
Trans. Roy. Soc., N.Zeal. Vol. 85, pt. 4, pp 647-658.

Battistini, Rene, 1960 beaches
Quelques aspects de la morphologie du littoral Mikea. (Côte sud-ouest de Madagascar).
Cahiers Oceanogr., C.C.O.E.C., 12(8):548-571.

beaches
Battistini, R., 1955.
Description du relief et des formations quaternaire du littoral breton entre Brignognan et Saint Pol de Leon (Finistère).
Bull. d'Info., C.C.O.E.C., 7(10):468-491, 9 pls.

beaches
Battistini, R., 1955.
Description du relief et des formations quaternaires du litoral occidental du Leon entre l'Anse des Blancs Sablons et l'Aber Benoit (Finistère).
Bull. d'Info., C.C.O.E.C. 7(2):64-78, 7 pls.

beaches
Beach Erosion Board, 1953.
Shore protection and design (preliminary, subject to revision). B.E.B. Bull., Special Issue, No. 2: 230 pp., 149 figs., plus app.
planning

beaches
Beach Erosion Board, 1948.
Proof test of water transparency method of depth determination. Eng. Notes No. 29:29 pp., 34 figs. data sheets (multilith).

beaches

Beasley, A.W., 1969.
Beach sands of the southern shore of Port Phillip Bay, Victoria, Australia.
Mem. natn. Mus. Victoria, 29:1-21.

beaches

Behrens, E. William, 1966.
Recent emerged beach in eastern Mexico.
Science, 152(3722):642-643.

beaches
Gulf of Mexico
sea level

beaches

Bellair, P., 1954.
Sur un échantillon de sable de la plage de Zemba.
Mém. Soc. Sci. Nat., Tunisie, 2:57-61.

beaches

Bellan-Santini, D., 1965.
Méthode de récolte et d'étude quantitative des peuplements sur substrat dur dans la zone d'agitation hydrodynamique.
Méthodes Quantitatives d'Étude du Benthos et Echelle Dimensionnelle des Benthontes: Colloque du Comité du Benthos (Marseille, Nov. 1963).
Comm. Int. Expl. Sci. Mer. Médit., Monaco, 23-24.

beaches

Ben-Avraham, Zvi, 1971.
Accumulation of stones on beaches by Codium fragile. Limnol. Oceanogr, 16(3): 553-554.

beaches

Bernard, Pierre, 1963
L'erosion de la plage de Coutainville et le problème de sa reconstitution.
Cahiers Océanogr., C.C.O.E.C., 15(4):219-223.

beaches

Bernard, Pierre, 1959.

Commentaires sur quelques relèves de la plage de Coutainville. Cahiers Océan. C.C.O.E.C. 11 (4): 212-218.

beaches

Berthois, L., 1950.
Sur la corrélation entre l'importance du volume des déplacements de l'air et l'évolution littorale. C.R. Acad. Sci., Paris, 230:2110-2111.

beaches

Berthois, L. and C., 1954.
Contribution a l'etude des formations sableuses du littoral de la norvege meridionale.
Trav. C.R.E.O. 1(4):13 pp., 4 textfigs.

beaches

Bertman, D.Ya., Yu.D. Shuisky and I.V. Shkarupo, 1971.
An experiment of studying planimetric-elevational changes of the beach as dependent on wind direction and force. (In Russian; English abstract). Okeanologiia 11(3): 450-459.

beaches, foraminifera

*Bhalla, S.N., 1969.

Recent foraminifera from Vishakapatnam beach sands and its relation to the known foram-geographical provinces in the Indian Ocean.
Bull. natn. Inst. Sci., India, 38(1): 376-392.

beaches

Bigarella, Joano José, 1965.
Sand-ridge structures from Paraná coastal plain.
Marine Geology Elsevier Publ. Co., 3(4):269-278.

beaches

Bigarella, J.T., 1954.
Nota sôbre os depósitos recentes do litoral sul-Brasileiro. Bol. Inst. Oceanol., São Paulo, 5: 233-236.

beaches

Bigarella, Joao Jose, Riad Salamuni and Zelia M. Pavao, 1959.

Contribuicao ao estudio dos sedimentos praiais recentes. 1. Praia de Ubatuba (Estado de Santa Catarina, Brasil). Universidade do Parana, Inst Geol. Geol. No. 1: 99 pp., English summary.

beaches

Billy, Cecile, 1964.
Étude minéralogique des sables côtiers de la Manche entre l'estuaire de la Dives et l'estuaire de la Somme.
Rev. Géogr. Phys. et Géol. Dynam., (2), 6(2):123-154.

beaches

Blackman, B., 1950.
Dredging at inlets on sandy coasts. Inst. Coastal Eng., Univ. Ext., Univ. Calif., Long Beach, 11-13 Oct. 1950:15 pp. (mimeographed).

beaches

Blake, G.J. 1968.
The rivers and the foreshore sediment of Pegasus Bay, South Island New Zealand.
N.Z. Jl. Geol. Geophys. 11 (1):225-235

beaches

Blanc, Jean-J., 1970
Recherches de géodynamique littorale appliquées à la plage du Prado (Marseille)
Téthys 1(4):1147-1166.

beaches

Blanc, J.-J., 1958.
Recherches de sédimentologie littorale et sous-marine en Provence occidentale.
Ann. Inst. Océan., Monaco, 35(1):1-140.

beaches

Blanc, J.J., 1956.
Recherches de sédimentologie littorale et sous-marine en Provence occidentale.
Bull. Mus. Hist. Nat., Marseille, 16:25-32.

beaches

Blanc, J.J., 1956.
Étude minéralogique des sables littoraux du Cap Lardier au Cap Couronne (Provence).
Bull. Mus. Hist. Nat., Marseille, 16:69-92.

beaches

Bloch, J.P. et J. Trichet, 1966.
Un exemple de grès de plage (côte Ligure italienne).
Marine Geol., 4(5):373-377.

beaches

* Bluck, B.J., 1969.
Particle rounding in beach gravels.
Geol. Mag. 106(1): 1-14.

beaches

Bluck, Brian J., 1967.
Sedimentation of beach gravels: examples from South Wales.
J. sedim. Petrol., 37(1):128-156.

beaches

*Boekschoten, G.J., 1966.
Shell borings of sessile epibiontic organisms as palaeoecological guides (with examples from the Dutch coast).
Palaeogr., Palaeoclimatol. Palaeoecol., 2:333-379.

reprint

beaches

Boekschoten, G.J., 1962
Beachrock at Limani Chersonisos, Crete.
Geol. en Mijnbouw, 41:3-7.

beaches

Boldyrev, V.L., 1960

[The beach barrier profile analysis] Izv. Vses. Geogr. Obshch., 92(5): 456-460.

beaches

Boldyrev, V.L., 1958.
[Degradation processes of coastal embankments as illustrated by the Kerchensky Strait coasts.]
Trudy Inst. Okeanol., 28:85-92.

beaches

Boldyrev, V.L., 1957.
[Type of coast with ancient slump- and slide phenomena.] Trudy Inst. Okeanol., 21:118-132.

beaches

Boltovskoy, Esteban, 1963
The littoral foraminiferal biocenoses of Puerto Deseado (Patagonia, Argentina).
Contrib. Cushman Found. Foram. Res., 14(2): 58-70.

Also in:
Centro Invest. Biol. Mar., Estacion Puerto Deseardo, Contrib. Cient., No. 3:

beaches

*Boon, John D., III, 1968.
Trend surface analysis of sand tracer distribution on a carbonate beach, Binini, B.W.I.
J. Geol., 76(1):71-87.

beaches

Borreswara Rao, C., and E.C. LaFond, 1956.
Sand sorting on the east coast beaches.
Current Science 25(3):77-79.

beaches

Bourcart, J., ???
Sur le substratum des plages. Actes du IV Congrès Int. du Quarternaire, Rome-Pise, Aout-Sept. 1953:3-8.

beaches

Bourcart, Jacques, 1959.
Les calcaires éocènes de la Manche et leur contribution aux sables des plages.
Eclogae Geol. Helv., 51(3):505-508.

Also in:
Trav. Lab. Géol. Sous-Marine, 10 (1960).

beaches

Bourcart, J., 1955.
Quelques remarques sur les littoraux actuels pour la compréhension des littoraux fossiles.
Bull. Soc. Géol., France (6) 5:571-576.

beaches

Bourcart, J., 1952.
Les frontières de l'océan. Albin Michel, Paris, 317 pp., 77 textfigs.

beaches
Bourcart, J., 1950.
Le littoral Breton, du Mont Saint-Michel au Finistere. Bull. d'Info., CO.E.C., 2(1):21-39, 2 figs.

beaches
Bourcart, M.J., 1949
Etude de la plage de Gavres. Bull. Info., Com. Centr. Oceanogr. et d'Etude des Cotes, Service Hydro. de la Marine, No.5, Mai, 1949, 9-11, Figs.1-3 (mimeographed).

beaches
Bourcart, J., 1948.
Le Quaternaire des grèves de Roscoff (Finistère). Bull. Soc. Géol., France, 5e ser., 18:181-197, 3 textfigs.

beaches
Bourcart, J., 1948.
II. Etude de la plage de gavies. 7 pp. (mimeographed), 4 figs. (multilith).

beaches
Bourcart, Jacques, Mlle. Claude Lalou, et Lucien Mallet, 1961.
Sur la présence d'hydrocarbures du type benzo 3,4 pyrene dans les vases côtières et les sables de la plage de la rade de Villebfranche (Alpes Maritimes).
C. R. Acad. Sci., Paris, 252:640-644.

Also in:
Trav. Lab. Géol. Sous-Marine, 10(1960).

beaches
Bowen, A.J., and D.L. Inman, 1966.
Budget of littoral sands in the vicinity of Point Arguello, California.
Tech.Memo., U.S. Army Coast.Eng.Res.Center, No.19: 41 pp. (multilithed).

beaches
Bowler, J.M., 1966.
The geology and geomorphology. Port Philip survey 1957-1963.
Mem. nat. Mus. Vict., 27:19-67.

beaches
Brafield, Alan E., 1965.
Quelques facteurs affectant la teneur en oxygène des eaux interstitielles littorales.

Vie et Milieu, Bull. Lab. Arago, (B)16(2-B):889-897.

beaches
Brafield, A.E., 1964.
The oxygen content of interstitial waters in sandy shore.
J. Animal Ecology, 33(1):97-116.

beaches
Brambati, Antonio, 1968.
Caratteristiche morfologiche e sedimentologiche della Costa Adriatica da Venezia a Monfalcone.
Studi trentini Sci. nat. (A) 45(2): 188-223. Also in: Raccolta Dat. Oceanogr. Comm. ital. Oceanogr. Consig. naz. Ricerche (A)(20)(1969)

beaches
Bratus, O.S., 1965.
Material composition of the beaches of the Crimea Peninsula. (In Russian).
Doklady Akad. Nauk, SSSR, 165(2):399-402.

beaches
Bratus, O.S., 1965.
The granulometric composition of arenaceous beach-deposits of the Crimaea. (In Russian).
Doklady, Akad. Nauk, SSSR, 163(2):431-434.

beaches
Bretschneider, Charles, 1971.
Coastal engineering practice(18). In: Impingement of man on the oceans, D.W. Hood, editor, Wiley Interscience: 489-501.

beaches
Broggi, J.A., 1952.
Migracion de arenas a lo largo de la costa peruana. Bol. Soc. Geol., Peru, 24: 25 pp., 2 pls., 1 textfig.

beaches
Brown, A.C., 1964.
Food relationships on the intertidal sandy beaches of the Cape peninsula.
S. African J. Sci., 60(2):35-41.

beaches
Bruun Per M., 1967
By-passing and back-passing off Florida
J. WatWays Harb. Div. Am. Soc. civ. Engrs. 93 (WW2): 101-125.

beaches
Bruun, P., 1954.
Coast erosion and the development of beach profiles. B.E.B. Tech. Memo. No. 44:82 pp., 25 figs.

beaches
Bruun, P., 1954.
Use of small-scale experiments with equilibrium profiles in studying actual problems and developing plans for coastal protection.
Trans. Amer. Geophys. Union 35(3):445-452, 11 textfigs.

beaches
Budanov, V.I., 1957.
[On the dynamics of the eastern shores of the Azov Sea.] Trudy Inst. Okeanol., 21:187-191.

beaches
Budanov, V.I., 1954.
[Procedures for the field investigation of seashores.] Trudy Inst. Oceanol. 10:169-178.

RT 3120 in Bibl. Transl. Rus. Sci. Tech. Lit. 25

beaches
Byerly, John Robert, 1963.
The relationship between watershed geology and beach radioactivity.
Beach Erosion Bd., Techn. Memo. (135):32 pp.

C

beaches
Caldwell, Joseph M. 1966.
Coastal processes and beach erosion.
J. Boston Soc. Civ. Engrs. 53 (2): 142-157.

beaches
Caldwell, J.M., 1949.
Beach erosion. Sci. Mon. 59:229-235.

beaches
* Carr, A.P., 1969.
Size grading along a pebble beach: Chesil Beach, England.
J. sed. Petr., 39(1): 297-311.

beaches
Carr, A.P., R. Gleason and A. King, 1970.
Significance of pebble size and shape in sorting by waves.
Sediment. Geol. 4 (2): 89-101

beaches
Carrier, G.F., and H.P. Greenspan, 1958.
Water waves of inite amplitude on a sloping beach. J. Fluid Mech., 4(1):97-109.

benthos
Caspers, Hubert, 1968.
La macrofaune benthique du Bosphore et les problémes de l'infiltration des eléments méditerranéens dan la mer Noire.
Rapp.P.-V.Reun.Comm.int.Explor.Sci.Méditerranée, 19(2):107-115.

beaches
Caulet, Jean-Claude, 1963.
Etude des plages entre Arzew et Port aux Poules.
Cahiers Océanogr., C.C.O.E.C., 15(9):617-637.

beaches
Chamberlain, Theodore, 1968.
The littoral sand budget, Hawaiian islands.
Pacific Science 22(2): 161-183.
Also in: Contrib. Hawaii Inst. Geophys. 1968.

beaches
Chamley, Hervé, Claude Froget et Léo Pastouret 1971.
Observations sur les plages quaternaires de l'Esterel. Relations avec les plages quaternaires des Maures.
C.r. hebd. Séanc. Acad. Sci. Paris (D) 273 (23): 2199

Beaches
Cherry, John A., 1966.

Sand movement along equilibrium beaches north of San Francisco.

J. Sed. Petr. 36(2): 341-352.

abstract

beaches
Christiansen, Sofus, 1958.
Bølgekraft og kystretning. Eksempal på kystudformning i det Sydfynske Øhav.
Geografisk Tidsskrift, 57:23-37.

beaches
Cooray, P.G. 1968.
A note on the occurrence of beachrock along the west coast of Ceylon.
J. sedim. Petrol. 38(2): 650-654.

beaches
Cotton, C.A., 1952.
Cyclic resection of headlands by marine erosion.
Geol. Mag., 89(3):221-225.

beaches
Croker, Robert A., 1970
Intertidal sand macrofauna from Long Island, New York.
Chesapeake Sci., 11 (3): 134-137.

beaches
Curray, Joseph R., and David G. Moore, 1964.
Holocene regressive littoral sand, Costa de Nayarit, Mexico.
In: Developments in Sedimentology, L.M.J.U. van Straaten, Editor, Elsevier Publishing Co., 1:76-82.

D

beaches
Darling, John M., 1966.
Study of pilot beaches in New England for the improvement of coastal storm warning.
CERC Bull. and Summary Rept of Res.Prog., 1965-1966:30-46.

beaches
Debouteville, C.D., 1955.
Sur la circulation des eaux marines et des eaux phréatiques continentales dans les plages des mers à marées.
C.R. Acad. Sci., Paris, 240(5):555-557.

beaches
Delamare-Deboutteville, C., 1960.
Biologie des eaux souterraines littorales et continentales.
Hermann, Paris, 740 pp., 254 textfigs., 70 N.F.

Reviewed by:
D.J. Zinn, 1962. Int. Revue Ges. Hydrobiol., 47(3):495.

beaches
Derijard Raoul 1966.
Contribution à l'étude du peuplement des sédiments sablo-vaseux intertidaux compactés ou fixés par la végétation de la région de Tuléar (République malgache).
Études malgaches, Univ. Madagascar, 17:1-94.

Beaches
Dexter, Deborah M., 1969.
Structure of an intertidal sandy-beach community in North Carolina.
Chesapeake Sci., 10(2): 93-98.

beaches
Dolan, Robert, John C. Ferm, David S. McArthur and James M. McCloy, 1969.
Measurements of beach process variables, outer banks, North Carolina.
Coastal Studies Ser, Louisiana State Univ. Press, 24: 79 pp.

beaches
Dorjes, Jurgen, 1971.
Der Golf von Gaeta (Tyrrhenisches Meer). IV. Das Makrobenthos und seine küstenparallele Zonierung. Senckenbergiana maritima 3(1): 203-246.

"beaches"
Doty, M.S., 1957.
Rocky intertidal surfaces. Ch. 18 in: Treatise on Marine Ecology and Paleoecology, Vol. 1.
Mem., Geol. Soc., Amer., 67:535-586.

beaches
Duboul-Razavet, C., et A. Monaco, 1966.
Étude minéralogique des sables du littoral Catatan espagnol.
Vie Milieu (B)17(1):217-241.

beaches
Duncan, John R., Jr., 1964.
The effects of water table and tide cycle on swash-backwash sediment distribution and beach profile development.
Marine Geology, 2(3):186-197.

beaches
Duplaix, S., 1950.
Origine des sables de deux coupes de plages des environs de la Pointe de la Coubre (Charante-Maritime) et leurs variations minérologiques en fonction des conditions météorologiques.
Bull. Soc. Géol., France, 5e ser., 20:239-244, 2 textfigs.

beaches
Duplaix, S., 1949
Etude des mineraux lourds de la plage de Gavres. Bull. Info., Com. Centr. Oceanogr. et d'Etude des Cotes, Service Hydro. de la Marine, No.5, Mai 1949, 11-15, Fig.4, Table (mimeographed)

E

beaches
Eagleson, P.S., and B. Glenne and J.A. Dracup, 1963.
Equilibrium characteristics of sand beaches.
J. Hydraulics Div., Proc. Amer. Soc. Civil. Eng. 89(1):35-57.

beaches
Eaton, R.O., 1950.
Littoral processes on sandy coasts. Inst. Coastal Eng., Univ. Ext., Univ. Calif., Long Beach, 11-13 Oct. 1950:42 pp., (mimeographed).

beaches
*Edelman, T., 1967.
Systematic measurements along the Dutch coast.
Proc. 10th Conf. Coast. Engng. Tokyo, 1966, 1:489-501.

beaches
Einstein, H.A., 1950.
Estimating quantities of sediment supplied by streams to a coast. Inst. Coastal Eng., Univ. Ext., Univ. Calif., Long Beach, 11-13 Oct. 1950: 6 pp. (mimeographed).

beaches
Einstein, Hans Albert, 1948
Movement of beach sands by water waves.
Trans. Am. Geophys. Un., 29(5):653-655.

beaches
Eisma, D., 1968.
Composition origin and distribution of Dutch coastal sands between Hoek van Holland and the island of Vlieland.
Netherlands J. Sea Res. 4(2): 123-267.

beaches
Eisma, D., 1965.
Eolian sorting and roundness of beach and sand dunes.
Netherland J. Sea. Res., 2 (4):541-555.

Beaches
Eisma, D., H. A. Das, D. Hoede, J. G. van Raaphorst and J. Zonderhuis, 1966.
Iron and trace elements in Dutch coastal sands.
Netherlands J. Sea Res., 3(1):68-94.

beaches
Elhai, H., 1952.
Morphologie du littoral breton entre Dahouet et Saint Cast. Bull. d'Info., C.C.O.E.C. 4(1)):389-402, 7 pls.

beaches
Emery, K.O., 1961.
A simple method of measuring beach profiles.
Limnol. & Oceanogr., 6(1):90-93.

beaches
Emery, K.O., 1955.
Grain size of marine beach gravels. J. Geol. 63(1):39-49.

beaches
Emery, K.O., 1945.
Tranportation of marine beach sand by flotation.
J. Sed. Petr. 15:84-87.

beaches
Emery, K.O., 1945.
The entrapment of air in beach sand. J. Sed. Petr 15:39-49.

beaches
Emery, K.O., and D.C. Cox, 1956.
Beach rock in the Hawaiian Islands. Pacific Sci., 10(4):382-402.

beaches
Emery, K.O., and J.F. Gale, 1951.
Swash and swash mark. Trans. Am. Geophys. Union 32(1):31-36, 4 textfigs.

beaches
Emery, K. O., and D. Neev, 1960
Mediterranean beaches of Israel. Min. Agric., Div. Fish., Israel, Sea Fish Res. Sta. Bull. No. 28: 1-24. Also, on same cover: Ministry of Develop., Israel, Geol. Surv., Bull No. 26: 1-24.

beaches
Emery, K.O., and R.E. Stevenson, 1950.
Laminated beach sand. J. Sed. Petr. 20(4):220-223, 2 textfigs.

beaches
Emery, K. O. and J. F. Foster, 1948
Water tables in marine beaches. J. Mar. Res. 7(3):644-654, 3 text figs.

beaches
Emery, K.O., J.I. Tracy, Jr. and H.S. Ladd, 1954.
Geology of Bikini and nearby atolls. Bikini and nearby atolls: Pt. 1. Geology.
Geol. Survey Prof. Pap. 260-A:1-265, 64 pls., 11 charts

beaches
Ertel, H., 1968.
Ausgleich von Störungen der Strandlinie an Flachküsten.
Acta hydrophysica 13(1): 5-10.

beaches
Ertel, H. und D. Hardke 1971.
Die Geschwindigkeit des Sedimenttransports längs der Strandlinie als Problem der Mechanik nichtstarrer Punktsysteme.
Acta hydrophisica 15(4): 249-258

beaches
Escoffier, F.F., 1954.
Travelling forelands and the shoreline processes associated with them. Bull. B.E.B. 8(4):11 pp., Fig. 4.

beaches
Etchichury, M.C., y J.R. Remiro 1971
Las arenas de la costa de la republica oriental del Uruguay en el tramo comprendido entre Nueva Palmira y la Barra del Chui.
Revta Mus. argent. Cienc. nat. Bernardino Rivadavia, Geol. 7(2): 153-195

F

beaches
Faure, Gérard, et Lucien Montaggioni 1971.
Étude lithologique et bionomique des calcaires littoraux sub-actuels de la région de Saint-Leu (Île de la Réunion, Océan Indien)
Téthys Suppl. 1: 281-295.

beaches
Fierro, Giliano, et Gian-Camillo Cortemiglia, 1965.
Observations sur le rôle de defense joue par la plage submergée de la Spiaggia delle Stella, Riviera ligure occidentels.
Cahiers Oceanogr., C.C.O.E.C., 17(10):715-726.

beaches
Fischer, Th., 1906.
Küstenstudien und Reiseeindrücke aus Algerien.
Zeits. Gesell. Erdkunde, 554-576, 12 photos.

beaches
Fischer, T. 1887.
Küstenstudien aus Nordafrika.
Petermanns Mitt. 33:1-13, 33-34, map.

beaches
Fischer, Th., 1895.
Zur Entwickelungsgeschichte des Küstens.
Petermanns Mitt. 31:409.

beaches
Fisher, R.L., 1955.
Cuspate spits of St. Lawrence Island Alaska.
J. Geol. 63:133-142.

beaches
Flemming, N.C., 1964.
Tank experiments on the sorting of beach material during cusp formation.
J. Sed. Petrol., 34(1):112-122.

beaches
Flinsch, H., 1953.
The effect of waves on a sand beach.
Proc. Minnesota Int. Hydraulics Conv., Sept. 1-4, 231-234.

beaches
Flouriot, Jean, 1967.
Le littoral marocain de Sale à Sidi Bou Knadel, etude morphologique.
Cah. océanogr., 19(1):17-39.

beaches
Forster, G.R., 1954.
Preliminary note on a survey of Stoke Point rocks with self-contained diving apparatus. J.M.B.A. 33:341-344.

beaches
France, Institut Geographique national, 1964.
La Photographie aérienne et l'étude des dépôts prélittoraux.
Etudes de Photo-Interprétation, Clos-Arceduc, Paris, No. 1:53 pp.

beaches
Frey, Robert W. and Taylor V. Mayou, 1971.
Decapod burrows in Holocene barrier island beaches and washover fans, Georgia. Senckenbergiana maritima 3(1): 53-77.

beaches
Fridman, R., 1957.
Sur des amas periodiques de materiel plus grossier à la surface de certaines plages sableuses.
Bull. d'Info., C.C.O.E.C., 9(9):508-510.

beaches
Friedman, Gerald M. 1967.
Dynamic processes and statistical parameters compared for size frequency distribution of beach and river sands.
J. Sed. Petr. 37 (2): 327-354.

beaches
Fuller, A.O., 1962.
Systematic fractionation of sand in the shallow marine and beach environment off the South African coast.
J. Sed. Petr., 32(3):602-606.

G

beaches
Gabis, V., 1955.
Les galets exotiques de la côte charentaise.
Bull. Soc. Géol., France, (6) 5:472-488.

beaches
Gadbois, P., and C. Laverdière, 1954.
Esquise géographique de la region de Floeberg Beach, nord de l'Ile Ellesmere.
Geogr. Bull., Canada, No. 6:17-44, 10 textfigs.

beaches
Gadow, Sibylle, 1971.
Der Golf von Gaeta (Tyrrhenisches Meer). 1. Die Sedimente. Senckenbergiana maritima 3(1): 103-133.

beaches
Galvin, Cyril J. Jr. 1968.
Breaker type classification on three laboratory beaches.
J. geophys. Res. 73(12): 3651-3659.

beaches
Galvin, C.J., Jr., and P.S. Eagleson, 1965.
Experimental study of longshore currents on a plane beach.
U.S. Army Coastal Engin. Res. Center, Techn. Memo., No. 10:80 pp.

beaches
Gameson, A.L.H., E.B. Pike and M.J. Barrett 1968.
Some factors influencing the concentration of coliform bacteria on beaches.
Rev. int. Océanogr. Méd. 9: 255-280.

benthos
Gamulin-Brida, Héléna, et Gordan Karaman, 1968.
Contribution aux recherches des biocoenoses benthiques de l'Adriatique méridionale-quelques particularites des biocoenoses du Golfe de Bokakotorska.
Rapp.P.-V. Réun.Comm.int.Explor.Sci.Méditerranée, 19(2):79-81.

beaches
Gangadharam, E.V., 1965.
Ilmenite from east coast beach sands.
Current Science, 34(9):286-287.

beaches
Garopati, P.N., and G. Chondrasekhara Rao, 1962.
Ecology of the interstitial fauna inhabiting the sandy beaches of Waltair coast.
J. Mar. Biol. Assoc., India, 4(1):44-57.

beaches
Gauld, D.T., and J.B. Buchanan, 1956.
The fauna of sandy beaches in the Gold Coast.
Oikos, 7(2):293-301.

beaches
Gee, Herbert C., 1965.
Beach nourishment from offshore sources.
J. Waterways and Harbors Div., Proc. Amer. Soc., Civil Engineers, 91(WW3):1-5.

beaches
Germaneau, J., 1967.
Contribution à l'étude dynamique des sables littoraux.
Trav. Cent. Rech. Étud océanogr., 7(2):5-17.

beaches
Germanovitch, D.E., and E.K. Zabieline, 1957.
Recherches geomorphologiques et lithologiques sur les rives de la mer d'Okhotsk, aux abords d'Okhotsk. Trav. Inst. Okean., Leningrad., 34:

not yet in MBL (Mar. 1958).

beaches
Gierloff, H. G., 1958.
Der Küstenschelf von El Salvador im Zusammenhang mit der Morphologie und Geologie des Festlandes.
Deutsche Hydrogr. Zeits., 11(6):240-246.

beaches
Gomoiu, M.T., 1963.
L'analyse granulométrique des sables de quelques plages de la Mer Noir (côte roumaine).
Rapp. Proc. Verb., Reunions, Comm. Int. Expl. Sci Mer Méditerranée, Monaco, 17(2):123-131.

beaches
Gorsline, D.S., 1966.
Dynamic charecteristics of west Florida Gulf coast beaches.
Marine Geol., 4(3):187-206.

beaches
Gorsline, D.S., 1964.
Beach studies in West Florida, U.S.A.
In: Developments in Sedimentology, L.M.J.U. van Straaten, Editor, Elsevier Publishing Co., 1: 144-147.

beaches
Götzinger, G., 1942.
Strandstudien an der westistrischen Adria. (Ergeb. der Chem. Österreichischen Adriaforschung)
Mitt. Geogr. Ges., Wien, 85:197-205.

beaches
Grant, A.S., and F.P. Shepard, 1946.
Effect of type of wave breaking on shore processes.
Bull. G.S.A. 57:1252.

beaches
Grant, U. S., 1948.
Influence of the water table on beach aggredation and degradation. J. Mar. Res. 7(3): 655-660, 1 textfig.

beaches
Granthem, K.N., 1953.
Wave run-up on sloping structures. Trans. Amer. Geophys. Union 34(5):720-724, 4 textfigs.

beaches
Great Britain, Hydraulics Research Board, 1957.
Hydraulics research, 1956, 54 pp.

beaches
Greene, H. Gary 1970.
Microrelief of an Arctic beach.
J. sedim. Petrol. 40(1): 419-427.

beaches
Greslou, L., 1964.
Méthodes d'investigation sur modèles réduits des differents problèmes d'hydraulique côtière.
La Houille Blanche, 19(6):706-717.

beaches
Guilcher, A., 1954.
Morphologie littorale et sous-marine. Presses Univ. France, 216 pp.

beaches
Guilcher, A., 1954.
Morphologie littorale du calcaire en Mediterranee occidentale (Catalogue et environs d'Alger).
Bull. Assoc. Geogr. Francais, No. 241-242:50-58.

beaches
Guilcher, A., 1955.
Géomorphologie de l'extremité septentrionale du Banc Farsan (Mer Rouge). Rés. Sci. Camp. Calypso. 1. Camp. 1951-1952 en Mer Rouge.
Ann. Inst. Océan., Monaco, 30:56-97, 9 textfigs., 20 pls.

beaches
Guilcher, A., 1949.
Observations sur les croissants de plage [beach cusps]. Bull. Soc. Géol., France, ser. 5, 19: 15-30, 2pls.

beaches
Guilcher, A., and F. Joly, 1954.
Recherches sur la morphologie de la côte atlantique du Maroc. Inst. Sci. Cherifien, 140 pp., 27 figs., 14 photographic plates.

Cited in Bull. d'Info., C.C.O.E.C., 6(8).

beaches
Guilcher, A., P. Vallantin, J.P. Angrand and P. Galloy, 1957.
Les cordons litteraux de rade de Brest.
Bull. d'Info., C.C.O.E.C., 9(1):21-54.

beaches
Guillou, R.B., and J.J. Glass, 1957.
A reconnaissance study of the beach sands of Puerto Rico.
Geol. Survey Bull., 1042-I:273-305.

H

beaches
Haase, G., and H. Richter, 1957.
Fossile Böden im Löss an der Schwarzmeerküste bei Constante. Petermanns Geogr. Mitt., 101(3): 161-173.

beaches
Hall, J. V., and W. J. Herron, 1950.
Test of nourishment of the shore by offshore deposition of sand, Long Beach, New Jersey.
Tech. Memo., B.E.B., No. 17:32 pp., 14 pls. (multilithed).

beaches
Hamada, T., 1951.
Breakers and beach erosion. Rept. Transport. Tech. Res. Inst., Rept. No. 1:165 pp., 20 photos., 39 textfigs.

beaches
Hand Bryce M. 1967.
Differentiation of beach and dune sands using settling velocities of light and heavy minerals.
J. Sed. Petrol. 37 (2): 514-520.

beaches
Handin, J.W., 1951.
The source, transportation, and deposition of beache sediment in Southern California. B.E.B. Tech. Rept. No. 22:112 pp., 12 pls.

beaches
Handin, J.W., 1950.
The geological aspects of coastal engineering.
Inst. Coastal Eng., Univ. Ext. Univ. Calif., Long. Beach, 11-13 Oct. 1950:4 pp. (mimeographed).

beaches
Handin, J. W., and T. C. Ludwick, 1950.
Accretion of beach behind a detached breakwater.
Tech. Memo., B.E.B., No. 16:13 pp. & 3 figs. (multilithed).

S.I.O. Contrib. No. 455.

beaches
Hansen, H. J., 1947.
Beach erosion studies in Florida. Univ. Florida Eng. Exp. Sta. Bull. No. 16: 68 pp.

beaches
Harris, R.L., 1954.
Restudy of test-shore nourishment by offshore deposition of sand, Long Branch, New Jersey.
Tech. Memo., B.E.B., No. 62:1-18, 8 figs.

beaches
Harris, Roger P., 1972.
The distribution and ecology of the interstitial meiofauna of a sandy beach at Whitsand Bay, East Cornwall. J. mar. biol. Ass. U.K. 52(1): 1-18.

beaches
Harris, T.F.W., 1961.
The nearshore circulation of water.
Marine studies off the Natal Coast, C.S.I.R. Symposium, No. S2:18-30. (Multilithed).

beaches
Harrison, W., and R. Morales Alamo, 1964.
Dynamic properties of immersed sand at Virginia Beach, Virginia.
U.S. Army Coastal Engng Res. Center, Techn. Memo., No. 9:52 pp. (multilithed).

beaches
Harrison, W., N.A. Pore and D.H. Tuck, Jr. 1965.
Predictor equations for beach processes and responses.
J. geophys. Res. 70(24): 6105-6109.

beaches
Harrison, W., and Kenneth A. Wagner, 1964.
Beach changes at Virginia Beach, Virginia.
U.S. Army, Coastal Engin. Res. Center, Misc. Paper, No. 6-64:25 pp. (multilithed).

beaches
Hatai, Kotora, 1959
Coastal geology of the Kii peninsula. Rec. Oceanogr. Wks., Japan, Spec. No. 3: 165-166.

beaches
Hatai, Kotora, and Minoru Saito, 1963.
Selective accumulation of beach gravels.
Rec. Oceanogr. Wks., Japan, n.s., 7(1):95-99.

beaches
Hayami, Shoitiro, 1958
Type of breakers, wave steepness and beach slope.
Coastal engineering in Japan, 1: 21-24;

beaches
Healy, J.J., 1953.
Wave damping effect on beaches. Proc. Minnesota Int. Hydraulics Conv., Sept. 1-4, 1953:213-220, 9 textfigs.

beaches
Hedgpeth, J.W., 1957.
Sandy beaches. Ch. 19 in: Treatise on Marine Ecology and Paleoecology, Vol. 1.
Mem. Geol. Soc., Amer., 67:587-608.

benthos
Henriksson, Rolf 1969.
Influence of pollution on the bottom fauna of the Sound (Öresund)
Oikos 20(2): 507-523.

beaches
Herbich, J.B., R.M. Sorensen and J.H. Willenbrock 1963.
Effect of berm on wave run-up on composite beaches.
J. Waterways Harbor Div., Proc. Amer. Soc. Civil Eng. 89(WW2):55-72.

beaches
Hernández-Pacheco, Francisco, e Isodoro Asensio Amor, 1965.
Sobre la formación artificial de una playa en la margin occidental de la ría del Eo.
Bol. Inst. Español Oceanogr., No. 118:38 pp.

beaches
*Herron, William J., and Robert L. Harris, 1967.
Littoral bypassing and beach restoration in the vicinity of Port Hueneme, California.
Proc. 10th Conf. Coast. Engng. Tokyo, 1966, 1:651-675.

beaches
Ho, D.V., and R.E. Meyer, 1962.
Climb of a bore on a beach. Part 1. Uniform beach slope.
J. Fluid Mech., 14(2):305-318.

beaches
Hodgson A.V., and W.B. Scott, 1970.
The identification of ancient beach sands by the combination of size analysis and electron microscopy.
Sedimentology 14 (1/2): 67-75.

beaches
Hodgson, W.A., 1966.
Coastal processes around the Otago Peninsula.
New Zealand J. Geol. Geophys., 9(1/2):76-90.

beaches
Hom-ma, Masashi, Kiyoshi Horikawa, and Choule Sonu, 1960
A study of beach erosion at the sheltered beaches of Katase and Kamakura, Japan.
Coastal Eng., Japan, 3: 101-122.

beaches
Hommeril, Pierre, et Claude Larsonneur, 1963.
Quelques effets morphologiques du gel intense de l'hiver 1963 sur le littoral Bas-Normand.
Cahiers Océanogr., C.C.O.E.C., 15(9):638-650.

beaches
Horton, D.F., 1948.
Shore effects of coastal structures. Mil. Eng. 40:402-405.

beaches
Hours, R., W.D. Nesteroff and V. Romanovsky, 1954
Méthode d'étude de l'évolution des plages par traceurs radio-actifs. Trav. C.R.E.O. 1(11):1-7.

beaches
Hours, R., W.D. Nesteroff and V. Romanovsky, 1955.
Utilisation d'un traceur radioactif dans l'étude de l'évolution d'une plage. C.R. Acad. Sci., Paris, 240(18): 1798-1800.

beaches
Hoyle, J.W., and G.T. King, 1955.
The longitudinal stability of beaches.
J. Inst. Municipal Eng. 82(5):181-191.

beaches
Hoyt, J.H., 1962.
High-angle beach stratification, Sapelo Island, Georgia.
J. Sed. Petr., 32:309-311.

beaches
Hoyt, John H., and Vernon J. Henry, Jr., 1964.
Development and geologic significance of soft beach sand.
Sedimentology, 3(1):44-51.

beaches,
Hoyt, John H., and Robert J. Weimer, 1963.
Comparison of modern and ancient beaches, central Georgia coast.
Bull. Am. Ass. Petrol. Geol., 47:529-531.
Also in:
Collected Reprints, Mar. Inst., Univ. Georgia, 4 (1965).

beaches
Hsu, K.J., 1960
Texture and mineralogy of the Recent sands of the Gulf Coast. J. Sed. Petr., 30(3): 380-403.

beaches a
*Hsu, Shih-Ang, 1971.
Measurement of shear stress and roughness length on a beach. J. geophys. Res., 76(12): 2880-2885.

beaches
Huang, Tun-Yow, 1961.
Smaller Foraminifera from beach sands at Tanmenkang, Pachao-tao, Penghu.
Proc. Geol. Soc., China, (4):83-90.

beaches
Hume, James D., and Marshall Schalk, 1964.
The effects of ice-push on Arctic beaches.
Amer. J. Sci., 262(2):267-273.

beaches
*Husmann, Siegfried, 1967.
Klassifizierung mariner, brackiger und limnischer Grundwasserbiotope.
Helgoländer wiss. Meeresunters. 16(3):271-278.

seashore
Hylander, C. J., 1950.
Sea and Shore. Macmillam Co., N.Y., 242 pp.,

I

beaches
Ichiye, T., 1953.
Some remarks on the non-linear theory of shallow water waves on a sloping beach. Ocean. Mag., Tokyo, 4(4):159-166.

beaches
Ingle, James C., 1966.
The movement of beach sand.
Elsevier, 231 pp. $ 14.50.

beaches
Ingle, James C., Jr., 1966.
The movement of beach sand; an analysis using fluorescent grains.
Develop. Sedimentol., Elseiver Publ. Co., 221 pp.

beaches
Ingle, James C., Jr., 1962.
Tracing beach sand movement by means of fluorescent dyed sand.
Shore and Beach, October, 1962: 6 pp.

beaches
Inman, D.L., 1953.
Areal and seasonal variations in beach and nearshore sediments at La Jolla, California.
B.E.B. Tech. Memo. No. 39:82 pp., 27 figs., numerous tables.

beaches
Inman, D. L., 1950.
Report on beach study in the vicinity of Mugu Lagoon, California. B.E.B. Tech. Memo. 14:47 pp., 27 figs. (multilithed).

S.I.O. Sub. Geol. Rept. No. 5.

beaches
Inman, D.L., and R.A. Bagnold, 1963.
21. Beach and nearshore processes. 2. Littoral processes.
In: The Sea, M.N. Hill, Editor, Interscience Publishers, 3:529-553.

beaches
Inman, Douglas L. and J. Filloux, 1960
Beach cycles related to tide and local wind wave regime. J. of Geol. 68(2): 225-231.

beaches
Inman, D.L., W.R Gayman, and D.C Cox, 1963
Littoral sedimentary processes on Kauai, a subtropical high island.
Pacific Science, 17(1):106-130.

beaches
Inman, D.L., and G.S. Rusnak, 1956.
Changes in sand level on the beach and shelf at La Jolla, California. B.E.B. Tech. Memo., 82:30 pp.

beaches a
Ionin, A., Y. Pavlidis y O. Avello, 1969.
Sobre la estructura geologica y geomorphologica de la zona costera y sedimentos de la ensenada de la Broa.
Serie Transform. Naturaleza, Inst. Oceanol. Acad. Cienc. Cuba, 8:1-21.

beaches
Ishihara, Tojiro, Yuichi Iwagaki and Masashi Murakami, 1958
On the investigation of beach erosion along the north coast of Akashi Strait. Coastal Engineering in Japan, 1: 97-109.

beaches
Ishihara, Tojiro, and Toru Sawaragi, 1964.
Stability of beaches using groins.
Proc. 9th Conf. Coastal Eng., Lisbon, June 1964, 299-309. Abstract in: Bull. Disaster Prevention Res. Inst., Kyoto Univ., 15(4):58.

beaches
Ishihara, Tojiro, Toru Sawaragi and Tetsuo Amano, 1959
Fundamental studies on the dynamics of sand drifts. Coastal Eng., Japan, 2: 35-52.

beaches
Iwagaki, Yuichi, and Hideaki Noda, 1963.
On the scale effect in two-dimensional beach processes.
Geophysical Papers dedicated to Professor Kenzo Sassa, 131-135.

beaches
Iwagaki, Yuichi, and Toru Sawaragi, 1962
A new method for estimation of the rate of littoral sand drift.
Coastal Engineering in Japan, 5:67-79.

beaches
Iwagaki, Y., and T. Sawaragi, 1958.
Experimental study on the equilibrium slopes of beaches and sand movement by breakers. Coastal engineering in Japan, 1:75-84.

J

beaches
Jansen, K.P., 1971.
Ecological studies on intertidal New Zealand Sphaeromatidae (Isopoda: Flabellifera).
Mar. Biol. 11(3): 262-285.

beaches
*Jansson, Bengt-Owe, 1967.
The availability of oxygen for the interstitial fauna of sandy beaches.
J. exp. mar. Biol. Ecol., 1(2):123-143.

beaches, plants
Jeanrenaud, Elena, Maria Truscă et C. Frăsinel, 1968.
Le régime hydrique et l'interrélation de la respiration chez quelques plantes du littoral de la Mer Noire aux environs de la Station Agigea-Constanța.
In: Lucrările Sesiunii Științifice a Stațiunii de Cercetări Marine "Prof. Ioan Borcea", Agigea, (1-2 Noiembre 1966), Volum Festiv, Iași, 1968: 159-184.

beaches
Johnson, Ralph Gordon, 1967.
Salinity of interstitial water in a sandy beach.
Limnol. Oceanogr., 12(1):1-7.

beaches
Jolliffe, I. P., 1964.
An experiment designed to compare the relative rates of movement of different sizes of beach pebble.
Proc. Geol. Assoc., Great Britain, 75(1):67-86.

beaches
Jolliffe, I P., (undated reprint)
The use of tracers to study beach movements; and the measurement of littoral drift by a fluorescent technique.
Revue de Géomorphologie Dynamique, unnumbered pp.

beaches
Judge, C.W. 1970.
Heavy minerals in beach and stream sediments as indicators of shore processes between Monterey and Los Angeles, California.
Techn. Memo. Coast. engng Res. Center 33:44 pp. (multilithed).

beaches
Jutson, J.T., 1949.
The shore platform of Point Lonsdale, Victoria.
Proc. Roy. Soc., Victoria, 61(n.s.):105-111, 6 textfigs., Pl. 13.

beaches
Jutson, J. T., 1949.
The shore platforms of Lorne, Victoria. Proc. Roy. Soc., Victoria, 61(n.s.):43-59, 17 textfigs. Pls. 5-7.

K

beaches
Kadib, Abdel-Latif, 1964.
Calculation procedure for sand transport by wind on natural beaches.
U.S.Army, Coastal Engineering Res. Cent., Misc. Paper, No. 2-64:25 pp. (multilithed).

beaches
Kaëib, Abdel-Latif, 1963.
Beach profile as affected by vertical walls.
Beach Erosion Bd., Techn. Memo., No. 134:41 pp. and tables.

beaches
Karo, H.A., 1958.
Basic surveys for beach and harbor studies.
Int. Hydrogr. Rev., 35(2):21-32.

beaches
Keary, Raymond, 1969.
Variations in the biogenic carbonate content of beach sands on the South Coast of Ireland.
Scient. Proc. R. Dublin Soc. (A) 3(18):193-202

beaches
*Keary, Raymond, 1967.
Biogenic carbonate in beach sediments of the west coast of Ireland.
Scient.Proc.R.Dublin Soc. (A)3(7):75-85.

beaches
Keary, R., 1965.
A note on the beach sands of the Cois Fharrige coast.
Irish Nat. J., 15(2):40-43.

beaches
Kempin, E.T., 1953.
Beach sand movement at Cottesloe, Western Australia. J.R. Soc., West. Australia, Inc., 37:35-58, 14 textfigs.

beaches
Kidson, C., & A.P. Carr, 1962
Marking beach materials for tracing experiments.
J. Hydraulics Div., Proc. Amer. Soc. Civil. Eng., 88(HY4): 43-60.

beaches
King, C.A., 1961.
Beaches and coasts.
Edward Arnold., Ltd., London, 403 pp., 65 shill.

beaches
King, C.A.M., 1951.
Depth of disturbance of sand on sea beaches by waves. J. Sed. Petr. 21(3):131-140, 7 textfigs.

beaches
King, C.A.M., and W.W. Williams, 1949
The formation and movement of sand bars by wave action. Geogr. Jour., 113:70-85; 16 textfigs.

beaches,
King, L., 1952.
The Durban beach problem. S. Afr. J. Sci. 48: 314-318.

beaches
Klein, G. de V., 1964.
Sedimentary facies in Bay of Fundy intertidal zone, Nova Scotia, Canada.
In: Developments in Sedimentology, L.M.J.U. van Straaten, Elsevier Publishing Co., 1:193-199.
Editor

beaches
Koh, Ryuji, 1969.
Beach erosion and Quaternary sea level.
Coast. Engng, Japan, 12: 121-128.

sea level
Quaternary

beaches
*Koldijk, W.S., 1968.
On environment-sensitive-size parameters.
Sedimentology, 10(1):57-69.

beaches a
*Komar, Paul D., 1971.
The mechanics of sand transport on beaches.
J. geophys. Res., 76(3): 713-721.

beaches
*Komar, Paul D. and Douglas L. Inman, 1970.
Longshore sand transport on beaches. J. geophys Res., 75(30): 5914-5927.

beaches
Komukai, Ryoshichi, 1959
On the marine geology of beach erosion in Omori-hama vicinity, Hakodate City.
H. O. Pub., Japan, 943:582 pp.

beaches
Komokai, R., 1959.
On the marine geology of beach erosion in Omori-hama vicinity, Hakodate City.
Pub. 943, Bull. Hydrogr. Office, Japan, 13(2):217-582.

beaches
Krumbein, W.C., 1959.
The "sorting out" of geological variables illustrated by regression analysis of factors controlling beach firmness.
J. Sed. Petr., 29(4):575-587.

beaches
Krumbein, W.C., 1957.
A method for specification of sand for beach fills
B.E.B. Tech. Memo., 102:81 pp.

beaches
Krumbein, W.C., 1954.
Statistical significance of beach sampling methods. B.E.B. Tech. Memo. No. 50:1-33, 5 figs.

beaches
Krumbein, W.C., 1953.
Statistical designs for sampling beach sand.
Trans. Amer. Geophys. Union 34(6):857-868, 2 textfigs.

beaches
Krumbein, W.S., 1950.
Littoral processes in lakes. Inst. Coastal Eng., Univ. Ext., Univ. Calif., Long Beach, 11-13 Oct. 1950:9 pp. (mimeographed).

beaches
Krumbein, W.C., and W.R. James, 1965.
A lognormal size distribution model for estimating stability of beach fill material.
U.S. Army, Coast Eng. Res. Center, Techn. Memo. No. 16: 17pp.

beaches
Krumbein, W.C., and L.E. Ohsiek, 1950.
Pulsational transport of sand by shore agents.
Trans. Am. Geophys. Union 31(2):216-220, 6 textfigs.

beaches
Kuenen, Ph. H., 1948.
The formation of beach cusps. J. Geol. 56(1): 34-40, 3 textfigs., 2 pls.

beaches
Kühl, Heinrich, und Hans Mann, 1966.
Anderungen im Chemismus des Interstitialwassers am Strand von Cuxhaven während einer Tide.
Helgoländer wiss Meeresunters, 13(3):238-245.

beaches
Kulm, L.D., and John V. Byrne 1967.
Sediments of Yaquina Bay, Oregon.
In: Estuaries G.H. Lauff editor. Publs Am. Ass. Advmt. Sci. 83: 226-238

beaches
Kulm, L.D., K.F. Scheidegger, J.V. Byrne and J.J. Spigai, 1968.
A preliminary investigation of the heavy mineral suites of the coastal rivers and beaches of Oregon and northern California. Ore Bin, 30(9): 165-180. Also in: Coll. Reprints, Dept. Oceanogr. Univ. Oregon, 7(1968).

L

beaches
LaFond, E.C., and R. Prasada Rao, 1956.
Erosion of the beach at Uppada. Port. Eng., India, 5:4-9.

beaches
Larras, J., 1957.
Plages et côtes de sable. Collection du Laboratoire National d'Hydraulique, Eyrolles, Paris, 117 pp.

Beaches
Lates, M., and A. Spătaru, 1970
Consequences of using the seaside beaches as sand source. (In Romanian)
Hidrotehnica, Romania, 15(2):87-91

beaches
LeFur, A., 1952.
Note sur la détermination du gradient des plages par examen de la houle sur les photographies aériennes. Ann. Hydrogr., Paris, 4th ser., 3:167-176, Pls. 35-38.

beaches
Lefur, A., 1951.
Note sur la détermination du gradient des plages par examen de la houle sur les photographies aériennes. Ann. Hydrogr., Paris, 2(723) :167-176.

beaches
LeFur, A., 1950.
Sur la determination du gradient des plages par examen de la houle sur les photographies aériennes. Bull. Info., C.C.O.E.C., 2 Ann., (4):125-137, 2 figs. (mimeographed).

beaches
Lewy, H., 1946.
Water waves on sloping beaches. Bull. Amer. Math. Soc. 52:737-775.

beaches
Lhermitte, P., and J. Germain, 1955.
[Littoral drift along shingle beaches; application to Port des Galets (Reunion Island).]
Mem. Soc. Hydrotechnic., France, 2:162-168.
Also in: Houille Blanche 10:658-669.

beaches
Longinov, V.V., 1961
[Some data concerning surfzone regime on the sand beach with smooth outer slope.]
Trudy Okeanogr. Komissii, Akad. Nauk, SSSR, 8:136-157.

beaches
Longuinov, V.V., 1958.
Quelques résultats de la mesure des pressions d'ondes de fond horizontales dans la zone maritime littorale.
Bull. d'Info., C.C.O.E.C., 10(9):531-540.

beaches
Longuinov, V.V., 1958.
[Some observations of horizontal wave-pressure in the near-bottom water layer of the coastal zone under natural conditions.] Trudy Inst. Okeanol., 28:100-156.

beaches
Longinov, V.V., 1950.
[Reshaping of beaches on the North Caucasian coast as a result of the August 1949 storm.] Priroda 7:55.

beaches
Longinov, V. V., 1948.
[Relation between the course of waves and the corresponding maximum rate of displacement of alluvium from sea-bottom slopes along the shore.] Izvest Akad. Nauk SSSR, Ser. Geogr. Geofiz., 12:362-368.

beaches
Longinov, V.V., and O. K. Leontov, 1951.
On the question of the dynamics of the profile of a sandy beach. (In Russian).
Trudy, Inst. Okeanol., Akad. Nauk, SSSR, 6:59-69.

beaches
Longinov, V.V., and A.D. Pasechnik, 1953.
Basic laws of growth of the profile of shingle beaches. (In Russian).
Trudy Inst. Okeanol., Akad. Nauk, SSSR, 7:135-15[illegible]

beaches
Loniunov, V.V., 1956.
[Some methodological problems in the study of the dynamics of the coastal zone.] Trudy Inst. Okeanol 19:144-155.

beaches
Lopez-Benito, Manuel, 1966.
Variación estacional del contenido de materia orgánica en las arenas de la playa de Areiño (Ría de Vigo).
Inv. Pesq., Barcelona, 30:233-246.

M

beaches
Mahadevan, C., and R. Prasada Rao, 1958
Causes of the growth of sand spit north of Godavari confluence.
Andhra Univ. Mem., Oceanogr., 2: 69-74.

beaches
Mahadevan, C., and A. Srirama Das, 1954.
Effects of high waves on the formation of coastal black sand deposits. Andhra Univ. Ocean. Mem. 1:57-62.

beaches
*Maloney, Neil J., 1967.
Textural characteristics of coastal and shallow marine sands of eastern Venezuela.
Boln Inst. Oceanogr., Univ. Oriente, 5(1/2):54-66.

beaches
Maloney, Neil J., 1965.
Particle density effects in estimating median diameters of sands.
Bol. Inst. Oceanogr., Univ-Oriente, 4(1):184-190.

beaches
Mancini, Renato, ed Enea Occella, 1967.
Natura e possibilità di concentrazione di minerali pesanti contenuti nelle sabbie marine lungo la linea di costa nella zona di Nettuno.
Boll. Ass. Mineraria Subalpina 4(4): 597-632. Also in: Progr. Ricerca Risorse mar. Fondo mar. (B)(33)(1969). (Comm. ital. Oceanogr. Consig. naz. Ricerche)

beaches
Maquet, G., 1955.
Minéralogie des sables de plage de la côte tunisienne de Ras Engela à Porto-Farina.
Bull. Soc. Sci. Nat., Tunisie, 8(1/2):57-64.

beaches
Martin, W.R.B., and A.M. Long, 1960.
Heavy mineral content and radioactive counts of beach sands west of Oreti River mouth to Blue Cliffs, Southland, New Zealand.
New Zealand J. Geol. and Geophys., 3(3):400-409.

beaches
Martins, Luiz B., 1965.
Significance of skewness and kurtosis in environmental interpretation.
J. sed. Petr., 35(3):768-770.

beaches
*Matthews, Barry, 1967.
Late Quaternary Marine fossils from Frobischer Bay (Baffin Island, N.W.T., Canada).
Palaeogr. Palaeoclimatol. Palaeoecol., 3(2):243-263.

beaches
Masry, David, 1970.
Ecological study of some sandy beaches along the Israeli Mediterranean coast, with a description of the interstitial harpacticoids (Crustacea Copepoda).
Cah. Biol. mar. 11 (3): 224-258.

beaches
Mauchline, John, Angela M. Taylor and Eric B. Ritson, 1964.
The radioecology of a beach.
Limnology and Oceanography, 9(2):187-194.

beaches
Mauriello, Louis J. 1968.
Beach rehabilitation by hopper dredge.
J. WatWays Harb. Div. Am. Soc. civ. Engrs. 94 (WW2): 175-188.

Mc

beaches
McCabe Robert A. 1970.
Beach behavior, north shore, Long Island Sound.
J. WatWays Harb. Div. Am. Soc. civ. Engrs 96 (WW4): 787-794

beaches
McCann, S.B. and F.A. Bryant 1970.
Beach processes and shoreline changes, Kouchibouguac Bay, New Brunswick.
Marit. Sed. 6 (3): 116-117 (Halifax)

beaches
McFarland, William N., 1963.
Seasonal plankton productivity in the surfzone of a south Texas beach.
Publ. Inst. Mar. Sci., Port Aransas, 9:77-90.

beaches
McIntyre, D.D., 1959
The hydraulic equivalence and size distribution of some mineral grains from a beach.
J. Geol., 67(3): 278-301.

beaches
McIntire, William G., and James P. Morgan, 1963.
Recent geomorphic history of Plum Island, Massachusetts and adjacent coasts.
Louisiana State Univ. Studies, Coast Studies Ser. (8):44 pp. (multilithed).

beaches
*McLean, Roger F., 1967.
Measurements of beachrock erosion by some tropical marine gastropods.
Bull. mar. Sci., Miami, 17(3):551-561.

beaches
*McLean, Roger F., 1967.
Origin and development of ridge-furrow systems in beachrock in Barbados, West Indies.
Mar. Geol., 5(3):181-193.

beaches
McMaster, R.L., 1960
Mineralogy as an indicator of beach sand movement along the Rhode Island shore. J. Sed. Petr., 30(3): 404-413.

beaches
*McLean, R.F., and R.M. Kirk, 1969.
Relationships between grain size, size-sorting, and foreshore slope on mixed sand-shingle beaches.
N.Z. Jl. Geol. Geophys. 12:138-155.

beaches
Mei, C.C., G.A. Tlapa and P.S. Eagleson 1968.
An asymptotic theory for water waves on beaches of mild slope.
J. geophys. Res. 73(14): 4555-4560

beach
Mero, John L., 1966.
Review of mineral values on and under the ocean.
Exploiting the Ocean, Trans. 2nd Ann. Mar. Techn. Soc. Conf., June 27-29, 1966, 61-78.

beaches
Mero, John L., 1965.
The mineral resources of the sea.
Elsevier Oceanography Ser., 312 pp.

beaches
Meyer, R.E., and R.B. Turner, 1967.
Some three-dimensional effects in surf.
J. geophys. Res., 72(10):2513-2518.

beaches
Miller, Robert L., and John M. Zeigler, 1964.
A study of sediment distribution in the zone of shoaling waves over complicated bottom topography.
In: Papers in Marine Geology, R.L. Miller, Editor, Macmillan Co., N.Y., 133-153.

beaches
Milling, Marcus E., and E. William Behrens, 1966.
Sedimentary structures of beach and dune deposits; Mustang Island, Texas.
Publs. Inst. mar. Sci., Univ. Texas, Port Aransas, 11:135-148.

beaches, organic matter in
Minas, Monique, 1965.
La substance organique et le calcaire dans deux types de vasières littorales de la région de Tuléar.
Rec. Trav. Sta. Mar., Endoume, hors sér., Suppl., No. 4:57-70. (Trav. Sta. Mar., Tuléar).

beaches
*Minch, John A., 1967.
Stratigraphy and structure of the Tijuana-Rosarito beach area, northwestern Baja California, Mexico.
Bull. geol Soc., Am., 78(9):1155-1178.

beaches
Minikin, R.R., 1948-1949.
Coast protection: a survey of beach stability.
Dock Harb. Auth. 29:165-169, 193-198, 232-236, 251-256, 281-285, 311-314.

beaches
Miyazaki, M., 1951.
[On the stable form of the sea bottom near the shore.] J. Ocean. Soc., Japan (Nippon Kaiyo Gakkaisi)6(3):157-159, 1 textfig.

beaches
Moberly, Ralph Jr. 1968.
Loss of Hawaiian littoral sand.
J. sed. Petr. 38(1):17-34.

beaches
Moberly, Ralph, Jr., L. David Bauer, Jr., and Anne Morrison, 1965.
Source and variation of Hawaiian littoral sands.
J. sed. Petr., 35(3):589-598.

Mokyevsky, O.B., 1960 beaches
Geographical zonation of marine littoral types. Limnol. & Oceanogr., 5(4):389-396.

beaches
Montaggioni, Lucien 1971.
Premières observations sur la répartition granulométrique et minéralogique des sables volcaniques littoraux de l'île de la Réunion (Océan Indien).
Téthys Suppl. 1:299-324.

beaches
Moore, J. G., 1947.
The determination of the depths and extinction coefficients of shallow water by air photography using colour filters. Phil. Trans. Roy. Soc. London, Ser. A, Math. Phys. Sci. 240(816): 163-217.

beaches
*Müller, German, and Gerd Tietz, 1966.
Recent dolomitization of Quaternary biocalcarenites from Fuerteventura (Canary Islands).
Contr. Mineral. Petrol., 13:89-96.

beaches
*Munk, W., and W. Wimbush, 1969.
A rule of thumb for wave breaking over sloping beaches. (In Russian; English abstract).
Okeanologiia, 9(1):71-75.

N

beaches
Nagata, Yutaka, 1964.
Deformation of temporal pattern of orbital wave velocity and sediment transport in shoaling water in breaker zone and on foreshore.
J. Oceanogr. Soc., Japan, 20(2):57-70.

beaches
Nagata, Y., 1960
Applications of electro-magnetic current meter for beach problems. Geophys. Notes. 13(1): No. 10 from Proc. 6th Conf. Coast. Eng. Japan, 1959: 45-48. No copy of reprint available for Geophys. Notes.

beaches
Nair R.R. 1971.
Beachrock and associated carbonate sediments on the Fifty Fathom Flat, a submarine terrace on the outer continental shelf off Bombay.
Proc. Indian Acad. Sci. (B) 73(3): 148-154

beaches
Nelson, Campbell S., and K. A. Rodgers 1969
Algal stabilisation of Holocene conglomerates by micritic high-magnesium calcite, southern New Caledonia.
N.Z. Jl mar freshw. Res. 3(3): 395-408.

beaches
Nevesskii, E.N., 1958.
Some results of the study of litoral sediment strata by means of vibrating-piston core samplers. (In Russian).
Biull., Okeanogr. Komissiia, Akad. Nauk, SSSR, (1):64-
not seen

beaches
Nevessky, E.N., 1958.
[The Anapa beach history based on analysis of littoral and coastal deposits.]
Trudy Inst. Okeanol., 28:14-22.

beaches
Nevessky, E.N., 1957.
[History of the development of the Anapa beach.]
Trudy Inst. Okeanol., 21:165-174.

beaches
Nevessky, E.N., 1957.
[On the relation of the coastal line contours to the geologic structure of abraded coasts.]
Trudy Inst. Okeanol., 21:175-186.

beaches
Nevesskii, E.N., 1957.
[The exploration of littoral deposits carried out with the aid of the vibro-piston tube.]
Doklady Akad. Nauk, SSSR, 112(3):418-

beaches
Newell, N.D., E. G. Purdy, and J. Imbrie 1960.
Bahamian oölitic sand.
J. Geol. 68(5): 481-497.

beaches
*Nichols, Robert L., 1968.
Coastal geomorphology, McMurdo Sound, Antarctica.
J. Glaciol., 7(51):449-478.

beaches
Nichols, R. L., 1949.
Recent shoreline changes at Shirley Gut, Boston Harbor. J. Geol. 57:85-89.

beaches
*Nicholson, D.S., 1967.
Distribution of economic minerals in South Island west coast beach sands.
N.Z.Jl. Sci., 10(2):447-

beaches
Nicholson, D.S., and H.E. Fyfe, 1958.
Borehole survey of North Island iron sands from New Plymouth to Kaipara Harbour.
N.Z.J. Geol. & Geophys., 1(4):617-634.

beaches
Nicholson, D.S., W.T. Shannon and T. Marshall, 1966.
Separation of ilmenite, zircon and monazite from Westport beach sands.
New Zealand J. Sci., 9(3):586-598.

beaches
Nievlessky, E.N., 1958.
Étude des sédiments marins litteraux à l'aide du tube à pistonvibreur.
Bull. d'Info., C.C.O.E.C., 10(6):309-323.

beach
Nonn, H., 1953.
Morphologie du littoral breton entre Saint-Cast et Dinard. Bull. d'Info., C.C.O.E.C. 5(1):30-46, Pls. 3-8. (mimeographed).

beaches
Norris, Robert M., 1964.
Dams and beach-sand supply in southern California In: Papers in Marine Geology, R.L. Miller, editor, Macmillan Co., N.Y., 154-171.

beaches
Norris, R.M., 1956.
Crescentic beach cusps and barkhan dunes.
Bull. Amer. Assoc. Petr. Geol., 40(1681-1686.

beaches
Norris, R.M., 1952.
Recent history of a sand spit at San Nicolas Island, California. J. Sed. Petr. 22:224-228.

O

beaches
Oakley, Horace Roy, and Edward Arthur Dyer, 1966.
Investigation of sea outfalls for Tyneside sewage disposal.
Inst. Civil Eng., London, Proc., 33:201-230.

beaches
O'Brien, Morrough P., 1969.
Equilibrium flow areas of inlets on sandy coasts.
J. WatWays Harb. Div. Am. Soc. civ. Engrs. 95(WW1): 43-52.

USA, west
beaches
inlets

beaches
Olson, F.C.W., 1952.
Shore developments of a bay. Pap. Ocean. Inst. Florida State Univ. Studies No. 7:28-31, 1 textfig.

beaches
Orsson, I., and W. Blake, Jr., 1961.
Problems of radiocarbon dating of raised beaches based on experience in Spitsbergen.
Norsk Geogr. Tids., 18(1/2):47-64.

beaches
Ottmann, Francois, 1965.
Introduction a la Geologie Marine et Littorale.
Masson et Cie, Paris, 259 pp. 47 F

beaches
Ottmann, Francis, Ramon Nobrega, Paulo Nobrega Coutinho, and Silvia Pericles B. de Oliveira, 1959.
Estudo topografico e sedimentologico de um perfil da Praia de Piedade-Recife-Pernambuco. Trabalhos Inst. Biol. Marit. e Oceanogr., Recife, 1(1):19-38.

beaches
Ottmann, J.-M., and F. Ottmann, 1961.
Sur le rapport C/N dans les sédiments litteraux
C.R. Acad. Sci., Paris, 252(15):2277-2279.

beaches
Otvos, Ervin G., Jr., 1965.
Sedimentation-erosion cycles of single tidal periods on Long Island Sound beaches.
J. sed. Petr., 35(3):604-609.

beaches
Otvos, Ervin G., Jr., 1965.
Types of rhomboid beach surface patterns.
Amer. J. Sci., 263(3):271-276.

beaches
Otvos, Ervin G., Jr., 1964.
Observation of beach cusp and beach ridge formation in the Long Island Sound.
J. Sed. Petr., 34(3):554-560.

beaches
Owens, E.H. 1971.
The restoration of beaches contaminated by oil in Chedabucto Bay, Nova Scotia.
Manuscript Rept. Ser., Mar. Sci. Br., Dept. Energy, Mines, Resources, Ottawa, 19: 75pp. (multilithed)

beaches
Owens, E.H. and S.B. McCann, 1970.
The role of ice in the Arctic beach environment with special reference to Cape Ricketts, southwest Devon Island, Northwest Territories, Canada.
Am. J. Sci. 268(5):397-414

P

beaches
Palausi, Guy, 1964.
Observations sur une plage de sable calcaire.
Rec. Trav. Sta. Mar. d'Endoume, Bull., 31(47):235-239.

beaches
*Panin, N., 1967.
Structure des dépôts de plage sur la côte de la Mer Noire.
Mar. Geol. 5(3):207-219.

beaches
Paregrine, D.H., 1967.
Long waves on a beach.
J. fluid Mech., 27(4):815-827.

beaches
Parran, A., 1890.
Communication sur les dunes et les plages du littoral algérien. C.R. Soc. Géol. Fr. 5:21-22.

beaches
Pavlidis, Yu.A., and F.A. Shcherbakov, 1965.
The lemination of beaches. Sediment drift and the genesis of the heavy mineral fields in the coastal zone of the sea. (In Russian; English abstract).
Trudy Inst. Okeanol., Akad. Nauk, SSSR, 76:126-136.

beaches
Pearson, E.A., Editor, 1960.
Proceedings of the First International Conference on Waste Disposal in the Marine Environment, Univ. California, 1959, Pergamon Press, 569 pp.

beaches
Pelletier, B.R., 1966.
Development of submarine physiography in the Canadian Arctic and its relation to crustal movements.
In: Continental drift, G.D. Garland, editor, Spec. Publs. R. Soc. Can., No. 9:77-101.

beaches
Perdikis, H.S., 1961
Behavior of beach fills in New England.
J. Waterways & Harbors Div., Proc. Amer. Soc. Civil Eng., 87(WW1): 75-110.

benthos
Peres, Jean-Marie, 1968.
Rapport du Comité benthos (1964-1966).
Rapp. P.-V. Réun. Comm. int. Explor. Sci. Méditerranée, 19(2):35-75.

beaches, minerals of
Petelin, V. P., 1964
The basic types of beach concentrates of heavy minerals in the Pacific Ocean basin. (In Russian)
Okeanologiia, Akad. Nauk, SSSR, 4(6):1052-1058.

beaches
Piasecki, Dionizy, 1963/1964
A graphical method of estimating the ranges of original sea shores. (In Polish; English summary).
Zeszyty Geogr., Wyzsza Szkola Pedagog., Gdansk., 5/6:75-102.

beaches
Pichon, Mireille, 1962.
Note préliminaire sur la répartition et le peuplement des sables fins et des sables vaseux non-fixés, de la zone intertidale, dans la région de Tuléar.
Ann. Fac. Sci. Techniques, 221-235.
Also in:
Rec. Trav. Sta. Mar., Endoume, hors series, Suppl. (1).
Trav. Sta. Mar. Tuléar, Republique Malgache.

beaches
Pierson, W.J., jr., 1951.
The accuracy of present wave forecasting methods with reference to problems in beach erosion on the New Jersey and Long Island coasts. B.E.B. Tech. Memo. No. 24:76 pp., 27 figs.

benthos
Pignatti, Sandro, 1968.
Recherches sur la productivité de la végétation benthique dans le golfe de Trieste.
Rapp. P.-V. Réun. Comm. int. Explor. Sci. Méditerranée, 19(2):209-211.

beaches
*Pilkey, Orrin H., Robert W. Morton, and John Luternaurer, 1967.
The carbonate fraction of beach and dune sands.
Sedimentology, 8(4):311-327.

beaches
Pilkey, Orrin H., and D. M. Richter, 1964.
Beach profiles of a Georgia barrier island.
Southeastern Geology, 6(1):11-19.
also in:
Collected Reprints, Mar. Inst., Univ. Georgia 4 (1965.)

beaches
Pimenta, J., 1956.
Cycles sédimentaires sur un littoral en déformation tectonique. C.R. Acad. Sci. Paris, 243(2):168-170.

beaches
Pollock, Leland W. and William D. Hummon, 1971.
Cyclic changes in interstitial water content, atmospheric exposure, and temperature in a marine beach. Limnol. Oceanogr. 16(3): 522-535.

beaches
Poncet, Jacques, 1961
Existence de "formes festonnées" prelittorales en certains points de la côte algerienne, mise en evidence de "courants d'arrachement" Cahiers Océanogr., C.C.O.E.C. 13(1): 32-37.

beaches
Popov, B.A., 1962
Formation of the underwater beach profiles caused by moderate tides. New investigations of sea coasts and reservoirs.
(In Russian).
Trudy Okeanogr., Komissii, Akad. Nauk, SSSR, 12:54-66.

beaches
Popov, S.M., 1960.
[The influence of the form of the coastline on dynamic wind regimes.]
Izv., Akad. Nauk, SSSR, Ser. Geofiz., (7):1072-1076.

Translation:
Eng. Ed. Pergamon Press (7): 714-716

beaches
Prenant, Marcel, 1963.
Études écologiques sur les sables intercotidaux. II. Distribution des granulométries sur les plages bretonnes exposées au large.
Cahiers Biol. Mar., Roscoff, 4(4):353-397.

beaches
Prenant, M., 1960.
Etudes ecologiques sur les sables intercotidaux 1. Questions de methode granulometrique - Application a trois annees de la baie de Quiberon.
Cahiers de Biol. Mar., Roscoff, 1(3):295-340.

beaches
Prenant, M., 1955.
Position des maxime caratéristiques dans les graphiques granulometriques de frequence établis pour des sables de la zone des marées sur les côtes bretonnes. C.R. Acad. Sci., Paris, 241(16):1062-1064.

beaches
Price, W.A., 1953.
The low energy coast and its new shoreline types in the Gulf of Mexico. Actes du IV Congres, ~~xxxx~~ Assoc. Int. Etude du Quaternaire (INQUA), Rome-Pise:3-8.

beaches
Price, W.A., 1951.
Barrier island, not "offshore bar". Science 113:487-488.

beaches
Pruitt, E.L., 1961
Beach processes.
Nav. Res. Rev., June: 14-19.

beaches
Pugh, D.C., 1953.
Étude mineralogique des plages Picardes et Flamandes. Bull. d'Info., C.C.O.E.C. 5(6):245-276, 16 figs.

beaches
Pytkowicz, Ricardo M., 1971.
Sand-seawater interactions in Bermuda beaches.
Geochim. cosmochim. Acta. 35(5): 509-515.

Q

R

beaches
Rao, C.B., 1957.
Beach erosion and concentration of heavy mineral sands. J. Sed. Petr., 27(2):143-147.

beaches
Rao, R. Prasada, and C. Mahadevan, 1958
Evolution of Visakhapatnam Beach.
Andhra Univ. Mem. Oceanogr., 2:33-47.

beaches
Reymont, Richard Arthur 1971.
Minor sed.-structures and shell orientations on a tidal beach (Bay of Arcachon, France)
Palaeogeogr. Palaeoclimatol. Palaeoecol. 9(4): 265-275

beaches
Rector, R.L., 1954.
Laboratory study of equilibrium profiles of beaches. B.E.B. Tech. Memo. No. 41:1-38, 13 figs.

beaches
Reineck, H.E., 1967.
Layered sediments of the tidal flats, beaches, and shelf bottoms of the North Sea.
In: Estuaries, G.H. Lauff, editor, Publs Am. Ass. Advmt Sci., 83:191-206.

beaches
Reinhard, H., 1950.
Zerstörungen durch Sturmfluten an der Küste des Darss und Zingst. Natur und Volk 80(1/2):32-41.

beaches
Revelle, R., and K.O. Emery, 1957.
Chemical erosion of beach rock and exposed reef rock. Geol. Survey Prof. Paper 260T:699-709.

beaches
Rex, Robert W., 1964.
Arctic beaches, Barrow, Alaska.
In: Papers in Marine Geology, R.L. Miller, Editor, Macmillan Co., N.Y., 384-400.

benthos
Reys, Jean-Pierre, 1968.
Quelques données quantitatives sur les biocoenoses benthiques du golfe de Marseille.
Rapp. P.-V. Réun. Comm. int. Explor. Sci. Méditerranée, 19(2):121-122.

beaches
Rivière, A., 1959.
Sur la notion de dispersion littorale.
Comptes Rendus, Acad. Sci., Paris, 249(19):1920-1922.

beaches
Rivière, A., 1957.
Études littorales. Bull. d'Info., C.C.O.E.C., 9(8):436-456, 2 pls.

beaches
Rivière, A., 1955.
[Recent researches on littoral oceanography; new solutions to some technical problems.]
Mem. Soc. Hydrotech., France, 2:150-161.

Also in: Houille Blanche 10:646-657.

beaches
Rivière, André et Solange Vernhet, 1967.
Interprétation sédimentologique de la radioactivité naturelle des plages de la côte atlantique
C. r. hebd. Séanc. Acad. Sci., Paris, (D) 264(15): 2184-2187.

beaches
Riviere, André, et Solange Vernhet 1966.
Signification sédimentologique de la radioactivité naturelle des plages du Golfe de Lion.
C. r. hebd. Séanc. Acad. Sci. Paris (D) 262 (4): 440-443

beaches
Rivière, André, et Solange Vernhet, 1965.
Études littorales. Contributions a l'étude des rivages du Golfe du Lion. Signification sédimentologique des radioactivités naturelles.
Cah. océanogr., 18(10):857-900.

beaches
Rivière, A., and S. Vernhet, 1962.
Les structures de plage à caractère périodique et leur role dans la morphologie littorale.
In: Océanographie Géologique et Géophysique de la Méditerranée Occidentale, Centre National de la Recherche Scientifique, Villefranche sur Mer, 4 au 8 avril 1961, 73-82.

beaches
Robinson, A.H.W., 1961
The hydrography of Start Bay and its relationship to beach changes at Hallsands.
Geogr. J., 127(1): 63-77.

beaches
Rogers, J.J.W., & H.C. Adams, Jr., 1959
The mineralogy and texture of beach sands of Galveston Island, Texas.
J. Sediment. Petrol., 29(2):207-211.

beaches
Roseau, M., 1951.
[Wave motion of the sea at a beach.] C.R. Acad. Sci., Paris, 231:1212-1214; 232:211-213, 303-306.

beaches
Rouville, ?, 1952.
Variations des plages. Bull. d'Info., C.O.E.C., 3(1):5.

beaches
Rouvillois Armelle, 1967.
Observations morphologiques sédimentologiques et écologiques sur la plage de la Ville Ger, dans l'estuaire de la Rance.
Cah. océanogr., 19(5): 375-389

beaches
Russell, C.H., and C. Inglis, 1953.
The influence of a vertical wall on a beach in front of it. Proc. Minnesota Hydraulics Conv. Sept. 1-4, 1953:221-226, 4 textfigs.
Int.

beaches
Russell, Richard J., 1969.
Where most grains of very coarse sand and fine gravel are deposited.
Sedimentology, 11 (1/2): 31-38.

beaches
Russell, Richard J., 1967.
River plains and sea coasts.
Univ. Calif. Press, 173 pp.

S

beaches
Sastri, C.S., and V. Sivaramakrishnan, 1968.
Study of the radioactivity of separated minerals of beach sands of Manavalakurichi, Madras State.
Current Science, 37(19):550.

beaches
Sawaragi, Toru, 1963.
Consideration on the applicability of experimental results to the analysis of prototype beaches.
Coastal Engineering in Japan, 6:21-27.

beaches
Saville, T., jr., and J.M. Caldwell, 1953.
Accuracy of hydrographic surveying in and near the surf zone. B.E.B. Tech. Memo. No. 32:28 pp., 8 figs.

beaches
Saville, T., jr., 1950.
Model study of sand moving along an infinitely long, straight beach. Trans. Am. Geophys. Union 31(4): 555-565, 6 textfigs.

beaches
Scatizzi, P., 1940.
Profili di spiaggia. [Profile des Strandes.]
Acta Pontif. Acad. Sci. 4:73-80.

beaches
Schiffman, Arnold, 1965.
Energy measurements in the swash-surf zone.
Limnology and Oceanography, 10(2):225-260.

beaches
Schlee, John, Elazar Uchupi and J.V.A. Trumbull, 1964.
Statistical parameters of Cape Cod beach and eolian sands.
U.S. Geol. Survey Prof. Paper, 501-D:D118-D122.

beaches, interstitial spaces
Schmidt, Peter 1969.
Die quantitative Verteilung und Populationsdynamik des Mesopsammons am Gezeiten-Sandstrand der Nordsee-Insel Sylt. II Quantitative Verteilung und Populationsdynamik einzelner Arten.
Int. Revue ges. Hydrobiol. 54 (1):95-174

beaches
Schou, A., 1945(1950).
Det marine Forland. Geografiske Studien over Danske Fladkystlandskabers dannelse og Formudvikling samt traek af disse omraaders Kulturgeografi. Folia Geogr. Danica. IV (Medd. Skalling Lab. 9):236 pp., 85 textfigs., 1 fold-in.

beaches
Schrom, Heinrich, 1966.
Verteilung einiger Gastrotrichen im oberen Eulitoral eines nordadriatischen Sandstrandes.
Veroff. Inst. Meeresforsch., Bremerh., Sonderband II:95-101.

beaches
Schuster, R., 1962.
Das marine Litoral als Lebenraum terrestricher Kleinarthropoden.
Int. Revue Ges. Hydrobiol., 47(3):359-412.

beaches
Schwartz, Maurice L. 1967.
Littoral zone tidal-cycle sedimentation.
J. Sed. Petr. 37(2): 677-709.

beaches
Schwartz, Maurice L., 1966.
Fluorescent tracer: transport in distance and depth in beach sands.
Science, 151 (3711):701-702.

beaches
Shen, M.C., and R.E. Meyer, 1963.
Climb of a bore on a beach. 2. Non-uniform beach slope.
J. Fluid Mechanics, 16(1):108-112.

beaches
Shen, M.C., and R.E. Meyer, 1963.
Climb of a bore on a beach. 3. Run-up.
J. Fluid Mechanics, 16(1):113-125.

beaches
Shepard, Francis P., 1964.
Criteria in modern sediments useful in recognizing ancient sedimentary environments.
In: Deltaic and shallow marine deposits, Developments, L.M.J.U. van Straaten, editor, Elsevier Publishing Co., 1:1-25
in Sedimentology

beaches
Shepard, F.P., 1950.
Beach cycles of southern California. B.E.B. Tech. Memo. No. 20:26 pp., 15 figs.

beaches
Shepard, F. P., 1950.
Longshore-bars and longshore-troughs. B.E.B. Tech. Memo. 15:32 pp., 19 figs. (multilithed).

S.I.O. Sub. Geol. Rept. No. 6.

beaches
Shepard, F.P., 1950.
Photography related to investigations of shore processes. Photogram. Eng. 16:756-769.

beaches
Shepard, F. P., 1949.
Dangerous currents in the surf. Physics Today 2: 20-29.

beaches
Shepard, F.P., D.L. Inman, and R.L. Fisher, 1951
Marine beaches of the Unites States. (Abstract).
Bull. G.S.A. 62:1477.

beaches
Shepard, F.P., & R. Young, 1961
Distinguishing between beach and dune sands.
J. Sediment. Petrol., 31(2): 196-214.

beaches
*Shinohara, Kinji, and Tooichiro Tsubaki, 1967.
Model study on the change of shoreline of sandy beach by the off shore breakwater.
Proc.10th Conf.Coast.Engng.Tokyo, 1966, 1:550-563.

beaches
Shinohara, K., and T. Tsubaki, 1959.
Laboratory study of sand movement and equilibrium profiles of beaches.
Coastal Engineering in Japan, 2:29-34.

beaches
Shinohara, Kinji, Toichiro Tsubaki, Masuo Yoshitaka and Chiuki Agemori, 1958
Sand transport along a model sandy beach by wave action.
Coastal Engineering in Japan, 1: 111-130.

beaches
Shuiskii Iu. D., 1969.
Regeneration of modern sea-beach placers as connected with the dynamics of the underwater sand slope. (In Russian).
Dokl. Akad. Nauk SSSR 189(5): 1111-1114

beaches
Shuliak, B.A., and V.L. Boldyrev, 1966.
On the problem of beach bar formation processes. (in Russian).
Okeanologiia, Akad. Nauk, SSSR, 6(1):109-116.

beaches
Shumway, George, David G. Moore and George B. Dowling, 1964.
Fairway Rock in Bering Strait.
In: Papers in Marine Geology, R.L. Miller, Editor, Macmillan Co., N.Y., 401-407.

beaches
Sibul, O.J., and E.G. Tickner, 1955.
A model study of the run-up of wind generated waves on levees with slopes of 1:3 and 1:6.
B.E.B. Tech. Memo. No. 67:19 pp., 7 figs.

beaches
Siesser, W.G., 1971
Mineralogy and diagenesis of some South African coastal and marine carbonates.
Marine Geol., 10(1): 15-38

beaches
Simonov, A.I., 1959.
Some peculiarities of the hydrological conditions along tidal beaches (according to an example of Cuban conditions).
Trudy Gosud. Okeanogr. Inst., 45:51-62.

beaches
Sitarz, Jean A., 1963.
Contribution à l'étude de l'évolution des plages à partir de la connaissance des profiles d'équilibre.
Trav. Centre de Recherches et d'Etudes Océanogr. 5(2,3,4):199 pp.

beaches
Sitarz, Jean A., 1960
Côtes Africaines. Étude des profils d'équilibre de plage. Trav. C.R.E.O., 3(4): 43-62.

beaches
Sonu, Choule J., and Johannes L. Van Beek, 1971.
Systematic beach changes on the outer banks North Carolina.
J. Geol. 79(4): 416-425.

beaches
Sourie, R., 1954.
Contribution à l'étude écologique des côtes rocheuses du Sénégal.
Mém. Inst. Francais d'Afrique Noire No. 38: 342 pp., 23 pls.

beaches
Stanley, Kirk W., and Howard J. Grey, 1966.
Spray on paint stripes to determine the direction of beach drifting.
J. Geol. 74(3): 257-261.

beaches
Steele, John H., and I. E. Baird 1968.
Production ecology of a sandy beach.
Limnol. Oceanogr. 13(1): 14-25

beaches
Steinberg, M., 1959.
Quelques précisions sur les croissants de plage et les sinuosités dans la région de Saint-Aygulf.
C.R. Acad. Sci., Paris, 248(22):3190-3192.

beaches
Stephenson, T.A., and A., 1961.
Life between tide marks in North America.
NA Vancouver Island, I.
Journal of Ecology, 49(1):1-29.

beaches
Stetson, H.C., 1954.
A preliminary investigation of shifting beach profiles. Coastal Geogr. Conf., 18 Feb. 1954:57-62, 92 textfigs.

beaches
Stoddart, D.R., and J.R. Cann, 1965.
Nature and origin of beach rock.
J. Sed. Petr., 35(1):243-273.

beaches
Stomianko, Pawel, 1960
[Analysis of the littoral transport by using marked luminescent sands as tracers. Seacoast investigations on Hal Peninsula.] Prace Instytutu Morskiego 1. Hydrotechnika, No. 4: 146 pp. multilithed or mimeographed.

beaches
Strahler, Arthur N. 1966.
Tidal cycle of changes in an equilibrium beach, Sandy Hook, New Jersey.
J. Geol. 74(3): 247-268.

beaches
Summerhayes, C.P. 1969.
Marine environments of economic mineral deposition around New Zealand: a review.
J. mar. Techn. Soc. 3(2): 57-66.

beaches
*Summerhayes, C.P., 1967.
Marine environments of economic mineral deposition around New Zealand: A review.
N.Z. Jl mar. Freshwat.Res., 1(3):267-282.

beaches, fauna of
Swedmark, Bertil, 1964.
The interstitial fauna of marine sand.
Biol. Rev., 39(1):1-42.

T

beaches
Tanner, William F. 1971.
Growth rates of Venezuelan beach ridges.
Sediment. Geol. 6(3): 215-220.

beaches
Tanner, William F., 1958.
The equilibrium beach.
Trans. Amer. Geophys. Union, 39(5):889-891.

beaches
Tanner, William F., and John D. Bates, 1965.
Submerged beach on a zero-energy coast.
Southeast. Geol., Duke Univ., 7(1):19-24.

beaches
Terry, Richard D., editor, 1966.
Ocean Engineering, 1. Energy sources and energy conversion, waste conversion and disposal. 2. Undersea construction, habitation, and vehicles; recreation.
Western Periodicals Co., North Hollywood, Calif.
Vol. 3:431 pp. (multilithed).

beaches
Timme, Richard C., and Dwight D. Pollard, 1965.
Long-term flourescent sand tracer studies.
Ocean Sci. and Ocean Eng., Mar. Techn. Soc., Amer. Soc. Limnol. Oceanogr., 2:1162-1166.

beaches
Tomczak, G., 1952.
Der Einfluss den Küstengestalt und des vorgelagerten Meeresbodens auf den windbesingten Anstau des Wassers an der deutschen Nordseeküste zwischen Ems und Elbe. Deutsche Hydrogr. Zeits. 5(5/6):277-284, 4 textfigs.

beaches
Trask, P.D., 1959.
Beaches near San Francisco, California, 1956-1957.
Beach Erosion Bd., Techn. Mem., No. 110:1-89.

beaches
Trask, P.D., 1956.
Changes in configuration of Point Reyes Beach, California, 1955-1956. B.E.B. Tech. Memo., 91:62 pp.

beaches
Trask, P.D., 1952.
Source of beach sand at Santa Barbara, California as indicated by mineral grain studies. B.E.B. Tech. Memo. No. 28:24 pp., 6 figs. (multilithed).

beaches
*Trevallion, A., A.D. Ansell, P. Sivadas and B. Narayanan, 1970.
A preliminary account of two sandy beaches in South West India. Mar. Biol., 6(3): 268-279.

beaches

Tricart, Jean, 1962
Étude générale de la desserte portuaire de la "Sasca". 11. Les sites portuaires, leurs caracteristiques morphodynamiques et leurs possibilites d'amenagement.
Cahiers Oceanogr., C.C.O.E.C., 14(3):146-161.

beaches

Tricart, J., 1959
Problemes geomorphologiques du littoral oriental du Bresil.
Cahiers Ocean.,C.C.O.E.C., 11(5):276-308.

beaches

Tricart, J., 1957.
Aspects et problèmes géomorphologiques du littoral de la Côte d'Ivoire.
Bull. I.F.A.N., (A), 19(1):1-20.

beaches

Tricart, J., 1951.
Études sur le faconnement des galets marins.
Proc. 3rd Int. Congress Sedimentol, Netherlands, 1951:245-255, 5 textfigs.

beaches

Trichet, Jean, 1965.
Essai d'explication de l'origine des gres de plage. Cas des gres de plage coralliens.
Comptes Rendus herb. Seance, Acad. Sci., Paris, 261(16):3176-3178.

beaches

Thrumbull, James V.A., and John C. Hathaway 1968.
Dark mineral accumulations in beach and dune sands of Cape Cod and vicinity.
U.S. Geol. Surv. prof. Pap. 600-B: B178-B184

beaches

Twenhofel, W.S., 1952.
Recent shore-line changes along the Pacific coast of Alaska. Am. J. Sci. 250:523-548.

U

beaches

Uchio, Takayasu, 1962
Recent Foraminifera thanatocoenoses of beach and nearshore sediments along the coast of Wakayama-Ken, Japan.
Publ. Seto Mar. Biol. Lab., 10(1) (Article 8) 133-144.

beaches

Udintsev, G.B., 1951.
[On ancient forms of sea abrasion.]
Trudy Inst. Okeanol., 6:78-82.

beaches

U.S.A.,U. S. Army Coastal Engineering Research Center, 1964.
A pictorial history of selected structures along the New Jersey coast.
Misc. Paper., No. 5-64:99 pp.

beaches

U.S.A., Army Coastal Engineering Research Center, 1964.
Land against the Sea.
Misc. Paper, No. 4-64:43 pp.

beaches

Ursell, F., 1952.
Edge waves on a sloping beach. Proc. R. Soc., London, A, 214:79-97, 3 textfigs.

beaches

Upchurch, Sam B. 1971.
Discriminant analysis of Bermuda carbonate strand line sediment.
Bull. Geol. Soc. Am. 83(1): 87-94

V

beaches

Vacelet, E., 1960.
Note preliminaire sur la faune infusorienne des "sables à Amphioxus" de la Baie de Marseille.
Rec. Trav. Sta. Mar. d'Endoume, 33(20):53-57.

beaches

* Van der Lingen, G.J., and Peter B. Andrews, 1967.
Hoof-print structures in beach sand.
J. sed. Petr. 39(1): 350-357.

beach sand formation

Van Overbeck, J. and R.E. Crist, 1947
The role of a tropical green alga in beach sand formation. Am. J. Bot. 34:299-300.

beaches

van Straaten, L.M.J.U., 1954.
Sedimentology of recent tidal flat deposits and the Psammites du Condroz (Devonian).
Geol. en Mijnbouw, (NW Ser.) 16e Jaargang:25-27, 2 pls., 15 textfigs.

beaches

Vasilchikov,N.V.,Yu. A. Pavlidis and N.P. Slovinsky-Sidak,1966.
On the vanadium titanomagnetic sea-beach placers in the Far East. (In Russian;English abstract).
Okeanologiia,Akad. Nauk,SSSR,6(5):823-829.

beaches

Vennin, J., 1951.
Étude sédimentologique de la rade de Safi.
Bull. Sci. Comm. Local Océan. Études Côtes du Maroc, 2nd ser., No. 1:37-43, figs.

beaches

*Venzo, G.A., e A. Brambati, 1968.
Evoluzione e difesa delle coste dell'alto Adriatico da Venezia a Trieste.
Riv. ital. Geotecn., 1968(3): 3-19.
Reprinted: Comm. Studio Oceanograf. Limnol., (a) (11)

beaches

Vernhet, Solange, 1966.
Signification sédimentologique de la radioactivité naturelle des plages du Golfe du Lion.
Comptes Rendus, Acad. Sci. Paris, 262(4):440-443.

radioactivity
Mediterranean, west
bottom sediments
beaches

beaches

Vernhet, S., 1955.
Influence de faibles courbures du littoral sur l'erosion des rivages sableux. Interprétation de l'allure sinueuse de caractère plus ou moins périodiques du tracé des grandes plages. C.R. Acad. Sci., Paris, 240(3):336-338.

beaches

Vesper, William H. 1967.
Behavior of beach fill and borrow area at Sherwood Island State Park, Westport, Connecticut.
Techn. Mem. U.S. Army Engng Res. Center 20: 25 pp. (multilithed)

beaches

Vesper, William H., 1965.
Behavior of beach fill and borrow area at Seaside Park, Bridgeport, Connecticut.
U.S. Army Coastal Engin. Res. Center, Techn. Memo., No. 11:24 pp.

beaches

Vladimirov, A.T., 1950.
[Formation of scalloped shorelines.] Priroda 7:54.

beaches

Vogt, H., A. Gomes and J. Tricart, 1960
Note preliminaire sur la morphologie du cordon littoral actuel entre Tramandai et Torres, Rio Grande do Sul, Bresil.
Cahiers Oceanogr., 12(7):453-460.

beaches

Vollbrecht, K., 1954.
Ueber die Natur des Sedimentgleichgewichter im Litoral. Geofis. Pura Appl. 28:159-170.

BEACHES

Vollbrecht, K., 1954.
Zur Frage der Sedimentbewegung im litoralen Gürtel.
ACTA Hydrophysica 2(2):43-80, 11 textfigs.

beaches

Völpel, F., 1960.
Die Bildung charakteristischer Sedimente in der küstennahen Flachsee.
Deutsche Hydrogr. Zeits., 13(6):290-307.

W

beaches

Walger, Eckart,1966.
Untersuchungen zum Vorgang der Transportsonderung Von Mineralen am Beispeil Von Strandsanden der westlichen Ostsee.
Meyniana, 16:55-106.

beaches

Warnke, Detlef A. 1969.
Beach changes at the location of landfall of Hurricane Alma.
SEast. Geol. 10(4): 189-200

beaches

Warnke, D.A., V. Goldsmith, P. Grose and J.J. Holt, 1966.
Drastic beach changes in a low-energy environment caused by Hurricane Betsy.
J. geophys. Res., 71(8):2013-2016.

beaches
hurricanes, effect of
USA, east

beaches

Warnke, Detlef A., and Donald K. Stauble 1971.
An application of reflected-light differential-interference microscopy: beach studies in eastern Long Island.
Sedimentology 17 (1/2): 103-114.

beaches

Warthin, A.S., Jr., 1959
Ironshore in some West Indian islands.
Trans. N.Y. Acad. Sci., Ser. II. 21(8): 649-652.

beaches

Watts, G.M., 1958.
Behavior of beach fill and borrow area at Harrison County, Mississippi
Beach Erosion Bd., Tech. Memo., No. 107:1-14.

beaches

Watts, G.M., 1954.
Laboratory study of effect of varying wave periods on beach profiles. B.E.B. Tech. Memo. No. 53:1-19, 8 figs.

beaches
Watts, G.M., and R.P. Dearduff, 1954.
Laboratory study of effect of tidal action on wave-formed beach profiles. Tech. Memo., B.E.B., 52:1-21, 8 figs.

Beaches
*Wellman, H.W., 1967.
Tilted marine beach ridges at Cape Turakirae, N. Z.
J. Geosci., Osaka City Univ., 10:123-129.

beaches
Wells, D.R., 1967.
Beach equilibrium and second-order wave theory.
J. geophys. Res., 72(2):497-504.

beaches
Wensink, J.J., and J.P. Bakker, 1951.
Five types of fine tidal flat sands from the sub-soil of Barradeel, N.W. Friesland, Netherlands. Proc. 3rd Int. Congress Sedimentol., Netherlands, 1951:273-279, 3 textfigs.

beaches
West, R.G., and B.W. Sparks, 1960
Coastal interglacial deposits of the English Channel. Phil. Trans., R. Soc., London, (B), 243 (701): 95-133.

beaches
*Westcott, M.F., and L.G. Parry, 1968.
Magnetic properties of some beach sand ilmenite particles.
J. geophys. Res., 73(4):1269-1277.

beaches
Westheide, W. 1968.
Zur quantitativen Verteilung von Bakterien und Hefen in einem Gezeitenstrand der Nordseeküste.
Mar. Biol. 1(4): 336-347.

beaches
Wicker, C.F., 1950.
A case history of the New Jersey coastline. Inst Coastal Eng., Univ. Ext., Univ. Calif., Long Beach, 11-13 Oct. 1950:26 pp. (mimeographed), 4 figs. (multilithed).

beaches
Wiegel, Robert L., 1964.
Oceanographical engineering.
Prentice-Hall Series in Fluid Mechanics, 532 pp.

beaches
Wiegel, R.L., 1953.
Waves, tides, currents and beaches: glossary of terms and of standard symbols.
Council on Wave Research, Berkeley, California, 113 pp.
list

beaches
Wiegel, R.L., D.A. Patrick and H.L. Kimberley, 1954.
Wave, longshore current and beach profile records for Santa Margarita River, Oceanside, California 1949. Trans. Amer. Geophys. Union 35(6):887-896.

beaches
Williams, A.T. 1971.
An analysis of some factors involved in the depth of disturbance of beach sand by waves.
Mar. Geol. 11 (3): 145-158

beaches
Williams, W.E., 1961
Waves on a sloping beach. Proc. Cambridge Phil. Soc. (Math. Phys. Sci.), 57(1): 160-165.

beaches
Williams, W.W., and C.A.M. King, 1950.
Observations faites sur la plage des Karentes.
Bull. d'Info., C.O.E.C., 2(10):363-368, 2 textfigs

beaches
*Wobber, Frank J., 1967.
The orientation of Donax on an Atlantic coast beach.
J. sedim. Petrol. 37(4):1233-1251.

Literaure not too well known.

beaches
Wunderlich, Friedrich, 1971.
Der Golf von Gaeta (Tyrrhenisches Meer). II. Strandaufbau und Stranddynamik. Senckenbergiana maritima 3(1): 135-183.

XYZ

beaches
Yalin, M. Selim, 1963
A model shingle beach with permeability and drag forces reproduced.
Int. Assoc. Hydr. Res. Congress, London, 1963:169-175.

beaches, chemistry of
Yamamoto, Sakujiro, 1959
Studies of beach-sands as chemical resources On sand-irons in beach-sands along the coastline of Tottori Prefecture.
Rec. Oceanogr. Wks., Japan, Spec. No. 3: 141-144.

beaches
Yasso, Warren E., 1966.
Heavy mineral concentration and sastrugi-like deflation furrows in a beach salcrete at Rockaway Point, New York.
J. Sed. Petr., 36(3):836-838.

beaches
Yasso, Warren E., 1966.
Formulation and use of fluorescent tracer coatings in sediment transport studies.
Sedimentology, 6(4):287-301.

beaches
Yegorov, E.N., 1957.
[On the stability of the contours of irregular coasts with a wave resultant normal to the coast line.] Trudy Inst. Okeanol., 21:107-117.

beaches
Zebrowski, George, 1965.
Micro-flora of the turbulent eulittoral zone. (Abstract).
Ocean Sci. and Ocean Eng., Mar. Techn. Soc., Amer. Soc. Limnol. Oceanogr., 1:96.

beaches
Zeigler, John M., 1964.
Some modern approaches to beach studies.
In: Oceanography and marine biology, Harold Barnes, Editor, George Allen & Unwin, Ltd., 2: 77-95.

beaches
Zeigler, J.M., C.R. Hayes and S.D. Tuttle, 1959.
Beach changes during storms on outer Cape Cod, Massachusetts.
J. Geology, 67(3):318-376.

beaches
Zeigler, J.M., and F.C. Ronne, 1957.
Time-lapse photography helps study changes in shore line. Research Reviews (April 1957):1-6.

beaches
Zeigler, John M., and Sherwood D. Tuttle, 1961
Beach changes based on daily measurement of four Cape Cod beaches.
J. Geology, 69(5):583-599.

beaches
Zeigler, John M., Sherwood D. Tuttle, Herman J. Tasha and Graham S. Giese, 1965.
The age and development of the Provincelands hook, outer Cape Cod, Massachusetts.
Limnol. and Oceanogr., Redfield Vol., Suppl. to 10:R298-R311.

beaches
Zenkovitch, V.P., 1965.
Buts et principaux axes de recherches des etude sur les zones maritimes littorales.
Cahiers Oceanogr., C.C.O.E.C., 17(9):605-623.

beaches
Zenkovich, V.P., 1957.
[On the origin of the offshore bars and lagoons.] Trudy Inst. Okeanol., 21:3-39.

beaches
Zenkovich, V.P., 1957.
[On the selection of shore-debris at the tip of marine spits.] Trudy Inst. Okeanol., 21:133-136.

beaches
Zenkovitch, V.P., 1956.
Etude de la dynamique du littoral. Essais de Geogr., Akad. Nauk, SSSR, 104-116.

beaches
Zenkovich, V.P., 1954.
[Cadastre of the shores of the USSR.] Trudy Inst. Okeanol., 10:35-43.

beaches
Zenkovich, V.P., 1954.
[Dynamical classification of ocean shores.] Trudy Inst. Okeanol., 10:112-134.

beaches
Zenkovitch, V.P., 1954.
[Some results and principal tasks in the study of seashores.]
Trudy Inst. Oceanol. 10:5-20.
RT3070 Bibl. Transl. Rus. Sci. Tech. Lit. 25.

beaches
Zenkovich, V.P., 1952.
[Double coastal beach ridges and barrier beaches.] Priroda, 41(2):113-

beaches
Zenkovich, V.P., 1949.
[Certain factors in the formation of sea benches] Dok. Akad. Nauk, USSR, 65:53-56.

beaches
Zenkovich, W.P., 1949.
[Deep bench as an indicator of a collapse of the shore.] Trudy Inst. Okeanol., 4:160-164.

beaches
Zenkovich, V.P., 1948.
[On the formation of beach cusps.] Trudy Inst. Okean. 2:35-48.

beaches
Zenkovich, V.P., 1948.
[Currents of coastal deposition of the Caucasus littoral of the Black Sea.] Dok. Akad. Nauk SSSR 60:263-266.

beaches

Zenkovich, V.P., 1948.
[Forms of accumulation of gravel alluvium on the Caucasus coast of the Black Sea.] Dok. Akad. Nauk SSSR. 60:641-644.

beaches

Zenkovich, V.P., 1946.
[On the study of littoral dynamics.] Trudy Inst. Okean. 1:11-122.

beaches

Zenkovitch, V.P., and A.M. Idanov, 1960.
[Why are the beaches of the Black Sea disappearing?]
Priroda, (10):51-54.

beaches

Zenkovitch, V. P., O. K. Leontiev, And E. N. Nevessky, 1960.
The influence of the eustatic post-glacial transgression on the development of the coastal zone of the USSR seas. Morsk. Geol., Doklady Sovetsk. Geo., 21 Sess., Int. Geol. Congress. 154-163.

beaches

Zhadanav, A.M., 1951.
[Determination of the power equivalent of wave motion on seashore.] Izvest. Akad. Nauk, Ser. Geogr. Geofiz., 15(1):51-56.

beaches

Zhadanov, A.M., 1951.
[Determination of flow power in shore deposits by direct observations.] Izvest. Akad. Nauk, Ser. Geogr. Geofiz., 15(2):81-90.

beaches

Zhdanov, A.M., 1958.
Wearing of beach gravel by waves. (In Russian). Biull., Okeanogr. Komissiia, Akad. Nauk, SSSR, (1):81-

not seen

beaches

Zhivago, A.V., and V.V. Petrikeev, 1949.
[Method of determining modifications of coastal relief of the sea bottom after storms.] Izvest. Akad. Nauk, USSR, Ser. Geogr., Geofiz., 13(2): 151-153.

beaches, biology of

beaches, amphipods in

*Croker, Robert A., 1968.
Distribution and abundance of some intertidal sand beach amphipods accompanying the passage of two hurricanes.
Chesapeake Sci., 9(3):157-162.

beaches, benthos

Amanieu, M., 1969.
Recherches écologiques sur les faunes des plages abritées de la région d'Arcachon.
Helgoländer wiss. Meeresunters. 19(4):455-567

beaches, benthos

Plawen, Luitfried Salvini, 1968.
Zur Kenntnis des Mesopsammales der Nordadria. 1. Die für den Meeresteil neuen Gruppen und Arten.
Thalassia Jugoslavica 4:11-17.

beaches benthos

Schrom, Heinrich, 1968.
Zur Kenntnis des Mesopsammals der Nordadria II: die Sande in der Umgebung von Rovinj und ihre Faunenmerkmale.
Thalassia Jugoslavica, 4:31-35

beaches, biology of

Fenchel, Tom, and Bengt-Owe Jansson, 1966.
On the vertical distribution of the microfauna in the sediments of a brackish-water beach.
Ophelia, 3:161-177.

beaches, biology

Johnson, Ralph Gordon, 1965.
Temperature variation in the infaunal environment of a sand flat.
Limnology and Oceanography, 10(1):114-120.

beaches, biological

Pichon, M., 1964.
Apercu preliminaire des peuplements sur sable et sables vaseux, libres ou couverts par des herbiers de phanerogames marines, de la region de Nosy-Be.
Cahiers, O.R.S.T.R.O.M., Oceanographie, 11(4):5-15.

beaches (biology of)

Remane, Adolf, und Erich Schulz, 1964.
Die Strandzonen des Roten Meeres und ihre Tierwelt. Wissenschaftliche Ergebnisse einer Forschungsreise von A. Remane und E. Schulz nach dem Roten Meer, E. 1.
Kieler Meeresf., 20(1):5-17.

beaches, biomass

*McIntyre, A.D., 1970
The range of biomass in intertidal sand, with special reference to the bivalve Tellina tenuis. J. mar. biol. Ass., U.K., 50(3): 561-575.

beaches, ciliates of

Petran Adriana, 1968.
Sur l'écologie des ciliés psammobiontes de la mer Noire (littoral roumain).
Revue Roumaine Biol. (Zool.) 13(6): 441-445.

beaches, calcium

beaches, calcium

Minas, Monique, 1965.
La substance organique et le calcaire dans deux types de vasières littorales de la région de Tuléar.
Rec. Trav. Sta. Mar., Endoume, hors sér., Suppl., No. 4:57-70. (Trav. Sta. Mar., Tuléar).

beaches, chemistry

beaches, chemistry of

Kühl, Heinrich and Hans Mann, 1966.
Chemische Untersuchungen am Sandstrand von Cuxhaven und Helgoland.
Veröff. Inst. Meeresforsch., Bremerh., Sonderband II:67-76.

beach cusps

beach cusps

Arbey, F., 1961.
Études littorales sur la côte des Maures. 1. Les croissants de plage (beach cusps).
Cahiers Océanogr., C.C.O.E.C., 13(6):380-396.

beach cusps

Cloud, Preston E., Jr., 1966.
Beach cusps: response to plateau's rule?
Science, 154(3751):890-891.

beach cusps

Egorov, E.N., 1953.
Observations on beach cusps. (In Russian).
Trudy Inst. Okeanol., Akad. Nauk, SSSR, 7:117-125.

cusps

Flemming, N.C., 1964.
Tank experiments on the sorting of beach material during cusp formation.
J. Sed. Petrol., 34(1):112-122.

beach cusps

Kuenen, Ph. H., 1948.
The formation of beach cusps. J. Geol. 56(1): 34-40, 3 textfigs., 2 pls.

beach cusps

Longuet-Higgins, M.S., & D.W. Parkin, 1962
Sea waves and beach cusps.
Geogr. J., 128(2): 194-200.

beach cusps

Russell, Richard J., and William G. McIntire, 1965.
Beach cusps.
Geol. Soc., Amer., Bull., 76(3):307-320.

beaches, ecology of

Panikkar, Bhaskara, and S. Rajan, 1970
Observations on the ecology of some sandy beaches of the southwest coast of India.
Proc. Indian Acad. Sci. (B) 71(6):247-260

beach, effect of

Nagashima, Hideki 1971.
Reflection and breaking of internal waves on a sloping beach.
J. oceanogr. Soc. Japan, 27(1):1-6.

beaches, effect of

Tuck, E.O., and Li-San Hwang 1972
Long wave generation on a sloping beach.
J. fluid Mech. 51(3):449-461

beach equilibrium

beach equilibrium

Clos-Arceduc, Albert, 1962
La demonstration de la loi de Lewis et les formes d'equilibre des plages.
Cahiers Oceanogr., C.C.O.E.C., 14(3):162-170.

beach erosion

beach erosion

Aki, K., 1953.
Beach erosion in Japan. Proc. Minnesota Int. Hydraulics Conv., Sept. 1-4, 1953, 227-230.

beach erosion

Bruun, Per, and J.A. Purpura, 1965.
Emergency methods to combat beach erosion: problem of controlling natural forces.
The Dock and Harbour Authority, 45(534):391-396.

beach erosion

Hayami, Shoitiro, Toijiro Ishihara and Yuichi Iwagaki, 1953.
Some studies on beach erosions.
Disaster Prevention Res. Inst., Bull., No. 5:29 pp

Also in:
Papers on Oceanogr. and Hydrol., (1949-1962), Geophys. Inst., Kyoto Univ., Contrib. No. 4:(1963)

beach erosion

Japan, Maritime Safety Board, 1956.
Beach erosion of Omari-hama, Hakodate City, Hokkaido, Japan.
Publ. No. 943, Bull. Hydrogr. Off., 13(1):215 pp. 211 textfigs., photos.

beaches, erosion of

Japan, The Organizing Committee of the Tenth Conference on Coastal Engineering, 1966.
Outline of Coastal Engineering in Japan: guide book to the Tenth Conference on Coastal Engineering, Tokyo, Japan, 161pp.

beach erosion

Kubo, Masafumi, and Nobuyuki Iwasa, 1961
Beach erosion and protection works in Imazu-Sakano Beach.
Coastal Eng., Japan, 4:103-118

beach erosion

LaFond, E.C., and R. Prasada Rao, 1954.
Beach erosion cycles near Waltair on Bay of Bengal.
Andhra Univ. Ocean. Mem. 1:63-77, 10 textfigs.

1

beach erosion
Mogi, A., and H. Iwasaki, 1963.
Beach erosions along the Ninomiya coast, Kanagawa Prefecture, and the Tokai Village coast, Ibaraki Prefecture. (In Japanese; English abstract).
Hydrogr. Bull., Tokyo, No. 73 (Publ. No. 981):11-16.

beach erosion
Munk, W.H., and M.A. Traylor, 1947.
Refraction of ocean waves: a process linking underwater topography to beach erosion. J. Geol. 55(1):1-25, Fig. 17, Table/ 8.

beach erosion
Paterson, D.E., 1956.
[Beach erosion at Durban, South Africa]
B.E.B. Bull. 10(1):11-20, chiefly figs.

beaches, effect of

beaches, effect of
Amein, Michael, 1966.
A method for determining the behavior of long waves climbing a sloping beach.
J. geophys. Res., 71(2):401-410.

beach (barrier), effect of
Ramanadham, R., K.R.G.K. Murty and B.S. Reddy 1971.
The influence of barrier beach on sediment transportation in the Kakinada Bay.
Pure appl. Geophys. 89: 192-196.

beaches, effect of
*Taira, Keisuke, and Yutaka Nagata, 1968.
Experimental study of wave reflection by a sloping beach.
J. oceanogr., Soc., Japan, 24(5):242-252.

beaches, fauna of

beaches, fauna of
Bush, Louise F. 1966.
Distribution of sand fauna in beaches at Miami, Florida.
Bull. mar. Sci. 16 (1): 58-75.

beaches, microfauna of
*Fenchel, Tom, Bengt-Oew Jansson and Wolf von Thun, 1967.
Vertical and horizontal distribution of the metazoan microfauna and of some physical factors in a sandy beach in the northern part of the Øresund.
Ophelia, 4:227-243.

beaches, fauna of
Jansson, Bengt-Owe, 1966.
Microdistribution of factors and fauna in marine sandy beaches.
Veröff. Inst. Meeresforsch., Bremerh., Sonderband II:77-84.

beaches, fauna of
McIntyre, A.D., 1968.
The meiofauna and macrofuana of some tropical beaches.
J. Zool. Lond., 156(3):377-392.

beaches (macrofauna)
*Morgens, J.F.C., 1967.
The macrofauna of an unstable beach discussed in relation to beach profile, texture and a progression in shelter from wave action.
Trans. R. Soc., N.Z., Zool., 9(10): 141-155.

beaches, macrofauna
*Morgens, J.F.C., 1967.
The macrofauna of an excellently sorted isolated beach at Kaikoura: and certain tidal observations
Trans. R. Soc., N.Z., Zool., 9(11):157-167.

beaches, flora
*Steele, J.H., A.L.S. Munro and G.S. Giese, 1970.
Environmental factors controlling the epipsammic flora on beach and sublittoral sands. J. mar. biol. Ass., U.K., 50(4): 907-918.

beach gradients

beach gradients
Williams, W.W., 1947
The determination of gradients on enemy-held beaches. Geograph. Jour. CIX (1/3): 76-90, 6 textfigs., 6 photos.

beach lamination
Clifton, H. Edward, 1969.
Beach lamination: nature and origin.
Marine Geol., 7(6): 553-559.

beaches, lists of spp.

beaches, lists of spp.
Renaud-Debyser, Jeanne, et Bernard Salvat, 1963.
Eléments de prospérité des biotopes des sédiments meubles intertidaux et écologie de leurs populations en microfaune et macrofaune.
Vie et Milieu, 14(3):463-550.

beaches, lists of spp.
Salvat, B., 1962.
Faune des sédiments meubles intertidaux du Bassin d'Arcachon.
Cahiers Biol. Mar., Roscoff, 3(3):219-244.

beaches, microfauna of
Ax Peter, und Renate Ax 1970
Das Verteilungsprinzip des subterranen Psammon am Übergang Meer-Süsswasser.
Mikrofauna des Meeresbodens 1: 51 pp.

beaches, mineralogy
Jurgaitis, A., 1970.
On the mineralogy and petrography of beach gravel and pebbles of the Southeastern Baltic Sea coast. (In Russian; English and German abstracts).
Baltica, Lietuv. TSR Mokslu Akad Geogr. Skyr. INQUA Taryb. Sek., Vilnius, 4: 249-257

beaches, minerals
Lincius, A., and A. Ugincius, 1970
Baltic Sea beach sands at Sventoji Harbour and their heavy minerals. (In Russian; English and German abstracts)
Baltica, Lietuv. TSR Mokslu Akad. Geogr. Skyr. INQUA Taryb. Sek., Vilnius, 4: 273-284

beach penetration

beach penetration
Allen, W.A., E.B. Mayfield and H.L. Morrison, 1957.
Dynamics of a projectile penetrating sand.
J. Appl. Phys., 28(3):370-376.

beaches, petrography
Jurgaitis, A., 1970.
On the mineralogy and petrography of beach gravel and pebbles of the Southeastern Baltic Sea coast. (In Russian; English and German abstracts).
Baltica, Lietuv. TSR Mokslu Akad Geogr. Skyr. INQUA Taryb. Sek., Vilnius, 4: 249-257

beach sands

beach sands
*Marvin, Ursula B., and Marco T. Einaudi, 1967.
Black, magnetic spherules from Pleistocene and Recent beach sands.
Geochim. cosmochim. Acta. 31(10):1871-1884.

beaches, surface

beaches (surf zone)
Schuster-Dieterichs, O., 1956.
Die Makrofauna am sandigen Brandungsstrand von El Salvador (mittelamerikanische Pazifikküste).
Senckenbergiana Biol. 37(1/2):1-56.

beach trafficability

beach trafficability
Glenn, A. H., and C. C. Bates, 1950.
The meteorological and oceanographic aspects of geophysical prospecting. Geophys. 15(2):427-456, 2 textfigs.

bearing

bearing, ship's
Ekman, V.W., 1953.
Studies on ocean currents. Results of a cruise on board the "Armauer Hansen" in 1930 under the leadership of Bjørn Helland-Hansen. Geograf. Pub. 19(1):106 pp. (text), 122 pp. (92 pls.).

Beaufort scale

Beaufort wind scale
Japan, Kobe Marine Observatory, 1956.
[On the Beaufort wind scale.] J. Ocean., Kobe, (2)

Beaufort scale
Kinsman Blair 1969
Who put the wind speeds in Admiral Beaufort's force scale?
Oceans 2(2) 15-25 (popular)

Beaufort scale

Beaufort scale, modifications of
Verploegh, G., 1954.
Nieuwe aequivalenten voor de Beaufortschaal geldend voor waarnemingen op zee. K. Nederl. Met. Inst., Wetenschap. Rapp. W.R. 54-003 (L4-009):1-24 (mimeographed).

Beaufort scale
Verploegh, G., 1956.
The equivalent velocities for the Beaufort estimates of the wind force at sea.
Medel. Verhandel. K. Nederl. Met. Inst., 66:33 pp

Beaufort scale
Wachter, H., 1960.
Was Begründet die Stufen der Beaufort-Skala?
Der Seewart, 21(3):85-101.

Beaufort scale
Walden Hans 1969.
Probleme bei der Festlegung von Äquivalentwerten zwischen gemessenen Windgeschwindigkeiten und geschätzten Beaufort-Stufen.
Einzelveröffentlichungen, dt Wetterdienst Seewetteramt, 65: 30 pp. tables, figs.

Beaufort scale
Walden, Hans, 1965.
Die Windgeschwindigkeits-Äquivalente der Beaufortgrade nach Beobachtungen deutscher Bordwetterwarten.
Deutscher Wetterdienst, Seewetteramt, Einzelveroffenthichungen, No. 117:47 pp. 12 tables, 11 figs.

bedding

bedding
Seibold, Eugen, 1963
Geological investigation of near-shore sand-transport-examples of methods and problems from the Baltic and North seas.
Progress in Oceanography, M. Sears, Edit., Pergamon Press, London, 1:1-70.

bedding
*Wunderlich, Friedrich, 1967.
Die Entstehung von "convolute bedding" an Platenrändern.
Senckenberg. leth., 48(3/4):345-349.

bedrock

bedrock

Horikawa, Kiyoshi and Tsuguo Sunamura, 1970.
A study on erosion of coastal cliffs and of submarine bedrocks. Coastal Engng Japan 13: 127-139.

bedrock

*Kranck, Kate, 1967.
Bedrock and sediments of Kouchibougauc Bay, New Brunswick.
J. Fish.Res.Bd.Can., 24(11):2241-2265.

bedrock (submarine)

Somerton, W.H., S.H. Ward, and M.S. King, 1963
Physical properties of MOHOLE test site basalt.
J. Geophys. Res., 68(3):849-856.

bedrock

Spjeldnaes, N. 1971.
Mesozoic (?) bedrock exposed on the bottom of the Barents Sea.
Mar. Geol. 11(4): M47-M50.

behavior

Brock, Vernon E. and Robert H. Riffenburgh, 1960 behavior
Fish schooling: a possible factor in reducing predation.
J. du Cons., 35(3): 307-317.

behavior

Stefan, M., 1958.
[Physiological analysis of the interrelation between the gas exchange and shoal behaviour of certain marine and freshwater species.]
Zool. Zhurn., 27(2):222-228.

benches

*Belderson, R.H., N.H. Kenyon and A.H. Stride, 1970.
10-km wide views of Mediterranean deep-sea floor
Deep-Sea Res., 17(2): 267-270.

benthic parameters

Driscoll, Egbert G. 1969.
Analysis of variance of benthic parameters.
Antarctic Jl, U.S. 4(5): 188-189

benthos: SEE ALSO Bottom fauna

Benthos

See also: bottom fauna

A

benthos

Amanieu M. 1969.
Recherches Ecologiques sur les faunes des plages abritées de la région d'Arcachon.
Helgoländer wiss. Meeresuntersuch. 19(4): 455-557.

benthos

Arnaud, Patrick, 1965.
Nature de l'étagement du benthos marin algal et animal dans l'Antarctique.
C.R. Acad. Sci, Paris, 261:265-266.

benthos

Auffret, G.A., L. Berthois et L. Cabioch 1971.
Aperçu sur la bathymétrie, la sédimentologie et les peuplements benthiques au large de Roscoff d'après la carte bathymétrique au 1/40.000e et les photographies sous-marines.
Cah. Biol. mar. 12(4): 497-504

benthos

Augarde, Jacques, et Roger Molinier 1968
Contribution à l'Etude Ecologique des peuplements marins superficiels: étude des facteurs hydrodynamiques à proximité de la surface.
Bull. Mus. Hist. nat. Marseille 28: 27-43.

benthos (biomass)

Australia, Commonwealth Scientific and Industrial Research Organization, Division of Fisheries and Oceanography, 1963.
Oceanographical observations in the Pacific Ocean in 1960, H.M.A.S. Gascoyne, Cruise G 3/60.
Oceanographical Cruise Report, No. 6:115 pp.

benthos

Azouz, Abderrazak, 1968.
Contribution à l'étude de l'étage circalittoral du golfe de Tunis.
Rapp. P.-V. Réun. Comm. int. Explor. Sci. Méditerranée, 19(2):125-127.

B

benthos

Bacescu, Mihai, Marian Traian Gomoiu et Elena Dumitrescu, 1968.
Quelques considérations sur la dynamique des organismes de la zone médio littorale sableuse en mer Noire.
Rapp. P.-V. Comm. Réun. int. Explor. Sci. Méditerranée, 19(2):117-119.

benthos

Bakaev, V.G., editor, 1966.
Atlas Antarktiki, Sovetskaia Antarktichеskaia Ekspeditsiia. 1.
Glabnoe Upravlenie Geodesii i Kartografii MG SSSR, Moskva-Leningrad, 225 charts.

benthos

Bakus, Gerald J., 1966.
Some relationships of fishes to benthic organisms on coral reefs.
Nature, 210(5033):280-284.

benthos

*Barham, E.G., N.J. Ayer, Jr., and R.E. Boyce, 1967.
Macrobenthos of the San Diego Trough: photographic cenus and observations from bathyscaphe Trieste.
Deep-Sea Res., 14(6):773-784.

benthos

Barham, Eric G., Nathan G. Ayer Jr. and Robert E. Boyce, 1966.
Mega-benthic fauna of the San Diego Trough: photographic study from bathyscopl "Trieste". (Abstract only).
Second Int. Oceanogr. Congr., 30 May-9 June 1966. Abstracts, Moscow: 20-21.

benthos

Barnard, J. Laurens and John R. Grady 1968
A biological survey of Bahia de los Angeles, Gulf of California, Mexico. 1. General account.
Trans. San Diego Soc. nat. Hist. 15(5): 51-66.

benthos

Bas, Carlos, 1968.
Le peuplement benthique du plateau continental de la province de Tarragona (mer Catalane). Note préliminaire.
Rapp. P.-V. Réun. Comm. int. Explor. Sci. Méditerranée, 19(2):129-132.

benthos

Batham, E.J., 1969.
Benthic ecology of Glory Cove, Stewart Island.
Trans. R. Soc. N.Z., Biol. Sci., 11(5): 73-81

benthos

Battaglia Bruno, 1967.
Genetic aspects of benthic ecology in brackish waters.
In: Estuaries, G.H. Lauff, editor, Publs Am. Ass. Advmt. Sci., 83:574-577.

Beliaev / Belyaev G.M.

benthos

Belyaev, G.M., 1969.
Study of sea floor fauna conducted by the Soviet Union during the IGY and IGC.
Annls int geophys. Year, 46: 210-216

benthos

Beliaev, G.M., 1962.
The study of bottom fauna of seas and oceans carried out by the Soviet scientists. (In Russian)
Mezhd. Geofiz. Komitet, Prezidiume, Akad. Nauk, SSSR, Rezult. Issled. Programme Mezhd. Geofiz. Goda, Okeanol. Issled., No. 7:97-104.

benthos biomass

Belyayev, G.M., 1960
[The quantitative distribution of the bottom fauna in the north-western Bering Sea.]
Trudy Inst. Okeanol., 34: 85-104.

benthos

Belyaev, G.M., 1958.
Some patterns in the quantitative distribution of bottom fauna in the Antarctic (In Russian).
Inform. Biull. Sovetsk. Antarkt. Exped., 1(3): 43-44.

Translation:
Scripta Tecnica, Inc., Elsevier Publ. Co., 118-121.

benthos

Belyayev, G.M., and N.G. Vinogradova, 1961.
[Quantitative distribution of bottom fauna in the northern half of the Indian Ocean.]
Doklady Akad. Nauk, SSSR, 138(5):1191-1194.

OTS-61-11147-18 JPRS:8807:13-14.

benthos

*Bellan, G., 1967.
Pollution et peuplements benthiques sur subtrat meuble dans la région de Marseille. 2. L'ensemble portuaire Marseillais.
Rev. intern Océanogr. Med., 8:51-95.

benthos

Bellan, G., 1965.
Une méthode de tri de la "Microfaune annélidienne"
Méthodes Quantitatives d'Étude du Benthos et Echelle Dimensionnelle des Benthontes: Colloque du Comité du Benthos (Marseille, Nov. 1963).
Comm. Int. Expl. Sci. Mer. Médit., Monaco, 33-34.

benthos

Bellan-Santini, Denise, 1970.
Methodologie pour l'Etude qualitative et quantitative des peuplements de substrat dur.
Thalassia Jugoslavica, 6: 129-134.

benthos

Bellan-Santini, D., 1965.
Méthode de récolte et d'étude quantitative des peuplements sur substrat dur dans la zone d'agitation hydrodynamique.
Méthodes Quantitatives d'Étude du Benthos et Echelle Dimensionnelle des Benthontes: Colloque du Comité du Benthos (Marseille, Nov. 1963).
Comm. Int. Expl. Sci. Mer. Médit., Monaco, 23-24.

benthos, deep

Bernard, Francis, 1960.
La vie marine en grande profondeur.
Bull. Soc. Zool., France, 85(4):257-274.

benthos

Beyer, Fredrik, 1968.
Zooplankton, zoobenthos, and bottom sediments as related to pollution and water exchange in the Oslofjord.
Helgoländer wiss. Meeresunters. 17: 496-509.

benthos

Bird, Samuel O., 1970.
Shallow-marine and estuarine benthic molluscan communities from area of Beaufort, North Carolina.
Bull. Am. Ass. Petrol. Geol. 54(9): 1651-1676

benthos

Bodeanu, Nicolae, 1968.
Recherches sur le microphytobenthos du littoral roumain de la mer Noire.
Rapp.P.-V.Réun.Comm.int.Explor.Sci.Mediterranée, 19(2):205-207.

benthos

Bourcier, Michel, 1970(1971).
Étude quantitative du macrobenthos de la baie de Cassis (zone sud du Cap Canaille).
Téthys 2(3): 633-638

benthos

Bourcier, M., et C. Poizat, 1969.
Particularités de la répartition numérique du macrobenthos dans les parages de Cassis.
Recl Trav. Stn mar. Endoume 46(62): 187-191

benthos

*Brawn, V.M., D.L. Peer and R.J. Bentley, 1968.
Caloric content of the standing crop of benthic and epibenthic invertebrates of St. Margaret's Bay, Nova Scotia.
J.Fish.Res.Bd.Can., 25(9):1803-1811.

Benthos

Broch, Hjalmar, 1961.
Benthonic problems in Antarctic and Arctic waters
Sci. Res. Norwegian Antarctic Exped., 1927-1928 et seq., financed by Consul Lars Christensen, Det Norske Videnskaps-Akad., Oslo, No. 38:32 pp.

benthos

Bruce, J.R., J.S. Colman, and N.S. Jones, Editors, 1963.
Marine fauna of the Isle of Man and its surrounding seas. 2nd Edition.
Liverpool Mar. Biol. Comm., Mem., No. 36:307 pp.

benthos

Brunel, Pierre, 1970.
Aperçu sur les peuplements d'invertébrés marins des fonds meubles de la baie de Gaspé.
Naturaliste can. 97(6): 679-710. (1956-60)

benthos

Brunel, Pierre 1961 (reprinted 1964)
Éléments d'écologie du benthos marin.
Cah. Inf. Stn biol. mar. Grande-Rivière 6: 25pp. (multilithed).

benthos

Bullivant, J.S., 1959
Photographs of the bottom fauna in the Ross Sea. N.Z.J. Sci. 2(4): 485-497.

benthos

Bullivant, John S., and John H. Dearborn, 1967.
The fauna of the Ross Sea. 5. general accounts, station lists and benthic ecology.
Bull. N.Z. Dept. Scient. indust. Res., 176:1-77. (mem. N.Z. oceanogr. Inst. 32).

C

benthos

Cabioch, Jacqueline 1969.
Les fonds de maerl de la baie de Morlaix et leur peuplement végétal.
Cah. Biol. mar. 10(2): 139-161

benthos

Cabioch, Louis, 1968.
Contribution à la connaissance des peuplements benthiques de la Manche occidentale.
Cah. Biol. mar. 9 (5-Suppl.): 493-720

benthos

Cabioch, L., 1961.
Étude de la répartition des peuplements benthiques au large de Roscoff.
Cahiers Biol. Mar., Roscoff, 2(1):1-40.

benthos

California, Humboldt State College, 1964.
An oceanographic study between the points of Trinidad Head and Eel River.
State Water Quality Control Bd., Resources Agency California, Sacremento, Publ., No. 25:136 pp.

benthos

Carey, Andrew G., Jr. 1967.
Energetics of the benthos of Long Island Sound. I. Oxygen utilization of sediment.
Bull. Bingham oceanogr. Coll. 19(2): 136-144.

benthos

Carey, Andrew G. Jr. 1965.
Preliminary studies of animal sediment interrelationships off the central Oregon coast.
Ocean Sci. Ocean Eng., Mar. Techn. Soc.-Am. Soc. Limnol. Oceanogr. 1:100-110

benthos

Carpine, Christian, 1965.
Quelques observations sur la faune bathyale dans le Canal de Corse.
Rapp. Proc. Verb., Réunions, Comm. Int. Expl. Sci., Mer Méditerranée, Moneco, 18(2):83.

benthos

Carpine, Christian, et Gaston, Fredj, 1965.
Méthodes de prélèvements et de tri en bionomie benthique.
Méthodes Quantitatives d'Étude du Benthos et Echelle Dimensionnelle des Benthontes: Colloque du Comité du Benthos (Marseille, Nov. 1963).
Comm. Int. Expl. Sci. Mer. Médit., Monaco, 29-31.

benthos

Carriker, Melbourne R., 1967.
Ecology of estuarine benthic invertebrates: a perspective.
In: Estuaries, G.H. Lauff, editor, Publs Am. Ass. Advmt Sci., 83:442-487.

benthos

Chipman, Walter, and Jean Thommeret 1970.
Manganese content and the occurrence of fallout ^{54}Mn in some marine benthos of the Mediterranean.
Bull. Inst. océanogr. Monaco, 69(1402): 15pp

benthos

Christomanos, Anast., and D. Giannitsis, 1962.
Dredging ewsults from the oceanographic cruise "Tithys" during August-September 1961.
Research Proc., Mar. Lab., Greece, 1(2):7-11.

benthos

Chukhchin, V.D., 1963.
Quantitative distribution of benthos in the eastern part of the Mediterranean Sea (In Russian
Trudy Sevastopol Biol. Sta., 16:215-223.

benthos

Conover, John T., 1962
Algal crusts and the formation of lagoon sediments.
The Environmental Chemistry of Marine Sediments, Proc. Symp., Univ., R.I., Jan. 13, 1962
Occ. Papers, Narragansett Mar. Lab., No. 1: 69-76.

benthos

Copeland, B.J., and Robert S. Jones, 1965.
Community metabolism in some hypersaline waters.
Texas J. Sci., 17(2):188-205.

Benthos

Coull, Bruce C., 1970.
Shallow water meiobenthos of the Bermuda platform.
Oecologia, Berl. 4: 325-357.

benthos

Coull, Bruce C., and Sidney S. Herman, 1970
Zoogeography and parallel level-bottom communities of the meiobenthic Harpacticoida (Crustacea, Copepoda) of Bermuda.
Oecologia, Berl. 5: 392-399.

benthos

Craig, G.Y., and N.S. Jones, 1966.
Marine benthos, substrate and palaeoecology.
Palaeontology, 9(1):30-38.

benthos

Crnkovic, Drago, 1964
Action de chalutage sur les populations benthiques dans la region des canaux de l'Adriatique nord. (In Jugoslavian; French resume).
Acta Adriatica, 11(1):47-57.

D

benthos

Day, J.H., 1969.
A guide to marine life on South African shores.
Univ. Cape Town, A.A. Balkema, 300pp.

benthos

Day, John H., John G. Field and Mary Potts Montgomery 1971.
The use of numerical methods to determine the distribution of the benthic fauna across the continental shelf of North Carolina.
J. anim. Ecol. 40: 93-125

benthos

de Gaillande, Daniel, 1970.
Peuplements benthiques de l'herbier de Posidonia oceanica (Delile), de la pelouse à Caulerpa prolifera Lamouroux et du large du Golfe de Gabes.
Téthys 2(2): 373-384

benthos

De Gaillande, Daniel, 1970
Note sur les peuplements de la zone centrale du Golfe de Gabès (Campagne Calypso 1965)
Téthys 2(1): 131-138

benthos

Dell, R.K., 1965.
Merine biology.
In: Anterctice, Trevor Hatherton, editor, Methuen & Co., Ltd., 129-152.

benthos

*Desai, B.N., and M. Krishman Kutty, 1967.
Studies on the benthic fauna of Cochin backwater.
Proc.Indian Acad.Sci., (B)66(4):123-142.

benthos

Diaz-Piferrer, M., 1969.
Distribution of the marine benthic flora of the Caribbean Sea.
Carib. J. Sci., 9(3/4): 151-178

bedrock

Dmitriev, L.V., A.G. Sharaskin and M.M. Farafonov, 1969.
Bedrock of the rift zones of the Indian Ocean cherts and some cherts in geochemistry. (In Russian).
Dokl. 185(2): 444-446

benthos

Dorjes, Jurgen, 1971.
Der Golf von Gaeta (Tyrrhenisches Meer). IV. Das Makrobenthos und seine küstenparallele Zonierung. Senckenberg. mar. maritima 3(1): 203-246.

benthos a

*Dörjes, Jürgen, Sibylle Gadow, Hans-Erich Reineck and Indra Bir Singh, 1970.
Sedimentologie und Makrobenthos der Nordergründe und der Aufsenjade (Nordsee). Senckenberg marit. 2: 31-59.

benthos

*Dörjes, Jürgen, Sibylle Gadow, Hans-Erich Reineck and Indra Bir Singh, 1969.
Die Rinnen der Jade (Südliche Nordsee). Sedimente und Makrobenthos. Senckenberg. marit. [1] 50: 5-62.

benthos

*Dörjes, Jürgen, Sibylle Gadow, Hans-Erich Reineck und Indra Bir Singh, 1969.
Die Rinnen der Jade (südliche Nordsee). Sedimente und makrobenthos. Senckenberg. Maritima 50: 5-62.

benthos

*Driscoll, Egbert G., 1967.
Attached epifauna - substrate relations. Limnol. Oceanogr., 12(4):633-641.

benthos

*Durand, J.R., 1967.
Étude des poissons benthiques du plateau continental congolais. 3. Étude de la repartition de l'abondance et des variations saisonnières. Cah. ORSTOM, Sér. Océanogr., 5(2):3-68.

benthos

Dybern, Bernt I. 1967.
Settlement of sessile animals on eternit slabs in two polls near Bergen. Sarsia 29: 137-150.

E

benthos

Edmondson, W.T., editor, 1966.
Ecology of Invertebrates, Marine Biology III, Proceeding of the Third International Interdisciplinary Conference. New York Acad. Sci., 313 pp. ($7.00).

Benthos

Emery, K. O., and J. Hülsemann, 1961 (1962).
The relationships of sediments, life and water in a marine basin. Deep-Sea Res., 8(3/4):165-180.

benthos

*Estcourt, I.N., 1967.
Distributions and associations of benthic inverebrates in a sheltered water soft-bottom environment (Marlborough Sounds, New Zealand). N.Z. Jl mar. Freshwat. Res., 1(3):352-370.

F

benthos

*Fager, E.W., 1969.
Production of stream benthos: A critique of the method of assessment proposed by Hynes and Coleman (1968). Limnol. Oceanogr., 14(5): 766-770.

benthos

Fausto Filho, José, Henry Ramos Matthews e Hermínia de Holanda Lima 1966.
Nota preliminar sôbre a fauna dos bancos de lagostas no Ceará. Arquivos Estação Biol. Marinha, Univ. Fed. Ceará 6(2): 127-130.

benthos

Fedorova, A.F., Kilezhenko, V.P., 1963
The radioactivity of bottom organisms in the Norwegian Sea. (In Russian). Okeanologiia, Akad. Nauk, SSSR, 3(1):123-126.

benthos

Fenchel, Tom, 1969
The ecology of marine microbenthos. IV Structure and function of the benthic ecosystem, its chemical and physical factors and the microfauna communities with special reference to the ciliated Protozoa.
Ophelia, 6: 1-182.

benthos

Ferreira-Correia, M.M., e F. Pinheiro-Vieira, 1969
Terceira contribuição ao inventário das algas marinhas bentônicas do nordeste Brasileiro.
Arq. Ciên. Mar, Fortaleza, Ceará, Brasil, 9(1): 21-26.

benthos

Ferrero, Letizia, 1961
Ricerche fisico-chimiche e biologiche sui La Laghi Salmastri Pontini in relazione alla produttivita. II. Il Lago di Paola (Sabaudia). Ricerche quantitative sulla fauna bentonica. Boll. Pesca, Piscicolt. e Idrobiol., n.s., 16(2):173-203.

benthos

Field, John G., 1971.
A numerical analysis of changes in the soft-bottom fauna along a transect across False Bay, South Africa. J. exp. mar. Biol. Ecol. 7(3): 215-253.

benthos a

Field, J.G., 1970.
The use of numerical methods to determine benthic distribution patterns from dredgings in False Bay. Trans. roy. Soc. S Afr. 39(2): 183-200

benthos

Field, J.G. and F.T. Robb, 1970.
Numerical methods in marine ecology. 2. Gradient analysis of rocky shore samples from False Bay.
Zool. Africana 5(2): 191-210

benthos

Filatova, Z.A., and N.G. Barsanova, 1964.
The communities of bottom fauna of the western part of the Bering Sea. Investigations of bottom fauna and flora of Far-Eastern Seas and the Pacific Ocean. (In Russian; English abstract). Trudy, Inst. Okeanol., Akad. Nauk, SSSR, 69:6-97.

benthos

Filatova, Z.A., and R.J. Levenstein, 1961.
[Quantitative distribution of deep-sea bottom fauna in the northeastern Pacific.] Trudy Inst. Okeanol., Akad. Nauk, SSSR, 45:190-213.

benthos

Filatova, Z.A., and A.A. Neiman, 1963
The biocoenosis of the bottom fauna in the Bering Sea. (In Russian). Okeanologiia, Akad. Nauk, SSSR, 3(6):1079-1084.

benthos

Fishelson, L., 1971.
Ecology and distribution of the benthic fauna in the shallow waters of the Red Sea. Mar. Biol. 10(2): 113-133.

benthos

Forster, G.R., 1961.
An underwater survey on the Lulworth Banks. J.M.B.A., U.K., 41(1):157-160.

benthos

Frankenberg, Dirk, 1965.
Variability in marine benthic communities off Georgia. (abstract).
Ocean Sci. and Ocean Eng., Mar. Techn. Soc., Amer. Soc. Limnol. Oceanogr., 2:1111.

benthos

Fricke, H.W. 1970
Beobachtungen über Verhalten und Lebensweise des im Sand lebenden Schlangensternes Amphioplus sp.
Helgoländer wiss. Meeresunters. 21(1/2): 124-133.

G

benthos

Gaidash, Yu. K. 1971.
Effect of metallurgic and metal working industry wastes on macrozoobenthos. (In Russian).
Gidrobiol. Zh. 7(4): 66-70

benthos

Gallardo, V.A. and J.C. Castillo 1969.
Quantitative benthic survey of the infauna of Chile Bay (Greenwich I., South Shetland Is.)
Gayana 16: 1-17.

benthos a

Gamulin-Brida, Helena, 1969.
Problems of benthic bionomics in the Adriatic Sea. (In Jugoslavian; English and Italian abstracts)
Thalassia Jugoslavica 5: 71-82

benthos

Gamulin-Bride, H., 1965.
Contribution aux recherches bionomiques sur les fonds Coralligènes au large de l'Adriatique moyen. Rapp. Proc. Verb., Réunions, Comm. Int. Expl. Sci., Mer Méditerranee, Monaco, 18(2):69-74.

benthos

Gamulin-Brida, Hélène, 1965.
Proposition pour le classification des organismes benthiques selon le grandeur.
Méthodes Quantitatives d'Étude du Benthos et Echelle Dimensionnelle des Benthontes: Colloque du Comité du Benthos (Marseille, Nov. 1963), Comm. Int. Expl. Sci. Mer. Médit., Monaco, 39-40.

benthos

Gamulin-Brida, Helena, 1965.
Contribution aux recherches sur la bionomie benthique de le baie de Porto Paone (Naples, Italie.) Repertition des biocoenoses benthiques. Pubbl. stsz. zool., Napoli, 34(3):376-500.

benthos

Gamulin-Brida, Helena, 1965.
Biocoenose des fonds vaseux au large de l'Adriatique moyenne, (In Jugoslairan; French resumé). Acta Adriatica, 10(10):27pp.

benthos

Gamulin-Brida, Helena, 1964.
Contribution aux recherches bionomiques sur les fonds vaseux du large de l'Adriatique moyenne. (In Jugoslavian; French resume). Acta Adriatica, 11(1):85-89.

benthos
Gamulin-Brida, Hélène, 1963.
Note préliminaire sur les recherches bionomiques dans l'Adriatique méridionale.
Rapp. Proc. Verb., Reunions, Comm. Int. Expl. Sci. Mer Méditerranée, Monaco, 17(2):85-92.

benthos
Gändefors, Dag, and Lars Orrhage 1968.
Patchiness of some marine bottom animals: a methodological study.
Oikos, 19(2): 311-321.

benthos
Gershanovich, D. E., and A. A. Neiman, 1964.
Bottom sediments and bottom fauna in the East ChinaSea. (In Russian).
Okeanologiia, Akad. Nauk, SSSR, 4(6):1089-1095.

benthos
Gilat, Eliezer, 1969.
Study of an ecosystem in the coastal waters of the Ligurian Sea. III Macrobenthic communities.
Bull. Inst. océanogr., Monaco 69 (1396): 76pp.
Also: IAEA Radioactivity in the Sea, Publ. 27.

benthos
Ginés, Hno., y R. Margalef, editores, 1967.
Ecologia marina.
Fundación La Salle de Ciencias Naturales, Caracas Monografía 14: 711 pp.

benthos
*Glémarec, Michel, 1969.
La "Grande Vasière" aperçu bionomique.
C.r.hebd.Séanc.,Acad.Sci.,Paris,268(1):155-157; 268(2):401-404.

benthos
Glemarec, Michel, 1965.
La faune benthique dans la partie méridionale du massif armoricain.
Cahiers Biol. Mar., Roscoff, 6(1):51-66.

benthos
Glémarec, Michel, 1964.
Bionomie benthique de la partie orientale du Golfe du Morbihan.
Cahiers de Biol. Mar., Roscoff, 5:33-96.
Also in:
Trav. Sta. Biol., Roscoff, n.s., 15. 1964.

benthos
Gilat, E., 1963.
The macrobenthic animal communities of the Israeli continental shelf in the Mediterranean.
Rapp. Proc. Verb., Reunions, Comm. Int. Expl. Sci. Mer Méditerranée, Monaco, 17(2):103-106.

benthos
*Golikov, A.N., 1968.
Distribution and variability of long-lived benthic animals as indicators of currents and hydrological conditions.
Sarsia, 34:199-208.

benthos
Golikov, A.N. and O.A. Skarlato, 1971.
Some results of diving hydrobiological investigations of the Posiet Bay (The Sea of Japan). (In Russian; English abstract). Gidrobiol. Zh. 7(5): 32-37.

benthos
*Golikov, A.N., and O.A. Scarlato, 1968.
Vertical and horizontal distribution of biocoenoses in the upper zones of the Japan and Okhotsk seas and their dependence on the hydrological system.
Sarsia, 34:109-116.

benthos
Golikov, A.N., and O.A. Scarlato, 1965.
Hydrobiological explorations in the Pasjet Bay with diving equipment. Fauna of the seas of the northwest Pacific. (In Russian).
Issled. Fauni Morei, Zool. Inst., Akad. Nauk, SSSR, 3(11):5-21.

benthos
Gomoiu, M.T., 1965.
Distribution of sand areas and their biocoenosis in the Romanian Black Sea coast.
Trav. Mus. Hist. nat. "Grigore Antipa", 8(1): 291-299

benthos
Govberg, L.I., 1965.
Faunistic complexes inthe near-coastal deposits of the northwest Black Sea. (In Russian).
Okeanologiia, Akad. Nauk, SSSR, 5(5):870-876.

benthos
*Griggs, G.B., A.G. Carey, Jr. and L.D. Kulm, 1969.
Deep-sea sedimentation and sédiment - fauna interaction in Cascadia Channel and on Cascadia Abyssal Plain. Deep-Sea Res., 16(2): 157-170.

benthos
Guérin, J.-P. 1970
Étude expérimental de l'établissement d'un peuplement de substrat meuble à partir de larves méroplanctoniques.
Cah. biol. mar. 11(2): 167-185.

benthos
Guille, Alain, 1970.
Les communautés benthiques des substrats meubles du plateau continental au large de Banyuls-sur-Mer.
C. r. hebd. Séanc. Acad. Sci. Paris (D) 271 (1): 189-192

benthos
Guille, Alain, et Jacques Soyer, 1970.
Bionomie benthique du plateau continental de la côte Catalane française.
Vie Milieu, 21 (1B): 137-147.

benthos
Guinther, Eric B., 1971.
Ecologic observations on an estuarine environment at Fanning Atoll.
Pacific Sci. 25(2): 249-259.

benthos
Gussev, A. V. and F. A. Pasternak, 1958.
Some remarks concerning the bottom fauna of Antarctic waters.
Doklady, Akad. Nauk USSR 123(5):841-844.

H

benthos
Hadzi, Jovan, 1964.
Genetic relationship between pelagic and benthal organisms. (In Jugoslavian; English summary).
Acta Adriatica, 11(1):121-126.

benthos
Hagerman, Lars, 1966.
The macro- and microfauna associated with Fucus Serratus L., with some ecological remarks.
Ophelia, 3:1-43.

benthos
Hanks, Robert W., 1964.
A benthic community in the Sheepscot River Estuary, Maine.
U.S.F.W.S. Fish. Bull., 63(2):343-353.

benthos
Harada, Eiji, and Tetsuya Narita, 1964.
Ecological aspects of the research of bottom animals of the deep sea with reference to the collections of JEDS-6. (French resumé).
La Mer. 1(2):62-65.
Also in: Repts. of JEDS, 4. Dec. 1963.

benthos
Harmelin, Jean-Georges, Denise Bellan-Santini, Charles-François Boudouresque, Thérèse Le Campion-Alsumard, Leung Tack Kit et Georges Salen 1970.
Étude expérimentale de la colonisation des surfaces vierges naturelles en eau pure et en eau polluée, dans la région marseillaise. 1. Conditions de l'expérience.
Téthys 2(2): 329-334

benthos
Harrison, W., and Marvin L. Wass, 1965.
Frequencies of infaunal invertebrates related to water content of Chesapeake Bay sediments.
Southeast Geol., 6(4):177-187.

benthos
Hempel, G., 1970.
Die biologischen Arbeiten auf der "Atlantischen Kuppenfahrt en 1967" des F.S. Meteor.
Meteor Forsch.-Ergebnisse (D) 7: 1-2.

benthos
Henriksson Rolf, 1965.
The bottom fauna in polluted areas of the Sound.
Oikos, 19(1): 111-125.

benthos
Hessler, Robert R., and Howard L. Sanders, 1967.
Faunal diversity in the deep-sea.
Deep-Sea Research, 14(1):65-78.

benthos
Hesthagen, Ivar H., 1970
On the near-bottom plankton and benthic invertebrate fauna of the Josephine Seamount and the Great Meteor Seamount.
Meteor Forsch.-Ergebnisse (D) 8: 61-70.

benthos
Hinschberger, Félix, Anne Saint-Requier et Anne Toulemont, 1966.
Recherches sédimentologiques et écologiques sur les fonds sous-marins dans les parages de la chaussée de Sein (Finistère).
Rev. Trav. Inst. Pêches marit., 31(4):425-452.

benthos
Holme, N.A., 1962.
Benthos in Antarctic waters. (Abstract for Symposium Antarctic Biol., Paris, 2-8 Sept. 1962);
Polar Record, 11(72):332-333.
Also: SCAR Bull. #12.

benthos
Holme, N.A., 1961
The bottom fauna of the English Channel.
J. Mar. Bio. Ass. U.K., 41(2):397-462.

benthos
Horikoshi, Masuoki, 1962
Bird's eye view of the studies of the benthos in Japan. (In Japanese; English abstract).
J. Oceanogr. Soc., Japan, 20th Ann. Vol., 707-723.

benthos
Horikoshi, Masuoki, 1962.
Distribution of benthic organisms and their remains at the entrance of Tokyo Bay, in relation to submarine topography, sediments and hydrography.
Nat. Sci. Rept., Ochanomizu Univ., 13(2):47-122.

benthos
Horikoshi, Masuoki, 1962
Distribution of benthic organisms and their remains at the entrance of Tokyo Bay. (Co-operative survey at the entrance of Tokyo Bay in 1959 - Part 7). (In Japanese; English abstract).
J. Oceanogr. Soc., Japan, 20th Ann. Vol., 146-154.

benthos
Horikoshi, Masuoki, 1961.
Distribution of benthic organisms and their remains at the entrance of Tokyo Bay in relation to submarine topography, sediments and hydrology. (Abstract).
Tenth Pacific Sci. Congr., Honolulu, 21 Aug.-6 Sept., 1961, Abstracts of Symposium Papers, 371-372.

benthos
Hughes, R.N. and M.L.H. Thomas 1971.
Classification and ordination of benthic samples from Bedeque Bay, an estuary in Prince Edward Island, Canada. Mar. Biol. 10(3): 227-235.

benthos
Hughes, Roger N., and Martin L.H. Thomas 1971.
The classification and ordination of shallow-water benthic samples from Prince Edward Island, Canada.
J. exp. mar. Biol. Ecol. 7(1): 1-39.

benthos
Hunkins, Kenneth, 1959.
The floor of the Arctic Ocean.
Trans. A.G.U., 40(2):159-162.

benthic fauna
Hunkins, Kenneth L., Maurice Ewing, Bruce C. Heezen, and Robert J. Menzies, 1960.
Biological and geological observations of the first photographs of the Arctic Ocean deep-sea floor.
Limnol. & Oceanogr., 5(2):154-161.

benthos
Hurley, D.E., 1965.
Benthic ecology of Milford Sound.
New Zealand, Sept. Sci. Industr. Res., Bull., 157:79-89.
Also:
New Zealand Oceanogr. Inst., Memoir, No. 17.

benthos
Huvé, A., P. Huvé and J. Picard, 1963.
Aperçu préliminaire sur le benthos littoral de la côte rocheuse adriatique italienne.
Rapp. Proc. Verb., Réunions, Comm. Int. Expl. Sci Mer Méditerranée, Monaco, 18(2):93-102.

I

J

benthos
Jones, M.E., 1961.
A quantitative evaluation of the benthic fauna off Point Richmond, California.
Univ. California, Publ. Zool., 67(3):219-320.

K

benthos
Kalinenko, V.O., 1952.
Bacteria and invertebrates inhabiting the sea-bed. (In Russian).
Mikrobiologiia, 21(4):555-

benthos
Kempf, Marc, 1970.
A Plataforma continental de Pernambuco (Brasil): Nota preliminar sôbre a natureza do fundo.
Trabhs oceanogr. Univ. Fed. Pernambuco 9/11: 111-124.

Benthos
*Kempf M., 1970.
Notes on the benthic bionomy of the N-NE Brazilian shelf. Mar. Biol., 5(3): 213-224.

benthos
Kikuchi, Taiji, 1966.
An ecological study on animal communities of the Zostera marina belt in Tomioka Bay, Amakusa, Kyushu.
Publ. Amakusa Mar. biol. Lab., 1(1):1-106.

benthos
Kirsteuer, Ernst 1968.
Bemerkungen zu den Aufsammlungen Evertebrata im Nordwesten von Madagaskar, unter besonderer Berücksichtigung der Nemertini, Ergebnisse der Österreichischen Indo-Westpazifik-Expedition 1959/60.
Zool. Anz. 180 (3/4): 165-177.

benthos
Kiseleva, M.I. and M.I. Shcherbina 1971.
Effect of the soda plant industrial wastes on some benthic invertebrates of the Black Sea. (In Russian)
Gidrobiol. Zh. 7(4): 62-66.

benthos
Kisseleva, Marta Ivanovna, 1968.
Le développement du benthos dans les mers du bassin méditerranéen.
Rapp. P.-V. Réun. Comm. int. Explor. Sci. Méditerranée, 19(2):103-105.

benthos, lists of spp.
Kiseleva, M.I., 1963.
Qualitative and quantitative distribution of the benthos in the Aegean Sea. (In Russian).
Trudy Sevastopol Biol. Sta., 16:192-200.

benthos
Kiseleva, M.I., 1961.
[The nature and quantitative distribution of the benthos in the Dardenelles region of the Aegean Sea.]
Trudy Sevastopol Biol. Sta., (14):135-146.

benthos
Kitamori, R., 1963.
Studies on the benthos communities of littoral areas in the Seto-Inland Sea and the adjacent waters. (In Japanese; English abstract).
Bull. Naikai Reg. Fish. Res. Lab., 21:1-90.

Abyssal benthos
Knudsen, Jørgen, 1961.
The bathyal and abyssal Xylophaga (Pholadidae, Bivalvia).
Galathea Rept., 5:163-209.

Benthos
Knudsen, Jørgen, 1961.
The bathyal and abyssal Xylophaga (Pholadidae, Bivalvia).
Galathea Rept., 5:163-209.

benthos
Kobiakova, Z.I., 1959.
The benthos in the northern part of the Tartary Strait and its influence on the feeding of fish. (In Russian).
Izv. Tikhookean. Nauchno-Issled. Inst. Ribn. Khoz i Okeanogr., 47:50-61.

benthos
Kosaka, Masaya, Masahiro Ogura, Kaoru Daimon and Kenichi Miyasita, 1971.
Studies on the benthic communities in Shimizu Harbour and Orido Bay. 1. Phase in summer. (In Japanese; English abstract). J. Tokai Univ. Coll. mar. Sci. Technol. 5:9-25.

benthos
Kosler, A. 1968.
Distributional patterns of the eulittoral fauna near the isle of Hiddensee (Baltic Sea, Rigz).
Mar. Biol. 1(4): 266-268.

benthos
Kühl, Heinrich, 1966.
Die Abfluss der Elbe im Jahre 1965 und seine Wirkung auf Salzgehalt, Plankton und Bewuchsbildung bei Cuxhaven.
Veröff. Inst. Meeresforsch. Bremerh., 10(2):61-70.

Benthos
Kühlmorgen-Hille, Georg, 1965.
The effect of the severe winter 1962/63 on the bottom fauna of Kiel Bay.
Ann. Biol., Cons. Perm. Int. Expl. Mer, 1963, 20: 98-99.

benthos
Kühlmorgen-Hille, Georg, 1963
Quantitative Untersuchungen der Bodenfauna in der Kieler Bucht und ihre jahreszeitlichen Veränderungen.
Kieler Meeresf., 19(1):42-66.
German and English abstract

benthos
Kurian, C.V., 1971.
Distribution of benthos on the south west coast of India. (Portuguese abstract). In: Fertility of the Sea, John D. Costlow, editor Gordon Breach, 1: 225-239.

benthos
Kuznetzov, A.P., 1970.
On the study of the Jermolinoxaja Inlet (Kandalaksha Bay, White Sea) as an eco-system. I. Bottom fauna. (In Russian; English abstract).
Trudy Inst. Okeanol. Akad. Nauk SSSR 88: 98-112.

benthos
Kuznetsov, A.P., 1970.
On the trophic structure and zonation of the bottom fauna distribution in the Azov and Baltic seas. (In Russian; English Abstract).
Trudy Inst. Okeanol. Akad. Nauk SSSR 88: 81-97.

benthos
Kuznetsov, A.P., 1970.
On the distribution of trophic groupings of bottom invertebrates in the Barents Sea. (In Russian; English abstract)
Trudy Inst. Okeanol. Akad. Nauk SSSR 88: 5-80.

benthos
Kusnetsov, A.P., 1964.
Distribution of the sea bottom fauna in the western part of the Bering Sea and the trophic zonation. Investigations of bottom fauna and flora of the Far-Eastern Seas and the Pacific Ocean. (In Russian).
Trudy Akad. Nauk, SSSR, 69:98-177.

benthos
Kuznetsov, A.P., 1961
[Data on the quantitative distribution of the bottom fauna in the Kamchatka Bay.]
Trudy Inst. Okeanol., 46:103-123.

benthos
Kuznetzov, A.P., 1961
[On ecology of some mass forms of benthos from Eastern Kamchatka and northern Kuril Islands.]
Trudy Inst. Okeanol., SSSR, 46: 85-97.

benthos
Kuznetsov, A., 1960.
[Data concerning quantitative distribution of bottom fauna of the bed of the Atlantic.]
Doklady Akad. Nauk, SSSR, 130(6):1345-1348.

benthos

Kuznetzov, A.P., 1959

Distribution of bottom fauna in the Kronotski Bay.

Trudy Inst. Okeanol., 36: 105-122.

benthos

Kuznetzov, A.P., 1959

Distribution of the bottom fauna in the North Kuril waters.

Trudy Inst. Okeanol., 36: 236-258.

benthos

Kuznetsov, A., 1956.

Data on the quantitative distribution of bottom fauna on the floor of the Atlantic Ocean.

Doklady Akad. Nauk, SSSR, 130(6):1345-1348.

OTS $0.50.

L

benthos

*Laakso, Mikko, 1968.

The bottom fauna in the surroundings of Helsinki.

Ann.Zool.Fenn.5(3):262-264.

benthos

Laakso, Mikko, 1965.

The bottom fauna in the surroungings of Helsinki.

Annales Zool. Fennici, 2:19-37.

benthos

*LaFond, E.C., 1967.

Movements of benthic organisms and bottom currents as measured from the bathyscaph Trieste.

In: Deep-sea photography, J.B.Hersey, editor, Johns Hopkins Oceanogr.Studies, 3:295-302.

benthos

LeDanois, Ed., 1948.

Les profondeurs de la mer. Trente ans de recherches sur la faune sous-marine au large des côtes de France. Payot, Paris, 303 pp., 8 pls., 56 textfigs.

benthos

Ledoyer, Michel 1969

Écologie de la faune vagile des biotopes méditerranéens accessibles en scaphandre autonome. V Étude des phénomènes nycthéméraux. Les variations nycthémerales des populations animales dans les biotopes. VI. Les pêches à la lumière. VII. Les variations nycthémerales des gaz dissous au sein des biotopes.

Téthys 1(2): 291-308; 309-320; 321-340.

benthos

Ledoyer, Michel, 1969.

Aperçu sur la faune vagile de quelques biotopes de substrat dur de Méditerranée orientale; Comparaison avec les mêmes biotopes en Méditerranée occidentale.

Téthys 1(2): 281-290.

benthos

Ledoyer, Michel, 1969.

La faune vagile des sables fins des hauts niveaux (S.F.H.N.) signification bionomique de ce biotope vue sous l'angle de la faune vagile.

Téthys 1(2): 275-280.

benthos

Ledoyer, Michel, 1966.

Ecologie de la faune vagile des biotopes méditerranéens accessibles en Scaphandre autonome. I. Introduction: données analytiques sur les biotopes de substract dur.

Recl. Trav. Stn. mar. Endoume, 40(56):103-149.

benthos

*Le Loeuff, P. et A. Intes, 1969.

Premières observations sur la faune benthique du plateau continental de côte d'Ivoire. Cah. O.R.S.TO.M. sér. Océanogr., 7(4): 61-66.

benthos

*Le Loeuff, P. et A. Intes, 1968.

La faune benthique du plateau continental de Côte d'Ivoire. Récoltes au chalut. Abondance Répartition. Variations saisonnières (mars 1966-février 1967). Doc. Centre Rech. océanog Abidjan, 025: 78 pp.

benthos

Leloup, E., Editor, with collaboration of L. Van Meel, Ph. Polk, R. Halewyck and A. Gryson, undated.

Recherches sur l'ostreiculture dans le Bassin de Chasse d'Ostende en 1962.

Ministere de l'Agriculture, Commission T.W.O.Z., Groupe de Travail - "Ostreiculture", 58 pp.

benthos

Leppäkoski, Erkki, 1969.

Transitory return of the benthic fauna of the Bornholm Basin after extermination by oxygen insufficiency.

Cah. Biol. mar. 10(2):163-172

benthos

Lewis, John B., 1965.

A preliminary description of some marine benthic communities from Barbados, West Indies.

Canadian J. Zool., 43(6):1049-1074.

benthos

Lie, Ulf, 1969.

Standing crop of benthic infauna in Puget Sound and off the coast of Washington.

J.Fish.Res.Bd.Can., 26(1):55-62.

benthos

*Lie, Ulf, 1968.

A quantitative study of benthic infauna in Puget Sound, Washington, U.S.A. in 1963-1964.

Fisk Dir.Skr.Ser.Havunds, 14(5):2 29-521.

benthos

Lisitzin, A.P., 1966.

Processes of Recent sedimentation in the Bering Sea. (In Russian).

Inst. Okeenol., Kom. Osad. Otdel Nauk o Zemle, Isdatel, Nauka, Moskva, 574 pp.

benthos

Littlepage, Jack L., and John S. Pearse, 1962

Biological and oceanographic observations under an Antarctic ice shelf.

Science, 137(3531):679-681.

benthos

Longhurst, Alan R., 1964.

A review of the present situation in benthic synecology.

Bull. Inst. Océanogr., Monaco, 63(1317):54 pp.

benthos

López López, Miguel A., y Taizo Okuda, 1968.

Algunas observaciones sobre características físico- químicas de los sedimentos y distribución de la fauna macro bentónica de la laguna Grande del Obispo (Venezuela).

Bol. Inst. Oceanogr. Univ. Oriente, Venezuela, 7(1): 107-127.

benthos

*Luksenas, Ju. K., 1969.

Biocoenoses of bottom invertebrates of the southern Baltic Sea and their trophic groups. (In Russian; English abstract). Okeanologiia, 9(6): 1078-1086.

benthos

*Lukshenas, Yu, K., 1967.

Zoogeographical complexes of bottom invertebrates in the southern Baltic Sea. (In Russian; English abstract).

Okeanologiia, Akad. Nauk, SSSR, 7(4):665-671.

benthos

Luss, V.J., and A.P. Kuznetsov, 1961

Data on the quantitative distribution of the bottom fauna in the Korpho Karaginsky region of the Bering Sea.

Trudy Inst. Okeanol., 46: 124-139.

M

benthos

*Mahadevan, S., and K. Nagappan Nayar, 1967(1968).

Underwater ecological observations in the Gulf of Mannar off Tuticorin. Vii. General topography and ecology of the rocky bottom.

J.mar.biol.Ass., India, 9(1):147-163.

benthos

*Makarov, Yu.N. and B.S. Averin, 1968.

On the quantitative distribution of zoobenthos in the shelf waters of the mozambique Strait. (In Russian; English abstract).

Okeanologiia, Akad.Nauk, SSSR, 8(6):1074-1077.

benthos

Makkaveeva, Elena Borisovna, 1968.

Rapports entre les composants végétaux et animaux dans les biocoenoses d'herbier des mers du bassin mediterranéen.

Rapp.P.-V.Réun.Comm.int.Explor.Sci.Méditerranée, 19(2):101-102.

benthos

Margalith Pinhas 1969

A contribution to the ecology of the marine microflora of the Gulf of Naples

Pubbl. Staz. Zool. Napoli, 37(2): 210-217.

benthos

Mars, P., 1963.

Les faunes et la stratigraphie du Quaternaire méditerranéen.

Rapp. Proc. Verb., Réunions, Comm. Int. Expl. Sci., Mer Méditerranée, Monaco, 17(3):1029-1044.

benthos

*Marshall, Nelson, 1967.

Some characteristics of the epibenthic environment of tidal shoals.

Chesapeake Sci., 8(3);155-169.

benthos

Marumo, Ryuzo, 1962

Research on deep-sea animals by the "Ryofu Maru". (In Japanese; English abstract).

Inf. Bull. Planktology, Japan, No. 8:14-22.

benthos

Marumo, Ryuzo, and Yutaka Kawarada, 1963

Research on deep-sea animals by the "Ryofu Maru" III. (In Japanese; English abstract).

Info. Bull. Plankt., Japan, No. 10:1-4.

benthos

Massé, Henri, 1971.

Etude quantitative de la macrofaune de peuplements des sables fins infralittoraux II. La Baie du Prado (Golfe de Marseille.

Téthys 3(1): 113-158.

benthos

Massé, Henri 1970 (1971).

Contribution à l'étude de la macrofaune de peuplements des sables fins infra-littoraux des côtes de Provence. I. La baie de Bandol.

Téthys 2(4): 783-820.

benthos

*Massé, Henri, 1968.

Sur la productivité des peuplements marins benthiques.

Cah.Biol.,mar.,9(4):363-372.

benthos

Masse, H., 1961.

Note préliminaire sur le peuplement des sables grossiers et fins graviers de l'étage infra-littoral aux environs de Marseille.

Rec. Trav. Sta., Mar. Endoume, 23(37):31-35.

Mc

benthos
McCauley, James E., 1964.
A preliminary report of the benthic animals collected on the USCGC NORTHWIND cruise during 1962.
U.S.C.G. Oceanogr. Rept., No. 1(CG373-1):17-22.

benthos
McErlean, A.J., 1964.
Characteristics of Macoma baltica populations in the middle Patuxent estuary.
Chesapeake Science, 5(4):200-208.

benthos
McIntyre, A.D., 1964
Meiobenthos of sub-littoral muds.
Jour. Mar. Biol. Assoc., U.K., 44(3):665-674.

benthos
McIntyre, A.D., 1956.
The use of trawl, grab and camera in estimating marine benthos. J.M.B.A., 35(2):419-430.

benthos
*McIntyre, A.D., and A. Eleftheriou, 1968.
The bottom fauna of a flatfish nursery ground.
J. mar. biol. Ass., U.K., 48(1):113-142.

benthos
McKnight, D. G., 1969.
An outline distribution of the New Zealand shelf fauna: benthos survey, station list, and distribution of Echinoidea.
Bull. N.Z. Dept. scient. industr. Res. 195: 1-91. Also: Mem. N.Z. oceanogr. Inst. 47.

benthos
*McKnight, D.G., 1968.
Features of the benthic ecology of chalky and preservation inlets.
N.Z. Jl. mar. Freshwat. Res., 2(4):716-720.

benthos
McLaughlin, P.A., 1963.
Survey of the benthic invertebrate fauna of the eastern Bering Sea.
U.S.F.W.S. Spec. Sci. Rept., Fish., No. 401:75 pp.

benthos, biomass of
McNulty, J.K., R.C. Work and H.B. Moore, 1962.
Some relationships between the infauna of the level bottom abd the sediments in south Florida.
Bull. Mar. Sci., Gulf and Caribbean, 12(3):322-332.

benthos, abyssal
Menzies, R.J., 1962.
On the food and feeding habits of abyssal organisms as exemplified by isopods.
Int. Revue Ges. Hydrobiol., 47(3):339-358.

benthos
Mileikovsky, S.A. 1970.
The temperature of water as a factor governing the reproduction and spawning of the bottom invertebrates in the shelf sea. (In Russian; English abstract)
Trudy Inst. Okeanol. Akad. Nauk SSSR 88: 113-149.

benthos
*Mileikovsky, S.A., 1967.
Shoals and mass aggregations of marine shallow-water free-moving bottom invertebrates and their biological significance. (In Russian; English abstract).
Okeanologiia, Akad. Nauk, SSSR, 7(4):655-664.

benthos, abyssal
Mileikovsky, S.A., 1961.
On the character and the nature of abyssal populations of eurybathic benthic invertebrate species with a pelagic development as noted in the Polychaeta Euphrosyne borealis Oersted 1843 found in the northern part of the Atlantic Ocean.
Okeanologiia, 1(4):679-687.

benthos
Milovidova, N. Yu., 1969.
Biomass and macrozoobenthic reserves in the bays of the northeastern Black Sea. (In Russian).
Gidrobiol. Zh., 5(1): 43-46.

benthos
Mokeivski, O.B., 1954.
Quantitative distribution of deep bottom fauna of Sea of Japan. (In Russian).
Trudy Inst. Okeanol., Akad. Nauk, SSSR, 8:147-163.

benthos
Mommaerts, J.P., 1969.
Données sur l'écologie de l'estuaire du Tamar (Plymouth).
Bull. K. Belg. Inst. Nat. Wet. 45(22): 1-24

benthos
Monniot, Claude, 1965.
Les "blocs à microcosmus" des fonds chalutables de la région de Banyuls-sur-Mer.
Vie et Milieu, Bull. Lab. Arago, (B)16(2-B):819-849.

benthos
Monniot, Claude, et Françoise Monniot 1968.
Les ascidies de grandes profondeurs récoltées par le navire océanographique américain Atlantis II (première note).
Bull. Inst. océanogr. Monaco 67 (1379): 45 pp.

benthos
Moore, H.B., 1931.
The muds of the Clyde Sea area. III. Chemical and physical conditions; rate and nature of sedimentation; and fauna.
J. Mar. Biol. Assoc., U.K., 17(2):325-358.

benthos
Moskalev, L.I. and F.A. Pasternak, 1970.
On quantitative distribution of bottom fauna in the south-eastern part of the Gulf of Mexico and waters off Cuba. (In Russian; English and Spanish abstracts). Okeanol. Issled Rezult. Issled. Mezhd. Geofiz. Proekt. 20: 128-143.

benthos
Mulicki, Zygmunt, and Ludwik Żmudziński, 1969.
Stock of zoobenthos in the southern Baltic (in the 1956-1957 period). (In Polish. English abstract).
Prace morsk Inst. Ryback (A) 15: 77-101

benthos
Müller, G.J., 1968.
Ergebnisse einer Tauchexpedition im Randgebiet des Phyllophora-Feldes im Schwarzen Meer.
Revue Roumaine Biol. (Zool.) 13(6): 425-431.

N

benthos
Nagoshi, Makoto, 1969.
On the benthos in the bays of Kumano Sea, with special reference to polychaetes.
Rept. Fac. Fish. Pref. Univ. Mie, 6(3): 97-107.

benthos
*Nakai, Zinziro, Masaya Kosaka, Fumio Hayashida, Tadashi Kubota and Masahiro Ogura, 1967.
Interrelationship between natural production of benthos and stability of the bottom. I. The condition in winter. (In Japanese; English abstract).
J. Coll. mar. Sci. Techn., Tokai Univ., (2):161-177.

benthos
Naumov, V.M., 1968.
Ecology of macrobenthos in Indian shelf waters of the Bay of Bengal. (In Russian).
Trudy, Vses. Nauchno-Issled. Inst. Morsk. Ribn. Okeanogr (VNIRO) 64, Trudy Azovo-Chernomorsk. Nauchno-Issled. Inst. Morsk. Ribn. Khoz. Okeanogr. (AscherNIRO), 28: 220-242.

benthos
Neyman, A.A., 1971.
Bottom population of the shelves of the Indian Ocean. (In Russian).
Vses. nauchno-issled. Inst. morsk. ribn. Khoz. Okeanogr. VNIRO, Trudy, 72: 56-64.

benthos
*Neyman, A.A., 1970.
Quantitative distribution of bottom fauna over the shelf in the Great Australian Bay and off New Zealand. Okeanologiia, 10(3): 517-520.
(In Russian; English abstract)

benthos
*Neyman, A.A., 1969.
Some data on the bottom fauna of the northern Indian Ocean shelves. (In Russian; English abstract). Okeanologiia, 9(6): 1071-1077.

benthos
Neyman, A.A., 1969.
On the distribution of trophic groupings of benthos on the shelf in different geographical zones. (In Russian).
Trudy Vses. nauchno-issled. Inst. morsk. rybn. Okean (VNIRO) 65: 282-295

benthos
Neyman, A.A., 1969.
Benthos of the west Kamchatka shelf (In Russian).
Trudy Vses. nauchno-issled. Inst. morsk. rybn. Khoz. Ocean (VNIRO) 65: 223-232

benthos
Neyman, A.A., 1968. (Neiman)
Characteristics of the bottom population on the shelf of western and southern Australia. (In Russian).
Trudy, Vses. Nauchno-Issled. Inst. Morsk. Ribn. Okeanogr (VNIRO) 64, Trudy Azovo-Chernomorsk. Nauchno-Issled. Inst. Morsk. Ribn. Khoz. Okeanogr. (AscherNIRO), 28: 204-209.

benthos
Neyman, A.A. 1967.
On limits of the application of the "trophic group" concept in benthos studies. (In Russian; English abstract).
Okeanologiia 7(2): 195-202.

benthos
Neyman, A.A., 1965.
Quantitative distribution of the benthos in the West Kamchatka Shelf and some problems involved with the methods of exploration. (In Russian).
Okeanologiia, Akad. Nauk, SSSR, 5(6):1052-1059.

benthos
Neyman, A.A., 1965.
Some regularities of the quantitative distribution of benthos on North Pacific shelves (In Russian).
Trudy vses. nauchno-issled. Inst. morsk. ryb. Khoz. Okeanogr. (VNIRO), 57:447-451.

benthos
Neyman, A.A., 1964.
Age of clams and the utilization of benthos by flatfishes in the southeastern Bering Sea. (In Russian).
Trudy, Vses. Nauchno-Issled. Inst. Morsk. Ribn. Choz. i Okeanogr. (VNIRO), 52:199-204.
Also: Izv. Tichookeansk. Nauchno-Issled. Inst. Morsk. Ribn. Choz. i Okeanogr. (TINRO), 53:199-204.

benthos

Neiman, A.A., 1961
[Some general laws concerned with the quantitative distribution of benthos in the Bering Sea.]
Okeanologiia, Akad. Nauk, SSSR, (2):294-304.

benthos

Neiman, A.A., 1961.
[The vertical distribution of zoogeographical complexes of the bottom fauna in the shelf and the upper horizons of the slope in the eastern part of the Bering Sea.]
Okeanologiia, Akad. Nauk, SSSR, 1(6):1073-1078.

benthos

Nesis, K.N., 1965.
Biocoenoses and biomass of benthos in the Newfoundland-Labrador region. Investigations in line with the programme of the International Geophysical Year, 2. (In Russian).
Trudy Vses. Nauchno-Issled. Inst. Morsk. Ribn. Choz. i Okeanogr. (VNIFO), 57:453-489.

benthos

Nesis, K.N., 1962.
Pacific elements in northwestern Atlantic benthos.
Sovetskie Riboch. Issledov., v Severo-Zapadnoi Atlant. Okeana, VNIRO-PINRO, Moskva, 83-98.

In Russian; English summary

benthos

Nesis, K.N., 1962.
Soviet investigations of benthos in Newfoundland-Labrador fishing area.
Sovetskie Riboch. Issledov. v Severo-Zapadnoi Atlant. Okeana, VNIRO-PINRO, Moskva, 219-225.

In Russian; English summary

benthos

Neumann, A. Conrad, 1965.
Processes of recent carbonate sedimentation in Harrington Sound, Bermuda.
Bull. Mar. Sci., 15(4): 987-1035.

benthos

Neumann, A. Conrad, 1963.
Processes of recent carbonate s edimentation in Harrington Sound, Bermuda.
Mar. Sci. Center, Lehigh Univ., 130 pp. (Unpublished manuscript).

benthos

Nikitin, V.N., 1964.
Quantitative distribution of bottom macrofauna in the Black Sea. Investigations of bottom fauna and flora of Far-Eastern Seas and the Pacific Ocean. (In Russian).
Trudy, Akad. Nauk, SSSR, 69:285-329.

benthos

Nikitin, V.N., 1962.
[Quantitative distribution of bottom macrofauna in the Black Sea near the shores of the Caucasus.]
Doklady Akad. Nauk, SSSR, 143(4):968-971.

benthos

Nishimura, Saburo, 1968.
The zoogeographical aspects of the Japan Sea. IV.
Publs Seto mar. biol. Lab., 15(5):329-352.

benthos

Noskova, E.D., 1965.

Investigations in Kursk and Bislinsk bays.
Atlantich. Nauchno-Issled. Inst. Ribn. Khoz. i Okeanogr. (ATLANTNIRO), Kaliningrad, 14:126 pp.

O

benthos

Okada, Yaichiro, Ichitaro Sakamoto, Ryohei Amano and Yoshiaki Tominaga 1966.
Preliminary report of the benthic biological survey in Suruga Bay.
J. Fac. Oceanogr. Tokai Univ. (1): 135-166.

benthos

Orel, Giuliano, e Barbara Mennea 1969.
I popolamenti bentonici di alcuni tipi di fondo mobile del Golfo di Trieste.
Pubbl. Staz. Zool. Napoli 37 (2 Suppl.): 261-276.

benthos

Oshima, Kazuo, 1963.
Ecological study of Usu Bay, Hokkaido Bay, Japan, 1. Bottom materials and benthic fauna.
(In Japanese; English abstract).
Bull. Hokkaido Reg. Fish. Res. Lab., Fish. Ag., No. 27:32-51.

P

benthos

Paranzan, P., 1962.
Prime notizie sulle biocenosi bentoniche del Mar Grande di Taranto.
Pubbl. Staz. Zool., Napoli, 32 (Suppl.):123-132.

benthos

Parenzan, Pietro, 1960
Il Mar Piccolo di Taranto.
Giovanni Semerano, Editore, Roma, 254 pp.

benthos

Parenzan, Pietro, 1960.
Su un tipo non ancora descritto del Mediterraneo: il "fondo a Cidaridi" di "Bocca Piccola" nel mare di Capri.
Thalassia Jonica, Ist. Speriment. Talassogr., 3:83-99.

benthos

*Parker, Robert H., 1969.
Megafaunal facies, estuary to shelf edge, surrounding the Gulf of Mexico. Trans. Gulf Coast Ass. geol. Socs, 19 : 503.

benthos a

Pasternak, F.A., 1970.
Marine biology of the East Atlantic continental region. In: The geology of the East Atlantic continental margin, 1. General and economic papers, ICSU/SCOR Working Party 31 Symposium, Cambridge 1970, Rept. No. 70/13:67-77

benthos

*Pasternak, F.A., 1968.
Botom fauna studies at the greatest depths in the Romanche Trench made from the Research Vessel Akademik Kurchatov. (In Russian; English abstract).
Okeanologiia, Akad. Nauk, SSSR, 8(2):312-316.

benthos

Pasternak, F.A., and A.V. Gusev, 1960

[Benthonic research.]
Arktich. i Antarkt. Nauchno-Issled. Inst., Sovetsk. Antarkt. Exped., Mezhd. Geofiz. God, Vtoraia Morsk. Exped., "Ob", 1956-1957, 7: 126-142.

benthos

*Pamatmat, Mario.M., 1968.
Ecology and metabolisim of a benthic community on an intertidal flat.
Int. Revue ges. Hydrobiol., 53(2):211-298.

benthos

Paul, L. J., 1966.
Observations on past and present distribution of mollusc beds in Ohiwa Harbour, Bay of Plenty.

New Zealand J. Sci., 9(1):30-40.

benthos a

*Pearson, T.H., 1971.
The benthic ecology of Loch Linnhe and Loch Eil, a sea-loch system on the west coast of Scotland. III. The effect on the benthic fauna of the introduction of pulp mill effluent. J. exp. mar. Biol. Ecol., 6(3): 211-233.

benthos

Pearson T.H. 1970.
The benthic ecology of Loch Linnhe and Loch Eil, a sea-loch system on the west coast of Scotland. 1. The physical environment and distribution of the macrobenthic fauna.
J. exp. mar. Biol. Ecol. 5(1): 1-34.

benthos

Peer, D.L., 1963
A preliminary study of the composition of benthic communities in the Gulf of St. Lawrence.
Fish. Res. Bd., Canada, Mss Rept. Ser. (Ocean and Limnol.), No. 145:24 pp. (multilithed).

benthos

*Pérès, Jean-Marie, 1968.
Un précurseur de l'etude du benthos de la Mediterranée: Louis Ferdinand, comte de Marsilli.
Bull. Inst. océanogr., Monaco, No. special 2: 369-376.

benthos

Pérès, J.M., 1967.
The Mediterranean benthos.
Oceanogr. Mar. Biol., Ann. Rev., H. Barnes, editor, George Allen and Unwin, Ltd., 5:449-533.

benthos

Pérès, J.M., 1965.
Aperçu sur les résultats de deux plongées effectuées dans le ravin de Puerto-Rico par le bathyscaphe Archimede.
Deep-Sea Res., 12(6):883-891.

benthos

Pérès, J.M., 1961.
Rapport sur les travaux récents concernant le benthos de la Mediterranee et de ses dependances.
Rapp. Proc. Verb., Réunions, Comm. Int. Expl. Sci. Mer Méditerranée, Monaco, 16(2):377-424.

benthos

Pérès, J.M. et J. Picard 1963
Note préliminaire générale sur le benthos littoral de la région de Tuléar (Madagascar).
Annls malgaches, Univ. Madagascar, 1: 145-151.

benthos

Pérès, J.M., and J. Picard, 1962.
Note préliminaire générale sur le benthos littoral de la region de Tuléar (Madagascar).
Ann. Fac. Sci. Techn., Madagascar, 145-151.

Also in:
Rec. Trav. Sta. Mar., Endoume, Marseille, Fasc., Hors Ser., Suppl. (1).
Trav. Sta. Mar. Tuléar (République Malgache)*
Int. Indian Ocean Exped., 145-151.

benthos

Pfitzenmeyer, H.T., and K.G. Drobeck, 1963.
Benthic survey for populations of soft-shelled clams, Mya arenaria, in the lower Potomac River, Maryland.
Chesapeake Science, 4(2):67-74.

benthos

*Picard, J., 1967.
Essai de classement des grands types de peuplements marins benthiques tropicaux, d'après les observations effectuées dans les parages de Tuléar (S.-W. de Madagascar).
Recl. Trav.Stn.mar.Endoume, hors série, Suppl.6: 3-24.

benthos
Picard, J., 1965.
Propositions pour une subdivision des benthontes en fonction de la taille.
Méthodes Quantitatives d'Étude du Benthos et Echelle Dimensionnelle des Benthontes: Colloque du Comité du Benthos (Marseille, Nov. 1963), Comm. Int. Expl. Sci. Mer. Médit., Monaco, 7-13.

benthos
Picard, J., 1961.
Distribution et particularités des peuplements benthiques des côtes de Provence entre le Bec de l'Aigle et l'Ile des Embiez.
Rapp. Proc. Verb., Réunions, Comm. Int. Expl. Sci. Mer Méditerranée, Monaco, 16(2):425-428.

benthos
Pichon, Mireille, 1967.
Contribution à l'étude des peuplements de la zone intertidale sur sables fins et sables vaseux non fixés dans la région de Tuléar.
Annales Univ. Madagascar 5: 171-214

BENTHOS
*Pichon, Michel M., 1967.
Caractères généraux des peuplements benthiques des récifs et lagons de l'Ile Maurice (Ocean Indien).
Cah. O.R.S.T.O.M., Ser. Océanogr., 5(4):31-45.

benthos
Plante, Raphaël 1963.
Note préliminaire sur la répartition et les peuplements des substrats solides dans la région de Tuléar.
Annls malgaches, Univ. Madagascar 1: 181-200.

benthos
Poizat, Claude, 1969.
Note sur la répartition des sédiments au débouché des calanques du littoral des massifs de Marseilleveyre Puget et Devenson (Marseille à Cassis)
Recl Trav. Stn mar. Endoume, 46 (62): 193-212

benthos
Popham, J.D. and D.V. Ellis 1971.
A comparison of traditional cluster and Zürich-Montpellier analyses of infaunal pelecypod associations from two adjacent sediment beds.
Mar. Biol. 8(3): 260-266.

benthos
Por, F.D., and Ruth Lerner-Seggev, 1966.
Preliminary data about the benthic fauna of the Gulf of Elat (Aqaba), Red Sea.
Israel J. Zool., 15(2):38-50.

benthos
Propp, M.V., 1962.
Underwater observations of sublittoral of the Barents Sea. (In Russian).
Trudy Okeanogr. Komissii, 14:73-75.

benthos
Pushkin, A.F. 1968.
The benthos in Chisha Bay. (In Russian)
Trudy Murmansk. Morsk. Biol. Inst. 17 (21): 48-57.

Q

R

benthos
Rankin, John S., Jr., Kerry B. Clark and Charles S. Biernbaum, 1969.
Weddell Sea benthic studies.
Antarctic J., U.S.A., 4(4):97.

benthos
Rankin, John S. Jr., Kerry B. Clark and Bruce W. Found 1968.
Zonation of the Weddell Sea benthos.
Antarct. J. U.S.A. 3(4): 85-86.

benthos
Renaud-Debyser, J., 1963.
Recherches écologiques sur la faune interstitielle des sables du Bassin d'Arcachon.
P.V. Soc. Linneene de Bordeaux, 99:1-8.

Also in:
Bull. Sta. Biol., Arcachon, No. 16(1964).

benthos
Reys, J.P., 1965.
Remarques sur les prélèvements quantitatifs du benthos de substrats meubles.
Méthodes Quantitatives d'Étude du Benthos et Echelle Dimensionelle des Benthontes: Colloque du Comité du Benthos (Marseille, Nov. 1963), Comm. Int. Expl. Sci. Mer. Médit., Monaco, 15-17.

benthos
Reyss, Daniel, 1964.
Contribution à l'étude du Rech Lacaze-Duthiers, vallée sous-marine des côtes du Roussillon.
Vie et Milieu, Bull. Lab. Arago, 15(1):1-46.

benthos
Riedl, Rupert, und Helmut Forstner, 1968.
Wasserbewegung im Mikrobereich von benthos.
Sarsia, 34:163-188.

benthos
Rieman, Franz, 1966.
Die interstitielle Fauna im Elbe-Aestuar: Verbreitung und Systematik.
Arcl. Hydrobiol. 3(1/2) (Suppl. 31):1-279.

benthos
Robins, M.W., and M.H. Thurston, 1969.
The ecology of Swanage Bay.
Rept. Underwat. Ass. 4: 67-74.

benthos
Rodríguez, Gilberto, 1967.
Las comunidades bentónicas.
In: Ecología marina, Monogr. Fundación La Salle de Ciencias Naturales, Caracas, 14:563-600.

S

benthos
Salvat, Bernard, 1971.
Evaluation quantitative totale de la faune benthique de la bordure lagunaire d'un atoll de Polynésie française.
C.r. hebd. Séanc., Acad. Sci. Paris (D) 272 (2): 211-214

benthos
Salvat, B. et J. Renaud-Mornant, 1969.
Étude écologique du macrobenthos et du meiobenthos d'un fond sableux du lagon de Mururoa (Tuamotu-Polynésie). Cah. Pacifique 13: 159-179.

benthos
Sanders, Howard L., 1969.
Benthic marine diversity and stability-time hypothesis.
Diversity and stability in ecological systems, Brookhaven Symposia in Biology, 22: 71-80

benthos
Sanders, Howard L., 1965.
Time, latitude, and structure of marine benthic communities.
Anais Acad. bras. Cienc., 37(Supl.):83-86.

benthos
Sanders, H.L. and R.R. Hessler, 1969.
Ecology of the deep-sea benthos. Science, 163(3874): 1419-1424.

benthos
Sanders, Howard L., and Robert R. Hessler, 1962.
Priapulus atlantisi and Priapulus profundus.
Two new species of priapulids from bathyal and abyssal depths of the North Atlantic.
Deep-Sea Res., 9(2):125-130.

benthos
Sanders, H.L., R.R. Hessler and G.R. Hampton, 1965.
An introduction to the study of deep-sea benthic faunal assemblages along the Gay Head-Bermuda transect.
Deep-Sea Res., 12(6):845-867.

benthos
Savilov, A.I., 1961
[Ecologic characteristics of the bottom communities of invertebrates in the Okhotsk Sea.]
Trudy Inst. Okeanol., SSSR, 46: 3-84.

benthos
Scaccini, Andrea, 1967.
Dati preliminari sulle zoocenosi bentoniche e sulla biomassa in una zona dell'alto e medio Adriatico.
Note Lab. Biol. mar. Pesca-Fano, Univ. Bologna, 2(3):25-56.
Reprinted: Comm. Studio Oceanograf. Limnol., Programma Ricerca Risorse mar. Fondo mar., (B) (3). 1958.

benthos
Schopf, Thomas J.M., and John B. Colton, Jr., 1966.
Bottom temperature and faunal provinces: continental shelf from Hudson Canyon to Nova Scotia. (Abstract only).
Biol. Bull., 131(2):406.

benthos
Schulz, Sigurd, 1969.
Benthos und Sediment in der Mecklenburger Bucht.
Beitr. Meereskunde 24-25: 15-55

benthos
Schurin, A.T., 1967.
The effect of manganese on the distribution of bottom invertebrates of the Baltic.
Annls.biol.Copenh. (1965)22:73.

benthos
Schwenke, Heinz, 1970.
Untersuchungen über den Aufbau des Phytals auf mobilen Substraten am Beispiel der westlichen Ostsee.
Thalassia Jugoslavica, 6: 169-180

benthos
Segerstråle, Sven G., 1965.
On the salinity conditions off the south coast of Finland since 1950, with comments on some remarkable hydrographical and biological phenomena in the Baltic area during this period.
Commentationes Biolog., Soc. Sci. Fennica, 28(7):28 pp.

benthos, fluctuations of
Segerstråle, Sven G., 1960
Fluctuations in the abundance of benthic animals in the Baltic area.
Soc. Sci. Fennica, Comm. Biol., 23(9): 19 pp.

benthos
Semenov, V.N., 1965.
The quantitative distribution of the bottom fauna on the shelf and the upper continental slope in the Gulf of Alaska. (In Russian).
SOVETSK. RIBOKHOZ. ISSLED. Severo-Vostochn. Chasti Tikhogo Okeana, 4 (Vses. Nauchno-Issled. Inst., VNIRO, Trudy, 58: Tikhookean. Nauchno-Issled. Inst., TINRO, Trudy, 53): 49-77.

benthos
Semenov, V.N., 1964.
Quantitative distribution of benthos on the shelf of the southeastern Bering Sea (the Bristol Bay, the coast of Alaska and Unimak Island). (In Russian).
Trudy, Vses. Nauchno-Issled. Inst. Morsk. Ribn. Choz. i Okeanogr. (VINRO), 52:177-184.

Also:
Izv. Tichookeansk. Nauchno-Issled. Inst. Morsk. Ribn. Choz. i Okeanogr. (TINRO), 53:177-184.

benthos
Shepherd, S.A., and H.B.S. Womersley, 1970.
The sublittoral ecology of West Island, South Australia. 1. Environmental features and the algal ecology.
Trans. R. Soc. S. Aust. 94: 105-137

benthos
Shevtsov, V.V., 1964.
Quantitative distribution and trophic groups of benthos in the Gulf of Alaska. (In Russian).
Trudy, Vses. Nauchno-Issled. Inst. Morsk. Ribn. Choz. i Okeanogr. (VINRO), 52:161-176.

Also:
Izv. Tichookeansk. Nauchno-Issled. Inst. Morsk. Ribn. Choz. i Okeanogr. (TINRO), 53:161-176.

benthos
Shevtsov, V.V., 1964
Some data on the quantitative distribution of the bottom fauna in the Gulf of Alaska. (In Russian).
Gosudarst. Kom. Sov. Ministr., SSSR, Ribn. Choz., Trudy VNIRO, 49; Izv. TINRO, 51:107-112.

benthos
*Skalkin, V.A., 1970.
Characteristics of some groupings of benthos in Aniva Bay. (Sea of Okhotsk). (In Russian; English abstract).
Zool.Zh.49(9):1405-1407.

benthos
Smith, K.L., Jr., K.A. Burns and J.M. Teal, 1972.
In situ respiration of benthic communities in Castle Harbor, Bermuda. Mar. Biol. 12(3): 196-199.

benthos
Sokolova, M.N., 1968.
On the relation between the trophic groups of deep-sea macrobenthos and the composition of bottom sediments. (In Russian; English abstract
Okeanologiia, Akad. Nauk, SSSR, 8(2):179-191.

benthos, abyssal
Sokolova, M.N., 1964.
On some regularities involved with the distribution of food groupings in the abyssal benthos. (In Russian).
Okeanologiia, Akad. Nauk, SSSR, 4(6):1079-1088.

benthos, deep
Sokolova, M.N., 1959.
On the distribution of deep-water bottom animals in relation to their feeding habits and the character of sedimentation.
Deep-Sea Research, 6(1):1-4.

benthos
Sokolova, M.N., 1956.
On the distribution regularities of deep sea benthos. The influence of the macrorelief and distribution of suspension upon the adaphic groups of bottom invertebrates.
Dokl. Akad. Nauk, SSSR, 110(4):692-695.

benthos (deep), nutrition
Sokolova, M.N., 1956.
The nutrition of deep water benthos. Nutrition of Laetmatonice producta V. Wyvillei McIntosh.
Doklady Akad. Nauk, SSSR, 110(6):1111-1114.

benthos (biomass)
Sokolova, M.B., and F.A. Pasternak, 1962.
Quantitative distribution of bottom fauna in the northern part of the Arabian Sea and the Bay of Bengal. (In Russian).
Doklady, Akad. Nauk, SSSR, 144(3):645-648.

Translation:
Soviet Oceanography, Issue 3:15-18. (1964).
(Scripta Tecnica, Inc., for AGU).

benthos
Southward, A.J., 1963
The distribution of some plankton animals in the English Channel and approaches. III. Theories about long-term biological changes, including fish.
J. Mar. Biol. Assoc., U.K., 43(1):1-29.

benthos
Squires, Donald F., David L. Pawson and Larry W. yeater, 1966.
Benthic invertebrate collections.
Antarctic J., U.S., 1(4):128

benthos
Stanley, Daniel J., and Gilbert Kelling, 1968.
Photographic investigation of sediment texture, bottom current activity, and benthonic organisms in the Wilmington Submarine Canyon.
U.S.C.G. Oceanogr. Rept. 22 (CG 373-22): 95 pp.

benthos
*Stephens, K., R.W. Sheldon and T.R. Parsons, 1967.
Seasonal variations in the availability of food for benthos in a coastal environment.
Ecology, 48(5):852-855.

benthos
Stripp, Konrad, 1969.
Das Verhältnis von Makrofauna und Meiofauna in den Sedimenten der Helgoländer Bucht.
Veröff. Inst. Meeresforsch. Bremerh. 12(2): 143-148

benthos
Stripp, Konrad, 1969.
Die Assoziationen des Benthos in der Helgoländer Bucht.
Veröff. Inst. Meeresforsch. Bremerh. 12(2):95-141

benthos
Stripp, Konrad, 1969.
Jahreszeitliche Fluktuationen von Makrofauna und Meiofauna in der Helgoländer Bucht.
Veröff. Inst. Meeresforsch. Bremerh. 12(2):65-94

benthos
Stripp, Konrad, und Sebastian A. Gerlach, 1969.
Die Bodenfauna im Verklappungsgebiet von Industrieabwässern nordwestlich von Helgoland.
Veröff. Inst. Meeresforsch. Bremerh. 12(2): 149-156.

benthos
Suyehiro, Tasuo, Yaichiro Okada, Masuoki Horokoshi and Eiji Iwai, 1962.
VI. A brief note on the benthic animals on the fourth cruise of the Japanese Expedition of Deep Seas (JEDS-4).
Oceanogr. Mag., Japan Meteorol. Agency, 13(2): 149-152.

JEDS Contrib. No. 34.

T

benthos
Thiel, Hjalmar, 1970.
Bericht über die Benthosuntersuchungen während der "Atlantischen Kuppenfahrten 1967" von F.S. Meteor.
Meteor Forsch.-Ergebnisse (D) 7: 23-42.

benthos
Thorson, Gunnar, 1966.
Some factors influencing the recruitment and establishment of marine benthic communities.
Neth. J. Sea Res., 3(2):267-293.

benthos
Tommasi, Luiz Roberto, 1970.
Observações sôbre a fauna bêntica do complexo estuarino-lagunar de Cananéia. Bolm Inst. oceanogr. São Paulo, 19: 43-56.

benthos
*Toriyama, Masahiro and Seiji Kudo, 1969.
On the small-trawl fishery in the Nobeoka-Wan-I. The Environmental factors of fishing ground (In Japanese; English abstract). Bull. Nansei reg. Fish. Res. Lab., 2: 105-130.

benthos
Tortonese, E., 1962.
Recenti ricerche del bentos in ambienti litorali del Mare Ligure.
Pubbl. Staz. Zool., Napoli, 32 (Suppl.):99-116.

benthos
True, Merrill A., 1970.
Étude quantitative de quatre peuplements sciaphiles sur substrat rocheux dans la région marseillaise.
Bull. Inst. océanogr. Monaco 69 (1401): 48 pp.

benthos
True, M.A., 1965.
Dispositif pour récolte total du peuplement sur substrat dur
Méthodes Quantitatives d'Étude du Benthos et Echelle Dimensionnelle des Benthontes: Colloque du Comité du Benthos (Marseille, Nov. 1963), Comm. Int. Expl. Sci. Mer. Médit., Monaco, 25-27.

benthos
Tsalkina, A.V., 1969.
On the characteristics of the epifauna of the west Kamchatka shelf. (In Russian).
Trudy vses. nauchno-issled. Inst. morsk. rybn. khoz. Okean (VNIRO) 65: 248-257.

benthos
Tulkki, Paavo, 1965.
Disappearence of the benthic fauna from the basin of Bornholm (southern Baltic) due to oxygen deficiency.
Cahiers Biol. mer, 6(4):455-463.

benthos
Turner, Charles H., Earl E. Ebert and Robert R. Given, 1966.
The marine environment in the vicinity of the Orange County Sanitation District's ocean outfall.
California Fish and Game, 52(1):28-41.

U

benthos
Uschakov, P.V., 1963.
Quelques particularités de la bionomique benthique de l'Antarctique de l'Est.
Cahiers de Biol. Mar., Roscoff, 4(1):81-89.

benthos
Ushakov, P.V., 1962.
Some characteristics of the distribution of bottom fauna off the coast of east Antarctica. (In Russian).
Inform. Biull., Sovetsk. Antarkt. Exped., 40:8-13.

Translation
Scripta Tecnica for AGU, 4(5):287-292. 1964.

benthos
Ushakov, P.V., 1962.
Bottom fauna of Far Eastern Seas and its composition and the regularity of its distribution.
(In Russian).
Sbornik Doklad. na II Plenume, Komissii Ribochoz. Issled. Zapad. Chasti Tichogo Okeana, Moskva, 73-82.

benthos
Ushakov, P.V., 1958
Investigations of the bottom fauna of the Far Eastern Seas of the USSR.
Proc. Ninth Pacific Sci. Congr., Pacific Sci. Assoc., 1957, 16(Oceanogr.):210-216.

benthos
Neiman, A.A., 1961
Some general laws concerned with the quantitative distribution of benthos in the Bering Sea.
Okeanologiia, Akad. Nauk, SSSR, (2):294-304.

benthos
Neiman, A.A., 1961.
The vertical distribution of zoogeographical complexes of the bottom fauna in the shelf and the upper horizons of the slope in the eastern part of the Bering Sea.
Okeanologiia, Akad. Nauk, SSSR, 1(6):1073-1078.

benthos
Nesis, K.N., 1965.
Biocoenoses and biomass of benthos in the Newfoundland-Labrador region. Investigations in line with the programme of the International Geophysical Year, 2. (In Russian).
Trudy Vses. Nauchno-Issled. Inst. Morsk. Ribn. Choz. i Okeanogr. (VNIFO), 57:453-489.

benthos
Nesis, K.N., 1962.
Pacific elements in northwestern Atlantic benthos.
Sovetskie Riboch. Issledov., v Severo-Zapadnoi Atlant. Okeana, VNIRO-PINRO, Moskva, 83-98.

In Russian; English summary

benthos
Nesis, K.N., 1962.
Soviet investigations of benthos in Newfoundland-Labrador fishing area.
Sovetskie Riboch. Issledov. v Severo-Zapadnoi Atlant. Okeana, VNIRO-PINRO, Moskva, 219-225.

In Russian; English summary

benthos
Neumann, A. Conrad, 1965.
Processes of recent carbonate sedimentation in Harrington Sound, Bermuda.
Bull. Mar. Sci., 15(4): 987-1035.

benthos
Neumann, A. Conrad, 1963.
Processes of recent carbonate sedimentation in Harrington Sound, Bermuda.
Mar. Sci. Center, Lehigh Univ., 130 pp. (Unpublished manuscript).

benthos
Nikitin, V.N., 1964.
Quantitative distribution of bottom macrofauna in the Black Sea. Investigations of bottom fauna and flora of Far-Eastern Seas and the Pacific Ocean. (In Russian).
Trudy, Akad. Nauk, SSSR, 69:285-329.

benthos
Nikitin, V.N., 1962.
Quantitative distribution of bottom macrofauna in the Black Sea near the shores of the Caucasus.
Doklady Akad. Nauk, SSSR, 143(4):968-971.

benthos
Nishimura, Saburo, 1968.
The zoogeographical aspects of the Japan Sea. IV.
Publs Seto mar. biol. Lab., 15(5):329-352.

benthos
Noskova, E.D., 1965.
Investigations in Kursk and Bislinsk bays.
Atlantich. Nauchno-Issled. Inst. Ribn. Khoz. i Okeanogr. (ATLANTNIRO), Kaliningrad, 14:126 pp.

O

benthos
Okada, Yaichiro, Ichitaro Sakamoto, Ryohei Amano and Yoshiaki Tominaga 1966.
Preliminary report of the benthic biological survey in Suruga Bay.
J. Fac. Oceanogr. Tokai Univ. (1): 135-166.

benthos
Orel, Giuliano, e Barbara Mennea 1969.
I popolamenti bentonici di alcuni tipi di fondo mobile del Golfo di Trieste.
Pubbl. Staz. Zool. Napoli 37 (2 Suppl.): 261-276.

benthos
Oshima, Kazuo, 1963.
Ecological study of Usu Bay, Hokkaido Bay, Japan, 1. Bottom materials and benthic fauna.
(In Japanese; English abstract).
Bull. Hokkaido Reg. Fish. Res. Lab., Fish. Ag., No. 27:32-51.

P

benthos
Paranzan, P., 1962.
Prime notizie sulle biocenosi bentoniche del Mar Grande di Taranto.
Pubbl. Staz. Zool., Napoli, 32 (Suppl.):123-132.

benthos
Parenzan, Pietro, 1960
Il Mar Piccolo di Taranto.
Giovanni Semerano, Editore, Roma, 254 pp.

benthos
Parenzan, Pietro, 1960.
Su un tipo non ancora descritto del Mediterraneo: il "fondo a Cidaridi" di "Bocca Piccola" nel mare di Capri.
Thalassia Jonica, Ist. Sperimont. Talassogr., 3:83-99.

benthos
*Parker, Robert H., 1969.
Megafaunal facies, estuary to shelf edge, surrounding the Gulf of Mexico. Trans. Gulf Coast Ass. geol. Socs, 19 : 503.

benthos a
Pasternak, F.A., 1970.
Marine biology of the East Atlantic continental region. In: The geology of the East Atlantic continental margin, 1. General and economic papers, ICSU/SCOR Working Party 31 Symposium, Cambridge 1970, Rept. No. 70/13:67-77

benthos
*Pasternak, F.A., 1968.
Botom fauna studies at the greatest depths in the Romanche Trench made from the Research Vessel Akademik Kurchatov. (In Russian; English abstract).
Okeanologiia, Akad. Nauk, SSSR, 8(2):312-316.

benthos
Pasternak, F.A., and A.V. Gusev, 1960
Benthonic research.
Arktich. i Antarkt. Nauchno-Issled. Inst., Sovetsk. Antarkt. Exped., Mezhd. Geofiz. God, Vtoraia Morsk. Exped., "Ob", 1956-1957, 7: 126-142.

benthos
*Pamatmat, Mario.M., 1968.
Ecology and metabolisim of a benthic community on an intertidal flat.
Int. Revue ges. Hydrobiol., 53(2):211-298.

benthos
Paul, L. J., 1966.
Observations on past and present distribution of mollusc beds in Ohiwa Harbour, Bay of Plenty.
New Zealand J. Sci., 9(1):30-40.

benthos a
*Pearson, T.H., 1971.
The benthic ecology of Loch Linnhe and Loch Eil, a sea-loch system on the west coast of Scotland. III. The effect on the benthic fauna of the introduction of pulp mill effluent. J. exp. mar. Biol. Ecol., 6(3): 211-233.

benthos
Pearson T.H. 1970.
The benthic ecology of Loch Linnhe and Loch Eil, a sea-loch system on the west coast of Scotland. 1. The physical environment and distribution of the macrobenthic fauna.
J. exp. mar. Biol. Ecol. 5 (1): 1-34

benthos
Peer, D.L., 1963
A preliminary study of the composition of benthic communities in the Gulf of St. Lawrence.
Fish. Res. Bd., Canada, Mss Rept. Ser. (Ocean and Limnol.), No. 145:24 pp. (multilithed).

benthos
*Pérès, Jean-Marie, 1968.
Un précurseur de l'etude du benthos de la Mediterranée: Louis Ferdinand, comte de Marsilli.
Bull. Inst. océanogr., Monaco, No. special 2: 369-376.

benthos
Pérès, J.M., 1967.
The Mediterranean benthos.
Oceanogr. Mar. Biol., Ann. Rev., H. Barnes, editor, George Allen and Unwin, Ltd., 5:449-533.

benthos
Pérès, J.M., 1965.
Aperçu sur les résultats de deux plongées effectuées dans le ravin de Puerto-Rico par le bathyscaphe Archimede.
Deep-Sea Res., 12(6):883-891.

benthos
Pérès, J.M., 1961.
Rapport sur les travaux récents concernant le benthos de la Mediterranee et de ses dépendances.
Rapp. Proc. Verb., Réunions, Comm. Int. Expl. Sci. Mer Méditerranée, Monaco, 16(2):377-424.

benthos
Pérès, J.M. et J. Picard 1963
Note préliminaire générale sur le benthos littoral de la région de Tuléar (Madagascar).
Annls malgaches, Univ. Madagascar, 1: 145-151.

benthos
Pérès, J.M., and J. Picard, 1962.
Note préliminaire générale sur le benthos littoral de la region de Tuléar (Madagascar).
Ann. Fac. Sci. Techn., Madagascar, 145-151.

Also in:
Rec. Trav. Sta. Mar., Endoume, Marseille, Fasc., Hors Ser., Suppl. (1).
Trav. Sta. Mar. Tuléar (République Malgache)
Int. Indian Ocean Exped., 145-151.

benthos
Pfitzenmeyer, H.T., and K.G. Drobeck, 1963.
Benthic survey for populations of soft-shelled clams, Mya arenaria, in the lower Potomac River, Maryland.
Chesapeake Science, 4(2):67-74.

benthos
*Picard, J., 1967.
Essai de classement des grands types de peuplements marins benthiques tropicaux, d'après les observations effectuées dans les parages de Tuléar (S.-W. de Madagascar).
Recl. Trav. Stn. mar. Endoume, hors série, Suppl. 6: 3-24.

V

benthos
Vaissiere, Raymond, et Gaston Fredj, 1964.
Contributions à l'étude bionomique de la Méditerranée occidentale (côte du Var et des Alpes maritimes - côte occidental de Corse). 5. Étude photographique préliminaire de l'étage bathyal dans la région de Saint-Tropez (ensemble A).
Bull. Inst. Océanogr., Monaco, 64(1323):70 pp.

benthos
*Vatova, Aristocle, 1967.
La fauna bentonica della costa occidentale dell' alto Adria.
Arch. Oceanogr. Limnol., 15(Suppl.):159-167.

benthos
Vervoort, W., 1971.
Zoological exploration of the continental shelf of Surinam. II. Hydrogr. Newsletter, R. Netherlands Navy, Spec. Publ. 6: 37-50.

benthos
Vidal, Annie 1967.
Étude des fonds rocheux circa-littoraux le long de la côte du Roussillon.
Vie Milieu (B) 18 (1B): 167-219

benthos
*Vinogradov, K.A., editor, 1967.
Biology of the northwestern Black Sea. (In Russian).
Naukova Dumka, Kiev, 268 pp.

benthos
Vinogradov, K.A., and V.P. Zakutsky, 1966.
Bottom biocoenoses in the western half of the Black Sea. (In Russian: English abstract).
Okeanologiia, Akad. Nauk, SSSR, 6(2):340-343

benthos
Vinogradov, L.G., 1963
Marine bottom biocoenoses and the application of the data on the distribution in fish-scouting operations. (In Russian).
Sovetsk. Ribokh. Issled. B Severo-Vostokh. Chasti Tikhogo Okeana, VNIRO 48, TINRO 50(1): 135-144.

benthos
*Vinogradov, L.G. and A.A. Neyman, 1969.
Bottom fauna on the shelf of the eastern sea of Okhotsk and some biological features of Kamchatka king crab. (In Russian; English abstract). Okeanologiia, 9(2): 329-340.

benthos
Vinogradov, L.G., and A.A. Neyman, 1965.
On the distribution of zoogeographical complexes of bottom invertebrates in the southern Bering Sea. (In Russian).
Sovetsk. Ribokhoz. Issled. Severo-Vostochn. Chasti Tokhogo Okeana, 4 (Vses. Nauchno-Issled. Inst., VNIRO, Trudy, 58: Tikhookean. Neuchno-Isaled. Inst., TINRO, Trudy, 53):45-48.

benthos
Vinogradov, L.G., and A.A. Neyman, 1965.
Zoogeographical complexes, trophic zones and marine bottom biocoenoses. Investigations in line with the programme of the International Geophysical Year, 2. (In Russian).
Trudy, Vses. Nauchno-Issled. Inst. Morsk. Ribn. Choz. i Okeanogr. (VNIRO), 57:425-445.

benthos, abyssal
Vinogradova, N.G., 1962
Some problems of the study of deep-sea bottom fauna.
J. Oceanogr. Soc., Japan, 20th Ann. Vol., 724-741.

benthos
Vinogradova, N.G., 1954.
Material on quantitative counts of the bottom fauna of certain bays of the Okhotsk and Bering seas. (In Russian).
Trudy Inst. Okeanol., Akad. Nauk, SSSR, 9:136-158.

Benthos
*Volva, G.N., 1970.
On fauna and ecology of some representatives of benthos in saltish basins of the Primorye (the Sea of Japan). Gidrobiol. Zh., 6(3): 17-22.

(In Russian; English abstract)

benthos
Vozhinskaya, V.B., 1964.
The bottom flora of Sakhalin. Investigations of bottom fauna and flora of Far-Eastern Seas and the Pacific Ocean. (In Russian).
Trudy, Akad. Nauk, SSSR, 39:330-340.

W

benthos, shallow
Wells, Harry W., and Mary Jane Wells, 1964.
The calico scallop community in North Carolina.
Bull. Mar. Sci., Gulf and Caribbean, 14(4):561-593.

benthos
Whitehouse, John W., and Brian G. Lewis, 1966.
The separation of benthos from stream samples by flotation with carbon tetrachloride.
Limnol. and Oceanogr., 11(1):124-126.

benthos
Wigley, Roland L., and K.O. Emery, 1967.
Benthic animals, particularly Hyalinoecia (Annelida) and Ophiomusium (Echinodermata), in sea-bottom photography from the continental slope.
In: Deep-sea photography, J.B. Hersey, editor, Johns Hopkins Oceanogr. Studies, 3:235-249.

benthos
Wigley, Roland L., and A.D. McIntyre, 1964
Some quantitative comparisons of offshore meiobenthos and macrobenthos south of Martha's Vineyard.
Limnology and Oceanography, 9(4):485-493.

benthos
Wiktor, Josef, 1962.
Quantitative and qualitative investigations of the Szczecin Firth bottom fauna. II.
Prace Morsk. Inst., Ryback., Oceano.-Ichtiol., Gdyni, 11(A):81-112.

In Polish, with Polish, English and Russian summary

XYZ

benthos
Young, D.K. and D.C. Rhoads, 1971.
Animal-sediment relations in Cape Cod Bay, Massachusetts. 1. A transect study. Mar. Biol. 11(3): 242-254.

benthos
Zakutsky, V.P., 1965.
On the concentration of some bottom and near-bottom organisms in the near-surface layer of the Black and Azov seas. (In Russian).
Okeanologiia, Akad. Nauk, SSSR, 5 (3):495-497.

benthos
Zakutsky, V.P., 1964.
The density of the microbenthos in the northwestern part of the Black Sea. (In Russian).
Okeanologiia, Akad. Nauk, SSSR, 4(4):684-686.

benthos
Zakutsky, V.P., 1963
The regularities in the distribution of the bottom fauna in the northern part of the Black Sea. (In Russian).
Okeanologiia, Akad. Nauk, SSSR, 3(6):1085-1087.

benthos
Zakutsky, V.P., 1963
The zoobenthos resources of the Black Sea. (In Russian).
Okeanologiia, Akad. Nauk, SSSR, 3(3):504-505.

benthos
Zakesskaya, N.T. 1969.
Distribution of benthos in South Shelikhov Bay (Okhotsk Sea). (In Russian)
Trudy vses. nauchno-issled. Inst. morsk. rybn. khoz. Okean (VNIRO) 65: 33-247.

benthos, quantitative
Zarenkov, N.A., W.W. Lieu, and N.T. K'An, 1963.
A general characteristic of quantitative distribution of plankton and benthos in the Tonkin Bay and the adjoining parts of the South China Sea. (In Russian).
Doklady, Akad. Nauk, SSSR, 148(6):1389-1391.

benthos
Zatsepin, V.I., and Z.A. Filatova, 1961
[Cyprina islandica (L.) (Mollusca Bivalvia), its geographical distribution and role in the bottom fauna communities.]
Trudy Inst. Okeanol., 46:201-216.

benthos
*Zatsepin, V.I. and L.A. Rittich, 1968.
The quantitative distribution of the main trophi gropps of the bottom invertebrates in the Barent Sea. (In Russian; English abstract). Trudy polyar. nauchno-issled. Inst. morsk. rby. Khoz. Okeanogr. (PINRO) 23: 527-545.

benthos
Zavodnik, Dugan, 1968.
Dynamisme annual de quelques zoocoenoses des fonds meubles dans les environs de Rovinj (Adriatique du Nord).
Rapp. P.-V. Réun. Comm. int. Explor. Sci. Méditerranée, 19(2):97-99.

benthos
Zavodnik, Duran,
Sur l'étude des peuplements du système phytal. Méthodes Quantitatives d'Étude du Benthos et Echelle Dimensionelle des Benthontes: Colloque du Comité du Benthos (Marseille, Nov. 1963).
Comm. Int. Expl. Sci. Mer. Médit., Monaco, 7-13.

benthos
Zenkevitch, L.A., 1958
Investigations of bottom fauna of the Far Eastern Seas and adjacent part of the Pacific. (Abstract).
Proc. Ninth Pacific Sci. Congr., Pacific Sci. Assoc., 1957, 16 (Oceanogr.):208.

benthos, abyssal
Zenkevitch, L.A., 1958.
[Points on biology in the study of the ocean depths.]
Vestnik, Akad. Nauk, SSSR, (3):48-

benthos, abyssal
Zenkevitch, L.A., 1958
Study of abyssal bottom fauna in the northwest part of the Pacific by Vitiaz. (Abstract).
Proc. Ninth Pacific Sci. Congr., Pacific Sci. Assoc., 1957, 16 (Oceanogr.):209.

benthos

Zenkevitch, L.A., & J.A. Birstein, 1960

On the problem of the antiquity of the deep-sea fauna.
Deep-Sea Res., 7(1); 10-23.

benthos

Zenkevitch, L.A., and Z. A. Filatova 1971
Some interesting zoological records from the region of the Peru-Chile Trench. (In Russian; English abstract)
Trudy Inst. Okeanol. P.P. Shirshova Akad Nauk SSSR 89: 77-80

benthos

Zmudzinski, Ludwik, 1965.
The resources of zoobenthos in the eastern part of Gdansk Bay.
Ann. Biol., Cons. Perm. Int. Expl. Mer, 1963, 20:99-100.

benthos

Ziegelmeier, E., 1961
Investigations on the bottom fauna in the shelf region off SW Iceland and SE Greenland during the research cruises with "Anton Dohrn" in the IGY. Contribution to Special IGY Meeting, 1959.
Cons. Perm. Int. Expl. Mer. Rapp. Proc. Verb. 149:209.

benthos, abyssal

benthos, abyssal

Clarke, Arthur H. 1969.
Diversity and composition of abyssal benthos.
Science 166 (3908): 1033-1034

benthos, abyssal

Okutani, T., 1964.
Report on the archibenthal and abyssal gastropod Mollusca mainly collected from Sagami Bay and adjacent waters by the R.V. Soya-Maru during the years 1955-1963.
J. Fac. Sci., Univ. Tokyo, (2), 15(3):371-447.

benthos, abyssal

Sanders, Howard L. and Robert R. Hessler, 1969
Diversity and composition of abyssal benthos.
Science, 166(3908): 1034.

benthos, abyssal

Vinogradova, N.G., 1959.
The zoogeographical distribution of the deep-water bottom fauna in the abyssal zone of the ocean.
Deep-Sea Res., 5(3):205-208.

benthos, abyssal

Vinogradova, N.G., 1956.
Some regularity in the vertical distribution of the abyssal bottom fauna of the World Ocean.
Dokl. Akad. Nauk, SSSR, 110(4):684-687.

abyssal benthos

Wolff, Torben, 1964.
Life in the ocean six miles down.
New Scientist, 24(414):241-244/

benthos, biomass

benthos, biomass

Arntz, Wolf E. 1971.
Biomasse und Produktion des Makrobenthos in den tieferen Teilen der Kieler Bucht im Jahr 1968.
Kieler Meeresforsch. 27(1): 36-72.

benthos, biomass

Băcescu, M.T. Gomoiu, N. Bodeanu, A. Petran, G. Müller și V. Manea, 1965.
Studii asupra variației vieții marine în zona litorală nisipoasă de la nord de Constanța (Cercetări efectuate în anii 1960-61 la puncte fixe situate în dreptul stațiunii Mamaia).
In: Ecologie marină, M. Băcescu, redactor, Edit. Acad. Republ. Pop. Române, București, 1: 7-138.

benthos, biomass

Băcescu, M., M.T. Gomoiu, N. Bodeanu, A. Petran, G.I. Müller și V. Chirilă, 1965.
Dinamica populațiilor animale și vegetale din zona nisipurilor fine de la nord de Constanța în condițiile anilor 1962-1965.
In: Ecologie marină, M. Băcescu, redactor, Edit. Acad. Republ. Pop. Române, București, 2:7-167.

benthos, biomass

Băcescu, M., G. Müller, H. Skolka, A. Petran, V. Elian, M.T. Gomoiu, N. Bodeanu și S. Stănescu, 1965.
Cercetări de ecologie marină în sectorul predeltaic în condițiile anilor 1960-1961.
In: Ecologie marină, M. Băcescu, redactor, Edit. Acad. Republ. Pop. Române, București, 1: 185-344.

benthos, biomass

Ellis, Derek V., 1967.
Quantitative benthic investigations. I. Satellite Channel biomass summaries and major taxon rank orders, February 1965-May 1967.
Tech. Rep. Fish. Res. Bd., Can., 25: 49 pp. (mimeographed).

benthos, biomass

Emery, K.O., Arthur S. Merrill, and James V.A. Trumbull, 1965.
Geology and biology of the sea floor as deduced from simultaneous photographs and samples.
Limnology and Oceanography, 10(1):1-21.

benthos, biomass

Knox, G.A., 1961.
The study of marine bottom communities.
Proc. R. Soc., New Zealand, 89(1):167-182.

Also in:
Collected Mar. Reprints, Edward Percival Mar. Lab., 1951-1964.

benthos, biomass

*Massé, H., 1968.
Etude des variations de la biomasse dans une biocoenose infralittorale de substrat meuble.
Recl Trav. Stn mar. Endoume, 43(59):25-31.

benthos biomass

Moore, Hilary B., Leon T. Davies, Thomas H. Fraser, Robert H. Gore and Nelia R. Lopez 1968.
Some biomass figures from a tidal flat in Biscayne Bay, Florida.
Bull. mar. Sci. Miami 18(2): 261-279

benthos biomass

Moskalev, L.I. and F.A. Pasternak, 1970.
On quantitative distribution of bottom fauna in the south-eastern part of the Gulf of Mexico and waters off Cuba. (In Russian; English and Spanish abstracts).
Okeanol. Issled. Rez. Issled. Mezhd. Geofiz. Proekt. Mezhd. Geofiz. Kom. Prezid. Akad. Nauk SSSR 20: 128-143.

benthos, biomass

. Murina V.V., V.D. Chujchin, O. Gomez y G. Suarez, 1969.
Distribución cuantitativa de la macrofauna bentonica del sublitoral superior de la plataforma cubana (region noro-occidental).
Serie Oceanol., Inst. Oceanol., Acad. Cienc. Cuba, 6: 1-14.

benthos, quantitative

Neymann, A.A., 1965.
Some regularities of the quantitative distribution of benthos on the North Pacific shelves. Investigations in line with the programme of the International Geophysical Year, 2. (In Russian).
Trudy, Vses. Nauchno-Issled. Inst. Morsk. Ribn. Choz. i Okeanogr., (VNIRO), 57:447-451.

benthos, quantitative

Neiman, A.A., 1965.
Some data on the quantitative distribution of benthos in the shelf coastal area of Australia. (In Russian).
Okeanologiia, Akad. Nauk, SSSR, 5(1):142-146.

benthos, biomass

Nesis, K.N., 1965.
Biocoenoses and biomass of benthos in the Newfoundland-Labrador region. (In Russian).
Trudy vses. nauchno-issled. Inst. morsk. ryb. Khoz. Okeanogr. (VNIRO), 57:453-489.

benthos biomass

Peer, D.L. 1970.
Relation between biomass, productivity, and loss to predators of a marine benthic polychaete Pectinaria hyperborea.
J. Fish. Res. Bd. Can. 27 (12): 2143-2153.

benthos, biomass

Pequenat, Willis E., 1968.
Distribution of epifaunal biomass on a sublittoral rock-reef.
Pacif. Sci., 22(1):37-40.

benthic biomass

Richards, Norman J., and Sarah W. Richards, 1965.
Effect of decalcification procedures on the dry weights of benthic invertebrates.
Limnol. Oceanogr., 10(3):469-471.

benthos biomass

Rowe, Gilbert T., 1971.
Benthic biomass and surface productivity. (Portuguese and English abstracts). In: Fertility of the Sea, John D. Costlow, editor, Gordon Breach, 2: 441-454.

benthos, biomass

Rowe, Gilbert T., 1971.
Benthic biomass in the Pisco, Peru upwelling.
Investigación pesq. 35(1): 127-135.

benthos biomass

Rowe, Gilbert T. and David W. Menzel, 1971.
Quantitative benthic samples from the deep Gulf of Mexico with some comments on the measurement of deep-sea biomass. Bull. mar. Sci. Miami 21(2): 556-566.

benthos, biomass

* Scaccini, Andrea, 1968.
Dati preliminari sulle zoocenosi bentoniche e sulla biomassa in una zone dell'alto e medio Adriatica.
Note Lab. Biol. mar. Pesca, Fano, Univ. Bologna, 2(3): 25-56.

Benthos (meioe) Biomass

*Sokolova, M.N., 1970.
Weight characteristics of meiobenthos from different parts of the deep-sea trophic regions of the Pacific Ocean. (In Russian; English abstract). Okeanologiia, 10(2): 348-356.

benthos, quantitative

Vilenkin, B. J., 1965.
On the interpertation of bottom quantitative data of benthos samples. (In Russian).
Okeanologiia, Akad. Nauk, SSSR, 5(1):128-133.

benthos, coastal

Paine, R.T., 1971.
A short-term experimental investigation of resource partitioning in a New Zealand rocky intertidal habitat. Ecology 52(6): 1097-1106.

benthos colonization

benthos, colonization
Puzanov, I.I., 1965.
Successive stages of Mediterranization of Black Sea fauna: new data. (In Russian).
Gidrobiol. Zhurn., Akad. Nauk. Ukrain., SSR, 1(2):54.

benthos dispersal of

benthos, dispersal of
Mileikovsky, S.A., 1966.
The range of dispersal of the pelagic larvae of bottom invertebrates by ocean currents and its distributional role on the example of Gastropoda and Lamellibranchia. (In Russian; English abstract).
Okeanologiia, Akad. Nauk, SSSR, 6(3):482-492.

benthos, effect of

benthos, effect of
Banse, Karl, Frederic H. Nichols and Dora R. May 1971.
Oxygen consumption by the sea bed. III. On the role of the macrofauna at three stations.
Vie Milieu Suppl. 22(1):32-52.

benthos, effect of
Laughton, A.S., 1963.
18. Microtopography.
In: The Sea, M.N. Hill, Editor, Interscience Publishers, 3:437-472.

benthos, effect of
Rhoads D.C. and D.K. Young 1970.
The influence of deposit-feeding organisms on sediment stability and community trophic structure.
J. mar. Res. 28(2):150-178.

benthos, infauna

benthos, infauna
Angel, Heather H., and Martin V. Angel 1967.
Distribution pattern analysis in a marine benthic community.
Helgoländer wiss. Meeresunters. 15(1/4): 445-454.

benthos (infauna)
Crisp, D.J. and R. Williams, 1971.
Direct measurement of pore-size distribution on artificial and natural deposits and prediction of pore space accessible to interstitial organisms. Mar. Biol. 10(3): 214-226.

benthos (Infauna)
Glémarec, Michel 1971.
L'endofaune du plateau continental Nord-Gascogne: étude des facteurs écologiques.
Vie Milieu Suppl. 22(1):94-108

benthos, infauna
Lie, Ulf, and James C. Kelley, 1970.
Benthic infauna communities off the coast of Washington and in Puget Sound: identification and distribution of the communities.
J. Fish. Res. Bd, Can., 27(4): 621-651

benthos (infauna)
Lie, Ulf, 1969.
The logarithmic series and the lognormal distribution applied to benthic infauna from Puget Sound, Washington, U.S.A.
Fisk Dir. Skr. Ser. HavUnders. 15(3): 234-245.

benthos (in fauna)
Lie, Ulf, and Dale S. Kisker, 1970.
Species composition and structure of benthic infauna communities off the coast of Washington.
J. Fish. Res. Bd, Can., 27(12): 2273-2282.

benthos infaunal
McKnight D.G., 1969.
Infaunal benthic communities of the New Zealand continental shelf.
N.Z. Jl mar. freshw. Res. 3(3): 409-444.

benthos (infauna), effect of
Young, David K. 1971.
Effects of infauna on the sediment and seston of a subtidal environment.
Vie Milieu Suppl. 22(2): 557-571

benthos intertidal

benthos (intertidal)
*Green, Roger H., Katherine D. Hobson and Stuart L. Santos, 1967.
Analysis of invertebrate distribution in the intertidal zone of Barnstable Harbor. (Abstract only).
Biol. Bull., mar. biol. lab., Woods Hole, 133(2): 454-455.

benthos
Sanders, Howard L., 1963.
The deep-sea benthos.
(Special Issue on Marine Biology), AIBS Bull., 13(5):61-63.

benthos, lists of species

benthos, lists of spp
Azouz, Abderrazak, 1966.
Étude des peuplements et des possibilités d'ostreiculture du Lac de Bizerte.
Inst. Nat. Sci. tech. Océanogr. Pêche, Salammbo, Ann. 15: 69 pp.

benthos, lists of spp
Bacescu, M., 1965.
Méthodes de la recherche du benthos en Mer Noire et importance des prélèvements directs en scaphandre autonome des échantillons de benthos pour l'étude quantitatives.
Méthodes Quantitatives d'Étude du Benthos et Echelle Dimensionnelle des Benthontes: Colloque du Comité du Benthos (Marseille, Nov. 1963), Comm. Int. Expl. Sci. Mer. Médit., Monaco, 48-62.

benthos, lists of spp.
*Bacescu, M., E. Dumitrescu, M.T. Gomoiu, et A. Petran, 1967.
Eléments pour la caractérisation de la zone sedimentaire médio-littorale de la Mer Noire.
Trav. Mus. Hist. nat. "Grigore Antipa," 7:1-14.

benthos, lists of spp.
Băcescu, M., M.T. Gomoiu, N. Bodeanu, A. Petran, G.I. Müller si V. Chirile, 1965.
Dinamica populaţiilor animale şi vegetale din zona nisipurilor fine de la nord de Constanta în condiţiile anilor 1962-1965.
In: Ecologie marină, M. Băcescu, redactor, Edit. Acad. Republ. Pop. Romîne, Bucuresti, 2:7-167.

benthos, lists of spp.
Băcescu, Mihai, Elena Dumitrescu, Amélie Marcus, Gabriela Paladian, et Rudolf Mayer 1963.
Données quantitatives sur la faune pétricole de la mer Noire à Agigea (Secteur Roumain) dans les conditions spéciales de l'année 1961.
Trav. Mus. Hist. Nat. "Grigore Antipa" 4: 131-155.

benthos, lists of spp.
Băcescu, M., G. Müller, H. Skolka, A. Petran, V. Elian, M.T. Gomoiu, N. Bodeanu şi S. Stănescu, 1965.
Cercetări de ecologie marină în sectorul predeltaic în condiţiile anilor 1960-1961.
In: Ecologie marina, M. Băcescu, redactor, Edit. Acad. Republ. Pop. Romîne, Bucureşti, 1:185-344.

Benthos, lists of spp
Bagge, Pauli, 1969.
Effects of pollution on estuarine ecosystems. I. Effects of effluents from wood-processing industries on the hydrography, bottom and fauna of Saltkällefjord (W. Sweden). II The succession of the bottom fauna communities in polluted estuarine habitats in the Baltic-Skagerak region.
Merentutkimuslait. Julk. 228: 130 pp.

benthos, lists of spp.
Bellan, G., 1967.
Pollution et peuplements benthiques sur substrat meuble dans la région de Marseille. I. Le secteur de Cortiou.
Rev. intern. Océanogr. Méd., 6/7:53-87.

benthos, lists of spp.
Bellan-Santini, Denise, 1970.
Salissures biologiques de substrats vierges artificiels immergés en eau pure, durant 26 mois, dans la région de Marseille (Méditerranée nord-occidentale). 1. Étude qualitative.
Téthys 2(2): 335-356.

benthos, lists of spp.
Bellan-Santini 1969.
Contribution à l'étude des peuplements infralittoraux sur substrat rocheux (Étude qualitative et quantitative de la frange supérieure) Rec. Trav. Sta. mar. Endoume 47(63): 294 pp.

benthos, lists of spp.
Black, J. Hope 1971.
9. Benthic communities. Port Phillip Bay Survey 2.
Mem. natn. Mus. Victoria 32: 129-170

benthos, lists of spp.
Bourcier, Michel, 1969.
Écoulement des boues-rouges dans le canyon de la Cassidaigne (décembre 1968)
Téthys 1(3): 779-782.

benthos, lists of spp.
Bourcier, M., 1968.
Étude du benthos du plateau continental de la baie de Cassis.
Rec. Trav. Sta. mar. Endoume, 44(60): 63-108

benthos, lists of spp.
Brunel, Pierre, 1970.
Catalogue d'invertébrés benthiques du Golfe Saint-Laurent recueillis de 1951 à 1966 par la Station de Biologie Marine de Grande-Rivière.
Trav. Pêch., Québec 32: 54 pp.

benthos, lists of spp.
Brunel, Pierre, 1964
Inventaire taxonomique des invertébrés benthiques marine du Golfe Saint-Laurent.
Rapp. Ann., 1963, Sta. Biol. Mar., Grande-Rivière: 39-44.

benthos, lists of spp.
Brunel, P., 1963.
Inventaire taxonomique des invertébrés marins du Golfe Saint-Laurent.
Rapp. Ann., 1962, Sta. Biol. Mar., Grande Rivière 81-89.

benthos, lists of spp.
Buchanan, John B., 1963.
The bottom fauna communities and their sediment relationships off the coast of Northumberland.
Oikos, 14(2):154-175.

benthos, lists of spp.
Buyanov, N.I., 1966.
Content of potassium in sea water and marine organisms of the White Sea. (In Russian).
Mater. Ribokhoz. Issled. severn. Basseina, Poliarn. Nauchno-Issled. Proektn. Inst. Morsk. Ribn. Khoz. Okeanogr. (PINRO), 7:183-191.

benthos, lists of spp.
Carpine, Christian, 1964.
La côte de l'Esterel, de la pointe des Lions à la pointe de l'Aiguille (région A2). Contributions à l'étude bionomique de la Méditerranée occidentale (côte du Var et des Alpes Maritimes, côte occidentale de Corse). Fasc. 3.
Bull. Inst. Océanogr., Monaco, 63(1312A,B):52 pp chart

benthos, lists of spp
Cerame-Vivas, M.J., and I.E. Gray, 1966.
The distributional pattern of benthic invertebrates of the continental shelf off North Carolina.
Ecology, 47(2):260-270.

benthos, lists of spp.
Christomanos, Anastas A., 1963
Dredging results from the oceanographic cruis "TithysII" during October 1962.
Res. Proc. Mar. Lab., Suppl. Thalassina Epist. Phylla, 1(6):27-38.

benthos, lists of spp.
Cory, Robert L., 1967.
Epifauna of the Patuxent River estuary, Maryland, for 1963 and 1964.
Chesapeake Sci., 8(2):71-89.

Crosnier, A., 1964. benthos, lists of spp.
Fonds de peche le long des cotes de la Republique Federale du Camaroun.
Cahiers, O.R.S.T.R.O.M., No. special:133 pp.

benthos, lists of spp.
Day, J.H., J.G. Field and M.J. Penrith, 1970.
The benthic fauna and fishes of False Bay, South Africa.
Trans. R. Soc. S.Afr. 39(1): 108 pp.

benthos, lists of spp.
Ellis, Derek V., 1967.
Quantitative benthic investigations. II. Satellite Channel species data, February 1965-May 1967.
Tech. Rep. Fish. Res. Bd. Can., 35: numerous pp. (unnumbered) (mimeographed).

benthos, lists of spp.
Dean, David and Harold H. Haskin, 1964
Benthic repopulation of the Raritan River estuary following pollution abatement.
Limnology and Oceanography 9(4):551-563.

Benthos, lists of spp.
Dragovich, Alexander, and John A. Kelly, Jr., 1964.
Ecological observations of macro-invertebrates in Tampa Bay, Florida.
Bull. Mar. Sci., Gulf and Caribbean, 14(1):74-102

benthos, lists of spp.
Fredj, Gaston, 1964.
La région de Saint-Topez: du cap Taillet au cap de Saint-Tropez. Contributions à l'étude bionomique de la Méditerranée occidentale (Côte du Var et des Alpes Maritimes, côte occidentale de Corse). Fasc. 2.
Bull. Inst. Océanogr., Monaco, 63(1311A,B):55 pp, charts

benthos, lists of spp.
Gamulin-Brida, Helena, 1967.
The benthic fauna of the Adriatic.
Oceanogr. Mar. Biol., Ann. Rev., H. Barnes, editor George Allen and Unwin, Ltd., 5:535-568.

benthos, lists of spp.
Febvre-Chevalier, Colette 1969.
Étude bionomique des substrats meubles dragables du Golfe de Fos.
Téthys 1 (2): 421-476.

benthos, list of species
Gamulin-Brida, Helena, 1962
Biocenoses du littoral plus profond (circa-littoral) dans les canaux de l'Adriatique moyenne. (In Jugoslavian; French abstract)
Acta Adriatica, Inst. Oceanogr. i Ribarst., Split, 9(7):3-196.

benthos, lists of spp.
Gilat, Eliezer, 1964.
The macrobenthonic invertebrate communities on the Mediterranean continental shelf of Israel.
Bull. Inst. Océanogr., Monaco, 62(1290):46 pp.

benthos, lists of spp.
Grindley, J.R., and B.F. Kensley, 1966.
Benthonic marine fauna obtained off the Orange River mouth by the Diamond Dredgr Emerson-K.
Cimbebasis, S.W.A., No. 16:14 pp.

benthos (lists of spp) a
Guerin-Ancey, Odile, 1970.
Étude des intrusions terrigènes fluviatiles dans les complexes récifaux: délimitation et dynamique des peuplements des vases et des sables vaseux du chenal postrécifal de Tuléar (S.W. de Madagascar).
Rec. Trav. Sta. mar. Endoume, hors sér suppl. 10: 3-46.

benthos, lists of spp. a
Guille, Alain, 1970.
Bionomie benthique du plateau continental de la côte catalane française. II Les communautés de la macrofaune.
Vie Milieu 21 (1B): 149-280

~~benthos, lists of spp~~
Guille, Alain et Jacques Soyer, 1968
La faune benthique des substrats meubles de Banyuls-sur-Mer: premières données qualitatives et quantitatives.
Vie Milieu 19 (2-B): 323-359

benthos lists of spp.
Haefner, Paul A. Jr. 1967.
Hydrography of the Penobscot River (Maine) estuary.
J. Fish. Res. Bd., Can., 24(7): 1553-1571

benthos, lists of spp.
Haertel, Lois, and Charles Osterberg, 1967.
Ecology of zooplankton, benthos and fishes in the Columbia River estuary.
Ecology, 48(3):459-472.

benthos, lists of sp
Heydemann, Berndt, 1963.
Deiche der Nordseeküste als besonderer Lebensraum - Ökologische Untersuchungen über die Arthropoden-Besiedlung.
Die Küste, 11:90-130.

benthos, lists of spp
Holme, N.A., 1967.
Changes in the bottom fauna of Weymouth Bay and Poole Bay following the severe winter of 1962-1963.
J. mar. biol. Ass., U.K., 47(2):397-405.

benthos, lists of spp.
Holme, N.A., 1966.
The bottom fauna of the English Channel, Part II.
J. mar biol. Assoc., 46(2):401-493.

benthos (list of species)
Inoue, A., 1963.
Study on productivity of Kasaoka Bay. (In Japanese; English abstract).
Bull. Nakai Reg. Fish. Res. Lab., 20:1-116.

benthos (infauna), lists of spp.
Keith, Don C., and Neil C. Hulings, 1965.
A quantitative study of selected nearshore infauna between Sabina Pass and Bolivar Point, Texas.
Publ. Inst. Mar. Sci., Port Aransas, 10:33-40.

benthos lists of spp.
Kensler, Craig B., 1965.
Distribution of crevice species along the Iberian Peninsula and northwest Africa.
Vie et Milieu, Bull. Lab. Arago, (B)16(2-B):851-887.

benthos.lists of spp.
Kikuchi, Taiji, 1968.
Faunal list of the Zostera marina belt in Tomioka Bay, Amakusa, Kyushu.
Publ. Amakusa mar. biol. Lab., 1(2):163-192.

benthos, lists of spp.
King, R.J., J. Hope Black and Sophie C. Ducker 1971 a
Intertidal ecology of Port Phillip Bay with systematic list of plants and animals.
Mem. natn. Mus. Victoria 32: 93-128

benthos, lists of spp.
Kiseleva, M.I., 1964.
Some data on the benthos of the Adriatic Sea. (In Russian).
Trudy Sevastopol Biol. Sta., 7:28-38.

benthos, lists of spp.
Knox, G.A., 1951.
A rock bottom fauna from 80 fathoms off Banks Peninsula.
Rec. Canterbury (N.Z.) Mus., 6(1):41-51.
Also in:
Collected Mar. Reprints, Edward Percival Mar. Lab., 1951-1964.

benthos, lists of spp.
Kramp, P.L., 1963
Summary of the zoological results of the "Godthaab" Expedition 1928.
Medd. om Grønland, 81(7):115 pp.

benthos, lists of spp.
Kühlmorgen-Hille, Georg, 1965.
Qualitative und quantitative Veränderungen der Bodenfauna der Kieler Bucht in den Jahren 1953-1965.
Kieler Meeresforsch., 21(2):167-191.

benthos, lists of spp.
Lacroix, G., and J. Bergeron, 1963.
Liste préliminaire des invertébrés du Banc de Bradelle, 1962.
Rapp. Ann., 1962. Sta. Biol. Mar., Grande Rivière, 59-67.

benthos, lists of spp. a
Ledoyer, Michel, 1970.
Additions à la liste des invertébrés benthiques recueillis dans le Golfe Saint-Laurent (Baie de Chaleurs).
Rapp. ann. 1969, Serv. Biol. Québec 37-43

benthos, lists of spp.
Ledoyer, Michel, 1968.
Écologie de la faune vagile des biotopes méditerranéens accessibles en scaphandre autonome (Région de Marseille principalement). IV. Synthèse de l'étude écologique.
Rec. Trav. Sta. mar. Endoume, 44 (60): 125-295.

benthos, lists of spp.
*Le Loeuff,P., et A. Intes,1968.
La faune benthique du plateau continental de Côte d'Ivoire récoltés au chalut: abondance, répartition,variations saisonnières,mars-1966 - février-1967.
Doc.Sci.Prov.Centre Recherches ocean.,Abidjan. 025:78 pp., app. (mimeographed).

benthos, lists of spp.
L'Herroux, Michel, 1970.
Peuplements des sables fins en baie de Saint-Brieuc (Manche).
Tethys, 2(1): 41-88

Lumare, Feto, 1968 benthos, lists of spp.
Osservazioni sulle zoocenosi caratteristiche dei fondi da pesca a strascico dall' Arcipelago Toscano a La Spezia.
Progr. ricerca Risorse mar. Fondo mar. Comm. ital. Oceanogr. Consig. naz. Ricerche (B)(85)(1968)

benthos, lists of spp.
McIntyre, A.D., 1961.
Quantitative differences in the fauna of boreal mud associations.
J. Mar. Biol. Assoc., U.K., 41(3):599-616.

benthos,lists of spp.
Menzies,R.J., O.H. Pilkey, B.W. Blackwelder, D. Dexter,P. Huling and L. McCloskey,1966.
A submerged reef off North Carolina.
Int. Rev. ges. Hydrobiol., 51(3):393-431.

benthos,lists of spp.
*Morgans, J.F.C.,1967.
The macrofauna of an excellently sorted isolated beach at Kaikoura: and certain tidal observations
Trans. R. Soc., N.Z. Zool., 9(11):157-167.

benthos,lists of spp.
*Morgans,J.F.C.,1967.
The macrofauna of an unstable beach discussed in relation to beach profile, texture and a progression in shelter from wave action.
Trans. R. Soc., N.Z.,Zool., 9(10):141-155.

benthos, lists of spp.
Morgans, J.F.C., 1962.
The benthic ecology of False Bay. II. Soft and rocky bottoms observed by diving and sampled by dredging and the recognition of grounds.
Trans. R. Soc., S. Africa, 36(4):287-334.

benthos, lists of spp.
Nakai, Zinziro, Masaya Kosaka, Masahiro Ogura, Katsuya Takahashi and Tin Yet Ho, 1970.
Interrelationship between natural production of benthos and stability of the bottom. II The condition in summer. (In Japanese; English abstract).
J. Coll. mar. Sci. Technol., Tokai Univ. 4:121-136.

benthos, lists of spp.
Newell, N.D., J. Imbrie, E.G. Purdy and D.L. Thurber, 1959.
Organism communities and bottom facies, Great Bahama Bank.
Bull. Amer. Mus. Nat. Hist., 117(4):183-228.

benthos, lists of spp.
Neyman, A.A., 1963
Quantitative distribution of benthos and food supply of demersal fish in the eastern part of the Bering Sea. (In Russian).
Sovetsk. Ribokh. Issled. B Severo-Vostokh. Chasti Tikhogo Okeana, VNIRO 48, TINRO 50(1): 145-206.

benthos,lists of spp.
O'Gower,A.K., and J.W. Wacasey,1967.
Animal communities associated with Thalassia, Diplantheria, and sand beds in Biscayne Bay. I. Analysis of communities in relation to water movements.
Bull. mar. Sci., Miami, 17(1):175-210.

benthos, lists of spp.
*Okutani, Takashi, 1969.
Synopsis of bathyal and abyssal megalo-invertebrates from Sagami Bay and south off Boso Peninsula trawled by the R/V Soyo-Maru.
Bull. Tokai reg. Fish. Res. Lab., 57: 1-61.

benthos, lists of spp.
*Olivier,Santiago Raul,Ricardo Bastida y Maria Rosa Torti,1968.
Resultados de las campañas oceanográficas Mar del Plata 1-5: Contribucion al trazado de una carta bionómica del área de Mar del Plata. Las asociaciones del sistema litoral entre 12 y 70m de profundedal.
Bol.Inst.Biol.mar,Mar del Plata,16:1-85.

benthos, lists of spp.
Parker, Robert H., 1964.
Zoogeography and ecology of macro-invertebrates of Gulf of California and continental slope of western Mexico.
In: Marine geology of the Gulf of California, a symposium, Amer. Assoc. Petr. Geol., Memoir, T. van Andel and G.G. Shor, Jr., Editors,3:331-376

benthos, lists of spp.
Parker, Robert H., 1964.
Zoogeography and ecology of some macroinvertebrates, particularly mollusks, in the Gulf of California and the continental slope off Mexico.
Vidensk. Medd., Dansk Naturh. Foren., 126:1-178.

Pearson, T.H. 1971. benthos, lists of spp.
Studies on the ecology of the macrobenthic fauna of Lochs Linnhe and Eil, west coast of Scotland. II. Analysis of the macrobenthic fauna by comparison of feeding groups.
Vie Milieu Suppl. 22(1): 53-91

benthos, lists of spp.
Penrith, M.-L., and B.F. Kensley, 1970 a
The constitution of the intertidal fauna of rocky shores of South West Africa. I. Lüderitzbucht.
Cimbebasia (A). (9). 191-239

benthos, lists of spp.
Pérès, J.M., et J. Picard, 1964.
Nouveau manual de bionome benthique de la Mer Méditerranée. Edition revue et augmenté.
Rec. Trav. Sta. Mar. d'Endoume, Bull., 31(47):5-137.

benthos, lists of spp.
Pichon,Michel,1966.
Note sur la faune des substrats sablo-vaseux infralittoraux de la Baie d'Ambaro (côte nord-ouest de Madagascar).
Cah. ORSTOM,Sér.Océanogr., 4(1):79-94.

benthos,lists of spp.
*Pichon,Mireille,1967.
Contribution à l'étude des peuplements de la zone intertidale sur rables fins et sables vaseux non fixés dans la région de Tuléar.
Recl.Trav.Stn.mar.Endoume,hors sér,Suppl.7:57-100

benthos,lists
*Plante,Raphaël,1967.
Étude quantitative du benthos dans la région de Nosy-Bé:note préliminaire.
Cah.ORSTOM,Sér.Oceanogr.,5(2):95-108.

benthos, lists of spp.
Poizat, Claude 1969
Le débouché des calanques entre Marseille et la Ciotat. étude des peuplements et des sédiments.
Recl Trav. Stn mar. Endoume 45(61): 367-436

benthos, lists of spp.
Propp, M.V. 1966.
Benthic communities of Laminaria and Lithothamnion in the upper sublittoral of the eastern Murmansk. (In Russian)
Trudy murmansk. biol. Inst. 11(15):92-114.

benthos, lists of spp.
*Reys,Jean-Pierre, et Simone Reys,1968.
Répartition quantitative du benthos de la région de Tuléar.
Recl.Trav.Stn.mar.Endoume,hors sér,Suppl.5:71-86.

benthos lists of spp.
Richards, Sarah W., and Gordon A. Riley 1967.
The benthic epifauna of Long Island Sound.
Bull. Bingham oceanogr. Coll. 19(2): 89-135.

benthos, lists of spp.
Rodrigues da Costa, Henrique, 1962.
Nota preliminar sobre a fauna de substrato duro no litoral dos estados do Rio de Janeiro e Guanabara.
Univ. Brasil., Fac. Nac. Filos., Centro Estudos Zool., Avulso, No. 15:10 pp.

benthos, lists of spp.
Sanders, H.L., P.C. Mangelsdorf, Jr., and G.R. Hampson, 1965.
Salinity anf faunal distribution in the Pocasset River, Massachusetts.
Limnol. and Oceanogr., Redfield Vol., Suppl. to 10:R216-R229.

benthos ,lists of spp.
Seno,Jiro, and Tatsuyoshi Masuda, 1966.
Biological specimens collected by otter trawling.
J. Tokyo Univ., Fish (Spec-ed.) 8(2):257-269.

benthos, lists of spp. a
Shepherd, S.A. and Jeanette E. Watson, 1970.
The sublittoral ecology of West Island, South Australia. 2 The association between hydroids and algal substrat.
Trans. R. Soc. S. Aust. 94: 139-146.

benthos, lists of spp.
Skalkin, V.A., 1960.
Benthos of Terpeniya Bay, its importance in feeding and on the distribution of the yellowfin flounder. (In Russian).
Izv., Tikhookean, Nauchno-Issled. Inst. Ribn. Khoz. i Okeanogr., 46:145-187.

benthos, list of spp.
Thomassin, Bernard, 1969.
Peuplements de deux biotopes de sables coralliens sur le grand récif de Tuléar, sud-ouest de Madagascar.
Rec. Trav. Sta. mar. Endoume, hors sér. Suppl. 9: 59-133

benthos,lists of spp.
*Tommasi,Luiz Roberto,1967.
Observações preliminares a fauna bêntica de sedimentos moles da Braie de Santos e regiões vizinhas.
Bolm Inst.oceanogr.S. Paulo,16(1):43-65.

benthos, lists of spp.
*Turner, Charles H., Earl E. Ebert and Robert R. Give, 1968.
The marine environment offshore from Point Loma, San Diego, County.
Fish. Bull. Dept. Fish Game, Calif. 140:185 pp.

benthos, lists of spp.
Turner, Charles H., Earl E. Ebert and Robert R. Given, 1965.
Survey of the marine environment offshore of San Elijo Lagoon.
California Fish & Game, 51(2):81-112.

benthos, lists of spp.
University of Southern California, Allan Hancock Foundation, 1965,
An oceanographic and biological survey of the southern California mainland shelf.
State of California, Resources Agency, State Water Quality Control Board, Publ. No. 27:232 pp. Appendix, 445 pp.

benthos, lists of spp.
Uschakov, P.V. 1970.
Observations sur la répartition de la faune benthique du littoral guinéen.
Cah. biol. mar. 11(4):435-457.

benthos, lists of spp.
Vaissière, Raymond, et Gaston Fredj, 1963.
Contribution à l'étude de la faune benthique du plateau continental de l'Algérie.
Bull. Inst. Océanogr., Monaco, 60(1272A/B), 83 pp

benthos, lists of spp.
*Vamvakas, Constantin-Nicolas E., 1970
Peuplements benthiques des substrats meubles du sud de la Mer Égée.
Téthys 2(1): 89-130

benthos, lists of spp.
Vinogradov, L.G., and A.A. Neyman, 1965.
Zoogeographical complexes, trophic zones and marine bottom biocenoses. (In Russian).
Trudy vses. nauchno-issled. Inst. morsk. ryb. Khoz. Okeanogr. (VNIRO), 57:425-445.

benthos, lists of spp.
Županović, Šime, 1969
Contribution à l'étude de la faune bentique de la dépression de Jabuka. (In Jugoslavian; French and Italian abstracts).
Thalassia Jugoslavica, 5:477-493

benthos, mass occurrence
Ziegelmeier, E., 1970
Über Massenvorkommen verschiedener makrobenthaler Wirbelloser während der Wiederbesiedlungsphase nach Schädigungen durch "katastrophale" Umwelteinflüsse.
Helgolander wiss. Meeresunters. 21(1/2):9-20

benthos (deep), feeding types
Sokolova, M.N., 1965.
On the irregularity in the distribution of nutrition groupings of deep-sea benthos in connection with the irregularity of sedimentation. (In Russian).
Okeanologiia, Akad. Nauk, SSSR, 5 (3):498-506.

benthos, littoral (lists of spp)
Băcescu, M.T. Gomoiu, N. Bodeanu, A. Petran, G. Müller, și V. Manea, 1965.
Studii asupra variației vieții marine în zona litorală nisipoasă de la nord de Constanța (Cercetări efectuate în anii 1960-1961 la punctele fixe situate în dreptul stațiunii Mamaia).
In: Ecologie marină, M. Băcescu, redactor, Edit. Acad. Republ. Pop. Romîne, București, 1: 7-138.

benthos, littoral
Ushakov, P.V., 1965.
The bionomic features of the near-coastal zone in the Guinea Republic (West Africa). (In Russian).
Okeanologiia, Akad. Nauk, SSSR, 5 (3):507-517.

benthos, microbiology of
Brisou, J., C. Tysset et Y de Rautlin de la Roy, 1964.
Étude sur la microbiologie du benthos.
Comm. Int. Expl. Sci., Mer Mediterranee, Symp. Pollution Mar., Microorgan. Prod. Petrol, Monaco 1964:115-121.

benthos, micro
Fenchel, Tom 1971.
The reduction-oxidation properties of marine sediments and the vertical distribution of the microfauna.
Vie Milieu Suppl. 22(2):509-521

benthos (microflora)
Hargrave, Barry T. 1970.
The effect of a deposit-feeding amphipod on the metabolism of benthic microflora.
Limnol. Oceanogr. 15(1):21-30.

benthos, numerical
Rodríguez, Gilberto y Anrés Eloy Esteves, 1966.
Ch. 4. Organismos benthonicos.
Estudios hidrobiologicos en el Estuario de Maracaibo, Inst. Venezolanus de Invest. Cient. 83-92.

Benthos, physiology of
*Theede, H., A. Ponat, K. Hiroki and C. Schlieper, 1969.
Studies on the resistance of marine bottom invertebrates to oxygen-deficiency and hydrogen sulphide. Marine Biol., 2(4): 325-337.

benthos, productivity
Arntz, Wolf E. 1971.
Biomasse und Produktion des Makrobenthos in den tieferen Teilen der Kieler Bucht im Jahr 1968.
Kieler Meeresforsch. 27(1):36-72.

benthos, productivity of
*Massé, Henri, 1968.
Evaluation de la production d'un peuplement benthique.
C.r. hebd. Séanc., Acad. Sci. Paris, (D), 266(7):713-715.

benthos, productivity
Peer, D.L. 1970.
Relation between biomass, productivity, and loss to predators of a marine benthic polychaete Pectinaria hyperborea.
J. Fish. Res. Bd. Can. 27(12):2143-2153.

benthos, quantitative
Chukhchin, V.D., 1964.
Quantitative data on the benthos of the Tyrrhenian Sea. (In Russian).
Trudy Sevastopol Biol. Sta., 7:48-50.

benthos, quantitative
*Dirk Frankenberg and Robert J. Menzies, 1968.
Some qunatitative analyses of deep-sea benthos off Peru.
Deep-Sea Res., 15(5):623-626.

benthos, quantitative
Grobov, A.G., and Michaylov, 1968.
Quantitative distribution of benthos in shelf waters of the northwestern Indian Ocean. (In Russian).
Trudy, Vses. Nauchno-Issled. Inst. Morsk. Ribn. Okeanogr (VNIRO) 64, Trudy Azovo-Chernomorsk. Nauchno-Issled. Inst. Morsk. Ribn. Khoz. Okeanogr. (AscherNIRO), 28: 196-203.

benthos, quantitative
Streltsov, V.E., 1966.
Quantitative distribution of the Polychaeta in the southern Barents Sea. (In Russian).
Trudy murmansk. biol. Inst., 11(15):71-91.

benthos, quantitative (data only)
Ellis, Derek V. 1968.
Quantitative benthic investigations. III. Locality and environmental data for selected stations (mainly from Satellite Channel, Straits of Georgia and adjacent inlets (February 1965-December 1967).
Techn. Rept. Fish. Res. Bd Can. 59: unnumbered pp. (mimeographed).

benthos, radioactivity of
Martin, DeCourcey, Jr., 1958
The uptake of radioactive wastes by benthic organisms.
Proc. Ninth Pacific Sci. Congr., Pacific Sci. Assoc., 1957, 16(Oceanogr.):167-169.

benthos, shallow
Fager, Edward W., 1968.
A sand-bottom epifaunal community of invertebrates in shallow water.
Limnol. Oceanogr., 13(3):448-464.

benthos vertical distribution
Patrick, Arnaud, 1966.
Remarques sur la répartition verticale du benthos du précontinent Antarctique. (Abstract only).
Second Int. Oceanogr. Congr., 30 May-9 June 1966. Abstracts, Moscow:11-12.

benthos, vertical migrations of
Zakutsky, V.P., 1968.
Diurnal rhythm of vertical migrations of bentho-hyponeuston organisms. (In Russian).
Gidrobiol. Zh., 4(6):55-59.

benthos, zonation of
*Rowe, Gilbert T. and Robert J. Menzies, 1969.
Zonation of large benthic invertebrates in the deep-sea off the Carolinas. Deep-Sea Res., 16 (5): 531-537.

berms
*Panin, N., 1967.
Structure des dépôts de plage sur la côte de la Mer Noire.
Mar. Geol. 5(3):207-219.

Bernard cells
Stevenson, Robert E., 1969.
From space: new tool for studying ocean weather.
Ocean Industry, 4(3):51-55.

Bertalanffy function
Hohendorf, Kurt, 1966.
Eine Diskussion der Bertalanffy-Funktionen und ihre Anwendung zur Charakterisierung des Wachstums von Fischen.
Kieler Meeresforsch., 22(1):70-97.

berthing, tankers

berthing, tankers
Great Britain, Department of Scientific and Industrial Research, 1962.
Hydraulics Research, 1961: the report of the Hydraulics Research Board with the report of the Director of Hydraulics Research (Wallingford), 96 pp.

beta-plane ocean
Mofjeld, H.O. and Maurice Rattray, Jr., 1971.
Free oscillations in a beta-plane ocean.
J. mar. Res. 29(3): 281-305.

BIBLIOGRAPHY

BIBLIOGRAPHIES
Here are entered not only proper bibliographies, but also citations for papers having helpful references to a subject.

Consult introduction for "arrangement."
Not all subheadings are entered on cards.

bibliographies
Boduszynska-Borowikowa, Maria, Editor, 1963.
Bibliografia oceanograficzna za rok 1962.
Polska Akad. Nauk, Komitet Badan Morza, Sopot, 109 pp. (mimeographed).

bibliographies
Commissione Internazionale per l'Esplorazione Scientifica del Mare Mediterraneo-Delegazione Italiana, 1956.
Bibliographia Oceanographica, Vol. 18(1945)

bibliographies
Cordes, Eilhard, 1970.
Die Literaturerschliessung in der Meereskunde.
Ergänzhft Dt. hydrogr. Z (A (8°)) 10: 43 pp.

bibliography
Dunod, Editeur, 1961.
Table de matieres. III. Science naturelles. 2. Physique du globe. Meteorologie-oceanographie physique.
La Nature - Science Progress, 89(Suppl.)(3320): 560.

Bibliographies (Oceanic Coordinate Index)
Gillenwaters, T. R., 1965.
Oceanic information - its collection and dissemination. (abstract).
Ocean Sci. and Ocean Eng., Mar. Techn. Soc.,-Amer. Soc. Limnol. Oceanogr., 2:1198-1204.

bibliographies
United Nations, F.A.O., 1959.
Current bibliography for aquatic sciences and fisheries, 2(1):1-94.

Acoustics - see noise

bibliography, acoustics
American Institute of Physics, 1949.
Cumulative index to acoustical literature and patents; 1938-1948 listed in the Journal of the Acoustical Society of America, Volumes 11-20.
J. Acoust. Soc., America:395 pp.

bibliography, acoustics
Backus, R.H., 1958.
Sound production by marine animals.
USN J. Underwater Acoustics, 8(2):191-202.

NAVEXOS -P-970
not in MBL library
MS owns

bibliographies, acoustics
Kellogg, W.N., 1953.
Bibliography of the noises made by marine organisms. Amer. Mus. Novitates No. 1611:5 pp.

bibliography, acoustics
Moulton, James M., 1962.
References dealing with animal acoustics, particularly marine forms. Second compilation, 78 pp., (mimeographed).
MS owns

bibliographies acoustics
Moulton, James M., 1959? (undated)
References dealing with animal acoustics, particularly of marine forms.
1266 references (mimeographed)

bibliography, acoustics
Schevill, W.E., R.H. Backus and J.B. Hersey, 1962
Sound production in marine animals. Ch. 14, Sect. IV. Transmission of energy within the sea. In: The Sea, Vol. 1, Physical Oceanography, Interscience Publishers, 540-566.
(Mss received September 1960)
WHOI Contrib. No. 1162

air-sea interface

bibliography, air-sea interface
Deacon, E.L., and E.K. Webb, 1962
Small-scale interactions. Ch. 3, Sect. II. Interchange of properties between sea and air. In: The Sea, Interscience Publishers, Vol. 1, Physical Oceanography, pp. 43-87.
(Mss received July 1960)

bibliography, air-sea interface
Panara, R., 1959.
Annotated bibliography on evaporation measurements.
Meteorol. Abstr., & Bibliogr., 10(8):1234-1262.

bibliography, air-sea interface
Sinha, E.Z., 1962.
Annotated bibliography on energy exchange. Ocean-atmosphere.
Meteorol. and Geoastrophys. Abstr., 13(5):1352-1429.

areal

Africa

bibliography, areal (Africa)
Hart, T. John, and Ronald I. Currie, 1960.
The Benguela Current.
Discovery Repts., 31:123-298.

bibliographies, areal (Africa, northwest)
Maurin, Cl. et Y. Aldebert 1970.
Aperçu bibliographique.
Rapp. P.-v. Réun. Cons. int. Explor. Mer 159: 272-289

bibliography, areal (Africa)
Sourie, R., 1954.
Contribution à l'étude écologique des côtes rocheuses du Sénégal.
Mém. Inst. Francais d'Afrique Noire, No. 39: 342 pp., 23 pls.

Antarctic

bibliography, Antarctic
Anonymous, 1963.
Recent Polar literature.
Polar Record, 11(74):639-663.

bibliography, areal (Antarctic)
Mackintosh, N.A., 1940.
Nomenclature of the major divisions of the Southern seas.
Spec. Art., Publ. Sci., Assoc. Ocean. Phys., No. 8:93-94, Figs. 6-7.

bibliographies areal (Antarctic)
Slaucitajs, L., 1954.
Aktivität argentiniens in Antarktisforschung 1902-1953. Geofisica Pura e Applicata, Milano, 29:226-229.

bibliography, areal (Antarctic)
USN Hydrographic Office, 1957.
Oceanographic atlas of the Polar seas. 1. Antarctic. H.O. Pub., No. 705:70 pp.

Arctic

bibliography - areal (Arctic)
Adrov, M.M., 1960
Further investigations of the hydrologic regime of the Barents Sea. (English Resume).
Soviet Fish. Invest. in North European Seas, VNIRO, Moscow, 1960: 9-22.

bibliographies, areal (Arctic)
Anonymous, 1963.
Recent Polar literature.
Polar Record, 11(74):630-663.

bibliography, Arctic
Anon., 1937
Report on the activities of Norges Svalbard og Ishavsundersøkelser 1927-1936.
Skrifter om Svalbard og Ishavet, No.73, 125 pp., 25 textfigs., 2 pls., and 1 pl.

bibliography, areal (Arctic)
Defant, A., 1936.
Bericht über die ozeanographischen Untersuchungen des Vermessungsschiffes "Meteor" in der Danemarkstrasse und in der Irmingersee.
Dritter Bericht Sitzber. Preuss. Akad. Wiss., Phys.-Math Kl. 19:13 pp., 2 pls.

bibliography, areal (Arctic)
Dunbar, M.J., 1951.
Eastern Arctic waters. Fish. Res. Bd., Canada, Bull. No. 88:131 pp., 32 textfigs.

bibliographies areal (Arctic)
Dutilly, A., 1945
Bibliography of Bibliographies on the Arctic. Catholic Univ. Washington, D.C. 47 pp. #1.00.

Bibliography - areal (Arctic)
Gershanovich, D.E., 1963
Bottom relief of the fishing grounds and some features of the Bering Sea geomorphology. (In Russian)
Sovetsk. Ribokh. Issled. B Severo-Vostokh. Chasti Tikhogo Okeana, VNIRO 48, TINRO 50(1): 13-76.

bibliography, areal (Arctic)
Lee, Arthur, 1963.
The hydrography of the European Arctic and subarctic seas.
In: Oceanography and Marine Biology, H. Barnes, Editor, George Allen and Unwin, 1:47-76.

bibliography areal Arctic
Lisitsin, A.P., 1959
[Bottom sediments of the Bering Sea.] Trudy Inst. Okeanol. 29: 65-187. (In Russian)

bibliography, areal (Arctic)
Maldonado-Koerdell, M., 1962
La Exploracion del oceano artico, con especial referencia al extremo norte del continente americano.
Bol. Bibliograf. de Geofisica y Oceanogr. Amer 2(Oceanogr.):365-488.

bibliography, areal (Arctic)
Tremaine, Marie, 1962.
Arctic bibliography.
Arctic, 15(4):247-250

bibliography areal (Arctic)
U.S.N. Hydrographic Office, 1958
Oceanographic atlas of the Polar seas. Pt. II. Arctic. H.O. Publ. No. 705: 139 pp., charts.

bibliography, areal (Arctic)
U.S. Navy Hydrographic Office, 1946.
Ice Atlas of the Northern Hemisphere. H.O. 550

bibliography areal (Argentina)
Popovici, Z., and A.E. Riggi, 1948
Los Estudios de Hidrobiologia en la Argentina, sus relaciones con el plan del superior gobierno de la nacion y sus proyecciones futuras. Mus. Argentino Ciencias Nat. "Bernadino Rivadavia", Miscelanea No.1:171 pp.

Atlantic

bibliographies, Atlantic
Bulatov, R.P., 1964.
Circulation of the waters of the Atlantic Ocean and adjoining seas. Bibliographic index (1638-1962). (In Russian).
Mezhd Geofiz. Kom., Presid. Akad. Nauk,SSSR, Moskva, 115 pp.

bibliography,Atlantic,tropical
Keehn,Pauline A.,1967.
Bibliography on oceanography of the tropical Atlantic.
Spec.Bibliogr.Oceanogr.,Contrib.Am.Met.Soc.,5:87 pp.

bibliography areal (Atlantic, northern)
Stefansson, Unnstein, 1962.
North Icelandic waters.
Rit Fiskideildar, Reykjavik, 3:269 pp.

bibliographies, North Atlantic
United Nations, Food and Agriculture Organization, Fisheries Resources and Exploitation Division, Biological Data Section, 1968.
North Atlantic bibliography and citation index.
FAO Fish. Techn. Pap., 54: unnumbered pp. (mimeographed). (Restricted distribution).

Australia

bibliography areal (Australia)
Dall, W., and W. Stephensen, 1953.
A bibliography of the marine invertebrates of Queensland. Univ. Queensland, Dept. Zool., Papers 1(2):21-47.

bibliography, areal (Australia)
Nicholls, A. G., 1948.
Marine biology in Western Australia. J. Roy. Soc. Western Australia 33:151-162.

Australasia

bibliography, areal (Australasia)
Knox, G.A., 1963.
The biogeography and intertidal ecology of the Australasian coasts.
In: Oceanography and Marine Biology, H. Barnes Editor, George Allen and Unwin, 1:341-404.

bibliography, areal (Baltic)
Segerstråle, Sven G., 1964
Literature on marine biology in the Baltic area published in the years 1953-1962.
Soc. Sci. Fennica, Comment. Biol., 27(3):44 pp.

bibliography, areal (Baltic)
Slaucitajs, L., 1947
Ozeanographie des Rigaischen Meerbusens. Teil 1, Statik. Contrib. Baltic Univ. No.45, Pinneberg, 110 pp., 69 text figs.

bibliographies, areal (Bay of Biscay)
Vanney, J.R. 1970.
Bibliographie analytique des fonds du Golfe de Gascogne.
Notes, Mém. Doc. Lab. Géomorphol. Écol. Pratique Hautes Études, Paris 17: 90 pp. (mimeographed).

bibliographies, Bermuda
Moore, H.B., undated
Ecological guide to Bermuda inshore waters. 48 pp. (duplicated), 18 figs. (ozalid).

Black Sea

bibliographies, Black Sea.
Bacescu, Mihai, 1965.
Bibliographie Roumaine de la Mar Noire.
Comm. Nat. Republique Populaire Roumaine pour l'UNESCO, Bucarest, 121 pp.

bibliography, areal (Black Sea)
Caspers, H., 1951.
Quantitative Untersuchungen über die Bodentierwelt des Schwarzen Meeres im bulgarischen Küstenbereich. Arch. f. Hydrobiol. 45(1/2):192 pp., 66 textfigs.

bibliographies
Ivanoff, S.N., 1940.
Bibliografia del Mar Nero. Cernomorsk. Nauk Inst. 1(42):1-268.

3515 references

bibliography, areal (Black Sea)
Laking, Phyllis N. 1970.
Bibliography of scientific literature on the Black Sea.
Ref. Woods Hole Oceanogr. Inst. 70-32:278 pp. (multilithed) (Unpublished manuscript). Copies may be purchased at cost (U.S.$5.00) from the Woods Hole Oceanographic Institution.

bibliography
Neumann, G., 1944.
Das Schwarze Meer. Zeitsch. Gesellschaft f. Erdkunde zu Berlin. 1944 (3/4):92-114, 25 text figs.

bibliography, areal (Black Sea)
Pova, E.A., 1946.
Problèmes de physiologie animale dans la Mer Noire. Bull. Inst. Océan., Monaco, No. 903:43 pp., 21 pp.

bibliographies areal (Black Sea)
Rojdestvensky, A.V. 1971.
Le deversement dans la mer Noire des eaux de la mer de Marmara.
Cah. océanogr. 23(3):283-289

bibliographies, areal (Black Sea)
Thufas, Valer 1969.
Marea Neagră note de curs, 1. Hidrologie R.S.R.
Univ. București, Fac. Geol. Geogr. 122 pp. (multilithed)

bibliography,Black Sea
*Vinogradov,K.A., editor, 1967.
Biology of the northwestern Black Sea. (In Russian).
Naukova Dumka,Kiev,268 pp.

bibliography, Brazil
Nomura,Hitoshi,1966.
Bibliografia sobre os recursos maronhos do Brasil.
Bolm Estud.Pesca,SUDENE,Recife,6 (2,3,4):41-58; 31-44;51-73.

bibliography, Brazil
Nomura,Hitoshi,1965.
Bibliografia sobre os recursos naminhos do Brasil.
Bol. Estadao Biol. Mar. Univ., Ceara,7:53 pp.

Canada

bibliography, areal (Canada)
Bishop, Y., N.M. Carter, D.Gailus, W.E. Ricker, and J.M. Speirs, 1957.
Index and list of titles, publications of the Fisheries Research Board of Canada, 1901-1954.
Fish. Res. Bd., Canada, Bull. 110:209 pp.

bibliographies areal (Canada)
Canada, Fisheries Research Board, 1962.
Publications and reports.
Ann. Rept. Fish. Res. Bd., Canada, 1961-1962: 173-196.

bibliography, areal (Canada)
Fraser, C. M., 1939(1940).
Oceanography in British Columbia. Proc. Sixth Pacific Sci Congr., 3:20-33.

bibliography, areal (Canada)
Hachey, H.B., 1961.
Oceanography and Canadian Atlantic waters.
Fish. Res. Bd., Canada, Bull., No. 134:120 pp.

bibliographies, areal, Canada
LeBlond,Robert,1967.
Le Saint-Laurent: orientation bibliographique.
Cah.Geogr.Québ., 11(23):419-464.

bibliographies,Canada
Roy, Jean-Marie, Gerard Beaulieu et Claire Talbot 1967.
Index des contributions du Department des Pêcheries.
Minist. Industr. Commerce, Québec 54 pp.

bibliographies, areal, Canada
Tully, John P., 1958.
Canadian Pacific Oceanography since 1953.
Proc. Ninth Pacific Sci. Congr., Pacific Sci. Assoc., 1957, Oceanogr., 16:6-13.

bibliography, Cape Cod region
Yentsch,Anne E., Melbourne R. Carriker, Robert H. Parker and Victor A. Zullo,1966.
Marine and estuarine environments, organisms and geology of the Cape Cod region: an indexed bibliography, 1665-1965.
Marine Biological Laboratory, Woods Hole,178 pp.

Caribbean

bibliographies, areal (Caribbean)
Bayer, Frederick M., Editor, 1963.
Regional bibliography for 1962.
Bull. Mar. Sci., Gulf and Caribbean, 13(4):578-659.

bibliographies, areal (Caribbean)
U.S.A. Bulletin of Marine Sciences, Gulf and Caribbean, 1964.
Regional bibliography for 1963.
Bull. Mar. Sci., Gulf and Caribbean, 14(4):663-749.

bibliographies, areal (Caribbean)
USA, University of Miami Press, 1962.
Regional bibliography for 1961.
Bull. Mar. Sci., Gulf and Caribbean, 12(4):702-756.

bibliography- English Channel
Armstrong, F.A.J., and E.I. Butler, 1962.
Hydrographic surveys off Plymouth in 1959 and 1960.
J. Mar. Biol. Assoc., U.K., 42(2):445-463.

bibliography areal (Germany)
Boehnecke, G., and Collaborators, 1948
German Oceanographic Work, 1934-1945.
Trans. Am. Geophys. Union, 29(1):59-68.

bibliography, areal (Germany)
Hansen, W., 1952.
Physikalische Ozeanographie in Deutschland 1945-1952. Assoc. Océanogr. Phys., Pros. Verb. No. 5:90-103.

bibliography, areal (Gulf of Mexico)
Geyer, R.A., 1950? (undated).
Bibliography of oceanography, marine biology, geology, geophysics, and meteorology of the Gulf of Mexico. Humble Oil Refining Co. (multilith) (Pages not numbered serially).

bibliography, areal (Gulf of Mexico)
Galtsoff, P.S., ed., 1954.
Gulf of Mexico, its origin, waters and marine life. Fish. Bull., Fish and Wildlife Service, 55:1-604, 74 textfigs.

bibliography, areal (Gulf of Mexico)
Geyer, R.A., 1950.
Bibliography of the Gulf of Mexico.
Texas J. Sci. 2(1):44-92.

bibliography, areal (Gulf of Mexico)
Geyer, R.A., 1948
Annotated bibliography of the Gulf of Mexico and contiguous area. Geophys. Res. Div., Expl. Dept., Humble Oil and Refining Co., 37 pp. (multilith)

bibliography, areal (Gulf of Mexico)
Gunter, G., 1952.
Historical changes in the Mississippi River and the adjacent marine environment. Publ. Inst. Mar Sci. 2(2:121-139.

bibliography, areal (Gulf of Mexico)
Murray, Grover E., 1966.
Salt structures of Gulf of Mexico Basin- A review.
Bull. Amer. Assoc. Petr. ol., 50(3-1):439-

bibliographies, Gulf of Mexico
United States, National Oceanographic Data Center 1970
CICAR bibliography on meteorology, climatology and physical/chemical oceanography.
Vol. 1: 350pp.
Vol. 2: 614pp.

bibliography, areal (Gulf of Mexico)
Vaughn, T.W., 1940.
The classification and nomenclature of the submarine features of the Gulf of Mexico and the Caribbean Sea. Assoc. d'Ocean. Phys., Spec. Art., Publ. Sci. No. 8:61-76, 1 fold-in.

bibliographies, areal (India)
Kanakasabapathi, K. 1970.
Bibliography of contributions from Central Marine Fisheries Research Institute, 1947-1969.
Bull. cent. mar. Fish. Res. Inst. 19: 73pp. (mimeographed)

bibliographies, areal (India)
Lal Mohan, R.S., D.B. James and K.K. Appukuttan 1969.
Bibliography of the Indian Ocean 1931-1961: a supplement to the "Partial Bibliography".
Bull. cent. mar. Fish. Res. Inst. Mandapam Camp 11:176pp.

bibliographies areal (India)
George, P.C. and P. Vedavyasa Rao 1967
An annotated bibliography of the biology and fishery of the commercially important prawns of India.
Proc. Symp. Crustacea, Ernakulam, Jan 12- 1965:1521-1547

bibliographies, areal (India)
George, P.C. and P. Vedavyasa Rao 1967.
An annotated bibliography of the biology and fishery of the edible crabs of India.
Proc. Symp. Crustacea, Ernakulam, Jan. 12- 1965:1548-1555.

Indian Ocean

bibliographies, Indian Ocean
Angot, D., Compiler, 1970.
Bibliographie des travaux scientifiques marins interessant Madagascar (quatrième liste), Centre ORSTOM de Nosy-Bé.
Imprimerie Nationale, Tananarive, 27pp.

bibliographies, Indian Ocean
Angot, M., 1967.
Bibliographie des travaux scientifiques marins interessant Madagascar.
Imprimerie Nationale, Tanarive, 24 pp.

bibliography Indian Ocean
France, Service Hydrographique de la Marine, 1963
Bibliographie des travaux scientifiques français dans l'océan indien figurant dans les publication du Service Hydrographique de la Marine ou émanant de marins et d'hydrographes.
Ann. Hydrogr., Paris, (4), 11:337-348.

bibliography areal (Indian Ocean)
Muromtsev, A.M., 1959
[The basis for the hydrology of the Indian Ocean.] Gidrometeoizdat, Leningrad, 437 pp.

bibliographies, areal (Indian Ocean)
Murty, V. Shiramachandra, D.C.V. Easterson and A. Bastian Fernando 1969.
Bibliography of the Indian Ocean, 1968, with a supplement for 1962-1967.
Bull. Centr. Mar. Fish. Res. Inst. Mandapam Camp, India 5: 146pp. (mimeographed).

bibliographies areal (Indian Ocean)
Samuel, Parkash, 1965.
Bibliography on physical oceanography of the Indian Ocean.
Amer. Meteorol. Soc., Meteorol. Geoastrophys. Abstr., Spec. Bibliogr., Oceanogr., Contrib., No. 2:122 pp. (multilithed).

bibliographies, areal (Indian Ocean)
Yentsch, Anne E., Compiler, 1962
A partial bibliography of the Indian Ocean. International Indian Ocean Expedition, U.S. Program in Biology, Woods Hole Oceanographic Institution, 390 pp.

Japan

bibliography, areal (Japan)
Japan, National Committee for the International Geophysical Coordination, Science Council of Japan, 1962
X. Oceanography.
Japan. Contrib. Int. Geophys. Year and Int. Geophys. Cooperation, 4:142-150.

bibliographies, areal (Japan)
Japan, Japanese National Commission for UNESCO, 1961
Oceanographic papers in Japan (Annotated bibliography). III: 180 pp. (multilithed).

bibliographies, areal (Japan)
Japan, Japanese National Commission for UNESCO, 1958.
Oceanographic papers in Japan (annotated bibliography)(1939-1957) II:217 pp.

bibliographies, areal (Japan)
Japan, Japanese National Commission for UNESCO, 1957.
Oceanography papers in Japan (Annotated bibliography) (1873-1938):235 pp.

bibliographies, areal (Japan)
Japan, Science Council of Japan, 1961.
Contents of previous issues of the Records of Oceanographic Works in Japan. 1. New Series. Published by the Japanese National Commission for UNESCO.
Rec. Oceanogr. Wks., Japan, 6(1):133-144.

Jugoslavia

bibliographies areal (Jugoslavia)
Jugoslavia, Godionjica Instituta za Oceanografiju i Ribarstvo, Split, 1961.
Spomen-Kniga, 1930-1960:35 pp.

bibliographies, Madagascar
Angot, M., 1966.
Bibliographie des travaux scientifiques marins interessant Madagascar.
Bull. Madagascar, (239):1-64.

Mediterranean

bibliography, areal (Mediterranean)
Cannicci, Gabriella, 1959
Considerazioni sulla possibilità di stabilire "indicatori ecologici" nel plancton del Mediterraneo. Boll. Pesca, Piscicolt, e Idrobiol. n.s., 14(2): 164-188.

bibliographies areal (Mediterranean)
Commissione Internazionale per l'Esplorazione Scientifica del Mare Mediterraneo-Delegazione Italiana 1956
Bibliographia Vol. 23(1950)

bibliography, areal (Mediterranean)
Lacombe, H., 1960
Bibliographie concernant la Méditerranée. (Physique de la mer). Comm. Int. Expl. Sci. Medit., Monaco, Rapp. Proc. Verb. des Reunions, 15(3): 221-226.

bibliographies, areal (Mediterranean)
Navarro, F. de P., 1943.
Operaciones oceanograficas en la Bahia de Palma de Mallorca (1932-1934). Registro y notas. Notas y Res. (2) 116:1-31.

bibliography, areal (Mediterranean)
Navarro, F. de P., 1942.
Bibliografía para un catalogo de la Fauna y Flora del Mar de Baleares. Trab. Inst. Español.Ocean. No. 15:93 pp.

bibliography, Mediterranean, west
Segre, A.G., 1961.
2. Considerazioni geologiche, commento alle anomalie gravimetriche e ai dati geomagnetici sull' Isola di Pantelleria.
Boll. Geofis. Theorica ed Applicata, Trieste, 3(12):273-287.

New Zealand

bibliography, areal (New Zealand)
Freed, D., 1963.
Bibliography of New Zealand marine biology, 1769-1899.
New Zealand Dept. Sci. Indus. Res., Bull., 148:46 pp.
Also: New Zealand Oceanogr. Inst., Mem., No. 16.

bibliography areal (New Zealand)
New Zealand Oceanographic Committee, 1955.
Bibliography of New Zealand Oceanography, 1949-1953. Geophys. Mem., N.Z. D.S.I.R., 4:19 pp.

bibliographies areal (North Sea)
Model, Fritz, 1964.
Geophysikalische Bibliographie von Nord- und Ostsee. 1. Chronologische Titelaufzahlung.
Deutsches Hydrogr. Inst., Hydrogr. Dokumentation 1961 pp. (mimeographed) (Unpublished manuscript).

bibliography, North Sea

Stocks, T., 1950.
Die Tiefenverhältnisse des Europäischen Nordmeeres. Deutsche Hydro. Zeits. 3(1/2):93-100, Pl. 6.

Norway

bibliographies, areal (Norway)

Brattström, Hans and Hans Tambs-Lyche, 1962.
Publications on marine fauna and flora off western Norway, 1949-1960.
Sarsia, Univ. i Bergen, 7:29-45.

bibliographies, areal (Norway)

Hoel, A., 1952.
Norsk Ishavsfangst. En fortegnelse over Litteratur. Medd. Norsk Polarinst. No. 69:23 pp.

Anon., 1935. bibliographies, areal (Pacific)
Bibliografiia Dal'nevostochnogo kraia 1890-1931, T.I. Fizicheskaia geografiia. Moscow.
(Bibliography of Far Eastern Regions, 1890-1931. V. 1. Physical Geography.)

bibliography (Pacific)

Dobrovolsky, A.D., Editor, 1968.
Hydrology of the Pacific Ocean. (In Russian).
P.P. Shirsnor Inst. Okeanol., Akad. Nauk, Isdatel "Nauka," Moskva, 524 pp.

bibliography, Pacific

Furon, Raymond, 1956(1953).
Importance paleogeographique des mouvements de subsidence du Pacifique central.
Proc. 8th Pacific Sci. Congr., Geol., Geophys., Meteorol., (Nat. Res. Counc., Philippines), 2A: 891-900.

bibliography- areal (Pacific, northeast)

Plachotnik, A.F., 1962.
Hydrology of the northeastern part of the Pacific Ocean (literature survey). Investigations on the program of the International Geophysical Year.
Trudy Vses. Nauchno-Issled. Inst. Morsk. Ribn. Chos. i Okean., VNIRO, 46:190-201.

In Russian

bibliography areal (Pacific, northwest)

Uda, M., 1939(1940).
A sketch of the recent development of hydrographical researches in the sea adjacent to Japan. Proc. Sixth Pacific Sci. Congr., 3:44-72.

bibliographies areal Pacific, southwest

Krebs, Betty N., 1964.
A bibliography of the oceanography of the Tasman and Coral seas, 1860-1960.
New Zealand Dept. Sci. Industr. Res., Bull., 156:27 pp.

Also:
New Zealand Oceanogr. Inst., Mem. (24).

bibliography Pacific, west

Hydrographic Office, 1946
References on the physical oceanography of the Western Pacific. H.O. Pub. 238, 174 pp.

bibliographies, areal (Poland)

Poland Instytut Geografii, Polska Akademia Nauk, 1957.
Bibliografia geografii Polskiej, 1952-1953:99 pp.

bibliography, areal (Polynesia)

Salvat, Bernard, 1970.
Les activités du Muséum National d'Histoire Naturelle en Polynésie française (sciences de la mer)
Cah. Pacifique, 14: 255-269.

institutions, France
Polynesie
bibliography, French Polynesia

bibliographies, Sénégal

Sénégal, Centre de Recherches Océanographiques de Dakar-Thiaroye, 1971
Liste des publications du Centre de Recherches Océanographiques de Dakar-Thiaroye à partir de 1966.
D.S.P. No. 34: unnumbered pp. (mimeographed)

South America

bibliography, Latin America

Emilsson, Ingvar, 1962.
Contribução à bibliografia Latino-Americana.
Setor Gráfico, Instituto Oceanográfia, São Paulo, Brasil, Outubro, 1962:16 pp.

bibliographies, S. America, west

Gilmartin, Malvern, 1964
Compilacion bibliografica sobre la oceanografia de las aguas litorales de Colombia, Ecuador y Peru, con especial referencia al fenomeno "El Nino".
Bol. Cient. Tecn. Inst. Nac., Pesca del Ecuador, Guayaquil, 1(1):15 pp.

bibliography, areal (Strait of Gibraltar)

DeBuen, R., 1927.
Résultats des investigations espagnoles dans le détroit de Gibraltar. Cons. Perm. Int. Expl. Mer, Rapp. Proc. Verb. 44:60-91, Textfigs. 29-50.

Bibliographies, Thailand

Thailand, Kasetsart University, Committee on Fishery Leterature, 1965.
First guide to the literature of the aquatic sciences and fisheries in Bangkok, Thailand.
Kasetsart Univ. Fish. Res., Bulll, No. 2:46 pp.

United States

bibliography, areal (U.S.A.)

California, Department of Fish and Game, Marine Research Committee, 1956.
Cooperative Oceanic Fisheries Investigations, Progress Report, 1 April 1955-30 June 1956:44 pp

bibliography, areal, U.S.A.

Colton, John B., Jr., 1964.
History of oceanography in the offshore waters of the Gulf of Maine.
U.S.F.W.S. Spec. Sci. Rept., Fish., No. 496:18 pp

bibliography, areal U.S.A.

Livingstone, Robert, Jr., 1965.
A preliminary bibliography with KWIC index on the ecology of estuaries and coastal areas of the eastern United States.
Spec. Scient. Rep. U.S. Fish. Wildl. Serv. Fish. No. 509 352 pp. (multilithed).

bibliography, areal (U.S.A.)

Nichols, Maynard M., 1962.
Bibliography of the physical, chemical and geological oceanography of Chesapeake Bay.
Virginia Inst., Mar. Sci., Glocester Pt., Spec. Sci. Rept. No. 34:43 pp. (mimeographed).

bibliography, areal, U.S.A.

Princeton University, Department of Civil Engineering, River and Harbor Program, 1958 (21 July).
Bibliography: Currents along the Eastern Seaboard of the United States and Coast Erosion. 6 pp. (mimeographed).

bibliography, areal (U.S.A)

Scruton, P.C., 1956.
Oceanography of Mississippi delta sedimentary environments. Bull. Amer. Assoc. Petr. Geol., 40(12):2864-2952.

bottom sediments - areal U.S.A.

Shepard, F.P., & R.R. Lankford, 1959
Sedimentary facies from shallow borings in lower Mississippi delta.
Bull. Amer. Assoc. Petrol. Geol., 43(9): 2051-2067.

bibliography, areal (U.S.A)

United States, National Academy of Sciences -National Research Council, 1962
Oceanography 1960-1970. 11. A history of oceanography: a brief account of the development of oceanography in the United States, 28 pp.

U.S.S.R

bibliographies, areal (USSR)

Anon., 1961
Soviet bibliography on oceanography.
Compiled from Refer. Zhurnal, Feb. 1960-June 1961. Listed in Techn. Transl., 6(10):753.

bibliographies areal (USSR)

Office of Technical Services, 1960

Soviet bibliography on oceanography.
U.S. Joint Publ. Res. Service, 205 E 42nd St., Suite 300, New York 17.

Bibliographies areal (USSR)

SSSR, Mezhduvedomstvennii Geofizicheskii Komitet pri Presidiume Akademii Nauk, SSSR, 1961.
[Bibliographic index to the literature in the Russian language, 1960.]
Mezhd. Geofiz. God, 124 pp.

bibliography, areal (U.S.S.R)

SSSR, Okeanograficheskoi Komissii, Akademia Nauk, 1959.
[Literature. Problems of the Caspian Sea.]
Trudy Okeanogr. Komissii, Akad. Nauk, SSSR, 5: 423-432.

bibliography, beaches

bibliography, beaches

Dolen, Robert and James McCoy, 1965.
Selected bibliography on beach factures and related nearshore processes.
Louisiana State Univ. Studies, Coastal Studies Ser. 11:59 pp.

biology

Bibliographies, biological

Algae through whales

bibliography, neurosecretion

Antheunisse, L.T., 1963.
Neurosecretory phenomena in the zebra mussel Dreissena polymorpha Pallas.
Arch. Nederl. Zool., 15(3):237-314.

bibliography

Briseno C., B., 1952.
Secreciones internas de los invertebrados.
Rev. Soc. Mexicana Hist. Nat. 13:1-21.

bibliographies

Douglas, B., T.T. Macan and J.C. Mackereth, 1956.
Abstracts of papers in the field of freshwater biology published in British journals in 1954.
Hydrobiologia 8:93/4):300-322.

BIBLIOGRAPHY

Geyer, R.A., 1948
Mass mortality of aquatic life from natural causes - bibliography of Geophys. Res. Div., Expl. Depart., Humble Oil and Refining Co. 15 pp. (multilith).

bibliography

Habe, T., H. Utinomi and D. Miyadi, 1957.
Some contributions on the taxonomy of marine animals by Japanese systematic zoologists.
Proc. UNESCO Symp., Phys. Ocean., Tokyo, 1955: 202-203.

bibliography

Jørgensen, C. Barker, 1955.
Quantitative aspects of filter feeding in invertebrates. Biol. Rev. 30(4):391-354.

bibliography
Kinne, Otto, 1963
The effects of temperature and salinity on marine and brackish water animals. In: Oceanography and Marine Biology, H. Barnes, Edit., George Allen & Unwin, 1:301-340.

bibliography
Krogh, A., 1931.
Dissolved substances as food of aquatic organisms Cons. Perm. Int. Expl. Mer, Rapp. Proc. Verb. 75: 7-36.

bibliography
Lebour, M.V., 1947
Notes on the Inshore Plankton of Plymouth JMBA 26(4):527-547, 1 textfig.

bibliography
Monod, Th., and M. Nickles, 1952.
Notes sur quelques Xylophages et Pétricoles marins de la côte ouest africaine.
I.F.A.N. Catalogues 8:7-68, 151 textfigs.

bibliography
Redfield, A. C., 1933.
The evolution of the respiratory function of the blood. Quart. Rev. Biol. 8(1):31-57, 6 textfigs.

WHOI Contrib. No. 13.

bibliographies
Schwartz, Frank J., 1961.
A bibliography of external forces on aquatic organisms.
Chesapeake Biol. Lab., Solomons, Maryland, Contrib. No. 168:85 pp. (multilithed)

bibliography
Schwenke, Heinz, 1960
Neuere Erkenntnisse uber die Beziehungen zwischen den Lebensfunktionen mariner Pflanzen und dem Salzgehalt des Meer-und Brackwassers. Kieler Meersf., 16(1): 28-37.

bibliography
Vinogradov, A.P., 1953 (translation).
The elementary chemical composition of marine organisms. Mem. Sears Found. Mar. Res. 2:1-647.

Translated by Julia Efron and J.K. Seltow with bibliography by V.W. Odum.

bibliography, learning
Wells, M. J., 1965.
Learning by Marine invertebrates.
In: Advances in Marine Biology, Sir Frederick S. Russell, editor, Academic Press, 3:1-62.

algae

bibliographies, algae

See also: diatoms
silicoflagellates

bibliography, algae, economic importance (25pp
Boney, A. D., 1965.
Aspects of the biology of the seaweeds of economic importance.
In: Advances in Marine Biology, Sir Frederick S. Russell, Academic Press, 3:105-253.

bibliographies, algae
Canada, Nova Scotia Research Foundation, 1970.
Selected bibliography on algae 10: 167 pp.

bibliographies, algae
Canada, Nova Scotia Research Foundation, 1968.
Selected bibliography on algae, 9:170 pp.

bibliographies, algae
Canada, Nova Scotia Research Foundation, 1967.
Selected bibliography on algae, 8:116 pp.

bibliography, algae
Canada, Nova Scotia Research Foundation, 1966.
Selected bibliography on algae, No.7:105 pp.

bibliographies, algae
Canada, Nova Scotia Research Foundation, 1965.
Selected bibliography on algae, No. 6:217 pp. $5.00.

(Earlier at $2.50)
Halifax, Nova Scotia

bibliography, algae
Dawson, E.Y., 1947
A guide to the Literature and Distributions of the Marine Algae of the Pacific Coast of North America. Mem. S. Calif. Acad. Sci., 3(1):1-134.

bibliography, algae
Dixon, P.S., 1963
The Rhodophyta: some aspects of their biology. In: Oceanography and Marine Biology, H. Barnes, Edit., George Allen & Unwin, 1: 177-196.

bibliography, algae
Humm, Harold J., 1962.
Bibliographic data useful in the study of marine algae.
Virginia Inst. Mar. Sci., Gloucester Pt., Spec. Sci. Rept., No. 29:11 pp. (mimeographed).

bibliography, algae (Laminaria)
Kain, Joanna M. (Mrs. N.S. Jones) 1971
Synopsis of biological data on Laminaria hyperborea.
FAO Fish. Synopsis 87: numerous pp.

bibliography, algae
Levring, T., 1947.
Submarine daylight and the photosynthesis of marine algae. Medd. Oceanogr. Inst., Göteborg 14 (Göteborgs Kungl. Vetenskaps- och Vitterhets-samhälles Handl., Sjätte Följden, Ser.B 5(6)): 1-89, 32 textfigs.

bibliography, algae
Lewin, Ralph A., Editor, 1962.
Physiology and biochemistry of algae.
Academic Press, New York and London, 929 pp.

bibliography, kelp (algae)
North, Wheeler J. editor, 1971.
The biology of giant kelp beds (Macrocystis) in California.
Beihefte Nova Hedwigia 32: 600 pp.

bibliography, algae
Papenfuss, G.F., 1968.
A history, catalogue, and bibliography of Red Sea benthic algae.
Israel J. Bot. 17(1): 1-118.

bibliographies, algae
Parke, Mary, and Peter S. Dixon, 1968.
Checklist of British marine algae - second revision.
J. mar. biol. Ass., U.K., 48(3):783-832.

bibliography, algae
Provosoli, Luigi, 1958
Nutrition and ecology of protozoa and algae.
Ann. Rev. Microbiol., 12:279-308.

bibliography, algae
Saoane-Cambra, Juan, 1965.
Estudios sobre las algas bentónicas en la costa sur de la Peninsula Ibérica (litoral de Cádiz)
Inv. pesq., Barcelona, 29:3-216.

bibliography, algae
Scagel, Robert F., 1961.
Marine plant resources of British Columbia.
Fish. Res. Bd., Canada, Bull. No. 127:1-89.

bibliography, algae
Scottish Seaweed Research Association, 1950.
Annual report for 1950:35 pp.

bibliography, phycology (algae)
Silva, P.C., 1970.
Phycological literature for 1963.
Phycologia 7 (3/4): 255-367.

bibliography, algae
Silva, P.C. compiler 1967
Phycological literature for 1962.
Phycologia 4 (4): 201-302

bibliography, algae
Symoens, J. -J., 1964.
Un siecle de recherches belges sur la floristique et l'ecologie des algues.
Bull. Soc. Roy. Botan., Belgique, 95:153-191.

annelids

bibliography, annelids
Åkesson, Bertil, 1962.
The embryology of Tomopteris helgolandica (Polychaeta).
Acta Zoologica, 43(2/3):135-199.

bibliography, annelids
Bellan, G., 1964.
Contribution à l'étude systématique, bionomique et écologique des annelides polychètes de la Méditerranée.
Rec. Trav. Sta. Mar., d'Endoume, 49(33):1-371.

bibliographies, annelids
Dales, R.P., 1957.
Pelagic polychaetes of the Pacific Ocean.
Bull., S.I.O., 7(2):99-168.

bibliography, annelids
Dales, R. Phillips, 1956.
An annotated list of the pelagic Polychaeta.
Ann. Mag. Nat. Hist., (12) 0(100):289-304.

bibliographies, annelids
Fauchald, K. 1963.
Nephtyidae (Polychaeta) from Norwegian waters.
Sarsia 13: 1-32.

bibliography, annelids
Friedrich, H., 1950
Vorkommen und Verbreitung der pelagischen Polychaeten im Atlantischen Ozean. Auf Grund der Fänge der "Meteor" Expedition. Kieler Meeresforschungen 7(1):5-23, 6 text figs.

bibliographies, annelids
Hartman, Olga 1965.
Catalogue of the polychaetous annelids of the world. Supplement 1960-1965 and index.
Allan Hancock Found. Publ. Occ. Paper 23: 197 pp.

Imajima, Minoru, 1967 bibliography, annelids
The Syllidae (polychaetous annelids) from Japan. VI. Distribution and literature.
Publs Seto Mar. biol. Lab., 14(5):351-368.

Japan
annelids, lists of species
bibliography, annelids

bibliographies, annelids
Kirkegaard, J.B. 1969.
A quantitative investigation of the central North Sea Polychaeta.
Spolia zool. Mus. haun. 29: 285 pp.

bibliography, annelids
Rullier, F., 1963.
Les annélides polychètes du Bosphore, de la Mer de Marmara et de la Mer Noire, en relation avec celles de la Méditerranée.
Rapp. Proc. Verb., Réunions, Comm. Int. Expl. Sci., Mer Méditerranée, Monaco, 17(2):161-260.

antibiotics

Bibliography, antibiotic
Aubert, M., J. Aubert, M. Gauthier et S. Daniel, 1966.
Origine et nature des substances antibiotiques présentes dans le milieu marin.
1. Étude bibliographique et analyse des travaux.
2. Méthodes d'étude et techniques experimentales.
Revue Int. Oceanogr. Médicale, C.E.R.B.O.M., 1:9-26; 27-34.

bacteria

bibliography, bacteria
Sieburth, John McN., 1960.
Soviet, aquatic bacteriology: a review of the past decade.
Quart. Rev. Biol., 35(3):179-205.

bibliography, bacteria
Vargues, H., 1962.
Contribution à l'étude du caractère halophile chez les bactéries isolées du milieu marin.
Bull. Inst. Océanogr., Monaco, 59(1231):176 pp.

bibliography, bacteria
Zobell, C.E., 1950.
Annotated bibliography on the ecology of marine bacteria. Comm. N.R.C., Div. Geol., Geogr. 10:31-54.

benthos

bibliography, benthos
Bacescu, Mihai 1971.
Rapport sur les travaux récents concernant le benthos de la Méditerranée et des mers dépendantes (1966-1968).
Rapp. P.-v. Réun. Comm. int. Explor. scient. Mer Médit. 20(2): 37-74

bibliography, benthos
Edmondson, W.T., editor, 1966.
Ecology of Invertebrates, Marine Biology, III. Proceeding of the Third International Interdisciplinary Conference.
New York Acad. Sci., 313 pp. ($7.00).

bibliography, benthos
Kitamori, R., 1963.
Studies on the benthos communities of littoral areas in the Seto-Inland Sea and the adjacent waters. (In Japanese; English abstract).
Bull. Naikai Reg. Fish. Res. Lab., 21:1-90.

bibliography, benthos
McIntyre, A.D. 1970
Bibliography on methods of studying the marine benthos.
FAO Fish. Techn. Pap. 98: 96 pp.

bibliography, benthos
McLaughlin, P.A., 1963.
Survey of the benthic invertebrate fauna of the eastern Bering Sea.
U.S.F.W.S. Spec. Sci. Rept., No. 401:75 pp.
Fish.,

bibliography, benthos,
Parker, Robert H., 1964.
Zoogeography and ecology of some macroinvertebrates, particularly mollusks, in the Gulf of California, and the continental slope off Mexico.
Vidensk. Medd., Dansk Naturh. Foren., 126:1-178.

bibliography, benthos
Peres, Joan-Marie, 1968.
Rapport du Comité benthos (1964-1966).
Rapp.P.-V.Réun.Comm.int.Explor.Sci.Méditerranée, 19(2):35-75.

bibliography, benthos
Pérès, J.M., 1961.
Rapport sur les travaux récents concernant le benthos de la Méditerranée et de sa dépendances.
Rapp. Proc. Verb., Réunions, Comm. Int. Expl. Sci. Mer Méditerranée, Monaco, 16(2):377-424.

bibliography, benthos
Ushakov, P.V., 1958
Investigations of the bottom fauna of the Far Eastern Seas of the USSR.
Proc. Ninth Pacific Sci. Congr., Pacific Sci. Assoc., 1957, 16(Oceanogr.):210-216.

bibliography, benthos
Vinogradova, N.G., 1962
Some problems of the study of deep-sea bottom fauna.
J. Oceanogr. Soc., Japan, 20th Ann. Vol., 724-741.

bioluminescence

bibliography, bioluminescence
Harvey, E.N., 1952.
Luminescent organisms. Amer. Sci. 40(3):468-481, 13 textfigs.

bibliography
Harvey, E.N., 1952.
Bioluminescence. Academic Press, Inc., N.Y., 650 ppl

bibliography, bioluminescence
Nicol, J.A.C., 1962.
Animal luminescence. In: Advances in comparative physiology and biochemistry, O. Lowenstein, Edit., 217-273. Academic Press, New York and London, 392 pp.

birds

bibliography birds (albatrosses)
Frings, Hubert, Mable Frings, and Carl Frings, 1966.
An annotated bibliography on North Pacific albatrosses.
Pacif. Sci., 20(3):312-337.

bibliography, bird migration
Griffin, Donald R., 1969.
The physiology and geophysics of bird navigation.
Q. Rev. Biol. 44(3): 255-276

bibliography, birds
Hays, Helen, and Robert W. Risebrough 1972
Pollutant concentrations in abnormal young terns from Long Island Sound.
The Auk 89 (1): 19-35

bibliography, birds
Løvenskiold, Herman L., 1963.
Avifauna Svalbardensis.
Norsk Polarinstitutt, Skrifter, No. 129:460 pp.

Bibliographies, biological.

Borers, see: marine borers

bibliographies, Brachiopoda
Dawson, Elliot W. 1971.
A reference list and bibliography of the Recent Brachiopoda of New Zealand.
J. Roy. Soc. New Zealand 1 (2): 159-174.

cephalopods

bibliography, cephalopods
Adam, W., 1962.
Cephalopodes de l'Archipel du Cap Vert, de l'Angola et du Mozambique.
Trab. Cent. Biol. Pisc., No. 32:9-64.

Also in:
Mem. Junta Invest. Ultram., (2), No. 33:

bibliography, cephalopods
Mangold-Wirz, K., 1963.
Biologie des Céphalopodes benthiques et nectoniques de la Mer Catalane.
Vie et Milieu, Suppl., 13:1-295.

chaetognath

bibliography, chaetognaths
Fowler, G.H., 1906.
The Chaetognatha of the Siboga Expedition with a discussion of the synonymy and distribution of the group. Rept. Siboga Exped. 21:1-88, 3 pls., 6 charts.

bibliographies chaetognaths
Russell, F.S., 1939.
Chaetognatha. Fiches d'Ident. Zooplancton, Cons. Perm. Int. Expl. Mer, 1:4 pp., 12 textfigs.

bibliography, chaetognaths
Thomson, J. M., 1947
The Chaetognatha of South-eastern Australia.
Counc. Sci. & Ind. Res., Australia, Bull. No.222, (Div. Fish. Rept. 14), 43 pp., 8 text figs.

bibliography, chaetognaths
Tokioka, Takase, 1962
The outline of the investigations made on chaetognaths in the Indian Ocean. (In Japanese and English).
Info. Bull. Planktology, Japan, No. 8:5-11.

bibliographies, chaetognaths
Tokioka, T., 1939.
Chaetognaths collected chiefly in the bays of Sagami and Suruga with some notes on the shape and structure of the seminal vesicles.
Rec. Oceanogr. Wks., Japan, 10(2):123-150.

bibliography, chaetognaths
von Ritter-Záhony, R., 1911
Vermes. Chaetognathi. Das Tierreich 29:34 pp., 16 text figs.

owned by MS

Bibliographies, biological

chordates, see: tunicates

cirripeds

bibliography cirripeds
Feyling-Hansen, R.W., 1953.
The barnacle Balanus balanoides (Linne 1766) in Spitzbergen. Skr. Norsk. Polarinst. No. 98:1-64, 15 textfigs., 9 pls.

Coelenterates

Bibliography, coelenterates
Berrill, N. J., 1949.
Developmental analysis of Scyphomedusae.
Biol. Rev. 24(4):393-410, 7 textfigs.

bibliography, coelenterates
Daniel, R., and A. Daniel, 1963.
On the siphonophores of the Bay of Bengal. 1. Madras coast.
J. Mar. biol. Ass., India 5(2): 185-220.

bibliography, coelenterates
Edwards, C., 1965.
The hydroid and the medusa Bougainvillia principis, and a review of the British species of Bougainvillia.
J. Mar. biol. Ass., U.K., 46(1):129-152.

bibliography coelenterates
Fewkes, J. W., 1884.
III. Acalephs. Bibliography to accompany "Selections from embryological monographs" compiled by Alexander Agassiz, Walter Faxon, and E.L. Mark". Bull. M.C.Z. 11(10):209-238.

bibliography, coelenterates
Hodgson, M.M., 1950.
A revision of the Tasmania Hydroida. Papers and Proc., Roy. Soc., Tasmania, 1949:1-65, 93 textfigs.

bibliography, coelenterates
Kramp, P.L., 1965.
The Hydromedusae of the Pacific and Indian Oceans. Dana Report No. 63: 162 pp.

bibliography, coelenterates
Kramp, P.L., 1961.
Synopsis of the medusae of the world.
Jour. Mar. Biol. Assoc., U.K., 40:1-469.

bibliography, medusae (freshwater), coelenterates
Matthews, Donald C., 1966.
A comparative study of Craspedacusta sowerbyi and Calpasoma dactyloptera life cycles.
Pacific Science, 20(2):246-259.

bibliographies, coelenterates
Milliman, John D., 1965.
An annotated bibliography of recent papers on corals and coral reefs.
Atoll. Res. Bull., No. 111:1-58.

bibliography, coelenterates
Rees, W.J., 1957
Evolutionary trends in the classification of capitate hydroids and medusae.
Bull. Brit. Mus. (N.H.), Zool., 4(9):453-534, Pls., 12-13, 58 textfigs.

bibliography, coelenterates
Roxas, H.A., 1936.
A fresh-water jelly-fish in the Philippines.
Philippine J. Sci. 60:37-44.

bibliography, coelenterates
Russell, F.S., 1953
The medusae of the British Isles. Anthomedusae, Leptomedusae, Limnomedusae, Trachymedusae and Narcomedusae. Cambridge University Press. 530 pp., 35 pls, 319 text figs.

bibliographies, coelenterates
Russell, F.S., 1950.
Hydromedusae, Family:Corynidae. Fiches d'Ident. Zooplancton, Cons. Perm. Int. Expl. Mer, 29:4 pp 8 textfigs.

bibliographies, coelenterates
Russell, F.S., 1950.
Hydromedusae, Family: Tubulariidae (contd.), Family: Margelopsidae. Fiches d'Ident. Zooplancton, Cons. Perm. Int. Expl. Mer, 28:3 pp., 5 text figs.

bibliographies, coelenterates
Russell, F.S., 1939.
Hydromedusae, Family: Tubulariidae. Fiches d'Ident. Zooplancton, Cons. Perm. Int. Expl. Mer, 2:4 pp., 7 textfigs.

bibliography, coelenterates
Thiel, Hjalmar, 1962
Untersuchungen über die Strobilisation von Aurelia aurita Lam. an einer Population der Kieler Förde.
Kieler Meeresf., 18(2):198-230.

bibliography, siphonophores (coelenterates)
Totton, A. K., and H. E. Bargmann, 1965
A synopsis of the Siphonophora.
Trustees of the British Museum, 230 pp. 153 textfigs and 40 pls.

bibliography, coelenterates
Vannucci, M., and W.J. Rees, 1961
A revision of the Genus Bougainvillia (Anthomedusae).
Bol. Inst. Oceanogr., Univ. Sao Paulo, 11(2):57-100.

Crustacea

bibliography, Crustacea
Cushing, D.H., 1951.
The vertical migration of Crustacea. Biol. Rev. 26(2):158-192.

bibliography, Crustacea
Gordon, Isabella 1966
Crustacea - general considerations.
Mém. Inst. fond. Afr. Noire 77: 27-86.

bibliography, Crustacea
Green, J., 1965.
Chemical embryology of the Crustacea.
Biol. Rev., 40(4):580-599.

bibliography, Crustacea
Knowles, F.G.W., and D.B. Carlisle, 1956.
Endocrine control in the Crustacea.
Biol. Rev. 31(4):396-473.

bibliography, Crustacea
Kurata, Hiroshi, 1962
Studies on the age and growth of Crustacea.
Bull. Hokkaido Reg. Fish. Res. Lab., No. 24: 1-115.

bibliography, crustaceans
Renaud, L., 1949.
Le cycle des réserves organiques chez les Crustacés Décapodes. Ann. Inst. Océan. 24(3): 259-357, 78 tables, 33 textfigs.

bibliography, Crustacea
Sars, G.O., 1913-1918
An account of the Crustacea of Norway with short descriptions and figures of all the species. Vol.VI. Copepoda Cyclopoida. Pts. I & II. Oithonidae, Cyclopinidae, Cyclopidae (part):1-32, pls.1-16. Pts.III & IV. Cyclopidae (continued):33-56, pls.17-32. Pts.V & VI. Cyclopidae (continued):57-80, pls.33-48. Pts.VII & VIII. Cyclopidae (concluded), Ascomyzontidae: 81-104, pls. 49-64. Pts.IX & X. Ascomyzontidae (concluded) Acontiophoridae, Myzopontiidae, Dyspontiidae, Artotrogidae, Cancerillidae, 105-140, pls. 65-80. Pts.XI & XII. Clausidiidae, Lichomol-
(over)

bibliography, Crustacea
Sars, G.O., 1903
An account of the Crustacea of Norway with short descriptions and figures of all species. Vol.IV. Copepoda Calanoida. Pts. 1 & 2. Calanidae, Eucalanidae, Paracalanidae, Pseudocalanidae, AEtideidae (part):28 pp. 16 pls. Pts.III & IV AEtideidae (concluded), Euchaetidae, Phaennidae, 29-48, pls.17-32, Pts.V & VI. Scolecithricidae, Diaixidae, Stephidae, Tharybidae, Pseudocyclopiidae, 49-73, pls.33-48. Pts.VII & VIII. Centropagidae Diaptomidae, 74-96, pls.49-64, Pts.IX & X Temoridae, Metridiidae, Heterorhabdidae (part), 97-120, pls.65-80. Pts. XI & XII.
(over)

bibliography, Crustacea
Sars, G.O., 1899-1900
An account of the Crustacea of Norway with short descriptions and figures of all the species. Vol.III Cumacea, Pts.I and II Cumidae, Lampropidae (part):24 pp., 16 pls. Pts.III and IV Lampropidae(concluded), Platyaspidae, Leuconidae:25-40, pls.17-32, Pts.V and VI. Diastylidae:41-68, pls.33-48, Parts VII and VIII Pseudocumidae, Nannastacidae, Campylaspidae:69-92, pls.49-64. Pts. IX and X. Anatomy, development, supplement: 65-110, Pls.65-72.

bibliography, Crustacea
Sars, G.O., 1896-1899
An account of the Crustacea of Norway with short descriptions and figures of all species. Vol II Isopoda. Pts.I,II. Apseudidae, Tanaidae:1-40, 16 pls. Pts.III,IV. Anthuridae, Gnathiidae, Aegidae, Cirolanidae, Limnoriidae:41-80, pls.17-32, Pts.V,VI. Idotheidae, Arcturidae, Asellidae, Ianiridae, Munnidae, 31-116, Pls.33-48. Pts.VII,VIII Desmosomidae, Munnopsidae (part):117-144, pls. 49-64. Pts.IX,X. Munnopsidae (concluded), Ligiidae, Trichoniscidae, Oniscidae (part): 145-184, Pls.65-83. Pts.X,XII. Oniscidae (concluded), Bopyridae, Dajidae:185-232,pls.
(over)

bibliography, Crustacea
Serene, R., Tran Van Duc and Nguyen Van Luom, 1958
Eumedoninae du Viet-Nam. (Crustacea) (avec un bibliographie de la sous-famille).
Treubia, 24(2):135-242.
Also:
Contrib. Inst. Océanogr., Nhatrang, No. 38.

bibliographies, Crustacean physiol.
Sindermann, Carl J., 1971.
Internal defenses of Crustacea: a review.
Fish.Bull. nat. mar. fish. Serv. NOAA 69(3): 455-489

bibliography, Crustacea
Wolff, T., 1947.
Liste over K.H. Stephensens videnskablige Afhandlinger. Vidensk. Medd. 110:VII-XV.

amphipods

bibliography, amphipods
Barnard, J. Laurens, 1962
South Atlantic abyssal amphipods collected by R.V. Vema.
Abyssal Crustacea, Vema Research Series, Columbia Univ. Press, New York and London, No. 1:1-78.

bibliography, amphipod
Bowman, Thomas E., and Lanelle W. Peterson, 1965.
Bibliography and list of new genera and species of amphipod crustaceans described by Clarence R. Shoemaker.
Crustaceana 9(3):309-316.

bibliography, amphipods
Charniaux-Cotton, H., 1957.
Croissance, régénération et déterminisme endocrinien des caractères sexuels d'Orchestia gammarella (Pallas)Crustacé Amphipode.
Ann. Sci. Nat., Zool. et Biol. Animale, (11) 19: 411-559.

bibliography, amphipod
Chevreux, E., and L. Fage, 1925
9. Amphipodes. Faune de France. 488 pp., 438 textfigs. Paul Lechevalier, Paris. (Office Central de Faunistique)

owned by M.S.

bibliography, amphipods
Clemens, H.P., 1950.
Life cycle and ecology of Gammarus fasciatus Say.
Franz Theodore Stone Inst. Hydrobiol. Contrib. No. 12:63 pp., 22 textfigs.

bibliography, amphipods
Crawford, G.I., 1937.
A review of the Amphipod Genus Corophium, with notes on the British species. J.M.B.A., n.s., 21:589-630, 4 textfigs.

bibliography, amphipods
Dunbar, M.J., 1954.
The amphipod Crustacea of Ungava Bay, Canadian eastern Arctic. J. Fish. Res. Bd., Canada, 11(6):709-798, 42 textfigs.

bibliography, amphipods
Hurley, D.E., 1955.
Pelagic amphipods of the suborder Hyperiidea in New Zealand. 1. Systematics.
Trans. R. Soc., New Zealand, 83(1):119-194, 18 textfigs.

bibliography, Ampelisca
Kaim-Malka, Richard A., 1970.
Contribution à l'étude de quelques espèces du genre Ampelisca (Crustacea-Amphipoda) en Méditerranée. II.
Téthys 1(4): 927-976.

bibliography, amphipods
Nagata, Kizo, 1966.
Studies on marine gammaridean Amphipoda of the Seto Inland Sea. IV.
Publ. Seto Mar. Biol. Lab., 13(5):327-348.

bibliography amphipods
Segerstrale, Sven. G., 1959.
Synopsis of data on the crustaceans Gammarus locusta, Gammarus oceanicus, Pontoporeia affinis and Corophium volutator (Amphipod Gammaridea).
Soc. Sci. Fennica, Comm. Biol. 20(5):23 pp.

bibliography, amphipods
Sexton, E.W., and D.M. Reid, 1951.
The life history of the multiform species, Jassa falcata (Montegu) (Crustacea Amphipoda) with a review of the bibliography of the species.
J. Linn. Soc., London, 42(283):29-91, Pls. 3-40.

Bibliography, amphipod
Shoemaker, C.R., 1955.
Amphipoda collected at the Arctic Laboratory, Office of Naval Research, Point Barrow, Alaska, by E.G. MacGinitie.
Smithsonian Misc. Coll., 128(1): 1-78.

bibliography, amphipods
Sivaprakasam, T. E., 1970.
Amphipoda from the east coast of India. 2. Grammaridea and Caprellidea.
J. Bombay nat. hist. Soc. 67(2):153-170.

Cladocera

bibliography, cladocerans
Baker, H.M., 1938.
Studies of the Cladocera of Monterey Bay. Proc. Calif. Acad. Sci., ser. 4, 23(23):311-365, Pls. 26-31.

bibliography, cladocerans
Casanova, Jean-Paul, 1967.
Penilia avirostris Dana en Méditerranée occidentale; sa valeur d'indicateur écologique.
Annls Fac. Sci., Marseille, 41: 95-119

indicator species
Penilia avirostris
bibliography, for Penilia

bibliographies, cladocerans
Rammner, W., 1939.
Cladocera. Fiches d'Ident. Zooplancton, Cons. Perm. Int. Expl. Mer, 3:4 pp., 4 textfigs.

Copepods

bibliography, copepods
Anraku, Masateru, 1963.
Feeding habits of planktonic copepods (a review). (In Japanese; English abstract).
Info. Bull. Plankt., Japan, No. 9:10-35.

bibliographies, copepods
Bernard, Michelle 1970.
Quelques aspects de la biologie du copepod pelagique Temora stylifera en Méditerranée: Essai d'écologie expérimentale.
Pelagos, Alger 11: 196 pp.

bibliography, copepods, parasitic
Boquet, C., and J.H. Stock, 1963.
Some recent trends in work on parasitic copepods. In: Oceanography and Marine Biology, H. Barnes, Editor, George Allen and Unwin, 1:289-300.

bibliography, copepods
Davis, C. C., 1949
A preliminary revision of the Monstrilloida, with descriptions of two new species. Trans. Am. Micros. Soc. 68(3):245-255, 1 pl. with 10 figs

bibliography copepods
Davis, C. C., 1949
The pelagic copepoda of the northeastern Pacific ocean. Univ. Washington Publ. Biol., 14:1-118, 15 pls.

bibliographies, copepods
Farran, G.P., 1948.
Copepoda, Sub-order: Calanoida, Family: Metridiae, Genus: Metridia. Fiches d'Ident. Zooplancton, Cons. Perm. Int. Expl. Mer, 14:4 pp., 7 textfigs.

bibliographies, copepods
Farran, G. P., 1948.
Copepoda, Sub-order: Calanoida, Family: Centropagidae, Genus: Centropages. Fiches d'Ident. Zooplancton, Cons. Perm. Int. Expl. Mer, 11:4 pp. 5 textfigs.

Bibliographies, copepods
Farran, G.P., 1948.
Copepoda, Sub-order: Calanoida, Family: Metridiae, Genus Pleuromamma. Fiches d'Ident. Zooplancton, Cons. Perm. Int. Expl. Mer, 17:4 pp., 5 textfigs.

bibliographies, copepods
Farran, G. P., 1948.
Copepoda, Sub-order: Calanoida, Family: Heterorhabdidae, Genus: Heterorhabdus. Fiches d'Ident. Zooplancton, Cons. Perm. Int. Expl. Mer, 16:4 pp. 7 textfigs.

bibliographies, copepods
Farran, G.P., 1948.
Copepoda, Sub-order: Calanoida, Family: Heterorhabdidae, Genus: Heterostylites, Hemirhabdus, Mesorhabdus, Disseta. Fiches d'Ident., Zooplancton, Cons. Perm. Int. Expl. Mer, 15:4 pp., 5 textfigs.

bibliographies, copepods
Farran, G.P., 1948.
Copepoda, Sub-order: Calanoida, Family: Candaciidae, Genus: Candacia. Fiches d'Ident. Zooplancton Cons. Perm. Int. Expl. Mer, 13:4 pp., 6 textfigs.

bibliographies, copepods
Farran, G.P., 1948.
Copepoda, Sub-order Calanoida, Family: Acartiidae, Genus: Acartia. Fiches d'Ident. Zooplancton, Cons. Perm. Int. Expl. Mer, 12:4 pp., 7 textfigs.

bibliography, copepods
Fish, A.G., 1962.
Pelagic copepods from Barbados.
Bull. Mar. Sci., Gulf and Caribbean, 12(1):1-38.

bibliography, copepods
Giron, Françoise, 1963.
Copépodes de la Mer d'Alboran (Campagne du "Président Théodore-Tissier", juin 1957).
Rev. Trav. Inst. Pêches Marit., 27(4):355-402.

bibliography, copepods
Grice, George D., 1961
Calanoid copepods from equatorial waters of the Pacific Ocean.
Fish. Bull., U.S. Fish & Wildlife Service (Fish. Bull. 186), 61: 172-246

bibliographies, copepods
Klie, W., 1943.
Copepoda I, Sub-order: Harpacticoida. Fiches d'Ident. Zooplancton, Cons. Perm. Int. Expl. Mer, 4:4 pp., 5 textfigs.

bibliography, copepods
van Breemen, P.J., 1908
VIII. Copepoden. Nordisches Plankton Lieferung 7., 263 pp., 251 text figs.

bibliography, copepods
Vervoort, Willem, 1964
Free-living Copepoda from Ifaluk Atoll in the Caroline Islands with notes on related species.
Bull. U.S. Nat. Mus. 236: 431 pp.

bibliography, copepods
Verwoort, W., 1952.
Copepoda, Sub-order: Calanoida, Family: Aetideidae, key to genera and references. Cons. Perm. Int. Expl. Mer, Fiches d'Ident. No. 4: 4 pp.

Cumacea

bibliography
Stebbing, T.R.R., 1913
Crustacea Cumacea (Sympoda). Das Tierreich, 39:1-210, 137 text figs.

owned by MS

Decapods

bibliographies, penaeids (decapods)
Allen, Donald M., and T.J. Costello, 1969.
Additional references on the biology of shrimp, family Penaeidae.
Fish. Bull. U.S. Dept. Comm. 68(1): 101-134

bibliography, decapods
Banner, Albert H., and Dora May Banner, 1960.
Contributions to the knowledge of the alpheid shrimp of the Pacific Ocean. V. The Indo-Pacific members of the Genus Athanas.
Pacific Science, 14(2):129-155.

bibliographies, decapod
Bergeron, Julien 1965.
Bibliographie du homard (Homarus americanus Milne-Edwards et Homarus gammarus (L.)).
Cah. Inf. Stn biol. mar., Grande Rivière 34: 81 pp. (multilithed)

bibliography, decapods
Boschi, Enrique E., and Victor Angelescu, 1962
Descripción de la morphologia externa e interna del langostino con algunos aplicaciones de indole taxonómica y biológica. Hymenopeneus mülleri (Bate), Crustacea fam. Penaeidae.
Bol. Inst. Biol. Mar., Mar del Plata, Argentina, No. 1:1-73.

bibliographies, decapods
Chin, Edward, and Donald M. Allen, 1959.
A list of references on the biology of the shrimp (Family Peneidae).
U.S.F.W.S. Spec. Sci. Rept., Fish., No. 276:143 p

bibliography, decapod
Coffin, Harold B., 1960.
The ovulation, embryology and developmental stages of the hermit crab (Pagurus samuelis Stimpson).
Walla Walla Coll. Publ., Dept. Biol. Sci., and Biol. Sta., No. 25:30. pp.

bibliographies
Corrivault, G. -W., and J. H. Tremblay, 1948. Contribution à la biologie du homard (Homarus americanus Milne-Edwards) dans la Baie-des-Chaleurs et le Golfe Saint-Laurent. Contrib. Sta. Biol. du St. Laurent, No. 19, 222 pp., 39 textfigs.

bibliographies, decapod
Dawson, C.E., 1954.
A bibliography of the lobster and the spiny lobster, families Homaridae and Palinuridae.
Fla. State Bd Conserv. 86 pp.

bibliography, decapods
Fize, A., and R. Serene, 1957
Les Hapalocarcinides du Viet-Nam.
Mem. Inst. Oceanogr., Nhatrang, 202 pp., 18 pls.

bibliography, decapod
George, M.J., 1969.
Genus Metapenaeus Wood-Mason & Alcock 1891.
Bull. cent. mar. Fish. Res. Inst., India, 14: 75-125 (mimeographed)

bibliography, decapods
George, M.J., 1969
Systematics-taxonomic consideration and general distribution.
Bull. cent. mar. Fish. Res. Inst., India, 14: 5-45. (mimeographed).

bibliographies, decapod
George, P.C., and P. Vedavyasa Rao 1967
An annotated bibliography of the biology and fishery of the edible crabs of India.
Proc. Symp. Crustacea, Ernakulam, Jan. 12-1965, 5: 1548-1555.

bibliographies, decapod
George, P.C. and P. Vedavyasa Rao 1967.
An annotated bibliography of the biology and fishery of the commercially important prawns of India.
Proc. Symp. Crustacea, Ernakulam, Jan. 12-1965, 5: 1521-1547.

bibliography decapods
Gordan, J., 1956.
A bibliography of pagurid crabs exclusive of Alcock 1905. Bull. Amer. Mus. Nat. Hist., 108(3): 253-352.

bibliographies, decapod
Guinot, Danièle, 1970 (1971).
Recherches préliminaires sur les groupements naturels chez les Crustacés décapodes brachyoures. VIII. Synthèse et bibliographie
Bull. Mus. nat. Hist. nat. (2) 42(5): 1063-1090

bibliographies, decapods
Guinot, Danièle 1966.
La faune carcinologique (Crustacea Brachyura) de l'Océan Indien occidental et de la Mer Rouge.
Mém. Inst. fond. Afr. noire 77: 235-352.

bibliography, decapods
Gunter, G., 1962.
Specific names of the Atlantic American white shrimp.
Gulf Res. Rept., 1(3):107-114.

bibliographies decapod
Gurney, R., 1939.
Bibliography of the larvae of decapod Crustacea
Ray Soc., London, 123 pp.

bibliography, decapod
Holthuis, L.B., 1959.
The Crustacea Decapoda of Suriname (Dutch Guiana).
Zool. Verhandl., Rijksmus. Nat. Hist., Leiden, No. 44:290 pp., 16 pls.

bibliography, decapod
Holthuis, L. B., and G. Gottlieb, 1958.
An annotated list of the decapod Crustacea of the Mediterranean coast of Israel with an appendix listing the Decapoda of the eastern Mediterranean.
Bull. Res. Counc., Israel (B) 7 (1-2): 126 pp. (Reprinted in Bull. Sea Fish Res. Sta. No. 18.

bibliography decapods
India, Central Inland Fisheries Institute, 1963.
Information on prawns from Indian waters: synopsis of biological data.
Indo-Pacific Fish. Council, Proc., 10th Sess., Seoul, Korea, 10-25 Oct., 1962, FAO, Bangkok, Sect. II:124-133.

bibliography, decapod
Jones, S., compiler, 1969.
Prawn fisheries of India.
Bull. cent. mar. Fish. Res. Inst., India, 14:303 pp. (mimeographed).

bibliography, decapod
Joubert, Leonie, S., 1965.
A preliminary report on the penaeid prawns of Durban Bay.
S. African Assoc. Mar. Biol. Res., Oceanogr. Res. Inst., Invest. Rept., No. 11:32 pp.

bibliographies, decapod
Kenster, Craig B., 1967.
An annotated bibliography of the marine spiny lobster Jasus verreauxi (H. Milne Edwards) (Crustacea Decapoda, Palinuridae).
Trans. R. Soc., N.Z. Zool., 8(19):207-210.

bibliography decapod
Kubo, I., 1949
Studies on the Peneids of Japanese and adjacent waters. J. Tokyo Coll., Fisheries 36(1):1-467, 160 text figs.

bibliography, decapod
Kunju, M. Mydeen, 1969.
Genera Solenocera Lucas 1850, Atypopenaeus Alcock 1905, Hippolysmata Stimpson 1860, Palaemon Weber 1795, and Acetes M. Edwards 1830.
Bull. cent. mar. Fish. Res. Inst., India, 14: 159-177 (mimeographed)

bibliography, decapod
Mohamed, K.H. 1969
Genus Penaeus Fabricius 1798.
Bull. cent. mar. Fish. Res. Inst., India, 14: 49-75. (mimeographed)

bibliographies, decapod
Monod, Théodore 1966
Crevettes et crabes de la côte occidentale d'Afrique.
Mém. Inst. fond. Afr. noire 77: 103-234

bibliography, decapod
Panouse, J.-B. 1946
Recherches sur les phénomènes humoraux chez les crustacés. L'adaptation chromatique et la crossance ovarienne chez la crevette Leander serratus. Ann. de l'Inst. Océan., Monaco 23(2):65-147

bibliography, decapod
Pike, R.B., 1947
Galathea. Liverpool Biol. Soc. Mem. 34 in Proc. Liverpool Biol. Soc. 55:1-138, 20 plates.

bibliography, decapod
Ramamurthy, S., and M.S. Muthu, 1969.
Prawn fishing methods.
Bull. cent. mar. Fish. Res. Inst., India, 16:235-257.

bibliographies, decapod
Rao, P. Vedavyasa 1969
Genus Parapenaeopsis Alcock 1901.
Bull. cent. mar. Fish. Res. Inst. India 14: 126-158 (mimeographed).

bibliography, decapod
Rauk, A.A., and W. Dall, 1965.
Littoral Penaeinee (Crustacea Decapoda) from northern Australia, New Guinea and adjacent waters.
Verhandel. K. Nederlandse Akad. Wetescheppen afd. Natuurkunde (2) 56(3): 116 pp., 13 pls.

bibliography, decapod
Sakai, Tune, 1965
The crabs of Sagami Bay collected by His Majesty the Emperor of Japan.
Meruzen Co., Ltd., Tokyo, 206 pp, 100 pls. + 92 pp. (Japanese)

bibliography, decapod
Scattergood, L. W., 1949.
A bibliography of lobster culture. U. S. Fish and Wildlife Service, Spec. Sci. Rept. 64, 26pp.

bibliography, decapod
Scattergood, L. W., 1949.
Translations of foreign literature concerning lobster culture and the early life history of the lobster. Spec. Sci. "ept., Fishery, "o. 6: 173 pp. (Mimeographed).

bibliography, decapods
Scrivener, J.C. and T. H. Butler 1971.
A bibliography of shrimps of the family Pandalidae emphasizing economically important species of the genus Pandalus.
Techn. Rept. Fish. Res. Bd., Can. 241: 42 pp (multilithed).

Bibliographies, decapod
Silas E.G. 1967.
On the taxonomy, biology and fishery of the spiny lobster Jasus lalandei frontalis (H. Milne Edwards) from St. Paul and New Amsterdam Islands in the southern Indian Ocean, with an annotated bibliography on species of the genus Jasus Parker.
Proc. Symp. Crustacea, Ernakulam, Jan. 12-15, 1965, 4: 1466-1520

bibliographies, decapod
Sims. Harold W., Jr., 1966.
An annotated bibliography of the spiny lobsters. Families Palinuridae and Scyllaridae.
Florida State Bd., Conserv., Techn. Ser. No. 48: 84 pp.

bibliographies, decapod
Teinsongrusmee Banchong 1967.
A bibliography of systematic and biology of shrimps (Peneidae)
Contrib. Mar. Fish. Lab. Bangkok 6:103pp.

bibliography, decapod
Vatova, A., 1943.
I Decapodi della Somalia. Thalassia 6(2):1-37, 5 pls.

bibliographies, decapod.
Wear, Robert G. 1970.
Notes and bibliography on the larvae of xanthid crabs.
Pacific Science 24 (1): 84-89.

bibliography, decapods (color change)
Chessard-Bouchard, Colette, 1965.
L'adaptation chromatique chez les Natantia (Crustacés decapodes).
Cahiers Biol. Mar., 6:469-576.

bibliography, decapod (color change)
Humbert, Chantal, 1965.
Etude experimentale du role de l'organe X (paro distalis) dans les changements de couleur et la mue de la crevette Palaemon serratus.
Trav. Inst. scient. cherif., Ser. Zool. No. 32, 96 pp.

bibliographies, decapod (tagging)
Neal, Richard A. 1969.
Methods of marking shrimp.
FAO Fish. Rept. 3 (57) (FRm/57.3 (Trm)): 1149-1165 (mimeographed)

isopods

Bibliography, isopods
Menzies, Robert J., 1962
The isopods of abyssal depths in the Atlantic Ocean.
Abyssal Crustacea, Vema Research Series. Columbia Univ. Press, New York and London, No. 1:79-206.

ostracods

bibliographies, ostracods
Benson, Richard H. 1959.
Arthropoda 1. Ecology of recent ostracodes of the Todos Santos Bay region, Baja California, Mexico.
Univ. Kansas, Paleontol. Contr., 1-80, 11 pls. 20 figs.

bibliography, ostracods
Bonaduce, Gioacchino, 1964.
Contributo alla conoscenza e correlazione sistematica nel l'ambito della Famiglio Polycopida (Ostracoda, Cladocopa) con particolare referimento alle relaziona tra parte molle e carapace.
Pubbl. Staz. Zool., Napoli, 34:160-184.

bibliography, ostracods
Hanai, T., 1959
Studies on the ostracoda from Japan-historical review. J. Fac. Sci. Tokyo Un. Sec. II. Vol. 11(4): 419-439.

bibliographies, ostracods
Klie, W., 1944.
Ostracoda II. Family: Conchoeciidae. Fiches d'Ident. Zooplancton, Cons. Perm. Int. Expl. Mer, 6:4 pp., 7 textfigs.

bibliographies, ostracods
Klie, W., 1944.
Ostracoda I. Family: Cyprinidae. Fiches d'Ident. Zooplancton, Cons. Perm. Int. Expl. Mer, 5:4 pp. 6 textfigs.

bibliographies, ostracod
Levinson, S.A., 1962.
Bibliography and index to new genera and species of Ostracoda for 1958-1959.
Micropaleontology, 8(1):77-105.

pycnogonids

bibliography pycnogonids
Hedgpeth, J.W., 1948
The Pycnogonida of the western North Atlantic and the Caribbean. Proc. U.S. Nat. Mus. 97:157-342, figs.4-53, charts 1-3.

schizopods

bibliography
Banner, A.H., 1949.
A taxonomic study of the Mysidacea and Euphausiacea (Crustacea) of the northeastern Pacific. Pt. III. Euphausiacea. R. Canadian Inst., Trans. 28:1-63, Pls. 1-4.

bibliography, schizopods
Einarsson, H., 1945
Euphausiacea. I. Northern Atlantic species. Dana Rept. No. 27: 185 pp., 84 text figs.

bibliography, schizopods
Fage, L., 1941
Mysidacea. Lophogastrida 1. Dana Rept. No. 19, 52 pp., 51 text figs.

bibliographies, schizopods
Marr, J.W.S., 1962.
The natural history and geography of the Antarctic krill (Euphausia superba Dana).
Discovery Repts., 32:33-464, Pl. 3.

bibliographies, schizopods
Nouvel, H., 1950.
Mysidacea, Fam. Mysidae (Suite 8): Mysinae (Suite 6): Mysini (Suite 2), Heteromysini; Mysidellina. Fiches d'Ident. Zooplancton, Cons. Perm. Int. Expl. Mer., 27:4 pp., Figs. 332-369.

bibliographies, schizopods
Nouvel, H., 1950.
Mysidacea. Fam. Mysidae (Suite 7): Mysinae (Suite 5): Mysini (Suite 1). Fiches d'Ident. Zooplancton Cons. Perm. Int. Expl. Mer, 26:3 pp., Figs. 303-331.

bibliographies, schizopods
Nouvel, H., 1950.
Mysidacea. Fam. Mysidae (Suite 6): Mysinae (Suite 4): Mysini (part.). Fiches d'Ident. Zooplancton, Cons. Perm. Int. Expl. Mer, 25:4 pp., Figs. 268-

bibliographies schizopods
Nouvel, H., 1950.
Mysidacea. Fam. Mysidae (Suite 5): Mysinae (Suite 3): Leptomysini. Fiches d'Ident. Zooplancton, Cons. Perm. Int. Expl. Mer, 24:4 pp., Figs. 222-267.

bibliographies, schizopods
Nouvel, H., 1950.
Mysidacea. Fam. Mysidae (Suite 4): Mysinae (Suite 2): Erythropini, (Suite 2): Calyptommini. Fiches d'Ident. Zooplancton, Cons. Perm. Int. Expl. Mer, 23:3 pp., Figs. 197-221.

bibliographies, schizopods
Nouvel, H., 1950.
Mysidacea. Fam. Mysidae (Suite 3): Mysinae (Suite 1): Erythropini (cont.). Fiches d'Ident. Zooplancton, Cons. Perm. Int. Expl. Mer, 22:4 pp., Figs. 165-196.

bibliographies, schizopods
Nouvel, H., 1950.
Mysidacea. Fam. Mysidae (Suite 2): Mysinae (part.): Erythropini (part.). Fiches d'Ident. Zooplancton, Cons. Perm. Int. Expl. Mer, 21:4 pp., Figs. 104-149.

bibliography, schizopods
Nouvel, H., 1950.
Mysidacea. Fam. Mysidae (Suite 1): Siriellinae, Gastrosaccinae. Fiches d'Ident. Zooplancton, Cons. Perm. Int. Expl. Mer, 20:4 pp., Figs. 55-103.

bibliographies, schizopods
Nouvel, H., 1950.
Mysidacea. Fam. Lophogastridae, Eucopiidae, Petalophthalmidae, Mysidae (part.): Boreomysinae. Fiches d'Ident. Zooplancton, Cons. Perm. Int. Expl. Mer, 19:4 pp., 54 figs.

bibliographies, schizopods
Nouvel, H., 1950.
Mysidacea. Généralités et bibliographie général. Fiches d'Ident. Zooplancton, Cons. Perm. Int. Expl. Mer, 18: 6 pp., 2 figs.

Bibliography, stomatopods
*Manning, Raymond B., 1969
Stomatopod Crustacea of the Western Atlantic.
Stud. Trop. Oceanogr., Univ. Miami, 8:380 pp.

bibliographies, schizopods
Pauli, V.L., 1957.
Determination of mysids of the Black Sea-Azov basin. Trudy Sevastopol. Biol. Stan. 9:113-166.

bibliographies schizopods
Ponomareva, L.A., 1963.
Euphausiids of the North Pacific: their distribution and ecology.
Akad. Nauk. SSSR., Inst. Okeanol.

Translation:
Israel Program Sci. Transl. TT 65-50098. 1966. 154 pp. $5.00.

bibliography, schizopods
Tattersall, W.M., 1951.
A review of the Mysidacea of the United States National Museum. Bull. U.S.Nat. Mus. 201:1-292, figs.

tanaids

bibliography, tanaids
Lang, K., 1950.
Contribution to the systematics and synonymies of the Tanaidacea. Ark. Zool. 42(3)(18):14 pp.

diatoms

bibliographies diatoms
Birnhak, Bruce I., Patricia V. Donelly and Richard P. Saunders 1967.
Studies on Guinardia flaccida (Castracane) Peragallo.
Fla. Bd. Conserv. Mar. Lab. St. Petersburg, Leaflet Ser. Phytoplankton 1 (3-3): 23pp.

bibliography, diatoms
Hendy, N. Ingram, 1964
An introductory account of the smaller algae of British coastal waters. V. Bacillariophyceae (Diatoms).
Her Majesty's Stationary Office, 317 pp., 45 pls.

bibliography, diatoms
Hendey, N. Ingram, 1959.
The structure of the diatom cell wall as revealed by the electron microscope.
J. Quekett Micros. Club, (4)5(6):147-175.

bibliography, diatoms
Lewin, Joyce C., and Robert R. L. Guillard, 1963
Diatoms.
Annual Review of Microbiology, 17:373-414.

bibliographies, diatoms
Lohman, K.E., 1957.
Diatoms. Annotated bibliography of marine paleoecology. In: Treatise on marine ecology and paleoecology, Vol. 2. Paleoecology.
G.S.A. Mem., 67:731-736.

bibliography, diatoms
Mann, A., 1907
Report on the diatoms of the Albatross voyages in the Pacific Ocean, 1888-1904.
Contrib. U. S. Nat. Herb. 10(5):221-419, Pls. XLIV-LIV.

bibliography diatoms
Simonsen, Reimer, 1962
Untersuchungen zur Systematik und Ökologie der Bodendiatomeen der westlichen Ostsee.
Inter. Rev. Gesamt. Hydrobiol., System. Beihefte 1:145 pp.

bibliography, diatoms
Zanon, D. V., 1949
Diatomee di Buenos Aires (Argentina)
Atti Accad. Naz. Lincei, Memorie, Cl. Sci. fis., mat. e. nat., ser. 7, 11(3):59-151, 2 pls.

bibliographies, dinoflagellates (fossil)
Downie, Charles, and William Antony S. Sarjent, 1964.
Bibliography and index of fossil dinoflagellates and acritarchs.
Geol. Soc., Amer., Mem., 94:180 pp.

bibliography dinoflagellate
Graham, H. W., 1942
Studies in the morphology, taxonomy, and ecology of the Peridiniales. Sci. Res. Cruise VII of the Carnegie, 1928-1929---Biol. III(542): 129 pp., 67 figs.

bibliographies, dinoflagellate
Halim, Youssef 1960
Étude quantitative et qualitative du cycle écologique des dinoflagellés dans les eaux de Villefranche-sur-Mer (1953-1955).
Ann. Inst. océanogr. Monaco 38: 123-232

bibliography, dinoflagellate
Matzenauer, L., 1933
Die Dinoflagellaten des indischen Ozeans (mit Ausnahme der Gattung Ceratium.) Bot. Arch. 35:437-510, 77 text figs., 2 charts.

bibliography, dinoflagellate
Paulsen, O., 1949
Observations on dinoflagellates. (Ed. J. Grøntoed) Kongl. Dansk. Videnskab. Selsk., Biol. Skr. 6(4):67 pp., 30 text figs.

bibliographies, dinoflagellate
Ryther, J.H., 1955.
Ecology of autotrophic marine dinoflagellates with reference to red water conditions. In: Luminescence of Biological Systems, F.H. Johnson, Edit., AAAS:387-414, 5 textfigs.

bibliography, dinoflagellates
Steidinger, Karen A., and Jean Williams 1970
Dinoflagellates.
Mem. Hourglass Cruises, Mar. Res. Lab., Fla. Dept. Nat. Res. 2:1-251.

bibliographies, dinoflagellate
Steidinger, Karen A., Joanne T. Davis and Jean Williams 1967
A key to the marine dinoflagellate genera of the west coast of Florida.
Techn. Ser. Fla Bd. Conserv. 52:45pp.

bibliography faecal pellets
Arakawa Kohman Y. 1970.
Scatological studies of the Bivalvia (Mollusca)
Adv. mar. Biol., F.S. Russell and Maurice Yonge, editors, Academic Press 8: 307-436.

bibliography, echinoderms
James, D.B., and R.S. Lal Mohan, 1969.
Bibliography of the echinoderms of the Indian Ocean.
Bull.cent.mar.Fish.Res.Inst.15:41 pp. (mimeographed).

bibliographies, echinoderms
Plessis, J. 1970
Note bibliographique sur les ouvrages récents (depuis 1960) traitant des échinodermes du Pacifique
Cah. Pacifique 14: 311-334.

ecology

bibliographies, ecological
Hedgpeth, J.W., Ed., 1957.
Treatise on Marine Ecology and Paleoecology, Vol. 1. Mem., Geol. Soc., Amer., 67:1296 pp.

bibliography, ecological
Knox, G.A., 1960.
Littoral ecology and biogeography of the southern oceans.
Proc. R. Soc., London, (B), 152(949):577-624.

fertilizers

bibliography, fertilizers
Mortimer, C.H., and C.F. Hickling, 1954.
Fertilizers in Fish Pond. A review and bibliography. Colonial Office, Fishery Publ., No. 5:4-155.

fish: see also fisheries

bibliographies, biological
fish, see also: fisheries

bibliography fishes
Asano, Hirotoshi, 1962
Studies on the congrid eels of Japan.
Bull. Misaki Mar. Biol. Inst., Kyoto Univ., No. 1:1-143.

bibliographies, fishes
Atz, James W. 1971
Dean bibliography of fishes 1968
Am. Mus. Nat. Hist. N.Y., 512 pp.

bibliography, fishes
Blache, J., 1964.
Note préliminaire dur les larves leptocéphales d'apodes du Golfe de Guinée.
Trav. Centre Océanogr., Pointe-Noire, Cahiers, O.R.S.T.R.O.M., Océanogr., Paris, 5:5-55.

bibliography fishes
Blackburn, M., 1950
A biological study of the anchovy, Engraulis australis (White), in Australian waters.
Australian J. Mar. and Freshwater Res. 1 (1): 3, 84, 5 pls., 11 textfigs.

bibliography, fishes
Davies, David H., 1961.
Shark research in Natal.
Marine Studies off the Natal Coast, C.S.I.R. Symposium, No. S2:81-88. (Multilithed).

bibliography fishes
deSylva, Donald P., 1963.
Systematics and life history of the Great Barracude Sphyraena barracuda (Walbaum).
Stud. trop. Oceanogr., Miami, 1:viii 179 pp., 32 tables, 36 figs. $2.50.

bibliography - fishes
Eales, J. Geoffrey, 1967.
A bibliography of the eels of the genus Anguilla
Tech.Rep.Fish.Res.Bd.Can., 28:171 pp. (mimeographed).

bibliographies (fishes)
France, Museum National d'Histoire Naturelle, 1961.
Bibliographie recente relative aux poissons dans le Pacifique.
Cahiers du Pacifique, Foundation Singer-Polignac, Mus. Nat. Hist. Nat., No. 3:131-144.

bibliography, fishes
Fuster de Plaza, M.L., and E.E Boschi, 1961
Areas de Migración y ecología de la anchoa Lycengraulis olidus (Günther) en las aguas Argentinas (Pisces, fam. Engraulidae).
Contrib., Inst. Biol. Mar., Mar del Plata, Argentina, No. 1:58 pp.

bibliography, fishes
Hubbs, C.L., and A.B. Rechnitzer, 1952.
Report on experiments designed to determine effects of underwater explosions on fish life.
Calif. Fish and Game 38(3):333-365.

bibliography, fishes
Iwai, Tamotsu, Izumi Nakamura and Kiyomatsu Matsubara, 1965.
Taxonomic study of the tunas. (In Japanese; English abstract).
Misaki Mar. Biol. Inst., Kyoto Univ., Spec. Rept. No.2:51 p.

bibliography, fishes
Jones, S., and P. Bensam, 1968.
An annotated bibliography on the breeding habits and development of fishes of the Indian Ocean.
Bull.Cent.mar.Fish.Res.Inst.3:154 pp. (mimeographed).

bibliographies, fishes
Klawe, Wifold L., and Makoto Peter Miyake, 1967.
An annotated bibliography on the biology and fishery of the skipjack tuna, Katsuwonus pelamis of the Pacific Ocean.
Bull.Inter-Am. trop. Tuna Comm., 12(4):363 pp.

bibliography, fishes
Matsui, I., 1952.
[Studies on the morphology, ecology and pond-culture of the Japanese eel (Anguilla japonica Temminck & Schlegel)] J. Simonoseki Coll. Fish. 2(2):1-245, 3 pls., 85 textfigs. (English summary).

bibliographies, fishes
Mohr. E. W.
Bibliographie der Alters- und Wachtums-Bestimmung bei Fischen.
Jour. du. Cons., 2:236-258

bibliographies, fish
Nakamura, Izumi, Tamotsu Iwai and Kiyomatsu Matsubara 1968.
A review of the sailfish, spearfish, marlin and swordfish of the world. (In Japanese)
Spec. Rept. Misaki mar. biol. Inst. Kyoto Univ. 4:95pp.

bibliography, fishes
Reintjes, John W., 1964.
Annotated bibliography on biology of menhadens and menhaden-like fishes of the world.
Fish. Bull., U.S.F.W.S., 63(3):531-549.

bibliographies, fish
Robinson, P.F., 1961?
A bibliography of papers dealing with the oyster toadfish, Opsanus tau.
Nat. Res. Inst., Chesapeake Biol. Inst. Contrib. No. 183.

bibliographies, fishes
Shimada, B.M., 1951.
An annotated bibliography on the biology of Pacific tunas. U.S.F.W.S., Fish. Bull. No. 58:58pp

bibliographies, fish and fisheries
Kelts, Lora I., and Janet I. Bressler 1971.
Fish and fisheries literature resources: an annotated bibliography.
Trans. Am. Fish. Soc. 100 (2):403-422.

bibliographies, fishery
Scattergood, L.W., 1957.
English translations of fishery literature, additional listings.
USFWS Spec. Sci. Rept., Fish., 227:66 pp.

bibliography, fishery
Union of South Africa, Division of Fisheries, Department of Commerce and Industries, 1961.
Fisheries Research in Natal Waters.
Marine Studies off the Natal Coast, C.S.I.R. Symposium, S2:89-117 (Multilithed).

Bibliography, fishery
United States, U.S. Fish and Wildlife Service, Bureau of Commercial Fisheries, 1969.
List of Special Scientific Reports and Special Scientific Report -- Fisheries of the U.S. Fish and Wildlife Service.
Fish. Leaflet, 624: 52pp

bibliography, fish anatomy
Jones, F.R. Harden, and N.B. Marshall, 1953.
The structure and functions of the teleostean swimmbladder. Biol. Rev. 28:16-83, 7 textfigs.

bibliography, fish behavior
Fleming, R.H., and T. Laevastu (compilers), 1956
The influence of hydrographic conditions on the behavior of fish (a preliminary literature survey). F.A.O. Fish. Bull. 9(4):181-196.

Bibliographies, biological

fish culture, see also: sea farming

bibliography, fish culture
Shelbourne, J.E., 1964.
The artificial propagation of marine fish.
In: Advances in Marine Biology, F.S. Russell, editor, Academic Press, 2:1-83.

bibliography fish detection
Cushing, D.H., Finn Devold, J.C. Marr, and H. Kristjonsson, 1952.
Some modern methods of fish detection: echo sounding, echo ranging and aerial scouting.
F.A.O. Fish. Bull. 5(3/4):95-119, 9 textfigs.

bibliography,
Cushing, D.H., and I.D. Richardson, 1955.
Echo sounding experiments on fish.
Min. Agric. Fish., Fish. Invest. (2)18(4):1-34.

bibliographies, fish anomalies
Dawson, C.E. 1971.
A bibliography of anomalies of fishes.
Gulf Res. Repts 3(2):215-239.

bibliography, fish detection
Le Gall, J., 1952.
La detection des bancs de poissons. Rapp. Proc. Verb., Cons. Perm. Int. Expl. Mer, 132:65-71, 4 textfigs.

bibliography, detection, fish
Shibata, Keishi 1970.
Study on details of ultrasonic reflection from individual fish.
Bull. Fac. Fish. Nagasaki Univ. 29: 1-82

fish diseases

Bibliography, fish diseases
Margolis, L., 1970.
A bibliography of parasites and diseases of fishes of Canada: 1879-1969.
Techn. Rept., Fish. Res. Bd Can., 185: 38pp. (mimeographed)

bibliography, fish diseases
Nigrelli, Ross, and George D. Ruggieri, 1965.
Studies on virus diseases of fishes, spontaneous and experimentally induced cellular hypertrophy (lymphocystis disease) in fishes of the New York aquarium, with a report of new cases and an annotated bibliography (1874-1965).
Zoologica, N.Y. Zool. Soc., 50(2):83-95.

bibliography, fish diseases
Oppenheimer, C., and C.L. Kesteven, 1953.
Disease as a factor in natural mortality of marine fish. FAO Fish. Bull. 6(6):215-222.

bibliographies, electric fishing
Halsband, Egon, and Inge Halsband 1970.
Bibliographie über die Electrofischerei und ihre Grundlagen. II.
Arch. Fischereiwiss. 21 (1):1-72

bibliography, fish flour
Bunn, Joseph B., 1959.
The chemistry of fish flour: a bibliography.
U.S.A. Chem. Warfare Labs., CWL Spec. Publ. 2-22:22 pp.

Bibliography, fish parasites
Margolis, L., 1970.
A bibliography of parasites and diseases of fishes of Canada: 1879-1969.
Techn. Rept., Fish. Res. Bd Can., 185: 38pp. (mimeographed)

bibliography, fish tagging
Fridriksson, A., 1952.
Marking of fish in Europe during 1927-1951.
Rapp. Proc. Verb., Cons. Perm. Int. Expl. Mer, 132:55-64, 4 textfigs.

bibliography, fish tagging
Paulik, G.J., 1963
Estimates of mortality rates from tag recoveries.
Biometrics, Biometric Soc., 19(1):28-57.

foraminifera

bibliography, foraminifera
Stainforth, R.M., 1960.
Estado actual de las correlaciones transatlanticas del Oligo-Mioceno por medio de foraminferos planktonicos.
Memoria Tercer Congreso Geologico Venezolano, 1: 382-406.

bibliographies, foraminifera
Todd, Ruth, 1964.
Recent literature on the Foraminifera.
Contrib. Cushman Found. Foram. Res., 15(1):39-44

fouling

bibliographies, fouling
DePalma, John R. 1968.
An annotated bibliography of marine fouling for marine scientists and engineers.
J. Ocean Techn. 2(3):33-44.

bibliography
Pyefinch, K. A., 1950.
Notes on the ecology of ship -fouling organisms.
J. Animal Ecol. 19(1):29-35.

bibliography
Woods Hole Oceanographic Institution, 1952.
Marine fouling and its prevention. U.S. Naval Inst., 388 pp., textfigs.

WHOI Contrib. No. 580.

fungi

bibliography
Johnson, T.W., and S.P. Meyers, 1957.
Literature on halophilus and halolimnic fungi.
Bull. Mar. Sci., Gulf and Caribbean, 7(4):330-359.

bibliography, fungi
Sparrow, Frederick K., Jr., 1960. (2nd Edit.).
Aquatic Phycomycetes.
Univ. Michigan Press, 1187 pp.

bibliographies, insects (marine)
Roth, James C., and Sikko Parma, 1970
A Chaoborus bibliography.
Bull. entomol. Soc. Am. 16(2):100-110

larvae

bibliography, larvae
Thorson, G., 1950.
Reproductive and larval ecology of marine bottom invertebrates. Biol. Rev. 25(1):1-45, 6 textfigs.

bibliography, larvae
Thorson, G., 1944.
Reproduction and larval development of Danish marine bottom invertebrates with special reference to the planktonic larvae in the sound (Øresund). Medd. Komm. Danmarks Fiskeri- og Havundersøgelser, Serie: Plankton, 4(1):523 pp., 199 textfigs.

mammals

Bibliographies, biological

mammals, see also: whales

bibliography, mammals
Clarke, Robert, 1957.
Migrations of marine mammals.
Norwegian Whaling Gaz. (11):609-630.

bibliography, mammals
Clarke, R., 1954.
Whales and seals as resources of the sea.
Norsk Hvalfangsttåd, 43 Årg (9):489-510.

bibliographies, mammals
Kenyon, K.W., and V.B. Schaeffer, 1955.
The seals, sea-lions and sea otter of the Pacific coast. Descriptions, life history notes photographs and drawings. Fish and Wildlife Circular 32:1-34.

bibliographies, mammals
Nishiwaki, M. 1965
Whales and pinnipeds. (In Japanese)
Univ. Tokyo Press 439 pp.

bibliography, mammals
Oppenheimer, Gerald J., 1960
Reference source for marine mammology.
USFWS Spec. Sci. Rept., Fish., No. 361:9 pp.

bibliography, mammals
Scheffer, V.B., and D.W. Rice, 1963.
A list of the marine mammals of the world.
U.S.F.W.S., Spec. Sci. Rept., Fish., No. 431: 12 pp.

bibliography, mammals
*Todd, Ethel I., 1968.
Books and articles on marine mammals.
U.S. Fish Wildl. Serv. Bur. Comm. Fish. Circ. 299: 14 pp.

bibliographies, mammals (dolphins-porpoises)
Whitfield, William K., Jr., 1971.
An annotated bibliography of dolphin and porpoise families Delphinidae and Platanistidae.
Spec. scient. Rept. Dept. Nat. Resources, Fla. 26: 104 pp.

marine borers

bibliographies, marine borers
Clapp, William F., (posthumous) and Roman Kenk, 1963
Marine borers: an annotated bibliography.
Office of Naval Research, Dept. of the Navy, ACR-74:1136 pp.

bibliographies, marine borers
Clapp, W.F., and R. Kenk, 1956, 1957.
Marine borers, a preliminary bibliography.
U.S. Library Congress, Tech. Info. Div., 1:358 pp 2:355 pp

bibliographies, marine borers
Clapp, W.F., and R. Kenk, 1956.
Marine borers: a preliminary bibliography.
Library of Congress, Tech. Info. Div., 346 pp. (mimeographed).

bibliography, meiofauna
Hulings, Neil C., and John S. Gray 1971
A manual for the study of meiofauna.
Smithson. Contrib. Zool. 78:83 pp.

microorganisms

bibliography, micro-organisms
Wood, E.J.F., 1963
Heterotrophic micro-organisms in the oceans.
In: Oceanography and Marine Biology, H. Barnes Edit., George Allen & Unwin, 1:197-222.

molluscs

bibliography, molluscs
Allen, J.A., 1963
Ecology and functional morphology of molluscs.
In: Oceanography and Marine Biology, H. Barnes Edit., George Allen & Unwin, 1:253-288.

bibliographies, molluscs
Baughman, J.L. 1948
An annotated bibliography of oysters with pertinent materials on mussels and other shellfish and an appendix on pollution.
Texas A and M Foundation, 794 pp.

bibliography, molluscs
Bernard, F., 1967.
Prodrome for a distributional check-list and bibliography of the recent marine Mollusca of the west coast of Canada.
Tech. Rep. Fish. Res. Bd., Can., 2:261-pp. (mimeographed).

bibliography, molluscs
Engel, H., and C.J. Van Eeken, 1962
Red Sea Opisthobranchia from the coast of Israel and Sinai. Contributions to the knowledge of the Red Sea, No. 22.
Sea Fish. Res. Sta., Haifa, Israel, Bull. No. 30: 15-34.

bibliographies, molluscs
Galtsoff, Paul S., 1964.
The American oyster, Crassostrea virginica Gmelin.
Fishery Bulletin, 64:480 pp.

bibliography, molluscs
Pasteur-Humbert, C., 1962.
Les mollusques marins testacés du Maroc. Catalogue non critique.
II. Les lamellibranches et les scaphopodes.
Trav. Inst. Sci., Chérifien, Zool., 28:180 pp.

bibliography, molluscs
Potts, W.T.W., 1967.
Excretion in the molluscs.
Biol. Rev., Cambridge Phil. Soc., 42(1):1-41.

bibliography, molluscs
Rees, C.B., 1950.
The identification and classification of Lamellibranch larvae. Hull Bull. Mar. Ecol. 3(19):73-104, 5 pls., 4 textfigs.

bibliography, molluscs
Scattergood, L. W., and C.C. Taylor, 1950.
The mussel resources of the North Atlantic region. Pt. 1. The survey to discover the locations and areas of the North Atlantic mussel-producing beds. Pt. 2. Observations on the biology and the methods of collecting and processing the mussel. Pt. 3. Development of the fishery and the possible need for conservation measures. Fishery Leaflet 364:30 pp., 7 figs. (multilith)

bibliography, molluscs
Sivalingam, S., 1962.
Bibliography of pearl oysters.
Bull. Fish. Res. Sta., Ceylon, 13: 21 pp.

bibliography, molluscs
Tesch, J. J., 1948
The Thecosomatous pteropods. II. The Indo-Pacific. Dana Rept. No. 30: 45 pp., 34 text figs., 3 pls.

bibliography, molluscs
Tesch, J. J., 1946
The Thecosomatous Pteropods. Dana Rept. No. 28:82 pp., 34 text figs., 8 pls.

bibliography, molluscs
Tesch, J.J., 1913
Mollusca. Pteropoda. Das Tierreich. 36:154 pp., 108 text figs.

bibliography, molluscs (Nautilus)
Toriyama, Ryuzo, Tadashi Sato, Takashi Hamada, and Pumwarn Komalarjun, 1965.
Nautilus pompilius drifts on the west coast of Thailand.
Japan. J. Geol. Geogr., Trans., 36(2/4):149-161.

bibliography, molluscs
Van der Spoel, S., 1967.
Euthecosomata, a group with remarkable developmental stages (Gastropoda, Pteropoda).
J. Noorduijn en Zoom, N.V., Gorinchem, 375 pp.

bibliography, molluscs
Yonge, C.M., 1947
The pallial organs in the Aspidobranch Gastropoda and their evolution throughout the Mollusca. Phil. Trans. Roy. Soc. London, Ser. B Biol. Sci. 232(591):443-518, 40 textfigs., 1 pl.

bibliography, molluscs (cultures)
Ranson, Gilbert, et Mlle. Pavetias, 1965.
Les huîtres biologie-culture: bibliographie.
Bull. Inst. océanogr., Monaco, 67(1388): 51 pp.

nannoplankton

bibliographies, nannoplankton (calcareous)
Loeblich, Alfred R., Jr., and Helen Tappan 1971.
Annotated index and bibliography of the calcareous nannoplankton VI.
Phycologia 10(4): 315-339.

bibliographies, nannoplankton
Loeblich, Alfred R. Jr., and Helen Tappan, 1970.
Annotated index and bibliography of the calcareous nannoplankton V.
Phycologia, 9(2): 157-174.

bibliography, nannoplankton
Leoblich, Alfred R., Jr., and Helen Tappan, 1966.
Annotated index and bibliography of the calcareous nannoplankton.
Phycologia, 5(2/3):81-216.

neuston

Bibliography, neuston
Zaitsev, Ju. P., 1968.
La neustonologie marine: objet, méthodes, réalisations principales et problèmes.
Pelagos, 8:1-48.

parasites

bibliography, parasites
Boschma, H., 1959.
Ellobiopsidae from Tropical West Africa.
Atlantide Rept., Sci. Res., Danish Exped., Coasts of Tropical West Africa, 1945-1946, 5:145-175.

bibliography, parasites (hosts)
Dollfus, R. Ph., 1963.
Liste des coelenteres marins, palearctiques et indicus ou ont été trouves des trematodes digenetiques.
Bull. Inst. Pêches Marit., Maroc., (9-10):33-57.

bibliography, parasites
Ormières, René, 1964.
Recherches sur les sporozoaires parasites des tuniciers.
Vie et Milieu, Bull. Lab. Arago, Univ. Paris, 15(4): 823-946.

bibliographies, parasites
Sindeman, Carl J. 1968.
Bibliography of oyster parasites and diseases.
Spec. scient. Rep. U.S. Fish. Wildl. Serv. Fish. 563: 13 pp.

peridinians

Bibliographies, biological

peridinians, see: dinoflagellates

plankton

bibliography, plankton
Bernard, M. F., 1938
Recherches récentes sur la densité du plancton méditerranéen. Rap. Proc. Verb des Réunion, Comm. Int. l'Expl. Sci. de la Méditerranée, n.s., XI:289-300.

bibliographies, plankton
Beyer, F. 1954.
Om metoder som brukes ved undersökelse av dyr eplanktonets fordeling og om deres begrensning.
12 pp. (mimeographed).

Bibliography, plankton
Brunel, Jules, 1962.
Le phytoplancton de la Baie des Chaleurs.
Contrib. Ministère de la Chasse et des Pêcheries, Province de Québec, No. 91: 365 pp.

bibliography, plankton
Cassie, R.M., 1963
Microdistribution of plankton. In: Oceanography and Marine Biology, H. Barnes, Edit., George Allen & Unwin, 1:223-252.

bibliography, plankton (chemistry)
*Corner, E.D.S., and C.B. Cowey, 1968.
Biochemical studies on the production of marine zooplankton.
Biol. Rev., 43(4):393-426.

bibliography, plankton
Fraser, J.H., 1961
The oceanic bathypelagic plankton of the North-East Atlantic and its possible significance to fisheries.
Dept. Agric. & Fish., Scotland, Marine Research (4):1-48.

bibliography plankton a
Furnestin, Marie-Louise, 1970.
Rapport sur le plancton.
Rapp. P. v. Réun. Cons. int. Explor. Mer 159: 90-115.

bibliography, plankton (Mediterranean)
*Furnestin, Marie-Louise, 1968.
Le zooplancton de la Mediterranee (bassin occidental). Essai de synthese.
J. Cons. perm. int. Explor. Mer, 32(1):25-69.

bibliographies plankton
Furnestin, Marie-Louise 1968.
Rapport sur les travaux concernant la planctonologie méditerranéenne publiés entre octobre 1964 et octobre 1966.
Rapp. P.-v. Réun. Comm. int. Explor. scient. Mer Méditerranée 19(3):335-366

bibliographies, plankton
Hasle, Grethe Rytter 1960.
Phytoplankton and ciliate species from the tropical Pacific.
Skr. Norske Videnskaps-Akad., Oslo, I. Mat.-Nat. Kl. 1960(2): 1-50.

bibliography plankton
Motoda, Sigeru, 1959
Devices of simple plankton apparatus.
Mem., Fac. Fish., Hokkaido Univ., 7(1/2): 73-94, 8pls.

bibliographies, plankton
Motoda, S., 1954.
On plankton research in Japan with an annotated bibliography.
Symp. Mar. &Fresh-water Plankton, Indo-Pacific, Bangkok, Jan. 25-26, 1954, FAO & UNESCO:57-70.

bibliography plankton
Rose, M., 1925.
Contribution à l'étude de la biologie du plankton. Le problème des migrations verticales journalières. Arch. Zool. expér. et gén. 64: 387-542, 41 textfigs.

bibliography plankton
Schulz, B., and A. Wulff, 1929
Hydrographie und Oberflächen plankton des westlichen Barentsmeeres im Sommer 1927.
Ber. deutschen wissensch. Komm. F. Meeresforsch. n.s. 4(5):232-372, 13 tables, 25 text figs.

bibliography plankton
Sproston, N. G., 1949.
A preliminary survey of the plankton of the Chu-San region with a review of the relevant literature. Sinensia 20:58-161.

bibliography, plankton, sampling
Tranter, D.J., editor, 1968.
Zooplankton sampling.
UNESCO, 174 pp.

bibliography, plankton
Tregouboff, G., 1956.
Rapport sur les travaux concernant le plancton Méditerranéen publiés entre Novembre 1952 et Novembre 1954.
Rapp. Proc. Verb., Comm. Int. Expl. Sci., Mer Méditerranee, 13:65-100

bibliography plankton
Tregouboff, G., 1952.
Rapport sur l'activité planctonologique actuelle des laboratoires maritimes de la Méditerranée occidentale.
Rapp. Proc. Verb., Comm. Int. Expl. Sci., Mer Medit., 12:53-77.

bibliography, plankton
Wiborg, K.F., 1954.
Investigations on zooplankton in coastal and offshore waters in western and northwestern Norway with special reference to the copepods.
Repts. Norwegian Fish. Mar. Invest. 11(1):1-246, 102 textfigs.

bibliography, plankton
Wimpenny, R.S., 1952.
Plankton. Rapp. Proc. Verb., Cons. Perm. Int. Expl. Mer, 132:28-35.

bibliography plankton
Yamazi, I., 1956.
Plankton investigations in inlet waters along the coast of Japan. XIX. Regional characteristics and classification of inlet waters based on the plankton communities. Publ. Seto Mar. Biol. Lab., 5(2)(9):157-196, Pls. 16-23

Pogonophora

bibliography, Pogonophora
Southward, E.C., 1963
Pogonophora. In: Oceanography and Marine Biology. H. Barnes, Edit., George Allen & Unwin, 1:405-428.

poisonous animals

bibliography, poisonous animals
De Clerq, Monique, 1964.
Aperçu sur les recherches scientifiques effectuées dans le domaine sw la toxicologie marine. Les animaux marins toxicophores.
L'Année Biol., Féd. Française, Soc. Sci. Nat., (4), 3(9/10):429-479.

bibliography, poisonous animals
Halstead, Bruce W., (with chemical sections by Donovan A. Courville), 1965.
Poisonous and venomous marine animals of the world. 1. Invertebrates.
U. S. Government Printing Office, 994 pp.

bibliography, poisonous animals
Nishimura, Saburo, 1965.
Droplets from the plankton net. XX. "Sea stings" caused by Creseis acicula Rang (Mollusca: Pteropoda) in Japan.
Publ. Seto Mar. Biol. Lab., 13(4):287-290.

bibliography, poisonous animals
Phillips, C., and W.H. Brady, 1953.
Sea pests. Poisonous or harmful sea life of Florida and the West Indies. Spec. Publ., Mar. Lab., Univ. Miami, Univ. Miami Press, 78 pp., 7 pls. (75 cents).

bibliography, poisonous animals
Russell, Findlay E., 1965.
Marine toxins and venomous and poisonous marine animals.
In: Advances in Marine Biology, Sir Frederick S. Russell, editor, Academic Press, 3:255-384.

bibliography, poisonous animals
Valette, E., 1969.
Les organismes marins toxiques et venimeux.
Bull. Cent. Etud. rech. sci., Biarritz, 7(3): 511-528

polychaetes

Bibliographies, biological

polychaets, see: annelids

bibliographies, priapulids a
Van der Land, J., 1970.
Systematics, zoogeography and ecology of the Priapulida.
Zool. Verhandel., Leiden, 112: 118 pp.

primary productivity

bibliography, primary productivity
Doty, Maxwell S., 1963.
A bibliography of articles pertinent to primary productivity
Proc. Conf., Primary Production Measurements, Marine and Freshwater, Univ. Hawaii, Aug. 21-Sept. 6, 1961, U.S. Atomic Energy Comm., Div. Techn. Inf., TID-7633:184-212.

bibliography primary productivity
Ketchum, B.H., 1954.
Mineral nutrition of phytoplankton.
Ann. Rev. Plant Physiol. 5:55-74.

bibliography, primary productivity
Oppenheimer, Carl H., editor, 1966.
Phytoplankton, Marine Biology II, Proceedings of the Second International Interdisciplinary Conference.
New York: Acad. Sci., 369 pp. ($8.00).

bibliography primary productivity
Provasoli, L., 1963
8. Organic regulation of phytoplankton fertility. In: The Sea, M.N. Hill, Edit., Vol. 2 (II) Fertility of the oceans, Interscience Publishers, New York and London 165-219.

bibliography, primary productivity
Steemann Nielsen, E., 1964.
Recent advances in measuring and understanding marine primary production.
Suppl. to J. Ecology, 52, and J. Animal Ecology, 33:119-130.
British Ecol. Soc., Jubilee Symposium

bibliography, primary productivity
Strickland, J.D.H., 1961.
Measuring production of marine phytoplankton.
Bull. Fish. Res. Bd., Canada, No. 122:172 pp.

bibliography, primary productivity
Yentsch, C.S., 1963.
Primary production.
In: Oceanography and Marine Biology, H. Barnes, Editor, George Allen and Unwin, 1:157-175.

bibliography, productivity (methods)
De Angelis, Costanzo, 1957.
Metodi di ricerca sulla produttivita del mare.
Boll. Pesca, Piscicolt. e Idrobiol., 12(2):159-211.

protozoa

Bibliographies, biological

protozoa, see also: radiolaria
tintinnids
silicoflagellates

bibliography, protozoans
Borror, Arthur Charles, 1963.
Morphology and ecology of the benthic ciliated Protozoa of Alligator Harbor, Florida.
Arch. Protistenkunde, 106(4):465-534.

bibliography protozoa
Provosoli, Luigi, 1958
Nutrition and ecology of protozoa and algae.
Ann. Rev. Microbiol., 12: 279 - 308

radiolaria

bibliographies, radiolaria
Campbell, A.S., and E.A. Holm, 1957.
Radiolaria. Annotated bibliography of marine paleoecology. In: Treatise on marine ecology and paleoecology, Vol. 2, Paleoecology.
G.S.A. Mem., 67:737-744.

red water

bibliography, red tides
Hayes, H.L., and T.S. Austin, 1951.
The distribution of discolored sea water.
Texas J. Sci. 3(4):530-541, 1 textfig.

bibliography, red tides
Hutton, R.F., 1956.
An annotated bibliography of red tides occurring in the marine waters of Florida.
Q.J. Fla. Acad. Sci., 19(2/3):123-186.

rotifers

bibliography, rotifers
Berzins, Bruno, 1960
Rotatoria VI. Order: Monogononta, (1) Sub-Order Floscularíaceae, (i) Family: Testudinellidae, Genera: Testudinealla, Filinia, Hexarthra, (ii) Family: Conchildae, Genus: Conochilus. (2) Sub-Order: Collothecaceae, Family: Collothecidae, Genus Collotheca.
Fiches d'Ident., Cons. Perm. Int. Expl. Mer, Zoopl. Sheet, 89: 4 pp.

bibliography, rotifers
Berzins, Bruno, 1960
Rotatoria, V. Order: Monogononta, Sub-Order: Ploima, (i) Family: Asplanchnidae, Genus: Asphanchna. (ii) Family: Synchaetidae, Genera: Ploesoma, Polyarthra.
Fiches d'Ident., Cons. Perm. Int. Expl. Mer, Zool. Sheet, 88: 4 pp.

bibliography, rotifers
Berzins, Bruno, 1960
Rotatoria IV. Order: Monogononta, Sub-order: Ploima, Family: Brachionidae (Cont.), Genera: Brachionus, Kellicottia, Argonotholca, Notholca, Pseudonotholca, Euchlanis, Tripleuchlanis.
Fiches d'Ident., Cons. Perm. Int. Expl. Mer, Zool. Sheet, 87: 5 pp.

bibliography, rotifers
Berzins, Bruno, 1960
Rotatoria III. Order: Monogonenta (sic), Sub-Order: Ploima, Family: Brachionidae; Genus: Keratella.
Fiches d'Ident., Cons. Perm. Int. Expl. Mer, Zoopl. Sheet, 86: 4 pp.

bibliography, rotifers
Berzins, Bruno, 1960
Rotatoria II. Order: Monogononta, Sub-order: Ploima, Family: Trichocercidae, Genus: Trichocerca.
Fiches d'Ident., Cons. Perm. Int. Expl. Mer, Zoopl. Sheet, 85: 3 pp.

bibliography, rotifers
Berzins, Bruno, 1960
Rotatoria I, Order Monogononta, Sub-order: Ploima, Family: Synchaetidae, Genus: Synchaeta
Fiches d'Ident., Cons. Perm. Int. Expl. Mer, Zoopl., Sheet 84: 7 pp.

bibliographies, sea farming
Hickling, C.F. 1970.
Estuarine fish farming.
Adv. mar. Biol., F.S. Russell and Maurice Yonge, editors, Academic Press, 8:119-213

Bibliographies, biological,

For seaweeds, see: algae

bibliography, self-purification
Paoletti, A., 1970.
Facteurs biologiques d'autoepuration des eaux de mer: points clairs et points obscurs d'une question discutée.
Rev. int. Océanogr. méd., 18-19:33-68

bibliographies, silicoflagellates
Glezer, Z.I. 1966.
Silicoflagellatophyceae. (In Russian)
Flora sporovykh rastenii SSSR, Acad. Nauk, Inst. IM. V.A. Komarova, 7.
Translation: Israel Program for Scientific Translations, Jerusalem, 1970, 363 pp.

bibliographies, silicoflagellates
Hanna, G.D., 1957
Silicoflagellata. Annotated bibliography of marine paleoecology. In: Treatise on Marine Ecology and Paleoecology, Vol. 2, Paleoecology.
G.S.A. Mem., 67:745-746.

Symbiosis

bibliography, symbiosis
Mansueti, Romeo J., 1963.
Symbiotic behavior between small fishes and jellyfishes with new data on that between the stomateid, Peprilus alepidotus, and the Scyphomedusa, Chrysaora quinquecirrha.
Copeia, (1):40-80.

tintinnids

bibliography, tintinnids
Entz, G., Jr., 1909
Studien über organisation und biologie der Tintinniden. Arch. f. Protistenkunde 15:93-226, Pls. 8-21, text figs.

bibliography, tintinnids
Kofoid, C. A. and A. S. Campbell, 1929
A conspectus of the marine and fresh-water Ciliata belonging to the suborder Tintinnoinea, with descriptions of new species principally from the Agassiz expedition to the eastern tropical Pacific, 1904-1905. Univ. Calif. Publ. Zool. 34:1-403, 697 text figs.

tunicates

bibliography, tunicates
Berrill, N.J., 1950.
The Tunicata with an account of the British species. Ray Soc. No. 133:354 pp., 120 textfigs.

bibliographies, tunicates
Buckman, A., 1945.
Appendicularia I-III. Fiches d'Ident. Zooplancton Cons. Perm. Int. Expl. Mer, 7-8 pp., 16 textfigs.

bibliography, tunicates
Fraser, J. H., 1949.
The distribution of Thaliacea (Salps and Doliolids) in Scottish Waters 1920 to 1939.
Scottish Home Dept., Fish. Div., Sci. Invest. 1949(1):44 pp., 16 textfigs.)

bibliographies, tunicates
Fraser, J. H., 1947.
Thaliacea - II. Family: Doliolidae. Fiches d'Ident Zooplancton, Cons. Perm. Int. Expl. Mer, 10:4 pp., 13 textfigs.

bibliographies, tunicates
Fraser, J.H., 1947.
Thaliacea I. Family: Salpidae. Fiches d'Ident. Zooplancton, Cons. Perm. Int. Expl. Mer, 9:4 pp., 17 textfigs.

bibliography, tunicates
Godeaux, J., 1957-1958.
Contribution à la connaissance des Thaliacés. Embryogénèse et blastogénèse du complexe neural. Constitution et développement du stolon prolifère.
Ann. Soc. Roy. Zool. Belgique, 88:5-285.

bibliography, tunicates
Ihle, J.E.W., 1912
Tunicata. Salpae I. Desmomyaria. Das Tierreich. 32:66 pp., 68 text figs.

owned by MS

bibliography, tunicates
Neumann, G., 1913
Tunicata Salpae II: Cyclomyaria et Pyrosomida. 40:36 pp., 19 text figs.
Das Tierreich

owned by MS

bibliography, tunicates
Tokioka, Takasi, 1960
Studies on the distribution of appendicularians and some thaliaceans of the North Pacific, with some morphological notes.
Publ. Set. Mar. Biol. Lab., 8(2) (27):351-443 tables.

Vitamins

bibliographies, vitamins

Carlucci, A.F., and S.B. Silbernagel, 1966.
Bioassay of seawater. II. Methods for the determination of concentrations of dissolved Vitamin B_1 in seawater.
Can. J. Microbiol., 12(6):1079-1089.

whales

bibliography, whales

Scheffer, V.B., and J.W. Slipp, 1948.
The whales and dolphins of Washington State with a key to the Cetaceans of the west coast of North America. Am. Mid. Nat. 39(2):257-337, 50 textfigs.

bibliography, whales

Slijper, E.J., 1962.
Whales. Hutchinson of London, 475 pp.

bibliography, whales (diseases of)

Stolk, A., 1962
Tumours in whales. III Granuloma malignum (Hodgkin's disease) in the fin whale, Balaenoptera physalus.
Proc. Kon. Nederl. Akad. van Wetenschap, Amsterdam, C, 65(3):250-268.

bibliography (whales)

Van Heel, W.H.D., 1962.
Sound and Cetacea.
Netherlands J. Sea Research, 1(4):407-507.

zooplankton

bibliography, zooplankton

Furnestin, Marie-Louise, 1971.
Rapport sur les travaux concernant la planctonologie méditerranéenne publiés entre octobre 1966 et octobre 1968. Méditerranée occidentale.
Rapp. P.-v. Réun. Comm. int. Explor. scient. Mer Médit. 20(2):115-139.

bibliography, zooplankton

Ghirardelli, Elvezio, 1971.
Rapport sur les travaux concernant la planctonologie méditerranéenne publiés entre octobre 1966 et octobre 1968. Mer Adriatique.
Rapp. P.-v. Réun. Comm. int. Explor. scient. Mer Médit. 20(2):141-146

Bibliography, zooplankton

Skolka, Vidor-Hilarius, 1971
Rapport sur les travaux concernant la planctonologie méditerranéenne publiés entre octobre 1966 et octobre 1968.
Rapp. P.-v. Réun. Comm. int. Explor. scient. Mer Médit. 20(2):147-166

Boats

Bibliography, boat

filed under name of vessel

bibliography, ships (polar)

Morley, J.P., 1963.
Polar ships and navigation in the Antarctic.
British Antarctic Survey, Bull., (2):1-25.

bibliography "Albatross"

Townsend, C.H., 1901.
Dredging and other records of the United States Fish Commission Steamer "Albatross" with bibliography relative to the work of the vessel.
U.S. Comm. Fish. Fisheries Rept., 1900:387-562, 7 pls.

bibliography, "Eltanin"

Sandved, K.G., 1966.
USNS Eltanin: four years of research.
Antarctic J., U.S., 1(4):164-174.

bibliography "Meteor"

Defant, A., 1941.
Abschliessender Bericht über die wissenschaftlichen Ergebnisse der Deutschen Atlantischen Expedition des Forschungs- und Vermessungsschiffes "Meteor", 1925-1927. Jahrbuch Preuss. Akad. Wiss., 1941:5 pp.

books

bibliographies

Emery, K.O., and Evelyn Sinha, 1967.
Oceanographic books of the world, 1957-1966.
Mar. Techn. Soc., TP 2:57 pp. $2.00.

bibliography, bubbles

Blanchard, Duncan C., 1963.
The electrification of the atmosphere by particles from bubbles in the sea.
Progress in oceanography, 1:71-202.

Chemistry

bibliography chemistry

Argentina, Servicio de Hidrografía Naval, 1959.
Química del agua del mar.
Servicio de Hidrografía Naval, Argentina, Publ., H. 604: 140 pp.

bibliography, chemistry

Bader, Richard G., 1962
Some experimental studies with organic compounds and minerals.
The Environmental Chemistry of Marine Sediments, Proc. Symp., Univ. R.I., Jan. 13, Occ. Papers, Narragansett Mar. Lab., No. 1:42-57.

bibliography, chemistry

Berrit, G., and B. Dussart, 1950.
Dosage dans les eaux naturelles des composés minéraux du phosphore (bibliographie). Circ., C.R.E.O., R.T.B., No. 4:11 pp.

bibliography, chemistry

Bruevich, S.W., editor, 1966.
Chemistry of the Pacific Ocean.
Inst. Okeanol., Akad. Nauk. SSSR, Isdatel, Nauka, Moskva, 358 pp.

bibliography, chemistry

Buch, K., 1952.
The cycle of nutrient salts and marine production
Rapp. Proc. Verb., Cons. Perm. Int. Expl. Mer, 132:36-46, 4 textfigs.

bibliography, chemistry

Buch, K., 1950.
Die Kohlensäurefaktoren des Meereswassers.
Cons. Perm. Int. Expl. Mer, Rapp. Proc. Verb. 67:51-88, figs.

bibliography, chemistry

Cheeseman, D.F., W.L. Lee and P.F. Zagalsky, 1967.
Carotenoproteins in invertebrates.
Biol. Rev., Cambridge Phil. Soc., 42(1):131-160.

bibliography, chemistry (drugs)

Der Marderosian, Ara Harold 1969
Biodynamic agents from marine sources as potential drugs.
J. mar. techn. Soc. 3(3):61-84

bibliographies, chemistry

Duursma, E.K., 1961
Dissolved organic carbon, nitrogen and phosphorus in the sea.
Netherlands J. Sea Res., 1(1/2):1-147.

bibliography, chemistry

Gillam, W.S., and J.W. McCutchan, 1961
Demineralization of saline waters. Current desalination processes and research hope for solution of our impending water crisis.
Science, 134(3485):1041-1048.

bibliography, chemistry

Grasshoff, Klaus, 1962
Untersuchungen über die Sauerstoffbestimmung im Meerwasser, 2.
Kieler Meeresforschungen, 18(2):151-160.

bibliography Chemistry

Hood, D.W., 1963
Chemical oceanography. In: Oceanography and Marine Biology, H. Barnes, Edit., George Allen & Unwin, 1:129-155.

bibliography, chemistry

Ishibashi, M., 1952.
Recent advances in chemical oceanography in Japan. List of titles in the symposium with references. Proc. Seventh Pacific Sci. Congr., Met. Ocean., 3:303-304.

bibliography, chemistry

Koyama, Tadashiro, 1962
Organic compounds in sea water.
J. Oceanogr. Soc., Japan, 20th Ann. Vol., 563-576.

bibliographies, chemistry (phosphate)

Olsen, Sigurd 1967.
Recent trends in the determination of orthophosphate in water.
In: Chemical environment in the aquatic habitat, H.L. Golterman and R.S. Clymo, editors, Proc. I.B.P. Symp. Amsterdam, Oct. 1966: 63-105

bibliography, chemistry

Richards, F.A., 1957.
Some current aspects of chemical oceanography.
In: Physics and chemistry of the earth, Pergamon Press, 2:77-128.

bibliography, chemistry (lacks titles)

Riley, J.P., 1965.
Analytical chemistry of sea water.
In: Chemical oceanography, J.P. Riley and G. Skirrow, editors, Academic Press, 2:295-424.

bibliography, chemistry

Saunders, George W., Francesco B. Trama and Roger W. Bachmann, 1962
Evaluation of a modified C14 technique for shipboard estimation of photosynthesis in large lakes.
Great Lakes Res. Div., Inst. Sci. and Techn., Univ. Michigan, Publ. No. 8:61 pp.

bibliography chemistry

Seibold, E., 1962.
Untersuchungen zur Kalkfällung und Kalklösung am Westrand der Grand Bahama Bank.
Sedimentology, 1(1):50-74.

bibliography, chemistry (sea water)

Soyer, J., 1963.
Contribution à l'étude des effets du mercure et de l'argent dans l'eau de mer.
Vie et Milieu, 14(1):1-36.

bibliography, chemistry

Thomsen, H., 1952.
The milieu. Rapp. Proc. Verb., Cons. Perm. Int. Expl. Mer, 132:21-27.

bibliography, chemistry
Toll, R., J.-M[a] Valles, and F. Saiz, 1952.
Sur l'utilisation directe d'une eau de mer quelconque comme eau auxiliaire pour la détermination de la chlorinité des eaux marins.
Vie et Milieu, Suppl. 2, Océan. Medit., Jour. Etudes, Lab. Arago:282-291.

bibliographies, chemistry
United States, National Oceanographic Data Center 1970.
CICAR bibliography on meteorology, climatology and physical/chemical oceanography.
Vol. 1: 380 pp.
Vol. 2: 614 pp.

bibliographies, chemistry
Vallentyne, J.R., 1957.
The molecular nature of organic matter in lakes and oceans, with lesser reference to sewage and terrestiral soils. J. Fish. Res. Bd., Canada, 14(1):33-82.

bibliographies, chemistry (organic compounds)
Wagner, Frank S. Jr. 1969.
Composition of the dissolved organic compounds in sea water: a review.
Contrib. mar. Biol., Port Aransas 14:115-153.

coral reefs

bibliographies, coral reefs
Davis, W.M., 1923.
The coral reef problem. Amer. Geogr. Soc., Spec. Publ. 9:596 pp.

bibliography, coral reefs
Ladd, H. S., and J.I. Tracy, jr., 1949.
The problem of coral reefs. Sci. Month. 69(5): 297-305, figs.

bibliographies, coral reefs
Ranson, G., 1958.
Coraux et récifs coralliens (bibliographie).
Bull. Inst. Océan., Monaco, No. 1121:80 pp.

bibliography, coral reefs
Stoddart, David R., editor, 1966.
Reef studies at Addu Atoll, Maldive Islands.
Atoll Res. Bull., 116:122 pp. (mimeographied).

bibliography, coral reefs
Tayama, R., 1952.
Coral reefs of the South Seas. Publ. No. 941, Bull. Hydrogr. Office 11:292 pp., (English resume, pp. 184-292) with Appendix 1 of 133 pp., of photos and Appendix 2 of 18 charts.

bibliography, coral reefs
Wiens, Herold J., 1961.
The role of mechanical abrasion in the erosion of coral reefs and land areas.
Proc. Ninth Pacific Sci. Congr., Pacific Sci. Assoc., 1957, 12(Geol.-Geophys.):361-366.

bibliography, coral reefs
Wiens, Herold J., 1961.
The evolution and destruction of atoll land.
Proc. Ninth Pacific Sci. Congr., Pacific Sci. Assoc., 1957, 12(Geol.-Geophys.):367-376.

bibliography, corrosion

bibliography, corrosion
Cuba, Academia de Ciencias de Cuba 1970.
Corrosión 1. Teoría de la corrosión.
Serie Inform. Cient. 39:135 pp. (mimeographed).

bibliography, corrosion
Keehn, Pauline A., 1967.
Bibliography marine corrosion.
Spec. Bibliogr. Oceanogr., Contrib. Am. Met. Soc., 4: 158 pp.

currents

bibliographies, currents
Associations d'Oceanographie Physique, 1957.
Bibliography on generation of currents and changes of surface level in oceans, seas and lakes by wind and atmospheric pressure, 1726-1955. Publ. Sci., 18:83 pp.

bibliography, currents (turbidity)
Kuenen, Ph. H., and F.L. Humbert, 1964
Bibliography of turbidity currents and turbidites. In: Turbidites, A.H. Bouma and A. Brouwer, Editors. Developments in Sedimentology, Elsevier Publishing Co., 3:222-246.

bibliography, currents
Montgomery, R.B., 1962
Equatorial Undercurrent observations in review.
J. Oceanogr. Soc., Japan, 20th Ann. Vol., 487-498.

bibliographies, currents
Sinha, Evelyn Z., Compiler, 1962.
Annotated bibliography on waves and currents.
Meteorol. Geoastrophys. Abstr., 13(9):2689-2741.

bibliography, current ripples
Dzulynski, Stanislaw, and John E. Sanders, 1962.
Current marks on firm mud bottoms.
Transactions, Connecticut Acad. Arts and Sci., 42 :57-96.

bibliographies, currents
Timonova, V.V., and I.I. Soskin, Ed., 1955.
Collected works on methods of studying ocean currents and tidal phenomena.
Trudy Gosud. Okean. Inst., 30(42):1-290.

Bibliographies, data
United States, National Oceanographic Data Center, 1961.
Reference sources for oceanographic station data. (Provisional).
Catalogue Ser., Publ., C-1:unnumbered pp. (multilithed).

bibliography, currents
Wiegel, R.L., and J.W. Johnson, 1960
Ocean currents, measurement of currents and analysis of data.
Waste disposal in the marine environment, Pergamon Press: 175-245.

bibliography, diffusion
Okubo, Akira, 1962
Horizontal diffusion from an instantaneous point-source due to oceanic turbulence.
Chesapeake Bay Inst., Techn. Rept. (Ref. 62-22) 32: 124 pp.

bibliography, diffusion
Okubo, Akira, 1962
A review of theoretical models of turbulent diffusion in the sea.
Chesapeake Bay Inst., Techn. Rept. 30 (Ref. 62-20): 105 pp. (multilithed).

bibliographies, diffusion
Okubo, Akira, 1962
A review of theoretical models for turbulent diffusion in the sea.
J. Oceanogr. Soc., Japan, 20th Ann. Vol., 286-320.

estuaries

bibliography, estuaries
Crance, Johnie H. 1969.
A selected bibliography of Alabama estuaries.
Bull. Alabama mar. Resources 2:21 pp.

bibliography, estuaries
Fisher, Leo, J., 1960.
An annotated bibliography of flushing and dispersion in tidal waters.
U.S. Navy Hydrogr. Off., Spec. Publ., SP-33:34 p

bibliography, estuarine
Rochford, D.J., 1951.
Studies in Australian estuarine hydrology.
Australian J. Mar. and Freshwater Res. 2(1):1-116, 1 pl., 7 textfigs.

bibliographies, estuaries
U.S.A. Corps of Engineers, U.S. Army Committee on Hydraulics 1965
Bibliography on tidal hydraulics.
Supplementary material compiled from May 1959 to May 1965. Tidal flow in rivers and harbors.
Rept. No. 2, Suppl. No. 4 (ES816): 218 pp. (multilithed)

evaporites

bibliography, evaporites
Cramer, Harvey Ross 1969
Evaporites, a selected bibliography.
Bull. Am. Ass. Petr. Geol. 53(4): 982-1011

expeditions

bibliography, expeditions ("Carnegie")
Crow, R.M., 1946.
Scientific results of Cruise VII of the Carnegie 1928-1929 --- Oceanography IV. The work of the Carnegie and suggestions for future scientific cruises. VIII. Complete bibliography of Cruise VII of the Carnegie. Publ. Carnegie Inst., Washington, 571:107-110.

bibliography, expeditions ("Albatross")
Riccardi, R., 1956.
La spedizione oceanografica dell'"Albatross" intorno al mondo. Boll. Pesca, Piscic. e Idrobiol., n.s., 9(2):300-309.

fisheries

bibliographies, fisheries
Allen, E.J., 1926.
A selected bibliography of marine bionomics and fishery investigation. J. du Cons. 1:77-96.

bibliographies, fisheries
Current Bibliography for Fisheries Science, 1958.
prep. by Biol. Branch, Fisheries Division
FAO/58/12/8947 Rome Nov. 1958

bibliographies, fisheries
Current Bibliography for Fisheries Science, 1958.
Biol. Branch, Fisheries Division.
FAO/58/11/8430 [2.1(2)]

bibliographies, fisheries
Blanco, G.J., and H.R. Montalban, 1953.
A bibliography of Philippine fishes and fisheries
Phil. J. Fish. 1(2):107-130.

bibliography, fisheries
Cushing, D.H., 1955.
Production and a pelagic fishery.
Fish. Invest., Min. Agric., Fish., & Food, (2), 18(7):1-104.

bibliography, fisheries
F.A.O. Fisheries Division, 1953.
Improving the fisheries contribution to World food supplies. F.A.O. Fish. Bull. 6(5):159-192, 2 textfigs.

bibliographies, fisheries
Longhurst, A.R., 1961.
Report on the Fisheries of Nigeria, 1961.
Federal Fisheries Service, Ministry of Economic Development, Lagos, numerous pp. (mimeographed).

bibliographies
Murty, V. Sriramachandra, D.C.V. Easterson, A. Bastian Fernando, K.K. Appukuttan and K.M.S. Ameer Hamsa, 1968. Bibliography of marine fisheries and oceanography of the Indian Ocean, 1962-1967
Bull. cent. mar. Fish. Res. Inst. 1: 208pp. (mimeographed).

bibliography, fisheries
Office Scientifique et Technique des Pêches Maritimes, 1948.
Bibliographie analytique des Publications de l'Office Scientifique et Technique des Pêches Maritimes. Notes et Rapports, n.s., No. 3:72 pp.

bibliography, fisheries
Redeke, H.C., 1927.
River pollution and fisheries.
Rapp. Proc. Verb., Cons. Perm. Int. Expl. Mer, 43:50 pp.

bibliographies, fisheries
Scattergood, L.W., 1958.
English translations of fishery literature, additional listings, 1958. USFWS Spec. Sci. Rept. Fish., 264:33 pp.

bibliographies, fisheries
Scattergood, L.W., 1954.
Bibliographic sources for fishery students and biologists
Trans. Amer. Fish. Soc., 83:20-37.

bibliographies, fisheries
Scattergood, L.W., 1954.
Bibliographic sources for fishery students and biologists. Trans. Amer. Fish. Soc., 1953:20-37.

bibliography(translations), fisheries
Sette, O.E., et al., 1954.
Progress in Pacific Ocean Fishery Investigations 1950-53. Spec. Sci. Rept.: Fish. No. 116:75 pp., 29 textfigs.

bibliographies, fisheries
United Nations, Food and Agricultural Organization, Fisheries Branch, 1960.
Current bibliography for aquatic sciences and fisheries. Explanation of coverage and arrangement, 3(1):Ex. 1-Ex. 88.

bibliographies, fisheries
U.S. Fish and Wildlife Service, 1955.
Fishery publications Index, 1920-1954. Circular 36:254 pp. $1.50 Superintendent of Documents

bibliographies, fisheries
van Campen, W., and E.E. Hoven, 1956.
Tunas and tuna fisheries of the world. An annotated bibliography, 1930-1953.
U.S.F.W.S. Bull. (Fish. Bull. 111) 57:173-249.

bibliographies, geochemistry
Graf, Donald L., 1960
Geochemistry of carbonate sediments and sedimentary carbonate rocks. IV B. Bibliography.
Ill. State Geol. Survey, Circular 309: 55 pp.

Bibliography

bibliography, geodesy
Heisakanen, W.A., 1955.
Intercontinental connection of geodetic systems.
Int. Hydrogr. Rev. 23(1):141-156, 7 textfigs.

Bibliographies, geographical areas

See: Bibliographies, areal

GEOLOGY

bibliography, geological
Boillot, Gilbert, 1964.
Géologie de la Manche occidentale fonds rocheux, dépôts quaternaires, sédiments actuels.
Ann. Inst. Océanogr., Monaco, 42(1):1-219.

bibliography, geological
Erimesco, P., 1963.
The expanding ocean floor.
Bull. Inst. Pêches Marit. Maroc, (9-10):3-31.

Bibliography, geological
Fairbridge, Rhodes W., 1965.
The Indian Ocean and the status of Gondwanaland.
Progress in Oceanography, 3:83-136.

abstract

bibliography, geology
Geyer R.A., 1950? (undated).
Bibliography of oceanography, marine biology, geology, geophysics and meteorology of the Gulf of Mexico. Humble Oil Refining Co. (multilith. Pages not numbered serially).

bibliography, geological
Geyer, R. A., 1948
Annotated bibliography of marine geophysical and geological surveys. Bull. G.S.A., 59(7):671-696.

bibliographies, geology
Hoel, A., and J. Norvik, 1962.
Glaciological bibliography of Norway.
Norsk Polarinst., Skrifter, No. 126:242 pp.

includes papers on sea as well as land.

bibliographies, geological
Ladd, H.S., Chairman, 1946
Report of the Committee on Marine Ecology as related to Palaentology, 1945-1946, 101 pp.

bibliography, geology
Ma, T.Y.H., 1948.
Origin of submarine canyons. Bull. Ocean. Inst. Taiwan, No. 4:37-46.

bibliography, geology
Pratje, O., 1951.
Die Erforschung des Meeresbodens. Geol. Rundschau 39(1):152-176, 1 textfig.

bibliography, geological
Stride, A.H., 1963
The geology of some continental shelves.
In: Oceanography and Marine Biology, H.Barnes Edit. George Allen & Unwin, 1:77-88.

bibliographies, geological
Terry, R.D., 1956.
Bibliography of marine geology and oceanography, California coast. Calif. Div. Mines, Spec. Rept. 43:

bibliographies, geological
Terry, R.D., and G.V. Chilingar, 1955.
Selected list of Soviet references on marine geology and oceanography. The Compass 32:220-227.

bibliography, geological
Thom, E.M., M. Hooker, and R.R. Dunaven, 1950.
Bibliography of North American geology 1948.
Geol. Survey Bull. 968:309 pp.

bibliographies geology
Thom, E. M., M. Hooker, and R.R. Dunaven, 1949
Bibliography of North American Geology.
Geol. Survey Bull. 958:658 pp.

bibliography (450 titles), geological
Udintsev, G.B., 1959.
Investigation of the relief of seas and oceans.
Akad. Nauk, SSSR, Itogi Nauki, Dostizheniia Okeanol. 1. Uspechi b Izuchenii Okean. Glubin (Biol. i Geol.):27-90.

bibliography geological
Waterways Experiment Station, 1949.
Subsurface exploration and sampling of soils for civil engineering purposes. Ed. M.J. Hvorslev 521 pp.

bibliography geological
Whitehouse, U. Grant, Lela M. Jeffrey, and James B. Debbrecht, 1959
Differential settling tendencies of clay minerals in saline waters. Seventh National Conference on Clays and Clay Minerals, Pergamon Press, 1-79.

Also in:
Contrib. Oceanogr. & Meteorol., A.& M. Coll. of Texas, Vol. 5, Contrib. No. 144.

beaches

Bibliography, beaches
Dolan, Robert, and James McCoy, 1965.
Selected bibliography on beach features and related near shore processes.
Lousiana State Univ. Studies, Coastal Studies Series, 11:59pp.

bibliography, beaches
Ellis, C.W., 1962.
Marine sedimentary environments in the vicinity of the Norwalk islands, Connecticut.
State Geol. Nat. Hist. Survey, Connecticut, Bull. No. 94:1-89.

bibliography, bottom sediments
Bezrukov, P.L., and A.P. Lisitsin, 1962.
The study of bottom sediments.(In Russian; English abstract).
Mezhd. Geofiz. Komitet, Prezidiume, Akad. Nauk, SSSR, Rezult. Issled. Programme Mezhd. Geofiz. Goda, Okeanol. Issled., No. 7:49-83.

bibliography bottom sedim.
Francis-Boeuf, C., 1948.
Sur la possibilité de concevoir une physiologie des sédiments marins. Bull. Mus. Hist. Nat., Marseille, 8(1):37-46.

bibliography, bottom sediments
Guilcher, Andre, 1964.
Present-time trends in the study of Recent marine sediments and in marine physiography.
Marine Geology, 1(1):4-15.

bibliography, bottom sediments
Hosokawa, Iwao, 1962
A review of chemical studies on marine sediments (chiefly on shallow-water deposits).
(In Japanese).
J. Oceanogr. Soc., Japan, 20th Ann. Vol., 541-562.

bibliographies, bottom sediments
Lapierre, Francis 1970.
Répartition des sédiments sur le plateau continental du Golfe de Gascogne. Intérêt des minéraux lourds. Trav. Cent. Rech. Études océanogr. ns. 10 (1/2/3): 127 pp.

bibliographies, bottom sediments, (minerals)
Mancini, Renato, 1969.
Relazione conclusiva sul lavoro di raccolta di documentazione in campo di risorse minerarie del fondo del mare. Progr. Ricerca Risorse mar Fondo mar, Comm. ital. Oceanogr. (B) 44: 164 pp.

bibliography, bottom sediments
Wiseman, J.D.H., 1946.
Marine sediments and related oceanographical subjects - an investigation into German developments. British Intelligence Objectives Sub-Committee, 32 Bryanston Square, London, W1, BIOS Trip No. 2485, August 1945:34 pp. (multilithed).
Final Rept. No. 1368, Item No. 2, 1

bibliography, geology, bottom sediments
Zheleznova, A.A., 1962.
On the suspension effect in connection with the pH determination of sea sediments. Review of the literature). (In Russian).
Trudy Inst. Okean., Akad. Nauk, SSSR, 54:83-99.

coasts

Bibliography, coasts
Dolan, Robert, and James McCoy, 1965.
Selected bibliography on beach features and related near shore processes.
Lousiana State Univ. Studies, Coastal Studies Series, 11:59pp.

bibliographies, coasts
Beach Erosion Board, 1953.
Shore protection planning and design. (Preliminary, subject to revision). B.E.B. Bull., Special Issue No. 2:230 pp., 149 figs., plus app.

bibliography, coasts
Fairbridge, R.W., 1952.
Marine erosion. Proc. Seventh Pacific Sci. Congr Met. Ocean., 3:347-359, 1 textfig.

bibliography, coasts
Francis-Boeuf, C., 1948.
Comportement des vases fluvio-marines vis-à-vis de l'oxygène dissous dans le milieu extérieur. Rev. Inst. français du Petrole et Ann. des Combustible liquides 3(5):109-113, 4 textfigs.

bibliography, coasts
Gibert, A., 1960
Observation des mouvements du sable sous l'eau au moyen de l'argent 110. Ministerio das Obras Publicas, Laboratorio Nacional de Engenharia Civil, Lisboa, Memoire, No. 143: 19 pp.

bibliography, coasts
Guilcher, A., 1967.
Morphologie et sédimentologie littorales et sous-marines.
Rapp. Nat. Trav. français, 1963-1966, Com. Nat. français Géodes.- Geophys., 244-255.

bibliography, coasts
Johnson, J.W., 1956.
Nearshore sediment movement.
Bull. Amer. Assoc. Petr. Geol., 40(9):2211-2232.

bibliography, coasts
Lorenzen, J.M., 1956 (nach 1940 mss)
Gedanken zur Generalplanung im nordfriesischen Wattenmeer. Die Küste, 5:9-48.

bibliographies, coasts
Mississippi River Commission, Corps of Engineers, War Department, 1947
List of publications, 12 pp. and ii. Waterways Experimental Station, Vicksburg Mississippi.

bibliography, coasts
Oren, O. H., and H. Steinetz, 1959.
Regional bibliography of the Mediterranean coast of Israel and the adjacent Levant countries.
Sea Fish. Res. Sta., Haifa, Israel, Bull. No. 22: 32 pp.

bibliography, coasts
Seibold, Eugen, 1963
Geological investigation of near-shore sand-transport-examples of methods and problems from the Baltic and North seas.
Progress in Oceanography, M. Sears, Edit., Pergamon Press, London, 1:1-70.

bibliography, coasts
Zenkovitch, V. P., 1960.
Fondements principaux d'une theorie sur la formation des structures d'accumulation dans la zone littorale.
Cahiers Ocean., C.C.O.E.C., 12(3):162-183.

bibliographies, deltas
Scruton, P.C., 1960.
Delta building and the deltaic sequence.
In: Recent sediments, northwest Gulf of Mexico, 1951-1958. Amer. Assoc. Petr. Geol., Tulsa, 82-102, with consolidated bibliography, pp. 368-381.

methods

bibliography, geology (methods)
*Richards, Adrian F., 1967.
Basic literature of marine geotechnique and related fields.
In: Marine geotechnique, A.F. Richards, editor, Univ. Illinois Press, 319-323.

minerals, etc.

bibliography, minerals
Manheim, F.T., 1965.
Manganese-iron accumulations in the shallow marine environment.
Narragansett Mar. Lab., Univ. Rhode Island, Occ. Publ., No. 3:217-276.

oil formation

bibliography, oil formation
Hodgson, G.W., Brian Hitchon and Kazuo Taguchi, 1964.
The water and hydrocarbon cycles in the formation of oil accumulations.
In: Recent researches in the fields of hydrosphere, atmosphere and nuclear geochemistry, Ken Sugawara festival volume, Y. Miyake and T. Koyama, editors, Maruzen Co., Ltd., Tokyo, 217-242.

bibliographies, rocks
Cann, J.R. and T. Simkin, 1971.
A bibliography of ocean-floor rocks. Phil. Trans. R. Soc. Lond. (A) 268(1192): 737-743.

sedimentary

bibliography, sedimentation
Clos-Arceduc, Albert, 1963
Étude simultanée des seiches et de l'alluvionnement.
Cahiers Océanogr., C.C.O.E.C., 15(1):53-56.

bibliography, sedimentation
Goldberg, E., editor, 1965.
Sedimentation: Annotated bibliography of foreign literature for 1959 to 1964, Survey No.1:192 pp. Israel Program Scientific Translation for U.S. Dept. Agriculture and National Science Foundation. ($5.00 from Clearing house for Federal Scientific and Technical Information, Springfield, Va., 22151).

bibliographies sedimentation
Terry, R.D., and G.V. Chilingar, 1956.
Selected list of Russian references of sedimentology. Trans. Amer. Geophys. Union, 37(2):245-254.

bibliographies, sedimentation
U.S.A. Corps of Engineers, U.S. Army, Committee on Hydraulics 1965.
Bibliography on tidal hydraulics. Supplementary material compiled from May 1959 to May 1965. Tidal flows in rivers and harbors.
Rept. No. 2, Suppl. No. 4 (ES816): 218 pp. (multilithed).

bibliography, geomagnetism
Allan, T.D., 1969
A review of marine geomagnetism.
Earth-Sci. Rev. 5(4): 217-254.

Geophysical (microseisms separately)

bibliography, geophysics
*Beloussov, V.V., 1968.
The earth's crust and upper mantle of the oceans. (In Russian; English abstract).
Roz. Issled. Mezhdunarod. Geofiz. Proekt., Mezhduvedomst. Geofiz. Kom., Akad. Nauk, SSSR, 253 pp.

bibliography, geophysics
Geyer, R. A., 1950? (undated).
Bibliography of oceanography, marine biology, geology, geophysics and meteorology in the Gulf of Mexico. Humble Oil Refining Co. (Multilith. Pages not numbered serially).

bibliographies, geophysics
Model, Fritz, 1962.
Geophysikalische Bibliographie von Nord- und Ostsee, 2 vols., unnumbered pp. (Unpublished manuscript).

bibliography, geophysical
Rabbitt, M. C., V.L. Skitsky, and S.T. Vesselowsky, 1950.
Geophysical abstracts 139 October-December 1949 (Numbers 11442-11678). Geol. Survey Bull. 966-D: 253-329.

bibliographies, geophysical
Schmidt, Peter, Compiler, 1970.
Zur Geschichte der Geologie, Geophysik, Mineralogie und Paläontologie: Bibliographie und Repertorium für die Deutsche Demokratische Republik.
Veröff. Bibliothek Bergakad., Freiburg, 40: 134 pp.

bibliography, geophysics
Talwani, M., 1964.
A review of marine geophysics.
Marine Geology, 2(1/2):29-80.

bibliography, geophysical
Uyeda, Seiya, 1962
Recent geophysical investigations of the ocean bottom. (In Japanese; English abstract)
J. Oceanogr. Soc., Japan, 20th Ann. Vol., 64-79.

bibliography, geophysics
Weibull, W., 1955.
Sound explorations.
Repts. Swedish Deep-Sea Exped., 1947-48, Bottom Invest., 4(1):3-31, photos. of records.

gravity

bibliography, gravity
Coster, H. P., 1945.
The gravity field of the western and central Mediterranean.
Proefschrift. Bij J. B. Wolters' Uitgevers-maatschappij, n. v., Groningen-Batavia, 1945: 57 pp.

bibliography, gravity (abstracts)
Kneissl, M., 1962
Literatur über das Europäische Gravimeter-Eichsystem.
Deutsche Geodatische Kommission, Bayer. Akad. Wiss., 80 pp. (multilithed)

Bibliographies, hydrography

See: Bibliographies, areal
Bibliographies, physical oceanography

ice

bibliographies, ice
Bradford, J.D. and S.M. Smirle 1970.
Bibliography on northern sea ice and related subjects.
Marine Operations, Ministry of Transport and Marine Sciences Branch, Department of Energy, Mines and Resources, Canada, 188 pp.

bibliography, ice
Great Britain, The Glaciological Society, 1964.
Glaciological literature.
J. Glaciol., 5(37):135-140.

bibliography, ice
Hoel, A., 1952.
Norsk Ishavsfangst en fortegnelse over litteratur
Medd. Norsk PolarInst. No. 69:23 pp.

bibliography, ice
Kingery, W.D., Editor, 1962.
Summary report - Project Ice Way.
Air Force Surveys in Geophysics, No. 145:210 pp.

bibliographies - ice
Kusunoki, K., 1952.
Bibliography of sea ice in Japan (for the years 1892-1950). J. Jap. Soc. Snow & Ice 13(4):125-128.

bibliography-ice
Nazarov, V.S., 1962
Ice of Antarctic waters. Oceanology, X section of IGY Program. (In Russian: English summary)
Rezult. Issled. Programme Mezhd. Geofis. Gode Mezhd. Geofiz Komitet, Prez. Akad. Nauk, SSSR 72 pp.

bibliography - ice
Palosuo, E., 1953.
A treatise on severe ice conditions in the central Baltic. Fennia 77(1):1-130, 55 textfigs.

bibliography-ice
Palusuo, E., 1953.
A treatise on severe ice conditions in the central Baltic. Merent. Julk. No. 156:1-130, 46 textfigs., 9 charts.

bibliography- ice
Pearce, D.C., 1951.
A bibliography on snow and ice. Canada, Nat. Res. Counc., 69 pp. (75 cents).

bibliography-ice
Pounder, E.R., 1962
The physics of sea-ice. Sect. VII. In: The Sea, Vol. 1, Physical Oceanography, Interscience Publishers, 826-838.
(Mss received June 1960)

bibliography-ice
U.S. Navy Hydrographic Office, 1945.
Bibliography on ice of the northern Hemisphere.
H.O. Pub. 240:1-179, i-xii.

ms

individuals

Bibliographies, individuals

Arranged under their name, not that of the author preparing the bibliography.
See also: "Biographies" in following section

bibliographies - Bacescu
Pora, Eugène A. 1968.
Mihai C. Băcescu à 60 ans.
Revue Roumaine Biol. (Zool.) 13(6):347-362.

bibliographies, Barbour
Bigelow, H.B., 1952.
Thomas Barbour. Biogr. Mem., Nat. Acad. Sci., 27:13-45.

bibliographies-Berg
Madsen, Bent Lauge, and Gunnar Nygaard 1970.
Professor Kaj Berg - 70 years old.
Hydrobiologia, Netherlands 35(2):345-351.

bibliographies, Berkeley
Arai, Mary Needler 1971.
Publications of Edith and/or Cyril Berkeley.
J. Fish. Res. Bd Can. 28(10):1365-1372.

bibliographies, Birstein
Chindonova, Yu. G., S.I. Ljovushkin and N.A. Zarenkov 1972.
Yakov Avadievich Birstein (7th April 1911-8th July 1970).
Crustaceana 22(1):103-112.

Bibliographies, Bjerknes
Bergeron, Tor, Olaf Devik, and Carl Ludvig Godske, 1962.
Vilhelm Bjerknes, March 14, 1862-April 19, 1951.
Geofys. Publik., Geophysica Norvegica, 24:7-25.

Bjerknes bibliography 27-38.

bibliography, Boschma
Klaauw, C.J., 1964.
Hilbrand Boschma.
Zool. Mededel., 39:ix-xlix.

bibliographies, Bowman
Carter, G.F., 1950.
Isaiah Bowman, 1878-1950. Ann. Assoc. Am. Geogr. 40(4):335-350.

bibliographies, Bowman
Wrigley, G.M., 1951.
Isaiah Bowman. Geogr. Rev. 41(1):7-65, 2 figs.

bibliographies - Bückmann
Schmidt, Ulrich 1970
Professor Dr. Adolf Bückmann zum siebzigsten Geburtstag.
Ber. dt. wissenschaft. Komm. Meeresforsch. n.f. 21:3-15.

bibliographies, D'Ancona
Tonolli, Vittorio, 1966.
Umberto D'Ancona (1896-1964).
Hydrobiologia, Acta Hydrobiol. Hydrogr. et Protistol., 27(1/2):260-273.

bibliography, Davies
Alexander, Anne J., and others, 1966
David H. Davies
S. Afr. Assoc. Mar. Biol. Res., Bull., No. 6:7-22.

biographies, Davies
bibliography

bibliographies, Dawson
Abbott, Isabella A., 1966.
Elmer Yale Dawson (1918-1966).
J. Phycology, 2(4):129-132.

bibliographies, Dawson
Garth, John S., 1967.
E. Yale Dawson.
Bull. S. Calif. Acad. Sci., 66(3):149-160.

bibliographies, de Buen
Chile, Universidad de Chile, Instituto de Biologia, 1963.
Fernando de Buen y Lozano (1895-1962).
Montemar, (3):91-98.

bibliographies, Defant
Wüst, Georg, 1964.
Albert Defant zum 80 Geburtstag.
Beiträge zur Physik Atmosph., 38(2):58-68.

bibliography, Dixon
Atkins, W.R.G., 1954.
Henry Horatio Dixon 1869-1953. Obituary Notices, Fellows Roy. Soc. 9:79-97, 1 pl.

bibliographies, Dobrovolsky
Mamaev, O.I., 1967.
60th birthday of Professor Aleksey Dmitrievich Dobrovolsky. (In Russian).
Okeanologiia, Akad. Nauk, SSSR, 7(4):738-742.

bibliography, Dohrn
Kühn, A., 1950.
Anton Dohrn und die Zoologie seiner Zeit. Pubbl. Staz. Zool., Napoli, Suppl. 1950:205 pp.

bibliography, Ehrenbaum
Lüling, K.-H., 1949.
Ernst Ehrenbaum. Ber. Deutschen Wiss. Komm. f. Meeresf., n.f., 11(4):435-442.

bibliographies, Ekman
Berg, Kaj, 1966.
Sven Ekman 31 - V - 1876---2 - 11 - 1964.
Hydrobiologia, Acta Hydrobiol., Hydrogr. et Protistol., 27(1/2):274-286.

bibliography, Emery
Clements, Thomas, 1963.
Dr. Kenneth Orris Emery an appreciation. In: Essays in Marine Geology in honor of K.O. Emery, Univ. S. California Press:ix-xi.

bibliography Ercegovic
Alfirevic, Slobodan 1970
Le Docteur Ante Ercegović (1895-1969), sa vie et son oeuvre. (In Jugoslavian and French)
Acta adriatica 13(5):23pp.

Bibliography, Ercegovic
Pucher-Petkovic, T., 1970.
Ante Ercegović (1895-1969) in memoriam.
Revue algol. n.s. 10(1):3-7.

bibliography, Fabricius
Tuxen, S.L., 1967.
Bibliographie von I. C. Fabricius.
Zoologische Anzeiger, 178(3/4):174-185.

bibliographies, Gilson
Van Straelen, V., 1948.
Gustave Gilson (1859-1944). Notice biographique avec liste bibliographique. Bull. Mus. Roy. Hist. Nat., Belgique, 24(1):21 pp., 1 pl.

bibliography Hentschel
Caspers, H., 1949.
Ernst Hentschel, 1876-1945. Arch. Hydrobiol. 62:490-499.

bibliography, Hentschel
Caspers, H., 1949.
Ernst Hentschel. Ber. Deutschen Wiss. Komm. f. Meeresf., n.f., 11(4):449-456.

Bibliographies, Hustedt
Behre, Karl, 1967.
Dr. Friedrich Hustedt, 80 Jahre alt.
Abh. naturw. Verein., Bremen, 37(2):97-108.

bibliography, Hustedt
Behre, K. 1970.
Friedrich Hustedts Leben und Werk.
Beihefte Nova Hedwigia 31: XI-XXII.

bibliographies, Järnefelt
Purasjoki, K.J., 1965.
Heikki Järnefelt, in memoriam.
Hydrobiologie, 25(3/4):571-578.

bibliographies, Järnefelt
Ryhänen, Reino, 1965.
Heikki Arvid Järnefelt, 23.6.1891-22. 10. 1963.
Arch. Hydrobiol., 61(2):249-256.

bibliography, Jespersen
Taning, A.V., 1952.
Poul Jespersen, 18 Marts 1891-20 December 1951.
Vidensk. Medd. fra Dansk Naturh. Foren. 114: 37-48.

bibliographies, Jewett
Buckley, O.E., 1952.
Frank Baldwin Jewett. Biogr. Mem., Nat. Acad. Sci. 27:239-264.

bibliographies - Kändler
Hempel, Gotthilf 1970.
Professor Dr. Rudolf Kändler zum siebzigsten Geburtstag.
Ber. dt. wissenschaft. Komm. Meeresforsch. n.f. 21:16-26.

bibliography, Einar Koefoed
Koefoed, Einar, 1963.
The scientific publications of Einar Koefoed.
Fiskeridirekt. Skrift., Ser. Havundersøgelser, 13(6):9-10.

bibliography, Kolosvery
Zullo, Victor A., William A. Newman and Arnold Ross 1972.
Kolosváry Gábor (Gabriel von Kolosváry) 1901-1968.
Crustaceana 22(1):96-102.

bibliography, Krogh
Spärck, R., 1949.
August Krogh. Liste over August Kroghs publikationer. Vidensk. Medd. Dansk Nat. Hist. För., København, 111:v-xxx

bibliography, Legendre
LeGrand, Y., 1954.
René Legendre (1880-1954). Bull. Inst. Océan., Monaco, No. 1044:29 pp.

bibliography, Maury
Brown, Ralph Minthorne, 1944.
Bibliography of Commander Mathew Fontaine Maury including a biographical sketch.
Bull. Va polytech. Inst., 37(12):46 pp.

bibliographies, Myers
Anon., 1970.
Annotated chronological bibliography to the publications of George Sprague Myers (to the end of 1969).
Proc. Calif. Acad. Sci. (4) 38:19-52.

bibliographies, Navarro
Lozano Cabo, Fernando, 1961.
D. Francisco de Paula Navarro Martin.
Bol. R. Soc. Esp. Hist. Nat., (B), 59:231-242.

bibliography, Petit
France, Laboratoire Arago, Banyuls-sur-Mer, 1964.
Volume jubilaire dédié à Georges Petit.
Vie et Milieu, Lab. Arago, Suppl., No. 17:i-liv.

bibliography, Ranson
Ranson, M. Gilbert, 1943
Titres et travaux scientifiques. Paris: Masson et Cie, Editeurs, Libraries de l'Academie de Medecine, 120 Boulevard Saint Germain. 88 pp.

bibliographies, Raymond
Stetson, H.C., 1953.
Memorial to Percy Edward Raymond (1879-1952).
Proc. G.S.A., Ann. Rept. for 1952:121-126, Pl. 11

bibliographies, Rollefsen
Sunnanå, Klaus 1969
[Gunnar Rollefsen]
Fisk Dir. Skr. Ser. HavUnders. 15(3):103-111

bibliographies, Schmidt
Bruun, A. Fr., 1934.
The life and work of Professor Johannes Schmidt.
Riv. di Biol. 16(1):3-22, 3 textfigs.

ms

bibliography (Scholander)
Scholander, Susan Irving, compiler, 1965.
Bibliography of written works by P.F. Scholander et al.
Hvalradets Skrifter, 48:7-14.

bibliography, individuals (Schulz)
Kalle, K., 1949.
Bruno Schulz. Ber. Deutschen Wiss. Komm. f. Meeresf., n.f., 11(4):443-449.

bibliographies, Sewell
Roonwal, M.L., 1963(publ.1967).
The late Lieut-Col.R.B.S. Sewell (1880-1964): an appreciation, with a complete list of his scientific writings.
Rec. Indian Mus. 60(3/4):327-336.

bibliographies, Skopintsev
USSR, Akademia Nauk, 1962
Boris Alexandrovich Skopintsev (in connection with the investigator's sixtieth birthday and the thirty-fifth year of his scientific activities). (In Russian).
Okeanologiia, Akad. Nauk, SSSR, 2(6):1120-1126.

bibliographies, Stockmann
Dobrobolsky, A.D., 1969.
Vladimir Borisobich Shtockman, 1909-1969. (In Russian).
Okeanologiia, 9(1): 5-14.

bibliographies, Stocks
Model, F., 1964.
In memoriam: Theodor Stocks.
Deutsche Hydrogr. Zeits., 17(1):41-45.

bibliographies Stschapova
Kureeva, M.S., 1957.
[Tat'iana Federovna Stschapova.] Trudy Inst. Okean. 23:5-14.

bibliography, Sverdrup
Revelle, R. R., and W. H. Munk, 1948.
Harald Ulrik Sverdrup - an appreciation (with bibliography). J. Mar. Res. 7(3):127-138.

bibliographies Thompson
Calman, W.T., 1948
Sir D'Arcy Thompson, C.B., F.R.S. Nature 162(4107):93-94.

bibliography Thompson
Thomas, B. D., 1958
Thomas Gordon Thompson. J. Mar. Res., 17: 11-22.

bibliography-Timonov
*Dobrvol'sky, A.D. and V.G. Bukhteev, 1970.
On V.V. Timonov's life-work. Okeanologiia, 10(5): 740-744. (in Russian; English abstract)

bibliographies, Trask
Gilluly, James, 1963.
Memorial to Parker Davies Trask.
Bull. Geol. Soc., Amer., 74(1):P13-P20.

biographies, Tregouboff
Fenaux, R., 1969.
A la mémoire de G. Tregouboff (1886-1969).
Trav. Sta. zool. Villefranche-sur-Mer, 29: 1-8.

bibliography, Vercelli
Anon., 1954.
Pubblicazioni di Francesco Vercelli.
Arch. Oceanografia e Limnol. 9(1/2):3-8.

bibliographies, Vercelli
Anon., 1952.
Francesco Vercelli. Bol. Pesca, Piscicult., Idrobiol., n.s., 7(2):246-251.

bibliography, Vercelli
Morelli, C., 1952.
In memoria di Francesco Vercelli. Osservatorio Geofis., n.s., Pubbl. No. 24: 14 pp.

bibliography
Böhnecke, G., 1944.
Professor Dr. Wattenberg. Ann. Hydr., usw., 72:291-293.

bibliography, Vercelli
Picotti, M., 1953.
Francesco Vercelli. La Ricerca Sci., A, 23(2): 277-282.

Ist. Talassogr., Trieste, Pubbl. 291.

bibliographies, individuals (Zenkevitch)
Belyaev, G.M. and S.A. Mileikovsky, 1971.
In memoriam of Academician Lev Alexandrovitch Zenkevitch. Mar. Biol. 11(4): 299-305.

institutions

bibliography institutions
Chevey, M. B., 1939(1940).
Travaux de l'Institut Océanographique de l'Indochine. Proc. Sixth Pacific Sci. Congr., 3: 38-42.

bibliography, institutions
Gemulin, Tomo, Miroslav Nikolic i Dusan Zovodnik 1964.
70 Godina Bioloskog Instituta u Rovinju 1891-1961 (In Jugoslavian; English abstract)
Thalassia Jugoslavica, 2(5/6):26 pp.

Bibliography institutions
Gibert, A., F. Abecasis, M. Goncalves Ferreira, J. Reis Carvalho and S. Cordeiro, 1960.
Tracing undersea sand movement with radioactive silver. Lab. Nacional Engenharia Civil, Ministerio das Obras Publicas, Lisbon, Tech. Paper, No. 150:
Reprint from 2nd UN Geneva Conf., Pergamon Press P/1820.

bibliographies, institutions
Høisaeter, Tore 1967.
Publications from the biological stations of the Bergens Museum and the University of Bergen, 1892-1967.
Sarsia 29:97-136.

bibliography, institutions
Nicolae, Ionescu, 1964.
Cercetari efectuarte in ultimi 20 de ani la litoralul Rominesc al Mării Negre de catre statiunea maritima de cercetari piscicole Constanta.
Bulet. Inst. Cercetari si Proiectari Piscicole, 23(2/3):135-155.

bibliography, institutions
Smithsonian Institution, 1947
A list and index of the publications of the United States National Museum (1875-1946).
Bull. U.S. Nat. Mus. 193:1-306.

owned by MS

bibliography, station Marine de Tulear
Thomassin Bernard A. 1971.
Revue bibliographique des travaux de la station marine de Tuléar (République Malgache) 1961-1970.
Téthys Suppl. 1:3-50

bibliography institutions
Utinomi, H., 1952.
List of publications of researches carried out at the Seto Marine Biological Laboratory during the period from 1922 to 1951 (exclusive of papers not referred to marine biology). Publ. Seto Mar. Biol. Lab., Kyoto Univ., 2(2):357-400.

bibliography, institutions
Vera-Cruz, D., 1960
Transporte sólido em costas arenosas.
Laboratorio Nacional de Engenharia Civil, Ministerio das Obras Publicas, Lisboa, Memoria, No. 145:1-12.

bibliography, institutions
Vodianitskii, V.A., 1963.
On the ninetieth year of the Sevastopol Biological Station A.O. Kovalevskii. (In Russian).
Trudy Sevastopol Biol. Sta., 16:3-25.

bibliography, institutions
Wüst, G., C. Hoffmann, C. Schliepor, R. Kändler, J. Krey and R. Jaeger, 1956.
Das Institut für Meereskunde der Universität Kiel nach seinem Wiederaufbau. Kieler Meeresf., 12(2):127-153.

internal waves

bibliographies, internal waves
Bernard, P. 1955.
Travaux recents sur les ondes internes.
Bull. d'Info. C.C.O.E.C. 7(1): 37-44.

bibliography, internal waves
Bernard, P., 1939.
Les "mardes" internes. Ann. Inst. Océan. 22(3): 145-192, 19 textfigs.

bibliography, internal waves
Cox, C.S., 1962
Internal waves, Part II. Ch. 21, Sect. V, Waves. In: The Sea, Vol. 1, Physical Oceanography, Interscience Publishers, 752-763.

bibliography, internal waves
Ichiye, Takashi, 1966.
Annotated bibliography on internal waves.
J. oceanogr. Soc., Japan, 22(5):201-222.

bibliographies internal waves
Ichiye, Takashi, 1962.
Annotated bibliography on internal waves.
Florida State Univ., Oceanogr. Inst., Techn. Rept No. 1:13 pp. (Unpublished manuscript)

bibliography, internal waves
McNeil, Mary, 1964
Internal waves (a bibliographic study).
Lockheed California Company, Systems Research Division, LR 17669: 98 pp.

bibliography, international commissions
Chapman, William McLeod, 1968.
The theory and practice of international fisheries commissions and bodies.
Proc. Gulf Carib. Fish. Inst. 20th Ann. Sess. 77-105

bibliography
France, Comité National Française Géodésique et Géophysique, 1957.
Participation française à l'Année Géophysique International (par P. Legay)
Bibliography of principal publications on physical oceanography in France, 1957.
C. R. Assemblée Générale du Comité Nat. Français U. Géod. et Géophys. Int., Paris, 16-21:143-159.

bibliography, international
Marson, Frank M., and Janet R. Terner, 1963
United States IGY bibliography, 1953-1960.
IGY World Data Center A, IGY General Rept. No. 18:391 pp.

bibliography international organizations
Takenouti, Y., 1960
X. Oceanography. Japanese Contrib. I.G.Y., 1957/8, 2: 124-145.

light

bibliographies, light
Das, S.M., 1954.
Submarine illumination in relation to phytoplankton. Sci. and Culture 19(11):528-534.

bibliography, light
Jerlov, N.G., 1963.
Optical oceanography.
In: Oceanography and Marine Biology, H. Barnes, Editor, George Allen and Unwin, 1:89-114.

bibliography, light
Sasaki, Tadayoshi, 1962
Optical oceanography. (In Japanese).
J. Oceanogr. Soc., Japan, 20th Ann. Vol., 377-385.

bibliography-light
Strickland, J.D.H., 1958.
Solar radiation penetrating the ocean, a review of requirement, data and methods of measurement, with particular reference to photosynthetic productivity. J. Fish. Res. Bd., Canada, 15(3): 453-493.

bibliography-light
Thorson, Gunnar, 1964
Light as an ecological factor in the dispersal and settlement of larvae of marine bottom invertebrates.
Ophelia, Mar. Biol. Lab., Helsingør, 1(1): 167-208.

limnology

bibliographies limnology
Meglis, A., Compiler, 1961
Bibliography on limnology.
Meteor. & Geoastrophys. Abs., 12(3): 559-619.

man in the sea

bibliography man-in-the-sea
Dugan, James, 1965.
Manned undersea stations. (abstract).
Ocean Sci. and Ocean Eng., Mar. Techn. Soc.,-Amer. Soc. Limnol. Oceanogr., 1:652-656.

bibliography, man-in-the-sea
Weltman, Gershom, and Glen H. Egstrom, 1968.
Measuring work effectiveness underwater.
J. Ocean Techn., 2(4):43-47.

bibliographies, marine matters
Woodburn, Kenneth D. 1965.
A discussion and selected annotated references of subjective or controversial marine matters.
Techn. Ser. Fla. Bd. Conserv. 46:50pp.

bibliographies, marine resources

Ingle, Robert M., 1969.
Selected references concerning Florida's marine resources. 1.
Spec. scient. Repts. Mar. Res Lab, Fla. Bd Conserv., St. Petersburg, 24: 117 pp.

materials

bibliography, materials
*Tuthill, A.H., and C.M. Schillmoller, 1967.
Guidelines for selection of marine materials.
J. Ocean Tech., 2(1):6-36.

meteorology

bibliography, meteorology
American Meteorological Society, 1950.
Meteorological abstracts and bibliography, 1(3): 147-214.

Special feature in this issue: A selective annotated bibliography on cloud physics and artificial precipitation.

bibliography, meteorology
Baum, W.A., 1947
An annotated bibliography and methodology for microclimatic purposes. QM Res. & Devel. Environmental, Protection Series Rept., 124 Pt.1. Oct. 1947, 98 pp.

bibliography, meteorology
Geyer, R. A., 1950?(undated).
Bibliography of oceanography, marine biology, geology, geophysics and meteorology of the Gulf of Mexico. Humble Oil Refining Co. (Multilith. Pages not numbered serially.)

bibliography, meteorology
Kramer, M.P. (compiler), 1954.
Selective annotated bibliography on general oceanographic meteorology. Met. Abstr. Bull. 5: 75-123.

bibliography, meteorology
Magin, G.B., Jr., and L.E. Randall, 1960.
Review literature on evaporation suppression.
Geol. Surv. Prof. Pap., 272-C:53-69.

bibliography, meteorology
Malkus, Joanne S., 1962
Large-scale interactions. Ch. 4, Sect. II. Interchange of properties between sea and air In: The Sea, Interscience Publishers, Vol 1, Physical Oceanography, 88-294.

bibliographies, meteorology
Nupen, W., and M. Rigby, 1956.
An annotated bibliography on tropical cyclones, hurricanes and typhoons.
Meteorol. Abstr. & Bibl., 7(9):1113-1163.

bibliography, meteorology
Polli, S., 1949.
100 anni di osservazioni meteorologiche eseguite a Trieste, 1841-1940. Pt. 4. La velocità del vento. Ist. Talass., Trieste, Publ. 234:42-87.

bibliography, meteorology
Royal Meteorological Society, 1949.
Bibliography of Meteorological literature. 6(5):231-302.

bibliography, meteorology
Royal Meteorological Society, 1947
Bibliography of meteorological literature. 5(10):229-284.

bibliography, meteorology
Royal Meteorological Society, 1939
Bibliography of meteorological literature 4(6):295-351.

bibliographies, meteorology
Royal Meteorological Society.
Bibliography of Meteorological Literature (a serial)

(prepared by the R. Meteorological Society with the collaboration of the Meteorological Office)

bibliographies, meteorology
Schneider, M., (Editor), 1961(1963).
Agrarmeteorologische Bibliographie 1961.
Bibliogr., Deutscher Wetterdienstes, (13):253 pp

bibliography, meteorology
Ultrasonic Corporation, 1948.
Contributions to the study of Natural fog. Final Rept., Pt. II on (Navy Fog Dispersal Project) June 30, 1948, 1v. incl. illus, diagra 279 refs. Unclassified. (Listed as item U4951 in Technical Information Pilot of 31 Aug. 1949)

microseisms

bibliography, microseisms
Darbyshire, J., 1962
Microseisms. Ch. 20, Sect. V. Waves. In: The Sea, Vol. 1, Physical Oceanography, Interscience Publishers, 700-719.

bibliography, microseisms
Gutenberg, B., and F. Andrews, 1952.
Bibliography of microseisms. 2nd ed., enlarged and revised. Sci. Rept. No. 1:94 pp. (mimeographed).

navigation

bibliography, navigation (polar)
Morley, J.P., 1963.
Polar ships and navigation in the Antarctic.
British Antarctic Survey Bull., (2):1-25.

bibliography, oceanography from space
United States, Ocean Engineering Information Service, 1970.
Oceanography from space and aircraft, an annotated bibliography.
Ocean. Eng. Info. Ser., La Jolla, Calif., 2, xv+ 79 pp.

paleoecology

bibliographies, paleoecology
ZoBell, C.E., et al., 1957.
Annotated bibliography of marine paleoecology. In: Treatise on Marine Ecology and Paleoecology, Vol. 2.
Mem. Geol. Soc., Amer., 67:691-1032.

particulate matter

bibliography, particulate matter
Jerlov, N.G., 1953.
Particle distribution in the ocean.
Repts. Swedish Deep-sea Exped., 1947-1948, Phys. Chem., 3(3):73-97, 25 textfigs.

bibliography, particulate matter
Parsons, T.R., 1963.
Suspended organic matter.
Progress in Oceanography, 1:205-239.

bibliography, particulate matter
Riley, Gordon A. 1970.
Particulate and organic matter in sea water.
Adv. mar. Biol. F.S. Russell and Maurice Yonge, editors Academic Press 8:1-118.

bibliography, pesticides, effect of
Risebrough, R.W., 1971.
Chlorinated hydrocarbons. (10) In: Impingement of man on the oceans, D.W. Hood, editor, Wiley Interscience: 259-286.

physical oceanography

bibliographies, oceanography
Anonymous, 1963.
The literature in meteorology and hydrology published in 1962. (In Russian).
Meteorol. i Gidrol., No. 4:54-57.

bibliography, physical oceanography
France, Comité National Français de Géodésie et Géophysique, 1960

Section d'Océanographie Physique.
C.R., Com. Nat. Francais, Géod. et Géophys., Année 1959:133-148.

bibliography, oceanography
Geyer, R. A., 1950(undated).
Bibliography of oceanography, marine biology, geology, geophysics and meteorology of the Gulf of Mexico. Humble Oil Refining Co. (multilith. Pages not numbered serially).

bibliographies, physical oceanography
Hela, Ilmo, 1965.
Bibliographical classification of physical oceanography. (In Finnish and English.
Merentutkimuslait. Julk. 226:45 pp.

bibliographies, physical oceanography
Lacombe, Henri 1971.
Références bibliographiques des travaux concernant l'océanographie physique de 1966 à 1968.
Rapp. P.-v. Réun. Comm. int. Explor. scient. Mer Médit. 20 (2): 109-112.

bibliographies physical oceanography
Lacombe, H., Secrétaire, 1964.
Section d'Océanographie physique. Rapport de la Section d'Océanographie physique.
Com. Nat. Français de Géodes. et Géophys., Comptes Rendus - Année 1963:161-176.

bibliographies, physical oceanography
Rice, M.L., Compiler, 1958.
Recent literature on dynamic and physical oceanography. (Supplement). Meteorol. Abstr. & Bibl., 9(6):731-773.

bibliographies, physical oceanography
Rice, M.L., Compiler, 1958.
Recent literature on physico-chemical oceanography. Meteorol. Abstr.& Bibl., 9(5):598-631.

bibliographies, physical oceanography
U.S. Navy Hydrographic Office, 1953.
Oceanographic and marine meteorological bibliography. Annales Hydrographiques, Ser. 1:1-41, 1848-1878; Ser. 2: 1-36, 1879-1916; Ser. 3:1-18, 1917-1946. H.O. Misc. 15257-3: numerous pp. (multilithed).

pollution

Bibliographies, pollution

See also: Bibliographies, wastes

bibliographies, pollution
Baughman, J.L. 1948.
An annotated bibliography of oysters with pertinent materials on mussels and other shellfish and an appendix on pollution.
Texas A+M Research Foundation, 794 pp.

bibliography, pollution, (oil)

Bertrand, A.R.V., J. Briant, A. Castela, P. Degobert, C. Gatellier, M. Masson, J.-L. Dutin et J. Pottier 1971.
Prévention et lutte contre la pollution au cours des opérations de forage et de production en mer.
Revue Inst. français Pétrole 26(9): 757-847.

bibliography, pollution

*Corner, E.D.S., A.J. Southward and E.C. Southward, 1968.
Toxicity of oil-spill removers ('detergents') to marine life: an assessment using the intertidal barnacle Elminius modestus.
J. mar. biol. Ass., U.K., 48(1):29-47.

bibliography, pollution

Devèze, L., 1965.
Rapport sur les travaux publiés durant la période 1962-64.
Rapp. P.-v. Réun., Comm. int. Explor. scient. Mer Méditerr., 18(3):591-598.

bibliography, pollution

Hays, Helen, and Robert W. Risebrough 1972
Pollutant concentrations in abnormal young terns from Long Island Sound.
The Auk 89 (1): 19-35

bibliographies, pollution (thermal)

Naylor, E. 1965.
Effects of heated effluents upon marine and estuarine organisms.
Adv. mar. Biol. Sir Frederick S. Russell, editor, Academic Press 3:63-103.

bibliographies, pollution (oil)

Nelson-Smith, A. 1970
The problem of oil pollution of the sea.
Adv. mar. Biol., F.S. Russell and Maurice Yonge, editors, Academic Press 8:215-306.

bibliography, pollution

Redeke, H.C. 1927.
River pollution and fisheries.
Rapp. P.-v. Cons. perm. int. Explor. Mer, 43:50 pp.

bibliography, pollution

Smithsonian Institution, Smithsonian Information Exchange 1971.
Environmental pollution: a guide to current research
CCM Inform. Corp. 851 pp.

bibliographies, pollution

U.S.A. Corps of Engineers, U.S. Army, Committee on Hydraulics 1965.
Bibliography on tidal hydraulics. Supplementary material compiled from May 1959 to May 1965. Tidal flows in rivers and harbors.
Rept. No. 2, Suppl. No. 4 (ES 816): 218 pp. (multilithed).

bibliography - pollution

ZoBell, C.E., 1963.
The occurrence, effects and fate of oil polluting the sea.
Int. J. Air Water Pollution, 7(2/3):173-199.

Paper presented at International Conference on Water Pollution Research, London, 3-7 September, 1962.

radioactivity

Bibliography, radioactivity

Hiyama, Yoshio, and Ryushi Ichikawa, 1958.
A measure on level of strontium-90 concentration in sea water around Japan at the end of 1956.
Proc. Ninth Pacific Sci. Congr., Pacific Sci. Assoc., 1957, Oceanogr., 16:141-145.

bibliography, radioactivity

Joseph, Joachim 1971.
Rapport du Comité de radioactivité marine.
Rapp. P.-v. Réun. Comm. int. Explor. scient. Mer Médit. 20(3): 167-168

bibliography, radioactivity

Templeton, W.L., 1965.
Ecological aspects of the disposal of radioactive wastes to the sea.
In: Ecology and the Industrial Society, Fifth Symposium of the British Ecological Society, Blackwell, Oxford, 65-97.

bibliography

Thuronyi, G., 1956.
Annotated bibliography on natural and artificial radioactivity in the atmosphere.
Met. Abstr. Bibl., A.M.S., 7(5):616-657.

saline water

bibliographies, saline water conversion

U.S. Department of the Interior, Office of Saline Water 1965.
Bibliography of saline water conversion literature.
Research and Development Progress Report No. 146: 262 + 46 pp.

salinity

bibliography, salinity

Cox, Roland A., 1963
The salinity problem.
Progress in Oceanography, M. Sears, Edit., Pergamon Press, London, 1:243-261.

bibliography, salinity

Buljan, M., 1953.
The fluctuations of salinity in Adriatic.
"Hvar" Repts. 2(2):1-63, 73 textfigs.

sea bottom

sea level

bibliographies, sea level

France, Centre National de la Recherche Scientifique, 1964.
Hydrologie et dynamique des mers: niveau moyen (bibliographie). Océanographie Physique.
Année Géophysique Internationale, Participation Française, (10), (2):49 pp.

bibliography, sea level

Lisitzin, E., 1963.
Mean sea level.
In: Oceanography and Marine Biology, H. Barnes, Editor, George Allen and Unwin, 1:27-45.

bibliography, seiches

Crystal, G., 1905.
Hydrodynamical theory of seiches. Trans. R. Soc. Edinburgh, 41(3):599-649.

seismology

bibliographies, seismology

Hodgson, J.H., and F.E. Langill, 1963
Bibliography of seismology. Items 15346-15888. July - December, 1962.
Publications of the Dominion Observatory, Ottawa, 22(12):331-360.

bibliographies, seismology

Hodgson, J.H., and F.E. Langill, 1962
Bibliography of seismology. Items 14535-15345. January-June, 1962.
Publications of the Dominion Observatory, Ottawa, 22(11):289-327.

bibliography seismics

Leenhardt, Olivier, 1967.
Bibliography on marine seismics.
Spec. Bibliogr. Oceanogr., Contrib. Am. Met. Soc. 3:95pp

ships

Bibliographies, ships,

See: Bibliographies boats

bibliographies, sea bottom

Duncan, John K., 1960.
A selective bibliography of the environmental controls on object stability on the sea bottom.
U.S. Navy Hydrogr. Off., Spec. Publ., SP-32:121pp

bibliographies, seismology

Smith, W.E.T., 1957.
Bibliography of seismology. No. 18, Items 9133-9381, July to December 1955. Publ. Dominion Observ., Ottawa, No. 18:385-409.

bibliography, seismology

Smith, W.E.T., 1956.
Bibliography of seismology. No. 17, Item 8948-9132
January-June 1955.
Publ. Dominion Obs., Ottawa, 14:363-382.

bibliographies, seismology

Smith, W.E.T., 1956.
Bibliography of seismology. No. 16 Item 8773-8947
July-December 1954. Publ. Dominion Observ., Ottawa, 14:339-360.

bibliography, seismology

Smith, W.E.T., 1955.
Bibliography of seismology, No. 14, Items 8302-8546, July to December 1953. Publ. Dominion Observatory, Ottawa, 14:281-308.

slicks

bibliography, slicks

Leonard, J.M., 1959.
Biological oil slicks. Part 1. Literature examination.
NRL Memo. Rept., 955:24 pp.

submersibles

bibliography, submersibles

Terry, Richard D., 1966.
The deep submersible.
Western Periodicals Co., North Hollywood, Calif. 456 pp.

surges

bibliography, surges

Groen, P., and G.W. Groves, 1962
Surges. Ch. 17, Sect. 5. Waves. In: The Sea, Vol. 1, Physical Oceanography, Interscience Publishers, 611-646.
(Mss received June 1960)

bibliography, surges

Kramer, M.P., 1955.
Annotated bibliography on storm surges.
Meteorol. Abstr. & Bibliogr. 6(3):370-390.

temperature

bibliography, temperature

Böhnecke, G., and G. Dietrich, 1951.
Monatskarten der Oberflächentemperatur für die Nord- und Ostsee und die angrenzenden Gewässer.
Deutsches Hydro. Inst. No. 2336:13 charts, 3 tables.

bibliography, temperature
*Dmitriev, A.A., 1967.
Thermal conditions in the ocean (Soviet studies for the last 50 years). (In Russian; English abstract).
Fisika Atmosfer. Okean., Izv. Akad. Nauk, SSSR, 3(10) 1035-1043.

bibliography, temperature
Margalef, R., 1955.
Temperatura, dimensiones y evolución.
Publ. Inst. Biol. Aplic. 19:13-94.

bibliography, thermal microstructure
Skudrzyk, E.J., 1963.
Thermal microstructure in the sea and its contribution to sound level fluctuations.
In: Underwater acoustics, V.M. Albers, Edit., Plenum Press, N.Y., 199-233.

tides

bibliographies, tidal hydraulics
U.S.A. Corps of Engineers, U.S. Army, Committee on Hydraulics 1965.
Bibliography on tidal hydraulics. Supplementary material compiled from May 1959-May 1965.
Rept. No. 2, Suppl. No. 4 (ES816): 215 pp. (multilithed).

bibliography, tidal hydraulics
U.S. Waterways Experiment Station, 1954.
Bibliography on tidal hydraulics.
Committee on Tidal Hydraulics, Vicksburg, Miss., 201 pp.

bibliographies, tides
Association d'Océanographie Physique, UGGI, 1957
Bibliography on tides, 1940-1954. Publ. Sci., No. 17: 63 pp.

bibliography, tides
Association d'Océanographic Physique, 1939
Bibliography on tides and certain kindred matters. Fifth Installment. Publ. Sci. No.6, 19 pp.

bibliography tides
Lacomb, H., 1950.
Le Mascaret. Analyse bibliographique. Bull. Info. C.C.O.E.C., 2 Ann., (4):138-151 (mimeographed).

bibliography, tides
Rossiter, J.R., 1963.
Tides.
In: Oceanography and Marine Biology, H. Barnes, Editor, George Allen and Unwin, 1:11-25.

bibliography, tides
Schönfeld, J.C., 1951.
Propagation of tides and similar waves. Staatsdrukkerij- en Uitgeverijbedrijf, s-'Gravenhage: 232 pp., 118 textfigs., 12 loose plates.

bibliography, tides
Union Géodésique et Géophysique International, 1939.
Bibliography on tides and certain kindred matters. (Fifth Installment). Publ. Sci., Assoc. Océan. Phys., No. 6:20 pp

bibliography, tides
Union Géodésique et Géophysique International, 1936.
Bibliography on tides and certain kindred matters. (Fourth installment). Publ. Sci., Assoc. Océan. Phys. No. 3:88 pp.

bibliography
Union Géodésique et Géophysique Internationale, 1932.
Tidal bibliography (third installment).
Publ. Sci., Assoc. Océan. Phys., No. 2:17 pp.

bibliography, tides
Wilson, Basil W., 1960
The prediction of hurricane storm tides in New York Bay.
B.E.B. Tech. Memo. No. 120: 107 pp.

bibliographies translations (sedimentation)
Goldberg, E., editor 1965
Sedimentation: annotated bibliography of foreign literature for 1959 to 1964, Survey No. 1: 192 pp.
Israel Program Scient. Translation, U.S. Dept. Agricult. Nat. Sci. Found. $5.00

bibliographies, translations
Scattergood, L.W. 1949.
Translations of foreign literature concerning lobster culture and the early life history of the lobster.
Spec. scient. Rept. Fish., USFWS 6:173 pp (mimeographed)

bibliographies, translations (fisheries)
Scattergood, L.W. 1958
English translations of fishery literature, additional listings, 1958.
USFWS Spec. Scient. Rept., Fish. 264: 33 pp.

bibliographies, translations
U.S.A., Department of Commerce, Office of Technical Services, 1962.
Oceanography - OTS selective bibliography, 24 pp.
OTS SB 497 $0.10
listed in Techn. Transl., 1963 9(1):38-39.

bibliography, transparency
Gonzalez Sabariegos, M. L., 1949.
Introduccion al estudio de la transparencia del agua del mer. Bol. Inst. Español Ocean. No. 21: 29 pp., 6 textfigs.

tsunamis

bibliographies tsunamis
Cuellar, M.P., 1953.
Annotated bibliography on tsunamis.
B.E.B. Tech. Memo. No. 30:69 pp. (multilithed).

bibliographies, tsunamis
Grigorash, Z.K., 1964.
Tsunamis: annotated bibliography of Russian and foreign languages, 1726-1962. (In Russian).
Mezhd. Geofiz. Kom., Prezidiume Akad. Nauk, SSSR, Izdatel, "Nauka", Moskva, 110 pp.

bibliographies, tsunamis
United States, U.S. Coast and Geodetic Survey, 1964.
Annotated bibliography on tsunamis.
Union Géodés. et Géophys. Int., Monogr., No. 27: 249 pp.

turbidites

bibliography, turbidites, etc.
Kuenen, Ph. H., 1964
Deep-sea sands and ancient turbidites.
In: Turbidites, A.H. Bouma and A. Brouwer, Editors, Developments in Sedimentology, Elsevier Publishing Co., 3:3-33.

bibliography, turbidites
Kuenen, Ph. H., and F.L. Humbert, 1964
Bibliography of turbidity currents and turbidites. In: Turbidites, A.H. Bouma and A. Brouwer, Editors. Developments in Sedimentology, Elsevier Publishing Co., 3:222-246.

bibliographies, turbidity currents
Kuenen, Ph. H., 1960 (30 Nov.)
Bibliography of turbidity currents and turbidities. 17 pp. legal size paper (mimeographed).

Typhoons

bibliography, typhoons
Hatakeyama, H., 1954.
Recent researches on typhoons in Japan.
Pap. Met. Geophys., Tokyo, 5(2): 101-113, 7 textfigs.

underwater television

bibliography, underwater television
Barnes, H., 1963.
Underwater television.
In: Oceanography and Marine Biology, H. Barnes, Editor, George Allen and Unwin, 1:115-123.

upwelling

BIBLIOGRAPHY
Brongersma-Sanders, M., 1948
The importance of upwelling water to vertebrate palaeontology and oil geology. Ver. Koninklijke Nederland Akad. Wetenshappen, afd. Naturkunde, Tweede Sectie, Deel XLV(4):112 pp., 7 text figs.

vertical motion

bibliography, vertical motion
Chekotillo, K.A., 1966.
Vertical water motions in the ocean. (In Russian; English abstract).
Rez. Issled. Mezhd. Geofiz. Proekt. Mezhd. Geofiz. Komitet Presidiume Akad. Nauk SSSR, Okeanol., 17:76pp.

bibliography, volcanism
McBirney, A.R. 1971
Oceanic volcanism: a review.
Rev. Geophys Space Phys. 9(3): 523-556

wastes

Bibliographies, wastes
See also: Bibliographies, pollution

bibliography, wastes
Friedrich, Hermann, und Hans Lüneberg, 1962
Beiträge zu einer Bibliographie "Abwasser in Meerwasser".
Veröffentlichungen Inst. Meeresf., Bremerhave 8(1):37-52.

bibliography, wastes
Johnson, J.W., 1960
The effect of wind and wave action on the mixing and dispersion of wastes.
Waste Disposal in the Marine Environment, Pergamon Press: 328-343.

waves

Bibliographies, waves
See also: Bibliographies, internal waves

bibliography, waves
Barber, N.F., and M.J. Tucker, 1962
Wind waves. Ch. 19, Sect. V. Waves. In: The Sea. Vol. 1, Physical Oceanography, Interscience publishers, 664-699.
(Mss received June 1960)

bibliography, waves (ripples)
Cox, C.S., 1962
Ripples. Ch. 21, Sect. V. Waves. In: The Sea, Vol. 1, Physical Oceanography, Interscience Publishers, 720-730.
(Mss received July 1960)

bibliography, waves
Kramer, M.P., 1954.
Pt. II. Selective annotated bibliography on wind waves and swell. Met. Abstr. & Bibl. 5(11): 1300-1337.

bibliographies, waves
Rice, M.L., Compiler, 1958.
Recent literature on waves, currents and swell. Meteorol. Abstr. & Bibl., 9(4):472-503.

bibliographies, waves
Sinha, Evelyn Z., Compiler, 1962
Annotated bibliography on waves and currents. Meteorol. Geoastrophys. Abstr., 13(9): 2689-2741.

bibliography, waves
Sirotov, K.M., 1962.
The study of sea waves. (In Russian). Mezhd. Geofiz. Komitet, Prezidiume, Akad. Nauk, SSSR, Rezult. Issled. Programme Mezhd. Geofiz. Goda, Okeanol. Issled., No. 7:105-113.

bibliography, waves
U.S. Naval Oceanographic Office and the Earth Sciences Division, National Academy of Sciences-National Research Council, 1963.
Ocean wave spectra.
Proc. Conf., Easton, Md., May 1-3, 1961, Prentice-Hall, Inc., viii + 357 pp.

bibliography, waves
Wiegel, R.L., and J.W. Johnson, 1950.
Elements of wave theory. Inst. Coastal Eng., Univ. Ext., Univ. Calif., Long Beach, 11-13 Oct. 1950:24 pp. (mimeographed), 17 figs. (multilith-ed).

bibliography, waves
Wilson, Basil W., 1959
The energy problem in the mooring of ships exposed to waves. Permanent Int. Assoc., Navigation Congresses, Bull. No. 50: 1-65. Presented at Princeton Univ. Conf. on "Berthing and Cargo Handling in Exposed Locations", 21-23 Oct. 1958. Also in: Contributions in Oceanography and Meteorology, A&M College of Texas, Vol. 5, Contr. No. 129.

bibliography, wood protection
Trussel, P.C., and E.B. Gareth Jones 1970.
Protection of wood in the marine environment.
Int. Biodetn. Bull. 6 (1): 3-7.

bibliography, zooplankton (Canadian)
Shih, C.-T., A.J.G. Figueira and E.H. Grainger 1971.
A synopsis of Canadian marine zooplankton.
Bull. Fish. Res. Bd. Can. 176: 264 pp.

biocoenose
Boudouresque, Charles François, 1970.
Recherches sur les concepts de biocoenoses et de continuum au niveau de peuplements benthiques sciaphiles.
Vie Milieu 21 (1B): 103-136.

biocoenosis

p. 132

biocoenoses
Chassé, C. 1971.
Distribution qualitative et quantitative des peuplements littoraux des sédiments meubles au long d'un gradient édaphique synthétique quantifié: l'instabilité (côtes de France, Manche et Atlantique).
Vie Milieu Suppl. 22 (2): 657-675.

biocoenoses
Hertweck, Gunther, 1971.
Der Golf von Gaeta (Tyrrhenisches Meer). V. Abfolge der Biofaziesbereiche in den Vorstrand- und Schelfsedimenten. Senckenbergiana maritima 3(1): 247-276.

biodeterioration
Jones, E.B. Gareth, and T. Le Campion-Alsumard 1970.
The biodeterioration of polyurethane by marine fungi.
Int. Biodetn. Bull. 6 (3): 119-124.

biogeochemistry
SEE CHEMISTRY

Biogeochemistry.
See: Chemistry

biogeography

biogeography
Beklemischev, C.W., 1965.
Biogeographical division of the upper layer of the Pacific pelagial and its dependence on the currents and water mass distribution. (In Russian; English abstract).
Okeanolog. Issled., Rezult. Issled Programme Mezhd. Geofiz. Goda, Mezhd. Geofiz. Komitet Presidiume, Akad. Nauk, SSSR, No. 13:123-127.

biogeography
Bernard, Michelle, 1967.
Recent advances in research on the zooplankton of the Mediterranean Sea.
Oceanogr. Mar. Biol., Ann. Rev., H. Barnes, editor, George Allen and Unwin, Ltd., 5:231-255.

biogeography
Bogorov, B.G., 1958
Biogeographical regions of the plankton of the North-Western Pacific Ocean and their influence on the deep sea. Deep-Sea Research, Vol. 5 (2): pp 149-161.

biogeography
Ginés, Hno., 7 R. Margalef, editores, 1967.
Ecología marina.
Fundación La Salle de Ciencias Naturales, Caracas Monografía 14: 711 pp.

biogeography
Lindberg, G.U. 1970.
Paradoxical conclusions of biogeography in the light of recent data of oceanic geology and geophysics. (In Russian; English abstract).
Zool. Zh. 49 (11): 1605-1613.

biogeography
Margalef, Ramón, 1967.
Biogeografía historica.
In: Ecología marina, Monogr. Fundación La Salle de Ciencias Naturales, Caracas, 14:356-376.

biogeography
McGowan, J.A. 1971.
Oceanic biogeography of the Pacific.
In: Micropalaeontology of oceans, B.M. Funnell and W.R. Riedel, editors, Cambridge Univ. Press, 3-74.

biographies

Biographies

This heading is misleading, for it merely contains, for the most part, short notices concerning an individual scientist.

This section is arranged alphabetically by the scientist concerned not by the author of the article.

biographies, Abe
Motoda, S., 1971.
Dr. Tohru H. Abé. Bull. Plankt. Soc. Japan 18(1): 1-2.

Abelson, P.H.

biography Abelson
Campbell, F.L., 1962
Philip Hauge Abelson, New Editor of Science. Science, 137(3526):267-268.

Aikawa, H.

biographies, Aikawa
Motoda, S., 1962
In memoriam Dr. Hiroaki Aikawa. (In Japanese and English)
Info. Bull. Planktology, Japan, No. 8: 1-2.

biographies, Aikawa
Tanaka, Otohiko, 1962
The late Dr. Hiroaki Aikawa. (In Japanese).
Info. Bull., Planktology, Japan, No. 8: 2.

Albert 1er
Prince of Monaco

biographies, Albert 1er, Prince de Monaco
Albert 1er, Prince de Monaco 1966.
La carrière d'un navigateur.
Editions des Archives du Palais Principier, Monaco, 238 pp.

biographies, Albert 1er de Monaco
Crovette, Arthur, 1965.
Le Prince Albert 1er "chef et propagateur de l'océanographie".
Colloque Internat., Hist. Biol. Mar., Banyuls-sur-Mer, 2-6 sept. 1963, Suppl., Vie et Milieu, No. 19:291-304.

biographies, Albert 1er de Monaco
Damien, Raymond, 1964
Albert 1er, Prince Souverain de Monaco.
Institut de Valois, Villemomble-Seine, près Paris, 515 pp.

biography, Prince Albert 1er, Monaco
Shadewaldt, H., 1965.
La croisière du Prince Albert 1er de Monaco en 1901 et la découverte de l'anaphylaxie.
Colloque Internat., Hist. Biol. Mar., Banyuls-sur-Mer, 2-6 sept. 1963, Suppl., Vie et Milieu, No. 19: 305-313.

Aldrovandi

biographies, Ulysse Aldrovandi
Grmek, Mirko Drazen, et Daniele Guinot, 1965.
Les crabes chez Ulysse Aldrovandi: un aperçu critique sur la carcinologie du XVIe siecle.
Colloque Internat., Hist. Biol. Mar., Banyuls-sur-Mer, 2-6 sept., 1963, Suppl., Vie et Milieu, No. 19:45-64.

biographies, Alexandrowicz

Andersen, Torben 1971
List of publications.
Fortegnelse Medd. Grønland, 130 pp.

biographies, Alexandrowicz

Bone, Q., 1971.
Professor J.S. Alexandrowicz. J. mar. biol. Ass. U.K. 51(4): 1007-1011.

Alpatov, Mikhail Grigorievich

biographies -Alpatov

Anon., 1965.
Mikhail Grigorievich Alpatov (1910-1965).
Okeanologiia, Akad. Nauk, SSSR, 5(4):765-766.

Arctowski, Prof. H.

biographies, Arctowski

Burdeckl, F., 1957.

In memoriam: Henryk Arctowski
Notos, Wea. Bur. South Africa, 6(3/4):126

biographies, Arctowski

Kosiba, Aleksander, 1959.
Dzialalność naukowa Profesora Henryka Arctowskiego.
Acta Geoph. Polonica, 7(3/4):250-281.

biographies, Arctowski

Kosiba, A., and S. Zych, 1959.

Prof. Henryk Arctowski. (15.VII.1871-21II.1958).
Przeglad Geofiz. Warsaw, 14(XII) (2):83-90.

biographies, Arctowski

Tison, L.J., 1960.
H. Arctowski et l'Expedition Antarctique de la "Belgica".
Acta Geophys. Polonica, 7(3/4):296-297.

Aristotle

biographies, Aristotle

Mangold, Katharina, et Georges Petit, 1965. Aristote et la biologie marine: les céphalopodes.
Colloque Internat., Hist. Biol. Mar., Banyuls-sur-Mer, 2-6 sept., Suppl., Vie et Milieu, No. 19:11-20. (1963)

Arne, Paule

biographies, Arne

France, Institut Scientifique et Technique des Pêches Maritimes, 1963.
A la memoire de Paule Arne 1878-1963.
Rev. Trav. Inst. Pêches Marit., 27(4):353.

Arrhenius, S.

biography, Arrhenius

Sweden, K Svenska Vetenskapsakedemiens Orasbok, Bilaga for 1959.
Svante Arrhenius, till 100 årsminnet av hans födelse, 115 pp.
Olander, Arne, Arrhenius och den elektrolytiska dissociationsteorin, 5-33.
Arrhenius, Olaf. Släkten Arrhenius, 35-41.
Arrhenius, Olaf. Arrhenius, det. första kvarteklet 43-64
Arrhenius-Wold, Anna-Lisa. Svante Arrhenius och utvecklingsoptismen, 65-100.
Arrhenius, G.O.S., Svante Arrhenius: contributions to earth science and cosmology

Atkins, W.G.R.

biographies - Atkins

Cooper, L.H.N., 1960
W.R.G. Atkins, CBB.E., O.B.E.(Mil.), Sc.D., F.R.I.C., F. INST. P., F.R.S., 1884-1959. J.M.B.A., 39:153-154.

biographies, Atkins

Russell, F.S., 1959.
Dr. W.R.G. Atkins.
Nature, 183:1228.

Audouin, Jean-Victor

biographies Audouin

*Théodoridès, Jean, 1968.
Les débuts de la biologie marine en France: Jean-Victor Audouin et Henri-Milne Edwardsm 1826-1829.
Bull. Inst. océanogr., Monaco, No. spécial 2: 417-437.

Baas Becking, L.G.M.

biography, Baas-Becking

Robertson, R.N., 1963.
Obituary - Dr. L.G.M. Baas-Becking.
Australian J. Sci., 26(1):15-16.

biography, Bacescu

Popescu-Goy, Aurelian, 1968.

Mihai Bacescu a L'aniversaire de ses 60 ans.

Trav. Mus. Hist. nat. "Grigore Antipa" 9:2-13.

biographies, Bacescu

Posa, Eugène A., 1968.
Mihai C. Băcescu à 60 ans.
Revue Roumaine Biol. (Zool.) 13(6): 347-362.

Beche, Alexander Dallas

biographies, Alexander Dallas Beche

Sinclair, Bruce 1966.
Early research at the Franklin Institute: the investigation into the causes of steam boiler explosions, 1830-1837.
Franklin Inst. Phila., 26+48 pp.

Barbour, T.

biographies (Barbour)

Romer, Alfred S., 1964.
Thomas Barbour.
Syst. Zool., 13(4):227-234.

biographies, Barbour

Bigelow, H.B., 1952.
Thomas Barbour. Biogr. Mem., Nat. Acad. Sci., 27: 13-45.

Barnard, K.H.

biographies, K.H. Barnard

Gordon, Isabella, 1966.
Dr. Keppel Harcourt Barnard.
Crustaceana, 10(2):219-221.

biographies, Barnard, K.H.

Grindley, John R., 1964.
Dr. K. H. Barnard.
Nature, 204(4959):625-626.

Bates, C.C.

biography, Bates

Fisk, H.N., 1954.
Charles Carpenter Bates, recipient of President's award. Bull. Amer. Assoc. Petr. Geol. 38(7):1637-1639.

Beklemishev, V.N.

biographies, V.N. Beklemishev

Smirnov, E.S., 1963.
Vladimire Nikolaevich Beklemishev. 5X1890-4 IX 1962.
Zool. Zhurnal, Akad. Nauk, SSSR, 42(2):314-317.

biography, V.N. Beklemishev

Bruce-Chwatt, L.J., 1962.
Prof. Vladimir Nikolaievitch Beklemishev.
Nature, 196(4857):812-813.

biographies, Beklemishev

Moshkovsky, Sh. D., 1963.
Vladimir Nikolaevitch Beklemishev 1890-1962 - Obituary.
Biull. Mosk. Obshch. Isp. Prirodi, n.s., Otdel Biolog., 68(2):157-160.

In Russian

Bellon, U.L.

biography - Bellon

Bellon, U.L., 1954.
Historia naturel del atun, Thunnus thynnus (L.), Ensayo de sintesis (precidida de la necrologia del autor.) Bol. Inst. Espanol Oceanogr. 67:1-88.

Bennett, Rawson II

biography, Bennett

Anon. 1968.
Rawson Bennett II 1905-1968.
J. acoust. Soc. Am. 43(5): 1190.

biographies - Berg

Madsen, Bent Lauge, and Gunnar Nygaard 1970
Professor Kaj Berg - 70 years old.
Hydrobiologia, Netherlands 35(2): 345-351

biographies, Berkeley

Stevenson, J. Cameron 1971.
Edith and Cyril Berkeley - an appreciation.
J. Fish. Res. Bd. Can. 28(10): 1360-1364

Berlinskii, N.A.

biographies Berlinskii

SSSR, Akademia Nauk, 1964. (In Russian).
N.A. Berlinskii, 1910-1964.
Okeanologiia, Akad. Nauk, SSSR, 4(2):364-365.

Bigelow, H.B.

biographies, Henry B. Bigelow

Bigelow, Henry B. 1964.
Memories of a long and active life.
The Cosmos Press, Cambridge, 41 pp.

biographies, Bigelow

France, Service Hydrographique de la Marine 1968.
Décès du Docteur Henry Bryant BIGELOW.
Cah. oceanogr. 20 (3): 186-187.

biographies, Bigelow

Graham, Michael 1968.
Obituary - Henry Bryant Bigelow (3 October 1879 - 11 December 1967).
Deep-Sea Research 15 (2): 125-132.

biographies, Birstein

Mileikovsky, S.A. and R.R. Makarov, 1971.
In memoriam of Professor Yakov Avad'evitch Birstein. Mar. Biol. 10(4):281-284.

biographies--Birstein

Mileikovsky, S.A. and R.R. Makarov, 1971.
Professor Yakov Avad'evich Birstein. (In Russian). Okeanologiia 11(3): 557-558.

biographies - Bjerkan
Rollefsen, G. 1970.
Paul Bjerkan (1874-1968).
J. Cons. int. Explor. Mer, 33 (2): 124-125.

biographies, Birstein
Anon, 1971.
In memoriam of L.A. Zenkevitch and J.A. Birstein.(In Russian). Trudy Inst. Okeanol. P.P. Shirshov 92: 5-8.

biographies, Birstein
Chindonova, Yu. G., S.I. Ljovushkin and N.A. Z-arenkov 1972.
Yakov Avadievich Birstein (7th April) 1911-8th July 1970).
Crustaceana 22(1):103-112.

Bjerknes

biographies, V.F.K. Bjerknes
Anon., 1951.
Vilhelm F.K. Bjerknes. Met. Mag. 80(949)(M.O.548) :181-184, 1 pl.

biographies, Bjerknes
Bergeron, T., 1952.
Vilhelm Bjerknes och hans livsverk. Svenska Fysikersamfondets Årsbok, Kosmos, 1952:9-24,

Also:
Met. Inst., K. Univ. Uppsala Medd. No. 30.

biographies, Bjerknes
Bergeron, Tor, Olaf Devik, and Carl Ludvig Godske, 1962.
Vilhelm Bjerknes, March 14, 1862-April 19, 1951.
Geofys. Publik., Geophysica Norvegica, 24:7-25.

biographies, Bjerknes
C.K.M.D., 1951.
Professor V.F.K. Bjerknes. Q.J. Roy. Met. Soc. 77(333):529-530.

biographies, Bjerknes
Gold, E., 1951.
Obituary: V.F.K. Bjerknes. Weather 6(7):216-217.

biographies, Bjerknes
Gold, E., 1951.
Prof. V.F.K. Bjerknes, For. Mem. R.S. Nature 167 (4256):838-839.

biographies, Bjerknes
Hrgian, A.H., 1962
Wilhelm Bjerknes (to the 100th anniversary of his birth). (In Russian).
Meteorol. i Gidrol., (12):46-48.

biographies
Sverdrup, H.U., 1951?
1. A la memorie de Vilhelm Bjerknes (1862-1951), C.R. Symposium sur la circulation generale des ~~xxxxxxxxxxxxxxxxxxxxxxxxxxxxxxxxxxxx~~ oceans et de l'atmosphere, Assoc. Meteorol., ~~xxxx~~,U.G.G.I., 1951:1-8.

biographies, Bjerknes
Sverdrup, H.U., 1951.
Vilhelm Bjerknes in memoriam. Tellus 3(4):217-221.

Blegvad, H.

biography, Blegvad
Poulsen, E.M., 1951.
Harald Blegvad, 1886-1951. J. du Cons. 18(1): 3-6.

Boeck A.

biographies, A. Boeck
Rollefsen, Gunnar 1966.
Norwegian fisheries research.
Fisk. Dir. Skr. Ser. Havunder. 14(6):1-36

Bogorov, B.G. V.G.

biographies, Bogorov
Anon., 1971.
Veniamin Grigorievich Bogorov. (In Russian).
Okeanologiia 11(4):762-763.

biographies, B.G. Bogorov
Bezrukov, P.L., 1965.
On the occasion of Prof. B.G. Bogorov's 60th birthday.(In Russian).
Okeanologiia, Akad. Nauk. SSSR, 5(3):573-574.

Translation:
Scripta Tecnica for AGU, 147-148

Bogucki, Mieczysla

biographies, Bogucki
Klekowski, R.Z. 1967.
Prof. Dr. Mieczyslaw Bogucki, 1884-1965.
Int. Revue ges. Hydrobiol. 52(4): 647-648.

Böhnecke, Günther

biography Böhnecke
Humphrey, G.F., 1966.
Dr. Günther Böhnecke.
Dt. hydrogr. Z., 19(5): 234-235.

Bohr, N.

biographies
Weisskopf, Victor F., 1963
Niels Bohr, 1885-1962.
Nucleus, Dunod, Edit., 4(2):94-96.

Boschma, H.

biography
Klaauw, C.J., 1964.
Hilbrand Boschma.
Zool. Mededel., 39(ix-xlix)

Bouquet de la Grye, Anatole

biographies Anatole Bouquet de la Grye
Gougenheim, André, 1967.
Deux ingénieurs hydrographes du XIXe siècle: precurseurs en matière de dynamique des mers.
Cah. Océanogr., 19(7):539-547.

biographies - Urbain Dortet de Tessan
biographies Anatole Bouquet de la Grye

Bourcart, Jacques

biographies-Bourcart
Anon., 1965.
In memoriam, Professor Jacques Bourcart 1892-1965
Marine Geology, Elsevier Publ. Co., 3(4):321.

biographies, Bourcart
Boillot, Gilbert, 1966.
L'oeuvre océanographique de J.Bourcart (1891-1965).
Cah. Biol. Mar., 7(4):463-465.

biographies, Bourcart
Carruthers, J.N., 1965.
Prof. Jacques Bourcart.
Nature, 20?(5000):915.

biographies, Bourcart
Casso, A.R. 1967
Notre maître et ami le professeur Jacques Bourcart membre de l'Institut.
Revue Géogr. phys. Géol. dyn., 9(3):173-176

biographies, Bourcart
Gougenheim, André, 1965.
Déces du Professeur Jacques Bourcart.
Cahiers Océanogr., C.C.O.E.C., 17(8):521-524.

biographies,
Lalou, Claude, 1969
Jacques Bourcart (1891-1964).
Rapp. P.-v. Réun. Commn int. Explor. scient. Mer Mediterr. 19(4): 597-598

biographies Bourcart
Lalou, Cl., 1966.
Jacques Bourcart, géologue du pay niçois.
Rev. Intern. Océanogr. Med. 3:55-57.

Bowman

biographies, Bowman
Carter, G.F., 1950.
Isaiah Bowman, 187801950. Ann. Assoc. Am. Geogr. 40(4):335-350.

biographies Isaiah Bowman
Murphy, R.C., 1943.
The new president of the American Association for the Advancement of Science.
Scientific Monthly, 56:570-573.

biography, Bowman
Ogilvie, A. G., 1950.
Isaiah Bowman; an appreciation. Geogr. J. 115(4/6):226-230, portrait.

biography, Bowman
Stamp, L. D., 1950.
Dr. Isaiah Bowman. Nature 165(4188):175.

biographies, Bowman
Wrigley, G.M., 1951.
Isaiah Bowman. Geogr. Rev. 41(1):7-65, 2 figs.

Breitfuss

biographies, Breitfuss
Anon., 1951.
Leonid Lvovich Breitfuss. Polar Record 6(41):124-125.

Brekhovskii, L.M.

biographies, Brekhovskikh
Anon., 1969.
L.M. Brekhovskikh, member of the Academy of Sciences of the USSR. (In Russian)
Akustichesk. Zh. 15(2):310.
Translation: Soviet Phys.-Acoust. 15(2): 273

biographies, Brekhovskii
Anon., 1967.
L.M. Brekhovskii (on this 50th birthday)(In Russian).
Akust.Zh.13(3):470-472.

Bronk, D.W.

biographies, Bronk
Lee, M.O., 1951.
Detlev W. Bronk, scientist. Science 113(2928): 143.

Brooks, C.E.P.

biography, C.E.P. Brooks
Rigby, M., 1958.
Dr. C.E.P. Brooks, Bull. A.M.S., 39(1):40-41.

Bruevich, S.V.

biographies, Bruevich

Vinogradov, A.P., et al., 1971.
Semen Wladimirovich Brujewicz. (In Russian).
Okeanologiia 11(4):764-765.

biographies, S.V. Bruevich

Zenkevitch, L.A., B.G. Bogorov & V.B. Stockmann 1965.
Semen Vladimirovich Bruevich (in connection with his 50th anniversary of scientific activities). (In Russian).
Okeanologiia, Akad. Nauk, SSSR, 5(5):931-932.

Brunel, Henry Marc

biographies, Brunel

Donovan, D.T. 1967.
Henry Marc Brunel: The first submarine geological survey and the invention of the gravity corer.
Mar. Geol. 5(1): 5-14.

Bruun, A.F.

biography, Bruun

Fraser, F.C., 1962.
Dr. A. F. Bruun.
Nature, 193(4822):1231-1232.

biographies, Bruun

Spärck, R., 1962.
Anton Frederik Bruun, 14 December 1901-13 December 1961.
J. du Cons., 27(2):121-123.

biographies, Bruun

SSSR, Akad. Nauk, 1963
Anton Frederik Bruun (obituary). (In Russian).
Okeanologiia, Akad. Nauk, SSSR, 3(3): 566.

biographies - Bückmann

Schmidt, Ulrich 1970.
Professor Dr. Adolf Bückmann zum siebzigsten Geburtstag.
Ber. dt. Wissenschaft. Komm. Meeresforsch. n.f. 21:3-15.

Calman, W.T.

biographies, Calman

Fage, L., 1953.
Notices nécrologiques sur Theodor Mortensen et William Thomas Calman. Bull. Soc. Zool., France, 77(5/6):319-323.

biography, Calman

Gordon, I., 1952.
Dr. W.T. Calman, C.B., F.R.S. Nature 170(4332): 780-781.

biographies, Calman

Peacock, A.D., 1953.
William Thomas Calman. Year Book, R.Soc., Edinburgh, 1953(session 1951-1952):12-14.

Carruthers, J.N.

biographies- Carruthers

Böhnecke, G., 1965.
Dr. J.N. Carruthers - 70th birthday.
Deutsche Hydrogr. Zeits., 18(2):76-77.

biographies - Carruthers

Lumby, J.R., 1965.
Dr. J.N. Carruthers- 70th birthday.
Deutsche Hydrogr. Zeits., 18(2):76.

Chapman, Sydney

Biographies, Chapman

Fanselau, G. 1968.
Sydney Chapman zum 80 Geburtstag.
Beitr. Geophys. 77(1): 1-4.

biographies, Chapman

Needler, A.W.H., 1970.
Wilbert McLeod Chapman.
J. Fish. Res. Bd, Can. 27 (12): 2333-2335.

biographies, Chapman

Roedel, Philip M., 1971.
In memoriam - Wilbert McLeod Chapman and Milner Baily Schaefer.
Fish. Bull. U.S. Dept. Comm. 69(1): 1-2

Clark, R.S.

biographies, Clark

Ritchie, J., 1950.
Dr. R.S. Clark. Nature 166(4234):1055.

biographies, Clark

Russell, E.S., 1951.
Robert Selbie Clark, 1882-1950. J. du Cons. 17 (2):99-101, photo.

Clemens, W.A.

biography, W.A. Clemens

Hoar, W.S., 1964.
W.A. Clemens - an appreciation.
J. Fish. Res. Bd., Canada, 21(5):869-872.

Conklin, E.G.

biographies, Conklin

Harvey, E.N., 1953.
Edwin Grant Conklin:1863-1952. Science 118(3052): 703-705.

biographies, Conklin

Waddington, C.H., 1952.
Prof. E.G. Conklin. Nature 170(4338):1046.

Cope

Cope

Romer, Alfred S., 1964.
Cope versus Marsh.
Systematic Zoology, 13(4):201-207.

Corkan, R.H.

biographies, Corkan

Anon., 1952.
Dr. R.H. Corkan. Liverpool Obs. and Tidal Inst., Ann. Rept., 1952:5-10.

biographies, Cousteau

Mowbray, Beverly, 1971.
Captain Jacques Cousteau.
Frontiers, Acad. nat. Sci., Phila., 35 (3): 2-6

Biographies, Coutiere

Chace, Fenner A. Jr., et J. Forest, 1970.
Henri Coutière: son oeuvre carcinologique, avec un index pour son mémoire de 1899 sur les Alpheidae.
Bull. Mus. natn. Hist. nat. (2) 41 (6): 1459-1486.

Cox, Roland Arthur

biographies - Cox

Culkin, F. 1968.
Roland Arthur Cox (1923-1967).
J. Cons. int. Explor. Mer 31 (3): 289-290.

biographies - Cox

Deacon, G.E.R. 1968.
Dr. Roland Arthur Cox (17 February - 19 March 1967).
Deep-Sea Res. 15 (2): 135-136.

Cromwell, Townsend

biographies, Cromwell

Montgomery, R.B., 1959.
Townsend Cromwell, 1922-1958. Obituary.
Limnol. & Oceanogr., 4(2):228.

Crossland, C.

biographies, Crossland

Bruun, A. Fr., 1943.
Cyril Crossland, April 19th, 1878-January 7th, 1943, in memoriam. Vidensk. Medd. fra Dansk Naturh. Foren. 106:xii-xvi.

Cushman, J.A.

biography, Cushman

Waters, J. A., 1949.
Joseph Augustine Cushman (1881-1949). Bull. Am. Assoc. Pet. Geol., 33(8):1457-1465, 1 fig.

Cuvier

biographies, Cuvier

Coleman, William, 1965.
Les organismes marins et l'anatomie comparée dite expérimentale: l'oeuvre de Georges Cuvier.
Colloque Internat., Hist. Biol. Mar., Banyuls-sur-Mer, 2-6 sept., 1963, Suppl., Vie et Milieu, No. 19:225-238.

Cviić, Vlaho

biographies - Cviić

Cviić, Vlaho (posthumous) 1965.
Primary organic production, distribution, and reproduction of bacteria in the middle Adriatic euphotic zone.
Acta Adriatica 10 (9): 21 pp.

Dahl, K.

biographies, Dahl

Sømme, S., 1952.
Knut Dahl, 1871-1951. J. du Cons. 18(2):101-103.

Dakin, W. J.

biography, Dakin

Marshall, A.J., 1950.
Prof. W.J. Dakin. Nature 165(4202):751-752.

D'Ancona, Umberto

biographies, D'Ancona

Battaglia, Bruno, 1965.
Ricordo di Umberto D'Ancona (9 maggio 1896-24 agosto 1964).
Arch. Oceanogr. e Limnol., 14(1):I-XI.

biographies, d'Ancona

Kosswig, C., 1966.
Umberto d'Ancona, 9.5.1896-24.8.1964.
Int. Rev. geo. Hydrobiol., 51(2):385-386.

biographies - D'Ancona

Maldura, C., 1964
Umberto D'Ancona
Boll. Pesca, Piscicolt. Idrobiol. n.s., 19(2): 297-298.

biographies, Umberto D'Ancona

Pantin, C.F.A., 1965.
Prof. Umberto D'Ancona.
Nature, 205(4967):128-129.

biographies, D'Ancona

Tonolli, Vittorio, 1966.
Umberto D'Ancona (1896-1964).
Hydrobiologia, Acta Hydrobiol. Hydrogr. et Protistol., 27(1/2):260-273.

Dannevig, A.N.

biographies, Dannevig
Ruud, Johan T., 1961.
Alf Nicolai Dannevig, 1886-1960.
J. du Conseil, 26(3):241-242.

Darby de Thiersant

biographies - Darby de Thiersant
Huard, Pierre, et Daniele Guinot, 1965.
Les crabes de Chine dans une serie d'aquerelles provenant de Darby de Thiersant.
Colloque Internat., Hist. Biol. Mar., Banyuls-sur-Mer, 2-6 sept. 1963, Suppl., Vie et Milieu, No. 19:35-43.

Davenport, C.B.

biographies, Davenport
Parker, G.H., 1944
Charles Benedict Davenport (1866-1944) Year Book, Am. Phil. Soc., 1944:358-362.

owned by M.S.

Davies, David H.

Biographies, Davies
Alexander, Anne J., and others 1966
David H. Davies.
S. Afr. Assoc. Mar. Biol. Res., Bull., No. 6: 7-22.

biographies, Davies
bibliography

Dawson, E. Yale

biographies, Dawson
Abbott, Isabella A., 1966.
Elmer Yale Dawson (1918-1966).
J. Phycology, 2(4):129-132.

biographies, Dawson
Dixon, Peter S. 1967.
Elmer Yale Dawson 1918-1966.
Int. Revue ges. Hydrobiol. 52(4): 651-652.

biographies, Dawson
Dorst, Jean, 1967.
Le Docteur E. Yale Dawson.
Noticias de Galepagos, UNESCO, Breixelles, 3-4.

biographies, Dawson
Garth, John S., 1967.
E. Yale Dawson.
Bull. S. Calif. Acad. Sci., 66(3):149-160.

biographies, Dawson
Silva, Paul C., 1967.
E. Yale Dawson (1918-1966).
Phycologia 6(4):218-236.

biographies, Day
Anon. 1971.
Vice Admiral Sir Archibald Day.
Int. hydrogr. Rev. 48(1): 7-9.

de Buen

biographies, de Buen
Balech, Enrique, 1963
Dr. Fernando de Buen Lozano (1895-1962?)
Bol. Inst. Biol. Mar., Mar del Plata, Argentina, No. 2:31-33.

biography, de Buen
Chile, Universidad de Chile, Instituto de Biologia, 1963.
Fernando de Buen y Lozano (1895-1962).
Montemar, (3):91-98.

biography, de Buen
Rodriquez, R.G., 1948
Odon de Buen y Del Cos 1863-1945. J. du Cons. 15(2):135-136.

Defant, A.

biographies, Defant
Wüst, G., 1964.
Albert Defant achtzig Jahre alt.
Die Naturwiss., 51(13):301-302.

biographies, Defant
Wüst, Georg, 1964.
Albert Defant zum 80 Geburtstag.
Beiträge zur Physik Atmosph., 37(2):58-68.

de Lacaze-Duthiers

biographies Lacaze-Duthiers
*Petit, Georges, 1968.
Henri de Lacaze-Dutiers (1821-1901) et ses "carnets" intimes.
Bull. Inst. océanogr., Monaco, No. special 2: 453-465.

biographies, Deryugin
Ushakov, P.V., 1971.
Konstantin Mikhailovich Deryugin (8.III.1878-27.XII. 1938). (In Russian). Okeanologiia 11(5) 926-930.

Desbrosses, P.

biographies, Desbrosses
France, Institute Scientifique et Technique des Pêches Maritimes, 1963.
À la mémoire de Pierre Desbrosses.
Rev. Trav. Inst. Pêches Marit., 27(2):117.

biographies, Desbrosses
Furnestin, J., 1964.
Pierre Desbrosses, 1902-1962.
Jour. du Conseil, 28(3):325-326.

biographies, Dietz
Wilson, J.T., 1971.
Fourth presentation, Walter H. Bucher Medal to Robert S. Dietz for original contributions to the basic knowledge of the earth's crust.
Eos, 52(7): 540-541

Disney, Lindsay P.

Dixon, H.H.

biography, Dixon
Atkins, W.R.G., 1954.
Henray Horatio Dixon, 1869-1953. Obituary Notices Fellows Roy. Soc. 9:79-97, 1 pl.

Dobrovolsky, A.D.

biographies, Dobrovolsky
Mamaev, O.I., 1967.
60th birthday of Professor Aleskey Dmitrievich Dobrovolsky. (In Russian).
Okeanologiia, Akad. Nauk, SSSR, 7(4):738-742.

Dobson, A.T.A.

biographies, Dobson
Graham, Michael, 1964
A.T.A. Dobson, 1885-1962.
J. du Conseil, 29(1):3-5.

Dohrn, A.

biographies, Dohrn
Heuss, Theodor, 1948.
Anton Dohrn.
Rainer Wunderlich Verlag Herman Leins.
Stuttgart und Tübingen, 448 pp.

biographies-Anton Dohrn
Kühn, A., 1950.
Anton Dohrn und die Zoologie seiner Zeit. Pubbl. Staz. Zool., Napoli, Suppl. 1950:205 pp.

Dohrn, Reinhard

biographies, Dohrn
Gamulin-Bride, H., 1963.
Reinhard Dohrn. Toute une vie consacrée à la Station Zoologique de Naples.
Biološki Glasnik, Hrvatsko Prirod. Društvo, Zagreb, 16(3-4):A11-A12.

Doodson, A.T.

biographies - Doodson
Gougenheim, A. 1968.
Le docteur A.T. Doodson (1890-1968).
Cah. océanogr. 20(4): 269-271.

biographies - Doodson
Proudman, J., 1960
A.T. Doodson, C.B.E., D. Sc., F.R.S., Hon. F.R.S.E. An appreciation.
Liverpool Obs. and Tidal Inst., Ann. Rept., 1960: 7-9.

biographies - Doodson
Rossiter, J.R. 1968.
A.T. Doodson, C.B.E., D. Sc., F.R.S., Hon. F.R.S.E. (1890-1968).
Deep-Sea Res. 15(2): 133-134.

d'Orbigny

biographies - A. d'Orbigny
Torlais, J., 1965.
Alcide d'Orbigny, voyageur et ethnologue.
Colloque Internat., Hist. Biol. Mar., Banyuls-sur-Mer, 2-6 sept., 1963, Suppl., Vie et Milieu, No. 19:219-223.

Dortet de Tessan

biographies-Urbain Dortet de Tessan
Gougenheim, André, 1967.
Deux ingénieurs hydrographes du XIXe siècle: precurseurs en matière de dynamique des mers.
Cah. Océanogr., 19(7):539-547.

Dostoevski, A.A.

biography - Dostoevski
Belov, S.V., 1963.
A.A. Dostoevski in memoriam. (In Russian).
Izvest. Vses. Geograf. Obsches., 95(4):371-373.

Edgell, Sir John

biographies, Edgell
Anon., 1963
Obituary - Vice Admiral Sir John Edgell (1880-1962).
Int. Hydrogr. Rev., 40(2):7-9.

biographies, Edgell
Deacon, G.E.R. 1963.
John Augustine Edgell, 1880-1962.
Biogr. Mem. Fellows R. Soc. 9: 87-90.

Ehrenbaum, E.

biography, Ehrenbaum
Lüling, K.-H., 1949.
Ernst Ehrenbaum. Ber. Deutschen Wiss. Komm. f. Meeresf., n.f., 11(4):435-442.

biographies - Ehrenberg
Engelmann, Gerhard 1969.
Christian Gottfried Ehrenberg, ein Wegbereiter der deutschen Tiefseeforschung.
Dt. hydrogr. Z. 22(4): 145-157.

Einarsson, Hermann

biographies, Einarsson
Bertelsen, E., 1968.
Hermann Einarsson, 1913-1966.
J. Cons. perm. int. Explor. Mer, 32(1):3-6.

Eklund, C.R.

biography - Eklund
Siple, P.A., 1963.
Carl R. Eklund (1909-1962) Obituary.
Arctic, 16(2):147-148.

Ekman, Sven

biographies, Sven Ekman
Berg, Kaj, 1966.
Sven Ekman 31 - V - 1876---2 - II - 1964.
Hydrobiologie, Acte Hydrobiol., Hydrogr. et Protistol., 27(1/2):274-286.

biographies, Sven Ekman
Pejler, Birger, 1964.
Sven Ekman zum Gedächtnis 1876-1964.
Int. Revue Ges. Hydrobiol., 49(4):637-638.

Ekman, V.W.

biographies, V.W. Ekman
Hansen, W., 1954.
Vagn Walfrid Ekman. Deutsche Hydrogr. Zeits. 7(1/2):65-66.

biographies, V.W. Ekman
Kullenberg, B., 1954.
Vagn Walfrid Ekman, 1874-1954. J. du Cons. 20(2): 140-143.

Emery, K.O.

biography, Emery
Clements, Thomas, 1963.
Dr. Kenneth Orris Emery: an appreciation. In: Essays in Marine Geology in honor of K.O. Emery, Univ. S. California Press, ix-xi.

biography Ercegovic
Alfirevic, Slobodan 1970
Le Docteur Ante Ercegović (1895-1969); sa vie et son oeuvre. (In Jugoslavian and French)
Acta adriatica 13(5):23pp.

biographies - Ercegovic
Pucher-Petkovic, T. 1970.
Ante Ercegovic (1895-1969) in memoriam.
Revue algol. n.s. 10(1):3-7.

Evgenov, N.I.

biographies - Evgenov
Gakkel, Ya. Ya., 1964
Nikolai Ivanovich Evgenov (1888-1964).
Okeanologiia, Akad. Nauk, SSSR, 4(5):929-931.

Fabricius, I.C.

biography, Fabricius
Tuxen, S.L., 1967.
Bibliographie von I.C. Fabricius.
Zoologische Anzeiger, 178(3/4): 174-185.

Fage, Louis

Biographies, Fage
Bernard F., 1965.
Louis Fage (1883-1964).
Journal du conseil, 29(3):233-236.

biographies Fage
DeBeer, Gavin, 1965.
Baptiste-Louis Fage.
Proc. linn. Soc., London, 176:223-224.

Biographies, Louis Fage
Drach, P., 1965.
Louis Fage (30 Septembre 1883-28 Mai 1964).
Deep-Sea Research, 12(3):280-283.

biographies - Fage
Forest, J., 1965.
Louis Fage, 1883-1964.
Crustaceana, 9(2):213-219.

biographies, Louis Fage
Gordon, Isabella, 1964.
Prof. Louis Fage.
Nature, 204(4965):1253-1254.

biographies, Fage
Lacombe, H., 1964.
Décès du Professor Louis FAGE.
Cahiers Oceanogr., C.C.O.E.C., 16(7):499-500.

biographies, Fage
Pérès, J.M., 1964.
À la mémoire du Professor Louis Fage.
Rec. Trav. Sta. Mar. d'Endoume, Bull., 31(47):1.

biographies, Fage
Petit, G., 1964.
Louis Fage (1883-1964).
Vie et Milieu, Bull. Lab. Arago, 15(1):227-229.

biographies - Louis Fage
Vachon, M., 1964.
Louis Fage 1883-1964. Notice biographique et bibliographique.
Bull. Mus. Nat. Hist. Nat., (2) 36(4):423-440.

biography - Fage
Zenkevich, L.A., 1965.
In memory of our foreign colleague. [Louis Fage]
Okeanologiia, Akad. Nauk, SSSR, 5(6):1127.

Farran, G.P.

biography - Farran
Went, A.E.J., 1950.
George Philip Farran 1876-1949. J. du Cons. 16(2) 160-163.

Field, Richard M.

biographies - Field
Hess, H.H., 1962.
Richard Montgomery Field.
Trans. Amer. Geophys. Union, 43(1):1-3.

Fisher, W.K.

biographies - Fisher
Blinks, L.R., 1954.
Prof. W.K. Fisher. Nature 173(4393):60.

Fleming, J.A.

biography - Fleming
Scott, W.E., 1956.
John Adam Fleming, 1877-1956. J. Geophys. Res., 61(4):589-592.

biographies - Fleming
Vestine, E.H., 1956.
John Adam Fleming. Trans. Amer. Geophys Union 37 (5):531-533.

Forbes, Alexander

biographies, Forbes
Fenn, Wallace O., 1966.
Alexander Forbes (1882-1965).
Am. Philosoph. Soc., Year Book 1965:140-145.

Forbes, Edward

Biographies - Forbes
Merriman, Daniel, 1965.
Edward Forbes - Manxman.
Progress in Oceanography, 3:191-206.

biography - Forbes
Pantin, C.F.A., 1958.
Edward Forbes and the modern Athens.
Perspectives in Biology & Medecine, 1(2):198-210.

Fox, H. Munro

biographies - Fox
Anon 1967.
Harold Munro Fox, 1889-1967, Editor Biological Reviews 1926-1967.
Biol. Rev. 42(2):161-162.

biographies, Fox
Wells, G.P., 1967.
Professor H. Munro Fox.
Nature, Lond., 213(5080):974.

Franklin, B.

Biographies - Franklin
Laktionov, A. Ph., 1961.
Benjamin Franklin and some problems of oceanography.
Okeanologiia, Akad. Nauk, SSSR, 1(5):951-955.

Frengelli, Dr. J.

biographies - Frengelli
Argentina, Instituto Antartico Argentino, 1958.
Fallecimiento del Dr. Joaquin Frengelli.
Bol. Inst. Antartico Argentino, 1(4):25.

Fridriksson, Arni

biographies, Fridriksson
*Olafsson, David, 1967.
Dr. Arni Fridriksson, 22 December 1898-16 October 1966.
J. Cons. perm. int. Explor. Mer, 31(2):139-144.

Fuglister, F.C.

biographies, Fuglister
Stommel, Henry, 1969.
Frederick C. Fuglister.
Deep-Sea Res., Suppl. 16: 1-3.

Gakkel, Yakov Yakovlevich

biographies, Gakkel
Anonymous, 1966.
Yakov Yakovlevich Gakkel.
Polar Record, 13(83):217-218.

biographies, Gakkel'
Armstrong, Terrance, 1966.
Yakov Yakovlerich Gakkel' (1901-1915).
Arctic, 19(2):211-212.

biographies - Gakkel
Belov, M.I., and P.S. Voronov, 1966.
Ya. Ya. Gakkel, in memoriam. (In Russian).
Izv. Vses. Geogr. Obshch., 98(3):269-271.

biographies, Gakkel
Gordienko, P.A., 1966.
Yakov Yakovlevich Gakkel, 1901-1965. (In Russian).
Okeanologiia, Akad. Nauk, SSSR, 6(3):564-566.

Galileo

biographies - Galileo
Belloni, Luigi, 1965.
La vessie natatoire des poissons selon Galilée et son école et comme paradigme du mécanisme biologique.
Colloque Internat., Hist. Biol. Mar., Banyuls-sur-Mer, 2-6 sept 1963, Suppl. Vie et Milieu, No. 19:65-81.

Garstang

biographies - Garstang
Hardy, A.C., 1951.
Walter Garstang 1868-1949. J.M.B.A. 29(3):561-566, 1 photo.

biography - Garstang
Hardy, A. C., 1950.
Walter Garstang. Proc. Linn. Soc., London, 162(1):99-105.

biographies - Garstang
Hardy, A.C., 1950.
Walter Garstang, 1868-1949. J. du Cons. 17(1):7-12.

Gilson, G.

biographies, Gilson
*Capart, André, 1968.
Naissance de l' océanographie en Belgique. Un précusseur: le Professor Gustave Gilson (1859-1944).
Bull. Inst. Océanogr., Monaco, No. Spécial 2: 311-316.

biography - Gilson
Koch, H.J., 1943
Gustave Gilson 1859-1944. J. du Cons. 15(2):132-134.

biographies - Gilson
Van Straelen, V., 1948.
Gustave Gilson (1859-1944). Notice biographique avec liste bibliographique. Bull. Mus. Roy. Hist Nat., Belgique, 24(1):21 pp., 1 pl.

biography, Goreau
Anon, 1970.
Professor Thomas F. Goreau.
Nature, Lond. 227(5257): 534.

Gran, H.H.

biography - Gran
Braarud, T., 1956
Minnetale over H.H. Gran.
Det Norske Videnskaps-Akad., Oslo, Aarbok, 1955: 53-57.

Granqvist, Gunnar

biographies, Granqvist
Lisitzin, E., 1966.
Gunner Granqvist (1888-1965).
J. Cons.perm.int. Explor. Mer, 30(3):279-280.

Green, Arda A.

biographies - Green
Colowick, S.P., 1958.
Arda Alden Green, protein chemist. Science 128(3323):519-521.

Grøntved, Julius

biographies, Grøntved
Steemann Nielsen, G., 1968.
Julius Grøntved, 1899-1967.
Skrift. Danmarks Fiskeri Havandersøgelser, 28:5-6.

Gubin, F.A

biographies, Gubin
Ivanenkov, V. N., E. M. Sjuzjumcv, and K. D. Sabinin, 1964.
F. A. Gubin, 1926-1964. (In Russian).
Okeanologiia, Akad. Nauk., SSSR, 4(6):1126.

Gurney, R.

biographies - Gurney
Calman, W.T., 1950.
Dr. Robert Gurney. Nature 165(4198):587-588.

biography - Gurney
Hardy, A.C., 1950.
Robert Gurney. Proc. Linn. Soc., London, 162(1):118-121.

Haeckel

biographies - Ernst Haeckel
Uschmann, Georg, 1965.
Die Beiträge Ernst Haeckels und seiner schüler zur Entwicklung der marinen Zoologie. (Resume).
Colloque Internat., Hist. Biol. Mar., Banyuls-sur-Mer, 2-6 sept., 1963, Suppl. Vie et Milieu, No. 19:259-260.

Haenke

biographies - Haenke
Schadewaldt, H., 1965.
Thaddeus Haenke (1761-1817) médecin et naturaliste autruchien et ses observations pendant la circumnavigation espagnole de Malaspina (1789-1793).
Colloque Internat., Hist. Biol. Mar., Banyuls-xxxx sur-Mer, 2-6 sept. 1963, Suppl., Vie et Milieu, No. 19:99-121.

Hale, Herbert Matthew

biographies - Hale
Mitchell, F.J. 1965.
Obituary and bibliography of Herbert Mathew Hale.
Rec. S.Austral. Mus. 15(1): 1-8.

Hansen, H. J.

biography - Hansen
Stephensen, K., 1937.
H.J. Hansen. 10 August 1855-26 Juli 1936.
H. J. Hansen as a carcinologist. Vidensk Medd. fra Dansk naturh. Foren. 100:V-XII.

Harrison, R. G.

biographies - Harrison
Nicholas, J. S., 1960.
Ross Granville Harrison, Experimental Embryologist.
Science 131(3397):337-339.

Harvey, E.N.

biographies - Harvey
Johnson, Frank H., 1960.
Edmund Newton Harvey, 1887-1959.
Arch. Biochem. Biophys., 87(2):i-iii.

Heilbrun, L.V.

biographies - Heilbrun
George, J.C., 1959
The late professor L.V. Heilbrun. J. Animal. Morph. Physiol. 6(2): 123.

biographies - Heilbrun
Steinbach, H. B., 1960
L. V. Heilbrun, General physiologist.
Science, 131 (3398):397-399.

Heinrici

biography - Heinrichi
Bückmann, A., 1949.
Carl Heinrichi. Ber. Deutschen Wiss. Komm. f. Meeresf., n.f., 11(4):429-434.

biography
Hjort, J., 1948
Carl Heinrici 1874-1944. J. du Cons. 15(2):130-131.

Heldt, H.

biographies - Heldt
Desbrosses, P., 1957.
Henri Heldt, 1891-1956. J. du Cons., 23(1):3-5.

Helland-Hansen

biographies - Helland-Hansen
Devik, O., 1956.
Bjørn Helland-Hansen. Festskrift til Professor B. Helland-Hansen fra Venner og Kolleger ved Chr. Michelsens Institutt, Bergen, 1956:1-23.

Oversikt over Professor Dr. B. Helland-Janens avhandlinger og artikler, 237-241.

biographies - Helland Hansen
Dietrich, G., 1958.
Bjørn Helland-Hansen. Deutsche Hydrogr. Zeits., 11(1):36-37.

biographies - Helland-Hansen
Mosby, H., 1958.
Bjørn Helland-Hansen, 1877-1957. J. du Cons., 23(3):321-323.

biographies - Helland-Hansen
Tait, J.B., 1958.
Prof. Bjørn Helland-Hansen. Nature 181(4607):453-454.

biographies, Hensen
Porep, R. 1970.
Der Physiologe und Planktonforscher Victor Hensen (1835-1924): Sein Leben und sein Werk.
Kieler Beiträge zur Geschichte der Medezin und Pharmazie, 9 (R. Herrlinger, F. Kudlien und G.E. Dann, Neumünster: Wachholz, 147 pp. 19.80 D.M. (not seen)

Hentschel, E.

biography - Hentschel
Caspers, H., 1949.
Ernst Hentschel. Ber. Deutschen Wiss. Komm. f. Meeresf., n.f., 11(4):449-456.

biography - Hentschel
Caspers, H., 1949.
Ernst Hentschel, 1876-1945. Arch. Hydrobiol. 62:490-499.

biographies, Herdman
Deacon, G.E.R., 1968
Dr. H.F.P. Herdman, O.B.E.
Geogr. J. 134: 174-175
Also in: Coll. Repr., Nat. Inst. Oceanogr. Wormley, 16 (651).

biographies, Hess
Anon., 1969.
Professor Harry Hammond Hess.
Nature, Lond., 224 (5217): 393.

Hidaka, Koji

biographies - Hidaka
Stockmann, W.V., 1964
In connection with the jubilee of Dr. Koji Hidaka. (In Russian).
Okeanologiia, Akad. Nauk, SSSR, 4(1):189-190.

biographies - Hidaka
Yoshida, Kozo, 1964.
Appreciation (Professor Hidaka).
In: Studies on Oceanography dedicated to Professor Hidaka in commemoration of his sixtieth birthday, i-ii.

Hill, Maurice N.

biographies - Hill
Davies, D. 1966.
Dr. Maurice N. Hill, F.R.S.
Nature, Lond. 209 (5030):1287.

Hjort

biography, Hjort
Andersson, K. A., 1949.
Johan Hjort, 1869-1948. Hydrobiologia, Acta Hydrobiol., Limnol., et Protistol., 2(1):97-99.

biographies, Hjort
Andersson, K. A., 1949.
Johan Hjort 1869-1948. J. du Cons. 16(1):3-8.

biography
Maurice, H.G., 1948
Prof. Johan Hjort, For. Mem. R.S.
Nature 162(4124):765-766.

biographies, Hjort
Rollefsen, Gunner 1966
Norwegian fisheries research.
Fisk. Dir. Ske. Ser. Hovunder. 14 (1):1-36

biography, Hjort
Russell, E.S., 1948
Prof. Johan Hjort, For. Mem. R.S. Nature 162 (4124):764-765.

biography
Ruud, J. T., 1948.
Johan Hjort. In Memoriam. Nor. Whaling Gazette, Nov. 1948, 15 pp.

Hoffman, C.

biographies, Hoffman
Levring, Tore, 1960.
In memoriam Curt Hoffman.
Bot. Mar., 1(3/4):65.

Huntsman, A.G.

biographies, A.G. Huntsman
Hart, J.L., 1965.
A.G. Huntsman - an appreciation.
J. Fish. Res. Bd., Canada, 22(2):255-258.

Hustedt, F.

biography, Hustedt
Behre, K. 1970.
Friedrich Hustedts Leben und Werk.
Beihefte Nova Hedwigia 31: XI-XXII.

Biographies, Hustedt
Behre, Karl., 1967.
Dr. Friedrich Hustedt, 80 Jahre alt.
Abh. naturw. Verein., Bremen, 37(2): 97-108.

biographies-Hustedt
Simonsen, Reimer, 1968.
Friedrich Hustedt zum Gedächtnis.
Int. Revue ges. Hydrobiol. 53(4): 651-653.

biographies, Hustedt
Subrahmanyan, R., 1968.
Obituary - Dr. Friedrich Hustedt.
J. mar. biol. Ass., India, 10(2):370-371

biographies, Hutchinson
Edmondson, Yvette H., 1971.
Some components of the Hutchinson legend.
Bibliography of G. Evelyn Hutchinson. Doctoral dissertations completed under the direction of G. Evelyn Hutchinson. Limnol. Oceanogr. 16(2): 157-163; 164-169; 169-172.

biographies, Hutchinson
Kohn, Alan J., 1971.
Phylogeny and biogeography of Hutchinsonia: G.E. Hutchinson's influence through his doctoral students. Limnol. Oceanogr. 16(2): 173-179.

Isachenko,

biographies - Isachenko
Kriss, A.E., 1949.
[B.L. Isachenko-founder of marine microbiology.]
Priroda (11):86-

biographies - Isachenko
Savich, V.P., 1949.
[In memory of Akademician B.L. Isachenko, 1871-1948.] Priroda (11):82-

biographies, Iselin
Carruthers, J.N. 1971.
The late Dr. Columbus O'Donnell Iselin. (1904-1971).
Deep-Sea Res. Oceanogr. 23 (6): 494-496

Ivlev, V.S.

biographies, Ivlev
Ivlev, V.S., 1966.
The biological productivity of waters.
J. Fish. Res. Bd., Can., 23(11):1727-1759.

Jacobsen, J.P.

biographies - Jacobsen
Knudsen, M., 1948
Jacob Peter Jacobsen 1877-1946. J. du Cons. 15(2):154-156.

Jarke, Joachim

biography, Jarke
Schotte, W., 1966.
In memoriam Dr. Joachim Jarke.
Dt. hydrogr. Z., 19(5): 232-234.

biography, Jarke

Järnefelt, Heikki

biographies, Järnefelt
Purasjoki, K.J., 1965.
Heikki Järnefelt, in memoriam.
Hydrobiologia, 25(3/4):571-578.

biographies, Järnefelt
Ryhänen, Reino, 1965.
Heikki Arvid Järnefelt, 23.6.1891-22. 10. 1963.
Arch. Hydrobiol., 61(2):249-256.

Järvi, T. H.

biographies - Järvi
Halme, Erkki, 1961.
Toivo Henrik Järvi, 1877-1960.
J. du Conseil, 26(3):238-240.

Jeffries, Sir H.

biographies - Jeffries
Stoneley, R., 1961.
Sir Harold Jeffries, an appreciation.
In: The Earth Today, A.H. Cook and T.F. Gaskell, Editors, pp. ix-xi.
Also: Geophys. Jour., R. Astron. Soc., Vol. 4.

Jensen, Adolph S.

biographies - Jensen
Thorson, G., 1954.
Adolf S. Jensen, 23 Maj 1866-29 August 1953, mindetale ved Dansk Naturhistorisk Forening's møde den 9 Oktober 1953. Vidensk. Medd. Dansk Naturhist. Foren., København, 116: v-xi.

Jespersen, P.

biographies - Jespersen
Grøntved, J., 1952.
Dr. phil. Poul Jespersen. 18 Marts 1891-20 December 1951. Naturh. Tid. 16(5/6):88-90.

biographies - Jespersen
Russell, F.S., 1952.
Poul Jespersen, 1891-1951. J. du Cons. 18(2): 104-106.

biography - Jespersen
Taning, A.V., 1952.
Poul Jespersen, 18 Marts 1891-20 December 1951.
Vidensk. Medd. fra Dansk Naturh. Foren., 114: 37-48.

Jewett, F.B.

biography - Jewett
Buckley, O.E., 1952.
Frank Baldwin Jewett. Biogr. Me., Nat. Acad. Sci 27:239-264.

biographies - Jewett
Gill, Frank, 1949.
Dr. Frank B. Jewett. Nature 164(4181):1032.

Jordan David Starr

biographies, Jordan
Hubbs, Carl L., 1964.
David Starr Jordan.
Syst. Zool., 13(4):195-200.

Jurva, R.

biography - Jurva
Granqvist, G., 1954.
Risto Jurva (1888-1953). J. du Cons. 20(1):6-7.

Kemp, S.W.

biographies - Kemp
Hardy, A.C., 1947
Stanley Wells Kemp, 1882-1945. JMBA 26(3):219-234.

biographies - Kemp
Hardy, A.C., 1946.
Stanley Wells Kemp, 1882-1945. J.M.B.A. 26:219-234, 1 pl.

Kendall, Percy Fry

biographies Kendall
*Carruthers, James N., 1968.
Some marine-geological speculations by a British contemporary of Prince Albert I: Professor Percy Fry Kendall F.R.S. (1856-1936).
Bull. Inst. océanogr., Monaco, No. spécial 2: 175-187.

Khmyznikov Pavel Konstantinovich

biographies, Khmyznikov
Lappo, S.D., 1966.
Pavel Konstantinovich Khmyznikov. The 70th anniversary. (In Russian).
Okeanologiia, Akad. Nauk, SSSR, 6(3):561-563.

Knipovich, N.M.

biographies - N. M. Knipovich
Alexeev, A. I., 1964.
A prominent explorer of the seas of the USSR. (In Russian).
Okeanologiia, Akad. Nauk, SSSR, 4(6):1123-1125.

biographies - Knipovich
Laktionov, A.F., 1962
N. M. Knipovich, - his life and scientific activities (dedicated to the Centenary of his birth). (In Russian).
Okeanologiia, Akad. Nauk, SSSR, 2(4):758-766.

Knudsen, M.

biographies - Knudsen
Anon., 1950.
Professor Martin Knudsen (1871-1949). Int. Hydr. Rev. 27(2):9-10.

biography - Knudsen
Thomsen, H., 1950.
Martin Knudsen, 1871-1949. J. du Cons. 16(2):155-159.

biographies - Kofoid
Kirby, H., 1950.
Charles Atwood Kofoid. Hydrobiol., Acta Hydrobiol. Limnol., Protistol., 2(4):383-385.

Koczy, Friedric Frans

biographies - Koczy
Bader, Richard G., Nils Jerlov, Berta Karlik and Anitra Thorhaug 1967.
Friedrich Frans Koczy.
Nature, Lond. 215 (5100):563.

biographies - Koczy
Emiliani, Cesare, 1967.
Friedrich Trans Koczy.
Studies Trop. Oceanogr., Miami, 5:V.

biography, Koczy
Gorshine, Donn S., 1967.
In Memoriam: Friedrich Frens Koczy 1914-1967.
Marine Geol., 5(5/6):557.

Koefoed, E.

biography, Einar Koefoed
Rasmussen, Birger, 1963.
Einar Koefoed and his work as zoologist during the practical-scientific fishery investigations in northern waters from 1923-38.
Fiskeridirekt. Skrift., Ser. Havundersøgelser, 13(6):11-19.

biography, Einar Koefoed
Rollefsen, Gunnar, 1963.
Forward to Contributions given in honor of Einar Koefoed's 60 years service in Norwegian fisheries research, 1902-1962.
Fiskeridirekt. Skrift., Ser. Havundersøgelser, 13(6):7-9.

biographies, Kokubo
Motoda, S., 1971.
Dr. Seiji Kokubo (1889-1971). Bull. Plankt. Soc. Japan 18(1): 3-4.

Kolesnikov, Arkady Georgievich

biographies, Kolesnikov
Beliaev, V.I., 1968.
Arkady Georgievich Kolesnikov (On his 60th birthday). (In Russian).
Fisika Atmosfer. Okean., Izv. Akad. Nauk, SSSR, 4(1):111.

biographies, Kolosvary
Zullo, Victor A., William A. Newman and Arnold Ross 1972.
Kolosváry Gábor (Gabriel von Kolosváry) 1901-1968.
Crustaceana 22(1):96-102.

biographies - Kozlova
Anon. 1970.
Olga Georgievna Kozlova. (In Russian)
Okeanologiia 10(3):565.

biographies, Kriss
Krassilnikov, N.A., 1968.
Anatolii Evseevich Kriss (on his 60th anniversary).
Mikrobiologiia, 37(6):1131.

Krogh, A.

biographies - Krogh
Berg, K., 1950.
A Krogh in memoriam. Hydrobiol., Acta Hydrobiol., Limnol., Protistol., 2(4):380-382, 1 photo.

biography - Krogh
Drinker, C., 1950
August Krogh 1874-1949. Science 112 (2900): 105-107

biographies - August Krogh
Spärck, R., 1949.
August Krogh. Liste over August Kroghs publikationer. Vidensk. Medd. Dansk. Nat. Hist. Før., København, 111:V-XXX.

Laktionov, Alexandre Fedorovitch

biographies, Laktionov
Anonymous, 1966.
Aleksendr Fedorovich Laktionov.
Polar Record, 13(83):218.

biographies - Laktionov
Baskakov, A.G., 1965.
Alexandre Fedorovitch Laktionov (1899-1965).
Okeanologiia, Akad. Nauk, SSSR, 5(4):763-764.

biographies, Lea
Ruud, J.T., 1971.
Einar Lea (1887-1969). J. Cons. int. Explor. Mer 33(3): 303-307.

Le Danois, Edouard

biographies - Le Danois
Anon., 1968.
Dr. Edouard Le Danois
Nature, Lond. 219 (5157): 987.

biographies - Le Danois
Letaconnoux, M.R. 1970.
Edouard Le Danois (1887-1968).
J. Cons. int. Explor. Mer. 33 (2): 122-123.

Le Gall, J. J.

biographies - Le Gall
France, C.C.O.E.C., 1954.
Jean Joseph LeGall. Bull. d'Info., C.C.O.E.C., 6(3):87-88.

biography - Le Gall
Desbresses, P., 1954.
Jean le Gall (1894-1954). J. du Cons. 20(1):3-5.

biography - Le Gall
France, Institut Scientifique et Technique des Pêches Maritimes, 1955.
Jean Le Gall (1894-1954). Rev. Trav. Inst. Pêches Marit., 19(1):5-6.

Legendre, R.

biography - Legendre
LeGrand, Y., 1954.
René Legendre (1880-1954). Bull. Inst. Océan., Monaco, No. 1044:29 pp.

Leim, A. H.

biography - Leim
Anon., 1962
Alexander Henry Leim, 1897-1960.
J. Fish. Res. Bd., Canada, 19(3):512-514.

biographies - Leim
Canada, Fisheries Research Board of Canana, 1962
Alexander Henry Leim, 1897-1960.
J. Fish. Res. Bd., Canada, 19(3):511-514.

Lidz, Louis

biographies - Lidz
Emiliani, C. 1969.
In memoriam: Mr. Louis Lidz 1939-1967.
Sedimentology 9 (3):264.

Liebert, F.

biography - Liebert
Tesch, J. J., 1950.
Francois Liebert 1882-1948. J. du Cons. 16(2): 164-165.

Lillie, F. R.

memoires - Lillie
Moore, C.R., 1948
Frank Rattray Lillie (1870-1947). Anat. Rec. 101(1):1-4.

Limbaugh, C.

biographies - Limbaugh
Simpson, Donald A., 1962
One in a million. Reminiscing about a promising young scientist who accomplished much in his brief career. [Conrad Limbaugh.]
Aquarium Journal, May 1962, 4 pp.

biographies, Lenin
Bogorov, B.G., 1970.
V.I. Lenin and oceanology. (In Russian). Oceanologiia, 10(5): 913.

Linneus, C.

biographies - Linneus
Hindle, E., 1957.
Carl Linneus as a traveller. Geogr. J., 123(4): 510-512.

Lomonosov, M.V.

biographies - Lomonosov
Anonymous, 1961.
[250 years since the birth of M.V. Lomonosov.]
Vestnik, Akad. Nauk, SSSR, 31(12):114-116.

biographies - Lomonosov
Baranov, N.A., 1961.
[Progressive ideas of M.V. Lomonosov in biology.]
Izv. Akad. Nauk SSSR, Ser. Biol., 1961(6):
English Edit., (6):931-942.

biographies - Lomonosov
Gelman, N.S., 1961.
[M.V. Lomonosov and modern biological chemistry.]
Izv. Akad. Nauk, SSSR, Ser. Biol., (6):
English Edit. (6):943-947.

biography-Lomonosov
Kvasov, D.G., 1961.
M.V. Lomonosov and physiology as a science. (In Russian).
Fiziolog Zhurn, SSSR imeni I.M. Sechenova, 47 (12):
English Edit. (undated)
Sechenov Physiol. J. USSR, 47(12):1-5.

biographies-Lomonosov
Laktionov, A.F., 1961.
M.V. Lomonosov and geography (250-anniversary of his birthday).
Izv. Vses. Geograf. Obshch., 93(6):461-478.

biographies-Lomonosov
Perevalov, V.A., 1961.
M.V. Lomonosov's works in oceanography.
Okeanologiia, Akad. Nauk, SSSR, 1(6):1120-1125.

Louis Ferdinand

biographies, Louis Ferdinand
*Pérès, Jean-Marie, 1968.
Un précurseur de l'etude du benthos de la Mediterranée: Louis Ferdinand, comte de Marsilli.
Bull. Inst. océanogr., Monaco, No. special 2: 369-376.

Lubbock, Sir John

biographies-Lubbock
Pumphrey, R.J., 1959.
The forgotten man: Sir John Lubbock.
Science, 129(3356):1087-1092.

biographies, Lund
Gunter, Gordon, and Gordon Marsh, 1970.
Elmer Julius Lund.
Contrib. mar. Sci., Port Aransas, 15: 3pp. (unnumbered).

Magid, Ahmad Ibn

biographies Magid
*Aleem, Anwar Abdel, 1968.
Ahmad Ibn Magid, Arab navigatior of the XV Century and his contribution to marine sciences.
Bull. Inst. océanogr., Monaco, No. special 2: 565-580.

Makarov, S.O.

biographies Makarov
*Soloviv, A. I., 1968.
S.O. Makarov and the significance of his research in oceanography.
Bull. Inst. oceanogr., Monaco, No. special 2: 615-625.

Mark, E.L.

biography-Mark
Kornhauser, S.J., 1948
Edward Laurens Mark (May 30, 1847 - December 16, 1946) Anat. Rec. 102(3):275-277.

Marr, James William Slesser

biographies, J.W.S. Marr
Deacon, G.E.R., 1965.
Dr. J.W.S. Marr.
Nature, 207(4995):351-352.

biographies-Marr
Roberts, B.B., 1966.
Dr. James William Slesser Marr.
Polar Record, 12(82):94-97.

Marsh

biographies, Marsh
Romer, Alfred S., 1964.
Cope versus Marsh.
Syst. Zool., 13(4):201-207.

Marsigli, L.F.

biographies-Marsigli
Olson, F.C.W., and M.C. Olson, 1958.
Luigi Ferdinando Marsigli, the last father of oceanography.
J. Florida Acad. Sci., 21(3):227-234.

biographies, Maslov
Marty, Yu.Yu., 1968
Nikolai Antonovich Maslov and his role in the study and development of fidh resources of northern seas. (in Russian). Trudy polyar. nauchno-issled. Inst. morsk. ryb. Khoz. Okeanogr (PINRO) 23: 5-20.

Matthews, D.J.

biographies-Matthews
Carruthers, J.N., 1956.
Donald John Matthews 1873-1956. J. du Cons. 22(1):1-8.

biography-Matthews
Carruthers, J.N., 1956.
Mr. D.J. Matthews. Nature 177(4505):410.

Maurice, H.G.

biographies-Maurice
Anon. (GCLB), 1951.
Henry Gascoyen Maurice. Polar Record 6(41):126-127.

biographies
Dobsen, A.T.A., 1950.
Henry Gasgoyne Maurice 1874-1950. J. du Cons. 17(1):3-6.

Maury

biographies-Maury
Anon. 1873.
M.F. Maury
Hydrogr. Mitt. 1:49.

biographies-Maury
Anon. 1873.
Captain M.F. Maury.
Nature 9:390

biography, Maury
Brown, Ralph Minthorne, 1944.
Bibliography of Commander Mathew Fontaine Maury including a biographical sketch.
Bull. Va polytech. Inst., 37(12):46 pp.

biographies-Maury
Canfield, N.L. 1953.
M.F. Maury and the World Meteorological Organization 1853-1953. H.O. Pilot Charts, Sept. 1953

biographies-Maury
Caskie, Jacqueline Amber 1928.
Life and letters of Maury.
Richmond.

biographies-Maury
Cromie, William J., 1964.
The first American oceanographer.
Proc. U.S. Naval Inst., 90(4):56-69.
Matthew Fontaine Maury

biographies, Maury
Deacon, G.E.R., 1964
Matthew Fontaine Maury, U.S.N.
J. R. Nav. Sci. Serv., 19:140-145
Also in:
Collected Reprints, Nat. Inst. Oceanogr., Wormley, 12. 1964.
A review of:
Williams, Frances Leigh, 1963. Matthew Fontaine Maury, Scientist of the Sea, Rutgers, Univ. Press xx+ 720 pp. $10.00.

biographies-Maury
Williams, Frances Leigh, 1963
Matthew Fontaine Maury, Scientist of the Sea. Rutgers Univ. Press, New Brunswick, N.J., 720 pp.
Review: Deacon, G.E.R. and Margaret B. Deacon, 1964, Science and the sea. Geographical Journal. 130(1):120-123.

biographies, Maury
*Leighly, John, 1968.
M.F. Maury in his time.
Bull. Inst. océanogr., Monaco, No. special 2: 147-161.

biographies-Maury
Lewis, C.L., 1927.
Matthew Fontaine Maury, the pathfinder of the seas. U.S. Naval Inst., 264 pp.

biographies-Maury
Lyman, J. 1948.
The centennial of pressure pattern navigation.
Proc. U.S. Naval Inst. 74 (509):

biographies-Maury
Maury, M.F., 1858
The physical geography of the sea.
[illegible] ed. Harper & Bros, 360 pp.

biographies-Maury
Maury, M.F., 1844
Remarks on the Gulf Stream and currents of the sea. Amer. J. Sc. & Art 47:161-181

biographies-Maury
Maury-Corbin, Diana Fontaine 1888.
A life of M.F. Maury.
London

biographies-Maury
Neumayer, G. 1872.
Die Erforschung des Südpolargebietes.
Z. ges. Erdkunde 7:120 (146)

biography-Maury
Römer, E., 1953.
M.F. Maury: zur hundertsten Wiederkehr des Tages der internationalen maritimen Konferenz zu Brüssel am 23. August 1853. Der Seewart 14(5): 1-4.

biography-Maury
Schumacher, A., 1953.
Matthew Fontaine Maury und die Brüsseler Konferenz 1853. Deutsch. Hydrogr. Zeits. 6(2):87-93.

biographies-Maury
Römer, E. 1941.
75 Jahre atlantisches Seekabel.
Der Seewart 10:119

biographies-Maury
Schott, G. 1895.
Die Verkehrswege der transozeanischen Segelschiffahrt in der Gegenwart.
Z. ges. Erdkunde 30:17

biographies-Maury
Wayland, John W. 1930.
The pathfinder of the seas.
Richmond.

Mayor Alfred Goldsborough

biographies, Mayor
Betz, Joseph J., 1965.
Pioneer biologist (Alfred Goldsborough Mayor).
Sea Frontiers, 11(5):286-295.

Merz, Alfred

biographies- Merz
Wüst, G., 1965.
Zum Gedenken an des Ausfahrt des Forschungs- und Vermessungsschiffes Meteor und an den Todestag von Alfred Merz vor vierzig Jahren.
Deutsche Hydrogr. Zeits 18(4): 178-179.

Messina Angelina Rose

biographies, Messina
Ellis, Brooks F., 1969.
Angelina Rose Messina (1910-1968)
Micropaleontology 15(2(: 133-134.

Miklukho-Maclay

biographies- Miklukho-Maclay
Mamaev, O.I., 1957.
[N.N. Miklukho-Maclay - an oceanographer.]
Izv. Vses. Geograf. Obsh. 89(3):255-258.

Milne Edwards, Henri

biographies, H-Milne Edwards
*Théodoridès, Jean, 1966.
Les débuts de la biologie marine en France: Jean-Victor Audouin et Henri-Milne Edwards, 1826-1829.
Bull. Inst. océanogr., Monaco, No. spécial 2: 417-437.

Mineev Vladimir Arefonovitch

biographies -Mineev
Anon., 1965.
Vladimir Arefonovitch Mineev (1935-1965).
Okeanologiia, Akad. Nauk, SSSR, 5(4):765.

Morris L.

biographies - Morris
Robinson, A.H.W., 1958.
Lewis Morris, an early Welsh hydrographer.
J. Inst. Navigation, 9(4):344-355.

Mortensen

biography- Mortensen
Broch, H., 1953.
Ole Theodor Mortensen. K. Norske Videnskabers Sleskabs, Trondheim, Medd. 25:60-64.

biographies - Mortensen
Fage, L., 1953.
Notices nécrologiques sur Theodor Mortensen et William Thomas Calman. Bull. Soc. Zool., France, 77(5/6):319-323.

biographies, Henry Nottidge Moseley
Schlee, Susan 1971.
Her Majesty's fauna on the bounding main.
Nat. Hist. Am. Mus. Nat. Hist. 80(8):76-80. (popular)

biographies, Myers
Walford, Lionel A., 1970.
On the natural history of George Sprague Myers. Proc. Calif. Acad. Sci. (4)38:1-18.

Nansen, F.

biography - Nansen
Brinkmann, August, Jr., 1961
Fridtjof Nansen som zoolog.
Naturen, (7/8):387-421.

biographies- Nansen
Dobrovolsky, A.D., 1961.
[F. Nansen (to the centenary of the explorer's birth).]
Okeanologiia, Akad. Nauk, SSSR, 1(5):956-957.

biographies - Nansen
Isachsen, F., 1961.
Fridtjof Nansen og polarforskningen.
Norsk Geogr. Tids., 18(1/2):33-38.

biography- Nansen
Knuth, E., 1948
Fridtjof Nansen og Knud Rasmussen en Slaegts-studie. Gyldendal, Copenhagen.

biographies - Nansen
Kosiba, Aleksander, 1962
Fridtjof Nansen (1861-1930). (In Polish; English summary).
Przeglad Geofiz., Warszawa, VII(XV) (2):75-83

biographies-Nansen
Lasukhin, A.A., 1962.
[The jubilee of Fridtjof Nansen.]
Izv., Akad. Nauk, SSSR, Ser. Geograf., (1):161-162.

biographies, Nansen
Rouch, J., 1958.
An unpublished letter from Nansen.
Int. Hydrogr. Rev., 35(2):137-145.

- biographies - Nansen
Volkov, N.A., 1961.
[Fridtjof Nansen - the great polar explorer and humanist (100-anniversary of his birthday).]
Izv. Vses. Geograf. Obshch., 93(6):479-486.

Nares, J.D.

biographies -Nares
Anon., 1957.
Vice-Admiral John D. Nares, D.S.O., R.N.
Int. Hydrogr. Rev., 34(1):9-10.

biographies -Nares
Edgell, J.A., 1957.
Vice-Admiral J.D. Nares (1877-1957). Deep-Sea Re 4(2):147.

Navarro Martin, F.

biographies - Navarro
Lozano Cabo, Fernando, 1961.
Francisco de Paula Navarro Martin. Nota biografica.
Bol. Inst. Espanol Oceanogr., 103:v-xvii.

biographies- Navarro
Lozano Cabo, Fernando, 1961.
D. Francisco de Paula Navarro Martin.
Bol. R. Soc. Esp. Hist. Nat. (B), 59:231-242.

Nellemose, W.

biography - Nellemose
Maurice, H.G., 1948
Wilhelm Nellemose 1890-1944. J. du Cons. 15(2):151-153.

Nelson, T.C.

biographies- Nelson
Carriker, Melbourne R., 1961
Thurlow Christian Nelson, marine biologist.
Limnol. & Oceanogr. 6(1): 79.

biography- Nelson
Carriker, M. R., 1960
Thurlow Christian Nelson, marine biologist.
Science, 132(3443): 1875-1876.

Newell G.E.

biographies, Newell
Anon., 1968.
Professor G.E. Newell.
Nature, Lond., 219 (5168): 1291

biographies, Nordenskiold
Kish, George, 1968.
Adolf Erik Nordenskiold (1832-1901) Polar explorer and historian of cartographer.
Geogr. J., Lond., 134(4):487-500.

Orr, A.P.

biographies - Orr
Yonge, C.M., 1962.
Dr A.P. Orr: an obituary.
Nature, 196(4856):719.

Orton, J.H.

biographies- Orton
Bruce, J.R., 1953.
Prof. J.H. Orton, F.R.S. Nature 171(4354):634.

biographies- Orton
Colman, J.S., 1953.
James Herbert Orton.
Mar. Biol. Sta., Port Erin, Ann. Rept. for 1952, (No. 65):5.

biographies - Orton
Turner, Ruth D., 1962
James H. Orton: his contributions to the field of fossil and Recent mollusks.
Revista Mus. Argentino Ciencias Nat. "Bernardino Rivadavia", Ciencias Zool., 8(7):89-99.

biographies, Panin
Anon., 1970.
K.I. Panin. (In Russian). Izv. Tichookean, nauchno-issled. Inst. Ribn. choz. Okeanogr. 70: 3-4.

Pantin C.F.A.

Biographies, Pantin
Anon., 1967.
Professor C.F.A. Pantin.
Nature, Lond., 213 (5075):447-448.

biographies- Pantin
Russell F.S. 1967.
Carl Frederick Abel Pantin 1899-1967: obituary.
J. mar. biol. Ass. U.K. 47 (2): 255-258.

Papanin, I.D.

biographies - Papanin
Anon., 1964.
Ivan Dmitrievich Papanin (in his seventieth year).
Okeanologiia, Akad. Nauk, SSSR, 4(6):1121-1122.

Paulsen, O.V.

biography - Paulsen
Jessen, K., 1947.
Ove Vilhelm Paulsen, 22 Marts 1874-29 April 1947.
Bot. Tidsskr. 42(2):136-140.

biographies - Paulsen
Steeman Nielsen, Einar, 1949.
Ove Paulsen 1874-1947. J. du Cons. 16(1):14-15.

Pavillard, J.

biographies - Pavillard
Tregouboff, G., 1963.
A la memoire de Jules Pavillard (1868-1961).
Rapp. Proc. Verb., Reunions, Comm. Int. Expl. Sci., Mer Mediterranee, Monaco, 17(2):449-454.

biography - Pavillard
Tregouboff, G., 1962.
A la memoire de Jules Pavillard (1868-1961).
Bull. Inst. Océanogr., Monaco, 59(1247):16 pp.

Pekina, Z.M.

biography - Pekina
Volkov, E.T., and E. Yu. Mikhasenok, 1962
Z.M. Pekina, the compiler of the most complete bibliographical material covering the history and development of Russia's marine book. (In Russian).
Okeanologiia, Akad. Nauk, SSSR, 2(5):954-958.

Pelseneer, P.

biographies - Pelseneer
Adam, W., and E. Leloup, 1947
Paul Pelseneer (1863-1945). Notice biographique. Bull. Mus. roy. d'Hist. Nat. de Belgique 23(1):45 pp., 2 pls.

biographies - Paul Pelseneer
Brien, Paul, 1965.
La biologie marine dans l'oeuvre zoologique de Paul Pelseneer (1863-1945).
Colloque Internat., Hist. Biol. Mar., Banyuls-sur-Mer, 2-6 sept. 1963, Suppl., Vie et Milieu, No. 19:275-289.

biographies - Pelseneer
Caullery, M., 1945.
Paul Pelseneer. C.R. Acad. Sci., Paris, 220:770.

Poey,

biographies, Petelin
Anon, 1971.
Veniaman Petrovich Petelin. (In Russian)
Okeanologiia 11(1):182

Peter the Great

biographies Peter the Great
Buzo, S. A., 1948.
Peter the Great, the first Russian oceanographer. Nauk. Zhizn., No. 5:42.

Petit, G.

biographies - Petit
France, Laboratoire Arago, Banyuls-sur-Mer, 1964.
Volume jubilaire dédié à George Petit.
Vie et Milieu, Lab. Arago, Suppl., No. 17:1-liv.

Pettersson, Hans

biographies, Pettersson (Hans)
Deacon, G.E.R., 1966.
Hans Pettersson, 1888-1966, Elected For. Mem. R.S. 1966.
Biogr. Mem. Fellows R. Soc., 12:405-421.

biographies, Pettersson
Ohlon, Sven Em. 1967.
Hans Pettersson 1888-1966 minnestal vid Kungl. Samhällets årshögtid 1967.
Årsb. VetenskVitterhSamh. Göteborg. 1967:41-50.

Pettersson, O.

biography - Pettersson
Thompson, D'Arcy, 1948
Otto Pettersson 1848-1941. J. du Cons. 15(2):121-125.

biographies - Pike
Rice, Dale W. 1970
Gordon C. Pike 1922-1968.
J. Mammal. 51(2): 434-435.

biographies - Poey
Jaume, M.L., 1955.
Poey, padre espiritual de los naturalistas cubanos
Mem. Soc. Cubano, Hist. Nat., 22(2):93-96.

Pratje, O.

biographies - Pratje
Anon., 1953.
Prof. Otto Pratje. Nature 171(4345):241.

biographies - Pratje
Böhnecke, G., 1952.
In Memoriam. Otto Pratje. Deutsche Hydrogr. Zeits. 5(5/6):286.

Preobrazhensky, V.

biography - Preobrazhensky
Goina and Logoina, K.S., 1961.
V.V. Preobrazhensky.
Meteorol. i Gidrol., (7):56-57.

Ramalho, A. de M.

biographies - Ramalho
Candeias, Alberto, 1961.
Alfredo de M. Ramalho, 1894-1959.
J. du Conseil, 26(3):235-237.

Ranson, H. G.

biography - Ranson
Ranson, M. Gilbert, 1943
Titres et travaux scientifiques. Paris: Masson et Cie, Editeurs, Libraries de l'Academie de Medecine, 120 Boulevard Saint-Germain. 88 pp.

Rasmussen, K.

biographies - Rasmussen
Knuth, E. 1948.
Fridtjof Nansen og Knud Rasmussen en Slægtsstudie.
Gyldendal, Copenhagen

Raymond, P.E.

biographies - Raymond
Stetson, H.C., 1953.
Memorial to Percy Edward Raymond (1879-1952).
Proc. G.S.A., Ann. Rept., for 1952:121-126, Pl. 11.

Redeke, H.C.

biography - Redeke
Tesch, J.J., 1948
H.C. Redeke. J. du Cons. 15(2):141-143.

Rees, C.B.

biographies - Rees
Lucas, C.E., 1957.
Obituary: C.B. Rees (1914-1956). J. du Cons., 22(3):267-269.

Rees, W.J.

biographies, Rees
Harding, J.P., 1967.
Dr. W.J. Rees.
Nature, Lond., 216(5116):728-729.

Revelle, Roger

biographies - Revelle
Anonymous, 1963.
Roger Revelle.
Scientific American, 209(3):24-25.

Richard, J.

biographies - Richard
Caullery, M., 1945.
Jules Richard. C.R. Acad. Sci., Paris, 220:768.

Ritter, Carl

biographies, Ritter
Troll, Carl, 1959.
The work of Alexander von Humboldt and Carl Ritter: a centernary address.
Advancement of Science, 16(64):441-452.

Rogalla, Ernst-Hellmuth

biographies - Rogalla
Tomczak, G. 1965.
In memoriam Dr. Ernst-Hellmuth Rogalla.
Dt. hydrogr. Z. 18(4): 179-181.

biographies, Rollefsen
Sunnanå, Klaus 1969
[Gunnar Rollefsen]
FiskDir. Skr. Ser. HavUnders. 15(3):103-111

biographies, Rollafsen
Went, Arthur E.J. 1969.
An appreciation. [Rollefsen]
FiskDir. Skr. Ser. HavUnders. 15(3):112-113.

Romer, A. S.

biographies, Alfred Sherwood Romer
Patterson, Bryan, 1965.
Alfred Sherwood Romer, Presient Elect.
Science, 147(3660):891-891.

Rossby, C.-G.

biographies - Rossby
Anon., 1953.
Professor Carl-Gustaf Rossby, Symons Medallist, 1953. Weather 8(6):168, Pl. 21.

biographies - Rossby
Bergeron, Tor, 1958.
Carl-Gustaf Rossby Minnesteckning.
K. Vetenskaps-Societetens Årsbok, 1958:17-23.

biographies - Rossby
Bolin, B., 1957.
Carl-Gustaf Rossby in memoriam. Tellus, 9(3): 257-258.

biographies - Rossby
Byers, H.R., 1958.
Carl-Gustaf Arvid Rossby (1898-1957).
Deep-Sea Res., 5(1):83-84.

biographies - Rossby
Flohn, H., 1958.
C.-G. Rossby. Meteorol. Rundsch., 11(1):1-2.

biographies - Rossby
Mason, B.J., 1957.
Professor Carl-Gustaf Rossby. Weather 12(11):351.

biographies - Rossby
Willett, H.C., 1958.
C.-G. Rossby, leader of modern meteorology.
Science 127(3300):686-687.

Rouch, le Commandant

biographies - Rouch
Cousteau, J.-Y, 1957.
Le Commandant Rouch. Les Amis du Musée Océanographique, Monaco, No. 41:1-20.

Runnström

biography - Runnström
Svärdson, G., 1966.
In memoriam Sven Runnström.
Hydrobiologia, 28(3/4):589-592.

bibliography - Runnström
Svärdson, G., 1966.
In memoriam Sven Runnström.
Hydrobiologia, 28(3/4):589-592.

Russell, E.S.

biographies - Russell
Graham, M., 1954.
E.S. Russell (1887-1954). J. du Cons. 20(2):135-139.

biographies, Ruud
Braarud, T., 1971.
Johan T. Ruud (1903-1970). J. Cons. int. Explor Mer 33(3): 308-311.

Ryabchikov, P.I.

biographies Ryabchikov
Anon., 1966.
Pyotr Ivanovich Ryabchikov. (In Russian).
Okeanologiia, Akad. Nauk, SSSR, 6(5):923.

Sabine, Edward

biographies, Sabine
Georgi, J., 1958
Edward Sabine, ein grosser Geophysiker des 19 Jahrhunderts. Deut. Hydro. Zeit., 11(6): 225-239.

Sachkova, Alexandra Ivanova

biographies - Sachkova
Anon., 1965.
Alexandra Ivanova Sachkova.
Okeanologiia, Akad. Nauk, SSSR, 5(6):1126.

biographies - Sachkova

Sars, George Ossian

biographies - Sars, G.O.
Sars, Michael
Rollefsen, Gunnar 1966.
Norwegian fisheries research.
Fisk. Dir. Skr. Ser. Havunder. 14(1):1-36.

Sars, Michael

biographies, Michael Sars
*Sivertsen, Erling, 1968.
Michael Sars, a pioneer in marine biology, with some aspects from the early history of biological oceanography in Norway.
Bull. Inst. Océanogr., Monaco, No. Special 2: 439-452.

Savage, Robt. Ed.

biographies - Savage
Wimpenny, R.S., 1958.
Robert Edward Savage, 1887-1957. J. du Cons., 23(3):324-326.

biographies - Savilov
Zenkevich, L.A., Z.A. Filatova, N.G. Vinogradova, L.I. Moskalev, et al. 1970.
Anatoly Ivanovich Savilov. (In Russian)
Okeanologiia 10(3): 563-564.

biographies, Schaefer
Roedel, Philip M., 1971.
In memoriam - Wilbert McLeod Chapman and Milner Baily Schaefer.
Fish. Bull. U.S. Dept. Comm. 69(1):1-2

Schmidt, J.

biographies - Schmidt
Bruun, A. Fr., 1934.
The life and work of Professor Johannes Schmidt.
Riv. di Biol. 16(1):3-22, 3 textfigs.

ms

biography - Schmidt
Regan, C.T., 1933.
Johannes Schmidt (1877-1933). J. du Cons. 8(2):145-160.

biography - Schmidt
Thompson, D'A. W., 1934.
Obituary notice, Johannes Schmidt. Proc. Roy. Soc., Edinburgh, 53(4):370-373.

Schott, Gerhard

biographies - Schott
Dunbar, Max J., 1961
Obituary. Gerhard Schott.
Geographical Review, 51(4):590.

Schulz, B.

biography - Schulz
Castens, G., 1944.
Bruno Schulz. Petermanns Geogr. Mitt., 90 Jahrg. 1944 (7/8):203-204.

biography - Schulz
Kalle, K., 1949.
Bruno Schulz. Ber. Deutschen Wiss. Komm. f. Meeresf., n.f., 11(4):443-449.

Schumacher, Arnold

biographies - Schumacher
Böhnecke, G., 1959.
Gedenktage-Arnold Schumacher. Deutsche Hydrogr. Zeits., 12(1): 33.

Schweigger, Erwin

biographies-Schweigger
Anon., 1965.
Fallecimiento del Dr. Erwin Schweigger.
Bol. Corp. Nac. Fertilizantes, Peru, 3(8):2-4.

Seco Serrano, E.

biography, Seco
Gómez Gallego, Julian, 1967.
Edmundo Seco Serrano (27-2-1921-14-4-1965)
Bol. Inst. español Oceanogr., 132:5-7.

Sewell, R.B.S.

biographies - Sewell
Roonwal, M.L., 1962 (publ. 1967).
The late Lieut-Col. R.B.S. Sewell (1880-1964): an appreciation, with a complete list of his scientific writings.
Rec. Indian Mus. 60(3/4):327-336.

Sexton, Mrs. E.W.

biographies - Sexton
Russell, F.S., 1959.
Mrs. E.W. Sexton.
Nature, 183:790.

Sheard, Keith

biographies - Sheard
Sadasavan, Tampi P.R. 1965.
Dr. Keith Sheard.
J. mar. biol. Ass. India 7(2): 453.

Shoemaker, Clarence R.

biographies - Shoemaker
Bowman, Thomas E. and Lanell W Peterson 1965
Bibliography and list of new genera and species of amphipod crustaceans described by Clarence R. Shoemaker.
Crustaceana 9(3): 309-316.

Shcherbakov, D.I.

biographies, Shcherbakov
Anon., 1966.
Dmitrii Ivanovich Shcherbakov. (In Russian).
Geofiz. Biull., Mezhd. Geofiz. Kom., Prezid. Akad. Nauk. SSSR, No. 17:162-163.

Shuleikin, V.V.

biographies, V.V. Shuleikin
Anon., 1965.
V.V. Shuleikin (On the 70th anniversary of his birth). (In Russian).
Meteorol. i Gidrol., (3):61.

biographies, Shuleikin, V.V.
Dobroklonskii, S.V., and G.E. Kononkova, 1965.
Akademician V.V. Shuleikin's 70th Birthday.
Fisika Atmosferi i Okeana, Izv. Akad. Nauk SSSR, 1(2):236-238.

Siedlecki, M.

biographies - Siedlecki
Russell E.S. 1948.
Michel Siedlecki 1873-1940
J. du Cons 15(2): 144-146

Sysoev, N.N.

biographies - Sysoyev
Bogorov, V.G., 1964.
Nikolay Nikolayevich Sysoyev.
Okeanologiia, 4(4):740-741.

Abstract: Soviet Bloc. Res. Geophys. Astron. Space, No. 94:41.

Skopintsev, B.A.

biographies - Skopintsev
USSR, Akademia Nauk, 1962
Boris Alexandrovich Skopintsev (in connection with the investigator's sixtieth birthday and the thirty-fifth year of his scientific activities). (In Russian).
Okeanologiia, Akad. Nauk, SSSR, 2(6):1120-1126

biographies - Skopintsev
SSSR, Akademii Nauk, 1962.
Boris Aleksandrovich Skopintsev.
Okeanologiia, Akad. Nauk, SSSR, 2(6):1120-1126.

Abstr, in:
Soviet Bloc Res. Geophys., Astron. & Space, 52: 21.

Smetanin, D.A.

biographies. Smetanin
Tovaricheii, Gruppa, 1964
Dmitri Alekseevich Smetanin 1928-1962 (In Russian)
Chemistry of the waters and sediments of the seas and oceans.
Trudy Inst. Okeanol., Akad. Nauk. SSSR, 67: 3-6.

Smith, R Adm. E. H.

Anon., 1962. biographies, Smith
R. Adm. Edward H. Smith, U.S.C.G. Obituary.
Polar Record, 11(72):345.

Smith, J.L.B.

Greenwood, P.H., 1968. biographies, Smith
Professor J.L.B. Smith.
Nature Lond., 217(5129):690-691.

Sømme, S.

biographies- Sømme
Rosseland, Leiv, 1962.
Sven Sømme, (1904-1961).
J. du Cons., 27(2):119-120.

Soule, Floyd M.

biographies, Soule
Anon, 1969.
Capt. Floyd M. Soule, USUGR (Ret.) (1901-1968.
Oceanogr. Rept., U.S.C.G.19 (CG373-19)ii-iii.

biographies - Soule
Barnes, Clifford A. 1969.
Obituary - Floyd Melville Soule.
Deep-Sea Res. 16(5): 399-403.

Spärck, Ragnar

Korringa, P., 1967 biographies, Spärck
Regnar Sparck, 1896-1965.
J. Cons. Perm. int. Explor. Mer, 31(1):3-4.

Wolff, Torben, 1965. biographies, Spärck
Prof. R. Spärck.
Nature, 208(5011):625.

biographies - Spärck
Zenkevich, L.A., 1966.
In memory of Prof. Ragnar Spärck. (in Russian).
Okeanologiia, Akad. Nauk. SSSR, 6(1):189.

Stejneger, Leonhard

biographies (Stejneger)
Schmitt, Waldo L., 1964.
Leonhard Stejneger.
Syst. Zool., 13(4):243-249.

Stephensen, K.H.

Bruun, A. Fr., 1947. biographies - Stephensen
K.H. Stephensen, Nature, 160:82.

biography- Stephensen
Bruun, A. Fr., 1947.
Knud Hensch Stephensen. 29 September 1882-13 Marts 1947. Vidensk. Medd. 110:V-VIII.

Steven, G.A.

biographies - Steven
Wilson, D.P., 1959.
Dr. G.A. Steven, F.R.S.E. Obituary.
J. Mar. Biol. Assoc., U.K., 38(1):1-2.

Stetson, H.C.

Hough, H.L., 1956. biographies - Stetson
Henry Crosby Stetson. In Memoriam. Deep-Sea Res. 3(4):291-293.

biographies- Stetson
Trask, P.D., 1957.
Memorial to Henry Crosby Stetson (1900-1955).
Proc. Vol. Geol. Soc., Amer., Ann. Rept., 1956: 171-174.

biographies - Stetson
Trask, P.D., 1956.
Henry Crosby Stetson (1900-1955).
Bull. Amer. Assoc. Petr. Geol., 40(5):1050-1051.

Stockmann, Vladimir Borisovich

biographies- Stockmann
Anon., 1968.
Vladimir Borisovich Stockmann. (deceased).(In Russian).
Fisika Atmosfer. Okean., Izv. Akad. Nauk, SSSR, 4(9):1012-1013.

biographies, Stockman
Anon, 1968.
Professor V.B. Stockman (obituary).
Nature, Lond., 220(5162):102.

biographies, Stockman
Anon., 1968.
Vladimir Borisovich Stockman.
Okeanologiia, Akad. Nauk, SSSR, 8(4):771.

biographies, Stockmann
Dobrowolsky, A.D., 1969.
Vladimir Borisovich Shtockman, 1909-1969. (In Russian).
Okeanologiia, 9(1): 5-14.

~~biography, Stockman~~
Hidaka, K. 1968.
Obituary: Professor W.B. Stockman.
(In Japanese).
J. Oceanogr. Soc. Japan, 25(1): 53-54

Stocks, T.

biographies - Stocks
Model, F., 1964.
In memoriam: Theodor Stocks.
Deutsche Hydrogr. Zeits., 17(1):41-45.

biographies, Strickland
Edmondson, Yvette H., 1970.
J.D.H. Strickland - 1920-1970. Limnol. Oceanogr. 15(6): 971.

biographies, Strickland
Parsons, T.R. 1971.
John Douglas Hipwell Strickland, 1920-1970.
J. Fish. Res. Bd. Can. 28(4): 599.

biographies, Strickland
Provasoli, Luigi, 1971.
Tribute to John D.H. Strickland. Mar. Biol. 9(1): 1-3.

Sugawara Ken

biographies-Ken Sugawara
Kogema, Tedashiro, 1964.
A short biography of Dr. Ken Sugawara.
In: Recent researches in the fields of hydrosphere, atmosphere and nuclear geochemistry, Ken Sugawara festival volume, Y. Miyake and T. Koyama, editors, Maruzen Co., Ltd., Tokyo, i-iv.

Sund, O.

Rollefsen, G., 1948 biography- Sund
Oscar Sund, 1883-1943. J. du Cons. 15(2): 147.

Sverdrup, H.U.

biographies - Sverdrup
Anon., 1958.
Harald Ulrik Sverdrup. Polar Record 9(58):57-58.

biographies
Deacon, G.E.R., 1958.
Professor Harald Ulrik Sverdrup (1888-1958).
Weather 13(1):17.

biographies - Sverdrup
Liestöl, Olav, 1958.
Harald Ulrik Sverdrup 1888-1957.
J. Glaciol., 3(23):226.

biographies- Sverdrup
Böhnecke, G., 1957.
In memoriam. H.U. Sverdrup. Deutsche Hydrogr. Zeits., 10(5):202-203.

biographies - Sverdrup
Fjeldstad, J. E., 1959
Harald Ulrik Sverdrup, 15 November 1888 - 21 August 1957.
Norsk Polarinst., Medd. 82: 23 pp.

biographies- Sverdrup
Fjeldstad, J.E., 1958.
Harald Ulrik Sverdrup, 1888-1957. J. du Cons., 23(2):147-150.

biographies- Sverdrup
Fleming, R.H., 1951.
Thirteenth award of the William Bowie medal.
Trans. A.G.U. 32(3):337-338.

biography - Sverdrup
France, Service Central Hydrographique, 1957.
H.U. Sverdrup. Bull. d'Info., C.C.O.E.C., 9(8): 4102 & b.

biographies- Sverdrup
Mosby, H., 1957
Harald Ulrik Sverdrup, 1888-1957.
Tellus 9(4):429-431.

biographies- Sverdrup
Munk, W.H., 1957.
Harald Ulrik Sverdrup (1888-1957). Deep-Sea Res., 4(4):289-290.

biography- Sverdrup
Revelle, R. R., and W. H. Munk, 1948.
Harald Ulrik Sverdrup - an appreciation (with bibliography). J. Mar. Res. 7(3):127-138.

Biographies - Sverdrup
Säger, Günther, 1961.
Vier Jahrzehnte Wirken im Dienste der Ozeanographie: Prof. H. U. Sverdrup 15.11. 1888-21. 8. 1957.
Geographische Berichte, 19(2):77-84.

BIOGRAPHIES - Sverdrup

Sager, Günther, 1957.
In memoriam Prof. Dr. Harald Ulrik Sverdrup.
Zeits. f. Meteorol., 11(9):257-259.

biographies - Sverdrup

Shuleikin, V.V., 1958.
Harald Ulrik Sverdrup, on the anniversary of his death. (In Russian).
Izv., Akad. Nauk, SSSR, Ser. Geofiz., (9):1151-1152.

English Edit., (9):667-668.

biographies - Sverdrup

Wordie, J.M., 1957.
Prof. H.U. Sverdrup. Nature 180(4594):1023.

Taning, A.V.

biographies - Tåning

Fridriksson, A., 1959.
Å. V. Tåning, 1890-1958, Obituary
J. du Conseil, 24(2):211-214.

biographies - Tåning

Hansen, P., 1959
Direktør, Dr. Phil. Åge Vedel Tåning
27 Juli 1890 - 26. September 1958.
Naturhistorisk Tidende 23(1-2):15-18.

Tarasov, Nikolai Ivanovich

Anon., 1965. biographies-Tarasov
Nikolai Ivanovich Tarasov (1905-1965). (In Russian).
Okeanologiia, Akad. Nauk, SSSR, 5(6):11251126

Tesch, J. J.

biographies-Tesch

Havinga, B., 1955.
Johan Jacob Tesch 1877-1954. J. du Cons. 20(3): 251-253.

Thompson, Sir D'Arcy W.

biography-Thompson

Anonymous, 1949.
Sir D'Arcy W. Thompson, M.O. 501. Met. Mag. 77(915):211.

Mere announcement.

biography-Thompson

Chambers, R., 1949
Sir D'Arcy Wentworth Thompson, C.B., R.R.S. (1860-1948). Sci.109:138-139, 151

biographies-Thompson

Clark, R. S., 1949.
D'Arcy Wentworth Thompson. 1860-1948 J. du Cons. 16(1): 9-13.

Thompson, Harold

biographies - Thompson

Tait, J.B., 1958.
Harold Thompson, 1890-1957. J. du Cons., 23(2): 151-154.

Thompson, Thomas G.

biographies-Thompson

Barnes, Clifford A., and Francis A. Richards 1962
Thomas Gordon Thompson, 1888-1961.
J. Mar. Res., 20(1):1-2.

biography - Thompson

Thomas, B.D., 1958
Thomas Gordon Thompson. J. Mar. Res., 17: 11-22

Thompson W.F.

biographie, W.F. Thompson

Van Cleve, R., 1966.
W.F. Thompson, 1888-1965.
J. Fish Res. Bd. Can., 23(11):1790-1793.

Thomson, Sir C. Wyville

Merriman, Daniel and Mary, 1958 biographies - Thomson
Sir C. Wyville Thomson's letters to Staff-Commander Thomas H. Tizard, 1877-1881. J. Mar. Res., 17:347-374.

biographies, Thorson

Anon. 1971.
Professor Gunnar Thorson.
Nature, Lond. 230 (5296): 607.

biographies, Thorson

Møller Christensen, Aage, 1971.
Gunnar Thorson 1906-1971.
Ophelia 9(1): 1-9

biographies, Timonev

Alekin, O.A., D.L. Laikhtman, L.F. Titov, B.D. Rusanov, K.K. Deryugin, V.I. Tyuryakov, E.I. Seryakov, L.A. Zhukov, et al.
Vsevolod Vsevolodovich Timonov. In memory of V.V. Timonov. (In Russian). Okeanologiia, 9(5): 917-918.

biographies-Timonov

*Dobrvol'sky, A.D. and V.G. Bukhteev, 1970.
On V.V. Timonov's life-work. Okeanologiia, 10(5): 740-744. (In Russian; English abstract)

Tissier, T.

biography-Tissier

LeDanois, Ed., 1948
Théodor Tissier 1866-1944. J. du Cons. 15(2):126-129.

Tonolli, V.

biographies - Tonolli

Rzóska, Julian 1967.
Professor V. Tonolli
Nature, Lond. 214 (5092): 1063.

Toscane

biographies Grand-Ducale de Toscane

Zanobio, Bruno, 1965.
Études de biologie marine à la cour Grand-Ducale de Toscane.
Colloque Internat., Hist. Biol. Mar., Banyuls-sur-Mer, 2-6 sept 1963, Suppl. Vie et Milieu, No. 19:83-98.

Trask, P.D.

biographies - Trask

Bailey, Thomas, L., 1962.
Parker Davies Trask (1899-1961).
Bull. Amer. Assoc. Petr. Geol., 46(5):714-717.

biographies - Trask

Gilluly, James, 1963.
Memorial to Parker Davies Trask.
Bull. Geol. Soc., Amer., 74(1):P13-P20.

biographies Uda

Yoshida, Kozo, 1969.
Appreciation [M. Uda]. Bull. Jap. Soc. Fis Oceanogr. Spec. No. (Prof. Uda Commem. Pap.). i-ii.

Ussachev, P.I.

biographies - Ussachev

Bogorov, B.G., 1963.
Pyotr (Peter) Ivanovich Ussachev. Biological Investigations of the Ocean. (In Russian).
Trudy Inst. Okeanol., Akad. Nauk, SSSR, 71:3-4.

Vaughn, T.W.

biographies - Vaughn

Thomas, H.D., 1952.
Dr. Thomas Wayland Vaughn. Nature 169(4305):734-735.

Vening-Meinesz, F.A.

biography - Vening Meinesz

Collette, B.J., 1966.
In memoriam Professor Dr. Ir. Felix Andries Vening Meinesz.
Geol. en Mijnbouw, 45 (9):285-290.

biographies - Vening Meinesz

Heiskanen, W.A., 1967.
Summary of Vening Meinesz's life.
Proc. First mar. Geod. Symp., 1966, 4.

biographies - Meinesz

Manten, 1966.
In memoriam F.A. Vening Meinesz.
Tectonophysics, 3(5):369-373.

biographies - Vening Meinesz

Bruins, G. J. 1967.
Professor F.A. Vening Meinesz.
Nature, Lond., 215 (5096): 109.

Vercelli, F.

biographies - Vercelli

Anon., 1953.
Prof. Francisco Vercelli. Nature 171(4345):241.

biographies - Vercelli

Anon., 1952.
Francesco Vercelli. Bol. Pesca, Piscicult., Idrobiol., n.s, 7(2):246-251.

biography - Vercelli

Morelli, C., 1952.
In memoria di Francesco Vercelli. Osservatorio Geofis., Trieste, n.s., Pubbl. No. 24: 14 pp.

biography - Vercelli

Picotti, M., 1953.
Francesco Vercelli. La Ricerca Sci., A, 23(2): 277-282.

Ist. Talassogr., Trieste, Pubbl. 291.

Vernadsky, V.I.

biographies - Vernadsky

Wolfson, V.I., 1964.
V.I. Vernadsky - the founder of a new branch of science - marine chemistry. (In Russian).
Okeanologiia, Akad. Nauk, SSSR, 4(2):193-204.

Vinogradov, B.S.

biographies - Vinogradov
Gureyev, A.A., 1960
B.S. Vinogradov (Obituary). Zool. Zhurn., Acad. Nauk, SSSR 39(2): 313

biographies - Vinogradov
Novikov, G.A., 1961.
In memoriam: Boris S. Vinogradov (25 III 1891 - 10 XII 1958)
Trudy Zool. Inst., Akad. Nauk, SSSR, 29:7-21.
Morphologiia i ekologiia pozvonochnix Zhivotnix.

von Humboldt

biography-von Humboldt
Kellner, L., 1963.
Alexander von Humboldt.
Oxford Univ. Press, N.Y., 247 pp. $5.75.
Reviewed:
Woodford, A.O., 1963. Science, 140(3570):973.

biography-von Humboldt
Kellner, L., 1960
Alexander von Humboldt and the history of international scientific collaboration.
Scientia, 54(8): 252-256.

biographies- von Humboldt
Schmidt, Herman, 1960.
De Terra, Helmut: Alexander von Humboldt und seine Zeit. Wiesbaden, F.A. Brockhaus 1956, 278 S u. 24 Bildtafl.
Naturwissenschaften, 47(1):24.

biographies - Alexander von Humboldt
Théodoridès, Jean, 1965.
Alexander von Humboldt et la biologie marine.
Colloque Internat., Hist. Biol. Mar., Banyuls-sur-Mer, 2-6 sept., 1963, Suppl., Vie et Milieu, No. 19:131-162.

biographies, von Humboldt
Troll, Carl, 1959.
The work of Alexander von Humboldt and Carl Ritter: a centernary address.
Advancement of Science, 16(64):441-452.

biography Von Humboldt
Wust, Georg, 1959.
Alexander von Humboldts Stellung in der Geschichte der Ozeanographie.
Alexander von Humboldt, Studien zu seiner universalen Geistehaltung, herausgegeben von Joachim H. Schultze. Festschrift zur Alexander von Humboldt-Feier veranstaltet aus Anlass der 100. Wiederkehr seines Todestages vom Humboldt-Komitee der Bundesrepublik Deitschland in Berlin am 18 und 19 Mai 1959. Verlag Walter de Guyter & Co. Berlin W35. pp. 90-104.

Wallace, Alfred R.

biographies-Wallace
Pantin, C.F.A., 1960.
Alfred Russel Wallace: his pre-Darwinian essay of 1855.
Proc. Linnean Soc., London, 171 Session, 1958-59(2):139-153.

Wattenberg, H.

biography-Wattenberg
Buch, K., 1945.
Hermann Wattenberg. Züge aus seinem Forschertum.
Kieler Meeresforschungen 6:7-10.

biography- Wattenberg
Böhnecke, G., 1944.
Professor Dr. Wattenberg. Ann. Hydr., usw., 72: 291-293.

Biographies, Wegener
Loewe, F. 1970.
Alfred Wegener - his life and work
Austral. Met. Mag. 18(4):177-190

Werenskiold, W.

biographies-Werenskiold
Isachsen, F., 1953.
Werner Werenskiold. Norsk Geogr. Tidsskr. 14: 1-5.

Wesenberg-Lund, C.

biographies-Wesenberg-Lund
Berg, K., 1956.
C. Wesenberg-Lund. Hydrobiologia, Acta Hydrobiol Hydrogr., et Protistol., 8(1/2):1-15.

biographies- Wesenberg-Lund
Thorson, Gunnar 1964.
Elise Wesenberg-Lund, 25 april 1896 - 19 juli 1969.
Videnox. Meddr. dansk naturh. Foren. 132:220-226.

Wexler, Harry

biographies, Harry Wexler
Chapman, Sydney, 1965.
Harry Wexler, 1911-1962, excellent and devoted student of our wonderful atmosphere.
Bull. Amer. Meteorol. Soc., 46(5):227-240.

Wilkens, Sir H. E.

biographies-Wilkins
Balchen, B., 1958.
Sir Hubert Wilkens 1888-1958, Obituary
Arctic J. Arctic Inst. N. Amer. 11(4):259.

biographies - Wilkins
Roberts, B.B., 1959.
Sir George Hubert Wilkins.
Polar Record, 9(62):489-491.

biographies - Wilkins
Wood, Walter A., 1959.
George Hubert Wilkins.
Geogr. Rev., 49(3):419-416.

Wilson, C. B.

biographies-Wilson
Schmitt, W.L., 1941.
Charles Branch Wilson (Obituary notice). Science 94(2442):258-259.

Winge, Øjvind

biographies-Winge
Beven, E.A., 1965.
Øjvind Winge.
Proc. Linn. Soc., London, 176:228-229.

Witting, R.J.

biography-Witting
Granqvist, G., 1948
Rolf Johan Witting 1879-1944. J. du Cons. 15(2):137-140.

BIOGRAPHIES WOLFLE
Abelson, Philip H., 1970.
Dael Wolfle at AAAS. Science, 168(3939): 1529.

biographies- Wood
Lucas, C.E., B.B. Parrish and B.B. Rae, 1970.
Henry Wood (1894-1969). J. Cons. int. Explor. Mer, 33(2): 119-121.

biographies, Wunsch
Press, Frank, and Henry Stommel 1971
Tenth presentation James B. Macelwane Award to Carl I. Wunsch in recognition of significant contributions to the geophysical sciences by a young scientist of outstanding ability.
Eos, 52(7): 539-540

Wüst, G.

biographies, Wüst
Joseph, J., 1970
Gedenktage. Prof. Dr. Wüst 80 Jahre.
Dt. hydrogr. Z. 23(1): 31.

biographies - Wüst
Stocks, Th., 1960.
Georg Wüst und seine Stellung in der neueren Ozeanographie.
Petermanns Geogr. Mitteil., 104(4):292-293.

Zariquiey Alvarez, Ricardo

biographies Zariquiey Alvarez
Anon., 1965.
[Don Ricardo Zariquiey Alvarez.]
Invest. Pesq., Barcelona, 28:1.

biographies, Zariquiey Alvarez
Gordon, Isabella, 1965.
Dr. Ricardo Zariquiey Alvarez.
Nature, 207(4999):806-807.

biographies Zariquiey Alvarez
Holthuis, L.B., 1965.
Dr. Ricardo Zariquiey Alvarez, 3 January 1897-27 January 1965.
Crustaceana, 9(1):105-110.

Yanulov, K.P.

biographies - Yanulov
Anon., 1966.
Konstantin Pavlovich Yanulov. (In Russian).
Okeanologiia, Akad. Nauk, SSSR, 6(5):924.

Zaitsev, A.A.

biographies - Zaitsev
Anon., 1966.
Anatoly Alexandrovich Zaitsev. (In Russian).
Okeanologiia, Akad. Nauk, SSSR, 6(5):925.

Zenkevitch, L.A.

biographies Zenkevitch
Anon, 1971.
In memoriam of L.A. Zenkevitch and J.A. Birstein. (In Russian). Trudy Inst. Okeanol. P.P. Shirshov 92: 5-8.

biographies, Zenkevich
Anon, 1970.
In memory of Lev Aleksandrovich Zenkevich (1889-1970). (In Russian). Gidrobiol. Zh., 6(6): 128-130.

biographies - Zenkevich
Anon. 1969.
Lev Aleksandrovich Zenkevich (on his 80th birthday). (In Russian)
Gidrobiol. Zh. (4): 5-8.

biography - Zenkevich
Anon., 1960.
[L.A. Zenkevitch (on his 70th birthday)]
Trudy Inst. Okeanol., 41:3-7.

biographies, Zenkevich

Beliaev, G.M., 1970.
Lev Aleksandrovich Zenkevich (1889-1970).
Zool. Zh. 49 (12):1891-1893.

biographies, Zenkevitch

Belyaev, G.M. and S.A. Mileikovsky, 1971.
In memoriam of Academician Lev Alexandrovitch Zenkevitch. Mar. Biol. 11(4): 299-305.

biographies - Zenkevich

Bogorov, V.G. 1969.
Prof. L.A. Zenkevich (On his 80th anniversary). (In Russian).
Zool. Zh. 48(8): 1261-1263.

biography-Zenkevich

Bogorov, B.G., 1964
Leo Alexandrovich Zenkevitch. (In Russian)
Investigations of bottom fauna and flora of Far - Eastern Seas and the Pacific Ocean.
Trudy Akad. Nauk, SSSR, 69: 3-5.

biographies, L.A. Zenkevich.

Pavlov, A.V., 1965.
The awarding of Lenin Prize to corresponding member, Academy of Sciences of USSR, L. A. Zenkewitsch. (In Russian).
Izv., Akad. Nauk, SSSR, Ser. Biol., (4):592-598.

biographies L. A. Zenkevich

SSSR, Akademia Nauk, 1964.
Lev Alexandrovich Zenkevich. (In Russian).
Okeanologiia, Akad. Nauk, SSSR., 4(6):937-938.

Zhdanov Alexander Mikhailovich

biographies, Zhdanov

Anon.,1968.
Alexander Mikhailovich Zhdanov.
Okeanologiia,Akad.Nauk,SSSR, 8(4):772-774.

Zorell, F.

biography-Zorell

Kalle, Kurt, 1956.
In memoriam. Franz Zorell. Deutsche Hydrogr. Zeits. 9(3):149-150.

Zubov, N.N.

biography-Zubov

Bruns, E., 1962.
In Memoriam Nikolai Nikolajewitsch Subov.
Beiträge z. Meereskunde, (6):5-6.

biography-Zubov

Dobrovolsky, V.A., 1961
[N.N. Zubov, one of the most prominent oceanographers. (24 March 1885-11 November 1960).]
Okeanologiia, Akad. Nauk, SSSR, (2):355-359.

biographies - Zubov

Kostrits, I. B., 1961.
[On the perpetuation of N.N. Zubov's memory.]
Izv. Vses. Geograf. Obshch., 93(5) :460.

biographies-Zubov

Lagutin, B.L., and A.M. Muromtsev, and A.A. Iushchak, 1961.
[In memory of N.N. Zubov.]
Meteorol. i Gidrol., (5):59-60.

"biolithosores"

"biolithosores"

Hommeril, Pierre, 1962
Étude locale (Gouville-sur-Mer,Manche) de la retinue des sédiments par deux polychaetes sedentaires: Sabellaria alveolata (Hermelle) et Lanice conchilega.
Cahiers Océanogr., C.C.O.E.C., 14(4):245-257.

biological assay

biological assay

Smayda, Theodore J., 1964.
Enrichment experiments using the marine centric diatom Cyclotella nana (clone 13-1) as an assay organism.
Narragansett Mar. Lab., Univ. Rhode Island, Occ. Publ. No. 2:25-32.

biological changes, long term

biological changes, long term

Southward, A.J., 1963
The distribution of some plankton animals in the English Channel and approaches. III. Theories about long-term biological changes, including fish.
J. Mar. Biol. Assoc., U.K., 43(1):1-29.

Biological indicators

See: indicator species

biological noise

See "noise, biological"

Biological noise

See: noise, biological

biological O_2 index

biological oceanography

Blackburn,Maurice,1966.
Biological oceanography of the eastern tropical Pacific: summary of existing information.
Spec. Scient.Rep.U.S.Fish.Wildl.Serv.,Fish., No. 540:18pp.

(Not at all complete)

biological problems

biological problems

Merriman, D., 1949
Biological problems of the ocean. Sci. Mon. 68(1):12-16.

biological resources

Bogdanov, A.S., 1971.
Problems of use of marine biological resources. (In Russian; English abstract). Okeanologiia 11(5): 842-847.

biological stations

Biological Variations

biological variations

Cushing, D.H. 1966
Biological and hydrographical changes in British seas during the last thirty years
Biol. Rev. 41 (2): 221-258

Bioluminescence

A

luminescence

Allen, W.E., 1939
"Phosphorescence" in the sea. Nautical Gazette, 2 pp., (26 Aug.)

bioluminescence

Anctil, M. and C.G. Gruchy, 1970.
Stimulation and photography of bioluminescence in lanternfishes (Myctophidae).
J. Fish. Res. Bd., Can., 27(4): 826-829.

bioluminescence

Angel, M.V. 1968
Bioluminescence in planktonic halocyprid ostracods
J. mar. biol. Ass. U. K. 48(1): 255-257.

luminescence

Araki, S., 1950.
The effect of light on the luminous solution of Cypridina hilgendorfii. Annotat. Zool. Japon. 23(3):98-102.

bioluminescence

Antiomkin, A.S., V.S. Filimonov, E.P. Baldina y V.N. Gruze, 1969.
Resultados preliminares sobre zooplancton y su luminiscencia en la región oriental del Mar Caribe.
Serie Oceanologica, Inst. Oceanol. Acad. Cienc., Cuba, 2:1-11.

B

bioluminescence

Backus, R.H., C.S. Yentsch and A. Wing, 1961.
Bioluminescence in the surface waters of the sea.
Nature, 192(4802):518.

bioluminescence

Baird, W., 1843.
Note on the luminous appearance of the sea with descriptions of some of the entomostracens insects by which it is occasioned. The Zoologist (Newman) 1:55-61.

luminescence

Bajkov, A.D. and T.W. Robinson, 1947
The dinoflagellates, their distribution, occurrence and toxicity. Aero Medical Laboratory, Army Air Forces, Air Material Command Engineering Division, Memo. Rept., Ser. No.TSEAA-691-3G, Expenditure Order No. 691-12, ADB/ml, 7 pp. (ozalid), 2 photos. 13 June, 1947

bioluminescence

Baguet, Fernand and James Case, 1971.
Luminescence control in Porichthys (Teleostei): excitation of isolated photophores. Biol. Bull mar. biol. Lab., Woods Hole, 140(1): 15-27.

bioluminescence

Barnes, Anthony T. and James F. Case, 1972.
Bioluminescence in the mesopelagic copepod, Gaussia princeps (T. Scott). J. exp. mar. Biol. Ecol. 8(1): 53-71.

bioluminescence

Bityukov, E.P., 1971.
Bioluminescence in the wake current in the Atlantic Ocean and Mediterranean Sea. (In Russian; English abstract). Okeanologiia, 11(1): 127-133.

bioluminescence

*Bityukov,E.P.,1968.
Characteristics of the daily rhythm of bioluminescence of Noctiluca miliaris (Flagellata, Peridinea). (In Russian;English abstract).
Zool.Zh.,47(1):36-40.

bioluminescence
*Bityukov, E.P., V.P. Rybasov and V.G. Shaida, 1967.
Annual variations of the bioluminescence field intensity in the neritic zone of the Black Sea. (In Russian;English abstract).
Okeanologiia,Akad.Nauk,SSSR, 7(6):1089-1099.

bioluminescence
Bode, V.C., R. DeSa and J.W. Hastings, 1963
Daily rhythm of luciferin activity in Gonyaulax polyedra.
Science, 141(3584):913-915.

bioluminescence
Boden, Brian P., and Elizabeth M. Kampa, 1964
Planktonic bioluminescence.
In: Oceanography and Marine Biology, Harold Barnes, Editor, George Allen & Unwin, Ltd. 2:341-373.

bioluminescence
Boden, Brian P. and Elizabeth M. Kampa, 1958
Lumière, bioluminescence et migrations de la couche diffusante profonde en Méditerranée occidentale. Vie et Milieu 9(1): 1-10.

bioluminescence
Boden, B.P., and E.M. Kampa, 1956.
Records of bioluminescence in the ocean.
Pacific Science 11(2):229-235.

bioluminescence
Boden, Brian P., Elizabeth M. Kampa and James M. Snodgrass, 1966.
Measurements of spontaneous bioluminescence in the sea.
Nature, 208(5015):1078-1080.

bioluminescence
Bolin, R.L., 1961.
The function of the luminous organs of deep-sea fishes.
Proc. Ninth Pacific Sci. Congr., Pacific Sci. Assoc., 1957, Fish., 10:37-39.

bioluminescence(Instrumentation)
Bourbeau, Frank, 1966.
Undersea photometer for marine biological studies
Undersea Techn., 7(9):39-45

bioluminescence
Bousfield, E.L., and W.L. Klawe, 1963.
Orchestoidea gracilis, a new beach hopper (Amphipoda: Talitridae) from Lower California, Mexico, with remarks on its luminescence.
Bull. S. California Acad. Sci., 62(1):1-8.

bioluminescence
Bowman Thomas E. 1967.
Bioluminescence in two species of pelagic amphipods.
J. Fish. Res. Bd. Can. 24(3): 687-688.

luminescence.
Burkenroad, M. D., 1943.
A possible function of bioluminescence.
J. Mar. Res., 5:161-164.

bioluminescence
Burkholder, Paul R., and Lillian M., 1958.
Studies on B vitamins in relation to productivity of the Bahia Fosforescente, Puerto Rico.
Bull. Mar. Sci., Gulf and Caribbean, 8(3):201-223.

C

bioliminescence
*Carpenter,J.H., and H.H. Seliger,1968.
Studies at Oyster Bay in Jamaica,West Indies. 2. Effects of flow patterns and exchange on bioluminescent distributions.
J.mar.Res.,26(3):256-272.

bioluminescence
Chace, F.A., jr., 1940
Plankton of the Bermuda Oceanographic Expeditions. IX. The Bathypelagic Caridean Crustacea. Zoologica, 25(2):117-209, 64 text figs.

owned by MS

bioluminescence
Chang, J.J., 1954.
Analysis of the luminescent response of the ctenophore, Mnemiopsis leidyi, to stimulation.
J. Cell. Comp. Physiol. 44(3):365-394.

bioluminescence
Chang, Joseph Jin, and Frank H. Johnson, 1959
The influence of pressure, temperature and urethane on the luminescent flash of Mnemiopsis leidyi. Biol. Bull., 116(1): 1-14.

bioluminescence
Chase,Aurin M.,1966.
Activity and inhibition of Cypridina luciferase; quantitative measurement; analysis of inhibition by urea; and some effects of sodium and potassium ions.
In: Bioluminescence in progress, F.H. Johnson and Y. Haneda,editors,115-136.

bioluminescence
Chase, A.M., F.S. Hurst and H.J. Zeft, 1959.
The effect of temperature on the non-luminescent oxidation of Cypridina luciferin.
J. Cell. and Comp. Physiol., 54(1):115-125.

bioluminescence
Christomanos, A.A., 1957.
[The biochemiluminescence of sea organisms.]
Ethnika Hellenika Okeanographika Etaireia Thalassina Phylla (Marine Pages), Apio 3:21-38.

bioluminescence
Church Ron 1970.
Bioluminescence: the sea's living light
Oceans Mag. 3(6):20-29. (popular)

bioluminescence
Clarke, George L., 1967.
Light in the sea.
Oceanology int., 2(7):40-42. (Popular)

bioluminescence
Clarke, G.L., 1961
The conditions of light in the sea with special reference to bioluminescence. Symposium on the radiant energy in the sea, Helsinki, 4-5 Aug. 1960.
U.G.G.I., Monogr., No. 10:101-103.

bioluminescence
Clarke, George L., 1958
The measurement of bioluminescence in the sea.
Proc. Ninth Pacific Sci. Congr., Pacific Sci. Assoc., 1957, 16(Oceanogr.):239-240.

bioluminescence
Clarke, George L., and Richard H. Backus, 1964.
Interrelations between the vertical migration of deep scattering layers, bioluminescence, and changes in daylight in the sea.
Bull. Inst. Oceanogr., Monaco, 64(1318):36 pp.

bioluminescence
Clarke, G.L., and R.H. Backus, 1956.
Measurements of light penetration in relation to vertical migration and record of luminescence of deep-sea animals. Deep-Sea Res. 4(1):1-14.

bioluminescence
Clarke, G.L. and L. R. Breslau, 1960.
Studies of luminescent flashing in Phosphorescent Bay, Puerto Rico, and in the Gulf of Naples using a portable bathyphotometer. Bul. l'Inst. Ocean., 57 (1171): 1-32.

bioluminescence
Clarke, G. L., and Lloyd R. Breslau, 1959.
Measurements of bioluminescence off Monaco and northern Corsica.
Bull. Inst. Ocean., Monaco, No. 1147:31 pp.

bioluminescence
Clarke, G.L., R.J. Conover, C.N. David and J.A.C. Nicol, 1962
Comparative studies of luminescence in copepods and other pelagic marine animals.
J. Mar. Biol. Assoc., U.K., 42(3):541-564.

bioluminescence
Clarke, G.L., and E.J. Denton, 1962
Light and animal life, Ch. 10, Sect. IV. Transmission of energy within the sea. In: The Sea, Interscience Publishers, Vol. 1, Physical Oceanography, 456-468.

bioluminescence
Clarke, G.L., and C.J. Hubbard, 1959.
Quantitative records of the luminescent flashing of oceanic animals at great depths.
Limnol. & Oceanogr., 4(2):163-180.

bioluminescence
Clarke, George L., and Mahlon G. Kelly, 1965.
Measurements of diurnal changes in bioluminescence from the sea surface to 2,000 meters using a new photometric device.
Limnol. and Oceanogr., Redfield Vol., Suppl. to 10:R54-R66.

bioluminescence
Clarke, George L., and Mahlon G. Kelly, 1964.
Variation in transparency and in bioluminescence on longitudinal transects in the Western Indian Ocean.
Bull. Inst. oceanogr., Monaco, 64(1319):20 pp.

bioluminescence
Clarke, M.R., 1965.
Large light organs on the dorsal surfaces of the squids Ommastrephes pteropus, 'Symplecteuthis oualaniensis' and 'Dosidicus gigas'.
Proc. Malac. Soc., Lond., 36:319-321.
Also in:
Collected Reprints, Nat. Inst. Oceanogr., Vol. 13.

bioluminescence
Cormier, M.J., 1962.
Studies on the bioluminescence of Renilla reniformis II. Requirement for 3', 5'-diphospho adenosine in the luminescent reaction.
J. Biol. Chem., 237:2032-2037.

bioluminescence
Cormier, Milton J., and Leon S. Dure, 1963.
Studies on the bioluminescence of Balanoglossus biminiensis extracts. 1. Requirement for hydrogen peroxide and characteristics of the system. 2. Evidence for the peroxidase nature of balanoglossid luciferase.
J. biol. Chem., 238(2):785-789; 790-793.
Also in:
Collected Reprints, Mar. Inst. Univ. Georgia, 4. 1965.

bioluminescence
Cormier, M.J., and C.B. Eckroade, 1962.
Studies on the bioluminescence of Renilla reniformis III. Some biochemical comparisons of the system to other Renilla species and determinations of the spectral energy distribution.
Biochim. Biophys. Acta, 64:340-344.

bioluminescence
Cormier, Milton J., and Kazuo Hori, 1964.
Studies on the bioluminescence of Renilia reniformis. 4. Non-enzymatic activation of Renilia luciferin.
Biochim.-Biophys. Acta. 86:99-104.

bioluminescence
Cormier,M.J., K. Hori, and P. Kreiss,1965.
Studies on the bioluminescence system of the sea Pansy, Renillia reniformis.
In: Bioluminescence in progress,F.H. Johnson and Y. Haneda,editors,Princeton Univ. Press. 349-362.

bioluminescence
Cphetomanos, A.A., 1957.
The biochemiluminescence of sea organisms. (In Greek)
Thalassina Phylla, 3:38 pp.

bioluminescence
Crane, Jules M., Jr. 1968.
Bioluminescence in the batfish Dibranchus atlanticus.
Copeia 1968(2): 410-411.

D

luminescent organs
Dahl, F., 1893.
Pleuromma, ein Krebs mit Leuchtorgan. Zool. Anz. 16(415):104-109.

bioluminescence
Dales, R. Phillips 1971.
Bioluminescence in pelagic polychaeta
J. Fish. Res. Bd. Can. 28 (10): 1487-1489.

bioluminescence
Davenport, D., and J.A.C. Nicol 1955
Luminescence in Hydromedusae.
Proc. R. Soc., London, B, 144(916):399-411.

bioluminescence
Dennell, R., 1955.
Observations on the luminescence of bathypelagic Crustacea Decapoda of the Bermuda area.
J. Linn. Soc., London, 42(287):393-406.

phosphorescence
Derbek, F., 1909.
Observations on the phosphorescence of the Okhotsk Sea. Ezhegodnik Zoologicheskago Muzeia Imperatorskoi Akad. Nauk, 14(3/4):29-32.

Translations may be purchased from the Scientific Translations Center, Library of Congress.

luminescence
Derbek, F.A., 1909.
Nabliūdeniīa nad svecheniem Okhotskogo Moriā. (Phosphorescence in the Sea of Okhotsk). Ezhegodn Zool. Muz., Akad. Nauk, 14(3/4):xxix-xxxii.

bioluminescence
Dörrbeck, F., 1947.
Leuchtendes Meer. Ein ungewöhnliches Phänomen im Ochotskischen Meer. Universum 2(3):53-54, 3 figs.

bioluminescence
*Doyle, Jean R., and R.H. Kay,1967.
Some studies on the bioluminescence of the euphausids Meganyctiphanes norvegica and Thysanoessa raschii.
J. mar. biol. Ass., U.K., 47(3):555-563.

bioluminescence
Dure, Leon S., and Milton J. Cormier, 1964.
Studies on the bioluminescence of Balanoglossus biminiensis extracts. 3. a kinetic comparison of luminescent and non luminescent peroxidation reactions and a proposed mechanism for peroxidase action.
J. biol. Chem., 239(7):2351-2359.

E

bioluminescence
Eckert,Roger,1966.
Excitation and luminescence in Noctiluca miliaris.
In: Bioluminescence in progress,F.H. Johnson and Y. Haneda,editors,Princeton Univ.Press. 269-300.

bioluminescence
Eckert, Roger, 1966.
Subcellular sources of luminescence in Noctiluca. Science, 151(3708):349-352.

bioluminescence
Eckert, Roger, 1965.
Bioelectric control of bioluminescence in the dinoflagellate Noctiluca. I. Specific nature of triggering events.
Science, 147(3662):1140-1142.

bioluminescence
Eckert, Roger, 1965.
Bioelectric control of bioluminescence in the dinoflagellate Noctiluca. II. Asynchronous flash initiation by a propagated triggering potential.
Science, 147(3662):1142-1145.

F

G

bioluminescence
Ganapati, P. N., D. G. V. Prasada Rao & M. V. Lakshmana Rao, 1959.
Bioluminescence in Visakhapatnam Harbour.
Current Sci., 28(6): 246-247.

bioluminescence
Giglioli, E.H., 1869-1870.
La fosforescenza del mare. Note press durante un viaggio di circumnavigazione. Atti. R. Acad. Sc., Torino, 5:485-505.

bioluminescence
Gitelson I.I., L.A. Levina, A.P. Shevyrnogov, V.S. Filimonov, L.S. Antemkin, R.N. Utiushev and Iu. A. Zagorodnyi, 1970.
The measurement of bioluminescence at maximum depths. (In Russian)
Dokl. Akad. Nauk, SSSR, 191 (3): 689-692

bioluminescence
Glahn, W., 1943.
Meerleuchten in Atlantischen Ozean. Seewart 12(2):17-25, 6 figs.

bioluminescence
Gold, Kenneth, 1965.
A note on the distribution of luminescent dinoflagellates and water constituents in Phosphorescent Bay, Puerto Rico.
Ocean Sci. and Ocean Eng., Mar. Techn. Soc.,-Amer. Soc. Limnol. Oceanogr., 1:77-80.

bioluminescence
Gotto, R.V., 1963.
Luminescent ophiuroids and and associated copepods.
The Irish Naturalists J., 14(7):137-139.

bioluminescence
Great Britain, Meteorological Office, 1961
The marine observers' log - Jan. Feb. March.
Marine Observer, 31(191): 6-23.

H

bioluminescence
Haneda, Yata, 1961.
Observations on luminescence of the sea. (Abstract).
Tenth Pacific Science Congress, Honolulu, 21 Aug.-6 Sept., 1961, Abstracts of Symposium Papers, 158-159.

bioluminescence
Haneda, Y., 1958.
Studies on luminescence in marine snails.
Pacific Science 12(2):152-156.

bioluminescence
Haneda, Y., 1958.
Studies on the luminous organisms found in the waters adjacent to the Pacific coasts of Japan. Report. II. Rec. Oceanogr. Wks., Japan, Spec. No. 2:171-174.

bioluminescence
Haneda, Y., 1957.
Studies on the luminous organisms found in the waters adjacent to the Pacific coasts of Japan.
Rec. Oceanogr. Wks., Japan, (Spec. No.):97-102.

bioluminescence
Haneda, Y., 1953.
Observation on some marine luminous organisms of Hachijo Island, Japan. Rec. Ocean. Wks., Japan, n.s., 1(1):103-106, 2 textfigs.

bioluminescence
Haneda, Yata, and Frank H. Johnson, 1962.
The photogenic organs of Parapriacanthus Beryciformes Franz and other fish with the indirect type of luminescent system.
J. Morphology, 110(2):187-198.

bioluminescence
Haneda, Y., F.H. Johnson, Y. Masuda, Y. Saiga, O. Shimomura, H.-C. Sie, N. Sugiyama and I. Takatsuki, 1961.
J. Cell. & Comp. Physiol., 57(1):55-62.
Crystalline luciferin from live Cypridina.

bioluminescence
Haneda, Yata, Frank H. Johnson, and Edward H. C. Sie,, 1959.
Luciferin and luciferase extracts of a fish, Apogon ellioti and their luminescent cross-reactions with those of a crustacean, Cypridina hilgendorfii.
Sci. Rept., Yokosuka Mus. No. 4:13-17.

bioluminescence
Haneda, Y., and T. Tokioka, 1954.
Droplets from the plankton net. 15. Records of a caudate form of Pegea confoederata from the Japanese waters, with some notes on its luminescence.
Publ. Seto Mar. Biol. Lab., 3(3):369-371.

bioluminescence
Haneda, Yata, and Frederick I. Tsuji, 1971.
Light production in the luminous fishes Photoblepharon and Anomalops from the Banda Islands. Science, 173(3992): 143-145.

bioluminescence
Haneda, Yata, Frederick I. Tsuji and Noboru Sugiyama 1969.
Luminescent systems in apogonid fishes from the Philippines.
Science 165 (3889): 188-190.

bioluminescence
Hardy, A.C., and R.H. Kay, 1964
Experimental studies of plankton luminescence.
Experimental studies of plankton luminescence.
Jour. Mar. Biol. Assoc., U.K., 44(2):435-484.

bioluminescence
Harvey, E.N., 1956.
Evolution and bioluminescence.
Quart. Rev. Biol., 31(4):270-287.

luminescence
Harvey, E.N., 1952.
Luminescent organisms. Amer. Sci. 40(3):468-481, 13 textfigs.

bioluminescence
Harvey, E.N., 1952.
Bioluminescence. Academic Press, Inc., N.Y., 650 pp.

bioluminescence
Harvey, E.Newton, and S.P. Marfey, 1958.
Fluorescence, phosphorescence and bioluminescence in the ctenophore Mnemiopsis leidyi. (Abstract).
Biol. Bull., 115(2):336-337.

bioluminescence
Hastings, J.W., 1971.
Light to hide by: ventral luminescence to camouflage the silhouette. Science, 173(4001): 1016-1017.

bioluminescence
Hastings, J.W. and George Mitchell, 1971.
Endosymbiotic bioluminescent bacteria from the light organ of pony fish. Biol. Bull. mar. biol. Lab. Woods Hole 141(2): 261-268.

bioluminescence
Hastings, J. Woodland, and Beatrice M. Sweeney, 1958.
A persistent diurnal rhythm of luminescence in Gonyaulax polyedra.
Biol. Bull., 115(3):440-458.

bioluminescence
Hastings, J.W., and B.M. Sweeney, 1957.
The luminescent reaction in extracts of the marine dinoflagellate, Gonyaulax polyedra. J. Cell. Comp. Physiol., 49(2):209-226.

bioluminescence
Hastings, J. Woodland, and Vernon C. Bode, 1961.
Ionic effects upon bioluminescence in Gonyaulax extracts.
Light and Life, W.D. McElroy and Bentley Glass, Edits., Johns Hopkins Univ. Press, 294-

bioluminescence
Hastings, J.W., Marcie Vergin, and R. De Sa, 1966.
Scintillons: the biochemistry of dinoflagellate bioluminescence.
In: Bioluminescence in Progress, F.H. Johnson and Y. Haneda, editors, Princeton Univ. Press, 301-329.

bioluminescence
Herring, P.J., 1967.
Luminescence in marine amphipods.
Nature, Lond., 214:1260-1261. Also in: Collected Reprints, Nat. Inst. Oceanogr., Wormley, 15 (1967).

bioluminescence
Hilder, B., 1955.
Radar and phosphorescence at sea. Nature 176 (4473):174-175.

bioluminescence
Hubbs, Carl L., Tamotsu Iwai and Kiyomatsu Matsubara 1967.
External and internal characters, horizontal and vertical distribution, luminescence and food of the dwarf pelagic shark, Euprotomicrus bispinatus.
Bull. Scripps Inst. Oceanogr. 10:1-64.

I

bioluminescence
Ivanov, B.G., 1969.
On the luminescence of the Antarctic krill (Euphausia superba). Okeanologiia 9(3): 505-506.
(In Russian; English abstract)

bioluminescence
Iwai, Tamotsu, 1960.
Luminous organs of the deep-sea squaloid shark, Centroscyllium ritteri Jordan and Fowler.
Pacific Science 14(1): 51-54.

J

bioluminescence
Johnson, Frank H., Edward H.-C. Sie and Y. Haneda, 1961
In: Light and Life, W.D. McElroy and Bentley Glass, Editors.
Johns Hopkins Press, 206-218.

bioluminescence
Johnson, F. H., H. Eyring and J. J. Chang, 1959.
Reaction rate control of light emission in bioluminescent systems.
Faraday Soc. Discussions, No. 27:191-198.

bioluminescence
Johnson F.H., O. Shimomura and Y. Saiga, 1962
Action of cyanide on Cypridina luciferin
J. Cell. and Comp. Physiol., 59(3):265-272.

bioluminescence
Johnson, F.H., Osamu Shimomura and Yo Saiga, 1961.
Luminescence potency of the Cypridina system.
Science, 134(3492):1755-1756.

bioluminescence
Johnson, Frank H., Osamu Shimomura, Yo Saiga, Lewis C. Gershman, George T. Reynolds and John R. Waters, 1962
Quantum efficiency of Cypridina luminescence, with a note on that of Aequorea.
J. Cell. and Comp. Physiol., 60(1):85-104.

bioluminescence
Johnson, F.H., H.D. Stachel, O. Shimomura and Y. Haneda, 1966.
Partial purification of the luminescence system of a deep-sea shrimp, Hoplophorus gracilorostris.
In: Bioluminescence in progress, F.H. Johnson and Y. Haneda, editors, Princeton Univ. Press, 523-532.

bioluminescence
Johnson, F.H., H.D. Stachel, E.C. Taylor and O. Shimomura, 1966.
Chemiluminescence and fluorescence of Cypridina luciferin and of some new indole compounds of dimethylsulfoxide.
In: Bioluminescence in progress, F.H. Johnson and Y. Haneda, editors, Princeton Univ. Press, 67-82.

bioluminescence
Johnson, F.H., N. Sugiyama, O. Shimomura, Y. Saiga, and Y. Haneda, 1961
Crustalline luciferin from a luminescent fish, Parapriacanthus beryciformes.
Proc. Nat. Acad. Sci., Vol. 47(4): 486-489.

K

Kalle, Kurt, 1960 bioluminescence
Die rätselhafte und "unheimliche" Naturerscheinung des "explodierenden" und des "rotierenden" Meeresleuchtens - eine Folge lokaler Seeleben? Deutsche Hydrogr. Zeits., 13 (2):49-77.

bioluminescence
Kampa, E.M., and B.P. Boden, 1957.
Light generation in a sonic scattering layer.
Deep-Sea Res., 4(2):73-92.

bioluminescence
Karabashev, G.S., 1969.
On the photometric technique for studying bioluminescence in the sea. (In Russian; English abstract). Okeanologiia, 9(6): 1100-1107.

bioluminescence
Kay, R.H., 1966.
The inhibition of optically stimulated bioluminescence in Meganyctiphanes norgvegicus.
In: Some contemporary studies in marine science, H. Barnes, editor, George Allen & Unwin, Ltd., 421-427.

bioluminescence
Kay, R.H., 1965.
Light-stimulated and light-inhibited bioluminescence of the euphausid Meganyctiphanes norvegica (G.O. Sars).
Proc. R. Soc., London, (B), 162(988):365-386.

bioluminescence
*Kelly, Mahlon G., 1968.
The occurrence of dinoflagellate bioluminescence at Woods Hole.
Biol. Bull., mar. biol. Lab., Woods Hole, 135(2):279-295.

bioluminescence
Kelly, Mahlon G., and Steven Katona, 1966.
An endogenous diurnal rhythm of bioluminescence in a natural population of dinoflagellates.
Biol. Bull., 131(1):115-126.

bioluminescence
Kielhorn, W.V., 1952
The biology of the surface zone zooplankton of a Boreo-Arctic Atlantic Ocean area. J. Fish Res. Bd., Canada 9 (5): 223-264, 13 text figs.

bioluminescence
Kishi, Y., T. Goto, Y. Hirata, O. Shimomura and F.H. Johnson, 1966.
The structure of Cypridina luciferin.
In: Bioluminescence in progress, F.H. Johnson and Y. Haneda, editors, Princeton Univ. Press, 89-114.

luminescence
Klenova, M.V., V.N. Florovskaja and N.M. Vikhrenko, 1956.
An attempt to a bituminological luminescence of the sea bottom. Dokl. Akad. Nauk, SSSR, 109(4): 846-848.

bioluminescence
Kornicker, L.S., and Charles E. King, 1965.
A new species of luminescent Ostracoda from Jamaica, Wes Indies.
Micropaleontology, 11(1):105-110.

bioluminescence
Koshtojantz, H.S., 1953.
[The relation of biological luminosity of Noctiluca to the composition of the reactive group of albumens and the gaseous exchange.]
Dok. Akad. Nauk, SSSR 91(5):1229.

bioluminescence
Kriegen, Neil, and J. W. Hastings 1968.
Bioluminescence: pH activity profiles of related luciferase fractions.
Science 161 (3841): 586-589.

bioluminescence
Kuznetsov, A.P., 1963.
The luminescence of Atlantic Ocean waters.
Priroda, 52(10):102-104.
Trans. cited:
Tech. Transl., 1964, 11(4):317.
JPRS:22251
OTS
TT-63-41349.

L

bioluminescence
Lochhead, J.H., 1954.
On the distribution of a marine cladoceran, Penilia avirostris Dana (Crustacea, Branchiopoda), with a note on its reported bioluminescence. Biol. Bull. 107(1):92-105.

M

bioluminescence
Marfey, S.P., L.C. Craig and E.N. Harvey, 1958
Fractionation of Cypridina luciferin and its benzoyl derivative. (Abstract).
Biol. Bull., 115(2):339.

bioluminescence
Margalef, Ramón, 1961
Hidrografía y fitoplancton de un área marina de la costa meridional de Puerto Rico.
Inv. Pesq., Barcelona, 18:38-96.

bioluminescense = (Peridinium bahamense)
Margalef, Ramon, 1957.
Fitoplancton de las costas de Puerto Rico.
Inv. Pesq., Barcelona, 6:39-52.

bioluminescence
Mauchline, J., 1958-59.
The biology of the euphausiid crustacean Meganyctiphanes norvegica (M. Sars).
Proc. R. Soc., Edinburgh, (B), 67(2):141-179.

bioluminescence
Mauchline, John, and the late Leonard R. Fisher 1969.
The biology of euphausids.
Adv. mar. Biol. F.S. Russell and Maurice Yonge, editors, Academic Press 7: 454pp $17.50

Mc

bioluminescence
McAllister, D.E. 1967.
The significance of ventral bioluminescence in fishes.
J. Fish. Res. Bd. Can. 24 (3): 537-554

bioluminescence
McElroy, W.D., and B.L. Streller, 1954.
Bioluminescence. Bacteriol. Rev. 18(3):177-194.

bioluminescence
Merrett, N.R., J. Badcock and P.J. Herring, 1971.
Observations on bioluminescence in a scopelarchid fish, Benthalbella. Deep-Sea Res. 18(12): 1265-1267.

bioluminescence, effect of
Mitsugi, Shinsuke, Takeo Koyama, Shin'ichi Yajima and Chosei Yoshimuta, 1966.
Behavior of fish schools against luminous substances in waters. (In Japanese; English abstract).
Bull. Tokai reg. Fish. Res. Lab., No. 48: 29-36.

bioluminescence
Musya, K., 1934.
On the luminous phenomena that accompanied the Great Sanriku Tunami in 1933.
Bull. Earthquake Res. Bull., Suppl. Vol. 1:87-111.

N

bioluminescence
Nares, J.D., 1957.
Sudden uprising of fish and phosphorescence in the Arabian Sea. Mar. Obs., 27(176):93-94.

bioluminescence
Neshyba, Steve 1967.
Pulsed light stimulation of marine bioluminescence.
Limnol. Oceanogr. 12(2): 222-235.

bioluminescence
Nicol, J.A.C., 1967.
The luminescence of fishes. In: Aspects of Marine Zoology, N.B. Marshall, editor, Symp. Zool. Soc., Lond., 19: 27-55.

bioluminescence
Nicol, J.A.C., 1963
Luminescence in animals.
Endeavour, Imperial Chemical Industries, London, 22:37-41.

bioluminescence
Nicol, J.A.C., 1962.
Animal luminescence.
In: Advances in comparative physiology and biochemistry, O. Lowenstein, Edit. Academic Press New York and London, 217-273.

bioluminescence
Nicol, J.A.C., 1962.
Bioluminescence.
Proc. R. Soc., London, (A) 265(1322):355-359.

bioluminescence
Nicol, J.A.C., 1961.
Luminescence and vision in marine animals. (Abstract).
Tenth Pacific Sci. Congr., Honolulu, 21 Aug.-6 Sept., 1961, Abstracts of Symposium Papers, 161.

bioluminescence
Nicol, J.A.C., 1958.
Observations on luminescence in Noctiluca.
J.M.B.A., U.K., 37(3):535-550.

bioluminescence
Nicol. J.A.C., 1958.
Observations on luminescence in pelagic animals.
J.M.B.A., U.K., 37(3):705-752.

bioluminescence
Nicol, J.A.C., 1955.
Observations on luminescence in Renilla (Pennatulacea). J. Exp. Biol. 32(2):299-320.

O

P

bioluminescence
*Petersson, Göran, 1968.
Studies on photophorus in the Euphausiacea.
Sarsia, 36:1-39.

luminescence
Plate, L., 1906.
Pyrodinium bahamense n.g., n.sp., die leucht-Peridinee des "Feuersees" von Nassau, Bahamas.
Arch. Protistenk. 7:411-429.

luminescence
Plate, L., 1889.
Observations on Noctiluca miliaris Suriray, and the sea luminosity produced by it. Ann. Mag. Nat. Hist., 6th ser., 3(13):22-28.

luminescence
Puissé gur, C., 1946.
Les Étres vivants lumineux. Sci. et Vie, Vol. LXX (348), translated by R. Widmer, David Taylor Model Basin Translation 221, dated May 1927 (as Luminous Living Organisms), 17 pp., 9 text figs.

Q

bio luminescence
*Quayle, D.B.
Paralytic shellfish poisoning in British Colombia.
Bull. Fish Res. Bd. Can., 168: 68 pp.

R

bioluminescence
Reynolds, George T., J.W. Hastings, Hidemi Sato and A. Randolph Sweeney, 1966.
The identity and photon yield of scintillons of Gonyaulax polyedra. (Abstract only).
Biol. Bull., 131(2):403.

bioluminescence
Roper, C.F.E., 1963.
Observations on bioluminescence in Ommastrephes pteropus (Steenstrup, 1855) with notes on its occurrence in the family Ommastrephidae (Mollusca: Cephalopoda).
Bull. Mar. Sci., Gulf and Caribbean, 13(2):343-353.

bioluminescence
*Rudjakov, J.A., 1968.
Bioluminescence potential and its relation to the concentration of luminescent plankton. (In Russian; English abstract).
Okeanologiia, Akad. Nauk, SSSR, 8(5):888-894.

bioluminescence
*Rudyakov, Yu. A., 1967.
On the methods of studying sea bioluminescence. (In Russian; English abstract).
Okeanologiia, Akad. Nauk, SSSR, 7(4):728-737.

bioluminescence
*Rudjakov, J.A., and N.M. Voronina, 1967.
Plankton and bioluminescence in the Red Sea and Gulf of Aden. (In Russian; English abstract).
Okeanologiia, Akad. Nauk, SSSR, 6(7):1076-1088.

S

bioluminescence

Seliger, H.H., J.H. Carpenter, M. Loftus, W.H. Biggley and W.D. McElroy, 1971. Bioluminescence and phytoplankton successions in Bahía Fosforescente, Puerto Rico. Limnol. Oceanogr. 16(4): 608-622.

bioluminescence

Seliger, H.H., J.H. Carpenter, M. Loftus and W.D. McElroy, 1970. Mechanisms for the accumulation of high concentrations of dinoflagellates in a bioluminescent bay. Limnol. Oceanogr., 15(2): 234-245.

bioluminescence

*Seliger, H.H., and W.G. Fastie, 1968. Studies at Oyster Bay in Jamaica, West Indies. 3. Measurement of underwater-sunlight spectra. J. mar. Res., 26(3):273-280.

bioluminescence

Seliger, H.H., W.G. Fastie and W.D. McElroy, 1960. Bioluminescence in Chesapeake Bay. Science, 133(3454):699-700.

bioluminescence

*Seliger, H.H., and W.D. McElroy, 1968. Studies at Oyster Bay in Jamaica, West Indies. 1. Intensity patterns of bioluminescence in a natural environment. J. mar. Res., 26(3):244-255.

bioluminescence

Shimomura, Osamu, John R. Beers, and Frank H. Johnson, 1964. The cyanide activation of Odontosyllis luminescence. J. Cell. and Comp. Physiol, 64(1):15-22.

bioluminescence

*Shimomura, Osamu, and Frank H. Johnson, 1968. Chaetopterus photoprotein: crystallization and cofactor requirements for bioluminescence. Science, 159(3820):1239-1240.

bioluminescence

Shimomura, O., and F.H. Johnson, 1966. Partial purification of the Chaetopterus luminescence system. In: Bioluminescence in progress, F.H. Johnson and Y. Haneda, editors, Princeton Univ. Press, 495-522.

bioluminescence

*Shimomura, Osamu, Frank H. Johnson and Takashi Masugi, 1969. Cypridina bioluminescence: light-emitting oxyluciferin-luciferase complex. Science, 164(3885):1299-1300.

bioluminescence

Shimomura, Osamu, Frank H. Johnson, and Yo Saiga, 1963. Extraction and properties of halistaurin, a bioluminescent protein from the hydromedusan Halistaura. J. Cell. Comp. Physiol., 62(1):9-16.

bioluminescence

Shimomura, Osamu, Frank H. Johnson, and Yo Saiga, 1963. Further data on the bioluminescent protein, aequorin. J. Cell. Comp. Physiol., 62(1):1-8.

bioluminescence

Shimomura, Osamu, Frank H. Johnson and Yo Saiga, 1963. Microdetermination of calcium by aequorin luminescence. Science, 140(3573):1339-1340.

bioluminescence

Shimomura, Osamu, Frank H. Johnson, and Yo Saiga, 1963. Partial purification and properties of the Odontosyllis luminescence system. J. Cell. Comp. Physiol., 61(3):275-292.

bioluminescence

Shimomura, Osamu, Frank H. Johnson and Yo Saiga, 1962. Extraction, purification and properties of aequorin, a bioluminescent protein from the luminous hydromedusan, Aequorea. J. Cell. and Comp. Physiol., 59(3):223-240.

bioluminescence

Shimomura, O., F.H. Johnson and Yo Saiga, 1961. Purification and properties of Cypridina luciferase. J. Cell. and Comp. Physiol., 58(2):113-124.

bioluminescence

Shoemaker, H.H., 1957. Observations of bioluminescence in the Atlantic fish (Porichthys porosissimum). Science, 126(3283):1112.

bioluminescence

Sie, Edward H.-C., W.D. McElroy, Frank M. Johnson and Y. Haneda, 1961. Spectroscopy of the Apogon luminescent system and of its cross reaction with the Cypridina system. Archives Biochem. and Biophys., 93(2): 286-291.

bioluminescence

Soli, Giogio, 1966. Bioluminescent cycle of photosynthetic dino-flagellates. Limnol. Oceanogr., 11(3):355-363.

Luminescence

Sommer, H., Whedon, W.F., Kofoid, C.A., and A. Stohler, 1937. Relation of paralytic shellfish poison to certain plankton organisms of the genus Gonyaulax. Arch. Path. Vol. 24, pp. 537-559.

bioluminescence

Stachel, H.D., E.C. Taylor, O. Shimomura and F.H. Johnson, 1966. Synthesis and properties of some indole derivatives related to Cypridina luciferin. In: Bioluminescence in progress, F.H. Johnson and Y. Haneda, editors, Princeton Univ. Press, 51-66.

bioluminescence

Strel'nikov, I.D., 1923. [Contributions to a knowledge of the Kara Sea] Izdanie Soveta Inst. pod direktora Instituta Nikolaia Morozova 6:71-80.

HO Transl. 310(Dec. 1955). Transl. by L.G. Robbins. Edited by C.S. Jenner, jr.

bioluminescence

Stukalin, M.V., 1934. [Bioluminescence of the Okhotsk Sea.] Vestnik Dal'nevostochnogo Filiala Akad. Nauk, SSSR, 9:134-137.

Translations may be purchased from the Scientific Translation Center, Library of Congress.

bioluminescence

Stukalin, M.V., 1934. [Bioluminescence of the Okhotsk Sea.] Vestnik Dal'nevostochnogo Filiale Akad. Nauk SSSR (9):137-139.

RT-1045 Bibl. Trans. Rus. Sci. Tech. Lit. 9.

bioluminescence

Sweeney, Beatrice M., 1963. Bioluminescent dinoflagellates. Biol. Bull., 125(1):177-181.

bioluminescence

Sweeney, Beatrice M., and G.B. Bouck, 1966. Crystal-like particles in luminous and non-luminous dinoflagellates. In: Bioluminescence in progress, F.H. Johnson and Y. Haneda, editors, Princeton Univ. Press, 331-348.

bioluminescence

Sweeney, B.M., and J.W. Hastings, 1957. Characteristics of the diurnal rhythm of luminescence in Gonyaulax poledra. J. Cell. Comp. Physiol., 49(1):115-128.

bioluminescence

Swift, Elijah, and W. Rowland Taylor, 1967. Bioluminescence and chloroplast movement in the dinoflagellate Pyrocystis lunula. J. Phycology, 3(2):77-81

T

luminescence

Tarasov, N.I., 1960. [On the stationary observations of the sea luminescence.] Biull. Okeanogr. Komissii, Akad. Nauk, SSSR, (6): 82-83.

bioluminescence

Tarasov, N.I., 1953. [Luminescence at sea.] Priroda 42(9):96.

bioluminescence

Tarasov, N.I., 1956. [Bioluminescence of the sea.] Akad. Nauk. SSSR, Nauchno Popul. Ser.:124 pp.

Abstr.: Meteorol. Abstr. & Bibl., 9(1):21.

bioluminescence

Tarasov, P. (or is it N?) I., 1957. [More on sea luminescence.] Priroda (12):101.

bioluminescence

Tarasov, N.I., and I.I. Gitelzon 1961. Comprehensive investigations of luminescence in the sea during scientific expeditions. (In Russian) Biull. Okeanogr. Komiss. Akad. Nauk SSSR 8: 75-80.

bioluminescence

Taylor W. Rowland, H.H. Selinger, W.G. Fastie and W.D. McElroy, 1966. Biological and physical observations on a phosphorescent bay in Falmouth Harbor, Jamaica, W.I. J. Mar. Res., 24(1):28-43.

bioluminescence

Terada, T., 1934. Luminous phenomena accompanying destructive sea-waves (tunami). Bull. Earthquake Res. Inst., Suppl. Vol. 1:25-35.

bioluminescence

Terio, B., 1964. Possibili interrelazioni tra bioluminescenza e fluorescenza di materiali fotosensi bili presenti nelle pinna e sui parapodi dei Tomopteridi. Osservazioni su Tomopteris (Johnstonella) nationalis Apstein. Atti Soc. Peloritana, Sci. Fis. Mat. e Nat., Messina, 10(2):127-138.

bioluminescence

Tett, P.B., 1971.
The relation between dinoflagellates and the bioluminescence of sea water. J. mar. biol. Ass., U.K., 51(1): 183-206.

bioluminescence

Tett, P.B., 1969.
The effects of temperature upon the flash-stimulated luminescence of the euphausiid Thysanoessa raschi.
J.mar.biol.Ass.,U.K., 49(1):245-258.

luminescence

Thompson, J.V., 1830.
On the luminosity of the ocean with descriptions of some remarkable species of Luminous animals (Pyrosoma pigmaea and Sapphirina indicator), and particularly four new genera, Noctiluca, Cynthia, Lucifer, and Podopsis of the Schizopoda. (Addenda to Memoir 1. Addendum to Memoir 2) Zoological researches and illustrations on natural history of nondescript or imperfectly known animals in a series of memoirs, Vol. 1, Mem. 3: 110 pp., 14 pls.

bioluminescence

Timofeeva, V.A., 1951.
[Distribution of luminosity in a violently agitated medium.] Dok. Akad. Nauk, 76:677-680.

bioluminescence

Timofeeva, V.A., 1951.
[The problem of the distribution of luminosity in the sea.] Dok. Akad. Nauk 76:831-834.

bioluminescence

Titschack, Herbert, 1964.
Untersuchungen ueber das Leuchten der Seefeder Veretillum cynomorium.
Vie et Milieu, Bull. Lab. Arago, 15(3):547-563.

bioluminescence

Tsuji, F.I., and Y. Haneda, 1966.
Chemistry of the luciferases of Cypridina hilgendorfi and Apogon ellioti.
In: Bioluminescence in progress, F.H.Johnson and Y. Haneda, editors, Princeton Univ. Press, 137-150.

bioluminescence

Tsuji, Frederick I., Richard V. Lynch, III and Yata Haneda, 1970.
Studies on the bioluminescence of the marine ostracod crustacean Cypridina serrata.
Biol.Bull.mar biol.Lab., Woods Hole, 139(2):386-401.

bioluminescence

Turner, R.J., 1966.
Marine bioluminescence.
Mar. Obs., 36(211):20-29.

U

bioluminescence

USN Hydrographic Office, 1957.
Oceanographic atlas of the Polar seas. 1. Antarctic. H.O. Pub., No. 705:70 pp.

V

luminescence

Venza, F., 1943.
Fosforescenze marine. Riv. Cult. Mar. 8(7/8): 128-132, (9/12):56-58.

bioluminescence

Verplegh, George, 1965.
The phosphorescent wheel.
Dt. hydrogr. Z. 21(4):152-162.

bioluminescence

Vinogradov, M.E., I.I. Gitelzon and Yu.I. Sorokin, 1970.
The vertical structure of a pelagic community in the tropical ocean. Mar. Biol., 6(3): 187-194.

bioluminescence

Voitov, V.I., A.A. Egorova and N.I. Tarasov, 1960.
On the luminescence of a free Black Sea bacterium, Bacterium issatchenkoi Egorova. (In Russian). Doklady, Akad. Nauk, SSSR, 132(6):1424-1426.

W

XYZ

bioluminescence

Yentsch, C.S., R.H. Backus, and Asa Wing, 1964.
Factors affecting the vertical distribution of bioluminescence in the euphotic zone.
Limnology and Oceanography, 9(4):519-524. (1964)

bioluminescence

Yentsch, Charles S. and J.C. Laird 1968.
Seasonal sequence of bioluminescence and the occurrence of endogenous rhythms in oceanic waters off Woods Hole, Massachusetts.
J. mar. Res. 26(2): 127-133.

bioluminescence, effect of

bioluminescence, effect of

Semenchenko, I.V., and A.V. Snytkina, 1961.
[The study of the spectral brightness at sea from an aircraft.]
Okeanologiia, Akad. Nauk, SSSR, 1(5):856-859.

biomass

Biomass

For particular groups of organisms, information on biomass will be found in the volumes concerned with these groups.

Biomass - here often loosely interpreted

biomass

Allen, K. Radway, 1971.
Relation between production and biomass.
J. Fish. Res. Bd Can. 28(10): 1573-1581.

biomass

Banse, Karl, 1962
Chemie und Autökologie bei produktions-biologischen Untersuchungen des Planktons.
Kieler Meeresf., 18(3) (Sonderheft):132-135.

biomass

Bogorov, B.G., 1957
Regularities of plankton distribution in north-west Pacific. Proc. Unesco Symp., Phys. Ocean., Tokyo, 1955: 260-276.

biomass

Bogorov, B.G., 1956
Unification of plankton research. Colloque Int. Biol. Mar., Sta. Biol. Roscoff, 24 Juin-4 Juillet 1956. Ann. Biol. 33(7-8): 299-315.

biomass

Bogorov, B.G., 1934.
[On the term "biomass" of a plankton organism".]
Biull. Vses. Nauchno-Issled. Inst. Morskogo Ribnogo Chosia. i Okeanogr., No. 2:26-29.

biomass

Bogorov, B.G., 1934.
[On the plankton organisms biomass.]
Biull., Vses. Nauchno-Issled. Inst. Morskogo Ribnogo Chosia. i Okeanogr., No. 1:1-18. (English summary.)

biomass

Bogorov, B.G., 1934.
Seasonal changes in biomass of Calanus finmarchicus in the Plymouth area in 1930. J.M.B.A. 19(2):585-612, 8 textfigs.

biomass

Bogorov, V.G., and M. Ye. Vinogradov, 1961
[Certain peculiarities of the plankton biomass distribution in the surface waters of the Indian Ocean in winter 1959/60.] (In Russian; English abstract).
Okeanolog. Issled., Mezhd. Komitet Proved. Mezhd. Geofiz. Goda, Prezidiume, Akad. Nauk, SSSR, (4):66-71.

biomass

Bogorov, B.G., and M. Ye. Vinogradov, 1961
[Certain esculiaritipe (sic!) of the plankton mass distribution in the surface waters of the Indian Ocean in winter 1959/60.]
Okeanol. Issledov., Mezhd. Komit., Mezhd. Geofiz. God, Presidiume Akad. Nauk, SSSR: 66-71

biomass

Chislenko, L.L., 1961.
[The role of Harpacticoida in mezobenthos biomass of some biotopes in the phytal of the White Sea.]
Zool. Zhurn., 40(7):983-996.

biomass

Della Croce, Norberto, 1962
Zonazione zooplanctonica nel Golfo di Napoli. (Italian and English abstracts).
Pubbl. Staz. Zool., Napoli. 32(Suppl.): 368-379.

Biomass

Emery, K. O., and J. Hülsemann, 1961 (1962).
The relationships of sediments, life and water in a marine basin.
Deep-Sea Res., 8(3/4):165-180.

biomass

Ganapati, P.N., and D.V. Subba Rao, 1958.
Quantitative study of plankton off Lawson's Bay, Waltair.
Proc. Indian Acad. Sci., (B), 48(4):189-209.

biomass

Greze, V.N., 1963
Specific traits noted in the structure of the pelagial in the Ionian Sea. (In Russian).
Okeanologiia, Akad. Nauk, SSSR, 3(1):100-109.

biomass (micro)

Hagmeier, Erik, 1964.
Zum Gehalt an Seston und Plankton im Indischen Ozean zwischen Australien und Indonesien.
Kieler Meeresf., 20(1):12-17.

biomass

Holme, N.A., 1953.
The biomass of the bottom fauna in the English Channel off Plymouth. J.M.B.A. 32(1):1-49, 7 textfigs.

biomass, microbial

Holm-Hansen, Osmund, 1970.
Determination of microbial biomass in deep ocean water. Occ. Pap. Inst. mar. Sci., Alaska, 1: 287-300.

biomass, microbial

Holm-Hansen, Osmund, 1969.
Determination of microbial biomass in ocean profiles. Limnol. Oceanogr., 14(5): 740-747.

biomass

Japan Meteorological Agency, 1962
Report of the Oceanographic observations in the sea east of Honshu from August to September, 1960. (In Japanese).
Res. Mar. Meteorol. and Oceanogr., July-Dec., 1960, Japan Meteorol. Agency, No. 28:21-29.

biomass

Kamšhilov, M.M., 1953.
Biomass of the zooplankton of the White Sea.
Dokl. Akad. Nauk, SSSR, 92(5):1057-1060.

Abstr. Biol. Abstr. 29(2):2791.

biomass

Kamshilov, M.M., 1953.
Biomass of the zooplankton of the White Sea.
Dokl. Akad. Nauk, SSSR, 92(5):1057-1060.

biomass

Kanaeva, I.P., 1965.
On the quantitative distribution of plankton in the Atlantic. Investigations in line with the programme of the International Geophysical year, 2. (In Russian).
Trudy, Vses. Nauchno-Issled. Inst. Morsk. Ribn. Choz. i Okeanogr. (VNIRO), 57:333-343.

biomass

Khromov, N.S., 1965.
Some data on plankton in the Dakar-Freetown area. Investigations in line with the programme of the International Geophysical Year, 2. (In Russian).
Trudy, Vses. Nauchno-Issled. Inst. Morsk. Ribn. Choz. i Okeanogr. (VNIRO), 57:393-404.

biomass

Khromov, N.S., 1965.
On the quantitative distribution of plankton in the northwestern Caribbean Sea and the Gulf of Mexico. Investigations in the line of the programme of the International Geophysical Year, 2. (In Russian).
Trudy, Vses. Nauchno-Issled. Inst. Morsk. Ribn. Choz. i Okeanogr. (VNIRO), 57:381-391.

biomass

Klumov, S.K., 1961
Plankton and the feeding of the whalebone whales (Mystioceti).
Trudy Inst. Okeanol., 51:142-156.

biomass (micro)

Krey, Johannes, 1964
Die mittlere Tiefenverteilung von Seston, Mikrobiomasse und Detritus im nördlichen Nordatlantik.
Kieler Meeresforsch., 20(1):18-29.

Biomass, micro-

Krey, Johannes, 1961.
Beobachtungen uber den Gehalt an Mikrobiomasse und Detritus in der Kieler Bucht 1958-1960.
Kieler Meeresf., 17(2):163-175.

biomass (micro-)

Krey, Johannes, und Karl-Heinz Szekielda, 1966.
Gesamtkohlenstoff und Mikrobiomasse in der Ostsee im Mai 1962.
Kieler Meeresforsch., 22(1):64-69.

biomass

Kriss, A.E., 1956.
The amount of microbial population and biomass at different depths of seas and oceans.
Dokl. Akad. Nauk, SSSR, 111(6):1356-1358.

biomass

Kuzmina, A.I., 1959

Some data on the spring and summer phytoplankton of the North Kuril waters.
Trudy Inst. Okeanol., 36: 215-229.

biomass

Legand, M., and H. Rotschi, 1962.
Bilan des recherches océanographiques en mer du Corail.
Cahiers Océanogr., C.C.O.E.C., 14(10):703-718.

biomass

Loobny-Herzyk, E.A., 1959

Distribution of zooplankton in the Kronotski Bay.
Trudy Inst. Okeanol., 36: 92-100.

biomass

Margalef, Ramon, 1962
Comunidades naturales.
Publicacion Especial, Inst. Biol. Marina, Univ. de Puerto Rico, Mayaquez, 469 pp.

biomass

Margalef, Ramon, y Enrique Morales, 1960

Fitoplancton de las costas de Blanes (Gerona) de julio de 1956 a junio de 1959.
Investigacion Pesquera, Barcelona 16:3-32.

biomass

Margineanu, C., and G.H. Serpoianu. 1961
Le développement du zooplancton marin au littoral roumain dans les conditions thermiques spécifiques à l'hiver 1960/1961.
(In Roumainian; French and Russian summaries)
Buletinul, Inst. Cercetări si Proiectări Piscicole, Roumania, 22(3):17-26.

biomass

McIntyre, A.D., 1961.
Quantitative differences in the fauna of boreal mud associations.
J. Mar. Biol. Assoc., U.K., 41(3):599-616.

biomass

*Nakai, Zinziro, Tadashi Kubota, Takeshi Mizushimia and Yukio Yamada, 1969.
Abundance and distribution of plankton in the waters around the Shimizu Harbour. 1. Diurnal and vertical change of biomass in summer season. (In Japanese; English abstract).
J. Coll. mar. Sci. Technol., Tokai Univ. (3):35-66.

biomass

Nikolaev, I.I., 1961.
Some general aspects involved with the distribution and the biology of mass fauna and flora species in the Baltic Sea.
Okeanologiia, Akad. Nauk, SSSR, 1(6):1046-1058.

biomass

Nikolaev, I., 1957.
Russian observations in the central Baltic between Klaipede and the Estonian Islands.
Ann. Biol., Cons. Perm. Int. Expl. Mer, 12:113, Table, 9.

biomass

Paasche, E., 1960.
Phytoplankton distribution in the Norwegian Sea in June, 1954, related to hydrography and compared with primary production data.
Fiskeridirektoratets Skr., Ser. Havundersøgelser, 12(11):1-77.

biomass

Patten, Bernard C., 1961

Negentropy flow in communities of plankton.
Limnol. & Oceanogr. 6(1): 26-30.

negative entropy = negentropy

biomass

Renaud-Debyser, J., and B. Salvat, 1963.
Le calcul des biovolumes dans l'étude des chaînes alimentaires de la faune endogénée des sédiments intertidaux.
Comptes Rendus, Acad. Sci., Paris, 256(12):2712-2714.
— meubles

biomass

Romanova, N.N., 1956.
Variations of the biomass of higher Crustacea in the northern Caspian Sea, as observed in the course of several years. Dokl. Akad. Nauk, SSSR, 109(2):393-396.

biomass

Savilov, A.I., 1961
Ecologic characteristics of the bottom communities of invertebrates in the Okhotsk Sea.
Trudy Inst. Okeanol., SSSR, 46: 3-84.

biomass

Semina, G.I., 1959

Distribution of phytoplankton in Kronotsk Bay.
Trudy Inst. Okeanol., 36: 73-91.

biomass

Seshadri, B., 1958
Seasonal variations in the total biomass and total organic matter of the plankton in the marine zone of the Vellar Estuary.
J. Zool. Soc., India, Calcutta, 9(2): 183-191.

biomass (benthos)

Shurin, A.T., 1960

Data on the fauna of the Russian bays and conditions for its dissemination.
Trudy Vses. Nauchno-Issled. Inst. Morsk. Ribnogo Chozia. i Okeanogr., 42: 37-60.

biomass

Štirn, Jože, 1969.
Prilog poznavanju kvantiteta i raspodjele pelagijske biomase u Sjevernom Jadranu.
Thalassia Jugoslavica 5: 361-367.

biomass

Timokina, A.F., 1965.
The distribution of zooplankton in different water masses of the Norwegian Sea in the spring and autumn of 1959. Investigations in line with the programme of the International Geophysical Year, 2 (In Russian).
Trudy Vses. Nauchno-Issled. Inst. Morsk. Ribn. Choz. i Okeanogr. (VNIRO), 57:405-424.

biomass

Vives, Francisco, y Manuel López-Benito, 1958
El fitoplancton de la Ría de Vigo y su relación con los factores térmicos y energéticos.
Inv. Pesq., Barcelona, 13:87-124.

biomass, micro

Zeitschel, B., 1969
Productivity and microbiomass in the tropical Atlantic in relation to the hydrographical conditions (with emphasis on the eastern area)
Actes Symp. Oceanogr. Ressources halieut. Atlant. trop., Abidjan, 20-28 Oct. 1966, UNESCO, 69-84.

plankton biomass

Zelikman, E.A., and M.M. Kamshilov, 1960
[Dynamics of the plankton biomass over many years in the southern part of the Barents Sea and the factors which determine this.]
Trudy Murmansk. Morsk. Biol. Inst., 2(6): 68-113.

biomass

Zhukova, A.I., 1955.
[Biomass of microorganisms in bottom sediments of the northern Caspian.] Mikrobiologiia 24(3):321-324.

biomass: data only

biomass (data only)

Australia, Commonwealth Scientific and Industrial Research Organization, 1963.
Coastal investigations at Port Hacking, New South Wales, 1960.
Div. Fish. and Oceanogr., Oceanogr. Sta. List, No. 52:135 pp.
A.D. Crooks, compiler

biomass, microbial

Holm-Hansen, Osmund, 1971.
Determination of microbial biomass in deep ocean profiles. (Portuguese abstract). In: Fertility of the Sea, John D. Costlow, editor, Gordon Breach, 1: 197-207.

biostratigraphy

biostratigraphy

Bartlett, Grant A., 1969.
Cretaceous biostratigraphy of the Grand Banks of Newfoundland.
Marit. Sediments 5(1):4-14.

biostratigraphy

Beckmann, J.P., 1971.
The Foraminifera of sites 68 to 75. Initial Repts Deep Sea Drilling Project 8: 713-725.

biostratigraphic

Blow, W.H., 1970.
Validity of biostratigraphic correlations based on the Globigerinacea.
Micropaleontology, 16(3): 257-268.

biostratigraphy

Bolli, H.M., J.E. Boudreaux, Cesare Emiliani, W.W. Hay, R.J. Hurley and J.I. Jones, 1968.
Biostratigraphy and paleotemperatures of a section cored on the Nicaragua Rise, Caribbean Sea.
Bull. geol. Soc. Am., 79(4): 459-470.

biostratigraphy

Bramlette, M.N. and W.R. Riedel, 1971.
Observations on the biostratigraphy of pelagic sediments. In: Micropaleontology of oceans, B.M. Funnell and W.R. Riedel, editors, Cambridge Univ. Press, 665-668.

biostratigraphy

Bukry, D., 1971.
Coccolith stratigraphy, Leg 8, Deep Sea Drilling Project. Initial Repts Deep Sea Drilling Project 8: 791-807.

biostratigraphy

Bukry, David, Robert G. Douglas, Stanley A. Kling and Valeri Krasheninnikov 1971.
Planktonic microfossil biostratigraphy of the northwestern Pacific Ocean.
Initial Reports, Deep Sea Drilling Project, Glomar Challenger 6: 1253-1300

Biostratigraphy

Kennett, James P., 1970.
Pleistocene paleoclimates and foraminiferal biostratigraphy in subantarctic deep-sea cores.
Deep-Sea Res., 17(1): 125-140.

biostratigraphy

Moore, T.C., 1971.
Radiolaria. Initial Repts Deep Sea Drilling Project, 8: 727-775.

biostratigraphy

Morin, Ronald W. 1971.
Late Quaternary biostratigraphy of cores from beneath the California Current.
Micropaleontology 17(4): 475-491.

biotites

Richards, J.R., and W.F. Willmott, 1970
K-Ar age of biotites from Torres Strait.
Aust. J. Sci., 32(9): 369-370

bioturbation

bioturbation

Diester, Liselotte, 1972.
Zur spätpleistozänen und holozänen Sedimentation im zentralen und östlichen Persischen Golf.
Meteor Forschungsergebnisse (C) 8: 37-64.

bioturbation

Frey, Robert W. and Taylor V. Mayou, 1971.
Decapod burrows in Holocene barrier island beaches and washover fans, Georgia. Senckenbergiana maritima 3(1): 53-77.

bioturbation

Piper, David J.W., and Neil F. Marshall, 1969
Bioturbation of Holocene sediments on La Jolla Deep Sea Fan, California.
J. Sedim. Petrol., 39(2): 601-606.

bioturbation

Reineck, Hans-Erich and Indra Bir Singh, 1971.
Der Golf von Gaeta (Tyrrhenisches Meer). III. Die Gefüge von Vorstrand- und Schelfsedimenten.
Senckenbergiana maritima 3(1): 185-201.

bioturbation

Stanley, Daniel J. 1971.
Bioturbation and sediment failure in some submarine canyons.
Vie Milieu Suppl. 22(2): 541-555.

bioturbation

Winston, Judith E., and Franz E. Anderson 1971.
Bioturbation of sediments in a northern temperate estuary.
Marine Geol. 10(1):39-49.

bipolarity

bipolarity

Birstein, J.A., 1960.
[The family Ischnomesidae (Crustacea, Isopoda, Asellota) in the northwestern part of the Pacific and the problems of amphiboreal and bipolar distribution of the deep sea fauna.]
Zool. Zhurn., 39(1):3-28.

bipolarity

Brodskii, K.A., 1960.
Zoogeographical zones in the southern part of the Pacific Ocean and the bipolar distribution of some calanoids. Biology of the seas. (In Russian).
Trudy Okeanogr. Komissii, Akad. Nauk, SSSR, 10(4):8-13.

bipolarity

Brodsky, K. A., 1959.
[About the bipolar distribution of some planktons.]
Biull. Sovetsk. Antarkt. Exped., No. 6:35-39.

bipolarity

Dell, R.K., 1965.
Marine biology.
In: Antarctica, Trevor Hatherton, editor, Methuen & Co., Ltd., 129-152.

bipolarity

Derjugin, K., 1927.
La distribution bipolaire des organismes marins.
Bull. Inst. Océan., Monaco, No. 495:23 pp.

bipolar species

Nesis, K.N., 1961.
[The pathways and time of formation of a broken up areal in amphiboreal species of marine bottom animals.]
Okeanologiia, Akad. Nauk, SSSR, 1(5):893-903.

Birds

birds

See also: guano birds

A

birds

Aldcroft, D.S., P.J. Cowan and R.J. Kennedy 1969.
Ross's gull at Cape Clear Island. a bird new to Ireland.
Ir. Nat. J., 16(8):237.

birds

Aldrich, John W., 1970.
Review of the problem of birds contaminated by oil and their rehabilitation.
Resource Publ. Bur. Sport Fish. Wildl., U.S. Fish Wildl. Serv. 87:23pp.

birds

Alexander, W.B., 1928
Birds of the Ocean. 428 pp., 87 pls.
G.P. Putnam's Sons, N.Y. - London

birds

Alexander, W.B., et.al. (joint letter with 15 authors).
The Families and Genera of the Petrels and their names.
Ibis. 108(1965):401-405.

birds

Amerson, A. Binion Jr., 1969.
Ornithology of the Marshall and Gilbert Islands.
Atoll Res. Bull. 127: 348pp (multilitted)

birds

Anderson, William, 1970.
The California least tern breeding in Alameda and San Mateo counties.
Calif. Fish Game 56(2): 136-137

birds

Anon., 1969.
The care and feeding of birds which seek shelter on ships at sea.
Mar. Obs. 39(226) (M.O. 814): 193-195

birds

Arsenjev, V.A., 1960

[Observations on the sea animals and birds of the Antarctic.] Arktich. i Antarkt. Nauchno-Issled. Inst. Sovetsk. Antarkt. Exped., Mezhd. Geofiz. God, Vtoraia Morsk. Exped., "Ob", 1956-1957, 7: 85-96.

birds

Ashmole, Myrtle J., 1970
Feeding of western and semipalmated sandpipers in Peruvian winter quarters.
Auk 87(1): 131-135

BIRDS

Ashmole, N. P , 1965.
Adaptive variation in the breeding regime of a tropical sea bird.
Proc. Nat. Acad. Sci., 53(2): 311-318.

BIRDS

Ashmole, N. P. and S. H. Tovar, 1968.
Prolonged parental care in royal terns and other birds. Auk, 85: 90-100.

birds

Ashmole, M.J. and N.P. Ashmole, 1968.
The use of food samples from sea birds in the study of seasonal variation in the surface fauna of tropical oceanic areas.
Pacif. Sci., 22:1-10.

birds

*Ashmole, N. Philip, and Myrtle J. Ashmole, 1967.
Comparative feeding ecology of sea birds of a tropical oceanic island.
Bull. Peabody Mus. N.H., Yale Univ. 24:1-131.

birds

Ashmole, N.P. and M.J. Ashmole, 1967.
Comparative feeding ecology of sea birds of a tropical oceanic island.
Yale Univ., Peabody Mus. Nat. Hist. Bull. 24:131 pp. 11 figs.

birds

Ashmole, M. J. and N. P. Ashmole 1967.
Notes on the breeding season and food of the red-footed booby (Sula sula) on Oahu, Hawaii.
Andea 55: 265-267.

B

BIRDS

Bailey, N. and W. R. P. Bourne, 1963.
Some records of petrels handled in the northern Indian Ocean.
J. Bombay Nat. Hist. Soc.

birds

Bailey, R.S., 1967.
Migrant waders in the Indian Ocean.
Ibis, 109:437-439.

birds

Bailey, Roger, 1966.
The sea-birds of the southeast coast of Arabia.
Ibis, 108:224-264.

BIRDS

*Bailey, R. S., R. Pocklington and P. R Willis, 1968.
Storm-petrels, Oceanodroma sp. in the Indian Ocean.
Ibis, 110: 27-34.

birds

Bang, B.G., 1965.
Anatomical adaptations for olfaction in the Snow Petrel.
Nature, 205 (4970):513-515.

BIRDS

Barth, Edvard K. 1967.
Egg dimensions and laying dates of Larus marinus, L. argentatus, L. fuscus and L. canus.
Nytt Mag. Zool, 15:1-34. Also: Medd. Zool. Mus. Oslo 81.

birds

Beck, J.R. 1969.
Food, moult and age of first breeding in the cape pigeon, Daption capensis Linnaeus.
Bull. Br. Antarct. Surv. 21: 33-44.

birds

Beck, J.R., and D.W. Brown 1971.
The breeding biology of the black-backed storm-petrel Fregetta tropica.
Ibis 113: 73-90

birds

*Beer, C.G., 1969.
Laughing gull chicks: recognition of their parents' voices. Science, 166(3908): 1030-1032.

birds

Belopolsky, L.O., 1957.
[Certain adaptive peculiarities in marine colonial birds in the Arctic.] Zool. Zhurnal, 36(3):432-442

birds

Bengston, Sven-Axel, 1971.
Breeding success of the arctic tern, Sterna paradisaea (Pontoppidan), in the Kongsfjord area, Spitsbergen in 1967. Norw. J. Zool. 19(1): 77-82.

birds (soaring of)

*Berger, Martin und Charlotte, 1968.
Das Meeressegeln des Eissturmvogels. (Fulmarus glacialis).
J. Ornithologie, 109 (4):418-420.

birds

Berndt, Rudolf, und Ute Rahme, 1968.
Erstnachweis der Ringschnabelmöwe (Larus delawarensis) in Europa.
J. Ornithologie, 109 (4):438-440.

birds

Berry, R.J. and P.E. Davis, 1970.
Polymorphism and behaviour in the arctic skua (Stercorarius parasiticus (L.)).
Proc. R. Soc. (B) 175 (1040): 255-267.

BIRDS

*Billings, S M., 1968.
Homing in Leach's petrels.
The Auk, 85: 36-43.

birds

Binford, Laurence C., 1971.
Seabirds of the Galapagos
Pacific Discovery 24(1): 23-27. (popular)

birds

Boecker, Maximilien 1968.
Zur Tagesaktivität der Seeschwalben
J. Ornithologie, 109(1): 62-66

birds

Bongiorno, Salvatore F., 1968.
Egg puncturing behavior in laughing gulls.
Auk, 85(4):697-699.

birds

Bourne W.R.P., 1965.
Observations of an encounter between birds and floating oil.
Nature, Lond., 219(5154): 632.

birds

Bourne, W.R.P., 1967.
Subfossil petrel bones from the Chatham Islands.
Ibis, 109:1-7.

birds

Bourne, W.R.P., 1967.
Long-distance vagrancy in the Petrels.
Ibis, 109:141-167.

birds

Bourne, W.R.P., 1966.
International collaboration in the analysis of seabird observations.
Sea Swallow, 18:9-11.

birds

Bourne, W.R.P., 1963.
"A Review of Oceanic Studies of the Biology of Seabirds".
Proc., 11th International Ornithological Congress, 831-854.

birds
Bourne, W.R.P., J.D. Parrack and G.R. Potts 1967.
Birds killed in the Torrey Canyon disaster.
Nature, Lond. 215 (5106): 1123-1125.

birds
Bourne, W.R.P. and J. Warham, 1966.
"Geographical variation in the Giant Petrels of the Genus Macronectes."
Ardea, 54(½).

birds
Bowen, Vaughn T., and Geoffrey D. Nichols 1968.
An egret observed on St. Paul's Rocks, equatorial Atlantic Ocean.
Auk, 85(1):130-131.

birds
Boyd, John C., and William J.L. Sladen, 1971.
Telemetry studies of the internal body temperatures of Adélie and Emperor penguins at Cape Crozier, Ross Island, Antarctica.
Auk, 88(2): 366-380

birds
Brackbill, Hervey, 1971.
Herring and ring-billed gulls paired or courting in Maryland in January and February.
Auk, 88(2): 438-440

birds
Brown, R.G.B. 1968.
Sea birds in Newfoundland and Greenland waters April-May 1966.
Can. Field-Nat. 82 (2): 88-102.

birds
Brown, R.G.B., 1967.
Sea birds off Halifax, March 1967.
Can.Fld.Nat. 81(4):276-278.

birds
Brown, R.G.B., 1967.
Breeding success and population growth in a colony of Herring and Lesser Black-backed Gulls Larus argentatus and Larus fuscus.
Ibis, 109:502-515.

birds
Brown, R.G.B., 1967.
Species isolation between the Herring Gull Larus argentatus and Lesser Black-backed Gull L. fuscus.
Ibis, 109:310-317.

birds
Brown, R.G.B. and D.E. Baird, 1965.
Social factors as possible regulators of Puffinus gravis numbers.
Ibis, 107:249-251.

birds
Brown, R.G.B., Eric L. Mills, Anne H. Mills and Roger Pocklington, 1968.
Identifying sea-birds of the northwest Atlantic Ocean: an introduction for ships at sea.
Inst. Oceanogr., Dalhousie Univ., Halifax, Canada, 37 pp. (mimeographed).

birds
Brun, Einar 1971.
Breeding distribution and population of cliff-breeding sea-birds in Sör-Varanger, North Norway.
Astarte 4 (2): 53-60

birds
Brun, Einar 1971
Census of puffins (Fratercula arctica) on Nord-Fugløy, Troms.
Astarte 4 (2): 41-45

birds
Buckley, P.A., and F.G. Buckley, 1970
Color variation in the soft parts and down of Royal Tern chicks
Auk 87(1): 1-13

birds, lists of spp.
Buden, Donald W., and Albert Schwartz, 1969.
Reptiles and birds of the Cay Sal Bank, Bahama islands.
Q Jl Fla. Acad. Sci., 31(4):290-320.

birds
Burton, P.J.K., 1971.
Comparative anatomy of head and neck in the Spoon-billed Sandpiper, Eurynorhynchus pygmeus and its allies.
J. Zool. Lond., 163 (2): 145-165

birds
*Burton, R.W., 1968.
Agonistic behaviour of the brown skua, Catharacta skua lönnbergi (Mathews).
Bull. Brit. antarct. Surv., (16):15-39.

C

birds
Campbell, Kenneth E., Jr., 1971.
First report of Sandwich terns in Peru.
Auk 88(3): 676.

birds
Campbell, R. Wayne, 1968.
Status of breeding herring gulls at Bridge Lake, British Columbia.
Can. Fld Nat. 82(3):217-219.

birds
Campbell, R. Wayne, and Theed Pearse 1968.
Notes on a twenty-year old glaucous-winged gull.
Bird Band. 39(3): 226-227.

gulls (herring)
Carey, Francis G., and Knut Schmidt-Nielsen, 1962
Secretion of iodide by the nasal gland of birds.
Science, 137(3533):866-867.

BIRDS
Cheke, A. S., 1964.
Notes on Calonectris laucomelas and other sea birds seen from S.S. Cattay, between Taiwan and Yokohama, 20-22 August, 1963.
Misc. Reps. Yamashina Inst. Orn. Zool., 4; 2 (22).

birds
Clapp, R.B. and F.C. Sibley, 1967.
New records of birds from the Phoenix and Line Islands.
Ibis, 109:122-125.

BIRDS
Clapp, R. B. and F. C. Sibley, 1966.
Longevity records of some central Pacific seabirds.
Bird-Banding, 37(3): 193-197.

birds
*Cline, David R., Donald B. Siniff, and Albert W. Erickson, 1969.
Summer birds of the pack ice in the Weddell Sea, Antarctica.
The Auk, 86(4): 701-716

birds
Coulson, J.C., G.R. Potts, I.R. Deans and S.M. Fraser, 1968.
Mortality of shags and other sea birds caused by paralytic shellfish poison.
Nature, Lond., 220(5162):23-24.

birds
Coulson, J.B., G.R. Potts, and Jean Horobin 1969.
Variation in the eggs of the shag (Phalacrocorax aristotelis).
Auk 86(2): 232-245

birds
Crowell, E.M., and Sears Crowell 1946.
The displacement of terns by gulls at the Weepecket islands, Massachusetts.
Bird-Banding 17 (1): 1-10.

D

birds
Davies, D.H., 1958.
The predation of sea-birds in the commercial fishery. Union S. Africa, Dept. Comm. & Indust., Div. Fish., Invest. Rept., No. 31:16 pp.

birds
Davies, D.H., 1954.
The South African pilchard (Sardinops ocellata): Bird predators, 1953-54.
Dept. Commerce & Industry, Div. Fish., Investig. Rept. No. 18:1-32.

Reprint from "Commerce & Industry", Jan. 1955.

birds
Davis, W. Marvin, 1971.
A specimen of the little gull from northern Mississippi.
Auk, 88(2): 437-438

birds
de Schauensee, R.M., 1963.
Yellow-headed blackbirds at sea in the Atlantic Ocean.
The Auk, 80(4):549.

birds
Devillers, Pierre, Guy McCaskie and Joseph R. Jehl Jr. 1971.
The distribution of certain large gulls (Larus) in southern California and Baja California.
Calif. Birds 2(1): 11-26.

birds
Donaldson, Grace, 1968.
Bill color changes in adult roseate terns
Auk, 85(4):662-668.

birds
Douglas, D.S., 1966.
Secretion of thiocyanate ion by the nasal gland of the Adelie Penguin.
Nature, 209(5028):1150-1151.

birds
Duncan, K.W., 1969.
The food of the black shag (Phalacrocorax carbo novae-hollandiae) in Otago inland waters.
Trans. R. Soc. N.Z., Biol. Sci. 11 (2): 9-23.

E

birds
Eliassen, E. and I. Hjelmtvedt, 1959.
The loss of water in wind drifted migratory birds.
Univ. Bergen Arbok, 1958, Natur. Rekke, 11: 20 pp.

BIRDS
Erard C., G. Jarry et F. Larigauderie, 1967.
Sur les observations d'oiseaux de mer en Méditerranée.
L'Oiseau, 37(4): 336-338.

birds
Erdman, Donald S., 1967.
Sea birds in relation to game fish schools off Puerto Rico and the Virgin Islands.
Carib. J. Sci., 7(1/2):79-85.

F

BIRDS
Falla, R. A., 1965.
Distribution of Hutton's shearwater in New Zealand.
Notornis 12: 66-70.

birds
Ferguson, A., 1971.
Notes on the breeding of the common scoter, Melanitta nigra L. in Ireland.
Ir. Nat. J., 17(2): 29-31.

birds
Firth, Roderick, Jr. 1971.
Roseate tern breeds during its second year.
Bird-Banding 42(4): 300-301.

BIRDS
Fisher, H. I., 1968.
The "two-egg clutch" in the Laysan albatross.
Auk, 85: 134-136.

birds
Fisher, H.I., 1967.
Body weights in Laysan Albatrosses.
Diomedea immutabilis.
Ibis, 109:373-382.

birds
Fisher, J., and R.M. Lockley, 1954.
Sea-birds. An introduction to the natural history of the sea-birds of the North Atlantic.
Houghton, Mifflin, Boston, xvi 320 pp., illus., pla., $6.00.

birds
Forbes, H.O. 1914
Notes on Molina's Pelican (Pelecanus thagus). Ibis., Ser. 10, II:403-420.

birds
Fordham, R.A. 1965.
Dispersion and dispersal of the Dominican Gull in Wellington, New Zealand.
Proc. N.Z. Ecol. Soc. 15: 40-50 (not seen)

birds
Furphy, J.S., F.D. Hamilton and O.J. Merne 1971.
Seabird deaths in Ireland, autumn 1969.
Ir. Nat. J. 17(2): 34-40.

G

birds
Gamarra Dulanto, Luis, 1964.
Las fluctuaciones en la poblacion de las aves guaneras en la costa Peruana.
Bol. Corp. Nacional de Fertilizantes, Lima, Peru, 2(12):3-12.

birds
Gill, Douglas E., William J.L. Sladen and Charles E. Huntington, 1970.
A technique for capturing petrels and shearwaters at sea.
Bird-Banding, 41 (2): 111-113

birds
Gill, F.B., 1967.
Birds of Rodriguez Island (Indian Ocean).
Ibis, 109:383-390.

birds
Gill, Frank B., 1967.
Observations on the pelagic distribution of sea birds in the western Indian Ocean.
Proc. U.S. nat. Mus., 123(3605):1-33.

BIRDS
*Gibson, J. D., 1967.
The wandering albatross (Diomedea exulans): results of banding and observations in New South Wales coastal waters and the Tasman Sea.
Notornis, 14 (2): 47-57.

birds
Gilmore, Raymond M., 1971.
Observations on marine mammals and birds off the coast of southern and central Chile, early winter 1970.
U.S. Antarctic Jl. 6(1): 10-11

birds
Gobeil, Robert E., 1970.
Arterial system of the herring gull (Larus argentatus). J. Zool., Lond. 160 (3): 339-354

birds
Goethe, Friedrich, 1970.
Märzaspekt des Vogellebens in der südlichen Nordsee.
Ber. dt. wiss. Komm. Meeresforsch. 21 (1/4): 430-443.

birds
Gordon, M.E., 1955.
Summer ecology of oceanic birds off southern New England. Auk 72:138-147, 1 textfig.

BIRDS
Gould, P. J., 1967.
Nocturnal feeding of Sterna furcata and Puffinus pacificus.
Condor, 69: 529.

BIRDS
Grive, J., 1964.
The skua, the nearest animal to the pole.
Sci. Progrès Nature, 331: 274-276.

birds
Gwynn, A.M., 1968.
The migration of the Arctic Tern.
Austral. Bird Bander 6(4): 71-75. (not seen)

H

birds
~~Hvalfangernes Assuranceforening, 1936:~~
Hansen, H. E., editor
Atlas over Antarktis of Sydishavet utgitt av Hvalfangernes Assuranceforening i anledning av Foreningens 25-aars Jubileum. Red. av........H.E. Hansen. 1 pl., 17 pp. incl. col. charts. 41.5x35 cm. Sandefjord, Norge

birds
Harrington, H.J. and I.C. McKellar, 1958
A radiocarbon date for penguin colonization of Cape Hallett, Antarctica. N.Z.J. Geol. Geophys. 1(3): 571-576.

birds
Harris, Michael P., 1970.
Breeding ecology of the swallow-tailed gull, Creagrus furcatus.
Auk 87 (2): 215-243.

birds
Harris, M.P. 1969.
Breeding seasons of sea-birds in the Galapagos Islands.
J. Zool. Lond., 159 (2): 145-165

birds
*Hatch, Jeremy J., 1970.
Predation and piracy by gulls at a ternery in Maine. Auk. 87(2): 244-254.

birds
Hays, Helen 1971.
Roseate tern, Sterna dougalli, banded on Atlantic coast recovered on Pacific.
Bird-Banding 42 (4): 295.

birds
Hays, Helen, and Robert W. Risebrough 1972
Pollutant concentrations in abnormal young terns from Long Island Sound.
The Auk 89 (1): 19-35

birds
Heppleston, P.B., 1970.
Anatomical observations on the bill of the oystercatcher (Haematopus ostralegus occidentalis) in relation to feeding behaviour.
J. Zool. Lond. 161 (4): 519-524.

birds (penguins)
*Herbert, C., 1967.
A timed series of embryonic developmental stages of the Adélie penguin (Pygoscelis adeliae) from Signy Island, South Orkney Islands.
Bull. Brit. Antarct. Surv., 14:45-67.

birds
Hildén, Olavi, 1971.
Occurrence, migration and colour phases of the Arctic skua (Stercorarius parasiticus) in Finland.
Annot. Zool. Fennici 8: 223-230

birds
Hillis, J.P. 1971.
Sea-birds scavenging at trawlers in Irish waters.
Ir. Nat. Jl 17(4): 129-132.

birds

Hjelmtveit, Ingvar, 1969.
On migratory birds at Ocean Weather Station "M".
Årbok Univ. i Bergen mat. naturv. Ser. 1969(4): 47 pp.

birds

Hjelmtveit, I., 1961.
Observations of birds at Ocean Weather Station "M".
Marine Observer, 31(191):38-41.

birds

Hubbs, Carl L., Arthur L. Kelly and Conrad Limbaugh (deceased 1960), 1970.
Diversity in feeding by Brandt's cormorant near San Diego.
Calif. Fish Game 56(3): 156-165

I

birds

Ingolfsson, Agnar 1969.
Sexual dimorphism of large gulls (Larus spp.).
Auk 86(4): 732-737.

birds

Ingolfsson, A. 1969.
Behavior of gulls robbing eiders.
Bird Study 16: 45-52 (not seen)

birds

Isenmann, P., 1970.
Contribution à la biologie de reproduction du petrel des neiges (Pagodroma nivea Forster).
Le problème de la petite et de la grande forme.
L'Oiseau, Rev. franç. Ornithol., 40 (no. spéc.): 99-134.
Also: Expéd. polaires franç. 318 (Mission Paul-Emile Victor)

birds

Isenmann, P., F. Lacan et J. Prévost, 1969.
Notes sur une breve visite à Cape Hunter et Cape Denison (Commonwealth Bay, King George V Land, Antarctique).
L'Oiseau, Rev. franç. Ornithol., 39 (no. spéc.): 2-10
Also: Expéd. polaires franç., Missions Paul-Emil Victor, 312.

J

birds (lists of spp)

Japan, Hokkaido University, Faculty of Fisheries 1957.
Data record of oceanographic observations and exploratory fishing, No. 1:247 pp.

birds

Japan, Science Council, National Committee for IIOE, 1966.
General report of the participation of Japan in the International Indian Ocean Expedition.
Rec. Oceanogr. Wks., Japan, n.s. 8(2): 133 pp.

birds

Jehl, Joseph R. Jr. 1971.
A hybrid glaucous x herring gull from San Diego.
Calif. Birds 2(1): 27-32.

birds

Jenkins, J. 1967.
Unusual records of birds at sea.
Notornis 14: 153.

birds

Jenkins, J. 1967.
Sightings of Kermadec petrels at sea.
Notornis 14: 113.

birds

Jensen, Albert C., 1971.
Land birds at sea.
Sea Frontiers, 17(1): 48-56

birds

Jordan S., Romulo, 1957.
El fenomeno de las regurgitaciones en el guanay (Phalacrocorax bougainvilli Lesson) y un metodo para estimar le ingestion diaria. Bol. C.A.G., 33(12):5-18.

birds

Jordán, Rómulo, y Humberto Fuentes, 1966.
Las poblaciones de aves guaneras y su situación actual.
Inst. del Mar, Peru, Informe (10):31 pp.

birds

Jouanin, C. and F.B. Gill, 1967.
Recherche du Petrel de Barau (Pterodroma baraui)
Oiseau et la Revue Francaise d'Ornithologie, 37:1-19.

K

birds

Kadlec, John A., and William H. Drury, Jr. 1969.
Loss of bands from adult herring gulls.
Bird Band. 40(3): 216-221

birds

*Kadlec, John A., and William H. Drury, 1968.
Structure of the New England herring gull population.
Ecology, 49(4):644-676.

BIRDS

Kale, H. W., 1963.
Occurrences of the greater shearwater along the southern Atlantic and Gulf coasts of the U.S.A.
Oriole, 28: 1-4.

BIRDS

King, W. B., 1967.
Sea birds of the tropical Pacific Ocean.
Prelim. Smithsonian Ident. Manual.
Smithsonian Inst., Washington. 126 pp.

birds

King, W.B., G.E. Watson, and P.J. Gould, 1967.
"An application of automatic data processing to the study of seabirds" 1. Numerical Coding.
Proc. U.S. Nat. Mus., 123(3609):1-29.

birds

Kinsky, F.C. 1968
An unusual bird mortality in southern North Island (New Zealand) April 1965.
Notornis 15: 143-155 (not seen)

birds

Kinsky, F.C., 1963.
The southern black-backed gull (Larus dominicanus) Lichtenstein.
Rec. Dom. Mus., New Zealand, 4(14):149-219.

birds

Kooyman, G.L., C.M. Drabek, R. Elsner and W.B. Campbell 1971.
Diving behavior of the Emperor penguin Aptenodytes forsteri.
Auk 88 (4): 775-795.

birds

Korotkevich, E.S., 1958.
Observations on birds during the first wintering of the Soviet Antarctic Expedition, 1956-1957.
Inform. Biull. Sovetsk. Antarkt. Exped., 1 (3): 83-84.

Translation:
Scripta Tecnica, Inc., Elsevier Publ.Co., 149-152.

birds

Kurochin, E.N., 1963.
Distribution of some sea birds in the North Pacific. (In Russian; English summary).
Zool. Zhurn., Akad. Nauk, SSSR, 42(8):1223-1231.

BIRDS

Kuroda, N., 1964.
Land birds at sea, a review.
Misc. Rep. Yamashina Inst. Orn. 4(2): 22.

BIRDS

Kuroda, N., 1963. A winter sea-bird census between Tokyo and Kushiro.
Misc. Rep. Yamashina Inst. Orn. 3(4): No. 19.

birds

Kuroda, Naghisa, 1960.
Analysis of sea bird distribution in the northwest Pacific Ocean.
Pacific Science, 14(1):55-67.

birds

Krylov, V.I. 1968
Marine mammals and birds in the Bellinghausen Station area. (In Russian).
Inform. Biull. Sovetsk. Antarkt. Exped. 71: 68-70.
Translation: Scripta Tecnica for AGU 7(3): 218-219

L

birds

Lacan, F., J. Prévost et M. van Beveren, 1969.
Etude des populations d'oiseaux de l'Archipel de Pointe Géologie de 1965 à 1968.
L'Oiseau, Rev. franç. Ornithol. 39 (no. spéc.): 11-32
Also: Expéd. polaires franç., Missions Paul-Emil Victor, 312

birds

Lavalle, J.A. de, 1917
Informe preliminar sobre la causa de la mortalidad de las aves ocurrida en el mes de marzo del presente año. Mem. Compañia Admin. Guano, VIII:61-84.

BIRDS
*Le Morvan P., J. L. Mougin et J. Prévost, 1967. Écologie du Skua antarctique (Stercorarius skua maccormicki) dans l'archipel de Pointe Géologie (Terre Adélie). L'Oiseau, 37(3): 193-220.

birds
Lloyd, H., 1963. Catharacta skua Brünnich sighted in North Pacific. Canadian Field Naturalist, 77(2):127.

birds
Loustau-Lalanne, P. 1963. Sea and shore birds of the Seychelles. Seychelles Soc., Occas. Publ. 2:

birds
Lostau-Lalanne, P. 1962. Land birds of the Granitic Islands of the Seychelles. Seychelles Soc. Occas. Publ. 1:

birds
Løvenskiolı, Herman L., 1963. Avifauna Svalbardensis. Norsk Polarinstitutt Skrifter, No. 129:460 pp.

birds
Loy, W., 1962. Ornithological profile from Iceland to Antarctica. Gerfaut, 52:626-640.

M

birds
Mattox, William G., 1970. Bird-banding in Greenland. Arctic 23(4): 217-228

birds
Maugin, J.-L., 1967. Étude écologique des deux éspèces de Fulmars: le Fulmar Atlantique (Fulmarus glacialis) et le Fulmar Antarctique (Fulmarus glacialoides). l'Oiseau et la Rev. Franc. d'Ornith, 37:57-103.

Mc

birds
McCaskie, Guy, 1970. The occurrences of four species of Pelecaniformes in the southwestern United States. Calif. Birds 1(4): 117-142.

birds (nos.)
McGary, James W., and Joseph J. Graham, 1960. Biological and oceanographic observations in the central north Pacific, July-September 1958. U.S.F.W.S. Spec. Sci. Rept., Fish., No. 358: 107 pp.

birds
McHugh, J.L., 1955. Distribution of black-footed albatross, Diomedea nigripes, off the west coast of North America, 1949 and 1950. Pacific Science 9(4):375-381.

birds
McNeil, Raymond, and Jean Burton, 1971. First authentic North American record of the British storm petrel (Hydrobates pelagicus). Auk 88(3): 671-672

birds
Meischke, M. H. A., 1967. Notes on sea birds 23. Observations of salt excretion by Pacific gull. Ardea, 55: 269.

birds
Metcalf, W.G., 1966. Observations of migrating Great Shearwaters Puffinus gravis off the Brazilian coast. Ibis, 108:138-140.

BIRDS
Mills, E. L., 1968. Observations of the ringed storm-petrel off the north-west coast of South America. The Condor, 70(1): 87-88.

birds
Mills, E.L., 1967. Bird records from Southwestern Ecuador. Ibis, 109:534-538.

birds
Moore, H.B., 1951. The seasonal distribution of oceanic birds in the western North Atlantic. Bull. Mar. Sci., Gulf and Caribbean 1(1):1-14, 10 textfigs.

birds
Moore, H.B., 1941 Notes on Bermuda Birds. Proc. Limn. Soc., N.Y. 1941:52-53.

owned by M.S.

birds
Moore, H.B., 1941 Notes on the Distributions of Oceanic Birds in the North Atlantic 1937-1941. Proc. Limn. Soc., N.Y., 1941:52-53.

owned by M.S.

birds
Moore, H.B., 1938 Notes on birds seen in the North Atlantic. Brit. Birds No.2, Vol.32:47.

owned by M.S.

birds
Morse, Douglass H., 1971. Great horned owls and nesting seabirds. Auk, 88(2): 426-427

BIRDS
Mörzer Bruyns, W. F. J., 1967. Notes on sea birds 24. Black-capped petrels (Pterodroma hasitata) in the Atlantic. Ardea, 55: 270.

birds
Morzer Bruyns, W.J. F., and K.H. Voers, 1967. Notes on sea-birds. 19. Black-capped Petrels (Pterodroma hasitata) in the Catibbean. Ardea, 55: 144-145.

birds
Motoda, S., and T. Fujii, 1956. Report from the "Oshoru maru" on oceanographic and biological investigations in the Bering Sea and northern North Pacific in the summer of 1955. 1. Programme of investigations and records of eye observations of sea birds and marine mammals. Bull. Fac. Fish., Hokkaido Univ., 6(4):280-297.

birds
Mougin, J.-L., 1968. Notes sur le cycle reproducteur et la mue du manchot Adélie (Pygoscelis adeliae) dans l'Archipel de Pointe Géologie (Terre Adélie). L'Oiseau, Rev. franç. Ornithol. 38 (spéc. no.): 89-94.
Also: Écologie des Oiseaux antarctiques, CNFRA

birds
Mougin, J-L., 1968. Étude écologique de quatre espèces de petrels antarctiques. L'Oiseau et R.F.O. 38 (spéc. no.): 1-52.
Also: Écologie des Oiseaux antarctiques, CNFRA.

birds
*Mougin, Jean-Louis, 1967. Etude ecologique des deux especes de fulmars, le fulmar atlantique (Fulmarus glacialis) et le fulmar antarctique (Fulmarus glacialoides). L'Oiseau et R.F.O., 37(1/2):57-103.

birds
Mougin, J.L., J. Prévost et M. Van Beveren, 1969. Note sur le baguage des oiseaux de l'Ile de la Possession (Archipel Crozet) de 1960 à 1968. L'Oiseau, Rev. franç. Ornithol., 39 (no. spéc.): 51-57.
Also: Direction Labs. scient. Territoires Terres australes et antarctiques franç. Sept. 1970

birds
Moynihan, M., 1956. Notes on the behaviour of some North American gulls. Behaviour, 10(1/2):126-178.

birds
Murphy, Robert Cushman, 1967. Distribution of North Atlantic pelagic birds. Ser. Atlas, Mar. Environment, 14:8 pls.

BIRDS
Murphy, R. C., 1963. Systematics and distribution of Antarctic petrels. Actual. scient. Symp. S.C.A.R. Biol. Antart., Paris, 1312:

birds
Murphy, Robert Cushman, 1962. The oceanic life of the Antarctic. Scientific American, 207(3):187-210.

N

birds
Napier, R.B., 1968. Erect-crested and rockhopper penguins inter breeding in the Falkland islands. Bull. Brit. antarct. Surv., (16):71-72.

birds

Nelson, J.B. 1969.
The breeding ecology of the red footed booby in the Galapagos.
J. Anim. Ecol. 38(1): 181-198.

birds

Nelson J.B. 1965.
Breeding behaviour of the swallow-tailed gull in the Galapagos.
Behaviour 30(2/3): 146-174

birds

Nelson, J.B., 1967.
The breeding behaviour of the White Booby
Sula dactylatra.
Ibis, 109:194-231.

birds

Nelson, J. B., 1966.
Population dynamics of the gannet (Sula bassana) at the Bass Rock with comparative information from other Sulidae.
J. Anim. Ecol., 35: 443-470.

birds

Nelson, J.B., 1967.
Colonial and cliff nesting in the Gannet.
Ardea, 55: 60-90.

birds

Norderhaug, Magnar. 1970.
The present status of the pink-footed goose (Anser fabalis brachyrhynchus) in Svalbard.
Årbok norsk Polarinst., 1969: 55-69.

birds

Norris, A. Y., 1965.
Observations of sea birds in the Tasman Sea and in New Zealand waters in October and November, & November, 1962.
Notornis, 12: 80-105.

birds

Novatti, Ricardo, 1962.
Distribucion pelagica de aves en el Mar de Weddell.
Contr. Inst. Antarctico Argentino, (67):22 pp.

birds

Novatti, Ricardo, 1960
Observaciones sobre aves oceanicas en el Mar de Weddell durante el verano 1959-1960.
Contr. Inst. Antartico Argentino, No. 53: 19 pp.

O

birds

Orians, Gordon H., 1969.
Age and hunting success in the brown pelican (Pelecanus occidentalis).
Anim. Behav. 17(2): 316-319.

birds

Ozawa, Keijiro, and Yoshihiko Nakamura, 1966.
Observations of birds in the Indian Ocean (II.)
J. Tokyo Univ. Fish., (Spec. edit.) 8(2):17-36.

birds

Ozawa, Keijiro, Tatsumi Yamada, Masatoshi Kira and Takehiko Shimizu, 1968.
Observations of sea-birds in the southern Ocean (III). J. Tokyo Univ. Fish., 9(2): 51-100.

P

birds

Peterson, Roger Tory, and George E. Watson, 1971.
Franklin's gull and bridled tern in southern Chile.
Auk 88 (3): 670-671

birds

Phillips, J.H., 1963.
The pelagic distribution of the Sooty Shearwater Procellaria grisea.
Ibis, 105(3):340-353.

birds

Pitman, C.R.S. 1967.
Sea fowl observations on a voyage, Capetown to London, 23 January to 8 February 1967.
Bull. Brit. Orni. Col. Club 87: 117-120.

birds

Pocklington, R., 1965.
Birds seen on Coco I., Cavgados Cavajos Shoals, I.O.
Ibis, 107: 387.

birds

Pocklington, R., 1965.
A Summary of birds seen over the Western Indian Ocean August-November, 1963.
Ibis, 107:385-386.

birds

*Potthoff, Thomas and William J. Richards, 1970.
Juvenile bluefin tuna, Thunnus thynnus (Linnaeus) and other scombrids taken by terns in the Dry Tortugas, Florida. Bull. mar. Sci., 20(2): 389-413.

birds

*Poulin, Jean M., 1968.
Croissance du jeune fou de bassan (Sula bassana) pendant sa periode pré-envol (île Bonaventura, Québec).
Naturaliste Can., 95(5):1131-1143.

penguins

* Poulter, Thomas C., 1969.
Sonar of penguins and Fur seals.
Proc. Calif. Acad. Sci., 36(13): 363-380.

birds

Purrington, Robert D. 1970
Nesting of the Sooty Tern in Louisiana.
Auk 87(1): 159-160

Q

R

birds

Ramamurthy V.D. 1970.
Antibacterial activity of the marine blue-green alga Trichodesmium erythraeum in the gastro-intestinal contents of the sea gull Laurus brunicephalus.
Mar. Biol. 6(1): 74-76.

birds

Rea, S.C. 1969.
Plastic device causes gull mortality.
Wilson Bull 81: 105-106 (not seen)

birds

Rees, E.I.S., 1963.
Marine birds in the Gulf of St. Lawrence and Str. of Belle Isle during November.
Stud. Stas. Fish. Res. Bd. Canada, (836)98-107.
(reprinted from Canadian Field-Naturalist,)77(2).

birds

Rees, E.I.S., 1963.
Marine birds in the Gulf of St. Lawrence and Strait of Belle Isle during November.
Canadian Field Naturalist, 77(2):98-107.

Birds

Richdale, L.E.
Supplementary notes on the diving petrel.
Trans and Proc. Roy. Soc. New Zealand, Vol. 75, Pt. 1, pp. 42-53, Pl. 4

birds

Risebrough, R.W., D.B. Menzel, D.J. Martin and H.S. Olcott 1967.
DDT residues in Pacific sea birds: a persistent insecticide in marine food chains.
Nature, Lond. 216: 589

birds

Robertson W.B., Jr. 1969.
Transatlantic migration of juvenile sooty terns.
Nature, Lond., 222 (5194): 632-634

birds

Rowan, M.K., 1965.
Regulation of sea-bird numbers.
Ibis, 107(1):54-60.

birds, anat.-physiol.

Russell, F.S., 1958.
Salt excretion in marine birds.
Nature, 182:1755.

S

birds

Sage, B.Z. 1968.
Ornithological transects in the North Atlantic.
Ibis 110: 1-16

birds

Salomonsen, Finn, 1971.
Recoveries in Greenland of birds ringed abroad.
Medd. Grønland 191 (2): 1-52.

birds (terns)

*Salomonsen, Finn., 1967.
Migratory movements of the Arctic tern (Sterna paradisaea Pontoppidan) in the Southern Ocean.
Kongelige Danske Videnskabernes Selskab, Biologiske Meddelelser 24(1):42 pp.

birds

Salomonsen, F., 1955.
The food production in the sea and the annual cycle of Faeroese marine birds. Oikos 6(1):92-100.

birds

Salt, George W. and Daniel E. Willard, 1971.
The hunting behavior and success of Forster's tern. Ecology 52(6): 989-998.

birds

Sanger, Gerald A., 1970.
The seasonal distribution of some seabirds off Washington and Oregon, with notes on their ecology and behavior.
Condor 72(3): 339-357

birds

Sanger, Gerald A., 1965.
Observations of wildlife off the coast of Washington and Oregon in 1963, with notes on the Laysan albatross (*Diomedea immutabilis*) in this area.
Murrelet, 46(1): 1-6.

birds

Schaefer, Milner B., 1970.
Men, birds and anchovies in the Peru Current - dynamic interactions.
Trans. Am. fish. Soc. 99(3):461-467.

birds

Schreiber, Ralph W. 1968.
Seasonal population fluctuations of herring gulls in central Maine.
Bird-Band., 39(2):81-106.

birds

Scott, J. Michael 1971.
Interbreeding of the glaucous-winged gull and western gull in the Pacific Northwest.
Calif. Birds 2(4):129-133

birds

Sdobnikov, V.M. 1971
Causes for breaks in breeding of Arctic birds. (In Russian; English abstract).
Zool. Zh. 50(5): 734-740

birds

Shuntov, V.P., 1968.
Some regularities in distribution of albatrosses (Tubinares, Diomedeidae) in the northern Pacific. (In Russian; English abstract
Zool. Zh., 47(7): 1054-1064

birds

Shuntov, V.P., 1968.
A quantitative record of sea birds in the eastern part of the Indian Ocean. (In Russian; English abstract).
Okeanologiia, Akad. Nauk, SSSR, 8(3):494-501.

birds

Sibley, F.C. and R.B. Clapp, 1967.
"Distribution and dispersal of Central Pacific Lesser Frigatebirds *Fregata ariel*".
Ibis, 109:328-337.

birds

Sibson, R.B. 1967.
Long-tailed skua ashore at Muriwai (N.Z.).
Notornis 14(2):79-81.

birds

Sick H., 1965.
Vogelwanderungen im kontinentalen Südamerika
Vogelwarte 24: 219-243 (not seen)

birds

Sladen, W.J.L. 1965.
Ornithological research in Antarctica.
BioScience 15(4): 264-268.

birds

Smith, Gordon C., 1969.
Inland Sea gulls.
Sea Frontiers, 15(1):12-20. (popular).

birds

Snow, D.W. and B.K. Snow, 1967.
The breeding cycle of the Swallow-tailed Gull *Creagrus furcatus*.
Ibis, 109:14-24.

birds

Solianik, G.A. 1969.
Some bird observations on Bouvet Island.
Soviet Antarctic Exped. 13: 34-37.

birds

Soper, M.F. 1967.
Some portraits of Kermadec sea birds.
Notornis 14: 114-121.

BIRDS

Spellerberg, I. F., 1965.
Brown skua, *Cattaracta skua lannbergi*, in the Antarctic.
Ibis, 107(1): 106.

birds

Stonehouse, B., 1967.
Feeding behaviour and diving rhythms of some New Zealand Shags, Phalacrocoracidae.
Ibis, 109:600-605.

birds

Stresemann, Erwin, and Vesta Stresemann 1970.
Über Mauser und Zug von *Puffinus gravis*.
J. Orn. 111 (3/4): 378-393.

birds

Stromar, L., 1967
Vierjährige Silbermövenberingung, *Larus argentatus* Pontopp., auf den Inselchen Makan und Bobara in Dalmatien. (In Serbo-Croatian; German summary)
Larus 19 (1965): 133-144 (not seen)

birds

Summerhayes C.P. 1969.
Seabirds seen in the northern Tasman Sea in winter.
N.Z. Jl mar. freshwat. Res. 3(4): 560-570.

birds

Swales, M.K. 1965.
The sea birds of Gough Island.
Ibis 107(1): 17-42; 215-229.

birds

Szijj, L.J., 1967.
Notes on the winter distribution of birds in the Western Antarctic and adjacent Pacific waters.
Auk, 84(3):366-378.

T

birds

Threlfall, William, 1968.
The food of three species of gulls in Newfoundland.
Can. Fld. Nat., 82(3):176-180.

birds

Tickell, W.L.N., 1970.
The great albatrosses.
Scient. Am. 223(5): 84-103.

birds

Ticknell, W.L.N., and J.D. Gibson 1968
Movements of wandering albatrosses *Diomedea exulans*.
Emu 68: 7-20.

birds

Tovar S., Humberto, 1969.
Areas de reproduccion y distribucion de las aves marinas en el litoral peruano.
Bol. Inst. Mar. Peru 1(10): 523-546

birds

Trawa, G. 1970.
Note préliminaire sur la vascularisation des membres des Sphéniscidés de Terre Adélie.
L'Oiseau, Rev. franç. Ornithol. 40 (no. spéc.): 142-156.

Also: Expéd. polaires franç., Missions Paul-Emil Victor 318

birds

Trimble, B. 1968.
Aberrant Wilson's petrel on the Newfoundland Banks.
Auk 85: 130

birds

Tuck, G.S. 1971.
Establishing the identity of a bird observed at sea.
Mar. Obs. (Met. O.839) 41 (233): 105-106.

birds

Tuck, Leslie M., 1968.
Laughing gulls (*Larus atricilla*) and black skimmers (*Rynchops nigra*) brought to Newfoundland by hurricane.
Bird Band. 39(3): 200-208.

birds

Tuck, G.S., 1970.
Sea-birds on boardships.
Mar. Obs. 40 (230) (Met. O.827): 180-182.

U

birds

Udvardy, Miklos, D.F., 1963.
Data on the body temperature of tropical sea and water birds.
The Auk, 80(2):191-194

V

birds
Vermeer, Kees, 1970.
Colonies of double-crested cormorants and white pelicans in Saskatchewan.
Can. Field-Nat., 84 (1):39-42

birds
Vermer, Kees, 1967.
Common terns (Sterno hirundo) nesting at Miquelon Lake, Alberta.
Can. Fld.Nat., 81(4):274-276.

birds
von Schmidt, Katherine, 1968.
Note on the mating behavior of the least tern, Sterna albifrons.
Auk, 85(4):694.

birds
Voisin, Jean-François, 1970
Some notes about birds and mammals in Svalbard, summer 1969.
Arbok norsk Polarinst. 1969: 107-115.

birds
Voisin, J.F. 1969.
L'albatros hurleur Diomedea exulans & l'Ile de la Possession.
L'Oiseau, Rev. franç. Ornithol. 39 (no. spéc.): 82-106
Also: Direction Labs. scient. Territoires Terres australes et antarctiques franç. Sept. 1970

birds
Voous, K.H. 1967.
Notes on sea birds. 22. Leach's storm petrel in the Caribbean.
Ardea 55: 268-269.

birds
Voous, K.H. 1967.
Notes on sea birds. 21. Leach's storm petrel in the tropical Pacific.
Ardea 55: 268.

birds
Voous, K.H. 1967.
Notes on sea birds. 19. Black-capped petrels (Pterodroma hasitata) in the Caribbean.
Ardea 55: 144-145.

birds
Voous, K.H. and J. Wattel, 1963.
Distribution and migration of the Greater Shearwater.
Ardea, 51(2-4):143-157.

birds
Vuilleumier, F., 1964.
Les oiseaux d'une traversee de L'Atlantique nord.
Nos Oiseaux, 27(8):239-245.

W

birds
Warham, J., 1964.
Marked Sooty Shearwaters Puffinus griseus in the Northern Hemisphere.
Ibis, 106(3):390-391.

birds
Watson, George E., 1971.
Molting Greater shearwaters (Puffinus gravis) off Tierra del Fuego.
Auk 88 (2): 440-442.

birds
Watson, George 1969.
A handbook to the birds of the Antarctic.
Antarctic J., U.S., 4(5): 196-197.

birds
Watson, George E., 1965.
Preliminary Smithsonian identification manualr seabirds of the Tropical Atlantic Ocean.
Smithsonian Institution, Washington, D.C. 108 pp.

birds
Wattel, J., 1965.
Preliminary report on bird observations during the ocean passages of H. NL. M.S. Luymes and Snellius (Luy-Snel Cruise)
Hydrogr. Newsl., Neth., 2(1):25-33.

birds
Wilbur, Henry M., 1969.
The breeding biology of Leach's petrel, Oceanodroma leucorhoa.
Auk, 86 (3): 433-442.

birds
Williams, Anthony J. 1971.
Laying and nest-building behaviour in the larger auks (Aves, Alcidae).
Astarte 4 (2): 61-67.

birds
*Wurster, C.F., Jr., and D.B. Wingate, 1968.
DDT residues and declining reproduction in the Bermuda petrel.
Science, 159(3818):979-981.

XYZ

birds
*Yeates, G.W., 1968.
Studies on the Adélie penquin at Cape Royds 1964-65 and 1965-66.
N.Z. Jl mar. Freshwat. Res., 2(3):472-486.

birds
Zapata, Abel, R.P. 1967.
Observaciones sobre aves de Puerto Deseado, Provincia de Santa Cruz.
Contrib. cient. Centro Invest. Biol. mar., Puerto Deseado, Argentina 44:
Also: El Hornero 10(4): 351-378.

birds
Zuta, Salvador, y Jorge Mejía, 1968.
Informe preliminar del Crucero Unanue 6708, 24 agosto-25 de Setiembre, invierno 1967.
Informe, Inst. Mar Peru 25: 1-23

bird counts
McGary, J.W., E.C. Jones and T.S. Austin, 1956.
Mid-Pacific oceanography. IX. Operation NORPAC.
Spec. Sci. Repts.: Fish. 168:127 pp.

birds, effect of

birds, effect of
Blakemore, L.C. and H.S. Gibbs 1968.
Effects of gannets on soil at Cape Kidnappers, Hawkes Bay.
N.Z. Jl Sci. 11 (1): 54-62.

birds, effect of
Dare, P.J., 1970
The movements of oyster catchers Haematopus ostralegus L. visiting or breeding in the British Isles.
Fish. Invest. Min. Agric. Fish. Food, (2) 25 (9):

birds, effect of
*Golovkin, A.N., 1967.
The influence of sea colonial birds on phytoplankton growth. (In Russian; English abstract).
Okeanologiia, Akad. Nauk, SSSR, 7(4):672-682.

birds, effect of
Heatwole, Harold 1971
Marine-dependent terrestrial biotic communities on some cays in the Coral Sea.
Ecology 52 (2): 363-366

birds, effect of
Mironov, O.G., 1965.
On the role of sea birds in transmitting radioactive substances from the sea to the land. (In Russian).
Okeanologiia, Akad. Nauk, SSSR, 5(4):715-718.

birds, effect of
Moran, Sh. and L. Fishelson, 1971.
Predation of a sand-dwelling mysid crustacean Gastrosaccus sanctus by plover birds (Charadriidae). Mar. Biol. 9(1): 63-64.

birds, guanay

birds (guanay)
Barreda O., Mario, 1959.
Recuperacion de guanayes (Phalacorax bougainvillii) caquecticos en cautiverio. Estudio de su ingestion y deyeccion.
Bol. C.A.G., 35(4):9-22.

birds, guanays
Flores, Luis, Oscar Guillén y Rogelio Villanieva 1966.
Informe preliminar del Crucero de invierno 1965 (Máncora-Morro Sama).
Inst. Mar, Peru, Informe, No.11:1-34 (multilithed).

birds, guanay
Jordan, Romulo, 1967.
The predation of guano birds on the Peruvian anchovy (Engraulis ringens Jenyns).
Rep. Calif. Coop. Oceanic Fish. Invest. 11:105-109.

birds (guanay)
Jordan, S., Romulo, 1959.
El fenomeno de las regurgitaciones en el guanay (Phalacrocorax bougainvillii L.) y un metodo para estimar la ingestion diaria.
Bol. C.A.G., 35(4):22-40.

birds, guanay
Jordan, Romulo, 1958.
El fenomeno de las regurgitaciones en el huanay (Phalacrocorax bougainvillii Lesson) y un metodo para estimar la ingestion diaria.
Bol. C.A.G. 34(1):9-19.

birds, guanay
Mejia, Jorge, y Luis Alberto Poma E., 1966.
Informe preliminar del Crucero de otoño 1966. (Cabo Blanco-Ilo).
Inst. Mar, Peru, Informe No.13:31pp (multilithed).

birds (guano)
Rand, R.W., 1963.
The biology of guano-producing birds. 5. Composition of colonies on the South West African islands.
Repub., S. Africa, Div. Sea Fish., Invest. Rept., No. 46:26 pp.
Reprint: Commerce & Industry, June 1963.

birds, guano

Rand, R.W., 1960

The biology of guano-producing sea-birds. 3. The distribution, abundance and feeding habits of the cormorants Phalacrocoracidae off the west coast of the Cape Province. Union S. Africa. Dep. Comm. & Ind., Div. Fish., 42: 32 pp.

birds

Vogt, William, 1964.
Informe sobre las aves guaneras. (Conclusion).
Bol. Corp. Nacional de Fertilizantes, Lima, Peru, 2 (10):5-40.

bird migration

Griffin, Donald R., 1969.
The physiology and geophysics of bird navigation.
Q. Rev. Biol. 44(3): 255-276

birds, parasites of

birds, parasites of
*Threlfall, William, 1968.
Studies on the helminth parasites of the American herry gull (Larus argentatus Pont.) in Newfoundland.
Can. J. Zool., 46(6):1119-1126.

birds, physiol.
Schmidt-Nielsen, Knut, C. Barker Jörgensen and Humio Osaki, 1958.
Extrarenal salt excretion in birds.
Amer. J. Physiol., 193(1):101-108.

birds, physiol.
Schmidt-Nielsen, K., and W.J.L. Sladen, 1958.
Nasal salt secretion in the Humboldt penguin.
Nature 181(4617):1217-1218.

bituminous rocks

Brongersma-Sanders Margaretha 1971.
Origin of major cyclicity of evaporites and bituminous rocks: an actualistic model.
Mar. Geol. 11 (2): 123-144

black tides

Ruellan F. 1968.
La "marée noire" en Bretagne.
Rev. gén. Sci. pures appl. 75 (7/8): 195-208. (not seen)

black tides

Furon, R. 1968.
La "marée noire" de 1967 et la pollution permanente des mers.
Rev. gén. Sci. pures appl. 75 (5): 157-160. (not seen)

bladders

bladders, effect of
*Weston, D.E., 1967.
Sound propagation in the presence of bladder fish.
In: Underwater Acoustics, V.M. Albers, editor, Plenum Press, 2: 55-88.

Blake event
*Smith, Jerry D. and John H. Foster, 1969.
Geomagnetic reversal in Brunhes normal polarity epoch.
Science 163(3867):565-567.

blisters

"blisters"

Strasburg, Donald W., and Heeny S.H. Yuen, 1960

Progress in observing tuna underwater at sea.
J. du Cons. 26(1): 80-93.

blocks, crustal

blocks, crustal
*Morgan, W. Jason, 1968.
Rises, trenches, great faults, and crustal blocks.
J. geophys. Res., 73(6):1959-1982.

block faulting

Barrett, D.L., and F. Aumento 1970.
The Mid-Atlantic Ridge near 45° N. XI
Seismic velocity, density and layering of the crust.
Can. J. Earth Sci. 7(4): 1117-1126.

block faulting

Heinrichs, D.F., 1970.
More bathymetric evidence for block faulting on the Gorda Rise. J. mar. Res., 28(3): 330-335.

blood

blood

Baron, J.C., 1968.
Étude préliminaire sur le sang de deux espèces de sardinelles (Sardinella aurita C.V., Sardinella eba C.V.).
Document scient. provisoire, Centre Océanogr., Côte d'Ivoire, ORSTOM, 029:48 pp. (mimeographed).

blood, invertebrate
Beck, A.B., and K. Sheard, 1949.
Copper and nickel content of the blood of the West Australian crayfish, Panulirus longipes (Milne Edwards), and of sea weed. Australian J. Exp. Biol. Med. Sci. 27:307-312.

blood, vertebrate (without hemoglobin)
Hureau, J.C., 1964.
Sur la probable identité des deux espèces du genre Chaenichthys de la famille Chaenichtyidae (Poissons a "sang blanc").
Bull. Mus. Nat. Hist. Nat., (2)36(4):450-456.

blood

Lenfant, Claude, 1969.
Physiological properties of blood of marine mammals.
In: The biology of marine mammals, H.T. Andersen, editor, Academic Press, 95-116

blood

Martsinkevitch, L. D., 1958.

Cell contents of blood of whiteblooded fishes in the Antarctic.
Inform. Biull. Sovetsk. Antarkt. Exped., (3):67-68.

blood
Redfield, A.C., 1933.
The evolution of the respiratory function of the blood. Quart.Rev. Biol. 8(1):31-57, 6 textfigs.

"blood"
Ruud, J.T., 1954.
Vertebrates without erythrocytes and blood pigment. Nature 173(4410):848--850.

blood (fishes)
Urist, Marshall R., and Karel A. Van de Putte, 1967.
Comparative biochemistry of the blood of fishes: identification of fishes by the chemical composition of serum. In: Sharks, skates, and rays, Perry W. Gilbert, Robert F. Mathewson and David P. Rall, editors, Johns Hopkins Univ. Press, 271-285.

blood, types

blood groups
Cushing, John E., 1964.
The blood groups of marine animals.
In: Advances in Marine Biology, F.S. Russell, editor, Academic Press, 2:85-131.

blood types
Cushing, John E., 1961.
Blood typing of marine animals.
Naval Research Reviews, May 1961:8-11.

blood types
Cushing, John, 1961.
Immunogenetic concepts in marine populations.
Tenth Pacific Sci. Congr., Honolulu, 21 Aug.-6 Sept., 1961, Abstracts of Symposium Papers, 182.

blood groups
Cushing, John E., Nora Lee Calaprice and Gary Trump, 1963
Blood group reactive substances in some marine invertebrates.
Biol. Bull., 125(1):69-80.

blood types
Cushing, John E., Kazuo Fujino and Nora Calaprice, 1963
The Ju blood typing system of the sperm whale and specific soluble substances.
Sci. Repts., Whales Res. Inst., No. 41:67-77.

blood types
Fujino, Kazuo, 1963
Intra-uterine selection due to maternal-fetal incompatibility of blood types in whales.
Sci. Repts., Whales Res. Inst., No. 14:53-65.

blood types
Fujino, K., 1962.
Blood types of species of Antarctic whales.
American Naturalist, 96(889):205-210.

blood types
Hildeman, W.H., 1961.
Immunogenetic studies of poikilothermic animals. (Abstract).
Tenth Pacific Sci. Congr., Honolulu, 21 Aug.-6 Sept., 1961, Abstracts of Symposium Papers, 183-184.

blood types
Ridgway, George J., 1961.
Special immunological techniques for the study of marine populations. (Abstracts).
Tenth Pacific Sci. Congr., Honolulu, 21 Aug.-6 Sept., 1961, Abstracts of Symposium Papers, 184-185.

blood types
Sprague, L.M., and J. R. Holloway, 1962.
Studies of the erythrocyte antigens of the skipjack (Katsuwonus pelamis).
American Naturalist, 96(889):233-238.

blood types
Suzuki, A., 1962.
On the blood types of yellowfin and bigeye tuna.
American Naturalist, 96(889):239-

blooms

Blooms

See also: under name of organism concerned in subsequent section

blooms (spp. unidentified)
Flemer, David A., 1969.
Continuous measurement of in vivo chlorophyll of a dinoflagellate bloom in Chesapeake Bay.
Chesapeake Sci., 10(2): 99-103.

blooms

VI, Jun 1971.
Mercury pollution of sea and fresh water: its accumulation into water blooms.
Rev. int. Océanogr. Méd. 22-23: 79-128.

BOATS: see SHIPS

Boats

In a separate portion of catalogue.

boat trucks

boat trucks
Silverman, Maxwell, and Roy D. Gaul, 1965.
The concept of portability applied to future oceanographic ship operation.
Ocean Sci. and Ocean Eng., Mar. Techn.Soc., Amer. Soc. Limnol. Oceanogr., 1: 384-397.

body temperatures
Boyd, John C. and William J.L. Sladen 1971.
Telemetry studies of the internal body temperatures of Adélie and Emperor penguins at Cape Crozier, Ross Island, Antarctica.
Auk 88 (2): 366-380.

BOMEX

BOMEX
Anon. 1969.
Doing something about the weather: BOMEX - a massive sea air study - may bring new insight into tropical weather.
Ocean Industry 4 (7): 65-68 (secondary sources only)

Bomex
Broecker, W.S., and T.-H. Peng, 1971.
The vertical distribution of radon in the Bomex area.
Earth planet. Sci. Letts 11 (2): 99-108.

BOMEX
Landis, Robert C., 1971.
Early BOMEX results of sea surface salinity and Amazon River water. J. phys. Oceanogr. 1(4): 278-281.

BOMEX
McAlister, E.D., William McLeish and Ernst A. Corduan 1971.
Airborne measurements of the total heat flux from the sea during BOMEX.
J. geophys. Res. 78 (19): 4172-4180.

BOMEX
Miyake, Mikio, Mark Donelan and Yasushi Mitsuta 1970
Airborne measurement of turbulent fluxes.
J. geophys. Res. 75 (24): 4506-4518.

BOMEX
Montgomery, Suzanne, 1969.
BOMEX aims for improved weather prediction.
UnderSea Techn., 10(5): 80-83.

BOMEX
Pond, S., G.T. Phelps, J.E. Paquin, G. McBean and R.W. Stewart 1971.
Measurements of the turbulent fluxes of momentum, moisture and sensible heat over the ocean.
J. atmos. Sci. 28 (6): 901-917

books

A

Books

Warning: Title of a book published in Great Britain and the U.S.A. more or less simultaneously often have entirely different titles!

books
Abbott, R. Tucker, 1962.
Sea shells of the world.
Golden Press, New York.

books
Adams, William Mansfield, 1970.
Tsunamis in the Pacific: proceedings of international symposium on tsunamis and tsunami research, Hawaii, Oct. 7-10, 1969. East-West Center, Honolulu 521 pp.
$15.00

books
Ahrens, L.H., editor 1968
Origin and distribution of the elements.
International Series of Monographs in Earth Sciences 30: 1178 pp.

books
Akademiia Nauk, SSSR, 1953.
Ocherki po gigrografii rek SSSR. 234 pp., figs., maps. $5.75.

A. Buschke, 80 East 11th Street, N.Y.3

books
Albers, Vernon M., editor, 1967.
Underwater Acoustics, Plenum Press, 2: 416 pp.

books
Albers, Vernon M., 1965.
Underwater acoustics handbook II.
The Pennsylvania State University Press, 356pp.
$12.50.

books
Albers, V.M., Editor, 1962
Underwater acoustics.
Plenum Press, New York, 354 pp.

books
Albert 1er, Prince of Monaco 1966.
La carrière d'un navigateur.
Editions des Archives du Palais Princier, Monaco, 238 pp.

books
Alekin, O.A., 1966.
Chemistry of the ocean. (In Russian).
Hydrometeoizdat., Leningrad, 248 pp. (not seen).

books
Alexander, Lewis M., editor, 1967.
The law of the sea.
Ohio State Univ. Press, 321 pp.

books
Alexander, W.B., 1928
Birds of the Ocean. 428 pp., 87 pls. G.P. Putnam's Sons, N.Y. - London

owned by M.S.

books
Allen, John R.L., 1968.
Current ripples. N.-Holland, 433 pp.
Hfl. 108 or £12.12s or $30.00 (not seen)

books
Aubert, M. et J. Aubert, 1969.
Océanographie médicale.
Gauthier-Villars, 298 pp. (not seen)

books
Australia, Bureau Meteorology, 1960
Antarctic meteorology (Melbourne, February, 1959).
Pergamon Press, N.Y.: 483 pp.

B

books
Băcescu, M., redactor responsabil, 1965.
Ecologie marină.
Edit. Acad. Republ. Pop. Române, Bucuresti, 1: 344 pp.; 2: 293 pp.

books
Badgley, Peter C., Leatha Miloy and Leo Childs, editors, 1969.
Oceans from space.
Gulf Publishing Co., Houston, Texas, 234 pp.

books
Barber, N.F., 1969.
Water waves. Wykeham Publ. Ltd., 142 pp.
20 shillings.

books
Barnes, R.D., 1963.
Invertebrate Zoology.
W.B. Saunders Co., Philadelphia and London, xii and 632 pp.

books
Barrington, E.J.W., 1965.
The biology of Hemichordata & Protochordata.
W.H. Freeman and Company, San Francisco, 176 pp. (paperback). $2.50.

books
Bascom, Willard, 1964
Waves and beaches: the dynamics of the ocean surface.
Anchor Books, Science Study Series, Doubleday & Co., Inc., New York, 267 pp. $1.45.
(popular paperback)

books
Bascom, Willard, 1961.
A hole in the bottom of the sea: the story of the MOHOLE Project.
Doubleday, New York, 352 pp., 53 figs., 8 pls.

Abstr. in: Geoscience Abstr., 3(9): 8.

books
Batchelor, G.K., 1967.
An introduction to fluid dynamics.
Cambridge Univ. Press., 615 pp.

books
Bayer, Frederick M., and Harding Owre, 1968.
The free living invertebrates.
Macmillan $11.95 (not seen).

books
Beagleole, J.C., Editor, 1961.
The journals of Captain James Cook on his voyages of discovery. Vol. 2. The voyage of the "Resolution" and "Adventure", 1772-1775.
Cambridge University press, clxx 1021 pp.

Reviewed by R. Firth, 1961 in Nature 192(4808): 1107-1108.

books
Behrman, Daniel, 1969.
The new world of oceans: men and oceanography.
Little, Brown and Co., 436 pp.

books
Bergman, P.G., and A. Yaspan, editors 1968
Physics of sound in the sea. I. Transmission
Gordon and Breach, 262 pp. $16.00

book
Bernhard, C.G., editor, 1966.
The functional organization of the compound eye.
Pergamon Press, 591 pp.

books
Berezkin, V., 1947.
Dinamika moria. Gidrometeorologicheskoe Izd-vo, Svordlovsk, 683 pp.

books
Bertin, Leon, 1956.
Eels: a biological study. Cleaver-Hume Press, Ltd., London, 200 pp., 55 illus., 9 pls.

books
Berzin, A.A. 1971.
Whales (In Russian; English abstract)
Tikhookean. nauchno-issled. Inst. ribn. Khoz. Okeanogr., Isdatel. "Pishchevaia Promishlennost" Moskva, 364 pp.

books
Bigelow, H. B., and W.T. Edmondson, 1947.
Wind waves at sea, breakers and surf. H.O. 602: xii and 177 pp., 57 textfigs., 24 pls.

Reviewed: J. du Cons. 16(2):242-243 by N.F. Barber

books
Bigelow, H. B., I.P. Farfante, and W.C. Schroeder 1949.
Fishes of the western North Atlantic: Lancelets, Cyclostomes, Sharks. Mem. Sears Found. Mar. Res. 576 pp.

Reviewed: Science 111:95 by L. P. Schultz.
J. du Cons. 16(3):397-398 by A. Fr. Bruun.

books
Bigelow, H.B., and W.C. Schroeder, 1953.
Fishes of the Gulf of Maine. First revision.
Fish. Bull. 74, Fish. Bull., Fish. and Wildlife Service, 53:1-577, 288 textfigs.

books
Birstein, J.A., 1963.
Deep water isopods of the north-west Pacific. (In Russian).
Izd. AN, USSR, Moscow, 216 pp.

Reviewed by:
Jusakin, O.G., 1963. Zool. Zhurn., 42(8):1278-1280. (In Russian).

books
Blackwelder, Richard E., 1963.
Classification of the animal kingdom.
Southern University Press, 94 pp. $7.00.

books
Blair, Carvel Hall, and Willits Dyer Ansel, 1968.
A guide to fishing boats and their gear.
Cornell Maritime Press, Inc. Cambridge, Md., 21613, xii + 142 pp. $5.00

books
Blanchard, Duncan C., 1967.
From raindrops to volcanoes: adventures with sea surface meteorology.
Doubleday, 180 pp. ($1.25 paper back; $4.95 hard cover)

books
Bogorov, V.G., 1955.
The ocean. Moscow, Militärverlag, 143 pp.
Cited in Petermanns Geogr. Mitt., 100(4)

books
Bolin, B. (Editor) 1960
The atmosphere and the sea in motion. The Rossby Memorial Volume. Rockefeller Inst. Press with Oxford U. Press, London: 509 pp.

Review by J. Paton, 1960, J. Atmos. Terr. Phys 19(3/4): 293.

books
Boltovskoy, Esteban, 1965.
Los foraminiferos recientes: biologia, metodos de estudio, aplicacion oceanografica.
Eudeba, Editorial Universitaria de Buenos Aires, 510 pp.

books
Boolootian, Richard A., editor, 1966.
The physiology of Echinodermata.
John Wiley & Sons, Inc., 772 pp. $45.00.

books
Boolootian, Richard A., and June Thomas 1967.
Marine biology; a study of life in the sea.
Holt Library of Science, 112 pp.

books
Borgstrom, Georg, Editor, 1962
Fish as food, Vol. 2. Nutrition, sanitation and utilization.
Academic Press, 778 pp., $25.00.

books
Borradaille, L.A., and F.A. Potts, 1958.
The invertebrata. Cambridge Univ. Press, 3rd edit 795 pp.

books
Bourcart, J., 1959
Problèmes de géologie sous-marins. Masson & Cie, éditeurs, Paris. (Collection "Evolution des Sciences")

Review: in Rev. Geog. Phys. Geol. Dyn. 2nd ser. 2(1): 67, by J.H. Brown.

books
Bourcart, J., 1952.
Les frontières de l'océan. Albin Michel, Paris, 317 pp., 77 textfigs.

books
Boyd, Waldo T., 1968.
Your career in oceanography. Julian Messner, 219 pp. $3.95.

books
Brahtz, John F. editor, 1968.
Ocean engineering: goals, environment, technology.
John Wiley and Sons, 720 pp. $17.95.

books
Brancanzio, Peter J., and A.G.W. Cameron, Editors 1964.
The origin and evolution of atmospheres and oceans.
John Wiley & Sons, Inc., 314 pp.

Proc. Conf. Goddard Inst. for Space Studies, NASA 8-9 Apr. 1963.

books
Bretschneider, Charles L., 1965.
Ocean wave spectra.
Prentice-Hall, Inc., Englewood Cliffs, N.J., 357 pp. $12.95

books
Briggs, Peter, 1970.
Laboratory at the bottom of the World (Antarctica).
David McKay Co., Inc., N.Y., 180 pp. $5.95. (not seen)

books
Briggs, Peter, 1969.
Mysteries of our world
David McKay Co., Inc., N.Y. 240 pp. $5.95 (not seen)

books
Briggs, Peter, 1969
Rivers in the sea.
Weybright and Talley, N.Y. 126 pp. $5.50. (not seen)

books
Briggs, Peter, 1968.
The great global rift
Weybright and Talley, N.Y., 128 pp. $5.50.

books
Briggs, Peter, 1967.
Water - the vital essence.
Harper Row, N.Y., 223 pp., $5.95. (not seen)

books
Bayer, Frederick M., and Harding Owre, 1968.
The free living invertebrates.
Macmillan $11.95 (not seen).

books
Beagleole, J.C., Editor, 1961.
The journals of Captain James Cook on his voyages of discovery. Vol. 2. The voyage of the "Resolution" and "Adventure", 1772-1775.
Cambridge University press, clxx 1021 pp.

Reviewed by R. Firth, 1961 in Nature 192(4808): 1107-1108.

books
Behrman, Daniel, 1969.
The new world of oceans: men and oceanography.
Little, Brown and Co., 436 pp.

books
Bergman, P.G., and A. Yaspan, editors 1968
Physics of sound in the sea. 1. Transmission
Gordon and Breach, 262 pp. $16.00

book
Bernhard, C.G., editor, 1966.
The functional organization of the compound eye.
Pergamon Press, 591 pp.

books
Berezkin, V., 1947.
Dinamika moria. Gidrometeorologicheskoe Izd-vo, Svordlovsk, 683 pp.

books
Bertin, Leon, 1956.
Eels: a biological study. Cleaver-Hume Press, Ltd., London, 200 pp., 55 illus., 9 pls.

books
Berzin, A.A. 1971.
Whales (In Russian; English abstract)
Tikhookean. nauchno-issled. Inst. ribn. Khoz. Okeanogr., Isdatel. "Pishchevaia Promishlennost" Moskva, 364 pp.

books
Bigelow, H. B., and W.T. Edmondson, 1947.
Wind waves at sea, breakers and surf. H.O. 602: xii and 177 pp., 57 textfigs., 24 pls.

Reviewed: J. du Cons. 16(2):242-243 by N.F. Barber

books
Bigelow, H. B., I.P. Farfante, and W.C. Schroeder 1949.
Fishes of the western North Atlantic: Lancelets, Cyclostomes, Sharks. Mem. Sears Found. Mar. Res. 576 pp.

Reviewed: Science 111:95 by L. P. Schultz.
J. du Cons. 16(3):397-398 by A. Fr. Bruun.

books
Bigelow, H.B., and W.C. Schroeder, 1953.
Fishes of the Gulf of Maine. First revision.
Fish. Bull. 74, Fish. Bull., Fish. and Wildlife Service, 53:1-577, 288 textfigs.

books
Birstein, J.A., 1963.
Deep water isopods of the north-west Pacific. (In Russian).
Izd. AN, USSR, Moscow, 216 pp.

Reviewed by:
Jusakin, O.G., 1963. Zool. Zhurn., 42(8):1278-1280. (In Russian).

books
Blackwelder, Richard E., 1963.
Classification of the animal kingdom.
Southern University Press, 94 pp. $7.00.

books
Blair, Carvel Hall, and Willits Dyer Ansel, 1968.
A guide to fishing boats and their gear.
Cornell Maritime Press, Inc. Cambridge, Md., 21613, xii + 142 pp. $5.00

books
Blanchard, Duncan C., 1967.
From raindrops to volcanoes: adventures with sea surface meteorology.
Doubleday, 180 pp. ($1.25 paper back; $4.95 hard cover)

books
Bogorov, V.G., 1955.
The ocean. Moscow, Militärverlag, 143 pp.

Cited in Petermanns Geogr. Mitt., 100(4)

books
Bolin, B. (Editor) 1960

The atmosphere and the sea in motion. The Rossby Memorial Volume. Rockefeller Inst. Press with Oxford U. Press, London: 509 pp.

Review by J. Paton, 1960, J. Atmos. Terr. Phys 19(3/4): 293.

books
Boltovskoy, Esteban, 1965.
Los foraminiferos recientes: biologia, metodos de estudio, aplicacion oceanografica.
Eudeba, Editorial Universitaria de Buenos Aires, 510 pp.

books
Boolootian, Richard A., editor, 1966.
The physiology of Echinoderms.
John Wiley & Sons, Inc., 772 pp. $45.00.

books
Boolootian, Richard A., and June Thomas 1967.
Marine biology: a study of life in the sea.
Holt Library of Science, 112 pp.

books
Borgstrom, Georg, Editor, 1962
Fish as food, Vol. 2. Nutrition, sanitation and utilization.
Academic Press, 778 pp., $25.00.

books
Borradaille, L.A., and F.A. Potts, 1958.
The invertebrata. Cambridge Univ. Press, 3rd edit 795 pp.

books
Bourcart, J., 1959

Problèmes de géologie sous-marins. Masson & Cie, éditeurs, Paris. (Collection "Evolution des Sciences")

Review: in Rev. Geog. Phys. Geol. Dyn. 2nd ser. 2(1): 67, by J.H. Brown.

books
Bourcart, J., 1952.
Les frontières de l'océan. Albin Michel, Paris, 317 pp., 77 textfigs.

books
Boyd, Waldo T., 1968.
Your career in oceanography. Julian Messner, 219 pp. $3.95.

books
Brahtz, John F. editor, 1968.
Ocean engineering: goals, environment, technology.
John Wiley and Sons, 720 pp. $17.95.

books
Brancanzio, Peter J., and A.G.W. Cameron, Editors 1964.
The origin and evolution of atmospheres and oceans.
John Wiley & Sons, Inc., 314 pp.

Proc. Conf. Goddard Inst. for Space Studies, NASA 8-9 Apr. 1963.

books
Bretschneider, Charles L., 1965.
Ocean wave spectra.
Prentice-Hall, Inc., Englewood Cliffs, N.J., 357 pp. $12.95

books
Briggs, Peter, 1970.
Laboratory at the bottom of the World (Antarctica).
David McKay Co., Inc., N.Y., 180 pp. $5.95. (not seen)

books
Briggs, Peter, 1969.
Mysteries of our world
David McKay Co., Inc., N.Y. 240 pp. $5.95 (not seen)

books
Briggs, Peter, 1969
Rivers in the sea.
Weybright and Talley, N.Y. 126 pp. $5.50. (not seen)

books
Briggs, Peter, 1968.
The great global rift
Weybright and Talley, N.Y., 128 pp. $5.50.

books
Briggs, Peter, 1967.
Water - the vital essence.
Harper Row, N.Y., 223 pp., $5.95. (not seen)

books
Brodal, Alf, and Ragnar Fänge, Editors, 1963.
The biology of Myxine.
Universitetsforlaget, Oslo, 588 pp.

books
Brown, M.E., Editor, 1957.
The physiology of fishes. 1. Metabolism.
Academic Press, Inc., N.Y., 447 pp.

books
Bruns, E., 1962.
Ozeanologie, Band II. Ozeanometrie 1.
Deutsch. Verlag Wiss., Berlin, 494 pp.

books
Bruns, Erich, 1958.
Ozeanologie. 1. Einführung in die Ozeanologie. Ozeanographie, Verb. Deutscher Verlag der Wissenschaften, Berlin, 420 pp., 145 figs. 41 DM.

Reviewed by Jens Smed, 1961. J. du Cons., 26(3): 354-355.

books
Bruns, E., 1955.
Handbuch der Wellen der Meere und Ozeane. VEB Deutscher Verlag der Wissenschaft, Berlin, 255 pp

books
Bruun, A.Fr., et Al., 1956.
The Galathea Deep Sea Expedition, 1950-1956.
MacMillan Co., N.Y., 296 pp.

books
Bruun, P., & F. Gerritsen, 1960
Stability of coastal inlets.
North Holland Publ. Co., Amsterdam: 123 pp.

Rev. in Science Progress, 49(194): 377
Okeanologiia 1(3):566 by V. P. Zenkovich and V.N. Mikhailov

books
Buchsbaum, Ralph and Lorus J. Milne, 1960
The Lower Animals - living invertebrates of the world. Doubleday and Co., N.Y., 303 pp., 315 photos.

(Magnificent photographs, some by D.P. Wilson)

books
Bullock, Theodore Holmes and G. Adrian Horridge, 1965.
Structure and function in the nervous systems of invertebrates.
W.H. Freeman and Company, San Francisco & London, 2 vols., 1719 pp., $75.00.

books
Burke, William T., 1969.
Towards a better use of the ocean.
Contemporary legal problems in ocean development.
Almqvist and Wiksell, Stockholm (Stockholm Internat. Peace Res. Inst. Monogr.) 231 pp.

books
Buzzati-Traverso, A.A., Editor, 1958
Perspectives in marine biology. Berkeley, U. of California Press, xvi plus 621 pp.

Review by: G.L. Kestevin, 1960. Limn. Ocean. 5(2): 237-242.
Review by: Ralph J. Johnson, U. of Chicago, 1959
Evolution 13(1):149

C

books
Cameron, Roderick, 1961.
Shells. G.P. Putnam's Sons, New York.

books
Camp, Leon, Richard Stern and B.M. Brown 1970.
Underwater acoustics.
Wiley-Interscience, 317 pp. $17.50

books
Capurro, Luis R.A. 1970.
Oceanography for practicing engineers.
Barnes and Noble, Inc. 181 pp. $4.95.

books
Carson, R.L., 1962.
Under the sea wind. A naturalists picture of ocean life. Simon and Shuster, N.Y., 314pp.

Popular.

books
Carthy, J. D., 1958
An introduction to the behaviour of invertebrates. Geo. Allen & Unwin, Ltd. London, 380 pp., 148 fig. 4 plates. Rev. by Albert Collier (1960) in: Trans. Amer. Fish. Soc., 89(3): 317-318.

books
Carthy, J.D., 1956.
Animal navigation.
Chas. Scribner's Sons, N.Y., 151 pp.

books
Chamberlain, Barbara Blau, 1964.
These fragile outposts - a geological look at Cape Cod, Marthas Vineyard and Nantucket.
The Natural Histroy Press, Garden City, New York, 327 pp.

books
Chapin, H., and F.G. Walton Smith, 1952.
The ocean river: the story of the Gulf Stream.
New York, Scribners, 325 pp., illus., $3.50.

Reviewed by H. Stommel, Science 116(3016):436.

books
Chapman, V.J., 1960.
Salt marshes and salt deserts of the world.
Leonard Hill, Ltd., London; Interscience Publishers, New York, 392 pp.

book
Christy, Francis T. Jr., and Anthony Scott 1965.
The common wealth in ocean fisheries; some problems of growth and economic allocation.
Resources for the Future, Inc., Johns Hopkins Univ. Press, 281 pp.

books
Clark, Ailsa M., 1962.
Starfishes and their relations.
British Museum (Natural History), 119 pp., illus. 11 shillings.

books
Clarke, G.L., 1954.
Elements of ecology. John Wiley & Sons, Inc., 534 pp.

books
Coker, R.E., 1947
This Great and Wide Sea. Univ. N. Carolina Press, 325 pp., 23 textfigs. 91 pls.

books
Cole, R.H., 1948.
Underwater explosions. Princeton Univ. Press, ix+437 pp.

Reviewed: Nature 166(4234):1045, W.G. Penney.

books
Colman, J.S., 1950.
The sea and its mysteries. London, G. Bell & Sons Ltd., 285 pp., 1 fold-in.

books
Cook, A.H., and T.F. Gaskell, Editors, 1961.
The Earth Today. Royal Astronomical Society, distributed by Interscience Publishers, Inc., N.Y., 404 pp.

Originally published as Geophysical Journal, R. Astron. Soc., Vol. 4.

books
Copeland, Herbert F., 1956.
The classification of lower organisms.
Pacific Books, Palo Alto, California, 302 pp.

books
Costlow, John D. 1971
The fertility of the sea.
Gordon and Breach Science Publishers,
Vol. 1: 308 pp.
2: 309-622

books
Cotter, Charles H., 1965.
The physical geography of the oceans.
American Elsevier Publ. Co., N.Y., 317 pp.

books
Coulson, C.A., 1952.
Waves: a mathematical account of the common types of wave motion. Oliver and Boyd, Ldt., London xii+159 pp., Price:7s6d.

books
Cowen, Robert C., 1960
Frontiers of the sea: the story of oceanographic exploration.
Victor Gollancz, Ltd., London, 307 pp.

Reviewed by GAT in Marine Observer, 1961, 31(193):149-150.
This is a popular book written by a non-oceanographer.

Reviewed by: Carruthers, J. N., 1961
J. Inst. Navigation, 14(4): 507-509

books
Cromie, William J., 1962.
Exploring the secrets of the sea.
Prentice-Hall, 300 pp., $5.95.

Reviewed by E. D. Goldberg, Science, 139(3557): 823.

books

Cushing, D.H. 1968.
Fisheries biology, a study in population dynamics.
Univ. Wisconsin Press, 200 pp. $7.50.

D

books

Daly, R.A., 1942.
The floor of the ocean. New light on old mysteries.
Univ. N. Carolina Press, 177 pp., figs.

books

Darwin, Sir George Howard, 1962.
The tides and kindred phenomena in the solar system.
W.H. Freeman & Co., San Francisco, 378 pp., $2.75.

a reprint
Reviewed by:
Blitzer, L., 1963, Amer. J. Physics, 31(1):70-71.

books

Daugherty, Charles Michael, 1961.
Searchers of the sea: pioneers in oceanography.
Viking Press, N.Y., 160 pp. (illustrated by Don Miller). $3.00.

eview in Veliger 4(2):118
By J.W. Hedgpeth

books

Davydov, L.K., and N.G. Konkina, 1958.
General hydrology. (In Russian).
Gidrometeoizdat, Leningrad, 488 pp.

Rev. in:
Appl. Mech. Rev., 13(7):#3748.

books

Day, J.H., 1969.
A guide to marine life on South African shores.
Univ. Cape Town, A.A. Balkema, 300 pp.

books

Dean, J.R., 1966.
History of oceanography. A century of oceanography.
Brown, Son and Ferguson, Glasgow, 128 pp. 42 s.

books

Deckert, K., and K. Gunther, 1956.
Creatures of the deep sea. Allen & Unwin, 40 Museum Street, London W.C. 1, England, 222 pp.

books

Defant, Albert, 1961.
Physical oceanography.
Pergamon Press, New York & London, Vol. 1 xvi+729 pp., Vol. 2 viii+598 pp.
£10 10s per set. ($35.00)

Reviewed by: J.C. Swallow, 1961 in Nature 192(4798):101.

Reviewed by J. B. Hersey, Science, 134(3488):1412

books

Degens, Egon T., 1965.
Geochemistry of sediments: a brief survey.
Prentice-Hall, Inc., 342 pp.

books

Delamare-Deboutteville, C., 1960.
Biologie des eaux souterraines littorales et continentales.
Hermann, Paris, 740 pp., 254 textfigs.
70 N.F.

Rev. by D.J. Zinn, 1962. Int. Revue Ges. Hydrobiol., 47(3):495.

books

Demel, Kazimierz, 1964.
Morze
Panstwowe Wydawnictwo Naukowe, Warsawa, 197 pp.

books

Dickinson, C.I., 1963.
British seaweeds.
London, Eyre and Spottiswoode, 232 pp. (25 s.)

Reviewed:

Newton, L., 1963. Nature, 199(4892):417.

books

Dietrich, G., and K. Kalle, 1957.
Allgemeine Meereskunde. Gebrüder Borntraeger, Berlin, 492 pp.

books

Djounkovski, N.N., and P.K. Bojitch, 1959.
La houle et son action sur les côtes et les ouvrages côtiers. Edition Eyrolles, Paris, 404 pp., 233 textfigs. - Gauthier-Villers, Paris (translation from the Russian)
Reviewed by C. Hensen in:
Deutsche Hydrogr. Zeits., 13(6):307-309.

books

Dobrovolsky, A.D., Editor, 1968.
Hydrology of the Pacific Ocean. (In Russian).
P.P. Shirshor Inst. Okeanol., Akad. Nauk, Isdatel "Nauka", Moskva, 524 pp.

books

Dronkers, J. J. 1964.
Tidal computations in rivers and coastal waters.
John Wiley + Sons, Inc. 518 pp. $17.00

books

Dubach, Harold W., and Robert W. Taber 1968.
Questions about the oceans.
Nat. Oceanogr. Data Center Gen. Ser. Publ. G-13: 121 pp. (paper). 55 cents

books

Düing, Walter 1970.
The monsoon regime of the currents in the Indian Ocean.
Int. Indian Oceanogr. Monogr. 1: 68 pp. $7.50.

books

Dunbar, M.J. 1968.
Ecological development in polar regions: a study in evolution.
Prentice-Hall, Inc. 119 pp. $4.95

books

Dunbar, M.J., Editor, 1963.
Marine distributions.
Univ. Toronto Press, viii+ 110 pp. $5.00

books

Dunn, G.E., and B.I. Miller, 1960.
Atlantic hurricanes.
Louisiana State Univ. Press, Baton Rouge, xx 326 pp., $10.00.

Reviewed by Alkire, H.L., 1962, Chesapeake Science, 3(1):50.

Also, Review by Paton, J., 1961
J. Atmos. Terres. Physics, 20(2/3): 220

books

du Plat, Joan, 1965
Marine archaeology
Hutchinson, London.

books

Duvanin, A.I., 1960.
Tides in the sea.
Gidrometeorologich. Isdatel., Leningrad, 390 pp.

E

books

Earl, G.E., and N. Peter, 1961.
Marine meteorology. The maritime Press, Ltd., London, 122 pp.

Reviewed by: L.B.P., 1962. Mar. Obs., 32(195):45-46.

books

Eckart, Carl, 1960

Hydrodynamics of oceans and atmospheres.
Pergamon Press, London, 290 pp.

books

Edmondson, W.T., Editor, 1966.
Ecology of Invertebrates, Marine Biology III, Proceeding of the Third International Interdisciplinary Conference.
New York Acad. Sci., 313pp. ($7.00).

books

Eguchi, Motoki, 1968.
The hydrocorals and scleractinian corals of Sagami Bay collected by His Majesty, The Emperor of Japan.
Marguen Co., Ltd. numerous pp. and plates.

books

Eicher, Don L. 1968.
Geologic time.
Prentice-Hall, Inc. 149 pp. $5.95

books

Ellis, C.B., 1954.
Fresh water from the ocean. For cities, industry and irrigation. Ronald, 240 pp., figs., About $5.00.

books

Emery, K.O. 1969.
A coastal pond studied by oceanographic methods.
American Elsevier Publishing Company, Inc. 80pp. $5.50.

books

Emery, K.O., 1960.
The sea off Southern California. A modern habitat of petroleum. Wiley, N.Y., 366 pp. $12.50.

Reviewed by P.E. Cloud in Science 131(3408):121; by Ph. H. Kuenen in Deep-Sea Res. 7
by J. W. Hedgpeth in Limnol. &Ocean., 5(3):347 - 348

books

Emery, K.O., and Evelyn Sinha, 1967.
Oceanographic books of the world, 1957-1966.
Mar. Techn. Soc., TP 2:57 pp. $2.00.

books
Ericson, David B., and Goesta Wollin, 1964.
The deep and the past.
Alfred A. Knopf, New York, 292 pp.

F

books
Fairbridge, Rhodes W., editor, 1966.
The encyclopedia of oceanography.
Reinhold Book Division, 1056 pp. $25.00.

book
Felsenbaum, A.I., 1960
[Theoretical basis and methods for calculating currents in the sea.]
Inst. Okeanol., Akad. Nauk, SSSR, 123 pp.

books
Fish, Marie Poland, and William H. Mowbray, 1970
Sounds of western North Atlantic fishes: a reference file of biological underwater sounds.
Johns Hopkins Univ. Press., 207 pp. $12.50

books
Fisher, J., and R.M. Lockley, 1954.
Sea-birds. An introduction to the natural history of the sea-birds of the North Atlantic.
Houghton, Mifflin, Boston, xvi 320 pp., illus., pls., $6.00.

books
Fogg, G.E., 1965.
Algal cultures and phytoplankton ecology.
The University of Wisconsin Press, Madison and Milwaukee, 126 pp., $5.50.

books
Fomin, L.M., 1964.
The dynamic method in oceanography.
American Elsevier Publ., Co., Inc., N.Y., 221 pp.

books
Fraser, J.H., 1962
Nature adrift: the story of marine plankton.
G.T. Foulis, & Co., London, 178 pp., 45 s.
Dufours Editions, Philadelphia. $8.95.
Reviewed:
A.J. Southward in J. Mar. Biol. Assoc., U.K., 43(1):279.

books
Freudenthal, Hugo D., editor, 1968.
Drugs from the sea.
Marine Techn. Soc., 297 pp. $12.00 (paperbound). (not seen).

books
Friedrich, Hermann 1969.
Marine biology: an introduction to its problems and results. Gwynne Vevers, translator.
Univ. Washington Press 474 pp.

Translation of: Meeresbiologie, Gebruder Borntraeger, Berlin, 1965.

books
Friedrich, Hermann, 1965.
Meeresbiologie, eine Einführung in die Probleme und Ergebenisse.
Gebrüder Born traeger, Berlin-Nikolassee, 436 pp.

book
Fritch, F.E., 1935
The structure and reproduction of the Algae. I. XVIII 791. Cambridge, England

books
Fujita, Hiroshi, 1962.
Mathematical theory of sedimentation analysis.
Physical Chemistry Monographs, Academic Press, New York and London, 315 pp.

books
Funnell, B.M., and W.R. Riedel, editors, 1971.
The micropalaeontology of oceans: proceedings of the symposium held in Cambridge from 10 to 17 September 1967 under the title "micropalaeontology of marine bottom sediments."
Cambridge Univ. Press. $55.00

G

books
Galtsoff, P.S., F.E. Lutz, P.S. Welch and J.G. Needham, 1937.
Culture methods for invertebrate animals.
Comstock Publ. Co., Ithaca, 590 pp., $4.00

book
Garland, G.D., editor, 1968.
Continental drift.
Univ. Toronto Press, $5.95 (not seen).

books
Gerjuoy, E., E.Yaspan and J.M. Major, editors 1968.
Physics of sound in the sea-2
Reverberation. 3. Reflection from submarines and surface vessels.
Gordon and Breach, 214 pp. $13.00

books
Gilbert, Perry W., Editor, 1963.
Sharks and Survival.
D.C. Heath Co., Boston, 578 pp.

books
Gilbert, Perry W., Robert F. Mathewson and David P. Rall, 1967.
Sharks, skates and rays.
Johns Hopkins Univ. Press, 624 pp. $15.00.

books
Glukhovsky, B. Kh., 1966.
Investigation of wind drifting waves. (In Russian).
Gidrometeoisdat, 284 pp.

books
Gorsky, N., 1961
The sea; friend and foe.
Foreign Languages Publishing House, Moscow, 266 pp.

books
Gorsky, N.N., 1960.
[The mystery of the oceans.]
The Publishing House of the USSR Academy of Sciences, Popular Science Series, Moscow.
Reviewed by: Baskakov, G.A., 1961. Okeanologiia, 1(5):948-950.

books
Gotto, R.V., 1969.
Marine animals
American Elsevier Publ. C., Inc. 96 pp. $4.25
(popular)

books
Graham, Michael, 1956.
Sea fisheries: their investigation in the United Kingdom. Edward Arnold, 488 pp., 112 drawings, 12 plates £5 5 s.

books
Grassé, P.P., R. Poisson and O. Tuzet, 1961.
Zoologie. I. Invertébrés. Précis de sciences biologiques publiés sous la direction du Pr. P.P. Grassé.
Masson et Cie, Editeurs, Paris, 919 pp.

books
Graustein, Jeannette E. 1967.
Thomas Nuttall naturalist - explorations in America 1808-1841.
Harvard Univ. Press 481 pp.

books
Green, J., 1961
A biology of Crustacea.
H.F. & G. Wetherby Ltd., London: 180 pp.

book
Greenberg, Jerry, 1968.
Fishmen fear-shark.
Seahawk Press, $2.00 (not seen).

books
Groen, P., 1951.
De Wateren der Wereldzee. C. de Boer Jr., Amsterdam, 300 pp., 148 figs., 40 photos.

books
Gullion, Edmund A., editor, 1968.
Uses of the sea. Prentice-Hall, Inc., 202 pp., $4.95.

H

books
Haigh, K.R. 1968.
Cable ships and submarine cables
The Trinity Press, Worcester and London, 416 pp.

books
Halstead, Bruce W. 1970.
Poisonous and venomous marine animals of the world. 3. Vertebrates (continued).
U.S. Govt. Printing Office, 1006 pp.

books
Halstead, Bruce W., 1967.
Poisonous and venomous marine animals of the world. II. Vertebrates.
U.S. Govt. Printing Off., 1070 pp.

books
Halstead, Bruce W., (with chemical sections by Donoven A. Courville), 1965.
Poisonous and venomous marine animals of the world. I. Invertebrates.
U. S. Government Printing Office, 994 pp.

books
Halstead, Bruce W., 1959
Dangerous marine animals. Cornell Maritime Press, Cambridge, Maryland. 146 pp. Reviewed by John P. Wise, 1960. Trans. Amer. Fish. Soc. 89(3): 318.

books
Hardy, Sir Alister, 1959.
The open sea: its natural history. II. Fish and fisheries.
Collins, London, 322 pp.

books
Hardy, Alister C., 1956.
The open sea: The world of plankton.
Collins, London, 335 pp., 103 textfigs., 24 pls. 30 shillings.

books
Harrison, Richard J., and Judith E. King, 1965.
Marine mammals.
Hutchinson & Co. Ltd., London; Hillary House Publishers Ltd., N.Y., 192 pp. $3.00.

books
Harvey, E.N., 1952.
Bioluminescence. Academic Press, Inc., N.Y., 650 pp.

book
Harvey, H.W., 1955.
The chemistry and fertility of sea water.
Cambridge University Press, 224 pp.

books
Harvey, H. W., 1945
Recent advances in the chemistry and biology of sea water. vii 164 pp, 29 figs. Cambridge Univ Press, London: MacMillan Co., New York

books
Harvey, H.W., 1928
Biological chemistry and physics of sea water. 194 pp., 65 textfigs. University Press, Cambridge.

owned by M.S.

books
Hedgpeth, J.W., Editor, 1957.
Treatise on Marine Ecology and Paleoecology, Vol. 1. Mem., Geol. Soc., Amer., 67:1296 pp.

books
Hegner, R.W., 1932
College Zoology. 713 pp., 568 textfigs.
MacMillian Co., N.Y.

owned by M.S.

books
Hela, Ilmo, and Taivo Laevastu, 1962
Fisheries hydrography; how oceanography and meteorology can and do serve fisheries.
Fishing News (Books), Ltd., 110 Fleet St., London EC1, 137 pp. £2 15 0

books
Herdman, W. A., 1923.
Founders of Oceanography and their work.
Arnold, London.

books
Hertwig, R., 1912
A manual of Zoology. Translated and edited by J.S. Kingley. 606 pp., 621 textfigs. Henry Holt and Co., N.Y.

owned by M.S.

books
Heuss, Theodor, 1948.
Anton Dohrn.
Rainer Wunderlich Verlag Herman Leins.
Stuttgart und Tübingen, 448 pp.

books
Hill, M.N., Editor, 1963.
The sea: ideas and observations on progress in the study of the seas. Vol. 3. The earth beneath the sea. History. Interscience Publishers New York and London, 963 pp.

books
Hill, M.N., Editor, 1963
The Sea: ideas and observations on progress in the study of the seas. Vol. 2. The composition of sea water. Comparative and descriptive oceanography, Interscience Publishers, New York and London, 554 pp.

books
Hill, Maurice N., Gen. Edit., 1962
The Sea; ideas and observations on progress in the study of the seas. Vol. 1, Physical Oceanography.
Interscience Publishers, 864 pp.

(See chapters listed individually under authors)

books
Hinds, Harold R. and Wilfred A. Hathaway, 1968.
Wildflowers of Cape Cod.
Chatham Press, Inc., 172 pp.

books
Hood, Donald W., editor 1971
Impingement of man on the oceans.
Wiley-Interscience, 738 pp.

book
Hood, Donald W., editor, 1970.
Symposium on organic matter in natural waters held at the University of Alaska, September 2-4, 1968.
Occ. Publ. Inst. mar. Sci. Alaska, 1: 625 pp. $10.00 (Paper cover $8.00).

books
Hopkins, David M., editor, 1968.
The Bering Land Bridge.
Stanford Univ. Press, 495 pp. $18.50. (not seen).

books
Horne, R.A., 1969.
Marine chemistry: the structure of water and the chemistry of the hydrosphere.
Wiley-Interscience, 568 pp. $19.95.

books
Horton, J.W., 1957.
Fundamentals of sonar. U.S. Naval Inst., Annapolis, 387 pp.

books
Hoult, David P., editor, 1969.
Oil on the sea.
Plenum Press, 114 pp.

books
Howe, H.V., 1962.
Ostracod taxonomy, 1962.
Louisiana State Univ. Press, 366 pp., $10.00.

Reviewed by:
A.G. Fisher, 1962, American Scientist, 50(3):314A

book
Huet, Marcel, 1970.
Traité de pisciculture. 4th edit., Editions, Ch. de Wyngaert, Bruxelles, XXIV + 718 pp.
$19.00

books
Hutchinson, G. Evelyn, 1967.
A treatise on limnology. II. Introduction to lake biology and the limnoplankton.
John Wiley & Sons, Inc. 1115 pp.

books
Hutchinson, G. Evelyn, 1966.
A treatise on limnology. Introduction to lake biology and the limnoplankton.
John Wiley & Sons, Inc. 992 pp. $29.00.

books
Hutchinson, G.E., 1957.
A treatise on limnology. 1. Geography, physics, and chemistry. J. Wiley & Sons, Inc., N.Y., 1015 pp.

books
Hylander, C.J., 1950.
Sea and Shore. Macmillan Co., N.Y., 242 pp.

books
Hyman, L.H., 1940
The investebrates: Protozoa through Ctenophora. 726 pp., 221 textfigs.
McGraw Hill Book Company, Inc., N.Y.

owned by M.S.

I

books
Ingle, James C., 1966.
The movement of beach sand.
Elsevier, 231 pp. $14.50

books
Ippen, Arthur T., 1966.
Estuary and coastline hydrodynamics. McGraw Hill Book Co, Inc., 744 pp.

books
Ivanov, A.V., 1963.
Pogonophora.
Consultants Bureau, N.Y., 479 pp. $12.50.

Tranlated from the Russian by D.B. Carlisle with additional material by Eve C. Southward.

First published in 1960 as one of the series "Fauna of the USSR", Zool. Inst., Akad. Nauk, SSSR.

J

books
Jackson, Daniel F., Editor, 1964.
Algae and Man.
Plenum Press, New York, 434 pp.

mostly freshwater

books
Jenkin, Penelope M., and John E. Harris, 1962.
Animal hormones: a comparative survey. Part 1. Kinetic and metabolic hormones.
International Monographs on Pure and Applied Biology, Pergamon Press, Vol. 6:310 pp.

books
Johnson, Frank H., and Yata Haneda, editors, 1966.
Bioluminescence in progress. Proceedings of the Luminescence Conference, Japan Society for the Promotion of Science.
National Science Foundation, Sept. 12-16, 1965.
Princeton University Press, 650 pp.

books
Johnson, T. W., Jr., and Frederick K. Sparrow, Jr., 1961
Fungi in oceans and estuaries.
J. Cramer, Publ., Weinheim, 668 pp. $30.00

Review by: Emerson, R. 1962
Science, 137(3531): 662-663

books
Jones F.R. Harden 1968.
Fish migration.
Edward Arnold Ltd. 325pp.

K

books
Keegan, Hugh L. and W.V. MacFarlane, Editors 1963
Venomous and poisonous animals and noxious plants of the Pacific region.
Pergamon Press, 456 pp.

books
Keen, M.J. 1968.
An introduction to marine geology
Pergamon Press Ltd, 218 pp.

books
Kellner, L., 1963.
Alexander von Humboldt.
Oxford University Press, N.Y., 247 pp. $5.75.

Reviewed: Woodford, A.O., 1963, Science, 140 (3570):973.

books
Kellogg, Winthrop N., 1961.
Porpoises and sonar. Univ. Chicago Press, 171 pp. $4.50.

Reviewed by H.O. Bull in:
Science, 134(3483):938-939.

books
King, C.A.M., 1962.
Oceanography for geographers.
Edward Arnold Publisher, London, 337 pp.

In the USA the title is:
An introduction to Oceanography.
McGraw Hill Book Company, New York, San Francisco 337 pp. (1963)

books
King, C.A., 1961.
Beaches and coasts.
Edward Arnold, Ltd., London, 403 pp., 65 shill.

Reviewed by H. Caspers, 1962.
Int. Rev. Ges. Hydrobiol., 47(3):493.

books
Kingsbury, John M. 1969.
Seaweeds of Cape Cod and the islands.
Chatham Press, Inc. 212 pp

books
Kinsman, Blair 1964.
Wind waves: their generation and propagation on the surface of the ocean.
Prentice-Hall, Inc., Englewood Cliffs, N.J., 640pp.
$26.60.

books
Kiselev, I.A. 1969
Plankton of the seas and continental waters. 1. Introduction and general problems of the planktology. (In Russian)
Izdatel. "Nauka", Leningradskoe Otdelenie, 657 pp

book
Klenova, M.V., 1960
Geology of the Barents Sea. Izdatel. Akad. Nauk, SSSR, Moscow, 1960, 367 pp.

books
Klenova, M.V., 1948.
Geologiia moria. Moskva, 495 pp., 3 pls.

books
Klenova, M.V., 1948.
Geologiia moria: uchebnae posobie dlia geograficheskikh fakul'tetov universitetov i pedagogicheskikh institutov. [Geology of the sea: textbook for geographical faculties of universities and pedagogical institutes.] Moskva, Gos- uchebnopedagogich. izd-vo Ministerstva prosverschenuia RSFSR, 1948:945 pp., maps.

books
Kraus, E.B. 1972.
Atmospheric ocean interactions.
Oxford Univ. Press 265 pp.

books
Kriss, A.E., 1961.
Meeresmikrobiologie, Tiefseeforschungen.
Gustav Fischer Verlag, Jena, 570 pp., 147 figs., 134 tables. 98.10 DM.

Reviewed:
DeLey, J., 1962.
Hydrobiologia, 20(4):382-383.
Translation from the Russian!

REVIEWED: Shewan, J.M., 1962
Nature, 195(4843): 743.

books
Krümmel, O., 1911
Handbuch de Ozeanographie. Stuttgart, 2nd Edition. 1:1-526;2:1-766.

books
Kuenen, Ph. H., 1955.
Realms of water. John Wiley & Sons, Inc., 320 pp.

books
Kuenen, P.H., 1950.
Marine geology. New York, John Wiley & Sons, 568 pp. ($7.50).

Reviewed: Science 114(2953):135-137, by W.H. Twenhoffel.

books
Kumé, Matazo, and Katsuma Dan, editors, 1968.
Invertebrate embryology. Translated from Japanese by Jean C. Dan, NOLIT Publ. House, Belgrade, 605 pp.

L

books
Lacombe H. 1968.
Les énergies de la mer.
Presses Universitaires de France 128 pp

books
Lacombe, Henri, 1965.
Cours d'oceanographie physique. (Theories de la circulation. Houles et vagues).
Gauthier-Villars, Paris, 392 pp.

books
Lane, F.W., 1960.
Kingdom of the octopus.
Sheridan House, New York, 300 pp.

books
Lange, Rolf, editor, 1969.
Chemical oceanography. Scandinavian University Books, Oslo, Norway, United States Distribution Office: Box 142, Boston, Massachusetts, 02113, 152 pp.

books
Lanyon, W.E., and W.N. Tavolga, Edit., 1960.
Animal sounds and communications.
Amer. Inst. Biol. Sci., Publ. No. 7:443 pp., 12-in L.P. Record, $9.50.

Reviewed:
H. Caspers, 1962. Int. Revue Ges. Hydrobiol., 47(3):487.

books
Lauff, George H., editor 1967.
Estuaries.
Publs Am. Ass. Advmt Sci., 83:757 pp.

books
Lebour, M. V., 1930.
The planktonic diatoms of northern seas.
Ray Society. London.

books
Leim, A.H., and W.B. Scott, 1966.
Fishes of the Atlantic coast of Canada.
Bull. Fish. Res. Bd Can., 155:485 pp. $8.50.

books
Leip, Hans, 1958 (English translation)

The River in the sea. G. P. Putnam's Sons, New York, 223 pp.

Reviewed by Leonard Outhwaite in:
Geogr. Rev., 49(1):138-139.

books
Leont'ev, O.K., 1963.
A short course of marine geology. (In Russian).
Isdatel. Moskovsk. Univ., 1963, 463 pp.

books
Leontiev, O.K., and A.I. Khalikov, 1965
Formation of the shores of the Caspian Sea.
(In Russian).
Inst. Geograf., Akad. Nauk, Azerbaid. SSR, Baku, 205 pp.

books
Levring, Tor, Heinz A. Hoppe and Otto J. Schmid, 1969
Marine algae: a survey of research and utilization.
Cram, De Gruyter & Co., Hamburg, 421 pp. (140 DM)
[Botanica Marina Handbooks]

books
Lewin, Ralph A., Editor, 1962.
Physiology and biochemistry of algae.
Academic Press, New York and London, 929 pp.

books
Lilly, John C., 1961.
Man and dolphin.
Doubleday, Garden City, N.Y., 312 pp. $4.95

Reviewed:
H.O. Bull, 1961. Science 134(3483):938-939.

books
Lineaweaver, Thomas H. III and Richard H. Backus 1970.
The natural history of sharks.
J.P. Lippincott Company, Phila. + N.Y., and André Deutsch Ltd., England, 265 pp.
$6.95; £2.75.

books
Lockwood, A.P.M., 1968.
Aspects of the physiology of Crustacea.
W.H. Freeman, $9.00 (not seen).

books
Lombard, A., 1956.
Geologie sedimentaire. Les series marines.
Masson et Cie, Paris, 722 pp.

Reviewed in: Amer. J. Sci., 256(5):365-367.

books
Long, John E., Editor, 1964.
Ocean sciences.
U.S. Naval Institute, 304 pp., $10.00

M

books
Mackintosh, N.A., 1965.
The stocks of whales.
Fishing News (Books) LTD, 232 pp.
Fisherman's Library, Buckland Foundation
£ 27 6
$7.25

books
MacMillan, D.H., 1966.
Tides.
American Elsevier Publishing Co., Inc. 240 pp., $9.50.

books
Marmer, H.A., 1930
The Sea. New York, 312 pp., illus.

books
Marshall, N.B. editor 1967.
Aspects of marine zoology.
Symp. Zool. Soc. London 19: 270 pp.

books
Martin, Dean F. 1968.
Marine geochemistry. 1. Analytical methods.
Marcel Dekker, Inc., 280 pp. $5.75.

Mc

books
McConnaughey, Bayard H., 1970.
Introduction to marine biology.
C.V. Mosby Co., St Louis, 449 pp. $11.00.

books
McCormick, Harold W., and Tom Allen with CAPT William E. Young, 1963.
Shadows in the sea: the sharks, skates and rays.
Chilton Books, Div. Chilton Publishers, 415 pp.

books
McLellan, Hugh J., 1965.
Elements of physical oceanography.
Pergamon Press, 150 pp. ($9.50).

books
Menard, H.W., 1964.
Marine geology of the Pacific.
McGraw-Hill Book Company, 271 pp.

book
Mertens, Lawrence E. 1970.
In-water photography, theory and practice.
Wiley-Interscience, 391 pp.

book.
Mileikovski, S.A. 1967.
History of plankton research in the Pacific Ocean.
In: Biologiia Tichogo Okeana Plankton.
Isdatel Nauka Moskva, 7-26

books
Miller, Robert L., Editor, 1964.
Papers in Marine Geology: Shepard Commemorative Volume.
Macmillan Co., N.Y., Collier-Macmillan Ltd., London, 531 pp.

books
Millott, N., editor, 1967.
Echinoderm biology.
Symp. Zool. Soc., Lond., Academic Press, 20: 240 pp. $11.00

books
Miyake, Yasuo, and Tadashiro Koyama, Editors, 1964.
Recent researches in the fields of hydrosphere, atmosphere and nuclear geochemistry. Ken Sugawara Festival Volume, Maruzen Co., Ltd., Tokyo, 404 pp

books
Moore, Hilary B., 1958.
Marine ecology.
John Wiley and Sons, New York 493 pp.

Reviewed in: Deep-Sea Res., 6(1):75-76.

books
Mordukhai-Boltovskoy, F.D., 1960.
Caspian fauna in the Azov-Black Sea basin.
(In Russian).
Akad. Nauk, SSSR, Inst. Biol. Vodochran., 286 pp.

Reviewed by L.A. Zenkevich, 1963. Zool. Zhurn., 42(3):473-476.

books
Morgan, R., 1956.
World sea fisheries. Methuen & Co., Ltd., 307 pp

books
Morton, J.E., 1958.
Molluscs. Hutchinson & Co., London, 232 pp.

Rev. by C.M. Yonge in Nature 183(4653):3.

books
Mosby, Håkon, 1960
Surrounding seas: Ch. 2 in: A geography of Norden. J.W. Cappelens Forlag, Oslo: 18-26, maps 6 and 7.

books
Mosetti, Ferruccio, 1964
Oceanografia.
Del Bianco Editore - Udine, 462 pp.

books
Mott, N.F., 1952.
Elements of wave mechanics. Cambridge Univ. Press, ix + 156 pp., 21 shillings.

books
Mouromtzev, A.M., 1956.
[L'océan mondial.] Bibliothèque Scientifique Populaire, Ed. Hydrometeorol., Leningrad, 1956, 86 pp., 24 figs.

books
Mouton, M.W., 1952.
The continental shelf. Martinus Nijhoff, The Hague, 367 pp.

"critical and comprehensive study of the juridical position of the continental shelf and of the questions concerning the utilization of the sea covering it and of its soil and subsoil beyond the limits of the territorial waters."

books
Munk, Walter H., and Gordon J.F. MacDonald, 1960.
The rotation of the earth: a geophysical discussion. Cambridge University Press, xix + 323 pp., $13.50.

Reviewed by Edgar W. Woolard in:
Trans. Amer. Geophys. Union, 42(3):307.

books
Muromtsev, A.M., 1958
[The basis for the hydrology of the Pacific Ocean.]
Gidrometeoizdat., Lengrad, 431 pp.

books
Muromtsev, A.M., 1959.
[The basis for the hydrology of the Indian Ocean.]
Gidrometeoizdat., Leningrad, 437 pp.

books
Murray, Grover E., 1961.
Geology of the Atlantic and Gulf Coastal province of North America. Harper, New York, xvii 692 pp., illus., $24. (Harpers Geoscience Ser.)

Reviewed by: Monroe, W.H., 1962.
Science, 135(3508):1057.

Reviewed by: LeGrand, H. E., 1962.
Trans. Amer. Geophys. Union, 43(1): 34

Reviewed by: Thomas, H. K., 1962. The Atlantic and Gulf Coast Geosyncline
Nature, 196(4849): 4 - 5.

books
Murray, J. and J. Hjort, 1912
The Depths of the Ocean. A general Account of the modern science of oceanography based largely on the scientific researches of the Norwegian steamer MICHAEL SARS in the North Atlantic. 821 pp., 575 textfigs.
Macmillan and Co. Ltd., London.

owned by M.S.

N

books
Nazarov, V.S., and A.M. Muromiev, 1954.
[Oceanography] Edited by Moscow-Transport, 166 pp., 68 figs.

books
Neumann, G., 1968.
Ocean currents.
Elsevier Publishing Co., 352 pp. $20.00.

Books
Neumann, Gerhard, and Willard J. Pierson, 1966.

Principles of Physical Oceanography,

Prentice-Hall, Inc., Englewood Cliffs, N. J., 545 pp.

books
Newsom, John D., editor, 1968.
Proceedings of the Marsh and Estuary Management Symposium, Louisiana State Univ., July 19-20, 1967: 250 pp.

books
Nicol, J.A. Colin, 1960.
The biology of marine animals.
Sir Isaac Pitman and Sons, Ltd., London, xi 707 pp., 294 textfigs.

Reviewed by D.P. Wilson, 1961
J.Mar. Biol. Assoc., U.K., 41(1):222

books
Nicholl, G.W.R., 1960

Survival at sea. Adlard Coles, Ltd., 166 pp. 25 shillings.

Reviewed in:
Marine Observer, 1961, 31(191): 45-46.

books
Nikolsky, G.V., 1963.
The ecology of fishes.
Academic Press, London, 352 pp.

Reviewed by:
H.G. Vevers, 1964, Chemistry & Industry, (31): 1378-1379.

books
Nishiwaki, M. 1965.
Whales and pinnipeds. (In Japanese)
Univ. Tokyo Press 439 pp.

O

books
Odum, E.P., 1959.
Fundamentals of ecology. 2nd Edit., 546 pp., 160 figs.
W.B. Saunders, Phila. & London, $7.50.

Reviewed by E. Nørgaard, 1959, Oikos, 10(2): 290-291.

books
Officer, C.B., 1958.
Introduction to the theory of sound transmission with applications to the ocean.
McGraw-Hill, 310 pp., 191 figs.

books
Olson, Theodore A., and Frederick F. Burgess 1967.
Pollution and marine ecology.
Interscience Publishers, XV + 364 pp.
96 shillings.

books
Ommanney, F.D., 1949.
The Ocean. 238 pp. Oxford Univ. Press.

books
Oppenheimer, Carl H., editor, 1966.
Phytoplankton, Marine Biology II, Proceedings of the Second International Interdisciplinary Conference.
New York Acad. Sci., 369 pp. ($8.00).

Books
Osokin, S.D., 1964.
A course in navigation. (In Russian).
Hydrographic Service of the Soviet Navy, Vols. 1-6, 1958-1963.
Reviewed in: Okeanologiia 4(4):737-738.

books
Ottmann, Francois, 1965.
Introduction a la Geologie Marine et Littorale.
Masson et Cie, Paris, 259 pp. 47 F

books
Oura, Hirobumi, editors, 1967.
Physics of snow and ice.
Inst. Low Temp.Sci., Hokkaido Univ., 711 pp.

P

books
Parsons, James 1962.
The green turtle and man.
Univ. Florida Press, 126 pp.

books
Pax, F., Editor, 1962.
Meeresprodukte. Ein Handwörterbuch der marinen Rohrstoffe.
Gebrüder Borntraeger, Berlin-Nikolassee, 259 pp.
471 pp., illust. DM78

Reviewed:
Hoppe, Heinz A., 1963. Botanica Marina, 5(4):128.

books
Pearson, E.A., Editor, 1960

Waste disposal in the marine environment (Berkeley, July 22-25, 1959)
Pergamon Press, N.Y.: 569pp.

"books"
Percier, Albert 1967.
Océanographie et technique des pêches maritimes.
Bull. Cent. Etud. Rech. scient., Biarritz 6(3): 353-659.

books
Pérès, J.M., 1961.
Oceanographie biologique et biologie marine. 1. La vie benthique.
Paris, Presses Universitaires de France, 541 pp

Reviewed:
ED 1963. Arch. Oceanografia e Limnol., 13(1): 152-153.

books
Pérès, J.M., and L. Deveze, 1963.
Océanographie biologique et biologie marine. "Euclide", Presses Universitaires de France, Paris, 514 pp.

books
Peterson, Mendel, 1965.
History under the sea: a handbook for underwate exploration.
Smithsonian Institution, Washington, D.C., 108 pp. 56 pls. $3.00.

books
Picard, J., 1961.
11000 Meter unter dem Meeresspiegel. Die Tauchfahrten des Bathyskaphs "Trieste", F.A. Brockhaus, Wiesbaden.

Reviewed by Th. Stocks, 1962, Deutsche Hydrogr. Zeits., 15(1):38.

books
Pickard, G.L., 1964.
Descriptive physical oceanography: an introduction.
Macmillan Co., New York, 199 pp. $4.50.

books
Porep, R. 1970.
Der Physiologe und Planktonforscher Victor Hensen (1835-1924): Sein Leben und sein Werk.
Kieler Beiträge zur Geschichte der Medizin und Pharmazie, 9 (R. Herrlinger, F. Kudlien und G.E. Dann), Neumünster: Wachholtz, 147 pp. 19.80 D.M. (not seen)

books
Prescott, G.W. 1968
The algae: a review.
Houghton Mifflin Co., Boston, 436 pp.

books
Price, Don. K., 1965
The scientific estate.
Harvard University Press, 323 pp.

books
Proudman, J., 1953.
Dynamical oceanography. Methuen, London: Wiley, N.Y., 409 pp., figs., $8.50.

Reviewed by H. Stommel in Science 118(3065):365.

books
Purchon, R.D. 1968.
The biology of the Mollusca.
Pergamon Press, 560 pp.

Q

R

books

Reitt, Helen, and Beatrice Moulton, 1968.
The Scripps Institution of Oceanography: first fifty years.
The Ward Ritchie Press, 217 pp.

books

Rankama, K., and Th. G. Sahama, 1950.
Geochemistry. Univ. Chicago Press, 912 pp., 44 textfigs.

books

Ray, C., and E. Ciampi, 1958.
The underwater guide to marine life.
Nicholas Kaye, Ltd., London, i-xii, 1-338, 16 pls

Reviewed:
Nature, 183(4665):852, by D.P. Wilson
Science Progress, 47(186):384 by D.B. Carlisle.

books

Ray, Dixie Lee, Ed., 1960
Marine Boring and fouling organigsms.
Univ Washington Press, 480, pp. $8.50

books

Raymont, John E.G., 1963.
Plankton and productivity in the oceans.
International Series of Monographs on Pure and Applied Biology, Pergamon Press, Ltd., 649 pp.

books

Rees, W.J., editor, 1966.
The Cnidaria and their evolution.
Symp. Zool. Soc., Lond., Academic Press, 16: 449 pp. $17.50

books

Reid, G.K., 1961

Ecology of inland waters and estuaries.
Reinhold Publ. Corp., N.Y.: 375 pp.

books

Richards, Adrian F., editor, 1967.
Marine Geotechnique.
Univ. Illinois Press, 327 pp. $8.95.

books

Richter, V.G., 1965.

Methods of study of the newest and recent tectonics of the shelf zone of seas and oceans.

Isdatel. "Nedra", Moscow, 244 pp.

books

Ricketts, E.F., and J. Calvin, 3rd Edit., revised by J.W. Hedgpeth, 1962.
Between Pacific tides. Stanford University Press, 516 pp., $8.75.

Reviewed by Smith, R., 1963. Veliger, 5(3):123-124

books

Robinson, A.H.W., 1962.
Marine cartography in Britain.
Leicester Univ. Press, 222 pp., 42 pls., 30 figs 5 5s.

Reviewed by:
G.S. Ritchie, 1963. J. Inst. Navigation, 16(1): 140-141.

books

Rogers, Howard, 1968.
Marine corrosion.
George Newnes Ltd., $7.50.

books

Roll, H.U., 1965.
Physics of the Marine Atmosphere.
International Geophysical Series, Academic Press 7:426 pp. $15.00

books

Romanovsky, V., C. Francis-Boeuf and J. Bourcart, 1953.
La Mer., Librairie Larousse, 503 pp., 16 color pls 870 textfigs. 5200 fr., $13.50.

Reviewed by F.G. Walton Smith, Ecology 38(2):370

books

Rongel, A., 1944
Elementos Oceanografia. D.H. 21, Marinha do Brasil, Hidrografia 1944 pp. figs.

no references

books

Ross, David A. 1970.
Introduction to oceanography.
Appleton-Century-Crofts, Div. Meredith Corp.
384 pp.

books

Ross R.D. 1968.
Industrial waste disposal.
Reinhold Book Corp., 340pp + I-XII
(not seen)

books

Rouch, J., 1961.
Les marées. Ed. Payot, Paris, 232 pp., 27 figs.

books

Rouch, J., 1959?
Les Océans. Armand Colin, No. 320:224 pp., 14 figs.

Reviewed by J. Bourcart in:
Rev. Géogr. Phys., Géol. Dyn., 1959 (2) 2(1):68.

books

Rouch, J., 1943.
Traité d'Océanographie physique. Sondages.
Payot, Paris, 256 pp., 100 textfigs.

books

Roumania, Comité d'État des Eaux, Conseil des Ministres, Republique Socialiste de Roumania, 1966
Contributions roumaines à l'étude de la Mer Noire.
Bucerest, 112 pp.

books

Rounsefell, G.A., and W.H. Everhart, 1953.
Fishery Science. xii 444 pp., John Wiley & Sons Inc., N.Y.

Reviewed:
J. du Cons. 20(1):102-105 by C.E. Lucas.

books

Runcorn, S.K., Editor, 1962.
Continental drift. International Geophysics Series, Vol. 3, Academic Press, New York and London, 331 pp., $12.00.

Reviewed by: Straley, H.W., III, 1963. J. Geol., 71(5):661-662.

books

Runcorn, S.K., 1960

Methods and techniques in geophysics. Vol. I.
Interscience Publishers, Inc., New York: 374 pp.

books

Russell, Frederick S.
The Medusae of the British Isles. 2. Pelagic Scyphozoa with a supplement to the first volume of the Hydromedusae
Cambridge Univ. Press. £7

books

Russell, F.S., editor, 1963.
Advances in Marine Biology, Academic Press, London and New York, 1:410 pp.

books

Russell, R.C.H., and D.H. Macmillan, 1953.
Waves and tides. Hutchinson's Sci. Tech. Publ. London and N.Y., 348 pp., 17 pls., 100 plus figs. $3.75.

for mariners and yachtsmen, not professional physicist or geophysicist.

books

Russell, Richard J., 1967.
River plains and sea coasts.
Univ. Calif. Press, 173 pp.

books

Russell-Hunter, W.D. 1968.
A biology of lower invertebrates.
MacMillan Co. 181 pp.

S

books

Sager, G., 1960
Gezeiten und Schiffahrt.
Fachbuch, Leipzig, 170pp.

Reviewed by:
Böhl, 1961 in Beitr. Meeres., 1:61-62.

books

Scheffer, V.B., 1958.
Seals, sea lions and walruses. A review of the Pinnipedia.
Stanford Univ. Press, Stanford, California, 179pp

book

Schmitt, Waldo, 1965.
Crustaceans.
Ann Arbor Sci. library, Univ. Michigan Press, 204 pp. Paperback, $1.95.

books

Schulejkin, W.W., 1960 (translated into German by Erich Bruns).
Theorie der Meereswellen.
Akademie - Verlag, Berlin, 158 pp., 49 textfigs.
Reviewed by Hans Walden in:
Deutsche Hydrogr. Zeits., 13(6):309-311.

books

Sears, M., Editor, 1961

Oceanography. Invited lectures presented at the International Oceanographic Congress held in New York - 1959.
Publ. Amer. Assoc. Adv. Sci., 67: 654 pp.

books

Shepard, F.P., 1959
The earth beneath the sea.
Johns Hopkins Press, Baltimore: 275 pp.

books

Schott, G. 1935
Geographie des indischen und stillen Ozeans. Hamburg 413 pp., 37 pls., 114 text figs.

books

Schott, G., 1942
Geographie des Atlantischen Ozeans im auftrage der Deutschen Seewarte vollständig neu bearbeitete dritte auflage. 438 pp., 141 text figs., 27 plates. C. Boysen. Hamburg.

Owned by MS

books

Shuleikin, V.V., 1953
[Molecular physics of the sea] Pt. 8, Sect. 10-17. Fizika Morye, 3rd Ed., Moscow, pp. 786-819.

OTS/ $1.25

books

Shepard, Francis P., and Robert F. Dill 1966
Submarine canyons and other sea valleys
Rand McNally + Co., 351 pp

books

Shokalskii, Iu. M., 1959.

Okeanographia. 2nd edition. Gidrometeorolicheskoe Isdatelstbo, Leningrad, 537 pp.

No references added after original edition in 1917.

(The author died in 1940 at the age of 84)

book

Schokalsky, J. 1917
Okeanografia (in Russian). Petrograd, xiii and iv, 1-614, maps illus.

books

Shuleikin, V.V., 1959.
Kratkii Kurs Fiziki Moria.
Gidrometeorol. Izdatel, Leningrad, 275 pp.

books

Shuleikin, V.V., 1949.
Ocherki po fiziki moria. Izd-vo Akad. Nauk, USSR, Moskau, 334 pp.

books

Schulejkin, W.W., 1960.
Theorie der Meereswellen. (Transl. E. Bruns). Akademie-Verlag, Berlin,

Rev., Bruns, E., 1961, Beitr. Meeres., 1:62-63.

books

Sinclair, Bruce, 1966.
Early research at the Franklin Institute: the investigation into the causes of steam boiler explosions, 1830-1837.
Franklin Inst. Phila., 26+48 pp.

books

Singer, S.F. editor, 1970
Global effects of environmental pollution.
Reidel, Dordrecht, Holland, 215 pp. (not seen)

books

Slijper, E.J., 1962.
Whales. Transl. A.J. Pomerans.
Hutchinson & Co., Ltd., London, 475 pp. 63 s.

Rev.
L.H. Matthews, 1962, Nature, 196(4853):404.

Dutch edition, 1958.

Also: Basic Books, Inc., New York, 1962

books

Smith, J.L.B., 1956.
The search beneath the sea. Henry Holt & Co., 260 pp. $3.95. in re Latimeria.

books

Sneschinskii, W.A., 1954.
[Handbuch der praktischen Ozeanographie]
Leningrad, 671 pp., 375 figs., 39 pls.

Reviewed by Gustav Thiel in Deutsche Hydrogr. Zeits., 9(4):199-200.

books

Somma, A., 1952.
Elementi di Meteorologia ed Oceanografia. Pt. II. Oceanografia. Casa Editrice Dott. Antonio Milani, Padova, Italia, xvii 758 pp., 322 textfigs., maps, tables, 4500 lire.

Books

Southward, A. J., 1965.

Life on the Sea - shore.

Harvard Univ. Press., 153 pp. $3.00

books

Spar, J., 1962.
Earth, sea and air. A survey of the geophysical sciences.
Addison-Wesley Publishing Co., Reading, 152 pp., $2.95.

books

Sparrow, Frederick K., Jr., 1960 (2nd Edition).
Aquatic Phycomycetes.
Univ. Michigan Press, 1187 pp.

books

Spiegler, K.S., 1962.
Salt water purification.
John Wiley and Sons, New York, London, 167 pp.

Reviewed:
Kalle, K., 1963. Deutsche Hydrogr. Zeits., 16(2):92-93.

book

Spern, Philip, 1966.
Fresh water from saline waters: The political, social, engineering and economic aspects of desalination.
Pergamon Press, 34 pp.

books

Steele, J.H. editor 1970.
Marine foodchains.
Univ. Calif. Press, 552 pp. $13.50

books

Steers, J.A., 1951.
The sea coast. Collins, London, 276 pp., $3.75

Reviewed: J. Geol. 63(1):94-97.

books

Sténuit, Robert (translated by Morris Kemp), 1966
The deepest days.
Coward-McCann, Inc., New York, 222 pp.

books

Stewart Harris B., Jr., 1966
Deep Challenge.
D. Van Nostrand Co., Inc. 202 pp $5.95

books

Stewart, Harris B., Jr., 1964.
The global sea. D. van Nostrand, Co., Princeton, N.J., $1.45 (paper bound).

books

Stoker, J.J., 1957.
Water waves: the mathematical theory with applications. Interscience Publ. Inc., N.Y., 567 pp.
Rev. in Physics Today 11(8):28.

books

Stoker, J.J., 1957.
Water waves. Interscience Publ. 595 pp. $12.00

books

Stommel, Henry, 1959.
The Gulf Stream. Univ. Calif. Press & Cambridge Univ. Press, 202 pp.
Reviewed Fr. Model in Deutsche Hydrogr. Zeits., 12(4):213-214.

books

Stommel, H., 1945
Science of the Seven Seas. 208 pp. numerous textfigs. Cornell Maritime Press, N.Y.

owned by M.S.

books

Strakhov, N.M., editor, 1966.
Geochemistry of Silica. (In Russian).
Isdatel. "Nauka", Moskva, 421 pp.

books

Sverdrup, H.U., 1952.
Havlaere. Fabritius & Sønners Forlag, Oslo, 110 pp., 43 textfigs.

books

Sverdrup, H.U., 1942
Oceanography for meteorologists. New York Prentice-Hall. 246 pp.

books

Sverdrup, H. U., M.W. Johnson and R.H. Fleming 1942.
The oceans: their physics, chemistry and general biology.
Prentice-Hall, x + 1087 pp., 265 figs., 7 charts.

T

books
Tait, J.B., 1952.
Hydrography in relation to fisheries. (Buckland lectures for 1938). Edward Arnold & Co., London: lxii & 106 pp., frontispiece and 19 figs.

Reviewed: J. du Cons. 19(2):211-213 by J.N. C-

books
Tannehill, I.R., 1956.
Hurricanes.
Princeton Univ. Press, 308 pp.

books
Tannehill, I.R., 1945.
Weather around the world. Princeton Univ. Press, 200 pp., 55 textfigs.

books
Tarling, D.H., and M.P. Tarling, 1970.
Continental drift.
Bell, 112 pp. 30s.

books
Tatarski, V.I., 1961.
Wave propagation in a turbulent medium.
McGraw-Hill Book Co., Inc., 257 pp., xivpp 6 pp., bibliography, 2 pp. appendix, and 22 pp notes and remarks. $9.75.

Reviewed in: Proc. IRE, 49(10):1590 by R. Bolgiano, Jr.

books
Tavolga, William N., editor, 1967.
Marine Bio-acoustics.
Pergamon Press, 2:353 pp.

books
Taylor, W. R., 1960.
Marine algae of the Eastern Tropical and Subtropical coasts of the Americas.
Univ. Michigan Press, Ambassador Books, Ltd., Toronto, x+870 pp., 14 textfigs., 80 pls. $19.50.

Reviewed by:
A.J. Lernatowicz, Limnol. & Oceanogr., 6(1):99-100.

books
Taylor, W. R., 1937.
Marine algae of the northeastern coast of North America. Univ. Michigan Press, 427 pp., 60 pls.

books
Terry, Richard D., 1966.
The deep submersible.
Western Periodicals Co.,
North Hollywood, Calif. 456 pp.

books
Tolstoy, Ivan, and C.S. Clay, 1966.
Ocean acoustics: Theory and equipment in underwater sound.
McGraw-Hill Book Co. 345 pp., $15.50.

books
Trask, P. D., ed., 1950.
Applied sedimentation. John Wiley Sons, 707 pp.

books
Traung, Jan-Olof, 1967.
Fishing boats of the world: 3
Fishing News (Books) Ltd., 648+XLII pp.
£4 15 0 ($23.50) (not seen)

books
Tricker, R.A.R., 1964.
Bores, breakers, waves and wakes; an introduction to the study of waves on water.
American Elsevier Publishing Co., Inc., 250 pp. $6.50.

books
Tucker, D.G., 1967.
Sonar in fisheries - a forward look.
Fishing News (Books) Ltd., 136 pp.

books
Tucker, D.G., 1966.
Underwater observations using sonar.
Fishing News (Books) Ltd. 144 pp.

books
Turekian Karl K. 1968.
Oceans.
Prentice-Hall, Inc., 120 pp. $2.00 (paper), $5.95 (cloth).

U

books
United States, National Academy of Sciences, 1969.
Eutrophication: causes, consequences, correctives; proceedings of a symposium, 661 pp.

books
United States, Underwater Society of America, 1962.
Underwater Yearbook, 1962. Garrand Press, Champaign, Ill., 70 pp., $5.00

books
Urick, Robert J., 1967.
Principles of underwater sound for engineers.
McGraw-Hill Book Co. 204 pp., $17.50.

V

books
Vallaux, C., 1933
Géographie générales des mers. Paris, 796 pp.

books
Vaucher, Charles, 1960 (translated by James Hogarth).
Sea Birds. Oliver & Boyd, Ltd., Edinburgh & London 254 pp., 255 photos.

Reviewed in:

Nature 188(4757):1139 by Frances Pitt, 1960. Sea birds.

books
*Vinogradov, K.A., editor, 1967.
Biology of the northwestern Black Sea. (In Russian).
Naukova Dumka, Kiev, 268 pp.

books
Vinogradov, M.E. 1968.
Vertical distribution of the oceanic plankton. (In Russian).
Izdatel. "Nauka", Moskva, 319 pp.

books
Von Arx, William S., 1962.
Introduction to physical oceanography.
Addison-Wesley Publ. Co., Reading, Mass., 422 pp. $15.00

Reviewed in: Undersea Technology, 3(4):11.

W

books
Walford, L.A., 1958.
Living resources of the sea. Opportunities for research and expansion.
Ronald Press Co., N.Y., xv 321 pp.

Reviewed by:
Cl. Delamare Deboutteville in: Vie et Milieu, 11(1):137-139.
R.S. Wimpenny in: Deep-Sea Research, 5(4):315-316.
J.P. Vine in: Science, 129(3356):1123

books
Waterman, T.H., Editor, 1961.
The physiology of Crustacea. Vol. 2. Sense organs, integration and behavior.
Academic Press, Inc., xiv+ 681 pps. $20.00.

books
Waterman, Talbot, H., 1960, Editor
The physiology of crustacea. Vol. I. Metabolism and growth. Academic Press, N.Y., 670 pp.

Review by: Milton Fingerman, 1961
Trans. Amer. Fish. Soc., 90(1): 88

books
Waterman, T.H., Edit., 1960, 1961.
The physiology of Crustacea. Vols. 1 and 2.

Reviewed by Joel W. Hedgpeth in AIBS Bull, 12(1): 40.

books
Weyl, Peter K., 1970.
Oceanography: an introduction to the marine environment. John Wiley and Sons, Inc., New York, xvii + 535 pp. $12.50.

books
Wheeler, Alwyne, 1969.
The fishes of the British Isles and North-West Europe. Michigan State University Press, 613 pp.

books
editors,
Whittard, W.F., and R. Bradshaw, 1965.
Submarine geology and geophysics.
Colston Papers, Butterworth's, London, No. 17:464 pp., 39 pls. $21.00.

books
Wickstead John H. 1965.
An introduction to the study of tropical plankton.
Hutchinson Tropical Monographs, 160 pp.

books
Wiegel, Robert L., 1964
Oceanographical engineering.
Prentice-Hall Series in Fluid Mechanics, 532 pp.

books
Wiens, H. J., 1962
Atoll environment and ecology.
Yale University Press, 532 pp.

books
Wiese, V.I., 1939.
Moria Sovetskoi Arktiki. Leningrad, Glavsevmorput Press, 2nd ed., 565 pp., illus., ports., maps.

LC G 630.R8W5

books
Wilbur, Karl M., and C.M. Yonge, editors, 1966.
Physiology of Mollusca.
Academic Press, Vol. 2:645 pp. $22.00

books
Wilbur, Karl M., and C.M. Yonge, Editors, 1964.
Physiology of Mollusca.
Academic Press xiii 473 pp. 114 s 6d

books
Wildt, R., editor, 1968.
Physics of sound in the sea. IV.
Acoustic properties of wakes.
Gordon and Breach 124pp. $17.50

books
Wilkes, C., 1845
Narrative of the United States Exploring Expedition during the years 1838-1842.
Philadelphia, 5 vols.

books
Williams, Jerome, 1962
Oceanography; an introduction to the marine sciences. Little Brown & Co., 242 pp.

books
Williams, Jerome, John J. Higginson and John D. Rohrbough, 1968.
Sea & Air: the naval environment.
U.S. Naval Institute, 338 pp. $11.50.

books
Wimpenny, R.S., 1966.
The plankton of the sea.
American Elsevier Publishing Co., 426 pp.
$16.00.

book
Wood, E.J. Ferguson, 1968.
Dinoflagellates of the Caribbean Sea and adjacent areas.
Univ. Miami Press, 143pp.

books
Wood, E.J. Ferguson, 1965.
Marine microbial biology.
Modern Biological Studies, Chapman & Hall, Ltd., London, Reinhold Publishing Corp., N.Y., 243 pp.

books
Woods Hole Oceanographic Institution, 1952.
Marine fouling and its prevention. U.S. Naval Inst. 388 pp., textfigs.

books
Woods, J.D., and J.N. Lythgoe, editors, 1971.
Underwater science: an introduction to experiments by divers.
Oxford Univ. Press 330pp.

book
Wooster, Warren S., editor, 1970.
Scientific Exploration of the South Pacific: proceedings of a symposium held during the ninth general meeting of the Scientific Committee on Oceanic Research, June 18-20, 1967 at the Scripps Institution of Oceanography, La Jolla, California.
Nat. Acad. Sci., 257pp.

XYZ

books
Yasso, Warren E., 1965
Oceanography: a study of inner space Holt, Rinehart and Winston, Inc. 176 pp.

Books
Yonge, C.M., 1960
Oysters. Collins, St. James's Place, London, 209 pp.

books
Yonge, C.M., 1949.
The Sea Shore. Collin's, St. James Place, London 311 pp., 61 color pls., 62 black-and-white pls., 88 textfigs.

books
Young, J.Z., 1964.
A model of the brain.
Oxford University Press, vii 348 pp. 35 shill.

books
Zenkevitch, L., 1963.
Biology of the Seas of the U.S.S.R.
S. Botcharskaya, Translator, George Allen and Unwin, Ltd. London, 955 pp.
Reviewed by C.M. Yonge, 1963, Nature, 200(4907): 617-618.

books
Zenkevitch, L., 1947 (2nd Edition).
Biology of the seas of the U.S.S.R.
Translation (1963) by S. Botcharskaya, Interscience Publishers, 955 pp.

books
Zenkovich, V.P., 1962
The basis of the concept on the development of sea coasts. Publishing House, Akademia Nauk, SSSR, 710 pp.

Reviewed by:
A.D. Dobrovolsky, 1962, Okeanologiia, 2(5): 950-953.

books
Zhukovskii, G.R., 1953.
Okeanografia dlia sudovoditelei. Leningrad-Moscow, Gos-izs-vovodnogo Transporta, 390 pp., supplements, $6.00.

A. Buschke
80 E. 11th Street
New York 3.

books
Zubov, N.N., 1947.
Dinamicheskaia okeanologiia. Moscow, 430 pp.

books
Zubov, N.N., 1945.
L'dy Arktiki. Glavsevmorput, 359 pp., Moscow.

borderland

borderland
Dehlinger, Peter, R.W. Couch, D.A. McManus and Michael Gemperle, 1970.
II. Regional observations. 4. Northeast Pacific Structure. In: The sea: ideas and observations on progress in the study of the seas, Arthur E. Maxwell, editor, Wiley-Interscience 4(2): 133-189.

borderland
Gorsline, Donn S., David E. Drake and Peter W. Barnes, 1968.
Holocene sedimentation in Tanner Basin, California continental borderland.
Bull. geol. Soc. Am., 79(6): 659-674

borderlands
*Hoshino, Michihei, 1969.
On the relationship between the distribution of epicenter and the submarine topography and geology. (In Japanese; English abstract).
J. Coll. mar. Sci. Technol. Tokai Univ., (3):1-10.

borderlands
Krause, Dale C., 1965.
Tectonics, bathymetry and geomagnetism of the southern continental borderland west of Baja California, Mexico.
Geol. Soc., Amer., Bull., 76(6):645-647.

borderland
Shor, G.G., Jr., and R.W. Raitt, 1958.
Seismic studies in the southern California borderland.
Congreso Geol. Int., 20 Sesion, Mexico, 1956, Sec. 9 (Geofis. Aplicada)(2):253-259.

bore-hole
Ewing, J., C. Windisch, and M. Ewing, 1970.
Correlation of horizon A with JOIDES bore-hole results. J. geophys. Res., 75(29): 5645-5653.

borers

borers
Amaral, S.F. do, 1956.
Attack of the marine borers, Teredo sp., Bankia sp., and Martesia striata Linne in pine wood boards treated with insecticudes.
Arqu. Inst. Biol., 23:1-19.

borers
Anderson, J.W., and D.J. Reish, 1967.
The effects of varied dissolved oxygen concentrations and temperature on the wood-boring isopod genus Limnoria.
Mar. Biol., 1(1): 56-59.

borers
Barnard, J. L., 1959.
(3) Generic partition in the amphipod family Cheluridae, marine wood borers.
Pacific Naturalist 1(3/4):2-12 5 figs. look very good.

borers
Balasubramanyan, R., 1965.
Protection of wooden hulls of fishing boats against marine borers.
Proc. Symposium on Marine Paints, New Delhi, 1964: 120-123.

borers
Balasubramanyan, R., and T.R. Menon, 1963.
Destruction of boat building timbers by marine organisms in the Port of Cochin. 1. Raft tests.
J. Mar. biol. Ass., India, 5(2):294-310.

borers
Barnard, J.L., and D.J. Reish, 1957.
First discovery of marine wood-boring copepods.
Science, 125(3241):236.

borers

Beckman, Carolyn and Robert Menzies, 1960.
The relationship of reproductive temperature and the geographical range of the marine wood-borer Limnoria tripunctata, Biol. Bull., 118(1): 9-16.

borers

Board, P.A., 1970.
Some observations on the tunnelling of shipworms.
J. Zool. Lond., 161(2): 193-201.

borers, effect of

Boer, Peter 1971.
Harpacticid copepods (Crustacea) living in wood infested by Limnoria from northwestern France.
Bull. zool. Mus. Amsterdam, 2(8): 63-72.

borers

Bohn, A., and C.C. Walden 1970
Survey of marine borers in Canadian Atlantic waters.
J. Fish. Res. Bd., Can., 27(6): 1151-1154

borers

Brunel, P., 1963.
Les isopodes xylophages Limnoria japonica et L. lignorum dans le Golfe Saint-Laurent: notes sur leur distribution et leurs cilies, ostracodes et copepodes commensaux.
Crustaceana, 5(1):35-46

borers

*Carriker, Melbourne R., Dirk Van Zandt and Garry Charlton, 1967.
Gastropod Urosalpinx: pH of accessory boring organ while boring.
Science, 158(3803):920-922.

borers

Chanley, Paul E., 1965.
Larval development of a boring clam, Barnea truncata.
Chesapeake Science, 6(3):162-164.

borers

Clapp, William F., (posthumous) and Roman Kenk, 1963
Marine borers: an annotated bibliography.
Office of Naval Research, Dept. of the Navy, ACR-74:1136 pp.

marine borers

Clapp, W.F., and R. Kenk, 1956.
Marine borers: a preliminary bibliography.
Library of Congress, Tech. Info. Div., 346 pp. (mimeographed).

borers

Clapp, W.F., and R. Kenk, 1956, 1957.
Marine borers, a preliminary bibliography.
U.S. Library Congress, Tech. Info. Div., 1:358 pp, 2:355 pp

borers

Daniel, A., 1963.
Factors influencing the settlement of marine foulers and borers in tropical seas.
Recent Advances in Zoology, Zoological Survey of India, 363-382.

borers

Daniel, A., 1958.
Notes on the distribution and seasonal variation of marine borers of the Madras area.
J. Madras Univ., (B), 28(2/3):115-128.

borers

DePalma, John R., 1963.
Marine fouling and boring organisms off Fort Lauderdale, Florida.
Naval Oceanogr. Off., IMR No. 0-70-63:28 pp.

borers

*Deschamps, P., 1968.
Quelques cas de biocoenose dans la faune des bois immergés en milieu marin.
Trav. Cent. Rech. Etud océanogr., n.s., 7(3/4):37-42.

Borers

Deschamps, Paul, 1965.
Fracteurs de repartition des perforants des bois immerges.
Trav. Centre de Recherches et d'Etudes Oceanogr. n.s., 6(1/4):397-400.

borers

Drisko, R.W., and H. Hochman, 1957.
Amino acid content of marine borers.
Biol. Bull., 112(3):325-329.

borers

Edmondson, C.H., 1962.
Teredinidae, ocean travelers.
Occ. Papers, Bernice P. Bishop Mus., 23(3):45-59

borers

Eltringham, S.K., 1965.
The effect of temperature upon the boring activity and survival of Limnoria (Isopoda).
J. Appl., Ecol., 2(1):149-157.

boring organisms

Eltringham, S.K., 1965.
The respiration of Limnoria (Isopoda) in relation to salinity.
Jour. Mar. Biol. Assoc., U.K., 45(1):145-152.

borers, physiol. of

Eltringham, S.K., 1964
Blood concentrations of Limnoria (Isopoda) in relation to salinity.
Jour. Mar. Biol. Assoc., U.K., 44(3):675-683.

borers

Eltringham, S.K., 1961.
The effect of salinity upon the boring activity and survival of Limnoria (Isopoda).
J. Mar. Biol. Assoc., U.K., 41(3):785-797.

borers

Eltringham, S.K., 1961
Wood-boring activity of Limnoria (Isopoda) in relation to oxygen tension.
Nature, 190(4775): 512-513.

borers

Eltringham, S.K., and A.R. Hockley, 1961.
Migration and reproduction of the wood-boring isopod, Limnoria, in Southampton Water.
Limnol. & Oceanogr., 6(4):467-482.

borers

Evans, John W. 1970.
Marine borer activity in test boards operated in the Newfoundland area during 1967-68.
J. Fish. Res. Bd, Can. 27(6): 201-203.

borers

Ganapati, P.N., and M.V. Lakshmana Rao, 1961.
The marine wood borer Xylophaga from Bay of Bengal.
Current Science, 30(12):464-465.

borers

Ganapati, P. N., & M. V. Lakshmana Rao, 1959.
Incidence of marine borers in the mangroves of the Godavari estuary.
Current Science, 38(8): 332

borers

Goreau, T. F. Goreau, N. I. Goreau and C.M. Yonge 1972
On the mode of boring in Fungiacava eilatensis (Bivalvia: Mytilidae).
J. Zool. Lond 166(1):55-60

borers

Griffin, F.J., 1964
Marine borers.
Chemistry and Industry 22(May):916-917

borers

Hart, C.W. Jr., N. Balakrishnan Nair and Dabney G. Hart, 1967.
A new ostracod (Ostracoda: Entocytheridae) commensal on a wood-boring marine isopod from India.
Notulae Naturae, 409:11 pp.

borers

Hidaka, T., and K. Saito, 1957.
Studies on the cellulose decomposing bacteria in the digestive organs of the ship-worm (Teredo navalis sp.). 2.
Mem. Fac. Fish., Kagoshima Univ., 5:172-177.

borers

Hurley, D.E., 1959.
The growth of Teredo (Bankia australis Calman) in Otago Harbour.
New Zealand J. Sci., 2(3):323-338.
Also in:
Collected Reprints, Portobello Mar. Biol. Sta., Vol. 2 (1961).

borers

Jeffrey, S.W., and Kazuo Shibata, 1969.
Some spectral characteristics of chlorophyll from Tridacna crocea zoozanthellae.
Biol. Bull., mar. biol. Lab., Woods Hole, 136(1):54-62.

borers

John, P.A., 1970.
Observations on the boring activity of Sphaeroma terebrans Spence Bate, a wood boring isopod.
Zool. Anz. 185(5/6):379-387

borers

John, P.A., 1967.
Biological analyses of Sphaeroma attack on timber treated with conventional wood preservatives.
Proc. Symp. Crustacea, Ernakulam, Jan. 12-15, 1965, 4: 1269-1273

borers
Johnson, F.H., and D.L. Ray, 1962.
Simple method of harvesting Limnoria from nature.
Science, 135(3506):795.

borers
Jones, Leslie T., 1963
The geographical and vertical distribution of British Limnoria (Crustacea:Isopoda).
J. Mar. Biol. Assoc., U.K., 43(3):589-603.

borers
Karande, Ashok A., and M.I. Jethmalani, 1965.
Natural resistance of Tetrameles nudiflora R.Br. to attack by teredinid wood borers - a preliminary note.
Proc. Symposium on Marine Paints, New Delhi, 1964: 124-126.

borers
Kristensen, Else, 1969.
Attacks by Teredo navalis L. in inner Danish waters in relation to environmental factors.
Vidensk. Medd. dansk naturh. Foren. 132: 199-210

borers
Kudinova-Pasternak, R.K., 1964.
Lethal action of high temperature on Teredo navalis L., (Mollusca, Bivalvia, Teredinidae). (In Russian; English abstract).
Zool. Zhurn., 43(7):1074-1076.

borers
Kudinova-Pasternak, R.K., 1962.
Effect of the marine water of a decreased salinity and of various temperature upon the larvae of Teredo navalis L.
Zool. Zhurnal, Akad. Nauk, SSSR, 41(1):49-57.

borers
Kudinova-Pasternak, R.K., 1957.
On the penetration possibility of Teredo navalis into the Caspian Sea. Zool. Zhurnal 36(6):847-851.

suggests one month quarantine in Volga-Don Canal from Black Sea to Caspian Sea.

borers
Kussakin, O.G., 1963.
Some data on the systematics of the family Limnoriidae (Isopoda) from northern and Far Eastern seas of the USSR.
Crustaceana, 5(4):281-292.

borers
Levine, Summer N., 1967.
Desalination and ocean technology.
Dover Publ. $4.00 (not seen).

borers
Li, S.-T., 1961
Depredation of timber in marine construction. 1. Marine borers (with particular reference to their distribution in U.S.A. waters).
Dock & Harbour Authority, 42(489):85-89.

borers
Liu, D., and P.M. Townsley, 1968.
Glucose metabolism in the caecum of the marine borer Bankia setacea.
J. Fish. Res. Bd, Can. 25(5):853-861

borers
Liu, D.L., and C.C. Walden, 1970.
Enzymes of glucose metabolism in the caecum of the marine borer Bankia setacea.
J. Fish. Res. Bd, Can., 27(6):1141-1146

borers
McCoy-Hill, M., 1967.
Protection against marine borer. 3.
Dock Harbour Auth., 48(566):239-244.

Borers
McCoy-Hill, M., 1965.
Marine wood preservation.
Trav. Centre de Recherches et d'Etudes Oceanogr. n.s., 6(1/4):389-396.

borers
Menzies, R.J., 1957.
The marine borer family Limnoridae (Crustacea, Isopoda).
Bull. Mar. Sci., Gulf and Caribbean, 7(2):101-200

borers
Menzies, Robert J., John Mohr and Carrol M. Wakeman, 1963.
The seasonal settlement of wood-borers in Los Angeles-Long Beach harbors.
Wasmann J. Biol., Univ. San Francisco, 21(2):97-120.

borers
Menzies, Robert J., and Donald J. Robinson, 1960.
Informe sobre los isopodes taladradores marinos colectados en el oriente de Venezuela.
Soc. Ciencias Nat. de la Salle, Mem., 20(56): 132-137.

Contrib. No. 1, Estacion de Investigaciones Marinas de Margarita, Fundacion La Salle de Ciencias Naturales.

borers
Meyers, S.P., and E.S. Reynolds, 1957.
Incidence of marine fungi in relation to marine borers. Science, 126(3280):969.

boring organisms (shipworms)
Monod, Th., and M. Nickles, 1952.
Notes sur quelques Xylophages et Pétricoles marins de la côte ouest africaine.
I.F.A.N. Catalogues 8:7-68, 151 textfigs.

borers
Nair, N. Balakrishnan, 1967.
Seasonal settlement of marine fouling and wood boring crustaceans at Cochin Harbour, south-west coast of India.
Proc. Symp. Crustacea, Ernakulam, Jan. 12-15, 1965, 4:1254-1268

borers
Nair, N.B., 1966.
Vertical distribution of marine wood boring animals in Cochin Harbor, south west coast of India.
Hydrobiologia, Acta Hydrobiol., Hydrogr. et Protistol., 27(1/2):248-259.

borers
Nair, N. Balakrishnan, 1965.
Marine timber-boring organisms of the Indian coast. Report on a collection from the south-east coast of India, with notes on distribution in the Indo-Pacific area.
J. Bombay Nat. Hist. Soc., 62(1):120-131.

borers
Nair, N. Balakrishnan, 1965.
Seasonal settlement of Marine wood boring animals at Cochin Harbour, south-west Coast of India.
Int. Revue Geo. Hydrobiol., 59(3):411-420.

borers
Nair, N.B., 1964.
Ecology of marine fouling and wood boring organisms of western Norway.
Sarsia, 8:1-88.

borers
Nair, N. Balakrishnan, 1963.
Shipworms from the Publicat Lake, East Coast, Madras State.
Current Science, 32(5):215-216.

borers
Nair, N. Balakrishnan, 1962.
Ecology of marine fouling and wood-boring organisms of western Norway.
Sarsia, 8:88 pp., 9 pls.

borers
Nair, N. Balakrishnan, 1961.
Some aspects of the marine borer problem in India.
J. Sci. & Industr. Res., 20(A)(10):584-590.

borers
Nair, N. Balakrishnan, 1960
The marine timber-boring molluscs and crustaceans of western Norway.
Årbok, Univ. Bergen, Naturv. Rekke, 1959(1): 23 pp.

borers
Nair, N.B., 1957.
The shipworms of South India with a note on the breeding season. J. Bombay Nat. Hist. Soc., 54(2):344-357.

borers
Nair, N. Balakrishnan, and O.N. Gurumani, 1957.
A new shipworm, Teredo (Teredora) vattanansis, from the east coast of India.
Ann. Mag. Nat. Hist., 10(111):174-176.

borers
Nair, N.B., and O.N. Gurumaui, 1957.
Timber boring molluscs of the Indian coast. 1. Report on a collection from Tondi and Adirampatnam, East Coast. J. Bombay Nat. Hist. Soc., 54(3):663-673.

borers
Nair, N. Balakrishnan and M. Saraswathy, 1971.
The biology of wood-boring teredinid molluscs.
Adv. mar. Biol. 9: 335-509.

borers
Neumann, A. Conrad, 1966.
Observations on coastal erosion in Bermuda and measurements of the boring rate of the sponge, Cliona lampa.
Limnol. and Oceanogr., 11(1):92-108.

borers
Petukhova, T.A., 1963.
Settlement of larvae of fouling organisms and marine borers in the region of Gelendjik and Novorossysk (Black Sea). (In Russian; English summary).
Trudy Inst. Okeanol., Akad. Nauk, SSSR, 70:151-156.

borers
Pillai, N. Krishna, 1967.
The role of Crustacea in the destruction of submerged timber.
Proc. Symp. Crustacea, Ernakulam, Jan. 12-15, 1965, 4: 1274-1283.

Borers

Quayle, D.B., 1965.
Dispersal of introduced marine wood borers in British Columbia waters.
Trav. Centre de Recherches et D'Etudes Oceanogr. n.s., 6(1/4):407-412.

borers

Rajagopal, A.S., 1964.
Two new species of marine borers of genus Nausitora (Mollusma; Teredinidae) from West Bengal, India.
J. Bombay Nat. Hist. Soc., 61(1):108-118.

borers

Rajagopalaiengar, A.S., 1961 (publ.1966).
Fuller description of a recently described species of th e marine borer Bankia(Neogankia) roomwali Rajagopalaiengar from West Bengal, India.
Rec.Indian Mus.59(4):449-454.

borers

Rancurel, P., 1965.
Description de la prodissoco nque de Teredo thomsoni Tryon et de Bankia anechoensis Roch.
Cahiers, O.R.S.T.R.O.M. - Océanogr., 3(1):101-105.

borers

Rancurel, P., 1964.
Presence de Teredo dicroa Roch 1929 en côte d'Ivoire.
Cahiers O.R.S.T.R.O.M., Océanogr., 11(4):127-133.

borers

Ray, Dixie Lee, Ed., 1960.
Marine boring and fouling organisms.
Univ. Washington Press, 480, pp. $850

borers

Ray, Dixie Lee, 1959.
An integrated approach to some problems of marine biological deterioration: destruction of wood in sea water.
Twentieth Ann. Biol. Coll., Marine Biol., Oregon State Coll., Apr. 3-4, 1959:70-87.

borers

Ray, D. L., and D. E. Stuntz, 1959.
Possible relation between marine fungi and Limnoria attack on submerged wood.
Science 129(3341):93-94.

borers

Reish, D.J., and W.M. Hetherington III, 1969.
The effects of hyper- and hypo-chlorinities on members of the wood-boring genus Limnoria.
Marine Biol., 2(2):137-139.

Borers

Roch F. and L.N. Santhakumaran 1967.
Notes on the Teredinidae from the Lagoon of Venice (Italy).
Boll. Pesca, Piscic. Idrobiol., ns. 12(1):37-48

borers

Ryabchikov, P.I., and G.G. Nikolaeva, 1963.
Settlement on wood of the larvae of Teredo navalis L. (Teredinidae, Mollusca) in the Black Sea. (English summary).
Trudy Inst. Okeanol., Akad. Nauk, SSSR, 70:179-185.
In Russian.

borers

Ryabchikov, P.I., I.N. Soldatova and S.E. Yesakova, 1961.
The first stage in the settlement of shipworms in the Sea of Azov. (In Russian).
Trudy Inst. Okeanol., Akad. Nauk, SSSR, 49:147-155.

USN-HO-TRANS 178
M. Slessers 1963
P.O. 32672

borers

Ryabchikov, P.I., I.N. Soldatova, S.E. Tesakova and T.A. Petukhova, 1963.
The beginning invasion of some species of ship worm teredo (Fam. Teredinidae, Mollusca) into the Sea of Azov. (In Russian; English summary).
Trudy Inst. Okeanol., Akad. Nauk, SSSR, 70:157-178.

Borers

Sampaio Franco, E.J., 1965.
Sur l'aptitude de quelques bois africains à l'emploi dans les travaux maritimes et la construction navale.
Trav. Centre de Recherches et d'Etudes Océanogr. n.s., 6(1/4):457-467.

borers

Santhakumaran, L.N., 1971.
Two new records of marine wood-borers (Mollusca: Teredinidae) from Bombay waters.
Current Sci. 40 (8): 199-200

borers

Santhakumaran, L.N., 1969.
Destruction of timber by crustacean wood borers in the lagoon of Venice.
Boll. Mus. Civ. Storia nat. Venezia, 19:7-11

borers

Saraswathy, M., 1964.
Shipworms from the Cochin Harbour, south-west coast, India.
J. Mar. biol. Ass., India, 6(2):309-310.

Borers

Schultz, Liselotte, 1961.
Verbreitung und Verbreitungsmoglichkeiten der Bohrmuschel Teredo navalis L. und ihr Vordringen in den NO-Kanal bei Kiel.
Kieler Meeresf., 17(2):228-236.

bottom motion, effect of

Slatkin, Montgomery W., 1971.
Long waves generated by ground motion.
J. fluid Mech. 48(1):81-90

borers

Soldatova, I.N., 1963.
Influence of v arying salinity on some physiological processes of the Black Sea bivalve mollusc Teredo pedicellata Quatr. (In Russian).
Trudy Inst. Okeanol., Akad. Nauk, SSSR, 70:186-196.

borers

Soldatova, I.N., 1961.
The effect of variable salinity conditions on the bivalve mollusk Teredo navalis L. (In Russian).
Trudy Inst. Okeanol., Akad. Nauk, SSSR, 49:162-179.

USN-HO-TRANS 179
M. Slessers 1963
P.O. 32672

borers

Soldatova, I.N., T.A. Lukasheva and I.N. Iljin 1967.
On the ecology of the bivalve mollusc Teredo navalis in the Azov Sea. (In Russian)
Trudy Inst. Okeanol. 85: 185-189

Borers

Southwell, C.R., C.W. Hummer, Jr., B.W. Forgeson, T.R. Price, T.R. Sweeney, and A.L. Alexander, 1965.
Natural resistance of woods to marine borer and other biological deterioration in tropical environments.
Trav. Centre de Recherches et d'Etudes Oceanogr. n.s., 6(1/4):419-432.

borers

*Strömberg, Jarl-Ove, 1967.
Segmentatation and organogenesis in Limnoria lignorum (Rathke)(Isopoda).
Arkiv Zoologi 20(5):91-139.

borers

Thomas, M.L.H. 1967.
Experiments in the control of shipworm Teredo sp. using bis (tri-N-butyltin) oxide.
Techn. Rept. Fish. Res. Bd. Can. 21:13pp., 10 tables (mimeographed)

borers

Townsley, P.M., R.A. Richy and P. C. Trussell, 1965.
The occurrence of protoporphyrin and myoglobin in the marine borer, Bankia setacea (Tryon).
Canadian J. Zool., 43(1):167-172.

borers

Turner, Ruth D. 1966.
A survey and illustrated catalogue of the Teredinidae (Mollusca: Bivalvia)
Mus. Comp. Zool. Harvard Univ., 265pp.

borers

Turner, R.D. and A.C. Johnson, 1968.
Biological studies in marine wood borers.
Ann. Repts. for 1968, Amer. Malacol. Union: 14-16. Also in: Contrib. Univ. Puerto Rico, Dept. mar. Sci. 8 (1968-1969).

borers

United States Navy Hydrographic Office, 1959.
Climatological and oceanographic atlas for mariners. Vol. 1.
North Atlantic Ocean., 182 charts.

borers

Walden, C.C, I.V.F. Allen and P.C. Trussell 1967.
Estimation of marine-borer attack on wooden surfaces
J. Fish. Res. Bd, Canada 24 (2): 261-272.

borers

Walden, C.C , and P.C. Trussell, 1965.
Sonic examination of marine piles.
Dock Harb. Auth., 46(535): 13-17.

borers (algae)

Wilkinson, M. and E.M. Burrows, 1972.
The distribution of marine shell-boring green algae. J. mar. biol. Ass. U.K. 52(1): 59-65.

borers, commensals with

BORERS, commensals with
Kirkegaard, J.B. and L.N. Santhakumaran, 1967
On a new species of annelid associate of marine wood borers; Cirriformia limnoriala n.sp. (Polychaeta)
Vidensk. Meddr dansk naturh. Foren. 130: 213-216

borers, commensals with
commensals

borers, resistance

borers
Anon., 1957.
Research against the Teredo. Ship and Boat Builder, 10(3):99.

marine borers
Ayers, J.C., 1954.
A method for rendering wood resistant to marine borers. Bull. Mar. Sci., Gulf and Caribbean, 3(4):297-304, 4 textfigs.

marine borers, resistence to
Woods, R.P., 1957.
Resistence of timbers to marine borers. Dock & Harbour Authority 38(441):101-104.

borers
Carpine, C., 1960.
Révision des connaissances sur la biologie de la prévention des Xylophages marins.
Bull. Inst. Océanogr., Monaco, 57(1187):1-50.

borer prevention
Muri, S., 1945.
Utility of wood impregnated with coal tar creosote oil for wooden ships to prevent the attack of borers. Physiol. Ecol. Contrib. Otsu Hydrobiol. Sta., Kyoto Univ., 35:1-18, textfigs.

borers
Vershinskii, N.V., 1958.
Method for controlling ship worms of the family Teredinidae by electric current. (In Russian).
Biull. Okeanogr. Komissiia, Akad. Nauk, SSSR, (1):60-
not seen

bores

bores
Abbot, M.R., 1956.
A theory of the propagation of bores in channels and rivers. Proc. Cambridge Phil. Soc., 52(2): 344-366.

bores
Chatley, H., 1950.
The Hangchow bore. Dock and Harbour Auth. 30(351):262-264.

bores
Dalton, F.K., 1951.
Fundy's prodigious tides and Petitcodiais tidal bore. J.R. Astron. Soc., Canada, 45(6):225-231.

bores
Fukui, Yoshiro, Makoto Nakamura, Hidehiko Shiraishi and Yasuo Sasaki, 1963.
Hydraulic study of tsunami.
Coastal Engineering in Japan, 6:67-82.

bores
Ho, D.V., and R.E. Meyer, 1962.
Climb of a bore on a beach. Part 1. Uniform beach slope.
J. Fluid Mech., 14(2):305-318.

bores
Howe, M.S. 1971.
On wave scattering by random inhomogeneities, with application to the theory of weak bores.
J. fluid Mech. 45(4): 785-804.

bores
Miller, Robert L. 1968.
Experimental determination of run-up of undulant and fully developed bores.
J. geophys. Res. 73(14): 4497-4510.

bores
Sandover, J.A., and C. Taylor, 1962.
Cnoidal waves and bores.
La Houille Blanche, 17(3):443-455 (English); 456-465 (French).

bores
Shen, M.C., and R.E. Meyer, 1963.
Climb of a bore on a beach. 2. Non-uniform beach slope.
J. Fluid Mechanics, 16(1):108-112.

bores
Shen, M.C., and R.E. Meyer, 1963.
Climb of a bore on a beach. 3. Run-up.
J. Fluid Mechanics, 16(1):113-125.

bores
Smith, F.G. Walton, 1969.
Ebb and flow
Sea Frontiers 15(2):86-96 (popular)

bores
Tricker, R.A.R., 1964.
Bores, breakers, waves and wakes: an introduction to the study of waves on water.
American Elsevier Publishing Co., Inc., 250 pp. $6.50.

bottom

See also: sea floor

bottom artefacts
Medcof, J., 1963.
Puzzling clay tubes from the sea bottom.
Canadian Field Naturalist, 77(4):214-219.

bottom configuration
Clay, C.S., and Peter A. Rona, 1963
On the existence of bottom corrugations in the Blake-Bahama Basin.
J. Geophys. Res., 69(2):231-234.

bottom, effect of

Bottom, effect of

See also: bottom topography, effect of

bottom, effect of
Abramson, H.N., and C.L. Bretschneider, 1954.
Some observations concerning the analysis of surface waves when the bottom is non rigid.
B.E.B. Tech. Memo. No. 46(App. 1):1-11, 3 figs.

With App. 2 - Summary of wave data, Pure Oil Structure A.

bottom, effect of
Bruun, Per 1967.
Bottom roughness: rivers, tidal inlets, ocean.
J. OceanTechn. 1(2):1-18.

bottom, effect of
Carrier, G.F., 1970.
Stochastically driven dynamical systems.
J. fluid Mech. 44(2): 249-264

bottom, effect of
Crnković, Drago, 1970.
A contribution to the study of biological and economic problems of trawling in the channels of the north-eastern Adriatic. (In Jugoslavian; English abstract).
Thalassia Jugoslavica 6: 5-90.

bottom, effect of
Gade, H. G., 1958.
Effects of a non rigid, impermeable bottom on plane surface waves in shallow water
J. Mar. Res., 16(2):61-82.

bottom, effect of
Gezentsvei, A.N., 1949.
The influence of the bottom on the distribution of temperature in water bodies.
Trudy Inst. Okeanol., 4:80-102.

bottom, effect of
Kornhauser, E.T., and N.P. Raney, 1955.
Attenuation in shallow-water propagation due to an absorbing bottom. J. Acoust. Soc., Amer., 27(4):689-691.

bottom, effect of
Kozlov V.F., 1969.
The influence of bottom relief on geostrophic currents in the Pacific Ocean. Okeanologiia, 9(4): 608-615.
(In Russian; English abstract)

bottom, effect of
Lehman, R.S., 1954.
Developments in the neighborhood of the beach of surface waves over an inclined bottom.
Pure and Appl. Math. 7(3):393-441.

bottom, effect of
Leipper, Dale F., 1955.
Sea temperature variations associated with tidal currents in stratified shallow water over an irregular bottom.
J. Mar. Res., 14(3):234-252.

bottom, effect of
Lysanov, Yu.P., 1967.
Time fluctuations of sound signals scattered by the ocean bottom. (In Russian).
Akust.Zh., 13(3):401-405.

bottom, effect of

Mei, C.C., and B. Le Méhauté, 1966.
Note on the equations of long waves over an uneven bottom.
J. geophys. Res., 71(2):393-400.

bottom, effect of

Reid, R.O., and K. Kajiura, 1957.
On the damping of gravity waves over a permeable sea bed. Trans. A.G.U., 38(5):662-666.

bottom, effect of

Robinson, A.R., and Sulochana Gadgil, 1970.
Time-dependent topographic meandering.
Geophys. fluid Dyn. 1 (4): 411-438.

bottom, effect of

Schmidt, P.B., 1971.
Monostatic and bistatic backscattering measurements from the deep ocean bottom.
J. acoust. Soc. Am. 50(1-2): 326-331

bottom, effect of

Schulman, Elliott E., and Pearn P. Niiler, 1970
Topographic effects on the wind-driven ocean circulation.
Geophys. fluid Dyn., 1 (4): 439-462

bottom, effect of

*Seifer, A.D., and M.J. Jacobson, 1968.
Ray transmission in an underwater acoustic duct with a bottom having curvature.
J. acoust. Soc. Am., 44(4):1103-1114.

bottom, effect of

Stockmann, W.B., 1949.
An investigation of the influence of wind and bottom relief on the resulting circulation and the distribution of mass in a non-uniform (baroclinic) ocean or sea. Trudy Inst. Okeanol., 3:3-65.

bottom, effect of

Tareiv, B.A., 1964
Internal baroclinic waves in the course of flowing around the irregularities of bottom contours and their effect on sedimentation processes. (In Russian).
Okeanologiia, Akad. Nauk, SSSR, 4(5):915.

bottom, effect of

Trites, R.W. 1956
The ocean floor and bottom movement.
Trans. R. Soc., Canada (3) 50: 83-91.

bottom, effect of

Wigley, Roland L., and Bruce R. Burns 1971.
Distribution and biology of mysids (Crustacea, Mysidacea) from the Atlantic coast of the United States in the NMFS Woods Hole collection.
Fish. Bull. U.S. NOAA 69(4): 717-746

bottom, effect of

Williams, A.B., 1958.
Substrates as a factor in shrimp distribution.
Limnol. & Oceanogr., 3(3):283-290.

bottom, effect of

Zhitkovsky Yu. Yu. and Yu. P. Lysanov, 1969.
On some peculiarities of the Fresnel diffraction of sound upon rough sea surface and bottom. (In Russian).
Fizika Atmosfer. Okean. Izv. Akad. Nauk, SSSR, 5(9): 982-985.

bottom, effect of

Zhukovets, A.M., 1963.
The influence of bottom roughness on wave motion in a shallow body of water. (In Russian).
Izv., Akad. Nauk, SSSR, Ser. Geofiz., (10):1561-1570.

Translation:
(10):943-948.

bottom, effect of

Zverev, S.M., 1960(1961)
Recordings of water waves in regions of the boundary of the shadow zone, which is generated by the ocean bottom.
Izv., Akad. Nauk, SSSR, Ser. Geofiz., (8): 1173-1186.

English Edition:
(8):777-785.

bottom, facies

bottom facies

*Reineck, Hans-Erich, Jürgen Dörjes, Sibylle Gadow und Günther Hertweck, 1968.
Sedimentologie, Faunenzonierung und Faziesabfolge vor der Ostküste der inneren Deutschen Bucht.
Senckenberg. leth., 49(4):261-309.

Deutsche Bucht
bottom sediments
bottom facies

bottom fauna, SEE ALSO, benthos

Bottom fauna

See also: benthos

bottom fauna

Aleem, A.A., 1956.
Quantitative underwater study of benthic communities inhabiting kelp beds off California.
Science, 123(3188):183.

bottom fauna

Bekman, M. Yu, 1952
Materials for the quantitative characterization of the Black Sea bottom fauna at Karadag. Trudy Karadags'ka Biol. Stantsiya, 12: 50-67. Rev. in Techn. Transl. 4(9): 510.

bottom fauna

Beliaev, G.M., and P.V. Ushakov, 1957.
Certain regularities in the quantitative distribution of bottom fauna in Antarctic waters.
Dokl. Akad. Nauk, SSSR, 112(1):137-140.

bottom fauna

Beliaev, G.M., and N.G. Vinogradova, 1961.
The quantitative distribution of bottom fauna in the northern part of the Indian Ocean.
Doklady Akad. Nauk, SSSR, 138(5):1191-1194.

bottom fauna

Bigelow, H.B., and W.C. Schroeder, 1939.
Notes on the fauna above mud bottoms in deep water in the Gulf of Maine. Biol. Bull. 46(3): 305-324, 8 textfigs.

bottom fauna

Bougis, P., 1950.
Methode pour l'etude quantitative de la microfaune des fond marins (Meiobenthos). Vie et Milieu 1(1):23-37, 3 textfigs.

bottom fauna

Caspers, H., 1938(1939).
Die Bodenfauna der Helgolander Tiefer Rinne.
Helgolander Wissensch. Meeresuntersuch., 2(1): 112 pp., 33 textfigs.

bottom fauna

Ewing, M., A. Vine, and J.L. Worzel, 1946
Photography of the Ocean Bottom. J. Optical Soc., America, 36(6):307-321, 18 textfigs.

bottom fauna (data only)

Granqvist, G., 1955.
The summer cruise with M/S Aranda in the northern Baltic, 1954. Merent. Julk., No. 166:56 pp.

bottom fauna

Holme, N.A., 1953.
The biomass of the bottom fauna in the English Channel off Plymouth. J.M.B.A. 32(1):1-49, 7 textfigs.

bottom fauna

Inman, D.L., 1950.
Submarine topography and sedimentation in the vicinity of Mugu Submarine Canyon, California.
B.E.B. Tech. Memo., No. 19:45 pp., 10 figs.

S.I.O. Sub. Geol. Rept. 10
S.I.O. Tech. Memo. 473.

bottom fauna and flora, lists of spp.

Jacquette, R., 1962.
Étude des fonds de maerl de Méditerranée.
Rec. Trav., Sta. Mar., Endoume, Bull., 26(41):141-235.

bottom fauna

Lee, R.E., 1944
A quantitative survey of the invertebrate bottom fauna in Menemsha Bight. Biol. Bull. 86(2):83-97, 5 charts, 3 graphs

bottom fauna

Nikitin, V.N., 1961.
Quantitative distribution of the bottom fauna in the north-western part of the Black Sea.
Doklady Akad. Nauk, SSSR, 138(5):1198-1201.

bottom fauna

Raymont, J.E.G., 1950.
A fish culture experiment in an arm of a sea loch. IV. The bottom fauna of Kyle Scotnish. Proc. Roy. Soc., Edinburgh, Sect. B, 64, Pt. 1, No. 4:65-108, 8 textfigs.

Bottom fauna

Smidt, E.L.B., 1944
Biological Studies of the Invertebrate Fauna of the Harbor of Copehagen. Vidensk. Medd. fra Dansk naturh. Foren. 107:235-316, 23 text figs.

owned by MS

bottom animals, deep

Sokolova, M. N., 1959.

On the distribution of deep-water bottom animals in relation to their feeding habits and the character of sedimentation.
Deep-Sea Res., 6(1):1-4.

bottom fauna

Sokolova, M.N., 1958.
Ernährung des Wirbellosen auf dem Tiefseeboden.
Trudy Inst. Okeanol., 27:123-153.

bottom fauna
Spärck, R., 1931
Some quantitative investigations on the bottom fauna at the west coast of Italy, in the Bay of Algiers, and at the coast of Portugal. Rept. on the Danish Oceanogr. Expeds. 1908-1910 to the Mediterranean and adjacent seas, III(7): 1-11.

owned by M.S.

bottom fauna
Stauffer, R. C., 1937.
Changes in the invertebrate community of a lagoon after the disappearance of the eel grass. Ecol. 18(3):427-431.

bottom fauna
Thorson, G., 1950.
Reproductive and larval ecology of marine bottom invertebrates. Biol. Rev. 25(1):1-45, 6 textfigs.

bottom fauna, lists of spp.
Ziegelmeier, E., 1959
Untersuchungen der Bodenfauna auf den Schelfgebieten vor Südostgrönland und vor SW Island im Spätwinter und Spätsommer 1958. Deutsch. Hydrogr. Zeits., Ergänzungsheft, Reihe B, 4(3):99-102.

bottom fauna, abyssal
Zenkevich, L.A., 1957.
The distribution of sea floor fauna in the north-west Pacific. Proc. UNESCO Symp., Phys. Ocean., Tokyo, 1955: 238-245.

bottom fauna, deep

bottom fauna, deep
Belyaev, G. M., 1960
Some regularities in the quantitative distribution of the bottom fauna in the western Pacific. Trudy Inst. Okeanol., 41:98-105.

bottom fauna, deep
Belyaev, G. M., and M.N. Sokolova, 1960.
Investigation of the bottom fauna in the Mariana Trench. Trudy Inst. Okeanol., 41:123-127.

bottom fauna, deep
Belyaev, G. M., N. G. Vinogradova, and Z. A. Filatova, 1960.
Investigation of the bottom fauna in the deep-water trenches of the southern Pacific. Trudy Inst. Okeanol., 41:106-122

bottom fauna, deep
Birstein, J. A., and M. N. Sokolova, 1960.
Bottom fauna of the Bougainville Trench. Trudy Inst. Okeanol., 41:128-131.

bottom fauna, deep
Filatova, Z. A., 1960.
On the quantitative distribution of the bottom fauna in the central Pacific. Trudy Inst. Okeanol., 41:85-97.

bottom fauna(deep)
Kiilerich, A., 1957.
Galathea-Ekspeditionens arbejde i Phillipinergrave. Ymer, 1957(3):200-222.

bottom fauna, deep
Sokolova, M.N., 1959
Some peculiarities in the ecology of the deep-water bottom invertebrates. Akad. Nauk, SSSR, Itogi Nauk, Dostizhenila Okeanol. 1. Uspechi b Izuchenii Okean. Glubn (Biol. i Geol.):188-203.

bottom fauna, deep
Vinogradov, M.E., 1961.
The sources of alimentation of deep-water fauna (on the decomposition rate of dead Pteropoda). Doklady Akad. Nauk, SSSR, 138(6):1439-1442.

bottom fauna, deep
Vinogradova, N.G., 1960
On the problem of geographic distribution of the deep water fauna in the Antarctic waters. Okean. Issle., IGY Com., SSSR:108-112.

bottom fauna, deep
Vinogradova, N. G., 1959.
(Zoogeographical abysses of the ocean (bottom fauna)) Akad. Nauk SSSR, Itogi Nauki, Dostighenila Okeanol. 1. Uspechi b izuchenii Okean. Glubin (Biol. i Geol.) 148-165.

bottom fauna, deep
Vinogradova, N.G., 1958.
Die Vertikalverteilung der Tiefseebodenfauna des Ozeans. Trudy Inst. Okeanol., 27:87-122.

bottom fauna, deep
Vinogradova, N.G., Ia. A. Birshtein and M.E. Vinogradov, 1959
Vertical zonation in the distribution of the deep-water fauna. Akad. Nauk., SSSR, Itogi Nauk, Dostizhenila Okeanol. 1. Uspechi b Izuchenii Okean. Glubin (Biol. i Geol.):166-187.

bottom fauna, deep
Zenkevich, L.A., G.M. Beliaev, Ia A. Birshtein, Z.A. Filatova, 1959.
Qualitative and quantitative data of the ocean fauna characteristic of deep water. Akad. Nauk, SSSR, Itogi Nauki, Dostizhenila Okeanol. 1. Uspechi b Izuchenii Okean. Glubin (Biol. i Geol.):106-147.

bottom fauna, deep
Zenkevich, L.A., & J.A. Birstein, 1961
On the geological antiquity of the deep-sea bottom fauna. Okeanologiya, 1: 110-124.

bottom fauna, deep
Zenkevitch, L.A., J.A. Birstein and G.M. Beliaev, 1955.
The bottom fauna of the Kurile-Kamchatka Trench. Trudy Inst. Oceanol., 12:345-381.

bottom fauna, deep
Zenkevich, L.A., and Z.A. Filatova, 1958.
Allgemeine Charakteristik der quantitativen Verteilung der Bodenfauna der ferőstlichen Meere der USSR und des Nord-Westlichen Teiles des Stillen Ozeans. Trudy Inst. Okeanol., 27:158-160

bottom flow

bottom flow
Dietrich, G., 1956.
Überstromung des Island-Färöer-Rückens in Bodennahe nach Beobachtungen mit dem Forschungsschiff "Anton Dohrn", 1954/55. Deutsche Hydrogr. Zeits. 9(2):79-89.

bottom flow
Heezen, Bruce C., and Charles Hollister, 1964.
Deep-sea current evidence from abyssal sediments. Marine Geology, 1(2):141-174.

bottom friction

bottom friction, effect of
*Hasselmann, Klaus, and J.I. Collins, 1968.
Spectral dissipation of finite-depth gravity waves due to turbulent bottom friction. J.mar.Res., 26(1):1-12.

bottom friction, effect of
Heaps, N.S., 1966.
Wind effects on the water in a narrow two-layered lake. 1. Theoretical analysis. Phil. Trans. R. Soc., (A), 259: 393-416. (1102)

bottom friction, effect of
*Iwagaki, Yuichi, and Yoshito Tsuchiya, 1967.
Linear damping of oscillatory waves due to bottom friction. Proc. 10th Conf.Coast.Engng,Tokyo, 1966,1:149-174

bottom friction
Iwagaki, Yuichi, and Tadao Kakinuma, 1963.
On the bottom friction factor of the Akita coast. Coastal Engineering in Japan, 6:83-91.

bottom friction, effect of
Iwagaki, Yuichi, Yoshito Tsuchiya and Masayuki Sakai, 1965.
Basic studies on the wave damping due to bottom friction. Coastal Engin., Japan, 8:37-49.

bottom friction
Kagan, B.A., 1971.
On the bottom friction in the one-dimensional tidal flow. (In Russian; English abstract). Fisika Atmos. Okean. Izv. Akad. Nauk SSSR 7(11): 1190-1200.

bottom friction
Mosby, H., 1950.
Experiments on bottom friction. Univ. Bergen Aarbok, 1949(2), Naturvitenskapelig rekke 10: 1-12, 8 textfigs.

bottom friction
Mosby, H., 1949.
Experiments on bottom friction. Proc. Verb., Assoc. d'Océan. Phys., Union Géodés. et Géophys. Internat. 4: 133.

bottom friction, effect of
Savage, R.P., and J.C. Fairchild, 1953.
Laboratory study of wave energy losses by bottom friction and percolation. B.E.B. Tech. Memo. No. 31:25 pp.

bottom friction
Titov, V.B., and A.S. Osadchy, 1967.
On the study of the structure of the bottom friction layer at the shallow water near shore. (In Russian). Fisika Atmosfer.Okean.Izv.Akad.Nauk,SSSR,3(10): 1119-1124.

Bottom, heat flow

See: heat flow

bottom irregularities

bottom irregularities

Volovov, V.I., 1968,
Recording of irregularities on the ocean bottom.
(In Russian)
Zh. Akustich. 14(4): 610.

Translation: Soviet Phys.-Acoust. 14(4): 509-510.

bottom, intrusions

Blanton, Jackson, 1971.
Exchange of Gulf Stream water with North Carolina Shelf Water in Onslow Bay during stratified conditions. Deep-Sea Res., 18(2): 167-178.

bottom lithology

bottom lithology

Palmer, Harold D., 1964
Marine geology of Rodriguez Seamount.
Deep-Sea Res., 11(5):737-756.

bottom nomenclature

bottom nomenclature

Koczy, F.F., 1954.
A survey on deep-sea features taken during the Swedish deep-sea expedition. Deep-Sea Res. 1(3): 176-184, 14 textfigs.

Bottom samples

United States, Department of Commerce, Environmental Sciences Services Administration, 1965

International Indian Ocean Expedition, USC&GS Ship Pioneer - 1964.
Vol. 1. Cruise Narrative and scientific results 139 pp.
Vol. 2. Data report: oceanographic stations, BT observations, and bottom samples, 183 pp.

bottom sampling

Riedl, R., 1955.
Aufsammlung tiefer Meeresböden in abgegrenzten Schichten und Flächen. Arch. f. Hydrobiol., 51(2):189-208, 6 textfigs.

bottom-reflection losses

Morris, Halcyon E., 1970.
Bottom-reflection-loss model with a velocity gradient.
J. acoust. Soc. Am., 48 (5-2): 1198-1202

bottom-sea interface

bottom-water interface

Schindler, J.E. and K.R. Honick, 1971.
Oxidation-reduction determinations at the mud-water interface. Limnol. Oceanogr. 16(5): 837-840.

bottom-sea interface

Sternberg, Richard W., and Joe S. Creager 1965
An instrument system to measure boundary-layer conditions at the sea floor.
Marine Geology 3(6): 475-482.

bottom rocks (volcanic)

Japan, Maritime Safety Board, 1958.
[On the igneous rocks found from the submarine banks of adjacent Sea of Japan.]
Hydrogr. Bull., Publ., (981) 55:29-36.

bottom sediments
see also: Submarine geology

Bottom sediments

See also: submarine geology
sedimentation, etc.
and in later years, subheadings under bottom sediments

A

bottom sediments

Agassiz, A., 1902.
1. Preliminary report and list of stations. With remarks on the deep-sea deposits by Sir John Murray. Reports on the scientific results of the expedition to the Tropical Pacific, in charge of Alexander Agassiz, by the U.S. Fish Commission Steamer "Albatross", from August, 1899, to March, 1900, ---- Mem. M.C.Z. 26(1):1-114, 19 pls.

bottom sediments

Agassiz, A., 1902
Report on the scientific results of the expedition of the Albatross to the tropical Pacific, 1899-1900. Preliminary report and list of stations with the remarks on the deep-sea deposits, by Sir John Murray.
Mem. MCZ 26(1):114 pp., illus., pls., charts.

bottom sediments (data)

Agassiz, A., 1892.
II. General sketch of the Expedition of the "Albatross", from February to May, 1891. Reports on the dredging operations of the west coast of Mexico and in the Gulf of California in charge of Alexander Agassiz, carried out by the U.S. Fish Commission Steamer "Albatross", Lieut. Commander Z.L. Tanner, U.S.N., commanding.
Bull. M.C.Z. 23(1):1-90, 22 pls.

bottom sediments

Aksenov, A.A., A.S. Ionin and F.A. Shcherbakov, 1964
New data on the structure of strata in the Recent near-shore deposits. (In Russian).
Okeanologiia, Akad. Nauk, SSSR, 4(5):842-849.

bottom sediments

Alarcón, Elías, 1970.
Descripción oceanográfica preliminar del Golfo de Arauco.
Bol. Cient. Inst. Fomento pesq., Chile, 13: 51 pp.

bottom sediments

Alfirevic, Slobodan, 1964.
La couverture sedimentologique de la region des canaux en Adriatique et les facteurs qui agissent sur sa formation. (In Jugoslavian; French resume).
Acta Adriatica, 11(1):9-17.

bottom sediments

Alfvievic, S., 1960.
[Results of the morphological and geological researches of marine sediments in the channels of the archipelago of Zadar.]
Hidrografski Godisnjak, Split, 1959(3):67-104.

bottom sediments

Allen, J.R.L., 1964.
The Nigerian continental margin: bottom sediments submarine morphology and geological evolution.
Marine Geology, 1(4):289-332.

bottom sediments

* Anderson, Franz E., 1968.
Seaward terminus of the Vashon continental glacier in the Strait of Juan de Fuca.
Marine Geol. 6(6): 419-438.

bottom sediments

Anon., 1950.
Exploracion oceanografica del Africa Occidental. Campana del "Maraspina" en enero de 1950 en aguas del Sahara desde Punto Durnford a Cabo Barbas.
Bol. Inst. Espanol Ocean. No. 38:12 pp., 2 figs.

bottom sediments

Anon., 1950.
Table 1. Direct current measurement in the Omura Bay for several stations occupied by a cutter.
Res. Mar. Met., Ocean. Obs., Tokyo, July-Dec. 1947, No. 2:243.

bottom sediments

Anon., 1949.
Campañas del "Xauen" en 1947 y 1948 en el mar de Alboran y en el estrecho de Gibraltar. Registro de operaciones. Bol. Inst. Español Ocean. 18: 1-53.

bottom sediments(data)

Anon., 1949.
Exploracion oceanografica del Africa Occidental. Campañas del "Malaspina" en 1947 y 1948 en aguas del Sahara desde Cabo Juby a Punta Durnford. Registro de operaciones. Bol. Inst. Español Ocean. No. 23: 28 pp., 2 textfigs.

bottom sediments

Anon., 1939 (?)
Synopsis of the Hydrographical investigations (July-December 1939). Semiannual Rept. Oceanogr. Invest. (July-Dec. 1939).
Imp. Fish. Expt. Sta., Tokyo, No.65, 147 pp.

bottom sediments

Anon., 1885.
Rapport sur les sondages exécutés par "Le Bruat" entre la Nouvelle Caledonie et l'Australie.
Ann. Hydrogr., Paris, (2) 7:

not seen

bottom sediments

Anon., 1882.
Sondages. Ann. Hydrogr., Paris, (2)4:89-123.

bottom sediments

Anon., 1882.
Sondages de l'aviso "Le Travailleur" dans le Golfe de Gascogne. Ann. Hydrogr., Paris, (2)4: 354-398.

bottom sediments

Antoine, John, 1968.
The structure, sediments and possible evolution of the Gulf of Mexico. Trans. Fifth Carib. Geol. Conf., St. Thomas, V.I., 1-5 July, 1968, Peter H. Mattson, editor. Geol. Bull. Queens Coll., Flushing 5: 8. (Abstract only).

bottom sediments

Aroutiounov, I.M., 1957.
[Constitution des sediments du nord de la mer Caspienne] Trav. Inst. Ocean., Leningrad, 34:

bottom sediments

Arrhenius, G., 1952.
Sediment cores from the East Pacific.
Repts. Swedish Deep-sea Expedition, 1947-1948, 5(1):227 pp., textfigs. with app. of pls.

bottom sediments

Avilov, I.K., 1965.
Some data on the bottom topography and grounds of the West African shelf. (In Russian).
Trudy vses. nauchno-issled. Inst. morsk. ryb. Khoz. Okeanogr. (VNIRO), 57:235-259.

bottom sediments

Avilov, I.K., 1965.
Relief and bottom sediments of the shelf and continental slope of the northwestern Atlantic. (In Russian).
Trudy vses. nauchno-issled. Inst. morsk. ryb. Khoz. Okeanogr. (VNIRO), 57:173-234.

bottom sediments
Avilov, I.K., 1962.
Peculiarities of the topographic structure and of the bottom sediments in the region southeast of Greenland. (In Russian).
Trudy Vses. Nauchno-Issledov. Inst. Morsk. Ribn. Chos. i Okean., VNIRO, 46:57-61.

bottom sediments
Avilov, I.K., 1958.
[A new texture from marine sediments.]
Doklady Akad. Nauk, SSSR, 118(6):1155-1157.

bottom sediments
Avilov, I.K., 1956.
[Thickness of contemporary deposits and post-glacial history of the White Sea.]
Gosudarst.Okean. Inst., Trudy, Leningrad, 31(43):5-57.

B

bottom sediments
Băcescu, M., G. Müller, H. Skolka, A. Petran, V. Elian, M.T. Gomoiu, N. Bodeanu și S. Stănescu, 1965.
Cercetări de ecologie marină în sectorul predeltaic în condițiile anilor 1960-1961.
In: Ecologie marină, M. Băcescu, redactor, Edit. Acad. Republ. Pop. Române, București, 1:185-344.

bottom sediments
Bader, R.G., 1954.
The role of organic matter in determining the distribution of pelecypoda in marine sediments.
J. Mar. Res. 13(1):32-47, 4 textfigs.

Bader, R.G., and V.J. Henry, 1958 bottom sediments
Marine sediments of Prince of Wales Strait and Amundsen Gulf, West Canadian Arctic. J. Mar. Res., 17:35-52.

bottom sediments
Bagnold, R.A., 1962.
Auto-suspension of transported sediment: turbidity currents.
Proc. R. Soc., London, (A) 265(1322):315-319.

bottom sediments
Bailey, J.W., 1856.
On some specimens of deep sea bottom from the Sea of Kamchatka. Am. J. Sci., ser. 2,2:284-285.

bottom sediments
Ballan, Y., A. Klingebiel et M. Pujos 1970
Premiers résultats d'une étude lithologique des sédiments superficiels de la plateforme littorale au large de Biscarrosse.
Cah. océanogr. 22(7): 727.

bottom sediments
Bandy, O.L., 1963
Miocene-Pliocene boundary in the Philippines as related to Late Tertiary stratigraphy of deep-sea sediments.
Science, 142(3597):1290-1292.

bottom sediments
Bannister, F.A., and M.H. Hey, 1936.
Some crystalline components of the Weddell Sea deposits. Discovery Repts. 13:60-76.

bottom sediments
Barghorn, E.S., & R.L. Nichols, 1961
Sulfate-reducing bacteria and pyritic sediments in Antarctica.
Science, 134(3473): 190.

bottom sediments
Barkovskaya, M.G., 1961.
The regularity of Recent bottom sediment distribution on the Black Sea shelf USSR. Dynamics of the coastal zone of the Black Sea. (In Russian).
Trudy Inst. Okeanol., Akad. Nauk, SSSR, 53:123-148.

bottom sediments
Barnard, J.L., 1962.
Benthic marine exploration of Bahia de San Quintin, Baja California, 1960-1961. No. 6. General.
Pacific Naturalist, 3(6/7):251-274.

bottom sediments
Barnard, J. Laurens, and John R. Grady 1968
A biological survey of Bahia de los Angeles, Gulf of California, Mexico. General account.
Trans. San Diego Soc. nat. Hist. 15 (6): 51-66.

bottom sediments
Berner, Ulrich, 1965.
Untersuchungen an Sedimenten vom Südausgang des Grossen Belts.
Mey-niana, Kiel, 15:1-28.

bottom sediments
Beach Erosion Board, 1953.
Japanese research in physical oceanography, 1948-1950. Bull. B.E.B. 7(1):26-35.

bottom sediments
Beall, A.O., Jr. and A.G. Fischer, 1969.
Sedimentology. Initial Reports of the Deep Sea Drilling Project, Glomar Challenger 1: 521-593.

bottom sediments
Beigbeder, Yvonne et Marie Moulinier, 1966.
Fonds sédimentaires et foraminifères dans la baie de Saint-Brieuc.
C.R. herbd. Séanc., Acad. Sci., Paris (d)263(4): 324-327.

bottom sediments
Bekker, Iu. R., N.S. Krylov and V.Z. Negrutsa 1970.
Geology of hyperborean sediments from the Aidovy islands in the Barents Sea. (In Russian).
Dokl. Akad. Nauk SSSR 193 (6): 1349-1352.

bottom sediments
Belderson, R.H., N.H. Kenyon and A.H. Stride, 1970.
Holocene sediments of the continental shelf west of the British Isles. In: The geology of the East Atlantic continental margin. 2. Europe, ICSU/SCOR Working Party 31 Symposium, Cambridge 1970, Rept. 70/14: 157-170.

bottom sediments
*Bellaiche, Gilbert, 1968.
Applications des méthodes radioactives à l'étude des transits sédimentaires: cas du golfe de Fréjus.
Cah. océanogr., 20(10):879-884.

bottom sediments
Belov, N.A., 1955.
[Results of the study of bottom deposits.]
Material. Nabluid. Nauch.-Issledov. Dreifuius., 1950/51, Morskoi Transport, 1:466-474; 476-532.
David Knauss, translator, A.M.S.

bottom sediments
Bennett, Lee C., Jr., and Samuel M. Savin, 1963.
The natural history of the Hardangerfjord. 6. Studies of the sediments of parts of the Ytre Samlafjord with the seismic profiler.
Sarsia, 14:79-94.

bottom sediments
Bennett, Richard H., George H. Keller, and Roswell F. Busby 1970.
Mass property variability in three closely spaced deep-sea sediment cores.
J. sedim. Petrol. 40(3): 1038-1043.

bottom sediments
Beresnev, A.F. and V.M. Kovylin, 1970.
The relief of the basement and the distribution of thicknesses of bottom sediment in the central part of the Japan Sea. (In Russian; English abstract). Okeanologiia, 10(1): 113-116.

bottom sediments
Berg, L.S., 1947.
Classification of marine sediments. Izvest. Vsesojuzn Geogr.-Obshch. SSSR 79(3):275-279.

bottom sediments
Bergeron, J., O. Leenhardt and C. Veysseyre, 1963.
De l'utilisation du "mud penetrator" dans les études des sédiments immergés superficiel.
Comptes Rendus, Acad. Sci., Paris, 256(24):5179-5181.

bottom sediments
Bernard, F., J. Lecal, and R. Codinet, 1950.
Étude des sédiments marins au large d'Alger. 1. Teneur en azote et carbone organique. Bull. Inst. Océan., Monaco, No. 963:10 pp., 4 textfigs., 1 fold-in.

bottom sediments
Berner, Robert A. 1970.
Pleistocene sea levels possibly indicated by buried black sediments in the Black Sea.
Nature, Lond. 227 (5259): 700.

bottom sediments
Berner, Robert A., 1967.
Diagenesis of iron sulfide in Recent marine sediments.
In: Estuaries, G.H. Lauff, editor, Publs Am. Ass. Advmt Sci., 83:268-272.

bottom sediments
Berrit, G.R., 1955.
Étude des teneurs en manganese et en carbonates de quelques carottes sedimentaires Atlantiques et Pacifiques. Göteborgs K. Vetenskaps- och Vitterhets-Samhalles Handl., Sjätte Följden, (B), 6(12):61 pp.
Also:
Medd. Oceanografiska Inst., Göteborg No. 23.

Berthois, Leopold

bottom sediments
Berthois, Leopold, 1966.
Hydrologie et sédimentologie dans le Kangerdlugssuaq (fjord à la côte ouest du Groenland).
C. r. hebd. séanc., Acad. Sci., Paris, (D), 262 (13):1400-1402.

Bottom Sediments
Berthois, Leopold, 1965.
Essai de correlation entre la sedimentation actuelle sur le bord externe des plateformes continentales et la dynamique fluviale.
Progress in Oceanography, 3:49-62.

bottom sediments
Berthois, L., 1957.
Recherches sur les sédiments de la Mer du Nord et la Mer d'Irlande.
Rev. Trav. Inst. Pêches Marit., 21(4):485-554.

bottom sediments
Berthois, L., 1956.
Déplacement des aires d'envasement dans l'estuaire de la Loire. C.R. Acad. Sci., Paris, 243(18):1343-1345.

bottom sediments
Berthois, L., 1954/55.
Recherches sur le tassement libre des sédiments et ses conséquences en géologie.
Rev. Fac. Ciências, Univ. Lisboa, (2a)(C)4(2): 371-385.

bottom sediments
Berthois, L., 1949.
Remarques sur la statistique granulometrique dans l'etude des sables. C.C.R.E.O., R.T.B. 3:

bottom sediments
Berthois, L.C., 1946.
Inventaire des échantillons lithologiques recueillis au cours des campagnes scientifiques du Prince Albert 1 de Monaco. Bull. Inst. Océan., Monaco, No. 898:1-31.

bottom sediments
Berthois, L. 1946
Recherches sur les sédiments du plateau continental atlantique. Ann. de l'Inst. Océan., 23(1): 1-63.

Berthois, L. 1939 bottom sediments
Contributions a l'etude des sédiments de la Méditerranée occidentale. Ann. de l'Inst. Océan, Monaco, 20(1): 1-50

bottom sediments
Berthois, L., and M. Aubert, 1950.
Nouvelle méthode d'étude des sables marins. C.R. Acad. Sci., Paris, 230:1304-1306, 1 textfig.

bottom sediments
Berthois, L. and C., 1955.
Étude lithographique des sédiments récoltés par le "President Theodore Tissier" en rade de Brest et en Manche. Rev. Trav. Inst. Pêches Marit., 19(4):467-499.

bottom sediments
Berthois, Léopold, et Suzanne Bouillé, 1965.
Sur le fractionnement des sédiments meubles.
C.R. Acad. Sci., Paris, 260(3):937-939.

bottom sediments
Berthois, L., R. Brenot, F. Auffret et Marie-Henriette Du Buit, 1969.
La sédimentation dans la région de la Bassurelle (Manche oriental): étude hydraulique, bathymétrique, dynamique et granulométrique.
Trav. Cent. Rech. Etud. océanogr. ns. 8(3): 13-29

bottom sediments
Berthois, L., A. Crosnier et Y. Le Calvez, 1968.
Contribution à l'étude sédimentologique du plateau continental dans la Baie de Biafra.
Cah. ORSTOM, sér. Océanogr., 6(3/4): 55-86.

bottom sediments
Berthois, L., and J. Furnestin, 1938.
Étude des sédiments dragués par le Président Théodore-Tissier. Rev. Trav. Off. Pêches Marit. 11(3):381-424.

bottom sediments
Berthois, L., and A. Guilcher, 1961
Étude de sédiments et fragments de roches fragues sur le banc Porcupine et à ses abords (Atlantique du Nord-est).
Revue des Travaux de l'Inst. des Pêches Marit. 25:355-385.

bottom sediments
Berthois, L., and Y. Le Calvez, 1959.
Deuxième contribution à l'étude de la sédimentation dans le Golfe de Gascogne
Rev. Trav. Inst. Pêches Marit., 23(3):323-376.

bottom sediments
Besnard, W., 1951.
Resultados científicos do cruzeiro do "Baependi" e do "Vega" à Ilha da Trinidade. Contribução para conhecimento da plataforma insular da Ilha da Trinidada. Bol. Inst. Paulista Ocean. 2(2):37-48, 5 pls.

bottom sediments
Beyer, Fredrik, 1968.
Zooplankton, zoobenthos, and bottom sediments as related to pollution and water exchange in the Oslofjord.
Helgoländer wiss. Meeresunters., 17: 496-509.

Bezrukov P.L.

bottom sediments
Bezrukov, P.L., 1964.
Sediments in the northern and central regions of the Indian Ocean. Investigations in the Indian Ocean, 33rd Voyage of E/S "Vitiaz". (In Russian).
Trudy Inst. Okeanol., Akad. Nauk, SSSR, 64:182-201.

Translation:
NIOT/73

bottom sediments
Bezrukov, P.L., 1964.
Sediments of the northern and central parts of the Indian Ocean. Investigations in the Indian Ocean (33rd Voyage of E/S "Vitiaz"). (In Russian).
Trudy Inst. Okeanol., Akad. Nauk, SSSR, 64:182-201.

bottom sediments
Bezrukov, P.L., 1962.
Some problems of zonation of sediment types in the World Ocean. Marine geology and the dynamics of coasts. (In Russian).
Trudy Okeanogr. Komissii, Akad. Nauk, SSSR, 10(3):3-8.

bottom sediments
Bezroukov, P.L., 1961
[Investigation of sediments in the northern Indian Ocean.] (In Russian; English abstract).
Okeanolog. Issled., Mezhd. Komitet Proved. Mezhd. Geofiz. Goda, Prezidiume, Akad. Nauk, SSSR, (4):76-90.

bottom sediments
Besrukov, P. L., 1960.
Sedimentation in the northwestern Pacific Ocean.
Morsk. Geol., Doklady Sovetsk. Geol., 21 sess., Int. Geol. Kongress, 45-58.
(English abstract)

bottom sediments
Bezrukov, P.L., 1960
[The bottom sediments of the Okhotsk Sea.]
Trudy Inst. Okeanol., 32: 15-95.

bottom sediments
Bezrukov, P.L., 1958.
Most recent investigations of bottom sediments in Far Eastern Seas and in the northwestern part of the Pacific Ocean. Oceanographic investigations in the northwest part of the Pacific Ocean.
Trudy Okeanogr. Komissii, Akad. Nauk, SSSR, 3:32-36.

In Russian

bottom sediments
Bezrukov, P.L., 1958
Sediments of the trenches in the northwestern Pacific.
Proc. Ninth Pacific Sci. Congr., Pacific Sci. Assoc., 1957, 16(Oceanogr.):204-207.

bottom sediments
Bezrukov, P.L., 1957.
[On deep-water deposits of the Idzu, Bonin, Marianas and Ryuku ocean depressions.]
Dokl. Akad. Nauk, SSSR, 114(2):387-390

bottom sediments
Bezrukov, P.L., 1955.
[Bottom deposits of the Kurile-Kamchatka Trench.]
Trudy Inst. Oceanol., 12:97-129.

bottom sediments
Bezrukov, P.L. and A.P. Lisitzin, 1969.
Bottom sediments.
Annls int. geophys. Year, 46:145-184.

bottom sediments
*Bezrukov, P.L., and A.P. Lisitzin, 1967.
Soviet investigations of bottom sediments in the oceans. (In Russian; English abstract).
Okeanologiia, Akad. Nauk, SSSR, 7(5):828-838.

bottom sediments
Bezrukov, P.L., and A.P. Lisitsin, 1962.
The study of bottom sediments. (In Russian; English abstract).
Mezhd. Geofiz. Komitet, Prezidiume, Akad. Nauk, SSSR, Rezult. Issled. Programme Mezhd. Geofiz. Goda, Okeanol. Issled., No. 7:49-83.

bottom sediments
Bezrukov, P.L., and A.P. Lisitsin, 1960
[Classification of bottom sediments in recent marine basins.] Trudy Inst. Okeanol., 32: 3-14.

bottom sediments
Besrukov, P.L., and I.O. Murdmaa, 1959
[Bottom sediments of the North Kuril area.]
Trudy Inst. Okeanol., 36: 169-190.

bottom sediments.
Bezrukov, P. L., and V.P. Petelin, 1962.
Data on the sediments of the deep-water trenches in the western part of the Pacific Ocean. Marine geology and the dynamics of coasts. (In Russian).
Trudy Okeanogr., Komissii, Akad. Nauk, SSSR, 10(3):66-69.

bottom sediments
Bezrukov, P.L., and B.P. Petelin, 1960.
[Manual on collection and preliminary treatment of samples of marine deposits.]
Trudy Inst. Okeanol., Akad. Nauk, SSSR, 44:81-111.

bottom sediments
Bhattacharya, S.K. and S.T. Ghotankar, 1969.
The estuary of the Hooghly. Bull. natn. Inst. Sci., India, 38(1): 25-32.

bottom sediments
Bigarella, João José, 1949.
Contribuição ao estudo da planície sedimentar da parte norte da Ilha de Santa Catarina. Arquivos de Biol. e Tecnologia, Inst. Biol. e Pesq. Tecnol., Secret. Agricult., Indust. e Comercio, Estado do Parana, Brazil, 4(16):107-140.

Biggs, Robert B., 1967. bottom sediments
The sediments of Chesapeake Bay. chemistry or In: Estuaries, G.H. Lauff, editor, Publs Am. Ass. Advmt Sci., 83:239-260.

bottom sediments
Blaik, M., and M.A. Ransome, 1962
Multiple reflections in deep-sea seismic reflection profiles. (Abstract).
J. Geophys. Res.. 67(9):3543.

Blanc, Jean J.

Bottom sediments
Blanc, J.J., 1969.
Sedimentary geology of the Mediterranean Sea.
In: Oceanography and marine biology: an annual review, H. Barnes, editor, George Allen & Unwin Ltd., 6: 377-454.

Blanc, Jean J., 1964. bottom sediments
Vases bathyales et sables détritiques au large de Marseille.
Rec. Trav. Sta. Mar. d'Endoume, Bull., 31(47):203-229.

Blanc, Jean J., 1964. bottom sediments
Recherches géologiques et sédimentologiques. Res. sci camp. "Calypso".
Ann. Inst. Océanogr., Monaco, n.s., 41:219-270.

bottom sediments
Blanc, J.J., 1958.
Sédimentologie sous-marine du détroit Siculo-Tunisien. Ann. Inst. Océan., Monaco, 34:91-126.

bottom sediments
Blanc, J.-J., 1958.
Recherches de sédimentologie littorale et sous-marine en Provence occidentale.
Ann. Inst. Océan., Monaco, 35(1):1-140.

bottom sediments
Blanc, J.J., 1958.
Recherches géologiques et sédimentologiques. Campagne 1955 en Méditerranée Nord-Orientale. Ann. Inst. Océan., Monaco, 34:157-212.

bottom sediments
Blanc, J.J., 1958.
Résultats scientifiques des campagnes de la Calypso. VIII. Campagne 1955 en Méditerranée Nord-Orientale. 1. Recherches géologiques et sédimentologiques. Ann. Inst. Océan., Monaco, 34:157-212.

bottom sediments
Blanc, Jean et Colerine Froget, 1967.
Recherches de géologie marine et sédimentologie. Campagne de la Calypso en Méditerranée orientale (quatrième mission, 1964).
Annls Inst. Oceanogr., n.s. 45(2): 257-292.

bottom sediments
Blanc-Vernhet, L., 1963.
Note préliminaire sur les Foraminifères des fonds détritiques côtiers et de la vase terrigène côtière dans la baie de Marseille.
Rec. Trav. Sta. Mar., Endoume, Bull., 30(45):83-93.

bottom sediments
Blanc-Vernhet, L., 1958.
Les milieux sédimentaires litteraux de la Provence occidentale (côte rocheuse). Relations entre la microfaune et la granulométrie du sédiment. Bull. Inst. Océan., Monaco, No. 1112: 45 pp.

bottom sediments
Blazhchishin, A.I., V.M. Litvin, L. Lukosevicius, and M.V. Rudenko, 1970.
New data on the bottom topography and sedimentary structure in the central area of the Baltic Sea. (In Russian; English and German abstracts)
Baltica, Lietuv. TSR Mokslu Akad. Geogr. Skyr. INQUA Tarybi Sek., Vilnius, 4: 145-168

bottom sediments
Bockel, M., 1962.
Projet d'une étude des propriétés acoustiques du fond de la mer par le bateau océanographique "Origny".
In: Océanographie Géologique et Géophysique de la Méditerranée Occidentale, Centre National de la Recherche Scientifique, Villefranche sur Mer, 4 au 8 avril 1961, 59-62.

bottom sediments
Boeggild, O.B., 1912.
The deposits of sea-bottom.
Rept. Danish Oceanogr. Exped., 1908-1910, 1:255-269, 1 pl.

bottom sediments
*Bogdanov, Yu. A., B.A. Koshelev and A.V. Soldatov, 1967.
Sediments of the Romanche Trench. (In Russian; English abstract).
Okeanologiia, Akad. Nauk, SSSR, 7(4):633-635.

bottom sediments
Bogorov, B.G., S.M. Brujewicz, M.V. Fedosov, and G.B. Udintzev 1961
Methods of investigation of oceans in the USSR
Ann. Int. Geophys. Year, 11:311-316.

Boillot, Gilbert

bottom sediments
Boillot, Gilbert, 1965.
Organogenic gradients in the study of neritic deposits of biological origin: The example of the western English Channel.
Marine Geology, 3(5):359-367.

bottom sediments
Boillot, Gilbert, 1963.
Sur la fosse de la Hague (Manche centrale).
Comptes Rendus, Acad. Sci., Paris, 257(25):3963-2966.

bottom sediments
Boillot, G., 1961.
La répartition des fonds sous-marins dans la Manche occidentale.
Cahiers Biol. Mar., Roscoff, 2(2):187-208.

bottom sediments
Boillot, Gilbert, 1961.
La répartition des sédiments en Baie de Morlaix et en Baie de Siec.
Cahiers de Biol. Marine, Roscoff, 2(1):53-66.

Also in:
Trav. Lab. Géol. Sous-Marine, 10(1960).

bottom sediments
Boillot, Gilbert, 1960.
La répartition des fonds sous-marins au large de Roscoff (Finistère).
Cahiers de Biol. Marine, 1:3-23.

Also in:
Trav. Lab. Géol. Sous-Marine, 10(1960).

bottom sediments
Boillot, Gilbert, and Yolande Le Calvez, 1961.
Étude de l'Eocène au large de Roscoff (Finistère) et au sud de la Manche occidentale.
Revue de Géogr. Phys. Géol. Dynamique (2), 4(1):15-30.

Also in:
Travaux, Lab. Géol. Sous-Marine, Sta. Oceanogr., Villefranche, 11(1961-1962).

bottom sediments
Bonder Constantin, Vasile Roventa, Pompiliu Besnea, Ilie Decu, et Anastase Dumitrascu, 1969.
La granulométrie de la couche superficielle du fond sous-marin de la mer Noire en face du littoral roumain.
Stud. Hidraul. 19:47-59
Abstract in: Rapp. P.-v. Reun. Commn int. Explor. scient. Mer Méditerr., 19(4): 627-628.

bottom sediments
Bordovsky, O.K., 1961.
Diagenetic changes in bottom sediments from the central Pacific Ocean. (Abstract).
Tenth Pacific Sci. Congr., Honolulu, 21 Aug.-6 Sept., 1961, Abstracts of Symposium Papers, 366.

bottom sediments
Bordorsky, O.K., 1957.
The bituminosity of deposits of the western part of the Bering Sea.
Dokl. Akad. Nauk, SSSR, 113(6):1321-1323.

bottom sediments
Bordovsky, O.K., 1956.
Study of the consistence of some recent marine sediments. Trudy Inst. Okeanol., 17:137-140.

bottom sediments
Borghetti, G., 1959.
Caratteristiche mecannico-fisiche di alcuni sedimenti recenti del Golfo di Genova.
Geofis. e Meteorol., 7(1/2):29-36.

bottom sediments
Borley, J.O., 1923.
The marine deposits of the southern North Sea.
Min. Agricult. Fish., Fish. Invest. (2), 4(6): 1-73, 14 charts, 9 figs.

bottom sediments
Bostnikov, V.S. and I.M. Belousov 1968
Preliminary results of geological investigations in the western Indian Ocean. (In Russian).
Trudy Vses. Nauchno-Issled. Inst. Morsk. Ribn. Okeanogr. (VNIRO) 64, Trudy Azovo-Chernomorsk Nauchno-Issled. Inst. Morsk. Ribn. Khoz. Okeanogr. (AscherNIRO) 28: 48-64.

bottom sediments
Bose, S.K., 1962
Wave propagation in marine sediments and water saturated soils.
Geofisica Pura e Applicata, 52(2):26-40.

bottom sediments
Bossolasco, M., I. Dagnino, A. Elena e C. Palau 1969.
Variazioni delle spiagge e sedimentazione di materiali ferromagnetici.
Geofisica Met. 18(5/6):107-120.

bottom sediments
Botteron, Germain, 1958.
Etude de sédiments recoltés au cours de plongées avec le bathyscaphe "Trieste" au large de Capri.
Bull. Lab. Géol., Mineralog. Geophys., et du Musée Geol., Univ. Lausanne, No. 124: 19 pp.

bottom sediments

Botteron, G., 1958

Etude de sédiments récoltés au cours de plongées avec le bathyscaphe "Trieste" au large de Capri Bull. Soc. Vaudoise Sci. Nat., 67 fas 2 (298): 73-92.

bottom sediments

Bouma, Arnold H., 1968.
Distribution of minor structures in Gulf of Mexico sediments.
Trans. Gulf Coast Ass. Geol. Socs 18: 26-33
Also in: Contr. Oceanogr. Texas A+M. Univ. 12 (383).

bottom sediments

Bouma, A.H., 1965.
Sedimentary characteristics of samples collected from some submarine canyons.
Marine Geology Elsevier Publ. Co., 3(4):291-320.

bottom sediments

Bouma, Arnold H. and William E. Sweet, Jr., 1969
The upper sediment column of the Gulf of Mexico presented in a new type of map. Trans. Gulf Coast Ass. geol. Socs, 19: 171-174.

Bourcart, Jacques

bottom sediments

Bourcart, Jacques, 1960.
Galets anciens dans les cinérites pliocènes du cap d'Ail.
Bull. Soc. Géol., France, (7), 2:38-40.

Also in:
Trav. Lab. Géol. Sous-Marine, 10(1960).

bottom sediments

Bourcart, J., 1960

Les sables profonds de la baie de Nice. Comm Int. Expl. Sci. Mer Mediterranée, Rapp. Proc. Verb., Monaco, 15(3): 323.

bottom sediments

Bourcart, J., 1960

Sur les propriétés physiques des vases profondes de la Méditerranée occidentale. C.R. Acad. Sci. 251(19): 1951-1953.

bottom sediments

Bourcart, J., 1959.

Hypothèses sur le mode de transport et de dépôt des sédiments en Méditerranée occidentale.
Comptes Rendus, Acad. Sci., Paris, 249(18):1783-1784.

bottom sediments

Bourcart, J., 1959.

Sur la répartition des sédiments en Méditerranée occidentale.
Comptes Rendus, Acad.,Sci., Paris, 249(17):1699-1700.

bottom sediments

Bourcart, Jacques, 1958.

Les sediments precontinentaux profonds dans le Golfe de Genes.
Deep-Sea Res., 5:215-221.

Reprinted in:
Trav. Lab. Geol. Sous-Marine, 8, 1958.

bottom sediments

Bourcart, J., 1957.
Géologie sous-marine de la Baie de Villefranche.
Ann. Inst. Océan., Monaco, 23(3):137-199.

bottom sediments

Bourcart, J., 1957.
Géologie sous-marine de la Baie de Villefranche.
Ann. Inst. Océan., Monaco, 33(3):137-200.

bottom sediments

Bourcart, J., 1957.
Premier rapport sur les fonds meubles de la Baie du Mont Saint-Michel. Bull. d'Indo., C.C.O.E.C., 9(6):325-336.

bottom sediments

Bourcart, J., 1955.
Les sables profonds de la Mediterranee.
Arch. Sci., Geneva, 8(1):5-13.

bottom sediments

Bourcart, J., 1955.
Étude des échantillons récoltés en juillet-août 1946 par le "President Théodore Tissier".
Rev. Trav. Inst. Pêches Marit., 19(4):447-464.

bottom sediments

Bourcart, J., 1955.
Recherches sur le plateau continental de Banyuls-sur-Mer.
Vie et Milieu, Bull. Lab. Arago, 6(4):435-524.

bottom sediments

Bourcart, J., 1954.
Les vases de la Méditerranée et leur mécanisme de dépôt. Deep-Sea Res. 1(3):126-130.

bottom sediments

Bourcart, J., 1954.
Description des echantillons recoltés par l' "Amiral Mouchez" sur la cote marocaine entre Port Lyautey et Casablanca.
Bull. d'Info., C.C.O.E.C. 6(5):207-211, 1 pl.

bottom sediments

Bourcart, J., 1953.
Sur la nécessité d'étudier ler propriétés d'ensemble des sediments actuels pour en déterminer le facies. Rev. Inst. Français du Pétrole et Annales des Combustibles Liquides, 8:100-101.

bottom sediments

Bourcart, J., 1953.
Suggestions sur l'origine de quelques sediments mediterraneens. Bull. d'Info., C.C.O.E.C. 5:211-219.

bottom sediments

Bourcart, J., 1952.
Sables "Néritiques" à 2750 m. de profondeur au large de Bougie (Algérie). C.R. Acad.Sci., Paris, 236:738-740.

bottom sediments

Bourcart, J., 1952.
Les frontières de l'ocean. Albin Michel, Paris, 317 pp., 77 textfigs.

sedimentation

Bourcart, J., 1950.
Le littoral Breton, du Mont Saint-Michel au Finistere. Bull. d'Info., C.O.E.C., 2(1):21-39, 2 figs.

bottom sediments

Bourcart, J., 1948.
Rapport sur une mission oceanographique sur le plateau continental marocain. Bull. Sci., C.O.E.C., Maroc, 2:22-36, 5 figs.

bottom sediments

Bourcart, J., 1947.
Sur les rechs, sillons sous-marins du Plateau Continental des Albères (Pyrénées Orientales). C. R. Acad. Sci. 224:1175-1177.

bottom sediments

Bourcart, J., 1947.
La sédimentation dans La Manche. C. R. Sess. Extr. Belg. Geol., Sept. 1946:14-43.

bottom sediments

Bourcart, J., 1947.
La répartition des sédiments dans la zone néritique. C.R. Soc. Geol. Fr. 1947, 16:325-327.

bottom sediments

Bourcart, J., 1947
Sur les vases du plateau continental français. C.R. Acad. Sci. Paris 225:137-139.

bottom sediments

Bourcart, J., 1945.
Sediments quartennaires sur la greve de la region de Roscoff (Finistere). C.R. Acad. Sci., Paris, 221(10/13):357-359.

bottom sediments

Bourcart, J., 1939.
Essai d'une définition de la vase des estuaires.
C.R. Acad. Sci., Paris, 209:542-543.

bottom sediments

Bourcart, J., 1939.
Sur les vases des estuaires de la Manche. C.R.S. Soc. Geol., France, 73-74.

bottom sediments

Bourcart, Jacques, et Gilbert Boillot, 1960.
Étude des dépots flandriens de l'anse Duguesclin pres de Cancale (Ille-et-Vilaine).
Bull. Soc. Géol., France, (7), 2:45-49.

Also in:
Trav. Lab. Géol. Sous-Marine, 10(1960).

bottom sediments

Bourcart, J., & G. Boillot, 1960

La répartition des sédiments dans la baie du Mont-Saint-Michel.
Rev. Geogr. Phys. et Geol. Dyn., 3(4): 189-199.

bottom sediments

Bourcart, Jacques, Gilbert Boillot, Jacques-Yves Cousteau, Maurice Gennesseaux, Eloi Klimer and Claude Lalou, 1958.

Les sédiments profonds au large de la côte niçoise.
C. R. Acad. Sci., Paris, 247(1):116-118.

Reprinted in:
Trav. Lab. Geol. Sous-Marine, 8, 1958.

bottom sediments

Bourcart, J., and C. Francis-Boeuf, 1949.
Nature et composition de quelques boues marines prelevees au large de Monaco. Bull. Sta. Ocean., Salammbo, No. 43:4 pp., tables.

bottom sediments

Bourcart, Jacques, M. Gennesseaux and E. Klimek, 1961.
Les canyons sous-marins de Banyuls et leur remplissage sédimentaire.
Comptes Rendus, Acad. Sci., Paris, 253(1):19-24.

bottom sediments

Bourcart, Jacques, Maurice Gennesseaux, and Eloi Klimek, 1961.
Sur le remplissage des canyons sous-marine de la Méditerranée française.
Comptes Rendus, Acad. Sci., Paris, 252:3693-3698.

Also in:
Travaux, Lab. Géol. Sous-Marine, Sta. Oceanogr., Villefranche, 11(1961-1962).

bottom sediments
Bourcart, Jacques, Maurice Gennesseaux, et Éloi Klimek, 1960.
Écoulements profonds de sables et de galets dans la grande vallée sous-marine de Nice.
C.R. Acad. Sci., Paris, 250:3761-3765.

Also in:
Trav. Lab. Géol. Sous-Marine, 10(1960).

bottom sediments
Bourcart, Jacques, Maurice Gennesseaux, Éloi Klimek, et Yolande Le Calvez, 1960.
Les sédiments des vallées sous-marines au large dans le Golfe de Gênes.
C.R. Acad. Sci., Paris, 251:1443-1445.

Also in:
Trav. Lab. Géol. Sous-Marine, 10(1960).

bottom sediments
Bourcart, J., and J. Jean-Jaquet, 1946.
Sur la répartition des sédiments dans la Baie du Mont-Saint-Michel. C. R. Acad. Sci., Paris, 222(26):1507-1508.

bottom sediments
Bourcart, Jacques, Mlle. Claude Lalou et Lucien Mallet, 1961.
Sur la présence d'hydrocarbures du type benzo-3.4 pyrène dans les vases côtières et les sables de la plage de la rade de Villefranche (Alpes Maritimes).
C. R. Acad. Sci., Paris, 252:640-644.

Also in:
Trav. Lab. Géol., Sous-Marine, 10(1960).

bottom sediments
Bourcart, Jacques, Francois Ottmann, and Jeanne-Maria Ottmann-Richard, 1958.
Premiers resultats de l'etude des carottes de la Baie des Anges, Nice.
Rev. Geograph. Phys. et Geol. Dynamique, (2) 1(3):167-173.

Also reprinted in:
Trav. Lab. Geol. Sous-Marine, 8, 1958.

bottom sediments
Bourcart, Jacques, and Pedro Roa-Morales, 1958
Les dépôts vaseux de la Rance maritime et du Mont-Saint-Michel.
Bull. Soc. Geol. de France, (6) 7:545-552.

Also reprinted in:
Trav. Lab. Geol. Sous-Marine, 8, 1958.

bottom sediments
Bourcart, Jacques, et Michel Siffre, 1958.
Le Quarternaire marin du pays niçois. Bull. Soc. Géol., France, 8:715-730.

bottom sediments
Bourcier, M., 1968.
Étude du benthos du plateau continental de la baie de Cassis.
Rec. Trav. Sta. mar. Endoume, 44(60): 63-108

Bouysse, Philippe

bottom sediments
Bouysse, Philippe et Jean-René Vanny, 1966.
La Baie de la Vilaine: étude sédimentologique et morphologique.
Cahiers Oceanogr., 18(4):319-341.

bottom sediments
Bowie, D.K., A.O. Fuller and W.G. Siesser, 1970.
The marine sediments of False Bay.
Trans. roy. Soc. S Afr. 39(2): 149-161

bottom sediments
Bowman, A., 1928.
The Moray Firth. Physical characteristics.
Cons. Perm. Int. Expl. Mer, Rapp. Proc. Verb., 52:39-42, 1 fig.

bottom sediments
Boychenko, I.G., 1961.
[Bottom topography of the Karaginsky Bay.]
Trudy Inst. Okeanol., Akad. Nauk, SSSR, 50:3-20.

bottom sediments
*Boynagryan, V.R., 1968.
Lithological peculiarities of surface sediments of the underwater shore slope in the southeastern Baltic Sea. (In Russian; English abstract).
Okeanologiia, Akad. Nauk, SSSR, 8(6):1036-1048.

bottom sediments
Braarud, T. and J. T. Ruud, 1932
The "Øst" Expedition to the Denmark Strait 1929. I. Hydrography. Hvalrådets Skr., No. 4:44 pp., 19 text figs.

bottom sediments
Bradner, Hugh, 1963
Probing sea-bottom sediments with microseismic noise.
J. Geophys. Res., 68(6):1788-1791.

bottom sediments
Brambati, Antonio, 1969
Sedimentazione recente nelle Lagune di Marano e di Grado (Adriatico settentrionale).
Studi trentini Sci. nat. (A) 46(1): 142-239. Also in: Raccolta Dati Oceanogr. Comm. ital. Oceanogr. Consig. natz Ricerche (A)(21)(1969)

bottom sediments
Brambati, Antonio, 1969.
Caratteristiche sedimentologiche dei depositi sabbiosi dei corsi d'acqua tributari dell'Adriatico settentrionale tra Venezia e Trieste e loro relazioni con la linea delle risorgive.
Studi trentini Sci. nat. (A) 46(1): 11-29. Also in: Raccolta Dati Oceanogr. Comm. ital. Oceanogr. Consig. natz Ricerche. (A) (20)(1969).

bottom sediments
Brambati, Antonio, e Giulio Antonio Venzo, 1967.
Recent sedimentation in the northern Adriatic Sea between Venice and Trieste.
Studi Trentini Sci. nat., (A), 44(2):202-274.
Reprinted: Comm. Studio Oceanogr. Limnol. (a) (4) 1968.

bottom sediments
Brambati, Antonio, e Maria Luisa Zucchi, 1969
Relazioni tra granulometrie e distribuzione dei Molluschi nei sedimenti recenti dell'Adriatico settentrionale tra Venezia e Trieste
Studi trentini Sci. nat. (A) 46(1): 30-40. Also in: Raccolta Dati Oceanogr. Comm. ital. Oceanogr. Consig. natz Ricerche (A)(20) (1969)

bottom sediments
Brattström, H., and E. Dahl, 1951.
Reports of the Lund University Chile Expedition, 1948-49. 1. General account, lits of stations, hydrography. Lunds Universitets Årsskrift, [K. Fysiograf. Sällskapets Handl., n.f., 61(8);] n.f., Avd. 2, 46(8):86 pp.

bottom sediments
Brennecke, W., 1921.
Die ozeanographischen Arbeiten der Deutschen Antarktischen Expedition, 1911-1912.
Arch. Deutschen Seewarte 39(1):1-216, 15 pls., 41 textfigs.

bottom sediments
Brennecke, W., 1915.
Oceanographischen Arbeiten S.M.S. "PLanet" im westlichen Stillen Ozean 1912-1913. Ann. Hydr., usw., 43:145-158.

bottom sediments (data)
Brennecke, W., 1910.
Lotungen des Kabeldampfers "Grossherzod von Oldenburg" im östliche Teil des Nordatlantischen Ozeans. Arch. Deutschen Seewarte 33(3):1-20, 1 textfig., 2 pls.

bottom sediments
Brodie, J.W., 1952.
Features of the sea floor west of New Zealand. N.Z. J. Sci. Tech., Sect. B, 33(5):373-384, 6 textfigs.

bottom sediments
Broggi, J.A., 1952.
Migracion de arenas a lo largo de la costa peruana. Bol. Soc. Geol., Peru, 24: 25 pp., 2 pls., 1 textfig.

bottom sediments
Brongersma-Sanders, Margaretha 1967.
Barium in pelagic sediments and in diatoms.
Proc. K. ned. Akad. Wet. (B) 70(1): 93-99

bottom sediments
Brouardel, J., and L. Fage, 1953.
Variation en mer de la teneur en oxygène dissous au proche voisinage des sédiments.
C.R. Acad. Sci., Paris, 237:1605-1607.

Bruevich, S.V.

bottom sediments
Bruevich, S.V., 1957.
Fresh water deposition below the Recent deposits of the Black Sea. (In Russian).
Doklady, Akad. Nauk, SSSR, 84(3):575

bottom sediments
Bruevich, S.V., 1949.
[Rate of formation of bottom deposits in the Caspian Sea.] Izvest. Akad. Nauk, USSR, Ser. Geogr., Geofiz. 13:9-32.

bottom sediments
Bruun, A.Fr., J.W. Brodie and C.A. Fleming, 1955.
Submarine geology of Milford Sound, New Zealand.
N.Z.J. Sci. Tech., B, 36:397-410, 7 textfigs.

bottom sediments
Buffington, E.C., and D.G. Moore, 1963.
Geophysical evidence on the origin of gullied submarine slopes, San Clemente, California.
J. Geology, 71(3):356-370.

Burckle, L.H.

bottom sediments
*Burckle, Lloyd, H., John Ewing, Tsunemasa Saito and Robert Leyden, 1967.
Tertiary sediment from the East Pacific Rise.
Science, 157(3788):537-540.

bottom sediments
Burckle, Lloyd H., and Tsunemasa Saito, 1966.
An Eocene dredge haul from the Tuamotu Ridge.
Deep-Sea Res., 13(6):1207-1208.

bottom sediments
Burke, Kevin 1967.
The Yallahs Basin: a sedimentary basin southeast of Kingston, Jamaica.
Marine Geol. 5(1):45-60.

bottom sediments

Burkholder, Paul R., and Lillian M., 1958.
Studies on B vitamins in relation to productivity of the Bahia Fosforescente, Puerto Rico.
Bull. Mar. Sci., Gulf and Caribbean, 8(3):201-223.

bottom sediments

Byrne, John V. and K. O. Emery, 1960
Sediments of the Gulf of California. Bull. Geol. Soc., Amer., 71(7): 983-1010.

C

bottom sediments

Cabioch, Louis, 1968.
Contribution à la connaissance des peuplements benthiques de la Manche occidentale.
Cah. Biol. mar. 9 (5-Suppl.): 493-720

bottom sediments

Cailleux, A., 1959.
Sur les galets dragués à 4255m. de profondeur entre les Acores et Brest.
Comptes Rendus, Acad. Sci. Paris, 249(13): 1128-1129.

bottom sediments

Callame, B., 1962.
Observations sur les échanges par diffusion entre les sédiments et l'eau de mer qui les recouvre.
In: Océanographie Géologique et Géophysique de la Méditerranée Occidentale, Centre National de la Recherche Scientifique, Villefranche sur Mer, 4 au 8 avril 1961, 83-87.

bottom sediments

Callame, B., and J. Debyser, 1954.
Observations sur les mouvements des diatomées à la surface des sédiments marins de la zone inter-cotidale. Vie et Milieu, 2:120-157.

Contr. C.R.E.O. Vol. 5

~~bottom sediments~~

Calmels, Augusto P., y Sergio P. Andreoli, 1968.
Contribución al conocimiento oceanográfico del area exterior de la Ria de Bahia Blanca.
Contrib. Inst. Oceanogr., Univ. Nac. Sur, Bahia Blanca 1: 46 pp. (mimeographed).

bottom sediments (data only)

Canada, Department of Mines and Technical Surveys, 1960.
Tidal and oceanographic survey of Hudson Strait, August and September, 1959. Data Record, 55 pp. (multilithed.)

bottom sediments (data)

Capart, A., 1951.
Liste des stations. Rés. Sci., Expéd. Océan. Belge dans les Eaux Côtières Africaines de l'Atlantique Sud (1948-1949) 1: 65 pp., 2 pls.

bottom sediments

Carzballo Muziotti, Luis Felipe, 1968
Sedimentos recientes de la Bahia de Mochima.
Bol. Inst. Oceanogr., Univ. Oriente 7(2): 45-60.

bottom sediments

Caralp, M., A. Klingebiel, A. Lamy et autres, 1968.
Etude micropaléontologique, sédimentologique et géochimique de quelques carottes de sédiments récents du Golfe de Gascogne.
Bull. Inst. Géol. Bassin d'Aquitaine, 5:3-73. (not seen)

bottom sediments

Carriker, Melbourne R., 1967.
Ecology of estuarine benthic invertebrates: a perspective.
In: Estuaries, G.H. Lauff, editor, Publs Am. Ass. Advmt Sci., 83:442-487.

bottom sediments

Carsola, A.J., 1954.
Recent marine sediments from Alaskan and North-west Canadian Arctic.
Bull. Amer. Assoc. Petr. Geol. 38(7):1552-1586, 8 textfigs.

bottom sediments

Carsola, A.J., 1954.
Recent marine sediments from Alaskan and north-west Canadian Arctic.
Bull. Amer. Assoc. Petr. Geol. 38(3):377-404.

bottom sediments

Carver, Robert E., 1971.
Holocene and late Pleistocene sediment sources, continental shelf off Brunswick, Georgia.
Jl sedim. Petrol. 41 (2): 517-525.

bottom sediments

Caspers, H., 1951.
Quantitative Untersuchungen über die Bodentierwelt des Schwarzen Meeres im bulgarischen Küstenbereich. Arch. f. Hydrobiol. 45(1/2):192 pp., 66 textfigs.

bottom sediments

Castany, G., and F. Ottman, 1957.
Le quarternaire marin de la Méditerranée occidentale.
Rev. Géogr. Phys., Géol. Dynam., (2)1(1):46-55.

bottom sediments

Chamley, Hervé, 1965.
Observations sur quelques sédiments marins prélevés près des côtes de Terre Adélie.
Rec. Trav. Sta. Mar., Endoume, Bull., 52(36):215-228.

Also:
Expéd. Polaires Françaises, Missions Paul-Emile Victor, No. 275.

bottom sediments

Charcot, J.B., 1931-1932.
Rapport préliminaire sur la campagne du "Pourquoi Pas?" en 1931. Ann. Hydrogr., Paris, (3)11:57-139.

bottom sediments

Charcot, J.B., 1931-1932.
Rapport préliminaire sur la campagne du "Pourquoi Pas?" en 1930. Ann. Hydrogr., Paris, (3)11:1-54.

bottom sediments

Charcot, J.B., 1927-1928.
Rapport preliminaire sur la campagne du "Pourquoi Pas?" en 1926. Ann. Hydrogr., Paris, (3)8:1-96, figs.

bottom sediments

Charcot, J.B., 1925-1926.
Rapport préliminaire sur la campagne du "Pourquoi Pas?" en 1925. Ann. Hydrogr., Paris, (3)7:191-389, figs.

bottom sediments

Charcot, J.B., 1925-1926.
Rapport préliminaire sur la campagne du "Pourquoi-Pas?" en 1924. Ann. Hydrogr., Paris, (3) 7:1-96, figs.

bottom sediments

Charcot, J.B., 1923-1924.
Rapport préliminaire sur la campagne du "Pourquoi-Pas?" en 1923. Ann. Hydrogr., Paris, (3) 6:1-89, figs.

bottom sediments

Chesterman, W.D. and H.K. Wong, 1971.
Bottom sediment distribution in certain inshore waters of Hong Kong. Int. hydrogr. Rev. 48(2): 51-69.

bottom sediments

Chidambaram, K., and A.D.I. Rajendran, 1951.
On the hydro-biological data collected on the Wadge Bank early in 1949. J. Bombay Nat. Hist. Soc. 49(4):738-748.

bottom sediments

Chidambaram, K., A.D.I. Rajandran, and A.P. Valsan, 1951.
Certain observations on the hydrography and biology of the pearl bank, Tholayviam Paar off Tuticorin in the Gulf of Manaar in 1949. (April)
J. Madras Univ., Sect. B, 21(1):48-74, 2 figs.

bottom sediments

Chilingar, G.V., 1958.
Some data on diagenesis obtained from Russian literature: a summary.
Geochimica et Cosmochimica Acta 13(2/3):213-217.

bottom sediments

Chilingar, G.V., 1956.
Black Sea and its sediments: a summary.
Bull. Amer. Assoc. Petr. Geol., 40(11):2765-2769.

bottom sediments

Chu, S. P., and H. Y. Shin, 1949.
The variation of certain chemical constituents of biological importance and some other properties of sea water in Chiaochow Bay, August, 1948, to May, 1949. (Contrib. Fish. Res. Inst., Dept. Fish., Nat. Univ. Shantung, No. 2) Sci. and Tech. in China 2(3):54-56.

bottom sediments

Ciabatti, Mario, e P. Polantoni, 1967.
Ricerche sui fondali antistanti il delta del Po.
G. Geol. 34: 109-120. Also in: Raccolta Dati Oceanogr. Comm. ital Oceanogr. Consig. naz. Ricerche (A) (15) (1968).

bottom sediments.

Coleman, James M., Sherwood M. Gagliano and James E. Webb, 1964.
Minor sedimentary structures in a prograding distributary.
Marine Geology, 1(3):240-258.

bottom sediments

Collette, B.J., J.I. Ewing, R.A. Lagaay and M. Truchan, 1969.
Sediment distribution in the oceans: the Atlantic between 10° and 19° N.
Marine Geol., 7(4): 279-345

bottom sediments

Collin, A.E., and M.J. Dunbar, 1964
Physical oceanography in Arctic Canada.
In: Oceanography and Marine Biology, Harold Barnes, Editor, George Allen & Unwin, Ltd., 2:45-75.

bottom sediments

Cone, R.A., N.S. Neidell, and K.E. Kenyon, 1963.
The natural history of the Hardangerfjord. 5. Studies of the deep-water sediments with the continuous seismic profiler.
Sarsia, 14:61-78.

bottom sediments
Conolly, John R., 1970
Sedimentary history of the continental margin of Australia.
Trans. N.Y. Acad. Sci (2)32 (3): 364-380

bottom sediments
*Conolly, John R., 1969.
Western Tasman sea floor.
N.Z. Jl Geol. Geophys. 12:310-343.

Cooper, L.H.N., 1960 bottom sediments
Some theorems and procedures in shallow-water oceanography applied to the Celtic Sea. JMBA 39:155-171.

bottom sediments
Cooper, L.H.N., 1951.
Chemical properties of the sea water in the neighborhood of the Babadie Bank. J.M.B.A. 30(1): 21-26, 2 textfigs.

bottom sediments
Cooper, L.H.N., 1960
Some theorems and procedures in shallow-water oceanography applied to the Celtic Sea. J. Mar. Biol. Ass'n U.K., 39(2): 155-171.

bottom sediments
Correns, C.W., 1950.
Faktoren der Sedimentbildung erläutet an Kalk- und Kieselsedimenten. Deutsche Hydro. Zeits. 3(1/2):83-88, 2 textfigs.

bottom sediments
Costa, Suzanne, 1960
Recherches sur les fonds à Halarachnion spatulatum de la baie de Marseille. Vie et Milieu, 11(1): 1-68.

bottom sédiments
* Cotton de Bennetot, Michelle, 1969.
Etude sédimentologique et morphologique de l'estuaire du Goayen (Finistère).
Cah. océanogr., 21(4): 355-377.

bottom sediments
Cotton de Bennetot, Michelle, Andre Guilcher et Anne Saint-Requier, 1965.
Morphologie et sédimentologie de l'Aber Benoît (Finistère).
Cahiers Océanogr., C.C.O.E.C., 17(6): 377-387.

bottom sediments
Crance, Johnie H. 1971.
Description of Alabama estuarine areas-Cooperative Gulf of Mexico Estuarine Inventory.
Bull. Alabama mar. Resources 6: 85pp.

bottom sediments
Crary, A. P., 1961
Marine-sediment thickness in the eastern Ross Sea area, Antarctica.
Bull. Geol. Soc. Amer., 72(5): 787-790.

bottom sediments
Creager, J.S., and D.A. McManus, 1964.
Notes on bottom sediments of the Chukchi Sea.
U.S.C.G. Oceanogr. Rept., No. 1(373-1):23-24.

Crozier, W. D., 1960 bottom sediments
Black magnetic spherules in sediments. J. Geophys. Res., 65 (9): 2971-2978.

bottom sediments
Cruickshank, Michael, 1963.
Ocean exploitation.
Undersea Technology, 4(10):16-17.

bottom sediments
Cuesta Urcelay, J., 1943.
Toponimia de los fondos del Cantabrico. (A puntes para la carta de pesca). Notas y Resumenes Inst. Espanol Oceanogr. 113:1-3, 3 maps.

bottom sediments
Cullen, David J. 1967.
The submarine geology of Foveaux Strait.
Bull. N.Z. sci. ind. Res. 184 (Mem. N.Z. Oceanogr. Inst. 33): 1-69.

bottom sediments
Curtis, R., G. Evans, D.J.J. Kinsman and D.J. Shearman, 1963.
Association of dolomite and anhydrite in the Recent sediments of the Persian Gulf.
Nature, 197(4868):679-680.

D

bottom sediments
Damuth, John E., and Rhodes W. Fairbridge, 1970.
Equatorial Atlantic deep-sea arkosic sands and ice-age aridity in tropical South America.
Bull. Geol. Soc. Am., 81 (1): 189-206

bottom sediments
da Nóbrega Coutinho, Paulo, e Jáder Onofre de Morais 1970
Distribucion de los Sedimentos en la plataforma continental norte y nordeste del Brasil.
Arq. Ciên. Mar, Fortaleza, Ceará, Brasil 10 (1): 79-90.

bottom sediments
Davies, David K., and W. Richard Moore 1970.
Dispersal of Mississippi sediment in the Gulf of Mexico.
J. sedim. Petrol. 40 (1): 339-353.

bottom sediments
Davies, T.A. and E.J.W. Jones, 1971.
Sedimentation in the area of Peake and Freen Deeps (Mid-Atlantic Ridge). Deep-Sea Res. 18(6) 619-630.

bottom sediments
DeBuen, R., 1927.
Résultats des investigations espagnoles dans le détroit de Gibraltar. Cons. Perm. Int. Expl. Mer, Rapp. Proc. Verb. 44:60-91, Textfigs. 29-50.

bottom sediments
de Buen, R., 1926. Analyse de douze échantillons de fonds marins provenant d'un sondage au trépan exécuté dans la ria de Vigo et considérations sur la genèse des rias. Bull. Inst. Océan., No. 474: 16 pp., 1 textfig.

bottom sediments
de Buen, O., 1916.
Première campagne de l'Institut espagnol d'Océanographie, dans la Mediterranée. Liste des stations et des operations. Bull. Inst. Océan., Monaco, No 314:23 pp.

bottom sediments
de Castro Barros, Aldemir, e Skapti Jonsson 1967.
Prospecção de camarões na região estuarina do Rio São Francisco.
Bol. Estud. Pesca, SUDENE, Recife 7(2):9-29

bottom sediments
de Gaillande, Daniel, 1970.
Peuplements benthiques de l'herbier de Posidonia oceanica (Delile), de la pelouse à Caulerpa prolifera Lamouroux et du large du Golfe de Gabes.
Tethys 2(2): 373-384

bottom sediments
de la Cruz S., Alfredo, 1966.
Estudios de plancton en la plataforma sur de Cuba.
Centre Invest. pesq., Cuba, 22:54 pp.

bottom sediments
Demel, K., and Mulicki, Z., 1954.
Quantitative investigations on the biological bottom productivity. Repts. Sea Fish Inst., Gdynia, No. 7:75-126, 14 textfigs.

bottom sediments
Demenitskaya, R.M. and K.L. Hunkins, 1970.
II. Regional observations. 6. Shape and structure of the Arctic Ocean (Feb. 1969).
In: The sea: ideas and observations on progress in the study of the seas, Arthur E. Maxwell, editor, Wiley Interscience 4(2): 223-249.

bottom sediments
De Miro Orell, Manuel, 1968.
Caracteristicas generales de los sedimentos recientes de los fondosmarinos de Venezuela.
Memoria Soc. Cienc. nat. La Salle, 28(79):97-137.

bottom sediments
Dietz, R.S., 1954.
Marine geology of northwestern Pacific: description of Japanese Bathymetric Chart 6901.
Bull. G.S.A. 65(1):1199-1224, 6 charts

bottom sediments
Dietz, Robert S., A.J. Carsola, E.C. Buffington, and Carl J. Shipek, 1964
Sediments and topography of the Alaskan shelves.
In: Papers in Marine Geology, R.L. Miller, Editor, Macmillan Co., N.Y., 241-256.

bottom sediments
Dietz, R.S., A.J. Carsola, and E.C. Lafond, 1948.
Topography and sediments of the Bering and Chukchi seas. (Abstract). Bull. G.S.A. 59:1318.

submarine geology
Dietz, R.S., K.O. Emery, and F.P. Shepard, 1942
Phosphorite deposits on the sea floor off southern California. Bull. G.S.A. 53: 815-848.

bottom sediments
*Dingle, R.V., 1970.
Quaternary sediments and erosional features off the north Yorkshire coast, western North Sea.
Marine Geol., 9(3):M17-M22.

bottom sediments
Dmitrenko, O.I., 1962.
The estimation of molecular capacity value as an indicator of the diagenetic transformations in colloidal sea sediments. (In Russian)
Trudy Inst. Okean., Akad. Nauk, SSSR, 54:147-157

bottom sediments
Dmitrenko, O.I., 1962.
The influence of some factors on the bound water content in the sea sediments. (In Russian)
Trudy Inst. Okeanol., Akad. Nauk, SSSR, 54:135-146.

bottom sediments
Dobson, M.R., W.E. Evans and K.H. James 1971.
The sediment on the floor of the southern Irish Sea.
Mar. Geol. 11(1):27-69.

bottom sediments
Dolet, M., P. Giresse, et Cl. Larsonneur, 1965.
Sédiments et sédimentations dans la baie du Mont Saint-Michel.
Bull. Soc. Linnéenne, Normandie, (10),6:51-65.

bottom sediments
*Dörjes, Jürgen, Sibylle Gadow, Hans-Erich Reineck and Indra Bir Singh, 1970.
Sedimentologie und Makrobenthos der Nordergründe und der Aufsenjade (Nordsee). Senckenberg marit. 2: 31-59.

bottom sediments
Dörjes, Jürgen, Sibylle Gadow, Hans-Erich Reineck and Indra Bir Singh, 1969.
Die Rinnen der Jade (Südliche Nordsee).
Sedimente und Makrobenthos. Senckenberg. marit. [1] 50: 5-62.

bottom sediments
*Dörjes, Jürgen, Sibylle Gadow, Hans-Erich Reineck und Indra Bir Singh, 1969.
Die Rinnen Jade (südliche Nordsee).
Sedimente und makrobenthos.
Senckenberg. maritima 50: 5-62.

bottom sediments
Drake, A.A., Jr., 1962.
Preliminary geologic report on the 1961 U.S. Expedition to Bellingshausen Sea, Antarctica.
Science, 135(3504):671-672.

bottom sediments
Duane, David B., and Edward P. Meisburger, 1969
Geomorphology and sediments of the nearshore continental shelf Miami to Palm Beach, Florida.
Techn. Memo. Coast. Engng Res. Cent. U.S. Army Corp Engrs. 29: 47+ pp.

bottom sediments
Düing, Walter, 1965.
Strömungsverhältnisse im Golf von Neapel.
Pubbl. Staz. Zool., Napoli, 34:256-316.

bottom sediments
Dunbar, M.J., 1951.
Eastern Arctic waters. Fish. Res. Bd., Canada, Bull. No. 88:131 pp., 32 textfigs.

bottom sediments
Duncan, John R., and L.D. Kulm, 1970.
Mineralogy, provenance, and dispersal history of late Quaternary deep-sea sands in Cascadia Basin and Blanco Fracture Zone off Oregon.
J. sedim. Petrol. 40(3): 874-887.

bottom sediments
Duplaix, S., and A. Cailleux, 1957.
Sur quelques sables des fonds de 1850 à 4270 m. de la Méditerranée. C.R. Acad. Sci., Paris, 244 (1):104-106.

bottom sediments
Duursma, E.K., 1970.
Organic chelation of ^{60}Co and ^{65}Zn by leucine in relation to sorption by sediments. Occ. Pap. Inst. mar. Sci., Alaska, 1: 387-397.

Bottom sediment
Dyer, K.R. 1970
The distribution and movement of sediment in the Solent, southern England.
Mar. Geol. 11(3): 175-187.

bottom sediments
Dyer, K.R., 1970.
Sediment distribution in Christchurch Bay, S. England. J. mar. biol. Ass., U.K., 50(3): 673-682.

E

bottom sediments
*Earley, Charles F., and H.G. Goodell, 1968.
The sediments of Card Sound, Florida.
J. sedim. Petr., 38(4):985-999.

bottom sediments
Edge, Billy L., and Paul G. Mayer, 1971
Dynamic structure-soil-wave model for deep water.
J. WatWays Harb. Div. Am. Soc. civ. Engrs. 97(WW1):167-184

bottom sediments
Edgerton, Harold E., 1965.
Sub-bottom penetration in Boston Harbor, 2.
J. Geophys. Res., 70(12):2931-2933.

bottom sediments
Einsele, Gerhard, und Friedrich Werner, 1968.
Zusammensetzung, Gefüge und mechanische Eigenschaften rezenter Sedimente vom Nildelta, Roten Meer und Golf von Aden.
Meteor Forschungsergebn (C) 1: 21-42

bottom sediments
El Wakeel, Saad K., 1964.
Recent bottom sediments from the neighborhood of Alexandria, Egypt.
Marine Geology, 2(1/2):137-146.

bottom sediments
Emelianov, E.M., 1961.
[Recent data on sediments of the Mediterranean.]
Doklady Akad. Nauk, SSSR, 137(6):1437-1440.

Emery, K.O.

bottom sediments
Emery, K.O. 1971.
Bottom sediment map of Malacca Strait.
Techn. Bull. Econ. Comm. Asia Far East 4:149-152.

Bottom sediments
Emery, K.O., 1969.
Distribution pattern of sediments on the continental shelves of western Indonesia.
Technical Bull. Econ. Comm Asia Far East, Comm. Coordination Joint Prospect. Mineral Res. Asian Offshore Areas, Geol. Surv., Japan, 2:79-82.

bottom sediments
Emery, K.O., 1965.
Geology of the continental margin off eastern United States.
In: Submarine Geology and Geophysics, Colston Papers, W.F. Whittard and R. Bradshaw, editors, Butterworth's, London, 1-17.

bottom sediments
Emery, K.O., 1956.
Sediments and water of Persian Guld.
Bull. Amer. Assoc. Petr. Geol., 40(10):2354-2383.

bottom sediments
Emery, K.O., 1954.
Some characteristics of southern California sediments. J. Sed. Petr. 24(1):50-59.

bottom sediments
Emery, K.O., 1953.
Some surface features of marine sediments made by animals. J. Sed. Petr. 23(3):202-204, 4 photos.

bottom sediments
Emery, K.O., 1952.
Continental shelf sediments of southern California. Bull. G.S.A. 63(11):1105-1108, 1 fold-in.

bottom sediments
EMERY, K.O., & Y.K. BENTOR, 1960
The continental shelf of Israel.
Israel, Ministry of Agric., Div. Fish. Sea Fish. Res. Sta., Haifa, Bull. No. 28: 25-40
is same publication as
Israel, Ministry of Development, Geol. Surv. Bull., No. 26: 25 - 40.

bottom sediments
Emery, K.O., W.S. Butcher, H.R. Gould, and F.P. Shepard, 1952.
Submarine geology off San Diego, California.
J. Geol. 60(6):511-548, 15 textfigs.

bottom sediments
Emery, K.O., D.S. Gorsline, E. Uchupi and R.D. Terry, 1957.
Sediments of three bays of Baja California: Sebastian Viscaino, San Critobal and Todos Santos. J. Sed. Petr., 27(2):95-115.

bottom sediments
Emery, K.O., Bruce C. Heezen and T.D. Allan, 1966.
Bathymetry of the eastern Mediterranean Sea.
Deep-Sea Res., 13(2):173-192.

bottom sediments
Emery, K.O., & J. Hülsemann, 1961
Stratification in recent sediments of Santa Barbara Basin as controlled by organisms and water character.
J. Geology, 69(3): 279-290.

bottom sediments
Emery, K.O., Arthur S. Merrill and J. V.A. Trumbull, 1965.
Geology and biology of the sea floor as deduced from simultaneous photographs and samples.
Limnology and Oceanography, 10(1):1-21.

bottom sediments
Emery, K.O. and John D. Milliman, 1970.
Quaternary sediments of the Atlantic Continental Shelf of the United States. Quaternaria 12: 3-18.

bottom sediments
Emery, K.O., and Hiroshi Niino, 1963.
Sediments of the Gulf of Thailand and adjacent continental shelf.
Bull.Geol.Soc.,Am.,74:451-554.

bottom sediments
Emery, K.O., and H. Niino, 1961.
Sediments of shallow portions of East China Sea and South China Sea. (Abstract).
Proc. Ninth Pacific Sci. Congr., Pacific Sci. Assoc., 1957, Geol. and Geophys., 12:40.

bottom sediments
Emery, K.O., and S.C. Rittenberg, 1952.
Early diagenesis of California basin sediments in relation to origin of oil. Bull Amer. Soc. Petr. Geol. 36(5):735-806, 30 textfigs.

bottom sediments
Emery, K.O., and S.C. Rittenberg, 1951.
Basin sediments off southern California.(Abstract
Bull. G.S.A. 62:1501-1502.

bottom sediments
Emery, K.O. and David A. Ross 1968.
Topography and sediments of a small area of the continental slope south of Martha's Vineyard.
Deep-Sea Res. 15(4): 415-422.

bottom sediments
Enequist, P., 1949.
Studies on the soft-bottom amphipods of the Skagerak. Zool. Bidrag, Uppsala, 28:297-492, 37 textfigs., 5 charts.

Ericson, David B

bottom sediments
Ericson, D.B., M. Ewing, and B.C. Heezen, 1952.
Turbidity currents and sediments in the North Atlantic. Bull. Amer. Assoc. Petr. Geol. 36(3): 489-511, 4 textfigs.

bottom sediments
Ericson, D.B., and G. Wollin, 1956.
Correlation of six cores from the equatorial Atlantic and the Caribbean. Deep-Sea Res. 3(2): 104-125.

bottom sediments
Escande, L., 1949.
Ondulations de sable des modèles réduits et dunes du désert. C.R. Acad. Sci., Paris, 229: 613-615.

bottom sediments
Escande, L., 1949.
Similitude des ondulations de sable des modeles reduits et des dunes du desert. C.R. Acad., Pari 229:701-702.

bottom sediments
Etchichury, Maria C., and Joaquin R. Remiro, 1960
Muestras de fondo de la plataforma continental comprendida entre los paralelos 34 y 36 30 de latitud sur y los meridianos 53 10 y 56 30 de longitud oeste.
Revista Argentino Ciencias Nat. "Bernardino Rivadavia", Ciencias Geol., 6(4):199-263.

bottom sediments
Evans G. 1966.
The recent sedimentary facies of the Persian Gulf region.
Phil. Trans. R. Soc. (A) 259(1099): 291-

bottom sediments
Evans, R.D., A. Kip and E. Moberg, 1938.
The radium and radon content of Pacific Ocean waters and sediments. Amer. J. Sci., (5), 36:241-259.
, life

bottom sediments
Evison, F.F., 1963.
Thickness of the earth's crust in Antarctica and the surrounding oceans: a reply.
Geophysical J., London, 7(4):469-476.

bottom sediments
Evison, F.F., C.E. Ingham, R.H. Orr and J.H. Le Fort, 1960
Thickness of the earth's crust in Antarctica and the surrounding oceans. Geophys. J., R. Astron. Soc., 3(3): 289-306.

Ewing, John I.

Ewing, John I. 1968?
History of the ocean basins as recorded in the sediments
Selected Papers from the Governor's Conference on Oceanography, October 11 and 12, 1967, at the Rockefeller University, New York, N.Y.: 144-155.

bottom sediments
Ewing, J., J. Antoine and M. Ewing, 1960
Geophysical measurements in the western Caribbean Sea and in the Gulf of Mexico.
J. Geophys. Res., 65(12): 4087-4126.

bottom sediments
Ewing, John I., Robert E. Houtz and William J. Ludwig, 1970.
Sediment distribution in the Coral Sea. J. Geophys. Res., 75(11): 1963-1972.

bottom sediments
Ewing, John I., and J.E. Nafe, 1963.
5. The unconsolidated sediments.
In: The Sea, M.N. Hill, Editor, Interscience Publishers, 3:73-84.

bottom sediments
Ewing, John I., J. Lamar Worzel and Maurice Ewing, 1962.
Sediments and oceanic structural hisotry of the Gulf of Mexico.
J. Geophys. Res., 67(6):2509-2527.

bottom sediments
Ewing, John, J.L. Worzel, Maurice Ewing and Charles Windisch, 1966.
Ages of Horizon A and the oldest Atlantic sediments: coring at an outcrop of Horizon A establishes it as a buried abyssal plain of Upper Cretaceous age.
Science, 154(3753):1125-1132.

Ewing, Maurice

bottom sediments
Ewing, Maurice and John Ewing, 1965.
The sediments of the Argentine Basin.
Anais Acad. bras. Cienc., 37 (Supl):31-61.

bottom sediments
Ewing, Maurice and John Ewing, 1964.
Continental drift and ocean-bottom sediments. (Abstract).
Geol. Soc., Amer., Special Paper, No. 76:55-56.

bottom sediments
Ewing, Maurice and John Ewing, 1964
Distribution of oceanic sediments.
In: Studies in Oceanography dedicated to Professor Hidaka in commemoration of his sixtieth birthday, 525-537.

bottom sediments
Ewing, Maurice, John I. Ewing, and Manik Talwani, 1963.
Sediment distribution in the oceans: the Mid-Atlantic Ridge.
Bull. Geol. Soc., Amer., 75(1):17-36.

bottom sediments
Ewing, Maurice, and John Ewing, 1963
Sediments at proposed LOCO drilling sites.
J. Geophys. Res., 68(1):251-256.

bottom sediments
Ewing, Maurice, John I. Ewing and Manik Talwani, 1964.
Sediment distribution in the oceans: the Mid-Atlantic Ridge.
Geol. Soc., Amer., Bull., 75(1):17-36.

bottom sediments
*Ewing, M., R. Houtz, and J. Ewing, 1969.
South Pacific sediment distribution. J. geophys. Res., 74(10): 2477-2493.

bottom sediments
Ewing, Maurice, Xavier Le Pichon and John Ewing, 1966.
Crustal structure of the mid-ocean ridges.
4. Sediment distribution in the South Atlantic Ocean and the Cenozoic history of the Mid-Atlantic Ridge.
J. Geophys. Res., 71(6):1611-1636.

bottom sediments
Ewing, Maurice, William J. Ludwig and John I. Ewing, 1964
Sediment distribution in the oceans: The Argentine Basin.
J. Geophys. Res., 69(10):2003-2032.

F

submarine geology
Fairbridge, R.W., 1947
Coarse sediments on the edge of the continental shelf. Amer. J. Sci.245:146-153.

bottom sediments
Fan, Pow-Foong, and Ross R. Grunwald 1971.
Sediment distribution in the Hawaiian Archipelago
Pacific Sci. 25(4): 454-488

bottom sediments
Farleigh, D.R.P., and Sir Claude Inglis, 1962
The behaviour and control of the Karnafuli Estuary, East Pakistan.
Proc. Conf. on Civil Engineering Problems Overseas, Inst. Civil Eng., London, 317-337.

bottom sediments
Fernandez del Riego, A., 1951.
Determinación del carbónico de los fondos de la Ria de Vigo, calculo del carbonato disuelto y consequencias geobiologicas. Bol. Inst. Español Ocean. No. 44:17 pp.

bottom sediments
Feldhausen, Peter H., 1970.
Ordination of sediments from the Cape Hatteras continental margin.
Mathem. Geol. 2(2): 113-129

bottom sediments
Fierro, Giuliano, 1969.
Répartition des sédiments dans la région des Bouches de Bonifacio
Rapp. P.-v. Reun. Commn int. Explor. scient. Mer Mediterr., 19(4): 645-647.

bottom sediments
Fierro, Giuliano, Franco Miglietta e Giovani Battista Piacentino, 1969.
Biologia delle Secche della Meloria. III. I sedimenti superficiali delle Secche e delle aree limitrofe dalla foce dell'Arno a Punta Fortullino.
Boll. Pesca Piscic. Idrobiol. 24 (2): 115-149.

bottom sediments
Fisher, R.L., G.L. Johnson and B.C. Heezen 1967.
Mascarene Plateau, western Indian Ocean.
Bull. geol. Soc. Am. 78: 1247-1266.

bottom sediments
Fisher, Robert L., 1961.
Middle America Trench: topography and structure. Bull. Geol. Soc., Amer., 72(5):703-720.

bottom sediments
*Flemming, N.C., and A.H. Stride, 1967.
Basal sand and gravel patches with separate indications of tidal current and storm-wave paths, near Plymouth.
J. mar. biol. Ass., U.K., 47(2):433-444.

bottom sediments (data)
Flint, J.M., 1905.
A contribution to the oceanography of the Pacific. Bull. U.S. Nat. Mus. 55:1-61.

bottom sediments
Forest, Jacques, 1959.
Campagne de la Calypso dans le golfe de Guinée et aux îles Principe, São Tomé, Annobon (1956). Ann. Inst. Océanogr., Monaco (Rés. Sci. Camp. "Calypso", 4):3-35.

bottom sediments
France, Laboratoire National d'Hydraulique, 1958.
Estuaire de la Gironde. Port autonome de Bordeaux Étude sur modèle réduit de la région du Bec d'Ambès et des Îles. Ser. A, Mars, 1958:90 pp.

bottom sediments
Francis-Boeuf and V. Romanovsky, 1946
Les dépôts fluviomarins entre le Rio Pongo et la Rio Nunez (Côte de la Guinée francaise). C.R.S. de la Soc. Géologique de France, 18 Nov. 1946, pp. 291-292.

bottom sediments
Francis-Boeuf, C., and V. Romanovsky, 1946.
L'envasement du port de Honfleur. Bull. Inst. Océan., Monaco, No. 901:8 pp., 5 figs.

bottom sediments
Bredrikson, K., 1956.
Cosmic spherules in deep-sea sediments. Nature 177(4497):32-33.

bottom sediments
Fridman, Melle Ruth, 1960.
Les sediments recents de l'Anse de l'Aiguillon et de ses limites marines à l'ouest.
Cahiers Océanogr., C.C.O.E.C., 12(4):268-274.

bottom sediments
Fuller, A.O., 1961
Size distribution characteristics of shallow marine sands from the Cape of Good Hope, South Africa.
J. Sediment. Petrol., 31(2): 256-261.

bottom sediments
Fukai, R., and S. Yamazaki, 1957.
[On the distribution of bottom sediments in Katsuragi Bay, Mie Prefecture.]
Bull. Tokai Reg. Fish. Res. Lab., No. 17:73-82.

bottom sediments
Fuller, A.O., 1962.
Systematic fractionation of sand in the shallow marine and beach environments off the South African coast.
J. Sed. Petr., 32(3):602-606.

G

bottom sediments
Galliher, E.W., 1932.
Sediments of Monterey Bay, California. Mining in California, Jan. 1932:42-79, 17 textfigs., 1 chart.

bottom sediments
Galtsoff, P.S., ed., 1954.
Gulf of Mexico, its origin, waters and marine life. Fish. Bull. Fish and Wildlife Service, 55:1-604, 74 textfigs.

bottom sediments
Gamazhenkov, V.S., 1962
The dynamics of coastal alluvium in the Gagra Sound.
Okeanologiia, Akad. Nauk, SSSR, 2(2):284-292.

bottom sediments
Gardner, James V. 1970.
Submarine geology of the western Coral Sea.
Bull. geol. Soc. Am. 81 (9): 2599-2614.

bottom sediments
Garland, C.F., 1952.
A study of water quality in Baltimore Harbor. Chesapeake Biol. Lab. Publ. 96:1-132, figs.

bottom sediments
Gaskell, T.F., and J.C. Swallow, 1953.
Seismic refraction experiments in the Indian Ocean and Mediterranean Sea. Nature 172(4377): 535-537, 2 textfigs.

bottom sediments
Gealy, E.L., 1971.
Saturated bulk density, grain density and porosity of sediment cores from the western equatorial Pacific: Leg 7, Glomar Challenger, Initial Repts. Deep Sea Drill. Proj. 7(2): 1081-1104.

bottom sediments
Gealy, E.L., 1971.
Sound velocity, elastic constants and related properties of marine sediments in the western equatorial Pacific: Leg 7, Glomar Challenger. Initial Repts. Deep Sea Drill. Proj. 7(2): 1105-1160A.

bottom sediments
Gennesseaux, M., 1962.
Travaux du Laboratoire de Géologie Sous-Marine concernant les grands carottages effectués sur le précontinent de la region Niçoise.
In: Océanographie Géologique et Géophysique de la Méditerranée Occidentale, Centre National de la Recherche Scientifique, Villefranche sur Mer, 4 au 8 avril 1961, 177-181.

bottom sediments
Gennesseaux, M., and Mme Le Calvez, 1960
Affleurement sous-marin de vases pliocènes dans la Baie des Anges (Nice). C.R. Acad. Sci. Paris, 251 (19): 2064-2066.

bottom sediments
Germaneau, J., 1968.
Caractères de la sédimentation dans l'estuaire de la Seine.
Bull. Inst. géol. Bassin d'Aquitaine 5: 140-167. (not seen)

Gershanovich, D.E.

bottom sediments
Gershanovich, D.E., 1967.
New data on geomorphology and Recent sediments of the Bering Sea and the Gulf of Alaska.
Marine Geol. 6(4): 281-296.

Gershanovich, D.E., 1965. bottom sediments
Chief results of marine geological surveys in the Bering Sea and in the Gulf of Alaska. (In Russian; English abstract).
Okeanolog. Issled., Rez. Issled. po Programme Mezhd. Geofiz. Goda, Mezhd. Geofiz. Komitet, Presidiume, Akad. Nauk, SSSR, No.13:189-198.

bottom sediments
Gershanovich, D.E., 1965.
Recent bottom sediment strength and sedimentation rate in the Bering Sea. Investigations in line with the programme of the International Geophysical Year, 2. (In Russian).
Trudy, Vses. Nauchno-Issled. Inst. Morsk. Ribn. Choz. i Okeanogr., (VNIRO), 57:261-269.

bottom sediments
Gershanovich, D.E., 1964.
Bottom sediments in the central and eastern Bering Sea. (In Russian).
Trudy, Vses. Nauchno-Issled. Inst. Morsk. Ribn. Choz. i Okeanogr. (VNIRO), 52:31-81.

Also:
Izv. Tichookeansk. Nauchno-Issled. Inst. Morsk. Ribn. Choz. i Okeanogr. (TINRO), 52:31-81.

bottom sediments
Gershanovich, D.E., 1962.
Relief and Recent sediments on the Bering Sea shelf. (In Russian).
Trudy Vses. Nauchno-Issledov. Inst. Morsk. Ribn. Chos. i Okean., VNIRO, 46:164-185.

bottom sediments
Gershanovitch, D.E., 1960
[Observations on bottom sediments in the course of the research cruise of the submarine "Severyanka".]
Biull. Okeanograf. Komissii. Akad. Nauk, SSSR, (6): 37-38.

bottom sediments
Gershanovitch, D. E., 1960.
[Recent shelf deposits in marginal seas of northeast Asia.] Morsk. Geol., Doklady Sovetsk. Geol., 21 Sess., Int. Geol. Congress, 116-122.
(English abstract)

bottom sediments
Gershanovich, D.E., 1958.
[Facies of modern sediments in the north-western part of the Okhotsk Sea.] Doklady Akad. Nauk, SSSR 118(2):355-358.

bottom sediments
Gershanovich, D.E., 1958.
Bottom deposits of sea straits. (In Russian).
Priroda, 47(7):97-
not seen

bottom sediments
Gershanovich, D.E., V.S. Kotenev, and V.N. Novikov, 1964.
Relief and Bottom sediments in the Gulf of Alaska. (In Russian).
Trudy, Vses. Nauchno-Issled. Inst. Morsk. Ribn. Choz. i Okeanogr., 52:83-133.
Also:
Izv. Tichookeansk. Nauchno-Issled. Inst. Morsk. Ribn. Choz. i Okeanogr. (TINRO), 52:83-133.

bottom sediments
Gershanovich, D. E., and A. A. Neiman. 1964.
Bottom sediments and bottom fauna in the East China Sea. (In Russian).
Okeanologiia, Akad. Nauk, SSSR, 4(6):1089-1095.

bottom sediments
*Gibson, Thomas G., and John Schlee, 1967.
Sediments and fossiliferous rocks from the eastern side of the Tongue of the Ocean, Bahamas.
Deep-Sea Res., 14(6):691-702.

bottom sediments
Giermann, Günter, 1961.
Erläuterungen zur bathymetrischen Karte der Strasse von Gibraltar.
Bull. Inst. Océanogr., Monaco, 58(1218A,B):1-28, 5 charts.

bottom sediments
Gilluly, James, 1964.
Atlantic sediments, erosion rates and the evolution of the continental shelf: some speculations.
Geol. Soc., Amer., Bull., 75(6):483-492.

bottom sediments
Gilvary, J.J., 1961.
The origin of ocean basins and continents.
Nature, 190(4781):1048-1053.

bottom sediments
Giorgetti, F., 1967.
Nota sui sedimenti costieri lungo la falesia a nord e nord ovest di Trieste.
Boll. Soc. adriatica Sci., Trieste, 55:12-17.
Reprinted: Comm. Studio Oceano raf. Limnol., Programma Ricerca Risorse mar. Fondo mar., (B)(10). 1968

bottom sediments
*Giraud, Bernard, Bruce C. Heezen, Wladimir, D. Nesteroff, et Germain Sabatier, 1968.
Les feldspaths dans les sédiments marins polaires
C.r.hebd.Séanc.,Acad.Sci.,Paris,(D)267(25):2099-2100.

bottom sediments
Giresse, Pierre 1969.
Carte sédimentologique des fonds sous-marins du delta de l'Ogoove.
Cah. océanogr. 21(10): 965-994.

bottom sediments
*Giresse, Pierre, Louis Dangeard et Pierre Hommeril, 1967.
Sur la présence de dépôts marins d'age dunkerquien le long des côtes normandes et sus quelques autres dépôts flandriens.
C.r.hebd.Séanc.Acad.Sci.,Paris,(1),265(25):1887-1890.

bottom sediments
Glangeard, L., 1939.
Rôle de la suspension tourbillonnaire et du roulement sur le fond dans la formation des sédiments actuels de l'estuaire girondin entre Bourdeaux et la Pointe de Grave. C.R. Acad. Sci., Paris, 208:1595-1597, 2 textfigs.

bottom sediments
Glangeaud, Louis, Jean Alinat, Christian Agarate, Olivier Leenhardt et Guy Pautot 1967
Les phénomènes ponto-plio-quaternaires dans la Méditerranée occidentale d'après les données de Géomède 1.
C.r. hebd. Séanc. Acad. Sci. Paris (D) 264(2): 208-211.

bottom sediments
Glangeaud, Louis, et Jean-Pierre Rehault, 1968.
Evolution ponto-plio-quaternaire du golfe de Genês.
C.r.hebd.Séanc.,Acad.Sci.Paris,(D)266(2):60-63.

bottom sediments
Glémarec, Michel, 1969.
Le plateau continental nord-Gascogne et la grande Vasière: étude bionomique.
Revue Trav. Inst. Pêches marit., 33(3): 301-310

bottom sediments
*Goldberg, Edward D. and John J. Griffin, 1970.
The sediments of the northern Indian Ocean.
Deep-Sea Res., 17(3): 513-537.

bottom sediments
Gomez de Llarena, J., 1955.
Algunos datos sobre los sedimentos recogidos por el "Xauen" en su campaña del otoño de 1952 (X-528). Bol. Inst. Español Ocean 69:1-9, 4 pls.

bottom sediments
Gómez de Llarena, J., 1952.
Observaciones sobre los sedimentos de las costas de Galicia (Campañas del "Xauen" on 1949 y 1950). Appendice: Analisis mineralogico de algunas muestras. By. J. Pérez Mateos. Bol. Inst. Español Ocean. No. 52:24 pp., 6 pls., 1 textfig.

bottom sediments
Gómez del Llarena, J., 1950.
Observaciones sobre los sedimentos recogidos entre los Cabos Juby y Bojador. Bol. Inst. Español Ocean. No. 21:29 pp.

bottom sediments
Gómez, de Llarena, J., 1950.
Exploracion oceanografica del Africa occidental. Observaciones sobre los sedimentos recogidos entre los Cabos Juby y Bojador. Bol. Inst. Español Ocean., No. 29:23 pp., 8 pls.

bottom sediments
Goodell, H.G., 1964?
Marine geology of the Drake Passage, Scotia Sea and South Sandwich Trench.
USNS Eltanin Mar. Geol. Cruises, 1-8, Florida State Univ., Sedimentol. Lab., 263 pp. (mimeographed).

bottom sediments
Goodell, H.G., N.D. Watkins, T.T. Mather and S. Koster 1968.
The Antarctic glacial history recorded in sediments of the Southern Ocean.
Palaeogr. Palaeoclimatol. Palaeoecol. 5(1): 41-62.

bottom sediments
Gor'kova, I.M., 1958
[Structure formation in marine sediments.] DAN, SSSR, 123(2): 343-345.
Translation NIOT/27 in MBL.

Gorshkova, T.I.

bottom sediments
Gorshkova, T.I., 1960.
[Sediments of the Baltic Sea.]
Trudy Vses. Nauchno-Issled. Inst. Morsk. Ribnogo Chozia. i Oceanogr., (VNIRO), 42:19-36.

bottom sediments
Gorshkova, T.I., and I.V. Solyankin, 1961
The grounds, water masses and regime of the Norwegian-Greenland Basin and adjacent areas. Contribution to Special IGY Meeting, 1959.
Cons. Perm. Int. Expl. Mer, Rapp. Proc. Verb. 149:44-45.

Gorsline, D.S.

bottom sediments
Gorsline, D.S., 1967.
Contrasts in coastal bay sediments on the Gulf and Pacific coasts.
IN: Estuaries, G.H. Lauff, editor, Publs Am. Ass. Advmt Sci., 83:219-225.

bottom sediments
Gorsline, D.S., 1963.
Bottom sediments of the Atlantic shelf and slope off the southern United States.
J. Geology, 71(4):422-440.

bottom sediments
Gorsline, D. S., 1957.
The relation of bottom sediment type to water motion; Sebastian Viscaino Bay, Baja California, Mexico.
Rev. Géo. Phys. Géol. Dyn. 2nd Ser. 1(2):83-92.

bottom sediments
Gorsline, Donn S., David E. Drake and Peter W. Barnes, 1968.
Holocene sedimentation in Tanner Basin, California continental borderland.
Bull. geol. Soc. Am., 79(6): 659-674

bottom sediments
Gorsline, D.S., and R.A. Stewart, 1962.
Benthic marine exploration of Bahia de San Quintin, Baja California, 1960-61 marine and Quarternary geology.
Pacific Naturalist, 3(8):283-319.

bottom sediments
Got, H., A. Guille, A. Monaco et J. Soyer, 1968.
Carte sédimentologique du plateau continental au large de la Côte Catalane française (P.-O.).
Vie Milieu 19(2-B): 273-290

bottom sediments
Gouleau, Dominique, 1968.
Etude hydrologique et sedimentologique de la Baie de Bourgneuf.
Trav. Lab. Geol. mar., Fac. Sci., Nantes, 187 pp. (multilithed).

bottom sediments
Gouleau, Dominique, et Juan Soriano, 1969.
Étude sédimentologique et géotechnique d'une carotte de la Baie de Bourgneuf.
Cah. océanogr., 21(1):57-70.

bottom sediments
Graindor, Maurice J., 1958.
Sur la présence de gros blocs encastrés dans les sables du golfe normanno-breton.
C. R. Acad. Sci., Paris, 247(25):2402-2404.

bottom sediments
Greenman, N.N., and R.J. LeBlanc, 1956.
Recent marine sediments and environments of northwest Gulf of Mexico.
Bull. Amer. Assoc. Petr. Geol., 40(5):813-847.

bottom sediments
Griffiths, J.C., 1951.
Size versus sorting in some Caribbean sediments.
J. Geol. 59(3):211-243, 13 textfigs.

bottom sediments
Grovel, Alain P. 1970.
Étude d'un estuaire dans son environnement le Blavet maritime et la region de Lorient.
Trav. Lab. Géol. mar. Fac. Sci. Nantes (C.N.R.S. A.O. 48-52): 122 pp. (multilithed).

bottom sediments
Gudkov, M.P., 1956.
Veränderung der Sedimente des nördlichen Kaspischen Meeres. Priroda 1956(1):91-93.

bottom sediments
Guilcher, Andre, 1964.
Present-time trends in the study of Recent marine sediments and in marine physiography.
Marine Geology, 1(1):4-15.

bottom sediments
Gunter, Gordon, and Gordon E. Hall, 1963.
Biological investigations of the St. Lucie Estuary (Florida) in connection with Lake Okeechobee discharges through the St. Lucie Canal.
Gulf Res. Repts., Ocean Springs, Miss., 1(5):189-307.

H

bottom sediments
Hachey, H.B., 1931.
Biological and oceanographic conditions in Hudson Bay. 2. Report on the Hudson Bay Fisheries Expedition of 1930. A. Open water investigations with the S.S. Loubyrne.
Contr. Canadian Biol. Fish., n.s., 6(23):465-471, 1 fig.

bottom sediments
Hales, Anton L., 1970.
II. Regional observations. 9. Eastern Continental Margin of the United States (July 1968). Part 2: A review. In: The sea: ideas and observations on progress in the study of the seas, Arthur E. Maxwell, editor, Wiley-Interscience 4(2): 311-320.

Hamilton, Edwin L.

bottom sediments
Hamilton, E.L., 1960.
Ocean basin ages and amounts of original sediments.
J. Sed. Petr., 30(3):370-379.

bottom sediments
Hamilton, E. L., 1959.
Thickness and consolidation of deep-sea sediments
Bull. Geol. Soc. Amer., 70(11):1399-1424.

bottom sediments
Hamilton, E.L., 1956.
Sunken islands of the mid-Pacific mountains.
Mem. G.S.A., 64:97 pp.

bottom sediments
Hamilton, E.L., and H.W. Menard, 1956.
Density and porosity of sea-floor surface sediments off San Diego, California.
Bull. Amer. Assoc. Petrol. Geol., 40(4):754-761.

bottom sediments
Hamilton, N. and A.I. Rees 1970.
Magnetic fabric of sediments from the shelf at La Jolla (California).
Marine Geol. 9(2): M6-M11.

bottom sediments
Hansen, K., 1964.
Lagoon sediments in Greenland.
In: Developments in sedimentology, L.M.J.U. van Straaten, Editor, Elsevier Publishing Co., 1:165-169.

bottom sediments
Hansen, Kaj., 1944.
1. Introduction and the bottom deposits. Investigations of the geography and natural history of the Praestø Fjord, Zealand. Folia Geogr. Danica 3(1):46 pp., 15 textfigs.

bottom sediments
Haranath, P.B.V. and P. Tiruvenganna Rao, 1958
Spectrographic examination of ocean bed sediments off Visakhapatnam coast. Jour. Sci. Ind. Res. Vol. 17B #9. Sept. '58 p. 354-356.

bottom sediments
Harder, P., Ad. S. Jensen, and D. Laursen, 1949.
The marine quaternary sediments in Disko Bugt.
Medd. Om Grønland 149(1): 85 pp., 18 textfigs., 8 pls., 1 table.

bottom sediments
Harrison, W., M.P. Lynch and A.G. Altschaeffl, 1964.
Sediments of lower Chesapeake Bay, with emphasis on mass properties.
J. Sed. Petr., 34(4):727-755.

bottom sediments
Hart, T. John, and Ronald I. Currie, 1960
The Benguela Current.
Discovery Repts., 31: 123-298.

estuarine deposits
Hartley, P.H.T., and G.M. Spooner, 1938
The ecology of the Tamar Estuary.
1. Introduction. JMBA XXII:501-508.

owned by M.S.

bottom sediments
Hartmann, M., H. Lange, E. Seibold und E. Walger 1971.
Oberflächensedimente im Persischen Golf und Golf von Oman. I. Geologisch-hydrologischer Engebnisse.
Meteor-Forsch-Engebn. (C) 4:1-76

bottom sediments
Hata, K., and H. Mii, 1955.
Observations on the tidal flats in Uranouchi Bay, Kochi Prefecture, Shikoku, Japan.
Rec. Ocean. Wks., Japan, 2(1):168-182, 2 pls.

bottom deposits
Haughton, S.H., 1956.
Phosphatic-glauconitic deposits off the west coast of South Africa.
Annals of South Africa Mus., 42(4)(16):329-334.

bottom sediments
Hawkins, James W., Jr., and John T. Whetten, 1969
Graywacke matrix minerals: hydrothermal reactions with Columbia River sediments.
Science, 166(3907): 868-870.

bottom sediments
Hayes, Dennis E., 1966.
A geophysical investigation of the Peru-Chile Trench.
Marine Geol., 4(5):309-351.

Hayes F.R.

bottom sediments
Hayes Miles O., 1967.
Relationship between coastal climate and bottom sediment type on the inner continental shelf.
Mar. Geol., 5(2):111-132.

bottom sediments
Hays, James D., 1967.
Quaternary sediments of the Antarctic Ocean.
Progress in Oceanography, 4:117-131.

bottom sediments
Hays, James D., Tsunemasa Saito, Neil D. Opdyke and Lloyd H. Burckle, 1969.
Pliocene-Pleistocene sediments of the equatorial Pacific: their paleomagnetic, biostratigraphic, and climatic record.
Bull. geol. Soc. Am., 80(8): 1481-1514.

Heezen, Bruce C.

bottom sediments
Heezen, B.C., E.T. Bunce, J.B. Hersey and M. Tharp, 1964.
Chain and Romanche Fracture zones.
Deep-Sea Research, 11(1):11-33.

bottom sediments
Heezen, Bruce C., Bill Glass and H. W. Menard, 1966.
The Manihiki Plateau.
Deep-Sea Research, 13(3):445-458.

bottom sediments
Heezen, Bruce C., and G. Leonard Johnson, 1969
Mediterranean undercurrent and microphysiography west of Gibraltar.
Bull. Inst. océanogr., Monaco, 67 (1380): 95 pp.

bottom sediments
*Henry, Vernon J., Jr., and John H. Hoyt, 1968.
Quaternary paralic and shelf sediments of Georgia.
SEast. Geol., Duke Univ., 9(4):195-214.

bottom sediments
Herman, Yvonne, 1963
Cretaceous, Paleocene, and Pleistocene sediments from the Indian Ocean.
Science, 140(3573):1316-1317.

bottom sediments
*Hertweck, Günther, und Hans-Erich Reineck, 1969.
Sedimentologie der Meeresbodensenke NW von Helgoland (Nordsee).
Senckenberg. maritima 50:153-164.

Bottom Sediments
*Hickel, W., 1969.
Sedimentbeschaffenheit und Bakteriengehalt im Sediment eines zukünftigen Verklappungsgebietes von Industrieabwässern nordwestlich Helgolands. Helgoländer wiss. Meeresunters, 19(1): 1-20

bottom sediments
Hinschberger, Félix, 1964.
La répartition des fonds sousmarins dans le vestibule du Goulet de Brest.
C.R. Acad. Sci., Paris, 258(26):6497-6499.

bottom sediments
Hinschberger, Félix, Anne Saint-Requier et Anne Toulemont, 1966.
Recherches sédimentologiques et écologiques sur les fonds sous-marins dans les parages de la chaussée de Sein (Finistère).
Rev. Trav. Inst. Peches marit., 31(4):425-452.

bottom sediments
Hishida, Kozo, 1962
On the drifting sands in Tsuiyama Harbour.
J. Oceanogr. Soc., Japan, 18(1):16-21.

bottom sediments
Hishida, K., 1950.
Hydrography of the mouth of Kumihama Bay.
Ocean. Mag., Tokyo, 2(2):67-68, 2 textfigs.

bottom sediments
Hoang, Ngoc, C., 1957.
Les sédiments des portions peu profondes des mers de Chine orientale and meridionale. Rev. Geog. Phys........ Geol. Dynam 2 Sér. 1(4):263-267.

bottom sediments
Hoeg, S., und G. Schellenberger 1971.
Messung der wellenbedingten Sedimentaufwirbelung in der küstennahen Zone.
Acta hydrophysica 16(2):45-82

bottom sediments
*Hollister, Charles D., and Bruce C. Heezen, 1967.
The floor of the Bellingshausen Sea.
In: Deep-sea photography, J.B. Hersey, editor, Johns Hopkins Oceanogr. Studies, 3:177-189.

bottom sediments
Holtedahl, H., 1959.
Geology and paleontology of Norwegian Sea bottom cover.
J. Sedimentary Petrology, 29(1):16-29.

bottom sediments
Holtedahl, H., 1956.
On the Norwegian continental terrace, primarily outside Møre-Romsdal: its geomorphology and sediments. With contributions on the Quarternary geology of the adjacent land and on the bottom deposits of the Norwegian Sea.
Aarbok, Univ. Bergen, Natur Rekke, 14:1-209.

bottom sediments
Holtedahl, H., 1952.
A study of the topography and the sediments of the continental slope west of Møre, W. Norway.
Univ. i Bergen, Årbok 1950, Naturvidensk. rekke, No. 5:1-58, 13 textfigs.

bottom sediments
Hommeril, P., 1965.
Répartition des sédiments sous-marins autour desîles anglo-normandes.
Comptes Rendus, Acad. Sci., Paris, 260(1):231-234.

bottom sediments
Hommeril, P., and M. Rioult, 1965.
Étude de la fixation des sédiments meubles par deux algues marines: Rhodothamniella floridula (Dillwyn) J. Feldm. et Microcoleus chtonoplastes Thur.
Marine geology, Elseirer Publ. Co., 3(1/2):131-155.

Bottom Sediments
Hommeril, P., and M. Rioult, 1962.
Phenomènes d'erosion et de sedimentation marines entre Sainte-Honorine-des-Pertes et Port-en-Bessin (Calvados). Role de Rhodothamniella floridula dans la retenue des sediments fins.
Cahiers Océanogr., C.C.O.E.C., 14(1):25-45.

bottom sediments
Hood, Donald W., editor 1971
Impingement of man on the oceans.
Wiley-Interscience, 738 pp.

bottom sediments
*Horikawa, Kiyoshi and Akira Watanabe, 1970.
Turbulence and sediment concentration due to wave. Coastal Engng Japan 13: 15-24.

bottom sediments
Horrer, Paul L., 1962
Oceanographic studies during Operation "Wigwam". Physical oceanography of the test area.
Limnol. and Oceanogr., Suppl. to Vol. 7: vii-xxvi.

bottom sediments
Hosino, M., and Y. Momose, 1953.
[On bottom sediments adjacent seas to North Hokkaido.] Publ. 981, Hydrogr. Bull., Tokyo, Spec. Number, No. 12:43-53, 4 figs.

Bottom sediments
Hoshino, Michihei, 1970.
The sediments of the continental shelf.
Kaiyokagaku (Mar. Sci. Mon.) 2(1). Also in: Coll. Repr. Coll. mar. Sci. Techn. Tokai Univ. 4: 287-292. (In Japanese; English abstract)

bottom sediments
*Houbolt, J.J.H.C., 1968.
Recent sediments in the Southern Bight of the North Sea.
Geologie Mijnb. 47(4):245-273.

bottom sediments
Houbolt, J.J.H.C., 1957.
Surface sediments of the Persian Gulf near the Qatar Peninsula. Mouton & Co., Den Haag, 113 pp., 31 pls., fold-ins.

bottom sediments
Hough, J.L., 1956.
Sediment distribution in the southern oceans around Antarctica. J. Sed. Petr., 26(4):301-306.

bottom sediments
Hough, J. L., 1950.
Pleistocene lithology of Antarctic ocean-bottom sediments. J. Geol. 58:254-260.

bottom sediments
Hough, J. L., 1948.
Pleistocene deposits of the Ross Sea and the southeastern Pacific Ocean. (Abstract). Bull. G.S.A. 59:1331.

submarine geology
Hough, J. L., 1942
Sediments of Cape Cod Bay, Massachusetts, J. Sed. Petrology, 12(1):10-30; textfigs. 1-9; tables 1-4.

bottom sediments
Hough, J. L., 1940.
Sediments of Buzzards Bay, Massachusetts.
J. Sed. Petr., 10(1):19-32, Figs. 1-4, Tables 1-3.

bottom sediments
Houtz, R.E., and J.I. Ewing, 1963
Detailed sedimentary velocities from seismic refraction profiles in the Western North Atlantic.
J. Geophys. Res., 68(18):5233-5258.

bottom sediments
Hoyt, John H., B.L. Oostdam and David D. Smith, 1969.
Offshore sediments and valleys of the Orange River (South and SouthWest Africa).
Marine Geol. 7(1). 64-84.

bottom sediments
Huang, Ter-Chien, and H.G. Goodell 1967
Sediments of Charlotte Harbor, southwestern Florida.
J. sedim. Petrol. 37(2):449-474

bottom sediments
Hubert, J.F., 1962
Dispersal patterns of Pleistocene sands on the North Atlantic deep-sea floor.
Science, 136(3514):383-384.

bottom sediments
Hülsemann, J., and K.O. Emery, 1961.
Stratification in Recent sediments of Santa Barbara as controlled by organisms and water character.
J. Geol., 69(3):279-290.

bottom sediments
Hunkins, Kenneth, 1959.
The floor of the Arctic Ocean.
Trans. A.G.U., 40(2):159-162.

bottom sediments
Hunkins, Kenneth L., Maurice Ewing, Bruce C. Heezen and Robert J. Menzies, 1960.
Biological and geological observations of the first photographs of the Arctic Ocean deep-sea floor.
Limnol. & Oceanogr., 5(2):154-161.

bottom sediments
Hunter, W., and D.W. Parkin, 1960
Cosmic dust in recent deep-sea sediments.
Proc. Roy. Soc., London, Ser. A., 255(1282): 382-397.

bottom sediments
*Hyne, Norman J., and H. Grant Goodell, 1967.
Origin of the sediments and the submarine geomorphology of the inner continental shelf off Choctawhatchee Bay, Florida.
Marine Geol., 5(4):299-313.

bottom sediments
Iijima, Azuma, 1960.
The bottom sediments of the Japan and Kuril Trenches collected by the Ryofu Maru during the Japanese Deep Sea Expeditions in 1959. Repts. of JEDS, 1:225-232.
Also in Oceanogr. Mag., 11(2):225-232.

bottom sediments
Iijima, Azuma, and Hideo Kagami, 1962
Cainozoic tectonic development of the continental slope, northeast of Japan. (In Japanese English abstract).
Repts. JEDS, Deep-Sea Res., Comm., Japan Soc. Promotion of Science, 3:561-577.
JEDS Contrib. No. 26.

bottom sediments
Il'in, A.V., and I.I. Shurko 1968.
Some peculiar features of distribution of psephytic material in sediments of the North Atlantic. (In Russian).
Dokl. Akad. Nauk SSSR 179(2): 447-450

bottom sediments
*Ingram, Roy L., 1968.
Vertical profiles of modern sediments along the North Carolina coast.
SEast, Geol., Duke Univ., 9(4):237-244.

bottom sediments,
Ingram, Roy L., 1965.
Facies maps based on the megascopic examination of modern sediments.
J. sed. Petr., 35(3):619-625.

bottom sediments
Indutkina, A.I., 1962
The magnitude of marine sediments in the eastern sector of the Ross Sea in the Antarcti
Okeanologiia, Akad. Nauk, SSSR, 2(2):379.

bottom sediments (shallow)
Inman, D.L., 1953.
Areal and seasonal variations in beach and nears shore sediments at La Jolla, California.
B.E.B. Tech. Memo. No. 39:82 pp., 27 figs., numerous tables.

bottom sediments
Inman, D.L., 1952.
Measures for describing the size distribution of sediments. J. Sed. Petr. 22(3):125-145, Figs. 1-9.

sedimentation
Inman, D. L., 1949.
Sorting of sediments in the light of fluid mechanics. (Abstract). Bull. G.S.A. 60(2, pt. 2) 1940.

sedimentation
Inman, D. L., 1949.
Sorting of sediments in the light of fluid mechanics. J. Sed. Petr. 19:51-70.

bottom sediments
Ionin, A.S., and F.A. Scherbakov, 1961.
The lamination of the near-coastal marine sediments in the southern part of the Black Sea.
Okeanologiia, Akad. Nauk, SSSR, 1(5):866-870.

J

bottom sediments
Jacobsen, N. Kingo, 1961.
Iagttagelser fra Saadil Fjord området. En foreløbig meddelse.
Geografisk Tidsskr., 60:54-73.

bottom sediments
Jakobsen, A.B., 1962.
Vadehavets sedimentomsaetning belyst ved kvantitative målinger.
Medd. Skalling-Lab., 17:87-103.
Reprinted from:
Geogr. Tidsskr., 60:87-103.

bottom sediments
Jakobsen, B., 1961.
Vadehavets sediment om saetning belyst ved kvantitative målinger.
Geografisk Tidsskr., 60:87-103.

bottom sediments (data)
Japan, Central Meteorological Observatory, 1951.
Table 11. Oceanographical observations taken in Nagasaki Harbor: physical and chemical data for stations occupied by R.S.S. "Asakaze-maru".
Res. Mar. Met. Ocean. Obs., Jan.-June 1949, No. 5:70-74.

bottom sediments (data)
Japan, Central Meteorological Observatory, 1951.
Table 9. Oceanographical observations taken in the Goto-nada and the Tsushima Straits. (A) Physical and chemical data for stations occupied by the fishing boat "Daikoku-maru": (B) ------- R.S.S. "Umikaze-maru".
Res. Mar. Met. Ocean. Obs., Jan.-June, 1949, No. 5:61-67.

bottom sediments (data)
Japan, Central Meteorological Observatory, 1951.
Table 8. Oceanographical observations taken in the Ariake Sea; physical and chemical data for stations occupied by R.S.S. "Umikaze-maru".
Res. Mar. Met. Ocean. Obs., Jan.-June, 1949, No. 5:50-60.

bottom sediments
Japan, Japanese Hydrographic Office, 1962
[Results of analyses for the bottom sediments lately dredged.]
Hydrogr. Bull., Tokyo, (Publ. No. 981,) No. 70: 39-48.

bottom sediments
Japan, Maritime Safety Board, 1958.
[On the bottom configuration and sediments in the adjacent sea of Mogami Bank, Japan Sea.]
Hydrogr. Bull. (Publ. 981), No. 55:37-53.

bottom sediments
Japan, Maritime Safety Board, 1957.
[Grain-analysis of sandy sediments (especially about Emery-tube analysis).] Hydrogr. Bull., (Publ. 981) (54):1-7.

bottom sediments
Japan, Maritime Safety Board, 1957.
[Inner part of Sagami Bay. (On the change of coast and sea bottom).] Hydrogr. Bull. (Publ. 981)(45): 39-49.

bottom sediments
Japan, Maritime Safety Board, 1954.
[On the bottom topography and sediments in the north-east part of the Sea of Japan.] Publ. 981, Hydrogr. Bull., Maritime Safety Bd., Tokyo, No. 42 121-137, 10 textfigs.

bottom sediments
Japan, Tokai Regional Fisheries Research Laboratory, 1952.
Report oceanographical investigation (Jan.-Dec. 1950, collectde), No. 74:143 pp.

bottom sediments (data)
Japan, Tokai Regional Fisheries Research Laboratory, 1951.
Report, Oceanographical investigation (Jan. 1943-Dec. 1944 collected), No. 72:105 pp., 1 fig.

bottom sediments
Jarke, Joachim, 1960.
Staubfall auf dem Schwarzen Meer.
Deutsche Hydrogr. Zeits., 13(5):225-229.

bottom sediments
Jarke, J., 1958.
Sedimente und Mikrofaunen im Bereich der Grenzschwelle zweier ozeanischer Räume, dargestellt an einem Schnitt uber den Island-Färöer-Rücken (nordatlantischer Ozean Rosengarten-europäisches Nordmeer).
Geol. Rundschau, 47(1):234-249.

bottom sediments
Jarke, J., 1956.
Eine neue Bodenkarte der südlichen Nordsee.
Deutsche Hydrogr. Zeits. 9(1):1-9.

bottom sediments
Jarke, J., 1956.
Neue Ergebnisse zur Bodenbedeckung der Deutschen Buscht.
Deutsche Hydrogr. Inst., Ozeanogr., 1956(22): 369-375.

bottom sediments
Jarke, J., 1951.
Die sedimentation in den schleswig-holsteinischen Förden. Skr. Naturwiss. Vereins f. Schleswig-Holstein, Karl-Gripp-Festschrift, 25:204-220, 2 textfigs.

bottom sediments
Jérémine, G., 1950
Contribution à la connaissance lithologique de la Grande Salvage. Bull. Inst. Ocean. Monaco No. 969: 10 pp., 1 textfig.

bottom sediments
Jones, D.T., 1954.
The characteristics of some Lower Palaeozoic marine sediments. Proc. R. Soc., London, A, 222(1150):327-332.

bottom sediments
Jordan, G.F., 1962
Redistribution of sediments in Alaskan bays and inlets.
Geogr. Rev., New York, 52(4):548-558.

Originally presented 10th Pacific Science Congress, Hawaii, Aug. 21-Sept. 6, 1961.

bottom sediments
Jouze, A., 1961.
Stratigraphy of sediments and paleogeographical conditions of sediment accumulation in the North Pacific Ocean. (Abstract).
Tenth Pacific Sci. Congr., Honolulu, 21 Aug.-6 Sept., 1961, Abstracts of Symposium Papers, 374-375.

bottom sediments
Just, Jean 1970.
Marine biological investigations of Jørgen Brønlund Fjord, North Greenland: physiographical and bathygraphical survey methods and lists of stations.
Meddr Grønland 184(5): 42 pp.

K

Bottom sediments
Kagami, Hideo, 1961.
Sources of shelf sediments off Sakata, Yamagata, Japan.
Japanese J., Geol. and Geogr., 32(3/4):411-420.

bottom sediments
Kagami, Hideo, 1961.
Specific gravity distribution of sediment and its meaning to the depositional environments on the continental shelf off Kamogawa, Pacific coast of Japan. (Abstract).
Tenth Pacific Sci. Congr., Honolulu, 21 Aug.-6 Sept., 1961, Abstracts of Symposium Papers, 375

bottom sediments
Kagami, Hideo, 1959.
Preliminary report on the shelf sediments off Kamo Yamagata, Japan.
Rec. Ocean. Wks., Japan., n.s., 60-67.

bottom sediments
Kagami, Hideo, and Azuma Iijima, 1960.
On the bottom sediments off Onagawa and Kushiro, the adjacent continental slope of Japan Trench. Repts. of JEDS, 1:233-242.

Also in Oceanogr. Mag. 11(2):233-242.

Bottom sediments
Kane, Henry E., 1966
Sediments of Sabine Lake, the Gulf of Mexico and adjacent water bodies, Texas - Louisiana.
J. Sed. Petr., 36(2):608-619.

bottom sediments
*Karlovac, Jozica, 1967.
Étude de l'écologie de la Sardine, Sardina pilchardus Walb., dans la phase planctonique de sa vie en Adriatique moyenne.
Acta adriat., 13(2):1-109.

bottom sediments
Kashin, Iu. S., 1956. (deceased)
[Investigations of "galechnix" deposits along the "Kavkazhsk" coast of the Black Sea between Gelendzhik and Tuapse. Studies of sea coasts and reservoirs.]
Trudy Okeanogr. Komissii, Akad. Nauk, SSSR, 1:73-76.

bottom sediments
Karlova, O., 1948/1949.
Le Parapenaeus longirostris (H. Lucas) de la haute Adriatique. Acta Adriatica 3(12):14 pp., 1 chart.

bottom sediments
Kenry, R. 1970.
Coastal climate and shelf-bottom sediments: a comment.
Mar. Geol. 8(5): 363-365.

sédimentation
Keller, W. O., and R. Foley, 1949.
Missouri river sediments in river water, ocean water, and sodium oxalate solution.
J. Sed. Petr. 19:78-81.

bottom sediments
Kempf, Marc, Janness Markus Mabesoone and Ivan de Medeiros Tinoco, 1970.
Estudo da plataforma continental na área do Recife (Brasil). I. Generalidades sôbre o fundo. Trabhs oceanogr. Univ. Fed. Pernambuco 9/11: 125-148.

bottom sediments
Kempf, Marc, Paulo Nobrega Coutinho and Jáder Onofre Morais, 1970.
Plataforma continental do norte e nordeste do Brasil. Nota preliminar sôbre a natureza do fundo. Trabhs oceanogr. Univ. Fed. Pernambuco 9/11: 9-26.

submarine geology
Keulegan, G.H., 1948
An experimental study of submarine sand bars. Tech. Rept. No.3, Beach Erosion Board, 39 pp., 20 text figs.

bottom sediments
Kiilerich, A., 1957.
Galathea-Ekspeditionens arbejde i Phillipiner-graven. Ymer 1957(3):200-222.

bottom sediments
King, Lewis H., 1967.
Use of a conventional echo-sounder and textural analysis in delineating sedimentary facies. Scotian Shelf.
Can. J. Earth Sci., 4(4): 691-705

bottom sediments
Kitching, J.A., S.J. Lilly, S.M. Lodge, J.F. Sloane, R. Bassindale and F.J. Ebling, 1952.
The ecology of the Lough Ine Rapids with special reference to water currents. J. Ecology 40(1): 179-201.

bottom sediments
Klein, George De Vries, 1966.
Relating directional current structures of modern sediments to the direction and velocity of tidal current systems, Five Islands tidal flat complex, Nova Scotia.
Marit. Sediments, 2(1):19-21. (mimeographed)

bottom sediments
Klein, George DeVries, and John E. Sanders, 1964.
Comparison of sediments from Bay of Fundy and Dutch Wadden Sea tidal flats.
J. Sed. Petr., 34(1):18-24.

Klenova, M.V.

bottom sediments
Klenova, M.V., 1960.
[Geology of the Barents Sea.]
Isdatel Akad. Nauk, SSSR, Moscow, 1960, 367 pp.

translation of Chs. 2, pp. 64-82
4 103-111
5 112-227
8 310-338

bottom sediments
Klenova, M. V., 1958.
[VIII Investigation of material composition of Recent sediments of submarine slopes in the Antarctic]. Opisanie Exped. DIE "Ob" 1955-1956, MGG. Trudy Kompacksnoi Antarkt. Exped., Acad. Nauk, SSSR:145-161.

submarine geology
Klenova, M.V., 1938
Sedimente des Motovskij Busens. (Zur Frage über die Komplexuntersuchung der recenten Meeressedimente). Trans. Inst. Mar. Fish. and Oceanogr., USSR, 5:3-63/ (German resume)

bottom sediments
Klenova, M.V., and D.E., Gershanovich, 1953.
[Deep sea facies of deposits of the Sea of Japan]
Dokl. Akad. Nauk, SSSR, 89:937-940.

bottom sediments
Klenova, M.V. and K.P. Savelevz, 1967.
Chart of the terrain of the northern part of the Atlantic Ocean. (In Russian)
Atlant. nauchno-issled. Inst. rybn. khoz. Okeanogr. (AtlantNIRO). Materialy Konferentsii po Rezul'tatam Okeanologischeskikh Issledovanii v Atlanticheskom Okeane, 245-250.

bottom sediments
Klenova, M.V., and K.P. Savelyeva, 1966.
Sediment maps of the North Atlantic. (In Russian; English abstract)
Okeanol. Issled., Mezhd. Geofiz. Kom., Presid., Akad. Nauk, SSSR, No. 15:124-130.

bottom sediments (data)
Klepikova, V.V., Edit., 1961
Third Marine Expedition on the D/E "Ob", 1957-1958. Data.
Trudy Sovetskoi Antarktich. Exped., Arktich. i Antarktich. Nauchno-Issled. Inst., Mezhd. Geofiz. God, 22:1-234 pp.

bottom sediments
Knight, R.J., 1971.
Distributional trends in the Recent marine sediments of Tasiujaq Cove of Ekalugad Fiord, Baffin Island, N.W.T.
Marit. Sed. 7(1): 1-18.

bottom sediments
Knox, G.A., 1957.
General account of the Chatham Islands 1954 Expedition.
N. Zealand D.S.I.R. Bull. (Mem. N.Z. Ocean. Inst. 2) 122:1-17.

bottom sediments
Koh, Ryuji. 1966.
Littoral drift along Iwafune Port.
Coast. Engng. Japan 9: 127-136.

bottom sediments
Koldewijn, Bernard William, 1958.
Sediments of the Paria-Trinidad Shelf.
Repts. Orinoco Shelf Exped., 3:109 pp., 11 pls., fold-in maps.
Geol. Inst., Groningen, Publ.

bottom sediments
Kolpack, Ronald L., 1967.
Surface sediments of Drake Passage.
Antarctic Jl., U.S.A., 2(5):183.

bottom sediment
Komukai, R., 1956.
On the bottom topography and sediments in the western passage of Tugaru Strait.
Rept. Hydrogr. Off., 31(4):45 pp.

bottom sediments
Kornicker, Louis S., and William R. Bryant 1968.
Sedimentation on continental shelf of Guatamala and Honduras.
Mem. Am. Ass. Petrol. Geol. 11: 244-257.
Also in: Contr. Oceanogr. Texas A.+M. Univ. 12 (Contr. 393).

bottom sediments
*Kotenev, B.N., 1968.
Marine geological studies in the vicinity of Iceland. (In Russian; English abstract).
Okeanologiia, Akad. Nauk, SSSR, 8(6):1049-1052.

bottom sediments
Kovach, Robert, and Frank Press, 1961.
A note on ocean sediment thickness from surface wave dispersion.
J. Geophys. Res., 66(9):3073-3074.

Kovylin, V.M.

bottom sediments
Kovylin, V.M., 1966.
Results of seismic research in the southwestern part of the abyssal basin of the Japan Sea. (In Russian; English abstract).
Okeanologiia, Akad. Nauk, SSSR, 6(2):294-305.

bottom sediments
Kovylin, V.M., 1964.
A study of the structure of sedimentary strata in the Mediterranean. (In Russian).
Okeanologiia, Akad. Nauk, SSSR, 4(1):81-85.

bottom sediments
Kovylin, V.M., 1961.
[Recent data on the thickness of bottom sediments of the Indian Ocean.]
Doklady Akad. Nauk, SSSR, 136(4):924-926.

bottom sediments
Kovylin, V.M., B.J. Karp and R.B. Shaiakhmetov 1966.
The structure of the earth-crust and of the sedimentary rock-mass of the Sea of Japan from seismic data. (In Russian)
Dokl. Akad. Nauk SSSR 168(5):1048-1051.

Bottom sediments
Kozlova, O.G., 1970.
Biostratigraphy of marine bottom sediments from the equatorial Indian Ocean. Okeanologiia, 10(3): 479-487.
(In Russian; English abstract)

bottom sediments
*Kranck, Kate, 1967.
Bedrock and sediments of Kouchibouguac Bay, New Brunswick.
J. Fish.Res.Bd., Can., 24(11):2241-2265.

bottom sediments
Krause, Dale C., 1961
Geology of the sea floor east of Guadalupe Island.
Deep-Sea Res., 8(1): 28-38.

bottom sediments
Kravitz, Joseph H. and Frederick H. Sorensen 1970
Sedimentological reconnaissance survey of Kane Basin.
Marit. Sed., 1(6):17-20

bottom sediments (data)
Kube, I., and F. Asada, 1957.
A quantitative study on crustacean bottom epifauna in Tokyo Bay. J. Tokyo Univ. Fish., 43(2) 249-289.

bottom sediments
Kuenen, Ph. H., 1959.
Transport and sources of marine sediments.
Geol. en Mijnbouw (nw. ser.) 21:191-196

bottom sediments
Kuenen, Ph. H., 1946
Rate and mass of deep-sea sedimentation.
Am. J. Sci., 244(8):563-572.

bottom sediments
Kuenen, Ph. H., and H.W. Menard, 1952.
Turbidity currents and non-graded deposits.
J. Sed. Petrol. 22:83-96.

bottom sediments
Kullenberg, B., 1952.
On the salinity of the water contained in marine sediments. Göteborgs K. Vetenskaps- och Vitterhets-Samhälles Handl., Sjätte Följden, Ser. B, 6(6):38 pp., 9 textfigs.
(Med. Ocean. Inst., Göteborg 21).

submarine geology
Kullenberg, B., 1945
(The New Swedish Core Sampler.) Dansk Geologisk Forening, Bd 10, Hft.5:557-560. (Dansk Geofysisk Forening, Meddelelse No.3) Translated from the Swedish by H.U. Sverdrup and F.P. Shepard, 4 pp., 1 fig. (mimeographed). 1945.

bottom sediments
Kulp, J.L., and L.E. Turner, 1952.
Extension of C14 age method. Rev. Sci. Instr. 23(6):296-297.

bottom sediments
Kulp, J.L., H.L. Volchok, H.D. Holland, and D.B. Ericson, 1952.
Thick source alpha activity of some north Atlantic cores. J. Mar. Res. 11(1):19-28, 2 textfigs.

bottom sediments
Kulp, J.L., and D.R. Carr, 1952.
Surface area of deep-sea sediments. J. Geol. 60(2):148-159, 2 textfigs.

submarine geology
Kulp, J.L., and D.R. Carr, 1949.
Surface area of deep-sea sediments. (Abstract). Bull. G.S.A. 60(2, pt. 2), 1902.

bottom sediments
Kuno, H., R.L. Fisher and N. Nasu, 1956.
Rock fragments and pebbles dredged near Jimmu Seamount, northwestern Pacific. Deep-Sea Res., 3(2):126-133.

L

bottom sediments
Laakso, Mikko, 1965.
The bottom fauna in the surroundings of Helsinki.
Annales Zool. Fennici, 2:19-37.

bottom sediments
Ladd, H.S., 1948
The value of Marine bottom samples.
J. Coast and Geodetic Survey, 1:83-84.

bottom sediments
Laevastu, T., and O. Mellis, 1955.
Extraterrestrial material in deep-sea deposits.
Trans. Amer. Geophys. Union, 36(3):385-388.

bottom sediments
Lavrov, V.M., 1964.
Deep-sea sediment cores of the North Atlantic basin of the Atlantic Ocean. Investigation of marine bottom sediments and suspended matter. (In Russian: English abstract).
Trudy Inst. Okeanol., Akad. Nauk, SSSR, 68:136-156.

bottom sediments
Lefevre, S., E. Leloup, and L. Van Meal, 1956.
Observations biologiques dans le port d'Ostende. K. Belg. Inst. Naturwet., Verhandl., [Inst. R. Sci. Nat. Belg., Mem.,] 133:157 pp.

bottom sediments
LaFond, E.C., 1954.
Physical oceanography and submarine geology of the seas to the west and north of Alaska. Arctic 7(2):93-101, 11 textfigs.

submarine geology
LaFond, E.C., 1940
Sand movements near the beach in relation to tides and waves. Sixth Pac. Sci. Congr. Calif. 1939, Proc.2:795-799.

submarine geology
LaFond, E.C., 1938
Relationship between sea level and sand movement. Science 88:112-113.

bottom sediments
Lalou, C., 1954.
Sur les formes cristallines observée dans les voiles calcaires par cultures bactériennes à partir des vases noires de Villefranche-sur-Mer (Alpes Maritimes).
C.R. Acad. Sci., Paris, 238:2329-2330.

bottom sediments
Lalou, C., 1954.
Sur la précipitation expérimentale de la calcite dans les vases de la baie de Villefranche-sur-mer. C.R. Acad. Sci., Paris, 238:603-605.

bottom sediments
Lapina, N. N., and H. A. Belov, 1960.
[Conditions for the formation of bottom sediments in the northern ice ocean].
Morsk. Geol., Doklady Sovetsk. Geol., 12th Sess. Internat. Geological Congress, 88-93.

bottom sediments
*Larsen, Birger, 1968.
Sediments from the Central Philippine Trench.
Galathea Rept., 9:7-21.

bottom sediments
Larsonneur, Claude, et Michel Rioult, 1969.
Le Bathonien et le Jurassique supérieur en Manche centrale.
C. r. hebd. Séanc. Acad. Sci., Paris (D):268(22): 2645-2648

bottom sediments
Larsonneur, Claude, 1965.
Recherches sédimentologiques et géologiques en Manche centrale.
Rev. Trav. Inst. Pêches Marit., 29(2):225-242.

bottom sediments
Lavrov, V.M. 1967.
Bottom sediments and relief of the North Atlantic. (In Russian).
Atlant. nauchno-issled. Inst. rybn. Khoz. Okeanogr. (AtlantNIRO). Materialy Konferentsii po Resul'tatem Okeanologicheskikh Issledovanii v Atlanticheskom Okeane, 251-257.

Laughton, A.S.

bottom sediments
Laughton, A.S., 1959.
The sea floor.
Science Progress, 47(186):230-249.

bottom sediments
Laughton, A. S., 1957.
Exploring the deep ocean floor.
J. R. Soc. of Arts, 106:36-56.
In: NIO Collected Reprints, Vol. 6.

bottom sediments, acoustic properties
Laughton, A.S., 1957.
Sound propagation in compacted ocean sediments.
Geophysics 22(2):233-260.

bottom sediments
Lavrov, V.M., 1964.
Bottom sediments of the North Africa Basin of the Atlantic Ocean. (In Russian).
Trudy, Inst. Okeanol. Akad. Nauk, SSSR, 64:136-156.
Translation:
John B. Southard
100 E. Palisade Ave. (Apt. B31)
Englewood, New Jersey

bottom sediments

Lavrov, V.M. and K.P. Savelieva, 1971.
Deep-sea sediments of the Guinea Basin. (In Russian; English abstract). Okeanologiia 11(4): 656-667.

bottom sediments

Lawson, A.C., 1950.
Sea bottom off the coast of California. A review of Special Paper 31, Geological Society of America, by Shepard and Emery, 1941. Bull. G.S.A. 61:1225-1242, 1 fig.

bottom sediments

Lebedev, L.I., 1963
The facial zones and the thickness of the Neo-Caspian sediments in the mid-Caspian sea area. (In Russian).
Okeanologiia, Akad. Nauk, SSSR, 3(6):1029-1038.

bottom sediments

LeClaire, Lucien, Jean-Pierre Caulet et Philippe Bouysse, 1965.
Prospection sédimentologique de la marge continentale nord-africaine.
Cahiers Océanogr., C.C.O.E.C., 17(7):467-479.

submarine geology

LeDanois, Ed., 1948.
Les profundeurs de la mer. Trente ans de recherches sur la faune sous-marine au large des côtes de France. Payot, Paris, 303 pp., 8 pls., 56 textfigs.

bottom sediments

Lefevre, P., and G. Lucas, 1955.
Étude de quelques sédiments marins des environs de Concarneau (Finistère). Bull. Inst. Océan., Monaco, No. 1062:1-35, 1 fold-in, 6 textfigs.

bottom sediments

Leonyjev, O.K., M.E. Bachtina and T.A. Dobrynina, 1959
[Nearshore sediments study of the N-W Caspian Sea. Questions of studies of marine coasts.]
Trudy Oceanogr., Komissii Akad. Nauk, SSSR, 4:18-30.

bottom sediments

Lewis, M.S. and J.D. Taylor 1966
Marine sediments and bottom communities of the Seychelles.
Phil. Trans. R. Soc. (A) 259 (1099): 279-290.

benthos

Lewis, M.S. and J.D. Taylor 1966.
Marine sediments and bottom communities of the Seychelles.
Phil. Trans. R. Soc. (A) 259 (1099): 279-290.

bottom sediments

L'Herroux, Michel, 1970.
Peuplements des sables fins en baie de Saint-Brieuc (Manche).
Téthys, 2(1): 41-88

Lisitzin, A.P.

bottom sediments

Lisitzin, A.P., 1966.
Processes of Recent sedimentation in the Bering Sea. (In Russian).
Inst. Okeanol., Kom. Osad. Otdel Nauk o Zemle, Isdatel, Nauka, Moskva, 574 pp.

bottom sediments

Lisitzin, A.P., 1962.
Data on Antarctic sediments. Marine geology and the dynamics of coasts. (In Russian).
Trudy Okeanogr. Komissii, Akad. Nauk, SSSR, 10(3):70-78.

bottom sediments

Lisitzin, A.P., 1961.
Bottom sediments of the Antarctic (Abstracts).
Tenth Pacific Sci. Congr., Honolulu, 21 Aug.-6 Sept., 1961, Abstracts of Symposium Papers, 316.

bottom sediments

Lisitzin, A.P., 1960

Bottom sediments of the eastern Antarctic and the southern Indian Ocean.
Deep-Sea Res., 7(2): 89-99.

bottom sediments

Lisitsin, A.P., 1960

[Marine geological research.] Arktich. i Antarktich. Nauchno-Issled. Inst., Sovetsk. Antarkt. Exped., Mezhd. Geofis, God, Vtoraia Morsk. Exped., "Ob", 1956-1957, 7: 7-43. (In Russian)

bottom sediments

Lisitzin, A. P., 1960.
Sedimentation in southern parts of the Indian and Pacific Oceans. Morsk. Geol., Doklady Sovetsk. Geol., 21 Sess., Int. Geol. Kongress, 69-87.
(English abstract)

bottom sediments

Lisitsin, A.P., 1959

[Bottom sediments of the Bering Sea.] Trudy Inst. Okeanol. 29: 65-187. (In Russian)

bottom sediments

Lisitsyn, A.P., 1958.
[On the types of marine deposits connected with ice activity.] Doklady Akad. Nauk, SSSR, 118(2): 373-376.

bottom sediments

Lisitzin, A.P., 1956.
[Mechanical analysis of marine sediments.]
Trudy Inst. Okeanol., 19:262-287.

bottom sediments

Lisitzyn, A.P., and A.V. Fotiev, 1956.
[The volume weight of modern sediments of the Bering Sea.] Doklady Akad. Nauk, SSSR, 109(1):75-78.

bottom sediments

Lisitsin, A.P., I.E. Mikhaltsev, N.N. Sysoev and G.B. Udintsev, 1957.
[New data on the thickness and bedding conditions of soft deposits in the northwestern part of the Pacific.] Doklady Akad. Nauk, SSSR, 115(6): 1107-1110.

bottom sediments

Lisitsin, A.P., and A.V. Zhivago, 1958.
[Bottom relief and sediments of the southern part of the Indian Ocean. 1.]
Izv., Akad. Nauk, SSSR, Ser. Geofiz., (2):9-21.

OTS $0.50.

bottom sediments

Lisitsin, A.P., A.V. Zhivago, 1958.
[Bottom relief and sediments of the southern part of the Indian Ocean. 2.]
Izv., Akad. Nauk, SSSR, Ser. Geofiz., (3):22-36.

OTS $0.50.

bottom sediments

Lisitzin, A.P., and A.V. Zhivago, 1958

[VII. Submarine geology]
Opisanie Exped. D/E "Ob" 1955-1956, Mezhd. Geofiz. God, Trudy Kompaeksnoi Antarkt. Exped., Akad. Nauk, SSSR: 103-144.

bottom sediments

Lisitsin, A. P., and A. V. Zhivago, 1958.

[The relief and sediments in the southern Indian Ocean.]
Izv. Akad Nauk SSSR, Ser. Geogr. (2):9-21
(3):22-36

bottom sediments

Litvin, V.M., and V.D. Rvachev, 1962
A study of bottom relief and types of ground in the fishing areas of Labrador and Newfoundland. (In Russian; English summary).
Sovetskie Ribochoz. Issledov. Severo-Zapad. Atlant. Okeana, VNIRO PINRO, 99-111.

bottom sediments

Logan, Brian W., 1961.
Cryptozoon and associate stromatolites from the Recent, Shark Bay, Western Australia.
J. Geology, 69(5):517-533.

bottom sediments

Lohse, E.A., 1955.
A theoretical curve for statistical analysis of sediments. J. Sed. Petrol. 25(4):293-296.

bottom sediments

Longère, Paul et Didier Dorel 1970.
Étude des sédiments meubles de la vasière de la Gironde et des régions avoisinantes.
Rev. Trav. Inst. Pêches marit. 34 (2): 233-256

bottom sediments

Lorin J., 1968
Contribution à l'étude des transits sédimentaires dans la partie orientale du pertuis breton et de la baie de l'Aiguillon.
Bull. Inst. Géol. Bassin d'Aquitaine 5: 111-139. (not seen)

bottom sediments

Loring, D.H., and D.J.G. Nota, 1966.
Sea-floor conditions around the Magdalen Islands in the southern Gulf of St. Lawrence.
J. Fish. Res. Bd., Canada, 28(8):1197-1207.

sediments

Loring, D.H., D.J.G. Nota, W.D. Chesterman and H.K. Wong, 1970.
Sedimentary environments on the Magdalen Shelf, southern Gulf of St. Lawrence.
Mar. Geol. 8(5): 337-354

bottom sediments

Lynts, George W., 1966.
Relationship of Sediment-size distribution to ecologic factors in Buttonwood Sound, Florida Bay.
J. Sed. Petrol., 36(1):66-74.

bottom sediments

Louderback, G.D., 1940.
San Francisco Bay sediments.
Proc. Sixth Pacific Sci. Congr., 1939, 2:783-793.

bottom sediments

Loughnan, F.C. & D.C. Craig, 1962
A preliminary investigation of the recent sediments off the east coast of Australia.
Austral. J. Mar. & Freshwater Res., 13(1): 48-56.

bottom sediments
Lowenstam, H., and S. Epstein, 1957.
On the origin of sedimentary aragonite needles of the Great Bahama Bank. J. Geol., 65(4):364-375.

bottom sediments
Lowman, S.W., 1949.
Sedimentary facies in Gulf Coast.
Bull. Amer. Assoc. Petr. Geol. 33(12):1939-1997, 35 textfigs.

bottom sediments
Lucas, G., J. Lang et C. Godard, 1969.
Étude sédimentologique de quelques échantillons prélevés dans le Golfe de Gabès.
Rec. Trav. Stn mar. Endoume 46 (62): 253-260

bottom sediments
Lucas, G., and P. Lefevre, 1956.
Contribution à l'étude de quelques sédiments marins et de récifs d'Hermelles de la baie du Mont-Saint-Michel. Rev. Trav. Inst. Pêches Marit. 20:85-112.

bottom sediments
Ludwick, John C., 1964.
Sediments in northeastern Gulf of Mexico.
In: Papers in Marine Geology, R.L. Miller, Editor, Macmillan Co., N.Y., 204-238.

bottom sediments
*Ludwig, William J., Dennis E. Hayes and John I. Ewing, 1967.
The Manila Trench and West Luzon Trough. I. Bathymetry and sediment distribution.
Deep-Sea Res., 14(5):533-544.

bottom sediments
Ludwig, W.J., R.E. Houtz and M. Ewing, 1971.
Sediment distribution in the Bering Sea: Bowers Ridge, Shirshov Ridge and enclosed basins.
J. geophys. Res. 76(26): 6367-6375.

bottom sediments
Lüneburg, Hans 1969.
Sedimenthabitus und Sedimentdynamik in den Prielrinnen von Büsum und im Hörnum-Tief vor Sylt (Deutsche Bucht).
Veröff. Inst. Meeresforsch. Bremerhaven 11 (2): 137-164.

bottom sediments
*Lüneburg, Hans, 1967.
Eigenschaften und Verteilung der Sedimente am Hornsriff.
Veröff. Inst. Meeresforsch. Bremerh. 10(3):187-208.

bottom sediments
Lüneburg, H., 1964.
Origin and significance of iron-oolitic sand-grains in the sediments of the Weser estuary.
Marine Geology, 1(1):106-110.

bottom sediments
Lydall, A.K., R.A. Gees, 1967.
Near-shore sediment distribution around Cape Sable Island.
Marit. Sed., Halifax, N.S., 3(2/3):65-66.

M

bottom sediments
*Mabesoone, Jannes M., e Ivan M. Tinoco, 1965/1966.
Shelf off Alagoas and Sergipe (northeastern Brazil). 2. Geology.
Trabhs Inst. Oceanogr., Univ. Fed. Pernambuco, Recife (7/8):151-186.

bottom sediments
Machado, L. de B., 1951.
Resultados científicos do cruzeiro do "Baependi" e do "Vega" à Ilha de Trindade. Oceanografia física. Contribuição para o conhecimento das características físicas e químicas das águas.
Bol. Inst. Paulista Ocean. 2(2):95-110, 5 pls.

bottom sediments
MacNeil, F.S., 1954.
Organic reefs and banks and associated detrital sediments. Amer. J. Sci. 252(7):383-401.

bottom sediments(data)
Maeda, H., 1953.
Studies on Yosa-Naikai. 3. Analytical investigations on the influence of the River Noda and the benthonic communities. J. Shimonoseki Coll. Fish. 3(2):141-149, 3 textfigs.

bottom sediments(data)
Maeda, H., 1953.
Studies on Yosa-Naikai. 2. Considerations upon the range of the stagnation and the influences by the River Noda and the open sea.
J. Shimonoseki Coll. Fish. 3(2):133-140, 2 textfigs.

bottom sediments
Magliocca, Argeo, and Arnaldo S. Kutner, 1965.
Bottom sediments of Flamengo Bay, Ubatuba. (Abstract).
Anais Acad. bras. Cienc., 37(Supl.):326

bottom sediments
Mahadevan, C., and P.J.N.S.R., Ajanayuho, 1954.
Studies on dredges sands off Visakapatnam Harbour. Andhra Univ. Ocean. Mem. 1:51-56, 2 textfigs.

bottom sediments
Mahadevan, C., and M. Poonachandra Rao, 1954.
Studies of ocean floor sediments off the east coast of India. Andhra Univ. Ocean. Mem., 1:1-35, 3 textfigs., 3 charts.

bottom sediments
Mahadevan, C., and M. Subba Rao, 1955.
Marine sediments off Kalingapatam on the east coast of India. Current Sci. 24(12):412-413.

bottom sediments
Maiev, E.G., 1961.
[The thickness of Recent deposits and the rate of sedimentation in the southern area of the Caspian Sea.] Okeanologiia, Akad. Nauk, SSSR, 1(4):658-663.

bottom sediments
Malkin, D.H., and D.J. Echols, 1948
Marine sedimentation and oil accumulation; II. Regressive marine offlap and overlap-offlap.
Bull. Am. Assoc. Pet. Geol. 32:252-261.

bottom sediments
Maloney, Neil J., 1965
Geomorphology of the central coast of Venezuela.
Bol. Inst. oceanogr., Univ. Oriente, 4(2): 246-265.

bottom sediments
Marin, Jean 1971.
Étude physico-chimique de l'estuaire du Belon.
Rev. Trav. Inst. Pêches marit. 35 (2): 109-156

bottom sediments
*Marlowe, James I., 1968.
Unconsolidated marine sediments in Baffin Bay.
J. sedim. Petr., 38(4):1065-1078.

bottom sediments
Masicka, Halina, 1964.
Application of echograms to determination of various kinds of sea bottom sediments. (In Polish; English summary).
Przeglad Geofiz., Warszawa, 9(17)(1):77-84.

bottom sediments
Masse, J.P. 1970.
Contribution à l'Etude de la cartographie sédimentaire du plateau continental Sénégalais.
Rapp. P.v. Réun. Cons. int. Explor. Mer. 159: 12-14.

bottom sediments
Matsudaira, Y., and T. Kawamoto, 1953.
[On the subsidence of ground level at Magasaki region. (II). Analytical data for boring cores and submarine muds.] Umi to Sora 3(30):31-39, 3 textfigs.

bottom sediments
Matthews, L.H., 1934.
The marine deposits of the Patagonian Continental Shelf. Discovery Repts. 9:177-205, Pls. 2-14.

bottom sediments
Maxwell, W.G.H. 1970.
The sedimentary framework of Moreton Bay, Queensland.
Aust. J. mar. Freshwat. Res. 21(2): 71-88

bottom sediments
McCone, Alistair.W., 1966.
The Hudson River Estuary; hydrology, sediments and pollution.
Geogr. Rev., 56(2):175-189.

bottom sediments
McDougall, J.C., 1961
Ironsand deposits offshore from the west coast, North Island, New Zealand.
New Zealand J. Geol. Geophys., 4(3):283-300.

bottom sediments
McIver, R.D., 1971.
Organic geochemical analyses of frozen samples from dsdp leg 8 cores. Inétial Repts. Deep Sea Drilling Project, 8: 871-872.

bottom sediments
McKinney, Thomas F. and Gerald M. Friedman 1970.
Continental shelf sediments of Long Island, New York.
J. sedim. Petrol. 40 (1): 213-248.

bottom sediments
McManus, D.A., and J.S. Creager, 1963.
Physical and sedimentary environments on a large spitlike shoal.
J. Geology, 71(4):498-512.

bottom sediments
McMaster, R.L., 1962.
Petrography and genesis of Recent sediments in Narragansett Bay and Rhode Island Sound, R.I.
J. Sedimentary Petrography, 32(3):484-501.

bottom sediments
McMaster, R.L., 1960.
Sediments of Narragansett Bay system and Rhode Island Sound, Rhode Island.
J. Sed. Petr., 30(2):249-274.

bottom sediments
McMeaty, Robert L., and Thomas P. Lachance, 1969.
Northwestern African continental shelf sediments.
Marine Geol. 7 (1): 57-67.

bottom sediments
McMullen, R.M., 1966.
Surface sediments on the Grand Banks of Newfoundland: a progress report.
Maritime Sediments, Halifax, 2(3):131-132

bottom sediments
Mellis, Otto, 1960
Gesteinsfragmente im Roten Ton des Atlantischen Ozeans (N 29° 21', W 58° 09', Tiefe 5 450 M). Medd. Oceanogr. Inst., Göteborg, 28 (Göteborgs Kungl. Vetenskaps-och Vitterhets-Samhälles Handl., Sjätte Följden, B, 8(6)): 3-18.

Mellis, O. 1952. bottom sediments
Replacement of plagioclase by orthoclase in deep sea deposits. Nature 169(4302):624, 1 textfig.

submarine geology
Mellis, O., 1948.
7. The coarse-grained horizons in the deep-sea sediments from the Tyrrhenian Sea. Medd. Ocean. Inst., Göteborg, 15 (Göteborgs Kungl. Vetenskaps-och Vitterhets- Samhalles Handlingar, Sjätte Följden, Ser. B, 5(13)):47-72, Figs. 8-11.

bottom sediments
Melnik, V.I., 1970.
Geological researches in the Gulf of Mexico and the Caribbean Sea. (In Russian; English and Spanish abstracts). Okeanol. Issled. Rez. Issled. Mezhd. Geofiz. Proekt. Mezhd. Geofiz. Kom. Prezid. Akad. Nauk SSSR 20: 22-41.

Menard, H.W.

bottom sediments
Menard, H.W., 1953.
Pleistocene and Recent sediment from the floor of the northeastern Pacific Ocean. Bull. G.S.A. 64(11):1279-1294, 5 textfigs.

bottom sediments
Mencher, Ely, R.A. Copeland and H. Payson Jr. 1968.
Surficial sediments of Boston Harbor, Massachusetts.
J. sedim. Petrol. 38(1):79-86.

bottom sediments
Menzies, R.J., O.H. Pilkey, B.W. Blackwelder, D. Dexter, P. Huling and L.McCloskey, 1966.
A submerged reef off North Carolina.
Int. Rev. ges. Hydrobiol., 51(3):393-431.

bottom sediments
Mihailescu, Nicolae, 1969.
Étude sédimentologique des graviers se trouvant sur le littoral de la Mer Noire entre Constantza et Vama Veche (Roumanie).
Rapp. P.-v. Reun. Commn int. Explor. scient. Mer Mediterr., 19(4): 623-626.

bottom sediments
Millard, N.A.H., and K.M.F. Scott, 1954.
The ecology of South African estuaries. VI. The Milnerton Estuary and the Diep River, Cape.
Trans. R. Soc., S. Africa, 34(2):279-324, 8 textfigs.

bottom sediments
Miller, R.L., and J.M. Zeigler, 1958
A model relating dynamics and sediment pattern in equilibrium in the region of shoaling waves, breaker zone and foreshore.
J. Geol., 66(4): 417-441.

bottom sediments
Mitsushio, Hiromi, 1964.
Bottom sediments near the Tomioko peninsula. Amakusa, Kumamoto Prefecture, Kyushu, Japan. (In Japanese; English abstract).
Science Repts., Fac. Sci., Kyushu Univ., Geol., 6(3):167-187.

bottom sediments
Moberly, R., Jr., and F.W. McCoy Jr. 1966
The sea floor north of the eastern Hawaiian Islands.
Marine Geology, 4(1):21-45

bottom sediments
Mogi, Akio, and Takahiro Sato, 1964.
Topography and sediment in the southern part of the Japan Trench.
J. Oceanogr. Soc. Japan, 20(2):51-56.

bottom sediments
Mogi, Akio, and Takahiro Sato, 1958.
On the bottom configuration and sediments in the adjacent sea of Mogami Bank, Japan Sea.
Hydrogr. Bull., Maritime Safety Bd., Japan, (Publ. 981):37-53.

bottom sediments
Moign, Annik, 1965.
Contribution à l'étude littorale et sous-marine de la Baie du Roi (Spitsberg - 79° N).
Cahiers Océanogr., C.C.O.E.C., 17(8):543-563.

bottom sediments
*Moign, Annik, et André Guilcher, 1967.
Une flèche littorale en milieu périglaciaire arctique: la flèche de Sars (Spitzberg).
Norois 14(56):549-568.

bottom sediments
Monaco, A. 1967
Étude sédimentologique et minéralogique des dépôts quaternaires du plateau continental et des rechs du Roussillon.
Vie Milieu (B) 18(1B):33-62.

bottom sediments
Moncure, Richard, and Maynard Nichols, 1968.
Characteristics of sediments in the James River estuary, Virginia.
Spec. scient. Rept. Virginia Inst. mar. Sci. 52: 40 pp (multilithed)

Bottom sediments
Monyuskho, A.M., 1962.
Post-sedimentation changes in the marine clays around the Baku Archipelago. (In Russian).
Doklady, Akad. Nauk, SSSR, 145(5):1118-1120.

Translation:
Earth Sci. Sect. (Amer. Geol. Inst.), 145(1-6): 134-136. 1964.

bottom sediments
*Morelock, Jack, 1969.
Shear strength and stability of continental slope deposits, western Gulf of Mexico.
J. geophys. Res., 74(2):465-482.

bottom sediments
Morovic, D., 1951.
Composition mécanique des sédiments au large de l'Adriatique. The M.V. "Hvar" cruises.
Researches into Fish. Biol. Izvjesca (Repts.) 3(1):3-18, 1 fold-in, 3 textfigs.

Moore, David G.

bottom sediments
Moore, David G., 1963.
Geological observations from the bathyscaph "Trieste" near the edge of the continental shelf off San Diego, California.
Geol. Soc., Amer., Bull., 74(8):1057-1062.

bottom sediments
Moore, D.G., and P.C. Scruton, 1957.
Minor internal structures of some recent unconsolidated sediments. Bull. Amer. Assoc. Petr. Geol., 41:2723-2751.

bottom sediments
Moore, David G., and George Shumway, 1959.
Sediment thickness and physical properties: Pigeon Point Shelf, California.
J. Geophys. Res., 64(3):367-374.

bottom sediments
Moore, Joseph E., and Donn S. Gorsline, 1960
Physical and chemical data for bottom sediments South Atlantic coast of the United States, M/V Theodore N. Gill Cruises 1-9. USFWS Spec. Sci. Rept., Fish., No. 366: 84 pp.

bottom sediments
Moore, J. Robert, 1968.
Recent sedimentation in northern Cardigan Bay, Wales.
Bull. Br. Mus. nat. Hist., Mineral., 2(2):21-131.

bottom sediments
Morgans, J.F.C., 1962.
The benthic ecology of False Bay, II. Soft and rocky bottoms observed by diving and sampled by dredging and the recognition of grounds.
Trans. R. Soc., S. Africa, 36(4):287-334.

bottom sediments
Moss, A.J., 1963.
The physical nature of common sandy and pebbly deposits. II.
Amer. J. Sci., 261(4):297-343.

bottom sediments
Mosby, H., 1938
Svalbard waters. Geophys. Publ. 12(4): 85 pp., 34 text figs.

bottom sediments
Moskalenko, V.N., 1963.
New data on the structure of the sedimentary layer in the Mediterranean. (In Russian).
Doklady, Akad. Nauk, SSSR, 152(6):1457-1460.

Translation:
Soviet Oceanography, Issue 3:40-42. (1964); (Scripta Tecnica, for AGU).

bottom sediments
Moss, A.J., 1962.
The physical nature of common sandy and pebbly deposits. Part 1.
Amer. J. Sci., 260:337-373.

bottom sediments
Mulicki, Z., 1957.
Ecology of the more important Baltic invertebrates. Prace Morsk. Inst. Ryback., Gydni, No. 9: 313-379.

bottom sediments
*Müller, German, 1966.
Grain size, carbonate content, and carbonate mineralogy of Recent sediments of the Indian Ocean off the eastern coast of Somalia.
Naturwissenschaften, 53(21):547-550.

bottom sediments
Müller, G., 1959.
Die Rezenten Sedimente im Golf von Neapel. 1. Die Sedimente des Golfes von Pozzuoli.
Pubbl. Sta. Zool., Napoli, 31(1):i-xxvii.

bottom sediments
*Müller, German, und Jens Müller, 1967.
Mineralogisch-sedimentpetrographische und chemische Untersuchungen an einem Bank-Sediment (Cross-Bank) der Florida Bay, USA.
Neues Jb. Miner. Abh., 106(3):257-286.

bottom sediments
Müller, German, Hans-Erich Reineck und Walter Staesche, 1968.
Mineralogisch-sedimontpetrographische Untersuchungen an Sedimenten der Deutschen Bucht (südöstliche Nordsee).
Senckenberg. leth., 49(4):347-365.

bottom sediments
Muraour, P., and G. Hollande, 1951.
Étude préliminaire de quelques sediments de la baie de Castiglione. Trav. C.O.E.C., Algerie.
Bull. Aquic. et Pêches, Castiglione, n.s., 3: 161-186, 8 figs., 3 pls.

N

bottom sediments
Nair, R.R. and Abraham Pylee, 1969.
Size distribution and carbonate content of the sediments of the western shelf of India. Bull. Inst. Sci., India, 38(1): 411-420.
instr.

bottom sediment
Nasu, K., 1960
Oceanographic investigation in the Chukchi Sea during the summer of 1958.
Sci. Repts., Whales Res. Inst., 15: 143-157.

bottom sediments
Nasu, Noriyuki, 1961.
The provenance of coarse sediments recently discovered, on the shelf and the trench slopes off the Japanese Pacific coast. (Abstract).
Tenth Pacific Sci. Congr., Honolulu, 21 Aug.-6 Sept., 1961, Abstracts of Symposium Papers, 381-382.

bottom sediments
Nasu, Noriyuki, Hideo Kagami and Junsuke Chujo, 1962
Submarine geology at the entrance of Tokyo Bay (cooperative survey at the entrance of Tokyo Bay). (In Japanese; English abstract).
J. Oceanogr. Soc., Japan, 20th Ann. Vol., 98-120.

bottom sediments
Nasu, Noriyuki, and Takahiro Sato, 1962.
VII. Geological results in the Japanese Deep Sea Expedition in 1961 (JEDS-4).
Oceanogr. Mag., Japan Meteorol. Agency, 13(2): 166.

JEDS Contrib. No. 35.

bottom sediments
Nasu, N., and Y. Saito, 1958.
Shelf sediments of the Gulf of Kumano, Japan, 1.
Rec. Oceanogr. Wks., Japan, (Spec. No. 2):205-210.

bottom sediments
Navarro, F. de P., 1947.
Exploración oceanografía del Africa occidental desde el Cabo Ghir al Cabo Judy: resultados de las campañas del "Malaspina" y del "Xauen" en Mayo 1946. Trab. Inst. Español Ocean. No. 20:40 pp., 9 figs.

Bottom Sediments
Nayudu, Y. Rammohanroy, 1965.
Petrology of submarine volcanics and sediments in the vicinity of the Mendicino Fracture Zone.
Progress in Oceanography, 3:207-220.

bottom sediments
Naylor, D., 1965.
Pleistocene and post-Pleistocene sediments in Dublin Bay.
Sci. Proc., R. Dublin Soc., (A), 2(11):175-188.

bottom sediments
Nayudu, Y.R., 1962
Origin and distribution of deep-sea sand-silt layers in Northeast Pacific.
Bull. Amer. Asoc. Petrl. Geol., 46(2):273-274.

Pacific, northeast

bottom sediments
Neeb, G.A., 1941.
De verspreiding van de Tambora- asch op den zeeboden. Handl., Nat. Geneesk. Congres, Utrecht, 28 259-262.

bottom sediments
Nekritz, Richard, 1963.
Bottom sediment distribution in the Mediterranean Sea.
U.S. Naval Oceanogr. Off., IMR No. 0-55-63:10 pp.

bottom sediments
Nellen, Walter, 1970.
2.9 Sediment Chemische, mikrobiologische und planktologische Untersuchungen in der Schlei im Hinblick auf deren Abwasserbelastung. Kieler Meeresforsch 26(2): 144-149.

bottom sediments
Neprochnov, Yu. P., 1961.
Sediment thickness of the Arabian Sea basin. (In Russian).
Doklady, Akad. Nauk, SSSR, 139(1):177.

Translation:
Consultants Bureau for Amer. Geol. Inst., (Earth Sciences Section only), 139(1-6):719-720. (1963)

bottom sediments
Neprochnov, Yu.P. and I.N. Elnikov, 1969.
Results of the seismic studies of structure of the Black Sea sediments along the Yalta-Sinop section. (In Russian; English abstract).
Okeanologiia, 9(5): 823-833.

bottom sediments
Neprochnov, Yu. P., and M.F. Mikhno, 1961.
Data on the structure of the sedimentary strata of the deep-water depression of the Black Sea in the region of Sochi. (In Russian).
Doklady, Akad. Nauk, SSSR, 137(5):1209-1211.

Translation:
Consultants Bureau for Amer. Geol. Inst., 137 (1-6):427-429. (1962).

Nesteroff, Wladimir D.

bottom sediments
*Nesteroff, W.D., 1966.
Les dépôts marins actuels de la feuille de Frejus-Cannes.
Bull. Carte géol. France, 61(278):203-211.

bottom sediments
Nesteroff, Wladimir, 1965.
Recherches sur les sédiments marins actuels de la région d'Antibes.
Ann. Inst. Océan., Monaco, 43(1):1-136.

bottom sediments
Nesteroff, Wladimir D., 1960
Les sédiments marins entre l'Esterel et l'embouchure du Var. Rev. Géogr. Phys. et Géol. Dynamique, (2) 3(1): 17-28.

bottom sediments
Nesteroff, Wladimir D., and Bruce C. Heezen, 1960
Les dépôts de courantes de turbidité, le flysch et leur signification tectonique.
C. R. Acad. Sci., Paris, 250:3690-3692.

Also in:
Trav. Lab. Géol. Sous-Marine, 10(1960).

bottom sediments
Nesteroff, W.D., and P. Roa-Morales, 1957.
Recherches sur les sédiments marins des bouches de Bonifacio.
Rev. Géol. Phys., Géol. Dyn., (2), 1(2):79-82.

bottom sediments
Nesteroff, Wladimir D., Germain Sabatier et Bruce C. Heezen, 1964.
Les minéraux argileux, le quartz et le calcaire dans quelques sédiments de l'océan Arctique.
C.r.heb.Séanc.Acad.Sci.Paris, 258:991-993.

bottom sediments
Netherlands Hydrographer, 1965.
Some oceanographic and meteorological data of the southern North Sea.
Hydrographic Newsletter, Spec. Issue No. to Vol. 1: numerous pp. not sequentially numbered.

bottom sediments
Nettleton, L.L., 1952.
Sedimentary volumes in Gulf Coastal Plain of the United States and Mexico. Pt. IV. Geophysical aspects. Bull. G.S.A. 63(12-1):1221-1228, 4 text-figs.

bottom sediments
Neumann, Georg, und Günter Bublitz, 1969.
Seegrunduntersuchungen im westlichen Teil der Oder-Bucht.
Beitr. Meereskunde 24-25: 81-109

bottom sediments
Nevessky, E.N., 1961.
Some data on the post-glacial evolution of the Karkinitsky Gulf and the accumulation of bottom debris
Trudy Inst. Okeanol., Akad. Nauk, SSSR, 48:88-103.

bottom sediments
Newell, N.D., J. Imbrie, E.G. Purdy and D.L. Thurber, 1959.
Organism communities and bottom facies, Great Bahama Bank.
Bull. Amer. Mus. Nat. Hist., 117(4):183-228.

Niino, Hiroshi

bottom sediments
Niino, H., 1957.
Sediments on three submarine banks as promising fishery grounds (sediments on the Umitaka, Yoneyama and Senjo-sho).
Rec. Ocean. Wks., Japan, (Suppl.):58-62.

bottom sediments
Niino, Hiroshi, 1956(1953).
Bottom characters of the continental shelf around the Japanese islands.
Proc. 8th Pacific Sci. Congr., Geol. Geophys., Meteorol., 2A:901-909.

bottom sediments
Niino, H., 1952.
Summary of the recherches on the submarine configuration and bottom deposits of the neighboring seas of Japan, 1939-1948.
Proc. Seventh Pacific Sci. Congr., Met. Ocean., 3:214-216.

bottom sediments
Niino, H., 1952.
The bottom character of the banks and submarine valleys on and around the continental shelf of the Japanese islands. J. Tokyo Univ., Fish., 38(3):391-410, 8 pls.

bottom sediments
Niino, H., 1952.
On the bottom character of some new banks in the waters neighboring the Japanese islands.
J. Tokyo Univ., Fish., 38(3):411-426.

bottom sediments
Niino, H., 1950.
[On the bottom deposits on the banks at the mouth of Wakasa Bay and on the adjacent continental shelf.] J. Tokyo Univ. Fish. 37(1):274 pp., 3 pls. 65 textfigs.

bottom sediments
Niino, H., 1950.
Bottomdeposits at the mouth of Wakasa Bay, Japan, and on the adjacent continental shelf.]
J. Sed. Petr. 20(1):37-54, 7 textfigs.

submarine geology
Nino, Hiroshi, 1948.
Sediments of Oki Bank in the Sea of Japan.
J. Sed. Petr. 18:79-85.

bottom sediments
Niino, H., & K.O. Emery, 1961
Sediments of shallow portions of East China Sea and South China Sea.
Bull. Geol. Soc. Amer., 72(5): 731-762.

bottom sediments
Niino, H., and T. Kumagori, 1956.
Preliminary report on the new bank at the southern end of Emperor Sea Mount.
J. Tokyo Univ. Fish., 42(1):101-102.

bottom sediments
Niino, H., N. Nasu and R.F. Parker, 1956.
A coastal survey of the Gulf of Mexico by Japanese fishing boats in 1936 and 1937.
J. Tokyo Univ. Fish., 42(1):93-99.

bottom sediments
Niino, H., and K. Ozawa, 1956.
A study on the visible figures of the bottom which resembles the so-called DSL.
J. Tokyo Univ. Fish., 42(2):165-168.

bottom sediments
Nikolayev, N.I., 1946.
[Les types génétiques des dépots continentaus récents.] Bull Soc. Nat., Moscou, Sect. Geol., 21(4):25-64.

Ninkovich, Dragoslav

bottom sediments
Ninkovich, Dragoslav 1968.
Pleistocene volcanic eruptions in New Zealand recorded in deep-sea sediments.
Earth Planet. Sci. Letts. 4(2):89-102.

bottom sediments
Ninkovich, Dragoslav, Bruce C. Heezen, John R. Conolly and Lloyd H. Burckle, 1964
South Sandwich tephra in deep-sea sediments.
Deep-Sea Res., 11(4):605-619.

bottom sediments
Nir, Yaacov, 1968.
Sédiments marins récents sur la plate-forme continentale du littoral israëlien de la mer Méditerranée.
Rapp. P.-v. Réun. Commn int. Explor. scient. Mer Mediterr., 19(4): 637-638.

bottom sediments
*Nobrega Coutinho, P., 1965/1966.
Contribuição a sedimentologia e microfauna da Baía de Sepetiba (Estado do Rio de Janeiro). 1. Sedimentos.
Trabhs Inst.Oceanogr.,Univ.Fed.Pernambuco,Recife (7/8):115-121.

bottom sediments
Nota, D.J.G., 1971.
Morphology and sediments off the Marowijne River, Eastern Surinam Shelf. Hydrogr. Newsletter, R. Netherlands Navy, Spec. Publ. 6: 31-35.

bottom sediments
Nota, D.J.G., 1958.
The sediments of the western Guiana shelf.
Medadel. Landbouwhogeschool Wagenigen Nederland, 58(2):1-98.
Also:
Geol. Inst. Publ.

O

bottom sediments
Ogaki, K., H. Kosugi and H. Tomabeti, 1953.
Submarine topography and distribution of bottom sediment on deep sea fishing ground of Hokkaido.
Bull. Hokkaido Regional Fish. Res. Lab., Fish. Agency, No. 9:1-16, 2 pls., 26 textfigs.

bottom sediments
Olausson, Eric, 1961
Remarks on some Cenozoic core sequences from the Central Pacific, with a discussion of the role of coccolithophorids and Foraminifera in carbonate deposition.'
Medd. Oceanogr. Inst., Göteborg, 29: 35 pp.
Also:
Göteborgs Kungl. Vetenskaps-och Vitterhets-Samhalles Handl., Sjatte Följden, (B), 8(10):

bottom sediments
Olausson, Eric, 1961.
Studies of deep-sea cores. Sediment cores from the Mediterranean Sea and the Red Sea.
Repts. Swedish Deep-Sea Exped., 1947-1948, 8(4): 337-391.

bottom sediments
Oliff, W.D., C.D. Berrisford, W.D. Turner, J.A. Ballard and D.C. McWilliam 1967.
The ecology and chemistry of sandy beaches and nearshore submarine sediments of Natal. 1. Pollution criteria for sandy beaches in Natal. 2. Pollution criteria for nearshore sediments of the Natal coast. Water Research 1(2): 115-146.

bottom sediments
*Olivier, Santiago Raul, Ricardo Bastida y Maria Rosa Torti, 1968.
Resultados de las campañas oceanográficas mar del Plata 1-5: Contribución al trazado de una carta bionómica del área de Mar del Plata. Las asociaciones del sistema litoral entre 12 7 70 m de profundedal.
Bol.Inst.Biol.mar.Mar del Plata, 16:1-85.

bottom sediments
Orlenok, V.V., 1971.
Structure and thickness of sediments in northern Atlantic according to seismic data. (In Russian; English abstract). Usloviia sediment. Atlantich. Okeana, Rez. Issled. Mezhd. Geofiz. Proekt. Mezhd. Geofiz. Kom. Prezid. Akad. Nauk SSSR: 271-296.

bottom sediments
Ottmann, Francois, 1959.
Estudo das Amostras do fundo recolhidas pelo N. E. "Almirante Saldanha" na regiao da emboucadura do Rio Amazonas. Missao de Diretoria de Hidrografia e Navegacao por ocasiao do Ano Geofisico International (Dezembro de 1958). Trabalhos. Inst. Biol. Marit. e Oceanogr., Recife 1(1): 77-106.

Bottom sediments
Ottmann, F., L. Bastaroux, A. Moche et G. Moulin, 1970.
Etude sédimentologique, géochimique et géotechnique des vases estuariennes (Vasière de Méan, près St-Nazaire, estuaire de la Loire).
C.R. Trav. 1967, 1968, Lab. Géol. Univ Nantes, 111 pp. (multilitted)

bottom sediments
Ottmann, Francois, et Paulo Nobrega Coutinho, 1963
Études sédimentologiques dans le port de Recife (Brasil).
Cahiers Océanogr., C.C.O.E.C., 15(3):161-169.

bottom sediments
Ottmann, François and Jeanne Marie Ottmann, 1970.
Estudo das amostras recolhidas na região de Cabo Frio (Brasil) Pelo Noc. Almirante Saldanha.
Trabhs. oceanogr. Univ. Fed. Pernambuco 9/11: 67-78.

bottom sediments
Ottmann, Francois, and Jeanne-Marie Ottmann, 1960
Estudo da Barra das Jangadas. 4. Estudo dos sedimentos. (In Portuguese; English and French resumes).
Trabalhos. Inst. Biol. Marit. e Oceanogr., Universidade do Recife, Brasil, 2(1):219-233.

bottom sediments
Ottmann, François, et Jeanne-Marie Ottmann, 1959.
Les sediments de l'embouchure du Capibaribe, Recife Bresil. Trabalhos Inst. Biol. Marit. e Oceanogr., Recife, 1(1):51-69.

bottom sediments
Ottmann, François, y Carlos M. Urien, 1965.
Observaciones preliminares sobre la distribución de los sedimentos en la zona externa del Rio de la Plata.
Anais Acad. bras. Cienc., 37(Supl.):283-288.

bottom sediments
Otto, L. 1971.
The frequency distribution of the current speed at the Netherlands lightvessels and its possible influence on the composition of sediments in the southern North Sea.
Geol. Mijnb. 50(3):475-478.

bottom sediments
Over, C.D., 1950.
On the interpretation of climatic variations as revealed by a study of samples from an equatorial Atlantic core. Centenary Proc. R. Met. Soc. 211-215. (deep sea

P

bottom sediments
Papazachos, Basil C., 1964.
Dispersion of Rayleigh waves in the Gulf of Mexico and Caribbean Sea.
Bull. Seismol. Soc., Amer., 54(3):909-925.

bottom sediments
Parfait, 1883.
Mission scientifique du Talisman.
Ann. Hydrogr., Paris, (2) 5:259-311, figs.

bottom sediments
Paris, J., 1954.
Contribution à la connaissance de la "zone nord der cannalots". Vie et Milieu, Bull. Lab. Arago, 5(4):469-512.

bottom sediments
Parker, Robert H., 1964.
Zoogeography and ecology of some macroinvertebrates, particularly mollusks, in the Gulf of California and the continental slope off Mexico.
Vidensk. Medd., Dansk Naturh. Foren., 126:1-178.

bottom sediments
*Partheniades, Emmanuel, and John F. Kennedy, 1967.
Depositional behavior of fine sediment in a turbulent fluid motion.
Proc. 10th Conf. Coast. Engng. Tokyo, 1966, 1:707-729.

Bottom Sediments
Patrick, Ruth, 1962.
Discussion of paper by Charles E. Renn, "Ecology and fine-grained sediments".
J. Geophys. Res., 67(4):1509-1510.

bottom sediments
Pemelun, V.P., 1956.
[Laboratory concentration of the sand-silt fraction.] Trudy Inst. Okeanol., 19:288-293.

bottom sediments
Pequegnat, Willis E., William R. Bryant and John E. Harris, 1971.
Carboniferous sediments from Sigsbee Knolls, Gulf of Mexico.
Bull. Am. Ass. Petrol. Geol., 55(1): 116-123

bottom sediments
Pérès, J.-M., 1959.
Observations en bathyscaphe de l'instabilité des vases bathyales méditerranéenes.
Rec. Trav. Sta. Mar., d'Endoume, 29(17):3-4.

bottom sediments
Perry, P.B., 1961
A study of the marine sediments of the Canadian Eastern Arctic Archipelago.
Fish. Res. Bd., Canada, Manuscript Rept. Ser., (Oceanogr. and Limnol.), No. 89: 80 pp. (multilithed).

Petelin, V.P.

bottom sediments
Petelin, B.P., 1961.
Choosing of methods for the mineralogical analysis of the sand-silt fractions of bottom sediments.
Trudy Inst. Okeanol., Akad. Nauk, SSSR, 50:170-183.

bottom sediments
Petelin, V.P., 1961
A new method of water mechanical analysis in marine sediments.
Okeanologiya, 1: 143-147.

bottom sediments
Petelin, V.P., 1960
[Bottom sediments in the western part of the Pacific Ocean.]
Okean. Issle., IGY Comm., SSSR:45-60.

bottom sediments
Petelin, V.P., 1959
[Bottom sediments of the Kronotski Bay.]
Trudy Inst. Okeanol., 36: 21-31.

bottom sediments
Petelin, B.P., 1957.
The relief of the floor and the bottom deposits in the north-west Pacific.
Proc. UNESCO Symp., Phys. Ocean., Tokyo, 1955: 225-237.

bottom sediments
Petzall, Wolf, 1967.
Sedimentación marina.
In: Ecología marina, Monogr. Fundación La Salle de Ciencias Naturales, Caracas, 14:35-66.

Pettersson, Hans

bottom sediments
Pettersson, H., 1960.
Poussière d'étoile.
Scientia, 95(12):367-369.

bottom sediments
Pettersson, H., 1955.
Résumé des résultats de la croisière de l'"Albatross" en Méditerranée.
Trav. Lab. Géol. Sous-Marine, 6:1-5.

bottom sediments
Petterssen, H., 1954.
Zur Bildung des Roten Tones.
Arch. Met., Geophys., Bioklim., A, 7:477-481.

bottom sediments
Pettersson, H., 1952.
Prelevements de sediments superficiels en Mer Tyrrhenienne. Bull. d'Info., C.C.O.E.C. 4(1): 7-10, 1 fig.

bottom sediments
Pevear, David R., and Orrin H. Pilkey, 1966.
Phosphorite in Georgia continental shelf sediments.
Geol. Soc., Am., Bull., 77(8):849-858.

Phleger, Fred B

bottom sediments
Phleger, Fred B, 1965.
Sedimentology of Guerrero Negro Lagoon, Baja California, Mexico.
In: Submarine geology and geophysics, Colston Papers, W.F. Whittard and R. Bradshaw, editors, Butterworth's, London, 17:205-235.

bottom sediments
Phleger, F. B, and G.C. Ewing, 1962.
Sedimentology and oceanography of coastal lagoons of Baja California, Mexico.
Bull. Geol. Soc., Amer., 73(2):145-182.

Bottom Sediments
Phipps, C.V.G., and L.H. King, 1969.
Chemical, mineralogical and textural variations in sediments from the Scotian Shelf.
Maritime Sediments 5(3): 101-112

bottom sediments
Pickard, G.L., 1961.
Oceanographic features of inlets in the British Columbia mainland coast.
J. Fish. Res. Bd., Canada, 18(6):907-999.

Pigorini, Bruno

~~bottom sediments~~
Pigorini, Bruno, 1969.
Provenance et dispersion des sédiments récents de la Mer Adriatique
Rapp. P.-v. Réun. Commn int. Explor. scient. Mer Mediterr., 19(4): 633-635.

bottom sediments
Pigorini, Bruno 1968.
Sources and dispersion of Recent sediments of the Adriatic Sea.
Mar. Geol. 6(3):187-229.

bottom sediments
*Pigorini, Bruno, 1967.
Aspetti sedimento logici del Mare Adriatico.
Mem. Soc. ital. Sci. nat. Mus. Civ. Stor. nat. Milano, 16(3):131-199.

bottom sediments
Pimm, A.C., R.E. Garrison and R.E. Boyce 1971.
Sedimentology synthesis: lithology, chemistry and physical properties of sediments in the northwestern Pacific Ocean.
Initial Repts Deep Sea Drilling Project, Glomar Challenger, 6:1131-1252.

bottom sediments
Pirie, Robert Gordon, 1965.
Petrology and physical-chemical environment of bottom sediments of the Rivière Bonaventure-Chaleur Bay area, Quebec, Canada.
Rept. B.I.O. 65-10:182 pp. (multilithed).

bottom sediments
*Poinsard, F., et J.P. Troadec, 1967.
La radiale de Pointe-Noire.
Cah. ORSTOM, Sér. Océanogr., 5(1):69-84.

bottom sediments
Poizat, Claude 1969.
Note sur la répartition des sédiments au débouché des calanques du littoral des massifs de Marseilleveyre, Puget et Devenson (Marseille à Cassis).
Recl. Trav. Stn mar. Endoume 46(62): 193-212.

bottom sediments
*Poizat, Claude, 1968.
Repartition des sediments au débouché des calanques du littoral des massifs de Marseille veyre et Puget (Marseille-Cassis).
C.r. hebd. Séanc., Acad. Sci., Paris, (D)267(8):831-834.

Bottom sediments
*Portmann, J.E., 1970.
The effect of China clay on the sediments of St. Austell and Mevagissey bays. J. mar. biol. Ass., U.K., 50(3): 577-591.

bottom sediments
Powers, M.C., 1953.
A new roundness scale for sedimentary particles.
J. Sed. Petr. 23(2):117-119.

bottom sediments
Pratje, O., 1952.
Die Erfahren bei der Gewinnung von rezenten, marinen Sedimenten in der letzten 25 Jahren.
Mitt. Geogr. Gesellsch., Hamburg, 50:118-197, 40 textfigs., Pls. 13-14.

bottom sediments
Pratje, O., 1950.
Die Bodenbedeckung des Englischen Kanals und die maximalen Gezeitenstromgeschwindigkeit.
Deutsch. Hydro. Zeits. 3(3/4):201-205.

bottom sediments
Pratje, O., 1949.
Die Bodenbedeckung der nordeuropaischen Meere.
Handbuch Seefisch. Nordeuropas 1(3):

bottom sediments
Pratje, O., 1949.
Die Bodenbedeckung des nordeuropäischen Meere.
Seefischerei Nordeuropas 1(3):23 pp., 3 charts.
Handb.

bottom sediments
Pratje, O., 1948
Die Schwankungen der Sedimentzusammensetzung auf engstem Raum und die Untersuchungsmethoden. German Hydro. Jour. 1(5/6):169-175, 1 text fig.

bottom sediments
Pratje, O., 1948
Die Boden bedeckung der südlichen und mittleren Odsee und ihre Bedeutung für die Ausdeutung fosseler Sedimente. Ger. Hydr. J. 1(2/3):45-61, 6 text figs.

submarine geology
Pratje, O., 1939
Die Sedimentation in der südlichen Ostsee. Ann. d. Hydrogr. 67:209-221.

bottom sediments
Pratje, O. (deceased) and F. Schüler, 1952.
Bodenkartierung des Seegebietes Hoofden (südliche Nordsee) mit Hilfe von Grundproben und Echogrammen. Deutsche Hydrogr. Zeits. 5(4):189-196, 6 textfigs., Chart 6 (in back pocket).

bottom sediments
Pratt, R.M., 1962.
The ocean bottom.
Science, 138(3539):492-495.

bottom sediments
Pravdić, V., 1970.
Surface charge characterization of sea sediments.
Limnol. Oceanogr., 15(2): 230-233.

bottom sediments
Price, W.A., & L.S. Kornicker, 1961
Marine and lagoonal deposits in clay dunes, Gulf Coast, Texas.
J. Sediment. Petrol., 31(2): 245-255.

bottom sediments
Priddy, R.R., R.M. Crisler, and Hu Burgord, 1954.
Sediments of parts of Mississippi Sound. (Abstract)
Bull. G.S.A. 65(2):1366-1367.

submarine geology
Putnam, J.A., K.J. Bermel, and J.W. Johnson, 1947
Suspended matter sampling and current observations in the vicinity of Hunters Point, San Francisco Bay. Trans. Am. Geophys. Un. 28(5):742-746, 4 text figs.

Q

R

Bottom sediments
Rao, Ch. Madhusudana and P.S.N. Murty, 1969.
Studies on the shelf sediments off the Madras coast. Bull. natn. Inst. Sci., India, 38(1): 442-448.

~~bottom sediments~~
Rao, G. Prabhakar, 1969.
Sediments of the near-shore region off Neendakara-Kayankulam coast and the Ashtamudi and Vatta estuaries, Kerala, India. Bull. natn. Inst. Sci., India, 38(1): 513-551.

bottom sediments
Rasmussen, K., 1957.
Investigations on marsh soils and Wadden Sea sediments in the Tønder region.
Medd. Skalling-Lab., 15:147-170.

bottom sediments
*Reboucas, Aldo C.,1965/1966.
Sedimentos da Baia de Tamandare Pernambuco.
Trabhs Inst.Oceanogr.,Univ.Fed.Pernambuco,Recife, (7/8):187-206.

bottom sediments
Reed, J.J., 1951.
Marine sediments near Sumner, Canterbury, New Zealand. N.Z. J. Sci. Tech., Sect. B, 33(2):129-137, 4 textfigs.

bottom sediments
Reed, J.J., and N. de B. Hornibrook, 1952.
Sediments from the Chatham Rise. 1. Petrology. 2. Recent and fossil microfaunas.
N.Z.J. Sci. & Tech., B, 34(3):173-188.

bottom sediments
Reed, J.J., and A.E. Leopard, 1954.
Sediments of Cook Strait. (N.Z. Ocean. Comm. Publ. No. 17). N.Z.J.Sci. Tech., Sect. B, 36(1):14-24, 3 figs.

bottom sediments
Reimnitz, Erk, Neil F. Marshall, 1965.
Effects of the Alaska earthquakes and tsunami on recent deltaic sediments.
Jour. Geophys. Res., 70(10):2363-2376.

abstract

earthquakes, effect of
tsunamis, effect of
deltas

Reineck, Hans-Erich

bottom sediments
Reineck, Hans Erich 1968.
Sedimengefüge in Golf von Neapel.
Pubbl. Staz. Zool. Napoli 36:112-137.

bottom sediments
Reineck, Hans-Erich, Jürgen Dörjes, Sibylle Gadow und Günther Hertweck 1968.
Sedimentologie, Faunenzonierung und Faziesabfolge vor der Ostküste der inneren Deutschen Bucht.
Senckenberg leth. 49(4):261-309.

bottom sediments
*Reineck,H.E., and Indra B. Singh,1967.
Primary sedimentary structures in the Recent Sediments of the Jade,North Sea,
Mer. Geol., 5(3):227-235.

bottom sediments, intertidal
Renaud-Debyser, J., and B. Salvat, 1963.
Le calcul des biovolumes dans l'étude des chaînes alimentaires de la faune endogée des sédiments meubles intertidaux.
Comptes Rendus, Acad. Sci., Paris, 256(12):2712-2714.

Bottom Sediments
Renn, Charles E., 1962.
Ecology and fine-grained sediments.
J. Geophys. Res., 67(4):1505-1507.

bottom sediments
Revelle, R., 1950.
1940 E.W. Scripps cruise to the Gulf of California. 5. Sedimentation and oceanography: survey of field observations. Mem. G.S.A., 43:1-6.

bottom sediments
Revelle, R.R. 1944
Scientific results of the cruise VII of the Carnegie during 1928-1929 under command of Captain J.P. Ault. Oceanography II. 1 Marine bottom samples collected in the Pacific Ocean by the Carnegie on its seventh cruise.
Publ. Carnegie Inst. Washington 556: 1-180.

submarine geology
Revelle, R.R., 1939
Sediments of the Gulf of California. Bull. G.S.A. 50:170-171.

bottom sediments
Revelle, R., and F.P. Shepard, 1939
Sediments off the California Coast.
Recent Marine Sediments, a symposium: 245-282. (A.A. P.G., Tulsa Oklahoma)

bottom sediments
Reyes,F Eduardo,1967.
Carta batilitologica de Valparaiso.
Rev.Biol.mar.Chile,13(1):59-69.

bottom sediments
Riccardi, R., 1946.
Lezioni di oceanografia. Ed. Perrele, 147 pp., 16 figs., 7 pls.

Bottom sediments
Rivière, André, Solenge Vernhet, François Arby et Marc Rivière, 1966.
Sur les terrains récents des côtes Atlantiques.
C. R. hebd. Séanc., Acad. Sci., Paris, (D) 262 (1):5-8.

bottom sediments
Riviere, A., and L. Visse,
Origin of the minerals of marine sediments.
Bull. Soc. Geol., France, (e) 4(7-9):

In Trans. Month. 3(9):3390.

Bottom sediments
Roa Morales, Pedro, and Francois Ottmann, 1961.
Primer estudo topografico y geologico del Golfo de Cariaco.
Bol. Inst. Oceanograf., Univ. Oriente, Cumana, Venezuela, 1(1):5-20.

bottom sediments
Rochford, D.J., 1951.
Summary to date of the hydrological work of the Fisheries Division, C.S.I.R.O. Proc. Indo-Pacific Fish. Counc., 17-28 Apr. 1950, Cronulla, N.S.W., Australia, Sects. II-III:51-59.

bottom sediments
Rodolfo, Kelvin S., 1969.
Sediments of the Andaman Basin, northeastern Indian Ocean.
Marine Geol. 7(5):371-402.

bottom sediments
Romankevich, E.A., and I.M. Urbanovich 1971
Organic matter (nitrogen carbohydrates) of the suspended matter, bottom sediments and interstitial water of the Peru-Chile region. (In Russian; English abstract)
Trudy Inst. Okeanol. P.P. Shirshova Akad. Nauk SSSR 89: 106-117.

bottom sediments
Rona, Peter A., and C.S. Clay, 1966.
Continuous seismic profiles of the continental terrace off southeast Florida.
Bull. Geol. Soc. Amer., 77(1):31-44.

bottom sediments
Ronov, A.B. 1968.
Distribution areas of the main genetic groups of modern sediments in the World Ocean. (In Russian).
Dokl. Akad. Nauk SSSR 179(3): 701-704.

bottom sediments
Rosenberg-Herman, Yvonne, 1965.
Étude des sédiments quaternaires de la Mer Rouge
Ann. Inst. Océanogr., Monaco, 42(3):339-415, 12 pls.

bottom sediments
Rosfelder, A., 1955.
Carte provisoire au 1/500,000 de la marge continental algérienne. Note de présentation.
Publ. Service Carte Géol., Algérie, n.s., Bull., Bull., No. 5, Trav. des Coll., 1954:57-106, 6 figs., 1 pl., 1 map.

bottom sediments
Ross, David A. 1970
Source and dispersion of surface sediments in the Gulf of Maine-Georges Bank area.
J. sedim. Petrol. 40(3): 906-920.

bottom sediments
Ruck, K.-W., 1952.
Seegrundkartierung der Lübecker Bucht.
Die Küste 1(2):55-57.

bottom sediments
Ruellan, F., and Y. Beigbeder, 1963.
Quelques observations préliminaires sur la répartition des sédiments sous-marins en baie de Saint-Brieuc (Côtes su Nord).
Comptes Rendus, Acad. Sci., Paris, 256(7):1566-1569.

bottom sediments
Ruellan, Francis, Yvonne Beibeder et André Dagorne, 1967
Répartition des fonds sédimentaires détritiques dans la partie méridionale du Golfe Normand-Breton (au sud du parallèle de 48°46'48"-54G20').
C. r. hebd. Séanc. Acad. Sci. Paris, (D) 264(12): 1550-1553

bottom sediments
Rumeau, J.L. and J.R. Vanney 1969
Caractères géochimiques et origine des sédiments récents du plateau continental atlantique dans le nord du Golfe de Gascogne.
Bull. Centre Rech. Pau-SNPA 3(1): 125-146.

bottom sediments (oolites)
Rusnak, G.A., 1960
Some observations of Recent oolites. J. Sed. Petr., 30(3): 471-480.

bottom sediments
Rvachev, V.D., 1964
Bottom contour and sediments on the Georges and Banquereau banks. (In Russian).
Materialy Ribochozh. Issled. Severn. Basseina, Gosudarst. Kom. Ribn. Chozh. SNCH,SSSR,Poliarn Nauchno-Issled. i Proektn. Inst. Morsk. Ribn. Chozh i Okeanogr., N.M. Knipovich (PINRO),2: 78-87.

bottom sediments
Rvachev, V.D., 1963
The topography and bottom sediments of the shelf area in southwestern Greenland. (In Russian).
Okeanologiia, Akad. Nauk. SSSR, 3(6):1046-1055.

bottom sediments
Ryan, J.D., 1953.
The sediments of Chesapeake Bay. Maryland Bd. Nat Res., Dept. Geol., Mines and Water Res. Bull. 12: 120 pp.

Abstr.: Chem. Abstr. 48:38681.

bottom sediments
Ryan, William B.F., Fifield Workum, Jr., and J.B. Hersey, 1964.
Sediments on the Tyrrhenian Sea abyssal plain. (Abstract).
Geol. Soc., Amer., Special Paper, No. 76:141-142.

S

bottom sediments
Saemundsson, B., 1946.
Description of Faxa Bay.
Rapp. Proc. Verb., Cons. Perm. Int. Expl. Mer, 26-29, 3 textfigs.

bottom sediments
Saidova, H.M. 1971.
On the Recent sediments near the Pacific coast of South America. (In Russian; English abstract)
Trudy Inst. Okeanol. P.P. Shirshova Akad. Nauk SSSR 89: 139-145

bottom sediments
Saito, Tsunemasa, Maurice Ewing and Lloyd H. Burckle, 1966.
Tertiary sediment from the Mid-Atlantic Ridge.
Science, 151(3714):1075-1079.

bottom sediments
Saks, V.N., 1950.
On the rate of accumulation of contemporary marine sediments. Priroda, 1950(6):24-33, diagr., sketch map.

bottom sediments
Salnikov, N.E., 1965.
Investigaciones de deconomía pesquera en el Golfo de Mexico y del Mar Caribe. (In Russian; Spenish abstract).
Sovetsk.-Cub. Ribokhoz. Issled., VNIRO:Tsentr. Ribokhoz. Issled. Natsional. Inst. Ribolovste Republ. Cuba. 93-179.

bottom sediments
Sarmiento, R., and R.A. Kirby, 1962
Recent sediments of Lake Maracaibo.
J. Sed. Petr., 32(4):698-724.

bottom sediments
Sato, T., 1962
A scoriaceous bed in Tokyo Kaiwan. (In Japanese; English abstract).
Hydrogr. Bull. (Publ. No. 981), Tokyo, No. 72: 28-28.

bottom sediments
Sato, T., 1959.
On the bottom sediments off the coast of Nojiro (XII).
Publ. 981, Hydrogr. Bull., No. 60:45-51.

bottom sediments
Sato, T., and M. Hoshino, 1963.
Bottom sediments dredged from Sagami Wan (Sagami Bay). (In Japanese; English abstract).
Hydrographic Bull., Tokyo, No. 73(Publ. No. 981): 6-10.

Bottom Sediments
Scaccini, A., e C. Piccinetti, 1969.
Il fondo del mare da Falconara a Tortoreto con annessa carta di pesca.
Programma particolare Risorse mar. Fondo mar. Consig. naz. Ricerche, 146 pp.

bottom sediments
Schäfer, W., 1954.
Dehnungsrisse unter Wasser im merrischen Sediment
Senck. Lett. 35(1/2):87-99, 12 textfigs.

bottom sediments
Scherbakov, Ph. A., 1959
[Lithological study of nearshore sediments in Anadyr-gulf (Bering Sea). Questions of studies of marine coasts.]
Trudy Okeanogr., Komissii Akad. Nauk SSSR, 4:31-43.

bottom sediments(data)
Schott, G., and P. Perlewitz, 1906.
Lotungen I.N.M.S. "Edi" und das Kabeldampfers "Stephen" im westlichen Stillen Ozeans.
Arch. Deutschen Seewarte 29(2):1-38, 4 pls.

bottom sediments
Schott, G., and B. Schulz, 1914.
Die Forschungsreise: S.M.S. "Möwe" im Jahre 1911.
Arch. Deutschen Seewarte 37(1):1-80, 8 pls., 16 textfigs.

bottom sediments
Schott, W., 1952.
On the sequence of deposits in the Equatorial Atlantic Ocean. Medd. Ocean. Inst. Göteborg, 18:15 pp., 3 textfigs.

Also:
Göteborgs K. Vetenskaps- och Vitterhets-Samhälles Handl., Sjätte Foljden B 6(2):15 pp., 3 textfigs.

bottom sediments
Schott, W., 1950.
Die flächenhafte Verteilung der Meeressedimente im atlantischen Ozean. Deutsche Hydro. Zeits. 3(1/2):89-93, 1 fig.

bottom sediments
Schüler, F., 1952.
Untersuchungen über die Mächtigkeit von Schlickschichten mit Hilfe des Echographem.
Deutsche Hydrogr. Zeits. 5(5/6):220-231, 9 textfigs.

bottom sediments
Schuster, O., 1952.
Die Vareler Rinne im Jadebusen. Die Bestandteile und das Gefüge einer Rinne im Watt. Abhandl. Senckenbergischen Naturforsch. Gesellshaft 486: 1-38, 14 textfigs.

bottom sediments
Scruton, P.C., 1960.
Delta building and the deltaic sequence.
In: Recent sediments, northwest Gulf of Mexico, 1951-1958. Amer. Assoc. Petr. Geol. Tulsa, 82-102, with consolidated bibliography, pp. 368-381

bottom sediments
Scruton, P.C. (undated)
Sediments of the eastern Mississippi delta.
Finding American Shorelines, 21-50.

bottom sediments

Seetaramaswamy, A., 1969.

Some aspects of the sediments off the Krishna Delta and Nizampatam Bay, East Coast of India. Bull. natn. Inst. Sci., India, 38(1): 428-435.

Seibold, Eugen

bottom sediments

Seibold, Eugen, 1961.
Der Boden der Ozeane und die Erdgeschichte.
Die Naturwiss., 48(9):319-324.

bottom sediments

Selariu, Octavien, Jeanne Mares-Marinesco, et Mariette Pauliuc, 1969.
Contribution à l'étude des dépôts marins quaternaires de la plateforme continentale de la mer Noire dans le secteur roumain
Rapp. P.-v. Réun. Commn int. Explor. scient. Mer Mediterr., 19(4): 629-631.

bottom sediments

Sharma, G.D., 1971.
5. Sediments. In: Impingement of man on the oceans, D.W. Hood, editor, Wiley Interscience: 169-188.

bottom sediments

Sharma G.D., and D. C. Burrell, 1970.
Sedimentary environment and sediments of Cook Inlet, Alaska.
Bull. Am. Ass. Petrol. Geol. 54(4): 647-654.

bottom sediments

Shearer, James M. 1967.
Bottom sediments of Port au Port Bay.
Marit. Sed., Halifax, 3(1): 5. (multilitted).

Shepard, F.P.

bottom sediments

Shepard, Francis P., 1964.
Ungraded sand and gravel deposits in submarine canyons. (Abstract).
Geol. Soc., Amer., Special Paper, No. 76:149-150.

bottom sediments

Shepard, F. P., 1956.
Marginal sediments of Mississippi delta.
Bull. Amer. Assoc. Petr. Geol., 40(11):2537-2633.

bottom sediments

Shepard, F.P., 1952.
Sediment distribution on the East Asiatic continental shelves. (Abstract). Proc. Seventh Pacific Sci. Congr., Met. Ocean. 3:209-210.

Abstract of: Shepard, F.P., K.O. Emery, and H.R. Gould, 1949 in: Allan Hancock Foundation Publications, Occ. Papers, No. 9:1-64.

bottom sediments

*Shepard, F.P., R.F. Dill and Ulrich von Rad, 1969.
Physiography and sedimentary processes of La Jolla Submarine Fan and Fan-Valley, California.
Bull.Am.Ass.Petrol.Geol., 53(2):390-420.

sediments

Shepard, F. P., and K. O. Emery, 1946.
Bottom-sediment charts of the Asiatic continental shelves. Bull. G. S. A. 57:1229-1230.

submarine geology

Shepard, F. P., K.O. Emery, and H.R. Gould, 1949
Distribution of sediments on East Asiatic Continental Shelf. Allan Hancock Found. Publ., Occ. Paper. No. 9:38 pp., 3 charts, 26 figs.

submarine geology

Shepard, F.P., and E.C. LaFond, 1940
Sand movements along the Scripps Institution pier. Am. J. Sci. 238:272-285.

submarine geology

Shepard, F.P., and G.E. MacDonald, 1938
Sediments of Santa Monica Bay, California
Bull. Am. Assoc. Pet. Geol. 22:201-216.

bottom sediments

Shepard, F.P., F. B Phleger and T.H. Van Andel, Editors, 1960
Recent sediments, northwest Gulf of Mexico.
Amer. Assoc. Petrol. Geol., Editors, 394 pp.

bottom sediments

Shepard, F.P., F.B. Phleger, & Tj. H. Van Andel, Editors, 1960

Recent sediments, northwest Gulf of Mexico.
Symposium, Prof. 51, Amer. Petrol. Inst. Tulsa: 394 pp.
Rev. in Amer. J. Sci., 259(7):551-554.

bottom sediments

Shepard, F.P., and G.A. Rusnek, 1957.
Texas Bay sediments.
Publ. Inst. Mar. Sci., Port Aransas, 4(2):5-13.

bottom sediments

Shor, George G., Jr., 1964.
Structure of the "ering(s) Sea and the Aleutian Ridge.
Marine Geology, 1(3):213-219.

bottom sediments

Shukri, N.M., 1945.
Bottom deposits of the Red Sea. Nature 155 (3932):306.

bottom sediments

Shurko, I. I. 1971.
The features of Late Quaternary sedimentation in the Argentine Trough (In Russian)
Dokl. Akad. Nauk SSSR 198(3): 696-698

bottom sediments

Simmons, E.G., 1957.
An ecological survey of the upper Laguna Madre of Texas. Publ. Inst. Mar. Sci., Port Aransas, 4(2):156-200.

bottom sediments

Singer, S. Fred, 1967.
Zodiacal dust and deep-sea sediments.
Science, 156(3778):1080-1083.

bottom sediments

Siyazuki, K., 1951.
[Studies on the foul-water drained from factories 1. On the influence of foul-waters drained from factories by the coast on the water of Mitaziri Bay.] Contr. Simonoseki Coll. Fish. Pt. 1: 4 textfigs. (English summary).

bottom sediments

*Skornyakova, N.S., and V.P. Petelin, 1967.
Sediments of the central part of the southern Pacific. (In Russian; English abstract).
Okeanologiia, Akad. Nauk, SSSR, 7(6):1005-1019.

bottom sediments

Smith, D.Taylor, and W.N. Li, 1966.
Echo-sounding and sea-floor sediments.
Marine Geol., 4(5):353-364.

bottom sediments

Snegovskii, S.S. 1971.
On sedimentary deposits of the South-Okhotsk abyssal depression. (In Russian)
Dokl. Akad. Nauk SSSR, 196(1): 87-90.

bottom sediments

Soot-Ryen, T., 1925.
The Folden Fiord. Zoological, hydrographical and quarternary geological observations made in the Folden Fiord during the summer of 1923. Introduction. Tromsø Mus. Skr. 1(1):11 pp., 5 figs., 1 chart (fold-in).

bottom sediments

South Africa, Fisheries and Marine Biological Survey Division, 1952.
Twenty-first annual report for the year ended December 1949. Station list R.S. "Africana".
Commerce and Industry: 51 pp., 6 charts.

bottom sediments (data)

South Africa, Division of Fisheries, 1950.
Station list - R.S. "Africana II", R.V. "Shipa" and P.B. "Palinurus". 22nd Ann. Rept.:21-169, 2 charts.

bottom sediments

Springer, S., and H.R. Bullis, 1952.
Exploratory shrimp fishing in the Gulf of Mexico, 1950-51. Fishery Leaflet 406:34 pp., 9 textfigs. (multilithed).

bottom sediments

Stanley, Daniel J., and Gilbert Kelling, 1968.
Photographic investigation of sediment texture, bottom current activity, and benthonic organisms in the Wilmington Submarine Canyon.
U.S.C.G. Oceanogr. Rept. 22 (CG 373-22): 95 pp.

bottom sediments

*Stanley, Daniel J., Donald J.P. Swift and Horace G. Richards, 1967.
Fossiliferous concretions on Georges Bank.
J. Sedim. Petrol. 37(4):1070-1083.

Bottom Sediments

Starik, I. E., and A. P. Jarkov, 1961.
The rapidity of accumulation of sediments in the Indian Ocean using the radio-carbon method.
Doklady Akad. Nauk, SSSR, 136(1): 203-205

bottom sediments

Sternberg, R.W., 1968.
Friction factors in tidal channels with differing bed roughness
Marine Geol. 6(3): 243-260

Stetson, Henry C.

bottom sediments

Stetson, H.C., 1953.
The sediments of the western Gulf of Mexico. Pt. I. The continental terrace of the western Gulf of Mexico: its surface sediments, origin and development. P.P.O.M. 12(4):1-45, 18 text figs.

submarine geology

Stetson, H. C., 1949.
The sediments and stratigraphy of the east coast continental margin; Georges Bank to Norfolk Canyon. P.P.O.M. 11(2):60 pp., 5 textfigs.

submarine geology
Stetson, H. C., 1941
Oceanography. Chapter in Geology, 1888-1938, Fiftieth Anniversary Volume. Geol. Soc. Amer., pp. 43-69.

submarine geology
Stetson, H.C., 1938
The sediments of the continental shelf off the eastern coast of the United States. P.P.O.M. 5(4):1-48, 15 textfigs.

owned by M.S.

submarine geology
Stetson, H.C. and F.B. Phleger, jr. 1947
Oceanography as related to Petroleum Geology. Bull. Am. Assoc. Petr. Geol., 31(1):175-178.

bottom sediments
Stewart, H.B., jr., 1956.
Contorted sediments in a modern coastal lagoon explained by laboratory experiments. Bull. Amer. Assoc. Petr. Geol., 40(1):153-161.

bottom sediments
Stewart, R.A., and D.S. Gorsline, 1962.
Recent sedimentary history of the St. Joseph Bay, Florida.
Sedimentology, 1(4):256-286.

bottom sediments
Stewart, Richard A., Orrin H. Pilkey and Bruce W. Nelson, 1965.
Sediments of the northern Arabian Sea.
Marin Geol., 3(6): 411-427.

bottom sediments
Stocks, Th., 1961
Eine neue Tiefenkarte der Deutschen Bucht.
Berichte zur deutschen Landskunde, 27(2): 280-305.

bottom sediments
Stocks, T., 1956.
Der Boden der südlichen Nordsee. 2. Beitrag. Eine neue Tiefenkarte der südlichen Nordsee.
Deutsche Hydrogr. Zeits., 9(6):265-280.

bottom sediments
Stocks, T., 1955.
Der Steingrund bei Helgoland.
Deutsche Hydrogr. Zeits. 8(3):112-118.

bottom sediments
Storr, John F., 1964.
Ecology and oceanography of the coral-reef tract Abaco Island, Bahamas.
Spec. Papers, Geol. Soc., Amer., 79:98 pp.

bottom sediments
Stride, A.H., 1963
Current-swept sea floors near the southern half of Great Britain.
Quart. Jour. Geol. Soc., London, 119:175-199.

Also in:
Collected Reprints, N.I.O., Vol. 11(458).1963.

bottom sediments
Steuer, A., 1935.
The fishery grounds near Alexandria. 1. Preliminary report. Ministry of Finance, Egypt, Coastguard and Fisheries Service, Fisheries Research Directorate, Memoirs and Notes No. 8:1-18, 4 textfigs., 2 charts.

bottom sediments
Strakhov, N., 1946.
[Types historicogéologiques de sédimentation.]
Bull. Akad. Nauk, SSSR, Geol., 2:39-71, Fig. 3.

submarine geology
Stubbings, H.G., 1939.
The marine deposits of the Arabian Sea. An investigation into their distribution and biology. John Murray Exped., 1933-34, Sci. Repts., 3(2):31-158, 4 pls., 4 charts, 5 textfigs.

bottom sediments
Subba Rao, M., 1964.
Some aspects of continental shelf sediments off the east coast of India.
Marine Geology, 1(1):59-87.

bottom sediments
Sutton, G.H., and H. Berckhemer and J.E. Nafe, 1957.
Physical analysis of deep-sea sediments.
Geophysics 22(4):779-812.

bottom sediments
Sverev, S.M., 1961.
[On the structure of the sedimentary stratum of several parts of the Pacific Ocean on the basis of seismic reflected wave data.]
Izv., Akad. Nauk, SSSR, Ser. Geol., 1961(2):80-86

Engl. Abstr. in:
OTS Soviet-bloc, Res. Geophys., Astron., and Space, 1961(10):11-12.

Bottom sediments
Swift, Donald J.P., Bernard R. Pelletier, Anil K. Lyall and James A. Miller, 1969.
Sediments of the Bay of Fundy – a preliminary report.
Maritime Sediments 5 (3): 95-100

bottom sediments
Sysoev, N.N., V.M. Kovylin, G.N. Lunarsky, Yu. P. Neprochnov, I.F. Mikhno, G.B. Udintsev and S.M. Zverev, 1961.
On the relations between the thickness of sediments and the structures in the northwestern Pacific Ocean. (Abstract).
Tenth Pacific Sci. Congr., Honolulu, 21 Aug.-6 Sept., 1961, Abstracts of Symposium Papers, 387.

T

bottom sediments
*Taft, William H., Frank Arrington, Allen Haimovitz, Catheryn MacDonald and Charles Woolheater, 1968.
Lithification of modern marine sediments at Yellow Bank, Bahamas.
Bull.mar.Sci., 18(4):762-828.

bottom sediments
Takemura, Y., and Sagara, J., 1961
[Bottom characters of pearling beds in the Arafura Sea II. Bottom characters of pearling beds from viewpoint of the size distribution of bottom sediments.]
Bull. Tokai Reg. Fish. Res. Lab., No. 29: 41-46.

bottom sediments (data)
Tanaka, Otohiko, Haruhiko Irie, Shozi IIzuka and Fumihiro Koga, 1961
The fundamental investigation on the biological productivity in the north-west of Kyushu. 1. Rec. Oceanogr. Wks., Japan, Spec. No. 5: 1-58.

bottom sediments (mud)
Tanita, S., S. Kato, and T. Okuda, 1950.
Studies on the environmental conditions of shellfish fields. 1. In the case of Hakodate Harbor. Bull. Fac. Fish., Hokkaido Univ., 1(1):18-27 7 textfigs. (In Japanese; English summary).

bottom sediments
Tanita, S., and T. Okuda, 1956.
[Fundamental investigation on the marine resources of Matsushima Bay. IV. Seasonal successions of water properties, bottom materials and benthos.]
Bull. Tohoku Reg. Fish. Res. Lab., (6):106-134.

bottom sediments
Tanner, W.F., 1956.
Size and roundness in sediments. Bull. G.S.A. 67(4):535-536.

bottom sediments
Tanner, W.F., R.G. Evans and C.W. Holmes, 1963.
Low-energy coast near Cape Romano, Florida.
J. Sed. Petrol., 33(3):713722.

bottom sediments
Tatara, K., 1955.
Preliminary note on the bottom topography of the "Damo" area. Rec. Ocean. Wks., Japan, 2(3):71-78.

bottom sediments
Terwindt, J.H.J. 1971.
Litho-facies of inshore estuarine and tidal-inlet deposits.
Geol. Mijnb. 50 (3): 515-526.

bottom sediments
Thomas, Charles W., 1959.
Lithology and zoology of an Antarctic Ocean bottom core.
Deep-Sea Res., 6(1):5-15.

bottom sediments
Thoulet, M. J., 1926.
Relations entre la composition des sediments sous-marins et les conditions des eaux superficielles. Bull. Inst. Ocean., Monaco, No. 470:28 pp

bottom sediments
Thoulet, J., 1910.
Etude des fonds marins de la Baie de la Seine.
Ann. Hydrogr., Paris, 53-78.

not seen.

bottom sediments
*Tooma, Samuel G., Jr., and Harry Iredale II, 1968.
Oceanography in the Channel Islands area off southern California, September and October 1965.
Tech. Rept., U.S. Naval Oceanogr. Off., TR-203:50pp. (multilithed).

bottom sediments
Toombs, R.B., 1956.
Some characteristics of Bute Inlet sediments.
Trans. R. Soc., Canada, (3)50:59-65.

bottom sediments
Trask, P.D., 1953.
The sediments of the western Gulf of Mexico. Pt. II. Chemical studies of sediments of the western Gulf of Mexico. P.P.O.M. 12(4):46-120, 21 text figs.

sedimentation
Trask, P. D., F. B Phleger, and H. C. Stetson, 1947.
Recent changes in sedimentation in the Gulf of Mexico. Science 106:460-461.

bottom sediments
Tregouboff, G., 1956.
Rapport sur les travaux concernant le plancton Méditerranéen publiés entre Novembre 1952 et Novembre 1954.
Rapp. Proc. Verb., Comm. Int. Expl. Sci., Mer Méditerranée, 13:65-100

bottom sediments
Tressler, Willis L., 1960.
Oceanographic observations at IGY Wilkes station Antarctica.
Trans., Amer. Geophys. Union, 41(1) :98-104.

bottom sediments
Tressler, Willis L., 1960
Oceanographic and hydrographic observations at Wilkes IGY Station, Antarctica.
J. Washington Acad. Sci., 50(5):1-13.

bottom sediments
Tressler, W. L. 1941
Geology and biology of North Atlantic deep-sea cores between Newfoundland and Ireland. 4. Ostracoda.
U.S. Geol. Surv. Prof. Pap. 196 C: C95-C106, 4 pls, 1 fig.

bottom sediments
Tromeur, J.Y., 1938-39.
Mission hydrographique du Saloum, 1930-1931.
Ann. Hydrogr., Paris, (3)16:5-33, figs.

bottom sediments
*Tsuchi, Ryuichi, 1966.
Report of geologic survey by the Umitaka maru, International Indian Ocean Expedition in the winter of 1963-1964. (In Japanese; English abstradt).
J. Tokyo Univ. Fish. (Spec.ed.) 8(2):71-81.

bottom sediments
Tsujita, T., 1953.
A marine ecological study on the Bay of Omura.
J. Ocean. Soc., Japan, 9(1):23-32, 6 textfigs.

U

bottom sediments
Uchupi, Elazar, 1966.
Structural framework of the Gulf of Maine.
J. geophys. Res., 71(12):3013-3028.

bottom sediments
Uchupi, Elazar, and K.O. Emery, 1968.
Structure of continental margin off Gulf coast of United States.
Bull. Am. Ass. Petr. Geol. 52 (7): 1162-1193.

bottom sediments
Uchupi, E., and R. Gaal, 1963.
Sediments of the Palos Verdes shelf.
In: Essays in Marine Geology in honor of K.O. Emery, Thomas Clements, Editor, Univ. Southern California Press, 171-189.

bottom sediments
Udintsev, G.B., Yu.P. Neprochnov, and V.M. Kovylin, 1965.
Thickness of the sedimentary layer and structure of the Earth's crust in seas and oceans based on the results of seismo-acoustic measurements. (In Russian; English abstract).
Okeanolog. Issled., Roz. Issled. po Programme Mezhd. Geofiz. Goda, Mezhd. Geofiz. Komitet, Prezidiume Akad. Nauk, SSSR, No. 13:181-188.

bottom sediments
United States Naval Oceanographic Office 1965.
Oceanographic atlas of the North Atlantic Ocean. V.
U.S. Oceanogr. Off. Publ. 700: 71 pp. (quarto)

bottom sediments
*Urien, Carlos Maria, 1967.
Los sedimentos modernos del Rio de la Plata exterior.
Bol. Servicio Hidrografia Naval, Armada Argentina, 4(2):113-213.

bottom sediments
Urien, Carlos M., 1966.
Distribución de los sedimentos modernos del rio de la Plata superior.
Bol. Serv. Hidrogr. Naval, Argentina, 3(3):197-203.

bottom sediments
Urry, W.D., 1948
Marine sediments and Pleistocene chronology.
Trans. N. Y. Acad. Sci., Ser. II 10:63-69.

bottom sediment
U.S.N. Hydrographic Office, 1958
Oceanographic atlas of the Polar seas. Pt. II. Arctic. H.O. Publ. No. 705: 139 pp., charts.

bottom sediments
U.S. Navy Hydrographic Office, 1957.
Operation Deep Freeze II, 1956-1957. Oceanographic results. H.O. Tech. Rept. 29:155 pp., (multilithed).
survey

V

bottom sediments
Van Andel, Tjeerd H., 1965.
Morphology and sediments of the Sahul Shelf, northwestern Australia.
Trans. N.Y. Acad. Sci., (2)28(1):81-89.

bottom sediments
Van Andel, Tjeerd H., 1964.
Recent marine sediments of Gulf of California.
In: Marine sediments of Gulf of California, a symposium, Amer. Assoc. Petr. Geol., Memoir, T. van Andel and G.G. Shor, Jr., editors, 3:216-310.

bottom sediments
Van Andel, Tjeerd H., Vaughan T. Bowen, Peter L Sachs and Raymond Siever, 1965.
Morphology and sediments of a portion of the Mid-Atlantic Ridge.
Science, 148 (3674):1214-1216.

Bottom sediments
Van Andel, T.J., and H. Postma, 1955.
Recent sediments in the Gulf of Paria.
Bull. Amer. Assoc. Petr. Geol. 39(10):2091-2094.

bottom sediments
Van Andel, T.J., and H. Postma, 1954.
Recent sediments of the Gulf of Paria.
Verh. K. Nederl. Akad. Wetensch., Afd. Natuurk., 20(5):245 pp., 4 fold-in maps.

bottom sediments (data)
Van Goethem, C., 1951.
Études physique et chimique du milieu marin.
Rés. Sci., Exped. Océan. Belge dans les Eaux Africaines de l'Atlantique Sud (1948-1949) 1:1-151, 1 pl.
Côtières

bottom sediments
Vannet, J.-R., and A. Guilcher, 1963.
La répartition des fonds sous marins dans le "Mor Bras" (Morbihan). Résultats préliminaires.
Comptes Rendus, Acad. Sci., Paris, 256(14):3170-3171.

bottom sediments
Van Straaten, L.M.J.U., 1965.
Sedimentation in the north-western part of the Adriatic Sea.
In: Submarine geology and geophysics, Colston Papers, W.F. Whittard and R. Bradshaw, editors, Butterworth's, London, 143-160. 143-171

bottom sediments
Van Straaten, L.M.J.U., Editor, 1964.
Deltaic and shallow marine deposits. Proceedings of the Sixth International Sedimentological Congress, the Netherlands and Belgium, 1963. Elsevier, Amsterdam, 464 pp.

Review with list of papers presented:
Marine Geology, 1(2):189-191.

bottom sediments
Van Straaten, L.M.J.U., 1959.
Minor structures of some littoral and neritic sediments. Geol. en Mijnbouw (nw. ser.) 21:197-216.

bottom sediments
Van Straaten, L.M.J.U., 1954.
Composition and structure of recent marine sediments in the Netherlands.
Leidse Geol. Mededal., 19:1-110.

bottom sediments
Vasiliev, G.D., y Iu. A. Torin, 1965.
Carastica oceanografica pesquera y biologica del Golfo de México y del Mar Caribe. (In Russian: Spanish abstract).
Sovetsk. -Cub. Ribokhoz. Issled., VNIRO: Tsentr. Ribokhoz. Issled. Natsional. Inst. Ribolovsta Republ. Cuba, 241-266.

bottom sediments
Vasiliev, B.I., and N.P. Vasilkovskii 1971.
Discovery of marine Miocene deposits on continental slope of Peter the Great Gulf (Sea of Japan). (In Russian)
Dokl. Akad. Nauk SSSR 198(5): 1195-1198.

bottom sediments
Vause, J.E., 1959.
Underwater geology and analysis of Recent sediments off the northwest Florida coast.
J. Sed. Petr., 29(4):555-563.

bottom sediments
Venkatarathnam, K., 1969.
Studies on the sediments of the continental shelf off the regions - Visakhapatnam-Pudimadaka and Pulicat Lake-Penner River confluence along the East Coast of India. Bull. natn. Inst. Sci., India, 38(1): 472-482.

bottom sediments
Vigneaux, M., J.C. Dumon et C. Latouche 1968
Étude sédimentologique et géochimique d'une carotte de vases marines du golfe de Gascogne.
Bull. Inst. géol. Bassin d'Aquitaine 5: 74-83.

bottom sediments
*Villegas, L., H. Trujillo y E. Löhre, 1967.
Informe sobre las investigaciones exploratorias realizadas en la zona de Chiloe-Guitecas, durante febrero y marzo de 1967, con el B/I Carlos Darwin.
Publ. Inst. Fomento Pesquero, Chile, 32: 19 pp. (mimeographed).

bottom sediments
Vincent, Joaquín Ros 1966.
Contribución al estudio de la materia orgánica en los sedimentos marinos de la cuenca occidental del Mediterráneo.
Boln Inst. esp. Oceanogr. 127:31pp.

Vinogradova, P.S., 1964. bottom sediments
Bottom sediments of the Norwegian Sea. (In Russian).
Trudy, Poliarn. Nauchno-Issled. i Proektn. Inst. Morsk. Ribn. Choz. i Okeanogr. im N.M. Knipovicha, 16:111-132.

bottom sediments
Vinogradova, P. S., 1959.
Some peculiarities of the sediment forming in the fjords of the Eastern Murman.
Trudy Pol. Nauch. Issle. Inst. Moskogo, 11:26-34.

Vinogradova, P.S., and V.M. Litvin, 1960 bottom sediments
Studies on the bottom relief and sediments in the Barents and Norwegian seas. Soviet Fish. Invest., North European Seas, VNIRO, PNIRO, Moscow. 1960:97-110.

bottom sediments
Vlassov, A. Ya., and G.V. Kovalenko, 1962.
Effect of compression on the natural remanent magnetization of benthic deposits of the Atlantic Ocean. (In Russian).
Izv. Akad. Nauk, SSSR, Ser. Geofiz., (5):639-643.
English Edit., (5):415-417.

bottom sediments
Vollbracht, K., 1953.
Zur Quarzachsenregelung sandiger Sedimente.
Acta Hydrophysica 1(2):61-88.

bottom sediments (data)
von Bonde, C., 1950.
Twentieth annual report for the year ended December 1948. Commerce and Industry, Apr. 1950: 412 pp., 12 charts.

bottom sediments (data)
von Bonde, C., 1949
Nineteenth Annual Report for the year ended December 1947. Division of Fisheries, Dept. Commerce and Industries. Commerce & Industry, S. Africa, Apr. 1949:415-476, 4 charts.

bottom sediments
Von Huene, Roland, George G. Shor, Jr., and Erk Reimnitz, 1967.
Geological interpretation of seismic profiles in Prince William Sound, Alaska.
Bull.geol.Soc.Am., 78(2):259-268.

bottom sediments
Vuletic, A., 1953.
Structure geologique du fond du Malo et du Veliko Jesero, sur l'Ile de Mjlet.
Acta Adriatica 6(1):3-63, 16 textfigs.

W

bottom sediments
Waldichuk, M., 1953.
Oceanography of the Strait of Georgia. III. Character of the bottom. Prog. Repts., Pacific Coast Stas., Fish. Res. Bd., Canada, 95:59-63, 4 textfigs.
aerial photos showing silt-laden fresh water.

bottom s ediments
Wangersky, Peter J., 1958.
The sea as a clue to the earth's history.
Yale Scientific Magazine, 33(3):6 pp.

bottom sediments
Weibull, W., 1947.
The thickness of ocean sediments measured by a reflexion method. Göteborgs K. Vetenskaps- och Vitterhets Samhalles Handl., Sjatte Foljden, B, 5(4):1-17.
Medd. Ocean. Inst., Göteborg 12

bottom sediments
*Werner, Friedrich, 1968.
Gefügeanalyse feingeschichteter Schlicksedimente der Eckernförder Bucht (westliche Ostsee).
Meyniana, 18:79-105.

bottom sediments
Weydert, Pierre, 1970.
Interprétation granulométrique d'un modèle actuel: les ensembles sédimentaires récifaux de la baie de Tuléar (Sud-ouest de Madagascar).
C.-r. hebd. Séanc., Acad. Sci. Paris (D) 271(20): 1748-1751

bottom sediments, bacteria
*Weyland, Horst, 1967.
Über die Verbreitung von Abwasserbakterien im Sediment des Weserästuars.
Veröff. Inst. Meeresforsch. Bremerh. 10(3):173-182.

bottom sediments
Whittard, W.F., 1962.
Geology of the western approaches of the English Channel, a progress report.
Proc. R. Soc., London, (A), 265(1322):395-406.

Bottom sediments
Wigley, Roland L., 1961.
Bottom sediments of Georges Bank.
J. Sed. Petr., 31(2):165-188.

bottom sediments
Williams, Joseph D., 1963.
The petrology and petrography of sediments from the Sigsbee blanket, Yucatan Shelf, Mexico.
Texas A. & M., Dept. of Oceanogr. and Meteorol., Ref., No. 63-12T:60 pp.

bottom sediments
Williams, H.F., 1951.
The Gulf of Mexico adjacent to Texas. Texas J. Sci. 3(2):237-250, 11 textfigs.

bottom sediments
Wimberley, C.S., 1955.
Marine sediments north of Scripps submarine canyon, La Jolla, California. J. Sed. Petr. 25(1):24-37, 10 textfigs.

bottom s ediments
Wimberley, Stanley, 1963.
Sediments of the mainland shelf near Santa Barbara, California.
In: Essays in Marine Geology in honor of K.O. Emery, Thomas Clements, Editor, Univ. Southern California Press, 191-201.

bottom sediments
Wiseman, J.D., & W.R. Riedel, 1960
Tertiary sediments from the floor of the Indian Ocean.
Deep-Sea Res., 7(3): 215-217.

Wüst, G., 1942. bottom sediments.
Die morphologischen und ozeanographischen Verhältnisse des Nordpolarbeckens. Veröffentlichungen des Deutsches Wissenschaftlichen Instituts zu Kopenhagen. Reihe 1: Arktis. Herausgegeben von Prof. Dr. Phil. Hans Freibold, No. 6: 21 pp., 7 textfigs., 1 pl. (Gebrüde Borntrager, Berlin-Zehlendorf, 1942).

bottom sediments
Yakuwa, R., 1946.
Über die Bodentemperaturen in den verschieden Bodenarten in Hokkaido. Geophys. Mag., Tokyo, 14(1):1-12, 2 figs.

bottom sediments
Yamazi, I., 1952?
Plankton investigations of inlet waters alon the coast of Japan. VI. The plankton of Nanao Bay.
Seto Mar. Biol. Lab. Contr. No. 191:309-319, 11 textfigs.

bottom sediments
Yemelyanov, Ye. M., 1961.
New data on Medditerranean sea deposits. (In Russian).
Doklady, Akad. Nauk, SSSR, 137(6):1437-1440.
translation:
Consultants Bureau for Amer. Geol. Inst., 137 (1-6):432-434. (1962).

bottom sediments
Yermakov, Yu. G., 1969.
The influence of tectonics on the distribution of the upper Mesozoic and Cenozoic deposits in the northern coastal region of the Black Sea. (In Russian)
Geotekton., Akad Nauk SSSR 1969(4):119-121. Translation: Geotectonics, Scripta Technica for AGU (1970): 281-283

bottom sediments
Yoshikawa, T., 1953.
Some consideration on the continental shelves around the Japanese Islands.
Nat. Sci. Rept., Ochanomizu Univ., 4(1):138-150, 9 textfigs.

bottom sediments
Young, D.K. and D.C. Rhoads, 1971.
Animal-sediment relations in Cape Cod Bay, Massachusetts. 1. A transect study. Mar. Biol. 11(3): 242-254.

bottom sediments
Zabel, M., 1962.
Ölsuche im Küstenvorfeld dabei Einsatz von Funkortungsverfahren vie Lorac und Decca.
Der Seewart, 23(5):177-180.

bottom sediments
Zeigler, John M., 1964.
The hydrography and sediments of the Gulf of Venezuela.
Limnology and Oceanography, 9(3):397-411.

bottom sediments
Zeigler, John M., and Ramon Perez Mena, 1960.
Distribucion de sedimentation en el Golfo de Venezuela.
Memoria Tercer Congreso Geologico Venezolano, 2: 895-904.

bottom sediments
Zen, E-An, 1959.
Mineralogy and petrography of marine bottom sediment samples off the coast of Peru and Chile.
J. Sed. Petr., 29(4):513-539.

bottom sediments
Zen, E-an, 1957.
Preliminary report on the mineralogy and petrology of some marine bottom s amples of the coast of Peru and Chile.
Amer. Mineralog., 42:889-903.

bottom sediments
Zenkevitch, N.L., 1961.
[New data on the bottom topography of the northeastern Pacific.]
Trudy Inst. Okeanol., Akad. Nauk, SSSR, 45:5-21.

bottom sediments
Zenkovich, V.P., 1969.
The near-shore shallows of western Cuba and their sediments. (In Russian; English abstract)
Okeanologiia, 9(2): 256-270.

bottom sediments
Zhivago, A.V., 1960
[Geomorphological research.] Arktich. i Antarktich. Nauchno-Issled. Inst. Mezhd. Geofiz. God Sovetsk. Antarkt. Exped., Vtoraia Morsk. Exped. "Ob", 1956-1957, 7: 44-72. (In Russian)

bottom sediments
Zhivago, A.V., 1960.
[On Easter Island.] Nauka i Zhizn' (7):44-49.

bottom sediments
Zhivago, A. V., and A. P. Lisitsin, 1958.
Bottom relief and the deposits of the Southern Ocean.
Inform. Biull. Sovetsk. Antarkt. Exped., (3):21-22.

bottom sediments
Zhivago, A.V., and A.P. Lisitsin, 1957.
[New data on the bottom relief and the sediments of east Antarctic seas.]
Izv., Akad. Nauk, SSSR, Ser. Geofiz., (1):19-35.
OTS $0.50

bottom sediments
Zverev, S.M., 1960.
Dynamic peculiarities of multiple reflected "water" waves in the ocean and their use in the determination of elastic waves in sediments. (In Russian).
Izv., Akad. Nauk, SSSR, Ser. Geofiz., (1):24-36.
English Transl., (1):12-20.

bottom sediments, abyssal

bottom sediments, abyssal
Bezrikov, P.L., 1962
The uneven distribution of abyssal oceanic sediments.
Okeanologiia, Akad. Nauk, SSSR, 2(1):9-25.
Abstracted in: Soviet Bloc Res. Geophys., Astron. and Space, 1962(35): 14-15.
(OTS61-11147-35 JPRS13739)

bottom sediments, abyssal
Clarke, R.H., 1968.
Burrow frequency in abyssal sediments.
Deep-Sea Res., 15(3):397-400.

bottom sediments
Emelyanov, E.M., and Shimkus, K.M., 1963
New data on the abyssal deposits related to the novoeuxenite rocks of the Black Sea. (In Russian).
Okeanologiia, Akad. Nauk, SSSR, 3(3):482-494.

bottom sediments (abyssal)
Emelianov, E.M., and K.M. Shimkus, 1962.
On the problem of variability of abyssal sediments in the Black Sea (In Russian).
Okeanologiia, Akad. Nauk, SSSR, 2(6):1040-1049.

bottom sediments, abyssal
Heath, G. Ross, and Theodore C. Moore, Jr., 1965.
Subbottom profile of abyssal sediments in the Central Equatorial Pacific.
Science, 149(3685):744-746.

bottom sediments, abyssal
Heezen, Bruce C., and A.S. Laughton, 1963.
14. Abyssal plains.
In: The Sea, M.N. Hill, Editor, Interscience Publishers, 3:312-364.

bottom sediments, abyssal
Il'in, A.V., I.I. Shurko, D.S. Nikolaev, and E.I. Efimova, 1967.
On the problem of stratification of abyssal deposits of the equatorial Atlantic Zone. (In Russian).
Dokl., Akad. Nauk, SSSR, 176(2):438-442.

bottom sediment accumulation
Carwile, R.H., and Faure, G., 1971.
Strontium isotope ratios and base metal content in a core from the Atlantis II Deep, Red Sea.
Chem. Geol. 8(1): 15-23

bottom sediments, accumulation of
Moore, T.C., 1971.
Radiolaria. Initial Repts Deep Sea Drilling Project, 8: 727-775.

bottom sediments, acoustic properties

bottom sediments, acoustic properties
Bennett, G.S., 1962.
Measurement of bottom acoustical characteristics.
J. Acoust. Soc., Amer., 34(10):1660-1661.

bottom sediments, acoustic properties
* Bouling, C., J. Bonnan, J. Greffard et J. Revel, 1969.
Traitement sur ordinateur de quelques données relatives à la propagation du son dans les sédiments marins recents.
Cah. océanogr., 21(2): 173-191.

bottom sediments, acoustic measurements
*Buchan, S., F.C.D. Dewes, D.M. McCann and D. Taylor Smith, 1967.
Measurements of the acoustic and geotechnical properties of marine sediment cores.
In: Marine geotechnology, A.F. Richards, editor, Univ. Illinois Press, 65-92.

bottom sediments, acoustic properties
Chassefière, B. et O. Leenhardt 1967.
Mesures acoustiques et mécaniques dans les vases.
Vie Milieu (B) 18 (1B): 1-32.

bottom sediments, acoustic properties
Chesterman, W. Deryck, J.M.P. St. Quinton Y. Chan and H.R. Matthews, 1967.
Acoustic surveys of the sea floor near Hong Kong.
Int. hydrogr. Rev., 44(1):35-54.

bottom sediments, acoustic properties
Fry John C. 1962
A low-velocity sediment layer in the Pacific. (Abstract).
J. Geophys. Res., 67(9):3559.

bottom sediments,
Fry, John C., and Russell W. Raitt, 1961.
Sound velocities at the surface of deep sea sediments.
J. Geophys. Res., 66(2):589-598.

bottom sediments, acoustic properties
* Greffard, Jacques, 1969.
Contribution à l'étude des relations entre la propagation du son et la nature physico-chimique des sédiments marins.
Cah. oceanogr., 21(2): 145-171.

bottom sediments, acoustic properties
Hamilton, Edwin L., 1971.
Prediction of in-situ acoustic and elastic properties of marine sediments.
Geophysics 36(2): 266-284.

bottom sediments, acoustic properties
Hamilton, Edwin L. 1970.
Sound channels in surficial marine sediments.
J. Acoust. Soc. Am. 48 (5-2):1296-1298

bottom sediment, acoustic properties
Hamilton, Edwin L., 1970.
Sound velocity and related properties of marine sediments, North Pacific. J. geophys. Res., 75(23): 4423-4446.

bottom sediments, acoustical properties
Hamilton, Edwin L., 1965.
Sound speed and related physical properties of sediments from experimental Mohole (Guadalupe Site).
Geophysics, 30(2):257-261.

bottom sediments, acoustics
Hamilton, Edwin L., 1963
Sediment sound velocity measurements made in situ from bathyscaph Trieste.
J. Geophys. Res., 68(21):5991-5998.

bottom sediments, acoustic properties
Hamilton, E.L., 1956.
Low sound velocities in high-porosity sediments.
J. Acoust. Soc., Amer., 28(1):16-19.

bottom sediments, acoustic properties of
*Hampton, Lloyd D., 1967.
Acoustic properties of sediments.
J. Acoust. Soc., Am., 42(4):882-890.

bottom sediments, acoustic properties of
Horton, C.W., 1956.
Acoustic properties of the ocean bottom at Bahia Todos Santos, Lower California. J. Acoust. Soc., Amer., 28(6):1243-1246.

bottom sediments, acoustic properties
Houtz, Robert, John Ewing and Xavier Le Pichon 1968.
Velocity of deep-sea sediments from sonobuoy data.
J. geophys. Res. 73 (8): 2615-2641.

bottom sediments, acoustic properties
Japan, Maritime Safety Board, 1954.
[Discrimination of bottom characteristics by listening sounds.] Publ. No. 981, Hydrogr. Bull., Maritime Safety Bd., Tokyo No. 42:139-141, 7 textfigs.

bottom sediments, acoustics
Kronengold, M. and J. M. Loewenstein 1968
A system for the study of underwater sound propagation.
Mar. Sci. Instrument. 4: 642-651.

BOTTOM SEDIMENTS, acoustics
*Le Pichon, Xavier, John Ewing and Robert E. Houtz, 1968.
Deep-sea sediment velocity determination made while reflection profiling.
J. geophys. Res., 73(8):2597-2614.

bottom sediments, acoustic properties
*McElroy, E.G., and A. DeLoach, 1968.
Sound speed and attenuation from 15 to 1500 kHz, measured in natural sea-floor sediments.
J. acoust. Soc. Am., 44(4):1148-1150.

bottom sediments, acoustic properties
McManus, Dean A., 1970.
Comparison of three methods of measuring or estimating sonic velocity in sediments. Initial Repts. Deep Sea Drilling Project, Glomar Challenger 5: 545-549.

bottom sediments, acoustical properties
Migikos, Jean-Pierre 1971.
Propriétés mécaniques et acoustique des sables marins. II. Propagation d'une onde acoustique et mécanismes de pertes liés à la presence du fluide saturant.
Cah. océanogr. 23 (9): 801-826

bottom sediments, acoustic properties
Migikos, J.-P., 1971
Lettre à l'éditeur à propos des propriétés mécaniques et acoustiques des sables marins.
Cah. oceanogr. 23 (10): 981-982

bottom sediments, acoustical properties
Migikos, Jean-Pierre, 1971.
Propriétés mécaniques et acoustiques des sables marins. I. Mecanismes de pertes dans la matrice et relations contrainte-deformation.
Cah. océanogr. 23(8): 709-732.

bottom sediments, acoustic properties
Moore, David G., 1964.
Acoustic-reflection reconnaissance of continental shelves: eastern Bering and Chukchi seas.
In: Papers in Marine Geology, R.L. Miller, editor Macmillan Co., N.Y., 319-362.

bottom sediments, acoustics
Moore, David G., 1960
Acoustic-reflection studies of the continental shelf and slope off southern California. Bull. Geol. Soc., Amer., 71: 1121-1136.

~~Bottom sediments~~, acoustics in
Schirmer, Florian, 1970.
Schallausbreitung im Schlick.
Dt. hydrogr. Z. 23(1): 24-30

bottom sediments, acoustics
Shumway, George, 1960.
Sound speed and absorption studies of marine sediments by a resonance method. II.
Geophysics, 25(3):659-682.

bottom sediments, acoustics
Shumway, George, 1958.
Sound velocity vs. temperature in water-saturated sediments.
Geophysics, 23(3):494-505.

bottom sediments, acoustics
Shumway, G., 1956.
Resonance chamber sound velocity and attenuation measurements in sediments. Geophysics 21(2):305-319.

bottom sediments, acoustic properties of
Stocks, T., 1952.
Schallweicher und schallharter Boden, ein Bereich des Stoller Grundes (Kieler Bucht) in Beziehung zur Geologie. Deutsches Hydrogr. Zeits. 5(2/3): 149-153, 3 textfigs.

Bottom sediments, acoustics
Stoll Robert D. and Perez M. Bryan 1970
Wave attenuation in saturated sediments.
J. acoust. Soc. Am., 47 (5-2): 1440-1447

bottom sediments, acoustic properties
Ulonska, Arno, 1968.
Versuche zur Messung der Schallgeschwindigkeit und Schalldämpfung im Sediment in situ.
Dt. hydrogr. Z., 21(2):49-58.

Bottom sediments, aeolian

bottom sediments age determination

bottom sediments, age
Cifelli, Richard, 1970.
Age relationships of Mid-Atlantic Ridge sediments.
Spec. Pap. Geol. Soc. Am. 124: 47-69.

bottom sediments, age
Dibner, V.D., V.A. Basov, A.A. Gerke, M.F. Soloviyeva, G.P. Sosipatrova and N.I. Shul'gina, 1970.
Age of the pre-Quaternary deposits from the sediments of the Barents Sea shelf. (In Russian; English abstract). Okeanologiia, 10(4): 670-68

bottom sediments, age determination

bottom sediments, age determination
Arrhenius, G., G. Kjellberg, and W.F. Libby, 1951.
Age determination of Pacific chalk ooze by radiocarbon and titanium content. Tellus 3: 222-229.

bottom sediments, age determination
Baranov, V.J., and L.R. Kuzmina, 1954.
Ionium method of the age determination of sea sediments. Dokl. Akad. Nauk, SSSR, 97:483-

bottom sediments, age determination
Dymond, Jack, 1969.
Age determination of deep-sea sediments: a comparison of three methods.
Earth Planet. Sci. Lett. 6(1): 9-14.

bottom sediments, age determination
Emery, K.O., and E.E. Bray, 1962.
Radiocarbon dating of California Basin sediments.
Bull. Amer. Assoc. Petrol. Geol., 46(10):1839-1856.

bottom sediments, age determination
Kharkar, D.P., D. Lal, N. Narsappaya, & B. Peters, 1958
Dating of ocean sediments.
C.R. Congr. Assoc. Sci. Pays Ocean Indien, 3, Tananarive, 1957, Sec. C: 39-42.
Abs. in Nucl. Sci. Abs., 15(13): #17134.

bottom sediments, age determination
Kroll, V.S., 1954.
On the age-determination in deep-sea sediments by radium measurements. Deep-Sea Res. 1(4):211-215.

bottom sediments, age determination of
Lal, D., 1962
Cosmic ray produced radionuclides in the sea.
J. Oceanogr. Soc., Japan, 20th Ann. Vol., 600-614.

bottom sediments, age determination
Naidu, A.S., 1969.
Radiocarbon date of an oolitic sand collected from the shelf off the East Coast of India.
Bull. natn. Inst. Sci., India, 38(1): 467-471.

Bottom Sediments, age determination
Olsson, Ingrid U., and K. Goata Eriksson, 1965.
Remarks on C14 dating of shell material in sea sediments.
Progress in Oceanography, 3:253-266.

abstract

bottom sediments, aeolian
Rex, R.W., and E.D. Goldberg, 1962
Insolubles. Ch. 5, Sect. II. Interchange of properties between sea and air. In: The Sea, Interscience Publishers, Vol. 1, Physical Oceanography, 295-304.

sediments, air-borne
Windom, Herbert L. 1970.
Contribution of atmospherically transported trace metals to South Pacific sediments.
Geochim. cosmochim. Acta, 34 (4): 509-514

bottom sediments, anaerobic
Soutar, A., 1971.
Micropalaeontology of anaerobic sediments and the California current. In: Micropalaeontology of oceans, B.M. Funnell and W.R. Riedel, editors, Cambridge Univ. Press, 223-230.

bottom sediments, anoxic

bottom sediments, anoxic
Drever, James I., 1971.
Magnesium-iron replacement in clay minerals in anoxic marine sediments. Science, 172(3990): 1334-1336.

bottom sediments, anoxic
Fujita, Yuji, Tadataka Taniguti, Shoji Iizuka and Buhei Zenitani 1967
Microbiological studies on shallow marine areas. IV On the liberation and accumulation of sulfides in mud sediment, and its relation to the formation of anoxic layer. (In Japanese; English abstract)
Bull. Fac. Fish. Nagasaki Univ. 24: 79-88.

bottom sediments, anoxic
Mengies, Robert J., Gilbert T. Rowe and Larry P. Atkinson 1968
A small basin containing anoxic sediments - Lookout Bight, North Carolina.
Int. Revue ges. Hydrobiol. 53(1): 77-81.

bottom sediments, ash
Anders, E., and D. N. Limber, 1959.
Origin of the Worzel deep-sea ash.
Nature, 184(4679): 44-45.

bottom sediments, authigenic
bottom sediments, authigenic
Pratt, Richard M., 1968.
Atlantic continental shelf and slope of the United States, physiography and sediments of the deep-sea basin.
Prof.Pap.U.S.geol.Surv., 529B:B1-B44, fold in chart.

bottom sediments, bacteriology of
bottom sediments, bacteria
Bader, R.G., 1961.
Bacterial activity in sediments of shallow marine bays.
Trans. Amer. Fish. Soc., 90(3):342.
Is review of:
Oppenheimer, C.H., 1960. Geochimica et Cosmochimica Acta, 19:244-260.

bottom sediments bacteria in
Bansemir, Klaus 1969.
Bakteriologische Untersuchungen von Wasser und Sedimenten aus dem Gebiet der Island-Färöer Schwelle.
Ber. dt. wiss. Komm. Meeresforsch. 20 (3/4): 282-287.

bottom sediments, bacteria
Bianchi, A.J.M., 1971.
Distribution de certaines bactéries hétérotrophes aérobies dans les sédiments profonds de Méditerranée Orientale. Mar. Biol. 11(2): 106-117.

bottom sediments, bacteria
*Bianchi, A.J.M., 1968.
Un milieu de culture permettant d'évaluer la population bactérienne des sédiments marins et certaines potentialités fonctionnelles de ses composants.
Recl.Trav.Stn.mar.Endoume, 43(59):345-349.

bottom sediments, bacteria
Deveze, L., 1964.
Activités bactériennes et matières organiques dans les sediments marins.
Ann. Inst. Pasteur, 107(Suppl., No.3):123-135.

bottom sediments, bacteria in
Deveze, L., 1963.
Quelques remarques sur la distribution verticale des bactéries dans des sédiments méditerranéens littoraux et profonds.
Rapp. Proc. Verb., Réunions, Comm. Int., Expl. Sci., Mer Méditerranée, Monaco, 17(3):695-700.

bottom sediments, bacteria
Hayes, F. R., and E. H. Anthony, 1959.
Lake water and sediment. VI. The standing crop of bacteria in lake sediments and its place in the classification of lakes.
Limnol. & Oceanogr., 4(3):299-315.

bottom sediments, bacteria
Hickel, W., 1969.
Sedimentbeschaffenheit und Bakteriengehalt im Sediment eines zukünftigen Verklappungsgebietes von Industrieabwässern nordwestlich Helgolands.
Helgoländer wiss. Meeresunters. 19(1): 1-20

bottom sediments, bacteria
Kriss, A.E., and E.A. Rukina, 1952.
Microorganisms in sea-bed deposits in ocean regions. (In Russian).
Izv., Akad. Nauk, SSSR, Ser. Biol., (6):67-

bottom sediments, bacteria
Krumbein, W.E. 1971.
Sedimentmikrobiologische Untersuchungen I. Über die Abhängigkeit der Keimzahl von der Korngrösse.
Vie Milieu Suppl. 22(1): 253-264.

bottom sediments, bacteria
Lagarde, Edmond, 1964.
Méthode d'estimation du pouvoir dénitrifiant des eaux et des sédiments marins.
Vie et Milieu, Bull. Lab. Arago, 15(1):213-218.

bottom sediments, bacteriology of
Morita, R.Y., and C.E. ZoBell, 1955.
Occurrence of bacteria in pelagic sediments collected during the Mid-Pacific Expedition.
Deep-Sea Res. 3(1):66-73.

bottom sediments, bacteria
Nedwell, D.B. and G.D. Floodgate, 1971.
The seasonal selection by temperature of heterotrophic bacteria in an intertidal sediment.
Mar. Biol. 11(4): 306-310.

bottom sediments, bacteria in
Oppenheimer, C.H., 1960.
Bacterial activity in sediments of shallow marine bays.
Geochim. et Cosmochim. Acta, 19(4):244-260.

bottom sediments, bacteria
Peroni, C., 1971.
Heterotrophic level of microorganisms.
CNEN Rept. RT/BIO (70)-11, M. Bernhard, editor: 69-87.

bottom sediments, bacteria
Psenin, L.N., 1959.
Nitrogen-fixing bacteria in the off-shore bottom deposits of the Black Sea. (In Russian).
Doklady, Akad. Nauk, SSSR, 129(4):930-933.

bottom sediments, bacteria
Rittenberg, S.C., 1940
Bacteriological analysis of some long cores of marine sediments. J. Mar. Res.3:191-201.

bottom sediments, bacteria
Rozenberg, L.A., 1962
Microbiological characteristics of the sediments and waters of the Mediterranean Sea.
Okeanologiia, Akad. Nauk, SSSR, 2(1):108-117
Abstracted in: Soviet Bloc Res., Geophys., Astron., and Space, 1962(35): 21-22.
(OTS61-11147-35 JPRS13739)

bottom sediments, bacteria
Tysset, Camille, Jean Brisou et Richard Moreau, 1963.
Contribution à l'étude des groupements physiologiques microbiens des vases de L'avant-port d'Alger.
Archives, Inst. Pasteur, Algérie, 41(3/4):105-114

bottom sediments, bacteria
Tysset, Camille, Jean Brisou et Richard Moreau, 1962.
Etude bactériologique complémentaire des boues du Port d'Alger.
Archives, Inst. Pasteur, Algérie, 41(1/2):19-40.

bottom sediments, bacteria
Velankar, N.K., 1957.
Bacteria isolated from sea-water and marine mud off Mandapam (Gulf of Manaar and Palk Bay).
Indian J. Fish., 4(1):208-227.

bottom sediments, bacteria
Walls, Nancy W., 1967.
Bacteriology of Antarctic region waters and sediments.
Antarctic Jl., U.S.A., 2(5):192-193.

bottom sediments bacteria
ZoBell, C.E., 1943
Influence of Bacterial Activity on Source Sediment. Oil Weekly, 26 Apr. 1943:1-12.

bottom sediments, bacteria
ZoBell, C.E., 1939
Occurrence and activity of bacteria in marine sediments. Recent Marine Sediments, a symposium: 416-427 (H.A.P.G., Tulsa Oklahoma)

bottom sediments (bacteria)
Zobell, C.E., 1938
Studies on the bacterial flora of marine bottom sediments. J. Sed. Petrol. 8:10-18.

bottom sediments, bacteria
Zobell, Claude E. and Joseph F. Prokop 1966.
Microbial oxidation of mineral oils in Barataria Bay bottom deposits.
Z. allg. Mikrobiol. 6(3):143-162.

bottom sediments bacteria
Zobell, C.E., and S.C. Rittenberg, 1948.
Sulfate-reducing bacteria in marine sediments.
J. Mar. Res. 7(3):602-617.

bottom sediments, banding (color)
Turner, Ralph R., 1971.
The significance of color banding in the upper layers of Kara Sea sediments.
Oceanogr. Rept. U.S. Cst Gd 36 (CG 373-36): 36pp.

sediments balance
Fleming, George, 1970.
Sediment balance of Clyde estuary.
J. Hydraul. Div. Am. Soc. civ. Engrs, 96 (HY11) 2219-2230

bottom sediments, bearing capacity
bottom sediments, bearing strength
Moore, David G., 1962.
Bearing strength and other physical properties of some shallow and deep-sea sediments from the North Pacific.
Bull. Geol. Soc., Amer., 73(9):1163-1166.

bottom sediments, bearing capacity
Richards, A.F., 1961.
Investigations of deep-sea sediment cores. 1. Shear strength, bearing capacity and consolidation.
U.S. Navy Hydrogr. Off., Techn. Rept., TR-63: 70 pp.

bottoms sediments, bioclastic fragments
Poizat, Claude, 1970.
Hydrodynamisme et sédimentation dans le Golf de Gabès (Tunisie).
Téthys 2 (1): 267-296.

BOTTOM SEDIMENTS, (BIOGENOUS)
Berger, Wolfgang H., 1970.
Biogenous deep-sea sediments: fractionation by deep-sea circulation.
Bull. geol. Soc. Am., 81 (5): 1385-1402.

bottom sediments, biogenous
Bezrukov, P.L., 1971.
Zonation of biogenous sedimentation in the oceans. In: Micropalaeontology of oceans, B.M. Funnell and W.R. Riedel, editors, Cambridge Univ. Press, 219-221.

bottom sediments, biogenous
Zaitseva, E.D., 1954.
Biogenetic elements in the bottom muds of Bering Sea sediments. Dokl. Akad. Nauk, SSSR, 98(6): 1005-1006.

Transl. Con. 3(9):ATS-43J15R

Bottom sediments, biology of

Bottom sediments, biology
See also: benthos
bottom sediments, bacteria of, etc.

bottom sediments, biology
Berggren, W.A., 1969.
Micropaleontologic investigations of Red Sea cores - summation of synthesis of results.
In: Hot brines and Recent heavy metal deposits in the Red Sea, E.T. Degens and D.A. Ross, editors, Springer-Verlag, New York, Inc., 329-335.

bottom sediments, biology
Berthois, Léopold, René Battistini et Alain Crosnier, 1964
Recherches sur le relief et la sédimentologie du plateau continental de l'extreme sud de Madagascar.
Cahiers Océanogr., C.C.O.E.C., 16(8):637-655.

bottom sediments, biology of
Berthois, Léopold, et Yolande Le Calvez, 1961
Étude dynamique de la sedimentation dans la baie de Sangarea (République de Guinée).
Cahiers Océan., C.C.O.E.C., 13(10):694-714.

bottom sediments, biology
Blanc-Vernet, L., H. Chamley et C. Froget, 1969
Analyse paléoclimatique d'une carotte de Méditerranée nord-occidentale. Comparaison entre les résultats de trois études: foraminifères, ptéropodes, fraction sédimentaire issue de Continent.
Palaeogeogr. Palaeoclimatol. Palaeoecol., 6 (3): 215-235.

bottom sediments, biology
Buchanan, John B., 1963.
The bottom fauna communities and their sediment relationships off the coast of Northumberland.
Oikos, 14(2):154-175.

bottom sediments, biology
Buchanan, J.B., 1958.
The bottom fauna communities across the continental shelf off Accra, Ghana (Gold Coast).
Proc. Zool. Soc., London, 130(1):1-56.

bottom sediments, biology of
Curry, D., E. Martini, A.J. Smith and W.F. Whittard, 1962.
The geology of the western approaches of the English Channel. 1. Chalky rocks from the upper reaches of the continental slope.
Phil. Trans., R. Soc., London, (B), 245(724): 267-290.

Bottom Sediments, biology of
Eklund, Melvin W., and Frank Poysky, 1965.
Clostridium botulinum Type F from marine sediments.
Science, 149(3681):306.

bottom sediments, biological remains
Emiliani, Cesare, and John D. Milliman, 1966.
Deep-sea sediments and their ecological record.
Earth-Sci. Rev., Elsevier Publ. Co., 1:105-132.

bottom sediments, biological
Evamy, B.D., and D.J. Shearman, 1965.
The development of over growths from echinoderm fragments.
Sedimentology, 5(3):211-233.

bottom sediments, biology of
Galhano, M. Helena, 1966.
Primeiras pesquisas sobre crustáceos intersticiais dos sedimentos marinhos de Portugal.
Publcoes Inst. Zool. Dr. Ausguste Nobre, Fac. Ciências Porto, 96:9-15.

bottom sediments biology
Ganapati, P.N., and P. Satyavati, 1958
Report on the Foraminifera in bottom sediments in the Bay of Bengal off the east coast of India. Andhra Univ. Mem., Oceanogr., 2: 100-127.

bottom sediments, biology of
Gennesseaux, Maurice, et Yolande Le Calvez, 1960
Affleurement sous-marin de vases pliocenes dans la baie des Anges (Nice).
C.R. Acad. Sci., Paris, 251:2064-2066.

Also in:
Trav. Lab. Géol. Sous-Marine, 10(1960).

bottom sediments, biology
Gilat, Eliezer, 1964.
The macrobenthonic invertebrate communities on the Mediterranean continental shelf of Israël.
Bull. Inst. Océanogr., Monaco, 62(1290):46 pp.

bottom sediments, biology of
Ginsburg, R.N., R. Michael Lloyd, K.W. Stockman, and J.S. McCallum, 1963.
22. Shallow-water carbonate sediments.
In: The Sea, M.N. Hill, Editor, Interscience Publishers, 3:554-582.

bottom sediments, biology of
Groot, J.J., C.R. Groot, M. Ewing, L. Burckle and J.R. Conolly, 1967.
Spores, pollen, diatoms and provenance of the Argentine Basin sediments.
Progress in Oceanography, 4:179-217.

bottom sediments, biology
Horikoshi, Masuoki, 1962.
Distribution of benthic organisms and their remains at the entrance of Tokyo Bay, in relation to submarine topography, sediments and hydrography.
Nat. Sci. Rept., Ochanomizu Univ., 13(2):47-122.

bottom sediments, biology
Isaacs, John D., 1968.
General features of the ocean. Ch. 6 in: Ocean engineering: goals, environment, technology, J.F. Brahtz editor, John Wiley & Sons, 157-201.

bottom sediments, biology of
Kanwisher, John, 1962
Gas exchange of shallow marine sediments.
The Environmental Chemistry of Marine Sediments, Proc. Symp., Univ. R.I., Jan. 13, 1962, Occ. Papers, Narragansett Mar. Lab., No. 1: 13-19.

bottom sediments, biology of
Kolosvary, G., 1963.
Madreporien und Balaniden aus rezenten Mittelmeersedimenten.
Int. Rev. Ges. Hydrobiol., 48(1):175

bottom sediments, biology of
Koreneva, E.V., 1961.
Distribution of spores and pollen of terrestrial plants in bottom sediments of the Pacific Ocean. (Abstract).
Tenth Pacific Sci. Congr., Honolulu, 21 Aug.-6 Sept., 1961, Abstracts of Symposium Papers, 145-146.

bottom sediments, biology of
Mukhina, V.V., 1966.
Siliceous organisms (diatoms, Silicoflagellata, Radiolaria, sponge spicules) in suspension and in the surface layer of the bottom sediments of the Indian Ocean (using data obstained on the 36th cruise of the R/V "Vityaz". (In Russian; English abstract).
Okeanologiia, Akad. Nauk, SSSR, 6(5):807-816.

bottom sediments, biology of
Neiheisel, James, 1965.
Source and distribution of sediments at Brunswick Harbor and Vicinity, Georgia.
U.S. Army Coastal Eng. Res. Center, Techn. Memo., No. 12:49 pp.

bottom sediments, biology of
Nötzold, Tilo, 1965.
Faziell-ökologische Aussagen auf Grund von Pflanzenfossilien aus dem Riss-Wurm Interglazial von Klien-Klütz-Höved in der Lübecker Bucht.
Beitrage Meeresk., Berlin (12/14);119-128.

bottom sediments, biology of
Nötzold, Tilo, 1965.
Pflanzenfossilien aus einem submarinen Torf der Mecklenburger Bucht.
Beiträge Meeresk., Berlin, (12/14):74-77.

bottom sediments, biology of
Parenzan, Pietro, 1959
Il fondo a deposito neritico di Vàvara.
Thalassia Jonica, 2:109-126.

bottom sediments, biology of
Pires de Carvalho, Maria da Gloria, 1970.
Componentes biogênicos dos sedimentos da plataforma dos estados do Espírito Santo e Rio de Janeiro.
Publ. Inst. Pesquisas Marinha, Brasil, 46: 14 pp.

bottom sediments, biology
Powers, M.C., and B. Kinsman, 1953.
Shell accumulations in underwater sediments and their relation to the thickness of the traction zone. J. Sed. Petr. 23(4):229-234, 5 textfigs.

bottom sediments biology
Rusnek, G.A., 1954.
Some remarks on the accumulation of shells in sediments. J. Sed. Petr. 24(4):283-285.

bottom sediments biological
Schneider, Eric D., and Bruce C. Heezen, 1966.
Sediments of the Caicos Outer Ridge, The Bahamas.
Bull. geol. Soc. Am., 77(12):1381-1398.

bottom sediments, biology
Sokolova M.N. 1968.
On the relation between the trophic groups of deep-sea macrobenthos and the composition of bottom sediments. (In Russian; English abstract)
Okeanologiia, 8(2):179-191.

bottom sediments, biology of
Wood, E.J. Ferguson, 1962
The microbiology of estuaries.
The Environmental Chemistry of Marine Sediments, Proc. Symp., Univ. R.I., Jan. 13, 1962, Occ. Papers, Narragansett Mar. Lab., No. 1: 20-26.

Bottom sediments, biology of

Subheadings follow:-

annelids	nannoplankton
cephalopod beaks	ostracods
coccolithophorids	palynomorphs
diatoms (+ oozes)	phytoplankton
echinoderms	pollen
foraminifera	pteropods
fungi	radiolaria
molluscs	sponge spicules
	spores

The forthcoming index on marine organisms should also be consulted for the several groups.

bottom sediments, biology, algae
Perkins, Ronald D., and Susan D. Halsey 1971.
Geologic significance of microboring fungi and algae in Carolina shelf sediments.
J. sedim. Petrol. 41(3): 843-853

bottom sediments, biology of (annelids)
Fager, Edward W., 1964.
Marine sediments: effects of tube-building polychaete.
Science, 143(3604):356-359.

bottom sediments biology (annelids)
Rhoads, Donald C. 1967
Biogenic reworking of intertidal and subtidal sediments in Barnstable Harbor and Buzzards Bay, Massachusetts.
J. Geol. 75(4):461-476.

bottom sediments, biology of (Bryozoa)
*Rucker, James Bivin, 1967.
Paleoecological analysis of cheilostome Bryozoa from Venezuela-British Guiana shelf sediments.
Bull. mar. Sci., Miami, 17(4):787-839.

bottom sediments, biology (cephalopod beaks)
Beliaev, G.M., 1962
The beaks of the cephalopods in the bottom sediments of the ocean.
Okeanologiia, Akad. Nauk, SSSR, 2(2):311-326.

bottom sediments, biology of, coccoliths
Bottom Sediments, Coccolithophorids
Bartolini, Carlo, 1970.
Coccoliths from sediments of the western Mediterranean.
Micropaleontology, 16(2):129-150

bottom sediments, biology (coccolithophorids)
Black, M., 1971.
The systematics of coccoliths in relation to the palaeontological record. In: Micropaleontology of oceans, B.M. Funnell and W.R. Riedel, editors, Cambridge Univ. Press, 611-624.

bottom sediments, biology of (coccolithophorids)
Black, M., 1964.
Cretaceous and Tertiary coccoliths from Atlantic seamounts.
Paleontology, 7(2):306-316.

bottom sediments, coccolithophorids (lists of spp)
Bukry, D., 1971.
Coccolith stratigraphy, Leg 8, Deep Sea Drilling Project. Initial Repts Deep Sea Drilling Project 8: 791-807.

bottom sediments, biology of (coccolithophores)
Bukry, David, 1971
Discoaster evolutionary trends.
Micropaleontology, Vol. 17(1):43-52

bottom sediments, coccolithophorids
Bukry, David, 1971.
Coccolith stratigraphy Leg 7, Deep Sea Drilling Project. Initial Repts Deep Sea Drill. Proj. 7(2): 1513-1528.

bottom sediments biology (coccolithophorids, lists of spp.)
Bukry, David, 1970.
Coccolith age determinations leg 4, deep sea drilling project. Initial Reports of the Deep Sea Drilling Project, Glomar Challenger, 4: 375-381.

bottom sediments, biology (coccolithophorids, lists of spp.)
Bukry, David, 1970.
Coccolith age determinations leg 2, deep sea drilling project. Initial Reports of the Deep Sea Drilling Project, Glomar Challenger, 2: 349-355.

bottom sediments, biology coccolithophorids, lists of spp.
Bukry, David, and M.N. Bramlette 1970.
Coccolith age determination leg 5, Deep Sea Drilling Project.
Initial Repts Deep Sea Drilling Project, Glomar Challenger 5: 487-494

bottom sediments, biology (coccolithophorids, lists of spp.)
Bukry, David and M.N. Bramlette, 1970.
Coccolith age determinations leg 3, deep sea drilling project. Initial Reports of the Deep Sea Drilling Project, Glomar Challenger 3: 589-611.

bottom sediments, biology of (coccolithophorids)
Cohen, C.L.D. 1965.
Coccoliths and discoasters from the bottom sediments of the Adriatic.
Rapp. P.-v. Réun. Comm. int. Explor. scient. Mer Méditerr. 18(3): 957.

Bottom sediments, coccolith
Cohen, C.L.D., 1965.
Coccoliths and discoasters from Adriatic bottom sediments.
Leidse Geol. Mededelingen. 35: 1-44, 25 pls.

bottom sediments biology of (coccolithophorids, lists of spp.)
Gartner, Stefan, Jr., 1970.
Coccolith age determinations leg 5, deep sea drilling project. Initial Repts. Deep Sea Drilling Project, Glomar Challenger 5: 495-500.

bottom sediments biology of (coccolithophorids, lists of spp.)
Gartner, Stefan, 1970.
Coccolith age determinations leg 3, deep sea drilling project. Initial Reports of the Deep Sea Drilling Project, Glomar Challenger 3: 613-627.

bottom sediments coccolithophorids
Geitzenauer, Kurt R., 1972.
The Pleistocene calcareous nannoplankton of the subantarctic Pacific Ocean. Deep-Sea Res. 19(1): 45-60.

bottom sediments, coccolithophorids, lists of spp.
Haq, B. and J.H. Lipps, 1971.
Calcareous nannofossils. Initial Repts Deep Sea Drilling Project 8: 777-789.

bottom sediments, biology of (coccolithophorids, lists of spp.)
Hay, W.W., 1970.
Calcareous nannofossils from cores recovered on leg 4. Initial Reports of the Deep Sea Drilling Project, Glomar Challenger, 4: 455-501.

bottom sediments, coccolithophorids
*McIntyre, Andrew, 1969.
The Coccolithophorida in Red Sea sediments.
In: Hot brines and Recent heavy metal deposits in the Red Sea, E.T. Degens and D.A. Ross, editors, Springer-Verlag, New York, Inc., 299-305.

bottom sediments, biology of
McIntyre, Andrew, Allan W.H. Bé, and David Krinsley, 1964.
Coccoliths and the Plio-Pleistocene boundary in deep-sea sediments. (Abstract).
Geol. Soc., Amer., Special Paper, No. 76:113.

bottom sediments, biology (coccolithophorids)
McIntyre, Andrew, Allan W.H. Bé, and Michael B. Roche, 1970
Modern Pacific Coccolithophorida: a paleontological thermometer.
Trans. N.Y. Acad. Sci. (2) 32(6): 720-731.

bottom sediments, coccoliths
Shumenko, S.I., and M.G. Ushakova, 1967.
An electron-microscopic investigation of coccoliths in the bottom deposits of the Pacific. (In Russian).
Dokl. Acad. Nauk, SSSR, 176(4):932-934.

bottom sediments, biology (coccolithophorids)

Uschakova, M.G., 1971.
Coccoliths in suspension and in the surface layer of sediment in the Pacific Ocean. In: Micropalaeontology of oceans, B.M. Funnell and W.R. Riedel, editors, Cambridge Univ. Press, 245-251.

bottom sediments, Coccoliths
Ushakova, M.G., 1966.
The biostratigraphic role of the Coccolithophoridae in the bottom sediments of the Pacific Ocean. (in Russian).
Okeanologiia, Akad. Nauk, SSSR, 6(1):136-143.

bottom sediments, biology (decapods)
Hartnoff, R.G., 1963
The biology of Manx spider crabs.
Proc. Zool. Soc., London, 141(3):423-496.

bottom sediments biology of, diatoms

bottom sediments, diatom oozes
Beliaeva, T.V., 1963.
The composition and distribution of the diatom algae in the surface layer of the Pacific Ocean sediments. (In Russian).
Okeanologiia, Akad. Nauk, SSSR, 3(4):684-696.

bottom sediments, diatoms
Belyaeva, (Sechkina), T.V., 1961
Diatoms in the upper layer of sediments of the north-western Pacific
Trudy Inst. Okeanol., 46:231-246.

bottom sediments, diatoms
Belyaeva (Sechkina), T.V., 1961
Diatoms in the upper layer of sediments of the Sea of Japan.
Trudy Inst. Okeanol., 46:247-262.

bottom sediments, diatom ooze
Brodie, J.W., 1965.
Oceanography.
In: Antarctica, Trevor Hatherton, editor, Methuen & Co., Ltd., 101-127.

bottom sediments, diatoms
Burckle, Lloyd H., Jessie H. Donahue, James D. Hays and Bruce C. Heezen, 1966.
Radiolaria and diatoms in sediments of the southern oceans.
Antarctic J., United States, 1(5):204.

bottom sediments, biology (diatoms)
Calvert, S.E., 1966.
Origin of diatom-rich, Varved sediments from the Gulf of California.
J. Geol., 74(5-1):546-565.

bottom sediments, diatoms
Calvert, S.E., 1964.
Factors affecting distribution of laminated diatomaceous sediments in Gulf of California.
In: Marine geology of the Gulf of California, a symposium, Amer. Assoc. Petr. Geol., Memoir, T. van Andel and G.G. Shor, Jr., editors, 3:311-330.

submarine geology, biology (diatoms)
Florin, M-B., 1948
9. Diatomeae in submarine cores from the Tyrrhenian Sea. Medd. Ocean. Inst., Göteborg, 15 (Göteborgs Kungl. Vetenskaps-och Viterhets Samhälles Handlingar, Sjätte Foljden, Ser. B 5(13):80-88.

bottom sediments, biology (diatoms)
Frenguelli, Joaquin and Hector A. Orlando, 1959
Analisis de algunas muestras del Pleistoceno del fondo del Mar. Mediterraneo envidas por el Lamont Geological Observatory, Secretaria de Marina, Servicio de Hidrograficia Naval, Argentina, Publico, H. 1014: 1-7.

bottom sediments, diatoms
Groot, J.J., C.R. Groot, M. Ewing, L. Burckle and J.R. Conolly, 1967.
Spores, pollen, diatoms and provenance of the Argentine Basin sediments.
Progress in Oceanography, 4:179-217.

bottom sediments, diatomaceous
Gross, M.Grant, 1967.
Concentration of minor elements in diatomaceous sediments of a stagnant fjord.
In: Estuaries, G.H. Lauff, editor, Publs Am. Ass. Advmt Sci., 83:273-282.

bottom sediments, biology of
Gross, M. Grant, 1964.
Heavy-metal concentrations of diatomaceous sediments, in a stagnant fjord. (Abstract).
Geol. Soc., Amer., Special Paper, No. 76:69.

bottom sediments, biology (diatoms)
Itihara, M., 1954.
Bottom sediments of Osaka Bay. 1. Mechanical analysis and frustules of diatom.
J. Inst. Polytechn., Osaka City Univ., G, 2:89-100.

bottom sediments, diatoms
Kanaya, T., 1971.
Some aspects of pre-Quaternary diatoms in the oceans. In: Micropaleontology of oceans, B.M. Funnell and W.R. Riedel, editors, Cambridge Univ. Press, 545-565.

bottom sediments, diatoms
Kanaya, Taro, 1961.
Characteristics and distribution of diatom thanatocoenoses in Pacific deep-sea cores.(abstract).
Tenth Pacific Sci. Congr., Honolulu, 21 Aug.-6 Sept., 1961, Abstracts of Symposium Papers, 375-376.

bottom sediments, diatoms
Kobayashi, Kazuo, Kazuhiro Kitazawa, Taro Kanaya and Toyosaburo Sakai, 1971.
Magnetic and micropaleontological study of deep-sea sediments from the west-central equatorial Pacific. Deep-Sea Res. 18(11): 1045-1062.

bottom sediments, biology of (diatoms)
*Kozlova, O.G., and V.V. Mukhina, 1967.
Diatoms and Silicoflagellates in suspension and floor sediments of the Pacific Ocean.
Int. Geol. Rev., 9(10):1322-1342.

(Translated from: Geokhimiya Kremnezema, NK 65-12 (51):192-218 (1966).

bottom sediments, diatoms
Kozlova, O.G., and V.V. Mukhina, 1966.
Diatoms and silicoflagellates in suspension and in the bottom sediments of the Pacific Ocean. (In Russian).
In: Geochemistry of silica, N.M. Strakhov, editor, Isdatel. "Nauka", Moskva, 192-218.

bottom sediments, biology (diatoms)
Mann, A., 1907
Report on the diatoms of the Albatross voyages in the Pacific Ocean, 1888-1904.
Contrib. U. S. Nat. Herb. 10(5):221-419, Pls. XLIV-LIV.

bottom sediments, diatoms
Mölder, Karl, 1943.
Studien über die Ökologie und Geologie der Boden-diatomeen in der Pojo-Bucht.
Annales Botanici Soc. Zool.-Bot. Fennicae Vanamo 18(2):204 pp.

bottom sediments, diatoms
Mukhina, V.V., 1966.
On the problem of the boundary surface between the sediments referred to the Quaternary and Tertiary periods in the Pacific Ocean (from the analysis of diatom data). (In Russian).
Okeanologiia, Akad. Nauk, SSSR, 6(1):122-135.

bottom sediments, diatoms in
Mukhina, V.V., 1963.
The biostratigraphic disintegration of bottom sediments at Station 3802 in the equatorial zone of the Pacific Ocean. (In Russian).
Okeanologiia, Akad. Nauk, SSSR, 3(5):861-869.

bottom sediments, diatoms
Nayudu, Y.R., and B.J. Enbysk, 1964.
Bio-lithology of northeast Pacific surface sediments.
Marine Geol., 2(4):310-342.

bottom sediments, biol. (diatoms)
Oshide, T., 1954.
On the diatoms from the bottom of Mutsu-Bay.
Studies from Geol. & Minerol. Inst., Tokyo Univ. Education (Mem. Vol. Prof. K. Kawada):159-165.

bottom sediments, biol. (diatoms)
Riedel, W.R., 1954.
The age of the sediment collected at Challenger (1875) Station 225 and the distribution of Ethmodiscus rex (Rattray). Deep-Sea Res. 1(3): 170-175, 1 pl.

bottom sediments, biology of (diatoms)
Round, F.E. 1968.
The phytoplankton of the Gulf of California. I. The distribution of phytoplanktonic diatoms in cores.
J. exp. mar. Biol. Ecol. 2(1): 64-86.

bottom sediments, diatoms
Schwarzenholz, Wilhelm, 1965.
Die Diatomeenflora in den Stechrohrkernen aus der Mecklenburger Bucht.
Beiträge Meeresk., Berlin, (12/14):85-118.

bottom sediments, diatoms
Simonsen, Reimer, 1962
Untersuchungen zur Systematik und Ökologie der Bodendiatomeen der westlichen Ostsee.
Inter. Rev. Gesamt. Hydrobiol., System. Beihefte 1:145 pp.

bottom sediments, biology of
Wood, E.J. Ferguson, 1963.
A study of the diatom flora of fresh sediments of the south Texas bays and adjacent waters.
Publ. Inst. Mar. Sci., Port Aransas, 9:237-310.

Zhuze, A.P.
Zhuse
Gzuze
Jousé

Zhuze

Also: Zhuse, Gzuze, Jousé

bottom sediments, diatoms
Jousé, A.P., 1971.
Diatoms in Pleistocene sediments from the northern Pacific Ocean. In: Micropalaeontology of oceans, B.M. Funnell and W.R. Riedel, editors, Cambridge Univ. Press, 407-421.

Zhuze bottom sediments, diatoms
Gzuse, A.P., 1963
The problems of stratigraphy and paleogeography in the northern part of the Pacific Ocean according to the data of diatom analysis. (In Russian).
Okeanologia, Akad. Nauk. SSSR, 3(6):1017-1028.

bottom sediments biology (diatoms)
Zhuze, A.P., 1960
[Diatoms in the surface layer of the Bering Sea sediments.] Trudy Inst. Okeanol., 32: 171-205.

bottom sediments biology (diatoms)
Jousé, A.P., 1960
Les diatomées des dépôts de fond de la partie nord-ouest de l'Océan Pacifique. Deep-Sea Research 6(3): 187-192.

bottom sediments, diatoms
Zhuze, A.P., 1958.
[Diatomaceae in bottom deposits in the northwestern part of the Pacific.]
Doklady Akad. Nauk, SSSR, 125(4):891-894.

bottom sediments (diatoms
Zhuze, A.P., 1954.
On the Tertiary age diatoms in the bottom sediments of the far eastern waters. (In Russian).
Trudy Inst. Okeanol., Akad. Nauk, SSSR, 9:119-135

bottom sediments, diatoms in
Zhuse, A.P., O.G. Kozlova and V.V. Mukhina, 1967.
Species composition and zonal distribution of diatoms in the surface layer of Pacific Ocean sediments. (In Russian).
Dokl., Akad.Nauk, SSSR., 172(5):1183-1186.

Zhuze bottom sediments, diatoms in
Jouse, A.P., G.S. Koroleva and G.A. Nagaeva, 1963.
Stratigraphical and paleogeographical investigations in the Indian sector of the Southern Ocean. (In Russian; English abstract).
Rez. Issled. Programme Mezhd. Geofiz. Goda, Oceanolog. Issled., Akad. Nauk, SSSR, No. 8: 137-161.

bottom sediments, diatoms
Zhuze
Jouse, A.P., G.S. Koroleva, G.A. Nagaeva, 1962
Diatoms in the surface layer of sediment in the Indian sector of the Antarctic. Investigations of marine bottom sediments. (In Russian; English summary).
Trudy Inst. Okeanol., Akad. Nauk, SSSR, 61:19-92.

bottom sediments, diatoms
Jouse, A.P., O.G. Kozlova and V.V. Muhina, 1971.
Distribution of diatoms in the surface layer of sediment from the Pacific Ocean. In: Micropalaeontology of oceans, B.M. Funnell and W.R. Riedel, editors, Cambridge Univ. Press, 263-269.

bottom sediments, coccolithophorids
Bukry, David 1971.
Coccolith stratigraphy leg 6, Deep Sea Drilling Project.
Initial Repts. Deep Sea Drilling Project, Glomar Challenger 6: 965-1004.

bottom sediments, coccolithophorids
Bukry, David and M.N. Bramlette, 1970.
Coccolith age determinations leg 5, deep sea drilling project. Initial Repts. Deep Sea Drilling Project, Glomar Challenger 5: 487-494.

bottom sediments, coccolithophorids
Bukry, David and M.N. Bramlette, 1969.
Coccolith age determinations leg 1, deep sea drilling project. Initial Reports of the Deep Sea Drilling Project, Glomar Challenger 1: 369-387.

bottom sediments, coccolithophorids
Gartner, Stefan, Jr., 1970.
Coccolith age determinations leg 5, deep sea drilling project. Initial Repts. Deep Sea Drilling Project, Glomar Challenger 5: 495-500.

bottom sediments, coccolithophorids
McIntyre, A. and R. McIntyre, 1971.
Coccolith concentrations and differential solution in oceanic sediments. In: Micropalaeontology of oceans, B.M. Funnell and W.R. Riedel, editors, Cambridge Univ. Press, 253-261.

bottom sidiments, coccolithophorids
Uschakova, M.G., 1971.
Coccoliths in suspension and in the surface layer of sediment in the Pacific Ocean. In: Micropalaeontology of oceans, B.M. Funnell and W.R. Riedel, editors, Cambridge Univ. Press, 245-251.

bottom sediments, dinoflagellates
Wall, D., 1971.
The lateral and vertical distribution of dinoflagellates in Quaternary sediments. In: Micropalaeontology of oceans, B.M. Funnell and W.R. Riedel, editors, Cambridge Univ. Press, 399-405.

bottom sediments, dinoflagellates
Williams, D.B. 1971.
The occurrence of dinoflagellates in marine sediments.
In: Micropalaeontology of oceans, B.M. Funnell and W.R. Riedel, editors, Cambridge Univ. Press, 231-243.

bottom sediments, biology of, discoasters
Bottom sediments, discoasters
Cohen, C.L.D., 1965.
Coccoliths and discoasters from Adriatic bottom sediments.
Leidse Geol. Mededelingen 35: 1-44, 25 pls.

bottom sediments, biology of (discoasters)
Cohen, C.L.D. 1965.
Coccoliths and discoasters from the bottom sediments of the Adriatic.
Rapp. P.-v. Réun., Comm. int. Explor. Scient. Mer Méditerr. 18(3): 957

bottom sediments, biology of
Wray, John L., and C. Howard Ellis, 1965.
Discoaster extinction in neritic sediments, northern Gulf of Mexico.
Bull. Amer. Assoc. Petr. Geol., 49(1):98-99.

bottom sediments, biology (echinoderms)
Jafar, S.A., 1970 bottom sediments, biology
A new species of holothurian sclerite from the Pleistocene of the Arabian Sea.
Micropaleontology 16(2): 233-234
[echinoderms]

bottom sediments, biology facal pellets
Kornicker, L.S., and E.G. Purdy, 1957.
A Bahamian faecal-pellet sediment.
J. Sediment. Petrol., 27(2):126-128.

bottom sediments, fish debris
Helms, Phyllis B. and W.R. Riedel, 1971.
Skeletal debris of fishes. Initial Repts Deep Sea Drill. Proj. 7(2): 1709-1720.

bottom sediments, fish debris
Soutar, Andrew, 1967.
The accumulation of fish debris in certain California coastal sediments.
Rep. Calif. Coop. Oceanic Fish. Invest. 11:136-139.

bottom sediments biology, Foraminifera
bottom sediments, foraminifera (planktonic
Adegoke, O.S., T.F.J. Dessauvagie and C.A. Kogbe, 1971
Planktonic foraminifera in Gulf of Guinea sediments.
Micropaleontology 17(2): 197-215

bottom sediments, foraminifera
Androsova, V.P., 1962.
Foraminifera of the bottom sediments in the western part of the Polar basin. (In Russian).
Trudy Vses. Nauchno-Issledov. Inst. Morsk. Ribn. Chos. i Okean., VNIRO, 46:102-117.

bottom sediments, foraminifera
Bandy, Orville L., 1964.
Foraminiferal biofacies in sediments of the Gulf of Batabano, Cuba, and their geological significance.
Bull. Amer. Assoc. Petr. Geol., 48(10):1666-1679.

bottom sediments, biology of (Foraminifera)
Bandy, Orville L., James C. Ingle, Jr., and Johanna M. Resig 1965.
Modification of foraminiferal distribution by the Orange County outfall, California.
Ocean Sci. Ocean Engng. Mar. Techn. Soc. Am.-Limnol. Oceanogr. 1: 54-76.

bottom sediments, foraminifera
Barash, M.S., 1971.
The vertical and horizontal distribution of planktonic Foraminifera in Quaternary sediments of the Atlantic Ocean. In: Micropaleontology of oceans, B.M. Funnell and W.R. Riedel, editors, Cambridge Univ. Press, 433-442.

bottom sediments, foraminifera (pelagic) (lists of spp.)
Beckmann, J.P., 1971.
The Foraminifera of sites 68 to 75. Initial Repts Deep Sea Drilling Project 8: 713-725.

bottom sediments, Foraminifera
Belyaeva, N.V., 1970.
The distribution of planktonic Foraminifera in sediment of the Atlantic Ocean. (In Russian; English asbtract). Okeanologiia, 10(6): 1016-1027.

bottom sediments, Foraminifera
Beliaeva, N., 1968.
Planktonic Foraminifera in deposits of the Atlantic. (In Russian).
Dokl. Akad. Nauk, SSSR, 183(2):445-448.

bottom sediments, biology of (Foraminifera)
Belyaeva N.V. 1968.
Quantitative distribution of planktonic foraminiferal tests in Recent sediments of the Pacific Ocean. [In Russian; English abstract].
Okeanologiia 8(1): 111-115.

bottom sediments biology, foraminifera
Beljaeva, N.V., 1960
Distribution of Foraminifera in the western part of the Bering Sea. Trudy Inst. Okeanol., 32: 158-170.

bottom sediments, biology (Foraminifera)
Beliaeva, N.V., and Kh. M. Saidova, 1965.
Relations between the benthic and planktonic Foraminifers in the uppermost layers of Pacific sediments. (In Russian).
Okeanologiia, Akad. Nauk, SSSR, 5(6):1010-1014.

Translation:
Scripta Tecnica, Inc., for AGU, 5(6):56-59.

bottom sediments, foraminifera
Berggren, W.A., 1969.
Biostratigraphy. Cenozoic foraminiferal faunas.
Initial Reports of the Deep Sea Drilling Project, Glomar Challenger 1: 594-623.

bottom sediments, Foraminifera
*Berthois, Leopold, Yolande Le Calvez et Alain Crosnier, 1968.
Répartition des foraminifères dans les sédiments du plateau continental camerounais, ESSAI d'écologie.
C.r. hebd. Séanc., Acad. Sci., Paris (D), 266 (7):660-662.

bottom sediments, Foraminifera
Berthois, L., A. Crosnier, et Y. Le Calvez, 1968.
Contribution a l'étude sédimentologique du plateau continental dans la Baie de Biafra.
Cah. ORSTOM, Océanogr., 6 (3/4): 55-86.

bottom sediments, foraminifera
Blanc-Vernet, L., 1969.
Contribution à l'étude des foraminifères de Méditerranée: relations entre la microfaune et le sédiment; biocoenoses actuelles thanatocoenoses pliocènes et quaternaires.
Rec. Trav. Sta. mar. Endoume 48 (64): 315 pp.

bottom sediments, foraminifera
Blow, W.H. 1971.
Deep Sea Drilling Project, leg 6, Foraminifera from selected samples.
Initial Repts. Deep Sea Drilling Project, Glomar Challenger 6: 1013-1026.

bottom sediments, foraminifera
Blow, W.H., 1969.
Deep sea drilling project, leg 1. Foraminifera from selected samples. Initial Reports of the Deep Sea Drilling Project, Glomar Challenger 1: 392-397.

Bottom sediments, foraminifera (benthonic)
Bock, Wayne D., 1971.
Paleoecology of a section cored on the Nicaragua Rise, Caribbean Sea.
Micropaleontology 17(2): 181-196

bottom sediments, Foraminifera
Boltovskoy, Estaban, 1966.
Depth at which Foraminifera can survive in sediments.
Contr. Cushman Lab. foramin. Res., 43-45.

bottom sediments, biology of (Foraminifera)
Burmistrova I.I. 1967.
Modern distribution of Foraminifera and stratigraphy of the Late Quaternary sediment in the Barents Sea. (In Russian; English abstract).
Okeanologiia 7(2): 302-308.

bottom sediments, biology of (Foraminifera)
Buzas, Martin A., 1969.
Foraminiferal species densities and environmental variables in an estuary. Limnol. Oceanogr., 14(3): 411-422.

bottom sediments, foraminifera
Caralp, M., A. Lamy et M. Pujos, 1970.
Contribution à la connaissance de la distribution bathymétrique des foraminifères dans le Golfe de Gascogne.
Revista Española Micropaleontologia, 2(1): 55-84.

bottom sediments, Foraminifera
Caralp, Michelle, et Michel Pujos, 1970.
Les foraminifères benthiques et planctoniques d'une carotte atlantique (golfe de Gascogne): variations climatiques du milieu marin au cours du Quaternaire récent.
Bull. Soc. géol. France (7) 12(1): 114-119.

~~bottom sediments~~ foraminifera
*Chatterjee, B.P. and M.N. Gururaja, 1969.
Foraminifera off Mangalore coast, south India.
Bull. natn. Inst. Sci., India, 38(1): 393-397.

bottom sediments, foraminifera
Chiji, Manzo, and Isao Konda, 1970.
Depth distribution of foraminiferal assemblage in the bottom sediments around Okushiri Island, North Japan Sea. (In Japanese; English abstract).
Bull. Osaka Mus. nat. Hist. 23: 35-50.

bottom sediments, biology
Cita, M.B., and M.A. Chierici, 1963
Crociera Talassografica Adriatica 1955.
V. Ricerche sui Foraminiferi contenuti in 18 carote prelavate sul fondo del mare Adriatico.
Arch. Oceanogr. & Limnol., 12(3):297-360.

bottom sediments, Foraminifera
Curry, D., J.W. Murray and W.F. Whittard, 1965.
The geology of the western approaches of the English Channel. III. The Globigerina silts and associated rocks.
In: Submarine geology and geophysics, Colston Papers, W.F. Whittard and R. Bradshaw, editors, Butterworth's, London, 17:239-261.

bottom sediments, foraminifera
Douglas, Robert G., 1971.
Cretaceous Foraminifera from the northwestern Pacific Ocean: leg 6, deep sea drilling project.
Initial Repts. Deep Sea Drilling Project, Glomar Challenger 6: 1027-1053.

bottom sediments, foraminifera
Douglas, Robert G. and Samuel M. Savin, 1971.
Isotopic analyses of planktonic Foraminifera from the Cenozoic of the northwest Pacific, leg 6. Initial Repts. Deep Sea Drilling Project, Glomar Challenger 6: 1123-1127.

bottom sediments, foraminifera
El-Wakeel, Saad K., Hosni F. Abdou and Saad D. Wahby, 1970.
Foraminifera from bottom sediments of Lake Maryut and Lake Manzalah, Egypt. Bull. Inst. Oceanogr. Fish., Cairo, 1: 429-448.

bottom sediments, foraminifera
Ericson, David B., Maurice Ewing and Gosta Wollin, 1964
The Pleistocene Epoch in deep-sea sediments.
A complete time scale dates the beginning of the first ice age at about 1 1/2 million years ago.
Science, 146(3645):723-732.

bottom sediments, biology of
Eriksson, K. Gösta, and Ingrid U. Olson, 1963.
Some problems in connection with C14 dating of tests of Foraminifera.
Bull. Geol. Inst., Univ. Uppsala, 42:1-13.

Also:
Publ., Inst. Quarternary Geol., Univ. Uppsala, Octavo Ser., No. 7.

bottom sediments, Foraminifera
Ewing, Maurice, Tsunemasa Saito, John I. Ewing and Lloyd H. Burckle, 1966.
Lower Cretaceous sediments from the northwest Pacific.
Science, 152(3723):751-755.

bottom sediments chem.
Gallignani, P., ed E. Rabbi, 1968.
Distribuzione del ferro libero nei sedimenti marini antistanti il delta del Po.
Also in: Raccolta Dati Oceanogr. P. Fest., (2) 24: 447-478. Comm. ital. Oceanogr. Consig. naz. Richerche (A)(5)(1969)

bottom sediments, coccolithophorids
Geitzenauer, Kurt R., 1969.
The Pleistocene Coccolithophoridae of the Southern Oceans.
Antarctic J., U.S., 4(5): 176-177.

bottom sediments, foraminifera
Gomoiu, M.T., 1965.
Une formule simple de calcul des foraminifères vivants dans un prélèvement sableux.
Méthodes Quantitatives d'Étude di Benthos et Echelle Dimensionnelle des Benthontes: Colloque du Comité du Benthos (Marseille, Nov. 1963), Comm. Int. Expl. Sci. Mer. Médit., Monaco, 45-48.

bottom sediments, molluscs
Gorbeng, L.I. 1970.
Distribution of molluscs in Holocene sediments from the White Sea. (In Russian; English abstract).
Okeanologiia 10(5): 837-846.

bottom sediments, foraminifera
Haman, Drew 1971.
Foraminiferal assemblages in Tremadoc Bay, North Wales, U.K.
J. foram. Res. 1(3): 126-143.

bottom sediments, foraminifera
Hooper, Kenneth, 1968.
Benthonic foraminiferal depth assemblages of the Continental shelf off eastern Canada.
Marit. Sed., 4(3): 96-99.

bottom sediments, foraminifera
Huang, Tunyow 1971.
Foraminiferal trends in the surface sediments of Taiwan Strait.
Techn. Bull. Econ. Comm. Asia Far East, 4:23-61

bottom sediments, Foraminifera
Jarke, Joachim, 1961
Beobachtungen über Kalkauflösung an Schalen von Mikrofossilien in Sedimenten der westlichen Ostsee.
Deutsche Hydrogr. Zeits., 14(1):6-11.

bottom sediments, Foraminifera
Jarke, Joachim, 1961
Die Beziehungen zwischen hydrographischen Ver-Hältnissen, Faziesentwicklung und Foraminiferen-verbreitung in der heutigen Nordsee als Vorbild für die Verhältnisse während der Miocän-Zeit.
Meyniana, 10:21-36.

In Okeanogr., Deutsch. Hydrogr. Inst., 1961.

bottom sediments, foraminifera
Kafescioglu, Ismail A. 1971.
Specific diversity of planktonic foraminifera on the continental shelves as a paleobathymetric tool.
Micropaleontology 17(4):455-470.

bottom sediments, foraminiferal
Keller, George H., and Adrian F. Richards, 1967.
Sediments of the Malacca Strait, Southeast Asia.
J.sedim. Petrol., 37(1):102-127.

bottom sediments, foraminifera
Kennett, James P., 1969.
Foraminiferal studies of Southern Ocean deep-sea cores.
Antarctic J., U.S. 4(5): 178-179

bottom sediments, Foraminfera
*Kennett, James P., 1968.
Latitudinal cariation in Globigerina pachyderma (Ehrenberg) in surface sediments of the southwest Pacific Ocean.
Micropaleontology, 14(3):305-318.

bottom sediments, foraminifera
Kim, Bong Kyun, Kim, Sung Woo and Kim, Joung Ja, 1970
Foraminifera in the bottom sediments off the southwestern coast of Korea.
Tech. Bull. Econ. Comm. Asia Far East Commit. Co-ordinat. Joint Prospect. miner. Resources Asian Offshore Areas 3:147-163

bottom sediments, biology (diatoms)
Kolbe, R.W., 1957.
Fresh-water diatoms from Atlantic deep-sea sediments. Science 126(3282):1053-1056.

bottom sediments, Foraminifera
Korneva, F.R., 1966.
Foraminifera distribution in the surface layer of sediments of the Eastern Mediterranean Sea. (In Russian; English abstract).
Okeanologiia, Akad. Nauk, SSSR, 6(5):817-822.

bottom sediments, foraminifera
Krasheninnikov, V. A. 1971.
Cenozoic Foraminifera.
Initial Repts. Deep Sea Drilling Project, Glomar Challenger 6: 1055-1068.

bottom sediments, biological content
Kustanowich, S., 1963.
Distribution of planktonic foraminifera in surface sediments of the south-west Pacific Ocean.
New Zealand J. Geol., Geophys., 6(4):534-565.

bottom sediments biology (foraminifera)
Levy, Alain 1967.
Contribution à l'étude des foraminifères des vechs du Roussillon et du plateau continental de bordure.
Vie Milieu (B) 18(1B): 63-102.

bottom sediments, Foraminifera
Lutze, Gerhard F., 1968.
Siedlungs - Strukturen rezenter Foraminiferen.
Meyniania, 18:31-34.

bottom sediments, Foraminifera
*Lutze, G.F., 1968.
Jahresgang der Foraminiferen-Fauna in der Bottsand-Lagune.
Meyniania, 18:13-30.

bottom sediments, Foraminifera
Lutze, Gerhard F., 1965.
Zur Foraminiferen-Fauna der Ostsee.
Mey-niana, Kiel, 15:75-142.

bottom sediments, foraminifera (planktonic)
Lynts, George W., 1971.
Analysis of the planktonic foraminiferal fauna of core 6275, Tongue of the Ocean, Bahamas.
Micropaleontology 17(2):152-166

bottom sediments, Foraminifera
Lynts, George W., 1966.
Variation of foraminiferal standing crop over short lateral distances in Buttonwood Sound, Florida Bay.
Limnol. Oceanogr., 11(4):562-566.

bottom sediments, Foraminifera
Marlowe, J.I., 1965.
Probable Tertiary sediments from a submarine canyon off Nova Scotia.
Marine Geology, Elsevier Publ. Co., 3(4):263-268.

bottom sediments, foraminifera
McRoberts, Jill H.E., 1968.
Post-glacial history of Northumberland Strait based on benthic Foraminifera.
Marit. Sed. 4(3):88-95

Bottom sediments, foraminifera
Morin, Ronald W., Fritz Theyer and Edith Vincent 1970.
Pleistocene climates in the Atlantic and Pacific oceans: a reevaluated comparison based on deep-sea sediments. Science, 169(3943): 365-366.

bottom sediments, Foraminifera
Murray, John W., 1969
Recent foraminifers from the Atlantic continental shelf of the United States.
Micropaleontology 15(4):401-419.

bottom sediments, Foraminifera
*Murray, John W., 1968.
Living foraminifera of lagoons and estuaries.
Micropaleontology, 14(4):435-455.

bottom sediments, foraminifera
Nayudu, Y.R., and B.J. Enbysk, 1964.
Bio-lithology of northeast Pacific surface sediments.
Marine Geol., 2(4):310-342.

bottom sediments, foraminifera
Nesteroff, Wladimir D., Solange Duplaix, Jacqueline Sauvage, Yves Lancelot, Frédéric Melieres et Edith Vincent 1968.
Les dépôts récents du canyon de Cap Breton.
Bull. Soc. géol. France (7) 10: 218-252.

bottom sediments, foraminifera
Norris, Robert M., 1964.
Sediments from Chatham Rise.
New Zealand Dept. Sci. Ind. Res., Bull., No. 159;
New Zealand Oceanogr. Inst., Memoir, No. 16:39pp.

bottom sediments, foraminifera
Parker, Frances L., 1971.
Distribution of planktonic Foraminifera in Recent deep-sea sediments. In: Micropalaeontology of oceans, B.M. Funnell and W.R. Riedel, editors, Cambridge Univ. Press, 289-307.

bottom sediments, biology of
Parker, Frances L., 1962.
Planktonic foraminiferal species in Pacific sediments.
Micropaleontology, 8(2):219-254.

bottom sediments, foraminifera
Pazotka von Lipinski, Günter, und Friedrich Wigand, 1969.
Foraminiferen aus dem Holozän der Doggerbank.
Beitr. Meereskunde, 24-25: 130-174

Bottom sediments, foraminifera
Rao, M. Subba and D. Vedantam, 1969.
Distribution of Foraminifera in the shelf sediments off Visakhapatnam. Bull. natn. Inst. Sci., India, 38(1): 491-501.

bottom sediments, foraminifera
Reiss, Z., P. Merling-Reiss and S. Moshkovitz 1971.
Quaternary planktonic Foraminiferida and nannoplankton from the Mediterranean continental shelf and slope of Israel.
Israel J. Earth-Sci. 20(4): 141-177

bottom sediments, Foraminifera
*Richter, Gotthard, 1967.
Faziesbeiriche rezenter und subrezenter Wattensedimente nach ihren Foraminiferen-Gemeinschaften.
Senckenberg. leth., 48(3/4):291-335.

bottom sediments, foraminifera
Rouvillois, Armelle, 1970.
Biocoenose et taphocoenose de foraminifères sur le plateau continental Atlantique au large de l'Île d'Yeu.
Revue Micropaléontol. 13(3): 188-204.

bottom sediments, Foraminifera

Ruddiman, William F., 1969.
Recent planktonic Foraminifera: dominance and diversity in North Atlantic surface sediments.
Science, 164(3884):1164-1167.

bottom sediments, foraminifera

Sachs, K.N., Jr., 1969.
Report on larger Foraminifera from sites 4 and 5. Initial Reports of the Deep Sea Drilling Project, Glomar Challenger 1: 398-399.

bottom sediments, biology (foraminifera)

Saidova, H.M., 1964.
Distribution of bottom Foraminifera and stratigraphy of sediment in the north-eastern Pacific. Investigation of marine bottom sediments and suspended matter. (In Russian: English abstract)
Trudy Inst. Okeanol., Akad. Nauk, SSSR, 68:84-119.

bottom sediments, foraminifera

Saidova, Kh. M., 1961.
[Quantitative distribution of bottom Foraminifera in the northeastern Pacific.]
Trudy Inst. Okeanol., Akad. Nauk, SSSR, 45:65-71.

bottom sediments biology (foraminifera)

Saidova, Kh. M., 1960
[Distribution of Foraminifera in the bottom sediments of the Okhotsk Sea.] Trudy Inst. Okeanol., 32: 96-157.

bottom sediments biology (foraminifera)

Saidova, Ch. M., 1960. Stratigraphy of sediments and paleogeography of the northeast sector of the Pacific Ocean according to sea-bottom foraminifers. Morsk. Geol., Doklady Sovetsk. Geol., 21 Sess., Inst. Geol. Kongress, 59-68. (English abstract)

bottom sediments, diatoms

Sechkina, T.V., 1959.
Diatoms in long cores of bottom sediments in the Sea of Japan. (In Russian).
Doklady, Akad. Nauk, SSSR, 126(1):171-174.

bottom sediments, biology (foraminifera)

Seiglie, George A., and Pedro J. Bermudez, 1963.
Distribucion de los foraminiferos del Golfo de Cariaco.
Bol., Inst. Oceanogr., Univ. de Oriente, Venezuela, 2(1):5-87.

bottom sediments, Foraminifera

Sen Gupta, B.K. and R.M. McMullen, 1969.
Foraminiferal distribution and sedimentary facies on the Grand Banks of Newfoundland.
Can. J. Earth Sci, 6 (3): 475-487

bottom sediments biology (foraminifera)

Shenton, E.H., 1957.
A study of the foraminifera and sediments of Matagorda Bay, Texas.
Trans., Gulf Coast Assoc. Geol. Socs., 7:135-150.

bottom sediments, Foraminifera

*Siddiquie, H.N., 1967.
Recent sediments in the Bay of Bengal.
Marine Geol. 5(4):249-291.

bottom sediments, foraminigera

Smith, A. Barrett, 1961.
A comparison of the distributions of living planktonic Foraminifera and Globigerina-rich sediment. (Abstract).
Tenth Pacific Sci. Congr., Honolulu, 21 Aug.-6 Sept., 1961, Abstracts of Symposium Papers, 386-387.

bottom sediments, foraminfera

*Tinoco, Ivan M., 1965/1966.
Contribuição a sedimentologia e microfauna da Baia de Sepetiba (Estado do Rio de Janeiro). 2. Foraminiferos.
Trabhs Inst. Oceanogr., Univ. Fed. Pernambuco, Recife, (7/8):123-126.

Bottom sediments, foraminifera

Vedantam, D. and M. Subba Rao, 1970.
Recent foraminifera from of Pentakota, east coast of India.
Micropaleontology 16(3): 325-344.

bottom sediments, Foraminifera

Vilks, G., 1968.
Foraminiferal study of the Magdalen Shallows, Gulf of St. Lawrence.
Marit. Sed., 4(1):14-21.

bottom sediments, biology (foraminifera)

Walton, William R., 1964.
Ecology of benthonic Foraminifera in the Tampa-Sarasota area, Florida.
In: Papers in Marine Geology, R.L. Miller, Editor, Macmillan Co., N.Y., 429-454.

bottom sediments, biology (fossils)

bottom sediments, fossils

Rade, J., 1966(1965).
Fossiliferous Cretaceous sediments in the vicinity of the Gulf of Carpentaria, Queensland.
Proc. R. Soc. Qd., 77:5-10.

bottom sediments, biology (fossils)

Schäfer, Wilhelm 1968.
Actuopaläontologische Beobachtungen. 8. Zur Unterscheidung von Salz- und Süsswasser-sedimenten Spurenfossilien und Marken.
Natur Mus., Frankf. 98(12): 496-506.

bottom sediments, fossiligerous

Singer, R., and A.O. Fuller, 1962.
The geology and description of a fossiliferous deposit near Zwartklip in False Bay.
Trans. R. Soc., S. Africa, 36(4):205-211.

bottom sediments, fungi

Bottom sediments, biology of (fungi)

Ahearn, Donald G., Frank J. Roth, Jr., and Samuel P. Meyers 1962.
A comparative study of marine and terrestrial strains of Rhodotorula.
Canadian J. Microbiol., 8:121-132.

bottom sediments biology of (fungi)

Gaertner, Alwin 1968.
Eine Methode des quantitativen Nachweises niederer mit Pollen köderbarer Pilze im Meerwasser und im Sediment.
Mar. Mykologie, Veröff. Inst. Meeresforsch. Bremerh., Sonderabend 8: 75-89

bottom sediments, fungi

Borut, S.Y., and T.W. Johnson, Jr., 1962.
Some biological observations on fungi in estuarine sediments.
Mycologia, 54(2):181-193.

bottom sediments, biology, fungi

Perkins, Ronald D., and Susan D. Halsey 1971.
Geologic significance of microboring fungi and algae in Carolina shelf sediments.
J. sedim. Petrol. 41 (3): 843-853

bottom sediments, fungi

Weyland, H., 1969.
Actinomycetes in North Sea and Atlantic Ocean sediments.
Nature, Lond., 223 (5205): 858.

bottom sediments, infusorians

Burkovsky, I.V., 1969.
Quantitative data on the distribution of psammophilic infusorians according to depth and the type of bottom sediment in the intertidal and sublittoral zones of Velikaya Salma (White Sea, Kandalaksha Bay). (In Russian; English abstract). Okeanologiia, 9(5): 874-880.

bottom sediments, microbenthos

*Fenchel, Tom., 1967.
The ecology of marine microbenthos. 1. The quantitative importance of ciliates as compared with metazoans in various types of sediments.
Ophelia, 4:121-137.

bottom sediments, microfauna of

Nicholls, M.M., and R.L. Ellison, 1967.
Sedimentary patterns of microfauna in a coastal plain estuary.
In: Estuaries, G.H. Lauff, editor, Publs Am. Ass Advmt Sci., 83:283-288.

bottom sediments, biology (microfossils)

bottom sediments, microfossils

Akers, W.H., 1965.
Pliocene-Pleistocene boundary, northern Gulf of Mexico.
Science, 149(3685):741-742.

bottom sediments, biology (microfossils)

Lisitzin A.P. 1971.
Distribution of carbonate microfossils in suspension and in bottom sediments.
In: Micropaleontology of Oceans, B.M. Funnell and W.R. Riedel, editors, Cambridge Univ. Press, 197-218.

Bottom sediments, microfossils

Lisitzin, A.P., 1971.
Distribution of siliceous microfossils in suspension and in bottom sediments. In: Micropalaeontology of oceans, B.M. Funnell and W.R. Riedel, editors, Cambridge Univ. Press, 173-195.

bottom sediments, biology (microorganisms)

McNulty, J.K., 1961
Ecological effects of sewage pollution in Biscayne Bay, Florida; sediments and their distribution of benthic and fouling microorganisms.
Bull. Mar. Sci., Gulf & Caribbean, 11(3): 394-447.

bottom sediments biology (microorganisms)

Zhukova, A.I., 1955.
[Biomass of microorganisms in bottom sediments of the northern Caspian.] Mikrobiologiia 24(3):321-324.

bottom sediments micropaleontology

Ericson, David B., and Gosta Wollin, 1959.
Micropaleontology and Lithology of Arctic sediment cores.
Sci. Studies, Fletcher's Ice Island, T-3, Vol. 1, Air Force Cambridge Res. Center, Geophys. Res. Pap., No. 63:50-58.

Bottom sediments, biology of (molluscs)

See also: bottom sediments, biology of (pteropods)

bottom sediments, biology, molluscs

Clifton, H. Edward, 1971.
Orientation of empty pelecypod shells and shell fragments in quiet water.
J. Sedim. Petrol. 41(3): 671-682.

bottom sediments, biology (molluscs)
*Emery, K.O., 1968.
Positions of empty pelecypod valves on the continental shelf.
J. sedim. Petr., 38(4):1264-1269.

bottom sediments, biology (molluscs)
Graf, I.E., 1956.
Die Fährten von Littorina littorea Linné (Gastr.) in verschieden Sedimenten.
Senckenberg. Lethea 37(3/4):305-317.

bottom sediments, molluscs
Greensmith, J.T., and E.V. Tucker, 1969.
The origin of Holocene shell deposits in the Chenier Plain facies of Essex (Great Britain)
Marine Geol. 7(5): 403-425

bottom sediments biology (molluscs)
Habe, Tadashige, 1959.
Pelecypod shell remains in Ariake Bay.
Rec. Oceanogr. Wks., Japan, Spec. No. 3:69-74.

bottom sediments, biology (molluscs)
Nevesskaya, L.A., 1959
[The Mollusca-complexes of the Upper Quarternary nearshore sediments of Anapa-region (Black Sea). Questions of studies of marine coasts.]
Trudy Oceanogr., Komissii Akad. Nauk, SSSR, 4:132-145.

bottom sediments, fungi
Smith, Louis DS., 1968.
The clostridial flora of marine sediments from a productive and from a non-productive area.
Can. J. Microbiol., 14(12):1301-1304.

bottom sediments, biology of (molluscs)
Van Straaten, L.M.J.U., 1966.
Micro-malacological investigation of cores from the southern Adriatic Sea.
Proc. K.ned. Akad. Wet., (B), 69 (3):429-445.

bottom sediments biology (molluscs)
Van Straaten, L.M.J.U., 1950.
Environment of formation and facies of the Wadden Sea sediments. Pt. 1. The mollusc fauna and the distribution of mollusc shells in the sediments. Waddensymposium, Tijdschr. Kon. Nederl. Aardrijkskundig Genootschap:94-108, 5 figs., 2 photos.

bottom sediments, biology of, nannoplankton

bottom sediments, nannoplankton
Hay, W.W., 1971.
Preliminary dating by fossil calcareous nannoplankton, deep sea drilling project, leg 8.
Initial Repts Deep Sea Drilling Project, 8: 809-818.

bottom sediments, nannoplankton
Hay, William W., 1971.
Preliminary dating by fossil calcareous nannoplankton deep sea drilling project: leg 6.
Initial Repts. Deep Sea Drilling Project, Glomar Challenger 6: 1005-1012.

bottom sediments, nannoplankton
Hay, W.W., 1969.
Preliminary dating by fossil calcareous nannoplankton deep sea drilling project, leg 1.
Initial Reports of the Deep Sea Drilling Project, Glomar Challenger 1: 388-391.

bottom sediments, nannoplankton
Martini, Erlend, 1965.
Mid-Tertiary calcareous nannoplankton from Pacific Deep-Sea cores.
In: Submarine geology and geophysics, Colston Papers, W.F. Whittard and R. Bradshaw, editors, Butterworth's, London, 17:393-410.

bottom sediments, biology, ostracods

bottom sediments, biology of (Ostracoda)
Ascoli, Piero, 1964.
Preliminary ecological study on Ostracoda from bottom cores of the Adriatic Sea.
Pubbl., Staz. Zool., Napoli, 33(Suppl.):213-246.

bottom sediments, ostracods
Diebel, Kurt, 1965.
Postglaziale Süsswasser-Ostracoden des Stechrohrkorns MB6 (Ostsee).
Beiträge Meeresk., Berlin, (12/14):11-17.

bottom sediments, ostracods
*Hazel, Joseph E., 1968.
Pleistocene ostracode Zoogeography in Atlantic coast submarine canyons.
J. Paleontol., 42(5):1264-1271.

bottom sediments, ostracods
Masoli, M., 1967.
Ostracodi recenti dell'Adriatico settentrionale tra Venezia e Trieste.
Memorie Mus Tridentino Sci. nat., 17(1):5-71.
Reprinted: Comm. Studio Oceanograf. Limnol. (A) (7). 1968

bottom sediments, biology
McKenzie, K.G., and F.M. Swain, 1967.
Recent Ostracoda from Scammon Lagoon, Baja, California.
J. Paleont., 41(2):281.

bottom, sediments, ostracods
Puri, H.S., 1971.
Occurrence of ostracodes in Bottom sediments.
In: Micropaleonology of oceans, B.M. Funnell and W.R. Riedel, editors, Cambridge Univ. Press, 353-358.

bottom sediments, ostracods
Swain, F.M., 1971.
Pliocene ostracodes from deep-sea sediments in the southwest Pacific and Indian Ocean. In: Micropaleontology of oceans, B.M. Funnell and W.R. Riedel, editors, Cambridge Univ. Press, 597-599.

bottom sediments, ostracods
Swain, F.M., 1971.
Pleistocene ostracoda from deep-sea sediments in the southeastern Pacific Ocean. In: Micropaleontology in oceans, B.M. Funnell and W.R. Riedel, editors, Cambridge Univ. Press, 487-492.

bottom sediments, biology of
Swain, F.M., and J.M. Gilby, 1967.
Recent Ostracods from Corinto Bay, western Nicaragua and their relationship to some other assemblages of the Pacific coast.
J. Paleont., 41(2):306-334.

bottom sediments, paleofauna
Curry, D., 1962
A Lower Tertiary outlier in the Central English Channel, with notes on the beds surrounding it.
Quart. J., Geol., London, 118:177-205.

Abstr. in:
J. Mar. Biol. Assoc., U.K., 43(2):568.

bottom sediments, biology, palynology

bottom sediments, biology of (palynomorphs)
Cross, Aureal T. Gary G. Thompson and James B. Zaitzeff 1966.
Source and distribution of palynomorphs in bottom sediments, southern part of Gulf of California.
Mar. Geol. 4(6): 467-524.

bottom sediments, palynology of
*Groot, J.J. and C.R. Groot, 1971.
Horizontal and vertical distribution of pollen and spores in Quaternary sequences. In: Micropaleontology of oceans, B.M. Funnell and W.R. Riedel, editors, Cambridge Univ. Press, 493-504.

bottom sediments, biology (phytoplankton)
Kashkin, N.I., 1964.
On the winter deposits of phytoplankton in the sublittoral. Regularity of the distribution of oceanic plankton. (In Russian; English abstract).
Trudy, Inst. Okeanol., Akad. Nauk, SSSR, 65:49-57.

bottom sediments, biology (palynology)
Kondratiene, O., A.I. Blaghchishin and E.M. Emelianov 1970
Compound [Components] and distribution of spores and pollen in the bottom sediments in the central and southeastern parts of the Baltic Sea. (In Russian, English and German abstracts)
Baltica, Lietuv. TSR Mokslu Akad. Geogr. Skyr. INQUA Taryb. Sex Vilnius 4: 181-195

bottom sediments, palynology of
Koreneva, E.V., 1971.
Spores and pollen in Mediterranean Bottom sediments. In: Micropaleontology of oceans, B.M. Funnell and W.R. Riedel, editors, Cambridge Univ. Press, 361-371.

bottom sediments biol. (palynology)
Muller, Jan, 1959.
Palynology of Recent Orinoco delta and shelf sediments. Reports of the Orinoco Shelf Expedition, Volume 5.
Micropaleontology, 5(1):1-32.

bottom sediments, biology (palynology)
Shatilova, I.I. 1968.
Pollen and spores from bottom deposits of the Black Sea. (In Russian).
Dokl. Akad. Nauk SSSR 179 (5): 1196-1199.

bottom sediments biology (palynology)
Traverse, Alfred, and Robert N. Ginsburg, 1966.
Palynology of the surface sediments of Great Bahama Bank, as related to water movement and sedimentation.
Marine Geol., 4(6): 417-459.

bottom sediments, biology of (phytoplankton)
Round, F.E., 1967.
The phytoplankton of the Gulf of California. I. Its composition, distribution and contribution to the sediments.
J. exp. mar. Biol., Ecol., 1(1):76-97.

bottom sediments, biology (plankton)
Bandy, Orville L., and Richard E. Casey 1969.
Major late Cenozoic planktonic datum planes, Antarctica to the tropics.
Antarctic J, U.S. 4 (5): 170-171.

bottom sediments, biology, pollen

bottom sediments, biology (pollen)

bottom sediments, pollen
Behre, Karl-Ernst, und Burchard Menke, 1969.
Pollenanalytische Untersuchungen an einem Bohrkern der südlichen Doggerbank.
Beitr. Meereskunde 24-25: 122-129

bottom sediments, pollen

Beug, Hans-Jürgen 1971.
Über die Pollenführung von Sedimentproben aus dem Persischen Golf und dem Golf von Oman.
Meteor-Forsch.-Engebnisse (C) 7:1-3

bottom sediments, biology of (pollen)

Boulouard, C., et H. Delauze 1966.
Analyse polynoplanctonologique de sédiments prélevés par le bathyscaphe Archimède dans la fosse de Japon.
Mar. Geol. 4(6): 461-466.

bottom sediments, pollen, etc.

Groot, Johan J., 1963
Palynological investigation of a core from the Biscay Abyssal Plain.
Science, 141(3580):522-523.

bottom sediments, biology of (pollen)

Groot, Johan J., and Catharina R. Groot 1966.
Pollen spectra from deep-sea sediments as indicators of climatic changes in southern South America.
Mar. Geol. 4(6): 525-537.

bottom sediments, biology of (pollen)

Groot, Catherine R., and Johan J. Groot, 1964.
The pollen flora of Quarternary sediments beneath Nantucket Shoals.
Amer. J. Sci., 262(4):488-493.

bottom sediments, pollen

Groot, J.J., C.R. Groot, M. Ewing, L. Burckle and J.R. Conolly, 1967.
Spores, pollen, diatoms and provenance of the Argentine Basin sediments.
Progress in Oceanography, 4:179-217.

bottom sediments, pollen-spores

*Horowitz, A., 1966.
Tropical and northern pollen and spores in Recent sediments from the Bay of Elat (Aqaba).
Israel J. Earth Sci., 15(3):125-130.

bottom sediments, biology (pollen)

Koroneva, K.B., 1961.
[The study of marine sediment cores from the Sea of Japan by applying the spore pollen method.]
Okeanologiia, Akad. Nauk, SSSR, 1(4):651-657.

bottom sediments, pollen-spores

Koreneva, E.V., 1957.
[Spore-pollen analysis of the bottom deposits of the Okhotsk Sea] Trudy Inst. Okeanol., 22:221-251

bottom sediments, pollen

Krog, Harald, 1965.
Ergebnisse pollenanalytischer Untersuchungen von 2 Torfkernen aus der Mecklenburger Bucht.
Beiträge Meeresk., Berlin, (12/14):60-61.

submarine geology pollen

Larsson, C., 1948.
8. Examination of pollen grains in three cores from the Tyrrhenian Sea. Medd. Inst. Ocean., Göteborg, 15 (Göteborgs Kungl. Vetenskaps- och Vitterhets Samhalles Handlingar, Sjätte Följden, Ser. B, 5(13)):73-79, Fig. 12.

bottom sediments, biology of (pollen)

Livingstone, D.A., 1964.
The pollen flora of submarine sediments from Nantucket Shoals.
Amer. J. Sci., 262(4):479-487.

bottom sediments, pollen

Lubliner-Mianowska, Karolina, 1965.
Die Pollenanalyse einer Stechrohr-Probe aus der Mecklenburger Bucht.
Beiträge Meeresk., Berlin, (12/14):62-73.

bottom sediments, pollen

Schulz, Horst, 1965.
Pollenanalytischer Beitrag zur Entwicklungsgeschichte der Mecklenburger Bucht.
Beiträge Meeresk., Berlin, (12/14):78-84.

bottom sediments, pollen

*Traverse, Alfred, and Robert N. Ginsburg, 1967.
Pollen and associated microfossils in the marine surface sediments of the Great Bahama Bank.
Rev. Palaeobot. Palynol. 3(1/4):243-254.

bottom sediments, biology (pollen)

Vronskii, V.A., and D.G. Panov, 1963.
Composition and distribution of pollen and spores in the surface layer of Mediterranean Sea sediments. (In Russian).
Doklady, Akad. Nauk, SSSR, 153(2):447-449.

bottom sediments pollen

Zagwijn, W.H., and H.J. Veenstra, 1966
A pollen-analytical study of cores from the Outer Silver Pit, North Sea.
Marine Geol., 4(6): 539-551.

bottom sediments, biology of pteropods

bottom sediments, pteropods

*Chen, C., 1971.
Occurrence of pteropods in pelagic sediments. (Abstract only). In: Micropalaeontology of oceans, B.M. Funnell and W.R. Riedel, editors, Cambridge Univ. Press, 351.

bottom sediments, pteropods

bottom sediments, pteropods

*Chen, Chin, 1969.
Pteropods in the hot brine sediments of the Red Sea.
In: Hot brines and Recent heavy metal deposits in the Red Sea, E.T. Degens and D.A. Ross, editors, Springer-Verlag, New York, Inc., 313-316.

bottom sediments, biology of (pteropods)

Chen, Chin 1968.
Pleistocene pteropods in pelagic sediments.
Nature, Lond., 219 (5159): 1145-1149.

bottom sediments, biology (pteropods)

Chen, Chin, 1964.
Pteropod ooze from Bermuda pedestal.
Science, 144(3614):60-62.

bottom sediments, pteropods

Curry, D., 1971.
The occurrence of pre-Quaternary pteropods.
In: Micropaleontology of oceans, B.M. Funnell and W.R. Riedel, editors, Cambridge Univ. Press, 595.

bottom sediments, pteropods

*Herman, Y., 1971.
Vertical and horizontal distribution of pteropods in Quaternary sequences. In: Micropaleontology of oceans, B.M. Funnell and W.R. Riedel, editors, Cambridge Univ. Press, 463-486.

bottom sediments, biology of Radiolaria

bottom sediments, Radiolaria

Benson, Richard N., 1964.
Preliminary report on Radiolaria in Recent sediments of the Gulf of California.
In: Marine geology of the Gulf of California, a symposium, Amer. Assoc. Petr. Geol., Memoir, T. van Andel and G.G. Shor, Jr., editors, 3:398-400.

bottom sediments, radiolaria

Burckle, Lloyd H., Jessie H. Donahue, James D. Hays and Bruce C. Heezen, 1966.
Radiolaria and diatoms in sediments of the southern oceans.
Antarctic J., United States, 1(5):204.

bottom sediments, radiolaria

Foreman, Helen P., 1971.
Cretaceous radiolaria, Leg 7, DSDP. Initial Repts Deep Sea Drill. Proj. 7(2): 1673-1693.

bottom sediments, Radiolaria

*Goll, Robert M., 1969.
Radiolaria: the history of a brief invasion.
In: Hot brines and Recent heavy metal deposits in the Red Sea, E.T. Degens and D.A. Ross, editors, Springer-Verlag, New York, Inc., 306-312.

bottom sediments, radiolaria

Goll, Robert M., and Kjell R. Bjørklund 1971
Radiolaria in surface sediments of the North Atlantic Ocean.
Micropaleontology 17(4): 434-454

bottom sediments, radiolaria

Hays, James D., 1965.
Radiolaria and Late Tertiary and Quaternary history of Antarctic seas.
In: Biology of Antarctic seas, II.
Antarctic Res. Ser., Amer. Geophys. Union, 5:125-184.

bottom sediments, biology of

Hays, James D., and Bruce Heezen, 1964.
Quarternary stratigraphy of Antarctic Ocean bottom sediments based on Radiolaria. (Abstract).
Geol. Soc., Amer., Special Paper, No. 76:76-77.

bottom sediments, biology (radiolaria)

Kling, Stanley A 1971.
Radiolaria: leg 6 of the Deep Sea Drilling Project.
Initial Repts Deep Sea Drilling Project, Glomar Challenger 6:1069-1117

bottom sediments, radiolaria

Kobayashi, Kazuo, Kazuhiro Kitazawa, Taro Kanaya and Toyosaburo Sakai, 1971.
Magnetic and micropaleontological study of deep-sea sediments from the west-central equatorial Pacific. Deep-Sea Res. 18(11): 1045-1062.

bottom sediments, biology of.

Nakaseko, Kojiro 1959.
On superfamily Liosphaericae (Radiolaria) from sediments in the sea near Antarctica. On Radiolaria from sediments in the sea near Antarctica.
Biol. Res. Japan Antarctic Res. Exped. Spec. Publ. Seto Mar. Biol. Lab., 13 pp.

bottom sediments, radiolaria
Nayudu, Y.R., and B.J. Enbysk, 1964.
Bio-lithology of northeast Pacific surface sediments.
Marine Geol., 2(4):310-342.

bottom sediments, biology (radiolaria)
Nigrini, C.A. 1971.
Radiolarian zones in the Quaternary of the equatorial Pacific Ocean.
In: Micropalaeontology of oceans. B.M. Funnell and W.R. Riedel, editors, Cambridge Univ. Press, 443-461.

bottom sediments, biology of (Radiolaria)
Nigrini, Catherine 1967.
Radiolaria in pelagic sediments from the Indian and Atlantic oceans.
Bull. Scripps Inst. Oceanogr. 11: 106 pp.

bottom sediments, radiolaria
Petrushevskaya, M.G., 1971.
Radiolaria in the plankton and Recent sediments from the Indian Ocean and Antarctic. In: Micropalaeontology of oceans, B.M. Funnell and W.R. Riedel, editors, Cambridge Univ. Press, 319-329.

bottom sediments, radiolaria
Petrushevskaya, M.G., 1971.
Spumellarian and nassellarian radiolaria in the plankton and bottom sediments of the central Pacific. In: Micropalaeontology of oceans, B.M. Funnell and W.R. Riedel, editors, Cambridge Univ. Press, 309-317.

bottom sediments, radiolaria
Petrushevskaia, M.G., 1966.
Radiolaria in the plankton and in the bottom sediments. (In Russian).
In: Geochemistry of silica, N.M. Strakhov, editor, Isdatel. "Nauka", Moskva, 219-245.

bottom sediments, radiolaria
Reshetnjak, V.V., 1971.
Occurrence of Phaeodarian radiolaria in Recent sediments and Tertiary deposits. In: Micropalaeontology of oceans, B.M. Funnell and W.R. Riedel, editors, Cambridge Univ. Press, 343-349.

bottom sediments, radiolaria
Riedel, W.R., 1971.
Cenozoic radiolaria from the western tropical Pacific, Leg 7. Initial Repts. Deep Sea Drill. Proj., 7(2): 1529-1672.

bottom sediments, radiolaria
Riedel, W.R., 1971.
The occurrence of pre-Quaternary radiolaria in deep-sea sediments. In: Micropalaeontology of oceans, B.M. Funnell and W.R. Riedel, editors, Cambridge Univ. Press, 567-594.

Bottom sediments, radiolaria
Riedel, William R., 1958
Radiolaria in Antarctic sediments.
Repts. B.A.N.Z. Antarctic Res.Exped., 1929-31 B, 6(10):219-255.

bottom sediments, biology (radiolaria)
Riedel, W.R., 1951.
Number of Radiolaria in sediments. Nature 167 (No. 4237):75.

bottom sediments, radiolaria
Riedel, W.R. and J.D. Hays, 1969.
Cenozoic radiolaria from leg 1. Initial Reports of the Deep Sea Drilling Project, Glomar Challenger 1: 400-402.

bottom sediments, Radiolaria
*Siddiquie, H.N., 1967.
Recent sediments in the Bay of Bengal.
Marine Geol., 5(4):249-291.

bottom sediments, biology silicoflagellates

bottom sediments, bilogy of (silicoflagellates)
Kozlova, O.G. and V.V. Mukhina 1967.
Diatoms and silicoflagellates in suspension and flosh sediments of the Pacific Ocean.
Int. Geol. Rev. 9(10): 1322-1342.

Translated from: Geokhimiya Kremnegema NK 65-12 (51): 192-218 (1966).

bottom sediments, silicoflagellates
Kozlova, O.G., and V.V. Mukhina, 1966.
Diatoms and silicoflagellates in suspension and in the bottom sediments of the Pacific Ocean. (In Russian).
In: Geochemistry of silica, N.M. Strakhov, editor, Isdatel. "Nauka", Moskva, 192-218.

bottom sediments, silicoflagellates
Mandra York T., 1969.
Silicoflagellates: a new tool for the study of Antarctic Tertiary climates
Antarctic J., U.S. 4(5): 172-174

bottom sediments, silicoflagellates
Martini, Erlend, 1971.
Neogene silicoflagellates from the equatorial Pacific. Initial Repts Deep Sea Drill. Proj. 7(2): 1695-1708.

bottom sediments, biology, sponge spicules

bottom sediments, sponge spicules
Koltun, V.M., 1966.
Sponge spicules in the surface layer of the sediments in the southern part of the Indian Ocean. (In Russian).
In: Geochemistry of silica, N.M. Strakhov, editor, Isdatel. "Nauka", Moskva, 262-283.

bottom sediments, biology, spores

bottom sediments, spores
Groot, J.J., C.R. Groot, M. Ewing, L. Burckle and J.R. Conolly, 1967.
Spores, pollen, diatoms and provenance of the Argentine Basin sediments.
Progress in Oceanography, 4:179-217.

bottom sediments, biology of (spores)
Vronskic, V.A., and D.G. Panov, 1963.
Composition and distribution of pollen and spores in the surface layer of Mediterranean Sea sediments. (In Russian).
Dokl. Akad. Nauk, SSSR, 153(2): 447-449.

bottom sediments, statoliths
Enbysk, Betty J. and F.I. Linger, 1966.
Mysid statoliths in shelf sediments off northwest North America.
J. sedim. Petrol., 36(3):839-840.

bottom sediments, yeasts
See also: fungi

Bottom sediments, biology of (yeasts)
See: bottom sediments, biology of (fungi)

bottom sediment boundaries

bottom sediment boundaries
Pilkey, Orrin H. and Dirk Frankenberg, 1964.
The relict-recent sediment boundary on the Georgia continental shelf.
Bull. Georgia Acad. Sci., 22(1):37-40.
Also in:
Collected Reprints, Mar. Inst., Univ. Georgia, 4 (1965.)

bottom sediments, brackish water

bottom sediments, brackish waters
Overbeck, Jürgen, 1964.
Der Fe/P-Quotient des Sediments als Merkmal des Stoffumsatzes in Brackwässern.
Helgol. Wiss. Meeresunters., 10(1/4):430-447.

bottom sediment budget

sediment budget
Pierce, J.W. 1969.
Sediment budget along a barrier island chain.
Sediment. Geol. 3(1): 5-16

bottom sediments caloric content

bottom sediments, caloric content
Hughes, Roger N. 1969.
Appraisal of the iodate-sulphuric acid wet-oxidation procedure for the estimation of the caloric content of marine sediments.
J. Fish. Res. Bd. Can., 26(7): 1959-1964

bottom sediments-cementation
MacIntyre, I.G., E.W. Mountjoy and B.F. d'Anglejan 1968.
An occurrence of submarine cementation of carbonate sediments off the west coast of Barbados, W.I.
J. sedim. Petrol. 38(2): 660-664.

bottom sediments, characteristics of
Accerboni, E., et F. Mosetti, 1966.
Mesures de résistivité électrique in situ sur le fond de la mer.
Boll. Geofis. teor. appl., 8(32):243-246.

bottom sediment, characteristics
Das, M.M. 1971.
Longshore sediment transport rates: a compilation of data.
Misc. Pap. Coast. Engng Res. Cent, U.S. Army Corps Engrs. 1-71: 75 pp.

bottom sediments, character
Ekman, S., 1947.
Ueber die Festigkeit der mariner Sedimente als Factor der Tierverbreitung, ein Beitrag zur Associationsanalyse. Zool. Bidrag., Uppsala, 25:1-20.

bottom sediment, characteristic
Nafe, J.E., and C.L. Drake, 1957.
Variations with depth in shallow and deep water marine sediments of porosity, density and the velocities of compressional and shear waves.
Geophysics 22(3):523-552.

bottom sediments, characteristics of
Slabaugh, W.H., and Arthur D. Stump, 1964
Surface areas and porosity of marine sediments.
J. Mar. Res., 69(22):4773-4778.

bottom sediments chemistry of

Bottom sediments, chemistry

See also: the subheading to individual chemicals or groups of chemicals which follow.
(Papers, where multiple chemicals are considered are under the heading: bottom sediments, chemistry)
See also: under CHEMISTRY and its subheadings.

A

bottom sediments, chemistry
Ackman, R.G., R.F. Addison, P.J. Ke and J.C. Sipos, 1971.
Examination of bottom deposits, Long Harbour, Newfoundland, for elemental phosphorus and for fluorides.
Tech Rept. Fish. Res. Bd, Canada, 233: 19pp.

bottom sediments, chemistry of
Alexandrov, A.P., and A.P. Reznikov, 1964
Minor elements in the sediments of the Black Sea. (In Russian).
Okeanologiia, Akad. Nauk, SSSR, 4(4):651-653.

bottom sediments, chemistry
Amin, B.S., D.P. Kharkar and D. Lal, 1966.
Cosmogenic ^{10}Be and ^{26}Al in marine sediments.
Deep-Sea Res., 13(5):805-824.

bottom sediments, chemistry
Amirkhanov, Kh. I., K.S. Magataev and E. Brandt, 1957.
Isotope dating of sedimentary minerals.
Doklady Akad. Nauk, SSSR, 117(4):675-677.

bottom sediments, chemistry
Angino, Ernest, 1966
Uranium and thorium in Antarctic glacial marine sediments.
Oceanogr. Mag., Jap. Met. Soc., 18(1/2):57-62.

bottom sediments, chemistry
Angino, Ernest E., 1966
Geochemistry of Antarctic pelagic sediments.
Geochimica et Cosmochimica Acta, 30(9):939-961.

bottom sediments, chemistry
Angino, Ernest E., and Robert S. Andrews, 1968.
Trace element chemistry, heavy minerals and sediment statistics of Weddell Sea sediments.
J. sedim Petrol. 38(2): 634-642.

bottom sediments, chemistry (organic)
Angino, Ernest E., and Lela M. Jeffrey, 1964.
Identification of sterols and fatty acids in Recent marine sediments. (Abstract).
Geol. Soc., Amer., Special Paper, No. 76:6.

bottom sediments, chemistry of
Anikouchine, William A., 1967.
Dissolved chemical substances in compacting marine sediments.
J. geophys. Res., 72(2):505-509.

bottom, sediments, chemistry
Arrhenius, Gustaf, 1966.
Sedimentary record of long-period phenomena.
In: Advances in earth science, P.M. Hurley, editor, M.I.T. Press, 155-174.

bottom sediments, chemistry of
Avilov, I.K., 1965.
Relief and bottom sediments of the shelf and continental slope of the Northwestern Atlantic. Investigations in line with the programme of the International Geophysical Year, 2. (In Russian).
Trudy, Vses. Nauchno-Issled. Inst. Morsk. Ribn. Choz. i Okeanogr. (VNIRO), 57:173-234.

B

bottom sediments, chemistry
Băcescu, M., M.T. Gomoiu, N. Bodeanu, A. Petran. G.I. Müller si V. Chirila, 1965.
Dinamice populațiilor animale și vegetale din zone nisipurilor fine de la nord de Constanța în condițiile anilor 1962-1965.
In: Ecologie marină, M. Băcescu, redactor, Edit. Acad. Republ. Pop. Române, București, 2: 7-167.

bottom sediments, chemistry
Bader, Richard G., 1962
Some experimental studies with organic compounds and minerals.
The Environmental Chemistry of Marine Sediments. Proc. Symp., Univ. R.I., Jan. 13, 1962, Occ. Papers, Narragansett Mar. Lab., No. 1: 42-57.

bottom sediments, chemistry
Bader, R.G., 1956.
The lignin fraction of marine sediments.
Deep-Sea Res. 4(1):15-22.

bottom sediments, chemistry of
Bader, R.G., 1955.
Carbon and nitrogen relations in surface and subsurface marine sediments.
Geochimica and Cosmochimica Acta 7:205-211.

bottom sediments, chemistry of
Bader, R.G., 1954.
Use of factors for converting carbon or nitrogen to total sedimentary organics. Science 120: 709-710.

bottom sediments, chemistry
Bandy, Orville L., and Kelvin S. Rodolfo, 1964
Distribution of Foraminifera and sediments, Peru-Chile Trench area.
Deep-Sea Res., 11(5):817-837.

bottom sediments, chem.
Barker, John L., Jr., and Edward Anders, 1968.
Accretion rate of cosmic matter from iridium and osmium contents of deep-sea sediments.
Geochim. cosmochim. Acta 32 (6): 627-645.

bottom sediments
Barrett, Peter J., 1966.
Effects of the 1964 Alaskan earthquake on some shallow-water sediments in Prince William Sound, southeast Alaska.
J. Sedim. Petrol., 36(4):992-1006.

bottom sediments, chemistry
*Baturin, G.N., A.V. Kochenov, and E.S. Trimonis, 1969
On the composition and origin of iron-ore sediments and hot brines in the Red Sea. Okeanologiia 9(3): 442-451.
In Russian; English abstract

bottom sediments
Beasley, A.W., 1966.
Bottom sediments. Port Philip survey 1957-1963.
Mem. nat. Mus. Vict., 27:69-105.

bottom sediments, chemistry
Welcner, John H., and Gordon E. Fogg, 1964.
Chlorophyll derivatives and carotenoids in the sediments of two English lakes.
In: Recent researches in the fields of hydrosphere, atmosphere and nuclear geochemistry, Ken Sugawara festival volume, Y. Miyake and T. Koyama, editors, Maruzen Co., Ltd., Tokyo, 39-48.

bottom sediments, chemistry
*Berner, Robert A., 1968.
Migration of iron and sulfur within anaerobic sediments during early diagenesis.
Am. J. Sci., 267(1):19-42.

bottom sediments, chemistry
Biggs, Robert B. 1967.
The sediments of Chesapeake Bay.
In: Estuaries, G.H. Lauff, editor, Publs. Am. Ass. Advmt Sci. 83: 239-260.

bottom sediments, chemistry of
Billings, Gale K. and Paul C. Ragland, 1968
Geochemistry and mineralogy of the Recent reef and lagoonal sediments south of Belize (British Honduras).
Chem. Geol. 3(2): 135-153.

bottom sediments, chemistry
*Bischoff, James L., 1969.
Red Sea geothermal brine deposits: their mineralogy, chemistry and genesis.
In: Hot brines and Recent heavy metal deposits in the Red Sea, E.T. Degens and D.A. Ross, editors, Springer-Verlag, New York, Inc., 368-401.

bottom sediments, chemistry of
*Blumer, Max, and W.J. Cooper, 1967.
Isoprenoid acids in Recent sediments.
Science, 158(3807): 1463-1464.

bottom sediments, chemistry
Bonatti, Enrico, D.E. Fisher, Oiva Joensuu, and H.S. Rydell, 1971.
Postdepositional mobility of some transition elements, phosphorus, uranium and thorium in deep-sea sediments.
Geochim. cosmochim. Acta, 35 (2): 189-201.

bottom sediments, chemistry
Bordovskiy, O.K., 1961
On chemistry of sediments in the Indian Ocean. (In Russian; English abstract.)
Okeanolog. Issled., Mezhd. Komitet Proved. Mezhd. Geofiz. Goda, Prezidiume, Akad Nauk, SSSR, (4):91-99.

bottom sediments, chemistry
Bordovsky, O.K., 1960
The chemistry of the sediments of the central Pacific.
Trudy Inst. Okeanol., 42: 107-116.

bottom sediments, chemistry
Bordorsky, O.K., 1957.
Humic substances in the deposits of the western part of the Bering Sea.
Dokl. Akad. Nauk, SSSR, 113(1):157-160.

bottom sediments, chemistry
Bordovskii, O.K., and B.A. Smirnov, 1962.
Structural-group composition of the oil fractions of bitumens from Bering Sea bottom sediments. (In Russian).
Geokhimiya, (11):1009-1013.

Translation:
Geochemistry, (11):1145-1150.

---

bottom sediments, chemistry
Boström, Kurt and David E. Fisher 1971.
Volcanogenic uranium, vanadium and iron in Indian Ocean sediments.
Earth planet. Sci. Letts 11(2): 95-98.

---

bottom sediments, chemistry
Boström, K., and M.N.A. Peterson, 1969.
The origin of aluminum-poor ferromanganoan sediments in areas of high heat flow on the East Pacific Rise.
Marine Geol., 7(5): 427-447.

---

bottom sediments, chemistry
Boström, Kurt, M.N.A. Peterson, Oiva Joensuu and David E. Fisher, 1969.
Aluminum-poor ferromanganoan sediments on active oceanic ridges. J. geophys. Res., 74(12) 3261-3270.

---

bottom sediments, chemistry
Bourcart, J., C. Lalou and L. Mallet, 1961
Sur la presence d'hydrocarbures du type benzo 3.4. pyrène dans les vases côtières et les sables de la plage de la rade de Villefranche (Alpes Maritimes).
C.R. Acad. Sci., Paris, 252(5): 640-644.

---

bottom sediments
Bowler, J.M., 1966.
The geology and geomorphology. Port Philip survey 1957-1963.
Mem. nat. Mus. Vict., 27:19-67.

---

bottom sediments, chemistry
Bruevich, S.W., editor, 1966.
Chemistry of the Pacific Ocean.
Inst. Okeanol., Akad. Nauk. SSSR, Isdatel. Nauka, Moskva, 358 pp.

---

bottom sediments
Bruevich, S.V., 1957.
(Fresh water depositions below the recent deposits of the Black Sea). DAN 84(3):575.

---

bottom sediments, chemistry
Brujewicz, S.W., 1957.
On certain chemical features of waters and sediments in north-west Pacific.
Proc. UNESCO Symp., Phys. Oceanogr., Tokyo, 1955: 277-292.

---

bottom sediments, chemistry
Bruyevich, S.V., 1956.
[The chemistry of the sediments of the Okhotsk Sea.] Trudy Inst. Okeanol., 17:41-132.

---

submarine geology-sediment chemistry
Bruevich, S.V., 1947.
[The chemical composition of solutions of the Caspian Sea.] Pt. 1. Gidrokhim Materiely13:19 pp.

---

bottom sediments, chemistry
Brujewicz, S.W., and E.G. Vinogradova, 1946
General features of Sedimentation in the Caspian Sea (according to the distribution of carbonates, Fe, Mor and P in Sea deposit CR (Doklady) Acad. Sci, URSS LII(9):789-792.

---

bottom sediments, chemistry
Brujewicz, S.W., and E.D. Zaitseva, 1961.
Some chemical features of sediments and interstitial solutions of the northwest Pacific Ocean.
Tenth Pacific Sci. Congr., Honolulu, 21 Aug.-6 Sept., 1961, Abstracts of Symposium Papers, 367.

---

bottom sediments, chemistry
Bruevich, S.W., and E.D. Zaitzeva, 1960
[The chemistry of sediments of the northwestern Pacific.]
Trudy Inst. Okeanol., 42: 3-88.

---

bottom sediments, chemistry
Burkholder, P.R., and L.M., 1956.
Microbiological assay of Vitamin B12 in marine solids. Science 123(3207):1071-1073.

C

bottom sediments, chemistry
California, Humboldt State College, 1964.
An oceanographic study between the points of Trinidad Head and Eel River.
State Water Quality Control Bd., Resources Agency California, Sacremento, Publ., No. 25:136 pp.

---

bottom sediments, chemistry
Carroll, Dorothy, 1964.
Ion-exchange capacity of sediments from the experiemental Mohole, Guadalupe site.
J. Sed. Petr., 34(3):537-542.

---

bottom sediments, chemistry
Chave, Keith E., and F.T.Mackenzie, 1961.
A statistical technique applied to the geochemistry of pelagic muds.
J. Geology, 69(5):572-582.

---

bottom sediments,chemistry
Chester,R.,1965.
Elemental geochemistry of marine sediments.
In:Chemical oceanography,J.P.Riley and G. Skirrow,editors,Academic Press,2:23-80.

---

bottom sediments,chemistry of
*Chester,R., and M.J.Hughes,1967.
A chemical technique for the separation of ferromanganese minerals, carbonate minerals and adsorbed trace elements from pelagic sediments.
Chem. Geol., 2(3):249-262.

---

bottom sediments, chemistry
Chester, R., and M.J. Hughes, 1966.
The distribution of manganese, iron and nickel in a North Pacific deep-sea clay core.
Deep-Sea Res., 13(4):627-634.

---

bottom sediments, chemistry
Choe, Sang, 1971.
Studies on marine sediments of the Korean Seas. 1. Concentrations and distributions of some geochemical elements in sediments from the sea off Eastern Korea. J. oceanogr. Soc. Korea 6(1): 1-15. (in Korean; English abstract)

---

bottom sediments, chemistry
Cabatti, M., P. Colantoni ed E. Rabbi, 1968.
Ricerche oceanografiche nell'alto Adriatico antistante il delta del Po: crociera estiva 1966.
G. Geol., (2) 34: 479-430. Also in: Raccolta Dati Oceanogr., Comm. ital. Oceanogr. Consig. naz. Ricerche, (A) (17) (1969).

---

bottom sediments (pelagic clay) chemistry of
Cronan, D.S., 1969.
Inter-element associations in some pelagic deposits.
Chem. Geol. 5 (2): 99-106

---

bottom sediments, chemistry
Crosnier, A., 1964.
Fonds de peche le long des cotes de la Republique Federale du Camaroun.
Cahiers, O.R.S.T.R.O.M., No. special:133 pp.

D

bottom sediments, chem.
Debyser, J., 1957.
La sédimentation dans le bassin d'Archachon.
Bull. C.E.R.S., Biarritz, 1(3):405-418.

---

bottom sediments, chemistry
Degens, Egon T., 1965.
Geochemistry of sediments: a brief survey.
Prentice-Hall, Inc., 342 pp.

---

bottom sediments, chemistry
Degens, E.T., and S. Epstein, 1962.
Relationship between O18/O16 ratios in coexisting carbonates, cherts and diatomites.
Bull. Amer. Assoc. Petr. Geol., 46(4):534-542.

---

bottom sediments, chemistry
Degens, Egon T., Johannes H. Reuter and Kenneth N.F. Shaw, 1964.
Biochemical compounds in offshore California sediments and sea waters.
Geochimica et Cosmochimica Acta, 28(1):45-66.

---

bottom sediments, chemistry
de las Heras, A.R., R. Lopez Costa, F. Cabano Ruegas, L. Rodriguez Molino and J.R. Besada Rial, 1956.
Análisis de fondos de la Bahía de Cádiz.
Bol. Inst. Espanol, Oceanogr., 75:20 pp.

---

bottom sediments, chemistry
Delauze, Henri, 1966.
Résultats d'analyses de prélèvements en Mer Ionienne, en Mer Méditerranée et dans la Fosse des Kouriles.
Cahiers Océanogr., C.C.O.E.C., 18(8):715-720.

---

bottom sediments,chemistry
Dietrich,R., W. Höhnk und W.D. Manzel,1966.
Studien zur Chemie ozeanischer Bodenproben. IV. Vergleichende Untersuchungen über die Beziehungen zwischen physikalischen und chemischen Grössen in Sedimenten des Indischen Ozeans und denen der Norddsee und der Aussenweser.
Veröff.Inst. Meeresforsch. Bremerh., 10(2):121-136.

---

bottom sediments,chemistry.
Dietrich, R., W. Höhnk und W.-D. Manzel, 1965.
Studien zur Chemie ozeanischer Bodenproben. III. Beziehungen zwischen physikalischen und chemischen Grössen in den Böden der Nordsee und der Aussenweser.
Veröffentlichungen, Inst. Meeresforschung, Bremerhaven, 9(2):242-278.

---

bottom sediments, chemistry
Drever, James I., 1971.
Chemical and mineralogical studies, site 66.
Initial Repts Deep-Sea Drill. Proj. 7(2): 965-975.

---

bottom sediments, chemistry
Duursma, Egbert Klaas 1967.
The mobility of compounds in sediments in relation to exchange between bottom and supernatant water.
In: Chemical environment in the aquatic habitat, Proc. I.B.P. Symp. Amsterdam, Oct 1966 : 288-299.

E

bottom sediments
Edgerton,H., E.Seibold,K.Vollbrecht und F. Werner,1966.
Morphologische Untersuchungen am Mittelgrund (Eckernförder Bucht, west liche Ostsee).
Meyniana, 16:37-50.

bottom sediments, chemistry
El Wakeel, S.K., and J.P. Riley, 1961.
Chemical and mineralogical studies of deep-sea sediments.
Geochimica et Cosmochimica Acta, 25(2):110-146.

bottom sediments, chemistry
Emery, K.O., 1964.
Sediments of Gulf of Aqaba (Eilat).
In: Papers in Marine Geology, R.L. Miller, Editor, Macmillan Co., N.Y., 257-273.

bottom sediments, chemistry
Emery, K.O., and Jobst Hulsemann, 1963
Submarine canyons of southern California. Part 1. Topography, water, and sediments.
Allan Hancock Pacific Expeditions, 27(1): 1-80.

bottom sediments,chemistry
Emery,K.O., and Hiroshi Niino,1963.
Sediments of the Gulf of Thailand and adjacent continental shelf.
Bull. Geol. Soc., Am., 74:451-554.

bottom sediments, chemical analyses
Emery, K.O., J.I. Tracy, jr., and H.S. Ladd, 1954.
Geology of Bikini and nearby atolls. Bikini and nearby atolls:Pt. 1, Geology.
Geol. Surv. Prof. Pap. 260-A:1-265, 64 pls., 11 charts

bottom sediments,chemistry
*Ernst,Wolfgang,1967.
Dünnschichtchromatographische Trennung und quantitative Bestimmung von Galacturon- und Glucuronsäure in Polysaccharidfraktionen des Meeresbodens.
Veröff. Inst. Meeresforsch. Bremerh. 10(3):183-185.

bottom sediments,chemistry
Ernst.W., 1966.
Nachweis,Identifizierung und quantitative Bestimmug von Uronsäuren in Polysaccharidfraktionen mariner Sedimente der Deutschen Bucht.
Veröff,Inst. Meeresforsch. Bremerh.,10(2):81-92.

F

bottom sediments
Faas,Richard W., 1966.
Paleoecology of an Arctic estuary.
Arctic, 19(4)343-348.

bottom sediments, chemistry
Febres, German, y Gilberto Rodriguez, 1966.
Ch. 3. Sedimentos.
Estudios hidrobiologicos en el Estuario de Maracaibo, Inst. Venezolano de Invest. Cient., 66-82. (Unpublished manuscript).

bottom sediments, chemistry
Feki, Mohamed, 1960.
Étude de quelques sédiments marins actuels recoltes au large des côtes de Tunisie.
Sta. Océanogr., Salammbô, Ann., No. 12:33 pp.

Note: Inside title page author is called Monsieur Feki Mohamed.

bottom sediments, chemistry
*Fujita, Yuji, Tadataka Taniguti and Buhei Zenitani, 1967.
Microbiological studies on shallow marine areas. III. On relation of the heterotrophic bacteria to the changes in carbon dioxide, oxygen consumption, organic acid and sulfides in the mud sediment. (In Japanese; English abstract).
Bull. Fac. Fish., Nagasaki Univ., 23: 187-196.

bottom sediments,chemistry
Fukai, Rinnosuke, 1965.
Chemical composition of shallow-water sediments in the Bay of Roquebrune.
Bull. Inst. Océanogr., Monaco,65(1337):15 pp.

G

bottom sediments, chemistry
Gadel, François, et Guy Cahet, 1968.
Etude des acides gras de sédiments récents: comparaison de milieux lagunaires et marins (Roussillon).
C.r. hebd. Séanc. Acad. Sci., Paris (D) 266 (20): 2040-2042.

bottom sediments, chemical
Gambarian, M.E., 1962.
A contribution to the technic of assaying the rate of destruction of organic matter in benthic sediments of deep reservoirs.
Mikrobiologiia, Akad. Nauk, SSSR, 31(5):895-898.

bottom sediments, chemistry
Gassaway, John D. 1970.
Mineral and chemical composition of sediments from the Straits of Florida.
J. sedim. Petrol. 40 (4): 1136-1146.

bottom sediments, chemistry
Gershanovich, D.E., 1962.
New data on Recent sediments of the Bering Sea. (In Russian).
Trudy Vses. Nauchno-Issledov. Inst. Morsk. Ribn. Chos i Okean., VNIRO, 46:128-164.

bottom sediments, chemistry
Gershanovich, D.E., 1956.
[Silica, calcium carbonate and organic carbon in deep-water sediments of the Japan Sea.]
Trudy Gosudarst. Inst. Okean., 31(43):72-79.

bottom sediments, chemistry
Gershanovich, D.E., V.S. Kotenev, and V.N. Novikov, 1964.
Relief and bottom sediments in the Gulf of Alaska. (In Russian).
Trudy, Vses. Nauchono-Issled. Inst. Morsk. Ribn. Choz. i Okeanogr., 52:83-133.

Also:
Izv. Tichookeansk. Nauchno-Issled. Inst. Morsk. Ribn. Choz. i Okeanogr. (TINRO), 52:83-133.

bottom sediments, chemistry of
*Girogetti, F., F. Mosetti e G. Macchi, 1968.
Caratteristiche morphologiche. Fisiche e chimiche del fondo marino del Golfo di Trieste nell' area compresa entro la Congiungente Punta Grossa-Bocche di Primero.
Boll. Soc. adiat. Sci. nat., 6(1):3-21.

bottom sediments, chemistry (data)
Glagdeva, M.A., 1961.
Regularities in the distribution of chemical elements in Recent sediments of the Black Sea. (In Russian).
Doklady Akad. Nauk, SSSR, 136(1):195-198.

English Edit., 1962, 136(1-6):1-4

bottom sidements, chemistry
Gogate S.S., V.N. Sastry, T.M. Krishnamoorthy and R. Viswanathan, 1970
Chemistry of shelf sediments on the west coast of India.
Current Sci., 39 (8): 171-174.

bottom sediments, chemistry
Goldberg, E.D., 1961
Chemical and mineralogical aspects of deep-sea sediments.
Physics and Chemistry of the Earth, Pergamon Press, 4(8): 281-302.

bottom sediments, chemistry
Goldberg, E.D., 1958
Determination of opal in marine sediments. J. Mar. Res., 17: 178-182.

bottom sediments, chemistry
Goldberg, E.D., and G.O.S. Arrhenius, 1958.
Chemistry of Pacific pelagic sediments.
Geochimica et Cosmochimica Acta 13(2/3):153-212.

bottom, sediments, chemistry
Goldberg, Edward D., and Minoru Koide, 1962.
Geochronological studies of deep-sea sediments by the ionium/thorium method.
Geochimica et Cosmochimica Acta, 26:417-450.

Bottom sediments, chemistry
Goldberg, Edward D., and Minoru Koide, 1962.
Ionium/thorium geochronology and the depth distribution of members of the U238 series in sediments of the Indian Ocean (Abstract).
J. Geophys. Res., 67(4):1638.

bottom sediments, chemistry
*Goodell,H. Grant,1967.
The sediments and sedimentary geochemistry of the southeastern Atlantic shelf.
J. Geol., 75(6):665-692.

bottom sediments, chemistry
Gorbunova, Z. N., 1960
[Highly dispersed minerals in the sediments of the Indian Ocean.] Doklady Akad. Nauk, SSSR, 134(4): 935-938.

bottom sediments, chemistry
Gorbunova, Z.N., 1960
[The composition of clayey minerals in different horizons of sediments of the Indian Ocean.]
Doklady Akad. Nauk, SSSR, 134(5): 1201-1203.

Bottom sediments, chemistry
Gorshkova, T.I. 1967.
Sedimentation in the Norwegian Sea. (Carbonate and organic matter in the sediments of the Norwegian Sea). (In Russian).
Atlant. nauchno-issled. Inst. rybn. Khoz. okeanogr. (AtlantNIRO). Materialy Konferentsii po Resul'tatam Okeanologicheskikh Issledovanii v Atlanticheskom Okeane, 300-308.

bottom sediments, chemistry
Gorshkova, T.I., 1965.
Sediments in the Norwegian Sea. (In Russian; English abstract).
Okeanolog. Issled.,Rezult. Issled. Programme Mezhd. Geofiz. Goda, Mezhd. Geofiz. Komitet Presidiume, Akad. Nauk, SSSR, No. 13:212-224.

bottom sediments, chemistry
Gorshkova, T.I., 1965.
Chlorophyll and carotinoids in the sediments of the Baltic Sea and Gulf of Riga. Investigations in line with the programme of the international Geophysical Year, 2. (In Russian).
Trudy, Vses. Nauchno-Issled. Inst. Morsk. Ribn. Choz. i Okeanogr.,(VNIRO), 57:313-323.

bottom sediments, chemistry
Gorshkova, T. I., 1960
Bottom sediments of the western Polar Basin (on materials of the drifting "North Pole I" station, 1937-1938.).
Soviet Fish. Invest., North European Seas, VNIRO, PNIRO, Moscow, 1960: 111-127.

bottom sediments, chemistry

Gorshkova, T.I., 1960

[Chemical composition of the soil solutions in the Norwegian Sea] Okean. Issle., IGY Com., SSSR: 113-116.

bottom sediments, chemistry of

Gorshkova, T.I., 1959.
[Carbonate and organic matter in the sediments of the middle and southern Caspian.]
Trudy VNIRO, 38:142-151.

bottom sediments, chemistry

Gorshkova, T.I., 1957.
[Sedimentation in the Caspian Sea.]
Trudy Vses. Gidrobiol. Obshsh., 8:68-99.

bottom sediments, chemistry

Gorshkova, T. I., 1960.
[(Sediments of the Norwegian Sea.) Morsk. Geol., Doklady Sovetsk. Geol., 21 Sess., Int. Geol. Congress, 132-139. English abstract

bottom sediments, chemical

Gorshkov, T.I., 1957.
[The chemical composition of interstitial waters of the Baltic Sea] Dokl. Akad. Nauk, SSSR, 113(4):863-865.

bottom sediments, chemistry

Gorshkova, T.N., 1956.
[Natural humidity, carbonates and organic matter of sediments as indicators of the conditions of ocean formation.] Trudy Inst. Okeanol., 17:141-147.

bottom sediments, chemistry

Greenfield, Leonard J., Robert D. Hamilton and Catherine Weiner 1970.
Nondestructive determination of protein, total amino acids, and ammonia in marine sediments.
Bull. mar. Sci. 20(2): 289-304.

bottom sediments, chemistry

Gucluer, Sevket M., and M. Grant Gross, 1964.
Recent marine sediments in Saanich Inlet, a stagnant marine basin.
Limnology and Oceanography, 9(3):359-376.

H

bottom sediments, chemistry

Hamada, Shichiro, and Ritsuko Hamada 1966.
The state of sediments in Omura Bay at the time of the Akashio occurrence. (In Japanese; English abstract).
Bull. Seikai reg. Fish. Res. Lab., 34: 149-159

bottom sediments, chemistry of

Hamaguchi, Hiroshi, Masumi Osawa, and Naoki Onima, 1962.
The vertical distribution of minor elements in core samples from the Japan Trench. (IN Japanese English abstract).
Repts. JEDS, Deep-Sea ~~Comm.~~ Res., Japan Soc., Promotion of Science, 3:1-2.

Originally published:
Nippon Kagaku Zasshi, 82:691-693.

JEDS Contr. No. 21.

bottom sediments, chemistry of

Hamaguchi, H., M. Tatsumoto and J. Itaya, 1954.
Chemical investigations of deep-sea deposits. XIX. J. Chem. Soc., Japan, Pure Chem. Sec., 75: 119-121.

bottom sediments, chemistry

*Hartmann, Martin, 1969.
Investigations of Atlantis II Deep samples taken by the FS Meteor.
In: Hot brines and Recent heavy metal deposits in the Red Sea, E.T. Degens and D.A. Ross, editors, Springer-Verlag, New York, Inc., 204-207.

bottom sediments, chemistry of

Hartmann, Martin, 1964.
Zur Geochemie von Mangan und Eisen in der Ostsee.
Meyniana, 14:3-20.

bottom sediments, chemistry

Hartmann, Martin, und Heimo Nielsen, 1969.
$\delta^{34}S$-Werte in rezenten Meeressedimenten und ihre Deutung am Beispiel einiger Sedimentprofile aus der westlichen Ostsee
Geol. Rundschau 58 (3): 621-655

bottom sediments, chemistry of

Haskin, Larry, and Mary A. Gehl, 1962.
The rare-earth dsitribution in sediments.
J. Geophys. Res., 67(6):2537-2541.

bottom sediments, chemistry of

Hata, Yoshihiko, 1963
Microbial activities in the development of reducing conditions in marine sediments (Preliminary report). (In Japanese; English abstract).
J. Shimonoseki Univ., Fish., 12(2/3):1-10.

bottom sediments, chemistry

*Hendricks, Ruth L., Fredric B. Reisbick, Edwin J. Mahaffey, D. Blair Roberts and Melvin N.A. Peterson, 1969.
Chemical composition of sediments and interstitial brines from the Atlantis II, Discovery and Chain deeps.
In: Hot brines and Recent heavy metal deposits in the Red Sea, E.T. Degens and D.A. Ross, editors, Springer, Verlag, New York, Inc., 407-440.

bottom sediments, chemistry

Heye, Dietrich, 1970.
A system for detection of ionium thorium and protactinium to date deep-sea cores.
Geochim. cosmochim. Acta, 34 (3): 389-397.

bottom sediments, chemistry

Heye, Dietrich, 1969.
Uranium, thorium, and radium in ocean water and deep-sea sediments
Earth planet. Sci. Letters 6(2): 112-116.

bottom sediments, chemistry

Hirst, D.M., 1962.
The geochemistry of modern sediments from the Gulf of Paria. I. The relationship between the mineralogy and the distribution of major elements.
Geochimica et Cosmochimica Acta, 26:309-334.

bottom sediments, chem.

Holland, H.D., and J.L. Kulp, 1954. chemistry
The transport and deposition of uranium, ionium and radium in rivers, oceans and ocean sediments.
Geochim. Cosmochim. Acta 5:197-213.

Abstr. Chem. Abstr. 11855a, 1954.

bottom sediments, chemistry

Holmes, Charles W., 1969.
In the Gulf of Mexico ---- geochemical exploration produces exciting results: sea floor samples lead to a surprising discovery of zircon ium, titanium and heavy minerals off the Texas coast.
Ocean Industry, 4(6):49-52. (popular)

bottom sediments, chem

Horowitz, Arthur, 1970.
The distribution of Pb, Ag, Sn, Tl and Zn in sediments on active oceanic ridges.
Marine Geol. 9(4): 241-259.

bottom sediments, chemical

Hoshino, M., and Y. Ichihara, 1960

[Submarine topography and bottom sediments off Kumanonada and Enshu-nada.] J. Oceanogr. Soc., Japan, 16(2): 41-46.

bottom sediments, chemistry

*Hoskins, Charles M., 1968.
Magnesium and strontium in mud fraction of Recent carbonate sediment, Alacrán Reef, Mexico.
Bull. Am. Ass. Petr. Geol., 52(11):2170-2177.

bottom sediments, chemistry

Hosokawa, Iwao, 1962
A review of chemical studies on marine sediments (chiefly on shallow-water deposits). (In Japanese).
J. Oceanogr. Soc., Japan, 20th Ann. Vol., 541-562.

bottom sediments, chemistry

Hosokawa, I.S. Okabe and S. Hamada, 1959.
[Oceanographical studies in the sediments in the East China Sea. 1. On the contents of organic carbon and total nitrogen.] J. Oceanogr. Soc., Japan, 15(2): 43-47.

bottom sediment, chemistry

Hosokawa, I., S. Okabe and S. Hamada, 1959.
[Oceanographical studies on sediments of East China Sea. II. On carbonate contents and soluble silicate.] J. Oceanogr. Soc., Japan, 15(3): 107-110.

bottom sediments, chemistry

Hunt, John M., Earl E. Hays, Egon T. Degens and David A. Ross, 1967.
Red Sea: detailed survey of hot-brine areas.
Science, 156(3774):514-516.

bottom sediments, chemistry of

Hurley, P.M., B.C. Heezen, W.H. Pinson and H.W. Fairbairn, 1963.
K-Ar age values in pelagic sediments of the North Atlantic.
Geochimica et Cosmochimica Acta, 27(4):393-399.

bottom sediments, chem.

Hutchinson, G.E., R.J. Benoit, W.B. Cotter, and P.J. Wangersky, 1955.
On the nickel, cobalt and copper contents of deep-sea sediments. Proc. Nat. Acad. Sci. 41(3): 160-162.

I

bottom sediments, chemistry

Imai, Y., 1960

[Oceanographical studies on the behaviour of chemical elements. 1. On iron and manganese in the sediments around "Shiome" (current rips) in Urado Bay.]
J. Oceanogr. Soc., Japan, 16(3): 134-138.

bottom sediments, chemistry of

Imai, Y., 1960

[Oceanographical studies on the behaviour of chemical elements. 2. Iron and aluminium distribution in Urado Bay - On the concentration of those elements in the "Shiome" zone, or the rip tide.]
J. Oceanogr. Soc., Japan, 16(4): 167-171.

bottom sediments, chemistry o
Ingle, R.M., A.R. Ceurvels and R. Leinacker, 1955
Chemical and biological studies of the muds of Mobile Bay.
Rept. to Div. Seafoods, Alabama Dept. Conservatio :1-14.

bottom sediments, chemistry
Isaeva, A.B., 1964.
Geochemical study of the sediments in the northern part of the Indian Ocean. Investigations in the Indian Ocean (33rd Voyage of E/S "Vitiaz"). (In Russian).
Trudy Inst. Okeanol., Akad. Nauk, SSSR, 64:227-235.

Bottom sediments, chemistry
Ishibashi, M., & T. Hara, 1959.
A systematic analysis of potassium, rubidium and cesium and its application to sea-muds.
Bull. Inst. Chem. Res. Kyoto Univ., 37(3):185-190.

bottom sediments, chemistry of
Ishibashi, M., and S. Ueda, 1956.
Chemical studies on the ocean (51-55). Chemical studies of the shallow water deposits.
Bull. Inst. Chem. Res., Kyoto Univ., 34(3): 117-141.

bottom sediments, chemistr
Ishibashi, M., and S. Ueda, 1956.
Chemical studies on the ocean. 59. Chemical studies of the shallow water deposits. 12. On th chemical constituents of the shallow-water deposits along the sea coasts of Okayama Prefecture. Bull. Inst. Chem. Res., 34(5):235-239.

bottom sediments, chemistry
Ishibashi, M., and S. Ueda, 1956.
Chemical studies on the ocean. 60. Chemical studies of the shallow-water deposits. 13. On the chemical constituents of the shallow-water deposits along the sea coasts of Yamaguchi and Shimane Prefectures. Bull. Inst. Chem. Res., 34(5):240-244.

bottom sediments, chemistry
Ishibashi, Masayoshi, Shunzo Ueda and Yoshikazu Yamamoto, 1970.
The chemical composition and the cadmium, chromium and vanadium contents of shallow-water deposits in Tokyo Bay. (In Japanese; English abstract)
J. Oceanogr. Soc. Japan 26(4): 189-194

bottom sediments, chemistry
Ishibashi, Masayoshi, Shunzo Ueda, and Yoshikazu Yamamoto, 1964
Studies on the utilization of the shallow water deposits (continued). On the cobalt and nickel contents of the shallow-water deposits.
Rec. Oceanogr. Wks., Japan, 7(2):37-42.

bottom sediments, chemistry
Ishibashi, M., S. Ueda and Y. Yamamoto, 1956.
Chemical studies on the ocean. 61. Chemical studies of the shallow water deposits. 14. On the chemical constituents of the shallow-water deposits along the sea-coasts of Tottori and Hyogo Prefectures. Bull. Inst. Chem. Res., 34(5):245-249.

bottom sediments, chemistry
Ishibashi, M., T. Yamamoto and F. Morii, 1962.
Chemical studies on the shallow-sea-water deposits. (Cont.).
Rec. Oceanogr. Wks., Japan, n.s., 6(2):163-168.

J

bottom sediments (calcareous)
Jacquette, R., 1962.
Etude des fonds de maerl de Méditerranée.
Rec. Trav. Sta. Mar., Endoume, Bull., 26(41):141-

bottom sediments, chemistry
*Jerbo, Allan, 1967.
Geochemical and strength aspects of Bothnian clay sediments.
In: Marine geotechnique, A.F. Richards, editor, Univ. Illinois Press, 177-186.

bottom sediments, chemistry
Jitts, H.R., 1959.
The adsorption of phosphate by estuarine bottom deposits.
Australian J. Marine & Freshwater Res., 10(1): 7-21.

K

bottom, sediments, chemistry
Kamatani, Akiyoshi, and Chikayoshi Matsudaira, 1966.
Extraction and determination methods of organic acids in sea water and marine sediment.
J. Oceanogr. Soc. Japan, 22(3):87-92.

bottom sediments, chemistry
Kamatani, Akiyoshi, and Chikayoshi Matsudaira, 1963.
On the fertility of Matsushima Bay. 1. Some chemical properties on the marine deposits.
Tohoku J. Agric. Res., 14(2):83-91.

bottom sediments, chemistry
Kaplan, I.R., and S.C. Rittenberg, 1963.
23. Basin sedimentation and diagenesis.
In: The Sea, M.N. Hill, Editor, Interscience Publishers, 3:583-619.

bottom sediments, chemistry
Kato, Kenji, 1961
Some aspects on biochemical characteristics of sea water and sediments in Mochima Bay, Venezuela.
Bol. Inst. Oceanogr., Univ. de Oriente, Venezuela 1(2):343-358.

bottom sediments, chemistry of
Kato, K., 1951.
Chemical studies on marine deposits. 4. On chemical composition of bottom samples of sea of the northwest of Hokkaido Island. Bull. Fac. Fish., Hokkaido Univ., 2(2):134-144, 12 textfigs.

submarine geology, chem
Kato, K., and T. Okuda, 1949.
Chemical studies on the marine deposits. II. On the catalytic action of the marine deposits.
J. Fish., Hakodate, Japan, No. 54:13-20, 4 textfigs.

bottom sediments, chemical
Keller, George H., and Adrian F. Richards, 1927.
Sediments of the Malacca Strait, Southeast Asia.
J. sedim. Petrol., 37(1):102-127.

bottom sediments, chemistry
Kharkar, D.P., D. Lal and B.L.K. Somayajulu, 1963.
Investigation in marine environments using radioisotopes produced by cosmic rays.
Radioactive Dating, Int. Atomic Energy Agency, Vienna, 175-186.

bottom sediments, chemistry
Knowles, G., R.W Edwards and R. Briggs, 1962
Polarographic measurement of the rate of respiration of natural sediments.
Limnol. and Oceanogr., 7(4):481-484.

bottom sediments, chemistry
Kochenov, A. V., G. N. Baturin, S. A. Kovaleva, Ye. M. Yemel'yanov and K. M. Shimkus, 1965.
Uranium and organic matter in the sediments of the Black and Mediterranean seas.
Geokhimiia, 1965(3):302-313.
Translation AGI $4.35

bottom sediments
Kolp, Otto, 1966.
Die Sedimente der westlichen und südlichen Ostsee und ihre Darstellung.
Beitr. Meereskunde, 17-18:9-60.

bottom sediments, chemistr
Kolp, O., 1957.
Die Schwermineralanteile veschiedener Meeresbodenarten der Beltsee. Ann. f. Hydrogr., 8:37-47

Bottom Sediments, Chemistry
Koutyurin, V. M., A. P. Lisitsin, 1961.
[Vegetable pigments in the suspended material and in bottom sediments of the Indian Ocean]
Mezhd. Kom. Mezhd. Geofiz. Goda, Presidiume Akad. Nauk, SSSR, Okeanol. Issled., (3):90-116.

bottom sediments, chemical anal.
Koyama, T., 1953.
Measurement and analysis of gases in sediments.
J. Earth Sci., Nagoya Univ., 1(2):107-118.

bottom sediments, chemistry
Kramer, J.R. 1961
Chemistry of Lake Erie.
Proc. Fourth Conf., Great Lakes Research Great Lakes Res. Div., Inst. Sci. & Tech. Univ. Michigan, Publ., (7):27-56.

bottom sediments, chemistry
Krause, Dales C., 1964.
Lithology and sedimentation in the southern continental borderland.
In: Papers in Marine Geology, R.L. Miller, Editor, Macmillan Co., N.Y., 274-318.

bottom sediments, chemistry
Krumbein, W.C., and R.M. Garrels, 1952.
Origin and classification of chemical sediments in terms of pH and oxidation-reduction potentials
J. Geol. 60(1):1933, 8 textfigs.

bottom sediments, chemistry
Ku, Teh-lung, 1964.
Ratios A_{U234}/A_{U238} of deep-sea sediments (Abstract).
Trans. Amer. Geophys. Union, 45(1):120.

bottom sediments, chemistry
Kulikov, N.N., 1970.
Distribution of main chemical components in bottom sediments of the Kara Sea. (In Russian)
Izv. Vses. Geograf. Obshch. 102:220-233

bottom sediments, chemistry
Kutyurin, E.M., and A.P. Lisitsin, 1962
[Vegetative pigments present in suspensions and bottom sediments in the western part of the Indian Ocean. Communication. 2. Quantitative distributions and qualitative contents of pigments present in suspensions]
Mezhd. Geofiz. Komitet, Prezidiume Akad. Nauk, SSSR, Rezult. Issled. Programme Mezhd. Geofiz. Goda, Okeanol. Issled., No. 5:112-129.

bottom sediments, chemistry
Kuznetsov, Yu.A., Z.N. Simonyak, A.P. Lisitsyn and M.S. Frenkel', 1968.
Uranium and radium in the surface layer of oceanic sediments. (In Russian)
Geokhimiya 1968(3): 323-333.
Translation: Geochem. int. 5(2): 306-311

L

Laevastu, Taivo, 1958 bottom sediments, chemistry
The occurrence of pigments in marine sediments. J. Mar. Res., 17:325-334.

bottom sediments, chemistry
Lalou, Claude, 1963.
Contribution à l'étude de la pollution des sédiments de la Méditerranée occidentale par les benzopyrènes.
Rapp. Proc. Verb., Réunions, Comm. Int., Expl. Sci. Mer Méditerranée, Monaco, 17(3):711-718.

bottom sediments, chemistry
Landergren, Sture, 1964.
On the geochemistry of deep-sea sediments.
Repts. Swedish Deep-Sea Exped., 10 (Spec. Invest., 5):61-148.

bottom sediments chemistry
Landergren, S., 1954.
On the geochemistry of the North Atlantic sediment core No. 238. Repts. Swedish Deep-Sea Exped. 1947-48, Sediment Cores, N. Atlantic Ocean, 7(2):125-148.

submarine geology, chemistry
Landergren, S., 1948
6. On the geochemistry of Mediterranean sediments. Preliminary report on the distribution of beryllium, boron, and the ferrides in three cores from the Tyrrhenian Sea. Medd. Ocean. Inst., Göteborg 15 (Göteborgs Kungl. Vetenskaps-och Vitterhets -Samhälles Handlingar, Sjätte Följden, Ser. B 5(13)):34-46, figs.6-7.

bottom sediments chemistry
Leclaire Lucien 1968
Contribution à l'étude de la relation entre le carbone et l'azote de la matière organique contenue dans les boues et vases du plateau continental algérien.
C. r. hebd. Séanc. Acad. Sci., Paris, (D) 266(20): 2049-2051

sediments
Leclaire, Lucien, 1963
Facteurs d'évolution d'une côte sablonneuse rectiligne très ouverte. Etude préliminaire à l'implantation d'un port de pêche et de plaisance.
Cahiers Océan., C.C.O.E.C., 15(8):540-556.

bottom sediments chemistry
Le Petit, J., et R. Matheron, 1963.
Contribution à l'étude des rapports existant entre certains facteurs physico-chimiques et la flore bactérienne dans un milieu sédimentaire lagunaire.
Rapp. Proc. Verb., Réunions, Comm. Int., Expl. Sci., Mer Méditerranée, Monaco, 17(3):687-694.

bottom sediments, chemistry
Lepore, Nevio, 1962.
Determinazione del carbonio organico e dei carbonati in sedimenti marini.
Archiv. di Oceanogr. Limnol., 12(3):275-295.

Also:
Ist. Sper. Talass., Trieste, Pubbl., No. 398.

bottom sediments, chemistry
Levinson, A.A., and John C. Ludwick, 1966.
Speculation on the incorporation of boron into argillaceous sediments.
Geochimica et Cosmochimica Acta, 30(9): 855-861

bottom sediments, chemistry
Lewis, G., and E.D. Goldberg, 1956.
X-ray fluorescence determination of barium, titanium and zinc in sediments.
Analyt. Chem., 28(8):1282-1285.

bottom sediments, chemistry
* Lidz, Louis, Walter B. Charm, Mahlon M. Ball, and Sylvia Valdes, 1969.
Marine basins off the coast of Venezuela.
Bull. Mar. Sci., 19(1): 1-17.

bottom sediments, chemistry
Lisitzin, A.P., 1966.
Processes of Recent sedimentation in the Bering Sea. (In Russian).
Inst. Okeanol., Kom. Osad. Otdel Nauk o Zemle, Isdatel, Nauka, Moskva, 574 pp.

bottom sediments, chemistry
Lisitzin, A.P., V.V. Serova, I.B. Zverinskaya, V. Lukashin, Z.N. Gorbunova, V.V. Gordeev, V.V. Zhurensko, A.M. Pchelintsev, Ju.I. Belyaev, N. I. Popov, O.V. Shishkina, N.M. Morozov, A.P. Jouse, O.G. Kozlova and V.V. Mukhina. 1971 Geochemical, mineralogical, and paleontological studies. Initial Repts. Deep Sea Drilling Project, Glomar Challenger 6: 829-960.

bottom sediments, chemistry
López López, Miguel A., y Laigo Okuda, 1968.
Algunas observaciones sobre características físico-químicas de los sedimentos y distribución de la fauna macro bentónica de la laguna Grande del Obispo (Venezuela).
Bol. Inst. Oceanogr. Univ. Oriente, Venezuela, 7(1): 107-127.

bottom sediments, chemistry
*Loring, D.H., and D.J.G. Nota, 1968.
Occurrence and significance of iron, manganese, and titanium in glacial marine sediments from the estuary of the St. Lawrence River.
J.Fish.Res.Bd.Can., 25(11):2327-2347.

bottom sediments, chemistry
Lüneberg, H., 1964.
Beiträge zur Sedimentologie der Weser- und Elbästuare.
Helgoländer Wiss. Meeresuntersuch., 10(1/4):217-230.

M

bottom sediments, chemistry
*Macchi, G., 1968.
Sulla composizione chimica dei sedimenti recenti del Golfo di Trieste.
Boll.Soc.adriat.Sci.nat. 56(1):22-41.

bottom sediments, chemical
Magliocca, A., & A.S. Kutner, 1965.
Sedimentos de fundo da Enseada do Flamengo-Ubatuba. (In Portuguese; English summary).
Contrib. Avulsas, Inst. Oceanogr., São Paulo, (Oceanogr. Fisica), No. 8:14 pp.

bottom sediments, chemistry of
Mallet, L., et Mme. Le Theule, 1961.
Recherche du benzo 3,4 pyrène dans les sables vaseux marins des régions côtières de la Manche et de l'Atlantique.
C.R. Acad. Sci., Paris, 252(4):565-657.

bottom sediments, chemistry
Mallet, L., L.V. Perdriau et J. Perdriau, 1963.
Pollution par les hydrocarbures polybenzéniques du type benzo-3,4 pyrène de la region occidentale de l'Ocean Glacial Artique.
Comptes Rendus, Acad. Sci., Paris, 256(16):3487-3489.

bottom sediments, chemistry of
Malyuga, D.P., 1949.
On the question of the content of cobalt, nickel, copper and other elements of the iron family in the deposits of the Black Sea.
Dokl. Akad. Nauk, SSSR, 67:1057-1060.

bottom sediments, chemistry
Manheim, F.T., 1970.
The diffusion of ions in unconsolidated sediments.
Earth planet. Sci. Letts. 9(4): 307-309.

bottom sediments, chemistry (plant pigments)
Margalef, Ramón, 1961
Hidrografía y fitoplancton de un área marina de la costa meridional de Puerto Rico.
Inv. Pesq., Barcelona, 18:38-96.

bottom sediments, chemistry of
Marshall, N., Editor, 1962.
Symposium on the environmental chemistry of marine sediments.
Occ. Publ., Grad. School Oceanogr., Univ. R.I., (1):85 pp.

bottom sediments, chemistry
Matsudaira, Yasuo, Haruyuki Koyama and Takuro Endo, 1961
Hydrographic conditions of Fukuyama Harbor.
J. Fac. Fish. and Animal Husbandry, Hiroshima Univ., 3(2):247-296.

Mc

bottom sediments, chemistry
McCone, Alistair W. 1967.
The Hudson River Estuary: sedimentary and geochemical properties between Kingston and Haverstraw, New York.
J. Sedim. Petrol. 37(2): 475-486.

bottom sediments, chemistry
McCrone, A.W., 1966.
Sediments from Long Island Sound (New York) physical and chemical properties reviewed.
J. Sed. Petr. 36(1):234-236.

bottom sediments, chemistry of
Meade, Robert H., 1964.
Mechanics of aquifer system. Removal of water and rearrangement of particles during compaction of clayey sediments. Review.
U.S. Geol. Survey Prof. Paper, 497-B:1-22.

bottom sediments, chemistry
Melnik, V.I., 1970.
Geological researches in the Gulf of Mexico and the Caribbean Sea. (In Russian; English and Spanish abstracts). Okeanol. Issled. Rezult Issled. Mezhd. Geofiz. Proekt. 20: 22-41.

bottom sediments, chemistry
Miller, A.R., C.D. Densmore, E.T. Degens, J.C. Hathaway, F.T. Manheim, P.F. McFarlin, R. Pocklington and A. Jokela, 1966.
Hot brines and recent iron deposits in deeps of the Red Sea.
Geochimica et Cosmochimica Acta, 30(3):341-360.

bottom sediments, chemistry
Mitsunobu, T., 1957.
Chemical investigations of deep-sea deposits. 24-25. The contents of copper and zinc in sea sediments. J. Chem. Soc., Japan, 78(3):405-415.

bottom sediments, chemistry
Miyake, Yasuo, and Yukio Sugimura, 1961
Ionium-thorium chronology of deep-sea sediments of the Western North Pacific Ocean.
June 9, 1961, 133(3467):1-3
(Abstract).

Also in:
Repts. of JEDS, Vol. 2(1961)

bottom sediments, chemistry
Moore, H.B., 1931.
The muds of the Clyde Sea area. III. Chemical and physical conditions; rate and nature of sedimentation; and fauna.
J. Mar. Biol. Assoc., U.K., 17(2):325-358.

bottom sediments, chemistry
Moore, H.B., 1930.
The muds of the Clyde Sea area. 1. Phosphate and nitrogen contents.
J. Mar. Biol. Assoc., U.K., 16(2):595-607.

bottom sediments, chemistry
Moore, Joseph E., and Donn S. Gorsline, 1960
Physical and chemical data for bottom sediments South Atlantic coast of the United States, M/V Theodore N. Gill Cruises 1-9. USFWS Spec. Sci. Rept., Fish., No. 366: 84 pp.

bottom sediments, chemistry of
Moore, Robert III, 1963
Bottom sediment studies, Buzzards Bay, Massachusetts.
J. Sed. Petr., 33(3):511-558.

bottom sediments, chemistry
Morii, Hideaki, Ryoiti Kanazu and Tadanobu Fukuhara, 1965.
Studies on the bottom muds in the seas of pearl farms. II. Vertical and seasonal variations of some constituents at each depth in the mud layers. (In Japanese; English abstract).
Bull. Fac. Fish., Nagasaki Univ., No. 19:81-84.

bottom sediments, chemistry
Morii, Hideaki, Ryoiti Kanazu and Tadanobu Fukuhara, 1965.
Studies on the bottom muds in the seas of pearl farms. 1. Seasonal variations of some constituents in the upper mud layers at the areas adjoining to the Haiki-Strait of the Sasebo and the Omura bays. (In Japanese; English abstract).
Bull. Fac. Fish., Nagasaki Univ., No. 19:74-80.

bottom sediments, chemistry
Mortimer, C.H., 1971.
Chemical exchanges between sediments and water in the Great Lakes - speculations on probable regulatory mechanisms. Limnol. Oceanogr. 16(2): 387-404.

bottom sediments chemistry
Mortimer, C.H., 1949.
Seasonal changes in chemical conditions near the mud surface in two lakes in the English Lake District. Verhandl. Int. Verein. f. Theoret. u. angewandte Limnol. 10:353-356, 1 fig.

bottom sediments
Moskalenko, V.N., 1966.
New data on the structure of sedimentary strata and solid basement in the Levant Sea (seismic profiling). (In Russian; English abstract).
Okeanologiia, Akad. Nauk, SSSR, 6(6):1030-1040.

bottom sediments, chemistry
Müller German, 1967
The HCl-soluble iron, manganese, and copper contents of Recent Indian Ocean sediments off the east coast of Somalia.
Mineralium Deposita, 2:54-61

N

bottom sediments, chemistry
Nayudu, Y. Rammohanroy, 1969.
Biolithology and chemistry of surface sediments in the subantarctic Pacific Ocean.
Antarctic J., U.S., 4(5): 180-181

bottom sediments, chemistry
*Neev, David, and K.O. Emery, 1967.
The Dead Sea: depositional processes and environments of evaporites.
Bull. Geol. Surv., Israel, 41:146 pp.

bottom sediments, chemistry
Nelson, Bruce W., 1962
Important aspects of estuarine sediment chemistry for benthic ecology.
The Environmental Chemistry of Marine Sediments, Proc. Symp., Univ. R.I., Jan. 13, 1962, Occ. Papers, Narragansett Mar. Lab., No. 1: 27-41.

bottom sediments, chemistr
Nicholls, G.D., 1963.
Environmental studies in sedimentary geochemistry.
Science Progress, 51(201):12-31.

bottom sediments, chemistry
Nienaber, James H., 1963.
Shallow marine sediments offshore from the Brazos River, Texas.
Publ. Inst. Mar. Sci., Port Aransas, 9:311-372.

bottom sediments.
Ninkovich, D., and B.C. Heezen, 1967.
Physical and chemical properties of volcanic glass shards from Pozzuolana ash, Thera Island, and from upper and lower ash layers in eastern Mediterranean deep sea sediments.
Nature, Lond., 213(5076):582-584.

bottom sediments, chemistry
Nota, D.J.G., and D.H. Loring, 1964.
Recent depositional conditions in the St. Lawrence River and Gulf - a reconnaissance survey.
Marine Geology, 2(3):198-235.

O

bottom sediments, chemistr
Okuda, Taizo, 1960
Metabolic circulation of phosphorus and nitrogen in Matsushima Bay (Japan) with special reference to exchange of these elements between sea water and sediments. (Portuguese, French resumés)
Trabalhos, Inst. Biol. Maritima e Oceanogr., Universidade do Recife, Brasil, 2(1):7-153.
Resumé pp. 147-149

bottom sediments, chemistry
Okuda, Taizo, Jose Benítez Alvarez, José Rafael Gómez 1965.
Caracteristicas quimicas de los sedimentos de la laguna y rio Unare.
Bol. Inst. Oceanogr. Univ. Oriente 4(1): 108-122.

bottom sediments, chemistry of
Okuda, Taizo, y Jose R. Gomez, 1964.
Distribucion del carbono y nitrogeno organicos de los sedimentos en la region nororiental de Venezuela.
Bol. Inst. Oceanogr., Univ. Oriente, Venezuela, 3(1/2):91-105.

bottom sediments, chemistr
Okuda, T., and S. Sato, 1955.
Fundamental investigations on the marine resources of Matsushima Bay. 1. On the bottom materials of Matsushima Bay.
Bull. Tohoku Reg. Fish. Res. Lab. (4):187-207.

bottom sediments, chemistry
Oppenheimer, C., and L.S. Kornicker, 1958.
Effect of the microbial production of hydrogen sulfide and carbon dioxide in the pH of Recent sediments.
Publ. Inst. Mar. Sci., 5:5-15.

bottom sediments, chemistry of
Osterberg, C., L.D. Kulm, and J.V. Byrne, 1963.
Gamma emitters in marine sediments near the Columbia River.
Science, 139(3558):916-917.

bottom sediments, chemistry of
Ostroumov, E.A., and I.I. Volkov, 1964.
Separation of indium and gallium from manganese, nickel, cobalt and zinc with the help of cinnamic acid in application to the study of sea deposits. Chemistry of the waters and sediments of the seas and oceans. (In Russian; English abstract).
Trudy Inst. Okeanol., Akad. Nauk, SSSR, 67:141-150.

bottom sediments, chemistry
Ottman, J.M., 1960.
Essai de détermination qualitative et quantitative de quelques constituants de la matière organique dans un sédiment marin.
Rev. Géogr. Phys., Géol. Dyn., (2), 3(1):49-52.

bottom sediments, chemistry
Ottmann, J.-M., and F. Ottmann, 1961.
Sur le rapport C/N dans les sédiments litteraux.
C.R. Acad. Sci., Paris, 252(15):2277-2279.

bottom sediments, chemistry.
Overbeck, Jürgen, 1964.
Der Fe/P-Quotient des Sediments als Merkmal des Stoffumsatzes in Brackwässern.
Helgol. Wiss. Meeresunters., 10(1/4):430-447.

P

bottom sediments, chemistry of
Pakhomova, A.S., 1959.
About the chemical composition of the suspended matter and bottom sediments of the Volga delta and the northern part of the Caspian Sea. (In Russian).
Trudy Gosud. Okeanogr. Inst., 45:117-144.

bottom sediments, chemistry
Palacas, James G., Vernon E. Swanson and George W. Moore, 1966.
Organic geochemistry of three North Pacific deep-sea sediment samples.
U.S. Geol. Survey Prof. Paper, 550-C;C102-C107.

bottom sediments, chemistry
Parker, P.L., 1964.
The biogeochemistry of the stable isotopes of carbon in a marine bay.
Geochim. e Cosmochim. Acta, 28(7):1155-1164.

bottom sediments (chemistry)
Pelletier, B.R., 1968.
Submarine physiography, bottom sediments, and modern sediment transport in Hudson Bay.
In: Earth Science Symposium on Hudson Bay, Otta February, 1968, Peter J. Hood, editor, GSC Pap. Geol. Surv. Can., 68-53:100-136

bottom sediments, chemistry
Perdriau, Jacques, 1964.
Pollution marine par les hydrocarbures cancérigènes - type benzo-3.4 pyrène - incidences biologiques. (Suite et fin).
Cahiers Océanogr., C.C.O.E.C., 16(3):205-229.

bottom sediments, chemistry of
Pettersson, H., 1959.
Manganese and nickel on the ocean floor.
Geochim. et Cosmochim. Acta, 17(3/4):209-213.

bottom sediments, chemistry
Piper, David Zink 1971.
The distribution of Co, Cr, Cu, Fe, Mn, Ni and Zn in Framvaren, a Norwegian anoxic fjord.
Geochim. cosmochim. Acta 35 (6): 531-550.

bottom sediments, chemistry
Plunkett, M.A., 1957.
The qualitative determination of some organic compounds in marine sediments. Deep-Sea Res., 4(4):259-262.

bottom sediments, chemistry
Price, N.B. and P.L. Wright 1970.
Variation in the mineralogy and chemistry of sediments from the south-western Barents Sea. In: The geology of the East Atlantic continental margin. 2. Europe, ICSU/SCOR Working Party 31 Symposium, Cambridge 1970, Rept. 70/14: 17-31.

Q

R

bottom sediments, chemistry
Rae, K.M., and R.G. Bader, 1959.
Clay-mineral sediments as a reservoir for radioactive materials in the sea.
Proc. Gulf and Caribbean Fish. Inst., 12th Sess., Nov., 1959:55-61.

bottom sediments, chemistry
*Rashid, M.A., and L.H. King, 1969.
Molecular weight distribution measurements on humic and fulvic acid fractions from marine clays on the Scotian Shelf.
Geochim. cosmochim. Acta, 33(1):147-151.

bottom sediments, chemistry
Richard, J. M., 1955.
Recherches sur le dosage du carbone organique dans les sediments marins. Resume du diplome d'etudes super., Paris, 355-358. (multilithed).
In: Trav. Sta. Zool., Villefranche-sur-mer, 15 (in MBL)

bottom sediments, chemistry
Riley, J.P., 1967.
The hot saline waters of the Red Sea bottom and their related sediments.
Oceanogr. Mar. Biol., Ann. Rev., H. Barnes, editor, George Allen and Unwin, Ltd., 5:141-157.

bottom sediments, chemistry
Riley, J.P. and G. Skirrow, editors, 1965.
Chemical oceanography.
Academic Press, Vol. 2:508 pp.

Bottom sediments, chemistry
Roa Morales, Pedro, and Francois Ottmann, 1961.
Primer estudo topografico y geologico del Golfo de Cariaco.
Bol. Inst. Oceanograf., Univ. Oriente, Cumana, Venezuela, 1(1):5-20.

bottom sediments, chemistry
Romankevich, E.A., 1962
[Organic substance in the surface layer of bottom sediments.]
Mezhd. Geofiz. Komitet. Prezidiume Akad. Nauk SSSR, Rezult. Issled. Programme Mezhd. Geofiz. Goda, Okeanol. Issled., No. 5:67-111.

bottom sediments, chemistry of
Rittenberg, S.C., K.O. Emery and W.L. Orr, 1955.
Regeneration of nutrients in sediments of marine basins. Deep-Sea Res. 3(1):23-45.

bottom sediments (chemistry)(data)
Rochford, D.J., 1953.
Analysis of bottom deposits in eastern and south-western Australia, 1951 and records of twenty-four hourly hydrological observations at selected stations in eastern Australian estuarine systems 1951. Oceanographical station list of investigations made by the Division of Fisheries, Commonwealth Scientific and Industrial Research Organization, Australia, 13:68 pp.

bottom sediments (chemistry
Rochford, D.J., 1951.
Studies in Australian estuarine hydrology.
Australian J. Mar. and Freshwater Res. 2(1):1-116, 1 pl., 7 textfigs.

bottom sediments, chemistry
*Romankevich, E.A., 1968.
Organic carbon and nitrogen distribution in sediments of the Pacific Ocean. (In Russian; English abstract).
Okeanologiia, Akad. Nauk, SSSR, 8(5):825-838.

bottom sediments, geochemistry of
Rotschi, H., 1954.
Quelques considerations geochimiques sur les sediments de l'ocean Pacifique.
Bull. d'Info., C.C.O.E.C., 6(2):55-76.

bottom sediments, chemistry
Rotschi, H., 1951.
Etude des teneurs en fer, manganèse et nickel de quelques carottes des grands fonds. Cahiers du Centre de Recherches et d'Études Océanographiques No. 4:2-22, 8 textfigs.

bottom sediments, chemistry
*Rozhdestvensky, A.V., 1967.
Granulometric and chemical studies of the near-shore sands and bottom sediments in the Bulgarian Black Sea coastal area. (In Russian; English abstract).
Okeanologiia, Akad. Nauk, SSSR, 7(6):1020-1024.

S

sediments
Sachet, M.-H., 1962
Geography and land ecology of Clipperton Island.
Atoll. Res. Bull., 86:115 pp.

bottom sediments, chemistry
Salomon, Milton, 1962
Soil Chemistry - a tool for the analysis of marine sediments.
The Environmental Chemistry of Marine Sediments, Proc. Symp., Univ. R.I., Jan. 13, 1962, Occ. Papers, Narragansett Mar. Lab., No. 1: 5-12.

bottom sediments, geochemistry
Savin, Samuel M., and Samuel Epstein, 1970.
The oxygen and hydrogen isotope geochemistry of ocean sediments and shales.
Geochim. Cosmochim. Acta, 34(1): 43-63.

bottom sediments, chemistry
*Schott, Wolfgang, und Ulrich von Stackelberg, 1965.
Über rezente Sedimentation im Indischen Ozean, ihre Bedeutung für die Entstehung kohlenwasserstoffhaltiger Sedimente.
Erdöl Kohle Erdgas Petrochem., 18(12):945-950.
Also in: Coll. Repr., Int. Indian Ocean Exped., 4(1967).

bottom sediments, chemistry
Seibold, E., 1962.
Untersuchungen zur Kalkfällung und Kalklösung am Westrand der Grand Bahama Bank.
Sedimentology, 1(1):50-74.

bottom sediments, chemistry
*Sevastyanov, V.F., and I.I. Volkov, 1967.
Redistribution of chemical elements in redox processes in the bottom sediments of the oxygen zone of the Black Sea. (In Russian; English abstract).
Trudy Inst. Okeanol., 83:115-134.

bottom sediments, chemistry
*Sevastyanov, V.F., and I.I. Volkov, 1967.
Redistribution of chemical elements in the oxidized layer of the sediments in the process of iron-manganese nodule formation in the Black Sea. (In Russian; English abstract).
Trudy Inst. Okeanol., 83(135-152.

bottom sediments, chemistry
Shcherbakov, F.A., 1958.
[Certain conditions of concentration of heavy minerals in littoral marine alluvia.]
Doklady Akad. Nauk, SSSR, 118(2):384-386.

bottom sediments, chemistry
Shabarova, N.T., 1955.
[The biochemical composition of deep-water marine mud deposits (ocean bottoms).] Biokhimiya 20(2): 146-151.
Translations may be purchased from Associated Technical Services, P.O. Box 271, East Orange, New Jersey for $8.00 (RJ-366).

bottom sediments, chemistry of
Shabarova, N.T., 1954.
[The organic matter of marine deposits.]
Uspekhi Sovremennoi Biol., 37(2):203-208.
Translations may be purchased for $13.40 (RJ-367) from Associated Technical Services, P.O. Box 271, East Orange, New Jersey.

bottom sediments, chemistry
Shapiro, Joseph, W.T. Edmondson and David E. Allison, 1971.
Changes in the chemical composition of sediments of Lake Washington, 1958-1970. Limnol. Oceanogr 16(2): 437-452.

Shishkina, O.V.

bottom sediments, chemistry, redox pot.
Shishkina, O.V., 1961.
The redox potential of the upper ten-meter layer of Quarternary deposits of the Black Sea. (In Russian).
Doklady, Akad. Nauk, SSSR, 139(4):1218-1220.
Translation: Consultants Bureau for Amer. Geol. Inst. (Earth Sci. Section only), 139(1-6):698-700.

bottom sediments, chemistry
Shiskana, O.V., 1959
[Chemical composition of interstitial waters of the Pacific Ocean.] Trudy Inst. Okeanol., 33: 126-145.
(index in English is incorrect)

bottom sediments, chemistry
*Sillén, Lars Gunnar, 1967.
Gibbs phase rule and marine sediments.
In: Equilibrium concepts in natural water systems, Werner Stumm, editor, Adv. Chem. Ser. 67: 57-69.

bottom sediments, chemistry
Skornyakova, N.S., 1961.
[Sediments in the northeastern Pacific.]
Trudy Inst. Okeanol., Akad. Nauk, SSSR, 45:22-64.

bottom sediments, chemistry
Smales, A.A., D. Mapper and A.J. Wood, 1957.
The determination by radioactivation of small quantities of nickel, cobalt and copper in rocks marine sediments and meteorites. Analyst 82(971) :75-88.

bottom sediments, chemistry
*Sreekvmaran, C., K.C. Pillai and T.R. Folsom, 1968.
The concentrations of lithium, potassium, rubidium and cesium in some western American rivers and marine sediments.
Geochim. cosmochim. Acta, 32(11):1229-1234.

bottom sediments, chemistry of
Starikova, N.D., and O.G. Jablokova, 1964.
Method of determination of ammonium and organic nitrogen in solid and liquid parts of marine sediments. Chemistry of the waters and sediments of the seas and oceans. (In Russian; English abstract).
Trudy Inst. Okeanol., Akad. Nauk, SSSR, 67:157-164.

bottom sediments, chemistry
*Stephens, K., R.W. Sheldon and T.R. Parsons, 1967.
Seasonal variations in the availability of food for benthos in a coastal environment.
Ecology, 48 (5):852-855.

bottom sediments, geochemistry
Strakhov, N.M., and I.L. Nesterova, 1968
Effects of volcanism on the geochemistry of marine deposits in the Sea of Okhotsk. (In Russian)
Geokhimiya Osadochnykh Porod i Rud, N.M. Strakhov, editor, Izdatel. Akad. Nauk SSSR, 223-259
Translation in Geochemistry int. 5(3):644-666

bottom sediments, chemistry
Stretta Etienne, 1959.
Position du Rio Capibaribe dans l'ensemble hydrogeologique du Bassin de Recife (Bresil).
Trabalhos, Inst. Biol. Marit. e Geogr., Recife, 71-76.

bottom sediments, chem.
Sugimura, Y., 1969.
Geochronological studies of deep sea sediment by means of natural radioactive nuclides. (In Japanese).
J. oceanogr. Soc. Japan, 25(5):261-268

bottom sediments, chemistry
Swanson, Vernon E., James G. Palacas and Alonza H. Love 1967.
Geochemistry of deep-sea sediment along the 160°W. meridian in the North Pacific Ocean.
Prof. Pap. U.S. geol. Surv. 575B: B137-B144.

T

bottom sediments, chemistry
Tanaka, S., K. Sakamoto, J. Takagi and M. Tsuchimoto 1968.
Aluminum-26 and beryllium-10 in marine sediments.
Science 160 (3834): 1348-1349.

bottom sediments, chemistry
Tanita, S., S. Kato, and T. Okuda, 1950
Studies on the environmental conditions of shellfish fields. 1. In the case of Hakodate Habbor.
Bull. Fac. Fish., Hokkaido Univ., 1(1):18-27, 7 textfigs. (In Japanese; English summary).

bottom sediments, chemistry
Tatsumoto, M., 1957.
Chemical investigations of deep-sea deposits. 22. The contents of cobalt and nickel in sea sediments. (2). 3. The contents of tin and lead in sea sediments. J. Chem. Soc., Japan, 78(1): 38-48.

bottom sediments, chem
Tatsumoto, M., 1956.
[Chemical studies on the deep-sea deposits. 21. The contents of cobalt and nickel in sea sediments.] J. Chem. Soc., Japan, 77(11):1637-1642.

bottom sediments, chemistry
Thommeret, Jean, et Yolande Thommeret, 1967.
Répartition des teneurs en carbone 14 naturel dans divers constituants de la biophase d'un sédiment superficiel de la Méditerranée occidentale.
Cah. océanogr., 19(6):495-504.

bottom sediments, chemistry
Thompson, Geoffrey, 1968.
Analyses of B, G_2, Rb and K in two deep-sea sediment cores, consideration of their use as paleoenvironmental indicators.
Marine Geol., 6(6): 463-477.

bottom sediments, chemistry
Thompson, S.O., and O.A. Schaeffer, 1964
Search for chlorine 36 in Pacific red clays. (Abstract).
Trans. Amer. Geophys. Union, 45(1):119.

bottom sediments, chem.
Tietjen, John H. 1968.
Chlorophyll and phaeo-pigments in estuarine sediments.
Limnol. Oceanogr. 13(1): 189-192.

bottom sediments, chem.
Tikhomireva, Ye. S., 1961.
The geochemistry of shale-bearing deposits of the Baltic basin. (In Russian).
Doklady Akad. Nauk, SSSR, 136(5):1209-1212.

English Edit., 1962, 136(1-6):17-19.

bottom sediments, chem
Till, Roger, 1970.
The relationship between environment and sediment composition (geochemistry and petrology) in the Bimini lagoon, Bahamas.
J. sedim. Petrol. 40(1): 367-385

bottom sediments, chemistry
Tooms, J.S., C.P. Summerhayes and D.S. Cronan, 1969.
Geochemistry of marine phosphate and manganese deposits. Oceanogr. Mar. Biol. Ann. Rev., H. Barnes, editor, George Allen and Unwin, Ltd., 7: 49-100.

bottom sediments, chemistry
Turekian, Karl K. 1967.
Estimates of the average Pacific deep-sea clay accumulation rate from material balance calculations.
Progress in Oceanography 4: 227-244.

bottom sediments, chemistry
Turekian, Karl K., 1965.
Some aspects of the geochemistry of marine sediments.
In: Chemical oceanography, J.P. Riley and G. Skirrow, editors, Academic Press, 2:81-126.

bottom sediments, chemistry
Turekian, K.K., 1964.
The geochemistry of the Atlantic Ocean basin.
Trans. N.Y. Acad. Sci., (2), 26(3):312-330.

U

bottom sediments, chemistry
Uchupi, Elazar, and K.O. Emery, 1963
The continental slope between San Francisco, California and Cedros Island, Mexico.
Deep-Sea Research, 10(4):397-447.

bottom sediments, chem
Ueda, S., 1957.
[Chemical studies on the ocean. 68. Chemical studies of the shallow-water deposits. 21. Vanadium and Chromium contents of the shallow-water deposits (1)(2).]
J. Oceanogr. Soc., Japan, 13(3):93-98; 99-106.

bottom sediments, chemistry
Ueda, S., 1957.
Chemical studies on the ocean. 65. Chemical studies of the shallow-water deposits (18). On the chemical constituents of the shallow water deposits along the sea-coasts of Osaka, Kyoto and Fukui Prefectures, and the generalized consideration of the shallow-water deposits in Japan proper. J. Ocean. Soc., Japan, 13(1):7-16.

bottom sediments, chemistry
Ueda, S., 1956.
Chemical studies on the ocean. LXIV. Chemical studies of the shallower water deposits. 17. On the chemical of the shallower-water deposits along the sea-coasts of Miyagi, Fusushima and Ibaraki Prefectures. J. Ocean. Soc., Japan, 12(3):89-92.

bottom sediments, chemistry
Ueda, S., 1956.
Chemical studies on the ocean. LXIII. Chemical studies of the shallow-water deposits along the sea-coasts of Chiba, Kanagawa and Shizuoka Prefectures. J. Ocean. Soc., Japan, 12(3):85-88.

Bottom sediments, chemistry
Ueda, S., 1956.
Chemical studies on the ocean. LXII. Chemical studies of the shallow-water deposits. 15. On the chemical constituents of the shallow-water deposits along the coast of Wakayama and Mie Prefectures. J. Ocean. Soc., Japan, 12(3):81-84.

bottom sediments
Ulonska, Arno, und Joachim Jarke, 1966.
Ein Gerät zur in situ- Messung der Schallgeschwindigkeit im marinen Sedimenten.
Dt. hydrogr. Z. 19(3):113-120.

bottom sediments, chemical
University of Southern California, Allan Hancock Foundation, 1965.
An oceanographic and biological survey of the southern California mainland shelf.
State of California, Resources Agency, State Water Quality Control Board, Publ. No. 27:232 pp. Appendix, 445 pp.

V

bottom sediments, chemistry
Van der Weijden, R.D. Shuiling and H.A. Das 1970.
Some geochemical characteristics of sediments from the North Atlantic Ocean.
Mar. Geol. 9(2): 81-99.

bottom sediments, chemistry
Varques, H., 1957.
Quelques aperçus concernant des phénomènes d'ammonification dans les boues marines provenant de la baie d'Alger.
Bull. Soc. Hist. Nat., Afrique du Nord, 48(56): 373-377.

bottom sediments, chemistry of
Vernhet, S., 1956.
Étude chimique et minéralogique de quelques sédiments méditerranéens de moyenne et grande profondeur. C.R. Acad. Sci., Paris, 242(8):1049-1052.

W

bottom sediments, chemistry
Watts, J.C.D., 1957.
The chemical composition of the bottom deposits from the Sierra Leone River estuary.
Bull. Inst. Francais d'Afrique Noire (A) 19(3): 1020-1029.

Bottom sediments, chemistry
Wiseman, John D. H., 1965.
Calcium and Magnesium carbonate in some Indian Ocean sediments.
Progress in Oceanography, 3:373-383.

sedimentation, chem
Wiseman, J.D.H., M.W. Strong and H.J.M. Bowen, 1956.
Marine organisms and biogeochemistry. 1. The rates of accumulation of nitrogen and calcium carbonate on the equatorial Atlantic floor. 2. Marine iron bacteria as rock forming organisms 3. The biogeochemistry of strontium.
Adv. Sci., 12(49):579-588.

XYZ

bottom sediments, chemistry
Yalkovsky, Ralph, 1967.
Signs test applied to Caribbean Core A 176-6.
Science, 155(3768):1408-1409.

bottom sediments, chemistry
*Yamamoto, Yoshikazu, 1968.
The chemical composition of shallow-water deposits of Nanao Bay, Japan.
J. oceanogr. Soc., Japan, 24(3):94-102.

bottom sediments, chemistry
*Yamamoto, Yoshikaza, 1968.
Minor elements in shallow-water deposits from Nanao Bay, Japan.
J. oceanogr., Soc., Japan, 24(4):160-166.

bottom sediments, chemistry
Yamamoto, H., 1959.
[Oceanographical studies on Japanese inlets. VIII. Differences between Urado Bay and Uranouchi Bay with reference to the chemical composition of bottom mud (Part 1) sediment in Urado Bay.] J. Oceanogr. Soc., Japan, 15(3): 99-102.

bottom sediments, chemistry
Yamomoto, H., and Y. Imai, 1959.
[Oceanographical studies on Japanese inlets IX. Differences between Urado Bay and Uranouchi Bay with reference to the chemical composition of bottom mud (Part 2) sediment in Uranouchi Bay.] J. Oceanogr. Soc., Japan, 15 (3): 103-105.

bottom sediments, chemistry
*Young, David K., 1968.
Chemistry of southern Chesapeake Bay sediments.
Chesapeake Sci., 9(4):254-260.

bottom sediments, chemistry
Young, Edward J., 1968.
Spectrographic data on cores from the Pacific Ocean and the Gulf of Mexico.
Geochim. cosmochim. Acta. 32(4): 466-471

bottom sediments, chem
Zaitseva, E.D., 1962.
Exchangeable cations in the sediments of the Black Sea. (In Russian; English summary).
Trudy Inst. Okeanol., Akad. Nauk, SSSR, 54:58-82

bottom sediments, chemistry
Zaitseva, E.D., 1960.
[On the problem of the capacity of the exchange and composition of cations in the sediments of the northwest part of the Pacific Ocean and Far Eastern Seas.]
Trudy Okeanogr. Komissii, Akad. Nauk, SSSR, 10(2): 47-55

bottom sediments, chemistry
Zaitseva, E.D., 1959
[Exchangeable cations capacity of the Far Eastern sea sediments.] Trudy Inst. Okeanol., 33: 146-164.

bottom sediments, chemistry
Zaitseva, E.D., 1957.
[Exchange capacity of kations of bottom deposits in the northwestern part of the Pacific.]
Dokl. Akad. Nauk, SSSR, 113(5):1106-1110.

bottom sediments, chemistry
Zaitzeva, E. D., 1956.
[Exchangeable ammonium in the sediments of the Pacific Ocean.] Dokl. Akad. Nauk, SSSR, 111(1): 144-147.

bottom sediments, chemistry
Zhelesnova, A.A., 1964.
The suspension effect in pH determination of sea sediments. (In Russian: English abstract).
Chemistry of the waters and sediments of the seas and oceans.
Trudy Inst. Okeanol., Akad. Nauk, SSSR, 67:135-140.

bottom sediments, chemistry
Zheleznova, A.A., 1962.
On the suspension effect in connection with the pH determination of sea sediments. Review of literature). (In Russian).
Trudy Inst. Okean., Akad. Nauk, SSSR., 54:83-99.

bottom sediments, chemistry
Zheleznova, A.A., and O.V. Shishkina, 1964.
Oxidizing-reduction and pH of sediments in the northern part of the Indian Ocean. Investigations in the Indian Ocean (33rd Voyage of E/S "Vitiaz"). (In Russian).
Trudy Inst. Okeanol., Akad. Nauk, SSSR, 64:236-249.

bottom sediments, chemistry
Zhukova, A.I., and M.V. Fedosov, 1961.
[The role of microorganisms in the upper layer of bottom sediments in the transformation of organic substances in shallow sea basin.]
Okeanologiia, (3):450-455.

bottom sediments, chemistry (alkaline earths)
Rao, S.R., and S.M. Shah, 1971.
Alkaline earth elements in marine sediments off Tarapur coast.
Current Sci., 40(2):25-27.

bottom sediments, chemistry, aluminum

bottom sediments, chemistry (aluminum)

bottom sediments, chemistry of
Wasson, John T., B. Alder and H. Oeschger, 1967.
Aluminum-26 in Pacific sediments: implications.
Science, 155(3761):446-448.

bottom sediments, amino acids

bottom sediments, chemistry (amino acids)

bottom sediments, chemistry of
Bajor, M., et B.M. Van der Weide, 1967.
Effets de la diagenèse sur la distribution des amino-acides dans les sédiments.
Bull. Centre Rech. PAU-SNPA, 1(1):173-186.

bottom sediments, chemistry (amino acids)
Clarke, R.H., 1967.
Amino-acids in Recent sediments of South-east Devon, England.
Nature, Lond., 213(5080):1003-1005.

bottom sediments, chemistry (amino acids)
Connan, J., 1967.
Signification géochimique de l'extraction des amino-acides des sédiments.
Bull. Centre Rech. PAU-SNPA, 1(1):165-171.

bottom sediments, chemistry of
Erdman, J.G., E.M. Mareth and W.E. Hanson, 1956.
Survival of amino acids in marine sediments.
Science 124(3230):1026.

sediments, chemistry of (amino acids)
Hoshino, Michihei, 1962
Amino acids in sediment from the Japan Trench.
J. Oceanogr. Soc., Japan, 18(1):1-3.

bottom sediments, chemistry (amino acids)
Starikova, N.D. and L.I. Korzhikova, 1969.
Amino acids in the Black Sea. Okeanologiia, 9(4): 625-636.
(In Russian; English abstract)

bottom sediments, chemistry (amino acids)
* Stevenson, F.J., and C.-N. Cheng, 1969.
Amino acid levels in the Argentine Basin sediments: correlation with Quaternary climatic changes.
J. sed. Petr., 39(1): 345-347.

bottom sediments, chemistry (arsenic

bottom sediments, chemistry (arsenic)
Boström, Kurt, and Sylvia Valdes 1969.
Arsenic in ocean floors.
Lithos, 2 (4): 351-360

bottom sediments, chem. (arsenic)
Bülow, B.Fr.v., 1935.
Arsengehalt ozeanischer Bodenproben der atlantischen "Meteor"-Expedition. Ann. Hydrogr., usw., Jahrg. 63 (Heft 10):395-397.

bottom sediments, chemistry
Pilipchuk, M.F., and V.F. Sevastianov 1968.
The arsenic from modern sediments of the World Ocean. (In Russian)
Dokl. Akad. Nauk SSSR 179(3): 697-700.

bottom sediments barium

bottom sediments, chemistry (barium)

bottom sediments, chemistry (barium)
Brongersma-Sanders, Margaretha, 1967.
Barium in pelagic sediments and in diatoms.
Proc. K. Ned. Akad. Wet., (B), 70 (1): 93-99.

bottom sediments, chemistry (barium)
Joensuu, Oiva, and Eric Olausson, 1964.
Barium content in deep-sea cores and its relationship to organic matter. (Abstract).
Geol. Soc., Amer., Special Paper, No. 76:87-87.

bottom sediments, chemistry (barium)
Turekian, K.K., and E.H. Tausch, 1964.
Barium in deep-sea sediments of the Atlantic Ocean.
Nature, 201(4920):696-697.

bottom sediments, chemistry (beryllium)

bottom sediments, chemistry
Goel, P.S., D.P. Kharkar, D. Lal, N. Narsappaya, B. Peters and V. Yatirajam, 1957.
The beryllium-10 concentration in deep-sea sediments.
Deep-Sea Res., 4(3):202-210.

bottom sediments, chem. (beryllium)
Tatsumoto, M., and M. Osawa, 1957.
Chemical investigations of deep-sea deposits. XXVI. The content of beryllium.
J. Chem. Soc., Japan, 78(4):502-508.

bottom sediments, chemistry (boron)

bottom sediments chemistry
Boon, John D. III, and William G. MacIntyre 1968.
The boron-salinity relationships in estuarine sediments of Rappahannock River, Virginia.
Chesapeake Sci. 9(1): 21-26

bottom sediments, chemistry (boron)
Gulyaeva, L., 1948.
On the content of boron in modern marine oozes.
Dokl. Akad. Nauk, SSSR, 60:833-835.

bottom sediments, chemistry
Krasintseva, V.V., and O.V. Shishkina, 1959.
A contribution to the problem of boron distribution in marine sediments. (In Russian).
Doklady Akad. Nauk, SSSR, 128(4):815-818.

bottom sediments, chemistry (boron)
*Yamamoto, Yoshikazu, 1968.
On the boron content of the shallow-water deposits.
J. oceanogr., Soc., Japan, 24(5):195-202.

bottom sediments, chemistry - bromide
Williams, Harold H., and Robert C. Harriss 1970.
Chloride and bromide in carbonate rocks in relation to the chemical history of sea water.
Can. J. Earth Sci., 7(6):1539-1555

bottom sediments, chemistry (cadmium)

bottom sediments, chemistry, cadmium
Ishibashi, Masayoshi, Shungo Ueda and Yoshikazu Yamamoto, 1970.
The chemical composition and the cadmium, chromium and vanadium contents of shallow-water deposits in Tokyo Bay. (In Japanese; English abstract)
J. oceanogr. Soc. Japan 26(4): 189-194

bottom sediments, chemistry (cadmium)
Ishibashi, Masayoshi, Shunzo Ueda and Yoshikazu Yamamoto, 1962
Studies on the utilization of shallow-water deposits (continued). On the cadmium content of shallow-water deposits.
Rec. Oceanogr. Wks., Japan, n.s., 6(2):169-176.

bottom sediments, chemistry (cadmium)
Mullin, J.B., and J.P. Riley, 1957.
The occurrence of cadmium in seawater and in marine organisms and sediments. J. Mar. Res., 15(2):103-122.

bottom sediments, chemistry, caesium-137
Auffret, J.P., P. Germain, P. Guegueniat, et Y. Lemosquet 1971
Étude expérimentale de la fixation du caesium 137 par certains sédiments de la Manche.
Cah. océanogr. 23(10): 935-955.

bottom sediments, calcareous

bottom sediments, chemistry (calcium and calcium compounds)

bottom sediments, calcium carbonate
Anwar, Y.M. and M.A. Mohamed, 1970.
The distribution of calcium carbonate in Continental Shelf sediments of Mediterranean Sea north of the Nile Delta in U.A.R. Bull. Inst. Oceanogr. Fish., U.A.R. 1: 449-460.

bottom sediments chemistry (calcium carbonate)
Arrhenius, G., 1952.
Sediment cores from the East Pacific.
Repts. Swedish Deep-sea Exped., 1947-1948, 5(1): 227 pp. textfigs., with appendix of pls.

bottom sediment, calcareous
Asensio Amor, Isidoro, y Pedro Balle Cruellas, 1967.
Primeros resultados del estudio sedimentológico de los fondos marinos del archipiélago Balear.
Bol. Inst. español Oceanogr., 130: 18pp.

bottom sediments, chemistry, calcium carbonate
Gealy, E.L. 1971.
Carbon-carbonate content of sediments from the western equatorial Pacific: Leg 7, Glomar Challenger.
Initial Repts Deep Sea Drill. Proj. 7 (2): 845-862.

bottom sediments, chemistry (calcium carbonate)
Gomoiu, Agripina and M.-T. Gomoiu, 1968.
Calcium carbonate content of sandy marine sediments from the beaches of the Black Sea Romanian littoral.
Revue Roumaine Biol. (Zool.) 13(6): 385-392.

bottom sediments, calcareous
Guilcher, André, 1964.
La sédimentation sous-marine dans la partie orientale de la Rade de Brest, France.
In: Developments in Sedimentology, L.M.J.U. van Straaten, Editor, Elsevier Publishing Co., 1:148-156.

bottom sediments - chemistry (calcium carbonate)
Guilcher, André, Anne Saint-Requier et François Doumenge 1966.
Les teneurs des sédiments en carbonate de calcium dans le lagon de Tahiti et dans les lagons derrière des barrières coralliennes.
C.R. hebd. Séanc. Acad. Sci., Paris, (D) 263 (1): 25-27.

bottom sediments, calcareous
Hommeril, Pierre, 1970.
Dynamique du transport des sédiments calcaires dans la partie nord du golfe normand-breton.
Bull. Soc. géol. France (7) 12(1): 31-41

bottom sediments, chemistry (calcium)
Hoshino, Michihei, and Ryoji Higano, 1961
Calcium carbonate content of deep sea sediment in Equatorial Region.
J. Oceanogr. Soc., Japan, 17(2):61-67.

English abstract

bottom sediments, chem. calcareous
Hoskins, Cortez W., 1964.
Molluscan biofacies in calcareous sediments, Gulf of Batabano, Cuba.
Bull. Amer. Assoc. Petr. Geol., 48(10):1680-1704.

bottom sediments, calcium carbonate
Kennett, James P., 1966.
Foraminiferal evidence of a shallow calcium carbonate solution boundary, Ross Sea, Antarctica
Science, 153(3732):191-193.

bottom sediments, chemistry (calcium)
Koczy, F., and H. Titze, 1956.
Eine rasche Prägisionbestimmung von Kalk in kleinen Sedimentmengen. Zeits. f. Anal. Chem., 150(2):100-110.

bottom sediments, chemistry, calcium carbonate
Lisitzin, Alexander P., 1970.
Sedimentation and geochemical considerations.
In: Scientific exploration of the South Pacific, W.S. Wooster, editor, Nat. Acad. Sci., 89-132.

bottom sediments, calcareous
*Michard, Gil, 1971.
Theoretical model for manganese distribution in calcareous sediment cores. J. geophys. Res., 76(9): 2179-2186.

Bottom sediments, chemistry, calcium carbonate
Yong Ahn Park and Moo Young Song, 1971.
Sediments of the continental shelf off the south coasts of Korea. J. oceanogr. Soc. Korea 6(1): 16-24.

bottom sediments, calcium carbonate
Pratt, R.M., and Bruce C. Heezen, 1964
Topography of the Blake Plateau.
Deep-Sea Res., 11(5):721-728.

bottom sediments chemistry (calcium carbonate)
Purdy, E.G., 1963.
Recent calcium carbonate facies of the Great Bahama Bank. 1. Petrography and reaction groups.
J. Geology, 71(3):334-355.

bottom sediments, chemistry (calcium carbonate)
Rao, M.S., 1958.
Distribution of calcium carbonate in the shelf sediments off east coast of India.
J. Sed. Petr., 28(3):274-285.

bottom sediments - chemistry (calcium carbonate)
Shinn, Eugene A. 1968.
Burrowing in Recent lime sediments of Florida and the Bahamas.
J. Paleontol. 42(4): 879-892

bottom sediments, chemistry, calcium, carbonate
Smith, Stephen V. 1971.
Budget of calcium carbonate, southern California continental borderland.
J. sedim. Petrol. 41(3): 798-808.

sedimentation, calcium carbonate
Turekian, K.K., 1963.
Rates of calcium carbonate deposition by deep-sea organisms, molluscs and the coral-algae associations
Nature, 197(4864):277-278.

bottom sediments, chemistry, calcium, carbonate

Vallier, Tracy L., 1970.
Carbon carbonate results. Initial Repts. Deep Sea Drilling Project, Glomar Challenger 5: 431-440.

bottom sediments, chemistry, calcium carbonate

Venkatarathnam, K., and William B.F. Ryan 1971.
Dispersal patterns of clay minerals in the sediments of the eastern Mediterranean Sea.
Mar. Geol. 11 (4): 261-282.

bottom sediments carbohydrates

bottom sediments, chemistry (carbohydrates)

bottom sediments, chemistry (carbohydrates)

Artem'yev V.E 1970.
Carbohydrates in bottom sediments of the Central Pacific. (In Russian; English abstract)
Okeanologiia 10 (4): 656-661.

bottom sediments, carbohydrates

*Artemyev, V.E., 1969.
Carbohydrates in the bottom sediments of the Kurile-Kamchatka Trench. (In Russian; English abstract). Okeanologiia, 9(2): 250-255.

bottom sediments, chemistry (carbohydrates)

Artemyev, V.E., L.N. Krayushkina and E.A. Romankevich, 1971.
Determination of carbohydrates in ocean floor sediments. (In Russian; English abstract).
Okeanologiia 11(6): 1125-1128.

bottom sediments, chemistry (carbohydrates)

Modzeleski, Judith E., William A. Laurie and Bartholomew Nagy 1970.
Carbohydrates from Santa Barbara Basin sediments: gas chromatographic-mass spectrometric analysis of trimethylsilyl derivatives.
Geochim. Cosmochim. Acta, 35 (8): 825-838.

bottom sediments, chemistry, carbon

Boyce, R.E. and G.W. Bode, 1972.
Carbon and carbonate analyses, leg 9. Initial Repts Deep Sea Drilling Project 9: 797-816.

bottom sediments, chemistry carbon isotopes

Sackett, William M., Walter R. Eckelmann, Michael L. Bender and Allan W.H. Be, 1965.
Temperature dependence of carbon isotope composition in marine plankton and sediments.
Science, 148(3667):235-237.

bottom sediments, chemistry (carbon 14)

Thommeret, Jean, et Yolande Thommeret 1969.
Le carbon 14 dans les sédiments de surface de la Méditerranée occidentale.
Rapp. P.-v. Réun. Commn int. Explor. scient. Mer Méditerr. 19 (5): 937-939.

Bottom sediments, carbonate

bottom sediments, chemistry (carbonates)

bottom sediments, chemistry (carbonate)

Anwar, Yehia M. and Mamdouh Abdel-Maksoud Mohamed, 1970.
The distribution of calcium carbonate in continental shelf sediments of the Mediterranean Sea north of the Nile delta in U.A.R. Bull. Inst. Oceanogr. Fish., Cairo, 1:451-460.

bottom sediments, carbonate

Bartlett, Grant A., and Robert G. Greggs, 1969.
Carbonate sediments: oriented lithified samples from the North Atlantic.
Science, 166 (3906): 740-741

bottom sediments, chemistry (carbonate)

Barusseau, Jean-Paul 1971.
Distribution granulométrique des sédiments bioclastiques carbonatés entre les Îles de Ré et d'Oléron et le plateau de Rochebonne.
Cah. océanogr. 23 (8): 687-707.

bottom sediments, carbonate

Berner, Robert A., 1966.
Diagenesis of carbonate sediments: interaction of magnesium in sea water with mineral grains.
Science, 153(3732): 188-191.

bottom sediments, chemistry (carbonates)

Bruevich, S.V., 1946.
[Carbonates in the bottom deposits of the Caspian Sea.] Dok. Akad. Nauk, USSR, 54:149-152.

bottom sediments, chemistry (carbonates)

Bruevich, S.V., and E.G. Vinogradova, 1946.
[General features of sedimentation in the Caspian Sea according to the distribution of carbonates, Fe, Mn, and P in sea deposits.] Dok. Akad. Nauk, USSR, n.s., 52:789-792.

bottom sediments, chem. (Carbonate carbon)

Calvert, S. E., 1966.

Accumulation of diatomaceous silica in the sediments of the Gulf of California.

Geol. Soc., Am., Bull., 77(6):569-596.

sedimentation, carbonate

Chave, Keith E., 1962
Processes of carbonate sedimentation.
The Environmental Chemistry of Marine Sediments, Proc. Symp., Univ. R.I., Jan. 13, 1962, Occ. Papers, Narragansett Mar. Lab., No. 1: 77-85.

bottom sediments, carbonate

Chave, Keith E., 1962
Processes of carbonate sedimentation.
The Environmental Chemistry of Marine Sediments, Proc. Symp., Univ. R.I., Jan. 13, 1962, Occ. Papers, Narragansett Mar. Lab., No. 1:77-85.

bottom sediments, carbonate

Cloud, Preston E., 1965.
Carbonate precipitation and dissolution in the marine environment.
In: Chemical oceanography, J.P.Riley and G. Skirrow, editors, 2:127-158.

bottom sediments, chemistry (carbonate)

Degens, Egon T., 1965.
Geochemistry of sediments; a brief survey.
Prentice-Hall, Inc., 342 pp.

bottom sediments, chemistry (carbonate)

Del Riego, A.F., 1951.
Determinación del carbónico de los fondos de la Ria de Vigo. Cálculo del carbonato disuelto y consequencias geobiológicas. Bol. Inst. Español Ocean., No. 44:17 pp.

bottom sediments, chemistry (carbonate)

Donahue, Jack, 1968.
Recent carbonate sediments in the Bahamas: a review. (Abstract only). Trans. Fifth Carib. Geol. Conf., St. Thomas, V.I., 1-5 July 1968, Peter H. Mattson, editor. Geol. Bull. Queens Coll., Flushing, 5: 47.

bottom sediments, carbonate

Emelianov, E.M., A.P. Lisitsyn and B.A. Koshelev, 1971.
Distribution and composition of carbonates in the upper bed of the Atlantic Ocean bottom sediments. (In Russian)
Dokl. Akad. Nauk SSSR, 196 (1): 207-210.

sedimentation (carbonate)

Evans, G., D.J.J. Kinsman and D.J. Shearman, 1964.
A reconnaissance survey of the environment of Recent carbonate sedimentation along the Trucial coast, Persian Gulf.
In: Developments in Sedimentology, L.M.J.U. van Straaten, Editor, Elsevier Publishing Co., 1: 129-135.

bottom sediments, chemistry (carbonate)

Fairbridge, R.W., 1955.
Warm marine carbonate environments and dolomitization. Tulsa Geol. Soc. Digest., 23:39-48.

bottom sediments, chemistry (carbonate)

Fernandez del Riego, A., 1951.
Determinación del carbónico de los fondos de la Ria de Vigo; cálculo del carbonato disuelto y consequencias geobiologicas. Bol. Inst. Español Ocean. No. 44:17 pp.

BOTTOM SEDIMENTS, CARBONATE

Friedman, Gerald M., 1968.
Geology and geochemistry of reefs, carbonate sediments and waters, Gulf of Aquaba (Elat), Red Sea.
J. Sedim. Petrol. 8(3): 895-919.

BOTTOM SEDIMENTS, CARBONATE

Gavish, Eliezer, and Gerald M. Friedman, 1969.
Progressive diagenesis in Quaternary to Late Tertiary carbonate sediments: sequence and time scale.
J. Sedim. Petrol., 39(3): 980-1006.

bottom sediments, carbonate

Gevirtz, Joel L., and Gerald M. Friedman, 1966.
Deep-sea carbonate sediments of the Red Sea and implications on marine lithification.
J. Sed. Petr. 36(1):143-151.

bottom sediments, carbonate

Ginsburg, R.N., R. Michael Lloyd, K.W. Stockman, and J.S. McCallum, 1963.
22. Shallow-water carbonate sediments.
In: The Sea, M.N. Hill, Editor, Interscience Publishers, 3:554-582.

bottom sediments, chemistry (carbonate)

Gershkova, T.I., 1967.
Sedimentation in the Norwegian Sea. (Carbonate and organic matter in the sediments of the Norwegian Sea). (In Russian).
Atlant. nauchno-issled. Inst. rybn. khoz. okeanogr. (AtlantNIRO). Materialy Konferentsii po Resul'tatam Okeanologischeskikh Issledovanii v Atlanticheskom Okeane, 300-308.

bottom sediments, carbonates, in
Gorshkova, T.I., 1965.
Carbonates in the sediments of the Norwegian-Greenland Sea basin as indicators of the distribution of water masses. (In Russian).
Trudy vses. nauchno-issled. Inst. morsk. ryb. Khoz. Okeanogr. (VNIRO), 57:297-312.

(bottom sediments, carbonates)
Gorshkova, T.I., 1965.
Sediments of the Norwegian Sea. (In Russian; English abstract).
Okeanolog. Issled., Rez. Issled. po Programme Mezhd. Geofiz. Goda, Mezhd. Geofiz. Komitet, Prezidiume, Akad. Nauk, SSSR, No.13:212-224.

bottom sediments, chemistry (carbonate)
Graf, Donald L., 1960
Geochemistry of carbonate sediments and sedimentary carbonate rocks. IV A. Isotopic composition-chemical analyses. Ill. State Geol. Survey, Circular 308: 42 pp.

bottom sediments, chemistry (carbonate)
Graf, Donald L., 1960
Geochemistry of carbonate sediments and sedimentary carbonate rocks. IV B. Bibliography. Ill. State Geol. Survey, Circular 309: 55 pp.

Graf, Donald L., 1960 bottom sediments, chemistry
Geochemistry of carbonate sediments and sedimentary carbonate rocks. III Minor element distribution. Illinois State Geol. Survey, Circ. 301:71 pp.

bottom sediments (carbonate)
Gross, M. Grant, 1965.
Carbonate deposits on Plantagenet Bank near Bermuda.
Geol. Soc., Am., Bull., 76(11):1283-1290.

bottom sediments (calcium carbonate
Guilcher, André, Léopold Berthois, Francois Dowmenge, et Alain Miche, 1965.
Les teneurs des sédiments en carbonate de calcium dans les lagons de Maupihea et de Bora-Bora (îles de la Société)
Comptes Rendus hebd. Séanc., Acad. Sci., Paris 261(23):5177-5179.

bottom sediments carbonate
Harris, William H., and R.K. Matthews, 1968.
Subaerial diagenesis of carbonate sediments: efficiency of the solution-reprecipitation process.
Science, 160(3823):77-79.

bottom sediments, chemistry
Harriss, Robert C., and Orren H. Pilkey, 1966.
Interstitial waters of some deep marine carbonate sediments.
Deep-Sea Res., 13(5):967-969.

bottom sediments, carbonates
Hodgson, W.A., 1966.
Carbon and oxygen isotope ratios in diagenetic carbonates from marine sediments.
Geochim. cosmochim. Acta, 30(12):1223-1233.

bottom sediments-chemistry (carbonate)
Hoskin, Charles M., and Richard V. Nelson 1969.
Modern marine carbonate sediment, Alexander Archipelago, Alaska.
J. sedim. Petrol. 39 (2): 581-590.

bottom sediments, carbonate
*Hülsemann, Jobst, 1967.
The continental margin off the Atlantic coast of the United States: carbonate sediments, Nova Scotia to Hudson Canyon.
Sedimentology, 8(2):121-145.

sediments, carbonate
Jansen, John F., and Yasushi Kitano, 1963.
The resistance of Recent marine carbonate sediments to solution.
J. Oceanogr. Soc., Japan, 18(4):208-219.

bottom sediments chem. (carbonate)
Karke, J., 1949.
Die Entstehungsmöglichkeiten von Eisenbikarbonat und Eisenkarbonat in rezenten Flachseesedimenten.
Deutsch. Hydro. Zeit. 2(6):286-291.

bottom sediments, carbonate
Kendall, Christopher G. St.C. and Patrick A. d'E. Skipwith, 1969.
Holocene shallow-water carbonate and evaporite sediments of Khor al Bazam, Abu Dhabi southwest Persian Gulf.
Bull. Am. Ass. Petr. Geol., 53(4): 841-869.

bottom sediments, carbonate mineralogy
Kier, Jerry S., and Orrin H. Pilkey 1971
The influence of sea level changes on sediment carbonate mineralogy, Tongue of the Ocean, Bahamas.
Mar. Geol. 11 (3): 189-200.

bottom sediments (carbonate)
Kinsmann, D.J.J., 1964.
The Recent carbonate sediments near Halat el Bahrani, Trucial coast, Persian Gulf.
In: Developments in Sedimentology, L.M.J.U. van Straaten, Editor, Elsevier Publishing Co., 1:185-192.

bottom sediments, chemistry (carbonates)
Lalou, C., 1962.
Formation des carbonates dans les vases prises à différents niveaux dans une carotte du fond de la baie de Villefranche-sur-Mer.
Trav. Centre Recherches et d'Études Océan., 4(4):11-18.

bottom sediments, chemistry (carbonates)
Lalou, C., 1957.
Étude expérimentale de la production de carbonates par les bactéries des vases de la Baie de Villefranche-sur-Mer. Ann. Inst. Océan., Monaco, 33(4):202-267.

bottom sediments, chemistry (carbonates)
Lalou, C., 1957.
Étude expérimental de la production de carbonates par les bactéries des vases de la Baie de Villefranche-sur-mer.
Ann. Inst. Océan., Monaco, 23(4):201-267.

bottom sediments, carbonate
Lewis, Michael Samuel, 1969.
Sedimentary environments and unconsolidated carbonate sediments of the fringing coral reefs of Mahé, Seychelles.
Marine Geol., 7(2): 95-127.

bottom sediments, chemistry (carbonate)
Lisitsyn, A.P., and V.P. Petelin, 1953.
Contemporary carbonate sediments of cold-water seas.
Moskovskoe Obshchestvo Ispytatelei Prirody, Biulleten, (otdel geol. 28(2), 58(2):82-83.

bottom sediments, carbonate
Maiklem, W.R., 1970.
Carbonate sediments in the Capricorn Reef complex, Great Barrier Reef, Australia.
J. sedim. Petrol., 40(1): 55-80

bottom sediments, chemistry (carbonate)
Maxwell, W.G.H., R.W. Day, & P.J.G. Fleming, 1961
Carbonate sedimentation on the Heron Island Reef, Great Barrier Reef.
J. Sediment. Petrol., 31(2): 215-230.

bottom sediments, carbonate
Milliman, John D., 1966.
Submarine lithification of carbonate sediments.
Science, 153(3739):994-996.

bottom sediments, carbonate
Milliman, John D., Orrin H. Pilkey and Blake W. Blackwelder, 1968.
Carbonate sediments on the continental shelf, Cape Hatteras to Cape Romain.
SEast, Geol., Duke Univ., 9(4):245-267.

bottom sediments, carbonate
Milliman, John D., David A. Ross and Teh-Lung Ku, 1969.
Precipitation and lithification of deep-sea carbonates in the Red Sea.
J. sedim. Petrol. 39(2): 724-736.

bottom sediments, carbonate
Nair, R.R. 1971.
Beachrock and associated carbonate sediments on the Fifty Fathom Flat, a submarine terrace on the outer continental shelf off Bombay.
Proc. Indian Acad. Sci. (B) 73 (3): 148-154

bottom sediments, carbonate
Nayudu, Y.R., 1964
Carbonate deposits and paleoclimatic implications in the northeast Pacific Ocean.
Science, 146(3643):515-517.

Bottom sediments, carbonate
Neumann, A. Conrad, 1965.
Processes of Recent carbonate sedimentation in Harrington Sound, Bermuda.
Bull. Mar. Sci., 15(4):987-1035.

sedimentation, carbonate
Neumann, A. Conrad, 1963.
Processes of recent carbonate sedimentation in Harrington Sound, Bermuda.
Mar. Sci. Center, Lehigh Univ., 130 pp. (Unpublished manuscript).

bottom sediments, carbonate
Neumann, A. Conrad, Conrad D. Gebelein and Terence P. Scoffin, 1970
The composition, structure and erodability of subtidal mats, Abaco, Bahamas.
J. sedim. Petrol., 40(1): 274-297.

bottom sediments, chemistry (carbonate)
Pettersson, H., 1948
Three sediment cores from the Tyrrhenian Sea. Medd. Ocean. Inst., Göteborg 15 (Göteborgs Kungl. Vetenskaps-och Vitterhets-Samhälles Handlingar. Sjätte Följden, Ser.B 5(13)):94 pp.

bottom sediments, carbonate
Pilkey, Orrin H., 1964.
The size distribution and mineralogy of the carbonate fraction of United States South Atlantic shelf and slope sediments.
Marine Geology, 2(1/2):121-136.

bottom sediments, carbonate
Pilkey, Orrin H., 1964.
Carbonate fraction of the U.S. South Atlantic shelf and upper slope sediments. (Abstract).
Geol. Soc., Amer., Special Paper, No. 76:130-131.

bottom sediments, chemistry, carbonate
Pimm, A.C., 1971.
Carbon carbonate results. Initial Repts. Deep Sea Drilling Project, Glomar Challenger 6: 739-752.

bottom sediments, chemistry (carbonates)
Pimm, A.C., 1970.
Carbon carbonate results, leg 4. Initial Reports of the Deep Sea Drilling Project, Glomar Challenger, 4: 307-314.

bottom sediments, chemistry (carbonate)
Pimm, A.C., 1970.
Carbon carbonate results, leg 3. Initial Reports of the Deep Sea Drilling Project, Glomar Challenger, 3: 495-507.

bottom sediments, carbonate
Pratt, Richard M., 1968.
Atlantic continental shelf and slope of the United States- physiography and sediments of the deep-sea basin.
Prof. Pap. U.S. geol. Surv., 529B:B1-B44, fold in chart.

bottom sediments carbonate
Pratt, R.M., 1966.
The Gulf Stream as a graded river.
Limnol. and Oceanogr., 11(1):60-67.

bottom sediments (calcium carbonate)
Purdy, E.G., 1963.
Recent calcium carbonate facies of the Great Bahama Bank. 2. Sedimentary facies.
J. Geology, 71(4):472-497.

bottom sediments, chemistry
Pytkowicz, Ricardo M. 1970.
On the carbonate depth in the Pacific.
Geochim. Cosmochim. Acta 34(7): 836-839.

bottom sediments, chemistry, carbonate
Roberts, H. H., 1971.
Mineralogical variation in lagoonal carbonates from North Sound, Grand Cayman Island (British West Indies)
Sediment. Geol. 6(3): 201-213.

bottom sediments, carbonate
*Robertson, Eugene C., 1967.
Laboratory consolidation of carbonate sediment.
In: Marine geotechnique, A.F. Richards, editor, Univ. Illinois Press, 118-127.

bottom sediments chemistry (carbonate)
Rucker, James B., 1965.
Carbonate mineralogy of sediments of Exuma Sound Bahamas.
J. Sed. Petr. 35(1): 68-72

bottom sediments, carbonate
Sanders, John E., and Gerald M. Friedman, 1969.
Position of regional carbonate/noncarbonate boundary in nearshore sediments along a coast: possible climatic indicator.
Bull. geol. Soc. Am., 80(9): 1789-1796

bottom sediments, carbonate
Schmalz, Robert F., 1967.
Kinetics and diagenesis of carbonate sediments.
J. sedim. Petrol., 37(1):60-67.

bottom sediments (carbonate)
Schmalz, R.F, and K.E Chave, 1963
Factors affecting saturation in ocean waters off Bermuda.
Science, 139(3560):1206-1207.

bottom sediments, calcium carbonate
Schneider, Eric D., and Bruce C. Heezen, 1966.
Sediments of the Caicos Outer Ridge, The Bahamas.
Bull. geol. Soc. Am., 77(12):1381-1398.

bottom sediments, carbonate
Scoffin, Terence P., 1970.
The trapping and binding of subtidal carbonate sediments by marine vegetation in Bimini Lagoon, Bahamas.
J. sedim. Petrol. 40(1): 249-273

bottom sediments, carbonate
Shinn, Eugene A., 1969.
Submarine lithification of Holocene carbonate sediments in the Persian Gulf.
Sedimentology, 12(1/2): 109-144

bottom sediments, chem. (carbonate)
*Biever, Raymond and Miriam Kastner, 1967.
Mineralogy and petrology of some Mid-Atlantic Ridge sediments.
J. mar. Res., 25(3):263-278.

bottom sediments, carbonate
Swinchatt, J.P., 1969.
Algal boring: a possible depth indicator in carbonate rocks and sediments.
Bull. geol. Soc. Am., 80(7): 1391-1396.

bottom sediments, carbonate.
Swinchatt, Jonathan P., 1965.
Significance of constituent composition, texture and skeletal breakdown in some Recent carbonate sediments.
J. Sed. Petr., 35(1):71-90.

bottom sediments, carbonate
Taft, William H., and John W. Harbaugh, 1964.
Modern carbonate sediments of southern Florida, Bahamas and Espiritu Santo Island, Baja California a comparison of their mineralogy and chemistry.
Stanford Univ. Publ., Geol. Sci., 8(2):133 pp.

bottom sediments, carbonate
*Thompson, Geoffrey, V.T. Bowen W.G. Melson and Richard Cifelli, 1968.
Lithified carbonates from the deep-sea of the equatorial Atlantic.
J. sedim. Petr., 38(4):1305-1312.

bottom sediments, chemistry (carbonate)
Vallier, Tracy L., 1970.
Carbon carbonate results. Initial Repts. Deep Sea Drilling Project, Glomar Challenger 5: 431-440.

bottom sediments, carbonate
Van der Linden, Willem J.M., 1967.
A textural analysis of Wellington Harbour sediments.
N.Z. Jl. mar. Freshwat. Res. 1(1): 26-37.

bottom sediment carbonate formations
*Vasil'Chikov, N.V., E.M. Natkinas and L.E. Shterenberg, 1971.
Origin of carbonate formations in the Kalamitskiy Bay (The Black Sea). (In Russian; English abstract). Okeanologiia, 11(1): 85-89.

bottom sediments, carbonate
Venzo, Giulio Antonio, e Sergio Stefanini, 1967
Distribuzione dei carbonati nei sedimenti di spiaggia e marini dell'Adriatico settentrionale tra Venezia e Trieste.
Studi Trentini Sci nat. (A), 44(2):178-201.
Reprinted: Comm. Studio Oceanograf. Limnol. (A), (3). 1968.

bottom sediments, carbonate
Von der Borch, Christopher, 1965.
The distribution and preliminary geochemistry of modern carbonate sediments of the Coorong area, South Australia.
Geochimica et Cosmochimica Acta, 29(7):781-799.

bottom sediments, chemistry (carbonate)
Wangersky, Peter J. and Oiva I. Joensuu 1967
The fractionation of carbonate deep-sea cores.
J. Geol. 75(2): 148-177.

bottom sediments, carbonate
Weber, Jon N., 1967.
Factors affecting the carbon and oxygen isotopic composition of marine carbonate sediments. I. Bermuda.
Am. J. Sci., 265(7): 586-608

bottom sediments, carbonate
*Weber, Jon N., and R.F. Schmalz, 1968.
Factors affecting the carbon and oxygen isotopic composition of marine carbonate sediments. III. Eniwetok Atoll.
J. sedim. Petr., 38(4):1270-1279.

bottom sediments, carbonate
*Weber, Jon N., and Peter M.J. Woodhead, 1969.
Factors affecting the carbon and oxygen isotopic composition of marine carbonate sediments. II. Heron Island, Great Barrier Reef, Australia.
Geochim. cosmochim. Acta, 33(1):19-38.

Bottom sediments, chemistry (carbonate)
Wiseman, J.D.H., 1959.
The relation between paleotemperature and carbonate in an equatorial Atlantic pilot core.
J. Geology, 67(6):685-690.

bottom sediments, chemistry (carbonate)
Wiseman, J.D.H., C. Emiliani, R. Yalkovsky, 1959.
The relationship between paleotemperatures and carbonate content in a deep-sea core: A discussion.
J. Geol., 67(5):572-576.

bottom sediments, chem

Yalkovsky, R., 1957.
The relationship between the paleotemperature and carbonate content in a deep-sea core.
J. Geol., 65(5):480-496.

bottom sediments, chemistry (carbon dioxide)

bottom sediments, chemistry
Moore, G.W., C.E. Robinson, and H.D. Nygren, 1962
Electrode determination of the carbon dioxide content of sea water and deep-sea sediment.
U.S. Geol. Survey, Prof. Paper 450-B:83-85.

bottom sediments carotenoids

bottom sediments, chemistry (carotenoids)

bottom sediments, carotenoids
Gorshkova, T.I., 1965.
Chlorophyll and carotinoids in the sediments of the Baltic Sea and the Gulf of Riga. (In Russian).
Trudy vses. nauchno-issled. Inst. morsk.ryb. Khoz. Okeanogr. (VNIRO), 57:313-328.

bottom sediments, chemistry (catalytic action)
Kato, K., and T. Okuda, 1949.
Chemical studies on the marine deposits. II. On the catalytic action of marine deposits. J. Fish Hakodate, Japan, No. 54:13-20, 4 textfigs.

bottom sediments, chemistry chloride
Williams, Harold H., and Robert C. Harriss, 1970.
Chloride and bromide in carbonate rocks in relation to the chemical history of ocean water.
Can. J. Earth Sci., 7(6):1539-1555

bottom sediments, chemistry (chlorins)

bottom sediments, chemistry of
Baker, B.L., and G.W. Hodgson 1968.
Chlorins in Recent marine sediments on the North Atlantic continental shelf off Nova Scotia (Canada).
Chem Geol., 3(2):119-133.

bottom sediments chlorophyll

bottom sediments, chemistry (chlorophyll)

bottom sediments, chemistry (chlorophyll)
Brongersma-Sanders, M., 1951.
On conditions favouring the preservation of chlorophyll in marine sediments.
Proc. Third World Petroleum Congr., Sect. 1:

submarine geology chem. (chlorophyll)
Jastrebova, L.A., 1938
Chlorophyll in Meeressedimenten. Trans. Inst. Mar. Fish. and Oceanogr. USSR, 5: 189-224, (German resume).

bottom sediments, chemist
Orr, W.L., K.O. Emery, and J.R. Grady, 1958.
Preservation of chlorophyll derivatives in sediments off southern California.
Bull. Amer. Assoc. Petr. Geol., 42(5):925-962.

bottom sediments, chemistry (chlorophyll)
Orr, W.L., and J.R. Grady, 1957.
Determination of chlorophyll derivatives in marine sediments. Deep-Sea Res., 4(4):263-271.

bottom sediments chemistry (chlorophyll)
Sugimura, Yukio, 1961
Geochemical studies on Recent sediments. IV. Chlorophyll degradation products in sruface muds from the Kagoshima Bay, Japan.
J. Oceanogr.Soc., Japan, 17(1):10-14.

bottom sediment, chemistry chlorophyll
Tietjen, John H. 1968
Chlorophyll and phaeo-pigments in estuarine sediments
Limnol. Oceanogr. 13(1): 189-192

bottom sediments, chemistry (chromium)
Turekian, Karl K., 1967.
Estimates of the average Pacific deep-sea clay accumulation rate from material balance calculations.
Progress in Oceanography, 4:227-244.

bottom sediments, chemistry (cobalt)
Turekian, Karl K., 1967.
Estimates of the average Pacific deep-sea clay accumulation rate from material balance calculations.
Progress in Oceanography, 4:227-244.

bottom sediments, chemistry (copper)

bottom sediments, chemistry of
Ishibashi, M., S. Ueda, Y. Yamamato and F. Morii 1958.
Studies on the utilization of the shallow-water deposits (continued). On the copper content of the shallow-water deposits at the seacoasts of the Kii Peninsula and other districts.
Rec. Oceanogr. Wks., Japan, Spec. No. 2:153-166.

bottom sediment chemistry, copper
Müller German, 1967
The HCl-soluble iron, manganese, and copper contents of Recent Indian Ocean sediments off the east coast of Somalia.
Mineralium Deposita, 2:54-61

bottom sediments, chemistry (copper)
Turekian, Karl K., 1967.
Estimates of the average Pacific deep-sea clay accumulation rate from material balance calculations.
Progress in Oceanography, 4:227-244.

bottom sediments, chemistry (denitrification)
Goering, J.J. and M.M. Pamatmat, 1971.
Denitrification in sediments of the sea off Peru. Investigación pesq. 35(1): 233-242.

bottom sediments, chem. (denitrification)
Lagarde, Edmond, 1962.
Contribution à l'étude du métabolisme de l'azote minéral au milieu marin. Microflore dénitrifiante de certaines zones littorales méditerranéennes.
Pubbl. Staz. Zool., Napoli, 32 (Suppl.):490-496.

bottom sediments, chemistry (fatty acids)

bottom sediments, chemistry (fatty acids)
*Cooper, W.J., and Max Blumer, 1968.
Linear, iso and anteiso fatty acids in Recent sediments of the North Atlantic.
Deep-Sea Res., 15(5):535-540.

bottom sediments, chemistry (fatty acids)
Farrington, John W., and James G. Quinn 1971
Fatty acid diagenesis in recent sediment from Narragansett Bay, Rhode Island.
Nature, Lond. 230(11): 67-69.

bottom sediments, chemistry of
Leo, Richard F., and Patrick L. Parker, 1966.
Branched chain fatty acids in sediments.
Science, 152(3722):649-650.

bottom sediments, chemistry (fatty acids)
Parker Patrick L. 1967
Fatty acids in Recent sediment.
Contrib. mar. Sci. Port Aransas 12:135-142.

bottom sediments, chemistry
Rhead, M.M., G. Eglinton, G.H. Draffan and P.J. England 1971.
Conversion of oleic acid to saturated fatty acids in Severn Estuary sediments.
Nature, Lond., 232 (5309): 327-330

bottom sediments, chemistry, fatty acids
Sever, Julia R., and Pat Haug 1971
Fatty acids and hydrocarbons in Surtsey sediment.
Nature, Lond., 234 (5330): 447-450

bottom sediments chemistry
Welte Dietrich H., and Götz Ebhardt, 1968.
Distribution of long chain n-paraffins and n-fatty acids from the Persian Gulf.
Geochim. cosmochim. Acta, 32(4): 465-466

bottom sediments, chemistry (fatty alcohols)

bottom sediments, chemistry of (fatty alcohols)
Sever, Judy, and P.L. Parker, 1969
Fatty alcohols (normal and isoprenoid) in sediments.
Science, 164(3883):1052-1054.

bottom sediments, chemistry, fatty acids
Simoneit, Bernd R. and A.L. Burlingame, 1971.
Further preliminary results on the higher weight hydrocarbons and fatty acids in the DSDP cores, legs 5-8. Initial Repts Deep Sea Drilling Project, 8: 873-900.

bottom sediments chemistry, fatty acids
Simoneit, Bernd R. and A.L. Burlingame, 1971.
Some preliminary results on the higher weight hydrocarbons and fatty acids in the deep sea drilling project cores, Legs 5-7. Initial Repts Deep-Sea Drill. Proj. 7(2): 889-912.

bottom sediments, chemistry, ferrides

Landergren, S., 1948
6. On the geochemistry of Mediterranean sediments. Preliminary report on the distribution of beryllium, boron, and the ferrides in three cores from the Tyrrhenian Sea. Medd. Ocean. Inst., Göteborg 15 (Göteborgs Kungl. Vetenskaps-och Vitterhets -Samhälles Handlingar, Sjätte Följden, Ser. B 5(13)):34-46, figs.6-7.

bottom sediments, chemistry ferrous sulphide

Ostroumov, E.A., and V.M. Shilov 1958.
The distribution of ferrous sulfide and hydrogen sulfide in the bottom sediments of the northwest Pacific Ocean. (In Russian)
Trudy Inst. Okeanol. 27: 77-86.

bottom sediments chemistry, fluorine

bottom sediments - chemistry (fluorine)

Shishkina, O.V. 1966.
Fluorine in oceanic sediments and their pore solutions. (In Russian)
Geokhimya (2): 236-243.
Translation: Geochem. int. Ann Arbor 3(1): 152-159

bottom sediments, chemistry (fluorine)
*Shishkina, O.V., G.A. Pavlova and V.S. Bykova, 1967.
The fluorine distribution in the interstitial water and in the bottom sediments of the Black Sea. (In Russian; English abstract).
Trudy Inst. Okeanol., 83:83-98.

bottom sediments, chemistry - fulvic acid
*Rashid, M.A., and L.H. King, 1969.
Molecular weight distribution measurements on humic and fulvic acid fractions from marine clays on the Scotian Shelf.
Geochim. cosmochim. Acta, 33(1):147-151.

bottom sediments chemistry, gallium

bottom sediments chemistry (gallium)

Ishibashi, M., T. Shigematsu, Y. Nishikawa, K. Hirake, 1961
Gallium content of sea water, marine organisms sediments, and other materials related to the ocean.
J. Chem. Soc., Japan, 82(9):1141-1142.

bottom sediments, chemistry of
Ishibashi, Masayoshi, Shunzo Ueda and Yoshikazu Yamamoto, 1966.
On the GALLIUM content of the shallow-water deposits.
J. oceanogr. Soc., Japan, 22(5):197-200.

bottom sediments, chemistry (halogens)

bottom sediments chem. (gases)
Olson, F.C.W., & B. Wilder, 1961
Gases in bottom sediments.
Bull. Mar. Sci. Gulf & Carib., 11(2): 207-209.

bottom sediments, chemistry (gases)
Reeburgh, William S., 1969.
Observations of gases in Chesapeake Bay sediments
Limnol. Oceanogr., 14(3): 368-375.

bottom sediments, chemistry (gas hydrates)

bottom sediments, gashydrates
Stoll, Robert D., John Ewing and George M. Bryan, 1971.
Anomalous wave velocities in sediments containing gas hydrates. J. geophys. Res., 76(8): 2090-2094.

bottom sediments, chemistry
Gulyayeva, L.A., and E.S. Itkina, 1962.
Halogens in marine and fresh-water sediments.
Geokhimia, (6):524-528.

Translation: Scripta Technica, (6):610-615.

(In Russian)

bottom sediments, chemistry (heavy metals)

bottom sediments, chemistry (heavy metals)
Brooks, R.R. and B.E. Quinn 1971
Heavy metals in stream sediments of the Port Pegasus area of Stewart Island.
N.Z. Jl Sci. 14 (1): 25-30.

bottom sediments, metals
Gross, M. Grant, 1964.
Heavy-metal concentrations of diatomaceous sediments in a stagnant fjord. (Abstract).
Geol. Soc., Amer., Special Paper, No. 76:69.

bottom sediments humic acids
Biber, V.A., and N.S. Bogolyubova, 1952.
The humic acids of estuary mud and their biological activity. (In Russian).
Doklady, Akad. Nauk, SSSR, 82(6):939.

bottom sediments, chemistry humic acid
Bordovskiy, O.K. 1965.
Accumulation and transformation of organic substance in marine sediments. 1. Summary and introduction. 2. Sources of organic matter in marine basins. 3. Accumulation of organic matter in bottom sediments. 4. Transformation of organic matter in marine sediments.
Marine Geology, Elsevier Publ. Co., 3(½):3-4; 5-31; 33-82; 83-114.

bottom sediments, chemistry humic acid
Kasatochkin, V.I. O.K. Bordovski, N.K. Larina and K.T. Chernikinskaia, 1965.
On the chemical nature of humic acids in the Indian Ocean bottom sediments. (In Russian).
Dokl. Akad. Nauk, SSSR, 179(3): 690-693.

bottom sediments, chemistry (humus)
Kato, K., 1956.
Chemical investigations on marine humus in bottom sediments. Mem. Fac. Fish., Hokkaido Univ. 4(2):91-209.

bottom sediments chem. (humus)
Kato, K., 1952.
Chemical studies on marine deposits. VI. On the organic constituents of marine humus of shallow water deposits. Bull. Fac. Fish., Hokkaido Univ., 3(2):121-127.

bottom sediments, chem. (humus)
Kato, K., 1951.
[Chemical studies on marine deposits. III. On the distribution of marine humus contents in the sea northwest of Hokkaido.] Bull. Fac. Fish., Hokkaido Univ., 1(2):10-30, 5 textfigs.

submarine geology, chem. (humus)
Kato, K., and T. Ishizuka, 1949.
Chemical studies on the marine deposits. I. On the distribution of humus contents in the marine deposits in Mutsu Bay. J. Fish., Hakodate, Japan, No 54:7-12, 4 textfigs. [In Japanese with English summary.]

bottom sediments, chemistry, humic acid
*Rashid, M.A., and L.H. King, 1969.
Molecular weight distribution measurements on humic and fulvic acid fractions from marine clays on the Scotian Shelf.
Geochim. cosmochim. Acta, 33(1):147-151.

bottom chemistry, Chemistry, hydrocarbons

bottom sediments, chemistry (hydrocarbons)

bottom sediments, chemistry (hydrocarbons)
Bray, E.E., and E.D. Evans, 1962.
Composition of sediment hydrocarbons reflect varying environment in the sea.
J. Mar. Biol. Assoc., 4(1):10-22.

bottom sediments, chemistry (hydrocarbons)
Dunton, M.L., and J.M. Hunt, 1962.
Distribution of low molecular-weight hydrocarbons in Recent and ancient sediments.
Bull. Amer. Assoc. Petrol. Geol., 46(12):2246-2248.

bottom sediments, chemistry, hydrocarbons
Khain, V.E., L.E. Levin, and L.I. Tuliani 1971.
The volume of sedimentary stratum and predicted reserves of hydrocarbons in the World Ocean trough system. (In Russian)
Dokl. Akad. Nauk SSSR, 200 (5): 1201-1202

bottom sediments - chemistry (hydrocarbons)
Mallet, Lucien, et Marie-Louise Priou, 1967.
Sur la rétention des hydrocarbures polybenzéniques du type benzo-3.4 pyrène par les sédiments, la faune et la flore marines de la baie de Saint-Malo.
C.r. hebd. Séanc. Acad. Sci. Paris (D) 264 (7): 969-971.

submarine geology, chem. (hydrocarbons)
Novelli, G.D., 1943
Bacterial oxidation of hydrocarbons in marine sediments. Proc. Soc. Exp. Biol. and Med. 52:133-134.

bottom sediments, chemistry (hydrocarbons)
Oro, J., D.W. Nooner, A. Zlatkis, S. A. Wikstrom and E. S. Barghoorn, 1965.
Hydrocarbons of biological origin in sediments about two billion years old.
Science, 148(3666):77-79.

bottom sediments, chemistry, hydrocarbons
*Schott, Wolfgang, und Ulrich von Stackelberg, 1965.
Über rezente Sedimentation im Indischen Ozean, ihre Bedeutung für die Entstehung kohlenwasserstoffhaltiger Sedimente.
Erdöl Kohle Ergas Petrochem., 18(12):945-950.
Also in: Coll. Repr., Int. Indian Ocean Exped., 4(1967).

bottom sediments, chemistry, hydrocarbons
Sever, Julia R., and Pat Haug 1971
Fatty acids and hydrocarbons in Surtsey sediment.
Nature, Lond., 234 (5330): 447-450

bottom sediments, chemistry, hydrocarbons
Simoneit, Bernd R. and A.L. Burlingame, 1971.
Further preliminary results on the higher weight hydrocarbons and fatty acids in the DSDP cores, legs 5-8. Initial Repts Deep Sea Drilling Project, 8: 873-900.

bottom sediments, chemistry, hydrocarbons

Simoneit, Bernd R. and A.L. Burlingame, 1971. Some preliminary results on the higher weight hydrocarbons and fatty acids in the deep sea drilling project cores, Legs 5-7. Initial Repts Deep-Sea Drill. Proj. 7(2): 889-912.

bottom sediments, chemistry (hydrocarbons)

Smith, P.V., jr., 1952. The occurrence of hydrocarbons in recent sediments from the Gulf of Mexico. Science 116 (3017):437-439.

bottom sediments, chemistry of

Stevens, N.P., E.E. Bray and E.D. Evans, 1956. Hydrocarbons in sediments of Gulf of Mexico. Bull. Amer. Assoc. Petr. Geol., 40(5):975-983.

bottom sediments, chemistry hydrogen sulphide

Ostroumov, E.A., and V.M. Shilov 1958. The distribution of ferrous sulfide and hydrogen sulfide in the bottom sediments of the northwest Pacific Ocean. (In Russian) Trudy Inst. Okeanol. 27: 77-86.

bottom sediments, chemistry (iodine)

Bojanowski, Ryszard, and Stefania Paslawska, 1970

On the occurrence of iodine in bottom sediments and interstitial waters of the southern Baltic Sea. Acta geophys. polon. 18(3/4):277-286

bottom sediments, iron

bottom sediments, chemistry (iron)

bottom sediments, iron ores

Bochert, H., 1965. Formation of marine sedimentary iron ores. In: Chemical oceanography, J.P. Riley and G. Skirrow, editors, Academic Press, 2:159-204.

bottom sediments, chemistry (iron)

Brodskaia, N.G., 1957. [Forms of iron in the modern sediments of the Okhotsk Sea.] Doklady Akad. Nauk, SSSR, 114(1): 165-

bottom sediments, chemistry

Brodskaya, N.G., and T.G. Martova, 1957. [Forms of iron in the modern sediments of the Okhotsk Sea.] Dokl. Akad. Nauk, SSSR, 114(2):165-168.

bottom sediments, chemistry (iron)

Cronan, D.S., T.H. van Andel, G. Ross Heath, M.G. Dinkelman, R.H. Bennett, David Bukry, Santiago Charleston, Ansis Kaneps, K.S. Rodolfo and R.S. Yeats, 1972. Iron-rich basal sediments from the eastern equatorial Pacific: leg 16, Deep Sea Drilling Project. Science 175(4017): 61-63.

bottom sediments, chemistry (iron)

Debyser, J., and P.-E., Rouge, 1956. Sur l'origine du fer dans les eaux interstitielles des sédiments marins actuels. C.R. Acad. Sci., Paris, 243(25):2111-2113.

bottom sediments, chemistry (iron)

Gennesseaux, Marice, 1960. Migration expérimentale du fer dans les sédiments marins. C.R. Acad. Sci., Paris, 251(22):2564-2565.

Also in: Trav. Lab. Géol. Sous-Marine, 10(1960).

bottom sediments, chemistry (iron)

Isayeva, A.B. 1971. Relation between titanium and iron in the sediments of the Indian Ocean. (In Russian). Geokhimya 1971 (3):310-317. Translation: Geochem. int. Am. geol. Inst. 8(2): 186-193.

bottom sediments, iron formation

*James, Harold L., 1969. Comparison between Red Sea deposits and older ironstone and iron-formation. In: Hot brines and Recent heavy metal deposits in the Red Sea, E.T. Degens and D.A. Ross, editors, Springer-Verlag, New York, Inc., 525-532.

bottom sediments, chemistry (iron)

Matheron, Robert, 1966. Teneurs en fer total des Sédiments de la lagune du Brusc (Var). Bull. Inst. Océanogr., Monaco, 66 (1375):8pp.

bottom sediments, chemistry (iron)

Mokievskaia, V.V., 1960. [On the problem of the existence of iron in its marine and interstitial waters of the Black Sea. Chemistry of the sea.] Trudy Komissii, Akad. Nauk, SSSR, 10(2):21-29. (Okeanogr.)

bottom sediments, chemistry, iron

Müller German, 1967. The HCl-soluble iron, manganese, and copper contents of Recent Indian Ocean sediments off the east coast of Somalia. Mineralium Deposita 2:54-61

bottom sediments, chemistry (iron)

Nevessky, E.N. and F.A. Shcherbakov, 1969. Peculiarities of iron accumulation and distribution in the sediments from the Kandalaksha Bay. Okeanologiia, 9(4): 649-660. (In Russian; English abstract)

bottom sediments, iron

*Nicelich, Rinaldo, Ferruccio Mosetti e Giuseppe Macchi, 1967. Sul significato dell suscettivita magnetica dei sedimenti di fondo marino. Atti Ist veneto Sci., 125:227-243.

bottom sediments, chemistry (iron)

Ostroumov, E.A., 1955. [Iron in the bottom sediments of the Okhotsk Sea.] Doklady Akad. Nauk, SSSR, 102(1):129-132.

bottom sediments, chemistry (iron)

Strakhov, N.M., 1958. [Forms of iron in sediments of the Black Sea.] Doklady Akad. Nauk, SSSR, 118(4):803-806.

Transl. listed in: Techn. Transl., 1962, 8(4):209.

bottom sediments, chemistry (iron)

Strakhov, N.M., 1948. [Distribution of iron in the sediments of lake and sea systems and the factors controlling it.] Izvest. Akad. Nauk, USSR, Geol., 4:30-50.

BOTTOM SEDIMENTS, CHEMISTRY

Turner, Ralph R. and Robert C. Harriss, 1970. The distribution of non-detrital iron and manganese in two cores from the Kara Sea. Deep-Sea Res., 17(3): 633-636.

bottom sediments, chemistry (iron)

von der Borch, C.C., W.D. Nesteroff and J.S. Galehouse, 1971. Iron-rich sediments cored during leg 8 of the deep sea drilling project. Initial Repts Deep Sea Drilling Project, 8: 829-835.

bottom sediments, chemistry (iron)

von der Borch, C.C. and R.W. Rex, 1970. Amorphous iron oxide precipitates in sediments cored during leg 5, deep sea drilling project. Initial Repts. Deep Sea Drilling Project, Glomar Challenger 5: 541-544.

bottom sediments, chemistry, isotope ratios

Carwile, R.H., and Faure, G., 1971. Strontium isotope ratios and base metal content in a core from the Atlantis II Deep, Red Sea. Chem. Geol. 8(1): 15-23

bottom sediments, chemistry (lead)

*Chow, Tsaihwa J., 1968. Lead isotopes of the Red Sea region. Earth Planet. Sci. Letters, 5(3):143-147.

bottom sediments, chemistry (lead)

Chow, T.J., 1958

Lead isotopes in sea water and marine sediments. J. Mar. Res., 17: 120-127.

bottom sediments, chemistry (lead)

Chow, T.J., and C.C. Patterson, 1962. The occurrence and significance of lead isotopes in pelagic sediments. Geochim. et Cosmochim. Acta, 26:263-308.

bottom sediments (pelagic), chemistry of

Chow, Tsaihwa J., and C.C. Patterson, 1961. The occurrence and significance of lead isotopes in pelagic sediments. (Abstract). Tenth Pacific Sci. Congr., Honolulu, 21 Aug.-6 Sept., 1961, Abstracts of Symposium Papers, 368.

bottom sediments, chemistry (lead)

Chow, Tsaihewa J., and M. Tatsumoto, 1964. Isotopic composition of lead in the sediments near Japan Trench. In: Recent researches in the fields of hydrosphere, atmosphere and nuclear geochemistry, Ken Sugawara festival volume, Y. Miyake and T. Koyama, editors, Maruzen Co., Ltd., Tokyo, 179-183.

bottom sediments, chemistry

*Cooper, J.A., and J.R. Richards, 1969. Lead isotope measurements on sediments from Atlantic II and Discovery deep areas. In: Hot brines and Recent heavy metal deposits in the Red Sea, E.T. Degens and D.A. Ross, editors, Springer-Verlag, New York, Inc., 499-511.

bottom sediments, chem (lead)

Ishibashi, Masayoshi, Shunzo Ueda and Yoshikazu Yamamoto, 1960

Studies on the utilization of the shallow-water deposits (Cont.). On the lead content of the shallow-water deposits. Rec. Oceanogr. Wks., Japan, Spec. No. 4: 111-122.

bottom sediments
chemistry (lead)

Ishibashi, Masayoshi, Sunzo Ueda and Yoshikazu Yamamoto, 1959.
Studies on the utilization of the shallow-water deposits (continued). On the zinc content of the shallow-water deposits.
Rec. Oceanogr. Wks., Japan, Spec. No. 3:123-133.

---

bottom sediments,
chemistry of (lead)

Lakshman, S.V.J., and P. Thruvenganna Rao, 1962.
Quantitative spectrochemical analysis of lead in ocean bed sediments.
J. Sci. Industr. Res., 21B(4):174-176.
India

---

bottom sediments, chemistry (lithium)

---

bottom sediments, lithium

Cheinikov, B.I., V.E. Karasev, and G.A. Krainikov, 1967.
Distribution of lithuim in phreatic waters of the sediments of the Pacific. (In Russian).
Dokl. Akad. Nauk, SSSR, 176(2):432-433.

---

bottom sediments, chemistry
Magnesium carbonate

Wiseman, John D.H., 1965.
Calcium and magnesium carbonate in some Indian Ocean sediments.
Progress in Oceanography, 3:373-383.

abstract

---

bottom sediments
chemistry, manganese

bottom sediments, manganese

Bender, Michael L., 1971.
Does upward diffusion supply the excess manganese in pelagic sediments? J. geophys. Res., 78(18): 4212-4215.

---

bottom sediments, chemistry (manganese)

Bonatti, Enrico, 1971.
Manganese fluctuations in Caribbean sediment cores due to postdepositional remobilization.
Bull. mar. Sci. Miami 21(2): 510-518.

---

bottom sediments, chemistry, manganese

Boström, Kurt 1970.
Deposition of manganese rich sediments during glacial periods.
Nature, Lond. 226 (5246): 629-630.

---

bottom sediments, chemistry
(manganese)

Chainikov, V.I., 1969.
On the source of manganese in bottom sediments of the Pacific. (In Russian)
Dokl. Akad. Nauk SSSR 187(4): 909-912

---

bottom sediments, chemistry
(manganese)

#Gorshkova, T.I., 1966.
Manganese in the bottom sediments of the Soviet northern seas and its biological significance. (In Russian).
Trudy vses. nauchno-issled. Inst. morsk. ryb. Khoz. Okeanogr. (VNIRO), 60:89-102.

---

bottom sediments, chemistry
(manganese)

Li, Yuan-Hui, James Bischoff and Guy Mathieu 1969.
The migration of manganese in the Arctic Basin sediment.
Earth Planet. Sci. Letts 7(3): 265-270.

---

bottom sediments, chemistry (manganese)

Li, Yuan-Hui, James Bischoff and Andrew Parker 1969.
The migration of manganese in the Arctic Basin sediment.
Earth Planet. Sci. Lett. 7(3): 265-270.

---

bottom sediments, chemistry, manganese

Lynn, D.C., and E. Bonatti 1965.
Mobility of manganese in diagenesis of deep-sea sediments.
Mar. Geol. 3(6): 457-474.

---

bottom sediments, chemistry, manganese

Michard, Gil, 1971.
Theoretical model for manganese distribution in calcareous sediment cores. J. geophys. Res., 76(9): 2179-2186.

---

bottom sediments chem. (manganese)

Moore, T.C., Jr., and G.R. Heath, 1966.
Manganese nodules topography and thickness of Quaternary sediments in the Central Pacific.
Nature, Lond., 212(5066):983-985.

---

bottom sediments, chemistry, manganese

Müller, German, 1967.
The HCl-soluble iron, manganese, and copper contents of Recent Indian Ocean sediments off the east coast of Somalia.
Mineralium Deposita 2: 54-61

Bottom sediments, chemistry

Murty, P.S.N., Ch. M. Rao, and C.V.G. Reddy, 1968
Manganese in the shelf sediments off the west Coast of India.
Current Sci. Sept. 5 1968 37(17): 481-483
Also in: Coll. Repr. Nat. Inst. Oceanogr. New Delhi, 1 (1963-1968).

---

bottom sediments chemistry
(manganese)

Rao, M. Subba, 1962.
Manganese in the shelf sediments off east coast of India.
Proc. Indian Acad. Sci., (A), 56(5):274-284.

---

bottom sediments, chemistry (manganese)

Rao, N.V.N. Durgaprasada, M. Rama Murty and M. Poornachandra 1970.
Manganese in the bottom sediments of the eastern part of the Bay of Bengal.
Current Sci. 39 (10): 225-226

---

bottom sediments, chemistry
(manganese)

Rotschi, H., 1951.
Étude des teneurs en fer, manganèse et nickel de quelques carottes des grands fonds. Cahiers du Centre de Recherches et d'Études Océanographiques No. 4:2-22, 8 textfigs.

---

bottom sediments, chemistry, MANGANESE

Turner, Ralph R. and Robert C. Harriss, 1970.
The distribution of non-detrital iron and manganese in two cores from the Kara Sea. Deep-Sea Res., 17(3): 633-636.

---

bottom sediments, chemistry, manganese

Turekian, Karl K., 1967.
Estimates of the average Pacific deep-sea clay accumulation rate from material balance calculations.
Progress in Oceanography, 4:227-244.

---

bottom sediments - Chemistry, manganese

Wangersky, Peter J., and Oiva Joensuu, 1964
Strontium, magnesium and manganese in fossil foraminiferal carbonates.
J. Geol., 72(4):477-483.

---

bottom sediments, chemistry (manganese)

Wolfe, L.A., and Harry Zeitlin 1970
An x-ray fluorescence spectroscopic method for the determination of total manganese in rocks and marine sediments.
Anal. Chim. Acta 51: 349-354

---

bottom sediments, chemistry (mercury)

---

bottom sediments, chemistry
(mercury)

Boström, Kurt and David E. Fisher 1969.
Distribution of mercury in East Pacific sediments.
Geochim. Cosmochim. Acta, 33(6): 743-745

---

bottom sediments, chemistry (mercury)

De Groot, A.J., J.J.M. De Goeij and C. Zegers, 1971.
Contents and behaviour of mercury as compared with other heavy metals in sediments from the rivers Rhine and Ems.
Geol. Mijnb. 50 (3): 393-398.

---

bottom sediments, chemistry

Hanya, Takahisa, Ryoshi Ishiwatari and Hisako Ichikuni, 1963.
The mechanism of removal of mercury from sea water to bottom muds in Minamata Bay. (In Japanese; English abstract).
J. Oceanogr. Soc., Japan, 19(2):94-100.

---

bottom sediments, chemistry, mercury

Huggett, Robert J., Michael E. Bender and Harold D. Slone 1971.
Mercury in sediments from three Virginia estuaries.
Chesapeake Sci. 12 (4): 280-282.

---

bottom sediments, chemistry
(mercury)

Jernelöv, Arne, 1970.
Release of methyl mercury from sediments with layers containing inorganic mercury at different depths. Limnol. Oceanogr., 15(6): 958-960.

---

bottom sediments,
metals

bottom sediments, chemistry,
metalliferous

Dasch, E. Julius, Jack R. Dymond and G. Ross Heath 1971.
Isotopic analysis of metalliferous sediment from the East Pacific Rise.
Earth Planet. Sci. Letts. 13 (1): 175-180

---

bottom sediments, metalliferous

Tooms, J.S., and M. Rugheim, 1969.
Additional metalliferous sediments in the Red Sea.
Nature, Lond., 223(5213): 1356-1359.

---

bottom sediments, minerals

Yong Ahn Park and Moo Young Song, 1971.
Sediments of the continental shelf off the south coasts of Korea. J. oceanogr. Soc. Korea 6(1): 16-24.

bottom sediments, chemistry (molybdenum)

bottom sediments, chemistry
Filipchuk, M.F., and I.I. Volkov, 1966.
Distribution of molybdenum in modern sediments of the Black Sea. (In Russian).
Doklady, Akad. Nauk, SSSR, 167(5):1143-1146.

bottom sediments, chemistry (molybdenum)
Sugawara, K., S. Okabe and M. Tanaka, 1961
Geochemistry of molybdenum in natural waters.
J. Earth Sci., Nagoya Univ., 9(1):114-128.

bottom sediments, chemistry (molybdenum)
Yamamoto, Yohsikazu, 1961
[Chemical studies on the ocean. 87. Molybdenum content of shallow-water deposits.]
J. Oceanogr. Soc., Japan, 17(1):15-20.

bottom sediments, chemistry of (molybdenum)
Yamamoto, Y., 1960
[Chemical studies on the ocean - 86. Molybdenum content of the shallow-water deposits (1).]
J. Oceanogr. Soc., Japan, 16(4): 163-166.

bottom sediments, chemistry (nickel)

bottom sediments, chemistry (nickel)
*Beasley, Thomas M., and Edward E. Held, 1969.
Nickel-63 in marine and terrestrial biota, soil and sediment.
Science, 164(3884):1161-1163.

bottom sediments, chemistry of
Bonner, Francis T., and Alzira Soares, 1964.
Nickel content of ocean core samples. (Abstract).
Trans. Amer. Geophys. Union, 45(1):118.

bottom sediments, chemistry (nickel)
Pettersson, H., and H. Rotschi, 1952.
The nickel content of deep-sea deposits.
Geochimica et Cosmochimica Acta 2:81-90, 4 text-figs.

bottom sediments, chem (nickel)
Smales, A.A., and J.D.H. Wiseman, 1955.
Origin of nickel in deep-sea sediments. Nature 175(4454):464-465.

bottom sediments, chemistry, nickel
Turekian, Karl K., 1967.
Estimates of the average Pacific deep-sea clay accumulation rate from material balance calculations.
Progress in Oceanography, 4:227-244.

bottom sediments, chemistry (nickel)
Yamakoshi, Kazuo, and Yuji Tazawa 1971.
Deposition of extraterrestrial nickel in marine sediments.
Nature, Lond. 233 (5321): 542-543.

bottom sediments, chemistry (nitrogen and its compounds)

bottom sediments- chemistry nitrogen
Arrhenius, G., 1952.
Sediment cores from the East Pacific.
Repts. Swedish Deep-Sea Exped., 1947-1948, 5(1): 227 pp., textfigs., with appendix of plates.

bottom sediments, Chemistry nitrogen
Arrhenius, O., 1948.
5. The concentration of nitrogen in samples of cores 13, 15, and 16. Medd. Ocean. Inst., Göteborg, 15 (Göteborgs Kungl. Vetenskaps- och Vitterhets- Samhälles Handlingar, Sjätte Följden Ser. B, 5(13)):30-33, Fig. 5.

bottom sediments, Chemistry, nitrogen
Bernard, F., J. Lecal, et R. Godinet, 1950.
Etudes des sédiments marins au large de L'Alger.
1. Teneur en azote et carbone organique. Bull. Inst. Océan. Monaco, No. 963: 10 pp., 4 textfigs. 1 fold-in.

bottom sediments, chem (nitrogen compounds)
Degens, Egon T., 1970.
Molecular nature of nitrogenous compounds in sea water and recent marine sediments. Occ. Pap. Inst. mar. Sci., Alaska, 1: 77-106.

bottom sediments, chem. (nitrogen)
Kawasaki, Nariko, 1962.
On the easily soluble nitrogen from the mud in Matsushima Bay. (In Japanese; English abstract)
Bull. Tohoku Reg. Fish. Res. Lab., No. 21:87-92.

bottom sediments, chemistry Nitrogen (organic, ammonia nitrate in bottom sediments)
Rittenberg, S.C., K.O. Emery and W.L. Orr, 1955.
Regeneration of nutrients in sediments of marine basins. Deep-Sea Res. 3(1):23-45.

bottom sediments, chemistry, nitrogen
Wiseman, J. D. H., and H. Bennett, 1940.
The distribution of organic carbon and nitrogen in sediments from the Arabian Sea. John Murray Exped., 1933-34, Sci. Repts. 3(4):193-221, 2 textfigs.

bottom sediments, chemistry (nitrogen compounds)
Yoshida, Yoichi, and Masao Kimata 1970.
Microbiological and chemical studies in the seas of Hiuchi and Bingo. II On the metabolic activities of inorganic nitrogen compounds by the microorganisms as a whole. (In Japanese; English abstract)
J. oceanogr. Soc. Japan 26 (1): 5-10.

bottom sediments, chemistry oils (hydrocarbonic fraction)
Smirnov, B.A., 1969.
On the biochemical aspects of the formation of the hydrocarbonic fractions of oils in the Recent marine sediments. (In Russian; English abstract).
Okeanologiia, 9(5): 791-795.

bottom sediments chemistry, oils (mineral)

bottom sediments, chemistry (oils)

bottom sediments - chemistry (mineral oils)
Zobell, Claude E., and Joseph F. Prokop 1966.
Microbial oxidation of mineral oils in Barataria Bay.
Z. allg. Mikrobiol. 6 (3): 143-162.

bottom sediments, chemistry (organic acids)

Bottom sediments, chem
Cahet, Guy, et François Gadel, 1970.
Dynamique saisonnière des acides gras II effets de la diagenèse dans les dépôts marins et lagunaires.
Cah. océanogr. 22 (10): 1033-1066

bottom sediments organic carbon

bottom sediments, chemistry (organic carbon)

bottom sediments, chemistry (organic carbon)
Del Riego, A.F., 1956.
El contenido en carbono orgánico en los sedimentos de la ría de Vigo. Algunos datos sobre la relación carbononitrógeno.
Bol. Inst. Espanol Oceanogr., 78:27 pp.

bottom sediments, chemistry
Eckelmann, W.R., W.S. Broecker, D.W. Whitlock, and J.R. Allsup, 1962.
Implications of carbon isotopic composition of total organic carbon of some recent sediments and ancient oils.
Bull. Amer. Assoc. Petrol. Geol., 46(5):699-704.

bottom sediments, chemistry
El Wakeel, S.K., and J.P. Riley, 1957.
The determination of organic carbon in marine muds. J. du Cons., 22(2):180-183.

sediments, chemistry (organic carbon)
Fernández del Riego, A., 1956.
El contenido en carbóno orgánico en los sedimentos de la ría de Vigo. Algunos datos sobre la relación carbono-nitrógeno.
Bol. Inst. Espanol Oceanogr., 78:29 pp.

(bottom sediments, organic carbon in)
Gorshkova, T.I., 1965.
Sediments of the Norwegian Sea. (In Russian; English abstract).
Okeanolog. Issled., Rez. Issled. po Programme Mezhd. Geofiz. Goda, Mezhd. Geofiz. Komitet, Prezidiume, Akad. Nauk, SSSR, No.13:212-224.

bottom sediments, chemistry (organic carbon)
Lisitsyn, A.P., 1955.
[The distribution of organic carbon in sediments of the Bering Sea.] Doklady Akad. Nauk, SSSR, 103(2):299-302.

translation: Associated Technical Services.

bottom sediments, chemistry (organic)
Welte, Dietrich H. und Götz Eb hardt 1968.
Die Verteilung höherer, geradkettiger Paraffine und Fettsäuren in einigem Sedimentprofil aus dem Persischen Golf.
Meteor Forschungsergebn. (C) 1: 43-52

bottom sediments, chemistry
Gross, M. Grant, 1967.
Organic carbon in surface sediment from the northeast Pacific Ocean.
Int. J. Oceanol. Limnol., 1(1):46-54.

bottom sediments, chemistry (organic carbon)
Okuda, Taizo, 1964.
Some problems for the determination of organic carbon in marine sediments.
Bol. Inst. Oceanogr., Univ. Oriente, Venezuela, 3(1/2):106-117.

bottom sediments, organic carbon
Oshima, Kazuo, 1963.
Ecological study of Usu Bay, Hokkaido, Japan. 1. Bottom materials and benthic fauna. (In Japanese; English summary).
Bull. Hokkaido Reg. Fish. Res. Lab., Fish. Ag., No. 27:32-51.

bottom sediments, organic carbon
Sackett, William L., 1964.
The depositional history and isotopic organic carbon composition of marine sediments.
Marine Geology, 2(3):173-185.

bottom sediments, organic carbon
Sackett, William M., Walter R. Eckelmann, Michael L. Bender and Allan W.H. Be, 1965.
Temperature dependence of carbon isotope composition in marine plankton and sediments.
Science, 148(3667):235-237/

bottom sediments, organic carbon
Vikhrenko, N.M., 1964.
Organic carbon in the bottom sediments of the northern Atlantic. (In Russian).
Okeanologiia, Akad. Nauk, SSSR, 4(3):437-446.

bottom sediments, chemistry (organic matter)

bottom sediments, chemistry (organic matter)
Anderson, D.Q., 1939.
Distribution of organic matter in marine sediments and its availability for further decomposition.
J. Mar. Res. 2(3):225-235, Textfig. 67, 2 tables.

bottom sediments, organic matter
Bordovsky, O.K., 1969.
Organic matter in the recent bottom sediment of the Caspian Sea. (In Russian; English abstract). Okeanologiia, 9(6): 996-1006.

BOTTOM SEDIMENTS, organic matter in
Bordovsky, O.K., 1968.
Organic matter in the glacial-marine sediments of the eastern Antarctic. (In Russian; English abstract).
Okeanologiia, Akad. Nauk, SSSR, 8(1):69-77.

bottom sediments, organic matter
Bordovskiy, O.K., 1965.
Accumulation and transformation of organic substance in marine sediments. 1. Summary and introduction. 2. Sources of organic matter in marine basins. 3. Accumulation of organic matter in bottom sediments. 4. Transformation of organic matter in marine sediments.
Marine Geology, Elsevier Publ. Co., 3(½):3-4; 5-31; 33-82; 83-114.

bottom sediments, chemistry (organic matter)
Bordovsky, O.K., 1960
Organic matter in the recent sediments of the Bering Sea. Trudy Inst. Okeanol, 42: 89-106.

bottom sediments, organic matter
Bordovskii, O.K., 1957.
The composition of organic matter in the Recent sediments of the Bering Sea. (In Russian).
Doklady, Akad. Nauk, SSSR, 116(3):443.

bottom sediments, organic matter
Buchanan, J.B. and M.R. Longbottom, 1970.
The determination of organic matter in marine muds: the effect of the presence of coal and the routine determination of protein. J. exp. mar. Biol. Ecol., 5(2): 158-169.

bottom sediments, organic content
Chalmers, G.V., and A.K. Sparks, 1959.
An ecological survey of the Houston ship channel and adjacent bays.
Publ. Inst. Mar. Sci., 6:213-250.

bottom sediments, organic matter
Conover, John T., 1962
Algal crusts and the formation of lagoon sediments.
The Environmental Chemistry of Marine Sediments, Proc. Symp., Univ., R.I., Jan. 13, 1962
Occ. Papers, Narragansett Mar. Lab., No. 1: 69-76.

bottom sediments, organic matter
Degens, Egon T., 1964.
Über biogeochemische Umsetzungen in Frühstadium der Diagenese.
In: Developments in Sedimentology, L.M.J.U. van Straaten, Editor, Elsevier Publishing Co., 1:81-92.

bottom sediments, organic matter in
Devèze, L., 1964.
Activités bacteriennes et matières organiques dans les sédiments marins.
Ann. Inst. Pasteur (3 Suppl.):123-135.

bottom sediments, organic matter
*Drozodova, T.V., A.V. Kochenov and G.N. Baturin, 1967.
Some compositional characteristics of organic matter in modern marine sediments. (In Russian).
Geokhimiya (10):1088-1093. Abstract in Geochem. Int. 4(5):979.

bottom sediments, chemistry (organic matter)
Florovskaya, V.N. and Yu.N. Gurskiy 1966
Organic matter in the pelagic sediments of the Black Sea. (In Russian)
Geokhimia (1): 123-128.
Translation: Geochem. int., Ann Arbor 3(1):78-83

bottom sediments, chemistry (organic carbon)
Froelich, Philip, Bruce Golden and Orrin H. Pilkey 1971.
Organic carbon in sediments of the North Carolina continental rise.
SEast Geol. 13(2):91-97.

bottom sediments, chemistry (organic matter)
Galkin, L.M., and I.B. Mizandrontsev 1971
Distribution of organic matter multistage decomposition products in bottom sediments (In Russian)
Dokl. Akad. Nauk, SSSR 19? (2): 423-425

bottom sediments, chemistry (organic matter)
Gershanowich, D.E. 1965.
New data on the accumulation of organic matter in Recent sediments in the northernmost part of the Pacific Ocean. (In Russian)
Okeanologiia 5(2): 298-303.

bottom sediments - organic, matter
Gorshkova, T.I. 1967.
Sedimentation in the Norwegian Sea. (Carbonate and organic matter in the sediments of the Norwegian Sea). (In Russian). Atlant. nauchno-issled. Inst. rybn. khoz. okeanogr. (AtlantNIRO). Materialy Konferentsii po Resul'tatam Okeanologicheskikh Issledovanii v Atlanticheskom Okeane, 300-308.

bottom sediments, organic matter
Gorshkova, T.I., 1962.
Organic matter in sediments of the Baltic Sea.
Trudy Vses. Nauchno-Issledov. Inst. Morsk. Ribn. Chos. i Okean., VNIRO, 46:117-123.

In Russian

bottom sediments, organic matter
Gorshkova, T.I., 1962.
Organic matter in the sediments of the Norwegian Sea and conditions for its accumulation.
Trudy Vses. Nauchno-Issled. Inst. Morsk. Ribn. Chos. i Okean., VNIRO, 46:38-57.

In Russian

bottom sediments, chem (organic matter)
Gorshkova, T.I., 1960.
Conditions for the accumulation of organic matter in marine sediments.
Trudy Okeanogr. Komissii, Akad. Nauk, SSSR, 10 (2):56-60.

bottom sediments, organic matter in
Gorshkova, T.I., 1959.
Carbonate and organic matter in the sediments of the middle and southern Caspian.
Trudy VNIRO, 38:142-151.

bottom sediments, organic matter in
Gorshkova, T.I., 1955.
Organic matter in the sediments of the Sea of Azov and Taganrogsk Gulf. Trudy VNIRO, 31:95-122

bottom sediments, organic content
Gudkov, M.P., and T.I. Gorshkova, 1959
The change in the upkeep of organic matter in the sediments of the northern Caspian in connection with the sinking of its level. Trudy VNIRO 38: 88-105.

bottom sediments, organic matter
Hamilton, R.D., and L.J. Greenfield, 1965.
Observations on the entrapment of organic matter within the particle structure of calcareous sediments.
Nature, 207(4997):627-628.

bottom sediments, organic matter
Kamatani, A., and C. Matsudaira, 1964.
On the fertility of Matsushima Bay. II. The mineralization of organic matter in marine mud.
Tohoku J. Agricult. Res., 15(3):279-294.

bottom sediments, organic matter in
Kato, K., 1955.
Accumulation of organic matter in marine bottom and its oceanographic environment in the sea to the southeast of Hokkaido.
Bull. Fac. Fish., Hokkaido Univ., 6(2):125-151, 15 textfigs.

bottom sediments, organic matter in
Kato, K., 1955.
Regional distribution of organic matter in bottom sediments in the vicinity of Okushiri Strait, Hokkaido.
Bull. Fac. Fish., Hokkaido Univ., 6(1):19-24, 1 table, 6 textfigs.

bottom sediments, organic matter
Koshimizu, Naobumi, 1961
On the distribution of organic matter in the sea-water and in the bottom sediments of Maizuru Bay.
Bull. Maizuru Mar. Obs., No. 7:(1-6).
Also in:
Oceanographical Magazine, Tokyo, 12(1):(1-6), 1960.

bottom sediments, organic matter
Koshimizu, N., 1960.
On the distribution of organic matter in the sea-water and in the bottom samples of Maizuru Bay.
Oceanogr. Mag., Tokyo, 12(1):1-6.

bottom sediments, chemistry (organic matter)

Longbottom, M.R., 1970.
The distribution of Arenicola marina (L.) with particular reference to the effects of particle size and organic matter of the sediments.
J. exp. mar. Biol. Ecol., 5(2): 138-157.

bottom sediments chemistry (organic matter)

Lux, R., 1938.
Sur la teneur en matière organique des sediments marins à Monaco et à Concarneau.
Bull. Inst. Océan., Monaco, 760:1-20.

bottom sediments, organic content

Magliocca, A., & A.S. Kutner, 1964.
Conteúdo orgânico dos sedimentos de fundo da Cananéia, São Paulo. (In Portuguese; English abstract).
Contrib. Avulsas, Inst. Oceanogr., São Paulo (Oceanogr. Fisica), No. 7:14 pp.

bottom sediments chem. (organic matter)

Mohamed, A.F., 1949
The distribution of organic matter in sediments from the Northern Red Sea. Am. J. Sci. 247(2):116-127.

Bottom Sediments, Organic matter

Murty, P.S.N., C.V.G. Reddy and V.V.R. Varadachari, 1969.
Distribution of organic matter in the marine sediments off the west coast of India.
Proc. Nat. Inst. Sci., India (B) 35(5):377-384

bottom sediments, organic matter

Neiheisel, James, 1965.
Source and distribution of sediments at Brunswick Harbor and vicinity, Georgia.
U.S. Army Coastal Eng. Res. Center, Techn. Memo., No. 12:49 pp.,

bottom sediments, organic matter in

Orr, W.L., and K.O. Emery, 1956.
Composition of organic matter in marine sediments: Preliminary data on hydrocarbon distribution in basins of southern California. Bull. G.S.A., 67(9):1247-1258.

bottom sediments, chemistry (organic matter)

Rao, M. Subba, 1960.
Organic matter in marine sediments off east coast of India.
Bull. Amer. Assoc. Petr. Geol., 44(10):1705-1713.

bottom sediments, organic matter in

Richard, Mlle., 1955.
Résultat de recherches sur la matière organique.
Trav. Lab. Géol. Sous-Marine, 6:27-28.

bottom sediments, organic content

Richards, F.A., and A.C. Redfield, 1954.
A correlation between the oxygen content of sea water and the organic content of marine sediments.
Ltr. to Edit., Deep-Sea Res. 1(4):279-281, 2 textfigs.

bottom sediments, organic matter

Romankevich, E.A., 1962
Organic substance in the surface layer of bottom sediments.
Mezhd. Geofiz. Komitet. Prezidiume Akad. Nauk. SSSR, Rezult. Issled. Programme Mezhd. Geofiz. Goda, Okeanol. Issled., No. 5:67-111.

bottom sediments, organic matter

Romankevich, E.A., 1961.
Organic substance in the surface layer of bottom sediments in the western Pacific Ocean. (Abstract)
Tenth Pacific Sci. Congr., Honolulu, 21 Aug.-6 Sept., 1961, Abstracts of Symposium Papers, 384-385.

bottom sediments, organic matter

Romankevich, E.A., 1960.
Organic matter in the bottom sediments of the Pacific Ocean, east of Kamchatka.
Trudy Okeanogr. Komissii, Akad. Nauk, SSSR, 10(2):39-47.

bottom sediments, organic matter

Romankevich, E.A., 1959
Organic matter in the marine sediments of Kronotski Bay.
Trudy Inst. Okeanol., 36: 32-39.

bottom sediments, organic matter in

Romankevich, E.A., 1957.
Organic matter in the bottom sediments columns from the northwest Pacific.
Doklady, Akad. Nauk, SSSR, 116(3):447.

bottom sediments, organic matter

Romankevich, E.A., and V.E. Artemyev, 1969.
The composition of organic matter of the sediments from the Kurile-Kamchatka Trench. (In Russian; English abstract). Okeanologiia, 9(5): 796-806.

bottom sediments, organic matter

Ros, Joaquin, 1962
Contribution à l'étude des substances "humiques" des vases marines.
In: Océanographie Géologique et Géophysique de la Méditerranée Occidentale. Centre National de la Recherche Scientifique, Villefranche sur Mer, 4 au 8 avril, 1961, 19-28.

bottom sediments, organic carbon

*Seigle, George A., 1968.
Foraminiferal assemblages as indicators of high organic carbon content in sediments and of polluted waters.
Bull. Am. Ass. Petr. Geol. 52(11):2231-2241.

bottom sediments, organic content

Shabarova, N.T., 1954.
Organic matter of marine deposits.
Uspekhi Sovremennoi Biol. 37(2):203-208.

RJ 367 in List 37 of
Bibl. Transl. Russ. Sci. Tech. Lit.

bottom sediments, chemistry (organic matter)

Shirai, Toru 1968.
Submarine geological study of the bottom sediments of the adjacent seas of the Japanese islands with special reference to the distribution of organic matter in sediments.
Mem. Fac. Sci., Kyoto Univ. Geol. Mineral. 34(2): 147-173.

bottom sediments, chemistry (organic)

Simoneit, Bernd R. and A.L. Burlingame, 1971.
Some preliminary results on the higher weight hydrocarbons and fatty acids in the deep sea drilling project cores, Legs 5-7. Initial Repts Deep-Sea Drill. Proj. 7(2): 889-912.

bottom sediments, chem (organic matter)

Starikova, N.D., 1962.
Investigation of qualitative composition of organic matter dissolved in sea and ocean sediments. (In Russian; English summary).
Trudy Inst. Okeanol., Akad. Nauk, SSSR, 54:22-30.

bottom sediments, chemistry (organic matter)

Starikova, N.D., 1961.
Organic matter in the liquid phase of marine and oceanic sediments.
Trudy Inst. Okeanol., Akad. Nauk, SSSR, 50:130-169.

bottom sediments, organic matter in

Starikova, N.D., 1961.
Organic matter in ground solutions and its distribution over the strata of marine and ocean sediments.
Doklady Akad. Nauk, SSSR, 149(6):1423-1426.

bottom sediments organic matter

Starikova, N.D., 1960.
Some ideas on the organic matter in interstitial water of the sediments in the Black Sea and the Sea of Azov.
Trudy, Comm. Oceanogr., Akad. Nauk, SSSR, 10(2): 30-38.

bottom sediments, organic matter

Starikova, N.D., 1960.
Some data on the organic matter of the liquid phase of the sediments in the Black Sea and Sea of Azov.
Trudy Okeanogr. Komissii, Akad. Nauk, SSSR, 10(2):30-38.

bottom sediments, chemistry (organic matter)

Starikova, N.D., 1959
Organic matter of the liquid phase of the sediments of the North West Pacific. Trudy Inst. Okeanol., 33: 165-177.

bottom sediments, organic content

Starikova, N.D., 1956.
The organic matter of the liquid phase of modern deposits of the Okhotsk Sea.
Dokl Akad. Nauk, SSSR, 108(5):892-894.

bottom sediments chemistry (organic matter)

Starikova, N.D., 1956.
Organic substance in the deep deposits of the Bering Sea.
Akad. Nauk, SSSR, Dokl., 106(3):519-522.

bottom sediments organic matter

Shabarova, N.T., 1954.
The organic matter of marine deposits.
Uspekhi Sovremennoi Biol., 37(2):203-208.

Translation may be purchased for $13.40 from Associated Technical Services, P.O. Box 271, East Orange, New Jersey (RJ-367).

Bottom sediments, organic matter

Vichrenko, N.M., 1967.
Contents and composition of the organic matter in the sediments of the northern part of the Atlantic Ocean. (In Russian).
Atlant. nauchno-issled. Inst. rybn. Khoz. okeanogr. (AtlantNIRO). Materialy Konferentsii po Rezul'tatam Okeanologischeskikh Issledovanii v Atlanticheskom Okeane 309-316

bottom sediments, organic matter

Vikhrenko, N.M., 1964.
Organic matter in the surface sediment layer of the northern part of the Atlantic Ocean. Investigation of marine bottom sediments and suspended matter. (In Russian; English abstract).
Trudy Inst. Okeanol., Akad. Nauk, SSSR, 67:120-135.

bottom sediments, organic matter

Vinogradova, T.L., 1969.
Organic matter in the upper layer of the sediments on the west Kamchatka shelf and in Shelikhov Bay. (In Russian).
Trudy Vses. nauchno-issled. Inst. morsk. rybn. Okean (VNIRO) 65: 267-281

bottom sediments, chem. (oxygen)
Francis-Boeuf, C., 1947
Sur la teneur en oxygène dissous du milieu intéreur des vases fluvio-marines. C.R., Paris, 225:392-394.

bottom sediments, chemistry (oxidation-reduction)

bottom sediments, chemistry
Callame, Bernard 1968.
Contribution à l'étude des potentiels d'oxydo-reduction dans les sédiments marins.
Cah. océanogr. 20 (4): 305-319.

bottom sediments, chemistry (oxidation-reduction)
Fenchel, Tom 1971.
The reduction-oxidation properties of marine sediments and the vertical distribution of the microfauna.
Vie Milieu Suppl. 22 (2): 509-521

bottom sediments, chem (oxidation-reduction)
Fomina, L.S., 1962.
The oxidative-reductive processes in the bottom sediments of the southwest Pacific.
Trudy Inst. Okeanol., Akad. Nauk, SSSR, 54:158-169.

In Russian; English summary

bottom sediments, chemistry
Hayes, F.R., B.L. Reid and M.L. Cameron, 1958.
Lake water and sediment. 2. Oxidation-reduction relations at the mud-water interface.
Limnol. & Oceanogr., 3(3):308-317.

bottom sediments, chemistry
Romankevich, E.A., and N.V. Petrov, 1961.
Oxidation-reduction potential Eh and pH of sediments in the northeastern Pacific.
Trudy Inst. Okeanol., Akad. Nauk, SSSR, 45:72-85.

bottom sediments, chemistry (O₂ consumption)
Francis-Boeuf, C., 1947.
Données sur la consommation d'oxygène in vitro de quelques vases fluvio-marines. C.R. Acad. Sci. 225:1083-1084.

bottom sediments, chemistry (O₂ consumption)
Francis-Boeuf, C., 1947.
Production et consommation d'oxygène par la pellicule superficielle des vases fluvio-marines. C.R. Acad. Sci. 225:820-822.

bottom sediments, oxygen consumption

Pamatmat, Mario M., 1971.
Oxygen consumption by the seabed IV. Shipboard and laboratory experiments. Limnol. Oceanogr. 16(3): 536-550.

bottom sediments, chemistry oxygen diffusion
Calleme, Bernard, 1967.
Sur la diffusion de l'oxygène à l'intérieur des sédiments marine.
Trav. Cent. Rech. Etud. océanogr., 7(2):25-29.

bottom sediments, chemistry oxygen utilization
Carey, Andrew G. Jr. 1967.
Energetics of the benthos of Long Island Sound. 1. Oxygen utilization of sediment.
Bull. Bingham oceanogr. Coll. 19 (2): 136-144.

bottom sediments, chemistry, paraffins

bottom sediments, chemistry (paraffins)

See also: chemistry, paraffins

bottom sediments, chemistry
Evans, E.D., G.S. Kenny, W.G. Meinschein and E.E. Bray, 1957.
Distribution of n-paraffins and separation of saturated hydrocarbons from recent marine sediments. Analyt. Chem., 29(12):1858-1861.

bottom sediments, chemistry
#Van der Weide, B., 1967.
Evolution des n-paraffines par traitement thermique de sédiments marins récents.
Bull. Centre Rech. PAU-SNPA, 1(1):161-164.

bottom sediments, chemistry (pigments)

bottom sediments, pigments

Baker, Earl W., 1970.
Tetrapyrrole pigments. Initial Reports of the Deep Sea Drilling Project, Glomar Challenger, 4: 431-438.

bottom sediments, pigments (photosynthetic)
Fenchel, Tom, and Birthe J. Straarup 1971.
Vertical distribution of photosynthetic pigments and the penetration of light in marine sediments.
Oikos 22 (2): 172-182.

bottom sediments, chemistry, phosphatase

Ayyakkannu, K. and D. Chandramohan, 1971.
Occurrence and distribution of phosphate solubilizing bacteria and phosphatase in marine sediments at Porto Novo. Mar. Biol. 11(3): 201-205.

bottom sediments, phosphatic

bottom sediments, chemistry (phosphorus and its compounds)

bottom sediments, chemistry phosphorus
Arrhenius, G., 1952.
Sediment cores from the East Pacific.
Repts. Swedish Deep-sea Exped., 1947-1948, 5(1): 227 pp., textfigs., appendix with plates.

bottom sediment, chemistry (phosphates)
Balasubrahmanyan, K., 1962
Studies in the ecology of the Vellar Estuary. 2. Phosphates in the bottom sediments.
J. Zool. Soc., India, Calcutta, 13(2):166-169.

Also in: Annamalai Univ. Mar. Biol. Sta. Porto Novo, S. India, Publ., 1961-1962.

bottom sediments, chemistry (phosphate)

Baturin, G.N., 1971.
Formation of phosphate sediments and water dynamics. (In Russian; English abstract).
Okeanologiia 11(3): 444-449.

sediments, chemistry of (phosphorus)
Bezrukov, P.L., 1957.
On phosphorus distribution in the sediments of the Okhotsk Sea. Doklady Akad. Nauk, SSSR, 113(1):142.

bottom sediments, chem
Bezrukov, P.L., and E.A. Ostroumov, 1957.
On phosphorus distribution in the sediments of the Okhotsk Sea. Dokl. Akad. Nauk, SSSR, 113(1):142-145.

bottom sediments, chemistry phosphorus
Bonatti, Enrico, D.E. Fisher, Olva Joensuu, and H.S. Rydell, 1971.
Postdepositional mobility of some transition elements, phosphorus, uranium and thorium in deep-sea sediments.
Geochim. cosmochim. Acta, 35 (2): 189-201.

bottom sediments, chemistry phosphorus
Chester, R., 1965.
Elemental geochemistry of marine sediments.
In: Chemical oceanography, J.P. Riley and G. Skirrow, editors, Academic Press, 2:23-80.

bottom sediments, chemistry (phosphate)
Cvijc, V., 1956.
Activity of bacteria in the liberation of phosphate from the sea sediments in bottom water.
Acta Adriatica 8(4):1-31.

bottom sediments, chemistry of
Datzko, V.G., 1948.
Phosphates in the bottom deposits of the Sea of Azov. Dokl. Akad. Nauk, SSSR., 59:275-277.

bottom sediments, chemistry phosphorus
Feuillet, Michelle 1971.
Etude du phosphore dans les sédiments ostréicoles du bassin des Chasses des Sables d'Olonne.
Revue Trav. Inst. Pêches marit. 35 (4): 443-453.

bottom sediments, phosphatic
Gorsline, Donn S., and Donald B. Milligan, 1963.
Phosphatic deposits along the margin of the Pourtalès Terrace, Florida.
Deep-Sea Res., 10(3):259-262.

bottom sediments, chem. (phosphate)
Macpherson, L.B., N.R. Sinclair and F.R. Hayes, 1958.
Lake water and sediment. III. The effect of pH on the partition of inorganic phosphate between water and oxidized mud or its ash.
Limnol. & Oceanogr., 3(3):318-326.

bottom sediments, chemistry (phosphorus)
Marini, Liana Bisi, 1962
La determinazione del fosforo nei sedimenti Adriatici.
Arch. Oceanogr. e Limnol., 12(3):267-274.

bottom sediments, chemistry (phosphorus)
Marino, Liana Bisi, 1962
La determinazione del fosforo nei sedimenti Adriatici.
Archiv. di Ocean. Limnol., XII (3):267-274
Istit. Sper. Talass., Trieste, Pubbl. No. 397.

bottom sediments, phosphorus

Murty, P.S.N. and C.V.G. Reddy, 1969.

Distribution of phosphorus in the marine sediments off the East Coast of India. Bull. natn. Inst. Sci., India, 38(1): 405-410.

Bottom sediments, Chem

Murty, P.S.N., C.V.G. Reddy and V.V.R. Varadachari, 1968
Distribution of total phosphorus in the shelf sediments off the west coast of India.
Proc. Nat. Inst. Sci. India (B) 34 (3): 134-141
Also in: Coll. Repr. Nat. Inst. Oceanogr. New Delhi, (1963-1968).

bottom sediments, chemistry (phosphorus)

*Murty, P.S.N., C.V.G. Reddy and V.V. Varadachari, 1968.
Distribution of total phosphorus in the shelf sediments off the west coast of India.
Proc. Nat. Inst. Sci., India, (B) 34(3):134-141.

bottom sediments, phosphorus

Naidu, A.S., and Y.L. Dora, 1967.
Geochemical behavior of phosphorus in the non-detrital sedimentary cycle - a review.
Bull. Dept. Mar. Biol. Oceanogr. Univ. Kerala, 3:33-40

bottom sediments, chemistry

Pomeroy, Lawrence, E.E. Smith, and Carol M. Grant, 1965.
The exchange of phosphates between estuarine water and sediments.
Limnology and Oceanography, 10(2):167-172.

sedimentary rocks, chemistry of

Ronov, A.B., and G.A. Korzina, 1960.
Phosphorus in sedimentary rocks.
Geokhimiia (8)

Translated as Geochemistry(8):805-829.

bottom sediments, phosphates

*Siddiquie, H.N. and A.N. Chowdhury, 1969.

The distribution of phosphates in some samples of the shelf sediments of the West Coast of India. Bull. Natn. Inst. Sci., India, 38(1): 483-490.

bottom sediments chemistry (phosphate)

Stephensen, W., 1951.
Preliminary observations upon the release of phosphate from estuarine mud. Proc. Indo-Pacific Fish. Counc., Cronulla N.S.W., Australia, Sects. II-III:184-189, 1 textfig.
17-28 Apr. 1950.

bottom sediments, chemistry (proactinium)

Sackett, W. M., 1960

Proactinium-231 content of ocean water and sediments. Science, 132(3441): 1761-1762.

bottom sediments chemistry (phosphate)

Tooms, J.S., C.P. Summerhayes and D.S. Cronan, 1969.
Geochemistry of marine phosphate and manganese deposits. Oceanogr. Mar. Biol. Ann. Rev., H. Barnes, editor, George Allen and Unwin, Ltd., 7: 49-100.

bottom sediments, chemistry, plutonium

Wong, Kai M. 1971.
Radiochemical determination of plutonium in sea water, sediments and marine organisms.
Analyt. Chim. Acta 56:355-364

Bottom sediments Chemistry, radioisotopes

bottom sediments chemistry (radium, radioisotopes, radioactivity, etc.)

Bottom sediments, chemistry, radioactivity

See also: Chemistry, radioactivity

Bottom sediments, chemistry

Agemirov, A. Sh., 1963.
Geochemical balance of radioactive elements in the Black Sea Basin. (In Russian)
Geokhimiya, (6):612-614.

Translations:
Geochemistry (6):630-633.

bottom sediments, chemistry

Arrhenius, G., M.N. Bramlette and E. Picciotto, 1957.
Localization of radioactive and stable heavy nuclides in ocean sediments. Nature 180:85-86.

bottom sediments, chemistry of

Arrhenius, G., and E.D. Goldberg, 1955.
Distribution of radioactivity in pelagic clays.
Tellus 7(2):226-231.

bottom sediments, chemi

Baranov, V.I., and L.A. Kuzmina, 1957.
Radiochemical analysis of deep-sea sediments in connection with the determination of the rate of sediment accumulation.
Int. Conf. Radioisotopes in Sci. Res., Paris, UNESCO/NS/RIC/222:1-16.

bottom sediments, radioactivity of

Baranova, D.D., 1966.
A comparative study of sorption and desorption of various radionuclides by marine shallow water sediments. (Abstract only).
Second Int. Oceanogr. Congr., 30 May-9 June 1966. Abstracts, Moscow:18-19.

bottom sediments, chemistry

Baranova, D.D., and G.G. Polikarpov, 1968.
Sorption and desorption of radionuclides by shallow-water sediments of the Black Sea. (In Russian; English abstract).
Okeanologiia, Akad. Nauk, SSSR, 8(3):427-430.

bottom sediments, sorption by

Baranova, D.D., and G.G. Polikarpov, 1965.
Sorption of strontium-90 and cesium-137 in the aleurite silts of the Black Sea. (In Russian).
Okeanologiia, Akad. Nauk, SSSR, 5(4):646-648.

bottom sediments, chemistry (radionuclides)

Bhandari, N., S.G. Bhat, S. Krishnaswamy and D. Lal 1971.
A rapid beta gamma coincidence technique for determination of natural radionuclides in marine deposits.
Earth planet. Sci. Letts 11(2): 121-126

bottom sediments, chemistry

Biscaye, Pierre E. and E. Julius Dasch, 1971.
The Rubidium, strontium, strontium-isotope system in deep-sea sediments: Argentine Basin.
J. geophys. Res. 76(21): 5087-5096.

bottom sediments, radioactivity of

Blanc, Jean Joseph, 1962.
Remarques sur divers types de sédiments sous-marins de Méditerranée et leur radioactivité.
In: Océanographie Géologique et Géophysique de la Méditerranée Occidentale, Centre National de la Recherche Scientifique, Villefranche sur Mer, 4 au 8 avril, 1961, 171-176.

bottom sediments, radioactivity of

Cerrai, E., B. Schreiber, C. Triulzi and L. Tassi-Pelati, 1964.
Contribution of Ce^{144} -Pr^{144} and Pm^{147} to the radioactivity of upper layers of coastal sediments of the Ligurian Sea.
Ist. Lombardo, Accad. Sci. e Lett. Rendiconti (B)

bottom sediments, radioactivity

Cutshall, N., and C. Osterberg, 1964.
Radioactive particles in sediment from the Columbia River.
Science, 144(3618):536-537.

Bottom sediments, chemistry

Duursma, E.K., and C.J. Bosch, 1970.
Theoretical, experimental and field studies concerning diffusion of radioisotopes in sediments and suspended particles of the sea. B. Methods and experiments.
Netherl. J. Sea Res. 4 (4): 395-469.

bottom sediments, chemistry, radioactivity

Duursma, E.K. and M.G. Gross, 1971.
Marine sediments and radioactivity. Radioactivity in the marine environment, U.S. Nat. Acad. Sci., 1971: 147-160.

bottom sediments, chemistry

Duursma, E.K., and C. Hoede 1967.
Theoretical, experimental and field studies concerning molecular diffusion of radioisotopes in sediments and suspended solid particles of the sea. A. Theories and mathematical calculations.
Neth. J. Sea Res. 3(3): 423-457

bottom sediments, radioactivity of

Fuiro, Giuliano, Federico Bedarida e Graziella Ricciardi, 1966
Misure di radioattività nei sedimenti del Mar Tirreno.
Atti Accad. ligure 22(1): 232-244

bottom sediments, radiation of

Gealy, E.L., 1971.
Natural gamma radiation of sediments from the western equatorial Pacific: Leg 7, Glomar Challenger. Initial Repts Deep Sea Drill. Proj. 7(2): 1037-1080.

bottom sediments, radioactivity of

Got, Henri, 1971.
Répartition du radium, thorium, potassium dans les sédiments du plateau continental catalan
C. r. hebd. Séanc. Acad. Sci. Paris (D) 272 (16): 2147-2150.

bottom sediments, chemistry (radioactivity)

Got, Henri 1968.
La radioactivité naturelle des sédiments de la baie de Banyuls.
Cah. océanogr. 20 (3): 225-235.

bottom sediments, radioactive

Gross, M.Grant, 1966.
Distribution of radioactive marine sediment derived from the Columbia River.
J. geophys. Res., 71(8):2017-2021.

bottom sediments, chem (radioactivity)
Jaffe, G., and J.M. Hughes, 1953.
The radioactivity of bottom sediments in Chesapeake Bay. Trans. Amer. Geophys. Union 34(4):539-542, 2 textfigs.

bottom sediments, radioactivity
Jefferies, Douglas F., 1968.
Fission-product radionuclides in sediments from the north-east Irish Sea.
Helgoländer wiss. Meeresunters. 17: 280-290.

bottom sediments, radioactivity of
Jennings, David, Norman Cutshall and Charles Osterberg, 1965.
Radioactivity: detection of gamm-ray emission in sediments in situ.
Science, 148(3672):948-950.

bottom sediments, chem. (radioactivity)
Mahadevan, C., and V. Aswathanarayana, 1954.
Radioactivity of sea floor sediments off the east coast of India. Andhra Univ. Ocean. Mem. 1: 36-50, 4 textfigs.

bottom sediments, radioactivity of
McCone, Alistair W., 1966.
The Hudson River Estuary: hydrology, sediments and pollution.
Geogr. Rev., 56(2):175-189.

bottom sediments Chem (radioactivity)
Niino, Hiroshi, 1960
Study on the natural radioactivity of bottom deposits. II.
Rec. Oceanogr. Wks., Japan, Spec. No. 4: 165-170.

bottom sediments, chemistry (radioactivity)
Niino, Hiroshi, 1959
On the natural radioactivity of the marine sediments (1.).
Rec. Oceanogr. Wks., Japan, Spec. No. 3: 163-164.

bottom sediments, radioactivity
Osterberg, C., L.D. Kulm and J.V. Byrne, 1963.
Gamma emitters in marine sediments near the Columbia River.
Science, 139(3558):916-917.

Bottom sediments chemistry (radioactivity)
Piggot, C.S., and W.D. Urry, 1942.
Radioactivity of ocean sediments. IV. The radium content of sediments of the Cayman Trough.
Amer. J. Sci., 240:1-12.

bottom sediments, chemistry (radium)
Revelle, R.R., and C.S. Piggot, 1944.
1. Marine bottom samples collected in the Pacific Ocean by the "Carnegie" on its seventh cruise. II Radium content of ocean bottom sediments.
Publ. Carnegie Inst., Washington, 556:5-196, 47 textfigs.

bottom sediments, chemistry (radioactivity)
Romankevich, E.A., P.L. Bezrukov, V.I. Baranov and L.A. Khristianova 1966.
Stratigraphy and absolute age of deep-sea sediments in the western Pacific. (In Russian; English abstract)
Ref. Issled. meghd. Geofiz. Proekt. Okeanol. Meghd. Geofiz. Komitet, Presid Akad. Nauk SSSR 14: 165pp.

bottom sediments, chem. (radium)
Sanderman, L.A., and C.L. Utterback, 1941.
Radium content of ocean bottom sediments from the Arctic Ocean, Bering Sea, Alaska peninsula and the coasts of southern Alaska and western Canada. J. Mar. Res. 4(2):132-141, 2 figs.

bottom sediments, chemistry (radioactivity)
Schofield, J.C. 1967.
Origin of radioactivity at Niue Island
N.Z. Jl Geol. Geophys. 10(6): 1362-1371.

bottom sediments, chemistry (radioactivity)
Schreiber, B. 1965.
Études sur la radioactivité du plancton et des sediments côtiers de la mer Ligurienne.
Rapp. P.-v. Réun. Comm. int. Explor. scient. Mer Medit. 18(3): 883-892.

bottom sediments, radioactivity
Schreiber, B., L. Pelati, E. Cerrai e C. Triulzi, 1964.
Gross beta radioactivity of litoral sediments of the Ligurian Sea.
Energia Nucleare, 11(11):616-624.

bottom sediments, radioactivity of
*Senin, Yu. M., and A.I. Sherstnev, 1967.
Total B-activity of the Recent bottom sediments from the northwestern African shelf. (In Russian; English abstract.
Okeanologiia, Akad. Nauk, SSSR, 7(4):628-632.

bottom sediments, radioactivity of
Sherstniov, A.I., T.I. Gorshkova and N.A. Kuznetsova, 1963.
The radioactivity of the bottom sediments in the Baltic Sea, its bays, and the characteristics of some elements of their composition. (In Russian).
Atlantich. Nauchno-Issled. Inst. Ribn. Khoz. i Okeanogr. (ATLANTNIRO), Trudy, 10:209-212.

bottom sediments, chemistry (radioactivity)
Somayajulu, B.L.K., and E.D. Goldberg 1966.
Thorium and uranium isotopes in seawater and sediments.
N.Z. Jl Geol. Geophys. 10(6): 1362-1371.

sediments, radioactivity
Starik, I. Ye., A. P. Lisitsyn, and Yu. V. Kuznet-sov, 1962.
Concerning the mechanism for removal of radium from sea water and its accumulation in sea and ocean sediments. (In Russian).
Antarktika, Moscow, 1962:70-133.

Abstracted in:
Soviet Bloc Res., Geophys., Astron and Space, 96: 47.

bottom sediments, radioactivity of
Starik, I. Ye., D.S. Nikolayev, Yu. V. Kuznetsov and V.K. Legin, 1961.
The radioactivity of Black Sea sediments. (In Russian)
Doklady, Akad. Nauk, SSSR, 139(6):1456-1459.

Translation:
Consultants Bureau for Amer. Geol. Inst. (Earth Sciences section only), 139(1-6):704-707. (1963).

bottom sediments, radioactivity of
Starik, I.E., D.S. Nikolaev, Iu. V. Kuznetsov, and V.K. Legin, 1961
[A comparative study of the radioactivity of Azov and Black Sea sediments.]
Doklady Akad. Nauk, SSSR, 139(2):456-460.

bottom sediments, chemistry (radium)
Ueda, S., 1957.
[Chemical studies on the ocean. 66-67. Chemical studies of the shallow-water deposits. 19-20. Radium content of the shallow water deposits. (1-2).] J. Oceanogr. Soc., Japan, 13(2):61-66; 67-72.

bottom sediments, chemistry (radioactivity)
Urry, W. D., 1949.
Radioactivity of ocean sediments. VI. Concentrations of the radioelements in marine sediments of the southern hemisphere. Amer. J. Sci., 247(4):257-275, 4 textfigs.

bottom sediments, chemistry (radioactivity)
Urry, W.D., 1948.
Radioactivity of ocean sediments. VII. Rate of deposition of deep sea sediments. J. Mar. Res. 7(3):618-634, 6 textfigs.

bottom sediments chemistry (radioactivity)
Urry, W.D., and C.S. Piggot, 1942.
Radioactivity of ocean sediments. V. Concentration of the radioelements and their significance in red clay. Amer. J. Sci. 240:93-103.

bottom sediments, chemistry (radioactivity)
Utterback, C. L., and L. A. Sanderman, 1948.
Radium analyses of marine sediments in northern Pacific and adjacent waters. J. Mar. Res. 7(3): 635-643.

bottom sediments
Vorhhet, Solange, 1966.
Signification sédimentologique de la radioactivité naturelle des plages du Golfe du Lion.
Comptes Rendus, Acad. Sci. Paris, 262(4):440-443.

bottom sediments radioactivity
Whittard, W.F. 1962.
Geology of the western approaches of the English Channel, a progress report.
Proc. R. Soc. London (A) 265 (1322): 395-406.

bottom sediments, chemistry (rare earths)

bottom sediments, chemistry (rare earths)
Balashov, Yu.A., Yu.A. Bogdanov, I.O. Murdmaa and V.I. Chernysheva, 1971.
Rare-earth elements in the lagoonal sediments of the Saint Paul Island. (In Russian; English abstract). Okeanologiia, 11(1): 71-77.

bottom sediments Chem (rare earths)
Haskin, L., & M.A. Gehl, 1962
The rare-earth distribution in sediments.
J. Geoph. Res., 9(1):2537-2542.

bottom sediments, chem (rare earths)
Ostroumov, E.A., 1953.
[Rare earths in deep-sea deposits of the Black Sea.] Dokl. Akad. Nauk, SSSR, 91:1175-1178.

bottom sediments, chem. (rare earths)
Ostroumov, E.A., A.A. Astmanuna, and T.G. Shokhor, 1956.
[Method for the determination of rare earths in marine sediments.] Trudy Inst. Okeanol., 19:297-303

bottom sediments, chemistry
Wildeman, Thomas R., and Larry Haskin, 1965.
Rare-earth elements in ocean sediments.
J. Geophys. Res., 70(12):2905-2910.

bottom sediments, rare metals
Krauskopf, K.B., 1956.
Factors controlling the concentrations of thirteen rare metals in sea-water.
Geochim. Cosmochim. Acta, 9:1-32B.

bottom sediments, chemistry (selenium)
Sokolova, E.G., and M.F. Pilipchuk 1970.
Selenium in Recent sediments of the Black Sea. (In Russian)
Dokl. Akad. Nauk SSSR 193 (3): 692-696

bottom sediments, chemistry silicon

bottom sediments, chemistry (silicon and its compounds)

bottom sediments, chemistry, silica
Arrhenius, G., 1952.
Sediment cores from the East Pacific.
Repts. Swedish Deep-sea Exped., 1947-1948, 5(1): 227 pp., textfigs., with appendix.

bottom sediments, siliceous
*Chester, R., and H. Elderfield, 1968.
The infrared determination of opal in siliceous deep-sea sediments.
Geochim. cosmochim. Acta, 32(10):1128-1140.

bottom sediments, silicon
Dangeard, L., and M. Rioult, 1965.
Le domaine de la géologie marine et ses frontières: confrontation de l'océanographie et du géologie.
In: Submarine geology and geophysics, Colston Papers, W.F. Whittard and R. Bradshaw, editors, Butterworth's, London, 17: 93-105.

bottom sediments, silica
Emelianov, E.M., 1966.
Distribution of authigenous silica in the suspension and in Recent sediments of the Mediterranean Sea. (In Russian).
In: Geochemistry of silica, N.M. Strakhov, editor, Isdatel. "Nauka", Moskva, 284-294.

bottom sediments, chemistry (silica)
Fanning, Kent A., and David R. Schink, 1969.
Interaction of marine sediments with dissolved silica. Limnol. Oceanogr., 14(1): 59-68.

bottom sediments, silicious
Keleda, G.A., 1966.
Main features in the evolution of silicious sediments. (In Russian).
In: Geochemistry of silica, N.M. Strakhov, editor, Isdatel, "Nauka", Moskva, 371-393.

bottom sediments, silica
Lisitzin, A.P., 1966.
Distribution of silica in Quaternary sediments owing to climatic zonality of the geological past. (In Russian).
In: Geochemistry of silica, N.M. Strakhov, editor, Isdatel. "Nauka", Moskva, 321-370.

bottom sediments, siliceous
Lisitzin, A.P., 1966.
Main regularities in the distribution of Recent siliceous sediments and their relationship with climatic zonation. (In Russian).
In: Geochemistry of silica, N.M. Strakhov, editor, Isdatel "Nauka", Moskva, 90-191.

bottom sediments, siliceous
Ramsay, A.T.S. 1971.
Occurrence of biogenic siliceous sediments in the Atlantic Ocean.
Nature, Lond. 233 (5315): 115-117.

bottom sediments, silica
Shurko, I.I., 1966.
Amorphous silica in the sediments of the North Atlantic. (In Russian).
In: Geochemistry of silica, N.M. Strakhov, Editor, Isdatel. "Nauka", Moskva, 295-300.

bottom sediments, chemistry
silicate detritus
Siever, Raymond and Miriam Kastner, 1967.
Mineralogy and petrology of some Mid-Atlantic Ridge sediments.
J. mar. Res., 25(3):263-278.

bottom sediments, siliceous
Zhuze, A.P., 1966.
Siliceous sediments in Recent and ancient lakes. (In Russian).
In: Geochemistry of silica, N.M. Strakhov, editor, Isdatel. "Nauka", Moskva, 301-318.

bottom sediments, sorption

bottom sediments, chemistry (sterols)

bottom sediments, chem (sterols)
Attaway, David and P.L. Parker, 1970.
Sterols in Recent marine sediments. Science, 169(3946): 674-675.

bottom sediments, chemistry
Schwendinger, Richard B., and J. Gordon Erdman, 1964
Sterols in Recent aquatic sediments.
Science, 144(3626):1575-1576.

bottom sediments, strontium

bottom sediments, chemistry (strontium)

bottom sediments, chemistry (strontium)
Dasch, E. Julius, 1969.
Strontium isotopes in weathering profiles, deep-sea sediments, and sedimentary rocks.
Geochim. cosmochim. Acta, 33(12): 1521-1552.

bottom sediments, chemistry (strontium)
Dasch, E. Julius, F. Allan Hills and Karl K. Turekian, 1966.
Strontium isotopes in deep-sea sediments.
Science, 153(3733):295-297.

Bottom Sediments, Chem.
Ishibashi, Masayoshi, Shunzo Ueda and Yoshikazu Yamamoto, 1969.
The contents of strontium and strontium-90 in shallow water deposits (In Japanese; English abstract).
J. oceanogr. Soc. Japan 25 (5): 233-238

bottom sediments, strontium
*Murthy, V. Rama, and E. Beiser, 1968.
Strontium isotopes in ocean water and marine sediments.
Geochim. cosmochim. Acta, 32(10):1121-1126.

bottom sediments sulfur compounds

bottom sediments, chemistry (sulfur and its compounds)

bottom sediments, chemistry (sulphur)
Berner, Robert A., 1964.
Distribution and diagenesis of sulphur in some sediments from the Gulf of California.
Marine Geology, 1(2):117-140.

bottom sediments, chemistry (sulphides)
Fenchel, T.M. and R.J. Riedl, 1970.
The sulfide system: a new biotic community underneath the oxidized layer of marine sand bottoms. Marine Biol., 7(3): 255-268.

bottom sediments, chemistry
Fujita, Yuji, Tadataka Taniguti, Shoji Iizuka and Bunkei Zenitani 1967.
Microbiological studies on shallow marine areas. IV On the liberation and accumulation of sulfides in mud sediment, and its relation to the formation of anoxic layer. (In Japanese; English abstract)
Bull. Fac. Fish. Nagasaki Univ. 24: 79-88.

bottom sediments, chemistry (sulphur)
Hartmann, Martin, und Heimo Nielsen, 1969.
δ^{34}S-Werte in rezenten Meeressedimenten und ihre Deutung am Beispiel einiger Sedimentprofile aus der westlichen Ostsee
Geol. Rundschau 58 (3): 621-655

bottom sediments, chemistry (sulphides)
Hata, Yoshihiko, 1965.
Microbial production of sulfides in marine and estuarine sediments. (In Japanese; English abstract).
J. Shimonoseki Univ., Fish., 14(2):37-83.

bottom sediments, chem- (sulphur)
Kaplan, I.R., K.O. Emery and S.C. Rittenberg, 1963.
The distribution and isotopic abundance of sulphur in Recent marine sediments off southern California.
Geochimica et Cosmochimica Acta, 27(4):297-331.

bottom sediments, chemistry-sulphur compounds
Kaplan, I.R., and T.A. Rafter, 1964.
Transformations of sulphur compounds in the sediments of Milford Sound.
New Zealand Dept. Sci. Ind. Res., Bull., No. 157:
New Zealand Oceanogr. Inst., Memoir, No. 17:73-76.

bottom sediments, chemistry
*Kaplan, I.R., R.E. Sweeney and Arie Nissenbaum, 1969.
Sulfur isotope studies on Red Sea geothermal brines and sediments.
In: Hot brines and Recent heavy metal deposits in the Red Sea, E.T. Degens and D.A. Ross, editors, Springer-Verlag, New York, Inc., 474-498.

bottom sediments, chemistry of (sulphur)
Koyama, T., and K. Sugawara, 1953.
Sulphur metabolism in bottom muds and related problems. (1). J. Earth Sci., Nagoya Univ., 1(1):24-34.

bottom sediments, chemistry (sulfur)
Ostroumov, E.A., 1957.
[Sulphur compounds in the bottom deposits of the Okhotsk Sea.] Trudy Inst. Okeanol., 22:139-157.

bottom sediments, chemistry of (sulfur)
Ostroumov, E.A., 1953.
[Different forms of combined sulfur in sediments of the Black Sea.] Trudy Inst. Okean. 7:80-90.

Abstr. in Chem. Abstr. 4346b, 1955.

bottom sediments, chemistry (sulfur)
Ostroumov, E.A., and L.S. Fomina, 1960
[On sulphuric compounds in the sediments of the northwest Pacific.] Trudy Inst. Okeanol., 32: 206-214.

bottom sediments, chem. H_2S
Ostroumov, E.A., and V.M. Shilov, 1956.
[Iron sulphide and hydrogen sulfide in bottom sediments of the northwestern part of the Pacific.] Dokl. Akad. Nauk, SSSR 106(3):501-504.

bottom sediments, chemistry (sulfur)
*Ostroumov, E.A., and I.I. Volkov, 1967.
Geochemical behaviour of sulfur in the bottom sediments of the Pacific. (In Russian; English abstract).
Trudy Inst. Okeanol., 83:68-82.

bottom sediments, chemistry, H_2S
Ostroumov, E.A., and I.I. Volkov, 1960
[Forms of sulphur compounds of bottom deposits in the Pacific Ocean near New Zealand] Trudy Inst. Okeanol, 42: 117-124.

sediments, chemistry (sulfur)
Ostroumov, E.A., I.I. Volkov and L.S. Fomina, 1961.
[Distribution of sulphur compounds in the bottom sediments of the Black Sea.]
Trudy Inst. Okeanol., Akad. Nauk, SSSR, 50:93-129.

bottom sediments, chemistry (Sulfate)
Shiskina, O.V., 1959
[Sulphate in the interstitial waters of the Black Sea.] Trudy Inst. Okeanol., 33: 178-193.

(index in English is incorrect)

bottom sediments, chemistry of
Sugawara, K., T. Koyama and A. Kozawa, 1954.
Distribution of various forms of sulphur in lake-, river- and sea-muds. II.
J. Earth Sci., Nagoya Univ., 2(1):1-4.

bottom sediments, chemistry of
Sugawara, K., T. Koyama and A. Kozawa, 1953.
Distribution of various forms of sulphur in lake rive and sea-muds. J. Earth Sci., Nagoya Univ., 1(1):17-23.

bottom sediments chem (sulfides)
Tomiyama, T., and Y. Kojima, 1953.
The errors and their removal in the determination of sulfide of bottom deposits.
Bull. Jap. Soc. Sci. Fish. 18(12):29-32.

bottom sediments, chem
Volkov, I.I., 1962.
The state of hydrogen sulphide in the water and sediments of the Black Sea.
Trudy Inst. Okeanol., Akad. Nauk, SSSR, 54:39-46.
In Russian; English summary

bottom sediments, chemistry
Volkov, I.I., 1961.
[Iron sulphide, their interrelations and transformations in the sediments of the Black Sea.]
Trudy Inst. Okeanol., Akad. Nauk, SSSR, 50:68-92.

bottom sediments, chemistry
Volkov, I.I., 1961.
[On free hydrogen sulphide and some products of its transformation in the sediments of the Black Sea.]
Trudy Inst. Okeanol., Akad. Nauk, SSSR, 50:29-67.

bottom sediments, chemistry
Volkov, I.I., 1960.
[Distribution of free hydrogen sulphide in the sediments of the Black Sea.]
Doklady Akad. Nauk, SSSR, 134(3):676-679.

bottom sediments, chemistry
Volkov, I.I., 1959
[Determination of different sulfuric compounds in sea sediments.] Trudy Inst. Okeanol., 33: 194-208.

bottom sediments, chemistry of (sulphates)
Volkov, I.I., and E.A. Ostroumov, 1960
[Distribution of the sulphates in the bottom sediments of the Pacific.]
Okean. Issle., IGY Com. SSSR:61-70.

bottom sediments, chemistry
Volkov, I.I., and E.A. Ostroumov, 1957.
[Concretions of iron sulphide in Black Sea deposits]
Doklady Akad. Nauk, SSSR, 116(4):645-648.

bottom sediments, chem.
Volkov, I.I., and E.A. Ostroumov, 1957.
[Determination of thiosulphate in silt waters of the Black Sea deposits.] Dokl. Akad. Nauk, SSSR, 114(4):853-855.

bottom sediments, chemistry
Watts, J.C.D., 1960
Sea-water as the primary source of sulfate in tidal swamp soils from Sierra Leone. Nature 186(721): 308-309.

bottom sediments, chemistry
Watts, J.C.D., 1960
Sulfate absorption by muds of relatively high organic content from the Sierra Leone River Estuary. Bull. de l'I.F.A.N. (2) 22(4):1153-1158.

bottom sediments, chemistry (tantalum)

bottom sediments, chemistry
Hamaguchi, Hiroshi, Rokuro Kuroda and Yoshichika Watanabe, 1963.
Tantalum contents of deep-sea sediments.
J. Chem. Soc. (Nippon Kagaku Zasshi), 84(9): 723-726. (Japan)

bottom sediments chemistry, thorium
Angino, Ernest, 1966
Uranium and thorium in Antarctic glacial marine sediments.
Oceanogr. Mag., Jap. Met. Soc., 18(1/2): 57-62.

bottom sediments, chemistry thorium
Antal, Paul S., 1966.
Diagenesis of thorium isotopes in deep-sea sediments.
Limnol. Oceanogr., 11(2):278-292.

bottom sediments, chemistry (thorium)
Bernat, Michel, and Edward D. Goldberg, 1969.
Thorium isotopes in the marine environment.
Earth Planet Sci. Letters, 5(5):308-312.

bottom sediments chemistry, thorium
Bonatti, Enrico, D.E. Fisher, Olva Joensuu and H.S. Rydell, 1971.
Postdepositional mobility of some transition elements, phosphorus, uranium and thorium in deep-sea sediments.
Geochim. Cosmochim. Acta, 35(2): 189-201.

Bottom sediments, chemistry (Thorium)
Koczy, F. F., 1962.
The mechanism of thorium enrichment in deep-sea sediments. (Abstract).
J. Geophys. Res., 67(4):1643-1644.

bottom sediments, chemistry (thorium)
Koczy, F.F., 1961
Ratio of thorium-230 to thorium-232 in deep-sea sediments.
Science, 134(3493): 1978-1979.

bottom sediments, chemistry (thorium)
Koczy, F. F., 1949.
Thorium in sea water and marine sediments. Geol. Fören. Förhandl. 71:238-242.

bottom sediments, chemistry (thorium)
Picciotto, E., and S. Wilgain, 1954.
Thorium determination of deep-sea sediments.
Nature 173(4405):632-633.

bottom sediments chemistry tin
Chester, R., 1965.
Elemental geochemistry of marine sediments.
In: Chemical oceanography, J.P. Riley and G. Skirrow, editors, Academic Press, 2:23-80.

bottom sediments, chemistry (tin)
Hill, Patrick Arthur, and Andrew Parker 1970.
Tin and zirconium in the sediments around the British Isles: a preliminary reconnaissance.
Econ. Geol. 65(4): 409-416.

bottom sediments, chemistry tin
Tatsumoto, M., 1957.
Chemical investigations of deep-sea deposits. 22. The contents of cobalt and nickel in sea sediment (2). 23. The contents of tin and lead in sea sediments. J. Chem. Soc., Japan, 78(1):38-48.

bottom sediments chemistry tin.
Wilson, Thomas A., 1965.
Offshore mining paves the way to ocean mineral wealth.
Engineering and Mining Jour., 166(6): 124-132.

bottom sediments, chemistry titanium
Arrhenius, G., 1952
Sediment cores from the East Pacific.
Repts. Swedish Deep-sea Exped., 1947-1948, 5(1): 227 pp., textfigs., with appendix of plates.

bottom sediments, chemistry titanium
Arrhenius, G., G. Kjellberg, and W.F. Libby, 1951.
Age determination of Pacific chalk ooze by radiocarbon and titanium content. Tellus 3: 222-229.

bottom sediments, chemistry titanium
Chester, R., 1965.
Elemental geochemistry of marine sediments.
In: Chemical oceanography, J.P. Riley and G. Skirrow, editors, Academic Press, 2:23-80.

bottom sediments, chemistry, titanium
Holmes, Charles W., 1969.
In the Gulf of Mexico ---- geochemical exploration produces exciting results: sea floor samples lead to a surprising discovery of zirconium, titanium and heavy minerals off the Texas coast.
Ocean Industry, 4(6):49-52. (popular)

bottom sediments, chemistry (titanium)
Isayeva, A.B. 1971.
Relation between titanium and iron in the sediments of the Indian Ocean. (In Russian).
Geokhimya 1971 (3):310-317.
Translation: Geochem. int. Am. geol. Inst. 8(2): 186-193.

titanium
Koczy, F.F., 1950.
Zur Sedimentation und Geochemie im aequatorischen Atlantischen Ozean. Medd. Oceanografiska Inst. Göteborg 17 (Göteborgs Kungl. Vetenskaps- och Vitterhets Samhälles Handlingar, Sjätte Följden Ser. B, 6(1)):44 pp., 17 textfigs.

bottom sediments, chemistry, titanium
Kullenberg, B., 1953.
Absolute chronology of deep-sea sediments and the deposition of clay on the ocean floor.
Tellus 5(3):302-305.

bottom sediments, chemistry of (titanium)
Ostroumov, E.A., 1956.
Titanium distribution in the deposits of the Okhotsk Sea. Geokhim. 1:90-96.

bottom sediments, chem. (titanium)
Ostroumov, E.A., 1956.
Titanium in deposits of the Okhotsk Sea.
Dokl. Akad. Nauk, SSSR, 107(3):444-447.

bottom sediments, chemistry (titanium)
Pilipchuk, M.F., 1965.
Titanium distribution in modern sediments of the Black Sea. (In Russian).
Dokl. Akad. Nauk SSSR 180(3): 715-718

bottom sediments - chemistry, titanium
Yalkovsky, Ralph, 1967.
Signs test applied to Caribbean Core A 176-6.
Science, 155(3768):1408-1409.

bottom sediments, chemistry, trace

bottom sediments, chemistry (trace elements)

bottom sediments, chemistry trace elements
Angino, Ernest E., and Robert S. Andrews, 1968.
Trace element chemistry, heavy minerals and sediment statistics of Weddell Sea sediments.
J. sedim. Petrol. 38(2): 634-642.

bottom sediments, chemistry trace elements
Arrhenius, Gustaf, 1967.
Deep-sea sedimentation: a critical review of U. S. work.
Trans. Am. geophys. Un., 48(2):604-631.

bottom sediments, chemistry
*Bender, Michael L. and Cynthia Schultz, 1969.
The distribution of trace metals in cores from a traverse across the Indian Ocean.
Geochim. cosmochim. Acta, 33(2):292-297.

bottom sediments, chemistry of (iodine)
Bennett, J.H., and O.K. Manuel, 1968.
On iodine abundances in deep-sea sediments.
J. geophys. Res., 73(6):2302-2303.

bottom sediments, chemistry trace elements
Brooks, R.R., B.J. Presley and I.R. Kaplan 1968.
Trace elements in the interstitial waters of marine sediments.
Geochim. cosmochim. Acta 32(4): 397-414.

bottom sediments, chemistry, trace elements
Chester, R., 1965.
Elemental geochemistry of marine sediments.
In: Chemical oceanography, J.P. Riley and G. Skirrow, editors, Academic Press, 2:23-80.

bottom sediments, chemistry, trace elements
Chester, R. and M.J. Hughes, 1969.
The trace element geochemistry of a North Pacific pelagic clay core. Deep-Sea Res., 16(6): 639-654.

bottom sediments, chemistry, trace elements
*Chester, R., and M.J. Hughes, 1967.
A chemical technique for the separation of ferro-manganese minerals, carbonate minerals and adsorbed trace elements from pelagic sediments.
Chem. Geol., 2(3):249-262.

bottom sediments, chemistry trace elements
*Chester, R., and R.G. Messiha-Hanna, 1970.
Trace element partition patterns in North Atlantic deep-sea sediments.
Geochim Cosmochim. Acta, 34(10):1121-1128.

bottom sediments, chemistry
Kulbicki, G., and J.L. Rumeau, 1967.
Influence du milieu sur les teneurs en éléments-traces de vases marines du Golfe de Gascogne.
Bull. Centre Rech. PAU-SNPA, 1(1):125-141.

Bottom Sediments, Chemistry (trace elements)
Landergren, Sture and Liva Joensuu, 1965.
Studies on trace element distribution in a sediment core from the Pacific Ocean.
Progress in Oceanography, 3:179-189.

abstract

bottom sediments, trace elements
*Nichols, G.D., 1967.
Trace elements in sediments: an assessment of their possible utility as depth indicators.
Marine Geol., 5(5/6):539-555.

bottom sediments, chemistry (trace elements)
Turekian, Karl K., and John Imbrie 1966
The distribution of trace elements in deep-sea sediments of the Atlantic Ocean.
Earth Planet. Sci. Letts 1 (4): 161-168.

bottom sediments, trace elements
Venkatarathnam, K. and V.V.S.S. Tilak, 1969.
Trace elements in the sediments of the continental shelf off certain parts of East Coast of India. Bull. natn. Inst. Sci., India, 38(1): 463-466.

bottom sediments chem. (trace elements)
Young, E., 1954.
Trace elements in recent marine sediments. (Abstract). Bull. G.S.A. 65(2):1329.

bottom sediments, chemistry (tungsten)

bottom sediments, chemistry
Pilipchuk, M.F., and I.I. Volkov, 1966.
Tungsten in modern sediments of the Black Sea. (In Russian).
Doklady, Akad. Nauk, SSSR, 167(2):430-433.

bottom sediments, chemistry, uranium

bottom sediments, chemistry (uranium)

bottom sediments, chemistry uranium
Angino, Ernest, 1966.
Uranium and thorium in Antarctic glacial marine sediments.
Nankyoku Shiryo (Antarctic Rec.), 15(1/2): 57-62. [Laznoga Mag., Jap. Met. Soc., 15(1/2):57-62]

bottom sediments chemistry, uranium
Baturin, G.N. 1971.
Uranium in ooze solutions of deposits in the southeastern part of the Atlantic Ocean. (In Russian).
Dokl. Akad. Nauk SSSR 198(5): 1186-1188

bottom sediments, chemistry (uranium)
Baturin, G.N., 1969.
Uranium in the surface sediment layer of the northwestern Indian Ocean. (In Russian; English abstract). Okeanologiia, 9(6): 1031-1037.

bottom sediments, chemistry (uranium)
Baturin, G.N., 1968.
Geochemistry of uranium in the Baltic.
Geokhimiya, 1968(3): 377-381
Translation: Geochem. int. 5(2): 344-345.

bottom sediments, chemistry (uranium)
Baturin, G.N., A.V. Kochenov and Yu. M. Senin 1971.
Uranium concentration in Recent ocean sediments in zones of rising currents. (In Russian).
Geokhimya 1971 (4): 456-462.
Translation: Geochim. int. Am. geol. Inst. 8(2): 281-286

bottom sediments, chemistry uranium
Bertine, K.K., L.H. Chan, and K.K. Turekian 1970.
Uranium determinations in deep-sea sediment and natural waters using fission tracks.
Geochim. cosmochim. Acta, 34(6): 641-648.

bottom sediments chemistry, uranium
Bonatti, Enrico, D.E. Fisher, Olva Joensuu and H.S. Rydell, 1971.
Postdepositional mobility of some transition elements, phosphorus, uranium and thorium in deep-sea sediments.
Geochim. cosmochim. Acta, 35 (2): 189-201.

chemistry, uranium
Boström, Kurt and David E. Fisher 1971.
Volcanogenic uranium, vanadium and iron in Indian Ocean sediments.
Earth planet. Sci. Letts 11(2): 95-98.

bottom sediments, chemistry, uranium
Fisher, David E. and Kurt Boström, 1969.
Uranium rich sediments on the East Pacific Rise.
Nature, Lond., 224 (5214): 64-65

bottom sediments, chemistry (uranium)
Goldsztein, Marcel, and Joaquin Ros, 1963.
Sur la teneur en uranium des sediments du bassin occidental de la Mediterranee.
Bull. Inst. Oceanogr., Monaco, 60(1267):19 pp.
Also:
Int. Atomic Energy Agency, Radioactivity in the Sea, Publ., No. 3.

bottom sediments, chemistry (uranium)
Goldzstein, M., et J. Ros, 1963.
Sur la teneur en uranium des sédiments du bassin occidental de la Méditerranée.
Rapp. Proc. Verb., Réunions, Comm. Int., Expl. Sci., Mer Méditerranée, Monaco, 18(3):1021-1028.

bottom sediments, chemistry (uranium)
*Kochenov, A.V., and G.N. Baturin, 1967.
Uranium distribution in the bottom sediments of the Aral Sea. (In Russian; English abstract).
Okeanologiia, Akad. Nauk, SSSR, 7(4):623-627.

bottom sediments, chemistry uranium
Kochenov, A. V., G. N. Baturin, S. A. Kovaleva, Ye. M. Yemel'yanov and K. M. Shimkus, 1965.
Uranium and organic matter in the sediments of the Black and Mediterranean seas.
Geokhimiia, 1965(3):302-313.
Translation AGI $4.35

bottom sediments chemistry, uranium
Kuznetsov, Yu. A., Z. N. Simonyak, A. P. Lisitsyn and M.S. Frenklikh, 1968.
Uranium and radium in the surface layer of oceanic sediments. (In Russian)
Geokhimya 1968(3): 323-333.
Translation: Geochem. int. 5(2): 306-311

bottom sediments, chemistry, uranium
Legin, V.K., Yu. V. Kuznetsov and K.F. Lazarev, 1966.
Forms of uranium occurrence in marine sediments. (In Russian).
Geokhimiya, (5): 606-608.
Abstract in: Geochem. Int. 3(3): 468
Translation: AGI. $1.20.

bottom sediments, chemistry, uranium
Mo, Tin, Barbara C. O'Brien and A.D. Suttle, Jr., 1971.
Uranium: further investigation of uranium content of Caribbean cores P6304-8 and P6304-9.
Earth Planet. Sci. Letts 10 (2): 175-178.

bottom sediments, chemistry, uranium
Rydell, Harold and David E. Fisher, 1971.
Uranium content of Caribbean core P6304-9.
Bull. mar. Sci., Coral Gables, 21(4):787-789.

bottom sediments, chemistry, uranium
Strom, K. M., 1948.
A concentration of uranium in black muds.
Nature 162:922.

bottom sediments, chemistry, uranium
Sugimura, Y., and T. Sugimura, 1962.
Uranium in recent Japanese sediments.
Nature, 194(4828):568-569.

bottom sediments, chemistry (vanadium)

bottom sediments, chemistry, vanadium
Boström, Kurt and David E. Fisher 1971.
Volcanogenic uranium, vanadium and iron in Indian Ocean sediments.
Earth planet. Sci. Letts 11(2): 95-98.

vanadium
Ishibashi, Masayoshi, Shungo Ueda and Yoshikazu Yamamoto, 1970.
The chemical composition and the cadmium, chromium and vanadium contents of shallow-water deposits in Tokyo Bay. (In Japanese; English abstract)
J. oceanogr. Soc. Japan 26(4): 189-194

bottom sediments, chemistry (vanadium)
Ostroumov, E.A., and O.M. Siline, 1952.
On certain regularity in the distribution of vanadium in modern marine deposits.
Dokl. Akad. Nauk, SSSR, 86:365-367.

bottom sediments, chemistry vanadium
Ueda, S., 1957.
Chemical studies on the ocean. 68. Chemical studies of the shallow-water deposits. 21. Vanadium and Chromium contents of the shallow-water deposits. (2)(2).
J. Oceanogr. Soc., Japan, 13(3):93-98;99-106.

bottom sediments, chemistry, vanadium
Vasilchikov, N.V., Yu, A. Pavlidis and N.P. Slovinsky-Sidak, 1966.
On the vanadium titanomagnetic sea-beach placers in the Far East. (In Russian; English abstract).
Okeanologiia, Akad. Nauk, SSSR, 6(5):823-829.

bottom sediments, chemistry (vitamins)

Bottom sediments, chem. (vitamins)
Kurata, Akira, 1970.
Vitamin B_{12} transudation from marine bottom muds. (In Japanese; English abstract).
J. oceanogr. Soc., Japan, 26 (2): 81-86.

bottom sediments, chemistry vitamins
Kurata, Akira, 1969.
Vitamin B_{12} content in bottom muds. (In Japanese, English abstract)
J. oceanogr. Soc. Japan, 25(2): 103-108.

bottom sediments, vitamins
Ohwada, Konichi, and Nobuo Taga, 1969.
Distribution of vitamin B_{12}, thiamine and biotin in marine sediments.
J. oceanogr. Soc., Japan, 25(3): 123-136.

bottom sediments, chemistry (zinc)

bottom sediments, chemistry (zinc)
Belova, I.V. 1970.
Zinc in Recent Black Sea sediments (In Russian).
Dokl. Akad. Nauk SSSR 193 (2): 433-436.

bottom sediments, chemistry (zinc)
Ishibashi, Masayoshi, Shunzo Ueda, Yoshikazu Yamamoto, 1959
Studies on the utilization of the shallow-water deposits (Cont.). On the zinc content of the shallow water deposits. Rec. Oceanogr. Wks., Japan, Spec. No. 3: 123-134.

bottom sediments, chemistry (zirconium)
Glazoleva, M.A. 1970.
Zirconium in Recent sediments of the Black Sea. (In Russian).
Dokl. Akad. Nauk SSSR 193(1): 184-187.

bottom sediments, chemistry (zirconium)
Hill, Patrick Arthur, and Andrew Parker 1970.
Tin and zirconium in the sediments around the British Isles: a preliminary reconnaissance.
Econ. Geol. 65 (4): 409-416.

bottom sediments, chemistry (zirconium)
Holmes, Charles W. 1971.
Zirconium on the continental shelf - possible indicator of ancient shoreline deposition.
Prof. Pap. U. S. Geol. Surv. 750-C: C7-C12.

bottom sediments, chronology of

bottom sediments, chronology of
Goldberg, E.D. and Koide, M., 1958
Ionium-thorium chronology in deep sea sediments of the Pacific. Science Vol. 128 No. 3330 p. 1003.

bottom sediments, chronology
Kullenberg, B., 1953.
Absolute chronology of deep-sea sediments and the deposition of clay on the ocean floor.
Tellus 5(3): 302-305.

bottom sediments, chronology
Lapierre, Francis, 1966.
Répartition et chronologie relative des sédiments sableux dans le golfe de Gascogne.
C.R. hebd. Séanc. Acad. Sci., Paris, (D):263: 1044-1047.

bottom sediments, chronology
Pettersson, H., 1950.
The chronology of the ocean floor. Advancement Sci. 7(25):72-75.

Bottom Sediments
Rona, Elizabeth, L.K. Akers, John E. Noakes, and Irwin Supernaw, 1965.
Geochronology in the Gulf of Mexico, Part 1.
Progress in Oceanography, 3:289-295.

bottom sediments classification

bottom sediments, classification of
Elliott, R.G., 1965.
A classification of subaqueous sedimentary structures based on rheological and kinemetical parameters.
Sedimentology, 5(3):193-209.

bottom sediments classification
Leclaire, Lucien, 1968.
Contribution à l'étude des sédiments marins non consolidés. Les bases d'une nouvelle classification.
C.r.hebd.Séanc.Acad.Sci., Paris, (D), 266(6):563-565.

bottom sediments, clastic
Masse, Jean-Pierre, 1970.
Contribution à l'Etude des sédiments bioclastiques actuels du complexe récifal de l'ile de Nossi-Bé (N.W. de Madagascar).
Rec. Trav. Sta. mar. Endoume, hors sér. Suppl. 10: 229-251

bottom sediments, classification
Panov, D.G., 1956.
On the genetic classification of the world ocean bottom. Dokl. Akad. Nauk, SSSR, 108(6):1061-1064.

bottom sediments, classification
Shepard, F.P., 1954.
Nomenclature based on sand- silt-clay ratios.
J. Sed. Petrol. 24:151-158.

bottom sediments, classification
Zubov, N.N., 1950.
Classification of sedimentary rocks and bottom deposits by their physical composition. Iz. v-s Geograf. Obshch. 82(2:191-198.

FDD 175T18
GG 202/2

bottom sediments, clays

bottom sediments, clay
Arrhenius, G., and E.D. Goldberg 1955.
Distribution of radioactivity in pelagic clays.
Tellus 7(2): 226-231.

bottom sediments, clay
Athearn, W.D., 1957.
Comparison of clay from the continental shelf off Long Island with the Gardiners clay. J. Geol. 65(4):448-449.

bottom sediments, clays
Powers, M.C., 1957.
Adjustment of land derived clays to the marine environment. J. Sed. Petr., 27(4):355-372.

bottom sediments, clay
Sastry, A.V.R., M. Poornachandra Rao and C. Mahadevan, 1958
Differential thermal analysis of plastic clays from the deltaic regions of Godavari and Krishna Rivers.
Andhra Univ. Mem. Oceanogr., 2:

bottom sediments, clay
Yermolayev, M.M. 1948.
Lithogenesis of plastic clay sediments in seas. (In Russian)
Izv. Akad. Nauk SSSR Ser. Geol. (1):121-138

bottom sediments, clay minerals
Berry, Richard W., and William D. Johns, 1966.
Mineralogy of the clay-sized fractions of some North Atlantic-Arctic Ocean bottom sediments.
Geol. Soc., Amer., Bull., 77(2):183-196.

bottom sediments, clay minerals
Dietz, R.S., 1942.
Clay minerals in recent sediments.
Amer. Minerol., 37:219-220.

bottom sediments, clay minerals
Gorbunova, Z.N., 1966.
Distribution of clay minerals in the sediments of the Indian Ocean. (In Russian; English abstract).
Okeanologiia, Akad. Nauk, SSSR, 6(2):267-275.

bottom sediments, clay minerals
Gorbunova, Z.N., 1962.
Clayey and other high-dispersed minerals in the sediments of the Bering Sea. (In Russian).
Okeanologiia, Akad. Nauk, SSSR, 2(6):1024-1034.

bottom sediments, clay minerals
Griffin, G.M., 1962
Regional clay-mineral facies--products of weathering intensity and current distribution in the northeastern Gulf of Mexico.
Geol. Soc. Amer. Bull., 73(6):737-768.

bottom sediments
Grim, R.E., R.S. Dietz, and W.F. Bradley, 1949.
Clay-mineral composition of some sediments from the Pacific Ocean of the California coast and Gulf of California. Bull. G.S.A. 60(11):1785-1808, 22 textfigs.

bottom sediments, clay minerals
Jacobs, Merian B., 1970.
Clay mineral investigations of Cretaceous and Quaternary deep-sea sediments of the North American Basin.
J. sedim. Petrol. 40(3): 864-868.

bottom sediments, clay minerals
Kobayashi, Kazuo, Kaoru Oinuma and Toshio Sudo, 1961.
20. Clay minerals in the core sample of deep sea sediment from Japan Trench. (English abstract).
Repts. JEDS, 2(16):

bottom sediments, clay minerals
Kobayashi, K., K. Oinuma and T. Sudo, 1960.
Clay mineralogical study on recent marine sediments (I),(II).
Oceanogr. Mag., Tokyo, 11(2):215-224.

bottom sediments, clay minerals
Murray, H.H., and J.L. Harrison, 1956.
Clay mineral composition of recent sediments from Sigsbee Deep. J. Sed. Petr. 26(4):363-368.

bottom sediments, clay minerals
Naidu, A.S., D.C. Burrell and D.W. Hood 1971.
Clay mineral composition and geologic significance of some Beaufort Sea sediments.
J. sedim. Petrol. 41(3): 691-694.

bottom sediments clay minerals
Oinuma, Kaoru, Kazuo Kobayashi, and Toshio Sudo, 1960.
Clay mineralogical study on recent marine sediments (II)-Sample JEDS-2-C2. (abstract). Repts. of JEDS, 1:223.

Also in Oceanogr. Mag. 11(2):223.

bottom sediments, clay minerals
Yeroshchev-Shak, V.A., 1962
Clayey minerals in the sediments of the Atlantic Ocean.
Okeanologiia, Akad. Nauk, SSSR, 2(1):98-105.

Abstracted in: Soviet Bloc Res., Geophys. Astron. and Space, 1962(35):20-21
(OTS61-11147-35 JPRS13739)

bottom sediments, clay minerals
Eroshchev-Shak, V.A., 1961.
Illite in the sediments of the Atlantic.
Doklady Akad. Nauk, SSSR, 137(4):951-953.

bottom sediments, coastal

bottom sediments, littoral
Adams, Rodney D., 1968.
Distribution of littoral sediment on the windward coast, St. Vincent, W.I. Trans. Fifth Carib. Geol. Conf., St. Thomas, V.I., 1-5 July, 1968, Peter H. Mattson, editor. Geol. Bull. Queens Coll., Flushing 5: 55-59.

bottom sediments, coastal
Inman, D.L., W.R Gayman, and D.C Cox, 1963
Littoral sedimentary processes on Kauai, a subtropical high island.
Pacific Science, 17(1):106-130.

bottom sediments, coastal
Reineck, Hans-Erich, 1962
Reliefgüsse ungestörter Sandproben.
Zeits. für Pflanzenernährung, Düngung, Bodenkunde 99(144) (2/2):151-153.

bottom sediments color of

bottom sediments, color of
Nayudu, Y.R., and B.J. Enbysk, 1964.
Bio-lithology of northeast Pacific surface sediments.
Marine Geol., 2(4):310-342.

bottom sediments, color
*Pantin, H.M., 1969.
The appearance and origin of colours in the muddy marine sediments around New Zealand.
N.Z. Jl.Geol.Geophys. 12:51-66.

bottom sediments, color
* Stanley, Daniel J., 1969.
Atlantic continental shelf and slope of the United States-color of marine sediments.
Prof. Pap., U.S. Geol. Surv. 529-D: D1-D15., foldin.

bottom sediments, compaction of

bottom, sediments, compaction of
Meade, Robert H., 1966.
Factors influencing the early stages of the compaction of clays and sands- review.
J. Sedim. Petrol., 36(4):1085-1101.

bottom sediments, compaction of
Meade, Robert H., 1964.
Mechanics of aquifer system. Removal of water and rearrangement of particles during the compaction of clayey sediments. Review.
U.S. Geol. Survey Prof. Paper, 497-B:1-22.

bottom sediments, compaction of
Parasnis, D.S., 1960
The compaction of sediments and its bearing on some geophysical problems.
Geophys. J., R. Astron. Soc., 3(1): 1-28.

bottom sediments, compactness
McMaster, Robert L., 1962.
Seasonal variability by compactness in marine sediments: a laboratory study.
Bull. Geol. Soc., Amer., 73(5):643-646.

bottom sediments, conductivity of
bottom sediments, consolidation of

bottom sediments, consolidation of
*Bryant, William R., Paul Cernock and Jack Morelock, 1967.
Shear strength and consolidation characteristics of marine sediments from the western Gulf of Mexico.
In: Marine geotechnique, A.F. Richards, editor, Univ. Illinois Press, 41-62.

bottom sediments, consolidation of
*McClelland, Bramlette, 1967.
Progress of consolidation in delta front and prodelta clays of the Mississippi River.
In: Marine geotechnique, A.F. Richards, editor, Univ. Illinois Press, 22-40.

bottom sediments, consolidation of
Richards, A.F., 1961
Investigations of deep-sea sediment cores. 1. Shear strength, bearing capacity and consolidation.
U.S. Navy Hydrogr. Off., Techn. Rept., TR-63: 70 pp.

bottom sediments, coral detritus
Deneufbourg, G., 1971.
Étude géologique du port de Papeete Tahiti (Polynésie Française). Cah. pacifiques 15: 75-82.

bottom sediments (data included)
bottom sediments, data only

bottom sediments, data only
Canada, University of British Columbia, Institute of Oceanography, 1962.
Data report #20 of sediment grain size analyses, 1960-1961, 12 pp.

bottom sediments, data only
Canada, University of British Columbia, Institute of Oceanography, 1963.
Sediment grain-size analyses, 1951, 1960, 1962.
Data Report, No. 22: 6 pp. (Multilithed)

bottom sediments (data only)
India, Naval Headquarters, Office of Scientific Research & Development, 1958-1959.
Indian Oceanographic Stations List, Serial Nos. 3-5 (June 1958; Jan.&June 1959): unnumbered pp. (mimeographed).

bottom sediments (data only)
Republic of China, Chinese National Committee on Oceanic Research, Academia Sinica, 1968.
Oceanographic Data Report of CSK, 2: 126 pp.

bottom sediments (data only)
U.S. Navy Hydrographic Office, 1960.
Oceanographic observations, Arctic waters, Task Force Five and Six, summer-autumn 1956, USS Requisite (AGS-18), USS Eldorado (AGC-11), USS Atka (AGB-3) and USCGC Eastwind (WAGB-279).
U.S. Navy Hydrogr. Off., Techn. Rept., TR-58: 89 pp.

bottom sediments, decomposition of
bottom sediments, deep sea

bottom sediments, deep-sea
*Broecker, Wallace S., David L. Thurber, John Goddard, Teh-Lung Ku, R.K. Matthews and Kenneth J. Mesolella, 1968.
Milankovitch hypothesis supported by precise dating of coral reefs and deep-sea sediments.
Science, 159(3812): 297-300.

bottom sediments, deep-sea
Davis, G.H., 1960
Thickness and consolidation of deep-sea sediments: a discussion. Bull. Geol. Soc., Amer., 71(11): 1727-1728.

bottom sediments, deep-sea
Dietz, Robert S., and John C. Holden, 1966.
Deep-sea deposits in but not on the continents.
Bull. Amer. Assoc. Petr. Geol., 50(2): 351-362.

bottom sediments, deep-sea
Dymond, Jack R., 1966.
Potassium-argon geochronology of deep-sea sediments.
Science, 152(3726): 1239-1241.

bottom sediments, deep-sea
El Wakeel, S.K., and J.P. Riley, 1961.
Chemical and mineralogical studies of deep-sea sediments.
Geochimica et Cosmochimica Acta, 25(2): 110-146.

bottom sediments, deep-sea
Emilianí, C., 1963
Deep-sea sediments.
United States National Report, 1960-1963. Thirteenth General Assembly, I.U.G.G.
In: Trans. Amer. Geophys. Union, 44(2): 495-498.

bottom sediments, deep-sea
Emiliani, C., 1960
Deep-sea sediments. Trans. Am. Geophys. Un. 41(2): 275-279.

bottom sediments, deep-sea
Emiliani, Cesare, and John D. Milliman, 1966.
Deep-sea sediments and their geological record.
Earth-Sci. Rev., 1 (2/3): 105-132.

bottom sediments, deep-sea
Ericson, D.B., W.S. Broecker, J.L. Kulp and G. Wallin, 1956.
Late-Pleistocene climates and deep-sea sediments.
Science 124(3218): 385-389.

bottom sediments, deep-sea
Ericson, D.B., M. Ewing, and B.C. Heezen, 1951.
Deep-sea sands and submarine canyons. Bull. G.S.A 62(8): 961-966, 1 textfig.

bottom sediments, deep-sea
Ericson, B.B., Maurice Ewing, Gösta Wollin and Bruce C. Heezen, 1961.
Atlantic deep-sea sediment cores.
Bull. Geol. Soc., Amer., 72(2): 193-286.

bottom sediments, deep-sea
Eriksson, K. Gösta, 1967.
Some deep-sea sediments in the western Mediterranean Sea.
Progress in Oceanography, 4: 267-280.

bottom sediments, clay
Ermolaev, M.M., 1950.
Lithogenesis of plastic clay marine sediments.
Izvest. Akad. Nauk, USSR, Ser. Geol., 1: 121-138.

Bottom sediments, deep-sea
*Foster, John H. and Neil D. Opdyke, 1970.
Upper Miocene to Recent magnetic stratigraphy in deep-sea sediments. J. geophys. Res., 75 (23): 4465-4473.

bottom sediments, deep-sea
Fredriksson, K., and L.R. Martin, 1963.
The origin of black spherules found in Pacific islands, deep-sea sediments and Antarctic ice.
Geochim. et Cosmochim. Acta, 27(3): 245-248.

bottom sediments, deep-sea
Hamilton, Edwin L., 1961.
On the consolidation and thickness of deep-sea sediments. (Abstract).
Proc. Ninth Pacific Sci. Congr., Pacific Sci. Assoc., (Geol.-Geophys.): 340.

bottom sediments, deep-sea
Kermabon, A., C. Gehin, P. Blavier and B. Tonarelli, 1969.
Acoustic and other physical properties of deep-sea sediments in the Tyrrhenian Abyssal Plain.
Marine Geol. 7(2): 129-145.

bottom sediments, deep
Kuenen, Ph. H., 1964
Deep-sea sands and ancient turbidites.
In: Turbidites, A.H. Bouma and A. Brouwer, Editors, Developments in Sedimentology, Elsevier Publishing Co., 3: 3-33.

bottom sediments, deep
Kukla, J., 1969.
The cause of the Holocene climatic change.
Geologie Mijnb., 48(3): 307-334.

bottom sediments, deep
Miyake, Y., and Y. Sugimura, 1961.
Lionium-thorium chronology of deep-sea sediments of the western North Pacific Ocean.
Science, 133 (3467): 1823-1824.

bottom sediments, deep-sea
*Morgenstein, Maury, 1967.
Authigenic cementation of scoriaceous deep-sea sediments west of the Osciety Ridge, South Pacific.
Sedimentology 9(2): 105-118.

bottom sediments, deep
Ninkovich, Dragoslav, Bruce C. Heezen, John R. Conolly and Lloyd H. Burckle, 1964.
South Sandwich tephra in deep-sea sediments.
Deep-Sea Res., 11(4): 605-619.

bottom sediments, deep-sea
Ninkovich, Dragoslav, Neil Ophyke, Bruce C. Heezen and John H. Foster, 1966.
Paleomagnetic Stratigraphy, rates of deposition and tephrachronology in North Pacific deep-sea sediments.
Earth Planet. Sci. Letters, 1(6): 476-498.

bottom sediments, deep-sea
Öpik, E.J., 1955.
Cosmic sources of deep-sea deposits. Nature 176(4489): 926-927.

bottom sediments, deep-sea
Sackett, William M., 1964.
Deep-sea sediment accumulation rates. (Abstract only).
Geol. Soc., Amer., Spec. Geol. Soc., Amer. Paper, No. 76: 142.

bottom sediments, deepwater
Stehman, Charles F., 1970.
Eocene deep water sediment from the northeast Providence Channel, Bahamas.
Marit. Sediments, 6(2): 65-67.

bottom sediments, diagenesis
Tageeva, N. V., 1960.
Study of diagenesis of marine sediments.
Morsk. Geol., Doklady Sovetsk. Geol., 21 Sess. Int. Geol. Congress, 140-153.

bottom sediments, dispersion of

bottom sediments, dispersion of
*Murray, Stephen Patrick, 1967.
Control of grain dispersion by particle size and wave state.
J. Geol., 75(5): 612-634.

bottom sediments, dispersal of
*Needham, H.D., D. Habib and B.C. Heezen, 1969.
Upper carboniferous palynomorphs as a tracer of red sediment dispersal patterns in the northwest Atlantic.
J. Geol. 77(1): 113-120.

bottom sediments, effect of

bottom sediments, effect of
Cullen, D.J., 1962
The influence of bottom sediments upon the distribution of oysters in Foreaux Strait, New Zealand.
New Zealand. J. Geol. & Geoph., 5(2):271-275.

bottom sediments, effect of
Brandt, H., 1960.
Factors affecting compressional wave velocity in unconsolidated marine sand sediments.
J. Acoust. Soc., Amer., 32(2):171-179.

bottom sediments, effect of
Gade, H.G., 1959.
Notes on the effect of elasticity of bottom sediments to the energy dissipation of surface waves in shallow water.
Arch. Math. Natur., BLV, No. 3

bottom sediments, effect of
Hallam, A., 1965.
Environmental causes of stunting in living and fossil marine invertebrates.
Paleontology, 8(1):132-155.

bottom sediments, effect of
Hoshino, M., 1956.
On the relation between abnormal distribution of blue mud and fishing ground.
J. Ocean. Soc., Japan, 12(4):103-108.

bottom sediments, effect of
Imai, Y., 1961
Oceanographical studies on the behaviour of chemical elements. V. On the influence of land water and sediment upon the quantityof inorganic phosphate contained in the sea water of Urado Bay. (In Japanese; English abstract).
J. Oceanogr. Soc., Japan, 17(3):157-160.

bottom sediments, effect of
Johnson Ralph Gordon, 1971.
Animal-sediment relations in shallow water benthic communities.
Mar. Geol. 11 (2): 93-104.

bottom sediments, effect of
*Nakai, Zinziro, Masaya Kosaka, Fumio Hayashida, Tadashi Kubota and Masahiro Ogura, 1967.
Interrelationship between natural production of benthos and stability of the bottom. I. The condition in winter. (In Japanese; English abstract).
J.Coll.mar.Sci.Techn.,Tokai Univ., (2):161-172.

bottom sediments, effect of
McNulty, J.K., R.C. Work and H.B. Moore, 1962.
Some relationships between the infauna of the level bottom and the sediment in south Florida.
Bull. Mar. Sci., Gulf and Caribbean, 12(3):322-332.

bottom sediments, effect of
Okuda, T., 1955.
On the soluble nutrients in bay deposits. III. Examinations on the diffusion of soluble nutrients to sea water from mud.
Bull. Tohoku Reg. Fish. Res. Lab., (4):215-242.

bottom sediments, effect of
Okuda, T., 1955.
On the soluble nutrients in bay deposits. IV. An experiment of the behavior of phosphate-phosphorus between mud and sea water.
Bull. Tohoku Reg. Fish. Res. Lab., (5):79-81.

bottom sediments, effect of
Scheltema, Rudolf S., 1961.
Metamorphosis of the veliger larvae of Nassarius obsoletus (Gastropoda) in response to bottom sediment.
Biol. Bull., 120(1):92-109.

bottom sediments, effect of
Warren, P.J., and R.W. Sheldon, 1968.
Association between Pandalus borealis and fine-grained sediments off Northumberland.
Nature, Lond., 217(5128):579-580.

bottom sediments, elastic properties
Hamilton, Edwin L., 1971.
Prediction of in-situ acoustic and elastic properties of marine sediments.
Geophysics 36 (2): 266-284.

bottom sediments, elasticity of
Hamilton, Edwin L., 1970.
Elastic properties of marine sediments. J. geophys. Res., 76(2): 579-604.

bottom sediments electrical conductivity

bottom sediments, electrical conductivity of
*Bannister, Peter R., 1968.
Determination of the electrical conductivity of the sea bed in shallow waters.
Geophysics 33(6):995-1003.

bottom sediments, electrical conductivity
Hutt, Jeremy R., and Joseph W. Berg, Jr. 1968
Thermal and electrical conductivity of sandstone rocks and ocean sediments
Geophysics 33 (3): 489-500.

bottom sediments, electrical conductivity of
Teramoto, Toshihiko, 1971.
Estimation of sea-bed conductivity and its influence on velocity measurements with towed electrodes
J. oceanogr. Soc. Japan, 27 (1): 7-18

bottom sediments, estuarine

bottom sediments, estuarine
Francis - Boeuf, C., 1946.
Recherches sur le milieu fluviomarin et les dépôts d'estuaire. Annales Inst. Océanogr. 23(3):

bottom sediments, estuarine
Guilcher, André, 1967.
Origin of sediments in estuaries.
In: Estuaries, G.H. Lauff, editor, Publs Am. Ass. Advmt Sci., 83:149-157.

bottom sediments estuarine
Kestner, Friedrich Julius Theodor, 1961
Short-term changes in the distribution of fine sediments in estuaries.
Proc. Inst. Civil Eng., 19:185-208.

bottom sediments, estuarine
Klein, George deVries, 1967.
Comparison of Recent and ancient tidal flat and estuarine sediments.
In: Estuaries, G.H. Lauff, editor, Publs Am. Ass. Advmt Sci., 83:207-218.

bottom sediments, estuarine
McMaster, Robert L., 1967.
Compactness variability of estuarine sediments: an in situ study.
In: Estuaries, G.H. Lauff, editor, Publs Am. Ass. Advmt Sci., 83:261-267.

bottom sediments (estuaries)
Postma, H., 1967.
Sediment transport and sedimentation in the estuarine environment.
In: Estuaries, G.H. Lauff, editor, Publs Am. Ass. Advmt Sci., 83:158-179.

bottom sediments evaporites

bottom sediments, evaporite
Kendall, Christopher G. St.C. and Patrick A. D'E. Skipwith, 1969.
Holocene shallow-water carbonate and evaporite sediments of Khor al Bazam, Abu Dhabi southwest Persian Gulf.
Bull. Am. Ass. Petr. Geol., 53 (4): 841-869.

bottom sediment flows

Bottom sediment flows
Courtois, G., et G. Sauzay 1966.
Les méthodes de bilan des taux de comptage de traceurs radioactifes appliquées à la mesure des débits massiques de charriage.
Le Houille blanche (3):279-289.

bottom sediments, freshwater

bottom sediments, freshwater
Frenkel, Lerry, and H.F. Thomas, 1966.
Evidence of freshwater lake deposits in Block Island Sound.
J. Geol., 74(2):240-242.

Bottom sediments, geochemistry
See: bottom sediments, chemistry

bottom sediments glacial

bottom sediments, glacial
Brodie, J.W., 1965.
Oceanography.
In: Antarctica, Trevor Hatherton, editor, Methuen & Co., Ltd., 101-127.

bottom sediments, glacial
Costello, W.R., 1970.
River channel and tidal bar deposits Gaspe Nord, Quebec.
Marit. Sediments, 6 (2): 68-71

bottom sediment, glacial
Dangeard, L., and M. Rioult, 1965.
Le domaine de la géologie marine et ses frontières: confrontation de l'océanographie et du géologie.
In: Submarine geology and geophysics, Colston Papers, W.F. Whittard and R. Bradshaw, editors, Butterworth's, London, 93-105.

bottom sediments, glacial
Griggs, G.B., and L.D. Kulm, 1969.
Glacial marine sediments from the northeast Pacific.
J. sedim. Petrol. 39 (3): 1142-1148.

bottom sediments, glacial
Harbison, Reginald N., 1968
Possible moraine deposits in the Gulf of Maine
Marit. Sediments 5 (1): 19-21

bottom sediments, glacial
Pratt, Richard M., 1968.
Atlantic continental shelf and slope of the United States - physiography and sediments of the deep-sea basin.
Prof.Pap.U.S.geol.Surv., 529B:B1-B44, fold in chart.

bottom sediments, glacio-marine
Sharma, Ghanshyam D., 1970.
Sediment-seawater interaction in glaciomarine sediments of southeast Alaska.
Bull. geol. Soc. Am. 81(4):1097-1106.

bottom sediments, glacial
*Stanley, Daniel J., 1968.
Reworking of glacial sediments in the North West Arm, a fjoid -like inlet on the southeast coast of Nova Scotia.
J. sedim. Petr., 38(4):1224-1241.

bottom sediments, glacial marine
Warnke, Detlef A., 1970.
Glacial erosion, ice rafting, and glacial-marine sediments: Antarctica and the Southern Ocean.
Am. J. Sci., 269(3): 276-294.

bottom sediments, graded

bottom sediments, graded bedding
Warme, John E. 1967.
Graded bedding in the Recent sediments of Mugu Lagoon, California.
J. sedim. Petrol. 37(2): 540-547.

bottom sediments, grain size

bottom sediments, grain size
Allen, George P., 1971.
Relationship between grain size parameter distribution and current patterns in the Gironde Estuary (France)
J. sedim. Petrol. 41(1): 74-88.

bottom sediments, grain size
Athearn, William D., 1967.
Estimation of relative grain size from sediment clouds.
In: Deep-sea photography, J.B. Hersey, editor, Johns Hopkins Oceanogr. Studies, 3:173-176.

bottom sediments, grain size
Boyce, R.E., 1972.
Grain size analyses, leg 9. Initial Repts Deep Sea Drilling Project 9: 779-796.

bottom sediments, clay minerals
Carroll, Dorothy, 1964.
Chlorite in sediments off the Atlantic coast of the United States. (Abstract).
Geol. Soc., Amer., Special Paper, No. 76:239-240

bottom sediments, grain size
Curray, J.R., 1956.
Dimensional grain orientation studies of recent coastal sands.
Bull. Amer. Assoc. Petr. Geol., 40(10):2440-2456.

bottom sediments, grain size
Davis, John C., 1970.
Information contained in sediment-size analyses.
Mathem. Geol. 2(2): 105-112

bottom, sediments, grain size
Donahue, Jessie G., Robert C. Allen and Bruce C. Heezen, 1966.
Sediment size distribution profile on the continental shelf off New Jersey.
Sedimentology, 7(2):155-159.

bottom sediments grain size
Dörjes, Jürgen, Sibylle Gadow, Hans-Erich Reineck and Indra Bir Singh, 1970.
Sedimentologie und Makrobenthos der Nordergründe und der Aufsenjade (Nordsee). Senckenberg marit. 2: 31-59.

bottom sediments, grain size
Folk, Robert L., 1966.
A review of grain-size parameters.
Sedimentology, 6(2):73-93.

bottom sediments, grain size
Fuller, A.O., 1962.
3. Size distribution characteristics of shallow marine sands from the Cape of Good Hope, South Africa.
J. Sed. Petrol., 1961, 31(2):256-261.
Reprinted in:
Univ. Cape Town, Oceanogr. Dept., Publ. 3-6. 1962

bottom sediments, grain size
Gadow, Sibylle, 1971.
Der Golf von Gaeta (Tyrrhenisches Meer). 1. Die Sedimente. Senckenbergiana maritima 3(1): 103-133.

bottom sediments, grain size
Gealy, E.L., 1971.
Grain size of sediments from the western equatorial Pacific: Leg 7, Glomar Challenger. Initial Repts Deep Sea Drill. Proj. 7(2): 1027-1036.

bottom sediments, grain size
Ginsburg, R.N., 1956.
Environmental relationships of grain size and constituent particles in some south Florida carbonate sediments.
Bull. Amer. Assoc. Petr. Geol., 40(10):2384-2427.

bottom sediments, grain size
Higashikawa, Seiji, 1970.
Analyses of bottom sediments of the East China Sea. (In Japanese; English abstract)
Mem. Fac. Fish., Kagoshima Univ. 19: 91-102

bottom sediments, grain size
Hulsey, J.D., 1961
Relations of settling velocity of sand sized spheres and sample weight.
J. Sediment. Petrol., 31(1): 101-112.

bottom sediments, grain size
Kagami, Hideo, 1962
Modal size distribution of the bottom sediment at the entrance of Tokyo Bay (Cooperative survey at the entrance of Tokyo Bay in 1959-Part 5). (In Japanese; English abstract).
J. Oceanogr. Soc., Japan, 20th Ann. Vol., 121-135.

bottom sediments, grain size
Kakinuma, Tadao, 1961
Size-frequency distribution of nearshore sediments at Ohsu beach along the Atsumi Bay coast.
Proc. 8th Conf. Coastal Eng., Japan, 156-160.
Also in:
Papers in Oceanogr. and Hydrol., Geophys. Inst., Kyoto Univ., (1949-1962).

Bottom sediments, grain size
Kagami, Hideo, 1961.
Modal analysis of marine sediments in the southern part of Tokyo Bay. Natural classification of clastic sediments.
Japanese J., Geol. and Geogr., 32(3/4):521-532.

bottom sediments, grain size
Krzeminska, Aleksandra, 1961
Granulometric and mineral composition of sands from a so-called Koszalin-Bay cross-section. (In Polish; English summary).
Prace Inst. Morsk., Gdansk, (1) Hydrotech., II Sesja Naukowa Inst. Morsk., 20-21 wrzesnia 1960:39-52; English summary, p. 144. (mimeographed).

bottom sediments grain size
Kuenen, Ph.H. and F. L. Humbert, 1970.
Grain size of turbidite ripples.
Sedimentology 13(3/4):253-261

Bottom Sediments, grain size
Lisitsin, A. P., 1962.
[Distribution and contents of suspensions of the Indian Ocean waters. Communication 3. Comparison of granulometric contents of suspensions and bottom sediments.]
Mezhd. Geofiz. Komitet, Prezidiume Akad. Nauk, SSSR, Rezult. Issled. Programme Mezhd. Geofiz. Goda, Okeanol. Issled., No. 5:130-139.

bottom sediments, grain size
Lisitsyn, A.P., V.F. Kanaev, 1956.
[Mechanical analysis of coarse materials in marine conditions.] Trudy Inst. Okeanol., 19:252-261.

bottom sediments grain size, effect of
Longbottom, M.R., 1970.
The distribution of Arenicola marina (L.) with particular reference to the effects of particle size and organic matter of the sediments.
J. exp. mar. Biol. Ecol., 5(2): 138-157.

bottom sediments, grain size
McMaster, Robert L., John D. Milliman and Asaf Ashraf, 1971.
Continental shelf and upper slope sediments off Portuguese Guinea, Guinea, and Sierra Leone West Africa.
J. sedim. Petrol. 41(1): 150-158.

bottom sediment, grain size
Miller, R.L., 1956.
Trend surfaces: their application to analysis and description of environments of sedimentation.
1. The relation of sediment-size parameters to current-wave systems and physiography.
J. Geol., 64(5):425-446.

bottom sediments, grain size
*Mishra, Sharad K., 1968.
Granulometric studies of Recent sediments in the Firth of Tay region (Scotland).
Sediment. Geol., 2(3):191-200.

bottom sediments, grain size
Müller, G., 1964.
Die Korngrössenverteilung in den rezenten Sedimenten des Golfes von Neapel.
In: Developments in Sedimentology, L.M.J.U. van Straaten, Editor, Elsevier Publishing Co., 1: 282-292.

bottom sediments, grain size
Petelin, V.P., 1958.
Simplified method for calculating the granulometric coefficients of bottom sediments. (In Russian).
Biull. Okeanogr. Komissia, Akad. Nauk, SSSR, (1):52-

bottom sediments, grain size
Petelin, V.P., 1956.
[A speedier method of determining medians and quartiles.]
Biull. Moskovskogo Obshch. Ispitatelei Prirody, Geol. Div., 31(1):95-97.

bottom sediments, grain size
*Rozhdestvensky, A.V., 1967.
Granulometric and chemical studies of the near-shore sands and bottom sediments in the Bulgarian Black Sea coastal area. (In Russian; English abstract).
Okeanologiia, Akad. Nauk, SSSR, 7(6):1020-1024.

bottom sediments, grain size
Sahu, Basanta K., 1965.
Sieving characteristics of symmetrical tabular grains with any roundness value.
J. sed. Petr., 35(3):763-765.

bottom sediments, grain size
Sakamoto, Ichitaro and Jun Yamada, 1969
Sedimentological study of submarine delta at the mouth of the Kiso River (mechanical analyses of bottom sediment). (In Japanese; English Abstract.
J. Fac. Fish. Pref. Univ. Mie, 8(1):17-140

bottomsediments, grain size
Park Won Cheon, and Sang Yong Kim, 1968.
Submarine topography and grain size distribution of sediments in the southern part of the East Sea of Korea. (In Korean; English abstract).
Bull. Fish. Res. Develop. Agency, Korea, 3:105-118

bottom sediments, grain size
Pimm, A.C., 1971.
Grain size results and composition of the sand fraction, leg 6. Initial Repts. Deep Sea Drilling Project, Glomar Challenger 6: 709-737.

bottom sediments, grain size
Pimm, A.C., 1970.
Leg 4 grain size results. Initial Reports of the Deep Sea Drilling Project, Glomar Challenger 4: 295-305.

bottom sediments, grain size
Pimm, A.C., 1970.
Grain size analysis, leg 3. Initial Reports of the Deep Sea Drilling Project, Glomar Challenger 3: 475-494.

bottom sediments, grain size
Rosfelder, Andre, 1960.
Contributions à l'analyse texturale des sediments. Thèses presentes pour obtenir le grade de docteur ès-sciences naturelles, 10 Dec. 1960, Univ. Alger, No. 11:356 pp. (multilithed).

bottom sediments, grain size
Schlee, John, and Jacqueline Webster, 1967
A computer program for grain-size data.
Sedimentology, 8(1):45-53.

bottom sediments, grain size
Shumway, G., and K. Igelman, 1960
Computed sediment grain surface areas. J. Sed. Petr., 30(3): 486-489.

bottom sediments, grain size
Swift, Donald J.P., and W. Richard Boehmer 1972.
Brown and gray sands on the Virginia shelf: color as a function of grain size.
Bull. geol. Soc. Am. 83(3):877-884.

bottom sediments, grain size
Trask, P.D., 1959
Effect of grain size on strength of mixtures of clay, sand and water.
Bull. Geol. Soc. Amer., 70(5):569-580.

bottom sediments, grain size
Uchio, Takayasu, 1968.
Influence of the River Shinano on Foraminifera and sediment grain-size distributions.
In: Papers in Marine Geology, R.L. Miller, Editor, Macmillan Co., N.Y., 411-428.

bottom sediments, grain size
Uchio, Takayasu, 1962
Influence of the River Shinano on Foraminifera and sediment grain size distributions. (In Japanese; English abstract).
J. Oceanogr. Soc., Japan, 20th Ann. Vol., 15-24.

bottom sediments, grain size
Uchio, Takayasu, 1962
Influence of the River Shinano on Foraminifera and sediment grain size distributions.
Publ. Seto Mar. Biol. Lab., 10(2) (Art.18): 363-392.

bottom sediments, grain size
Vallier, Tracy L., 1970.
Grain size analysis. Initial Repts. Deep Sea Drilling Project, Glomar Challenger 5: 421-430.

bottom sediments, grain size
Werner, Friedrich, 1964.
Sedimentkorne aus den Rinnen der Kieler Bucht.
Meyniana, 14:52-65.

bottom sediments, grain size
Wunderlich, Friedrich, 1970.
Korngrössenverschiebung durch Lanice conchilega (Pallas). Senckenberg marit. 2: 119-125.

bottom sediments, grain size
Wildo, Pat, 1965.
Estimates of bottom current velocities from grain size measurements for sediments from the Monterey deep-sea fan.
Ocean Sci. and Ocean Eng., Mar. Techn. Soc.,-Amer. Soc. Limnol. Oceanogr., 2:718-727.

bottom sediments, clay minerals
Yerochtchev-Shak, V.A., 1961
[Kaolinite in the bottom sediments of the Atlantic.]
Doklady Akad. Nauk, SSSR, 137(3):695-697.

bottom sediments, clay minerals
Yerochtchev-Shak, V.A., 1961
Illite in the bottom sediments of the Atlantic.
Doklady Akad. Nauk, SSSR, 137(4):951-953.

bottom sediments, gravels
Schwarzacker, W., & K. Hunkins, 1961
Analysis of gravels from the Arctic Ocean floor.
IGY Bull., N.A.S., 47: 1-7.

bottom sediments gravels
Veenstra, H.J., 1969.
Gravels of the southern North Sea.
Marine Geol. 7(5): 449-464.

bottom sediments, ice-rafted

bottom sediments, ice-rafted
*Lepley, Larry K., 1966.
Submarine geomorphology of eastern Ross Sea and Sulzberger Bay, Antarctica.
Tech. Rept., U.S. Naval Oceanogr. Off., TR-172:34 pp. (multilithed).

bottom sediments(rafted)
Nechaev, A.P., 1961.
[Ice drift sediments accumulation in the Far East.]
Izv. Vses. Geograf. Ser., 93:76-78.

bottom sediments, ice-rafted
Needham, H.D., 1962
Ice-rafted rocks from the Atlantic Ocean off the coast of the Cape of Good Hope.
Deep-Sea Res., 9(5):475-486.

Bottom sediments interstitial water

bottom sediments, interstitial water
Brujewicz, S.B., 1956.
[Vertical distribution of biogenic elements in sediments solutions (interstial water) of the Okhotsk Sea.] Dokl. Akad. Nauk, SSSR, 111(2):391-392.

bottom sediments, interstitial water
Brujewicz, S.W., 1946
Salinity of Sediment Solution of the Caspian Sea. Comptes Rendus (Doklady) de l'Acad. d. Sci. URSS, LIV (3):239-242.

bottom sediments, interstitial water
Bruevich, S.V., and E.G. Vinogradova, 1947.
[The chemical composition of sediment solutions of the Caspian Sea. I. The northern Caspian (1932). II. The northern, middle, and southern Caspian (1935, 1936, 1940).] (English summaries)
Gidrokhim Mat. 8:129-186.

bottom sediments, interstitial water
Brujewicz, S. W. and E. G. Vinogradova, 1946
[Biogenic elements in the sediment solutions of the northern, middle and southern parts of the Caspian Sea.] Comptes Rendus (Doklady) Acad. Sci. URSS, LIV (5):419-422.

bottom sediments, interstitial water
Lissitzin, A.P., 1956.
[Water content of bottom deposits of the western part of the Bering Sea.] Dokl. Akad. Nauk, SSSR, 107(3):455-458.

bottom sediments interstitial water
Modierskaya, V.V., 1960
[The question of the existence of iron in marine interstitial water of the Black Sea.]
Trudy, Comm. Oceanogr., Akad. Nauk, SSSR, 10 (2): 21-29.

bottom sediments, interstitial water
Rullier, F., 1957.
Teneur en air et en eau interstitiel des sables marins et son influence sur les conditions d'habitat. C.R. Acad. Sci., Paris, 245(10):936-938.

bottom sediments, interstitial water
Tageeva, N.V., 1960
[Water migration in marine sediments on their diagenesis] Doklady Akad. Nauk, SSSR, 134(4): 917-919.

bottom sediment layers
*Houtz, R. and R. Meijer, 1970.
Structure of the Ross Sea shelf from profiler data.
J. geophys. Res., 75(32):6592-6597.

bottom sediments, light penetration

bottom sediments, light penetration
Fenchel, Tom, and Birthe J. Straarup 1971.
Vertical distribution of photosynthetic pigments and the penetration of light in marine sediments.
Oikos 22 (2): 172-182.

bottom sediments, light penetration
Gomoiu, M.-T. 1967.
Some quantitative data on light penetration.
Helgoländer wiss. Meeresunters. 15 (1/4): 120-127.

bottom sediments, lithology of

bottom sediments, lithology of
Keen, M.J., 1963
The magnetization of sediment cores from the eastern basin of the North Atlantic Ocean.
Deep-Sea Res., 10(5):607-622.

bottom sediments, lithology
Nota, D.J.G., and D.H. Loring, 1964.
Recent depositional conditions in the St. Lawrence River and Gulf - a reconnaissance survey.
Marine Geology, 2(3):198-235.

bottom sediments, lithology of
Pavlidis, Y. A., 1964.
Peculiar features of the lithology of the nearshore deposits in the Kuril Islands. (In Russian).
Okeanologiia, Akad. Nauk, SSSR, 4(6):1044-1051.

bottom sediments, magnetic particles

bottom sediments, magnetic particles
Grjebine, Torvy, 1965.
Spherules magnetiques dans les sédiments de la Mediterranée.
Bull. Inst. Océanogr., Monaco, 65(1338):12 pp.

bottom sediments, magnetic orientation of

bottom sediments, magnetic, orientation of
Nozharov, Peter B., 1968.
Comments on the influence of fluid motion upon the magnetic orientation of sediements.
Pure Appl. Geophys., 70:81-87.

bottom sediments, magnetism of

bottom sediments, magnetism of
Belshe, John C., 1962
Magnetic properties of some sediment cores from the deep ocean. (Abstract)
J. Geophys. Res., 67(9):3542.

bottom sediments, magnetism
Grjebine, T., 1965.
Sphérules magnétiques dans les sédiments de la Méditerranée.
Rapp. P.-v. Réun., Comm. int Explor scient. Mer Méditerr., 18(3):959-961.

bottom sediments, magnetism
King, R.F., 1955.
The remanent magnetism of artificially deposited sediments. Mon. Not. R. Astr. Soc., Geophys. Suppl., 7(3):115-134.

bottom sediments, magnetism
King, R.F., and A.I. Rees, 1966.
Detrital magnetism in sediments; an examination of some theoretical models.
J. geophys. Res., 71(2):561-571.

bottom sediments, magnetism
Niino, Hiroshi 1971.
A study of the sediments and magnetics across the continental shelf between Borneo and Malaya peninsula.
Techn. Bull. Econ. Comm. Asia Far East 4: 143-147.

bottom sediments, magnetism of
*Poutiers, Jacques, 1969.
Sur la susceptibilité magnétique des sables littoraux de la Baie des Anges.
C.r.hebd.Séanc., Acad.Sci., Paris, 268(3):485-488.

bottom sediments, magnetism
Rees, A.I. 1970.
Magnetic properties of some canyon sediments.
Mar. Geol. 9 (2): M12-M16.

bottom sediments, magnetic remanence
Rees, A.I., 1961
The effect of water currents on the magnetic remanence and anistropy of susceptibility of some sediments.
Geophys. J., 5(3):235-251.

bottom sediments, mechanical properties
Mizikos, Jean-Pierre 1971.
Propriétés mécaniques et acoustique des sables marins. II. Propagation d'une onde acoustique et mécanismes de pertes liés à la présence du fluide saturant.
Cah. océanogr. 23 (9): 801-826

bottom sediments, mechanical properties
Mizikos, J.-P., 1971
Lettre à l'éditeur à propos des propriétés mécaniques et acoustiques des sables marins.
Cah. océanogr. 23 (10): 981-982

bottom sediments, mechanical properties
Mizikos, Jean-Pierre, 1971.
Propriétés mécaniques et acoustiques des sables marins. I. Mécanismes de pertes dans la matrice et relations contrainte-deformation.
Cah. océanogr. 23(8): 709-732

bottom sediments, metamorphosis
Esquevin, J., 1969.
Influence de la composition chimique des illites sur leur cristallinité.
Bull. Centre Rech. Pau-SNPA 3(1): 147-153

bottom sediments, mica
Berthois, L., 1962.
Étude du comportement hydraulique du mica.
Sedimentology, 1(1):40-49.

bottom sediments microstructure

bottom sediments, microstructure
*Bowles, Frederick A., 1968.
Microstructure of sediments: investigations with ultrathin sections.
Science, 159(3820):1236-1237.

Bouguer anomalies
Grushinskii, N.P. and N.B. Sazhina 1969
Some considerations about Bouguer anomalies on the oceans. (In Russian)
Dokl. Akad. Nauk SSSR 184 (2): 331-333.

bottom sediments, minerals in
See also individual "minerals" under "minerals"

Bottom sediments, minerals
See also: minerals

bottom sediments, minerals
Alexina, I.A., 1963.
A contribution to the description of the mineralogic composition of the coarse aleuritic fraction of the bottom-set beds of the north-western part of the Pacific. (In Russian).
Doklady, Akad. Nauk, SSSR, 149(6):1420-1423.

bottom sediments, mineralogy
Alexina, I.A., 1962
Mineralogy of the macroaleuritic fraction in the sediments of Kronotzk and Avachinck bays. Investigations of marine bottom sediments. (In Russian; English summary).
Trudy Inst. Okeanol., Akad. Nauk, SSSR, 61: 104-154.

bottom sediments, minerals
*Aoki, Hitoshi, Takeshi Yoshihara and Michihei Hoshino, 1967.
Geology of the Suruga Bay. I. Submarine distribution of gravels. (In Japanese; English abstract).
J.Coll.mar.Sci.Techn., Tokai Univ., (2):85-92.

bottom sediments, mineralogy of
Avilov, I.K., 1965.
Relief and bottom sediments of the shelf and continental slope of the Northwestern Atlantic. Investigations in line with the programme of the International Geophysical Year, 2. (In Russian).
Trudy, Vses. Nauchno-Issled. Inst. Morsk. Ribn. Choz. i Okeanogr. (VNIRO), 57:173-234.

bottom sediments, mineralogy of
Avilov, I.K., 1965.
Some data on the bottom topography and grounds of the West African shelf. Investigations in line with the programme of the International Geophysical Year, 2. (In Russian).
Trudy, Vses. Nauchno-Issled. Inst. Morsk. Ribn. Choz. i Okeanogr., (VNIRO), 57:235-259.

bottom sediments, minerals
Berner, Robert A., 1964.
Stability fields of iron minerals in anaerobic marine sediments.
J. Geol., 72(6):826-834.

bottom sediments, minerals
Berry, Richard W., and William D. Johns, 1966.
Mineralogy of the clay-sized fractions of some North Atlantic-Arctic Ocean bottom sediments.
Geol. Soc., Amer., Bull., 77(2):183-196.

bottom sediments, minerals
*Berthois, Léopold, et Pedro Roa, 1967.
Interprétation des analyses de minéraux lourds par pourcentages pondérés, dans la région des Feröer-Écosse- Hebrides.
C.r.hebd. Séanc., Acad. Sci., Paris, (D)265(3): 196-199.

Biggs, Robert B., 1967. bottom sediments, mineralogy or
The sediments of Chesapeake Bay.
In: Estuaries, G.H. Lauff, editor, Publs Am. Ass. Advmt Sci., 83:239-260.

bottom sediments, mineralogy
Berthois, Léopold, et Gérard Auffret, 1968.
Contribution a l'étude des conditions de sédimentation dans la rade de Brest.
Cah. océanogr., 20(10):893-920.

bottom sediments, minerals
Berthois, Léopold et Philippe Bois, 1969.
Le cours inférieur et l'estuaire de la rivière du Château en période d'étiage, Iles de Kerguelen. Étude hydraulique, sédimentologique et chimique.
Cah. océanogr. 21(8): 727-771.

bottom sediments, mineralogy of
Biscaye, Pierre E., 1965.
Mineralogy and sedimentation of Recent deep-sea clay in the Atlantic Ocean and adjacent seas and oceans.
Bull. Geol. Soc., Amer., 76(7):803-832.

bottom sediments (deep), minerals in
Biscaye, Pierre E., 1964.
Mineralogy and sedimentation of Atlantic and Antarctic Ocean deep-sea sediments. (Abstract).
Geol. Soc., Amer., Special Paper, No. 76:16.

bottom sediments, minerals
Blanc, J.J., 1963.
Petit guide et tableaux pour la determination des mineraux dans les sediments.
Rec. Trav. Sta. Mar., Endoume, 29(44):75-86.

bottom sediments, minerals
Blazhchishin, A.I., and M.M. Usonis 1970.
Sedimentation characteristics in the southeastern Baltic Sea on the data of mineralogical analysis (In Russian; English and German abstracts).
Baltica, Lietuv. TSR Mokslu Akad. Geogr. Skyr. INQUA Taryb. Sek., Vilnius, 4: 115-144

bottom sediments, minerals
Bonatti, E., 1963.
Zeolites in Pacific pelagic sediments.
Trans. N.Y. Acad. Sci., (2) 25(8):938-948.

bottom sediments, minerals
Bordovsky, O.K., 1956.
[Some data on the argillaceous minerals of sediments of the northwest part of the Pacific Ocean]
Trudy Inst. Okeanol., 17:133-136.

bottom sediments, minerals in
Brückner, W.D., and H.J. Morgan, 1964.
Heavy mineral distribution on the continental shelf off Accra, Ghana, West Africa.
In: Developments in Sedimentology, L.M.J.U. van Straaten, Editor, Elsevier Publishing Co., 1:54-61.

bottom sediments, minerals
Carroll, Dorothy, 1964.
Ion-exchange capacity of sediments from the experimental Mohole, Guadalupe site.
J. Sed. Petr., 34(3):537-542.

bottom sediments, minerals
Chamley, Herve, 1964.
Remarques sur les minéraux argileux des sédiments fluviatiles et marins de la région du Bas-Rhône.
Rec. Trav. Sta. Mar., Endoume, 34(50):263-280.

bottom sediments, minerals
Chamley, H., 1963.
Contribution à l'étude minéralogique et sédimentologique de vases méditerranéenes.
Rec. Trav. Sta. Mar., Endoume, 29(44):91-195.

bottom sediments, minerals
Chamley, Hervé, 1963.
Sur la nature des minéraux argileux de vases méditerranéenes (Méditerranée centrale et nord-orientale).
Rapp. Proc. Verb., Réunions, Comm. Int. Expl. Sci., Mer Méditerranée, Monaco, 17(3):1055-1060.

bottom sediments, minerals
Chassefière B. 1969.
Étude minéralogique de la fraction lourde de sédiments de la région de Thau (Hérault).
Vie Milieu (B) 20(1): 37-50.

bottom sediments, mineralogy of
Chave, Keith E., 1962
Factors influencing the mineralogy of carbonate sediments.
Limnology and Oceanography, 7(2):218-223.

bottom sediments, minerals
Cherry, John., 1965.
Sand movement along a portion of the Northern California coast.
U.S. Army Coast Eng. Res. Center, Techn. Memo. No. 14:125 pp.

bottom sediments, minerals
Chou, J.T. 1971.
Recent sediments in the shallow part of the South China Sea.
Proc. Geol. Soc., China 14:99-117

bottom sediments, mineralogy
Cook, H.E., R.W. Rex, W.A. Eklund and B. Murray, 1971.
X-ray mineralogy studies, Leg 7. Initial Repts Deep-Sea Drill Proj. 7(2): 913-963.

bottom sediments, mineralogy
Cook, H.E. and I. Zemmels, 1971.
X-ray mineralogy studies - leg 8. Initial Repts Deep Sea Drilling Project, 8: 901-950.

bottom sediments, minerals in
Damiani, L. Favretto and G.L. Morelli, 1964.
Le argille della fossa mesoadriatica.
Archiv. Oceanogr. e Limnol., 13(2):187-196.

bottom sediments, mineralogy
Drever, James I., 1971.
Chemical and mineralogical studies, site 66.
Initial Repts Deep-Sea Drill. Proj. 7(2): 965-975.

bottom sediments, minerals
Duplaix, S., 1953.
Étude minéralogique des sédiments sableux des nappes alluviales anciennes du Gave de Pau et de l'Adour.
Bull. Soc., Géol., France, (6), 3:369-375.

bottom sediments minerals
Duplaix, S., 1958
Etude mineralogique des niveaux sableus des carottes prelevées sur le fond de la Mediterranée. Repts. Swedish Deep-Sea Exped., 8 (Sediments cores from the Mediterranean Sea and the Red Sea, No. 2): 139-166.

bottom sediments, minerals
Duplaix, Solange, et Gilbert Boillot 1968.
Sur la minéralogie et l'origine des sables siliceux contenus dans les sédiments de la Manche occidentale.
Rev. géogr. phys. Géol. dynam. (2) 10(2): 147-161.

bottom sediments, minerals
Duplaix, S., and A. Cailleux, 1952.
Étude minéralogique et morphoscopique de quelques sables des grands fonds de l'océan Atlantique occidental. Göteborgs K. Vetenskaps- och Vitterhets-Samhälles Handl., Sjätte Följden, Ser. B, 6(5):1-27, 37 textfigs.

Medd. från Oceanogr. Inst., Göteborg, 20:

bottom sediments, minerals
*Duplaix, Solange, et Maurice Gennesseaux, 1968.
Les minéraux lourds des formations Tertiaires et Quaternaires de la région niçoise (Alpes-Maritimes).
Revue Géogr. phys. Géol. dynam. (2)10(4):353-374.

bottom sediments, mineral in
Duplaix, Solange, et Maurice Gennesseaux, 1967.
Les minéraux lourds des sables du Var, du Paillon et de la Roya et les dépôts sous-marins de la Mer de Ligurie.
Cah. océanogr., 19(3):219-236.

bottom sediments, minerals
Duplaix, Solange, et Vladimir D. Nesteroff, 1959.
Recherches sur les mineraux lourds du littoral du golfe de la Napoule et du golfe Juan. Bull. Soc. Geol., France, (7): 1: 107-111.

bottom sediments, minerals in
Duplaix, S., W.D. Nesteroff et B.C. Heezen, 1965
Minéralogie comparée des sédiments du Tage (Portugal) et de quelques sablés profonds de la plaine abyssale correspondante.
Deep-Sea Research, 12(2):211-217.

bottom sediments, minerals
Edwards, Dennis S., and H. Grant Goodell, 1969.
The detrital mineralogy of ocean floor surface sediments adjacent to the Antarctic Peninsula, Antarctica.
Marine Geol., 17(3): 207-234

bottom sediments, mineralogy
El Wakeel, S.K., and J.P. Riley, 1961.
Chemical and mineralogical studies of deep-sea sediments.
Geochimica et Cosmochimica Acta, 25(2):110-146.

bottom sediments, minerals
Eroshchev-Shak, V.A., 1964.
Clay minerals of the Atlantic Ocean. Oceanographic investigations in the Atlantic Ocean and Black Sea. (In Russian).
Trudy Morsk. Gidrofiz. Inst., Akad. Nauk Ukrain. SSR, 30:116-136.

bottom sediments, minerals in
Eroshchev-Shak, V.A., 1963.
On the distribution of argillaceous minerals in deep-water sediments of the Atlantic.
(In Russian; English abstract).
Rez. Issled., Programme Mezhd. Geofiz. Goda, Okeanolog. Issled., Akad. Nauk, SSSR, No. 8:125-135.

Sediments, minerals
Eroschev-Shak, V.A., 1962
[On zoning of argillaceous minerals distribution in deposits of the Atlantic Ocean.]
Trudy Inst. Okeanol., Akad. Nauk, SSSR, 56: 59-69.

bottom sediments, minerals

Esteoule, Jr., J. Esteoule-Choux, M. Melguen et E. Seibold, 1970.
Sur la présence d'attapulgite dans des sédiments récents du nord-est du Golfe Persique. C.r. hebd. Séan. Acad. Sci. Paris, 271: 1153-1156.
Also in: Recueil. Trav. Groupe d'Etude Marge Continental, Lab. Géol. sous-marine Univ. Rennes 1.

bottom sediments, mineralogy of

*Etchichury, M.C., y J.R. Remiro, 1967.
Los sedimentos litorales de la Provincia de Santa Cruz entre Punta Dungeness y Punta Desengáno.
Revta Mus. argent. Cienc. nat. Bernardino Rivadavia Inst. nac. Invest., Geol., 6(8): 323-376.

bottom sediments, minerals in

Etchichury, M.C., and J.R. Remiro, 1963
La corriente de Malvinas y los sedimentos Pampeano-Patagonicos.
Communicaciones, Mus. Argentino, Ciencias Nat. "Bernardino Rivadavia" e Inst. Nacional Invest. Ciencias Nat., Ciencias Geol., 1(20): 1-11.

bottom sediments, minerals

Evans, G., and D.J. Sherman, 1964.
Recent celestine from the sediments of the Trucial coast of the Persian Gulf.
Nature, 202(4930): 385-386.

bottom sediments, minerals in

Fairbank, Nora Gladwin, 1962
Minerals from the eastern Gulf of Mexico.
Deep-Sea Res., 9(4): 307-338.

bottom sediments, mineralogy

Fairbank, N.G., 1956.
Methods and preliminary results in a study of minerals from the eastern Gulf of Mexico.
J. Sed. Petr., 26(3): 268-275.

bottom sediments, minerals

Favrette, Luciano, 1966.
Authigenic ferriferous aragonite from bottom sediments of the Adriatic Sea.
Mineralog. Mag., 35: 781-783.

minerals - pyrite

Frankel, J.J., 1964.
Recent Foraminifera filled and encrusted with pyrite from Durban Bay.
S. African J. Sci., 60(10): 299.

bottom sediments, minerals

Gadow, Sibylle, 1971.
Der Golf von Gaeta (Tyrrhenisches Meer). 1. Die Sedimente. Senckenbergiana maritima 3(1): 103-133.

bottom sediments, minerals

Gassaway, John D. 1970
Mineral and chemical composition of sediments from the Straits of Florida.
J. sedim. Petrol. 40(4): 1136-1146.

bottom sediments, minerals

Gershanovich, D.E., V.S. Kotenev, and V.N. Novikov, 1964.
Relief and bottom sediments in the Gulf of Alaska. (In Russian).
Trudy, Vses. Nauchno-Issled. Inst. Morsk. Ribn. Choz. i Okeanogr, 52: 83-133.

Also:
Izv. Tichookeansk. Nauchno-Issled. Inst. Morsk. Ribn. Choz. i Okeanogr. (TINRO), 52: 83-133.

bottom sediments, minerals

Giresse, Pierre, 1965.
Observations sur la présence de "glauconie" actuelle dans les sédiments ferrugineux peu profonds du bassin gabonais.
Comptes Rendus, Acad. Sci., Paris, 260(21): 5597-5600.

bottom sediments, mineralogy

Goldberg, Edward D., and John J. Griffin, 1964
Sedimentation rates and mineralogy in the South Atlantic.
J. Geophys. Res., 69(20): 4293-4309.

bottom sediments, minerals

Gorbunova, Z.N., 1963.
Minerals of the montmorillonite group in the sediments of the Pacific Ocean. (In Russian).
Okeanologiia, Akad. Nauk, SSSR, 3(5): 870-875.

bottom sediments, minerals

Gorbunova, Z.N., 1962
Clay and associated minerals of the Indian Ocean sediments. Investigations of marine bottom sediments. (In Russian: English summary).
Trudy Inst. Okeanol., Akad. Nauk, SSSR, 61: 93-103.

bottom sediments, minerals in

Gorbunova, Z.N., 1962
Clayey and other high-dispersed minerals in the sediments of the Bering Sea. (In Russian).
Okeanologiia, Akad. Nauk, SSSR, 2(6): 1024-1034.
Abstr. in:
Soviet Bloc Res. Geophys., Astron. & Space, 52: 24-25.

bottom sediments, minerals

Griffin, J.J., and E.D. Goldberg, 1963.
26. Clay-mineral distributions in the Pacific Ocean.
In: The Sea, M.N. Hill, Editor, Interscience Publishers, 3: 728-741.

bottom sediments, mineral

Hamilton, N., 1965.
Description of a bottom-sediment sample dredged from Halley Bay.
Brit. Antarctic Surv. Bull., No. 7: 47-52.

bottom sediments chamosite

Hardjosoesastro, Rachmat, 1971.
Note on chamosite in sediments of the Surinam shelf.
Geologie Mijnb. 50(1): 29-33

Bottom sediments, minerals in

Hardmann, Martin, and Ludwig Lohmann, 1965.
Untersuchungen an der heissen Salzlauge und am Sediment des Atlantis II-Tiefs im Roten Meer.
Meteor Forschungsergebn. (C) 1: 13-20.

bottom sediments, mineralogy

Heath, G. Ross, 1969.
Mineralogy of Cenozoic deep-sea sediments from the Equatorial Pacific Ocean.
Bull. geol. Soc. Am. 80(10): 1997-2018

bottom sediments, mineral

Hebara, Tashiyuki, 1965.
Role of minerals constituents in mechanical properties of the estuarine sediments along the Hiroshima Bay.
J. Sci., Hiroshima Univ., (C) 4(4): 429-454.

bottom sediments, minerals

Heezen, Bruce C., Wladimir D. Nesteroff, Agnes Oberlin et Germain Sabatier, 1965.
Découverte d'attapulgite dans les sédiments profonds du Golfe d'Aden et de la Mer Rouge.
C.R. Acad. Sci., Paris, 260(22): 5819-5821.

bottom sediments, minerals

*Herman, Yvoone and P.E. Rosenberg, 1969.
Mineralogy and micropaleontology of a goethite-bearing Red Sea core.
In: Hot brines and Recent heavy mineral deposits in the Red Sea, E.T. Degen and D.A. Ross, editors, Springer-Verlag, New York, Inc., 448-459.

bottom sediments, minerals in

Hubert, John F., and William J. Neal, 1967.
Mineral composition and dispersal patterns of deep-sea sands in the western North Atlantic petrologic province.
Bull. Geol. Soc., Am. 78(6): 749-772

bottom sediments mineralogy

Iijima, Azuma and Hideo Kagami, 1960

Lawsonite-and pumpellyite-rock fragments from deep-sea bottom off the coast of Sanriku, Northeastern Japan. Report on JEDS-2-S of the Japanese Deep-Sea Expedition.
Japan J. Geol. and Geogr., 31(2/4): 253-260.

bottom sediments, minerals

Imbrie, John., and Tjeerd H. Van Andel, 1964.
Vector analysis of heavy mineral data.
Geol. Soc., Amer., Bull., 75(11): 1131-1156.

bottom sediments, minerals

Japan, Maritime Safety Board, 1957.
[Mineralogical studies in recent marine sediments]
Hydrogr. Bull. (Publ. 981)(54): 8-38.

bottom sediments, mineral

Keller, George H., and Adrian F. Richards, 1967.
Sediments of the Malacca Strait, Southeast Asia.
J. Sedim. Petrol., 37(1): 102-127.

bottom sediments, minerals

Klingebiel, André, et Francis Lapierre, 1966.
Sur la répartition des sables recouvrant le plateau continental du golfe de Gascogne. Intérêt de l'étude des minéraux lourds.
C.R. herbd. Séanc., Acad. Sci., Paris, (D) 263 (17): 1195-1198.

bottom sediments, minerals

Klingebiel, André, Francis Lapierre et Claude Latouche, 1966.
Sur la nature et la répartition des minéraux argileux dans les sédiments récents du Golfe de Gascogne.
C.R. herbd. Séanc., Acad. Sci., Paris (D) 263(18): 1293-1294.

bottom sediments, minerals

Kobayashi, K., K. Oinuma and T. Sudo, 1964.
Clay mineralogy of Recent marine sediments and sedimentary rocks from Japan.
Sedimentology, Elsevier Publ. Co., 3(3): 233-239.

bottom sediments, minerals

Kobayashi, Kazuo, Kaoru Oinuma and Toshio Sudo, 1961.
Clay mineral composition of the cores samples collected in the Japan Trench. (Abstract).
Tenth Pacific Sci. Congr., Honolulu, 21 Aug.-6 Sept., 1961, Abstracts of Symposium Papers, 376

bottom sediments, mineralogy of

Kobayashi, K., K. Oinuma and T. Sudo, 1960.
Clay mineralogical study on Recent marine sediments. (1) Samples JEDS-1-R', JEDS-R-R".
Oceanogr. Mag., Japan, 11(2): 215-222.

bottom sediments, mineralogy of

Krzeminska, Aleksandra, 1961
Granulometric and mineral composition of sands from a so-called Koszelin-Bay cross-section. (In Polish; English summary).
Prace Inst. Morsk., Gdansk, (1) Hydrotech. II. Sesja Naukowa Inst. Morsk., 20-21 wrzesnia 1960: 39-52; English summary, p. 144. (mimeographed).

bottom sediments, minerals
Kwon, Nak-Yon, and Sang-Jung Park 1970.
A study of submarine topography and characteristic of bottom sediment in the central part of the Eastern Sea of Korea. (In Korean; English abstract).
Bull. Fish. Res. Dev. Agency, Pusan, 5: 107-125.

bottom sediments, minerals
Lisitzin, A.P., V.V. Serova, I.B. Zverinskaya, V. Lukashin, Z.N. Gorbunova, V.V. Gordeev, V.V. Zhurensko, A.M. Pchelintsev, Ju.I. Belyaev, N. I. Popov, O.V. Shishkina, N.M. Morozov, A.P. Jouse, O.G. Kozlova and V.V. Mukhina. 1971 Geochemical, mineralogical, and paleontological studies. Initial Repts. Deep Sea Drilling Project, Glomar Challenger 6: 829-960.

bottom sediments, mineral
Mallik, T.K., 1969.
Heavy minerals of shelf sediments between Vishakhapatnam and the Penner delta, eastern coast of India. Bull. natn. Inst. Sci., India, 38(1): 502-512.

bottom sediments, mineralogy of
Marlowe, J.I., 1966.
Mineralogy as an indicator of long-term current fluctuations in Baffin Bay.
Canadian J. Earth Sci., 3(2):191-202.

bottom sediments, minerals
McAllister, Raymond F., Jr., 1964.
Clay minerals from west Mississippi delta marine sediments.
In: Papers in Marine Geology, R.L. Miller, editor Macmillan Co., N.Y., 457-473.

bottom sediments, minerals
McCrone, A.W., 1966.
Sediments from Long Island Sound (New York) physical and chemical properties reviewed.
J. Sed. Petr. 36(1):234-236.

bottom sediments, mineralogy of
McMaster, Robert L., and Louis E. Garrison, 1966.
Mineralogy and origin of southern New England shelf sediements.
J. Sedim. Petrol., 36(4):1131-1142.

bottom sediments, mineralog
Mero, John L., 1965.
The mineral resources of the sea.
Elsevier Oceanogr. Ser., 312 pp.

bottom sediments, minerals
Nachtigall, K.H., 1962.
Über die Regelung von Langquarzen in aquatisch sedimentierten Sanden.
Meyniana, 12:9-24.

bottom sediments, minerals in
Nesteroff, Wladimir D., Bruce C. Heezen and Germain Sabatier, 1963.
Repartition des minéraux argileux dans les sédiment profonds de l'océan Indien.
Comptes Rendus, Acad. Sci., Paris, 257(4):941-943.

bottom sediments, minerals in
Nesteroff, W.D., G. Sabatier et B. C. Heezen, 1963.
Les mineraux argileux dans les sédiments du bassin occidental de la Méditerranée. (Sommaire).
Rapp. Proc. Verb., Réunions, Comm. Int., Expl. Sci., Mer Méditerranée, Monaco, 17(3):1005-1007.

bottom sediments, mineral
Neumann, A. Conrad, 1965.
Processes of Recent carbonate sedimentation in Harrington Sound, Bermuda.
Bull. Mar. Sci., 15(4):987-1035.

bottom sediments, minerals
Nevessky, E.N., 1957.
[Some data on the composition of the clay minerals in the bottom deposits of the Okhotsk Sea.]
Trudy Inst. Okeanol., 22:158-163.

bottom sediments, mineralogy of
Nevesky, E.N., 1951.
The change of the mineralogical constitution of near-shore sea alluvia and their dependence on depth and bottom relief. (In Russian).
Trudy Inst. Okeanol., Akad. Nauk, SSSR, 6:99-104

bottom sediments minerals
Nicolich, R., e F. Mosetti, 1967.
Magnetite content as a method of sediment analysis.
Boll. Soc adriat. Sci., Trieste, 55:39-45.

bottom sediments, mineralogy
Nota, D.J.G., and D.H. Loring, 1964.
Recent depositional conditions in the St. Lawrence River and Gulf - a reconnaissance survey.
Marine Geology, 2(3):198-235.

bottom sediments, minerals in
Oinuma, Kaoru, and Kazuo Kobayashi, 1962
Clay minerals in the sediments of the entrance of Tokyo Bay. (Cooperative survey at the entrance of Tokyo Bay in 1959 - Part 6). (In Japanese; English abstract).
J. Oceanogr. Soc., Japan, 20th Ann. Vol., 136-145.

bottom sediments, mineralogy of
Perez Mateos, Josefina, 1958.
Mineralogia de la fraccion "arena" de los sedimentos marinos que se estudian.
Boll. Inst. Español Oceanogr., Madrid, No. 90: 29-35.

bottom sediments, minerals
Petelin, V.P., 1957.
[Mineralogy of the sand-alevrite fraction of the bottom sediments of the Okhotsk Sea.]
Trudy Inst. Okeanol., 22:77-138.

bottom sediments, minerals
Petelin, V.P., 1954.
On the finding of autogenous glauconite in contemporary marine sediments. (In Russian).
Trudy Inst. Okeanol., Akad. Nauk, SSSR, 8:220-228.

bottom sediments, minerals in
Petelin, V.P., and I.A. Aleksina, 1965.
Division of the Pacific into mineralogical provinces. (In Russian; English abstract).
Okeanolog. Issled., Rezult. Issled. Programme Mezhd. Geofiz. Goda, Mezhd. Geofiz. Komitet Presidiume, Akad. Nauk, SSSR, No. 13:199-204.

bottom sediments mineralogy of
Peterson, M.N.A., and Edward D. Goldberg, 1962
Feldspar distributions in South Pacific pelagic sediments.
J. Geophys. Res., 67(9):3477-3492.

bottom sediments, minerals
Peterson, M.N.A., and J.J. Griffin, 1964
Volcanism and clay minerals in the south-eastern Pacific.
J. Mar. Res., 22(1):13-21.

bottom sediments, minerals
Peterson, M.N.A., V. Rama Murthy, and J.F. Evernden, 1964.
Studies on acidic pyroclastic material from the southern East Pacific Rise. (Abstract).
Trans. Amer. Geophys. Union, 45(1):115.

bottom sediments, minerals
Pilkey, Orrin H., 1964.
Distribution of heavy minerals in sediments of the United States South Atlantic continental shelf and slope. (Abstract).
Geol. Soc., Amer., Special Paper, No. 76:254.

bottom sediments, mineralog
Pilkey, Orrin H., 1964.
Mineralogy of the fine fraction in certain carbonate cores.
Bull. Mar. Sci., Gulf and Caribbean, 14(1):126-139

bottom sediments, mineral.
Pilkey, Orrin H., and Blake W. Blackwelder 1968.
Mineralogy of some Recent marine, terrigenous and carbonate sediments.
J. Sedim. Petrol. 38(3): 799-810.

bottom sediments, mineralogy
Pilkey, Orrin H., and David Noble, 1966.
Carbonate and clay mineralogy of the Persian Gulf
Deep-Sea. Res., 13(1):1-16.

bottom sediments, mineralogy of
Pilkey, Orrin H., and James B. Rucker, 1966.
Mineralogy of Tongue of the Ocean sediments.
J. mar. Res., 24(3):276-285.

bottom sediments, minerals in
Pomerancblum, Malvina, 1966.
The distribution of heavy minerals and their hydraulic equivalents in sediments of the Mediterranean continental shelf of Israel.
J. Sed. Petr., 36(1):162-174.

bottom sediments, mineralogy of
Pratt, W.L., 1963.
Glauconite from the sea floor off Southern California.
In: Essays in Marine Geology in honor of K.O. Emery, Thomas Clements, Editor, Univ. Southern California Press, 97-119.

bottom sediments, mineralogy
Price, N.B. and P.L. Wright 1970.
Variation in the mineralogy and chemistry of sediments from the south-western Barents Sea. In: The geology of the East Atlantic continental margin. 2. Europe, ICSU/SCOR Working Party 31 Symposium, Cambridge 1970, Rept. 70/14: 17-31.

bottom sediments, minerals
Rao, M. Subba 1963.
Clay mineral composition of shelf sediments off the east coast of India.
Proc. Indian Acad. Sci., (A), 58(1):6-15.

bottom sediments, minerals in
Rateev, M.A., 1952.
[Clay minerals in bottom sediments of the Black Sea.] Dokl. Akad. Nauk, SSSR, 83:287-290.

bottom sediments, minerals in
Rateeva, M. A., 1952.
[Clay minerals in bottom sediments of the Aral Sea] Dokl. Akad. Nauk, SSSR, 86:997-1000.

bottom sediments, minerals
*Rateev, M.A., Z.N. Gordunova, A.P. Lisitsin and G.I. Nosov, 1968.
Climatic zonality of the argillaceous minerals in the World Ocean sediments. (In Russian; English abstract).
Okeanolog. Issled., Rez. Issled. Mezhd. Geofiz. Proekt. Mezhd. Geofiz. Komit., Pres. Akad. Nauk, SSSR, 18:283-311.

~~bottom sediments, stratification~~
Reineck, H.E., 1967.
Layered sediments of the tidal flats, beaches, and shelf bottoms of the North Sea.
In: Estuaries, G.H. Lauff, editor, Publs Am. Ass. Advmt Sci., 83:191-206.

bottom sediments, mineralogy
Rex, R.W., 1970.
X-ray mineralogy studies, leg 3. Initial Reports of the Deep Sea Drilling Project, Glomar Challenger 3: 509-581.

bottom sediments, minerals
Rex, R.W., 1970.
X-ray mineralogy studies - leg 2. Initial Reports of the Deep Sea Drilling Project, Glomar Challenger 2: 329-346.

bottom sediments, minerals
Rex, R.W. and B. Murray, 1970.
X-ray mineralogy studies, leg 4. Initial Reports of the Deep Sea Drilling Project, Glomar Challenger, 4: 325-369.

bottom sediments, mineral
Roberts, H. H., 1971.
Mineralogical variation in lagoonal carbonates from North Sound Grand Cayman Island (British West Indies)
Sediment. Geol. 6 (3): 201-213.

bottom sediments, minerals
Ross, David A., 1971.
Sediments of the northern Middle America Trench.
Bull. geol. Soc. Am. 82 (2): 303-322.

bottom sediments, minerals
Ross, David A. 1970.
Atlantic continental shelf and slope of the United States - heavy minerals of the continental margin from southern Nova Scotia to northern New Jersey.
Prof. Pap. U.S. Geol. Surv. 529 G: G1-G40.

bottom sediments, minerals
Ross, David A., 1967.
Heavy-mineral assemblages in the nearshore surface sediments of the Gulf of Maine.
Prof. Pap. U.S. geol. Surv., 575-C: C77-C80.

bottom sediments, minerals
Ryan, William B.F., Fifield Workum, Jr., and J.B. Hersey, 1965
Sediments on the Tyrrhenian Abyssal Plain.
Geol. Soc. Am., Bull., 76 (11): 1261-1282.

bottom sediments, minerals
Seibold, Eugen, 1963
Geological investigation of near-shore sand-transport-examples of methods and problems from the Baltic and North seas.
Progress in Oceanography, M. Sears, Edit., Pergamon Press, London, 1:1-70.

bottom sediments, minerals
Sevon, W.D., 1966.
Sediment variation on Farewell Spit, New Zealand.
New Zealand J. Geol Geophys., 9(1/2):60-75.

bottom sediments, mineralogy of
Shcherbakov, F.A., and J.A Pavlidis, 1962
The specific features in the distribution of heavy minerals in the sea coastal zone. (In Russian).
Okeanologiia, Akad. Nauk, SSSR, 2(4):651-663.

bottom sediments, minerals in
Shchurko, I.I., 1966.
Terrigenic-volcanogenic minerals in modern sediments of the North Atlantic. (In Russian)
Dokl., Akad. Nauk, SSSR, 171(2):461-464.

bottom sediments, minerals
Soldatov, A.V. and I.O. Murdmaa, 1970.
Mineral composition of bottom sediments from the Romanche Trench. Okeanologiia, 10(3): 488-495. (In Russian; English abstract)

bottom sediments, minerals
Stoicovici, E., and M.T. Gomoiu, 1968.
Data on composition of sediments from meso-littoral area of the Black Sea.
Trav. Mus. Hist. nat. "Grigore Antipa" 8(1): 301-310.

bottom sediments, minerals
Vallentyne, J.R., 1963.
Isolation of pyrite spherules from Recent sediments.
Limnology and Oceanography, 8(1):16-30.

chiefly freshwater

bottom sediments, minerals
Vogel, D.E., 1965.
Thin sections for determining of the composition of the light-mineral fraction of unconsolidated sediments.
Geol. en Mijnbouw, 44(2):64-65.

bottom sediments, minerals
Volkov, I.I., A.G. Rozanov, and T.A. Iagodinskaia, 1971.
Pyrite microconcretions in sediments of the Black Sea. (In Russian).
Dokl. Akad. Nauk 197 (1): 195-198

botton sediments, minerals in
Volkov, P.A., 1965.
Experiments on the mechanism of sorting of heavy minerals in the littoral zone.
Sediment drift and the genesis of the heavy mineral fields in the coastal zone of the sea. (In Russian: English abstract).
Trudy Inst. Okeanol., Akad. Nauk, SSSR, 76:137-151.

bottom sediments, heavy mineral
Von der Borch and R. L. Oliver, 1965.
Comparison of heavy minerals in marine sediments with mainland rock outcrops along the coast of Antarctica between longitudes 40°E and 150°E.
Sedim. Geol. 2(1): 77-80

bottom sediments, mineralogy
White, Stan M., 1970
Mineralogy and geochemistry of continental shelf sediments of the Washington-Oregon coast.
J. Sedim. Petrol. 40 (1): 58-54

bottom sediments, minerals
Windom, Herbert L., William J. Neal, and Kevin C. Beck, 1971.
Mineralogy of sediments in three Georgia estuaries
J. Sedim. Petrol. 41 (2): 497-504.

bottom sediments, minerals
Yeroshchev-Shak, V.A., 1963.
Complex method of determining clay minerals in Atlantic Ocean sediments. (In Russian).
Trudy Morsk. Gidrofiz. Inst., Akad. Nauk, Ukrain. SSR, 28:102-107.

Translation:
Soviet Oceanography, Issue 2, 1963 series.
Scripta Tecnica, Inc., for AGU, 71-75. 1965.

bottom sediments, minerals
Zen, E-an, 1957.
Preliminary report on the mineralogy and petrology of some marine bottom samples of the coast of Peru and Chile. Amer. Mineralog., 42:889-903.

bottom sediments muddy

bottom sediments, mud
Biggs, Robert Bruce, 1963.
Deposition and early diagenesis of modern Chesapeake Bay muds.
Mar. Sci. Center, Lehigh Univ., 119 pp. (Unpublished manuscript).

bottom sediments, mud
Boswell, P.G.H., 1961
Muddy sediments. Geotechnical studies for geologists, engineers and soil scientists.
W. Heffer & Sons, Ltd., Cambridge, 140 pp.

bottom sediments, mud
Choquette, Philip W., 1968.
Marine diagenesis of shallow marine lime-mud sediments: insights from δO^{18} and δC^{13} data.
Science, 161(3846):1130-1132.

bottom sediments, mud
Damodarn, R., and C. Hridayanathan 1966.
Studies on the mud banks of the Kerala Coast.
Bull. Dept. mar. biol. Oceanogr. Univ. Kerala 2:61-68

bottom sediments, mud
Dobson, Max R., 1965.
Black mud in the outer Thames estuary.
Dock Harb. Auth., 46(535):18-21.

bottom sediments, mud
Dora, Y.L., R. Damodaran and V. Jos Anto, 1968.
Texture of the Narakat mud bank sediments.
Bull. Dept. mar. Biol. Oceanogr. Univ. Kerala, 4: 1-10.

bottom sediment, muds
Fox, D. L., and L. J. Anderson, 1941.
Pigments from marine muds. Proc. Nat. Acad. Sci. 27(7):333-337

bottom sediments, mud
Fox, D.L., and C.H. Oppenheimer, 1954.
The riddle of sterol and carotenoid metabolism in muds of the ocean floor. Arch. Biochem. Biophys. 51(2):323-328.

bottom sediment, mud
Gamulin-Brida, Helena, Slobodan Alfirević et Drago Crnković 1971.
Contribution à la biologie de la faune endogée des fonds vaseux de l'Adriatique avec égard spécial à l'espèce Nephrops norvegicus (L.).
Vie Milieu Suppl. 22 (2): 637-655

bottom sediments mud
Hamada, S., and I. Hosokawa, 1959
[On the muddy sediments of the East China Sea. J. Oceanogr. Soc., Japan,] 15(2): 39-42.

bottom sediments, mud
Hinschberger, F., et A. Saint-Réquier, 1970
Les sédiments vaseux au large du Finistère.
Norois 17 (67): 401-403.

bottom sediments, mud
Japan, Maritime Safety Board, 1958.
[On the depth in the mud zone.] Hydrogr. Bull. (Publ. 981), No. 55:54-72, pls.

bottom sediments, mud
Komudai, Ryoshichi, Kunio Sugiura and Fuhio Takabe, 1958.
[On the depth in the mud zone.]
Hydrogr. Bull., Mar. Safety Agency, Tokyo, (Publ. 981):54-72.

bottom sediments, mud
Nesteroff, W. D., A. Hinterlechner and G. Sabatier, 1959.
Sur la composition de quelques vases mediterraneenes.
Bull. Soc. Franc. Miner. Crist., 81:72-73.
Also in:
Trav. Lab. Geol. Sous-Marine, 8, 1958.

bottom sediments, mud
Owaki, N., 1963
A note on depth when the bottom is soft mud.
Int. Hydrogr. Rev., 40(2):41-43.

bottom sediments, mud
Pinot, J. P., 1961
Les accumulations vaseuses littorales au sud de l'ile-grande.
Cahiers Oceanogr., C.C.O.E.C., 13(7):462-484.

bottom sediments, mud.
*Reineck, Hans-Erich, Wolfgang Friedrich Gutman und Günther Hertweck, 1967.
Das Schlickgebiet südlich Helgoland als Beispiel rezenter Schelfablagerungen.
Senckenberg. leth., 48(3/4):219-275.
pp. 258-259.

bottom sediments, mud
Riedl, R., 1961.
Études des fonds vaseux de l'Adriatique. Methodes et résultats.
Rec. Trav. Sta. Mar., Endoume, 23(37):161-169.

bottom sediments, mud
Rigomier, D., J. Dupuis et P. Jambu 1971.
Les phénomènes de maturation du sol et leur incidence sur les modifications de l'activité biologique dans une séquence d'assèchement de vases marines (Prés salés et polders de l'Anse de l'Aiguillon).
Rev. Géogr. phys. Géol. dyn. (2) 8 (2): 101-122

bottom sediments, mud
*Seki, H., J. Skelding and T.R. Parson, 1968.
Observations on the decomposition of a marine mud.
Limnol. Oceanogr., 13(3):440-447.

bottom sediments, mud
Shimp, N.F., J. Witters, Paul Edwin Potter and J.A. Schleicher 1969.
Distinguishing marine and freshwater muds.
J. geol. 77 (5): 566-580.

bottom sediments, nannoplankton
Reiss, Z., P. Merling-Reiss and S. Moshkovitz 1971.
Quaternary planktonic Foraminiferida and nannoplankton from the Mediterranean continental shelf and slope of Israel.
Israel J. Earth-Sci. 20 (4): 141-177

bottom sediments, near shore
bottom sediments, near-shore
Parker, Francis L., 1964.
Criteria in modern sediments useful in recognizing ancient sedimentary environments.
In: Deltaic and shallow marine deposits, Developments in Sedimentology, L.M.J.U. van Straaten, editor, Elsevier Publishing Co., 1:1-25.

bottom sediments, nearshore
Sanders, John E., and Gerald M. Friedman, 1969.
Position of regional carbonate/noncarbonate boundary in nearshore sediments along a coast: possible climatic indicator.
Bull. geol. Soc. Am., 80 (9): 1789-1796

bottom sediments, near-shore
Shepard, F.P., 1964.
Criteria in modern sediments in recognizing ancient sedimentary environments.
In: Developments in Sedimentology, L.M.J.U. van Straaten, Editor, Elsevier Publishing Co., 1:1-25.

bottom sediments, oil-bearing
Ma, Ting-Ying H., 1970.
Development of grabens and age of the contained oil-bearing sediments as a basis for re-assessment of past geological deductions.
Acta geol. Taiwanica, 13: 59-82

bottom sediments paleomagnetism
bottom sediments, paleomagnetism
Harrison, C.G.A., 1966.
The paleomagnetism of deep sea sediments.
J. geophys. Res., 71(12):3033-3043.

bottom sediments, palaeomagnetic inclination, error.
Irving, E., 1967.
Evidence for palaeomagnetic inclination error in sediment.
Nature, Lond., 213(5075):483-484.

bottom sediments, paleomagnetism of
Linkova, T.I., and A.P. Lisitsyn 1971.
Paleomagnetic investigations of bottom sediments of the Indian Ocean sector of the Antarctic. (In Russian).
Dokl. Akad. Nauk SSSR 199 (6): 1409-1412

bottom sediments, paleomagnetism
Opdyke, N.D., 1970.
1. General observations. 5. Paleomagnetism.
In: The sea: ideas and observations on progress in the study of the seas, Arthur E. Maxwell, editor, Wiley-Interscience, 4(1): 157-182.

bottom sediments, paleomagnetic
Watkins, N.D., and H.G. Goodell, 1967.
Paleomagnetic studies of deep-sea sediments from Antarctic seas.
Antarctic Jl., U.S.A., 2(5):182.

bottom sediments, pelagic
bottom sediments, pelagic
Angino, Ernest E., 1966.
Geochemistry of Antarctic pelagic sediments.
Geochimica et Cosmochimica Acta, 30 (9): 939-961.

bottom sediments, pelagic (chemistry of)
Arrhenius, G., 1963.
25. Pelagic sediments.
In: The Sea, M.N. Hill, Editor, Interscience Publishers, 3:655-727.

Bottom Sediments, Pelagic
Arrhenius, Gustaf, and Enrico Bonatti, 1965.
Neptunism and vulcanism in the ocean.
Progress in Oceanography, 3:7-22.

abstract

bottom sediments, pelagic
Arrhenius, G., and E. Bonatti, 1961.
The formation of pelagic sediments. (Abstract).
Tenth Pacific Sci. Congr., Honolulu, 21 Aug.-6 Sept., 1961, Abstracts of Symposium Papers, 366.

bottom sediments pelagic
Berg, L.S., 1948.
[The formation of pelagic deposits.]
Vsesoyuznogo Geograf. Obsch. 80(3):272.

bottom sediments, pelagic
Berger, Wolfgang H., and G. Ross Heath 1968.
Vertical mixing in pelagic sediments.
J. mar. Res. 26 (2): 134-143.

bottom sediments, pelagic
Bonatti, E., 1963.
Zeolites in Pacific pelagic sediments.
Trans. N.Y. Acad. Sci., (2) 25(8):938-948.

bottom sediments, pelagic
Bramlette, M.N. and W.R. Riedel, 1971.
Observations on the biostratigraphy of pelagic sediments. In: Micropaleontology of oceans, B.M. Funnell and W.R. Riedel, editors, Cambridge Univ. Press, 665-668.

bottom sediments, pelagic
Chen, Chin 1968.
Pleistocene pteropods in pelagic sediments.
Nature, Lond. 219 (5159): 1145-1149.

bottom sediments, pelagic
Chester, R., 1965.
Elemental geochemistry of marine sediments.
In: Chemical oceanography, J.P. Riley and G. Skirrow, editors, Academic Press, 2:23-80.

bottom sediments, pelagic (chemistry)
Corliss, John B., 1971.
The origin of metal-bearing submarine hydrothermal solutions. J. geophys. Res. 76(33): 8128-8138.

bottom sediments, pelagic
Emiliani, Cesare, and John D. Milliman, 1966.
Deep-sea sediments and their geological record.
Earth-Sci. Rev., 1 (2/3):105-132.

bottom sediments, pelagic
Ewing, Maurice, Xavier Le Pichon and John Ewing, 1966.
Crustal structure of the mid-ocean ridges.
4. Sediment distribution in the South Atlantic Ocean and the Cenozoic history of the Mid-Atlantic Ridge.
J. Geophys. Res., 71(6):1611-1636.

Bottom sediments, Pelagic
Fairbridge, Rhodes W., 1965.
The Indian Ocean and the status of Gondwanaland.
Progress in Oceanography, 3:83-136.

abstract

bottom sediments, pelagic
Hamaguchi, Hiroshi, Masumi Osawa and Naoki Onima, 1962
The vertical distribution of minor elements in core samples from the Japan Trench. (In Japanese; English abstract).
Repts. JEDS, Deep-Sea Res. Comm., Japan Soc., Promotion of Science, 3:1-2.
Originally published:
Nippon Kagaku Zasshi, 82:691-693 (1961).
JEDS Contrib. No. 21.

bottom sediments, pelagic
Hamilton, Edwin L., 1967.
Marine geology of abyssal plains in the Gulf of Alaska.
J. geophys. Res., 72(16):4189-4213.

bottom sediments, pelagic
Jones, E.J., J. Ewing and M. Truchan, 1971.
Aleutian Plain sediments and lithospheric plate motions. J. geophys. Res. 76(33): 8121-8127.

Bottom Sediments, pelagic
Ku, Teh-Lung, 1965.
An evaluation of the U234/U235 method as a tool for dating pelagic sediments.
J. Geophys. Res., 70(14):3457-3474.

bottom sediments, pelagic
Kuenen, Ph. H., 1964
Deep-sea sands and ancient turbidites.
In: Turbidites, A.H. Bouma and A. Brouwer, Editors, Developments in Sedimentology, Elsevier Publishing Co., 3:3-33.

bottom sediments, pelagic
Landergren, Sture, 1964.
On the geochemistry of deep-sea sediments.
Repts. Swedish Deep-Sea Exped., 10(Spec. Invest. 5):61-148.

bottom sediments, pelagic
Le Pichon, Xavier, 1966.
Étude de la Dorsale Médio-Atlantique. IV.
Distribution des sediments non consolidés sur la Dorsale Medio-Atlantique: histoire géologique de la Dorsale durant l'ère tertiaire.
Cahiers Océanogr., C.C.O.E.C., 18(8):669-713.

bottom sediments pelagic
Peterson M.N.A., and Edward D. Goldberg, 1962
Feldspar distributions in South Pacific pelagic sediments
J. Geophys. Res., 67(9):3477-3492.

bottom sediments, pelagic
Revelle, R., M. Bramlette, G. Arrhenius and E.D. Goldberg, 1955.
Pelagic sediments of the Pacific. In: Crust of the Earth, Eric Poldervaart, Ed., G.S.A. Spec. Paper, 62:221-235.

bottom sediments, pelagic
Rex, R.W., and E.D. Goldberg, 1958.
Quartz contents of pelagic sediments of the Pacific Ocean. Tellus, 10(1):153-159.

bottom sediments, pelagic
*Rex, R.W., J.K. Syers, M.L. Jackson and R.N. Clayton, 1969.
Eolian origin of quartz in soils of Hawaiian islands and in Pacific pelagic sediments.
Science, 163(3864):277-279.

bottom sediments, pelagic (paleontology of)
Riedel, W.R., 1963.
33. The preserved record: paleontology of pelagic sediments.
In: The Sea, M.N. Hill, Editor, Interscience Publishers, 3:866-887.

bottom sediments, pelagic
*Scholl, David W., Roland von Huene and James B. Ridlon, 1968.
Spreading of the ocean floor: undeformed sediments in the Peru-Chile Trench.
Science, 159(3817):869-871.

bottom sediments, sound velocity in
*Schreiber, B. Charlotte, 1968.
Sound velocity in deep sea sediments.
J. geophys. Res., 73(4):1259-1267.

bottom sediments, pelagic
Smith, Stephen V., Joseph A. Dygas and Keith E. Chave, 1968.
Distribution of calcium carbonate in pelagic sediments.
Marine Geol., 6(5): 391-400

bottom sediments, pelagic
Zubakov, V.A., 1966.
A comparison of the radiometric scale of the continental Pleistocene with the chronological diagrams of bathypelagic bottom sediments and the curve of solar radiation. (In Russian).
Dokl., Akad. Nauk, SSSR, 171(5):1153-1155.

bottom sediments, mud-water interface
See: water - bottom interface
bottom - sea interface

bottom sediment - water interface
Gieskes, Joris M., 1970.
Einige Beobachtungen über Lösungsvorgänge am Boden des Ozeans.
Meteor Forschungsergebn. (A) 8:12-17

sediment-seawater interface
Sharma, Ghanshyam D., 1970.
Sediment-seawater interaction in glaciomarine sediments of southeast Alaska.
Bull. geol. Soc. Am. 81(4): 1097-1106.

bottom sediments, physical properties
Also See: bottom sediments, mechanical analysis

Bottom sediments, physical properties
See also: bottom sediments, acoustic properties
bottom sediments, bearing capacity
bottom sediments, caloric content
Etc.

bottom sediments, physical characteristics
Biggs, Robert B., 1957.
The sediments of Chesapeake Bay.
In: Estuaries, G.H. Lauff, editor, Publs Am. Ass. Advmt Sci., 83:239-260.

bottom sediments, physical properties
*Buchan, S., F.C.D. Dewes, D.M. McCann and D. Taylor Smith, 1967.
Measurements of the acoustic and geotechnical properties of marine sediment cores.
In: Marine geotechnology, A.F. Richards, editor, Univ. Illinois Press, 65-92.

bottom sediments, physical properties
Cook, F.M., 1972.
Physical properties synthesis. Initial Repts Deep Sea Drilling Project 9: 945-946.

bottom sediments, physical properties
Hamilton, Edwin L., 1965.
Sound speed and related physical properties of sediments from experimental Mohole (Guadalupe Site).
Geophysics, 30(2):257-261.

bottom sediments, physical properties of
Hamilton, E.L., G. Shumway, H.W. Menard and C.J. Shipek, 1956.
Acoustic and other physical properties of shallow water sediments off San Diego.
J. Acoust. Soc., Amer., 28(1):16-19.

bottom sediments, physical properties
Inderbitzen, A.L., 1970.
Empirical relationships between mass physical properties for recent marine sediments off southern California.
Marine Geol. 9(5):311-329

bottom sediments, physical properties
Inderbitzen A.L. 1968.
A study of the effects of various core samples on mass physical properties in marine sediments.
J. Sedim. Petrol. 38(2): 473-489.

bottom sediment property
Inderbitzen, A.L., and F. Simpson 1971.
Relationships between bottom topography and marine sediment properties in an area of submarine gullies.
J. Sedim. Petrol. 41 (4): 1126-1133.

bottom sediments, physical properties
Keller, George H., 1968.
Shear strength and other physical properties of sediments from some ocean basins: Proc. Conf. on Civil Engineering in the Oceans, Sept. 1968, Amer. Soc. Civil Engr., 391-417. Also in: Coll. Reprints, Atlantic-Pacific Oceanogr. Labs. ESSA, 1968.

bottom sediments,physical properties
*Keller,George H., and Richard A. Bennett,1970.
Variations in the mass physical properties of selected submarine sediments.
Marine Geol., 9(3):215-223.

bottom sediments, physical properties
McCrone,A.W.,1966.
Sediments from Long Island Sound (New York) physical and chemical properties reviewed.
J. Sed. Petr. 36(1):234-236.

bottom sediments, physical properties
Nafe, J.E., and C.L. Drake, 1963.
29 Physical properties of marine sediments.
In: The Sea, M.N. Hill, Editor, Interscience Publishers, 3:794-815.

bottom sediments, physical properties
Keller, George H. and R.H. Bennett, 1968.
Mass physical properties of submarine sediments in the Atlantic and Pacific basins: Proc. 23 Intern. Geol. Cong. Prague, 8: 33-50. Also in: Coll. Reprints, Atlantic-Pacific Oceanogr. Labs., ESSA, 1968.

bottom sediments, physical properties
Nafe, J.E., and C.L. Drake, 1957.
Variation with depth in shallow and deep water marine sediments of porosity, density and the velocities of compressional and shear waves.
Geophysics 22(3):523-552.

bottom sediments, physical properties
Richards, Adrian F., 1962
Investigations of deep-sea sediment cores. II. Mass physical properties.
U.S. Navy Hydrogr. Off., Techn. Rept., TR-106:146 pp.

bottom sediments, physical properties
*Ross, David A., 1971.
Mass physical properties and slope stability of sediments of the northern Middle America Trench.
J. geophys. Res., 76(3): 704-712.

sediments, physical properties
Waskom, J.D., 1958.
Roundness as an indicator of environment along the coast of panhandle Florida.
J. Sed. Petr. 28(3):351-360.

bottom sediments, pelagic
Heath, G. Ross and Ralph Moberly, Jr., 1971.
Noncalcareous pelagic sediments from the western Pacific: Leg 7, Deep Sea Drilling Project.
Initial Repts Deep Sea Drill. Proj. 7(2): 987-990.

bottom sediments, petrology of

bottom sediments, petrology of
Fuller, M.D., C.G.A. Harrison and Y.R. Nayudu, 1966.
Magnetic and petrologic studies of sediment found above basalt in experimental Mohole core EM7.
Bull. Amer. Assoc. Petr. Geol., 50(3):566-573.

bottom sediment ponding

bottom sediment ponding
Horsey, J.B., 1965.
Sediment ponding in the deep-sea.
Geol. Soc. Am. Bull., F6(11):1251-1260.

bottom sediments, ponding
Ryan, William B.F., Fifield Workum Jr. and J.B. Hersey 1965
Sediments on the Tyrrhenian Abyssal Plain.
Bull. geol. Soc. Am. 76(11): 1261-1282

bottom sediments, ponded
Van Andel, Tjeerd H., and Paul D. Komar, 1969.
Ponded sediments of the Mid-Atlantic Ridge between 22° and 23° North Latitude.
Bull. Geol. Soc. Am. 80(7): 1163-1190

bottom sediment, porosity

bottom sediments, porosity
Bennett, Richard H., and Douglas N. Lambert 1971.
Rapid and reliable technique for determining unit weight and porosity of deep-sea sediments.
Mar. Geol. 11(3): 201-207

bottom sediments, porosity of
Crisp, D.J. and R. Williams, 1971.
Direct measurement of pore-size distribution on artificial and natural deposits and prediction of pore space accessible to interstitial organisms. Mar. Biol. 10(3): 214-226.

bottom sediment-porosity
Inderbitzen, Anton L., 1965.
An investigation of submarine slope stability.
Ocean Sci. and Ocean Eng., Mar. Techn. Soc., Amer. Soc. Limnol. Oceangor., 2:1309-1344.

bottom sediments, post-glacial
Medvedev, V.S., and E.N. Nevesskii 1971.
New data on sedimentation in the White Sea during late-post glacial period. (In Russian)
Dokl. Akad. Nauk SSSR 200 (1): 179-181

bottom sediments, red

bottom sediments, red
Hinge Carsten und Dieter Meischner 1968
Gibt es regenten Rot-Sediment in der Adria?
Mar. Geol. 6(1): 53-71

bottom sediments, reef

bottom sediments, reef
Ladd, Harry S., Joshua I. Tracey, Jr., and M. Grant Gross, 1967.
Drilling on Midway Atoll, Hawaii.
Science, 156(2778):1088-1094.

bottom sediments, relict

bottom sediments, relict
*Emery, K.O., 1968.
Relict sediments on continental shelves of the world.
Bull.Am.Ass.Petrol.Geol., 52(3):445-464.

bottom sediments, relict
Swift, Donald J.P. Daniel J. Stanley and Joseph R. Curray, 1971.
Relict sediments on continental shelves: a reconsideration.
J. Geol. 79(3): 322-346

bottom sediments, resistivity

bottom sediments, resistivity
Boyce, Robert E 1968
Electrical resistivity of modern marine sediments.
J. geophys. Res. 73(14): 4759-4766.

bottom sediments, reworking of

Bottom sediments, reworking of

See also: burrows, burrowing, etc.

bottom sediments
Glagoleva, M.A., 1960.
The influence of bottom organisms on the distribution of the elements in Recent sediments of the Black Sea.
Doklady Akad. Nauk, SSSR, 135(5):1233
135(1-6)1091-1093 (1961) in English

bottom sediment, reworking of
Haven, Dexter S., and Reinaldo Morales-Alamo, 1965.
The use of flourescent particles in the study of sediment mixing by invertebrates. (abstract).
Ocean Sci. and Ocean Eng., Mar. Techn. Soc., Amer. Soc. Limnol. Oceanogr., 2:736a.

bottom sediments, reworking of
Rhoads, Donald C., 1963.
Rates of sediment reworking by Yoldia limulata in Buzzards Bay, Massachusetts and Long Island Sound.
J. Sed. Petr., 33(3):723-727.

bottom sediments, reworking of
Rhoads, D.C. and D.K. Young, 1971.
Animal-sediment relations in Cape Cod Bay, Massachusetts. II. Reworking by Molpadia oolitica (Holothuroidea). Mar. Biol. 11(3): 255-261.

bottom sediments, sandy

bottom sediments, sands (volcanic origin)
Deneufbourg, G., 1971.
Étude géologique du port de Papeete Tahiti (Polynésie Française). Cah. pacifiques 15: 75-82.

bottom sediments, sandy
*Dill, Robert F., 1967.
Effects of explosive loading on the strength of sea-floor sands.
In: Marine geotechnique, A.F. Richards, editor, Univ. Illinois Press, 291-303.

bottom sediments, sandy
*Harrison, W., and A.M. Richardson, 1967.
Plate-load tests on sandy marine sediments, lower Chesapeake Bay.
In: Marine geotechnique, A.F. Richards, editor, Univ. Illinois Press, 274-290.

bottoms sediments, Sand
Kuenen, Ph. H., 1964.
Pivotability studies of sand by a shape-sorter.
In: Developments in Sedimentation, L.M.J.U. van Straaten, Editor, Elsevier Publishing Co., 1:207-215.

bottom sediments, sand
Kuenen, Ph. H., 1960.
Experimental abrasion of sand grains.
Rept. Int. Geol. Congress, 21st Session, Norden, 1960 (10 Submarine Geology):50-53.

bottom sediments, sand

Kuenen, Ph. H., 1959.
Sand its origin, transportation, abrasion and accumulation. Alex. L. du Toit Memorial Lectures No. 6. Geol. Soc. S. Africa, Annexure to Vol. 62:33 pp.

bottom sediments, sand

Menard, H. W., jr., 1949.
Synthesis of sand mixtures. J. Sed. Petr. 19 (2):71-77, Fig. 1.

bottom sediments, sand (carbonate)

Moore, Clyde H., Jr., 1969.
Factors controlling carbonate sand distribution in the shallow shelf environment: illustrated by the Texas Cretaceous. Trans. Gulf Coast Ass. geol. Socs. 19 : 507.

bottom sediments, sand

Rex, R.W., and S.V. Margolis, 1969.
Surface features on sand grains from Antarctic continental shelf and deep-sea cores.
Antarctic J., U.S., 4(5): 168-170.

bottom sediments, sand

Scheidegger, K.F., L.D. Kulm and E.J. Runge 1971.
Sediment sources and dispersal patterns of Oregon continental shelf sands.
J. sedim. Petrol. 41 (4): 1112-1120.

bottom sediments, sand

Wang, Chao-Siang, 1961.
Sand fraction study of the shelf sediments off the China coast.
Proc. Geol. Soc., China, (4):33-49.

bottom sediments, sand

Wang, Chao-Siang, 1961.
Sand fraction study of Chinese-Asiatic shelf sediments. (Abstract).
Tenth Pacific Sci. Congr., Honolulu, 21 Aug.-6 Sept., 1961, Abstracts of Symposium Papers, 389.

bottom sediments, seismic waves

Hamilton, E.L., H.P. Bucker, D.L. Keir and J.A. Whitney 1970.
Velocities of compressional and shear waves in marine sediments determined in situ from a research submersible.
J. geophys Res. 75: 4039-4049.

bottom sediments, sandforms

Reineck, Hans-Erich, Indra Bir Singh and Friedrich Wunderlich, 1971.
Einteilung der Rippeln und anderer mariner Sandkörper. Senckenberg. marit. 3(1): 93-101.

bottom sediments, shallow

bottom sediments, shallow

Rasmussen, K., 1957.
Investigations on marsh soils and Wadden Sea sediments in the Tøndre region.
Medd. Skalling Lab., 15:147-170.

bottom sediments, shear strength of

bottom sediments, shear strength

*Bryant, William R., Paul Cernock and Jack Morelock, 1967.
Shear strength and consolidation characteristics of marine sediments from the western Gulf of Mexico.
In: Marine geotechnique, A.F. Richards, editor, Univ. Illinois Press, 41-62.

bottom sediments, shear strength

Bryant, William, Adrian F. Richards and George H. Keller, 1969.
Shear strength of sediments measured in place near the Mississippi Delta compared to measurements obtained from cored material. Trans. Gulf Coast Ass. geol. Socs. 19 : 267.

bottom sediments, shear strength

Bryant, William R., and Charles S. Wallin, 1968
Stability and geotechnical characteristics of marine sediments, Gulf of Mexico.
Trans. Gulf Coast Ass. Geol. Socs, 18: 334-356.
Also in: Oceanogr. Texas A&M Univ., 12 (Contr. 392) Contr.

bottom sediments, cohesion)

Holmes, C.W., and H.G. Goodell, 1964.
The prediction of strength in the sediments of St. Andrew Bay, Florida.
J. Sed. Petr., 34(1):134-143.

bottom sediments, shear strength of

Moore, D.G., and A.F. Richards, 1962
Conversion of "relative shear strength" measurements by Arrhenius on east Pacific deep-sea cores to conventional units of shear strength.
Geotechnique, March, 1962, Inst. Civil Engineers, London, 55-59.

bottom sediments, shear strength

Richards, A.F., 1961
Investigations of deep-sea sediment cores. 1. Shear strength, bearing capacity and consolidation.
U.S. Navy Hydrogr. Off., Techn. Rept., TR-63: 70 pp.

bottom sediments, slope stability

Inderbitzen, Anton L., 1965.
An investigation of submarine slope stability.
Ocean Sci. and Ocean Eng., Mar. Techn. Soc., Amer. Soc. Limnol. Oceanogr., 2:1309-1344.

sediments, solidification of

sediments, solidification of

Heveski, E.N., 1949.
Solidification of shells and calcium carbonate sands on several portions of the shores of the Black and Caspian Seas. Trudy Inst. Okeanol., 4: 172-178.

Bottom sediments, sound in

See: bottom sediments, acoustic properties

bottom sediments, sorting of

bottom sediments, sorting

Cherry, John., 1965.
Sand movement along a portion of the Northern California coast.
U.S. Army Coast Eng. Res. Center, Techn. Memo; No. 14:125 pp.

bottom sediments, sorting of

Giresse, P., 1967
Mécanismes de répartition des minéraux argileux des sédiments marins actuels sur le littoral sud du Cotentin.
Marine Geol., 5(1): 61-69.

bottom sediments, sorting of

Ippen, A.T., and P.S. Eagleson, 1955.
A study of sediment sorting by waves shoaling on a plane beach. B.E.B. Tech. Memo. 63:1-83.

bottom sediments, sorting of

Volkov, P.A., 1965.
Experiments on the mechanism of sorting of heavy minerals in the littoral zone.
Sediment drift and the genesis of the heavy mineral fields in the coastal zone of the sea. (In Russian;English abstract).
Trudy Inst. Okeanol., Akad. Nauk, SSSR, 76:137-151.

bottom sediments, sources of

bottom sediments, sources

Kulm, L.D., and John V. Byrne, 1967.
Sediments of Yaquina Bay, Oregon.
In: Estuaries, G.H. Lauff, editor, Publs Am. Ass. Advmt Sci., 83:226-238.

bottom sediments, spherules

Fredriksson, Kurt, 1961.
Origin of black spherules from Pacific islands, deep-sea sediments and Antarctic ice. (Abstract).
Tenth Pacific Sci. Congr., Honolulu, 21 Aug.-6 Sept., 1961, Abstracts of Symposium Papers, 370-371.

sediments, stratification

See: bottom sediments, layering of

bottom sediments, stratigraphy of

Davidova, N.N., R.N. Jinoridze, D.D. Kvasov, G. Masicke and E.A. Spiridonova, 1970.
New data on the stratigraphy of the bottom deposits of the southern Baltic. (In Russian; English and German abstracts.
Baltica, Lietuv. TSR Mokslu Akad., Geogr. Skyr. INQUA Taryb. Sek. 4: 33-48. (Vilnius,

sediments, stratification

Lavrov, V.M., 1962
Features of sediments stratification of the North-Eastern Atlantic.
Trudy Inst. Okeanol., Akad. Nauk, SSSR, 56: 15-22.

bottom sediments, stratigraphy

Maev, E.G., 1961. (also Mayev)
A contribution to the stratigraphy of deep-water deposits of the South Caspian.
Doklady Akad. Nauk, SSSR, 136(6):1428-1431.

bottom sediments, stratigraphy

Mayev, Ye. G., 1961.
The stratigraphy of the deep-sea deposits of the southern Caspian.
Doklady Akad. Nauk, SSSR, 136(6):1428.

English Edit. (1962), 136(1-6):147-149.

bottom sediments, stratification

Stubbings, H.G., 1939.
Stratification of biological remains in marine deposits. John Murray Exped., 1933-34, Sci. Repts. 3(3):159-192, 4 textfigs.

bottom sediments, strength of

bottom sediments, strength of

Rucker, James B., Newell T. Stiles and Roswell F. Busby, 1967.
Sea-floor strength observations from the DRV ALVIN in the Tongue of the Ocean, Bahamas.
SEast. Geol., Duke Univ., 8(1):1-8.

bottom sediments, strength

Gershenovich, D.E., 1965.
Recent bottom sediment strength and sedimentation rate in the Bering Sea. (In Russian).
Trudy vses. nauchno-issled. Inst. morsk. ryb. Khoz. Okeanogr. (VNIRO), 57:261-269.

bottom sediments, surface

Allen, George P., Patrice Castaing and Andre Klingebiel, 1971.
Preliminary investigation of the surficial sediments in the Cap-Breton Canyon (Southwest France) and the surrounding continental shelf
Mar. Geol. 10 (5): M27-M32.

bottom sediments, surface

Sarnthein, Michael, 1971.
Oberflächensedimente im Persischen Golf und Golf von Oman. II Quantitative Komponentenanalyse der Grobfraktion.
Meteor-Forsch.-Engebn. (C) 5: 1-113.

Sediments, suspended
See also: Suspended particles + matter

bottom sediments, suspensions

Chien, N., 1952.
The efficiency of depth-integrating suspended-sediment sampling. Trans. Amer. Geophys. Union 33(5):693-698, 2 textfigs.

sediment suspensions

Fukushima, Hisoa and Masakazu Kashiwamura, 1959
Field investigations of suspended sediment by the use of bamboo samplers. Coastal Eng., Japan, 2: 53-57.

bottom sediments, suspended

*Halliwell, A.R., and B.A. O'Connor, 1967.
Suspended sediment in a tidal estuary.
Proc. 10th Conf. Coast. Engng. Tokyo, 1966, 1:687-706.

bottom sediments, suspension of

Iwagaki, Yuichi, Yoshito Tsuchiya and Yoichire Yano, 1965.
Some experiments on the influence of size frequency distributions of sediments on their suspension.
Annuals, Disaster Prevention Res. Inst. Kyoto, Univ., No. 8:353-362.

bottom sediments, suspensions

Schubel, J.R. 1968
Suspended sediment of the northern Chesapeake Bay
Techn. Rept. Chesapeake Bay Inst. 68-2: 264 pp. (multilithed)

Bottom sediments, temperature

bottom sediments, temperature

Vatova, A., 1961.
Sur les temperatures des fonds vaseux de la Lagune de Venise.
Rapp. Proc. Verb., Réunions, Comm. Int. Expl. Sci., Mer Mediterranee, Monaco, 16(3):787-788.

Bottom sediments, terrigenous

bottom sediments, terrigenous

Barkovskaya, M.G., 1961.
General laws governing the distribution of terrigenic material in the shoreline zone USSR of the Black Sea. Dynamics of the coastal zone of the Black Sea. (In Russian).
Trudy Inst. Okeanol., Akad. Nauk, SSSR, 53:64-94.

bottom sediments, terrigenous

Bunce, Elizabeth T., M. G. Langseth, R.L. Chase and M. Ewing 1967.
Structure of the western Somali Basin.
J. geophys. Res. 72 (10): 2547-2555.

bottom sediments, terriginous

Pratt, Richard M., 1968.
Atlantic continental shelf and slope of the United States- physiography and sediments of the deep-sea basin.
Prof. Pap. U.S. geol. Surv., 529B: B1-B44, fold in chart.

bottom sediments, terrigenous

Pratt, R.M., 1966.
The Gulf Stream as a graded river.
Limnol. and Oceanogr., 11(1):60-67.

bottom sediments, terrigenous

Weser, Oscar E., 1970.
Lithologic summary. Initial Repts. Deep Sea Drilling Project, Glomar Challenger 5: 569-620.

Bottom sediments, Tertiary

bottom sediments, Tertiary

Bryant, William R., and Thomas E. Pyle, 1965.
Tertiary sediments from Sigsbee Knolls, Gulf of Mexico.
Bull. Amer. Assoc. Petr. Geol., 49(9):1517-1518.

bottom sediments, texture

Alfirevic, S., 1961.
Influence des facteurs géomorpholigiques, hydrophysiques et biologiques sur la sèlection granulometrique des sediments dans les chenaux de l'Adriatique.
Rapp. Proc. Verb., Réunions, Comm. Int. Expl. Sci., Mer Mediterranee, Monaco, 16(3):749-755.

bottom sediments, texture

Andreieff, Patrick, Philippe Bouysse, Jean-Jacques Chateauneuf, Alain L'Homer et Georges Scolari, 1971.
La couverture sédimentaire meuble du plateau continental externe de la Bretagne meridionale (Nord du golfe de Gascogne).
Cah. océanogr. 13(4): 343-381

bottom sediments, texture

Angino, Ernest E., and Robert S. Andrews 1968.
Trace element chemistry, heavy minerals, and sediment statistics of Weddell Sea sediments.
J. sedim. Petrol. 38(2): 634-642.

bottom sediments, texture

Asensio Amor, Isidoro, y Pedro Balle Cruellas, 1967.
Primeros resultados del estudio sedimentológico de los fondos marinos del archipiélago Balear.
Bol. Inst. español Oceanogr., 130: 18pp.

bottom sediments, texture

Asensio Amor, Isidoro, y Pedro Balle Cruellas, 1967.
Consideraciones sedimentológicos sobre los fondos marinos frente a la Costa Brava Catalana.
Bol. Inst. español Oceanogr. 129: 11 pp.

bottom sediments, texture

Avilov, I.K., 1965.
Relief and bottom sediments of the shelf and continental slope of the Northwestern Atlantic. Investigations in line with the programme of the International Geophysical Year, 2. (In Russian).
Trudy, Vses. Nauchno-Issled. Inst. Morsk. Ribn. Choz. i Okeanogr. (VNIRO), 57:173-234.

bottom sediments, texture

Avilov, I.K., 1965.
Some data on the bottom topography and grounds of the West African shelf. Investigations in line with the programme of the International Geophysical Year, 2. (In Russian).
Trudy, Vses. Nauchno-Issled. Inst. Morsk. Ribn. Choz. i Okeanogr., (VNIRO), 57:235-259.

bottom sediments, texture

Barusseau, Jean Paul, 1970.
Etude granulométrique des sédiments du plateau de Chardonnière (Ile d'Oleron); intérêt de l'analyse des modes des courbes de fréquences.
Cah. océanogr. 22(5): 439-455

bottom sediments, texture

Barusseau, Jean-Paul, 1967.
Etude granulométrique préliminaire des sédiments détritiques non cohesifs des pertuis charantais.
Cah. océanogr., 19(4):311-328.

bottom sediments, texture

Berthois, L., 1969.
Contribution à l'étude sédimentologique du Kangerdlugssuaq, Côte ouest du Groenland.
Meddr Grønland 187(1): 135 pp.

bottom sediments, texture

Berthois, Léopold, René Battistini et Alain Crosnier, 1964
Recherches sur le relief et la sédimentologie du plateau continental de l'extreme sud de Madagascar.
Cahiers Océanogr., C.C.O.E.C., 16(8):637-655.

bottom sediments, texture

Berthois, Léopold, et André Gendre, 1967.
Recherches sur le comportement hydraulique des particules sédimentaires.
Cah. océanogr., 19(2):95-123.

bottom sediments, texture

Berthois, Léopold, et Yolande Le Calvez, 1961
Étude dynamique de la sedimentation dans la baie de Sangarea (République de Guinée).
Cahiers Océan., C.C.O.E.C., 13(10):694-714.

bottom sediments, texture

*Bruevich, S.W., and E.Z. Kulik, 1967.
Physical properties of sediments in the Pacific Ocean. (In Russian; English abstract).
Trudy Inst. Okeanol., 83:153-165.

bottom sediments, texture

Carey, Andrew G., Jr. 1965.
Preliminary studies of animal sediment interrelationships off the central Oregon coast.
Ocean Sci. Ocean Engng., Mar. Tech. Soc. — Am. Soc. Limnol. Oceanogr. 1: 100-110

Bottom sediments, texture

Conaghan, P. J., 1966.
Sediments and sedimentary processes in Gladstone Harbor, Queensland.
Univ. Queensland Papers, Dept. Geol., 6 (1): 7 - 52.

bottom sediments, texture

Conti, S., and G. Fiero, 1961.
Studio dei movimenti e delle migrazioni dei depositi litorali di alcuni bassi fondali del mar Ligure, in rapporto alla batimetria, alla granulometria e al moto ondoso.
Rapp. Proc. Verb., Reunions, Comm. Int. Expl. Sci., Mer Mediterranee, Monaco, 16(3):735.

bottom sediments, texture

Crosnier, A., 1964.
Fonds de peche le long des cotes de la Republique Federale du Camaroun.
Cahiers, O.R.S.T.R.O.M., No. special:133 pp.

bottom sediments, texture

Davidson, C., and N.J. Keen, 1963.
Size analyses of turbidity current sediment.
Nature, 197(4863):372-373.

bottom sediments, texture
Dora, Y.L., R. Damodaran and V. Jos Anto 1968.
Texture of the Narakal mud bank sediments.
Bull. Dept. mar. Biol. Oceanogr. Univ. Kerala, 4: 1-10.

bottom sediments, texture
Emery, K.O., and Jobst Hülsemann, 1963
Submarine canyons of southern California.
Part 1. Topography, water, and sediments.
Allan Hancock Pacific Expeditions, 27(1): 1-80.

bottom sediments, texture
Febres, German, y Gilberto Rodriguez, 1966.
Ch. 3. Sedimentos.
Estudios hidròbiologicos en el Estuario de Maracaibo, Inst. Venezolano de Invest. Cient. 66-82. (Unpublished manuscript).

bottom sediments, texture
Feki, Mohamed, 1960.
Étude de quelques sédiments marins actuels recoltés au large des côtes de Tunisie.
Sta. Océanogr., Salammbo, Ann., No. 12:33 pp.

Note:
Inside title page author is called Monsieur Feki Mohamed.

bottom sediments, texture
Fierro, G., e R. Passega, 1965.
Studio sedimentologico di 32 carote prelevate nel Nord. Tirreno.
Atti XIV Conuegno Annuale, Assoc. Geogis. Ital., Roma, 18-20 Feb., 1965:223-234.

bottom sediments, textures
Gardet, Monique, 1961
Notes preliminaires à l'étude des sediments marins du bassin de la Tour Rouge.
Les Cahiers du Centre d'Études et de Recherches de Biologie et d'Océanographie (C.E.R.B.O.M.), (2): 33-39.

bottom sediments, texture
Gershanovich, D.E., 1962.
New data on Recent sediments of the Bering Sea. (In Russian).
Trudy Vses. Nauchno-Issledov. Inst. Morsk. Ribn. Chos. i Okean., VNIRO, 46:128-164.

bottom sediments, texture
*Girogetti, F., F. Mosetti e G. Macchi, 1968.
Caratteristiche morphologiche. Fisiche e chimiche del fondo marino del Golfo di Trieste nell' area compresa entro la Congiungente Punta Grossa-Bocche di Primero.
Boll. Soc. adiat. Sci. nat., 6(1):3-21.

bottom sediments, texture
Guilcher, André, 1964.
La sédimentation sous-marine dans la partie orientale de la Rade de Brest, France.
In: Developments in Sedimentology, L.M.J.U. van Straaten, Editor, Elsevier Publishing Co., 1:148-156.

bottom sediments, texture
Hamada, Shichiro and Ritsuko Hamada 1966.
The state of sediments in Omura Bay at the time of the Akashio occurrence. (In Japanese; English abstract).
Bull. Seikai reg. Fish. Res. Lab. 34: 149-159

bottom sediments, texture
Hamilton, Edwin L., 1964
Consolidation characteristics and related properties of sediments from experimental Mohole (Guadalupe site).
J. Geophys. Res., 69(20):4257-4269.

bottom sediments, texture
Hamilton, N., 1965.
Description of a bottom-sediment sample dredged from Halley Bay.
Brit. Antarctic Surv. Bull., No. 7:47-52.

bottom sediments, texture
Hishida, Kozo, Katsumi Tanioka, Takeo Yasuoka Toshio Wakabayashi, 1961
On the blocking of the mouth of the River Yura
Bull. Maizuru Mar. Obs., No. 7:31-44, (51-64)

bottom sediments, texture
Hoshino, M., and Y. Ichihara, 1960
[Submarine topography and bottom sediments off Kumanonada and Enshu-nada.]
J. Oceanogr. Soc., Japan, 16(2): 41-46.

bottom sediments, texture
Iwabuti, Y., 1962
[Continental slope and outer continental shelf off Zyoban.]
Hydrogr. Bull., Tokyo, (Publ. No. 981), 70:33-38.

bottom sediments, texture
*James, Noel P., and Daniel J. Stanley, 1968.
Sable Island Bank off Nova Scotia: Sediment dispersal and Recent history.
Bull. Am. Ass. Petr. Geol., 52(11):2208-2230.

bottom sediments, texture
Jones, Anthony S.G., 1971.
A textural study of marine sediments in a portion of Cardigan Bay (Wales).
J. sedim. Petrol. 41 (2): 505-516.

Bottom sediments, texture
Kagami, Hideo, 1961.
Submarine sediments off Sakata, Yamagata, Japan.
Japanese J., Geol. and Geogr., 32(3/4):397-412.

bottom sediments, texture
Keller, George H., and Adrian F. Richards, 1967.
Sediments of the Malacca Strait, Southeast Asia.
J. sedim. Petrol., 37(1):102-127.

bottom sediments, texture
Kepinski, A. 1968
Korngusammensetzung wasserführender Bodenschichten.
Acta hydrophysica 12 (3): 101-114

bottom sediments, texture
Lüneberg, H., 1964.
Beiträge zur Sedimentologie der Weser- und Elbästuare.
Helgoländer Wiss. Meeresuntersuch., 10(1/4):217-230.

bottom sediments, texture
Magliocca, A., & A.S. Kutner, 1965.
Sedimentos de fundo da Enseada do Flamengo-Ubatuba. (In Portuguese; English summary).
Contrib. Avulsas, Inst. Oceanogr., São Paulo, (Oceanogr. Fisica), No. 8:14 pp.

bottom sediments, texture
Mathieu, Robert, 1968.
Les Sediments du plateau continental du Maroc Atlantique entre Dar-Bou-Azza et Mohammedia.
Bull. Inst. Pêches Marit. Maroc. 19: 65-76.

bottom sediments, texture
McKenzie, Kenneth G., 1964.
The ecologic associations of an ostracode fauna from Oyster Harbour, a marginal marine environment near Albany, Western Australia.
Pubbl., Staz. Zool., Napoli, 33(Suppl.):421-461.

bottom sediments, texture
Meisburger, Edward P., and David B. Duane, 1971.
Geomorphology and sediments of the inner continental shelf, Palm Beach to Cape Kennedy, Florida.
Techn. Memo. Coast. Engng Res. Cent., U.S. Army Engrs, 34: 111 pp. (multilithed)

bottom sediments, texture
Nichols, Maynard M., 1964.
Characteristics of sedimentary environments in Moriches Bay.
In: Papers in Marine Geology, R.L. Miller, Editor, Macmillan Co., N.Y., 363-383.

bottom sediments, texture
Pantin, H.M., 1966.
Sedimentation in Hawke Bay.
New Zealand Dept. Sci. Ind. Res., Bull., 171:1-64.
Also: New Zealand Oceanogr. Inst. Mem. No. 28:1-64.

bottom sediments, texture
Yong Ahn Park and Moo Young Song, 1971.
Sediments of the continental shelf off the south coasts of Korea. J. oceanogr. Soc. Korea 6(1): 16-24.

bottom sediments, texture of
Pelletier, B.R. 1970.
Sedimentological sampling and results from the diver lock-out facility of the submersible Shelf Diver, Bay of Fundy, Nova Scotia.
Marit. Sed. Halifax 6(3): 102-109.

bottom sediments, texture
Pichon, Michel, 1966.
Note sur la faune des substrats sablo-vaseux infralittoraux de la Baie d'Ambaro (côte nord-ouest de Madagascar).
Cah. ORSTOM, Sér. Océanogr., 4(1):79-94.

bottom sediments, texture
Prenant, Marcel, 1963.
Études écologiques sur les sables intercotidaux.
II. Distribution des granulométries sur les plages bretonnes exposées au large.
Cahiers Biol. Mar., Roscoff, 4(4):353-397.

bottom sediments, texture
*Rossi, S., F. Mosetti e B. Cescon, 1968.
Morfologia e natura del fondo nel Golfo di Trieste (Adriatico settentrionale fra Punta Tagliamento e Punta Salvore).
Boll. Soc. adriat. Sci., 56(2):187-206.

bottom sediments, texture
Ryan, William B.F., Fifield Workum Jr. and J.B. Hersey 1965
Sediments on the Tyrrhenian Abyssal Plain.
Bull. geol. Soc. Am. 76 (11): 1261-1282.

bottom sediments, texture
Sato, T., and M. Tamaki, 1962
Bottom sediments of the Aomori Bay. (In Japanese; English abstract).
Hydrogr. Bull., (Publ. No. 981), Tokyo, No. 72: 16-25.

bottom sediments, texture
Schneider, Eric D., and Bruce C. Heezen, 1966.
Sediments of the Caicos Outer Ridge, The Bahamas.
Bull. geol. Soc. Am., 77(12):1381-1398.

bottom sediments, texture
Seibold, Eugen, 1963
Geological investigation of near-shore sand-transport-examples of methods and problems from the Baltic and North seas.
Progress in Oceanography, M. Sears, Edit., Pergamon Press, London, 1:1-70.

bottom sediments, texture
*Siddiquie, H.N., 1967.
Recent sediments in the Bay of Bengal.
Marine Geol., 5(4):249-291.

bottom sediments, texture
*Summerhayes, C.P., 1969.
Recent sedimentation around northernmost New Zealand.
N.Z. Jl Geol. Geophys. 12:172-207.

bottom sediments, texture
Uchupi, Elazar, and K.O. Emery, 1963
The continental slope between San Francisco, California and Cedros Island, Mexico.
Deep-Sea Research, 10(4):397-447.

bottom sediments, texture
University of Southern California, Allan Hancock Foundation, 1965.
An oceanographic and biological survey of the southern California mainland shelf.
State of California, Resources Agency, State Water Quality Control Board, Publ. No. 27:232 pp. Appendix, 445 pp.

bottom sediments, texture
Van der Linden, Willem J.M., 1969.
Off-shore sediments, north-west Nelson, South Island, New Zealand.
N.Z. Jl Geol. Geophys. 12: 87-103.

bottom sediments, texture
Van der Linden, Willem J.M., 1967.
A textural analysis of Wellington Harbour sediments.
N.Z. Jl. mar. Freshwat. Res. 1(1):26-37.

bottom sediments, texture
Van Straaten, L.M.J.U., 1960.
Transport and composition of sediments.
Verhandel. Kon. Nederl., Geol. Mijnb. k. Gen., Geol. Ser. (Symposium Ems-Estuarium(Nordsee), 19:279-292.

bottom sediments, texture
Veenstra, H.J., 1965.
Geology of the Dogger Bank area, North Sea.
Marine Geology, Elsevier Publ. Co., 3(4):245-262.

bottom sediments, texture
Volkov, P.A., 1963
The hydraulic characteristics of shell sediments. (In Russian).
Okeanologiia, Akad. Nauk, SSSR, 3(4):680-683.

bottom sediments, texture
Weiler, R.R., and A.A. Mills, 1965.
Surface properties and pore structure of marine sediments.
Deep-Sea Res., 12(4):511-529.

bottom sediment texture
Whitmarsh, R.B., 1971.
Interpretation of long range sonar records obtained near the Azores. Deep-Sea Res., 18(4): 433-440.

bottom sediments, thermal conductivity

Bottom sediments, thermal conductivity
See also: heat flow

bottom sediments, thermal conductivity
Hutt, Jeremy R., and Joseph W. Berg, Jr. 1968.
Thermal and electrical conductivities of sandstone rocks and ocean sediments.
Geophysics 33(3): 489-500.

bottom sediments, thermal conductivity
Negi, Janardan G., and Rishi Narain Singh, 1967.
On heat transfer in layered oceanic sediments.
Earth Planet. Sci. Letters 2(4): 335-336

bottom sediments, thermal conductivity
Ratcliffe, E.H., 1960.
The thermal conductivities of oceanic sediments.
J. Geophys. Res., 65(5):1535-1542.

bottom sediments, thermal con.
*Sclater, J.G., C.E. Corry and V. Vacquier, 1969.
In situ measurement of the thermal conductivity of ocean-floor sediments.
J. Geophys. Res., 74(4):1070-1081.

bottom sediments, thermal conductivity of
Uyeda, Seiya, Ki-iti Horai, Masashi Yasui and Hideo Akabatsu, 1962.
IX. Heat flow measurements over the Japan Trench during the JEDS-4.
Oceanogr. Mag., Japan Meteorol. Agency, 13(2):185-189.
JEDS Contrib. No. 37.

bottom sediments, thermo magnetic properties

bottom sediments, thermo magnetic properties
*Schwartz, E.J., 1968.
Thermo-magnetic properties of banded manganiferous sediment from the Mid-Atlantic Ridge.
Can. J. earth Sci., 5(6):1517-1518.

bottom sediments, thermal properties
Chmelik, Frank B. 1969.
Thermal properties of marine sediments.
Trans. Marine Temperature Measurements Symposium, Mar. Tech. Soc., June 1969. 275-288.
Also in: Contr. Oceanogr. Texas A. M. Univ. 12 (Contr. 380)

bottom sediments, thickness of

bottom sediments, thickness of
Bezrukov, P.L., 1961
[Investigation of sediments in the northern Indian Ocean.]
Okeanol. Issledov., Mezhd. Komit., Mezhd. Geofiz. God, Presidium Akad. Nauk, SSSR: 76-90

sediment thickness
*Conolly, John R., and C.C. von der Borch, 1967.
Sedimentation and physiography of the sea-floor south of Australia.
Sedimentary Geol., 1(2):181-220.

bottom sediments, thickness of
Drake, Charles L., 1966.
Recent investigations of the continental margin of eastern United States.
In: Continental margins and island arcs, W.H. Poole, editor, Geol. Surv. Pap., Can., 66-15:33-47.

bottom sediments, thickness of
Edgerton, H.E., et O. Leenhardt 1966.
Mesures d'épaisseur de la vase sur les fortes pentes du precontinent.
C. r. hebd. Séanc. Acad. Sci. Paris (D) 262 (19): 2005-2007.

bottom sediments, thickness
Ewing, John, and Maurice Ewing, 1967.
Sediment distribution on the mid-ocean ridges with respect to spreading of the sea floor.
Science, 156 (3782):1590-1592.

bottom sediments (thickness)
Hamilton, Edwin L., 1967.
Marine geology of abyssal plains in the Gulf of Alaska.
J. geophys. Res., 72(16):4189-4213.

bottom sediments, thickness of
Kovach, R., and F. Press, 1961
A note on ocean sediment thickness from surface wave dispersion.
J. Geophys. Res., 66(9):3073-3074.

bottom sediments, thickness of
Kovylin, V.M., 1961
[Multichannel acoustic network for seismo-acoustic measurements in the ocean.]
Okeanol. Issledov., Mezhd. Komit., Mezhd. Geofiz. God, Presidiume Akad. Nauk, SSSR,: 100-109.

sediment thickness
Kroenke, Loren W., 1965.
Seismic refraction studies of sediment thickness around the Hawaiian Ridge.
Pacific Science, 19(3):335-338.

sediment, thickness of
Moore, David G., and George Shumway, 1959.
Sediment thickness and physical properties: Pigeon Point Shelf, California.
J. Geophys. Res., 64(3):367-374.

bottom sediment thickness
Malahoff, Alexander and G.P. Woollard, 1970.
II. Regional observations. 3. Geophysical studies of the Hawaiian Ridge and Murray Fracture Zone. In: The sea: ideas and observations on progress in the study of the seas, Arthur E. Maxwell, editor, Wiley-Interscience 4(2): 73-131.

bottom sediments, thickness of
Orlenok, V.V., 1971.
Structure and thickness of sediments in northern Atlantic according to seismic data. (In Russian; English abstract). Okeanol. Issled Rezult. Issled. Mezhd. Geofiz. Proekt. 21: 271-296.

bottom sediments, thickness
Shor, G.G., Jr., H.W. Menard, and R.W. Raitt, 1970.
II. Regional observations. 1. Structure of the Pacific Basin. In: The sea: ideas and observations on progress in the study of the seas, Arthur E. Maxwell, editor, Wiley-Interscience 4(2): 3-27.

bottom sediments, thickness of
Wong, How-Kin, and Edward F.K. Zarudzki, 1969.
Thickness of unconsolidated sediments in the eastern Mediterranean Sea.
Bull. geol. Soc. Am. 80(12): 2611-2614

bottom sediments, thin sections
Wood, G.V., 1970.
Sediment thin section data. Initial Reports of the Deep Sea Drilling Project, Glomar Challenger 4: 315-323.

bottom sediments, thin sections
Wood, G.V., 1970.
Sediment thin section data, leg 2. Initial Reports of the Deep Sea Drilling Project, Glomar Challenger 2: 323-328.

bottom sediments, thin sections
Wood, G.V. 1969.
Sediment thin section data.
Initial Repts Deep Sea Drilling Project, Glomar Challenger, 1: 345-353.

Sediment, transport
see also: Transport, sediments

Bottom sediments, transport of
See: Transport, sediments

bottom sediments, turbidites
Francis, T.J.G. 1971
Effect of earthquakes on deep-sea sediments.
Nature, Lond. 233 (5315): 98-102.

bottom sediments (varved)
Gross, M. Grant, Sevket M. Gucluer, Joe S. Creager and William A. Dawson, 1963.
Varved marine sediments in a stagnant fjord.
Science, 141(3584):918-919.

bottom sediments volcanic ash
Nayudu, Y. Rammohanroy, 1964.
Volcanic ash deposits in the Gulf of Alaska and problems of correlation of deep-sea ash deposits.
Marine Geology, 1(3):194-212.

bottom sediments, volcanic ash
Mellis, O., 1955.
Volcanic ash-horizons in deep-sea sediments from the eastern Mediterranean. Deep-Sea Res. 2(2): 89-92, 3 textfigs.

bottom sediments, volcanic in origin
Kawamura, B., and T. Sato, 1963.
Scoria-fall and the northeastward of Miyake Sima (Miyake Island), in 1962. (In Japanese; English abstract).
Hydrogr. Bull., Tokyo, No. 73(Publ. No. 981):1-5.

bottom sediments, volume
Gilluly, James, 1969.
Oceanic sediment volumes and continental drift.
Science, 166(3908): 992-993.

Bottom sediments, water content
See: bottom sediments, interstitial water

Bottom sediment-water interface

sediment-water interface
Bieri, R., M. Koide and E.D. Goldberg, 1967.
Geophysical implications of the excess helium found in Pacific waters.
J. geophys. Res., 72(10):2497-2511.

bottom sediments
Hayes, F.R., 1964
The mud-water interface.
In: Oceanography and Marine Biology, Harold Barnes, Editor, George Allen & Unwin, Ltd., 2: 121-145.

bottom sediments
Hayes, F. R., and M. A. MacAulay, 1959.
Lake water and sediment. V. Oxygen consumed in water over sediment cores.
Limnol. & Oceanogr., 4(3):291-298.

bottom sediments
Hayes, F.R., and J. E. Phillips, 1958
Lake water and sediment. IV Radiophosphorus equilibrium with mud, plants; and bacteria under oxidized and reduced conditions. Limnol. & Oceanogr. 3(4):459-475.

bottom sediments (exchange of materials - lakes)
Mortimer, C.H., 1941-1942.
The exchange of dissolved substances between mud and water in lakes. J. Ecol. 29(2):280-329, 21 textfigs.; 30(1):147-201, Figs. 22-46.

Bottom sediments, x-rays of

bottom sediments, x-rays of
Bouma, Arnold, H., 1964.
Notes on X-ray interpretation of marine sediments
Marine Geol., 2(4):278-309.

Bottom tectonics
See: tectonics

bottom terrain
Avilov, I.K. 1967.
Relief and terrain of areas of mineral resources in the northwest and southeast Atlantic. (In Russian)
Atlant. nauchno-issled. Inst. rybn. Khoz. okeanogr. (AtlantNIRO). Material. Konferentsii Resul'tatam Okeanolog. Issledovan. Atlant. Okeane, 277-289.

Bottom terrain
Il'in, A.V. 1967.
New data on the morphology of the bottom of the Atlantic Ocean. (In Russian).
Atlant. nauchno-issled. Inst. rybn. Khoz. okeanogr. (AtlantNIRO). Materialy Konferentsii po Resul'tatam Okeanologischeskikh Okean... Issledovanii v Atlanticheskom
255-276

Bottom topography

bottom topography
Adams, R.D., 1957.
Exploration of the deep ocean floor. Nature 180(4590):778-780.

bottom topography
*Agapova, G.V., and A.D. Yempolsky, 1967.
On the possibility of using spectral characteristics for the analysis of underwater relief (Shatsky Rise). (In Russian; English abstract).
Okeanologiia, Akad. Nauk, SSSR, 7(4):703-710.

bottom topography
Atlasov, I.P., V.A. Vakar, V.D. Dibner, B. Kh. Yegiazarov, A.B. Zimkin and B.S. Romanovich, 1964
A new tectonic chart of the Arctic. (In Russian).
Doklady, Akad. Nauk, SSSR, 156(6):1341-1342.

Translation:
T412R
Library, National Research Council,
Sussex Street, Ottawa, Canada

bottom topography
Aubert, M., et A. Thomas, 1964.
Levé des fonds marins côtiers de la Baie des Anges.
Cahiers C.E.R.B.O.M., Nice, (1):7-12.

bottom topography
Agassiz, A., 1892.
II. General sketch of the Expedition of the "Albatross" from February to May, 1891. Reports on the dresging operations of the west coast of Mexico and in the Gulf of California, in charge of Alexander Agassiz, carried out by the U.S. Fish Commission Steamer "Albatross", Lieut. Commander Z.L. Tanner, U.S.N., commanding.
Bull. M.C.Z. 23(1):1-90, 22 pls.

bottom topography
Avilov, I.K., 1965.
Some data on the bottom topography and grounds of the West African shelf. (In Russian).
Trudy vses. nauchno-issled. Inst. morsk. ryb. Khoz. Okeanogr. (VNIRO), 57:235-259.

bottom topography
Barnes, C.A., 1954.
Some problems in laying a submarine power line. (Abstract). Bull. G.S.A. 65(2):1332.

bottom topography (Bays)
Behrens, E. William, 1963.
Buried Pleistocene river valleys in Aransas and Baffin bays, Texas.
Publ. Inst. Mar. Sci., Port Aransas, 9:7-18.

bottom topography
Beloussov, I.M., 1970.
Certain geomorphological peculiarities of the southern part of the Gulf of Mexico and the region of Cuba. (In Russian; English and Spanish abstracts). Okeanol. Issled. Rezult. Issled. Mezhd. Geofiz. Proekt. 20: 11-21.

bottom topography
Beloussov, V.V., 1968.
The earth's crust and upper mantle of the oceans. (In Russian; English abstract).
Rez. Issled. Mezhdunarod. Geofiz. Proekt., Mezhduvedomst. Geofiz. Kom., Akad. Nauk, SSSR, 255 pp.

bottom topography
Belousov, I.M., 1964.
On the bottom topography of the northwestern Indian Ocean. (In Russian).
Mezhdunarodni Geologicheskii Kongress 22 Sessiya, Doklady Sovetski Geologov, Problema 16, Geologiya dna Okeanov i Moryu, Moscow, 30-40.
Translation:
National Lending Library for Science and Technology
Boston Spa, Yorkshire, England

bottom topography
Belousov, I.M., V.F. Kanaev and N.A. Marova, 1964.
The bottom relief in the northern part of the Indian Ocean. (In Russian).
Doklady, Akad. Nauk, SSSR, 155(5):1174-1177.

bottom topography
Belousov, I.M., A.I. Ioffe, and V.P. Smilga, 1968.
Least squares method used to restore the sea bottom relief at sites defying measurements.
(In Russian).
Dokl. Akad. Nauk, SSSR, 182(1):89-92.

bottom topography

Beresnev, A.F. and V.M. Kovylin, 1970.
The relief of the basement and the distribution of thicknesses of bottom sediment in the central part of the Japan Sea. (In Russian; English abstract). Okeanologiia, 10(1): 113-116.

submarine topography

Bezrukov, P.L., 1957.
(New data on the regularities of the morphology of submarine relief).
DAN 116(5):841.

bottom topography

Blazhchishin, A.I., V.M. Litvin, L. Lukosevicius and M.V. Rudenko, 1970.
New data on the bottom topography and sedimentary structure in the central area of the Baltic Sea. (In Russian; English and German abstracts)
Baltica, Lietuv. TSR Mokslu Akad. Geogr. Skyr. INQUA Taryb. Sek., Vilnius, 4: 145-163

bottom topography

Bogorov, G.V. and A.V. Il'in, 1971.
On the methods for quantitative description of bottom relief irregularities. (In Russian; English abstract). Okeanologiia 11(2): 326-333.

bottom topography

Boillot, Gilbert, 1963.
Sur la fosse centrale de la Manche.
Comptes Rendus, Acad. Sci., Paris, 257(26):4199-4202.

bottom topography

Boillot, Gilbert, 1963.
Sur une nouvelle fosse de la Manche occidentale, la fosse du Pluteus.
Comptes Rendus, Acad. Sci., Paris, 257(22):3448-3451.

bottom topography

Boillot, G., 1961.
Sur une nouvelle fosse de la Manche occidentale, la "fosse de l'île Vierge".
Comptes Rendus, Acad. Sci., Paris, 253(1):156-158

bottom topography

Bossolasco, M., I. Dagnino, A. Elena e C. Palau 1969.
Variazioni delle spiagge e sedimentazione di materiali ferromagnetici.
Geofis. Met. 18 (5/6): 107-120.

bottom topography

Bowin, C.O., A.J. Nalwalk and J.B. Hersey, 1966.
Serpentinized peridotite from the north wall of the Puerto Rico Trench.
Geol. Soc., Amer., Bull., 77(3):257-270.

bottom topography

Bowler, J.M., 1966.
The geology and geomorphology. Port Philip survey 1957-1963.
Mem. nat. Mus., Vict., 27:19-67.

bottom topography

Boychenko, I.G., 1961.
Bottom topography of the Karaginsky Bay.
Trudy Inst. Okeanol., Akad. Nauk, SSSR, 50:3-20.

bottom topography

Braarud, T. and J. T. Ruud, 1932
The "Øst" Expedition to the Denmark Strait 1929. I. Hydrography. Hvalrådets Skr., No. 4:44 pp., 19 text figs.

bottom topography

Brenot, Roger, 1963.
Sur un point de topographie dans le détroit de Sicile (parages de l'Ile Julia). (By title).
Rapp. Proc. Verb., Réunions, Comm. Int. Expl. Sci Mer Méditerranée, Monaco, 17(3):1061.

Paper published in:
Rev. Trav. Inst. Pêches Marit., 26(3):369. 1962

bottom topography

Carbonnel, J.-P., 1962.
Une science et un art: la topographie sous-marine.
La Nature, (3331):475-481.

topography

Carsola, A.J., 1954.
Microrelief in Arctic sea floor.
Bull. Amer. Assoc. Petr. Geol. 38(7):1587-1601.

topography

Chenoweth, P.A., 1962.
Comparison of the ocean floor with the lunar surface.
Bull. Geol. Soc., Amer., 73(2):199-210.

bottom topography

Chernysheva, V.I., 1963.
Olivine basalt in the region of the northern ending of the underwater Hawaya mountain range. (In Russian).
Doklady, Akad. Nauk, SSSR, 151(6):1433-1436.

bottom topography

Conolly, J.R., and M. Ewing, 1967.
Sedimentation in the Puerto Rico Trench.
J. sedim. Petrol., 37(1):44-59.

bottom topography

Cooper, L. H. N., and D. Vaux, 1949.
Cascading over the continental slope of water from the Celtic Sea. J.M.B.A. 28(3):719-750, 14 textfigs.

bottom topography

Curl, H., Jr., 1957.
Changes in bottom topography off Alligator Harbor since 1889. Q.J. Florida Acad. Sci., 20(3):205-208.

bottom topography

D'Arrigo, A., 1951.
Le variazione morfologiche del fondo marino nel Mediterraneo, ricostruibili della carte nautiche dell'ammiragliato inglese e del l'istituto idrografico della Marina Italiana intercorso dal 1824-1924. Atti Convegno Int. Met. Marit., Genova 20-22/IX/1951:96-101.

bottom topography

Dietrich, Gunter 1965.
New hydrographical aspects of the northwest Atlantic.
ICNAF Spec. Publ. 6: 29-51.

bottom topography

Dietrich, Gunter, 1964.
Oceanic Polar Front Survey in the North Atlantic.
In: Research in Geophysics. Solid Earth and Interface Phenomena, 2:291-308.

bottom topography

Dietz, Robert S., A.J. Carsola, E.C. Buffington, and Carl J. Shipek, 1964
Sediments and topography of the Alaskan shelves.
In: Papers in Marine Geology, R.L. Miller, Editor, Macmillan Co., N.Y., 241-256.

topography

Dietz, R.S., and G. Shumway, 1961.
Arctic Basin geomorphology.
Bull. Geol. Soc., Amer., 72:1319-1330.

bottom topography

Donnelly, Thomas W., 1965.
Sea-bottom morphology suggestive of post-pleistocene tectonic activity of the eastern greater Antilles.
Geol. Soc., Am., Bull., 76(11):1291-1294.

bottom topography

Dubrovin, L.I., 1962.
Bottom relief in the Lazarev Station region. (In Russian).
Sovetsk. Antarktich. Expedits., Inform. Biull., (32):27-28.

Translation:

(Scripta Tecnica, Inc., for AGU) 4(1):45-46.

bottom topography

Elliott, F.E., W.H. Myers and W.L. Tressler, 1955.
A comparison of the environmental characteristics of some shelf areas of eastern United States. J. Washington Acad. Sci. 45(8):248-259, 4 figs.

A flagrant case of scientific plagiarism

bottom topography

Elmendorf, C.H., and B.C. Heezen, 1957.
Oceanographic information for engineering submarine cable systems. Bell Syst. Tech. J., 36(5): 1047-1094.

bottom topography

Emery, K.O., 1966.
The Atlantic continental shelf and slope of the United States: geologic background.
Prof. Pap. U.S. geol. Surv., 529-A:A1-A23.

bottom topography

Emery, K.O., 1965.
Geology of the continental margin off eastern United States.

In: Submarine Geology and Geophysics, Colston Papers, W.F. Whittard and R. Bradshaw, editors, Butterworth's, London, 17: 1-17.

bottom topography

Emery, K.O., 1951.
Topography of the Arctic Basin. Proc. Alaskan Sci. Conf., Bull. Nat. Res. Coun. 122:81.

bottom topography

Erimesco, P., 1961.
Geophysics of ocean basins.
Bull. Inst. Pêches Mar. Maroc, 6:3-56.

bottom topography

Fisher, Robert L., 1961.
Middle America Trench: topography and structure.
Bull. Geol. Soc. Amer., 72(5):703-720.

bottom topography

Fonselius, Stig H., 1962
Hydrography of the Baltic deep basins.
Fishery Board of Sweden, Ser. Hydrogr., Rept. 13: 40 pp.

bottom topography

*Forsbergh, Eric D., 1969.

On the climatology, oceanography and fisheries of the Panama Bight. (In Spanish and English).
Bull. Inter-Am. Trop. Tuna Comm. 14(2): 385 pp.

bottom topography

Fujiwara, Kenzo, 1971.
Soundings and submarine topography of the glaciated continental shelf in Lützow-Holm Bay, East Antarctica. (In Japanese; English abstract). Antarct. Rec., Repts. Japan Antarctic Res. Exped. 41: 81-103.

bottom topography

Fukuoka, Jiro, 1962
An analysis of hydrographical condition along the Tsushima warm current in the Sea of Japan. Rec. Oceanogr. Wks., Japan, n.s., 6(2):9-30.

bottom topography

Gamulin-Brida, Hélène, 1963.
Note préliminaire sur les recherches bionomiques dans l'Adriatique méridionale.
Rapp. Proc. Verb., Reunions, Comm. Int. Expl. Sci., Mer Méditerranée, Monaco, 17(2):85-92.

bottom topography

Garner, D.M., 1964.
The hydrology of Milford Sound.
New Zealand Dept. Sci. Ind. Res., Bull., No. 157: New Zealand Oceanogr. Inst., Memoir, No. 17:25-33

bottom topography

*Gaynanov, A.G., E.N. Isaev, and G.B. Udintsev, 1968.
Magnetic anomalies and bottom morphology of island arcs in the Northwestern Pacific. (In Russian; English abstract).
Okeanologiia, Akad. Nauk, SSSR, 8(6):1017-1024.

bottom topography

Gennesseau, M., 1964.
L'évolution des fonds sous-marins de la Baie des Anges et le delta du Var.
Cahiers, C.E.R.B.O.M., Nice, (1):29-45.

bottom topography

Gershanovich, D.E., 1965.
On the geomorphological pattern of the remotest part of the Northeastern Pacific. (In Russian).
Trudy vses. nauchno-issled. Inst. morsk. ryb. Khoz. Okeanogr. (VNIRO), 57:271-283.

bottom topography

Gershanovich, D.E., 1965.
On the geomorphological pattern of the remotest part of the northeastern Pacific. Investigations in line with the programme of the International Geophysical Year, 2. (In Russian).
Trudy, Vses. Nauchno-Issled. Inst. Morsk. Ribn. Choz. i Okeanogr. (VNIRO), 57:271-283.

bottom topography

Gershanovich, D.E., V.S. Kotenev, and V.N. Novikov, 1964.
Relief and bottom sediments in the Gulf of Alaska. (In Russian).
Trudy, Vses, Nauchno-Issled. Inst. Morsk. Ribn. Choz. i Okeanogr., 52:83-133.

Also:
Izv. Tichookeansk. Nauchno-Issled. Inst. Morsk. Ribn. Choz. i Okeanogr. (TINRO), 52:83-133.

bottom topography

Gibson, William M., 1960

Submarine topography in the Gulf of Alaska.
Bull. Geol. Soc., Amer., 71(7): 1087-1108.

bottom topography

Glangeaud, Louis, Christian Agarate, Gilbirt Bellaiche et Guy Pantot, 1965.
Morphotectonique de la terminaison sous-marine orientale des maures et de l'Esterel.
Comptes Rendus herbdom. Seanc., Acad. Sci., Paris, 261(22):4795-4798.

bottom topography

Grabovsky, N.A., R. Kh. Greku, A.P. Metalnikov, 1961.
[Some geomorphological aspects of the bottom relief in the Atlantic Ocean along the thirteenth meridian from the North Polar Circle to the southern tropic.]
Okeanologiia, Akad. Nauk, SSSR, 1(5):860-865.

bottom topography

Gudelis, V., 1970.
Main features of geology and bottom topography of the mid-Baltic Sea. (In Russian, English and German abstracts)
Baltica, Lietuv. TSR Mokslu Akad, Geogr. Skyr. INQUA Taryb. Sek., Vilnius, 4: 103-113.

bottom topography

Guilcher, Andre, 1964.
Present-time trends in the study of Recent marine sediments and in marine physiography.
Marine Geology, 1(1):4-15.

bottom topography

Hachey, H.B., 1961.
Oceanography and Canadian Atlantic waters.
Fish. Res. Bd., Canada, Bull. No. 134:120 pp.

bottom topography

Hamilton, E.L., 1963
Sea floor relief.
United States National Report, 1960-1963. Thirteenth General Assembly, I.U.G.G.
In: Trans. Amer. Geophys. Union, 44(2):493-494.

bottom topography

Heezen, B.C. 1966.
Physiography of the Indian Ocean.
Phil. Trans. R. Soc. (A) 259 (1099): 137-149.

bottom topography

Heezen, Bruce C., and Charles D. Hollister, 1967.
Physiography and bottom currents in the Bellingshausen Sea.
Antarctic Jl. U.S.A., 2(5):184-185.

bottom topography

Heezen, Bruce C., and G. Leonard Johnson, 1969
Mediterranean undercurrent and microphysiography west of Gibraltar.
Bull. Inst. océanogr., Monaco, 67 (1380): 95 pp.

bottom topography

Heezen, Bruce C., and G. Leonard Johnson, 1963.
A moated knoll in the Canary Passage.
Deutsche Hydrogr. Zeits., 16(6):269-272.

bottom topography

Heezen, Bruce C., and H.W. Menard 1963.
12. Topography of the deep-sea floor.
In: The Sea, M.N. Hill, Editor, Interscience Publishers, 3:233-280

and H.W. Menard
bot top

bottom topography

Hess, H.H., 1947
Drowned ancient islands of the Pacific basin. Int. Hydr. Rev. 24:81-91, 8 figs.

Same article as in Am. Jour. Sci.

bottom topography

Hinschberger, Félix, 1963.
Un problème de morphologie sous-marine: la Fosse d'Ouessant.
Norois, 10(39):217-233.

bottom topography

Horrer, Paul L., 1962
Oceanographic studies during Operation "Wigwam". Physical oceanography of the test area.
Limnol. and Oceanogr., Suppl. to Vol. 7: vii-xxvi.

bottom topography

Hoshino, Michihei, 1970.
Submerged topography in shallow sea. (In Japanese). Bull. Coast. Oceanogr. 8(1):
Also in: Coll. Repr. Coll. mar. Sci. Techn. Tokai Univ. 4: 275-284

bottom topography

Hoshino, Michihei 1969
On the relationship between the distribution of epicenters and the submarine topography and geology. (In Japanese; English abstract)
J. Coll. Mar. Sci. Technol. Tokai Univ., 3: 1-10

bottom topography

Hoshino, Michihei, 1961.
Some problems of submarine topography in the seas adjacent to Japan. (Abstract).
Tenth Pacific Sci. Congr., Honolulu, 21 Aug.-6 Sept., 1961, Abstracts of Symposium Papers, 372.

bottom topography

Hunkins, Kenneth, 1959.

The floor of the Arctic Ocean.
Trans. A.G.U., 40(2):159-162.

bottom topography

*Hyne, Norman J., and H. Grant Goodell, 1967.
Origin of the sediments and the submarine geomorphology of the inner continental shelf off Choctawhatchee Bay, Florida.
Marine Geol., 5(4):299-313.

bottom topography

Inderbitzen, A.L., and F. Simpson 1971.
Relationships between bottom topography and marine sediment properties in an area of submarine gullies.
J. Sedim. Petrol. 41 (4): 1126-1133.

bottom topography

Ilyin, A.V., 1971.
Main features of geomorphology of Atlantic bottom. (In Russian; English abstract).
Usloviia sediment. Atlantich. Okeana, Rez. Issled. Mezhd. Geofiz. Proekt. Mezhd. Geofiz. Kom. Prezid. Akad Nauk SSSR: 107-246.

bottom topography

Ilyin, A.V., 1961.
[Bottom topography of the Kamchatka Bay.]
Trudy Inst. Okeanol., Akad. Nauk, SSSR, 50:21-28.

bottom topography

Illin, A.V., 1957.
New data on the bottom relief in the region of the Komandorski islands.
Doklady, Akad. Nauk, SSSR, 116(3):397-

bottom topography

Ivanov, M.M., 1963.
The relationship between the magnetic field in the Atlantic Ocean and the bottom relief. (In Russian).
Geomagnet. i Aeron., 3(4):781-783.

English abstract in:
Soviet Bloc Res. Geophys. Astron., Space, 1963, (68):25-26.

topography

Japan, Maritime Safety Board, 1954.
[On the bottom topography and sediments in the north-east part of the Sea of Japan.] Publ. 981, Hydrogr. Bull., Maritime Safety Bd., Tokyo, No. 42:121-137, 10 textfigs.

bottom topography

Kaplin, P.A., and A.S. Ionin, 1962.
Results of the direct study of the bottom relief in coastal areas of the sea and some problems of submarine geomorphological investigation. Methods and results of submarine investigations. (In Russian).
Trudy, Okeanogr. Komissii, Akad. Nauk, SSSR, 14: 45-62.

bottom topography

King, Lewis H., 1967.
Use of a conventional echo-sounder and textural analyses in delineating sedimentary facies: Scotian Shelf.
Can. J. Earth Sci., 4(4): 691-708

topography

Klenova, M.V., 1960.
[Geology of the Barents Sea.]
Isdatel., Akad. Nauk, SSSR, Moscow, 1960, 367 pp

translatio of:
Ch. 2, pp. 64-82
Ch. 4, pp. 103-111
Ch. 5, pp. 112-227
Ch. 8, pp. 310-338.

bottom topography

Klenova, M.V., V.M. Lavrov, & V.K. Nikolaeva, 1962
The distribution of suspension in the Atlantic as connected with its bottom topography.
Doklady Akad. Nauk, SSSR, 144(5): 1153-1155.

bottom topography

Kolp, Otto, 1965.
Paläogeographische Ergebnisse der Kartierung des Meeresgrundes des westlichen Ostsee zwischen Fehmarn und Arkona.
Beiträge Meeresk., Berlin, (12/14):19-59.

bottom topography

*Kotenev, B.N., 1968.
Marine geological studies in the vicinity of Iceland. (In Russian; English abstract).
Okeanologiia, Akad. Nauk, SSSR, 8(6):1049-1052.

bottom topography

*Kranck, Kate, 1967.
Bedrock and sediments of Kouchibouguac Bay, New Brunswick.
J. Fish. Res. Bd., Can., 24(11):2241-2265.

bottom topography

Krause, Dale C., 1967.
Sea floor relief.
Trans. Am. geophys. Un., 48(2):631-638.

bottom topography

Krause, Dale C., 1961
Geology of the sea floor east of Guadalupe Island.
Deep-Sea Res., 8(1): 28-38.

bottom topography

Krishnan, M.S., 1960
The mid-ocean ridges.
Proc. Nat. Inst. Sci., India, (A), 26(Suppl.1) 195-218.

bottom topography

Kuenen, Ph. H., 1962.
Giant rivers under the sea.
Sea Frontiers, 8(4):237-244.

bottom topography

Kwon, Nak-Yon, and Sang-Jung Park 1970.
A study of submarine topography and characteristic of bottom sediment in the central part of the Eastern Sea of Korea. (In Korean; English abstract).
Bull. Fish. Res. Dev. Agency, Pusan, 5: 107-125.

bottom topography

Laborel, J., J. M. Pérès, J. Picard & J. Vacelet, 1961
Étude directe des fonds des parages de Marseille de 30 à 300 m. avec la soucoupe plongeante Cousteau.
Bull. Inst. Océan., Monaco, 1206: 16 pp.

bottom topography

LaFond, E.C., 1954.
Physical oceanography and submarine geology of the seas to the west and north of Alaska. Arctic 7(2):93-101, 11 textfigs.

bottom topography

Laughton, A.S. 1966.
The Gulf of Aden.
Phil. Trans. R. Soc. (A) 259(1099):150-171.

bottom topography

Lisitzin, A.P., 1966.
Processes of Recent sedimentation in the Bering Sea. (In Russian).
Inst. Okeanol., Kom. Osad. Otdel Nauk o Zemle, Isdatel, Nauka, Moskva, 574 pp.

topography

Lisitsin, A.P., and A.V. Zhivago, 1958.
[Bottom relief and sediments of the southern part of the Indian Ocean. 2.]
Izv., Akad. Nauk, SSSR, Ser. Geofiz., (3):22-36.

topography

Lisitsin, A.P., and A.V. Zhivago, 1958.
[Bottom relief and sediments of the southern part of the Indian Ocean. 1.]
Izv., Akad. Nauk, SSSR, Ser. Geofiz., (2):9-21.

OTS $0.50.

bottom topography

Lisitzin, A.P., and A.V. Zhivago, 1958

[VII. Submarine geology]
Opisanie Exped. D/E "Ob" 1955-1956, Mezhd. Geofiz. God, Trudy Kompaeksnoi Antarkt. Exped., Akad. Nauk, SSSR: 103-144.

bottom topography

Litvin, V.M. 1965.
On the origin of the bottom relief of the Norwegian Sea. (In Russian)
Okeanologiia 5(4): 692-700.

bottom topography

Litvin, V.M. 1964.
Bottom configuration of the Norwegian Sea. (In Russian).
Trudy Poliarn. Nauchno-Issled. Proektn. Inst. Morsk. Ribn. Khoz. Okeanogr. N.M. Knipovicha 16:89-109.

bottom topography

Litvin, V.M., and S.A. Bodnar 1964.
Some data on the relief and bottom of fishing grounds in East Greenland. (In Russian).
Material. Ribokhoz. Issled. Severn. Basseina, Poliarn. Nauchno-Issled. Proektn. Inst. Morsk. Ribn. Khoz. Okeanogr. N.M. Knipovicha, PINRO, Gosud. Proizvodst. Komm. Ribn. Khoz. SSSR 4: 106-111.

bottom topography

Litvin, V.M., and V.D. Rvachev, 1962.
A study of bottom relief and types of ground in the fishing areas of Labrador and Newfoundland.
Sovetskie Riboch. Issledov. v Severo-Zapadnoi Atlant. Okeana, VNIRO-PINRO, Moskva, 99-111.

bottom topography

Longhurst, Alan R., 1963.
The bionomics of the fisheries resources of the eastern tropical Atlantic.
Great Britain, Colonial Office, Fish. Bull., (20):66 pp.

bottom topography

Ludwig, W.J., R.E. Houtz and M. Ewing, 1971.
Sediment distribution in the Bering Sea: Bowers Ridge, Shirshov Ridge and enclosed basins.
J. geophys. Res. 76(26): 6367-6375.

bottom topography

Ma, T.L.H., 1957.
Reef corals used for proving the occurrence of shift in crustal masses and the equator and submarine features used to prove the sudden total displacement of the solid earth shell.
Proc. UNESCO Symp., Phys. Ocean., Tokyo, 1955: 220-224.

bottom topography

Mal'tsev, V.M., 1959.
Bottom relief of the Cape Agulhas Basin. (B). (In Russian).
Inform. Biull. Sovetsk. Antarkt. Exped., 2( ):

Translation:
Scripta Tecnica, Inc., Elsevier Publ. Co., 100-101.

bottom topography

Marlowe, J.I., 1965.
Probable Tertiary sediments from a submarine canyon off Nova Scotia.
Marine Geology, Elsevier Publ. Co., 3(4):263-268.

bottom topography

*Marova, N.A., 1967.
Physiographic map of the western Pacific. (In Russian; English abstract).
Okeanologiia, Akad. Nauk, SSSR, 7(4):710-718.

bottom topography

Marova, N.A., 1966.
Bottom topography of the Indian Ocean in the Java Trench area. (In Russian; English abstract).
Okeanologiia, Akad. Nauk, SSSR, 6(3):464-474.

bottom topography

McDonald, Martin F. and Eli Joel Katz, 1969.
Quantitative method for describing the regional topography of the ocean floor. J. geophys. Res 74(10): 2597-2607.

bottom topography

Menard, H.W., 1967.
Sea floor spreading, topography and the second layer.
Science, 157(3791):923-924.

bottom topography

Mikhailov, D.V., and V.P. Goncharov, 1969.
Geomorphology of the floor of the Mediterranean Basin. (In Russian)
Izv. Akad. Nauk SSSR, Geogr. Ser. (2): 25-37

bottom topography

*Misawa, Yoshihumi, Michihei Ho Shino, Hitoshi Aoki and Kei Amma, 1969.
On the submarine topography of the northern Shikoku Basin. (In Japanese; English abstract).
J.Coll.mar.Sci.Technol., Tokai Univ., (3):11-22.

bottom topography

Misawa, Yoshifumi, and Takeshi Yoshiwara 1968.
Submarine topography of the Suruga Bay. (In Japanese; English abstract).
Fossa Magna, Preprint Ann. Meet. Geol. Soc. Japan, 1968.
Also in: Coll. Repr. Coll. mar. Sci. Technol.

bottom topography

Mogi, Akio, 1962
Subm--ine topography of the adjacent seas of Japan Islands. (In Japanese).
J. Oceanogr. Soc., Japan, 20th Ann. Vol., 52-63.

bottom topography

Moign, Annik, 1965.
Contribution à l'étude littorale et sous-marine de la Baie du Roi (Spitsberg - 79° N).
Cahiers Océanogr., C.C.O.E.C., 17(8):543-563.

bottom topography

Moore, David G., 1963.
Geological observations from the bathyscaph "Trieste" near the edge of the continental shelf off San Diego, California.
Geol. Soc. Amer., Bull., 74(8):1057-1062.

bottom topography

Mosby, H., 1938
Svalbard waters. Geofysiske Publ. 12(4): 85 pp., 34 textfigs.

bottom topography

Mosetti, Ferruccio, e Carlo d'Ambrosi 1968.
Cenni sulle vicissitudini costiere dell'alto Adriatico dedotte dalla attuale morfologia del fondo marino.
Atti Memorie Comm. Grotte "Eugenio Boegan", Trieste 6:19-31.

bottom topography, nomenclature

Muromtsev, A.M., 1959.
Nomenclature of basic relief forms in the Pacific Ocean. (In Russian).
Trudy Gosud. Okeanogr. Inst., 48:86-94.

bottom topography

Murray, H.W., 1947
Topography of the Gulf of Maine, field season of 1940. Bull. G.S.A. 58:153-196 plus 12 pls.

bottom topography

Nasu, Noryuki, and Takahiro Sato, 1962.
VII. Geological results in the Japanese Dee Sea Expedition in 1961 (JEDS-4).
Oceanogr. Mag., Japan Meteorol. Agency, 13(2): 166.

bottom topography

Nishijima, S., K. Yamazato and S. Kamura, 1969.
Notes on the characteristics of precious coral grounds in the Ryukyu islands. (In Japanese; English abstract). Bull. Jap. Soc. fish. Oceanogr. Spec. No. (Prof. Uda Commem. Pap.): 291-297.

bottom, topography

Nota, D.J.G., 1971.
Morphology and sediments off the Marowijne River, Eastern Surinam Shelf. Hydrogr. Newsletter, R. Netherlands Navy, Spec. Publ. 6: 31-35.

bottom topography

Novozhilov, L.V., 1967.
Statistic characteristics of the mesoform's relief of the floor of the world ocean.
Izv. Vses. Geograf. Obshch., 100(6):488-493.

bottom topography

Ostenso, Ned A., 1964.
Structure of the Arctic Ocean Basin. (Abstract).
Geol. Soc., Amer., Special Paper, No. 76:314.

bottom topography

Ostrovskii, A.B., 1966.
On the structure of overdeepenings of river valleys at the Black Sea shore of the Caucasus.
Doklady Akad. Nauk SSSR, 167(6): 1362-1364.

bottom topography

Ozawa, Keijiro and Tsotomu Isouchi, 1966.
Continuous echo sounding and bottom configuration in the eastern Indian Ocean.
J. Tokyo Univ. Fish., (Spec.ed.)8(2):55-69.

bottom topography

Park Won Cheon, and Sang Yong Kim, 1968.
Submarine topography and grain size distribution of sediments in the southern part of the East Sea of Korea. (In Korean; English abstract).
Bull. Fish. Res. Develop. Agency, Korea, 3:105-118

bottom topography

Pickard, G.L. 1967.
Some oceanographic characteristics of the larger inlets of southeast Alaska.
J. Fish. Res. Bd Can. 24(7): 1475-1506.

Pratt, R.M., 1965. bottom topography
Ocean-bottom topography: the divide between the Sohm and Hatteras Abyssal plains.
Science, 148(3677):1598-1599.

bottom topography

Puri, Harbans S., Gioacchino Bonaduce and John Malloy, 1964.
Ecology of the Gulf of Naples.
Pubbl. Staz. Zool., Napoli, 33(Suppl.):87-199.

bottom topography

Rvachyov, V.D., 1962.
Bottom relief and grounds of the West-Greenland fishing banks. (In Russian).
Nauchno-Technich. Biull., PINRO, Murmansk, No. 4 (22):31-33.

bottom topography, effect of

Rzheplinsky, D.G., 1970.
The effect of Coriolis force and bottom topography on wind currents around oceanic islands. (In Russian; English abstract).
Fisika Atmosfer. Okean. Izv. Akad. Nauk SSSR, 6(7): 718-727

circulation

bottom topography

Sachet, M.-H., 1962
Geography and land ecology of Clipperton Island.
Atoll Res. Bull., 86:115 pp.

bottom topography

Saks, V.N., 1958(1960).
[Some considerations on the geological history of the Arctic.]
Problemy Severa, (1):

translations:
Problems of the North, (1):70-90.

bottom topography

Sato, K., 1962
On the survey for sinking of sea bottom in Niigata. (In Japanese; English abstract).
Hydrogr. Bull. (Publ. No. 981), Tokyo, No. 72: 9-15.

bottom topography

Sato, R., 1960.
On the determination of crustal structure by the dispersion of surface waves. III.
Geophys. Notes, Tokyo, 13(1):No. 26.

Reprinted from Zisin, Ser. II, 12(4):

bottom relief

Shtokman, V.B., 1949.
[Influence of the relief of the bottom on the direction of ocean currents.] Priroda 38(1):10-23.

bottom topography

Sclater, J.G., F.J.W. Jones and S.P. Miller 1970.
The relationship of heat flow, bottom topography and basement relief in Peake and Freen deeps, northeast Atlantic.
Tectonophysics 10(1/3): 283-300.

bottom topography

Spiess, F.N. and John D. Mudie, 1970.
1. General observations. 7. Small scale topographic and magnetic features. In: The sea: ideas and observations on progress in the study of the seas, Arthur E. Maxwell, editor, Wiley-Interscience 4(1): 205-250.

bottom topography

Stepanov, V.N., 1960

[Types of bottom relief in seas.]
Biull. Oceanogr. Komm., (5): 48-53.

bottom topography

Sverdrup, H.U., 1941
The influence of bottom topography on ocean currents. pp. 66-75 in Contrib. Appl. Mechanics and Related Subjects, Theodore von Karman Anniv. Vol., Pasadena, California Inst. Tech. 337 pp.

Bottom Topography

Tarassov, B. V., 1961. ref.
[New features characteristic of the bottom relief in the Arctic Ocean.]
Probl. Arkt. i Antarkt., (8):89-90.

bottom topography

Torphy, S.R., and J.M. Zeigler, 1957.
Submarine topography of the Eastern Channel, Gulf of Maine. J. Geol., 65(4):433-441.

bottom topography

Uchupi, Elazar, 1966.
Topography and structure of Northeast Channel, Gulf of Maine.
Bull. Am. Assoc. Petr. Geol., 50(1):165-167.

bottom topography

Uchupi, Elazar, and K.O. Emery, 1963
The continental slope between San Francisco, California and Cedros Island, Mexico.
Deep-Sea Research, 10(4):397-447.

bottom topography

Windsor G.B., 1969.
IGY investigations of submarine topography.
Annls int. geophys. Year, 46:131-147

bottom topography

Udintsev, G.B., 1965.
New data on the bottom topography of the Indian Ocean. (In Russian).
Okeanologiia, Akad. Nauk, SSSR, 5(6):993-998.

bottom topography

Udintsev, G. B., 1963
New charts of the bottom relief in the Pacific Ocean. (In Russian).
Okeanologiia, Akad. Nauk, SSSR, 3(1):169-175.

bottom topography

Udintzev, G.B., 1962.
Relief of the ocean and the question of tectonics (In Russian). Marine Geology and the Dynamics of coasts.
Trudy Okeanogr. Komissii, Akad. Nauk, SSSR, 10(3):38-44.

bottom topography

Udintsev, G.B., 1962.
Studies of the submarine relief. (In Russian; English abstract).
Mezhd. Geofiz. Komitet, Presidiume, Akad. Nauk, SSSR, Rezult. Issled. Programme Mezhd. Geofiz. Goda, Okeanol. Issled., No. 7:33-48.

bottom topography

Udintsev, G.B., 1961.
Evolutionary features in the structure of the Pacific Ocean floor expressed in the submarine relief. (Abstract).
Tenth Pacific Sci. Congr., Honolulu, 21 Aug.-6 Sept., 1961, Abstracts of Symposium Papers, 388-389.

bottom topography

Udintsev, G.B., 1955.
[Bottom relief of the Okhotsk Sea.]
Trudy Inst. Okeanol., 13:5-15.

bottom topography

van Andel, Tjeerd H., and John J. Veevers, 1965.
Submarine morphology of the Sahul Shelf, northwestern Australia.
Bull. Geol. Soc., Amer., 76(6):695-700.

bottom topography

Veenstra, H.J., 1965.
Geology of the Dogger Bank area, North Sea.
Marine Geology, Elsevier Publ. Co., 3(4):245-262.

bottom topography

*Verzhbitsky, E.V., E.N. Isaev, and A.A. Shreyder, 1968.
Relationship between magnetic heterogeneity and bottom structure in the northwestern Indian Ocean.
Okeanologiia, Akad. Nauk, SSSR, 8(6):1025-1035.

bottom topography

Voelz, E., 1961.
Neue Untersuchungen des Bodenreliefs in der Grönland See.
Der Seewart, 22(4):148-157.

bottom topography

Volkov, P., 1961
[New explorations of the floor relief in the Greenland Sea.]
Morskoi Flot, 21(3): 35-37
Abs. in Techn. Transl. 6(9): 658.

bottom topography

Waldichuk, M., 1953.
Oceanography of the Strait of Georgia. III. Character of the bottom. Prog. Repts., Pacific Coast Stas., Fish. Res. Bd., Canada, 95:59-63, 4 textfigs.

bottom topography

Wiborg, K.F., 1944
The production of zooplankton in a landlocked fjord, the Nordåsvatn near Bergen, in 1941-42, with special reference to the copepods... (Repts. Norwegian Fish. and Mar. Invest.) 7(7):83 pp., 40 text figs.

bottom topography

Yo, K., 1953.
[Submarine topography of the adjacent sea off Tokati.] Publ. 981, Hydrogr. Bull., Tokyo, Spec. Number 12 No. 12:62-82, 19 textfigs.

bottom topography

Zatonsky, L.K., 1964.
The new data concerning the bottom topography of the Indian Ocean. Investigations in the Indian Ocean (33rd Voyage of the E/S "Vitiaz"). (In Russian).
Trudy Inst. Okeanol., Akad. Nauk, SSSR, 64:158-181.

bottom topography

Zenkevitch, N.L., 1961.
[New data on the bottom topography of the northeastern Pacific.]
Trudy Inst. Okeanol., Akad. Nauk, SSSR, 45:5-21.

bottom relief

Zenkovich, V.P., 1950.
[Preservation of low-relief forms on the bottom of a deep sea.] Dok. Akad. Nauk, USSR, 73:67-69.

bottom topography

Zhivago, A.V., 1960.
Outlines of Southern Ocean geomorphology. (Abstract).
Tenth Pacific Sci. Congr., Honolulu, 21 Aug.-6 Sept., 1961, Abstracts of Symposium Papers, 321.

bottom topography

Zhivago, A.V., 1960

[Geomorphological research.] Arktich. i Antarktich. Nauchno-Issled. Inst. Mezhd. Geofiz. God Sovetsk. Antarkt. Exped., Vtoraia Morsk. Exped. "Ob", 1956-1957, 7: 44-72. (In Russian)

bottom topography

Zhivago, A.V., and A.P. Lisitsin, 1957.
[New data on the bottom relief and the sediments of East Antarctic seas.]
Izv., Akad. Nauk, SSSR Ser. Geofiz., (1):19-35.

OTS $0.50

bottom topography

Zhivago, A.V., O.N. Vinogradov, G.M. Braslavskaya and N.A. Timofeeva, 1965.
New map of the relief of sea bottom in the southern part of the Indian Ocean. (In Russian).
Izv., Akad. Nauk, Ser. Geograf., (2):23-28.

bottom topography, effect of

Bottom topography, effect of

See also: bottom, effect of

bottom topography, effect of

Arakawa, H., 1959

Coast effect on typhoon movement. Meteorol & Geophys., Tokyo, 9(3/4): 123-126.

bottom topography, effect of

Boyer, Don L., 1971.
Rotating flow over long shallow ridges.
Geophys. fluid Dynam. 2(2): 165-184

bottom topography, effect of

Clarke, R.A. and N.P. Fofonoff, 1969.
Oceanic flow over varying bottom topography.
J. mar. Res., 27(2): 226-240.

bottom topography, effect of

Felsenbaum, A.I., 1956.
[An extension of Ekman's theory to the case of a non-uniform wind and an arbitrary bottom relief in a closed sea.] Dokl. Akad. Nauk, SSSR, 109(2): 299-302.

bottom topography, effect of

Fomin, L.M., 1969.
V.B. Stockman's method of total flows for variable depth ocean. (In Russian; English abstract).
Okeanologiia, 9(1):119-124.

bottom topography, effect of

Fukuoka, J., 1958.
On the variation of the Kuroshio current south of Honshu (Japanese Main Land). Geophys. Mag., Tokyo, 28(3):343-355.
Also in: Collected Reprints, Oceanogr. Lab., Meteorol. Res. Inst., Tokyo, June 1960.
(In English).

bottom topography, effect of

Furuhashi, Kenzo, 1961
On the distribution of chaetognaths in the waters off the south-eastern coast of Japan. (JEDS-3).
Publ. Seto Mar. Biol. Lab., 9(1) (2):17-30.

Also in: Repts. of JEDS, Vol. 2(1961)

bottom topography, effect of

Furuhashi, Kenzo, 1961
On the possible segregation found in the copepod fauna in the deep waters off the southeastern coast of Japan. (JEDS-3).
Publ. Seto Mar. Biol. Lab., 9(1)(L):1-15.

Also in: Repts. of JEDS, Vol. 2(1961)

bottom topography, effect of

Görtler, H., 1941.
Einfluss der Bodentopographie auf Strömungen über der rotierenden Erde. Z. angew. Math. Mech. 21:279-303.

bottom topography, effect of

Greenspan, H.P., 1963.
A note concerning topography and inertial currents.
J. Mar. Res., 21(3):147-154.

bottom topography, effect of

*Holland, William R., 1967.
On the wind-driven circulation in an ocean with bottom topography.
Tellus, 19(4):582-600.

bottom topography, effect of

Hsueh, Y., 1967.
On the effect of bottom topography and variable wind stress on ocean water movements.
J. geophys. Res., 72(16):4101-4107.

bottom topography, effect of

Ivanov, Iu. A., and V.M. Kamenkovich, 1959.
Bottom relief as the main factor responsible for the non-zonal course of the Antarctic Circumpolar Current.
Doklady Akad. Nauk, SSSR, 128(6):1167-1170.

bottom topography, effect of

Jacobs, S.J., 1964.
On stratified flow over bottom topography.
J. Mar. Res., 22(3):223-235.

bottom topography, effect of

Johns, B. 1968.
Some effects of topography on The tidal flow in a river estuary.
Geophys. J. R. astron. Soc. 15(5):501-507.

bottom topography, effect of

Johnson, J.A., C.B. Fandry and L.M. Leslie 1971.
On the variation of ocean circulation produced by bottom topography.
Tellus 23(2):113-121.

bottom topography, effect of

Kajiura, J., 1953.
On the influence of bottom topography on ocean currents. J. Ocean. Soc., Japan, 9(1):1-14, 3 textfigs.

bottom topography, effect of

Kamenkovich, V.M., 1960

[Antarctic circumpolar current as influenced by bottom relief] Doklady Akad. Nauk, SSSR, 134 (5): 1076-1078.

bottom topography, effect of

Kamenkovich, V.M., 1956.
[On the influence of the relief of the bottom on a pure drift current in a homogeneous shoreless sea.] Izv. Akad. Nauk, SSSR, Ser. Geofiz., (10): 1182-

bottom topography, effect of

Kamenkovich, V.M., and V.A. Mitrofanov 1971.
On a case of bottom relief influence on currents in the ocean. (In Russian).
Dokl. Akad. Sci. SSSR 199(1):78-81.

bottom topography

Kanaev, V. Ph., and G.B. Udintsev, 1960.
[Study of bottom relief during oceanographic expeditions.]
Trudy Inst. Okeanol., Akad. Nauk, SSSR, 44:3-53.

bottom topography, effect of

Koblents, Ya. P., 1960.
[Influence of the Antarctic shelf relief on the development of discharge glaciers.]
Inform. Bull., Soviet Antarctic Exped., 21:10-15.

bottom topography, effect of

Moore, D.R., 1969.
Construction of f/H maps and comparison of measured deep-sea motions with the f/H map for the North Atlantic Ocean.
Techn. Rept. Chesapeake Bay Inst., 51 (Ref. 69-4). 61 pp.

bottom topography, effect of

Nakano, M., 1955.
On a problem concerning the vertical circulation of sea water produced by winds, with special reference to its bearing on submarine geology and submarine topography.
Rec. Ocean. Wks., Japan, 2(2):68-81.

bottom topography, effect of

Neumann, Gerhard, 1960.
On the effect of bottom topography on ocean currents. Deutsche Hydrogr. Zeits., 13(3):132-141.

bottom topography, effect of

Oi, M., 1956.
Deflection of an ocean current due to a submarine ridge.
Pap. Oceanogr. Inst., Fla. State Univ., 22(2): 7-12.

(Studies)

bottom topography, effect of

Orlanski Isidoro, 1969.
The influence of bottom topography on the stability of jets in a baroclinic fluid.
J. atmos. Sci., 26(6):1216-1232

bottom topography, effect

Rhines, P.B. 1969.
Slow oscillations in an ocean of varying depth. 1. Abrupt topography. 2 Islands and seamounts.
J. fluid Mech. 37(1): 161-189; 191-205

bottom topography, effect of

Roberts, James A., and Chen-Wu Chien, 1965.
The effects of bottom topography on the refraction of the tsunami of 27-28 March 1964: the Crescent City case.
Ocean Sci. and Ocean Eng., Mar Techn. Soc.,-Amer. Soc. Limnol. Oceanogr., 2:707-716.

bottom topography, effect of

Robinson, A., and H. Stommel, 1959
Amplification of transient response of the ocean to storms by the effect of bottom topography.
Deep-Sea Res., 5(4):312-314.

bottom topography, effect of

Ryzhkov, Yu. G., and L. A. Koveshnikov, 1963.
Genesis of convergence and divergence zones above sharp changes in the slope of the ocean floor. (In Russian).
Izv., Akad. Nauk, SSSR, Ser. Geofiz., (6):953-959

Translation (AGU): (6):585-588.

bottom topography, effect of

Shtokman, V.B., 1948.
Effect of bottom topography on the direction of the transport of water set up by the wind or the mass-field in a non-homogeneous ocean.
Dok. Akad. Nauk, SSSR 59(5):889-892.

Translation: E.R. Hope - Canada Defence Sci. Info. DRB - 25 Oct. 1952. T57R.

bottom topography, effect of

Shtokman, V.B., 1947.
[Effect of bottom topography on the direction of currents in the sea.] Priroda 11:10-23.

T57R

bottom topography, effect of

Shtokman, V.B., 1949.
[Effect of bottom topography on the direction of currents in the sea.] Priroda, 38(11):10-23.

RT-2108 Bibl. Transl. Rus. Sci Tech. Lit. 15.

bottom topography, effect of

Shuliak, B.A., 1961.
Determination of wave-flow parameters based on the parameters of the periodic bottom structures formed by the flow. (In Russian).
Doklady, Akad. Nauk, SSSR, 137(3):580-583.

Translation:
Consultant's Bureau for Amer. Geol. Inst., Earth Sci. Sect., 137(1-6):397-399. (1962).

bottom topography, effect of

Shulekin, V.V., 1967.
Calculation of vertical currents in a sea with a complex bottom relief. (In Russian)
Dokl. Akad. Nauk, SSSR, 175(3): 575-551

РАСЧЕТ ВИХРЕВЫХ ТОКОВ В МОРЕ СО СЛОЖНЫМ РЕЛЬЕФОМ ДНА.

bottom topography, effect of

Sokolova, M.N., 1956.
[On the distribution regularities of deep sea benthos. The influence of the macrorelief and distribution of suspension upon the adaphic groups of bottom invertebrates.]
Dokl. Akad. Nauk, SSSR, 110(4):692-695.

bottom topography, effect of

Takano, Kenzo, 1966.
A possible effect of the bottom topography on the general circulation in an ocean.
J. oceanogr. Soc., Japan, 22(6): 264-273.

bottom topography, effect of

Tareyev, B.A., 1971.
Gradient-vorticity waves on a shelf. Fizika Atmosfer. Okean. Izv. Akad. Nauk SSSR 7(4): 431-436. (In Russian; English abstract).

bottom topography, effect of

Tareev, B.A., 1965.
Internal baroclinic waves in the course of flowing about the roughness of the ocean floor and their influence upon the process of sedimentation in the depth. (In Russian).
Okeanologiia, Akad. Nauk, SSSR, 5(1):45-51.

bottom topography, effect of

Tickner, E.G., 1960

Effects of reefs and bottom slopes on wind set-up in shallow water.
Beach Erosion Bd., Tech. Memo., No. 122: 1-20.

bottom topography
U.S.A., National Academy of Sciences, 1961.
Arctic Basin seismic studies from IGY Drifting Station Alpha - Hunkins, K.
Trans. Amer. Geophys. Union, IGY Bull., No. 46: 1-5.

Original report prepared by Lamont Geological Observatory:
Seismic studies of the Arctic Ocean floor, 1960.
Geophys. Res. Directorate, Air Force, Cambridge Res. Lab., AFCRC-TN-60. 257.

bottom topography, effect of
Vajk, Raoul, 1964.
Correction of gravity anomalies at sea for submarine topography.
J. Geophys. Res., 69(18):3837-3844.

bottom topography, effect of
Veronis, George, 1966.
Rossby waves with bottom topography.
J. mar. Res., 24(3):338-349.

bottom topography, effect of
Warren, Bruce A., 1963.
Topographic influences on the path of the Gulf Stream.
Tellus, 15(2):167-183.

topography, effect of
*Welander, Pierre, 1969.
Effects of planetary topography on the deep-sea circulation.
Deep-Sea Res., Suppl. 16: 369-391.

bottom topography, effect of
Zubov, N. N., 1959. ?
[The influence "barichesk" relief on sea level and on the phenomenon of "vergentsii".]
Meteorol. i Gidiol. (6):52-54.

bottom topography, micro
Sanders, J.E., K.O. Emery and Elazar Uchupi, 1969
Microtopography of five small areas of the continental shelf by side-scanning sonar.
Bull. geol. Soc. Am., 80(4): 561-572

bottom water

bottom water
When named such as Antarctic Bottom Water, SEE: Regional Cumulation

bottom water
Castens, G., 1934.
Das Bodenwasser und die Gliederung des atlantischen Ozeans. Ann. Hydrogr., usw., Jahrg. 62 (Heft 5):185-191.

bottom water
Codispoti, L.A., and F.A. Richards 1968.
Micronutrient distributions in the East Siberian and Laptev seas during summer 1963.
Arctic 21(2): 67-83.

bottom water
Deacon, G.E.R. 1959.
The Antarctic Ocean.
Science Progress. 47(188): 647-660

bottom water
de Maio, A., and L. Trotti, 1961.
Sur la formation d'eau de fond et d'eau profonde dans la Mer Méditerranée. (Résumé).
Rapp. Proc. Verb., Réunions, Comm. Int. Expl. Sci., Mer Méditerranée, Monaco, 16(3):595.

Complete Paper in:
Cahiers Océanogr., C.C.O.E.C., 1961, 13(4):277-233.

bottom water
Fofonoff, N.P., 1956.
Some properties of sea water influencing the formation of Antarctic bottom water. Deep-Sea Res. 4(1):32-35.

bottom water
Furnestin, J., et Ch. Allain, 1963.
La formation de l'eau de fond "Algéro-atlantique" en hiver sur la côte algérienne et sa progression vers le centre du bassin occidental. (Résumé).
Rapp. Proc. Verb., Réunions, Comm. Int., Expl. Sci., Mer Méditerranée, Monaco, 17(3):913-916.

bottom water
Gordon, Arnold L., 1966.
Potential temperature, oxygen and circulation of bottom water in the Southern Ocean.
Deep-Sea Res., 13(6):1125-1138.

bottom water
Gordon, Arnold L., Paul J. Grim and Marcus Langseth, 1966.
Layer of abnormally cold bottom water over southern Aves Ridge.
Science, 151(3717):1525-1526.

bottom water
Klepikov, V. V., 1960.
[On bottom waters in the Weddell Sea.]
Inform. Bull., Soviet Antarctic Exped., 18:21-24

bottom water
Koczy, F.F., 1950.
Die bodennahe Wasserschichten der Tiefsee.
Naturwiss. 15:360.

bottom water
Koczy, F., and B. Szabo, 1962
Renewal time of bottom water in the Pacific and Indian oceans.
J. Oceanogr. Soc., Japan, 20th Ann. Vol., 590-599.

bottom water
Krause, D.C., 1962
Renewal of bottom water in the Sangihe Basin, East Indies, and bottom temperatures of the Recorder Expedition.
Austral. J. Mar. & Freshwater Res., 13(1): 57-60.

bottom water
Metcalf, W.G., 1955.
On the formation of bottom water in the Norwegian basins. Trans. Amer. Geophys. Union, 36(4):596-600, 2 textfigs.

bottom water formation
Mosby, Håkon, 1967.
Bunnvannsdannelse i havet.
Fridtjof Nansen Minneforelesningen, Norske Videnskaps Akademi, 3:29 pp.

bottom water
Mosby, Hakon, 1962
Recording the formation of bottom water in the Norwegian Sea.
Proc. Symposium on Mathematical-Hydrodynamical Methods of Phys. Oceanogr., Sept. 1961, Inst. Meeresk., Hamburg, 289-296.

bottom water
Nansen, F., 1912.
Das Bodenwasser und die Abkühlung des Meeres.
Int. Rev. Ges. Hydrobiol., u. Hydrogr., 5(1):1-42.

bottom water
*Rehrer, R., A.C. Jones and M.A. Roessler, 1967.
Bottom water drift on the Tortugas grounds.
Bull. mar. Sci., Miami, 17(3):562-575.

bottom water
Reid, Joseph L. and Ronald J. Lynn, 1971.
On the influence of the Norwegian-Greenland and Weddell seas upon the bottom waters of the Indian and Pacific oceans. Deep-Sea Res, 18(11): 1063-1068.

bottom water
Riley, G.A., 1951.
Oxygen, phosphate, and nitrate in the Atlantic Ocean. Bull. Bingham Ocean. Coll. 8(1):126 pp., 33 pp.

Bottom Water
Shannon, L.V., and M. van Rijswijck, 1969.
Physical oceanography of the Walvis Ridge region.
Investl Rep. Div. Sea Fish. S Afr. 70: 1-19.

bottom water
Timofeyev, V.T., 1957
[Formation of the bottom waters of the central part of the Arctic Basin.] Problemy Arktiki, 1: 29-33.

T 349R-Jan. 1961-E.R. Hope.

bottom water
Van Riel, P.M., 1956.
Oceanographic results. Part 5. The bottom water. Ch. II. Temperature. Snellius Exped., 2(5):59 pp.

bottom water
Van Riel, P.M., 1943.
The bottom water. Introductory remarks and oxygen content. Snellius Exped., Vol. 2, Pt. 5, Ch. 1:77pp., 34 textfigs.

bottom-water formation
Worthington, L.V., 1955.
A new theory of bottom-water formation.
Deep-Sea Res. 3(1):82-87.
Caribbean

bottom water
Wüst, Georg, 1963
On the stratification and the circulation in the cold water sphere of the Antillean-Caribbean basins.
Deep-Sea Res., 10(3):165-187.

bottom water (data)
Wüst, Georg, 1961
Das Bodenwasser und die Vertikalzirkulation des Mittellandischen Meeres. 3. Beitrag zum mittelmeerischen Zirkulationsproblem.
Deutsche Hydrogr. Zeits., 14(3):81-92.

bottom water
Wüst, Georg, 1957.
Die Stromgeschwindigkeiten, besonders im Tiefenund Bodenwasser.
Wiss. Ergebn. dt. atlant. Exped. 'Meteor', 6(2): 321-351.

Translation: USN Oceanogr. Off. TRANS 348. (M. Slessers). 1967.

bottom water
Wüst, G., 1957.
Stromgeschwindigkeiten und Strommengen in den Tiefen des Atlantischen Ozeans unter besonderer Berücksichtigung des Tiefen- und Bodenwassers. Quantitative Untersuchungen zur Statik und Dynamik des Atlantischen Ozeans.
Wiss. Ergebn. Deutschen Atlant. Exped., "Meteor", 1925-1927, 6(2):262-420, 36 figs.

bottom water
Wüst, G., 1943.
Der subarktische Bodenstrom in der westatlantischen Mulde. Ann. Hydr., usw., 71(4/6):249-255.

bottom water
Wüst, G., 1941.
Relief und Bodenwasser im Nordpolarbecken. Zeitschr. d. Ges. f. Erdkunde, Berlin, 1941: 163-180.

bottom water
Wüst, G., 1938.
Bodentemperatur und Bodenstrom in der atlantischen, indischen, und pazifischen Tiefsee. Gerlands Beiträge zur Geophysik, 54(1):1-8, 1 map.

bottom water
Wüst, Georg, 1933.
Das Bodenwasser und die Gliederung der atlantischen Tiefsee.
Wiss. Egebn. dt. atlant. Exped. 'Meteor', 6(1): 1-106.

Translation: USN Oceanogr. Off. TRANS 340. (M. Slessers). 1967.

bottom water
Wüst, Georg (with Arnold Gordon), 1964
Stratification and circulation in the Antil-lean-Caribbean basins. 1. Spreading and mixing of the water types with an oceanographic atlas.
Vema Research Series, Columbia Univ. Press, New York and London, No. 2:201 pp.

bottom water
Wyrtki, Klaus, 1961
The thermohaline circulation in relation to the general circulation in the oceans.
Deep-Sea Res., 8(1):39-64.

bottom water
Yakuwa, R., 1946.
Über die Bodentemperaturen in den verschieden Bodenarten in Hokkaido. Geophys. Mag., Tokyo, 14(1):1-12, 2 figs.

botulism

botulism
Cote, Alfred J., 1967.
Botulism/Selmonella on fisheries 'menu': National Fisheries Institute - will fisheries be government controlled?
Oceanology Int., 2(2):41-43.

Botulism
Eklund, Melvin W., and Frank Poysky, 1965.
Clostridium botulinum Type F from marine sediments.
Science, 149(3681):306.

Bouguer anomalies

Bouguer anomalies
Bowin, Carl O. 1968
Geophysical study of the Cayman Trough.
J. geophys. Res. 73 (16): 5159-5173.

Bouguer anomalies
Davey, F.J. 1970.
Bouguer anomaly map of the north Celtic Sea and entrance to the Bristol Channel.
Geophys. J. R. astron Soc. 22 (3): 277-282.

Bouguer anomalies
Harrison, J.C., and R.E. von Huene, 1965.
The surface-ship gravity motor as a tool for exploring the geological structure of continental shelves.
Ocean Sci. and Ocean Eng., Mar. Techn. Soc.,- Amer. Soc. Limnol. Oceanogr., 1:414-431.

boulders

boulders
Laughton, A.S., 1963.
18. Microtopography.
In: The Sea, M.N. Hill, Editor, Interscience Publishers, 3:437-472.

boundary conditions

boundary conditions
*Blandford, Robert R., 1971.
Boundary conditions in homogeneous ocean models. Deep-Sea Res. 18(7): 739-751.

boundary conditions, thermal
Haney, Robert L., 1971.
Surface thermal boundary condition for ocean circulation models. J. phys. Oceanogr. 1(4): 241-248.

boundary conditions
Ivanov, Y.A., 1965.
The role of boundary conditions and advection in the formation and distribution of extreme values of oceanological characteristics according to depth. (In Russian).
Okeanologiia, Akad. Nauk, SSSR, 5(1):40-44.

Boundary currents

See also: currents, boundary

boundary currents
Baker, D. James Jr., 1971.
A source-sink laboratory model of the ocean circulation.
Geophys. fluid Dyn. 2 (1): 17-29

boundary currents
Warren, Bruce A., and Arthur D. Voorhis, 1970.
Velocity measurements in the deep western boundary current of the Pacific.
Nature, Lond., 228 (5274): 849-850

boundary friction

boundary friction
Schmitz, H.P., 1962
Über die Zufuhr von Turbulenzinergie durch Grenzflächen-reibung und labile Schichtung.
Deutsche Hydro. Zeits., 15(3):114-124.

boundary layers

boundary layer, effect of
Abbott, M.R., 1960
Boundary layer effects in estuaries. J. Mar. Res., 18(2): 83-100.

boundary layers (waves)
Bye, John A.T., 1966.
The wave boundary layer. (Abstract only).
Trans. Am. Geophys. Un., 47(3):478.

boundary layers
*Charney, J.G., 1969.
What determines the depth of the planetary boundary layer in a neutral atmosphere? (In Russian; English abstract).
Okeanologiia, 9(1): 143-145.

boundary layer planetary
Dvoryaninov, G.S., 1971.
The effect of stratification on the planetary boundary layer in the ocean. (In Russian; English abstract). Fisika Atmosfer. Okean., Izv Akad. Nauk SSSR, 7(3): 327-336.

boundary-layer flow
Faller, Alan J., 1962
The instability of Ekman boundary-layer flow and its application as a possible explanation of 'Langmuir' circulation cells in the ocean. (Abstract).
J. Geophys. Res., 67(9):3556.

boundary layers
Grosbh, C.E., and S.J. Lukasik, 1965.
Ocean wave boundary layers. (summary).
Symposium, Diffusion in oceans and fresh waters, Lamont Geol. Obs., 31 Aug.-2 Sept., 1964, 149-150.

boundary layers
Harper, J.F., 1963.
On boundary layers in two-dimensional flow with vorticity.
J. Fluid Mechanics, 17(1):141-153.

boundary layers
Hart, J.E., 1971.
A possible mechanism for boundary layer mixing and layer formation in a stratified fluid.
J. phys. Oceanogr. 1(4): 258-262.

boundary layers
*Ichiye, Takeshi, 1967.
Upper ocean boundary-layer flow determined by dye diffusion.
Physics Fluids 10 (9-2):S270-S277.

boundary zone
Ishida, Masami, Tsuneyoshi Suzuki, Noritatsu Sano, Ichiro Saito and Seikichi Mishima, 1960.
[On the detection of the boundary zone.]
Bull. Fac. Fish., Hokkaido Univ., 10(4):291-302.

boundary layers (coastal)
Janowitz, Gerald S. 1970.
The coastal boundary layers of a lake when horizontal and vertical Ekman numbers are of different orders of magnitudes.
Tellus 22 (6): 585-596.

BOUNDARY LAYERS
Johnson, J.A., 1970.
Oceanic boundary layers. Deep-Sea Res., 17(3): 455-465.

boundary layers,wave
*Jonsson,Ivar G.,1967.
Wave boundary layers and friction factors.
Proc. 10th Conf.Coast.Engng,Tokyo, 1966,1:126-148.

boundary layers
Kao, S.-K., 1960.
Stationary flow in the planetary boundary layer with an inversion and a sea breeze.
J. Geophys. Res., 65(6):1731-1736.

boundary layers
Kelly, R.E. 1970.
Wave induced boundary layers in a stratified fluid.
J. fluid Mech. 42(1): 139-150.

boundary layers
Leikhmen,D.L.,1966.
Dynamics of atmospheric and oceanic boundary layers taking into account their interaction and nonlinear effects. (In Russian;English abstract).
Fisika Atmosferi i Okeana,Izv. Akad. Nauk,SSSR. 2(10):1017-1025.

boundary layers
Nece, Ronald E., and J. Dungan Smith 1970
Boundary shear stress in rivers and estuaries.
J. WatWays Harb. Div. Am. Soc. civ. Engrs 96 (WW2): 335-358.

boundaries, geological (Würm/Flandrian)
Olausson, Eric 1969.
On the Würm-Flandrian boundary in deep-sea cores.
Geol. Mijnb. 48(3): 349-361.

boundary layers,inertial
Pedlosky, Joseph, 1965.
A necessary condition for the existence of an inertial boundary layer in a baroclinic ocean.
J. Mar. Res., 23(2):69-72.

boundaries, geological (Pliocene/Pleistocene)
Phillips, J.D., W.A. Berggren, A. Bertels and D. Wall 1968.
Paleomagnetic stratigraphy and micropaleontology of three deep sea cores from the central North Atlantic Ocean.
Earth Planet. Sci. Letts. 4(2): 118-130.

boundary layers
Schubauer, G.B., and W.G. Spangenberg, 1960.
Forced mixing in boundary layers.
J. Fluid Mech., 8(1):10-32.

boundary layers
Stern, Melvin E., 1963.
Trapping of low frequency oscillations in an equatorial "boundary layer".
Tellus, 15(3):246-250.

boundaries, deep-sea
Sternberg, R.W., 1970.
Field measurements of the hydrodynamic roughness of the deep-sea boundary. Deep-Sea Res., 17(3) 413-420.

boundary layers
Trenin, V.P., 1961.
Calculation of the mean gradient in boundary layers. (In Russian).
Trudy Gosud. Okeanogr. Inst., 63:95-103.

boundary layers
Vager, B.G., and B.A. Kagan, 1969.
Dynamics of a turbulent boundary layer in a tidal current. (In Russian; English abstract).
Fisika Atmosfer. Okean.,Izv. Akad. Nauk, SSSR, 5(2):168-179.

boundary layers
Van Dyke, M., 1962.
Higher approximations in boundary-layer theory. 2. Approaches to leading edges.
J. Fluid Mechanics, 14(4):481-495.

boundary layers, benthic
Wimbush, Mark and Walter Munk, 1970.
1. General observations. 19. The benthic boundary layer. In: The sea: ideas and observations on progress in the study of the seas, Arthur E. Maxwell, editor, Wiley-Interscience 4(1): 731-758.

boundary layers, thermal
Wu, Jin, 1971.
An estimation of oceanic thermal-sublayer thickness. J. phys. Oceanogr. 1(4): 284-286.

boundary mixing
Wunsch, Carl, 1970.
On oceanic boundary mixing. Deep-Sea Res., 17(2): 293-301.

Boundary layers, turbulent
Zaslavsky, M.M. 1970.
On the theory of surface wave generation by a turbulent boundary layer. (In Russian; English abstract).
Fizika Atmosfer. Okean., Izv. Akad. Nauk SSSR 6(6): 627-634.

boundaries-biogeographical

boundaries, biogeographical
Beklemishev, C.W., and H.J. Seimina, 1956.
On the boundary of the biogeographical boundary between the boreal and tropical pelagic region of the western North Pacific.
Dokl. Akad. Nauk, SSSR, 108(6):1057-1060.

boundaries
Hortig,F.J.,1967.
Jurisdictional,administrative, and technical problems related to the establishment of California coastal and offshore boundaries.
In: The law of the sea, L.M. Alexander,editor, 230-241.

boundaries geological

Cretaceous-Tertiary boundary
Berggren, W.A. 1971.
Paleogene planktonic foraminiferal faunas on Legs I-IV (Atlantic Ocean), JOIDES Deep-Sea Drilling Program - a Synthesis.
Proc. II Plankton. Conf., Roma 1970, A. Farinacci, editor, 57-77.

boundaries, Tertiary
*Berggren, W.A., 1971.
Tertiary boundaries and correlations. In: Micropaleontology of oceans, B.M. Funnell and W.R. Riedel, editors, Cambridge Univ. Press, 693-809.

boundaries,Brunhes/Matuyama
* Goodell,H.G., and N.D. Watkins,1968.
The paleomagnetic stratigraphy of the Southern Ocean: 20° West to 160° East Longitude.
Deep-Sea Res., 15(1):89-112.

boundaries,Matuyama/Gauss
*Goodell,H.G., and N.D. Watkins,1968.
The paleomagnetic stratigraphy of the Southern Ocean: 20° West to 160° East longitude.
Deep-Sea Res., 15(1):89-112.

boundaries,Pleistocene/Recent
*Frerichs,William E.,1968.
Pleistocene-Recent boundary and Wisconsin glacial stratigraphy in the northern Indian Ocean.
Science, 159(3822):1456-1458.

boundaries, Pliocene-Pleistocene
Hays, J.D. and W.A. Berggren, 1971.
Quaternary boundaries and correlations. In: Micropaleontology of oceans, B.M. Funnell and W.R. Riedel, editors, Cambridge Univ. Press, 669-691.

boundaries, Plio-Pleistocene
Kennett, James P. and Kurt R. Geitzenauer, 1969
Relationships between Globorotalia truncatulinoides and G. tosaensis in a Pliocene-Pleistocene deep-sea core from the South Pacific. Trans. Gulf Coast Ass. geol. Socs, 19: 613.

boundaries, Neogene
Lamb, James L., 1969.
Planktonic foraminiferal datums and late Neogene epoch boundaries in the Mediterranean, Caribbean and Gulf of Mexico. Trans. Gulf Coast Ass. geol Socs, 19: 559-578.

boundary layer, bottom
Weatherly, Georges L., 1972
A study of the bottom boundary layer of the Florida Current. J. phys. Oceanogr. 2(1): 54-72.

boundary layer, coastal
Csanady, G.T., 1972.
The coastal boundary layer in Lake Ontario. Part I: the spring regime. J. phys. Oceanogr. 2(1): 41-53.

boundaries(salt vs fresh-water)
den Hartog, C. 1971.
The border environment between the sea and the fresh water, with special reference to the estuary.
Vie et Milieu Suppl. 22(2): 739-751.

boundaries, wavy

boundaries, wavy
* Davis, Russ E., 1969.
On the high Reynolds number flow over a wavy boundary.
J. fluid Mech. 36(2): 337-346.

bow-riding

bow-riding
Lang, Thomas G. 1966.
Hydrodynamics analysis of cetacean performance.
In: Whales dolphins and porpoises, K.S. Norris, editor, Univ. Calif. Press, 410-432.

bow-riding

Yuen, Heeny S.H., 1961
Bow wave riding of dolphins. Science 134 (3484): 1011-1012.

brachiopods

brachiopods, lists of spp.
Dell, R.K., 1964.
A list of Mollusca and Brachiopoda collected by N.Z.O.I. from Milford Sound.
New Zealand Dept. Sci. Ind. Res., Bull., No. 157; New Zealand Oceanogr. Inst., Memoir, No. 17:91-92

brachiopods
Foster, Merrill W. 1969.
Antarctic and subantarctic brachiopods
Antarctic J. U.S. 4 (5): 191-192.

brachiopods

McCammon, Helen M., 1971.
Behavior in the brachiopod Terebratulina septentrionalis (Couthouy). J. exp. mar. Biol. Ecol., 6(1): 35-45.

brachiopods, physiology
McCammon, Helen M., 1965.
Filtering currents in brachiopods measured with a thermistor flowmeter.
Ocean Sci. and Ocean Eng., Mar. Techn. Soc.,-Amer. Soc. Limnol. Oceanogr., 2:772-779.

brackishness

brackish waters
Turmel, J.M., 1956.
Ecologie des prés salés: Morphologie, circulation et salinité des eaux.
Bull. Lab. Marit. Dinard, 42:41-49.

" brackishness"
Zambriborshch, F.S., 1966.
"Brackishness" of the northwestern part of the Black Sea and fishes populating it. (In Russian).
Gidrobiol. Zhurn., 2(1): 10-17
English abstract
Black Sea
"brackishness"

brackish water, effect of

Lange, Rolf, 1969.
Marine biology and chemistry. In: Chemical Oceanography, Rolf Lange, editor, Universitetsforlaget, Oslo: 35-46.

bradycardia

bradycardia
Elsner, R.W., David W. Kenney and Kent Burgess, 1966.
Diving bradycardia in the trained dolphin.
Nature, 212 (5060):407-408.

bradycardia
Irving, Laurence 1966.
Elective regulation of the circulation in diving animals.
In: Whales, dolphins and porpoises, K.S. Norris, editor, Univ. Calif. Press 381-396.

bradycardia
Murdaugh, H. Victor, Jr., 1966.
Adaptations to diving in the harbor seal; cardiac output during diving.
Am. J. Physiol., 210(1):176-180.

breakers

breakers
See also: waves, breaking

breakers

Akedo, K., 1960
Surf and breaker at the coast from Osaka to Kobe.
Bull. Kobe Mar. Obs., No. 167(16):39-46.

breakers

Barber, Norman Frederick, 1950
Ocean waves and swell.
The Institution of Civil Engineers, Maritime and Waterways Engineering Division, Session 1949-1950:3-22.

breakers
Bruun, Per, 1963
Longshore currents and longshore troughs.
J. Geophys. Res., 68(4):1065-1078.

breakers
Castanho, J., 1966.
Rebentacão das ondas e transporte litoral.
Lab. Nac. Engenharia Civil, Memoria, Liboa, No. 275-278 pp.

English synopsis (2 pp).

breakers
Collins, J.I., 1964
The effect of currents on the mass transport of progressive waves.
J. Geophys. Res., 69(6):1051-1056.

breakers

Ertel, H. 1971.
Eine Betrachtung zur geomorphologisch wirksamen Arbeit der Brandungswellen an Flachküsten.
Acta hydrophys. 16(1): 5-10.

breakers
Ertel, H., 1960.
Condición para la ruptura de las crestas de olas marinas.
Gerlands Beitr. f. Geophysik, 69(2):114-118.

breakers
Galvin, Cyril J., 1969.
Breaker travel and choice of design wave height.
J. WatWays Harb. Div. Am. Soc. civ. Engrs 95 (WW2): 175-200.

breakers
Galvin, Cyril J., Jr. 1968.
Breaker type classification on three laboratory beaches
J. geophys. Res. 73(12): 3651-3659.

breakers
Galvin, C.J., Jr., and P.S. Eagleson, 1965.
Experimental study of longshore currents on a plane beach.
U.S. Army Coastal Engin. Res. Center, Techn. Memo., No. 10:80 pp.

breakers
Groen, P., and R. Dorrestein, 1958.
Zeegolven.
K. Nederl. Meteorol. Inst., Opstellen op Oceanografisch en Maritiem Meteorol. Gebied, 11: 98 pp.

breakers
Groen, P., and M.P.H. Weenink, 1950.
Two diagrams for finding breaker characteristics along a straight coast. Trans. Am. Geophys. Union 31(3):398-400, 2 textfigs.

breakers
Hamada, T., 1951.
Breakers and beach erosion. Rept. Transport. Tech. Res. Inst., No. 1:165 pp., 20 photos., 39 textfigs.

breakers
Harris, T.F.W., 1961.
The nearshore circulation of water.
Marine Studies off the Natal Coast, C.S.I.R. Symposium, No. S2:18030. (Multilithed).

breakers

Hayami, Shoitiro, 1958
Type of breakers, wave steepness and beach slope.
Coastal engineering in Japan, 1: 21-24.

breakers
Hayami, Shoitiro, 1955.
Mechanism of wave breaking. (II). (In Japanese).
Proc., 2nd Conf. Coastal Engin., Japan, 1955:13-15.

Also in:
Papers on Oceanogr. and Hydrol. (1949-1962), Geophys. Inst., Kyoto Univ., Contrib. No. 7.

breakers
Hayami, Shoitiro, 1954.
Mechanism of wave breaking. (In Japanese).
Proc. 1st Conf. Coastal Engin., Japan, 1954:35-43.

Also in:
Papers on Oceanogr. and Hydrol., (1949-1962) Geophys. Inst., Kyoto Univ., Contrib. No. 6. (1963).

breakers
Hayashi, Taizo, and Masatoro Hattori, 1964.
Thrusts exerted upon composite-type breakwaters by the action of breaking waves.
Coastal Engineering in Japan, 7:65-84.

breaker

Hayashi, Taizo, and Masataro Hattori, 1958.

Pressure of the breaker against a vertical wall.
Coastal Engineering in Japan, 1: 25-38.

breakers

Hom-ma, Masashi, Kiyoshi Horikawa, and Choule Sonu, 1960

A study on beach erosion at the sheltered beaches of Katase and Kamakura, Japan.
Coastal Eng., Japan, 3:101-122.

breakers.
Iversen, H.W., 1952.
Laboratory study of breakers. Proc. N.B.S. Semicent. Symp., Gravity Waves, June 18-20, 1951, Nat. Bur. Stand. Circ. 521:9-32, 14 textfigs.

breakers

Iwagaki, Y., and T. Sawaragi, 1958.
Experimental study on the equilibrium slopes of beaches and sand movement by breakers. Coastal Engineering In Japan, 1:75-84.

breakers (surf)

Jones, W.M., 1950.
Progress in hydrography and physical oceanography N.Z. Dept. Sci. Ind. Res., Gephys. Conf., 1950, Paper No. 11: 2 pp. (mimeographed).

breakers

Jordaan, J.M., Jr., 1961.
Basic model studies of nearshore wave action. Marine Studies off the Natal Coast, C.S.I.R. Symposium, No. S2:118-134 (multilithed).

breakers

Kühlmann, D.H.H. 1970
Studien über physikalische und chemische Faktoren in kubanischen Riffgebieten.
Acta hydrophys. 15(2):105-152.

breakers

Larras, J., 1952.
[Experimental research on breaking waves.]
An. Ponts Chauss. 122:525-542.

Abstr.: Appl. Mech. Rev. 6:261.

breakers

*LeMéhauté, Bernard, Robert C.Y. Koh and Li-San Hwanz, 1968.
A synthesis on wave run-up.
J. WatWays Harb.Div.Am.Soc.civ.Engrs., 94(WW1): 77-92.

breakers

Mason, M.A., 1952.
Some observations of breaking waves. Proc. N.B.S Semicent. Symp., Gravity Waves, June 18-20, 1951 Nat. Bur. Stand. Circ. 521:215-220, 3 textfigs.

breakers

*Mitsuyasu, Hisashi, 1967.
Shock pressure of breaking wave.
Proc. 10th Conf.Coast.Engng.Tokyo, 1967, 1:268-283.

breakers

Miyazaki, M., 1952.
Mathematical studies of surf and breakers - nonlinear wave theory in shallow water of constant slope. II. Ocean. Mag., 4(1):13-21, 8 textfigs.
, Tokyo,

breaker zone

Nagata, Yutaka, 1964.
Deformation of temporal pattern of orbital wave velocity and sediment transport in shoaling water in breaker zone and on foreshore.
J. Oceanogr. Soc., Japan, 20(2):57-70.

breakers

*Nakamura, Makoto, Hidehiko Shiraishi and Yasuo Sasaki, 1967.
Wave decaying due to breaking.
Proc. 10th Conf. Coast. Engng. Tokyo, 1966, 1:234-253.

breakers

O'Brien, M. P., 1949.
The causes of plunging and spilling breakers.
Bull. B.E.B., 3(3):7-10, 1 fig.

(First issued as Tech. Rept. HE-116-192, Fluid Mechanics Lab., Univ. Calif.)

breakers

Shteinbach, B.V., 1962.
On the method of observation of storm waves on a shallow coast. (In Russian).
Trudy Gosud. Okeanogr. Inst., 66:121-126.

breakers

Sitarz, Jean A., 1963.
Contribution à l'étude de l'évolution des plages à partir de la connaissance des profiles d'équilibre.
Trav. Centre de Recherches et d'Etudes Océanogr. 5(2,3,4):199 pp.

breakers

Stoker, J.J., jr., 1949.
The breaking of waves in shallow water. Ann. N. Y. Acad. Sci., 51(3):360-375, 12 textfigs.

breakers

Takenouti, Y., and K. Nanasawa, 1961
[About "GUNKAN NAMI" (warship wave) which reaches on the Imabari Beach as significant breakers.]
J. Oceanogr. Soc., Japan, 17(2):80-90.

In Japanese, English abstract.

breakers

Tricker, R.A.R., 1964.
Bores, breakers, waves and wakes: an introduction to the study of waves on water.
American Elsevier Publishing Co., Inc., 250 pp. $6.50.

breakers

University of California, Department of Engineering, 1949(1947).
Forecasting breakers and surf on a straight beach of infinite length. Bull. B.E.B., 3(4):23-32, 2 figs.+2 figs.

Tech. Rept. HE-116-13
HE-116-67

breakers

Vera-Cruz, D., 1962
Ondas na rebentação.
Lab. Nacional de Engenhara Civil, Memoria, Lisbon, No. 199: 16 pp.

breakers

*Wilson, W. Stanley, 1968.
On the origin of certain breakers off the island of Aruba.
Tech.Rep.Chesapeake Bay Inst. (Ref.68-13) 43:27 pp. (multilithed).

breakers, effect of

breakers, effect of

Entel H. und D. Hartke 1971.
Die Geschwindigkeit des Sedimenttransports längs der Strandlinie als Problem der Mechanik nichtstarrer Punktsysteme.
Acta hydrophysica 15(4): 249-258.

breakers, effect of

Miller, R.L., and J.M. Zeigler, 1958

A model relating dynamics and sediment pattern in equilibrium in the region of shoaling waves, breaker zone and foreshore.
J. Geol., 66(4): 417-441.

breaker zone

breaker zone

Hodgson, W.A., 1966.
Coastal processes around the Otago Peninsula.
New Zealand J. Geol. Geophys., 9(1/2):76-90.

breaker zone

Ingle, James C., Jr., 1966.
The movement of beach sand: an analysis using fluorescent grains.
Develop. Sedimentol., Elsevier Publ., Co., 221 pp.

breaker zone

Ingle, James C., Jr., 1962.
Tracing beach sand movement by means of fluorescent dyed sand.
Shore and Beach, October 1962, 6 pp.

breaker zone

Inman, D.L., and N. Nasu, 1956.
Orbital velocity associated with wave action near the breaker zone. B.E.B. Tech. Memo., 79: 70 pp.

breaker zone

Prasad, Rao, R., 1954.
Swash and breaker zone. Current Sci. 24:46-47.

breaker zone

Schwartz, Maurice L. 1967.
Littoral zone tidal-cycle sedimentation.
J. sedim. Petrol. 37(2): 677-709.

breakwaters

breakwaters

Beach Erosion Board, 1953.
Shore protection planning and design. (Preliminary, subject to revision). B.E.B. Bull., Special Issue, No. 2:230 pp., 149 figs., plus app.

Breakwaters

Bruun, Per, 1953.

Breakwaters for coastal protection. Hydraulic principles in design.
XVIII Int. Navig. Congr., Rome, SII-Q1:5-39.

breakwaters

de Carvalho, J.J.R., and D. Vera-Cruz, 1962
On the stability of rubble-mound breakwaters.
Ministerio das Obras Publicas, Lab. Nacional de Engenharia Civil, Techn. Paper, 200:633-658.

Reprinted? from Coastal Engineering.

breakwaters

Evans, J.T., 1955.
Pneumatic and similar breakwaters. Proc. R. Soc. London, A, 231:457-466.

breakwaters

Great Britain, Department of Scientific and Industrial Research, 1961.
Waves and sea defences.
Hydraulics Research, 1960:62-69.

breakwaters

Hayashi, Taizo, and Masatoro Hattori, 1964.
Thrusts exerted upon composite-type breakwaters by the action of breaking waves.
Coastal Engineering in Japan, 7:65-84.

breakwaters
Hayashi, Taizo and Masataro Hattori, 1961
Stability of the breakwater against sliding due to pressure of breaking waves.
Coastal Eng., Japan, 4:23-33.

breakwaters
Hedar, P.A., 1953.
Design of rock-fill breakwaters. Proc. Minnesota Int. Hydraulics Conv., Sept. 1-4, 1953:241-260, 30 textfigs.

breakwaters
Hom-ma, Masashi, and Kiyoshi Horikawa, 1961
A study of submerged breakwaters.
Coastal Eng., Japan, 4: 85-102.

breakwaters
Hom-ma Masashi, Kiyoshi Horikawa and Hiromasa Mochizuki, 1964.
An experimental study on floating breakwaters.
Coastal Engineering in Japan, 7:85-94.

breakwaters
Iribarren Cavalilles, R., and C. Nogales y Olano, 1953.
Nueva confirmacion de la formule para el calculo de los diques de escollera. Rev. Obras Publ., Jan. 1953:3 pp.

breakwaters
Kaplan, K., 1952.
Notes on determination of stable underwater breakwater slopes. Bull. B.E.B. 6(3):20-22.

breakwaters
Kurihara, Michiaki, 1965.
On the study of pneumatic breakwater in Japan.
Coastal Engin., Japan, 8:71-83.

breakwaters
Larras, J., 1952.
[The turning of the swell around jetties.]
An. Ponts Chaus. 122:517-523.

Abstr. Appl. Mech. Rev. 6:166.(1953)

breakwaters
Muraki, Yoshio, 1966.
Field investigation on the oscillation of breakwater caused by wave action.
Coast. Engng Japan, 9:97-106.

breakwaters
Ross, C.W., 1957.
Model tests on a triple-bulkhead type of floating breakwater.
Beach Erosion Bd., Tech. Memo., 99:30 pp.

breakwaters, effect of
Shiraishi, Naobumi, Atsushi Numata and Naoki Hase, 1960
The effect and damage of submerged by breakwater in Niigata coast.
Coastal Eng., Japan, 3: 89-99.

breakwaters
Stoker, J.J., 1952.
Theory of floating breakwaters in shallow water. (Abstract). Proc. N.B.S. Semicent. Symp., Gravity Waves, June 18-20, 1951, Nat. Bur. Stand. Circ. 521:33.

"breakwaters"
Taylor, G., 1955.
The action of a surface currents used as a breakwater. Proc. R. Soc., London, A, 231(1187):466-478.

breakwaters, effect of
Wada, Akira, 1964.
On the disturbed waves in a bay sheltered by a breakwater.
Coastal engineering in Japan, 7:31-44.

breakwaters
Wiegel, R.L., H.W. Shen and J.D. Camming, 1962.
Hovering breakwater.
J. Waterways and Harbor Div., 88(WW2):23-50.

breakwaters, effect of

breakwaters, effect of
*Fukuuchi, Hiromasa and Yoshiyuki Ito, 1967.
On the effect of breakwater against tsunami.
Proc. 10th Conf. Coast. Engng., Tokyo, 1966, 2:821-839

breakwaters, effect of
Green, J.L., 1961
Pneumatic breakwaters to protect dredges.
J. Waterways & Harb. Div., Proc. Amer. Soc. Civil Eng., 87(WW2): 67-88.

breakwaters, effect of
Hayashi, Taizo, 1965.
Virtual mass and the damping factor of the breakwater during rocking, and the modification of their effect of the expression of the thrusts exerted upon breakwaters by the action of breaking waves.
Coastal Engin., Japan, 8:105-117.

breakwaters, effect of
Hayashi, Taizo, Masataro Hattori, Tokutaro Kano and Masujiro Shirai 1966.
Hydraulic research on the closely spaced pile breakwater
Coast. Engng Japan 9:107-117.

breakers, effect of
Ingle, James C., Jr., 1962.
Tracing beach sand movement by means of fluorescent dyed sand.
Shore and Beach, October 1962, 6 pp.

breakwater, effect of
Kajiura, K., 1963.
27. Effects of a breakwater on the oscillations of bay water. (In Japanese; English abstract).
Bull. Earthquake Res. Inst., Tokyo, 41(2):403-418

breakwaters, effect of
Kato, Juichi, Toshifumi Noma, Yukio Uekita and Seiya Hagino, 1969.
Damping effect of floating breakwater.
J. WatWays Harb. Div. Am. Soc. civ. Engrs. 95(WW3): 337-344

breakwaters, effect of
Longuet-Higgins, M.S. 1967.
On the wave-induced difference in mean sea level between the two sides of a submerged breakwater.
J. mar. Res. 25(2): 148-153.

breakwaters, effect of
*Momoi, Takao, 1967.
A long wave around a breakwater (case of perpendicular incidence).II.
Bull. Earthq. Res. Inst., Tokyo Univ., 45(3):749-783.

breakwaters, effect of
Momoi, Takeo 1967.
A long wave around a breakwater (case of perpendicular incidence). I.
Bull. Earthq. Res. Inst. 45(1): 91-136.

breakwaters, effect of
Sato, Shoji and Isao Irie, 1970.
Variation of topography of sea-bed caused by the construction of breakwaters. Coastal Engng Japan 13: 141-152.

breccias

breccias
Cann, J.R., and F.J. Vine 1966.
An area on the crest of the Carlsberg Ridge: petrology and magnetic survey.
Phil. Trans. R. Soc. (A) 259 (1099): 198-217.

Breccias
Nayudu, Y. Rammohanroy, 1965.
Petrology of submarine volcanics and sediments in the vicinity of the Mendicino Fracture Zone.
Progress in Oceanography, 3:207-220.

breezes
Yoshino, Itaru 1971.
On the land and sea breeze of Osaka Bay. (In Japanese; English abstract)
Umi to Sora 46 (3/4): 99-111.

brines, concentration of

brines
Brewer, P.G., T.R.S. Wilson, J.W. Murray, R.G. Munns and C.D. Densmore 1971.
Hydrographic observations on the Red Sea brines indicate a marked temperature increase.
Nature, Lond. 231 (5297): 37-38

brines, geothermal
Craig, H., 1966.
Isotopic composition and origin of the Red Sea and Salton Sea geothermal brines.
Science, 154(3756):1544-1548.

brine
*Delevaux, M.H., B.R. Doe and G.F. Brown, 1967.
Preliminary lead isotope investigations of brine from the Red Sea, galena from the Kingdom of Saudi Arabia, and galena from United Arab Republic (Egypt).
Earth Planet Sci. Letters, 3(2):139-144.

bryozoans
Emschermann, P., 1971.
Lozomespilon perezi - ein Entoproctenfund im Mittelatlantik. Überlegungen zur Benthosbesiedlung der Grossen Meteorbank. Mar. Biol. 9(1): 51-62.

brine
Heine, A.J., 1968.
Brine in the McMurdo Ice Shelf, Antarctica.
N.Z. Jl. Geol. Geophys. 11(4): 829-839.

brines
Hunt, John M., Earl E. Hays, Egon T. Degens and David A. Ross, 1967.
Red Sea: detailed survey of hot-brine areas.
Science, 156(3774):514-516.

brines

*Lerman, Abraham, 1967.
Model of chemical evolution of a chloride lake-The Dead Sea.
Geochim. cosmochim Acta, 31(12):2309-2330.

brines

Manheim, F.T. and F.L. Sayles, 1970.
Brines and interstitial brackish water in drill cores from the deep Gulf of Mexico.
Science, 170(3953): 57-61.

brines, effect of

Raup, Omer B., 1970.
Brine mixing: an additional mechanism for formation of basin evaporites.
Bull. Am. Ass. Petrol. Geol. 54(12): 2246-2259.

brine, effect of

*Redfield, Alfred C. and Irving Friedman, 1969.
The effect of meteoric water, melt water and brine on the composition of Polar Sea water and of the deep waters of the ocean.
Deep-Sea Res., Suppl. 16: 197-214.

brines

Thompson, T.G., and K.H. Nelson, 1956.
Concentration of brines and deposition of salts from sea water under frigid conditions.
Am. J. Sci., 254(4):227-238.

broaching

broaching

Grim, O., 1963
Surging motion and broaching tendencies in a severe irregular sea.
Deutsche Hydrogr. Zeits., 16(5):201-231.

brood strength

brood strength

Carruthers, J.N., A.L. Lawford, and V.F.C. Veley, 1951.
Variations in brood strength in the North Sea haddock in the light of relevant wind conditions.
Nature 168:317-319.

brood strength

Carruthers, J.N., A.L. Lawford and V.F.C. Veley, 1951.
Fishery hydrography: brood strength fluctuations in various North Sea fish, with suggested method of prediction. Kieler Meeresf. 8(1):5-15.

brood strength

Graham, M., 1951.
Letter to the editor. Winds and brood strength.
J. du Cons. 18(1):73.

brucite

brucite

Schmalz, R.F., 1965.
Brucite in carbonate secreted by the red alga Goniolithon sp.
Science, 149(3687):993-996.

brucite

Weber, Jon N., and John W. Kaufman, 1965.
Brucite in the calcareous alga Goniolithon.
Science, 149(3687)996-997.

Brunhes epoch

*Smith, Jerry D., and John H. Foster, 1969.
Geomagnetic reversal in Brunhes normal polarity epoch.
Science 163(3867): 565-567.

Bryozoa

bryozoa

Androsova, E.I., 1965.
Bryozoans of the orders Cyclostomata and Ctenostomata of the northern part of the Sea of Japan. Fauna of the seas of the northwest Pacific. (In Russian).
Issled. Fauni Morei, Zool. Inst., Akad. Nauk, SSSR, 3(11):72-114.

bryozoa

Bobin, Genevieve, 1970.
Loxosoma fishelsoni n. sp. (Entoprocta: Loxosomatidae) du Golfe d'Akaba.
Israel J. Zool. 19(2): 111-121

bryozoa

Bobin, Geneviève, 1968.
Loxosomella museriensis n. sp. entoprocte Loxosomatidae de Mer Rouge.
Israel J. Zool. 17(4): 175-189.

bryozoa

Cook, Patricia L., 1968.
Bryozoa (Polyzoa) from the coast of tropical Africa.
Atlantide Rept., 10:115-262.

Bryozoa

Cook, Patricia L. 1966.
Some "sand fauna" Polyzoa (Bryozoa) from eastern Africa and the northern Indian Ocean.
Cah. Biol. mar. Roscoff 7(2): 207-223

bryozoa

Gordon, D.P., 1967.
A report of the ectoproct Polyzoa of some Auckland shores.
Tane, J. Auckland Univ. Fld Club 13: 43-76

Bryozoa

Gostilovskaia, M.G. 1968.
Bryozoa in Chesk Bay, Barents Sea. (In Russian).
Trudy Murmansk Morsk Biol. Inst. 17(21): 58-73

Bryozoa

Harmelin, Jean-Georges 1969.
Bryozoaires des grottes sous-marines obscures de la région de Marseillaise: faunistique et écologie.
Tethys 1(3): 793-806.

Bryozoa

Jebram, Diethardt, 1970.
Zur Bryozoen-Fauna der deutschen Meeresgebiete und Brackwässer.
Kieler Meeresforsch. 25(2): 336-347
I. Neue Funde

Bryozoa, checklists

Powell, N.A., 1969.
A checklist of Indo-Pacific Bryozoa in the Red Sea.
Israel J. Zool. 18(4): 357-362.

Bryozoa

Powell, N.A. 1969.
Indo-Pacific Bryozoa new to the Mediterranean coast of Israel.
Israel J. Zool. 18(2/3): 157-168

Bryozoa

Powell, N.A., 1971.
The marine Bryozoa near the Panama Canal.
Bull. mar. Sci. Coral Gables, 21(3): 766-778.

bryozoa

*Powell, N.A., and G.D. Crowell, 1967.
Studies on Bryozoa (Polyzoa) of the Bay of Fundy region. I. Bryozoa from the intertidal zone of Minas Basin and Bay of Fundy.
Cah. Biol. mar., 8(4):331-347.

Bryozoa

*Rucker, James Bivin, 1967.
Paleoecological analysis of cheilostome Bryozoa from Venezuela-British Guiana shelf sediments.
Bull. mar. Sci., Miami, 17(4):757-839.

Bryozoa

Ryland, J.S., 1967.
Polyzoa.
Oceanogr. Mar. Biol., Ann. Rev., H. Barnes, editor, George Allen and Unwin, Ltd., 5:343-369.

Bryozoa

Schopf, Thomas, J.M., 1968.
Ectoprocta, Entoprocta, and Bryozoa.
Syst. Zool., 17(4):470-471.

Bryozoa

*Schopf, Thomas J.M., and Frank T. Manheim, 1967.
Chemical composition of Ectoprocta (Bryozoa).
J. Paleont., 41(5):1197-1225.

Bryozoa

Soule, John D., and Dorothy F. Soule 1969.
Three new species of burrowing bryozoans (Ectoprocta) from the Hawaiian islands
Occ. Pap. Calif. Acad. Sci. 78: 9pp.

Bryozoa

Soule, John D., and Dorothy F. Soule, 1968.
Perspectives on the Bryozoa: Ectoprocta question.
Syst. Zool., 17(4):468-470.

Bryozoa

Teal, John M., 1967.
Is an ectoproct possible?
Nature Lond., 216(5121):1239-1240.

bubbles

bubbles

Abe, T., 1959.
A supplementary note on the foaming of sea water (2) - On the mean thickness of the membrane surrounding a bubble of sea water. Umi to Sora (33(6):97-101.

bubbles

Abe, T., 1955.
A study on the foaming of sea water. A note on the analogy between the coagulation process of colloidal particles and that of bubbles in foam layer of sea water.
Pap. Meteorol. Geophys., Tokyo, 6(1):57-62.

bubbles

Abe, T., 1955.
A study on the foaming of sea water (1). On the mechanism of decay of bubbles and their size distribution in foam layer of sea water.
Pap. Met. Geophys., Tokyo, 5(3/4):240-247, 6 textfigs.

bubbles

Abe, T., 1955.
A study on the foaming of sea water. On the mechanism of the decay of bubbles and their size distribution in foam layer of sea water. Rec. Ocean. Wks., Japan, 2(1):1-6, 2 textfigs.

bubbles
Abe, T., 1953.
A study on the foaming of sea water. On the mechanism of the decay of foam layer of sea water. Rec. Ocean. Wks., Japan, n.s., 1(2):18-24, 5 textfigs., 1 photo.

bubbles
Astu, G., 1959.
Quelques aspects de l'écume océanique.
Bull. du Cen. Études Rech. Sci., 2(3):357-367.

bubbles
Baylor, E.R., W.H. Sutcliffe and D.S. Hirschfeld, 1962
Adsorption of phosphates onto bubbles.
Deep-Sea Res., 9(2):120-124.

Bubbles
Barbour, Richard T., 1966.
Interaction of bubbles and bacteria in the formation of organic aggregates in sea-water.
Nature, 211 (5046):257-258.

bubbles
*Barnes, C., and D.V. Anderson, 1968.
Damped oscillations of a bubble in an extended liquid.
J. acoust. Soc. Am., 43(3):639.

bubbles
Blanchard, Duncan C., 1964
Sea-to-air transport of surface active material.
Science, 146(3642):396-397.

bubbles
Blanchard, Duncan C., 1963.
The electrification of the atmosphere by particles from bubbles in the sea.
Progress in Oceanography, 1:71-202.

bubbles
Blanchard, D.C., 1956.
Air bubbles in water. Amer. J. Phys., 24(3):177-178.

bubbles
Blanchard, D.C., 1954.
Bursting of bubbles at an air-water interface.
Nature 173(4413):1048.

bubbles
Blanchard, D.C., and A.H. Woodcock, 1957.
Bubble formation and modification in the sea and its meteorological significance. Tellus 9(2):145-158.

bubbles
Bulson, P.S., 1963.
Bubble breakwaters with intermittant air supply - abstracts from report on some experimental studies.
Dock & Harbour Authority, 44(514):129-134.

bubbles
Bulson, P.S., 1961
Currents produced by an air curtain in deep water. Report on recent experiments at Southampton.
Dock & Harb. Auth., 42(487): 15-22.

bubbles
Day, J.A., 1967.
Bursting air bubbles studies by the time exposure technique.
Nature, Lond., 216(5120):1097-1099.

bubbles
Duncan, A.R., and N.M. Pantin, 1969.
Evidence for submarine geothermal activity in the Bay of Plenty.
N.Z. Jl mar. freshwat. Res. 3(4). 602-606

bubbles
*Eller, Anthony, 1968.
Force on a bubble in a standing acoustic wave.
J. acoust. Soc. Am., 43(1):170-171.

bubbles
Eller, Anthony, and H.G. Flynn, 1965.
Rectified diffusion during nonlinear pulsations of cavitation bubbles.
J. Acoust. Soc., Amer., 37(3):493-503.

bubbles
Friedman, B., 1950.
Theory of underwater explosion bubbles. Comm. Pure Applied Math. 3:177-199.

bubbles
Garrett, William D., 1968.
The influence of monomolecular surface films on the production of condensation nuclei from bubbled sea water.
J. geophys. Res., 73(16):5146-5150.

BUBBLES
*Garrett, William D., 1967.
Stabilization of air bubbles at the sea-surface by surface active material.
Deep-Sea Res., 14(6):661-672.

bubbles
*Garrett, W.D., 1967.
Damping of capillary waves at the air-sea interface by oceanic surface-active material.
J. mar. Res., 25(3):279-291.

bubbles
*Gathman, Stuart, and Eva Mae Trent, 1968.
Space charge over the open ocean.
J. atmos. Sci., 25(6):1075-1079.

bubbles, effect of
Haeske, H., 1956.
Experimental determination of the damping of pulsating air bubbles in water in the frequency range 100 to 300 kc/s. Acustica 6:266-275.

bubbles
Hayami, S., and Y. Toba, 1958.
Drop production by bursting of air bubbles on the sea surface. 1.
J. Oceanogr. Soc., Japan, 14(4):145-150.

bubbles
Hempleman, H.V., 1968.
Bubble formation and decompression sickness.
Rept. Underwat. Ass. 1968: 63-65

bubbles
Kanwisher, John W., 1963
On the exchange of gases between the atmosphere and the sea.
Deep-Sea Research, 10(3):195-207.

bubbles
Kientzler, C.F., A.B. Arons, D.C. Blanchard and A.H. Woodcock, 1954.
Photographic investigation of the projection of droplets by bubbles bursting at a water surface.
Tellus 6(1):1-7, 3 textfigs.

bubbles
Knelma, K., N. Dombrowski, and D.M. Neivitt, 1954.
Mechanism of the bursting of bubbles. Nature 173(4397):261, 2 textfigs.

bubbles
Knudsen, W.C., 1961.
Elimination of secondary pressure pulses in offshore exploration.
Geophysics, 26(4):425-436.

bubbles
Komabayashi, M., T. Gonda and K. Isono, 1964.
Life time of water drops before breaking and size distribution of fragment droplets.
J. Meteorol. Japan, (2), 42 (5):330-340.
Also in:
Collected Papers on Sciences of Atmosphere and Hydrosphere, Water Res. Lab., Nagoya Univ., 1964, 2

bubbles
Langlois, W.E., 1963
Similarity rules for isothermal bubble growth.
J. Fluid Mech., 15(1):111-118.

bubbles
Leonard, J.H., and G. Houghton, 1961.
Effect of mass transfer on the velocity of rise of bubbles in water.
Nature, 190(4777):687-688.

bubbles
Liebermann, L., 1957.
Air bubbles in water. K. Appl. Phys., 28(2: 205-211.

bubbles
Loye, D.P. and W.F. Arndt, 1948
Sheet of Air bubbles as an acoustical screen for underwater noise. J. Acous. Soc., America, 20(2):143-145, 1 text fig.

bubbles
Ludlam, F.H., 1954.
Large air bubbles in water. Weather 9(6):169, 6 figs.

bubbles
Mason, B.J., 1954.
Bursting of air bubbles at the surface of the sea. Nature 174:470.

bubbles
McCartney, B.S., and B. McK. Bary, 1965.
Echo-sounding on probable gas bubbles from the bottom of Saanich Inlet, British Columbia.
Deep-Sea Research, 12(3):285-294.

bubbles
Medwin, Herman, 1970.
In situ acoustic measurements of bubble populations in coastal ocean waters. J. geophys. Res., 75(3): 599-611.

bubbles
Menard, H. W., 1950.
Transportation of sediments by bubbles. J. Sed. Petr. 20(2):98-106, 2 textfigs.

bubbling

Menzel, David W., 1966.
Bubbling of sea water and the production of organic particles: a re-evaluation.
Deep-Sea Res., 13(5):963-966.

bubbles

Meyer, E., and E. Skudrzyk, 1953.
[Acoustical properties of gas bubbles in water.] (in German). Akust.Beheift No. 3:434-440.

Abstr. Appl. Mech. Rev. 7:3040(1954)

seemingly incomplete reference - cited from Trans. A.G.U.

bubbles

Miyake, Y., 1951.
The possibility and the allowable limit of formation of air bubbles in the sea. Pap. Met. Geophys., Tokyo, 2(1):95-101, 3 textfigs.

bubbles

Miyake, Y., and T. Kuwabara, 1944.
[The scale of bubbles and their durability.] (In Japanese). Umi to Sori 24:291-295.

bubbles

Monahan, E.C., 1968.
Sea spray and whitecaps.
Oceanus, 14(3):21-24. (not technical).

bubbles

Plesset, M.S., 1949.
The dynamics of cavitation bubbles. J. App. Mech Sept. 1949:277-282, 8 textfigs.

bubbles

Ramsey, W.L., 1962.
Bubble growth from dissolved oxygen near the sea surface.
Limnology & Oceanogr., 7(1):1-7.

bubbles

Ramsey, W.L., 1962
Dissolved oxygen in shallow near-shore water and its relation to possible bubble formation.
Limnol. and Oceanogr., 7(4):453-461.

bubbles

Steinberg, M.S. 1967.
Resonance scattering of sound by a small gaseous object of arbitrary form.
J. acoust. Soc. Am. 41(5): 1352-1357.

bubbles

Sugewara, K., 1965.
Exchange of chemical substance between air and sea.
In: Oceanography and Marine Biology, H. Barnes, editor, George Allen and Unwin, 3:59-77.

bubbles

Sutcliffe, William H., Jr., Edward R. Baylor and David W. Menzel, 1963
Sea surface chemistry and Langmuir circulation.
Deep-Sea Res., 10(3):233-243.

bubbles

Toba, Toshiaki, 1962.
Drop production of bursting air bubbles on the sea surface. III. Study by use of a wind flume.
J. Meteorol. Soc., Japan, (2), 40(1):63-64.

bubbles

Toba, Yoshiaki, 1961
Drop production by bursting of air bubbles on the sea surface. III. Study by use of a wind flume (short report).
J. Oceanogr. Soc., Japan, 17(4):169-178.

bubbles

Toba, Y., 1961
Drop production by bursting of air bubbles on the sea surface (III). Study by use of a wind flume.
Mem. Coll. Sci., Univ. Kyoto, 29(3):313-344.

bubbles

Toba, Y., 1959.
[Drop production by bursting of air bubbles on the sea surface. II. Theoretical study of the shape of floating bubbles]
J. Oceanogr. Soc., Japan, 15(3):121-130.

bubbles

Toba, Y., 1958.
Observation of sea water droplets by filter paper.
J. Ocean. Soc. Japan, 14(4):151-154.

bubbles

Wangersky, Peter J., 1965.
The organic chemistry of sea water.
Am. Scient., 53(3):358-374.

bubbles

Woo Seng H., and Paul R. Pasley, 1967.
Nonlinear oscillation of a bubble.
J. acoust. Soc. Am., 42(1):114-120

bubbles

Woodcock, A.H., 1962
Solubles. Ch. 6, Sect. II, Interchange of properties between sea and air. In: The Sea, Interscience Publishers, Vol. 1, Physical Oceanography, 305-312.
(mss received October 1960)

bubbles

Woodcock, A.H., C.F. Kientzler, A.B. Arons, and D.C. Blanchard, 1953.
Giant condensation nuclei bursting from bubbles.
Nature 172(4390):1144-1145, 2 textfigs.

bubbles

Wyman, J., jr., P.F. Scholander, G.A. Edwards, and L. Irving, 1952.
On the stability of gas bubbles in sea water.
J. Mar. Res. 11(1):47-62, 5 textfigs.

bubbles, chemistry of

bubbles, chemistry of

Komabayashi, M., 1964.
Primary fractionation of chemical components in the formation of submicron spray drops from seasalt solution.
J. Meteorol., Japan (2), 42 (5):309-316.

Also in:
Collected papers on Sciences of Atmosphere and Hydrosphere, Water Res. Lab., Nagoya Univ., 1964, 2.

bubbles, effect of

bubbles, effect of

Albers, Vernon M., 1965.
Underwater acoustics handbook II.
The Pennsylvania State University Press, 356 pp. $12.50.

bubbles, effect of

Blanchard, Duncan C., and Lawrence Syzdek 1970.
Mechanism for the water-to-air transfer and concentration of bacteria.
Science 170(3958):626-628.

bubbles, effect of

Carlucci, A.F., and P.M. Williams, 1965.
Concentrations of bacteria from sea water by bubble scavenging.
J. Cons. perm. Int. Explor. Mer. 30(1):28-33.

bubbles, effect of

Davis, C.E., 1955.
Scattering of light by an air bubble in water.
J. Optical Soc., Amer., 45(7):572-581.

bubbles, effect of

Fox, F.E., S.R. Curley and G.S. Larson, 1955.
Phase velocity and adsorption measurements in water containing air bubbles.
J. Acoust. Soc., Amer., 27:534-539.

bubbles, effect of

Gibson, Frederick W., 1970.
Measurement of the effect of air bubbles on the speed of sound in water.
J. acoust. Soc. Am., 48(5-2): 1195-1197

bubbles, effect of

Glotov, V.P., P.A. Kolobaev and G.G. Neuimin, 1961.
[Investigation of the scattering of sound by bubbles generated by an artificial wind in sea water and the statistical distribution of bubble sizes.]
Akusticheskii Zhurnal, 7(4):421-427.

English translation: Soviet Physics, Acoustics, 7(4):341-345.

bubbles, effect of

Glotov, V.P., and Yu. P. Lysanov, 1965.
Influence of a non uniform distribution of air bubbles on the reflection of sound from the surface layer of the ocean. (In Russian).
Akustich. Zhurm., 11(4):492-495.

Translation:
Soviet Phys. Acoust. 11(4):421-423.

bubbles, effect of

Hsieh, D.-Y., and M.S. Plesset, 1961.
Theory of rectified diffusion of mass into gas bubbles.
J. Acoust. Soc., Amer., 33(2):206-215.

bubbles, effect of

Ishida, Masami, and Noritatsu Sano, 1962.
On reflection of ultrasound from discontinuous boundaries in the sea. 1. Approximate treatment of ultrasound from the group of small reflecting bodies in the sea. (In Japanese; English abstract)
Bull. Fac. Fish., Hokkaido Univ., 12(4):279-292.

bubbles, effect of

Klotov, V.P., P.A. Kolobaev, and G.G. Neuimin 1962.
Investigation of the scattering of sound bubbles generated by an artificial wind in sea water and the statistical distribution of bubble size.
Akust. Zhurn., 7(4):
Translation: Soviet Physics-Acoustics, 7(4): 341-345.

bubbles, effect of

Macpherson, J.D., 1957.
The effect of gas bubbles on sound propagation in water. Proc. Phys. Soc., London, B 70(445):85-92

bubbles, effect of

MacIntyre, Ferren 1970.
Geochemical fractionation during mass transfer from sea to air by breaking bubbles.
Tellus 22(4): 451-462.

bubbles, effect of

Welsby V.G., and M.H. Safar, 1969/1970.
Acoustic non-linearity due to micro-bubbles in water.
Acustica 22(3): 177-182

bubbles, effect of

Williams, G.P., 1961
Winter water temperatures and ice prevention by air bubbling.
Dock & Harbour Authority, 42(490):111-115.

bubble pulse

bubble pulse

Blaik, Maurice, and Ermine A. Christian, 1965.
Neaur-surface measurements of deep explosions. 1. Pressure pulses from small charges.
J. Acoust. Soc., Amer., 38(1):50-56.

bubble pulses

Hovem, Jens M. 1970.
Deconvolution for removing the effects of the bubble pulse of explosive charges.
J. acoust. Soc. Am. 47(1-2): 281-284

bubbling, effect of

bubbling

Barber, Richard T., 1966.
Organic aggregates formed from dissolved organic material by bubbling. (Abstract only).
Second Int. Oceanogr. Congr., 30 May-9 June 1966. Abstracts, Moscow:19.

bubbling

*Batoosingh, Edward, Gordon A. Riley and Barbara Keshwar, 1969.
An analysis of experimental methods for producing particulate organic matter in sea water by bubbling. Deep-Sea Res., 16(2): 213-2 .

"buffer domain"

buffer domain

Momoi, Takao, 1965.
A long wave in the vicinity of an estuary.
2. An analysis by the method of the buffer domain.
Bull. Earthquake Res. Inst., 43(3):459-498.

"buffer domain"

Momoi, Takao, 1965.
A long wave in the vicinity of an estuary (1.).
An analysis by the method of buffer domain.
Bull. Earthquake Res. Inst., Tokyo, 43(2):291-316

"buffer domain"

Momoi, T., 1965.
The method of the "buffer domain" in water with a step bottom.
Bull. Earthquake Res. Inst., Tokyo, 43(2):269-289

buoyancy mechanisms

buoyancy

Aleyev, Yu. G. 1966.
Buoyancy and hydrodynamic function of the body of nektonic animals. (In Russian; English abstract).
Zool. Zh. 44(4): 575-584

buoyancy

Baldridge, H. David 1970.
Sinking factors and average densities of Florida sharks as a function of liver buoyancy.
Copeia, 1970(4): 744-754.

buoyancy

Bidder, A.M., 1962.
Use of the tentacles, swimming and buoyancy control in the pearly nautilus.
Nature, 196(4853):451-454.

buoyancy

Clarke, Malcolm R. 1970.
Function of the spermaceti organ of the sperm whale.
Nature, Lond., 228(5274): 873-874.

buoyancy

Corner, E.D.S., E.J. Denton and G.R. Forster, 1969.
On the buoyancy of some deep-sea sharks.
Proc. R. Soc. (B) 171(1025): 415-429.

buoyancy

Denton, E.J., 1963
Buoyancy mechanisms of sea creatures.
Endeavour, Imperial Chemical Industries, London, 22: 3-8.

buoyancy

Denton, E.J., 1962.
Some recently discovered buoyancy mechanisms in marine animals.
Proc. R. Soc., London, (A), 265(1322):366-370.

buoyancy

Denton, E.J., 1960.
The buoyancy of marine animals.
American Scientist, 203(1):118-129.

buoyancy

Denton, E.J. and J.B. Gilpin-Brown, 1971.
Further observations on the buoyancy of Spirula.
J. mar. biol. Ass. U.K. 51(2): 363-373.

buoyancy

Denton, E.J., and J.B. Gilpin-Brown, 1966.
On the buoyancy of the pearly nautilus.
J. mar. biol. Assoc. U.K., 46(3):723-759.

buoyancy

Denton, E.J., and J.B. Gilpin-Brown, 1959.
Buoyancy of cuttlefish.
Nature, 184(4695):1330-1331.

buoyancy

Denton, E.J., J.B. Gilpin-Brown and J.V. Howarth, 1967.
On the buoyancy of Spirula spirula.
J. mar. biol. Ass. U.K., 47(1):181-191.

buoyancy

Denton, E.J., J.B. Gilpin-Brown and T.I. Shaw, 1969.
A buoyancy mechanism found in cranchid squid.
Proc. R. Soc. (B) 174(1036): 271-279.

buoyancy (fish)

Denton, E.J., and N.B. Marshall, 1958.
The buoyancy of bathypelagic fishes without a gass-filled swimbladder.
J. mar. biol. Ass., U.K., 37(3):753-768.

buoyancy

Gross, Seymour, 1966.
A low-cost buoyant element for deep-submergence applications.
Undersea Techn., 7(3):23-27.

buoyancy

Smayda, Theodore J., 1966.
The importance of buoyancy to marine plants and animals.
Maritimes, Univ. R.I., 10(2):1-3, 11.

buoyant jets

Tamai, Nobuyuki 1969.
Surface discharge of horizontal buoyant jets.
Coast. Engng. Japan, 12: 159-177.

buoyant sphere

Larsen, Lawrence H., 1969.
Oscillations of a neutrally buoyant sphere in a stratified fluid. Deep-Sea Res 16(6): 587-603.

buoys, buoy stations, buoy, ice, buoys, drifting, etc.

SEE ALSO: anchor stations; see also under: instrumentation

Buoys

See also: instrumentation, buoys

buoys

Akamatsu, Hideo, 1970.
The ocean data station buoys of the Japan Meteorological Agency. Oceanogrl Mag. 22(2): 75-94.

buoys, (drift)

Alekseev, A.P., and A.J. Kisljakov, 1964.
Buoys drift in the Norwegian and Greenland seas. (In Russian).
Materialiy Ribochoz. Issled. Severn. Basseina, Gosudarst. Kom. Rib. Choz., SNCH, SSSR, Poliarn. Nauchno-Issled. i Proektn. Inst. Morsk. Ribn. Choz. i Okeanogr., N.M. Knipovich (PINRO), 2:103-104.

buoys

Anon., 1969.
Giant and pygmy buoys.
Oceanol. Int., 4(1):31-33. (popular).

buoys

Anon., 1968.
Monster buoys for Britain.
Hydrospace, 1(4):46-47.

buoys

Anon., 1968.
ODESSA: small buoys with big voices.
Ocean Indust. 3(10):55-59.

buoy stations
Aubert, M., 1965.
Projet de station autometrique d'exploration océanographique.
Cahiers, C.E.R.B.O.M., 2:99-101.

"buoy stations"
Baskakov, G.A., and N.F. Kudryavtsev, 1957.
Long-operating automatic station for observations of currents on the high seas, V. SB.
(In Russian).
Problemy Arktiki, (Morsk. Transp.), (2):93-96.
RZhGeofiz 6/58-4000

buoys
Berteaux, H.O., and N.P. Fofonoff, 1967.
Oceanographic buoys gather data from surface to sea floor.
Oceanol. int., 2(5):39-42. (popular).

buoy stations
Boguslavsky, S.G., and Yu. M. Beljakov, 1966.
Peculiarities of the dynamics of the Subantarctic Intermediate Current in the Atlantic Ocean.
(In Russian; English abstract).
Fisika Atmosferi i Okeana, Izv. Akad. Nauk, SSSR, 2(10):1082-1088.

buoys
Borovikov, P.A., 1965.
A system of continuous measurements of physical parameters at sea. (In Russian).
Okeanologiia, Akad. Nauk, SSSR, 5(3):566-568.

buoys
Brekhovskikh, Academician L.M., K.N. Fedorov, L.M. Fomin, M.N. Koshlyakov and A.D. Yampolsky, 1971.
Large-scale multi-buoy experiment in the tropical Atlantic. Deep-Sea Res. 18(12): 1189-1206.

buoys
Buckley, C. Peter, 1967.
The ideal buoy concept.
Trans. 2nd Int. Buoy Techn. Symp. 449-472

buoys
Bullen, L.G., and H. Castelliz, 1969.
Automatically deployed oceanographic buoys.
Oceanol. int., 69, 2: 6 pp., 8 figs.

buoys, drifting
Chaplygin, Ye. I., and Yu. K. Alekseyev, 1957.
Guide to observations on currents. Chap. 6. Study of marine currents and ice drifts with the aid of automatic drifting radio beacons.
(In Russian).
Rudovodstvo po Nablyudeniyam nad Techeniyami, Moscow, 152-176.
OTS, $0.75

buoys
Clark, J., 1965.
Buoy watching in the North Atlantic.
Mar. Obs. (M.O. 764), 35(209):130-132.

buoys
Danilevskaya, M.V., 1966.
The U.S. buoy for ocean studies is anchored. (In Russian)
Okeanologiia, Akad. Nauk, SSSR, 6(3):559.

buoys
Daubin, Scott C., 1962
The General Motors deep-sea oceanographic buoy system.
Proc. 2nd Interindustrial Oceanogr. Symposium Lockheed Aircraft Corp., 19-26.

buoys
Deberaux, R.F., J.D. Isaacs and F.D. Jennings, 1969.
Long distance telemetry of environmental data for the North Pacific buoy study. Oceanol. int. 69, 3: 8 pp., 13 figs.

buoys
Devereux, Robert F., J.W. Petre and Ralph F. Kosic 1968.
Real time oceanography: The monster buoy.
Mar. Sciences Instrument. 4:472-484.

buoy stations
Dubovic, R.A., 1961.
Comparison of observations from ship drifting and anchored with observations at buoy station. (In Russian).
Meteorol. i Gidrol., (10):50-53.

buoys, data from
Evans, Martha W., Richard A. Schwartzlose and John D. Isaacs 1968.
Data from deep moored instrument stations.
Scripps Inst. Oceanogr. Ref. 68-17: 145pp.
(unpublished manuscript)

buoys, ice
Evgenov, N.I., 1935.
[On ice buoys cast in Chukchi Sea and the eastern part of the East-Sibirian Sea from the ice-breaker Krasin in 1934.] Buill. Arktich. Inst., Leningrad, 5(3/4):93.

buoys
Ewing, J.A., 1969.
Some measurements of the directional wave spectrum. J. mar. Res., 27(2): 163-171.

buoys
Fofonoff, N.P. 1967.
Current measurements from moored buoys.
Trans. 2nd int. Buoy Techn Symp. 409-418.

buoys
Fofonoff, N.P., and F. Webster 1971.
Current measurements in the western Atlantic.
Phil. Trans. R. Soc. (A) 270: 423-436

buoys
Frassetto, R., 1968.
Progress on a compact multipurpose and economical buoy system having a low hydrodynamic and aerodynamic drag.
J. Ocean Techn., 2(2):17-21.
Reprinted: Comm. Studio Oceanograf. Limnol., (A), (12)

buoys
Frassetto, R., 1963.
A miniaturized system for long term recording of temperature microstructure from anchored buoys having an accuracy of measurement of the order of 0.01° C. (Summary).
Rapp. Proc. Verb., Réunions, Comm. Int., Expl. Sci., Mer Méditerranée, Monaco, 17(3):889-891.

buoys
Garner, D.M., 1969.
Vertical surface acceleration in a wind-generated sea.
Dt. hydrogr. Z. 22(4):163-168.

buoys
Gaul, R.D. and N.L. Brown 1967
A comparison of wave measurements from a free-floating wave meter and the monster buoy.
Trans. 2nd int. Buoy Symp. Techn 473-494.

buoys
Giraytys, James 1967.
Utilization of buoys from a user's standpoint.
Trans. 2nd int. Buoy Technol. Symp. 531-538.

buoys
Glass, C.J. 1967.
Factors influencing the management of the study of the feasibility of national data buoy systems.
Trans. 2nd Int. Buoy Techn. Symp. 539-555.

buoy stations
Gorodenskii, N.B., N.F. Kudriavtsev and V.G. Labeish, 1961.
Investigations by means of models of the effect of currents and swell at autonomous stations in observing currents. Methods for investigating oscillations of sea level and currents with the help of autonomous stations of long duration. Methods of oceanological investigations, a collection of papers. (In Russian).
Trudy Arktich. i Antarktich. Nauchno-Issled. Inst., 210:13-22.

buoys
*Grant, D.A., 1968.
Current, temperature, and depth data from a moored oceanograpic buoy.
Can. J. earth Sci., 5(5):1261-

buoys
Holmes, John F. 1967.
Tuned spherical buoy.
Trans. 2nd Int. Buoy Techn. Symp. 495-502.

buoys
Inyutkina, A.I., 1964
Electronic buoys for data collection concerned with the Labrador Current. (In Russian
Okeanologiia, Akad. Nauk, SSSR, 4(3):546-

buoys
Isaacs, John D., 1967.
Remarks on some present and future buoy developments.
Trans. 2nd Int. Buoy Techn Symp, 503-524

buoy stations
Karnaushenko, N.N., 1965.
A buoy with neutral buoyancy for measuring currents from large expedition ships. (In Russian).
Trudy Morsk. Gidrofiz. Inst., Akad. Nauk, SSSR, 31:105-111.

buoys
Kudriavtsev, N.F., 1965.
On the experience of employing long-term autonomic stations for the exploration of sea currents. (In Russian).
Okeanologiia, Akad. Nauk, SSSR, 5(3):534-541.

buoy stations
Kudriavtsev, N.F., 1964
On the method for the calculation of dynamic loads conditioned by the effect of currents and waves on the elements of autonomic stations. (In Russian).
Okeanologiia, Akad. Nauk, SSSR, 4(1):142-145.

buoys
Lee, A.J. 1969.
Maff hydrographic buoy study. On the use of moored current meter networks in the shelf seas around Britain.
Oceanol. int. 69 (2): 5pp.

buoys
Metzler, A.R., 1962
An untended digital data acquisition system.
Proc. 2nd Interindustrial Oceanogr. Symposium, Lockheed Aircraft Corp., 3-10.

buoys
Millard, R.C., Jr., 1971.
Wind measurements from buoys: a sampling scheme.
J. geophys. Res. 76(24): 5819-5828.

buoys
MINAS, H.J., A. TRAVERS, M. Travers et S. Maestrini, 1968.
Première utilisation à Villefranche-sur-mer de la Bouée Laboratoire du COMEXO pour l'étude de la distribution du microplancton et de certains facteurs écologiques.
Rec. Trav. Sta. mar. Endoume 44 (60): 13-45.

buoys
*Morrow, Bertan, and Wen F. Chang, 1967.
Determination of the optimum scope of a moored buoy.
J. Ocean Tech., 2(1):37-42.

buoys
Munske, Richard E., Editor, 1964.
Buoys, buoys and more buoys.
Undersea Technology, 5(3):11-14.

buoys
Nan-niti, Tosio, Akimitsu Fujiki and Hideo Akamatsu, 1968.
Micro-meteorological observations over the sea (1).
J. oceanogr. Soc., Japan, 24 (6): 281-294

buoy stations anchor buoy
Nan-niti, Tosio, Hideo Akamatsu and Toshisuke Nakai, 1964.
A further observation of a deep current in the east-north-east sea of Torishima.
Oceanogr. Mag., Tokyo, 16(1/2):11-19.

buoys (navigational)
*O'Connell, J.M., and J.W. Petre, 1969.
Operational experience with the prototype large navigational buoy. Oceanol. int., 69, 2: 6 pp., 4 figs.

buoy stations
Ovchinnikov, I.M., 1961
[Circulation of waters in the northern Indian Ocean during the winter monsoon.] (In Russian; English abstract).
Okeanolog. Issledov., Mezhd. Komitet Proved. Mezhd. Geofiz. Goda, Prezidiume Akad. Nauk, SSSR, (4):18-24.

buoys
Petre, J.W., and R. Devereux 1968.
The monster buoy - a progress report for 1967.
Mar. Sci. Instrument. 4:434-443.

buoys
Picard, Jacques 1965.
French "King-size" buoy - La Bouée Laboratoire.
GeoMar. Techn. 1(9):27-28.

buoys
*Rebaudi, Roberto, S., 1967.
Sistema de boyas para medir corrientes sobre la plataforma submarina.
Bol. Servicio Hidrografia, Naval, Armada Argentina, 4(2):225.

buoys, moored
Richardson, W.S., P.B. Stimson, and C.H. Wilkins 1963.
Current measurements from moored buoys.
Deep-Sea Res., 10(4):369-388.

buoy stations
Romanov, Yu. A., 1961
[Dynamic method as applied to the Equatorial Indian Ocean.] (In Russian; English abstract).
Okeanolog. Issled., Mezhd. Komitet Proved. Mezhd. Geofiz. Goda, Prezidiume, Akad. Nauk, SSSR (4):25-30.

buoys
Romanovsky, V., 1963.
Le courantographe american Richardson. (Resume).
Rapp. Proc. Verb., Reunions, Comm. Int., Expl. Sci., Mer Mediterranee, Monaco, 17(3):887.

buoys
*Rudnick, Philip, 1969.
Wave directions from a large spar buoy.
J.mar.Res., 27(1):7-23.

buoys
Ruff, Ronald E. 1968.
Self-contained oceanographic data acquisition buoy.
Mar. Sci. Instrument., 4: 485-489.

buoy stations
Shirey, V.A., 1961.
Methods of observations on currents by buoy.
Meteorol i Gidrol., (9):48-50.
(In Russian).

buoy stations
Siedler, G., and G. Krause, 1964
An anchored vertically moving instrument and its application as parameter follower.
Trans. 1964 Buoy Techn. Symposium, Mar. Techn. Symposium, 24-25 Mar. 1964, Washington, 483-488.

Buoys
Skornjakov, V. M., 1961.
(Lightened construction of tied buoys.)
Meteorol. i Gidrol., (7):37.

buoy stations
Sobchenko, E.A. 1964.
Experience using the stations on buoys at small depths with the changing of sea level. (In Russian).
Materiali, Ribokhoz. Issled. Severn. Basseina Poliarn. Nauchno-Issled. Proektn. Inst. Morsk. Rib. Khoz. Okeanogr. N.M. Knipovich PINRO Gosud. Proizvodst. Komm. Ribn. Khoz. SSSR 4: 112-114.

buoys
Stas, I.I., 1962.
Performance of hydrographic buoy stations. (In Russian).
Trudy Morsk. Gidrofiz. Inst., 26:65-69.

Translations:
Scripta Tecnica, Inc., for Amer. Geophys. Union, 26:49-52. (1964).

buoys
Svendsen, Harald 1971.
Investigation of the Norwegian coast current off Egersund September 1968.
Rept. Geophys. Inst. (A) Phys. Oceanogr. Bergen 28: 15 pp., 13 figs. (multilithed).

moored buoys
Vershinsky, N.V., and P.A. Borovikov, 1965.
On the calculation of stations with automatically variable depth. Electronic instruments for oceanographic investigations. (In Russian; English abstract).
Trudy Inst. Okeanol., Akad. Nauk, SSSR, 74:85-89.

buoy stations
Vinogradov, V.V., 1962
Some experience involved with the establishment of buoy stations in the ocean.
Okeanologiia, Akad. Nauk, SSSR, 2(2):346-352.

buoys
Webster, Ferris 1967.
A scheme for sampling deep-sea currents from moored buoys.
Trans. 2nd int. Buoy Techn. Symp. 419-431.

anchored buoy stations
Zikov, I.D., 1963.
Radiotransmitter set for the autonomous buoy stations. (In Russian).
Meteorol. i Gidrol., (4):47-48.

buoy, stations
Istoshin, Yu.V. and E.M. Sauskan, 1968.
On the counter currents of the Kuroshio. (In Russian; English abstract).
Akad. Nauk, SSSR, 8(6):949-959.

buoys, cost effectiveness
Marechal, N., Edward U. Graham, 1969.
Cost effectiveness study of international copperative logistics support for buoy data collection arrays. Oceanol. int., 69, 2: 7 pp.

buoy data

data from buoys

Evans, M.W., A.L. Moore, R.A. Schwartzlose and A.M. Tubbs, 1971
Data from deep-moored instrument stations: volume 3.
SIO Ref. 71-2: 85 pp. (multilithed)

buoy data

Grant, D.A., 1969.
Current, temperature, and depth data from a moored oceanographic buoy: reply to G.T. Needler and R.W. Stewart.
Can. J. Earth Sci. 6(3): 523

buoy data

Needler, G.T., 1969.
Current, temperature and depth data from a moored oceanographic buoy: discussion
Can. J. Earth Sci. 6(3): 521-522.

buoy data

Stewart R.W., 1969
Current, temperature and depth data from a moored oceanographic buoy.
Can. J. Earth Sci., 6(3): 522.

buoy lines

Nath, John H., 1971.
Dynamic response of taut lines for buoys.
J. mar. techn. Soc. 5(6): 44-46

buoys, telemetering

Kebe, Hans-Werner 1971.
Eine ferngesteuerte Messboje mit Datenspeicherung für refraktionsseismische Untersuchungen auf See.
Meteor-Forsch.-Ergebnisse (C) 6: 14-20.

burrowing

burrows

Bajard, Jacques 1966
Figures et structures sédimentaires dans la zone intertidale de la partie orientale de la Baie du Mont-Saint-Michel.
Rev. Géogr. Phys. Géol. Dyn. (2) 8(1): 39-111.

burrows

Clarke, R.H., 1968.
Burrow frequency in abyssal sediments.
Deep-Sea Res., 15(3):397-400.

burrows
bottom sediments, abyssal

burrows

Donahue, Jack 1971.
Burrow morphologies in north-central Pacific sediments.
Mar. Geol 11 (1): M1-M7.

burrows

Frey, Robert W. and Taylor V. Mayou, 1971.
Decapod burrows in Holocene barrier island beaches and washover fans, Georgia. Senckenbergiana maritima 3(1): 53-77.

burrows

Goreau, T. F. Goreau, N.I. Goreau and C.M. Yonge 1972
On the mode of boring in Fungiacava eilatensis (Bivalvia: Mytilidae).
J. Zool. Lond 166(1): 55-60

burrows

Rice, A.L. and C.J. Chapman, 1971.
Observations on the burrows and burrowing behaviour of two mud-dwelling decapod crustaceans, Nephrops norvegicus and Goneplax rhomboides. Mar. Biol. 10(4): 330-342.

burrowing (molluscs)

Breum, Ole 1970
Stimulation of burrowing activity by wave action in some marine bivalves.
Ophelia 8: 197-207.

burrowing, effect of

Howard, James D., 1969.
X-ray radiography for examination of burrowing in sediments by marine invertebrate organisms.
Sedimentology, 11 (3/4): 249-258

burrowing of organisms

Kuenen, Ph. H., 1961
Some arched and spiral structures in sediments.
Geologie en Mijnbouw, 40:71-74.

Also:
Publikatie, Geol. Inst., Groningen, No. 140

burrowing

Payne, Robert R., John R. Conolly and William H. Abbott 1971.
Turbidite muds within diatom ooze off Antarctica: Pleistocene sediment variation defined by closely spaced piston cores.
Bull. geol. Soc. Am. 83 (2): 481-486.

burrowing

Phillips, Philip J. 1971.
Observations on the biology of mudshrimps of the genus Callianassa (Anomura: Thalassinidea) in Mississippi Sound.
Gulf Res. Repts 3(2): 165-196.

burrowing

Röder, Heinrich, 1971.
Gangsysteme von Paraonis fulgens Levinsen 1883 (Polychaeta) in ökologischer, ethologischer und aktuopaläontologischer Sicht. Senckenbergiana maritima 3(1): 3-51.

burrowing

Shinn, Eugene A. 1968.
Burrowing in Recent lime sediments of Florida and the Bahamas.
J. Paleontol. 42(4): 879-894

burrowing

Soule, John D. and Dorothy F. Soule 1969
Three new species of burrowing bryozoans (Ectoprocta) from the Hawaiian islands.
Occ. Pap. Calif. Acad. Sci. 78: 9 pp.

burrowing

Stanley, Steven M. 1969.
Bivalve mollusk burrowing aided by discordant shell ornamentation.
Science 166 (3905): 634-635.

burrowing

Trueman, E.R. and A.D. Ansell, 1969.
The mechanisms of burrowing into soft substrata by marine animals. Oceanogr. Mar. Biol. Ann. Rev., H. Barnes, editor, George Allen and Unwin, Ltd., 7: 315-366.

burrowing

*Werner, Friedrich, 1968.
Gefügeanalyse feingeschichteter Schlicksedimente der Eckernförder Bucht (westliche Ostsee).
Meyniana, 18:79-105.

burrowers, effect of

Warme, J.E., T.B. Scanland, and N.F. Marshall, 1971.
Submarine canyon erosion: contribution of marine rock burrowers. Science 173(4002): 1127-1129.

business

business

Ridler, K.E.W., 1968.
Exploration is only valuable if it leads either to profitable business or to the meeting of social needs.
Hydrospace, 1(4):16-18.

by-passing

by-passing

Bruun, Per M. 1967
By-passing and back-passing off Florida.
J. WatWays Harb. Div. Am. Soc. civ. Engrs 93 (WW2): 101-128.

byssal threads

*Van Winkle, W., Jr., 1970.
Effect of environmental factors on byssal thread formation.
Marine Biol., 7(2):143-148.

Cables

cables

*Bengelsdorff, Edgar, 1967.
Ein neues hochfestes Seilkabel für ozeanographische "in-situ - Geräte"
Beitr. Meereskunde, 21:72-75.

cables

Buchanan, George R., 1971
Three dimensional analysis of a partially buried cable.
Ocean Engng 2(2): 83-90

cables

Buchanan, George R., and Robert L. Ho 1970.
Analysis of an underwater cable.
Ocean Engng 1(6): 617-630.

cable angle
FRANCE, Service Hydrographique de la Marine, 1964
Résultats d'observations. Campagne internationale d'observations dans le détroit de Gibraltar (15 mai-15 juin 1961). Mesures de courant d'hydrologie et de météorologie effectuées à bord de la "Calypso".
Cahiers Océanogr., C.C.O.E.C., 16(1):23-94.

cables
Gay, S.M., 1968.
Computer analysis and design of undersea cable systems.
UnderSea Techn., 9(10):43, 48-49.

cables
Haigh, K.R., 1968.
Cable ships and submarine cables.
The Trinity Press, Worcester and London 416pp.

cables
Jack, J.S., W.H. Leech and H.A. Lewis, 1957.
Route selection and cable laying for the Transatlantic cable system.
Bell. System Tech. J., 36(1):293-326.

cables
Johnson, G. Leonard, and Bruce C. Heezen 1969.
Natural hazards to submarine cables.
Ocean Engng 1(5): 535-553.

cables
Marskuev, W.I., 1965.
Cables for deep water television units and the peculiarities of their use. Electronic instruments for oceanographic investigations. (In Russian; English abstract).
Trudy Inst. Okeanol., Akad. Nauk, SSSR, 74-40-46.

cables
Munitz, Allan J., 1966.
Analyzing failures of ocean communications cable.
Undersea Techn., 7(5):45, 47-49.

cables
Peffenberger, J.C., E.A. Capadona and R.B. Siter, 1966.
Dynamic testing of cables.
Exploiting the Ocean, Trans. 2nd Mar. Techn. Soc. Conf., June 27-29, 1966, 485-523.

cables
Richardson, William S. 1967.
Buoy mooring cables, past, present and future.
Trans. 2nd int. Buoy Technol. Symp. 15-18.

cables
Stimson, Paul B., 1967.
Deep-Sea mooring cables.
Trans 2nd Int. Buoy Techn. Symp., 115-118

cables
Walsh, Don. K., 1966
Underwater electrical cables & connectors engineered as a single requirement.
Exploiting the Ocean, Trans. 2nd Mar. Techn. Soc. Conf., June 27-29, 1966, 469-484.

Cable breaks

cables, breaking of
Bogdanov, K.T., and Yu. A. Ivanov, 1958.
On causes of cable break during taking of deep water series of bathymeters. (In Russian).
Meteorol. i Gidrol., (8):49-50.

Cables (damage)
Borgorodsky, M. M., 1961.
[Method of discovery of inner damage in the cable of distant installations in sea conditions.]
Meteorol. i Gidrol., (7):38-39.

cable breaks
Burke, Kevin 1967.
The Yallahs Basin: a sedimentary basin southeast of Kingston, Jamaica.
Marine Geol. 5(1):45-60.

cable length
Darling, Robert C. 1966.
Cable length determinations for deep-sea oceanographic operations.
Techn. Bull. ESSA, C.G.S, No 30: 6pp.

Cable breaks
Krause, Dale C., William C. White, David J. W. Piper, and Bruce C. Heezen, 1970.
Turbidity currents and cable breaks in the western New Britain Trench.
Bull. geol. Soc. Am., 2153-2160.

cables, shape (catenation, breaking, etc)

cable trajectory
Carruthers, J.N., 1960
On determining the shape of trawls and towing cables.
Fishing News, (2468): 6.
In NIO Collected Reprints, 1960

cable shape
Carruthers, J.N., 1954.
On the instrumental measurement of line shape under water. Deutsche Hydrogr. Zeits. 7(1/2): 22-35.

cable catenary
Kumasawa, N., 1963.
Theoretical study on the motion of long-line gear in water. 1. Shape of the main line under constant steady current.
J. Tokyo Univ. Fish., 49(1):1-24.

cables concatenation of
Miyazaki, Yoshio, 1970.
Studies on approximate formulas for tension and configuration of the towing rope. II. Method of calculation. (In Japanese; English abstract).
Bull. Jap. Soc. scient. Fish. 36(1): 58-67

cables concatenation of
Miyazaki, Yoshio 1970.
Studies on approximate formulas for tension and configuration of the towing rope. I. Method of approximations. (In Japanese; English abstract)
Bull. Jap. Soc. scient. Fish. 36(1): 48-57.

cable concatenation
Miyazaki, Yoshio 1970.
The configuration and tension of a rope and a plane net set in a uniform stream.
J. Tokyo Univ. Fish. 56(1/2): 49-117.

cable shape
Paka, V.T., V.P. Makushkin, M.F. Naumenko and K.I. Chigrakov, 1964
Deep sinking of counters in the course of the movement of the ship. (In Russian).
Okeanologiia, Akad. Nauk, SSSR, 4(1):128-131.

Cable concatenation
Tapley, I., and H. Costello 1971.
The configuration of a mooring cable during deployment of an automatic mooring system: a numerical analysis.
J. Can. Aeronautics Space 17: 221-228.

cables, catenation of
Walton, T.S., and Polachek, 1960.
Calculation of transient motion of submerged cables.
Math. of Computation, 14(69):27-46.

cable, shape of
Wood, W.W., 1962.
The fall of a towed cable.
Proc. R. Soc., London, A, 269(1337):205-218.

cables, effect of
Donovan, John E., 1970.
Triboelectric noise generation in some cables commonly used with underwater electroacoustic transducers.
J. acoust. Soc. Am. 48(3-2): 714-724.

Calcification

calcification
*Angell, Robert W., 1967.
Test recalcification in Rosalina floridana (Cushman).
Contrib. Cushman Fdn. foramin. Res., 18(4):176-177.

calcification
Chockalingam, S., 1971.
Studies on enzymes associated with calcification of the cuticle of the hermit crab Clibanarius olivaceous. Mar. Biol. 10(2): 169-182.

calcification
*Malone, Philip G., and J. Robert Dodd, 1967.
Temperature and salinity effects on calcification rate in Mytilus edulis and its paleoecological implications.
Limnol. Oceanogr., 12(3):432-436.

calibration

calibration, international
Bogoyavlensky, A.N., 1965.
Reproducibility and errors in results of hydrochemical determinations in the oceans. 1. Results of the second series of international intercalibration of methods for chemical determinations at sea. Methods of marine hydrochemical investigations at sea. (In Russian).
Trudy Inst. Okeanol., Akad. Nauk, SSSR, 79:34-48.

calibration
Doty, M.S., 1962
Analysis of the productivity data from the September Honolulu Intercalibration Trials. The Xth Pacific Science Congress.
Okeanologiia, Akad. Nauk, SSSR, 2(3):543-553.

calibration, sound

*Trott, W.James, 1968.
International round robin calibration in underwater sound.
J. acoust. Soc. Am., 44(4):1158-1159.

Calming the sea

calming the sea

Ashton, E.W.S., and J.K. O'Sullivan, 1949.
Effect of rain in calming the sea. Nature, 164 (4164):320-321.

calming the sea

Barnaby, C. F., 1949.
Effect of rain in calming the sea. Nature 164 (4179):968.

calming the sea

Sainsbury, G. L., and I. C. Cheeseman, 1950
Effect of rain in calming the sea. Nature 166 (4210): 79

"caloricity"

Shushkina, E.A., V.I. Kuz'micheva and L.A. Ostapenko, 1971.
Energetic equivalents of body mass, respiration and caloricity of the Japan Sea mysids. (In Russian; English abstract). Okeanologiia 11(6): 1065-1074.

Calorific value

calorific value

Comita, G.W., S.M. Marshall and the late A.P. Orr, 1966.
On the biology of Calanus finmarchicus, XIII. Seasonal change in weight, calorific value and organic matter.
J. mar. biol. Ass., U.K., 46(1):1-17.

calorific values

Comita, G.W., and D.W. Schindler, 1963
Calorific values of Microcrustacea.
Science, 140(3574):1394-1395.

calorific value

Copenhagen, W. J. and L. D., 1949
Variation in the phytoplankton of Table Bay, October 1934 to October 1935. With a note on the calorific value of Chaetoceros spp. Trans. Roy. Soc. S. Africa, 32(2):113-123, 2 text figs.

calorimetry

Ostapenya, A.P., L.M. Sushchenya, N.N. Khmeleva, 1967.
Calorimetry of plankton from the tropical zones of the ocean. (In Russian; English abstract).
Okeanologiia, Akad. Nauk, SSSR, 7(6):1100-1107.

caloric content

*Pandian, Thavamani J., 1967.
Changes in chemical composition and caloric content of developing eggs of the shrimp Crangon crangon.
Helgoländer wiss. Meeresunters., 16(3):216-224.

caloric content

*Pandian, Thavamani J., and Karl-Heinz Schumann, 1967.
Chemical composition and caloric content of egg and zoea of the hermit crab Eupagurus bernhardus
Helgoländer wiss Meeresunters. 16(3):225-230.

camouflage

Denton, Eric 1971.
Reflectors in fishes.
Scient. Am. 224 (1): 65-72.

camouflage

Hastings, J.W., 1971.
Light to hide by: ventral luminescence to camouflage the silhouette. Science, 173(4001): 1016-1017.

Canals

canals

Aron, William I., and Stanford H. Smith 1971.
Ship canals and aquatic ecosystems.
Science 174 (4004): 13-30.

canals

Lamoen, J., 1949.
Tides and current velocities in a sea level canal. Engineering 168(4357):97-99, 3 figs.

canals

Sekerzh-Zenkovich, T. Ia., 1959.
[Peculiar determination problems on the expansion of free tidal waves in a canal of variable depth.]
Izv. Akad. Nauk. SSSR, Ser Geofiz. (10):1460-1467.

canals, effect of

Topp, Robert W., 1969.
Interoceanic sea-level canals; effects on the fish faunas.
Science 165 (3900): 1324-1327.

Capes

capes

*Dolan, Robert, and John C. Ferm, 1968.
Crescentic land forms along the Atlantic coast of the United States.
Science, 159(3815):627-629.

capes

Hoyt, John H., and Vernon J. Henry, Jr. 1971
Origin of capes and shoals along the southeastern coast of the United States.
Bull. geol. Soc. Am., 82(6): 59-66.

capes

Dolan, Robert, 1971.
Coastal landforms: crescentic and rhythmic.
Bull. geol. Soc. Am. 82(1): 177-180

Carapace

carapace

Digby, Peter S.B. 1968.
Mobility and crystalline form of the lime in the cuticle of the shore crab, Carcinus maenas.
J. Zool. Lond. 154 (3): 273-286.

Carcenogens,

See also: under chemistry

Carcenogens

See also: Chemistry, benzo-3,4 pyrene
chemistry, carcenogens

carcenogens

de Lima-Zanghi, Carmen, 1968.
Bilan des acides gras du plancton marin et pollution par le benzo-3.4 pyrène.
Cah. océanogr., 20(3):203-216.

careers

Fanning, Odum, 1969.
Opportunities in oceanographic careers.
Universal Publ. Distribut. Co. $1.95.
(not seen)

Cartography

cartography

Beaton, Robert J., 1967.
Impact of cartography on international relations.
Int. hydrogr. Rev., 44(1):167-176.

cartography

*Beigbeder, Yvonne (posthumous) et Fernand Verger, 1967.
Essai de cartographie de la géomorphologie dynamique de la Pointe d'Arçay, en Vendée.
Revue Géogr. phys. Géol. dyn. (2)9(5):409-414.

cartography

Brudhikrai, L.J., 1947
Development of hydrographic work in Siam from the beginning up to the present. Int. Hydr. Rev. 24:48-53.

cartography

Chinese delegation, 1947
Hydrographic work of China, between the years 1937 and 1947. Int. Hydr. Rev. 24:196.

cartography

Dahlgren, P., H. Richter, 1947
Swedish Hydrography (1644-1944). Summary of Sveriges Sjökarta. The nautical chart of Sweden. Contributions to the history of Swedish Hydrography. Int. Hydr. Rev. 24:186-195.

cartography

Day, A., 1953.
Navigation and hydrography. J. Inst. Navigation 6(1):1-14.

cartography

Moore, J.G., 1947
The determination of the depths and extinction coefficients of shallow water by air photography using colour filters. Philos. Trans. R. Soc., London, ser.A Math. Phys. Sci. No.816, V. 240:163-217.

cartography

*Oren, Oton Haim, 1968.
Jews in cartography and navigation (from the XIth to the beginning of the XVth century).
Bull. Inst. océanogr., Monaco, No. spécial 2: 189-197.

cartography

Robinson, A.H.W., 1962.
Marine cartography in Britain.
Leicester Univ. Press, 222 pp., 42 pls., 30 figs.

Reviewed by:
G.S. Ritchie, 1963. J. Inst. Navigation, 16(1): 140-141.

cartography

Stassinopoulos, A.C., 1947
Report on the first-order triangulation of Greece. Int. Hydr. Rev. 24:208-209, 3 tables.

cascading

cascading
Boden, B.P., and E.M. Kampa, 1953.
Winter cascading from an oceanic island and its biological implication. Nature 171:426-427.

cascading
Bougis, P., and M. Ruivo, 1954.
Sur un descente d'eaux superficielles en profondeur (cascading) dans le sud du Golfe du Lion. Bull. d'Info., C.C.O.E.C. 6(8):147-154, 2 pls.

Cascading
Cooper, L. H. N., and D. Vaux, 1949.
Cascading over the continental slope of water from the Celtic Sea. J.M.B.A. 28(3):719-750, 14 textfigs.

cascading
Stefánsson, Unnsteinn, Larry P. Atkinson and Dean F. Bumpus, 1971.
Hydrographic properties and circulation of the North Carolina Shelf and slope waters. Deep-Sea Res., 18(4): 383-420.

catabolism
Budd, J.A., 1969.
Catabolism of trimethylamine by a marine bacterium, Pseudomonas NCMB 1154. Marine Biol., 4(3): 257-266.

catalogues

catalogues
Rossiter, J.R., 1961
Catalogue of published mean sea level data (1807-1958).
Union Géodes. et Géophys. Int. Assoc. Océanogr Phys., Publ. Sci., No. 23: 64 pp.
Monographie No. 13.

catalogues
Snyder, H. George, and A. Fleminger, 1965.
A catalogue of zooplankton samples in the marine invertebrate collections of Scripps Institution of Oceanography.
SIO Ref., No. 65-14:140 pp., 48 charts. (Unpublished manuscript).

catalogues
United States, National Academy of Sciences 1965
Catalogue of data in World Center A-Oceanography. Supplement No.3, data received during the period 30 June 1965: 99 pp.

Catalogues
United States, National Academy of Sciences 1965
Catalogue of data in World Center A. oceanography. Supplement No. 2, data received during the period 1 July - 31 December 1964: 77 pp.

Catalogues
United States, National Academy of Sciences 1965
Catalogue of publications in World Data Center A - oceanography - received 1 July 1957 - 31 December 1964.
World Data Center A, Washington, D.C.: 135 pp.

catalogs
United States, National Oceanographic Data Center, 1965.
Reference sources for oceanographic station data Catalog Ser., Publ. C-1 (2 vols): unnumbered pp. $4.00

Catamaran

catamarans
Anon., 1969.
Duplus - a catamaran for severe seas.
Ocean Industry, 4(3):28-29.

catamarans
Anon. 1968.
Catamarans for research.
Hydrospace 1(3): 16-18.

catamaran
Hamlin, Cyrus, 1965.
The catamaran as a seagoing work platform. (abstract).
Ocean Sci. and Ocean Eng., Mar. Techn. Soc.,-Amer. Soc. Limnol. Oceanogr., 2:1127-1144.

catamaran
Mavor, James W., Jr., 1966.
Ten months with ALVIN, a rundown of lesson, limitations, capabilities.
Geo-Mar. Techn., 2(2):8-18.

catamaran
Vershinsky, N.V., and V.P. Nikolaev, 1967.
Floating laboratory on a pontoon-catamaran. (In Russian).
Okeanologiia, Akad. Nauk, SSSR, 7(1):187-188.

catastrophic burial
McKnight, D.G., 1969.
A recent, possibly catastrophic burial in a marine molluscan community. N.Z. Jl mar. Freshwat. Res., 3(1): 177-179. Also in: Coll. Repr. N.Z. Oceanogr. Inst.

catch variability

catch variability
Bridger, J.P., 1956.
On day and night variation in catches of fish larvae. J. du Cons. 22(4):42-57.

catches, day vs. night
Isaacs, John D., 1964
Night-caught and day-caught larvae of the California sardine.
Science, 144(3622):1132-1133.

Cathodic protection

Cathodic protection
Cornet, I, T.W. Pross, Jr., and B.C. Bloom, 196[illegible]
Current requirements for cathodic protection of disks rotating in salt water - a mass transfer analysis.
Trav. Centre de Recherches et d'Etudes Oceanogr. n.s., 6(1/4):175-181.

Cathodic protection
Cotton, J.B., and D.C. Moore, 1965.
Use of platinised titanium anodes for cathodic protection.
Trav. Centre de Recherches et d'Etudes Oceanogr. n.s., 6(1/4):183-189.

Cathodic protection
Frinken, H., and J. Kubisch, 1965.
Uber die Beinflussung von Edalstahlen durch verschiedene Methoden des kathodischen Schutzes.
Trav. Centre de Recherches et d'Etudes Oceanogr. n.s., 6(1/4):191-196.

Cathodic protection
Guillen Rodrigo, Miguel A., 1965.
Influence de la temperature et de l'agitation sur la formation du depot magnesium-calcaire d ans l'acier protege cathodiquement dans l'eau de mer.
Trav. Centre de Recherches et d'Etudes Oceangr. n.s., 6(1/4):197-203.

Cathodic protection
Heuze, Bernard, 1965.
La relativite du potentiel dans les procedes speciaux d'equipement et de surveillance pour la protection cathodique des ouv rages a la mer.
Trav. Centre de Recherches et d'Etudes Oceanogr. n.s., 6(1/4):255-265.

Cathodic protection
Lowe, R.A., and C. Richardson, 1965.
Some developments in marine cathodic protection with particular reference to ships.
Trav. Centre de Recherches et d'Etudes Oceanogr. n.s., 6(1/4):267-270.

Cathodic protection
Morgan, J.H., 1965.
Instrumentation and automation in marine cathodic protection.
Trav. Centre de Recherches et d'Etudes Oceanogr. n.s., 6(1/4):251-253.

Cathodic protection
Pourbaix, M., 1965.
Protection cathodique en presence d'eau de mer.
Trav. Centre de Recherches et d'Etudes Oceanogr. n.s., 6(1/4):271-273.

Cathodic protection
Schnock, A., 1965.
La protection cathodique des coques de navires par "trailing anode".
Trav. Centre de Recherches et d'Etudes Oceanogr. N.s., 6(1/4):205-208.

Cathodic protection
Souske, R., 1965.
Etat actuel de la protection cathodique par le zinc en France dans les ouvrages maritimes.
Trav. Centre de Recherches et d'Etudes Oceanogr. n.s., 6(1/4):285-288.

cathodic protection
Wood, Herbert T., 1971.
Cathodic protection of wire rope.
J. mar. techn. Soc. 5(3): 31-34.

cavitation

cavitation
Barnes, H.L., 1956.
Cavitation as a geological agent. Am. J. Sci. 254(8):493-505.

cavitation
Cary, Boyd B., 1967.
Nonlinear losses induced in spherical waves.
J. acoust. Soc. Am. 42(1): 88-92

cavitation
*Dubosset, M., et M. LaVergne, 1968.
Calcul de la cavitation due aux explosions sous-marines à faible profondeur.
Acustica, 2 0(5):289-298.

cavitation
Eller, Anthony, and H.G. Flynn, 1965.
Rectified diffusion during nonlinear pulsations of cavitation bubbles.
J. Acoust. Soc., Amer., 37(3):493-503.

cavitation, effect of
Furduev, A.V., 1966.
Undersurface cavitation as a source of noise in the ocean. (In Russian; English abstract).
Fisika Atmosferi i Okeana, 2(5):523-533.

cavitation
Gilbarg, D., 1957.
Free-streamline theory and steady-state cavitation.
In: Symposium, Naval Hydrodynamics, F.S. Sherman, Ed., Nat. Res. Counc. Nat. Acad. Sci., Publ., 281-291.

cavitation
Gilbarg, D., and J. Serrin, 1950.
Free boundaries and jets in the theory of cavitation. J. Math. and Phys. 29(1):5 textfigs.

cavitation
Hansen, Paule G., and Eric G. Barham, 1962.
Resonant cavity measurements of the effects of "red water" plankton on the attenuation of underwater sound.
Limnology & Oceanography, 7(1):8-13.

cavitation
Jorgensen, D.W., 1961.
Noise from cavitating submerged water jets.
J. Acoust. Soc., Amer., 33(10):1334-1338.

cavitation
Kogarko, B.S., 1964.
Unsteady one-dimensional motion of a liquid with a cavitation appearing and developing. (In Russian).
Doklady, Akad. Nauk, SSSR, 155(4):779-782.

cavitation
Konstantinov, V.A., 1946.
Vliianie Chisla Reinoldsa na Otryvnoe Obtekanie. [Influence of Reynolds number on the separation (cavitation) flow.] Izvest. Akad. Nauk. SSSR, Otdel Tech. Nauk, No. 10:1355-1373.

Translations DTMB No. 233 by G. Weinblum.

cavitation
Kornfeld and Suvorav, 1944.
[On the destructive action of cavitation.]
[J. Physics, USSR.] 7(3):171-181.

cavitation
Numachi, F., 1951.
Summary report on the research of cavitational phenomena. Abstract Notes and Data concerning Subjects at 6th Int. Conf. Ship Tank Superintendents, Trans. Tech. Res. Inst.:32-38.

cavitation
Plesset, M.S., 1949.
The dynamics of cavitation bubbles. J. App. Mech Sept. 1949:277-282, 8 textfigs.

cavitation
Rasmussen, R.E.H., 1949.
Experiments on flow cavitation in water mixed with air. Trans. Danish Acad. Tech. Sci., A.T.S., 1949 (1):60 pp.

cavitation, effect of
Richardson, E.G., 1960.
The transmission of sound through a wake.
Proc. Physics Soc., London, 76(1):25-32.

Cavitation
Richardson, E. G., 1950.
Cavitation in liquids. Endeavour 9(35):149-153, 6 textfigs.

cavitation
Rusby, J.S.M., 1970.
The onset of sound wave distortion and cavitation in water and sea water. J. Sound Vibration 13(3): 257-267.

cavitation
Sette, D., 1967.
Research on cavitation nuclei.
In: Underwater acoustics, V.M. Albers, editor, Plenum Press, 2: 139-160.

cavitation
Shal'nev, K.K., 1950.
[The cavitation property of hydrodynamical circuits.] Doklady Akad. Nauk, SSSR, 72(4):645-

cavitation
*Tuthill, A.H., and C.M. Schillmoller, 1967.
Guidelines for selection of marine materials.
J. Ocean Tech., 2(1):6-36.

cays
Stoddart, D.R. 1971
Sea-level change and the origin of sand cays: radiometric evidence.
J. mar. biol. Ass. India 11 (1/2): 44-58

cell division

cell division
Thomas, William H. 1966.
Effects of temperature and illuminance on cell division rates of three species of tropical oceanic phytoplankton.
J. Phycol. 2: 17-22.

cell size

cell size
Migita, Seiji, 1969.
Seasonal variation of cell size in Skeletonema costatum and Melosira moniliformis. (In Japanese; English abstract)
Bull. Fac. Fish. Nagasaki Univ. 27: 9-17.

cell size
Semina, H.J., 1971.
Oceanic conditions affecting the cell size of phytoplankton. In: Micropalaeontology of oceans, B.M. Funnell and W.R. Riedel, editors, Cambridge Univ. Press, 89

cell size
Semina, H.J., 1969.
The size of phytoplankton cells along 174°W in the Pacific Ocean. Okeanologiia 9(3): 479-487.
(In Russian; English abstract)

cells size
*Semina, H.J., 1968.
Water movement and the size of phytoplankton cells.
Sarsia, 34:267-272.

cell size
Semina, H.I. and V.V. Aratskaia 1970.
Main pycnocline, cell size, and the distribution patterns of phytoplankton species (In Russian).
Dokl. Akad. Nauk SSSR. 191(2): 449-452.

cellulose, decomposition of
Maciejowska, Modesta, 1969.
The effect of physical and chemical factors of the sea environment upon decomposition of cellulose in brackish waters. (In Polish; English abstract).
Prace morsk. Inst. Ryback (A) 15: 33-54

cell wall
Swift, Elijah, and Charles C. Remsen, 1970.
The cell wall of Pyrocystis spp. (Dinococcales)
J. Phycol. 6(1): 79-86.

cement

cement
Mero, John L., 1965.
The mineral resources of the sea.
Elsevier Oceanogr. Ser., 312 pp.

cementation

cementation
De Groot, K, 1969.
The chemistry of submarine cement formation at Dohat Hussain in the Persian Gulf.
Sedimentology, 12(1/2): 63-68

cementation
Garrison, Robert E., John L. Luternauer, Edwin V. Grill, Robert D. MacDonald and James W. Murray, 1969.
Early diagenetic cementation of Recent sands, Fraser River delta, British Columbia.
Sedimentology, 12(1/2): 27-46.

cementation
Land, Lynton S. and Thomas F. Goreau 1970
Submarine lithification of Jamaican reefs.
J. sedim. Petrol. 40(1): 457-462.

cementation, authigenic
*Morgenstein, Maury, 1967.
Authigenic cementation of scoriaceous deep-sea sediments west of the Society Ridge, South Pacific.
Sedimentology 9(2):105-118.

cementation
Taylor, J. C.M., and L.V. Illing, 1969.
Holocene intertidal calcium carbonate cementation, Qatar, Persian Gulf.
Sedimentology, 12(1/2): 69-107.

charts
Robinson, A.H.W., 1965.
The mapping of hydrographic data in atlases: a plea for morphological maps.
Int. Hydrogr. Rev., 42(1): 63071.

charts
Robinson, A.H.W., 1962.
Marine cartography in Britain. A history of the sea chart to 1855.
Oxford (Leicester) Univ. Press, 222 pp.

Reviewed by Th. Stocks, 1963, Deutsche Hydrogr. Zeits., 16(2):93-94.

charts
Ruddock, A., 1961.
The earliest original English seaman's rutter and pilot's charts.
J. Inst. Navigation, 14(4):409-431.

chart datum level
Sager, G., 1960.
Das Seekartennull der europäischen Küsten.
Vermessungstechnik, 8(7):199-204.

charts
Schott, G., 1942.
Die Grundlagen einer Weltkarte der Meeresstromungen. Ann. Hydr. 70:329-340.

charts
*Viglieri, Alfredo, 1968.
La carte générale bathymétrique des océans établie par S.A.S. le Prince Albert 1er.
Bull. Inst. océanogr., Monaco, No. spécial 2: 243-253.

charts
Weyl, R., 1949.
Eine neue Teifenkarte der Caribischen See und ihre tektonische Ausdeutung. Petermanns Geogr. Mitt., 93 Jahrgang, 173-174, 1 textfig.

chart
Zetler, B.D., 1948
A seismic sea wave travel time chart.
J. Coast and Geodetic Survey 1:56-58, 1 fig.

Charting

Charting
Friedman, Bob 1967.
Fish charts ocean floor.
Sea Frontiers 13(2): 98-101.

Check lists

check list
Dawson, E.Y., 1947
A guide to the Literature and Distributions of the Marine Algae of the Pacific Coast of North America. Mem. S. Calif. Acad. Sci. 3(1):1-134.

Chemical stimulation
McLeese, D.W., 1970.
Detection of dissolved substances by the American lobster (Homarus americanus) and olfactory attraction between lobsters.
J. Fish. Res. Bd. Can., 27(8): 1371-1378

Chemistry

CHEMISTRY

Subheadings are arranged alphabetically.

See also: BOTTOM SEDIMENTS, chemistry
INSTRUMENTATION, chemistry
METHODS, chemistry
phytoplankton, chemistry
zooplankton, chemistry, etc.
(The two last headings are in the Organismal Index.)

chemistry
Abe, T., 1935
Seasonal variation in the chemical elements of seawater at the mouth of Sinagawa-wan. (In Japanese). J. Ocean. 8:137-153, tables.

chemistry
Achmet'eva, E.A., 1962.
Peculiarities of the hydrochemical regime of the Danish Strait.
Trudy Vses. Nauchno-Issledov. Inst. Morsk. Ribn. Chos. i Okean., VNIRO, 46:68-73.

In Russian

chemistry
Alekin, O.A., 1966.
Chemistry of the ocean. (In Russian).
Hydrometeoizdat., Leningrad, 248 pp. (not seen).

chemistry
Alekin, O.A., 1948.
Obshchaia gidrokhimiia; khimiia prviodnykh ved. (General hydrochemistry; chemistry of natural waters). Gidromat. izdvo, Leningrad, 206 pp.

chemistry
Anonymous, 1964.
Chemistry and the oceans.
Chemical and Engineering News, 42(22):48 pp.

chemistry
Armstrong, E.F., and L.M. Miall, 1947.
Raw materials from the sea. Chem. Pub., 196 pp., 1947.

chemistry
Armstrong, F.A.J., and E.I. Butler, 1960.
Chemical changes in sea water off Plymouth during 1959.
J.M.B.A., U.K., 39(3):425-428.

chemistry
Armstrong, F. A. J., and E. I. Butler, 1959.
Chemical changes in sea water off Plymouth during 1957.
J. Mar. Biol. Ass. U.K. 38(1):41-45.

chemistry
Balbi, R., 1951.
Physical and chemical characteristics of sea water at the Lido of Venice as compared with the mineral waters of Salsomaggiore. Arch. Opped. al Mare 3:127-132.

Abstr. Chem. Abstr., 1952:2355f.

chemistry
Baranov, I.V., 1946.
Hydrochemical characteristics of the northern part of the Onezhskaya Guba. Nauchnyi Biull. No. 13:19-20.

chemistry
Barnes, C.A., and T.G. Thompson, 1938.
Physical and chemical investigations in Bering Sea and portions of the North Pacific Ocean.
Univ. Washington Publ. Ocean. 3:35-79, and Appen.

chemical oceanography
Barnes, H., 1955.
Chemical aspects of oceanography.
R. Inst. Chem., Lectures Monogr. & Repts., 1955(4):

chemistry
Blinov, L.K., 1956
Hydrochemistry of the Aral Sea.
Mono. Gidrokhim. Aral'sk Morya, Leningrad: 252 pp.
Listed in Techn. Transl., 6(10): 754.

chemistry
Bogojavlensky, A.N., 1955.
Chemistry of the Kurile-Kamchatka Trench waters.
Trudy Inst. Oceanol., 12:161-176.

chemistry
Braarud, T. and A. Klem, 1931
Hydrographical and chemical investigations in coastal waters off Norway. Hvalrådets Skrifter, No. 1:

chemistry
*Brewer, Peter G., and Derek W. Spencer, 1969.
Note on the chemical composition of the Red Sea brines.
In: Hot brines and Recent heavy metal deposits in the Red Sea, E.T. Degens and D.A. Ross, editors, Springer-Verlag, New York, Inc., 174-179.

chemistry
Brujewicz, S.W., 1962.
Investigation of chemical stratification in the bottom layer and water mass of the Black Sea.
Trudy Inst. Okean., Akad. Nauk, SSSR, 54:31-38.

In Russian

chemistry
Bruevich, S.V., 1960.
Bringing into perspective the development of chemical oceanography.
Trudy Okeanogr. Komissii, Akad. Nauk, SSSR, 10(2):3-12.

chemistry
Bruevich, S.V., 1960
Hydrochemical investigations on the White Sea.
Trudy Inst. Okeanol, 42: 199-254.

chemistry
Bruevich, S. V., 1948.
Elementary composition of water in the world ocean. Trudy Inst. Okean., 2:21-25.

chemistry
Bruevich, S.V., 1947.
The chemical composition of sediment solutions of the Caspian Sea. Pt. 1. Gidrokhim Materialy 13:19 pp.

chemistry
Bruevich, S. V., A.N. Bogoyavlensky and V.V. Mokievskaya, 1960
Hydrochemical features of the Okhotsk Sea
Trudy Inst. Okeanol, 42: 125-198.

CHEMISTRY
Bruevich, S.V., and E.G. Vinogradova, 1947.
[The chemical composition of sediment solutions of the Caspian Sea. I. The northern Caspian (1932). II. The northern, middle, and southern Caspian (1935, 1936, 1940).] (English summaries) Gidrokhim Mat. 8:129-186.

chemistry
Buch, K., 1948.
Amnesomsättningen i Skärgårdsvattnen.
Skärgårdsboken:136-146, Helsingfors.

chemistry
Buljan, M., 1954.
Influence of deep submarine volcanisms upon the chemistry of sea water. Comm. Int. Expl. Sci. Mer Méditerranée, Rapp. Proc. Verb., n.s., 12: 135-142.

chemistry
Carritt, D.E., 1962
Use of anion-exchange resins in the analysis of sea water. (Abstract).
J. Geophys. Res., 67(9):3548.

chemistry
Carritt, D.E., 1958
Analytical chemistry in oceanography. J. Chem. Educ., 35: 119-122.

chemistry
Charcot, J.B., 1931-1932.
Rapport préliminaire sur la campagne du "Pourquoi Pas?" en 1931. Ann. Hydrogr., Paris, (3)11:57-139

chemistry
Charcot, J.B., 1931-1932.
Rapport préliminaire sur la campagne du "Pourquoi Pas?", en 1930. Ann. Hydrogr., Paris, (3)11:1-54.

chemistry
Charcot, J.B., 1923-1924.
Rapport préliminaire sur la campagne du "Pourquoi-Pas?" en 1923. Ann. Hydrogr., Paris, (3) 6:1-89, figs.

chemistry
Chosen. Fishery Experiment Station, 1938.
Research in the neighboring sea of the Korean Gulf on board the R.M.S. Misago-maru in Feb-Mar 1933. Ann. Rept. Hydrogr. Obs., Chosen Fish. Exp. Sta., 8(IIA):7-22.

chemistry
Chosen. Fishery Experiment Station, 1938.
Oceanographical investigations during Jun-Jul 1933, in the Japan Sea offshore along the east coast of Tyosen on board the R.M.S. Misago-maru. (In Japanese). Ann. Rept. Hydrogr. Obs., Chosen Fish. Exp. Sta, 8:11, 23-29, 38-60, 62-65.

chemistry
Chosen. Fishery Experiment Station, 1938.
Oceanographic investigations in the Japan Sea made during Aug. and Oct. 1933, on board the R.M.S. Misago-maru. (In Japanese; English headings for tables and charts). Ann. Rept. Hydrogr. Obs., Chosen Fish. Exp. Sta. 8:77-113, tables, charts.

chemistry
Chosen. Fishery Experiment Station, 1938.
Hydrographic observations during Aug. 1933, on board the R.M.S. Misago-maru in the southwest part of the Japan Sea. (In Japanese; English abstract). Ann. Rept. Hydrogr. Obs., Chosen Fish. Exp. Sta., 8:iii-iv; 67-76.

chemistry
Chosen. Fishery Experiment Station, 1936-1938.
Monthly observations from Urusaki to Kawaziri Misaki during the year 1932-1933. Ann. Rept. Hydrogr. Obs., Chosen Fish. Exp. Sta., 7:94-113; 8:139-153.

chemistry
Chosen. Fishery Experiment Station, 1936.
Hydrographical investigations off the shore along the east coast of Tyosen, on board the Misago-maru in Oct. 1932. (In Japanese; English title and headings for tables.) Ann. Rept. Hydrogr. Obs., Chosen Fish. Exp. Sta., 7:86-93.

chemistry
Cooper, L.H.N., 1962.
Chemistry of the sea.
Proc. R. Soc., London, (A) 265(1322):371-380.

Chemistry
Cox, Roland A., 1959.
The chemistry of sea water.
The New Scientist, 6:518-521.

chemistry
Darmois, R., and J. Darmois, 1949.
Déterminations physicochimiques sur l'eau méditerranéenne dans la région de Monaco.
C. R. Acad. Sci., Paris, 228:417-418.

Abstr. in Chem. Abstr. 1949:4058.

chemistry
Deacon, G.E.R., 1934.
Die Nordgrenzen antarktischen und subantarktischen Wassers im Weltmeer. Ann. Hydrogr., usw., Jahrg. 62(Heft 3):129-136.

chemistry
Dimaksian, A.M., 1960
[On the radioactive methods of some hydrological regime elements measurement.]
Meteorol. i Gidrol., (12): 32-36.

chemistry
Fedosov, M.V., and R.L. Davidovich, 1963
Some peculiarities of the hydrochemical regime of the Bering Sea. (In Russian).
Sovetsk. Ribokh. Issled. B Severo-Vostokh. Chasti Tikhogo Okeana, VNIRO48, TINRO 50(1): 77-84.

chemistry
Fedosov, M.V., and I.A. Ermachenko, 1962.
Conditions for the formation of the hydrochemical regime and primary productivity of the Norwegian and Greenland seas. (In Russian)
Trudy Vses. Nauchno-Issledov. Inst. Morsk. Ribn. Chos. i Okeanogr., VNIRO, 46:18-37.

chemistry
Garrels, R.M., and M.E. Thompson, 1962
A chemical model for sea water at 25°C and one atmosphere total pressure.
Amer. J. Sci., 260(1):57-66.

chemistry
Goodman, J.R., J.H. Lincoln, T.G. Thompson and F.A. Zeusler 1942.
Physical and chemical investigations: Bering Sea, Bering Strait, Chukchi Sea during the summers of 1937 and 1938.
Publ. Univ. Washington, Oceanogr. 3(4):105-169, 37 maps.

Chemistry
Graham, H.W. and E.G. Moberg 1944.
Chemical results of the last cruise of the Carnegie. Chemistry 1. Scientific results of Cruise VII of the Carnegie during 1928-1929 under command of Captain J.P. Ault
Carnegie Publ. Washington, 562: 58pp, 7 maps.

chemistry
Grainger, E. H., 1959.
The annual oceanographic cycle at Igloolik in the Canadian Arctic. I. The zooplankton and physical and chemical observations.
J. Fish. Res. Bd. Canada, 16(4): 453-501.

chemistry
Gran, H.H., and T. Braarud, 1935
A quantitative study of the phytoplankton in the Bay of Fundy and the Gulf of Maine (including observations on hydrography, chemistry, and turbidity).
J. Biol. Bd., Canada, 1(5):279-467, 69 text figs.

chemistry
Greece, National Hellenic Oceanographic Society, 1957
Chemistry of sea water. Marine Pages (Thalassina Phylla), No. 8:115-129.

chemistry
Groen, P., 1967.
The waters of the sea.
D. Van Nostrand Co., Ltd. 328 pp. $9.00

chemistry
Harvey, H. W., 1945
Recent advances in the chemistry and biology of sea water. vii 164 pp., 29 figs., Cambridge Univ. Press, London:MacMillan Co., N.Y.

chemistry
Harvey, H.W., 1928
Biological chemistry and physics of sea water. 194 pp., 65 textfigs. University Press, Cambridge.

owned by M.S.

chemistry
Hindman, J.C., L.J. Anderson, and E.G. Moberg, 1949.
A precise determination of chlorinity of sea water using the Ag-AgCl indicator electrode.
J. Mar. Res. 8(1):30-35.

chemistry
Hood, Donald W., 1963.
Some chemical aspects of the marine environment.
Great Lakes Res. Div., Inst. Sci. and Techn., Univ. Michigan, Publ. 10:91-111.

Chemistry
Horne, R.A. 1969.
Marine Chemistry: the structure of water and the chemistry of the hydrosphere.
Wiley-Interscience, 568 pp. $19.95.

chemistry
Ishibashi, M., 1957.
Quantitative distribution of chemical elements in the sea water. Proc. UNESCO Symp., Phys. Ocean., Tokyo, 1955:175-188.

chemistry
Ishibashi, M., 1952.
Recent advances in chemical oceanography in Japan. List of titles in the symposium with references. Proc. Seventh Pacific Sci. Congr., Met. Ocean., 3:303-304.

chemistry
Ishibashi, M., M. Shinagawa, and T. Suzube, 1943.
[Chemical research concerning the ocean. XVIII.] Umi to Sora 23:317-337.

Chemistry
Ivanenkov, V.N., 1967.
On the quality of the hydrochemical observations. (In Russian)
Atlant. nauchno-issled. Inst. rybn. khoz. okeanogr. (AtlantNIRO). Materialy Konferentsii po Rezul'tatam Okeanologicheskikh Issledovanii v Atlanticheskom Okeane 101-105.

Chemical oceanography
Japan, Kobe Marine Observatory, Oceanographical Section, 1964.?
Researches in chemical oceanography made at K.M.O. for the recent ten years (1954-1963). (In Japanese).
Weather Service Bull., Japan. Meteorol. Agency, 31(7):188-191.

Also in:
Bull. Kobe Mar. Obs., No. 175. 1965.

chemistry
Kalle, K., 1948.
Physik und Chemie des Meeres. Geophys. Dieterich'sche Verlagsbuchhandlung, Wiesbaden, Pt. 2, Vol. 164-177.

chemistry
Konovalov, G.S., 1957.
[On the investigation of micro-elements in natural waters.]
Gidrochim. Materialy, 26:19.

chemistry
Laktionov, A.F., 1943.
Gidrokhimicheskiye usloviya v more Lattevykh. [Chemical conditions in the Laptev Sea.] Probl. Arkt. 1:107-128.

chemistry
Lange, Rolf 1969.
Chemical oceanography.
Universitetsforlaget, Oslo, 152 pp.

chemistry
Lebedintsev, A.A., 1892.
[Preliminary account of chemical investigations in the Black Sea and Sea of Azov in the summer of 1891] Zap. Novoross. Obschsh. Estestvoisp. 16(2)

chemistry
Legendre, R., 1947.
Les industries chemiques de la mer. Bull. Inst. Océan., Monaco, No. 922:15 pp.

chemistry
Lyman, J. and R. H. Fleming 1940
Composition of sea water. J. Mar. Res., III, pp. 134-146

chemistry
Majori, L., 1969.
Ricerche sull'inquinamento delle acque del mare dell'alto Adriatico.
Prog. Ricerca Risorse mar. Fondo mar. Comm. ital. Oceanogr. Consig. naz. Ricerche, (B)(26): 1-40

chemistry
Manuelli, A., 1914.
Ricerche de chimica talassografica sul rapportò fra vari sali nell'acque de mare. Ann. Chim. Appl. (2):132-133.

chemistry
Matida, Y., 1951.
On the chemical composition of sea water in the Tokyo Bay. Contr. Cent. Fish. Sta., Japan, 1948-1949 (138):6 pp.

Chemistry
McCone, Alistair W. 1967.
The Hudson River Estuary: sedimentary and geochemical properties between Kingston and Haverstraw, New York.
J. sedim. Petrol. 37(2): 475-486.

chemicals
McIlhenny, W.F., 1966.
The oceans: technology's new challenge.
Chem. Engng., (Nov. 7):247-254.

chemistry
Minas, H.J., 1961.
Étude comparée de quelques facteurs physicochimiques des eaux portuaires et des eaux du golfe de Marseille.
Rapp. Proc. Verb., Réunions, Comm. Inst. Expl. Sci., Mer Méditerranée, Monaco, 16(3):663-667.

chemistry
Miyake, Y., 1939.
Chemical studies of the western Pacific Ocean. 1. The chemical composition of the oceanic salts. Pts. 1, 2. Bull. Chem. Soc., Japan, 14(2):29-34; 55-58.

chemistry
Mokievskaya, V.V., & D.A. Smetanin, 1962
Chemical investigations in the Pacific Ocean
The Xth Pacific Science Congress.
Okeanologiia, Akad. Nauk, SSSR, 2(3):464-468

chemistry
Musina, A.A., and E.V. Belisheva, 1960.
[Peculiarities of the hydrochemistry of Arctic seas.]
Trudy Okeanogr. Komissii, Akad. Nauk, SSSR, 10(2):61-68.

chemistry
Nishikawa, T., T. Okuno, M. Maeda and Y. Ogata, 1939.
Ocean water taken in various parts of the world. (In Japanese).
J. Soc. Chem. Ind., Japan, 42(Suppl.),:71-72.

chemistry
Okubo, I., 1963.
14. On the chemical elements of sea water in the Kuril Trench taken by the bathyscaphe "Archimede" (Étude chimique de l'eau relevée de la fosse des Kouriles par le bathyscaphe français "Archimede"). (In Japanese; French resumé).
La Mer, 1(1):3-6.

Reprinted in:
Bull. Hakodate Mar. Obs., (10):3-6.

chemistry
Ommanney, F. D., 1949.
The Ocean. 238 pp. Oxford Univ. Press.

chemistry
Phillips, F. C., 1947.
Oceanic salt deposits. Quart. Rev. Chem. Soc. 1:91-111.

chemistry
Picotti, M., 1955.
Unita di misura nella chimica talassografica. Arch. Ocean. e Limnol., 10(3):201-215.

chemistry
Picotti, M., 1941.
Unita di mesura della chimica talassograffica. (Mem. CCLXXXVI), Arch. Ocean. Limn., Anno 1, Fasc. 1, 1941XIX:5-29.

chemistry
Picotti, M., 1941.
Composizione chimica dell'aequa marina desunta della concentrazione dei constituenti predomanti. (Memoria CCLXXVII), Arch. di Ocean. e Limn., Anno I, Fasc. 2- 1941XIX:89-102.

Pirie, Robert Gordon, 1965.
Petrology and physical-chemical environment of bottom sediments of the Rivière Bonaventure-Chaleur Bay area, Quebec, Canada.
Rept. B.I.O. 65-10:182 pp. (multilithed).

chemistry
Rakestraw, N. W., 1949.
The conception of alkalinity of excess base of sea water. J. Mar. Res. 8(1):14-20.

Chemistry
Rakestraw, N.W., 1933
Studies on the biology and chemistry of the Gulf of Maine. I. Chemistry of the waters of the Gulf of Maine in August 1932. Biol. Bull., 64:149-158

chemical oceanography
Richards, Francis A., 1967.
Chemical oceanography.
Trans. Am. geophys. Un., 48(2):595-604.

chemical oceanography
Richards, F.A., 1957.
Some current aspects of chemical oceanography. In: Progress in physics and chemistry of the earth, Pergamon Press, Ltd., 2:77-128.

chemistry
Robinson, R. J., and F. W., Knapman, 1941.
The sodium-chlorinity ratio of ocean waters from the northeast Pacific. J. Mar. Res., 4 (2):142-152.

chemistry
Rochford, D., 1947
The preparation and use of Harvey's reduced strychnine reagent in oceanographical chemistry. Australia, Coun. Sci. Ind. Res., Bull. No.220, 37 pp.

chemistry
Rotschi, Henri, Michel Angot, Michel Legand and H.R. Jitts, 1959
"Orsom III", Resultats de la croisiere BOUSSOLE 2. Chimie, productivite et zooplancton et resultats "production primaire" de la croisiere 56-5.
O.R.S.T.O.M., Inst. Francais d'Oceanie, Rapp. Sci., No. 13: 1-83.

chemistry
Rotschi, H., B. Wauthy and M. LeGand, 1961
Orsom III. Résultats de la croisière "Epi" 2eme partie. Chimie et biologie.
O.R.S.T.O.M., I.F.O., Rapp. Sci., (23):4-111.

chemistry
Schulz, B., 1934.
Hydrographische Untersuchungsfahrten in die Nord und Ostsee mit dem R.F.D. "POSEIDON" 1931/33.
Ann. Hydrogr., usw., Jahrg. 62(Heft 3):116-119.

chemistry
Sinjukov, V.V., 1964.
On the influence of volcanic eruption on the chemistry of oceanic waters. (In Russian).
Okeanologiia, Akad. Nauk, SSSR, 4(4):644-650.

chemistry
Skopintsev, B.A., 1962.
Recent work on the chemistry of the sea. (In Russian).
Trudy Morsk. Gidrofiz. Inst., 25:82-109.

chemistry
Skopintsev, B.A., 1948.
[Investigation of acid products of hydrolyzed organic substances in natural waters]
Dok. Akad. Nauk SSSR 62:243-246.

chemistry
Skopintsev, B.A., F.A. Gubin, R.V. Vorobieva and O.A. Vershinina, 1958.
Percentage of the chief components of salt composition in waters of the Black Sea and some questions of water transfer. Heat of the sea. Chemistry of the sea. (In Russian).
Trudy Morsk. Gidrofiz. Inst., 13:89-112.

chemistry
Slinn, D.J., 1960.
Chemical constituents of sea water off Port Erin during 1959.
Ann. Rept. Mar. Biol. Sta., Univ. Liverpool, 72: 21-24.

chemistry
Smetanin, B.A., 1962.
Some features of meridional distribution of the chemical characteristics in the Pacific.
Trudy Inst. Okeanol., Akad. Nauk, SSSR, 54:3-21.
In Russian; English summary

chemistry
Smetanin, D.A., 1958.
[Hydrochemie im Gebiet des Kurilen-Kamtschatka tiefsee Grabens. 1. Einige Frage der Hydrologie und Chemie der unteren subarktischen Gewasser in Gebiet des Kurilen-Kamtschatka Grabens.] Trudy Inst. Okeanol., 27:22-54.

chemistry
SSSR, Akademia Nauk, Institute of Oceanology, 1966.
Chemistry of the Pacific Ocean. (In Russian).
Isdatal Nauke, Moskoa, 358 pp.

chemistry
Sugawara, Ken, 1969.
Scope of the chemistry of the sea. Bull. Jap. Soc. fish. Oceanogr. Spec. No. (Prof. Uda Commem. Pap.): 7-9.

chemistry
Thompson, T.G., and C.A. Barnes, 1951.
Physical and chemical oceanography of the Gulf of Alaska and the Aleutian Islands. Proc. Alaskan Sci. Conf., Bull. Nat. Res. Coun. 122:82.

chemistry
Thompson, E. F., and H. C. Gilson, 1937.
Chemical and physical investigations. Introduction. John Murray Exped., 1933-34, Sci. Repts. 2(2):15-20.

chemistry
Thompson, T. G., and R. J. Robinson 1932
Chemistry of the Sea. Physics of the Earth. Vol. 5, Oceanography, pp. 95-203. Nat. Res. Council Bull., No. 85. Washington, D. C.

chemistry
*Volostnykh, B.V., V.F. Rambayev, and L.N. Lebedeva, 1967.
On hydrochemical study of the Atlantic Ocean waters. (In Russian; English abstract).
Trudy Inst. Okeanol., 83(51-62.

chemistry
Voronokoff, P.P., 1954.
[Characteristic features of the chemical composition of surface waters in various geographical regions.] Dokl. Akad. Nauk, SSSR, 94(2):293.

chemistry
Wattenberg, H., 1943.
Zur Chemie des Meereswassers, neuere Untersuchungen über gelöste Gase. Z. anorg. allgem. Chem. 251:71-75.

chemistry
Wattenberg, H., 1943.
Zur Chemie des Meerwassers. Über die in Spuren vorkommende Elemente.
Zeits. Anorg. Chem., 251:86-91.

chemistry
Wattenberg, H., 1941.
Über die Grenzen zwischen Nord- und Ostseewasser.
Ann. Hydr., usw., 69:265-279.

chemistry
Wattenberg, H., 1940.
Der hydrographisch-chemische Zustand der Ostsee im Sommer 1939. Ergebnisse der 'Triton'-Fahrt von 27 Juli bis 10 August, 1939. Ann. Hydrogr. usw. 68:185-194.

chemistry
Wattenberg, H., 1939.
Einige Ergebnisse der Untersuchungsfahrten mit dem Reichsforschungsdampfer Poseidon in der westlichen Ostsee, 1938. Ber. Deutsch. Wiss. Komm. Meeresf. 9(n.s.):541-560.

chemistry
Wattenberg, H., 1938.
Zur Chemie des Meerwassers.
Zeits. Anorg. Chem., 236:339-360.

chemistry
Wattenberg, H., 1937.
IV. Die chemischen Arbeiten auf der "Meteor"-Fahrt, Februar bis Mai 1937. Ann. Hydr., usw., 1937:17-22, 3 figs.

alkalinity
Wattenberg, H., and H. Wittig., 1938.
Ueber die Bestimmung der Titrationsalkalinität des Seewassers. Kieler Meeresf. 3:238-262.

chemistry, sea water
Wood, P.C., 1961.
The principles of water sterilisation by ultraviolet light and theri application in the purification of oysters.
Min. Agric. Fish. Food, Fish. Invest. (2)13(6): 48 pp.

chemistry
Yegorova, V.A., 1957.
[Hydrochemical investigations in the coastal zone of the northeastern part of the Black Sea.]
Trudy Inst. Okeanol., 21:137-164.

chemistry
Zaitsev, G.N., and M.V. Fedosov, 1959.
[Vertical displacement and formation of the hydrochemical regime of the upper water layers of the middle and southern Caspian.]
Trudy VNIRO, 38:134-

chemistry
Zorell, F., 1933.
Beiträge zur Kenntniss der Alkalinität des Meerwassers. Ann. Hydrogr., usw., Jahrg. 61(Heft I/II):18-22.

acid-iron waste
Chemistry - acid-iron waste
Arnold, E.L., jr., and W. F. Royce, 1950.
Observations on the effect of acid-iron waste disposal at sea on animal populations. Spec. Sci Rept., Fisheries, No. 11: 12 pp., 2 figs. (multilithed).

Chemistry acids, organic (yellow)
*Ghassemi, Masood and R.F. Christman, 1968.
Properties of the yellow organic acids of natural waters.
Limnol. Oceanogr., 13(4):583-598.

chemistry, acrolein
Dahlberg, Michael D. 1971
Toxicity of acrolein to barnacles (Balanus eburneus).
Chesapeake Sci. 12(4): 282-284

chemistry, acrylic acid
Glombitza, K.-W. und R. Heyser, 1971.
Antimikrobielle Inhaltsstoffe in Algen. 4. Mitteilung. Wirkung der Acrylsäure auf Atmung und Makromolekülsynthese bei Staphylococcus aureus und Escherichia coli. Helgoländer wiss. Meeresunters 22(3/4): 442-453.

Chemistry - actinomycin, effect of
Greenhouse, Gerald A., Richard O. Hynes and Paul R. Gross, 1971.
Sea urchin embryos are permeable to actinomycin
Science, 171(3972): 686-689.

adenine
Chemistry, adenine
Litchfield, Carol D., and Donald W. Hood, 1966.
Microbiological assay for organic compounds in seawater. II. Distribution of adenine, uracil, and threonine.
Appl. Microbiol., 14(2):145-151.

Chemistry, adenine
Rosenberg, E., 1964.
Purine and pyrimidines in sediments from the experimental Mohole.
Science, 146(3652):1680-1681.

adenosine
Chemistry, adenosine triphosphate
Hamilton, Robert D., and Osmund Holm-Hansen 1967
Adenosine triphosphate content of marine bacteria.
Limnol. Oceanogr. 12(2):319-324.

chemistry, adenosine triphosphate
Holm-Hansen, Osmund, and Charles R. Booth, 1966.
The measurement of adenosine triphosphate in the ocean and its ecological significance.
Limnol. Oceanogr., 11(4):510-519.

Chemistry - adenylic acid
Iida, Atsushi, Isao Araki and Kiichi Murata 1961.
Biochemical studies on muscle of sea animals. I. On adenylic acid in the boiled liquor with the muscle of scallop, Pecten yessoensis. (In Japanese; English abstract).
Bull. Fac. Fish., Hokkaido Univ. 12(2): 151-159.

activity coefficients

chemistry - activity coefficients
Berner, Robert A., 1965.
Activity coefficients of bicarbonate, carbonate and calcium ions in sea water.
Geochimica et Cosmochimica Acta, 29(8):947-965.

chemistry - activity coefficients
Platford, R.F., 1965.
The activity coefficient of sodium chloride in seawater.
J. Mar. Res., 23(2):55-62.

Chemistry - activity coefficient
Platford, R.F., and Thomas Defoe, 1965.
The activity coefficient of sodium sulfate.
J. Mar. Res., 23(2):63-68.

aequorin

chemistry, aequorin
Shimomura, Osamu, Frank H. Johnson and Yo Saiga, 1963
Microdetermination of calcium by aequorin luminescence.
Science, 140(3573):1339-1340.

chemistry - agglutinins
Smith, A.C. and R.A. Goldstein, 1971.
"Natural" agglutinins against sea urchin sperm in the hemolymph of the crab Cardisoma guanhumi.
Marine Biol., 8(1): 6.

aggregates

chemistry - Aggregates, organic
Barber, Richard T., 1966.
Interaction of bubbles and bacteria in the formation of organic aggregates in sea-water.
Nature, 211 (5046):257-258.

chemistry, aggregates (organic)
Siegel, Alvin, and Barbara Burke 1965.
Sorption studies of cations on "bubble produced organic aggregates" in sea water.
Deep-Sea Res., 12 (6): 789-796.

air chemistry

chemistry, air
*Egorov, V.V., T.N. Znigalovskaya and S.G. Malakhov, 1970.
Microelement content of surface air above the continent and the ocean. J. geophys. Res., 75 (18): 3650-3656.

albumin

chemistry, albumen
Hagmeier, Erik, 1961.
Plankton-Äquivalente Auswertungen von chemischen und mikroskopischen Analysen.
Kieler Meeresf., 17(1):32-47.

chemistry, albumen
Krey, Johannes, 1961.
Beobachtungen uber den Gehalt an Mikrobiomasse und Detritus in der Kieler Bucht 1958-1960.
Kieler Meeresf., 17(2):163-175.

chemistry, albumen
Krey, Johannes, 1958
Chemical determinations of net plankton with special reference to equivalent albumin content. J. Mar. Res., 17:312-324.

chemistry, albumen
Krey, J., D. Hantschmann, and St. Wellershaus, 1959.
Der Sestongehalt entlang eines Schniffes von Kap Farvel bis zur Flämischen Kappe im April und September 1958.
Deutsch. Hydrogr. Zeits., Ergänzungsheft Reihe, B (4) No. 3:73-80.

chemistry, albumen
*Krishnamurthy, K., 1967.
Some aspects of chemical composition of plankton.
Kieler Meeresforsch., 23(2):99-104.

chemistry, albumen
*Lenz, Jürgen, Heinz Schöne und Bernt Zeitschel, 1967.
Planktonologische Beobachtungen auf einem Schnitt durch die Nordsee von Cuxhaven nach Edinburgh.
Kieler Meeresforsch., 23(2):92-98.

Chemistry, alcohols
Nonaka, Junsaku, and Chiaki Koizumi 1964
Component fatty acids and alcohols of Euphausia superba lipid by gas-liquid chromatography. (In Japanese; English abstract)
Bull. Jap. Soc. scient. Fish. 30(8): 630-634.

chemistry, aldehydes
Corwin, James F., 1969.
Volatile oxygen-containing organic compounds in sea water: determination. Bull. mar. Sci., 19(3): 504-509.

algin

chemical composition - algae
Black, W.A.P., and E.T. Dewar, 1949.
Correlation of some of the physical and chemical properties of the sea with the chemical composition of the algae. J.M.B.A. 28(3):673-699, 14 textfigs.

chemistry - algin
Davis, F. W., 1950.
Algin from sargassum. Science 111:150.

alginates

Chemistry, alginates
Oguro, Miki, 1961.
Studies on soluble alginates. II. The pH of soluble alginates, including lithium alginate, sodium alginate, potassium alginate and ammonium alginate. (In Japanese; English abstract).
Bull. Fac. Fish., Hokkaido Univ., 12(1):88-92.

alkali

Chemistry - alkali
Cooper, J. A. and J. R. Richards, 1966.
Isotopic and alkali measurements from the Vema Seamount of the South Atlantic Ocean.
Nature, 210(5042):1245-1246.

alkaline earths

chemistry - alkaline earths/chlorinity
Billings, Gale K., Owen P. Bricker, Fred T. Mackenzie and Albert L. Brook 1969.
Temporal variation of alkaline earth element/chlorinity ratios in the Sargasso Sea.
Earth Planet. Sci. Letters, 6(3): 231-236.

chemistry - alkaline earths
Voipio, Aarno, 1959
On the alkaline-earth metal and magnesium contents of sea water.
Suomen Kemistilehti, B, 32: 61-65.

Alkalinity

Chemistry, alkalinity (data)
Acara, A., 1961.
Contribution on the variation of alkalinity in the southeastern Black Sea.
Rapp. Proc. Verb., Réunions, Comm. Int. Expl. Sci., Mer Méditerranée, Monaco, 16(3):699-703.

Chemistry - alkalinity
*Akiyama, Ysutomu, Takeshi Sagi, Takeshi Yura and Yoshisuke Maeda, 1968.
Onthe distribution of pH in situ and total alkalinity in the western North Pacific Ocean.
Oceanogrl Mag., 20(1):1-8.

chemistry, alkalinity (data)
Aleksandrovskaya, N.B., M.V. Kales and E.A. Yole 1966.
Peculiarities of the hydrological regime of the Baltic Sea in 1964.
Annls. biol., Copenh., 21:43-45.

Chemistry, alkalinity
Anderson, D.H., and R.J. Robinson, 1946.
Rapid electrometric determination of the alkalinity of sea water using a glass electrode.
Ind. Eng. Chem., Anal. Edit., 18:767.

chemistry, alkalinity (data)
Argentina, Servicio de Hidrografia Naval, 1960.
Operacion Oceanografica Malvinas (Resultados preliminares).
Servicio de Hidrografia Naval, Argentina, Publ., H. 606: numerous unnumbered pp.

chemistry, alkalinity (data)
Argentina, Servicio de Hidrografia Naval, 1959.
Operacion Oceanografica Atlantico Sur. Resultados preliminares.
Servicio de Hidrografia Naval, Argentina, Publ., H. 608: numerous unnumbered pp.

chemistry, alkalinity (data)
Argentina, Servicio de Hidrografia Naval, 1959.
Operacion Oceanografica Drake. 1. Resultados preliminares.
Servicio de Hidrografia Naval, Argentina, Publ., H. 613: numerous unnumbered pp.

chemistry, alkalinity (data)
Argentina, Servicio de Hidrografia Naval, 1959.
Operacion Oceanografica Meridiano. Resultados preliminares.
Servicio de Hidrografia Naval, Argentina, Publ., H. 617: numerous unnumbered pp.

chemistry, alkalinity(data)
Argentina, Servicio de Hidrografia Naval, 1959. Trabajos oceanograficos realizados en la campana Antartica 1958/1959. Resultados preliminares. Servicio de Hidrografia Naval, Argentina, Publ., H. 616: 127 pp.

chemistry, alkalinity(data)
Argentina, Servicio de Hidrografia Naval, 1959. Trabajos oceanograficos realizados en la campana Antartica, 1957-1958. Resultados preliminares. Servicio de Hidrografia Naval, Argentina, Publ., H. 615: numerous unnumbered pp.

chemistry - alkalinity(data)
Argentina, Servicio de Hidrografia Naval, 1959. Trabajos oceanograficos realizados en la campana Antartica, 1955-1956. Servicio de Hidrografia Naval, Argentina, Publ., H. 620: numerous unnumbered pp.

Chemistry alkalinity
Ben-Yaakov, Sam, 1971. A multivariable regression analysis of the vertical distribution of TCO_2 in the eastern Pacific. J. geophys. Res. 76(30): 7417-7431.

chemistry, alkalinity
Bienati, Norberto L., y Rufino A. Comes 1970. Variación estacional de la composicion fisico-quimica del agua de mar en Puerto Paraiso, Antartida occidental. Contrib. Inst. Antart. Argentino 130: 45pp.

chemistry, alkalinity
Bogoyavlensky, A.N., 1959
[Hydrochemical investigations.] Arktich. i Antarktich. Nauchno-Issled. Inst., Mezhd. Geofiz. God, Sovetsk. Antarkt. Eksped., 5: 159-172.

chemistry - alkalinity(monthly means)
Bose, B.B., 1956. Observations on the hydrology of the Hooghly Estuary. Indian J. Fish., 1(3):101-118.

chemistry - alkalinity
*Brosin, Hans-Jürgen, und Dietwart Nehring, 1968. Der Aquatoriale Unterstrom im Atlantischen Ozean auf 29°30'W im September und Dezember 1966. Beiträge Meeresk., 22:5-17.

chemistry, alkalinity(data)
Bruneau, L., N.G. Jerlov, F.F. Koczy, 1953. Physical and chemical methods. Repts. Swedish Deep-sea Exped., 1947-1948, Phys. Chem., 3(4):101-112, i-iv, 7 textfigs., 2 tables.

chemistry, alkalinity
Burkholder, Paul R., 1960
Distribution of some chemical values in Lake Erie. Limnological survey of western and central Lake Erie, 1928-1929. USFWS Spec. Sci. Rept., Fish., No. 334 :71-110.

chemistry, alkalinity(data)
Chesapeake Bay Institute, 1954. Choptank Bay spring cruise, 28 April-1 May 1952. Ref. 54-1:37 pp.

chemistry, alkalinity (data)
Dragovich, Alexander, John H. Finucane, John A. Kelly, Jr., and Billie Z. May, 1963. Counts of red-tide organisms, Gymnodinium breve, and associated oceanographic data from Florida west coast, 1960-61. U.S. Fish and Wildlife Service, Spec. Sci. Repts. Fish., No. 455:1-40.

chemistry, alkalinity and chloride
Dutta, N., J.C. Malhotra and B.B. Bose, 1954. Hydrology and seasonal fluctuations of the plankton in the Hooghly estuary. Symp. Mar. Fresh-water Plankton, Indo-Pacific, Bangkok, Jan. 25-26, 1954, FAO-UNESCO:35-47.

chemistry, alkalinity
Dyrssen, David, 1969. The carbonate system. In: Chemical oceanography, Rolf Lange, editor, Universitetsforlaget, Oslo: 59-64.

chemistry, alkalinity
*Dyrssen, David, et al., 1967. Analysis of sea water in the Uddevalla Fjord system 1. Rept. Chem. Sea Water, Univ. Göteborg, 4:1-8 pp. (mimeographed).

chemistry, alkalinity
Dyrssen, David, and Lars Gunner Sillén, 1967. Alkalinity and total carbonate in sea water. A plea for p-T-independent data. Tellus, 19(1):113-121.

chemistry - alkalinity (specific)
Egorova, V.A., and A.A. Zheleznova, 1963 Specific alkalinity in the surface layer of the Mediterranean Sea on the basis of summer observations of 1959-1960. (In Russian). Okeanologiia, Akad. Nauk, SSSR, 3(4):653-665.

chemistry, alkalinity(data)
El-Sayed, Sayed Z., and Enrique F. Mandelli, Primary production and standing crop of phytoplankton in the Weddell Sea and Drake Passage. In: Biology of Antarctic seas, II. Antarctic Res. Ser., Am. Geophys. Union, 5:87-106

Chemistry, alkalinity
Fonselius, S.H. 1968. Observations at Swedish lightships and in the central Baltic 1966. Annls biol. Copenh. 1966, 23:74-78.

Chemistry, alkalinity (data)
Franco, Paolo 1970. Oceanography of northern Adriatic Sea. I. Hydrologic features: cruises July-August and October-November 1965. Archo Oceanogr. Limnol. 16 (Suppl.):1-93.

alkalinity
Franco, Paolo, 1967. Condizioni idrologiche e produttività primaria nel Golfo di Venezia - nota preliminare. Arch. Oceanogr. Limnol. 15(1): 69-83.

chemistry, alkalinity
Galtsoff, P.S., ed., 1954. Gulf of Mexico, its origin, waters and marine life. Fish. Bull., Fish and Wildlife Service, 55:1-604, 74 textfigs.

chemistry - alkalinity
Gast, J.A., and T.G. Thompson, 1958. Determination of alkalinity and borate concentration in sea water. Anal. Chem., 30(9):1549-1551.

chemistry - alkalinity(data)
Gololobov, Ya. K. 1963
Hydrochemical characteristics of the Aegean Sea during autumn of 1959. (In Russian) Mezhd. Geofiz. Komitet, Prezidiume, Akad. Nauk SSSR, Rezult. Issled. Programme Mezhd. Geofiz. Goda, Okeanol. Issled., (8):90-96.

chemistry, alkalinity
Gorsline, Donn S., 1963. Oceanography of Apalachicola Bay, Florida. In: Essays in Marine Geology in honor of K.O. Emery, Thomas Clements, Editor, Univ. Southern California Press, 69-96.

Chemistry - alkalinity
Gripenberg, S., 1961
Alkalinity and boric acid content of Barents Sea water. Contribution to Special IGY Meeting, 1959. Cons. Perm. Int. Expl. Mer. Rapp. Proc. Verb 149:31-37.

Chemistry - alkalinity
Gripenberg, Stina, 1960
On the alkalinity of Baltic waters. J. du Conseil 26(1): 5-20.

chemistry - alkalinity
Gupta, R.Sen. and Abraham Pylee, 1969.
Specific alkalinity in the northern Indian Ocean during the south west monsoon. Bull. natn. Inst. Sci., India, 38(1): 324-333.

Chemistry - alkalinity (data)
Halim, Youssef, 1960. Étude quantitative et qualitative du cycle écologique des dinoflagellés dans les eaux de Villefranche-sur-Mer (1953-1955). Ann. Inst. Océanogr. Monaco 38: 123-232.

chemistry - alkalinity/chlorinity
Ivanenkov, V.N., 1964. Distribution of PCO_2, pH, alk/cl in the northern part of the Indian Ocean. Investigations in the Indian Ocean (33rd voyage of E/S "Vitiaz"). Trudy Inst. Okeanol., Akad. Nauk, SSSR, 64:128-143.

chemistry - alkalinity
Japan, Kobe Marine Observatory 1967. Report of the oceanographic observations in the sea south of Honshu and in the Seto-Naikai from May to June 1964. (In Japanese) Bull. Kobe Mar. Obs. No. 178: 37-

chemistry - alkalinity (data)
Kaleis, M.V., and N.B. Aleksandrovskaya, 1965. Hydrological regime of the Baltic Sea in 1963. Ann. Biol., Cons. Perm. Int. Expl. Mer, 1963, 20:70-72.

chemistry-alkalinity

Keeling, Charles D., and Bert Bolin, 1968. The simultaneous use of chemical tracers in oceanic studies. II. A three-reservoir model of the North and South Pacific oceans. Tellus, 20(1): 17-54

chemistry - alkalinity

Kirillova, E.P., 1965. The regime of chlorine, alkalinity and oxygen in the mouth of the Iuzhnii Bug. (In Russian). Trudy, Gosudarst. Okeanogr. Inst., No. 83:158-171.

chemistry - alkalinity

Kitamura, Hiroyuki, and Takeshi Sagi, 1965. On the chemical elements in the sea south of Honshu, Japan. II. (In Japanese; English abstract) Bull. Kobe Mar. Obs., No. 174:39-55.

chemistry - alkalinity

Knull, James R. and Francis A. Richards, 1969. A note on the sources of excess alkalinity in anoxic waters. Deep-Sea Res., 16(2): 205-212.

Chemistry - alkalinity

Koczy, F.F., 1956. The specific alkalinity. Deep-Sea Res. 3(4):279-288.

chemistry, alkalinity

Koroleff, Folke, 1965. The results of alkalinity intercalibration measurements in Copenhagen. UNESCO, Techn. Papers, Mar. Sci., No. 3:9-11. (mimeographed).

chemistry - alkalinity(data)

Kramer J.R. 1961 Chemistry of Lake Erie. Proc. Fourth Conf., Great Lakes Research. Great Lakes Res. Div., Inst. Sci. & Tech. Univ. Michigan. Publ., (7):27-56.

chemistry - alkalinity

Leloup, E., editor, 1966. Recherches sur l'ostreiculture dans le bassin d'Ostende en 1964. Minist. Agricult. Comm. T.W.O.Z., Groupe de Travail "Ostreiculture": 58pp.

chemistry - alkalinity

Leloup, E., Editor, with collaboration of L. Van Meel, Ph. Polk, R. Halewyck and A. Gryson, undated. Recherches sur l'ostreiculture dans le Bassin de Cahsse d'Ostande en 1962. Ministere de l'Agriculture, Commission T.W.O.Z., Groupe de Travail - "Ostreiculture", 58 pp.

chemistry, alkalinity

Lemasson, L., et Y. Magnier, 1966. Résultats des observations scientifiques de Le Dunkerquoise sous le commandement du Capitaine de Corvette Brosset. Croisière Hunter. Cah. ORSTOM, Sér. Océanogr., 4(1):3-78.

Chemistry - alkalinity

Lisitzin, A.P., 1966. Processes of Recent sedimentation in the Bering Sea. (In Russian). Inst. Okeanol., Kom. Osad. Otdel Nauk o Zemle, Isdatel, Nauka, Moskva, 574 pp.

chemistry, alkalinity

Matsue, Yoshiyuki, Yuzo Komaki and Masaaki Murano, 1957

On the distribution of minute nutrients in the North Japan Sea; March-April, 1956. Bull. Japan Sea Reg. Fish. Res. Lab., 6: 316-320. Abstr. in: Rec. Res., Fac. Agric., Tokyo Univ., Mar. 1958, 7(81): 58.

Chemistry, alkalinity

Matsue, Yoshiyuki, Yuzo Komaki and Masaaki Murano, 1957

On the distribution of minute nutrients in the North Japan Sea, during the close of August, 1955. Bull. Japan Sea Reg. Fish. Res. Lab., 6:121-127 Abstr. in: Rec. Res., Fac. Agric., Tokyo Univ., Mar. 1958. 7(80): 57-58.

chemistry, alkalinity

Minas, H.J., 1963. La relation alcalinité-chlorosité de certaines eaux côtiers du littoral de la région Marseillaise Rapp. Proc. Verb., Réunions, Comm. Int., Expl. Sci., Mer Méditerranée, Monaco, 17(3):855-856.

Chemistry - alkalinity

Minas, H.J., and M.J. Pizarro, 1962. Quelques observations hydrologiques sur les eaux au large des côtes orientales de la Corse. Bull. Inst. Océanogr., Monaco, 59(1232):1-23.

chemistry - alkalinity

Mokiyevskaya, V.V., 1961 [Some peculiarities of hydrochemistry in the northern Indian Ocean.] (In Russian; English abstract). Okeanolog. Issled., Mezhd. Komitet Proved. Mezhd. Geofiz. Goda, Prezidiume, Akad. Nauk, SSSR, (4):50-61.

alkalinity (data)

Nehring, D. and H.-J. Brosin 1968 Ozeanographische Beobachtungen im äquatorialen Atlantik und auf dem Patagonischen Schelf während der 1. Südatlantik Expedition mit dem Fischereiforschungsschiff Ernst Haeckel von August bis Dezember 1966. Geod. Geoph. Veröff. 4(3): 93pp.

Chemistry, alkalinity (data)

Nehring, Dietwart, und Karl-Heinz Rohde 1967. Weitere Untersuchungen über anormale Ionenverhältnisse in der Ostsee. Beitr. Meereskunde 20:10-33.

chemistry, alkalinity

Norina, A.M., 1965. Hydrochemical characteristics of the northern part of the Barents Sea. (In Russian). Trudy, Gosudarst. Okeanogr. Inst., No. 83:243-271.

chemistry-alkalinity (specific)

Oren, O.H., 1962 A note on the hydrography of the Gulf of Eylath. Contributions to the knowledge of the Red Sea No. 21. Sea Fish. Res. Sta., Haifa, Israel, Bull. No. 30: 3-14.

chemistry, alkalinity

Park, Kilho 1968. Alkalinity and pH off the west coast of Oregon. Deep-Sea Res. 15(2): 171-183

chemistry - alkalinity

Park, Kilho 1967. Chemical features of the subarctic boundary near 170°W. J. Fish. Res. Bd. Can. 24(5): 899-908.

chemistry, alkalinity

Park, K., M. Oliphant and H. Freund, 1963. Conductometric determination of alkalinity of sea water. Analyt. Chem., 35(10):1549.

Ref. in: Chem. Titles, 1963(18):73.

Chemistry, alkalinity

*Park, P. Kilho, George R. Webster, and Roy Yamamoto, 1969. Alkalinity budget of the Columbia River. Limnol. Oceanogr. 14(4):559-567.

chemistry, alkalinity

Picotti, Mario, 1965. La crociera idrografico-talassografica di Capo Matapan. Pubbl. Commissione Ital. Comitato int. Geofis., Ser. IGC, 42: 63 pp.

chemistry, alkalinity

Puri, Harbans S., Gioacchino Bonaduce and John Malloy, 1964. Ecology of the Gulf of Naples. Pubbl. Staz. Zool., Napoli, 33(Suppl.):87-199.

Chemistry, alkalinity

Richards, Francis A., 1968. Chemical and biological factors in the marine environment. Ch. 8 in: Ocean engineering: goals, environment, technology, J.F. Brahtz, editor, John Wiley & Sons, 259-303.

chemistry - Alkalinity

Rotschi, Henri, 1965. Le pH et l'alcalinité des eaux profondes de la Fosse des Hebrides et du Bassin des Fidji. Progress in Oceanography, 3:301-310.

Chemistry, alkalinity(data)

Rotschi, Henri, 1960 Orsom III, Resultats des Croisières diverses de 1959. Océanographie physique. Centre d'Océanogr., Inst. Francais d'Océanie Rapp. Sci., No. 17: 59 pp.

chemistry - alkalinity

Rotschi, H., 1959 Chimie, "Orsom III", Resultats de la Croisière BOUSSOLE. O.R.S.T.O.M., Inst. Français d'Océanie, Rapp Sci. 13: 3-60.

chemistry, alkalinity

Schulz, B., 1930. Die Alkalinität des Oberflächenwassers der Nordsee und des Nordatlantik. Cons. Perm. Int. Expl. Mer, Rapp. Proc. Verb., 67:91-92.

chemistry, alkalinity(data)
Schulz, B., 1922.
Hydrographischen Beobachtungen inbesondere über die Kohlensäure in der Nord- und Ostsee im Sommer 1921. (Forschungsschiffe "Poseidon" und "Skagerak"). Arch. Deutschen Seewarte 40(2):1-44, 2 textfigs., 4 pls.

chemistry, alkalinity (data)
Seibold, E., 1962.
Untersuchungen zur Kalkfällung und Kalklösung am Westrand der Grand Bahama Bank.
Sedimentology, 1(1):50-74.

chemistry, alkalinity
Sen Gupta, R., and Abraham Pylee 1968.
Specific alkalinity in the northern Indian Ocean during the south west monsoon.
Bull. natn. Inst. Sci. India 38:324-333.

alkalinity
Škrivanić, Ante, 1969.
Hydrographic and biotical conditions in the north Adriatic-1: Hydrochemistry and some factors influencing hydrography. (Jugoslavian and Italian abstracts).
Thalassia Jugoslavica, 5: 315-328.

chemistry, alkalinity
Svansson, Artur, 1960(1958)
Observations at Swedish light vessels.
Ann. Biol., Cons. Perm. Int. Expl. Mer, 15: 43-45.

chemistry, alkalinity
Traganza, Eugene D., and Barney J. Szabo 1967.
Calculation of calcium anomalies on the Great Bahama Bank from alkalinity and chlorinity data.
Limnol. Oceanogr. 12(2): 281-286.

chemistry, alkalinity
Turekian, Karl K., 1971
2. Rivers, tributaries and estuaries. In: Impingement of man on the oceans, D.W. Hood, editor, Wiley Interscience, 9-73.

chemistry, alkalinity
Turekian, Karl K., Donald F. Shutz, Peter Bower and David G. Johnson, 1967.
Alkalinity and strontium profiles in Antarctic waters.
Antarctic Jl. U.S.A., 2(5):186-188.

chemistry, alkalinity
Vatova, A. 1965
Les conditions hydrographiques de la Mar Piccolo de Tarante pendant l'année 1963.
Rapp. P.-v. Réun. Commn int. Explor. scient. Mer Méditerr. 18(3): 653-655.

chemistry, alkalinity
Vatova, A., 1944.
Osservazioni idrografiche periodiche nell'alto Adriatico (1937-1944).
Boll. Pesca, Piscic. Idrobiol. 3(2):247-277.

chemistry - alkalinity (data)
Villanueva, Sebastian F., Alberto Gomez, Aldo Orlando y Andres J. Lusquiños 1969.
Datos y resultados de las campañas pesquerias "Pesqueria VII", 16 de febrero al 1° de marzo de 1968.
Publ. Ser. Inf. técn. Proyecto Desarollo Pesq. Mar del Plata (10/VII): unnumbered pp.

chemistry - alkalinity(data)
Vize, V. Iu., 1933
[Some data on the winter hydrological regime of the Petschora Sea.] (German summary).
Arctica. Vses. Arktich. Inst., 1:99-114.
Reprint in MBL library.

chemistry, alkalinity
Waldichuk, Michael, 1965.
Water exchange in Port Moody, British Columbia, and its effect on waste disposal.
J. Fish. Res. Bd., Canada, 22(3):801-822.

chemistry, alkalinity
Ward, Ronald W., Valerie Vreeland, Charles H. Southwick and Anthony J. Reading, 1965.
Ecological studies related to plankton productivity in two Chesapeake Bay estuaries.
Chesapeake Science, 6(4):214-225.

chemistry, alkalinity
West, L.E. and R.J. Robinson 1941.
Potentiometric analysis of sea water. II. Determination of titration alkalinity.
J. mar. Res. 4(1): 33-41.

chemistry, alkalinity
Wittig, H., 1940
Über die Verteilung des Kalziums und der Alkalinität in der Ostsee. Kieler Meeresforschungen, III:460-496, 6 textfigs.

alkalinity (data only)

chemistry, alkalinity (data only)
Aragno, Federico, Alberto Gomez, Aldo Orlando, y Andres J. Lusquiños 1968.
Datos y resultados preliminares de las campañas pesquerias "Pesqueria I" (12 de agosto al 8 setiembre de 1966).
Publ. (Ser. Informes técn.) Mar del Plata, Argentina 10(1): 1-159.

chemistry - alkalinity (data only)
Argentina, Secretaria de Marina, Servicio de Hidrografia Naval, 1961.
Trabajos oceanograficos realizados en la campana Antartica 1959/1960. Resultados preliminares
Publico, H. 623:unnumbered pp.

chemistry - alkalinity (data only)
Argentina, Secretaria de Marina, Servicio de Hidrografia Naval, 1961.
Operacion oceanografica, Vema- Canepa 1, Resultados preliminares.
Publico H. 628:30. pp.

chemistry, alkalinity (data only)
Conseil International pour l'Exploration de la Mer, 1968.
Cooperative synoptic investigation of the Batlic 1964. 5.U.S.S.R.
ICES Oceanogr., Data Lists: 173 pp. (multilithed)

chemistry - alkalinity(data only)
Conseil Permanent International pour l'Exploration de la Mer, 1957.
Bulletin hydrographique pour l'année 1953:167 pp.

chemistry - alkalinity (data only)
Fonselius, S.H., 1967.
Hydrographical observations at Swedish lightships and in the central Baltic, 1965.
Annls. biol. Copenh. (1965)22:54-56.

chemistry - alkalinity (data only)
Granqvist, G., 1955.
The summer cruise with M/S Aranda in the northern Baltic, 1954. Merent. Julk., No. 166:56 pp.

chemistry - Alkalinity (data only)
Japan, Japan Meteorological Agency, 1970
The results of marine meteorological and oceanographical observations. (The results of the Japaneses Expedition of Deep Sea (JEDS-11); January-June 1967 41: 332 pp.

chemistry - alkalinity (data only)
Japan, Japanese Oceanographic Data Center, 1969.
U.M. Schokalsky, USSR, April 28 - June 3, 1967, Northwest of the North Pacific Ocean. Prelim. Data Rept. CSK (KDC Ref. 90K014) 121: 41 pp. (multilithed).

chemistry - alkalinity (data only)
Japan, Japanese Meteorological Agency, 1967.
The results of marine meteorological and oceanographical observations, July-December, 1964, 36: 367 pp.

chemistry - alkalinity (data only)
Japan, Japanese Oceanographic Data Center, 1968.
Ryofu Maru, Marine Division, Japan Meteorological Agency, January 11-February 24, 1967, West of the North Pacific Ocean.
Prelim. Data Rept. CSK (KDC 49K040)82:52 pp.

chemistry - alkalinity (data only)
Japan, Maritime Safety Agency 1966
Results of oceanographic observations in 1963.
Data Rept. Hydrogr. Obs. Ser. Oceanogr. Publ. 792(3): 74pp.

chemistry - alkalinity (data only)
Postma, H., 1959.
Chemical results and a survey of water masses and currents. Tables: oxygen, hydrogen ion, alkalinity and phosphate.
Snellius Exped., Eastern Part of the East Indian Archipelago, 1929-1930, Chem Res., 4:35 pp., 1 fig.

chemistry - alkalinity (data only)
Rotschi, H., Ph. Hisard, L. Lemasson, Y. Magnier, J. Noel et B. Piton, 1967.
Resultats des observations physico-chimiques de la croisiere "Alize" du N.O. Coriolis.
Rapp. scient., Off. Rech. scient. tech. Outre-Mer, Noumea, 2:56 pp. (mimeographed).

chemistry - alkalinity (data only)
Rotschi, H., Ph. Hisard, L. Lemasson, Y. Magnier, J. Noel, et B. Piton, 1966.
Resultats des observations physico-chimiques de la croisiere "Alize".
Centre ORSTROM Noumea Rapp. Sci., No. 28:56 pp. (mimeographed).

chemistry - alkalinity (data only)
Rotschi, Henri, Yves Magnier, Maryse Tirelli, et Jean Garbe, 1964.
Résultats des observations scientifiques de "La Dunkerquoise" (Croisière "Guadalcanal").
Océanographie, Cahiers O.R.S.T.R.O.M., 11(1):49-154.

chemistry - alkalinity (data only)
Sweden, Havsfiskelaboratoriet, Lysekil, Hydrografiska Avd., Göteborg, 1967.
Hydrographical data, January-June 1967, R.V. Skagerak.
Meddn. Havsfiskelab., Lysekil, Hydrogr. Avd., Göteborg, 30: numerous pp. (mimeographed).

chemistry - alkalinity (data only)
Valdez, Alberto J., Alberto Gomez, Aldo Orlando y Andres J. Lusquinos 1968
Datos y resultados de las campañas pesqueras "Pesqueria IV" (7 de junio al 4 de julio 1967).
Publ. (Ser. Informes técn.) Mar del Plata, Argentina 10(4): 1-159.

chemistry - alkalinity (data only)
Zvereva, A.A., Edit., 1959.
[Data, 2nd Marine Expedition, "Ob", 1956-1957.]
Arktich. i Antarkt. Nauchno-Issled. Inst., Mezhd. Geofiz. God, Sovetsk. Antarkt. Exped., 6: 1-387.

alkali reserves

chemistry, alkali (data) (excess)
Franco, Paolo, 1962
Condizioni fisiche e chimiche delle acque lagunari nel Porto-Canale di Malamocco. 1. Giugno 1960-Giugno 1961.
Arch. Oceanogr. e Limnol., 12(3):225-255.

alkalinity reserve (data)
Berthois, Léopold, 1965.
Remarques sur les propriétés physico-chimiques des eaux marines dans le Golfe de Gascogne, en mai 1964.
Rev. Trav. Inst. Pêches marit., 29(4):383-392.

chemistry - alkali reserves
Bogoyavlenskii, A.N., 1955.
[Chemical characteristics of the water in the region of the Kurile-Kamchatka Trench.]
Trudy Inst. Okeanol., 12:161-176.

chemistry alkalinity (specific) /salinity
Park, Kilho, 1966.
Columbia River plume identification by specific alkalinity.
Limnol. and Oceanogr., 11(1):118-120.

alkanes

chemistry - alkanes
Koons, Charles B., George W. Jamieson and Leon S. Ciereszko, 1965.
Normal alkane distributions in marine organisms; possible significance to petroleum origin.
Bull. Amer. Assoc. Petr. Geol., 49(3):201-304.

alloys

chemistry - alloys (aluminum)
Anon., 1967.
Aluminum alloys keyed for deep ocean applications: profile - Aluminum Company of America oceanographic program.
Undersea Techn., 8(4):20-21.

aluminosilicates

chemistry - aluminosilicates
Arrhenius, G., 1954.
Origin and accumulation of aluminosilicates in the ocean. Tellus 6(3):215-220, 1 textfig.

chemistry, aluminosilicates
Helgeson, Harold C., 1971.
Kinetics of mass transfer among silicates and aqueous solution.
Geochim. cosmochim. Acta 35 (5): 421-469

chemistry, aluminium
Amin, B.S., D.P. Kharkar and D. Lal, 1966.
Cosmogenic ^{10}Be and ^{26}Al in marine sediments.
Deep-Sea Res., 13(5):805-824.

chemistry, aluminum
Atkinson, Larry P., and Unnsteinn Stefansson, 1969.
Particulate aluminum and iron in sea water off the southeastern coast of the United States.
Geochim. cosmochim. Acta 33(11): 1449-1453

chemistry - aluminum
Chester, R., 1965.
Elemental geochemistry of marine sediments.
In: Chemical oceanography, J.P. Riley and G. Skirrow, editors, Academic Press, 2:23-80.

chemistry - aluminum
Haendler, H.M., and T.G. Thompson, 1939.
Determination and occurrence of aluminum in sea water. J. Mar. Res., 2:12-16.

chemistry - aluminium
Hashitani, H., and K. Yamamoto, 1959.
Simultaneous determination of traces of iron and aluminium in sea water.
J. Chem. Soc., Japan, 80(7):727-731.

Abstr. in Anal. Abstr., 7(6):#2498.

chemistry - aluminium
Hood, D.W., 1963
Chemical oceanography. In: Oceanography and Marine Biology, H. Barnes, Edit., George Allen & Unwin, 1:129-155.

chemistry - aluminium
Imai, Yoshihiko, 1961
Oceanographical studies on the behaviour of chemical elements, III. Iron and aluminium distribution in Susaki Bay with the concentration of those elements in the "Shiome" zone.
J. Oceanogr. Soc., Japan, 17(1):48-53.

chemistry - aluminium
Imai, Yoshihiko, 1961
Oceanographical studies on the behaviour of chemical elements. IV. Distribution of iron and aluminium in Uranouchi Bay - particularly on concentration of those elements in the "Shiome" zone of the tide rip.
J. Oceanogr. Soc., Japan, 17(2):96-100.

chemistry - aluminium
Imai, Y., 1960
[Oceanographical studies on the behaviour of chemical elements. 2. Iron and aluminium distribution in Urado Bay - On the concentration of those elements in the "Shiome" zone, or the rip tide.]
J. Oceanogr. Soc., Japan, 16(4): 167-171.

chemistry - aluminium
Ishibashi, M., and T. Fujinaga, 1952.
Chemical studies on the ocean. 28 Determination of aluminium in sea water. J. Chem. Soc., Pure Chem. Sect., Japan, 73:783-785.

chemistry - aluminum
Ishibashi, M., and T. Kamai, 1952.
Chemical studies in the ocean. 24. Determination of aluminum in sea water. J. Chem. Soc., Japan, 73:380-382.

Abstr. Chem. Abstr. 2554a, 1953.

chemistry - aluminum
Ishibashi, M., and K. Motojima, 1952.
Chemical studies in sea water. 25. Aluminum content of sea water. J. Chem. Soc., Japan, 73:491-493.

Abstr.: Chem. Abstr. 2554b, 1953.

chemistry - aluminium (data only)
Japan, Hokkaido University, Faculty of Fisheries, 1967.
Data record of oceanographic obsergations and exploratory fishing, 11:383 pp.

chemistry - aluminum
Joyner, Timothy, 1964.
The determination and distribution of particulate aluminum and iron in the coastal waters of the Pacific northwest.
J. Mar. Res., 22(3):259-268.

chemistry - aluminum
Okabe, Shiro, Yoshimasa Toyota and Takafumi Murakami, 1967.
Major aspects of chemical oceanography. (In Japanese; English abstract).
J. Coll. mar. Sci. Techn., Tokai Univ., (2):39-56.

chemistry - aluminum
Sackett, W., and G. Arrhenius, 1962.
Distribution of aluminium species in the hydrosphere. 1. Aluminium in the ocean.
Geochimica et Cosmochimica Acta, 26:955-968.

chemistry - aluminum
Simons, L.H., P.H. Monaghan and M.S. Taggert, 1953.
Aluminum and iron in Atlantic and Gulf of Mexico waters.
Analyt. Chem., 25:989-990.

chemistry - aluminum
Toyota, Yoshimasa, and Shiro Okabe 1967.
Vertical distribution of iron, aluminum, silicon and phosphorus in particulate matter collected in the western North Pacific, Indian and Antarctic oceans. (In Japanese; English abstract).
J. oceanogr. Soc Japan 23 (1): 1-9.

chemistry - Aluminium
Toyota, Yoshimasa, and Shiro Okabe, 1967.
Vertical distribution of iron, aluminum, silicon, and phosphorus in particulate matter collected in the western North Pacific, Indian, and Antarctic oceans. (In Japanese; English abstract)
J. oceanogr. Soc. Japan, 23 (1)
Also in: Coll. Repr. Coll. mar. Sci. Techn., Tokai Univ., 1967-68, 3: 335-343.

chemistry - aluminum
Toyota, Yoshimasa and Shiro Okabe, 1967.
Surface distribution of iron, aluminum, silicon and phosphorus in the western North Pacific and eastern Indian Oceans. (In Japanese).
J. Coll. mar. Sci. Techn., Tokai Univ., (2):227-229.

aluminum

aluminum

Atkinson, Larry P., and Unnsteinn Stefánsson, 1969.
Particulate aluminum and iron in sea water of the southeastern coast of the United States.
Geochim. cosmochim. Acta, 33: 1449-1453.

Chemistry aluminium (particulate)

Feely, R.A., W.M. Sackett and J.E. Harris, 1971
Distribution of particulate aluminum in the Gulf of Mexico. J. geophys. Res, 76(24): 5893-5902.

chemistry, aluminum

Yalkovsky, Ralph, 1967.
Signs test applied to Caribbean Core A 176-6.
Science, 155(3768):1408-1409.

aluminium-26

chemistry, aluminium-26

Singer, S. Fred, 1967.
Zodiacal dust and deep-sea sediments.
Science, 156(3778):1080-1083.

chemistry - aluminum 26

Tanaka, S., K. Sakamoto, J. Takagi and M. Tsuchimoto 1968.
Aluminum-26 and beryllium-10 in marine sediments.
Science 160 (3834): 1348-1349.

chemistry, Aluminum-26

Wasson, John T., B. Alder and H. Oeschger, 1967.
Aluminum-26 in Pacific sediments: implications.
Science, 155(3761):446-448.

aluminum, effect of

aluminum, effect of

Hanks, Robert W., 1965.
Effect of metallic aluminum particles on oysters and clams.
Chesapeake Science, 6(3):146-149.

amino acids

chemistry, amino acids

Allen, W.V. 1971.
Amino acid and fatty acid composition of the Dungeness crab (Cancer magister)
J. Fish. Res. Bd. Can. 28(8): 1191-1195

chemistry amino acids

Andrews, P. and P.J. LeB. Williams, 1971.
Heterotrophic utilization of dissolved organic compounds in the sea. III. Measurement of the oxidation rates and concentrations of glucose and amino acids in sea water. J. mar. biol. Ass., U.K., 51(1): 111-125.

chemistry, amino-acids

Bajor, M., et B.M. Van der Weide, 1967.
Effets de la diagenèse sur la distribution des amino-acides dans les sédiments.
Bull. Centre Rech. PAR-SNPA, 1(1):173-186.

amino acids

Belser, W.L., 1963
9. Bioassay of trace substances. In: The Sea, M.N. Hill, Edit., Vol. 2 (II) Fertilit of the oceans, Interscience Publishers, New York and London, 220-231.

chemistry, amino acids

*Besnier, Vincent, 1969.
Action des ultrasons sur l'extraction des acides aminés libres d'une algae planctonique.
C.r. hebd. Séanc. Acad. Sci., Paris, (1) 268(11): 1505-1507.

chemistry, amino acids

Bishop, A.D., and L.R. Louden, 1965.
Separation and identification of amino acids in Galveston and Baffin bays, Texas.
Ocean Sci. and Ocean Eng., Mar. Techn. Soc.,-Amer. Soc. Limnol. Oceanogr., 2:1104-1108.

amino acids

Boffi, V. e A. Lucarelli, 1963.
3. Biochimica oceanografica.
Rapp. Attività Sci. e Tecn., Lab. Studio della Contaminazione Radioativva del Mare, Fiascherino, La Spezia, (maggio 1959-maggio 1962), Comit. Naz. Energia Nucleare, Roma, RT/BIO (63), 8:41-56. (multilithed).

chemistry - amino acids

Bohling, H. 1970.
Untersuchungen über freie gelöste Aminosäuren in Meerwasser.
Mar. Biol. 6(3): 213-225.

chemistry, amino acids

*Chau, Y.K., L. Chuecas and J.P. Riley, 1967.
The component combined amino acids of some marine phytoplankton species.
J. mar. biol. Ass., U.K., 47(3):543-554.

chemistry, amino-acids

Chau, Y.K., and J. P. Riley, 1966.
The determination of amino-acids in sea water.
Deep-Sea Res., 13(6):1115-1124.

chemistry, amino acids

Clarke, R.H., 1967.
Amino-acids in Recent sediments of South-east Devon, England.
Nature, Lond., 213(5080):1003-1005.

~~chemistry, amino-acids~~

Connan, J., 1967.
Signification géochimique de l'extraction des amino-acides des sédiments.
Bull. Centre Rech. PAU-SNPA, 1(1):165-171.

chemistry, amino-acids

Cowey, C.B., and E.D.S. Corner, 1966.
The amino-acids composition certain unicellular algae and of the faecal pellets produced by Calanus finmarchicus when feeding on them.
In: Some contemporary studies in marine biology, H. Barnes, editor, George Allen & Unwin, Ltd., 225-231.

Chemistry - amino acids

Cowey, C.B., and E.D.S. Corner, 1963
On the nutrition and metabolism of zooplankton II. The relationship between the marine copepod Calanus helgolandicus and particulate material in Plymouth sea water, in terms of amino acid composition.
J. Mar. Biol. Assoc., 43(2):495-511.

chemistry, amino acids

Daumas, Raoul, et Hubert J. Ceccaldi, 1965.
Contribution à l'étude biochimique d'organismes marins. 1 Acides amines libres et proteiques chez Beroe ovata (Eschscholtz), Ciona intestinalis (L.), Cymbulia peroni (De Blainville) et Rhizostoma pulmo (Agassiz).
Rec. Trav. Sta. Mar. Endoume, 38(54):3-14.

Also in:
Trav. Sta. Zool., Villefranche-sur-Mer, 25.

Chemistry, amino acids

Degens, E.T., J.M. Hunt, J.H. Reuter and W.E. Reed 1964.
Data in distribution of amino acids and oxygen isotopes in petroleum brine waters of various geologic ages.
Sedimentology, Elsevier Publ. Co. 3(3): 199-225

chemicals amino acids

Degens, Egon T., Johannes H. Reuter, and Kenneth N.F. Shaw, 1964.
Biochemical compounds in offshore California sediments and sea waters.
Geochimica et Cosmochimica Acta, 28(1):45-66.

Chemistry - amino acids

*Dietrich, R., W. Höhnk und W.-D. Manzel, 1967.
Studien zur Chemie ozeanischer Bodenproben.
V. Der Nachweis von freien Aminosäuren und von Kohlenhydraten in der organischen Subtanz der Sedimente des Indischen Ozeans.
Veröff. Inst. Meeresforsch. Bremerh. 10(3):141-148.

Chemistry - amino acids

Donnelly, Patricia V., M. A. Burklew and Rose A. Overstreet 1967
Amino acids and organic nitrogen content in Florida Gulf coast waters and in artificial cultures of marine algae.
Prof. Pap. Ser. Florida Bd. Conserv. Mar. Lab. St. Petersburg, 9: 90-97.

Chemistry, amino acids

Hall, K.J., W. C. Weimer and G. Fred Lee 1970.
Amino acids in an estuarine environment
Limnol. Oceanogr. 15(1): 162-164.

chemistry - amino acids

*Hobbie, John E., Claude C. Crawford and Kenneth L. Webb, 1968.
Amino acid flux in an estuary.
Science, 159(3822):1463-1464.

chemistry, amino acids

Jackson, Togwell A., 1971.
Preferential polymerization and absorption of L-optical isomers of amino acids relative to D-optical isomers on kaolinite templates.
Chem. Geol. 7 (4): 295-306

chemistry, amino acids

Jeffries, H. Perry 1969
Seasonal composition of temperate plankton communities: free amino acid.
Limnol. Oceanogr. 14(1): 41-52.

Chemistry, amino acids

Johannes, R. E., and Kenneth L. Webb, 1965.
Release of dissolved amino acids by marine zooplankton.
Science, 150(3692):76-77.

chemistry, amino acids

Kubota, Minoru, Koichi Hagiya and Shigeru Kimura, 1971.
Amino acid composition and solubization of fish egg shell membrane. (In Japanese; English abstract). J. Tokyo Univ. Fish. 57(2-1): 87-94.

chemistry, amino acids
Lee, Byung D. and Pyung Chin 1969
The amino acids of Penaeus japonicus Bate during the developmental stages. (In Korean; English abstract).
Publ. mar. Lab. Pusan Fish. Coll. 2:51-54

chemistry - AMINO ACIDS
Litchfield, Carol D. and J.M. Prescott, 1970.
Analysis by dansylation of amino acids dissolved in marine and freshwaters. Limnol. Oceanogr., 15(2): 250-256.

Chemistry amino acids
Palmork, Karsten H., 1963.
Studies of the dissolved organic compounds in the sea.
Fisheridirekt. Skrift., Ser. Havundersøgelser, 13(6): 120-125.

chemistry, amino acids
Palmork, Karsten H., 1963.
The use of 2,4 dinitro-1-fluorobenzene in the separation and identification of amino acids from sea water.
Acta Chem. Scand., 17(5):1456-1457.

chemistry-amino acids
Park, Kilho, W.T. Williams, J.M. Prescott, and D.W. Hood, 1963.
Amino acids in Redfish Bay, Texas.
Publ. Mar. Sci., Port Aransas, 9:59-63.

chemistry, amino acids
Park, Kilho, W.T. Williams, J.M. Prescott and D.W. Hood, 1962
Amino acids in deep-sea water.
Science, 138(3539):531-532.

chemistry - amino acids
Parsons, T.R., K. Stephens and J.D.H. Strickland, 1961
On the chemical composition of eleven species of marine phytoplankters.
J. Fish. Res. Bd., Canada, 18(6):1001-1016.

chemistry - amino-acids
Pocklington, Roger 1971.
Free amino-acids dissolved in North Atlantic Ocean waters.
Nature, Lond., 230(5293): 374-375

chemistry, amino acids
Regnault, M., 1971.
Acides aminés libres chez les larves de Crangon septemspinosa (Caridea). Variation de leur taux de l'éclosion à la métamorphose. Leur rôle au cours du développement et leur importance dans la nutrition. Mar. Biol. 11(1): 35-44.

chemistry - amino acids
Riley, J.P. and D.A. Segar, 1970.
The seasonal variation of the free and combined dissolved amino acids in the Irish Sea. J. mar. biol. Ass., U.K., 50(3): 713-720.

chemistry - amino acids
Rucker, James B., 1965.
Amino acids in calcareous marine skeletons.
Canadian J. Zool., 43(2):351-356.

chemistry - amino acids
Schaefer, Heinz, 1965.
Isoberung von gelosten organischen Verbindungen aus dem Meerwasser unter besonderer. Berücksichtigung der Aminosäuren.
Helgoländer wiss Meeresunters. 12(3):239-252.

chemistry - amino acids
Siegel, Alvin, and Egon T. Degens, 1966.
Concentration of dissolved amino acids from saline waters by ligand-exchange chrmoatography.
Science, 151(3714):1098-1101.

chemistry, amino acids
Srinivasagam, R.T., J.E. Raymont, C.F. Moodie and J.K.B. Raymont, 1971.
Biochemical studies on marine zooplankton. X. The amino acid composition of Euphausia superba, Meganyctiphanes norvegica and Neomysis integer.
J. mar. biol. Ass. U.K. 51(4): 917-925.

chemistry - amino acids
Starikova, N.D. and L.I. Korzhikova, 1969.
Amino acids in the Black Sea. Okeanologiia, 9 625-636.
(In Russian; English abstract)

chemistry, amino acids
Stephens, Grover C. and Barbara B. North, 1971.
Extrusion of carbon accompanying uptake of amino acids by marine phytoplankton. Limnol. Oceanogr. 16(5): 752-757.

chemistry, amino acids
Takagi, Mitsuzo, Atsushi Iida, Hanako Murayama and Suga Soma, 1970.
Free amino acid composition of seven species of shellfish muscle (In Japanese; English abstract)
Bull. Fac. Fish. Hokkaido Univ. 21(2): 128-132

chemistry, amino acids
*Takagi, Mitsuzo, Keiichi Oishi, and Ayako Okumura, 1967.
Free amino acid composition of some species of marine algae. (In Japanese; English abstract).
Bull. Jap. Soc. scient. Fish., 33(7):669-673.

chemistry, amino acids
Tatsumoto, Mitsunobu, W.T. Williams, J.M. Prescott, and D.W. Hood, 1961
Amino acids in samples of surface sea water.
J. Mar. Res., 19(2):89-95.

chemistry, amino acids
*Wada, Koji, 1967.
Studies on the mineralization of the calcified tissue in mollusco. XIV. Modification of the amino acid pattern of proteins in the extra pallial fluid during the process of formation and mineralization of nacreous conchiolin in some bivalves.
Bull. Jap. Soc., scient. Fish. 33(7):613-617.

chemistry, amino acids
*Webb, Kenneth L., and R.E. Johannes, 1967.
Studies of the release of dissolved free amino acids by marine zooplankton.
Limnol. Oceanogr., 12(3):376-382.

chemistry, amino acids
Webb, K.L., R.E. Johannes and S.J. Coward, 1971.
Effects of salinity and starvation on release of dissolved free amino acids by Dugesia dorotocephala and Bdelloura candida (Platyhelminthes, Turbellaria). Biol. Bull. mar. biol. Lab. Woods Hole 141(2): 364-371.

amino acids
Zlobin, V.S. and M.F. Perlyuk, 1970.
Amino acids and some products of the hydrolysis of nucleic acids in the flora of the sea. (In Russian)
Poliarn nauchno-issled. Proektn. Inst. morsk. Ribn. Khoz. Okeanogr. N.M. Knipovicha (PINRO) 14: 183-202

amino acids, effect of
Jackson, Togwell A. and James L. Bischoff, 1971.
The influence of amino acids on the kinetics of the recrystallization of aragonite to calcite.
J. Geol. 79(4): 493-497

Ammonia

chemistry, ammonia
Atkins, W.R.G., 1957.
The direct estimation of ammonia in sea water, with notes on nitrate, copper, zinc and sugars.
J. du Cons. 22(3):271-277.

chemistry - ammonia
Auninsh, E.A., 1966.
Biogenic elements in the Gulf of Riga. (In Russian).
Trudy, Gosudarst. Okeanogr. Inst., No. 83:172-206.

chemistry, ammonia
*Bada, Jeffrey L., and Stanley L. Miller, 1968.
Ammonium ion concentration in the primitive ocean.
Science, 159(3813):423-425.

chemistry - ammonia
Bal, D.V., L.B. Pradhan, and K.G. Gupte, 1946
A preliminary record of some of the chemical and physical conditions in waters of the Bombay harbour during 1944-45. Proc. Indian Acad. Sci. Sect. B24(2): 60-73, 4 figs.

chemistry - ammonia
Bandel, W., 1940.
Phytoplankton- und-Nahrstoffgehalt der Ostsee im Gebiet der Darsser Schwelle. Internat. Rev. ges. Hydrobiol. u. Hydrogr., 40:249-304.

chemistry ammonia
Beers, John R., 1964.
Diurnal variation of ammonia in the Sargasso Sea off Bermuda. (Abstract).
Rept., Assoc. Island Mar. Labs., Caribbean, Fifth Meet., Lerner Mar. Lab., Bimini, Nov. 6-8, 1963: 1.

chemistry ammonia
Beers, John R., and Alison C. Kelly, 1965.
Short-term variation of ammonia in the Sargasso Sea off Bermuda.
Deep-Sea Res., 12(1):21-25.

Chemistry - ammonia
Braarud, T., and A. Klem, 1931.
Hydrographical and chemical investigations in the coastal waters off Møre and in the Romsdalfjord.
Hvalrådets Skrifter, 1931(1):88 pp., 19 figs.

Chemistry - ammonia
Bruevich, S.W., A.N. Bogoyavlensky, and V.V. Mokievskaya, 1960
[Hydrochemical features of the Okhotsk Sea.]
Trudy Inst. Okeanol., 42: 125-198.

Chemistry, ammonia
Brujewicz, S. W. and E. G. Vinogradova, 1946
[Biogenic elements in the sediment solutions of the northern, middle and southern parts of the Caspian Sea.] Comptes Rendus (Doklady) Acad. Sci. URSS, LIV (5):419-422.

Chemistry - ammonia
Buljan, M., 1951.
A modification of Teorell's method for determining small quantities of ammonia.
Arh. Hem. Farm. 23:119-122.

Abstr: J.M.B.A. 33(2):561.

Chemistry - ammonia
Buljan, M., 1951.
Note on a method for determination of ammonia in sea water. J.M.B.A. 30:277-280.

Chemistry - ammonia
*Choe, Sang, Tai Wha Chung and Hi-Sang Kwak, 1968.
Seasonal variations in nutrients and principal ions contents of the Han River water and its water characteristics. (In Korean; English abstract).
J. oceanogr. Soc., Korea, 3(1):26-38.

Chemistry - ammonia
Cooper, L.H.N., 1948
Particulate ammonia in sea water, JMBA, 27 (2):322-325.

ammonia
Corcoran, E.F., and James E. Alexander, 1963.
Nutrient, chlorophyll and primary production studies in the Florida Current.
Bull. Mar. Sci., Gulf and Caribbean, 13(4):527-541.

Chemistry - ammonia
Defant, A., G. Böhnecke, H. Wattenberg, 1936.
I. Plan und Reiseberichte die Tiefenkarte das Beobachtungsmaterial. Die Ozeanographischen Arbeiten des Vermessungsschiffes "Meteor" in der Dänemarkstrasse und Irmingersee während der Fischereischutzfahrten 1929, 1930, 1933 und 1935. Veroffentlichungen des Instituts für Meereskunde, n.f., A. Geogr.-naturwiss. Reihe, 32:1-152 pp., 7 text figs., 1 plate.

Chemistry - ammonia (data)
Dragovich, Alexander, John H. Finucane, John A. Kelly, Jr., and Billie Z. May, 1963.
Counts of red-tide organisms, Gymnodinium breve, and associated oceanographic data from Florida west coast, 1960-61.
U.S. Fish and Wildlife Service, Spec. Sci. Repts. Fish., No. 455:1-40.

Chemistry - ammonia
Fedosov, M.V., and E.G. Vinogradova, 1955.
[Hydrological and hydrochemical regime, primary production in the Azov Sea and forecasting change. Fundamental basis of the hydrochemical regime in the sea of Azov.] Trudy VNIRO, 31(1):9-34.

Chemistry, ammonia
Fonselius S.H. 1968.
Observations at Swedish lightships and in the central Baltic 1966.
Annls biol. Copenh. 1966, 23: 74-78.

Chemistry - ammonia
Føyn, E., 1950.
Ammonia determination in sea water. J. du Cons. 16(2):175-178, 3 textfigs.

Chemistry - ammonia
Frega, F., 1967.
Hidrografia de la ria de Vigo, 1962, con especial referencia a los compuestos nitrogens.
Investigacion pesq., 31(1):145-149.

Chemistry - Ammonia
Fraga, F., 1960.
Variación estacional de la materia orgánica suspendida y di disuelta en la Ria de Vigo.
Inv. Pesq., Barcelona, 17:127-140.

Chemistry, ammonia
Fraga, F. y A. Ballester 1966.
Distribución vertical del nitrogeno organico e inorganico en la fosse de Cariaco y su relación con el fosforo total.
Memoria Soc. Cienc. nat. La Salle, Venezuela 26 (75): 274-282.

Chemistry - ammonia
Goering, J.J., R.C. Dugdale and D.W. Menzel, 1964
Cyclic diurnal variations in the uptake of ammonia and nitrate by photosynthetic organisms in the Sargasso Sea.
Limnology and Oceanography, 9(3):448-451.

Chemistry, ammonia
Guarrera, S.A., 1950.
Estudios hidrobiologicos en el Rio de la Plata.
Rev. Inst. Nac., Invest. Cienc. Nat., Ciencias Botanicas 2(1):62 pp.

Chemistry - ammonia
Hatzikakidis, A., 1950.
[Chemical and microbiological study of marine waters.] ΠΡΑΚΤΙΚΑ ΕΛΛΗΝΙΚΟΥ ΥΔΡΟΒΙΟΛΟΓΙΚΟΥ ΙΝΣΤΙΤΟΥΤΟΥ
4(1):103-120.

Ammonia
Hollan, Eckard, 1970.
Eine physikalische Analyse kleinräumiger Änderungen chemischer Parameter in den tiefen Wasserschichten der Gotlandsee.
Kieler Meeresforsch. 25(2): 255-267

Chemistry - ammonia
Holm-Hansen, Osmund, J.D.H. Strickland and P.M. Williams, 1966.
A detailed analysis of biologically important substances in a profile off southern California.
Limnol. Oceanogr., 11(4):548-561.

Chemistry, ammonia
Hung Tsu-Chang 1970
Photo- and radiation effects on inorganic nitrogen compounds in pure water and sea water near Taiwan in the Kuroshio Current.
Bull. Inst. Chem. Acad. Sinica 18: 37-56

Chemistry - ammonia (data)
Japan, Central Meteorological Observatory, 1951.
Table 6. Oceanographical observations taken in the Akashi-seto, the Yura-seto and the Kii Suido; physical and chemical data for stations occupied by R.M.S. "Shumpu-maru".
Res. Mar. Met. Ocean. Obs., Jan.-June 1949, No. 5:40-47.

Chemistry, ammonia
Japan, Kobe Marine Observatory 1967.
Report of the oceanographic observations in the sea south of Honshu and in the Seto-Naikai from May to June 1964. (In Japanese).
Bull. Kobe Mar. Obs. 176: 37-

Chemistry - ammonia
Japan, Kobe Marine Observatory, Oceanographical Section, 1964.?
Researches in chemical oceanography made at K.M.O. for the recent ten years (1954-1963). (In Japanese).
Weather Service Bull., Japan. Meteorol. Agency, 31(7):188-191.

Also in:
Bull. Kobe Mar. Obs., No. 175. 1965.

Chemistry ammonia
Toussot-Dubrin, Jacques and Abdelmajid Kedir, 1970
Photosensitized oxidation of ammonia by singlet oxygen in aqueous solution and in seawater.
Nature, Lond, 227(5259): 700-701.

Chemistry - ammonia
Kändler, Rudolf, 1963
Hydrographische Untersuchungen über die Abwasserbelastung der Flensburger Förde.
Kieler Meeresforsch., 19(2):142-157.

Chemistry - ammonia (data)
Kato, Kenji, 1966.
Geochemical studies on the mangrove region of Cananéia, Brazil. 1. Tidal variations of water properties.
Bolm Inst. Oceanogr., S. Paulo, 15(1):13-20.

Chemistry, ammonia
Kato, Kenji, 1961.
Oceanographical studies in the Gulf of Cariaco.
1. Chemical and hydrographical observations in January, 1961.
Bol. Inst. Oceanograf., Univ. Oriente, Cumana, Venezuela, 1(1):49-72.

Chemistry, ammonia
Konnov, V.A., 1965.
Determination of ammonia in sea water. Methods of marine hydrochemical investigation. (In Russian).
Trudy, Inst. Okeanol., Akad. Nauk, SSSR, 79:11-13.

ammonia
Kriss, A.E., 1949.
The role of microorganisms in the accumulation of hydrogen sulphide, ammonia and nitrogen in the depth of the Black Sea. Priroda 1949(6):35-46.

Chemistry, ammonia
Krogh, A., 1934
A method for the determination of ammonia in water and air. Biol. Bull. 67(1): 126-131, 1 text fig.

chemistry, ammonia
Lagarde, E., 1963.
Metabolisme de l'azote minéral en milieu marin.
Vie et Milieu, 14(1):37-54.

chemistry, ammonia
Lagarde, E., and P. Forget, 1961.
Contribution à l'étude du metabolisme de l'azote mineral en milieu marin. Étude de souche pure de bacteries dénitrificatrices et réductrices des nitrates.
Rapp. Proc. Verb., Réunions, Comm. Int. Expl. Sci. Mer Méditerranée, Monaco, 16(2):245-250.

chemistry ammonia
Lisitzin, A.P., 1966.
Processes of Recent sedimentation in the Bering Sea. (In Russian).
Inst. Okeanol., Kom. Osad. Otdel Nauk o Zemle, Isdatel, Nauka, Moskva, 574 pp.

chemistry ammonia
MacIsaac, Jane J., 1967.
Ammonia determinations by two methods in the northeast equatorial Pacific Ocean.
Limnol. Oceanogr., 12(3):552-554.

chemistry - ammonia
MacIsaac, J.J. and R.K. Olund, 1971.
An automated extraction procedure for the determination of ammonia in seawater. Investigación pesq. 35(1): 221-232.

chemistry, ammonia (data)
Marumo, Ryuzo, Editor, 1970.
Preliminary report of the Hakuho Maru Cruise KH-69-4, August 12-November 13, 1969, The North and Equatorial Pacific Ocean.
Ocean Res. Inst. Univ. Tokyo, 68pp.

chemistry - ammonia
Matsue, Y., 1950.
[The variation of titratable base in sea water by the growth of diatoms.] J. Ocean. Soc., Tokyo, 6(1):32-38, 2 textfigs. (In Japanese with English abstract).

chemistry - ammonia
Menzel, David W., and Jane P. Spaeth, 1962
Occurrence of ammonia in Sargasso Sea waters and in rain water.
Limnology and Oceanography, 7(2):159-162.

chemistry, ammonia
Morii, Hideaki, Yusho Akishige, Ryoichi Kanazu and Tadanobu Fukuhara 1967.
Chemical studies of the Kuroshio and adjacent region. I. On the nutrient salts and the dissolved oxygen. (In Japanese; English abstract).
Bull. Fac. Fish., Nagasaki Univ. 22:91-103

chemistry - ammonia
#Newell, B.S., 1967.
The determination of ammonia in sea water.
J. mar. biol. Ass., U.K., 42(2):271-280.
7

chemistry - ammonia
Nümann, W., 1954.
Der Nährstoffhaushalt in der/Adria. nordöstlichen
Thalassia 5(2):1-68, 12 textfigs.

chemistry, ammonia
Ohwada, Mamoru, and Katsumi Yamamoto 1966.
Some chemical elements in the Japan Sea.
Oceanogr. Mag. Japan. Met. Soc. 18(1/2): 31-37

chemistry, ammonia
#Okuda, Taizo, 1967.
Vertical distribution of inorganic nitrogen in the equatorial Atlantic Ocean.
Boln Inst. Oceanogr., Univ. Oriente, 5(½):67-83.

chemistry, ammonia
Okuda, Taizo, 1962
Physical and chemical oceanography over continental shelf between Cabo Frio and Vitoria (Central Brazil).
J. Oceanogr. Soc., Japan, 20th Ann. Vol., 514-540.

chemistry, ammonia
Okuda, Taizo, 1960
Chemical oceanography in the South Atlantic Ocean, adjacent to north-eastern Brazil. (Portuguese and French resumés).
Trabalhos, Inst. Biol. Marit. e Oceanogr., Universidade do Recife, Brasil, 2(1):155-174.

chemistry, ammonia
Okuda, Taizo, 1960
Metabolic circulation of phosphorus and nitrogen in Matsushima Bay (Japan) with special reference to exchange of these elements between sea water and sediments. (Portuguese, French resumé)
Trabalhos, Inst. Biol. Maritima e Oceanogr., Universidade do Recife, Brasil, 2(1):7-153.
Resumé pp. 147-149

chemistry - Ammonia
Okuda, Taizo, and Lourinaldo Barreto Cavalcanti, 1963.
Algumas condicoes oceanograficas na area nordeste de Natal. (Septembro 1960).
Trab. Inst. Oceanogr., Univ. do Recife, 3(1): 3-25.

chemistry, ammonia
Okuda, Taizo, Angel José Garcia, José Benetoz Alvarez, 1965.
Variacion estacional de las elementos nutritions en el agua de la laguna y el rio Unare.
Bol. Inst. Oceanogr., Univ. Oriente, 4(1): 123-135.

chemistry - ammonia (data)
Platt, Trevor, and Brian Irwin, 1968.
Primary productivity measurements in St. Margaret's Bay, 1967.
Techn. Rept., Fish. Res. Bd., Can., 77: 123 pp.

chemistry, ammonia (data)
Postma, H., 1966.
The cycle of nitrogen in the Wadden Sea and adjacent areas.
Neth. J. Sea Res., 3(2):186-221.

chemistry ammonia
Prochazkova, L., 1960.
Einfluss der Nitrate und Nitrite auf die Bestimmung des organischen Stickstoff und Ammoniums im Wasser.
Arch. f. Hydrobiol., 56(3):179-185.

chemistry ammonia
Rakestraw, N. W., and A. Hollaender, 1936.
Photochemical oxidation of ammonia in sea water.
Science, 84(2185):442-443.

chemistry - ammonia
Redfield, A.C., and A.B. Keys, 1938.
The distribution of ammonia in the waters of the Gulf of Maine. Biol. Bull. 74(1):83-92, 6 textfigs.

chemistry, ammonia
Rheinheimer, Gerhard 1967.
Ökologische Untersuchungen zur nitrifikation in Nord- und Ostsee.
Helgoländer wiss. Meeresunters. 15(1/4): 243-252.

chemistry, ammonia
Richards, Francis A., Joel D. Cline, William W. Broenkow and Larry P. Atkinson, 1965.
Some consequences of the decomposition of organic matter in Lake Nitinat, an anoxic fjord.
Limnol. and Oceanogr., Redfield Vol., Suppl. to, 10:R185-R201.

chemistry, ammonia
Richards, Francis A., and Richard A. Kletsch, 1964.
The spectrophotometric determination of ammonia and labile amino compounds in fresh and seawater by oxidation to nitrite.
In: Recent researches in the fields of hydrosphere, atmosphere and nuclear geochemistry, Ken Sugawara festival volume, Y. Miyake and T. Koyama, editors, Maruzen Co., Ltd., Tokyo, 65-81.

chemistry, ammonia
Riley, J.P., 1953.
The spectrophotometric determination of ammonia in natural waters with particuler reference to sea water.
Analyt. Chim. Acta, 9(6):575-589.

chemistry, AMMONIA
Robertson, A.J., 1907.
On hydrographical investigations in the Faroe-Shetland Channel and the northern part of the North Sea during the years 1904-1905. Fish. Bd., Scotland, North Sea Fish. Invest. Comm., Northern Area, Rept. No. 2:1-140, 13 textfigs., 15 pls.

chemistry, ammonia
Robinson, R.J., and H.E. Wirth, 1934.
Free ammonia, albuminoid nitrogen and organic nitrogen in the waters of the Pacific Ocean off the coasts of Washington and Vancouver Island.
J. du Cons., 9(2):187-195.

chemistry, ammonia
Robinson, R.J., and H.E. Wirth, 1934.
Report on the free ammonia, albuminoid nitrogen and organic nitrogen in the waters of the Puget Sound during the summers of 1931 and 1932.
J. du Cons., 9(1):15-27.

chemistry ammonia
Ryther, John H., D.W. Menzel and Nathaniel Corwin, 1967.
Influence of the Amazon River outflow on the ecology of the western tropical Atlantic. I. Hydrography and nutrient chemistry.
J. mar. Res., 25(1):69-83.

chemistry, ammonia
Sagi, T., 1969.
The ammonia content in sea water in the western North Pacific Ocean. Oceanogrl Mag., 21(2): 113-119.

chemistry ammonia

Scaccini Cicatelli, Marta, 1968.
Un anno di osservazioni idrologiche in una stazione fissa nelle acque dell'Adriatico occidentale presso Fano.
Note Lab. Biol. mar. Pesca, Fano, Univ. Bologna 2(9): 181-228.

Chemistry, ammonia (data)

Scaccini Cicatelli, Marta 1967.
Distribuzione stagionali dei sali nutritivi in una zona dell' alto e medio Adriatico.
Bull. Pesca, Piscic. Idrobiol. n.s. 12(1): 49-82.

Chemistry ammonia

Schachter, D., 1954.
Contribution à l'étude hydrographique et hydrologique de l'étang de Berre (Bouches-du-Rhone).
Bull. Inst. Océan., Monaco, No. 1048:20 pp.

chemistry Ammonia

Schott, Friedrich, and Manfred Ehrhardt, 1970.
On fluctuations and mean relations of chemical parameters in the northwestern North Sea.
Kieler Meeresforsch. 25(2): 272-278

Chemistry, ammonia

Seelkopf, Carl, and Luis Boscan F., 1960

Hydrochemische Untersuchungen im Maracaibo-See.
Deutsche Hydrogr. Zeits., 13(4): 174-180.

Chemistry, ammonia

Sen Gupta, R. 1968
Inorganic nitrogen compounds in ocean stagnation and resupply.
Science 160 (3830): 884-885.

Chemistry, ammonia

Siyazuki, K., 1951.
[Studies on the foul-water drained from factories. I. On the influence of foul-waters drained from factories by the coast on the water of Mitaziri Bay.] Contr. Simonoseki Coll. Fish. No. 1:155-158, 4 textfigs. (English summary).

Chemistry, ammonia

Smayda, Theodore J., 1966.
A quantitative analysis of the phytoplankton of the Gulf of Panama. III General ecological conditions and the phytoplankton dynamics at 8o 45'N, 79o 23'W from November 1954 to May 1957.
Inter-Amer. Trop. Tuna Comm., Bull., 11(5): 355-612.

chemistry ammonia

Solov'eva, N.F., 1959.
[Hydrochemical analyses of the Aral Sea in 1948.]
Akad. Nauk, SSSR, Lab. Ozerovedeniya, Trudy, 8:
5 Sept. 61 JPRS:10054
OTS-SLA $3.60 61-28414

chemistry, ammonia (data)

Soot-Ryen, T., 1947.
Hydrographical investigations in the Tromsø-district 1934-1938 (Tables).
Tromsø Mus. Aarsheft., Naturhist. Avd. No. 33, 66(1943)(1): numerous pp. (unnumbered).

Chemistry, ammonia (data)

Soot-Ryen, T., 1938.
Hydrographical investigations in the Tromsø district, in 1931.
Tromsø Mus. Aarsheft., Naturhist. Avd. No. 10, 54(2):1-6, 41 pp. of tables.

chemistry, ammonia

Spencer, C.P., 1957.
The oxidation of ammonia in the sea (abstract).
Ann. Rept. Challenger Soc., 3(9):21.

Chemistry, ammonia

Spencer, C.P., 1956.
The bacterial oxidation of ammonia in the sea.
J.M.B.A., 35(3):621-630.

Chemistry ammonia

Steemann-Nielsen, Einar, 1951
The marine vegetation of the Isefjord. A study on ecology and production, Medd. Komm. Danmarks Fiskeri-og Havundersøgelser. Ser. Plankton. 5(4); 114pp., 46 text figs.

Chemistry ammonia

Strickland, J.D.H., and K.H. Austin, 1959
The direct estimation of ammonia in sea water with notes on "reactive" iron, nitrate and inorganic phosphorus.
J. du Conseil, 24(3):446-451.

chemistry, ammonia

Sugiyama, Teruyuki, Yoshio Miyake and Kuniyasu Fujisawa, 1971.
On the distribution of inorganic nutrients along the coastal waters of Okayama Prefecture (1970).
Bull. Fish. Exp. Sta. Okayama Pref. (1970): 22-28. (In Japanese)

Chemistry, ammonia

Thomas, William H., 1966.
On denitrification in the northeastern tropical Pacific Ocean.
Deep-Sea Res., 13(6):1109-1114.

chemistry ammonia

Uyeno, Fukuzo, 1964.
Relationships between production of foods and oceanographical condition of sea water in pearl farms. II. On the seasonal changes of sea water constituents and of bottom condition, and the effect of bottom cultivation. (In Japanese; English abstract)
J. Fac. Fish., Pref. Univ. Mie, 6(2):145-169.

ammonia

Vaccaro, Ralph F., 1965.
inorganic nitrogen in sea water.
In: Chemical Oceanography, J.P. Riley and G. Skirrow, editors, Academic Press, 1:365-408.

chemistry ammonia

Vaccaro, Ralph F., 1963.
Available nitrogen and phosphorus and the biochemical cycle in the Atlantic off New England.
J. Mar. Res., 21(3):284-301.

chemistry, ammonia

Vaccaro, R. F., 1962.
The oxidation of ammonia in sea water.
Journal du Conseil, 27(1):3-14.

chemistry, ammonia

Vatova, A., 1955.
Il dossaggio dell'azoto ammoniacale nell'acque di mare con l'elettrofotometro Elko II.
Nova Thalassia 2(4):1-22.

chemistry, ammonia

Wattenberg, H., 1937.
Medthoden zur Bestimmung von Phosphat, Silicat, Nitrat, und Ammoniak im Seewasser. Rapp. Proc. Verb. 103(1):1-26.

Chemistry, ammonia

Watts, J.C.D., 1958.
The hydrology of a tropical West African estuary.
Bull. Inst. Francais, Afrique Noire, 20(3):697-752.

chemistry ammonia

Won, Chong Hun, 1964.
Tidal variations of chemical constituents of the estuary water at the Luva bed in the Nack-Dong River from Nov. 1962 to Oct. 1963. (In Korean: English abstract).
Bull. Pusan Fish. Coll., 6(1):21-32.

chemistry, ammonia

Won, Chong Hun, 1963.
Distribution of chemical constituents of the estuary water in Gwang-yang Inlet. (In Korean; English abstract).
Bull. Fish. Coll., Pusan Nat. Univ., 5(1):1-10.

chemistry ammonia concentration, effect of

Thomas, William H., 1970.
Effect of ammonium and nitrate concentration on chlorophyll increases in natural tropical Pacific phytoplankton populations. Limnol. Oceanogr., 15(3): 386-394.

ammonia (data only)

chemistry, ammonia (data only)

Australia, Commonwealth Scientific and Industrial Research Organization, 1964.
Oceanographical observations in the Indian Ocean in 1961, H.M.A.S. Diamantina, Cruise Dm 3/61.
Div. Fish. and Oceanogr., Oceanogr. Cruise Rept., No. 11:215 pp.

Chemistry - ammonia, (data only)

Brasil, Marinha do Brasil, Diretoria de Hidrografia e Navegação, 1963
Operação "TRIDENTE I", Estudo das condições oceanográficas entre o Rio de Janeiro e o Rio da Prata, durante o inverno (Agosto-Septembro), ano de 1962.
DG-06-XV:unnumbered pp. (mimeographed).

chemistry, ammonia (data only)

Brasil, Diretoria de Hidrografia e Navegacao, 1961
Estudo das condições oceanográficas nas proximidades do Rio de Janeiro durante o mês de Dezembro.
DG-06-XIII:mimeographed sheets.

chemistry, ammonia (data only)

Brasil, Marinha do Brasil, Diretoria Hidrografia e Navegacao, 1960.
Estudo das condicoes oceanograficas na regiao profunda a Nornordeste de Natal, Estado do Rio Grande do Norte.
DG-06-XI(Sept. 1960):unnumbered pp. (mimeographed).

chemistry ammonia (data only)

Brasil, Marinha do Brasil, Diretoria de Hidrografia e Navegação, 1960.
Estudo das condições oceanográficas sobre a plataforma continental, entre Cabo-Frio e Vitoria, durante o outono (abril-maio).
DG-06-X(junho):unnumbered pp. (mimeographed).

chemistry, ammonia (data only)

Brasil, Marinha do Brasil, Diretoria de Hidrografia e Navegacao, 1959.
Levantamento oceanografico da costa nordeste.
DG-06-IX:unnumbered pp. (mimeographed).

chemistry - ammonia (data only)

Conseil International pour l'Exploration de la Joint Skagerak Expedition 1966. 1. Oceanographic stations, temperature-salinity-oxygen content: 2. Oceanographic stations, chemical observations. ICES Oceanogr. Data lists: 250 pp; 209 pp.

chemistry - ammonia (data only)

Hela, Ilmo, and Folke Koroleff, 1958
Hydrographical and chemical data collected in 1957 on board the R/V Aranda in the Barents Sea.
Merent. Julk., No. 179: 67 pp.

chemistry, ammonia (data only)

Japan, Hokkaido University, Faculty of Fisheries 1970.
Data record of oceanographic observations and exploratory fishing 13:406pp.

chemistry ammonia (data only)

Japan, Hokkaido University, Faculty of Fisheries, 1968.
Data record of oceanographic observations and exploratory fishing. No. 12:420 pp.

chemistry - ammonia (data only)

Japan, Hokkaido University, Faculty of Fisheries, 1968.
The Oshoro Maru cruise 23 to the east of Cape Erimo, Hokkaido, April 1967.
Data Record Oceanogr. Obs. Expl. Fish., 12: 115-169.

chemistry - ammonia (data only)

Japan, Hokkaido University, Faculty of Fisheries, 1968.
The Oshoro Maru cruise 21 to the Southern Sea of Japan, January 1967.
Data Record Oceanogr. Obs. Explor. Fish. 12: 1-97; 113-119.

chemistry - ammonia (data only)

Japan, Hokkaido University, Faculty of Fisheries, 1967.
The Oshoro Maru cruise 16 to the Great Australian Bight November 1965-February 1966.
Data Record Oceanogr. Obs. Explor. Fish., Fac. Fish., Hokkaido Univ. 11: 1-97; 113-119.

chemistry ammonia (data only)

Japan, Hokkaido University, Faculty of Fisheries, 1967.
The Oshoro Maru cruise 18 to the east of Cape Erimo, Hokkaido, April 1966.
Data Record Oceanogr. Obs. Explor. Fish. 11: 121-164.

chemistry ammonia (data only)

Japan, Japan Meteorological Agency, 1971.
The results of marine meteorological and oceanographical observations, 45: 338 pp.

chemistry, ammonia (data only)

Japan, Japan Meteorological Agency, 1959.
The results of marine meteorological and oceanographical observations, July-December, 1958, No. 24:289 pp.

chemistry, ammonia (data only)

Japan, Japanese Oceanographic Data Center, 1967.
Ryofu Maru, Marine Division, Japan Meteorological Agency, Japan September 13-17, 1966, eastern Sea of Japan.
Prelim. Data Rept. CSK (KDC Ref. No. 49K033) 59:11 pp. (multilithed)

chemistry, ammonia (data only)

Platt, Trevor and Brian Irwin 1971.
Phytoplankton production and nutrients in Bedford Basin, 1969-70.
Techn. Rept. Fish. Res. Bd. Can. 247:172 pp. (multilithed).

chemistry - ammonia (data only)

\# Scaccini Cicatelli, Marta, 1967.
Distribuzione stagionale dei sali nutrivi in una zone dell'alto e medio Adriatico.
Boll. Pesca, Piscic. Idrobiol., n.s.,22(1): 49-82.

chemistry ammonia (data only)

Sweden, Havsfiskelaboratoriet, Lysekil 1971.
Hydrographical data January-June 1970, R.V. Skagerak, R.V. Thetis and TV252, 1970.
Meddn. Hydrogr. avd. Göteborg, 104: unnumbered pp. (multilithed).

chemistry ammonia (data only)

Sweden, Havsfiskelaboratoriet, Lysekil 1970.
Hydrographical data 1966, R.V. Skagerak, R.V. Thetis.
Meddn Hydrogr. avd. Göteborg 85:255pp.

chemistry ammonia (data only)

Sweden, Havsfiskelaboratoriet, Lysekil 1970.
Hydrographical data, R.V. Skagerak, R.V. Thetis.
Hydrogr. avd. Göteborg
Meddn. 84: 296 pp. (multilithed).

chemistry, ammonia (effect of)

Zgurovskaya, L.N., and N.G. Kustenko 1968
The action of ammonia nitrogen on cell division, photosynthesis and the accumulation of pigments in Skeletonema costatum (Grev.), Chaetoceros sp. and Prorocentrum micans. (In Russian)(English abstract).
Okeanologiia 8(1):116-125.

anions

chemistry, anions

Nehring, Dietwart, und Karl-Heinz Rohde 1967.
Weitere Untersuchungen über anormale Ionenverhältnisse in der Ostsee.
Beitr. Meereskunde 20:10-33.

chemistry, anions

Riley, J.P., 1965.
Analytical chemistry of sea water.
In: Chemical oceanography, J.P. Riley and G. Skirrow, editors, Academic Press, 2:295-424.

chemistry, anion content

Nicholls, G.D., 1965.
The geochemical history of the oceans.
In: Chemical oceanography, J.P. Riley and G. Skirrow, editors, Academic Press, 2:277-294.

anoxia

chemistry anoxia

Hood, Donald W., editor 1971
Impingement of man on the oceans.
Wiley-Interscience, 738 pp.

chemistry, anoxic waters

Knull, James R. and Francis A. Richards, 1969.
A note on the sources of excess alkalinity in anoxic waters. Deep-Sea Res., 16(2): 205-212.

chemistry, anoxic environments

Richards, Francis A., 1971.
Anoxic versus oxic environments. (?) In: Impingement of man on the oceans, D.W. Hood, editor, Wiley Interscience: 201-217.

chemistry, anoxia

Richards, F.A., 1965.
Anoxic basins and fjords.
In: Chemical oceanography, J.P. Riley and G. Skirrow, editors, Academic Press, 1:611-645.

chemistry, antibiotics

See: antibiotics

antimony

chemistry, antimony

Schutz, D.F., and K.K. Turekian, 1964.
The distribution of selenium, antimony, silver, cobalt and nickel in sea water. (Abstract).
Trans. Amer. Geophys. Union, 45(1):118.

chemistry antimycin

Finucane, John H. 1969
Antimycin as a toxicant in a marine habitat.
Trans. Am. Fish. Soc., 98(2): 288-292

chemistry - apparent oxygen utilization

Pytkowicz, Ricardo M., 1971.
On the apparent oxygen utilization and the preformed phosphate in the oceans. Limnol. Oceanogr. 16(1): 39-42.

arabinose (data)

Chemistry - arabinose

Anderson, William W., and Jack W. Gehringer 1959.
Physical oceanography, biological and chemical data, South Atlantic coast of the United States. M/V. Theodore N. Gill Cruise 7.
USFWS Spec. scient. Rept. Fish. 278: 277 pp.

chemistry- arabinose (data only)

Anderson, William W., and Jack W. Gehringer, 1959.
Physical oceanographic, biological and chemical, South Atlantic coast of the United States, M/V Theodore N. Gill cruise 9.
USFWS. Spec. Sci. Rept., Fish. No. 313:226 pp.

chemistry- arabinose (data only)
Anderson, W.W., and J.W. Gehringer, 1959.
Physical oceanographic, biological and chemical data South Atlantic coast of the United States, M/V Theodore N. Gill Cruise 8.
USFWS Spec. Sci. Rept., Fish., No. 303:227 pp.

chemistry- arabinose (data only)
Anderson, William W., and Jack W. Gehringer, 1958.
Physical oceanographic, biological and chemical data, south Atlantic coast of the United States, M/V Theodore N. Gill Cruise 6.
USFWS Spec. Sci. Rept., Fish., No. 265:99 pp.

chemistry - arabinose (data only)
Anderson, W.W., and J.W. Gehringer, 1958.
Physical oceanographic, biological and chemical data, South Atlantic coast of the United States, M/V Theodore N. Gill Cruise 5.
USFWS Spec Sci. Rept., Fish., 248:220 pp.

chemistry- arabinose (data only)
Anderson, W.W., and J.W. Gehringer, 1957.
Physical oceanographic, biological and chemical data, South Atlantic coast of the United States, Theodore N. Gill Cruise.
USFWS Spec. Sci. Rept., Fish., No. 210:208 pp.

chemistry - arabinose (data only)
Anderson, W.W., and J.W. Gehringer, 1957.
Physical oceanographic, biological and chemical data, South Atlantic coast of the United States, M/V Theodore N. Gill cruise 4.
USFWS Spec. Sci. Rept., Fish., 234:192 pp.

chemistry - arabinose (data only)
Anderson, W.W., J.W. Gehringer and E. Cohen, 1956
Physical oceanographic, biological and chemical data, South Atlantic coast of the United States, Theodore N. Gill, Cruise 2.
Spec. Sci. Rept., Fish. No. 198:270 pp.

chemistry, arabinose (data)
Anderson, W.W., J.W. Gehringer and E. Cohen, 1956
Physical oceanographic, biological and chemical data, south Atlantic coast of the United States, M/V Theodore N. Gill, Cruise 1.
USFWS Spec. Sci. Rept., Fish., No. 178:160 pp.

chemistry arabinose
Dragovich, A., 1963.
Hydrology and plankton of coastal waters at Naples, Florida.
Q.J. Florida Acad. Sci., 26(1):22-47.

argon

chemistry - argon

Benson, Bruce B., and Peter D.M. Parker, 1961

Nitrogen/argon and nitrogen isotope ratios in aerobic sea water.
Deep-Sea Res., 7(4):237-253.

chemistry, argon

Bieri, Rudolf H. 1971.
Dissolved noble gases in marine waters.
Earth Planet. Sci. Letts, 10(3): 329-333

chemistry - argon (data)

Bieri, Rudolf H., Minoru Koide and Edward D. Goldberg 1968.
Noble gas contents of marine waters.
Earth Planet. Sci. Letts 4(5): 329-340.

chemistry, argon
Bieri, Rudolf H., Mineru Koide and Edward D. Goldberg, 1966.
The noble gas contents of Pacific seawaters.
J. geophys. Res., 71(22):5243-5265.

chemistry - argon
Bieri, Rudolph, Minoru Koide and Edward D. Goldberg, 1964
Noble gases in sea water.
Science, 146(3647):1035-1037.

chemistry - argon
*Craig, H., and R.F. Weiss, 1968.
Argon concentrations in the ocean: a discussion.
Earth Planet. Sci. Letters, 5(3):175-183.

argon
*Craig, H., R.F. Weiss and W.B. Clarke, 1967.
Dissolved gases in the equatorial and South Pacific Ocean.
J. geophys. Res., 72(24):6165-6181.

chemistry, argon-40
Dalrymple, G. Brent, and James G. Moore 1968.
Argon-40: excess in submarine pillow basalts from Kilauea Volcano, Hawaii.
Science 161 (3846): 1132-1135.

chemistry - argon
Dymond, Jack, 1970.
Excess argon in submarine basalt pillows.
Bull. geol. Soc. Am., 81(4): 1229-1232.

chemistry, argon
Fisher, David E., 1971.
Incorporation of Ar in East Pacific basalts.
Earth Planet. Sci. Letts. 12 (3): 321-324.

chemistry - argon
Klots, C.E, and B.B. Benson, 1963
Solubilities of nitrogen, oxygen and argon in distilled water.
J. Mar. Res., 21(1):48-57.

chemistry, argon
König, H., H. Wänke, G.S. Bien, N.W. Rakestraw and H.E. Suess, 1964.
Helium, neon and argon in the oceans.
Deep-Sea Research, 11(2):243-247.

chemistry - argon
Koyama, Tadashiro, 1958.
A new method of direct determination of argon and nitrogen.
J. Earth Sci., Nagoya Univ., 6(1):1-11.

Also in:
Collected Papers on Sciences of Atmosphere and Hydrosphere, 1958-1963, 1 (1964).

chemistry - argon
Mazor, E., G.J. Wasserburg, and H. Craig, 1964.
Rare gases in Pacific Ocean water.
Deep-Sea Res., 11(6):929-932.

chemistry - argon

Murray, C.N. and J.P. Riley, 1970.
The solubility of gases in distilled water and sea water - III. Argon. Deep-Sea Res., 17(1): 203-209.

chemistry - argon
Oana, S., 1957.
Bestimmung des Argons im besonderen Hinblick auf gelöste Gase in natürlichen Wassern.
J. Earth Sci., Nagoya Univ., 5(2):103-124.

chemistry, argon
Rakestraw, N. W., and V.M. Emmel, 1938.
The solubility of nitrogen and argon in sea water
J. Phys. Chem. 42(9):1211-1215, 2 textfigs.

chemistry - argon
Richards, F. A., 1965.
Dissolved gases other than carbon dioxide.
In: Chemical oceanography, J.P. Riley and G. Skirrow, editors, Academic Press, 1:197-225.

chemistry - argon

Richards, Francis A., and Bruce B. Benson, 1961

Nitrogen/argon and nitrogen isotope ratios in two anaerobic environments, the Cariaco Trench in the Caribbean Sea and Dramsfjord, Norway.
Deep-Sea Res., 7(4): 254-264.

chemistry, argon

Weiss, R.F., 1971.
The effect of salinity on the solubility of argon in seawater. Deep-Sea Res., 18(2): 225-230.

chemistry - Argon

Weiss, R.F., 1970.
The solubility of nitrogen, oxygen and argon in water and seawater. Deep-Sea Res., 17(4): 721-735.

chemistry, Arochlor®

Nimmo, D.R., R.R. Blackman, A.J. Wilson, Jr. and J. Forester, 1971.
Toxicity and distribution of Aroclor® 1254 in the pink shrimp Penaeus duorarum. Mar. Biol. 11(3): 191-197.

Arsenates

Chemistry- arsenates
Denigès, G., 1920.
Réaction de coloration extrêmement sensibles des phosphates et des arseniates.
C.R. Acad. Sci., Paris, 171:802-804.

Arsenic

chemistry - Arsenic
Atkins, W.R.G., and E.G. Wilson, 1927
The phosphorus and Arsenic compounds of sea-water. J.M.B.A. 14: 609-614.

chemistry, arsenic
Boström, Kurt and Sylvia Valdes 1969
Arsenic in ocean floors.
Lithos, 2 (4): 351-360

chemistry - astacin carotenoids (data only)
Saloman, Carl H., and John L. Taylor 1968.
Hydrographic observations in Tampa Bay, Florida, and the adjacent Gulf of Mexico, 1965-1966.
Data Rept. U.S. Fish Wildl. Serv., Bur. Comm. Fish. 6 cards (microfiche).

chemistry of atmosphere
Barger, W.R. and W.D. Garrett, 1970.
Surface active organic material in the marine atmosphere. J. geophys. Res., 75(24): 4561-4566.

chemistry, atmospheric
Junge, C.E., 1956.
Recent investigations in air chemistry.
Tellus 8(2):127-139.

auxins

chemistry - auxins
Mowat, Joyce A., (née Bentley), 1964.
A survey of results on the occurrence of auxins and gibberellins in algae.
Botanica Marina, 8(1):149-155.

chemistry, Baltic water
Kremling, Klaus, 1970.
Untersuchungen über die chemische Zusammensetzung des Meerwassers aus der Ostsee. II. Frühjahr 1967- Frühjahr 1968.
Kieler Meeresforsch. 26(1): 1-20.

barium

chemistry - barium (data)
Bolter, Ernst, Karl K. Turekian and Donald F. Schutz, 1964.
The distribution of rubidium, cesium and barium in the oceans.
Geochimica et Cosmochimica Acta, 28(9):1459-1466.

chemistry - barium
Bowen, H.J.M., 1956.
Strontium and barium in sea water and marine organisms.
J.M.B.A. 35(3):451-460.

chemistry, barium
Brongersma-Sanders, Margaretha, 1967.
Barium in pelagic sediments and in diatoms.
Proc. K. ned. Akad. Wet., (B), 70 (1): 93-99.

chemistry - barium
Chester, R., 1965.
Elemental geochemistry of marine sediments.
In: Chemical oceanography, J.P. Riley and G. Skirrow, editors, Academic Press, ":23-80.

chemistry - barium
Chow, Tsaihwa J., and E.D. Goldberg, 1960
On the marine geochemistry of barium. Geochim. Cosmochim. Acta., 20(3/4): 192-198.

chemistry, barium
Chow, T.J., and C.C. Patterson, 1966.
Concentration profiles of barium and lead in Atlantic waters off Bermuda.
Earth Planet. Soil Letters, 1(6):397-400.

chemistry, barium
Desai, M.V.M., Elizabeth Koshy and A.K. Ganguly, 1968.
Stability of barium in sea-water in presence of dissolved organic matter.
Current Sci., 38(5): 107-108.

chemistry - barium
Ishibashi, Masayoshi, Koichi Emi, Takeshi Kusaka and Miysuyuki Mitooka, 1960
Basic studies on the analysis and the separation of boric acid (2): On the complex precipitate of barium boro manno saccharate and barium boro aravotrihydroxyglutarate. Rec. Oceanogr. Wks., Japan, Spec. No. 4: 95-110.

chemistry - barium
Mauchline, J., and W.L. Templeton, 1966.
Strontium, calcium and barium in marine organisms from the Irish Sea.
J. Cons. perm int. Expl. Mer, 30(2):161-170.

chemistry - barium
Pilkey, Orrin H., and H.G. Goodell, 1963
Trace elements in Recent mollusk shells.
Limnol. and Oceanogr., 8(2):137-148.

chemistry — barium
Szabo, Barney J., and Oiva Joensuu, 1967.
Emission spectrographic determination of barium in sea water using a cation exchange concentration procedure.
Environment. Sci. Techn., 1(6):499-502.

chemistry, barium
Turekian, Karl K., 1971
2. Rivers, tributaries and estuaries. In: Impingement of man on the oceans, D.W. Hood, editor, Wiley Interscience, 9-73.

chemistry - barium
Turekian, Karl K. 1968.
Deep-sea deposition of barium, cobalt and silver.
Geochim. cosmochim. Acta 32 (1): 603-612.

chemistry, barium
Turekian, Karl K., 1967.
Estimates of the average Pacific deep-sea clay accumulation rate from material balance calculations.
Progress in Oceanography, 4:227-244.

chemistry, barium
Turekian, Karl K., and David G. Johnson, 1966.
The barium distribution in sea water.
Geochim. Cosmochim. Acta, 30(11):1153-1174.

chemistry - barium
Turekian, Karl E., and Donald F. Schutz, 1965.
Trace element economy in the oceans.
Narragansett Mar. Lab., Univ. Rhode Island, Occ. Publ., No. 3:41-89.

chemistry, barium
Turekian, Karl K., Donald F. Schultz and David Johnson, 1966.
The distribution of Sr, Ba, Co, Ni, and Ag in ocean water profiles of the Pacific sector of the Antarctic seas.
Antarctic J., United States, 1(5):224.

chemistry - barium
Turekian, K.K., and E.H. Tausch, 1964.
Barium in deep-sea sediments of the Atlantic Ocean.
Nature, 201(4920):696-697.

chemistry barium
Von Engelhardt, W., 1936.
Die Geochemie des Bariums.
Chem. Erde, 10:187-246.

cited in:
Chem. Abstr., 1936, 30:33716.

chemistry, barium.
Wolgemuth, K., and W.S. Broecker 1970
Barium in sea water
Earth Planet. Sci. Letts, 8(5): 372-378.

barium sulphate

chemistry - barium sulphate
Burton, J.D., N.J. Marshall, and A.J. Phillips, 1968.
Solubility of barium sulphate in sea water.
Nature, Lond., 217(5131):834-835.

chemistry - barium sulphate
Mero, John L., 1965.
The mineral resources of the sea.
Elsevier Oceanogr. Ser., 312 pp.

base

chemistry - base, total
Baas Becking, L.G.M., 1956.
Biological processes in the estuarine environment.
IX Observations on total base. Proc. K. Nederl. Akad. Wetensk., Amsterdam, B, 59(5): 408-420

chemistry - base (excess)
Cooper, L.H.N., 1933.
Chemical constituents of biological importance in the English Channel, November 1930 to January 1932. II. Hydrogen ion concentration, excess base, carbon dioxide and oxygen. J.M.B.A., 18: 729-753.

chemistry - base
Matsue, Y., 1950.
The variation of titratable base in sea water by the growth of diatoms. J. Ocean. Soc., Tokyo, 6(1):32-38, 2 textfigs. (In Japanese with English abstract).

benzo-3.4 pyrene

chemistry, benzo-3.4 pyrene
de Lima-Zanghi, Carmen 1968.
Bilan des acides gras du plancton marin et pollution par le benzo-3.4 pyrene.
Cah. océanogr. 20 (3): 205-216.

chemistry - benzo-3.4 pyrene
Mallet, Lucien, 1967.
Pollutions marines par les hydrocarbures polycondenses du type benzo-3.4 pyrène des côtes nord et ouest de France; leur incidence sur le milieu biologique et en particulier sur le plancton.
Cah. océanogr., 19(3):237-243.

chemistry - benzo-3.4 pyrene
Mallet, Lucien, et Marie-Louise Priou 1967.
Sur la rétention des hydrocarbures polybenzéniques du type benzo-3.4 pyrène par les sédiments, la faune et la flore marines de la baie de Saint-Malo.
C.r. hebd. Séanc. Acad. Sci. Paris (D) 264(7): 969-971.

chemistry, benzo-3,4 pyrene
Mallet, L., L.V. Perdriau et J. Perdriau, 1963.
Pollution par les hydrocarbures polybenzéniques du type benzo-3,4 pyrène de la region occidentale de l'Ocean Glacial Arctique.
C.R. Acad. Sci., Paris, 256(16):3487-3489.

chemistry, benzo-3,4 pyrene
Mallet, Lucien, et Jacques Sardou, 1964.
Recherche de la présence de l'hydrocarbure polybenzénique benzo 3-4 pyrène dans le milieu planctonique de la région de la Baie de Villefranche.
Comm. Inst. Expl. Mer. Médit., Symp. Pollut. par Microorgan. Prod. Petrol., Monaco, 1964:331-334.

Also in:
Trav. Sta. Zool., Villefranche-sur-Mer, 24. (1964)

chemistry - benzo 3-4 pyrene
Piccinetti, C., 1967.
Diffusione dell'idrocarburo cancerigeno benzo 3-4 pirene nell'alto e medio Adriatico.
Arch. Oceanogr. Limnol., 15(Suppl.):169-183.

beryllium

chemistry, beryllium
Amin, B.S., D.P. Kharkar and D. Lal, 1966.
Cosmogenic ^{10}Be and ^{26}Al in marine sediments.
Deep-Sea Res., 13(5):805-824.

chemistry - beryllium
Ishibashi Masayoshi, Taitito Fujinaga, Tooru Kuwamoto, 1960
Chemical studies on the ocean (82). Chemical Studies on the seaweeds. Quantitative determination of beryllium in seaweeds. Rec. Oceanogr. Wks., Japan, Spec. No. 4: 87-90.

chemistry - berylium
Ishibashi, M. T. Shigematsu and Y. Nishikawa, 1956.
On the amount of berylium in sea water.
Bull. Inst. Chem. Res., Kyoto Univ., 34(4):210-213.

chemistry - beryllium
Kharkar, D.P., D. Lal and B.L.K. Somayajulu, 1963.
Investigation in marine environments using radioisotopes produced by cosmic rays.
Radioactive Dating, Int. Atomic Energy Agency, Vienna, 175-186.

chemistry - beryllium
Landergren, S., 1948
6. On the geochemistry of Mediterranean sediments. Preliminary report on the distribution of beryllium, boron, and the ferrides in three cores from the Tyrrhenian Sea. Medd. Ocean. Inst., Göteborg 15 (Göteborgs Kungl. Vetenskaps-och Vitterhets -Samhälles Handlingar Sjätte Följden, Ser. B 5(13)):34-46, figs. 6-7.

chemistry - Beryllium
Ostroumov, E.A., and I.I. Volkov, 1964.
A new method for gravimetric determination of beryllium and the separation from manganese, nickel, cobalt and zinc with the help of cinnamic acid. Chemistry of the waters and sediments of the seas and oceans. (In Russian; English abstract).
Trudy Inst. Okeanol., Akad. Nauk, SSSR, 67:151-156.

chemistry, beryllium
Silker, W.B., D.E. Robertson, H.G. Rieck, Jr. R.W. Perkins and J.M. Prospero 1968.
Beryllium-7 in ocean water.
Science 161 (3844): 879-880.

chemistry - beryllium
Tatsumoto, M., and M. Osawa, 1957.
Chemical investigations of deep-sea deposits. CCVI. The content of Beryllium.
J. Chem. Soc., Japan, 78(4):502-508.

beryllium-10

chemistry - beryllium-10
Burton, J.D., 1965.
Radioactive nuclides in sea water, marine sediments and marine organisms.
In: Chemical oceanography, J.P. Riley and G. Skirrow, editors, Academic Press, 2:425-475.

chemistry - beryllium-10
Goel, P.S., D.P. Kharkar, D. Lal, N. Narsappaya, B. Peters, and V. Yatirajam, 1957.
The beryllium-10 concentration in deep-sea sediments.
Deep-Sea Res., 4(3):202-210.

chemistry, beryllium-10
Somayajulu, B.L.K. 1967.
Beryllium-10 in a manganese nodule.
Science 156 (3779): 1219-1220.

chemistry - beryllium-10
Tanaka, S., K. Sakamoto, J. Takagi and M. Tsuchimoto, 1968.
Aluminum-26 and beryllium-10 in marine sedimen
Science, 160(3834):1348-1349.

bicarbonate

chemistry - bicarbonate
Berner, Robert A., 1965.
Activity coefficients of bicarbonate, carbonate and calcium ions in sea water.
Geochimica et Cosmochimica Acta, 29(8):947-965.

chemistry - Bicarbonate (data only)
India, Naval Hdqtrs., Office of Scientific Research and Development, 1957.
Indian Oceanographic Stations List No. 1/57: 5 pp. (mimeographed)

chemistry - bicarbonate
Matida, Y., (undated in English).
On the chemical composition of the sea water in the Tokyo Bay. J. Ocean. Soc., Tokyo, (Nippon Kaiyo Gakkaisi) 5(2/4):105-110, 3 figs.
(In Japanese; English summary).

chemistry, bicarbonate
MacKenzie, Fred T., and Robert M. Garrels, 1966.
Silica-bicarbonate balance in the ocean and early diagenesis.
J. Sedim. Petrol., 36(4):1075-1084.

bioassay

chemistry - bioassay
Carlucci, A.F., and S.B. Silbernagel 1967.
Bioassay of seawater. IV. The determination of dissolved biotin in seawater using ^{14}C uptake by cells of Amphidinium carteri.
Can. J. Microbiol. 13(8): 979-986.

chemistry, biochemistry
Iida, Atsushi, Isao Arai, Kiichi Murata and Keiichi Oishi 1961.
Biochemical studies on the muscle of sea animals. II. On color-evaluation of hotate-kaibashira adductor muscle of scallop (Pecten yessoensis). (In Japanese; English abstract)
Bull. Fac. Fish. Hokkaido Univ. 12 (3): 239-245.

chemistry, biochemistry
Khailov, K.M. 1965.
Prospects in the development of marine dynamic biochemistry. (In Russian)
Okeanologiia 5(1):3-13.

chemistry, biochemistry
Vinogradova, Z. A. 1964
Some biochemical aspects related to comparative plankton studies with regard to the Black, Azov and Caspian seas. (In Russian)
Okeanologiia 4(2): 232-242.

chemistry, biochemistry
Webb, D.A. 1937.
Studies on the ultimate composition of biological material. II. Spectrographic analyses of marine invertebrates, with special reference to the chemical composition of their environment.
Sci. Proc. R. Dublin Soc. n.s. 21 (47): 505-539.

chemistry, biodegradability
Foret-Montando, Paule, 1971.
Evolution dans le temps de la toxicité des détergents issus de la pétroléochimie: Etude réalisée sur Scolelepis fuliginosa (polychète sédentaire).
Téthys 3(1): 173-182

biotin

Chemistry biotin
Carlucci, A.F., 1970
Vitamine B_{12}, Thiamine, and biotin
Bull. Scripps. Inst. Oceanogr. 17:23-31

chemistry - biotin
Carlucci, A.F., and S.B. Silbernagel 1967.
Bioassay of seawater. IV. The determination of dissolved biotin in seawater using ^{14}C uptake by cells of Amphidinium carteri.
Can. J. Microbiol. 13 (8): 979-986.

Chemistry - biotin
Carlucci, A.F., S.B. Silbernagel and P.M. McNally, 1969.
Influence of temperature and solar radiation on persistence of vitamin B_{12}, Thiamine and biotin in sea water.
J. Phycol., 5(4): 302-305.

Chemistry biotin
Litchfield, Carol D., and Donald W. Hood, 1965.
Microbiological assay for organic compounds in seawater. 1. Quantitative assay procedures and biotin distribution.
Appl. Microbiol., 13(6):886-894.

Also in:
Texas A&M Univ., Contrib. Oceanogr., Meteorol., 10. 1965/1966.

bitumens

Chemistry - bitumens
Bordovskiy, O.K., 1965.
Accumulation and transformation of organic substance in marine sediments. 1. Summary and introduction. 2. Sources of organic matter in marine basins. 3. Accumulation of organic matter in bottom sediments. 4. Transformation of organic matter in marine sediments.
Marine Geology, Elsevier Publ. Co., 3(1/2):3-4; 5-31; 33-82; 83-114.

chemistry, bitumen
Lisitzin, A.P., Yu.A. Bogdanov and L.I. Ovchinnikova, 1967.
Some results of bituminous suspension studies from the Pacific Ocean. (In Russian; English abstract).
Okeanologiia, Akad. Nauk, SSSR, 7(1):120-129.

Blood pigments

Chemistry - blood pigments
Eliassen, E., 1954.
The physiology of the vascular system of invertebrates. 1. A monograph on the blood-pigments; their physical-chemical qualities, distribution and physiological importance. Univ. i Bergen, Aarbok., Naturvitensk. rekke, 1953(11):1-65, 26 textfigs.

Chemistry, blood
Redfield, A. C., and E.N. Ingalls, 1933.
The oxygen dissociation curves of some bloods containing hemocyanin. J. Cell. & Comp. Phys. 3(2):169-202, 20 textfigs.

Blood proteins

Chemistry - Blood proteins
Wieser, Wolfgang, 1965.
Electrophoretic studies on blood proteins in an ecological series of isopod and amphipod species. Jour. Mar. Biol. Assoc., U.K., 45(2):507-523.

borate

Chemistry - borate
Gast, J.A., and T.G. Thompson, 1958.
Determination of alkalinity and borate concentration in sea water. Anal. Chem., 30(9):1549-1551.

Chemistry borate
Keeling, Charles D., and Bert Bolin, 1968
The simultaneous use of chemical tracers in oceanic studies. II. A three-reservoir model of the North and South Pacific oceans. Tellus, 20(1): 17-54

boric acid

Chemistry, boric acid
Buch, K., 1933.
On boric acid in the sea and its influence on the carbonic acid equilibrium. J. du Cons. 8(3):309-325, 2 textfigs.

Chemistry boric acid
Buch, K., 1933.
Der Borsäuregehalt des Meerwassers und seine Bedeutung bei der Berechnung des Kohlensäuresystems im Meerwasser. Ergängen zum Berichte über die im Frühjahr 1931 im Helsingfors von einer Arbeitskommission der Internationalen Meeresforschung ausgeführten Untersuchungen. Rapp. Proc. Verb. 85:71-75, 1 fold-in.

Chemistry boric acid
Coste, B., 1968.
Dosage de l'acide borique dans l'eau de mer. Recl Trav. Stn mar. Endoume, 43(59):65-80.

chemistry, boric acid
Culberson, C., D.R. Kester and R.M. Pytkowicz, 1967.
High-pressure dissociation of carbonic and boric acids in sea water.
Science, 157(3784):59-61.

Chemistry - boric acid
Culberson, C., and R.M. Pytkowicz, 1968.
Effect of pressure on carbonic acid, boric acid, and the pH in seawater.
Limnol. Oceanogr., 13(3):403-417.

chemistry, boric acid
Gripenberg, Stina, 1966.
Equilibria of the complexes formed by mannitol in sea water.
Commentat. physico-math., 32(1):3-38.

Chemistry - boric acid
Gripenberg, S., 1961
Alkalinity and boric acid content of Barents Sea water. Contribution to Special IGY Meeting, 1959.
Cons. Perm. Int. Expl. Mer. Rapp. Proc. Verb 149:31-37.

Chemistry - boric acid
Ishibashi, Masayoshi, Koichi Emi, Takeshi Kusaka and Miysuyuki Mitooka, 1960
Basic studies on the analysis and the separation of boric acid (2): On the complex precipitate of barium boro manno saccharate and barium boro aravotrihydroxyglutarate. Rec. Oceanogr. Wks., Japan, Spec. No. 4: 95-110.

Chemistry, boric acid
Miyake, Y., and H. Matui, 1939.
Boric acid content in sea water.
Geophys. Mag., Tokyo, 12:201-303.

Chemistry - boric acid
Noakes, John E., and Donald W. Hood, 1961
Boron-boric acid complexes in sea water.
Deep-Sea Res., 8(2):121-129.

chemistry, boric acid
Williams, P.M., 1966.
Complexes of boric acid with organic cis-diols in seawater.
Limnol. Oceanogr., 11(3):401-404.

boron

chemistry, boron
Brujewicz, S.W. and V.D. Korzh, 1971.
On boron interchange between the sea and the atmosphere. (In Russian; English abstract).
Okeanologiia 11(3): 414-422.

chemistry, boron
Harriss, Robert C., 1969.
Boron regulation in the oceans.
Nature, Lond., 223(5203):290-291

Chemistry - boron
Rakestraw, N.W., and H.E. Mahncke, 1935.
Boron content of sea water of the North Atlantic Coast. Ind. Chem. Eng., Anal. Ed., 7:425.

Chemistry - boron
Reynolds, Robert C., Jr., 1965.
boron and oceanic evolution: a reply.
Geochimica et Cosmochimica Acta, 29(8):1008-1009.

Chemistry - boron
Chester, R., 1965.
Elemental geochemistry of marine sediments.
In: Chemical oceanography, J.P. Riley and G. Skirrow, editors, Academic Press, 2:23-80.

Chemistry, boron
Culkin, Frederick, 1965.
The major constituents of sea water.
In: Chemical oceanography, J.P. Riley and G. Skirrow, editors, Academic Press, 1: 121-161.

chemistry, boron
Gassaway, John D., 1967.
New method for boron determination in sea water and some preliminary results.
Int. J. Oceanol. Limnol., 1(2):86-90.

Chemistry - boron
Greenhalgh, R., and J.P. Riley, 1962.
The development of a reproducible spectrophotometric curcumin method for determining boron, and its application to sea water.
Analyst, 87:970-976.

chemistry - boron
Gregor, Bryan, 1965.
Boron and oceanic evolution.
Geochimica et Cosmochimica Acta, 29(8):1007-1008.

Chemistry - boron
Gripenberg, Stina, 1960
On the alkalinity of Baltic waters. J. du Conseil 26(1): 5-20.

Chemistry, boron
Ichie, T., K. Tanioka, and T. Kawamoto, 1950.
Reports of the Oceanographical observations on board the R.M.S. "Yushio Maru" off Shionomisaki (Aug. 1949). Papers and Repts., Ocean., Kobe Mar. Obs., Ocean. Dept., No. 5:15 pp., 33 figs. (Odd atlas-sized pages - mimeographed).

chemistry, boron
Kawamoto, T., 1951.
On the distribution of boron in the sea. Mem., Kobe Mar. Obs., 9:9-14, 7 textfigs.

Chemistry, boron
Krasintseva, V.V., and O.V. Shishkina, 1959.
A contribution to the problem of boron distribution in marine sediments. (In Russian).
Doklady, Akad. Nauk, SSSR, 128(4):815-818.

chemistry, boron
Landergren, Sture, 1964.
On the geochemistry of deep-sea sediments.
Repts. Swedish Deep-Sea Exped., 10(Spec. Invest. 5):61-148.

Chemistry boron
Landergren, S., 1948
6. On the geochemistry of Mediterranean sediments. Preliminary report on the distribution of beryllium, boron, and the ferrides in three cores from the Tyrrhenian Sea. Medd. Ocean. Inst., Göteborg 15 (Göteborgs Kungl. Vetenskaps-och Vitterhets -Samhälles Handlingar, Sjätte Följden, Ser. B 5(13)):34-46, figs.6-7.

chemistry boron
Lerman, A., 1966.
Boron in clays and estimation of paleosalinities.
Sedimentology, 6(4);267-286.

chemistry, boron
Lewin, Joyce, 1966.
Boron as a growth requirement for diatoms.
J. Phycology, 2(4):160-163.

Chemistry - boron
Miyake, Y., and S. Sakurai, 1952.
Boron in sea water as an indicator for the water mass analysis. Umi to Sora 30(1/2):14-18, 6 text-

chemistry - boron (data)

Noakes, John E., and Donald W. Hood, 1961
Boron-boric acid complexes in sea water.
Deep-Sea Res., 8(2):121-129.

chemistry - boron

Reynolds, R.C., Jr., 1965.
The concentration of boron in Precambrian seas.
Geochimica et Cosmochimica Acta, 29(1):1-16.

chemistry - boron

Schwarcz, H.P., E.K. Agyei and C.C. McMullen, 1969.
Boron isotopic fractionation during clay adsorption from sea-water.
Earth Planet Sci. Lett. 6(1):1-5.

chemistry - boron

Yamamoto, Hiroshi, and Yoshihiko Imai, 1960.
Distribution of boron in Urado Bay. Especially about the variation of boron content accompanied by the dilution of the sea water.
Rec. Oceanogr. Wks., Japan, Spec. No. 4:123-128.

chemistry, boron/salinity

Boon, John D. III, and William G. MacIntyre 1968.
The boron-salinity relationships in estuarine sediments of the Rappahannock River, Virginia.
Chesapeake Science 9(1): 21-26.

bromide

chemistry, bromide

Andersen, Raymond J., Donald L. Graf and Blair F. Jones, 1966.
Calcium and bromide contents of natural waters.
Science, 153(3744):1637-1638.

chemistry - bromide

Brewer, Peter G., and Derek W. Spencer, 1969.
A note on the chemical composition of the Red Sea brines.
In: Hot brines and Recent heavy metal deposits in the Red Sea, E.T. Degens and D.A. Ross, editors, Springer-Verlag, New York, Inc., 174-179.

chemistry, bromides

*Morcos, Selim A., 1967.
The chemical composition of sea water from the Suez Canal region. I. The major anions.
Kieler Meeresforsch., 23(2):80-91.

bromide/chlorinity

chemistry, bromide/chlorinity

Morris, A.W., and J.P. Riley, 1966.
The bromide/chlorinity and sulphate/chlorinity ratio in sea water.
Deep-Sea Res., 13(4):699-705.

chemistry - bromine/chlorine

Kikkawa, Kyozo, and Shiko Shiga, 1965.
The modified chemical constituent of sea water intruding into the coastal aquifer.
Spec. Contrib., Geophys. Inst., Kyoto Univ., No. 5:7-16.

bromine

chemistry, bromine

Bruevich, S.W., 1960
[Hydrochemical investigations on the White Sea]
Trudy Inst. Okeanol., 42: 199-254.

bromine chemistry, bromine

Culkin, Frederick, 1965.
The major constituents of sea water.
In: Chemical oceanography, J.P. Riley and G. Skirrow, editors, Academic Press, 1: 121-161.

chemistry - bromine

Haslam, J., and R.O. Gibson, 1950.
The analytical chemistry of bromine manufacture III. Survey of the in-shore waters round the coasts of Great Britain. Analyst 75:357-370.

chemistry, bromine

Kato, K., 1949.
Studies on the direct separation of bromine from sea water. J. Fish., Hakodate, Japan, No. 54:1-6, 4 textfigs. [In Japanese with English summary.]

chemistry, bromine

Matida, Y., and N. Yamauchi, (undated in English)
On the distribution of bromine in the ocean.
J. Ocean. Soc., Tokyo, (Nippon Kaiyo Gakkaisi) 5(2/4):111-115.

(In Japanese; English summary).

chemistry - bromine

Selivanov, L.S., 1947.
[On the origin of chlorine and bromine in the salts of the ocean.] Bull. Volcan. Station, Kamchatka, 11:26-34.

chemistry, bromine

Thompson, T.G., and E. Korpi, 1942.
The bromine-chlorinity ratio of sea water. J. Mar. Res. 5(1):28-36.

chemistry, bromine

Vasilief, V.V., 1937.
[Content of bromine in Japanese sea.]
Zhurn. Prikl. Khim., 10(2)(7):1296-1301.

Transl. cited:
USFWS Spec. Sci. Rept., Fish., 227.

chemistry, bromine

Wells, Leslie E., 1967.
Better fuel from sea water.
Sea Frontiers, 13(5):269-273 (popular)

Chemistry, bromine

Winchester, John W., and Robert A. Duce, 1965.
Geochemistry of iodine, bromine, and chlorine in the air-sea sediment system.
Narragansett Mar. Lab., Univ. Rhode Island, Occ. Publ., No. 3:185-201.

buffers, of seawater / buffer capacity, of sea water

Chemistry - buffer capacity

Mitchell, P.H., and N.W. Rakestraw, 1933.
The buffer capacity of sea water.
Biol. Bull., 65(3):437-451.

Chemistry, buffers

Mitchell, P.H., and J.L. Solinger, 1934.
The effects of land drainage upon the excess bases of sea water.
Biol. Bull., 66(2):97-101.

buffers

Pytkowicz, Ricardo M., 1967.
Carbonate cycle and the buffer mechanism of recent oceans.
Geochim. cosmochim. Acta, 31(1):63-73.

Chemistry - buffer capacity

Richards, Francis A., 1968.
Chemical and biological factors in the marine environment. Ch. 8 in: Ocean engineering: goals, environment, technology, J.F. Brahtz, editor, John Wiley & Sons, 259-303.

Cadmium

chemistry - cadmium

Hiyama, Yoshio, and Makoto Shimizu, 1964
On the concentration factors of radioactive Cs, Sr, Cd, Zn, and Ce in marine organisms.
Rec. Oceanogr. Wks., Japan, 7(2):43-77.

chemistry, cadmium

Ishibashi, Masayoshi, Taitiro Fujinaga, Fuji Morii, Yoshihiko Kanchiku, and Fumio Kamiyama, 1964
Chemical studies on the ocean (Part 94). Chemical studies on the seaweeds (19). Determination of zinc, copper, lead, cadmium and nickel in seaweeds using dithizone extraction and polarographic method.
Rec. Oceanogr. Wks., Japan, 7(2):33-36.

chemistry - cadmium

Ishibashi, M., T. Shigematsu, M. Tabushi, Y. Nishikawa, & S. Goda, 1962
Determination of cadmium in sea-water.
J. Chem. Soc. Japan, 83(3): 295-297.

chemistry, cadmium

Ishibashi, Masayoshi, Shunzo Ueda and Yoshikazu Yamamoto, 1962
Studies on the utilization of shallow-water deposits (continued). On the cadmium content of shallow-water deposits.
Rec. Oceanogr. Wks., Japan, n.s., 6(2):169-176.

chemistry, cadmium

Maljković, Dubravka and Marko Branica 1971.
Polarography of seawater. II. Complex formation of cadmium with EDTA. Limnol. Oceanogr. 16(5): 779-785.

Chemistry, cadmium

Mullin, J.B., and J.P. Riley, 1957.
The occurrence of cadmium in seawater and in marine organisms and sediments. J. Mar. Res., 15(2):103-122.

chemistry, cadmium

Mullin, J.B., and J.P. Riley, 1954.
Cadmium in sea water. Nature 174:42.

Caesium

Chemistry, cesium (data)

Bolter, Ernst, Karl K., Turekian and Donald F. Schutz, 1964.
The distribution of rubidium, cesium and barium in the oceans.
Geochimica et Cosmochimica Acta, 28(9):1459-1466.

Chemistry - caesium

Bryan, G.W., 1963
The accumulation of 137 Cs by brackish water invertebrates and its relation to the regulation of potassium and sodium.
J. Mar. Biol. Assoc., U.K., 43(2):541-565.

Chemistry - caesium

Bryan, G.W., 1963
The accumulation of radioactive caesium by marine invertebrates.
J. Mar. Biol. Assoc., U.K., 43(2):519-539.

chemistry, caesium
Burovina, I.V., et al., 1963.
The content of lithium, sodium, potassium, rubidium and caesium in the muscles of marine animals of the Barents Sea and Black Sea.
Doklady, Akad. Nauk, SSSR, 149(2):413-415.

chemistry, cesium
Folson, T.R., and Katsuko Saruhashi, 1963
A comparison of analytical techniques used for determination of fallout cesium in sea water for oceanographic purpose.
J. Radiation Res., 4(1):39-53.

chemistry - cesium
Folsom, T.R., C. Feldman, and T.C. Rains, 1964.
Variation of cesium in the ocean.
Science, 144(2618):538-539.

chemistry, cesium
Harada, Y., 1943.
Chemical studies on the ocean. XVII. On the determinations of cesium in sea water and in bittern. J. Chem. Soc., Japan, 64:1049-1053.
Chem. Abstr., 1947, 41:3394.

chemistry, caesium
Hiyama, Yoshio, and Makoto Shimizu, 1964
On the concentration factors of radioactive Cs, Sr, Cd, Zn, and Ce in marine organisms.
Rec. Oceanogr. Wks., Japan, 7(2):43-77.

chemistry, cesium
Ishibashi, M., & T. Hara, 1959
A systematic analysis of potassium, rubidium and cesium and its application to sea-muds.
Bull. Inst. Chem. Res. Kyoto Univ., 37(3):185-190.

chemistry, cesium
Ishibashi, M., & T. Hara, 1959.
On concentrating rubidium and cesium from a large volume of aqueous solution.
Bull. Inst. Chem. Res. Kyoto Univ., 37(3):172-178.

chemistry - cesium
Ishibashi, M., & T. Hara, 1959.
On the amount of cesium dissolved in sea-water.
Bull. Inst. Chem. Res. Kyoto Univ., 37(3):179-184.

chemistry - caesium
Ishibashi, M., and T. Hara, 1955.
On the content of rubidium and caesium in sea-water.
Rec. Ocean. Wks., Japan, 2(1):45-48.

chemistry, cesium
Kautsky, Hans, 1970.
Auftreten und Veränderungen des Sr-90- und Cs-137-Gehaltes im Oberflächenwasser der Barentssee in den Jahren 1961 bis 1969
Dt. hydrogr. Z. 23(6):264-268

chemistry, caesium 137
Lucu, Čedomil, and Olga Jelisavčić, 1970
Uptake of ^{137}Cs in some marine animals in relation to temperature, salinity, weight and moulting.
Int. Revue ges. Hydrobiol. 55(5): 783-796.

chemistry, caesium
Mohanrao, G.J., and T.R. Folsom, 1963
Gamma-ray spectrometric determination of low concentrations of radioactive caesium in sea water by a nickel ferrocyanide method.
Analyst, 88(1043):105-108.

cesium
Morozov, N.P. 1968.
On the geochemistry of rare alkaline elements in the oceans and seas. (In Russian; English abstract)
Okeanologiia, 8(2): 216-224.

chemistry - cesium
Parchevsky, V.P., 1965.
Radionuclides of cesium, ruthenium and zirconium in animal and plant organisms of the Black Sea. (In Russian).
Okeanologiia, Akad. Nauk, SSSR, 5(5):856-862.

chemistry - caesium
Parshevskiy, V.P., and K.M. Khailov, 1968.
Formation of Ce^{141}, Ru^{106}, Cs^{137}, and Zn^{65} complexes with hydrophyl high-molecular combinations dissolved in sea water. (In Russian).
Okeanologiia, Akad. Nauk, SSSR., 8(6):1092.

chemistry cesium
Seto, Yoshio, 1965.
On the contribution of rubidium to the determination of radioactive cesium in deep-sea water.
J. Oceanogr. Soc., Japan, 21(5):202-205.

chemistry, caesium
Smales, A.A., and L. Salmon, 1955.
Determination by radioactivation of small amounts of rubidium and caesium in sea water and related materials of geological interest.
Analyst 80:37-50.
Anal. Abstr. 2(5):1141.

chemistry - cesium
Sreekumaran, C., K.C. Pillai and T.R. Folsom 1968.
The concentration of lithium, potassium, rubidium and cesium in some western American rivers and marine sediments.
Geochim. cosmochim. Acta 32 (11): 1229-1234.

chemistry, cesium-137
Volchok, H.L., V.T. Bowen, T.R. Folsom, W.S. Broecker, E.A. Schuert and G.S. Bien, 1971.
Oceanic distributions of radionuclides from nuclear explosions. Radioactivity in the marine environment, U.S. Nat. Acad. Sci., 1971: 42-89.

chemistry cesium
Yamagata, N., 1957.
Separation of tracer cesium by ion exchange chromatography. A preliminary study for the determination of cesium content in sea water.
J. Chem. Soc., Japan, 78(4):513-517.
Abstr. in Anal Abstr. 5(2):#716

Caesium 134

chemistry - caesium-134
Morgan, F., 1964.
The uptake of radioactivity by fish and shellfish. 1. 134 caesium by whole animals.
J. Mar. Biol. Assoc., U.K., 44(1):259-271.

Caesium 137

chemistry - cesium 137
Baptist, J.P., and T.J. Price, 1962.
Accumulation and retention of cesium 137 by marine fishes.
Fish. Bull. 206 from Fish. Bull., Fish and Wildlife Service, 62:177-187.

chemistry - cesium-137
Baranova, D.D., and G.G. Polikarpov, 1965.
Sorption of strontium-90 and cesium-137 in the aleurite silts of the Black Sea. (In Russian).
Okeanologiia, Akad. Nauk, SSSR, 5(4):646-648.

chemistry, cesium-137
Barinov, G.V., 1965.
The exchange of Ca^{45}, Cs^{137} and Ce^{144} between algae and the sea water. (In Russian).
Okeanologiia, Akad. Nauk, SSSR, 5(1):111-116.

cesium-137
Belyaev, L.I., L. I. Gedeonov, and G.V. Yakovleva, 1966.
Estimation of Sr^{90} and Cs^{137} content in the Black Sea. (In Russian; English abstract).
Okeanologiia, Akad. Nauk, SSSR, 6(4): 641-644

chemistry - cesium 137
Broecker, Wallace S., Erno R. Bonebakker and Gregory G. Rocco, 1966.
The vertical distribution of cesium 137 and strontium 90 in oceans. 2.
J. geophys. Res., 71(8):1999-2003.

chemistry, caesium 137
Broecker, W.S., and G.G. Rocco, 1963.
Direct comparison of radiocarbon and fallout measurements in sea water.
Symposium on oceanic mixing. In: Nuclear Geophysics, Proc. Conf., Woods Hole, Massachusetts, June 7-9, 1962.
NAS-NRC, Publ., No. 1075:150-151.
Also: Nuclear Sci. Ser. Rept., No. 38.

cesium (radioactive)
Bryan, G.W., 1965.
Ionic regulation in the squat lobster Galathea squamifera, with special reference to the relationship between potassium metabolism and the accumulation of radioactive cesium.
J. Mar. Biol. Assoc., U.K., 45(1):97-113.

chemistry, caesium-137
Bryan, G.W., 1961.
The accumulation of radioactive caesium in crabs.
J. Mar. Biol. Assoc., U.K., 41(3):551-575.

chemistry - caesium-137
Folsom, T.R., and G.J. Mohanrao, 1963.
Variation of Cs-137 in California coastal sea water. Radioactive tracers in oceanography.
Symposium 10th Pacific Sci. Congr., 1961.
Union Géodes. Géophys. Int., Monogr., 20:31-32.

chemistry, cesium-137
Folsom, T.R., and G.J. Mohanrao, 1962
Distribution of cesium-137 in the Pacific and Indian Oceans. (Abstract).
J. Geophys. Res., 67(9):3558.

chemistry, caesium-137.
Folsom, T.R., and G.J. Mahanrao, 1960.
Trend of caesium-137 in the effluent of a large city.
Nature, 188(4755):979-982.

chemistry, caesium-137
Fukai, R., and N. Yamagato, 1962.
Estimation of the levels of caesium-137 in sea-water by the analysis of marine organisms.
Nature, 194(4827):466.

chemistry, cesium-137
Gutknecht, John, 1965.
Uptake and retention of cesium 137 and zinc 65 by seaweeds.
Limnology and Oceanography, 10(1):58-66.

chemistry, cesium-137
Kupferman, S.L., 1971.
Cesium-137 in the North Atlantic measured by selective absorption in situ. J. mar. Res., 29(1): 11-18.

chemistry, cesium-137
Le Beyec, Y., A. Morel et P. Sligewicq 1971.
Note sur la distribution du césium-137 dans le bassin liguro-provençal.
Cah. océanogr. 23(9):859-869.

chemistry, cesium-137
Mironov, O. G., 1964.
On the problem of migration of strontium-90 and cesium-137 into the organism of man in the course of consuming some species of marine animals. (In Russian).
Okeanologiia, Akad. Nauk, SSSR, 4(6):1059-1061.

chemistry, caesium-137
Miyake, Yasuo, Katsuko Saruhashi, Yukio Katsuragi and Teruko Kanazawa, 1962
Penetration of 90 Sr and 137 Cs in deep layer of the Pacific and vertical diffusion rate of deep water.
Repts. JEDS, Deep-Sea Res. Comm., Japan Soc., Promotion of Science, 3:141-147.
Originally published (1962):
J. Radiation Res., 3(3):141-147.
JEDS Contrib. No. 25.

chemistry, cesium 137.
Miyake, Y., K. Saruhashi, Y. Katsuragi and T. Kanazawa, 1961.
Cesium 137 and strontium 90 in sea water.
Papers in Meteorol. & Geophys., Tokyo, 12(1): 85-88.

chemistry - cesium 137
Miyaki, Y., K. Sarneshi, Y. Katsuraji, T. Kanazawa and Y. Sugimura, 1964.
Uranium, radium, thorium, ionium, strontium 90 and cesium 137 in coastal waters of Japan.
In: Recent researchers in the fields of hydrosphere, atmosphere and nuclear geochemistry, Ken Sugawara festival volume, Y. Miyake and T. Koyama, editors, Maruzen Co., Ltd., Tokyo, 127-141.

chemistry, cesium137
Morgan, A., and G.M. Arkell, 1963.
Method for the determination of cesium-137 in sea water.
Health Physics, 9(8):857-862.

Ref. in: Chem. Titles, 1963 (18):83.

chemistry - caesium-137
Park, Kilho, Merilyn J. George, Yasuo Miyake, Katsuko Saruhashi, Yukio Katsuragi and Teruko Kanazawa, 1964.
Strontium-90 and caesium-137 in Columbia River plume, July 1964.
Nature, 208(5015):1084-1085.

chemistry - cesium-137
Polikarpov, G.G., 1961.
The role of detritus formation in the migration of strontium-90, cesium-137 and cerium-144. Experiments with the sea alga Cystoseira barbata.
Doklady Akad. Nauk, SSSR, 136(4):921-923.

English Edit., 1962, 136(1-6):11-13.

chemistry - Cesium 137
Rocco, Gregory G., and Wallace S. Broecker, 1963
The vertical distribution of Cesium 137 and Strontium 90 in the oceans.
J. Geophys. Res., 68(15):4501-4512.

chemistry - caesium-137
Schroeder, B.W., and R.D. Cherry, 1962.
Caesium-137 in the seas off the Cape of Good Hope.
Nature, 194(4829):669.

chemistry, cesium-137
Sreekumaran, C., S.S. Gogate, G.R. Doshi, V.N. Sastry and R. Viswanathan 1968.
Distribution of cesium-137 and strontium-90 in the Arabian Sea and Bay of Bengal.
Current Sci. 37(22): 629-631.

chemistry, cesium-137
Wolfe, Douglas A., 1971.
Fallout cesium-137 in clams (Rangia cuneata) from the Neuse River Estuary, North Carolina.
Limnol. Oceanogr. 16(5): 797-805.

chemistry - cesium-137
*Yakovleva, G.V., 1967.
The effect of trade-winds air transport on water pollution by strontium-90 and cesium-137 of the equatorial Atlantic. (In Russian; English abstract).
Okeanologiia, Akad. Nauk, SSSR, 7(4):617-622.

chemistry - Cesium-137
Yamagata, N., and S. Matsuda, 1959.
Cesium-137 in the coastal waters of Japan.
Bull. Chem. Soc., Japan, 32(5):497-502.

calcareous deposits

chemistry - calcium (deposition of)
Goreau, Thomas F., 1959
The physiology of skeleton formation in corals 1. A method for measuring the rate of calcium deposition by corals under different conditions. Biol. Bull., 116(1): 59-75.

chemistry - calcareous concretions
*Laborel, Jacques Louis, e Marc Kempf, 1965/1966.
Formações de vermetos e algas calcárias nas costas do Brasil.
Trabhs Inst. Oceanogr., Univ. Fed. Pernambuco, Recife, (7/8):33-50.

chemistry - calcareous deposits
Nesteroff, W.D., 1956.
Le substratum organique dans dépots calcaires, sa signficance. Bull. Soc. Géol., France (6) 6:381-390.

chemistry - Calcareous nodules
*Van Andel, Tjeerd H., G. Ross Heath, T.C. Moore and David F.R. McGeary, 1967.
Late Quaternary history, climate and oceanography of the Timor Sea, northwestern Australia.
Am. J. Sci., 265(9):737-758.

Calcification

chemistry - calcification
Chave, Keith E., and Bradner D. Wheeler, Jr., 1965.
Mineralogic changes during growth in the red alga, Clathromorphum compactum. Science, 147 (3658):621.

chemistry - calcification
Guilcher, A., 1957.
Formes de corrosion littorale du calcaire sur les cotes du Portugal.
Tijdschr. Konink. Nederl. Aardrijk. Gen., 74(3): 263-269.

chemistry - calcification
Lewin, Ralph A., Editor, 1962.
Physiology and biochemistry of algae.
Academic Press, New York and London, 929 pp.

chemistry calcification, corals
Pearse, Vicki Buchsbaum and Leonard Muscatine, 1971.
Role of symbiotic algae (Zooxanthellae) in coral calcification. Biol. Bull. mar. biol. Lab. Woods Hole, 141(2): 350-363.

Calcium

chemistry, calcium
Andersen, Raymend J., Donald L. Graf and Blair F. Jones, 1966.
Calcium and bromide contents of natural waters.
Science, 153(3744):1637-1638.

chemistry, calcium
Ayers, John C., David C. Chandler, George H. Lauf, Charles F. Powers and E. Bennett Henson, 1958
Currents and water masses of Lake Michigan.
Great Lakes Res. Inst., Publ. No. 3:169 pp.

chemistry, calcium (dissolved)
Berner, Robert A., 1969.
Chemical changes affecting dissolved calcium during the bacterial decomposition of fish and clams in sea water.
Marine Geol. 17(3): 253-274

chemistry calcium
Berner, Robert A., 1965.
Activity coefficients of bicarbonate, carbonate and calcium ions in sea water.
Geochimica et Cosmochimica Acta, 29(8):947-965.

chemistry, calcium (data)
Berthois, L., et A. Crosnier, 1966.
Étude dynamique de la sédimentation au large de l'estuaire de la Betsiboka
Cah. ORSTOM, Sér. océanogr. 4(2):49-130.

chemistry, calcium
Carpenter, J.H., 1957.
The determination of calcium in natural waters.
Limnol. & Oceanogr. 2(3):271-280.

chemistry, calcium
Carroll, J.J., L.J. Greenfield and R.F. Johnson, 1965.
The mechanism of calcium and magnesium uptake from sea water by a marine bacterium.
J. Cell. Comp. Physiol., 66(1):109-118.

chemistry, calcium
Chester, R., 1965.
Elemental geochemistry of marine sediments.
In: Chemical oceanography, J.P. Riley and G. Skirrow, editors, Academic Press, 2:23-80.

chemistry calcium
Chow, T.J., and T.G. Thompson, 1955.
Flame photometric determinations of calcium in sea water and marine organisms. Analyt. Chem. 27:910-913.

Chemistry - calcium
*Choe, Sang, Tai Wha Chung and Hi-Sang Kwak, 1968.
Seasonal variations in nutrients and principal ions contents of the Han River water and its water characteristics. (In Korean; English abstract).
J. oceanogr. Soc., Korea, 3(1):26-38.

Chemistry calcium
Chu, S. P., 1949
Experimental studies on the environmental factors influencing the growth of phytoplankton. Sci. & Tech. in China 2(3):37-52.

Chemistry - calcium
Culkin, Frederick, 1965.
The major constituents of sea water.
In: Chemical oceanography, J.P. Riley and G. Skirrow, editors, Academic Press, 1: 121-161.

chemistry, calcium
Culkin, F., and R.A. Cox, 1966.
Sodium, potassium, magnesium, calcium and strontium in sea water.
Deep-Sea Res., 13(5):789-804.

Chemistry - calcium
DeSousa, A., 1954.
The rapid determination of calcium and magnesium in sea water. Anal. Chim. Acta 11(3):221-224.

Abstr. Anal. Abstr. 2(1):207.

Chemistry - calcium
*Dyrssen, David, et al., 1967.
Analysis of sea water in the Uddevalla Fjord system 1.
Rept. Chem. Sea Water, Univ. Göteborg, 4:1-8 pp. (mimeographed).

chemistry - calcium
Gaarder, T., and R. Spärck, 1932.
Hydrographisch-biochemische Untersuchungen in norwegischen Austern-Pollen. Bergens Mus. Aarbok, Naturvidensk.-rekke, No. 1:5-144, 75 textfigs.

chemistry calcium
Goedecke, E., 1936.
Der Kalkgehalt im Oberflächenwasser der Unterelbe und Deutschen Bucht.
Arch. Deutschen Seewarte 55(1):1-37, 4 pls.

Chemistry - calcium
Ishibashi, Masayoshi, and Toshio Yamamoto, 1960
Chemical studies on the ocean (Pts. 78-79).
Chemical studies on the seaweeds (VI & VII).
The content of calcium, magnesium and phosphorus in seaweeds. Iron content in seaweeds.
Rec. Oceanogr. Wks., Japan, Spec. No. 4: 73-78; 79-86.

chemistry calcium
Ishihara, Tadashi, and Masato Yasuda, 1964.
Determination of calcium by chelatemetry using calcein. 1. Determination of calcium in drinking water and sea water. (In Japanese; English abstract.).
Bull., Fac., Fish., Nagasaki Univ., No. 17:110-123.

chemistry, calcium
Kato, Kenji, 1966.
Studies on calcium content in sea water. III. Calcium in the waters of Cananéia lagoon and its adjacent regions, State of Sao Paulo, Brazil.
Bolm Inst. Oceanogr., S. Paulo, 15(1):41-45.

chemistry, calcium
Kato, Konji, 1966.
Studies on calcium content in sea water. II. Distribution of calcium in the Atlantic water off south Brazil.
Bolm Inst. Oceanogr., S. Paulo, 15(1):29-39.

chemistry, calcium
Kato, Kenji, 1966.
Studies on calcium content in sea water. 1. Chelatometric determination of calcium in sea water.
Bolm Inst. Oceanogr., S. Paulo, 15(1):25-28.

chemistry - calcium
Kevern, Niles R., 1964.
Relative strontium and calcium uptake by green algae.
Science, 146(3650):1488.

Chemistry, calcium
Khan, Khan Umuardaraz, and Yoshio Hiyama, 1964
Mutual effect of Sr-Ca upon their uptake by fish and freshwater plants.
Rec. Oceanogr. Wks., Japan, 7(2):107-122.

chemistry - calcium
Kawaguti, S., and D. Sakumoto, 1948.
The effect of light on the calcium deposition of corals. Bull. Ocean. Inst., Taiwan, No. 4: 65-70.

Chemistry, calcium
Koczy, F.F., 1950.
Zur Sedimentation und Geochemie im aequatorischen Atlantischen Ozean. Medd. Oceanografiska Inst., Göteborg 17 (Göteborgs Kungl. Vetenskaps- och Vitterhets Samhälles Handlingar, Sjätte Följden Ser. B, 6(1):44 pp., 17 textfigs.

Chemistry - calcium (data)
Kramer J.R. 1961
Chemistry of Lake Erie.
Proc. Fourth Conf., Great Lakes Research Great Lakes Res. Div., Inst. Sci. & Tech. Univ. Michigan, Publ., (7):27-56.

chemistry, calcium
Lebedev, V.I., 1965.
Calcium content and some other compositional characteristics of the Precambrian Seas.
Geokhimiya (9):1154-1164.

Translation: Geochemistry, Int., Ann Arbor, 2(5):843-852.

chemistry, calcium
Lyakhin, Yu.I., 1971.
Calcium and magnesium in the western tropical Atlantic. (In Russian; English abstract).
Okeanologiia 11(4): 635-641.

Chemistry, calcium
Mameli, D., e F. Mosetti, 1968.
Some new investigations on the Ca and Mg content in sea water.
Boll. Soc. adriat. Sci., Trieste, 55:27-38.

chemistry - calcium
Mameli, D., e F. Mosetti, 1966.
Un nuovo metodo di marcatura delle acque di mare costiere: il contenuto in calcio e magnesio quale indicatore del mescolamento di acque continentali in acqua marina.
Boll. Geofis. teor. appl., 8(32):294-308.

Chemistry - calcium
Mauchline, J., and W.L. Templeton, 1966.
Strontium, calcium and barium in marine organisms from the Irish Sea.
J. Cons. perm. int. Expl. Mer, 30(2):161-170.

chemistry, calcium
Murray, J.W., 1966.
A study of the seasonal changes of the water mass of Christchurch Harbour, England.
Jour. mar. biol. Assoc., U.K., 46(3):561-578.

chemistry, calcium (data)
Nehring, Dietwart, und Karl-Heinz Rohde 1967
Weitere Untersuchungen über anormale Ionenverhältnisse in der Ostsee.
Beitr. Meereskunde 20:10-33

Chemistry, calcium
Okuda, Taizo, 1964.
Calcium and magnesium contents in the river and sea waters of a tropical area.
Bol. Inst. Oceanogr., Univ. Oriente, Venezuela, 3(1/2):118-135.

Chemistry, calcium
Oren, O.H., 1962
A note on the hydrography of the Gulf of Eylath. Contributions to the knowledge of the Red Sea No. 21.
Sea Fish. Res. Sta., Haifa, Israel, Bull. No. 30: 3-14.

Chemistry calcium
Pate, J. B., and R.J. Robinson, 1958
The (Ethylenedinitrilo) tetraacetate titration of calcium and magnesium in ocean waters. 1. Determination of calcium. J. Mar. Res., 17:390-402.

chemistry, calcium
Pedersen, E., 1947
Bestemmelse av kalsium i sjøvann. Rept. Norwegian Fish and Mar. Invest. 9(1):19 pp.

Chemistry - calcium
Rao, S.R., S.M. Shah and R. Viswanathan, 1968.
Calcium, strontium and radium content of molluscan shells.
J. mar. biol. Ass., India, 10(1): 159-165.

Chemistry, calcium (data)
Rial, J.R. esada and L.R. Molins, 1962.
Determinación complexométrica de los iones calcio y magnesio en el agua de mar y estudio de las variaciones de su concentración en las aquas de la Ría de Viga.
Bol. Inst. Español Oceanogr., 111:11 pp.

chemistry, calcium
Romanovsky, V., 1964.
Coastal effects of the Cape Sicié sewer outfall (French Mediterranean coast west of Toulon).
Air and Water Pollution, 8(10):557-589.

calcium
Sagi, Takeshi, 1969.
The concentration of calcium and the calcium chlorosity ratio in the western North Pacific Ocean.
Oceanogr. Mag., 21(1): 61-66.

Chemistry, calcium
Schachter, D., 1954.
Contribution à l'étude hydrographique et hydrologique de l'étang de Berre (Bouches-du-Rhone).
Bull. Inst. Océan., Monaco, No. 1048:20 pp.

chemistry, calcium
Seibold, E. 1962
Untersuchungen zur Kalkfällung und Kalklösung am Westrand der Grand Bahama Bank.
Sedimentology 1(1):50-74

chemistry, calcium
Skopintsev, B.A., R.V. Vorob'eva and L.A. Shtukovskaya, 1957.
[The determination of calcium and the amount of calcium and magnesium in sea water by a complexometric method.]
Gidrokhim. Materialy, 27:146-

chemistry, calcium
Sugawara, K., and N. Kawasaki, 1958.
Strontium and calcium distribution in western Pac Pacific, Indian and Antarctic Oceans.
Rec. Oceanogr. Wks., Japan, Spec. No. 2:227-242.

chemistry, calcium
Sugawara, Ken, Kikuo Terada, Satoru Kanamori, Nobuko Kanamori and Shiro Okabe, 1962
On different distribution of calcium, strontium, iodine, arsenic and Molydenum in the northwestern Pacific, Indian and Antarctic Oceans.
J. Earth Sciences, Nagoya Univ., 10(1):34-50.

chemistry, calcium
Swan, E.F., 1956.
The meaning of strontium-calcium ratios.
Deep-Sea Res., 4:71.

chemistry- calcium
*Tsunogai, Shizuo, Masakichi Nishimura and Syu Nakaya, 1968.
Calcium and magnesium in sea water and the ratio of calcium to chlorinity as a tracer of water masses.
J. oceanogr. Soc., Japan, 24(4):153-159.

chemistry, calcium
Turekian, Karl K., 1971
2. Rivers, tributaries and estuaries. In: Impingement of man on the oceans, D.W. Hood, editor, Wiley Interscience, 9-73.

chemistry, calcium
Uzumasa, Y., Y. Nasu, and T. Seo, 1960.
Flame photometric determination of strontium and calcium in natural waters.
J. Chem. Soc., Japan, 81(3):430-433.

chemistry, calcium
Valori, P., M. Talenti and F. Savoine, 1957.
Flame photometry in the analysis of natural waters. III. Determination of calcium. Ric. Sci., 27(6):1901-1914.

Abstr. in Anal. Abstr., 5(4):#1382.

chemistry -calcium
*Viswanathan, R., S.M. Shah and C.K. Unni, 1969
Atomic absorption and flüorometric analyses of seawater in the Indian Ocean. Bull. natn. Inst. Sci., India, 38(1): 284-288.

chemistry, calcium
Wattenberg, H. 1931
Beziehungen zwischen Kalkgehalt des Meerwassers und Plankton.
Rapp. Proc. Verb. Cons. Perm. int. Explor. Mer, 75: 67-79.

chemistry, calcium
Wattenberg, H., 1925.
Kalkauflösung und Wasserbewegung am Meeresboden.
Ann. Hydr., us., 63:387-391.

chemistry, calcium
Wattenberg, H. and E. Timmermann, 1936.
Ueber Sättigung des Seewassers an $CaCO_3$ und die anorganogene Bildung von Kalksedimenten. Ann. Hydr., usw., 64:23-31.

chemistry, calcium
Webb, D.A., 1938.
Strontium in sea water and its effect on calcium determinations.
Nature, 142(3599):751-752.

chemistry, calcium
Williams, Louis G., 1964.
Relative strontium and calcium uptake by green algae.
Science, 146(3650):1488.

chemistry, calcium
Wittig, H., 1940
Über die Verteilung des Kalziums und der Alkalinität in der Ostsee. Kieler Meeresforschungen, III:460-498, 6 textfigs.

chemistry, calcium
Won, Chong Hun, 1964.
Tidal variations of chemical constituents of the estuary water at the Lava bed in the Nack-Dong River from Nov. 1962 to Oct. 1963. (In Korean: English abstract).
Bull. Pusan Fish. Coll., 6(1):21-32

chemistry, calcium
Won, Chong Hun, 1963.
Distribution of chemical constituents of the estuary water in Gwang-Yang Inlet. (In Korean; English abstract).
Bull. Fish. Coll., Pusan Nat. Univ., 5(1):1-10.

chemistry - Calcium
Yamamoto, Hiroshi, and Yoshihiko Imai, 1959.
The hydrographic condition and variation of calcium in Susaki Bay.
Rec. Oceanogr. Wks., Japan, Spec. No. 3:117-121.

Calcium, anomalies

chemistry, calcium anomalies
Traganza, Eugene D., and Barney J. Szabo, 1967
Calculation of calcium anomalies on the Great Bahama Bank from alkalinity and chlorinity data.
Limnol. Oceanogr., 12(2):281-286.

Calcium-45

chemistry, calcium-45
Barinov, G.V., 1965.
The exchange of Ca^{45}, Cs^{137} and Ce^{144} between algae and the sea water. (In Russian).
Okeanologiia, Akad. Nauk, SSSR, 5(1):111-116.

Calcium carbonate

calcium carbonate
Alekin, O.A., Iu.I. Liakhin, 1968.
On the causes of oversaturation of sea water with calcium carbonate. (In Russian).
Dokl. Akad. Nauk, SSSR, 170(1):191-194.

chemistry - calcium carbonate
Alekin, O.A., and N.P. Moricheva, 1966.
On the saturation of Black Sea water with calcium carbonate. (In Russian).
Doklady, Akad. Nauk, SSSR, 167(2):423-425.

chemistry - calcium carbonate, precipitation of
Alekin, O.A., and N.P. Moricheva, 1961.
The precipitation of calcium carbonate from sea water by organisms. (In Russian).
Doklady Akad. Nauk, SSSR, 136(6):1454-1457.

English Edit., 1962, 136(1-6):20-22.

chemistry, calcium carbonate
Anon. 1968
Ocean bottom minerals.
Ocean Industry 3(6):61-73.

chemistry, calcium carbonate
Ben-Yaakov, S. and I.R. Kaplan 1971.
Deep-sea in situ calcium carbonate saturometry.
J. geophys. Res. 76(3): 722-731

chemistry - calcium carbonate
Blanchard, Richard L., 1965.
U 234/U238 ratios in coastal marine waters and calcium carbonates.
J. Geophys. Res., 70(160):4055-4061.

calcium carbonate
Bogorov, B.G., O.K. Bordovsky and M.E. Vinogradov 1966.
Biogeochemistry of the oceanic plankton. The distribution of some chemical components of the plankton in the Indian Ocean. (In Russian; English abstract).
Okeanologiia, Akad. Nauk, SSSR, 6(2):314-325.

chemistry, calcium carbonate
Bruevich, S. W. and E. G. Vinogradova, 1946
[Biogenic elements in the sediment solutions of the northern, middle and southern parts of the Caspian Sea.] Comptes Rendus (Doklady) Acad. Sci. URSS, LIV (5):419-422.

chemistry, calcium carbonate
Chu, S.P., and K.C. Young, 1949.
The variation with depth of certain nutrient salts for phytoplankton growth and some other properties of water in the fishing ground east of Chusan Islands in the Chinese East Sea.
Science (China) 31(6):181-182, 1 fig.

chemistry, calcium carbonate
Cloud, P.E., Jr., 1962
Behaviour of calcium carbonate in sea water.
Geochim. et Cosmochim. Acta 26: 867-884.

chemistry, calcium carbonate
Correns, C.W., 1950.
Faktoren der Sedimentbildung erläutet an Kalk- und Kieselsedimenten. Deutsche Hydro. Zeits. 3(1/2):83-88, 2 textfigs.

chemistry - calcium carbonate
Edmond, John M., and J.M.T.M. Gieskes, 1970.
On the calculation of the degree of saturation of sea water with respect to calcium carbonate under in situ conditions.
Geochim. cosmochim. Acta, 34(12):1261-1291.

chemistry calcium carbonate
Febres, German, y Gilberto Rodriguez, 1966.
Ch. 3. Sedimentos.
Estudios hidrobiologicos en el Estuario de Maracaibo, Inst. Venezolano de Invest. Cient., 66-82. (Unpublished manuscript).

Chemistry, calcium carbonate
Force, L.M. 1969.
Calcium carbonate size distribution on the west Florida shelf and experimental studies on the microarchitectural control of skeletal breakdown.
J. sedim. Petrol. 39 (3): 902-934.

chemistry, calcium carbonate, (solubility of)
Hawley, J. and R.M. Pytkowicz 1969.
Solubility of calcium carbonate in seawater at high pressures and 2°C.
Geochim. cosmochim. Acta 33 (12): 1557-1561.

Chemistry calcium carbonate
Kennett, James P., 1966.
Foraminiferal evidence of a shallow calcium carbonate solution boundary, Ross Sea, Antarctia
Science, 153(3732):191-193.

Chemistry, calcium carbonate
Kitano, Yasushi, 1964.
On factors influencing the polymorphic crystallization of calcium carbonate found in marine biological systems.
In: Recent researches in the fields of hydrosphere, atmosphere and nuclear geochemistry, Ken Sugawara festival volume, Y. Miyake and T. Koyama, editors, Maruzen Co., Ltd., Tokyo, 305-319.

Kitano, Yasushi, 1962. Chemistry calcium carbonate
The behavior of various inorganic ions in the separation of calcium carbonate from a bicarbonate solution.
Bull. Chem. Soc., Japan, 35(12):1973-1985.

Also in:
Collected Papers on Sciences of atmosphere and Hydrosphere, 1958-1963, Water Res. Lab., Nagoya Univ., 1(19)

Chemistry, Calcium carbonate
Kitano, Yasushi, and Donald W. Hood, 1965.
The influence of organic material on the polymorphic crystallization of calcium carbonate
Geochimica et Cosmochimica Acta, 29(1):29-42.

Chemistry, calcium carbonate
Kitano, Yasushi and Donald W. Hood, 1962.
Calcium carbonate crystal forms formed from sea water by inorganic processes.
J. Oceanogr. Soc., Japan, 18(3):141-145.

chemistry, calcium carbonate
Kitano, Yasushi, and Tamotsu Oomori 1971.
The coprecipitation of uranium with calcium carbonate.
J. oceanogr. Soc. Japan, 27(1): 34-42

Chemistry, calcium carbonate
Kuenen, Ph. H., 1941.
Het gehalte aan kalk en organische stof van de Indische deepzeeafzettingen. Nat. Geneesk Congr., Utrecht, Handl XXVIII:258-259.

Chemistry, Calcium carbonate
Li Yuan-Hui, Taro Takahashi, and Wallace S. Broecker, 1969.
Degree of saturation of $CaCO_3$ in the oceans.
J. geophys. Res., 74(23): 5507-5525.

Chemistry, calcium carbonate
Lisitsin, A.P., 1964.
Distribution and chemical composition of suspended matter in the waters of the Indian Ocean. (In Russian; English Abstract).
Rezult. Issled., Programme Mezhd. Geofiz. Goda, Mezhd. Geofiz. Kom., Presidume Akad. Nauk, SSSR, Okeanol., No. 10:136 pp.

Chemistry, calcium carbonate
Lyakhin, Yu.I., 1970.
Saturation of the Sea of Okhotsk water with calcium carbonate. (In Russian; English abstract). Okeanologiia, 10(6): 980-987.

Chemistry, calcium carbonate
MacIntyre, W.G., and R.F. Platford, 1964.
Dissolved calcium carbonate in the Labrador Sea.
J. Fish. Res. Bd., Canada, 21(6):1475-1480.

Chemistry - calcium carbonate
Maschhaupt, J.G., 1950.
Het Koolzure-kalkgehalte der Dollard-Gronden.
Waddensymposium, Tijdschr. Kon. Nederl. Aardrijkskundig Genootschap:114-121.

Chemistry, Calcium carbonate
Miyake, Y., 1957.
A study on the organic productivity and the solubility product CaCO3 in the ocean by means of the radiocarbon C14.
Int. Conf., Radioisotopes in Sci. Res., Paris, Sept., 1957, UNESCO/NS/RIC/138:1-4.

Chemistry, Calcium carbonate
Nishizawa, Tanzo and T. Muraki, 1940
Chemical studies in the sea adjacent to Palau. I. A survey crossing the sea from Palau to New Guinea. Kagaku Nanyo (Sci. of the South Sea) 2(3):1-7.

Chemistry, Calcium carbonate
Park, Kilho, 1964.
Electrolytic conductance of sea water: effect of calcium carbonate dissolution.
Science, 146(3640):56-57.

Chemistry -calcium carbonate
*Pilkey, Orrin H., Robert W. Morton and John Luternauer, 1967.
The carbonate fraction of beach and dune sands.
Sedimentology, 8(4):311-327.

chemistry, calcium carbonate
Pytkowicz, Ricardo M., 1971.
Sand-seawater interactions in Bermuda beaches.
Geochim. cosmochim. Acta. 35 (5): 509-515.

Chemistry, Calcium carbonate
Pytkowicz, Ricardo M., 1970.
On the carbonate depth in the Pacific Ocean.
Geochim. cosmochim. Acta, 34 (7): 836-839.

Chemistry - calcium carbonate
Pytkowicz, R.M., 1965.
Calcium carbonate saturation in the ocean.
Limnology and Oceanography, 10(2):220-225.

chemistry, calcium carbonate
Pytkowicz, R.M., 1964.
Calcium carbonate in the oceans. (Abstract).
Trans. Amer. Geophys. Union, 45(1):118.

Chemistry calcium carbonate
Pytkowicz, Ricardo M., 1963
Calcium carbonate and the in situ pH.
Deep-Sea Res., 10(5):633-638.

Chemistry, calcium carbonate
Pytkowicz, R.M., and D.N. Conners, 1964
High pressure solubility of calcium carbonate in seawater.
Science, 144(3620):840-841.

chemistry, calcium carbonate
Pytkowicz, R.M., A. Disteche and S. Distriche, 1965.
Calcium carbonate solubility in seawater at in situ pressures.
Earth Planet Sci. Letters, 2(5):430-432.

Chemistry, calcium carbonate
Rao, M.S., 1958.
Distribution of calcium carbonate in the shelf sediments off east coast of India.
J. Sed. Petr., 28(3):274-285.

chemistry - calcium carbonate
Robertson, A.J., 1907.
On hydrographical investigations in the Faroe-Shetland Channel and the northern part of the North Sea during the years 1904-1905. Fish. Bd., Scotland, North Sea Fish. Invest. Comm., Northern Area, Rept. No. 2:1-140, 13 text-figs., 15 pls.

Chemistry - calcium carbonate
Royer, L., 1945.
Au sujet d'une précipitation de carbonate de calcium observée dans la Mer Morte. C. R. Acad. Sci., Paris, 221(6/9):239-240

Chemistry, calcium carbonate (data)
Seibold, E., 1962.
Untersuchung zur Kalkfällung und Kalklösung am Westrand der Grand Bahama Bank.
Sedimentology, 1(1):50-74.

Chemistry, Calcium carbonate
Smith, C. L., 1941.
The solubility of calcium carbonate in tropical sea water. JMBA, 25(2):235-242.

Chemistry calcium carbonate
Smith, C. L., 1940.
The Great Bahama Bank. II. Calcium carbonate precipitation. J. Mar. Res., 3(2):171-189, 1 fig.

calcium carbonate
Smith, Stephen V., Joseph A. Dygas and Keith E. Chave, 1968.
Distribution of calcium carbonate in pelagic sediments.
Marine Geol., 6(5): 391-400

Chemistry, calcium carbonate (suspended)
Smith, S.V., R.J. Roy, H.G. Schiesser, G.L. Shepard and K.E. Chave 1971.
Flux of suspended calcium carbonate ($CaCO_3$), Fanning Island Lagoon.
Pacific Sci. 25 (2): 206-221.

Chemistry, calcium carbonate
Summerhayes, C.P., 1969.
Marine environments of economic mineral deposition around New Zealand: a review.
J. mar. techn. Soc. 3(2): 57-66

Chemistry, Calcium carbonate
*Summerhayes, C.P., 1967.
Marine environments of economic mineral deposition around New Zealand: A review.
N.Z. Jl mar. Freshwat.Res., 1(3):267-282.

chemistry calcium carbonate
Turekian, Karl K., 1967.
Estimates of the average Pacific deep-sea clay accumulation rate from material balance calculations.
Progress in Oceanography, 4:227-244.

Chemistry, calcium carbonate
Turekian, Karl K., 1965.
Some aspects of the geochemistry of marine sediments.
In: Chemical oceanography, J.P.Riley and G. Skirrow, editors, Academic Press, 2:81-126.

Chemistry calcium carbonate
Wattenberg, H., 1936.
Kohlensäure und Kalziumkarbonat im Meere.
Fortschr. Mineralog. Kristalog. Petrogr., 20:168-195.

Transl. cited:
USFWS Spec. Sci. Rept., Fish., 227.

Chemistry calcium carbonate
Wattenberg, H., 1931.
Ueber die Löslichkeit von CaCO3 im Meerwasser.
Naturwiss. 19(48):965.

transl. cited.
USFWS Spec. Sci. Rept., Fish., 227.

Chemistry
calcium carbonate
Wiseman, John D. H., 1965.
Calcium and magnesium carbonate in some Indian Ocean sediments.
Progress in Oceanography, 3:373-383.

Chemistry
calcium carbonate
Wiseman, J.D.H., M.W. Strong and H.J.M. Bowen, 1956.
Marine organisms and biogeochemistry. 1. The rates of accumulation of nitrogen and calcium carbonate on the equatorial Atlantic floor. 2. Marine iron bacteria as rock forming organisms. 3. The biogeochemistry of strontium.
Adv. Sci., 12(49):579-588.

chemistry, calcium carbonate
Yalkovsky, Ralph, 1967.
Signs test applied to Caribbean Core A 176-6.
Science, 155(3768):1408-1409.

calcium (data only)

Chemistry calcium (data only)
Australia, Commonwealth Scientific and Industrial Research Organization, 1964.
Oceanographical observations in the Indian Ocean in 1961, H.M.A.S. Diamantina, Cruise Dm 3/61.
Div. Fish. and Oceanogr., Oceanogr. Cruise Rept., No. 11:215 pp.

Chemistry calcium (data)
Dragovich, Alexander, John H. Finucane, John A. Kelly, Jr., and Billie Z. May, 1963.
Counts of red-tide organisms, Gymnodinium breve, and associated oceanographic data from Florida west coast, 1960-61.
U.S. Fish and Wildlife Service, Spec. Sci. Repts. Fish., No. 455:1-40.

Chemistry calcium (data only)
Granqvist, G., 1955.
The summer cruise with M/S Aranda in the northern Baltic, 1954. Merent. Julk., No. 166:56 pp.

Calcium carbonate budget

chemistry, calcium carbonate budget
Olausson, Eric, 1967.
Climatological geoeconomical and paleooceanographical aspects on carbonate deposition.
Progress in Oceanography, 4:245-265.

Calcium carbonate, precipitation of, Bottom deposits

calcium carbonate compensation
United States, Glomar Challenger Shipboard Scientific Party, 1971.
Leg 4 of the Deep Sea Drilling Project: coring reveals new information on the geologic history of the western Atlantic and Caribbean basins.
Science 172 (3989): 1197-1205.

Chemistry, calcium carbonate deposition
Wilbur, Karl M., and Norimitsu Watabe, 1967.
Mechanisms of calcium carbonate deposition in coccolithophorids and molluscs.
Studies Trop. Oceanogr., Miami, 5:133-154.

chemistry, calcium carbonate monohydrate
Duedal, J.W., and D.E. Buckley 1971.
Calcium carbonate monohydrate in seawater.
Nature, Phys. Sci. Lond. 234 (45): 39-40

Chemistry, calcium carbonate, precipitation of C14/C12
Broecker, Wallace S., and Taro Takahashi, 1966.
Calcium carbonate precipitation on the Bahama Banks.
J. Geophys. Res., 71(6):1575-1602.

chemistry calcium carbonate precipitation
Pytkowicz, Ricardo M., 1965.
Rates of inorganic calcium carbonate nucleation.
J. Geology, 73(1):196-199.

Chemistry calcium carbonate, precipitation of
Simkiss, Kenneth, 1964.
Variations in the crystalline form of calcium carbonate precipitated from artificial sea water.
Nature, 201(4918):492-493.

Chemistry, calcium carbonate, precipitation of
Simkiss, K., 1964
The inhibitory effects of some metabolites on the precipitation of calcium carbonate from artificial and natural sea water.
Journal du Conseil, 29(1):6-18.

Chemistry calciumcarbonate, precipitation of
Wells, A.J., and L.V. Illing, 1964.
Present day precipitation of calcium carbonate in the Persian Gulf.
In: Developments in Sedimentology, L.M.J.U. van Straaten, (Editor,) Elsevier Publishing Co., 1:429-435.

Chemistry - calcium carbonate saturation
Lyakhin, Yu. I., 1968.
Saturation of the Pacific waters with calcium carbonate. (In Russian; English abstract).
Okeanologiia, Akad. Nauk, SSSR, 8(1):58-68.

calcium-chlorinity ratio

Chemistry, calcium/chlorinity ratio
Berthois L 1969.
Contribution à l'étude sédimentologique du Kangerdlugssuaq, côte ouest du Groenland.
Meddr Grønland 187(1): 186pp.

Chemistry, Ca/Cl
*Dyrssen, David, et al., 1967.
Analysis of sea water in the Uddevalla Fjord system
Rept.Chem.Sea Water, Univ.Göteborg, 4:1-8 pp. (mimeographed).

chemistry, calcium chlorinity data)
Kato, Taxeo 1969.
On the distribution of the water mass produced by water-mixing and the change of chemical composition. (Ratio $Ca^{2+}:Cl^-$, $SO_4^{2-}:Cl^-$) - Oceanographical studies on the East China Sea, 3. (In Japanese; English abstract)
Umi to Sora, 44 (2/3): 55-80

Chemistry Ca/Cl
Mameli, D., e F. Mosetti, 1966.
Un nuovo metodo di marcatura delle acque di mare costiere: il contenuto in calcio e magnesio quale indicatore del mescolamento di acque continentali in acqua marina.
Boll. Geofis. teor. appl., 8(32):294-308.

Chemistry calcium-chlorinity ratio
May, Billie Z., 1964.
Surface calcium-chlorinity relationships.
Q.J. Florida Acad. Sci., 27(3):177-185.

Chemistry, Ca/Cl
*Riley, J.P., 1967.
The major cation/chlorinity ratios in sea water.
Chem. Geol., 2(3):263-269.

Chemistry, calcium/chlorinity ratio
Rohde, Karl-Heinz 1967
Untersuchungen über die Calcium-Chlor- und Magnesium-Chlor-Relationen in Flussmündungen und Bodden der westlichen Ostsee.
Beitr. Meereskunde 20: 34-42

calcium /chlorinity (data)
Rohde, Karl-Heinz, 1966.
Untersuchungen über die Calcium- und Magnesiumanomalie.
Beitr. Meereskunde, 19:18-31.

Chemistry calcium/chlorinity ratio
Sugi, Takeshi, 1969.
The concentration of calcium and the calcium chlorosity ratio in the western North Pacific Ocean.
Oceanogrl Mag., 21 (1): 61-66.

Chemistry, Ca/chlorinity
*Tsunogai, Shizuo, Masakichi Nishimura and Syu Nakaya, 1968.
Calcium and magnesium in sea water and the ratio of calcium to chlorinity as a tracer of water masses.
J. oceanogr. Soc., Japan, 24(4):153-159.

Calcium ions

Chemistry, calcium ions
Kikkawa, Kyozo, and Shiko Shiga, 1965.
The modified chemical constituent of sea water intruding into the coastal aquifer.
Spec. Contrib., Geophys. Inst., Kyoto Univ., No. 5:7-16.

Calcium metabolism

Chemistry, calcium metabolism
Chaisemartin, Claude, 1964.
Importance des gastrolithes dans l'économie du calcium chez Astacus pallipes Lereboullet. Bilan de l'exuviation.
Vie et Milieu, Lab. Arago, 15(2):457-474.

Calcium phosphate

Chemistry, calcium phosphate
Pytkowicz, R.M., and D.R. Kester, 1967.
Relative calcium phosphate saturation in two regions of the North Pacific Ocean.
Limnol. Oceanogr., 12(4):714-718.

Calcium precipitation

Chemistry, calcium precipitation
Kalinenko, V.O., 1949.
[Precipitation by bacteria of calcium in the sea.]
Trudy Inst. Okeanol., 3:200-215.

Calcium silicate

Chemistry, calcium silicate
Nordlii, O., 1950.
Precipitate in the sea-water samples for a study of plankton. J. du Cons. 16(3):310.

Chemistry calcium sulfate
*Kester, Dana R. and Ricardo M. Pytkowicz, 1969.
Sodium, magnesium and calcium sulfate ion-pairs in seawater at 25C. Limnol. Oceanogr., 14(5): 686-692.

Calcium uptake

Chemistry - calcium (uptake)
Boroughs, H.T., S.J. Townsley and R.W. Hiatt, 1957.
The metabolism of radionuclides by marine organisms. III. The uptake of calcium in solution by marine fishes. Limnol. Oceanogr., 2(1):28-32.

Chemistry calcium uptake
Goreau, T.F., and V.T. Bowen, 1955.
Calcium uptake by a coral. Science 122(3181): 1188-1189.

W-H.O.I. Contr. 786.

Chemistry - calcium uptake
Hsiao, S.C.T., and H. Boroughs, 1958.
The uptake of radioactive calcium by sea urchin eggs. 1. Entrance of Ca 45 into unfertilized cytoplasm.
Biol. Bull., 114(2):196-204.

Carbohydrate

Chemistry, carbohydrates
Antia, N.J., and C.Y. Lee, 1963.
Studies on the determination and differential analysis of dissolved carbohydrate in sea water.
Fish. Res. Bd., Canada, Mss Rept. Ser. (Oceanogr. & Limnol.), No. 168:75 pp., 18 tables.

Chemistry carbohydrate
Beers, John R., 1966.
Studies on the chemical composition of the major zooplankton groups in the Sargasso Sea.
Limnol. Oceanogr., 11(4):520-528.

Chemistry, carbohydrate (particulate)
Biggs, Robert B., and Carolyn D. Wetzel 1968.
Concentrations of particulate carbohydrate at the halocline in Chesapeake Bay.
Limnol. Oceanogr. 13(1): 169-171.

Chemistry, carbohydrates (data)
Boffi, V., e A. Lucarelli, 1963.
3. Biochimica oceanografica.
Rapp. Attività Sci. e Tecn., Lab. Studio della Contaminazione Radioattiva del Mare, Fiascherino, La Spezia, (maggio 1959-maggio 1962), Comit. Naz. Energia Nucleare, Roma, RT/BIO (63), 8:41-56. (multilithed).

Chemistry, carbohydrates (data only)
Collier, Albert, 1958.
Gulf of Mexico physical and chemical data from ALASKA cruises.
USFWS Spec. Sci. Rept., Fish., No. 249:417 pp.

Chemistry carbohydrates
*Dietrich, R., W. Hohnk und W.-D. Menzel, 1967.
Studien zur Chemie ozeanischer Bodenproben.
V. Der Nachweis von freien Aminosäuren und von Kohlenhydraten in der organischen Substanz der Sedimente des Indischen Ozeans.
Veröff. Inst. Meeresforsch. Bremerh. 10(3):141-148.

Chemistry - carbohydrates
Donnelly, Patricia V., and Mary A. Burklew, 1967.
Concentration of carbohydrates of the coastal waters of West Florida.
Prof. Pap. Ser. Florida Bd. Conserv. Mar. Lab. St. Petersburg 9: 85-89

Chemistry, carbohydrates
Donnelly, Patricia V., and Mary Ann Burklew, 1966.
Carbohydrate distribution in the northwest coastal waters of Florida.
Florida Bd. Conserv., St. Petersburg, Mar. Lab., Prof. Papers Ser., No. 8:39-42.

Chemistry, carbohydrate (data)
Dragovich, Alexander, 1961
Relative abundance of plankton off Naples, Florida, and associated hydrographic data, 1956-57.
USFWS Spec. Sci. Rept., Fish., No. 372:41 pp.

Chemistry, carbohydrates
Finucane, John H., and Alexander Dragovich 1959
Counts of red tide organisms, Gymnodinium brevis and associated oceanographic data from Florida west coast, 1954-1957.
USFWS Spec. scient. Rept. Fish. 289: 220 pp.

Chemistry - carbohydrates
Fogg, G.E., 1966.
The extracellular products of algae.
In: Oceanography and marine biology, H. Barnes, editor, George Allen & Unwin, Ltd., 4:195-212.

chemistry, carbohydrates
*Handa, Nobuhiko, 1967.
Identification of carbohydrates in marine particulate matters and their vertical distribution.
Rec. oceanogr. Wks. Japan, 9(1):65-73.

Chemistry, carbohydrates
*Handa, Nobuhiko, 1967.
The distribution of the dissolved and the particulate carbohydrates in the Kuroshio and its adjacent areas.
J. oceanogr. Soc., Japan, 23(3):115-123.

Chemistry - carbohydrate
Handa Nobuhiko 1966.
Distribution of dissolved carbohydrate in the Indian Ocean.
J. Oceanogr. Soc. Japan 22(2): 50-55.

Chemistry carbohydrates
Handa, N. and H. Tominaga, 1969.
A detailed analysis of carbohydrates in marine particulate matter. Marine Biol. 2(3): 228-235

Chemistry carbohydrates
Handa, N. and K. Yanagi, 1969.
Studies on water-extractable carbohydrates of the particulate matter from the northwest Pacific Ocean. Mar Biol., 4(3): 197-207.

Chemistry, carbohydrates
*Hussein, M. Fawzy, R. Boulus and F.M. Hanna, 1967.
Studies on the chemical composition of plankton of Lake Qarun. I. Seasonal variations in the protein, lipids and carbohydrate content of plankton.
Bull. Fac. Sci., Cairo Univ. Press, 40:121-131.

Chemistry, carbohydrate (data only)
Japan, Hokkaido University, Faculty of Fisheries, 1967.
The Oshoro Maru cruise 18 to the east of Cape Erimo, Hokkaido, April 1966.
Data Record Oceanogr. Obs. Explor. Fish. 11: 121-164.

Chemistry, carbohydrates
Marker, A.F.H., 1965.
Extracellular carbohydrate liberation in the flagellates Isochrysis galbana and Prymnesium parvum.
J. mar. biol. Ass., U.K., 45(3):755-772.

Chemistry carbohydrate
Marshall, S.N., and A.P. Orr, 1962
Carbohydrate as a measure of phytoplankton.
J. Mar. Biol. Assoc., U.K., 42(3):511-519.

Chemistry, carbohydrate
*Tokuda, Hiroshi, 1969.
Excretion of carbohydrate by a marine pennate diatom, Nitzschia closterium.
Rec. oceanogr. Wks. Japan, 10(1):109-122.

Chemistry - carbohydrates
*Kawahara, Hosaku, Yoshiaki Maita and Jiro Ishii, 1967.
Studies on the organic matter in sea water.
1. The distribution of carbohydrate in waters of southern Hokkaido. (In Japanese; English abstract).
Bull. Jap. Soc. scient. Fish., 33(9):825-933.

Chemistry - carbohydrate (suspension)
Marshall, Sheina M., and the late A.P. Orr, 1964
Carbohydrate and organic matter in suspension in Loch Striven during 1962.
Jour. Mar. Biol. Assoc., U.K., 44(2):285-292.

Chemistry carbohydrate
May, B.Z., 1960.
Stabilization of the carbohydrate content of sea water samples.
Limnol. and Oceanogr., 5(3):342-343.

chemistry, carbohydrates
Modzeleski, Judith E., William A. Laurie and Bartholomew Nagy 1970.
Carbohydrates from Santa Barbara Basin sediments: gas chromatographic-mass spectometric analysis of trimethylsilyl derivatives.
Geochim. cosmochim. Acta, 35 (8): 825-838.

Chemistry, carbohydrates, algal
Percival, Elizabeth, 1969.
Marine algal carbohydrates. In: Oceanography and marine biology: an annual review, H. Barnes, editor, George Allen & Unwin Ltd., 6: 137-161.

chemistry - carbohydrate content

Ramaswamy, T.S., 1953.
Carbohydrate and fat contents of fishes.
J. Madras Univ., B, 23(3):232-238.

chemistry, carbohydrate

Raymont, J.E.G., J.Austin and Eileen Linford, 1967.
The biochemical composition of certain zooplankton decapods.
Deep-Sea Research, 14(1):113-115.

Chemistry carbohydrate content

Raymont, J.E.G., and Robert J. Conover, 1961
Further investigations on the carbohydrate content of marine zooplankton.
Limnol. & Oceanogr., 6(2): 154-164.

chemistry, carbohydrate

Raymont, J.E.G., R.T. Srinivasagam and J.K.B. Raymont, 1971.
Biochemical studies on marine zooplankton - VIII. Further investigations on Meganyctiphanes norvegica (M. Sars). Deep-Sea Res. 18(12): 1167-1178.

chemistry - carbohydrate

Raymont, J.E.G., R.T. Srinivasagam and J.K. B. Raymont, 1969.
Biochemical studies on marine zooplankton. IV. Investigation on Meganyctiphanes norvegica (M. Sars.) Deep-Sea Res., 16(2): 141-156.

chemistry, carbohydrates

Schaefer, Heinz, 1965.
Isoberung von gelösten Kohlenhydraten aus dem Meerwasser.
Helgoländer wiss Meeresunters, 12(3):353-359.

chemistry - Carbohydrate

Seki, Humitake, 1965.
Studies on microbial participation to food cycle in the sea II. Carbohydrate as the only organic source in the microcosm.
Jour. Oceanogr. Soc., Japan, 29(6):278-285.

chemistry - carbohydrate

Walsh, Gerald E., 1966.
Studies of dissolved carbohydrate in Cape Cod waters. III. Seasonal variation in Oyster Pond and Wequaquet Lake, Massachusetts.
Limnol. Oceanogr., 11(2):249-256.

carbohydrate

Walsh, Gerald E., 1965.
Studies on dissolved carbohydrate in Cape Cod waters. II. Diurnal fluctuation in Oyster Pond
Limnol. Oceanogr., 10(4):577-582.

carbohydrate

Walsh, Gerald E., 1965.
Studies on dissolved carbohydrate in Cape Cod waters. 1. General survey.
Limnol. Oceanogr., 10(4):570-576.

chemistry, carbohydrate

Walsh, Gerald E., and Jane Douglass, 1966.
Vertical distribution of dissolved carbohydrate in the Sargasso Sea off Bermuda.
Limnol. Oceanogr., 11(3):406-408.

chemistry carbohydrates

Wangersky, Peter J., 1959.
Dissolved carbohydrates in Long Island Sound, 1956-1958.
Bull. Bingham Oceanogr. Coll., 17(1):87-94.

chemistry carbohydrate metabolism

Parvathy, K., 1971.
Carbohydrate metabolism of two crustaceans during starvation. Marine Biol., 8(1): 1-5.

Carbon

chemistry, carbon

Beers, John R., 1966.
Studies on the chemical composition of the major zooplankton groups in the Sargasso Sea.
Limnol. Oceanogr., 11(4):520-528.

chemistry - carbon

Berthois, Leopold, 1966.
Hydrologie et sédimentologie dans le Kangerd-lugssuaq (fjord à la côte ouest du Groenland).
C. r. hebd. séanc., Acad. Sci., Paris, (D), 262 (13): 1400-1402.

chemistry, carbon

Craig, H., 1971.
Son of abyssal carbon. J. geophys. Res. 76(21): 5133-5139.

chemistry - carbon (abyssal)

Craig, H., 1969.
Abyssal carbon and radiocarbon in the Pacific.
J. geophys. Res., 74(23): 5491-5506.

chemistry carbon (data)

Kabanova, Yu. G., 1964.
Primary production and biogenic elements content of the Indian Ocean waters. Investigations in the Indian Ocean (33rd Voyage of E/S "Vityaz"). (In Russian; English summary).
Trudy Inst. Okeanol., Akad. Nauk, SSSR, 64:85-93.

chemistry, carbon

Parsons, T.R., K. Stephens and J.D.H. Strickland, 1961
On the chemical composition of eleven species of marine phytoplankters.
J. Fish. Res. Bd., Canada, 18(6):1001-1016.

chemistry, carbon

Skopintsev, B.A., 1960.
Organic matter in waters of the sea. Marine chemistry, hydrology, oceanic geology. (In Russian).
Trudy Morsk. Gidrofiz. Inst., 19:3-20.

chemistry, carbon

Small, Lawrence F. and Donald A. Ramberg, 1971.
Chlorophyll a, carbon and nitrogen in particles from a unique coastal environment. In: Fertility of the Sea, John D. Costlow, editor, Gordon Breach, 2: 475-492.

chemistry, carbon (extrusion of)

Stephens, Grover C. and Barbara B. North, 1971.
Extrusion of carbon accompanying uptake of amino acids by marine phytoplankters. Limnol. Oceanogr. 16(5): 752-757.

Chemistry, carbon

Suess, H.E. and Edward Goldberg, 1971.
Comments on paper by H. Craig 'Abyssal carbon and radiocarbon in the Pacific'. J. geophys. Res. 76(21): 5131-5132.

chemistry - carbon (particulate)

Szekielda, Karl-Heinz 1967.
Some remarks on the influence of hydrographic conditions on the concentration of particulate carbon in sea water.
In: Chemical environment in the aquatic habitat, Proc. I.B.P. Sym. Amsterdam, Oct. 1966: 314-322

Carbon assimilation

chemistry, carbon assimilation

Curl, H.C. Jr., and L.F. Small, 1965
The significance of variations in photosynthetic carbon assimilation numbers in the marine environment. (Abstract).
Ocean Sci. and Ocean Eng., Mar. Techn. Soc., Amer. Soc. Limnol. Oceanogr., 1:299.

chemistry, carbon assimilation

Mandelli, Enrique F., and Paul R. Burkholder, 1966.
Primary productivity in the Gerlache and Bransfield straits of Antarctica.
J. Mar. Res., 24(1):15-27.

chemistry, carbon assimilation

Parsons, T.R., and J.D.H. Strickland, 1961(1962).
On the production of particulate organic carbon by heterotrophic processes in sea water.
Deep-Sea Res., 8(3/4):211-222.

Carbonate

chemistry carbonate

Alekin, O.A., 1957.
On the problem of stability of the carbonate system in natural waters. (In Russian).
Doklady, Akad. Nauk, SSSR, 117(6):1090-

chemistry, carbonates

Alekin, O.A., and N.P. Moricheva, 1958.
(Content of organic matter in natural waters as affected by carbonate system.) DAN 119(2):322.

carbonate

Antoine, J.W., and T.E. Pyle, 1970.
Crustal studies in the Gulf of Mexico.
Tectonophysics 10(5/6): 477-494

chemistry, carbonate

Argentina, Servicio de Hidrografia Naval, 1959.
Quimica del agua del mar.
Servicio de Hidrografia Naval, Argentina, Publ., H. 604:140 pp.

chemistry, carbonate

Berner, Robert A., 1965.
Activity coefficients of bicarbonate, carbonate and calcium ions in sea water.
Geochimica et Cosmochimica Acta, 29(8):947-965.

chemistry, carbonate

Berrit, G.R., 1955.
Étude des teneurs en manganese et en carbonates de quelques carottes sedimentaires Atlantiques et Pacifiques. Göteborgs K. Vetenskaps- och Vitterhets-Samhalles Handl., Sjätte Följden, (B), 6(12):61 pp.
Also:
Medd. Oceanografiska Inst., Göteborg, No. 23.

chemistry - carbonates

Berrit, G.R., 1955.
Étude des teneurs en manganese et en carbonates de quelques carottes sedimentaires atlantiques et pacifiques. Göteborgs K. Vetenskaps- och Vitterhets-Samhalles Handl. (7)B6(12):1-61.

Medd. Ocean. Inst., Göteborg 23.

Chemistry - carbonates
Bishev, L. L., 1955
[Results of hydrochemical investigations the Kubansk delta estuaries.] Trudy VNIRO 31: 145-150.

Chemistry - carbonate
Bogoyavlensky, A.N., 1959
[Hydrochemical investigations.]
Arktich. i Antarktich. Nauchno-Issled. Inst., Mezhd. Geofiz. God, Sovetsk. Antarkt. Eksped., 5: 159-172.

chemistry - carbonate system
Bruevich, S.W., editor, 1966.
Chemistry of the Pacific Ocean.
Inst. Okeanol., Akad. Nauk, SSSR, Isdatel. Nauka, Moskva, 358 pp.

Chemistry, carbonate
Buch, K., 1930.
Die Kohlensäurefaktoren des Meereswassers.
Cons. Perm. Int. Expl. Mer, Rapp. Proc. Verb. 67:51-85, figs.

Chemistry - carbonates
Chave, K.E., 1965.
Carbonates: association with orgaic matter in surface seawater.
Science, 148(3678):1723-1724.

Chemistry, carbonates
Chave, K.E., K.S. Deffeyes, P.K. Weyl, R.M. Garrels and M.E. Thompson, 1962.
Observations on the solubility of skeletal carbonates in aqueous solutions.
Science, 137(3523):33-34.

chemistry, carbonate
Chave, Keith E., and Robert F. Schmalz, 1966.
Carbonate-seawater interactions.
Geochim. Cosmochim. Acta, 30(10):1037-1048.

Chemistry, carbonate
Chidambaram, K., A.D.I. Isaac Rajandran, and A.P. Valsan, 1951.
Certain observations on the hydrography and biology of the pearl bank, Tholayviam Paar off Tuticorin in the Gulf of Manaar in April 1949.
J. Madras Univ., Sect. B, 21(1):48-74, 2 figs.

Chemistry, carbonate dissolution
Cloud, Preston E., 1965.
Carbonate precipitation and dissolution in the marine environment.
In: Chemical oceanography, J.P. Riley and G. Skirrow, editors, 2:127-158.

Chemistry, carbonate precipitation
Cloud, Preston E., 1965.
Carbonate precipitation and dissolution in the marine environment.
In: Chemical oceanography, J.P. Riley and G. Skirrow, editors, 2:127-158.

chemistry, carborate
Dyrssen, David, 1969.
The carbonate system. In: Chemical oceanography, Rolf Lange, editor, Universitetsforlaget, Oslo: 59-67.

Chemistry, carbonate (total)
*Dyrssen, David, et al., 1967.
Analysis of sea water in the Uddevalla Fjord system 1.
Rept. Chem. Sea Water, Univ. Göteborg, 4:1-8 pp. (mimeographed).

Chemistry, carbonate
Dyrssen, David, and Lars Gunnar Sillén, 1967.
Alkalinity and total carbonate in sea water. A plea for p-T-independent data.
Tellus, 19(1):113-121.

Chemistry carbonate
Goodell, H.G., and R.K. Garman, 1969.
Carbonate geochemistry of Superior deep test well, Andros Island, Bahamas.
Bull. Am. Ass. Petrol. Geol., 53(3): 513-536.

Chemistry, carbonate
Gripenberg, S., 1953.
Acidimetric carbonate analysis.
Medd. Ocean. Inst., Göteborg 22:1-32, 1 pl.

Göteborgs K. Vetenskaps- och Vitterhets-Salhälles Handl., Sjätte Följden, Ser. B, 6(9):

Chemistry, carbonates
Hathaway, John C., and Egon T. Degens 1969.
Methane-derived marine carbonates of Pleistocene age.
Science 165 (3894): 690-692.

Chemistry, carbonates
Khailov, K.M. 1971.
A phenomenon of the direct participation of carbonates soluble in sea water in biosynthesis and growth of Mytilus edulis. (In Russian; English abstract).
Okeanologiia 11(3): 494-500.

chemistry, carbonate
Irwin, M.L., 1965.
General theory of epeiric clear water sedimentation.
Bull. Amer. Assoc. Petr. Geol., 49(4):445-459.

Chemistry carbonate
Keeling, Charles D., and Bert Bolin, 1968
The simultaneous use of chemical tracers in oceanic studies. II. A three-reservoir model of the North and South Pacific oceans.
Tellus, 20(1): 17-54

chemistry, carbonate
Kramer, James R., 1964
Sea water: saturation with apatites and carbonates.
Science, 146(3644):637-638.

Chemistry, carbonates
Lalou, C., 1957.
Formations expérimentale des carbonates dans le mileau marin. Nouveaux résultats obtenus dans les cultures de longue durée.
Trav. Sta. Zool., Villefranche-sur-Mer, 16(5):

Is in:
Rev. Géo. Phys., Géol. Dyn., (2), 1(2):93-98.

Chemistry, carbonates
Lalou, C., 1963.
Étude expérimentale de la production de carbonates par les bactéries des vases de la Baie de Villefranche-sur-mer.
Ann. Inst. Océan., Monaco, 23(4):201-267.

Chemistry carbonates
Lalou, C., 1957.
Studies on bacterial precipitation of carbonates in sea water. J. Sed. Petr., 27(2):190-195.

chemistry carbonates
Lalou, Mlle., 1955.
La précipitation expérimentale des carbonates par les bactéries.
Bull. d'Info., C.C.O.E.C., 7(8):358-359.

chemistry, carbonate
Lange, Rolf 1969.
Chemical oceanography.
Universitetsforlaget, Oslo, 152 pp.

chemistry carbonates
Malone, Ph. G., and K.M. Towe, 1970.
Microbial carbonate and phosphate precipitates from sea water cultures.
Marine Geol., 9(5): 301-309.

Chemistry carbonate
Marlowe, James I. 1971
Dolomite, phosphorite, and carbonate diagenesis on a Caribbean seamount.
J. Sedim. Petrol. 41(3): 809-827.

chemistry, carbonate
Olausson, Eric, 1967.
Climatological geoeconomical and paleooceanographical aspects on carbonate deposition.
Progress in Oceanography, 4:245-265.

Chemistry, carbonate cycle
Pytkowicz, Ricardo M., 1967.
Carbonate cycle and the buffer mechanism of recent oceans.
Geochim. cosmochim. Acta, 31(1):63-73.

Chemistry carbonates
Revelle, R., and R. Fairbridge, 1957.
Carbonates and carbon dioxide. Ch. 10 in: Treatise on Marine Ecology and Paleoecology, Vol 1. Mem., Geol. Soc., Amer., 67:239-296.

Chemistry, bicarbonate
Rotschi, H., 1961.
Chimie.
O.R.S.T.O.M., I.F.O., Rapp. Sci., 23:4-45.

Chemistry carbonates
Schachter, D., 1954.
Contribution à l'étude hydrographique et hydrologique de l'étang de Berre (Bouches-du-Rhone).
Bull. Inst. Océan., Monaco, No. 1048:20 pp.

chemistry, carbonate
* Schmalz, Robert F., and Frederick J. Swanson, 1969.
Diurnal variations in the carbonate saturation of sea water.
J. sed. Petr., 39(1): 255-267.

Chemistry - carbonate
Skirrow, Geoffrey, 1965.
The dissolved gases - carbon dioxide.
In: Chemical oceanography, J.P. Riley and G. Skirrow, editors, Academic Press, 1:227-322.

chemistry, carbonates
Weber, Jon N. and Peter M.J. Woodhead, 1972
Temperature dependence of oxygen-18 concentration in reef coral carbonates. J. geophys. Res. 77(3): 463-473.

carbonate, biogenic

carbonate,biogenic
*Keary, Raymond, 1967.
Biogenic carbonate in beach sediments of the west coast of Ireland.
Scient.Proc.R.Dublin Soc.(A)3(7):75-85.

carbonate, data only

Chemistry, carbonate (total) (data only)
Japan Maritime Safety Agency 1966.
Results of oceanographic observations in 1963.
Data Rept. Hydrogr. Obs. Ser. Oceanogr. Publ. 792(3): 74 pp.

Carbonate(data only)
India, Naval Hdqtrs., Office of Scientific Research and Development, 1957.
Indian Oceanographic Stations List No. 1/57: 5 pp. (mimeographed)

Carbonate deposition of

chemistry - carbonate, deposition of
Lalou, C., 1954.
Sur un mécanisme bactérien possible dans la formation des dépôts de carbonates dépourvus d'organismes.
C.R.S. Soc. Geol., France, No. 14:369-371.

chemistry carbon, detrital
Strickland, J.D.H., Lucia Solórzano and R.W. Eppley, 1970
General Introduction, hydrography and chemistry.
Bull. Scripps Inst. Oceanogr. 17: 1-22.

Chemistry - Carbonate dissolution
*Gartner, Stefan, Jr., 1970.
Sea-floor spreading, carbonate dissolution level and the nature of horizon A. Science, 169(3950) 1077-1079.

Carbonate, effect of

Chemistry carbonate, effect of
Alekin, O.A., and N.P. Moricheva, 1958.
Content of organic matter in natural waters as affected by the carbonate system. (In Russian).
Doklady, Akad. Nauk, SSSR, 119(2):322-

Carbonate equilibria

Chemistry - carbonate equilibria
Deffeyes, Kenneth S., 1965.
Carbonate equilibria: a graphic and algebraic approach.
Limnol. Oceanogr., 10(3):412-426.

Carbonate, particulate

Chemistry - carbonate, particulate
Wangersky, Peter J., and Donald C. Gordon, Jr., 1965.
Particulate carbonate, organic carbon and Mn^{++} in the open ocean.
Limnol. Oceanogr., 10(4):544-550.

Chemistry - carbonates (precipitated)
Arrhenius, Gustaf, 1967.
Deep-sea sedimentation: a critical review of U.S. work.
Trans. Am. geophys. Un., 48(2):604-631.

chemistry - carbonate precipitation
Lalou, Mlle., 1955.
La précipitation expérimentale des carbonates par les bactéries. Trav. Lab. Géol. Sous-Marine, 6:28-29.

carbonate precipitation
Towe, K.M., and Philip G. Malone, 1970
Precipitation of metastable carbonate phases from seawater.
Nature, Lond., 226(5243): 348-349.

carbon-chlorophyll

Carbon/chlorophyll
Lorenzen, Carl J. 1968.
Carbon/chlorophyll relationships in an upwelling area.
Limnol. Oceanogr. 13(1): 202-204.

carbon-chlorophyll
Steele, J.H., and I.E. Baird, 1962.
Carbon-chlorophyll relations in cultures.
Limnol. and Oceanogr., 7(1):101-102.

Carbon cycle

Chemistry carbon cycle
Eriksson, E., and P. Welander, 1956.
On a mathematical model of the carbon cycle in nature. Tellus, 8(2):155-174.

CO_2

chemistry, carbon dioxide
Akiyama, T., 1969.
Carbon dioxide in the atmosphere and in sea water in the Pacific Ocean east of Japan.
Oceanogrl Mag., 21(2): 129-135.

chemistry, carbon dioxide
Akiyama Tsutomu 1969.
Carbon dioxide in the atmosphere and in sea water over the western North Pacific Ocean.
Oceanogrl Mag. 21(2):121-127.

Chemistry, carbon dioxide
Akiyama Tsutomu 1969
Carbon dioxide in the atmosphere and in sea water in the adjacent seas of Japan.
Oceanogrl Mag. 21(1):53-59.

chemistry, carbon dioxide
Akiyama, Tsutomu 1968.
Partial pressure of carbon dioxide in the atmosphere and in sea water over the western North Pacific Ocean.
Oceanogrl Mag. 20(2): 133-146.

chemistry, carbon dioxide
Baker, F.W.G., 1969.
Carbon dioxide.
Annls int. geophys. Year, 46: 191-194

Carbon dioxide 164

Chemistry carbon dioxide
Ben-Yaakov, Sam, 1971.
A multivariable regression analysis of the vertical distribution of TCO_2 in the eastern Pacific. J. geophys. Res. 76(30): 7417-7431.

chemistry - carbon dioxide
Berger, Rainer, and W.F. Libby, 1969
Equilibration of atmospheric carbon dioxide with sea water: possible enzymatic control of the rate.
Science, 164(3886): 1395-1397.

Chemistry carbon dioxide (data)
Beyers, Robert J., James L. Larimer, Howard T. Odum, Richard B. Parker and Neal E. Armstrong, 1963.
Directions for the determination of changes in carbon dioxide concentration from changes in pH.
Publ. Inst. Mar. Sci., Port Aransas, 9:454-489.

chemistry - carbon dioxide
Bolin, B., 1960
On the exchange of carbon dioxide between the atmosphere and the sea. Tellus, 12(3): 274-281

chemistry, carbon dioxide
Brannon, H.R., Jr., A.C. Daughtry, D. Perry, W.W. Whitaker and M. Williams, 1957.
Radiocarbon evidence on the dilution of atmospheric and oceanic carbon by carbon from fossil fuels. Trans. A.G.U., 38(5):643-650.

chemistry, carbon dioxide
Broecker, W., 1963
4. Radioisotopes and large scale oceanic mixing. In: The Sea, M.N. Hill, Edit., Vol. 2. The composition of sea water, Interscience Publishers, New York and London, 88-108.

chemistry, carbon dioxide
Broecker, Wallace S., Yuan-Hui Li and Tsung-Hung Peng, 1971.
Carbon dioxide - man's unseen artifact.(11)In: Impingement of man on the oceans, D.W. Hood, editor, Wiley Interscience: 287-324.

Chemistry carbon dioxide
Buch, Kurt, 1960
Dissoziation der Kohlensäure, Gleichgewichte und Puffersysteme. Handbuch der Pflanzenphysiologie, 5 (Edit. A. Pirson, Springer-Verlag, Berlin, Gottingen-Heidelberg):1-11

chemistry carbondioxide
Buch, Kurt, 1960
Kohlendioxyd im Meerwasser. Handbüch der Pflanzenphysiologie 5:47-61.
Edited by A. Pirson, Springer-Verlag, Berlin-Göttingen-Heidelberg.

Chemistry, carbon dioxide
Buch, K., 1952.
The cycle of nutrient salts and marine production. Rapp. Proc. Verb., Cons. Perm. Int. Expl. Mer, 132:36-46, 4 textfigs.

Chemistry, carbon dioxide
Buch, K., 1951.
Das Kohlensäure Gleichgewichtssystem im Meerwasser kritische Durchsicht und Neuberechnungen der Konstituenten. Havsforskningsinstitutets Skrift. No. 151:18 pp.

Chemistry, carbon dioxide
Buch, K., 1949.
Über den biochemischen Stoffwechsel in der Ostsee
Kieler Meeresforschungen 6:31-44, 6 textfigs.

Chemistry carbon dioxide
Buch, K., 1945.
Kolsyrejämvikten i Baltiska havet. Fennia 68(5): 1-208.

Chemistry, carbon dioxide
Buch, K., 1942.
Kohlensäure in Atmosphäre und Meer. Ann. Hydr., usw., 70:193-205.

Chemistry, carbon dioxide
Buch, K., 1942.
Kohlensäure in Atmosphäre und Meer. Forsch. u. Fortschr. 18:216-217.

Chemistry, carbondioxide
Buch, K., 1939.
Kohlensäure in Atmosphäre und Meer an der Grenze zum Arkticum. Acta Acad. Aboensis, Math., Phys., 11(12):41 pp., 3 textfigs.

Chemistry, carbon dioxide
Buch, K., 1939.
Beobachtungen über das Kohlensäure-gleichgewicht und über den Kohlensäure-austausch zurwischen Atmosphäre und Meer im Nordatlantischen Ozean. Acta Academiae Aboensis, Math. et Phys. XI.9:31 pp., 2 text figs.

Chemistry, carbon dioxide
Buch, K., 1938.
New determination of the second dissociation constant of carbonic acid in sea water. Acta Acad. Aboensis, Math. Phys., 11(5):17 pp., 1 textfig.

Chemistry, Carbon dioxide
Buch, K., 1934.
Beobachtungen über chemische Faktoren in der Nordsee, zwischen Nordsee und Island sowie auf dem Schelfgebiete nördlich von Island. Rapp. Proc. Verb. 89(3)13-31, 8 tables, 2 textfigs.

Chemistry, carbon dioxide
Buch, K., 1933.
Der Borsäuregehalt des Meerwassers und seine Bedeutung bei der Berechnung des Kohlensäuresystems im Meerwasser. Ergängen zum Bericht über die im Frühjahr 1931 in Helsingfors von einer Arbeitskommission der Internationalen Meeresforschung ausgeführten Untersuchungen. Rapp. Proc Verb. 85:71-75, 1 fold-in.

Chemistry, carbon dioxide
Buch, K., 1933.
On boric acid in the sea and its influence on the carbonic acid equilibrium. J. du Cons. 8(3):309-325, 2 textfigs.

Chemistry, carbon dioxide
Buch, K., and S. Gripenberg, 1938.
Jahreszeitlicher Verlauf der chemischen und biologischen Faktoren im Meerwasser bei Hangö im Jahre 1935. Havforskningsinstitutets Skrift No. 118:26 pp., 5 textfigs.

chemistry, carbon dioxide
Buch, K., H.W. Harvey, and H. Wattenberg, 1931
Die scheinbaren Dissoziationskonstanten der Kohlensäure in Seewasser verschiedenen Salzgehalts. Naturwiss. 19(37):773-

Trans cited FWS Spec. Sci. Rept., Fish. 227.

CO_2 1931

Chemistry, carbon dioxide
Buch, K., H.W. Harvey, H. Wattenberg, and S. Gripenberg, 1932.
Über das Kohlensäuresystem im Meerwasser. Bericht über die im Frühjahr 1931 im Helsingfors von einer Arbeitskommission der Internationale Meeresforschung ausgeführten Untersuchungen. Cons. Perm. Int. Expl. Mer, Rapp. Proc. Verb. 79:70 pp., 9 textfigs., 1 fold-in.

chemistry, carbon dioxide (air)
Büchen, Matthias 1971.
Engebnisse der CO_2-Konzentrationsmessung in der ozeannahen Luftschicht und im Oberflächenwasser während der Atlantischen Expedition 1969.
Meteor-Forschungsergebnisse (B) 7:55-70

Chemistry - carbon dioxide
Burkholder, Paul R., 1960.
Distribution of some chemical values in Lake Erie. Limnological survey of western and central Lake Erie, 1928-1929. USFWS Spec. Sci. Rept., Fish., No. 334:71-110.

Chemistry, carbon dioxide
Caspers, H., 1951.
Quantitative Untersuchungen über die Bodentierwelt des Schwarzen Meeres im bulgarischen Küstenbereich. Arch. f. Hydrobiol. 45(1/2):192 pp., 66 textfigs.

Carbon dioxide
*Choe, Sang, Tai Wha Chung and Hi-Sang Kwak, 1968.
Seasonal variations in nutrients and principal ions contents of the Han River water and its water characteristics. (In Korean; English abstract).
J. oceanogr. Soc., Korea, 3(1):26-38.

Chemistry, carbon dioxide
Cooper, L. H.N., 1933.
Chemical constituents of biological importance in the English Channel, November 1930 to January 1932. II. Hydrogen ion concentration, excess base, carbon dioxide and oxygen. J.M.B.A., 18: 729-753.

chemistry, carbon dioxide
Cooper, L.H.N., 1948.
The nutrient balance in the sea. Research 1:

Chemistry - carbon dioxide
Craig, H., 1957.
The natural distribution of radiocarbon and the exchange time of carbon dioxide between atmosphere and sea. Tellus 9(1):1-17.

chemistry carbon dioxide
Dingle, A.N., 1954.
The carbon dioxide exchange between the North Atlantic Ocean and the atmosphere. Tellus 6(4): 342-350, 4 textfigs.

Chemistry, carbon dioxide
Ercegovic, A., 1935.
Recherches sur l'alcalinité et l'équilibre de l'acide carbonique dans les eaux cotières de l'Atlantique oriental moyen. Acta Adriatica 1(7): 1-84.

Chemistry, carbon dioxide
Fiske, John D., Clinton E. Watson and Philip G. Coates, 1966.
A study of the marine resources of the North River.
Comm. Mass., Div. Mar. Fish., Monogr. Ser. 3:53 pp.

Chemistry, carbon dioxide
Galtsoff, P.S., ed., 1954.
Gulf of Mexico, its origin, waters and marine life. Fish. Bull., Fish and Wildlife Service, 55:1-604, 74 textfigs.

Chemistry, carbon dioxide
Hamm, R. E., and T.G. Thompson, 1941.
Dissolved nitrogen in the sea water of the northwest Pacific with notes on the total carbon dioxide and dissolved oxygen. J. Mar. Res., 4(1):11-27.

Chemistry carbon dioxide
Hanya, Takashisha and Ryoshi Ishiwatari, 1961
Carbon dioxide exchange between air and sea water. 1.
J. Oceanogr. Soc., Japan, 17(1):28-32.

Chemistry carbon dioxide
Hood, Donald W., editor 1971
Impingement of man on the oceans.
Wiley-Interscience, 738 pp.

Chemistry carbon dioxide
Hood, D.W., 1963
Chemical oceanography. In: Oceanography and Marine Biology, H. Barnes, Edit., George Allen & Unwin, 1:129-155.

Chemistry - carbon dioxide
*Hoover, Thomas E., and David C. Berkshire, 1969.
Effects of hydration on carbon dioxide exchange across an air-water interface.
J. geophys. Res., 74(2):456-464.

carbon dioxide
Ivanenkov, V.N., 1964.
Distribution of PCO2, pH, alk/cl in the northern part of the Indian Ocean. Investigations in the Indian Ocean (33rd voyage of E/S "Vitiaz"). (In Russian).
Trudy Inst. Okeanol., Akad. Nauk, SSSR, 64:128-143.

Chemistry, carbon dioxide
Kändler, R., 1930.
Beiträge zur Kenntnis über die Beziehungen zwischen Wasserstoffionen-Konzentration, freier Kohlensäure und Alkalinität im Meerwasser. Cons. Perm. Int. Expl. Mer, Rapp. Proc. Verb. 67: 89-90.

Chemistry carbon dioxide
Kanwisher, John W., 1963
On the exchange of gases between the atmosphere and the sea.
Deep-Sea Research, 10(3):195-207.

Chemistry, carbon dioxide
Kay, H., 1954.
Eine Mikromethode zur chemischen Bestimmung des organisch gebundenen Kohlenstoffs im Meerwasser. Kieler Meeresf. 10(1):26-36, Textfigs. 16-17.

Chemistry, carbon dioxide
Keeling, Charles D. 1968.
Carbon dioxide in surface ocean waters. 4 Global distribution.
J. geophys. Res. 73(14): 4543-4553.

Chemistry, carbon dioxide
Keeling, Charles D. 1965.
Carbon dioxide in surface waters of the Pacific Ocean. 2. Calculation of the exchange with the atmosphere.
J. geophys. Res. 70(24): 6099-6102.

Chemistry, carbon dioxide
Keeling, Charles D., and Lee S. Waterman 1968.
Carbon dioxide in surface ocean waters. 3. Measurements on Lusiad Expedition, 1962-1963.
J. geophys. Res. 73(14): 4529-4541.

Chemistry, carbon dioxide
Kelley, J.J. 1968.
Carbon dioxide in the seawater under the Arctic ice.
Nature, Lond. 218(5144): 862-864

Chemistry carbon dioxide
Kelley, J.J. and D.W. Hood, 1971.
Carbon dioxide in the Pacific Ocean and Bering Sea: upwelling and mixing. J. geophys. Res., 745-752.

chemistry, carbon dioxide
Kelley, J.J., L.L. Longerich and D.W. Hood, 1971.
Effect of upwelling, mixing, and high primary productivity on CO_2 concentrations in surface waters of the Bering Sea. J. geophys. Res. 76(36): 8687-8693.

Chemistry - carbon dioxide
Kitamura, Hiroyuki, and Takeshi Saga, 1964.
On the chemical elements in the sea south of Honshu, Japan. (In Japanese; English abstract).
Bull. Kobe Mar. Obs., No. 172:6-54.

chemistry - carbon dioxide
Kitching, J.A., S.J. Lilly, S.M. Lodge, J.F. Sloane, R. Bassindale, and F.J. Ebling, 1952.
The ecology of the Lough Ine Rapids with special reference to water currents. J. Ecology 40(1): 179-201.

Chemistry carbon dioxide
*Kühme, Heinrich, 1968.
Untersuchungen der Konzentration atmosphärischen Spurengase über dem Atlantik.
Meteor Forschungsergeb, (A)2:83-93.

Chemistry - carbon dioxide (data)
Lemasson, L., et Y. Magnier, 1966.
Résultats des observations scientifiques de Da Dunkerquoise sous le commandement du Capitaine de Corvette Brosset. Croisière Hunter.
Cah. ORSTOM, Sér. Océanogr., 4(1):3-78.

Chemistry - carbon dioxide assimilation
Lewin, Ralph A., Editor, 1962.
Physiology and biochemistry of algae.
Academic Press, New York and London, 9292 pp.

chemistry, carbon dioxide
Lyakhin, Yu.I., 1971.
Carbon dioxide exchange between the ocean and the atmosphere in the southeastern Atlantic.
Okeanologiia 11(1): 48-52. (In Russian; English abstract).

Chemistry carbon dioxide
Lyman, John and Jacob Verduin, 1961.
Changes in pH and total CO2 in natural waters.
Limnol. & Oceanogr., 6(1):80-83.

Chemistry carbon dioxide
Makkaveeva, N.S., 1965.
The hydrochemical investigations of the Korovinskaya Inlet. (In Russian).
Material. Ribochoz. Issled. Severn. Bassin. Poliarn. Nauchno-Issled. i Proekt. Inst. Morsk. Rib. Choz. i Okeanogr., N.M. Knipovich, 5:117-122.

Chemistry carbon dioxide (data)
Masuzawa, Jotaro, and Hideo Akamatsu, 1962.
1. Hydrographic observations during the JEDS-4.
Oceanogr. Mag., Japan Meteorol. Agency, 13(2): 122-130.

Chemistry carbon dioxide (data)
Matsudaira, Yasuo, 1964
Cooperative studies on primary productivity in the coastal waters of Japan, 1962-63. (In Japanese; English abstract).
Inform. Bull., Planktol., Japan, No. 11:24-73.

Chemistry carbon dioxide
MacIntyre, W.G., and R.F. Platford, 1964.
Dissolved calcium carbonate in the Labrador Sea.
J. Fish. Res. Bd., Canada, 21(6):1475-1480.

Chemistry carbon dioxide
Milburn, T.R., and L.C. Beadle, 1960
The determination of total carbon dioxide in water. J. Exp. Biol. 37(3): 444-460.

chemistry, carbon dioxide
*Miyake, Yasuo, and Yukio Sugimura, 1969.
Carbon dioxide in the surface water and the atmosphere in the Pacific, the Indian and the Antarctic Ocean areas.
Rec. oceanogr. Wks. Japan, 10(1):23-28.

Chemistry carbon dioxide
Miyake, Yasuo, and Yoshio Sugiura, 1962
Vertical distribution of CO2 in air and ocean near sea surface.
Rec. Oceanogr. Wks., Japan, Spec. No. 6:219-221.

chemistry, carbon dioxide
Murray, C.N. and J.P. Riley, 1971.
The solubility of gases in distilled water and sea water - IV. Carbon dioxide. Deep-Sea Res. 18(5): 533-541.

Chemistry carbon dioxide
Neumann, W., 1949.
Beiträge zur Hydrographie des Vrana-Sees (Insel Cherso), inbesondere Untersuchungen über organisches sowie anorganische Phosphor- und Stickstoffverbindungen. Nova Thalassia 1(6):17 pp.

Chemistry carbon dioxide
Park, Kilho, 1965.
Total carbon dioxide in sea water.
J. oceanogr. Soc., Japan, 21(2):54-59.

Chemistry - carbon dioxide
Park, P. Kilho, Louis I. Gordon, Stephen W. Hager, and Milton C. Cissell, 1969.
Carbon dioxide partial pressure in the Columbia River.
Science, 166 (3907): 867-868

Chemistry carbon dioxide
Park, Kilho, and Donald W. Hood, 1963
Radiometric field measurements of CO2 exchange from the atmosphere to the sea.
Limnol. and Oceanogr., 8(2):287-289.

Chemistry carbon dioxide
Postma, H., 1964.
The exchange of oxygen and carbon dioxide between the ocean and the atmosphere.
Netherlands J. Sea Res., 2(2):258-283.

chemistry, carbon dioxide
Quinn, J.A. and N.C. Otto, 1971.
Carbon dioxide exchange at the air-sea interface: flux augmentation by chemical reaction.
J. geophys. Res., 76(6): 1539-1549.

Chemistry - carbon dioxide
Rakestraw, N. W. and D. E. Carritt, 1948
Some seasonal chemical changes in the open ocean. J. Mar. Res. 7(3):362-369, 6 text figs.

Chemistry, Carbon dioxide
Redfield, A. C., 1934.
On the proportions of organic derivatives in sea water and their relation to the composition of plankton. In: James Johnstone Memorial Volume:176-192, 5 textfigs.

Chemistry - carbon dioxide
Revelle, R., and R. Fairbridge, 1957.
Carbonates and carbon dioxide. Ch. 10 in: Treatise on Marine ecology and Paleoecology, Vol. 1. Mem. Geol. Soc., Amer., 67:239-296.

Chemistry carbon dioxide
Revelle, R., and H.E. Suess, 1962
Gases, Ch. 7, Sect. II, Interchange of properties between sea and air. In: The Sea, Interscience Publishers, Vol. 1, Physical Oceanography, 313-321.

Chemistry - carbon dioxide
Revelle, R. and H.E. Suess, 1957.
Carbon dioxide exchange between atmosphere and ocean and the question of an increase of atmospheric CO2 during the past decades. Tellus 9(1): 18-27.

Chemistry - carbon dioxide
Richards, F.A., 1957.
Some current aspects of chemical oceanography.
In: Physics and chemistry of the earth, Pergamon Press, 2:77-128.

chemistry - carbon dioxide
Richards, Francis A., Joel D. Cline, William W. Broenkow and Larry P. Atkinson, 1965.
Some consequences of the decomposition of organic matter in Lake Nitinat, an anoxic fjord.
Limnol. and Oceanogr., Redfield Vol., Suppl. to 10:R185-R201.

chemistry carbon dioxide
Riley, J.P., 1965.
Analytical chemistry of sea water.
In: Chemical oceanography, J.P. Riley and G. Skirrow, editors, Academic Press, 2:295-424.

Chemistry carbon dioxide
Rotschi, Henri, 1965.
Le pH et l'alcalinité des eaux profondes de la Fosse des Hebrides et du Bassin des Fidji.
Progress in Oceanography, 3:301-310.

Chemistry, carbon dioxide
Rotschi, H., 1961.
Chimie.
O.R.S.T.O.M., I.F.O., Rapp. Sci., 23:4-45.

chemistry carbon dioxide

Rotschi, Henri, 1961(1962).
Oxygène, phosphate et gaz carbonique total en Mer de Corail.
Deep-Sea Res., 8(3/4):181-195.

chemistry, carbon dioxide (total) (data)

Rotschi, Henri, 1960
Orsom III, Resultats des Croisières diverses de 1959. Oceanographie physique.
Centre d'Océanogr., Inst. Français d'Océanie Rapp. Sci., No. 17: 59 pp.

chemistry carbon dioxide

Rotschi, Henri, 1959
Chimie, "Orsom III", Resultats de la Croisière BOUSSOLE.
O.R.S.T.O.M., Inst. Français d'Océanie, Rapp. Sci. 13: 3-60.

chemistry, carbon dioxide (data)

Rotschi, Henri, Michel Angot and Roger Desrosieres, 1960.
Orsom III, Resultats de la croisiere "Choiseul" 2ème partie. Chimie, productivité, phytoplancton qualitatif.
Rapp. Sci., Noumeau, No. 16:91 pp. (mimeographed).

chemistry, carbon dioxide (data)

Rotschi, Henri, Michel Angot, et Michel Legand, 1959
Orsom III. Resultats de la croisiere "Astrolabe" 2. Chimie, productivite et zooplankton
Rapp. Sci., Inst. Francais d'Oceanie, No. 9: 97 pp. (mimeographed).

chemistry carbon dioxide

Rouch, J., 1946
Traité d'Océanographie physique. L'eau de mer, Payot, Paris, 349 pp., 150 text figs.

chemistry carbon dioxide

Sargent, M.C., and J.C. Hindman, 1943
The ratio of carbon dioxide consumption to oxygen evolution in sea water in the light. J. Mar. Res. 5:131-135.

Chemistry, carbon dioxide

Sargent, M.C., and J.C. Hindman, 1943
The ratio of carbon dioxide consumption to oxygen evolution in sea water in the light. J. Mar. Res. V(2):131-135.

Chemistry carbon dioxide

Saruhashi, K., 1953.
On the total carbonaceous matter and hydrogen ion concentration in sea water - a study on the metabolism of natural water (1): Pap. Met. Geophys. 3(3):202-206, 2 textfigs.

chemistry, carbon dioxide

Saruhashi, K., 1952.
A study on pH and the total carbon dioxid in sea water. Umi to Sora 30(1/2):26-29, 2 textfigs.

chemistry, carbon dioxide

Schulek, E., J. Trompler, A. Endroi-Havas and L. Remport, 1961
Bestimmung sehr kleiner Mengen Kohlendioxyds und die analytische Anwendung des Verfahrens.
Analyt. Chim. Acta, 24(1): 11-19.

chemistry - carbon dioxide (data)

Schulz, B., 1923.
Hydrographische Untersuchungen besonders über den Durchlüftungszustand in der Ostsee in Jahre 1922. (Forschungsschiffe "Skagerak" und "Nautilus"). Arch. Deutschen Seewarte 41(1):1-64, 2 textfigs., 5 pls.

chemistry carbon dioxide (data)

Schulz, B., 1922.
Hydrographischen Beobachtungen inbesondere über die Kohlensäure in der Nord- und Ostsee im Sommer 1921. (Forschungsschiffe "Poseidon" und "Skagerak"). Arch. Deutschen Seewarte 40(2):1-44, 2 textfigs., 4 pls.

chemistry - carbon dioxide

*Seki, Humitake, 1967.
Effect of organic nutrients on dark assimilation of carbon dioxide in the sea.
Inf.Bull.Planktol.Japan, Comm.No.Dr.Y.Matsue, 201-205.

chemistry - carbon dioxide

Skirrow, Geoffrey, 1965.
The dissolved gases - carbon dioxide.
In: Chemical oceanography, J.P. Riley and G. Skirrow, editors, Academic Press, 1:227-322.

chemistry carbon dioxide

Slocum, G., 1955.
Has the amount of carbon dioxide changed significantly since the beginning of the twentieth century?
Monthly Weather Review, 83(10):225-231.

chemistry, carbon dioxide

Spencer, C.P., 1965.
The carbon dioxide system in sea water: a critical appraisal.
In: Oceanography and Marine Biology, H. Barnes, editor, George Allen and Unwin, 3:31-57.

chemistry, carbon dioxide

Subba Rao, D.V., 1965.
The measurement of total carbon dioxide in dilute tropical waters.
Australian J. Mar. Freshwater Res. 16(3):273-280.

Chemistry, carbon dioxide

Sugiura, Yoshio 1966.
The total carbon dioxide in the ocean.
Pap. Met. Geophys. Tokyo 16 (3/4):230-236.

chemistry carbon dioxide

Sugiura, Yoshio, E.R. Ibert and D.W. Hodd, 1963
Mass transfer of carbon dioxide across sea surfaces.
J. Mar. Res., 21(1):11-24.

chemistry, carbon dioxide

*Sugiura, Yoshio, and Hirozo Yoshimura, 1967.
Total carbon dioxide and its bearing on the dissolved oxygen in the Oyashio and in the frontal region of the Kuroshio.
Geochem.J., Japan, 1(3):125-130.

Chemistry, Carbon dioxide

Taft, Bruce A., and John A. Knauss 1967.
The Equatorial Undercurrent of the Indian Ocean as observed by the Lusiad Expedition.
Bull. Scripps Inst. Oceanogr. 9:1-163.

chemistry carbon dioxide

Takahashi, Taro, 1961
Carbon dioxide in the atmosphere and in Atlantic Ocean water. J. Geophys. Res., 66(2): 477-494.

chemistry, carbon dioxide

Teal, John M., 1967.
Biological production and distribution of pCO_2 in Woods Hole waters.
In: Estuaries, G.H. Lauff, editor, Publs Am. Ass. Advmt Sci., 83:336-340.

Chemistry - Carbon dioxide

Teal, John M., and John Kanwisher, 1966.
The use of pCO_2 for the calculation of biological production, with examples from waters off Massachusetts.
J. Mar. Res., 24(1):4-14.

chemistry, carbon dioxide

Traganza, Eugene D., 1967.
Dynamics of the carbon dioxide system on the Great Bahama Bank.
Bull. mar. Sci., Miami, 17(2):348-366.

chemistry - carbon dioxide

Voipo, Aarno, 1963.
The mass transfer coefficient of carbon dioxide between the gas and solution phase. II. The effect of electrolytes in the aqueous phase.
Suomen Kemistilehti (B), 36:79-81.

chemistry carbon dioxide

Voipio, Aarno, and Erkki Häsänen, 1963
The mass transfer coefficient of carbon dioxide between the gas and solution phase. III. The process in non-aqueous solvents.
Suomen Kemistilehti, (B) 36:187-190.

chemistry - carbon dioxide

Ward, Ronald W., Valerie Vreeland, Charles H. Southwick and Anthony J. Reading, 1965.
Ecological studies related to plankton productivity in two Chesapeake Bay estuaries.
Chesapeake Science, 6(4):214-225.

chemistry - carbon dioxide

Waterman, Lee S., 1965.
Carbon dioxide in surface waters.
Nature, 205(4976):1099-1100.

chemistry - carbon dioxide

Wattenberg, H., 1936.
Kohlensäure und Kalziumkarbonat im Meere.
Fortschr. Mineralog. Kristalog. Petrogr., 20:168-195.

Transl. cited:
USFWS Spec. Sci. Rept., Fish., 227.

chemistry, carbon dioxide

Yuan-Hui Li and Tien-Fung Tsui, 1971.
The solubility of CO_2 in water and sea water.
J. geophys. Res., 78(18): 4203-4207.

carbon dioxide/ ammonia

Chemistry - carbon dioxide/ammonia

Richards, Francis A., Joel D. Cline, William W. Broenkow and Larry P. Atkinson, 1965.
Some consequences of the decomposition of organic matter in Lake Nitinat, an anoxic fjord.
Limnol. and Oceanogr., Redfield Vol., Suppl. to, 10:R185-R201.

carbon dioxide anomalies

chemistry, carbon dioxide (anomalies)
Park, K., H.C. Curl, Jr., W.A. Glooschenko, 1967.
Large surface carbon dioxide anomalies in the North Pacific Ocean.
Nature, Lond., 215 (5099): 380-381.

Chemistry, carbon dioxide assimilation
Gabrielsen, E.K., and E. Steemann Nielsen, 1938.
Kohlensäureassimilation und Lichtqualität bei den marinen Planktondiatomeen (Vorlaufig Mitteilung).
Cons. Perm. Int. Expl. Mer, Rapp. Proc. Verb. 108(2):20-21, 2 textfigs.

CO_2, atmospheric

chemistry, carbon dioxide (atmospheric)
Bray, J. R., 1959
An analysis of the possible recent change in atmospheric carbon dioxide concentration.
Tellus, 11(2):220-230.

chemistry, carbon dioxide, atmospheric
Brown, Craig W., and Charles D. Keeling, 1965.
The concentration of atmospheric carbon dioxide in Antarctica.
J. geophys. Res., 70(24):6077-6085.

Chemistry - carbon dioxide (atmospheric)
Bruevich, S.V., and S.V. Lyutsarev, 1954.
Carbon dioxide content of the atmosphere above the Pacific and Indian Oceans and in the region of the Black Sea. (In Russian).
Doklady, Akad. Nauk, SSSR, 136(2):405-407.

Translations:
SLA Translations Center, John Crerar Library or Office of Technical Services, Dept. Commerce.

chemistry, carbon dioxide (data)
Bruneau, L., N.G. Jerlov, and F.F. Koczy, 1953.
Physical and chemical methods.
Repts. Swedish Deep-sea Exped., 1947-1948, Phys. Chem., 3(4):101-112, i-iv, 7 textfigs., 2 tables.

chemistry carbon dioxide
Buch, Kurt, 1960
Das Kohlendioxyd in der Atmospharischen Luft.
Handbuch der Pflanzenphysiologie, 5(A. Pirson, edit., Springer-Verlag, Berlin-Gottingen-Heidelberg):12-23

Chemistry, carbon dioxide (atmospheric)
Callendar, G.S., 1958.
On the amount of carbon dioxide in the atmosphere. Tellus, 10(2):243-248.

chemistry, carbon dioxide, atmospheric
Callendar, G.S., 1957.
The effect of fuel combustion on the amount of carbon dioxide in the atmosphere. Tellus 9(3):421-422.

chemistry, carbon dioxide (atmospheric)
Keeling, Charles D, Thomas B. Harris and Eugene M. Wilkins 1968.
Concentration of atmospheric carbon dioxide at 500 and 700 millibars.
J. geophys. Res. 73(14): 4511-4528.

chemistry, carbon dioxide (air)
Lutsarev, S.V., and S.W. Brujewicz, 1964.
Carbon dioxide content of the air over the Pacific and Indian Oceans and ocer the region of the Black Sea. Chemistry of the waters and sediments of seas and oceans. (In Russian: English abstract).
Trudy Inst. Okeanol., Akad. Nauk, SSSR, 67:7-40.

chemistry - carbon dioxide (atmospheric)
Miyake, M., and S. Matsuo, 1963.
A role of sea ice and sea water in the Antarctic on the carbon dioxide xyxle in the a tmosphere.
Papers in Meteorol. and Geophys., Tokyo, 14(2): 120-125.

chemistry, carbon dioxide (atmospheric)
Pales, Jack C, and Charles D. Keeling 1965.
The concentration of atmospheric carbon dioxide in Hawaii.
J. geophys. Res. 70(24): 6053-6076.

chemistry, carbon dioxide, atmospheric
Wilkins, E.M., 1961
Seasonal variations in atmospheric carbon dioxide concentration.
J. Geophys. Res., 66(4): 1314-1315.

Chemistry - carbon dioxide - carbonate system
Pytkowicz, Ricardo M., 1969.
The carbon dioxide-carbonate system at high pressures in the oceans. In: Oceanography and marine biology: an annual review, H. Barnes, editor, George Allen & Unwin Ltd., 6: 83-135.

chemistry - carbon dioxide-carbonate
Pytkowicz, Ricardo M., 1968.
The carbon dioxide-carbonate system at high pressures in the oceans. Oceanogr. Mar. Biol. Ann. Rev., 1968, 6: 83-135. Also in: Coll. Reprints, Dept. Oceanogr. Univ. Oregon, 7(1968).

CO_2: data only

Chemistry, carbon dioxide (total) (data only)
Japan, Japanese Meteorological Agency, 1967.
The results of marine meteorological and oceanographical observations, July-December, 1964, 36: 367 pp.

Chemistry, carbon dioxide (data only)
Japan, Japan Meteorological Agency, 1961
The results of marine meteorological and oceanographical observations, January-June 1960. The results of the Japanese Expedition of Deep-Sea (JEDS-2, JEDS-3), No. 27: 257 pp.

chemistry, carbon dioxide (data only)
Rotschi, Henri, Michel Angot, Michel Legand and Roget Desrosieres, 1961
Orsom III, Resultats de la Croisiere "Dillon", 2eme Partie. Chimie et Biologie.
ORSTOM, Inst. Francais d'Oceanie, Centre d'Oceanogr., Rapp. Sci., No. 19: 105 pp. (mimeographed).

chemistry, carbon dioxide (total) (data only)
Rotschi, H., Ph. Hisard, L. Lemasson, Y. Magnier, J. Noel et B. Piton, 1967.
Resultats des observations physico-chimiques de la croisiere "Alize" du N.O. Coriolis.
Rapp. scient., Off. Rech. scient. tech. Outre-Mer, Noumea, 2:56 pp. (mimeographed).

Chemistry carbon dioxide (data only)
Rotschi, Henri, Michel Legand and Roger Desrosieres, 1961
Orsom III, Croisières diverses de 1960, physique chimie et biologie. ORSTOM, Inst. Francais d'Océanie, Centre d'Océanogr., Noumea, Rapp. Sci., No. 20: 59 pp. (mimeographed).

chemistry, carbon dioxide (total) (data only)
Rotschi, H., Ph. Hisard L. Lemasson, Y. Magnier, J. Noel, et B. Piton, 1966.
Resultats des observations physico-chimiques de la croisiere "Alize".
Centre ORSTROM Noumea Rapp. Sci., No. 28:56 pp. (mimeographed).

Chemistry, carbon dioxide (data only)
Rotschi, Henri, Yves Magnier, Maryse Tirelli, et Jean Garbe, 1964.
Résultats des observations scientifiques de "La Dunkerquoise " (Croisière "Guadalcanal").
Océanographie, Cahiers O.R.S.T.R.O.M., 11(1):49-154.

carbon dioxide, effect of
Rasool, S.I. and S.H. Schneider, 1971.
Atmospheric carbon dioxide and aerosols: effects of large increases on global climate.
Science, 173(3992): 138-141.

carbon dioxide (fixation)

chemistry carbon dioxide (fixation)
Hammen, Carl S., 1964.
Carbon dioxide fixation in marine invertebrates: a review.
Narragansett Mar. Lab., Univ. Rhode Island, Occ. Publ., No. 2:48-50.

Chemistry, carbon dioxide (data) fixation
Rotschi, Henri, Michel Angot, Michel LeGand, and H.R. Jitt, 1959
Chimie, productivité et zooplancton. "Orsom III". Resultats de la Croisiere "Boussole". Resultats "production primaire' de la croisière 56-5.
Rapp. Sci. Inst. Francais d'Oceanie, Centre d'Oceanogr., No. 13:

chemistry - carbon dioxide, partial pressure of
Rudolf, Werner, 1971.
Eine Methode zur kontinuierlichen Analyse des CO_2-Partialdruckes im Meerwasser.
Meteor Forsch.-Ergebn. (B) 6:12-36.

chemistry, carbon dioxide (total) (data)
Franco, Paolo 1970.
Oceanography of northern Adriatic Sea. I. Hydrologic features: cruises July-August and October-November 1965.
Archo Oceanogr. Limnol. 16 (Suppl): 1-93.

Chemistry, carbon dioxide uptake
Australia, Commonwealth Scientific and Industrial Research Organization, 1961.
F.R.V. "Derwent Hunter".
C.S.I.R.O., Div. Fish. and Oceanogr., Rept., No. 32:56 pp.

chemistry carbon dioxide (uptake)
Australia, Marine Biological Laboratory, Cronulla, 1960.
F.R.V. "Derwent Hunter", scientific report of -- cruises 10-20/58 ----------.
C.S.I.R.O., Div. Fish. & Oceanogr., Rept., 30: 53 pp., numerous figs. (mimeographed).

for complete "title", see author card.

carbon, dissolved

chemistry carbon, dissolved (data)
Duursma, E.K., 1961.
Dissolved organic carbon, nitrogen and phosphorus in the sea.
Netherlands J. Sea Res., 1(1/2):1-147.

Carbon fixation

Chemistry
carbon fixation
Doty, Maxwell S., and Mikihiko Oguri, 1958
Primary production patterns in enriched areas.
Proc. Ninth Pacific Sci. Congr., Pacific Sci. Assoc., 1957, 16(Oceanogr.):94-97.

chemistry
carbon fixation (data)
Forsbergh, Eric D., William W. Broenkow, 1965.
Observaciones oceanograficas del oceano Pacific oriental recolectadas por el barco Shoyo Maru, octubre 1963-marzo 1964.
Comision Interamericana del Atun Tropical, Bol. 19(2):85-237.

chemistry, carbon dioxide uptake
Broecker, Wallace S., Yuan-Hui Li and Tsung-Hung Peng, 1971.
Carbon dioxide - man's unseen artifact. (11) In: Impingement of man on the oceans, D.W. Hood, editor, Wiley Interscience: 287-324.

carbonic acid

chemistry, carbonic acid
Buch, K., 1946.
[The carbonic acid equilibrium in the sea.] (In Finnish.) Suomen Kemistil., A, 19:4-9.

chemistry, carbonic acid
Culberson, C., D.R. Kester and R.M. Pytkowicz, 1967.
High-pressure dissociation of carbonic and boric acids in sea water.
Science, 157(3784):59-61.

Chemistry - carbonic acid
*Culberson, C., and R.M. Pytkowicz, 1968.
Effect of pressure on carbonic acid, boric acid, and the pH in seawater.
Limnol. Oceanogr., 13(3):403-417.

carbonic acid
Fox, C.J.J., 1909.
On the coefficients of absorption of nitrogen and oxygen in distilled water and sea water, and of atmospheric carbonic acid in sea water.
Trans. Faraday Soc., 5:68-87.

Chemistry, carbonic acid
Saruhashi, K., 1955.
On the equilibrium concentration ratio of carbonic acid substances dissolved in natural water. A study of metabolism in natural waters.
Pap. Meteorol. Geophys., Japan, 6(1):38-55.

Skirrow, Geoffrey, 1965. carbonic acid
The dissolved gases - carbon dioxide.
In: Chemical oceanography, J.P. Riley and G. Skirrow, editors, Academic Press, 1:227-322.

Carbon, inorganic

chemistry, carbon (inorganic)
Deuser, W.G. and J.M. Hunt, 1969.
Stable isotope ratios of dissolved inorganic carbon in the Atlantic. Deep-Sea Res. 16(2): 221-225.

chemistry, carbon, inorganic
Keeling, Charles D., and Bert Bolin, 1968
The simultaneous use of chemical tracers in oceanic studies. II. A three-reservoir model of the North and South Pacific oceans.
Tellus, 20(1): 17-54

chemistry, carbon (inorganic)
Ku, Teh-Lung, Yuan-Hui Li and Guy G. Mathieu 1969
Radium and inorganic carbon in Antarctic waters.
Antarctic Jl. U.S. 4(5):186-187.

chemistry, carbon, (inorganic)
Sackett, William M., and Willard S. Moore, 1966.
Isotopic variations of dissolved inorganic carbon.
Chem. geol., 1(4):323-328.

Carbon isotopes

Chemistry, carbon-14

See also: Chemistry, radiocarbon

Chemistry carbon 14
Arnold, J.R., and E.C. Anderson, 1957.
The distribution of Carbon-14 in nature.
Tellus 9(1):28-32.

Chemistry, Carbon - 14
Berger, Rainer, R.E. Taylor and W.F. Libby, 1966
Radiocarbon content of marine shells from the California and Mexican west coast.
Science, 153(3738):864-866.

Chemistry, carbon 14
Bolin, Bert and Henry Stommel, 1961
On the abyssal circulation of the World Ocean. IV. Origin and rate of circulation of deep ocean water as determined with the aid of tracers.
Deep-Sea Res., 8(2):95-110.

Chemistry, carbon-14
Burton, J.D., 1965.
Radioactive nuclides in sea water, marine sediments and marine organisms.
In: Chemical oceanography, J.P. Riley and G. Skirrow, editors, Academic Press, 2:425-475.

Chemistry - carbon 14
Cooper, L.H.N., 1956.
On assessing the age of deep oceanic water by Carbon 14. J.M.B.A., 35(2):341-354.

chemistry - carbon 13
Craig, H., 1970.
Abyssal carbon 13 in the South Pacific. J. geophys. Res., 75(3): 691-695.

chemistry, carbon isotopes
Degens, E.T., M. Behrendt, B. Gotthardt and E. Reppmann, 1968.
Metabolic fractionation of carbon isotopes in marine plankton. II. Data on samples collected off the coasts of Peru and Ecuador.
Deep-Sea Res., 15(1):11-20.

chemistry, carbon isotopes
Degens, E.T., R.R.L. Guillard, W.M. Sackett and J.A. Hellebust, 1968.
Metabolic fractionation of carbon isotopes in marine plankton. 1. Temperature and respiration experiments.
Deep-Sea Res., 15(1):1-9.

chemistry - carbon 13
Deuser, W.G. 1970.
Carbon-13 in Black Sea waters and implications for the origen of hydrogen sulfide.
Science 168(3939): 1575-1577.

chemistry - Carbon isotope fractionation
Deuser, W.G., and E.T. Degens, 1967
Carbon isotope fractionation in the system CO_2 (gas) - CO_2 (aqueous) - HCO_3^- (aqueous).
Nature, Lond., 215 (5105): 1033-1035

chemistry, carbon isotopes
Deuser, W.G., E.T. Degens and R.R.L. Guillard 1968
Carbon isotope relationships between plankton and sea water.
Geochim. cosmochim. Acta 32 (16): 657-660.

Chemistry
C14 uptake (data)
El-Sayed, Sayed Z., and Enrique F. Mandelli, 1965
Primary production and standing crop of phytoplankton in the Weddell Sea and Drake Passage.
In: Biology of Antarctic seas, II.
Antarctic Res. Ser., Am. Geophys. Union, 5:87-106.

Chemistry, carbon 14
Fonselius, S. and G. Östlund, 1959
Natural radiocarbon measurements on surface water from the North Atlantic and the Arctic Sea. Tellus, 11(1): 77-82.

chemistry, carbon -14 depletion (data)
Houtman, Th. J., 1966.
Repeat measurements of temperature, salinity, and carbon 14 depletion at an ocean station.
N.Z. J. Sci., 9 (2):457-471.

Chemistry, carbon 14
Knauss, John A., 1962
On some aspects of the deep circulation of the Pacific.
J. Geophys. Res., 67(10):3943-3954.

Chemistry, Carbon 14
Kulp, J.L., L.E. Tryon, W.E. Eckelmann, and W.A. Snell, 1952.
Lamont natural radiocarbon measurements. II.
Science 116(3016):409-414.

Chemistry Carbon 14
Kulp, L.J., L.E. Tryon, and H.W. Feely, 1952.
Trans. Amer. Geophys. Union 33(2):183-192.

Chemistry
carbon-14
Lasker, Reuben, 1960.
Utilization of organic carbon by a marine crustacean: analysis with carbon-14.
Science, 131(3407):1098-2000.

Chemistry, carbon 14
Legand, M., 1957.
Variations quantitatives du zooplancton récolté par l'Orsom III pendant la croisière EQUAPAC.
Off. Recherche Sci. et Tech. Outre-Mer, Inst. Français d'Océanie, Sect. Océanogr., Rapp. Sci., No. 2:31 pp., (Mimeographed).

Chemistry, carbon 14
Rafter, T.A., 1953.
The preparation of Carbon for C 14 age measurements. N.Z.J. Sci. Tech., B, 35(1):64-89, 11 text-figs.

Chemistry
carbon C^{13}
Rogers, M.A. and C.B. Koons, 1969.
Organic carbon C^{13} values from Quarternary marine sequences in the Gulf of Mexico: a reflection of paleotemperature changes. Trans. Gulf Coast Ass. geol. Socs, 19 : 529-534.

chemistry, carbon 14

Rubin, Meyer, and Dwight W. Taylor, 1963
Radiocarbon activity of shells from living clams and snails.
Science, 141(3581):637.

chemistry, carbon isotopes

Sackett, William M., Walter R. Eckelmann, Michael L. Bender and Allan W.H. Be, 1965.
Temperature dependence of carbon isotope composition in marine plankton and sediments.
Science, 148(3667):235-237.

chemistry, carbon-14

Servant, Jean 1966
La radioactivité de l'eau de mer.
Cahier océanogr. 18(4): 277-318.

Skirrow, Geoffrey, 1965. carbon isotopes
The dissolved gases - carbon dioxide.
In: Chemical oceanography, J.P. Riley and G. Skirrow, editors, Academic Press, 1:227-322.

chemistry, carbon 14

Somayajulu, B.L.K., D. Lal and S. Kusumgar, 1969
Man-made carbon-14 in deep Pacific waters: transport by biological skeletal material.
Science, 166(3911): 1397-1399.

chemistry 14C discrimination

Steemann Nielsen, 1955.
The interaction of photosynthesis and respiration and its importance for the determination of 14C-discrimination in photosynthesis.
Physiol. Plant., 8:945-953.

chemistry, 14-Carbon

Steemann Nielsen, E., and A.A.Al Kholy, 1956.
Use of 14 C-technique in measuring photosynthesis of phosphorus or nitrogen deficient algae.
Physiol. Plant. 9:144-153.

chemistry, carbon 14

Suess H.E., 1970.
The transfer of carbon 14 and tritium from the atmosphere to the ocean. J. geophys. Res., 75(12): 2363-2364.

chemistry, carbon 14

Thommeret, Jean, et Yolande Thommeret, 1967.
Répartition des teneurs en carbone 14 naturel dans divers constituants de la biophase d'un séd-iment superficiel de la Méditerranée occidentale.
Cah. océanogr., 19(6):495-504.

chemistry
Carbon 14

Thommeret, Jean, Yolande Thommeret et Jean Galliot, 1965.
Teneur en radiocarbone des eaux profondes et superficielles de l'océan Indian (Mer d'Oman).
Bull. Inst. océanogr., Monaco, 63(1347):8pp.

chemistry - carbon-14

Turekian, Karl K., 1965.
Some aspects of the geochemistry of marine sediments.
In: Chemical oceanography, J.P. Riley and G. Skirrow, editors, Academic Press, 2;81-126.

chemistry carbon 14

Ulubekova, M.V., and L.A. Kuzmina, 1953.
[Use of C14 in study of photo-reduction and chemosynthesis in green algae.]
Dokl. Akad. Nauk, SSSR, 93(5):915.

chemistry, carbon-14

Volchok, H.L., V.T. Bowen, T.R. Folsom, W.S. Broecker, E.A. Schuert and G.S. Bien, 1971.
Oceanic distributions of radionuclides from nuclear explosions. Radioactivity in the marine environment, U.S. Nat. Acad. Sci., 1971: 42-89.

chemistry, carbon isotopes

Williams, P.M., 1968.
Stable carbon isotopes in the dissolved organic matter of the sea.
Nature, Lond., 219 (5150): 152-153

carbon-14

Williams, P.M., J.A. McGowan and M. Stuiver, 1970
Bomb carbon-14 in deep sea organisms.
Nature, Lond., 227(5256): 375-376

carbon14/carbon12

chemistry
C14/C12

Broecker, Wallace S., 1963
C14/C12 ratios in surface ocean water. In: Nuclear Geophysics.
Nat. Acad. Sci.-Nat. Res. Council, Publ. No. 1075:138-149.

chemistry - C^{13}/C^{12}

*Williams, P.M. and L.I. Gordon, 1970.
Carbon-13: carbon-12 ratios in dissolved and particulate organic matter in the sea. Deep-Sea Res., 17(1): 19-27.

chemistry, carbon monoxide

Douglas, Everett, 1967.
Carbon monoxide solubilities in sea water.
J. phys. Chem. 71(6):1931-1933.

chemistry, carbon monoxide

Lamontagne, R.A., J.W. Swinnerton and V.J. Linnenbom, 1971.
Nonequilibrium of carbon monoxide and methane at the air-sea interface. J. geophys. Res. 76(21): 5117-5121.

chemistry - carbon monoxide, atmospheric

Seiler, W. and C. Junge, 1970.
Carbon monoxide in the atmosphere. J. geophys. Res., 75(12): 2217-2226.

chemistry - carbon monoxide

Swinnerton, J.W., V.J. Linnenbom, R.A. Lamontagne, 1970.
The ocean: a natural source of carbon monoxide.
Science, 167(3920): 984-986.

chemistry, carbon monoxide

Swinnerton, J.W., R.A. Lamontagne and V.J. Linnenbom, 1971.
Carbon monoxide in rainwater. Science 172 (3986): 943-945.

carbon, organic

chemistry, carbon, organic

Barber, Richard T., 1968.
Dissolved organic carbon from deep waters resists microbial oxidation.
Nature, Lond., 220(5164):274-275.

chemistry - organic carbon

Bernard, F., J. Lecal, and R. Codinet, 1950.
Études des sédiments marins au large de l'Alger.
1. Teneur en azote et carbone organique. Bull. Inst. Océan, Monaco, No. 963:10 pp., 4 textfigs., 1 fold-in.

chemistry
organic carbon

Bogorov, B.G., O.K. Bordovsky and M.E. Vinogradov 1966.
Biogeochemistry of the oceanic plankton. The distribution of some chemical components of the plankton in the Indian Ocean. (In Russian; English abstract).
Okeanologiia, Akad. Nauk, SSSR, 6(2):314-325.

chemistry
organic carbon

Camps, J.M., and E. Arias, 1963
Determinación del carbono orgánico presente en aguas de Castellón de la Plana.
Inv. Pesq., Barcelona, 23:125-131.

chemistry - carbon, organic

Camps, J.M., J. Selga and E. Arias, 1960
Determinacion del carbono organico en al agua de mar. Invest. Pesq. 16: 139-150.

chemistry - carbon, organic

Datzo, V.G., and V.E., 1950.
[Method of determination of organic carbon in natural waters.] Dokl. Akad. Nauk, SSSR, 73:337-339.

chemistry, carbon(organic) (budget)

Deuser, W.G., 1971.
Organic-carbon budget of the Black Sea. Deep-Sea Res. 18(10): 995-1004.

chemistry organic carbon

Duursma, E.K., 1960
Dissolved organic carbon, nitrogen and phosphorus in the sea. Netherlands J. Mar. Res. 1(1): 1-148.

chemistry
carbon (organic)

Duursma, E.K., 1959
Über gelöste organische Substanz auf einem Meridionalschnitt durch den nördlichen Nordatlantischen Ozean im April 1958.
Deutsche Hydrogr. Zeits., Ergänzungsheft, Reihe B, 4(3):69-72.

chemistry, organic carbon

Febres, German, y Gilberto Rodriguez, 1966.
Ch. 3. Sedimentos.
Estudios hidrobiologicos en el Estuario de Maracaibo, Inst. Venezolano de Invest. Cient., 66-82. (Unpublished manuscript).

chemistry, carbon (organic)

Fredericks, Alan D. and William M. Sackett, 1970.
Organic carbon in the Gulf of Mexico. J. geophy Res., 75(12): 2199-2206.

Chemistry, carbon (organic)
Gordon, Donald C., Jr. 1971.
Organic carbon budget of Fanning Island lagoon.
Pacific Sci. 25(2): 222-227.

chemistry, carbon organic
Gross, M.Grant, 1967.
Organic carbon in surface sediment from the northeast Pacific Ocean.
Int. J. Oceanol. Limnol., 1(1):46-54.

chemistry organic carbon
Holm-Hansen, Osmund, J.D.H. Strickland and P.M. Williams, 1966.
A detailed analysis of biologically important substances in a profile off southern California.
Limnol. Oceanogr., 11(4):548-561.

chemistry, organic carbon
Krey, Johannes, Peter H. Koske and Karl-Heinz Szekielda, 1965.
Produktionsbiologische und hydrographische Untersuchungen in der Eckernförder Bucht.
Kieler Meeresforsch., 21(2):135-143.

chemistry, organic carbon
Krogh, A. and A. Keys, 1934
Methods for the determination of dissolved organic carbon and nitrogen in sea water.
Biol. Bull., 67:132-144

chemistry organic carbon
Krylova, L.P., 1957.
Determination of carbon in the organic substances present in natural waters by means of dry combistion.
Gidrokhim. Materialy, 26:237.

chemistry organic carbon
Lasker, Reuben, 1960.
Utilization of organic carbon by a marine crustacean: analysis with carbon-14.
Science, 131(3407):1098-2000.

chemistry organic carbon
Lisitsin, A.P., 1964.
Distribution and chemical composition of suspended matter in the waters of the Indian Ocean. (In Russian); English Abstract.
Rezult. Issled., Programme Mezhd. Geofiz. Goda, Mezhd. Geofiz. Kom., Presidume Akad. Nauk, SSSR, Okeanol., No. 10:136 pp.

chemistry organic carbon
Lynn, W.R. and Won Tak Yang, 1960
The ecological effects of sewage in Biscayne Bay. Oxygen demand and organic carbon determinations. Bull. Mar. Sci. Gulf Caribb., 10 (4): 491-509.

chemistry carbon organic (fixation of)
Margalef, Ramón, 1964.
Modelos experimentales de poblaciones de fitoplancton: nuevas observaciones sobre pigmentos y fijación de carbono inorgánico.
Inv. Pesq. Barcelona, 26:195-203.

chemistry, carbon (organic)
McIver, R.D., 1971.
Organic geochemical analyses of frozen samples from dsdp leg 8 cores. Inétial Repts. Deep Sea Drilling Project, 8: 871-872.

chemistry carbon, organic
McKenzie, Kenneth G., 1964.
The ecologic associations of an ostracode fauna from Oyster Harbour, a marginal marine environment near Albany, Western Australia.
Pubbl., Staz. Zool., Napoli, 33(Suppl.):421-461.

chemistry organic carbon
Menzel, David W., 1964
The distribution of dissolved organic carbon in the Western Indian Ocean.
Deep-Sea Res., 11(5):757-765.

Chemistry, carbon (organic)
Menzel, David W. and John H. Ryther 1968.
Organic carbon and the oxygen minimum in the South Atlantic Ocean.
Deep-Sea Res. 15(3):327-337.

Chemistry carbon, organic
Menzel, David W., and Ralph F. Vaccaro, 1964
The measurement of dissolved organic and particulate carbon in seawater.
Limnology and Oceanography, 9(1):138-142.

chemistry carbon, organic
Okuda, Taizo, 1964.
Some problems for the determination of organic carbon in marine sediments.
Bol. Inst. Oceanogr., Univ. Oriente, Venezuela, 3(1/2):106-117.

chemistry carbon, organic
Okuda, Taizo, y Jose R. Gomez, 1964.
Distribucion del carbono y nitrogeno organico de los sedimentos en la region nororiental de Venezuela.
Bol. Inst. Oceanogr., Univ. Oriente, Venezuela, 3(1/2):91-105.

chemistry organic carbon
Richard, J. M., 1955.
Recherches sur le dosage du carbone organique dans les sediments marins. Resume du diplome d'etudes super., Paris, 355-358. (multilithed).
In: Trav. Sta. Zool., Villefranche-sur-mer, 15 (in MBL)

Chemistry, carbon, (organic)
Seki, Humitake, 1967.
Mineralization rate of organic carbon by microorganisms in the sea.
J. oceanogr. Soc., Japan, 23(1): 15-23.

carbon, organic
Skopintsev, B.A., 1965.
Calculation of the organic matter formation and oxidation in sea water. (In Russian; English abstract).
Okeanolog. Issled., Rez. Issled. po Programme Mezhd. Geofiz. Goda, Mezhd. Geofiz. Komitet, Prezidiume Akad. Nauk, SSSR, No. 13:96-107.

Chemistry, carbon (organic)
Skopintsev, B.A., N.N. Romenskaya and M.V. Sokolova 1968.
Organic carbon in the waters of the Norwegian Sea and the northeast Atlantic. (In Russian; English abstract)
Okeanologiia 8(2):225-234.

carbon, organic
Skopintsev, B.A., and S.N. Timofeeva, 1962.
Contents of organic carbon in the waters of the Baltic and North seas and in the subtropical and tropical Atlantic. (In Russian).
Trudy Morsk. Gidrofiz. Inst., 25:110-117.

Chemistry carbon, organic
Skopintsev, B.A., and S.N. Timofeeva, 1960.
Organic carbon in the waters of the northern part of the Black Sea.
Doklady Akad. Nauk, SSSR, 134(3):688-690.

carbon (organic)
Skopintsev, B.A., S.N. Timofeeva, A.F. Danilenko and M.V. Sokolova, 1967.
Organic carbon, nitrogen, phosphorus and their mineralization products in the Black Sea water. (In Russian; English abstract).
Okeanologiia, Akad. Nauk, SSSR, 7(3):457-469.

organic carbon
Skopintsev, B.A., S.N. Timofeeva and O.A. Vershinina, 1966.
Organic carbon in the waters of the near-equatorial and southern parts of the Atlantic Ocean and in the Mediterranean Sea. (In Russian ;English abstract).
Okeanologiia, Akad. Nauk, SSSR, 6(2):251-260.

Chemistry carbon (organic)
Starikova, N.D., 1970.
Patterns of the vertical distribution of dissolved organic carbon in the sea water and interstitial solutions. (In Russian; English abstract). Okeanologiia, 10(6): 988-1000.

chemistry, organic carbon
*Strathmann, Richard R., 1967.
Estimating the organic carbon content of phytoplankton from cell volume or plasma volume.
Limnol. Oceanogr., 12(3):411-418.

chemistry, carbon (organic)
Szekielda, Karl-Heinz 1971.
Organisch gelöster und partikulärer Kohlenstoff in einem Nebenmeer mit starken Salzgehaltsschwankungen (Ostsee).
Vie Milieu Suppl. 22(2): 579-612

Chemistry, carbon (organic)
*Szekielda, Karl-Heinz, 1968.
Vergleichende Untersuchungen über den Gehalt an organischem Kohlenstoff im Meerwasser und dem Kaliumpermanganatverbrauch.
J. Cons. perm. int. Explor. Mer. 32(1):17-24.

organic carbon
Uyeno, Fukuzo, 1964.
Relationships between production of foods and oceanographical condition of sea water in pearl farms. II. On the seasonal changes of sea water constituents and of bottom condition, and the effect of bottom cultivation. (In Japanese; English abstract)
J. Fac. Fish., Pref. Univ. Mie, 6(2):145-169.

chemistry, carbon, organic
Wangersky, Peter J., 1965.
The organic chemistry of sea water.
Am. Scient., 53(3):358-374.

carbon, organic
Wangersky, Peter J., and Donald C. Gordon, Jr., 1965.
Particulate carbonate, organic carbon and Mn^{++} in the open ocean.
Limnol. Oceanogr., 10(4):544-550.

Chemistry carbon, organic
*Williams, P.M., 1967.
Sea surface chemistry: organic carbon and organic and inorganic nitrogen and phosphorus in surface films and subsurface waters.
Deep-Sea Res., 14(6):791-800.

chemistry, carbon, organic (data)
Wilson, Ronald F., 1963.
Organic carbon levels in some aquatic ecosystems.
Publ. Inst. Mar. Sci., Port Aransas, 9:64-76.

chemistry organic carbon
Wilson, Ronald F., 1961
Measurement of organic carbon in sea water.
Limnol. & Oceanogr., 6(3): 259-261.

organic carbon
Wiseman, J.D.H., and H. Bennett, 1940.
The distribution of organic carbon and nitrogen in sediments from the Arabian Sea. John Murray Exped., 1933-34, Sci. Repts. 3(4):193-221, 2 textfigs.

chemistry, carbon (organic) (data only)
Ivanoff, A., A. Morel, Mme Vesin-Couffinhal, C. Amiel, J.P. Bethoux, J. Boutler, C. Copin-Montegut, P. Courau, P. Geistdoerfer, et F. Nyffeler, 1969.
Résultats des observations effectuées en Mer Méditerranée orientale et en mer Tyrrhénienne à bord du navire Amalthée en mars 1967.
Cah. océanogr. 21 (Suppl. 2): 245-263

chemistry, carbon (organic) (data only)
Japan, Hokkaido University, Faculty of Fisheries, 1967.
Data record of oceanographic observations and exploratory fishing, 11:383 pp.

chemistry carbon, organic (dissolved)
Ehrhardt, Manfred, 1969.
The particulate organic carbon and nitrogen, and the dissolved organic carbon in the Gotland Deep in May 1968.
Kieler Meeresforsch. 25(1): 71-80

chemistry, carbon, organic (dissolved)
*Ogura, Norio, 1970.
The relation between dissolved organic carbon and apparent oxygen utilization in the Western North Pacific. Deep-Sea Res., 17(2): 221-231.

carbon, organic (dissolved)
*Starikova, N.D., 1967.
Quantitative distribution of dissolved organic carbon in the Indian Ocean water. (In Russian; English abstract).
Trudy Inst. Okeanol., 83:38-50.

chemistry, carbon (organic) (dissolved)
Williams, P.M., H. Oeschger and P. Kinney 1969
Natural radiocarbon activity of the dissolved organic carbon in the north-east Pacific Ocean.
Nature, Lond. 224 (5216): 256-258

chemistry carbon, organic (particulate)
Adams, J.A., and I.E. Baird, 1965.
Chlorophyll a particulate organic carbon and zooplankton standing crop in the northern North Sea.
Ann. Biol., Cons. Perm. Int. Expl. Mer, 1963, 20:91-92.

chemistry, carbon, organic (particulate)
Ehrhardt, Manfred 1970
2.8. Partikulärer organischer Kohlenstoff und Stickstoff sowie gelöster organischer Kohlenstoff. Chemische, mikrobiologische und planktologische Untersuchungen in der Schlei im Hinblick auf deren Abwasserbelastung. Kieler Meeresforsch 26(2): 138-144.

chemistry, carbon (organic) (particulate)
Ehrhardt, Manfred 1969.
The particulate organic carbon and nitrogen and the dissolved organic carbon in the Gotland Deep in May 1968.
Kieler Meeresforsch. 25(1): 71-80

chemistry, carbon (organic) (particulate)
Fournier, Robert O., 1968.
Observations of particulate organic carbon in the Mediterranean Sea and their relevance to the deep-living Coccolithophorid Cyclococcolithus fragilis.
Limnol. Oceanogr., 13(4):693-696.

chemistry, carbon, organic (particulate)
Gordon, Donald C., Jr., 1970.
Some studies on the distribution and composition of particulate organic carbon in the North Atlantic Ocean. Deep-Sea Res., 17(2): 233-243.

chemistry, carbon (organic) (particulate)
Hobson, Louis A., 1971.
Relationships between particulate organic carbon and micro-organisms in upwelling areas off southwest Africa. Investigación pesq. 35(1): 195-208.

chemistry, carbon (particulate) (data only)
Japan, Hokkaido University, Faculty of Fisheries, 1968.
The Oshoro Maru cruise 23 to the east of Cape Erimo, Hokkaido, April 1967.
Data Record Oceanogr. Obs. Expl. Fish., 12: 115-169.

chemistry, carbon (oxidizable) (data only)
Krey, Johannes, Wolf Boje, Max Gillbricht und Jürgen Lenz 1971.
Planktologisch-chemische Daten der Meteor-Expedition in den Indischen Ozean 1964/65.
Meteor Forsch.-Ergebn. (D) 9:1-120.

carbon, particulate

chemistry carbon particulate (data only)
Australia, Commonwealth Scientific and Industrial Research Organization, 1965
Oceanographic observations in the Indian Ocean in 1963, H.M.A.S. Gascoyne, Cruise G 1/63.
Div. Fish. and Oceanogr., Oceanogr. Cruise Rept., No. 21:135 pp.

chemistry, carbon particulate (data only)
Australia, Commonwealth Scientific and Industrial Research Organiz-tion.
Oceanographic observations in the Indian Ocean in 1963, H.M.A.S. DIAMANTINA Cruise DM 3/63.
Div. Fish. Oceanogr., Oceanogr. Cruise Rept. No. 25: 147 pp.

chemistry, carbon (particulate)
Biggs, R.B. and D.A. Flemer, 1972
The flux of particulate carbon in an estuary.
Mar. Biol. 12(1): 11-17.

chemistry particulate carbon
*Duing, Walter, 1966.
Die Vertikalzirkulation in den Küstennahen Gewässern des Arabischen Meeres während der Zeit des Nordostmonuns.
"Meteor" Forschungsergebnisse (A)(3):67-83.

chemistry particulate carbon
Flemer, David A. and Robert B. Biggs 1971.
Particulate carbon: nitrogen relations in northern Chesapeake Bay.
J. Fish. Res. Bd, Can, 28(6): 911-918.

chemistry, carbon (organic)(particulate)
*Gordon, Donald C., Jr., 1971.
Distribution of particulate organic carbon and nitrogen at an oceanic station in the central Pacific. Deep-Sea Res, 18(11): 1127-1134.

chemistry carbon, particulate
Gordon, Donald C., Jr., 1970.
Faulty particulate organic carbon concentrations
Deep-Sea Res., 17(6): 1025.

chemistry, particulate carbon
Koske, Peter H., and Karl-Heinz Szekielda, 1965.
Zur Hydrographie des Nord-Ostsee Kanals.
Kieler Meeresforsch., 21(2):132-134.

chemistry, carbon (particulate) (data only)
Krey, Johannes, Wolf Boje, Max Gillbricht und Jürgen Lenz 1971.
Planktologisch-chemische Daten der Meteor-Expedition in den Indischen Ozean 1964/65.
Meteor Forsch.-Ergebn. (D) 9:1-120.

chemistry carbon particulate
Menzel, David W., and Ralph F. Vaccaro, 1964
The measurement of dissolved organic and particulate carbon in seawater.
Limnology and Oceanography, 9(1):138-142.

chemistry organic carbon, particulate
Mullin, Michael M., 1965.
Size fractionation of particulate organic carbon in the surface waters of the western Indian Ocean.
Limnol. Oceanogr., 10(4):610-611.

carbon, organic (particulate)
Mullin, Michael, M., 1965.
Size fractionation of particulate organic carbon in the surface waters of the western Indian Ocean.
Limnol. Oceanogr., 10(3):359-362.

chemistry carbon particulate
Newell, B.S., 1969.
Seasonal variations in the Indian Ocean along 110°E. II. Particulate carbon.
Austr. J. mar. Freshwat. Res., 20(1): 51-54

chemistry particulate organic carbon
Riley, Gordon A., Peter J. Wangersky and Denise Van Hemert, 1964
Organic aggregates in tropical and subtropical surface waters of the North Atlantic Ocean.
Limnology and Oceanography, 9(4):546-550.

chemistry carbon, particulate
Steele, J. H., and I. E. Baird, 1962.
Further relations between primary production, Chlorophyll and particulate carbon.
Limnol. & Oceanogr., 7(1):42-47.

chemistry particulate carbon

Steele, J.H., and I.E. Baird, 1961
Relations between primary production, chlorophyll and particulate carbon.
Limnol. & Oceanogr., 6(1): 68-78.

carbon, particulate
*Stephens, K., R.W. Sheldon and T.R. Parsons, 1967.
Seasonal variations in the availability of food for benthos in a coastal environment.
Ecology, 48 (5):852-855.

Chemistry - Carbon, particulate

SZEKIELDA, K.-H., 1969.
Le dosage du carbone particulaire dans l'eau de mer et son application dans le Golfe du Lion.
J. Cons. perm. int. Explor. Mer, 32(3): 318-343.

chemistry carbon, particulate
Sutcliffe, W.H., Jr., R.W. Sheldon and A. Prakash, 1970.
Certain aspects of production and standing stock of particulate matter in the surface waters of the northwest Atlantic Ocean.
J. Fish. Res. Bd, Can., 27 (11): 1917-1926.

chemistry, carbon (particulate)
Szekielda, Karl-Heinz 1971.
Organisch gelöster und partikulärer Kohlenstoff in einem Nebenmeer mit starken Salzgehaltsschwankungen (Ostsee).
Vie Milieu Suppl. 22 (2): 579-612

Chemistry, carbon, particulate
*Szekielda, K.H., 1968.
Le carbone particulaire dans les masses d'eau profondes.
Marine Biol., 2(1):71-72.

chemistry, carbon-phosphorus bond
Quin, Louis D., 1965.
The presence of compounds with a carbon-phosphorus bond in some marine invertebrates.
Biochemistry, 4(2):324-330.

Carbon reservoir

chemistry, carbon reservoir
*Weber, Jon N., 1967.
Possible changes in the isotopic composition of the oceanic and atmospheric carbon reservoir over geologic time.
Geochim. cosmochim. Acta, 31(12):2343-2352.

Carbon, surface and bottom

Chemistry carbon, surface and bottom (data)
Raghu Prasad, R., and P.V. Ramachandran Nair, 1963.
Studies on organic production. 1. Gulf of Manaar.
J. Mar. Biol. Assoc. India, 5(1):1-26.

Carbon total

carbon, total
Cooper, L.H.N., 1965.
Chemistry of the sea. 2. Organic.
Chemistry in Britain, 1965, 1:150-154.

Chemistry carbon (total)
Krey, Johannes, und Karl-Heinz Szekielda, 1966.
Gesamtkohlenstoff und Mikrobiomasse in der Ostsee im Mai 1962.
Kieler Meeresforsch., 22(1):64-69.

Carcinogens

chemistry carcenogens
Bourcart, Jacques, Mlle. Claude Lalou et Lucien Mallet, 1961.
Sur la présence d'hydrocarbures du type benzo 3.4 pyrène dans les vases côtières et les sables de la plage de la rade de Villefranche (Alpes Maritimes).
C.R. Acad. Sci., Paris, 252:640-644.

Also in: Trav. Lab. Géol. Sous-Marine, 10(1960).

Chemistry carcenogens
Mallet, L., et Mme Le Theule, 1961.
Recherche du benzo 3.4 pyrène dans les sables vaseux marins des régions côtières de la Manche et de l'Atlantique.
C.R. Acad. Sci., Paris, 252(4):565-567.

Chemistry - carcenogens
Niaussat, P., 1970.
Pollution, par biosynthèse in situ d'hydrocarbures cancérigènes, d'une biocoenose lagunaire reproduction in vitro de a phénomène.
Rev. intern. Océanogr. Méd. 17: 87-95

chemistry - carcenogens
*Pasquini, Pasquale, 1966.
Il benzo 3-4 pirene, idrocarburo cancerigeno, nell'ambiente marino.
Archo zool. ital., 51(2):747-774.

chemistry carcenogens
Perdriau, Jacques, 1964.
Pollution marine par les hydrocarbures cancérigènes - type benzo-3.4 pyrène - incidences biologiques. (Suite et fin).
Cahiers Océanogr., C.C.O.E.C., 16(3):205-229.

chemistry - carcenogens
Perdriau, Jacques, 1964.
Pollution marine par les hydrocarbures cancérigènes-type benzo-3.4 pyrène - incidences biologiques. (1.).
Cahiers Océanogr., C.C.O.E.C., 16(2):125-138.

Chemistry - Carcenogens
*Piccinetti, C., 1967.
Diffusione dell'idrocarburo cancerigeno benzo 3-4 pirene nell'alto e medio Adriatico.
Arch. Oceanogr. Limnol., 15(Suppl.):169-183.

Chemistry - carcenogens
*Scaccini-Cicatelli, Marta, 1965.
Studio dei fenomeni di accumulo del benzo 3-4 pirene nell'organismo di Tubifex.
Boll. Pesca, Piscic. Idrobiol., n.s., 20(2):245-250.

Carotene

Chemistry Carotene (data)

Forsbergh, Eric D., William W. Broenkow, 1965.
Observaciones oceanograficas del oceano Pacifio oriental recolectadas por el barco Shoyo Maru, octubre 1963-marzo 1964.
Comision Interamericana del Atun Tropical, Bol.

chemistry - carotene
Jeffrey, S.W., and F.T. Haxo 1968
Photosynthetic pigments of symbiotic dinoflagellates (zooxanthellae) from corals and clams.
Biol. Bull. mar. biol. Lab. Woods Hole 135 (1): 149-165

Chemistry - carotene
Taylor, W. Rowland, and Conrad D. Gebelein, 1966.
Plant pigments and light penetration in intertidal sediments.
Helgoländer wiss Meeresunters. 13(3):229-237.

Carotenoids

chemistry carotenoids
Belcner, John H., and Gordon E. Fogg, 1964.
Chlorophyll derivatives and carotenoids in the sediments of two English lakes.
In: Recent researches in the fields of hydrosphere, atmosphere and nuclear geochemistry, Ken Sugawara festival volume, Y. Miyake and T. Koyama, editors, Maruzen Co., Ltd., Tokyo, 39-48.

Chemistry - carotenoid content

*Campbell, S.A., 1969.
Seasonal cycles in the carotenoid content in Mytilus edulis. Marine Biol., 4(3): 227-232.

chemistry - cartenoids
*Czeczuga, B., 1970.
Some carotenoids in the jelly-fish Aurelia aurita (Scyphozoa: Discomedusae). Mar. Biol., 5(2): 141-144.

chemistry carotinoids
Establier, R., 1966.
Estudios sobre los carotênoides de animales y plantas marinos. III. Carotinoides contenidos en las pinnulas del atún, Thunnus thynnus (L.).
Inv. Pesq., Barcelona, 30:497-500.

chemistry, carotinoids
Establier, R., 1966.
Estudios sobre los carotinoides de plantas y animales marinos. II. Variaciones del contênido en carotinoides de los ovarios y hepatopâncreas del crustáceo Plesiopenaeus edwardsianus Johnson.
Inv. Perq., Barcelona, 30:223-232.

chemistry, carotinoids
Establier, R., 1966.
Estudio sobre los carotinoides de plantas y animales marinos. 1. Distribucion de carotinoides en el crustáceo Plesiopenaeus edwardsianus (Johns on).
Inv. Pesq., Barcelona, 30:207-222.

chemistry carotenoids
Establier, R., and R. Margalef, 1964
Fitoplancton e hidrografía de las costas de Cadiz (Barbate), de junio de 1961 a agosto de 1962.
Invest. Pesquera, Barcelona, 25:5-31.

chemistry carotenoids
Fisher, L.R., S.K. Kon and S.Y. Thompson, 1964
Vitamin A and carotenoids in certain invertebrates. VII. Crustacea: Copepoda.
Jour. Mar. Biol. Assoc., U.K., 44(3):685-692.

chemistry, carotenoids
Fisher, L.R., S.K. Kon, and S.Y. Thompson, 1952.
Vitamin A and carotenoids in certain invertebrates. J.M.B.A. 31(2):229-258, 6 textfigs.
1, Marine crustacea.

chemistry - carotenoids
Fox, D.L., and C.H. Oppenheimer, 1954.
The riddle of sterol and carotenoid metabolism in muds of the ocean floor. Arch. Biochem. Biophys. 51(2):323-328.

Chemistry, carotenoid
Goodwin, J.W., and D.L. Fox, 1955.
Some unusual carotenoids from two nudibranch slugs and a lamprid fish. Nature 175(4468):1086-1087, 1 textfig.

Chemistry - carotinoid pigments
Gorshkova, T.I., 1965.
Chlorophyll and carotinoids in the sediments of the Baltic Sea and Gulf of Riga. Investigations in line with the programme of the International Geophysical Year, 2. (In Russian).
Trudy, Vses. Nauchno-Issled. Inst. Morsk. Ribn. Choz. i Okeanogr., (VNIRO), 57:313-323.

Chemistry - caretenoids
*Ishikawa, Yusuke, Yoshio Miyake and Shigeki Yasuie, 1967.
Chromatophores and caretenoids in cultured karuma-prawn, Penaeus japonicus Bate, of different body colours. (In Japanese; English abstract).
Bull. Fish. Exp. Stn, Okayama Pref., 1966:18-24.

Chemistry, carotenoids
Komaki, S., 1957.
Seasonal variations in the concentrations of net-plankton pigments in sea water.
Bull. Hokkaido Reg. Fish. Res. Lab., No. 16:84-91.

chemistry, carotenoids
Lee, Kang-Ho and Chong-Bae Kim, 1971.
Pigments in marine fish and shell-fish. 1. Carotenoids in sea mussel, Mytilus edulis.
Bull. Pusan Fish. Coll. (nat. Sci.) 11(1): 57-62. (In Korean; English abstract).

Chemistry, Carotenoids
Lee, Welton, L. Barbara M. Gilchrist and R. Phillips Dales, 1967.
Carotenoid pigments in Sabella penicillus.
J. mar. biol. Ass., U.K., 47(1):33-37.

Chemistry carotenoid pigments
Lenel, Roland, 1965.
Nature et métabolisme des pigments carotenoides des téguments de Pachygrapsus marmoratus (Décapode Brachyoure).
Comptes Rendus, Acad. Sci., Paris, 261(4):1119-1122.

Chemistry carotenoids
Lewin, Ralph A., 1962.
Physiology and biochemistry of algae.
Academic Press, New York and London, 929 pp.

Chemistry carotenoids
Liaci, Lidia, 1964.
Pigmenti e steroli negli invertebrati marini.
Archivio Zool. Ital., 49:281-300.

Chemistry - carotenoids
Mandelli, Enrique F. 1969
Carotenoid interconversion in light dark cultures of the dinoflagellate Amphidinium klebsi.
J. Phycol. 5(4): 382-384

Chemistry carotinoids (data)
Margalef, Ramón, 1961
Hidrografía y fitoplancton de un área marina de la costa meridional de Puerto Rico.
Inv. Pesq., Barcelona, 18:38-06.

Chemistry carotenoids (data)
Margalef, Ramón, and Juan Herrera, 1963
Hidrografía y fitoplancton de las costas de Castellón, de julio de 1959 a junio de 1960.
Inv. Pesq., Barcelona, 22:49-109.

Chemistry carotenoids
Millott, N., and H.G. Vevers, 1955.
Carotenoid pigments in the optic cushion of Marthasterias glacialis (L.). J.M.B.A. 34(2): 279-287, 3 textfigs.

Chemistry carotenoids
Morales, E., y E. Arias, 1965.
Ecología del puerto de Barcelona y desarrollo de adherencias orgánicas sobre placas su mergidas.
Inv. Pesq., Barcelona, 28:49-79.

Chemistry carotenoids
Munoz Sardon, Filipe, 1962
Estudio del fitoplancton del Golfo de Cadiz en relacion con el regimen regional de vientos.
Inv. Pesq., Barcelona, 21:165-188.

Chemistry Carotenoids
Muñoz, Felipe, y Jose M. San Feliu, 1965.
Hidrografia y fitoplancton de las costas de Castellón de agosto de 1962 a julio de 1963.
Inv. Pesq., Barcelona, 28:173-209.

chemistry, carotenoids
Qasim, S.Z., and C.V. G. Reddy, 1967.
The estimation of plant pigments of Cochin backwater during the monsoon months.
Bull. mar. Sci., Miami, 17(1):95-110.

Chemistry carotenoids
Parsons, T.R., 1961
On the pigment composition of eleven species of marine phytoplankters.
J. Fish. Res. Bd., Canada, 18(6):1017-1025.

Chemistry carotenoid
Vinogradova, Z.A., G.K. Yatsenko and L.V. Antsupova, 1966.
On the study of the seasonal variability of plankton pigment composition in the northwestern part of the Black Sea. (In Russian; English abstract).
Okeanologiia, Akad. Nauk, SSSR, 6(5):853-860.

Chemistry cerotenoids
Ward, Ronald W., Velerie Vreeland, Cherles H. Southwick and Anthony J. Reading, 1965.
Ecological studies related to plankton productivity in two Chesapeake Bay estuaries.
Chesapeake Science, 6(4):214-225.

Carotenoproteins

chemistry, carotenoproteids
Ceccaldi, Hubert J., et Bernard H. Allemand, 1964
Carotenoproteides. 3. Extraction du pigment bleu de la carapace du homard Homarus gammarus (L.).
Rec. Trav. Sta. Mar., Endoume, 51(35):3-7.

Chemistry carotenoproteins
Cheeseman, D.F., W.L. Lee and P.F. Zagelsky, 1967.
Carotenoproteins in invertebrates.
Biol. Rev., Cambridge Phil. Soc., 42(1):131-160.

carotin

Chemistry carotin
Sushcheniya, L.M., and Z.Z. Finenko, 1964
On the study of plankton productivity in the tropical area of the Atlantic Ocean. (In Russian).
Okeanologiia, Akad. Nauk, SSSR, 4(5):866-872.

Catalytic activity

Chemistry, catalytic activity
Yamazi, I., 1955.
Plankton investigations in inlet waters along the coast of Japan. VIII. The plankton of Miyazu Bay in relation to the water movement.
Publ. Seto Mar. Biol. Labl, 4(2/3):269-284, 16 textfigs.

Chemistry catalytic activity
Yamazi, I., 1954.
Plankton investigations in inlet waters along the coast of Japan. XIV. The plankton of Turuga Bay on the Japan Sea coast.
Publ. Seto Mar. Biol. Lab. 4(1):115-126, 11 textfigs.

Chemistry catalytic activity
Yamazi, I., 1954.
Plankton investigation in inlet waters along the coast of Japan. 13. The plankton of Olama Bay on the Japan Sea coast.
Publ. Seto Mar. Biol. Lab. 4(1):103-114, 9 textfigs.

Cations

Chemistry, cations
Nehring, Dietwart, und Karl-Heinz Rohde 1967.
Weitere Untersuchungen über anormale Ionenverhältnisse in der Ostsee.
Beitr. Meereskunde 20: 10-33

Chemistry cations
Riley, J.P., 1965.
Analytical chemistry of sea water.
In: Chemical oceanography, J.P. Riley and G. Skirrow, editors, Academic Press, 2:295-424.

chemistry cation content
Nicholls, G.D., 1965.
The geochemical history of the oceans.
In: Chemical oceanography, J.P. Riley and G. Skirrow, editors, Academic Press, 2:277-294.

Chemistry cations
Shanklin, D.R., 1954.
Cation analysis of Woods Hole sea water. Nature 173(4393):82.

chemistry, cation exchange
Turekian, Karl K., 1971
2. Rivers, tributaries and estuaries. In: Impingement of man on the oceans, D.W. Hood, editor, Wiley Interscience, 9-73.

cellulose

Chemistry cellulose
Elroi, D., and B. Komarovsky, 1961
On the possible use of the fouling ascidian Ciona intestinalis as a source of vanadium, cellulose and other products.
Proc. Gen. Fis. Counc., Medit., Proc. and Techn. Papers, 6:261-267.

Technical Paper No. 37.

cerium

chemistry, cerium
*Carpenter, James H., and Virginia E. Grant, 1967.
Concentration and state of cerium in coastal waters.
J. mar. Res., 25(3):228-238.

Chemistry - cerium.
Fukai, Rinnosuke, and Lang Huynh-ngoc 1968.
Studies on the chemical behaviour of radionuclides in sea-water. 1. General considerations and study of precipitation of trace amounts of chromium, manganese, iron, cobalt, zinc and cerium.
IAEA Radioactivity in the sea, Publ. 22: 1-26

Chemistry cerium
Goldberg, Edward D., Minoru Koide, R.A. Schmitt and R.H. Smith, 1963
Rare-earth distribution in the marine environment.
J. Geophys. Res., 68(14):4209-4217.

Chemistry cerium
Hiyama, Yoshio, and Makoto Shimizu, 1964
On the concentration factors of radioactive Cs, Sr, Cd, Zn, and Ce in marine organisms.
Rec. Oceanogr. Wks., Japan, 7(2):43-77.

Chemistry - cerium-141
Parchevskiy, V.P., and K.M. Khailov, 1968.
Formation of Ce^{141}, Ru^{106}, Cs^{137}, and Zn^{65} complexes with hydrophyl high-molecular combinations dissolved in sea water. (In Russian)
Okeanologiia, Akad. Nauk, SSSR, 8(6):1092.

cerium-144

chemistry, cerium-144
Spitsyn, B.I., R.N. Bernovskaia, and N.I. Popov, 1969.
State of ultrasmall quantities of cerium-144 and yttrium-91 in sea water. (In Russian)
Dokl. Acad Nauk SSSR, 185(1): 111-114.

Chemistry, cerium-144
Barinov, G.V., 1965.
The exchange of Ca^{45}, Cs^{137} and Ce^{144} between algae and the sea water. (In Russian).
Okeanalgiia, Akad. Nauk, SSSR, 5(1):111-116.

Chemistry cerium 144
Bowen, Vaughan T., and Thomas T. Sugihara, 1965.
Oceanographic implications of radioactive fall-out distributions in the Atlantic Ocean: from 20°N to 25° S, from 1957 to 1961.
J. Mar. Res., 23(2):123-146.

Chemistry cerium-144
Polikarpov, G.G., 1961. formation
The role of detritus in the migration of strontium-90, cesium-137 and cerium-144. Experiments with the sea alga Cystoseira barbata.
Doklady Akad. Nauk, SSSR, 136(4):921-923.
English Edit., 1962, 136(1-6):11-13.

Cesium: SEE: CAESIUM

Cesium

See under: Caesium

Chemistry - cerium-144
Rice, T. R., and Virginia M. Willis, 1959.
Uptake, accumulation and loss of radioactive cerium-144 by marine planktonic algae.
Limnol. & Oceanogr., 4(3):277-290.

chemistry, changes in
Armstrong, F.A.J., and E.I. Butler, 1968.
Chemical changes in sea water off Plymouth during the years 1962 to 1965.
J. mar. biol. Ass., U.K., 48(1):153-160.

chelators

see also:
Chemistry: chelation process

Chemistry, chelators (organic)
Barber, Richard T. and John. H. Ryther, 1969.
Organic chelators: factors affecting primary production in the Cromwell Current upwelling.
J. exp. mar. Biol. Ecol., 3(2): 191-199.

chemistry chelators
Cooper, L.H.N., 1965.
Chemistry of the sea. "2. Organic.
Chemistry in Britain, 1965, 1:150-154.

Chemistry chelation
Duursma, E. K., and W. Sevenhuysen, 1966.
Note on chelation and solubility of certain metals in sea water at different PH values.
Netherlands J. Sea Res., 3(1):95-106.

Chemistry chelation processes
Hood, D.W., 1963
Chemical oceanography. In: Oceanography and Marine Biology, H. Barnes, Edit., George Allen & Unwin, 1:129-155.

chemistry, chelation
Johnston, R., 1964
Sea water, the natural medium of phytoplankton II. Trace metals and chelation, and general discussion.
J. Mar. Biol. Assoc., U.K., 44(1):87-109.

Chemistry - chelating agents, effect of
Jones, Galen E., 1964.
Effect of chelating agents on the growth of Escherichia coli in seawater.
J. Bacteriol, 87(3):483-499.

chemistry, chelators
Lewis, A.G., A. Ramnarine and M.S. Evans, 1971.
Natural chelators - an indication of activity with the calanoid copepod Euchaeta japonica. Mar. Biol. 11(1): 1-4.

chemical composition, fishes
Goldberg, Edward D., 1962
Oceanographic studies during Operation "Wigwam" Elemental composition of some pelagic fishes.
Limnol. and Oceanogr., Suppl. to Vol. 7: lxxii-lxxv.

Chemical constituents

CHEMISTRY
Chemicals, accumulation within plants and animals

See: Chemicals, concentration of

chemical constituents
Clarke, F.W., and W.C. Wheeler, 1922.
The inorganic constituents of marine invertebrates Second edition, revised and enlarged. U.S. Geol. Survey, Prof. Pap. 124:1-62.

Chemistry, chemical elements
Hosokawa, Iwao, Fumio Ohshima and Norihiko Kondo 1970
On the concentrations of the dissolved chemical elements in the estuary water of the Chikugogawa River. (In Japanese; English abstract).
J. oceanogr. Soc. Japan, 26(1): 1-5.

Chemistry, chemical environment
Golterman, H. L. and R.S. Clymo, editors 1967.
Chemical environment in the aquatic environment.
Proc. IBP Symp., Amsterdam, Oct. 1966: 322 pp.

coordination chemistry
*Martin, Dean F., 1967.
Coordination chemistry of the oceans.
In: Equilibrium concepts in natural water systems, Werner Stumm, editor, Adv. Chem. Ser. 67: 255-269.

Chemistry, methods

See: Methods, chemistry

chemical units
Helland-Hansen, B., J.P. Jacobsen, and T.G. Thompson, 1948.
Chemical methods and units. Rept. from Committee UGGI, Assoc. Ocean. Phys., Publ. Sci. No. 9:1-28

chemistry, citric acid
Creac'h, P. 1955.
Sur la presence des acides citrique et malique dans les eaux marines littorales.
C. r. hebd. Seanc. Acad. Sci. Paris 240(26): 2551-2552

Chemicals, concentration of

See also: Radioactivity, accumulation and concentration

Chemicals, concentration of or accumulation within plants and animals.

Chemistry, chemicals (concentration within organisms)

See also: chemistry, radioactivity (contamination)

Chemistry, concentration of chemicals within organisms
Amano, K., K. Yamada, M. Bito, A. Takase and S. Tanaka, 1955.
Studies on the radioactivity in certain pelagic fish. 1. Distribution of radioactivity in various tissues of fish. Bull. Japan. Soc. Sci. Fish., 20(10):907-915.

concentration within organisms
Anderson, J.W. and G.C. Stephens, 1969.
Uptake of organic material by aquatic invertebrates. VI. Role of epiflora in apparent uptake of glycine by marine crustaceans. Marine Biol., 4(3): 243-249.

chemistry, concentration of chemicals within organisms
Angino, Ernest E., John E. Simek and Jim A. Davis 1965.
Fixing of fallout material by floating marine organisms, Sargassum fluitans and S. natans.
Publ. Inst. Mar. Sci., Port Aransas, 10:173-178.

Chemistry accumulation of chemicals
Aubert, M., R. Chesselet, and D. Normann, 1962.
Mesures de radioactivité (α, β et γ) effectuées sur les échantillons de plancton et d'algues prelevés en Méditerranée dans la zone cotière de la ville de Nice.
Cahiers, Centre d'Études et de Recherches Biol. Océanogr. Med., (2):23-43. (Mimeographed)

concentration of chemicals within organisms
*Avargues M., J. Ancellin et A. Vilquinn,1968.
Recherches expérimentales sur l'accumulation des radionuclides par les organismes marins
Rev. intern. Océanogr. Med. 11:87-100.

Chemistry chemicals, concentration of
Avio, Carlo Maria, Miranda Rossi-Toretti and Luciano Lenzerini, 1964.
Captazione dello Sr 90 da parte di Chlorella vulgaris 1. Captazione dello Sr 90 in funzione dello sviluppo di una popolazione.
Pubbl. Staz. Zool., Napoli, 33(3):264-273.

freshwater

chemistry, accumulation of chemicals within organisms.
Bache, C.A., W.H. Gutenmann and D.J. Lisk 1971.
Residues of total mercury and methyl-mercuric salts in lake trout as a function of age.
Science 172 (3986): 951-952.

chemistry, concentration of chemicals within organisms
Baptist, John P., 1966
Uptake of mixed fission products by marine fishes.
Trans. Amer. Fish. Soc., 95(2):145-152.

concentration of chemicals within organisms
Baptist, J.P., and T.J. Price, 1962.
Accumulation and retention of cesium 137 by marine fishes.
Fish. Bull. 206 from Fish Bull., Fish and Wildlife Service, 62:177-187.

chemistry, concentration of chemicals
Barinov, H.V., 1966.
Kinetic regularities of accumulation and exchange of radionuclides by sea algae. (Abstract only).
Second Int. Oceanogr. Congr., 30 May-9 June 1966. Abstracts, Moscow: 21-22.

Chemistry - concentration of chemicals
Bagge, P., and Anneli Salo, 1967.
Biological detectors of radioactive contamination in the Baltic.
Rep. Inst. Radiation Phys., Helsinki, SFL-A9:43pp. (mimeographed).

concentration of chemicals
Barinov, G.V., 1965.
The exchange of Ca^{45}, and Cs^{137} and Ce^{144} between algae and the sea water. (In Russian).
Okeanologiia, Akad. Nauk, SSSR, 5(1):111-116.

chemicals, concentration by organisms
Barrington, E.J.W., 1957.
The distribution and significance of organically bound iodine in the ascidian Ciona intestinalis Linnaeus.
J. Mar. Biol. Assoc., U.K., 36(1):1-16.

concentration of chemicals within organisms
*Battani, M., M.D. Chambost et M. Leandri, 1968.
Etude de la contamination par l'iodide 131 et le fer 59 de Ciona intestinalis Linné (tunicier).
Rev. intern. Océanogr. Med. 11:71-86.

chemicals, concentration of
Bein, S.J., 1957.
The relationship of total phosphorus concentration in sea water to red tide blooms.
Bull. Mar. Sci., Gulf and Caribbean, 7(4):316-329

accumulation of chemicals within organisms
Bernhard, M. 1970.
The role of vertebrates in the uptake and loss of radioisotopes.
Rev. int. Océanogr. Med. 20: 101-123

concentration of chemicals within organisms
*Bernhard, M., 1968.
Research on the metabolism of some radioactive element in the marine environment.
Rev. intern. Océanogr. Med., 11:11-59.

Chemistry, concentration by organisms
Bernhard, M., preparator, 1965.
Studies on the radioactive contamination of the sea, annual report 1964.
Com. Naz. Energ. Nucleare, La Spezia, Rept., No. RT/BIO (65) 18:35 pp.

Chemistry concentration of chemicals
Bernhard, M., 1964.
Chemical composition and the radiocontamination of marine organisms.
Proc. Symp. Nuclear Detonations and Marine Radioactivity, S.H. Small, editor, 137-150.

concentration of chemicals within organisms
Bernhard, M., L. Rampi and A. Zattera 1971
First trophic level of the food chain.
CNEN Rept. RT/BIO (70)-11, M. Bernhard, editor: 23-40

chemicals, concentration within organisms
*Bernard, M. and A. Zattera, 1969.
A comparison between the uptake of radioactive and stable zinc by a marine unicellular alga.
Symp. Radioecology, Proc. Second Natn. Symp., Ann Arbor, Mich., 1967 (Conf. 670503): 389-398.

concentration within organisms
* Bhatt, Y.M., V.N. Sastry, S.M. Shah and T.M. Krichnamoorthy, 1968.
Zinc, mangenese and cobalt contents of some marine bivalves from Bombay.
Proc. nat. Inst. Sci., India (B), 34(6): 283-287.

Chemistry, concentration of chemicals within organisms
Boclet, Daniel, Marie-Louise Drugy, Gérard Lambert, Joaquin Ros, Jean Thommeret, 1962.
Essai d'analyse de la radioactivité de l'éponge "Hercina variabilis".
In: Océanographie, Géologique et Géophysique de la Méditerranée Occidentale, Centre National de la Recherche Scientifique, Villefranche sur Mer, 4 au 8 avril, 211-220.

chemicals, concentration of
Boroughs, Howard, 1958.
The metabolism of radiostrontium by marine fishes.
Proc. Ninth Pacific Sci. Congr., Pacific Sci. Assoc., 1957, 16(Oceanogr):146-151.

chemistry, concentration of chemicals within organisms
Boroughs, H., W.A. Chipman and T.R. Rice, 1957.
Ch. 8. Laboratory experiments on the uptake accumulation and loss of radionuclides by marine organisms. NAS-NRC Publ. 551:80-87.

chemistry, concentration of chemicals within organisms
Boroughs, H., and D.F. Reid, 1958.
The role of the blood in the transportation of strontium 90-yttrium 90 in teleost fish.
Biol. Bull., 115(1):64-73.

chemicals, concentration of
Boroughs, H.T., S.J. Townsley and R.W. Hiatt, 1957.
The metabolism of radionuclides by marine organisms. III. The uptake of calcium in solution by marine fishes.
Limnol. and Oceanogr., 2(1):28-32.

Chemistry, concentration of chemicals within organisms
Boroughs, H., S.J. Townsley and R.W. Hiatt, 1956.
The metabolism of radionuclides by marine organisms. II. The uptake, accumulation and loss of yttrium 91 by marine fish, and the importance of short-lived radionuclides in the sea.
Biol. Bull. 111(3):352-357.

Chemistry, concentration of chemicals within organisms
Boroughs, H., S.J. Townsley and R.W. Hiatt, 1956.
The metabolism of radionuclides by marine organisms. I. The uptake, accumulation and loss of strontium 89 by fishes. Biol. Bull. 111(3):336-351.

Chemistry, concentration of chemicals within organisms
Boss, W.R., 1954.
Accumulation of radioactivity in marine invertebrate and vertebrate animals. Meeting of Radiological Survey between Japan and U.S.A. (Dec. 1954).

Chemistry, chemicals concentration within organisms
Bowen, H.J.M. 1968.
The uptake of gold by marine sponges.
J. mar. biol. Ass. U.K. 48 (2): 275-277.

chemicals, concentration of
Bowen, H.J.M., 1956.
Strontium and barium in sea water and marine organisms.
J.M.B.A., 35(3):451-460.

chemistry, chemicals accumulated within organisms
Bowen, V.T., J.S. Olsen, C.L. Osterberg and J. Ravera, 1971.
Ecological interactions of marine radioactivity.
Radioactivity in the marine environment, U.S. Nat. Acad. Sci., 1971: 200-222.

chemicals, concentration of
Brooks, R.R., and M.G. Rumsby, 1967.
Studies on the uptake of cadmium by the oyster, Ostrea sinuata (Lamarck).
Aust. J. mar. Freshwat. Res., 18(1): 53-61

concentration of chemicals
Brooks, Robert R., and Martin G. Rumsby, 1965.
The biogeochemistry of trace element uptake by some New Zealand bivalves.
Limnol. Oceanogr., 10(4):521-527.

concentration of chemicals within organisms
*Bryan, G.W., 1969.
The absorption of zinc and other metals by the brown seaweed Laminaria digitata.
J. mar. biol. Ass., U.K., 49(1):225-243.

Chemistry, chemicals concentration within organisms
Bryan, G.W. 1968.
Concentrations of zinc and copper in the tissues of decapod crustaceans.
J. mar. biol. Ass. U.K. 48(2): 303-321.

Chemistry, concentration of chemicals within organisms
Bryan, G. W., 1964.
The accumulation of radio-caesium by marine and brackish water invertebrates.
Proc. Symp. Nuclear Detonation mar. Radioactivity Norwegian Defense Res. Establ.
Edited by S.H. Small, 85-93.
Abstract in Jour. Mar. Biol. Assoc., U.K., 45 (2), 559.

Chemistry concentrations of chemicals
Bryan, G.W., 1964
Zinc regulation in the lobster Homarus vulgaris 1. Tissue zinc and copper concentrations.
J. Mar. Biol. Assoc., U.K., 44(3):549-563.

Chemistry, concentration of chemicals within organisms
Bryan, G.W., 1963
The accumulation of 137 Cs by brackish water invertebrates and its relation to the regulation of potassium and sodium.
J. Mar. Biol. Assoc., U.K., 43(2):541-565.

Chemistry, concentration of chemicals within organisms
Bryan, G.W., 1963
The accumulation of radioactive caesium by marine invertebrates.
J. Mar. Biol. Assoc., U.K., 43(2):519-539.

Chemistry concentration of chemicals
Bryan, G.W., 1961.
The accumulation of radioactive caesium in crabs.
J. Mar. Biol. Assoc., U.K., 41(3):551-575.

Chemistry, concentration of chemicals within organisms
Bryan, G.W., and Eileen Ward, 1965.
The absorption and loss of radioactive and non-radioactive manganese by the lobster Homarus vulgaris.
Jour. Mar. Biol. Assoc., U.K., 45(1):65-95.

chemicals, concentration of
Bryan, G.W., and Eileen Ward, 1962.
Potassium metabolism and the accumulation of 137 cesium by decapod Crustacea.
J. Mar. Biol. Assoc., U.K., 42(2):199-242.

Chemistry concentration of chemicals
Burovina, I.V., V.V. Glazunov, V.G. Leont'yev, V.P. Nesterov, I.A. Skul'skiy, D.G. Fleyshman and M.N. Smith, 1963.
Lithium, sodium, potassium, rubidium and cesium in the muscles of marine organisms of the Barents and Black Seas.
Doklady, Akad. Nauk, SSSR, 149(2):413-415.
Translation AGI 149:170-172.

chemical concentration by organisms
Burton, J.D., 1965.
Radioactive nuclides in sea water, marine sediments and marine organisms.
In: Chemical oceanography, J.P. Riley and G. Skirrow, editors, Academic Press, 2:425-475.

Chemistry concentration by organisms
Buyanov, N.I., and V.P. Kilezhoenko, 1961.
Potassium radioactivity of some marine organisms in the southern part of the Barents Sea. (In Russian).
Materiali, Ribochoz. Issled. Severn. Basseina, Poliarn. Nauchno-Issled. i Proektn. Inst. Morsk. Rib. Choz. i Okeanogr. im. N.M. Knipovicha, PINRO, Gosud. Proizvodst. Komm. Ribn. Choz. SSSR. 4:142-148.

concentration within marine organisms
*Carey, A.G., W.G. Pearcy and C.L. Osterberg, 1966.
Artificial radionuclides in marine organisms in the northeast Pacific Ocean off Oregon.
Disposal of radioactive wastes into seas, oceans and surface waters, Int. Atomic Energy Agency, 303-319.
Also in: Coll. Repr., Dep. Oceanogr., Oregon State Univ. 5.

Carlisle, D.B., 1968. Concentration of chemicals
Vanadium and other metals in ascidians.
Proc. R. Soc. B. 171(1022): 31-42.

chemicals, concentration of
Carlisle, D.B., 1958.
Niobium in ascidians.
Nature, 181(4613):933.

Chemistry, concentration of chemicals
Carlucci, A.F. and S.B. Silbernagel 1967
Bioassay of seawater. IV The determination of dissolved biotin in seawater using ^{14}C uptake by cells of Amphidinium carteri.
Can. J. Microbiol. 13(8): 979-986.

chemicals, concentration of
Carroll, James J., and Leonard J. Greenfield, 1964.
Mechanism of calcium and magnesium uptake from the sea by marine bacteria. (Abstract).
Geol. Soc., Amer., Special Paper, No. 76:30.

Chemistry concentration of chemicals
Carroll, J.J., L.J. Greenfield and R.F. Johnson, 1965.
The mechanism of calcium and magnesium uptake from seawater by a marine bacterium.
J. Cell. Comp. Physiol., 66(1):109-115.

Chemistry concentration of chemicals by organisms
Cerrai, E., B., Schreiber, C. Triulzi and L. Tassi-Pelati, 1964.
Ra^{226}, Pm^{147}, Mn^{54} content in a plankton sample of the Tyrrhenian Sea.
Ist. Lombardo, Acad. Sci. e Lett., Rendiconti (B) preprint

Chemistry concentration of chemicals
Chave, Keith E., and Bradner D. Wheeler, Jr., 1965.
Mineralogic changes during growth in the red alga, Clathromorphum compactum.
Science, 147(3658):621.

concentration of chemicals in organisms
Cherry, R.D., I.H. Fericke, and L.V. Shannon, 1969.
Thorium-228 in marine plankton and sea-water.
Earth planet. Sci. Letts. 6(6): 451-456.

concentration of chemicals
Chesselet, Roger, et Claude Lalou, 1964.
Concentrations en radionuclides émetteurs gamma presentées par les Holothuries prelevées dans la zone côtière d'Antibes au Cap Ferrat en août 1962.
Bull. Inst. Océanogr., Monaco, 63(1305):16 pp.

Chemistry, chemicals concentration within organisms
Chipman, W.A. 1966.
Uptake and accumulation of chromium-51 by the clam, Tapes decussatus, in relation to physical and chemical form.
Disposal of Radioactive Wastes into Seas, Oceans and Surface Waters IAEA (SM-72/35): 571-582.

CONCENTRATION WITHIN ORGANISMS
Chipman, W.A., 1966.
Some aspects of the accumulation of ^{51}Cr by marine organisms.
In: Radioecological Concentration Processes, Pergamon Press, 931-942.

chemistry, concentration of chemicals within organisms
Chipman, W.A., 1959
Accumulation of radioactive materials by fishery organisms. Proc. Gulf Caribb. Fish. Inst. 11th Session Nov. 1958: 97-110.

CONCENTRATION within organ
Chipman, Walter, and Evelyn Schommers, 1968.
Role of surface-associated organisms in the uptake of radioactive manganese by the clam, Tapes decussatus.
IAEA Radioactivity in the Sea, Publ. 24:1-11.

CONCENTRATION WITHIN ORGAN.
Chipman, Walter, Evelyn Schommers and Mireille Boyer, 1968.
Uptake, accumulation and retention of radioactive manganese by the marine annelid worm, Hermione hystrix.
IAEA Radioactivity in the Sea, Pub. 25:1-16.

chemistry, concentration within organisms
Chipman, Walter, and Jean Thommeret 1970.
Manganese content and the occurrence of fallout ^{54}Mn in some marine benthos of the Mediterranean.
Bull. Inst. océanogr. Monaco, 69(1402):15pp

chemicals, concentration of
Christomanos, An.An., A. Dimitriades and V. Gardiki, 1962
Contribution to plancton chemistry. (In English; Greek summary).
Chemia Chronika, Greece, 27A:23-26.

chemicals, concentration of
Conover, R.J., 1961.
The turnover of phosphorus by Calanus finmarchicus.
J. Mar. Biol. Assoc., U.K., 41:484-488.

Chemistry, concentration of chemicals within organisms
Corcoran, E.F., and J.F. Kimball, Jr., 1963.
The uptake, accumulation and exchange of strontium-90 by open sea phytoplankton.
Radioecology, V. Schultz and A.W. Klement, Jr., editors, Proc. First. Nat. Symp., Radioecology, Sept., 1961, Reinhold Publ. Corp., and Amer. Inst. Biol. Sci., 187-191.

*Cross, Ford A., John M. Dean and Charles L. Osterberg, 1969.
The effect of temperature, sediment and feeding on the behavior of four radionuclides in a marine benthic amphipod.
Symp. Radioecology, Proc. Second Natn. Symp., Ann Arbor, Mich., 1967 (Conf.670503):450-461.

concentration of chemicals within organisms
*Cross, F.A., S.W. Fowler, J.M. Dean, L.F. Small and C.L. Osterberg, 1968.
Distribution of 65Zn in tissues of two marine crustaceans determined by autoradiography.
J. Fish Res. Bd., Can., 25(11):2461-2466.

chemistry, chemicals accumulated within organisms
Cross, Ford A., James N. Willis and John P. Baptist 1971.
Distribution of radioactive and stable zinc in an experimental marine ecosystem.
J. Fish. Res. Bd Can. 28 (11): 1783-1788

chemicals, concentration of
Culkin, F., and J.P. Riley, 1958.
The occurrence of gallium in marine organisms.
J. Mar. Biol. Assoc., U.K., 37:607-616.

Chemistry, concentration of chemicals within organisms
Davis, J.J., and R.F. Foster, 1958.
Bioaccumulation of radioisotopes through aquatic food chains.
Ecology, 39(3):530-535.

chemistry concentration of chemicals
de Loyola e Silva, Jayme, 1965.
Marine isopods from highly radioactive places (Guarapari, Espiritu Santo). (Abstract).
Anais Acad. bras. Cienc., 37 (Supl.):259.

Chemistry, chemicals concentration within organisms
Dittmar, Heinrich, und Klaus Vogel 1968.
Die Spurenelemente Mangan und Vanadium in Brachiopodenschalen in Abhängigkeit vom Biotop.
Chem. Geol. 3(2): 95-110

chemistry concentration of chemicals
Dodd, J. Robert, 1965.
Environmental control of strontium and magnesium in Mytilus.
Geochimica et Cosmochimica Acta, 29(5):385-398.

chemicals, concentration of
Dodd, J. Robert, 1964.
Environmentally controlled strontium and magnesium variation in Mytilus. (Abstract).
Geol. Soc., Amer., Special Paper, No. 76:46.

Chemistry - Concentration of chemicals
Duke, Thomas W., John P. Baptist, and Donald E. Hoss, 1966.
Bioaccumulation of radioactive gold as a sediment tracer in the estuarine environment.
Fish.Bull.,U.S.Fish and Wildlife, 65(2):427-436.

Chemistry concentration of chemicals within organisms
Edgington, David N., Solon A. Gordon, Michael M. Thommes and Luis R. Almodovar, 1970.
The concentration of radium, thorium, and uranium by tropical marine algae. Limnol. Oceanogr., 15(6): 945-955.

Chemistry concentration of chemicals
*Eisler, Ronald, and Melvin P. Weinstein, 1967.
Changes in metal composition of the quahaug clam, Mercenaria mercenaria, after exposure to insecticides.
Chesapeake Sci., 8(4):253-258.

chemistry chemicals concentration of
Elroi, D., and B. Komarovsky, 1961.
On the possible use of the fouling ascidian Ciona intestinalis as a source of vanadium, cellulose and other products.
Proc. Gen. Fis. Counc., Medit., Proc. and Techn. Papers, 6:261-267.
Technical Paper No. 37.

Chemistry concentration of chemicals
Federov, Anatol F., 1964.
Mathematical formulas for concentration coefficient study of radioactive material to sea biota.
Bull. Inst. Océanogr., Monaco, 63(1304):11 pp.

chemistry, chemicals concentration within organisms
Establier, R., 1969.
Contenido en cobre, hierro, manganeso y cinc de los ostiones (Crassostrea angulata) de las costas de Cádiz. Investigación pesq. 33(1): 335-343.

chemistry, chemicals, concentration within organisms
Establier, R. 1969.
Estudios del contenido en cobre del agua de mar y ostiores (Crassostrea angulata) de las costas de Cádiz. Investigación pesq. 33(1): 69-86.

Chemistry concentrations of chemicals within organisms
Ferguson, John Carruthers, 1971.
Uptake and release of free amino acids by starfishes. Biol. Bull. mar. biol. Lab. Woods Hole 141(1): 122-129.

Chemistry concentration within organisms
Fletcher, G.L., 1971.
Accumulation of yellow phosphorus by several marine invertebrates and seaweed.
J. Fish. Res. Bd. Can. 28(6): 793-796.

chemistry, chemicals, concentration in organisms
Folsom, T.R., D.R. Young and C. Sreekumaran, 1969.
An estimate of the response rate of albacore to cesium.
Symp.Radioecology, Proc.Second Natn.Symp.Ann Arbor,Mich., 1967 (Conf.670503):337-345.

chemistry, chemicals accumulated within organisms
Foster, R.F., I.L. Ophel and A. Preston, 1971.
Evaluation of human radiation exposure. Radioactivity in the marine environment, U.S. Nat. Acad. Sci., 1971: 240-260.

concentration of chemicals within organisms
Fowler, S.W., L.F. Small and J.M. Dean, 1971.
Experimental studies on elimination of zinc-65, cesium-137 and cerium-144 by euphausiids. Marine Biol., 8(3): 224-231.

chemicals, concentration within organisms
*Fowler, Scott W., Lawrence F. Small, and John Mark Dean, 1969.
Metabolism of zinc-65 in euphausiids.
Symp.Radioecology, Proc.Second Natn.Symp.,Ann Arbor,Mich., 1967 (Conf.670503): 399-411.

chemistry chemicals concentrated within organisms
Fraizier A. et A. Vilquin, 1971.
Etude expérimentale de l'élimination du ^{137}Cs chez le mulet Mugil chelo et la blennie Blennius pholis. Mar. Biol. 10(2): 154-156.

chemistry, chemicals concentration within organisms
Fukai, Rinnosuke 1968.
Distribution of cobalt in marine organisms.
IAEA Radioactivity in the sea, Publ. 23:1-19

chemicals, concentration of
Fukai, Rinnosuke, 1964.
A note on the strontium-90 content of marine organisms as an index for variations of strontium-90 in sea water.
Bull. Inst. Océanogr., Monaco, 63(1307):16 pp.

Chemistry, concentration of chemicals
Fukai, Rinnosuki, and Daniele Broquet, 1965.
Distribution of chromium in marine organisms.
Bull. Inst. Oceanogr., Monaco, 65(1336): 19 pp.

Chemicals, concentration by organisms
Fukai, R., H. Suzuki and K. Watanabe, 1962.
Strontium-90 in marine organisms during the period 1957-1961.
Bull. Inst. Océanogr., Monaco, 59(1251):16 pp.

Chemistry, Concentration of chemicals
Folsom, Theodore R., D.R. Young and L.E. Finnin 1965.
Sumcoincidence gamma-ray spectrometry in tracing cobalt-60 and silver-110 in marine organisms.
In: Radioisotope sample measurement techniques in medecine and biology, Int. Atom. Energy Agency 1965: 57-69.

Chemistry, concentration of chemicals
Godoy, Oswaldo T., e Rudolf Barth 1967.
Concentração de cobre na agua e sua influência sôbre o plâncton. (In Portuguese; English summary)
Publ. Inst. Pesq. marinha, Rio de Janeiro, 01) 1967: 16 pp. (multilithed).

chemistry concentration of chemicals
Geneslay, Raymond, 1967.
Les ressources minérales des océans.
Cah. Océanogr. 19(10):833-845.

Chemistry, chemicals, Concentration within organisms
Glooschenko, Walter A. 1969.
Accumulation of ^{203}Hg by the marine diatom Chaetoceros costatum.
J. Phycol. 5(3): 224-226.

chemistry concentration by organisms
Goldberg, Edward D., 1965.
Minor elements in sea water.
In: Chemical oceanography, J.P. Riley and G. Skirrow, editors, Academic Press, 1:163-196.

chemistry, concentration of chemicals
*Gustafson, P.F., S.S. Brar, D.M. Nelson and S.E. Muniak, 1967.
Radioactive ^{54}Mn and ^{65}Zn in euryhaline fish.
Can. J. Zool., 45(5):729-735.

chemistry concentration of chemicals
Gutknecht, John, 1965.
Uptake and retention of cesium 137 and zinc 65 by seaweeds.
Limnology and Oceanography, 10(1):58-66.

chemistry elements, concentration of
Gutknecht, John, 1963
Zn 65 uptake by benthic marine algae.
Limnology and Oceanography, 8(1):31-38.

chemistry chemical elements, concentration of
Gutknecht, John, 1962
Zn uptake by Ulva lactuca.
Limnology and Oceanography, 7(2):270-271.

chemistry concentration of elements
Gutknecht, J., 1961.
Mechanisms of radioactive zinc uptake by Ulva lactuca.
Limnol. & Oceanogr., 6(4):426-431.

chemistry
Concentration of Chemicals
Harris, Robert C., 1965
Trace element distribution in molluscan skeletal material. 1. Magnesium, iron, manganese and strontium. Bull. Mar. Sci., 15(2):265-273.

chemistry concentration of chemical
Harriss, Robert C., 1965.
Trace element regulation in the marine alga Caulerpa racemosa.
Ecology, 46(4):539-540.

chemistry, concentration of chemicals
Harriss, Robert C., and Orrin H. Pilkey, 1966.
Temperature and salinity control of the concentration of skeletal Na, Mn, and Fe in Dendraster excentricus.
Pacific Science, 20(2):238.

chemicals, concentration of
Hela, Ilmo, 1963.
Alternative ways of expressing the concentration factors for radioactive substances in aquatic organisms.
Bull. Inst. Océanogr., Monaco, 61(1280):1-8.

chemistry, concentration of chemicals within organisms
Hiyama, Y., and R. Ichikawa, 1957.
Up-take of strontium by marine fish from the environment. Rec. Oceanogr. Wks., Japan, n.s., 3(1):78-84.

chemistry
concentration of chemicals
Hiyama, Yoshio, and Junko Matsubara Khan, 1964
On the concentration factors of radioactive I, Co, Fe and Ru in marine organisms.
Rec. Oceanogr. Wks., Japan, 7(2):79-106.

chemistry
concentration of chemicals
Hiyama, Yoshio, and Makoto Shimizu, 1964
On the concentration factors of radioactive Cs, Sr, Cd, Zn, and Ce in marine organisms.
Rec. Oceanogr. Wks., Japan, 7(2):43-77.

chemistry concentration of chemicals
*Hobden, D.J., 1967.
Iron metabolism in Mytilus edulis. 1. Variation in total content and distribution.
J. mar. biol. Ass., U.K., 47(3):597-606.

chemicals, concentration within organisms
*Holtzman, Richard B., 1969.
Concentrations of the naturally occurring radionuclides ^{226}Ra, ^{210}Pb and ^{210}Po in aquatic fauna.
Symp. Radioecology, Proc. Second Natn. Symp., Ann Arbor, Mich., 1967 (Conf. 670503):535-546.

chemicals, accumulation of
Hoss, Donald E., 1964.
Accumulation of zinc-65 by flounder of the Genus Paralichthys.
Trans. Amer. Fish. Soc., 93(4):364-368.

chemistry, concentration of chemicals within organisms
Hsiao, S.C.T., and H. Boroughs, 1958.
The uptake of radioactive calcium by sea urchin eggs. 1. Entrance of Ca 45 into unfertilized cytoplasm. Biol. Bull. 114(2):196-204.

Chemistry, chemicals concentration within organisms
Ikuta, Kunio 1968.
Studies on accumulation of heavy metals in aquatic organisms. II. On accumulation of copper and zinc in oysters. III. On accumulation of copper and zinc in the parts of oysters. (In Japanese; English abstract).
Bull. Jap. Soc. Scient. Fish. 34(2): 112-116; 117-122.

Chemistry - concentration of chemicals
Ikuta, Kunio 1967.
Studies on accumulation of heavy metals in aquatic organisms I. On the copper content in oysters. (In Japanese, English abstract).
Bull. Jap. Soc. scient. Fish., 33(5): 8405-409.

chemistry concentration of chemicals
Iorgulescu, Adriana, M. Oncescu, O. Serbanescu, and Florica Porumb, 1965.
Active fission γ (gamma) elements identified in marine plankton.
Rev. Roum. Biol. Zool., 10(5):373-378.

Chemistry
concentration of chemicals
Ishibashi, Masayoshi, Toshio Yamamoto and Tetsuo Fukita, 1964
Chemical studies on the ocean (Part 92).
Chemical studies on the seaweeds (17). Cobalt content in seaweeds.
Rec. Oceanogr. Wks., Japan, 7(2):17-24.

chemistry
concentration of chemicals
Ishibashi, Masayoshi, Toshio Yamamoto, and Tetsuo Fujita, 1964
Chemical studies on the ocean (Part 93).
Chemical studies on the seaweeds (18). Nickel content in seaweeds.
Rec. Oceanogr. Wks., Japan, 7(2):25-32.

chemistry, accumulation of chemicals within organisms
*Jefferies, D.F. and C.J. Hewett, 1971.
The accumulation and excretion of radioactive caesium by the plaice (Pleuronectes platessa) and the thornback ray (Raia clavata). J. mar. biol. Ass. U.K. 51(2): 411-422.

Chemistry, chemicals Concentrated within organisms
Jelisavcic Olga, Zvonimir Kolar, Petar Strohal, Stjepan Keckeš and Stjepan Lulić 1969.
Gross beta activity in selected organisms from the north Adriatic.
Rapp. P.-v. Reun. Commn int. Explor. scient. Mer Mediterr. 19(5): 957-959.

Chemistry
concentration of chemicals (phosphorus)
Johannes, R.E., 1964.
Uptake and release of phosphorus by a benthic marine amphipod.
Limnology and Oceanography, 9(2):235-242.

chemistry
concentration of chemicals
Kalk, Margaret, 1963.
Cytoplasmic transmission of a vanadium compound in a tunicate oocyte visible with electron-microscopy.
Acta Embroylogiae et Morphologiae Experimentalis. 6:289-303.

chemistry concentration of chemicals
Kalk, Margaret, 1963.
Absorption of vanadium by tunicates.
Nature, 198(4884):1010-1011.

Chemistry, chemicals, Concentration within organisms
Keckeš, Stjepan, Bartolo Ozretić and Mirjana Krajnović 1969
Metabolism of Zn^{65} in mussels (Mytilus galloprovincialis Lam.). Uptake of Zn^{65}.
Rapp. P.-v. Reun. Commn int. Explor. scient. Mer Mediterr. 19(5): 949-952.

chemistry
concentration of chemicals
*Keckes, Stjepan, Zvonimir Pučar and Ljerka Marazović, 1967.
Accumulation of electrodialytically separated physico-chemical forms of Ru by mussels.
Int. J. Oceanol. Limnol., 1(4):246-253.

chemistry, concentration of chemicals
Keckeš, S., Z. Pučar and Lj. Marazović, 1966.
The influence of the physico-chemical form of ^{106}Ru on its uptake by mussels from sea water.
In: Radioecological concentration processes, Proc. Int. Symp., Stockholm, 25-29 Apr. 1966, Pergamon Press, 993-994.

Chemistry
concentration of chemicals
Kevern, Niles R., 1964.
Relative strontium and calcium uptake by green algae.
Science, 146(3650):1488.

freshwater

Chemistry
concentration of chemicals
Khan, Kahn Umuardaraz, and Yoshio Hiyama, 1964
Mutual effect of Sr-Ca upon their uptake by fish and freshwater plants.
Rec. Oceanogr. Wks., Japan, 7(2):107-122.

chemistry concentration by organisms
Kilezhenko, V.P., and V.N. Podymakhin, 1964.
Accumulation of trace elements by some organisms in the southern part of the Barents Sea. (In Russian).
Materialy Ribochoz. Issled. Severn. Basseina, Poliarn. Nauchno-Issled. i Proektn. Inst. Morsk. Rib. Choz. i Okeanogr. im N.M. Knipovicha, PINRO, Gosud. Proizvodst. Komm. Ribn. Choz. SSSR. 4:135-134.

Chemistry concentration of chemicals
Kokubu, N., and T. Hidaka, 1965.
Tantalum and niobium in ascidians.
Nature, 205(4975):1028-1029.

chemicals concentration of
Krishnamurthy, K., 1963
Phosphorus in plankton. 1.
J. Zool. Soc., India, Calcutta, 14(2):161-164.
Also in:
Annamalai Univ. Mar. Biol. Sta. Porto Novo, S. India, Publ., 1961-1962.

Chemicals concentration within organisms
*Kuenzler, Edward J., 1969.
Elimination and transport of cobalt by marine zooplankton.
Symp. Radioecology, Proc. Second Natn. Symp., Ann Arbor, Mich., 1967 (Conf. 670503):483-492.

*Kujala, Norman F., I. Lauren Larsen and Charles L. Osterberg, 1969.
Radiosotope measurements of the viscera of Pacific salmon.
Symp. Radioecology, Proc. Second Natn. Symp., Ann Arbor, Mich., 1967 (Conf. 670503): 440-449.

Chemistry, Concentration of chemicals
Kulebakina, L.G., and G.G. Polikarpov. 1967.
On algal radioecology of the Black Sea shelf. (In Russian; English abstract)
Okeanologiia 7(2): 279-286.

chemistry concentration by organisms
Lerman, Abraham, 1965.
Strontium and magnesium in water and in Crassostrea calcite.
Science, 150(3697):745-751.

Chemistry concentration of chemicals
Lewis, Gary B., and Allyn H. Seymour, 1965.
Distribution of zinc-65 in plankton from offshore waters of Washington and Oregon, 1961-1963.
Ocean Sci. and Ocean Eng., Mar. Techn. Soc.,-Amer. Soc. Limnol. Oceanogr., 2:956-967.

Chemistry Concentration, chemicals
Lockwood, A.P.M., 1964.
Activation of the sodium uptake system at high block concentration in the amphipod Gammarus duebeni.
J. Exp. Biol., 41:447-458.

chemicals, concentration of
Lopez-Benito, Manuel, 1963
Estudio de la composicion quimica del Lithothamnium calcareum (Aresch) y su aplicacion como corrector de terrenos de vultivo.
Inv. Pesq., Barcelona, 23:53-70.

chemistry, chemicals accumulated within organisms
Lowman, F.G., T.R. Rice and F.A. Richards, 1971.
Accumulation and redistribution of radionuclides by marine organisms. Radioactivity in the marine environment, U.S. Nat. Acad. Sci., 1971: 161-199.

concentration of chemicals within organisms
Lucu, Čedomil, and Olga Jelisavčić, 1970
Uptake of ^{137}Cs in some marine animals in relation to temperature, salinity, weight and moulting.
Int. Revue ges. Hydrobiol. 55(5): 783-796.

chemistry, chemicals, concentration within organisms
Lucu, Č., O. Jelisavčić, S. Lulić and P. Strohal, 1969.
Interactions of ^{233}P with tissues of Mytilus galloprovincialis and Carcinus mediterraneus.
Mar. Biol. 2(2): 103-104

chemistry concentration of chemicals within organisms
Lyubimov, A.A., A. Ya. Zesenko, and L.N. Leshchenko, 1970.
The effect of changing physical-chemical state of radioyttrium on its accumulation by marine organisms. (In Russian; English abstract).
Okeanologiia, 10(6): 1001-1008.

Chemistry accumulation of chemicals within organisms
Mauchline, John, and Angela M. Taylor, 1964.
The accumulation of radionucleides in the thornback ray, Raia clavata L., in the Irish Sea.
Limnology and Oceanography, 9(3):303-309.

chemicals, concentration of
Mauchline, John, Angela M. Taylor and Eric B. Ritson, 1964.
The radioecology of a beach.
Limnology and Oceanography, 9(2):187-194.

concentration of chemicals
Mauchline, J., and W.L. Templeton, 1966.
Strontium, calcium and barium in marine organisms from the Irish Sea.
J. Cons. perm. int. Expl. Mer, 30(2):161-170.

Chemistry concentration of chemicals
Mehran, A.R., and J.L. Tremblay, 1965.
Un aspect du metabolisme du zinc chez Littorina obtusata L., et Fucus edentata de la Pylaie.
Revue canadienne de Biologie, 24(3):157-161.

Chemistry concentration of chemicals
Michon, G., 1965.
Facteurs de concentration biologique en milieu marin. Application aux problèmes des pollutions radioactifs.
Cahiers du C.E.R.B.O.M., Nice, 2:125-145.

Chemistry concentration of chemicals
Mironov, O. G., 1964.
On the problem of migration of strontium-90 and cesium-137 into the organism of man in the course of consuming some species of marine animals. (In Russian).
Okeanologiia, Akad. Nauk, SSSR, 4(6):1059-1061.

Chemistry elements, concentration of
Mishima, Jiro, and Eugene P. Odum, 1963.
Excretion rate of Zn 65 by Littorina irrorata in relation to temperature and body size.
Limnology and Oceanography, 8(1):39-44.

Chemistry, concentration of chemicals within organisms
Miyake, Y., 1963
3. Artificial radioactivity in the sea. In: The Sea, M.N. Hill, Edit., Vol. 2. The composition of sea water, Interscience Publishers, New York and London, 78-87.

Chemistry, concentration of chemicals within organisms
Morgan, F., 1964.
The uptake of radioactivity by fish and shellfish 1. 134 caesium by whole animals.
J. Mar. Biol. Assoc., U.K., 44(1):259-271.

Chemistry, concentration of chemicals within organisms
Mori, Takajiro, Keishi Amano, Toshiharu Kawabata and Masamichi Saiki, 1957.
Studies on the accumulation of fission products in the fish. 1.
Pap. First Conf., Radioisotopes, Japan Atomic Industrial Forum:546-551.
Abstr. in:
Rec. Res., Fac. Agricult., Tokyo Univ., 7(91):62.

Chemistry, concentration of chemicals
Munda, Ivka, und Dušan Zavodnik 1967.
Algenbestände als Konzentrationen organische Materials im Meer.
Helgoländer wiss. Meeresunters. 15 (1/4): 622-629

concentration of chemicals by organisms
Nagaya, Yutaka, and Theodore R. Folsom, 1966.
Zinc-65 and other fallout nuclides in marine organisms of the California coast.
J. Radiation Res., 5(1):82-89.

Chemistry, chemicals, concentration within organisms
Nair, the late K.V.K., Y.M. Bhatt and G.R. Doshi 1969.
Fallout radioactivity in three brackish water molluscs from Kerala.
Current Sci. 38 (14): 332-333.

Chemistry, concentration of chemicals within organisms
Nakano, Eizo, Kayo Okazaki and Takashi Iwamatsu, 1963
Accumulation of radioactive calcium in larvae of the sea urchin Pseudocentrotus depressus.
Biol. Bull., 125(1):125-132.

chemicals concentrated by animals
Noddack, I., and W. Noddack, 1940.
Die Haufigkeiten der Schwermetalle in Meerestieren.
Ark. Zool., 32A(4):35 pp.

Chemistry concentration of chemicals by animals
Odum, H.T., 1951.
Notes on the strontium content of sea water, celertite radiolaria and strontianite snail shells.
Science, 114:211-213.

Chemistry concentration of chemicals
Oguri, Mikio, Naoko Takada and Ryushi Ichikawa, 1965.
Metabolism of radionuclides in fish. IV. Strontium-calcium discrimination in the renal excretion of fish.
Bull. Jap. Soc. Sci. Fish., 31(6):435-438.

Chemistry concentration of chemicals
Osterberg, Charles, 1965.
Radioactivity from the Columbia River.
Ocean Sci. and Ocean Eng., Mar. Techn. Soc.,-Amer. Soc. Limnol. Oceanogr., 2:968-979.

Chemistry, concentration of chemicals within organisms
Osterberg, Charles, Andrew G. Carey and Herbert Curl, 1963.
Acceleration of sinking rates of radionuclides in the ocean.
Nature, 200(4913):1276-1277.

chemicals accumulation of
Osterberg, Charles, June Pattullo and William Pearcy, 1964
Zinc-65 in euphausiids as related to Columbia River water off the Oregon Coast.
Limnology and Oceanography, 9(2):249-257.

concentration of chemicals within organisms
*Ozretic, B., and M. Krajnovic, 1968.
The turnover of Zn 65 during the early embryonal development of the sea urchin Paracentrotus lividus Lam.
Rev. Intern. Océanogr. Med., 11:101-106.

Chemistry concentration, chemicals
Paasche, E., 1964.
A tracer study of the inorganic carbon uptake during coccolith formation and photosynthesis in the coccolithophorid coccolithus huxleyi.
Physiol. Plant. Suppl. 3:82 pp.

chemicals, concentration of
Paoletti, A., 1966
Captation du strontium radioactif par des Algues microscopiques
Revue Int. Océanogr. Médicale, C.E.R.B.O.M. 1: 92-99

concentration of chemicals within organisms
Parchevsky, V.P., 1965.
Radionuclides of cesium, ruthenium and zirconium in animal and plant organisms of the Black Sea. (In Russian).
Okeanologiia, Akad. Nauk, SSSR, 5(5):856-862.

Chemistry, concentration of chemicals
Parchevskii, V.P. 1964.
Concentration of a mixture of fission products in some Black Sea organisms. (In Russian)
Trudy Sevastopol. Biol. Stants. Akad Nauk 15: 493-498.

chemistry concentration by organisms
Parchevskii, V.P., G.G. Polycarpov and I.S. Zaburunova, 1965.
Certain regularities in the accumulation of yttrium and strontium by marine organisms. (In Russian).
Doklady, Akad. Nauk, SSSR, 164(4):913-916.

chemistry, chemicals concentrated within organisms

Piro, A., M. Verzi e C. Papucci 1969. L'importanza dello stato fisico-chimico degli elementi per l'accumulo negli organismi marini. I. Lo stato chimico-fisico dello zinco in acque di mare. Pubbl. Staz. Zool. Napoli 37 (2 Suppl.): 295-310.

concentration of chemicals

Polikarpov, G.G., and A. Ya. Zesenko, 1965. On the possibility of using the concept of the "build-up factor" in the radio-ecological study of the seas. (In Russian). Okeanologiia, Akad. Nauk, SSSR, 5(6):1099-1100.

chemistry - concentration of chemicals

Parker, Patrick L., 1966. Movement of radioisotopes in a marine bay: cobalt-60, iron-59, manganese-54, zinc-65, soduim-22. Publs. Inst. mar. Sci., Univ. Texas, Port Aransas, 11:102-107.

chemistry - concentration of chemicals

Pearcy, William G., and Charles L. Osterberg, 1964. Vertical distribution of radionuclides as measured ~~by~~ in oceanic animals. Nature, 204(4957):440-441.

chemicals, concentration within organisms

Pentreath, R.J. and D.F. Jefferies, 1971. The uptake of radionuclides by I-group plaice (Pleuronectes platessa) off the Cumberland Coast, Irish Sea. J. mar. biol. Ass. U.K. 51(4): 963-976.

chemicals, concentration within organisms

Policarpov, G.C., 1961. [On the stability of accumulation coefficients of Sr 90, Y91 and Ce 144 in sea algae.] Doklady Akad. Nauk, SSSR, 140(5):1192-1194.

chemicals, concentration of by marine organisms

Polikarpov, G.G., 1961. [Material on coefficient of accumulation of P32, S35, Sr90, Y91, Cs137 and Ce144 in marine organisms.] Trudy Sevastopol Biol. Sta., (14):314-328.

chemistry concentration by organisms

Poulsen, V., 1963. Notes on Hyolithellus Billings, 1871, Class Pogonophora Johannson, 1937. Biol. Medd. K. Danske Vidensk. Selsk., 23(12): 15 pp.

chemistry, chemicals concentration within organisms

Preston, A. 1970. Concentrations of iron-55 in commercial fish species from the North Atlantic. Mar. Biol. 6(4): 345-349.

chemistry, chemicals concentration within organisms

Preston, Alan, 1965. The control of radioactive pollution in a North Sea oyster fishery. Helgoländer wiss. Meeresunters. 17: 269-279.

chemistry, chemicals, concentration within organisms

Preston, Eric M., 1971. The importance of ingestion in chromium-51 accumulation by Crassostrea virginica (Gmelin). J. exp. mar. Biol. Ecol., 6(1): 47-54.

concentration within organ.

Pringle, B.H., D.E. Hissong, E.L. Katz and S.T. Mulawka, 1968. Trace metal accumulation by estuarine molluscs. J. Sanit. Engng. Div. Am. Soc. Civ. Engrs. 94 (SA3): 455-475.

Concentration of chemicals within organisms

Pyle, Thomas E. and Thomas T. Tieh, 1970. Strontium, vanadium, and zinc in the shells of pteropods. Limnol. Oceanogr., 15(1): 153-154.

chemistry - concentration of chemicals

*Reish, D.J. and G.C. Stephens, 1969. Uptake of organic material by aquatic invertebrates. V. The influence of age on the uptake of glycine-C^{14} by the polychaete Neanthes arenaceodentata. Marine Biol., 3(4): 352-355.

chemistry concentration of radioactive elements

Roche, Jean, Simonne Andre and Italo Covelli, 1963. Sur la fixation de l'iode (131 I) par la moule (Mytilus galloprivincialis L.) et la nature des combinaisons iodées elaborées. Comp. Biochem. Physiol., 9(4):291-300.

chemicals concentrated within organisms

Romeril, M.G., 1971. The uptake and distribution of ^{65}Zn in oysters. Mar. Biol. 9(4): 347-354.

Chemistry, chemicals concentration within organisms

Samuels, E.R., M. Cawthorn, B.H. Lauer and B.E. Baker 1970. Strontium-90 and cesium-137 levels in tissues of fin whale (Balaenoptera physalus) and harp seal (Pagophilus groenlandicus). Can. J. Zool. 48(2): 267-269.

Chemistry concentration by organisms

Schreiber, B., 1963. Acantharia as "scavengers" for strontium and their role in the sedimentation of radioactive debris. In: Nuclear detonations and marine radioactivity the report of a symposium held at the Norwegian Defence Research Establishment, 16-20 September, 1963. Forsvarets Forskningsinstitutt, P.O. Box 25, Kjeller, Norge, 113-126. (multilithed).

chemistry Concentration of chemicals

Schreiber, Bruno, Enrico Cerrai, Cesare Triulzi and Laura Tassi-Pelati, 1964. Gross beta radioactivity, content of Sr^{90} and of other radionuclides in plankton collected in the Tyrrhenian Sea during the 1st oceanographic campaign of M/N "Bannock" (September-October) 1963. Ist. Lombardo, Accad. Sci. e Lett, Rendiconti, (B), 98:143-166.

concentration of elements within organisms

Segar, D.A., J.D. Collins and J.P. Riley, 1971. The distribution of the major and some minor elements in marine animals. J. mar. biol. Ass. U.K., 51(1): 131-136.

chemistry - accumulation within organisms

Seki, H., 1970. Bioaccumulation of radioisotopes by microorganisms in the sea. (In Japanese). J. oceanogr. Soc., Japan, 26(6): 367-372.

concentration of chemicals

*Seki, Humitake, and Nobuo Taga, 1967. Removal of radio nuclides at different multiplication phases of marine bacteria. J. oceanogr. Soc., Japan, 23(3):136-140.

chemicals accumulated within organisms

Shannon, L.V., and R.D. Cherry 1971. Radium-226 in marine phytoplankton. Earth planet. Sci. Letts 11(4): 339-343

Chemistry - concentration of chemicals

Sherstnev, A. I., and I.I. Baksheeva, 1962. Maintenance of common potassium and the radio-active isotope potassium-40 in the organs of some different fish from the west coast of Africa. (In Russian). Trudy, Baltiisk. Nauchno-Issled. Inst. Morsk. Ribn. Khoz. i Okeanogr. (BALTNIRO), 8:135-137.

concentration of chemicals

*Shimizu, Makoto, Lydia S. Ignatiades and Yoshio Hiyama, 1968. On the concentration factors of 198 Au in several marine organisms. Rec. oceanogr. Wks, Japan, n.s., 9(2):257-261.

Chemistry, chemicals concentrated within animals

Shimizu, Makoto, Takeshi Kajihara and Yoshio Hiyama 1970. Uptake of ^{60}Co by marine animals. Rec. oceanogr. Wks, Japan 10(2): 137-145.

Chemistry, accumulation of chemicals within organisms

Shuliene R. Yu. 1971. On the accumulation of radiocarbon by Gammarus pulex L. from water solution of sodium carbonate. (In Russian) Gidrob. Zh. 7(3): 83-84.

concentration of chemicals within organisms

Shuster, Carl N., Jr., and Benjamin H. Pringle, 1969. Trace metal accumulation by the American eastern oyster, Crassostrea virginica. Proc. nat. Shellfish. Ass., 59: 91-103

chemicals, concentration within organisms

*Simek, John E., J.A. Davis, C.E. Day, III, and Ernest E. Angino, 1969. Sorption of radioactive nuclides by Sargassum fluitans and S. natans. Symp. Radioecology, Proc. Second Natn. Symp., Ann Arbor, Mich., 1967 (Conf. 670503):505-508.

concentration of chemicals

*Small, Lawrence F., 1969. Experimental studies on the transfer of ^{65}Zn in high concentration by euphausiids. J. exp. mar. Biol. Ecol., 3(1):106-123.

Chemistry chemicals Concentration within organisms

Smith, J. David 1970. Tin in organisms and water in the Gulf of Naples. Nature, Lond. 225 (5227): 103-104

concentration by organisms

Stevenson, Robert A., 1965. Differences in trace element composition in the sea urchins Tripneustes esculentus (Leske) and Echinometra lucunter (L.). (Abstract). Ocean Sci. and Ocean Eng., Mar. Techn. Soc., Amer. Soc. Limnol. Oceanogr., 1:3 0.

concentration of elements by organisms
Stevenson, Robert A., and Sara Lugo Ufret, 1966.
Iron, manganese, and nickel in skeletons and food of the sea urchins Tripneustes esculentus and Echinometra lucunter.
Limnol. and Oceanogr., 11(1):11-17.

concentration of chemicals
Thommeret, M., 1963.
Aspect des recherches du C.S.M. dans le somaine des pollutions de la mer et de la concentration du C14 dans les organismes marins. (Résumé).
Rapp. Proc. Verb., Réunions, Comm. Int., Expl. Sci., Mer Méditerranée, Monaco, 17(3):953-955.

chemistry concentration of chemicals within organisms
Timet, Dubravko, Vladimir Mitin, Melita Herak and Dubravko Emanović, 1969.
Investigations in the biochemistry of the Lim Channel-oyster (Ostrea edulis L.) from the Lim Channel-IV: The concentration of calcium, magnesium and inorganic phosphorus in the extrapallial fluid of the oyster. (Jugoslavian and Italian abstracts)
Thalassia Jugoslavica, 5:415-420.

chemistry, concentration of chemicals within organisms
Timet, Dubravko, Vladimir Mitin, Melita Herak and Dubravko Emanović, 1969.
Investigations in the biochemistry of the oyster (Ostrea edulis L.) from the Lim Channel-III: The concentration of sodium and potassium in the extrapallial fluid of the oyster. Jugoslavian and Italian abstracts).
Thalassia Jugoslavica, 5:409-414

chemistry, concentrations of chemicals within organisms
Timet, Dubravko, Vladimir Mitin, Melita Herak and Dubravko Emanović, 1969.
Investigations in the biochemistry of the oyster (Ostrea edulis L.) from the Lim Channel-II: The concentration of calcium magnesium, and inorganic phosphorus in the tissues of the oyster. (Jugoslavian and Italian abstracts).
Thalassia Jugoslavica, 5:403-408

chemistry - concentrations of chemicals within organisms
Timet Dubravko, Vladimir Mitin, Melita Herak and Dubravko Emanović, 1969.
Investigations in the biochemistry of the oyster (Ostrea edulis L.) from the Lim Channel I: The concentration of sodium and potassium in the tissues of the oyster. (Jugoslavian and Italian abstracts).
Thalassia Jugoslavica, 5:395-401.

Chemistry, chemicals concentrated within organisms
Tsuruga, Hanato, 1968.
Some long-lived radionuclides in marine organisms on the Pacific coast of Japan.
J. radiation Res. 9-2: 63-72. Also in: Coll. Repr. 1968, Tokai reg. Fish. Res. Lab. B-491.

chemistry concentration of chemicals by marine organisms
Tsuruga, Hanato, 1965.
Sequential analysis of radionuclides in marine organisms.
Bull. Jap. Soc. Sci. Fish., 31(9):651-658.

chemistry concentration of chemicals
Tsuruga, H., 1962.
The uptake of radioruthenium by several kinds of littoral seaweed.
Bull. Jap. Soc. Sci. Fish., 28(12):1149-1154.

concentration of chemicals
*Umezu, Takeshi, and Masamichi Saiki, 1967.
Gross uptake of radionuclides ny marine micro-organisms in batch culture. (In Japanese; English abstract).
Bull. Naikai reg. Fish. Res. Lab., 24: 1-9.

Chemistry, chemicals, concentration within organisms
Vosjan, J.H. 1969.
Effect of chelation on the uptake and loss of yttrium-91 by Porphyra.
Neth. J. Sea Res. 4(3): 310-316.

Chemistry concentration of chemicals
Williams, Louis G., 1964.
Relative strontium and calcium uptake by green algae.
Science, 146(3650):1488.

freshwater

chemistry, chemicals concentration within organisms
Wolfe, Douglas A. 1970.
Levels of stable Zn and 65Zn in Crassostrea virginica from North Carolina.
J. Fish. Res. Bd. Can. 27(1): 47-57.

Concentration of chemicals
*Wolfe, Douglas A. and Claire L. Schelske, 1969.
Accumulation of fallout radioisotpoes by bivalve molluscs from the lower Trent and Neuse rivers.
Symp. Radioecology, Proc. Second Natn., Symp., Ann Arbor, Mich., 1967 (Conf. 670503):493-504.

chemistry uptake by organisms
Wright, Richard T., and John E. Hobbie, 1965.
The uptake of organic solutes by planktonic bacteria and algae. (abstract).
Ocean Sci. and Ocean Eng., Mar. Techn. Soc.,-Amer. Soc. Limnol. Oceanogr., 1:116-127.

concentration within organisms
*Yamamoto, Toshio, Tatsuo Fujita, and Tseunenobu Shigematsu, 1969.
Chemical studies on the seaweeds (24): Strontium content in seaweeds.
Rec. Oceanogr. Wks. Japan, 10(1):29-38.

chemistry - concentration of chemicals
*Yoshida, Tamao, Toshiharu Kawabata and Yoshiyuki Matsue, 1967.
Transference mechanism of mercury in marine environment.
J. Tokyo Univ. Fish., 53(½):73-84.

Chemistry concentration of chemicals
*Young, D.R., and T.R. Folsom, 1967.
Loss of Zn 65 from the California sea-mussel Mytilus californianus.
Biol. Bull., mar. biol. Lab., Woods Hole, 133(2):438-447.

chemistry, concentration of chemicals within organisms
Zattera, Antonio, e Michael Bernhard 1969.
L'importanza dello stato chimico-fisico degli elementi per l'accumulo negli organismi marini. II Accumulo di zinco stabile e radioattivo in Phaeodactylum tricornutum.
Pubbl. Staz. Zool. Napoli, 37 (2 Suppl.): 386-399.

concentration within organisms
Zesenko, A. Ya., A.A. Lyubimov, V.N. Ivanov, and L.N. Leshchenko, 1968.
Changes of the state of radionuclides in sea water and its influence on the accumulation by the sea organisms. (In Russian).
Okeanologiia, Akad. Nauk. SSSR, 8(6):1092.

Chemistry - CONCENTRATION OF CHEMICALS
Zlobin, V.S., 1968.
Dynamics of radiostrontium accumulation by some brown algae and the influence of sea water salinity on the coefficients of accumulation. (In Russian; English abstract).
Okeanologiia, Akad. Nauk, SSSR, 8(1):78-85.

chemistry, chemical environment
Goldberg, E.D., W.S. Broecker, M.G. Gross and K.K. Turekian, 1971.
Marine chemistry. Radioactivity in the marine environment, U.S. Nat. Acad. Sci., 1971: 137-146.

chemistry, chemical properties
Dussart, B.H. 1967.
Some comments on "integrative" and "specific" properties of the aquatic environment.
In: Chemical environment in the aquatic habitat, H.L. Golterman and R.S. Clymo, editors, Proc. I.B.P. Sym. Amsterdam, Oct. 1966: 24-29.

chitin

chitin
Blackwell, J., K.D. Parker and K.M. Rudall, 1965.
Chitin in pogonophore tubes.
J. mar. biol. Ass., U.K., 45(3):659-661.

Chemistry chitin synthesis
Dall, W., 1965.
Studies on the physiology of a shrimp Metapenaeus sp. (Crustacea:Decapoda:Penaeidae). IV. Carbohydrate metabolism.
Australian J. Mar. Freshwater Res., 16(2):163-180.

Chemistry chitin
Daste, Ph., 1956.
La décomposition de la chitine par les microorganismes.
L'Année Biol., (3), 32(11-12):473-488.

Chemistry, chitin
*Greze, I.I., 1967.
On the amount of chitin and calcite in shells of Amplipoda (Gammaridea).
Zool. Zh., 46(11):1655-1658.

chitinous shells
Harding, J.P., 1964.
Crustacean cuticle with reference to the ostracod carapace.
Pubbl. Staz. Zool., Napoli, 33(Suppl.):9-31.

Chemistry chitin
Hyman, Libbie H., 1966.
Further notes on the occurence of chitin in invertebrates.
Biol. Bull., 130(1):94-95.

Chemistry, chitin synthesis
Parvathy, K., 1970.
Blood sugars in relation to chitin synthesis during cuticle formation in Emerita asiatica.
Mar. Biol., 5(2): 108-112.

Chemistry, chitin
Raymont, J.E.G., J. Austin and Eileen Linford, 1967.
The biochemical composition of certain zooplankton decapods.
Deep-Sea Research, 14(1):113-115.

chemistry, chitin
Raymont, J.E.G., R.T. Srinivasagam and J.K.B. Raymont, 1971.
Biochemical studies on marine zooplankton - VIII. Further investigations on Meganyctiphanes norvegica (M. Sars). Deep-Sea Res. 18(12): 1167-1178.

chemistry - Chitin

*Raymont, J.E.G., R.T. Srinivasagam and J.K. B. Raymont, 1969.
Biochemical studies on marine zooplankton. IV. Investigation on Meganyctiphanes norvegica (M. Sars.) Deep-Sea Res., 16(2): 141-156.

chemistry, chitin

Seki, H., 1966.
Microbiological studies on the decomposition of chitin in the marine environment. (In Japanese). J. oceanogr. Soc., Japan, 22(5):236-240.

Chemistry chitin, decomposition of

Seki, Humitake, 1965.
Microbiological studies on the decomposition of chitin in marine environment IX. Rough estimation on chitin decomposition in the ocean. X. Decomposition of chitin in marine sediments. J. Oceanogr. Soc., Japan, 21(6):253-260;261-269.

chemistry-chitin

Seki, Humitake, and Nobuo Tago, 1965.
Microbiological studies on the decomposition of chitin in marine environment. VIII. Distribution of chitinoclastic bacteria in the pelagic and neretic waters.
J. Oceanogr. Soc., Japan, 21(4):174-187.

Chemistry Chitin

Seki, Humitake, and Nobuo Taga, 1965.
Micorbiological studies on the decomposition of chitin in marine environment. VI. Chitinoclastic bacteria in the digestive tracts of whales from the Antarctic Ocean.
Jour. Oceanogr. Soc., Japan 20(6):272-277.

chemistry, chitin

Takeda, Michio and Teruo Tomida 1969.
Chitin absorbent in thin-layer chromatography for separation of phenols and amino acids. (In Japanese; English abstract).
J. Shimonoseki Univ. Fish. 18(1):36-44.

Chemistry-chitin

Takeda, Michio, Teruo Tomida, Takashige Omura and Hiroshi Katsuura, 1965.
Thin-layer chromatography of saccharides obtained by degradation of chitin. 1. Separation of glucosamine, N-acetylglucosamine and chito-oligosaccharides. (In Japanese; English abstract
J. Shimonoseki Univ., Fish., 14(1):1-8.

Chloride

Chemistry chloride

Anderson, L. J., 1948.
Conductometric titration of chloride in sea water and marine sediments. ANalyt. Chem. 20(7): 618-619.

Chemistry, chloride

Brewer, Peter G., and Derek W. Spencer, 1969.
A note on the chemical composition of the Red Sea brines.
In: Hot brines and Recent heavy metal deposits in the Red Sea, E.T. Degens and D.A. Ross, editors, Springer-Verlag, New York, Inc., 174-179.

chemistry - chloride

Burkholder, Paul R., 1960.
Distribution of some chemical values in Lake Erie. Limnological survey of western and central Lake Erie, 1928-1929. USFWS Spec. Sci. Rept., Fish., No. 334 :71-110.

Chemistry chloride

Eriksson, Erik, 1960.
The yearly circulation of chloride and sulfur in nature; meteorological, geochemical and pedological implications. II.
Tellus, 12(1):63-109.

Chemistry, chloride

Honma, M., 1955.
Flame photometric determination of chloride in sea water. Analyt. Chem., 27 :1656-1659.

Chemistry chloride

Jones, H.B., and H. Baum, 1955.
Determination of chloride by automatic titration Analyt. Chem. 27(1):99-100.

chemistry chlorides

Koske, Peter H., 1964.
Über ein potentiometrisches Verfahren zur Bestimmung von Chloridkonzentration in Meerwasser Kieler Meeresf., 20(2):138-142.

Chemistry chloride (data)

Kramer J.R., 1961
Chemistry of Lake Erie.
Proc. Fourth Conf., Great Lakes Research, Great Lakes Res. Div., Inst. Sci. & Tech. Univ. Michigan, Publ., (7):27-56.

chlorides

*Lerman, Abraham, 1967.
Model of chemical evolution of a chloride lake-The Dead Sea.
Geochim. cosmochim Acta, 31(12):2309-2330.

chlorides

van Dam, L., 1940.
Estimation of chlorides in 1 cm^3 sea water sample by means of syringe pipettes. Treubia 17:473-477, 1 textfig.

chemistry, chlorinated hydrocarbons

Hays, Helen, and Robert W. Risebrough 1972
Pollutant concentrations in abnormal young terns from Long Island Sound.
The Auk 89 (1): 19-35

chlorine

Chemistry, chlorine

Bjoern-Andersen, 1912.
Exact determination of the chlorine in some samples of sea water from the Mediterranean. Rept. Danish Oceanogr. Exped., 1908-1910, 1:193-197.

chemistry, chlorine

*Choe, Sang, Tai Wha Chung and Hi-Sang Kwak 1968.
Seasonal variations in nutrients and principal ions contents of the Han River water and its water characteristics. (In Korean; English abstract).
J. oceanogr. Soc., Korea, 3(1):36-38.

Chemistry chlorine

Hoering, T.C., & P.L. Parker, 1961

The geochemistry of the stable isotopes of chlorine.
Geochim. et Cosmochim. Acta, 23 (3/4): 186-199.

chlorine ions

Kawabata, Hiroshi, and Shiko Shiga, 1966.
Correlations among tide, discharge, temperature and chemical contents at Kamegawa, Beppu City. Spec. Contrib., Geophys. Inst., Kyoto Univ., No. 5:25-32.

chemistry chlorine

Masurenkov, Iu. P., & S.I. Pakhomov, 1961
[A contribution to the geochemistry of chlorine] Doklady Akad. Nauk, SSSR, 139(2):453-455.

Chemistry, chlorine

Ruppin, E., 1911.
Bericht über das Verhältnis du Cl, SO3 und % Werte in einer Reihe von 14 Meerwassen Proben. Zeits. Anorg. Allgem. Chem., 69:232-246.

Transl. cited: USFWS Spec. Sci. Rept., Fish. 227

chlorine

Selivanov, L.S., 1947.
[On the origin of chlorine and bromine in the salts of the ocean] Bull. Volcan. Station, Kamchatka, 11:26-34.

Chemistry - chlorine

Winchester, John W., and Robert A. Duce, 1965.
Geochemistry of iodine, bromine, and chlorine in the air-sea-sediment system.
Narragansett Mar. Lab., Univ. Rhode Island, Occ. Publ., No. 3:185-201.

Chlorine 36

chemistry chlorine 36

Thompson, S.O., and O.A.Schaeffer, 1964
Search for chlorine 36 in Pacific red clays. (Abstract).
Trans. Amer. Geophys. Union, 45(1):119.

Chemistry chlorine, effect of

Hirayama, K. and R. Hirano, 1970.
Influences of high temperature and residual chlorine on marine phytoplankton. Marine Biol., 7(3): 205-213.

Chemistry, chlorine (effect of)

Zibrowius, Helmut, et Gérard Bellan 1969
Sur un nouveau cas de salissures biologiques favorisées par le chlore.
Téthys 1(2): 375-382.

chlorinity

A

Chemistry chlorinity

Akagawa, M., 1956.
[On the oceanographical conditions of the north Japan Sea (west off the Tsugaru Straits) in summer (Part 1).]
Bull. Hakodate Mar. Obs., No. 3:1-11
190-199.

Chemistry chlorinity

Akagawa, M., 1956.
On the oceanographical conditions of the North Japan Sea (west of the Tsugaru Straits) in summer (Part 2.).
Bull. Hakodate Mar. Obs., No. 3:5-11
1-7.

Chemistry chlorinity

Akagawa, M., 1954.
[On the oceanographical conditions of the north Japan Sea (west off the Tsugaru-Straits) in summer] J. Ocean. Soc., Japan, 10(4):189-199, 5 textfigs.

Chemistry, Chlorinity

Akiyama, Tsutomu, Takeshi Sagi, Takeshi Yura and Kan Kimura 1966.
On the distribution of in situ pH in the adjacent seas of Japan.
Oceanogrl Mag. 18(1/2): 83-90.

Chemistry - chlorinity

Anon., 1950.
[The results of harmonic analysis of tidal currents, water temperature and chemical components in the Kii-Suido.] J. Ocean., Kobe Obs., 2nd ser. 2(1):28-31, 2 figs.

chemistry chlorinity (data)
Angino, Ernest E., and Gale K. Billings, 1966.
Lithium content of sea water by atomic absorption spectrometry.
Geochim. et Cosmochim. Acta, 30(2):153-158.

chemistry chlorinity (data)
Angot, M., et R. Gerard, 1965.
Hydrologie de la région de Nosy-Bé: juillet à novembre 1963.
Cahiers, O.R.S.T.R.O.M. - Océanogr., 3(1):3-29.

chemistry, chlorinity (data)
Angot, M., et R. Gerard, 1965.
Hydrologie de la région de Nosy-Be: decembre 1963 a mars 1 964.
Cahiers, O.R.S.T.R.O.M. - Océanogr., 3(1):31-53.

chemistry chlorinity
Anon., 1951.
Bulletin of the Marine Biological Station of Asamushi 4(3/4): 15 pp.

chemistry, chlorinity
Aota, Masaaki 1968.
Study of the variation of oceanographic conditions northeast off Hokkaido in the Sea of Okhotsk. (In Japanese; English abstract)
Low Temp. Sci. (A)26: 351-356.

chemistry chlorinity
Australia, Marine Biological Laboratory, Cronulla, 1960.
F.R. V. "Derwent Hunter", scientific report of cruises 10-20/58 ---------.
C.S.I.R.O., Div. Fish. & Oceanogr., Rept., 30: 53 pp., numerous figs. (mimeographed).

chemistry chlorinity
Australia, C.S.I.R.O., Division of Fisheries and Oceanography, 1959.
Hydrological investigations from F.R.V. Derwent Hunter, 1957.
Oceanogr. Sta. List, No. 37:96 pp.

chemistry, chlorinity
Australia, C.S.I.R.O., Division of Fisheries and Oceanography, 1959.
Scientific reports of a cruise on H.M.A. Ships "Queenborough" and "Quickmatch", March 24-April 26, 1958. Rept. No. 24: 24 pp. (mimeographed).

B

chemistry chlorinity (data)
Băcescu, M., G. Müller, H. Skolka, A. Petran, V. Elian, M.T. Gomoiu, N. Bodeanu si S. Stănescu, 1965.
Cercetări de ecologie marină în sectorul predeltaic în condițiile anilor 1960-1961.
In: Ecologie marină, M. Băcescu, redactor, Edit. Acad. Republ. Pop. Române, București, 1: 185-344.

chemistry chlorinity
Bather, J.M., and J.P. Riley, 1954.
The chemistry of the Irish Sea. Part 1. The sulphate-chlorinity ratio. J. du Cons. 20(2): 145-152.

chemistry - chlorinity
Bather, J.M., and J.P. Riley, 1953.
The precise and routing potentiometric determination of the chlorinity of sea water. J. du Cons 18(3):277-286, 3 textfigs.

chemistry, chlorinity (data)
Berthois, L., et A. Crosnier 1966.
Étude dynamique de la sedimentation au large de la Betsiboka.
Cah. ORSTOM, Ser. Océanogr. 4(2): 49-130.

chlorinity
Bjerrum, N., 1904.
On the determination of chlorine in sea water and examination of the accuracy with which Knudsen's pipette measures a volume of sea water.
Medd. Komm. Havundersøgelser, Ser. Hydrogr., 1(3):11 pp.

chemistry, chlorinity
Brujewicz, S.W., 1964.
The methods of determination of chlorinity (salinity) in sea water. Chemistry of the waters and sediments of seas and oceans. (In Russian; English abstract).
Trudy Inst. Okeanol., Akad. Nauk, SSSR, 67:177-215.

chlorinity
Brun, Guy, 1967.
Étude écologique de l'estuaire du "Grand Rhone".
Bull. Inst. océanogr., Monaco, 66(1371):46 pp.

chlorinity
Buljan, Miljenko 1965
Some results of long-term hydrographic investigations at the Stončica Station (Middle Adriatic) (Preliminary Report).
Rapp. P.-v. Réun., Comm. int. Explor. scient., Mer Méditerr., 18(3):767-771.

chemistry, chlorinity
Buljan, M., 1948/1949.
Sur l'emploi de certains indicateurs pour le dosage de la chlorinité des eaux. Acta Adriatica 3(11):24 pp., 4 textfigs.

C

chemistry - chlorinity
California Academy of Sciences
California Division of Fish and Game
Scripps Institution of Oceanography
U. S. Fish and Wildlife Service — 1950
California Cooperative Sardine Research Program.
Progress Rept. 1950:54 pp., 37 text figs.

chemistry, chlorinity
Calorels, Augusto Pablo, y Hugo Alfredo Taffetani, 1969
Nuevos aportes al conocimiento oceanográfico de la ria de Bahía Blanca: Puerto de Ingeniero White.
Contrib. Inst. Oceanogr. Univ. Nac. Sur, Bahía Blanca, Argentina, 4: 35 pp.

chemistry chlorinity (data)
Capart, J.J., et F. Closset, 1964.
Expéditions Antarctiques Belges Mission IRIS - 1960, C.N.R.P.B. (1): Rapport de la section océanographique.
Bull. Inst. R. Sci. Nat. Belgique, 40(2):47 pp.

chemistry chlorinity
Carritt, D.E., 1963
5. Chemical instrumentation. In: The Sea, M.N. Hill, Edit., Vol. 2. The composition of sea water, Interscience Publishers, New York and London, 109-123.

chemistry chlorinity
Carritt, Dayton E., and James H. Carpenter, 1959.
The composition of sea water and the salinity-chlorinity-density problems. Nat. Acad. Sci.-Nat. Res. Counc., Publ. No. 600: 67-86.

chemistry - chlorinity
Castillejo, F. F., 1966.
Non-seasonal variations in the hydrological environment off Port Hackvej, Sydney.
C.S.I.R.O., Div. Fish. Oceanogr. Techn. Paper, No. 21:12 pp.

chemistry - chlorinity
Cavaliere, Antonio, 1963.
Studi sulla biologia e pesca di Xiphias gladius. II.
Boll. Pesca, Piscicolt. e Idrobiol., 6(2):143-170.

chemistry chlorinity
Chalmers, G.V., and A.K. Sparks, 1959.
An ecological survey of the Houston ship channel and adjacent bays.
Publ. Inst. Mar. Sci., 6:213-250.

chemistry chlorinity
Charnock, H., and J. Crease, 1966.
A salinity estimator based on measurement of conductivity ratio.
UNESCO, Techn. Papers, Mar. Sci., No. 4:App. B. (mimeographed)

chemistry chlorinity
Chau, Y.K., 1958
Some hydrological features of the surface waters of the Pearl River estuary between Hong Kong and Macau. Hong Kong Univ. Fish. J. (2): 37-41.

chlorinite
Codde, R.E.L., J. Lamoen, J.E.L. Verschave, 1953.
La chlorinité de l'Escaut Maritime.
XVIII Congr. Int. Navigation, Rome 1953, Sect.II. Navigation Maritime, Communication, 3:43-60, 2 figs., 4 tables.

chemistry chlorinity
Cox, Roland A., 1963
The salinity problem.
Progress in Oceanography, M. Sears, Edit., Pergamon Press, London, 1:243-261.

chemistry, chlorinity
Cox, R.A., 1962.
Report on the scientific results of the programme for comparing physical properties of sea water.
Rapp. Proc. Verb., Reunion, Cons. Perm. Int. Expl. Mer, 1961:94-98.

chlorinity chemistry, chlorinity
Cox, R.A., F. Culkin, R. Greenhalgh and J.P. Riley, 1962.
Chlorinity, conductivity and density of sea-water.
Nature, 193(4815):518-520.

Chemistry, chlorinity
Cox, R.A., F. Culkin and J.P. Riley 1967
The electrical conductivity/chlorinity relationship in natural sea water.
Deep-Sea Research 14(2): 203-220.

D

chemistry, chlorinity
de Quay, Alain, 1966.
Calcul théorique de la densité de l'eau de mer.
Cahiers Océanogr., C.C.O.E.C., 19(1):43-61.

chemistry chlorinity
Dorrestein, R., 1960.
On the distribution of salinity and of some other properties of the water in the Ems-Estuary.
Verh. Kon. Ned. Geol. Mijnb. k. Gen., Geol. Ser., 19:43-74.

chemistry chlorinity
Dugal, L.-P., 1932-1933.
Observations sur le chlore et l'oxygène dessous de l'estuaire du Saint-Laurent. Publ. Sta. Biol. Saint-Laurent, Contr. No. 4, 20 pp.

chlorinity
*Dyrssen, David, et al., 1967.
Analysis of sea water in the Uddevalla Fjord system 1.
Rept. Chem. Sea Water, Univ. Göteborg, 4:1-8 pp. (mimeographed).

E

Chemistry, chlorinity
Emery, K.O., 1954.
Source of water in basins off southern California. J. Mar. Res. 13(1):1-21, 6 textfigs.

Chemistry, chlorinity (data)
Endo, Takuo, 1965.
On primary production in the Seto Inland Sea. I. Primary production and hydrographic conditions. (In Japanese: English Abstract).
J. Fac. Fish., Animal Husbandry, Hiroshima Univ. 6(1):85-100.

F

Chemistry, chlorinity (data)
Febres, Germán, Gilberto Rodríguez y Andrés Eloy Esteves, 1966.
Ch. 2. Quimica del agua.
Estudios Hidrobiologicos en el Estuario de Maracaibo, Inst. Venezolano de Invest. Cient., 21-65. (Unpublished manuscript).

Chemistry
Chlorinity (data)
Forsbergh, Eric D., William W. Broenkow, 1965.
Observaciones oceanograficas del oceano Pacifico oriental recolectada por el barco Shoyo Maru, octubre 1963-marzo 1964.
Comision Interamericana del Atun Tropical, Bol.

Chemistry, chlorinity
Fujino, Kazuo, 1966.
Oceanographic observations on the Drifting Station ARLIS II, June-November 1964. (In Japanese; English abstract)
Low Temp. Sci. (A) 24:269-284.

chemistry, chlorinity
Fukai, R., 1953.
On the distribution of nutrient salts in the Equatorial North Pacific.
Bull. Chem Soc., Japan, 26(9):485-489, 5 textfigs.

Also reprinted in Bull. Tokai Reg. Fish. Lab. 7(Contr. B) as Contr. 113.

Chemistry, chlorinity
Fukuo, Yoshiaki, 1963.
Why is relative composition of chemical constituents in sea water maintained to be uniform.
Geophysical Papers dedicated to Professor Kenzo Sassa, 7-11.

Also in:
Spec. Contr., Geophys. Inst., Kyoto Univ., (1): 1-5.

chemistry
chlorinity
Fukuo, Yoshiaki, 1962
On the exchange of water and the productivity of a bay with special reference to Tanabe Bay (IV).
Rec. Oceanogr. Wks., Japan, Spec. No. 6:161-168.

chemistry
chlorinity
Fukuoka, Jiro, 1962
Characteristics of hydrography of the Japan Sea - in comparison withhydrography of the North Pacific -. (In Japanese; English abstract).
J. Oceanogr. Soc., Japan, 20th Ann. Vol., 180-188.

Chemistry, chlorinity
Fukuoka, J., and I. Tsuiki, 1954.
On the variation in the Kuroshio area (especially on the low chlorinity in summer).
Ocean. Mag., Tokyo, 6(1):15-23, 7 textfigs.

chemistry - chlorinity(data)
Fuse, Shin-ichiro, and Eiji Harada, 1960
A study on the productivity of Tanabe Bay. (3) Result of the survey in the summer of 1958.
Rec. Oceanogr. Wks., Japan, Spec. No. 4: 13-28.

Oceanographic conditions of Tanabe Bay.

chemistry, chlorinity
Fuse, S., I. Yamazi, and E. Harada, 1958.
A study on the productivity of the Tanabe Bay. (Part 1). 1. Oceanographic conditions of the Tanabe Bay. Results of the survey in the autumn of 1956.
Rec. Oceanogr. Wks., Japan, Spec. No. 2:3-9.

G

Chlorinity
Gololobov, Ya. K., 1963.
Hydrochemical characteristics of the Aegean Sea during the autumn of 1959. (In Russian; English abstract).
Rez. Issled. Programme Mezhd. Geofiz. Goda, Okeanolog. Issled., Akad. Nauk, SSSR, No. 8:90-96

chemistry, chlorinity
Grasshoff, K., 1965.
The results of the chlorinity and salinity intercalibration measurements during the informal meeting in Copenhagen.
UNESCO, Techn. Papers, Mar. Sci., 3:6-8. (mimeographed)

Chemistry, chlorinity (data)
Guillén, Oscar, y Francisco Vásquez 1966.
Informe preliminar del Crucero 6602 (Cabo Blanco - Arica).
Inst. del Mar, Peru, Informe (12):27 pp.

H

chemistry, chlorinity (data)
Halim, Youssef 1960.
Etude quantitative et qualitative du cycle écologique des dinoflagellés dans les eaux de Villefranche-sur-Mer (1953-1955).
Ann. Inst. Océanogr., Monaco 38:123-232.

Chemistry
chlorinity (some data)
Hanaoka, T., and A. Murakami, 1954.
On the submarine illumination of bay water.
Bull. Nakai Regional Fish. Res. Lab. 6:7-14, 6 textfigs.

chemistry, chlorinity
Hanzawa, M., and T. Tsuchida, 1954.
A report on the oceanographical observations in the Antarctic carried out on board the Japanese whaling fleet during the years 1946 to 1952.
J. Ocean. Soc., Japan, 10(3):99-111, 7 textfigs.

chemistry, chlorinity
Hermann, F., 1951.
High accuracy potentiometric determination of the chlorinity of sea water. J. du Cons. 17(3):223-230, 3 textfigs.

Chlorinity
Hirano, Toshiyuki, 1958.
The oceanographic study on the subarctic region of the northwestern Pacific Ocean. II. On the formation of the subarctic water systems. Bull. Tokai Reg. Fish. Lab., Tokyo, No. 15:57-69

chemistry - chlorinity
Hirano, Toshiyuki, 1957.
The oceanographic study on the subarctic region of the northwestern Pacific Ocean. 1. On the water systems in the subarctic region based upon the oceanographic survey by the R.V. Tenyo-maru in summer, 1954)
Bull. Tokai Reg. Fish. Res. Lab., Tokyo, No. 15: 39-55

Chemistry, chlorinity (data only)
Hirota, Reiichiro, 1970.
Some oceanographical conditions observed at a definite station off Aitsu Marine Biological Station (1967-1969) (In Japanese)
Calanus, Aitsu mar. biol. Sta, Kumamoto Univ., 2:47-54.

Chemistry, chlorinity (data)
Hisaoka, Minoru, Kazuhiko Nogami, Osamu Takeuchi, Masaya Suzuki and Hitomi Sugimoto, 1966.
Studies on sea water exchange in fish farm. 2. exchange of sea water in floating net.
Bull. Naikai reg. Fish. Res. Lab., No. 23:21-43. (In Japanese; English abstract)

Chemistry, chlorinity (data)
Horikoshi, Masuoki, 1962.
Distribution of benthic organisms and their remains at the entrance of Tokyo Bay, in relation to submarine topography, sediments and hydrography.
Nat. Sci. Rept. Ochanomizu Univ., 13(2):47-122.

Chemistry chlorinity
Hung, Tsu-Chang 1970
Photo- and radiation effects on inorganic nitrogen compounds in pure water and sea water near Taivan in the Kuroshio Current.
Bull. Inst. Chem. Acad. Sinica 18:37-56

I

Chemistry, chlorinity
Ichiye, T., 1956.
On the annual variation of chlorinities at the upper layer of the Kuroshio. Ocean. Mag., Tokyo, 7(2):87-93.

chemistry, chlorinity(profiles)
Ichiye, T., 1953.
On the variation of circulation. IV.
Ocean. Mag., Tokyo, 5(1):23-44, 9 textfigs.

chemistry, chlorinity
Ichiye, Takashi, 1952?
On the hydrographical condition in the Kuroshio region (1952). 1. Southern area of Honshu.
Bull. Kobe Mar. Obs., No. 163:1-30.

Chemistry
chlorinity
Ichie, Takashi, 1951
On the hydrography of the Kii-Suido (1951).
Bull. Kobe Mar. Obs., No. 164:253-278(top of page); 35-60(bottom of page).

Chemistry, chlorinity
Ichie, T., and S. Moriyasu, 1950.
Results of the oceanographical observations of the Miharo Seto and the Ondo Seto (Oct. 1950).
J. Ocean., Kobe Obs., 2nd ser., 1(4):1-14, 7 figs

Chemistry
chlorinity
Ichiye, Takashi, Sigeo Moriyasu and Hiroyuki Kitamura, 1951
On the hydrography near the estuaries.
Bull. Kobe Mar. Obs., No. 164: 349(top of page)-369; 53(bottom of page)-75.

Chemistry
chlorinity

Iida, Hayato, 1962
On the water masses in the coastal region of the south-western Okhotsk Sea.
J. Oceanogr. Soc., Japan, 20th Ann. Vol., 272-278.

Chemistry - chlorinity

Iizuka, Shoji, and Haruhiko Irie, 1966.
The hydrographic conditions and the fisheries damages by the red water occurred in Omura Bay in summer 1965 - II. The biological aspects of a dominant species in the red water. (In Japanese; English abstract).
Bull. Fac. Fish. Nagasaki Univ., No. 21:67-101.

Chemistry
chlorinity(data)

Inoue, A., 1963.
Study on productivity of Kasaoka Bay. (In Japanese; English abstract).
Bull. Naikai Reg. Fish. Res. Lab., 20:1-116.

Chemistry, chlorinity

Irie, Haruhiko, and Kentaro Hamashima, 1966.
The hydrographic conditions and the fisheries damages by the red water occurred in Omura Bay in summer 1965 - I. (In Japanese; English abstract).
Bull. Fac. Fish. Nagasaki Univ., No. 21:59-65.

J

chemistry
Chlorinity

Japan, Central Meteorological Observatory, 1955.
The results of marine meteorological and oceanographical observations. Pt. 1. Oceanography January-June, 1955. No. 16:120 pp.

Chemistry, chlorinity
(data only)

Japan, Fisheries Agency 1970.
The results of fisheries oceanographical observation, January-December 1967, 1701 pp.

Japan, Hakodate Marine Observatory

Chemistry, chlorinity

In the references that follow the pagination is erratic and often overlapping. Most references are to preliminary and/or data reports.

Chemistry, chlorinity

Japan, Hakodate Marine Observatory, 1969.
Report of the oceanographic observations in the sea south of Hokkaido and in the Sea of Okhotsk from October to November 1965 (In Japanese)
Bull. Hakodate mar. Obs. 14:16-21

Chemistry, chlorinity

Japan, Hakodate Marine Observatory, 1969.
Report of the oceanographic observations in the sea east of Hokkaido and the Kuril islands from May to June 1965. (In Japanese)
Bull. Hakodate mar. Obs. 14:12-17.

Chemistry chlorinity

Japan Hakodate Marine Observatory, 1969.
Report of the oceanographic observations in the sea east of Hokkaido and the Kuril islands, and in the Okhotsk Sea from July to September, 1965.
Bull. Hakodate mar. Obs. 14:3-15. (In Japanese)

Chemistry chlorinity

Japan, Hakodate Marine Observatory, 1969.
Report of the oceanographic observations in the sea east of Hokkaido and Tohoku District from February to March, 1965. (In Japanese)
Bull. Hakodate mar. Obs. 14:3-9

Chemistry - chlorinity

Japan, Hakodate Marine Observatory, 1967.
Report of the oceanographic observations of the sea southeast of Hokkaido, from June to July, 1964. (In Japanese).
Bull. Hakodate mar. Obs., 13:3-6.

Chemistry - chlorinity

Japan, Hakodate Marine Observatory, 1967.
Report of the oceanographic observations in the Okhotsk Sea and east of the Kurile islands and Hokkaido from October to November 1964. (In Japanese).
Bull. Hakodate mar. Obs., 13:20-28.

Chemistry - chlorinity

Japan, Hakodate Marine Observatory, 1967.
Report of the oceanographic observations in the sea east of Hokkaido and in the southern part of the Okhotsk Sea from May to June, 1964. (In Japanese).
Bull. Hakodate mar. Obs., 13:10-17.

Chemistry, chlorinity

Japan, Hakodate Marine Observatory, 1967.
Report of the oceanographic observations in the sea southeast of Hokkaido in February, 1964. (In Japanese).
Bull. Hakodate mar. Obs., 13:3-9.

chemistry, chlorinity

Japan, Hakodate Marine Observatory, 1967.
Report of the oceanographic observations in the Okhotsk Sea, east of the Kurile islands and Hokkaido and east of the Tohoku district from August to September, 1964. (In Japanese).
Bull. Hakodate mar. Obs., 13:7-19.

Chlorinity

Japan, Hakodate Marine Observatory, 1966.
Report of the oceanographic observations in the sea east of Honshu and Hokkaido in March 1963.
. . in the sea east of Tohoku district in May 1963.
. . in the northern part of the Japan Sea in June 1963.
. . in the sea southeast of Hokkaido in July 1963.
. . in the sea southeast of the Kurile Islands and Hokkaido from August to September 1963.
. . in the Okhotsk Sea in November 1963. (In Japanese)
Bull. Hakodate Mar. Obs. No. 12:3-8; 9-14; 15-18; 3-4; 5-11; 12-17.

Chemistry, chlorinity

Japan, Hakodate Marine Observatory, 1964.
Report of the oceanographic observations in the sea east of the Tohoku District from February to March 1962.
Report of the oceanographic observations in the Tsugaru Straits in April 1962.---in May 1962.---in June, 1962. from August to September, 1962. Report of the oceanographic observations in the sea south of Hokkaido in June 1962. Report of the oceanographic observations in the sea west of Tsugaru Straits, the Tsugaru Straits and South of Hokkaido in May, 1962. (In Japanese).
Bull. Hakodate Mar. Obs.,
No. 11: misnumbered pp.

Chemistry, chlorinity

Japan, Hakodate Marine Observatory, 1963.
Report of the oceanographic observations in the sea east of Tohoku District from February to March 1961, 3-8.
Report of the oceanographic observations in the sea east of Tohoku District in May 1961, 9-12.
Report of the oceanographic observations in the Japan Sea in June 1961:59-79.
Report of the oceanographic observations in the western part of Wakasa Bay from January to May 1961:80.
Report of the oceanographic observations in the sea south of Hokkaido in July 1961:3-4.
OVER
In Japanese

Chemistry
chlorinity

Japan, Hakodate Marine Observatory, Oceanographical Section, 1962
Report of the oceanographic observations in the sea east of Tohoku District in May, 1961. (In Japanese).
Res. Mar. Meteorol. and Oceanogr. Obs., Jan.-June, 1961, No. 29:9-12.

Chemistry
chlorinity

Japan, Hakodate Marine Observatory, Oceanographical Section, 1962
Report of the oceanographic observations in the sea east of Tohoku District from February to March, 1961. (In Japanese).
Res. Mar. Meteorol., and Oceanogr. Obs., Jan.-June, 1961, No. 29:3-8.

Chemistry
chlorinity

Japan, Hakodate Marine Observatory, 1962
Report of the oceanographic observations in the sea east of Tohoku District and in the Okhotsk Sea from August to September 1960. (In Japanese).
Results, Mar. Meteorol. and Oceanogr., July-Dec., 1960, Japan Meteorol. Agency, No. 28: 7-16.

Chemistry
chlorinity

Japan, Hakodate Marine Observatory, 1962
Report of the oceanographic observations in the sea south of Hokkaido in July, 1960. (In Japanese).
Res. Mar. Meteorol. and Oceanogr., July-Dec., 1960, Japan Meteorol. Agency, No. 28: 3-6.

Chemistry
chlorinity

Japan, Hakodate Marine Observatory, 1962
Report of the oceanographic observations in the sea west of Hiyama (Hokkaido) and in the sea east of Tohoku District in November 1960. (In Japanese).
Res. Mar. Meteorol. and Oceanogr., July to Dec 1960, Japan Meteorol. Agency, No. 28:17-20.

Chemistry
chlorinity

Japan, Hakodate Marine Observatory, 1961
Report of the oceanographic observations in the Okhotsk Sea and in the sea east of Tohoku District from May to June, 1959.
Bull. Hakodate Mar. Obs., (8):8-16.

Chemistry,
chlorinity

Japan, Hakodate Marine Observatory, 1961
[Report of the oceanographic observations in the sea east of Tohoku District from November to December 1959.]
Bull. Hakodate Mar. Obs., (8):15-19.

Chemistry, chlorinity

Japan, Hakodate Marine Observatory, 1961.
[Report of the Oceanographic observations in the sea east of Tohoku District and in the Okhotsk Sea from August to September 1959.]
Bull. Hakodate Mar. Obs., (8):6-14.

Chemistry
chlorinity

Japan, Hakodate Marine Observatory, 1961
[Report of the oceanographic observations in the sea east of Tohoku District from February to March 1959.]
Bull. Hakodate Mar. Obs., (8):3-7.

Chemistry, chlorinity

Japan, Hakodate Marine Observatory, 1961
[Report of the oceanographic observations in the sea south of Hokkaido from July to August 1959.]
Bull. Hakodate Mar. Obs., (8):3-5.

Chemistry - chlorinity

Japan, Hakodate Marine Observatory, 1957.
[Report of the oceanographic observations in the sea east of Tohoku District from February to March 1956.]
Bull. Hakodate Mar. Obs., No. 4:49-57
1-9

Chemistry chlorinity
Japan, Hakodate Marine Observatory, 1957
[Report of the oceanographic observations in the sea east of the Tohoku district in August 1956.]
Bull. Hakodate Mar. Obs., No. 4: 1-12.

Chemistry chlorinity
Japan, Hakodate Marine Observatory, 1957
[Report of the oceanographic observations in the sea east of Tohoku district from October to November 1956.]
Bull. Hakodate Mar. Obs., No. 4: 1-8.

Chemistry chlorinity
Japan, Hakodate Marine Observatory, 1957
[Report of the oceanographic observations in the Tsugaru Straits in July 1956.]
Bull. Hakodate Mar. Obs., No. 4: 13-21.

Chemistry chlorinity
Japan, Hakodate Marine Observatory, 1957.
[Report of the oceanographic observations of the Okhotsk Sea from April to May 1956.]
Bull. Hakodate Mar. Obs., No. 4:105-112.
1-8.

Chemistry chlorinity
Japan, Hakodate Marine Observatory, 1957.
[Report of the oceanographic observations in the sea east of Tohoku District from May to June 1956.] Bull. Hakodate Mar. Obs., No. 4:113-119.
9-15.

Chemistry chlorinity
Japan, Hakodate Marine Observatiory, 1956.
[Report of the oceanographical observations in the sea off the Sanriku District in Feb.-Mar., 1955.]
Bull. Hakodate Mar. Obs., (3):11-16.

Chemistry chlorinity
Japan, Hakodate Marine Observatory, 1956.
[Report of the oceanographical observations in the sea east off Tohoku District in May-June 1955]
Bull. Hakodate Mar. Obs., (3):11-16.

Chemistry chlorinity
Japan, Hakodate Marine Observatory, 1956.
[Report of the oceanographical observations east off Sanriku District in 1954.]
Bull. Hakodate Mar. Obs., (3):25-45.

Chemistry chlorinity
Japan, Hakodate Marine Observatory, 1956.
[Report of the oceanographical observations east off Tohoku District from August to September 1955.]
Bull. Hakodate Mar. Obs., (3):13-21.

Chemistry, chlorinity
Japan, Hakodate Marine Observatory, 1956.
[Report of the oceanographical observations east of Tohoku District in November 1955.]
Bull. Hakodate Mar. Obs., (3):11-12.

Chemistry-chlorinity
Japan, Hakodate Marine Observatory, 1956.
[Report on the oceanographical observation of the sea off Tohoku District made on board the R.M.S. Yoshio-maru in summer 1953 (1st paper).]
Bull. Hakodate Mar. Obs., No. 2:1-3.

Chemistry-chlorinity
Japan, Hakodate Marine Observatory, 1956.
[Report of the oceanographical observations in the sea off Tohoku District made on board R.M.S. Yushio-maru in spring and summer 1952. 1.]
Bull. Hakodate Mar. Obs., (3):205-215.
1-11.

Chemistry-chlorinity (data only)
Japan, Hokkaido University, Faculty of Fisheries, 1964
Data record of oceanographic observations and exploratory fishing, No. 8:303 pp.

Chemistry chlorinity(data)
Japan, Hokkaido University, Faculty of Fisheries, 1962
II. The "Oshoro Maru" cruise 48 to the Bering Sea and northwestern North Pacific in June-July 1961.
Data Record Oceanogr. Obs., Expl. Fish. 6: 22-149.

Chemistry chlorinity(data)
Japan, Japan Meteorological Agency, 1964.
Oceanographic observations.
Res. Mar. Met. & Oceanogr. Obs., No. 31:220 pp.

chemistry chlorinity
Japan, Japan Meteorological Agency, Oceanographical Section, 1962
Report of the oceanographic observations in the sea east of Honshu from February to March, 1961. (In Japanese).
Res. Mar. Meteorol. and Oceanogr. Obs., Jan.-June, 1961, No. 29:13-21.

Chemistry chlorinity
Japan,
Japan Meteorological Agency, 1962
Report of the Oceanographic observations in the sea east of Honshu from August to September, 1960. (In Japanese).
Res. Mar. Meteorol. and Oceanogr., July-Dec., 1960, Japan Meteorol. Agency, No. 28: 21-29.

chlorinity
Japan, Japan Meterrology Agency, 1959.
The results of marine meteorological and oceanographical observations, January-June 1958, No. 23:240 pp.

Japan, Kobe Marine Observatory

Chemistry chlorinity
Japan, Kobe Marine Observatory 1967.
Report of the oceanographic observations in the sea south of Honshu from October to November 1963. (In Japanese)
Bull. Kobe Mar. Obs. No. 178: 41-49.

chemistry, chlorinity
Japan, Kobe Marine Observatory 1967.
Report of the oceanographic observations in the sea South of Honshu from July to August 1963. (In Japanese).
Bull. Kobe Mar. Obs. No. 178: 31-40.

Chemistry, chlorinity
Japan, Kobe Marine Observatory 1967.
Report of the oceanographic observations in the sea south of Honshu and in the Seto-Naikai from May to June 1964. (In Japanese).
Bull. Kobe Mar. Obs. No. 178: 37-

chemistry, chlorinity
Japan, Kobe Marine Observatory 1967.
Report of the oceanographic observations in the sea south of Honshu from February to March 1964. (In Japanese)
Bull. Kobe Mar. Obs. No. 178: 27-

Chemistry-chlorinity
Japan, Kobe Marine Observatory, Oceanographical Section, 1964.
Report of the oceanographic observations in the Kuroshio and region east of Kyushu from October to November 1962. (In Japanese).
Res. Mar. Meteorol. and Oceanogr., Japan. Meteorol. Agency, 32: 41-50.
Also in: Bull. Kobe Mar. Obs., 175. 1965.

Chemistry - chlorinity
Japan, Kobe Marine Observatory, Oceanographical Section, 1964.
Report of the oceanographic observations in the sea south of Honshu from July to August, 1962. Res. Mar. Meteorol. and Oceanogr., Japan. Meteorol. Agency, 32: 32-40. (In Japanese).
Also in: Bull. Kobe Mar. Obs. 175.

Chemistry-chlorinity
Japan, Kobe Marine Observatory, Oceanographical Section, 1964.
Report of the oceanographic observations in the sea south of Honshu from February to March, 1963. Res. Mar. Meteorol. and Oceanogr., Japan Meteorol. Agency, 33: 27-32.
Also in: Bull. Kobe Mar. Obs., 175. 1965.

Chemistry chlorinity
Japan, Kobe Marine Observatory, Oceanographical Section, 1963.
Report of the oceanographic observations in the sea south of Honshu from February to March, 1962. (In Japanese).
Res. Mar. Met. & Ocean., J.M.A., 31:37-44.
Also in:
Bull. Kobe Mar. Obs., No. 173(4). 1964.

Chemistry chlorinity
Japan, Kobe Marine Observatory, Oceanographical Section, 1963
Report of the oceanographic observations in the sea south of Honshu from May to June, 1962. (In Japanese).
Res. Mar. Met. & Ocean., J.M.A., 31:45-49.
Also in:
Bull. Kobe Mar. Obs., No. 173(4):1964.

Chemistry, Chlorinity
Japan, Kobe Marine Observatory, 1963.
Report of the oceanographical observations in the sea south of Honshu from February to March 1961.
Report of the oceanographical observations in the cold water region off Enshu Nada in May 1961. (In Japanese).
Bull. Kobe Mar. Obs., 171(4):22-35.

Chemistry, chlorinity
Japan, Kobe Marine Observatory, 1963.
Report of the oceanographic observations in the sea south of Honshu from July to August and from the cold water region south of Enshu Nada October to November 1960. (In Japanese).
Bull. Kobe Mar. Obs., 171(3):36-52.

Chemistry chlorinity
Japan, Kobe Marine Observatory, Oceanographical Observatory, 1962
Report of the oceanographic observations in the cold water region off Enshu Nada in May, 1961. (In Japanese).
Res. Mar. Meteorol. and Oceanogr. Obs., Jan.-June, 1961, No. 29:28-35.

Chemistry chlorinity
Japan, Kobe Marine Observatory, Oceanographical Section, 1962
Report of the oceanographic observations in the cold water region off Kii Peninsula from October to November 1961. (In Japanese).
Res. Mar. Meteorol. & Oceanogr., No. 30:49-55.
Also in:
Bull. Kobe Mar. Obs., No. 173(3). 1964

Chemistry chlorinity
Japan, Kobe Marine Observatory, 1962
Report of the oceanographic observations in the cold water region south of Enshu Nada from October to November, 1960. (In Japanese). Res. Mar. Meteorol. and Oceanogr., July-Dec. 1960, Japan Meteorol. Agency, No. 28:43-51.

Chemistry chlorinity
Japan, Kobe Marine Observatory, Oceanographical Section, 1962
Report of the oceanographic observations in the sea south of Honshu from February to March, 1961. (In Japanese).
Res. Mar. Meteorol. and Oceanogr. Obs., Jan.-June, 1961, No. 29:22-27.

Chemistry chlorinity
Japan, Kobe Marine Observatory, Oceanographical Section, 1962
Report of the oceanographical observations in the sea south of Honshu from July to August 1961. (In Japanese).
Res. Mar. Meteorol. & Oceanogr., 30:39-48.

Also in:
Bull. Kobe Mar. Obs., No. 173(3) 1964.

Chemistry chlorinity
Japan, Kobe Marine Observatory, Oceanographical Section, 1962
Report of the oceanographic observations in the sea south of Honshu in May, 1960. (In Japanese).
Bull. Kobe Mar. Obs., No. 169(12):27-33.

Chemistry chlorinity
Japan, Kobe Marine Observatory, 1962
Report of the oceanographic observations in the sea south of Honshu from July to August, 1960. (In Japanese).
Res. Mar. Meteorol. and Oceanogr., July-Dec., 1960, Japan Meteorol. Agency, No. 28:36-42.

Chemistry chlorinity
Japan, Kobe Marine Observatory, Oceanographical Section, 1962
Report of the oceanographic observations in the sea south of Honshu in March, 1960. (In Japanese).
Bull. Kobe Mar. Obs., No. 169(12):22-33.

Chemistry chlorinity
Japan, Kobe Marine Observatory, Oceanographical Section, 1962
Report of the oceanographic observations in the sea south of Honshu from October to November, 1959. (In Japanese).
Bull. Kobe Mar. Obs., No. 169(11):44-50.

Chemistry chlorinity
Japan, Kobe Marine Observatory, Oceanographical Section, 1962
Report on the oceanographic observations in the sea south of Honshu from July to August, 1959.
Bull. Kobe Mar. Obs., No. 169(11):37-43.
(In Japanese)

Chemistry chlorinity
Japan, Kobe Marine Observatory, 1961
[Report of the oceanographic observations in the sea south of Honshu in March 1958.]
Bull. Kobe Mar. Obs., No. 167(21-22):30-36.

--from May to June, 1958(21-22):37-42
--from July to September, 1958(23-24):34-40
--from October to December, 1958(23-24):41-47
--from February to March,1959(25-26):33-47.

Chemistry-chlorinity
Japan, Kobe Marine Observatory, 1958.
[Report of the oceanographic observations in the sea south of Honshu from November to December 1957.]
J. Oceanogr., Kobe Mar. Obs., (2), 10(1):21-28.

in May 1957.] Ibid., 9(2):69-78.

in August, 1957.] Ibid., 9(2):79-86.

Chemistry, chlorinity
Japan, Kobe Marine Observatory, 1956?
[Report on the oceanographical observations south off Honshu in 1954.] J. Ocean., Kobe, (2)7(2):46-69.

Chemistry, chlorinity
Japan, Kobe Marine Observatory, 1956.
[Report of the oceanographic observations in the sea south of Honshu in August 1955.]
J. Ocean., Kobe (2)7(3):23-32.

Chemistry, chlorinity
Japan, Kobe Marine Observatory, 1956.
[Report of the oceanographical observations in the sea south off Honshu in May 1955.]
J. Ocean., Kobe (2)7(2):17-25.

pagination repeats that of previous article.

Chemistry, chlorinity
Japan, Kobe Marine Observatory, 1956.
[Report of the oceanographical observations in the sea south off Honshu in March 1955.]
J. Ocean., Kobe, (2)7(2):17-24.

Chemistry, chlorinity
Japan, Kobe Marine Observatory, 1955.
[The oultine of the oceanographical observation on the Southern Sea of Japan on board R.M.S. "Syunpu-maru" (Jan. 1955).]
J. Ocean., Kobe Mar. Obs., (2)6(1):1-19, 10 figs.

Chemistry, chlorinity
Japan, Kobe Marine Observatory, 1954.
[Results of oceanographical observations off Enshu Nada (June 1954).] J. Ocean. (2)5(6):1-6,

Chemistry, chlorinity
Japan, Kobe Marine Observatory, 1954.
[The outline of the oceanographical observations off Shionomisaki on board the R.M.S. "Syunpu-maru" (May 1954).] J. Ocean. (2)5(5):1-11, 14 figs

Chemistry, chlorinity
Japan, Kobe Marine Observatory, 1954.
[On the annual variation of water temperature and chlorinity in the Kii Suido.] J. Ocean. (2)5(3):1-7, 3 figs.

Chemistry, chlorinity
Japan, Kobe Marine Observatory, 1954.
[The results of the harmonic analysis to use the regular monthly oceanographic observation on board the R.M.S. "Syunpu-maru" in Osaka Wan, 1953]
J. Ocean. (2)5(2):1-15, 11 figs.

Chemistry, chlorinity
Japan, Kobe Marine Observatory, 1954.
[The results of the regular monthly oceanographical observations on board the R.M.S. "Syunpu-maru" in the Kii Suido and Osaka Wan.] J. Ocean. (2)5(1):1-19, 16 figs.

Chemistry, chlorinity
Japan, Kobe Marine Observatory, 1953.
[The results of the regular monthly oceanographical observations on board the R.M.S. "Syunpu-maru" in the Kii Suido and Osaka Wan.]
Jan. 1953 - J. Ocean. (2)4(1):1-10, 9 figs.
Feb. 1953 - J. Ocean. (2)4(2):1-9, 8 figs.,
Mar. 1953 - J. Ocean. (2)4(3):1-10, 9 figs.
Apr. 1953 - J. Ocean. (2)4(5):1-9, 9 figs.
May 1953 - J. Ocean. (2)4(6):1-12, 9 figs.
June 15-22, 1953 - J. Ocean. (2)4(6):13-22, 8 figs.
July, 1953 - J. Ocean. (2)4(8):1-25, 29 figs.
Aug. 1953 - (2)4(9):1-30, 27 figs. (J. Ocean.)
Sept. 1953 - J. Ocean. (2)4(10):1-13, 9 figs.
Oct. 1953 - J. Ocean. (2)4(11):1-21, 15 figs.
Nov. 11-18, 1953 - J. Ocean. (2)4(12):1-15, 12 figs.

Japan, Maizuru Marine Observatory

Chemistry chlorinity
Japan, Maizuru Marine Observatory 1969.
Report of the oceanographic observations in the western part of the Japan Sea in November 1965. (In Japanese).
Bull. Maizuru mar. Obs. 11: 98-103.

Chemistry chlorinity
Japan, Maizuru Marine Observatory, 1967.
Report of the oceanographic observations in the Japan Sea from October to November, 1964. (In Japanese).
Bull. Maizuru mar. Obs., 10:87-94.

Chemistry - chlorinity
Japan, Maizuru Marine Observatory, 1967.
Report of the oceanographical observations in the Japan Sea from August to September, 1964. (In Japanese).
Bull. Maizuru mar. Obs., 10:74-86.

Chemistry - chlorinity
Japan, Maizuru Marine Observatory, 1967.
Report of the oceanographic observations in the Japan Sea from May to June, 1964.
Bull. Maizuru mar. Obs., 10:65-76.
(In Japanese)

Chemistry - chlorinity
Japan, Maizuru Marine Observatory, 1965.
Report of the oceanographic observations in the central part of the Japan Sea from February to March, 1962.-----in the Japan Sea from June to July, 1962.-----in the western part of Wakasa Bay from January to April, 1962.----in the central part of the Japan Sea from September to October, 1962.---in the western part of Wakasa Bay from May to November, 1962.---in the central part of the Japan Sea in March, 1963.---in the Japan Sea in June, 1963.---in Wakasa Bay in July, 1963.---in the central part of the Japan Sea in October, 1963. (In Japanese).
Bull. Maizuru Mar. Obs., No.9:67-73;74-88;89-95;71-80;81-87;59-65;66-77;80-84;85-91.

Chemistry chlorinity
Japan, Maizuru Marine Observatory, 1963
Report of the oceanographic observations in the western part of Wakasa Bay from August to December, 1961. (In Japanese).
Bull. Maizuru Mar. Obs., No. 8:96-102.

Chemistry chlorinity
Japan, Maizuru Marine Observatory, 1963
Report of the oceanographic observations in the Wakasa Bay in August, 1961. (In Japanese).
Bull. Maizuru Mar. Obs., No. 8:89-95.

Chemistry chlorinity
Japan, Maizuru Marine Observatory, 1963
Report of the oceanographic observations in the western part of Wakasa Bay from January to May, 1961. (In Japanese).
Bull. Maizuru Mar. Obs., No. 8:80-90.

Chemistry chlorinity
Japan, Maizuru Marine Observatory, 1963
Report of the oceanographic observations in the central part of the Japan Sea in October 1961. (In Japanese).
Bull. Maizuru Mar. Obs., No. 8:78-88.

Chemistry chlorinity
Japan, Maizuru Marine Observatory, 1963
Report of the oceanographic observations in the western part of Wakasa Bay from June to December, 1960. (In Japanese).
Bull. Maizuru Mar. Obs., No. 8:76-83.

chemistry chlorinity
Japan, Maizuru Marine Observatory, 1963
Report of the oceanographic observations in the Wakasa Bay from July to August, 1960. (In Japanese).
Bull. Maizuru Mar. Obs., No. 8:69-75.

chemistry chlorinity
Japan, Maizuru Marine Observatory, 1963
Report of the oceanographic observations in the western part of Wakasa-Bay from January to June, 1960. (In Japanese)
Bull. Maizuru Mar. Obs., No. 8:68-79, 59.

chemistry chlorinity
Japan, Maizuru Marine Observatory, 1963
Report of the oceanographic observations in the central part of the Japan Sea in September 1960. (In Japanese).
Bull. Maizuru Mar. Obs., No. 8:60-68.

chemistry chlorinity
Japan, Maizuru Marine Observatory, 1963
Report of the oceanographic observations in the Japan Sea in June 1961. (In Japanese).
Bull. Maizuru Mar. Obs., No. 8:59-79.

chemistry chlorinity
Japan, Maizuru Marine Observatory, 1963
Report of the oceanographic observations in the Japan Sea from May to June, 1960. (In Japanese).
Bull. Maizuru Mar. Obs., No. 8:56-67.

chemistry chlorinity
Japan, Maizuru Marine Observatory, 1963
Report of the oceanographic observations in the central part of the Japan Sea in February, 1961. (In Japanese).
Bull. Maizuru Mar. Obs., No. 8:54-58.

chemistry chlorinity
Japan, Maizuru Marine Observatory, Oceanographical Section, 1962
Report of the oceanographic observations in the western part of Wakasa Bay from June to May, 1961. (In Japanese).
Res. Mar. Meteorol. and Oceanogr. Obs., Jan.-June, 1961, No. 29:80-90.

chemistry chlorinity
Japan, Maizuru Marine Observatory, 1962
Report of the oceanographic observations in the western part of Wakasa Bay from June to December 1960. (In Japanese).
Res. Mar. Meteorol. and Oceanogr., July-Dec., 1960, Japan Meteorol. Agency, No. 28:76-83.

chemistry chlorinity
Japan, Maizuru Marine Observatory, 1962
Report of the oceanographic observations in the Wasaka Bay from July to August, 1960. (In Japanese).
Res. Mar. Meteorol. and Oceanogr., July-Dec., 1960, Japan Meteorol. Agency, No. 28:69-75.

chemistry chlorinity
Japan, Maizuru Marine Observatory and Hakodate Marine Observatory, Oceanographical Sections, 1962
Report of the oceanographic observations in the Japan Sea in June 1961. (In Japanese).
Res. Mar. Meteorol. and Oceanogr. Obs., Jan.-June, 1961, No. 29:59-79.

chemistry chlorinity
Japan, Maizuru Marine Observatory, Oceanographical Section, 1962
Report of the oceanographic observations in the central part of the Japan Sea in February 1961. (In Japanese).
Res. Mar. Meteorol. and Oceanogr. Obs., Jan.-June, 1961, No. 29:54-58.

chemistry chlorinity
Japan, Meteorological Agency, 1962
Report of the Oceanographic Observations in the sea south and west of Japan from October to November, 1960. (In Japanese).
Res. Mar. Meteorol. and Oceanogr., July-Dec., 1960, Japan Meteorol. Agency, No. 28: 30-35.

chemistry chlorinity
Japan, Maizuru Marine Observatory, Oceanographical Observatory, 1961
[Report of the Oceanographic observations in the western part of Wakasa Bay from August to December, 1959.]
Bull. Maizuru Mar. Obs., No. 7:68-74.

chemistry chlorinity
Japan, Maizuru Marine Observatory, Oceanographical Section, 1961
[Report of the oceanographic observations off Kyoga-misaki and in the western part of Wakasa Bay from February to June, 1959.]
Bull. Maizuru Mar. Obs., No. 7:65-72, 61.

Chemistry chlorinity
Japan, Maizuru Marine Observatory, Oceanographical Section, 1961
[Report of the oceanographic observations in the sea north of San'in District in August, 1958.]
Bull. Maizuru Mar. Obs., No. 7: 64-68.

chemistry chlorinity
Japan, Maizuru Marine Observatory, Oceanographical Section, 1961
[Report of the oceanographic observations in the Wakasa Bay from July to August, 1959.]
Bull. Maizuru Mar. Obs., No. 7:62-67.

chemistry chlorinity
Japan, Maizuru Marine Observatory, Oceanographical Section, 1961
Report of the oceanographic observations in the Japan Sea from June to July, 1958.
Bull. Maizuru Mar. Obs., No. 7:60-67.

chemistry chlorinity
Japan, Maizuru Marine Observatory, Oceanographical Section, 1961
[Report of the oceanographic observations in the sea north of Kyoga-misaki from August to December 1958.]
Bull. Maizuru Mar. Obs., No. 7:60-63.

chemistry chlorinity
Japan, Maizuru Marine Observatory, Oceanographical Section, 1961
[Report of the oceanographic observations in the Japan Sea from June to July, 1959.]
Bull. Maizuru Mar. Obs., No. 7:57-64.

chemistry chlorinity
Japan, Maizuru Marine Observatory, Oceanographical Section, 1961
[Report of the oceanographic observations off Kyoga-misaki in the Japan Sea in April and May, 1958.]
Bull. Maizuru Mar. Obs., No. 7:55-60.

chemistry chlorinity
Japan, Maizuru Marine Observatory, Oceanographical Section, 1961
[Report of the oceanographic observations in the western part of Wakasa Bay from January to March, 1958.]
Bull. Maizuru Mar. Obs., No. 7:50-54.

chemistry chlorinity
Japan, Maizuru Marine Observatory, Oceanographical Section, 1961
[Report of the oceanographic observations off Kyoga-misaki in Japan Sea from October to December, 1957.]
Bull. Maizuru Mar. Obs., No. 7:29-36.

chemistry chlorinity
Japan, Maizuru Marine Observatory, 1958
[Report of the oceanographic observations off the east part of Sanin district in May and June 1956.]
Bull. Maizuru Mar. Obs., No. 6: 149-156. [45-52.]

chemistry chlorinity
Japan, Maizuru Marine Observatory, 1958
[Report of the oceanographic observations in the sea north of Sanin and Hokuriku districts in August 1957.]
Bull. Maizuru Mar. Obs., No. 6: 245-253. [107-115.]

chemistry chlorinity
Japan, Maizuru Marine Observatory, 1958
[Report of the oceanographic observations north off Kyoga-misaki and off the east part of the Sanin district in April to June, 1957.]
Bull. Maizuru Mar. Obs., No. 6: 234-243. [95-105.]

chemistry chlorinity
Japan, Maizuru Marine Observatory, 1958
[Report of the serial observations off Kyoga-misaki from May 1955 to April 1956.]
Bull. Maizuru Mar. Obs., No. 6: 95-104. [47-56.]

chemistry, chlorinity
Japan, Maizuru Marine Observatory, 1958
[Report of the oceanographic observations north of Sanin and Hokuriku districts in summer 1956]
Bull. Maizuru Mar. Obs., No. 6: 157-unnumbered. [53-85.]

chemistry - chlorinity
Japan, Maizuru Marine Observatory, 1958
[Report of the oceanographic observations off Kyoga-misaki in the Japan Sea in January and March, 1957.]
Bull. Maizuru Mar. Obs., No. 6: 129-137. [45-53.]

chemistry, chlorinity
Japan, Maizuru Marine Observatory, 1958
[Report of the oceanographic observations off Kyoga-misaki in Japan Sea in October and November, 1956.]
Bull. Maizuru Mar. Obs., No. 6: 37-44.

chemistry chlorinity
Japan, Maizuru Marine Observatory, 1956.
[Report of the oceanographical observations north off Sanin and Hokuriku districts in summer 1954.]
[Report of oceanographic observations north off Sanin and Hokuriku districts in summer 1955.]
Bull. Maizuru Mar. Obs., (5):85-94;49-56.

chemistry- chlorinity

Japan, Maizuru Marine Observatory, 1956

[Report of the oceanographic observations off Kyoga-misaki during the latter half of 1955.] Bull. Maizuru Mar. Obs., (5): 31-37.

chemistry- chlorinity

Japan, Maizuru Marine Observatory, 1956

[Report of the serial observations off Kyoga-misaki during the first half of 1955.] Bull. Maizuru Mar. Obs., (5): 27-32.

chemistry, chlorinity

Japan, Maizuru Marine Observatory, 1955. [Report of the oceanographical observations off San'in and Hokuri-ku (July-August, 1952).] Bull. Maizuru Mar. Obs., (4):49-63 (168-181).

chemistry, chlorinity

Japan, Maizuru Marine Observatory, 1955. [Report of the oceanographical observations (taken) off San'in and Hokuriku (July-August, 1953).] Bull. Maizuru Mar. Obs., (4):13-27. 217-231.

chemistry, chlorinity (data only)

Japan, Maritime Safety Agency, Tokyo 1964. Tables of results from oceanographic observations in 1961. Hydrogr. Bull. Tokyo (Publ. No. 981) No. 77: 82 pp.

chemistry chlorinity, hourly

Japan, Maritime Safety Board, 1955. Oceanographical state in the northern half of Ki-i Suido in May 1954.] Publ. No. 981, Hydrogr. Bull., Spec. No. 17:53-62.

chemistry chlorinity

Japan, Maritime Safety Board, 1959 Tables of results from Oceanographic observation in 1954 and 1955. Hydrogr. Bull., (Publ. 981), No. 58: 139 pp.

chemistry, chlorinity

Japan, Japan Meteorological Agency. 1965. The results of marine meteorological and oceanographical observations, July-December 1963, No. 34: 360 pp.

Japan, Nagasaki Marine Observatory

chemistry chlorinity

Japan, Nagasaki Marine Observatory 1971. Report of the oceanographic observations in the sea southward of Kyushu from October to November, 1965. (In Japanese). Oceanogr. Met. Nagasaki mar. Obs. 15: 78-82

chemistry, chlorinity

Japan, Nagasaki Marine Observatory, 1971. Report of the oceanographic observations in the sea southeast of Kyushu in September, 1965. (In Japanese) Oceanogr. Met. Nagasaki mar. Obs. 15: 75-77

chemistry - chlorinity

Japan, Nagasaki Marine Observatory 1971. Report of the oceanographic observations in the sea west of Japan from June to August, 1965. (In Japanese) Oceanogr. Met. Nagasaki Mar. Obs. 15: 59-74

Chemistry chlorinity

Japan, Nagasaki Marine Observatory, Oceanographic Section, 1965. Report of the oceanographic observations in the sea west of Japan from February to March 1963,-- from July to August 1963. Res. Mar. Meteorol. Oceanogr., Japan, Meteorol Agency, 33:39-58; 34:53-80.

Also in:- Oceanogr. Meteorol., Nagasaki, 15(227-228).

Chemistry chlorinity

Japan, Nagasaki Marine Observatory, Oceanographical Section, 1962 Report of the oceanographic observations in the sea west of Japan from April to May, 1961 (In Japanese). Res. Mar. Meteorol. and Oceanogr. Obs. Jan.-June, 1961, No. 29: 45-53.

Chemistry chlorinity

Japan, Nagasaki Marine Observatory, Oceanographical Section, 1962 Report of the oceanographic observations in the sea west of Japan from February to March, 1961. (In Japanese). Res. Mar. Meteorol. and Oceanogr. Obs., Jan.-June, 1961, No. 29:36-44.

chemistry chlorinity

Japan, Nagasaki Marine Observatory, 1962 Report of the oceanographic observations in the sea west of Japan from October to November, 1960. (In Japanese). Res. Mar. Meteorol. and Oceanogr., July-Dec., 1960, Japan Meteorol. Agency, No. 28:52-59.

chemistry chlorinity

Japan, Nagasaki Marine Observatory, Oceanographical Section 1960. Report of the oceanographic observation in the sea west of Japan from January to February, 1960. Report of the Oceanographic observations in the sea north-west of Kyushu from April to May, 1960. Results Mar. Meteorol. & Oceanogr., J.M.A., 27:42-50; 51-67.

Also in: Oceanogr. & Meteorol., Nagasaki Mar. Obs., (1961) 11 (202).

chemistry chlorinity

Japan, Nagasaki Marine Observatory, Oceanographical Section, 1960 [Report of the oceanographic observations in the sea west of Japan from June to July, 1959] Res. Mar. Meteorol. & Oceanogr., J.M.A., 26: 51-57. Also in: Oceanogr. & Meteorol., Nagasaki Mar. Obs., (1961), 11(200).

chemistry chlorinity

Japan, Nagasaki Marine Observatory, Oceanographical Section, 1960 [Report of the oceanographic observation in the sea west of Japan from January to February 1959.] Res. Mar. Meteorol. & Ocean., J.M.A., 25: 48-56. Also in: Oceanogr. & Meteorol., Nagasaki Mar. Obs. (1961) 11(199).

chemistry, chlorinity

Japan, Nagasaki Marine Observatory, 1959. Report of the oceanographic observations in the sea west of Japan from June to July 1958 and in the sea north west of Kyushu in October 1958. Results Mar. Meteorol. & Oceanogr., J.M.A., 24: 47-60.

Also in: Oceanogr. & Meteorol., Nagasaki Mar. Obs., 10(195). 1960.

chemistry, chlorinity

Japan, Nagasaki Marine Observatory, 1959. Report of the oceanographic observations in the sea west of Japan from January to February, 1958. Results Mar. Meteorol. & Oceanogr., J.M.A., 23: 43-49.

Also in: Oceanogr. & Meteorol., Nagasaki Mar. Obs., 10(194). 1960.

Chemistry chlorinity

Japan, National Committee for the International Geophysical Coordination, Science Council of Japan, 1961 The results of the Japanese oceanographic project for the International Geophysical Cooperation 1959: 1-65.

chemistry, chlorinity

Japan, Tokai Regional Fisheries Research Laboratory 1959 IGY physical and chemical data by the R.V. Soyo-maru, 25 July - 14 September 1958, 17 pp. (multilithed)

Chemistry, chlorinity

Japan, Tokai Regional Fisheries Research Laboratory, 1952. Report oceanographical investigation (Jan.-Dec. 1950, collected), No. 74:143 pp.

K

Chemistry, chlorinity (data)

Kato, Kenji, 1966. Studies on calcium content in sea water. III. Calcium in the waters of Cananéia lagoon and its adjacent regions, State of Sao Paulo, Brazil. Bolm Inst. Oceanogr., S. Paulo, 15(1):41-45.

Chemistry, chlorinity (data)

Kato, Kenji, 1966. Studies on calcium content in sea water. II. Distribution of calcium in the Atlantic water off south Brazil. Bolm Inst. Oceanogr., S. Paulo, 15(1):29-39.

Chemistry, chlorinity (data)

Kato, Kenji, 1966. Geochemical studies on the mangrove region of Cananéia, Brazil. II. Physico-chemical observations on the reduction states. Bolm Inst. Oceanogr., S. Paulo, 15(1):21-24.

chemistry chlorinity (data)

Kato, Kenji, 1966. Geochemical studies on the mangrove region of Cananéia, Brazil. 1. Tidal variations of water properties. Bolm Inst. Oceanogr., S. Paulo, 15(1):13-20.

Chemistry, chlorinity (data)

Kato, Takeo, 1969. On the distribution of the water mass produced by water-mixing and the change of chemical composition. (Ratio $Ca^{2+}:Cl^-$, $SO_4^{2-}:Cl^-$) - Oceanographical studies on the East China Sea, 3. (In Japanese; English abstract) Umi to Sora, 44(2/3): 55-80

chemistry, chlorinity

Kawai, Hideo, and Minoru Sasaki, 1961. An example of the short-period fluctuation of the oceanographic condition in the vicinity of the Kuroshio front. Bull. Tohoku Regional Fish. Res. Lab., No. 19:119-134.

chemistry, chlorinity

Kawarada, Y., 1953. [Plankton associations in Japan Sea and Tsugaru Straits in May 1950.] J. Ocean. Soc., Japan, 9(2):103-108, 5 textfigs.

chemistry, chlorinity
Kawarada, Y., 1953.
[On the plankton association in Japan Sea (1).]
J. Ocean. Soc., Japan, 9(2):95-102, 7 textfigs.

chemistry, chlorinity
Kirillova, E.P., 1965.
The regime of chlorine, alkalinity and oxygen in the mouth of the Iuzhnii Bug. (In Russian).
Trudy, Gosudarst. Okeanogr. Inst., No. 83:158-171.

chemistry, chlorinity
Kitamura, Hiroyuki, and Takeshi Sagi, 1964.
On the chemical elements in the sea south of Honshu, Japan. (In Japanese; English abstract).
Bull. Kobe Mar. Obs., No. 172:6-54.

chemistry
chlorinity
Koganei, Shoichi, and Masuoki Horikoshi, 1962
Hydrography at the entrance of Tokyo Bay (Cooperative survey at the entrance of Tokyo Bay in 1959 - Part 3). (In Japanese; English abstract).
J. Oceanogr. Soc., Japan, 20th Ann. Vol. 90-97.

chemistry, chlorinity
Kon, H., 1953.
On the distribution of phytoplankton in the northeastern part of the Pacific (Tohoku district) in autumn. (Oct.-Nov., 1951).
J. Ocean. Soc., Japan, 9(2):109-114, 4 textfigs.

chemistry, chlorinity
Koshimizu, N., 1952.
[The variation of the chlorinity and silicon content of the surface water between Maizuru and Nakhodoka] Bull. Maizuru Mar. Obs., (3):32-35

chemistry, chlorinity
Kubo, Tadashi, Yoshikazu Sato and Shigeshi Komaki 1967.
On the backflow of sea water into Lake Abashiri. I. An estimate of charged volume and the condition of sea water back-flowed into Lake Abashiri. (In Japanese; English abstract).
Bull. Hokkaido reg. Fish. Res. Lab. No. 32: 49-61.

chemistry, chlorinity
Kudô, Hideo, 1964.
A method for the determination of the concentration of sulfides in sea water and the concentration of sulfides in sea water in Matsushima Bay. (In Japanese; English abstract).
Bull. Tohoku Reg. Fish. Res. Lab., No. 24:1-8.

chemistry
chlorinity
Kusunoki, Kou, 1962
Hydrography of the Arctic Ocean with special reference to the Beaufort Sea.
Contrib. Inst., Low Temp. Sci., Sec. A, (17): 1-74.

chemistry, chlorinity
Kwiecinski, Bogdan, 1965.
The relation between the chlorinity and the conductivity in Baltic water.
Dee-Sea Res., 12(2):113-120.

L

chemistry, chlorinity
Lefevre, S., E. Leloup and L. Van Meal, 1956.
Observations biologiques dans le port d'Ostende.
K. Belg. Inst. Naturwet., Verhandl., [Inst. R. Sci. Nat. Belg., Mem.,] 133:157 pp.

chemistry
chlorinity
Leloup, E., Editor, with collaboration of L. Van Meel, Ph. Polk, R. Halewyck and A. Gryson, undated.
Recherches sur l'ostreiculture dans le Bassin de Chasse d'Ostende en 1962.
Ministere de l'Agriculture, Commission T.W.O.Z., Groupe de Travail- "Ostreiculture". 58pp.

chemistry, chlorinity
Leprevost, Alsedo, e João José Bigarella, 1949.
Notas preliminares sobre a composicao química de algumas águas do litoral paranaense.
Arquivos de Biologia e Tecnologia, Inst. Biol. e Pesquisas Tecnolog., Secret. Agricult., Indust. e Comercio, Estado do Parana, Brazil, 4 (13):73-86.

M

chemistry, chlorinity
Maeda, H., 1953.
The relation between chlorinity and silicate concentration of water observed in some estuaries.
J. Shimonoseki Coll. Fish. 3(2):167-180, 16 textfigs.

chemistry, chlorinity
Maeda, H., 1952.
The relation between chlorinity and silicate concentration of water observed in some estuaries.
Publ. Seto Mar. Biol. Lab., Kyoto Univ., 2(2): 249-255, 5 textfigs.

chemistry
chlorinity
Maeda Hiroshi, and Kaoru Takesue, 1961
The relation between chlorinity and silicate concentration of waters observed in some estuaries. V. Consideration upon the fitness of [Cl]-[SiO2] relation of the river water for the relation formula computed from the estuarine waters and upon the seasonal variation of the estimated constants.
Rec. Oceanogr. Wks., Japan, 6(1):112-119.

chemistry, chlorinity
Maeda, Sonosuke 1968.
On the cold water belt along the northern coast of Hokkaido in the Okhotsk Sea.
Umi to Sora 43(3):71-90

chemistry, chlorinity
Makimoto, H., H. Maeda and S. Era, 1955.
The relation between chlorinity and silicate concentration of water observed in some estuaries
Rec. Ocean. Wks., Japan, 2(1):106-112, 4 textfigs.

chemistry, chlorinity
Marumo, R., 1954.
[Relation between planktonological and oceanographical conditions of a sea area east of Kinkazan in winter.] J. Ocean. Soc., Japan, 10(2):77-84, 5 textfigs.

chemistry, chlorinity
Masuzawa, J., 1955.
Preliminary report on the Kuroshio in the Eastern Sea of Japan (Currents and water masses of the Kuroshio system III).
Rec. Ocean. Wks., Japan, 2(1):132-140, 5 textfigs

chemistry, chlorinity (data)
Masuzawa, J., M. Yasui and H. Akamatsu, 1960.
A hydrographic section across the Kurile Trench.
Repts. of JEDS, 1:165-172.

Also in Oceanogr. Mag. 11 (2) (same pagination)

chemistry, chlorinity
Matida, Y., (undated in English).
On the chemical composition of the sea water in the Tokyo Bay. J. Ocean. Soc., Tokyo, (Nippon Kaiyo Gakkaisi) 5(2/4):105-110, 3 figs.

(In Japanese; English summary)

chemistry
chlorinity (data)
Matsudaira, Yasuo, 1964
Cooperative studies on primary productivity in the coastal waters of Japan, 1962-63. (In Japanese; English abstract).
Inform. Bull., Planktol., Japan, No. 11:24-73.

Mc

chemistry - chlorinity
McKenzie, Kenneth G., 1964.
The ecologic associations of an ostracode fauna from Oyster Harbor, a marginal marine environment near Albany, Western Australia.
Pubbl., Staz. Zool., Napoli, 33(Suppl.):421-461.

chemistry - chlorinity
Millard, N.A.H., and K.M.F. Scott, 1954.
The ecology of South African estuaries. VI. Milnerton Estuary and the Diep River, Cape.
Trans. R. Soc., S. Africa 34(2):279-324, 8 textfigs.

chemistry, chlorinity
Minas, H.J., 1963.
La relation alcalinité-chlorosité de certaines eaux côtiers du littoral de la région Marseillaise
Rapp. Proc. Verb., Réunions, Comm. Int., Expl. Sci., Mer Méditerranée, Monaco, 17(3):855-856.

chemistry - chlorinity
Miyake, Y., 1939.
A new indicator for the determination of chlorinity in sea water.
Geophys. Mag., Tokyo, 12:299-300.

chemistry - chlorinity
Miyake, Y., 1938.
A new indicator for the determination of chlorinity in sea water. Geophys. Mag., Japan, 12:299-300.

chemistry - chlorinity
Miyake, Y., Y. Sugiura and K. Kameda, 1955.
On the distribution of radioactivity in the sea around Bikini Atoll in June 1954.
Pap. Met. Geophys., Tokyo, 5(3/4):253-262.

chemistry - chlorinity
Miyake, Y., Y. Sugiura and K. Kamada, 1953.
A study of the property of the coastal waters around Hachijo Island. Rec. Ocean. Wks., Japan, n.s., 1(1):93-99, 10 textfigs.

chemistry - chlorinity
Morcos, Selim A., and J.P. Riley, 1966.
Chlorinity, salinity, density and conductivity of sea water from the Suez Canal region.
Deep-Sea Res., 13(4):741-749.

chemistry - chlorinity
Mori, Isamu, and Haruhiko Irie, 1966.
The hydrographic conditions and the fisheries damages by the red water occurred in Omura Bay in summer 1965 - III. The oceanographic conditions in the offing region of Omura Bay during the term of the red water. (In Japanese; English abstract).
Bull. Fac. Fish. Nagasaki Univ., No. 21:103-113.

chemistry - chlorinity
Moriyasu, S., 1955.
On the variation of water temperature and chlorinity after heavy precipitation in the Osaka-Wan and the Kii-Suide. Ocean. Mag., Tokyo, 6(4): 178-180, 9 textfigs.

chemistry - chlorinity
Murakami, A., 1954.
[Oceanography of Kasaoka Bay in Seto Inland Sea.]
Bull. Nakai Regional Fish. Res. Lab. 6:15-57, 42 textfigs.

chemistry - chlorinity
Murray, J.W., 1966.
A study of the seasonal changes of the water mass of Christchurch Harbour, England.
Jour. mar. biol. Assoc., U.K., 46(3):561-578.

N

chemistry-chlorinity
Nakano, M., S. Unoki, M. Hanzawa, R. Marumo and J. Fukuoka, 1954.
Oceanographic features of a submarine eruption that destroyed the Kaiyo-maru No. 5. J. Mar. Res. 13(1):48-66, 10 textfigs.

chemistry-chlorinity
Nakayama, I., 1956.
[On the influence of the river water on the oceanographical condition in the western part of Wakase Bay.] Bull. Maizuru Mar. Obs., (5):28-30.

chemistry, chlorinity
Nannichi, T., 1951.
On the fluctuation of the Kuroshio and the Oyashio. Pap. Met. Geophys. 2(1):102-111, 6 textfigs.

chemistry-chlorinity
Nan-niti, Tosio, Hideo Akamatsu and Toshisuke Nakai, 1964.
A further observation of a deep current in the east-north-east sea of Torishima.
Oceanogr. Mag., Tokyo, 16(1/2):11-19.

chemistry-chlorinity
Nehring, Dietwart, und Karl-Heinz Rohde 1967.
Weitere Untersuchungen über anormale Ionenverhältnisse in der Ostsee.
Beitr. Meereskunde 20: 10-33

chemistry-chlorinity
Neumann, W., 1949.
Beiträge zur Hydrographie des Vrana-Sees (Insel Cherso), insbesondere Untersuchungen über organisches sowie anorganische Phosphor- und Stickstoffverbindungen. Nova Thalassia 1(6):17 pp.

chemistry
chlorinity
Nishizawa, Satoshi, Naoichi Inoue and Yoshio Akiba, 1959
Turbidity distribution in the subarctic water of the North Pacific in the summer of 1957.
Rec. Oceanogr. Wks., Japan, Spec. No. 3: 231-241.

chemistry, chlorinity
Nooter, K., et F. Liebregts, 1971
Une recherche sur le taux de chlorinité de l'eau interstitielle dans le lit de la Slack.
Bijdragen tot de Dierkunde 41 (1):23-30

O

chemistry- chlorinity (data)
Okabe, Shiro, Yoshimasa Toyota and Takafumi Murakami, 1967.
Major aspects of chemical oceanography. (In Japanese; English abstract).
J.Coll.mar.Sci.Techn., Tokai Univ., (2):39-56.

chemistry-chlorinity
Okitsu, T., T. Tokui and B. Tsubata, 1954.
Oceanographical observations off Asamushi during 1952. Bull. Mar. Biol. Sta., Asamushi, 7(1):21-25, 4 textfigs.

chemistry-chlorinity
Okuda, Taizo, 1964.
Calcium and magnesium contents in the river and sea waters of a tropical area.
Bol. Inst. Oceanogr., Univ. Oriente, Venezuela, 118-135.

chemistry
chlorinity
Okuda, Taizo, 1960
Metabolic circulation of phosphorus and nitrogen in Matsushima Bay (Japan) with special reference to exchange of these elements between sea water and sediments. Portuguese, French resumés)
Trabalhos, Inst. Biol. Maritima e Oceanogr., Universidade do Recife, Brasil, 2(1):7-153.

chemistry
chlorinity
Okuda, Taizo, and Ramon Nobrega, 1960
Estudoda Barra das Jangadas. 1. Distribuicão e movimento da clorinidade - quantidade de corrente. (In Portuguese; English and French resumés).
Trabalhos, Inst. Biol. Marit. e Oceanogr., Universidade do Recife, Brasil, 2(1):175-191.

P

chemistry
chlorinity
Patten, B.C., R.A. Mulford and J.E. Warinner, 1963
An annual phytoplankton cycle in the lower Chesapeake Bay.
Chesapeake Science, 4(1):1-20.

chemistry, chlorinity
Picotti, M., 1930.
Ricerche di oceanografia chimica. 1. Tabelle generali delle analisi clorometrische e dei temperatura, salinita e densita. Ann. Idrogr., 11(2):69-116.

translation cited:
USFWS Spec. Sci. Rept., Fish. 227.

chemistry, chlorinity
Pollak, M.J., 1954.
The use of electrical conductivity measurements for chlorinity determination. J. Mar. Res. 13(2):228-231.

chemistry, chlorinity (data)
Postma, H., 1966.
The cycle of nitrogen in the Wadden Sea and adjacent areas.
Neth. J. Sea Res., 3(2):186-221.

Q

R

chemistry, chlorinity
Reeburgh, W.S., 1966.
Some implications of the 1940 redefinition of chlorinity.
Deep-Sea Res., 13(5):975-976.

chemistry-chlorinity
Richards, F.A., 1957.
Some current aspects of chemical oceanography. In: Physics and chemistry of the earth, Pergamon Press, 2:77-128.

chemistry, chlorinity (data)
Riley, J.P., 1967.
The hot saline waters of the Red Sea bottom and their related sediments.
Oceanogr. Mar. Biol., Ann.Rev., H. Barnes, editor, George Allen and Unwin., Ltd., 5:141-157.

chemistry, chlorinity
Riley, J.P., 1965.
Analytical chemistry of sea water.
IN: Chemical oceanography, J.P. Riley and G. Skirrow, editors, Academic Press, 2:295-424.

chemistry, chlorinity (water samples)
Rittenberg, S.C., K.O. Emery and W.L. Orr, 1955.
Regeneration of nutrients in sediments of marine basins. Deep-Sea Res. 3(1):23-45.

chemistry-chlorinity
Rochford, D.J., 1966.
Hydrology. Port Philip survey 1957-1963.
Mem. nat. Mus. Vict., 27:107-118.

chemistry, chlorinity
Rochford, D.J., 1951.
Studies in Australian estuarine hydrology.
Australian J. Mar. and Freshwater Res. 2(1):1-116, 1 pl., 7 textfigs.

chemistry-chlorinity
Rohde, Karl-Heinz, 1966.
Untersuchungen über die Calcium-und Magnesiumanomalie.
Beitr. Meereskunde, 19:18-31.

chemistry, chlorinity (data)
Rodriguez, Gilberto, y Eduardo Ormeno, 1966.
Ch. 5. Zooplancton.
Estudios hidrobiologicos en el Estuario de Maracaibo, Inst. Venezolano de Invest. Cient. 93-121.

S

chemistry-chlorinity
Saruhashi, K., 1953.
The chlorinity determinations of sea water by a micro-analytical method.
Pap. Met. Geophys., Tokyo, 4 :90-92.

chemistry-chlorinity
Saruhashi, K., 1953.
The chlorinity determination of sea water by micro-analytical method.
Rec. Ocean. Wks., Japan, n.s., 1(2):52-54, 1 textfig.

chemistry, chlorinity (data)
Scaccini Cicatelli, Marta 1967
Distribuzione stagionali dei sali nutrivi in una zona dell'alto e medio Adriatico.
Boll. Pesca, Piscic. Idrobiol. n.s. 12(1):49-82.

chemistry-chlorinity
Scruton, P.C., 1956.
Oceanography of Mississippi delta sedimentary environments. Bull. Amer. Assoc. Petr. Geol. 40(12):2864-2952.

chemistry - chlorinity
Sebastian, Alfonso, Manual N. Llorca and Vitaliano B. Encina, 1965.
Oceanography of Lingayen Gulf.
Philipp. J. Fish., 7(1):13-33.

chemistry
chlorinity
Senta, Tetsuchi, Tokuzo Harada, Shigeki Yasuie and Mikio Azuma, 1968.
Seasonal variations of water temperature and chlorinity in Bisan-seto, central part of the Inland Sea in 1967. (In Japanese).
Bull.Fish.Exp.Sta., Okayama Pref. 42:96-104.

chemistry, chlorinity
Senta, Tetsushi, Shigeki Yasuie and Tokuzo Harada, 1967.
Comparisons between the seasonal variations of water temperature and chlorinity in Bisan-seto, central part of the Seto Inland Sea, in normal years and 1965-1966. (In Japanese).
Bull.Fish.Exp.Stn, Okayama Pref., 1966:54-60.

chemistry-chlorinity
Siyazuki, K., 1951.
[Studies on the foul-water drained from factories. 1. On the influence of foul-waters drained from factories by the coast on the water of Mitaziri Bay.] Contr. Simonoseki Fish. Coll. Pt. 1:155-158, 4 textfigs. (English summary).

Chemistry-chlorinity
Solov'eva, N.F., 1959.
[Hydrochemical analyses of the Aral Sea in 1948.]
Akad. Nauk, SSSR, Lab. Ozerovedeniya, Trudy, 8:3-22.

Chemistry-chlorinity
Sørensen, S.P.L., 1902.
Bestimmung des Chlor- und Salzgehalts.
Wiss. Meersunters., Kiel, n.f., 6:136-142.

Transl. cited:
USFWS Spec. Sci. Rept., Fish., 227.

Chemistry, chlorinity
Spencer, R.S., 1956.
Studies in Australian estuarine hydrology.
Australian J. Mar. & Freshwater Res. 7(2):193-253

Chemistry, chlorinity (data)
Sugawara, Ken, and Shiro Okabe, 1966.
Molybdenum and Vanadium determination of Umitaka-maru samples from her 1962-1963 and 1963-1964 cruises of the International Indian Ocean Expedition. (In Japanese; English abstract)
J. Tokyo Univ. Fish., (Spec. ed.) 8(2):165-171.

Chemistry - chlorinity (data)
Sugimoto, Hitomi, Minoru Hisaoka, Kazuhiko Nogami, Osamu Takeuchi and Masaya Suzuki, 1966.
Studies on sea water exchange in fish farm. 1. Exchange of sea water in fish farm surrounded by net. (In Japanese; English abstract).
Bull. Naikai reg. Fish. Res. Lab., No. 23:1-20.

chemistry, chlorinity
Sugiura, Yoshio, 1967.
The significance of the difference in conductometric chlorinity minus titrimetric chlorinity.
Rec. oceanogr. Wks, Japan, 9(1):55-64.

Chemistry chlorinity
Sugiura, Yoshio, 1963
Some chemico-oceanographical properties of the water of Suruga Bay. (In Japanese; English abstract).
J. Oceanogr. Soc., Japan, 18(4):193-199.

T

chemistry, chlorinity
Tanaka, Otohiko, Fumihiro Koga, Haruhiko Irie, Shozi Iizuka, Yosie Dotu, Keitaro Uchida, Satoshi Mito, Seiro Kimura, Osame Tabeta and Sadahiko Imai, 1962.
The fundamental investigation of the biological productivity in the north-western sea area of Kyushu. II. Study on plankton productivity in the waters of Genkai-Nada region.
Rec. Oceanogr. Wks., Japan, Spec. No. 6:1-20.
Also in:
Contrib. Dept. Fish., and Fish. Res. Lab., Kyushu Univ., No. 8.

chemistry, chlorinity
Tanita, S., K. Kato, and T. Okuda, 1951.
Studies on the environmental conditions of shell-fish-fields. In the case of Mureran Harbour.(2).
Bull. Fac. Fish., Hokkaido Univ., 2(3):220-230.

chemistry, chlorinity
Tanioka, K. 1966.
Oceanographical conditions of the Japan Sea (IV). (In Japanese; English abstract).
Umi to Sora 41(1/2): 50-57.

Chemistry, chlorinity
Thomas, I.M., and S.J. Edmonds, 1956.
Chlorinities of coastal waters in South Australia. Trans. R. Soc., S. Australia, 79:152-166.

Chemistry, chlorinity
Thompson, T. G., and E. Korpi, 1942.
The bromine-chlorinity ratio of sea water.
J. Mar. Res. 5(1):28-36.

Chemistry, chlorinity
Tokui, T., 1952.
On the dissolved oxygen and chlorinity of the water of Mutsu Bay during 1950.
Bull. Mar. Biol. Sta., Asamushi, 5(1/4):55-57.

Chemistry, Chlorinity
Toll, R., J.ma. Valles and F. Saiz, 1951.
Sobre la utilización de un agua de mar cualquiera como agua petrón auxiliar para la determinación de la chlorinidad de las aguas marinas.
Publ. Inst. Biol. Aplic. 9:169-179.

chemistry, chlorinity
Toll, R., J.Ma. Valles, 1951.
Sobre el empleo como disolución standard para la determinación de la clorinidad de las aguas del mar, de una disolución de cloruro, sódico de igual Cl/l litro que el del agua internacional.
Publ. Inst. Biol. Apl. 9:179-184.

Chemistry, chlorinity (data)
Torii, Tetsuya and Sadao Murata 1966.
The distribution of copper and zinc in the Indian and the Southern Ocean waters.
J. oceanogr. Soc. Japan 22(2): 56-60

Chemistry-chlorinity (data)
Torii, Tetsuya, and Sadao Murata, 1964.
Distribution of uranium in the Indian and the Southern Ocean waters.
In: Recent researches in the fields of hydrosphere, atmosphere and nuclear geochemistry, Ken Sugawara festival volume, Y. Miyake and T. Koyama editors, Maruzen, Co., Ltd., Tokyo, 321-334.

Chemistry, chlorinity
Traganza, Eugene D. and Barney J. Szabo 1967.
Calculation of calcium anomalies on the Great Bahama Bank from alkalinity and chlorinity data.
Limnol. Oceanogr. 12(2): 281-286

Chemistry, chlorinity
Tratet, Gérard, 1964.
Variations du phytoplancton à Tanger.
Trav. Inst. Sci., Cherifien, Rabat, Ser. Botan., (29):204 pp.

Chemistry, chlorinity
Tseu, W.S.L., 1953.
Seasonal variations in the physical environment of the ponds at the Hawaii Marine Laboratory and the adjacent waters of Kaneohe Bay, Oahu.
Pacific Science 7(3):278-290, 16 text figs.

Chemistry - chlorinity
Tsubata, B., and T. Numakunai, 1960.
Oceanographical conditions observed at definite station off Asamushi during 1958 and 1959.
Bull. Mar Biol. Sta., Asamushi, Tohoku Univ., 10(1):73-

chemistry, chlorinity
Tsujita, T., 1963.
The fishery oceanography of the East China Sea and Tsushima Strait. 2. On the characteristics and fluctuations of the cold water masses in view point of the production oceanography.
Bull. Seikai Reg. Fish. Res. Lab., 28:57-68.

Chemistry, Chlorinity
Tsujita, T., 1954.
[On the observed oceanographic structure of the Tsushima fishing grounds in winter and ecological relationship between the structure and the fishing conditions.] J. Ocean. Soc., Japan, 10(3):158-170, 12 textfigs.

U

Chemistry, chlorinity (data)
*Usuki, Itaru, 1967.
A record of hydrographic conditions of inshore water AROUND the vicinity of the Sado Marine Biological Station during 1964 to 1966.
Sci. Rep. Niigata Univ. (D)4:87-107.

Chemistry, chlorinity
Uyeno, Fukuzo, 1964.
Relationships between production of foods and oceanographical condition of sea water in pearl farms. II. On the seasonal changes of sea water constituents and of bottom condition, and the effect of bottom cultivation. (In Japanese; J. Fac. Fish., Pref. Univ. Mie, 6(2):145-169.

English abstract)

Chemistry, chlorinity
Uyeno, Fukuzo, 1966.
Nutrient and energy cycles in an estuarine oyster area.
J. Fish. Res. Bd., Can., 23(11):1635-1652.

V

Chemistry, chlorinity
Vardaro, Emilio, 1958.
Sul rapporto solfato-clorori nelle acque del Mar Piccolo, del Mar Grande e del Golfo di Taranto.
Thalassia Jonica, Ist. Sperimment. Talassogr., Taranto, 1:81-64.

Chemistry, chlorinity
Villalobos-Figueroa, Alejando, José A. Suarez-Caabro, Samuel Gomo, Guadalupe de la Lanza, Mauricio Aceves, Fernando Manrique and Jorge Cabrera, 1967.
Considerations on the hydrography and productivity of Alverado Lagoon, Vera Cruz, Mexico.
Proc. Gulf Caribb. Fish. Inst., 19th Sess., 75-85.

Chemistry, chlorinity
Voipio, Aarno, and Erkki Häsänen, 1962
Relationships between chlorinity, density and specific conductivity in Baltic waters.
Annales Acad. Sci. Fennicae, (A) (Chemica) (111): 18 pp.

W

Chemistry, Chlorinity
Watanabe, Nobuo 1967.
Chlorinity distribution and its change in the estuary of Kiso-Sansen. (In Japanese; English abstract.
Bull. coast. Oceanogr. 6(1).
Also in: Coll. Repr. Coll. mar. Sci. Techn. Tokai Univ., 1967-68, 3: 219-231

Chemistry - chlorinity (data)
*Watanabe, Nobuo, 1967.
Estimation of chlorinity change due to effluent from a hydraulic power plant. (In Japanese; English abstract).
J. Coll. mar. Sci. Techn., Tokai Univ., (2):1-19.

Chemistry, chlorinity
Watanabe, Nobuo 1965.
Oceanographic conditions around Miho Key and Shimizu Port - an example of transition in coastal micro-oceanography. (In Japanese; English abstract).
Bull. coast. Oceanogr. Japan 4(1): 11-22.

Reprinted in Coll. Repr. Coll. mar. Sci. Technol. Tokai Univ. 2(1966).

Chemistry chlorinity

Watanabe, N., 1962
On the fall of chlorinity of sea-water due to the flowing of land-water. (In Japanese, English summary).
Bull. Tokai Reg. Fish. Res. Lab., Tokyo, No. 32: 123-130.

Chemistry, chlorinity

Watanabe, N., 1958
On the mechanism of variations in water temperature and chlorinity of upper layer in the seas southwest off Kyushu and around Nansei-Shoto.
Bull. Tokai Reg. Fish. Res. Lab. No. 21: 15-24.

Chemistry, chlorinity

Watanabe, Nobuo, Toshiyuki Hirano, Rinnosuke Fukai, Kozi Matsumoto and Fumiko Shiokawa, 1957.
A preliminary report on the oceanographic survey in the "Kuroshio" area, south of Honshu, June-July 1955.
Rec. Oceanogr. Wks., Japan (Suppl.):197-

Chemistry, chlorinity

West, L.E., and R.J. Robinson, 1941.
Potentiometric analysis of sea water. 1. Determination of chlorinity. J. Mar. Res. 4(1):1-10.

Chemistry, chlorinity

Williams, Hulen B., Vincent Farrugia, Anthony Ekker and Elard L. Haden, 1964.
Chlorinity determinations in estuarine waters by physical methods.
J. Mar. Res., 22(2):190-196.

Chemistry chlorinity

Won, Chong Hun, 1964.
Tidal variations of chemical constituents of the estuary water at the lava bed in the Nack-Dong River from Nov. 1962-Oct. 1963. (In Korean; English abstract).
Bull. Pusan Fish. Coll., 6(1):21-32.

Chemistry, chlorinity

Won, Chong Hun, 1963.
Distribution of chemical constituents of the estuary water in Gwang-Yang Inlet. (In Korean; English abstract).
Bull. Fish. Coll., Pusan Nat. Univ., 5(1):1-10.

Chemistry, chlorinity

Won, Chong Hun, and Kil Soon Park, 1968.
Tidal variations of chlorinity and pH at the Yong-Ho Basin from Mar. 1 to Mar. 20, 1968. (In Korean; English abstract).
Bull. Pusan Fish. Coll. (Nat. Sci.), 8(2):103-111.

XYZ

Chemistry - chlorinity

Yamamoto, Hiroshi, and Yoshihiko Imai, 1959.
The hydrographic condition and variation of calcium in Susaki Bay.
Rec. Oceanogr. Wks., Japan, Spec. No. 3:117-121.

Chemistry, chlorinity

Yamanaka, Hajime, Noboru Anraku and Jiro Morita, 1965.
Seasonal and long-term variations in oceanographic conditions in the western North Pacific Ocean.
Rept. Nankai Reg. Fish. Res. Lab., No. 22: 35-70.

Chemistry, chlorinity

Yamanaka, I., 1951.
[On the hydrographical conditions of Japan Sea in spring and summer 1949. Pt. II.] J. Ocean. Soc., Japan, (Nippon Kaiyo Gakkaisi) 6(3):150-156, 4 textfigs.

Chemistry, chlorinity

Yamanaka, I., 1951.
[On the hydrographical conditions of Japan Sea in spring and summer 1949. Pt. I.] J. Ocean. Soc., Japan, (Nippon Kaiyo Gakkaisi) 6(3):143-149, 3 textfigs.

Chemistry, chlorinity

Yamazi, I., 1955.
Plankton investigations in inlet waters along the coast of Japan. XVI. The plankton of Tokyo Bay in relation to water movement.
Publ. Seto Mar. Biol. Lab., 4(2/3):285-309, Pls. 19-20, 22 textfigs.

Chemistry, chlorinity

Yamazi, I., 1953.
Plankton investigations of inlet waters along the coast of Japan. X. Plankton of Kamaisi Bay on the eastern coast of Tohoku District.
Publ. Seto Mar. Biol. Lab. 3(2)(Article 18):189-204, 16 textfigs.

Chemistry, Chlorinity

Yamazi, I., 1953.
Plankton investigations in inlet waters along the coast of Japan. IX. The plankton of Onagawa Bay on the eastern coast of Tohoku District.
Publ. Seto Mar. Biol. Lab. 3(2)(Article 17):173-187, 10 textfigs.

chemistry, chlorinity

Yamazi, I., 1953.
Plankton investigation in inlet waters along the coast of Japan. VII. The plankton collected during the cruises to the new Yamamoto Bank in the Sea of Japan. Publ. Seto Mar. Biol. Lab. 3(1): 75-108, 19 textfigs.

Chemistry - chlorinity

Yamazi, I., 1952.
Plankton investigations in inlet waters along the coast of Japan. III. The plankton of Imari Bay in Kyusyu. IV. The plankton of Nagasaki Bay and Harbour in Kyusyu; V. The plankton of Hiroshima Bay in the Seto-Naikai (Inland Sea).
Publ. Seto Mar. Biol. Lab., Kyoto Univ., 2(2): 289-304; 305-318; 319-330, 8 and 8 and 7 textfigs

chemistry, chlorinity

Yasui, Masashi, Takeo Yasuoka, Katsumi Tanioka and Osami Shiota 1967.
Oceanographic studies of the Japan Sea. I. Water characteristics.
Oceanogrl Mag. 19(2): 177-196

Chemistry, chlorinity

*Yasuoka, Takeo, 1967.
Hydrography of the Okhotsk Sea - (1).
Oceanogrl Mag., 19(1):61-72.

Chlorinity-defined

Chemistry
chlorinity - defined

Cox, R.A., 1965.
The physical properties of sea water.
In: Chemical oceanography, J.P. Riley and G. Skirrow, editors, Academic Press, 1: 73-120.

chlorinity, annual variation

chlorinity, annual variations

Koizumi, M., 1955.
Researches on the variations of oceanographic conditions in the region of the Ocean Weather Station "Extra" in the North Pacific Ocean. 1. "Normal" values and annual variations of oceanographic elements. Pap. Met., Geophys., Tokyo, 6(2):185-201.

Chemistry
chlorinity, annual variation

Moriyasu, Shigewo, 1951
[On the annual variation of water temperature in Osaka Wan.]
Bull. Kobe Mar. Obs., No. 164:337(top of page) -346; 43 (bottom of page)-52.

Chemistry, chlorinity, annual variation

Nagai, M., 1954.
[On the annual variation of water temperature and chlorinity at Gotonada.] Oceanogr., Meteorol., Nagasaki, 6(2-4): 129-136.

chlorinity (data only)

Chemistry
chlorinity (data only)

Akamatsu, Hideo, Tsutomu Akiyama and Tsutomu Sawara, 1965.
Preliminary report of the Japanese Expedition of Deep-Sea, the Tenth Cruise, 1965(JEDS-10).
Oceanogr. Mag., Tokyo, 17(1/2):49-68.

Chemistry, Chlorinity (data only)

Angot, M., et R. Gerard 1965.
Hydrologie de la région de Nosy-Bé: juillet à novembre 1963.
Cahiers ORSTOM, Océanogr. 3(4): 3-29.

Chemistry, Chlorinity (data only)

Anon., 1951.
[Report of the oceanographical observations on board the R.M.S. "Syunpu Maru" in the Kii-Suido (Oct. 1950).] J. Ocean., Kobe Obs., 2nd ser., 1(6):1-15, 7 figs.

Chemistry, Chlorinity (data only)

Anon., 1950.
Table 3. Oceanographical observations taken in the North Pacific Ocean east off Sanriku; physical and chemical data for stations occupied by "Oshoro-maru". Res. Mar. Met., Ocean. Obs., Tokyo, July-Dec. 1947, No. 2:8-12.

chemistry, chlorinity (data only)

Anon., 1950.
Table 2. Oceanographical observations taken in the Uchiura Bay; physical and chemical data for stations occupied by R.M.S. "Kuroshio-maru". Res. Mar. Met., Ocean. Obs., Tokyo, July-Dec. 1947, No. 2:3-7.

Chemistry, chlorinity (data only)

Anon., 1950.
Table 1. Oceanographical observations taken in the Uchiura Bay; physical and chemical data for stations occupied by a fishing boat belonging to Mori Fishery Assoc. Res. Mar. Met., Ocean. Obs., Tokyo, July-Dec., 1947, No. 2:1-2.

Chemistry, chlorinity (data only)

Anon., 1950.
[Report of the oceanographical observations from Fisheries Experimental Stations (Sept. 1950).]
J. Ocean., Kobe Obs., 2nd ser., 1(2):5-11.

Chemistry, chlorinity (data only)

Anon., 1950.
[The report of the oceanographical observations in the Bungo Suido and off Ashizuri-Misaki in August 1950.] J. Ocean., Kobe Obs., 2nd ser., 1(1):15-44, 7 figs., tables.

Chemistry, Chlorinity (data only)

Anon., 1949-1950.
[The results of the regular monthly oceanographical observations on board the R.M.S. "Syunpu Maru" and "Takatori Maru" in the Osaka Wan.]
J. Ocean., Kobe Obs., 2nd ser., 2(1):1-28.

chemistry, chlorinity (data only)
Australia, Commonwealth Scientific and Industrial Research Organization 1970.
Coastal investigations off Port Hacking, New South Wales in 1965.
Oceanogr. Sta. List. Div. Fish. Oceanogr. 85: 124 pp.

chemistry, chlorinity (data only)
Australia, Commonwealth Scientific and Industrial Organization, 1968.
Coastal investigations off Port Hacking, New South Wales, in 1964.
Oceanogr. Sta. List, 84: 49 pp. (multilithed)

chemistry, chlorinity (data only)
Australia, Commonwealth Scientific and Industrial Organization, 1968.
Coastal investigations off Port Hacking, New South Wales, in 1961.
Oceanogr. Sta. List, 81: 55 pp. (multilithed)

chemistry, chlorinity (data only)
Australia, Commonwealth Scientific and Industrial Research Organization, 1963.
Coastal hydrological investigations in the New South Wales tuna fishing area, 1963.
Div. Fish. and Oceanogr., Oceanogr. Sta. List, 53:81 pp.

chlorinity (data only)
Australia, Commonwealth Scientific and Industrial Research Organization, 1963.
Coastal investigations at Port Hacking, New South Wales, 1960.
Division Fish. and Oceanogr., Oceanogr. Sta. List, No. 52:135 pp.

chemistry, chlorinity(data only)
Australia, Commonwealth Scientific and Industrial Research Organization, 1963
Coastal hydrological investigations in eastern Australia, 1960.
Div. Fish. and Oceanogr., Oceanogr. Sta. List 51: 46 pp.

chemistry, chlorinity(data only)
Australia, Commonwealth Scientific and Industrial Research Organization, 1961
Coastal hydrological sampling at Rottnest Island, W.A., and Port Moresby, Papua, during the I.G.Y. (1957-58), and surface sampling in the Tasman and Coral seas, 1959.
Div. Fish. & Oceanogr., Oceanogr. Sta. List, 48:239 pp.

Chemistry, chlorinity(data only)
Australia, Commonwealth Scientific and Industrial Research Organization, 1961
Coastal investigations at Port Hacking, New South Wales, 1959.
Div. Fish. & Oceanogr., Oceanogr. Sta. List, No. 47: 135 pp.

chemistry, chlorinity(data only)
Australia, Commonwealth Scientific and Industrial Research Organization, 1961
Coastal hydrological investigations in the New South Wales tuna fishing area, 1959.
Div. Fish. & Oceanogr., Oceanogr. Sta. List, 46: 132 pp.

Chemistry, chlorinity (data only)
Australia, Commonwealth Scientific and Industrial Research Organization, 1960
Coastal hydrological investigations in eastern Australia, 1959.
Div. Fish. and Oceanogr., Oceanogr. Sta. List, 45: 24 pp.

Chlorinity(data only)
Australia, Commonwealth Scientific and Industrial Research Organization, 1960.
Coastal investigations at Port Hacking, New South Wales, 1958.
Oceanogr. Sta. List, Div. Fish. & Oceanogr., 42: 99 pp.

Chemistry, chlorinity (data only)
Australia, Commonwealth Scientific and Industrial Research Organization, 1960
Coastal hydrological investigations in south-eastern Australia, 1958. Oceanogr. Sta. List. Div. Fish. and Oceanogr., 60 pp.

chlorinity (data only)
Australia, C.S.I.R.O., Division of Fisheries and Oceanography, 1959
Coastal hydrological investigations in the New South Wales tuna fishing area, 1958.
Oceanogr. Sta. Lists, 38:96 pp.

Chemistry chlorinity (data only)
Australia, C.S.I.R.O., Division of Fisheries, 1959
Coastal hydrological investigations at Eden, New South Wales, 1957.
Oceanogr. Sta. List, 35: 36 pp.

chemistry, chlorinity (data only)
Australia, C.S.I.R.O., Division of Fisheries, 1959
Coastal hydrological investigations at Port Hacking, New South Wales, 1957.
Oceanogr. Sta. List, 34:72 pp.

Chemistry chlorinity (data only)
Australia, Commonwealth Scientific and Indistrial Research Organisation, 1957.
Estuarine hydrological investigations in eastern and south-western Australia.
Oceanogr. Sta. List, Div. Fish. and Oceanogr., 32:170 pp.

chemistry chlorinity(data only)
Australia, Commonwealth Scientific and Industrial Research Organization, 1957.
Estuarine hydrological investigations in eastern and south-western Australia, 1955.
Ocean. Station List, 29:93 pp.

Chemistry, chlorinity(data only)
Australia, Commonwealth Scientific and Industrial Research Organization, 1957
Onshore and oceanic hydrological investigations in eastern and southwestern Australia, 1955.
Oceanogr. Sta. List, 27:145 pp.

chemistry, Chlorinity(data only)
Australia, Commonwealth Scientific and Industrial Research Organization, 1957
Surface sampling in the Tasman and Coral Seas and the south-eastern Indian Ocean, 1956. Oceanogr. Sta. List, Div. Fish. & Oceanogr. 31:150 pp.

chemistry - chlorinity (data only)
Australia, Commonwealth Scientific and Industrial Research Organization, 1956.
Onshore hydrological investigations in eastern and south-western Australia, 1954. Compiled by D.R. Rochford.
Ocean. Sta. List. Invest. Div. Fish., 24:119 pp.

Chemistry - chlorinity(data only)
Australia, Commonwealth Scientific and Industrial Research Organization, 1954.
Onshore hydrological investigations in eastern and south-western Australia, 1953.
Ocean. Sta. List, Invest. Div. Fish., 18:64 pp.

Chemistry, chlorinity(data)
Australia, Commonwealth Scientific and Industrial Research Organization, 1953.
Onshore hydrological investigations in eastern and south-western Australia, 1951. Vol. 14:64 pp.

Chemistry - chlorinity (data)
Australia, Commonwealth Scientific and Industrial Research Organization, Melbourne, 1952.
Oceanographical station list of investigations made by the Division of Fisheries, C.I.R.O.
Vol. 5. Estuarine hydrological investigations in eastern Australia 1940-50;150 pp.
Vol. 6. Ibid.:137 pp.
Vol. 7. Ibid. :139 pp.
Vol. 8. Hydrological investigations in south-western Australia, 1944-50:152 pp.
Vol. 10. Records of twenty-four hourly hydrological observations at Shell Point, Georges River, New South Wales, 1942-50:134 pp.

Chemistry, chlorinity (data only)
Australia, Commonwealth Scientific and Industrial Research Organization, 1952.
Records of twenty-four hourly hydrological observations at Shell Point, Georges River, New South Wales, 1942-50. Compiled by D.R. Rochford.
Ocean. Sta. List., Invest. Fish. Div., 10:134 pp.

chemistry, chlorinity (data only)
Buljan, M., and M. Marinkovic, 1956.
Some data on hydrography of the Adriatic. (1946-1951). Acta Adriatica, 7(12):1-55.

chemistry, chlorinity (data only)
Buljan, Miljenko, and Mira Zore-Armanda 1966.
Hydrographic data on the Adriatic Sea collected in the period from 1952 through 1962.
Acta adriat. 12: 438 pp.

Chemistry, chlorinity (data only)
Cordini, I.R., 1955.
Contribucion al conocimiento del sector antartico Argentino. Inst. Antartico Argentino, Publ. 1: 277 pp., 82 textfigs., 56 pls.

Chemistry, chlorinity (data only)
France, Service Hydrographique de la Marine, 1963.
Resultats d'observations.
Cahiers Oceanogr., C.C.O.E.C., 15(8):576-588.

Chemistry, chlorinity(data only)
France, Service Hydrographique de la Marine, 1963
Résultats d'observations. Première campagne océanographique du "Commandant Robert Giraud", en Canal de Mozambique.
Cahiers Océanogr., C.C.O.E.C., 15(4):260-285.

Chemistry, Chlorinity (data only)
France, Service Hydrographique de la Marine, 1959.
Resultats d'observations océanographiques.
Cahiers Océanogr., C.C.O.E.C., 11(5):323-370.

Chemistry, chlorinity(data)
Franco, Paolo, 1962
Condizioni fisiche e chimiche delle acque lagunari nel Porto-Canale di Malamocco. 1. Giugno 1960-Giugno 1961.
Arch. Oceanogr. e Limnol., 12(3):225-255.

Chemistry, chlorinity (data)
Hasegawa, Y., M. Yokoseki, E. Fukuhara, and K. Terai, 1952.
On the environmental conditions for the culture of laver in Usu-Bay, Iburi Prov., Hokkaido.
Bull. Hokkaido Regional Fish. Res. Lab. 6:1-24, 8 textfigs., 9 tables.

Chemistry chlorinity(data only)
Hela, Ilmo, and Folke Koroleff, 1958
Hydrographical and chemical data collected in 1957 on board the R/V Aranda in the Barents Sea.
Merent. Julk., No. 179: 67 pp.

Chemistry, chlorinity (data only)
Hirota, Reiichiro, 1966.
Some oceanographical and meteorological conditions observed at a definite station off Mukaishima Marine Biological Station in 1965.
Contrib. Mukaishima Mar. Biol. Sta., No. 85:13pp.

chemistry chlorinity (data only)
Hirota, Reiichiro, 1965.
Some oceanographical and meteorological conditions observed at a definite station off Mukaishima Marine Biological Station in 1964. Contrib. Mukaishima Mar. Biol. Sta., No. 82:13pp.

Chemistry chlorinity(data)
Hirota, Reiichiro, 1961
Zooplankton investigations in the Bingo-Nada region of the Setonaikai (Inland Sea of Japan)
J. Sci. Hiroshima Univ., (B) Div. 1, 20:83-145
Also in:
Contrib., Mukaishima Mar. Biol. Sta., Hiroshima Univ., No. 67.

chemistry chlorinity (data only)
Hoshino, Zen-ichiro, 1970.
Oceanographical conditions observed at definite station off Asamushi during 1969.
Bull. mar. biol. Sta. Asamushi, 14(1):63 (foldout)

Chemistry, chlorinity(data only)
India, Naval Headquarters, Office of Scientific Research & Development, New Delhi, 1960.
Indian oceanographic stations list, Ser. No. 6 (Jan.):11 pp. (mimeographed).

Chemistry, chlorinity(data only)
Japan, Central Meteorological Observatory, 1956.
The results of marine meteorological and oceanographical observations (NORPAC Expedition Spec. No.) 17:131 pp.

Chemistry chlorinity(data only)
Japan, Central Meteorological Observatory, 1955.
The results of marine meteorological and oceanographical observations, January-June 1951.
No. 9:177 pp.

Chemistry, chlorinity (data only)
Japan, Central Meteorological Observatory, 1955.
The results of marine meteorological and oceanographical observations.
1. Oceanography, January-June 1954, No. 14:91 pp.
1. Oceanography, July-December 1954, No. 15:134 pp.

Chemistry, chlorinity(data only)
Japan, Central Meteorological Observatory, 1954.
The results of marine meteorological and oceanographical observations, Part 1, Oceanography, July-December, 1952, No. 12:138 pp.

chemistry, chlorinity (data only)
Japan, Central Meteorological Observatory, 1954.
The results of marine meteorological and oceanographical observations. 13(1):1-210.

Chemistry, chlorinity (data only)
Japan, Central Meteorological Observatory, 1953
The results of Marine meteorological and oceanographical observations. Jan.-June 1952, No. 11:362, 1 fig.

Chemistry, chlorinity (data only)
Japan, Central Meteorological Observatory, 1952
The results of Marine Meteorological and oceanographical observations, July - Dec. 1951, No. 10:310 pp., 1 fig.

Chemistry, chlorinity(data)
Japan, Central Meteorological Observatory, 1952.
The results of Marine meteorological and oceanographical observations. July - Dec. 1950. No. 8: 299 pp.

Chemistry chlorinity(data)
Japan, Central Meteorological Observatory, 1952.
The Results of Marine Meteorological and oceanographical observations. Jan. - June 1950. No. 7: 220 pp.

Chemistry - chlorinity(data)
Japan, Central Meteorological Observatory, 1951.
The results of marine meteorological and oceanographic observations. July - Dec. 1949. No. 6: 423 pp.

Chemistry, chlorinity (data only)
Japan, Central Meteorological Observatory, 1951.
Table 4. Oceanographical observations taken in Sagami Bay; physical and chemical data for stations occupied by R.M.S. "Asashio-maru". Res. Mar. Met. Ocean. Obs., Jan.-June, 1949, No. 5:30-37.

Chemistry, chlorinity (data only)
Japan, Central Meteorological Observatory, 1951.
Table 1. Oceanographical observations taken in the sea area south of Hokkaido; physical and chemical data for stations occupied by the R.M.S. "Yushio-maru". Res. Mar. Met. Ocean. Obs., Jan.-June 1949, No. 5:2-3.

Chemistry - chlorinity(data)
Japan, Central Meteorological Observatory, 1951.
Table 5. Oceanographical observations taken in the sea area to the south of the Kii Peninsula along "E"-line; physical and chemical data for stations occupied by R.M.S. "Chikubu-maru".
Res. Mar. Met. Ocean. Obs., Jan.-June, 1949, No. 5:37-40.

Chemistry chlorinity (data only)
Japan, Fisheries Agency, 1971
The results of fisheries oceanographical observation, January-December, 1968, 1729 pp.

Chemistry, chlorinity(data)
Japan, Fisheries Agency, Research Division, 1956.
Radiological survey of western area of the dangerous zone around the Bikini-Eniwetok Atolls, investigated by the "Shunkotsu maru" in 1956, Part 1:143 pp.

Chemistry chlorinity (data only)
Japan, Hokkaido University, Faculty of Fisheries, 1968.
Data record of oceanographic observations and exploratory fishing.
No. 12:420 pp.

Chemistry - Chlorinity (data only)
Japan, Hokkaido University, Faculty of Fisheries, 1966.
Data record of oceanographic observations and exploratory fishery, No 10:388 pp.

Chemistry - chlorinity (data only)
Japan, Hokkaido University, Faculty of Fisheries. 1965.
Data record of oceanographic observations and exploratory fishing, No. 9: 343 pp.

Chemistry. chlorinity(data only)
Japan, Hokkaido University, Faculty of Fisheries, 1963
Data Record of Oceanographic observations and exploratory fishing, No. 7:262 pp.

Chemistry chlorinity (data only)
Japan Meteorological Agency, 1968.
The results of marine meteorological and oceanographical observations, July-December, 1965, 38: 404 pp. (multilithed)

Chemistry, chlorinity (data only)
Japan, Japan Meteorological Agency, 1968.
The results of the Japanese Expedition of Deep Sea (JEDS-10).
Res.mar.met.oceanogr.,Observ.,Jan-June 1965:37: 385.

Chemistry, chlorinity (data only)
Japan, Japenese Meteorological Agency, 1967.
The results of marine meteorological and oceanographical observations, July-December, 1964, 36: 367 pp.

Chemistry, chlorinity (data only)
Japan, Japan Meteorological Agency, 1966.
The results of the Japanese Expedition of Deep Sea (JEDS-8).
Results mar.met.oceanogr.Obsns.Tokyo,35:328 pp.

Chemistry, chlorinity (data only)
Japan, Japan Meteorological Agency. 1965.
The results of marine meteorological and oceanographical observations, July-December 1963, No. 34: 360 pp.

Chemistry, chlorinity (data only)
Japan, Japan Meteorological Agency, 1964.
The results of marine meteorological and oceanographical observations, January-June 1963, No. 33:289 pp.

Chemistry chlorinity (data only)
Japan, Japan Meteorological Agency, 1964.
The results of the Japanese Expedition of Deep Sea (JEDS-5).
Res. Mar. Meteorol. and Oceanogr. Obs., July-Dec. 1962, No. 32:328 pp.

Chemistry chlorinity(data only)
Japan, Japan Meteorological Agency, 1962
The results of marine meteorological and oceanographical observations, July-December 1961, No. 30:326 pp.

chemistry
chlorinity(data only)
Japan, Japan Meteorological Agency, 1962
The results of marine meteorological and oceanographical observations, January-June 1961, No. 29: 284 pp.

chemistry,
chlorinity(data only)
Japan, Meteorological Agency, 1962
The results of marine meteorological and oceanographical observations, July-December, 1960, No. 28: 304 pp.

chemistry, chlorinity (data only)
Japan, Japan Meteorological Agency 1961.
The results of marine meteorological and oceanographical observations, January-June 1960.
The Results of the Japanese Expedition of Deep-Sea (JEDS-2, JEDS-3), No. 27: 257 pp.

chemistry Chlorinity(data only)
Japan, Japan Meteorological Agency, 1960.
The results of marine meteorological and oceanographical observations, July-December, 1959, No. 26:256 pp.

chlorinity (data only)
Japan, Japan Meteorological Agency, 1960.
The results of marine meteorological and oceanographical observations, Jan-June 1959, No. 25: 258 pp.

chemistry chlorinity(data only)
Japan, Japan Meteorological Agency, 1959.
The results of marine meteorological and oceanographical observations, July-December, 1958, No. 24:289 pp.

chemistry chlorinity(data only)
Japan, Japan Meteorological Agency, 1958
The results of marine meteorological and oceanographical observations, July-December 1957, No. 22: 183 pp.

chlorinity(data only)
Japan, Japan Meteorological Agency, 1958.
The results of marine meteorological and oceanographical observations, January-June, 1957, No. 21:168 pp.

chemistry, chlorinity (data only)
Japan, Japanese Meteorological Agency 1957
The results of marine meteorological and oceanographical observations, Jan.-June, 1956: 184 pp.
July-December, No. 20: 191 pp.

chemistry
Chlorinity(data only)
Japan, Japan Meteorological Agency, 1956.
The results of marine meteorological and oceanographical observations. Part 1. Oceanography, July-December, 1955. No. 18:90 pp.

chemistry chlorinity(data only)
Japan, Kagoshima University, Faculty of Fisheries, 1956.
Oceanographical observation made during the International Cooperation Expedition, EQUAPAC, in July-August, 1956, by M.S. Kagoshima-maru and M.S. Keitan-maru, 68 pp. (mimeographed).

chemistry
chlorinity(data only)
Japan, Kobe Marine Observatory, 1961
Data of the oceanographic observations in the sea south of Honshu from February to March and in May, 1959.
Bull. Kobe Mar. Obs., No. 167(27):99-108; 127-130;149-152;161-164;205-218.

chemistry - chlorinity
Japan, Kobe Marine Observatory, 1953?
[The results of the regular monthly oceanographical observations on board the R.M.S. "Syunpu-maru" in the Kii Suido and Osaka Wan (Oct. 1953)]
J. Ocean., Kobe Mar. Obs. (2nd ser.) 4(11):1-21, 15 figs.

Reports with similar titles in all other numbers of Vol. 4 this journal. These have not been entered in index.

chemistry, chlorinity (data only)
Japan Maritime Safety Agency 1966.
Results of oceanographic observations in 1963.
Data Rept. Hydrogr. Obs. Ser. Oceanogr. Publ. 792(3): 74 pp.

chemistry, chlorinity (data only)
Japan Maritime Safety Agency 1965.
Results of oceanographic observations in 1962.
Data Rept. Hydrogr. Obs. Ser. Oceanogr. (Publ. 792) Nov. 1965, 1: 65 pp.

chemistry - chlorinity (data only)
Japan, Maritime Safety Agency, 1964.
Tables of results from oceanographic observations in 1960.
Hydrogr. Bull., (Publ. No. 981), No. 75:86 pp.

chemistry, chlorinity (data only)
Japan Maritime Safety Board 1961.
Tables of results from oceanographic observation in 1959.
Hydrogr. Bull. Tokyo No. 68 (Publ. No. 981): 112 pp.

chemistry, chlorinity (data only)
Japan, Maritime Safety Board 1961
Tables of results from oceanographic observations in 1958.
Hydrogr. Bull. Tokyo No. 66 (Publ. No 981): 153 pp.

chemistry, chlorinity (data only)
Japan, Maritime Safety Board, 1960.
Tables of results from oceanographic observation in 1957.
Hydrogr. Bull. (Publ. No. 981), No. 64:103 pp.

chemistry chlorinity(data only)
Japan, Maritime Safety Board, 1956.
Tables of results from oceanographic observations in 1952 and 1953.
Hydrogr. Bull. (Publ. 981) (51):1-171.

chemistry chlorinity(data)
Japan, Maritime Safety Board, 1954.
[Tables of results from oceanographic observations in 1950.] Publ. No. 981, Hydrogr. Bull. Maritime Safety Bd., Spec. Number (Ocean. Repts.) No. 14:26-164, 5 textfigs.

chemistry
chlorinity(data only)
Japan, Maizuru Marine Observatory, 1963
Data of the oceanographic observations (1960-1961) (35-36):115-272.

chemistry chlorinity (data)
Karlovac, O., 1956.
Station list of the M.V. "Hvar" fishery-biological cruises, 1948-1949. Inst. Ocean. i Ribarstvo, Split, Repts., 1(3):177 pp.

chemistry, chlorinity(data only)
Katsuura, Hiroshi, Hideo Akamatsu, Tsutomu Akiyama, 1964.
Preliminary report of the Japanese Expedition of Deep-Sea, the Eighth Cruise (JEDS-8).
Oceanogr. Mag., Tokyo, 16(1/2):125-136.

chemistry chlorinity (data)
Lüneberg, H., 1939.
Hydrochemische Untersuchungen in der Elbmündung mittels Elektrocolorimeter.
Arch. Deutschen Seewarte 59(1):1-27, 8 pls.

chemistry chlorinity(data)
Maeda, H., 1953.
Studies on Yoda-Naikai. 2. Considerations upon the range of the stagnation and the influences by the River Noda and the open sea.
J. Shimonoseki Coll. Fish. 3(2):133-140, 2 textfigs.

chemistry chlorinity(data)
Maeda, H., 1953.
Studies on Yosa-Naikai. 3. Analytical investigations on the influence of the River Noda and the benthonic communities. J. Shimonoseki Coll. Fish. 3(2):141-149, 3 textfigs.

chemistry
chlorinity (data)
Masuzawa, Jotaro, and Hideo Akamatsu, 1962.
1. Hydrographic observations during the JEDS-4.
Oceanogr. Mag., Japan Meteorological Agency, 13(2):122-130.

chemistry
chlorinity(data)
Marinkovic-Roje, Marija, and Miroslav Nicolic 1961.
[Oceanographic researches in the areas of Rovinj and Limski Kanal from 1959-1961.] In Jugoslavian: English summary.
Hidrografskog Godisnjaka 1960, Split:61-67.

chemistry chlorinity(data)
Matsudaira, Yasuo, Haruyuki Koyama and Takuro Endo, 1961
[Hydrographic conditions of Fukuyama Harbor.]
J. Fac. Fish. and Animal Husbandry, Hiroshima Univ., 3(2):247-296.

chemistry chlorinity(data)
Menaché, M., 1954.
Étude hydrologique sommaire de la région d'Anjouan en rapport avec la pêche des coelacanthes.
Mém. Inst. Sci., Madagascar, A, 9:151-185, 34 textfigs.

chemistry, chlorinity (data)
Navarro, F. de P., F. Lozano, J.M. Navaz, E. Otero J. Sáinz Pardo and others, 1943.
La Pesca de Arrastre en los fondos del Cabo Blanco y del Banco Arguin (Africa Sahariana).
Trab. Inst. Español Ocean. No. 18:225 pp., 38 pls

chemistry, chlorinity (data only)

Nishihira, Moritaka 1968. Oceanographical conditions observed at definite station off Asamushi during 1967. Bull. mar. biol. Sta. Asamushi 13(2), 151

chemistry chlorinity(data)

Noakes, John E., and Donald W. Hood, 1961
Boron-boric acid complexes in sea water.
Deep-Sea Res., 8(2): 121-129.

Chemistry chlorinity(data)

Oren, O.H., 1962
A note on the hydrography of the Gulf of Eylath. Contributions to the knowledge of the Red Sea No. 21.
Sea Fish. Res. Sta., Haifa, Israel, Bull. No. 30: 3-14.

Chemistry, chlorinity(data)

Picotti, M., and A. Vatova, 1942.
Osservazioni fisiche e chemiche periodiche nell' Alto Adriatico (1920-1938). Thalassia 5(1):1-157.

Chemistry, chlorinity(data)

Reish, D.J., and H.A. Winter, 1954.
The ecology of Alamitos Bay, California, with special reference to pollution.
Calif. Fish and Game 40(2):105-121, 1 textfig.

Chemistry, chlorinity(data)

Rial, J.R. Besada, and L.R. Molins, 1962.
Determinación complexométrica de los iones calcio y magnesio en el agua de mar y estudio de las variaciones de su concentración en las aguas de la Ría de Vigo.
Bol. Inst. Español Oceanogr., 111:11 pp.

chemistry, chlorinity(data)

Rochford, D.J., 1953.
Analysis of bottom deposits in eastern and south-western Australia, 1951 and records of twenty-four hourly hydrological observations at selected stations in eastern Australian estuarine systems 1951. Oceanographical station list of investigations made by the Division of Fisheries, Commonwealth Scientific and Industrial Research Organization, Australia, 13:68 pp.

Chemistry, chlorinity (data)

Rochford, D.J., 1953.
Estuarine hydrological investigations in eastern and south-western Australia, 1951. Oceanographical station list of investigations made by the Division of Fisheries, Commonwealth Scientific and Industrial Research Organization, Australia, 12:111 pp.

chemistry chlorinity (data only)

* Scaccini Cicatelli, Marta, 1967.
Distribuzione stagionale dei sali nutrivi in una zona dell'alto e medio Adriatico.
Boll. Pesca, Piscic. Idrobiol., n.s., 22(1): 49-62.

Chemistry chlorinity ratios

Shesteperov, I.A., 1969.
Verification of Tsurikov's formula by the measurements in the sea. Okeanologiia, 9(4): 616-618.
(In Russian; English abstract)

chemistry, chlorinity (data)

Shimomura, T., and K. Miyata, 1957.
The oceanographical conditions of the Japan Sea and its water systems, laying stress on the summer of 1955.
Bull. Japan Sea Reg. Fish. Res. Lab., No. 6 (General survey of the warm Tsushima Current 1): 23-120.

Chemistry, chlorinity(data)

Tanak, Otohiko, Haruhiko Irie, Shozi IIzuka and Fumihiro Koga, 1961
The fundamental investigation on the biological productivity in the north-west of Kyushu. 1. Rec. Oceanogr. Wks., Japan, Spec. No. 5: 1-58.

Chemistry, chlorinity

Tanioka, Katsumi, 1962
The oceanographical conditions of the Japan-Sea. (1) On the annual variations of chlorinity. (In Japanese; English abstract).
Umi to Sora, Mar. Meteorol. Soc., Kobe, 38(3): 90-100.

Chemistry, chlorinity(data only)

Tsubata, B., 1958.
Oceanographical conditions observed at definite station off Asamushi during 1956.
Bull. Mar. Biol. Sta., Asamushi, 9(1):43.

Chemistry, chlorinity(data)

Tsubata, B., 1956.
Oceanographical observations off Asamushi during 1954. Bull. Mar. Biol. Sta., 8(1):43-48.

Chemistry, chlorinity(data)

Uda, Michitaka, and Makoto Ishino, 1960
Researches on the currents of Kuroshio.
Rec. Oceanogr. Wks., Japan, Spec. No. 4: 59-72.

Chemistry, chlorinity(data)

Uda, M., N. Watanabe and M. Ishino, 1956.
General results of the oceanographic surveys (1952-1955) on the fishing grounds in relation to the scattering layer. J. Tokyo Univ. Fish., 42(2):169-207.

chlorinity, effect of

Chemistry, chlorinity, effect of

Kojima, S., 1954.
On the fluctuation in the sardine catch of the western Japan Sea. 1. Surface temperature, clorinity and the sardine fishery.
Bull. Jap. Soc. Sci. Fish. 20(5):375-379, 3 textfigs.

chlorinity, effect of

Kwiecinski, Bogdan, 1965.
The sulfate content of Baltic water and its relation to the chlorinity.
Deep-Sea Res., 12(6):797-804.

Chemistry chlorinity, effect of

Uyeno, Fukuzo, 1961
Oceanographical and ecological studies on primary production of the sea, with special references to relationship between diatom production and temperature and chlorinity of water.
Rept., Fac. Fish., Pref. Univ., Mie, 4(1): 1-64.

Chemistry chlorinity, effect of

Uyeno, F., 1958.
Relation between diatom quantity, temperature and chlorinity in the neighboring sea near the coast of Honshu, Japan. Mem. Kobe Mar. Obs., 12(1):1-6.

Chemistry, chlorinity, effect of

Uyeno, Fukuzo, 1957.
Relation between diatom quantity, temperature and chlorinity in the neighboring sea south of Honshu.
Umi to Sora, 33(4/5):70-74.

chlorinity, effect of

Uyeno, F., 1957.
The variation of diatom communities and the schematic explanation of their increase in Osaka Bay in summer. III. Schematic explanation of increase of diatoms in relation to the distribution of chlorinity.
J. Oceanogr. Soc., Japan, 13(3):107-110.

Chlorinity, monthly means

Chemistry Chlorinity (monthly means)

Newell, B. S., 1966.
Seasonal changes in the hydrological and biological environments off Port Hacking, Sydney.
Australian J. Mar. freshw. Res., 17(1):77-91.

Chlorinity/silicate

Chemistry chlorinity/silicate

Sugiura, Yoshio, and Suphachai Chaitiamvong, 1964.
Relation of silicate concentration to dissolved oxygen amount in sea water collected in the northern frontal region of Kuroshio.
J. Oceanogr. Soc., Japan, 20(2):89=92.

Chlorinity surface

Chemistry, chlorinity, surface

Akasuka, K. (deceased), F. Uyeno, K. Mitani and M. Miyamura, 1960.
On the relation between the distribution of plankton and the annual changes of sea conditions in Ise Bay.
J. Oceanogr. Soc., Japan, 16(2):83-91.

Chemistry, chlorinity (surface) (data only)

Australia, C.S.I.R.O., Division of Fisheries, 1959.
Surface sampling in the Tasman and Coral Seas, 1957.
Oceanogr. Sta. List, 36:175 pp.

Chemistry, chlorinity, surface (data only)

Australia, Commonwealth Scientific and Industrial Research Organization, 1957.
Surface sampling in the Tasman and Coral seas, 1955. Oceanogr. Sta. List, 28:88 pp.

Chemistry chlorinity(surface)(data only)

Australia, Commonwealth Scientific and Industrial Research Organization, 1954.
Surface sampling in the Tasman Sea, 1954.
D.R. Rochford, Compiler.
Ocean Sta. List., Invest. Div. Fish., 25:79 pp.

Chemistry, chlorinity (surface)

Japan, Kobe Marine Observatory, 1955.
The result of the regular monthly surface observations on board the M.S. (Hayanami" in Osaka Bay
Jan. 14-15, 1954 - J. Ocean. (2)5(1):20-21.
Feb.-Apr., 1954 - J. Ocean. (2)5(4):1-5, 3 figs.
May 11, 1954 - J. Ocean. (2)5(5):12-13, 12 figs.

Chemistry, chlorinities, surface

Japan, Kobe Marine Observatory, 1953.
On the distribution of surface chlorinities from Japan to Antarctic whaling grounds. J. Ocean. (2)4(1):13-16, 3 figs.

Chemistry, chlorinity, surface

Japan, Kobe Marine Observatory, 1953.
On the annual deviation of the surface water temperature and chlorinity. J. Ocean. (2)4(1):17-21, 3 figs.

chemistry chlorinity surface
Kawai, H., and M. Sasaki, 1961
An example of the short-period fluctuation of the oceanographic condition in the vicinity of the Kuroshio front.
Bull. Tohoku Reg. Fish. Res. Lab., (19): 119-134.

chemistry chlorinity (surface-bottom) (data)
Kubo, I., and E. Asada, 1957.
A quantitative study on crustacean bottom epifauna in Tokyo Bay. J. Tokyo Univ., Fish., 43(2): 249-289.

chemistry chlorinity, surface
Moriyasu, S., 1958.
On the fluctuation of the Kuroshio south of Honshu. (2). (Seasonal variations of water temperature and chlorinity in the upper layer).
Mem. Kobe Mar. Obs., 12(2):1-18.

chemistry chlorinity, surface (data)
Nakamura, Akikazu, 1962
On the surface oceanographic observations across the North Pacific Ocean in winter, 1959-1960, and in summer and autumn, 1960.
Rec. Oceanogr. Wks., Japan, n.s., 6(2):31-57.

Chlorinity/temperature

See: temperature/chlorinity

Chlorellin

chemistry chlorellin
Pratt, R., T.C. Daniels, J.J. Eiler, J.B. Gunnison, W.D. Kumler, J.F. Oneto, L.A. Strait, H.A. Spoehr, G.J. Hardin, H.W. Milner, J.H.C. Smith, and H.H. Strain, 1944.
Chlorellin, an antibacterial substance from Chlorella. Science 99(2574):351-352.

Chlorophyll

chemistry chlorophyll
Alexander, James E., E.F. Corcoran, 1963
Distribution of chlorophyll in the Straits of Florida.
Limnology and Oceanography, 8(2):294-297.

chemistry chlorophyll
Alexander, J.E., J.H. Steele and E.F. Corcoran, 1962
The seasonal cycle of chlorophyll in the Florida Straits.
Proc. Gulf and Caribbean Fish. Inst., Inst. Mar. Sci., Univ. Miami, 14th Ann. Sess.: 63-67.

Chemistry Chlorophyll
Anderson, G.C., 1969.
Subsurface chlorophyll maximum in the northeast Pacific Ocean. Limnol. Oceanogr., 14(3): 386-391.

chemistry chlorophyll
Anderson, George C., 1964.
The seasonal and geographic distribution of primary productivity off the Washington and Oregon coasts.
Limnology and Oceanography, 9(3):284-302.

chemistry chlorophyll
Angot, Michel, et Robert Gerard, 1966.
Hydrologie et phytoplancton de l'eau de surface en avril 1965 a Nosy Be.
Cah. ORSTOM, Ser. Oceanogr., 4(1):95-136.

chemistry chlorophyll
Apollonio, Spencer, 1965.
Chlorophyll in Arctic sea ice.
Arctic, J. Arctic Inst., N. Amer., 18(2):118-122.

Chemistry chlorophyll
Aruga, Yusho, and Masami Monsi, 1962
Primary production in the northwestern part of the Pacific off Honshu, Japan.
J. Oceanogr. Soc., Japan, 18(2):85-94.

Chemistry chlorophyll
Atkins, W.R.G., and P.G. Jenkins, 1953.
Seasonal changes in the phytoplankton during the year 1951-52 as indicated by spectrophotometric chlorophyll estimations. J.M.B.A. 31(3):495-508, 8 textfigs.

chemistry chlorophyll
Atkins, W.R.G., and M. Parke, 1951.
Seasonal changes in the phytoplankton as indicated by chlorophyss estimation. J.M.B.A. 29(3): 609-618.

chlorophyll
Banse, K., 1956.
Produktionsbiologische Serienbestimmungen im südlichen Teil der Nordsee im März 1955.
Kieler Meersf., 12(2):166-179.

chemistry chlorophyll
Beers, John R. and Gene L. Stewart, 1971.
Micro-zooplankters in the plankton communities of the upper waters of the eastern tropical Pacific. Deep-Sea Res. 18(9): 861-883.

chlorophyll
Belcher, John H., and Gordon E. Fogg, 1964.
Chlorophyll derivatives and carotenoids in the sediments of two English lakes.
In: Recent researches in the fields of hydrosphere, atmosphere and nuclear geochemistry, Ken Sugawara festival volume, Y. Miyake and T. Koyama, editors, Maruzen Co., Ltd., Tokyo, 39-48.

chemistry chlorophyll (data)
Bennett, Edward B., and Milner B. Schaefer, 1960
Studies of physical, chemical and biological oceanography in the vicinity of the Revilla Gigedo Islands during "Island Current Survey" of 1957.
Bull. Inter-American Tropical Tuna Comm., 4(5): 219-317. (Also in Spanish).

chemistry, chlorophyll
Bogoyavlensky, A.N., 1959
[Hydrochemical investigations.]
Arktich. i Antarktich. Nauchno-Issled. Inst., Mezhd. Geofiz. God, Sovetsk. Antarkt. Eksped., 5: 159-172.

chemistry, chlorophyll
Brongersma-Sanders, M., 1951.
On conditions favouring the preservation of chlorophyll in marine sediments. Proc. Third World Petroleum Congr., Sect. 1:

chlorophyll
Burkholder, Paul R., Lillian M. Burkholder and Juan A. Rivero, 1959.
Chlorophyll A in some corals and marine plants.
Nature 183:1338-1339.

chemistry chlorophyll
Burkholder, Paul R., and John M. Sieburth, 1961.
Phytoplankton and chlorophyll in the Gerlache and Bransfield Straits of Antarctica.
Limnol. & Oceanogr., 6(1):45-52.

chemistry - chlorophyll
California Academy of Sciences
California Division of Fish and Game
Scripps Institution of Oceanography
U. S. Fish and Wildlife Service
1950
California Cooperative Sardine Research Program.
Progress Rept. 1950:54 pp., 37 text figs.

chemistry - chlorophyll (data)
Chesapeake Bay Institute, 1954.
Data report, Choptank River winter cruise, 7 December to 10 December 1952. Ref. 54-11:1-37, 1 fig.
(24)

chemistry chlorophyll (data)
Chesapeake Bay Institute, 1954.
Choptank River autumn cruise, 23 Sept-27 Sept 1954. Data Rept. 22, Ref. 54-9:37 pp., 1 fig.

chemistry, chlorophyll (data)
Chesapeake Bay Institute, 1954.
Choptank River Spring Cruise, 28 April-1May 1952.
Ref. 54-1:37 pp.

chemistry chlorophyll (data)
Chesapeake Bay Institute, 1952.
Data report, St. Mary's River cruise, June 19-July 18, 1951. Rept. 11, Ref. 52-19:115 pp.

chemistry chlorophyll (data)
Chesapeake Bay Institute, 1951.
Data report, Cruise VIII, Jan. 10, 1951-Jan. 23, 1951. Rept. No. 6:29 pp. (duplicated).

chemistry chlorophyll (data)
Chesapeake Bay Institute, 1951.
Data report, Cruise VII, 14 Oct. 1950-2 Nov. 1950. Rept. No. 5:41 pp. (duplicated).

chemistry - chlorophyll
Chesapeake Bay Institute, 1951.
Data report. Cruise VII. October 14, 1950-November 2, 1950. Rept. No. 2:41 pp. (duplicated).

chemistry chlorophyll
Chesapeake Bay Institute, 1950.
Data Report, Cruise V and VI, May 20, 1950-May 25, 1950, July 14, 1950-July 19, 1950. Rept. No. 4:51 pp. (mimeographed).

chemistry - chlorophyll (data)
Chesapeake Bay Institute, 1949.
Quarterly report, July 1, 1949-October 1, 1949.
Rept. No. 1:121 pp. (multilith).

chemistry, chlorophyll
Conover, S.A.M., 1954.
Observation on the structure of red tides in New Haven Harbor, Connecticut. J. Mar. Res. 13(1):145-155, 6 textfigs.

chemistry chlorophyll
Conover, S.A.M., 1956.
Oceanography of Long Island Sound, 1952-1954.
4. Phytoplankton. Bull. Bingham Oceanogr. Coll., 15:62-112.

chemistry, chlorophyll
Corcoran, E.F., and James E. Alexander, 1963.
Nutrient, chlorophyll and primary production studies in the Florida Current.
Bull. Mar. Sci., Gulf and Caribbean, 13(4):527-541.

chemistry, chlorophyll
Davis, P.S., 1957.
A method for the determination of chlorophyll in sea-water. C.S.I.R.O., Australia, Div. Fish. and Oceanogr., Rept., No. 7:8 pp. (mimeographed)

chemistry, Chlorophyll
Derenbach, Jens B., 1970.
Partikuläre Substanz und Plankton an Hand chemischer und biologischer Daten gemessen in den oberen Wasserschichten des Gotland-Tief im Mai 1968.
Kieler Meeresforsch. 25(2): 279-289

chemistry, chlorophyll
Dyrssen, David, 1969.
Stiochiometry and chemical equilibrium. In: Chemical oceanography, Rolf Lange, editor, Universitetsforlaget, Oslo: 47-59.

chemistry, chlorophyl
Edmondson, W. T., and Y.H. Edmondson, 1947.
Measurements of production in fertilized salt-water. J. Mar. Res. 6(3):228-246, Figs. 55-62.

chemistry
chlorophyll (data)
El-Sayed, Sayed Z., and Enrique F. Mandelli, 1965.
Primary production and standing crop of phytoplankton in the Weddell Sea and Drake Passage. In: Biology of Antarctic seas, II. Antarctic Res. Ser., Am. Geophys. Union, 5:87-106.

chemistry, chlorophyll
Establier, R., 1969.
Fitoplancton e hidrografía de la bahía de Cádiz de abril de 1965 a noviembre de 1967. Investigación pesq. 33(1): 97-118.

chemistry
chlorophyll
Establier, R., y R. Margalef, 1964
Fitoplancton e hidrografía de las costas de Cádiz (Barbate) de junio de 1961 a agosto de 1962.
Inv. Pesq., Barcelona, 25:5-31.

chemistry
chlorophyll
Finenko, Z.Z., 1964
The chlorophyll content in the plankton of the Black and Azov seas. (In Russian).
Okeanologiia, Akad. Nauk, SSSR, 4(3):462-468.

chemistry chlorophyll
Flemer, David A., 1969.
Continuous measurement of in vivo chlorophyll of a dinoflagellate bloom in Chesapeake Bay.
Chesapeake Sci., 10(2): 99-103.

chemistry,
chlorophyll
Forsbergh, Eric D., 1963
Some relationships of meteorological, hydrographic, and biological variables in the Gulf of Panama. (In English and Spanish).
Bull., Inter-American Tropical Tuna Comm., 7(1): 109 pp.

chemistry
Chlorophyll (data)
Forsbergh, Eric D., William W. Broenkow, 1965.
Observaciones oceanograficas del oceano Pacifico oriental recolectadas por el barco Shoyo Maru, octubre 1963-marzo 1964.
Comision Interamericana del Atun Tropical, Bol. 10(2): 85-237.

chemistry, chlorophyll
Gifford, Cameron E., and Eugene P. Odum, 1961.
Chlorophyll a content of intertidal zones on a rocky shore.
Limnol. & Oceanogr., 6(1):83-85.

chemistry, chlorophyll
Gillbricht, M., 1952.
Untersuchungen zur Produktionsbiologie des Planktons in der Kieler Bucht. 1. Die zeitliche und räumliche Verteilung des Planktons und die quantitativen Beziehungen zwischen Plankton-, Chlorophyll- und Sestonbestimmungen.
Kieler Meeresf., 9(2):177-189.

Translation available, Fisheries station, Aberdeen.

chemistry chlorophyll
Godnev, T.N., S.V. Kaishevich and G.V. Zakharich, 1950.
On the chlorophyll content of fresh-water plankton. Doklady Akad. Nauk, SSSR, 73(5):1041-

chemistry chlorophyll
Gorshkova, T.I., 1965.
Chlorophyll and carotinoids in the sediments of the Baltic Sea and Gulf of Riga. Investigations in line with the programme of the International Geophysical Year, 2. (In Russian).
Trudy, Vses. Nauchno-Issled. Inst. Morsk. Ribn. Choz. i Okeanogr., (VNIRO), 57:313-323.

chemistry, chlorophyll.
Graham, H. W., 1943.
Chlorophyll content of marine plankton.
J. Mar. Res., 5:153-160.

chemistry chlorophyll
Hagmeier, Erik, 1961.
Plankton-Äquivalente Auswertungen von chemischen und mikroskopischen Analysen.
Kieler Meeresf., 17(1):32-47.

chemistry chlorophyll
Harvey, H.W., 1953.
Synthesis of organic compounds and chlorophyll by Nitzschia closterium. J.M.B.A. 31(3):477-487, 4 textfigs.

chemistry
chlorophyll (data)
Herrera, Juan, y Ramón Margalef, 1963
Hidrografía y fitoplancton de la costa comprendida entre Castellón y la desembocadura del Ebro, de julio de 1960 a junio de 1961.
Inv. Pesq., Barcelona, 24:33-112.

chemistry
chlorophyll (data)
Herrara, Juan, and Ramon Margalef, 1961
Hidrografía y fitoplancton de las costas de Castellón de julio de 1958 a junio de 1959.
Inv. Pesq., Bacelona, 20:17-63.

chemistry, chlorophyll
*Hickel, Wolfgang, 1967.
Untersuchungen über die Phytoplanktonblüte in der westlichen Ostsee.
Helgoländer wiss. Meeresunters., 16(½):1-66.

chemistry chlorophyll
Holmes, R.W., M.B. Schaeffer and B.M. Shimada, 1957.
Primary production, chlorophyll and zooplankton volumes in the tropical eastern Pacific Ocean.
Bull. Inter-Amer. Trop. Tuna Comm., 2(4):129-156.

chemistry chlorophyll
Hoy, H. 1970.
Chlorophyll in the sea off South Africa.
Fish. Bull. S Afr. 6: 1-9.

chemistry - chlorophyll
Humphrey, G.F., 1970.
The concentrations of chlorophylls a and c in the south-west Pacific Ocean.
Aust. J. mar. Freshwat. Res. 21(1): 1-10.

chemistry, chlorophyll
Humphrey, G.F., 1966.
The concentration of chlorophylls a and c in the south-east Indian Ocean.
Aust. J. mar. Freshwat. Res., 17(2):135-145.

chemistry, chlorophyll
Humphrey, G.F., 1963.
Chlorophyll a and c in cultures of marine algae.
Australian J. Mar. Freshwater Res., 14(2):148-154.

chemistry, chlorophyll
Ichimura, Shun-ei, 1965.
A short review on primary production in the Kuroshio. (In Japanese; English abstract).
Inf. Bull. Planktol., Japan, No. 12:1-6.

chemistry, chlorophyll
Irie, Haruhiko, and Syozi Iizuka, 1966.
Studies of the oceanographic characteristics of Haiki Channel and the adjacent waters and of effects of closing of the channel on pearl frams. 1. Present status of plankton-biota and its preseumptive changes. (In Japanese; English abstract.)
Bull. Fac. Fish., Nagasaki Univ., No. 20:14-21.

chemistry, chlorophyll
Ivanoff, A., M. Bauer, J. Bortler, C. Pauet, G. Copin-Montegut, M. LeRoy et A. Morel, 1969.
Résultats des observations effectuées en mer Tyrrhénienne à bord du navire océanographique Calypso en juillet 1964.
Cah. océanogr. 21 (Suppl. 2): 193-202.

chemistry chlorophyll
Jacques, Guy 1968.
Étude du plancton de la région de Banyuls-sur-Mer. Variations saisonnières des pigments chlorophylliens de la couche superficielle (Point côtier, août 1965-août 1966).
Rapp. Proc.-Verb. Reun. Comm. int. Explor. scient. Mer Méditerranée, 19(3): 557-559.

chemistry chlorophyll
(data)
Jacques, G. 1967.
Aspects quantitatifs du phytoplancton de Banyuls-sur-Mer (Golfe du Lion) 1. Pigments et population phytoplanctoniques dans le Golfe du Lion en mars 1966.
Vie Milieu (B) 18 (2B): 239-271.

chlorophyll
Japan, National Working Group on Photosynthesis-Chlorophyll of IIOE, 1965.
Summary report on photosynthesis and chlorophyll in the eastern Indian Ocean observed by Japanese ships during IIOE. (In Japanese; English abstract
Inf. Bull. Planktol., Japan, No. 12:72-78.

chemistry, chlorophyll
Jastrebova, L.A., 1938
Chlorophyll in Meeressedimenten. Trans. Inst. Mar. Fish. and Oceanogr. USSR, 5:189-224, (German resume)

chemistry
chlorophyll
Jeffrey, S.W. 1968.
Pigment composition of Siphonales algae in the brain coral, Favia.
Biol. Bull. mar. biol. Lab. Woods Hole, 135(1): 141-148.

chemistry
chlorophyll

Jeffrey, S.W., and F.T. Haxo 1968.
Photosynthetic pigments of symbiotic dinoflagellates (zooxanthellae) from corals and clams.
Biol. Bull. mar. biol. Lab., Woods Hole, 135(1): 149-165.

chemistry, chlorophyll estimations

Jenkins, P.M., 1955 (with intro. by W.R.G. Atkins)
Seasonal changes in the phytoplankton as indicated by spectrophotometric clorophyll estimations 1952-53.
Pap. Mar. Biol. and Oceanogr., Deep-Sea Res., Suppl. to Vol. 3:58-67.

chemistry, chlorophyll (data)

Kalle, K., 1956.
Chemisch-hydrographische Untersuchungen in der inneren Deutschen Bucht. Deutsche Hydrogr. Zeits. 9(2):55-65.

chemistry chlorophyll

Kalle, K., 1951.
Meereskundlich-chemische Untersuchungen mit Hilfe des Pulfrich-Photometers von Zeiss. VII. Die Mikrobestimmungen des Chlorophylls und die Eigenfluoreszenz des Meerwassers. Deutsche Hydrogr. Zeits. 4(3):92-96, 2 text figs.

chemistry
chlorophyll

Kawarada, Y. and A. Sano 1968
Distribution of chlorophyll and phaeophytin in the western North Pacific
Oceanogrl Mag. 21(2): 137-146

chemistry, chlorophyll

Kay, H., 1954.
Untersuchungen zur Menge und Verteilung der organischen Substanz im Meerwasser.
Kieler Meeresf. 10(2):202-214, Figs. 18-25.

chemistry, chlorophyll

Kimball, J.F., Jr., Eugene F. Corcoran and E.J. Ferguson Wood, 1963.
Chlorophyll-containing microorganisms in the aphotic zone of the oceans.
Bull. Mar. Sci., Gulf and Caribbean, 13(4):574-577.

chemistry, chlorophylls

Komaki, S., 1957.
[Seasonal variations in the concentrations of net-plankton pigments in sea water.]
Bull. Hokkaido Reg. Fish. Res. Lab., No. 16:84-91.

chemistry, chlorophyll

Krey, J., 1956.
Die Trophie küstennaher Meeresgebiete.
Kieler Meeresf., 12(1):46-64.

chemistry
Chlorophyll

Krey, J., 1934
Die Bestimmung des Chlorophylls im Meerwasser. Schöpfproben. J. du Cons. 14:201-209.

chemistry
chlorophyll

Krey, Johannes, und Karl-Heinz Szekielda, 1966.
Gesamtkohlenstoff und Mikrobiomasse in der Ostsee im Mai 1962.
Kieler Meeresforsch., 22(1):62-69.

Kutyurin, V. M., 1959 chemistry, chlorophyll

Determination of chlorophyll content in sea water and the spectral analysis of phytoplankton pigments. Arctic and Antarctic Sci. Res. Inst., Soviet Antarctic Expeds., I.G. Y., 2nd Marine Exped. on the "Ob," 1956-1957, 5:173-175. Publisher: Morskoi Transport, Leningrad).

Arktich. i Antarktich. Nauchno-Issled. Inst., Mezhd. Geofiz. God, Sovetsk. Antarkt. Exsped., 5: 173 - 176

chemistry
chlorophyll.

Kutyurin, E.M., and A.P. Lisitsin, 1962
[Vegetative pigments present in suspensions and bottom sediments in the western part of the Indian Ocean. Communication. 2. Quantitative distributions and qualitative content of pigments present in suspensions.]
Mezhd. Geofiz. Komitet, Prezidiume Akad. Nauk, SSSR, Rezult. Issled. Programme Mezhd. Geofiz. Goda, Okeanol. Issled., No. 5: 112-129.

chemistry, chlorophyll

Léger, Guy, 1964.
Les populations phytoplanctoniques en mer Ligure (radiale Monaco-Calvi) en juin 1963.
Bull. Inst. Océanogr., Monaco, 64(1326):32 pp.

Chemistry
chlorophyll

Leloup, E. editor, 1966.
Recherches sur l'ostreiculture dans le bassin d'Ostende en 1964.
Minist. Agricult. Comm. T.W.O.Z., Groupe de Travail "Ostreiculture": 58pp.

chemistry
chlorophyll

Lenz, Jürgen, 1970.
4. Planktologie. 4.1. Seston, Chlorophyll- und Eiweissgehalt. 4.3. Zooplankton.
Chemische, mikrobiologische und planktologische Untersuchungen in der Schlei im Hinblick auf deren Abwasserbelastung. Kieler Meeresforsch 26(2): 180; 180-129; 203-213.

chemistry, chlorophyll

Lewin, Ralph A., Editor, 1962.
Physiology and biochemistry of algae.
Academic Press, New York and London, 929 pp.

chemistry, chlorophyll

Loftus, M.E. and J.H. Carpenter, 1971. A fluorometric method for determining chlorophylls a, b and c. J. mar. Res. 29(3): 319-338.

chemistry, chlorophyll

Lorenzen, Carl J., 1971.
Continuity in the distribution of surface chlorophyll. J. Cons. int. Explor. Mer. 34 (1): 18-23.

Chemistry, chlorophyll

Lorenzen, Carl J., 1970.
Surface chlorophyll as an index of the depth, chlorophyll content and primary productivity of the euphotic layer. Limnol. Oceanogr., 15 (3): 479-480.

chemistry, chlorophyll

Lorenzen, Carl J., 1967.
Vertical distribution of chlorophyll and phaeo-pigments: Baja California.
Deep-Sea Res., 14(6):735-745.

chemistry, chlorophyll

Lorenzen, Carl J., 1966.
A method for the continuous measurement of in vivo chlorophyll concentration.
Deep-Sea Res., 13(2):223-227.

chemistry, chlorophyll

Lorenzen, Carl J., 1965.
A note on the chlorophyll and phaeophytin content of the chlorophyll maximum.
Limnol. Oceanogr., 10(3):482-483.

chlorophyll (data)

*Maestrini, Serge, et Mariano Javier Pigarro, 1966
Contribution à l'étude de l'hydrologie et de la productivité primaire des eaux cotières de la région de Tuléar.
Recl. Trav. Stn. mar. Endoume, hors sér., Suppl. 5: 7-23.

chemistry, chlorophyll

Mandolli, Enrique F., 1965.
Conribucion al conocimiento de la produccion organica primaria en aguas sub-antarticas (Ocean Atlantico Sud-Occidental).
Anais Acad. bras. Cienc., 37(Supl.):399-407

Chemistry
chlorophyll

Margalef, Ramón, 1964.
Fitoplancton de las costas de Blanes (provincia de Garona, Mediterráneo Occidental), de julio de 1959 a junio de 1963.
Inv. Pesq., Barcelona., 26:131-164.

Chemistry,
chlorophyll (data)

Margalef, Ramón, 1961
Hidrografía y fitoplancton de un área marina de la costa meridional de Puerto Rico.
Inv. Pesq., Barcelona, 18:38-96.

Chemistry
chlorophyll

Margalef, Ramón, y Juan Herrera, 1964.
Hidrografía y fitoplancton de la costa comprendida entre Castellón y la desembocadura del Ebro, de julio de 1961 de julio de 1962.
Inv. Pesq., Barcelona, 26:49-90.

chemistry
chlorophyll (data)

Margalef, Ramón, and Juan Herrera, 1963
Hidrografía y fitoplancton de las costas de Castellón, de julio de 1959 a junio de 1960.
Inv. Pesq., Barcelona, 22: 49-109.

chemistry, chlorophyll

Marshall, N., 1956.
Chlorophylla a in the phytoplankton of coastal waters of the eastern Gulf of Mexico.
J. Mar. Res., 15(1):14-32.

chemistry chlorophyll (data)

Marumo, Ryuzo, Editor, 1970.
Preliminary report of the Hakuho Maru Cruise KH-69-4, August 12-November 13, 1969, The North and Equatorial Pacific Ocean.
Ocean Res. Inst. Univ. Tokyo, 68pp.

Chemistry - Chlorophyll

Matterne, Manfred, 1970
Vergleich zwischen Primärproduktion und Syntheseraten organischer Zellbestandteile mariner Phytoplankter.
Kieler Meeresforsch. 25(2):290-313.

Chemistry
Chlorophyll

Mazuzawa, Jotaro, Tsutomu Akiyama, Yutaka Kawarada and Tsutomu Sawara, 1970.
Preliminary report of the Ryofu Maru Cruise Ry7001 in January-March 1970. Oceanogrl Mag., 22(1): 1-25.

Chemistry
chlorophyll

McLaren, I.A., 1961
The hydrography and zooplankton of Ogac Lake, a landlocked fjord on Baffin Island.
Fish. Res. Bd., Canada, MSS Rept. Ser. (Biol.) No. 709: 167 pp. (multilithed).

Chemistry chlorophyll

McIntyre, A.D., and J.H. Steele, 1956.
Hydro-biological conditions in the Denmark Strait, May 1954.
Ann. Biol., Cons. Perm. Int. Expl. Mer, 11:20-25, Figs., 25-28.

chemistry, chlorophyll
Minas, H.J., and M.J. Pizarro, 1962.
Quelques observations hydrologiques sur les eaux au large des côtes orientales de la Corse.
Bull. Inst. Océanogr., Monaco, 59(1232):1-23.

chemistry chlorophyll
Morales, E., y E. Arias, 1965.
Ecología del puerto de Barcelona y desarrollo de adherencias orgánicas sobre placas sumergidas.
Inv. Pesq., Barcelona, 28:49-79.

chemistry chlorophyll
Munoz Sardon, Filipe, 1962
Estudio del fitoplancton del Golfo de Cadiz en relacion con el regimen regional de vientos.
Inv. Pesq., Barcelona, 21:165-188.

chemistry chlorophyll
Muñoz, F. y J.M. San Feliu 1969.
Hidrografía y fitoplancton de las costas de Castellón, de febrero a junio de 1967.
Invest. pesq. Barcelona 33(1): 313-334.

chemistry Chlorophyll,
Muñoz, Felipe, y Jose M. San Feliu, 1965.
Hidrografia y fitoplancton de las costas de Castellon de agosto de 1962 a julio de 1963.
Inv. Pesq., Barcelona, 28:173-209.

Chemistry chlorophyll
*Nakajima, Kohki, and Satoshi Nishizawa, 1968.
Seasonal cycles of chlorophyll and seston in the surface water of the Tsugaru Strait area.
Rec. oceanogr. Wks, Japan, n.s., 9(2):219-246.

chemistry, chlorophyll
Nehring, Dietwart, Sigurd Schulz und Karl-Heinz Rohde, 1969.
Untersuchungen über die Productivität der Ostsee. I. Chemisch-biologische Untersuchungen in der mittleren Ostsee und in der Beltsee im April/Mai 1967.
Beitr. Meeresk., 23:5-36

chemistry, chlorophyll
*Nellen, Walter, 1967.
Horizontale und vertikale Verteilung der Planktonproduktion im Golfe von Guinea und in angrenzenden Meeresgebieten während der Monate Februar bis Mai 1964.
Kieler Meeresforsch., 23(1):48-67.

chemistry chlorophyll
*Nemoto, Takahisa, 1968.
Chlorophyll pigments in the stomach of Euphausiids.
J. oceanogr., Soc., Japan, 24(5):253-260.

chemistry, chlorophyll (data)
Niemi, Åke, Heinrichs Skuja and Torbjörn Willén 1970
Phytoplankton from the Pojoviken-Tvärminne area, S. coast of Finland.
Mem. Soc. Fauna Flora Fennica 46: 14-28.

chemistry chlorophyll
Parsons, T.R., 1961
On the pigment composition of eleven species of marine phytoplankters.
J. Fish. Res. Bd. Canada, 18(6):1017-1025.

chemistry chlorophyll (data)
Parsons, T.R., 1960.
A data record and discussion of some observations made in 1958-1960 of significance to primary productivity.
Fish. Res. Bd., Canada, Manuscript Rept. Ser., (Oceanogr. & Limnol.); No. 81:19 pp. (multilithed).

chemistry chlorophyll
Patten, B.C., R.A. Mulford and J.E. Warinner, 1963
An annual phytoplankton cycle in the lower Chesapeake Bay.
Chesapeake Science, 4(1):1-20.

chemistry, chlorophyll (data)
Platt, Trevor, and Brian Irwin, 1968.
Primary productivity measurements in St. Margaret's Bay, 1967.
Techn. Rept., Fish. Res. Bd., Can., 77: 123 pp.

chemistry, chlorophyll
Postma, H., 1954.
Hydrography of the Dutch Waddensea. A study of the relations between water movement, the transport of suspended materials and the production of organic matter. Arch Néerl. Zool., 10(4):405-511, 55 textfigs.

chemistry, chlorophyll
Pratt, D. M., 1949.
Experiments in the fertilization of a salt water pond. J. Mar. Res. 8(1):36-59, 8 textfigs.

chemistry chlorophyll
Qasim, S.Z., and C.V.G. Reddy, 1967.
The estimation of plant pigments of Cochin backwater during the monsoon months.
Bull. mar. Sci., Miami, 17(1):95-110.

chemistry chlorophyll
Ramamurthy, S., 1953.
Measurement of diatom population by pigment extraction method. J. Madras Univ., B, 23(2): 164-173.

chemistry, chlorophyll
Rao, D.V. Subba and T. Platt, 1969.
Optimal extraction conditions of chlorophylls from cultures of five species of marine phytoplankton.
J. Fish. Res. Bd. Can. 26(6): 1625-1630

chlorophyll
Riley, G. A., 1939.
Plankton studies. II. The western North Atlantic, May-June, 1939. J. Mar. Res. 2(2):145-162, textfigs. 49-51, 4 tables.

chemistry chlorophyll determinations
Riley, G. A., 1937.
The significance of Mississippi River drainage for biological conditions in the northern Gulf of Mexico. J. Mar. Res. 1(1):60-74, Textfigs. 19-25.

chemistry chlorophyll
Riley, G.A., and S.A.M. Conover, 1956.
Oceanography of Long Island Sound, 1952-1954. 3. Chemical oceanography.
Bull. Bingham Oceanogr. Coll., 15:47-61.

chemistry chlorophyll
Rodhe, W., 1948
Environmental requirements of fresh-water plankton algae. Experimental studies in the ecology of phytoplankton. Symbolae Botanicae Upsalienses X(1):149 pp., 30 figs.

Chemistry - chlorophyll
Saijo, Yatsuka, 1969.
Chlorophyll pigments in the deep sea. Bull. Jap. Soc. fish. Oceanogr. Spec. No. (Prof. Uda Commem. Pap.): 179-182.

chemistry chlorophyll
Saijo, Y., and S. Ichimura, 1960
Primary production in the northwestern Pacific Ocean.
J. Oceanogr. Soc., Japan, 16(3): 139-145.

chemistry chlorophyll
Saijo, Y., S. Iizuka and O. Asaoka, 1969.
Chlorophyll maxima in Kuroshio and adjacent area.
Mar Biol., 4(3): 190-196.

chemistry chlorophyll (data)
Sakamoto, Ichitaro, and Kanau Matsuike, 1966.
A preliminary report on the primary productivity in the eastern Indian Ocean in winter- IIOE report on productivity, 1963-1964, Umitaka-maru-methods of experiments data, and outlines.
J. Tokyo Univ. Fish., (Spec. ed.) 8(2):173-226.

Chemistry, chlorophyll (data)
San Feliu, J.M. y F. Muñoz 1967.
Hidrografía y fitoplancton de las costas de Castellón, de mayo de 1965 a julio de 1966.
Invest. pesq. Barcelona 31(3): 419-461.

chemistry chlorophyll
Scripps Institution of Oceanography, 1949
Marine Life Research Program. Progress Report to 30 April 1949, 25 pp. (mimeographed), numerous figs. (photo + ozalid). 1 May 1949.

chemistry, chlorophyll
Shabarova, N.T., 1954.
Quantitative determination of chlorophyll in marine plants and contemporary deposits. Biokhim., 19(2):156.

chemistry chlorophyll
Slinn, D.J. and Gaenor Offlow (Mrs. M. Solly) 1969
Chemical constituents in sea water off Port Erin during 1968.
Ann. Rep. mar. biol. Sta. Port Erin 81 (1968): 31-

chemistry chlorophyll
*Sournia, Alain, 1967.
Rythme nycthéméral du rapport "intensité photosynthétique /chlorophylle " dans le plancton marin.
C.r.hebd. Séanc., Acad. Sci. Paris, (D)265(14): 1000-1003.

chemistry - chlorophyll
Spencer, C.P., 1964.
The estimation of phytoplankton pigments.
J. du Conseil, 28(3):327-334.

chemistry chlorophyll
Steele, J.H., 1964.
A study of production in the Gulf of Mexico.
J. Mar. Res., 22(3):211-222.

chemistry chlorophyll
Steele, J.H., 1962
Environmental control of photosynthesis in the sea.
Limnology and Oceanography, 7(2):137-150.

chemistry, chlorophyll
Steele, J.H., 1956.
Plant production on the Fladen Ground. J.M.B.A. 35(1):1-33.

chemistry, chlorophyll
Steele, John H., and I. E. Baird, 1965.
The chlorophyll a content of particulate organic matter in the northern North Sea.
Limnology and Oceanography, 10(2):261-267.

chemistry, chlorophyll
Steele, J.H., and I.E. Baird, 1962.
Carbon-chlorophyll relations in cultures.
Limnol. and Oceanogr., 7(1):101-102.

chemistry, Chlorophyll
Steele, J. H., and I. E. Baird, 1962.
Further relations between primary production, Chlorophyll and particulate carbon.
Limnol. & Oceanogr., 7(1):42-47.

chemistry, chlorophyll
Steele, J.H., and I.E. Baird, 1961
Relations between primary production, chlorophyll and particulate carbon.
Limnol. & Oceanogr., 6(1): 68-78.

chemistry, chlorophyll
Steele, J.H., I.E. Baird and R. Johnston, 1971.
Evidence of upwelling on Rockall Bank. Deep-Sea Res., 18(2): 261-268.

chemistry, chlorophyll
Steemann Nielsen, E., 1962.
On the maximum quantity of plankton chlorophyll per surface unit of a lake or the sea.
Int. Rev. Ges. Hydrobiol., 47(3):333-338.

chemistry, chlorophyll
Steemann Nielsen, E., 1961
Chlorophyll concentration and rate of photosynthesis in Chlorella vulgaris.
Physiol. Plantarum, 14:868-876.

chemistry, chlorophyll
Steemann Nielsen, E., 1940.
Die Productionsbedingungen des Phytoplanktons im Übergangsgebiet zwischen der Nord- und Ostsee.
Medd. Komm. Havundersøgelser, Ser. Plankton, 3(4):55 pp., 17 textfigs., 22 tables.

chemistry, chlorophyll
Stephens, K., R.W. Sheldon and T.R. Parsons, 1967.
Seasonal variations in the availability of food for benthos in a coastal environment.
Ecology, 48 (5):852-855.

chemistry, chlorophyll
Suschenia, L.M., 1961
[The chlorophyll content of plankton in the Aegean, Ionic and Adriatic seas.]
Okeanologiia, Akad. Nauk, SSSR, 1(6):1039-1045.

chemistry, chlorophyll
Sushcheniya, L.M., and Z.Z. Finenko, 1964
On the study of plankton productivity in the tropical area of the Atlantic Ocean. (In Russian).
Okeanologiia, Akad. Nauk, SSSR, 4(5):966-872.

chemistry, chlorophyll (data)
Szekielda, Karl-Heinz 1971.
Organisch gelösten und partikulärer Kohlenstoff in einem Nebenmeer mit starken Salzgehaltsschwankungen (Ostsee).
Vie Milieu Suppl. 22 (2): 579-612

chemistry, chlorophyll (data)
Tanaka, Otohiko, Haruhiko Irie, Shozi IIzuka and Fumihiro Koga, 1961
The fundamental investigation on the biological productivity in the north-west of Kyushu. 1. Rec. Oceanogr. Wks., Japan, Spec. No. 5: 1-58.

chemistry, chlorophyll
Tanaka, Otohiko, Fumihiro Koga, Haruhiko Irie, Shozi Iizuka, Yosie Dotu, Keitaro Uchida, Satoshi Mito, Seiro Kimura, Osame Tabeta, and Sadahiko Imai, 1962.
The fundamental investigation of the biological productivity in the north-western sea area of Kyushu. II. Study on plankton productivity in the waters of Genkai-Nada region.
Rec. Oceanogr. Wks., Japan, Spec., No. 6:1-20.
Also in:
Contrib., Dept. Fish., and Fish. Res. Lab., Kyushu Univ., No. 8.

chemistry - chlorophyll
Taylor, W. Rowland, and Conrad D. Bebelein, 1966.
Plant pigments and light penetration in intertidal sediments.
Helgoländer wiss Meeresunters. 13(3):229-237.

chemistry, chlorophyll
Tyler, John E., 1964
In situ detection and estimation of chlorophyll and other pigments - preliminary results.
Proc. Nat. Acad. Sci., 51(4):671-678.

chemistry, chlorophyll (data)
Uyeno, Fukuzo, 1966.
Nutrient and energy cycles in an estuarine oyster area.
J. Fish. Res. Bd., Can., 23(11):1135-1652.

chemistry - chlorophyll
Uyeno, Fukuzo, 1964.
Relationships between production of foods and oceanographical condition of sea water in pearl farms. II. On the seasonal changes of sea water constituents and of bottom condition, and the effect of bottom cultivation. (In Japanese;
J. Fac. Fish., Pref. Univ. Mie, 6(2):145-169.
English abstract)

chemistry, chlorophyll
Uyeno, Fuxugo, Kyoichi Kawaguchi, Nagao Terada and Tadashi Okada 1970.
Decomposition, effluent and deposition of phytoplankton in an estuarine pearl oyster area.
Rept. Fac. Fish. Prefect. Univ. Mie 7(1):7-41

chemistry, Chlorophyll
Vaccaro, Ralph F., 1963.
Available nitrogen and phosphorus and the biochemical cycle in the Atlantic off New England.
J. Mar. Res., 21(3):284-301.

chemistry, chlorophyll
Vedernikov, V.I., and E.G. Starodubtsev 1971.
Primary production and chlorophyl in the southeastern Pacific. (In Russian; English abstract).
Trudy Inst. Okeanol. P.P. Shirshova, Akad. Nauk SSSR 89: 33-42.
S. America, west

chemistry, chlorophyll
Vinberg, G.G., E.P. Muravleva and Z.Z. Finenko, 1964.
Some data on the chlorophyll content of the plankton and primary production in the Black Sea. (In Russian).
Trudy Sevastopol Biol. Sta., 7:212-220.

chemistry - chlorophyll
Vinogradova, Z.A., G.K. Yatsenko and L.V. Antsupova, 1966.
On the study of the seasonal variability of plankton pigment composition in the northwestern part of the Black Sea. (In Russian; English abstract).
Okeanologiia, Akad. Nauk, SSSR, 6(5):853-860.

chemistry - chlorophyll, s.b.c.
Ward, Ronald W., Valerie Vreeland, Charles H. Southwick and Anthony J. Reading, 1965.
Ecological studies related to plankton productivity in two Chesapeake Bay estuaries.
Chesapeake Science, 6(4):214-225.

chemistry - chlorophyll
Wauthy, B., R. Desrosières et J. Le Bourhis, 1967.
Importance presumée de l'ultraplancton dans les eaux tropicales oligotrophes du Pacifique central sud.
Cah. ORSTOM, Sér. Océanogr., 5(2):109-116.

chemistry, chlorophyll
Wellershaus, Stefan, 1964.
Die Schichtungsverhältnisse im Pelagial des Bornholmbeckens. Über den Jahresgang einiger biotischer Faktoren.
Kieler Meeresf., 20(2):148-156.

Chemistry, chlorophyll
Yentsch, C.S., 1957.
A non-extractive method for the quantitative estimation of chlorophyll in algal cultures.
Nature 179:1302-1304.

chemistry, chlorophyll
Yentsch, Charles S., and David W. Menzel, 1963
A method for the determination of phytoplankton chlorophyll and phaeophytin by fluorescence.
Deep-Sea Res., 10(3):221-231.

chemistry, chlorophyll
Yentsch, Charles S., and Carol A. Reichert, 1963
The effects of prolonged darkness on photosynthesis, respiration and chlorophyll in the marine flagellate, Dunaliella euchlora.
Limnology and Oceanograpy, 8(3):338-342.

chemistry, chlorophyll
Yentsch, C.S., and J.H. Ryther, 1959.
Relative significance of the net phytoplankton and nanoplankton in the waters of Vineyard Sound.
J. du Conseil, 24(2):231-238.

Chlorophyll "a"

chemistry, chlorophyll a
Adams, J.A., and I.E. Baird 1968
Investigations from Aberdeen in 1966: chlorophyll a and zooplankton standing crop.
Annls biol. Copenh. (1966) 23: 92-93.

chemistry - chlorophyll a
Adams, J.A., and I.E. Baird, 1967.
Scottish plankton investigations in the near northern seas 1965, chlorophyll a and zooplankton standing crop.
Annls. biol. Copenh. (1965)22:65-66.

chemistry, chlorophyll a
Adams, J.A., and I.E. Baird, 1966.
Investigations from Aberdeen 1964 in the near northern seas area.
Annls. biol., Copenh., 21:63-65.

chemistry chlorophyll a
Adams, J.A., and I.E. Baird, 1965.
Chlorophyll a, particulate organic carbon and zooplankton standing crop in the northern North Sea.
Ann. Biol., Cons. Perm. Int. Expl. Mer, 1963, 20:91-92.

chemistry - chlorophyll a
Adams, J. A., and I.E. Baird, 1964.
Chlorophyll a, particulate organic carbon and zooplankton.
Ann. Biol., Cons. Perm. Int. Expl. Mer, 1962, 19:63-65.

chemistry, chlorophyll a
*Angot, Michel, 1967.
Rapports entre la concentration en chlorophylle a, le taux d'assimilation du carbone et la valeur de l'energie lumineuse en eau tropicale littorale.
Cah. ORSTOM, Sér. Océanogr., 5(1):39-45.

chemistry chlorophyll a (data)
Apollonio, S., 1961.
The chlorophyll content of Arctic sea ice.
Arctic, 14(3):197-199.

chemistry, chlorophyll a
Australia, Marine Biological Laboratory, Cronulla, 1960.
F.R.V. "Derwent Hunter", scientific report of ----- cruises 10-20/58 -----.
C.S.I.R.O., Div. Fish. & Oceanogr., Rept., 30: 53 pp., numerous figs. (mimeographed).
See author card for complete "title"

chemistry chlorophyll a
Beers, John R. and Sidney S. Herman, 1969.
The ecology of inshore plankton populations in Bermuda. Part I. Seasonal variation in the hydrography and nutrient chemistry. Bull. mar. Sci., 19(2): 253-278.

chemistry chlorophyll a
Blackburn, Maurice, 1966.
Relationship between standing crops at successive trophic levels in the eastern Pacific.
Pacific Science, 20(1): 36-59.

chemistry - chlorophyll a
Burkholder, Paul R., Lillian M. Burkholder and Luis R. Almodóvar, 1967.
Carbon assimilation of marine flagellate blooms in neritic waters of southern Puerto Rico.
Bull. mar. Sci., Miami, 17(1):1-15.

chemistry, chlorophyll a (data)
Endo, Takuo, 1965.
On primary production in the Seto Inland Sea. I. Primary production and hydrographic conditions. (In Japanese; English abstract.)
J. Fac. Fish., Animal Husbandry, Hiroshima Univ. 6(1):85-100.

chemistry, chlorophyll a
English, T. Saunders, 1961.
Biological oceanography in the North Polar Sea from IGY Drifting Station Alpha, 1957-58.
Trans. Amer. Geophys. Union, 42(4):518-525.
Reprinted from:
"Some biological oceanographic observations in the Central North Polar Sea, Drift Station Alpha, 1957-1958".
Arctic Inst., North America, Res. Paper, No. 13.
Also:
Air Force Cambridge Res. Lab. Sci. Rept. (AFCRL-625), No. 15.

chemistry - chlorophyll a
Eppley, Richard W., and P. R. Sloan, 1966.
Growth rates of marine phytoplankton: correlation with light absorption by cell chlorophyll a
Physiologia Pl. 19:47-59.

chemistry chlorophyll a
Establier, R., and R. Margalef, 1964
Fitoplancton e hidrografia de las costas de Cadiz (Barbate), de junio de 1961 a agosto de 1962.
Invest. Pesquera, Barcelona, 25:5-31.

chemistry, chlorophyll a
Forsbergh, Eric D., 1969.
On the climatology, oceanography and fisheries of the Panama Bight. (In Spanish and English).
Inter-Am. Trop. Tuna Comm. 14(2): 385 pp.
Bull.

chemistry chlorophyll a (data)
Forsbergh, Eric D., and James Joseph, 1964
Biological production in the eastern Pacific Ocean. (In English and Spanish).
Inter-American Tropical Tuna Commission, Bull., 8(9):479-527.

chemistry chlorophyll a
Forsbergh, Eric D., and James Joseph. 1963
Phytoplankton production in the south-eastern Pacific.
Nature, 200(4901):87-88.

chemistry, chlorophyll a
*French, C. Stacy, 1967.
Changes with age in the absorption spectrum of chlorophyll a in a diatom.
Archiv für Mikrobiol., 59: 93-103.

chemistry, chlorophyll a
Griffiths, Raymond C., 1965.
A study of ocean fronts off Cape San Lucas, Lower California.
U.S.F.W.S. Spec. Sci. Repts., Fish., No. 499:54 p.

chemistry chlorophyll a
Hansen, Vagn Kr., 1961.
Danish investigations on the primary production and the distribution of chlorophyll a at the surface of the North Atlantic during summer.
Rapp. Proc. Verb., Cons. Perm. Int. Expl. Mer., 149:160-166.

chemistry, chlorophyll a
Holmes, R. W., 1962.
Oceanographic studies during Operation "Wigwam", Marine phytoplankton - areal surveys.
Limnol. and Oceanogr., Suppl. to Vol. 7:xxvii-xxviii.

chemistry chlorophyll a
Holmes, R.W., M.B. Schaefer and B.M. Shimada, 1958
Scope measurements of productivity, chlorophyll a and zooplankton volumes. U.S.F.W.S. Sp. Sci. Rept. Fisheries No. 279 pt. 2: 59-68.

chemistry chlorophyll-a
Ichimura, Shun-ei and Hiromu Kobayashi, 1964
Primary production in Tokyo Bay. (In Japanese; English abstract).
Inform. Bull., Planktol., Japan, No. 11:6-8.

chemistry, chlorophyll a
Japan, Science Council, National Committee for IIOE, 1966.
General report of the participation of Japan in the International Indian Ocean Expedition.
Rec. Oceanogr. Wks., Japan, n.s. 8(2): 133 pp.

chemistry, chlorophyll a
Kerr, J.D., and D.V. Subba Rao, 1966.
Extraction of chlorophyll a from Nitzschia closterium by grinding.
UNESCO, Monogr. Oceanogr. Methodol., 1:65-69.

chemistry chlorophyll-a
Krey, J., D. Hantschmann, and St. Wellershaus, 1959.
Der Sestongehalt entlang eines Schniffes von Kap Farvel bis zur Flämischen Kappe im April und September 1958.
Deutsch. Hydrogr. Zeits., Ergänzungsheft Reihe B (4) No. 3:73-80.

chemistry, chlorophyll a
Kovalevskaya, R.Z., and A.P. Ostapenya, 1966.
Observations of chorophyll a in the marine seston of the surface water. (In Russian; English abstract).
Okeanologiia, Akad. Nauk, SSSR, 6(5):849-852.

chemistry chlorophyll a
*Krishnamurthy, K., 1967.
Some aspects of chemical composition of plankton.
Kieler Meeresforsch., 23(2):99-104.

chemistry chlorophyll a
*Lenz, Jürgen, Heinz Schöne und Bernt Zeitschel, 1967.
Planktonologische Beobachtungen auf einem Schnitt durch die Nordsee von Cuxhaven nach Edinburgh.
Kieler Meeresforsch., 23(2):92-98.

chemistry chlorophyll a
Lenz, Jürgen, und Bernt Zeitschel 1968.
Zur Bestimmung des Extinktions-Koeffizienten für Chlorophyll a in Methanol.
Kieler Meeresforsch. 24(1): 41-50.

chemistry, chlorophyll a concentration
Malone, T.C., 1971.
Diurnal rhythms in netplankton and nannoplankton assimilation ratios. Mar. Biol. 10(4): 285-289.

chemistry - chlorophyll a
*Mandelli, Enrique F., 1967.
Enhanced photosynthetic assimilation ratios in Antarctic polar front (convergence) diatoms.
Limnol. Oceanogr., 12(3):484-491.

chemistry chlorophyll a
Mandelli, Enrique F., and Paul R. Burkholder, 1966.
Primary productivity in the Gerlache and Bransfield straits of Antarctica.
J. Mar. Res., 24(1):15-27.

chemistry chlorophyll a
Margalef, Ramón, 1964.
Modelos experimentales de poblaciones de fitoplancton: nuevas observaciones sobre pigmentos y fijación de carbono inorgánico.
Inv. Pesq., Barcelona, 26:195-203.

chemistry chlorophyll a (data)
Matsudaira, Yasuo, 1964
Cooperative studies on primary productivity in the coastal waters of Japan, 1962-63. (In Japanese; English abstract).
Inform. Bull., Planktol., Japan, No. 11:24-73.

chemistry chlorophyll a
McAllister, C.D., 1962
Data record, photosynthesis and chlorophyll a measurements at Ocean Weather Station "P", July 1959 to November 1961.
Fish. Res. Bd., Canada, Mss. Rept. Ser., (Oceanogr. and Limnol.), No. 126:14 pp. (multilithed).

chemistry, chlorophyll a
Motoda, Sigeru, Teruyoshi Kawamura, Tsuneyoshi Suzuki and Takashi Minoda, 1963.
Photosynthesis of a natural phytoplankton population mainly composed of a cold diatom, Thalassiosira hyalina, in Hakodate harbor, March, 1962.
Bull. Fac. Fish., Hokkaido Univ., 14(3):127-130.

chemistry, chlorophyll a
Niemi, Åke 1971.
Late summer phytoplankton of the Kimito Archipelago (SW coast of Finland).
Merentutkimuslait. Julk. 233:3-17

chemistry, chlorophyll "a"
Odum, H.T., W. McConnell & W. Abbott, 1958.
The chlorophyll "A" of communities.
Publ. Inst. Mar. Sci., 5:65-96.

chemistry - chlorophyll a
Orlando, Aldo Mario, Enrique F. Mandelli y Paul R. Burkholder, 1965.
El fitoplancton antártico y las variables físico-químicas del medio.
Bol Servicio Hidrograf naval, 5(3):201-212

chemistry, chlorophyll a
Owen, R.W., Jr., 1967.
Atlas of July oceanographic conditions in the northeast Pacific Ocean, 1961-64.
Spec. scient. Rep. U.S. Fish. Wildl. Serv., Fish., 549:85 pp.

Chemistry chlorophyll a
Owen, R.W., Jr., 1963.
Northeast Pacific albacore oceanography survey, 1961.
U.S.F.W.S. Spec. Sci. Rept., Fish., No. 444:1-35.

chemistry chlorophyll a
Parsons, T.R., K. Stephens and J.D.H. Strickland, 1961
On the chemical composition of eleven species of marine phytoplankters.
J. Fish. Res. Bd., Canada, 18(6):1001-1016.

chemistry chlorophyll a
Patterson, J., and T.R. Parsons, 1963
Distribution of chlorophyll a and degradation products in various marine materials.
Limnology and Oceanography, 8(3):355-356.

chemistry chlorophylla
Piton, B., et Y. Magnier, 1971.
Sur la détermination de la chlorophylle a dans l'eau de mer côtière tropicale.
Doc. scient. Cent. Nosy-Bé, Off. Rech. scient. techn. Outre-Mer 20: 15 pp. + figs. (mimeographed)

chemistry Chlorophyll a
Saijo, Yatsuka, and Kaoru Takesue, 1965.
Further studies on the size distribution of photosynthesizing phytoplankton in the Indian Ocean.
Jour. Oceanogr. Soc., Japan, 20(6):265-271.

chemistry, chlorophyll a
Small, Lawrence F. and Donald A. Ramberg, 1971.
Chlorophyll a, carbon and nitrogen in particles from a unique coastal environment. In: Fertility of the Sea, John D. Costlow, editor, Gordon Breach, 2: 475-492.

chemistry chlorophyll a
Stevenson, Merritt R. 1970.
On the physical and biological oceanography near the entrance of the Gulf of California, October 1966 - August 1967. (In English and Spanish)
Bull. int. Am. trop. Tuna Commn 14(3): 389-504.

chemistry chlorophyll a
Stroup, E.D. and J.H. Wood 1966.
Atlas of the distribution of turbidity, phosphate and chlorophyll in Chesapeake Bay, 1949-1951.
Chesapeake Bay Inst. Johns Hopkins Univ. Ref. 66-1: 193 pp. (Unpublished manuscript)

chemistry, chlorophyll A
Talling, J.F., and D. Driver, 1963
Some problems in the estimation of chlorophyll -A in phytoplankton.
Proc. Conf., Primary Production Measurements, Marine and Freshwater, Univ. Hawaii, Aug. 21-Sept. 6, 1961, U.S. Atomic Energy Comm., Div. Techn. Info., TID-7633:142-146.

chemistry, chlorophyll a
Travers, A., 1962.
Recherches sur le phytoplancton du Golfe de Marseille. II. Étude quantitative des populations phytoplantoniques du Golfe de Marseille.
Rec. Trav. Sta. Mar. Endoume, Bull., 28(40):70-140.

Chemistry chlorophyll a
Vaccaro, Ralph F., and John H. Ryther, 1960
Marine phytoplankton and the distribution of nitrite in the sea. J. du Cons., 35(3):260-271,

Chemistry chlorophyll a
Walsh, Gerald E., 1966.
Studies of dissolved carbohydrate in Cape Cod waters. III. Seasonal variation in Oyster Pond and Wequaquet Lake, Massachusetts.
Limnol. Oceanogr., 11(2):249-256.

Chemistry chlorophyll-a
Yone, Yasuo, Koji Takeshita, Otohiko Tanaka and Tetsuo Tomiyama, 1964
Primary productivity in Fukuoka Bay. (In Japanese; English abstract).
Inform. Bull., Planktol., Japan, No. 11:9-11.

chlorophyll "a", data only

chemistry, chlorophyll a (data only)
Ballester, Antonio, 1965(1967).
Tablas hidrograficas.
Memoria Soc. Cienc. nat. La Salle, 25(70/71/72):41-137.

chemistry chlorophyll a (data)
Ballester, A., 1965.
Hidrografia y nutriantes de la Fosa de Cariaco.
Informe de Progresso del Estudio Hidrografico de La Fosa de Cariaco, Fundacion, La Salle de Ciencias Naturales Estacion de Investigacion es Marinas de Margarita, Caracas, Sept. 1965. (mimeographed): 3-12.

chemistry - chlorophyll "a" (data only)
Blackburn, M., R.C. Griffiths, R.W. Holmes and W.H. Thomas, 1962.
Physical, chemical and biological observations in the eastern tropical Pacific Ocean: three cruises to the Gulf of Tehuantepec, 1958-1959.
U.S.F.W.S. Spec. Sci. Rept., Fish., No. 420:170pp.

chemistry, chlorophyll a (data only)
Japan, Hokkaido University, Faculty of Fisheries 1970.
Data record of oceanographic observations and exploratory fishing 13: 406 pp.

chemistry chorophyll a (data only)
Japan, Hokkaido University, Faculty of Fisheries, 1968.
Data record of oceanographic observations and exploratory fishing. No. 12:420 pp.

chlorophyll a (data only)
Love, C.M., 1966.
Physical, chemical, and biological data from the northeast Pacific Ocean: Columbia river effluent area, January-June 1963. 6. Brown Bear Cruise 326:13-23 June: CNAV Oshawa Cruise Oshawa-3:17-30 June.
Univ. Washington, Dept. Oceanogr., Tech. Rep., No. 134:230 pp. (Unpublished manuscript).

Chemistry chlorophyll a (data only)
McAllister, C.D., 1962
Data record. Photosynthesis and chlorophyll a measurements at Ocean Weather Station "P", July 1959 to November 1961.
Fish. Res. Bd., Canada, Mss. Rept. Ser. (Ocean. and Limnol.), No. 126:14 pp. (multilithed).

chemistry chlorophyll a (data only)
Owen, R.W., Jr., 1967.
Northeast Pacific albacore-oceanography data, 1962-64.
U.S. Fish Wildl. Serv., Data Rep., 15:47 pp. (microfiche - 1 card).

Chemistry, chlorophyll a (data only)
Saloman, Carl H., and John L. Taylor 1968.
Hydrographic observations in Tampa Bay, Florida and the adjacent Gulf of Mexico, 1965-1966.
Data Rept. U.S. Fish Wildl. Serv. Bur. Comm. Fish. 6 cards (microfiche).

chlorophyll a, diurnal variation
Wood, E.J. Ferguson, and Eugene F. Corcoran, 1966.
Diurnal variation in phytoplankton.
Bull. Mar. Sci., 16(3):383-403.

chlorophyll a and zooplankton
Adams, J.A., and I.E. Barid, 1966.
Investigations from Aberdeen 1964 in the near northern seas area.
Annls. biol., Copenh., 21:63-65.

Chlorophyll a+b

chemistry, chlorophyll a & b
Australia, Commonwealth Scientific and Industrial Organization, 1962.
Oceanographic observations in the Indian Ocean in 1961. H.M.A.S. Diamantina.
Div. Fish. Oceanogr., Cruise DM 1/61:88 pp.

Chlorophyll c.

chemistry chlorophyll c
*Wauthy, Bruno, et Jacques Le Bourhis, 1967.
Sur l'importance relative des chlorophylles a et c dans la composition pigmentaire du phytoplancton en zone tropicale oligotrope.
Cah. ORSTOM, Ser. Océanogr., 5(1):59-64.

Chemistry chlorophyll budget a

*Platt, T. and R.J. Conover, 1971.
Variability and its effect on the 24th chlorophyll budget of a small marine basin. Mar. Biol. 10(1): 52-65.

chlorophyll budget grazing

Chlorophyll concentration

Chemistry, chlorophyll concentrations
Banse, K., and George C. Anderson, 1967.
Computations of chlorophyll concentrations from spectrophotometric readings.
Limnol. Oceanogr., 12(4):696-697.

Chlorophyll (data only)

chemistry, chlorophyll (data only)
Aragno, Federico, Alberto Gomez, Aldo Orlando y Andres J. Lusquiños 1968.
Datos y resultados preliminares de las campañas pesqueria "Pesqueria I" (12 de agosto al 8 de setiembre de 1966)
Publ. (Ser. Informes técn.) Mar del Plata, Argentina 10(1):1-159.

chemistry chlorophyll (data only)
Australia, Commonwealth Scientific and Industrial Research Organization, 1968.
Oceanographical observations in the Pacific Ocean in 1963, H.M.A.S. Gascoyne, Cruise 63/63.
Oceanogr. Cruise Rept., Div. Fish. Oceanogr., 26: 134 pp.

chemistry chlorophyll (data only)
Australia, Commonwealth Scientific and Industrial Research Organization, 1967.
Oceanographical observations in the Indian Ocean in 1962, H.M.A.S. Diamantina Cruise Dm 4/62.
Div. Fish. Oceanogr., Oceanogr. Cruise Rep. 20: 138 pp.

chemistry chlorophyll (data only)
Australia, Commonwealth Scientific and Industrial Research Organization, 1967.
Oceanographical observations in the Pacific Ocean in 1962, H.M.A.S. Gascoyne, Cruise G 1/62.
Div. Fish. Oceanogr., Oceanogr. Cruise Rep. 13: 180 pp.

chemistry - chlorophyll (data only)
Australia, Commonwealth Scientific and Industrial Research Organization, 1967.
Oceanographical observations in the Pacific Ocean in 1961, H.M.A.S. Gascoyne Cruise G 3/61.
Div. Fish. Oceanogr., Oceanogr. Cruise Rep., 12: 126 pp.

chemistry - chlorophyll (data only)
Australia, Commonwealth Scientific and Industrial Research Organization, 1966.
Oceanographical observations in the Indian Ocean in 1962 H.M.A.S. Gascoyne Cruise G 1/62.
Div. Fish. Oceanogr., Oceanogr. Cruise Rep. No. 17: 151 pp.

chemistry, chlorophyll (data only)
Australia, Commonwealth Scientific and Industrial Organization 1965.
Oceanographical observations in the Indian Ocean in 1963, H.M.A.S. Diamantina, Cruise Dm 2/63.
Div. Fish. Oceanogr., Oceanogr. Cruise Rept. No. 24: 153 pp.

Chemistry - chlorophyll (data only)
Australia, Commonwealth Scientific and Industrial Organization 1965.
Oceanographical observations in the Indian Ocean in 1963, H.M.A.S. Diamantina, Cruise Dm 1/63.
Div. Fish. Oceanogr., Oceanogr. Cruise Rept., No. 23: 176 pp.

chemistry chlorophyll (data only)
Australia, Commonwealth Scientific and Industrial Research Organization, 1965.
Oceanographic observations in the Indian Ocean in 1963, H.M.A.S. Gascoyne, Cruise G 1/63.
Div. Fish. and Oceanogr., Oceanogr. Cruise Rept., No. 21:135 pp.

chemistry chlorophyll (data only)
Australia, Commonwealth Scientific and Industrial Research Organiz-tion.
Oceanographic observations in the Indian Ocean in 1963, H.M.A.S. DIAMANTINA Cruise DM 3/63.
Div. Fish. Oceanogr., Oceanogr. Cruise Rept. No. 25: 147 pp.

chemistry - chlorophyll (data only)
Australia, Commonwealth Scientific and Industrial Research Organization, 1964.
Oceanographical observations in the Indian Ocean in 1962, H.M.A.S. Diamantina, Cruise D m 2/62.
Oceanogr. Cruise Rept., Div. Fish. and Oceanogr. No. 15:117 pp.

chemistry - chlorophyll (data only)
Australia, Commonwealth Scientific and Industrial Research Organization, 1964.
Oceanographical observations in the Indian Ocean in 1961, H.M.A.S. Diamantina, Cruise Dm 3/61.
Div. Fish. and Oceanogr., Oceanogr. Cruise Rept., No. 11:215 pp.

chemistry chlorophyll (data only)
Australia, Commonwealth Scientific and Industrial Research Organization, 1963.
Coastal investigations at Port Hacking, New South Wales, 1960.
Div. Fish. and Oceanogr., Oceanogr. Sta. List, No. 52:135 pp.

chemistry chlorophyll (data only)
Australia, Commonwealth Scientific and Industrial Research Organization, 1963
Oceanographical observations in the Indian Ocean in 1961, H.M.A.S. Diamantina Cruise Dm 2/61.
Oceanogr. Cruise Rept., Div. Fish. and Oceanogr., No. 9:155 pp., 14 figs.

chemistry chlorophyll (data only)
Australia, Commonwealth Scientific and Industrial Research Organization, 1963.
Oceanographical observations in the Pacific Ocean in 1961, H.M.A.S. Gascoyne, Cruise G 1/61.
Oceanogr. Cruise Rept., Div. Fish. and Oceanogr., No. 8:130 pp., 12 figs.

chemistry chlorophyll (data only)
Australia, Commonwealth Scientific and Industrial Research Organization, Division of Fisheries and Oceanography, 1963.
Oceanographical observations in the Indian Ocean in 1960, H.M.A.S. Diamantina, Cruise Dm 2/60.
Oceanographical Cruise Report No. 3:347 pp.

chemistry, chlorophyll (data only)
Australia, Commonwealth Scientific and Industrial Research Organization, Division of Fisheries and Oceanography, 1962
Oceanographic observations in the Pacific Ocean in 1960, H.M.A.S. Gascoyne, Cruises G 1/60 and G 2/60.
Oceanographical Cruise Report No. 5: 255 pp.

chemistry Chlorophyll (data only)
Australia, Commonwealth Scientific and Industrial Organization, 1962.
Oceanographical observations in the Indian Ocean in 1959, H.M.A.S. Diamantina, Cruises Dm 1/59 and Dm 2/59.
Oceanogr. Cruise Rept., Div. Fish. and Oceanogr., No. 1:134 pp.

chemistry, chlorophyll (data only)
Australia, Commonwealth Scientific and Industrial Research Organization, 1961
Oceanic investigations in Eastern Australian waters, F.R.V. "Derwent Hunter", 1959.
Div. Fish. and Oceanogr., Oceanogr. Sta. List 48: 84 pp.

chemistry - chlorophyll (data only)
Australia, Commonwealth Scientific and Industrial Research Organization, 1960
Oceanic observations in Antarctic waters, M.V. Magga Dan, 1959.
Div. Fish. and Oceanogr., Oceanogr. Sta. List, 44: 78 pp.

chemistry chlorophyll (data only)
Australia, Commonwealth Scientific and Industrial Research Organization, 1960.
Oceanic investigations in eastern Australia, H.M.A. Ships Queenborough, Quickmatch and Warrego, 1958.
Div. Fish. and Oceanogr., Oceanogr. Sta. List, 43:57 pp.

chlorophyll (data only)
Australia, Commonwealth Scientific and Industrial Research Organization, 1960.
Coastal investigations at Port Hacking, New South Wales, 1958.
Oceanogr. Sta. List, Div. Fish. & Oceanogr., 42: 99 pp.

chemistry, chlorophyll (data only)
Australia, Commonwealth Scientific and Industrial Organization, 1960
Oceanic investigations in eastern Australian waters, F.R.V. Derwent Hunter, 1958. Oceanogr. Sta. List., Div. Fish. and Oceanogr., No. 41: 232 pp.

Chemistry Chlorophyll (data only)
Australia Commonwealth and Industrial Research Organization, 1957.
Onshore and oceanic hydrological investigations in eastern and south-western Australia, 1956.
Oceanogr. Sta. List, Div. Fish. & Oceanogr. 30:79pp.

chemistry chlorophyll (data only)
Chesapeake Bay Institute, 1954.
Data Report 23, Patuxent River Winter Cruise, 3 December- 7 December 1952. Ref. 54-10:1-44, 1 textfig.

chlorophyll (data only)
*Colton, John B., Jr., Robert R. Marak, Samuel R. Nickerson and Ruth R. Stoddard, 1968.
Physical, chemical and biological observations on the continental shelf, Nova Scotia to Long Island, 1964-1966.
U.S. Fish Wildl. Data Rept., 23: 195 pp. on 3 microfiche.

chemistry, chlorophyll (data only)

Japan, Hokkaido University, Faculty of Fisheries, 1968.
The Oshoro Maru cruise 23 to the east of Cape Erimo, Hokkaido, April 1967.
Data Record Oceanogr. Obs. Expl. Fish., 12: 115-169.

chemistry chlorophyll (data only)

Japan, Hokkaido University, Faculty of Fisheries, 1967.
The Oshoro Maru cruise 16 to the Great Australian Bight November 1965-February 1966.
Data Record Oceanogr. Obs. Explor. Fish., Fac. Fish., Hokkaido Univ. 11: 1-97; 113-119.

chemistry - chlorophyll (data only)
Japan, Hokkaido University, Faculty of Fisheries, 1967.
Data record of oceanographic observations and exploratory fishing, 11:383 pp.

chemistry - chlorophyll (data only)

Japan, Hokkaido University, Faculty of Fisheries, 1967.
The Oshoro Maru cruise 18 to the east of Cape Erimo, Hokkaido, April 1966.
Data Record Oceanogr. Obs. Explor. Fish. 11: 121-164.

chemistry, chlorophyll (data only)
Japan, Hokkaido University, Faculty of Fisheries, 1957
Data record of oceanographic observations and exploratory fishing, No. 1:247 pp.

chemistry - chlorophyll (data only)
McGary, James W., and Joseph J. Graham, 1960.
Biological and oceanographic observations in the central north Pacific, July-September 1958.
U.S.F.W.S., Spec. Sci. Rept., Fish., No. 358: 107 pp.

chlorophyll (data)
Nehring, D. and H. J. Brosin 1968
Ozeanographische Beobachtungen im äquatorialen Atlantik und auf dem Patagonischen Schelf während der 1. Südatlantik Expedition mit dem Fischereiforschungsschiff Walther Herwig von August bis Dezember 1966.
Geod. Geoph. Veröff. 4 (3): 93 pp.

chemistry, chlorophyll (data only)
Platt, Trevor, and Brian Irwin 1971.
Phytoplankton production and nutrients in Bedford Basin 1969-70.
Techn. Rept. Fish. Res. Bd. Can. 247: 172 pp. (multilithed)

chemistry, chlorophyll (data only)
Platt, Trevor, and Brian Irwin 1970.
Primary productivity measurements in St. Margaret's Bay 1968-1970.
Techn. Rept. Fish. Res. Bd. Can. 203: 67 pp. (multilithed).

chemistry, chlorophyll (data only)
Valdez, Alberto J., Alberto Gomez, Aldo Orlando y Andres J. Lusquiños 1968.
Datos y resultados de las campañas pesquería "Pesquería IV" (7 de junio al 4 de julio de 1967).
Publ. (Ser. Informes tecn.) Mar del Plata

chemistry, chlorophyll (data only)
Villanueva, Sebastian F., Alberto Gomez, Aldo Orlando y Andres J. Lusquiños, 1969.
Datos y resultados de las campañas pesquería "Pesquería III", 16 de febrero al 1° de marzo de 1968.
Publ. Ser. Inform. tecn. Proyecto Desarrollo Pesq., Mar del Plata (10/III): unnumbered pp.

chemistry - chlorophyll decomposition

Moreth, Clarice M. and Charles S. Yentsch, 1970.
The role of chlorophyllase and light in the decomposition of chlorophyll from marine phytoplankton. J. exp. mar. Biol. Ecol., 4(3): 238-249.

chemistry, chlorophyll (decomposition of)
Yentsch, Charles S. 1967.
The measurement of chloroplastic pigments - Thirty years of progress?
In: Chemical environment in the aquatic habitat, H. L. Golterman and R. S. Clymo, editors, Proc. I.B.P. Symp. Amsterdam, Oct 1966: 255-270.

chlorophyll derivatives

chemistry, chlorophyll derivatives
Orr, W.L., K.O. Emery and J.R. Grady, 1958.
Preservation of chlorophyll derivatives in sediments off southern California.
Bull. Amer. Assoc. Petr. Geol., 42(5):925-962.

chemistry chlorophyll derivatives

Sugimura, Yukio, 1961
Geochemical studies on Recent sediments. IV. Chlorophyll degradation products in surface muds from the Kagoshima Bay, Japan.
J. Oceanogr. Soc., Japan, 17(1):10-14.

chlorophyll, intraspecific variation
Margalef, Ramón, 1961.
Variaciones intraspecíficas de los pigmentos asimiladores en cloroficeas y fanerógamos acuaticos.
Inv. Pesq., Barcelona, 19:111-118.

chemistry, chlorophyll a, monthly means
Slinn, D.J., and Gaenor Offlow (Mrs. M. Solly), 1968.
Chemical constituents of sea water off Port Erin during 1967.
Ann. Rept. mar. biol Stn. Port Erin, 80:37-42.

chemistry, chlorophyll (particulate)
Small, Lawrence F., and Herbert Curl, Jr., 1968.
The relative contribution of particulate chlorophyll and river tripton to the extinction of light off the coast of Oregon.
Limnol. Oceanogr., 13(1):84-91.

chemistry - chlorophyll variations

Nival, P., 1971.
Problèmes posés par la conception d'un modèle de variations annuelles de la chlorophylle en un point d'une zone côtière (Villefranche-sur-Mer). Investigación pesq. 35(1): 351-360.

chemistry, chlorophyllase

Barrett, J. and S.W. Jeffrey, 1971.
A note on the occurrence of chlorophyllase in marine algae. J. exp. mar. Biol. Ecol. 7(3): 255-262.

chlorophyllase, effect of

Moreth, Clarice M. and Charles S. Yentsch, 1970.
The role of chlorophyllase and light in the decomposition of chlorophyll from marine phytoplankton. J. exp. mar. Biol. Ecol., 4(3): 238-249.

Cholesterol

chemistry, cholesterol
Kaneda, T., 1963.
The effects of marine animal oils and other marine products on cholesterol. (In Japanese).
Bull. Jap. Soc. Sci. Fish., 29(4):387-394.

chemistry cholesterol
Schwendinger, Richard B., and J. Gordon Erdman 1964
Sterols in Recent aquatic sediments.
Science, 144(3626):1575-1576.

chemistry, chromatophorotropins

Fingerman, Milton, Clelmer K. Bartell and Robert A. Krasnow, 1971.
Comparison of chromatophorotropins from the horseshoe crab Limulus polyphemus and the fiddler crab, Uca pugilator. Biol. Bull. mar. biol Lab. Woods Hole, 140(3): 376-388.

Chromium

chemistry - chromium
Chester, R., 1965.
Elemental geochemistry of marine sediments.
In: Chemical oceanography, J.P. Riley and G. Skirrow, editors, Academic Press, 2:23-80.

chemistry - CHROMIUM
Chipman, W.A., 1966.
Uptake and accumulation of Chromium-51 by the clam, Tapes decussatus, in relation to physical and chemical form.
Disposal of Radioactive Wastes into Seas, Oceans and Surface Waters, IAEA (SM-72/35): 571-582.

chemistry, CHROMIUM
Chipman, W.A., 1966.
Some aspects of the accumulation of ^{51}Cr by marine organisms.
In: Radioecological Concentration Processes, Pergamon Press, 931-942.

chemistry - chromium
Curl, H.C., Jr., Norman Cutschall and Charles Osterberg, 1965.
Uptake of chromium (III) by particles in sea water.
Nature, 205(4968):275-276.

chemistry chromium
Elderfield, H., 1970.
Chromium speciation in sea water.
Earth Planet. Sci. Letts., 9(1): 10-16.

chemistry, chromium

Fonselius, Stig H., 1970.
Some trace metal analyses in the Mediterranean, the Red Sea and the Arabian Sea.
Bull. Inst. océanogr., Monaco 69 (1407): 15 pp.
Also: IAEA Radioactivity in the Sea, Publ. 29.

chemistry, chromium

Fukai, R., 1967.
Valency state of chromium in seawater.
Nature, Lond., 213(5079):901.

Chemistry - chromium

Fukai, Rinnosuke, and Daniele Broguet, 1965.
Distribution of chromium in marine organisms.
Bull. Inst. Océanogr., Monaco, 65(1336):19 pp.

Chemistry, chromium

Fukai, Rinnosuke, and Lang Huynh-ngoc, 1968.
Studies on the chemical behaviour of radionuclides in sea-water. 1. General consideration and study of precipitation of trace amounts of chromium, manganese, iron, cobalt, zinc and cerium.
IAEA Radioactivity in the sea, Publ. 22:1-26.

chemistry, chromium

Fukai, Rinnosuke, and Daniele Vas 1969.
Changes in the chemical forms of chromium on the standing of sea-water samples.
J. oceanogr. Soc., Japan, 25(2): 109-111

chemistry - chromium

Fukai, Rinnosuke, and Daniele Vas, 1968
A differential method of analysis for trivalent and hexavalent chromium in sea water.
J. oceanogr. Soc., Japan, 23(6): 298-305

Chemistry chromium

Ishibashi, M., 1953.
Studies on minute elements in sea water.
Rec. Ocean. Wks., Japan, 1(1):88-92.
n.s.

chemistry, chromium

Ishibashi, M., and T. Shigematsu, 1950.
Determination of chromium in sea water.
Bull. Inst. Chem. Res., Kyoto Univ., 23:59-60.

Abstr.: Chem. Abstr. 11024g, 1952.

chemistry - chromium

Ishibashi, Masayoshi, Shunzo Ueda and Yoshikazu Yamamoto, 1970.
The chemical composition and the cadmium, chromium and vanadium contents of shallow-water deposits in Tokyo Bay. (In Japanese; English abstract)
J. oceanogr. Soc. Japan 26(4): 189-194

chemistry - chromium, effect of

Raymont, J.E.G., and J. Shields, 1963.
Toxicity of copper and chromium in the marine environment.
Int. J. Air Water Poll., 7(4/5):435-443.

chemistry - chromium

Turekian, Karl E., and Donald F. Schutz, 1965.
Trace element economy in the oceans.
Narragansett Mar. Lab., Univ. Rhode Island, Occ. Publ., No. 3:41-89.

chemistry, chromium

Ueda, S., 1957.
Chemical studies on the ocean. 68. Chemical studies of the shallow-water deposits. 21. Vanadium and Chromium contents of the shallow water deposits. (1)(2).
J. Oceanogr. Soc., Japan, 13(3):93-98; 99-106.

chemistry cobalt

Ishibashi, Masayoshi, Toshio Yamamoto and Tetsuo Fukita, 1964
Chemical studies on the ocean (Part 92). Chemical studies on the seaweeds (17). Cobalt content in seaweeds.
Rec. Oceanogr. Wks., Japan, 7(2):17-24.

chemistry cobalt

Ishibashi, Masayoshi, Shunzo Ueda, and Yoshikazu Yamamoto, 1964
Studies on the utilization of the shallow-water deposits (continued). On the cobalt and nickel contents of the shallow-water deposits.
Rec. Oceanogr. Wks., Japan, 7(2):37-42.

chemistry cobalt

Johnston, R., 1964
Sea water, the natural medium of phytoplankton. II. Trace metals and chelation, and general discussion.
J. Mar. Biol. Assoc., U.K., 44(1):87-109.

chemistry, cobalt

*Presley, B.J., R.R. Brooks and I.R. Kaplan, 1967.
Mangenese and related elements in the interstitial water of marine sediments.
Science, 158(3803):906-910.

Chromium-51

chemistry, chromium-51

Cutshall, Norman, Vermon Johnson and Charles Osterberg, 1966.
Chromium-51 in sea water: chemistry.
Science, 152(3719):202-203.

Chemistry - chromium-51

Osterberg, Charles, Norman Cutshall and John Cronin, 1965.
Chromium-51 as a radioactive tracer of Columbia River water at sea.
Science, 150(3703):1585-1586.

chemistry chromium-51

Osterberg, C., L.D. Kulm, and J.V. Byrne, 1963.
Gamma emitters in marine sediments near the Columbia River.
Science, 139(3558):916-917.

Chemistry chromium-51

Preston, Eric M., 1971.
The importance of ingestion in chromium-51 accumulation by Crassostrea virginica (Gmelin).
J. exp. mar. Biol. Ecol., 6(1): 47-54.

chemistry, chromium-51

Tennant, David A., and William O. Forster 1969.
Seasonal variation and distribution of ^{65}Zn, ^{54}Mn and ^{51}Cr in tissues of the crab Cancer magister Dana.
Health Physics 18: 649-657.
Also in: Coll. Repr. Dept. Oceanogr. Oregon State Univ. 9(2)(1970).

Chromolipoids

chemistry - chromolipoids

Liaci, Lidia, 1964.
Pigmenti e steroli neglo invertebrato marini.
Archivio Zool. Ital., 49:281-300.

chemistry coal-tar derivatives, effect of

Powell, N.A., C.S. Sayce and D.F. Tufts, 1970.
Hyperplasia in an estuarine bryozoan attributable to coal tar derivatives.
J. Fish. Res. Bd, Can., 27(11): 2095-2096.

Cobalamin

Chemistry cobalamines

Benghitsky, A.G., L.G. Gutveib, and M.N. Lebedeva, 1970
Synthesis of cobalamines with bacteria isolated from the digestive organs of Black Sea fish. (In Russian)
Gidrobiol. Zh. 6(5): 72-75

chemistry cobalamin

Droop, M.R., 1955.
A pelagic marine diatom requiring cobalamin.
J.M.B.A. 34(2):229-231.

chemistry, cobalamin

Dyrssen, David, 1969.
Stoichometry and chemical equilibrium. In: Chemical oceanography, Rolf Lange, editor, Universitetsforlaget, Oslo: 47-59.

Chemistry, cobalamins

Parker, Bruce C. 1969.
Influence of method for removal of seston on the dissolved organic matter II Cobalamins.
J. Phycol. 5(2): 124-127.

Cobalt

CHEMISTRY, COBALT

Ashton, S., R. Chester and L.R. Johnson 1972.
Uptake of cobalt from sea water by aeolian dust.
Nature, Lond., 235 (5338): 380-381

Chemistry cobalt

Bernhard, M., 1964.
Chemical composition and the radiocontamination of marine organisms.
Proc. Symp. Nuclear Detonations and Marine Radioactivity, S.H. Small, editor, 137-150.

chemicals, cobalt

* Bhatt, Y.M., V.N. Sastry, S.M. Shah and T.M. Krishnamoorthy, 1968.
Zinc, manganese and cobaly contents of some marine bivalves from Bombay.
Proc. nat. Inst. Sci., India (B), 34(6): 283-287.

Chemistry, cobalt

Burns, Roger G., 1965.
Formation of cobalt (III) in the amorphous FeOOH . H2O phase of manganese nodules.
Nature, 205(4975):999.

Chemistry cobalt

Chester, R., 1965.
Adsorption of zinc and cobalt on illite in sea-water.
Nature, 206(4987):884-886.

chemistry cobalt
Chester, R., 1965.
Elemental geochemistry of marine sediments.
In: Chemical oceanography, J.P. Riley and G. Skirrow, editors, Academic Press, 2:23-80.

Chemistry Cobalt
Duursma, E. K., and W. Sevenhuysen, 1966.
Note on chelation and solubility of certain metals in sea water at different PH values.
Netherlands J. Sea Res., 3(1):95-106.

Chemistry - COBALT
Fukai, Rinnosuke, 1968.
Distribution of cobalt in marine organisms.
IAEA Radioactivity in the Sea, Publ. 23:1-19.

Chemistry, cobalt
Fukai, Rinnosuke and Lang Huynh-ngoc 1968.
Studies on the chemical behaviour of radionuclides in sea-water. I. General consideration and study of precipitation of trace amounts of chromium, manganese, iron, cobalt, zinc and cerium.
IAEA Radioactivity in the sea, Publ. 22:1-26.

Chemistry - cobalt
Fukai, R., L. Huynh-Ngoc and D. Vas 1966.
Determination of trace amounts of cobalt in sea-water after enrichment with solid manganese dioxide.
Nature, Lond. 211 (5050): 726-729.

Chemistry cobalt
Hiyama, Yoshio, and Junko Matsubara Khan, 1964
On the concentration factors of radioactive I, Co, Fe and Ru in marine organisms.
Rec. Oceanogr. Wks., Japan, 7(2):79-106.

chemistry cobalt
Hood, D.W., 1963
Chemical oceanography. In: Oceanography and Marine Biology, H. Barnes, Edit., George Allen & Unwin, 1:129-155.

Chemistry - cobalt
Hutchinson, G.E., R.J. Benoit, W.B. Cotter and P.J. Wangersky, 1955.
On the nickel, cobalt and copper contents of deep-sea sediments. Proc. Nat. Acad. Sci. 41(3):160-162.

Chemistry, cobalt
Ishibashi, M., 1953.
Studies on minute elements in sea water.
Rec. Ocean. Wks., Japan, n.s., 1(1):88-92.

Chemistry, cobalt
Ostroumov, E.A., and I.I. Volkov, 1962.
Separation of titanium, zirconium and thorium from manganese, nickel, cobalt and zinc by means of cinnamic acid.
Trudy Inst. Okeanol., Akad. Nauk, SSSR, 54:170-181.
In Russian; English summary

Chemistry - cobalt
Parker, Patrick L., Ann Gibbs and Robert Lawler, 1963.
Cobalt, iron and manganese in a Texas bay.
Publ. Inst. Mar. Sci., Port Aransas, 9:28-32.

chemistry cobalt
Robertson, D.E., 1970.
The distribution of cobalt in oceanic waters.
Geochim. cosmochim. Acta 34(5): 553-567

Chemistry, cobalt
Rozhanskaya, L.I. 1966.
Some data on the cobalt content in the Red Sea and Gulf of Aden. (In Russian)
Gidrobiol. Zh. 2(2): 40-42

Chemistry, cobalt
Rozhanskaia, L.I., 1963.
Distribution of cobalt in the waters of the Black and Azov seas. (In Russian).
Trudy Sevastopol Biol. Sta., 16:467-471.

cobalt
Schutz, Donald F., and Karl K. Turekian, 1965.
The distribution of cobalt, nickel, and silver in ocean water profiles around Pacific Antarctica.
J. geophys. Res., 70(22):5519-5528.

chemistry, cobalt
Schutz, D.F., and K.K. Turekian, 1964.
The distribution of selenium, antimony, silver, cobalt and nickel in sea water. (Abstract).
Trans. Amer. Geophys. Union, 45(1):118.

Chemistry, cobalt
Tatsumoto, M., 1957.
Chemical investigations of deep-sea deposits. 22. The contents of cobalt and nickel in sea sediments (2). 23. The contents of tin and lead in sea sediments. J. Chem. Soc., Japan, 78(1):38-48.

Chemistry, cobalt
Tatsumoto, M., 1956.
Chemical studies on the deep-sea deposits. 21. The contents of cobalt and nickel in sea sediments (1). J. Chem. Soc., Japan, 77(11):1637-1642.

chemistry, cobalt
Topping, Graham, 1969.
Concentrations of Mn, Co, Cu, Fe, and Zn in the northern Indian Ocean and Arabian Sea. J. mar. Res., 27(3): 318-326.

chemistry, cobalt
Turekian, Karl K., 1971
2. Rivers, tributaries and estuaries. In: Impingement of man on the oceans, D.W. Hood, editor, Wiley Interscience, 9-73.

Chemistry, cobalt
Turekian, Karl K. 1968.
Deep-sea deposition of barium, cobalt and silver.
Geochim. Cosmochim. Acta 32(1): 603-612.

Chemistry, cobalt
Turekian, Karl K., and Donald F. Schutz, 1965.
Trace element economy in the oceans.
Narragansett Mar. Lab., Univ. Rhode Island, Occ. Publ., No. 3:41-89.

chemistry, cobalt
Turekian, Karl K., Donald F. Schultz and David Johnson, 1966.
The distribution of Sr, Ba, Co, Ni, and Ag in ocean water profiles of the Pacific sector of the Antarctic seas.
Antarctic J., United States, 1(5):224.

Chemistry - cobalt
Thompson, T.G. and Taivo Laevastu, 1960
Determination and occurrence of cobalt in sea water. J. Mar. Res. 18(3): 189-192.

Chemistry, cobalt
Veeh, H. Herbert, and Karl K. Turekian 1968.
Cobalt, silver and uranium concentrations of reef-building corals in the Pacific Ocean.
Limnol Oceanogr. 13(2): 304-308.

Chemistry cobalt
Weiss, Herbert V., and John A. Reed, 1960
Determination of cobalt in sea water. J. Mar. Res. 18(3): 185-188.

Chemistry cobalt
Yatsimirsky, K.B., E.M. Emel'yanov, V.K. Pavlova, and Ya. S. Savichenko, 1970.
Determination of microquantities of nickel and cobalt in small weighed portions of marine suspended matter (based on the data from the Baltic Sea and the Atlantic Ocean). (In Russian; English abstract). Okeanologiia, 10(6):1111-1116.

Chemistry, cobalt
Young, R.S., 1957.
The geochemistry of cobalt.
Geochimica et Cosmochimica Acta, 13(1):28-41.

Cobalt-60

Chemistry, cobalt-60
Folsom, Theodore R., D. R. Young and L.E. Finnin, 1965.
Sum coincidence gamma-ray spectrometry in tracing cobalt-60 and silver-110 in marine organisms.
In: Radioisotope sample measurement techniques in medecine and biology, Internat. Atom. Energy Agency, 1965: 57-69.

chemistry, cobalt-60
Parker, Patrick L., 1966.
Movement of radioisotopes in a marine bay: cobalt-60, iron-59, manganese-54, zinc-65, sodium-22.
Publs. Inst. mar. Sci. Univ. Texas, Port Aransas, 11:102-107.

Co-enzymes

chemistry, co-enzymes
Cooper, L.H.N., 1965.
Chemistry of the sea. 2. Organic.
Chemistry in Britain, 1965, 1:150-154.

Conchiolin

Chemistry, conchiolin
Wada, Koji, 1967.
Studies on the mineralization of the calcified tissue in molluscs. XIV. Modification of the amino acid pattern of proteins in the extrapallial fluid during the process of formation and mineralizations of nacreous conchiolin in some bivalves.
Bull. Jap. Soc. scient. Fish, 33(7):613-617.

Cu

Chemistry, copper
Alexander, J.E. and E.F. Corcoran 1967.
The distribution of copper in tropical seawater.
Limnol. Oceanogr. 12(2): 236-242.

Chemistry, copper
Atkins, W.R.G., 1957.
The direct estimation of ammonia in sea water, with notes on nitrate, copper, zinc and sugars.
J. du Cons., 22(3):271-277.

chemistry, copper
Atkins, W.R.G., 1953.
The seasonal variation in the copper content of sea water. J.M.B.A. 31(3):493-494, 1 textfig.

Chemistry, copper
Barnes, H., 1946.
The estimation in sea-water solutions of micro-quantities of copper by means of dithizone.
J.M.B.A. 26:303-311.

Chemistry, copper
Barnes, H., and K.A. Pyefinch, 1947.
Copper in diatoms. Nature, 160 (4055):97.

chemistry, copper
*Barth, Rudolf, Oswaldo T. Godoy e Gilberto Hauila, 1967.
Observacões em nanoplancton e concentracão de cobre na corrente do Brasil.
Inst. Pesquisas Marinha, Brasil, Publ. 003: 11 pp (multilithed).

chemistry, copper
Beck, A.B., and K. Sheard, 1949.
Copper and nickel content of the blood of the West Australian crayfish, Panulirus longipes (Milne Edwards) and of sea weeds. Australian J. Exp. Biol. Med. Sci. 27:307-312.

chemistry, copper
Bernhard, M., 1964.
Chemical composition and the radiocontamination of marine organisms.
Proc. Symp. Nuclear Detonations and Marine Radioactivity, S.H. Small, editor, 137-150.

Chemistry, copper
Bougis, P., 1962.
Le cuivre en ecologie marine. Problemi ecologici delle zone litorali del Mediterraneo 17-23 luglio 1961.
Pubbl. Staz. Zool., Napoli, 32 (Suppl.):497-514.

chemistry, copper
Bougis, P., 1962.
Données sur la teneur en cuivre des eaux de mer littorales des environs de Roscoff en août 1961.
Cahier de Biologie Marine, 3(3):317-323.

chemistry, copper, effect of
Bougis, P., 1959.
Sur l'effet biologique du cuivre en eau de mer.
Trav. Sta. Zool., Villefranche-sur-Mer, 18(13).

Reprinted from: C.R. Acad. Sci., Paris, 249(2): 326-328.

Chemistry, copper
Bregant, Davide 1965.
La distribuzione del Rame nel Tirreno.
Rapp. P.-v. Réun. Commn int. Explor. scient. Mer Méditerr. 18(3):729-731.

Chemistry, Copper
Bregant, Davide, 1961.
Elementi oligodinamici: Rame. Talassografia Adriatica, A.G.I. 1959-1960.
Consiglio Naz. delle Ricerche, Commissione Naz. Ital. per la Cooperazione Geofisica Internaz., Pubbl., No. 3:3-6.

Also:
Ist. Sperimentale Talassogr., Trieste, Pubbl., No. 372. (1961).

Chemistry, copper
Brooks, R.R.,
Trace elements in New Zealand coastal waters.
Geochimica et Cosmochimica Acta, 29(12): 1369-1370.

Chemistry, copper
Bryan, G.W. 1968.
Concentrations of zinc and copper in the tissues of decapod crustaceans.
J. mar. biol. Ass. U.K. 48(2): 303-321.

Chemistry copper
Bryan, G.W., 1964
Zinc regulation in the lobster Homarus vulgaris.
1. Tissue zinc and copper concentrations.
J. Mar. Biol. Assoc., U.K., 44(3):549-563.

Chemistry, copper, effect of
Bryan, G.W. and L.G. Hummerstone, 1971.
Adaptation of the polychaete Nereis diversicolor to estuarine sediments containing high concentrations of heavy metals. 1. General observations and adaptation to copper. J. mar. biol. Ass. U.K. 51(4): 845-863.

chemistry, copper
Chester, R., 1965.
Elemental geochemistry of marine sediments.
In: Chemical oceanography, J.P. Riley and G. Skirrow, editors, Academic Press, 2:23-80.

Chemistry, copper
Chierego, N., 1955.
La determinazione del rame in merteriale biologico e nelle acque. Arch. Ocean. e Limnol., 10(3): 198-199.

Pubbl. Ist. Sper. Talassogr., Trieste, No. 322.

Chemistry, copper
*Choe, Sang, Tai Wha Chung and Hi-Sang Kwak, 1968.
Seasonal variations in nutrients and principal ions contents of the Han River water and its water characteristics. (In Korean; English abstract).
J. oceanogr. Soc., Korea, 3(1):26-38.

Chemistry, copper
Chow, T.J., and T.G. Thompson, 1954.
Seasonal variations in the concentration of copper in the surface waters of San Juan Channel, Washington. J. Mar. Res. 13(3):233-244, 2 textfig

Chemistry, copper
Chow, T.J., and T.G. Thompson, 1952.
The determination and distribution of copper in sea water. 1. The spectrophotometric determination of copper in sea water. J. Mar. Res. 11(2): 124-138, 3 textfigs.

Chemistry, copper
Corcoran, E.F., and J.E. Alexander, 1964.
The distribution of certain trace elements in tropical sea water and their biological significance.
Bull. Mar. Sci., Gulf and Caribbean, 14(4):594-601.

Chemistry Copper
Decleir, W., J. Lemaire and A. Richard, 1970.
Determination of copper in embryos and very young specimens of Sepia officinalis. Mar. Biol., 5(3): 256-258.

chemistry, copper (data only)
Dragovich, Alexander, John H. Finucane and Billie Z. May, 1961.
Counts of red tide organisms, Gymnodinium breve, and a associated oceanographic data from Florida west coast, 1957-1959.
USFWS Spec. Sci. Rept., Fish., No. 369:175 pp.

chemistry copper (data)
Dragovich, A., and B.Z. May, 1962
Hydrological characteristics of Tampa Bay tributaries.
U.S.F.W.S. Fish. Bull. (205) 62: 163-176.

Chemistry - copper
Dragovich, Alexander, and James E. Sykes 1967.
Oceanographic atlas for Tampa, Florida, and adjacent waters of the Gulf of Mexico, 1958-61.
Circular Bur. Comm. Fish. U.S. Fish Wildl. Serv. 255: 466 pp. (quarto)

Chemistry Copper
Duursma, E. K., and W. Sevenhuysen, 1966.
Note on chelation and solubility of certain metals in sea water at different PH values.
Netherlands J. Sea Res., 3(1):95-106.

chemistry, copper
Establier, R. 1969.
Estudios del contenido en cobre del agua de mar y ostiones (Crassostrea angulata) de las costas de Cádiz. Investigación pesq. 33(1): 69-86.

Chemistry, copper
Finucane, John H., and Alexander Dragovich 1959
Counts of red tide organisms, Gymnodinium brevis and associated oceanographic data from Florida west coast, 1954-1957.
USFWS Spec. scient. Rept. Fish. No. 289: 220 pp.

chemistry, copper
Fonselius, Stig H., 1970.
Some trace metal analyses in the Mediterranean, the Red Sea and the Arabian Sea.
Bull. Inst. océanogr., Monaco, 69 (1407): 15 pp.
Also: IAEA Radioactivity in the Sea, Publ. 29.

Chemistry copper
Fonselius, Stig H., and Folke Koroleff, 1963.
Copper and zinc content of the water in the Ligurian Sea.
Bull. Inst. Oceanogr., Monaco, 61(1281):1-15.

chemistry, copper
Foster, P. and A.W. Morris, 1971.
The seasonal variation of dissolved ionic and organically associated copper in the Menai Straits. Deep-Sea Res., 18(2): 231-236.

chemistry, copper

Galtsoff, P.S., ed., 1954.
Gulf of Mexico, its origin, waters and marine life
Fish. Bull., Fish and Wildlife Service, 55:1-604, 74 textfigs.

Chemistry, copper

Galtsoff, P. S., 1943.
Copper content of sea water. Ecol., 24(2): 263-265.

chemistry copper

Godoy, Oswaldo T., e Rudolf Barth, 1967.
Concentração de cobre na agua e sua influência sôbre o plâncton. (In Portuguese; English summary)
Publicação, Inst. Pesquisas marinha, Rio de Janeiro, 01/1967: 16 pp. (multilithed)

Chemistry copper

Hood, D.W., 1963
Chemical oceanography. In: Oceanography and Marine Biology, H. Barnes, Edit., George Allen & Unwin, 1:129-155.

Chemistry copper

Hutchinson, G.E., R.J. Benoit, W.B. Cotter and P.J. Wangersky, 1955.
On the nickel, cobalt and copper contents of deep-sea sediments. Proc. Nat. Acad. Sci. 41(3): 160-162.

chemistry, copper

Ikuta, Kunio 1968.
Studies on accumulation of heavy metals in aquatic organisms. II On accumulation of copper and zinc in oysters III On accumulation of copper and zinc in the parts of oysters. (In Japanese; English abstract).
Bull. Jap. Soc. scient. Fish. 34 (2): 112-116; 117-122.

chemistry copper

Ishibashi, Masayoshi, Toshio Yamamoto and Fuji Morii, 1962
Chemical studies on the ocean (Pt. 85).
Chemical studies on the seaweeds (11).
Copper content in seaweeds.
Rec. Oceanogr. Wks., Japan, N.S., 6(2):157-168.

chemistry, copper

Igarashi, H., S. Komakai and M. Matsumura, 1957.
Preliminary report on copper content in the sea water and the marine algae of the inlet of Toyohama, Ishikari Bay.
Bull. Hokkaido Reg. Fish. Res. Lab., 16:92-99.

chemistry copper

Ishibashi, Masayoshi, Taitiro Fujinaga, Fuji Morii, Yoshihiko Kanchiku, and Fumio Kamiyama, 1964.
Chemical studies on the ocean (Part 94). Chemical studies on the seaweeds (19). Determination of zinc, copper, lead, cadmium and nickel in seaweeds using dithizone extraction and polarographic method.
Rec. Oceanogr. Wks., Japan, 7(2):33-63.

chemistry, copper

Ishibashi, M. S. Ueda, Y. Yamamoto and F. Morii, 1958.
Studies on the utilization of the shallow-water deposits (continued). On the copper content of the shallow-water deposits at the seacoasts of the Kii Peninsula and other districts.
Rec. Oceanogr. Wks., Japan, Spec. No. 2:153-166.

chemistry copper

Johnston, R., 1964
Sea water, the natural medium of phytoplankton II. Trace metals and chelation, and general discussion.
J. Mar. Biol. Assoc., U.K., 44(1):87-109.

Chemistry, copper

Kalle, K., and H. Wattenberg, 1938.
Über Kupfergehalt des Ozeanwassers. Naturwiss., 26:630-631.

Chemistry, copper

Kamada, M., and T. Onishi, 1964.
The copper and zinc contents of the river waters flowing into Ariake Sea and Yatusiro Bay. (In Japanese).
Sci. Repts., Kagoshima Univ., No. 13:17-27.

Chemistry, copper

Leisegang, E.C., and M.J. Orren 1966.
Trace element concentrations in the sea off South Africa.
Nature, Lond. 211 (5054): 1166-1167.

Chemistry copper

Meng-Chierego, N., and M. Picotti, 1960

Crociera Talassografica Adriatica, 1955. IV. La distribuzione del rame nelle acque dell' Adriatico. Archivio di Oceanografia e Limnol., 11(3): 421-424.

Chemistry copper

Meyer, H., 1938.
Die photometrische Bestimmung des Kupfers im Seewasser.
Ann. Hydrogr. u. Marit. Meteorol., 66:325.

Chemistry copper

Meyer, Y., 1939.
Die photometrische Bestimmung des Kupfers im Seewasser. Ann. Hydrogr., usw., 66:325.

Chemistry, copper

Mitsunobu, T., 1957.
[Chemical investigations of deep-sea deposits. 24-25. The contents of copper and zinc in sea sediments.] J. Chem. Soc., Japan, 78(3):405-415.

Chemistry, copper

Morita, Y., 1953.
Copper and zinc in Pacific waters.
Rec. Ocean. Wks., Japan, n.s., 1(2):49-51.

Chemistry, copper

Morita, Y., 1950.
Distribution of copper and zinc. IV. Copper and zinc contents of sea water. (In Japanese).
J. Chem. Soc., Japan, Pure Chem. Sect., 71:246-248.

Chem. Abstr., 1951, 45:4856.

Chemistry copper

Orren, M.J. 1967.
Trace elements (copper, iron and manganese) off the coast of South Africa.
Invest. Rep. Fish. mar. Biol. Surv. Div. Un. S. Afr. 59: 1-40.

Chemistry copper(data)

Picotti, Mario, 1960

Crociera Talassografica Adriatica 1955. III. Tabelle delle osservazioni fisiche, chimiche, biologiche e psammografiche.
Archivio di Oceanograf. e Limnol., 11(3): 371-377, plus tables.

copper Chemistry, copper

Picotti, M., 1956.
Les éléments ologodynamiques marins en relation avec la cuivre. Ist. Sperimentale Talassografico, Trieste, Pubbl. No. 333:181-184.
Also:
Rapp. Proc. Verb., Comm. Int. Expl. Sci., Mer Medite 13:181-184.

Chemistry - copper

Riley, G. A., 1937.
The significance of the Mississippi River drainage for biological conditions in the northern Gulf of Mexico. J. Mar. Res. 1(1):60-74, Textfigs 19-25.

Chemistry copper

Riley, J.P., and P. Sinhaseni, 1958.
The determination of copper in sea water, silicate rocks and biological materials. Analyst 83(986): 299-303.

Chemistry, copper

Rozhanskaya, L.I., 1965.
The concentration and distribution of copper in the waters of the Sea of Azov. (In Russian).
Okeanologiia, Akad. Nauk, SSSR, 5(6):983-986.

Chemistry, copper

Skopintsev, B.A., and T.P. Popova, 1960.
Some results of the determination of the content of iron, manganese and copper in the waters of the Black Sea. Chemistry of the Sea, hydrology, marine chemistry. (In Russian).
Trudy Morsk. Gidrofiz. Inst., 19:21-30.

chemistry, copper

Slowey, J. Frank, and Donald W. Hood, 1971.

Copper, manganese and zinc concentrations in Gulf of Mexico waters
Geochim. Cosmochim. Acta 35 (2): 121-138.

Chemistry - copper (organic complexed)

Slowey, J. Frank, Lela M. Jeffrey and D.W. Hood 1967.
Evidence for organic complexed copper in sea water.
Nature, Lond. 214 (5086): 377-378.

chemistry, copper

Spencer, Derek W., and Peter G. Brewer, 1969.
The distribution of copper, zinc and nickel in sea water of the Gulf of Maine and the Sargasso Sea.
Geochim. cosmochim. Acta. 33(3): 325-339.

Chemistry, Copper

*Steemann Nielsen, E. and S. Wium-Andersen, 1970.
Copper ions as poison in the sea and in fresh-water. Marine Biol., 6(2): 93-97.

Chemistry copper
Tichonov, M.K., and V.K. Chavoronkina, 1958.
On the question of the determination of copper in marine waters. (In Russian).
Trudy Morsk. Gidrofiz. Inst., 13:137-142.

Chemistry copper
Topping, Graham, 1969.
Concentrations of Mn, Co, Cu, Fe, and Zn in the northern Indian Ocean and Arabian Sea. J. mar. Res., 27(3): 318-326.

Chemistry, copper (data)
Torii, Tetsuya, and Sadao Murata 1966.
The distribution of copper and zinc in the Indian Ocean and the Southern Ocean waters.
J. oceanogr. Soc. Japan 22(2): 56-60.

Chemistry, copper
Uppström, Leif, 1968.
Analysis of boron in sea water by a modified circumin method.
Rept.Chem.Sea Water,Univ.Göteborg,4:8-13. (mimeographed).

Chemistry, copper
Wieser, Wolfgang, 1966.
Copper and the role of isopods in degradation of organic matter.
Science, 153(3731):67-69.

chemistry, copper
Williams, P.M., 1969.
The association of copper with dissolved organic matter in seawater. Limnol. Oceanogr., 14(1): 156-158.

chemistry, copper
Yatsimirsky, K.B., E.M. Emelyanov, V.K. Pavlova and Ya.S. Savichenko, 1971.
Determination of manganese and copper microquantities in small portions of marine suspended matter (from the Baltic Sea and the Atlantic Ocean). (In Russian; English abstract). Okeanologiia 11(4): 730-734.

Chemistry copper
Zuckerlandl, E., 1960
Hémocyanine et cuivre chez un crustacé décapode dans leurs rapports avec le cycle d'intermue.
Ann. Inst. Océanogr., Monaco, 38:1-122.

copper, effect of

Chemistry copper, effect of
Mandelli, E.F., 1969.
The inhibitory effects of copper on marine phytoplankton.
Contrib. mar. Sci., Port Aransas, 14: 47-57

Chemistry copper, effect of
Starr, T.J., and M.E. Jones, 1957.
The effect of copper on the growth of bacteria isolated from marine environments.
Limnol. Oceanogr., 2(1):33-36.

copper, effect of
Raymont, J.E.G., and J. Shields, 1963.
Toxicity of copper and chromium in the marine environment.
Int. J. Air Water Poll., 7(4/5):435-443.

copper (toxicity)

copper (toxicity)
Crisp, D.C., C.P. Spencer and D.H.A. Marr, 1957.
Toxicity of copper compounds in the sea (abstract)
Ann. Rept. Challenger Soc., 3(9):22.

Chemistry copper, toxicity of
Hueck, Hendrik J. and Dorothea M.M. Adema, 1968.
Toxicological investigations in an artificial ecosystem. A progress report on copper toxicity towards algae and daphniae.
Helgoländer wiss. Meeresunters. 17: 188-199

chemistry, cyanocobalamin (B12)
Provasoli, L., 1971.
Nutritional relationships in marine organisms.
In: Fertility of the Sea, John D. Costlow, editor, Gordon Breach, 2: 369-382.

Cytosine

cytosine
Rosenberg, E., 1964.
Purine and pyrimidines in sediments from the experimental Mohole.
Science, 146(3652):1680-1681.

DDT

Chemistry, DDT
Brewerton, H.V., 1969.
DDT in fats of Antarctic animals
N.Z. Jl Sci. 12 (2): 194-199.

Chemistry DDT
Butler, Philip A., 1966.
Fixation of DDT in estuaries.
Trans. 31st N.Amer.Wildl.Nat.Resources Conf., Pittsburgh 184-189.

Chemistry DDT residues
Cox, James L., 1971.
DDT residues in seawater and particulate matter in the California Current system.
Fish. Bull. nat. mar. Fish. Serv. 69 (2): 443-450

Chemistry DDT
Cox, James L. 1971.
Uptake, assimilation, and loss of DDT residues by Euphausia pacifica, a euphausiid shrimp.
Fish. Bull. nat. mar. fish. Serv. NOAA 69 (3): 627-633.

Chemistry - DDT
Cox, James L., 1970.
DDT residues in marine phytoplankton: increase from 1955 to 1969. Science, 170(3953): 71-73.

Chemistry DDT
Duffy, J.R., and D. O'Donnell, 1968.
DDT residues and metabolites in Canadian Atlantic coast fish.
J. Fish.Res.Bd.,Can., 25(1):189-195.

DDT
Ernst, W., 1969.
Stoffwechsel von Pesticiden in marinen Organismen. 1. Vorläufig Untersuchungen über die Umwandlung und Akkumulation von DDT-¹⁴C durch den Polychaeten Nereis diversicolor.
Veröff. Inst. Meeresforsch. Bremerhaven, 11 (2): 327-331.

chemistry DDT
Fougeras-Lavergnolle, Jean 1971.
Recherche des pesticides organochlorés dans les milieux littoraux.
Revue Trav. Inst. Pêches marit. 35(3): 367-371.

chemistry, DDT
Gillette, Robert, 1971.
DDT: in field and courtroom a persistent pesticide lives on. Science 174(4014): 1108-1110.

chemistry, DDT
Hays, Helen, and Robert W. Risebrough 1972
Pollutant concentrations in abnormal young terns from Long Island Sound.
The Auk 89 (1): 19-35

Chemistry DDT, effect of
Lowe, Jack I., 1965.
Chronic exposure of blue crabs, Callinectes sapidus, to sublethal concentrations of DDT.
Ecology, 46(6):899-900.

chemistry DDT
Shaw, Stanton B., 1972.
DDT residues in eight California marine fishes.
Calif. Fish Game 58(1): 22-26

chemistry, DDT
Woodwell, G.M., P.P. Craig and H.A. Johnson, 1971.
DDT in the biosphere: where does it go?
Science 174(4014): 1101-1107.

chemistry, DDT
Woodwell, George M., Charles F. Wurster, Jr., and Peter A. Isaacson, 1967.
DDT residues in an east coast estuary: a case of biological concentration of a persistent insecticide.
Science, 156(3776):821-823.

Chemistry DDT, effect of
*Wurster, Charles F., Jr., 1968.
DDT reduces photosynthesis by marine phytoplankton.
Science, 159(3822):1474-1475.

decalcification

decalcification
Jarke, Joachim, 1961
Beobachtungen über Kalkauflösung an Schalen von Mikrofossilien in Sedimenten der westlichen Ostsee.
Deutsche Hydrogr. Zeits., 14(1):6-11.

chemistry degradation
Foret-Montardo, Paule, 1971.
Evolution dans le temps de la toxicité des détergents issus de la pétroléochimie: Etude réalisée sur Scolelepis fuliginosa (polychète sédentaire).
Téthys 3 (1): 173-182

denitrification

Chemistry denitrification
*Barbaree, J.M., and W.J. Payne, 1967.
Products of denitrification by a marine bacterium as revealed by gas chromatography.
Marine Biol., 1(2):136-139.

Chemistry - denitrification
Goering, John J. 1968
Denitrification in the oxygen minimum layer of the eastern tropical Pacific.
Deep-Sea Res. 15(2): 157-164.

Chemistry denitrification
Goering, John J. and Joel D. Cline, 1970.
A note on denitrification in seawater. Limnol. Oceanogr., 15(2): 306-309.

chemistry, denitrification
Goering, J.J., and R.C. Dugdale, 1966.
Denitrification rates in an island bay in the equatorial Pacific Ocean.
Science, 154(3748): 505-506.

Chemistry denitrification
Goering, J.J. and M.M. Pamatmat, 1971.
Denitrification in sediments of the sea off Peru. Investigación pesq. 35(1): 233-242.

Chemistry denitrification
Lagarde, Edmond, 1964.
Méthode d'estimation du pouvoir dénitrifiant des eaux et des sédiments marins.
Vie et Milieu, Bull. Lab. Arago, 15(1): 213-218.

denitrification
Skopintsev, B.A., 1968.
A study of some reduction and oxidation processes in the Black Sea. (In Russian; English abstract).
Okeanologiia, Akad. Nauk, SSSR, 8(3): 412-426.

Chemistry, denitrification
Vaccaro, Ralph F. 1965.
Inorganic nitrogen in sea water.
In: Chemical oceanography, J.P. Riley and G. Skirrow, editors, Academic Press, 1: 365-408.

Chemistry, detergents, effect of
*Bellamy, D.J., P.H. Clarke, D.M. John, D. Jones, A. Whittick and T. Darke, 1967.
Effects of pollution from the Torrey Canyon, on littoral and sublittoral ecosystems.
Nature, Lond., 216(5121): 1170-1173.

chemistry detergents, effect of
Bellan, Gérard, Francis Carwille, Paule Foret-Montardo, Richard A. Kaim-Malka, Leung Tack Kit, 1969.
Contribution à l'étude de différents facteurs physicochimiques polluants sur les organismes marins. 1. Action des détergents sur la polychète Scolelepis fuliginosa (note préliminaire).
Téthys 1(2): 367-374

chemistry detergents, effect of
Bellan, Gérard, Donald J. Reish et Jean-Paul Foret 1971.
Action toxique d'un détergent sur le cycle de développement de la polychète Capitella capitata (Fab.).
C.r. hebd. Séanc. Acad. Sci. Paris (D) 272 (19): 2476-2479.

Chemistry - detergents, effect of
Bryan, G.W., 1969.
The effects of oil-spill removers ('detergents') on the gastropod Nucella lapillus on a rocky shore and in the laboratory. J. mar. biol. Ass. U.K., 49(4): 1067-1092.

Chemistry - detergents, effect of
Corner, E.D.S., A.J. Southward and E.C. Southward, 1968.
Toxicity of oil-spill removers ('detergents') to marine life: an assessment using the intertidal barnacle Elminius modestus.
J. mar. biol. Ass., U.K., 48(1): 29-47.

chemistry, detergents
Foret-Montardo, Paule, 1971.
Évolution dans le temps de la toxicité des détergents issus de la pétroléochimie: Étude réalisée sur Scolelepis fuliginosa (polychète sédentaire).
Téthys 3(1): 178-182

Chemistry detergents, effect of
Foret-Montardo, Paule 1970 (1971).
Étude de l'action des produits de base entrant dans la composition des détergents issus de la pétroléochimie vis-à-vis de quelques invertébrés benthiques marins.
Téthys 2(3): 567-614

Chemistry, detergents, effect of
Kühl, Heinrich und Hans Mann 1967.
Die Toxizität verschiedener Ölbekämpfungsmittel für See- und Süsswassertiere.
Helgoländer wiss. Meeresunters. 16(4): 321-327.

Chemistry detergent effect of
*Lacaze, Jean-Claude, 1967.
Étude de la croissance d'une algue planctonique en présence d'un détergent utilisé pour la destruction des nappes de pétrole en mer.
C.r. hebd. Séanc. Acad. Sci., Paris, (D), 265(20): 1489-1491.

Chemistry detergents, effect of
*Manwell, Clyde, and C.M. Ann Baker, 1967.
A study of detergent pollution by molecular methods: starch gel electrophoresis of a variety of enzymes and other proteins.
J. mar. biol. Ass., U.K., 47(3): 659-675.

detergents, effect of
Renzoni, A. 1971.
The influence of some detergents on the larval life of marine bivalve larvae. (Abstract only)
Rev. int. Océanogr. méd. 24: 50-52.

Chemistry - DETERGENTS, effect of
Wilson, Douglas P., 1968.
Temporary adsorption on a substrate of an oil-spill remover ('detergent'): tests with larvae of Sabellaria spinulosa.
J. mar. biol. Ass., U.K., 48(2): 183-186.

Chemistry - detergents, effect of
Wilson, Douglas P. 1968.
Long-term effects of low concentrations of an oil spill remover ('detergent'): studies with the larvae of Sabellaria spinulosa.
J. mar. biol. Ass. U.K. 48(1): 177-186.

deuterium

chemistry, deuterium
Arnason, Bragi, and Thorbjorn Sigurgeirsson 1968.
Deuterium content of water vapour and hydrogen in volcanic gas at Surtsey, Iceland.
Geochim. cosmochim. Acta 32(8): 807-813

Chemistry - deuterium
Berthois, L. 1969.
Contribution à l'étude sédimentologique du Kangerdlugssuaq, côte ouest du Groenland.
Meddr Grønland 187(1): 185pp.

Chemistry, deuterium
Berthois, Leopold, 1966.
Hydrologie et sedimentologie dans le Kangerd-lugssuaq (fjord a la cote ouest du Groenland).
C.r. hebd. seanc., Acad. Sci., Paris, (D), 262(13): 1400-1402.

Chemistry - deuterium
*Craig, H., 1969.
Geochemistry and origin of the Red Sea brines.
In: Hot brines and Recent heavy metal deposits in the Red Sea, E.T. Degens and D.A. Ross, editors, Springer-Verlag, New York, Inc., 208-242.

Chemistry, deuterium
Craig, H., 1961
Standard for reporting concentrations of deuterium and oxygen-18 in natural waters.
Science, 133(3467): 1833-1834.

chemistry, deuterium
Craig, H., and L.I. Gordon, 1965.
Isotopic oceanography: deuterium and oxygen 18 variations in the ocean and marine atmosphere.
Narragansett Mar. Lab., Univ. Rhode Island, Occ. Publ., No. 3: 277-374.

Chemistry deuterium
Crespi, H. L., & J. J. Katz, 1961
The determination of deuterium in biological fluids.
Anal. Biochim., 2(3): 274-279.

chemistry deuterium
Ehhalt, Dieter H. and H. Göte Östlund, 1970.
Deuterium in hurricane Faith 1966: preliminary results. J. geophys. Res., 75(12): 2323-2327.

Chemistry, deuterium
Friedman, I., 1953.
Deuterium content of natural waters and other substances. Geochim. et Cosmochim Acta, 4(1/2): 89-103.

Chemistry, deuterium
Friedman, I., D.R. Norton, D.B. Carter and A. C. Redfield, 1956.
The deuterium balance in Lake Maracaibo.
Limnol. & Oceanogr., 1(4): 239-246.

Chemistry, deuterium
Friedman, Irving, Alfred C. Redfield, Beatrice Schoen and Joseph Harris, 1964.
The variation of the deuterium content of natural waters in the hydrologic cycle.
Reviews of Geophysics, 2(1): 177-224.

chemistry, deuterium
Friedman, Irving, Beatrice Schoen and Joseph Harris, 1961.
The deuterium concentration in Arctic Sea ice.
J. Geophys. Res., 66(6):1861-1865.

chemistry, deuterium
Greene, C.H., and R.J. Voskuyl, 1939.
The deuterium-protium ratio. 1. The densities of natural waters from various sources.
J. Amer. Chem. Soc., 61:1342-1349.

chemistry
deuterium
Horibe, Yoshio and Mituko Kobayakawa, 1960
Deuterium abundance of natural waters. Geochim Cosmochim. Acta, 20(3/4): 273-283.

chemistry, deuterium
*Horibe, Yoshio, and Nobuko Ogura, 1968.
Deuterium content as a parameter of water mass in the ocean.
J. geophys. Res., 73(4):1239-1249.

chemistry
deuterium
Miyake, Y., and S. Matsuo, 1962
A note on the deuterium content in the atmosphere and the hydrosphere.
Papers Meteor. Geophys., 13(3-4):245-259.

chemistry deuterium
Sugiura, Yoshio, 1969.
Discussion of paper by Y. Horibe and N. Ogura 'Deuterium content as a parameter of water mass in the ocean'.
J.Geophys.Res., 74(6):1705-1707.

chemistry
deuterium
Woodcock, A.H., and Irving Friedman, 1963
The deuterium content of raindrops.
J. Geophys. Res., 68(15):4477-4483.

deuterium-hydrogen ratio

chemistry, deuterium-hydrogran ratio
Friedman, I., and A.H. Woodcock, 1957.
Determination of deuterium-hydrogrn ratios in Hawaiian waters. Tellus 9(4):553-556.

chemistry, deuterium/salinity
Ehhalt, D.H. 1969
On the deuterium-salinity relationship in the Baltic Sea.
Tellus 21(3):429-435.

deuterium oxide

chemistry deuterium oxide
Washburn, H.W., C.E. Berry and L.G. Hall, 1953.
Measurement of deuterium oxide concentration in water samples by the mass spectrometer.
Analyt. Chem., 25(1):130-134.

chemistry, diadinoxanthin
Jeffrey, S.W., and F.T. Haxo 1968.
Photosynthetic pigments of symbiotic dinoflagellates (zooxanthellae) from corals and clams.
Biol. Bull. mar. biol. Lab. Woods Hole 135(1): 149-165.

chemistry, Diadinoxanthin
Taylor, W. Rowland, and Conrad D. Gebelein, 1966.
Plant pigments and light penetration in intertidal sediments.
Helgolander wiss Meeresunters. 13(3):229-237.

chemistry, diagenesis
Thorstenson, Donald C., and Fred T. Mackenzie 1971
Experimental decomposition of algae in seawater and early diagenesis.
Nature, Lond. 234(5331): 543-545

Diatoxanthin
Taylor, W. Rowland and Conrad D. Gebelein, 1966.
Plant pigments and light penetration in intertidal sediments.
Helgolander wiss Meeresunters. 13(3):229-237.

chemistry, dieldrin
Epifanio, C.E., 1971.
Effects of dieldrin in seawater on the development of two species of crab larvae, Leptodius floridanus and Panopeus herbstii.
Mar. Biol. 11(4): 356-362.

chemistry-dieldrin
Lane, Charles E., and Robert J. Livingston 1970.
Some acute and chronic effects of dieldrin on the sailfin molly, Poecilia latipinna.
Trans. Am. fish. Soc. 99(3): 489-495

chemistry
dinoxanthin
Jeffrey, S.W., and F.T. Haxo 1968.
Photosynthetic pigments of symbiotic dinoflagellates (zooxanthellae) from coral and clams.
Biol. Bull. mar. biol. Lab. Woods Hole 135(1): 149-165.

chemistry, dispersants
Canevari, Gerard P., 1969.
The role of chemical dispersants in oil cleanup.
In: Oil on the sea, D.P. Hoult, editor, Plenum Press, 29-51.

chemistry-DNA
Derenbach, Jens B., 1970.
Partikulare Substanz und Plankton an Hand chemischer und biologischer Daten gemessen in den oberen Wasserschichten des Gotland-Tief im Mai 1968.
Kieler Meeresforsch. 25(2): 279-289

chemicals, effect of
Aubert, M. et D. Pesando 1971.
Télémédiateurs chimiques et équilibre biologique océanique. 2. Nature chimique de l'inhibiteur de la synthèse d'un antibiotique produit par une diatomée.
Rev. int. Océanogr. Méd. 21:17-22

Eh

chemistry
Eh (data)
Kramer, J.R., 1961
Chemistry of Lake Erie.
Proc. Fourth Conf., Great Lakes Research, Great Lakes Res. Div., Inst. Sci. & Tech., Univ. Michigan, Publ., (7):27-56.

chemistry Eh
Pirie, Robert Gordon, 1965.
Petrology and physical-chemical environment of bottom sediments of the Rivière Bonaventure-Chaleur Bay area, Québec, Canada.
Rept. B.I.O. 65-10-182 pp. (multilithed).

Chemistry, Eh (data)
Seibold, E., 1962.
Untersuchungen zur Kalkfällung und Kalklösung am Weßtrand der Great Bahama Bank.
Sedimentology, 1(1):50-74.

Chemistry - Eh
*Whitfield, M., 1969.
Eh as an operational parameter in estuarine studies.
Limnol. Oceanogr. 14(4):547-558.

chemistry, Eiweiss (data)
Szekielda, Karl-Heinz 1971.
Organisch gelöster und partikulärer Kohlenstoff in einem Nebenmeer mit starken Salzgehaltsschwankungen (Ostsee).
Vie Milieu Suppl. 22(2): 579-612

electrolytes

chemistry
electrolytic conductance
Park, Kilho, 1964.
Electrolytic conductance of sea water: effect of calcium carbonate dissolution.
Science, 146(3640): 56-57.

chemistry, electrolytic conductance
Park, Kilho, 1964
Partial equivalent conductance of electrolytes in sea water.
Deep-Sea Res., 11(5):729-736.

chemistry, electrolytic conductance
Connors, Donald N., and Kilho Park, 1967.
The partial equivalent conductance of electrolytes in seawater: a revision.
Deep-Sea Research, 14(4):481-484.

Chemistry
electrons, hydrated
Swallow, A.J., 1969
Hydrated electrons in seawater.
Nature Lond 222(5191): 369-370.

Elements

chemistry elements
(salts and their ions)
Bruevich, S.W., 1960
[Hydrochemical investigations on the White Sea]
Trudy Inst. Okeanol., 42: 199-254.

Chemistry elements
(cycle)
Golterman, H.L., 1960
Studies on the cycle of elements in fresh water. Acta Botanica Neerlandica, 9: 1-58.

Chemistry, elements, radioactive
*Miyake, Yasuo, Katsuko Saruhashi and Yukio Sugimura, 1968.
Biogeochemical balance of natural radioactive elements in the sea.
Rec.oceanogr.Wks, Japan, n.s., 9(2):180-187.

Chemistry, elements (geochemical)
Taylor, S.R., 1964.
Abundance of chemical elements in the continental crust: a new table.
Geochim. Cosmochim. Acta, 28:1273-1285.

chemistry - chemical elements
Vinogradov, A.P., 1953 (translation).
The elementary chemical composition of marine organisms. Mem. Sears Found. Mar. Res. 2:1-647.

Translated by J. Efron and J.K. Seltow with bibliography by V.W. Odum.

Enzymes

Chemistry, enzymes
Aubert, M., et C. Margat 1965.
Les facteurs probiotiques du milieu marin.
Cah. C.E.R.B.O.M. 18:43-48.

chemistry, enzymes, effect of
Berger, Rainer, and W.F. Libby, 1969
Equilibration of atmospheric carbon dioxide with sea water: possible enzymatic control of the rate.
Science, 164 (3886): 1395-1397.

chemistry, enzymes, nitrate reductase
Eppley, R.W., J.L. Coatsworth, and Lucia Solorzano, 1969.
Studies of nitrate reductase in marine phytoplankton. Limnol. Oceanogr., 14(2): 194-205.

chemistry Enzymes
Eppley, R.W., T.T. Packard and J.J. MacIs 1970.
Nitrate reductase in Peru current phytoplankton. Mar. Biol. 6(3): 195-199.

Chemistry enzymes
Fogg, G.E., 1966.
The extracellular products of algae.
In: Oceanography and marine biology, H. Barnes, editor, George Allen & Unwin, Ltd., 4:195-212.

Chemistry, enzymes
Gillen, R.G., 1971
The effect of pressure on muscle lactate dehydrogenase activity of some deep-sea and shallow-water fishes. Marine Biol., 8(1): 7-11.

Chemistry, enzymes
Hammen, C.S., 1969.
Lactate and succinate oxidoreductases in marine invertebrates. Mar Biol., 4(3): 233-238.

chemistry, enzymes
Jacobi, Günter, 1959
Salinitätswirkung des Seewassers auf die Enzymaktivität von Ulva lactuca. Kieler Meeresforsch., 15: 161-163.

chemistry, enzymes (zinc)
Wolfe, Douglas A. 1970.
Zinc enzymes in Crassostrea virginica.
J. Fish. Res. Bd Can. 27(1): 59-69

Chemistry equivalent volumes

chemistry, equivalent volumes (partial)
Duedall, Iver W., and Peter K. Weyl, 1967.
The partial equivalent volumes of salts in seawater.
Limnol. Oceanogr., 12(1):52-59.

esters

chemistry, esters
Jarvis, N.L., W.D. Garrett, M.A. Scheiman and C.O. Timmons, 1967.
Surface chemical characterization of surface-active material in seawater.
Limnol. Oceanogr., 12(1):88-96.

chemistry, esters (wax)
Lee, Richard F., Jed Hirota and Arthur M. Barnett, 1971.
Distribution and importance of wax esters in marine copepods and other zooplankton. Deep-Sea Res. 18(12): 1147-1165.

chemistry, wax esters
Lee, R.F., J.C. Nevenzel and G.-A. Paffenhöfer, 1971.
Importance of wax esters and other lipids in the marine food chain: phytoplankton and copepods. Mar. Biol. 9(2): 99-108.

chemistry, esters (wax)
Lee, Richard F., Judd C. Nevenzel, G.-A. Paffenhöfer and A.A. Benson 1970.
The metabolism of wax esters and other lipids by the marine copepod, Calanus helgolandicus.
J. Lipid Res. 11(3): 237-240
Also in: Coll. Repr. Scripps Inst. Oceanogr. 40.

ethylenediamine tetr-acetic acid

chemistry, ethylenediaminete traacetate (EDTA), effect of
Dingle, J.R. and J.A. Hines 1971.
Degradation of inosine 5'-monophosphate in the skeletal muscle of several North Atlantic fishes.
J. Fish. Res. Bd. Can. 28(8): 1125-1131

Chemistry - EDTA
Johnston, R., 1964
Sea water, the natural medium of phytoplankton. II. Trace metals and chelation, and general discussion.
J. Mar. Biol. Assoc., U.K., 44(1):87-109.

ethylenediamine tetra-acetic acid
Spencer, C.P., 1958.
The chemistry of ethylenediamine tetra-acetic acid in sea water. J.M.B.A., 37(1):127-144.

europium

chemistry europium
Goldberg, Edward D., Minoru Koide, R.A. Schmitt and R.H. Smith, 1963.
Rare-earth distribution in the marine environment.
J. Geophys. Res., 68(14):4209-4217.

fats

chemistry, fats
Fisher, L.R. (posthumous) and Zena D. Hosking, 1966.
Vitamin A and fat in the herring of the Blackwater estuary.
In: Some contemporary studies in marine biology, H. Barnes, editor, George Allen & Unwin, Ltd., 297-311.

fatty acids

chemistry fatty acids
Ackman, R.G., 1964.
Structural homogeneity in unsaturated fatty acids of marine lipids. A review.
J. Fish. Res. Bd., Canada, 21(2):247-254.

chemistry, fatty acids
Ackman, R.G., and C.A. Eaton 1971.
Investigation of the fatty acid composition of oils and lipids from the sand lance (Ammodytes americanus) from Nova Scotia waters.
J. Fish. Res. Bd. Can. 28(4): 601-606

chemistry, fatty acids
Ackman, R.G., and C.A. Eaton 1967.
Fatty acid composition of the decapod shrimp, Pandalus borealis, in relation to that of the euphausid, Meganyctiphanes norvegica.
J. Fish. Res. Bd Can. 24(2): 467-471.

chemistry, fatty acids
Ackman, R.G., C.S. Tocher and J. McLachlan 1968.
Marine phytoplankter fatty acids.
J. Fish. Res. Bd Can. 25(8): 1603-1620.

chemistry, fatty acids
Allen, W.V. 1971.
Amino acid and fatty acid composition of the Dungeness crab (Cancer magister)
J. Fish. Res. Bd. Can. 28(8): 1191-1195

chemistry, fatty acids
Beach, David H., Glenn W. Harrington and George G. Holz, Jr., 1970.
The polyunsaturated fatty acids of marine and freshwater cryptomonads.
J. Protozool. 17(3): 501-510

Chemistry - fatty acids
Chuecas, L., and J.P. Riley, 1966.
The component fatty acids of some sea-weed fats.
J. mar. biol. Ass., U.K., 46(1):153-159.

Chemistry, fatty acids
Collier, Albert 1967.
Fatty acids in certain plankton organisms.
In: Estuaries, G.H. Lauff, editor, Publs Am. Ass. Advmt Sci. 83: 353-360

Chemistry fatty acids
Cooper, L.H.N., 1965.
Chemistry of the sea. 2. Organic.
Chemistry in Britain, 1965, 1:150-154.

chemistry, fatty acids
Cooper, W.J., and Max Blumer, 1968.
Linear, iso and anteiso fatty acids in Recent sediments of the North Atlantic.
Deep-Sea Res., 15(5):535-540.

Chemistry, fatty acids
Culkin, F. and R.J. Morris, 1970.
The fatty acids of some cephalopods. Deep-Sea Res., 17(1): 171-174.

Chemistry, fatty acids
Culkin, F. and R. J. Morris 1970.
The fatty acids of some marine teleosts.
J. fish. Biol 2(2): 107-112.

chemistry, fatty acids
Culkin, F. and R.J. Morris, 1969.
The fatty acids of some marine crustaceans.
Deep-Sea Res., 16(2): 109-116.

chemistry, fatty acids
Farrington, John W., and James G. Quinn 1971
Fatty acid diagenesis in recent sediment from Narragansett Bay, Rhode Island.
Nature, Lond. 230(11): 67-69.

chemistry, fatty acids, polyunsaturated
Harrington, Glenn W., David H. Beach, Joyce E. Dunham and George G. Holz, Jr., 1970.
The polyunsaturated fatty acids of marine dinoflagellates.
J. Protozool. 17(2): 213-219.

chemistry, fatty acids (unsaturated)
Jarvis, N.L., W.D. Garrett, M.A. Scheiman and C.O. Timmons, 1967.
Surface chemical characterization of surface-active material in seawater.
Limnol. Oceanogr., 12(1):88-96.

chemistry, fatty acids
Jeffrey, Lela M., Nestor R. Bottino and Raymond Reiser, 1966.
The distribution of fatty acid classes in lipids of Antarctic euphausids.
Antarctic J., United States, 1(5):209.

Chemistry - fatty acids
Jeffries, H. Perry, 1970.
Seasonal composition of temperate plankton communities: fatty acids. Limnol. Oceanogr., 15(3): 419-426.

chemistry, fatty-acids
Hinchcliffe, P.R. and J.P. Riley, 1972.
The effect of diet on the component fatty-acid composition of Artemia salina. J. mar. biol. Ass. U.K. 52(1): 203-211.

chemistry, fatty acids
Kamazu, Ryoichi, Hideaki Morii and Tadanobu Fukuhara, 1969.
Studies on the little toothed whales in the west sea area of Kyushu. XVII. About higher branched chain fatty acids in head oil of little toothed whale. 1 and 2. (In Japanese; English abstract).
Bull. Fac. Fish. Nagasaki Univ. 28: 161-165, 167-171

Chemistry - fatty acids
Kates, M., 1966.
Lipid components of diatoms.
Biochimica et Biophysica Acta, 116(2):264-278.

chemistry, fatty acids
Leo, Richard F., and Patrick L. Parker, 1966.
Branched chain fatty acids in sediments.
Science, 152(3722):649-650.

chemistry, fatty acids
Lewis, R.W., 1969.
The fatty acid composition of Arctic marine phytoplankton and zooplankton with special reference to minor acids. Limnol. Oceanogr. 14(1): 35-40.

Chemistry, fatty acids
Lewis, R.W. 1967.
Fatty acid composition of some marine animals from different depths.
J. Fish. Res. Bd Can. 24(5): 1101-1115.

Chemistry, fatty acids
Margalef, Ramón 1967.
Un ejemplo de diversidad química.
Investigación pesq. Barcelona 3(3): 489-490.

chemistry, fatty acids
Maurer, Larry G., and P.L. Parker, 1967.
Fatty acids in sea grasses and marsh plants.
Contrib. mar. Sci., Port Aransas, 12: 113-119

chemistry, fatty acids
Meyers, Phillip A., and James G. Quinn, 1971.
Fatty acid-clay mineral association in artificial and natural sea water solutions
Geochim. Cosmochim. Acta 35(6): 628-632.

chemistry, fatty acids
Morris, R.J., 1971.
Variations in the fatty acid composition of oceanic euphausiids. Deep-Sea Res., 18(5): 525-529.

Chemistry, fatty acids
Nonaka, Junsaku, and Chiaki Koizumi 1964
Component fatty acids and alcohols of Euphausia superba lipid by gas-liquid chromatography. (In Japanese)
Bull. Jap. Soc. scient. Fish. 30(8): 630-634
English abstract

Chemistry, fatty acids
Parker, Patrick L. 1967.
Fatty acids in Recent sediment.
Contrib. mar. Sci. Port Aransas 12: 135-142.

chemistry fatty acids
Pugh, P.R., 1971.
Changes in the fatty acid composition of Coscinodiscus eccentricus with culture-age and salinity. Mar. Biol. 11(2): 118-124.

chemistry, fatty acids
Rhead, M.M., G. Eglinton, G.H. Draffan and P.J. England 1971.
Conversion of oleic acid to saturated fatty acids in Severn Estuary sediments.
Nature, Lond., 232 (5309): 327-330

Chemistry, fatty acids
Shatunovskii, M.I., 1970.
The composition of fatty acids of lipids of spring- and autumn-breeding Baltic herring roe, fingerlings and adults of the Gulf of Riga (the Baltic Sea).
Dokl. Akad. Nauk, SSSR, 195(4): 962-964

chemistry - fatty acids
Sipos, J.C., and R.G. Ackman, 1965.
Jellyfish (Cyanea capillata) lipids: fatty acid composition.
J. Fish. Res. Bd. Can., 25(8): 1561-1569.

chemistry, fatty acids
Slowey, J.F., L.M. Jeffrey and D.W. Hood, 1962.
The fatty-acid content of ocean water.
Geochimica et Cosmochimica Acta, 26(June):607-616.

chemistry, fatty acids
Tsuyuki, Hideo and Shingo Itoh, 1971.
Fatty acid components of Ganges River dolphin oil. Scient. Repts. Whales Res. Inst. 23: 141-147.

chemistry, Fatty acids
Tsuyuki, Hideo and Shingo Itoh, 1970.
Fatty acid components of black right whale oil by gas chromatography. Scient. Repts Whales Res. Inst., 22: 165-170.

chemistry, fatty acids
Ushakov, A.N., D.M. Vityuk, V.A. Vaver and L.D. Bergelson, 1966.
Fatty, acids in Black Sea water. (In Russian; English abstract).
Okeanologiia, Akad. Nauk, SSSR, 6(5):891-894.

chemistry, fatty acids
Welte, Dietrich H., and Götz Ebhardt, 1968.
Distribution of long chain n-paraffins and n-fatty acids from the Persian Gulf.
Geochim. Cosmochim. Acta, 32(4): 465-466

Chemistry, fatty acids
Welte, Dietrich H., und Götz Ebhardt 1968.
Die Verteilung höherer, geradkettiger Paraffine und Fettsäuren in einem Sedimentprofil aus dem Persischen Gulf.
Meteor Forschungsergebn. (C) 1: 43-52.

chemistry, fatty acids
Williams, P.M., 1965.
Fatty acids derived from lipids of marine origin.
J. Fish. Res. Bd., Canada 22(5):1107-1122.

Chemistry, Fatty acids
Yakovleva, K.K., 1969.
Dynamics of polyunsaturated fatty acids in certain species of fishes of the Black Sea. 1. The sea scorpion, Scorpaena porcus L. (In Russian)
Voprosy Ikhtiol. 9(4): (not seen)
Translation: 9(4): 597-602

chemistry - fatty acid amide
Nakamura, Takashi, and Masamichi Toyomizu, 1970
Studies on fatty acid amide in fishes. (In Japanese; English abstract).
Bull. Jap. Soc. scient. Fish. 36(6): 631-637.

chemistry - ferric hydroxide
Brewer, Peter G., and Derek W. Spencer, 1969.
A note on the chemical composition of the Red Sea brines.
In: Hot brines and Recent heavy metal deposits in the Red Sea, E.T. Degens and D.A. Ross, editors, Springer-Verlag, New York, Inc., 174-179.

chemistry, ferric hydroxide, effect of
Rosenthal, H., 1971.
Wirkungen von "Rotschlamm" auf Embryonen und Larven des Herings Clupea harengus. Helgoländer wiss. Meeresunters 22(3/4): 366-376.

ferritin

chemistry ferritin
Towe, K.M., H.A. Lowenstam and M.H. Nesson, 1963
Invertebrate ferritin: occurrence in Mollusca. Science, 142(3588):63-64.

ferro-manganese nodules

See also under:
manganese nodules

Chemistry, ferro-manganese nodules
See also: chemistry, ferrous sulphide nodules
chemistry, manganese nodules

chemistry, ferro-manganese nodules
Barnes, S.S., and J.R. Dymond, 1967.
Rates of accumulation of ferro-manganese nodules.
Nature, Lond, 213(5082):1218-1219

chemistry, ferromanganese nodules
Shterenberg, L.E. 1971.
Some aspects of ferromanganese nodules of the Gulf of Riga. (In Russian).
Dokl. Akad. Nauk SSSR 201(2): 457-460.

Chemistry ferro -manganese nodules
Skornyakova, N.S., and P.F. Andryushchenko, 1968.
Iron-manganese nodules from the central South Pacific Ocean. (In Russian; English abstract).
Okeanologiia, Akad. Nauk, SSSR, 8(5):865-877.

ferrous sulfate

chemistry, ferrous sulfate
Syazuki, K., 1953.
Studies on the foul-water drained from factories 2. On the harmful components and the water pollution of the foul-water drained from the metal plants. J. Shimonoseki Coll. Fish. 3(2):181-185, 2 textfigs.

ferrous sulphide nodules

chemistry, ferrous sulphide nodules
Volkov, I.I., 1964.
Formation and regularities and chemical composition in the Black Sea deposits. Chemistry of the waters and sediments of seas and oceans. (In Russian: English abstract).
Trudy Inst. Okeanol., Akad. Nauk, SSSR, 67:101-134.

fluorescence data

fluorescence (data)
Kalle, K., 1956.
Chemisch-hydrographische Untersuchungen in der inneren Deutschen Bucht. Deutsche Hydrogr. Zeits. 9(2):55-65.

fluoride

chemistry fluoride
Ackman, R.G., R.P. Addison, P.J. Ke and J.C. Sipos, 1971.
Examination of bottom deposits, Long Harbour, Newfoundland, for elemental phosphorus and for fluorides.
Tech. Rept. Fish. Res. Bd. Canada, 233: 19 pp.

Chemistry, fluoride
Anselm, C.D., and R.J. Robinson, 1951.
The spectrophotometric determination of fluoride in sea water. J. Mar. Res. 10(2):203-214, 6 textfigs.

chemistry, fluoride
Bewers, J.M., 1971.
North Atlantic fluoride profiles. Deep-Sea Res., 18(2): 237-241.

chemistry - fluoride
Brewer, P.G., D.W. Spencer and P.E. Wilkniss, 1970.
Anomalous fluoride concentrations in the North Atlantic. Deep-Sea Res., 17(1): 1-7.

chemistry, fluorides
Danielsen, M.E., 1954.
Photoelectric determination of small amounts of fluorides in aqueous solutions.
Univ. i Bergen, Aarbok, 1953, Naturv. Rekke, 9:

chemistry, fluoride
Greenhalgh, R., and J.P. Riley, 1963.
Occurrence of abnormally high fluoride concentrations at depth in the oceans.
Nature, 197(4865):371-372.

chemistry, fluorides
*Morcos, Selim A., 1967.
The chemical composition of sea water from the Suez Canal region. I. The major anions.
Kieler Meeresforsch., 23(2):80-91.

chemistry, fluoride
Riley, J.P., 1965.
The occurrence of anomalously high fluoride concentrations in the North Atlantic.
Deep-Sea Research, 12(2):219-220.

chemistry, fluorides
Thompson, T.G., and H.J. Taylor, 1933.
Determination and occurrence of fluorides in sea water.
Ind. Eng. Chem., Analyt. Edit., 5:87-89.

chemistry fluoride
Warner, Theodore B., 1971.
Normal fluoride content of seawater. Deep-Sea Res., 18(12): 1255-1263.

chemistry, fluoride
Wilkniss, P.E., T.B. Warner and R.A. Carr, 1971.
Some aspects of the geochemistry of F, Fe and Mn in coastal waters and in fresh-water springs on the southeast coast of Hawaii.
Mar. Geol. 11(4): M39-M46

chemistry, fluoride
Windom, Herbert L., 1971.
Fluoride concentration in coastal and estuarine waters of Georgia. Limnol. Oceanogr. 16(5): 806-810.

fluoride/chlorinity data

Chemistry, fluoride/chlorinity (data)
Greenhalgh, R., and J.P. Riley, 1963.
Occurrence of abnormally high fluoride concentrations at depth in the oceans.
Nature, 197(4865):371-372.

chemistry fluoride/chlorinity ratio
Kester, Dana R., 1971.
Fluoride chlorinity ratio of sea water between the Grand Banks and the Mid-Atlantic Ridge.
Deep-Sea Res. 18(11): 1123-1126.

fluorine

Chemistry, fluorine
Bochert, H., 1952.
Geochemistry of fluorine. Heidelberger Beitr. Mineral. Petrogr., 3(1):36-43.
Chem. Abstr., 1952, 46:11058.

chemistry, fluorine
Carpenter, Roy, 1969.
Factors controlling the marine geochemistry of fluorine.
Geochim. cosmochim. Acta, 33(10): 1153-1167.

Chemistry fluorine
Culkin, Frederick, 1965.
The major constituents of sea water.
In: Chemical oceanography, J.P. Riley and G. Skirrow, editors, Academic Press, 1: 121-161.

chemistry fluorine
Danielsen, M.E., and T. Gaarder, 1955.
Fluorine content of drinking water and food in western Norway, the Bergen District.
Arbok, Univ. i Bergen, Naturvitensk. rekke, No. 15:20 pp.

fluorine
Matida, Y., 1954.
On the source and the fate of fluorine in water in Tokyo Bay. J. Ocean. Soc., Japan, 10(2): 71-76, 1 textfig.

chemistry, fluorine
*Shishkina, O.V., G.A. Pavlova and V.S. Bykova, 1967.
The fluorine distribution in the interstitial water and in the bottom sediments of the Black Sea. (In Russian; English abstract).
Trudy Inst. Okeanol., 83:83-98.

chemistry, fluorine
Shishkina, O.V. 1966.
Fluorine in oceanic sediments and their pore solutions. (In Russian)
Geokhimya (2):236-249.
Translation: Geochem. int. Ann Arbor 3(1): 152-159.

chemistry, fluorine
Wilkniss, P.E. and D.J. Bressan, 1971.
Chemical processes at the air-sea interface: the behavior of flourine. J. geophys. Res., 76(3): 736-741.

chemistry, fluorine
Wilkniss, P.K., and V.J. Linnenbom, 1968.
The determination of fluorine in seawater by photon activation analysis.
Limnol. Oceanogr., 13(3):530-533.

chemistry, fractionation (geochemical)

MacIntyre, Ferren 1970.
Geochemical fractionation during mass transfer from sea to air by breaking bubbles.
Tellus 22(4): 451-462.

francolite

fucoxanthin

chemistry, fucoxanthin
Garside, C., and J.P. Riley, 1968.
The absorptivity of fucoxanthin.
Deep-Sea Res., 15(5):627.

Chemistry, fucoxanthin
Parsons, T.R., and J.D.H. Strickland, 1963.
Discussion of spectrophotometric determination of marine-plant pigments, with revised equations for ascertaining chlorophylls and carotenoids.
J. Mar. Res., 21(3):155-163.

fucoxanthin
Taylor, W. Rowland, and Conrad D. Gebelein, 1966.
Plant pigments and light penetration in intertidal sediments.
Helgoländer wiss Meeresunters. 13(3):229-237.

galacturonic acid

Chemistry, Galacturonic acid
*Ernst, Wolfgang, 1967.
Dünnschichtchromatographische Trennung und quantitative Bestimmung von Galacturon- und Glucuronsäure in Polysaccharidfraktionen des Meeresbodens.
Veröff. Inst. Meeresforsch. Bremerh. 10(3):183-185.

gallium

Chemistry, gallium
Chester, R., 1965.
Elemental geochemistry of marine sediments.
In: Chemical oceanography, J.P. Riley and G. Skirrow, editors, Academic Press.

chemistry - gallium
Culkin, F. and J.P. Riley, 1958
The occurrence of gallium in marine organisms
J. Mar. Biol. Ass. 37: pps. 607-616.

Chemistry gallium
Ishibashi, M., T. Shigematsu, Y. Nishikawa, K. Hirake, 1961
Gallium content of sea water, marine organisms sediments, and other materials related to the ocean.
J. Chem. Soc., Japan, 82(9):1141-1142.

chemistry, gallium
Ishibashi, Masayoshi, Shunzo Ueda and Yoshikazu Yamamoto, 1966.
On the GALLIUM content of the shallow-water deposits.
J. oceanogr. Soc. Japan, 22(5):197-200.

chemistry, gallium
Ostroumov, E.A., and I.I. Volkov, 1964.
Separation of indium and gallium from manganese, nickel, cobalt and zinc with the help of cinnamic acid in application to the study of sea deposits. Chemistry of the waters and sediments of the seas and oceans. (In Russian
Trudy Inst. Okeanol., Akad. Nauk, SSSR, 67:141-150.

gamma ray emitters

Chemistry, gamma ray emitters
Chesselet R., 1967.
Application en océanographie de la méthode de spectrométrie gamma in situ.
Rev. int. Océanogr. Méd. 5:5-21.

gases, atmospheric

Chemistry - gases, atmospheric
Fox, C.J., 1905.
On the determination of the atmospheric gases dis-solved in sea-water.
Publ. de Circ., Cons. Perm. Int. Expl. Mer, No. 21:1-24, 1 pl., 4 textfigs.

Chemistry, gases
Benson, Bruce B., 1965.
Some thoughts on gases dissolved in the oceans.
Narragansett Mar. Lab., Univ. Rhode Island, Occ. Publ., No. 3:91-107.

Chemistry gases, dissolved
Carritt, D.E., 1963
5. Chemical instrumentation. In: The Sea, M.N. Hill, Edit., Vol. 2. The composition of sea water, Interscience Publishers, New York and London, 109-123.

Chemistry - gases, dissolved.
Deacon, G.E.R., 1965.
Chemistry of the sea. 1. Inorganic.
Chemistry in Britain, 1:48-53.
Also in:
Collected Reprints, Nat. Inst. Oceanogr., 13.

chemistry, gases
Goldberg, Edward D., 1971.
3. Atmospheric transport. In: Impingement of man on the oceans, D.W. Hood, editor, Wiley Interscience: 75-88.

Chemistry - gases (dissolved)
Richards, Francis A., 1968.
Chemical and biological factors in the marine environment. Ch. 8 in: Ocean engineering: goals, environment, technology, J.F. Brahtz, editor, John Wiley & Sons, 259-303.

Chemistry - gases, effect of
Lebedev, V.L., and G.A. Safyanov, 1966.
The effect of solid and gaseous suspensions on the hydrostatic pressure of sea water. (In Russian; English abstract).
Okeanologiia, Akad. Nauk, SSSR, 6(6):1023-1029.

Chemistry gas exchange
Kanwisher, John W., 1963
On the exchange of gases between the atmo-sphere and the sea.
Deep-Sea Research, 10(3):195-207.

Chemistry gas exchange
Kanwisher, John, 1962
Gas exchange of shallow marine sediments.
The Environmental Chemistry of Marine Sediments, Proc. Symp., Univ. R.I., Jan. 13, 1962, Occ. Papers, Narragansett Mar. Lab., No. 1: 13-19.

chemistry, gas exchange
Revelle, R., and H.E. Suess, 1962
Gases, Ch. 7, Sect. II, Interchange of properties between sea and air. In: The Sea, Interscience Publishers, Vol. 1, Physical Oceanography, 313-321.

gases, noble

chemistry, gases (noble)
Bieri, Rudolf H., Mineru Koide and Edward D. Goldberg, 1966.
The noble gas contents of Pacific seawaters.
J. geophys. Res., 71(22):5243-5265.

gases, solubility of

Chemistry gases, solubility of
Carritt, D.E., 1954.
Atmospheric pressure changes and gas solubility.
Deep-Sea Res. 2(1):59-62, 1 textfig.

chemistry, gases (volcanic)
Arnason, Bragi, and Thorbjörn Sigurgeirsson 1968.
Deuterium content of water vapour and hydrogen in volcanic gas at Surtsey, Iceland.
Geochim. cosmochim. Acta 32(8): 807-813.

chemistry, gases (volcanic)
Björnsson, Sveinbjörn 1968.
Radon and water in volcanic gas at Surtsey Island.
Geochim. cosmochim. Acta 32(8): 815-820.

chemistry, gases (volcanic)
Sigvaldason, Gudmundur E. and Gunnlaugur Elísson 1968.
Collection and analysis of volcanic gases at Surtsey, Iceland
Geochim. cosmochim. Acta 32(8): 797-805.

geochemistry: see under "G" of main file

Chemistry, geochemistry

See: Geochemistry

germanium

chemistry germanium
Chester, R., 1965.
Elemental geochemistry of marine sediments.
In: Chemical oceanography, J.P. Riley and G. Skirrow, editors, Academic Press, 2:23-80.

Chemistry germanium
El Wardani, S.A., 1957.
On the geochemistry of germanium.
Geochimica et Cosmochimica Acta, 13(1):5-19.

Chemistry, germanium
Sokolov, V.S., and M.F. Pilipchuk, 1969.
Germanium in modern deposits of the Black Sea. (In Russian).
Dokl. Akad. Nauk SSSR, 185(3): 679-682

giberellic acid

Chemistry - gibberellic acid
Birch, A.J., and J. Winter, 1963.
A partial synthesis of 14C phyllocladene: some observations on the biosynthesis of gibberellic acid.
J. Chem. Soc., London, Paper No. 1051: 5547-5548.

Chemistry Gibberellic acid
Paster, Zvi and Bernard C. Abbott, 1970.
Gibberellic acid: a growth factor in the unicellular alga Gymnodinium breve. Science, 169(3945): 600-601.

chemistry - gibberellic acid
Ramamurthy, V.D. and R. Seshadri 1966.
Effects of gibberellic acid (GA) on laboratory cultures of Trichodesmium erythraeum (Ehr.) and Melosira sulcata (Ehr.).
Proc. Indian Acad. Sci. (B), 64 (3):146-151

gibberellins

chemistry, gibberellins, effect of
Bentley-Mowat, J.A., and S.M. Reid, 1969.
Effect of gibberellins, kinetin and other factors on the growth of unicellular marine algae in cultures.
Botanica mar. 12(1/4): 155-199

chemistry
gibberellins, effect of
Johnston, R., 1963
Effects of gibberellins on marine algae in mixed cultures.
Limnol. and Oceanogr., 8(2):270-275.

gibberellins
Mowat, Joyce A. (née Bentley), 1964.
A survey of results on the occurrence of auxins and gibberellins in algae.
Botanica marina, 8(1):149-155.

glucose

chemistry - glucose
Andrews, P. and P.J. LeB. Williams, 1971.
Heterotrophic utilization of dissolved organic compounds in the sea. III. Measurement of the oxidation rates and concentrations of glucose and amino acids in sea water. J. mar. biol. Ass., U.K., 51(1): 111-125.

chemistry, glucose
Takahashi, M. and S. Ichimura, 1971.
Glucose uptake in ocean profiles with special reference to temperature. Mar. Biol. 11(3): 206-213.

Chemistry, glucose
Vaccaro, Ralph F., Sonja E. Hicks, Holger W. Jannasch and Francis G. Carey, 1968.
The occurrence and role of glucose in seawater.
Limnol. Oceanogr. 13(2): 356-360

chemistry, glucose
Vaccaro, Ralph F., and Holger W. Jannasch, 1967.
Variations in uptake kinetics for glucose by natural populations in seawater.
Limnol. Oceanogr., 12(3):540-542.

glucuronic acid

chemistry, glucuronic acid
Ernst, Wolfgang, 1967.
Dünnschichtchromatographische Trennung und quantitative Bestimmung von Galacturon- und Glucuronsäure in Polysaccharidfraktionen des Meeresbodens.
Veröff. Inst. Meeresforsch. Bremerh. 10(3):183-185.

glycollic acid

chemistry glycollic acid
Droop, M.R., and Susanne McGill, 1966.
The carbon nutrition of some algae: the inability to utilize glycollic acid for growth.
Jour. mar. biol. Assoc., U.K., 46(3):679-684.

chemistry, glycollic acid
Fogg, G.E., 1966.
The extracellular products of algae.
In: Oceanography and marine biology, H. Barnes, editor, George Allen & Unwin, Ltd., 4:195-212.

gold

chemistry, gold
Anoshin, G.N., and E.M. Emelianov, 1969.
Gold content of the Atlantic Ocean igneous rocks. (Radioactivation analysis data). (In Russian).
Dokl. Akad. Nauk SSSR, 189(5): 1107-1110

chemistry, gold
Baur, E., 1942.
Sur la significance et le dosage de l'or des eaux marines. Bull. Inst. Océan., Monaco, No. 830:7 ppl

chemistry - gold
Bowen, H.J.M., 1968.
The uptake of gold by marine sponges.
J. mar. biol. Ass., U.K., 48(2):275-277.

chemistry gold
Caldwell, W.E., 1938.
Gold content of sea water. J. Chem. Educ., 15: 507-510.

chemistry - gold
Fukai, Rinnosuki, and W. Wayne Meinke, 1962
Activation analyses of vanadium, arsenic, molybdenum, tungsten, rhenium and gold in marine organisms.
Limnology and Oceanography, 7(2):186-200.

chemistry - gold
Haber, F., 1928.
Das Gold im Meer. Zeits. Ges. Erdk., Berlin, Ergänzungsheft, 2:3-12.

chemistry - gold
Haber, F., 1927.
Das Gold im Meerwasser. Zeits. Angew. Chem., 40: 303-314.

Chemistry, gold
*Harriss, Robert C., J.H. Crockett and M. Stainton, 1968.
Palladium, iridium and gold in deep-sea manganese nodules.
Geochim. cosmochim. Acta, 32(10):1049-1056.

chemistry, gold
Hummel, R.W., 1957.
Determination of gold in sea water by radioactivation analysis. Analyst 82(976):483-487.

chemistry, gold
Jaenicke, J., 1935.
Habers Forschungen über das Goldvorkommen im Meerwasser. Naturwiss. 23:57-63.

Chemistry, gold
Libby, Fred, 1969.
Searching for alluvial gold deposits off Nova Scotia.
Ocean Industry, Gulf Publishing Co., 4(1):43-46.

chemistry - gold
Peshchevitskii, B.I., G.N. Anoshin and A.M. Erenburg, 1965.
On the chemical varieties of gold in sea-water. (In Russian).
Doklady, Akad. Nauk, SSSR, 162(4):915-917.

chemistry gold
Polikarpov, Iu.A., and I.P. Ielovenko, 1970
Gold findings from off-shore and liman sands, the Black Sea north-western coast. (In Russian).
Dokl. Akad. Nauk SSSR 191(4): 905-908

chemistry - gold
Putnam, G.L., 1953.
The gold content of sea water. J. Chem. Educ. 30:576-579.

chemistry gold
Robineau, CH., and Ch. Desbois, 1955.
Etude sur la presence de l'or dans l'eau de mer.
Trav., C.R.E.O., 2(3):1-4.

chemistry, gold
Stark, W., 1943.
Über die Goldführung der Meere.
Helv. Chem. Acta, 26:424-441.

chemistry
gold
Weiss, H.V., and M.-G. Lai, 1963
Cocrystallization of ultramicro quantities of elements with 2-mercaptobenzimidazole.
Determination of gold in sea water.
Anal. Chim. Acta, 28(3):242-248.

guanine

chemistry
guanine
Rosenberg, E., 1964.
Purine and pyrimidines in sediments from the experimental Mohole.
Science, 146(3652):1680-1681.

hemocyanin.

chemistry, hemocyanin
Felsenfeld, G., 1954.
The binding of copper by hemocyanin.
J. Cell. Comp. Physiol. 43(1):23-38, 2 textfigs.

chemistry
hemocyanin
Melentis, John, and Anastasios A. Christomanos, 1970
Approach to the evolution and ontogenesis of hemocyanins.
Folia Biochim. Biolog. Graeca 7(3/4): 122-136

chemistry, hemocyanin
Redfield, A.C., 1950.
Hemocyanin, pp. 174-190 in: Copper metabolism, a symposium on animal, plant, and soil relationships. W.D. McElroy and B. Glass, editors. Johns Hopkins University Press, 443 pp.

chemistry
hemocyanin
Zuckerkandl, E., 1960
Hémocyanine et cuivre chez un crustacé décapode dans leurs rapports avec le cycle d'intermue.
Ann. Inst. Océanogr., Monaco, 38:1-122.

haemoglobin

Chemistry haemoglobin
Fox, H.M., 1957.
Haemoglobin in Branchiura. Nature 179(4565):873

Chemistry haemoglobin
Fox, H.M., 1957.
Haemoglobin in the Crustacea. Nature 179(4551): 148.

Chemistry, haemoglobin
Manwell, Clyde, E.C. Southward and A.J. Southheard, 1966.
Preliminary studies on haemoglobin and other proteins of the Pogonophora.
J. mar. biol. Ass., U.K., 46(1):115-124.

haemoglobin
Ruud, Johan T., 1965.
The ice fish.
Scientific American, 213(5):108-114.

Chemistry haemoglobin
Southward, Eve C., 1963.
Haemoglobin in barnacles.
Nature 200(4908):798-799.

halides

Chemistry, halides(geochemical)
Degens, Egon T., 1965.
Geochemistry of sediments: a brief survey.
Prentice-Hall, Inc., 342 pp.

halogens

Chemistry – halogens, geochemistry of
Correns, C.W., 1956.
The geochemistry of the halogens. In: Physics and Chemistry of the Earth, A.H. Ahrens, K. Rankama and S.K. Runcord, Eds., Pergamon Press, Ltd., 1(7):181-234.

Chemistry halogens
Giral, J., and F.A. Gila, 1923.
Sur l'emploi de chlorure de sodium comme étalon dans les dosages des halogènes de l'eau de mer.
C.R. Acad. Sci., Paris, 176(24):1729-1730.

Chemistry, halogens
Lewin, Ralph A., Editor, 1962.
Physiology and biochemistry of algae.
Academic Press, New York and London, 929 pp.

Chemistry, halogens
Suzuki, N., 1951.
Chemical studies on sea water (Pt. 1). On the determination of halogens (1). Bull. Fac. Fish., Hokkaido Univ., 1(2):69-80, 2 textfigs.

heavy metals

chemistry, heavy metals
Abrahamczik, E., 1938.
Zur Bestimmung geringer Menger von Schwermetallen in Wassern. Mikrochemie 25:228-233.

Cooper, L.H.N., 1956 chemistry, heavy metals
Some chemical and physical factors controlling the biological productivity of temperate and tropical oceanic waters. Proc. Eighth Pacific Sci. Congr., 3A:1157-1163.

Chemistry heavy-metals, effect of
Hood, Donald W., editor 1971
Impingement of man on the oceans.
Wiley-Interscience, 738 pp.

Chemistry, heavy metals
Koroleff, F., 1950.
Determination of traces of heavy metals in sea water by means of dithizone. 1. Metal-dithizone eqilibria. Merent. Julk., Havsforskningsinst. Skrift., No. 145:69 pp., 20 textfigs.

Chemistry, heavy metals
Nishikawa, Katsuo and Kenji Tabata, 1969.
Studies on the toxicity of heavy metals to aquatic animals and the factors to decrease the toxicity - III. Bull. Tokai reg. Fish. Res. Lab., 58: 233-241.
On the low toxicity of some heavy metal complexes to aquatic animals.
(In Japanese; English abstract)

chemistry - heavy metals
Noddack, I., and W. Noddack, 1940.
Die Haufigkeiten der Schwermetalle in Meerestieren.
Ark. Zool., 32A(4):35 pp.

Chemistry heavy metals
Tabata, Kenji, 1969.
Studies on the toxicity of heavy metals to aquatic animals and the factors to decrease the toxicity - IV. (In Japanese; English abstract). Bull. Tokai reg. Fish. Res. Lab., 58: 243-253.
On the relation between the toxicity of heavy metals and the quality of environmental water.

chemistry heavy metals
Tabata, Kenji, 1969.
Studies on the toxicity of heavy metals to aquatic animals and the factors to decrease the toxicity - II. (In Japanese; English abstract). Bull. Tokai reg. Fish. Res. Lab., 58: 215-232.
The antagonistic action of hardness components in water on the toxicity of heavy metal ions.

Chemistry, heavy metals
Tabata, Kenji, 1969.
Studies on the toxicity of heavy metals to aquatic animals and the factors to decrease the toxicity - I. (In Japanese; English abstract. Bull. Tokai Reg. Fish. Res. Lab., 58: 203-214.
On the formation and the toxicity of precipitate of heavy metals.

Chemistry heavy metals
Tabata, Kenji and Katsuo Nishikawa, 1969.
Studies on the toxicity of heavy metals to aquatic animals and the factors to decrease the toxicity - V. A trial to decrease the toxicity of heavy metal ions by the addition of complexing agents. (In Japanese; English abstract). Bull. Tokai reg. Fish. Res. Lab., 58: 255-264.

helium

chemistry, helium
Bieri, Rudolf H. 1971.
Dissolved noble gases in marine waters.
Earth Planet. Sci. Letts, 10(3): 329-333

Chemistry, helium (data)
Bieri, Rudolf H., Minoru Koide and Edward D. Goldberg.
Noble gas contents of marine waters.
Earth Planet. Sci. Letters, 4(5) 329-340

chemistry, helium
Bieri, Rudolf H., Mineru Koide and Edward D. Goldberg, 1966.
The noble gas contents of Pacific seawaters.
J. geophys. Res., 71(22):5243-5265.

chemistry helium
Bieri, Rudolph, Minoru Koide and Edward D. Goldberg, 1964
Noble gases in sea water.
Science, 146(3647):1035-1037.

chemistry helium
Craig, Harmon and W.B. Clarke, 1970
Oceanic ^{3}He: contribution from cosmogenic tritium.
Earth Planet. Sci. Letts 9(1): 45-48.

chemistry, helium
Craig, H. and R.F. Weiss, 1971.
Dissolved gas saturation anomalies and excess helium in the oceans.
Earth Planet. Sci. Letts 10(3): 289-296

Chemistry, helium
*Craig, H., R.F. Weiss and W.B. Clarke, 1967.
Dissolved gases in the equatorial and South Pacific Ocean.
J. geophys. Res., 72(24):6165-6181.

Chemistry - helium
König, H., H. Wänke, G.S. Bien, N.W. Rakestraw, and H.E. Suess, 1964.
Helium, neon and argon in the oceans.
Deep-Sea Research, 11(2):243-247.

chemistry - helium
Mazor, E., G.J. Wasserburg, and H. Craig, 1964.
Rare gases in Pacific Ocean water.
Deep-Sea Res., 11(6):929-932.

chemistry, helium
Rakestraw, N. W., C.E. Herrick, jr., and W.D. Urry, 1939.
The helium-neon content of sea water and its relation to the oxygen content. J. Am. Chem. Soc. 61:2806-2807.

chemistry, helium
Urry, W.D., 1935.
Further studies in the rare gases. III. The helium-neon content of ocean waters.
J. Amer. Chem. Soc., 57:657-659.

helium excess

Chemistry, helium excess
Bieri, R., M. Koide and E.D. Goldberg 1967.
Geophysical implications of the excess helium found in Pacific waters.
J. geophys. Res. 72 (10): 2497-2511.

chemistry, helium (excess)
Clarke, W.B., M.A. Beg, and Harmon Craig, 1969.
Excess ^{3}He in the sea: evidence for terrestrial primordial helium.
Earth Planet Sci Letters 6(3): 213-220

chemistry, heavy metals, effect of
Merlini, Margaret, 1971.
Heavy-metal contamination. (1) In: Impingement of man on the oceans, D.W. Hood, editor, Wiley Interscience: 461-486.

helium flux

helium flux
Suess, Hans E., and Heinrich Wanke, 1965.
On the possibility of a helium flux through the ocean floor.
Progress in Oceanography, 3:347-353.

chemistry helium isotopes
Weiss, R.F. 1970.
Helium isotope effect in solution in water and seawater. Science, 168(3928): 247-248.

Hexachlorocyclohexane

Chemistry, hexachlorocyclohexane
Werner, A.E., and M. Waldichuk, 1961.
Decay of Hexachlorocyclohexane in sea water.
J. Fish. Res. Bd., Canada, 18(2):287-289.

chemistry homogenates
Barnes, H., and D.M. Finlayson, 1963
Estimation of lactic acid in sea water solutions and homogenates.
Limnol. and Oceanogr., 8(2):292-294.

humic acid

chemistry, humus
Gadel, F., 1969.
Etude des substances humiques de sediments lagunaires et marins.
Vie Milieu (B)20(2): 221-255.

Chemistry humus water
Järnefelt, H., 1956.
Zooplankton und Humuswasser. Ann. Acad. Sci. Fennicae (A), 4 Biol., 31:3-14.

humus
Kato, K., 1956.
Chemical investigations on marine Humus in bottom sediments. Mem. Fac. Fish., Hokkaido Univ. 4(22):91-209.

Chemistry humus, marine
Kato, K., 1955.
Sedimentological examination on the influences of oceanographic environments upon marine humus distribution. Rec. Ocean. Wks., Japan, 2(3):53-62, 9 textfigs.

Chemistry humus
Kato, K., 1951.
Chemical studies on marine deposits. III. On the distribution of marine humus content in the sea to the northwest of Hokkaido. Bull. Fac. Fish., Hokkaido Univ., 1(2):10-30, 5 textfigs.

Chemistry humus
Kriss, A.E., and E.M. Markianovitch, 1959.
On the utilization of water humus in the sea by microorganisms.
Mikrobiologiia 28(3):399-406.

chemistry - humic substances
Ogura, N., 1967.
Humic substances in sea water. (A review). (In Japanese).
J. oceanogr. Soc., Japan, 23(3):141-147.

Chemistry humus
Oguro, Miki, 1961.
Chemical studies on the current rip in Ishikuri Bay caused by the Ishikari River flowing into that bay. 2. On the distribution of marine humus contents. (In Japanese; English abstract).
Bull. Fac. Fish., Hokkaido Univ., 12(1):81-87.

Chemistry humic substances, effect of
Prakash, A. and M.A. Rashid, 1968.
Influence of humic substances on the growth of marine phytoplankton: dinoflagellates.
Limnol. Oceanogr., 13(4):598-606.

chemistry, humic matter
Prakash, A., 1971.
Terrigenous organic matter and coastal phytoplankton fertility. (English and Portuguese abstracts). In: Fertility of the Sea, John D. Costlow, editor, Gordon Breach, 2: 351-368.

humus
Waksman, S.A., 1933.
On the distribution of organic matter in the sea bottom and the chemical nature of marine humus. Soil Sci. 36(2):125-147, 2 textfigs.

hydrocarbons

chemistry, hydrocarbons
Blumer, Max 1967
Hydrocarbons in digestive tract and liver of a basking shark.
Science 156 (3773): 390-391.

Chemistry hydrocarbons
Blumer, M., R.R. L. Guillard and T. Chase, 1971.
Hydrocarbons of marine phytoplankton. Marine Biol. 8(3): 183-189.

chemistry, hydrocarbons
Blumer, M., M.M. Mullin and R.R.L. Guillard 1970.
A polyunsaturated hydrocarbon (3, 6, 9, 12, 15, 18-heneicosahexaene) in the marine food web.
Mar. Biol. 6(3): 226-235.

Chemistry hydrocarbons
Boylan, D.B., and B.W. Tripp, 1971.
Determination of hydrocarbons in seawater extracts of crude oil and crude oil fractions.
Nature, Lond., 230 (5288): 44-47.

chemistry, hydrocarbons
Clark, Robert C., Jr., and Max Blumer, 1967.
Distribution of n-paraffins in marine organisms and sediments.
Limnol. Oceanogr., 12(1):79-87.

Chemistry hydrocarbons
Duursma, E.K., 1965.
The dissolved organic constituents of sea water. In: Chemical oceanography, J.P. Riley and G. Skirrow, editors, Academic Press, 1:433-475.

hydrocarbons
Hodgson, G.W., Brian Hitchon and Kazuo Taguchi, 1964.
The water and hydrocarbon cycles in the formation of oil accumulations.
In: Recent researches in the fields of hydrosphere, atmosphere, and nuclear geochemistry, Ken Sugawara festival volume, Y. Miyake and 217-242.

Chemistry, hydrocarbons
Jarvis, N.L., W.D. Garrett, M.A. Scheiman and C.O. Timmons, 1967.
Surface chemical characterization of surface-active material in seawater.
Limnol. Oceanogr., 12(1):88-96.

chemistry hydrocarbons
Mironov, O.G., 1969.
Microorganisms growing on hydrocarbons from the Black Sea.
Mikrobiologiia 38(4): 728-731.

Chemistry hydrocarbons
Rezak, Richard, Arnold H. Bouma and Lela M. Jeffrey, 1969
Hydrocarbons cored from knolls in southwestern Gulf of Mexico. Trans. Gulf Coast Ass. geol. Socs, 19 : 115-118.

chemistry, hydrocarbons
Swinnerton, J.W., and V.J. Linnenbom 1967.
Gaseous hydrocarbons in sea water: determination.
Science 156 (3778): 1119-1120.

chemistry, hydrocarbons
Lisitzin, A.P., Yu.A. Bogdanov and I.I. Ovchinnikova, 1967.
Some results of bituminous suspension studies from the Pacific Ocean. (In Russian; English abstract).
Okeanologiia, Akad. Nauk, SSSR, 7(1):120-129.

chemistry hydrocarbons
Youngblood, W.W., M. Blumer, R.L. Guillard and F. Fiore, 1971.
Saturated and unsaturated hydrocarbons in marine benthic algae. Marine Biol. 8(3): 190-201.

hydrocarbons aliphatic

Chemistry - hydrocarbons (aliphatic)
Nowlin, W.D., Jr., J.L. Harding and D.E. Amstutz, 1965.
A reconnaissance study of the Sigsbee knolls of the Gulf of Mexico.
J. Geophys. Res., 70(6):1339-1347.

hydrocarbons cancerigenic

chemistry, hydrocarbons, cancerigenic

See also: chemistry, benzo-3.4 pyrene

chemistry, hydrocarbons, cancerigenic
Greffard, Jacques, et Jean Meury, 1967.
Note sur la pollution en rade de Toulon par les hydrocarbures cancérigènes.
Cah. océanogr., 19(6):457-468.

Chemistry, hydrocarbons chlorinated
Hood, Donald W., editor 1971
Impingement of man on the oceans.
Wiley-Interscience, 738 pp.

chemistry, hydrocarbons (chlorinated)
Risebrough, R.W., 1971.
Chlorinated hydrocarbons. (10) In: Impingement of man on the oceans, D.W. Hood, editor, Wiley Interscience: 259-286.

Chemistry, hydrocarbons, effect of
Mironov, Oleg G., 1968.
Hydrocarbon pollution of the sea and its influence on marine organisms.
Helgoländer wiss. Meeresunters. 17: 335-339.

chemistry, hydrocarbon oxidation
Button, D.K., 1971.
Petroleum - biological effects in the marine environment. (14) In: Impingement of man on the oceans, D.W. Hood, editor, Wiley Interscience: 421-429.

hydrocarbon provinces

Chemistry - hydrocarbon provinces
Meyerhoff, A.A., 1967.
Future hydrocarbon provinces of Gulf of Mexico - Caribbean region.
Trans. Gulf Coast Geol. Assoc. Soc. 17th Ann. Meet.: 217-260.

1

hydrogen isotopes

chemistry - hydrogen isotopes

Culkin, Frederick, 1965.
The major constituents of sea water.
In: Chemical oceanography, J.P. Riley and G. Skirrow, editors, Academic Press, 1: 121-161.

chemistry, hydrogen isotopes

Savin, Samuel M., and Samuel Epstein, 1970.
The oxygen and hydrogen isotope geochemistry of ocean sediments and shales.
Geochim. Cosmochim. Acta, 34(1): 43-63.

hydrogen peroxide

chemistry, hydrogen peroxide

Van Baalen, C., and J.E. Marler 1966
Occurrence of hydrogen peroxide in sea water.
Nature, Lond. 211 (5052): 951.

H_2S

chemistry
hydrogen sulfide

Acara, A., and C. Erol, 1960
On the pollution of Golden Horn Estuary.
Rapp. Proc. Verb. Reun. Monaco, 15(3): 27-32.
Comm. Int. Expl. Sci. Mer. Med.

chemistry, hydrogen sulphide

Aleksandrovskaya, N.B., M.V. Kales and E.A. Yola, 1966.
Peculiarities of the hydrological regime of the Baltic Sea in 1964.
Annls. biol., Copenh., 21:43-45.

Andrusov, N.I., 1891. chemistry, hydrogen sulphide
Problem of a further study of the Black Sea and the surrounding countries. Ser. II. On the hydrogen sulphide fermentation in the Black Sea.
Zap. Russ. Akad. Nauk po fiz.-mat. otd. 1:1

not seen.

chemistry - hydrogen sulphide

Bacescu, M., 1961
Cercetări fizico-chimice şi biologice romînesti la Marea Neagră efectuate în perioda 1954-1959.
Hidrobiologia, Acad. Repub. Pop. Rom., (3): 17-46.

chemistry - hydrogen sulphide

Braarud, Trygve, Bjørg Føyn and Grethe Rytter Hasle, 1958
The marine and fresh-water phytoplankton of the Dramsfjord and the adjacent part of the Oslo-fjord, March - December 1951.
Hvalradets Skr., 43:102 pp.

chemistry, hydrogen sulphide

Brongersma-Sanders, M., 1947.
On the desirability of a research into certain phenomena in the region of upwelling along the coast of southwest Africa. Koninklijke Nederlandsche Akademie van Wetenschappen, Proc. 50(6):659-665.

chemistry - hydrogen sulphide

Chalmers, G.V., and A.K. Sparks, 1959.
An ecological survey of the Houston ship channel and adjacent bays.
Publ. Inst. Mar. Sci., 6:213-250.

chemistry, hydrogen sulfide

Copenhagen, W.J., 1953 (data)
The periodic mortality of fish in the Walvis Region. A phenomenon within the Benguela Current
Union of S. Africa, Div. Fish., Investig. Rept. No. 14:1-35, 9 pls.

chemistry, hydrogen sulfate

Deuser, W.G., 1970.
Carbon-13 in Black Sea waters and implications for the origin of hydrogen sulfide. Science, 168(3939): 1575-1577.

chemistry
hydrogen sulfide

Fonselius, Stig H., 1969.
Hydrography of the Baltic deep basins. III.
Rept., Fish. Bd., Sweden, Ser. Hydrogr., 23: 1-97.

chemistry
hydrogen sulphide

Fonselius, S.H. 1968.
Observations at Swedish lightships and in the central Baltic 1966.
Annls biol. Copenh. (1966) 23: 74-78.

chemistry - hydrogen sulfide

Fonselius, Stig H., 1967.
Hydrography of the Baltic deep basins II.
Rep. Fishery Bd., Swed., Hydrogr., 20:31 pp.

chemistry
hydrogen sulphide

Fonselius, Stig H., 1962
Hydrography of the Baltic deep basins.
Fishery Board of Sweden, Ser. Hydrogr., Rept. 13: 40 pp.

chemistry, hydrogen sulphide

Gaarder, T., and R. Spärck, 1932.
Hydrographisch-biochemische Untersuchungen in norwegischen Austern-Pollen. Bergens Mus. Aarbok, Naturvidensk.-rekke, No. 1:5-144, 75 textfigs.

chemistry, hydrogen sulfide

Gieskes, Joris M.T.M., and Klaus Grasshoff 1969.
A study of the variability in the hydrochemical factors in the Baltic Sea on the basis of two anchor stations, September 1967 and May 1968.
Kieler Meeresforsch. 25(1): 105-132

chemistry, hydrogen sulphide

Gilsen, K.K., 1918.
[Formation of hydrogen sulfide on the bottom of Onejakol Lake.] Bull. Akad. Nauk, SSSR, (6) 12(2): 2233.

Transl. cited:
USFWS Spec. Sci. Rept., Fish., 227.

chemistry, hydrogen sulphide

Golobolov, Y.K., 1953.
[The thickness of the oxygen-hydrogen sulfide layer in the Black Sea] Gidrokhim. Materialy, 21:3-9.

Abstr. Chem. Abstr. 11126e, 1954.

chemistry, hydrogen sulphide

Gololobov, I.K., and M.V. Pirogova, 1948.
[The upper boundary of the hydrogen sulfide region in the eastern portion of the Black Sea.] Dok. Akad Nauk, USSR, 63:179-182.

chemistry, hydrogen sulphide

Gucluer, Sevket M., and M. Grant Gross, 1964.
Recent marine sediments in Saanich Inlet, a stagnant marine basin.
Limnology and Oceanography, 9(3):359-376.

chemistry - hydrogen sulfide

Hata, Yoshihiko, 1965.
Microbial production of sulfides in marine and estuarine sediments. (In Japanese; English abstract).
J. Shimonoseki Univ., Fish., 14(2):37-83.

chemistry, hydrogen sulfide

Issatchenko, B.L., 1924.
Sur la fermentation sulphydrique dans la Mer Noire. C.R. Acad. Sci., Paris, 178:2204.

Chemistry
hydrogen sulphide

Ivannenkov, V.N., 1961
[The hydrogen sulphide contamination of the intermediate water layers of the Arabian sea and the Bay of Bengal.]
Okeanologiia, (3):443-449.

chemistry - Hydrogen sulfide

Kadota, Hajime, and Yuzaburo Ishida, 1968.
Evolution of volatile sulfur compounds from unicellular marine algae.
Bull. Misaki mar. biol. Inst. Kyoto Univ., 12: 35-48.

chemistry
hydrogen sulphide

Kaleis, M.V., N.B. Alexandrovskaya and E.A. Yula 1968.
Some peculiarities in the oceanographical regime of the Baltic in 1966.
Annls biol. Copenh. (1966) 23: 78-81.

Kändler, Rudolf 1971. hydrogen sulfide
Untersuchungen über die Abwasserbelastung der Untertrave.
Kieler Meeresforsch. 27(1): 20-29.

chemistry, hydrogen sulphide

Kandler, R., 1953.
Hydrographische Untersuchungen zum Abwasserproblem in den Buchten und Förden der Ostseeküste Schleswig-Holstein. Kieler Meeresf. 9(2):176-200, Pls. 9-14(18 figs.).

Chemistry-hydrogen sulphide

Kriss, A.E., 1949.
The role of microorganisms in the accumulation of hydrogen sulphide, ammonia and nitrogen in the depths of the Black Sea. Priroda 1949(6):35-46.

chemistry, hydrogen sulphide

Kaleis, M.V., and N.B. Aleksandrovskaye, 1965.
Hydrological regime of the Baltic Sea in 1963.
Ann. Biol., Cons. Perm. Int. Expl. Mer. 1963, 20:70-72.

chemistry, hydrogen sulphide

Kriss, A.E., and E.A. Rukina, 1949.
[Sur l'origine de l'hydrogène sulfuré dans la Mer Noire.] Microbiologija 18(4):332-345, 6 tables.

chemistry, hydrogen sulphide

Kriss, A.E., and E.A. Rukina, 1949.
[Les processus d'oxydoréduction dans la zone de la Mer Noire où les eaux sont chargées en H2S.
Microbiologija 18(5):402-415.

chemistry, hydrogen sulphide

Kriss, A.E., and E.A. Rukina, 1949.
[On the origin of hydrogen sulphide in the Black Sea.] Tr. Sevastop. Biol. Sta. 7:

Chemistry, hydrogen sulphide
Kriss, A.E., and E.A. Rukina, 1949.
[Oxidation and reduction processes in the hydrogen sulphide region of the Black Sea.]
Tr. Sevastop. Biol. St. 7:

Chemistry, hydrogen sulphide
Kriss, A.E., E.A. Rukina and A.S. Tikhonenko, 1956.
[Biomass of micro-organisms on the bottom of the hydrogen sulfide region, of the Black Sea.]
Dokl. Akad. Nauk, SSSR, 75:453-456.

Chemistry- Hydrogen sulphide
Koyama, Tadashiro, Nobuyuki Nakai and Eijire Kamata, 1965.
Possible discharge rate of hydrogen sulphide from polluted coastal belts in Japan.
J. Earth Sci., Nagoya Univ., 13(1):1-11.

Chemistry, hydrogen sulphide
Kudô, Hideo, 1964.
A method for the determination of the concentration of sulfides in sea water and the concentration of sulfides in sea water in Matsushima Bay. (In Japanese; English abstract).
Bull. Tohoku Reg. Fish. Res. Lab., No. 24:1-7.

Chemistry, hydrogen sulphide
Lefevre, S., E. Leloup, and L. Van Meal, 1956.
Observations biologiques dans le port d'Ostende.
K. Belg. Inst. Naturwet., Verhandl., [Inst. R. Sci. Nat. Belg., Mem.] 133:157 pp.

Chemistry, hydrogen sulphide (data)
Liubimova, E.M., 1959.
Vertical distribution of organic phosphorus in the Black Sea. Hydrometeorology. Hydrochemistry. (In Russian).
Trudy Morsk. Gidrofiz. Inst., 16:127-160.

Chemistry hydrogen sulphide
Naguib, Monir, 1961.
Studies on the ecology of Lake Quarun.
Kieler Meeresf., 17(1):94-131.

Chemistry, hydrogen sulfide
Nakai, Nobuyuki, and Mead LeRoy Jensen, 1960
Biogeochemistry of sulfur isotopes.
J. Earth Sciences, Nagoya Univ., 8(2): 181-196.

Chemistry, hydrogen sulphide
Neumann, G., 1944.
Das Schwarze Meer. Zeitsch. Gesellschaft f. Erdkunde zu Berlin. 1944 (3/4):92-114, 25 text figs.

Chemistry, hydrogen sulphide
Orr, A.P., 1947
An experiment in marine fish cultivation: II. Some physical and chemical conditions in a fertilized sea-loch (Loch Craiglin, Argyll).
Proc. Roy. Soc., Edinburgh, Sect.B, 63, Pt.1 (2):3-20,19 text figs.

owned by MS

Chemistry, hydrogen sulphide
Ostroumov, E.A., and V.M. Shilov, 1956.
[Iron sulfide and hydrogen sulfide in bottom sediments of the northwestern part of the Pacific.]
Dokl. Akad. Nauk, SSSR, 106(3):501-504.

Chemistry, hydrogen sulphide
Ostroumov, E.A., I.I. Volkov and L.S. Fomina, 1961.
[Distribution of sulphur compounds in the bottom sediments of the Black Sea.]
Trudy Inst. Okeanol., Akad. Nauk, SSSR, 50:93-129.

Chemistry, hydrogen sulphide
Özturgut, Erdogan 1965.
Hydrogen sulphide concentration in the Black Sea.
Rapp. P.-v. Réun. Commn int. Explor. scient. Mer Méditerr. 18(3): 745-747.

Chemistry hydrogen sulfide
Pieterse, F., and D.C. van der Post, 1967.
The pilchard of South West Africa (Sardinops ocellata): oceanographic conditions associated with red-tides and fish mortalities in the Walvis Bay region.
Investl Rept., Mar. Res. Lab., SWest Africa, 14: 1s5 pp.

Chemistry, hydrogen sulphide
Pova, E.A., 1946.
Problèmes de physiologie animale dans la Mer Noire. Bull. Inst. Océan., Monaco, No. 903:43 pp, 21 textfigs.

Chemistry hydrogen sulfide
Richards, Francis A., 1965.
Chemical observations in some anoxic, sulfide-bearing basins and fjords.
Proc. Int. Water Pollution Res. Conf. Tokyo, 1964, 215-243.

Chemistry - hydrogen sulfide
Richards, Francis A., Joel D. Cline, William W. Broenkow and Larry P. Atkinson, 1965.
Some consequences of the decomposition of organic matter in Lake Nitinat, an anoxic fjord.
Limnol. and Oceanogr., Redfield Vol., Suppl. to 10:R185-R201.

Chemistry- hydrogen sulphide
Riley, J.P., 1965.
Analytical chemistry of sea water.
In: Chemical oceanography, J.P. Riley and G. Skirrow, editors, Academic Press, 2:295-424.

Chemistry - hydrogen sulphide
Seelkopf, Carl, and Luis Boscan F., 1960
Hydrochemische Untersuchungen im Maracaibo-See.
Deutsche Hydrogr. Zeits., 13(4): 174-180.

Chemistry - hydrogen sulfide
Skopintsev, B.A., 1968.
A study of some reduction and oxidation processes in the Black Sea. (In Russian; English abstract).
Okeanologiia, Akad. Nauk, SSSR, 8(3):412-426.

Chemistry, hydrogen sulphide
Skopintsev, B.A., and E.V. Smirnov 1965.
Hydrogen sulphide in the abyssal waters of the Black Sea. (In Russian)
Okeanologiia Akad. Nauk SSSR 5(6): 969-982.

Chemistry hydrogen sulphide
Skopintsev, B.A., and E.V. Smirnov, 1962
The distribution of hydrogen sulphide in the Black Sea during the autumn of 1960.
Okeanologiia, Akad. Nauk, SSSR, 2(2):419-434.

Chemistry, hydrogen sulphide
Sorokin, Yu.I., 1971.
Experimental data on the oxidation rate of hydrogen sulphide in the Black Sea. (In Russian; English abstract). Okeanologiia 11(3): 423-431.

Chemistry - hydrogen sulphide
Sorokin, Yu.I., 1970.
Experimental studies of the rate and mechanism of oxidation of hydrogen sulphide in the Black Sea with the aid of sulphur-35. (In Russian; English abstract). Okeanologiia, 10(1): 51-62.

Chemistry - hydrogen sulphide
Sorokin, J.I., 1964.
On the primary production and bacterial activities in the Black Sea.
Journal du Conseil, 29(1):41-60.

Chemistry - hydrogen sulfide (data only)
Sweden, Havsfiskelaboratoriet, Lysekil, 1968.
Hydrographical data, July-December 1967, R.V. Skagerak.
Meddn Havsfiskelab., Lysekil, Hydrogr. Avd., Göteborg, 52: unnumbered pp. (multilithed).

Chemistry, hydrogen sulphide
Tambs-Lyche, H., 1957.
A simple method for quantitative determinations of H2S in sea water. Univ. i Bergen, Årbok 1956, Naturvit. rekke, (12):1-5.

Chemistry, hydrogen sulfide
Tambs-Lyche, H., 1954.
Notes on the hydrography of the Bolstadfjord, a land-locked fjord near Bergen.
Publ. Biol. Sta. 9 (Univ. i Bergen Aarbok, 1954, Naturv. rekke, No. 4):1-14, 6 textfigs.

Chemistry - hydrogen sulfide
Tanita, S., K. Kato, and T. Okuda, 1951.
Studies on the environmental conditions of shell-fish-fields. In the case of Muroran Harbour (2):
Bull. Fac. Fish., Hokkaido Univ. 2(3):220-230.

Chemistry, hydrogen sulphide
Tanita, S., S. Kato, and T. Okuda, 1950.
Studies on the environmental conditions of shell-fish fields. 1. In the case of Hakodate Harbor.
Bull. Fac. Fish., Hokkaido Univ., 1(1):18-27, 7 textfigs. (In Japanese; English summary).

Chemistry - hydrogen sulphide
Trimonis, E.S. and K.M. Shimkus, 1969.
Methods for collecting suspended matter from the sea water contaminated with suphuretted hydrogen. (In Russian; English abstract).
Okeanologiia 9(2): 358-360.

Chemistry, hydrogen sulphide
Ullyott, P. and Orhan Ilgaz 1946
The hydrography of the Bosphorus: an introduction.
Geol. Rev. 36(1): 44-66.

Chemistry hydrogen sulphide(data)
Volkov, I.I., 1961.
[On free hydrogen sulphide and some products of its transformation in the sediments of the Black Sea.]
Trudy Inst. Okeanol., Akad. Nauk, SSSR, 50:29-67.

Chemistry, hydrogen sulphide
Volkov, I.I., 1962.
The state of hydrogen sulphide in the water and sediments of the Black Sea.
Trudy Inst. Okeanol., Akad. Nauk, SSSR, 54:39-46.
In Russian; English summary

Chemistry - hydrogen sulphide
Volkov, I.I., 1960.
[Distribution of free hydrogen sulphide in the sediments of the Black Sea.]
Doklady Akad. Nauk, SSSR, 134(3):676-679.

chemistry, hydrogen sulphide
Wiborg, K.F., 1944
The production of plankton in a land locked fjord. The Nordåsvatn near Bergen in 1941-1947. With special reference to copepods. Fiskeridirektoratets Skrifter Serie Havundersøkelser (Rept. Norwegian Fish. and Marine Invest.) 7(7):1-83, Map.

chemistry, hydrogen sulphide
Zelinsky, 1893.
[The sulfhydric fermentation in the Black Sea.]
Russ. Fizika-Khim. Obshch. Zhurn., (25):298-303.

Transl. cited:
USFWS Spec. Sci. Rept., Fish., 227.

chemistry, hydrogen sulfide
Zelinski, N.D., 1893.
On the hydrogen sulphide fermentation in the Black Sea and the estuaries of Odessa. Zhurn. Russk. Fiz.-Khim. Obshch. 25:299-

hydrogen sulfide / ammonia

chemistry - hydrogen sulfide/ammonia
Richards, Francis A., Joel D. Cline, William W. Broenkow and Larry P. Atkinson, 1965.
Some consequences of the decomposition of organic matter in Lake Nitinat, an anoxic fjord.
Limnol. and Oceanogr., Redfield Vol., Suppl. to 10:R185-R201.

hydrogen sulphide (data only)

chemistry, hydrogen sulphide (data only)
Sweden, Havsfiskelaboratoriet, Lysekil, Hydrografiska Avd., Göteborg, 1967.
Hydrographical data, January-June 1967, R.V. Skagerak.
Meddn. Havsfiskelab., Lysekil, Hydrogr. Avd., Göteborg, 38: numerous pp. (mimeographed).

chemistry, hydrogen sulphide (data only)
Sweden, Havsfiskelaboratoriet, Lysekil, Hydrografiska Avd., Göteborg, 1967.
Hydrographical data, January-June 1967, R.V. Thetis.
Meddn. Havsfiskelab., Lysekil. Hydrogr. Avd., Göteborg, 41: numerous pp. (mimeographed).

chemistry, hydrogen sulfide, effect of
Dimov, I., 1964.
Influence de l'acide sulfhydrique sur la distribution verticale du zooplancton. (In Bulgarian; Russian and French summaries).
Izv. Inst. Ribiov'dstvo Ribolov, Varna, Isdatel. Bilgarskata Akad. Naukite, Sofia, 4:25-30.

chemistry, hydrogen sulfide, effect of
Williams, Peter M., and Arthur R. Coote, 1962
Mullin and Riley nitrate determination in the presence of hydrogen sulfide.
Limnology and Oceanography, 7(2):258.

hydroxides

chemistry, hydroxides (geochemical)
Degens, Egon T., 1965.
Geochemistry of sediments: a brief survey.
Prentice-Hall, Inc., 342 pp.

hydroxylamine

chemistry, hydroxylamine
Fiadeiro, M., L. Solórzano and J.D.H. Strickland, 1967.
Hydroxylamine in sea water.
Limnol. Oceanogr., 12(3):555-556.

hydroxyapatite

Chemistry hydroxyapatite
Kramer, James R., 1964
Sea water: saturation with apatites and carbonates.
Science, 146(3644):637-638.

hydroxyl ion (data only)

Chemistry, hydroxyl ions (data only)
Kvinge, Tor, 1963.
The "Conrad Holmboe" Expedition to East Greenland waters in 1923.
Årbok, Univ. i Bergen, Mat.-Naturv. Ser., (15): 44 pp.

Chemistry indicators, pigments
Margalef, Ramón, 1963
Modelos simplificados del ambiente marino para el estudio de la sucesión y distribución del fitoplancton y del valor indicador de sus pigmentos.
Invest. Pesquera, Barcelona, 23:11-52.

Chemistry indicators
Park, Kilho, June G. Pattullo and Bruce Wyatt, 1962
Chemical properties as indicators of upwelling along the Oregon coast.
Limnol. and Oceanogr., 7(3):435-437.

chemistry, indicators
Ramírez R., Boris, 1970.
Análisis de pigmentos vegetales marinos en un corte meridiano del Pacífico sur.
Inv. pesq. Barcelona 34(2): 309-318.

indium

Chemistry - indium
Barić, A. and M. Branica, 1969.
Behavior of indium in seawater (a preliminary study). Limnol. Oceanogr., 14(5): 796-798.

Chemistry, indium
Matthews, A.D., and J.P. Riley 1970.
Occurrence of indium in sea water and some marine sediments.
Nature, Lond. 225(5239): 1242.

Chemistry, indium
Ostroumov, E.A., and I.I. Volkov, 1964.
Separation of indium and gallium from manganese, nickel, cobalt and zinc with the help of cinnamic acid in application to study of sea deposits. Chemistry of the waters and sediments of the seas and oceans. (In Russian; English abstract).
Trudy Inst. Okeanol., Akad. Nauk, SSSR, 67:141-150.

chemistry, indole acetic acid
Chandramohan, D., 1971.
Indole acetic acid synthesis in sea.
Proc. Indian Acad. Sci. (B) 73(3): 105-109.

chemistry, inosine 5'-Monophosphate (IMP), effect of
Dingle, J.R., and J.A. Hines 1971.
Degradation of inosine 5'-monophosphate in the skeletal muscle of several North Atlantic fishes.
J. Fish. Res. Bd. Can. 28(8): 1125-1131

chemistry, intercalibration
Palmork, Karsten H., 1969.
Determination of reliability in marine chemistry by means of intercalibration and statistics.
In: Chemical oceanography, Rolf Lange, editor, Universitetsforlaget, Oslo: 91-103.

chemistry, invertebrates
Christomanos, Anast. A., Vula Gardiki und Olympia Vavatsi, 1955.
Uber das Eiweiss der Medusengallerte (Rhizostoma pulmo).
Folia Biochem. et Biol. Graeca, 2:25-32.

iodate

chemistry iodate
Petek, Milica and Marko Branica, 1969.
Hydrographical and biotical conditions in north Adriatic - III. Distribution of zinc and iodate. (Jugoslavian and Italian abstracts).
Thalassia Jugoslavica 5: 257-263.

chemistry, iodate
Yonehara, N., 1964.
The colorimetric determination of minute amounts of iodide and iodate in sea water by means of their catalytic effects.
Bull. Chem. Soc., Japan, 37(8):1101-1107.

iodide

chemistry - iodide
Tsunogai, S. and T. Sase, 1969.
Formation of iodide-iodine in the ocean. Deep-Sea Res., 16(5): 489-496.

chemistry, iodide
Yonehara, N., 1964.
The colorimetric determination of minute amounts of iodide and iodate in sea water by means of their catalytic effects.
Bull. Chem. Soc., Japan, 37(8):1101-1107.

iodine

chemistry, iodine
Barkley, R.A., and T.G. Thompson, 1960.
Determination of chemically combined iodine in sea water by amperometric and catalytic methods.
Anal. Chem., 32:154-158.

chemistry iodine
Barkley, R.A., & T.G. Thompson, 1960
The total iodine and iodate-iodine content of sea water.
Deep-Sea Res., 7(1): 24-34.

chemistry, iodine
Barrington, E.J.W., 1957.
The distribution and significance of organically bound iodine in the ascidian Ciona intestinalis Linnaeus.
J. Mar. Biol. Assoc., U.K., 36(1):1-16.

chemistry, iodine
Bruevich, S.W., 1960
[Hydrochemical investigations on the White Sea.]
Trudy Inst. Okeanol., 42: 199-254.

Chemistry iodine
Cooper, L.H.N., 1958
Oxidized iodine in sea water. Nature, 182:250-252

chemistry, iodine
Dean, G.A., 1963.
The iodine content of some New Zealand drinking waters with a note on the contribution from sea spray to the iodine in rain.
New Zealand Journal of Science, 6(2):208-214.

chemistry, iodine
Dubravcic, M., 1955.
Determination of iodine in natural waters.
Analyst 80:295-300.

Chemistry, iodine
Fukai, Rinnosuke 1965.
Remarks on the chemical problems in relation to the marine radioactivity
Rapp. P. v. Réun. Commn int. Explor. scient. Mer méditerr. 18(3): 861-863

Chemistry iodine
Hiyama, Yoshio, and Junko Matsubara Khan, 1964
On the concentration factors of radioactive I, Co, Fe and Ru in marine organisms.
Rec. Oceanogr. Wks., Japan, 7(2):79-106.

Chemistry, iodine
Ishibashi, Masayoshi, Koichi Emi, Tamotsu Matsumoto, and Takashi Teramoto, 1959
Basic Studies of the new method of separation of iodine from natural waters. Adsorption of iodine to Polyvinyl glycerin acetal. Rec. Oceanogr. Wks., Japan, Spec. No. 3: 103-108.

chemistry, iodine
Kappanna, A.N., and V. Sitakara Rao, 1962
Iodine content of marine algae from Gujarat Coast.
J.Sci. Industr. Res., (B), 21(11):559-560.

Chemistry -iodine
Kikkawa, Kyozo, and Shiko Shiga, 1965.
The modified chemical constituent of sea water intruding into the coastal aquifer.
Spec. Contrib., Geophys. Inst., Kyoto Univ., No. 5:7-16.

Chemistry, iodine
*Kuenzler, Edward J., 1969.
Elimation of iodine, cobalt, iron and zinc by marine zooplankton.
Symp. Radioecology, Proc. Second Natn. Symp., Ann Arbor, Mich., 1967 (Conf. 670503): 462-473.

Chemistry iodine
Miyake, Yasuo, and Shizuo Tsunogai, 1963
Evaporation of iodine from the ocean.
J. Geophys. Res., 68(13):3989-3993.

chemistry, iodine
Pastowska, Stefania, and Stanislaw Ostrowski 1968.
On the iodine content in maritime air of the southern Baltic Sea.
Acta geophys. pol. 16(2): 181-194

Chemistry iodine
Pavlova, G.A., and O.V. Shishkina, 1964.
The method of the determination of iodine in the interstitial waters. Chemistry of the waters and sediments of the seas and oceans. (In Russian; English abstract).
Trudy Inst. Okeanol., Akad. Nauk, SSSR, 67:165-176.

Chemistry iodate
Petek, Milica, and Marko Branica, 1969 (1968)
Hydrographic and historical conditions in North Adriatic. III Distribution of zinc and iodate.
Thalassia jugosl. 5 (1968)
Abstract in: Rapp. P.-v. Reun. Commn int. Explor. scient. Mer Méditerr., 19(4): 765.

chemistry, iodine
Reith, J.F., 1930.
Der Jodgehalt von Meerwasser.
Res. Trav. Chim., Pays-Bas, 49:142-150.

chemistry, iodine
Roche, Jean, 1959.
On some aspects of iodine biochemistry in marine animals.
Pubbl. Sta. Zool., Napoli, 31(Suppl.):176-189.

chemistry, iodine
Revel Jean 1969
Dosage de l'iode minéral total dans l'eau de mer par méthode automatique à l'autoanalyseur technicon.
Cah. océanogr. 21(3): 273-281.

chemistry, iodine
Roche, Jean, et Simonne Andre, 1965.
Sur la biochemie comparée du transport des iodures chez les vertébrés, les invertébrés et dans les algues.
Rivista Biol., Univ. Perugia, 58(1):3-25.

Chemistry, iodine
Schulz, B., 1930.
Die Beziehung zwischen Jod- und Salzgehalt des Meerwassers. Ann. Hydrogr., usw., (5):187-
Transl. cited:
USFWS Spec. Sci. Rept., Fish., 227.

chemistry, iodine
Shaw, T.I., and L.H.N. Cooper, 1957.
State of iodine in sea water. Nature 180(4579): 250

chemistry, iodine
Skopintsev, B.A., 1933.
[The determination of iodine in sea water.]
Trudy Gosud. Okeanogr. Inst., Leningrad, 3(3): 105-119.

chemistry, iodine
Skopintsev, B. A., and L.A. Michailouskaia, 1933
[Iodine in the White Sea water.]
Trudy Gosud. Okeanogr. Inst., Leningrad, 3(3):79-87.

chemistry, iodine
*Sugawara, Ken, 1967.
Retrospection and future problems of the study of iodine in sea water.
Rec. oceanogr. Wks. Japan, 9(1):23-36.

Chemistry, iodine
Sugawara, K., 1957.
The distribution of some minor bio-elements in Western Pacific waters. Proc. UNESCO Symp., Phys. Ocean., Tokyo, 1955:169-174.

chemistry, iodine
Sugawara, K., T. Koyama and K. Terada, 1955.
A new method of spectrophotometric determination of iodine in natural waters. Bull. Chem. Soc., Japan, 28(7):494-497.

chemistry, iodine
Sugawara, Ken, and Kikuo Terada, 1958.
Iodine distribution in the western Pacific Ocean (Abstract).
Proc. Ninth Pacific Sci. Congr., Pacific Sci. Assoc., 1957, Oceanogr., 16:200.
Published in:
J. Earth Sciences, Nagoya Univ., Japan, 5(2), 1957.

chemistry, iodine
Sugawara, K., and K. Terada, 1957.
Iodine distribution in the western Pacific.
J. Earth Sci., Nagoya Univ., 5(2):81-102.

chemistry, iodine
Sugawara, K., K. Terada, J.K. Johannesson, T.I. Shaw and L.H.N. Cooper, 1958.
Oxidised iodine in sea water. Nature 182(4630): 250-252.

chemistry iodine
Sugawara, Ken, Kikuo Terada, Satoru Kanamori, Nobuko Kanamori and Shiro Okabe, 1962
On different distribution of calcium, strontium iodine, arsenic and Molydenum in the north-western Pacific, Indian and Antarctic Oceans.
J. Earth Sciences, Nagoya Univ., 10(1):34-50.

chemistry, iodine
Tsunogai, Shizuo, 1971.
Iodine in the deep water of the ocean. Deep-Sea Res. 18(9): 913-919.

chemistry, iodine
Tsunogai, Shizuo, and Takashi Henmi 1971.
Iodine in the surface water of the ocean.
J. Oceanogr. Soc. Japan, 27(2): 67-72.

chemistry iodine
Voipio, A., 1961
The iodine content of Barents Sea water.
Contribution to Special IGY Meeting, 1959.
Cons. Perm. Int. Expl. Mer, Rapp. Proc. Verb. 149:38-39.

Chemistry, iodine
Winchester, John W., and Robert A. Duce, 1965.
Geochemistry of iodine, bromine, and chlorine in the air-sea sediment system.
Narragansett Mar. Lab., Univ. Rhode Island, Occ. Publ., No. 3:185-201.

iodine/chlorine
Chemistry iodine/chlorine
Kikkawa, Kyozo, and Shiko Shiga, 1965.
The modified chemical constituent of sea water intruding into the coastal aquifer.
Spec. Contrib., Geophys. Inst., Kyoto Univ., No. 5:7-16.

iodine vapor
Chemistry, iodine vapor
Barringer, A.R., 1967.
Detecting the ocean's food (and pollutants) from space.
Ocean Industry, 2(5): 25-34

chemistry, ions
Garrels R.M. and M.E. Thompson, 1962
A chemical model for sea water at 25 C and one atmosphere total pressure.
Amer. J. Sci.. 260(1):57-66.

chemistry
ions (complex)

Mangel, Marc S., 1971.
A treatment of complex ions in sea water.
Mar. Geol. 11 (2): M24-M26

chemistry
ions

Millero, Frank J., 1969.
The partial molal volumes of ions in seawater.
Limnol. Oceanogr., 14(3): 376-385.

ion concentrations

Almazov, A.M., 1956. chemistry, ion concentrations
On the proportions between ion concentrations in waters of open estuaries. (In Russian).
Doklady, Akad. Nauk, SSSR, 108(5):833-836.

chemistry ion concentrations

Nellen, Walter 1970.
2.5. Wasserstoffionenkonzentration. Chemische, mikrobiologische und planktologische Untersuchungen in der Schlei im Hinblick auf deren Abwasserbelastung. Kieler Meeresforsch 26(2): 128-130.

chemistry - ionic discharge

*Alekin, O.A., 1967.
On the significance for the ocean of ion discharge of rivers. (In Russian; English abstract
Okeanologiia, Akad. Nauk, SSSR, 7(4):555-560.

ionic ratios

chemistry, ion ratios
*Khlebovich, V.V., 1968.
Some peculiar features of the hydrochemical regime and the fauna of mesohaline waters.
Marine Biol., 2(1):47-49.

chemistry, ionic ratios
Kirsch, M., 1956.
Ionic ratios of some of the major components in river-diluted sea water in Bute and Knight Inlets, British Columbia.
J. Fish. Res. Bd., Canada, 13(3):273-289.

ionium

chemistry, ionium
Baranov, V.J., and L.R. Kuzmina, 1954.
[Ionium method of the age determination of sea sediments.] Dokl. Akad. Nauk, SSSR 97:483.

chemistry, ionium
Baranov, V.I., and Kuz'mina, L.A., 1954.
[An ionium method for determining the age of sea cores. The direct determination of ionium.]
Dokl. Akad. Nauk, SSSR, 92(3):483.

chemistry, ionium

Heye, Dietrich, 1970.
A system for detection of ionium thorium and protactinium to date deep-sea cores.
Geochim. cosmochim. Acta, 34 (3): 389-397.

chemistry, ionium
Holland, H.D., and J.L. Kulp, 1954.
The mechanism of removal of ionium and radium from the oceans. Geochim. Cosmochim. Acta 5:214-224.

Abstr: Chem. Abstr. 11855b, 1954.

chemistry, ionium
Holland, H.D., and J.L. Kulp, 1954.
The transport and deposition of uranium, ionium and radium in rivers, oceans and ocean sediments.
Geochim. Cosmochim. Acta 5:197-213.

Abstr.: Chem. Abstr. 11855a, 1954.

chemistry, ionium
Holland, H.D., and J.L. Kulp, 1954.
The transport and deposition of uranium, ionium and radium in rivers, oceans and ocean sediments.
Geochimica et Cosmochimica Acta 5:197-213.

chemistry, ionium
Holland, H.D., and L.J. Kulp, 1954.
The mechanism of removal of ionium and radium from the oceans. Geochimida et Cosmochimica Acta 5:214-224, 6 textfigs.

chemistry, ionium
Isaac, N., and E. Picciotto, 1953.
Ionium determination in deep-sea sediments.
Nature 171(4356):742-743.

chemistry - ionium
Miyaki, Y., K. Saruhashi, Y. Katsuraji, J. Kanazawa and Y. Sugimura, 1964.
Uranium, radium, thorium, ionium, strontium 90 and cesium 137 in coastal waters of Japan.
In: Recent researchers in the fields of hydrosphere, atmosphere and nuclear geochemistry, Ken Sugawara festival volume, Y. Miyake and T. Koyama, editors, Maruzen Co., Ltd., Tokyo, 127-141.

chemistry, ionium
Miyake, Yasuo, Katsuko Saruhashi and Yukio Sugimura, 1968.
Biogeochemical balance of natural radioactive elements in the sea.
Rec. oceanogr. Wks, Japan, n.s., 9(2):180-187.

chemistry, iridium
Barker, John L. Jr., and Edward Anders 1968.
Accretion rate of cosmic matter from iridium and osmium contents of deep-sea sediments
Geochim. cosmochim. Acta 32 (6): 627-645.

chemistry, iridium
*Harriss, Robert C., J.H. Crockett and M. Stainton, 1968.
Palladium, iridium and gold in deep-sea manganese nodules.
Geochim. cosmochim. Acta, 32(10):1049-1056.

iron

chemistry, iron
Armstrong, F.A. J., 1957.
The iron content of sea water. J.M.B.A., 36(3): 509-517.

chemistry, iron

Atkinson, Larry P., and Unnsteinn Stefansson 1969.
Particulate aluminum and iron in sea water off the southeastern coast of the United States.
Geochim. cosmochim. Acta 33(6): 1449-1453

chemistry, iron
*Aubert, M., J. Aubert, S. Daniel, J.-P. Gambarotta, L.-A. Romey, J.-Ph. Mangin, P.-F. Bulard, P. Irr et Y. Chevalier, 1967
Dynamique d'eaux résiduaires rejetées en bassin maritime.
Rev. intern. Océanogr., Med. 8:5-40.

chemistry, iron
Aubert, M., et J.P. Gambarotta, 1966.
Étude de la répartition du fer dans les eaux de surface au large des côtes des Alpes-Maritimes.
Rev. Intern. Océanog. Méd., 2:27-51.

chemistry, iron
Aubert, M., J.P. Gambarotta et F. Laumond, 1967.
Étude de la répartition du fer au large des côtes de Provence et de Corse: Étude de la dispersion des apports terrigènes.
Rev. int. Océanogr. Méd. 5:23-61.

chemistry, iron
*Berner, Robert A., 1968.
Migration of iron and sulfur within anaerobic sediments during early diagenesis.
Am. J. Sci., 267(1):19-42.

chemistry, iron
Bernhard, M., 1964.
Chemical composition and the radiocontamination of marine organisms.
Proc. Symp. Nuclear Detonations and Marine Radioactivity, S.H. Small, editor, 137-150.

chemistry, iron
Bonatti, Enrico, and Oiva Joensuu, 1966.
Deep-Sea iron deposit from the South Pacific.
Science, 154(3749):643-645.

chemistry, iron

Boström, Kurt and David E. Fisher 1971.
Volcanogenic uranium, vanadium and iron in Indian Ocean sediments.
Earth planet. Sci. Letts 11(2): 95-98.

chemistry, iron
Braarud, T., and A. Klem, 1931.
Hydrographical and chemical investigations in the coastal waters off Møre and in the Romsdalfjord.
Hvalrådets Skrifter, 1931(1):88 pp., 19 figs.

chemistry, iron
Brodskaya, N.G., and T.G. Martova, 1957.
[Forms of iron in the modern sediments of the Okhotsk Sea.] Dokl. Akad. Nauk, SSSR, 114(2):165-168.

chemistry, iron
Bruevich, S.V., and E.G. Vinogradska, 1946.
General features of sedimentation in the Caspian Sea according to the distribution of carbonates, Fe, Mn, and P in sea deposits. Dok. Akad. Nauk, SSSR, n.s., 52:789-792.

chemistry, iron

Bryan, G.W. 1968.
Concentrations of zinc and copper in the tissues of decapod crustaceans.
J. mar. biol. Ass. U.K. 48(2): 303-321.

chemistry, IRON

*Burton, J.D. and P.C. Head, 1970.
Observations on the analysis of iron in seawater with particular reference to estuarine waters.
Limnol. Oceanogr., 15(1): 164-167.

estuaries
iron

chemistry - iron
Chester, R., 1965.
Elemental geochemistry of marine sediments.
In: Chemical oceanography, J.P. Riley and G. Skirrow, editors, Academic Press, 2:23-80.

chemistry, iron

*Choe, Sang, Tai Wha Chung and Hi-Sang Kwak, 1968. Seasonal variations in nutrients and principal ions contents of the Han River water and its water characteristics. (In Korean; English abstract). J. oceanogr. Soc., Korea, 3(1):26-38.

chemistry
Iron (ferric)

Chow, Tsaihwa J., 1965. Chemical equilibrium of ferric iron in sodium chloride medium. Trav. Centre de Recherches et d'Etudes Oceanogr. n.s., 6(1/4):53-55.

chemistry, iron

Chu, S. P., 1949
Experimental studies on the environmental factors influencing the growth of phytoplankton. Sci. & Tech. in China 2(3):37-52.

chemistry
iron

Collins, Peter, and Harvey Diehl, 1960

Tripridyltriazine, a reagent for the determination of iron in sea water. J. Mar. Res., 18 (3): 152-156.

chemistry iron

Cooper, L.H.N., 1948
The distribution of iron in the waters of the western English Channel. JMBA 27(2):279-313, 2 text figs.

chemistry, iron

Cooper, L.H.N., 1948
Some chemical considerations on the distribution of iron in the sea. JMBA, 27(2): 314-321.

chemistry, iron

Cooper, L.H.N., 1939.
Phosphorus, nitrogen, iron, and manganese in marine zooplankton. J.M.B.A. 23:387-390.

chemistry, iron

Corcoran, E.F., and J.E. Alexander, 1964. The distribution of certain trace elements in tropical sea water and their biological significance. Bull. Mar. Sci., Gulf and Caribbean, 14(4):594-601.

chemistry, iron

Corcoran, E.F., and James E. Alexander, 1963. Nutrient, chlorophyll and primary production studies in the Florida Current. Bull. Mar. Sci., Gulf and Caribbean, 13(4):527-541.

chemistry, iron (data)

Donnelly, P.V., J. Vuille, M.C. Jayaswal, R.A. Overstreet, J. Williams, M.A. Burklew and R.M. Ingle, 1966.
A study of contributory chemical parameters to red tide in Apalachee Bay. Florida Bd., Conserv., St. Petersburg, Mar. Lab. Prof. Papers Ser. No. 8:43-83.

chemistry, iron

*Emilianov, E.M., N.B. Vlasenko and S.A. Orlova, 1968.
Some data on iron distribution in suspended matter from the Baltic. (In Russian; English abstract). Okeanologiia, Akad. Nauk, SSSR, 8(4):638-645.

chemistry, iron

Fukai, Rinnosuke, and Lang Huynh-ngoc 1968.
Studies on the chemical behaviour of radionuclides in sea-water. 1. General considerations and study of precipitation of trace amounts of chromium, manganese, iron, cobalt, zinc and cerium.
IAEA Radioactivity in the sea, Publ. 22:1-26

chemistry, iron

Garland, C.F., 1952.
A study of water quality in Baltimore Harbor. Chesapeake Biol. Lab. Publ. 96:1-132, figs.

chemistry, iron

Gennesseaux, Maurice, 1960.
Migration expérimentale du fer dans les sédiments marins. C.R. Acad. Sci., Paris, 251(22):2564-2565.

Also in: Trav. Lab. Géol. Sous-Marine, 10(1960).

chemistry, iron

Goldberg, E.D., 1952.
Iron assimilation by marine diatoms. Biol. Bull. 102(3):243-248.

chemistry
Iron

Harris, Robert C., 1965.
Trace element distribution in molluscan skeletal material. 1. Magnesium, iron, manganese and strontium. Bull. Mar. Sci., 15(2):265-273.

chemistry iron

*Hartmann, Martin, 1969.
Investigations of Atlantis II Deep samples taken by the FS Meteor.
In: Hot brines and Recent heavy metal deposits in the Red Sea, E.T. Degens and D.A. Ross, editors, Springer-Verlag, New York, Inc., 204-207.

chemistry, iron

Japan, Hokkaido University, Faculty of Fisheries, 1967.
Data record of oceanographic observations and exploratory fishing, 11:383 pp.

chemistry, iron

Hartmann, Martin, 1964.
Zur Geochemie von Mangan und Eisen in der Ostsee. Meyniana, 14:3-20.

chemistry, iron

Hashitani, H., and K. Yamamoto, 1959.
Simultaneous determination of traces of iron and aluminium in sea water. J. Chem. Soc., Japan, 80(7):727-731.

Abstr. in: Anal. Abstr., 7(6):#2498.

chemistry, iron

Head, P.C., 1971.
Observations on the concentration of iron in sea water, with particular reference to Southampton Water. J. mar. biol. Ass. U.K. 51(4): 891-903.

chemistry
iron

Hiyama, Yoshio, and Junko Matsubara Khan, 1964
On the concentration factors of radioactive I, Co, Fe and Ru in marine organisms.
Rec. Oceanogr. Wks., Japan, 7(2):79-106.

chemistry,
iron

Hood, D.W., 1963
Chemical oceanography. In: Oceanography and Marine Biology, H. Barnes, Edit., George Allen & Unwin, 1:129-155.

chemistry, iron

Ichie, T., K. Tanioka, and T. Kawamoto, 1950.
Reports of the oceanographical observations on board the R.M.S. "Yushio Maru" off Shionomisaki (Aug. 1949). Papers and Repts., Ocean., Kobe Mar. Obs., Ocean. Dept., No. 5:15 pp., 33 figs. (Odd atlas-sized pages - mimeographed).

chemistry
iron

Imai, Yoshihiko, 1961
Oceanographical studies on the behaviour of chemical elements. IV. Distribution of iron and aluminium in Uranouchi Bay - particularly on concentration of those elements in the "Shiome" zone of the tide rip.
J. Oceanogr. Soc., Japan, 17(2):96-100.

chemistry,
iron

Imai, Yoshihiko, 1961
Oceanographical studies on the behaviour of chemical elements, III. Iron and aluminium distribution in Susaki Bay with the concentration of those elements in the "Shiome" zone.
J. Oceanogr. Soc., Japan, 17(1):48-53.

chemistry
iron

Imai, Y., 1960

[Oceanographical studies on the behaviour of chemical elements. 2. Iron and aluminum distribution in Urado Bay - On the concentration of those elements in the "Shiome" zone, or the rip tide.]
J. Oceanogr. Soc., Japan, 16(4): 167-171.

chemistry
iron

Imai, Y., 1960

[Oceanographical studies on the behaviour of chemical elements. 1. On iron and manganese in the sediments around "Shiome" (current rips) in Urado Bay.]
J. Oceanogr. Soc., Japan, 16(3): 134-138.

Chemistry iron

Ishibashi, Masayoshi, and Toshio Yamamoto, 1960

Chemical studies on the ocean (Pts. 78-79). Chemical studies on the seaweeds (VI & VII). The content of calcium, magnesium and phosphorus in seaweeds. Iron content in sea weeds. Rec. Oceanogr. Wks., Japan, Spec. No. 4: 73-78; 79-86.

chemistry, iron(data)

Japan, Central Meteorological Observatory, 1951.
The results of marine meteorological and oceanographic observations. July - Dec. 1949. No. 6: 423 pp.

chemistry, iron(data)

Japan, Central Meteorological Observatory, 1951.
Table 6. Oceanographical Observations taken in the Akashi-seto, the Yura-seto and the Kii Suido; physical and chemical data for stations occupied by R.M.S. "Shumpu-maru".
Res. Mar. Met. Ocean. Obs., Jan.-June 1949, No. 5:40-47.

chemistry, iron

Jarke, J., 1949.
Die Entstehungsmöglichkeiten von Eisenbikarbonat und Eisenkarbonat in rezenten Flachseesedimenten. Deutsch. Hydro. Zeit. 2(6):286-291.

chemistry
iron

Johnston, R., 1964
Sea water, the natural medium of phytoplankton II. Trace metals and chelation, and general discussion.
J. Mar. Biol. Assoc., U.K., 44(1):87-109.

chemistry
iron

Joyner, Timothy, 1964.
The determination and distribution of particulate aluminum and iron in the coastal waters of the Pacific northwest.
J. Mar. Res., 22(3):259-268.

Chemistry, iron

Ketchum, B.H., A.C. Redfield, and J.C. Ayers, 1951.
The oceanography of the New York Bight. P.P.O.M. 12(1):1-46, 20 textfigs.

Reviewed: J. du Cons. 18(2):246-247 by D. Vaux.

Chemistry, iron

Koczy, F.F., 1950.
Zur Sedimentation und Geochemie im aequatorischen Atlantischen Ozean. Medd. Oceanografiska Inst., Göteborg 17 (Göteborgs Kungl. Vetenskaps- och Vitterhets Samhälles Handlingar, Sjätte Följden Ser. B, 6(1)):44 pp., 17 textfigs.

Chemistry, iron

Labeyrie, J., 1968.
Le fer dans la mer.
Rev. intern. Oceanogr. Med., 11:129-139.

Chemistry, iron

Laevastu, Taivo, and T. G. Thompson, 1958.
Soluble iron in coastal waters.
J. Mar. Res., 16(3):192-198.

chemistry, iron

Leisegang, E.C. and M.J. Orren, 1966
Trace element concentrations in the sea off South Africa.
Nature 211 (5054): 1166-1167.

chemistry, iron

Lewin, Joyce and Ching-Hong Chen, 1971.
Available iron: a limiting factor for marine phytoplankton. Limnol. Oceanogr. 16(4): 670-675.

Chemistry, iron

Lewis, G.J., jr., and E.D. Goldberg, 1954.
Iron in marine waters. J. Mar. Res. 13(2):183-197, 2 textfigs.

Chemistry, iron

Lisitsin, A.P., 1964.
Distribution and chemical composition of suspended matter in the waters of the Indian Ocean. (In Russian: English Abstract).
Rezult. Issled., Programme Mezhd. Geofiz. Goda, Mezhd. Geofiz. Kom., Presidume Akad. Nauk, SSSR, Okeanol., No. 10:136 pp.

Chemistry, iron

*López-Benito, M., 1969.
Distribución del hierro en la ría de Vigo.
Investigación pesq. 33(1): 119-141.

chemistry, iron

López-Benito, Manuel, 1967.
Determinación espectrofotométrica de hierro en el agua de mar con el reactivo de dipiridilo.
Investigación pesq., 31(1):17-31.

Chemistry, iron

López-Benito, Manuel 1967.
Estudio espectrofotométrico de diferentes reacciones analíticas formadoras de complejos del hierro y su aplicación al conocimiento del estado de este catión en el agua de mar.
Investigación pesq. Barcelona 31(3): 585-609.

Chemistry, iron (data)

Matsudaira, Yasuo, Haruyuki Koyama and Takuro Endo, 1961
[Hydrographic conditions of Fukuyama Harbor.]
J. Fac. Fish. and Animal Husbandry, Hiroshima Univ., 3(2):247-296.

Chemistry, iron

McDougall, J.C., 1961
Ironsand deposits offshore from the west coast, North Island, New Zealand.
New Zealand J. Geol. Geophys., 4(3):283-300.

chemistry, iron

Menzel, D.W., and J.H. Ryther, 1961
Nutrients limiting the production of phytoplankton in the Sargasso Sea, with reference to iron.
Deep-Sea Res., 7(4): 276-281.

Chemistry, iron

Menzel, David W., and Jane P. Spaeth, 1962
Occurrence of iron in the Sargasso Sea off Bermuda
Limnology and Oceanography, 7 :155-158.

Chemistry, iron

Mero, John L., 1965.
The mineral resources of the sea.
Elsevier Oceanogr. Ser., 312. pp.

Chemistry, iron

Miller, A.R., C.D. Densmore, E.T. Degens, J.C. Hathaway, F.T. Manheim, P.F. McFarlin, R. Pocklington and A. Jokela, 1966.
Hot brines and recent iron deposits in deeps of the Red Sea.
Geochimica et Cosmochimica Acta, 30(3):341-360.

Chemistry, iron

Mokyevskaya, V.V., 1962.
Methods of determination of iron in sea and interstitial water. (In Russian; English summary)
Trudy Inst. Okeanol., Akad. Nauk, SSSR, 54:115-122.

Chemistry, iron

Mokievskaia, V.V., 1960.
[On the problem of the existence of iron in the marine and interstitial waters of the Black Sea. Chemistry of the sea.]
Trudy Okeanogr. Komissii, Akad. Nauk, SSSR, 10(2):21-29.

chemistry, iron

Mokievskaja, V.V., 1959
[Geochemistry of iron in sea water.] Trudy Inst. Okeanol., 33: 114-125.

Chemistry, iron

Naguib, Monir, 1961.
Studies on the ecology of Lake Quarun.
Kieler Meeresf., 18(1):94-131.

Chemistry, iron

Neuman, W., 1949.
Beiträge zur Hydrographie des Vrana-sees (Insel Cherso), insbesondere Untersuchungen über organisches sowie anorganische Phosphor- und Stickstoffverbindungen. Nova Thalassia 1(6):17 pp.

Chemistry, iron

Nevessky, E.N. and F.A. Shcherbakov, 1969.
Peculiarities of iron accumulation and distribution in the sediments from the Kandalaksha Bay. Okeanologiia, 9(4): 649-660.
(In Russian; English abstract)

Chemistry, iron

Okabe, Shiro, Yoshimasa Toyota and Takafumi Murakami, 1967.
Major aspects of chemical oceanography. (In Japanese; English abstract).
J. Coll. mar. Sci. Techn., Tokai Univ., (2):39-56.

Chemistry, iron

Orren, M.J. 1967.
Trace elements (copper, iron and manganese) off the coast of South Africa.
Invest'l Rep. Fish. mar. Biol. Surv. Div. Un. S. Afr. 59: 1-40.

Chemistry, iron

Parker, Patrick L., Ann Gibbs and Robert Lawler, 1963.
Cobalt, iron and Manganese in a Texas bay.
Publ. Mar. Sci., Port Aransas, 9:28-32.

chemistry, iron

Pettersson, H., and H. Rotschi, 1952.
The nickel content of deep-sea deposits.
Geochimica et Cosmochimica Acta 2:81-90, 4 textfigs.

Chemistry, iron

Pilkey, Orrin H., and H.G. Goodell, 1963
Trace elements in Recent mollusk shells.
Limnol. and Oceanogr., 8(2):137-148.

iron

Rakestraw, N. W., H. E. Mahncke, and E. F. Beach, 1936.
Determination of iron in sea water. Ind. Eng. Chem., Analyt. Ed., 8:136-138.

Chemistry, iron

Richards, Francis A., Joel D. Cline, William W. Broenkow and Larry P. Atkinson, 1965.
Some consequences of the decomposition of organic matter in Lake Nitinat, an anoxic fjord.
Limnol. and Oceanogr., Redfield Vol., Suppl. to 10:R185-R201.

Chemistry, iron

Rodhe, W., 1948
Environmental requirements of fresh-water plankton algae. Experimental studies in the ecology of phytoplankton. Symbolae Botanicae Upsalienses X(1):149 pp., 30 figs.

Chemistry, iron

Rotschi, H., 1951.
Étude des teneurs en fer, manganèse et nickel de quelques carottes des grands fonds. Cahiers du Centre de Recherches et d'Études Océanographiques No. 4:2-22, 8 textfigs.

Chemistry, iron

Ryther, John H., D.W. Menzel and Nathaniel Corwin, 1967.
Influence of the Amazon River outflow on the ecology of the western tropical Atlantic. I. Hydrography and nutrient chemistry.
J. mar. Res., 25(1):69-83.

Chemistry, iron

Sananman, Michael, and Donald W. Lear, 1961.
Iron in Chesapeake Bay waters.
Chesapeake Science, 2(3/4):207-209.

chemistry, iron
Seiwell, G. E., 1935.
Note on iron analysis of Atlantic coastal waters.
Ecology 16(4):663-664.

chemistry, iron
Simons, L.H., P.H. Monaghan and M.S. Taggert, 1953.
Aluminum and iron in Atlantic and Gulf of Mexico waters.
Analyt. Chem., 25:989-990

Chemistry, iron
Skopintsev, B.A. and T.P. Popova 1960.
Some results of the determination of the content of iron, manganese and copper in the waters of the Black Sea. Chemistry of the sea, hydrology and marine geology. (In Russian)
Trudy morsk. Gidrofiz. Inst. 19:21-30.

chemistry iron
Smeyda, Theodore J., 1966.
A quantitative analysis of the phytoplankton of the Gulf of Panama. III General ecological conditions and the phytoplankton dynamics at 8o 45'N, 79o 23'W from November 1954 to May 1957.
Inter-Amer. Trop. Tuna Comm., Bull., 11(5): 355-612.

chemistry, iron
Smayda, T.J., 1957.
Phytoplankton studies in lower Narragansett Bay.
Limnol. & Oceanogr., 2(4):342-357.

chemistry, iron
Strakhov, N.M., 1958.
On the types of iron sediments of the Black Sea.
Doklady Akad. Nauk, SSSR, 118(4):803-806.

chemistry, iron
Strakhov, N.M., 1948.
Distribution of iron in the sediments of lake and sea systems and the factors controlling it.
Izvest. Akad. Nauk, USSR, Geol., 4:30-50.

Chemistry iron
Strickland, J.D.H., & K.H. Austin, 1959
The direct estimation of ammonia in sea water with notes on "reactive" iron, nitrate and inorganic phosphorus.
J. du Conseil, 24(3):446-451.

chemistry, iron
Subrahmanyan, R., and R. Sen Gupta, 1965.
Studies on the plankton of the east coast of India. II Seasonal cycle of plankton and factors affecting marine production with special reference to iron content of water.
Proc. Indian Acad. Sci., (B), 61(1):12-24.

chemistry, iron
Tanita, S., K. Kato and T. Okuda, 1951.
Studies on the environmental conditions of shell-fish-fields. In the case of Muroran Harbour (2).
Bull. Fac. Fish., Hokkaido Univ., 2(3):220-230.

chemistry, iron
Taylor W. Rowland, H.H. Selinger, W.G. Fastie and W.D. McElroy, 1966.
Biological and physical observations on a phosphorescent bay in Falmouth Harbor, Jamaica, W.I.
J. Mar. Res., 24(1):28-43.

Chemistry, iron
Topping, Graham, 1969.
Concentrations of Mn, Co, Cu, Fe, and Zn in the northern Indian Ocean and Arabian Sea. J. mar. Res., 27(3): 318-326.

chemistry, iron
Toyota, Yoshimasa and Shiro Okabe, 1967.
Surface distribution of iron, aluminum, silicon and phosphorus in the western North Pacific and eastern Indian Oceans. (In Japanese).
J. Coll. mar. Sci. Techn., Tokai Univ., (2):227-229.

Chemistry Iron
Toyota, Yoshimasa and Shiro Okabe, 1967.
Vertical distribution of iron, aluminum, silicon, and phosphorus in particulate matter collected in the western North Pacific, Indian, and Antarctic oceans. (In Japanese; English abstract)
J. Oceanogr. Soc. Japan, 23(1):1-9
Also in: Coll. Repr. Coll. mar. Sci. Techn., Tokai Univ., 1967-68, 3:335-343.

chemistry iron
Tranter, D.J., and B.S. Newell, 1963
Enrichment experiments in the Indian Ocean.
Deep-Sea Res., 10(1/2): 1-9.

Chemistry IRON
Turner, Ralph R. and Robert C. Harriss, 1970.
The distribution of non-detrital iron and manganese in two cores from the Kara Sea. Deep-Sea Res., 17(3): 633-636.

chemistry, iron
Wilkniss, P.E., T.B. Warner and R.A. Carr 1971.
Some aspects of the geochemistry of F, Fe and Mn in coastal waters and in fresh-water springs on the southeast coast of Hawaii.
Mar. Geol. 11 (4): M39-M46

Chemistry, iron
Wilson, Thomas A., 1965.
Offshore mining paves the way to ocean mineral wealth.
Engineering and Mining Jour., 166(6): 124-132.

Chemistry, iron (acid soluble)
Won, Chong Hun, 1963.
Distribution of chemical constituents of the estuary water in Gwang-Yang Inlet. (In Korean; English abstract).
Bull. Fish. Coll., Pusan Nat. Univ., 5(1):1-10.

chemistry, iron
Yalkovsky, Ralph, 1967.
Signs test applied to Caribbean Core A 176-6.
Science, 155 (3768):1408-1409.

Chemistry iron
Yamamoto, H., and Y. Imai, 1959
Oceanographical studies on Japanese inlets. X. On the distribution of iron and manganese in sea water in Urado Bay. J. Oceanogr. Soc., Japan, 15(4): 185-190.

chemistry iron
Yamamoto, Sakujiro, 1959
Studies of beach-sands as chemical resources. On sand-irons in beach-sands along the coastline of Tottori Prefecture.
Rec. Oceanogr. Wks., Japan, Spec. No. 3: 141-144.

Chemistry, iron
Zelenov, K.K., 1964.
Iron and manganese in exhalations of the submarine Banu Vukhu volcano (Indonesia).
(IN Russian).
Doklady, Akad. Nauk, SSSR, 155(6):1317-1320.

Chemistry, iron
Zelenov, K.R., 1958.
On iron in solution carried into the Okhotsk Sea by the hot springs of the Ebeko volcano (Paramushiro Island) Doklady Akad. Nauk, SSSR, 120(5):1089-1092.

Chemistry iron
Zenkevich, N.L., and N.S. Skornyakova, 1961.
Iron and manganese on the ocean floor.
Priroda, 1961(2):47-50.

OTS - Soviet-bloc Res. Geophys. Astron., and Space, No. 10(1960):7.

Chemistry, iron (data only)
Anon., 1950.
Table 14. Oceanographical observations taken in the Maizuru Bay; physical and chemical data for stations occupied by R.M.S. "Syunpu-maru". Res. Mar. Met., Ocean. Obs., Tokyo, July-Dec. 1947, No. 2:71-76.

Chemistry, iron (data only)
Ivanoff, A., A. Morel, Mme Vesin-Couffinhal, C. Amiel, J.P. Bethoux, J. Boutler, C. Copin-Montegut, P. Courau, P. Geistdoerfer, et F. Nyffler 1969.
Résultats des observations effectuées en mer Méditerranée orientale et en mer Tyrrhénienne à bord du navire Amalthée en mars 1967.
Cah. océanogr. 21 (Suppl 2): 245-263

Chemistry, iron (effect of)
Aubert, M., J.P. Gambarotta et F. Laumond 1968.
Rôle des apports terrigènes dans la multiplication du phytoplancton marin: cas particulier du fer.
Rev. int. Océan. Med. 12: 75-121

Chemistry iron, effect of
Hayward, J., 1968.
Studies on the growth of Phaeodactylum tricornutum. III. The effect of iron on growth.
J. mar. biol. Ass., U.K., 48(2):295-302.

Chemistry iron, effect of
Ryther, John H., and Dana D. Kramer, 1961
Relative iron requirement of some coastal and offshore algae.
Ecology, 42(2): 444-446.

iron metabolism

chemistry, iron metabolism
*Hobden, D.J., 1967.
Iron metabolism in Mytilus edulis. 1. Variation in total content and distribution.
J. mar. biol. Ass., U.K., 47(3):597-606.

Iron oxides

Chemistry iron oxides
Hunt, John M., Earl E. Hays, Egon T. Degens and David A. Ross, 1967.
Red Sea: detailed survey of hot-brine areas.
Science, 156(3774):514-516.

chemistry iron oxides (amorphous)
von der Borch, C.C. and R.W. Rex, 1970.
Amorphous iron oxide precipitates in sediments cored during leg 5, deep sea drilling project.
Initial Repts. Deep Sea Drilling Project, Glomar Challenger 5: 541-544.

iron, particulate

chemistry, iron (particulate)

Betzer, Peter R. and Michael E.Q. Pilson, 1971.
Particulate iron and the nepheloid layer in the western North Atlantic, Caribbean and Gulf of Mexico. Deep-Sea Res. 18(7): 753-761.

Chemistry, iron (particulate)

Betzer, P.R. and M.E.Q. Pilson 1970
Concentrations of particulate iron in Atlantic open-ocean water.
J. mar. Res. 28(2): 251-267.

Chemistry iron (particulate)

Kato, Kenji, 1957.
Particulate iron in Bering Sea.
Bull. Fac. Fish., Hokkaido Univ., 7(4):291-299.

Reprinted in 1959 in:
Plankton Invest., Hokkaido Univ.

Chemistry iron, particulate

Schaefer, M.B., and Y.M.M. Bishop, 1958.
Particulate iron in offshore water of the Panama Bight and in the Gulf of Panama. Limnol. & Oceanogr., 3(2):137-149.

chemistry, iron phorphyrin

Dyrssen, David, 1969.
Stiochiometry and chemical equilibrium. In: Chemical oceanography, Rolf Lange, editor, Universitetsforlaget, Oslo: 47-59.

Iron, radioactive

Chemistry iron (radioactive isotopes)

Kautsky, Hans, and Dieter E. Schmitt, 1962
Eine Bestimmungsmethode für die radioaktiven Isotope 55 Fe und 59 Fe in Meerwasser.
Deutsche Hydrogr. Zeits., 15(5):199-204.

Iron-55

chemistry, iron-55

Palmer, H.E., T.M. Beasley and Theodore R. Folsom, 1966.
Iron-55 in marine environment and in people who eat ocean fish.
Nature, 211(5055):983-985.

Chemistry, iron-55

Servant, Jean 1966.
La radioactivité de l'eau de mer.
Cah. Océanogr. C.C.O.E.C. 18(4): 277-318.

iron-59

chemistry, iron-59

Parker, Patrick L., 1966.
Movement of radioisotopes in a marine bay: cobalt-60, iron-59, manganese-54, zinc-65, soduim-22.
Publs. Inst. mar. Sci. Univ. Texas Fort Aransas, 11:102-107.

chemistry, iron-59

Tsukidate, Jun-ichi and Shunzo Suto, 1971.
Tracer experiments on the effect of micro-nutrients on the growth of Porphyra plants - 1. Iron-59 assimilation in relation to environmental factors. Bull. Tokai reg. Fish. Res. Lab. 64: 89-100.

iron sulfide

chemistry, iron sulfide

Berner, Robert A., 1967.
Diagenesis of iron sulfide in Recent marine sediments.
In: Estuaries, G.H. Lauff, editor, Publs Am. Ass. Advmt Sci., 83:268-272.

chemistry - iron sulphide

Volkov, I.I., 1957.
[Concretions of iron sulfide in Black Sea deposits]
Doklady Akad. Nauk, SSSR, 116(4):654-

chemistry - iron sulphide

Volkov, I.I., and E.A. Ostroumov, 1957.
[Concretions of iron sulfide in Black Sea deposits]
Doklady Akad. Nauk, SSSR, 116(4):645-648.

Chemistry - iron sulfide

Ostroumov, E.A., and V.M. Shilov, 1956.
[Iron sulfide and hydrogen sulfide in bottom sediments of the northwestern part of the Pacific]
Dokl. Akad. Nauk, SSSR, 106(3):501-504.

chemistry, iron sulphide

Ostroumov, E.A., and W.M. Schilov, 1958.
[Verteilung von Schwefeleisen und Schwefelwasserstoff in den Bodenablagerungen des Nord-Westlich-en Teiles des Stillen Ozeans.]
Trudy Inst. Okeanol., 27:77-86.

chemistry - iron transformation

Sorokin, Yu.I. and Yu.A. Bogdanov, 1971.
Transformation of iron in the process of bacterial decomposition of the plankton organic substance. (In Russian). Gidrobiol. Zh., 7(2): 106-107.

isoprenoid acids

chemistry, isoprenoid acids

Blumer, Max, and W.J. Cooper, 1967.
Isoprenoid acids in Recent sediments.
Science, 158(3807):1463-1464.

isotopes

Chemistry, isotopes

Isotopes are chiefly entered under the Chemical element

Chemistry, isotopes

Brodsky, A.E., and N.P. Radschenko, 1940.
[The isotope content of Arctic Seas and ice]
Acta Physicochemica URSS 13:145.

chemistry, isotopes

Brodskii, A.I., N.P. Radchenko and B.L. Smolenskaya, 1940.
[Die Isotopenzusammensetzung der arktischen Meere und Eise] Acta Physicochim., SSSR, 13(1):145-156.

Chemistry, isotopes

Brodskii, A.I., N.P. Radchenko and B.L. Smolenskaya, 1939.
Isotopic composition of Arctic waters, ices and glaciers. J. PHys. Chem., SSSR, 13:1494-1501.

Chem. Abstr., 1941;35:353.

Chemistry, isotopes

Cooper, J. A. and J. R. Richards, 1966.
Isotopic and alkali measurements from the Vema Seamount of the South Atlantic Ocean.

Nature, 210(5042):1245-1246.

chemistry, isotopes

Dansgaard, W., 1961
The isotopic composition of natural waters with special reference to the Greenland ice cap.
Medd. om Grønland, 165(2): 1-120.

chemistry, isotopes

Johnston, William H., 1962.
Some new applications of isotopes in oceanography. (by title only).
J. Geophys. Res., 67(9):3569.

Chemistry isotopes (produced by cosmic rays)

Lal, D., 1962
Cosmic ray produced radionuclides in the sea.
J. Oceanogr. Soc., Japan, 20th Ann. Vol., 600-614.

Chemistry, isotopes

Model, Fr., 1955.
Isotope und Ozeanographie. Ein Literaturbericht.
Deutsche Hydrogr. Zeits., 8(5):203-206.

Chemistry, isotope dating

See: Age determination (geological)

chemistry, isotope ratios

Deuser, Werner G., and Egon T. Degens, 1969.
δ^{18}/O^{16} and C^{13}/C^{12} ratios of fossils from the hot-brine deep area of the central Red Sea.
In: Hot brines and Recent heavy metal deposits in the Red Sea, E.T. Degens and D.A. Ross, editors, Springer-Verlag, New York, Inc., 336-347.

Chemistry isotopic composition, variation in volume

Menache, Maurice, 1966.
Variation de la masse volumique de l'eau en fonction de sa composition isotopique.
Cahiers Oceanogr., C.C.O.E.C., 18(6):477-496.

chemistry, isotope variations

Parker, Patrick L., 1971.
Petroleum - stable isotope ratio variations.(15)
In: Impingement of man on the oceans, D.W. Hood, editor, Wiley Interscience: 431-444.

Chemistry isotopic water

Wirth, H.E., T.G. Thompson and C.L. Utterback, 1935.
Distribution of isotopic water in the sea.
J. Amer. Chem. Soc., 57:400-404.

chemistry, ketones

Corwin, James F., 1969.
Volatile oxygen-containing organic compounds in sea water: determination. Bull. mar. Sci., 19(3): 504-509.

Chemistry - kinetin, effect of

Bentley-Mowat, J.A., and S.M. Reid, 1969.
Effect of gibberellins, kinetin and other factors on the growth of unicellular marine algae in cultures.
Botanica mar. 12(1/4): 185-199

krypton

chemistry, krypton

Bieri, Rudolf H., Mineru Koide and Edward D. Goldberg, 1966.
The noble gas contents of Pacific seawaters.
J. geophys. Res., 71(22):5243-5265.

krypton (data)

Bieri, Rudolf H., Mineru Koide and Edward D. Goldberg.
Noble gas contents of marine waters.
Earth Planet. Sci. Letters, 4(5) 329-340

chemistry krypton
Bieri, Rudolph, Minoru Koide and Edward D. Goldberg, 1964
Noble gases in sea water.
Science, 146(3647):1035-1037.

chemistry krypton
Mazor, E., G.J. Wasserburg, and H. Craig, 1964.
Rare gases in Pacific Ocean water.
Deep-Sea Res., 11(6):929-932.

lactic acid

chemistry, lactic acid
Barnes, H., and D.M. Finlayson, 1963
Estimation of lactic acid in sea water solutions and homogenates.
Limnol. and Oceanogr., 8(2):292-294.

chemistry, lanthanides
Graham, A.L., and G.D. Nicholls, 1969.
Massspectrographic determinations of lanthanide element contents in basalts.
Geochim. cosmochim. Acta 33(5):555-568

chemistry lanthanides
*Masuda, Akimasa, 1968.
Nature of the experimental Mohole basalt - redetermination of lanthanides.
J. geophys. Res., 73(16):5425-5428.

lead

Chemistry, lead
Boury, M., 1938.
Le plomb dans le milieu marin.
Trav. Rev. Off. Pêches Marit., 11:157-165.

Chemistry, lead
Chester, R., 1965.
Elemental geochemistry of marine sediments.
In: Chemical oceanography, J.P. Riley and G. Skirrow, editors, Academic Press, 2:23-80.

Chemistry lead
Chow, T.J., 1958
Lead isotopes in sea water and marine sediments. J. Mar. Res., 17: 120-127.

chemistry lead
Chow, Tsaihwa J. and John L. Earl, 1970.
Lead aerosols in the atmosphere: increasing concentrations. Science 169(3945): 577-580.

chemistry, lead
Chow, T.J., and C.C. Patterson, 1966.
Concentration profile of barium and lead in Atlantic waters off Bermuda.
Earth Planet. Sci. Letters, 1(6):397-400.

Chemistry lead
Chow, T.J., M. Tatsumoto and C.C. Patterson, 1962
Lead isotopes and uranium contents in experimental Mohole cores (Guadalupe site).
J. Sed. Petr., 32(4):866-869.

chemistry, lead
Gusyatskaya, E.V., and L.G. Longinova, 1955.
Spectrographic method of determining lead in natural waters. Izv. Akad. Nauk, SSSR, Ser. Fiz., 19(2):194-196.
Abstr. Anal. Abstr., 3(12):#3768.

chemistry - lead
Hood, Donald W., editor 1971
Impingement of man on the oceans.
Wiley-Interscience, 738 pp.

chemistry, lead
Ishibashi, Masayoshi, Taitiro Fujinaga, Fuji Morii, Yoshihiko Kanchiku, and Fumio Kamiyama, 1964.
Chemical studies on the ocean (Part 94). Chemical studies on the seaweeds (19). Determination of zinc, copper, lead, cadmium and nickel in seaweeds using dithizone extraction and polarographic method.
Rec. Oceanogr. Wks., Japan, 7(2):33-63.

Chemistry, lead
Ishibashi, Masayoshi, Shunzo Ueda and Yoshikazu Yamamoto, 1960
Studies on the utilization of the shallow-water deposits (Cont.). On the lead content of the shallow-water deposits. Rec. Oceanogr. Wks., Japan, Spec. No. 4: 111-122.

Chemistry, lead
Lakshman, S.V.J., and P. Tiruvenganna Rao, 1962.
Quantitative spectrochemical analysis of lead in ocean bed sediments.
J. Sci. Industr. Res., India, 21B(4):174-176.

chemistry, lead
Murozumi, M., T.J. Chow and C. Patterson, 1965.
Concentration of common lead in Greenland snows.
Narragansett Mar. Lab., Univ. Rhode Island, Occ. Publ., No. 3:213-215.

Chemistry, lead
Oversby, V.M., and P.W. Gast, 1968.
Oceanic basalt leads and the age of the earth.
Science, 162(3856):925-927.

chemistry, lead
Patterson, Clair, 1971.
Lead.(?)In: Impingement of man on the oceans, D.W. Hood, editor, Wiley Interscience: 245-258.

Chemistry, lead
Tatsumoto, M., 1957.
Chemical investigations of deep-sea deposits. 22. The contents of cobalt and nickel in sea sediment (2). 23. The contents of tin and lead in sea sediments. J. Chem. Soc., Japan, 78(1):38-48.

Chemistry, lead
Tatsumoto, M., and C.C. Patterson, 1963.
The concentration of common lead in sea water. Symposium on oceanic mixing. In: Nuclear Geophysics, Proc. Conf., Woods Hole, Massachusett, June 7-9, 1962.
NAS-NRC, Publ., No. 1075:167-175.
Also: Nuclear Sci. Ser. Rept., No. 38.

chemistry, lead
Turekian, Karl E., and Donald F. Schutz, 1965.
Trace element economy in the oceans.
Narragansett Mar. Lab., Univ. Rhode Island, Occ. Publ., No. 3:41-89.

Chemistry - lead
Ulrych, T.J., 1968.
Oceanic basalt leads and the age of the earth.
Science, 162(3856):928.

lead isotopes

Chemistry - lead isotopes
*Chow, Tsaihwa J., 1968.
Lead isotopes of the Red Sea region.
Earth Planet. Sci. Letters, 5(3):143-147.

Chemistry, lead isotopes
Chow, Tsaihwa J., 1965.
Radiogenic leads of Canadian and Baltic shield regions,
Narragansett Mar. Lab., Univ. Rhode Island, Occ. Publ., No. 3:169-184.

Chemistry - lead isotopes
Chow, Tsaiheva J., and M. Tatsumoto, 1964.
Isotopic compostition of lead in the sediments near Japan Trench.
In: Recent researches in the fields of hydrosphere, atmosphere and nuclear geochemistry, Ken Sugawara festival volume, Y. Miyake and T. Koyama, editors, Maruzen Co., Ltd., Tokyo, 179-183.

Chemistry lead isotopes
Cooper, J.A., and J.R. Richards, 1969.
Lead isotopes measurements on sediments from Atlantic II and discovery deep areas.
In: Hot brines and Recent heavy metal deposits in the Red Sea, E.T. Degens and D.A. Ross, editors, Springer-Verlag, New York, Inc., 499-511.

Chemistry - lead isotopes
*Delevaux, M.H., B.R. Doe and G.F. Brown, 1967.
Preliminary lead isotope investigations of brine from the Red Sea, galena from the Kingdom of Saudi Arabia, and galena from United Arab Republic (Egypt).
Earth Planet. Sci., Letters, 3(2):139-144.

Chemistry, lead isotopes
Oversby, Virginia M., and Paul W. Gast 1970.
Isotopic composition of lead from oceanic islands.
J. geophys. Res. 75(11):2097-2114

chemistry, lead isotopes
Reynolds, Peter H. and E. Julius Dasch, 1971.
Lead isotopes in marine manganese nodules and the Ore-lead growth curve. J. geophys. Res., 76(21):5124-5129.

Chemistry, lead-210
Shannon, L.V., R.D. Cherry and M.J. Orren 1970.
Polonium-210 and lead-210 in the marine environment.
Geochim. Cosmochim. Acta 34(6): 701-711.

chemistry, lead isotopes
Tatsumoto, Mitsunobu, 1969.
Lead isotopes in volcanic rocks and possible ocean-floor thrusting beneath island arcs.
Earth Planet. Sci. Letters 6(5): 369-376

chemistry, lead isotopes
Tatsumoto, M., 1966.
Genetic relations of oceanic basalts as indicated by lead isotopes.
Science, 153(3740):1094-1101.

chemistry, lead isotopes
*Ulrych, T.J., 1967.
Oceanic basalt leads: a new interpretation and an independent age for the earth.
Science, 158(3798):252-256.

lead-210

chemistry, lead 210

Goldberg, Edward S., 1962
Lead 210 in natural waters. (Abstract)
J. Geophys. Res., 67(9):3561.

chemistry, lead-210

Rama, M. Koide, and E.D. Goldberg, 1961.
Lead-210 in natural waters.
Science, 134(3472):98-99.

chemistry, lead-210

Servant, Jean, 1966.
La radioactivité de l'eau de mer.
Cahier Océanogr., C.C.O.E.C., 18(4):277-318.

leptocel

chemistry, leptocel

Fox, D.J., J.D. Isaacs, and E. F. Corcoran, 1952.
Marine leptocel, its recovery, measurement and distribution. J. Mar. Res. 11(1):29-46, 4 text-figs.

lipids

see also: chromolipoids

chemistry, lipids

Ackman, R.G., and C.A. Eaton 1971.
Investigation of the fatty acid composition of oils and lipids from the sand lance (Ammodytes americanus) from Nova Scotia waters.
J. Fish. Res. Bd, Can. 28(4):601-606

chemistry, lipids

Bogorov, B.G., O.K. Bordovsky and M.E. Vinogradov 1966.
Biogeochemistry of the oceanic plankton. The distribution of some chemical components of the plankton in the Indian Ocean. (In Russian; English abstract).
Okeanologiia, Akad. Nauk, SSSR, 6(2):314-325.

chemistry, lipids

Boucaud-Camou, E., 1971.
Constituants lipidiques du foie de Sepia officinalis. Marine Biol., 8(1): 66-69.

chemistry lipids

Duursma, E.K., 1965.
The dissolved organic constituents of sea water.
In: Chemical oceanography, J.P. Riley and G. Skirrow, editors, Academic Press, 1:433-475.

chemistry, lipids

*Hussein, M. Fawzy, R. Boulus and F.M. Hanna, 1967.
Studies on the chemical composition of plankton of Lake Qarun. I. Seasonal variations in the protein, lipids and carbohydrate content of plankton.
Bull. Fac. Sci., Cairo Univ. Press, 40:121-131.

chemistry, lipids

Jeffrey, Lela M., undated reprint.
Lipids in sea water.
J. Amer. Oil Chemists' Soc., 43(4):211-214.
Also in:
Texas A & M Univ., Contrib. Oceanogr. Meteorol. 10. 1965-1966.

chemistry lipids

Kates, M., 1966.
Lipid components of diatoms.
Biochimica et Biophysica Acta, 116(2):264-278.

chemistry, lipids

Lee, R.F., J.C. Nevenzel and G.-A. Paffenhöfer, 1971.
Importance of wax esters and other lipids in the marine food chain: phytoplankton and copepods. Mar. Biol. 9(2): 99-108.

chemistry, lipids

Lee, Richard F., Judd C. Nevenzel, G.-A. Paffenhöfer and A.A. Benson 1970.
The metabolism of wax esters and other lipids by the marine copepod, Calanus helgolandicus.
J. Lipid Res. 11(3):237-240
Also in: Coll. Repr. Scripps Inst. Oceanogr. 40.

chemistry lipids

Linford, Eileen, 1965.
Biochemical studies on marine zooplankton. II. Variations in the lipid content of some Mysidaces.
J. Cons. perm. int. Explor. Mer, 30(1):16-27.

chemistry, lipids

Lovern, J.A., 1964.
The lipids of marine organisms.
In: Oceanography and Marine Biology, Harold Barnes, Editor, George Allen & Unwin, Ltd., 2:169-191.

chemistry lipids

Morris, Robert J., 1971.
Seasonal and environmental effects on the lipid composition of Neomysis integer. J. mar. biol. Ass., U.K., 51(1): 21-31.

chemistry, lipids

Nonaka, Junsaku, and Chiaki Koizumi 1964.
Component fatty acids and alcohols of Euphausia superba lipid by gas-liquid chromatography. (In Japanese; English abstract)
Bull. Jap. Soc. scient. Fish. 30(8):630-634.

chemistry, lipids

Raymont, J.E.G., J. Austin and Eileen Linford, 1967.
The biochemical composition of certain zooplankton decapods.
Deep-Sea Research, 14(1):113-115.

chemistry, lipid

Raymont, J.E.G., R.T. Srinivasagam and J.K.B. Raymont, 1971.
Biochemical studies on marine zooplankton - VIII. Further investigations on Meganyctiphanes norvegica (M. Sars). Deep-Sea Res. 18(12): 1167-1178.

chemistry lipid

Raymont, J.E.G., R.T. Srinivasagam and J.K. B. Raymont, 1969.
Biochemical studies on marine zooplankton. IV. Investigation on Meganyctiphanes norvegica (M. Sars.) Deep-Sea Res., 16(2): 141-156.

chemistry lipids

Shatunovskii, M.I., 1970.
The composition of fatty acids of lipids of spring- and autumn-breeding Baltic herring roe, fingerlings and adults of the Gulf of Riga (the Baltic Sea).
Dokl. Akad. Nauk, SSSR, 195(4):962-964

chemistry lipids

Sipos, J.C., and R.G. Ackman, 1968.
Jellyfish (Cyanea capillata) lipids fatty acid composition.
J. Fish. Res. Bd., Can., 25(8):1561-1569.

chemistry, lipids

Williams, P.M., 1965.
Fatty acids from lipids of marine origin.
J. Fish. Res. Bd., Canada, 22(5):1107-1122.

lithium

chemistry, lithium (data)

Angino, Ernest E., and Gale K. Billings, 1966.
Lithium content of sea water by atomic absorption spectrometry.
Geochim. et Cosmochim. Acta, 30(2):153-158.

chemistry, lithium

Bardet, T., E. Tcharkirian and R. LaGrange, 1937
Dosage du lithium dans l'eau de mer.
C.R. Acad. Sci., Paris, 204:443-445.

chemistry, lithium

Borovik-Romanova, J.F., V.V. Korolev and Y.I. Kutsenko, 1954.
Spectroscopic determination of strontium and lithium in natural waters. Zhurn. Analit. Khim., 9(5):265.

chemistry, lithium

Burovina, I.V., et al., 1963.
The content of lithium, sodium, potassium, rubidium and caesium in the muscles of marine animals of the Barents Sea and Black Sea. (In Russian).
Doklady Akad. Nauk, SSSR, 149(2):413-415.

chemistry, lithuim

Chelnikov, B.I., V.E. Kerasev, and G.A. Krainikov, 1967.
Distribution of lothuim in phreatic waters of the sediments of the Pacific. (In Russian).
Dokl., Akad. Nauk, SSSR, 176(2):432-433.

chemistry lithium

Chow, T.J., and E.D. Goldberg, 1962
Mass spectrometric determination of lithium in seawater.
J. Mar. Res., 20(3):163-167.

chemistry, lithium

Fabricand, E.S. Imbimbo, M. E. Brey and J.A. Weston, 1966.
Atomic absorption for Li, Mg, K, Rb, and Sr in ocean waters.
J. Geophys. Res., 71(6):3917-3921.

chemistry, lithium

Ishibashi, M., and K. Kurata, 1939.
Determination of lithium in sea water and bittern
J. Chem. Soc., Japan, 60:1109-1111.

Chem. Abstr. 1940, 34:4314.

chemistry, lithium

Morozov, N.P., 1969.
Direct determinations of lithium and rubidium in the sea water and in the interstitial solutions with the flame-spectrophotometric technique. (In Russian; English abstract).
Okeanologiia 9(2): 353-358.

chemistry, lithium

*Morozov, N.P., 1968.
Lithium and rubidium in the waters of inland seas. (In Russian; English abstract).
Okeanologiia, Akad. Nauk, SSSR, 8(4): 612-615.

chemistry, lithium

Morozov, N.P. 1968.
On the geochemistry of rare alkaline elements in the oceans and seas.
(In Russian; English abstract)
Okeanologiia 8(2): 216-224.

chemistry, lithium

*Presley, B.J., R.R. Brooks and I.R. Kaplan, 1967.
Manganese and related elements in the interstitial water of marine sediments.
Science, 158(3803)906-910.

chemistry lithium

Riley, J.P., and M. Tongudai, 1964.
The lithium content of sea water.
Deep-Sea Res., 11(4):563-568.

chemistry, lithium

Sreekvmaran, C., K.C. Pillai and T.R. Folsom, 1968.
The concentrations of lithium, potassium, rubidium and cesium in some western American rivers and marine sediments.
Geochim. cosmochim. Acta, 32(11):1229-1234.

chemistry, lithium

Strock, L.W., 1936.
Zur Geochemie des Lithiums.
Nachr. Ges. Wiss., Göttingen, Math.-Phys. Kl. n.f., (4), 1(15):171-201.

chemistry, lithium

Thomas, B.B., and T.G. Thompson, 1933.
Lithium in sea water.
Science, 77:547-548.

chemistry, macromolecules

Aizatullin, T.A. and K.M. Khailov, 1970.
Kinetics of enzymic hydrolysis of macromolecules solved in sea water in the presence of bacteria.
Gidrobiol. Zh., 6(6): 49-55. (In Russian; English abstract)

chemistry - macromolecules

Erokhin, V.E., 1970.
On possibility of sorbtion accumulation of macromolecules solved in sea water by copepods Tigriopus brevicornis O.F. Müller and larvae of Balanus improvisus Darw. (In Russian).
Gidrobiol. Zh., 6(6): 94-98.

chemistry - macromolecules, organic

Khaylov, K.M., 1968.
Dissolved organic macromolecules in sea water (In Russian)
Geokhimiya (5): 595-603.
Translation: Geochemistry int. 5(3): 497-503

Magnesium

chemistry magnesium

Berthois, Leopold, 1966.
Hydrologie et sedimentologie dans le Kangerd-lugssuaq (fjord a la cote ouest du Groenland).
C. r. hebd. seanc., Acad. Sci., Paris (D), 262(13): 1400-1402.

chemistry, magnesium (data)

Berthois, L., et A. Crosnier 1966
Etude dynamique de la sédimentation au large de l'estuaire de la Betsiboka.
Cah. ORSTOM, Sér. Océanogr. 4(2): 49-130.

chemistry, magnesium

Carroll, J.J., L.J. Greenfield and R.F. Johnson, 1965.
The mechanism of calcium and magnesium uptake from sea water by a marine bacterium.
J. Cell. Comp. Physiol., 66(1):109-113.

Chemistry, magnesium

Chave, K.E., 1954.
Aspects of the biogeochemistry of magnesium. 1. Calcareous marine organisms. J. Geol. 62(3): 266-283.

chemistry, magnesium

Chester, R., 1965.
Elemental geochemistry of marine sediments.
In: Chemical oceanography, J.P. Riley and G. Skirrow, editors, Academic Press, 2:23-80.

chemistry magnesium

Culkin, Frederick, 1965.
The major constituents of sea water.
In: Chemical oceanography, J.P. Riley and G. Skirrow, editors, Academic Press, 1: 121-161.

chemistry, magnesium

Culkin, F., anf R.A. Cox, 1966.
Sodium, potassium, magnesium, calcium and strontium in sea water.
Deep-Sea Res., 13(5):789-804.

chemistry, magnesium

DeSousa, A., 1954.
The rapid determination of calcium and magnesium in sea water. Anal. Chim. Acta. 11(3):221-224.

Abstr. Anal. Abstr. 2(1):207.

chemistry, magnesium

Dodd, J. Robert, 1965.
Environmental control of strontium and magnesium in Mytilus.
Geochimica et Cosmochimica Acta, 29(5):385-398.

magnesium

*Dyrssen, David, et al., 1967.
Analysis of sea water in the Uddevalla Fjord system 1.
Rept. Chem. Sea Water, Univ. Göteborg, 4:1-8 pp. (mimeographed).

chemistry, magnesium

Fabricand, E.S. Imbimbo, N. E. Brey and J.A. Weston, 1966.
Atomic absorption for Li, Mg, K, Rb, and Sr in ocean waters.
J. geophys. Res., 71(6):3917-3921.

chemistry magnesium

Fesenko, N.G., 1957

[Contribution to the question concerning a direct trilon method of determining magnesium in water.] Gidrokhim. Materialy 27: 135.

chemistry, magnesium

Gilbert, F.C., and W.C. Gilpin 1951.
Production of magnesia from sea water and dolomite.
Research 4(8): 348-353.

Magnesium (data only)

Granqvist, G., 1955.
The cruise with M/S Aranda in the northern Baltic, 1954. Merent. Julk., No. 166:56 pp.

chemistry, Magnesium

Harris, Robert C., 1965.
Trace element distribution in molluscan skeletal material. 1. Magnesium, iron, manganese and strontium. Bull. Mar. Sci., 15(2):265-273.

ABSTRACT

chemistry, magnesium

Ishibashi, Masayoshi, and Toshio Yamamoto, 1960

Chemical studies on the ocean (Pts. 78-79).
Chemical studies on the seaweeds (VI & VII).
The content of calcium, magnesium and phosphorus in seaweeds. Iron content in seaweeds.
Rec. Oceanogr. Wks., Japan, Spec. No. 4: 73-78; 79-86.

chemistry magnesium

Kitano, Yasushi, and Tsuyako Furutsu, 1959.
The state of a small amount of magnesium contained in calcareous shells.
Bull. Chem. Soc., Japan, 33(1):1-4.

Also in:
Collected Papers on Sciences of Atmosphere and Hydrosphere, 1958-1963. 1(3). 1964.

chemistry magnesium (data)

Kramer J.R., 1961
Chemistry of Lake Erie.
Proc. Fourth Conf., Great Lakes Research Great Lakes Res. Div., Inst. Sci. & Tech. Univ. Michigan. Publ. (7):27-56.

chemistry, magnesium

Lerman, Abraham, 1965.
Strontium and magnesium in water and in Crassostrea calcite.
Science, 150(3697):745-751.

chemistry - magnesium

Lowenstam, Heinz A., 1964.
Coexistry calcites and aragonites from skeletal carbonetes of marine organisms and their strontium and magnesium contents.
In: Recent researches in the fields of hydrosphere, atmosphere and nuclear geochemistry, Ken Sugawara, festival volume, Y. Miyake and T. Koyama, editors, Maruzen Co, Ltd., Tokyo, 373-304.

chemistry magnesium

Lyakhin, Yu.I., 1971.
Calcium and magnesium in the western tropical Atlantic. (In Russian; English abstract).
Okeanologiia 11(4): 635-641.

chemistry, magnesium

Mameli, D., e F. Mosetti, 1967.
Some new investigations on the Ca and Mg content in sea water.
Boll. Soc. adriat. Sci., Trieste, 55:27-38.

chemistry, magnesium

Mameli, D., e F. Mosetti, 1966.
Un nuovo metodo di marcatura delle acque di mare costiere: il contenuto in calcio e magnesio quale indicatore del mescolamento di acque continentali in acqua marina.
Boll. Geofis. teor. appl., 8(32):294-308.

chemistry, magnesium

Milliman, John D., Manfred Gastner and Jens Müller 1971.
Utilization of magnesium in coralline algae.
Bull. geol. Soc. Am. 82(3):573-580

chemistry, magnesium

Murray, J.W., 1966.
A study of the seasonal changes of the water mass of Christchurch Harbour, England.
Jour. mar. biol. Assoc. U.K., 46(3):561-578.

chemistry, magnesium (data)

Nehring, Dietwart, und Karl-Heinz Rohde 1967.
Weitere Untersuchungen über anormale Ionenverhältnisse in der Ostsee.
Beitr. Meereskunde 20:10-33

chemistry, magnesium

Nishizawa, Tanzo and T. Muraki, 1940
Chemical studies in the sea adjacent to Palau. I. A survey crossing the sea from Palau to New Guinea. Kagaku Nanyo (Sci. of the South Sea) 2(3):1-7.

chemistry magnesium

Okuda, Taizo, 1964.
Calcium and magnesium contents in the river and sea waters of a tropical area.
Bol. Inst. Oceanogr., Univ. Oriente, Venezuela, 3(1/2):118-135.

chemistry, magnesium

Oren, O.H., 1962
A note on the hydrography of the Gulf of Eylath. Contributions to the knowledge of the Red Sea No. 21.
Sea Fish. Res. Sta., Haifa, Israel, Bull. No. 30: 3-14.

chemistry, magnesium

Pate, John B., and Rex J. Robinson, 1961
The (ethylenedinitrilo)-tetraacetate titration of calcium and magnesium in ocean waters. II. Determination of magnesium.
J. Mar. Res., 19(1): 12-20.

chemistry magnesium

Pilkey, Orrin H., and H.G. Goodell, 1963
Trace elements in Recent mollusk shells.
Limnol. and Oceanogr., 8(2):137-148.

chemistry, magnesium

Platford, R.F., 1965.
Activity coefficient of the magnesium ion in sea water.
J. Fish. Res. Bd., Canada, 22(1):113-116.

chemistry, magnesium

Pytkowicz, R.M., I.W. Duedall and D.N. Connors, 1966.
Magnesium ions: activity in seawater.
Science, 152(3722):640-642.

chemistry, magnesium

Rial, J.R. Besada, and L.R. Molins, 1962.
Determinación complexométrica de los iones calcio y magnesio en el agua de mar y estudio de las variaciones de su concentración en las aguas de la Ría de Viga.
Bol. Inst. Español/ Oceanogr., 111:11 pp.

chemistry, magnesium

Robertson, A.J., 1907.
On hydrographical investigations in the Faroe-Shetland Channel and the northern part of the North Sea during the years 1904-1905. Fish. Bd., Scotland, North Sea Fish. Invest. Comm., Northern Area, Rept. No. 2:1-140, 13 text-figs., 15 pls.

chemistry magnesium

Rodhe, W., 1948
Environmental requirements of fresh-water plankton algae. Experimental studies in the ecology of phytoplankton. Symbolae Botanicae Upsalienses X(1):149 pp., 30 figs.

chemistry, magnesium

Romanovsky, V., 1964.
Coastal effects of the Cape Sicié sewer outfall (French Mediterranean coast west of Toulon).
Air and Water Pollution, 8(10):557-589.

chemistry, magnesium

Schachter, D., 1954.
Contribution à l'étude hydrographique et hydrologique de l'étang de Berre (Bouches-du-Rhode).
Bull. Inst. Océan., Monaco, No. 1048:20 pp.

chemistry, magnesium

Skopintsev, B.A., R.V. Vorob'eva and L.A. Shtukovskaya, 1957.
[The determination of calcium and the amount of calcium and magnesium in sea water by a complexometric method.]
Gidrokhim. Materialy, 27:146-

chemistry, magnesium

Thompson, Mary E., 1966.
Magnesium in sea water: an electrode measurement.
Science, 153(3738):866-867.

chemistry, magnesium

*Tsunogai, Shizuo, Masakichi Nishimura and Syu Nakaya, 1968.
Calcium and magnesium in sea water and the ratio of calcium to chlorinity as a tracer of water masses.
J. oceanogr. Soc., Japan, 24(4):153-159.

chemistry magnesium

Voipio, Aarno, 1959
On the alkaline-earth metal and magnesium contents of sea water.
Suomen Kemistilehti, B, 32: 61-65.

chemistry, magnesium

Wattenberg, H., and E. Timmermann, 1938.
Die Löslichkeit von Magnesiumkarbonat und Strontiumkarbonat in Seewasser. Kieler Meeresf. 2: 81-94.

chemistry magnesium

Wangersky, Peter J., and Oiva Joensuu, 1964
Strontium, magnesium and manganese in fossil foraminiferal carbonates.
J. Geol., 72(4):477-483.

magnesium

Won, Chong Hun, 1964.
Tidal variations of chemical constituents of the estuary water at the Lava bed in the Nack-Dong River from Nov. 1962 to Oct. 1963. (In Korean: English abstract).
Bull. Pusan Fish. Coll., 6(1): 21-32.

tides, effect of
Korea
estuaries
temperature
P.H. Chlorinity
calcium
oxygen
silicate
ammonia
nitrite
nitrate

chemistry, magnesium

Won, Chong Hun, 1963.
Distribution of chemical constituents of the estuary water in Gwang-Yang Inlet. (In Korean; English abstract).
Bull. Fish. Coll., Pusan Nat. Univ., 5(1):1-10.

chemistry, magnesium

Yalkovsky, Ralph, 1967.
Signs test applied to Caribbean Core A 176-6.
Science, 155 (3768):1408-1409.

Magnesium/Chlorinity ratio

chemistry, magnesium/chlorine ratio

Berthois, L. 1969.
Contribution à l'étude sédimentologique du Kangerdlugssuaq, côte ouest du Groenland.
Meddr Grønland 187 (1): 185 pp.

chemistry, magnesium/chlorinity ratios

Mameli, D., e F. Mosetti, 1966.
Un nuovo metodo di marcatura delle acque di mare costiere: il contenuto in calcio e magnesio quale indicatore del mescolamento di acque continentali in acqua marina.
Boll. Geofis. teor. appl., 8(32):294-308.

chemistry, magnesium/chlorinity ratios

*Riley, J.P., and M. Tongudai, 1967.
The major cation/chlorinity ratios in sea water.
Chem. Geol., 2(3):263-269.

chemistry, magnesium/chlorinity ratio

Rohde, Karl Heinz 1967.
Untersuchungen über die Calcium-Chlor- und Magnesium-Chlor- Relationen in Flussmündungen und Bodden der westlichen Ostsee.
Beitr. Meereskunde 20:34-42

chemistry, magnesium / chlorinity (data)

Rohde, Karl-Heinz, 1966.
Untersuchungen über die Calcium-und Magnesiumanomalie.
Beitr, Meereskunde, 19:18-31.

chemistry, magnesium chlorinity

Viswanathan, R., S.M. Shah and C.K. Unni, 196

Atomic absorption and flurometric analyses of seawater in the Indian Ocean. Bull. natn. Inst. Sci., India, 38(1): 284-288.

Magnesium sulfate

chemistry, magnesium sulfate

Glotov, V.P., 1964.
Calculation of the relaxation time for the degree of dissociation of magnesium sulfate in fresh and sea water as a function of temperature. (In Russian).
Akustich. Zhurn., 10(1):40-47.

Translation:
Soviet Phys. Acoustics (Amer. Inst. Phys., Inc.) (1964), 10(1):33-38.

chemistry, magnesium sulfate

Kester, Dana R. and Ricardo M. Pytkowicz, 1969.
Sodium, magnesium and calcium sulfate ion-pairs in seawater at 25C. Limnol. Oceanogr., 14(5): 686-692.

chemistry, magnesium sulfate
*Kester, Dana R. and Ricardo M. Pytkowicz, 1968.
Magnesium sulfate association at 25C in synthetic seawater.
Limnol. Oceanogr., 13(4):670-674.

chemistry, magnesium sulphate
Pytkowicz, R.M. and R. Gates 1968.
Magnesium sulphate interactions in seawater from solubility measurements.
Science 161 (3842):690-691.

Chemistry, malic acid
Creac'h, P., 1955.
Sur la presence des acides citrique et malique dans les eaux marines littorales.
C.R. Acad. Sci., Paris, 240(26):2551-2552.

Manganese

chemistry, manganese
Bernhard, M., 1964.
Chemical composition and the radiocontamination of marine organisms.
Proc. Symp. Nuclear Detonations and Marine Radioactivity, S.H. Small, editor, 137-150.

chemistry, manganese
Berrit, G.R., 1955.
Étude des teneurs en manganese et en carbonates de quelques carrotes sedimentaires atlantiques et pacifiques.
Göteborgs K. Vetenskaps- och Vitterhets-Samhälles Handl. (7)B6(12):1-61.
Sjätte Följden
Medd. Ocean. Inst., Göteborg, 23.

chemistry, manganese
* Bhatt, Y.M., V.N. Sastry, S.M. Shah and T.M. Krishnamoorthy, 1968.
Zinc, manganese and cobalt contents of some marine bivalves from Bombay.
Proc. nat. Inst. Sci., India (B); 34(6): 283-287.

chemistry, manganese
Bonatti, Enrico, 1965.
Il manganese nei fondi oceanici.
Boll. Soc. Toscana Sci. Nat., (A), 72:21 pp.

chemistry, manganese
Bonatti, Enrico, and Oiva Joensuu, 1966.
Deep-Sea iron deposit from the South Pacific.
Science, 154(3749):643-645.

chemistry - manganese
Brewer, Peter G., and Derek Spencer, 1969.
A note on the chemical composition of the Red Sea brines.
In: Hot brines and Recent heavy metal deposits in the Red Sea, E.T. Degens and D.A. Ross, editors, Springer-Verlag, New York, Inc., 174-179.

chemistry, manganese
Bruevich, S.V., and E.G. Vinogradova, 1946.
General features of sedimentation in the Caspian Sea according to the distribution of carbonates, Fe, Mn, and P. in sea deposits. Dok. Akad. Nauk, SSSR, n.s., 52:789-792.

chemistry, manganese
Chainikov, V.I., 1969.
On the source of manganese in bottom sediments of the Pacific. (In Russian)
Dokl. Akad. Nauk SSSR 187(4):909-912

chemistry - MANGANESE
Chipman, Walter, and Evelyn Schommers, 1968.
Role of surface-associated organisms in the uptake of radioactive manganese by the clam, Tapes decussatus.
IAEA Radioactivity in the Sea, Publ. 24:1-11.

chemistry - MANGANESE
Chipman, Walter, Evelyn Schommers and Mireille Boyer, 1968.
Uptake, accumulation and retention of radioactive manganese by the marine annelid worm, Hermione hystrix.
IAEA Radioactivity in the Sea, Pub , 25:1-16.

chemistry - manganese
*Choe, Sang, Tai Wha Chung and Hi-Sang Kwak, 1968.
Seasonal variations in nutrients and principal ions contents of the Han River water and its water characteristics. (In Korean; English abstract).
J. oceanogr. Soc., Korea, 3(1):26-38.

chemistry, manganese
Cooper, L.H.N., 1939.
Phosphorus, nitrogen, iron, and manganese in marine zooplankton. J.M.B.A. 23:387-390.

chemistry, manganese
Dieulafait, L., 1883
Le manganese dans les eaux de mers actuelles et dans certains de leur depots. C.R. Acad. Sci., Paris 96, 718-721.

chemistry, manganese
Dietz, R.S., 1955.
Manganese deposits on the northeast Pacific floor. Calif. J. Mines & Geol. 51(3):
cited from Bull. d'Info.

chemistry, manganese
*Fomina, L.S., 1967.
On the determination of manganese of different valency in its combined presence. (In Russian: English abstract).
Trudy Inst. Okeanol., 83:99-114.

chemistry, manganese
Fukai, Rinnosuke 1965.
Remarks on the chemical problems in relation to the marine radioactivity.
Rapp. P.-v. Réun. Commn int. Explor. scient. Mer Méditerr. 18(3):861-863.

chemistry, manganese
Fukai, Rinnosuke, and Lang Huynh-ngoc 1968.
Studies on the chemical behaviour of radionuclides in sea-water. I. General consideration and study of precipitation of trace amounts of chromium, manganese, iron, cobalt, zinc and cerium.
IAEA Radioactivity in the Sea, Publ. 22:1-26.

chemistry, manganese
Goodell, H.G., and J.K. Osmond, 1966.
Marine geological investigations in the South Pacific Ocean.
Antarctic J., United States, 1(5):203.

chemistry manganese
Gorshkova, T.I., 1966.
Manganese in the bottom sediments of the Soviet northern seas and its biological significance. (In Russian).
Trudy vses. nauchno-issled. Inst. morsk. ryb. Khoz. Okeanogr. (VNIRO), 60:89-102.

chemistry Manganese
Harris, Robert C., 1965.
Trace element distribution in molluscan skeletal material. 1. Magnesium, iron, manganese and strontium. Bull. Mar. Sci., 15(2):265-273.

chemistry - manganese
*Hartmann, Martin, 1969.
Investigations of Atlantis II Deep samples taken by the FS Meteor.
In: Hot brines and Recent heavy metal deposits in the Red Sea, E.T. Degens and D.A. Ross, editors, Springer-Verlag, New York, Inc., 204-207.

chemistry, manganese
Hartmann, Martin, 1964.
Zur Geochemie von Mangen und Eisen in der Ostsee.
Meyniana, 14:3-20.

chemistry, manganese
Harvey, H.W., 1949.
On manganese in sea and fresh water. J.M.B.A., n.s., 28(1):155-164, textfigs.

chemistry, manganese
Harvey, H.W., 1947
Manganese and the growth of phytoplankton. JMBA 26(4):562-579, 2 text figs.

chemistry, manganese
Hood, D.W., 1963
Chemical oceanography. In: Oceanography and Marine Biology, H. Barnes, Edit., George Allen & Unwin, 1:129-155.

chemistry manganese
Imai, Y., 1960
[Oceanographical studies on the behaviour of chemical elements. 1. On iron and manganese in the sediments around "Shiome" (current rips) in Urado Bay.]
J. Oceanogr. Soc., Japan, 16(3): 134-138.

chemistry manganese
Ishibashi, Masayoshi, Tsunenobu Shigematsu, and Yasuharu Nishikawa, 1960
Determination of manganese in sea-water. Rec. Oceanogr. Wks., Japan, n.s., 5(2): 63-65.

chemistry manganese
Johnston, R., 1964
Sea water, the natural medium of phytoplankton II. Trace metals and chelation, and general discussion.
J. Mar. Biol. Assoc., U.K., 44(1):87-109.

chemistry, manganese
Koroleff, F., 1947.
Determination of manganese in natural waters.
Acta Chem. Scandinavica 1:503-506.

chemistry, manganese
Koczy, F.F., 1950.
Zur Sedimentation und Geochemie im aequatorischen Atlantischen Ozean. Medd. Oceanografiska Inst., Göteborg 17 (Göteborgs Kungl. Vetenskaps- och Vitterhets- Samhälles Handlingar, Sjätte Följden Ser. B, 6(1)): 44 pp., 17 textfigs.

chemistry, manganese

Krotov, B.P., 1951.
[Concerning genesis of submarine manganese deposits.] Dok. Akad. Nauk, SSSR, 77:93-95.

chemistry, manganese

Leisegang, E.C. and M.J. Orren 1966.
Trace element concentrations in the sea off South Africa.
Nature, Lond. 211 (5054):1166-1167.

chemistry manganese

Lisitsin, A.P., 1964.
Distribution and chemical composition of suspended matter in the waters of the Indian Ocean. (In Russian; English Abstract).
Rezult. Issled., Programme Mezhd. Geofiz. Goda. Mezhd. Geofiz. Kom. Presidume Akad. Nauk. SSSR. Okeanol., No 10:136 pp.

chemistry, manganese

Marchandise, H., 1958
Contribution to the study of sedimentary manganese deposits.
Trans. Internat. Geol. Cong., Mexico, (20): 107-118.
Listed in Tech. Trans., 2(2): 132

chemistry, manganese

Mokyevskaya, V.V., 1961.
Manganese in waters of the Black Sea. (In Russian).
Doklady, Akad. Nauk, SSSR, 137(6):1445-1447.

Translation:
Consultant's Bureau for Amer. Geol. Inst., Earth Sci. Sect., 137(1-6):251-253. 1962

chemistry - manganese

Murty, P.S.N., Ch.M.Rao and C.V.G.Reddy, 1968.
Manganese in the shelf sediments off the west coast of India.
Current Sci., 37(17):481-483.

chemistry, manganese

Orren, M.J. 1967.
Trace elements (copper, iron and manganese) off the coast of South Africa.
Invest. Rep. Fish. mar. biol. Surv. Div. Un. S. Afr. 59: 1-40.

chemistry, manganese

Ostroumov, E.A., and I.I. Volkov, 1962.
Separation of titanium, zirconium and thorium from manganese, nickel, cobalt and zinc by means of cinnamic acid.
Trudy Inst. Okeanol., Akad. Nauk, SSSR, 54:170-181.

In Russian; English summary

chemistry, manganese

Parker, Patrick L., Ann Gibbs and Robert Lawler, 1963.
Cobalt, iron and manganese in a Texas bay.
Publ. Mar. Sci., Port Aransas, 9:28-32.

chemistry, manganese

Pettersson, H., 1959.
Manganese and nickel on the ocean floor.
Geochim. et Cosmochim. Acta, 17(3/4):209-213.

chemistry, manganese

Pettersson, H., and H. Rotschi, 1952.
The nickel content of deep-sea deposits.
Geochimica et Cosmochimica Acta 2:81-90, 4 text-figs.

chemistry, manganese

Pilkey, Orrin H., and H.G. Goodell, 1963
Trace elements in Recent mollusk shells.
Limnol. and Oceanogr., 8(2):137-148.

chemistry, manganese

*Presley, B.J., R.R. Brooks and I.R. Kaplan, 1967.
Manganese and related elements in the interstitial water of marine sediments.
Science, 158(3803):906-910.

chemistry, manganese

Rao, M. Subba, 1962.
Manganese in the shelf sediments off east coast of India.
Proc. Indian Acad. Sci., (A), 56(5):274-284.

chemistry, manganese

Rona, Elizabeth, Donald W. Hood, Lowell Muse and Benjamin Buglio, 1962
Activation analysis of manganese and zinc in sea water.
Limnology and Oceanography, 7(2):201-206.

chemistry manganese

Ronov, A.B., and A.I. Ermishkina, 1959
[Distribution of manganese in sedimentary rocks.]
Geokhimiya, (3):

Geochemistry (3):254-278.

chemistry manganese

Skopintsev, B.A., and T.P. Popova, 1960.
Some results of the determination of the content of iron, manganese and copper in the waters of the Black Sea. Chemistry of the sea, hydrology, marine geology. (In Russian).
Trudy Morsk. Gidrofiz. Inst., 19:21-30.

chemistry, manganese

Slowey, J. Frank, and Donald W. Hood, 1971.
Copper, manganese and zinc concentrations in Gulf of Mexico waters
Geochim. Cosmochim. Acta 35 (2):121-138.

chemistry manganese

Spencer, Derek W. and Peter G. Brewer, 1971.
Vertical advection, diffusion and redox potentials as controls on the distribution of manganese and other trace metals dissolved in waters of the Black Sea. J. geophys. Res. 76(24): 5877-5892.

chemistry manganese

Thompson, T.G., and T.L. Wilson, 1935.
The occurrence and determination of manganese in sea water.
J. Amer. Chem. Soc., 57:233-236.

chemistry manganese

Topping, Graham, 1969.
Concentrations of Mn, Co, Cu, Fe, and Zn in the northern Indian Ocean and Arabian Sea. J. mar. Res., 27(3): 318-326.

chemistry manganese

Turekian, Karl E., and Donald F. Schutz, 1965.
Trace element economy in the oceans.
Narragansett Mar. Lab., Univ. Rhode Island, 6 Occ. Publ., No. 3:41-89.

chemistry manganese

Varentsov, I.M., and M.I. Stepanetz 1970.
Experiments involved in the modelling of processes of manganese lixiviation by sea water from volcanic materials of basal composition. (In Russian).
Dokl. Akad. Nauk SSSR 190 (3):679-682.

chemistry manganese

Wangersky, Peter J., 1963.
Manganese in ecology.
Radioecology, V. Schultz and A.W. Klement, Jr., Editors, Proc. First Nat. Symp., Radioecology, Sept., 1961, Reinhold Publ. Corp., and Amer. Inst. Biol. Sci., 499-408.

chemistry, manganese

Wangersky, Peter J., and Donald C. Gordon, Jr., 1965.
Particulate carbonate, organic carbon and Mn^{++} in the open ocean.
Limnol. Oceanogr., 10(4):544-550.

chemistry, manganese

Wangersky, P.J., and G.E. Hutchinson, 1958.
Manganese deposition and deep water movements in the Caribbean. Nature 181(4602):108-109.

chemistry, manganese

Wilkniss, P.E., T.B. Warner and R.A. Carr 1971.
Some aspects of the geochemistry of F, Fe and Mn in coastal waters and in fresh-water springs on the southeast coast of Hawaii.
Mar. Geol. 11 (4): M39-M46

chemistry, manganese

Yelkovsky, Ralph, 1967.
Signs test applied to Caribbean Core A 176-6.
Science, 155(3768):1408-1409.

chemistry - manganese

Yamamoto, H., and Y. Imai, 1959.
[Oceanographical studies on Japanese inlets. X. On the distribution of iron and manganese in sea water in Urado Bay.] J. Oceanor. Soc., Japan, 15(4): 185-190.

chemistry, manganese

Yatsimirsky, K.B., E.M. Emelyanov, V.K. Pavlova and Ya.S. Savichenko, 1971.
Determination of manganese and copper micro-quantities in small portions of marine suspended matter (from the Baltic Sea and the Atlantic Ocean). (In Russian; English abstract). Okeanologiia 11(4): 730-734.

chemistry manganese

Zelenov, K.K., 1964.
Iron and manganese in exhalations of the submarine Banu Vukhu volcano (Indonesia). (In Russian).
Doklady, Akad. Nauk, SSSR, 155(6):1317-1320.

Manganese-54

chemistry, manganese-54

Chipman, Walter, and Jean Thommeret 1970.
Manganese content and the occurrence of fallout ^{54}Mn in some marine benthos of the Mediterranean.
Bull. Inst. océanogr. Monaco, 69(1402):15pp

chemistry, manganese-54
Parker, Patrick L., 1966.
Movement of radioisotopes in a marine bay: cobalt-60, iron-59, manganese-54, zinc-65, soduim-22.
Publs. Inst. mar. Sci. Univ. Texas, Port Aransas, 11:102-107.

chemistry, manganese-54
*Pearcy, William G., and Charles L. Osterberg, 1968.
Zinc-65 and manganese-54 in albacore Thunnus alalunga from the west coast of North America.
Limnol. Oceanogr., 13(3):490-498.

chemistry, manganese-54
Tennant, David A., and William O. Forster 1969.
Seasonal variation and distribution of ^{65}Zn, ^{54}Mn and ^{51}Cr in tissues of the crab Cancer magister Dana.
Health Physics 18: 649-657.
Also in: Coll. Repr. Dept. Oceanogr. Oregon State Univ. 9(2) (1970).

manganese deposits

Chemistry manganese crusts
St. Kroll, V., 1955.
Radium in manganese crusts. Goteborgs K. Vetenskaps- och Vitterhets-Samhalles Handl., Sjätte Följden, (B), 6(12):10 pp., 9 textfigs.

Also:
Medd. Oceanografiska Inst., Göteborg, No. 28.

Chemistry, manganese deposits
Krotov, B.P., 1951.
Genesis of marine deposits of manganese.
Doklady Akad. Nauk, SSSR, 77(1):93-

Chemistry, manganese deposits
McKelvey, V.E., and Livingston Chase, 1966.
Selecting areas favorable for subsea prospecting
Exploiting the Ocean, Trans. 2nd Ann. Mar. Techn. Soc., Conf., June 27-29, 1966, 44-60.

manganese, effect of

Chemistry, manganese, effect of
Schurin, A.T., 1967.
The effect of manganese on the distribution of bottom invertebrates of the Baltic.
Annls. biol. Copenh. (1965) 22; 73.

chemistry, manganese, effect of
Takeda, Keiji, 1970.
Relative growth of a marine centric diatom Chaetoceros calcitrans f. pumilus (Paulsen) Takano in media containing various concentration of manganese. Bull. plankt. Soc., Japan, 17(2): 77-83.

Chemistry, manganese encrustations
Glasby, G.P., J.S. Tooms and J.R. Cann, 1971.
The geochemistry of manganese encrustations from the Gulf of Aden. Deep-Sea Res. 18(12): 1179-1187.

Chemistry, manganese fallout
Chipman, Walter, and Jean Thommeret, 1970.
Manganese content and the occurrence of fallout ^{54}Mn in some marine benthos of the Mediterranean.
Bull. Inst. océanogr. Monaco, 69 (1402): 15 pp. Also: IAEA, Radioactivity in the Sea 25.

Chemistry manganese minerals
Arrhenius, G., J. Mero and J. Korkisch, 1964
Origin of oceanic manganese minerals.
Science, 144(3615):170-172.

MANGANESE NODULES
SEE ALSO: FERROUS SULFIDE NODULES

Chemistry, manganese nodules
*Ahrens, L.H., J.P. Willis and C.O. Osthuizen, 1967.
Further observations on the composition of manganese nodules, with particular reference to some of the rarer elements.
Geochim. cosmochim. Acta, 41(11):2196-2180.

Chemistry, manganese deposition
Allen, J.A., 1960.
Manganese deposition on the shells of living molluscs.
Nature, 185(4709):336-337.

chemistry - manganese nodules
Andrews James E., 1971.
Abyssal hills as evidence of transcurrent faulting on North Pacific fracture zones.
Bull. geol. Soc. Am, 82(2): 463-470.

chemistry, manganese nodules
Andrushchenko, P.F., and N.S. Skornyakova, 1969.
Structure and mineral composition of iron-manganese nodules of the south Pacific Ocean. (In Russian; English abstract). Okeanologiia, 9(2): 282-294.

chemistry, manganese nodules.
Anon. 1968.
Ocean-bottom minerals
Ocean Industry 3(6): 61-73.

Chemistry manganese nodules
Arrhenius, G., 1963.
25. Pelagic sediments.
In: The Sea, M. N. Hill, Editor, Interscience Publishers, 3:655-727.

Chemistry, Manganese Nodules
Arrhenius, Gustaf, and Enrico Bonatti, 1965.
Neptunism and vulcanism in the ocean.
Progress in Oceanography, 3:7-22.

chemistry manganese, nodules
Ball, James 1967.
New concept for lifting nodules
Ocean Industry 2(6): 37-39

chemistry manganese nodules
Barnes, Steven S. 1967.
Minor element composition of ferro-manganese nodules.
Science 157 (3784): 63-65.

Chemistry manganese nodules
Bender, Michael L., Teh-Lung Ku and Wallace S. Broecker, 1966.
Manganese nodules: their evolution.
Science, 151(3708):325-328.

Chemistry manganese nodules
Bezrukov, P.L., 1963
Studies of the Indian Ocean during the 35th cruise of R/V 'Vityaz'. (In Russian).
Okeanologiia, Akad. Nauk, SSSR, 3(3):540-549.

Chemistry manganese nodules
Bezrukov, P.L., 1962
Distribution of ferro-manganese concretions at the floor of the Indian Ocean. (In Russian)
Okeanologiia, Akad. Nauk, SSSR, 2(6):1014-1019.
Abstr. in:
Soviet Bloc Res. Geophys., Astron & Space, 52: 24.

chemistry, manganese nodules
Bonatti, Enrico, and Oiva Joensuu, 1966.
Deep-Sea iron deposit from the South Pacific.
Science, 154(3749):643-645.

Chemistry, manganese nodules.
Bonatti, Enrico, and Y. Rammohanroy Nayudu 1965.
The origin of manganese nodules on the ocean floor.
Am. J. Sci. 263(1): 17-39.

chemistry, manganese nodules
Bonatti, Enrico, and Y. Rammohanroy Nayudu, 1964.
The origin of manganese nodules on the ocean floor.
Amer. J. Sci., 263(1):17-39.

Chemistry, manganese nodules
Buchowiecki, J., and R.D. Cherry 1968.
Thorium, radium and potassium in manganese nodules.
Chem. Geol. 3(2): 111-117.

Chemistry manganese nodules
Burns, Roger G., 1965.
Formation of cobalt (III) in the amorphous FeOOH . H2O phase of manganese nodules.
Nature, 205(4975):999.

Chemistry, manganese nodules
*Cheney, E.S., and L.D. Vredenburgh, 1968.
The role of iron sulfides in the diagenetic formation of iron-poor manganese nodules.
J. sedim. Petr., 38(4):1363-1365.

chemistry, manganese nodules
Cherdyntsev, V.V., N.B. Kadyrov and N.V. Novichkova 1971.
Origin of manganese nodules of the Pacific Ocean from radioisotope data. (In Russian).
Geokhimya 1971, (3): 339-354.
Translation: Geochem. int. Am. geol. Inst. 8(2): 211-225.

Chemistry, manganese nodules
Chester, R., 1965.
Elemental geochemistry of marine sediments.
In: Chemical oceanography, J.P. Riley and G. Skirrow, editors, Academic Press, 2:23-80.

Chemistry, manganese nodules
Chow, T.J., and C.R. McKinney, 1958.
Mass spectrometric determination of lead in manganese nodules. Anal. Chem., 30(9):1499-1503.

chemistry, manganese nodules, elements in
Cronan, D.S., 1969.
Inter-element associations in some pelagic deposits.
Chem. Geol. 5 (2): 99-106

chemistry - manganese nodules
Cronan, D.S. and J.S. Tooms, 1969.
The geochemistry of manganese nodules and associated pelagic deposits from the Pacific and Indian Oceans. Deep-Sea Res., 16(4): 335-359.

chemistry, manganese nodules
Cronan, D.S., and J.S. Tooms, 1968.
A microscopic and electron probe investigation of manganese nodules from the northwest Indian Ocean
Deep-Sea Res., 15(2):215-223.

chemistry, manganese nodules
Cronan, D.S., and J.S. Tooms, 1967.
Geochemistry of manganese nodules from the N.W. Indian Ocean.
Deep-Sea Research, 14(2):239-249.

chemistry, manganese nodules
Indian Ocean

chemistry, manganese nodules
Cronan, D.S., and J.S. Tooms, 1967.
Sub-surface concentrations of manganese nodules in Pacific sediments.
Deep-Sea Research, 14(1):117-119.

Chemistry, manganese nodules
Edgington, David N., and Edward Callender 1970.
Minor element geochemistry of Lake Michigan ferro-manganese nodules.
Earth Planet. Sci. Letts. 8 (2): 97-100.

Chemistry manganese oxide
Emery, K.O., 1966.
Geological methods for locating mineral deposits on the ocean floor.
Exploiting the Ocean, Trans. 2nd Ann. Mar. Techn. Soc. Conf., June 27-29, 1966, 24-43.

chemistry - manganese nodules
Finkelman, Robert B., 1970.
Magnetic particles extracted from manganese nodules: suggested origin from stony and iron meteorites. Science, 167(3920): 982-984.

Chemistry, manganese nodules
Fomina, L.S., 1966.
Accumulation and redistribution of rare-earth elements upon the formation of iron-manganese concretions in the ocean.
Doklady, Akad. Nauk, SSSR, 170(5):1181-1184.

chemistry, manganese nodules
Fomina, L.S., and I.I. Volokov, 1969.
Rare-earth elements in ferromanganesian concretions of the Black Sea. (In Russian)
Dokl. Acad. Nauk, SSSR, 185(1): 188-191.

chemistry, manganese nodules
Fredriksson, K., 1956.
Cosmic spherules in deep-sea sediments. Nature 177(4497):32-33.

chemistry manganese nodules
Gager, H.M., 1968.
Mössbauer spectra of deep-sea iron-manganese nodules.
Nature, Lond., 220(5171):1021-1023.

Chemistry manganese nodules
Glasby, G.P., and G.W. Hodgson 1971.
The distribution of organic pigments in marine manganese nodules from the northwest Indian Ocean.
Geochim. cosmochim. Acta. 35 (8): 845-851

chemistry, manganese nodules
Goldberg, E.D., 1961.
Chemical and mineralogical aspects of deep-sea sediments.
Physics and Chemistry of the Earth, Pergamon Press, 4(8):281-302.

Chemistry manganese nodules
Goldberg, Edward D., Minoru Koide, R.A. Schmitt and R.H. Smith, 1963
Rare-earth distribution in the marine environment.
J. Geophys. Res., 68(14):4209-4217.

Chemistry, manganese nodules
Goldberg, E.D., and E. Picciotto, 1955.
Thorium determination in manganese nodules.
Science 121(3147):613-614.

Chemistry, manganese nodules
Graham, J. W. and S. C. Cooper, 1959.
Biological origin of manganese-rich deposits of the sea floor.
Nature, 183(4667):1050-1051.

Chemistry, "manganese" nodules
Grjebine T. 1965.
Sphérules magnétiques dans les sédiments de la Méditerranée.
Rapp. P.-v. Réun. Commn int. Explor. scient. Mer Méditerr. 18(3): 959-961.

Chemistry - manganese nodules
Grjebine, Torvy, 1965.
Spherules magnétiques dans les sédiments de la Mediterranée.
Bull. Inst. Océanogr., Monaco, 65(1338):12 pp.

chemistry, manganese dioxide
Hamilton, E.L., 1956.
Sunken islands in the mid-Pacific mountains.
Mem. G.S.A., 64:97 pp.

Chemistry - manganese nodules
*Harriss, Robert C., J.H. Crockett, and M. Stainton, 1968.
Palladium, iridium and gold in deep-sea manganese nodules.
Geochim. cosmochim. Acta, 32(10):1049-1056.

chemistry, manganese nodules
Harriss, Robert C. and Arthur G. Troup, 1970.
Chemistry and origin of freshwater ferromanganese concretions. Limnol. Oceanogr., 15(5): 702-712.

Chemistry, manganese nodules
Harriss, Robert C., and Arthur G. Troup 1969.
Freshwater ferromanganese concretions: chemistry and internal structure.
Science 166 (3905): 604-606

chemistry manganese nodules
Hawkins, Larry K., 1969.
Visual observations of manganese deposits on the Blake Plateau. J. geophys. Res., 74(28): 7009-7017.

chemistry, manganese nodules
Heezen, Bruce C., Bill Glass and H.W. Menard 1966.
The Manihiki Plateau.
Deep-Sea Res. 13(3): 445-458.

Chemistry, manganese nodules
Herzenberg, C.L., and D.L. Riley 1969.
Interpretation of the Mössbauer spectra of marine iron-manganese nodules.
Nature, Lond., 224 (5216): 259-260.

chemistry, manganese nodules
Hurley, Robert J., 1966.
Geological studies of the West Indies.
In: Continental margins and island arcs, W.H. Poole, editor, Geol. Surv. Pap., Can., 66-15:139-150.

Chemistry, manganese nodules
Jedwab, Jacques 1970.
Les sphérules cosmiques dans les nodules de manganèse.
Geochim. cosmochim. Acta 34 (4): 447-457.

Chemistry manganese nodules
Kalinenko, V.O., O.V. Belopytova, and G.G. Nikolaeva, 1962
The bacteriogenic formation of ferro-manganese concretions in the Indian Ocean. (In Russian).
Okeanologiia, Akad. Nauk, SSSR, 2(6):1050-1059.
Abstr. in:
Soviet Bloc Res. Geophys. Astron., & Space, 52: 23-24.

chemistry manganese nodules
Krause, Dale C., 1964.
Lithology and sedimentation in the southern continental borderland.
In: Papers in Marine Geology, R.L. Miller, Macmillan Co., N.Y., 274-318.

chemistry - manganese nodules
Ku, Teh-Lung and Wallace S. Broecker 1969.
Radiochemical studies on manganese nodules of deep-sea origin. Deep-Sea Res., 16(6): 625-637.

chemistry, manganese nodules
Lakin, H.W., C.E. Thompson and D.F. Davidson, 1963
Tellurium content of marine manganese oxides and other manganese oxides.
Science 142(3599):1568-1569.

Chemistry, manganese nodules
La Que, F.L. 1971
Prospects for and from deep ocean mining
J. mar. techn. Soc. 5(2): 5-15

Chemistry manganese nodules
McFarlin, Peter F., 1967.
Aragonite vein fillings in marine manganese nodules.
J. sedim. Petrol., 37(1):68-72.

chemistry, manganese nodules
McIlhenny, W.F., 1966.
The oceans: technology's new challenge.
Chem. Engng., (Nov.7):247-254.

Chemistry manganese nodules
Mero, John L., 1966.
Review of mineral values on and under the ocean.
Exploiting the Ocean, Trans. 2nd Ann. Mar. Techn. Soc. Conf., June 27-29, 1966, 61-78.

chemistry manganese nodules
Mero, John L., 1961.
Sea-floor manganese nodules. (Abstract)
Tenth Pacific Sci. Congr., Honolulu, 21 Aug.-6 Sept., 1961, Abstracts of Symposium Papers, 378-379.

chemistry, manganese nodules
Moore, T.C., Jr., and G.R. Heath, 1966.
Manganese nodules, topography and thickness of Quaternary sediments in the Central Pacific.
Nature, Lond., 212(5066):983-985.

Chemistry manganese nodules
Morgenstein, Maury, 1971.
A study of the growth of two deep-sea manganese meganodules.
Pacific Sci. 25(3): 308-312

chemistry - manganese nodules
Morgenstein, Maury, and Murray Felsher 1971.
The origin of manganese nodules: a combined theory with special reference to palagonization.
Pacific Sci. 25(3): 301-307.

chemistry, manganese nodules
Nasu, Noriyuki, and Takahiro Sato, 1962.
VII. Geological results in the Japanese Deep Sea Expedition in 1961 (JEDS-4).
Oceanogr. Mag., Japan Meteorol. Agency, 13(2): 166.

Jeds Contrib. No. 35.

chemistry,
Manganese Nodules
Nayudu, Y. Rammohanroy, 1965.
Petrology of submarine volcanics and sediments in the vicinity of the Mendicino Fracture Zone
Progress in Oceanography, 3:207-220.

Chemistry, manganese nodules
Niino, H., 1955.
On a manganese nodule and Perotrochus dredged from the banks near the Izu Islands, Japan.
Rec. Ocean. Wks., Japan, 2(2):120-124, 2 pls.

chemistry, manganese nodules
Nikolayev, D.S., and E.I. Yefimova, 1963.
On the age of iron-manganese concretions from the Indian and Pacific oceans. (In Russian).
Geokhimia (7):678-688.

Translation:
Geochemistry (Scripta Tecnica, Inc.)(7):703-714.

chemistry
manganese nodules
Olausson, Eric, and Sulo Uusitalo, 1963
On the influence of seismic vibrations on sediments.
Comptes Rendus, Soc., Geol., Finlande, 35: 101-114.

chemistry, manganese nodules
Pettersson, H., 1955.
Manganese nodules and oceanic radium.
Pap. Mar. Biol. and Oceanogr., Deep-Sea Res., Suppl. to Vol. 3:335-345.

chemistry, manganese nodules
Pettersson, H., 1955.
Résumé des résultats de la croisière de l'"Albatross" en Méditerranée.
Trav. Lab. Géol. Sous-Marine, 6:1-5.

Chemistry - manganese nodules
Pratt, Richard M., 1971.
Lithology of rocks dredged from the Blake Plateau.
SEast. Geol. 13(1):19-38.

chemistry - manganese nodules
Pratt, Richard M., and Peter F. McFarlin, 1966.
Manganese pavements on the Blake Plateau.
Science, 151(3714):1080-1082.

chemistry, manganese nodules
*Price, N.B., 1967.
Some geochemical observations on manganese-iron oxide nodules from different depth environments.
Marine Geol., 5(5/6):511-538.

chemistry manganese nodules
Reynolds, Peter H. and E. Julius Dasch, 1971.
Lead isotopes in marine manganese nodules and the Ore-lead growth curve. J. geophys. Res., 76(21):5124-5129.

Riley, J.P., and Prapas Sinhaseni, 1958 manganese nodules
Chemical composition of three manganese nodules from the Pacific Ocean. J. Mar. Res., 17:466-482.

Chemistry - manganese nodules
*Rossmann, Ronald and Edward Callender, 1968.
Manganese nodules in Lake Michigan.
Science, 162(3858):1123-1124.

Chemistry manganese nodules
Sackett, William M., 1966.
Manganese nodules: thorium-230:protactinium-231 ratios.
Science, 154(3749):646-647.

chemistry, manganese nodules
Schoettle, Manfred, and Gerald M. Friedman 1971
Fresh water iron-manganese nodules in Lake George, New York.
Bull. geol. Soc. Am., 82(1): 101-110

Chemistry, manganese nodules
*Sevastyanov, V.F., and I.I. Volkov, 1967.
Redistribution of chemical elements in the oxidized layer of the sediments in the process of iron-manganese nodule formation in the Black Sea. (In Russian; English abstract).
Trudy Inst. Okeanol., 83(135-152.

Chemistry manganese nodules
Sevastionov, V.F., and I.I. Volkov, 1966.
The chemical composition of the ferromanganese concretions of the Black Sea. (In Russian).
Doklady, Akad. Nauk, SSSR, 166(3):701-704.

Chemistry
manganese nodules
Shipek, Carl J., 1960.
Photographic study of some deep-sea floor environments in the eastern Pacific.
Bull. Geol. Soc., Amer., 71(7):1067-1074.

Chemistry
manganese nodules
Skorniakova, N.S., P.F. Androoschenko, and L.S. Fomina, 1962
The chemical composition of the ferromanganese concretions in the Pacific Ocean.
Okeanologiia, Akad. Nauk, SSSR, 2(2):264-277.

chemistry
manganese nodules
Skornyakova, N.S., & N. L. Zenkevich, 1961
The distribution of the ferromanganese concretions in the surface sedimentary layers of the Pacific Ocean.
Okeanologiya, 1: 86-94.

Chemistry, manganese nodules
Smith, R.E., J.D. Gassaway and H.N. Giles 1968.
Iron-manganese nodules from Nares Abyssal Plain: geochemistry and mineralogy.
Science 161 (3843): 780-781.

chemistry, manganese nodules
Somayajulu, B.L.K. 1967.
Beryllium-10 in a manganese nodule
Science 156 (3779): 1219-1220

chemistry, manganese nodules
Somayajulu, B.L.K., G.R. Heath, T.C. Moore, Jr., and D.S. Cronan, 1971.
Rates of accumulation of manganese nodules and associated sediment from the equatorial Pacific.
Geochim. cosmochim. Acta 35(6): 621-624.

Chemistry
manganese nodules
Summerhayes, C.P., 1969.
Marine environments of economic mineral deposition around New Zealand: a review.
J. mar. techn. Soc. 3(2): 57-66

chemistry - manganese nodules

Summerhayes C.P. 1969.
Marine geology of the New Zealand subantarctic sea floor.
Bull. N.Z. scient. indust. Res. 190
(N.Z. Oceanogr. Inst. Mem. 50): 92 pp.

Chemistry, manganese nodules

Summerhayes, C.P. 1967.
Manganese nodules from the south-western Pacific.
N.Z. Jl. Geol. Geophys. 10(6): 1372-1381.

Chemistry manganese nodules

*Summerhayes, C.P., 1967.
Marine environments of economic mineral deposition around New Zealand: A review.
N.Z. Jl mar. Freshwat. Res., 1(3): 267-282.

chemistry, manganese nodules

Thomas, D.W., and Max Blumer, 1964.
Pyrene and fluoranthene in manganese nodules.
Science, 143(3601): 39.

chemistry manganese deposits

Tooms, J.S., C.P. Summerhayes and D.S. Cronan, 1969.
Geochemistry of marine phosphate and manganese deposits. Oceanogr. Mar. Biol. Ann. Rev., H. Barnes, editor, George Allen and Unwin, Ltd., 7: 49-100.

chemistry - manganese nodules

Turekian, Karl K., 1965.
Some aspects of the geochemistry of marine sediments.
In: Chemical oceanography, J.P. Riley and G. Skirrow, editors, Academic Press, 2: 81-126.

chemistry manganese nodules

Willis, J.P., & L.H. Ahrens, 1962
Some investigations on the composition of manganese nodules. with particular reference to certain trace elements.
Geochim. et Cosmochim. Acta, 26: 751-764.

Chemistry manganese nodules

Zenkevich, N.L., and N. S. Skornyakova, 1961.
Iron and manganese on the ocean floor.
Priroda, 1961(2): 47-50.

OTS - Soviet-bloc Research on Geophys., Astron., and Space, No. 10(1960): 7.

chemistry, manganese, oxidation of

Krumbein, W.E., 1971.
Manganese-oxidizing fungi and bacteria.
Naturwissenschaften 58(1): 56-57.

mannitol

chemistry, mannitol

Gripenberg, Stina, 1966.
Equilibria of the complexes formed by mannitol in sea water.
Commentat. physico-math., 32(1): 3-38.

Melanins

Chemistry, melanin

Liaci, Lidia, 1964.
Pigmenti e steroli negli invertebrati marini.
Archivio Zool. Ital., 49: 281-300.

melanoidines

chemistry melanoidines

Kalle, Kurt, 1962
Über die gelösten organischen Komponenten im Meerwasser.
Kieler Meeresf., 18(3) (Sonderheft): 128-131.

chemistry, mercaptans

Adams, Donald D., and Francis A. Richards, 1968.
Dissolved organic matter in an anoxic fjord, with special reference to the presence of mercaptans.
Deep-Sea Res., 15(4): 471-481.

Chemistry, Mercaptan

Kadota, Hajime, and Yuzaburo Ishida, 1968.
Evolution of volatile sulfur compounds from unicellular marine algae.
Bull. Misaki mar. biol. Inst. Kyoto Univ., 12: 35-48.

mercury

Chemistry, mercury

Bache, C.A., W.H. Gutenmann and D.J. Lisk 1971.
Residues of total mercury and methylmercuric salts in lake trout as a function of age.
Science 172 (3986): 951-952.

Chemistry, mercury

De Groot, A.J., J.J.M. De Goeij and C. Zegers, 1971.
Contents and behaviour of mercury as compared with other heavy metals in sediments from the rivers Rhine and Ems.
Geol. Mijnb. 50 (3): 393-398.

chemistry, mercury

Glooschenko, Walter A. 1969.
Accumulation of ^{203}Hg by the marine diatom Chaetoceros costatum.
J. Phycol. 5(3): 224-226

Chemistry mercury

Hanya, Takahisa, Ryoshi Ishiwatari and Hisako Ichikuni, 1963.
The mechanism of removal of mercury from sea water to bottom muds in Minamata Bay. (In Japanese; English abstract).
J. Oceanogr. Soc., Japan, 19(2): 94-100.

chemistry, mercury

Hays, Helen, and Robert W. Risebrough 1972
Pollutant concentrations in abnormal young terns from Long Island Sound.
The Auk 89 (1): 19-35

Chemistry mercury

Hosohara Kyoichi, 1962(1961)
Mercury content of deep sea water. (In Japanese; English summary).
Repts. JEDS, Deep-Sea Res. Comm., Japan Soc., Promotion of Science, 3: 1-5.

Original published (1961):
Nippon Kagaku Zasshi, 82: 1107-1108.
JEDS Contrib. No. 23.

chemistry mercury

Hosohara, Kyoichi, 1961
Mercury content of deep sea water. (In Japanese; English abstract).
Nippon Kagaku Zasshi, 82: 1107-1108.

Also in Repts. of JEDS, 3(23)

Chemistry mercury

Hosohara, K., 1961
Mercury content of deep sea water.
J. Chem. Soc., Japan, 82(8): 1107-1108.

Chemistry mercury

Hosohara, K., R. Kuroda, & H. Hamaguchi, 1961
[Photometric determination of mercury in sea water]
J. Chem. Soc. Japan, 82(3): 347-349 (Abs.p.A21)

Also: Repts. of JEDS, Vol. 2, Contr. 19: 1 - 4

chemistry (mercury)

Idyll, C.P. 1971.
Mercury and fish.
Sea Frontiers 17 (4): 230-240. (popular)

chemistry, methyl mercury

Jernelöv, Arne, 1970.
Release of methyl mercury from sediments with layers containing inorganic mercury at different depths. Limnol. Oceanogr., 15(6): 958-960.

chemistry, mercury

Smith, J. David, R.A. Nicholson and P.J. Moore 1971
Mercury in water of the tidal Thames.
Nature, Lond. 232 (5310): 393-394

chemistry, mercury

Stock, A., and F. Cucuel, 1934.
Die Verbreitung des Quecksilbers.
Naturwiss., 22: 390-393.

chemistry, mercury

Underdal, Bjarne, and Tore Håstein 1971.
Mercury in fish and water from a river and a fjord in the Kragerø region, South Norway.
Oikos 22(1): 101-105

chemistry, mercury

*Yoshida, Tamao, Toshiharu Kawabata and Yoshiyuki Matsue, 1967.
Transference mechanism of mercury in marine environment.
J. Tokyo Univ. Fish., 53(1/2): 73-84.

chemistry, mercury

Young, David R., 1971.
Mercury in the environment: a summary of information pertinent to the distribution of mercury in the Southern California Bight.
Southern Calif. Coastal Water Research Project, 31 pp. $1.70.

mercury, effect of

Chemistry Mercury, effect of

Akiyama, Akio, 1970.
Acute toxicity of two organic mercury compounds to the teleost, Oryzias latipes in different stages of development.
Bull. Jap. Soc. scient. Fish. 36 (6): 563-570

Chemistry, mercury, effect of

Soyer, J., 1963.
Contribution a l'etude des effets biologiques du mercure et de l'argent dans l'eau de mer.
Vie et Milieu, 14(1):1-36.

Chemistry - Mercury poisoning'

Selye, Hans, 1970.
Mercury poisoning: prevention by spironolactone.
Science, 169 (3947): 775-776.

mesothorium

Chemistry, mesothorium

Lazarev, K.F., S.M. Grashchenko, D.S. Nikolaev, and V.M. Drozhzhin, 1965.
The concentration of mesothorium-1 in Black Sea water. (In Russian).
Doklady, Akad. Nauk, SSSR, 164(4):910-912.

metallic salts

Chemistry metallic salts

Clarke, G.L., 1947
Poisoning and recovery in barnacles and mussels Biol. Bull., 92(1):73-91.

metals

chemistry, metals, (heavy)

Brooks, R.R., and B.F. Quin, 1971
Heavy metals in stream sediments of the Port Pegasus area of Stewart Island
N.Z. Jl Sci. 14 (1): 25-30

chemistry, metals

Davankov, A.B., V.M. Laufer, E.G. Azhazha, A.V. Gordiyevskiy and V.N. Kiryushov, 1962.
Experimental ectraction of uranium and other elements from the waters of the Atlantic Ocean. Ordzhonikidze, Izv. VUZ Tsvetnaya Metallurgiya (2):118-123.

Abstr. in:
Soviet-Bloc Res. Geophys., Astron., and Space, No. 36:12-13.

61-11147-36 JPRS:13931 (31 May 1962).

Chemistry, metals (heavy)

Grice, C. Fitzhugh, 1968.
The Red Sea's hot brines and heavy metals.
Ocean Industry, 3(3):52-53,55,57.

Chemistry, metals

Pillai, T.N.V., M.V.M. Desai, Elizabeth Mathew, S. Ganapathy and A.K. Ganguly 1971.
Organic materials in the marine environment and the associated metallic elements.
Current Sci. 40 (4): 75-81.

chemistry Methyl amines

Sakevich, A.I., 1970.
Detection of methyl amines in the culture of Stephanodiscus hantzschii Grun. Gidrobiol. Zh., 6(3): 98-100. (In Russian)

methane

chemistry, methane

*Atkinson, Larry P., and Francis A. Richards, 1967.
The occurrence and distribution of methane in the marine environment.
Deep-Sea Res., 14(6):673-684.

chemistry, methane

Lamontagne, R.A., J.W. Swinnerton and V.J. Linnenbom, 1971.
Nonequilibrium of carbon monoxide and methane at the air-sea interface. J. geophys. Res. 76(21): 5117-5121.

Chemistry micronutrients

Lewin, Ralph A., Editor, 1962.
Physiology and biochemistry of algae.
Academic Press, New York and London, 929 pp.

Chemistry, micronutrients

Richards, Francis A., 1968.
Chemical and biological factors in the marine environment.
Ch. 8 in: Ocean engineering: goals, environment technology, J. F. Brahtz, editor, John Wiley & Sons, 259-303.

chemistry, methane

Swinnerton, J.W., and V. J. Linnenbom, 1967.
Gaseous hydrocarbons in sea water: determination.
Science, 156(3778):1119-1120.

mineralization

Chemistry, mineralization

*Wade, Koji, 1967.
Studies on the mineralization of the calcified tissue in Mollusco. XIV. Modification of the amino acid pattern of proteins in the extra pallial fluid during the process of formation and mineralization of nacreous conchiolin in some bivalves.
Bull. Jap. Soc. scient. Fish, 33(7):613-617.

chemistry, mineral precipitates

Lowestam, H.A., 1964.
Mineral precipitates and their relation to the organic chemistry of marine invertebrates. (Abstract).
Geol. Soc., Amer., Special Paper, No. 76:105.

Minor chemical constituents

chemistry, minor constituents

Dyrssen, David, 1969.
Stiochiometry and chemical equilibrium. In: Chemical oceanography, Rolf Lange, editor, Universitetsforlaget, Oslo: 47-58.

Chemistry - minor elements

Goldberg, Edward D., 1965.
Minor elements in sea water.
In: Chemical oceanography, J.P. Riley and G. Skirrow, editors, Academic Press, 1:163-196.

chemistry, minor elements

Gross, M.Grant, 1967.
Concentration of minor elements in diatomaceous sediments of a stagnant fjord.
In: Estuaries, G.H. Lauff, (editor), Publs Am. Ass. Advmt Sci., 83:273-282.

minor chemical constituents

Miyake, Y., 1939.
Chemical studies of the western Pacific Ocean. 6. The vertical variation of minor constituents of the "Kurosio" region.
Bull. Chem. Soc., Japan, 14(10):467-471.

minor inorganic constituents

Riley, J.P., 1965.
Analytical chemistry of sea water.
In: Chemical oceanography, J.P. Riley and G. Skirrow, editors, Academic Press, 2:295-424.

Chemistry - minor elements

Sreekumaran, C., J.R. Naidu, S.S. Gogate, M. Rama Rao, G.R. Doshi, V.N. Sastry, S.M. Shah, C.K. Unni and R. Viswanathan 1968.
Minor and trace elements in the marine environment of the west coast of India.
J. mar. biol. Ass. India 10(1): 152-158.

chemistry molal volumes

Millero, Frank J., 1969.
The partial molal volumes of ions in seawater.
Limnol. Oceanogr., 14(3): 376-385.

molybdenum

Chemistry molybdenum

Bachmann, Roger W., and Charles R. Goldman, 1964
The determination of microgram quantities of molybdenum in natural waters.
Limnology and Oceanography, 9(1):143-146.

Chemistry, molybdenum

Brooks, R.R.,
Trace elements in New Zealand coastal waters.
Geochimica et Cosmochimica Acta, 29(12): 1369-1370.

chemistry, molybdenum

del Riego, A.F., 1962.
El molibdeno en el planeton de la ría de Vigo. Algunas determinaciones de nitrógeno y fósforo: su relación y contenida en materia organica y humedad.
Bol. Inst. Espanol Oceanogr., Madrid, 108:24 pp.

Chemistry, molybdenum

Ernst, T., and H. Hoermann, 1936.
Bestimmung von Vanadium, Nickel und Molybdän im Meerwassers. Nach. Ges. Wiss., Göttingen, Math.-Phys. Kl., n.s., (4) 1(16):205-208.

Chemistry molybdenum

Fukai, Rinnosuke, and W. Wayne Meinke, 1962
Activation analyses of vanadium, arsenic, molybdenum, tungsten, rhenium and gold in marine organisms.
Limnology and Oceanography, 7(2):186-200.

Chemistry molybdenum

Head, P.C. and J.D. Burton, 1970.
Molybdenum in some ocean and estuarine waters.
J. mar. biol. Ass. U.K., 50(2): 439-448.

Chemistry molybdenum

Imai, Yoshihiko, 1961
Oceanographical studies on the behaviour of chemical elements. VI. On the distribution of molybdenum in sea water in Uranouchi Bay and Susaki Bay.
J. Oceanogr. Soc., Japan, 17(4):201-204.

Chemistry, molybdenum

Imai, Yoshihiko, 1961
Oceanographical studies on the behavior of chemical elements. VII. On the distribution of molybdenum in sea water in Urado Bay.
J. Oceanogr. Soc., Japan, 17(4):205-207.

Chemistry, molybdenum

Ishibashi, M., 1953.
Studies on minute elements in sea water.
Rec. Ocean. Wks., Japan, n.s., 1(1):88-92.

Chemistry
molybdenum

Ishibashi, Masayoshi, Taitiro Fujinaga and Tooru Kuwamoto, 1962
Chemical studies on the ocean. Part 89. Fundamental investigation on the dissolution and deposition of molybdenum, tungsten and vanadium in the sea.
Rec. Oceanogr. Wks., Japan, Spec. No. 6: 215-218.

Chemistry Molybdenum blue

Levine, H., J.J. Rowe, and F.S. Grimaldi, 1955.
Molybdenum blue reaction and determination of phosphorus in waters containing arsenic, silicon and germanium. Analyt. Chem. 27(2):258-262.

Molybdenum

Okabe, Shiro, and Toyoko Morinaga 1968.
Vanadium and molybdenum in the rivers and estuary water which pour into the Suruga Bay, Japan. (In Japanese; English abstract).
Nippon Kagaku Zasshi, 89 (3).
Also in: Coll. Repr. Coll. mar. Sci. Techn., Tokai Univ. 1967-1968, 3:329-332.

Molybdenum

Okabe, Shiro, and Yoshimasa Toyota 1967.
Molybdenum contents in suspended matters in the sea water. (In Japanese; English abstract).
Nippon Kagaku Zasshi 88(5)
Also in: Coll. Repr. Coll. mar. Sci. Techn., Tokai Univ., 1967-68, 3:323-326.

chemistry, molybdenum

Okabe, Shiro, Yoshimasa Toyota and Toyoko Morinaga, 1970.
Pollution in coastwise waters of Japan. Kagaku-no-Ryoiki 24(1). Also in: Coll. Repr. Coll. mar. Sci. Techn. Tokai Univ. 4: 303-309.

chemistry
molybdenum

Pilipchuk, M.F., and I.I. Volkov, 1966.
Distribution of molybdenum in modern sediments of the Black Sea. (In Russian).
Doklady, Akad. Nauk, SSSR, 167(5):1143-1146.

chemistry, Molybdenum (data)

Sugawara, Ken, and Shiro Okabe, 1966.
Molybdenum and Vanadium determination of Umitaka-maru samples from her 1962-1963 and 1963-1964 cruises of the International Indian Ocean Expedition. (In Japanese; English abstract).
J. Tokyo Univ. Fish.,)Spec.ed.)8(2):165-171.

Chemistry, molybdenum

Sugawara, Ken and Shiro Okabe 1966.
Molybdenum and vanadium determination of Umitaka-maru samples from her 1962-1963 and 1963-1964 cruises of the International Indian Ocean Expedition. (In Japanese, English abstract).
J. Tokyo Univ. Fish. (Spec. Ed.) 8 (2): 25-31
Reprinted in: Coll. Repr. Coll. mar. Sci. Technol. Tokai Univ. 2 (1966): 25-31.

Chemistry, molybdenum

Sugawara, K., and S. Okabe, 1960.
Geochemistry of molybdenus in natural waters (1).
J. Earth Sci., Nagoya Univ., 8(1):93-107.

Chemistry
molybdenum

Sugawara, K., S. Okabe and M. Tanaka, 1961
Geochemistry of molybdenum in natural waters.
J. Earth Sci. Nagoya Univ., 9(1):114-128.

chemistry
molydenum

Sugawara, Ken, Kikuo Terada, Satoru Kanamori, Nobuko Kanamori and Shiro Okabe, 1962
On different distribution of calcium, strontium, iodine, arsenic and Molybdenum in the northwestern Pacific, Indian and Antarctic Oceans.
J. Earth Sciences, Nagoya Univ., 10(1):34-50.

chemistry
molybdenum

Yamamoto, Yohsikazu, 1961
[Chemical studies on the ocean. 87. Molybdenum content of shallow-water deposits.]
J. Oceanogr. Soc., Japan, 17(1):15-20.

Chemistry - molybdenum

Yamamoto, Y., 1960
[Chemical studies on the ocean - 86. Molybdenum content of the shallow-water deposits (1).]
J. Oceanogr. Soc., Japan, 16(4): 163-166.

chemistry, molybdenum

*Yamamoto, Toshio, Tetsuo Fujita, Tsunenobu Shigematsu and Masayoshi Ishibashi, 1968.
Chemical studies on the seaweeds. (23). Molybdenum content in seaweeds.
Rec. oceanogr. Wks., Japan, n.s., 9(2):209-217.

Myoglobin

Chemistry,
myoglobin

Townsley, P.M., R.A. Richy and P.C. Trussell, 1965.
The occurrence of protoporphyrin and myoglobin in the marine borer, Bankia setacea (Tryon).
Canadaian J. Zool., 43(1):167-172.

Chemistry,
neo-dinoxanthin

Jeffrey, S.W. and F.T. Haxo 1968.
Photosynthetic pigments from symbiotic dinoflagellates (zooxanthellae) from corals and clams.
Biol. Bull. mar. biol. Lab. Woods Hole 135(1): 149-165.

neofucoxanthin

neofucoxanthin

Taylor, W. Rowland, and Conrad D. Gebelein, 1966.
Plant pigments and light penetration in intertidal sediments.
Helgoländer wiss Meeresunters. 13(3):229-237.

neon

chemistry, neon

Bieri, Rudolf H. 1971.
Dissolved noble gases in marine waters.
Earth Planet. Sci. Letts, 10(3): 329-333

chemistry, neon

Bieri, Rudolf H., Mineru Koide and Edward D. Goldberg, 1966.
The noble gas contents of Pacific seawaters.
J. geophys. Res., 71(22):5243-5365.

Chemistry
neon

Bieri, Rudolph, Minoru Koide and Edward D. Goldberg, 1964
Noble gases in sea water.
Science, 146(3647):1035-1037.

chemistry, neon

Craig, H. and R.F. Weiss, 1971.
Dissolved gas saturation anomalies and excess helium in the ocean
Earth Planet. Sci. Letts 10 (3): 289-296

Chemistry, neon

*Craig, H., R.F. Weiss and W.B. Clarke, 1967.
Dissolved gases in the equatorial and South Pacific Ocean.
J. geophys. Res., 72(24):6165-6181.

chemistry, neon

König, H., H. Wänke, G.S. Bien, N.W. Rakestraw and H.E. Suess, 1964.
Helium, neon and argon in the oceans.
Deep-Sea Research, 11(2):243-247.

Chemistry
neon

Mazor, E., G.J. Wasserburg, and H. Craig, 1964.
Rare gases in Pacific Ocean water.
Deep-Sea Res., 11(6):929-932.

Chemistry, neon

Rakestraw, N. W., C.E. Herrick, jr., and W.D. Urry, 1939.
The helium-neon content of sea water and its relation to the oxygen content. J. Am. Chem. Soc. 61:2806-2807.

Chemistry
neon

Urry, W.D., 1935.
Further studies in the rare gases. III. The heliu-neon content of ocean waters.
J. Amer. Chem. Soc., 57:657-659.

Chemistry,
neo-peridinin

Jeffrey, S.W., and F.T. Haxo 1968.
Photosynthetic pigments of symbiotic dinoflagellates (zooxanthellae) from corals and clams.
Biol. Bull. mar. biol. Lab. Woods Hole 135 (1): 149-165.

Chemistry, neoxanthin

Jeffrey, S.W. 1968
Pigment composition of Siphonales algae in the brain coral, Favia.
Biol. Bull. mar. biol. Lab. Woods Hole 135(1): 141-148.

Nickel

Chemistry, nickel

Beck, A.B., and K. Sheard, 1949.
Copper and nickel content of the blood of the West Australian crayfish, Panulirus longipes (Milne Edwards) and of sea weeds. Australian J. Exp. Biol. Med. Sci. 27:307-312.

chemistry, nickel
Bernhard, M., 1964.
Chemical composition and the radiocontamination of marine organisms.
Proc. Symp. Nuclear Detonations and Marine Radioactivity, S.H. Small, editor, 137-150.

chemistry, nickel
Bonner, Francis T., and Alzira Soares Lourenço, 1965.
Nickel content of Pacific Ocean cores.
Nature, 207(5000):933-935.

chemistry, nickel
Bonner, Francis T., and Alzira Soares, 1964.
Nickel content of ocean core samples. (Abstract).
Trans. Amer. Geophys. Union, 45(1):118.

chemistry, nickel
Chester, R., 1965.
Elemental geochemistry of marine sediments.
In: Chemical oceanography, J.P. Riley and G. Skirrow, editors, Academic Press, 2:23-80.

chemistry, nickel
Corcoran, E.F., and J.E. Alexander, 1964.
The distribution of certain trace elements in tropical sea water and their biological significance.
Bull. Mar. Sci., Gulf and Caribbean, 14(4):594-601.

chemistry
Nickel
Duursma, E. K., and W. Sevenhuysen, 1966.
Note on chelation and solubility of certain metals in sea water at different PH values.
Netherlands J. Sea Res., 3(1):95-106.

chemistry, nickel
Ernst, T., and H. Hoermann, 1936.
Bestimmung von Vanadium, Nickel und Molybdän im Meerwassers. Nach. Ges. Wiss., Göttingen, Mat.-Phys. Kl., n.s., (4) 1(16):205-208.

chemistry, nickel
Hutchinson, G.E., R.J. Benoit, W.B. Cotter and P.J. Wangersky, 1955.
On the nickel, cobalt and copper contents of deep-sea sediments. Proc. Nat. Acad. Sci. 41(3): 160-162.

chemistry, nickel
Ishibashi, M., 1953.
Studies on minute elements in sea water.
Rec. Ocean. Wks., Japan, n.s., 1(1):88-92.

chemistry
nickel
Ishibashi, Masayoshi, Taitiro Fujinaga, Fuji Morii, Yoshihiko Kanchiku, and Fumio Kamiyama, 1964
Chemical studies on the ocean (Part 94).
Chemical studies on the seaweeds (19). Determination of zinc, copper, lead, cadmium and nickel in seaweeds using dithizone extraction and polarographic method.
Rec. Oceanogr. Wks., Japan, 7(2):33-36.

chemistry
nickel
Ishibashi, Masayoshi, Toshio Yamamoto, and Tetsuo Fujita, 1964
Chemical studies on the ocean (Part 93).
Chemical studies on the seaweeds (18). Nickel content in seaweeds.
Rec. Oceanogr. Wks., Japan, 7(2):25-32.

chemistry
nickel
Ishibashi, Masayoshi, Shunzo Ueda, and Yoshikazu Yamamoto, 1964
Studies on the utilization of the shallow-water deposits (continued). On the cobalt and nickel contents of the shallow-water deposits.
Rec. Oceanogr. Wks., Japan, 7(2):37-42.

chemistry, nickel
Laevastu, T., and T.G. Thompson, 1956.
The determination and occurrentce of nickel in sea water, marine organisms and sediments.
J. du Cons., 21(2):125-143.

chemistry, nickel
Ostroumov, E.A., and I.I. Volkov, 1962.
Separation of titanium, zirconium and thorium from manganese, nickel, cobalt and zinc by means of cinnamic acid.
Trudy Inst. Okeanol., Akad. Nauk, SSSR, 54:170-181.

In Russian; English summary

chemistry, nickel
Pettersson, H., 1959.
Manganese and nickel on the ocean floor.
Geochim. et Cosmochim. Acta, 17(3/4):209-213.

chemistry, nickel
Pettersson, H., and H. Rotschi, 1952.
The nickel content of deep-sea deposits.
Geochimica et Cosmochimica Acta 2:81-90, 4 textfigs.

chemistry nickel
Pettersson, H., and H. Rotschi, 1950.
Nickel content of deep-sea deposits. Nature 166: 308.

chemistry nickel
*Presley, B.J., R.R. Brooks and I.R. Kaplan, 1967.
Manganese and related elements in the interstitial water of marine sediments.
Science, 158(3803):906-910.

chemistry, nickel
Rotschi, H., 1951.
Étude des teneurs en fer, manganèse et nickel de quelques carottes de grands fonds. Cahiers du Centre de Recherches et d'Études Océanographiques No. 4:2-22, 8 textfigs.

nickel
Schutz, Donald F., and Karl K. Turekian, 1965.
The distribution of cobalt, nickel, and silver in ocean water profiles around Pacific Antarctica.
J. geophys. Res., 70(22):5519-5528.

chemistry, nickel
Schutz, D.F., and K.K. Turekian, 1964.
The distribution of selenium, antimony, silver, cobalt and nickel in sea water. (Abstract).
Trans. Amer. Geophys. Union, 45(1):118.

chemistry, nickel
Smales, A.A., and J.D.H. Wiseman, 1955.
Origin of nickel in deep-sea sediments. Nature 175(4454):464-465.

chemistry, nickel
*Spencer, Derek. W., and Peter G. Brewer, 1969.
The distribution of copper, zinc and nickel in sea water of the Gulf of Maine and the Sargasso Sea.
Geo chim. Acta . 33(3): 325-339.

chemistry, nickel
Tatsumoto, M., 1957.
Chemical investigations of deep-sea deposits. 22. The content of cobalt and nickel in sea sediments (2). 23. The contents of tin and lead in sea sediments. J. Chem. Soc., Japan, 78(1):38-48.

chemistry, nickel
Tatsumoto, M., 1956.
[Chemical studies on the deep-sea deposits. 21. The contents of cobalt and nickel in sea sediments] (1). J. Chem. Soc., Japan, 77(11):1637-1642.

chemistry, nickel
Turekian, Karl K., 1971
2. Rivers, tributaries and estuaries. In: Impingement of man on the oceans, D.W. Hood, editor, Wiley Interscience, 9-73.

chemistry, nickel
Turekian, Karl K., and Donald F. Schutz, 1965.
Trace element economy in the oceans.
Narragansett Mar. Lab., Univ. Rhode Island, Occ. Publ., No. 3:41-89.

chemistry, nickel
Turekian, Karl K., Donald F. Schultz and David Johnson, 1966.
The distribution of Sr, Ba, Co, Ni, and Ag in ocean water profiles of the Pacific sector of the Antarctic seas.
Antarctic J., United States, 1(5):L224.

~~chemistry, nickel-63~~
Beasley, Thomas M., and Edward E. Held, 1969.
Nickel-63 in marine and terrestrial biota, soil and sediment.
Science, 164(3884):1161-1163.

chemistry nickel
Yatsimirsky, K.B., E.M. Emel'yanov, V.K. Pavlova, and Ya. S. Savichenko, 1970.
Determination of microquantities of nickel and cobalt in small weighed portions of marine suspended matter (based on the data from the Baltic Sea and the Atlantic Ocean). (In Russian; English abstract). Okeanologiia, 10(6):1111-1116.

chemistry, nicotinic acid
Hiltz, Doris Fraser 1970.
Occurrence of trigonelline (N-methyl nicotinic acid) in the adductor muscle of a lamellibranch, the sea scallop (Placopecten magellanicus).
J. Fish. Res. Bd Can. 27(3):604-606

Niobium

chemistry, niobium
Carlisle, D.B., 1958.
Niobium in ascidians. Nature 181(4613):933.

chemistry, niobium
Carlisle, D.B., and L.G. Hummerstone, 1958.
Niobium in sea water. Nature 181(4616):1002-1003

chemistry, niobium
Kokubu, N., and T. Hidaka, 1965.
Tantalum and niobium in ascidians.
Nature, 205(4975):1028-1029.

niobium-95

chemistry, niobium-95
Pearcy, William G., and Charles L. Osterberg, 1964.
Vertical distribution of radionuclides as measured in oceanic animals.
Nature, 204(4957):440-441.

nitrate-radical

Chemistry, nitrates, etc.

See also: chemistry, nutrients

A

chemistry nitrate
Adrov, M.M., 1962.
Hydrological investigations off West Greenland.
Sovetskie Riboch. Issledov. v Severo-Zapadnoi Atlant. Okeana, VINRO-PINRO, Moskva, 137-153.

chemistry, nitrate
Anderson, G.C., T.R. Parsons and K. Stephens, 1969.
Nitrate distribution in the subarctic Northeast Pacific Ocean. Deep-Sea Res., 16(4): 329-334.

chemistry, nitrate
Anon., 1950.
Report of the oceanographical observations from Fisheries Experimental Stations (Sept. 1950).
J. Ocean., Kobe Obs., 2nd ser., 1(2):5-11.

chemistry nitrates
Armstrong, F.A.J., 1963
Determination of nitrate in water by ultraviolet spectrophotometry.
Analyt. Chem., 35:1292-1293.

abstract in:
J. Mar. Biol. Assoc., U.K., 44(1):273.

chemistry, nitrate
Armstrong, F.A.J., and E.C. LaFond, 1966.
Chemical nutrient concentrations and their relationship to internal waves and turbidity off southern California.
Limnol. Oceanogr., 11(4):538-547.

chemistry, nitrate
Atkins, W.R.G., 1957.
The direct estimation of ammonia, in sea water, with notes on nitrate, copper, zinc and sugars.
J. du Cons., 22(3):271-277.

chemistry, nitrate
Atkins, W.R.G., 1954.
Note on the use of diphenylbenzidine for the estimation of nitrate in sea water. J. du Cons. 20(2):153-155.

chemistry, nitrate
Atkins, W.G.R., 1932
Nitrate in sea water and its estimation by means of diphenylbenzidine. JMBA, 18:167-192

chemistry, nitrate
Auninsh, E.A., 1966.
Biogenic elements in the Gulf of Riga. (In Russian).
Trudy, Gosudarst. Okeanogr. Inst., No. 83:172-206.

B

chemistry nitrate
Băcescu, M., M.T. Gomoiu, N. Bodeanu, A. Petran, G.I. Müller si V. Chirila, 1965.
Dinamica populaţiilor animale şi vegetale din zona nisipurilor fine de la nord de Constanţa în condiţiile anilor 1962-1965.
In: Ecologie marină, M. Băcescu, redactor, Edit. Acad. Republ. Pop. Romane, Bucureşti, 2:7-167.

chemistry, nitrate (data)
Băcescu, M., G. Müller, H. Skolka, A. Petran, V. Elian, M.T. Gomoiu, N. Bodeanu si S. Stănescu, 1965.
Cercetări de ecologie marină în sectorul predeltaic în condiţiile anilor 1960-1961.
In: Ecologie marină, M. Băcescu, redactor, Edit. Acad. Republ. Pop. Romane, Bucureşti, 1:185-344.

chemistry, nitrate (data)
Ballester, Antonio, Enrique Arias, Antonio Cruzado, Dolores Blasco y José María Camps 1967.
Estudio hidrográfico de la costa catalana de junio de 1965 a mayo de 1967.
Invest. pesq. Barcelona 31(3): 621-662.

chemistry, nitrate
Bandel, W., 1940
Phytoplankton- und Nahrstoffgehalt der Ostsee im Gebiet der Darsser Schwelle. Internat. Rev. ges. Hydrobiol. u. Hydrogr., 40:249-304

chemistry, nitrate
Bardin, I.P., 1958.
Hydrological, hydrochemical, geological and biological studies, Research Ship "Ob", 1955-1957. Trudy Kompleks. Antarkt. Exped., Akad. Nauk, SSSR, Mezh. Geofiz. God, Gidrometeorol. Izdatel., Leningrad, 217 pp.

chemistry, nitrate
Barnes, H., 1950.
A modified 2:4 xylenol method for nitrate estimation. The Analyst 75(892):388-391.

chemistry, nitrate (some data)
Bernard, F., 1956.
Contribution à la connaissance du détroit de Gibraltar (Hydrographie et nannoplancton en juin 1954). Bull. Inst. Océan., Monaco, No. 1074:1-22.

chemistry, nitrate
Bernard, F., 1952.
Eaux atlantiques et mediterraneennes au large de l'Algerie. 1. Hydrographie, sals nutritifs et phytoplancton en 1950. Ann. Hydrogr., Monaco, n.s., 27(1):48 pp., 15 textfigs.

chemistry, nitrates
Bernard, F., 1948
Recherches préliminaires sur la fertilité marine au large d'Alger. J. du Cons. 15(3): 260-267, 5 text figs.

chemistry, nitrates
Bernard, M. F., 1938
Recherches récentes sur la densité du plancton méditerranéen. Rap. Proc. Verb des Réunion, Comm. Int. l'Expl. Sci. de la Méditerranée, n.s., XI:289-300.

chemistry, nitrate
Bernard, F., 1937.
Résultats d'une année de recherches quantitatives sur le phytoplancton de Monaco. Cons. Perm. Int. Expl. Mer, Rapp. Proc. Verb. 105:28-39, 8 textfig

chemistry, nitrate
Bigelow, H.B., and M. Leslie, 1930
Reconnaissance of the waters and plankton of Monterey Bay, July 1928. Bull. M.C.Z., 70(5):429-481, 43 text figs.

chemistry, nitrate
Bigelow, H.B., L.C. Lillick, and M. Sears, 1940.
Phytoplankton and planktonic protozoa of the offshore waters of the Gulf of Maine. Pt. 1. Numerical distribution. Trans. Am. Phil. Soc., n.s., 31(3):149-191, 10 textfigs.

chemistry, nitrate
Black, W.A.P., and E.T., 1949.
Correlation of some of the physical and chemical properties of the sea with the chemical composition of the algae. J.M.B.A. 28(3):673-699, 14 textfigs.

chemistry, nitrate
Bogoyavlensky, A.N., 1959
Hydrochemical investigations. Arktich. i Antarktich. Nauchno-Issled. Inst., Mezhd. Geofiz. God, Sovetsk. Antarkt. Eksped., 5: 159-172.

chemistry, nitrate
Bogoyavlenskii, A.N., 1955.
Chemical characteristics of the water in the region of the Kurile-Kamchatka Trench.
Trudy Inst. Okeanol., 12:161-176.

chemistry nitrate
Bogojavlensky, A.N. and O.V. Shishkina 1971.
On the hydrochemistry off Peru and Chile. (In Russian; English abstract)
Trudy Inst. Okeanol. P.P. Shirshova Akad. Nauk SSSR 89:91-105

chemistry, nitrate
Boto, R. G., 1945
Contribuicao para os estudios de oceanografia ao longo da costa de Portugal. Fosfatos e nitratos. Trav. St. Biol. Mar., Lisbonne, No. 49:102 pp., 57 figs.

chemistry, nitrate
Braarud, T., 1934
A note on the phytoplankton of the Gulf of Maine in the summer of 1933. Biol. Bull. 67(1):76-82.

chemistry, nitrate
Braarud, T., and A. Klem, 1931.
Hydrographical and chemical investigations in the coastal waters off Møre and in the Romsdalfjord. Hvalrådets Skrifter, 1931(1):88 pp., 19 figs.

chemistry nitrate (data)
Brazil, Diretoria de Hidrografia e Navegacão, 1957.
Ano Geofisico Internacional. Publicacão DG-06-II mimeographed.

chemistry, nitrate
Broenkow, William W., 1965.
The distribution of nutrients in the Costa Rica dome in the eastern tropical Pacific Ocean.
Limnology and Oceanography, 10(1):40-52.

Chemistry, nitrate

Bruevich, S.W., 1960

[Hydrochemical investigations on the White Sea.]
Trudy Inst. Okeanol., 42: 199-254.

Chemistry, nitrate

Bruevich, S.W., A.N. Bogoyavlensky, and V.V. Mokievskaya. 1960

Hydrochemical features of the Okhotsk Sea.
Trudy Inst. Okeanol., 42: 125-198.

Chemistry, nitrate (data)

Bruevich, S.V., and I.A. Chikina, 1933
[Hydrochemical observations in the northern part of the Kola Fjord (Barents Sea) in summer 1931.]
Trudy Gosud. Okeanogr. Inst., 3(3): 120-124.

Chemistry, nitrates

Brunelli, G., 1947.
Su alcune caratteristiche del Mare Mediteranneo.
Atti Accad. Lincei Rend. Cl. Sci. Fis. Mat. e Nat., Ser. 8A, 3(1/2):34-36.

Chemistry, nitrate

Bsharah, L., 1957.
Plankton of the Florida Current 5. Environmental conditions, standing crop, seasonal and diurnal changes at a station forty miles east of Miami.
Bull. Mar. Sci., Gulf & Caribbean, 7(3):201-251.

Chemistry, nitrate

Bucalossi, G., 1960.
Étude quantitative des variations du phytoplancton dans la baie d'Alger en fontion du milieu (novembre 1959 à mai 1960).
Bull. Inst. Océanogr., Monaco, 57(1189):1-40.

Chemistry, nitrate

Buch, K.,,1934.
Beobachtungen über chemische Faktoren in der Nordsee, zwischen Nordsee und Island sowie auf dem Schelfgebiete nördlich von Island. Rapp. Proc. Verb., 89(3):13-31, 8 tables, 2 textfigs.

Chemistry, nitrate

Buljan, M., 1953.
The nutrient salts in the Adriatic waters.
Acta Adriatica 5(9):3-15.

Chemistry, nitrate

Burns, R.B.,1967.
Chemical observations in the North Sea in 1965.
Annls biol. Copenh. (1965)22:29-31.

Chemistry, nitrate

Burns,R.B., and R. Johnston,1966.
Chemical observations from 1964 in the northern North Sea.
Annls.biol., Copenh., 21:29-36.

Chemistry, nitrate

Burns, R.B., and R. Johnston, 1965.
Chemical observations on the waters of the Scottish west coast and the western approaches.
Ann. Biol., Cons. Perm. Int. Expl. Mer, 1963, 20:

Chemistry, nitrate

Burns, R.B., and R. Johnston, 1965.
Chemical observations in the northern North Sea, 1963.
Ann. Biol., Cons. Perm. Int. Expl. Mer, 1963, 20:57-60.

Chemistry, nitrate (data)

Burns, R.B., and R. Johnston, 1964.
Chemical observations.
Ann. Biol., Cons. Perm. Int. Expl. Mer, 1962, 19:42-44.

Chemistry, nitrate

Burns, R.B., and R. Johnston, 1964.
Chemical and hydrographic survey of the North Channel and Clyde Sea area in April, 1962.
Ann. Biol., Cons. Perm. Int. Expl. Mer, 1962, 19:32-33.

Chemistry, nitrate (data)

Burns, R.B., and R. Johnston, 1963.
Chemical observations.
Ann. Biol., Cons. Perm. Int. Expl. Mer, 1961, 18:47-48.

C

Chemistry, nitrate (data)

California, Humboldt State College, 1964.
An oceanographic study between the points of Trinidad Head and the Eel River.
State Water Quality Control Bd., Resources Agency, California, Sacremento, Publ., No. 25:

Chemistry, nitrate

Carlucci, A.F. and Hazel R. Schubert, 1969.
Nitrate reduction in seawater of the deep nitrite maximum off Peru. Limnol. Oceanogr., 14(2): 187-193.

Chemistry, nitrate

Castillejo, F. F., 1966.
Non-seasonal variations in the Hydrological environment off Port Hackvej, Sydney.

C.S.I.R.O., Div. Fish. Oceanogr. Techn. Paper, No. 21:12 pp.

Chemistry, nitrate

Chauchan, V.D. 1967.
Some observations on chemical and physical conditions of the sea water at Port Okha.
In: Proc. Seminar, Sea, Salt and Plants, V. Krishnamurthy, editor, Bhavnagar, India 41-44.

Chemistry, nitrate

*Choe,Sang,Tai Wha Chung and Hi-Sang Kwak,1968.
Seasonal variations in nutrients and principal ions contents of the Han River water and its water characteristics. (In Korean;English abstract).
J.oceanogr.Soc.,Korea, 3(1):26-38.

Chemistry, nitrate

Chow, D.T.-W., and R.J. Robinson, 1953.
Polarographic determination of nitrate in sea water. J. Mar. Res. 12(1):1-12, 1 textfig.

chemistry,nitrate

Chow,Tsaihwa J., and Arnold W. Mantyla,1965.
Inorganic nutrient anions in deep ocean waters.
Nature,206(4982):383-385.

Also in:
Collected Reprints,Int. Indian Ocean Exped., UNESCO,3. 1966.

Chemistry, nitrate

Chromov, N.S., 1962.
Distribution and dynamics of the plankton and the food of sardines in the fishing areas off the west African coast. (In Russian).
Trudy Vses. Nauchno-Issled. Inst. Morsk. Ribn. Chos. i Okean., VNIRO, 46:214-235.

Chemistry, nitrate

*Codispoti, Louis A., 1968.
Some results of an oceanographic survey in the northern Greenland Sea, summer 1964.
Techn. Rept., U.S. Nav. Oceanogr. Off., TR202: 49 pp.

Chemistry, nitrate

Codispoti, L.A., and F. A. Rihcards, 1968.
Micronutrient distributions in the East Siberian and Laptev seas during summer 1963.
Artic, 21(2): 67-83

Chemistry, nitrate

Conover, S.A.M., 1954.
Observation on the structure of red tides in New Haven Harbor, Connecticut. J. Mar. Res. 13(1):145-155, 6 textefigs.

Chemistry, nitrate

Corcoran, E.F., and James E. Alexander, 1963.
Nutrient, chlorophyll and primary production studies in the Florida Current.
Bull. Mar. Sci., Gulf and Caribbean, 13(4):527-541.

Chemistry, nitrate (data)

Corwin, Nathaniel, and David A. McGill, 1963.
Nutrient distribution in the Labrador Sea and Baffin Bay.
U.S.C.G. Bull., No. 48:79-94; 95-153.

nitrate (data)

COSTE, B., et H.J. MINAS, 1968.
Production organique primaire et sels nutritifs au large des côtes occidentales Corso-Sardes en février 1966.
Rec. Trav. Sta. mar. Endoume, 44(60):89-61

Chemistry, nitrate (data)

Coste, Bernard, et Hans-Joachim Minas 1967.
Premières observations sur la distribution des taux de productivité et des concentrations en sels nutritifs des eaux de surface du Golfe de Lion.
Cah. Océanogr. 19(5): 417-429.

Chemistry, nitrate

Crehuet, Ramón Fernández y Maria Jesús del Val Cordón, 1960

Observaciones oceanograficas en la Bahia de Malaga (Marzo a marzo 1957). Bol. Inst. Esp. Ocean., 98: 1-29.

D

Chemistry, nitrate

Davidovitch, R.L., 1964.
Short chemical characteristics of the northwest Pacific Ocean. (In Russian).
Gosudarst. Kom. Sov. Ministr., SSSR, Ribn. Choz., Trudy VNIRO, 49, Izv. TINRO, 51:93-98.

Chemistry, nitrate (data)

Davidovich, R.L., 1963
Hydrochemical features of the southern and southeastern parts of the Bering Sea. (In Russian).
Sovetsk. Ribokh. Issled. B Severo-Vostokh. Chasti Tikhogo Okeana, VNIRO 48, TINRO 50(1): 85-96.

Chemistry, nitrate

Deacon, G. E. R., 1933
A general account of the hydrology of the South Atlantic Ocean. Discovery Repts. 7:173-238, pls.8-10.

Chemistry, nitrate

Defant, A., G. Böhnecke, H. Wattenberg, 1936.
I. Plan und Reiseberichte die Tiefenkarte das Beobachtungsmaterial. Die Ozeanographischen Arbeiten des Vermessungsschiffes "Meteor" in der Dänemarkstrasse und Irmingersee während der Fischereischutzfahrten 1929, 1930, 1933 und 1935. Veroffentlichungen des Instituts für Meereskunde, n.f., A. Geogr.-naturwiss. Reihe, 32:1-152 pp., 7 text figs., 1 plate.

Chemistry, nitrate (data)

De Queiroz Santos, Neuzon 1967.
Principais nutrientes e alguns dadas fisico-químicos do região lagunar de Cananéia.
Bolm Inst. Biol. mar. Univ. Fed. Rio Grande Norte, Brasil 4:1-14.

Chemistry nitrates

Dobrzhanskaya, M.A., 1954.
[Nitrates in the Black Sea.]
Dokl. Akad. Nauk, SSSR, 99(1):61

Chemistry, nitrate (data)

Dragovich, Alexander, 1961
Relative abundance of plankton off Naples, Florida, and associated hydrographic data, 1956-57.
USFWS Spec. Sci. Rept., Fish., No. 372:41 pp.

Chemistry, nitrate

Dragovich, Alexander, John H. Finucane, John A. Kelly, Jr., and Billie Z. May, 1963.
Counts of red-tide organisms, Gymnodinium breve, and associated oceanographic data from Florida west coast, 1960-61.
U.S. Fish and Wildlife Service, Spec. Sci. Repts. Fish., No. 455:1-40.

Chemistry nitrate

Dragovich, A., and B.Z. May, 1962
Hydrological characteristics of Tampa Bay tributaries.
U.S.F.W.S. Fish. Bull. (205) 62: 163-176.

Chemistry - nitrate

Dugdale, R.C. and Jane J. MacIsaac, 1971.
A computation model for the uptake of nitrate in the Peru upwelling region. Investigación pesq. 35(1): 299-308.

Chemistry, nitrate

*Duxbury, Alyn C. and Noel B. McGary, 1968.
Local changes of salinity and nutrients off the mouth of the Columbia River.
Limnol. Oceanogr., 13(4):626-636.

E

chemistry, nitrate

Eisma, D. and A.J. van Bennekom, 1971.
Oceanographic observations on the eastern Surinam Shelf. Hydrogr. Newsletter, R. Netherlands Navy, Spec. Publ. 6: 25-29.

Chemistry, nitrate (data)

El-Sayed, Sayed Z., and Enrique F. Mandelli, 1965.
Primary production and standing crop of phytoplankton in the Weddell Sea and Drake Passage.
In: Biology of Antarctic seas. II.
Antarctic Res. Ser., Am. Geophys. Union, 5:87-106.

Chemistry, nitrate

Eppley, Richard W., and James L. Coatsworth 1968.
Uptake of nitrate and nitrite by Ditylum brightwelli - kinetics and mechanisms.
J. Phycol. 4(2): 151-156.

chemistry nitrate

*Ewins, P.A., and C.P. Spencer, 1967.
The annual cycle of nutrients in the Menai Straits.
J. mar. biol. Ass., U.K., 47(3):533-542.

F

chemistry, nitrate (data)

Faganelli, A., 1961.
Primi risultati relativi alla concentrazione dei sali nutritivi nelle acque del Mar Mediterraneo centrale et mari adiacenti.
Rapp. Proc. Verb., Réunions, Comm. Int. Expl. Sci., Mer Méditerranée, Monaco, 16(3):675-686.

chemistry, Nitrate (data)

Faganelli, Armando, 1961.
Primi risultati relativi alla concentrazione dei sali nutritivi nelle acque del Mare Mediterraneo Centrale e mari adiacenti.
Archivio Oceanograf. e Limnol., 12(2):191-208.

Also:
Ist. Sperimentale Talassograf., Trieste, Pubbl. No. 377 (1961)

Chemistry, nitrate

Febres, Germán, Gilberto Rodríguez y Andrés Eloy Esteves, 1966.
Ch. 2. Química del agua.
Estudios Hidrobiologicos en el Estuario de Maracaibo, Inst. Venezolano de Invest. Cient., 21-65. (Unpublished manuscript).

Chemistry nitrate

Fedosov, M. V., 1955
[Chemical basis for the foodstuffs in the Azov Sea and prognosis of its change in connection with water construction projects in rivers.]
Trudy VNIRO 31: 35-61.

chemistry, nitrate

Fedosov, M.V., and E.G. Vinogradova, 1955.
[Hydrological and hydrochemical regime, primary production in the Azov Sea and forecasting change. Fundamental basis of the hydrochemical regime in the Azov Sea.] Trudy VNIRO, 31(1):9-34.

Chemistry, nitrate

Fedosov, M.V., and G.N. Zaitsev, 1960
[Water balance and the chemical regime in the Baltic Sea and its gulfs.]
Trudy Vses. Nauchno-Issled. Inst. Morsk. Ribnogo Chozia i Oceanogr. (VNIRO) 42: 7-14.

Chemistry, nitrate

Finucane, John H., and Alexander Dragovich, 1959.
Counts of red tide organisms, Gymnodinium brevis and associated oceanographic data from Florida west coast, 1954-1957.
USFWS Spec. scient. Rept., Fish. No. 289:220pp.

Chemistry
Nitrate (data)

Forsbergh, Eric D., William W. Broenkow, 1965.
Observaciones oceanograficas del oceano Pacific oriebtal recolectadas por el barco Shoyo Maru, octubre 1963-marzo 1964.
Comision Interamericana del Atun Tropical, Bol. 10(2): 85-237.

Chemistry, nitrate

Fraga, F., 1967.
Hidrografía de la ria de Vigo, 1962, con especial referencia a los compuestos nitrógens.
Investigación pesq., 31(1):145-149.

Chemistry, nitrate

Fukai, R., 1955.
Critical studies on the analytical methods for minor chemical constituents in sea water (Part 3) - Remarks on the method of estimation of nitrate-nitrogen by means of reduced strychnine reagent. J. Ocean. Soc., Japan, 11(1):19-23.

Also: Bull. Tokai Reg. Fish. Res. Lab., 11(B-152)

Chemistry nitrate

Fukai, R., 1953.
On the distribution of nutrient salts in the Equatorial North Pacific.
Bull. Chem. Soc., Japan, 26(9):485-489, 5 textfigs.

Also reprinted in: Bull. Tokai Reg. Fish. Lab. 7(Contr. B): Contr. 113.

chemistry, nitrate

Fukai, R., 1952.
Seasonal variations of minor chemical constituents in the waters of the Zunan-Kuroshio region. 1. On nitrogen-phosphorus content ratio with the behavior of nitrate and phosphate.
Bull. Chem. Soc., Japan, 25(5):323-325, 3 textfig

Also in:
Bull. Tokai Regional Fish. Res. Lab., Fish. Agency, (Contrib B)6, 1953.

G

Chemistry, nitrate

Gaarder, Karen Ringdal, 1938.
Phytoplankton studies from the Tromsø district, 1930-31. Tromsø Mus. Årshefter, Naturhist. Avd., 11, 55 (1):159 pp., 4 fold-in pls., 12 textfigs.

Chemistry, nitrate

Gaarder, T., and R. Spärek, 1932.
Hydrographisch-biochemische Untersuchungen in norwegischen Austern-Pollen. Bergens Mus. Aarbok, Naturvidensk.-rekke, No. 1:5-144, 75 textfigs.

Chemistry, nitrate

Gad, G., M. Knetsch, and H. Schlichting, 1948.
Ein Beitrag zur Nitratbestimmung im Wasser.
Gesundheits-Ingenieur 69(4/5):137.

Chemistry, nitrate

Galtsoff, P.S., ed., 1954.
Gulf of Mexico, its origin, waters and marine life. Fish. Bull., Fish and Wildlife Service, 55:1-604, 74 textfigs.

Chemistry nitrate

Ganapati, P.N., and D. Venkata Rama Sarma, 1958.
Hydrography in relation to the production of plankton off Waltair coast.
Andhra Univ. Mem. Oceanogr., 2:168-192.

Chemistry, nitrate

Garkavaia, G.P., and L.E. Pozdniakova 1968.
Data on the hydrochemical regime in Chesha Bay, Barents Sea. (In Russian)
Trudy Murmansk. Morsk. Biol. Inst. 17(21): 4-21.

Chemistry nitrate (data)

Gennesseaux, Maurice, 1960

L'oxygène, les phosphates et les nitrates dissous dans les eaux du bassin sud-ouest de la Méditerranée occidentale (des Iles Baléares à Gibraltar). Trav. C.R.E.O. ns. 3(4): 5- 22.

Also in: Trav. lab. Géol. Sous-Marine, 10 (1960).

Chemistry, nitrate

Gieskes, Joris M.T.M., and Klaus Grasshoff 1969. A study of the variability in the hydrochemical factors in the Baltic Sea on the basis of two anchor stations, September 1967 and May 1968. Kieler Meeresforsch. 25 (1): 105-132.

chemistry, nitrate

Goering, J.J., and R.C. Dugdale, 1966. Denitrification rates in an island bay in the equatorial Pacific Ocean. Science, 154(3748):505-506.

Chemistry nitrate

Goering, J.J., R.C. Dugdale and D.W. Menzel, 1964 Cyclic diurnal variations in the uptake of ammonia and nitrate by photosynthetic organisms in the Sargasso Sea. Limnology and Oceanography, 9(3):448-451.

chemistry, nitrate

Gololobov, Ya. K., 1963. Hydrochemical characteristics of the Aegean Sea during the autumn of 1959. (In Russian; English abstract). Rez. Issled. Programme Mezhd. Geofiz. Goda, Okeanolog. Issled., Akad. Nauk, SSSR, No. 8:90-96.

chemistry, nitrate

Griffiths, Raymond C., 1965. A study of ocean fronts off Cape San Lucas, Lower California. U.S.F.W.S. Spec. Sci. Rept., Fish., No. 499:54 pp

Chemistry, nitrate

Grøntved, J., 1952. Investigations on the phytoplankton in the southern North Sea in May 1947. Medd. Komm. Danmarks Fisk.- og Havundersøgelser, Plankton ser., 5(5): 1-49, 1 pl., 21 tables, 24 textfigs.

H

Chemistry nitrates

Halim, Youssef, 1960 Étude quantitative et qualitative du cycle écologique des dinoflagellés dans les eaux de Villefranche-sur-Mer. (1953-1955). Ann. Inst. Océanogr., Monaco, 38:123-232.

Chemistry nitrates

Harrison, G.A.F., 1962 The determination of nitrate in water. Talanta, 9(6): 533-535.

chemistry nitrate

Harvey, H.W., 1926. Nitrate in the sea. J. Mar. Biol. Assoc., U.K., 14:71-88.

Chemistry, nitrate

Hela, I., 1955. Ecological observations on a locally limited red tide bloom. Bull. Mar. Sci., Gulf and Caribbean, 5(4):269-291, 16 textfigs.

Chemistry, nitrate

Higaki, Shiro, and Zeitlin, 1962. Nitrate in Hawaiian waters. Limnol. & Oceanogr., 7(1):103.

Chemistry nitrate

Hollan, Eckard, 1970. Eine physikalische Analyse kleinräumiger Änderungen chemischer Parameter in den tiefen Wasserschichten der Gotlandsee. Kieler Meeresforsch. 25(2): 255-267

Chemistry, nitrate

Holm-Hansen, Osmund, J.D.H. Strickland and P.M. Williams, 1966. A detailed analysis of biologically important substances in a profile off southern California. Limnol. Oceanogr., 11(4):548-561.

chemistry, nitrate

Hulburt, Edward M., 1966. The distribution of phytoplankton, and its relationship to hydrography, between southern New England and Venezuela. J. Mar. Res., 24(1):67-81.

chemistry, Nitrate

Hung, Tsu-Chang 1970 Photo- and radiation effects on inorganic nitrogen compounds in pure water and sea water near Taiwan in the Kuroshio Current. Bull. Inst. Chem. Acad. Sinica 18: 37-56

chemistry nitrate (data)

Husby, David M., and Gary L. Hufford 1971. Oceanographic investigation of the northern Bering Sea and Bering Strait 8-21 June 1969. Oceanogr. Rept. U.S. Cst Gd 42 (CG 373-42): 55 pp.

I

chemistry, nitrate (data)

Ignatiades, Lydia, and Theano Becacos-Kontos 1969. Nutrient investigations in lower Saronicos Bay, Aegean Sea. Vie Milieu (B) 20 (1): 51-61.

Chemistry nitrate (data)

Ilyina, N.L., 1961 [Some characteristics of hydrochemistry of the Northern Part of the Norwegian and the Greenland seas (summer 1958).] (English abstract). Mezhd. Kom. Mezhd. Geofiz. Goda, Presidiume Akad. Nauk, SSSR, Okeanol. Issled., (3):151-161.

Chemistry nitrates

Isaeva, A.B., 1956. K metodike opredeleniya nitratov v morskoi vode difenilaminnym metodom pri pomoshchi fotoelektricheskogo kolorimetra. Trudy Inst. Okeanol., 19:

Chemistry, nitrate

Ivanenkov, V.N., and F.A. Gebin, 1960. Water masses and hydrochemistry of the western and southern parts of the Indian Ocean. Physics of the Sea. Hydrology. (In Russian). Trudy Morsk. Gidrofiz. Inst., 22:33-115.

Translation: Scripta Technica, Inc. for Amer. Geophys. Union, 27-99.

J

Chemistry, nitrate (data)

Jacobs, Stanley S., Peter M. Bruchausen and Edward B. Bauer, 1970 Cruises 32-36, 1968, hydrographic stations, bottom photographs, current measurements. Eltanin Repts Lamont-Doherty Geol. Obs., Nat. Sci. Found., U.S. Antarctic Res. Program, 460 pp. (multilithed)

Chemistry, nitrate

Japan, Kobe Marine Observatory, 1967. Report of the oceanographic observations in the sea south of Honshu from July to August 1965. (In Japanese). Bull. Kobe Mar. Obs. No. 178: 31-40

chemistry, nitrate

Japan Kobe Marine Observatory, 1967. Report of the oceanographic observations in the sea south of Honshu from February to March 1964. (In Japanese) Bull. Kobe Mar. Obs. No. 178: 27-

chemistry, nitrate

Japan, Kobe Marine Observatory, Oceanographical Section, 1964. Report of the oceanographic observations in the Kuroshio and region east of Kyushu from October to November 1962. (In Japanese) Res. Mar. Meteorol. and Oceanogr., Japan. Meteorol. Agency, 32: 41-50.

Also in: Bull. Kobe Mar. Obs., 175. 1965.

chemistry, nitrate

Japan, Kobe Marine Observatory, Oceanographical Section, 1964. Report of the oceanographic observations in the sea south of Honshu from February to March, 1963. Res. Mar. Meteorol. and Oceanogr., Japan Meteorol. Agency, 33: 27-32.

Also in: Bull. Kobe Mar. Obs., 175. 1965.

chemistry, nitrate

Japan, Kobe Marine Observatory, Oceanographical Section, 1964. Report of the oceanographic observations in the sea south of Honshu from July to August, 1962. Res. Mar. Meteorol. and Oceanogr., Japan. Meteorol. Agency, 32: 32-40. (In Japanese).

Also in: Bull. Kobe Mar. Obs. 175.

Chemistry, nitrate

Japan, Kobe Marine Observatory, Oceanographical Section, 1964.? Researches in chemical oceanography made at K.M.O. for the recent ten years (1954-1963). (In Japanese). Weather Service Bull., Japan. Meteorol. Agency. 31(7):188-191.

Also in: Bull. Kobe Mar. Obs., No. 175. 1965.

chemistry nitrate

Japan, Kobe Marine Observatory, Oceanographical Section, 1962
Report of the oceanographical observations in the sea south of Honshu from July to August 1961. (In Japanese).
Res. Mar. Meteorol. & Oceanogr., 30:39-48.

Also in:
Bull. Kobe Mar. Obs., No. 173(3). 1964

chemistry nitrate

Japan, Kobe Marine Observatory, Oceanographical Section, 1962
Report of the oceanographic observations in the sea south of Honshu in May, 1960. (In Japanese).
Bull. Kobe Mar. Obs., No. 169(12):27-33.

chemistry - nitrate

Japan, Science Council, National Committee for IIOE, 1966.
General report of the participation of Japan in the International Indian Ocean Expedition.
Rec. Oceanogr. Wks., Japan, n.s. 8(2): 133 pp.

chemistry, nitrates

Jayaraman, R., 1954.
Seasonal variations in salinity, dissolved oxygen and nutrient salts in the inshore waters of the Gulf of Manner and Palk Bay near Mandapam (S. India). Indian J. Fish. 1:345-364.

chemistry - nitrates

Jayaraman, R., 1951.
Observations on the chemistry of the waters of the Bay of Bengal off Madras City during 1948-1949. Proc. Indian Acad. Sci., b, 33(2):92-99, figs.

chemistry, nitrate

Jeffries, Harry P., 1962.
Environmental characteristics of Raritan Bay, a polluted estuary.
Limnol. and Oceanogr., 7(1):21-31.

chemistry nitrate

Jeffries, Harry P., 1962
The atypical phosphate cycle of estuaries in relation to benthic metabolism.
The Environmental Chemistry of Marine Sediments. Proc. Symp., Univ. R.I., Jan. 13, 1962 Occ. Papers, Narragansett Mar. Lab., No. 1: 58-67.

chemistry nitrate

Jones, P.G.W., 1971.
The southern Benguela Current region in February 1966: Part I. Chemical observations with particular reference to upwelling. Deep-Sea Res., 18(2): 193-208.

chemistry - nitrate

Jones, P. G. W., 1966.
The nitrate and phosphate content of the eastern equatorial Atlantic Ocean.
Bull. I.F.A.N.(A), 28(2):444-449.

K

chemistry nitrate

Kabanova, Yu. G., 1961
[Primary production and nutrients in the Indian Ocean.] (In Russian; English abstract).
Okeanolog. Issled., Mezhd. Komitet Proved. Mezhd. Geofiz. Goda, Prezidiume, Akad. Nauk. SSSR, (4):72-75.

chemistry nitrate

Kändler, Rudolf, 1963
Hydrographische Untersuchungen über die Abwasserbelastung der Flensburger Förde.
Kieler Meeresforsch. 19(2):142-157.

chemistry, nitrate (data)

Kato, Kenji, 1966.
Geochemical studies on the mangrove region of Cananéia, Brazil. 1. Tidal variations of water properties.
Bolm Inst. Oceanogr., S. Paulo, 15(1):13-20.

chemistry - nitrates

Kato, K., 1966.
Chemical investigations on the hydrographical system of Cananéia lagoon.
Bolm Inst. Oceanogr., S. Paulo, 15(1):1-12.

chemistry, nitrate

Kato, Kenji, 1961.
Oceanographical studies in the Gulf of Cariaco.
1. Chemical and hydrographical observations in January, 1961.
Bol. Inst. Oceanograf., Univ. Oriente, Cumana, Venezuela, 1(1):49-72.

chemistry nitrate (data)

Kato, Kenji, 1961
Some aspects on biochemical characteristics of sea water and sediments in Mochima Bay, Venezuela.
Bol. Inst. Oceanogr., Univ. de Oriente, Venezuela 1(2):343-358.

chemistry nitrate

Ketchum, B.H., 1947
The biochemical Relations between Marine Organisms and their Environment. Ecol. Monogr., 17:309-315, 5 textfigs.

chemistry nitrate, (surface)

Khan, M.A. and D.I. Williamson, 1970.
Seasonal changes in the distribution of chaetognatha and other plankton in the eastern Irish Sea. J. exp. mar. Biol. Ecol., 5(3): 285-303.

chemistry, nitrate

Kijowski, S., 1937.
Nieco danych o skladzie chemicznym wod Zatoki Gdanskiej. (Quelques données sur la composition chimique des eaux du Golf de Dantzig). Bull. Sta. Mar. Hel 1(1):33-41, 1 fig.

chemistry nitrate

*Kirkwood, L.F., 1967.
Inorganic phosphate, organic phosphorus, and nitrate in Australian waters.
Tech. Pap. Div.Fish Oceanogr.,CSIRO,Australia, 25: 18 pp.

chemistry, nitrate

Kitamura, H., 1958.
[On the nitrate-nitrogen in the Kuroshio. Chemical oceanography of the North Pacific. (II).]
Umi to Sora, 34(1):11-19.

chemistry, nitrate

Kitamura, Hiroyuki, and Takeshi Sagi, 1965.
On the chemical elements in the sea south of Honshu, Japan. II. (In Japanese;English abstract)
Bull. Hokkaido Mar. Obs., No. 174:39-55.

chemistry, nitrate

Kitamura, Hiroyuki, and Takeshi Saga, 1964.
On the chemical elements in the sea south of Honshu, Japan. (In Japanese; English abstract).
Bull. Kobe Mar. Obs., No. 172:6-54.

chemistry nitrate (data)

Klepikova, V.V., Edit., 1961
Third Marine Expedition on the D/E "Ob", 1957-1958. Data.
Trudy Sovetskoi Antarktich. Exped., Arktich. i Antarktich. Nauchno-Issled. Inst., Mezhd. Geofiz. God, 22:1-234 pp.

chemistry, nitrate

*Krishnamurthy,K., 1967.
The cycle of nutrient salts in Porto Novo(India) water.
Int.Rev.ges.Hydrobiol., 52(3):427-436.

chemistry, nitrates

Kruger, D., 1950.
Variations quantitatives des protistes marins au voisinage du Port d'Alger durant l'hiver 1949-1950. Bull. Inst. Océan., Monaco, No. 978:20 pp. 5 textfigs.

chemistry nitrate

Kusunoki, Kou, 1962
Hydrography of the Arctic Ocean with special reference to the Beaufort Sea.
Contrib. Inst., Low Temp. Sci., Sec.A,(17): 1-74.

L

chemistry, nitrate

Lagarde, E., and P. Forget, 1961.
Contribution à l'étude du métabolisme de l'azote mineral en milieu marin. Étude de souche pure de bacteries dénitrificatrices et réductrices des nitrates.
Rapp. Proc. Verb., Réunions, Comm. Int. Expl. Sci. Mer/ Mediterranée, Monaco, 16(2):245-250.

chemistry, nitrate

Lauzier, L., and C. Ouellet, 1943.
Détermination colorimétrique des nitrites et nitrates dans l'eau de mer. Ann. A.C.F.A.S. 9:94.

chemistry, nitrate

Lecal, J., 1952.
Répartition en profondeur des coccolithophorides en quelques stations méditerranéenes occidentales
Bull. Inst. Océan., Monaco, No. 1018:13 pp., 4 textfigs.

chemistry, nitrates

Leloup, E., Editor, with collaboration of L. Van Meel, Ph. Polk, R. Halewyck and A. Gryson, undated.
Recherches sur l'ostreiculture dans le Bassin de Chasse d'Ostende en 1962.
Ministere de l'Agriculture, Commission T.W.O.Z. Groupe de Travail - "Ostreiculture". 58 pp.

chemistry nitrate

Lillick, L.C., 1937
Seasonal studies of the phytoplankton off Woods Hole, Massachusetts. Biol. Bull. LXXIII (3):488-503, 3 text figs.

chemistry, nitrate

Lindquist, Armin, 1959
Studien über das Zooplankton der Bottensee II. Zur Verbreitung und Zusammensetzung des Zooplanktons.
Inst. Mar. Res., Lysekil, Ser. Biol., Rept. No. 11:136 pp.

chemistry, nitrates
López Costa, R., 1951.
Sôbre la determinación del nitrogeno nítrico en el agua de mar. 1. La resorcina como reactivo de los nitratos. Bol. Inst. Español. Ocean. 43:13 pp

chemistry, nitrate
Lund, J.W.G., 1950.
Studies on Asterionella formosa Hass. II. Nutrient depletion and the spring maximum. Pt. II. Discussion. J. Ecol. 38(1):15-35.

chemistry, nitrate
Lund, J.W.G., 1950.
Studies on Asterionella formosa Hass. II. Nutrient depletion and the spring maximum. Pt. I. Observations on Windermere, Esthwaite water and Blelham Tarn. J. Ecol. 38(1):1-14, 7 textfigs.

M

chemistry, nitrates
Marchesoni, V., 1954.
Il trofismo della Laguna Veneta e la vivificazione marina. III. Ricerche sulle variazioni quantitative del fitoplancton.
Arch. Oceanogr. e Limnol. 9(3):153-284.

chemistry, nitrates (data)
Margalef, R., J. Herrera, M. Steyaert et J. Steyaert, 1966.
Distributions et caractéristiques des communautés phytoplanctoniques dans la bassin Tyrrhenien de la Méditerranée en fonction des facteurs ambiants et à la fin de la stratification estivale de l'année 1963.
Bull. Inst. r. Sci. nat. Belg., 42(5):1-56.

chemistry, nitrate
Marshall, S.M. and A.P. Orr, 1948
Further experiments on the fertilization of a sea loch (Loch Craiglin). The effect of different plant nutrients on the phytoplankton. J.M.B.A. 27(2):360-379, 10 text figs.

chemistry, nitrate (data)
Marumo, Ryuzo, Editor, 1970.
Preliminary report of the Hakuho Maru Cruise KH-69-4, August 12-November 13, 1969, The North and Equatorial Pacific Ocean.
Ocean Res. Inst. Univ. Tokyo, 68pp.

nitrate
Matsue, Y., 1950.
[The variation of titratiable base in sea water by the growth of diatoms.] J. Ocean. Soc., Tokyo, 6(1):32-38, 2 textfigs. (In Japanese with English abstract).

chemistry, nitrate
Matsue, Yoshiyuki, Yuzo Komaki and Masaaki Murano, 1957
On the distribution of minute nutrients in the North Japan Sea, during the close of August, 1955.
Bull. Japan Sea Reg. Fish. Res. Lab., 6:121-127
Abstr. in:
Rec. Res. Fac. Agric., Tokyo Univ., Mar. 1958. 7(80): 57-58.

Mc

chemistry, nitrate
McGill, David A. 1965.
The relative supplies of phosphate, nitrate and silicate in the Mediterranean Sea.
Rapp. P.-v. Réun. Commn int. Explor. scient. Mer Méditerr. 18(3): 737-744.

chemistry, nitrate (data)
McGill, David A., and Nathaniel Corwin, 1965.
The distribution of nutrients in the Labrador Sea, summer 1964.
U.S.C.G. Rept., No. 10(373-10):25-33; 277-285.
[Oceanogr.

chemistry, nitrate
Miller, S.M., H.B. Moore and K.R. Kvammen, 1953.
Plankton of the Florida Current. 1. General conditions. Bull. Mar. Sci., Gulf and Caribbean 2(3): 465-485, 27 textfigs.

chemistry, nitrate
Minas, H.J., and M.J. Pizarro, 1962.
Quelques observations hydrologiques sur les eaux au large des côtes orientales de la Corse.
Bull. Inst. Océanogr., Monaco, 59(1232):1-23.

chemistry, nitrate
Miyake, Y., Y. Sugiura and K. Kamada, 1953.
A study of the property of the coastal waters around Hachijo Island. Rec. Ocean. Wks., Japan, n.s., 1(1):93-99, 10 textfigs.

chemistry, nitrate
Mokiyevskaya, V.V., 1961.
[Some peculiarities of hydrochemistry in the northern Indian Ocean.]
Okeanol. Issledov., Mezhd. Komit., Mezhd. Geofiz. God, Presidiume, Akad. Nauk, SSSR, :50-61.

chemistry, nitrate
Mokievskaja, V.V., 1959
[Nutrients salts in the upper layer of the Bering Sea.] Trudy Inst. Okeanol., 33: 87-113.

chemistry nitrate (data)
Moreira da Silva, Paulo, 1957
Oceanografia do triangulo Cabo-Frio-Trinidade-Salvador.
Anais Hidrograficos, Brasil, 16:213-308.

chemistry, nitrate
Morii, Hideaki, Yusho Arishige, Ryoichi Kanagu and Tadanobu Fukuhara 1967.
Chemical studies of the Kuroshio and adjacent region. 1. On the nutrient salts and the dissolved oxygen. (In Japanese; English abstract).
Bull. Fac. Fish. Nagasaki Univ. 22: 91-103

chemistry, nitrate
Morris, A.W., and J.P. Riley, 1963.
The determination of nitrate in sea-water.
Analytica Chimica Acta, 29(3):272-279.

chemistry, nitrate
Mullin, J.B., and J.P. Riley, 1955.
The spectrophotometric determination of nitrate in natural waters with particular reference to sea water.
Anal. Chim. Acta 12(5):464-480.
Anal. Abstr. 2(10):2869.

N

chemistry, nitrate (data)
Nagaya, Y., 1959
On the distribution of the nutrient salts in the Equatorial region of the Western Pacific, Jan.-Feb. 1958.
Publ. 981, Hydrogr. Bull., No. 59: 31-40.

chemistry, nitrate
Nehring, Dietwart, Sigurd Schulz und Karl-Heinz Rohde, 1969.
Untersuchungen über die Productivität der Ostsee. I. Chemisch-biologische Untersuchungen in der mittleren Ostsee und in der Bottensee im April/Mai 1967.
Beitr. Meeresk., 23:5-36.

nitrate chemistry, nitrate
Neumann, W., 1949.
Beiträge zur Hydrographie des Vrana-Sees (Insel Cherso), inbesondere Untersuchungen über organisches sowie anorganische Phosphor- und Stickstoffverbindungen. Nova Thalassia 1(6):17 pp.

chemistry nitrate (data)
Nishida, Keizo, 1958
[On the oceanographic conditions of the lower cold water in the Japan Sea.]
J. Fac. Fish. and Animal Husbandry, Hiroshima Univ., 2(1): 1-65.

chemistry nitrate
Norina, A.M., 1965
Hydrochemical characteristics of the northern part of the Barents Sea. (In Russian).
Trudy, Gosudarst Okeanogr. Inst., No. 83: 243-271.

chemistry, nitrate
Nümann, W., 1941
Der Nährstoffhaushalt in der nordöstlichen Adria. Thalassia 5(2):1-68, 12 figs.

O

nitrate (data)
Ogura, Norio, and Takahisa Hanya, 1967.
Ultraviolet absorption of the sea water, in relation to organic and inorganic matters.
Int. J. Oceanol. Limnol., 1(2):91-102.

chemistry, nitrate (data)
Okabe, Shiro, Yoshimasa Toyota and Takafumi Murakami, 1967.
Major aspects of chemical oceanography. (In Japanese; English abstract).
J. Coll. mar. Sci. Techn., Tokai Univ., (2):39-56.

chemistry, nitrate
*Okuda, Taizo, 1967.
Vertical distribution of inorganic nitrogen in the equatorial Atlantic Ocean.
Boln Inst. Oceanogr., Univ. Oriente, 5(½):67-83.

chemistry nitrate
Okuda, Taizo, 1962
Physical and chemical oceanography over continental shelf between Cabo Frio and Vitoria (Central Brazil).
J. Oceanogr. Soc., Japan, 20th Ann. Vol., 514-540.

chemistry nitrate
Okuda, Taizo, 1960
Chemical oceanography in the South Atlantic Ocean, adjacent to north-eastern Brazil. (Portuguese and French resumés).
Trabalhos, Inst. Biol. Marit. e Oceanogr., Universidade do Recife, Brasil, 2(1):155-174.

chemistry
nitrate

Okuda, Taizo, 1960
Metabolic circulation of phosphorus and nitrogen in Matshishima Bay (Japan) with special reference to exchange of these elements between sea water and sediments. (Portuguese, French resumés)
Trabalhos, Inst. Biol. Maritima e Oceanogr., Universidade do Recife, Brasil, 2(1):7-153.

Chemistry
nitrate (data)

Okuda, Taizo, and Lourinaldo Barreto Cavalcanti, 1963.
Algumas condicoes oceanograficas na area nordeste de Natal. (Septembro 1960).
Trab. Inst. Oceanogr., Univ. do Recife, 3(1): 3-25.

English summary

Chemistry, nitrate

Okuda, Taizo, Angel José Garcia, José Benetoz Alvarez 1965.
Variacion estacional de las elementos nutritiona en el agua de la lagune y el rio Unare.
Bol. Inst. Oceanogr., Univ Oriente, 4(1):123-13

Orr, A.P., 1947 chemistry nitrate
An experiment in marine fish cultivation: II. Some physical and chemical conditions in a fertilized sea-loch (Loch Craiglin, Argyll).
Proc. Roy. Soc., Edinburgh, Sect.B, 63, Pt.1 (2):3-20, 19 text figs.

Chemistry nitrate

Orr, A. P., 1933
Physical and chemical conditions in the sea in the neighborhood of the Great Barrier Reef. Brit. Mus. (N.H.) Great Barrier Reef Exped., 1928-29, Sci. Repts. 2(3):37-86, 7 text figs.

P

Chemistry, nitrate

Park, Kilho 1967.
Chemical features of the subarctic boundary near 170° W.
J. Fish. Res. Bd Can. 24(5):899-908.

chemistry, nitrate (data)

Parsons, T.R., 1960.
A data record and discussion of some observations made in 1958-1960 of significance to primary productivity.
Fish. Res. Bd., Canada, Manuscript Rept. Ser., (Oceanogr. & Limnol.), No. 81:19 pp. (multilithed).

chemistry
nitrate

Patten, B.C., R.A. Mulford and J.E. Warinner, 1963
An annual phytoplankton cycle in the lower Chesapeake Bay.
Chesapeake Science, 4(1):1-20.

Chemistry
nitrates

Phifer, L. D. and T. G. Thompson, 1937.
Seasonal variations in the surface waters of San Juan channel during the five year period, January 1931 to December 30, 1935. J. Mar. Res., 1(1):34-59, text figs. 12-18, 12 tables.

chemistry
nitrate (data)

Picotti, Mario, 1960

Crociera Talassografica Adriatica 1955. III. Tabelle delle osservazioni fisiche, chimiche, biologiche e psammografiche.
Archivio di Oceanograf. e Limnol., 11(3):371-377, plus tables.

chemistry, nitrate

Piro, A. 1971.
Chemical environmental factors in marine radiocontamination.
CNEN Rept. RT/BIO(70)-11, M. Bernhard, editor: 11-22.

chemistry - nitrate

Ponomarenko, L.S., 1966.
Hydrochemical conditions in the Norwegian and Greenland seas in the summer of 1964. (In Russian).
Mater. Ribokhoz. Issled. severn. Basseina, Poliarn. Nauchno-Issled. Proektn. Inst. Morsk. Ribn. Khoz. Okeanogr. (PINRO), 7:155-167.

chemistry - nitrate

Ponomarenko, L.S., 1966.
Hydrochemical conditions in the Norwegian and Greenland seas in summer (from the data of the June surveys 1954-1964). (In Russian).
Trudy, Poliarn. Nauchno-Issled. Proektn. Inst. Morsk. Ribn. Khoz. Okeanogr., N. M. Knipovich, PINRO, 18:125-145.

nitrate chemistry, nitrate

Posner, G.S., 1957.
The Peru Current.
Bull. Bingham Oceanogr. Coll., 16(2):106-153.

chemistry, nitrate (data)

Postma, H., 1966.
The cycle of nitrogen in the Wadden Sea and adjacent areas.
Neth. J. Sea Res., 3(2):186-221.

nitrates

Prasad, R. Raghu, 1956.
Further studies on the plankton of the inshore waters off Mandapam. Indian J. Fish., 3(1):1-42.

nitrate (average, monthly)

Prasad, R.R., 1954.
The characteristics of marine plankton at an inshore station in the Gulf of Mannar near Mandapam. Indian J. Fish. 1(1/2):1-36, 8 textfig

Chemistry, nitrate

Pratt, David M., 1965.
The winter-spring diatom flowering in Narragansett Bay.
Limnology and Oceanography, 10(2):175-184.

chemistry, nitrate

Pratt, D.M., 1949.
Experiments in the fertilization of a saltwater pond. J. Mar. Res. 8(1):36-59, 8 textfigs.

Q

chemistry, nitrate

Qasim, S.Z., and C.V.G. Reddy, 1967.
The estimation of plant pigments of Cochin backwater during the monsoon months.
Bull. mar. Sci., Miami, 17(1):95-110.

R

Chemistry, nitrate

Rakestraw, N. W., 1933.
Studies on the biology and chemistry of the Gulf of Maine. 1. Chemistry of the waters of the Gulf of Maine in August, 1932. Biol. Bull. 64(2): 149-158, 4 textfigs.

Chemistry, nitrate

Rakestraw, N. W., and H.P. Smith, 1937.
A contribution of the chemistry of the Caribbean and Cayman Seas. Bull. Bingham Ocean. Coll 6(1):1-41, 32 text-figs.

Chemistry, nitrate

Ramalho, A. de M., R. G. Boto, B. C. Goncalves, H. Vilela, 1936.
Observacoes oceanograficas. "Albacora", 1935-1936. Trav. St. Biol. Mar. Lisbonne, No.39: 25 pp., 2 figs.

Chemistry, nitrate (data)

Ramalho, A. de M., L. S. Dentinho, C. A. M. de Sousa, Fronteira, F. L. Mamede, and H. Vilela, 1935.
Observacoes oceanograficas. "Albacora" 1934. Trav. St. Biol. Mar., Lisbonne, No. 35: 35 pp., 1 fig.

Chemistry, nitrate

Ramamurthy, S. 1963.
Studies on the hydrological factors in the north Kanara coastal waters.
Indian J. Fish. (A) 10(1):75-93.

Chemistry, nitrate

Ramamurthy, S., 1954.
Hydrobiological studies in the Madras coastal waters. J. Madras Univ., B, 23(2):148-163, 3 textfigs.

chemistry, nitrate

Ramamurthy, V.D., K. Krishnamurthy and R. Seshadri, 1965.
Comparative hydrographical studies of the near shore and estuarine waters of Porto-Novo, S. India.
J. Annamalai Univ., 26:154-164.

Also in: Collected Reprints, Mar. Biol. Sta., Porto Novo, 1963/64.

chemistry, nitrate

Redfield, A.C., 1934.
On the proportion of organic derivatives in sea water and their relation to the composition of plankton. IN: James Johnstone Memorial Volume: 176-192, 5 textfigs.

Chemistry, nitrate

Rheinheimer, Gerhard 1967.
Ökologische Untersuchungen zur nitrification in Nord- und Ostsee.
Helgoländer wiss. Meeresunters. 15 (1/4): 243-252.

Chemistry
nitrate

Riley, Gordon A., 1959
Oceanography of Long Island Sound, 1954-1955.
Bull. Bingham Oceanogr. Coll., 17(1):9-30.

Chemistry, nitrate

Riley, G.A., 1951.
Oxygen, phosphate, and nitrate in the Atlantic Ocean. Bull. Bingham Ocean. Coll. 8(1):126 pp. 33 textfigs.

Chemistry, nitrates

Riley, G. A., 1941.
Plankton studies. IV. Georges Bank. Bull. Bingham Ocean. Coll. 2(7):73 pp.,

Chemistry, nitrate

Riley, G. A., 1939.
Plankton studies. II. The western North Atlantic, May-June, 1939. J. Mar. Res. 2(2):145-162, Textfigs. 49-51, 4 tables.

chemistry, nitrate
Riley, G.A. and S.A.M. Conover 1956.
Oceanography of Long Island Sound, 1952-1954. 3. Chemical oceanography.
Bull. Bingham Oceanogr. Coll. 15: 47-61.

Chemistry
nitrate
Riley, G. A., H. Stommel, and D. F. Bumpus, 1949
Quantitative ecology of the plankton of the western North Atlantic. Bull. Bingham Ocean. Coll. 12(3):169 pp., 39 text figs.

Chemistry, nitrate
Riley, G. A., and R. von Arx, 1949.
Theoretical analysis of seasonal changes in the phytoplankton of Husan Harbor, Korea. J. Mar. Res. 8(1):60-72, 5 textfigs.

Chemistry
nitrate
Riley, J.P., 1967.
The hot saline waters of the Red Sea bottom and their related sediments.
Oceanogr. Mar. Biol., Ann.Rev., H. Barnes, editor, George Allen and Unwin, Ltd., 5:141-157.

nitrate(water samples and sediments)
Rittenberg, S.C., K.O. Emery and W.L. Orr, 1955.
Regeneration of nutrients in sediments of marine basins. Deep-Sea Res. 3(1):23-45.

chemistry, nitrate
Robertson, A.J., 1907.
On hydrographical investigations in the Faroe-Shetland Channel and the northern part of the North Sea during the years 1904-1905. Fish. Bd., Scotland, North Sea Fish. Invest. Comm., Northern Area, Rept. No. 2:1-140, 13 textfigs., 15 pls.

Chemistry, nitrate
Robinson, R.J., and B.M.G. Zwicker, 1941.
Preparing nitrate-free sea water. Science 94(2427):25-26.

chemistry, nitrate
Rochford, D.J. 1969.
Seasonal variation in the Indian Ocean along 110°E. 1. Hydrological structure of the upper 500m.
Aust. J. mar. Freshwat. Res. 20(1):1-50.

Chemistry, nitrate
Rochford, D.J., 1966.
Hydrology. Port Philip survey 1957-1963.
Mem. nat. Mus. Vict., 27:107-118.

Chemistry, NITRATE
Rochford, D. J., 1966.
Some hydrological features of the eastern Arafura Sea and the Gulf of Carpenteria in August 1964.
Australian J. Mar. freshw. Res., 17(1):31-60.

chemistry, nitrate (data)
Rochford, D.J., 1953.
Estuarine hydrological investigations in eastern and south-western Australia, 1951. Oceanographical station list of investigations made by the Division of Fisheries, Commonwealth Scientific and Industrial Research Organization, Australia, 12:111 pp.

chemistry, nitrate
Rochford, D.J., 1952.
A comparison of the hydrological conditions of the eastern and western coasts of Australia.
Indo-Pacific Fish. Counc., Proc., 3rd meeting, 1-16 Feb. 1951, Sects. 2/3:61-68, 7 textfigs.

Chemistry, nitrate
Rochford, D.J., 1951.
Hydrology of the estuarine development. Proc. Indo-Pacific Fish. Cound., 17-28 Apr. 1950, Cronulla N.S.W., Australia, Sects. II-III:157-168, 10 textfigs., Table 6.

nitrate chemistry, nitrate
Rochford, D.J., 1951.
Studies in Australian estuarine hydrology.
Australian J. Mar. and Freshwater Res. 2(1):1-116, 1 pl., 7 textfigs.

Chemistry, nitrate
Rodhe, W., 1948
Environmental requirements of fresh-water plankton algae. Experimental studies in the ecology of phytoplankton. Symbolae Botanicae Upsalienses X(1):149 pp., 30 figs.

chemistry, nitrate
Rouch, J., 1946
Traité d'Océanographie physique. L'eau de mer. Payot, Paris, 349 pp., 150 text figs.

chemistry, nitrates
Rozanov, A.G., and V.S. Bykova, 1964. (data)
The distribution of nitrates and nitrites in water of the northern part of the Indian Ocean. Investigations in the Indian Ocean (33rd voyage of E/S "Vitiaz"). (In Russian).
Trudy Inst. Okeanol., Akad. Nauk, SSSR, 64:94-101.

Chemistry
nitrate
Rubinchik, E.E., 1959
[Data on the hydrochemistry of the southern Caspian.]
Trudy V.N.I.R.O., 38: 152-164.

chemistry, Nitrate
Ruud, J. T., 1932.
On the biology of southern Euphausiidae.
Hvalrådets Skrifter No. 2:1-105, 37 text figs.

Chemistry, nitrate
Ryther, John H., D.W. Menzel and Nathaniel Corwin, 1967.
Influence of the Amazon River outflow on the ecology of the western tropical Atlantic. I. Hydrography and nutrient chemistry.
J.mar.Res., 25(1):69-83.

S

chemistry nitrate
Sagi, Takeshi, 1970.
On the distribution of nitrate nitrogen in the western North Pacific Ocean. Oceanogrl Mag. 22(2): 63-74.

Chemistry
nitrate
Scaccini Cicatelli, Marta, 1966.
Un anno di osservazioni idrologiche in una stazione fissa nelle acque dell'Adriatico occidentale presso Fano.
Note Lab. Biol. mar. Pesca, Fano, Univ. Bologna 2(9): 181-228.

chemistry, nitrate (data)
Scaccini Cicatelli, Marta 1967
Distribuzione stagionali dei sali nutrivi in una zona dell'alto e medio Adriatico.
Boll. Pesca, Piscic. Idrobiol. n.s. 12(1):49-82.

Chemistry, nitrates
Schachter, D., 1954.
Contribution à l'étude hydrographique et hydrologique de l'étang de Berre (Bouches-du-Rhone).
Bull. Inst. Océan., Monaco, No. 1048:20 pp.

Chemistry
nitrate
*Sholkovitz, Edward R. and Joris M. Gieskes, 1971.
A physical-chemical study of the flushing of the Santa Barbara Basin. Limnol. Oceanogr. 16(3): 479-489.

Chemistry, Nitrate
Schott, Friedrich, and Manfred Ehrhardt, 1970.
On fluctuations and mean relations of chemical parameters in the northwestern North Sea.
Kieler Meeresforsch. 25(2): 272-278

Chemistry, nitrate
Sen Gupta, R., 1968.
Inorganic nitrogen compounds in ocean stagnation and nutrient resupply.
Science, 160(3830):884-885.

Chemistry, nitrate
Shimomura, T., and K. Miyata, 1957.
[The oceanographical conditions of the Japan Sea and its water systems, laying stress on the summer of 1955.]
Bull. Japan Sea Reg. Fish. Res. Lab., No. 6 (General survey of the warm Tsushima Current 1): 23-120.

Chemistry, nitrate
Skolka H.V., et M.-T. Gomoiu 1967.
Recherches océanologiques roumaines dans la mer Noire. (In Roumanian; French abstract).
Hydrobiologie, Bucuresti 8:15-30

Chemistry, nitrate
Slinn, D.J., 1966.
Chemical constituents in sea water off Port Erin during 1965.
Mar. biol. Sta., Univ. Liverpool. No. 78:29-33.

Chemistry
nitrate
Slinn D.J., 1961
Chemical constituents in sea water off Port Erin in 1960.
Ann. Rept. Mar. Biol. Sta., Univ. Liverpool 1960, 73:23-28.

Chemistry, nitrate
Slinn, D.J., and Winifred Chapman, 1965.
Chemical constituents in sea water off Port Erin in 1964.
Mar. Biol. Sta. Port Erin, Ann. Rept. (1964), No. 77:43-48.

Chemistry, nitrate
Slinn, D.J., and Winifred Chapman, 1964.
Chemical constituents in sea water off Port Erin during 1963. Liverpool,
Mar. Biol. Sta., Univ./PortErin, Ann. Rept., No. 76(1963):26-31.

chemistry, nitrate
Slinn, D.J., and Gaenor Offlow (Mrs. M. Solly) 1969.
Chemical constituents in sea water off Port Erin during 1968.
Ann. Rep. mar. biol. Sta. Port Erin 81 (1968):31-

chemistry, nitrate

Smayda, Theodore J., 1966.
A quantitative analysis of the phytoplankton of the Gulf of Panama. III General ecological conditions and the phytoplankton dynamics at 8o 45'N, 79o 23'W from November 1954 to May 1957.
Inter-Amer. Trop. Tuna Comm., Bull., 11(5): 355-612.

Chemistry, nitrate

Smetanin, D.A., 1960

[Some features of the chemistry of water in the central Pacific.]
Trudy Inst. Okeanol., 40: 58-71.

Chemistry, nitrate

Smetanin, D.A., 1959

[Hydrochemistry of the Kuril-Kamchatka deep-sea trench.] Trudy Inst. Okeanol., 33: 43-86.

chemistry nitrate (data)

Soot-Ryen, T., 1947.
Hydrographical investigations in the Tromsø-district 1934-1938 (Tables).
Tromsø Mus. Aarsheft., Naturhist.Avd. No. 33, 66(1943)(1): numerous pp. (unnumbered).

Chemistry, nitrate(data)

Soot-Ryen, T., 1938.
Hydrographical investigations in the Tromsø district in 1931. Tromsø Mus. Aarsheft., Naturhist. Avd. No. 10, 54(2):I-6 plus 41 pp. of tables.

chemistry, nitrate

Spencer, R.S., 1956.
Studies in Australian estuarine hydrology.
Australian J. Mar. & Freshwater Res. 7(2):193-253.

Chemistry, nitrate

Steele, J. H., and I. E. Baird, 1962.
Further relations between primary production, Chlorophyll and particulate carbon.
Limnol. & Oceanogr., 7(1):42-47.

chemistry, nitrate

Steele, J.H., I.E. Baird and R. Johnston, 1971.
Evidence of upwelling on Rockall Bank. Deep-Sea Res., 18(2): 261-268.

chemistry, nitrate

Steemann Nielsen, E., 1953.
The production of phytoplankton at the Faroe Isles, Iceland, East Greenland and in the waters around.
Medd. Komm. Danmarks Fisk- og Havundersøgelser, Ser. Plankton, 3(1):1-93, 5 textfigs.

chemistry, nitrate

Steemann-Nielsen, Einar, 1951
The marine vegetation of the Isefjord. A study on ecology and production. Medd. Komm. Danmarks Fiskeri-og Havundersøgelser. Ser. Plankton. 5(4); 114pp., 46 text figs.

chemistry, nitrate

Steemann Nielsen, E., 1940.
Die Produktionsbedingungen des Phytoplanktons im Übergangsgebiet zwischen der Nord- und Ostsee.
Medd. Komm. Havundersøgelser, Ser. Plankton, 3(4):55 pp., 17 textfigs., 22 tables.

Chemistry, nitrate

Steemann Nielsen, E., and E.A. Jensen, 1957.
Primary oceanic production. The autotrophic production of organic matter in the oceans.
Galathea Repts. No. 1:49-136.

chemistry nitrate

*Stefánsson, U., 1968.
Nitrate-phosphate relationships in the Irminger Sea.
J. Cons. perm. int. Explor. Mer, 32(2):188-200.

Chemistry, nitrate

Stefansson, Unstein, 1966.
Influence of the Surtsey eruption on the nutrient content of the surrounding seawater.
J. Mar. Res., 24(2):241-268.

Chemistry nitrate

Stefansson, Unnstein, and Francis A. Richards, 1963.
Processes contributing to the nutrient distributions off the Columbia River and Strait of Juan de Fuca.
Limnology and Oceanography, 8(4):394-410.

chemistry nitrate

*Stephens, K., R.W. Sheldon and T.R. Parsons, 1967.
Seasonal variations in the availability of food for benthos in a coastal environment.
Ecology, 48 (5):852-855.

chemistry nitrate

Strickland, J.D.H., & K.H. Austin, 1959
The direct estimation of ammonia in sea water with notes on "reactive" iron, nitrate and inorganic phosphorus.
J. du Conseil, 24(3): 446-451.

Chemistry, nitrate

Subrahmanyan, R., 1959.

Studies on the phytoplankton of the west coast of India. 1. Quantitative and qualitative fluctuation of the total phytoplankton crop, the zooplankton crop and their interrelationship, with remarks on the magnitude of the standing crop and production of matter and their relationship to fish landings. 2. Physical and chemical factors influencing the production of phytoplankton, with remarks on the cycle of nutrients and on the relationship of the phosphate content to fish landings.
Proc. Indian Acad. Sci., 1:113-252.

chemistry nitrate

Sugiura, Yoshio, 1963
Some chemico-oceanographical properties of the water of Suruga Bay. (In Japanese; English abstract).
J. Oceanogr. Soc. Japan, 18(4):193-199.

chemistry, nitrate

Sugiyama, Teruyuki, Yoshio Miyake and Kuniyasu Fujisawa, 1971.
On the distribution of inorganic nutrients along the coastal waters of Okayama Prefecture (1970).
Bull. Fish. Exp. Sta. Okayama Pref. (1970): 22-28. (In Japanese)

chemistry nitrate

Sverdlov, A.I., 1965.
Distribution of hydrochemical elements in the area of Faroe-Shetland Strait in November-January 1963-1964. (In Russian).
Materail. Ribochoz. Issled. Severn. Bassin. Poliarn. Nauchno-Issled. i Proekt. Inst. Morsk. Rib. Choz. i Okeanogr., N.M. Knipovich, 5:91-98.

Chemistry, Nitrates

Sykes, J.B. and A.D. Boney, 1970.
Seasonal variations in inorganic phytoplankton nutrients in the inshore waters of Cardigan Bay. J. mar. biol. Ass., U.K., 50(3): 819-827.

chemistry nitrate

*Szekielda, Karl-Heinz, 1968.
Ein chemisches Modell für den Auf-und Abbau organischen Materials und dessen Anwendung in der offenen See.
J. Cons. perm. int. Explor. Mer, 32 (2):18-187.

T

Chemistry nitrate(data)

Tanaka, Otohiko, Haruhiko Irie, Shozi IIzuka and Fumihiro Koga, 1961
The funadmental investigation on the biological productivity in the north-west of Kyushu. 1. Rec. Oceanogr. Wks., Japan, Spec. No. 5: 1-58.

Chemistry, nitrate

Thomas, William H., 1966.
On denitrification in the northeastern tropical Pacific Ocean.
Deep-Sea Res., 13(6):1109-1114.

Chemistry, nitrate

Thomsen, H., 1933.
The distribution of phosphate and nitrate in the North Sea in May 1932. Rapp. Proc. Verb., Cons. Perm. Int. Expl. Mer 85:42-46, 8 textfigs.

Chemistry, nitrate

Thomsen, H., 1931.
Nitrate and phosphate contents of Mediterranean water.
Rept. Danish Oceanogr. Exped., 1908-1910, 3(6): 14 pp., 2 textfigs., 1 pl.

Chemistry nitrate

Tranter, D.J., and B.S. Newell, 1963
Enrichment experiments in the Indian Ocean.
Deep-Sea Res., 10(1/2):1-9.

Chemistry, nitrate

Tully, J.P., and A.J. Dodimead, 1957.
Properties of the water in the Strait of Georgia, British Columbia and influencing factors.
J. Fish. Res. Bd., Canada, 14(3):241-319.

U

Chemistry nitrate (data)

University of Southern California, Allan Hancock Foundation, 1965.
An oceanographic and biological survey of the southern California mainland shelf.
State of California, Resources Agency, State Water Quality Control Board, Publ. No. 27:232 pp. Appendix, 445 pp.

Chemistry nitrate (data)

Uyeno, Fukuzo, 1966.
Nutrient and energy cycles in an estuarine oyster area.
J. Fish. Res. Bd., Can., 23(11):1635-1652.

Chemistry, nitrate

Uyeno, Fukuzo, 1964.
Relationships between production of foods and oceanographical condition of sea water in pearl farms. II. On the seasonal changes of sea water constituents and of bottom condition, and the effect of bottom cultivation. (In Japanese;
J. Fac. Fish., Pref. Univ. Mie, 6(2):145-169.

English abstract)

V

chemistry, nitrate

Vaccaro, Ralph F., 1965.
inorganic ni ogen in sea water.
In: Chemical Oceanography, J.P. Riley and G. Skirrow, editors, Academic Press, 1:365-408.

Vaccaro, Ralph F., 1963.
Available nitrogen and phosphorus and the biochemical cycle in the Atlantic off New England.
J. Mar. Res., 21(3):284-301.

Chemistry, nitrate

Vaccaro, Ralph F., and John H. Ryther, 1960
Marine phytoplankton and the distribution of nitrite in the sea. J. du Cons., 35(3):260-271.

Chemistry, nitrate

Vatova, A. 1965.
Les conditions hydrographiques de la Mar Piccolo de Taranto pendant l'année 1963.
Rapp. P.-v. Reun. Commn int. Explor. scient. Mer Méditerr. 18(3): 653-655.

Chemistry, nitrate

Vatova, A., 1956.
Elektrophotometrische Nitratbestimmung im Meerwasser mit dem Photometer "Elko II".
Deutsche Hydrogr. Zeits., 9(4):194-198.

Chemistry, nitrate

Vatova, A., 1956.
Il dosaggio dell'azote nitrico nell'acqua di mare con l'elettrofotometro Elko II.
Nova Thalassia 2(5):5-25.

Chemistry, nitrate

Van Goethem, C., 1951.
Étude physique et chimique du milieu marin.
Rés. Sci. Expéd. Océan. Belge dans les Eaux Côtières Africaines de l'Amerique Sud (1948-1949) 1:1-151, 1 pl.

Chemistry, nitrate (data)

Villanueva, Sebastian F., Alberto Gomez, Aldo Orlando y Andres J. Lusquinos 1969.
Datos y resultados de las campañas pesqueria "Pesqueria VII", 16 de febrero al 1° de marzo de 1968.
Publ. Ser. Inf. técn. Proyecto Desarrollo Pesq. Mar del Plata (10/VII): unnumbered pp.

chemistry, nitrate

Viorel, Chirilă, 1965.
Observații asupra condițiilor fizico-chimice ale mării la Mamaia, în anii 1959 și 1960.
In: Ecologie marină, M. Băcescu, redactor, Edit. Acad. Republ. Pop. Române, București, 1: 139-184.

Chemistry nitrate (data)

Voit, S.S., D.A. Aksenov, M.M. Bogorodsky, V.V. Sinukov and V.A. Vladimirzev, 1961
[Some peculiar features of water circulation in the Black Sea and the water regime in the Pre-Bosphorus area.]
Okeanologiia, Akad. Nauk, SSSR, 1(4):613-625.

W

Chemistry, nitrate

Wattenberg, H., 1937.
Methoden zur Bestimmung von Phosphat, Silicat, Nitrat, und Ammoniak im Seewasser. Rapp. Proc. Verb. 103(1):1-26.

Chemistry, nitrate

*Wauthy, B., R. Desrosières et J. Le Bourhis, 1967.
Importance presumée de l'ultraplancton dans les eaux tropicales oligotrophes du Pacifique central sud.
Cah. ORSTOM, Sér. Océanogr., 5(2):109-116.

chemistry, nitrate

Wattenberg, H., 1929.
7. Die Phosphat- und Nitrat-Untersuchungen der Deutschen Atlantischen Expedition aur V.S. "Meteor".
Rapp. Proc. Verb., Cons. Perm. Int. Expl. Mer, 53:90-94.

Transl. cited:
USFWS Spec. Sci. Rept., Fish. 227.

Chemistry, nitrate

Watts, J.C.D., 1958.
The hydrology of a tropical West African estuary.
Bull. Inst. Francais, Afrique Noire, 20(3):697-752.

chemistry, nitrates

Williams, Peter M., and Arthur R. Coote, 1962
Mullin and Riley nitrate determination in the presence of hydrogen sulfide.
Limnology and Oceanography, 7(2):258.

Chemistry, nitrate

Won, Chong Hun, 1964.
Tidal variation of chemical constituents of the estuary water at the Lava bed in the Nack-Dong River from Nov. 1962 to Oct. 1963. (In Korean; English abstract).
Bull. Pusan Fish. Coll., 6(1):21-32.

chemistry, nitrate

Won, Chong Hun, 1963.
Distribution of chemical constituents of the estuary water in Gwang-Yang Inlet. (In Korean; English abstract).
Bull. Fish. Coll., Pusan Nat. Univ., 5(1):1-10.

XYZ

chemistry, nitrate

Yoshida, Yoichi, Akira Kawai and Masao Kimata, 1967.
Studies on marine nitrifying bacteria (nitrite formers and nitrate formers). IV. On the nitrite formation of the marine nitrite formers (2). (In Japanese; English abstract).
Bull. Jap. Soc. scient. Fish. 33(4): 367-371.

Chemistry, nitrate

*Zlobin, V.S., N.G. Sapronetskaya and A.G. Alekseeva, 1968.
Distribution and regeneration of phosphates in the Norwegian Sea in 1965. (In Russian; English abstract).
Okeanologiia, Akad. Nauk, SSSR, 8(4):616-627.

chemistry, nitrate

Zuta, Salvador, y Oscar Guillen, 1970.
Oceanografia de las aguas costeras del Peru.
Bol. Inst. Mar., Peru, 2(5):157-324.

chemistry, nitrate

Zwicker, B.M.G., and R.J. Robinson, 1944.
The photometric determinations of nitrate in sea water with a strychnidine reagent. J. Mar. Res. 5(3):214-232, Figs. 52-58.

Chemistry, nitrate concentration, effect of

Thomas, William H., 1970.
Effect of ammonium and nitrate concentration on chlorophyll increases in natural tropical Pacific phytoplankton populations. Limnol. Oceanogr., 15(3): 386-394.

Chemistry, nitrate (data only)

Anderson, William W., and Jack W. Gehringer, 1959.
Physical oceanographic, biological and chemical, South Atlantic coast of the United States, M/V Theodore N. Gill cruise 9.
USFWS Spec. Sci. Rept., Fish. No. 313:226 pp.

Chemistry nitrate (data only)

Anderson, W.W., and J.W. Gehringer, 1959.
Physical oceanographic, biological and chemical data South Atlantic coast of the United States, M/V Theodore N. Gill Cruise 8.
USFWS Spec. Sci. Rept., Fish., No. 303: 227 pp.

chemistry nitrate (data only)

Anderson, William W., and Jack W. Gehringer, 1959.
Physical oceanography, biological and chemical data, South Atlantic coast of the United States. M/V Theodore N. Gill, Cruise 7.
USFWS Spec. Sci. Rept., Fish. No. 278:277 pp.

Chemistry nitrate (data only)

Anderson, William W., and Jack W. Gehringer, 1958.
Physical oceanographic, biological and chemical data, south Atlantic coast of the United States, M/V Theodore N. Gill Cruise 6.
USFWS Spec. Sci. Rept., Fish., No. 265:99 pp.

chemistry, nitrate (data only)

Anderson, W.W., and J.W. Gehringer, 1958.
Physical oceanographic, biological and chemical data, south Atlantic coast of the United States, M/V Theodore N. Gill cruise 5.
USFWSSpec. Sci. Rept., Fish., 248:220 pp.

Chemistry, nitrate (data only)

Anderson, W.W., and J.W. Gehringer, 1957.
Physical oceanographic, biological and chemical data, South Atlantic coast of the United States, M/V Theodore N. Gill, cruise 4.
USFWS Spec. Sci. Rept., Fish., 234:192 pp.

chemistry nitrate (data only)

Anderson, W.W., and J.W. Gehringer, 1957.
Physical oceanographic, biological and chemical data, South Atlantic coast of the United States, Theodore N. Gill cruise.
USFWSSpec. Sci. Rept., Fish., No. 210:208 pp.

Chemistry, nitrate (data only)

Anderson, W.W., J.W. Gehringer and E. Cohen, 1956
Physical oceanographic, biological and chemical data, South Atlantic coast of the United States, Theodore N. Gill, Cruise 2.
Spec. Sci. Rept., Fish., No. 198:270 pp.

Chemistry, nitrate (data only)

Anderson, W.W., J.W. Gehringer and E. Cohen, 1956.
Physical oceanographic, biological and chemical data, south Atlantic coast of the United States, M/V Theodore N. Gill, Cruise, 1.
USFWS Spec. Sci. Rept., Fish., No. 178:160 pp.

Chemistry, nitrate (data only)

Aragno, Federico J., Alberto Gomez, Aldo Orlando y Andres J. Lusquinos 1968.
Datos y resultados de las campañas pesqueria "Pesqueria III" (20 de febrero al 20 de marzo de 1967).
Publ. (Ser. Informes técn.) Mar del Plata, Argentina 10(3): 1-162.

Chemistry, nitrate (data only)
Aragno, Federico, Alberto Gomez, Aldo Orlando, y Andres J. Lusquinos 1968. Datos y resultados de las Campañas pesqueria "Pesqueria II" (9 de noviembre al 12 diciembre de 1966. Publ. (Ser. Informes tecn.) Mar del Plata, Argentina 10(2):1-129.

Chemistry, nitrate (data only)
Aragno, Federico, Alberto Gomez, Aldo Orlando y Andres J. Lusquinos 1968. Datos y resultados preliminares de las campañas pesqueria "Pesqueria I" (12 de agosto al 8 de setiembre de 1966. Publ. (Ser. Informes tecn) Mar del Plata, Argentina 10(1): 1-159.

Chemistry, nitrate (data only)
Australia, Commonwealth Scientific and Industrial Research Organization 1970. Coastal investigations off Port Hacking, New South Wales in 1965. Oceanogr. Sta. List. Div. Fish. Oceanogr. 85: 124 pp.

Chemistry, nitrate (data only)
Australia, Commonwealth Scientific and Industrial Research Organization 1968. Oceanographical observations in the Indian Ocean in 1965, H.M.A.S. Gascoyne, Cruise G 2/65. Oceanogr. Cruise Rept., Div. Fish. Oceanogr. 43:1-58.

Chemistry, nitrate (data only)
Australia, Commonwealth Scientific and Industrial Research Organization 1968. Oceanographical observations in the Pacific Ocean in 1964, H.M.A.S. Gascoyne, Cruise G6/64. Oceanogr. Cruise Rept. Div. Fish. Oceanogr. 42: 53 pp.

Chemistry, nitrate (data only)
Australia, Commonwealth Scientific and Industrial Research Organization 1968. Oceanographical observations in the Indian Ocean in 1964, H.M.A.S. Gascoyne, Cruise G5/64. Oceanogr. Cruise Rept., Div. Fish. Oceanogr. 41: 52 pp.

Chemistry, nitrate (data only)
Australia, Commonwealth Scientific and Industrial Research Organization 1968. Oceanographical observations in the Indian Ocean in 1964, H.M.A.S. Diamantina, Cruise Dm 5/64. Oceanogr. Cruise Rept. Div. Fish. Oceanogr. 40: 48 pp.

Chemistry nitrate (data only)
Argentina, Secretaria de Marina, Servicio de Hidrografia Naval, 1961. Trabajos oceanograficos realizados en la campana Antartica 1960/1961. Resultados preliminares. Publico, H. 629: unnumbered pp.

Chemistry nitrate (data only)
Argentina, Secretaria de Marina, Servicio de Hidrografia Naval, 1961. Operacion oceanografica, Vema - Canepa 1, Resultados preliminares. Publico, H. 628:30 pp.

Chemistry nitrate (data only)
Argentina, Secretaria de Marina, Servicio de Hidrografia Naval, 1961. Trabajos oceanograficos realizados en la campana Antartica 1959/1960. Resultados preliminares Publico, H. 623: unnumbered pp.

Chemistry nitrate (data only)
Argentina, Servicio de Hidrografia Naval, 1960. Operacion Oceanografica Malvinas (Resultados preliminares). Servicio de Hidrografia Naval, Argentina, Publ., H. 606: numerous unnumbered pp.

Chemistry nitrate (data only)
Argentina, Servicio de Hidrografia Naval, 1960. Operacion Oceanografica Drake II. Resultados preliminares. Servicio de Hidrografia Naval, Argentina, Publ., H. 14: numerous unnumbered pp.

Chemistry, nitrate (data only)
Argentina, Servicio de Hidrografia Naval, 1959. Operacion Oceanografica Meridiano. Resultados preliminares. Servicio de Hidrografia Naval, Argentina, Publ., H., 617: numerous unnumbered pp.

Chemistry nitrate (data only)
Argentina, Servicio de Hidrografia Naval, 1959. Trabajos oceanograficos realizados en la campana Antartica, 1958/1959. Resultados preliminares. Servicio de Hidrografia Naval, Argentina, Publ., H. 616: 127 pp.

Chemistry nitrate (data only)
Argentina, Servicio de Hidrografia Naval, 1959. Trabajos oceanograficos realizados en la campana Antartica, 1957-1958. Resultados preliminares. Servicio de Hidrografia Naval, Argentina, Publ., H. 615: numerous unnumbered pp.

Chemistry nitrate (data only)
Argentina, Servicio de Hidrografia Naval, 1959. Operacion Oceanografica Drake. 1. Resultados preliminares. Servicio de Hidrografia Naval, Argentina, Publ., H. 613: numerous unnumbered pp.

Chemistry, nitrate (data only)
Argentina, Servicio de Hidrografia Naval, 1959. Operacion Oceanografica Atlantico Sur. Resultados preliminares. Servicio de Hidrografia Naval, Argentina, Publ., H. 608: numerous unnumbered pp.

Chemistry nitrate (data only)
Argentina, Servicio de Hidrografia Naval, 1959. Operacion Oceanografica Cuenca. Resultados preliminares. Servicio de Hidrografia Naval, Argentina, Publ., H. 607: numerous unnumbered pp.

Chemistry, nitrate (data only)
Australia, Commonwealth Scientific and Industrial Research Organization, 1969. Oceanographical observations in the Indian Ocean in 1965 H.M.A.S. Diamantina Cruise Dm2/65. Div. Fish. Oceanogr. Oceanogr. Cruise Rept. 49: 57 pp. (multilithed)

Chemistry – nitrate (data only)
Australia, Commonwealth Scientific and Industrial Research Organization, 1969. Oceanographical observations in the Pacific Ocean in 1965 H.M.A.S. Gascoyne Cruise G3/65. Div. Fish. Oceanogr. Oceanogr. Cruise Rept. 44: 24 pp. (multilithed).

Chemistry nitrate (data only)
Australia, Commonwealth Scientific and Industrial Organization, 1968. Coastal investigations off Port Hacking, New South Wales, in 1961. Oceanogr. Sta. List, 81: 55 pp. (multilithed)

Chemistry nitrate (data only)
Australia, Commonwealth Scientific and Industrial Organization, 1968. Coastal investigations off Port Hacking, New South Wales, in 1964. Oceanogr. Sta. List, 84: 49 pp. (multilithed)

Chemistry, nitrate (data only)
Australia, Commonwealth Scientific and Industrial Research Organization, 1968. Oceanographical observations in the Pacific Ocean in 1963, H.M.A.S. Gascoyne, Cruise G3/63. Oceanogr. Cruise Rept., Div. Fish. Oceanogr., 26: 134 pp.

Chemistry, nitrate (data only)
Australia, Commonwealth Scientific and Industrial Organization 1967. Oceanographical observations in the Indian Ocean in 1965, H.M.A.S. Gascoyne, Cruise G5/65. Div. Fish. Oceanogr., Oceanogr. Cruise Rept. 46: 62 pp.

Chemistry nitrate (data only)
Australia, Commonwealth Scientific and Industrial Organization, 1967. Oceanographical observations in the Indian Ocean in 1964, H.M.A.S. Diamantina, Cruise Dm 2/64. Oceanogr. Cruise Rep. Div. Fish. Oceanogr. 36: 53 pp.

Chemistry, nitrate (data only)
Australia, Commonwealth Scientific and Industrial Organization 1967. Oceanographical observations in the Indian Ocean in 1962, H.M.A.S. Diamantina Cruise Dm 4/62. Div. Fish. Oceanogr., Oceanogr. Cruise Rept. 20: 138 pp.

Chemistry, nitrate (data only)
Australia, Commonwealth Scientific and Industrial Organization 1967. Oceanographical observations in the Pacific and Indian oceans in 1962, H.M.A.S. Gascoyne, cruises G 2/62 and G 3/62. Div. Fish. Oceanogr., Oceanogr. Cruise Rept. 16: 90 pp.

chemistry, nitrate (data only)
Australia, Commonwealth Scientific and Industrial Organization, 1967.
Oceanographical observations in the Pacific Ocean in 1962, H.M.A.S. Gascoyne, Cruise G 1/62.
Div. Fish. Oceanogr., Oceanogr. Cruise Rept. 13:180 pp.

chemistry, nitrate (data only)
Australia, Commonwealth Scientific and Industrial Organization 1965.
Oceanographical observations in the Indian Ocean in 1963, H.M.A.S. Diamantina, Cruise Dm 2/63.
Div. Fish. Oceanogr., Oceanogr. Cruise Rept. 24:153 pp.

chemistry, nitrate (data only)
Australia Commonwealth Scientific and Industrial Organization 1965.
Oceanographical observations in the Indian Ocean in 1963, H.M.A.S. Diamantina, Cruise Dm 1/63.
Div. Fish. Oceanogr., Oceanogr. Cruise Rept. 23:175 pp.

chemistry nitrate (data only)
Australia, Commonwealth Scientific and Industrial Research Organization, 1965.
Oceanographic observations in the Indian Ocean in 1963, H.M.A.S. Gascoyne, Cruise G 1/63.
Div. Fish. and Oceanogr., Oceanogr. Cruise Rept., No. 21:135 pp.

Chemistry, nitrate (data only)
Australia, Commonwealth Scientific and Industrial Research Organization, 1964.
Oceanographical observations in the Indian Ocean in 1961, H.M.A.S. Diamantina, Cruise, Dm 3/61.
Div. Fish. and Oceanogr., Oceanogr. Cruise Rept., No. 11:215 pp.

chemistry nitrate (data only)
Australia, Commonwealth Scientific and Industrial Research Organization. 1964.
Oceanographic observations in the Indian Ocean in 1962, H.M.A.S. Diamantina, Cruise Dm 1/62/
Oceanogr. Cruise Rept., Div. Fish Oceanogr., No. 10: 128 pp.

chemistry, nitrate (data only)
Australia, Commonwealth Scientific and Industrial Research Organizatoon. 1963
Coastal hydrological investigations in eastern Australia, 1960.
Div. Fish. and Oceanogr., Oceanogr. Sta. List 51: 46 pp.

chemistry nitrate (data only)
Australia, Commonwealth Scientific and Industrial Research Organization, 1963
Oceanographical observations in the Indian Ocean in 1961, H.M.A.S. Diamantina Cruise Dm 2/61.
Oceanogr. Cruise Rept., Div. Fish. and Oceanogr., No. 9: 155 pp., 14 figs.

chemistry, nitrate (data only)
Australia, Commonwealth Scientific and Industrial Research Organization, 1963.
Oceanographical observations in the Pacific Ocean in 1961, H.M.A.S. Gascoyne, Cruise G 1/61.
Oceanogr. Cruise Rept., Div. Fish. and Oceanogr., No. 8:130 pp., 12 figs.

chemistry nitrate (data only)
Australia, Commonwealth Scientific and Industrial Organization, 1962.
Oceanographical observations in the Indian Ocean in 1960, H.M.A.S. DIAMANTINA Cruise Dm 3/60.
Oceanogr. Cruise Rept., Div. Fish. and Oceanogr., No. 4:39 pp.

chemistry nitrate(data only)
Australia, Commonwealth Scientific and Industrial Research Organization, 1961
Coastal investigations at Port Hacking, New South Wales, 1959.
Div. Fish. & Oceanogr., Oceanogr. Sta. List, No. 47:135 pp.

chemistry, nitrate (data only)
Australia, Commonwealth Scientific and Industrial Research Organization, 1960.
Coastal investigations at Port Hacking, New South Wales, 1958.
Oceanogr. Sta. List, Div. Fish. & Oceanogr., 42:99 pp.

chemistry, nitrate (data only)
Australia, Commonwealth Scientific and Industrial Research Organization, 1967.
Oceanographical observations in the Pacific Ocean in 1964, H.M.A.S. Gascoyne, Cruise G 1/64.
Div. Fish. Oceanogr., Oceanogr. Cruise Rep. 32:66pp (multilithed).

chemistry nitrate (data only)
Australia, Commonwealth Scientific and Industrial Research Organisation, 1967.
Oceanographical observations in the Pacific Ocean in 1963, H.M.A.S. Gascoyne, Cruise G 4/63.
Div. Fish. Oceanogr., Oceanogr. Cruise Rep. 29:64 pp. (multilithed).

nitrate (data only)
Australia, Commonwealth Scientific and Industrial Research Organization, 1967.
Oceanographical observations in the Pacific Ocean in 1961, H.M.A.S. Gascoyne Cruise G 3/61.
Div. Fish. Oceanogr., Oceanogr. Cruise Rep., 12:126 pp.

chemistry nitrate (data only)
Australia, Commonwealth Scientific and Industrial Research Organization, 1966.
Oceanographical observations in the Indian Ocean in 1962 H.M.A.S. Gascoyne Cruise G 1/62.
Div. Fish. Oceanogr., Oceanogr. Cruise Rep. No. 17: 151 pp.

chemistry, nitrate (data only)
Australia, Commonwealth Scientific and Industrial Research Organization.
Oceanographic observations in the Indian Ocean in 1963, H.M.A.S. DIAMANTINA Cruise DM 3/63.
Div. Fish. Oceanogr., Oceanogr. Cruise Rept. No. 25: 147 pp.

Chemistry, nitrate (data only)
Australia, Commonwealth Scientific and Industrial Research Organization, 1964.
Oceanographical observation in the Indian Ocean in 1962, H.M.A.S. Diamantina, Cruise Dm 2?62.
Oceanogr. Cruise Rept., Div. Fish. and Oceanogr. No. 15:117 pp.

chemistry, nitrate (data only)
Australia, Commonwealth Scientific and Industrial Research Organization, 1963.
Coastal investigations at Port Hacking, New South Wales, 1960.
Div. Fish. and Oceanogr., Oceanogr. Sta. List, No. 52:135 pp.

chemistry nitrate (data only)
Australia, Commonwealth Scientific and Industrial Research Organization, 1960
Coastal hydrological investigations in eastern Australia, 1959.
Div. Fish. and Oceanogr., Oceanogr. Sta. List, 45: 24 pp.

chemistry nitrate (data only)
Australia, Commonwealth Scientific and Industrial Organization, 1960
Coastal hydrological investigations in south-eastern Australia, 1958. Oceanogr. Sta. List. Div. Fish. and Oceanogr., 60 pp.

chemistry nitrate (data only)
Australia, C.S.I.R.O., Division of Fisheries, 1959
Coastal hydrological investigations at Eden, New South Wales, 1957.
Oceanogr. Sta. List, 35: 36 pp.

Chemistry nitrate (data only)
Australia, C.S.I.R.O., Division of Fisheries, 1959
Coastal hydrological investigations at Port Hacking, New South Wales, 1957.
Oceanogr. Sta. List, 34:72 pp.

nitrate (data only)
Australia, Division of Fisheries and Oceanography, C.S.I.R.O., 1959.
Coastal hydrological investigations in the New South Wales tuna fishing area, 1958.
Oceanogr. Sta. Lists, 38:96 pp.

chemistry, nitrate (data only)
Australia, Commonwealth Scientific and Industrial Research Organization, 1957.
Estuarine hydrological investigations in eastern and south-western Australia, 1956.
Oceanogr. Sta. List, Div. Fish. and Oceanogr., 32:170 pp.

chemistry nitrate (data only)
Australia, Commonwealth Scientific and Industrial Research Organization, 1957.
Estuarine hydrological investigations in eastern and south-western Australia, 1955.
Ocean. Sta. List, 29:93 pp.

chemistry
Nitrate(data only)
Australia Commonwealth and Industrial Research Organization, 1957
Onshore and oceanic hydrological investigations in eastern and southwestern Australia, 1956.
Oceanogr. Sta. List., Div. Fish. & Oceanogr. 30:79pp.

chemistry, nitrate (data only)
Australia, Commonwealth Scientific and Industrial Research Organization, 1957.
Onshore and oceanic hydrological investigations in eastern and southwestern Australia, 1955.
Oceanogr. Sta. List, 27:145 pp.

Chemistry, nitrate(data only)
Australia, Commonwealth Scientific and Industrial Research Organization, 1956.
Onshore hydrological investigations in eastern and south-western Australia, 1954
Ocean. Sta. List, Invest. Fish. Div., 24:119 pp.

Chemistry, nitrate (data only)
Australia, Commonwealth Scientific and Industrial Research Organization, 1954.
Onshore hydrological investigations in eastern and south-western Australia, 1953.
Ocean. Sta. List, Invest. Div. Fish. 18:64 pp.

chemistry, nitrate (data only)
Australia, Commonwealth Scientific and Industrial Research Organization, 1953.
Onshore hydrological investigations in eastern and south-western Australia, 1951. Vol. 14:64 pp

chemistry, nitrate (data only)
Australia, Commonwealth Scientific and Industrial Research Organization, Melbourne, 1952.
Oceanographical station list of investigations made by the Fisheries Division of C.S.I.R.O.
Vol. 5. Estuarine hydrological investigations in eastern Australia 1940-50:150 pp.
Vol. 6. Ibid.:137 pp.
Vol. 7: Ibid.:139 pp.
Vol. 8. Hydrological investigations in south-western Australia, 1944-50:152 pp.
Vol. 10. Records of twenty-four hourly hydrological observations at Shell Point, Georges River, New South Wales, 1942-50:134 pp.

Chemistry, nitrate (data only)
Australia, Commonwealth Scientific and Industrial Research Organization, 1952.
Records of twenty-four hourly hydrological observations at Shell Point, Georges River, New South Wales. (Compiled by D.R. Rochford).
Ocean. Sta. List., Invest. Fish. Div., 10:134 pp.

chemistry, nitrate (data only)
Australia, Commonwealth Scientific and Industrial Research Organization, 1951.
Oceanographic station list of investigations made by the Fisheries Division, C.S.I.R.O.
Vol. 4. Onshore hydrological investigations in eastern Australia, 1942-50:114 pp.

chemistry nitrate (data only)
Ballester, A. 1965.
Hidrografia y nutrientes de la Fosa de Cariaco.
Informe de Progreso del Estudio Hidrografico de la Fosa de Cariaco, Fundación La Salle de Ciencias Naturales Estación de Investigaciones Marinas de Margarita, Caracas, Sept. 1965: 3-12. (mimeographed).

Chemistry, nitrate (data only)
Ballester, Antonio, 1965(1967).
Tablas hidrograficas.
Memoria Soc. Cienc. nat. La Salle, 25(70/71/72):41: 137.

chemistry nitrate (data only)
*Brazil, Diretoria de Hidrografia e Navegacao, 1968.
XXXI Comissao oceanografica noc Almirante Saldanha (14/11 a 16/12/66, DG26-VIII:249 pp.

Chemistry nitrate (data only)
Brasil, Diretoria de Hidrografia e Navegacao, 1968.
XXXV Comissao Oceanografica "Operacao Norte/Nordeste 1" noc Almirante Saldanha (14/9 a 16/12/1967.)
DG 26-XI: 600 pp.

Chemistry, nitrate (data only)
Brazil, Diretoria de Hidrografia e Navegação 1967.
XXXII Comissão Oceanográfica no Almirante Saldanha (4/3 a 3/5/67) DG 26-X: 411 pp.

Chemistry nitrate (data only)
Brasil, Marinha do Brasil, Diretoria de Hidrografia e Navegação, 1963
Operação "TRIDENTE I", Estudo das condições oceanográficas entre o Rio de Janeiro e o Rio da Prata, durante o inverno (Agosto-Septembro) ano de 1962.
DG-06-XV: unnumbered pp. (mimeographed).

chemistry, nitrate (data only)
Brasil, Diretoria de Hidrografia e Navegacao, 1961
Estudo das condicões oceanográficas nas proximidades do Rio de Janeiro durante o mês de Dezembro.
DG-06-XIII: mimeographed sheets.

chemistry, nitrate (data only)
Brazil, Diretoria de Hidrografia e Navegacao, 1958.
Ano Geofisico Internacional, Publ. DG-06-VII: mimeographed sheets.

Chemistry, nitrate (data only)
Brasil, Diretoria de Hidrografia e Navegacão, 1957.
Ano Geofisico Internacional. Publ. DG-06-IV :6 pp. (mimeographed)

chemistry, nitrate (data only)
Brasil, Marinha do Brasil, Diretoria de Hidrografia e Navegacao, 1960.
Estudo das condicoes oceanograficas na regiao profunda a Nornordeste de Natal, Estado do Rio Grande do Norte.
DG-06-XI(Sept. 1960): unnumbered pp. (mimeographed)

chemistry, nitrate (data only)
Brasil, Marinha do Brasil, Diretoria de Hidrografia e Navegação, 1960.
Estudo das condições oceanográficas sobre a plataforma continental, entre Cabo-Frio e Vitoria, durante o outono (abril-maio).
DG-06OX(junho): unnumbered pp. (mimeographed).

chemistry, nitrate (data only)
Brasil, Marinha do Brasil, Diretoria de Hidrografia e Navagação, 1959.
Levantamento oceanografico da costa nordeste.
DG-06-IX: unnumbered pp. (mimeographed).

chemistry nitrate (data)
Bregant, Davide 1967 (1969).
Distribuzione dei sali nutritivi nell'area delle Bocche di Bonifacio e del Golfo dell'Asinara – Crociera Bannock 1964.
Boll. Pesca Piscic. Idrobiol. n.s. 22(2): 113-120

chemistry, nitrate (data only)
Buljan, Miljenko, and Mira Zore-Armanda 1966.
Hydrographic data on the Adriatic Sea collected in the period from 1952 through 1962.
Acta adriat. 12: 438 pp.

Chemistry, nitrate (data only)
Canada, Fisheries Research Board, 1959.
Physical and chemical data record, coastal seaways project, November 12 to December 5, 1958.
MSS Rept. Ser. (Oceanogr. & Limnol.), No. 36: 120 pp.

Chemistry, nitrate (data only)
Canada, University of British Columbia 1969.
British Columbia inlet cruises 1968.
Data Rept. 28: 59 pp. (multilithed).

Chemistry nitrate (data only)
Carlsberg Foundation, 1937.
Hydrographical observations made during the "Dana" Expedition 1928-30 with an introduction by Helge Thomsen. Dana Rept. No. 12: 46 pp.

Chemistry, nitrate (data)
Choe, Sang 1969.
Phytoplankton studies in Korean waters. III. Surface phytoplankton survey of the north-eastern Korea Strait in May of 1967. (In Korean: English abstract)
J. oceanogr. Soc. Korea 4(1): 1-8.

Chemistry, nitrate (data only)
Collier, Albert, 1958.
Gulf of Mexico physical and chemical data from ALASKA cruises.
USFWS Spec. Sci. Rept., Fish., No. 249:417 pp.

Chemistry - nitrate (data only)
Conseil Permanent International pour l'Exploration de la Mer, 1967.
ICES Oceanographic data lists, 1960, No. 6: 295 pp. (multilithed).

Chemistry, nitrate (data only)
Conseil International pour l'Exploration de la Joint Skagerak Expedition 1966. 1. Oceanographic stations, temperature-salinity-oxygen content. 2. Oceanographic stations, chemical observations.
ICES Oceanogr. Data lists: 250 pp; 209 pp.

chemistry - nitrate (data)
Conseil Permanent International pour l'Exploration de la Mer, 1952.
Bulletin Hydrographique 1948:87 pp.

Chemistry, nitrate (data)
Conseil Permanent International pour l'Exploration de la Mer, 1950.
Bulletin hydrographique pour les années 1940-1946 avec un appendix pour les années 1936-1939. 190 pp.

chemistry - nitrate (data)
Conseil Permanent International pour l'Exploration de la Mer, 1939
Bulletin Hydrographique pour l'annee 1937: 106 pp., 4 maps., 4 figs.

Chemistry, nitrate (data only)
Conseil Permanent International pour l'Exploration de la Mer, 1956.
Bulletin hydrographique pour l'année 1952:164 pp.

chemistry - Nitrate-N (data)
Conseil Permanent International pour l'Exploration de la Mer, 1954.
Bulletin Hydrographique pour l'Année 1950: 114 pp., 5 charts.

chemistry, nitrate (data only)
Dragovich, Alexander, John H. Finucane and Billie Z. May, 1961.
Counts of red tide organisms, Gymnodinium breve, and associated oceanographic data from Florida west coast, 1957-1959.
USFWS Spec. Sci. Rept., Fish., No. 369:175 pp.

Chemistry
nitrate (data only)
Favorite, Felix, John W. Schantz and Charles R. Hebard, 1961
Oceanographic observations in Bristol Bay and the Bering Sea 1939-41 (USCGT Redwing).
U.S.F.W.S. Spec. Sci. Rept., Fish. No. 381: 323 pp.

Chemistry nitrate (data only)
France, Centre de Recherches et d'Etudes Océanographiques, 1960.
Stations hydrologiques effectuées par le "Passeur du Printemps" dans le cadre des travaux de l'Année Géophysique Internationale.
Travaux, C.R.E.O., 3(4):17-22.

Chemistry
nitrate (data only)
Grainger, E.H. 1971
Biological oceanographic observations in Frobisher Bay. 1. Physical, nutrient and primary production data, 1967-1971.
Techn. Rept. Fish. Res. Bd Can. 265: 75pp
(multilithed)

Chemistry, nitrate (data only)
Great Britain, Discovery Committee, 1957.
Station list, 1950-1951. Discovery Repts., 28: 299-398.

Chemistry, nitrate (data only)
Great Britain, Discovery Committee, 1937.
Station list, 1933-1935. Discovery Repts., 22:3-196, Pls. 1-4.

Chemistry, nitrate (data only)
Great Britain, Discovery Committee, 1942.
Station list, Discovery Repts., 4:3-230, Pls. 1-5.

Chemistry, nitrate (data only)
Great Britain Discovery Committee, 1931.
Station list. Discovery Repts., 3:3-132, Pls. 1-10.

Chemistry, nitrate (data only)
Great Britain, Discovery Committee, 1929.
Station List. Discovery Repts. 1:3-140.

Chemistry, nitrate (data only)
Guillén, Oscar, y Francisco Vasquez 1966.
Informe preliminar del Crucero 6602 (Cabo Blanco-Arica)
Informe, Inst. Mar. Peru No. 12: 27 pp.

Chemistry, nitrate (data only)
Hapgood, William, 1959.
Hydrographic observations in the Bay of Naples (Golfo di Napoli), January 1957-January 1958 (station lists).
Pubbl. Sta. Zool., Napoli, 31(2):337-371.

Chemistry, nitrate (data only)
Hela, Ilmo, and F. Koroleff 1958.
Hydrographical and chemical data collected in 1956 on board the R/V Aranda in the Baltic Sea.
Merent. Julk. 183:1-52.

Chemistry
nitrate (data only)
Hela, Ilmo, and Folke Koroleff, 1953
Hydrographical and chemical data collected in 1957 on board the R/V Aranda in the Barents Sea.
Merent. Julk., No. 179: 67 pp.

Chemistry, Nitrate (data only)
Hisard, P., F. Jarrige, et P. Rual, 1968.
Résultats des observations physico-chimiques de la croisière CYCLONE 2 du N.O. Coriolis.
Rapp. sci., O.R.S.T.O.M., Nouméa, 20:21 pp. (mimeographed).

Chemistry, nitrate (data only)
Hufford, Gary L., and James M. Seabrooke, 1970
Oceanography of the Weddell Sea in 1969 (IWSOE).
Oceanogr. Rept. U.S. Coast Guard 31 (CG373-31) 32 pp.

Chemistry, nitrate (data only)
Japan, Hokkaido University, Faculty of Fisheries 1970.
Data record of oceanographic observations and exploratory fishing 13: 406 pp.

Chemistry, nitrate (data only)
Japan, Japanese Oceanographic Data Center 1969.
Onlick, USSR July 21-August 20, 1968, northwest North Pacific Ocean.
Data Rept. CSK (KDC Ref 90K020) 191: 32 pp. (multilithed)

Chemistry nitrate (data only)
Japan, Hokkaido University, Faculty of Fisheries, 1968.
The Oshoro Maru cruise 23 to the east of Cape Erimo, Hokkaido, April 1967.
Data Record Oceanogr. Obs. Expl. Fish., 12: 115-169.

Chemistry, nitrate (data only)
Japan, Hokkaido University, Faculty of Fisheries, 1968.
Data record of oceanographic observations and exploratory fishing.
No. 12:420 pp.

Chemistry, nitrate (data only)
Japan, Hokkaido University, Faculty of Fisheries, 1968.
The Oshoro Maru cruise 21 to the Southern Sea of Japan, January 1967.
Data Record Oceanogr. Obs. Explor. Fish. 12:1-97; 113-119.

Chemistry, nitrate (data only)
Japan, Hokkaido University, Faculty of Fisheries, 1967.
The Oshoro Maru Cruise 18 to the east of Cape Erimo, Hokkaido, April, 1966.
Data Record Oceanogr. Obs. Explor. Fish. 11: 121-164.

Chemistry nitrate (data only)
Japan, Hokkaido University, Faculty of Fisheries, 1967.
Data record of oceanographic observations and exploratory fishing, 11:383 pp.

Chemistry
nitrate (data only)
Japan, Japan Meteorological Agency 1971.
The results of marine meteorological and oceanographical observations, July - December 1969, 46:270 pp.

Chemistry
nitrate (data only)
Japan, Japan Meteorological Agency, 1971.
The results of marine meteorological and oceanographical observations, 45: 338 pp.

Chemistry
nitrate (data only)
Japan, Japan Meteorological Agency, 1970.
The results of marine meteorological and oceanographical observations, July-December, 1968, 44:311 pp.

Chemistry
Nitrate (data only)
Japan, Japan Meteorological Agency, 1970
The results of marine meteorological and oceanographical observations. (The results of the Japanese Expedition of Deep Sea (JEDS-11); January-June 1967 41: 332 pp.

Chemistry, nitrate (data only)
Japan, Japan Meteorological Agency 1970:
The results of marine meteorological and oceanographical observations, July-December 1966, 40:336 pp.

Chemistry, nitrate (data only)
Japan Meteorological Agency, 1968.
The results of marine meteorological and oceanographical observations, July-December, 1965, 38: 404 pp. (multilithed)

Chemistry nitrate (data only)
Japan, Japan Meteorological Agency, 1966.
The results of the Japanese Expedition of Deep Sea (JEDS-8).
Results mer.met.oceanogr.Obsns.Tokyo, 35:328 pp.

Chemistry
nitrate (data only)
Japan, Japan Meteorological Agency, 1964.
The results of marine meteorological and oceanographical observations, January-June 1963, No. 33:289 pp.

Chemistry nitrate (data only)
Japan, Japan Meteorological Agency, 1964.
The results of the Japanese Expedition of Deep Sea (JEDS-5).
Res. Mar. Meteorol. and Oceanogr. Obs., July-Dec. 1962, No. 32:328 pp.

Chemistry
nitrate (data only)
Japan, Meteorological Agency, 1962
The results of marine meteorological and oceanographical observations, July-December, 1960, No. 28: 304 pp.

Chemistry, nitrate (data only)
Japan, Japan Meteorological Agency, 1961
The results of marine meteorological and oceanographical observations, January-June 1960. The results of the Japanese Expedition of Deep-Sea (JEDS-2, JEDS-3), No. 27: 257 pp.

Chemistry, nitrate (data only)
Japan, Japan Meteorological Agency, 1959.
The results of marine meteorological and oceanographical observations, July-December, 1958, No. 24:289 pp.

Chemistry nitrate (data only)
Japan, Japan Meteorology Agency, 1959.
The results of marine meteorological and oceanographical observations, January-June 1958, No. 23:240 pp.

Chemistry, nitrate (data only)
Japan, Central Meteorological Observatory, 1951.
The results of marine meteorological and oceanographic observations. July - Dec. 1949. No. 6: 423 pp.

Chemistry nitrate (data only)
Japan, Japanese Oceanographic Data Center, 1971. Kofu Maru, Hadkodate Marine Observatory, Japan Meteorological Agency, 27 June - 9 August 1968, East of Japan & Okhotsk Sea. Data Rept. CSK (KDC Ref. 49K084) 178: 32 pp. (multilithed).

Chemistry, nitrate (data only)
Japan, Japanese Oceanographic Data Center, 1969
Cape St. Mary, Fisheries Research Station, UK (Hong Kong). May 16 - 23, 1968, South China Sea. Prelim. Data Rept. CSK (KDC Ref. No. 74K010) 203: 15 pp. (multilithed).

Chemistry, nitrate (data only)
Japan, Japanese Oceanographic Data Center, 1969. Shumpu Maru, Kobe Marine Observatory, Japan Meteorological Agency, July 19-31, 1968, South of Japan.
Data Rept. CSK (KDC Ref. 49K085) 179: 9 pp.

Chemistry nitrate (data only)
Japan, Japanese Oceanographic Data Center, 1969
Takuyo, Hydrographic Division, Maritime Safety Agency, Japan. July 19-September 6, 1968, Kuroshio Extension Area. Prelim. Data Rept. CSK (KDC Ref. No. 49K082) 176: 27 pp. (multilithed).

Chemistry, nitrate (data only)
Japan, Japanese Oceanographic Data Center, 1968. Ryofu Maru, Marine Division, Japan Meteorological Agency, Japan, January 13-March 22, 1968, West of North Pacific Ocean.
Data Rept. CSK (KDC Ref. 49K069) 155: 70 pp. (multilithed).

Chemistry, nitrate (data only)
Japan, Japanese Oceanographic Data Center, 1968. Takuyo, Hydrographic Division, Maritime Safety Agency, July 12-August 30, 1967, Central Part of the North Pacific Ocean.
Prelim. Data Rept. CSK (KDC Ref. 49K053) 105:25 pp. (multilithed)

Chemistry nitrate (data only)
Japan, Japanese Oceanographic Data Center, 1968. Yang Ming, Chinese National Committee on Ocean Research, Republic of China, April 1-May 14, 1967, Surrounding waters of Taiwan.
Data Rept. CSK (KDC Ref. 21K004)102: 15 pp. (multilithed).

Chemistry, nitrate (data only)
Japan, Japanese Oceanographic Data Center, 1967. Yang Ming, Chinese National Committee on Oceanic Research, Republic of China, September 10-October 14, 1966, Adjacent Sea of Taiwan.
Prelim. Data Rept. CSK (KDC Ref. 21K003)67:17pp.

Chemistry, nitrate (data only)
Japan, Japanese Oceanographic Data Center, 1966. Koyo Maru, Shimonoseki University of Fisheries, October 26-29, south of Japan.
Prelim. Data Rept. CSK (KDC Ref. No. 49K039)66:7 pp. (multilithed).

Chemistry, nitrate (data only)
Japan, Japanese Oceanographic Data Center 1967.
Ryofu Maru, Marine Division, Japan Meteorological Agency, Japan, September 13-17, 1966, eastern Sea of Japan.
Prelim. Data Rept., CSK (KDC Ref. No. 49K033) 59:11 pp. (multilithed).

Chemistry, nitrate (data only)
Japan, Japanese Oceanographic Data Center 1967.
Nagasaki Maru, Faculty of Fisheries, Nagasaki University, Japan, June 16-21, 1966, southern Sea of Japan.
Prelim. Data Rept. CSK (KDC Ref. No. 49K032) 58:8 pp. (multilithed)

Chemistry, nitrate (data only)
Japan, Japanese Data Center, 1966.
Preliminary data report of CSK, Shumpu Maru, Kobe Marine Obsefvatory, Japan Meteorological Agency, February 19-24, 1966, Southern Sea of Japan, No. 30 (October 1966).
KDC Ref. No. 49K019:6 (Mulilithed).

Chemistry, nitrate (data only)
Japan, Japanese Oceanographic Data Center, 1966. Cape St. Mary, Fisheries Research Station, Hong Kong, October 2-October 10, 1965, South China Sea. Prelim. Data Rept., CSK, (KDC Ref. 74K001):20:12pp (multilithed)

Chemistry nitrate, (data only)
Japan, Japanese Oceanographic Data Center, 1966. U.M. Shokalski, USSR, July 16-August 18, 1965, E. & S. of Japan.
Prelim. Data Rept., CSK, (KDC Ref. 90K001), 23: 41 pp. (multilithed).

Chemistry nitrate (data only)
Japan, Japanese Oceanographic Data Center, 1966. Ryofu Maru, Japan Meteorological Agency, Japan, February 4-February 28, 1966, East of Japan, Prelim. Data Rept., CSK (KDC Ref. 49K017), 28: 17 pp. (multilithed).

Chemistry, nitrate (data only)
Japan, Japanese Oceanographic Data Center, 1966. Yang Ming, Chinese National Committee on Oceanic Research, Republic of China, August 10-October 13, 1965, Adjacent Sea of Taiwan.
Prelim. Data Rept., CSK (KDC Ref. 21K001), 22: 18 pp. (multilithed).

Chemistry nitrate (data only)
Japan, Maritime Safety Board, 1961
Tables of results from oceanographic observation in 1959.
Hydrogr. Bull. (Publ. No. 981), No. 68: 112 pp.

Chemistry, nitrate (data only)
Japan, Maritime Safety Board, 1961
Tables of results from oceanographic observations in 1958.
Hydrogr. Bull., Tokyo, No. 66 (Publ. No. 981): 153 pp.

Chemistry, nitrate (data only)
Japan, Tokyo University of Fisheries, 1966.
J. Tokyo Univ. Fish. (Spec.ed.)8 (3):1-44.

Chemistry, nitrate (data only)
Jarrige, F., J. Merle et J. Noel, 1968.
Résultats des observations physico-chimiques de la croisière CYCLONE 4 du N.O. Coriolis.
Rapp. sci., O.R.S.T.O.M. Noumea, 25:22 pp.

Chemistry nitrate (data only)
Katsuura, Hiroshi, Hideo Akamatsu, Tsutomu Akiyama, 1964.
Preliminary report of the Japanese Expedition of Deep-Sea, the Eighth Cruise (JEDS-8).
Oceanogr. Mag., Tokyo, 16(1/2):125-136.

Chemistry, nitrate (data only)
Love, C.M., 1966.
Physical, chemical, and biological data from the northeast Pacific Ocean: Columbia river effluent area, January-June 1963. 6. Brown Bear Cruise 326:13-23 June: CNAV Oshawa Cruise Oshawa-3:17-30 June.
Univ. Washington, Dept. Oceanogr., Tech. Rep., No. 134:230 pp. (Unpublished manuscript).

Chemistry Nitrate (data only)
Magnier, Y., H. Rotschi, B. Voituriez et J. Merle 1968.
Résultats des observations physico-chimiques de la croisière CYCLONE 3 du N.O. Coriolis.
Rapp. sci., O.R.S.T.O.M., Nouméa, 21:23 pp. (mimeographed)

Chemistry nitrate (data only)
Magnier, Yves, Pierre Rual et Bruno Voituriez, 1968.
Résultats des observations physico-chimiques de la croisière CYCLONE 6 du N.O. Coriolis.
Rapp. sci. O.R.S.T.O.M., Noumea, 27:23 pp. (mimeographed)

Chemistry, nitrate (data only)
Marvin, K.T., 1955.
Oceanographic observations in west coast Florida waters, 1949-52. U.S. Fish & Wildlife Service, Spec. Sci. Repts., No. 149:6 pp., 2 figs., numerous pp tables.

Chemistry nitrate (data only)
Minas, H.J. 1971.
Résultats de la campagne "Mediprod I" du Jean Charcot.
Cah. océanogr. 23 (Suppl. 1):93-144.

Chemistry nitrate (data only)
*Nasu, Keiji, and Tsugio Shimano, 1966.
The physical results of oceanographic survey in the south east Indian Ocean in 1963/64.
J. Tokyo Univ. Fish., (Spec.ed.) 8(2):133-164.

Chemistry, nitrate (data only)
Oren, O.H., 1967.
Croisière Chypre-04 dans la Méditerranée orientale, février-mars 1965. Résultats des observations hydrographiques.
Bull. Sea Fish. Res. Stn Israel, 47:37-54.

Chemistry, nitrate (data only)
Oren, Oton Haim, 1967.
Croisière Chypre-04 en Méditerranée orientale, février-mars, 1965: résultats des observations hydrologiques.
Cah. océanogr., 19(9):783-798.

Chemistry, nitrate (data only)
Oren, Oton Haim, 1966.
Croisière "Chypre-02" en Méditerranée orientale, juillet-août 1963. Résultats des observations hydrologiques.
Cah. océanogr., 18(Suppl.):1-17.

chemistry, nitrate (data only)
Piton, B., et Y. Magnier 1971.
Observations physico-chimiques faites par le Vauban le long de la côte nord-ouest de Madagascar de janvier à septembre 1970.
Doc. scient. Cent. Nosy-Bé, Off. Rech. scient. techn. Outre-Mer 21: unnumbered pp. (mimeographed)

Chemistry, nitrate (data only)
Platt, Trevor, and Brian Irwin 1971.
Phytoplankton production and nutrients in Bedford Basin, 1969-70.
Techn. Rept. Fish. Res. Bd, Can. 247: 172 pp. (multilithed).

Chemistry, nitrate (data)
Platt, Trevor, and Brian Irwin, 1968.
Primary productivity measurements in St. Margaret's Bay, 1967.
Tech. Rept. Fish. Res. Bd., Can., 77: 123 pp.

Chemistry, nitrate (data only)
Republic of China, Chinese National Committee on Oceanic Research, Academia Sinica, 1968.
Oceanographic Data Report of CSK, 2: 126 pp.

Chemistry, nitrate (data only)
Republic of China, China National Committee on Oceanic Research, Academia Sinica 1966.
Oceanographic data report of CSK 1: 123 pp.

Chemistry, nitrate (data)
Rochford, D.J., 1953.
Analysis of bottom deposits in eastern and south-western Australia, 1951 and records of twenty-four hourly hydrological observations at selected stations in eastern Australian estuarine systems 1951. Oceanographical station list of investigations made by the Division of Fisheries, Commonwealth Scientific and Industrial Research Organization, Australia, 13:68 pp.

Chemistry, nitrate (data only)
Rotschi, H., Ph. Hisard, L. Lemasson, Y. Magnier, J. Noel, et B. Piton, 1966.
Resultats des observations physico-chimiques de la croisiere "Alize".
Centre ORSTROM Noumea Rapp. Sci., No. 28:56 pp. (mimeographed).

Chemistry, nitrate (data only)
Rotschi, H., Ph. Hisard, L. Lemasson, Y. Magnier, J. Noel et B. Piton, 1967.
Resultats des observations physico-chimiques de la croisiere "Alize" du N.O. Coriolis.
Rapp. scient., Off. Rech. scient. tech. Outre-Mer, Noumea, 2:56 pp. (mimeographed).

Chemistry, nitrate (data only)
Rotschi, H., G. Pickard, P. Hisard et J. Cannevet, 1968.
Résultats des observations physico-chimiques de la croisière CYCLONE 5 du N.O. Coriolis.
Rapp. sci., O.R.S.T.O.M., Noumea, 26:23 pp.

chemistry nitrate (data only)
* Scaccini Cicatelli, Marta, 1967.
Distribuzione stagionale dei sali nutrivi in una zone dell'alto e medio Adriatico.
Boll. Pesca, Piscic. Idrobiol., n.s., 22(1): 49-82.

Chemistry nitrate (data only)
Sweden, Havsfiskelaboratoriet, Lysekil 1971.
Hydrographical data, January-June 1970 R.V. Skagerak, R.V. Thetis and TV252, 1970.
Meddn. Hydrogr. avd. Göteborg, 104: unnumbered pp. (multilithed).

Chemistry nitrate (data only)
Sweden, Havsfiskelaboratoriet, Lysekil 1970.
Hydrographical data 1966, R.V. Skagerak, R.V. Thetis.
Meddn Hydrogr. avd. Göteborg 85: 255 pp.

Chemistry nitrate (data only)
Sweden, Havsfiskelaboratoriet, Lysekil 1970.
Hydrographical data, R.V. Skagerak, R.V. Thetis.
Meddn. Hydrogr. avd. Göteborg 84: 296 pp. (multilithed).

Chemistry nitrate (data only)
United States, National Oceanographic Data Center, 1965.
Data report EQUALANT III.
Nat. Oceanogr. Data Cent., Gen. Ser., Publ. G-7: 339 pp. $5.00

Chemistry nitrate (data only)
United States, U.S. Coast Guard, 1965.
Oceanographic cruise USCGC Northwind: Chukchi, East Siberian and Laptev seas, August-September, 1963.
U.S.C.G. Oceanogr. Rept., No. 6(CG373-6):69 pp.

Chemistry nitrate (data only)
United States, National Oceanographic Data Center, 1964.
Data report EQUALANT II.
NODC Gen. Ser., Publ. G-5: numerous pp. (not serially numbered; loose leaf - $5.00).

Chemistry, nitrate (data only)
Valdez, Alberto J., Alberto Gomez, Aldo Orlando and Andres J. Lusquinos 1969.
Datos y resultados de las campañas pesquería "Pesquería V" (28 de agosto al 7 de setiembre de 1967).
Publ. (Ser. Informes técn.) Mar del Plata, Argentina 10(5): unnumbered pp.

Chemistry, nitrate (data only)
Valdez, Alberto J., Alberto Gomez, Aldo Orlando y Andres J. Lusquinos 1968.
Datos y resultados de las campañas pesquería "Pesquería IV" (7 de junio al 4 de julio de 1967.
Publ. (Ser. Informes técn.) Mar del Plata, Argentina 10(4): 1-159.

Chemistry nitrate (data only)
Zvereva, A.A., Edit., 1959.
[Data, 2nd Marine Exped., "Ob", 1956-1957.]
Arktich. i Antarkt. Nauchno-Issled. Inst., Mezhd. Geofiz. God, Sovetsk. Antarkt. Exped., 6: 1-387.

Chemistry nitrate depth distribution
Eppley, R.W., 1970.
Relationships of phytoplankton species distribution to the depth distribution of nitrate
Bull. Scripps Inst. Oceanogr. 17: 43-49

nitrate, effect of

Chemistry, nitrate, effect of
Procharzkova, L., 1960.
Einfluss der Nitrate und Nitrite aud die Bestimmung des organischen Stickstoffs und Ammoniums im Wasser.
Arch. f. Hydrobiol., 56(3):179-185.

Chemistry, nitrogen fixation
Brooks, Ralph H., Jr., Patrick L. Brezonik, Hugh D. Putnam and Michael A. Keirn, 1971.
Nitrogen fixation in an estuarine environment: the Waccasassa on the Florida Gulf Coast.
Limnol. Oceanogr. 16(5): 701-710.

nitrate (mean)(data)

Chemistry, nitrate (mean)(data)
Dragovich, A., 1963.
Hydrology and plankton of coastal waters at Naples, Florida.
Q.J. Florida Acad. Sci., 26(1):22-47.

Chemistry Nitrate (monthly means)
Newell, B. S., 1966.
Seasonal changes in the hydrological and biological environments off Port Hacking, Sydney.
Australian J. Mar. freshw. Res., 17(1):77-91.

Chemistry, nitrate, monthly means
Slinn, D.J., and Gaenor Offlow (Mrs. M. Solly), 1968.
Chemical constituents of sea water off Port Erin during 1967.
Ann. Rept. mar. biol Stn, Port Erin, 80:37-42.

nitrate-nitrite

Chemistry nitrate-nitrite
Slinn, S.J., 1962
Chemical constituents in sea water off Port Erin during 1961.
Mar. Biol. Sta., Univ. Liverpool, Port Erin, Ann. Rept., 74(1961):23-26.

nitrate/oxygen A.O.U.

Chemistry, nitrate/oxygen A.O.U.
Kitamura, Hiroyuki, and Takeshi Sagi, 1965.
On the chemical elements in the sea south of Honshu, Japan. II. (In Japanese; English abstract Bull. Kobe Mar. Obs., No. 174:39-55.

NO_3/PO_4^3

Chemistry, nitrate/phosphate ratio
Broenkow, William W., 1965.
The distribution of nutrients in the Costa Rica dome in the eastern tropical Pacific Ocean.
Limnology and Oceanography, 10(1):40-52.

chemistry, nitrate-phosphate correlations
Stefansson, Unstein, 1966.
Influence of the Surtsey eruption on the nutrient content of the surrounding seawater.
J. Mar. Res., 24(2):241-268.

chemistry nitrate-phosphate ratio, anomaly of
Cooper, L.H.N., 1938
Redefinition of the anomaly of the nitrate-phosphate ratio. JMBA, 23:179

Chemistry, nitrate reductase
Packard, T.T., Dolores Blasco, J.J. MacIsaac and R.C. Dugdale, 1971.
Variations of nitrate reductase activity in marine phytoplankton. Investigación pesq. 35(1): 209-219.

nitrate reduction

chemistry, nitrate reduction
*Fiadeiro, Manuel, and J.D.H. Strickland, 1968.
Nitrate reduction and the occurrence of a deep nitrite maximum in the ocean off the west coast of South America.
J. mar. Res., 26(3):187-201.

chemistry, nitrate reduction
Richards, F.A. and W.W. Broenkow, 1971.
Chemical changes, including nitrate reduction in Darwin Bay, Galapagos Archipelago, over a 2-month period, 1969. Limnol. Oceanogr. 16(5) 758-765.

chemistry, nitrate reduction
ZoBell, Claude, E., and Keith M. Budge, 1965.
Nitrate reduction by marine bacteria at increased hydrostatic pressures.
Limnology and Oceanography, 10(2):207-214.

chemistry, nitrate reductase
Eppley, R.W., T.T. Packard and J.J. MacIsaac 1970.
Nitrate reductase in Peru Current phytoplankton.
Mar. Biol. 6(3): 195-199.

nitrate/silicate

Chemistry, nitrate-silicate correlations
Stefansson, Unstein, 1966.
Influence of the Surtsey eruption on the nutrient content of the surrounding seawater.
J. Mar. Res., 24(2):241-268.

Chemistry, nitric oxide
*Kühme, Heinrich, 1968.
Untersuchungen der Konzentration atmosphärischen Spurengase über dem Atlantik.
Meteor Forschungsergeb, (A)2:83-93.

chemistry, nitrification
Carlucci, A.F. and P.M. McNally, 1969.
Nitrification by marine bacteria in low concentrations of substrate and oxygen. Limnol. Oceanogr., 14(5): 736-739.

Chemistry, nitrification
Fiala M. 1967.
Contribution à l'étude des eaux et des sédiments de l'Etang des Bages-Sigean (Aude). V Recherches sur les processus de nitrification (Etude préliminaire).
Vie Milieu (B)18 (2B): 227-238.

nitrite

chemistry, nitrite
Auninsh, E.A., 1966.
Biogenic elements in the Gulf of Riga. (In Russian).
Trudy, Gosudarst. Okeanogr. Inst., No. 83:172-206.

Chemistry, nitrite
Bata Nobus, 1969.
On the distribution of nitrite nitrogen in Osaka Bay. (In Japanese; English abstract).
Bull. Kobe mar. Obs., 181: 1-8.

Chemistry nitrite
Bal, D.V., L.B. Pradhan, and K.G. Gupte, 1946
A preliminary record of some of the chemical and physical conditions in waters of the Bombay harbour during 1944-45. Proc. Indian Acad. Sci. Sect. B24(2): 60-73, 4 figs.

Chemistry, nitrite (data)
Ballester, A. 1965.
Hidrografía y nutrientes de la Fosa de Cariaco.
Informe de Progreso del Estudio Hidrográfico de la Fosa de Cariaco, Fundación La Salle de Ciencias Naturales, Estación de Investigaciones Marinas de Margarita, Caracas, Sept. 1965: 3-12 (mimeographed)

chemistry, nitrite (data)
Ballester, Antonio, Enrique Arias, Antonio Cruzado, Dolores Blasco y José María Camps, 1967.
Estudio hidrográfico de la costa catalana de junio de 1965 a mayo de 1967.
Investigación, pesq. 31(3): 621-662

Chemistry, nitrite
Bandel, W., 1940.
Phytoplankton-und Nahrstoffgehalt der Ostsee in Gebiet der Darsser Schwelle. Internat. Rev. ges. Hydrobiol. u. Hydrogr., 40:249-304

nitrites
Barnes, H., 1954.
The estimation of nitrites.
Mem. Ist. Ital. Idrobiol., "Dott. Marco de Marchi" 8:73-79, 2 textfigs.

Chemistry nitrites
Barnes, H., and A.B. Folkard, 1951.
The determination of nitrites. Analyst 76(907): 599-603.

Chemistry nitrite
Bendschneider, K., and R.J. Robinson, 1952.
A new spectrophotometric method for the determination of nitrite in sea water. J. Mar. Res. 11(1):87-96, 5 textfigs.

Chemistry, nitrite
Bennett, Edward B., and Milner B. Schaefer, 1960
Studies of physical, chemical and biological oceanography in the vicinity of the Revilla Gigedo Islands during "Island Current Survey" of 1957.
Bull. Inter-American Tropical Tuna Comm., 4(5): 219-317. (Also in Spanish)

Chemistry nitrites
Bogoyavlenskii, A.N., 1955.
[Chemical characteristics of the water in the region of the Kurile-Kamchatka Trench.]
Trudy Inst. Okeanol., 12:161-176.

Chemistry nitrite
Braarud, T., and E. Føyn, 1951.
Nitrite in polluted sea-water. Observations from the Oslo Fjord, 1946-1948. Avhandl. Norske Videnskaps-Akad., Oslo, Math.-Naturvid. Kl., 1951 (3):24 pp., 9 textfigs.

Chemistry, nitrite
Braarud, T., and A. Klem, 1931.
Hydrographical and chemical investigations in the coastal waters off Møre and in the Romsdalfjord. Hvalrådets Skrifter, 1931(1):88 pp., 19 figs.

Chemistry, nitrite (data)
Brandhorst, Wilhelm, 1959.
Nitrification and denitrification in the eastern tropical North Pacific.
J. du Cons., 25 (1):3-20.

Chemistry, nitrite
Bruevich, S.V., 1954.
[Nitrites and nitrification in the sea.] Trudy Inst. Okeanol., 8:3-17.

Chemistry nitrite
Bruevich, S.W., A.N. Bogoyavlensky, and V.V. Mokievskaya, 1960
[Hydrochemical features of the Okhotsk Sea.]
Trudy Inst. Okeanol., 42: 125-198.

Chemistry nitrite
Burns, R.B., 1967.
Chemical observations in the North Sea in 1965.
Annls biol. Copenh. (1965)22:29-31.

Chemistry nitrite
*Carlucci, A.F. and Hazel R. Schubert, 1969.
Nitrate reduction in seawater of the deep nitrite maximum off Peru. Limnol. Oceanogr., 14(2): 187-193.

chemistry, nitrite
*Choe, Sang, Tai Wha Chung and Hi-Sang Kwak, 1968.
Seasonal variations in nutrients and principal ions contents of the Han River water and its water characteristics. (In Korean; English abstract).
J. oceanogr. Soc. Korea, 3(1):26-38.

Chemistry nitrite (data)
Corwin, Nathaniel, and David A. McGill, 1963.
Nutrient distribution in the Labrador Sea and Baffin Bay.
U.S.C.G., Bull., No. 48:79-94; 95-153.

chemistry, nitrite (data)

COSTE, B., et H.J. MINAS, 1968.
Production organique primaire et sels nutritifs au large des côtes occidentales Corso-Sardes en février 1966.
Rec. Trav. Sta. mar. Endoume, 44(60): 49-61

---

chemistry, nitrite (data)

De Queiroz Santos, Neugon 1967.
Principais nutrientes e alguns dados físico-químicos do região lagunar de Cananéia.
Bolm Inst. Biol. mar. Univ. Fed. Rio Grande Norte, Brasil 4:1-14.

Chemistry nitrite

Dobrzhanskaia, M.A., 1963
The amount and distribution of nitrite in the Black Sea. (In Russian).
Trudy Sevastopol Biol. Sta., 16:488-495.

---

chemistry nitrite(data)

Dragovich, Alexander, 1961
Relative abundance of plankton off Naples, Florida, and associated hydrographic data, 1956-'57.
USFWS Spec. Sci. Rept., Fish., No. 372:41 pp.

Chemistry nitrite

Dragovich, A., and B.Z. May, 1962
Hydrological characteristics of Tampa Bay tributaries.
U.S.F.W.S. Fish. Bull. (205) 62:163-176.

---

chemistry, nitrite

Dyer, W.J., 1946.
Colorimetric nitrite determination. J. Fish. Res. Bd., Canada, 6:414-418.

---

chemistry, nitrite

*Egorov, V.A., 1968.
Nitrites in Mediterranean waters. (In Russian; English abstract).
Okeanologiia, Akad. Nauk, SSSR, 8(4):628-637.

---

chemistry, nitrite

Eisma, D. and A.J. van Bennekom, 1971.
Oceanographic observations on the eastern Surinam Shelf. Hydrogr. Newsletter, R. Netherlands Navy, Spec. Publ. 6: 25-29.

Chemistry nitrite (data)

El-Sayed, Sayed Z., and Enrique F. Mandelli, 1965
Primary production and standing crop of phytoplankton in the Weddell Sea and Drake Passage.
In: Biology of Antarctic seas, II.
Antarctic Res. Ser., Am. Geophys. Union, 5:87-106

Chemistry, nitrite

Eppley, Richard W., and James L. Coatsworth 1968.
Uptake of nitrate and nitrite by Ditylum brightwellii - kinetics and mechanisms.
J. Phycol. 4(2): 151-156.

---

chemistry, nitrite

Febres, German, Gilberto Rodriguez y Andres Eloy Esteves, 1966.
Ch. 2. Quimica del agua.
Estudios Hidrobiologicos en el Estuario de Maracaibo, Inst. Venezolano de Invest. Cient., 21-65. (Unpublished manuscript).

---

chemistry, nitrite

Fedosov, M.V., and E.G. Vinogradova, 1955.
[Hydrological and hydrochemical regime, primary production in the Azov Sea and forecasting change. Fundamental basis of the hydrochemical regime in the Azov Sea.] Trudy VNIRO, 31(1):9-34.

Chemistry, nitrite

Fonselius, S.H. 1968.
Observations at Swedish lightships and in the central Baltic 1966.
Annls biol. Copenh. 1966 23:74-78.

---

chemistry
Nitrite (data)

Forsbergh, Eric D., William W. Broenkow, 1965.
Observaciones oceanograficas del oceano Pacifico oriental recolectadas por el barco Shoyo Maru, octubre 1963-marzo 1964.
Comision Interamericana del Atun Tropical, Bol. 10(2): 85-237.

---

chemistry - nitrite

Fraga, F., 1967.
Hidrografia de la ria de Vigo, 1962, con especial referencia a los compuestos nitrógens.
Investigacion pesq., 31(1):145-149.

---

chemistry, nitrite

Fraga F. y A. Ballester 1966.
Distribución vertical del nitrogeno organico e inorganico en la fosa de Cariaco y su relación con el fosforo total.
Memoria Soc Cienc. nat. La Salle 26(75): 274-282.

---

chemistry, nitrite

Gaarder, Karen Ringdal, 1938.
Phytoplankton studies from the Tromsø district, 1930-31. Tromsø Mus. Årshefter, Naturhist. Avd., 11, 55(1):159 pp., 4 fold-in pls., 12 textfigs.

---

chemistry, nitrite

Galtsoff, P.S., ed., 1954.
Gulf of Mexico, its orig in, waters and marine life. Fish. Bull., Fish and Wildlife Service, 55:1-604, 74 textfigs.

Chemistry, nitrite

Garkavaia, G.P., and L.E. Pogodniakova 1968.
Data on the hydrochemical regime in Chesha Bay, Barents Sea. (In Russian)
Trudy Murmansk. Morsk. Biol. Inst. 17(21): 4-21.

---

chemistry nitrite

Goering, John J., and David Wallen, 1967.
The vertical distribution of phosphate and nitrite in the upper one-half meter in the southeast Pacific Ocean.
Deep-Sea Research, 14(1):29-33.

Goodman, J.R., ~~J.H.Lincoln, T.G.Thompson,~~ and ~~F.A.Zeusler,~~ 1942.
Physical and chemical investigations: Bering Sea, Bering Strait, Chukchi Sea during the summers of 1937 and 1938. Univ. Washington Publ. Ocean., 3(4):105-169, 37 maps.

---

nitrite chemistry, nitrite

Guarrera, S.A., 1950.
Estudios hidrobiologicos en el Rio de la Plata.
Rev. Inst. Nac., Cienc. Nat., Ciencias Botanicas 2(1):62 pp.
(Invest.)

---

chemistry, nitrite

Hattori, Akihko and Eitaro Wada, 1971.
Nitrite distribution and its regulating processes in the equatorial Pacific Ocean. Deep-Sea Res. 18(6): 557-568.

---

chemistry, nitrite

Hatzikakidis, A., 1950.
[Chemical and microbiological study of marine waters.] ΠΡΑΚΤΙΚΑ ΕΛΛΗΝΙΚΟΝ ΥΔΡΟΒΙΟΛΟΓ. ΙΚΟΥ ΙΝΣΤΙΤΟΥΤΟΥ 4(1):103-120.

---

chemistry, nitrite

Hisard, Ph and B. Piton, 1969.
La distribution du nitrite dans le système des courants equatoriaux de l'océan pacifique, à 170°E. J. Cons. perm. int. Explor. Mer, 32(3): 303-317.

---

chemistry Nitrite

Hollan, Eckard, 1970.
Eine physikalische Analyse kleinräumiger Änderungen chemischer Parameter in den tiefen Wasserschichten der Gotlandsee.
Kieler Meeresforsch. 25(2): 255-267

---

chemistry nitrite (data only)

Hufford, Gary L. and James M. Seabrooke, 1970
Oceanography of the Weddell Sea in 1969 (IWSOE).
Oceanogr. Rept. U.S. Coast Guard 31 (CG373-31) 32 pp.

Chemistry, nitrite

Hulburt, Edward M., 1966.
The distribution of phytoplankton, and its relationship to hydrography, between southern New England and Venezuela.
J. Mar. Res., 24(1):67-81.

---

chemistry nitrite

Hung, Tsu-Chang 1970
Photo- and radiation effects on inorganic nitrogen compounds in pure water and sea water near Taiwan in the Kuroshio Current.
Bull. Inst. Chem. Acad. Sinica 18: 37-56

Chemistry nitrite

Hung, Tsu-Chang, and Ching-Wei Lee, 1967.
Research on chemical oceanography in the Kuroshio around Taiwan Island.
Bull. Inst. Chem., Acad. Sinica, Taiwan, 14:80-102.

Chemistry, nitrite (data)

Husby, David M., and Gary L. Hufford 1971.
Oceanographic investigation of the northern Bering Sea and Bering Strait 8-21 June 1969.
Oceanogr. Rept. U.S. Cst Gd 42 (CG373-42): 55 pp.

---

chemistry nitrite (data)

*Ishino Makoto, Keiji Nasu, Yoshimi Morita and Makoto Hamada, 1968.
Oceanographic conditions in the west Pacific Southern Ocean in summer of 1964-1965. J. Tokyo Univ. Fish., 9(1): 115-208.

chemistry, nitrites
Ivanenkov, V.N., and F.A. Gebin, 1960.
Water masses and hydrochemistry of the western and southern Indian Ocean. Physics of the Sea. Hydrology. (In Russian).
Trudy Morsk. Gidrofiz. Inst., 22:33-115.

Translation: Scripta Technica, Inc., for Amer. Geophys. Union, 27-99.

parts of

chemistry, nitrite
Japan, Kobe Marine Observatory 1967.
Report of the oceanographic observations in the sea south of Honshu and in the Seto-Naikai from May to June 1964.
(In Japanese)
Bull. Kobe Mar. Obs. 178:37-

chemistry, nitrite
Japan, Kobe Marine Observatory 1967.
Report of the oceanographic observations in the sea south of Honshu from July to August 1963. (In Japanese)
Bull. Kobe Mar. Obs. 178:31-40.

chemistry-nitrite
Japan, Kobe Marine Observatory, Oceanographical Section, 1964.
Report of the oceanographic observations in the Kuroshio and region east of Kyushu from October to November 1962. (In Japanese)
Res. Mar. Meteorol. and Oceanogr., Japan. Meteorol. Agency, 32: 41-50.

Also in: Bull. Kobe Mar. Obs., 175. 1965.

chemistry nitrite
Japan, Kobe Marine Observatory, Oceanographical Section, 1964.
Report of the oceanographic observations in the sea south of Honshu from February to March, 1963. Res. Mar. Meteorol. and Oceanogr., Japan Meteorol. Agency, 33: 27-32.

Also in: Bull. Kobe Mar. Obs., 175. 1965.

chemistry, nitrite
Japan, Kobe Marine Observatory, Oceanographical Section, 1964.
Report of the oceanographic observations in the sea south of Honshu from July to August, 1962. Res. Mar. Meteorol. and Oceanogr., Japan. Meteorol. Agency, 32: 32-40. (In Japanese).

Also in: Bull. Kobe Mar. Obs. 175!

chemistry - nitrite
Japan, Kobe Marine Observatory, Oceanographical Section, 1964.?
Researches in chemical oceanography made at K.M.O. for the recent ten years (1954-1963). (In Japanese).
Weather Service Bull., Japan. Meteorol. Agency, 31(7):188-191.

Also in:
Bull. Kobe Mar. Obs., No. 175. 1965.

chemistry nitrite
Japan, Kobe Marine Observatory, 1963.
Report of the oceanographical observations in the sea south of Honshu from February to March 1961.
Report of the oceanographical observations in the cold water region off Enshu Nada in May 1961. (In Japanese).
Bull. Kobe Mar. Obs., 171(4):22-35.

chemistry nitrite
Japan, Kobe Marine Observatory, Oceanographical Section, 1962
Report of the oceanographical observations in the sea south of Honshu from July to August 1961. (In Japanese).
Res. Mar. Meteorol. & Oceanogr., 30:39-48.

Also in:
Bull. Kobe Mar. Obs., No. 173(3). 1964.

chemistry nitrite
Japan, Kobe Marine Observatory, Oceanographical Section, 1962
Report of the oceanographic observations in the sea south of Honshu from February to March, 1961. (In Japanese).
Res. Mar. Meteorol. and Oceanogr. Obs., Jan.-June, 1961, No. 29:22-27.

chemistry nitrite
Japan, Kobe Marine Observatory, 1962
Report of the oceanographic observations in the sea south of Honshu from July to August 1960. (In Japanese).
Res. Mar. Meteorol. and Oceanogr., July-Dec., 1960, Japan Meteorol. Agency, No. 28:36-42.

chemistry nitrite
Japan, Nagasaki Marine Observatory 1971
Report of the oceanographic observations in the sea south of Kyushu in April, 1966 (In Japanese).
Oceanogr. Met. Nagasaki Mar. Obs. 18:53-57.

chemistry, nitrate
Kändler, Rudolf, 1963
Hydrographische Untersuchungen über die Abwasserbelastung der Flensburger Förde.
Kieler Meeresforsch., 19(2):142-157.

chemistry, nitrite
Kandler, R., 1953.
Hydrographische Untersuchungen zum Abwasserproblem in den Buchten und Förden der Ostseeküste Schleswig-Holstein. Kieler Meeresf. 9(2):176-200, Pls. 9-14(18 figs.).

chemistry, nitrite (data)
Kato, Kenji 1966
Geochemical studies on the mangrove region of Cananéia, Brazil. I. Tidal variations of water properties.
Bolm Inst. Oceanogr. S. Paulo 15(1):13-20

chemistry Nitrite
Kato, Kenji, 1961.
Oceanographical studies in the Gulf of Cariaco.
1. Chemical and hydrographical observations in January, 1961.
Bol. Inst. Oceanograf., Univ. Oriente, Cumana, Venezuela, 1(1):49-72.

chemistry nitrite (data)
Kato, Kenji, 1961
Some aspects on biochemical characteristics of sea water and sediments in Mochima Bay, Venezuela.
Bol. Inst. Oceanogr., Univ. de Oriente, Venezuela 1(2):343-358.

chemistry nitrite
Kay, H., 1954.
Untersuchungen zur Menge und Verteilung der organischen Substanz im Meerwasser.
Kieler Meeresf. 10(2):202-214, Figs. 18-25.

chemistry, nitrite
Kitamura, Hiroyuki, and Takeshi Sagi, 1964.
On the chemical elements in the sea south of Honshu, Japan. (In Japanese; English abstract).
Bull. Kobe Mar. Obs., No. 172:6-54.

nitrates
Laktionov, A.F., V.A. Shamontyev and A. V. Yanes, 1960
Oceanographic characteristics of the North Greenland Sea.
Soviet Fish., Invest., North European Seas, VNIRO, PNIRO, Moscow, 1960:51-65.

chemistry, nitrite
Lauzier, L., and C. Ouellet, 1943.
Détermination colorimétrique des nitrites et nitrates dans l'eau de mer. Ann. A.C.F.A.S. 9:94.

chemistry nitrites
Leloup, E., Editor, with collaboration of L. Van Meel, Ph. Polk, R. Halewyck and A. Gryson, undated.
Recherches sur l'ostreiculture dans le Bassin de Chasse d'Ostende en 1962.
Ministere de l'Agriculture, Commission T.W.O.Z., Groupe de Travail - "Ostreiculture", 58 pp.

chemistry nitrites
Lisitzin, A.P., 1966.
Processes of Recent sedimentation in the Bering Sea. (In Russian).
Inst. Okeanol., Kom. Osad. Otdel Nauk o Zemle, Isdstel, Nauka, Moskva, 574 pp.

chemistry - nitrite(N/m3)(data)
Lüneberg, H., 1939.
Hydrochemische Untersuchungen in der Elbmündung mittels Elektrokolorimeter.
Arch. Deutschen Seewarte 59(1):1-27, 8 pls.

chemistry, nitrites (data)
Margalef, R., J. Herrera, M. Steyaert et J. Steyaert, 1966.
Distributions et caracteristiques des communautés phytoplanctoniques dans le bassin Tyrrhenien de la Méditerranee en fonction des facteurs ambiants et à la fin de la stratification estivale de l'annee 1963.
Bull. Inst. r. Sci. nat. Belg., 42(5):1-56.

chemistry nitrite (data)
Marumo, Ryuzo, Editor, 1970.
Preliminary report of the Hakuho Maru Cruise KH-69-4, August 12-November 13, 1969, The North and Equatorial Pacific Ocean.
Ocean Res. Inst. Univ. Tokyo, 68 pp.

chemistry, nitrite
Matsue, Yoshiyuki, Yuzo Komaki and Masaaki Murano, 1957
On the distribution of minute nutrients in the North Japan Sea, during the close of August, 1955.
Bull. Japan Sea Reg. Fish. Res. Lab., 6:121-127
Abstr. in:
Rec. Res., Fac. Agric., Tokyo Univ., Mar. 1958, 7(80): 57-58.

chemistry, nitrite
Miyake, Y., Y. Sugiura and K. Kamada, 1953.
A study of the property of the coastal waters around Hachijo Island. Rec. Ocean. Wks., Japan, n.s., 1(1):93-99, 10 textfigs.

Chemistry, nitrite
Mokiyevskaya, V.V., 1961
[Some peculiarities of hydrochemistry in the northern Indian Ocean.] (In Russian; English abstract).
Okeanolog. Issled., Mezhd. Komitet Proved. Mezhd. Geofiz. Goda, Prezidiume, Akad. Nauk, SSSR, (4):50-61.

chemistry, nitrite
Neumann, W., 1949.
Beiträge zur Hydrographie des Vrana-Sees (Insel Cherso), inbesondere Untersuchungen über organisches sowie anorganische Phosphor- und Stickstoffverbindungen. Nova Thalassia 1(6):17 pp.

chemistry, nitrite
Nümann, W., 1941.
Der Nährstoffhaushalt in der nordöstlichen Adria. Thalassia 5(2):1-68, 12 textfigs.

chemistry nitrite
Noble, A., 1968.
Studies on sea-water off the North Kanara coast.
J. mar. biol. Ass., India, 10(2):197-223.

chemistry nitrite
Norina, A.M., 1965.
Hydrochemical characteristics of the northern part of the Barents Sea. (In Russian).
Trudy, Gosudarst Okeanogr. Inst., No. 83:243-271.

nitrite (data)
Okabe, Shiro, Yoshimasa Toyota and Takafumi Murakami, 1967.
Major aspects of chemical oceanography. (In Japanese; English abstract).
J.Coll.mar.Sci.Techn., Tokai Univ., (2):39-56.

chemistry, nitrite
*Okuda, Taizo, 1967.
Vertical distribution of inorganic nitrogen in the equatorial Atlantic Ocean.
Boln Inst. Oceanogr., Univ.Oriente, 5(½):67-83.

chemistry nitrite
Okuda, Taizo, 1962
Physical and chemical oceanography over continental shelf between Cabo Frio and Vitoria (Central Brazil).
J. Oceanogr. Soc., Japan, 20th Ann. Vol., 514-540.

chemistry nitrite
Okuda, Taizo, 1960
Chemical oceanography in the South Atlantic Ocean, adjacent to north-eastern Brazil. (Portuguese and French resumés).
Trabalhos, Inst. Biol. Marit. e Oceanogr., Universidade do Recife, Brasil, 2(1):155-174.

chemistry nitrite
Okuda, Taizo, 1960
Metabolic circulation of phosphorus and nitrogen in Matsushima Bay (Japan) with special reference to exchange of these elements between sea water and sediments. (Portuguese, French resumes)
Trabalhos, Inst. Biol. Maritima e Oceanogr., Universidade do Recife, Brasil, 2(1):7-153.

chemistry, nitrite
Okuda, Taizo, Angel José Garcia, José Benétoz Alvarez, 1965.
Variacion estacional de las elementos nitritions en al agua de la laguna y el rio Unare.
Bol. Inst. Oceanogr., Univ. Oriente, 4(1):123-135.

chemistry, nitrite
Oudot C., P. Hisard et B. Voituriez 1969.
Nitrite et circulation meridienne à l'Equateur dans l'océan Pacifique occidental.
Cah. ORSTOM Sér. Océanogr. 7(4): 67-82.

chemistry, nitrites
Phifer, L. D. and T. G. Thompson, 1937.
Seasonal variations in the surface waters of San Juan channel during the five year period, January 1931 to December 30, 1935. J. Mar. Res., 1(1):34-59, text figs. 12-18, 12 tables.

Chemistry nitrite
Pirro, A. 1971.
Chemical environmental factors in marine radiocontamination.
CNEN Rept. RT/BIO (70)-11, M. Bernhard, editor: 11-22.

chemistry, nitrite
Ponomarenko, L.S., 1966.
Hydrochemical conditions in the Norwegian and Greenland seas in the summer of 1964. (In Russian).
Mater. Ribokhoz. Issled. severn. Basseina, Poliarn. Nauchno-Issled. Proektn. Inst. Morsk. Ribn. Khoz. Okeanogr. (PINRO), 7:155-167.

chemistry, nitrite
Ponomarenko, L.S., 1966.
Hydrochemical conditions in the Norwegian and Greenland seas in summer (from the data of the June surveys 1954-1964). (In Russian).
Trudy, Poliarn. Nauchno-Issled. Proektn. Inst. Morsk. Ribn. Khoz. Okeanogr., N. M. Knipovich, PINRO, 10:125-145.

chemistry nitrite
Ponomarenko, L.S., and M.A. Istoshina, 1962
Hydrochemical studies in Newfoundland area. (In Russian; English summary).
Sovetskie Ribochoz. Issledov. Severo-Zapad. Atlant. Okeana, VNIRO-PINRO, 113-124.

chemistry nitrite
Pozdniakova, L.E., and V.N. Vinogradov, 1966.
Some peculiarities in the distribution of the hydrochemical elements in the southeastern Barents Sea in August-September 1959. (In Russian).
Trudy murmansk. biol. Inst., 11(15):140-156.

chemistry, nitrite (data)
Postma, H., 1966.
The cycle of nitrogen in the Wadden Sea and adjacent areas.
Neth. J. Sea Res., 3(2):186-221.

Chemistry nitrite
Proctor, Raphael R., Jr., 1962
Stabilization of the nitrite content of sea water by freezing.
Limnol. and Oceanogr., 7(4):479-480.

chemistry nitrite
Rakestraw, N. W., 1936.
The occurrence and significance of nitrite in the sea. Biol. Bull. 71(1):133-167, 12 textfig

chemistry nitrite
Rakestraw, N.W., 1933.
Studies on the biology and chemistry of the Gulf of Maine. 1. Chemistry of the waters of the Gulf of Maine in August, 1932. Biol. Bull. 64(2):149-158, 4 textfigs.

Chemistry, nitrite
Ramamurthy, S., 1965.
Studies on the plankton of the North Canara coast in relation to the pelagic fishery.
J.mar. biol. Ass.India, 7(1):127-149.

chemistry nitrite (data)
Ramamurthy, V.D., K. Krishnamurthy and R. Seshadri, 1965.
Comparative hydrographical studies of the near shore and estuarine waters of Porto-Novo, S. India.
J. Annamalai Univ., 26:154-164.

Also in; Collected Reprints, Mar. Biol. Sta., Porto Novo, 1963/64.

chemistry, nitrite
Redfield, A.C., and A.B. Keys, 1938.
The distribution of ammonia in the waters of the Gulf of Maine. Biol. Bull. 74(1):83-92, 6 textfigs.

Chemistry, nitrite
Rheinheimer, Gerhard 1967.
Ökologische Untersuchungen zur Nitrification in Nord- und Ostsee.
Helgoländer wiss. Meeresunters. 15 (1/4): 243-252.

Chemistry - nitrite
Richards, Francis A., and Richard A. Kletsch, 1964.
The spectrophometric determination of ammonia and labile amino compounds in fresh and seawater by oxidation to nitrite.
In: Recent researches in the fields of hydrosphere, atmosphere and nuclear geochemistry, Ken Sugawara festival volume, Y. Miyake and T. Koyama, editors, Maruzen Co., Ltd., Tokyo, 65-81.

Chemistry nitrite
Robertson, A.J., 1907.
On hydrographical investigations in the Faroe-Shetland Channel and the northern part of the North Sea during the years 1904-1905. Fish. Bd., Scotland, North Sea Fish. Invest. Comm., Northern Area, Rept. No. 2:1-140, 13 textfig., 15 pls.

chemistry, nitrites
Robinson, R.J., and T.G. Thompson, 1948
The determination of nitrites in sea water.
J. Mar. Res. 7(1):42-48.

chemistry, nitrites (data)
Rozanov, A.B., and V.S. Bykova, 1964.
The distribution of nitrates and nitrites in water of the northern part of the Indian Ocean. Investigations in the Indian Ocean (33rd voyage of E/S "Vitiaz"). (In Russian)
Trudy Inst. Okeanol., Akad. Nauk, SSSR, 64:94-101

chemistry nitrite
Rubinchik, E.E., 1959
[Data on the hydrochemistry of the southern Caspian.]
Trudy V.N.I.R.O., 38: 152-164.

chemistry, nitrite
* Scaccini, Cicatelli, Marta, 1968.
Un anno di osservazioni idrologiche in una Stazione fissa nelle acque dell'Adriatico Occodentale presso Fano.
Note Lab. Biol. mar. Pesca, Fano, Univ. Bologna 2(9): 181-228.

chemistry, nitrite (data)
Scaccini Cicatelli, Marta 1967.
Distribuzione stagionali dei sali nutrivi in una zona dell' alto e medio Adriatico.
Boll. Pesca Piscic. Idrobiol., n.s. 12(1): 49-82

chemistry - nitrites
Schachter, D., 1954.
Contribution à l'étude hydrographique et hydrologique de l'étang de Berre (Bouches-du-Rhone).
Bull. Inst. Océan., Monaco, No. 1048:20 pp.

chemistry, Nitrite
Schott, Friedrich, and Manfred Ehrhardt, 1970.
On fluctuations and mean relations of chemical parameters in the northwestern North Sea.
Kieler Meeresforsch. 25(2): 272-278

Chemistry, nitrite
Sen Gupta, R. 1968.
Inorganic nitrogen compounds in ocean stagnation and nutrient resupply.
Science 160 (3830): 884-885.

chemistry, nitrite (data)
Shimomura, T., and K. Miyata, 1957.
[The oceanographical conditions of the Japan Sea and its water systems, laying stress on the summer of 1955.]
Bull. Japan Sea Reg. Fish. Res. Lab., No. 6 (General survey of the warm Tsushima Current 1): 23-120.

chemistry nitrite
Slinn D.J. 1961
Chemical constituents in sea water off Port Erin in 1960.
Ann. Rept. Mar. Biol. Sta., Univ. Liverpool 1960, 73:23-28.

chemistry nitrite
Slinn, D.J., and Winifred Chapman, 1965.
Chemical constituents in sea water off Port Erin in 1964.
Mar. Biol. Sta. Port Erin, Ann. Rept. (1964), No. 77:43-48.

chemistry, nitrite
Slinn, D.J., and Winifred Chapman, 1964.
Chemical constituents in sea water off Port Erin during 1963.
Mar. Biol. Sta., Univ. Liverpool, Port Erin, Ann. Rept., No. 76(1963):26-31.

chemistry, nitrite
Slinn, D.J., and Eleanor Offlow (Mrs. M. Solly) 1969.
Chemical constituents in sea water off Port Erin during 1968.
Ann. Rep. mar. biol. Sta. Port Erin 81 (1968): 31-

chemistry, nitrate
Smith, F.G.W., R.H. Williams, and C.C. Davis, 1950.
An ecological survey of the sub-tropical inshore waters adjacent to Miami. Ecol. 31(1):119-146, 7 textfigs.

chemistry nitrite (data)
Soot-Ryen, T., 1947.
Hydrographical investigations in the Tromsø-district 1934-1938 (Tables).
Tromsø Mus. Aarsheft., Naturhist. Avd., No. 33, 66(1943)(1):numerous pages (unnumbered).

chemistry, nitrite(data)
Soot-Ryen, T., 1938.
Hydrographical investigations in the Tromsø district in 1931.
Tromsø Mus. Aarsheft., Naturhist. Avd., No. 10, 54(2):1-6 plus 41 pp. of tables.

chemistry, nitrite
Sugiyama, Teruyuki, Yoshio Miyake and Kuniyasu Fujisawa, 1971.
On the distribution of inorganic nutrients along the coastal waters of Okayama Prefecture (1970).
Bull. Fish. Exp. Sta. Okayama Pref. (1970): 22-28. (In Japanese)

chemistry nitrite(data)
Tanaka, Otohiko, Haruhiko Irie, Shozi Iizuka and Fumihiro Koga, 1961
The fundamental investigation on the biological productivity in the north-west of Kyushu. 1. Rec. Oceanogr. Wks., Japan, Spec. No. 5: 1-58.

Chemistry nitrite
Tanita, S., K. Kato and T. Okuda, 1951.
Studies on the environmental conditions of shell-fish-fields. In the case of Muroran Harbour (2).
Bull. Fac. Fish., Hokkaido Univ., 2(3):220-230.

Chemistry, nitrite
Thomas, William H., 1966.
On denitrification in the northeastern tropical Pacific Ocean.
Deep-Sea Res., 13(6):1109-1114.

chemistry nitrite
Tomo, Aldo P., 1970
Cadenas troficas observadas en la bahia de puerto Paraiso (Peninsula Antartica) en relación con las variaciones de la fertilidad de sus aguas.
Contrib. Inst. antart. argentino 131:14pp

chemistry, nitrites
Tully, J.P., and A.J. Dodimead, 1957.
Properties of the water in the Strait of Georgia British Columbia, and influencing factors.
J. Fish. Res. Bd., Canada, 14(3):241-319.

chemistry, nitrite
Udaya, Varma Thirupad, P., and C.V. Gangadhara Reddy, 1959
Seasonal variations of the hydrological factors of the Madras coastal waters.
Indian J. Fish., 6(2):298-305.

chemistry, nitrite (data)
Uyeno, Fukuzo, 1966.
Nutrient and energy cycles in an estuarine oyster area.
J. Fish. Res. Bd., Can., 23(11):1635-1652.

Chemistry, nitrite
Uyeno, Fukuzo, 1964.
Relationships between production of foods and oceanographical condition of sea water in pearl farms. II. On the seasonal changes of sea water constituents and of bottom condition, and the effect of bottom cultivation. (In Japanese; English abstract)
J. Fac. Fish., Pref. Univ. Mie, 6(2):145-169.

Chemistry, nitrite
Uyguner, B., 1957.
Le dosage du nitrite dans les eaux du Bisphore, Dardenelles et Trebizonde. Considérations sur le production biologique du nitrite et le cycle d'azote. Hidrobiologi, Istanbul Univ. Fen Fakült. (B) 4(2/3):50-61, 6 figs.

Chemistry, nitrite
Vaccaro, Ralph F., 1965.
inorganic nitrogen in sea water.
In: Chemical Oceanography, J.P. Riley and G. Skirrow, editors, Academic Press, 1:365-408.

Chemistry, nitrite (data)
Vaccaro, Ralph F., 1963.
Available nitrogen and phosphorus and the biochemical cycle in the Atlantic off New England.
J. Mar. Res., 21(3):284-301.

chemistry nitrite
Vaccaro, Ralph F., and John H. Ryther, 1960
Marine phytoplankton and the distribution of nitrite in the sea. J. du Cons., 35(3):260-271.

Chemistry, nitrite (data)
Villanueva, Sebastian F., Alberto Gomez, Aldo Orlando y Andres J. Lusquiños 1969.
Datos y resultados de las campañas pesqueria "Pesqueria VII" 16 de febrero al 1° de marzo de 1968.
Publ. Ser. Inf. tecn., Proyecto Desarrollo Pesq. Mar del Plata (10/VII): unnumbered pp.

chemistry, nitrite
Wattenberg, H., 1937.
IV. Die chemischen Arbeiten auf der "Meteor"-Fahrt Februar bis Mai 1937. Ann. Hydr., usw., 1937:17-22, 3 figs.

chemistry, nitrite

Watts, J. C. D., 1961.
Seasonal fluctuations and distribution of nitrite in a tropical West African estuary.
Nature, 191(4791):929.

chemistry nitrite

Watts, J.C.D., 1958.
The hydrology of a tropical West African estuary
Bull. Inst. Francais, Afrique Noire, 20(3):697-752.

Won, Chong Hun, 1964. chemistry, nitrite
Tidal variations of chemical constituents of the estuary water at the Lava bed in the Nack-Dong River from Nov. 1962 to Oct. 1963. (In Korean; English abstract).
Bull. Pusan Fish. Coll., 6(1):21-32.

nitrite chemistry, nitrite

Won, Chong Hun, 1963.
Distribution of chemical constituents of the estuary water in Gwang-Yang Inlet. (In Korean; English abstract).
Bull. Fish. Coll., Pusan Nat. Univ., 5(1):1-10.

nitrite

Wooster, Warren S., Tsaihwa J. Chow and Izadore Barrett, 1965.
Nitrite distribution in Peru Current waters.
J. Mar. Res., 23(3):210-221.

chemistry, nitrite

Zuta, Salvador, y Oscar Guillen, 1970.

Oceanografia de las aguas costeras del Peru.
Bol. Inst. Mar., Peru, 2(5):157-324.

nitrite (data only)

chemistry, nitrite (data only)

Anon., 1950.
Table 1. Oceanographical observations taken in the Uchiura Bay; physical and chemical data for stations occupied by a fishing boat belonging to Mori Fishery Assoc., Res. Mar. Met., Ocean. Obs., Tokyo, July-Dec. 1947, No. 2:1-2.

chemistry nitrate (data only)

Anonymous, 1947
Discovery investigations Station List 1937-1939. Discovery Report, XXIV: 197-422. pls IV-VI.

chemistry, nitrite (data only)

Aragno, Federico, Alberto Gomez, Aldo Orlando, y Andres J. Lusquinos 1968.
Datos y resultados de las campañas pesqueria "Pesqueria II" (9 de noviembre al 12 diciembre de 1966).
Publ. (Ser. Informes tecn.) Mar del Plata, Argentina 10(2):1-129.

chemistry, nitrite (data only)

Aragno, Federico, Alberto Gomez, Aldo Orlando y Andres J. Lusquinos 1968.
Datos y resultados preliminares de las campañas pesqueria "Pesqueria I" (12 de agosto al 8 de setiembre de 1966).
Publ. (Ser. Informes tecn.) Mar del Plata, Argentina 10(1): 1-159.

nitrite (data only)

Ballester, Antonio, 1965(1967).
Tablas hidrograficas.
Memoria Soc. Cienc. nat. La Salle, 25(70/71/72):41-137.

chemistry nitrite (data only)

Brasil, Diretoria de Hidrografia e Navegacao, 1968.
XXXV Comissao Oceanografica "Operacao Norte/Nordeste 1" noe Almirante Saldanha (14/9 a 16/12/1967).
DG 26-XI: 600 pp.

chemistry nitrite (data only)

*Brazil, Diretoria de Hidrografia e Navegação, 1968.
XXXI Commissao oceanografica noc Almirante Saldanha (14/11 a 16/12/66, DG26-VIII:249 pp.

chemistry, nitrite (data only)

Brazil, Diretoria de Hidrografia e Navegação 1967.
XXVII Comissão Oceanográfica No Almirante Saldanha (14/3 A 3/5/67), DG26-X: 411 pp

chemistry nitrite (data only)

Brasil, Marinha do Brasil, Diretoria de Hidrografia e Navegação, 1963
Operação, "TRIDENTE I", Estudo das condições oceanográficas entre o Rio de Janeiro e o Rio da Prata, durante o inverno (Agosto-Septembro), ano de 1962.
DG-06-XV: unnumbered pp. (mimeographed).

chemistry nitrite (data only)

Brasil, Diretoria de Hidrografia e Navegacao, 1961
Estudo das condições oceanográficas nas proximidades do Rio de Janeiro durante o mês de Dezembro.
DG-06-XIII: mimeographed sheets.

chemistry, nitrite (data only)

Brasil, Marinha do Brasil, Diretoria de Hidrografia e Navegacao, 1960.
Estudo das condicoes oceanograficas na regiao profunda a Nornordeste de Natal, Estado do Rio Grande do Norte.
DG-06-XI (Sept. 1960): unnumbered pp. (mimeographed)

chemistry, nitrite (data only)

Brasil, Marinha do Brasil, Diretoria de Hidrografia e Navegação, 1960.
Estudo das condições oceanograficas sobre a plataforma continental, entre Cabo-Frio e Vitoria, durante o outono (abril-maio).
DG-06-X (junho): unnumbered pp. (mimeographed).

chemistry, nitrate (data only)

Brasil, Marinha do Brasil, Diretoria de Hidrografia e Navegação, 1959.
Levantamento oceanografico da costa nordeste.
DG-06-IX: unnumbered pp. (mimeographed).

chemistry, nitrate (data only)

Canada, Fisheries Research Board, 1959.
Data record, Ocean Weather Station "P" (Latitude 50 00'N, Longitude 145 00'W) January 22 - July 11, 1958.
MSS Rept. Ser. (Oceanogr. & Limnol.), No. 31: 112 pp.

chemistry nitrite (data only)

Conseil International pour l'Exploration de la Joint Skagerak Expedition 1966. 1. Oceanographic stations, temperature-salinity-oxygen content. 2. Oceanographic stations, chemical observations.
ICES Oceanogr. Data lists: 250 pp; 209 pp.

chemistry nitrite (data only)

Conseil Permanent International pour l'Exploration de la Mer, 1956.
Bulletin hydrographique pour l'année 1952:164 pp.

nitrite (data only)

Conseil Permanent International pour l'Exploration de la Mer, 1952.
Bulletin Hydrographique, 1948:87 pp.

nitrate (data only)

Conseil Permanent International pour l'Exploration de la Mer, 1936.
Bulletin Hydrographique pour l'annee 1935: 105 pp., 4 maps, 4 figs.

chemistry nitrite (data only)

Côte d'Ivoire, Centre de Recherches Océanographiques, 1963.
Résultats hydrologiques effectuées au large de la Côte d'Ivoire de 1956-1963. Première partie: observations, 1956-1963.
Trav. Centre de Recherches Océanogr., Ministere de la Production Animale, Côte d'Ivoire, unnumbered pp. (multilithed).

chemistry nitrite (data only)

Dragovich, Alexander, John H. Finucane and Billie Z. May, 1961.
Counts of red tide organisms, Gymnodinium breve, and associated oceanographic data from Florida west coast, 1957-1959.
USFWS Spec. Sci. Rept., Fish., No. 369:175 pp.

chemistry, nitrite (data only)

Great Britain, Discovery Committee, 1945.
Station list, 1935-1937.
Discovery Rept., 24:3-196.

chemistry, nitrite (data only)

Great Britain, Discovery Committee, 1957.
Station list, 1950-1951.
Discovery Rept. 28:300-398.

Nitrite (data only)

Great Britain, Discovery Committee, 1957.
Station List 1950-51. Discovery Repts. 28:299-398.

chemistry, nitrite (data only)

Great Britain, Discovery Committee, 1947.
Station list, 1937-1939. Discovery Repts., 24: 198-422, Pls. 4-6.

chemistry, nitrite (data only)

Great Britain, Discovery Committee, 1945.
Station list, 1935-1937. Discovery Repts., 24: 3-196, Pls. 1-3.

chemistry, nitrite (data only)

Great Britain, Discovery Committee, 1942.
Station list, 1933-1935. Discovery Repts., 22:3-196, Pls. 1-4.

chemistry, nitrite (data only)

Great Britain, Discovery Committee, 1932.
Station list. Discovery Repts. 4:3-230, Pls. 1-5.

chemistry, nitrite (data only)
Great Britain, Discovery Committee, 1931.
Station list. Discovery Repts. 3:3-132, Pls. 1-10.

chemistry, nitrite (data only)
Great Britain, Discovery Committee, 1929.
Station list. Discovery Repts. 1:3-140.

chemistry nitrite (data only)
Hapgood, William, 1959.
Hydrographic observations in the Bay of Naples (Golfo di Napoli), January 1957-January 1958 (station lists).
Pubbl. Sta. Zool., Napoli, 31(2):337-371.

Chemistry nitrite (data only)
Hela, Ilmo, and Folke Koroleff, 1958
Hydrographical and chemical data collected in 1957 on board the R/V Aranda in the Barents Sea.
Merent. Julk., No. 179: 67 pp.

chemistry, nitrite (data only)
Hela, Ilmo, and F. Koroleff, 1958.
Hydrographical and chemical data collected in 1956 on board the R/V "Aranda" in the Baltic Sea.
Merent. Julk., No. 183:1-52.

chemistry, nitrite (data only)
India, Naval Headquarters, New Delhi, 1958.
Indian oceanographic station list, Ser. No. 2: unnumbered pp. (mimeographed).

chemistry, nitrite (data only)
India, Naval Headquarters, Office of Scientific Research & Development, 1957.
Indian Oceanographic Stations List, No. 1/57: unnumbered pp. (mimeographed).

chemistry, nitrite (data only)
Japan, Japan Meteorological Agency, 1964.
The results of marine meteorological and oceanographical observations, January-June 1963, No. 33:289 pp.

Chemistry nitrite (data only)
Japan, Japan Meterrology Agency, 1959.
The results of marine meteorological and oceanographical observations, January-June 1958, No. 23:240 pp.

chemistry, nitrite (data only)
Japan, Central Meteorological Observatory, 1952
The results of Marine Meteorological and oceanographical observations, July - Dec. 1951, No. 10:310 pp., 1 fig.

Chemistry, nitrite (data only)
Japan, Central Meteorological Observatory, 1952.
The Results of Marine Meteorological and oceanographical observations. Jan. - June 1950. No. 7: 220 pp.

chemistry, nitrite (data only)
Japan, Central Meteorological Observatory, 1951.
Table 6. Oceanographical observations taken in the Akashi-seto; the Yura-seto and the Kii Suido; physical and chemical data for stations occupied by R.M.S. "Shumpu-maru".
Res. Mar. Met. Ocean. Obs., Jan.-June 1949, No. 5:40-47.

Japan nitrite (data only)
Central Meteorological Observatory, Japan, 1951.
The results of marine meteorological and oceanographical observations, July-Dec. 1948, No. 4: vi+414 pp., of 57 tables, 1 map, 1 photo.

Chemistry, nitrite (data only)
Japan, Hokkaido University, Faculty of Fisheries 1970
Data record of oceanographic observations and exploratory fishing 13:406 pp.

Chemistry nitrite (data only)
Japan, Hokkaido University, Faculty of Fisheries, 1968.
The Oshoro Maru cruise 23 to the east of Cape Erimo, Hokkaido, April 1967.
Data Record Oceanogr. Obs. Expl. Fish., 12: 115-169.

Chemistry, nitrite (data only)
Japan, Hokkaido University, Faculty of Fisheries, 1968.
The Oshoro Maru cruise 21 to the Southern Sea of Japan, January 1967.
Data Record Oceanogr. Obs. Explor. Fish. 12: 1-97; 113-119.

Chemistry nitrite (data only)
Japan, Hokkaido University, Faculty of Fisheries, 1968.
Data record of oceanographic observations and exploratory fishing. No. 12:420 pp.

Chemistry, nitrite (data only)
Japan, Hokkaido University, Faculty of Fisheries, 1967.
Data record of oceanographic observations and exploratory fishing, 11:383 pp.

Chemistry nitrite (data only)
Japan, Hokkaido University, Faculty of Fisheries, 1967.
The Oshoro Maru cruise 18 to the east of Cape Erimo, Hokkaido, April 1966.
Data Record Oceanogr. Obs. Explor. Fish. 11: 121-164.

Chemistry nitrate (data only)
Japan, Hokkaido University, Faculty of Fisheries, 1967.
The Oshoro Maru cruise 16 to the Great Australian Bight November 1965-February 1966.
Data Record Oceanogr. Obs. Explor. Fish., Fac. Fish., Hokkaido Univ. 11: 1-97; 113-119.

Chemistry nitrite (data only)
Japan, Hokkaido University, Faculty of Fisheries, 1967.
The Oshoro Maru cruise 16 to the Great Australian Bight November 1965-February 1966.
Data Record Oceanogr. Obs. Explor. Fish., Fac. Fish., Hokkaido Univ. 11: 1-97; 113-119.

Chemistry nitrite (data only)
Japan, Japan Meteorological Agency, 1971.
The results of marine meteorological and oceanographical observations, 45: 338 pp.

Chemistry nitrite (data only)
Japan, Japanese Oceanographic Data Center, 1971.
Shumpu Maru, Kobe Marine Observatory, Japan Meteorological Agency, 16 July - 8 August 1970, south of Japan. Data Rept. CSK (KDC Ref. 49K127) 279: 7 pp. (multilithed).

chemistry, nitrite (data only)
Japan, Japanese Oceanographic Data Center, 1970.
Shumpu Maru, Kobe Marine Observatory, Japan Meteorological Agency, Feb. 21 - Mar. 11, 1970, South of Japan. Data Rept. CSK (KDC Ref. 49K122) 266: 4 pp. (multilithed).

Chemistry nitrite (data only)
Japan, Japanese Oceanographic Data Center, 1970
Orlick, USSR, Jan. 27 - Mar. 6, 1969, Northwest of the North Pacific Ocean. Data Rept. CSK (KDC Ref. 90K023) 227: 30 pp. (multilithed).

Chemistry - nitrite (data only)
Japan, Japanese Oceanographic Data Center, 1970.
Han Ra San, Fisheries Research and Development Agency, Republic of Korea, August 9 - 21, 1969, East of the Yellow Sea. Data Rept. CSK (KDC Ref. 24K033) 238: 17 pp. (multilithed).

Chemistry nitrite (data only)
Japan, Japanese Oceanographic Data Center, 1969,
Ryofu Maru, Marine Division, Japan Meteorological Agency, Japan. October 7 - November 9, 1968. East China Sea. Data Rept. CSK (KDC Ref. No. 49K449) 217: 39 pp. (multilithed).

Chemistry - Nitrite (data only)
Japan, Japanese Oceanographic Data Center, 1969
Cape St. Mary, Fisheries Research Station, UK (Hong Kong). May 16 - 23, 1968, South China Sea. Prelim. Data Rept. CSK (KDC Ref. No. 74K010) 203: 15 pp. (multilithed).

Chemistry, nitrite (data only)
Japan, Japanese Oceanographic Data Center 1969.
Iskatel, USSR, January 26 - March 11, 1968, northwest North Pacific Ocean.
Data Rept. CSK (KDC Ref 90K019) 187: 32 pp. (multilithed)

Chemistry nitrite (data only)
Japan, Japanese Oceanographic Data Center, 1969.
Shumpu Maru, Kobe Marine Observatory, Japan Meteorological Agency, July 19-31, 1968, South of Japan.
Data Rept. CSK (KDC Ref. 49K085) 179: 9 pp.

Chemistry nitrite (data only)
Japan, Japanese Oceanographic Data Center, 1969
Takuyo, Hydrographic Division, Maritime Safety Agency, Japan. July 19-September 6, 1968, Kuroshio Extension Area. Prelim. Data Rept. CSK (KDC Ref. No. 49K082) 176: 27 pp. (multilithed).

Chemistry, nitrite (data only)
Japan, Japanese Oceanographic Data Center 1969.
Hakuho Maru, Ocean Research Institute, University of Tokyo, Japan, May 15-June 8, 1968 southwest of Japan and the East China Sea.
Data Rept. CSK (KDC Ref. 49K078) 172: 7pp. (multilithed).

Chemistry nitrite (data only)
Japan, Japanese Oceanographic Data Center, 1969
U.M. Schokalsky, USSR, April 28 - June 3, 1967, Northwest of the North Pacific Ocean. Prelim. Data Rept. CSK (KDC Ref. 90K014) 121: 41 pp. (multilithed).

Chemistry - nitrite (data only)
Japan, Japanese Oceanographic Data Center, 1968.
Tae Baek San. Fisheries Research and Development Agency, Republic of Korea, February 16 - March 12, 1968, South of Korea.
Data Rept. CSK (KDC Ref. 24K022) 163: 12 pp. (multilithed)

Chemistry - nitrite (data only)
Japan, Japanese Oceanographic Data Center, 1968.
Han ra San, Fisheries Research and Development Agency, Republic of Korea, February 15- March 12, 1968, East of the Yellow Sea.
Data Rept. CSK (KDC Ref. 24K021) 162: 18 pp. (multilithed)

Chemistry- nitrite (data only)
Japan, Japanese Oceanographic Data Center, 1968.
Baek du San. Fisheries Research and Development Agency, Republic of Korea, February 17-March 12, 1968, West of the Japan Sea.
Data Rept. CSK (KDC Ref. 24K020) 161:23 pp. (multilithed)

Chemistry nitrite (data only)
Japan, Japanese Oceanographic Data Center, 1968.
Shumpu Maru, Kobe Marine Observatory, Japan Meteorological Agency, Japan, February 17-28, 1968, South of Japan.
Data Rept. CSK (KDC Ref. 49K071) 157: 6 pp. (multilithed).

Chemistry, nitrite (data only)
Japan, Japanese Oceanographic Data Center, 1968.
Ryofu Maru, Marine Division, Japan Meteorological Agency, Japan, January 13-March 22, 1968, West of North Pacific Ocean.
Data Rept. CSK (KDC Ref. 49K069) 155: 70 pp. (multilithed).

Chemistry, nitrite (data only)
Japan, Japanese Oceanographic Data Center, 1968.
Suro No. 1, Hydrographic Office, Korea, September 23-October 21, 1967, South of Korea.
Data Rept. CSK (KDC Ref. 24K019) 136: 13 pp. (multilithed).

Chemistry, nitrite (data only)
Japan, Japanese Oceanographic Data Center, 1968.
Nagasaki Maru, Faculty of Fisheries, Nagasaki University of Fisheries, Japan, November 7-29, 1967, East China Sea.
Data Rept. CSK (KDC Ref. 49K076) 133: 5 pp. (multilithed).

Chemistry - nitrite (data only)
Japan, Japanese Oceanographic Data Center, 1968.
Han Ra San, Fisheries Research and Development Agency, Korea, August 12-September 4, 1967, North of East China Sea.
Prelim. Data Rept. CSK (KDC Ref. 24K018) 117:12 pp. (multilithed).

Chemistry, nitrite (data only)
Japan, Japanese Oceanographic Data Center, 1968.
Chun Ma San, Fisheries Research and Development Agency, Korea, August 13-21, 1967, East of Yellow Sea.
Prelim. Data Rept. CSK (KDC Ref. 24K017) 116:16pp. (multilithed).

Chemistry nitrite (data only)
Japan, Japanese Oceanographic Data Center, 1968.
Baek Du San, Fisheries Research and Development Agency, Korea, August 12-25, 1967, West of Japan Sea.
Data Rept. CSK (KDC Ref. 24K016) 115: 26 pp. (multilithed)

Chemistry, nitrite (data only)
Japan, Japanese Oceanographic Data Center, 1968.
Takuyo, Hydrographic Division, Maritime Safety Agency, July 12-August 30, 1967, Central Part of the North Pacific Ocean.
Prelim. Data Rept. CSK (KDC Ref. 49K053) 105:25pp. (multilithed)

Chemistry nitrite (data only)
Japan, Japanese Oceanographic Data Center, 1968.
Yang Ming, Chinese National Committee on Ocean Research, Republic of China, April 1-May 14, 1967, Surrounding waters of Taiwan.
Data Rept. CSK (KDC Ref. 21K004) 102: 15 pp. (multilithed).

Chemistry nitrite (data only)
Japan, Japanese Oceanographic Data Center, 1968.
Orlick, USSR, February 5-March 11, 1967, Northwest of North Pacific Ocean.
Prelim. Data Rept. CSK (KDC Ref. 90K012) 94:29pp.

Chemistry nitrite (data only)
Japan, Japanese Oceanographic Data Center, 1967.
Bukhansan, Fisheries Research and Development Agency, Korea, July 16- August 9, 1966, East [of the] Yellow Sea.
Prelim. Data Rept., CSK (KDC Ref. 24K010) 69:17 pp.

Chemistry, nitrite (data only)
Japan, Japanese, Oceanographic Data Center, 1967
Baekdusan, Fisheries Research and Development Agency, Korea, July 14-28, 1966, West [of the] Japan Sea.
Prelim. Data Rept. CSK (KDC Ref. 24K009) 65:22 pp.

Chemistry, nitrite (data only)
Japan, Japanese Oceanographic Data Center, 1967.
Yang Ming, Chinese National Committee on Oceanic Research, Republic of China, September 10-October 14, 1966, Adjacent Sea of Taiwan.
Prelim. Data Rept. CSK (KDC Ref. 21K003) 67:17pp.

Chemistry, nitrite (data only)
Japan, Japanese Oceanographic Data Center 1967.
Ryofu Maru, Marine Division, Japan Meteorological Agency, Japan, September 13-17, 1966, eastern Sea of Japan.
Prelim. Data Rept. CSK (KDC Ref. 49K033) 59: 11 pp. (multilithed).

Chemistry, nitrite (data only)
Japan, Japanese Oceanographic Data Center 1967.
Nagasaki Maru, Faculty of Fisheries Nagasaki University, Japan, June 16-21, 1966, southern Sea of Japan.
Prelim. Data Rept. CSK (KDC Ref No. 49K032) 58: 8pp. (multilithed).

Chemistry nitrite (data only)
Japan, Japanese Oceanographic Data Center, 1966.
Ryofu Maru, Japan Meteorological Agency, Japan, February 4-February 28, 1966, East of Japan.
Prelim. Data Rept., CSK (KDC Ref. 49(K017), 28: 17 pp. (multilithed).

Chemistry - nitrite (data only)
Japan, Japanese Oceanographic Data Center, 1966.
Yang Ming, Chinese National Committee on Oceanic Research, Republic of China, August 10-October 13, 1965, Adjacent Sea of Taiwan.
Prelim. Data Rept., CSK (KDC Ref. 21K001), 22: 18 pp. (multilithed).

Chemistry nitrite (data only)
Japan, Japan Meteorological Agency, 1970.
The results of marine meteorological and oceanographical observations, July-December, 1968, 44:311 pp.

Chemistry, Nitrite (data only)
Japan, Japan Meteorological Agency, 1970
The results of marine meteorological and oceanographical observations. (The results of the Japaneses Expedition of Deep Sea (JEDS-11); January-June 1967 41: 332 pp.

Chemistry, nitrite (data only)
Japan, Japan Meteorological Agency 1970.
The results of marine meteorological and oceanographical observations, July-December 1966, 40: 336 pp.

Chemistry, nitrite (data only)
Japan, Japan Meteorological agency 1969.
The results of marine meteorological and oceanographical observations, Jan.-June 1966, 39: 349 pp.

Chemistry, nitrite (data only)
Japan, Maritime Safety Agency 1967.
Results of oceanographic observations in 1965.
Data Rept. Hydrogr. Obs. Ser. Oceanogr. (Publ. 792) (Oct. 1967) 5: 115pp.

Chemistry, nitrite (data only)
Japan, Maritime Safety Agency 1967.
Results of oceanographic observations in 1964.
Data Rept. Hydrogr. Obs. Ser. Oceanogr. Publ. 792 (4): 88pp.

Chemistry nitrite (data only)
Japan, Japan Meteorological Agency, 1964.
The results of the Japanese Expedition of Deep Sea (JEDS-5).
Res. Mar. Meteorol. and Oceanogr. Obs., July-Dec. 1962, No. 32:328 pp.

Chemistry nitrite (data only)
Japan, Japan Meteorological Agency, 1962
The results of marine meteorological and oceanographical observations, July-December 1961, No. 30:326 pp.

Chemistry nitrite (data only)
Japan, Japan Meteorological Agency, 1961
The results of marine meteorological and oceanographical observations, January-June 1960. The results of the Japanese Expedition of Deep-Sea (JEDS-2, JEDS-3), No. 27: 257 pp.

Chemistry nitrite (data only)
Japan, Japan Meteorological Agency, 1959.
The results of marine meteorological and oceanographical observations, July-December, 1958, No. 24:289 pp.

Chemistry, nitrite (data only)
Japan, Maritime Safety Agency 1966.
Results of oceanographic observations in 1963.
Data Rept. Hydrogr. Obs. Ser. Oceanogr., Publ. 792(3): 74 pp.

Chemistry nitrite (data only)
Japan, Maritime Safety Board, 1961
Tables of results from oceanographic observation in 1959.
Hydrogr. Bull. (Publ. No. 981), No. 68: 112 pp.

chemistry, nitrite (data only)
Japan, Maritime Safety Board, 1961
Tables of results from oceanographic observations in 1958.
Hydrogr. Bull., Tokyo, No. 66 (Publ. No. 981): 153 pp.

Chemistry nitrite (data only)
Japan, Tokyo University of Fisheries, 1966.
J. Tokyo Univ. Fish. (Spec.ed.)8(3):1-44.

chemistry, nitrite (data only)
Jarrige, F., J. Merle et J. Noel, 1968.
Résultats des observations physico-chimiques de la croisière CYCLONE 4 du N.O. Coriolis.
Rapp. sci., O.R.S.T.O.M., Noumea, 25:22 pp.

Chemistry nitrite (data only)
Katsuura, Hiroshi, Hideo Akamatsu, Tsutomu Akiyama, 1964.
Preliminary report of the Japanese Expedition of Deep-Sea, the Eighth Cruise (JEDS-8).
Oceanogr. Mag. Tokyo, 16(1/2):125-136.

nitrite (data only)
Korea, Republic of, Fisheries Research and Development Agency, 1971
Annual report of oceanographic observations, 19:717 pp.

chemistry nitrite (data only)
Korea, Fisheries Research and Development Agency, 1968.
Annual report of oceanographic observations, 16:691 pp.

Chemistry nitrite (data only)
Korea, Fisheries Research and Development Agency, 1967.
Annual report of oceanographic observations, 15 (1966):459 pp.

nitrate (data only)
Krauel, David P., 1969
Bedford Basin data report, 1967.
Techn. Rept. Fish. Res. Bd., Can., 120:84 pp (multilithed).

Chemistry Nitrite (data only)
Magnier, Y., H. Rotschi, B. Voituriez et J. Merle 1968.
Résultats des observations physico-chimiques de la croisière CYCLONE 3 du N.O. Coriolis.
Rapp. sci., O.R.S.T.O.M., Nouméa, 21:23 pp. (mimeographed)

Chemistry, nitrite (data only)
Oren, Oton Haïm, 1967.
Croisière Chypre-04 en Méditerranée orientale, février-mars, 1965: résultats des observations hydrologiques.
Cah. océanogr., 19(9):783-798.

Chemistry, nitrite (data only)
Oren, O.H., 1967.
Croisière Chypre-04 dans la Méditerranée orientale, février-mars 1965. Résultats des observations hydrographiques.
Bull. Sea Fish. Res. Stn Israel, 47:37-54.

Chemistry nitrite (data only)
Oren, Oton Haïm, 1966.
Croisière "Chypre-02" en Méditerranée orientale, juillet-août 1963. Résultats des observations hydrologiques.
Cah. océanogr., 18(Suppl.):1-17.

chemistry, nitrite (data only)
Piton, B., et Y. Magnier 1971.
Observations physico-chimiques faites par le Vauban le long de la côte nord-ouest de Madagascar de janvier à septembre 1970.
Doc. scient. Cent. Nosy-Bé, Off. Rech. scient. techn. Outre-Mer 21: unnumbered pp. (mimeographed)

Chemistry, nitrite (data only)
Republic of China, Chinese National Committee on Oceanic Research, Academia Sinica, 1968.
Oceanographic Data Report of CSK, 2: 126 pp.

Chemistry, nitrite (data only)
Republic of China, Chinese National Committee on Oceanic Research, Academia Sinica 1966.
Oceanographic Data Rept. CSK 1: 123 pp.

Chemistry nitrite (data only)
Rotschi, H., Ph. Hisard, L. Lemasson, Y. Magnier, J. Noel et B. Piton, 1967.
Resultats des observations physico-chimiques de la croisiere "Alize" du N.O. Coriolis.
Rapp. scient. Off. Rech. scient. tech. Outre-Mer, Noumea, 2:56 pp. (mimeographed)

Chemistry nitrite (data only)
Rotschi, H., Ph. Hisard, L. Lemasson, Y. Magnier, J. Noel, et B. Piton, 1966.
Resultats des observations physico-chimiques de la croisiere "Alize".
Centre ORSTROM Noumea Rapp. Sci., No. 28:56 pp. (mimeographed)

chemistry, nitrite (data only)
Rotschi, H., G. Pickard, P. Hisard et J. Cannevet, 1968.
Résultats des observations physico-chimiques de la croisière CYCLONE 5 du N.O. Coriolis.
Rapp. sci., O.R.S.T.O.M., Noumea, 26:23 pp.

Chemistry nitrite (data only)
* Scaccini Cicatelli, Marta, 1967.
Distribuzione stagionale dei sali nutrivi in una zone dell' alto e medio Adriatico.
Boll. Pesca, Piscic. Idrobiol., n.s., 22(1): 49-82.

Chemistry nitrite (data only)
Sweden, Havsfiskelaboratoriet, Lysekil 1971.
Hydrographical data January-June 1970, R.V. Skagerak, R.V. Thetis and TV252, 1970.
Meddn. Hydrogr. avd. Göteborg, 104: unnumbered pp. (multilithed).

Chemistry nitrite (data only)
Sweden, Havsfiskelaboratoriet, Lysekil 1970.
Hydrographical data 1966, R.V. Skagerak, R.V. Thetis.
Meddn Hydrogr. avd. Göteborg 85:255 pp.

Chemistry nitrite (data only)
Sweden, Havsfiskelaboratoriet, Lysekil 1970.
Hydrographical data, R.V. Skagerak, R.V. Thetis.
Meddn. Hydrogr. avd. Göteborg 84: 296 pp. (multilithed).

Chemistry nitrite (data only)
Tabata, S., C.D. McAllister, R.L. Johnston, D.G. Robertson, J.H. Meikle, and H.J. Hollister, 1961.
Data record. Ocean Weather station "P" (Latitude 50 00'N, Longitude 145 00"W), December 9, 1959 to January 19, 1961.
Fish. Res. Bd., Canada, MSS. Rept. Ser. (Ocean & Limnol.), No. 98:296 pp. (Multilithed).

Chemistry, nitrite (data only)
United States, National Oceanographic Data Center 1965.
Data report EQUALANT III.
Publ. Nat. Oceanogr. Data Center U.S.A. Gen. Ser. G-7:339 pp.

nitrite, effect of

Chemistry nitrite, effect of
Prochazkova, L., 1960.
Einfluss der Nitrate und Nitrite auf die Bestimmung des organischen Stickstoffs und Ammoniums im Wasser.
Arch. f. Hydrobiol., 56(3):179-185.

nitrite, inorganic

Chemistry nitrite, inorganic
Richards, Francis A., Joel D. Cline, William W. Broenkow and Larry P. Atkinson, 1965.
Some consequences of the decomposition of organic matter in Lake Nitinat, an anoxic fjord.
Limnol. and Oceanogr., Redfield Vol., Suppl. to 10:R185-R201.

nitrite maximum

nitrite maximum
*Fiadeiro, Manuel, and J.D.H. Strickland, 1968.
Nitrate reduction and the occurrence of a deep nitrite maximum in the ocean off the west coast of South America.
J. mar. Res., 26(3):187-201.

Chemistry, nitrite maximum (secondary)
Wooster, Warren S. 1967.
Further observations on the secondary nitrite maximum in the northern equatorial Pacific.
J. mar. Res. 25(2): 154-161.

chemistry, nitrite metabolism

Wada, Eitaro and Akihiko Hattori, 1971. Nitrite metabolism in the euphotic layer of the central North Pacific Ocean. Limnol. Oceanogr. 16(5): 766-772.

Chemistry, nitrite, monthly means

Slinn, D.J., and Gaenor Offlow (Mrs.M.Solly), 1968. Chemical constituents of sea water off Port Erin during 1967. Ann.Rept.mar.biol.Stn.Port Erin, 80:37-42.

nitrite/oxygen A.O.U.

Chemistry, nitrite/oxygen A.O.U.

Kitamura, Hiroyuki, and Takeshi Sagi, 1965. On the chemical elements in the sea south of Honshu, Japan. II. (In Japanese; English abstract) Bull. Kobe Mar. Obs., No. 174:39-55.

Chemistry nitrate/phosphate

*Stefánsson, U., 1968. Nitrate-phosphate relationships in the Irminger Sea. J. Cons.perm.int.Explor.Mer, 32(2):188-200.

N_2-element

chemistry nitrogen

Anon., 1950. The report of the oceanographical observations in the Bungo Suido and off Ashizuri-Misaki in August 1950. J. Ocean., Kobe Obs., 2nd ser., 1(1):15-44, 7 figs., tables.

chemistry, nitrogen

Argentina, Servicio de Hidrografia Naval, 1959. Quimica d l agua del mar. Servicio de Hidrografia Naval, Argentina, Publ., H. 604:140 pp.

Chemistry, nitrogen

Armstrong, F.A.J., and E.I. Butler, 1960. Chemical changes in sea water off Plymouth during 1958. J.M.B.A., 39(2):299-302.

Chemistry, nitrogen

Beers, John R., 1966. Studies on the chemical composition of the major zooplankton groups in the Sargasso Sea. Limnol. Oceanogr., 11(4):520-528.

chemistry nitrogen

Berthois, Leopold, 1966. Hydrologie et sedimentologie dans le Kangerd-lugssuaq (fjord a la cote ouest du Groenland). C.r. hebd. seanc., Acad. Sci., Paris, (D), 262(13): 1400-1402.

chemistry nitrogen

Bruevich, S.W., editor, 1966. Chemistry of the Pacific Ocean. Inst. Okeanol., Akad. Nauk, SSSR, Isdatel. Nauka, Moskva, 358pp.

Chemistry Nitrogen

Cooper, L.H.N., 1948. The nutrient balance in the sea. Research 1: 242-248, 2 textfigs.

Chemistry, nitrogen

Cooper, L.H.N., 1939. Phosphorus, nitrogen, iron, and maganese in marine zooplankton. J.M.B.A. 23:387-390.

chemistry, nitrogen

Cooper, L. H. N., 1937. The nitrogen cycle in the sea. J.M.B.A., XXII:183-204.

chemistry, nitrogen

Cooper, L. H. N., 1934. The determination of phosphorus and nitrogen in plankton. JMBA, XIX:755-759

chemistry, nitrogen

*Craig, H., R.F.Weiss and W.B. Clarke, 1967. Dissolved gases in the equatorial and South Pacific Ocean. J. geophys. Res., 72(24):6165-6181.

chemistry nitrogen

Cushing, D.H., and H.F. Nicolson, 1963. Studies on a Calanus patch. IV. Nutrient salts off the north-east coast of England in the spring of 1954. J. Mar. Biol. Assoc., U.K., 43(2):373-386.

Chemistry, nitrogen

Dragovich, Alexander, and James E. Sykes 1967. Oceanographic atlas for Tampa, Florida, and adjacent waters of the Gulf of Mexico, 1958-61. Circular, Fish Wildl. Serv. Bur. Comm. Fish. 255: 466pp. (quarto)

chemistry, nitrogen

Duursma, E.K., 1960

Dissolved organic carbon, nitrogen and phosphorus in the sea. Netherlands J. Mar. Res. 1(1): 1-148.

chemistry nitrogen

Duursma, E.K., 1959 Über gelöste organische Substanz auf einem Meridionalschnitt durch den nördlichen Nordatlantischen Ozean im April 1958. Deutsche Hydrogr. Zeits., Ergänzungsheft, Reihe B, 4(3):69-72.

chemistry nitrogen

Flemer, David A. and Robert B. Biggs 1971. Particulate carbon: nitrogen relations in northern Chesapeake Bay. J. Fish. Res. Bd, Can, 28(6): 911-918.

chemistry, nitrogen

Fox, C.J.J., 1909. On the coefficients of absorption of nitrogen and oxygen in distilled water and sea water, and of atmospheric carbonic acid in sea water. Trans. Faraday Soc., 5:68-87.

Chemistry, nitrogen

Føyn, E., 1951. Nitrogen determinations in sea water. Repts. Norwegian Fish. & Mar. Invest. 9(14):7 pp., 3 textfigs.

chemistry nitrogen

Fraga, F., 1966. Distribution of particulate and dissolved nitrogen in the Western Indian Ocean. Deep-Sea Research, 13(3):413-425.

abstract
Indian Ocean
nitrogen, particulate
nitrogen, dissolved

chemistry nitrogen

Fraga, F., 1959.

Determinación de nitrógeno orgánico suspendido y disuelto en el agua del mar. Inv. Pesq., 14:121-127.

Chemistry, nitrogen

Fraga, F., y A. Ballester 1966. Distribución vertical del nitrógeno orgánico e inorgánico en la fosa de Cariaco y su relación con el fósforo total. Memoria Soc. Cienc. nat. La Salle 26 (75): 274-282.

chemistry, nitrogen

Gomes, A.L., 1955. Possibilidades da pesca maritima no Brasil. Rev. Biol. Mar. Valparaiso, 6(1-3):6-20.

chemistry, nitrogen

Gilson, H. C., 1937 Chemical and Physical Investigations. The nitrogen cycle. John Murray Exped., 1933-34, Sci. Repts., 2(2):21-81, 16 text figs.

Chemistry, nitrogen

Hamm, R. E., and T.G. Thompson, 1941. Dissolved nitrogen in the sea water of the northwest Pacific with notes on the total carbon dioxide and dissolved oxygen. J. Mar. Res 4(1):11-27.

Chemistry, nitrogen

Harvey, H.W., 1957. Bio-assay of nitrogen available to two species of phytoplankton in an offshore water. J.M.B.A., 36(1):157-160.

chemistry nitrogen

Heather, R. C., & R. F. Rackham, 1959.

Oxidised nitrogen in waters and sewage effluents observed by ultra-violet spectrophotometry. Analyst, 84(1002):548-551.

chemistry nitrogen

Klots, C.E., and B.B Benson, 1963 Solubilities of nitrogen, oxygen and argon in distilled water. J. Mar. Res., 21(1):48-57.

chemistry nitrogen

Koyama, Tadashiro, 1958. A new method of direct determination of argon and nitrogen. J. Earth Sci., Nagoya Univ., 6(1):1-11.

Also in: Collected Papers on Sciences of Atmosphere and Hydrosphere, 1953-1963, 1 (1964).

Chemistry, nitrogen

Krey, J., 1942. Nährstoff und Chlorophylluntersuchungen in der Kieler Förde 1939. Kieler Meeresforsch. 4:1-17, textfigs.

chemistry, nitrogen

Knudsen, M., 1903. Ueber den Gebrauch von Stickstoffbestimmungen in der Hydrographie. Publ. de Circ. (Cons. Perm. Int. Expl. Mer.) No. 4:1-9.

Chemistry, nitrogen

Kriss, A.E., 1949.
The role of microorganisms in the accumulation of hydrogen sulphide, ammonia and nitrogen in the depths of the North Sea. Priroda, 1949(6):35-46.

Chemistry, nitrogen

Lagarde, E., 1963.
Métabolisme de l'azote minéral en milieu marin. Vie et Milieu, 14(1):37-54.

Chemistry, nitrogen (data)

Lee, Min Jai, Jae Hyung Shim and Chong Kyun Kim 1967.
Studies on the plankton of the neighboring seas of Korea. 1. On the marine conditions and phytoplankton of the Yellow Sea in summer. Repts Inst. mar. Biol. Seoul Nat. Univ. 1(6): 1-14.

Chemistry, nitrogen

Megia, T.G., and R.G. Lao, 1955.
A report on the O:N:P ratios of Philippine and adjacent waters. Philippine J. Fish., 3(1):55-61, 3 textfigs.

Chemistry nitrogen

Miyake Yasuo and Katsuko Saruhashi, 1967.
The geochemical balance of nutrient matter in the ocean.
Pap. Met. Geophys., Tokyo, 15(?): 89-94.

chemistry, nitrogen

Miyake, Yasuo and Eitaro Wada, 1971.
The isotope effect on the nitrogen in biochemical, oxidation-reduction reactions. Rec. oceanogr. Wks, Japan n.s. 11(1): 1-6.

Chemistry, nitrogen

Nümann, W., 1949.
Kolorimetrische Bestimmung von Silicat, Phosphor und Stickstoff im Meerwasser. Deutsch. Hydro. Zeitschr. 2(4):137-153.

chemistry nitrogen (gas)

Nutt, D.C., L.K. Coachman and P.F. Scholander, 1961.
Dissolved nitrogen in West Greenland waters. J. Mar. Res., 19(1):6-11.

Chemistry, nitrogen

*Okuda, Taizo, 1967.
Vertical distribution of inorganic nitrogen in the equatorial Atlantic Ocean. Boln Inst. Oceanogr., Univ. Oriente, 5(½):67-83.

Chemistry, nitrogen

Okuda, Taizo, 1962
Physical and chemical oceanography over continental shelf between Cabo Frio and Vitoria (Central Brazil). J. Oceanogr. Soc., Japan, 20th Ann. Vol., 514-540.

Chemistry nitrogen, inorganic

Okuda, Taizo, José Benítez and Esther Fernández A. 1969.
Vertical distribution of inorganic and organic nitrogen in the Cariaco Trench.
Bol. Inst. oceanogr. Univ. Oriente 8(1/2): 28-34.

Chemistry, nitrogen

Okuda, Taizo, Lourinaldo Cavalcanti, and Manoel Pereira Borba, 1960
Estudo da Barra das Jangadas. 3. Variação de nitrogenio e fosfato durante o ano. (In Portuguese; English and French abstracts) Trabalhos, Inst. Biol. Marit. e Oceanogr., Universidade do Recife, Brasil, 2(1):207-218.

Abstract p. 217-218

Chemistry, nitrogen

Okuda, Taizo, Angel José García, José Benítoz Alvarez, 1965.
Variacion estacional de las elementos nutritions en el ague de la laguna y el rio Unare. Bol. Inst. Oceanogr., Univ. Oriente, 4(1): 123-135.

Chemistry nitrogen

Parsons, T.R., K. Stephens and J.D.H. Strickland, 1961
On the chemical composition of eleven species of marine phytoplankters. J. Fish. Res. Bd., Canada, 18(6):1001-1016.

Chemistry nitrogen

Rakestraw, N. W., and V.M. Emmel, 1938.
The solubility of nitrogen and argon in sea water. J. Phy. Chem. 42(9):1211-1215, 2 textfigs.

Chemistry, nitrogen

Rakestraw, N. W., and V.M. Emmel, 1938.
The relation of dissolved oxygen to nitrogen in some Atlantic waters. J. Mar. Res. 1(3):207-216, Text-figs. 58-81.

Chemistry nitrogen

Rakestraw, N. W., and V.M. Emmel, 1937.
The determination of dissolved nitrogen in water. Ind. and Eng. Chem., Anal. Ed., 9:344-346.

Chemistry, nitrogen

Rakestraw, N. W. and V. M. Emmel, 1937
The determination of dissolved nitrogen in water. Ind. and Eng. Chem. (Anal. Ed.), 9: 344-346

Chemistry nitrogen

Revelle, R., and H.E. Suess, 1962
Gases, Ch. 7, Sect. II, Interchange of properties between sea and air. In: The Sea, Interscience Publishers, Vol. 1, Physical Oceanography, 313-321.

Chemistry nitrogen

Richards, F. A., 1965.
Dissolved gases other than carbon dioxide. In: Chemical oceanography, J.P. Riley and G. Skirrow, editors, Academic Press, 1:197-225.

Chemistry, nitrogen

Richards, Francis A., Joel D. Cline, William W. Broenkow and Larry P. Atkinson, 1965.
Some consequences of the decomposition of organic matter in Lake Nitinat, an anoxic fjord. Limnol. and Oceanogr., Redfield Vol., Suppl. to. 10:R185-R201.

chemistry nitrogen

Riley, J.P., 1965.
Analytical chemistry of sea water. In: Chemical oceanography, J.P. Riley and G. Skirrow, editors, Academic Press, 2:295-424.

chemistry nitrogen

Riley, J.P., and P. Sinhaseni, 1957.
The determination of ammonia and total inorganic nitrogen in sea water. J.M.B.A., 36(1):161-168.

chemistry nitrogen

Sen Gupta, R. 1969
Biochemical relationships and inorganic nitrogen equilibrium in semi-enclosed basins.
Tellus 21(2): 270-281

nitrogen chemistry, nitrogen

Skopintsev, B.A., 1960.
Organic matter in waters of the sea. Marine chemistry, hydrology, oceanic geology. (In Russian). Trudy Morsk. Gidrofiz. Inst., 19:3-20.

chemistry, nitrogen

Skopintsev, B.A., 1948.
The change in nitrogen and phosphorus content in terrigenous suspended particles in a water medium. Izvest. Akad. Nauk SSSR, Ser. Geogr. Geofiz. 12:107-118.

Chemistry nitrogen

Skopintsev, B.A., S.N. Timofeeva, A.F. Danilenko and M.V. Sokolova, 1967.
Organic carbon, nitrogen, phosphorus and their mineralization products in the Black Sea water. (In Russian; English abstract). Okeanologiia, Akad. Nauk, SSSR, 7(3):457-469.

chemistry, nitrogen

Small, Lawrence F. and Donald A. Ramberg, 1971.
Chlorophyll a, carbon and nitrogen in particles from a unique coastal environment. In: Fertility of the Sea, John D. Costlow, editor, Gordon Breach, 2: 475-492.

chemistry - nitrogen

Stickney, Alden P., 1968.
Supersaturation of atmospheric gases in the coastal waters of the Gulf of Maine.
Fish. Bull. Bur. Comm. Fish. U.S.F.W.S., 67(1): 117-123

chemistry nitrogen

Thomas, William H., 1966.
Surface nitrogenous nutrients and phytoplankton in the Northeastern tropical Pacific Ocean. Limnol. Oceanogr., 11(3):393-400.

Chemistry, nitrogen

Vaccaro, Ralph F., 1965.
Inorganic nitrogen in sea water. In: Chemical oceanography, J.P. Riley and G. Skirrow, editors, Academic Press, 1:365-408.

chemistry nitrogen

Vives, Francisco, y Manuel López-Benito, 1957
El fitoplancton de la Ría de Vigo desde julio de 1955 a junio de 1956. Inv. Pesq., Barcelona, 10: 45-146.

Chemistry, nitrogen

von Brand, Th., and N. W. Rakestraw, 1941.
The determinations of dissolved nitrogen in sea water. J. Mar. Res. 4(1):76-80.

chemistry, nitrogen
Voronkev, P.P., 1948.
[Basic characteristics of the hydrochemical regime of the coastal zone of Barents Sea in the central Murman region.] Trudy Dal'niye Zelentsy, 1948(1):39-101, 25 diagrams, sketch map.

chemistry, nitrogen
Weiss R.F. 1970.
The solubility of nitrogen, oxygen and argon in water and seawater.
Deep-Sea Res 17(4): 721-735

chemistry nitrogen
Wilkinson, D., 1963.
Nitrogen transformations in a polluted estuary.
Int. J. Air Water Pollution, 7(6/7):737-752.

chemistry nitrogen
*Williams, P.M., 1967.
Sea surface chemistry: organic carbon and organic and inorganic nitrogen and phosphorus in surface films and subsurface waters.
Deep-Sea Res., 14(6):791-800.

chemistry nitrogen
Wiseman, J.D.H., M.W. Strong and H.J.M. Bowen, 1956.
Marine organisms and biogeochemistry. 1. The rates of accumulation of nitrogen and calcium carbonate on the equatorial Atlantic floor. 2. Marine iron bacteria as rock forming organisms. 3. The biogeochemistry of strontium.
Adv. Sci., 12(49):579-588.

chemistry, nitrogen (inorganic)
Yoshida, Yoichi, and Masao Kimata 1969.
Studies on the marine microorganisms utilizing inorganic nitrogen compounds. IV On the liberation rates of inorganic nitrogen compounds from bottom muds to sea water. V. On the uptake or liberation of inorganic nitrogen compounds by the microorganisms as a whole in seawater. (In Japanese; English abstract)
Bull. Jap. Soc. scient. Fish. 35(3): 303-306.

N_2, albumoid

chemistry nitrogen, albuminoid
Robinson, R.J., and H.E. Wirth, 1934.
Free ammonia, albuminoid nitrogen and organic nitrogen in the waters of the Pacific Ocean off Washington and Vancouver Island.
J. du Cons., 9(1):187-195.

chemistry, nitrogen, albuminoid
Robinson, R.J., and H.E. Wirth, 1934.
Report on the free ammonia, albuminoid nitrogen and organic nitrogen in the waters of the Puget Sound area during the summers of 1931 and 1932.
J. du Cons., 9(1):15-27.

Nitrogen/argon ratio

nitrogen/argon ratio
Benson, Bruce B., and Peter D.M. Parker, 1961
Nitrogen/argon and nitrogen isotope ratios in aerobic sea water.
Deep-Sea Res., 7(4): 237-253.

nitrogen/argon ratio
Richards, Francis A., and Bruce B. Benson, 1961
Nitrogen/argon and nitrogen isotope ratios in two anaerobic environments, the Cariaco Trench in the Caribbean Sea and Dramsfjord, Norway.
Deep-Sea Res., 7(4): 254-264.

nitrogen assimilation

chemistry, nitrogen assimilation
Lewin, Ralph A., 1962.
Physiology and biochemistry of algae.
Academic Press, New York and London, 929 pp.

chemistry nitrogen assimilation
Vaccaro, Ralph F., 1965.
Inorganic nitrogen in sea water.
In: Chemical Oceanography, J.P. Riley and G. Skirrow, editors, Academic Press, 1:365-408.

nitrogen compounds

chemistry nitrogen compounds
Brandt, K., 1905.
On the production and the conditions of production in the sea. Cons. Perm. Int. Expl. Mer, Rapp. Proc. Verb. 3(1):12 pp.

chemistry nitrogen compounds
Burkholder, Paul R., 1960.
Distribution of some chemical values in Lake Erie.
Limnological survey of western and central Lake Erie, 1928-1929. USFWS Spec. Sci. Rept., Fish., No. 334:71-110.

chemistry, nitrogen compounds
Fraga, F., 1967.
Hidrografía de la ría de Vigo, 1962, con especial referencia a los compuestos nitrógens.
Investigación pesq., 31(1):145-149.

chemistry, nitrogen compounds
Giral, J., and O. Gomez Ibanez, 1930.
Determinations des composés azotés dans l'eau de mer. Cons. Perm. Int. Expl. Mer, Rapp. Proc. Verb. 67:93-99.

chemistry, nitrogen compounds
Gomez Ibanez, O., 1929.
Determinación del nitrogeno en sus formes amoniacal, nitroso, y nitrico en el agua de mar. Inst. Espanol Oceanogr., Notas y Res., 2nd ser., No. 36:24 pp., figs.

chemistry nitrogen compounds
Kitamura, H., 1962.
(Marine chemical studies on the phosphorus distributed in the North Pacific Ocean.)
Bull. Kobe Mar. Obs., No. 168:40-94.

chemistry, nitrogen compounds
Rakestraw, N. W. and D. E. Carritt, 1948
Some seasonal chemical changes in the open ocean. J. Mar. Res. 7(3):362-369, 6 text figs.

chemistry nitrogen compound
Rakestraw, N.W. and Th. v. Brand, 1947
Decomposition and Regeneration of Nitrogenous Organic Matter in Sea Water. VI. The effect of Enzyme poisons: Biol. Bull. 92(2):110-114, 1 text figs.

chemistry, nitrogen compounds (inorganic)
Sen Gupta, R., 1971.
Oceanography of the Black Sea: inorganic nitrogen compounds. Deep-Sea Res., 18(5): 457-475.

chemistry, nitrogen compounds
Von Brand, T., and N.W. Rakestraw 1941
Decomposition and regeneration of nitrogenous organic matter in sea water. IV Interrelationship of various stages; influence of concentration and nature of particulate matter.
Biol. Bull. mar. biol. Lab. Woods Hole 81(1): 63-69.

chemistry nitrogen compounds
von Brand, Th., and N.W. Rakestraw, 1940.
Decomposition and regeneration of nitrogenous organic matter in sea water. III. Influence of temperature and source and condition of water.
Biol. Bull. 79(2):231-236, 2 textfigs.

Reviewed: J. du Cons. 16(1):113-116, by H. Barnes

chemistry nitrogenous substances
Waksman, S.A., M. Hotchkiss, C.L. Carey, and Y. Hardman, 1938.
Decomposition of nitrogenous substances in sea water by bacteria. J. Bact. 35(5):477-486.

chemistry nitrogen compounds, regeneratio
Skopontzev, B.A., and E.S. Bruck, 1940.
[Contribution to the study of regeneration of nitrogen and phosphorus compounds in the course of decomposition of dead plankton.] C.R. Acad. Sci., Leningrad, 26:807-810.

nitrogen cycle

chemistry, nitrogen cycle
Acker, Robert F. 1967.
The why of marine microbiology.
[U.S.] Naval Res. Rev. (March): 10-17.

nitrogen cycle
Corner, E.D.S. and Anthony G. Davies, 1971.
Plankton as a factor in the nitrogen and phosphorus cycles in the sea. Adv. mar. Biol, 9: 101-204.

chemistry nitrogen cycle
Gilson, H. C., 1937
Chemical and Physical Investigations. The nitrogen cycle. John Murray Exped., 1933-34, Sci. Repts., 2(2):21-81, 16 text figs.

chemistry nitrogen cycle
Hamilton, R.D., 1964
Photochemical processes in the inorganic nitrogen cycle of the sea.
Limnology and Oceanography, 9(1):107-111.

chemistry, nitrogen cycle
Harris, Eugene, 1959.
The nitrogen cycle in Long Island Sound decreased.
Bull. Bingham Oceanogr. Coll., 17(1):31-65.

chemistry, nitrogen cycle
Ketchum, B.H., R.F. Vaccaro and Nathaniel Corwin, 1958
The annual cycle of phosphorus and nitrogen in New England coastal waters. J. Mar. Res., 17: 282-301.

Chemistry, nitrogen cycle
Matida, Y., 1953.
The cycles of phosphorus and nitrogen in Tokyo Bay. Bull. Japan. Soc. Sci. Fish. 19(4):429-434, 7 textfigs.

Chemistry - nitrogen cycle
Miyake, Yasuo, and Eitaro Wada, 1968.
The nitrogen cycle in the sea.
Rec. oceanogr. Wks., Japan, n.s., 9(2):197-208.

chemistry, nitrogen cycle
Postma, H., 1966.
The cycle of nitrogen in the Wadden Sea and adjacent areas.
Neth. J. Sea Res., 3(2):186-221.

Chemistry nitrogen cycle
Rheinheimer, Gerhard 1970.
3.4 Einfluss verschiedener Faktoren auf die Bakterienflora. 3.5. Bakterien und Stickstoffkreislauf. Chemische, mikrobiologische und planktologische Untersuchungen in der Schlei im Hinblick auf deren Abwasserbelastung. Kieler Meeresforsch 26(2): 161-168; 168-170.

Chemistry, nitrogen cycle
Uyguner, B., 1957.
Le dosage du nitrite dans les eaux du Bosphore, Dardenelles et Trebizonde. Considérations sur la production biologique du nitrite et le cycle d'azote. Hidrobiologi, Istanbul Univ. Fen Fakult. (B)4(2/3):50-61, 6 figs.

Chemistry, nitrogen cycle
Virtanen, A.I., 1952.
Molecular nitrogen fixation and nitrogen cycle in nature. Tellus 4(4):304-306.

Chemistry, nitrogen cycle
von Brand, Th., N.W. Rakestraw, and C.E. Renn, 1939.
Further experiments on the decomposition and regeneration of nitrogenous organic matter in sea water. Biol. Bull. 77(2):285-296, 3 textfigs

Reviewed: J. du Cons. 16(1):113-116 by H. Barnes

chemistry, nitrogen cycle
von Brand, Th., N.W. Rakestraw, and C.E. Renn, 1937.
The experimental decomposition and regeneration of nitrogenous organic matter in sea water.
Biol. Bull. 72(2):165-175, 2 textfigs.

Reviewed: J. du Cons. 16(1):113-116 by H. Barnes

Chemistry, nitrogen cycle
Von Brand, T., N.W. Rakestraw and J.W. Zabor 1942.
Decomposition and regeneration of nitrogenous organic matter in sea water. V. Factors influencing the length of the cycle; observations upon the gaseous and dissolved organic nitrogen.
Biol. Bull. mar. biol. Lab. Woods Hole 83(2): 273-282.

Chemistry, nitrogen cycle
Waksman, S.A., M. Hotchkiss, and C.L. Carey, 1933
Marine bacteria and their role in the cycle of life in the sea. II. Bacteria concerned in the cycle of nitrogen in the sea. Biol. Bull. 65:137-167, 1 pl.

nitrogen, data only

Chemistry nitrogen (data only)
Anon., 1950.
Table 14. Oceanographical observations taken in the Maizuru; physical and chemical data for stations occupied by R.M.S. "Syunpu-maru". Res. Mar. Met., Ocean. Obs., Tokyo, July-Dec. 1947, No. 2:71-76.

Chemistry, nitrogen (organic) (data only)
Conseil International pour l'Exploration de la Joint Skagerak Expedition 1966. 1. Oceanographic stations, temperature-salinity-oxygen content. 2. Oceanographic stations, chemical observations.
ICES Oceanogr. Data lists: 250 pp; 209 pp.

Chemistry nitrogen (data only)
Japan, Hokkaido University, Faculty of Fisheries, 1968.
The Oshoro Maru cruise 21 to the Southern Sea of Japan, January 1967.
Data Record Oceanogr. Obs. Explor. Fish. 12: 1-97; 113-119.

Chemistry, nitrogen (organic) (data only)
Japan, Hokkaido University, Faculty of Fisheries, 1967.
Data record of oceanographic observations and exploratory fishing, 11:383 pp.

Chemistry total nitrogen (data only)
Sweden, Havsfiskelaboratoriet, Lysekil 1971.
Hydrographical data January-June 1970 R.V. Skagerak, R.V. Thetis and TV252, 1970.
Meddn. Hydrogr. avd. Göteborg, 104: unnumbered pp. (multilithed).

Chemistry nitrogen deficiency
Thomas, William H., 1970.
On nitrogen deficiency in tropical Pacific oceanic phytoplankton: photosynthetic parameters in poor and rich water. Limnol. Oceanogr., 15 (3): 380-385.

Chemistry nitrogen (data only)
United States, U.S. Coast Guard, 1965.
Oceanographic cruise, USCGC Northwind: Chukchi, East Siberian and Laptev seas, August-September, 1963.
U.S.C.G. Oceanogr. Rept., No. 6(CG373-6):69 pp.

Chemistry nitrogen dioxide
*Kühme, Heinrich, 1968.
Untersuchungen der Konzentration atmosphärischen Spurengase über dem Atlantik.
Meteor Forschungsergeb, (A)2:83-93.

nitrogen, dissolved

nitrogen fixation

Chemistry nitrogen fixation
Dugdale, R.C., D.W. Menzel and J.H. Ryther, 1961
Nitrogen fixation in the Sargasso Sea.
Deep-Sea Res., 7(4): 297-300.

Chemistry, nitrogen fixation
Dugdale, Richard C., John J. Goering and John H. Ryther, 1964
High nitrogen fixation rates in the Sargasso Sea and the Arabian Sea.
Limnology and Oceanography, 9(4):507-510.

Chemistry, nitrogen fixation
Lewin, Ralph A., 1962.
Physiology and biochemistry of algae.
Academic Press, New York and London, 929 pp.

Chemistry nitrogen fixation
Sisler, F.D., and C.E. ZoBell, 1951.
Nitrogen fixation by sulfate-reducing bacteria indicated by nitrogen-argon ratio. Science 113 (2940):511-512.

chemistry, nitrogen fixation
Stewart, W.D.P., 1971.
Nitrogen fixation in the sea. In: Fertility of the Sea, John D. Costlow, editor, Gordon Breach 2: 537-564.

Chemistry, nitrogen fixation
Stewart, W.D.P., 1962.
Fixation of elemental nitrogen by marine blue-green algae.
Annals of Botany, 26(103):439-446.

Chemistry, nitrogen fixation
Vaccaro, Ralph F., 1965.
Inorganic nitrogen in sea water.
In: Chemical Oceanography, J.P. Riley and G. Skirrow, editors, Academic Press, 1:365-408.

Chemistry, nitrogen flow
Walsh, John J. and Richard C. Dugdale, 1971.
A simulation model of the nitrogen flow in the Peruvian upwelling system. Investigación pesq. 35(1): 309-330.

nitrogen isotope ratio

nitrogen isotope ratios
Benson, Bruce B., and Peter D.M. Parker, 1961
Nitrogen/argon and nitrogen isotope ratios in aerobic sea water.
Deep-Sea Res., 7(4): 237-253.

chemistry, nitrogen isotopes
Miyake, Yasuo, and Eitaro Wada, 1967.
The abundance ratio of 15N/14N in marine environments.
Rec. oceanogr. Wks, Japan, 9(1):37-53.

nitrogen isotope ratio
Richards, Francis A., and Bruce B. Benson, 1961
Nitrogen/argon and nitrogen isotope ratios in two anaerobic environments, the Cariaco Trench in the Caribbean Sea and Dramsfjord, Norway.
Deep-Sea Res., 7(4): 254-264.

nitrogen (organic)

Chemistry nitrogen, organic
Benétoz Alvarez, José, 1965.
Algunas okervaciones del nitrogeno organico en el agus de mar y la composicion quinice del majillon (Mytilus edulis) en la ria de Vigo. (Espana).
Bol. Inst. Oceanogr., liniv, Oriente, 4(1): 172-183.

Chemistry, nitrogen (organic)
Donnelly, Patricia V., Mary A. Burklew and Rose A. Overstreet 1967.
Amino acids and organic nitrogen content in Florida gulf coast waters and in artificial cultures of marine algae.
Prof. Pap. Ser. Fla. Bd Conserv. 9:90-97.

Chemistry, nitrogen, organic
Duursma, E.K., 1961
Dissolved organic carbon, nitrogen and phosphorus in the sea
Netherlands J. Sea Res., 1(1/2):1-147.

chemistry nitrogen, organic (particulate)
Ehrhardt, Manfred, 1970
2.8. Partikulärer organischer Kohlenstoff und Stickstoff sowie gelöster organischer Kohlenstoff. Chemische, mikrobiologische und planktologische Untersuchungen in der Schlei im Hinblick auf deren Abwasserbelastung. Kieler Meeresforsch. 26(2): 138-144.

Chemistry nitrogen, organic
*Fedosov, M.V., and I.A. Ermachenko, 1967.
Hydrochemical characteristics of water masses and water exchange between Iceland and Faroes (hydrographical surveys in 1960).
Rapp. P.-V. Réun. Cons. perm. int. Explor. Mer. 157:196.

Chemistry, nitrogen (organic)
Fraga, F., y A. Ballester 1966.
Distribución vertical del nitrogeno organico e inorganico en la fosa de Cariaco y su relación con el fosforo total.
Memoria Soc. Cienc. nat La Salle 26 (75): 274-282.

chemistry nitrogen (organic)
Fraga, F., and F. Vives, 1961.
La descomposicion de la materia organica nitrogenada en el mar.
Inv. Pesq., Barcelona, 19:65-79.

Chemistry nitrogen (organic)
Harvey, H.W., 1951.
Micro-determination of nitrogen in organic matter without distillation. The Analyst 76:657-660.

Chemistry nitrogen, organic (dissolved)
Holm-Hansen, Osmund, J.D.H. Strickland and P.M. Williams, 1966.
A detailed analysis of biologically important substances in a profile off southern California.
Limnol. Oceanogr., 11(4):548-561.

Chemistry, nitrogen (organic)
Krogh, A. and A. Keys, 1934
Methods for the determination of dissolved organic carbon and nitrogen in sea water.
Biol. Bull., 67:132-144.

Chemistry Nitrogen (organic)
Minkina, A. L., 1961.
[Organic nitrogen and phosphorus in the water masses of the Barents Sea.]
Mezhd. Kom. Mezhd. Geofiz. Goda, Presidium Akad. Nauk, SSSR, Okeanol. Issled., (3):162-171.

chemistry nitrogen, organic
Okuda, Taizo, José Benítez and Esther Fernández A. 1969.
Vertical distribution of inorganic and organic nitrogen in the Cariaco Trench.
Bol. Inst. oceanogr. Univ. Oriente 8 (1/2): 28-34.

Chemistry nitrogen, organic
Okuda, Taizo, y Jose R. Gomez, 1964.
Distribucion del carbono y nitrogeno organicos de los sedimentos en la region nororiental de Venezuela.
Bol. Inst. Oceanogr., Univ. Oriente, Venezuela, 3(1/2):91-105.

Chemistry nitrogen (organic)
Prochazkova, L., 1960.
Einfluss der Nitrate und Nitrite auf die Bestimmung des organischen Stickstoffs und Ammoniums im Wasser.
Arch. f. Hydrobiol., 56(3):179-185.

Chemistry nitrogen, organic
Robinson, R.J., and H.E. Wirth, 1934.
Free ammonia, albuminoid nitrogen and organic nitrogen in the waters of the Pacific Ocean off the coasts of Washington and Vancouver Island.
J. du Cons., 9(2):187-195.

Chemistry nitrogen, organic
Robinson, R.J., and H.E. Wirth, 1934.
Report on the free ammonia, albuminoid nitrogen and organic nitrogen in the waters of the Puget Sound area during the summers of 1931 and 1932.
J. du Cons., 9(1):15-27.

nitrogen, organic
Strickland, J.D.H., Lucia Solórzano and R.W. Eppley, 1970
General Introduction, hydrography and chemistry.
Bull. Scripps Inst. Oceanogr. 17: 1-22.

nitrogen, organic
Thomas, William H., Edward H. Renger and Anne N. Dodson, 1971.
Near-surface organic nitrogen in the Eastern Tropical Pacific Ocean. Deep-Sea Res., 18(1): 65-71.

Chemistry, nitrogen (organic)
von Brand, T., and N.W. Rakestraw, 1941.
The determination of dissolved organic nitrogen in sea water. J. Mar. Res., IV(1):76-80

Chemistry nitrogen (organic)
Yentsch, Charles S., and Ralph F. Vaccaro, 1958.
Phytoplankton nitrogen in the oceans.
Limnol. & Oceanogr., 3(4):443-448.

nitrogen, organic
Yoshida, Yoichi, and Masao Kimata, 1969.
Studies on the marine microorganisms utilizing inorganic nitrogen compounds.
IV. On the liberation rates of inorganic nitrogen compounds from bottom muds to sea water.
V. On the uptake or liberation of inorganic nitrogen compounds by the microorganisms as a whole in sea water. (In Japanese; English abstract)
Bull. Jap. Soc. scient. Fish., 35(3): 303-306; 307-10.

Chemistry, nitrogen (dissolved organic) (data only)
Japan, Hokkaido University, Faculty of Fisheries 1970.
Data record of oceanographic observations and exploratory fishing 13:406 pp.

Chemistry, nitrogen (dissolved organic) (data only)
Japan, Hokkaido University, Faculty of Fisheries, 1968.
Data record of oceanographic observations and exploratory fishing. No. 12:420 pp.

Chemistry, nitrogen (total) (data only)
Saloman, Carl H., and John L. Taylor 1968.
Hydrographic observations in Tampa Bay, Florida, and the adjacent Gulf of Mexico, 1965-1966.
Data Rept. U.S. Fish Wildl. Bur. Comm. Fish. 6 cards (microfiche)

nitrogen, particulate
Chemistry, nitrogen (particulate)
Ehrhardt, Manfred, 1969.
The particulate organic carbon and nitrogen, and the dissolved organic carbon in the Gotland Deep in May 1968.
Kieler Meeresforsch. 25(1): 71-80

Chemistry nitrogen, particulate
*Fraga, F., 1969.
Distribución del nitrógeno orgánico en el Océano Índico occidental. II. Investigación pesq. 33(1): 163-17

chemistry, nitrogen (particulate) (data only)
Japan, Hokkaido University, Faculty of Fisheries, 1968.
The Oshoro Maru cruise 23 to the east of Cape Erimo, Hokkaido, April 1967.
Data Record Oceanogr. Obs. Expl. Fish., 12: 115-169.

Chemistry Nitrogen, particulate
Fraga, F., 1966.
Distribution of particulate and dissolved nitrogen in the Western Indian Ocean.
Deep-Sea Research, 13(3):413-425.

chemistry, nitrogen (particulate)(data)
Postma, H., 1966.
The cycle of nitrogen in the Wadden Sea and adjacent areas.
Neth. J. Sea Res., 3(2):186-221.

Chemistry nitrogen, particulate
*Stephens, K. R.W. Sheldon and T.R. Parsons, 1967.
Seasonal variations in the availability of food for benthos in a coastal environment.
Ecology, 48 (5):852-855.

Chemistry nitrogen, particulate
Vaccaro, Ralph F., 1963.
Available nitrogen and phosphorus and the biochemical cycle in the Atlantic off New England.
J. Mar. Res., 21(3):284-301.

Chemistry, nitrogen (particulate)
von Brand, Th., 1938.
Quantitative determination of nitrogen in the particulate matter of the sea. J. du Cons. 12(2): 187-196.

Chemistry, nitrogen (particulate)

von Brand, Th., 1937.
Observations upon the nitrogen of the particulate matter in the sea. Biol. Bull. 72(1):1-6.

nitrogen/phosphorus ratio

Chemistry, nitrogen/phosphorus ratio

Armstrong, F.A.J., 1965.
Phosphorus.
In: Chemical oceanography, J.P. Riley and G. Skirrow, editors, Academic Press, 1: 323-364.

Chemistry, nitrogen/phosphorus ratios
*Ewins. P.A., and C.P. Spencer, 1967.
The annual cycle of nutrients in the Menai Straits.
J. mar. biol. Ass., U.K., 47(3):533-542.

Chemistry, nitrogen/phosphorus ratio

Okuda, Taizo, 1960
Chemical oceanography in the South Atlantic Ocean, adjacent to north-eastern Brazil. (Portuguese and French resumés).
Trabalhos, Inst. Biol. Marit. e Oceanogr., Universidade do Recife, Brasil, 2(1):155-174.

nitrogen, total

Chemistry, nitrogen, total
McKenzie, Kenneth G., 1964.
The ecologic associations of an ostracode fauna from Oyster Harbour, a marginal marine environment near Albany, Western Australia.
Pubbl., Staz. Zool., Napoli, 33(Suppl.):421-461.

Chemistry, nitrogen, total
Nichols, Maynard M., 1964.
Characteristics of sedimentary environments in Moriches Bay.
In: Papers in Marine Geology, R.L. Miller, Editor, Macmillan Co., N.Y., 363-383.

nitrogen, uptake

Chemistry - nitrogen uptake
Dugdale, R.C., and J.J. Goering, 1967.
Uptake of new and regenerated forms of nitrogen in primary productivity.
Limnol. Oceanogr., 12(2):196-206.

Chemistry, nitrogen uptake

Goering, J.J., D.D. Wallen and R.M. Nauman, 1970.
Nitrogen uptake by phytoplankton in the discontinuity layer of the eastern subtropical Pacific Ocean. Limnol. Oceanogr., 15(5): 789-796.

Chemistry, nitrogenous substances
Fogg, G.E., 1966.
The extracelluler products of algae.
In: Oceanography and marine biology, H. Barnes, editor, George Allen & Unwin, Ltd., 4:195-212.

nitrous oxide

Chemistry, nitrous oxide
Craig, H., and L.I. Gordon, 1963.
Nitrous oxide in the ocean and the marine atmosphere.
Geochimica et Cosmochimica Acta, 27(9):949-955.

Chemistry, nitrous oxide
Junge, C., B. Bockholt, K. Schütz and R. Beck 1971.
N_2O measurements in air and seawater over the Atlantic.
Meteor Forsch.-Engebn. (B) 6: 1-11.

chemistry, nitrous oxide
Junge, C. and J. Hahn, 1971.
N_2O measurements in the North Atlantic. J. geophys. Res. 76(33): 8143-8146.

noble gases
See also gases, noble

Chemistry, noble gases

See also: chemistry, gases

noble gases
Richards, F. A., 1965.
Dissolved gases other than carbon dioxide.
In: Chemical oceanography, J.P. Riley and G. Skirrow, editors, Academic Press, 1:197-225.

non-astacin, SEE ASTACIN

nonastacin carotenoids (data only)
Saloman, Carl H., and John L. Taylor 1968.
Hydrographic observations in Tampa Bay, Florida, and the adjacent Gulf of Mexico, 1965-1966.
Data Rept. U.S. Fish Wildl. Bur. Comm. Fish., 6 cards (microfiche)

nucleic acids

Chemistry, nucleic acids
Lewin, Ralph A., Editor, 1962.
Physiology and biochemistry of algae.
Academic Press, New York and Loneon, 929 pp.

chemistry, nucleic acids
Pillai, T.N.V., and A.K. Ganguly, 1970.
Nucleic acids in the dissolved constituents of sea-water.
Current Sci., 39 (22): 501-504.

chemistry Nuclide silver- 108m
Folsom, T.R., R. Grismore and D.R. Young 1970
Long-lived γ-ray emitting nuclide silver-108m found in Pacific marine organisms and used for dating.
Nature, Lond., 227 (5261): 941-943.

nucleotides

Chemistry nucleotides
Arai, Ken-ichi, Tomoe Furukawa and Tsuneyuki Saito, 1961.
III. Acid-soluble nucleotides in muscle of marine invertebrates. Acid-soluble nucleotides in foot and adductor muscle of seashells. (In Japanese; English abstract).
Bull. Fac. Fish., Hokkaido Univ., 12(1):66-70.

Chemistry nucleotides
Lewin, Ralph A., Editor, 1962.
Physiology and biochemistry of algae.
Academic Press, New York and London, 929 pp.

chemistry, nutrient cycle
Postma, H., 1971.
Distribution of nutrients in the sea and the oceanic nutrient cycle. (English and Portuguese abstracts). In: Fertility of the Sea, John D. Costlow, editor, Gordon Breach, 2: 337-349.

chemistry nutrient enrichment
Glooschenko, Walter A. and Herbert Curl, Jr. 1971.
Influence of nutrient enrichment on photosynthesis and assimilation ratios in natural North Pacific phytoplankton communities.
J. Fish. Res. Bd Can. 28 (5): 790-793.

chemistry nutrient regeneration
Calvert, S.E. and N.B. Price, 1971.
Upwelling and nutrient regeneration in the Benguela Current. Deep-Sea Res., 18(5): 505-523.

Nutrient salts
See nutrients in Main file

Chemistry, nutrients

See also: chemistry, nitrates
chemistry, phosphates, etc

Chemistry nutrients
Adrov, M.M., 1962
Hydrologic investigations off West Greenland. (In Russian; English summary).
Sovetskie Ribochoz. Issledov. Severo-Zapad. Atlant. Okeana, VNIRO-PINRO, 137-153.

chemistry, nutrients
*Aizatullin, T.A., 1967.
Formal chemical-kinetic characteristics of the process of regeneration of mineral combinations of biogenic elements. (In Russian; English abstract).
Trudy Inst. Okeanol., 83(20*37.

Chemistry nutrients
Anderson, George C., 1964.
The seasonal and geographic distribution of primary productivity off the Washington and Oregon coasts.
Limnology and Oceanography, 9(3):284-302.

Chemistry nutrients
Ansell, Alan D., J. Coughlan, K.F. Lander, and F.A. Loosmore, 1964.
Studies on the mass culture of Phaeodactylum. IV. Production and nutrient utilization in outdoor mass culture.
Limnology & Oceanography, 9(3):334-342.

chemistry, nutrients
Armstrong, F.A.J., and E.C. LaFond, 1966.
Chemical nutrient concentrations and their relationship to internal waves and turbidity off southern California.
Limnol. Oceanogr., 11(4):538-547.

Chemistry nutrients
Aruga, Yusho, and Masami Monsi, 1962
Primary production in the northwestern part of the Pacific off Honshu, Japan.
J. Oceanogr. Soc., Japan, 18(2):85-94.

Chemistry, nutrients
Bainbridge, V., 1960
The plankton of inshore waters off Freetown, Sierra Leone. Colonial Off., Fish. Publ., London, No. 13: 48 pp.

chemistry nutrients
Banse, Karl, 1960
Bemerkungen zu meereskundlichen Beobachtungen vor der Ostküste von Indien.
Kieler Meeresf., 16(1): 214-220.

chemistry, nutrients
Barnes, H., 1957.
Nutrient elements. Ch. 11 in: Treatise on Marine Ecology and Paleoecology, Vol. 1.
Mem., Geol. Soc., Amer., 67:297-344.

chemistry, nutrients
Bocacos-Kontos, Theano, and Lydia Ignatiades 1970.
Preliminary biological, chemical and physical observations in the Corinth Canal area.
Cah. océanogr. 22(3): 259-267.

chemistry nutrients
Beers, John R. and Sidney S. Herman, 1969.
The ecology of inshore plankton populations in Bermuda. Part I. Seasonal variation in the hydrography and nutrient chemistry. Bull. mar. Sci., 19(2): 253-278.

chemistry nutrients
Belser, W. L., 1959.
Bioassay of organic micronutrients in the sea.
Proc. Nat. Acad. Sci., 45(10):1533-1542.

chemistry, nutrients
Bernard, Francis, 1967.
Research on phytoplankton and pelagic Protozoa in the Mediterranean Sea from 1953-1966.
Oceanogr. Mar. Biol., Ann. Rev., H. Barnes, editor, George Allen and Unwin, Ltd., 5:205-229.

chemistry, nutrients
Bernard, Francis, 1958
Le courant Atlantique en Méditerranée.
Rapp. Proc. Verb., Comm. Int. Expl. Sci., Mer Medit., n.s., 14:97-100.

chemistry, nutrients
Bernard, F., 1939.
Variations du nanoplancton et des sels nutritifs en mer profonde: les eaux cotières de Monaco en 1938. Cons. Perm. Int. Expl. de la Mer, Rapp. et Proc. Verb. 109(3):51-59.

chemistry nutrients
Bogojavlensky, A.N. and O.V. Shishkina 1971.
On the hydrochemistry off Peru and Chile. (In Russian; English abstract)
Trudy Inst. Okeanol. P.P. Shirshova Akad. Nauk SSSR 89:91-105

chemistry, nutrients
Braarud, T., 1945
A phytoplankton survey of the polluted waters of inner Oslo Fjord. Hvalrådets Skrifter, No.28, 142 pp., 19 text figs., 17 tables.

owned by MS

chemistry, nutrients
Buch, K., 1952.
The cycle of nutrient salts and marine production. Rapp. Proc. Verb., Cons. Perm. Int. Expl. Mer, 132:36-46, 4 textfigs.

chemistry, nutrients
Buchanan, J.B., 1958.
The bottom fauna communities across the continental shelf off Accra, Ghana (Gold Coast).
Proc. Zool. Soc., London, 130(1):1-56.

chemistry, nutrients (organic)
Button, D.K. 1969.
Effect of clay on the availability of dilute organic nutrients to steady-state heterotrophic populations.
Limnol. Oceanogr. 14(1):95-100.

chemistry nutrients
*Choe, Sang, Tai Wha Chung and Hi-Sang Kwak, 1968.
Seasonal variations in nutrients and principal ions contents of the Han River water and its water characteristics. (In Korean; English abstract).
J.oceanogr.Soc., Korea, 3(1):26-38.

chemistry nutrient anions
Chow, Tsaihwa J., and Arnold W. Mantyla, 1965.
Inorganic nutrient anions in deep ocean waters.
Nature, 206 (4982):383-385.
Also in:
Collected Reprints, Int. Indian Ocean Exped., UNESCO, 3. 1966.

chemistry, nutrients
Codispoti, Louis A. 1968.
Some results of an oceanographic survey in the northern Greenland Sea, summer 1964.
Techn. Rept., U.S. Navy Oceanogr. Off. TR202: 49 pp.

chemistry, nutrients
Cooper, L.H.N., 1958.
Consumption of nutrient salts in the English Channel as a means of measuring production.
Rapp. Proc. Verb., Cons. Perm. Int. Expl. Mer, No. 144:35-37.

chemistry nutrients
Cooper, L.H.N., 1956
Some chemical and physical factors controlling the biological productivity of temperate and tropical oceanic waters. Proc. Eighth Pacific Sci. Congr., 3A:1157-1163.

chemistry, nutrients
Cooper, L.H.N., 1952.
Processes of enrichment of surface water with nutrients due to strong winds blowing on to a continental slope. J.M.B.A. 30:453-464, 4 text-figs.

chemistry nutrients
Coste, Bernard, 1971.
Les sels nutritifs entre la Sicile, la Sardaigne et la Tunisie.
Cah. océanogr. 23(1): 49-83.

nutrients
COSTE, B., et H.J. MINAS, 1968.
Production organique primaire et sels nutritifs au large des côtes occidentales Corso-Sardes en février 1966.
Rec. Trav. Sta. mar. Endoume, 44(60): 49-61

chemistry, nutrients
Countryman, Kenneth A. 1969.
Some summer oceanographic features of the Norwegian Sea, summer 1963.
Techn. Rept. U.S.N. Oceanogr. Off. TR-216: 35 pp.

chemistry, nutrients
Day, J.H., 1951.
The ecology of South African estuaries. 1. A review of estuarine conditions in general. Trans. R. Soc., South Africa, 33(1):53-91, 2 textfigs.

chemistry, nutrients
Deacon, G.E.R., 1965.
Chemistry of the sea. 1. Inorganic.
Chemistry in Britain, 1:48-53.
Also in:
Collected Reprints, Nat. Inst. Oceanogr., 13.

chemistry nutrients
Deacon, G.E.R., 1963
12. The Southern Ocean. In: The Sea, M.N. Hill, Edit., Vol. 2. (III) Currents, Interscience Publishers, New York and London, 281-296.

chemistry, nutrients
De Queiroz Santos, Neuzon 1967.
Principais nutrientes e alguns dados físico-químicos do região lagunar de Cananéia.
Bolm Inst. Biol. mar. Univ. Fed. Rio Grande Norte, Brasil 4:1-14.

nutrients
Dragovich, Alexander, John A. Kelly, Jr. and H. Grant Goodell, 1968.
Hydrological and biological characteristics of Florida's west coast tributaries.
Fish. Bull. U.S. Fish. Wildl. Serv., 66(3): 463-477.

chemistry, nutrients
Dugdale, Richard C. and John J. Goering, 1971.
A model of nutrient-limited phytoplankton growth. In: Impingement of man on the oceans, D.W. Hood, editor, Wiley Interscience: 589-600. (23)

chemistry nutrients
*Duke, T.W., and T.R. Rice, 1967.
Cycling of nutrients in estuaries.
Proc.Gulf Caribb. Fish.Inst., 19th.Sess.:59-67.

chemistry, nutrients
Emery, K.O., W.L. Orr and S.C. Rittenberg, 1955.
Nutrient budgets in the ocean. In: Essays in the Natural Sciences in Honor of Captain Allan Hancock, Univ., S. Calif., 299-309.

chemistry, nutrients
English, T. Saunders, 1961.
Biological oceanography in the North Polar Sea from IGY Drifting Station Alpha, 1957-58.
Trans. Amer. Geophys. Union, 42(4):518-525.
Reprinted from:
"Some biological oceanographic observations in the Central North Polar Sea, Drift Station Alpha 1957-1958".
Trans. Amer. Geophys. Union, 42(4):518-525.
Also:
Air Force Cambridge Res. Lab., Sci. Rept. (AFCRL-625), No. 15.

Chemistry, nutrients
*Ewins, P.A., and C.P. Spencer, 1967.
The annual cycle of nutrients in the Menai Straits.
J. mar. biol. Ass., U.K., 47(3):533-542.

Chemistry, nutrients
Eyster, Clyde, 1968.
Seawater as a source of plant nutrients.
Nature, Lond., 220(5164)260-261.

Chemistry, nutrients
Faganelli, A., 1961.
Primi risultati relativi alla concentrazione dei sali nutrivi nelle acque del Mar Mediterraneo centrale et mari adiacenti.
Rapp. Proc. Verb., Réunions, Comm. Int. Expl. Sci., Mer Méditerranée, Monaco, 16(3):675-686.

Chemistry, nutrients
Fedosov, M.V., 1965.
Conditions of formation of the primary food resources in the ocean. Investigations in line with the programme of the International Geophysical Year. (In Russian).
Trudy, Vses. Nauchno-Issled. Inst. Morsk. Ribn. Choz. i Okeanogr. (VNIRO), 57:145-160.

Chemistry nutrients
Fedosov, M.V., 1962.
Formation of the chemical elements for primary productivity in the North Sea. (In Russian).
Trudy Vses. Nauchno-Issledov. Inst. Morsk. Ribn. Chos. i Okean., VNIRO, 46:13-18.

Chemistry, nutrients
Fonselius, Stig H., 1962
Hydrography of the Baltic deep basins.
Fishery Board of Sweden, Ser. Hydrogr., Rept 13: 40 pp.

Chemistry nutrients
Føyn, Ernst, 1968.
Biochemical and dynamic circulation of nutrients in the Oslofjord.
Helgoländer wiss. Meeresunters. 17: 489-495

Chemistry, nutrients
Frega, Fernando, 1967.
El Agua marina.
In: Ecología marina, Monogr. Fundación La Salle de Ciencias Naturales, Caracas, 14:67-99.

chemistry, nutrients ("reserved nutrients")
Fukai, Rinnosuke 1966.
Comments on Dr. Sugiura's "reserved nutrients."
J. oceanogr. Soc. Japan 22(4): 17-19

chemistry, nutrients
Fukai, R., 1954.
On the distribution and variation of nutrient salts in the adjacent water to the Aleutian Islands in the North Pacific.
Bull. Chem. Soc., Japan, 27:402-408.

Chemistry nutrients
Gilmartin, Malvern, 1964.
The primary production of a British Columbia fjord.
J. Fish. Res. Bd., Canada, 21(3):505-538.

nutrients
Goering, John J., and David W. Menzel, 1965.
The nutrient chemistry of the sea surface.
Deep-Sea Res., 12(6):839-843.

Chemistry, nutrients
Gololobov, Ya. K., 1963.
Hydrochemical characteristics of the Aegean Sea during the autumn of 1959. (In Russian; English abstract).
Rez. Issled., Programme Mezhd. Geofiz. Goda, Oceanolog. Issled., Akad. Nauk, SSSR, No. 8:90-96.

chemistry, nutrients
Halldal, P., 1953.
Phytoplankton investigations from Weather Ship M in the Norwegian Sea, 1948-49 (including observations during the "Armauer Hansen" cruise, July 1949). Hvalrådets Skrifter No. 38:91 pp., 20 tables, 21 textfigs.

Chemistry nutrients
Hood, D.W., 1963
Chemical oceanography. In: Oceanography and Marine Biology, H. Barnes, Edit., George Allen & Unwin, 1:129-155.

Chemistry, nutrients
Ignatiades, Lydia, and Theano Becacos-Kontos 1969.
Nutrient investigations in lower Saronicos Bay, Aegean Sea.
Vie Milieu (B) 20(1): 51-61.

Chemistry nutrients
Jones, P.G.W. and A.R. Folkard, 1971.
Hydrographic observations in the eastern Irish Sea with particular reference to the distribution of nutrient salts. J. mar. biol. Ass., U.K., 51 (1): 159-182.

Chemistry nutrients
Jones, P.G.W., and A.R. Folkard, 1970.
Chemical oceanographic observations off the coast of north-west Africa, with special reference to the process of upwelling.
Rapp. P.-V. Réun. Cons. int. Explor. Mer 159: 38-60

chemistry, nutrients
Kabanova, Yu. G., 1964.
Primary production and biogenic elements content of the Indian Ocean waters. Investigations in the Indian Ocean (33rd Voyage of E/S "Vityaz"). (In Russian; English summary).
Trudy Inst. Okeanol., Akad. Nauk, SSSR, 64:85-93.

Chemistry, nutrients
Ketchum, Bostwick H., 1967.
Phytoplankton nutrients in estuaries.
In: Estuaries, G.H. Lauff, editor, Publs Am. Ass Advmt Sci., 83:329-335.

Chemistry, nutrients
Ketchum, B.H., 1962.
23. The regeneration of nutrients. Contributions to symposium on zooplankton production, 1961.
Rapp. Proc. Verb., Cons. Perm. Int. Expl. Mer, 153:142-147.

Chemistry, nutrients
Kinney, Patrick, Martin E. Arhelger and David C. Burrell 1970.
Chemical characteristics of water masses in the Amerasian Basin of the Arctic Ocean.
J. Geophys. Res. 75(21): 4097-4104.

Chemistry, nutrient
*Krishnamurthy, K., 1967.
The cycle of nutrient salts in Porto Novo (India) water.
Int. Rev. ges. Hydrobiol., 52(3):427-436.

chemistry nutrients
Leach, J.H., 1971.
Hydrology of the Ythan Estuary with reference to distribution of major nutrients and detritus
J. mar. biol. Ass., U.K., 51(1): 137-157.

Chemistry, nutrients
Lewin, Joyce C., and Robert R.L. Guillard, 1963.
Diatoms.
Annual Review of Microbiology, 17:373-414.

Chemistry, nutrients
Mandelli, Enrique F., 1965.
Contribucion al conocimiento de la produccion organica primaria en aguas sub-antarticas (Ocean Atlantico Sud-Occidental).
Anais Acad. bras. Cienc., 37(Supl.):399-407.

chemistry, nutrients
Margalef, R., J. Herrera, M. Steyaert et J. Steyaert, 1966.
Distributions et caractéristiques des communautés phytoplanctoniques dans le bassin Tyrrhenien de la Méditerranée en fonction des facteurs ambiants et à la fin de la stratification estivale de l'année 1963.
Bull. Inst. r. Sci. nat. Belg., 42(5):1-56.

Chemistry nutrients
Marin, Jean 1971.
Étude physico-chimique de l'estuaire du Belon.
Rev. Trav. Inst. Pêches marit. 35(2): 109-156

Chemistry nutrients
Matsudaira, C., 1940.
Seasonal changes of chemical composition of sea water and phytoplankton in the Bay of Ise.
Suisangakkaiho, 8(2):148-162.

Chemistry, nutrients
Maucha, R., 1949.
Einige Gedanken aur Frage des Nahrstoffhaushalts der Gewasser. Hydrobiol. 1(3):225-237.

Chemistry, nutrients
McGill, David A., 1969.
A budget for dissolved nutrient salts in the Mediterranean Sea.
Cah. océanogr. 21(6): 543-554

Chemistry, nutrients
Miyake, Yasuo, and Katsuko Saruhashi 1967.
The geochemical balance of nutrient matters in the oceans.
Pap. Met. Geophys. Tokyo 18(2): 89-94.

Chemistry nutrients
Mokievskaja, V.V., 1959
[Nutrients salts in the upper layer of the Bering Sea.] Trudy Inst. Okeanol., 33: 87-113.

Chemistry nutrients
Molinas, Marinella Poli, e Maria Vittoria Olm 1969.
Ulteriori osservazioni sull'apporto in sali nutritivi di alcuni corsi d'acqua sfocianti nell'Adriatico.
Note Lab. Biol. mar. pesca-Fano, Ist. Zool, Univ. Bologna, 3(3): 41-72

chemistry, nutrients
Muñoz, F., y J.M. San Feliu 1969.
Hidrografía y fitoplancton de las Costas de Castellón, de febrero a junio de 1967.
Invest. pesq. Barcelona 33 (1): 313-334.

chemistry
Nutrients
Mostert, S.A. 1970.
Analysis of particulate matter and nutrients in sea-water.
Fish. Bull. SAfr. 6: 34-35

chemistry, nutrients
Munk, W.H., and G.A. Riley, 1952.
Absorption of nutrients by aquatic plants.
J. Mar. Res., 11(2):215-240, 6 textfigs.

chemistry nutrients
Murakami, Akio 1969.
A balance sheet of the nitrogen as the plant nutrient in a laver culture ground. Bull. Nansei reg. Fish. Res. Lab. 2: 1-18.
(In Japanese; English abbtract).

chemistry
nutrients
Noble, A., 1968.
Studies on sea-water off the North Kanara coast.
J. mar. biol. Ass., India, 10(2): 197-223.

chemistry, nutrients
Noskova, E.D., 1965.
Investigations in Kursk and Bislinsk bays.
Atlantich. Nauchno-Issled. Inst. Ribn. Khoz. i Okeanogr. (ATLANTNIRO), Kaliningrad, 14:126 pp.

chemistry, nutrients
Ohle, Waldemar, 1964.
Interstitiallösungen der Sedimente, Nährstoffgehalt des Wassers und Primärproduktion des Phytoplanktons in Seen.
Helgol. Wiss. Meeresunters., 10(1/4):411-429.

nutrients
Okabe, Shiro, Yoshimasa Toyota and Takafumi Murakami, 1967.
Major aspects of chemical oceanography. (In Japanese; English abstract).
J. Coll. mar. Sci. Techn., Tokai Univ., (2):39-56.

chemistry
nutrients
Okuda, Taizo, 1960
Metabolic circulation of phosphorus and nitrogen in Matsushima Bay (Japan) with special reference to exchange of these elements between sea water and sediments. (Portuguese and French resumés)
Trabalhos, Inst. Biol. Maritima e Oceanogr., Universidade do Recife, Brasil, 2(1):7-153.

chemistry, nutrients
Okuda, T., 1955.
On the soluble nutrients in bay deposits. III. Examinations on the diffusion of soluble nutrients to sea water from mud.
Bull. Tohoku Reg. Fish. Res. Lab., (4):215-242.

chemistry, nutrients
Okuda, Taizo, and Lourinaldo B. Cavalcanti 1963.
Uma nota sobre os elementos nutritivos na agua intersticial dos sedimentos na area de Mangues da Barra das Jangadas (Brasil)
Trab. Inst. Oceanogr. Univ. do Recife 3(1): 27-31

chemistry, nutrients
Olmo, Maria Vittoria, e Marinella Poli Molinas 1970.
Un terzo anno di osservazioni sull'apporto in sali nutritivo di alcuni corsi d'acqua sfocianti nell'Adriatico.
Note Lab. Biol. Mar. Pesca-Fano, annesso Ist. Zool. Univ. Bologna 3(8): 177-224.

chemistry, nutrients
Petersen, G. Høpner, 1964.
The hydrography, primary production, bathymetry, and "tagsâq" of Disko Bugt, West Greenland.
Medd om Grønland, 159(10):45 pp.

chemistry, nutrients
Picotti, Mario, 1965.
La crociera idrografico-talassografica di Capo Matapan.
Pubbl. Commissione Ital. Comitato int. Geofis., Ser. IGC, 42:63 pp.

chemistry, nutrients
* Poli Molinas, Marinella, e Maria Vittoria Olmo, 1968.
L'apporto in sali nutritivi di alcuni corsi d'acqua sfocianti nell'Adriatico.
Note Lab. Biol. mar. Pesca, Fano, Univ. Bologna, 2(6): 85-116.

chemistry, nutrients
Reddy, C.V. Gangadhara, and V.N. Sankaranarayanan 1969.
Distribution of nutrients in the shelf waters of the Arabian Sea along the west coast of India.
Bull. natn. Inst. Sci., India 38(1): 206-220.

chemistry
nutrients
Redfield, A.C., B.H. Ketchum and F.A. Richards, 1963
2. The influence of organisms on the composition of sea-water. In: The Sea, M.N. Hill, Edit., Vol. 2. The composition of sea water, Interscience Publishers, New York and London, 26-77.

chemistry nutrients
Rheinheimer, Gerhard, 1970.
2.6 Ammoniak-, Nitrit-, Nitrat- und Phosphatgehalt. Chemische, mikrobiologische und planktologische Untersuchungen in der Schlei im Hinblick auf deren Abwasserbelastung. Kieler Meeresforsch 26(2): 130-132.

chemistry, nutrients
Richards, F.A., 1957.
Some current aspects of chemical oceanography.
In: Physics and chemistry of the earth, Pergamon Press, 2:77-128.

chemistry, nutrients
Riley, Gordon A. 1967.
Mathematical model of nutrient conditions in coastal waters.
Bull. Bingham oceanogr. Coll. 19(2): 72-80

chemistry
nutrients
Riley, Gordon A., 1959
Oceanography of Long Island Sound, 1954-1955
Bull. Bingham Oceanogr. Coll., 17(1): 9-30.

chemistry, nutrients
Rochford, D.J., 1952.
A comparison of the hydrological conditions of the eastern and western coasts of Australia.
Indo-Pacific Fish. Cound., Proc., 3rd meeting, 1-16 Feb. 1952, Sects. 2/3:61-68, 7 textfigs.

chemistry
nutrients
Rojdestvensky, A.V., 1968.
Sur les éléments biogènes dans la partie NW de la mer Noire. (In Russian; French abstract).
Izv. Nauchnoizsledovatel. Inst. ribn. Stopan. Okeanol. - Varna, 9:9-15

chemistry
nutrients
Ryther, John H., 1963
17. Geographic variations in productivity.
In: The Sea, M.N. Hill, Edit., Vol. 2. (IV) Biological Oceanography, Interscience Publishers, New York and London, 347-380.

chemistry nutrients
Ryther, John H., D.W. Menzel and Nathaniel Corwin, 1967.
Influence of the Amazon River outflow on the ecology of the western tropical Atlantic. I. Hydrography and nutrient chemistry.
J. mar. Res., 25(1):69-83.

chemistry
nutrients
Sankaranarayanan, V.N. and C.V. Gangadhara Reddy, 1969.
Nutrients of the northwestern Bay of Bengal.
Bull. natn. Inst. Sci., India, 38(1): 148-163.

chemistry, nutrients
Sawasaki, Michitaka, Shigeki Torii and Kunisuki Nakamura, 1965.
Studies on cultural grounds of laver in the flat reef of Moheji. Relation between nutrient constituents in the sea water and quality of laver. (In Japanese; English abstract).
Sci. Repts., Hokkaido Fish. Exp. Sta., (3):51-56.

nutrients
Scaccini Cicatelli, Marta, 1969
Un anno di osservazioni idrologiche in una stazione fissa nelle acque dell'Adriatico occidentale presso Fano.
Prog. Ricerca Risorse mar. Fondo mar., Comm. ital. Oceanogr. Consig. naz. Ricerche, (B)(40): 153-225

chemistry, nutrients
* Scaccini Cicatelli, Marta, 1968.
Un anno di osservazioni idrologiche in una stazione fissa nelle acque dell'Adriatico occidentale presso Fano.
Note Lab. Biol. mar. Pesca, Fano, Univ. Bologna. 2(9): 181-228.

chemistry
nutrients
* Scaccini Cicatelli, Marta, 1967.
Distribuzione stagionale dei sali nutrivi in una zona dell'alto e medio Adriatico.
Boll. Pesca, Piscic. Idrobiol., n.s., 22(1): 49-82.

Chemistry nutrients
Scaccini, Andrea, et Marta Scaccini-Cicatelli, 1969.
Premiers résultats sur la diffusion des sels nutritifs dans les eaux de l'Adriatique occidentale.
Rapp. P.-v. Réun. Commn int. Explor. scient. Mer Mediterr., 19(4): 753-754.

Chemistry, nutrients
Schelske, Claire L., and Eugene P. Odum, 1962.
Mechanisms maintaining high productivity in Georgia estuaries.
Proc., Gulf and Caribbean Fish. Inst., Inst. Mar. Sci., Univ. Miami, 14th Ann. Sess.:75-80.

Chemistry nutrients (organic)
*Seki, Humitake, 1967.
Effect of organic nutrients on dark assimilation of carbon dioxide in the sea.
Inf. Bull. Planktol. Japan, Comm. No. Dr. Y. Matsue, 201-205.

Chemistry nutrients
Smayda, Theodore J., 1966.
A quantitative analysis of the phytoplankton of the Gulf of Panama. III. General ecological conditions and the phytoplankton dynamics at 8o 45'N, 79o 23'W from November 1954 to May 1957.
Inter- Amer. Trop. Tuna Comm., Bull., 11(5): 355-612.

Chemistry (nutrients?)
Smirnov, E., N. Mijailov y V. Rossov 1967
Algunas características hidroquímicas del Mediterráneo americano.
Estudios, Inst. Oceanol. Acad. Cienc. Cuba, 2(1): 51-74. (not seen).

chemistry, nutrients
*Stefansson, Unnsteinn, 1968.
Dissolved nutrients and water masses in the northern Irminger Sea.
Deep-Sea Res., 15(5):541-575.

Chemistry, nutrients
Stefansson, Unstein, 1966.
Influence of the Surtsey eruption on the nutrient content of the surrounding seawater.
J. Mar. Res., 24(2):241-268.

Chemistry nutrients
Stefánsson, Unnsteinn and Larry P. Atkinson, 1971.
Nutrient-density relationships in the western North Atlantic between Cape Lookout and Bermuda.
Limnol. Oceanogr. 16(1): 51-59.

Chemistry nutrients
Stefánsson, Unnsteinn, Larry P. Atkinson and Dean F. Bumpus, 1971.
Hydrographic properties and circulation of the North Carolina Shelf and slope waters. Deep-Sea Res., 18(4): 383-420.

Chemistry nutrients
Strickland, J.D.H., R.W. Eppley y Blanca Rojas de Mendiola, 1969.
Poblaciones de fitoplancton, nutrientes y fotosíntesis en aguas costeras peruanas.
Bol. Inst. Mar, Peru 2(1): 1-45
(In Spanish and English)

Chemistry nutrients
Strickland, J.D.H., Lucia Solórzano and R.W. Eppley, 1970
General Introduction, hydrography and chemistry.
Bull. Scripps Inst. Oceanogr. 17: 1-22.

chemistry nutrients
Subrahmanyan, R., 1959.
Studies on the phytoplankton of the west coast of India. 1. Quantitative and qualitative fluctuation of the total phytoplankton crop, the zooplankton crop and their interrelationship, with remarks on the magnitude of the standing crop and production of matter and their relationship to fish landings. 2. Physical and chemical factors influencing the production of phytoplankton, with remarks on the cycle of nutrients and on the relationship of the phosphate content to fish landings.
Proc. Indian Acad. Sci., 1:113-252.

chemistry, nutrients
Sugiyama, Teruyuki, Yoshio Miyake and Kuniyasu Fujisawa, 1971.
On the distribution of inorganic nutrients along the coastal waters of Okayama Prefecture (1970).
Bull. Fish. Exp. Sta. Okayama Pref. (1970): 22-28. (In Japanese)

Chemistry nutrients
Sugiyama, Teruyuki, and Kazuo Ukida, 1969.
On the distribution of nutrients along the coastal waters Okayama Prefecture (1969).
(In Japanese). Bull. Fish. exp. Sta. Okayama, 1969: 115-120.

Chemistry, nutrients
Sykes, J.B. and A.D. Boney 1970.
Seasonal variations in inorganic phytoplankton nutrients in the inshore waters of Cardigan Bay.
J. mar. biol. Ass. U.K. 50 (3): 819-827.

chemistry, nutrients
*Szekielda, Karl-Heinz, 1968.
Ein chemisches Modell für den Auf-und Abbau organischen Materials und dessen Anwendung in der offenen See.
J. Cons. perm. int. Explor. Mer, 32(2):180-187.

nutrients
Taleb, Rachida, et Youcef Lalami 1970.
Étude des facteurs physico-chimiques et météorologiques au large de la baie d'Alger.
Bull. Soc. Hist. nat. Afr. nord, 61 (3/4): 121-153

Chemistry nutrients
Taniguti, Tadataka, 1965.
Microbiological studies on shallow marine areas. 1. On the release of nutrient salts and organic acids from sediment to water layer. (In Japanese; English abstract).
Bull. Fac. Fish., Nagasaki Univ., No. 19:91-99.

Chemistry - nutrients
Thayer, Gordon W. 1971.
Phytoplankton production and the distribution of nutrients in a shallow unstratified estuarine system near Beaufort, N.C.
Chesapeake Sci. 12 (4): 240-253.

Chemistry nutrients, nitrogenous
Thomas, William H., 1966.
Surface nitrogenous nutrients and phytoplankton in the northeastern tropical Pacific Ocean.
Limnol. Oceanogr., 11(3):393-400.

Chemistry, nutrients
Tomo, Aldo P. 1970.
Cadenas tróficas observadas en la bahía de Puerto Paraiso (Península Antártica) en relación con las variaciones de la fertilidad de sus aguas.
Contrib. Inst. antart. argentino, 131: 14 pp.

Chemistry nutrients
Vacelet, Eveline, 1969
Rôle des populations phytoplanctoniques et bactériennes dans le cycle du phosphore et de l'azote en mer et dans les flaques supralittorales du Golfe de Marseille.
Téthys 1(1): 5-118

Chemistry nutrients
Vacelet, E., 1963.
La richesse en éléments nutritifs en période estivale dans les cuvettes à salinité variable.
Rec. Trav. Sta. Mar., Endoume, (44)(29):11-17.

Chemistry nutrients
Varlet, F., 1958
Les traits essentials du régime côtise de l'Atlantique près d'Abidjan (Côte d'Ivoire).
Bull. d'I.F.A.N., (A), 20(4):1089-1102.

Chemistry nutrients, surface
Wooster, Warren S., Milner B. Schaefer and Margaret K. Robinson, 1967.
Atlas of the Arabian Sea for fishery oceanography.
Inst. Mar. Resources, Univ. Calif., San Diego, IMR Ref. 67-12: numerous pp. (unnumbered) (multilthed).

chemistry, nutrients
Zuta, Salvador, y Oscar Guillen, 1970.
Oceanografia de las aguas costeras del Peru.
Bol. Inst. Mar., Peru, 2(5):157-324.

Chemistry, nutrients, competition for
Hulburt, Edward M., 1970.
Competition for nutrients by marine phytoplankton in oceanic, coastal and estuarine regions.
Ecology, 51(3): 475-484.

Chemistry nutrient cycle
Kühl, Heinrich, und Hans Mann, 1962
Modellversuche zum Stoffhaushalt in Aquarien bei verschiedenem Salzgehalt.
Kieler Meeresf., 18(3) (Sonderheft):89-92.

Chemistry nutrient deficiency
Holmes, Robert W., 1966.
Light microscope observations on cytological manifestations of nitrate, phosphate, and silicate deficiency in four marine centric diatoms.
J. Phycology, 2(4):136-140.

nutrients, effect of

Chemistry nutrients, effect of
Castellví, Josefina, 1964
Un sencillo experimento para demostrar la influencia de la concentracion de elementos nitritivos sobre la calidad de los pigmentos de las algas.
Inv. Pesq., Barcelona, 25:157-160.

Chemistry nutrients (effect of)
Goldman, M.R. 1965.
Micronutrient limiting factors and their detection in phytoplankton populations.
Mem. Ist. Ital. Idrobiol. 18 (Suppl): 121-135.

Chemistry nutrients, effect of
Ryther, J. H., and R. R. L. Guillard, 1959.
Enrichment experiments as a means of studying nutrients limiting to phytoplankton production.
Deep-Sea Res., 6(1):65-69.

Chemistry nutrients, effect of
Thomas, William H., 1969.
Phytoplankton nutrient enrichment experiments off Baja California and in the eastern Equatorial Pacific Ocean.
J. Fish. Res. Bd, Can. 26(5): 1133-1145

Chemistry nutrient liberation
Okubo, A., 1956.
An additional note on the decomposition of sinking remains of plankton organisms and its relationship to nutrient liberation.
J. Ocean. Soc., Japan, 12(2):45-47.

nutrient limitations

Chemistry nutrient limitations
*Dugdale, R.C., 1967.
Nutrient limitation in the sea: dynamics, identification and significance.
Limnol. Oceanogr., 12(4):685-695.

Chemistry nutrient, limitations by
Steele, J.H., 1962
Environmental control of photosynthesis in the sea.
Limnology and Oceanography, 7(2):137-150.

nutrients, periodicity
Ballester, A. 1969.
Periodicidad en la distribucion de nutrientes en la Fosa de Cariaco.
Mem. Soc. Cienc. nat. La Salle, Venezuela, 29(83): 122-141

nutrients, regeneration of

Chemistry nutrients, regeneration of
Grill, Edwin V., and Francis A. Richards, 1964
Nutrient regeneration from phytoplankton decomposing in seawater.
J. Mar. Res., 22(1):51-69.

Chemistry nutrient regeneration
Johannes, R.E., 1965.
Influence of marine Protozoa on nutrient regeneration.
Limnol. Oceanogr., 10(3):434-442.

Chemistry nutrients, regeneration of
Kametani, Akiyoshi, 1968.
Regeneration of inorganic nutrients from diatom decomposition.
J. oceanogr. Soc. Japan, 25(2):63-74.

Chemistry nutrients, renewal of
Hulburt, Edward M., and Nathaniel Corwin, 1970.
Relation of the phytoplankton to turbulence and nutrient renewal in Casco Bay, Maine.
J. Fish. Res. Bd Can., 27(11): 2081-2090.

Chemistry nutrients, replenishment of
Bhavanarayana, P.V., and E.C. LaFond, 1957.
On the replenishment of some plant nutrients during the upwelling period on the east coast of India. Indian J. Fish., 4(1):75-79.

nutrition
Oppenheimer, Carl H., editor, 1966.
Phytoplankton, Marine Biology II, Proceedings of the Second International Interdisciplinary Conference.
New York Acad. Sci., 369pp. ($8.00).

chemistry, oils
Ackman, R.G., and C.A. Eaton, 1971.
Investigation of the fatty acid composition of oils and lipids from the sand lance (Ammodytes americanus) from Nova Scotia waters.
J. Fish. Res. Bd. Can. 28(4): 601-606

Chemistry OILS (fish)
Ueda, Tadashi, 1967.
Fatty composition of oils from 33 species of marine fish. (In Japanese; English abstract).
J. Shimonoseki Univ. Fish., 16(1):1-10.

chemistry, oleic acid
Rhead, M.M., G. Eglinton, G.H. Draffan and P.J. England 1971.
Conversion of oleic acid to saturated fatty acids in Severn Estuary sediments.
Nature, Lond., 232 (5309): 327-330

chemistry, organic
Cooper, L.H.N., 1965.
Chemistry of the sea. 2. Organic.
Chemistry in Britain, 1965, 1:150-154.

organic acids

chemistry, organic acids
Droop, M.R., 1966.
Organic acids and bases and the lag phase in Nannochloris oculata.
Jour. mar. biol. Assoc., U.K., 46(3):673-678.

chemistry, organic bases
Droop, M.R., 1966.
Organic acids and bases and the lag phase in Nannochloris oculata.
Jour. mar. biol. Assoc., U.K., 46(3):673-678.

Chemistry Organic Acids
Koyama, Tadashiro, and Thomas G. Thompson (posthumous), 1964.
Identification and determination of organic acids in sea water by partition chromatography.
Jour. Oceanogr. Soc., Japan, 20(5):209-220.

Chemistry organic acids
Quinn, James G. and Philip A. Meyers, 1971.
Retention of dissolved organic acids in sea-water by various filters. Limnol. Oceanogr. 16(1): 129-131.

organic compounds
also; SEE chemical concerned as carbon, organic phosphorous, organic.

Chemistry organic compounds
Corwin, James F., 1969.
Volatile oxygen-containing organic compounds in sea water: determination. Bull. mar. Sci., 19(3): 504-509.

Chemistry organic compounds
*Khailov, K.M., and Z.Z. Finenko, 1968.
The interaction of detritus with high-molecular components of dissolved organic matter in sea water.
(In Russian; English abstract).
Okeanologiia, Akad. Nauk, SSSR, 8(6):980-991.

Chemistry organic compounds
Schaefer, Heinz, 1965.
Isoberung von gelösten organischen Verbindungen aus dem Meerwasser unter besonderer Berücksichtigung der Aminosäuren.
Helgoländer wiss Meeresunters, 12(3):239-252.

Chemistry organic compounds
Skopintsev, B.A., 1958.
Investigation of the contents of suspended matter and colored organic compounds in the Azov and Black Sea. Heat of the sea, Chemistry of the Sea. (In Russian).
Trudy Morsk. Gidrofiz. Inst., 13:113-129.

chemistry, organic
Strickland, J.D. H., and Lucia Solórzano, 1966.
Determination of monoesterase hydrolysable phosphate and phosphomonoesterase activity in sea water.
In: Some contemporary studies in marine science H. Barnes, editor, George Allen & Unwin, Ltd., 665-674.

Chemistry organic compounds
Taniguti, Tadataka, 1964.
Consumption of dissolved oxygen by bacterial decomposition of various organic compounds added into inshore sea water.
Bull., Fac. Fish., Nagasaki Univ., No. 17:104-109
abstract, p. 104.

Chemistry, organic compounds
Wagner, Frank S. Jr. 1969.
Composition of the dissolved organic compounds in sea water: a review.
Contrib. mar. Biol. Port Aransas 14:115-153.

organic constituent

Chemistry organic constituents, lists of
Duursma, E.K., 1965.
The dissolved organic constituents of sea water.
In: Chemical oceanography, J.P. Riley and G. Skirrow, editors, Academic Press, 1:433-475.

Chemistry organic compounds
Harvey, H.W., 1953.
Synthesis of organic compounds and chlorophyll by Nitschia closterium. J.M.B.A. 31(3):477-487, 4 textfigs.

nutrients, effect of

Chemistry
nutrients, effect of

Castellví, Josefina, 1964
Un sencillo experimento para demostrar la influencia de la concentracion de elementos nitritivos sobre la calidad de los pigmentos de las algas.
Inv. Pesq., Barcelona, 25:157-160.

chemistry, nutrients (effect of)

Goldman, M.R. 1965.
Micronutrient limiting factors and their detection in phytoplankton populations.
Mem. Ist. Ital. Idrobiol. 18 (Suppl): 121-135.

Chemistry
nutrients, effect of

Ryther, J. H., and R. R. L. Guillard, 1959.
Enrichment experiments as a means of studying nutrients limiting to phytoplankton production.
Deep-Sea Res., 6(1):65-69.

Chemistry
nutrients, effect of

Thomas, William H., 1969.
Phytoplankton nutrient enrichment experiments off Baja California and in the eastern equatorial Pacific Ocean.
J. Fish. Res. Bd, Can. 26(5): 1133-1145

Chemistry
nutrient liberation

Okubo, A., 1956.
An additional note on the decomposition of sinking remains of plankton organisms and its relationship to nutrient liberation.
J. Ocean. Soc., Japan, 12(2):45-47.

nutrient limitations

Chemistry
nutrient limitations

*Dugdale, R.C., 1967.
Nutrient limitation in the sea: dynamics, identification and significance.
Limnol. Oceanogr., 12(4):685-695.

Chemistry
nutrient, limitations by

Steele, J.H., 1962
Environmental control of photosynthesis in the sea.
Limnology and Oceanography, 7(2):137-150.

nutrients, periodicity

Ballester, A. 1969.
Periodicidad en la distribucion de nutrientes en la Fosa de Cariaco.
Mem. Soc. Cienc. nat. La Salle, Venezuela, 29(83): 122-141

nutrients, regeneration of

Chemistry
nutrients, regeneration of

Grill, Edwin V., and Francis A. Richards, 1964
Nutrient regeneration from phytoplankton decomposing in seawater.
J. Mar. Res., 22(1):51-69.

Chemistry
nutrient regeneration

Johannes, R.E., 1965.
Influence of marine Protozoa on nutrient regeneration.
Limnol. Oceanogr., 10(3):434-442.

Chemistry
nutrients, regeneration of

Kamatani, Akiyoshi, 1969.
Regeneration of inorganic nutrients from diatom decomposition.
J. oceanogr. Soc. Japan, 25(2): 63-74.

Chemistry
nutrients, renewal of

Hulburt, Edward M., and Nathaniel Corwin, 1970.
Relation of the phytoplankton to turbulence and nutrient renewal in Casco Bay, Maine.
J. Fish. Res. Bd Can., 27(11): 2081-2090.

Chemistry
nutrients, replenishment of

Bhavanarayana, P.V., and E.C. LaFond, 1957.
On the replenishment of some plant nutrients during the upwelling period on the east coast of India. Indian J. Fish., 4(1):75-79.

nutrition

Oppenheimer, Carl H., editor, 1966.
Phytoplankton, Marine Biology II, Proceedings of the Second International Interdisciplinary Conference.
New York Acad. Sci., 369pp. ($8.00).

chemistry, oils

Ackman, R.G., and C.A. Eaton, 1971.
Investigation of the fatty acid composition of oils and lipids from the sand lance (Ammodytes americanus) from Nova Scotia waters.
J. Fish. Res. Bd, Can. 28(4): 601-606

Chemistry
OILS (fish)

Ueda, Tadashi, 1967.
Fatty composition of oils from 33 species of marine fish. (In Japanese; English abstract).
J. Shimonoseki Univ. Fish., 16(1):1-10.

chemistry, oleic acid

Rhead, M.M., G. Eglinton, G.H. Draffan and P.J. England 1971.
Conversion of oleic acid to saturated fatty acids in Severn Estuary sediments.
Nature, Lond., 232 (5309): 327-330

Cooper, L.H.N., 1965. chemistry, organic
Chemistry of the sea. 2. Organic.
Chemistry in Britain, 1965, 1:150-154.

organic acids

chemistry, organic acids

Droop, M.R., 1966.
Organic acids and bases and the lag phase in Nannochloris oculata.
Jour. mar. biol. Assoc., U.K., 46(3):673-678.

chemistry, organic bases

Droop, M.R., 1966.
Organic acids and bases and the lag phase in Nannochloris oculata.
Jour. mar. biol. Assoc., U.K., 46(3):673-678.

Chemistry
Organic Acids

Koyama, Tadashiro, and Thomas G. Thompson (posthumous), 1964.
Identification and determination of organic acids in sea water by partition chromatography.
Jour. Oceanogr. Soc., Japan, 20(5):209-220.

Chemistry
organic acids

Quinn, James G. and Philip A. Meyers, 1971.
Retention of dissolved organic acids in sea-water by various filters. Limnol. Oceanogr. 16(1): 129-131.

organic compounds

also:
SEE chemical concerned as carbon, organic
phosphorous, organic -

Chemistry
organic compounds

Corwin, James F., 1969.
Volatile oxygen-containing organic compounds in sea water: determination. Bull. mar. Sci., 19(3): 504-509.

Chemistry
organic compounds

*Khailov, K.M., and Z.Z. Finenko, 1968.
The interaction of detritus with high-molecular components of dissolved organic matter in sea water.
(In Russian; English abstract).
Okeanologiia, Akad. Nauk, SSSR, 8(6):980-991.

chemistry
organic compounds

Schaefer, Heinz, 1965.
Isoberung von gelösten organischen Verbindungen aus dem Meerwasser unter besonderer Berücksichtigung der Aminosäuren.
Helgoländer wiss Meeresunters, 12(3):239-252.

Chemistry
organic compounds

Skopintsev, B.A., 1958.
Investigation of the contents of suspended matter and colored organic compounds in the Azov and Black Sea. Heat of the sea, Chemistry of the Sea. (In Russian).
Trudy Morsk. Gidrofiz. Inst., 13:113-129.

chemistry, organic

Strickland, J.D. H., and Lucia Solórzano, 1966.
Determination of monoesterase hydrolysable phosphate and phosphomonoesterase activity in sea water.
In: Some contemporary studies in marine science H. Barnes, editor, George Allen & Unwin, Ltd., 665-674.

chemistry
organic compounds

Taniguti, Tadataka, 1964.
Consumption of dissolved oxygen by bacterial decomposition of various organic compounds added into inshore sea water.
Bull., Fac. Fish., Nagasaki Univ., No. 17:104-109

abstract, p. 104.

chemistry, organic Compounds

Wagner, Frank S. Jr. 1969.
Composition of the dissolved organic compounds in sea water: a review.
Contrib. mar. Biol. Port Aransas 14: 115-153

organic constituents

Chemistry
organic constituents, lists of

Duursma, E.K., 1965.
The dissolved organic constituents of sea water.
In: Chemical oceanography, J.P. Riley and G. Skirrow, editors, Academic Press, 1:433-475.

Chemistry
organic compounds

Harvey, H.W., 1953.
Synthesis of organic compounds and chlorophyll by Nitschia closterium. J.M.B.A. 31(3):477-487, 4 textfigs.

Chemistry
organic compounds
Kashiwada, Ken-ichi and Daiichi Kakimoto, 1962
Studies on organic compounds in natural water 1. On the folic acid content in Inland Water. (In Japanese; English abstract).
Memoirs, Fac. Fish., Kagoshima Univ., 11(2): 158-164.

Chemistry
organic compounds
Khaylov, K.M., 1962.
Some unknown organic substances of sea water. (In Russian).
Doklady, Akad. Nauk, SSSR, 147(5):1200-1203.

Chemistry
organic compounds
Koyama, Tadashiro, 1962
Organic compounds in sea water.
J. Oceanogr. Soc., Japan, 20th Ann. Vol., 563-576.

Chemistry organic compound?, effect of
Kristensen, Ingivar, 1964.
Hypersaline bays as an environment for young fish
Gulf and Caribbean, Fish. Inst., Proc. 16th Ann. Sess., 1963:139-142.

Chemistry organic constituents
Riley, J.P., 1965.
Analytical chemistry of sea water.
In: Chemical oceanography, J.P.Riley and G. Skirrow, editors, Academic Press, 2:295-424.

organic matter

Chemistry
organic matter
Adams, Donald D., and Francis A. Richards, 1968.
Dissolved organic matter in an anoxic fjord, with special reference to the presence of mercaptans.
Deep-Sea Res., 15(4):471-481.

Chemistry, organic matter
Alekin, O.A., and N.P. Moricheva, 1958.
Content of organic matter in natural waters as affected by the carbonate system. (In Russian).
Doklady, Akad. Nauk, SSSR, 119(2):322-

Chemistry, organic matter.
Anderson, D.Q., 1939.
Distribution of organic matter in marine sediments and its availability in further decomposition. J. Mar. Res. 2(3):225-235, Textfig. 67, 2 tables.

Chemistry
organic matter
Bader, R. G., D.W. Hood and J.B. Smith 1960.
Recovery of dissolved organic matter in sea-water and organic sorption by particulate material.
Geochimica et Cosmochimica Acta, 19(4):236-243.

Chemistry
organic matter
Banse, Karl, und Johannes Krey, 1962
Quantitative Aspekte des Kreislaufes der organischen Substanz im Meere.
Kieler Meeresf., 18(3) (Sonderheft): 97-106.

organic matter,
Baylor, E.R., and W.H. Sutcliffe, 1963.
Dissolved organic matter in seawater as a source of particulate food.
Limnology and Oceanography, 8(4):369-381.

Chemistry
organic matter
Blanchard, Duncan C., 1964
Sea-to-air transport of surface active material
Science, 146(3642):396-397.

chemistry, organic matter
Brouardel, Jean, et E. Rinck, 1958.
Appareillage employé en Méditerranée pour l'étude de la production de matière organique.
J. du Conseil, 24(1):10-15.

Chemistry, organic matter (dissolved)
Bruevich, S.W., editor, 1966.
Chemistry of the Pacific Ocean.
Inst. Okeanol., Akad. Nauk, SSSR, Isdatel. Nauka, Moskva, 358 pp.

Chemistry
organic matter
Bruevich, S.W., 1960
[Hydrochemical investigations on the White Sea.]
Trudy Inst. Okeanol., 42: 199-254.

Chemistry organic matter
Chave, K.E., 1965.
Carbonates: association with organic matter in surface seawater.
Science, 148(3678):1723-1724.

Chemistry
organic matter
Christomanos, Anast. A., Aphr. Dimitriadis, and Demetr. Giannitsis, 1962
Contribution ot the knowledge of dissolved organic matter in sea water. (In English; Greek summary).
Res. Proc., Mar. Lab., Athens, Greece, 1(4): 21-25.

Chemistry organic matter
Collier, A., S. Ray, and W. Magnitzky, 1950.
A preliminary note on naturally occurring organic substances in sea water affecting feeding of oysters. Science 111:151-152, 1 textfig.

Chemistry organic matter
Comita, G.W., S.M. Marshall and the late A.P. Orr, 1966.
On the biology of Calanus finmarchicus. XIII. Seasonal change in weight, calcrific value and organic matter.
J. mar. biol. Ass., U.K., 46(1):1-17.

chemistry, organic matter
Conover, Robert J., 1966.
Assimilation of organic matter by zooplankton.
Limnol. Oceanogr., 11(3):338-345.

ORganic matter, assimilation of
Conover, Robert J., 1965.
Assiliation of organic matter by zooplankton. (Abstract).
Ocean Sci. and Ocean Eng., Mar. Techn. Soc., Amer. Sci. Limnol. Oceanogr., 1:303.

chemistry, organic matter
Cruac'h P. 1955.
Quelques composants de la matière organique de l'eau de mer littorale. Helio-oxydation dans le milieu marin.
C.r. hebd. Seanc. Acad. Sci. Paris 241(4): 437-439.

Chemistry
organic matter(suspended)
Dal Pont, G., and B. Newell, 1963.
Suspended organic matter in the Tasman Sea.
Australian J. Mar. Freshwater Res., 14(2):155-165.

Chemistry
organic matter
Datsko, V.G., 1957
[Basic problems in the investigation of organic substances in natural waters.] Gidrokhim. Materialy 26: 7

Chemistry organic matter
Datzo, V.G., 1951.
[Vertical distribution of organic matter in the Black Sea.] Dokl. Akad. Nauk, SSSR, 77:1059-1062.

Chemistry organic matter
Datzo, V.G., 1939.
[Organic matter in water of certain seas.]
Dokl. Akad. Nauk, SSSR, 24(3):294-297.

Chemistry organic matter
DeBuen, R., 1937.
Déterminations physico-chimiques dans les eaux de la Baie de Santander pendant l'été de 1935.
Cons. Perm. Int. Expl. Mer, Rapp. Proc. Verb. 104:26-31, 6 textfigs.

chemistry, organic matter
Dobrzhanskaia, M.A. 1964.
Distribution of organic material in the waters of the Mediterranean Sea. (In Russian).
Trudy Sevastopol. Biol. Stants. Akad. Nauk 15:499-502.

Chemistry
organic matt -er
Dobrzhanskaia, M.A., 1963.
Amount and distribution of organic matter in the Black Sea (based on oxidizability). (In Russian).
Trudy Sevastopol Biol. Sta., 16:472-487.

Chemistry organic matter, dissolved
Dobrzhanskaya, M.A., 1956.
[Seasonal and 24-hour fluctuations in the content of dissolved organic matter in the Black Sea.]
Dokl. Akad. Nauk, SSSR, 111(2):462-465.

Chemistry
organic matter (dissolved)
Duursma, Egbert Klaas, 1963.
The production of dissolved organic matter in the sea, as related to the primary gross production of organic matter.
Netherlands J. Sea Res., 2(1):85-94.

Chemistry
organic matter, dissolved
Duursma, E.K., 1959
Über gelöste organische Substanz auf einem Meridionalschnitt durch den nördlichen Nordatlantischen Ozean im April 1958.
Deutsche Hydrogr. Zeits., Ergänzungsheft, Reihe B, 4(3):69-72.

Chemistry, organic matter
Ermachenko, I.A., 1965.
Conditions for the formation and decay of the organic matter in the ice areas of the Greenland and Barents seas. (In Russian).
Trudy vses. nauchno-issled. Inst. morsk. ryb. Khoz. Okeanogr. (VNIRO), 57:161-171.

chemistry — dissolved substances
Eucken, A., 1947.
Der Einfluss gelöste Stoffe auf die Konstitution des Wassers. Nachr. Akad. Wiss., Göttingen, Math.-Phys. Kl. 2:33.

Chemistry organic matter

Fedosov, M.V., and G.N. Zaitsev, 1960
[Water balance and the chemical regime in the Baltic Sea and its gulfs.]
Trudy Vses. Nauchno-Issled. Inst. Morsk. Ribnogo Chozia i Oceanogr. (VNIRO) 42: 7-14.

Chemistry organic matter

Fincham, A.A., 1969.
Organic matter in sand on Port Erin beach.
Ann. Rep., mar. biol. Sta., Port Erin, 81 (1968): 47-50

Chemistry organic matter

Fisher, A.G., 1964
Latitudinal variations in organic range. (In Russian).
Okeanologiia, Akad. Nauk, SSSR, 4(1):98-111.

Taken from the American Scientist, 1961, 49(1):50-74.

Organic Matter (total, dissolved, particulate)

Fraga, F., 1960.
Variación estacional de la materia orgánica suspendida y disuelta en la Ría de Vigo.
Influencia de la luz y la temperatura.
Inv. Pesq., Barcelona, 17:127-140.

Chemistry organic matter

Fraga, F., and F. Vives, 1961.
La descomposicion de la materia organica nitrogenada en el Mar.
Inv. Pesq., Barcelona, 19:65-79.

Chemistry organic matter

Fraga, F., and F. Vives, 1961.
Variación estacional de la materia orgánica en la Ría de Vigo.
Inv. Pesq., Barcelona, 20:65-71.

Chemistry organic matter

Gillbricht, M., 1959
Die gelöste organische Substanz in der Irmingersee im Spätwinter und Spätsommer 1958.
Deutsche Hydrogr. Zeits., Ergänzungsheft, Reihe B, 4(3):87-89.

Chemistry organic matter

Gillbricht, M., 1957.
Ein Verfahren zum oxydativen Nachweis von organischen Substanz im Seewasser.
Helgolander Wiss. Meeresunters., 6(1):76-83.

Chemistry organic matter

Gololobov, Ya. K., 1963.
Hydrochemical characteristics of the Aegean Sea during the autumn of 1959. (In Russian; English abstract).
Rez. Issled., Programme Mezhd. Geofiz. Goda, Okeanolog. Issled., Akad. Nauk, SSSR, No. 8:90-96.

Chemistry organic matter (data)

Gomez Ibáñez, O., 1928.
Contribución a la determinación de la materia orgánica contenida en el agua de mar.
Notas y Res., Inst. Espanol Ocean., 2nd ser., No. 26:10 pp.

Chemistry, organic matter

Gordon, Donald C., Jr., 1970.
A microscopic study of organic particles in the North Atlantic Ocean. Deep-Sea Res., 17(1): 175-185.

Chemistry organic matter

Gorgy, Samy, 1966.
Les pêcheries et le milieu marin dans le Secteur Méditerranéan de la Republique Arabe linie.
Rev. Trav. Inst. Pêches Marit, 30(1):25-

Chemistry organic material

Gorschkova, T.I., 1938
Organischer Stoff in den Sedimenten des Motovskij Busens. Trans. Inst. Mar. Fish. and Oceanogr. USSR, V:71-84 (German resume)

Chemistry Organic matter

Grall, Jean-René, 1966.
Détermination de la production de matière organique en Manche occidentale à l'aide du carbone 14.
C. R. hebd. séanc., Acad. Sci., Paris, (D) 262(24):2514-2517.

Chemistry organic matter

Hood, D.W., 1963
Chemical oceanography. In: Oceanography and Marine Biology, H. Barnes, Edit., George Allen & Unwin, 1:129-155.

Chemistry organic matter

Ichie, T., K. Tanioka, and T. Kawamoto, 1950.
Reports of the oceanographical observations on board the R.M.S. "Yushio Maru" off Shionomisaki (Aug. 1949). Papers and Repts., Ocean., Kobe Mar. Obs., Ocean. Dept., No. 5:15 pp., 33 figs. (Odd atlas-sized pages - mimeographed).

Chemistry - organic matter

Jannasch, Holger W., Kjell Eimhjellen, Carl O. Wirsen and A. Farmanfarmaian, 1971.
Microbial degradation of organic matter in the deep sea. Science, 171(3972): 672-675.

chemistry, organic material

Jarvis, N.L., W.D. Garrett, M.A. Scheiman and C.O. Timmons, 1967.
Surface chemical characterization of surface-active material in seawater.
Limnol. Oceanogr., 12(1):88-96.

organic matter

Jeffrey, Lela M., and D.W. Hood, 1958
Organic matter in sea water; an evaluation of various methods for isolation. J. Mar. Res., 17:247-271.

Chemistry organic matter

Kalle, Kurt, 1963.
Uber das Verhalten und die Herkunft der in den Gewässern und in der Atmosphäre vorhandenen himmelblauen Fluorescenz.
Deutsche Hydrogr. Zeits., 16(4):153-166.

Chemistry organic matter

Kalle, Kurt, 1962
Über die gelösten organischen Komponenten im Meerwasser.
Kieler Meeresf., 18(3) (Sonderheft):128-131.

Chemistry organic matter

Kawamoto, T., 1951.
On the soluble organic matter in sea water. Mem. Kobe Mar. Obs., 9:15-22, 8 textfigs.

Chemistry organic matter

Khailov, K.M., 1971.
The rate of utilization of dissolved organic matter of sea water by echinoderms and molluscs. (In Russian).
Dokl. Akad. Nauk SSSR 198(2):443-446.

Chemistry organic matter

Khailov, K.M., 1962
The utilization of the phenol extraction in the investigation of the organic complex in sea water. (In Russian).
Okeanologiia, Akad. Nauk, SSSR, 2(5):835-844.

Chemistry - organic matter

*Khailov, K.M., and Z.P. Burlakova, 1969.
Release of dissolved organic matter by marine seaweeds and distribution of their total organic production to inshore communities.
Limnol. Oceanogr., 14(4):521-527.

Chemistry organic matter (dissolved)

Khailov, K.M. and V.E. Erokhin, 1971.
On the utilization of dissolved organic matter by Tigriopus brevicornis and Calanus finmarchicus (In Russian; English abstract). Okeanologiia, 11(1): 117-126.

Chemistry - organic matter

*Khailov, K.M. and Z.Z. Finenko, 1968.
The interaction of detritus with high-molecular components of dissolved organic matter in sea water. (In Russian; English abstract).
Okeanologiia, Akad. Nauk, SSSR, 8(6):980-991.

Chemistry organic matter (dissolved)

Khailov, K.M. and Yu. A. Gorbenko, 1969.
On the role of periphytic communities in ecological metabolism in the sea. The interaction of the communities with dissolved organic matter of the sea water. (In Russian; English abstract). Okeanologiia, 9(5): 834-845.

Chemistry, organic matter

*Khailov, K.M., and Yu. A. Gorbenko, 1967.
On the role of the communities of periphytic microorganisms in the ecological metabolism of the sea: methods of studying the external metabolism of periphyton. (In Russian; English abstract).
Okeanologiia, Akad. Nauk, SSSR, 7(4):718-727.

Chemistry organic matter

Kinney, Patrick J., Theodore C. Loder and Joanne Groves, 1971.
Particulate and dissolved organic matter in the Amerasian Basin of the Arctic Ocean.
Limnol. Oceanogr. 16(1): 132-137.

Chemistry, organic matter

Kochenov, A. V., G. N. Baturin, S. A. Kovaleva, Ye. M. Yemel'yanov and K. M. Shimkus, 1965.
Uranium and organic matter in the sediments of the Black and Mediterranean seas.

Geokhimiia, 1965(3):302-313.

Translation AGI $4.35

organic matter

Korringa, P., 1956.
Hydrographical, biological and ostreological observations in the Knysna Lagoon and with notes on conditions in other South African waters.
S. Africa, Dept. Comm. & Fish., Div. Fish., Invest. Rept., No. 20:63 pp., 23 pls.

Chemistry organic matter
Koshimizu, N., 1960.
On the distribution of organic matter in the sea-water and in the bottom samples of Maizuru Bay.
Oceanogr. Mag., Tokyo, 12(1):1-6.
Also in: Bull. Maizuru Mar. Obs., 1961, (7):1-6

Chemistry organic matter
Krey, J., 1956.
Die Trophie küstennaher Meeresgebiete.
Kieler Meeresf., 12(1):46-64.

Chemistry, organic matter
*Kriss, A.G., 1968.
On the distribution of unstable and stable forms of organic matter in the water column of the World Ocean.
Int. Revue ges. Hydrobiol. 53(3):443-452.

Chemistry organic matter
Kriss, A.E., 1963.
On the distribution of unstable and stable forms of organic matter in the water mass of the world ocean. (In Russian; English abstract).
Mikrobiologiia, 32(1):103-112.

Chemistry organic matter, dissolved
Krogh, A., 1931.
Dissolved substances as food of aquatic organisms
Cons. Perm. Int. Expl. Mer, Rapp. Proc. Verb., 75 :7-36.

Chemistry organic matter
Kuenen, Ph. H., 1941.
Het gehalte aan kalk en organische stof van de Indische deepzeeafzettingen. Nat. Geneesk Congr., Utrecht, Handl. XXVIII:258-259.

Chemistry organic matter
Lopez-Benito, Manuel, 1966.
Variación estacional del contenido de materia orgánica en las arenas de la playa de Areiño (Ría de Vigo).
Inv. Pesq., Barcelona, 30:233-246.

Chemistry organic matter
Lopez-Benito, M., 1955.
Sobre determinación de materia organica en el agua del mar con permanganato potásico.
Invest. Pesq. 1:67-71.

Chemistry, organisms
Lunde, Gulbrand 1967.
Activation analysis of bromine, iodine and arsenic in oils from fishes, whales, phyto- and zooplankton of marine and limnetic biotopes.
Int. Revue ges. Hydrobiol. 52(2): 265-279.

Chemistry organic matter
Lux, R., 1938.
Sur la teneur en matière organique des sediments marins à Monaco et à Concarneau.
Bull. Inst. Océan., Monaco, 760:1-20

Chemistry organic matter
Margalef, D. Ramon, 1963.
El ecosistema pelagico de un area costera del Mediterraneo occidental.
Memorias, Real Acad., Ciencias y Artes de Barcelona, 35(1):3-48.

Chemistry organic matter (suspension)
Marshall, Sheina M., and the late A.P. Orr, 1964
Carbohydrate and organic matter in suspension in Loch Striven during 1962.
Jour. Mar. Biol. Assoc., U.K., 44(2):285-292.

organic matter, particulate
McAllister, C.D., T. R. Parsons and J.D.H. Strickland, 1960
Primary productivity at station "P" in the north-east Pacific Ocean. J. du Cons., 35(3):240-259.

organic particles
Menzel, David W., 1966.
Bubbling of sea water and the production of organic particles: a re-evaluation.
Deep-Sea Res., 13(5):963-966.

organic matter (particulate)
Menzel, David W., and John H. Ryther, 1964.
The composition of particulate organic matter in the western North Atlantic.
Limnology and Oceanography, 9(2):179-196.

Chemistry organic matter
Motoda, S., 1940.
Organic matter in sea water of Palao, South Sea.
Sapporo Nat. Hist. Soc. 16(2):100-104.

chemistry, organic matter, dissolved
Ogura, Norio, 1970.
On the presence of 0.1 - 0.5μ dissolved organic matter in seawater. Limnol. Oceanogr., 15(3): 476-479.

Chemistry organic matter
Ogura, Norio, 1965.
Method of concentration of dissolved organic substances from sea water by dialysis. (In Japanese; English abstract).
J. Oceanogr. Soc., Japan, 21(5):206-211.

organic matter
Oguro, Miki, 1961.
Chemical studies on the current rip in Ishikari Bay caused by the Ishikari River flowing into that bay. 1. On the soluble organic matter. (In Japanese; English abstract).
Bull. Fac. Fish., Hokkaido Univ., 12(1):71-80.

organic matter (data)
Ogura, Norio, and Takahisa Hanya, 1967.
Ultraviolet absorption of the sea water, in relation to organic and inorganic matters.
Int. J. Oceanol. Limnol., 1(2):91-102.

Chemistry organic matter
Orr, A. P., 1933
Physical and chemical conditions in the sea in the neighborhood of the Great Barrier Reef, Brit. Mus. (N.H.) Great Barrier Reef Exped., 1928-29, Sci. Repts. 2(3):37-86, 7 text figs.

Chemistry, organic matter
Pillai, T.N.V., M.V.M. Desai, Elizabeth Mathew, S. Ganapathy and A.K. Ganguly 1971.
Organic materials in the marine environment and the associated metallic elements.
Current Sci. 40(4): 75-81.

Chemistry organic matter
Plunkett, M.A., and N.W. Rakestraw, 1955.
Dissolved organic matter in the sea.
Pap. Mar. Biol. and Oceanogr., Deep-Sea Res., Suppl. to Vol. 3:12-14.

Chemistry organic matter
Provasoli, L., 1963
8. Organic regulation of phytoplankton fertility. In: The Sea, M.N. Hill, Edit., Vol.2. (II) Fertility of the oceans, Interscience Publishers, New York and London, 165-219.

Chemistry - dissolved organic matter
Ranson, M.G., 1936
- Le rôle de la matière dissoute dans l'eau et les théories de Pütter (suite). Bull. Mus. Nat. Hist. Nat. 2e série, 7:160-172.

dissolved organic matter
Ranson, M.G., 1935
Le rôle de la matière organique dissoute dans l'eau et les théories de Pütter. Bull. Mus. Nat. Hist. Nat., 2e série, 7:359-366.

Chemistry, organic matter (dissolved)
Ricci, E., 1956.
Teneur en matières organiques dissoutes dans le golfe de Tunis. Bull. Sta. Océan., Salammbo, 53:69-74.

Chemistry organic matter
Richard, Mlle, 1955.
Résultat de recherches sur la matière organique.
Bull. d'Info., C.C.O.E.C., 7(8):357-358.

organic matter
Richards, F.A., 1957.
Some current aspects of chemical oceanography.
In: Physics and chemistry of the earth, Pergamon Press, 2:77-128.

Chemistry - organic matter
Riley, G.A., 1956.
Oceanography of Long Island Sound, 1952-1954.
9. Production and utilization of organic matter.
Bull. Bingham Oceanogr. Coll., 15:234-244.

organic matter
Romankievitch, E.A., 1960.
[Organic matter in the sediments on the bottom of the Pacific Ocean east of Kamchatka.]
Trudy, Comm. Oceanogr., Akad. Nauk, SSSR, 10(2): 39-47.

Chemistry - organic matter
Ryther, J.H., and D. W. Menzel, 1965.
On the production, composition, and distribution of organic matter in the Western Arabian Sea.
Deep-Sea Research, 12(2):199-209.

Chemistry organic matter
Ryther, J.H., D.W. Menzel, E.M. Hulburt, C.J. Lorenzen and N. Corwin, 1971.
The production and utilization of organic matter in the Peru coastal current. Investigación pesq. 35(1): 43-59.
Originally published in: Anton Bruun Repts scient. Res. SE Pacific Exped. 4: 12 pp. 1970.

Chemistry, organic matter (dissolved)
Saunders, G.W., 1957.
Interrelations of dissolved organic matter and phytoplankton. Bot. Rev., 23(6):389-409.

Chemistry, organic matter
Schaefer, Heinz, 1964.
Beiträge zur Entsalzung mit Retardion 11A8.
Helgoländer Wiss. Meeresuntersuchungen, 11(3/4): 301-322.

chemistry, organic matter
Skopintsev, B.A., 1971.
Recent achievements in the studies of organic matter from the ocean water. Okeanologiia 11(6): 939-956. (In Russian; English abstract)

chemistry organic matter

Skopintsev, B.A., 1968.
A study of some reduction and oxidation processes in the Black Sea. (In Russian; English abstract). Okeanologiia, Akad. Nauk, SSSR, 8(3):412-426.

Chemistry organic matter

Skopintsev, B.A., 1966.
Some considerations on the distribution and state of organic matter in ocean water.
(In Russian; English abstract)
Okeanologiia, Akad. Nauk, SSSR, 6(3)441-450.

organic matter

Skopintsev, B.A., 1965.
Calculation of the organic matter formation and oxidation in sea water. (In Russian; English abstract).
Okeanolog. Issled., Rez. Issled. po Programme Mezhd. Geofiz. Goda, Mezhd. Geofiz. Komitet. Prezidiume Akad. Nauk, SSSR, No. 13:96-107.

Chemistry organic matter

Skopintsev, B.A., 1960.
Organic matter in waters of the sea. Marine chemistry, hydrology, oceanic geology. (In Russian).
Trudy Morsk. Geofiz. Inst., 19:3-20.

Chemistry organic matter

Skopintsev, B.A., 1957.
[On the identification of the colloidal fraction of the organic matter in natural waters.]
Gidrokhim. Materialy, 26:243-

Chemistry organic matter

Skopintzev, B.A., 1939.
Organic matter in the waters of the Barents, Polar and Kara seas.
Doklady, Akad. Nauk, SSSR,, 22:448-451.

Chemistry organic matter

Smetanin, D.A., 1959

[Hydrochemistry of the Kuril-Kamchatka deep-sea trench.] Trudy Inst. Okeanol., 33: 43-86.

Chemistry, organic matter

Sorokin, Iu. I., 1970
On the quantitative of the role played by bacterio-plankton in rotation of organic matter in tropical waters. (In Russian).
Dokl. Akad. Nauk, SSSR, 193(4): 923-925.

Chemistry, organic matter

Stephens, Grover C. 1967.
Dissolved organic material as a nutritional source for marine and estuarine invertebrates.
In: Estuaries, G.H. Lauff, editor, Publs Am. Ass. Advmt Sci. 83: 367-373.

organic matter

Štirn, Jože, 1969.
The distribution of the pelagic organic matter in the North Adriatic.

Rapp. P.-v. Réun. Commn int. Explor. scient. Mer Mediterr., 19(4): 755-758.

chemistry organic matter

Strickland, J.D.H., 1965.
Production of organic matter in the primary stages of the marine food chain.
In: Chemical oceanography, J.P. Riley and G. Skirrow, editors, Academic Press, 1:477-610.

chemistry - organic matter

Sukhoruk V.I., and E.Z. Samyshev, 1969.
Quantitative ratio of organic matter components in the waters of Guinea Bay.
Gidrobiol. Zh. 5 (3): 34-39

Chemistry ORGANIC MATTER

Szekielda, Karl-Heinz, 1970.
The liberated energy potentially available from oxidation processes in the Arabian Sea. Deep-Sea Res., 17(3): 641-646.

chemistry organic matter

Szekielda Karl-Heinz 1969.
La répartition de matériel organique devant les côtes.
C.r. hebd. Séanc. Acad. Sci., Paris, (D) 268 (19): 2323-2326.

Chemistry organic matter

*Szekielda, Karl-Heniz, 1968.
The transport of organic matter by the bottom water of the oceans.
Sarsia, 34:243-252.

chemistry - organic matter

*Szekielda, Karl-Heinz, 1968.
Ein chemisches Modell für den Auf-und Abbau organischen Materials und dessen Anwendung in der offenen See.
J.Cons.perm.int.Explor.Mer, 32(2):180-187.

Chemistry - organic matter

*Szekielda, Karl-Heinz, 1967.
Modellrechnungen fur die frigesetzte energie bei der Oxydation organischen Materials in Seewasser.
Dt.hydrogr.Z., 20(6):265-269.

chemistry organic matter

Taniguti, T., 1962.
A method of estimating the dissolved oxygen in the sea water polluted by organic matter.
Bull. Jap. Soc. Sci. Fish., 28(4):448-452.

Chemistry organic matter

Tanita, S., K. Kato, and T. Okuda, 1951.
Studies on the environmental conditions of shell-fish-fields. In the case of Muroran Harbour (2).
Bull. Fac. Fish., Hokkaido Univ. 2(3):220-230.

Chemistry organic matter

Teixeira, Clovis, and Miryam B. Kutner, 1962
Plankton studies in a mangrove environment.
1. First assessment of standing stock and principal ecological factors.
Bol. Inst. Oceanogr., Sao Paulo, Brasil, 12(3): 101-124.

chemistry, organic matter (dissolved)

Thomas, J.P., 1971.
Release of dissolved organic matter from natural populations of marine phytoplankton. Mar. Biol. 11(4): 311-323.

Chemistry organic matter

Traganza, Eugene D., 1969.
Fluorescence excitation and emission spectra of dissolved organic matter in sea water. Bull mar. Sci., 19(4): 897-904.

Chemistry, organic,matter (suspended)

Tsujita, T., 1953.
A preliminary study on naturally occurring suspended organic matter in waters adjacent to Japan J. Ocean. Soc., Japan, 8(3/4):113-125, 3 textfigs

chemistry, organic matter

Vallentyne, J.R., 1957.
The molecular nature of organic matter in lakes and oceans, with lesser reference to sewage and terrestrial soils. J. Fish. Res. Bd., Canada, 14(1):33-82.

chemistry, organic matter

Vasiliu, F., 1968.
Note préliminaire sur les matières organiques se trouvant dans les eaux roumaines de la Mer Noire en 1967.
Revue Roumaine Biol. (Zool.) 13(6): 473-478

Chemistry organic matter

Vikhrenko, M.M., 1962
[Luminescent-bituminological characteristic of the organic matter in cores from the North Atlantic Ocean.]
Trudy Inst. Okeanol., Akad. Nauk, SSSR: 56: 32-58.

Chemistry, organic matter

von Brand, Th., N.W. Rakestraw, and C.E. Renn, 1939.
Further experiments on the decomposition and regeneration of nitrogenous organic matter in sea water. Biol. Bull. 77(2):285-296, 3 textfigs.

Reviewed: J. du Cons. 16(1):113-116 by H. Barnes

Chemistry, organic matter

von Brand, Th., N.W. Rakestraw, and C.E. Renn, 1937.
The experimental decomposition and regeneration of nitrogenous organic matter in sea water.
Biol. Bull. 72(2):165-175, 2 textfigs.

Reviewed: J. du Cons. 16(1):113-116 by H. Barnes

Chemistry organic matter

Waksman, S.A., 1933.
On the distribution of organic matter in the sea bottom and the chemical nature of marine humus.
Soil Sci. 36(2):125-147, 2 textfigs.

Chemistry organic matter, Dissolved

Watt, W. D., 1966.
Release of dissolved organic material from the cells of phytoplankton populations.
Proc. R. Soc., (B) 164:521-551.

Chemistry organic matter

Wellershaus, Stefan, 1964.
Die Schichtungsverhältnisse im Pelagial des Bornholmbeckens. Über den Jahresgang einiger biotischer Faktoren.
Kieler Meeresf., 20(2):148-156.

chemistry organic matter

Williams, P.M., 1969.

The association of copper with dissolved organic matter in seawater. Limnol. Oceanogr., 14(1): 156-158.

chemistry
organic matter

Williams, P.M., 1968.
Stable carbon isotopes in the dissolved organic matter of the sea.
Nature, Lond., 219 (5150): 152-153

chemistry organic matter
Williams, P.M., and A. Zirino, 1964.
Scavenging of 'dissolved' organic matter from sea-water with hydrated metal oxides.
Nature, 204(4957):462-464.

Chemistry, organic matter
Zobell, C.E., 1940.
The effect of oxygen tension on the rate of oxidation of organic matter in sea water by bacteria.
J. Mar. Res. 3(3):211-223.

Chemistry, organic matter, decomposition of (effect of)

Menzel, David W., 1970.
The role of in situ decomposition of organic matter on the concentration of non-conservative properties in the sea. Deep-Sea Res., 17(4): 751-764.

Chemistry - organic matter, dissolved

Anderson, G.C. and R.P. Zeutschel, 1970.
Release of dissolved organic matter by marine phytoplankton in coastal and offshore areas of the northeast Pacific Ocean. Limnol. Oceanogr., 15(3): 402-407.

organic matter effect of

Chemistry organic matter, effect of
Bader, R.G., 1954.
The role of organic matter in determining the distribution of pelecypods in marine sediments.
J. Mar. Res. 13(1):32-47, 4 textfigs.

Chemistry, organic matter, effect of
Kitano, Yasushi, and Donald W. Hood, 1965.
The influence of organic material on the polymorphic crystallization of calcium carbonate.
Geochimica et Cosmochimica Acta, 29(1):29-42.

Chemistry - organic matter, effect of

Kitano, Y., N. Kanamori and A. Tokuyama, 1970.
Influence of organic matter on inorganic precipitation. Occ. Pap. Inst. mar. Sci., Alaska, 1: 413-447.

chemistry, organic matter (mineralization)
Seki, Humitake 1968.
Relation between production and mineralization of organic matter in Aburatsubo Inlet, Japan.
J. Fish. Res. Bd Can., 25 (4): 625-637.

organic matter particulate

Chemistry, organic matter, particulate
Menzel, David W., 1967.
Particulate organic carbon in the deep sea.
Deep-Sea Research, 14(2):229-238.

Chemistry Organic matter, recycling

Steele, J.H., editor, 1970.
Marine food chains.
Univ. Calif. Press, 552 pp. $13.50.

Chemistry organic matter, steady state
Matsudaira Chikayoshi, and Keinosuke Motohashi, 1970.
A study on the steady state of total organic matter in the ocean. (In Japanese; English abstract).
J. Oceanogr. Soc. Japan 26(6): 354-359.

organic matter suspended

Chemistry, organic matter, suspended
Bogdanov, Y.A., 1965.
The suspended organic matter in the waters of the Pacific Ocean. (In Russian).
Okeanologiia, Akad. Nauk, SSSR, 5(2):286-297.

chemistry, organic matter (terrigenous)
Prakash, A., 1971.
Terrigenous organic matter and coastal phytoplankton fertility. (English and Portuguese abstracts). In: Fertility of the Sea, John D. Costlow, editor, Gordon Breach, 2: 351-368.

chemistry organic molecules

Thorstenson, Donald C., 1970
Equilibrium distribution of small organic molecules in natural waters.
Geochim. Cosmochim. Acta, 34(7): 745-770

organisms

chemistry
organisms,

Alekin, O.A., & N.P. Moricheva, 1961

Calcium carbonate absorption by organisms from sea-water.
Doklady Akad. Nauk, SSSR, 136(6): 1454-1458.

chemistry, organisms
Ashworth, Raymond B., and Milton J. Cormier, 1967.
Isolation of 2, 6-dibromophenol from the marine hemichrodate, Balenoglossus biminiensis.
Science, 155(3769):1558-1559.

chemistry organisms
Benétoz Alvarez, José, 1965.
Algunas okervaciones del nitrogeno organico en el agus de mar y la composicion quinice del majillon (Mytilus edulis) en le ria de Vigo. (España).
Bol. Inst. Oceanogr., liniv. Oriente, 4(1):172-183.

organism, chemistry.

Bernhard, Michael, 1965.
γ-spectra of marine organisms.
Rapp. P.-v. Réun., Comm. int. Explor. scient., Mer Méditerr., 18(3):899-905.

radioactivity(marine organisms)

Chemistry organisms,

Cook, A.C., 1961

The carbon isotopic compositions of certain marine invertebrates and coals from the Australian Permian.
Geochim. et Cosmochim. Acta, 22(2-4): 289-290.

chemistry organisms (marine),

Costa, Rafael Lopez, Lorenzo Rodriguez Molins, Jose Ramon Basada Rial 1957

Estudios guimicos sobre el mejillon (Mytilus edulis) de la ria de Vige
Ist. Espanol de Ocean. No. 84-85-86-87 Dec. 1957

chemistry organisms (Marine)

Cubillos, R., 1958.

Determinacion cualitativa del fierro enconches y caparazones de algunas animales marinas de Montemar
Rev. Biol. Mar. (Valparaiso) 8(1,2,3):149-151.

chemistry, marine organisms
Daumas, Raoul, et Hubert J. Ceccaldi, 1965.
Contribution à l'étude biochimique d'organismes marins. 1 Acides amines libres et proteiques chez Beroe ovata (Eschscholtz), Ciona intestinalis (L.), Cymbulia peroni (De Blainville) et Rhizostoma pulmo (Agassiz).
Rec. Trav. Sta. Mar. Endoume, 38(54):3-14.

Also in:
Trav. Sta. Zool., Villefranche-sur-Mer, 25.

chemistry, organisms
Fried, George H., Carleton Ray, Jack Hiller, Steve Rabinow and William Antopol, 1967.
Alpha-glycerophosphate dehydrogenase and glucose-6-phosphete dehydrogenase in tissues of the Weddell seal.
Science, 155(3769):1560-1561.

Chemistry organisms,
Fukai, Rinnosuke, and W. Wayne Meinke, 1962
Activation analyses of vanadium, arsenic, molybdenum, tungsten, rhenium and gold in marine organisms.
Limnology and Oceanography, 7(2):186-200.

organisms, chemistry of
Hsaio, Sidney C., 1965.
Kinetic studies on alkaline phosphatase from echinoplutei.
Limnol. and Oceanogr., Redfield Vol., Suppl. to 10:R129-R136.

Chemistry organisms,

Imanishi, Noboru, 1959.
Studies on the inorganic chemical constituents of sea fishes (I).
Rec. Oceanogr. Wks., Japan. Spec. No. 3:135-139.

organisms chemistry of
Lanskaia, L.A., and T.I. Pshenin, 1961.
[Conteint of protein, fat, carbohydrate ash in some plankton algae of the Black Sea raised in culture.]
Trudy Sevastopol Biol. Sta., (14):292-302.

Chemistry, organic (organisms)
Miyazawa, Keisuke, Keiji Ito, and Fumio Matsumoto, 1970.
Occurrence of d-2-hydroxy-3-aminopropane sulfonic acid and 3-aminopropane sulfonic acid in a red alga, Grateloupia livida
Bull. Jap. Soc. scient. Fish., 36(1): 109-119

chemistry of organisms
Patton, S., G. Fuller, A.R. Loeblich, III and A.A. Benson, 1966.
Fatty acids of the "red tide" organism, Gonyaulax polyedra.
Biochimica et Biophysica Acta, 116:577-579.

organisms, biochemistry of
Quin, Louis D., 1964
2-aminoethylphosphonic acid in insoluble protein of the sea anemone Metridium dianthus.
Science, 144(3622):1133-1134.

organisms, biochemistry of
Raymont, G.E.G., J. Austin and Eileen Linford, 1964.
Biochemical studies on marine zooplankton. 1. The biochemical composition of Neomysis integer.
Jour. du Conseil, 28(3):354-363.

organisms, chemistry of
Sutcliffe, W.H., Jr., 1965.
Growth estimates from ribonucleic acid content in some small organisms.
Limnol. and Oceanogr., Redfield Vol., Suppl. to 10:R253-R258.

chemistry of organisms
*Tundisi, J., and S. Krishnaswamy, 1967.
Total phosphorus content of Neomysis integer.
Bolm Inst. Oceanogr., S. Paulo, 16(1):99-100.

chemistry, organisms
Vinogradov, A.P., 1953. (translation).
The elementary chemical composition of marine organisms. Mem. Sears Found. Mar. Res. 2:1-647.

Translated by J. Efron and J.K. Seltow with bibliography by V.W. Odum.

organisms, chemistry of
Vinogradov, A.P., 1933.
La composition chimique élementaire des organismes vivants et le système périodique des éléments chimiques. C.R. Acad. Sci., Paris, 197 (25):1673-1675.

Transl. cited:
USFWS Spec. Sci. Rept., Fish., 227.

orthophosphate

Chemistry orthophosphate
Sugaware, K., & S. Kanamori, 1961
Spectrophotometric determination of submicrogram quantities of orthophosphate in natural waters.
Bull. Chem. Soc. Japan, 34(2): 258-261.

Chemistry, osmium
Barker, John L. Jr., and Edward Anders 1968.
Accretion rate of cosmic matter from iridium and osmium contents of deep-sea sediments.
Geochim. Cosmochim. Acta 32 (6): 627-645.

Chemistry, osmium
Sharma, N.N. and J.M. Parekh, 1969.
Osmium in seawater - its determination.
Bull. natn. Inst. Sci., India, 38(1): 236-239.

chemistry, oxic environments
Richards, Francis A., 1971.
Anoxic versus oxic environments.(?)In: Impingement of man on the oceans, D.W. Hood, editor, Wiley Interscience: 201-217.

oxidation

Chemistry oxidation
Creac'h, P., 1955.
Quelques composants de la matière organique de l'eau de mer littorale. Hélio-oxydation dans le milieu marin. C.R. Acad. Sci., Paris, 141(4): 437-439.

OXIDATION
Szekielda, Karl-Heinz, 1970.
The liberated energy potentially available from oxidation processes in the Arabian Sea. Deep-Sea Res., 17(3): 641-646.

oxidation capacity

oxidation reduction

Chemistry, oxidation-reduction reactions
Miyake, Yasuo, and Eitaro Wada 1971.
The isotope effect on the nitrogen in biochemical, oxidation-reduction reactions.
Rec. oceanogr. Wks Japan n.s. 11 (1): 1-6.

Chemistry, oxidation-reduction (data)
Odum, Howard T., Rene P. Cuzon du Rest, Robert J. Beyers, and Clyde Allbaugh, 1963.
Diurnal metabolism, total phosphorus, Ohle anomaly and zooplankton diversity of abnormal marine ecosystems of Texas.
Publ. Inst. Mar. Sci., Port Aransas, 9:404-453.

Chemistry oxidation-reduction potential
Pirie, Robert Gordon, 1965.
Petrology and physical-chemical environment of bottom sediments of the Rivière Bonaventure-Chaleur Bay area, Quebec, Canada.
Rept. B.I.O. 65-10:182 pp. (multilithed).

Chemistry oxidation reduction
Rivière, A., and Solange Vernhet, 1959
États d'oxydo-reduction dans les milieux naturels. Technique de determination directe du rH. Quelques resultats en milieu lagunaire.
Cahiers Océan., C.C.O.E.C., 11(5):309-314.

Chemistry oxidation-reduction potential
Romankevich, E.A., and N.V. Petrov, 1961.
[Oxidation-reduction potential Eh and pH of sediments in the northeastern Pacific.]
Trudy Inst. Okeanol., Akad. Nauk, SSSR, 45:72-85.

chemistry, oxidation-reduction
Schindler, J.E. and K.R. Honick, 1971.
Oxidation-reduction determinations at the mud-water interface. Limnol. Oceanogr. 16(5): 837-840.

chemistry, oxidation-reduction potential
Skopintsev, B.A., N.N. Romenskaya and E.V. Smirnov, 1966.
New determinations of the oxidation-reduction potential in the Black Sea water. (In Russian; English abstract).
Okeanologiia, Akad. Nauk, SSSR, 6(5):799-806.

Chemistry oxidation-reduction potentials
Valentine, B., and C.S. McClesky, 1956.
Oxidation-reduction potentials of the water and sediments of Barataria Bay, Louisiana.
Bull. Mar. Sci., Gulf and Caribbean, 6(3):200-208

oxides

Chemistry oxides (geochemical)
Degens, Egon T., 1965.
Geochemistry of sediments: a brief survey.
Prentice-Hall, Inc., 342 pp.

oxidizability

Chemistry oxidizability
Ruppin, E., 1904.
Ueber die Oxydierbarkeit des Meerwassers durch Kaliumpermanganat.
Publ. de Circ., Cons. Perm. Int. Expl. Mer, No. 20:1-9.

oxidation capacity
Skopintzev, B.A., 1952.
[The oxidation capacity of waters of Black Sea and Azov Sea.] Dokl. Akad. Nauk, SSSR, 87:829-831.

O_2

A

Chemistry oxygen
Aasen, O., 1952.
The Lusterfjord herring and its environment.
Repts. Norwegian Fish. & Mar. Invest. 10(2): 63 pp., 3 pls., 20 textfigs.

Chemistry oxygen
Aasen, O., and E. Akyuz, 1956.
Further observations on the hydrography and occurrence of fish in the Black Sea.
Repts. Fish. Res. Center, Meat & Fish Off., Ser. Mar. Res., 1(6):5-34.

Chemistry oxygen
Aasen, O., and I. Artuz, 1956.
Some observations on the hydrography and occurrence of fish off the Turkish Black Sea coast 11 October - 1 November 1954.
Repts. Fish. Res. Center, Meat & Fish Off. Rept. 1:5-23.

Chemistry oxygen
Aasen, O., I. Artuz and E. Akyuz, 1956.
A contribution to the fishery investigations in the Sea of Marmora. Repts. Fish. Res. Center, Meat & Fish Off., Ser. Mar. Res., 1(2):5-31.

Chemistry oxygen
Aasen, O., I. Artuz and E. Akyuz, 1956.
Report on a survey of the Turkish Black Sea coast
Repts. Fish. Res. Center, Meat & Fish Off., Ser. Mar. Res., 1(5):5-29.

Chemistry oxygen
Acara, A., 1961
Poor water masses in the North Pacific Ocean. The description of the water mass. Pt. 1.
Hidrobiologi, Istanbul, (B), 5(3/4):97-112.

Chemistry oxygen
Acara, A., 1961
Poor water masses in the North Pacific Ocean. The distribution of the water mass. Pt. 2.
Hidrobiologi, Istanbul, (B), 5(3/4): 113-128.

Chemistry oxygen
Acara, A., and C. Erol, 1960
On the pollution of Golden Horn Estuary.
Rapp. Proc. Verb. Reun. Monaco, 15(3): 27-32. Comm. Int. Expl. Sci. Mer. Med.

Chemistry oxygen
Ackefors, Hans, 1965.
On the zooplankton fauna at Askö (The Baltic-Sweden).
Ophelia, 2(2): 269-280

Adrov, M.M.
Chemistry oxygen
Adrov, M.M., 1967
Dissolved oxygen in the waters of the Faroe-Islandic region. (In Russian).
Atlant. nauchno-issled. Inst. ribn. khoz. okeanogr. (AtlantNIRO). Materialy Konferentsii po Resul'tatam Okeanologicheskikh Issledovanii v. Atlanticheskom Okeane 53-63.

Chemistry oxygen
*Adrov, M.M., 1967.
Dissolved oxygen in the waters of the Iceland-Faroe Ridge area, June 1960.
Rapp. P.-V. Réun. Cons. perm.int. Explor.Mer. 157:184-195.

Chemistry oxygen
Adrov, M.M., 1966.
The problem of the practical use of the distribution of some oceanographic parameters. (In Russian).
Materiali, Sess. Uchen. Soveta PINRO Rez. Issled., 1964, Minist. Ribn. Khoz., SSSR, Murmansk, 148-162.

Chemistry oxygen
Adrov, M.M., 1962.
Hydrologic investigations off West Greenland.
SOVETSKIE Riboch. Issledov. v Severo-Zapadnoi Atlant. Okeana, VINRO-PINRO, Moskva, 137-153.
In Russian, English summary

Chemistry oxygen
Akagawa, M., 1956.
On the oceanographical conditions of the north Japan Sea (west off the Tsugaru Straits) in summer (Part 1)
Bull. Hakodate Mar. Obs., No. 3:1-11.
190-199.

Chemistry, oxygen
Akagawa, M., 1954.
On the oceanographical conditions of the north Japan Sea (west off the Tsugaru-Straits) in summer. J. Ocean. Soc., Japan, 10(4):189-199, 5 textfigs.

Chemistry oxygen
Akiyama, Tsutomu, 1968
The distribution of dissolved oxygen and phosphate phosphorus in the adjacent seas of Japan.
Oceanogr. Mag. 20(2): 147-172

Chemistry oxygen
Alander, H., 1951.
Swedish observations (Baltic). Ann. Biol. 7: 112-113, Textfigs. 3-7.

Chemistry oxygen
Alander, H., 1950.
Baltic area. Hydrography (Swedish observations).
Cons. Perm. Int. Expl. Mer, Ann. Biol. 6:157-158, Textfigs. 3-5.

Chemistry oxygen
Alander, H., 1949.
Swedish observations. Ann. Biol. 5:133-134, Figs. 3-6.

chemistry, oxygen
Alander, H., 1949
Hydrography. Ann. Biol., Int. Cons. 4: 139-141; text figs. 1-11.

Chemistry, oxygen
Alarcón, Elías, 1970.
Descripción oceanográfica preliminar del Golfo de Arauco.
Bol. Cient. Inst. Fomento pesq., Chile, 13: 51 pp.

Chemistry oxygen (data)
Aleksandrovskaya, N.B., M.V. Kales and E.A. Yola, 1966.
Peculiarities of the hydrological regime of the Baltic Sea in 1964.
Annls. Biol., Copenh., 21:43-45.

Chemistry oxygen
Aleksejev, A., B. Istoshin, W. Pachorukov, P. Myrland, H. Einarsson, F. Hermann and U. Stefansson, (1959) 1961.
International investigations on hydrographic conditions in the Norwegian Sea, June 1959.
Ann. Biol., Cons. Perm. Int. Expl. Mer, 16: 25.

Chemistry oxygen
Aleksejev, A., B. Istoshin, L. Ponomarenko, (1959) 1961
Hydrographic conditions in the Norwegian Sea, November-December 1959.
Ann. Biol., Cons. Perm. Int. Expl. Mer, 16: 26-28.

Chemistry oxygen
Allain, Ch., 1965.
L'hydrologie de la bordure atlantique nord-américaine du banc St- Pierre au cap Cod, en été 1962 (campagne de la "Thalassa", 19 juillet r 27 août 1962).
Rev. Trav. Inst. Pêches marit., 29(4):357-381.

Chemistry oxygen
Allain, Charles, 1964.
L'hydrologie et les courants du Détroit de Gibraltar pendant l'été de 1959.
Rev. Trav. Inst. Pêches Marit., 28(1):1-102.

Chemistry oxygen
Allen, J.A., 1955.
Solubility of oxygen in water. Nature 175(4445): 83.

Chemistry oxygen
Anand, S.P., C.B. Murty, R. Jayaraman and B.M. Aggarwa, 1969.
Distribution of temperature and oxygen in the Arabian Sea and Bay of Bengal during the monsoon season. Bull. natn. Inst. Sci., India, 38(1): 1-24.

Angot, Michel
chemistry oxygen
Angot, Michel, 1965.
Cycle annuel de l'hydrologie dans la région proche de Nosy-Be.
Cahiers, O.R.S.T.R.O.M. - Océanogr., 3(1):55-66.

Chemistry oxygen (data)
Angot, Michel, et Robert Gerard, 1966.
Caractères hydrologiques de l'eau de surface au Centre ORSTOM de Nosy Bé de 1962 à 1965.
Cah. ORSTOM, Océanogr., 4 (3): 37-53.

Chemistry, oxygen
Angot, Michel, et Robert Gerard, 1966.
Hydrologie et phytoplancton de l'eau de surface en avril 1965 a Nosy Be.
Cah. ORSTOM, Ser. Oceanogr., 4(1):95-136.

Chemistry, oxygen (data)
Angot, M. et R. Gerard, 1965.
Hydrologie de la région de Nosy-Be: juillet à novembre, 1963.
Cahiers, O.R.S.T.R.O.M. - Océanogr., 3(1):3-29.

oxygen (data)
Angot, M., et R. Gerard, 1965.
Hydrologie de la région de Nosy-Be; decembre 1963 à mars 1964.
Cahiers, O.R.S.T.R.O.M. - Océanogr., 3(1):31-53.

Chemistry, oxygen (data)
Angot, M., et R. Gerard, 1963.
Hydrologie de la région de Nosy-Bé: mars-avril-mai-juin 1963.
Trav. Centre Océanogr., Nosy-Bé, Cahiers, O.R.S.T.R.O.M., Océanogr., Paris, No. 6:255-283.

Chemistry oxygen
Anon., 1951
Bulletin of the Marine Biological Station of Asamushi 4(3/4): 15 pp.

Chemistry oxygen
Anon., 1950.
The results of harmonic analysis of tidal currents, water temperature and chemical components in the Kii-Suido. J. Ocean., Kobe Obs., 2nd ser., 2(1):28-31, 2 figs.

Chemistry oxygen
Anon., 1949-1950.
The results of the regular monthly oceanographical observations on board the R.M.S. "Syunpu Maru" and "Takatori Maru" in the Osaka Wan.
J. Ocean., Kobe Obs., 2nd ser., 2(1):1-28.

O_2
Anonymous, 1947
Discovery investigations Station List 1937-1939. Discovery Report, XXIV: 197-422. pls IV-VI.

Antonov, A.E.
Chemistry, oxygen
Antonov, A.E., 1963.
The change of the dissolved oxygen contents in the waters of the Gdansk and the Bornholm deeps in 1947-1961. (In Russian).
Atlantich. Nauchno-Issled. Inst. Ribn. Khoz. i Okeanogr. (ATLANTNIRO), Trudy, 10:10-14.

Chemistry oxygen
Antonov, A.E., 1959
Hydrographic situation in the North Sea, April 1957.
Biull. Okeanograf. Komissii, Akad. Nauk, SSSR, (3): 46-52.

Chemistry oxygen
Antonov, A.E., and O.S. Roudneva, 1965.
Peculiarities of hydrological and hydrochemical conditions in the south Baltic during the IGY. (In Russian; English abstract).
Okeanolog. Issled. Rezult. Issled. Programme Mezhd. Geofiz. Goda, Mezhd. Geofiz. Komitet Presidiume, Akad. Nauk, SSSR, no. 13:90-95.

Chemistry, oxygen

Austin, Thomas S., 1960

Oceanography of the east central equatorial Pacific as observed during expedition "Eastropic". U.S.F.&.W.S., Fish. Bull. 168, Vol. 60: 257-282.

Chemistry oxygen

Austin, Thomas S., 1958.

Variations with depth of oceanographic properties along the equator in the Pacific. Trans. A.G.U., 39(6):1055-1063.

Chemistry oxygen

Azouz, Abderrazak, 1966.
Étude des peuplements et des possibilités d'ostreiculture du Lac de Bizerte.
Inst. nat. Sci. Tech. Océanogr. Pêche, Salammbo, Ann. 15: 69 pp.

B

Chemistry oxygen

Bacescu, M., 1961
Cercetări fizico-chimice şi biologice romîneşti la Marea Neagra, efectuate în perioda 1954-1959.
Hidrobiologia, Acad. Repub. Pop. Rom., (3): 17-46.

Chemistry oxygen

Băcescu, M., M.T. Gomoiu, N. Bodeanu, A. Petran, G.I. Müller si V. Chirila, 1965.
Dinamica populaţiilor animale şi vegetale din zona nisipurilor fine de la nord de Constanţa în conditiile anilor 1962-1965.
In: Ecologie marină, M. Băcescu, redactor, Edit. Acad. Republ. Pop. Romane, Bucuresti, 2:7-167.

Chemistry oxygen (data)

Băcescu, M., G. Müller, H. Skolka, A. Petran, V. Elian, M.T. Gomoiu, N. Bodeanu şi S. Stănescu, 1965.
Cercetări de ecologie marină în sectorul predeltaic în conditiile anilor 1960-1961.
In: Ecologie marină, M. Băcescu, redactor, Edit. Acad. Republ. Pop. Romane, Bucureşti, 1: 185-344.

Chemistry, oxygen

Baggesgaard-Rasmussen and J.P. Jacobsen, 1930
Contribution to the hydrography of the waters round Greenland in the year 1925.
Medd. fra Komm. for Havundersøgelser, Serie: Hydrografi. 2(10):24 pp., 25 text figs.

Chemistry oxygen

Bailey, W.B., 1956.
On the oxygen of deep Baffin Bay water.
J. Fish. Res. Bd., Canada, 13:303-308.

Chemistry oxygen

Baksev, V.G., editor, 1966.
Atlas Anterktiki, Sovetskaia Antarktichkeskaia Ekspeditsiie.I.
Glebnoe Upravlenie Geodesii i Kartografii MG SSSR, Moskva-Leningrad, 225 Charts.

Chemistry oxygen (data)

Ballester, A., 1965.
Hidrografia y nutrientes de la Fosa de Cariaco.
Informe de Progresso del Estudio Hidrografico de la Fosa de Cariaco, Fundación La Salle de Ciencias Naturales Estación de Investigacion -es Marinas de Margarita, Caracas, Sept. 1965. (mimeographed):3- 12.

Chemistry oxygen (data)

Ballester Antonio Enrique Arias, Antonio Cruzado, Dolores Blasco y José María Camps, 1967.
Estudio hidrográfico de la costa catelana de junio de 1965 a mayo de 1967.
Investigación, pesq. 31(3): 621-662

Chemistry oxygen

Banse, Karl, 1968.
Hydrography of the Arabian Sea shelf of India and Pakistan and effects on demersal fishes.
Deep-Sea Res., 15(1):45-79.

Chemistry oxygen

Banse, Karl, 1960

Bemerkungen zu meereskundlichen Beobachtungen vor der Ostkuste von Indien.
Kieler Meeresf., 16(1): 214-220.

Chemistry, oxygen

Banse, K., 1959

On upwelling and bottom-trawling off the south west coast of India. J. Mar. Biol. Ass., India 1(1): 33-49.

Chemistry, oxygen

Banse, K., 1957.
Ergebnisse eines hydrographisch-produktionsbiologischen Längsschnitts durch die Ostsee im Sommer 1956. II. Die Verteilung von Sauerstoff, Phosphat und suspendierten Substanz.
Kieler Meeresf. 13(2):186-201.

Chemistry, oxygen

Barber, F.G., 1967.
A contribution to the oceanography of Hudson Bay.
Manuscript Rep. Ser., Dept. Energy, Mines, Resources, Can., No. 4:69 pp. (multilithed).

Chemistry, oxygen

Barber, F.G., and C.J. Glennie, 1964.
On the oceanography of Hudson Bay, an atlas presentation of data obtained in 1961.
Manuscr. Rep. Ser., mar. Sci. Br., Dept. Mines tech Surv., Ottawa, 1: numerous pp. (unnumbered) (multilithed).

Chemistry, oxygen (data)

Bardin, I.P., 1958.
Hydrological, hydrochemical, geological and biological studies, Research Ship "Ob", 1955-1957. Trudy Kompleks. Antarkt. Exped., Akad. Nauk, SSSR, Mezh. Geofiz. God, Gidrometeorol. Izdatel, Leningrad, 218 pp.

oxygen

Barkley, Richard A., 1968.

Oceanographic atlas of the Pacific Ocean.
University of Hawaii Press. 156 figures.

Chemistry, oxygen

Barnes, C.A., and E.E. Collias, 1958

Some considerations of oxygen utilization rates in Puget Sound, J. Mar. Res., 17:68-80.

Chemistry oxygen

Beck, B., K. Kalle and E. Rogalla, 1959

Die Schichtung im Sauerstoffgehalt im Jahreszeitlichen Wechsel zwischen Kap Farvel und der Flämischen Kappe. Deutschen Hydrogr. Zeits., Ergänzungsheft Reihe B (4°) No. 3: 66-69.

Also In: Deutsches Hydrogr. Inst., Ozeanogr., 1959(1960).

Chemistry oxygen

Bennett, Edward B., 1963.
An oceanographic atlas of the Eastern Tropical Pacific Ocean, based on data from EASTROPIC expedition. October-December 1955.
Bull. Inter-American Trop. Tuna Comm., 8(2):33-165.

Chemistry oxygen

Bennett, Edward B., and Milner B. Schaefer, 1960.
Studies of physical, chemical and biological oceanography in the vicinity of the Revilla Gigedo Islands during "Island Current Survey" of 1957.
Bull. Inter-American Tropical Tuna Comm., 4(5): 219-317. (Also in Spanish).

Chemistry oxygen

Ben-Yaakov, Sam, 1971.
A multivariable regression analysis of the vertical distribution of TCO_2 in the eastern Pacific. J. geophys. Res. 76(30): 7417-7431.

Chemistry oxygen

Bernard, F., 1948
Recherches préliminaires sur la fertilité marine au large d'Alger. J. du Cons. 15(3): 260-267, 5 text figs.

Chemistry oxygen (data)

Bernhard, Michel, 1963.
1. Introduzione. 2. Chimica oceanografica.
Rapp. Attività Sci. e Tecn., Lab. Studio della Contaminazione Radioattiva del Mare, Fiascherino, La Spezia (maggio, 1959-maggio 1962), Comit. Naz. Energia Nucleare, Roma, RT/BIO, (63), 8:7-39. (multilithed).

Berthois, Léopold

Chemistry, oxygen

Berthois, L., 1969.
Contribution à l'étude sédimentologique du Kangerdlugssuaq, Côte ouest du Groenland.
Medd. Grønland 187 (1): 185 pp.

Chemistry oxygen

Berthois, Leopold, 1966.
Hydrologie et sedimentologie dans le Kangerd-lugssuaq (fjord a la cote ouest du Groenland).
C. r. hebd. seanc., Acad. Sci., Paris (D), 262(13): 1400-1402.

Chemistry oxygen (data)

Berthois, Léopold, 1965.
Remarques sur les propriétés physico-chimiques des eaux marines dans le Golfe de Gascogne, en mai, 1964.
Rev. Trav. Inst. Pêches marit., 29(4):383-392.

Chemistry oxygen

Bessonov, N.M., 1964
On some features of variations of hydrochemical characteristics in fishery areas off Dakar and Takoradi. (In Russian).
Okeanologiia, Akad. Nauk, SSSR, 4(5):813-824.

Chemistry, oxygen

Bessonov, N.M. and V.N. Kochikov 1967.
On some variable oceanographic conditions off the productive regions of Dakar and Takograd. (In Russian)
Atlant. nauchno-issled. Inst. rybn. Khoz. okeanogr. (AtlantNIRO). Materialy Konferentsii po Resul'tatem Okeanologicheskikh Issledovanii v. Atlanticheskom Okeane, 94-100.

Chemistry, oxygen

Beyer, Fr., and E. Føyn, 1951.
Surstoffmangel i Oslofjorden. En kritisk situasjon for fjordens dybebestand. Naturen, 1951, No. 10:289-306, 9 textfigs.

Chemistry oxygen

Bezrukov, P.L., 1959

[Oceanographic investigations in the North-West Pacific, August-October, 1954.] Trudy Inst. Okeanol., 16: 70-97.

Chemistry, oxygen

Bibik, V.A. 1964

Dynamic characteristics of the water of the southeastern part of the Black Sea and the distribution of its oceanographic elements. (In Russian)

Trudy azov.-chernomorsk. nauchno-issled. Inst. morsk. ryb. khoz. okeanogr. 23: 23-31.

Chemistry oxygen (data)

Bien, George S., Norris W. Rakestraw and Hans E. Suess, 1963.

Radiocarbon dating of deep water of the Pacific and Indian oceans.

Bull. Inst. Oceanogr., Monaco, 61(1278):1-16.

chemistry, oxygen (data)

Bienati, Norberto L., y Rufino A. Comes 1970.

Variación estacional de la composicion fisico-quimica del agua de mar en Puerto Paraiso, Antartida occidental.

Contrib. Inst. Antart. Argentino 130: 45pp.

Chemistry oxygen

Biernacka, I., 1948

(Tintinnoinea in the Gulf of Gdansk and adjoining waters) Buiuletyn Morskiego Laboratorium Rybackiego w Gdyni dawniej-Stacji Morskiej w. Helu. No. 4:73-91, 4 text figs., 1 pl. with 21 figs. (Bull. Lab. Mar., Gdynia, formerly Bull. Sta. Hel.)

Chemistry, oxygen

Bigelow, H.B., and M. Leslie, 1930

Reconnaissance of the waters and plankton of Monterey Bay, July 1928. Bull. M.C.Z., 70(5):429-481, 43 text figs.

Chemistry, oxygen

Birkenes, E., and T. Braarud, 1954.

Phytoplankton in the Oslo Fjord during a "Coccolithus huxleyi-summer".

Avhandl. Norske Videnskaps.-Akad., Oslo. 1. Mat.-Naturvid. Kl., 1952(2):1-23, 1 textfig.

Chemistry oxygen

Bishev, L. L., 1955

[Results of hydrochemical investigations the Kubansk delta estuaries.] Trudy VNIRO 31: 145-150.

Chemistry oxygen

Bjerrum, N., 1904.

On the determination of oxygen in sea water. Medd. Komm. Havundersøgelser, Ser. Hydrogr., 1(5):13 pp.

Chemistry oxygen

Black, W.A.P., and E.T. Dewar, 1949.

Correlation of some of the physical and chemical properties of the sea with the chemical composition of the algae. J.M.B.A. 28(3):673-699, 14 textfigs.

Chemistry Oxygen (data)

Blackburn, M., 1962.

An oceanographic study of the Gulf of Tehuantepec U.S.F.W.S. Spec. Sci. Rept., Fish., No. 404:28 pp

Chemistry oxygen

Bogdanov, D.V., 1965.

Algunos rasgos de la oceanografía del Golfo de México y del Mar Caribe. (In Russian: Spanish abstract).

Sovetsk.-Cub. Ribokhoz. Issled., VNIRO: Tsentr. Ribokhoz. Issled. Natsional. Inst. Ribolovsta Republ. Cuba, 23-45.

Chemistry oxygen

Bogoyavlensky, A.N., 1963

On the study of oxygen distribution in sea water based on the investigations carried out in the Southern Ocean. (In Russian).

Okeanologiia, Akad. Nauk, SSSR, 3(2):271-277.

Chemistry, oxygen

Bogoyavlensky, A.N., 1959

[Hydrochemical investigations.] Arktich. i Antarktich. Nauchno-Issled. Inst., Mezhd. Geofiz. God, Sovetsk. Antarkt. Eksped., 5: 159-172.

Chemistry oxygen

Bogojavlensky, A. N., 1958.

Certain peculiarities of the oxygen, phosphorus and siliceous acid distribution in Antarctic waters. Inform. Biull. Sovetsk. Antarkt. Exped., (3):19-20.

Chemistry oxygen

Bogoiavlenskii, A.N., 1958.

[VI. Hydrochemical work.] Opisanie Exped. D/E. "Ob", 1955-1956, MGG, Trudy Kompleksnoi Antarkt. Exped., Akad. Nauk, USSR, 91-102.

Chemistry, oxygen

Bogoyavlenskii, A.N., 1955.

[Chemical characteristics of the water in the region of the Kurile-Kamchatka Trench.] Trudy Inst. Okeanol., 12:161-176.

Chemistry oxygen

Bogojavlensky, A.N. and O.V. Shishkina 1971.

On the hydrochemistry of Peru and Chile. (In Russian; English abstract)

Trudy Inst. Okeanol. P.P. Shirshova Akad. Nauk SSSR 89: 91-105

Chemistry, oxygen

Braarud, T., 1945

A phytoplankton survey of the polluted waters of inner Oslo Fjord. Hvalrådets Skrifter, No.28, 142 pp., 19 text figs., 17 tables.

Chemistry, oxygen

Braarud, T., 1939

Observations on the phytoplankton of the Oslo Fjord, March-April, 1937. Nytt Magasin for Naturvidenskapene, 80:211-218, 1 text fig.

Chemistry, oxygen

Braarud, T., 1934

A note on the phytoplankton of the Gulf of Maine in the summer of 1933. Biol. Bull. 67(1):76-82. (Contribution No.46 of the Woods Hole Oceanographic Institution)

Chemistry, oxygen

Braarud, T., and Adam Bursa, 1939

On the phytoplankton of the Oslo Fjord, 1933-1934. Hvalrådets Skr. No.19:1-63; 9 text figs. Reviewed: J. du. Cons. 14(3): 418-420. A.C. Gardiner.

Chemistry, oxygen

Braarud, T., and A. Klem, 1931.

Hydrographical and chemical investigations in the coastal waters off Møre and in the Romsdalfjord. Hvalrådets Skrifter, 1931(1):88 pp., 19 figs.

Chemistry, oxygen (data)

Braarud, T., K.R. Gaarder and J. Grøntved, 1953.

The phytoplankton of the North Sea and adjacent waters in May 1948. Rapp. Proc. Verb., Cons. Perm Int. Expl. Mer, 133:1-87, 29 tables, Pls. A-B, 18 textfigs.

Chemistry oxygen

Braarud, T. and J. T. Ruud, 1937

The Hydrographic conditions and aeration of the Oslo Fjord, 1933-1934. Hvalrådets Skr. No. 15:56 pp., 24 figs.

Reviewed: J. du Cons. XIV(3):406-408. J. N. Carruthers.

Chemistry oxygen (data)

Brandhorst, Wilhelm, 1959.

Nitrification and denitrification in the eastern tropical North Pacific. J. du Cons., 25(1):3-20.

Chemistry, oxygen

Brandhorst, W., 1955.

Hydrographie des Nord- Ostsee Kanal. Kieler Meeresf. 11(2):184-187.

Chemistry oxygen

Brandhorst, Wilhelm, y Hector Inostroza, 1965.

Descripcion grafica de las condiciones oceanograficas de aguas chilenas en base a datos de las expediciones "William Scoresby" y "Chiper". Inst. Fomento Pesq., Bol. Cient., No. 1(2):17-70.

Chemistry oxygen (data)

*Brewer, P.G., C.D. Densmore, R.Munns and R.J. Stanley, 1969.

Hydrography of the Red Sea brines.

In: Hot brines and Recent heavy metal deposits in the Red Sea, E.T. Degens and D.A. Ross, editors, Springer-Verlag, New York, Inc., 138-147.

Chemistry, oxygen

Broenkow, William W., 1965.

The distribution of nutrients in the Costa Rica dome in the eastern tropical Pacific Ocean. Limnology and Oceanography, 10(1):40-52.

Chemistry, oxygen

Brouardel, J., and L. Fage, 1953.

Variation en mer de la teneur en oxygene dissous au proche voisinage des sediments. C.R. Acad. Sci., Paris, 237:1605-1607.

Chemistry, oxygen

Brouardel, J., and J. Vernet, 1958.

Recherches expérimentales sur la variation, en Méditerranée, de la teneur en oxygène de l'eau au proche voisinage des sediments. Bull. Inst. Océan., Monaco, 1111:34 pp.

Chemistry oxygen

*Brosin, Hans-Jürgen, und Dietwart Nehring, 1968.

Der Aquatoriale Unterstrom im Atlantischen Ozean auf 29o30'W im September und Dezember 1966. Beiträge Meeresk., 22:5-17.

Bruevich, S.W.

Chemistry, oxygen (dissolved)

Bruevich, S.W., editor, 1966.

Chemistry of the Pacific Ocean. Inst. Okeanol., Akad. Nauk, SSSR, Isdatel. Nauka, Moskva, 358 pp.

Chemistry, oxygen

Bruevich, S.W., 1960

[Hydrochemical investigations on the White Sea.] Trudy Inst. Okeanol., 42: 199-254.

Chemistry oxygen

Bruevich, S.W., A.N. Bogoyavlensky, and V.V. Mokievskaya, 1960

[Hydrochemical features of the Okhotsk Sea.] Trudy Inst. Okeanol., 42: 125-198.

Chemistry, oxygen (data)

Bruevich, S.V., and I.A. Chikina, 1933
[Hydrochemical observations in the northern part of the Kola Fjord (Barents Sea) in summer 1931] Trudy Gosud. Okeanogr.Inst., 3(3): 120-124.

Chemistry, oxygen

Bryson, R.A., and V.E. Suomi, 1951. Midsummer renewal of oxygen within the hypolimnion. J. Mar. Res. 10(3):263-269, 6 textfigs.

Chemistry, oxygen

Bsharah, L., 1957. Plankton of the Florida Current 5. Environmental conditions, standing crop, seasonal and diurnal changes at a station forty miles east of Miami. Bull. Mar. Sci., Gulf & Caribbean 7(3):201-251.

Chemistry oxygen

Buch, K., 1949. Über den biochemischen Stoffwechsel in der Ostsee Kieler Meeresforschungen 6:31-44, 6 textfigs.

Chemistry, oxygen

Buch, K., 1939. Beobachtungen über das Kohlensäure-gleichgeweit und über den Kohlensäure-austausch zurwischen Atmosphäre und Meer im Nordatlantischen Ozean. Acta Academiae Aboensis, Math. et Phys. XI.9:31 pp., 2 text figs.

Chemistry oxygen

Buch, K., 1934. Beobachtungen über chemische Faktoren in der Nordsee, zwischen Nordsee und Island sowie auf dem Schelfgebiete nördlich von Island. Rapp. Proc. Verb. 89(3):13-31, 8 tables, 2 textfigs.

Chemistry oxygen

Buch, K., and S. Gripenberg, 1938. Jahreszeitlicher Verlauf der chemischen und biologischen Faktoren im Meerwasser bei Hangö im Jahre 1935. Havforskningsinstitutets Skrift No. 118:26 pp., 5 textfigs.

Chemistry oxygen

Buchoff, L.S., N.M. Ingber, and J.H. Brady, 1955. Colorimetric determination of low concentrations of dissolved oxygen in water. Anal. Chem., 27(9) 1401-1404.

Chemistry oxygen

Burkholder, Paul R., 1960. Distribution of some chemical values in Lake Erie. Limnological survey of western and central Lake Erie, 1928-1929. USFWS Spec. Sci. Rept., Fish., No. 334 :71-110.

Chemistry oxygen

Burns, R.B.,1967. Chemical observations in the North Sea in 1965. Annls biol. Copenh. (1965)22:29-31.

Chemistry oxygen

Burns, R.B., 1957. Chemical observations. Ann. Biol., Cons. Perm. Int. Expl. Mer, 12:68-71, Figs. 9-15.

Chemistry oxygen

Burns,R.B., and R. Johnston,1966. Chemical observations from 1964 in the northern North Sea. Annls.biol., Copenh., 21:29-36.

Chemistry, oxygen

Burns, R.B., and R. Johnston, 1965. Chemical observations on the waters of the Scottish west coast and the western approaches. Ann. Biol., Cons. Perm. Int. Expl. Mer, 1963, 20:

Chemistry, oxygen

Burns, R.B., and R. Johnston, 1965. Chemical observations in the northern North Sea, 1963. Ann. Biol., Cons. Perm. Int. Expl. Mer, 1963, 20:57-60.

Chemistry oxygen (data)

Burns, R.B., and R. Johnston, 1964. Chemical observations. Ann. Biol., Cons. Perm. Int. Expl. Mer, 1962, 19: 42-44.

Chemistry oxygen (data)

Burns, R.B., and R. Johnston, 1963. Chemical observations. Ann. Biol., Cons. Perm. Int. Expl. Mer, 1961, 18: 47-48.

Chemistry oxygen

Bursa, Adam S., 1961 The annual oceanographic cycle at Igloolik in the Canadian Arctic. II. The phytoplankton. J. Fish. Res. Bd., Canada, 18(4):563-615.

C

Chemistry, oxygen

Cairns, Alan A. 1967. The zooplankton of Tanquary Fjord, Ellesmere Island, with special reference to calanoid copepods. J. Fish. Res. Bd Can. 24(3):555-568.

Chemistry, oxygen

California Academy of Sciences
California Division of Fish and Game
Scripps Institution of Oceanography
U. S. Fish and Wildlife Service 1950
California Cooperative Sardine Research Program. Progress Rept. 1950:54 pp., 37 text figs.

Chemistry oxygen (data)

California, Humboldt State College, 1964. An oceanographic study between the points of Trinidad Head and the Eel River. State Water Quality Control Bd., Resources Agency, California, Sacremento, Publ., No. 25: 136 pp.

Chemistry oxygen

Callame, Bernard, 1965. Etude d'hydrologie cotière à la Rochelle-La Pallice. Cahiers Océanogr., C.C.O.E.C., 17(6):397-414.

Chemistry oxygen

Canada, Fisheries Research Board, 1959. Physical and chemical data record, coastal seaways project, November 12 to Dece,ber 5, 1958. MSS Rept. Ser. (Oceanogr. & Limnol.), No. 36:120 pp.

Chemistry oxygen

Cannon,Glenn A., 1966. Tropical waters in the western Pacific Ocean, August-September 1957. Deep-Sea Res., 13(6):1139-1148.

Chemistry oxygen (data)

Capurro, L.R.A., 1955. Expedicion Argentina al Mar de Weddell (diciembre 1954 a Enero de 1955). Ministerio de Marina, Argentina, Direccion Gen. de Navegacion e Hidrografia, 184 pp.

Chemistry oxygen

Carpenter, James H., 1966. New measurements of oxygen solubility in pure and natural water. Limnol. Oceanogr., 11(2):264-277.

Chemistry oxygen

Carruthers, J. N., S. S. Gogate, J. R. Naidu & T. Laevastu, 1959. Shorewards upslope of the layer of minimum oxygen off Bombay: its influence on marine biology, especially fisheries. Nature, London, 183:1084-1087.

Chemistry, oxygen

Caspers, H., 1951. Quantitative Untersuchungen über die Bodentierwelt des Schwarzen Meeres im bulgarischen Küstenbereich. Arch. f. Hydrobiol. 45(1/2):192 pp., 66 textfigs.

Chemistry oxygen

Castillejo, F. F., 1966. Non-seasonal variations in the hydrological environment off Port Hackvej, Sydney.

C.S.I.R.O., Div. Fish. Oceanogr. Techn. Paper, No. 21:12 pp.

Chemistry, oxygen

Cavaliere, Antonio, 1963. Studi sulla biologia e pesca di Xiphias gladius. II. Boll. Pesca, Piscicolt. e Idrobiol., 8(2):143-170.

Chemistry oxygen

Chalmers, G.V., and A.K. Sparks, 1959. An ecological survey of the Houston ship channel and adjacent bays. Publ. Inst. Mar. Sci., 6:213-250.

Chemistry oxygen (data)

*Charnell, Robert L., David W.K. Au and Gunter R. Seckel,1967. The Trade Wind Zone Oceanography Pilot Study 11: Townsend Cromwell cruises 4,5, and 6, May to July 1964. Spec. scient. Rep. U.S Fish Wildl. Serv., Fish. 553: 78 pp. (multilithed).

Chemistry, oxygen (data)

*Charnell, Robert L., David W.K. Au, and Gunter R. Seckel,1967. The Trade Wind Zone Oceanography Pilot Study IV: Townsend Cromwell cruises 11,12, and 13, December 1964 to February 1965. Spec. Scient. Rep. U.S. Fish. Wildl. Serv. Fish. 555: 78 pp. (multilithed).

Chemistry oxygen (data)

*Charnell,Robert L., David W.K. Au and Gunter R. Seckel,1967. The Trade Wind Zone Oceanography Pilot Study 111: Townsend Cromwell cruises 8,9, and 10, September to November 1964. Spec. Scient. Rep. U.S. Fish. Wildl. Serv. Fish. 554: 78 pp. (multilithed).

Chemistry oxygen

Chau, Y.K. and R. Abesser, 1958

A preliminary study of the hydrology of Hong Kong territorial waters. Hong Kong Univ. Fish J. (2): 43-57.

Chemistry, oxygen

Chauhan V.D., 1967.
Some observations on chemical and physical conditions of the sea water at Port Okha.
In: Proc. Seminar, Sea Salt and Plants, V. Krishnamurthy, editor, Bhavnagar, India 41-44

Chemistry, oxygen (data)

Chew, F., 1952.
Results of hydrographic and chemical investigations in the region of the "red tide" bloom on the west coast of Florida in November 1952. Bull. Mar. Sci., Gulf and Caribbean 2(4):610-625, 10 textfigs.

Chemistry, oxygen

Chidambaram, K., A.D.I. Rajandran, and A.P. Valsan, 1951.
Certain observations on the hydrography and biol-ogy of the pearl bank, Tholayyiam Paar off Tuti-corin in the Gulf of Manaar in April 1949. J. Madras Univ., Sect. B, 21(1): 48-74, 2 fig

chemistry, oxygen

Chippenfield, P.N.J., 1962.
Some thought on the discharge of industrial efflu-ents to tidal waters.
Chem. Indust., 38:1660-1666.

Chemistry, oxygen

Chirita, Viorel, 1969.
Observations sur l'oxygène dans les eaux marines sur le littoral roumain de la mer Noire
Rapp. P.-v. Réun. Commn int. Explor. scient. Mer Mediterr., 19(4): 771-774.

oxygen

*Choe, Sang, Tai Wha Chung and Hi-Sang Kwak, 1968.
Seasonal variations in nutrients and principal ions contents of the Han River water and its water characteristics. (In Korean; English abstract).
J. oceanogr. Soc., Korea, 3(1):26-38.

oxygen

*Choe, Sang, Tai Wha Chung and Hi-Sang Kwak, 1968.
Seasonal variations in primary productivity and pigments of downstream water of the Han River. (In Korean; English abstract).
J. oceanogr., Soc., Korea, 3(1):16-25.

Chemistry oxygen

Clowes, A.J., 1950.
An introduction to the hydrology of South African waters. Fish. & Mar. Biol. Surv. Div., Dept. Commerce & Industries, Union of South Africa, Investigational Rept. No. 12:42 pp., 20 fold-ins 14 figs.

Chemistry, oxygen

Clowes, A.J., and G.E.R. Deacon, 1935.
The deep-water circulation of the Indian Ocean. Nature 136:936-938, 4 textfigs.

Chemistry, oxygen

Codispoti, Louis A., 1968.
Some results of an oceanographic survey in the northern Greenland Sea, summer 1964.
Techn. Rept., U.S. Nav. Oceanogr. Off., TR202: 49 pp.

Chemistry
oxygen

Cooper, L.H.N., 1961
Vertical and horizontal movements in the ocean. Oceanography, Amer. Assoc. Adv. Sci. Publ. No. 67:599-621.

Chemistry oxygen

Cooper, L.H.N., 1952.
The physical and chemical oceanography of the waters bathing the continental slope of the Celtic Sea. J.M.B.A. 30:465-510, 15 textfigs.

Chemistry oxygen

Cooper, L.H.N., 1933.
Chemical constituents of biological importance in the English Channel, November 1930 to January 1932. II. Hydrogen ion concentration, excess base, carbon dioxide and oxygen. J.M.B.A., 18: 729-753.

Chemistry oxygen

Copeland, B.J., 1967
Environmental characteristics of hypersaline lagoons.
Contrib. mar. Sci., Port Aransas, 12: 207-215

Chemistry
oxygen

Copeland, B.J., and W.R. Duffer, 1964
Use of a clear plastic dome to measure gaseous diffusion rates in natural waters.
Limnology and Oceanography, 9(4):494-499.

Chemistry oxygen(data)

Copenhagen W.J., 1953.
The periodic mortality of fish in the Walvis Region. A phenomenon within the Benguela Current. Union of S. Africa, Div. Fish., Investig. Rept., No. 14:1-35, 9 pls.

Chemistry
oxygen (data)

Corwin, Nathaniel, and David A. McGill, 1963.
Nutrient distribution in the Labrador Sea and Baffin Bay.
U.S.C.G. Bull., No. 48:79-94; 95-153.

Chemistry, oxygen (data)

Coste, Bernard, et Hans-Joachim Minas 1967.
Premières observations sur la distribution des taux de productivité et des concentrations en sels nutritifs des eaux de surface du golfe du Lion.
Cah. oceanogr., 19(5): 417-429

Chemistry oxygen

Craig, R.E., 1956.
Hydrography, near northern seas and approaches.
Ann. Biol., Cons. Perm. Int. Expl. Mer, 11:33-36, Figs. 1-2.

Chemistry oxygen

Craig, R.E., 1954.
A first study of the detailed hydrography of some Scottish West Highland sea lochs (Lochs Inchard, Kanaird, ~~Laxx~~ and the Cairnbawn group).
Ann. Biol., Cons. Perm. Int. Expl. Mer, 10:16-19, Textfigs. 4-5.

oxygen

Craig, R.E., 1952.
Hydrography of the Firth of Forth in the spring.
Cons. Perm. Int. Expl. Mer, Ann. Biol. 8:98-99, Text-fig. 1.

Chemistry, oxygen

*Craig, H., R.F. Weiss and W.B. Clarke, 1967.
Dissolved gases in the equatorial and South Pacific Ocean.
J. geophys. Res., 72(24):6165-6181.

chemistry
oxygen

Crehuet, Ramón Fernández y Maria Jesús del Val Cordón, 1960

Observaciones oceanograficas en la Bahia de Malaga (Marzo 1955 a marzo 1957). Bol. Inst. Esp. Ocean., 98: 1-29.

Chemistry, oxygen (data)

Cromwell, T., 1954.
Mid-Pacific oceanography II. Transequatorial waters, June-August 1950, January-March 1951.
U.S.F.W.S. Spec. Sci. Rept. - Fish. No. 131:

oxygen Chemistry, oxygen

Cromwell, T., 1953.
Circulation in a meridional plane in the central equatorial Pacific. J. Mar. Res. 12(2):196-213, 9 textfigs.

Chemistry, oxygen (data)

Cromwell, T., 1951.
Mid-Pacific oceanography, January through March 1950. Spec. Sci. Rept., Fish., No. 54:9 pp., 17 figs., station data.

Chemistry, oxygen

Cronin, L. Eugene, Joanne C. Daiber and Edward M. Hulburt, 1962
Quantitative seasonal aspects of zooplankton in the Delaware River estuary.
Chesapeake Science, 3(2):63-93.

Chemistry
oxygen

Cubillos, M., Regina, 1960.
Estudio de las expediciones oceanográfico-pesqueras realizadas en Tarapacá y Antofagasta, 1954-1955.
Rev. Biol. Mar., 10(1/3):181-194.

Chemistry
oxygen

Currie, Ronald I., 1965.
The oceanography of the south-east Atlantic.
Anais Acad. bras. Cienc., 7(Supl.):11-22.
no abstract

D

Chemistry, oxygen

D'Anglejean, B.F., 1967.
Origin of marine phosphorites off Baja California, Mexico.
Marine Geol., 5(1): 15-44.

Chemistry
oxygen

Darbyshire, J., 1964
A hydrological investigation of the Agulhas Current.
Deep-Sea Res., 11(5):781-815.

Chemistry, oxygen

Davidovitch, R.L., 1964.
Short chemical characteristics of the northwest Pacific Ocean. (In Russian).
Gosudarst. Kom. Sov. Ministr., SSSR, Ribn. Choz. Trudy, VNIRO, 49, Izv. TINRO, 51:93-98.

Chemistry oxygen (data)

Davidovich, R.L., 1963
Hydrochemical features of the southern and southeastern parts of the Bering Sea. (In Russian).
Sovetsk. Ribokh. Issled. B Severo-Vostokh. Chasti Tikhogo Okeana, VNIRO 48, TINRO 50(1): 85-96.

Chemistry, oxygen

Day, J.H., 1951.
The ecology of South African estuaries. 1. A review of estuarine conditions in general. Trans R. Soc., S. Africa, 33(1):53-91, 2 textfigs.

Chemistry oxygen

Deacon, G. E. R., 1933
A general account of the hydrology of the South Atlantic Ocean. Discovery Repts. 7:173-238, pls.8-10.

Chemistry oxygen

Dean, David and Harold H. Haskin, 1964
Benthic repopulation of the Raritan River estuary following pollution abatement.
Limnology and Oceanography, 9(4):551-563.

Chemistry, oxygen

de Barros Machado, L., 1950.
Pequisas fisicas e quimicas do sistema hidrografico da regiao lagunar de Cananeia. 1. Cursos de aguas. Bol. Inst. Paulista Oceanogr. 1(1):45-67, 3 textfigs.

Chemistry, oxygen

DeBuen, R., 1937.
Déterminations physico-chimiques dans les eaux de la Baie de Santander pendant l'été de 1935. Cons. Perm. Int. Expl. Mer, Rapp. Proc. Verb. 104:26-31, 6 textfigs.

Chemistry, oxygen

DeBuen, O., 1925.
Croisière océanographique du Transport "Almirante Lobo". Cons. Perm. Int. Expl. Mer, Rapp. Proc. Verb. 37:33-57.

Chemistry, oxygen

de Buen, O., 1916.
Première campagne de l'Institut espagnol d'Océanographie dans la Mediterranée. Liste des stations et des operations. Bull. Inst. Océan., Monaco, No. 314:23 pp.

Defant, A., G. Böhnecke, H. Wattenberg, 1936.
I. Plan und Reiseberichte die Tiefenkarte das Beobachtungsmaterial. Die Ozeanographischen Arbeiten des Vermessungsschiffes "Meteor" in der Dänemarkstrasse und Irmingersee während der Fischereischutzfahrten 1929, 1930, 1933 und 1935. Veroffentlichungen des Instituts für Meereskunde, n.f., A. Geogr.-naturwiss. Reihe, 32:1-152 pp., 7 text figs., 1 plate.

chemistry, oxygen

De Groot, S.J., and H. Postma, 1968.
The oxygen content of the Wadden Sea.
Netherlands J. Sea Res, 4(1):1-10.

Chemistry, oxygen

De La Lanza Espino 1965.
Algunas caracteristicas hidrograficas del sistema litoral de Veracruz, Ver.
Anales Inst. Biol. Univ. Mex. 36 (1/2): 47-52.

chemistry oxygen

Demel, K., and Z. Mulicki, 1954.
Quantitative investigations on the biological bottom productivity. Rept. Sea. Fish. Inst., Gdynia, No. 7:75-126, 14 textfigs.

chemistry, oxygen (data)

De Queiroz Santos, Neuzon 1967.
Principais nutrientes e alguns dados fisico-quimicos da região lagunar de Cananela.
Bolm Inst. Biol. mar. Univ.Fed. Rio Grande Norte 4:1-14

Chemistry, Oxygen

Dietrich, Gunter, 1965
New hydrographica aspects of the northwest Atlantic.
ICNAF Spec. Publ. No. 6:29-51,

Chemistry oxygen

Dietrich, Gunter, 1964.
Oceanic Polar Front Survey in the North Atlantic.
In: Research in Geophysics. Solid Earth and Interface Phenomena, 2:291-308.

Chemistry, oxygen

Dietrich, G., 1963.
Die Meere.
Die Grosse Illustrierte Länderkunde, Bertelsmann Verlag, 2:1523-1606.

Chemistry oxygen

Dietrich, Günter, 1961
Some thoughts on the working-up of the observations made during the "Polar Front Survey" in the IGY 1958.
Rapp. Proc. Verb., Cons. Perm. Int. Expl. Mer, 149:103-110.

Chemistry oxygen

Dietrich, G., 1960
Temperatur-, Salzgehalts-u. Sauerstoff-Verteilung auf den Schnitten von F.F.S. ANTON DOHRN u. V.F.S. GAUSS im Internationalen Geophysikalischen Jahr 1957/1958.
Deut. Hydrogr. Zeit., Erganz. Reihe B (4): 103 pp.

Chemistry oxygen

Dietrich, G., 1939
Das Amerikanische Mittelmeer; ein meereskundlicher über blick. Zeitschr. Gesellschaft für Erdkunde zu Berlin. Jahrgang 1939 (3/4):108-130, 38 textfigs.

chemistry, oxygen

Dietrich, G., 1937.
Über Bewegung und Herkunft des Golfstromwassers. Veröff. Inst. Meereskunde, n.f., Ser. A. Geogr.-Naturwiss. Reihe. 33:53-91, 26 textfigs.

Chemistry, oxygen

Dobrazhanskaia, M.A. 1970.
Distribution of oxygen in the Red Sea and in the Gulf of Aden based on the investigations of the E/S Akademik A. Kovalevskii. (In Russian)
Biol. Moria, Kiev, 21: 41-65.

Chemistry, oxygen

Dobrzhanskaya, M.A., 1957.
Seasonal peculiarities in the vertical distribution of oxygen within the zone of photosynthesis in the Black Sea.
Doklady Akad. Nauk, SSSR, 115(4):755-758.

Chemistry oxygen

Dodimead, A.J., F. Favorite and T. Hirano, 1964.
Review of oceanography of the subarctic Pacific Region. Salmon of the North Pacific Ocean. II. Collected Reprints, Tokai Reg. Fish. Res. Lab., No. 2:187 pp.

Chemistry, oxygen

Doe, L. A. E., 1965.
Physical conditions on the shelf near Karachi during the post-monsoonal calm, 1964. Ocean Sci. and Ocean Eng., Mar. Sci. Techn. Soc. - Amer. Soc. Limnol. Oceanogr., 1: 278-292.

Chemistry, oxygen

Doe, L.A.E., 1955.
Offshore waters of the Canadian Pacific coast. J. Fish. Res. Bd., Canada, 12(1):1-34, 19 textfigs.

Chemistry, oxygen

Donguy, Jean Rene, et Michel Prive, 1964.
Les conditions de l'Atlantique entre Abidjan et l'Equateur.
Cahiers Oceanogr., C.C.O.E.C., 16(5):393-398.

Chemistry oxygen (data)

*Brainville, Gérard, 1968.
Le fjord du Saguenay: 1. Contribution à l'océanographie.
Le Naturaliste canadien, 95(4):809-855.

chemistry, oxygen

Duedall, Iver W. and Arthur R. Coote, 1972.
Oxygen distribution in the south Atlantic.
J. geophys. Res. 77(3): 496-498.

chemistry, oxygen

Dugal, L. -P., 1932-1933.
Observations sur le chlore et l'oxygène dessous de l'estuaire du Saint-Laurent. Publ. Sta. Biol. Saint-Laurent Contr. No. 4: 20 pp.,

Chemistry oxygen

Düing, Walter, 1965.
Strömungsverhältnisse im Golf von Neapel.
Pubbl. Staz. Zool., Napoli, 34:256-316.

Chemistry oxygen

*Düing, Walter, Klaus Grasshoff und Gunther Krause, 1967.
Hydrographische Beobachtungen auf einem Äquatorschnitt im Indischen Ozean.
"Meteor Forschungsergebnisse (A)(3):84-92.

Chemistry, oxygen

Dussart, B., and C. Francis-Boeuf, 1949
Technique du dosage de l'oxygene dissous dans l'eau basee sur le methode Winkler. Circ. C.R.E.O., 1(1):8 pp.

Chemistry, oxygen

Dutta, N., J.C. Malhotra and B.B. Bose, 1954.
Hydrology and seasonal fluctuations of the plankton in the Hooghly estuary.
Symp. Mar. Fresh-water Plankton, Indo-Pacific, Bangkok, Jan. 25-26, 1954, FAO-UNESCO:35-47.

Chemistry, oxygen (data)

Dybern, Bernt I, 1967.
Topography and hydrography of Kviturdvikpollen and Vågsbøpollen on the west coast of Norway.
Sarsia 30:1-27.

E

Chemistry oxygen

Edmondson, W.T., and Y.H. Edmondson, 1947.
Measurements of production in fertilized sea water. J. Mar. Res. 6(3):228-246, Figs. 55-62.

Chemistry, oxygen
Eggvin, J., 1933.
A Norwegian fat-herring fjord. An oceanographical study of the Eidsfjord. Repts. Norwegian Fish. Mar. Invest. 4(6):22 pp., 11 textfigs.

Chemistry oxygen (data)
*Ehrhardt, Jean-Paul, 1967.
Contribution a l'etude du planoton superficiel et Sub-Superficiel du Canal de Sardaigne et de la mer Sud-Tyrrhenienne: Campagne de l' Orbgny, du 15 Septembre au 19 Octobre 1963.
Cah. oceanogr. 19(8):657-686.

chemistry, oxygen
Eisma, D. and A.J. van Bennekom, 1971.
Oceanographic observations on the eastern Surinam Shelf. Hydrogr. Newsletter, R. Netherlands Navy, Spec. Publ. 6: 25-29.

Chemistry, oxygen
Elizarov, A.A. 1971.
Oceanographic surveys of the shelf and continental slope area of west India. (In Russian).
Vses. nauchno-issled. Inst. Morsk. ribn. Khoz. Oceanogr. VNIRO, Trudy 72:34-45

Chemistry - oxygen
Elizarov, A.A., 1968.
Preliminary results of oceanographic investigations of the west coast of India. (In Russian).
Trudy, Vses. Nauchno-Issled. Inst. Morsk. Ribn. Okeanogr (VNIRO) 64, Trudy Azovo-Chernomorsk. Nauchno-Issled. Inst. Morsk. Ribn. Khoz. Okeanogr. (AscherNIRO), 28: 94-101.

Chemistry, oxygen (data)
El-Sayed, Sayed Z., and Enrique F. Mandelli, 1965
Primary production and standing crop of phytoplankton in the Weddell Sea and Drake Passage.
In: Biology of Antarctic seas II.
Antarctic Res. Ser., Am. Geophys. Union, 5:87-106

Chemistry, oxygen
Emery, K.O., 1954.
Source of water in basins off southern California. J. Mar. Res. 13(1):1-21, 6 textfigs.

chemistry oxygen (data)
Emery, K.O., and Jobst Hülsemann, 1963
Submarine canyons of southern California.
Part 1. Topography, water, and sediments.
Allan Hancock Pacific Expeditions, 27(1): 1-80.

Chemistry, Oxygen
Emery, K. O., and J. Hülsemann, 1961 (1962).
The relationships of sediments, life and water in a marine basin.
Deep-Sea Res., 8(3/4):165-180.

Chemistry, oxygen (chiefly data)
Emilsson, I., 1956.
Relatorio e resultados fisico-quimicos de tres cruzeiros oceanograficos em 1956.
Contrib. Avulsas, Inst. Ocean., Univ. Sao Paulo, Ocean. Fisica, No. 1:1-70.

Chemistry oxygen (data)
Endo, Takuo, 1965.
On primary production in the Seto Inland Sea.
I. Primary production and hydrographic conditions. (In Japanese; English abstract).
J. Fac. Fish., Animal Husbandry, Hiroshima Univ. 6(1):85-100.

Chemistry, oxygen
English, T. Saunders, 1961.
Biological oceanography in the North Polar Sea from IGY Drifting Station Alpha, 1957-58.
Amer. Geophys. Union, Trans., 42(4):518-525.

Reprinted from:
"Some biological oceanographic observations in the Central North Polar Sea, Drift Station Alpha, 1957-1958".
Arctic Inst., North America, Res. Paper, No. 13.

Also:
Air Force Cambridge Res. Lab. Sci. Rept. (AFCRL-625) No. 15

Chemistry, oxygen
Epstein, I. M., 1959.
Determination of oxygen in a reservoir with an oxygen sounding device.
Trans. Hydrobiol. Soc. U.S.S.R., 9:379-386.

Chemistry oxygen
Ercegovic, A., 1940
Weitere Untersuchungen über einige hydrographische Verhältnisse und über die Phytoplanktonproduktion in den Gewässern der Östlichen Mitteladria. Acta Adriatica 2(3):95-134, 8 text figs.

oxygen
Ercegovic, A., 1934.
Température, salinité, oxygène et phosphates dans les eaux cotière de l'Adriatique occidental moyen. Acta Adriatica 1(5):1-51, 19 textfigs.

F

Chemistry - oxygen
Fedosov, M. V., 1955
Chemical basis for the foodstuffs in the Azov Sea and prognosis of its change in connection with water construction projects in rivers.
Trudy VNIRO 31: 35-61.

Chemistry oxygen
Fedosov, M.V., and N.A. Azova, 1965.
Hydrochemical constituents of the balance of biogenic matter in the Gulf of Alaska. (In Russian).
Sovetsk. Ribokhoz. Issled. Severo-Vostochn. Chesti Tikhogo Okeana, 4 (Vses. Nauchno-Issled. Inst. VNIRO Trudy 58:Tikhookean. Nauchno-Issled. Inst., TINRO Trudy 53):11-20.

Chemistry, oxygen
Fedosov, M.V., and L.A. Barsukova, 1959.
Gas regime of the water mass of the northern Caspian. Trudy VNIRO, 38:78-87.

Chemistry, oxygen
*Fedosov, M.V., and I.A. Ermachenko, 1967.
Hydrochemical characteristics of water masses and water exchange between Iceland and Faroes (hydrographical surveys in 1960).
Rapp. P.-V.Réun.Cons. perm. int. Explor. Mer, 157:196-

Chemistry, oxygen
Fedosov, M.V., and V.V. Volkovinsky, 1965.
The oxygen regime as an indicator of the primary productivity of marine waters. Investigations in line with the programme of the International Geophysical Year. (In Russian).
Trudy, Vses. Nauchno-Issled. Inst. Morsk. Ribn. Choz. i Okeanogr., (VNIRO), 57:131-144.

Chemistry oxygen (data)
Fieux, Michèle 1971.
Observations hydrologiques hivernales dans le Rech Lacaze-Duthiers (Golfe du Lion).
Cah. océanogr. 23(8):677-686.

Chemistry, oxygen
Filarski, J., 1958(1960)
Polish observations in the southern Baltic March 1958/February 1959. Cons. Perm. Int. l'Expl. Mer: Ann. Biol. 15: 45-47.

Chemistry, oxygen
Filarski, J., 1957.
Polish investigations in the southern Baltic.
Ann. Biol., Cons. Perm. Int. Expl. Mer, 12:111, Figs., 1-7.

oxygen (data)
Filarski, J., 1955.
Hydrographical conditions of the southern Baltic for the period from April 1952 to May 1953.
Prace Morskiego Inst. Ryback., Gydni, No. 8:255-282.

Chemistry oxygen
Fiske, John D., Clinton E. Watson and Philip G. Coates, 1966.
A study of the marine resources of the North River.
Comm. Mass., Div. Mar. Fish., Monogr. Ser. 3:53 pp.

Chemistry, oxygen
Flores P., Luis Alberto 1967.
Informe preliminar del crucero 6611 de la primavera de 1966 (Cabo Blanco-Punta Coles).
Informe, Inst. Mar, Peru 17:16 pp.

Chemistry oxygen
Flores, Luis, Oscar Guillén y Rogelio Villanieva, 1966.
Informe preliminar del Crucero de invierno 1965 (Máncora-Morro Sama).
Inst. Mar. Peru, Informe, No. 11:1-34 (multilithed)

Chemistry, oxygen
Flores P., Luis Alberto, y Louis A. Poma Elias 1967.
Informe preliminar del crucero 6608-09 de invierno 1966 (Mancora-Ilo).
Informe Inst. Mar, Peru 16:24 pp

Fonselius, Stig H.

chemistry, oxygen
Fonselius, Stig H. 1970.
On the stagnation and recent turnover of the water in the Baltic.
Tellus 22 (5): 533-544.

Chemistry oxygen
Fonselius, Stig H., 1969.
Hydrography of the Baltic deep basins. III.
Rept., Fish. Bd., Sweden, Ser. Hydrogr., 23: 1-97.

Chemistry, oxygen
Fonselius, S.H. 1968.
Observations at Swedish lightships and in the central Baltic 1966.
Annls biol. Copenh. 1966, 23:74-78.

Chemistry, oxygen
Fonselius, Stig H. 1967.
Hydrography of the Baltic deep basins. II
Rep. Fishery Bd, Swed., Hydrogr. 20:31 pp.

Chemistry, oxygen (data)

Fonselius, S.H., 1966.
Observations at Swedish lightships in the Central Baltic in 1964.
Annls biol., Copenh., 21:46-49.

chemistry, oxygen

Fonselius, S.H., 1965.
The oxygen analysis during the informal intercalibration meeting in Copenhagen.
UNESCO, Techn. Papers, Mar. Sci., No. 3:3-5. (mimeographed)

Chemistry oxygen

Fonselius, Stig H., 1965.
Observations at Swedish light vessels and in the Central Baltic.
Ann. Biol., Cons. Perm. Int. Expl. Mer, 1963, 20:66-68.

Chemistry oxygen

Fonselius, Stig H., 1962
Hydrography of the Baltic deep basins.
Fishery Board of Sweden, Ser. Hydrogr., Rept. 13: 40 pp.

Chemistry oxygen

Forsbergh, Eric D., 1963
Some relationships of meteorological, hydrographic, and biological variables in the Gulf of Panama. (In English and Spanish).
Bull., Inter-American Tropical Tuna Comm., 7(1): 109 pp.

chemistry
Oxygen (data)

Forsbergh, Eric D. William W. Broenkow, 1965.
Observanciones oceanograficas del oceano Pacific oriental recolectadas por el barco Shoyo Maru, octubre 1963-marzo 1964.
Comision Interamericana del Atun Tropical, Bol. 10(2): 85-237.

Chemistry, oxygen

Føyn, E., 1955.
Continuous oxygen recording in sea water.
Repts. Norwegian Fish. Mar. Invest. 11(3):1-8, 4 textfigs.

Chemistry, oxygen

Fox, C.J.J., 1909.
On the coefficients of absorption of nitrogen and oxygen in distilled water and sea water, and of atmospheric carbonic acid in sea water.
Trans. Faraday Soc., 5:68-87.

Chemistry oxygen

Francis-Boeuf, C., 1947.
Données sur la consommation d'oxygène in vitro de quelques vases fluvio-marines. C.R. Acad. Sci. 225:1083-1084.

Chemistry, oxygen

Francis-Boeuf, C., 1947.
Production et consommation d'oxygène par la pellicule superficielle des vases fluvio-marines
C.R. Adad. Sci. 225:820-822.

chemistry oxygen

Francis-Boeuf, C., 1947
Sur la teneur en oxygène dissous du milieu intéreur des vases fluors marines. C.R., Paris, 225:392-394.

chemistry, oxygen (data)

Franco Paolo 1970.
Oceanography of northern Adriatic Sea. I. Hydrologic features: cruises July-August and October-November 1965.
Archo Oceanogr. Limnol. 16 (Suppl.):1-93

Chemistry
oxygen (data)

Franco, Paolo, 1962
Condizioni fisiche e chimiche delle acque lagunari nel Porto-Canale di Malamocco. 1. Giugno 1960-Giugno 1961.
Arch. Oceanogr. e Limnol., 12(3):225-255.

Chemistry, oxygen

Freier, R., and G. Resch, 1956.
Über die Bestimmung des in Wasser gelösten Sauerstoffs. Zeits. f. Anal. Chem. 148(6):427-434.

Chemistry, oxygen (data)

Fuglister, F.C., 1963.
Gulf Stream '60.
Progress in Oceanography, 1:263-373.

Chemistry
oxygen

Fukai, R., 1958
On the deep circulation in the north-western North Pacific with reference to vertical distribution of dissolved oxygen.
Proc. UNESCO Symp. Phys. Oceanogr., Tokyo, 1955, 149-152.

Included in COLLECTED REPRINTS, 1958, Tokai Reg. Fish. Res. Lab., B-237.

Chemistry, oxygen

Fukai, R., 1957.
On the deep circulation in the northwestern North Pacific with reference to vertical distribution of dissolved oxygen.
Proc. UNESCO Symp., Phys. Ocean., Tokyo, 1955: 149-152.

Chemistry, oxygen

Fukai, R., 1953.
On the distribution of nutrient salts in the Equatorial North Pacific.
Bull. Chem. Soc., Japan, 26(9):485-489, 5 textfigs.

Also reprinted in: Bull. Tokai Reg. Fish. Lab. 7(Contr. B) as Contr. 113.

Chemistry, oxygen

Fukase, S., 1951.
On the distribution of dissolved oxygen in the Okhotsk Sea. Bull. Maizuru Mar. Obs. No. 2:26-29, 9 textfigs.

Chemistry oxygen

Fukuoka, Jiro, 1965 (1967).
Condiciones meteorologicas e hidrograficas de los mares adyacentes a Venezuela 1962-1963.
Memoria Soc. Cienc. nat. La Salle, 25(70/71/72):11-38.

Chemistry oxygen

Fukuoka, Jiro, 1965.
Hydrography of the adjacent sea. (1). The circulation in the Japan Sea.
J. Oceanogr., Soc., Japan, 21(3):95-

Chemistry oxygen

Fukuoka, Jiro, 1962.
Abyssal circulation in the Atlantic near the poles and abyssal circulation in the Pacific and other oceans in relation to the former.
J. Oceanogr. Soc., Japan, 18(1):5-12.

JEDS Contrib. No. 27.

Chemistry
oxygen

Fukuoka, Jiro, 1962.
Caracteristicas de las condiciones hidrograficas del Mar Caribe.
Memoria, Soc. Ciencias Nat. La Salle, Venezuela, 22(63):198-205.

Chemistry
oxygen

Fukuoka, Jiro, 1962
Characteristics of hydrography of the Japan Sea - in comparison with hydrography of the North Pacific -. (In Japanese; English abstract).
J. Oceanogr. Soc., Japan, 20th Ann. Vol., 180-188.

Chemistry, oxygen

Furnestin, Jean, 1963.
Teneur en oxygène des eaux de la Méditerranée Algéro-Tunisienne (supplément à l'étude hydrologique de février 1960 campagne du "Président-Théodore-Tissier").
Rev. Trav. Inst. Pêches Marit., 27(3):119-126.

Chemistry
oxygen

Furnestin, J., 1961.
Teneur en oxygène des eaux de la Méditerranée occidentale (supplement à l'étude hydrologique de juin-juillet 1957).
Rapp. Proc. Verb., Réunions, Comm. Int. Expl. Sci., Mer Méditerranée, Monaco, 16(3):583-584.

chemistry
oxygen

Furnestin, J., 1960

Teneur en oxygène des eaux de la Méditerranée occidentale (Supplément à l'étude hydrologique de juin-juillet 1957).
Rev. Trav. Inst. Pêches Marit., 24: 453-480.

G

Chemistry oxygen

Gaarder, Karen Ringdal, 1938.
Phytoplankton studies from the Tromsø district, 1930-31. Tromsø Mus. Årshefter, Naturhist. Avd., 11, 55 (1):159 pp., 4 fold-in pls., 12 textfigs.

Chemistry, oxygen (data)

Gaarder, T., 1927.
Die Sauerstoffverhältnisse im östlichen Teil des Nord Atlantischen Ozeans. Cruises of the "Armauer Hansen", No. 2. Geofys. Publ. 4:72 pp., 2 figs.

Gaarder, T., and R. Spärck, 1932.
Hydrographisch-biochemische Untersuchungen in norwegischen Austern-Pollen. Bergens Mus. Aarbok, naturvidensk.-rekke, No. 1:5-144, 75 textfigs.

Chemistry, oxygen

Gad, G., 1948.
Die titrimetrische Bestimmung des im Wasser gelösten Sauerstoffs ohne Verwendung jodhaftiger Reagenzien. Gesundheits Ingenieur 69(1):22.

Chemistry
oxygen

Gade, Herman G., 1963
Some hydrographic observations of the inner Oslofjord during 1959.
Hvalrådets Skrifter, No. 46:1-62.

Chemistry, oxygen

Gade, Herman G., 1962.
Further hydrographic observations in the Gulf of Cariaco, Venezuela. The circulation and water exchange.
Bol. Inst. Oceanogr., Univ. de Oriente, Venezuela, 1(2):359-395.

Chemistry, oxygen

Gade, Herman G., 1961.
Informe sobre las condiciones hidrograficas en el Golfo de Cariaco, para el periodo que empieza en mayo y termina en noviembre de 1960. (In English).
Bol. Inst. Oceanograf., Univ. Oriente, Cumana, Venezuela, 1(1):21-47.

Chemistry, oxygen

Gade, Herman G., 1961.
On some oceanographic observations in the southeastern Caribbean Sea and adjacent Atlantic Ocean with special reference to the influence of the Orinoco River.
Bol. Inst. Oceanogr., Univ. de Oriente, Venezuela, 1(2):287-342.

Chemistry, oxygen

Gallardo Yves 1970
Contribution à l'étude du Golfe de Guinée hydrologie et courants dans la région de l'Île Annobon.
Cah. océanogr. 22(3): 277-288

Chemistry, oxygen

Galtsoff, P.S., ed., 1954.
Gulf od Mexico, its origin, waters and marine life. Fish. Bull. Fish and Wildlife Service, 55:1-604, 74 textfigs.

Chemistry, oxygen

Gameson, A.L.H., and K.G. Robertson, 1955.
The solubility of oxygen in pure water and sea water. J. Appl. Chem. 5(9):502.

Chemistry, oxygen

Ganopati, P.N., and G. Chondrasekhara Rao, 1962.
Ecology of the interstitial fauna inhabiting the sandy beaches of Waltair coast.
J. Mar. Biol. Assoc., India, 4(1):44-57.

in sea water and interstitial water

chemistry oxygen(data)

Ganapati, P.N., E.C. LaFond and P.V. Bhavanarayana, 1956.
On the vertical distribution of chemical constituents in the shelf waters off Waltair.
Proc. Indian Acad. Sci., 44:68-71.

chemistry oxygen

Ganapati, P.N., and D. Venkata Rama Sarma, 1958.
Hydrography in relation to the production of plankton off Waltair coast.
Andhra Univ. Mem. Oceanogr., 2:168-192.

Chemistry, oxygen

Ganapati, P.N., and D.V. Subba Rao, 1958.
Quantitative study of plankton off Lawson's Bay, Waltair.
Proc. Indian Acad. Sci., (B), 48(4):189-209.

Chemistry, oxygen

Ganapati, P.N., and D.V. Subba Rao, 1957.
On upwelling and productivity of the waters off Lawson's Bay, Waltair. Current Science 26(11): 347-348.

Garg, J.N., C.B. Murty and R. Jayaraman, 1969.

Vertical distribution of oxygen in the Bay of Bengal and the Andaman Sea during February-March 1963. Bull. natn. Inst. Sci., India, 38(1) 40-48.

Chemistry, oxygen

Garkavaia, G.P., and L.E. Pogdniakova 1968.
Data on the hydrochemical regime in Chesha Bay, Barents Sea. (In Russian)
Trudy Murmansk. morsk. biol. Inst. 17(21): 4-21.

Chemistry, oxygen

Garland, C.F., 1952
A study of water quality in Baltimore Harbor.
Chesapeake Biol. Lab. Publ. 96:1-132, figs.

chemistry oxygen(data)

Gennesseaux, Maurice, 1960

L'oxygène, les phosphates et les nitrates dissous dans les eaux du bassin sud-ouest de la Méditerranée occidentale (des Îles Baléares à Gibraltar). Trav. C.R.E.O. ns. 3(4): 5-1622.

Also in:
Trav. Lab. Géol. Sous-marine, 10 (1960)

Chemistry oxygen

Genovese, S., 1952.
Osservazioni idrologiche esequite nella Tonnara del "Tono" (Milazzo) durante la Campagna di Pesca 1952. Bol. Pesca, Piscicult., Idrobiol., n.s., 7(2):196-200, 3 textfigs.

oxygen

Gieskes, Joris M.T.M., and Klaus Grasshoff 1969.
A study of the variability in the hydrochemical factors in the Baltic Sea on the basis of two anchor stations, September 1967 and May 1968.
Kieler Meeresforsch. 25(1): 105-132

Chemistry, oxygen

Gilmartin, M., 1962.
Annual cyclic changes in the physical oceanography of a British Columbia fjord.
J. Fish. Res. Bd., Canada, 19(5):921-974.

Chemistry, oxygen

Gilson, H. C., 1937
Chemical and Physical Investigations. The nitrogen cycle. John Murray Exped., 1933-34, Sci. Repts., 2(2):21-81, 16 text figs.

Glowinska, A.

Chemistry, oxygen

Glowinska, A., 1967.
Polish observations in the southern Baltic 1965.
Annls. biol. Copenh. (1965)22:51-54.

chemistry oxygen (data)

Glowinska, A., 1964.
Polish observations in the southern Baltic, 1962.
Ann. Biol., Cons. Perm. Int. Expl. Mer, 1962, 19: 48-49.

Chemistry, oxygen

Glowinska, Amalia, 1963.
Hydrologic conditions in the southern Baltic in the years 1951-1960. (In Polish; Russian and English summaries).
Prace Morsk. Inst. Ryback. w Gdyni, 12A:23-35.

chemistry oxygen(data)

Glowinska, A., (1959) 1961

Polish observations in the southern Baltic in 1959.
Ann. Biol., Cons. Perm. Int. Expl. Mer, 16: 61-65.

oxygen (data)

Glowinska, A., 1954.
[Hydrologic research in the southern Baltic in 1951.] Rept. Sea Fish. Inst., Gdynia, No. 7:159-190, 27 textfigs.

oxygen Chemistry, oxygen

Glowinska, A., 1949.
Observations off the Polish coast 1947 and 1948.
Ann. Biol. 5:131-133, 2 textfigs.

Chemistry, oxygen

Glowinska, A., 1950.
Hydrographical characteristics of the southern Baltic in the time from November 1948 to October 1949. Bull. Inst. Pêches Maritimes, Gydnia, No. 5:123-143, 3 textfigs. (In Polish; English summary).

Chemistry, oxygen O_2

Glowinska, A., 1949
The Hydrography of the Gulf of Gdansk.
Ann. Biol. Int. Cons., 4:142-143, text fig.12.

Chemistry, oxygen

Glowinska, A., 1948.
[Hydrographic conditions in the Gulf of Gdansk in the second part of 1946.] (In Polish, with English summary.) Bull. Lab. Mar., Gydnia, No. 4:171-185, 9 textfigs.

Chemistry, oxygen

Gololobov Ia. K. 1964.
Analysis of the chemical compounds in biological production in the Black Sea and some peculiarities of their formation. (In Russian).
Trudy azov. chernomorsk. nauchno-issled. Inst. morsk. ryb. khoz. okeanogr. 23:33-47.

chemistry oxygen(data)

Gololobov, Ya. K. 1963
Hydrochemical characteristics of the Aegean Sea during autumn of 1959. (In Russian).
Mezhd. Geofiz. Komitet, Prezidiume, Akad. Nauk SSSR, Rezult. Issled. Programme Mezhd. Geofiz. Goda, Okeanol. Issled., (8):90-96.

Chemistry, oxygen

Gololobov, Y.K., 1953.
[The thickness of the oxygen-hydrogen sulfide layer in the Black Sea.] Gidrochim. Materialy, 21:3-9.

Abstr. Chem. Abstr. 11126e, 1954.

Chemistry, oxygen

Goodman, J.R., J.H. Lincoln, T.G. Thompson and F.A. Zeusler 1942.
Physical and chemical investigations: Bering Sea, Bering Strait, Chukchi Sea during the summers of 1937 and 1938.
Publ. Oceanogr. Univ. Washington 3(4): 105-169.

Chemistry oxygen

Gordon, Arnold L., 1966.
Potential temperature, oxygen and circulation of bottom water in the Southern Ocean.
Deep-Sea Res., 13(6):1125-1138.

Chemistry, oxygen

Gorsline, Donn S., 1963.
Oceanography of Apalachicola Bay, Florida.
In: Essays in Marine Geology in honor of K.O. Emery, Thomas Clements, Editor, Univ. Southern California Press, 69-96.

Chemistry, oxygen

*Gostan, Jacques, 1968.
Conditions hydrologiques observées pendant l'été entre la Riviera et la Corse.
Cah. océanogr., 20(1):37-66.

Chemistry, oxygen (data only)

Goulet, Julien R., Jr., and Merton C. Ingham 1968.
Oceanic conditions in the northwestern Gulf of Guinea: Geronimo Cruise 3, 10 February to 21 April 1964
Data Report, U.S. Department of the Interior, Bur. Comm. Fish. 25: 1 film (microfiche)

Chemistry, oxygen
Graham, H.W., and E.G. Moberg 1944.
Chemical results of the last cruise of the Carnegie. Chemistry 1. Scientific results of Cruise VII of the Carnegie during 1928-1929 under command of Captain J. P. Ault.
Carnegie Publ. Washington 562:58pp.

chemistry, oxygen
Gran, H.H., 1927.
The production of plankton in the coastal waters off Bergen, March-April, 1922. Rept. Norwegian Fish. Mar. Invest. 3(8):74 pp., 8 textfigs.

chemistry, oxygen
Gran, H.H., and T. Braarud, 1935
A quantitative study of the phytoplankton in the Bay of Fundy and the Gulf of Maine (including observations on hydrography, chemistry, and turbidity).
J. Biol. Bd., Canada, 1(5):279-467, 69 text figs.

Chemistry oxygen
Grasshoff, Klaus, 1964.
Über ein neues Gerät zur Herstellung eines absoluten Standards für die Sauerstoffbestimmung nach der Winklermethode.
Kieler Meeresf., 20(2):143-147.

chemistry, oxygen
Greece, The National Hellenic Oceanographic Society, 1959
The oxygen content of the sea water.
Thalassina Phylla, Athens, 3(1):1-6.

chemistry oxygen
Greffard, J., and J.C. Braconnot, 1964.
Etude de la répartition des teneurs en oxygène dissous relevées au cours d'une campagne d'été effectuée dans la partie nord du bassin occidental méditerranéen. Comparison avec la répartition des salinités.
Cahiers Océanogr., C.C.O.E.C., 16(6):439-456.

Chemistry, oxygen
Griffiths, Raymond C., 1965.
A study of ocean fronts off Cape San Lucas, Lower California.
U.S.F.W.S. Spec. Sci. Rept., Fish., No. 499:54 pp.

chemistry, oxygen
Griffiths, V.S., and M.I. Jackman, 1962.
A polarographic study of dissolved oxygen. I.
Talanta, 9:205-212.

Chemistry, oxygen (data)
Grøntved, J., 1952.
Investigations on the phytoplankton in the southern North Sea in May 1947. Medd. Komm. Danmarks Fisk.- og Havundersøgelser, Plankton Ser. 5(5): 1-49, 1 pl., 21 tables, 24 textfigs.

Chemistry oxygen
*Guillén, Oscar, 1967.
Anomalies in the waters off the Peruvian Coast during March and April 1965.
Stud. trop. Oceanogr., Miami, 5:452-465.

Chemistry, oxygen
Guillén G., Oscar, y Luis Alberto Flores P., 1968.
Informe preliminar del Crucero 6702 del verano de 1967 (Cabo Blanco-Arica).
Informe, Inst. Mar. Peru 18:17pp.

chemistry, oxygen (data)
Guillén, Oscar, y Francisco Vásquez 1966.
Informe preliminar del Crucero 6602 (Cabo Blanco-Arica).
Inst. del Mar, Peru, Informe (12):27pp.

H

Chemistry oxygen
Haefner, Paul A., Jr., 1967.
Hydrography of the Penobscot River (Maine) estuary.
J. Fish. Res. Bd., Can., 24(7):1553-1571

chemistry, oxygen (data)
Halim, Youssef 1960.
Etude quantitative et qualitative du cycle écologique des dinoflagellés dans les eaux de Villefranche-sur-Mer (1953-1955).
Ann. Inst. océanogr. Monaco 38:123-232.

chemistry, oxygen
Hamm, R.E, and T.G. Thompson 1941.
Dissolved nitrogen in the sea water of the northeast Pacific with notes on the total carbon dioxide and dissolved oxygen.
J. mar. Res. 4(1):11-27.

chemistry, oxygen
Hanzawa, M., and T. Tsuchida, 1954.
A report on the oceanographical observations in the Antarctic carried out on board the Japanese whaling fleet during the years 1946 to 1952.
J. Ocean. Soc., Japan, 10(3):99-111, 7 textfigs.

chemistry, oxygen
Harper, E.L., 1953.
Semimicrodetermination of dissolved oxygen.
Anal. Chem. 25:187, 1 fig.

Chemistry, oxygen
Hart, T. John, and Ronald I. Currie, 1960
The Benguela Current.
Discovery Repts., 31: 123-298.

chemistry, oxygen (data)
Hasegawa, Y., M. Yokoseki, E. Fukuhara, and K. Terai, 1952.
On the environmental conditions for the culture of laver in usu-Bay, buri Prov., Hokkaido.
Bull. Hokkaido Regional Fish. Res. Lab. 6:1-24, 8 textfigs., 9 tables.

chemistry, oxygen
Hela, Ilmo, Eliezer Gilat and Jean-Claude Martin, 1964.
Study of an ecosystem in the coastal waters of the Ligurian Sea. 1. Hydrographic conditions.
Bull. Inst. Océanogr., Monaco, 63(1306):20 pp.

Chemistry oxygen
Herlinveaux, R.H., and J. P. Tully, 1961
Some oceanographic features of Juan de Fuca Strait.
J. Fish. Res. Bd., Canada, 18(6):1027-1071.

Chemistry Oxygen (data)
Hisaoka, Minoru, Kazuhiko Nogami, Osamu Takeuchi, Masaya Suzuki and Hitomi Sugimoto, 1966.
Studies on sea water exchange in fish farm. 2. exchange of sea water in floating net.
Bull. Naikai reg. Fish. Res. Lab., No. 23:21-43.

Chemistry Oxygen
Hollan, Eckard, 1970.
Eine physikalische Analyse kleinräumiger Änderungen chemischer Parameter in den tiefen Wasserschichten der Gotlandsee.
Kieler Meeresforsch. 25(2): 255-267

chemistry, oxygen
Holland, William R. 1971.
Ocean tracer distributions. 1. A preliminary numerical experiment.
Tellus 23 (4/5): 371-392

Chemistry oxygen
Holm-Hansen, Osmund, J.D.H. Strickland and P.M. Williams, 1966.
A detailed analysis of biologically important substances in a profile off southern California.
Limnol. Oceanogr., 11(4):548-561.

Chemistry oxygen
Hood, D.W., 1963
Chemical oceanography. In: Oceanography and Marine Biology, H. Barnes, Edit., George Allen & Unwin, 1:129-155.

chemistry, oxygen
Hülsemann, J., and K. O. Emery, 1961.
Stratification in Recent sediments of Santa Barbara Basin as controlled by organisms and water character.
J. Geol., 69(3):279-290.

Chemistry oxygen
Hung, Tsu-Chang, and Ching-Wei Lee, 1967.
Research on chemical oceanography in the Kuroshio around Taiwan Island.
Bull. Inst. Chem., Acad. Sinica, Taiwan, 14:80-102.

Chemistry oxygen
Hyber, Lothar, 1967.
Untersuchungen über Sauerstoffgehalt und Sauerstoffzehrung in der Adria im Bereich der Stadt Rovinj/Istria.
Thalassia jugosl., 3(1/6):195-199.

Chemistry oxygen
Ichiye, Takashi, 1962
On formation of the intermediate water in the northern Pacific Ocean.
Geofisica Pura e Applicata, 51(1):108-119.

Chemistry oxygen
Ichiye, T., 1954.
On the distribution of oxygen and their seasonal variations in the adjacent seas of Japan. II.
Ocean. Mag., Tokyo, 6(2):67-100, 25 textfigs.

Chemistry, oxygen
Ichiye, T., 1954.
On the distributions of oxygen and their seasonal variations in the adjacent seas of Japan. I. Preliminary report on the characteristics of oxygen distributions. Ocean. Mag., Tokyo, 6(2): 41-66, 20 textfigs.

Chemistry
oxygen

Ichiye, Takashi, 1952?
[On the hydrographical condition in the Kuroshio region (1952). 1. Southern area of Honshu.]
Bull. Kobe Mar. Obs., No. 163: 1-30.

Chemistry
oxygen

Ichie, Takashi, 1951
[On the hydrography of the Kii-Suido (1951).]
Bull. Kobe Mar. Obs., No. 164:253-278(top of page); 35-60 (bottom of page).

chemistry, oxygen

Ichiye, Takashi, 1950?
[On the distribution of oxygen in the ocean.]
Bull. Kobe Mar. Obs., No. 162:1-10.

chemistry, oxygen

Ichie, T., 1949.
Report on the oceanographical observations on board R.M.S. "Syunpu Maru" in the Akashi Seto and the Yura Seto in March 1949. Papers and Repts., Ocean., Kobe Mar. Obs., Ocean. Dept., No. 4:10 pp. numerous figs. and tables, (atlas-sized pages - mimeographed).

Chemistry
oxygen

Ichiye, Takashi, Sigeo Moriyasu and Hiroyuki Kitamura, 1951
On the hydrography near the estuaries.
Bull. Kobe Mar. Obs., No. 164: 349 (top of page)-369; 53 (bottom of page)-75.

chemistry, oxygen

Ichie, T., K. Tanioka, and T. Kawamoto, 1950.
Reports of the oceanographical observations on board the R.M.S. "Yushio Maru" off Shionomisaki (Aug. 1949). Papers and Repts., Ocean., Kobe Mar. Obs., Ocean. Dept., No. 5:15 pp., 33 figs. (Odd atlas-sized pages - mimeographed).

chemistry - oxygen

Iizuka, Syoji, and Haruhiko Irie, 1967.
Studies on the oceanographic characteristics of Haikai Channel and the adjacent waters and of the effects of closing of the channel on pearl farms. II Movement of sea water and amount of dissolved oxygen in the farm. (In Japanese; English abstract).
Bull. Fac. Fish. Nagasaki Univ., 22:1-14.

chemistry - oxygen

Iizuka, Shoji, and Haruhiko Irie, 1966.
The hydrographic conditions and the fisheries damages by the red water occurred in Omura Bay in summer 1965 - II. The biological aspects of a dominant species in the red water. (In Japanese; English abstract).
Bull. Fac. Fish. Nagasaki Univ., No. 21:67-101.

Chemistry
oxygen

Ilyina, N.L., 1961
[Some characteristics of hydrochemistry of the Northern Part of the Norwegian and the Greenland seas (summer 1958).] (English abstract).
Mezhd. Kom. Mezhd. Geofiz. Goda, Presidiume Akad. Nauk, SSSR, Okeanol. Issled., (3):151-161.

chemistry, oxygen

*Ingraham, W. James, Jr., 1968.
The geostrophic circulation and distribution of water properties off the coasts of Vancouver Island and Washington spring and fall 1963.
Fishery Bull. Fish Wildl. Serv. U.S. 66(2):223-250.

Chemistry oxygen

Ishino, M., 1955.
Hydrographic survey in the Equatorial Pacific Ocean, the South China Sea and the Formosa-Satsunan,- Kuroshio region.
Rec. Ocean. Wks., Japan, 2(1):125-131, 4 textfigs

chemistry
oxygen (data)

*Ishino Makoto, Keiji Nasu, Yoshimi Morita and Makoto Hamada, 1968.
Oceanographic conditions in the west Pacific Southern Ocean in summer of 1964-1965. J. Tokyo Univ. Fish., 9(1): 115-208.

chemistry oxygen

Irie, Haruhiko, and Kentaro Hamashima, 1966.
The hydrographic conditions and the fisheries damages by the red water occurred in Omura Bay in summer 1965 - I. (In Japanese; English abstract).
Bull. Fac. Fish. Nagasaki Univ., No. 21:59-65.

Chemistry
oxygen

Ivanenkov, V.N., and F.A. Gebin, 1960.
Water masses and hydrochemistry in the western and southern parts of the Indian Ocean. Physics of the Sea. Hydrology. (In Russian).
Trudy Morsk. Gidrofiz. Inst., 22:33-115.

Translation: Scripta Technica, Inc., for Amer. Geophys. Union, 27-99.

Chemistry oxygen (data)

Ivanenkov, V.N., V.R. Vintovkin and K. S. Shatzkov 1964.
Distribution of dissolved oxygen in the northern part of the Indian Ocean. Investigations in the Indian Ocean (33rd voyage of E/S "Vitiaz"). (In Russian).
Trudy Inst. Okeanol., Akad. Nauk, SSSR, 64:115-127.

J

chemistry, oxygen

Jacob, Otto Ernst, 1961.
Über die kurzfristige Veränderlichkeit in der Vertikalverteilung ausgewählter chemischer Faktoren in der östlichen und mittleren Ostsee im Sommer 1960.
Kieler Meeresf., 17(2):154-158.

chemistry, oxygen (data)

Jacobs, Stanley S., Peter M. Bruchausen and Edward B. Bauer 1970.
Cruises 32-36, 1968, hydrographic stations, bottom photographs, current measurements. Eltanin Repts, Lamont-Doherty Geol. Obs., Nat. Sci. Found. U.S. Antarctic Res. Program, 460 pp. (multilithed).

Chemistry oxygen

Jacobsen, J.P. 1916
Contribution to the Hydrography of the Atlantic. Researches from the M/S "Margrethe" 1913. Medd. fra Komm. for Havundersøgelser, Serie: Hydrografi, 2(5):23 pp., 7 figs.

Chemistry oxygen

Jacobsen, J.P., 1912.
The amount of oxygen in the water of the Mediterranean.
Rept. Danish Oceanogr. Exped., 1908-1910, 1:209-236, 6 pls.

Chemistry oxygen (data)

Jacobsen, J.P., 1908.
Der Sauerstoffgehalt des Meereswassers in den daenischen Gewässern innerhalb Skagens.
Medd. Komm. Havundersøgelser, Ser. Hydrogr., 1(12):23 pp., 5 pls.

Chemistry oxygen

Jacobsen, J.P., 1905.
Die Löslichkeit von Sauerstoff im Meerwasser durch Winklers Titriermethode bestimmt. Medd. Komm. Havundersøgelser, Ser. Hydrogr., 1(8):13 pp., 4 textfigs.

Chemistry oxygen

Jacobsen, J.P., and A.J.C. Jensen, 1925.
Examination of hydrographical measurements from the Research Vessels "Explorer" and "Dana" during the summer of 1924. Cons. Perm. Int. Expl. Mer, Rapp. Proc. Verb. 37:31-84, Figs. 8-52.

Chemistry, oxygen

Jacobsen, J.P., R.J. Robinson and T.G. Thompson, 1950.
A review of the determination of dissolved oxygen in sea water by the Winkler method.
Publ. Sci., Assoc. Ocean. Phys. No. 11:

Chemistry
Oxygen (data)

Jakleln, A., 1936
Oceanographic investigations in East Greenland waters in the summers of 1930-1932. Skr. om Svalbard og Ishavet No. 67:79 pp., Pl. 2, 28 text figs.

Japan, Hakodate Marine Observatory

chemistry
oxygen

Japan, Hakodate Marine Observatory, 1970.
Report of the oceanographic observations in the sea east of Honshu and Hokkaido, and in the Tsugaru Straits from April to May, 1966. (In Japanese)
Bull. Hakodate mar. Obs. 15:11-16

chemistry, oxygen

Japan, Hakodate Marine Observatory 1970
Report of the oceanographic observations in the sea east of Honshu and Hokkaido from February to March 1966. (In Japanese).
Bull. Hakodate mar. Obs. 15: 3-10.

chemistry
oxygen

Japan, Hakodate Marine Observatory, 1969.
Report of the oceanographic observations in the sea south of Hokkaido and in the Sea of Okhotsk from October to November 1965 (In Japanese)
Bull. Hakodate mar. Obs. 14:16-21

chemistry
oxygen

Japan, Hakodate Marine Observatory, 1969.
Report of the oceanographic observations in the sea east of Hokkaido and the Kuril islands from May to June 1965. (In Japanese)
Bull. Hakodate mar. Obs. 14: 12-17.

chemistry
oxygen

Japan, Hakodate Marine Observatory, 1969.
Report of the oceanographic observations in the sea east of Hokkaido and Tohoku District from February to March, 1965. (In Japanese)
Bull. Hakodate mar. Obs. 14:3-9

chemistry,
oxygen

Japan Hakodate Marine Observatory, 1969.
Report of the oceanographic observations in the sea east of Hokkaido and the Kuril islands and in the Okhotsk Sea from July to September, 1965.
Bull. Hakodate mar. Obs. 14: 3-15. (In Japanese)

chemistry, oxygen

Japan, Hakodate Marine Observatory, 1967.
Report of the oceanographic observations in the Okhotsk Sea and east of the Kurile islands and Hokkaido from October to November 1964. (In Japanese).
Bull. Hakodate mar. Obs., 13: 20-28.

chemistry oxygen

Japan, Hakodate Marine Observatory, 1967.
Report of the oceanographic observations in the sea east of Hokkaido and in the southern part of the Okhotsk Sea from May to June, 1964. (In Japanese).
Bull. Hakodate mar. Obs., 13:10-17.

chemistry, oxygen

Japan, Hakodate Marine Observatory, 1967.
Report of the oceanographic observations in the sea southeast of Hokkaido in February, 1964. (In Japanese).
Bull. Hakodate mar. Obs. 13:3-9.

chemistry oxygen

Japan, Hakodate Marine Observatory, 1967.
Report of the oceanographic observations in the Okhotsk Sea, east of the Kurile islands and Hokkaido and east of the Tohoku district from August to September, 1964. (In Japanese).
Bull. Hakodate mar. Obs., 13:7-19.

chemistry, oxygen

Japan, Hakodate Marine Observatory, 1966
Sea ice observation along the Okhotsk coast of Hokkaido from January to April 1963.
Sea ice conditions in the sea off Hokkaido from December 1962 to 1963.
Local sea ice conditions based on coastal observations of the Okhotsk Sea based on the broadcasting of U.S.S.R., 1962-1963. (In Japanese; English abstracts)
Bull. Hakodate Mar. Obs. No.12;1, 1-35,38-52; 53-67.

chemistry oxygen

Japan, Hakodate Marine Observatory, 1964.
Report of the oceanographic observations in the sea east of the Tohoku District from February to March 1962. Report of the oceanographic observations in the Tsugaru Straits in April 1962.---in May 1962.---in June, 1962. from August to September, 1962. Report of the oceanographic observations in the sea south of Hokkaido in June 1962. Report of the oceanographic observations in the sea west of Tsugaru Straits, the Tsugaru Straits and South of Hokkaido in May, 1962. (In Japanese).
Bull. Hakodate Mar. Obs., (I No. 11:misnumbered pp.

chemistry oxygen

Japan, Hakodate Marine Observatory, 1963.
Report of the oceanographic observations in the sea east of Tohoku District from February to March 1961:3-8.
Report of the oceanographic observations in the sea east of Tohoku District in May 1961:9-12.
Report of the oceanographic observations in the Japan Sea in June 1961:59-79.
Report of the oceanographic observations in the western part of Wakasa Bay from January to May 1961:80.
Report of the oceanographic observations in the sea south of Hokkaido in July 1961:3-4.
Bull. Hakodate Mar. Obs. 10: numerous pp. (pagination erratic). (In Japanese)

chemistry oxygen

Japan, Hakodate Marine Observatory, 1962
Report of the oceanographic observations in the sea east of Tohoku District and in the Okhotsk Sea from August to September 1960. (In Japanese).
Results. Mar. Meteorol. and Oceanogr., July-Dec., 1960, Japan Meteorol. Agency, No. 28: 7-16.

chemistry oxygen

Japan, Hakodate Marine Observatory, Oceanographical Section, 1962
Report of the oceanographic observations in the sea east of Tohoku District in May, 1961. (In Japanese).
Res. Mar. Meteorol. and Oceanogr. Obs., Jan.-June, 1961, No. 29: 9-12.

chemistry, oxygen

Japan, Hakodate Marine Observatory, Oceanographical Section, 1962
Report of the oceanographic observations in the sea east of Tohoku District from February to March, 1961. (In Japanese).
Res. Mar. Meteorol., and Oceanogr. Obs., Jan.-June, 1961, No. 29:3-8.

chemistry, oxygen

Japan, Hakodate Marine Observatory, 1962
Report of the oceanographic observations in the sea south of Hokkaido in July, 1960. (In Japanese).
Res. Mar. Meteorol. and Oceanogr., July-Dec., 1960, Japan Meteorol. Agency, No. 28:3-6.

chemistry, oxygen

Japan, Hakodate Marine Observatory, 1962
Report of the oceanographic observations in the sea west of Hiyama (Hokkaido) and in the sea east of Tohoku District in November 1960. (In Japanese).
Res. Mar. Meteorol. and Oceanogr., July-Dec., 1960, Japan Meteorol. Agency, No. 28:17-20.

chemistry, oxygen

Japan, Hakodate Marine Observatory, 1961
Report of the oceanographic observations in the Okhotsk Sea and in the sea east of Tohoku District from May to June, 1959.
Bull. Hakodate Mar. Obs., (8):8-16.

chemistry, oxygen

Japan, Hakodate Marine Observatory, 1961.
[Report of the oceanographic observations in the sea east of Tohoku District and in the Okhotsk Sea from August to September 1959.]
Bull. Hakodate Mar. Obs., (8):6-14.

chemistry oxygen

Japan, Hakodate Marine Observatory, 1961
[Report of the oceanographic observations in the sea east of Tohoku District from November to December 1959.]
Bull. Hakodate Mar. Obs., (8):15-19.

chemistry oxygen

Japan, Hakodate Marine Observatory, 1961.
[Report of the oceanographic observations in the sea east of Tohoku District from February to March 1959.]
Bull. Hakodate Mar. Obs., (8):3-7.

chemistry, oxygen

Japan, Hakodate Marine Observatory, 1957.
[Report of the oceanographic observations in the sea east of Tohoku District from February to March 1956.]
Bull. Hakodate Mar. Obs., No. 4:49-57. 1-9.

chemistry, oxygen

Japan, Hakodate Marine Observatory, 1957.
[Report of the oceanographic observations in the sea east of Tohoku District from May to June 1956.]
Bull. Hakodate Mar. Obs., No. 4:113-119. 9-15.

chemistry, oxygen

Japan, Hakodate Marine Observatory, 1957
[Report of the oceanographic observations in the sea east of the Tohoku district in August 1956.]
Bull. Hakodate Mar. Obs., No. 4: 1-12.

chemistry oxygen

Japan, Hakodate Marine Observatory, 1957
[Report of the oceanographic observations in the sea east of Tohoku district from October to November 1956.]
Bull. Hakodate Mar. Obs., No. 4: 1-8.

chemistry, oxygen

Japan, Hakodate Marine Observatory, 1956.
[Report of the oceanographical observations in the sea off the Sanriku District in Feb.-Mar., 1955.]
Bull. Hakodate Mar. Obs., (3):11-16.

chemistry, oxygen

Japan, Hakodate Marine Observatory, 1956.
[Report of the oceanographical observations east off Tohoky District from August to September 1955.]
Bull. Hakodate Mar. Obs., (3):13-21.

chemistry, oxygen

Japan, Hakodate Marine Observatory, 1956.
[Report of the oceanograhic observations east of Tohoku District in November 1955.]
Bull. Hakodate Mar. Obs., (3):11-12.

chemistry, oxygen

Japan, Hakodate Marine Observatory, 1956.
[Report on the oceanographical observation of the sea off Tohoku District made on board the R.M.S. Yoshio-maru in summer 1953 (1st paper).]
Bull. Hakodate Mar. Obs., No. 2:1-3.

chemistry, oxygen

Japan, Hakodate Marine Observatory, 1956.
[Report of the oceanographical observations in the sea off Tohoku District made on board the R.M.S. Yushio-maru in spring and summer 1952. 1.]
Bull. Hakodate Mar. Obs., (3):205-215. 1-11.

Japan Meteorological Agency

oxygen

Japan, Japan Meteorological Agency. 1965.
The results of marine meteorological and oceanographical observations, July-December 1963, No. 34: 360 pp.

chemistry oxygen

Japan, Japan Meteorological Agency, Oceanographical Section, 1962
Report of the oceanographic observations in the sea east of Honshu from February to March, 1961. (In Japanese).
Res. Mar. Meteorol. and Oceanogr. Obs., Jan.-June, 1961, No. 29:13-21.

chemistry oxygen

Japan, Japan Meteorological Agency, 1962
Report of the Oceanographic observations in the sea east of Honshu from August to September, 1960. (In Japanese).
Res. Mar. Meteorol. and Oceanogr., July-Dec., 1960, Japan Meteorol. Agency, No. 28:21-29.

chemistry oxygen

Japan, Japan Meteorological Agency, 1962
Report of the oceanographic observations in the sea south and west of Japan from October to November, 1960. (In Japanese).
Res. Mar. Meteorol. and Oceanogr., July-Dec., 1960, Japan Meteorol. Agency, No. 28:30-35.

chemistry oxygen

Japan, Japan Meterology Agency, 1959.
The results of marine meteorological and oceanographical observations, January-June 1958, No. 23:240 pp.

Japan Kobe Marine Observatory

chemistry, oxygen

Japan, Kobe Marine Observatory 1967.
Report of the oceanographic observations in the sea south of Honshu from October to November 1963. (In Japanese)
Bull. Kobe Mar. Obs. 178:41-49.

chemistry, oxygen

Japan Kobe Marine Observatory, 1967.
Report of the oceanographic observations in the sea south of Honshu and in the Seto-Naikai from May to June 1964 (In Japanese)
Bull. Kobe Mar. Obs. No. 178: 37-

chemistry, oxygen
Japan, Kobe Marine Observatory, 1967.
Report of the oceanographic observations in the sea south of Honshu from July to August 1965. (In Japanese).
Bull. Kobe Mar. Obs. No. 178: 31-40

Chemistry, oxygen
Japan, Kobe Marine Observatory, 1967.
Report of the oceanographic observations in the sea south of Honshu from February to March 1964. (In Japanese)
Bull. Kobe Mar. Obs. No. 178: 27-

Chemistry, oxygen
Japan, Kobe Marine Observatory, Oceanographical Section, 1964.?
Researches in chemical oceanography made at K.M.O. for the recent ten years (1954-1963). (In Japanese).
Weather Service Bull., Japan. Meteorol. Agency, 31(7):188-191.

Also in:
Bull. Kobe Mar. Obs., No. 175. 1965.

chemistry, oxygen
Japan, Kobe Marine Observatory, Oceanographical Section, 1964.
Report of the oceanographic observations in the Kuroshio and region east of Kyushu from October to November 1962. (In Japanese) Res. Mar. Meteorol. and Oceanogr., Japan. Meteorol. Agency, 32: 41-50.

Also in: Bull. Kobe Mar. Obs., 175. 1965.

chemistry, oxygen
Japan, Kobe Marine Observatory, Oceanographical Section, 1964.
Report of the oceanographic observations in the sea south of Honshu from February to March, 1963. Res. Mar. Meteorol. and Oceanogr., Japan Meteorol. Agency, 33: 27-32.

Also in: Bull. Kobe Mar. Obs., 175. 1965.

chemistry, oxygen
Japan, Kobe Marine Observatory, Oceanographical Section, 1964.
Report of the oceanographic observations in the sea south of Honshu from July to August, 1962. Res. Mar. Meteorol. and Oceanogr., Japan. Meteorol. Agency, 32: 32-40. (In Japanese).

Also in: Bull. Kobe Mar. Obs. 175.

chemistry, oxygen
Japan, Kobe Marine Observatory, Oceanographical Section, 1963
Report of the oceanographic observations in the sea south of Honshu from May to June, 1962. (In Japanese).
Res. Mar. Met. & Ocean., J.M.A., 31:45-49.

Also in:
Bull. Kobe Mar. Obs., No. 173(4):1964.

chemistry, oxygen
Japan, Kobe Marine Observatory, Oceanographical Section, 1963
Report of the oceanographic observations in the sea south of Honshu from February to March 1962. (In Japanese).
Res. Mar. Met. & Ocean., J.M.A., 31:37-44.

Also in:
Bull. Kobe Mar. Obs., No. 173(4):1964.

Chemistry, oxygen
Japan, Kobe Marine Observatory, 1963.
Report of the oceanographic observations in the sea south of Honshu from July to August and from the cold water region south of Enshu Nada October to November 1960. (In Japanese).
Bull. Kobe Mar. Obs., 171(3):36-52.

chemistry, oxygen
Japan, Kobe Marine Observatory, 1963.
Report of the oceanographical observations in the sea south of Honshu from February to March 1961. Report of the oceanographical observations in the cold water region off Enshu Nada in May 1961. (In Japanese).
Bull. Kobe Mar. Obs., 171(4):22-35.

chemistry, oxygen
Japan, Kobe Marine Observatory, Oceanographical Section, 1962
Report of the oceanographic observations in the cold water region off Kii Peninsula from October to November 1961. (In Japanese).
Res. Mar. Meteorol. & Oceanogr., No. 30:49-55.

Also in:
Bull. Kobe Mar. Obs., No. 173(3):1964

chemistry, oxygen
Japan, Kobe Marine Observatory, Oceanographical Observatory, 1962
Report of the oceanographic observations in the cold water region off Enshu Nada in May, 1961. (In Japanese).
Res. Mar. Meteorol. and Oceanogr. Obs. Jan.-June, 1961, No. 29:28-35.

chemistry, oxygen
Japan, Kobe Marine Observatory, 1962
Report of the oceanographic observations in the cold water region south of Enshu Nada from October to November, 1960. (In Japanese)
Res. Mar. Meteorol. and Oceanogr., July-Dec. 1960, Japan Meteorol. Agency, No. 28: 43-51.

chemistry, oxygen
Japan, Kobe Marine Observatory, Oceanographical Section, 1962
Report of the oceanographical observations in the sea south of Honshu from July to August 1961. (In Japanese).
Res. Mar. Meteorol. & Oceanogr., 30:39-48.

Also in:
Bull. Kobe Mar. Obs., No. 173(3). 1964.

chemistry, oxygen
Japan, Kobe Marine Observatory, Oceanographical Section, 1962
Report of the oceanographic observations in the sea south of Honshu from February to March, 1961. (In Japanese).
Res. Mar. Meteorol. and Oceanogr. Obs., Jan.-June, 1961, No. 29:22-27.

chemistry, oxygen
Japan, Kobe Marine Observatory, Oceanographical Section, 1962
Report of the oceanographic observations in the sea south of Honshu in March 1960. (In Japanese).
Bull. Kobe Mar. Obs., No. 169(12):22-33.

chemistry, oxygen
Japan, Kobe Marine Observatory, 1962
Report of the oceanographic observations in the sea south of Honshu from July to August, 1960. (In Japanese).
Res. Mar. Meteorol. and Oceanogr., July-Dec., 1960, Japan Meteorol. Agency, No. 28:36-42.

chemistry, oxygen
Japan, Kobe Marine Observatory, Oceanographical Section, 1962
Report of the oceanographic observations in the sea south of Honshu in May, 1960. (In Japanese).
Bull. Kobe Mar. Obs., No. 169(12):27-33.

chemistry, oxygen
Japan, Kobe Marine Observatory, Oceanographical Section, 1962
Report on the oceanographic observations in the sea south of Honshu from July to August 1959.
Bull. Kobe Mar. Obs., No. 169(11):37-43. (In Japanese)

chemistry, oxygen
Japan, Kobe Marine Observatory, Oceanographical Section, 1962
Report of the oceanographic observations in the sea south of Honshu from October to November, 1959. (In Japanese).
Bull. Kobe Mar. Obs., No. 169(11):44-50.

chemistry, oxygen
Japan, Kobe Marine Observatory, 1961
[Report of the oceanographic observations in the sea south of Honshu in March 1958.]
Bull. Kobe Mar. Obs., No. 167(21-22):30-36.

--from May to June, 1958(21-22):37-42
--from July to September, 1958(23-24):34-40
--from October to December, 1958(23-24):41-47
--from February to March, 1959(25-26):33-47.

oxygen chemistry, oxygen
Japan, Kobe Marine Observatory, 1958.
[Report of the oceanographic observations in the sea south of Honshu from November to December 1957.]
J. Oceanogr., Kobe Mar. Obs., (2) 10(1):21-28.

in May 1957.]
Ibid., 9(2):69-78.

in August 1957.]
Ibid., 9(2):79-86.

oxygen chemistry, oxygen
Japan, Kobe Marine Observatory, 1956.
[Report of the oceanographic observations in the sea south of Honshu in August 1955.]
J. Ocean., Kobe (2)7(3):23-32.

chemistry, oxygen
Japan, Kobe Marine Observatory, 1956.
[Report of oceanographical observations in the sea south off Honshu in May 1955.]
J. Ocean., Kobe, (2)7(2):17-25.

pagination repeats that of previous article.

chemistry, oxygen
Japan, Kobe Marine Observatory, 1956.
[Report of the oceanographical observations in the sea south off Honshu in March 1955.]
J. Ocean., Kobe (2)7(2):14-24.

chemistry, oxygen
Japan, Kobe Marine Observatory, 1956?
[Report on the oceanographical observations south off Honshu in 1954.] J. Ocean., Kobe, (2)7(2):46-69.

oxygen chemistry, oxygen
Japan, Kobe Marine Observatory, 1955.
[The outline of the oceanographical observation on the Southern Sea of Japan on board R.M.S. "Syunpu-maru" (Jan. 1955).]
J. Ocean., Kobe Mar. Obs., (2)6(1):1-19, 10 figs.

Chemistry, oxygen

Japan, Kobe Marine Observatory, 1954.
[The outline of the oceanographic observations in the southern area of Honshu on board the R.M.S. "Syunpu-maru".]
Aug.-Sept. 1954, - J. Ocean. (2)5(9):1-44.
Oct. 1954 - J. Ocean. (2)5(10):1-44.

Chemistry, oxygen

Japan, Kobe Marine Observatory, 1954.
[The results of the regular monthly oceanographical observations on board the R.M.S. "Syunpu-maru" in the Kii Suido and Osaka Wan.] J. Ocean. (2)5(1):1-19, 16 figs.

Chemistry, oxygen

Japan, Kobe Marine Observatory, 1954.
[The outline of the oceanographical observations off Shionomisaki on board the R.M.S. "Syunpu-maru" (May 1954).] J. Ocean. (2)5(5):1-11, 14 figs

Chemistry, oxygen

Japan, Kobe Marine Observatory, 1953?
[The results of the regular monthly oceanographical observations on board the R.M.S. "Syunpu-maru" in the Kii Suido and Osaka Wan (Oct. 1953)]
J. Ocean. Kobe Mar. Obs., 2nd ser., 4(11):1-21, 15 textfigs.

reports with similar titles in all other numbers of Vol. 4, this journal. These not entered in index.

Chemistry, oxygen

Japan, Kobe Marine Observatory, 1953.
[The results of the regular monthly oceanographical observations on board the R.M.S. "Syunpu-maru" in the Kii Suido and Osaka Wan.]
Jan. 1953 - J. Ocean. (2)4(1):1-10, 9 figs.
Feb. 1953. - J. Ocean. (2)4(2):1-9, 8 figs.
Mar. 1953 - J. Ocean. (2)4(3):1-10, 9 figs.
Apr. 1953 - J. Ocean. (2)4(5):1-9, 9 figs.
May 1953 - J. Ocean. (2)4(6):1-12, 9 figs.
June 15-22, 1953 - J. Ocean. (2)4(6):13-22, 8 figs.
July 1953 - J. Ocean. (2)4(8):1-25, 29 figs.
Aug. 1953 - J. Ocean. (2)4(9):1-30, 27 figs.
Sept. 1953 - (2)4(10):1-13, 9 figs. J. Ocean.
Oct. 1953 - J. Ocean. (2)4(11):1-21, 15 figs.
Nov. 11-18, 1953 - J. Ocean. (2)4(12):1-15, 12 figs.

Chemistry, oxygen

Japan, Kobe Marine Observatory, 1953.
[Diurnal variation of oxygen in the Bisan-Seto (Seto Inland Sea) (Sept. 1952).]
J. Ocean. (2)4(4):10-12, 3 figs.

Chemistry, oxygen

Japan, Kobe Marine Observatory, 1953.
On the chemical conditions in the northern part of Osaka Bay after heavy rain (July 1952).] J. Ocean. (2)4(4):1-9, 5 figs.

Japan, Maizuru Marine Observatory

Chemistry, oxygen

Japan, Maizuru Marine Observatory 1969.
Report of the oceanographic observations in the western part of the Japan Sea in November 1965. (In Japanese)
Bull. Maizuru mar. Obs. 11: 98-103.

Chemistry, oxygen

Japan, Maizuru Marine Observatory, 1967.
Report of the oceanographic observations in the Japan Sea from October to November, 1964. (In Japanese).
Bull. Maizuru mar. Obs., 10:87-94.

Chemistry, oxygen

Japan, Maizuru Marine Observatory, 1968.
Report of the oceanographical observations in the Japan Sea from August to September, 1964. (In Japanese).
Bull. Maizuru mar. Obs., 10:74-86.

Chemistry oxygen

Japan, Maizuru Marine Observatory, 1967.
Report of the oceanographic observations in the Japan Sea from May to June, 1964.
Bull. Maizuru mar. Obs., 10:65-76.
(In Japanese)

Chemistry, oxygen

Japan, Maizuru Marine Observatory, 1965.
Report of the oceanographic observations in the central part of the Japan Sea from February to March, 1962---in the Japan Sea from June to July, 1962.---in the western part of Wakasa Bay from January to April, 1962.---in the central part of the Japan Sea from September to October, 1962.---in the western part of Wakasa Bay from May to November, 1962.---in the central part of the Japan Sea in March, 1963.---in the Japan Sea in June, 1963.---in Wakasa Bay in July, 1963.---in the central part of the Japan Sea in October, 1963. (In Japanese).
Bull. Maizuru Mar. Obs., No. 9:67-73; 74-88; 89-95. 71-80; 81-87; 59-65; 66-77; 80-84; 85-91.

Chemistry oxygen

Japan, Maizuru Marine Observatory, 1963
Report of the oceanographic observations in the central part of the Japan Sea in February, 1961. (In Japanese).
Bull. Maizuru Mar. Obs., No. 8:54-58.

Chemistry oxygen

Japan, Maizuru Marine Observatory, 1963
Report of the oceanographic observations in the central part of the Japan Sea in October 1961. (In Japanese).
Bull. Maizuru Mar. Obs., No. 8:78-88.

Chemistry oxygen

Japan, Maizuru Marine Observatory, 1963
Report of the oceanographic observations in the central part of the Japan Sea in September 1960. (In Japanese).
Bull. Maizuru Mar. Obs., No. 8:60-68.

Chemistry oxygen

Japan, Maizuru Marine Observatory, 1963
Report of the oceanographic observations in the Japan Sea from May to June, 1960. (In Japanese).
Bull. Maizuru Mar. Obs., No. 8:56-67.

Chemistry oxygen

Japan, Maizuru Marine Observatory, 1963
Report of the oceanographic observations in the Japan Sea in June 1961. (In Japanese).
Bull. Maizuru Mar. Obs., No. 8:59-79.

Chemistry oxygen

Japan, Maizuru Marine Observatory, 1963
Report of the oceanographic observations in the Wakasa Bay in August, 1961. (In Japanese).
Bull. Maizuru Mar. Obs., No. 8:89-95.

Chemistry oxygen

Japan, Maizuru Marine Observatory, 1963
Report of the oceanographic observations in the Wakasa Bay from July to August, 1960. (In Japanese).
Bull. Maizuru Mar. Obs., No. 8:69-75.

Chemistry oxygen

Japan, Maizuru Marine Observatory, 1963
Report of the oceanographic observations in the western part of Wakasa Bay from January to May, 1961. (In Japanese).
Bull. Maizuru Mar. Obs., No. 8:80-90.

Chemistry oxygen

Japan, Maizuru Marine Observatory, 1963
Report of the oceanographic observations in the western part of Wakasa-Bay from January to June, 1960.
Bull. Maizuru Mar. Obs., No. 8:68-79, 59.

Chemistry oxygen

Japan, Maizuru Marine Observatory, 1963
Report of the oceanographic observations in the western part of Wakasa Bay from June to December, 1960. (In Japanese).
Bull. Maizuru Mar. Obs., No. 8:76-83.

Chemistry oxygen

Japan, Maizuru Marine Observatory, Oceanographical Section, 1962
Report of the oceanographic observations in the central part of the Japan Sea in February 1961. (In Japanese).
Res. Mar. Meteorol. and Oceanogr. Obs., Jan.-June, 1961, No. 29:54-58.

Chemistry, oxygen

Japan, Maizuru Marine Observatory and Hakodate Marine Observatory, Oceanographical Sections, 1962.
Report of the oceanographic observations in the Japan Sea in June, 1961. (In Japanese).
Res. Mar. Meteorol. and Oceanogr. Obs., Jan.-June, 1961, No. 29:59-79.

Chemistry oxygen

Japan, Maizuru Marine Observatory, Oceanographical Section, 1962
Report of the oceanographic observations in the western part of Wakasa Bay from June to May, 1961. (In Japanese).
Res. Mar. Meteorol. and Oceanogr. Obs., Jan.-June, 1961, No. 29:80-90.

Chemistry oxygen

Japan, Maizuru Marine Observatory, 1962
Report of the oceanographic observations in the central part of the Japan Sea in September 1960. (In Japanese).
Res. Mar. Meteorol. and Oceanogr., July-Dec., 1960, Japan Meteorol. Agency, No. 28:60-68.

Chemistry oxygen

Japan, Maizuru Marine Observatory, 1962
Report of the oceanographic observations in the Wasaka Bay from July to August, 1960. (In Japanese).
Res. Mar. Meteorol. and Oceanogr., July-Dec., 1960, Japan Meteorol. Agency, No. 28:69-75.

Chemistry oxygen

Japan, Maizuru Marine Observatory, 1962
Report of the oceanographic observations in the western part of Wakasa Bay from June to December 1960. (In Japanese).
Res. Mar. Meteorol. and Oceanogr., July-Dec., 1960, Japan Meteorol. Agency, No. 28:76-83.

Chemistry oxygen

Japan, Maizuru Marine Observatory, Oceanographical Section, 1961
[Report of the oceanographic observations in the Japan Sea from June to July, 1959.]
Bull. Maizuru Mar. Obs., No. 7:57-64.

Chemistry oxygen

Japan, Maizuru Marine Observatory, Oceanographical Observatory, 1961
[Report of the Oceanographic observations in the western part of Wakasa Bay from August to December, 1959.]
Bull. Maizuru Mar. Obs., No. 7:68-74.

Chemistry oxygen

Japan, Maizuru Marine Observatory, Oceanographical Section, 1961
[Report of the oceanographic obserbations in the sea north of Kyoga-misaki from August to December 1958.]
Bull. Maizuru Mar. Obs., No. 7:60-63.

chemistry
oxygen
Japan, Maizuru Marine Observatory, Oceanographical Section, 1961
[Report of the oceanographic observations in the Wakasa Bay from July to August, 1959.]
Bull. Maizuru Mar. Obs., No. 7:62-67.

chemistry
oxygen
Japan, Maizuru Marine Observatory, Oceanographical Section, 1961
[Report of the oceanographic observations in the western part of Wakasa Bay from January to March, 1958.]
Bull. Maizuru Mar. Obs., No. 7:50-54.

chemistry
oxygen
Japan, Maizuru Marine Observatory, Oceanographical Section, 1961
Report of the oceanographic observations in the Japan Sea from June to July, 1958.
Bull. Maizuru Mar. Obs., No. 7:60-67.

chemistry
oxygen
Japan, Maizuru Marine Observatory, Oceanographical Section, 1961.
[Report of the oceanographic observations in the sea north of San'in District in August, 1958.]
Bull. Maizuru Mar. Obs., No. 7: 64-68.

chemistry
oxygen
Japan, Maizuru Marine Observatory, Oceanographical Section, 1961
[Report of the oceanographic observations off Kyoga-misaki and in the western part of Wakasa Bay from February to June, 1959.]
Bull. Maizuru Mar. Obs., No. 7:65-72, 61.

chemistry
oxygen
Japan, Maizuru Marine Observatory, Oceanographical Section, 1961
[Report of the oceanographic observations off Kyoga-misaki in the Japan Sea in April and May, 1958.]
Bull. Maizuru Mar. Obs., No. 7:55-60.

chemistry
oxygen
Japan, Maizuru Marine Observatory, Oceanographical Section, 1961
[Report of the oceanographic observations off Kyoga-misaki in Japan Sea from October to December, 1957.]
Bull. Maizuru Mar. Obs., No. 7:29-36.

chemistry, oxygen
Japan, Maizuru Marine Observatory, 1958
[Report of the oceanographic observations north off Kyoga-misaki and off the east part of the Sanin district in April to June, 1957.]
Bull. Maizuru Mar. Obs., No. 6: 234-243.
[95-105.]

chemistry oxygen
Japan, Maizuru Marine Observatory, 1958
[Report of the oceanographic observations in the sea north of Sanin and Hokuriku districts in August 1957.]
Bull. Maizuru Mar. Obs., No. 6: 245-253.
[107-115.]

chemistry, oxygen
Japan, Maizuru Marine Observatory, 1958
[Report of the oceanographic observations north of Sanin and Hokuriku districts in summer 1956]
Bull. Maizuru Mar. Obs., No. 6: 157-unnumbered.
[53-85.]

chemistry, oxygen
Japan, Maizuru Marine Observatory, 1958
[Report of the oceanographic observations off Kyoga-misaki in the Japan Sea in January and March, 1957.]
Bull. Maizuru Mar. Obs., No. 6: 129-137.
[45-53.]

chemistry oxygen
Japan, Maizuru Marine Observatory, 1958
[Report of the oceanographic observations off the east part of Sanin district in May and June 1956.]
Bull. Maizuru Mar. Obs., No. 6: 149-156.
[45-52.]

chemistry
oxygen
Japan, Maizuru Marine Observatory, 1958
[Report of the oceanographic observations off Kyoga-misaki in Japan Sea in October and November, 1956.]
Bull. Maizuru Mar. Obs., No. 6: [37-44.]

chemistry, oxygen
Japan, Maizuru Marine Observatory, 1958
[Report of the serial observations off Kyoga-misaki from May 1955 to April 1956.]
Bull. Maizuru Mar. Obs., No. 6: 95-104.
[47-56.]

chemistry, oxygen
Japan, Maizuru Marine Observatory, 1956.
[Report of the oceanographic observations off Kayoga-misaki during the latter half of 1955.]
Bull. Maizuru Mar. Obs., (5):31-37.

chemistry
oxygen
Japan, Maizuru Marine Observatory, 1956
[Report of the serial observations off Kyoga-misaki during the first half of 1955.]
Bull. Maizuru Mar. Obs., (5): 27-32.

chemistry, oxygen
Japan, Maizuru Marine Observatory, 1956.
[Report of the oceanographical observations north off Sanin and Hokuriku districts in summer 1954]
[Report of the oceanographical observations north off Sanin and Hokuriku districts in summer 1955]
Bull. Maizuru Mar. Obs., (5):85-94; 49-56.

chemistry, oxygen
Japan, Maizuru Marine Observatory, 1955.
[Report of the oceanographical observations (taken) off San'in and Hokuriku (July-August, 1953).]
Bull. Maizuru Mar. Obs., (4):13-27.
217-231.

chemistry, oxygen
Japan, Maizuru Marine Observatory, 1955.
[Report of the oceanographical observations off San'in and Hokuri-ku (July and August 1952)]
Bull. Maizuru Mar. Obs., (4):49-63 (168-181).

Japan Maritime Safety Agency

chemistry
oxygen
Japan, Maritime Safety Board, 1959
Tables of results from Oceanographic observation in 1954 and 1955.
Hydrogr. Bull., (Publ. 981) No. 58: 139 pp.

Japan Nagasaki Marine Observatory

chemistry, oxygen
Japan, Nagasaki Marine Observatory, 1971.
Report of the oceanographic observations in the sea southeast of Kyushu in September, 1965. (In Japanese)
Oceanogr. Met. Nagasaki mar. Obs. 18: 75-77

chemistry, oxygen
Japan, Nagasaki Marine Observatory 1971.
Report of the oceanographic observations in the sea west of Japan from June to August, 1965. (In Japanese)
Oceanogr. Met. Nagasaki Mar. Obs. 18: 59-74

chemistry oxygen
Japan, Nagasaki Marine Observatory 1971
Report of the oceanographic observations in the sea south of Kyushu in April, 1966 (In Japanese).
Oceanogr. Met. Nagasaki mar. Obs. 18: 53-57.

oxygen
Japan, Nagasaki Marine Observatory 1971
Report of the oceanographic observations in the sea south of Kyushu in March 1966.
Oceanogr. Met. Nagasaki Mar. Obs. 18: 50-52
(In Japanese)

chemistry
oxygen
Japan, Nagasaki Marine Observatory 1971.
Report of the oceanographic observations in the sea west of Japan from January to February 1966. (In Japanese)
Oceanogr. Met. Nagasaki Mar. Obs. 18: 34-49.

oxygen
Japan, Nagasaki Marine Observatory 1971.
Report of the oceanographic observations in the sea west of Japan from June to August, 1967. (In Japanese)
Oceanogr. Met. Nagasaki mar. Obs. 18: 1-12.

chemistry
oxygen
Japan, Nagasaki Marine Observatory 1971.
Report of the oceanographic observations in the sea west of Japan from January to February 1968 (In Japanese)
Oceanogr. Met. Nagasaki mar. Obs. 18: 1-10.

chemistry, oxygen
Japan, Nagasaki Marine Observatory 1971.
Report of the oceanographic observations in the sea west of Japan from January to February 1967. (In Japanese)
Oceanogr. Met. Nagasaki Mar. Obs. 18: 1-10

chemistry, oxygen
Japan, Nagesaki Marine Observatory, Oceanographic Section, 1965.
Report of the oceanographic observations in the sea west of Japan from February to March 1963,--from July to August 1963.
Res. Mar. Meteorol. Oceanogr., Japan Meteorol Agency, 33:39-58; 34:53-80.

Also in:-Oceanogr. Meteorol., Nagesaki, 15 (227-228).

Chemistry oxygen
Japan, Nagasaki Marine Observatory, Oceanographical Section, 1962
Report of the oceanographic observations in the sea west of Japan from April to May, 1961 (In Japanese).
Res. Mar. Meteorol. and Oceanogr. Obs., Jan.-June, 1961, No. 29: 45-53.

Chemistry oxygen
Japan, Nagasaki Marine Observatory, Oceanographical Section, 1962
Report of the oceanographic observations in the sea west of Japan from February to March, 1961. (In Japanese).
Res. Mar. Meteorol. and Oceanogr. Obs., Jan.-June, 1961, No. 29:36-44.

Chemistry oxygen
Japan, Nagasaki Marine Observatory, 1962
Report of the oceanographic observations in the sea west of Japan from October to November, 1960. (In Japanese).
Res. Mar. Meteorol. and Oceanogr., July-Dec., 1960, Japan Meteorol. Agency, No. 28:52-59.

Chemistry oxygen
Japan, Nagasaki Marine Observatory, Oceanographical Section 1960.
Report of the oceanographic observation in the sea west of Japan from January to February, 1960. Report of the Oceanographic observation in the sea north-west of Kyushu from April to May, 1960.
Results Mar. Meteorol. & Oceanogr., J.M.A., 27: 42-50; 51-67.
Also in:
Oceanogr. & Meteorol., Nagasaki Mar. Obs., (1961) 11 (202).

Chemistry oxygen
Japan, Nagasaki Marine Observatory, Oceanographical Section, 1960
[Report of the oceanographic observations in the sea west of Japan from June to July, 1959]
Res. Mar. Meteorol. & Oceanogr., J.M.A., 26: 51-57.
Also in:
Oceanogr. & Meteorol., Nagasaki Mar. Obs., (1961), 11(200).

Chemistry oxygen
Japan, Nagasaki Marine Observatory, 1959.
Report of the oceanographic observations in the sea west of Japan from June to July 1958 and in the sea north west of Kyushu in October 1958.
Results Mar. Meteorol. & Oceanogr., J.M.A., 24: 47-60.
Also in:
Oceanogr. & Meteorol., Nagasaki Mar. Obs., 10(195): 1960.

Chemistry, oxygen
Japan, National Committee for the International Geophysical Year 1960
The results of the Japanese oceanographic project for the International Geophysical Year 1957/58: 198 pp.

Japanese Oceanographic Data Center

Chemistry, oxygen
Japan, Japanese Oceanographic Data Center 1970.
CSK atlas. 4. Winter 1967: 32 pp.

Chemistry, Oxygen
Japan, Science Council, National Committee for IIOE, 1966.
General report of the participation of Japan in the International Indian Ocean Expedition.
Rec. Oceanogr. Wks., Japan, n.s. 8(2): 133 pp.

Chemistry, oxygen
Japan, Tokai Regional Fisheries Research Laboratory, 1959.
IGY Physical and chemical data by the R. V. Soyo-maru, 25 July-14 September 1958.
17 pp. (multilithed).

Chemistry, oxygen
Jayaraman, R., 1954.
Seasonal variations in salinity, dissolved oxygen and nutrient salts in the inshore waters of the Gulf of Manner and Palk Bay near Mandapam (S. India). Indian J. Fish. 1:345-364.

Chemistry oxygen
Jayaraman, R., 1951.
Observations on the chemistry of the waters of the Bay of Bengal off Madras City during 1948-1949. Proc. Indian Acad. Sci., B, 33(2):92-99, figs.

Chemistry, oxygen
Jayaraman, R., C.P. Ramamritham and K.V. Sundararaman, 1959.
The vertical distribution of dissolved oxygen in the deeper waters of the Arabian Sea in the neighborhood of the Laccadives during the summer of 1959.
J. Mar. Biol. Assoc., India, 1(2):206-211.

Chemistry oxygen
Jayaraman, R., R. Viswanathan and S.S. Gogate, 1961.
Characteristics of sea water near the light house, Bombay.
J. Mar. Biol. Assoc., India, 3(1/2):1-5.

Chemistry, oxygen
Jeffries, Harry P., 1962.
Environmental characteristics of Raritan Bay, a polluted estuary.
Limnol. and Oceanogr., 7(1):21-31.

Chemistry, oxygen
Jensen, A.J.C., 1949.
Hydrography (Baltic). Danish investigations.
Ann. Biol. 5:131.

Chemistry, oxygen(data)
Jensen, A.J.C., 1944.
The hydrography of the Praestø Fjord.
Folia Geogr. Danica 3(2):47-55, 2 textfigs., 4 tables.

oxygen
Jerlov, N.G., 1956.
The Equatorial currents in the Pacific Ocean.
Repts. Swedish Deep-Sea Exped., 3(Phys. Chem. 6): 129-154.

Chemistry, oxygen
Jerlov, N.G., 1953.
The Equatorial Currents in the Indian Ocean.
Repts. Swedish Deep-sea Exped., 1947-1948, Phys. Chem., 3(5):115-125, 14 textfigs.

Chemistry oxygen
Johnston, R., 1956.
Chemical observations, northern North Sea.
Ann. Biol., Cons. Perm. Int. Expl. Mer, 11:41-43, Figs. 7-10.

Chemistry oxygen
Johnston, R., 1954.
Chemical observations.
Ann. Biol., Cons. Perm. Int. Expl. Mer, 10:84-86, Textfigs. 6-9.

Chemistry oxygen
Johnston, R., 1953.
Northern North Sea and approaches. Chemical observations.
Ann. Biol. Cons. Perm. Int. Expl. Mer, 9:99-103, Figs. 3-6.

Chemistry oxygen
Johnston, R., and R.B. Burns, 1958(1960).
Chemical observations in the North Sea.
Cons. Perm. Int. Expl. Mer, 15:38-40. (Ann. Biol.)

Chemistry oxygen
Johnston, R., and R.E. Craig, 1954.
Hydrographic conditions off the north coast of Scotland.
Ann. Biol., Cons. Perm. Int. Expl. Mer, 10:89-90, Fig. 12.

Chemistry oxygen
Jones, J.A., 1963.
Ecological studies of the southeastern Florida patch reefs. 1. Diurnal and seasonal changes in the environment.
Bull. Mar. Sci., Gulf and Caribbean, 13(2):282-307.

Chemistry, oxygen
*Jones, James I., 1967.
Significance of distribution of planktonic Foreminifera in the Equatorial Atlantic Undercurrent.
Micropaleontology, 13(4):489-501.

Chemistry oxygen
Jones, P.G.W., 1971.
The southern Benguela Current region in February 1966: Part I. Chemical observations with particular reference to upwelling. Deep-Sea Res., 18(2): 193-208.

K

Chemistry, oxygen (data)
Kaleis, M.V., and N.B. Aleksandrovskaye, 1965.
Hydrological régime of the Baltic Sea in 1963.
Ann. Biol., Cons. Perm. Int. Expl. Mer, 1963, 20:70-72.

chemistry, oxygen
Kaleis, M.V., N.B. Alexandrovskaya and E.A. Yula 1968
Some peculiarities in the oceanographical regime of the Baltic in 1966.
Annls biol. Copenh. 1966, 23: 78-81.

Chemistry, oxygen
Kalle, K., 1957.
Chemische Untersuchungen in der Irminger See im Juni 1955. Ber. Deutschen Wiss. Komm. f. Meeresf. n.f., 14(4):313-328.

Chemistry, oxygen
Kalle, K., 1943.
Die grosse Wasserumschichtung im Gotland-tief vom Jahre 1933-1934. Ann. Hydr., usw., 71:142-146, 6 textfigs.

Chemistry, oxygen
Kalle, K., 1939
V. Die chemischen Arbeiten auf der "Meteor" Fahrt Januar bis Mai 1938. Bericht über die zweite Teilfahrt des Deutschen Nordatlantischen Expedition des Forschungs-und Vermessungsschiffes "Meteor", Januar bis Juli, 1938. Ann. Hydro. u. Mar. Meteorol. 1939: 23-30, 6 text figs.

Chemistry oxygen
Kändler, Rudolf, 1963
Hydrographische Untersuchungen über die Abwasserbelastung der Flensburger Förde.
Kieler Meeresforsch., 19(2):142-157.

Chemistry, oxygen
Kändler, R., 1957.
Cruises with the R/V "Südfall",
Ann. Biol., Cons. Perm. Int. Expl. Mer, 12:109.

chemistry oxygen
Kandler, R., 1953.
Hydrographische Untersuchungen zum Abwasserproblem in den Buchten und Förden der Ostseeküste Schleswig-Holstein. Kieler Meeresf. 9(2):176-200, Pls. 9-14(18 figs.).

chemistry, oxygen
Kasturirangan, L.R., 1957.
A study of the seasonal changes in the dissolved oxygen of the surface waters of the sea on the Malabar coast. Indian J. Fish., 4(1):134-189.

chemistry, oxygen (data)
Kato, Takeo 1969.
On the distribution of the water mass produced by water-mixing and the change of chemical composition. (Ratio $Ca^{2+}:Cl^-$, $SO_4^{2-}:Cl^-$) - Oceanographical studies on the East China Sea, 3. (In Japanese; English abstract)
Umi to Sora 44(2/3): 55-80

chemistry, oxygen (data)
Kato, Kenji, 1966.
Geochemical studies on the mangrove region of Cananéia, Brazil. 1. Tidal variations of water properties.
Bolm Inst. Oceanogr., S. Paulo, 15(1):13-20.

chemistry, oxygen
Kato, K., 1966.
Chemical investigations on the hydrographic system of Cananéia lagoon.
Bolm Inst. Oceanogr., S. Paulo, 15(1):1-12.

chemistry, oxygen (data)
Kato, Kenji, 1966.
Geochemical studies on the mangrove region of Cananéia, Brazil. II. Physico-chemical observations on the reduction states.
Bolm Inst. Oceanogr., S. Paulo, 15(1):21-24.

chemistry, oxygen (data)
Kato, Kenji, 1966.
Studies on calcium content in sea water. II. Distribution of calcium in the Atlantic water off south Brazil.
Bolm Inst. Oceanogr., S. Paulo, 15(1):29-39.

chemistry, oxygen
Kato, Kenji, 1961.
Oceanographical studies in the Gulf of Cariaco.
1. Chemical and hydrographical observations in January, 1961.
Bol. Inst. Oceanograf., Univ. Oriente, Cumana, Venezuela, 1(1):49-72.

chemistry oxygen (data)
Kato, Kenji, 1961
Some aspects on biochemical characteristics of sea water and sediments in Mochima Bay, Venezuela.
Bol. Inst. Oceanogr., Univ. de Oriente, Venezuela 1(2):343-358.

chemistry, oxygen
Kawamoto, T., 1954.
[On the saturation percentage of the dissolved oxygen in the sea - on the case of the Tsushima and Liman cold-current region in the Japan Sea.] Spec. Publ. Sea Reg. Fish. Res. Lab., 3rd Anniv., 115-123.

chemistry oxygen (data)
Kawamura, Teruyoshi, 1966.
Distribution of phytoplankton populations in Sandy Hook Bay and adjacent areas in relation to hydrographic conditions in June 1962.
Tech. Pap. Bur. Sport Fish Wildl., U.S., (1):1-37. (multilithed).

chemistry, oxygen
Kay, H., 1954.
Untersuchungen zur Menge und Verteilung der organischen Substanz im Meerwasser.
Kieler Meeresf. 10(2):202-214, Figs. 18-25.

chemistry oxygen
Keeling, Charles D., and Bert Bolin, 1968
The simultaneous use of chemical tracers in oceanic studies. II. A three-reservoir model of the North and South Pacific oceans.
Tellus, 20(1): 17-54

chemistry, oxygen
Ketchum, B.H., J.C. Ayers, and R.F. Vaccaro, 1952.
Processes contributing to the decrease of coliform bacteria in a tidal estuary. Ecology 33(2): 247-258, 4 textfigs.

chemistry, oxygen
Ketchum, B.H., A.C. Redfield, and J.C. Ayers, 1951.
The oceanography of the New York Bight. P.P.O.M. 12(1):1-46, 20 textfigs.

Reviewed: J. du Cons. 18(2):246-247 by D. Vaux

chemistry, oxygen
Khimitsa, V.A., 1968.
On the vertical and horizontal distribution of oxygen along the west coast of Hindustani during the northeast monsoon. (In Russian).
Trudy, Vses. Nauchno-Issled. Inst. Morks. Ribn. Okeanogr (VNIRO) 64, Trudy Azovo-Chernomorsk. Nauchno-Issled. Inst. Morsk. Ribn. Khoz. Okeanogr. (AscherNIRO), 28: 162-180.

chemistry oxygen
Kijowski, S., 1937.
Nieco danych o skladzii chemicznym wod Zatoki Gdanskiej. (Quelques données sur la composition chimique des eaux du Golf de Dantzig). Bull. Sta. Mar. Hel 1(1):33-41.

chemistry oxygen
King, Joseph E., 1958
Variation in abundance of zooplankton and forage organisms in the Central Pacific in respect to equatorial upwelling.
Proc. Ninth Pacific Sci. Congr., Pacific Sci. Assoc., 1957, 16(Oceanogr.):98-107.

chemistry, oxygen
Kinney, Patrick, Martin E. Arhelger and David C. Burrell 1970.
Chemical characteristics of water masses in the Amerasian Basin of the Arctic Ocean.
J. geophys. Res. 75(21): 4097-4104.

chemistry oxygen
Kinsey, D.W., and Barbara E. Kinsey, 1967.
Diurnal changes in oxygen content of the water over the coral reef platform at Heron I.
Aust. J. mar. Freshwat. Res., 18(1): 23-34

chemistry oxygen
Kirillova, E.P., 1965.
The regime of chlorine, alkalinity and oxygen in the mouth of the Iuzhnii Bug. (In Russian).
Trudy, Gosudarst. Okeanogr. Inst., No. 83:158-171.

chemistry oxygen
Kitamura, Hiroyuki, 1959.
[A short term variation of the dissolved oxygen. Chemical oceanography of the North Pacific. V.] Umi to Sora, 34(3); 35(1):7-12.

chemistry oxygen
Kitamura, Hiroyuki, 1958.
[Oxygen distribution in the North Pacific Ocean. Chemical oceanography of the North Pacific. IV.] Umi to Sora, 33(12):71-75.
34(4)

chemistry oxygen
Kitamura, Hiroyuki, and Takeshi Sagi, 1964.
On the chemical elements in the sea south of Honshu, Japan. (In Japanese; English abstract).
Bull. Kobe Mar. Obs., No. 172:6-54.

chemistry, oxygen
Kitching, J.A., S.J. Lilly and S.M. Lodge, J.F. Sloane, R. Bassindale and F.J. Ebling, 1952.
The ecology of the Lough Ine Rapids with special reference to water currents. J. Ecology 40(1): 179-201.

chemistry oxygen
Klepikov, V.V., 1958(1960)

The origin and diffusion of Antarctic ocean-bed water. Problemy Severa, (1):

Translation in:
Problems of the North, 1: 321-333.

chemistry oxygen (data)
Klepikova, V.V., Edit., 1961
Third Marine Expedition on the D/E "Ob", 1957-1958. Data.
Trudy Sovetskoi Antarktich. Exped., Arktich. i Antarktich. Nauchno-Issled. Inst., Mezhd. Geofiz. God, 22:1-234 pp.

chemistry oxygen
Klots, C.E., and B.B Benson, 1963
Solubilities of nitrogen, oxygen and argon in distilled water.
J. Mar. Res., 21(1):48-57.

chemistry, oxygen (data)
Knudsen, M., 1911.
Danish hydrographical investigations at the Faroe Islands in the spring of 1910. Medd. Komm. Havundersøgelser, Ser. Hydrogr., 2(1):17 pp., 5 text-figs., 2 pls.

chemistry, oxygen
Kollmeyer, Ronald C., Robert M. O'Hagan, and Richard M. Morse, 1965.
Oceanography of the Grand Banks region and the Labrador Sea in 1964.
U.S.C.G. Oceanogr. Rept., No. 10(373-10):1-24; 34-285.

chemistry, oxygen
Korshun, M.A., and N.E. Gel'man, 1946.
[Apparatus for direct microdetermination of oxygen.] Zavofskaia Laboratoriia 12(4-5):500-502.

RT-1020 Bibl. Transl. Rus. Sci. Tech. Lit. 7.

chemistry, oxygen
Kostikova, A.N., 1965.
Hydrochemical characteristics of the Gulf of Alaska waters in the summer-autumn 1962. (In Russian).
Sovetsk. Ribokhoz. Issled. Severo- Vostochn. Chasti Tikhogo Okeana, 4 (Vses. Nauchno-Issled. Inst., VNIRO Trudy 58; Tikhookean. Nauchno-Issled. Inst., TINRO, Trudy, 53):21-33.

chemistry, oxygen

Kovalev, A.D., and V.I. Cherniavskii, 1965.
On the daily variability of the hydrologic elements in Tauisk inlet. (In Russian).
Izv., Tichookean. Nauchno-Issled. Inst. Ribn. Khoz. i Okeanogr., 59:39-47.

chemistry oxygen (data)

Kramer J.R. 1961
Chemistry of Lake Erie.
Proc. Fourth Conf., Great Lakes Research, Great Lakes Res. Div., Inst. Sci. & Tech. Univ. Michigan. Publ., (7):27-56.

Chemistry, oxygen

Kruger, D., 1950.
Variations quantitatives des protistes marins au voisinage du Port d'Alger durant l'hiver 1949-1950. Bull. Inst. Océan., Monaco, No. 978:20 pp., 5 textfigs.

chemistry oxygen

Kusunoki, Kou, 1962
Hydrography of the Arctic Ocean with special reference to the Beaufort Sea.
Contrib. Inst., Low Temp. Sci., Sec. A,(17): 1-74.

chemistry oxygen (data)

Kvinge, Tor, 1963.
The "Conrad Holmboe" Expedition to East Greenland waters in 1923.
Årbok, Univ. i Bergen, Mat.-Naturv. Ser., (15): 44 pp.

L

chemistry, oxygen

Khovansky, Y.A., 1962.
Some features of the shelf-water dynamics off the South West Africa coast. (In Russian).
Baltisk. Nauchno-Issled. Inst. Morsk. Ribn. Khoz. i Okeanogr. (BALTIRO), Trudy, 9:57-69.

Chemistry, oxygen

Kuenen, Ph.H., 1948
Influence of the earth's rotation on ventilation currents of the Moluccan deep-sea basins. Proc. Koninklijke Nederlandsche Akademie van Wetenschappen, LI(4):417-426, 4 text figs.

chemistry, oxygen

Kuznetsova, L.N., 1966.
On the vertical circulation of water masses in the northwestern North Atlantic. (In Russian).
Mater. Ribokhoz. Issled. severn. Basseina, Poliarn. Nauchno-Issled. Proektn. Inst. Morsk. Ribn. Khoz. Okeanogr. (PINRO), 7:129-136.

Laktionov, A.F., V.A. Shamontyev and A.V. Yanes, 1960 oxygen

Oceanographic characteristics of the North Greenland Sea.
Soviet Fish. Invest., North European Seas, VNIRO, PNIRO, Moscow, 1960:51-65.

chemistry, oxygen

Lefevre, S., E. Leloup, and L. Van Meal, 1956.
Observations biologiques dans le port d'Ostende.
K. Belg. Inst. Naturwet., Verhandl., [Inst. R. Sci. Nat. Belg., Mem.,] 133:157 pp.

chemistry, oxygen

Legand, M., 1957.
Variations quantitatives du zooplancton récolté par l'Orsom III pendant la croisière EQUAPAC.
Off. Recherche Sci. et Tech. Outre-Mer, Inst. Français d'Océanie, Sect. Océanogr., Rapp. Sci., No. 2:31 pp. (mimeographed).

chemistry oxygen

Leloup, E., Editor, with collaboration of L. Van Meel, Ph. Polk, R. Halewyck and A. Gryson, 1966.
Recherches sur l'ostreiculture dans le Bassin de Chasse d'Ostende en 1962.
Ministere de l'Agriculture, Commission T.W.O.Z. Groupe de Travail - "Ostreiculture", 58 pp.

chemistry, oxygen (data)

Lemmasson, L., et Y. Magnier, 1966.
Résultats des observations scientifiques de La Dunkerquoise sous le commandement du Capitaine de Corvette Brosset. Croisière Hunter.
Cah. ORSTOM, Sér. Océanogr., 4(1):3-78.

chemistry oxygen (data)

Leontyeva, V.V., 1960

[Some data on the hydrology of the Tonga and Kermadec Trenches.]
Trudy Inst. Okeanol., 40:72-82.

chemistry, oxygen

Leontieva, V.V., and A.E. Gamutilov, 1959

[The influence of the oceanic waters on the hydrological conditions of the Kronotski Bay.]
Trudy Inst. Okeanol., 36: 59-72.

chemistry OXYGEN

Lida, Michitaka, 1964.
Dissolved oxygen in the Pacific Ocean and adjacent seas as an important element to study the circulation and structure in the changing marine environment.
In: Recent researches in the fields of hydrosphere, atmosphere and nuclear geochemistry, Ken Sugawara festival volume, Y. Miyake and T. Koyama, editors, Maruzen Co., Ltd., Tokyo, 349-356.

chemistry oxygen

Lieber, M., and R. Eliassen, 1949.
The use of the amidol test for dissolved oxygen in sea water. J. Mar. Res. 8(2):107-119, 3 textfigs.

Chemistry oxygen

Lisitzin, A.P., 1966.
Processes of Recent sedimentation in the Bering Sea. (In Russian).
Inst. Okeanol., Kom. Osad. Otdel Nauk o Zemle, Isdatel, Nauka, Moskva, 574 pp.

Chemistry oxygen (data)

Littlepage, Jack L., 1965.
Oceanographic investigations in McMurdo Sound, Antarctica.
In: Biology of Antarctic seas, II.
Antarctic Res. Ser., Am. Geophys. Union, 5:1-37.

chemistry oxygen

*Lockerman, Robert C., 1968.
Some summer oceanographic features of the Laptev and East Siberian seas.
Tech. Rept., U.S. Naval Oceanogr. Off., TR-200: 50 pp. (multilithed).

Chemistry oxygen (data)

Longhurst, A.R., 1964.
The coastal oceanography of western Nigeria.
Bull. Inst. Français, Afrique Noire, (A), 26(2): 337-402.

Chemistry oxygen

Longhurst, Alan R., 1962.
A review of the oceanography of the Gulf of Guinea.
Bull. Inst. Français Afrique Noire, 24(3):633-663.

Chemistry oxygen

Lynn, W.R. and Won Tak Yang, 1960

The ecological effects of sewage in Biscayne Bay. Oxygen demand and organic carbon determinations. Bull. Mar. Sci. Gulf Caribb., 10 (4): 491-509.

M

Chemistry oxygen

Machado, L. de B., 1952.
Pesquisas fisicas e quimicas do sistema hidrografico da regiao Lagunar de Cananeia.
Bol. Inst. Ocean., Univ. Sao Paulo, 3(1/2):55-75, 4 graphs, 1 chart.

Chemistry oxygen

Machado, L. de B., 1951.
Resultados científicos do cruzeiro do "Baependi" e do "Vega" à Ilha de Trindada. Oceanografia física. Contribuição para o conhecimento das características físicas e químicas das águas.
Bol. Inst. Paulista Ocean. 2(2):95-110, 5 pls.

Chemistry oxygen

Machado, L. de B., 1950.
Pesquisas fisicas e quimicas do sistema hidrográfico da região lagunar de Cananéia. 1. Cursos de água. Bol. Inst. Paulista Ocean. 1(1):45-68.

Chemistry oxygen (data)

Maeda, H., 1953.
Studies on Yosa-Naikai. 3. Analytical investigations on the influence of the River Noda and the benthonic communities. J. Shimonoseki Coll. Fish 3(2):141-149, 3 textfigs.

chemistry oxygen (data)

Maeda, H., 1953.
Studies on Yosa-Naikai. 2. Considerations upon the range of the stagnation and the influences by the River Noda and the open sea.
J. Shimonoseki Coll. Fish. 3(2):133-140, 2 textfigs.

chemistry, oxygen (data)

*Maestrini, Serge, et Mariano Javier Pigarro, 1966
Contribution a l'étude de l'hydrologie et de la productivité primairé des eaux cotières de la région de Tuléar.
Recl. Trav. Stn. mar. Endoume, hors sér., Suppl. 5:7-23.

Chemistry oxygen (data)

Mahadeva, N., 1962
Preliminary report on the experiments with Marutoku-B net and a 45 cm x 90 cm net of No. 0 bolting silk off Trincomalee, Ceylon, during October 1960.
Indo-Pacific Fish. Council, FAO, Proc., 9th Sess., (2/3):17-24.

Chemistry, oxygen

Makkaveeva, N.S., 1965.
The hydrochemical investigations of the Korovinskaya Inlet. (In Russian).
Material. Ribochoz. Issled. Severn. Bassin. Poliarn. Nauchno-Issled. i Proekt. Inst. Morsk. Rib. Choz. i Okeanogr., N.M. Knipovich, 5:117-122.

Chemistry oxygen

Memeli, D., and F. Mosetti, 1965.
The significance of the dissolved oxygen determination in hydrological researches.
Boll. Geofis. Teor. ed Appl., 7(28):322-325.

German summary

Chemistry, oxygen

Mann, C.R., A.B. Grant and T.R. Foote, 1966.
Atlas of oceanographic sections, temperature-salinity-dissolved oxygen, Northwest Atlantic Ocean, Newfoundland Basin enf Gulf Stream, February, 1962-July, 1964.
Rept. B.I.O. 65-16:51 pp. (quarto) (multilithed).

Chemistry oxygen

Mao, H., and K. Yoshida, 1955.
Physical oceanography in the Marshall Islands area. Bikini and nearby atolls, Marshall Islands. Prof. Papers, Geol. Survey, 260-R:645-684, Figs. 179-218.

Margalef, Ramon

chemistry
oxygen(data)

Margalef, R., F. Cervignon and G. Yepez T., 1960
Exploracion preliminar de las caracteristicas hidrograficas y de la distribucion del fitoplancton en el area da la Isla Margarita (Venezuela).
Mem. Soc., Ciencias Nat. de la Salle, 22(57): 210-221.
Contribucion No. 2, Estacion de Investigaciones Marinas de Margarita, Fundacion La Salle de Ciencias Naturales.

chemistry
oxygen(data)

Margalef, Ramón, and Juan Herrera, 1963
Hidrografía y fitoplancton de las costas de Castellón, de julio de 1959 a junio de 1960.
Inv. Pesq., Barcelona, 22: 49-109.

chemistry
oxygen(data)

Marinkovic-Roje, Marija, and Miroslav Nicolic, 1961
[Oceanographic researches in the areas of Rovinj and Limski Kanal from 1959-1961.] In Jugoslavian: English summary.
Hidrografskog Godisnjaka 1960 Split:61-67.

Chemistry, oxygen

Marshall, S.M. and A.P. Orr, 1948
Further experiments on the fertilization of a sea loch (Loch Craiglin). The effect of different plant nutrients on the phytoplankton.
J.M.B.A. 27(2):360-379, 10 text figs.

chemistry oxygen (data)

Marukawa, H., and T. Kamiya, 1926.
Outline of the hydrographical features of the Japan Sea.
Annot. Oceanogr. Res., 1(1):1-7, tables, charts.

Chemistry, oxygen

Maslennikov, V.V., 1971.
Oceanographic surveys in the Andaman Sea and the northeastern Bay of Bengal. (In Russian)
Vses. nauchno-issled. Inst. morsk. Ribn. Khoz. Okeanogr. VNIRO, Trudy 72: 46-55

Masuzawa, Jotaro

Chemistry
oxygen (data)

Magazzu, Giuseppe, e Guglielmo Cavallaro 1969.
Rapporto sulle crociere di studio lungo le coste meridionali Calabresi e della Sicilia orientale (1967-1968). 1. Idrografia.
Programma Ricerca Risorse Mar. Fondo mar. (B) 45: 5-70.

oxygen (data)

Magazzu, Giuseppe, Guglielmo Cavallaro e Letterio Guglielmo 1969.
Considerazioni preliminari sulle condizioni chimico-fisiche e sullo zooplancton delle acque costiere fra C. po Milazzo e C. po D'Orlando (Messina)
Programma Ricerca Risorse mar. Fondo mar. (B) 45: 71-90

Chemistry, oxygen

Masuzawa Jotaro 1967.
An oceanographic section from Japan to New Guinea at 137°E in January 1967.
Oceanogrl Mag. 19(2): 95-118

Chemistry
oxygen

Masuzawa, Jotaro, 1965.
Water characteristics of the Kuroshio.
Oceanogr. Mag., Tokyo, 17(1/2):37-47.

chemistry - oxygen

Masuzawa, Jotaro, 1964.
A typical hydrographic section of the Kuroshio extension.
Oceanogr. Mag., Tokyo, 16(1/2):21-30.

Chemistry
oxygen

Masuzawa, Jotaro, 1962
The deep water in the western boundary of the North Pacific.
J. Oceanogr. Soc., Japan, 20th Ann. Vol., 279-285.
Also JEDS Contrib. No. 39

chemistry
oxygen (data)

Masuzawa, Jotaro, 1961
Preliminary report of the Japanese Expedition of Deep-Sea, the Third Cruise (JEDS-3).
Oceanogr. Mag., Tokyo, 12(2):207-218.

chemistry, oxygen

Masuzawa, J., 1955.
Preliminary report of the Kuroshio in the Eastern Sea of Japan (currents and water masses of the Kuroshio system III).
Rec. Ocean. Wks., Japan, 2(1):132-140, 5 textfigs

chemistry, oxygen

Masuzawa, J., 1950.
On the intermediate water in the southern Sea of Japan. Ocean. Mag., Tokyo, 2(4):137-144, 6 text-figs.

chemistry
oxygen(data)

Masuzawa, Jotaro, and Hideo Akamatsu, 1962.
1. Hydrographic observations during the JEDS-4.
Oceanogr. Mag., Japan Meteorol. Agency, 13(2): 122-130.

chemistry, oxygen

Mazuzawa, Jotaro, Tsutomu Akiyama, Yutaka Kawarada and Tsutomu Sawara, 1970.
Preliminary report of the Ryofu Maru Cruise Ry7001 in January-March 1970. Oceanogrl Mag., 22(1): 1-25.

chemistry oxygen (data)

Masuzawa, J., M. Yasui and H. Akamatsu, 1960.
A hydrographic section across the Kurile Trench.
Repts. of JEDS, 1:165-172.
Also in Oceanogr. Mag. 11 (2) (same pagination)

chemistry
oxygen (data)

Matsudaira, Yasuo, 1964
Cooperative studies on primary productivity in the coastal waters of Japan, 1962-63. (In Japanese; English abstract).
Inform. Bull., Planktol., Japan, No. 11:24-73.

chemistry
oxygen (data)

Matsudaira, Yasuo, Haruyuki Koyama and Takuro Endo, 1961
[Hydrographic conditions of Fukuyama Harbor.]
J. Fac. Fish. and Animal Husbandry, Hiroshima Univ., 3(2):247-296.

chemistry, oxygen

*Maximova, M.P., and A.M. Chernyakova, 1968.
Oxygen solubility in sea water. (In Russian; English abstract).
Okeanologiia, Akad. Nauk, SSSR, 8(5):912-919.

chemistry
oxygen (data)

McGary, James W., Edward D. Stroup, 1958
Oceanographic observations in the central North Pacific, September 1954-August 1955.
U.S.F.W.S. Spec. Sci. Rept., Fish., No. 252: 250 pp.

chemistry
oxygen

McGill, David A., 1964.
The distribution of phosphorus and oxygen in the Atlantic Ocean as observed during the I.G.Y., 1957-1958.
Progress in Oceanography, Pergamon Press, 2:127-211.

chemistry oxygen (data)

McGill, David A., and Nathaniel Corwin, 1965.
The distribution of nutrients in the Labrador Sea, Summer, 1964.
U.S.C.G. Oceanogr. Rept. No. 10(CG 373-10):25-33, 277-255.

chemistry oxygen

*McLain, Douglas R., 1968.
Oceanographic surveys of Traitors Cove, Revillagigedo Island, Alaska.
Spec. scient. Rep. U.S. Fish Wildl. Serv. Fish, 576: 15 pp. (multilithed).

chemistry
oxygen

McLaren, I.A., 1961
The hydrography and zooplankton of Ogac Lake, a landlocked fjord on Baffin Island.
Fish. Res. Bd., Canada, MSS Rept. Ser. (Biol.) No. 709: 167 pp. (multilithed).

chemistry oxygen

McLellan, H.J., 1957.
On the distinctness and origin of the slope water off the Scotian shelf and its easterly flow south of the Grand Banks.
J. Fish. Res. Bd., Canada, 14(2):213-239.

chemistry
oxygen

McLellan, H.J., and W.D. Nowlin, 1963
Features of deep water in the Gulf of Mexico.
J. Mar. Res., 21(3):233-245.

chemistry, oxygen

Megia, T.G., and R.G. Lao, 1955.
A report on the O:N:P ratios of Philippine and adjacent waters. Philippine J. Sci., 3(1):55-61, 3 textfigs.

chemistry oxygen

Menzies, R.J., O.H. Pilkey, B.W. Blackwelder, D. Dexter, P. Huling and L. McCloskey, 1966.
A submerged reef off North Carolina.
Int. Rev. ges. Hydrobiol., 51(3):393-431.

chemistry, oxygen

Metcalf, W.G., 1955.
On the formation of bottom water in the Norwegian basin. Trans. Amer. Geophys. Union, 36(4):596-600, 2 textfigs.

chemistry - oxygen

*Metcalf, W.G., and M.C. Stalcup, 1967.
Origin of the Atlantic Equatorial Undercurrent.
J. geophys. Res., 72(20):4959-4975.

Chemistry oxygen
Metcalf, W.G., A.D. Voorhis and M.C. Stalcup, 1962.
The Atlantic Equatorial Undercurrent.
J. Geophys. Res., 67(6):2499-2508.

Chemistry oxygen
Meyer, P.F., and K. Kalle, 1950.
Die biologische Umstimmung der Ostsee in den letzten Jahrzehnten - eine Folge hydrographische Wasserumschichtungen? Archiv f. Fischereiwiss., 2 Jahrg.,:1-9, 6 textfigs.

Chemistry oxygen
Millard, N.A.H., and K.M.F. Scott, 1954.
The ecology of South African estuaries. VI. Milnerton estuary and the Diep River, Cape. Trans. R. Soc., S. Africa, 34(2):279-324, 8 textfigs.

Chemistry oxygen (data)
*Minas, Hans Joachim, 1968.
A propos d'une remontée d'eaux "profondes" dans les parages du Golfe de Marseille (octobre 1964): conséquences biologiques.
Cah. océanogr., 20(8):647-674.

Chemistry oxygen
Minas, H.J., and B. Coste, 1964.
Étude de la structure hydrologique et de quelques aspects de la productivité de la zone euphotique en fin d'été au niveau d'une station fixe (Bouée-Laboratoire du Comexo) en rade de Villefranche s/mer.
Rec. Trac. Sta. Mar., Endoume, 34(50):133-155.

Chemistry, oxygen
Minas, H.J., and M.J. Pizarro, 1962.
Quelques observations hydrologiques sur les eaux au large des côtes orientales de la Corse.
Bull. Inst. Océanogr., Monaco, 59(1232):1-23.

Chemistry oxygen
Miwa, Katsutoshi, 1962
Report on dissolved nutrients investigations in the North Pacific and the Bering Sea during June-August 1960. (In Japanese; English summary)
Bull. Hokkaido Reg. Fish. Res. Lab., (25): 58-65.

Chemistry oxygen
Miyake, Y., 1948
The diurnal variation of dissolved oxygen in sea water consisting of different water masses. Geophys. Mag., Japan, 16(2/4):66-70, 6 text figs.

Chemistry, oxygen
*Miyake, Y., and K. Saruhashi, 1967.
A study of the dissolved oxygen in the ocean.
Pap. Met. Geophys., Tokyo, 17(3):210-217.

Chemistry, oxygen
Miyake, Y., and K. Saruhashi, 1957.
On the vertical distribution of the dissolved oxygen in the ocean. Proc. UNESCO Symp., Phys. Ocean., Tokyo, 1955:156-159.

Chemistry, oxygen
Miyake, Y., Y. Sugiura and K. Kameda, 1955.
On the distribution of radioactivity in the sea around Bikini Atoll in June 1954.
Pap. Met. Geophys., Tokyo, 5(3/4):253-263.

Chemistry oxygen
Miyake, Y., Y. Sugiura and K. Kamada, 1953.
A study of the property of the coastal waters around Hachijo Island. Rec. Ocean. Wks., Japan, n.s., 1(1):93-99, 10 textfigs.

Chemistry oxygen
Miyake, Yasuo, Yoshio Sugiura, Yukio Sugimura, Tsutomu Akiyama and Hibrozo Yoshimura, 1962.
II. Chemistry in the Japanese Expedition of the Deep Sea (JEDS-4).
Oceanogr. Mag., Japan Meteorol. Agency, 13(2): 131-132.

JEDS Contrib. 30.

Chemistry oxygen
Mokiyevskaya, V.V., 1961
[Some peculiarities of hydrochemistry in the northern Indian Ocean.] (In Russian; English abstract).
Okeanolog. Issled., Mezhd. Komitet Proved. Mezhd. Geofiz. Goda, Prezidiume, Akad. Nauk, SSSR, (4):50-61.

Chemistry, oxygen
Mokievskaja, V.V., 1959
[Nutrients salts in the upper layer of the Bering Sea.] Trudy Inst. Okeanol., 33: 87-113.

Chemistry, oxygen
Montgomery, H.A.C., N.S. Thom and A. Cockburn, 1964.
Determination of dissolved oxygen by the Winkler method and the solubility of oxygen in pure water and sea water.
J. Appl. Chem., 14(7):280-296.

Chemistry oxygen (data)
Morales, E., y E. Arias, 1965.
Ecología del puerto de Barcelona y desarrollo de adherencias orgánicas sobre placas sumergidas.
Inv. Pesq., Barcelona, 28:49-79.

Chemistry oxygen (data)
Moreira da Silva, Paulo, 1957
Oceanografia do triangulo Cabo-Frio-Trinidade-Salvador.
Anais Hidrograficos, Brasil, 16:213-308.

Chemistry - oxygen
Mori, Isamu, and Haruhiko Irie, 1966.
The hydrographic conditions and the fisheries damages by the red water occurred in Omura Bay in summer 1965 - III. The oceanographic conditions in the offing region of Omura Bay during the term of the red water. (In Japanese; English abstract).
Bull. Fac. Fish. Nagasaki Univ., No. 21:103-113.

Chemistry, oxygen
Morii, Hideaki, Yuako Akishige, Ryoichi Kanagu and Tadanobu Fukuhara 1967.
Chemical studies of the Kuroshio and adjacent region. I. On the nutrient salts and the dissolved oxygen. (In Japanese; English abstract)
Bull. Fac. Fish. Nagasaki Univ. 22:91-103.

Chemistry, oxygen
Morse, D.C., 1947
Some observations on seasonal variations in plankton population Patuxant River, Maryland 1943-1945. Bd. Nat. Res., Publ. No.65, Chesapeake Biol. Lab., 31, 3 figs.

Chemistry, oxygen
Mosby, Hakon, 1959
Deep water in the Norwegian Sea.
Geofys. Publ., Norge, 21(3): 1-62.

Chemistry oxygen
Mosby, H., 1938
Svalbard waters. Geophys. Publ. 12(4): 85 pp., 34 text figs.

Chemistry oxygen
Mostert, S.A., 1966.
Distribution of inorganic phosphate and dissolved oxygen in the south west Indian Ocean.
Repub. S. Africa, Dept. Comm. Indust., Div. Sea Fish. Invest. Rept., No. 54:1-23.

Chemistry oxygen
Motoda, S., 1940.
The environment and the life of massive reef coral, Goniastrea aspera Verrill, inhabiting the reef flat in Palao.
Palao Trop. Biol. Sta. Studies 2(1):61-104, 9 textfigs.

Chemistry oxygen
Motoda, S., 1940.
Comparison of the conditions of water in the bay, lagoon and open sea in Palao.
Palao Trop. Biol. Sta. Studies 2(1):41-48, 2 textfigs.

Chemistry oxygen
Motoda, S., 1940.
Organic matter in sea water of Palao, South Sea.
Trans. Sapporo Nat. Hist. Soc. 16(2):100-104.

Chemistry oxygen
Moya, Regina C., 1948.
Determinacion del oxigeno en el agua de mar.
Rev. Biol. Mar. 1(1):46-56.

Chemistry oxygen
Mulicki, Z., 1957.
[Ecology of the more important Baltic invertebrates.] Prace Morsk. Inst. Ryback., Gdyni, No. 9: 313-379.

Chemistry oxygen
Murakami, A., 1954.
[Oceanography of Kasaoka Bay in Seto Inland Sea.] Bull. Nakai Regional Fish. Res. Lab. 6:15-57, 42 textfigs.

Chemistry, oxygen
Muromtsev, A.M., 1962.
On the hydrology of the Suez, Red Sea and Aden Bay. (In Russian).
Meteorol. i Gidrol., (2):42-45.

Chemistry oxygen (data)
Muromtsev, A.M., 1959
[The basis for the hydrology of the Indian Ocean.] Gidrometeoizdat, Leningrad, 437 pp.

Chemistry, oxygen
Muromtsev, A.M., 1959
[Basic data for the hydrology of the Indian Ocean. Appendix II. Atlas of vertical section and charts of temperature, salinity, density and oxygen content.]
Gosud. Okeanogr. Inst. Glabnoe Upravlenie Gidrometeorol. Sluzhbi pri Sovete Ministrov SSSR, Gidrometoizdat, Leningrad, 112 pp.

Chemistry, oxygen (data)

Muromtsev, A.M., 1958
[The basis for the hydrology of the Pacific Ocean.]
Gidrometeoizdat., Leningrad, 431 pp.

Chemistry, oxygen

Muromtsev, A.M., 1958
[Basic data for the hydrology of the Pacific Ocean. Appendix II. Atlas of vertical sections and charts of temperature, salinity, density and oxygen content.]
Gosud. Okeanogr. Inst., Glabnoe Upravlenie Gidrometeorol. Sluzhbi pri Sovete Ministrov, SSSR, Gidrometeoizdat, Leningrad, 124 pp.

chemistry, oxygen

Murray, J.W., 1966.
A study of the seasonal changes of the water mass of Christchurch Harbour, England.
Jour. mar. biol. Assoc., U.K., 46(3):561-578.

N

Chemistry oxygen

Nagai, M., 1956.
[A note on the annual variation of dissolved oxygen in sea water (2).]
Ocean. & Meteorol., Nagasaki, Mar. Obs., 7(1/4): 127-132.

Chemistry, oxygen

Nagai, M., 1956.
[A note on the annual variation of dissolved oxygen in sea water.] Ocean. & Meteorol., Nagasaki Mar. Obs., 7(1/4):63-68.

Chemistry, oxygen

Nagai, M., 1955.
A note on the annual variation of dissolved oxygen in sea water. J. Ocean. Soc., Japan, 11(3): 127-132.

chemistry, oxygen

Nagai, M., 1955.
[A note on the annual variation of dissolved oxygen in sea water. Part 1] J. Ocean. Soc., Japan, 11(2):63-68.

Chemistry oxygen (data)

Nagaya, Y., 1959
On the distribution of the nutrient salts in the Equatorial region of the Western Pacific, Jan.-Feb., 1958.
Publ. 981, Hydrogr. Bull., No. 59: 31-40.

Chemistry oxygen

Nakano, M., S. Unoki, M. Hanzawa, R. Marumo and J. Fukuoka, 1954.
Oceanographic features of a submarine eruption that destroyed the Kaiyo-maru No. 5. J. Mar. Res. 13(1):48-66, 10 textfigs.

Chemistry oxygen

Nannichi, T., 1951.
On the fluctuation of the Kuroshio and/Oyashio. the
Pap. Met. Geophys. 2(1):102-111, 6 textfigs.

Chemistry, oxygen

Nan-niti, Tosio, Hideo Akanatsu and Toshisuke Nakai, 1964.
A further observation of a deep current in the east-north-east sea of Torishima.
Oceanogr. Mag., Tokyo, 16(1/2):11-19.

Chemistry oxygen

Nasu, Keiju, 1963
Oceanography and whaling ground in the Subarctic region of the Pacific Ocean.
Sci. Repts., Whales Res. Inst., No. 14:105-155.

Chemistry, oxygen

Nasu, K., 1960
Oceanographic investigation in the Chukchi Sea during the summer of 1958.
Sci. Repts., Whales Res. Inst., 15: 143-157.

Chemistry oxygen (data)

Navarro, F. de P., F. Lozano, J.M. Navaz, E. Otero, J. Sáinz Pardo and others, 1943.
La pesca de Arrastre en los fondos del Cabo Blanco y del Banco Arguin (Africa Sahariana).
Trab. Inst. Español Ocean. No.18:225 pp., 38 pls.

oxygen (data)

Nehring, D. and H.-J. Brosin 1968
Ozeanographische Beobachtungen im äquatorialen Atlantik und auf dem Patagonischen Schelf während der 1. Südatlantik Expedition mit dem Fischereiforschungsschiff Ernst Haeckel von August bis Dezember 1966.
Geod. Geoph. Veröff. 4(3):93pp.

oxygen

Nehring, Dietwart, Sigurd Schulz und Karl-Heinz Rohde, 1969.
Untersuchungen über die Produktivität der Ostsee [I. Chemisch-biologische Untersuchungen in der mittleren Ostsee und in der Bottensee im April/Mai 1967.]
Beitr. Meeresk., 23:5-36

oxygen

Nekrasova, V.A., and V.N. Stepanov, 1963.
Meridional hydrological profiles of the oceans from IGY data. (In Russian; English abstract).
Rez. Issled. Programme Mezhd. Geofiz. Goda, Okeanolog. Issled., Akad. Nauk, SSSR, No. 8:34-51.

Chemistry oxygen

Neumann, A. Conrad, 1965.
Processes of recent carbonate sedimentation in Harrington Sound, Bermuda.
Mar. Sci. Center, Lehigh Univ., 130 pp. (Unpublished manuscript).

Chemistry, oxygen

Neumann, A. Conrad, and David A. McGill, 1961 (1962).
Circulation of the Red Sea in early summer.
Deep-Sea Res., 8(3/4):223-235.

Chemistry oxygen

Neumann, W., 1949.
Beiträge zur Hydrographie des Vrana-Sees (Insel Cherso), insbesondere Untersuchungen über organisches sowie anorganisches Phosphor- und Stickstoffverbindungen. Nova Thalassia 1(6):17 pp.

Chemistry oxygen

Newell, B.S., 1959
The hydrography of the British East African coastal waters. II. Colonial Off., Fish. Publ. London, No. 12: 18 pp.

Chemistry, oxygen

Nichols, Maynard M., 1964.
Characteristics of sedimentary environments in Moriches Bay.
In: Papers in Marine Geology, R.L. Miller, Macmillan Co., N.Y., 363-383.

Chemistry, oxygen (data)

Nichols, M.M., and R.C. Barnes, 1964.
Shelf observations - hydrography: cruise of August 21-26, 1962.
Virginia Inst. Mar. Sci., Spec. Sci. Rept., (41): 23 pp. (multilithed).

chemistry, oxygen (data)

Niemi, Åke, Heinrichs Skuja and Torbjörn Willén 1970
Phytoplankton from the Pojoviken-Tvärminne area, S. coast of Finland.
Mem. Soc. Fauna Flora Fennica 46: 14-28.

Chemistry, oxygen

Nikiforov, E. G., E.V. Belysheva and N. I. Blinov, 1966.
On the structure of the water masses in the eastern part of the Arctic Basin. (In Russian).
Okeanologiia, Akad., Nauk, SSSR, 6(1):76-81.

Chemistry oxygen(data)

Nishida, Keizo, 1958
[On the oceanographic conditions of the lower cold water in the Japan Sea.]
J. Fac. Fish. and Animal Husbandry, Hiroshima Univ., 2(1): 1-65.

Chemistry, oxygen

Nishizawa, Tanzo and T. Muraki, 1940
Chemical studies in the sea adjacent to Palau. I. A survey crossing the sea from Palau to New Guinea. Kagaku Nanyo (Sci. of the South Sea) 2(3):1-7.

Chemistry oxygen (data)

Noakes, John E., and Donald W. Hood, 1961
Boron-boric acid complexes in sea water.
Deep-Sea Res., 8(2):121-129.

Chemistry oxygen

Noble, A., 1968.
Studies on sea-water off the North Kanara coast.
J. mar. biol. Ass., India, 10(2): 197-223.

Chemistry, oxygen

Norina, A.M., 1965.
Hydrochemical characteristics of the northern part of the Barents Sea. (In Russian).
Trudy, Gosudarst Okeanogr. Inst., No. 83:243-271.

Chemistry, oxygen

NORPAC Committee, 1960.
The NORPAC Atlas. Oceanic observations of the Pacific, 1955.

Chemistry oxygen

Novosalov, A.A., 1962
The study of the biochemical consumption of oxygen in the waters of the northern part of the Atlantic Ocean.
Okeanologiia, Akad. Nauk, SSSR, 2(1):84-91.

Abstracted in: Soviet Bloc Res., Geophys. Astron. and Space, 1962(35): 19. (OTS61-11147-35 JPRS13739)

Chemistry - oxygen

Nowlin, W.D., Jr., and H.J. McLellan, 1967.
A characterization of the Gulf of Mexico waters in winter.
J. mar. Res., 25(1):29-59.

chemistry, oxygen
*Nowlin, W.D., Jr., E.F. Paskausky and H.J. McLellan, 1969.
Recent dissolved-oxygen measurements in the Gulf of Mexico deep waters.
J. mar. Res., 27(1):39-44.

oxygen
Nusbaum, I., and H.E. Miller, 1952.
The oxygen resources of San Diego Bay.
Sewage and Industrial Wastes 24(12):1512-1527.

chemistry, oxygen (data)
Nutt, D.C., and L.K. Coachman, 1956.
The oceanography of Hebron Fjord, Labrador.
J. Fish. Res. Bd., Canada, 13(5):709-758.

O

chemistry, oxygen
Oceanographic Section, C.M.O., 1949.
Report on sea and weather observations on Antarctic Whaling Ground (1948-1949). Ocean. Mag., Tokyo, 1(3):142-173, 11 textfigs.

chemistry, oxygen (data)
Odum, Howard T., Rene P. Cuzon du Rest, Robert J. Beyers and Clyde Allbaugh, 1963.
Diurnal metabolism, total phosphorus, Ohle anomaly and zooplankton diversity of abnormal marine ecosystems of Texas.
Publ. Inst. Mar. Sci., Port Aransas, 9:404-453.

carbon
Ogura Norio, 1970
Dissolved organic carbon in the equatorial region of the central Pacific.
Nature, Lond. 227(5265):1335-1336

chemistry
oxygen
Ohwada, Mamoru, and Hisanori Kon, 1963
A microplankton survey as a contribution to the hydrography of the North Pacific and adjacent seas. (II). Distribution of the microplankton and their relation to the character of water masses in the Bering Sea and northern North Pacific Ocean in the summer of 1960.
Oceanogr. Magazine, Japan Meteorol. Agency, 14(2):87-99.

chemistry, oxygen
Ohwada, Mamoru and Futomi Ogawa 1966.
Plankton in the Japan Sea.
Oceanogr. Mag. 18(1/2):39-51.

chemistry, oxygen
Ohwada, Mamoru and Katsumi Yamamoto 1966.
Some chemical elements in the Japan Sea.
Oceanogr. Mag. 18(1/2):31-37.

chemistry, oxygen (data)
Okabe, Shiro, Yoshimasa Toyota and Takafumi Murakami, 1967.
Major aspects of chemical oceanography. (In Japanese; English abstract).
J. Coll. mar. Sci. Techn., Tokai Univ., (2):39-56.

chemistry, oxygen
Okitsu, T., T. Tokui and B. Tsubata, 1954.
Oceanographical observations off Asamushi during 1952. Bull. Mar. Biol. Sta., Asamushi, 7(1):21-25, 4 textfigs.

chemistry
oxygen
Okubo, Akira, 1959.
An estimation of the average compensation depth by a study of the vertical distribution of dissolved oxygen.
Rec. Oceanogr. Wks., Japan, n.s., 5(1):51-59.

chemistry
oxygen
Okubo, A., 1958.
The distribution of dissolved oxygen in the north-western part of the North Pacific Ocean in the aspect of physical oceanography. 1. General features of the oxygen distribution. (1).
Oceanogr. Mag., Tokyo, 10(1):137-156.

chemistry, oxygen
Okubo, A., 1957.
[A note on the super- and under-saturation of dissolved oxygen in surface waters (The distribution of dissolved oxygen in the sea east of Honshu, Japan.). 2.]
J. Ocean. Soc., Japan, 13(4):139-144.

chemistry, oxygen
Okubo, A., 1957.
Estimation of biological effects in the distribution of dissolved oxygen by use of isopycnic analysis. The distribution of dissolved oxygen in the sea east of Honshu, Japan, 1.
Ocean. Mag., Tokyo, 9(1):75-86.

Okuda, Teizo

chemistry oxygen
*Okuda, Taizo, 1967.
Vertical distribution of inorganic nitrogen in the equatorial Atlantic Ocean.
Boln Inst. Oceanogr., Univ. Oriente, 5(1/2):67-83.

chemistry
oxygen
Okuda, Taizo, José Rafael Gomez, José Benitez Alvarez and Angel José Garcia, 1965.
Condiciones hidrograficas de la laguna y rio Unere.
Bol. Inst. Oceanogr., Univ. Oriente, 4(1):60-107.

chemistry
oxygen
Okuda, Taizo, 1962
Physical and chemical oceanography over continental shelf between Cabo Frio and Vitoria (Central Brazil).
J. Oceanogr. Soc., Japan, 20th Ann. Vol., 514-540.

chemistry
oxygen
Okuda, Taizo, 1960
Chemical oceanography in the South Atlantic Ocean, adjacent to north-eastern Brazil. (Portuguese and French resumés).
Trabalhos, Inst. Biol. Marit. e Oceanogr., Universidade do Recife, Brasil, 2(1):155-174.

chemistry
oxygen
Okuda, Taizo, 1960
Metabolic circulation of phosphorus and nitrogen in Matsushima Bay (Japan) with special reference to exchange of these elements between sea water and sediments. (Portuguese, French resumés)
Trabalhos, Inst. Biol. Maritima e Oceanogr., Universidade do Recife, Brasil, 2(1):7-153.

chemistry
oxygen (data)
Okuda, Taizo, and Lourinaldo Barreto Cavalcanti, 1963.
Algumas condicoes oceanograficas na area nordeste de Natal. (Septembro 1960).
Trab. Inst. Oceanogr., Univ. do Recife, 3(1):3-25.

chemistry
oxygen
Okuda, Taizo, Lourinaldo Cavalcanti and Manoel Pereira Borba, 1960
Estudo da Barra das Jangadas. 2. Variacao do pH, Oxigenio dissolvido e consumo de permaganato durante o ano. (In Portuguese; English and French resumés)
Trabalhos, Inst. Biol. Marit. e Oceanogr., Universidade do Recife, Brasil, 2(1):193-205.

chemistry - oxygen
Omaly, Nicole, 1968.
Sur la repartition du zooplankton de la Baie d'Alger et la notion d'essai.
Pelagos, 9:29-72.

chemistry
oxygen (data)
Oren, O.H., 1962
A note on the hydrography of the Gulf of Eylath. Contributions to the knowledge of the Red Sea No. 21.
Sea Fish. Res. Sta., Haifa, Israel, Bull. No. 30: 3-14.

chemistry, oxygen
Orr, A.P., 1947
An experiment in marine fish cultivation: II. Some physical and chemical conditions in a fertilized sea-loch (Loch Craiglin, Argyll).
Proc. Roy. Soc., Edinburgh, Sect.B, 63, Pt.1 (2):3-20, 19 text figs.

chemistry, oxygen
Orr, A. P., 1933
Physical and chemical conditions in the sea in the neighborhood of the Great Barrier Reef. Brit. Mus. (N.H.) Great Barrier Reef Exped., 1928-29, Sci. Repts. 2(3):37-86, 7 text figs.

chemistry
oxygen
Orr, A.P., and F.W. Moorhouse, 1933.
(a) Variations in some physical and chemical conditions on and near Low Isles Reef.
(b) Temperature of the waters in the anchorage, Low Isles.
(c) Physical and chemical conditions in mangrove swamps.
Brit. Mus. (N.H.) Great Barrier Reef Exped., 1928-1929, Sci. Rept., 2(4):87-110.

chemistry
oxygen (data)
Orren, M.J., 1963.
Hydrological observations in the south west Indian Ocean.
Rept., S. Africa, Dept. Comm. Industr., Div. Sea Fish., Invest. Rept., No. 45:61 pp.

Reprinted from: Commerce & Industry, May, 1963.

chemistry, oxygen
Ottmann, Francois, and Taizo Okuda, 1961

Etude des conditions physico-chimiques des eaux de deux estuaires du Nord-est Bresilien.
Cahiers Oceanogr., C.C.O.E.C., 13(4): 234-242.

chemistry, Oxygen
Ottmann, François, et Jeanne-Marie Ottmann, 1959.
La marée de salinité dans le Capibaribe, Recife-Bresil.
Trabalhos Inst. Biol. Marit. e Oceanogr., Recife, 1(1):39-49.

chemistry
oxygen
Oulman, C.S., and E.R. Baumann, 1956.
A colorimetric method for determining dissolved oxygen. Sewage & Indr. Wastes 28:1461-1465.

Abstr. in Publ. Health Eng. Abstr., 37(7):26.

chemistry
oxygen
Owen, Robert W., Jr., 1968.
Oceanographic conditions in the northeast Pacific Ocean and their relation to the albacore fishery.
Fish. Bull. U.S. Fish. Wildl. Serv., 66(3): 503-526.

chemistry
oxygen
Owen, R.W., Jr., 1967.
Atlas of July oceanographic conditions in the northeast Pacific Ocean, 1961-64.
Spec.scient.Rep.U.S.Fish Wildl.Serv., Fish., 549:85 pp.

P

chemistry
oxygen
Paccagnini, Ruben N., and Alberto O. Casellas, 1961
Estudios y resultados preliminares sobre trabajos oceanograficos en el area del Mar de Weddell. (In Spanish; Spanish, English, French German and Italian resumes).
Contrib., Inst. Antartico Argentino, No. 64: 12 pp.

chemistry, oxygen
Park, Joo Suck 1968.
Chaetognaths and plankton in Korean waters. II. The distribution of Chaetognaths in the southern waters and their relation to the character of the water masses in the summer of 1967. (In Korean; English abstract)
Bull. Fish. Res. Develop. Agency, Korea 3:83-102.

chemistry, oxygen
Park, Kilho 1967.
Chemical features of the subarctic boundary near 170°W.
J. Fish. Res. Bd. Can. 24(5): 899-908.

chemistry
oxygen
Parker, Robert H., 1964.
Zoogeography and ecology of macro-invertebrates of Gulf od California and continental slope of western Mexico.
In: Marine geology of the Gulf of California, a symposium, Amer. Assoc. Petr. Geol., Memoir, T. van Andel and G.G. Shor, Jr., Editors, 3:331-376.

chemistry, oxygen
Patil, M.R., and C.P. Ramamirtham 1963.
Hydrography of the Laccadives offshore waters - a study of the winter conditions.
J. mar. biol. Ass. India 5(2):159-169.

chemistry, oxygen
Patil, M.R., C.P. Ramamirtham, P. Udaya Varma, C.P. Aravindakshan Nair and Per Myrland, 1964.
Hydrography of the west coast of India during the pre-monsoon period of the year 1962. 1. Shelf waters of Maharashtra and southwest Sourashtra coasts.
J. Mar. biol. Ass., India, 6(1):151-166.

chemistry
oxygen
Patten, B.C., R.A. Mulford and J.E. Warinner, 1963
An annual phytoplankton cycle in the lower Chesapeake Bay.
Chesapeake Science, 4(1):1-20.

chemistry
oxygen
Pearson, Erman A., and George A. Holt, 1960.
Water quality and upwelling at Grays Harbor entrance.
Limnol. & Oceanogr., 5(1):48-56.

chemistry
oxygen
Pellegrini, Liliane, et Max Pellegrini 1970
Observations sur la teneur en oxygène dissous des eaux de la lagune du Brusc.
Bull. Inst. Océanogr. Monaco 69(1406):23pp.

chemistry, oxygen
*Pelletier, B.R., 1968.
Submarine physiography, bottom sediments, and modern sediment transport in Hudson Bay.
In: Earth Science Symposium on Hudson Bay, Ottawa February, 1968, Peter J. Hood, editor, GSC Pap. Geol. Surv. Can., 68-53:100-136

chemistry
oxygen
Peterson, Clifford L., 1960.
The physical oceanography of the Gulf of Nicoya, Costa Rica, a tropical estuary.
Inter-American Tropical Tuna Comm., Bull., 4(4): 139-216.

chemistry, oxygen
Phifer, L. D. and T. G. Thompson, 1937.
Seasonal variations in the surface waters of San Juan channel during the five year period, January 1931 to December 30, 1935. J. Mar. Res., 1(1):34-59, text figs. 12-18, 12 tables.

chemistry, oxygen
Pickard, G.L. 1967.
Some oceanographic characteristics of the larger inlets of southeast Alaska.
J. Fish. Res. Bd. Can. 24(7): 1475-1506.

chemistry oxygen
Pickard, G.L., 1963.
Oceanographic characteristics of inlets of Vancouver Island, British Columbia.
J. Fish. Res. Bd., Canada, 20(5):1109-1144.

chemistry
oxygen
Pickard, G.L., 1961
Oceanographic features of inlets in the British Columbia mainland coast.
J. Fish. Res. Bd., Canada, 18(6):907-999.

chemistry oxygen
Pickard, G.L., 1954.
Oceanography of British Columbia mainland inlets. 4. Dissolved oxygen distribution. Fish. Res. Bd. Canada, Pacific Coast Stas., Prog. Repts., 99:9-11.

chemistry, oxygen
Picotti, Mario, 1965.
La crociera idrografico-talassografica di Capo Matapan.
Pubbl.Commissione Ital.Comitato int. Geofis., Ser. IGC, 42: 63 pp.

chemistry
oxygen(data)
Picotti, Mario, 1960
Crociera Talassografica Adriatica 1955. III. Tabelle delle osservazioni fisiche, chimiche, biologiche e psammografiche.
Archivio di Oceanograf. e Limnol., 11(3): 371-377, plus tables.

chemistry oxygen
Picotti, M. and A. Vatova, 1942.
Osservazioni fisiche e chimiche periodiche nell' Alto Adriatico (1920-1938). Thalassia V (1):157 pp., 8 tables, 11 figs.

chemistry
oxygen
*Pieterse, F., and D.C. van der Post, 1967.
The pilchard of South West Africa (Sardinops ocellata): oceanographic conditions associated with red-tides and fish mortalities in the Walvis Bay region.
Investl Rept., Mar. Res. Lab., SWest Africa, 14: 1s5 pp.

chemistry, oxygen
Pirie, Robert Gordon, 1965.
Petrology and physical-chemical environment of bottom sediments of the Rivière Bonaventure-Chaleur Bay area, Quebec, Canada.
Rept. B.I.O. 65-10:182 pp. (multilithed).

chemistry, oxygen
*Pizarro, Mariano Javier, 1967.
Distribución del oxígeno disuelto en la zona oeste de la convergencia subtropical del Atlantico sud.
Bolm.Inst.oceanogr.S. Paulo, 16(1):67-85.

chemistry oxygen (data)
Platt, Trevor, and Brian Irwin, 1968.
Primary productivity measurements in St. Margaret's Bay, 1967.
Techn.Rept.Fish.Res.Bd., Can., 77: 123 pp.

chemistry, oxygen
*Poinsard, F., et J.P. Troadec, 1967.
La radiale de Pointe-Noire.
Cah. ORSTOM, Sér. Océanogr., 5(1):69-84.

chemistry, oxygen
Pomeroy, A.S., 1965.
Notes on the physical oceanographic environment of the Republic of South Africa.
Naval Oceanogr. Res. Unit, 41 pp. (Unpublished manuscript).

chemistry, oxygen
Ponomarenko, L.S., 1966.
Hydrochemical conditions in the Norwegian and Greenland seas in the summer of 1964. (In Russian).
Mater. Ribokhoz. Issled. severn. Basseina, Poliarn. Nauchno-Issled. Proektn. Inst. Morsk. Ribn. Khoz. Okeanogr. (PINRO), 7:155-167.

chemistry - oxygen
Ponomarenko, L.S., 1966.
Hydrochemical conditions in the Norwegian and Greenland seas in summer (from the data of the June surveys 1954-1964). (In Russian).
Trudy, Poliarn. Nauchno-Issled. Proektn. Inst. Morsk. Ribn. Khoz. Okeanogr., N. M. Knipovich, PINRO, 18:125-145.

chemistry
oxygen
Ponomarenko, L.S., and M.A. Istoshina, 1962
Hydrochemical studies in Newfoundland area. (In Russian; English summary).
Sovetskie Ribochoz. Issledov. Severo-Zapad. Atlant. Okeana, VNIRO-PINRO, 113-124.

chemistry, oxygen
Popescu, Dumitru, et Maria Stadniciuc, 1968.
Mensurations quantitatives de la photosynthèse à différentes profondeurs dans les eaux de la Mer Noire, en surface du littoral d'Agigea. (In Roumanian; French abstract).
In: Lucrările Sesiunii Stiințifice a Stațiunii de Cercetări Marine "Prof. Ioan Borcea", Agigea, (1-2 Noiembrie 1966). Volum Festiv, Iași, 1968: 187-195.

chemistry
oxygen
Postma, H., 1964.
The exhange of oxygen and carbon dioxide between the ocean and the atmosphere.
Netherlands J. Sea Res., 2(2):258-283.

chemistry, oxygen
Pova, E.A., 1946.
Problèmes de physiologie animale dans la Mer Noire. Bull. Inst. Océan. Monaco, No. 903:43 pp., 21 textfigs.

chemistry, oxygen
Powers, Charles F., 1963.
Some aspects of the oceanography of Little Port Walter Estuary, Baranof Island, Alaska.
U.S.F.W.S. Fish. Bull., 63(1):143-164.

Chemistry, oxygen

Pozdniakova, L.E., and V.N. Vinogradov, 1966. Some peculiarities in the distribution of the hydrochemical elements in the southeastern Barents Sea in August-September 1959. (In Russian). Trudy murmansk. biol. Inst., 11(15):140-156

Chemistry, oxygen

Pritchard, D.W., and R.E. Bunce, 1959. Physical and chemical hydrography of the Magothy River. Chesapeake Bay Inst., Tech. Rept., (Ref. 59-2), 17:1-22. (Unpublished manuscript).

Chemistry, oxygen

Puri, Harbans S., Gioacchino Bonaduce and John Malloy, 1964. Ecology of the Gulf of Naples. Pubbl. Staz. Zool., Napoli, 33(Suppl.):87-199.

Chemistry, oxygen

Pytkowicz, R.M., and D.R. Kester 1966. Oxygen and phosphate as indicators for the deep intermediate waters in the northeast Pacific Ocean. Deep-Sea Research 13(3):393-379.

Chemistry oxygen

Qasim, S.Z., and C.V.G. Reddy, 1967. The estimation of plant pigments of Cochin backwater during the monsoon months. Bull.mar.Sci., Miami, 17(1):95-110.

R

Chemistry oxygen

Rae, B.B., R. Johnston and J.A. Adams, 1965. The incidence of dead and dying fish in the Moray Firth, September 1963. Jour. Mar. Biol. Assoc., U.K., 45(1):29-47.

Chemistry oxygen (data)

Raghu Prasad, R., and P.V. Ramachandran Nair, 1963. Studies on organic production. 1. Gulf of Manaar. J. Mar. Biol. Assoc., India, 5(1):1-26.

oxygen (data)

Rakestraw, Norris W., 1964. Some observations on silicate and oxygen in the Indian Ocean. In: Recent researches in the fields of hydrosphere, atmosphere and nuclear geochemistry, Ken Sugawara festival volume, Y. Miyake and T. Koyama, editors, Maruzen Co., Ltd., Tokyo, 243-255.

Chemistry oxygen

Rakestraw, N.W., 1947. Oxygen consumption in sea water over long periods. J. Mar. Res. 6(3):259-263.

Chemistry - oxygen

Rakestraw, N. W., 1933. Studies on the biology and chemistry of the Gulf of Maine. 1. Chemistry of the waters of the Gulf of Maine in August, 1932. Biol. Bull. 64(2):149-158, 4 textfigs.

Chemistry - oxygen

Rakestraw, N. W. and D. E. Carritt, 1948. Some seasonal chemical changes in the open ocean. J. Mar. Res. 7(3):362-369, 6 text figs.

Chemistry oxygen

Rakestraw, N.W., and V.M. Emmel, 1938. The relation of dissolved oxygen to nitrogen in some Atlantic waters. J. Mar. Res. 1(3):207-216, Text-figs. 78-81.

Chemistry oxygen

Rakestraw, N. W., C.E. Herrick, Jr., and W.D. Urry, 1939. The helium-neon content of sea water and its relation to the oxygen content. J. Am. Chem. Soc. 61:2806-2807.

Chemistry - oxygen

Rakestraw, N.W., P.D. Rudd, and M. Dole, 1951. Isotopic composition of oxygen in air dissolved in Pacific Ocean water as a function of depth. J. Amer. Chem. Soc. 73:2976.

Chemistry, oxygen

Rakestraw, N. W., and H.P. Smith, 1937. A contribution to the chemistry of the Caribbean and Cayman Seas. Bull. Bingham Ocean. Coll. 6(1) 1-41, 32 text-figs.

Chemistry oxygen (data)

Ramalho, A. de M., 1942. Observacoes oceanograficas. "Albacora", 1940. Trav. St. Biol. Mar., Lisbonne, No. 45: 96 XVI pp., 2 figs. (fold-in).

Chemistry, oxygen (data)

Ramalho, A. de M., 1941. Observacoes oceanograficas. "Albacora", 1930, 1938 e 1939. Trav. St. Biol. Mar., Lisbonne, No. 44:52 pp., 1 fold-in (with 8 figs.)

Chemistry, oxygen (data)

Ramalho, A. de M., B. C. Goncalves, R. G. Boto, Z. Vilela, 1938. Observacoes oceanograficas. "Albacora", 1937. Trav. St. Biol. Mar., Lisbonne, No. 43: 30 pp., 1 fig.

Chemistry oxygen (data)

Ramalho, A. de M., R. G. Boto, B. C. Goncalves, H. Vilela, 1936. Observacoes oceanograficas. "Albacora", 1935-1936. Trav. St. Biol. Mar. Lisbonne, No.39: 25 pp., 2 figs.

Chemistry oxygen (data)

Ramalho, A. de M., L. S. Dentinho, C. A. M. de Sousa, Fronteira, F. L. Mamede, and H. Vilela, 1935. Observacoes oceanograficas. "Albacora" 1934. Trav. St. Biol. Mar., Lisbonne, No. 35: 35 pp., 1 fig.

Chemistry - oxygen

Ramamurthy S. 1963. Studies on the hydrological factors in the north Kanara coastal waters. Indian J. Fish. (A) 10(1): 75-93

Ramamirtham, C.P.

Chemistry, oxygen (data)

Ramamirtham, C.P., and R. Jayaraman, 1960. Hydrographical features of the continental shelf waters off Cochin during the years 1958 and 1959. J. Mar. Biol. Assoc., India, 2(2):199-207.

Chemistry, oxygen

Ramamirtham, C.P., and M.R. Patil, 1965. Hydrography of the west coast of India during the premonsoon period of the year 1962. 2. In and offshore waters of the Konkan and Malabar coasts. J. mar. biol. Ass., India, 7(1):150-168.

Chemistry oxygen.

Ramamurthy, S., 1953. Seasonal changes in the hydrogen-ion concentration and the dissolved oxygen content of the surface waters of the Madras coast. J. Madras Univ., B, 23(1):52-60, 2 textfigs.

Chemistry oxygen

Ramamurthy, S., 1954. Hydrobiological studies in the Madras coastal waters. J. Madras Univ., B, 23(2):148-163, 3 textfigs.

Chemistry oxygen (data)

Ramamurthy, V.D., K. Krishnamurthy and R. Seshadri, 1965. Comparative hydrographical studies of the near shore and estuarine waters of Porto-Novo, S. India. J. Annamalai Univ., 26:154-164.

Also in: Collected Reprints, Mar. Biol. Sta., Porto Novo, 1963/64.

Chemistry oxygen

Ramsey, W.L., 1962. Dissolved oxygen in shallow near-shore water and its relation to possible bubble formation. Limnol. and Oceanogr., 7(4):453-461.

Chemistry, oxygen

Rao, D. Panakala, and R. Jayaraman 1970. On the occurrence of oxygen maxima and minima in the upper 500 meters of the north-western Indian Ocean. Proc. Indian Acad. Sci. (B), 71 (6): 230-246

Chemistry oxygen

Rao, D.S., 1967. A preliminary note on the fluctuation of some hydrographic properties in the Arabian Sea. J.mar.biol.Ass.India 9(2):426-430.

Chemistry oxygen

Rao, D.S., and N. Madhavan, 1964. On some pH measurements in the Arabian Sea along the west coast of India. J. Mar. Biol. Ass., India, 6(2):217-221.

Chemistry, oxygen

Rao, L.V. Gangadhara and R. Jayaraman 1966. Upwelling in the Minicoy region of the Arabian Sea. Current Sci. 35 (15): 378-380

Chemistry, oxygen (data)

Rao, T.S. Satyanarayana and V. Chalapathi Rao 1963. Studies on diurnal variations in the hydrobiological conditions of the Waltair coast. J. mar. biol. Ass. India 4 (1): 23-43.

Chemistry oxygen

Rattray, Maurice, Jr., Cuthbert M. Love, and Diane E. Heggerty, 1962
Distribution of physical properties below the level of seasonal influence in the Eastern North Pacific Ocean.
J. Geophys. Res., 67(3): 1099-1108.

Chemistry, oxygen

Redfield, A.C., 1955.
The hydrography of the Gulf of Venezuela.
Pap. Mar. Biol. and Oceanogr., Deep-Sea Res., Suppl. to Vol. 3:115-133.

Chemistry, oxygen

Redfield, A. C., 1948.
The exchange of oxygen across the sea surface.
J. Mar. Res. 7(3):347-361, 4 textfigs.

Chemistry oxygen

Redfield, A.C., 1942.
The processes determining the concentration of oxygen, phosphate and other organic derivatives within the depths of the Atlantic Ocean.
P.P.O.M. 9(2):1-22, 13 textfigs.

Chemistry, oxygen

Redfield, A. C., 1934.
On the proportions of organic derivatives in sea water and their relation to the composition of plankton. In: James Johnstone Memorial Volume: 176-192, 5 textfigs.

Chemistry oxygen

Redfield, A.C., B.H. Ketchum and F.A. Richards 1963
2. The influence of organisms on the composition of sea-water. In: The Sea, M.N. Hill, Edit., Vol. 2. The composition of sea water, Interscience Publishers, New York and London, 26-77.

Chemistry oxygen

Reichelt, W., 1941.
Die ozeanographischen Verhältnisse bis zur warmen Zwischeschicht an der antarktischen Eisgrenze im Südsommer 1936/37. Nach Beobachtungen auf dem Walfangmutterschiff "Jan Wellem" im Weddell-Meer. Archiv. Deutschen Seewarte 61(5):54 pp., 11 pls.

Chemistry, oxygen (data)

Reichelt, W., 1941.
Die ozeanographischen Verhältnisse bis zur warmen Zwischenschicht an der antarktischen Eisgrenze in Südsommer 1936/37. Nach Beobachtungen auf dem Walfang-Mutterschiff "Jan Wellem" im Weddell-Meer Arch. Deutschen Seewarte 61(5):1-54, 11 pls.

Reid, Joseph L., Jr.

oxygen

Reid, Joseph L., Jr., 1965.
Intermediate waters of the Pacific Ocean.
The Johns Hopkins Oceanogr. Studes, No. 2: 1-85.

Chemistry oxygen

Reid, Joseph L., 1964.
A transequatorial Atlantic oceanographic section in July 1963 compared with other Atlantic and Pacific sections.
J. Geophys. Res., 69(24):5205-5215.

Chemistry oxygen

Reid, Joseph L., Jr., 1962
Distribution of dissolved oxygen in the summer thermocline.
J. Mar. Res., 20(2):138-148.

Chemistry oxygen

Reid, Joseph L., Jr., 1960
Oceanography of the northwestern Pacific Ocean during the last ten years.
California C oop. Ocean. Fish. Invest., Rept. 7: 77-90.

Chemistry, oxygen (data)

Reish, D.J., and H.A. Winter, 1954.
The ecology of Alamitos Bay, California, with special reference to pollution.
Calif. Fish and Game 40(2):105-121, 1 textfig.

Chemistry oxygen

Revelle, R., and H.E. Suess, 1962
Gases, Ch. 7, Sect. II, Interchange of properties between sea and air. In: The Sea, Interscience Publishers, Vol. 1, Physical Oceanography, 313-321.

Chemistry oxygen

Rheinheimer, Gerhard, 1970.
2.4. Sauerstoffhaushalt. Chemische, mikrobiologische und planktologische Untersuchungen in der Schlei im Hinblick auf deren Abwasserbelastung. Kieler Meeresforsch 26(2): 126-128.

Chemistry oxygen

Riabikov, O.G., 1960.
[On the change in the oxygen regime of the waters in the Baltic sea in the vicinity of the Gotland deep.]
Trudy Vses. Nauchno-Issled. Inst. Morsk. Ribnogo Chozia i Okeanogr. (VNIRO), 42:15-18.

Richards, Francis A.

Chemistry oxygen

Richards, F. A., 1965.
Dissolved gases other than carbon dioxide.
In: Chemical oceanography, J.P. Riley and G. Skirrow, editors, Academic Press, 1:197-225.

Chemistry oxygen

Richards, F.A., 1957.
Some current aspects of chemical oceanography.
In: Physics and chemistry of the earth, Pergamon Press, 2:77-128.

Chemistry, oxygen

Richards, F.A., 1957.
Oxygen in the ocean. Ch. 9 in: Treatise on Marine Ecology and Paleoecology, Vol. 1.
Mem. Geol. Soc., Amer., 67:185-238.

Chemistry, oxygen

Richards, F.A., and N. Corwin, 1956.
Some oceanographic applications of recent determinations of the solubility of oxygen in sea water. Limnol. & Oceanogr., 1(4):263-267.

Chemistry oxygen

Richards, Francis A., Joel D. Cline, William W. Broenkow and Larry P. Atkinson, 1965.
Some consequences of the decomposition of organic matter in Lake Nitinat, an anoxic fjord.
Limnol. and Oceanogr., Redfield Vol., Suppl. to 10:R185-R201.

Chemistry, oxygen

Richards, F.A., and A.C. Redfield, 1954.
A correlation between the oxygen content of sea water and the organic content of marine sediments.
Ltr. to Edit., Deep-Sea Res. 1(4):279-281, 2 textfigs.

Chemistry, oxygen

Riel, P.M. van 1939 (1940)
The oxygen content of the bottom water in the inland seas of the Asiatic-Australian Archipelago.
Proc. Sixth Pacific Sc. Congr. 3: 197-200

Riley, Gordon A.

Chemistry oxygen

Riley, Gordon A., 1959
Oceanography of Long Island Sound, 1954-1955
Bull. Bingham Oceanogr. Coll., 17(1):9-30.

oxygen, Chemistry, oxygen

Riley, G.A., 1952.
Phytoplankton of Block Island Sound, 1949.
Bull. Bingham Ocean. Coll. 13(3):40-64, 6 textfigs.

Chemistry, oxygen

Riley, G.A., 1951.
Oxygen, phosphate, and nitrate in the Atlantic Ocean. Bull. Bingham Ocean. Coll. 8(1):126 pp., 33 textfigs.

Chemistry, oxygen

Riley, G.A. 1942.
The relationship of vertical turbulence and spring diatom flowerings.
J. mar. Res. 5(1): 67-87.

Chemistry oxygen

Riley, G. A., 1941.
Plankton studies. IV. Georges Bank. Bull. Bingham Ocean. Coll. 7(4):73 pp.

Riley, J.P.

Chemistry, oxygen (data)

Riley, J.P., 1967.
The hot saline waters of the Red Sea bottom and their related sediments.
Oceanogr. Mar. Biol., Ann.Rev., H. Barnes, editor, George Allen and Unwin, Ltd., 5:141-157.

Chemistry, oxygen

Riley, J.P., 1965.
Analytical chemistry of sea water.
In: Chemical oceanography, J.P. Riley and G. Skirrow, editors, Academic Press, 2:295-424.

Chemistry oxygen (water samples)

Rittenberg, S.C., K.O. Emery and W.L. Orr, 1955.
Regeneration of nutrients in sediments of marine basins. Deep-Sea Res. 3(1):23-45.

Chemistry, oxygen

Robles, Fernando, 1966.
Descripcion grafica de las condiciones oceanografica frente a la provincia de Tarapaca en base a los datos de la Operacion Oceanografica Marchile II.
Inst. Hidrogr. Armada, Chile, unnumbered pp.

Chemistry, oxygen

Roche, A., and A. Roubault, 1947.
Observations sur les courants superficiels de la Mer de Monaco. Bull. Inst. Océan., Monaco, No. 909: 11 pp., 2 textfigs.

Rochford, D.J.

chemistry, oxygen
Rochford, D.J. 1969.
Seasonal variation in the Indian Ocean along 110°E. 1. Hydrological structure of the upper 500m.
Austr. J. mar. Freshwat. Res. 20(1):1-50.

chemistry oxygen
Rochford, D.J., 1966.
Hydrology. Port Philip survey 1957-1963.
Mem. nat. Mus. Vict., 27:107-118.

chemistry oxygen
Rochford, D.J., 1964.
Hydrology of the Indian Ocean.
Australian J. mar. freshwater Res., 15(1):25-55.

Also in:
Collected Reprints, Int. Indian Ocean Exped., UNESCO, 3. 1966.

chemistry oxygen
Rochford, D.J., 1962.
Hydrology of the Indian Ocean. II. The surface waters of the south-east Indian Ocean and Arafura Sea, in the spring and summer.
Australian J. Mar. Freshwater Res., 13(2):226-251.

chemistry oxygen
Rochford, D.J., 1961.
Hydrology of the Indian Ocean. 1. The water masses in intermediate depths of the south-east Indian Ocean.
Australian J. Mar. Freshwater Res., 12(2):129-149.

chemistry oxygen (data)
Rochford, D.J., 1958.
Characteristics and flow paths of the intermediate depth waters of the southeast Indian Ocean.
J. Mar. Res., 17:483-504.

chemistry oxygen (data)
Rochford, D.J., 1953.
Analysis of bottom deposits in eastern and south-western Australia, 1951 and records of twenty-four hourly hydrological observations at selected stations in eastern Australian estuarine systems 1951. Oceanographical station list of investigations made by the Division of Fisheries, Commonwealth Scientific and Industrial Research Organization, Australia, 13:68 pp.

chemistry, oxygen (data)
Rochford, D.J., 1953.
Estuarine hydrological investigations in eastern and south-western Australia, 1951. Oceanographical station list of investigations made by the Division of Fisheries, Commonwealth Scientific and Industrial Research Organization, Australia, 12:111 pp.

chemistry oxygen
Rochford, D.J., 1952.
A comparison of the hydrological conditions of the eastern and western coasts of Australia.
Indo-Pacific Fish. Counc., Proc., 3rd meeting, 1-16 Feb. 1951, Sects. 2/3:61-68, 7 textfigs.

chemistry, oxygen
Rochford, D.J., 1951.
Hydrology of the estuarine development. Proc. Indo-Pacific Fish. Counc., 17-28 Apr. 1950, Cronulla N.S.W., Australia, Sects. II-III:157-168, 10 textfigs., Table 6.

chemistry oxygen
Rochford, D.J., 1951.
Studies in Australian estuarine hydrology.
Australian J. Mar. and Freshwater Res. 2(1):1-116, 1 pl., 7 textfigs.

chemistry, oxygen
Roden, Gunnar I., 1964.
Oceanographic aspects of Gulf of California.
In: Marine geology of the Gulf of California, a symposium, Amer. Assoc. Petr. Geol., Memoir, T. van Andel, and G.G. Shor, Jr., editors, 3:30-58

chemistry oxygen
Roden, Gunnar I., 1963
On sea level, temperature, and salinity variations in the Central Tropical Pacific and on Pacific Ocean Islands.
J. Geophys. Res., 68(2):455-472.

chemistry, oxygen
Roden, G.I., 1958.
Oceanographic and meteorological aspects of the Gulf of California. Pacific Science 12(1):21-45.

chemistry oxygen
Roden, G.I., and G.W. Groves, 1959
Recent oceanographic investigations in the Gulf of California.
J. Mar. Res., 18(1):10-35.

chemistry - oxygen
*Roger, C., 1967.
Contribution à la connaissance des euphausiacés du Pacificque équatorial.
Cah. ORSTOM, Sér. Océanogr., 5(1):29-37.

chemistry oxygen
Roger, C., 1966.
Étude sur quelques espèces d'euphausiacés de l'est de l'Océan Indien (110°E.).
Cah. ORSTOM, Sér. Océanogr., 4(4):73-103.

chemistry, oxygen
Romanovsky, V., 1964.
Coastal effect of the Cape Sicié sewer outfall (French Mediterranean, west of Toulon).
Air and Water Pollution, 8(10):557-589.
coast

chemistry, oxygen
Rosfelder, A., 1955.
Carte provisoire au 1/500,000 de la marge continental algérienne. Note de présentation.
Publ. Service Carte Geol., Algérie, n.s., Bull. No. 5, Trav. des Coll., 1954:57-106, 6 figs., 1 pl., 1 map.

Rotschi, Henri

chemistry Oxygen
Rotschi, Henri, 1965.
Le pH et l'alcalinite des eaux profondes de la Fosse des Hebrides et du Bassin des Fidji.
Progress in Oceanography, 3:301-310.

chemistry oxygen
Rotschi, Henri, 1961
Contribution française en 1960 à la connaissance de la Mer de Coral: Océanographique physique.
Cahiers Océanogr., C.C.O.E.C., 13(7):434-455.

chemistry oxygen
Rotschi, Henri, 1961.
Sur certaines propriétés chimiques des eaux equatoriales et tropicales du Pacifique.
Caractères generaux de la distribution de l'oxygène dissous.
Cahier Océanogr., C.C.O.E.C., 13(1):14-29.

chemistry oxygen
Rotschi, Henri, 1961(1962).
Oxygène, phosphate et gaz carbonique total en Mer de Corail.
Deep-Sea Res., 8(3/4):181-195.

chemistry oxygen
Rotschi, Henri, 1961
Influence de la divergence des Salomon sur la répartition de certaines propriétés des eaux.
C.R. Acad. Sci., Paris, 253:2559-2561.

chemistry oxygen (data)
Rotschi, Henri, 1960
Orsom III, Resultats des Croisières diverses de 1959. Oceanographie physique.
Centre d'Océanogr., Inst. Français d'Océanie, Rapp. Sci., No. 17: 59 pp.

chemistry oxygen
Rotschi, Henri, 1960
Récents progrès des recherches océanographiques entreprises dans le Pacifique Sud-Ouest.
Cahiers Oceanogr., C.C.O.E.C., 12(4):248-267.

chemistry oxygen
Rotschi, H., 1960.
Remarques sur la relation entre l'oxygène et le phosphore mineral dissous en Mer de Corail.
C. R. Acad. Sci., Paris, 250(13):2403-2405.

Resume in:
Cahiers du Pacifique, 1962(4):85.

chemistry oxygen
Rotschi, Henri, 1959
Chimie, "Orsom III", Resultats de la Croisière BOUSSOLE.
O.R.S.T.O.M., Inst. Français d'Océanie, Rapp. Sci. 13: 3-60.

chemistry, oxygen (data)
Rotschi, Henri, Michel Angot and Roger Desrosieres, 1960.
Orsom III, Resultats de la croisiere "Choiseul" phytoplancton qualitatif. Rapp. Sci., Noumeau, No. 16:91 pp. (mimeographed).
2ème partie. Chimie, productivité

chemistry oxygen (data)
Rotschi, Henri, Michel Angot, et Michel Legrand, 1959
Orsom III. Resultats de la croisiere "Astrolabe". 2. Chimie, productivite et zooplankton.
Rapp. Sci., Inst. Francais d'Oceanie, No. 9: 97 pp. (mimeographed).

chemistry oxygen (data)
Rotschi, Henri, Michel Angot, Michel LeGand, and H.R. Jitt, 1959
Chimie, productivité et zooplancton. "Orsom III". Resultats de la Croisiere "Boussole".
Resultats "production primaire" de la croisière 56-5.
Rapp. Sci. Inst. Francais d'Oceanie, Centre d'Oceanogr. No. 13:

chemistry oxygen
Rotschi, H., and L. Lemasson, 1967.
Oceanography of the Coral and Tasman seas.
Oceanogr. Mar. Biol. Ann. Rev., Harold Barnes, editor, George Allen and Unwin, Ltd., 5:49-97.

chemistry
oxygen(data)
Rotschi, Henri, et Yves Magnier, 1963
Resultats des observations scientifiques de La DUNKERQUOISE---------------.
Inst. Français d'Océanie, Rapp. Sci., O.R.S.T. R.O.M., Noumea, No. 24:67 pp.

chemistry
oxygen
Rotschi, H., B. Wauthy and M. LeGand, 1961
Orsom III. Résultats de la croisière "Epi" 2eme partie. Chimie et biologie.
O.R.S.T.O.M., I.F.O., Rapp. Sci., (23):4-111.

chemistry, oxygen
Rotthauwe, H.W., 1958.
Die Sauerstoffbestimmung in See- und Süsswasser mit Hilfe der Quecksilbertropelektrode und ihre Anwendung bei physiologischen Untersuchungen.
Kieler Meeresf., 14(1):48-63.

chemistry, oxygen
Rouch, J., 1950.
Le Canal de Panama. Bull. Inst. Océan., Monaco, No. 975:20 pp., 4 textfigs.

chemistry, oxygen
Rouch, J., 1946
Traité d'Océanographie physique. L'eau de mer. Payot, Paris, 349 pp., 150 text figs.

chemistry, oxygen
Rounsefell, George A., 1964.
Preconstruction study of the fisheries of the estuarine areas traversed by the Mississippi River-Gulf outlet project.
U.S.F.W.S. Fish. Bull., 63(2):373-393.

chemistry
oxygen (data)
Rozanov, A.G., and V.S. Bykova, 1964.
The distribution of nitrates and nitrites in eater of the northern part of the Indian Ocean. Investigations in the Indian Ocean (33rd voyage of E/S "Vitiaz"). (IN Russian).
Trudy Inst. Okeanol., Akad. Nauk, SSSR., 64:94-101

chemistry
oxygen
Rubinchik, E.E., 1959
[Data on the hydrochemistry of the southern Caspian.]
Trudy V.N.I.R.O., 38: 152-164.

chemistry, oxygen
Rumkówna, A., 1948
[List of the phytoplankton species occurring in the superficial water layers in the Gulf of Gdańsk] Bull. Lab. mar., Gdynia, No. 4: 139-141 with tables in back.

chemistry - oxygen
Rupp, S.M., 1969
Temporal variations of the deep water at ocean station P.
Tech. Rept. Chesapeake Bay Inst. 57: 38pp. (multilith)

chemistry oxygen
Ruud, B., 1926.
Quantitative investigations of plankton at Lofoten, March-April, 1922-1924. Preliminary report.
Rept. Norwegian Fish. Mar. Invest. 3(7):30 pp., 3 charts, 5 diagrams.

chemistry, oxygen
Ruud, J. T., 1932.
On the biology of southern Euphausiidae.
Hvalrådets Skrifter No. 2:1-105, 37 text figs.

chemistry oxygen
Ryther, John H., 1970.
Is the world's oxygen supply threatened?
Nature, Lond., 227(5256): 374-375

chemistry, oxygen
Ryther, John H., John R. Hall, Allan K. Pease, Andrew Bakun and Mark M. Jones, 1966.
Primary organic production in relation to the chemistry and hydrography of the western Indian Ocean.
Limnol. Oceanogr., 11(3):371-380.

S

chemistry, oxygen
Saelen, O.H., 1950.
The hydrography of some fjords in northern Norway: Balsfjord, Ulfsfjord, Grøtsund, Vengsøyfjord, and Malangen. Tromsø Mus. Årshefter, Naturhist. Afd., 38, 70(1):102 pp., 10 pls., 42 textfigs.

chemistry, oxygen
Salnikov, N.E., 1965.
Investigaciones de economía pesquera en el Golfo de México y del Mar Caribe. (In Russian; Spanish abstract).
Sovetsk.-Cub. Ribokhoz. Issled., VNIRO:Tsentr. Ribokhoz. Issled. Natsional. Inst. Ribolovsta Republ. Cuba, 93-179.

chemistry
oxygen
Sandoval Eliseo, 1970.
Distribución de los atunes en el primer trimestre del año en relación con las condiciones oceanográficas generales frente a Chile y Perú.
Bol. cient. Inst. Fomento pesq. Chile 4: 86pp.

chemistry
oxygen
San Feliu, J.M., y F. Muñoz, 1970.
Hidrografía y fitoplancton de las costas de Castellón, de julio de 1967 a junio de 1968.
Inv. pesq. Barcelona, 34(2): 417-449.

chemistry
oxygen
San Feliu, J.M., y F. Muñoz, 1965.
Hidrografía y plancton del puerto de Castellón de junio de 1961 a enero de 1963.
Inv. Pesq., Barcelona, 28:3-48.

chemistry, oxygen
*Sapozhnikov, V.V., 1967.
The application of the computation parameter "Performed phosphate" to elucidating the basic regularities in the distribution of phosphate and oxygen along the meridional sections in the Pacific and Atlantic oceans. (In Russian; English abstract).
Trudy Inst. Okeanol., 83:5-19.

chemistry, oxygen
Sapozhinkov, V.V., 1965.
Hydrochemical characteristics as applied to distinguishing the Cromwell Current. Methods of marine hydrochemical investigation. (In Russian).
Trudy Inst. Okeanol., Akad. Nauk, SSSR, 79:87-96.

chemistry, oxygen
Sargent, M.C., and J.C. Hindman, 1943
The ratio of carbon dioxide consumption to oxygen evolution in sea water in the light. J. Mar. Res. V(2):131-135.

oxygen
Scaccini Cicatelli, Marta, 1969
Un anno di osservazioni idrologiche in una stazione fissa nelle acque dell'Adriatico occidentale presso Fano.
Prog. Ricerca Risorse mar. Fondo mar. Comm. ital. Oceanogr. Consig. naz. Ricerche, (B)(40): 153-225

chemistry, oxygen
* Scaccini Cicatelli, Marta, 1968.
Un anno di osservazioni idrologiche in una stazione fissa nelle acque dell'Adriatico occidentale presso Fano.
Note Lab. Biol. mar. Pesca, Fano, Univ. Bologna 2(9): 181-228.

Oxygen
Schott, Friedrich, and Manfred Ehrhardt, 1970.
On fluctuations and mean relations of chemical parameters in the northwestern North Sea.
Kieler Meeresforsch. 25(2): 272-278

chemistry oxygen
Schulz, B., 1934.
Die Ergebnisse der Polarexpedition auf dem U-Boat "Nautilus". Ann. Hydrogr., usw., 62:147-152, Pl. 16 with 2 figs.

chemistry
oxygen (data)
Schulz, B., 1923.
Hydrographische Untersuchungen besonders über den Durchlüftungszustand in der Ostsee in Jahre 1922. (Forschungsschiffe "Skagerak" und "Nautilus").
Arch. Deutschen Seewarte 41(1):1-64, 2 textfigs., 5 pls.

chemistry, oxygen(data)
Schulz, B., 1922.
Hydrographischen Beobachtungen inbesondere über die Kohlensäure in der Nord- und Ostsee im Sommer 1921. (Forschungsschiffe "Poseidon" und "Skagerak"). Arch. Deutschen Seewarte 40(2):1-44, 2 textfigs., 4 pls.

chemistry, oxygen
Schulz, B., and A. Wulff, 1929
Hydrographie und Oberflächen plankton des westlichen Barentsmeeres im Sommer 1927.
Ber. deutschen wissensch. Komm. F. Meeresforsch. n.s. 4(5):232-372, 13 tables, 25 text figs.

chemistry, oxygen
Sebastian, Alfonso, Manual N. Llorca and Vitaliano B. Encina, 1965.
Oceanography of Lingayen Gulf.
Philipp. J. Fish., 7(1):13-33.

chemistry oxygen
Segerstråle, Sven G., 1965.
On the salinity conditions off the south coast of Finland since 1950, with comments on some remarkable hydrographical and biological phenomena in the Baltic area during this period.
Commentationes Biolog., Soc. Sci. Fennica, 28(7):28 pp.

chemistry, oxygen
Seiwell, H. R., 1939.
Atlantis cruise to the tropical North Atlantic January to March 1939. Trans. Am. Geophys. Union of 1939:417-422, 5 textfigs.

Chemistry, oxygen
Seiwell, H. R., 1939.
Die Verwendung der Verteilung von Sauerstoff auf die physische Ozeanographie des Karibischen Meeresgebietes. Gerlands Beiträge z. Geophys. 54(4):1-7, 3 textfigs.

Chemistry, oxygen
Seiwell, H. R., 1938.
Use of non-conservative properties of sea water in physical oceanographical problems. Nature 142:164.

Chemistry, oxygen
Seiwell, H. R., 1938.
Application of the distribution of oxygen to the physical oceanography of the Caribbean Sea region. P.P.O.M. 6(1):60 pp., 42 figs.

Reviewed: J. du Cons. 14(3): 410-411. G. Dietrich

Chemistry, oxygen
Seiwell, H. R., 1937.
Relationship of minimum oxygen concentration to density of the water column in the Western North Atlantic. Gerlands Beiträge z. Geophysik 50: 302-306, 1 textfig.

Chemistry, oxygen
Seiwell, H. R., 1937.
The oxygen-poor layers in the Western North Atlantic. J. du Cons. 12(3):277-283, 2 figs.

Chemistry, oxygen
Seiwell, H.R., 1937
The minimum oxygen concentration in the Western Basin of the North Atlantic. P.P.O.M. 5(3):1-23, 8 textfigs.

Chemistry, oxygen
Seiwell, H. R., 1937.
Consumption of oxygen in sea water under controlled laboratory condistions. Nature 140: 506, 1 textfig.

Chemistry, oxygen
Seiwell, H. R., 1934.
The distribution of oxygen in the western basin of the North Atlantic. P.P.O.M. 3(1): 86 pp.

Chemistry, oxygen
Seiwell, H. R., and G. E. Seiwell, 1938.
The sinking of decomposing plankton in sea water and its relationship to oxygen consumption and phosphorus liberation. Proc. Am. Phil. Soc. 78(3):465-481, 5 textfigs.

Chemistry oxygen (data)
Serpoianu, GH., 1967.
Considérations sur la pénétration des eaux méditerranéennes dans le bassin de la Mer Noire.
Hydrobiologie, Bucureşti, 8: 239-251

chemistry, oxygen (data)
Serpoianu Gheorghe, et Viorel Chirila 1969.
Observations sur les particularités hydrologiques des eaux de la mer Noire dans la couche où la vie commence à disparaître.
Rapp. P.-v. Réun. Commn int. Explor. scient. Mer Mediterr., 19(4): 689-692.

chemistry oxygen
Seryi, V.V., and V.A. Khimitza, 1963
On the hydrology and chemistry of the Gulf of Aden and the Arabian Sea. (In Russian).
Okeanologiia, Akad. Nauk, SSSR, 3(6):994-1003

Chemistry - oxygen
*Sevastyanov, V.F., and I.I. Volkov, 1967.
Redistribution of chemical elements in redox processes in the bottom sediments of the oxygen zone of the Black Sea. (In Russian; English abstract).
Trudy Inst. Okeanol., 83:115-134.

Chemistry, oxygen
Sewell, R.B. Seymour, and L. Fage, 1948
Minimum oxygen layer in the ocean. Nature 162 (4129):949-951.

chemistry, oxygen
Shannon, L.V., and M. van Rijswijck 1969.
Physical oceanography of the Walvis Ridge region.
Invest. Rept. Div. Sea Fish. S.Afr. 70:19pp.

Chemistry, oxygen (data)
Shimomura, T., and K. Miyata, 1957.
[The oceanographical conditions of the Japan Sea and its water systems, laying stress on the summer of 1955.]
Bull. Japan Sea Reg. Fish. Res. Lab., No. 6 (General survey of the warm Tsushima Current 1): 23-120.

Chemistry, oxygen
Shiokawa, Tsukasa, and Haruhiko Irie, 1966.
The hydrographic conditions and the fisheries damages by the red water occurred in Omura Bay in summer 1965 - IV. Mass-mortality of fishes resultant in the red water. (In Japanese; English abstract).
Bull. Fac. Fish. Nagasaki Univ., No. 21:115-129.

chemistry, oxygen
Simonov, A.I., 1965.
Basic facts of the formation of the daily oxygen regime in the syrface layer of nearshore waters. (In Russian).
Trudy, Gosudarst. Okeanogr., Inst. No. 83:56-71.

Chemistry oxygen
Siyazuki, K., 1951.
[Studies on the foul-water drained from factories 1. On the influence of foul-waters frained from factories by the coast on the water of Mitaziri Bay.] Contr. Simonoseki Coll. Fish. Pt. 1:155-158, 4 textfigs. (English summary).

chemistry oxygen (data)
Sjarif, Sjarmilah, 1959 (Miss!)
Seasonal fluctuations in the surface salinity along the coast of the southern part of Kalimantan (Borneo). Mar. Res., Indonesia, (Penjelidikan Laut di Indonesia), No. 4:29 pp.

chemistry, oxygen
Skolka, H.V., et M.-T. Gomoiu 1967.
Recherches océanologiques roumaines dans la mer Noire. (In Rumanian; French abstract).
Hydrobiologie, Bucureşti 8:15-30.

oxygen (data)
Skopintsev, B.A., 1965.
Investigation of the oxygen minimum layer in the North Atlantic during the autumn of 1959. (In Russian; English abstract).
Okeanolog. Issled., Rez. Issled. po Programme Mezhd. Geofiz. Goda, Mezhd. Geofiz. Komitet, Prezidiume Akad. Nauk, SSSR, No.13:108-114.

chemistry, oxygen
Skopintsev, B.A., and M.S. Ledovskoy, 1963
The oxygen of the Black Sea waters during 1959-1960. (In Russian).
Okeanologiia, Akad. Nauk, SSSR, 3(6):1004-1016

chemistry, oxygen
Skopintsev, B.A., and V. K. Zhavoronkina, 1962.
Results of the determination of dissolved oxygen of the subtropical and tropical sections of the North Atlantic Ocean in August-October 1959. (In Russian).
Trudy Morsk. Gidrofiz. Inst., 25:118-129.

Chemistry oxygen
Škrivanić, Ante, 1969.
Hydrographic and biotical conditions in the North Adriatic -1: Hydrochemistry and some factors influencing hydrography. (Jugoslavian and Italian abstracts).
Thalassia Jugoslavica, 5: 315-328.

chemistry, oxygen
Slaucitajs, L., 1947
Ozeanographie des Rigaischen Meerbusens. Teil 1, Statik. Contrib. Baltic Univ. No.45, Pinneberg, 110 pp., 69 text figs.

Chemistry, oxygen
Slinn, D.J., 1966.
Chemical constituents in sea water off Port Erin during 1965.
Mar. biol. Sta., Univ. Liverpool, No. 78:29-33.

Chemistry oxygen
Slinn, D.J., 1962
Chemical constituents in sea water off Port Erin during 1961.
Mar. Biol. Sta., Univ. Liverpool, Port Erin, Ann. Rept., 74(1961):23-26.

chemistry oxygen
Slinn D.J. 1961
Chemical constituents in sea water off Port Erin in 1960.
Ann. Rept. Mar. Biol. Sta., Univ. Liverpool 1960, 73:23-28.

chemistry oxygen
Slinn, D.J., 1959
Chemical constituents in sea water off Port Erin during 1958.
Ann. Rept., Mar. Biol. Sta., Univ. Liverpool 1958 (No.71): 24-28.

Chemistry, oxygen
Slinn, D.J., 1958.
Phosphate and oxygen in sea-water off Port Erin during 1957.
Ann. Rept., Mar. Biol. Sta., Univ. Liverpool, 70:27-30.

chemistry, oxygen
Slinn, D.J., 1956.
Phosphate and oxygen in sea-water off the Isle of Man, 1952-1954. Ann. Rept. Mar. Biol. Sta., Isle of Man, 68:30-38.

chemistry, oxygen
Slinn, D.J., and Winifred Chapman, 1965.
Chemical constituents in sea water off Port Erin in 1964.
Mar. Biol. Sta. Port Erin, Ann. Rept. (1964), No. 77:43-48.

oxygen chemistry, oxygen
Slinn, D.J., and Winifred Chapman, 1964.
Chemical constituents of sea water off Port Erin during 1963.
Mar. Biol. Sta., Univ. Liverpool, Port Erin, Ann. Rept., No. 76(1963):26-31.

chemistry oxygen
Slinn, D.J. and Gaenor Offlow (Mrs. M. Solly) 1969.
Chemical constituents in sea water off Port Erin during 1968.
Ann. Rep. mar. biol. Sta. Port Erin 81 (1968): 31-

chemistry, oxygen
Smeyde, Theodore J., 1966.
A quantitative analysis of the phytoplankton of the Gulf of Panama. III General ecological conditions and the phytoplankton dynamics at 8o 45'N, 79o 23'W from November 1954 to May 1957.
Inter-Amer. Trop. Tune Comm., Bull., 11(5): 355-612.

chemistry oxygen
Smetanin, D.A., 1961.
[Some features of the chemistry of waters in the northeastern Pacific based on observations made in 1958-1959.]
Trudy Inst. Okeanol., Akad. Nauk, SSSR, 45:130-141.

Chemistry
oxygen
Smetanin, D.A., 1960
[Some features of the chemistry of water in the central Pacific.]
Trudy Inst. Okeanol., 40: 58-71.

Chemistry, oxygen
Smetanin, D.A., 1959
[Hydrochemistry of the Kuril-Kamchatka deep-sea trench.] Trudy Inst. Okeanol., 33: 43-86.

Chemistry
oxygen
Smetanin, D.A., 1958
[Hydrochemistry in the region of the Kurile-Kamchatka Trench. 1. Certain questions of the hydrology and chemistry of the lower subarctic water in the region of the Kurile-Kamchatka Trench.]
Trudy Inst. Okeanol., 27: 22-54.

Chemistry, oxygen
*Smirnov, E.V., N.N. Mijailov y V.V. Rossov, 1967.
Algunas características hidroquímicas del Mediterráneo americano.
Estudios Inst. Oceanol., Acad. Ciencias, Cuba, 2(1): 51-74.

chemistry, oxygen
Smith, F.G.W., R.W. Williams, and C.C. Davis, 1950.
An ecological survey of the subtropical inshore waters adjacent to Miami. Ecol. 31(1):119-146, 7 textfigs.

chemistry, oxygen
Soeraatmadja, Rd. E., 1957.
The coastal current south of Java.
Penjelidikan Laut, Indonesia (Mar. Res.), No. 3: 41-55.

Chemistry oxygen
Solov'eva, N. F., 1959.
[Hydrochemical analyses of the Aral Sea in 1948.]
Akad. Nauk, SSSR, Lab. Ozerovedeniya, Trudy, 8: 3-22.

5 Sept. 61 JPRS:10054
OTS-SLA $3.60 61-28414

chemistry, oxygen(data)
Soot-Ryen, T., 1956.
Report on the hydrographical conditions in West-Finnmark, March-April, 1935.
Acta Borealia, Tromsø Mus., A. Scientia, No. 10: 1-37.

chemistry oxygen(data)
Soot-Ryen, T., 1947.
Hydrographical investigations in the Tromsø-district 1934-1938 (Tables). Tromsø Mus. Aarsheft., Naturhist. Avd. No. 33, 66(1943)(1): numerous pp. (unnumbered pages).

chemistry
oxygen (data)
Soot-Ryen, T., 1938.
Hydrographical investigations in the Tromsø district in 1931.
Tromsø Mus. Aarsheft., Naturhist. Avd. No. 10, 54(2):1-6, plus 41 pp. tables.

Chemistry, oxygen
Soule, F.M., A.P. Franceschetti and R.M. O'Hagen, 1963.
Physical oceanography of the Grand Banks region and the Labrador Sea in 1961.
U.S. Coast Guard Bull., No. 47:19-82.

Chemistry, oxygen
Sourie, R., 1954.
Contribution à l'étude écologique des côtes roches du Sénégal.
Mém. Inst. Français d'Afrique Noire No. 38: 342 pp., 23 pls.

chemistry
oxygen (data)
South Africa, Department of Commerce and Industries, Division of Fisheries, 1961
Thirtieth Annual Report, 1 April 1958 to 31 March 1959: 160 pp.

chemistry, oxygen
Spencer, R.S., 1956.
Studies in Australian estuarine hydrology.
Australian J. Mar. & Freshwater Res. 7(2):193-253.

Chemistry, oxygen
Stander, G.H., 1967.
The Benguela Current off South West Africa.
Fish. Bull., Misc. Contr. Oceanogr. Mar. Biol. S. Afr., 4:1-1-8.

chemistry, oxygen
Steele, J.H., 1957.
Notes on oxygen sampling on the Fladen Ground.
J. Mar. Biol. Assoc., U.K., 36(2):227-232.

Chemistry, oxygen
Steemann-Nielsen, Einar, 1951
The marine vegetation of the Isefjord. A study on ecology and production. Medd. Komm. Danmarks Fiskeri-og Havundersøgelser. Ser. Plankton. 5(4); 114pp., 46 text figs.

Chemistry oxygen
Steemann Nielsen, E., 1935.
The production of phytoplankton at the Faroe Isles, Iceland, East Greenland and in the waters around.
Medd. Komm. Danmarks Fisk- og Havundersøgelser, Ser. Plankton, 3(1):1-93, 5 textfigs.

Chemistry
oxygen
Stefansson, Unnstein, and Francis A. Richards, 1964.
Distributions of dissolved oxygen, density and nutrients off the Washington and Oregon coasts.
Deep-Sea Res., 11(3):355-380.

Chemistry
oxygen
Stefansson, Unnstein, and Francis A. Richards, 1963.
Processes contributing to the nutrient distributions off the Columbia River and Strait of Juan de Fuca.
Limnology and Oceanography, 8(4):394-410.

chemistry, oxygen
Stephensen, W., 1951.
Preliminary observations upon the release of phosphate from estuarine mud. Proc. Indo-Pacific Fish. Counc., 17-27 Apr. 1950, Cronulla, N.S.W., Australia, Sects. II-III:184-189, 1 textfig.

oxygen
Stickney, Alden P., 1968.
Supersaturation of atmospheric gases in the coastal waters of the Gulf of Maine.
Fish. Bull. Bur. Comm. Fish., U.S.F.W.S., 67(1): 117-123

Chemistry, oxygen
Storr, John F., 1964.
Ecology and oceanography of the coral-reef tract, Abaco Island, Bahamas.
Spec. Papers, Geol. Soc., Amer., 79:98 pp.

Chemistry, oxygen (data)
Stroup, E.D., 1954.
Mid-Pacific oceanography IV. Transequatorial waters, January-March 1952.
U.S.F.W.S. Spec. Sci. Rept. - Fish. No. 135: 1-52, 18 figs. (multilithed).

chemistry, oxygen
Suarez-Caabro, José 1965.
Datos meteorologicos, hidrograficos y planctonicos del litoral de Veracruz, Ver.
Anales Inst. Biol. Univ. Mexico 36 (1/2): 25-46.

chemistry
oxygen
Subrahmanyan, R., 1959.
Studies on the phytoplankton of the west coast of India. 1. Quantitative and qualitative fluctuation of the total phytoplankton crop, the zooplankton crop and their interrelationship, with remarks on the magnitude of the standing crop and production of matter and their relationship to fish landings. 2. Physical and chemical factors influencing the production of phytoplankton, with remarks on the cycle of nutrients and on the relationship of the phosphate content to fish landings.
Proc. Indian Acad. Sci., 1:113-252.

Chemistry
oxygen (data)
Sugimoto, Hitomi, Minoru Hisaoka, Kazuhiko Nogami Osamu Takeuchi and Masaya Suzuki, 1966.
Studies on sea water exchange in fish farm. 1. Exchange of sea water in fish farm surrounded by net. (In Japanese; English abstract).
Bull. Naikai reg. Fish. Res. Lab., No. 23:1-20.

chemistry, oxygen
Sugiura, Yoshio 1969.
The relationship between phosphate and oxygen concentrations in the upwelling waters and its oceanographical significance.
Oceanogr. Mag. 21 (1): 47-52.

chemistry oxygen

Sugiura, Yoshio, 1963
Some chemico-oceanographical properties of the water of Suruga Bay. (In Japanese; English abstract).
J. Oceanogr. Soc., Japan, 18(4):193-199.

chemistry oxygen

Sugiura, Y., 1954.
[On the diurnal variation of dissolved oxygen.(1)]
J. Ocean. Soc., Japan, 10(1):22-28, 2 textfigs.

chemistry oxygen

Sugiura, Y., 1954.
[On the diurnal variation of dissolved oxygen. (II).] J. Ocean. Soc., Japan, 10(2):65-70, 5 text -figs.

Chemistry oxygen

Sugiura, Y., 1953.
On the diurnal variations of oxygen content in surface layers of the hydrosphere.
Pap. Met. Geophys., Tokyo, 4:79-89.

Chemistry oxygen

Sugiura, Yoshio, and Hirozo Yoshimura, 1964.
Distribution and mutual relation of dissolved oxygen and phosphate in the Oyashio and the northern part of Kuroshio.
J. Oceanogr. Soc., Japan., 20(1):14-23.

Chemistry oxygen

Suzuki, T., and N. Sano, 1960.
[On the current rip in Ishikari Bay cause by the Ishikari River flowing into that bay.]
Bull. Fac. Fish., Hokkaido Univ., 11(3):132-161.

Chemistry oxygen

Svansson, Artur, 1964.
Observations at Swedish lightships and in the central Baltic.
Ann. Biol., Cons. Perm. Int. Expl. Mer, 1962, 19: 46-48.

Chemistry oxygen (data)

Svansson, Artur, 1963.
Hydrography of the Kattegat area. Swedish data.
Ann. Biol., Cons. Perm. Int. Expl. Mer, 1961, 18: 44-46.

chemistry oxygen

Svansson, Artur, 1960(1958)
Observations at Swedish light vessels.
Ann. Biol., Cons. Perm. Int. Expl. Mer, 15: 43-45.

Chemistry - oxygen

Sverdlov, A.I., 1965.
Distribution of hydrochemical elements in the area of Faroe-Shetland Strait in November-January 1963-1964. (In Russian).
Material. Ribochoz. Issled. Severn. Bassin. Poliarn. Nauchno-Issled. I Proekt. Inst. Morsk. Rib. Choz. i Okeanogr., N.M. Knipovich, 5:91-98.

Chemistry, oxygen

Sverdrup, H.U., 1953.
The currents off the coast of Queen Maud Land.
Norsk Geografisk Tidsskrift. 16(1/4):239-249, 5 textfigs.

Chemistry, oxygen

Sverdrup, H.U., 1938
On the explanation of the oxygen minima and maxima in the oceans. J. du Cons.13:163-172.

Chemistry, oxygen

Sverdrup, H. U., 1933
On vertical circulation in the ocean due to the action of the wind with application to conditions within the Antarctic Circumpolar Current. Discovery Repts. 7:139-169, 23 text figs.

Chemistry - oxygen

Sverdrup, H. U., and R. H. Fleming, 1941.
The waters of the coast of southern California, March to July 1937. Bull. S.I.O. 4(10):261-275, 66 textfigs.

Chemistry oxygen

*Szekielda, Karl-Heinz, 1968.
Ein chemisches Modell für den Auf-und Abbau organischen Materials und dessen Anwendung in der offenen See.
J. Cons.perm.int.Explor.Mer, 32(2):180-187.

Chemistry oxygen

Tabata, S., 1961
Temporal changes of salinity, temperature and dissolved oxygen content of the water at Station "P" in the northeast Pacific Ocean, and some of their determining factors.
J. Fish. Res. Bd., Canada, 18(6):1073-1124.

Chemistry oxygen

Tabata, S., 1960
Characteristics of water and variations of salinity, temperature and dissolved oxygen content of the water at ocean Weather Station "P" in the northeast Pacific Ocean.
J. Fish. Res. Bd., Canada, 17(3):353-370.

chemistry oxygen

Tabata, S., and G.L. Pickard, 1957.
The physical oceanography of Bute Inlet, British Columbia. J. Fish. Res. Bd., Canada, 14(4):487-520.

Chemistry oxygen

Taft, B.A., 1963.
Distribution of salinity and dissolved oxygen on surfaces of uniform potential specific volume in the South Atlantic, South Pacific and Indian oceans.
J. Mar. Res., 21(2):129-146.

Chemistry, oxygen

Taft, Bruce A., and John A. Knauss 1967.
The equatorial undercurrent of the Indian Ocean as observed by the Lusiad Expedition.
Bull. Scripps Inst. Oceanogr. 9:1-163.

Chemistry oxygen

Takahashi, Tadao, Masaaki Chaen and Soichi Ueda, 1960.
Report of the Kagoshima-maru IGY cruise, 1958.
Mem. Fac. Fish., Kagoshima Univ., 8: 82-86.

Chemistry oxygen (data)

Tanaka, Otohiko, Haruhiko Irie, Shozi IIzuka and Fumihiro Koga, 1961
The fundamental investigation on the biological productivity in the north-west of Kyushu. 1. Rec. Oceanogr. Wks., Japan, Spec. No. 5: 1-58.

Chemistry, oxygen (data)

Tanaka, Otohiko, Fumihiro Koga, Haruhiko Irie, Shozi Iizuka, Yosie Dotu, Keitaro Uchida, Satoshi Mito, Seiro Kimura, Osame Tabeta, and Sadahiko Imai, 1962.
The fundamental investigation of the biological productivity in the north-western sea area of Kyushu. II. Study on plankton productivity in the waters of Genkai-Nada region.
Rec. Oceanogr. Wks., Japan, Spec. No. 6:1-20.
Also in:
Contrib., Dept. Fish., and Fish. Res. Lab., Kyushu Univ., No. 8.

Chemistry oxygen

Taniguti, T., 1962.
A method of estimating the dissolved oxygen in the sea water polluted by organic matter.
Bull. Jap. Soc. Sci. Fish., 28(4):448-452.

Chemistry oxygen

Tanita, S., K. Kato, and T. Okuda, 1951.
Studies on the environmental conditions of shell-fish-fields. In the case of Muroran Harbour. (2).
Bull. Fac. Fish. Hokkaido Univ. 2(3):220-230.

Chemistry oxygen

Tanita, S., S. Kato, and T. Okuda, 1950.
Studies on the environmental conditions of shell-fish fields. 1. In the case of Hakodate Harbor.
Bull. Fac. Fish., Hokkaido Univ., 1(1):18-27, 7 textfigs. (In Japanese; English summary).

Chemistry, oxygen (data)

Tawara, Satoru, and Arao Tsuruta, 1966.
Water temperature relevant to distribution of zooplankton biomass in the tuna fishing ground in the eastern tropical Pacific. (In Japanese; English abstract).
J. Shimonoseki Univ., Fish., 14(3):199-213.

Chemistry oxygen (data)

Teixeira, Clovis, and Miryam B. Kutner, 1962
Plankton studies in a mangrove environment.
1. First assessment of standing stock and principal ecological factors.
Bol. Inst. Oceanogr., Sao Paulo, Brasil, 12 (3):101-124.

Chemistry oxygen (data)

Teixeira, C., J. Tundisi and M.B. Kutner, 1965.
Plankton studies in a mangrove environment. II. The standing stock and some ecological factors.
Bolm Inst. Oceanogr., S Paulo, 14(1):13-41.

chemistry, oxygen

Theisen, E. 1946
Tanafjorden. Enfinmarksfjords oceanografi. Fiskeridirektoratets skrifter Ser. Havundersøkelser. (Repts. on Norwegian Fishery and Marine Investigations) VIII (8):1-77, 23 text figs., 8 pls.

Chemistry, oxygen

*Thomas, Robert W., 1968.
Oceanography survey results off Point Arguello, California, January and November-December 1964.
Tech.Rept., U.S.Naval Oceanogr.Off., TR-201:53 pp. (multilithed).

Chemistry oxygen

Thompson, E. F., 1939.
Chemical and physical investigations. The exchange of water between the Red Sea and the Gulf of Aden over the "sill". John Murray Exped., 1933-34, Sci. Repts. 2(4):105-119, 10 textfigs.

chemistry, oxygen

Thompson, E.F., 1939.
Chemical and physical investigations. The general hydrography of the Red Sea. John Murray Exped., 1933-34, Sci. Repts. 2(3):83-103, 12 textfigs.

Chemistry, oxygen

Thompson, T.G., and R.J. Robinson, 1939.
Notes on the determination of dissolved oxygen in sea water. J. Mar. Res. 2(1):1-8.

chemistry, oxygen

Thomsen, H., 1952.
Hydrography. Danish observations. Cons. Perm. Int. Expl. Mer, Ann. Biol. 8:130, 1 textfig.

Chemistry, oxygen
Thomsen, H., 1951.
Hydrography (Baltic).
Ann. Biol., Cons. Perm. Int. Expl. Mer, 7:112, 2 textfigs.

Chemistry, oxygen
Thomsen, H., 1947
Hydrografiske undersøgelser i Limfjorden 1943. 1. Observations materialet. Publ. fra Det Danske Meteorologiske Institutet Meddelelser, No.9:1-104.

Chemistry, oxygen
Thomsen, H., 1938.
Hydrographical observations made during the DANA Expedition 1928-30. Dana Rept. 12:46 pp.

Chemistry, oxygen
Thomsen, H., and A.J.C. Jensen, 1950.
Baltic area. Hydrography. Monthly means of temperature and salinity at Christiansø, N of Bornholm.
Cons. Perm. Int. Expl. Mer, Ann. Biol. 6:156-157.

Chemistry, oxygen
Timofeev, N.A., 1968.
Oxygen and phosphorus distribution in the Red Sea. (In Russian).
Trudy, Vses. Nauchno-Issled. Inst. Morsk. Ribn. Okeanogr (VNIRO) 64, Trudy Azovo-Chernomorsk. Nauchno-Issled. Inst. Morsk. Ribn. Khoz. Okeanogr. (AscherNIRO), 28: 152-161.

Chemistry oxygen
Tokui, T., 1952.
On the dissolved oxygen and chlorinity of the water of Mutsu Bay during 1950.
Bull. Mar. Biol. Sta. 5(1/4):55-57.

Chemistry oxygen
Tolmachev, V.A., 1957.
On the seasonal fluctuations of dissolved oxygen at great depths of Lake Baikal.
Doklady Akad. Nauk, SSSR, 113(2):395-398.

Chemistry oxygen
Trites, R.W., 1960.
An oceanographical and biological reconnaissance of Kennebecasis Bay and the Saint John River estuary.
J. Fish. Res. Bd., Canada, 17(3):377-408.

Chemistry, oxygen
Trotti, L., 1954.
Report on the oceanographic investigations in the Ligurian and North Tirrenian Seas. Hydrography. Centro Talassografico Tirreno, Pubbl. 16: 1-39, 5 pls.

Chemistry, oxygen
Truesdale, G.A., A.L. Downing and G.F. Lowden, 1954.
The solubility of oxygen in pure water and sea-water. J. Appl. Chem., 5:53-62.

Chemistry, oxygen
Truesdale, G.A., and G. Knowles, 1956.
Some recent work on dissolved oxygen in natural waters.
J. du Cons. 21(3):263-267.

Chemistry, oxygen
Tseu, W.S.L., 1953.
Seasonal variations in the physical environment of the ponds at the Hawaii Marine Laboratory and the adjacent waters of Kaneohe Bay, Oahu.
Pacific Science 7(3):278-290, 16 text figs.

Chemistry, oxygen
Tsujita, T., 1953.
A marine ecological study on the Bay of Omura.
J. Ocean. Soc., Japan, 9(1):23-32, 6 textfigs.

Chemistry, oxygen
Tsubata, B., and T. Numakunai, 1960.
Oceanographical conditions observed at definite station off Asamushi during 1958 and 1959.
Bull. Mar. Biol. Sta., Asamushi, Tohoku Univ. 10(1):73-

Chemistry oxygen
Tsuchiya, Mizuki, 1968.
Upper waters of the intertropical Pacific Ocean.
Johns Hopkins Univ. Studies, 4:50 pp.

Chemistry, oxygen (data)
Tsuruta, Arao, Satoru Tawara and Tadashi Ueda 1967.
On the distribution of dissolved oxygen and nutrient salts in tuna fishing grounds of the Central Tropical Atlantic.
J. Shimonoseki Univ. Fish. 15(3): 295-305.

Chemistry, oxygen
Tully, J.P., and A.J. Dodimead, 1957.
Properties of the water in the Strait of Georgia, British Columbia, and influencing factors.
J. Fish. Res. Bd., Canada, 14(-):241-319.

U

Chemistry, oxygen
Uda, Michitake, 1963.
Oceanography of the subarctic Pacific Ocean.
J. Fish. Res. Bd., Canada, 20(1):119-179.

Chemistry oxygen
Uda, Michitaka, 1962
Subarctic oceanography in relation to whaling and salmon fisheries.
Sci. Repts., Whales Res. Inst., No. 16:105-119.

Chemistry oxygen (data)
Uda, Michitaka, and Makoto Ishino, 1960
Researches on the currents of Kuroshio.
Rec. Oceanogr. Wks., Japan, Spec. No. 4: 59-72.

Chemistry, oxygen (data)
Uda, M., N. Watanabe and M. Ishino, 1956.
General results of the oceanographic surveys (1952-1955) on the fishing grounds in relation to the scattering layer. J. Tokyo Univ. Fish., 42(2):169-207.

Chemistry, oxygen
Udaya Varma Thirupad, P., and C.V. Gangadhara Reddy, 1959
Seasonal variations of the hydrological factors of the Madras coastal waters.
Indian J. Fish., 6(2):298-305.

Chemistry, oxygen
Ullyott, P. and Orhan Ilgaz, 1946
The Hydrography of the Bosphorus: an Introduction. Geol. Rev., 36 (1), pp. 44-66

Chemistry, oxygen
United States, Department of Commerce, Environmental Sciences Services Administration, 1965
International Indian Ocean Expedition, USC&GS Ship Pioneer - 1964.
Vol. 1. Cruise Narrative and scientific results 139 pp.
Vol. 2. Data report: oceanographic stations, BT observations, and bottom samples, 183 pp.

Chemistry, oxygen
University of Southern California, Allan Hancock Foundation, 1965.
An oceanographic and biological survey of the southern California mainland shelf.
State of California, Resources Agency, State Water Quality Control Board, Publ. No. 27:232 pp. Appendix, 445 pp.

Chemistry oxygen (data)
Uriarte, L.B., 1932.
Premiers travaux du Laboratoire Oceanographique des Iles Canaries.
Rapp. Proc. Verb., Cons. Perm. Int. Expl. Mer, 67:65-88, 8 textfigs.

Chemistry oxygen (data)
U.S. Navy Hydrographic Office, 1957.
Operation Deep Freeze, II, 1956-1957. Oceanographic survey results. H.O. Tech. Rept., 29: 155 pp. (multilithed).

Chemistry oxygen (data)
Uyeno, Fukuzo, 1966.
Nutrient and energy cycles in an estuarine oyster area.
J. Fish. Res. Bd., Can., 23(11):1635-1652.

Chemistry, oxygen
Uyeno, Fukuzo, 1964.
Relationships between production of foods and oceanographical condition of sea water in pearl farms. II. On the seasonal changes of sea water constituents and of bottom condition, and the effect of bottom cultivation. (In Japanese;
J. Fac. Fish., Pref. Univ. Mie, 6(2):145-169.

V

Chemistry oxygen
Valdivia, Julie, y Oscar Guillén, 1966.
Informe preliminar del crucero de primavera 1965 (Cabo Blanco-Morro Sama).
Inst. Mar. Peru, Informe, No. 11:35-70.

Chemistry oxygen
Vallaux, C., 1943.
Signification et portée de l'oxygène dissous dans les eaux océanique. Bull. Inst. Océan., Monaco, No. 852:7 pp.

Chemistry, oxygen (data)
Van Goethem, C., 1951.
Étude physique et chimique du milieu marin.
Rés. Sci., Expéd. Océan. Belge dans les Eaux Côtières Africaines de l'Atlantique Sud (1948-1949) 1:1-151, 1 pl.

Chemistry - oxygen (data)
Van Riel, P. M., 1943.
The bottom water. Introductory remarks and oxygen content. Snellius Exped. Vol. 2, Pt. 5, Ch. 1:77 pp., 34 textfigs.

Chemistry, oxygen
Varlet, F., 1958
Les traits essentials du régime côtise de l'Atlantique près d'Abidjan (Côte d'Ivoire).
Bull. d'I.F.A.N., (A), 20(4):1089-1102.

Vatova, A.

Chemistry, oxygen
Vatova, A. 1965.
Les Conditions hydrographiques de la Mar Piccolo de Taranto pendant l'année 1963.
Rapp. P.-v. Réun. Comm. int. Explor. scient. Mer. Méditerr. 18(3): 653-655.

Chemistry, oxygen
Vatova, A., 1944.
Osservazioni idrografiche periodiche nell'alto Adriatico (1937-1944). Boll. Pesca, Piscic. Idrobiol. 3(2):247-277.

Chemistry, oxygen
Vatova, A., 1934.
L'anormale regime fisico-chimico dell' Alto Adriatico nel 1929 e le sue ripercussioni sulla fauna. Thalassia 1(8):49 pp., 3 fold-ins, 16 figs., tables.

Chemistry oxygen (data)
Vatova, A., and P.M. di Villagrazia, 1950.
Sulle condizioni idrographiche de Canal di Leme in Istria. Nova Thalassia 1(8):24 pp., graphs.

Chemistry, oxygen
Vatova, A., and P. Milo di Villagrazia, 1948.
Sulle condizioni chimicofisiche del Canale di Lema presso Rovigno d'Istria.
Bol. Pesca, Piscicol. Idrobiol., n.s., 3(1): 3(1):5-27, 2 textfigs., 5 pls.

Chemistry, oxygen
Vega Rodriguez, Filiberto, y Virgilio Arenas Fuentes 1965.
Resultados preliminares sobre la distribución del plancton y datos hidrográficos del Arrecife "La Blanquilla", Veracruz, Ver.
Anales Inst. Biol. Univ. Mexico 36(1/2): 53-59.

Chemistry, oxygen
Verstraete, J.M., 1970.
L'oxygène au large de Grand Bassam. Doc. scient Cent. Rech. océanogr., Côte d'Ivoire, ORSTOM, 1(3): 19-35.

Chemistry, oxygen
Villagrazia, P. M. de, 1947.
La fotocromia applicata all'analisi colorimetrica per il dosaggio approssimativo dell'ossigeno disciolto nelle acque dolci e marina. Boll. Pesca, Piscicolt. Idrobiol., n.s., 2(1):78-93.

Chemistry, oxygen
Villalobos-Figueroa, Alejandro, José A. Suarez-Caabro, Samuel Gomo, Guadalupe de la Lanze, Mauricio Aceves, Fernando Manrique and Jorge Cabrera, 1967.
Considerations on the hydrography and productivity of Alvarado lagoon, Vera Cruz, Mexico.
Proc.Gulf Caribb. Fish.Inst., 19th Sess., 75-85.

Chemistry, oxygen
Villagrazia, P. Milo de, 1950.
Ulteriore contributo alla conoscenza delle condizioni idrografiche e biologiche del Canali di Lema (Istria). Bol. Pesca, Piscicolt. Idrobiol. 5(2):225-247, 3 textfigs.

Chemistry
oxygen
Vinogradov, A.P., V.M. Kutyurin & I.K. Zadorozhnyi, 1959
[Isotope fractionation of atmospheric oxygen.]
Geokhimiya, 3: 241-253.

Pagination is for English edition of journal.

oxygen chemistry, oxygen
Vinogradov, M.E., 1953.
[Influence of zooplankton respiration on deficiency in oxygen content in various layers of water.]
Dokl. Akad. Nauk, SSSR, 82:637-639.

Chemistry, oxygen (data)
Viorel, Chiriló, 1965.
Observaţii asupra condiţiilor fizico-chimice ale mării la Mamaia, în anii 1959 şi 1960.
In: Ecologie marină, M. Băcescu, redactor, Edit. Acad. Republ. Pop. Române, Bucureşti, 1:139-184.

Chemistry
oxygen (data)
Vives, Francisco, y Manuel López-Benito, 1957
El fitoplancton de la Ría de Vigo desde julio de 1955 a junio de 1956.
Inv. Pesq., Barcelona, 10:45-146.

Chemistry
oxygen (data)
Voit, S.S., D.A. Aksenov, M.M. Bogorodsky, V.V. Sinukov and V.A. Vladimirzev, 1961
[Some peculiar features of water circulation in the Black Sea and the water regime in the Pre-Bosphorus area.]
Okeanologiia, Akad. Nauk, SSSR, 1(4):613-625.

Chemistry
oxygen
*Volkovinsky, V.V., 1967.
Application of the two-layer system to the computations of the primary productivity to sea water. (In Russian; English abstract).
Okeanologiia, Akad. Nauk, SSSR, 7(6):1037-1052.

Chemistry, oxygen
Voloshonsky, A.G. 1966
Hydrochemical conditions in the Faroe-Iceland Channel in November-December 1964. (In Russian).
Mater. Ribokhoz. Issled. severn. Basseina, Poliarn. Nauchno-Issled. Proektn. Inst. Morsk. Khoz. Okeanogr. (PINRO) 7:146-154.

Chemistry oxygen
Voronkev, P.P., 1948.
[Basic characteristics of the hydrochemical regime of the coastal zone of Barents Sea in the central Murman re ion.] Trudy Dal'niye Zelentsy, 1948(1):39-101, 25 diagrams, sketch map.

Chemistry
oxygen
Vorontsova, R. V., 1964.
The hydrochemical conditions in the southern part of the Davis Strait in the spring and summer of 1963. (In Russian).
Material. Sess. Uchen. Sov., PINRO, Rez. Issled. 1962-1963., Murmansk, 130-135.

Chemistry Oxygen
Vul'fson, V.I., 1970.
On the problem of free oxygen resources.
Okeanologiia 10(3):387-395. (In Russian; English abstract)

Chemistry, oxygen
Vul'fson, V.I., and M.D. Alekseeva, 1957.
[Use of a permanganate method in the determination of oxygen dissolved in water.]
Gidrokhim. Materialy, 26:226-

W

Chemistry, oxygen
Waldichuk, Michael, 1965.
Water exchange in Port Moody, British Columbia, and its effect on waste disposal.
J. Fish. Res. Bd., Canada, 22(3):801-822.

Chemistry, oxygen
Waldichuk, M., 1956.
Basic production of Trevor Channel and Alberni Inlet from chemical measurements.
J. Fish. Res. Bd., Canada, 13(1):7-20.

Chemistry, oxygen
Waldichuk, M., 1955.
Effluent disposal from the proposed pulp mill at Crofton, B.C. Fish. Res. Bd., Canada, Pacific Coast Stas., Prog. Repts., No. 102:6-9, 3 text-figs.

Chemistry, oxygen
Waldichuk, M., 1954.
Effect of pulpmill waste in Alberni Harbour.
Fish. Res. Bd., Canada, Pacific Coast Stas., Prog. Repts., No. 101:23-26, 2 textfigs.

Chemistry oxygen
Waldichuk, M., 1953.
Oceanography of the Strait of Georgia. IV. Dissolved oxygen distribution. Prog. Repts., Pacific Coastal Stas., Fish. Res. Bd., Canada, No. 96:6-10, 3 textfigs.

oxygen (dissolved)
Ward, Ronald W., Valerie Vreeland, Charles H. Southwick and Anthony J. Reeding, 1965.
Ecological studies related to plankton productivity in two Chesapeake Bay estuaries.
Chesapeake Science, 6(4):214-225.

Chemistry oxygen
Watanabe, Nobuo, Toshiyuki Hirano, Rinnosuke Fukai, Kozi Matsumoto and Fumiko Shiokawa, 1957.
A preliminary report on the oceanographic survey in the "Kuroshio" aream south of Honshu, June-July 1955.
Rec. Oceanogr. Wks., Japan (Suppl.):197-

Chemistry, oxygen
Wattenberg, H., 1939.
Die Entstehung der Sauerstoffarmen Zwischenschicht im Ozean. Ann. Hydr., usw., 67:257-266.

Chemistry, oxygen
Wattenberg, H., 1938.
Die Verteilung des Sauerstoffs und des Phosphats im Atlantischen Ozean. Wiss. Ergeb., Meteor, 9(1):132 pp.

Chemistry, oxygen
Watts, J.C.D., 1958.
The hydrology of a tropical West African estuary.
Bull. Inst. Francais, Afrique Noire, 20(3):697-752.

Chemistry, oxygen
Weiss, R.F., 1970.
The solubility of nitrogen, oxygen and argon in water and seawater. Deep-Sea Res., 17(4): 721-735.

Chemistry, oxygen
Wennekens, M.P., 1959
Water mass properties of the Straits of Florida and related waters.
Bull. Mar. Sci., Gulf and Caribbean, 9(1): 1-52.

Chemistry, oxygen
Weyl, Peter K., 1965.
On the oxygen supply of the deep Pacific Ocean.
Limnology and Oceanography, 10(2):215-219.

Chemistry oxygen
Wheatland, A.B., and L.J. Smith, 1955.
Gasometric determination of dissolved oxygen in pure and saline water as a check of titrimetric methods.
J. Appl. Chem., 5(3):144-148.

Chemistry, oxygen
Wiborg, K.F., 1944
The production of plankton in a land locked fjord. The Nordåsvatn near Bergen in 1941-1947. With special reference to copepods. Fiskeridirektoratets Skrifter Serie Havundersøkelser (Rept. Norwegian Fish. and Marine Invest.) 7(7):1-83, Map.

chemistry oxygen
Wiktorowie, Jozef i Krystyna, 1962
Some hydrological properties of the Pomeranian Bay water
Prace Morsk. Inst., Ryback., Gdyni. Oceanogr. Ictiol., 11(A):113-136.

chemistry oxygen
Wolf, G., 1961
Uber die hydrologischen Verhältnisse in der westlichen Ostsee im November 1953.
Beitr. Meeres., 1:39-47.

chemistry, oxygen
Won, Chong Hun, 1964.
Tidal variations of chemical constituents of the estuary water at the Lava bed in the Nang-Dong River from Nov.1962 to Oct. 1963. (In Korean: English abstract).
Bull. Pusan Fish. Coll., 6(1): 21-32

chemistry oxygen
Won, Chong Hun, 1963.
Distribution of chemical constituents of the estuary water in Gwang-Yang Inlet. (In Korean; English abstract).
Bull. Fish. Coll., Pusan Nat. Univ., 5(1):1-10.

chemistry oxygen
Wooster, W.S., 1951.
Distribution of oxygen and phosphate in the Arctic Sea. Proc. Alaskan Sci. Conf., Bull. Nat. Res. Coun. 122:81.

chemistry, oxygen
Wooster, W.S., and T. Cromwell, 1958.
An oceanographic description of the eastern tropical Pacific. Bull. S.I.O., 7(3):169-282.

chemistry oxygen
Wooster, W.S., and Malvern Gilmartin, 1961
The Peru-Chile Undercurrent.
J. Mar. Res., 19(3):97-122.

chemistry oxygen
Wooster, W.S., and J.L. Reid, Jr., 1963
11. Eastern boundary currents. In: The Sea, M.N. Hill, Edit., Vol. 2. (III) Currents, Interscience publishers, New York and London, 253-280.

chemistry oxygen
Wüst, Georg, 1963
On the stratification and the circulation in the cold water sphere of the Antillean-Caribbean basins.
Deep-Sea Res., 10(3):165-187.

chemistry oxygen
Wüst, Georg, 1961
[On the vertical circulation of the Mediterranean Sea.]
J. Geophys. Res., 66(10):3261-3272.

chemistry, oxygen
Wüst, G., and W. Brogmus, 1955.
Ozeanographische Ergebnisse einer Untersuchungsfahrt mit Forschungskutter "Südfall" durch die Ostsee Juni-Juli 1954 (anlässlich der totalen Sonnenfinsternis auf Öland).
Kieler Meeresf. 11(1):3-21, 8 textfigs.

chemistry oxygen
Wüst, Georg (with Arnold Gordon), 1964
Stratification and circulation in the Antillean-Caribbean basins. 1. Spreading and mixing of the water types with an oceanographi atlas.
Vema Research Series, Columbia Univ. Press, New York and London, No. 2: 201 pp.

chemistry, oxygen
Wyrtki, Klaus, 1964.
Upwelling in the Costa Rica dome.
U.S.F.W.S. Fish. Bull., 63(2):355-372.

chemistry oxygen (data)
Wyrtki, Klaus, 1961
The flow of water into the deep sea basins of the western South Pacific Ocean.
Australian J. Mar. & Freshwater Res., 12(1): 1-16.

XYZ

chemistry oxygen
Yamamoto, H., 1959.
Oceanographical studies on Japanese inlets. 5. Relations between variation of chemical compositions and hydrographic conditions in Uranouchi Bay. 6. On the consumption of dissolved oxygen in Urado Bay.
J. Oceanogr. Soc., Japan, 15(1):15-18, 19-22.

chemistry, oxygen
Yamazi, I., 1955.
Plankton investigations in inlet waters along the coast of Japan. XVI. The plankton of Tokyo Bay in relation to water movement.
Publ. Seto Mar. Biol. Lab., 4(2/3):285-309, Pls. 19-20, 22 textfigs.

chemistry oxygen(profiles)
Yamazi, I., 1953.
Plankton investigation in inlet waters along the coast of Japan. VII. The plankton collected during the druises to the new Yamamoto Bank in the Sea of Japan. Publ. Seto Mar. Biol. Lab. 3(1):75-108, 19 textfigs.

chemistry oxygen
Yamazi, I., 1952.
Plankton investigations in inlet waters along the coast of Japan. III. The plankton of Imari Bay in Kyusyu. IV. The plankton of Nagasaki Bay and Harbout in Kyusyu. V. The plankton of Hiroshima Bay in the Seto-Nakai (Inland Sea).
Publ. Seto Mar. Biol. Lab., Kyoto Univ., 2(2): 289-304; 305-318; 319-330, 8 and 8 and 7 textfigs

chemistry oxygen
Yamazi, I., 1952?
Plankton investigations in inlet waters along the coast of Japan. VI. The plankton of Nanao Bay.
Seto Mar. Biol. Lab. Contr. 191:309-319, 11 textfigs.

chemistry, oxygen
Yamazi, I., 1951.
Plankton investigations in inlet waters along the coast of Japan. II. The plankton of Hakodate Harbour and Yoichi Inlet in Hokkaido.
Publ. Seto Mar. Biol. Sta., Kyoto Univ., 1(4): 185-194, 3 textfigs.

chemistry oxygen
Yamazi, I., 1950.
Plankton investigations in inlet waters along the coast of Japan. 1. Publ. Seto Mar. Biol. Lab. 1(3):93-113, 14 textfigs.

chemistry oxygen (data)
*Yasuoka, Takeo, 1967.
Hydrography of the Okhotsk Sea (1).
Oceanogrl. Mag., 19(1):61-72.

chemistry oxygen (data)
Zeigler, John M., 1964.
The hydrography and sediments of the Gulf of Venezuela.
Limnology and Oceanography, 9(3):397-411.

chemistry, oxygen
Zlobin, V.S., M.F. Perlyuk and N.G. Sapronetskaya 1968.
Some regularities in the distribution of phosphates and oxygen in the Davis Strait, summer and winter 1965.
Annls biol, Copenh 1966, 23:13-20.

chemistry, oxygen
Zlobin, V.S., N.G. Sapronetskaya, and A.P. Alekseeva, 1968.
Some regularities in the formation of the hydrochemical regime and primary productivity of the Norwegian Sea. (In Russian).
Mater. Ribokhoz. Issled. Severn. Basseina, 11: 133-144.

chemistry, oxygen
Zuta, Salvador, y Oscar Guillen, 1970.
Oceanografia de las aguas costeras del Peru.
Bol. Inst. Mar., Peru, 2(5):157-324.

apparent oxygen production (AOP)
Stefansson, Unnsteinn, and Francis A. Richards, 1964.
Distributions of dissolved oxygen, density and nutrients off the Washington and Oregon coasts.
Deep-Sea Res., 11(3):355-380.

Oxygen (A.O.U.)

chemistry, oxygen (apparent oxygen utilization)
Alvarez-Borrego, Saul, and P. Kilho Park 1971.
AOU as indicator of water-flow direction in the central North Pacific.
J. Oceanogr. Soc. Japan 27(4):142-151

chemistry OXYGEN AOU
Broenkow, William W., 1965.
The distribution of nutrients in the Costa Rica dome in the eastern tropical Pacific Ocean.
Limnology and Oceanography, 10(1):40-52.

chemistry Oxygen AOU (data
Forsbergh, Eric D., William W. Broenkow, 1965.
Observaciones oceanograficas del oceano Pacific oriental recolecgadas por el barco Shoyo Maru, octubre 1963-marzo 1964.
Comision Interamericana del Atun Tropical, Bol. 10(2): 85-237.

Chemistry, oxygen (AOU)
Japan, Science Council, National Committee for IIOE, 1966.
General report of the participation of Japan in the International Indian Ocean Expedition.
Rec. Oceanogr. Wks., Japan, n.s. 8(2): 133 pp.

chemistry oxygen A.O.U.
Kitamura, Hiroyuki, and Takeshi Sagi, 1965.
On the chemical elements in the sea south of Honshu, Japan. II. (In Japanese; English abstract)
Bull. Kobe Mar. Obs., No. 174:39-55.

Chemistry, oxygen, AOU
*Ogura, Norio, 1970.
The relation between dissolved organic carbon and apparent oxygen utilization in the Western North Pacific. Deep-Sea Res., 17(2): 221-231.

Chemistry, oxygen (AOU)
Sugiura, Yoshio 1966.
Reply to Dr. Fukai's Comment.
J. oceanogr. Soc. Japan 22(1): 19.

Chemistry
Oxgen (A. O. U.)
Sugiura Yoshio, 1965.
Some chemico-oceanographic properties of the Kuroshio and its adjacent regions.
Proc. Symp., Kuroshio, Tokyo, Oct. 29, 1963, Oceanogr. Soc., Japan, andUNESCO, 12-13.

Chemistry, oxygen (A.O.U.)
Suguira, Yoshio, 1964.
Some chemico-oceanographical properties of the Kurishio and its adjacent regions.
In: Recent researches in the fields of hydrosphere, atmosphere and nuclear geochemistry, Ken Sugawara festival volume, Y. Miyake and T. Koyama, editors, Maruzen Co., Ltd., Tokyo, 49-63.

Chemistry
oxygen (A.O.U.)
Sugiura, Yoshio, and Suphachai Chaitiamvong, 1964.
Relation of silicate concentration to dissolved oxygen amount in sea water collected in the northern frontal region of Kuroshio.
J. Oceanogr. Soc., Japan, 20(2):89-92.

Chemistry oxygen (AOU)
*Sugiura, Yoshio, and Hirozo Yoshimura, 1967.
Total carbon dioxide and its bearing on the dissolved oxygen in the Oyashio and in the frontal region of the Kuroshio.
Geochem.J., Japan, 1(3):125-130.

O_2 - anomaly

Chemistry oxygen anomalies
Burt, Wayne V., W. Bruce McAlister and John Queen, 1959.
Oxygen anomalies in the surf near Coos Bay, Oregon.
Ecology, 40(2):305-306.

Chemistry, oxygen anomaly
Matsudaira, Y., T. Kawamoto and H. Kitamura, 1956
On the distribution of dissolved oxygen in the Kuroshio region. J. Fac. Fish. & Animal Husbandry Hiroshima Univ., 1(2):297-301.

Oxygen annual variation

Chemistry, oxygen, annual variation
Ichiye, T., 1954.
On the distribution of oxygen and their seasonal variations in the adjacent Seas of Japan. III.
Ocean. Mag., Tokyo, 6(3):101-131, 19 textfigs.

Chemistry oxygen, annual variation
Kamps*, L.F., R. Dorrestein and L. Otto, 1960.
Note on the annual variation of salinity, temperature and oxygen content in the Ems-Estuary.
Verh. Kon. Ned. Geol. Mijnb. k. Gen., Geol. Ser., 19:75-81.

Chemistry
oxygen, atmospheric
Machta, L. and E. Hughes, 1970.
Atmospheric oxygen in 1967 to 1970. Science, 168(3939): 1582-1584.

oxygen balance

Chemistry, oxygen balance
Kanwisher, John W., and Stephen A. Wainwright, 1967.
Oxygen balance in some reef corals.
Biol. Bull., mar.biol.Lab., Woods Hole, 133(2): 378-390.

Oxygen, bottom

Chemistry oxygen(bottom)
Brennecke, W., 1921.
Die ozeanographischen Arbeiten der Deutschen Antarktischen Expedition, 1911-1912.
Arch. Deutschen Seewarte 39(1):1-216, 41 textfigs 15 pls.

Chemistry, oxygen, bottom
Brouardel, J., and L. Fage, 1954.
Variation de la teneur en oxygène de l'eau au proche voisinage des sédiments. Deep-Sea Res. 1(2):86-94.

Chemistry oxygen, bottom
Glowinska, A., 1966.
Polish observations in the southern Baltic 1964.
Annls biol., Copenh., 21:42-43.

Chemistry, oxygen (bottom)
Glowinska, A., 1963.
Hydrography of the southern Baltic in 1963.
Ann. Biol., Cons. Perm. Int. Expl. Mer, 1963, 20:69-70.

Chemistry oxygen (bottom)(data)
Parker, Robert H., 1964.
Zoogeography and ecology of some macroinvertebrates, particularly mollusks, in the Gulf of California, and the continental slope off Mexico.
Vidensk. Medd., Dansk Naturh. Foren., 126:1-178.

oxygen-carbon dioxide
Culberson, Charles and Ricardo M. Pytkowicz, 1970.
Oxygen-total carbon dioxide correlation in the Eastern Pacific Ocean.
J. oceanogr. Soc. Japan, 26(2): 95-108

Oxygen-chlorinity relationships

Chemistry - oxygen-chlorinity relationships
Ichiye, T., 1954.
On the distribution of oxygen and their seasonal variations in the adjacent seas of Japan. II.
Ocean. Mag., Tokyo, 6(2):67-100, 25 textfigs.

Chemistry, oxygen/chlorinity
Ichie, Takashi, 1951
On the hydrography of the Kii-Suido(1951).
Bull. Kobe Mar. Obs., No. 164: 253-278(top of page); 35-60 (bottom of page).

Chemistry oxygen concentration
White, Warren B., 1971.
The westward extension of the low-oxyty distribution in the Pacific Ocean off the west coast of South America. J. geophys. Res. 76 (24): 5842-5851.

Oxygen consumption

oxygen consumption
Aizatullin, T.A., S.G. Kara-Murza and A.V. Leonov 1968.
Some peculiarities of oxygen consumption kinetics in sea-water samples. (In Russian).
Mater. Ribokhoz. Issled. Severn. Basseina, 12: 166-172.

Chemistry, oxygen consumption
Alyakrinskaya, I.O., 1966.
Experimental data on oxygen consumption in sea water polluted by petroleum. (in Russian).
Okeanologiia, Akad. Nauk. SSSR, 6(1):89-97.

chemistry, oxygen consumption
Banse, Karl, Frederic H. Nichols and Dora R. May 1971.
Oxygen consumption by the sea bed. III. On the role of the macrofauna at three stations.
Vie Milieu Suppl. 22(1):32-52.

Chemistry
oxygen consumption
Beyer, Fredrik, 1972
Om Vannutvekslingen i Oslofjorden og dens Betydning for faunaen.
Rapp. Inst. mar. Biol. Avd. A+C, 3: 25pp. (multilithed)

Chemistry
oxygen consumption
Craig, H., 1971.
The deep metabolism: oxygen consumption in abyssal ocean water. J. geophys. Res. 76(21): 5078-5086.

Chemistry, oxygen consumption
Krüger, F. 1970.
Untersuchungen über die Temperaturabhängigkeit des Sauerstoffverbrauchs von Crepidula fornicata (Mollusca: Prosobranchia).
Mar. Biol. 5(2): 145-153.

Chemistry
oxygen consumption
Kutty, M.N., 1969.
Oxygen consumption in the mullet Liza macrolepis with special reference to swimming velocity. Marine Biol., 4(3): 239-242.

chemistry, oxygen consumption
Kutty, M.N., G. Murugapoopathy and T.S. Krishnan, 1971.
Influence of salinity and temperature on the oxygen consumption in young juveniles of the Indian prawn Penaeus indicus. Mar. Biol. 11(2): 125-131.

Chemistry
oxygen consumption
Laevastu, Taivo, Harry Zeitlin and Moon Ki Song. 1965.
Notes on oxygen consumption in sea water.
Limnology and Oceanography, 10(1):144-146.

Chemistry - oxygen consumption
Lange, Rolf 1969.
Marine biology and chemistry.
In: Chemical oceanography, Rolf Lange, editor, Universitetsforlaget Oslo: 35-46

Chemistry, oxygen consumption
*Lee, Byung Don, and Pyung Chin (1970).
On the oxygen consumption of gill tissue of certain crabs in relation to difference inhabitat. (In Korean; English abstract).
Publ. Mar. Lab. Pusan Fish. Coll. 3:53-56.

Chemistry
oxygen consumption
Madanmohanrao, G., & K. Pampapathi Rao, 1962
Oxygen consumption in a brackish water crustacean, Sesarma plicatum (Latreille) and a marine crustacean, Lepas anserifera L.
Crustaceana, 4(1): 75-81.

Chemistry oxygen consumption
McFarland, W.N., and P.E. Pickens, 1965.
The effects of season, temperature and salinity on standard and active oxygen consumption of the grass shrimp, Palaemonetes vulgaris (Say).
Canadian Jour., Zool., 43(3):571-585.

Chemistry, oxygen consumption
McLeese, D.W., and J. Watson 1968.
Oxygen consumption of the spider crab (Chionoectes opilio) and the American lobster (Homarus americanus) at a low temperature.
J. Fish. Res. Bd Can., 25(8): 1729-1732.

Chemistry, oxygen consumption
Motoda, S., 1938.
O_2 consumption of plankton in the sea. Kagaku Nanyo (Sci. of the South Sea)1(2):33-34.

Chemistry, oxygen consumption
Motohashi, Keinosuke and Chikayoshi Matsudaira 1969.
On the relation between the oxygen consumption and the phosphate regeneration from phytoplankton decomposing in stored sea water.
J. oceanogr. Soc. Japan 25(5): 249-254.

Chemistry, oxygen ~~consumption~~
Pamatmat, Mario M. and Karl Banse, 1969.
Oxygen consumption by the seabed. II. In situ measurements to a depth of 180 m. Limnol. Oceanogr., 14(2): 250-259.

Chemistry
oxygen consumption
Skopintsev, B.A., 1962
On the biochemical consumption of oxygen in the waters of the Northern Atlantic. (In Russian).
Okeanologiia, Akad. Nauk. SSSR, 2(6):1009-1013.

Chemistry oxygen consumption
Sholkovitz, Edward R. and Joris M. Gieskes, 1971.
A physical-chemical study of the flushing of the Santa Barbara Basin. Limnol. Oceanogr. 16(3): 479-489.

Chemistry - oxygen consumption
Yamamoto, H., 1959.
[Oceanographical studies on Japanese inlets. 7. On the consumption of dissolved oxygen in Urado Bay (Pt. 2).] J. Oceanogr. Soc., Japan, 15(2): 61-63.

Chemistry oxygen content
Rougerie, F., 1969.
Sur un noyau à forte teneur en oxygène dans la partie inférieure du courant de Cromwell.
Cah. ORSTOM ser. Océanogr., 7(3): 21-28.

Oxygen data only
A

Chemistry oxygen (data only)
Akamatsu, Hideo, Tsutomu Akiyama and Tsutomu Sawara, 1965.
Preliminary report of the Japanese Expedition of Deep-Sea, the Tenth Cruise, 1965(JEDS-10).
Oceanogr. Mag., Tokyo, 17(1/2):49-68.

Chemistry - oxygen (data only)
Anderson, William W., and Jack W. Gehringer, 1959.
Physical oceanographic, biological and chemical, South Atlantic coast of the United States, M/V Theodore N. Gill cruise 9.
USFWS Spec. Sci. Rept., Fish. No.313:226 pp.

Chemistry oxygen (data only)
Anderson, W.W, and J.W. Gehringer, 1959.
Physical oceanographic, biological and chemical data South Atlantic coast of the United States, M/V Theodore N. Gill Cruise 8.
USFWS Spec. Sci. Rept., Fish., No. 303:227 pp.

Chemistry oxygen (data only)
Anderson, William W., and Jack W. Gehringer, 1959.
Physical oceanography, biological and chemical data, South Atlantic coast of the United States, M/V Theodore N. Gill, Cruise 7.
USFWS Spec. Sci. Rept., Fish. No. 278:277 pp.

Chemistry oxygen (data only)
Anderson, William W., and Jack W. Gehringer, 1958.
Physical oceanographic, biological and chemical data, south Atlantic coast of the United States, M/V Theodore N. Gill Cruise 6.
USFWS Spec. Sci. Rept., Fish., No. 265:99 pp.

Chemistry oxygen (data only)
Anderson, W.W., and J.W. Gehringer, 1958.
Physical oceanographic, biological and chemical data, South Atlantic coast of the United States, M/V Theodore N. Gill cruise 5.
USFWS Spec. Sci. Rept., Fish. 248:220 pp.

Chemistry oxygen (data only)
Anderson, W.W., and J.W. Gehringer, 1957.
Physical oceanographic, biological and chemical data, South Atlantic coast of the United States, Theodore N. Gill cruise.
USFWS Spec. Sci. Rept., Fish., No. 210:208 pp.

Chemistry oxygen (data only)
Anderson, W.W., and J.W. Gehringer, 1957.
Physical oceanographic, biological and chemical data, South Atlantic coast of the United States, M/V Theodore N. Gill cruise 4.
USFWS Spec. Sci. Rept., Fish., 234:192 pp.

Chemistry oxygen(data only)
Anderson, W.W., J.W. Gehringer and E. Cohen, 1956
Physical oceanographic, biological and chemical data, South Atlantic coast of the United States, Theodore N. Gill, Cruise 2.
Spec. Sci. Rept., Fish., No. 198:270 pp.

Chemistry oxygen (data only)
Anderson, W.W., J.W. Gehringer and E. Cohen, 1956
Physical oceanographic, biological and chemical data, south Atlantic coast of the United States, M/V Theodore N. Gill, Cruise 1.
USFWS Spec. Sci. Rept., Fish., No. 178:160 pp.

Chemistry, oxygen (data only)
Angot, M., et R. Gerard 1965.
Hydrologie de la région de Nosy-Bé: décembre 1963 à mars 1964.
Cahiers ORSTOM Océanogr. 4(1): 31-53.

Chemistry, oxygen (data only)
Angot, M., et R. Gerard 1965.
Hydrologie de la région de Nosy-Bé: juillet à novembre 1963.
Cahiers ORSTOM Océanogr. 3(1): 3-29.

Chemistry oxygen (data only)
Anon., 1951.
[Report of the oceanographical observations on board the R.M.S. "Syunpu Maru" in the Kii-Suido (Oct. 1950).] J. Ocean., Kobe Obs., 2nd ser., 1(6):1-15, 7 figs.

Chemistry oxygen (data only)
Anon., 1950.
Table 14. Oceanographical observations taken in the Maizuru Bay; physical and chemical data for stations occupied by R.M.S. "Syunpu-maru". Res. Mar. Met., Ocean. Obs., Tokyo, July-Dec. 1947, No. 2:71-76.

Chemistry oxygen(data only)
Anon., 1950.
Table 13. Oceanographical observations taken in the Nagasaki Harbour; physical and chemical data for stations occupied by a cutter. Res. Mar. Met. Ocean. Obs., Tokyo, July-Dec. 1947, No. 2:65-70.

Chemistry oxygen (data only)
Anon., 1950.
Table 10. Oceanographical observations taken in the sea northwest of Kyushu; physical and chemical data for stations occupied by the Ex "Fugen-maru". Res. Mar. Met., Ocean. Obs., Tokyo, July-Dec. 1947, No. 2:57-58.

Chemistry oxygen(data only)
Anon., 1950.
Table 3. Oceanographical observations taken in the North Pacific Ocean east off Sanriku; physical and chemical data for stations occupied by "Oshoro-maru". Res. Mar. Met., Ocean. Obs., Tokyo, July-Dec. 1947, No. 2:8-12.

Chemistry oxygen(data only)
Anon., 1950.
Table 2. Oceanographical observations taken in the Uchiura Bay; physical and chemical data for stations occupied by R.M.S. "Kuroshio-maru". Res. Mar. Met., Ocean. Obs., Tokyo, July-Dec. 1947, No. 2:3-7.

Chemistry oxygen (data only)
Anon., 1950.
Oceanographical observations taken in the North Pacific Ocean (east of Sanriku); physical and chemical data for stations occupied by R.M.S. "Yushio-maru" (continued). Res. Mar. Met., Ocean. Obs., Jan.-June 1947, No. 1:28037, Table 4.

Chemistry oxygen(data only)
Anon., 1950.
Table 5. Oceanographical observations taken in the North Pacific Ocean east off Tohoku District; physical and chemical data for stations occupied by R.M.S. "Ryofu-maru". Res. Mar. Met., Ocean. Obs., Tokyo, July-Dec. 1947, No. 2:22-43.

Chemistry oxygen (data only)
Anon., 1950.
[Report of the oceanograhical pbservations from Fisheries Experimental Stations (Sept. 1950).] J. Ocean., Kobe Obs., 2nd ser., 1(2):5-11.

Chemistry (organic) oxygen (data only)
Anon., 1950.
[The report of the oceanographical observations in the Bungo Suido and off Ashizuri-Misaki in August 1950.] J. Ocean., Kobe Obs., 2nd ser., 1(1):15-44, 7 figs., tables.

Chemistry, oxygen (data only)
Anon., 1950.
[The report of the oceanographical observations in the Bungo Suido and off Ashizuri-Misaki in August 1950.] J. Ocean., Kobe Obs., 2nd ser., 1(1): 18-44, 7 figs., tables.

Chemistry, oxygen (data only)
Aragno, Federico, Alberto Gomez, Aldo Orlando y Andres J. Lusquiños 1968.
Datos y resultados de la campaña pesquería "Pesquería II" (9 de noviembre al 12 diciembre de 1966).
Publ. (Ser. Inf. Tecn.), Mar del Plata 10(2): 1-129.

Chemistry, oxygen (data only)
Aragno, Federico J., Alberto Gomez, Aldo Orlando y Andres J. Lusquiños 1968.
Datos y resultados de la campaña pesquería "Pesquería III" (20 de febrero al 20 de marzo de 1967).
Publ. Ser. Inf. tecn. Mar del Plata 10(3): 1-162.

Chemistry, oxygen (data only)
Aragno, Federico, Alberto Gomez, Aldo Orlando y Andres J. Lusquiños 1968.
Datos y resultados preliminares de la campaña pesquería "Pesquería I" (12 de agosto al 8 de setiembre de 1966).
Publ. Ser. Inf. tecn. Mar del Plata 10(1): 1-159.

chemistry oxygen (data only)
Argentina, Secretaria de Marina, Servicio de Hidrografia Naval, 1961.
Trabajos oceanograficos realizados en la campana Antartica 1960/1961. Resultados preliminares.
Publico, H. 629:unnumbered pp.

chemistry oxygen (data only)
Argentina, Secretaria de Marina, Servicio de Hidrografia Naval, 1961.
Trabajos oceanograficos realizados en la campana Antartica 1959/1960. Resultados preliminares
Publico, H. 623:unnumbered pp.

chemistry oxygen (data only)
Argentina, Secretaria de Marina, Servicio de Hidrografia Naval, 1961.
Operacion oceanografica, Vema - Canepa 1, Resultados preliminares.
Publico, H. 628:30 pp.

Chemistry oxygen (data only)
Argentina, Servicio de Hidrografia Naval, 1960.
Operacion Oceanografica Drake II. Resultados preliminares.
Servicio de Hidrografia Naval, Argentina, Publ., H. 14: numerous unnumbered pp.

Chemistry oxygen (data only)
Argentina, Servicio de Hidrografia Naval, 1960.
Operacion Oceanografiaa Malvinas (Resultados preliminares).
Servicio de Hidrografia Naval, Argentina, Publ., H. 606: numerous unnumbered pp.

Chemistry oxygen (data only)
Argentina, Servicio de Hidrografia Naval, 1959.
Operacion Oceanografica Cuenca. Resultados preliminares.
Servicio de Hidrografia Naval, Argentina, Publ., H. 607: numerous unnumbered pp.

chemistry oxygen (data only)
Argentina, Servicio de Hidrografia Naval, 1959.
Operacion Oceanografica Atlantico Sur. Resultados preliminares.
Servicio de Hidrografia Naval, Argentina, Publ., H. 608: numerous unnumbered pp.

Chemistry oxygen(data only)
Argentina, Servicio de Hidrografia Naval, 1959.
Operacion Oceanografica Drake. 1. Resultados preliminares.
Servicio de Hidrografia Naval, Argentina, Publ., H. 613: numerous unnumbered pp.

Chemistry oxygen(data only)
Argentina, Servicio de Hidrografia Naval, 1959.
Trabajos oceanograficos realizados en la campana Antartica, 1958/1959. Resultados preliminares.
Servicio de Hidrografia Naval, Argentina, Publ., H. 616:127 pp.

Chemistry oxygen (data only)
Argentina, Servicio de Hidrografia Naval, 1959.
Trabajos oceanograficos realizados en la campana Antartica, 1957-1958. Resultados preliminares.
Servicio de Hidrografia Naval, Argentina, Publ., H. 615: numerous unnumbered pp.

Chemistry oxygen (data only)
Argentina, Servicio de Hidrografia Naval, 1959.
Trabajos oceanograficos realizados en la campana Antertica, 1955-1956.
Servicio de Hidrografia Naval, Argentina, Publ., H. 620: numerous unnumbered pp.

Chemistry oxygen(data only)
Austin, T.S., 1957.
Summary, oceanographic and fishery data, Marquesas Islands area, August-September 1956, (EQUAPAC). USFWS Spec. Sci. Rept., Fish., No. 217:186 pp.

Chemistry oxygen (data only)
Austin, T.S., 1954.
Mid-Pacific oceanography III. Transequatorial waters, August-October 1951.
U.S.F.W.S. Spec. Sci. Rept. - Fish. No. 131: numerous pp. (unnumbered), 13 figs. (multilithed)

Australia, Commonwealth Scientific and Industrial Research Organization

Chemistry, oxygen (data only)
Australia, Commonwealth Scientific and Industrial Research Organization 1970.
Coastal investigations off Port Hacking, New South Wales in 1965.
Oceanogr. Sta. List Div. Fish. Oceanogr. 85: 124 pp.

Chemistry, oxygen (data only)
Australia, Commonwealth Scientific and Industrial Research Organization, 1969.
Oceanographical observations in the Indian Ocean in 1965 H.M.A.S. Diamantina Cruise Dm2/65. Div. Fish. Oceanogr. Oceanogr. Cruise Rept. 49: 57 pp. (multilithed)

Chemistry, oxygen (data only)
Australia Commonwealth Scientific and Industrial Research Organization 1969.
Oceanographical observations in the Pacific Ocean in 1965, H.M.A.S. Gascoyne Cruise G3/65.
Div. Fish. Oceanogr. Oceanogr. Cruise Rept. 44: 24 pp. (multilithed).

Chemistry, oxygen (data only)
Australia, Commonwealth Scientific and Industrial Research Organization, 1969.
Oceanographical observations in the Indian Ocean in 1964 H.M.A.S. Diamantina Cruise DM4/64. Div. Fish. Oceanogr. Oceangr. Cruise Rept. 38: 65 pp. (multilithed).

Chemistry oxygen (data only)
Australia, Commonwealth Scientific and Industrial Organization, 1968.
Coastal investigations off Port Hacking, New South Wales, in 1964.
Oceanogr. Sta. List, 84: 49 pp, (multilithed)

chemistry oxygen (data only)
Australia, Commonwealth Scientific and Industrial Organization, 1968.
Coastal investigations off Port Hacking, New South Wales, in 1961.
Oceanogr. Sta. List, 81: 55 pp. (multilithed)

chemistry oxygen (data only)
Australia, Commonwealth Scientific and Industrial Research Organization, 1968.
Investigations by F.V. Degei in Queensland waters in 1965.
Div.Fish.Oceanogr., Oceanogr.Sta.Lists,73: 49 pp. (multilithed).

Chemistry oxygen (data only)
Australia, Commonwealth Scientific and Industrial Research Organization, 1968.
Investigations by F.R.V.Marelda on the eastern Australian tuna grounds in 1965.
Div.Fish.Oceanogr., Oceanogr., Sta.List,72: 58 pp. (multilithed).

Chemistry oxygen (data only)
Australia, Commonwealth Scientific and Industrial Research Organization, 1968.
Investigations by F.R.V.Investigator on the South Australian tuna grounds in 1965.
Div.Fish.Oceanogr., Oceanogr.Sta.List,70:21pp. (multilithed).

Chemistry oxygen (data only)
Australia, Commonwealth Scientific and Industrial Research Organization, 1968.
Investigations by F.R.V. Marelda on the eastern Australian tuna grounds in 1964.
Oceanogr.Sta.List,Div.Fish.Oceanogr.,68: 91 pp. (multilithed).

Chemistry oxygen (data only)
Australia, Commonwealth Scientific and Industrial Research Organization, 1968.
Investigations by F.R.V. Investigator on the South Australian tuna grounds in 1964.
Oceanogr.Sta.List,Div.Fish.Oceanogr.,67: 78 pp. (multilithed).

Chemistry, oxygen (data only)
Australia Commonwealth Scientific and Industrial Organization, 1968.
Investigations by F.R.V. Marelda on the eastern Australian tuna grounds in 1963
Div. Fish. Oceanogr., Oceanogr. Sta. List, 65: 144 pp.

Chemistry oxygen (data only)
Australia, Commonwealth Scientific and Industrial Organization, 1968.
Investigations by F.R.V. Investigator on the south Australian tuna grounds in 1963.
Oceanogr. Data List, Div.Fish.Oceanogr., 64: 95 pp. (multilithed).

chemistry oxygen (data only)
Australia, Commonwealth Scientific and Industrial Organization 1968.
Investigations by F.R.V. Marelda on the eastern Australian grounds in 1962.
Div. Fish. Oceanogr., Oceanogr. List. 61: 135 pp.

chemistry, oxygen (data only)
Australia, Commonwealth and Scientific and Industrial Research Organization 1968.
Investigations by F.R.V. Investigator on the South Australian tuna grounds in 1969.
Oceanogr. Sta. List, Div. Fish. Oceanogr. 60: 23 pp.

chemistry oxygen (data only)
Australia, Commonwealth Scientific and Industrial Organization, 1968.
Investigations by F.R.V. Derwent Hunter on the eastern Australian tuna grounds in 1962.
Div. Fish. Oceanogr., Oceanogr. List 59: 74 pp.

chemistry oxygen (data only)
Australia, Commonwealth Scientific and Industrial Research Organization, 1968.
Investigations by F.R.V. Weerutta on the south Australian tuna grounds in 1961.
Oceanogr. Sta. List, Div. Fish. Oceanogr., 55: 27 pp. (multilithed).

chemistry oxygen (data only)
Australia, Commonwealth Scientific and Industrial Organization, 1968
Investigations by F.R.V. Derwent Hunter on the eastern Australian tuna grounds in 1961.
Div. Fish. Oceanogr., Oceanogr. List 54: 235 pp.

chemistry, oxygen (data only)
Australia, Commonwealth Scientific and Industrial Organization 1967.
Oceanographical observations in the Indian Ocean in 1965, H.M.A.S. Gascoyne, Cruise G 5/65.
Div. Fish. Oceanogr. Oceanogr. Cruise Rep. 46: 62 pp.

chemistry, oxygen (data only)
Australia, Commonwealth Scientific and Industrial Organization 1968.
Oceanographical observations in the Indian Ocean in 1965, H.M.A.S. Gascoyne, Cruise G 2/65.
Oceanogr. Cruise Rept. 43: 1-58.

chemistry, oxygen (data only)
Australia, Commonwealth Scientific and Industrial Research Organization 1968.
Oceanographical observations in the Pacific Ocean in 1964, H.M.A.S. Gascoyne, Cruise G 6/64.
Oceanogr. Cruise Rept. 42: 53 pp.

chemistry, oxygen (data only)
Australia Commonwealth Scientific and Industrial Research Organization 1968
Oceanographical observations in the Indian Ocean in 1964, H.M.A.S. Gascoyne, Cruise G5/64.
Oceanogr. Cruise Rept., Div. Fish. Oceanogr. 41: 52 pp.

chemistry, oxygen (data only)
Australia, Commonwealth Scientific and Industrial Research Organization 1968.
Oceanographical observations in the Indian Ocean in 1964, H.M.A.S. Diamantina, Cruise Dm 5/64.
Oceanogr. Cruise Rept., Div. Fish. Oceanogr. 40: 48 pp.

chemistry oxygen (data only)
Australia, Commonwealth Scientific and Industrial Research Organization, 1968.
Oceanographical observations in the Pacific Ocean in 1963, H.M.A.S. Gascoyne Cruise G3/63.
Oceanogr. Cruise Rept., Div. Fish. Oceanogr., 26: 134 pp.

chemistry oxygen (data only)
Australia, Commonwealth Scientific and Industrial Research Organization, 1967.
Oceanographical observations in the Pacific Ocean in 1964, H.M.A.S. Gascoyne Cruise G4/64.
Oceanogr. Cruise Rept., Div. Fish. Oceanogr. 39: 39 pp.

chemistry, oxygen (data only)
Australia, Commonwealth Scientific and Industrial Organization, 1967.
Oceanographical observations in the Indian Ocean in 1964, H.M.A.S. Diamantina, Cruise Dm 2/64.
Oceanogr. Cruise Rep. Div. Fish. Oceanogr. 36: 53 pp.

chemistry oxygen (data only)
Australia, Commonwealth Scientific and Industrial Organization, 1967.
Oceanographical observations in the Indian and Pacific oceans in 1964, H.M.A.S. Gascoyne, Cruise G3/64.
Oceanogr. Cruise Rep. Div. Fish. Oceanogr., 35: 40 pp.

chemistry - oxygen (data only)
Australia, Commonwealth Scientific and Industrial Research Organization, 1967.
Oceanographical observations in the Indian Ocean in 1964, H.M.A.S. Gascoyne, Cruise G 2/64.
Div. Fish. Oceanogr., Oceanogr. Cruise Rep. 34: 46 pp. (multilithed)

chemistry oxygen (data only)
Australia, Commonwealth Scientific and Industrial Research Organization, 1967.
Oceanographical observations in the Pacific Ocean in 1964, H.M.A.S. Gascoyne, Cruise G 1/64.
Div. Fish. Oceanogr., Oceanogr. Cruise Rep. 32: 66 pp (multilithed).

chemistry oxygen (data only)
Australia, Commonwealth Scientific and Industrial Organization, 1967.
Oceanographic observations in the Pacific Ocean in 1963, H.M.A.S. Gascoyne, Cruise G 5/63.
Div. Fish. Oceanogr. Cruise Rep. 31: 57 pp.

chemistry, oxygen (data only)
Australia, Commonwealth Scientific and Industrial Research Organisation, 1967.
Oceanographical observations in the Pacific Ocean in 1963, H.M.A.S. Gascoyne, Cruise G 4/63.
Div. Fish. Oceanogr., Oceanogr. Cruise Rep. 29: 64 pp. (multilithed).

chemistry oxygen (data only)
Australia, Commonwealth Scientific and Industrial Research Organiz-tion.
Oceanographic observations in the Indian Ocean in 1963, H.M.A.S. DIAMANTINA Cruise DM 3/63.
Div. Fish. Oceanogr., Oceanogr. Cruise Rept. No. 25: 147 pp.

chemistry oxygen (data only)
Australia, Commonwealth Scientific and Industrial Research Organization, 1967.
Oceanographic observations in the Indian Ocean in 1963, H.M.A.S. Gascoyne, Cruise G 2/63.
Div. Fish. Oceanogr., Oceanogr. Cruise Rep. 22: 51 pp.

chemistry, oxygen (data only)
Australia, Commonwealth Scientific and Industrial Research Organization 1967.
Oceanographical observations in the Indian Ocean in 1962, H.M.A.S. Diamantina, Cruise Dm 4/62.
Div. Fish. Oceanogr. Oceanogr. Cruise Rept. 20: 138 pp.

chemistry, oxygen (data only)
Australia, Commonwealth Scientific and Industrial Research Organization, 1967.
Oceanographical observations in the Pacific Ocean in 1962 H.M.A.S. Gascoyne Cruise G 5/62.
Oceanogr. Cruise Rep. Div. Fish. Oceanogr., 19: 71 pp.

chemistry, oxygen (data only)
Australia, Commonwealth Scientific and Industrial Organization 1967.
Oceanographical observations in the Pacific and Indian oceans in 1962 H.M.A.S. Gascoyne, Cruises G 2/62 and G 3/62.
Div. Fish. Oceanogr., Oceanogr. Cruise Rept. 16: 90 pp.

chemistry, oxygen (data only)
Australia, Commonwealth Scientific and Industrial Research Organization, 1967.
Oceanographical observations in the Pacific Ocean in 1962, H.M.A.S. Gascoyne, Cruise G 1/62.
Div. Fish. Oceanogr., Oceanogr. Cruise Rep. 13: 180 pp.

chemistry oxygen (data only)
Australia, Commonwealth Scientific and Industrial Research Organization, 1967.
Oceanographical observations in the Pacific Ocean in 1961, H.M.A.S. Gascoyne Cruise G 3/61.
Div. Fish. Oceanogr., Oceanogr. Cruise Rep., 12: 126 pp.

chemistry - oxygen (data only)
Australia, Commonwealth Scientific and Industrial Research Organization, 1966.
Oceanographical observations in the Indian Ocean in 1962 H.M.A.S. Gascoyne Cruise G 1/62.
Div. Fish. Oceanogr., Oceanogr. Cruise Rep. No. 17: 151 pp.

chemistry, oxygen (data only)
Australia Commonwealth Scientific and Industrial Organization 1965.
Oceanographical observations in the Indian Ocean in 1963, H.M.A.S. Diamantina, Cruise Dm 2/63.
Div. Fish. Oceanogr. Oceanogr. Cruise Rept. No. 24: 153 pp.

chemistry oxygen (data only)
Australia, Commonwealth Scientific and Industrial Organization, 1965.
Oceanographical observations in the Indian Ocean in 1963, H.M.A.S. Diamantina, Cruise Dm 1/63.
Div. Fish. Oceanogr., Oceanogr. Cruise Rept., No. 23: 175 pp.

Chemistry
oxygen (data only)
Australia, Commonwealth Scientific and Industrial Research Organization, 1965.

Oceanographic observations in the Indian Ocean in 1963, H.M.A.S. Gascoyne, Cruise G 1/63.
Div. Fish. and Oceanogr., Oceanogr. Cruise Rept., No. 21: 135 pp.

Chemistry oxygen (data only)
Australia, Commonwealth Scientific and Industrial Research Organization, 1964.
Oceanographical observations in the Indian Ocean in 1962, H.M.A.S. Diamantina, Cruise Dm 2/62.
Oceanogr. Cruise Rept., Div. Fish. and Oceanogr. No. 15:117 pp.

Chemistry oxygen (data only)
Australia, Commonwealth Scientific and Industrial Research Organization, 1964.
Oceanographical observations in the Indian Ocean in 1961, H.M.A.S. Diamantina, Cruise Dm 3/61.
Div. Fish. and Oceanogr., Oceanogr. Cruise Rept. No. 11:215 pp.

Chemistry oxygen (data only)
Australia, Commonwealth Scientific and Industrial Research Organization, 1964.
Oceanographic observations in the Indian Ocean in 1962, H.M.A.S. Diamantina, Cruise Dm 1/62.
Oceanogr. Cruise Rept., Div. Fish. Oceanogr., No. 10: 128 pp.

Chemistry oxygen (data only)
Australia, Commonwealth Scientific and Industrial Research Organization, 1963.
Coastal hydrological investigations in the New South Wales tuna fishing area, 1963.
Div. Fish. and Oceanogr., Oceanogr. Sta. List, 53:81 pp.

A.D. Crooks, compiler

Chemistry oxygen (data only)
Australia, Commonwealth Scientific and Industrial Research Organization, 1963.
Coastal investigations at Port Hacking, New South Wales, 1960.
Div. Fish. and Oceanogr., Oceanogr. Sta. List, No. 52:135 pp.

Chemistry
oxygen (data only)
Australia, Commonwealth Scientific and Industrial Research Organization, 1963
Coastal hydrological investigations in eastern Australia, 1960.
Div. Fish. and Oceanogr., Oceanogr. Sta. List, 51: 46 pp.

Chemistry
oxygen (data only)
Australia, Commonwealth Scientific and Industrial Research Organization, 1963
Oceanographical observations in the Indian Ocean in 1961, H.M.A.S. Diamantina Cruise Dm 2/61.
Oceanogr. Cruise Rept., Div. Fish and Oceanogr. No. 9: 155 pp., 14 figs.

Chemistry oxygen (data only)
Australia, Commonwealth Scientific and Industrial Research Organization, 1963.
Oceanographical observations in the Pacific Ocean in 1961, H.M.A.S. Gascoyne, Cruise G 1/61.
Oceanogr. Cruise Rept., Div. Fish. and Oceanogr. No. 8:130 pp., 12 figs.

Chemistry
oxygen (data only)
Australia, Commonwealth Scientific and Industrial Research Organization, Division of Fisheries and Oceanography, 1963
Oceanographical observations in the Indian Ocean in 1960, H.M.A.S. Diamantina, Cruise Dm 2/ 60.
Oceanographical Cruise Report No. 3: 347 pp.

Chemistry
oxygen (data only)
Australia, Commonwealth Scientific and Industrial Organization, 1962.
Oceanographic observations in the Indian Ocean in 1961. H.M.A.S. Diamantina.
Div. Fish Oceanogr., Cruise DM 1/61:88 pp.

Chemistry
oxygen (data only)
Australia, Commonwealth Scientific and Industrial Research Organization, Division of Fisheries and Oceanography, 1962.
Oceanographic observations in the Pacific Ocean in 1960, H.M.A.S. Gascoyne, Cruises G 1/60 and G 2/60.
Oceanographical Cruise Report No. 5:255 pp.

Chemistry oxygen (data only)
Australia, Commonwealth Scientific and Industrial Organization, 1962.
Oceanographical observations in the Indian Ocean in 1960, H.M.A.S. DIAMANTINA Cruise Dm 3/60.
Oceanogr. Cruise Rept., Div. Fish. and Oceanogr., No. 4:39 pp.

Chemistry oxygen (data only)
Australia, Commonwealth Scientific and Industrial Organization, 1962.
Oceanographical observations in the Indian Ocean in 1960, H.M.A.S. Diamantina, Cruise Dm 1/60.
Oceanogr. Cruise Rept., Div. Fish. and Oceanogr. No. 2:128 pp.

Chemistry oxygen (data only)
Australia, Commonwealth Scientific and Industrial Organization, 1962.
Oceanographical observations in the Indian Ocean in 1959, H.M.A.S. Diamantina, Cruise Dm 1/59 and 2/59.
Oceanogr. Cruise Rept., Div. Fish. and Oceanogr., No. 1:134 pp.

Chemistry
oxygen (data only)
Australia, Commonwealth Scientific and Industrial Research Organization, 1961
Coastal investigations at Port Hacking, New South Wales, 1959.
Div. Fish. & Oceanogr., Oceanogr. Sta. List, No. 47: 135 pp.

Chemistry
oxygen (data only)
Australia, Commonwealth Scientific and Industrial Research Organization, 1961
Oceanic investigations in Eastern Australian waters, F.R.V. "Derwent Hunter", 1959.
Div. Fish. and Oceanogr., Oceanogr. Sta. List, 48: 84 pp.

Chemistry
oxygen (data only)
Australia, Commonwealth Scientific and Industrial Research Organization, 1961
Coastal hydrological investigations in the New South Wales tuna fishing area, 1959.
Div. Fish. & Oceanogr., Oceanogr. Sta. List, 46: 132 pp.

Chemistry
oxygen (data only)
Australia, Commonwealth Scientific and Industrial Research Organization, Division of Fisheries and Oceanography, 1960.

F.RV. "Derwent Hunter" Scientific reports of Cruises 1-8, 1958 57 pp. (Mimeographed)

Chemistry
oxygen (data only
Australia, Commonwealth Scientific and Industrial Research Organization, 1960

Coastal hydrological investigations in eastern Australia, 1959.
Div. Fish. and Oceanogr., Oceanogr. Sta. List, 45: 24 pp.

Chemistry oxygen (data only)
Australia, Commonwealth Scientific and Industrial Organization, 1960

Coastal hydrological investigations in south-eastern Australia, 1958. Oceanogr. Sta. List. Div. Fish. and Oceanogr., 60 pp.

Chemistry oxygen (data only)
Australia, Marine Biological Laboratory, Cronulla, 1960.
F.R.V. "Derwent Hunter", scientific report of --- cruises 10-20/58 -------.
C.S.I.R.O., Div. Fish. & Oceanogr., Rept., 30: 53 pp., numerous figs. (mimeographed).

For complete "title" see author card.

Chemistry oxygen (data only)
Australia, Commonwealth Scientific and Industrial Research Organization, 1960.
Oceanic investigations in eastern Australia, H.M.A. Queenborough, Quickmatch and Warrego, 1958.
Div. Fish. and Oceanogr., Oceanogr. Sta. List, 43:57 pp.
Ships

Chemistry oxygen (data only)
Australia, Commonwealth Scientific and Industrial Research Organization, 1960.
Coastal investigations at Port Hacking, New South Wales, 1958.
Oceanogr. Sta. List, Div. Fish. & Oceanogr., 42: 99 pp.

Chemistry oxygen (data only)

Australia, Commonwealth Scientific and Industrial Organization, 1960

Oceanic investigations in eastern Australian waters, F.R.V. Derwent Hunter, 1958. Oceanogr. Sta. List., Div. Fish. and Oceanogr. No. 41: 232 pp.

Chemistry oxygen (data only)
Australia, C.S.I.R.O., Division of Fisheries and Oceanography, Marine Biological Laboratory, Cronulla, 1959.
F.R.V. "Derwent Hunter", Scientific report of Cruise DH9/57, Aug. 19-25, 1957; Cruise DH10/57, Sept. 4-11, 1957; Cruise DH11/57, Sept. 18-21, 1957; Cruise DH12/57, Sept. 26-Oct. 11, 1957.
CSIRO, Div. Fis. & Ocean. Rept. No. 20:20 pp. (mimeographed)

Chemistry
oxygen (data only)
Australia, C.S.I.R.O., Division of Fisheries and Oceanography, 1959.

Hydrological investigations from F.R.V. Derwent Hunter, 1957.
Oceanogr. Sta. List, No. 37:96 pp.

(compiled by D. J. Rochford)

Chemistry oxygen (data only)
Australia, C.S.I.R.O., Division of Fisheries and Oceanography, 1959.
Scientific reports of a cruise on H.M.A. Ships "Queenborough" and "Quickmatch", March 24-April 26, 1958. Rept. 24: 24 pp. (mimeographed)

Chemistry oxygen (data only)
Australia, Division of Fisheries and Oceanography C.S.I.R.O., 1959.
Coastal hydrological investigations in the New South Wales tuna fishing area, 1958.
Oceanogr. Sta. List, 38:96 pp.

Chemistry
oxygen (data only)

Australia, C.S.I.R.O., Division of Fisheries, 1959

Coastal hydrological investigations at Eden, New South Wales, 1957.
Oceanogr. Sta. List, 35: 36 pp.

Chemistry oxygen (data only)
Australia, C.S.I.R.O., Division of Fisheries, 1959
Coastal hydrological investigations at Port Hacking, New South Wales, 1957.
Oceanogr. Sta. List, 34:72 pp.

Chemistry oxygen (data only)
Australia, C.S.I.R.O., Division of Fisheries and Oceanography, 1958/1959.
FVR "Derwent Hunter. Rept. No. 19:16 pp.
No. 21:16 pp.
(mimeographed)

Chemistry oxygen (data only)
Australia, Commonwealth Scientific and Industrial Research Organization, 1957.
Estuarine hydrological investigations in eastern and south-western Australia, 1957.
Oceanogr. Sta. List, Div. Fish. and Oceanogr., 32:170 pp.

Chemistry oxygen (data only)
Australia Commonwealth Scientific and Industrial Research Organization, 1957.
Onshore and oceanic hydrological investigations in eastern and southwestern Australia, 1956.
Oceanogr. Sta. List, Div. Fish. & Oceanogr. 30:79pp.

Chemistry oxygen(data only)
Australia, Commonwealth Scientific and Industrial Research Organization, 1957.
Estuarine hydrological investigations in eastern and south-western Australia, 1955.
Ocean. Station List, 29:93 pp.

Chemistry oxygen (data only)
Australia, Commonwealth Scientific and Industrial Research Organization, 1957.
Onshore and oceanic hydrological investigations in eastern and southwestern Australia, 1955.
Oceanogr. Sta. List, 27:145 pp.

Chemistry oxygen (data only)
Australia, Commonwealth Scientific and Industrial Research Organization, 1956.
Onshore hydrological investigations in eastern and south-western Australia, 1954. Compiled by D. R. Rochford.
Ocean. Sta. List., Invest. Div. Fish., 24:119 pp.

Chemistry oxygen (data only)
Australia, Commonwealth Scientific and Industrial Research Organization, 1954.
Onshore hydrological investigations in eastern and south-western Australia, 1953.
Ocean. Sta. List, Invest. Div. Fish. 18:64 pp.

Chemistry oxygen(data)
Australia, Commonwealth Scientific and Industrial Research Organization, 1953.
Onshore hydrological investigations in eastern and south-western Australia, 1951. Vol. 14:64 pp.

Chemistry oxygen (data only)
Australia, Commonwealth Scientific and Industrial Research Organization, 1952.
Records of twenty-four hourly hydrological observations at Shell Point, Georges River, New South Wales, 1942-1950. (Compiled by D.R. Rochford). Ocean. Sta. List, Invest. Fish. Div., 10:134 pp.

Chemistry oxygen (data)
Australia, Commonwealth Scientific and Industrial Research Organization, Melbourne, 1952.
Oceanographical station list of investigations made by the Fisheries Division, C.S.I.R.O.
Vol. 5. Estuarine hydrological investigations in eastern Asutralia, 1940-50:150 pp.
Vol. 6. Ibid.:137 pp.
Vol. 7. Ibid.:139 pp.
Vol. 8. Hydrological investigations in south-western Australia, 1944-50:152 pp.
Vol. 10. Records of twenty-four hourly hydrological observations at Shell Point, Georges River, New South Wales, 1942-50:134 pp

Chemistry oxygen(data)
Australia, Commonwealth Scientific and Industrial Research Organization, 1951.
Oceanographic station list of investigations made by the Fisheries Division, C.S.I.R.O.
Vol. 4. Onshore hydrological investigations in eastern Australia, 1942-50:114 pp.

B

Chemistry oxygen (data only)
Ballester, Antonio, 1965 (1967).
Tablas hidrograficas.
Memoria Soc.Cienc.nat.La Salle, 25(70/71/72):41-137.

Chemistry oxygen (data only)
Bang, N.D., and F.C. Pearce 1970.
Hydrological data Agulhas Current project, March 1969, R.V. Thomas B. Davie.
Data Rept. Inst. Oceanogr. Univ. Cape Town, 4: 26 pp. Also in: Collected Repr. Oceanogr. Inst. Univ Cape Town 9.

Chemistry oxygen (data only)
Berrit, G.R., R. Gerard, L. Lemasson, J.P. Rebert et L. Vercesi, 1968.
Observations océanographiques exécutées en 1967.
1. Stations hydrologiques; observations de surface at de fond; stations côtières.
Doc. sci. provis., Centre Recherches océanogr., Côte d'Ivoire, 026:133pp (mimeographed)

Chemistry oxygen (data only)
Berrit, G.R., R. Gerard and L. Vercesi, 1968.
Observations océanographiques executées en 1966.
II. Stations côtières d'Abidjan, Lome et Cotonou - observations de surface et de fond.
ORSTOM Centre Rech. Océanogr., Côte d'Ivoire, Document scient. provis. 017: 71 pp.

Chemistry oxygen (data only)
Blackburn, M., R.C. Griffiths, R.W. Holmes and W.H. Thomas, 1962.
Physical, chemical and biological observations in the eastern tropical Pacific Ocean: three cruises to the Gulf of Tehuantepec, 1958-1959.
U.S.F.W.S. Spec. Sci. Rept., Fish., No. 420: 170 pp.

chemistry, oxygen (data only)
Brasil, Diretoria de Hidrografia e Navegação 1970.
IV, V e VI Comissões oceanograficas no Almirante Saldanha (29/5 a 4/6/1957), (5/10 a 17/10/1957) (26/11 a 4/12/1957).
DG20-IV: 149 pp.

oxygen (data only)
Brasil, Diretoria de Hidrografia e Navegação, 1969
III Comissão oceanográfica no Almirante Saldanha (20/3 a 16/4/1957), DG 20-III: 73 pp.

Chemistry oxygen (data only)
Brasil, Diretoria de Hidrografia e Navegação, 1969.
II Comissão Oceanográfica no Almirante Saldanha (15 A 28/2/1957).
DG20-II:60 pp.

Chemistry oxygen (data only)
Brasil, Diretoria de Hidrografia e Navegacao, 1968.
XXXV Comissao Oceanografica "Operacao Norte/Nordeste 1" noc Almirante Saldanha (14/9 a 16/12/1967.)
DG 26-XI: 600 pp.

Chemistry oxygen (data only)
*Brazil, Diretoria de Hidrografia e Navegação, 1968.
XXXI Comissao oceanografica noc Almirante Saldanha (14/11 a ;6/12/66, DG26-VIII:249 pp.

chemistry, oxygen (data only)
Brazil, Diretoria de Hidrografia e Navegação 1967.
XXXII Comissão Oceanográfica Na Almirante Saldanha (14/3 a 3/5/67), DG 26-X: 411 pp.

Chemistry oxygen (data only)
Brasil, Marinha do Brasil, Diretoria de Hidrografia e Navegação, 1963
Operação, "TRIDENTE I", Estudo das condições oceanográficas entre o Rio de Janeiro e o Rio da Prata, durante o inverno (Agosto-Septembro) ano de 1962.
DG-06-XV:unnumbered pp. (mimeographed).

Chemistry oxygen (data only)
Brasil, Diretoria de Hidrografia e Navegação, Marinha do Brasil, 1962
Estudo das condições oceanográficas entre o Rio de Janeiro e o Rio da Prata, durante o outono (Maio de 1962).
DG-06-XIV:unnumbered pp., charts, (mimeographed)

Chemistry oxygen (data only)
Brasil, Diretoria de Hidrografia e Navegacao 1961
Estudo das condições oceanograficas nas proximidades do Rio de Janeiro durante o mês de Dezembro.
DG-06-XIII: mimeographed sheets.

Chemistry oxygen (data only)
Brasil, Diretoria de Hidrografia e Navegacao, Marinha do Brasil, 1961
Estudo das condições oceanográficas sobre a plataforma continental, entre Cabo-Frio e a Ponta do Boi, durante o mes de setembro (transição inverno-primervera).
DG-06-XII: unnumbered pp. (mimeographed).

Chemistry oxygen (data only)
Brasil, Marinha do Brasil, Diretoria de Hidrografia e Navegação, 1960.
Estudo das condições oceanográficas sobre a plataforma continental, entre Cabo-Frio e Vitoria, durante o outono (abril-maio).
DG-06-X(junho): unnumbered pp. (mimeographed).

Chemistry oxygen (data only)
Brasil, Marinha do Brasil, Diretoria de Hidrografia e Navegação, 1959.
Levantamento oceanografico da costa nordeste.
DG-06-IX:unnumbered pp. (mimeographed).

Chemistry oxygen (data only)
Brazil, Diretoria de Hidrografia e Navegacao, 1958.
Ano Geogisico Internacional , Publ. DG-06-VII: mimeographed sheets.

Chemistry oxygen (data only)
Brasil, Diretoria de Hidrografia e Navegacao, 1957.
Ano Geofisico Internacional. Publicacao DG-06-V: 3 pp., 18 figs., 3 pp. of tables.

Chemistry oxygen(data only)
Brazil, Diretoria de Hidrografia e Navegacao, 1957.
Ano Geofisico Internacional. Publ. DG-06-III and IV. (mimeographed)

Chemistry oxygen (data)
Brazil, Diretoria de Hidrografia e Navegacão, 1957.
Ano Geofisico Internacional. Publicacão DG-06-II mimeographed.

Chemistry, oxygen (data)
Brazil, Diretoria de Hidrografia e Navegacao, 1957.
Ano Geofisico Internacional, Publicacao DG-06-VI: unnumbered mimeographed pages.

Chemistry, oxygen (data only)
Brucks, J.T., M.C. Ingham and T.D. Leming 1968.
Oceanic conditions off Sierra Leone, 10 February to 2 March 1965 (part of Geronimo cruise 5).
Data Rept, U.S. Dept. Interior, Bur. Comm. Fish. 28: 1 film (microfiche)

Chemistry, oxygen (data only)
Brucks, John T., Merton C. Ingham and Thomas D. Leming 1968
Oceanic conditions in the northwestern Gulf of Guinea, 14 to 30 March 1965 (part of Geronimo cruise 5).
Data Rept U.S. Dept. Interior, Bur. Comm. Fish. 27: 1 film (microfiche)

Chemistry oxygen, (data)
Brujewicz, S.W., 1957.
On certain chemical features of waters and sediments in north-west Pacific. Proc. UNESCO Symp., Phys. Ocean., Tokyo, 1955:277-292.

oxygen(data)
Bruneau, L., N.G. Jerlov and F.F. Koczy, 1953.
Physical and chemical methods.
Repts. Swedish Deep-sea Exped., 1947-1948, Phys. Chem., 3(4): 101-112, i-iv, 7 textfigs., 2 tables

Chemistry - oxygen (data only)
Buljan, Miljenko, and Mira Zore-Armanda 1966.
Hydrographic data on the Adriatic Sea collected in the period from 1952 through 1962.
Acta adriat. 12: 438 pp.

Chemistry oxygen (data only)
Buljan, M., and M. Marinkovic, 1956.
Some data on hydrography of the Adriatic (1946-1951). Acta Adriatica, 7(12):1-55.

C

Chemistry oxygen (data only)
*Callame, B., 1968.
Étude d'hydrologie côtière à la Rochelle-la Pallice (Charante-Maritime) de 1964 à 1967.
Trav. Cent. Rech. Étud. océanogr., n.s., 7(3/4):43-52.

Chemistry oxygen (data Only)
Callaway, Richard J., and James W. McGary, 1959.
Northeastern Pacific albacore survey. 2. Oceanographic and meteorological observations.
USFWS Spec. Sci. Rept., Fish., No. 315:133 pp.

oxygen (data only)
Canada, Canadian Oceanographic Data Centre, 1969
Davis Strait and Northern Labrador Sea, August 27 to October 22, 1965.
1969 Data Record Series 7: 203 pp (multilithed)

Chemistry, oxygen (data only)
Canada, Canadian Oceanographic Data Centre 1969.
Ocean Weather Station 'P' North Pacific Ocean, December 3, 1967, to February 25, 1968.
Data Record Series 6: 116 pp.
(multilithed)

Chemistry, oxygen (data only)
Canada, Canadian Oceanographic Data Centre 1969.
East Greenland, Denmark Strait and Irminger Sea, January 16 to April 5, 1967.
1969 Data Rec. Ser. 4: 158 pp. (multilithed).

Chemistry oxygen (data only)
Canada, Canadian Oceanographic Data Centre, 1969.
Labrador and Irminger seas, March 12 to May 12, 1966.
1969 Data Record Ser. 1: 152 pp. (multilithed)

Chemistry oxygen (data only)
Canada, Canadian Oceanographic Data Centre, 1968.
Scotian shelf January 20 to January 27, 1968.
1968 Data Record Series, 7: 49 pp. (multilithed).

Chemistry oxygen (data only)
Canada, Canadian Oceanographic Data Center, 1968
Ocean Weather Station 'P', North Pacific Ocean, April 7 to July 6, 1967.
1968 Data Record Ser. 5:140 pp. (multilithed)

Chemistry, oxygen (data only)
Canada, Canadian Oceanographic Data Centre 1968.
Gulf of St. Lawrence, November 16 to November 27, 1967.
1968 Data Record Ser. 3: 81 pp. (multilithed).

Chemistry oxygen (data only)
Canada, Canadian Oceanographic Data Centre, 1967.
Ocean Weather Station 'P' North Pacific Ocean, October 25, 1966 to January 9, 1967.
1967 Data Record Ser. 8: 111 pp.

Chemistry oxygen (data only)
Canada, Canadian Oceanographic Data Centre, 1966.
Ocean Weather Station 'P' North Pacific Ocean, March 3 to June 2 1966.
1966 Data Record Series 11: 168 pp.

oxygen (data only)
Canada, Canadian Oceanographic Data Centre, 1967.
Ocean Weather Station "P", August 5 to October 31, 1966.
1967 Data Record Ser., 6:164 pp. (mimeographed).

Chemistry, oxygen (data only)
Canada, Canadian Oceanographic Data Centre 1966.
Ocean Weather Station 'P', North Pacific Ocean, December 11, 1965, to March 9, 1966.
1966 Data Record Series, No. 8, 144 pp (multilithed).

Chemistry, oxygen (data only)
Canada, Canadian Oceanographic Data Center, 1966.
Gulf Stream between Cape Cod and Bermuda, November 16 to December 15, 1964.
1966 Data Record Series, No. 7:59 pp. (multilithed).

Chemistry, oxygen (data only)
Canada, Canadian Oceanographic Data Centre 1966.
Ocean Weather Station 'P', North Pacific Ocean, September 17 to December 15, 1965.
1966 Data Record Series, No. 6: 170 pp (multilithed).

Chemistry oxygen (data only)
Canadian Oceanographic Data Centre, 1966.
Ocean Weather Station "P" North Pacific Ocean, April 17 to June 3, 1965.
1966 Data Record Ser. No. 3:150pp. (multilithed)

Chemistry oxygen (data only)
Canada, Canadian Oceanographic Data Centre, 1966.
Arctic, Hudson Bay and Hudson Strait, August 5 to October 4, 1962.
1966 Data Record Series, No. 4:247 pp.

Chemistry oxygen (data only)
Canada, Canadian Oceanographic Data Centre, 1966.
Western North Atlantic and Caribbean Sea, February 1 to February 27, 1965.
1966 Data Record Ser., No. 2:78 pp. (multilithed)

Chemistry oxygen (data only)
Canada, Canadian Oceanographic Data Centre, 1966
Arctic 1961, August 2 to October 12, 1961.
1966 Data Record Series, 322 pp.

Chemistry oxygen (data only)
Canada, Canadian Oceanographic Data Centre, 1965.
Data Record, Grand Banks to the Azores, June 1 to July 15, 1964.
1965 Data Record Series, No. 13:221 pp. (mimeographed).

Chemistry, oxygen (data only)
Canada, Canadian Oceanographic Data Centre 1965.
North Atlantic, east of Nova Scotia, south of the Grand Banks, March 5 to August 10, 1962.
1965 Data Record Series, No. 10: 226 pp.

Chemistry oxygen (data only)
Canada, Canadian Oceanographic Data Centre, 1965.
Data Record Gulf of St. Lawrence, July 23 to August 23, 1964.
1965 Data Record Ser., No. 9:262 pp. (multilith)

Chemistry oxygen (data only)
Canadian Oceanographic Data Centre, 1965.
Data record, Saguenay and Gulf of St. Lawrence, August 19 to August 30, 1963.
1965 Data Record Series, No. 12:123 pp. (multilithed).

Chemistry oxygen (data only)
Canada, Canadian Oceanographic Data Centre, 1965.
Baffin Bay, Smith Sound to Strait of Belle Isle September 4 to October 24, 1964.
1965 Data Record Ser., No. 11:165 pp.

Chemistry oxygen (data only)
Canada, Canadian Committee on Oceanography, 1965.
Ocean Weather Station "P", North Pacific Ocean, May 16 to August 12, 1964.
Canadian Oceanographic Data Centre, 1965 Data Record Series, No. 3:112 pp. (Unpublished manuscript).

chemistry, oxygen (data only)
Conseil, Permanent International pour l'Exploration de la Mer, 1968.
ICES Oceanographic Data Lists 1962(3): 27 pp.

chemistry oxygen (data only)
Conseil Permanent International pour l'Exploration de la Mer, 1968.
ICES Oceanographic Data Lists, 1962, 1:153 pp., 2:85 pp.

chemistry oxygen (data only)
Conseil International pour l'Exploration de la Mer, 1968.
ICES Oceanographic Data Lists, 1961, 10: 217 pp. (multilithed).

chemistry oxygen (data only)
Conseil International pour l'Exploration de la Mer, 1968.
ICES Oceanographic Data Lists 1961, 8:101 pp. (multilithed).

chemistry oxygen (data only)
Conseil Permanent International pour l'Exploration de la Mer, 1968.
ICES Oceanographic data lists 1961, No.5: 247 pp.

chemistry oxygen (data only)
Conseil Permanent International pour l'Exploration de la Mer, 1967.
ICES Oceanographic data lists, 1961, 4: 124 pp. (multilithed).

chemistry oxygen (data only)
Conseil Permanent International pour l'Exploration de la Mer, 1968
ICES Oceanographic Data Lists, 1960, 11: 178 pp.

chemistry oxygen (data only)
Conseil International pour l'Exploration de la Mer. 1968.
Cooperative synoptic investigation of the Baltic 1964. Sweden.
ICES Oceanogr. Data Lists 4:181 pp. (mimeographed)

chemistry, oxygen (data only)
Conseil International pour l'Exploration de la Mer, 1968.
Cooperative synoptic investigation of the Baltic 1964. 1. Finland.
ICES Oceanogr. Data lists: 82 pp. (multilithed).

chemistry, oxygen (data only)
Conseil, International pour l'Exploration de la Mer, 1968.
Cooperative synoptic investigation of the Batlic 1964. 3. Poland.
ICES Oceanogr., Data Lists: 266 pp. (multilitned).

chemistry, oxygen (data only)
Conseil Permanent International pour l'Exploration de la Mer 1966.
ICES oceanographic data lists 1961 (3): 166 pp.

chemistry oxygen (data only)
Conseil Permanent International pour l'Exploration de la Mer, 1967.
ICES Oceanographic data lists, 1960, No.8:227 pp. (multilithed).

chemistry, oxygen (data only)
*Conseil Permanent International pour l'Exploration de la Mer, 1967.
ICES oceanographic data lists, 1960, No. 7: 270 pp (multilithed).

chemistry oxygen (data only)
Conseil Permanent International pour l'Exploration de la Mer, 1967.
ICES Oceanographic data lists, 1960, No. 6: 295 pp. (multilithed).

chemistry, oxygen (data only)
Conseil Permanent International pour l'Exploration de la Mer 1967.
ICES oceanographic data lists 1959, No. 9: 235 pp.

chemistry oxygen (data only)
Conseil Permanent International pour l'Exploration de la Mer, 1967.
ICES Oceanographic data lists, 1958, 12: 111 pp. (multilithed).

chemistry, oxygen (data only)
Conseil Permanent International pour l'Exploration de la Mer 1967.
ICES oceanographic data lists, 1957, No. 9: 199 pp.

chemistry oxygen (data only)
Conseil, Permanent International pour l'Exploration de la Mer.
ICES oceanographic data lists, 1959, 7:203 pp. (mimeographed).

chemistry, oxygen (data only)
Conseil International pour l'Exploration de la Joint Skagerak Expedition 1966. 1. Oceanographic stations, temperature-salinity-oxygen content. 2. Oceanographic stations, chemical observations.
ICES Oceanogr. Data lists: 250 pp; 209 pp.

chemistry, oxygen (data only)
Conseil Permanent International pour l'Exploration de la Mer, 1966.
ICES Oceanographic Data Lists, 1961, No.2:177 pp (multilithed).

chemistry oxygen (data only)
Conseil Permanent International pour l'Exploration de la Mer, 1966.
ICES Oceanographic data lists, 1960, No.3:118 pp.
ICES Oceanographic data lists, 1960, No.4:190 pp.

chemistry oxygen (data only)
Conseil Permanent International pour l'Exploration de la Mer, 1966.
ICES Oceanographic data lists, 1960, No.2:166 pp.

chemistry oxygen, (data only)
Conseil Permanent International pour l'Exploration de la Mer, Service Hydrographique, 1966.
ICES Oceanographic data lists, 1957(8):75 pp.
ICES Oceanographic data lists, 1958(10):72 pp.
ICES Oceanographic data lists, 1960(1):198 pp.

chemistry oxygen, (data only)
Conseil Permanent International pour l'Exploration de la Mer. 1966.
ICES Oceanographic data lists, 1959, No.6:202pp.

chemistry oxygen (data only)
Conseil Permanent International pour l'Exploration de la Mer, 1965.
ICES Oceanographic Data Lists, 1959, No.4: 224 pp
1959, No.5: 158 pp

chemistry, oxygen (data only)
Conseil Permanent International pour l'Exploration de la Mer 1966.
ICES oceanographic data lists, 1958; No. 9: 175 pp.

chemistry oxygen (data only)
Conseil Permanent International pour l'Exploration de la Mer, 1965.
ICES oceanographic data lists, 1958, No. 6:199pp.

chemistry, oxygen (data only)
Conseil Permanent International pour l'Exploration de la Mer, 1965.
ICES oceanographic data lists, 1958, No. 7:192pp.
No. 8:286pp.

chemistry oxygen (data only)
Conseil Permanent International pour l'Exploration de la Mer, 1965.
ICES oceanographic data lists, 1958, No. 5:284 pp.

chemistry, oxygen (data only)
Conseil Permanent International pour l'Exploration de la Mer, 1965.
ICES oceanographic data lists 1952, No. 3: 214 pp.

chemistry oxygen (data only)
Conseil Permanent International pour l'Exploration de la Mer, 1964.
ICES oceanographic data lists, 1957, No. 3:167pp.
No. 4:178pp.
No. 5:255pp.
No. 6:160pp
1958, No. 2:157pp.
No. 3:174pp.
No. 4:241pp.
1959, No. 1:201pp.

chemistry oxygen (data only)
Conseil Permanent International pour L'Exploration de la Mer, Service Hydrographique, 1963.
ICES oceanographic data lists, 1958(1):1-259.

chemistry oxygen (data only)
Conseil Permanent International pour l'Exploration de la Mer, 1963.
ICES Oceanogr. Data Lists, 1957(2):353 pp. (Multilithed).

chemistry oxygen (data only)
Conseil Permanent International pour l'Exploration de la er, 1963.
ICES oceanographic data lists, 1957 (1):277 pp.

chemistry oxygen (data only)
Conseil Permanent International pour l'Exploration de la Mer, 1957.
Bulletin Hydrographique pour l'année 1953:167 pp.

chemistry oxygen (data only)
Conseil Permanent International pour l'Exploration de la Mer, 1956.
Bulletin hydrographique pour l'année 1952:164 pp.

chemistry oxygen (data only)
Great Britain, Discovery Committee, 1957.
Station list, 1950-1951.
Discovery Rept. 28:300-398.

chemistry oxygen (data only)
Great Britain, Discovery Committee, 1955.
Station list, R.R.S. "William Scoresby".
Discovery Repts., 26:212-258, Pls. 11-12.

chemistry, oxygen (data only)
Great Britain, Discovery Committee 1953.
Station list, R.R.S. "William Scoresby" 1950.
Discovery Repts 26:211-258, Pls. 11-12.

Chemistry oxygen (data only)
Great Britain, Discovery Committee, 1949.
Station list, R.R.S. William Scoresby, 1931-1938.
Discovery Repts., 25:143-280, Pls. 34-37.

Chemistry-oxygen (data only)
Great Britain, Discovery Committee 1947.
Station list, 1937-1939.
Discovery Repts 24:198-422, Pls. 4-6.

Chemistry oxygen (data only)
Great Britain, Discovery Committee, 1945.
Station list, 1935-1937.
Discovery Rept., 24:3-196.

Chemistry oxygen(data only)
Great Britain, Discovery Committee, 1942.
Station list, 1933-1935. Discovery Repts., 22:3-196, Pls. 1-4.

Chemistry oxygen (data only)
Great Britain, Discovery Committee, 1932.
Station List. Discovery Repts. 4:3-230, Pls. 1-5.

Chemistry oxygen (data only)
Great Britain, Discovery Committee, 1931.
Station list. Discovery Repts. 3:3-132, Pls. 1-10.

chemistry oxygen (data only)
Great Britain, Discovery Committee, 1929.
Station list. Discovery Repts. 1:3-140.

chemistry oxygen (data only)
Great Britain, Fisheries Laboratory, Lowestoft, 1957?
Research Vessel 'Ernest Holt', hydrographical observations, 1951:unnumbered pages.

chemistry oxygen (data only)
Great Britain, Fisheries Laboratory, Lowestoft, 1957.
Research Vessel 'Ernest Holt', hydrographical observations, 1955:86 pp.

Chemistry oxygen (data only)
Great Britain, Fisheries Laboratory, Lowestoft, 1956.
Research Vessel "Ernest Holt". Hydrographical observations, 192 pp. (multilithed).

chemistry oxygen (data only)
Great Britain, Fisheries Laboratory, Lowestoft, 1956.
Research Vessel, 'Ernest Holt', hydrographical observations, 1950:150 pp.

chemistry oxygen(data only)
Great Britain, Fisheries Laboratory, Lowestoft, 1954.
Research Vessel "Ernest Holt", hydrographical observations, 130 pp.

chemistry Oxygen(data only)
Great Britain, Fisheries Laboratory, Lowestoft, 1952.
Research vessel, "Ernest Holt", Hydrographical observations, 1952, 98 pp.

chemistry oxygen (data only)
Great Britain, Ministry of Agriculture, Fisheries and Food, 1953.
Fisheries Laboratory, Lowestoft, Research Vessel "Ernest Holt"; hydrographical observations.
Unnumbered paggs

chemistry oxygen (data only)
Great Britain, Ministry of Agriculture, Fisheries and Food, 1949.
Fisheries Laboratory, Lowestoft, Research Vessel "Ernest Holt"; hydrographical observations.
Unnumbered pages.

oxygen (data)
Gudkovich, Z.M., 1955.
[Results of a preliminary analysis of the deep-water hydrological observations.] Material. Nabluid. Nauch.-Issledov. Dreifuius., 1950/51, Morskoi Transport, 1:41-46; 48-170.

David Knauss translator, AMS-Astia Doc. 117133

H

chemistry oxygen (data only)
Hapgood, William, 1959.
Hydrographic observations in the Bay of Naples (Golfo di Napoli), January 1957-January 1958 (Station lists).
Pubbl. Sta. Zool., Napoli, 31(2):337-371.

chemistry, oxygen (data only)
Healey, D.A., and R.L.K. Tripe 1969.
Oceanographic observations at Ocean Station P (50°N, 145°W) 21 February to 9 April 1969.
Techn Rept. Fish. Res. Bd. Can. 145: 116 pp. (multilithed).

Chemistry oxygen (data only)
Hela, Ilmo, and Folke Koroleff, 1958
Hydrographical and chemical data collected in 1957 on board the R/V Aranda in the Barents Sea.
Merent. Julk., No. 179: 67 pp.

Chemistry oxygen (data only)
Hela, Ilmo, and F. Koroleff, 1958.
Hydrographical and chemical data collected in 1956 on board the R/V"Aranda" in the Baltic Sea.
Merent. Julk., No. 153:1-52.

Chemistry oxygen (data only)
Henrotte-Bois, Maurice, 1968.
Résultats de mesures faites à bord de l'Origny en mer Tyrrhenienne. Campagne internationale de l'OTAN (16 septembre-15 Octobre 1963).
Présentation de résultats.
Cah. océanogr. 20(Suppl.1):65-125.

chemistry oxygen (data only)
Henrotte-Bois, Maurice, 1969.
Résultats de mesures hydrologiques faites à bord de l'Origny: campagne "Mediterranean outflow" (septembre-octobre 1965).
Cah. oceanogr. 21 (Suppl. 1): 49-192.

oxygen(data only)
Herlinveaux, R.H. 1970.
Sapack 8/68 - oceanographic and biological observations.
Techn. Rept. Fish. Res. Bd. Can. 159: 60 pp.

oxygen (data only)
Hisard, P., F. Jarrige, et P. Rual, 1968.
Résultats des observations physico-chimiques de la croisière CYCLONE 2 du N.O. Coriolis.
Rapp. sci., O.R.S.T.O.M., Nouméa, 20:21 pp. (mimeographed).

chemistry oxygen (data only)
Holmes, Robert W., and Maurice Blackburn, 1960
Physical, chemical and biological observations in the eastern tropical Pacific Ocean, Scot Expedition, April-June 1958.
USFWS Spec. Sci. Rept. Fish., No. 345:106 pp.

oxygen (data only)
Hoshino, Zen-ichiro, 1970.
Oceanographical conditions observed at definite station off Asamushi during 1969.
Bull. mar. biol. Sta. Asamushi, 14(1):63 (foldout)

oxygen (data only)
Hufford, Gary L., and James M. Seabrooke, 1970
Oceanography of the Weddell Sea in 1969 (IWSOE).
Oceanogr. Rept. U.S. Coast Guard 31 (CG 373-31) 32 pp.

oxygen (data only)
Husby, David M., 1971.
Oceanographic investigations in the northern Bering Sea and Bering Strait, June-July 1968.
Oceanogr. Rept. U.S. Cst Gd 40 (CG 373-40): 49 pp.

Chemistry oxygen (data)
Husby, David M., and Gary L. Hufford 1971.
Oceanographic investigation of the northern Bering Sea and Bering Strait 8-21 June 1969.
Oceanogr. Rept. U.S. Cst Gd 42 (CG 373-42): 55 pp.

I

chemistry, oxygen (data only)
Ilahude, Abdul Gani 1971.
Oceanographic Station List 1963-1966.
Special Issue, Oceanogr. Cruise Rept. (1963-1966), Djakarta, 67 pp.

Chemistry, oxygen (data only)
India, Central Marine Fisheries Research Institute 1965.
Oceanographic Station List III.
Indian J. Fish. 12 (A+B): 237-457

Chemistry oxygen (data only)
India, Naval Headquarters, New Delhi, 1958.
Indian oceanographic station list, Ser. No. 2: unnumbered pp. (mimeographed).

Chemistry, oxygen (data only)
India, Central Marine Fisheries Research Institute, 1964.
Oceanographic Station List II
Indian J. Fish. (A) 11(2): 735-965

Chemistry oxygen (data only)
India, Central Marine Fisheries Research Institute, 1962.
Oceanographic station list.
Indian Jour. , Fish., (A), 9(1):213-431.

oxygen (data only)
India, Naval Headquarters, Office of Scientific Research & Development, 1958-1959.
Indian Oceanographic Stations List, Serial Nos. 3-5(June 1958; Jan. & June 1959): unnumbered pp. (mimeographed)

Chemistry oxygen (data only)
India, Naval Headquarters, Office of Scientific Research & Development, 1957.
Indian Oceanographic Stations List, No. 1/57: unnumbered pp. (mimeographed).

Chemistry oxygen (data only)
Indonesia, Institute of Marine Research, 1956.
Oceanographic station list, 1956.
Mar. Res. Indonesia, No. 2:49-55.

Chemistry oxygen (data only)
Indonesia, Lembaga Penjelidikan Laut, 1957.
Oceanographic station list 1957.
Penjel. Laut, Indonesia (Mar. Res.) No. 3:57-69.

Chemistry oxygen (data only)
Ingraham, W. James,Jr., 1967.
Distribution of physical-chemical properties and tabulations of station data, Washington and British Columbia coasts, October-November 1963.
Data Rep. U.S. Fish Wildl.Serv.,Bur.Comm.Fish. 2 cards (microfiche).

Chemistry oxygen (data only)
Ingham, Merton C., and Robert B. Elder, 1970.
Oceanic conditions off northeastern Brazil, February to March and October-November 1966.
Oceanogr. Rept. U.S. Cst Guard, 34 (CG373-34): 155 pp.

Chemistry, oxygen (data only)
Ingham, Merton C., Julien R. Goulet Jr. and John T. Brucks 1968.
Oceanic conditions in the northwestern Gulf of Guinea, Geronimo Cruise 4, 5 August to 13 October 1964.
Data Rept. U.S. Dept. Interior, Bur. Comm. Fish. 26: 1 film (microfiche)

chemistry, oxygen (data only)
Ivanoff, A., A. Morel, J. Boutler, G. Copin-Montegut, F. Varlet, P. Courau, P. Geistdoerfer, F. Nyffeler et J.P. Bethoux 1969.
Résultats des observations effectuées en Mer Méditerranée principalement dans le Détroit de Sicile à bord du navire océanographique Calypso en mai 1965 et juillet 1966.
Cah. océanogr. 21 (Suppl. 2): 203-244.

J

Chemistry oxygen (data only)
Japan, Central Meteorological Observatory, 1956.
The results of marine meteorological and oceanographic observations (NORPAC Expeditions Spec. No.)17:131 pp.

Chemistry
Oxygen (data only)
Japan, Central Meteorological Observatory, 1956.
The results of marine meteorological and oceanographical observations. Pt. 1. Oceanography January-June, 1955. No. 16:120 pp.

Chemistry oxygen(data only)
Japan, Central Meteorological Observatory, 1955.
The results of marine meteorological and oceanographical observations.
1. Oceanography, January-June 1954, No. 14:91 pp.
1. Oceanography, July-December 1954, No. 15:134 pp.

Chemistry oxygen (data only)
Japan, Central Meteorological Observatory, 1955.
The results of marine meteorological and oceanographical observations, January-June 1951. No. 9:177 pp.

Chemistry oxygen(data only)
Japan, Central Meteorological Observatory, 1954.
The results of marine meteorological and oceanographical observations. 13(1):1-210.

chemistry oxygen(data only)
Japan, Central Meteorological Observatory, 1954.
The results of marine meteorological and oceanographical observations. Part 1, Oceanograhy, July-December, 1952, No. 12:138 pp.

Chemistry oxygen(data only)
Japan, Central Meteorological Observatory, 1953
The results of Marine meteorological and oceanographical observations. Jan.-June 1952,No. 11:362, 1 fig.

Chemistry oxygen(data only)
Japan, Central Meteorological Observatory, 1952
The results of Marine Meteorological and oceanographical observations, July - Dec. 1951, No. 10:310 pp., 1 fig.

Chemistry oxygen(data only)
Japan, Central Meteorological Observatory, 1952.
The results of Marine meteorological and oceanographical observations. July - Dec. 1950. No. 8: 299 pp.

Chemistry oxygen(data only)
Japan, Central Meteorological Observatory, 1952.
The Results of Marine Meteorological and oceanographical observations. Jan. - June 1950. No. 7: 220 pp.

chemistry oxygen(data only)
Japan, Central Meteorological Observatory, 1951.
The results of marine meteorological and oceanographic observations. July - Dec. 1949. No. 6: 423 pp.

Chemistry oxygen (data only)
Japan, Central Meteorological Observatory, 1951.
Table 14. Oceanographical observations taken in Miyazu Bay: physical and chemical data for stations occupied by R.M.S. "Asanagi-maru".
Res. Mar. Met. Ocean. Obs., Jan.-June, 1949, No. 5:77-84.

Chemistry oxygen(data only)
Japan, Central Meteorological Observatory, 1951.
Table 9. Oceanographical observations taken in the Goto-nada and the Tsushima Straits. (A) Physical and chemical data for stations occupied by the fishing boat "Daikoku-maru": (B) ------- R.S.S. "Umikaze-maru".
Res. Mar. Met. Ocean. Obs., Jan.-June, 1949, No. 5:61-67.

Chemistry oxygen(data only)
Japan, Central Meteorological Observatory, 1951.
Table 8. Oceanographical observations taken in the Ariake Sea; physical and chemical data for stations occupied by R.S.S. "Umikaze-maru".
Res. Mar. Met. Ocean. Obs., Jan.-June, 1949, No. 5:50-60.

Chemistry oxygen(data only)
Japan, Central Meteorological Observatory, 1951.
Table 6. Oceanographical observations taken in the Akashi-seto, the Yura-seto and the Kii Suido; physical and chemical data for stations occupied by R.M.S. "Shumpu-maru".
Res. Mar. Met. Ocean. Obs., Jan.-June 1949, No. 5:40-47.

Chemistry oxygen(data only)
Japan, Central Meteorological Observatory, 1951.
Table 5. Oceanographical observations taken in the sea area to the south of the Kii Peninsula along "E"-line; physical and chemical data for stations occupied by R.M.S. "Chikubu-maru".
Res. Mar. Met. Ocean. Obs., Jan.-June, 1949, No. 5:37-40.

Chemistry oxygen(data only)
Japan, Central Meteorological Observatory, 1951.
Table 4. Oceanographical observations taken in Sagami Bay; physical and chemical data for stations occupied by R.M.S. "Asashio-maru".
Res. Mar. Met. Ocean. Obs., Jan.-June 1949, No. 5:30-37.

Chemistry, oxygen(data only)
Japan, Central Meteorological Observatory, 1951.
Table 3. Oceanographical observations taken in the North Pacific Ocean along "C" line; (A)Physical and chemical data for stations occupied by R.M.S. "Ukuru-maru";(B) --- R.M.S. Ikuna-maru"; (C) --- R.M.S. "Ryofu-maru";(D) --- R.M.S. Shinnan-maru";(E) --- R.M.S. "Kung-maru";(F) --- R.M.S. "Chikubu-maru";(G)"Ryofu-maru"; (H) --- "Ikuna-maru";(I) --- "Chikubu-maru";(J) ---- "Ukuru-maru"; (K) --- "Shinnan-maru";(L) --- ""Ukuna-maru";(M) --- "Ryofu-maru".
Res. Mar. Met. Ocean. Obs., Jan.-June 1949, No. 5:13-30.

chemistry oxygen(data only)
Japan, Central Meteorological Observatory, 1951.
Table 2. Oceanographical observations taken in the sea area along the east coast of Tohoku District. (A). Physical and chemical data for stations occupied by R.M.S. "Oyashio-maru": (B) Physical and chemical data for stations occupied by R.M.S. "Kuroshio-maru". Res. Mar. Met. Ocean. Obs., Jan.-June 1949, No. 5:3-12.

Chemistry oxygen(data only)
Japan, Central Meteorological Observatory, 1951.
Table 1. Oceanographical observations taken in the sea area south of Hokkaido; physical and chemical data for stations occupied by R.M.S. "Yushio-maru". Res. Mar. Met. Ocean. Obs., Jan.-June 1949, No. 5:2-3.

JAPAN Chemistry oxygen(data only)
Central Meteorological Observatory, Japan, 1951.
The results of marine meteorological and oceanographical observations, July-Dec. 1948, No. 4: vi+ 414 pp., 1 map, 1 photo. of 57 tables,

JAPAN Chemistry oxygen(data only)
Central Meteorological Observatory, Japan, 1951.
The results of marine meteorological and oceanographical observations, Jan.-June 1948, No. 3: 256 pp.

JAPAN Chemistry oxygen (data only)
Central Meteorological Observatory, 1949.
Report on sea and weather observations on Antarctic whaling ground (1948-1949). Ocean. Mag., Tokyo, 1(3):142-173, 11 textfigs.

JAPAN, Chemistry oxygen (data only)
Central Meteorological Observatory, Japan, 1950.
The results of marine meteorological and oceanographical observations, Jan.-June 1947, No. 1: 113 pp., 19 tables.

Chemistry oxygen (data only)
Japan, Fisheries Agency, Research Division, 1956.
Radiological survey of western area of the dangerous zone around the Bikini-Eniwetok Atolls, investigated by the "Shunkotsu maru" in 1956, Part 1:143 pp.

chemistry, oxygen (data only)
Japan, Hokkaido University, Faculty of Fisheries 1970.
Data record of oceanographic observations and exploratory fishing 13:406 pp.

Chemistry oxygen (data only)
Japan, Hokkaido University, Faculty of Fisheries, 1968.
Data record of oceanographic observations and exploratory fishing. No. 12:420 pp.

chemistry oxygen (data only)
Japan, Hokkaido University, Faculty of Fisheries, 1968.
The Oshoro Maru cruise 23 to the east of Cape Erimo, Hokkaido, April, 1967.
Data Record Oceanogr. Obs. Expl. Fish., 12: 115-169.

Chemistry, oxygen (data only)
Japan, Hokkaido University, Faculty of Fisheries, 1968.
The Oshoro Maru cruise 21 to the Southern Sea of Japan, January 1967.
Data Record Oceanogr. Obs. Explor. Fish. 12: 1-97; 113-119.

chemistry, oxygen (data only)
Japan, Hokkaido University, Faculty of Fisheries, 1967.
Data record of oceanographic observations and exploratory fishing, 11:383 pp.

Chemistry oxygen (data only)
Japan, Hokkaido University, Faculty of Fisheries, 1967.
The Oshoro Maru cruise 16 to the Great Australian Bight November 1965-February 1966.
Data Record Oceanogr. Obs. Explor. Fish., Fac. Fish., Hokkaido Univ. 11: 1-97; 113-119.

chemistry, oxygen (data only)
Japan, Hokkaido University, Faculty of Fisheries, 1967.
The Oshoro Maru cruise 18 to the east of Cape Erimo, Hokkaido, April, 1966.
Data Record Oceanogr. Obs. Explor. Fish. 11: 121-164.

chemistry, oxygen (data only)
Japan, Hokkaido University, Faculty of Fisheries 1961.
Data record of oceanographic observations and exploratory fishing, No. 5:391 pp.

Chemistry, oxygen (data only)
Japan, Hokkaido University, Faculty of Fisheries, 1960.
Data record of oceanographic observations and exploratory fishing, No. 4:221 pp.

Chemistry, Oxygen (data only)
Japan, Hokkaido University, Faculty of Fisheries, 1959.
Data record of oceanographic observations and exploratory fishing, No. 3:296 pp.

Chemistry oxygen (data only)
Japan, Hokkaido University, Faculty of Fisheries 1957.
Data record of oceanographic observations and exploratory fishing, No. 1:247 pp.

Chemistry oxygen (data only)
Japan, Hokkaido University, Faculty of Fisheries, 1955.
Correction of data presented in "Hydrographic data obtained principally in the Bering Sea by Training Ship 'Oshoro Maru' in the summer of 1955" published in January 1956: 8 pp.

Chemistry oxygen (data only)
Japan, Japanese Meteorological Agency, 1966.
The results of the Japanese Expedition of Deep Sea (JEDS-8).
Results mar. met. oceanogr. Obsns. Tokyo, 35:328 pp.

Japan Japan Meteorological Agency

Chemistry oxygen (data only)
Japan, Japan Meteorological Agency 1971.
The results of marine meteorological and oceanographical observations, July - December 1969, 46:270 pp.

Chemistry oxygen (data only)
Japan, Japan Meteorological Agency, 1971.
The results of marine meteorological and oceanographical observations, 45: 338 pp.

Chemistry oxygen (data only)
Japan, Japan Meteorological Agency, 1970.
The results of marine meteorological and oceanographical observations, July-December, 1968, 44:311 pp.

Chemistry Oxygen (data only)
Japan, Japan Meteorological Agency, 1970
The results of marine meteorological and oceanographical observations. (The results of the Japaneses Expedition of Deep Sea (JEDS-11); January-June 1967 41: 332 pp.

Chemistry, oxygen (data only)
Japan, Japan Meteorological Agency 1970?
The results of marine meteorological and oceanographical observations, July-December 1966, 40:336 pp.

Chemistry, oxygen (data only)
Japan Meteorological Agency 1969.
The results of marine meteorological and oceanographical observations, Jan.-June 1966, 39:349 pp. (multilithed).

Chemistry oxygen (data only)
Japan Meteorological Agency, 1968.
The results of marine meteorological and oceanographical observations, July-December, 1965, 38: 404 pp. (multilithed)

Chemistry oxygen, (data only)
Japan, Japanese Meteorological Agency, 1967.
The results of marine meteorological and oceanographical observations, July-December, 1964, 36: 367 pp.

Chemistry oxygen (data only)
Japan, Japan Meteorological Agency. 1965.
The results of marine meteorological and oceanographical observations. July-December 1963. No. 34: 360 pp.

Chemistry, oxygen (data only)
Japan, Japan Meteorological Agency, 1964.
The results of marine meteorological and oceanographical observations, January-June 1963, No. 33:289 pp.

Chemistry oxygen (data only)
Japan, Japan Meteorological Agency, 1964.
The results of the Japanese Expedition of Deep Sea (JEDS-5).
Res. Mar. Meteorol. and Oceanogr. Obs., July-Dec. 1962, No. 32:328 pp.

Chemistry oxygen (data only)
Japan, Japan Meteorological Agency, 1964.
Oceanographic observations.
Res. Mar. Met. & Oceanogr. Obs., No. 31:220 pp.

Chemistry oxygen (data only)
Japan, Japan Meteorological Agency, 1962
The results of marine meteorological and oceanographical observations, July-December 1961, No. 30:326 pp.

Chemistry oxygen (data only)
Japan, Japan Meteorological Agency, 1962
The results of marine meteorological and oceanographical observations, January-June 1961, No. 29: 284 pp.

Chemistry oxygen (data only)
Japan, Meteorological Agency, 1962
The results of marine meteorological and oceanographical observations, July-December, 1960, No. 28: 304 pp.

Chemistry oxygen (data only)
Japan, Japan Meteorological Agency, 1960.
The results of marine meteorological and oceanographical observations, July-December, 1959, No. 26:256 pp.

Chemistry oxygen (data only)
Japan, Japan Meteorological Agency, 1960.
The results of marine meteorological and oceanographical observations, Jan.-June 1959, No. 25: 258 pp.

Chemistry oxygen (data only)
Japan, Japan Meteorological Agency, 1959.
The results of marine meteorological and oceanographical observations, July-December, 1958, No. 24:289 pp.

Chemistry oxygen (data only)
Japan, Japan Meteorological Agency, 1958
The results of marine meteorological and oceanographical observations, July-December 1957, No. 22: 183 pp.

Chemistry oxygen (data only)
Japan, Japan Meteorological Agency, 1958.
The results of marine meteorological and oceanographical observations, January-June 1957, No. 21:168 pp.

chemistry, oxygen (data only)

Japan, Japanese Meteorological Agency 1957. The results of marine meteorological and oceanographical observations, Jan.-June, 1956: 184 pp. July-December, No. 20: 191 pp.

Chemistry Oxygen (data only)

Japan, Japan Meteorological Agency, 1956.

The results of marine meteorological and oceanographical observations. Part 1. Oceanography, July-December, 1955. No. 18: 90 pp.

Japan Japanese Oceanographic Data

Chemistry oxygen (data only)

Japan, Japanese Oceanographic Data Center, 1971. Shumpu Maru, Kobe Marine Observatory, Japan Meteorological Agency, 16 July - 8 August 1970, south of Japan. Data Rept. CSK (KDC Ref. 49K127) 279: 7 pp. (multilithed).

Chemistry oxygen (data only)

Japan, Japanese Oceanographic Data Center, 1971. Takuyo, Hydrographic Department, Maritime Safety Agency, Japan, 10-30 July 1970, south & east of Japan. Data Rept. CSK (KDC Ref. 49K126) 278: 19 pp. (multilithed).

Chemistry oxygen (data only)

Japan, Japanese Oceanographic Data Center, 1971. Seifu Maru, Maizuru Marine Observatory, Japan Meteorological Agency, 7 May - 14 June 1969, Southeast of Taiwan & East China Sea. Data Rept. CSK (KDC Ref. 49K107) 242: 6 pp. (multilithed).

Chemistry oxygen (data only)

Japan, Japanese Oceanographic Data Center, 1971. Kofu Maru, Hadkodate Marine Observatory, Japan Meteorological Agency, 27 June - 9 August 1968, East of Japan & Okhotsk Sea. Data Rept. CSK (KDC Ref. 49K084) 178: 32 pp. (multilithed).

Chemistry oxygen (data only)

Japan, Japanese Oceanographic Data Center, 1970 Ji Ri San, Fisheries Research and Development Agency, Republic of Korea, Feb. 18 - 26, 1970, West of the Japan Sea. Data Rept. CSK (KDC Ref. 24K038) 272: 16 pp. (multilithed).

Chemistry oxygen (data only)

Japan, Japanese Oceanographic Data Center, 1970. Han Ra San, Fisheries Research and Development Agency, Republic of Korea, Feb. 11 - 24, 1970, East of the Yellow Sea. Data Rept. CSK (KDC Ref. 24K037) 271: 14 pp. (multilithed).

Chemistry oxygen (data only)

Japan, Japanese Oceanographic Data Center, 1970. Baek Du San, Fisheries Research and Development Agency, Republic of Korea, Feb. 4 - 20, 1970, South of Korea & East China Sea. Data Rept. CSK (KDC Ref. 24K036) 270: 15 pp. (multilithed).

Chemistry Oxygen (data only)

Japan, Japanese Oceanographic Data Center, 1970 Seifu Maru, Maizuru Marine Observatory, Japan Meteorological Agency, Feb. 6 - Mar. 16, 1970, Japan Sea. Data Rept. CSK (KDC Ref. 49K124) 268: 8 pp. (multilithed).

Chemistry oxygen (data only)

Japan, Japanese Oceanographic Data Center, 1970. Shumpu Maru, Kobe Marine Observatory, Japan Meteorological Agency, Feb. 21 - Mar. 11, 1970, South of Japan. Data Rept. CSK (KDC Ref. 49K122) 266: 4 pp. (multilithed).

Chemistry oxygen (data only)

Japan, Japanese Oceanographic Data Center, 1970 Chofu Maru, Nagasaki Marine Observatory, Japan Meteorological Agency, Jan. 19 - 21, 1970, East China Sea. Data Rept. CSK (KDC Ref. 49K123) 267: 5 pp. (multilithed).

Chemistry oxygen (data only)

Japan, Japanese Oceanographic Data Center, 1970 Takuyo, Hydrographic Department, Maritime Safety Agency, Japan, Feb. 12 - Mar. 7, 1970. South and East of Japan. Data Rept. CSK (KDC Ref. 49K120) 264: 13 pp. (multilithed).

Chemistry oxygen (data only)

Japan, Japanese Oceanographic Data Center, 1970. Kaiyo, Hydrographic Department, Maritime Safety Agency Japan, Nov. 7 - 29, 1969, South of Japan. Data Rept. CSK (KDC Ref. 49K116) 257: 10 pp. (multilithed).

Chemistry, OXYGEN (DATA ONLY)

Japan, Japanese Oceanographic Data Center, 1970. Suro No. 3, Hydrographic Office, Republic of Korea, July 23 - August 19, 1969, South of Korea Data Rept. CSK (KDC Ref. 24K035) 240: 19 pp. (multilithed).

Chemistry, oxygen (data only)

Japan, Japanese Oceanographic Data Center 1970. Han Ra San, Fisheries Research and Development Agency, Republic of Korea, August 9-21, 1969 east of the Yellow Sea. Data Rept CSK (KDC Ref. No. 24K033) 238: 17 pp. (multilithed).

Chemistry OXYGEN (DATA ONLY)

Japan, Japanese Oceanographic Data Center, 1970. Takuyo, Hydrographic Department, Maritime Safety Agency, Japan, August 11 - September 2, 1969, South and East of Japan. Data Rept. CSK (KDC Ref. 49K112) 229: 21 pp. (multilithed).

Chemistry Oxygen (data only)

Japan, Japanese Oceanographic Data Center, 1970 Shumpu Maru, Kobe Marine Observatory, Japan Meteorological Agency, July 18 - August 14, 196 South of Japan. Data Rept. CSK (KDC Ref. 49K11 (228:) 7 pp. (multilithed).

Chemistry, oxygen (data only)

Japan, Japanese Oceanographic Data Center, 1970 Orlick, USSR, Jan. 27 - Mar. 6, 1969, Northwest of the North Pacific Ocean. Data Rept. CSK (KDC Ref. 90K023) 227: 30 pp. (multilithed).

Chemistry, oxygen (data only)

Japan, Japanese Oceanographic Data Center, 1970. Oshoro Maru, Faculty of Fisheries, Hokkaido University, Japan, November 5, 1968 - January 20, 1969, Southwest of the North Pacific Ocean. Data Rept. CSK (KDC Ref. 49K104) 216: 13 pp. (multilithed).

Chemistry oxygen (data only)

Japan, Japanese Oceanographic Data Center, 1970 G. Nevelskov, USSR, Oct. 28, 1968 - Jan. 14 1969, Northwest of the North Pacific Ocean. Data Rept. CSK (KDC Ref. 90K024) 209: 55 pp. (multilithed).

Chemistry, oxygen (data only)

Japan, Japanese Oceanographic Data Center 1969. Seifu Maru, Maizuru Marine Observatory, Japan Meteorological Agency, Japan, February 8-26, 1969, Japan Sea. Data Rept. CSK (KDC Ref. No. 49K103) 215: 7 pp. (multilithed).

Chemistry oxygen (data only)

Japan, Japanese Oceanographic Data Center, 1969 Cape St. Mary, Fisheries Research Station, UK (Hong Kong). May 16 - 23, 1968, South China Sea. Prelim. Data Rept. CSK (KDC Ref. No. 74K010) 203: 15 pp. (multilithed).

Chemistry oxygen (data only)

Japa Japanese Oceanographic Data Center, 1970. Suro No. 3, Hydrographic Office, Republic of Korea October 12 - November 8, 1968, South of Korea. Data Report CSK (KDC Ref. 24K027) 201: 15 pp. (multilithed).

Chemistry - oxygen (data only)

Japan, Japanese Oceanographic Data Center 1970. Suro No. 3, Hydrographic Office, Republic of Korea, July 22 - August 7, 1968 South of Korea. Data Rept. CSK (KDC Ref. 24K029) 197: 17 pp. (multilithed).

Chemistry - oxygen (data only)

Japan, Japanese Oceanographic Data Center 1969. Orlick, USSR, July 21 - August 20, 1968 northwest North Pacific Ocean. Data Rept. CSK (KDC Ref. No 90K020) 191: 32 pp. (multilithed).

Chemistry, oxygen (data only)

Japan, Japanese Oceanographic Data Center 1969. Iskatel, USSR, January 26 - March 11, 1968, northwest North Pacific Ocean. Data Rept. CSK (KDC Ref. 90K019) 187: 32 pp. (multilithed).

Chemistry, oxygen (data only)

Japan, Japanese Oceanographic Data Center, 1969. Tenyo Maru, Shimonoseki University of Fisheries, Japan, August 31 - September 2, 1968, East China Sea. Prelim. Data Rept. CSK (KDC Ref. No. 49K092) 186: 5 pp. (multilithed).

Chemistry, oxygen (data only)

Japan, Japanese Oceanographic Data Center, 1969. Seifu Maru, Maizuru Marine Observatory, Japan Meteorological Agency. August 6 - 10, 1968, Japan Sea. Prelim. Data Rept. CSK (KDC Ref. No. 49K087) 181: 9 pp. (multilithed).

Chemistry
oxygen (data only)

Japan, Japanese Oceanographic Data Center, 1969

Takuyo, Hydrographic Division, Maritime Safety Agency, Japan. July 19-September 6, 1968, Kuroshio Extension Area. Prelim. Data Rept. CSK (KDC Ref. No. 49K082) 176: 27 pp. (multilithed).

chemistry, oxygen (data only)

Japan, Japanese Oceanographic Data Center 1969.
Hakuho Maru, Ocean Research Institute, University of Tokyo, Japan, May 15-June 8, 1968, Southwest of Japan and the East China Sea.
Data Rept. CSK (KDC Ref. 49K078) 172: 7 pp. (multilithed).

Chemistry
oxygen (data only)

Japan, Japanese Oceanographic Data Center, 1969.

Cape St. Mary, Fisheries Research Station, UK (Hong Kong). January 6 - 14, 1968, South China Sea. Prelim. Data Rept. CSK (KDC Ref. No. 74K009) 165: 15 pp. (multilithed).

Chemistry
OXYGEN (DATA ONLY)

Japan, Japanese Oceanographic Data Center, 1970
OCEANOGRAPHIC VESSEL No. 1, Hydrographic Department, Royal Thai Navy, Thailand, December 14 1967 - January 18, 1968, Gulf of Thailand. Data Rept. CSK (KDC Ref. 86K011) 148: 21 pp. (multilithed).

Chemistry, oxygen (data Only)

Japan, Japanese Oceanographic Data Center, 1969.
Ryofu Maru, Marine Division, Japan Meteorological Agency, Japan. October 7 - November 9, 1968. East China Sea. Data Rept. CSK (KDC Ref. No. 49K449) 217: 39 pp. (multilithed).

Chemistry oxygen (data only)

Japan, Japanese Oceanographic Data Center, 1969. Researcher I, Bureau of Coast & Geodetic Survey, Manila, Philippines. April 20 - July 19, 1968. North & east of Philippine Region. Data Rept. CSK (KDC Ref. No. 66K001) 202: 49 pp. (Multilithed)

Chemistry, oxygen (data only)

Japan, Japanese Oceanographic Data Center, 1969.
Chofu Maru, Nagasaki Marine Observatory, Japan Meteorological Agency, August 4 - 5, 1968, East China Sea. Prelim. Data Rept. CSK (KDC Ref. No. 49K086) 180: 5 pp. (multilithed).

Chemistry, oxygen (data only)

Japan, Japanese Oceanographic Data Center, 1969.
Shumpu Maru, Kobe Marine Observatory, Japan Meteorological Agency, July 19-31, 1968, South of Japan.
Data Rept. CSK (KDC Ref. 49K085) 179: 9 pp.

Chemistry
oxygen (data only)

Japan, Japanese Oceanographic Data Center, 1969.

Keiten Maru, Faculty of Fisheries, Kagoshima University, Japan. April 23 - May 12, 1968, East China Sea & southeast of Taiwan. Data Rept. CSK (KDC Ref. No. 49K079) 173: 10 pp.

Chemistry, oxygen (data only)

Japan, Japanese Oceanographic Data Center, 1969.
Takuyo, Hydrographic Division, Maritime Safety Agency, May 14-25, 1968, South of Japan.
Data Rept. CSK (KDC Ref. 49K077) 171: 16 pp.

Chemistry oxygen (data only)

Japan, Japanese Oceanographic Data Center, 1969.
Seifu Maru, Maizuru Marine Observatory, Japan Meteorological Agency, February 17-24, 1968, Japan Sea.
Data Rept. CSK (KDC Ref. 49K073) 159: 7 pp.

Chemistry oxygen (data only)

Japan, Japanese Oceanographic Data Center, 1969.
Kofu Maru, Hakodate Marine Observatory, Japan Meteorological Agency, February 7-26, 1968, east of Japan.
Data Rept. CSK (KDC Ref. 49K070) 156: 9 pp.

Chemistry, oxygen (data only)

Japan, Japanese Oceanographic Data Center, 1969
Tenyo Maru, Shimonoseki University of Fisheries Japan. August 23 - 26, 1967, South of Kyushu. Prelim. Data. Rept. CSK (KDC Ref. No. 49K431) 144: 8 pp. (multilithed).

Chemistry
oxygen (data only)

Japan, Japanese Oceanographic Data Center, 1969.

Koyo Maru, Shimonoseki University of Fisheries, Japan, October 29 - 31, 1967, East China Sea and Southeast of Taiwan. Prelim. Data Rept. CSK (KDC Ref. No. 49K067) 132: 6 pp. (multilithed).

Chemistry, oxygen (data only)

Japan, Japanese Oceanographic Data Center, 1969.

Umitaka Maru, Tokyo University of Fisheries, Japan, November 1, 1967- February 23, 1968, West of the North Pacific Ocean. Prelim. Data Rept. CSK (KDC Ref. No. 49K066) 131: 13 pp. (multilithed).

Chemistry oxygen (data only)

Japan, Japanese Oceanographic Data Center, 1969.

Orlick, USSR, July 29 - September 3, 1967, Northwest of the North Pacific Ocean. Prelim. Data Rept. CSK (KDC Ref. No. 90K015) 122: 31 pp. (multilithed).

Chemistry oxygen (data only)

Japan, Japanese Oceanographic Data Center, 1969.

U.M. Schokalsky, USSR, April 28 - June 3, 1967, Northwest of the North Pacific Ocean. Prelim. Data Rept. CSK (KDC Ref. No. 90K014) 121: 41 pp. (multilithed).

Chemistry OXYGEN (data only)

Japan, Japanese Oceanographic Data Center, 1968.
Tae Baek San, Fisheries Research and Development Agency, Republic of Korea, Frbruary 16- March 12, 1968, South of Korea.
Data Rept. CSK (KDC Ref. 24K022) 163: 12 pp. (Multilithed)

Chemistry oxygen (data only)

Japan, Japanese Oceanographic Data Center, 1968.
Han ra San, Fisheries Research and Development Agency, Republic of Korea, February 15- March 12, 1968, East of the Yellow Sea.
Data Rept. CSK (KDC Ref. 24K021) 162: 18 pp. (multilithed)

Chemistry, oxygen (data only)

Japan, Japanese Oceanographic Data Center, 1968.
Baek du San. Fisheries Research and Development Agency, Republic of Korea, February 17-March 12, 1968, West of the Japan Sea.
Data Rept. CSK (KDC Ref. 24K020) 161: 23 pp. (multilithed)

Chemistry oxygen (data only)

Japan, Japanese Oceanographic Data Center, 1968.
Chofu Maru, Nagasaki Marine Observatory, Japan Meteorological Agency, Japan, February 2-4, 1968, East China Sea.
Data Rept. CSK (KDC Ref. 49K072) 158: 5 pp. (multilithed).

Chemistry oxygen (data only)

Japan, Japanese Oceanographic Data Center, 1968.
Shumpu Maru, Kobe Marine Observatory, Japan Meteorological Agency, Japan, February 17-28, 1968, South of Japan.
Data Rept. CSK (KDC Ref. 49K071) 157: 6 pp. (multilithed).

Chemistry oxygen (data only)

Japan, Japanese Oceanographic Data Center, 1968.
Ryofu Maru, Marine Division, Japan Meteorological Agency, Japan, January 13-March 22, 1968, West of North Pacific Ocean.
Data Rept. CSK (KDC Ref. 49K069) 155: 70 pp. (multilithed).

Chemistry, oxygen (data only)

Japan, Japanese Oceanographic Data Center, 1968.
Takuyo, Hydrographic Division, Maritime Safety Agency, Japan, February 19-March 10, 1968, south and east of Japan.
Data Rept. CSK (KDC Ref. 49K068) 154: 14 pp. (multilithed).

Chemistry oxygen (data only)

Japan, Japanese Oceanographic Data Center, 1968, Fisheries Research Vessel No. 2, Department of Fisheries, Thailand, November 1- December 10, 1967, Gulf of Thailand and South China Sea.
Data Rept. CSK (KDC Ref. 68K005) 139: 29 pp. (Multilithed)

Chemistry oxygen (data only)

Japan, Japanese Oceanographic Data Center, 1968.
Cape St. Mary, Fisheries Research Station, Hong Kong, September 19-October 11, 1967, South China Sea.
Prelim. Data Rept. CSK (KDC Ref. 74K008) 138: 16 pp. (multilithed).

Chemistry oxygen (data only)

Japan, Japanese Oceanographic Data Center, 1968.
Jalanidhi, Naval Hydrographic Office, Indonesia, October 4-19, 1967, South China Sea.
Data Rept. CSK (KDC Ref. 42K002) 137: 15 pp. (multilithed).

Chemistry oxygen (data only)

Japan, Japanese Oceanographic Data Center, 1968.
Suro No. 1, Hydrographic Office, Korea, September 23-October 21, 1967, South of Korea.
Data Rept. CSK (KDC Ref. 24K019) 136: 13 pp. (mutlilithed).

Chemistry, oxygen (data only)

Japan, Japanese Oceanographic Data Center, 1968.
Chofu Maru, Nagasaki Marine Obse.vatory, Japan Meteorological Agency, Japan, July 3-4, 1967, Aug. 28-29, 1967, Oct. 6-7, 1967, Nov. 9-10, 1967.
Data Rept. CSK (KDC Refs. 49K319, 49K320, 49K321, 49K322) 134: 18 pp. (multilithed).

Chemistry oxygen (data only)

Japan, Japanese Oceanographic Data Center, 1968.
Nagasaki Maru, Faculty of Fisheries, Nagasaki University of Fisheries, Japan, November 7-29, 1967, East China Sea.
Data Rept. CSK (KDC Ref. (49K076) 133: 5 pp. (multilithed).

Chemistry oxygen (data only)
Japan, Japanese Oceanographic Data Center, 1968. Chofu Maru, Nagasaki Marine Observatory, Japan Meteorological Agency, Japan, October 5-6, 1967, East China Sea.
Data Rept. CSK (KCD Ref. 49K065) 130: 5 pp. (multilithed).

Chemistry oxygen (data only)
Japan, Japanese Oceanographic Data Center, 1968. Oceanographic Vessel, No. 1, Hydrographic Department, Royal Thai Navy, Thailand, August 2-11, 1967, Gulf of Thailand.
Data Rept. CSK (KDC Ref. 86K004) 127:9 pp. (multilithed)

Chemistry, oxygen (data only)
Japan, Japanese Oceanographic Data Center, 1968. Oceanographic Vessel, No 1, Hydrographic Department, Royal Thai Navy, Thailand, June 7-12, 1967, Gulf of Thailand.
Data Rept. CSK (KDC Ref. 86K003) 126:9 pp. (multilithed)

Chemistry oxygen (data only)
Japan, Japanese Oceanographic Center, 1968. Oceanographic Vessel No. 1, Hydrographic Department, Royal Thai Navy, Thailand, March 18-23, 1967, Gulf of Thailand.
Data Rept. CSK (KDC Ref. 86K002) 125:9 pp. (multilithed)

Chemistry, oxygen (data only)
Japan, Japanese Oceanographic Data Center, 1968. Oceanographic Vessel No. 1, Hydrographic Department, Royal Thai Navy, Thailand, January 18-24, 1967, Gulf of Thailand.
Data Rept. CSK (KDC Ref. 86K001) 124:9 pp. (multilithed)

Chemistry oxygen (data only)
Japan, Japanese Oceanographic Data Center, 1968. Burudjulasad, Naval Hydrographic Office, Indonesia, August 25-September 5, 1967, South China Sea.
Data Rept. CSK (KDC Ref. 42K001) 119: 11 pp. (multilithed).

Chemistry, oxygen (data only)
Japan, Japanese Oceanographic Data Center, 1968. Han Ra San, Fisheries Research and Development Agency, Korea, August 12-September 4, 1967, North of East China Sea.
Prelim. Data Rept. CSK(KDC Ref. 24K018) 117:12 pp. (multilithed).

Chemistry oxygen (data only)
Japan, Japanese Oceanographic Data Center, 1968. Chun Ma San, Fisheries Research and Development Agency, Korea, August 13-21, 1967, East of Yellow Sea
Prelim. Data Rept. CSK(KDC Ref. 24K017) 116: 16 pp.
(multilithed)

Chemistry oxygen (data only)
Japan, Japanese Oceanographic Data Center, 1968. Baek Du San, Fisheries Research and Development Agency, Korea, August 12-25, 1967, West of Japan Sea.
Data Rept. CSK (KDC Ref. 24K016) 115: 26 pp. (multilithed)

Chemistry, oxygen (data only)
Japan, Japanese Oceanographic Data Center, 1968. Hakuho Maru, Ocean Research Institute, University of Tokyo, Japan, September 6-20, 1967, South of Japan.
Data Rept. CSK (KDC Ref. 49K059) 111: 9 pp. (multilithed).

Chemistry oxygen (data only)
Japan, Japanese Oceanographic Data Center, 1968. Seifu maru, Maizuru Marine Observatory, Japan Meteorological Agency, Japan, August 2-24, 1967, Japan Sea.
Data Rept. CSK (KDC Ref. (49K058) 110: 9 pp. (multilithed).

Chemistry oxygen (data only)
Japan, Japanese Oceanographic Data Center, 1968. Takuyo, Hydrographic Division, Maritime Safety Agency, July 12-August 30, 1967, Central Part of the North Pacific Ocean.
Prelim. Data Rept. CSK(KDC Ref. 49K053) 105:25 pp. (multilithed)

Chemistry oxygen (data only)
Japan, Japanese Oceanographic Data Center, 1968. Yang Ming, Chinese National Committee on Ocean Research, Republic of China, April 1-May 14, 1967, Surrounding waters of Taiwan.
Data Rept. CSK (KDC REF. 21K004) 102: 15 pp. (multilithed).

Chemistry oxygen (data only)
Japan, Japanese Oceanographic Data Center, 1968. Kofu Maru, Hakodate Marine Observatory, Japan Meteorological Agency, Japan, May 17-19, 1967, East of Japan.
Data Rept. CSK (KDC Ref. 49K049) 98: 6pp. (multilithed).

Chemistry oxygen (data only)
Japan, Japanese Oceanographic Data Center, 1968. Orlick, USSR, February 5-March 11, 1967, Northwest of North Pacific Ocean.
Prelim. Data Rept. CSK (KDC Ref. 90K012) 94:29 pp.

Chemistry oxygen (data only)
Japan, Japanese Oceanographic Data Center, 1968. Bering Strait, U.S. Coast Guard, January 14-18, 1967, East of Japan.
Data Rept. CSK (KDC Ref. 31K007) 91: 7 pp. (mutlilithed).

Chemistry oxygen (data only)
Japan, Japanese Oceanographic Data Center, 1968. Seifu Maru, Maizuru Marine Observatory, Japan Meteorological Agency, February 10-March 2, 1967, Japan Sea.
Prelim. Data Rept. CSK(KDC 49K044) 86:14 pp. (multilithed).

Chemistry oxygen (data only)
Japan, Japanese Oceanographic Data Center, 1968. Ryofu Maru, Marine Division, Japan Meteorological Agency, January 11-February 24, 1967, West of the North Pacific Ocean.
Prelim. Data Rept. CSK (KDC 49K040) 82:52 pp.

Chemistry oxygen (data only)
Japan, Japanese Oceanographic Data Center, 1968. U.M. Schokalsky, USSR, July 20-August 23, 1966, North-west of North Pacific Ocean.
Prelim. Data Rept. CSK(KDC Ref. 90K010) 75:41 pp. (multilithed).

Chemistry oxygen (data only)
Japan, Japanese Oceanographic Data Center, 1968. G. Nevelskoy, USSR, July 13-September 17, 1966, Northwest of North Pacific Ocean.
Prelim. Data Rept. CSK (KDC Ref. 90K008) 73: 32 pp (multilithed).

Chemistry oxygen (data only)
Japan, Japanese Oceanographic Data Center, 1967. Nagasaki Maru, Faculty of Fisheries, Nagasaki University, June 13-17, 1967, South of Japan.
Prelim. Data Rept. CSK (KDC Ref. No. 49K061) 113:6pp.

Chemistry oxygen (data only)
Japan, Japanese Oceanographic Data Center, 1967. Chofu Maru (NMO), Jan. 13-14, 1967; Mar. 19-20, 1967, Apr. 15-15, 1967, May 11-12, 1967, Japan Meteorological Agency, South-East of Yakushima.
Prelim. Data Rept. CSK (KDC Ref. Nos. 49K315, 49K316, 49K317, 49K318) 104: 18 pp.

Chemistry, oxygen (data only)
Japan, Japanese Oceanographic Data Center, 1967. Chofu maru, Nagasaki Marine Observatory, Japan Meteorological Agency, Japan, May 17-18, 1967, East China Sea.
Prelim. Data Rept. CSK(KDC Ref. 49K051) 100: 5pp.

Chemistry, oxygen (data only)
Japan, Japanese Oceanographic Data Center, 1967. Shumpu maru, Kobe Marine Observatory, Japan Meteorological Agency, Japan, 13-14, 1967, South of Japan.
Prelim. Data Rept. CSK (KDC Ref. 49K050) 99:5pp.

Chemistry, oxygen (data only)
Japan, Japanese Oceanographic Data Center, 1967. Nagasaki maru, The Faculty of Fisheries, Nagasaki University, Japan, January 19-22, 1967, East China Sea.
Prelim. Data Rept. CSK (KDC Ref. 49K046) 88: 7pp.

Chemistry, oxygen (data only)
Japan, Japanese Oceanographic Data Center, 1967. Oshoro maru, the Faculty of Fisheries, Hokkaido University, Japan, January 15-February 1, 1967, South of Japan.
Prelim. Data Rept. CSK (KDC Ref. 49K045) 87: 9pp.

Chemistry, oxygen (data only)
Japan, Japanese Oceanographic Data Center, 1967. Chofu maru, Nagasaki Marine Observatory, Japan Meteorological Agency, Japan, January 20-February 22, 1967, East China Sea.
Prelim. Data Rept. CSK (KDC Ref. 49K043) 85: 11 pp.

Chemistry oxygen (data only)
Japan, Japanese Oceanographic Data Center, 1967. Shumpu maru, Kobe Marine Observatory, Japan Meteorological Agency, Japan, February 26-27, 1967, south of Japan.
Prelim. Data Rept. CSK(KDC Ref. 49K042) 84: 6 pp.

Chemistry, oxygen (data only)
Japan, Japanese Oceanographic Data Center, 1967. Kofu Maru, Hakodate Marine Observatory, Japan Meteorological Agency, February 4-March 7, 1967, East of Japan.
Prelim. Data Rept. CSK (KDC Ref. No. 49K041) 83:15pp.

Chemistry, oxygen (data only)
Japan, Japanese Oceanographic Data Center, 1967. Cape St. Mary, Fisheries Research Station, Hong Kong, June 1-8, 1966, South China Sea.
Prelim. Data Rept. CSK (KDC Ref. 74K003) 77:12 pp.

chemistry, oxygen (data only)
Japan, Japanese Oceanographic Data Center, 1967. Bering Strait, U.S. Coast Guard, July 8-14, 1966 South-east of Japan.
Prelim. Data Rept., CSK(KDC Ref. 31K005), 71:11 pp.

oxygen
Japan, Japanese, Oceanographic Data Center, 1967. Suro No. 1, Hydrographic Division, Korea, October 12-November 2, 1966, south of Korea.
Prelim. Data Rept., CSK (KDC Ref. 24K011) 70 12 pp.

Chemistry oxygen (data only)
Japan, Japanese Oceanographic Data Center, 1967. Bukhansan, Fisheries Research and Development Agency, Korea, July 16- August 9, 1966, East of the Yellow Sea.
Prelim. Data Rept., CSK (KDC Ref. 24K010) 69 17 pp.

Chemistry oxygen (data only)
Japan, Japanese, Oceanographic Data Center, 1967. Baekdusan, Fisheries Research and Development Agency, Korea, July 14-28, 1966, West of the Japan Sea. Prelim. Data Rept. CSK (KDC Ref. 24K009) 65:22 pp.

Chemistry oxygen (data only)
Japan, Japanese Oceanographic Data Center, 1966. Koyo Maru, Shimonoseki University of Fisheries, October 26-29, south of Japan. Prelim. Data Rept. CSK (KDC Ref. No. 49K038):66:7pp. (multilithed).

Chemistry, oxygen (data only)
Japan, Japanese Oceanographic Data Center, 1967. Kagoshima maru, the Faculty of Fisheries, Kagoshima University, Japan, August 5-14, 1966, East China Sea. Prelim. Data Rept. CSK (KDC Ref. 49K031) 57: 9pp.

chemistry - oxygen (data only)
Japan, Japanese Oceanographic Data Center 1967. Ryofu maru, Marine Division, Japan Meteorological Agency, Japan, September 13-17, 1966, eastern Sea of Japan. Prelim. Data Rept. CSK (KDC Ref. No. 49K033) 59: 11 pp. (multilithed)

chemistry, oxygen (data only)
Japan, Japanese Oceanographic Data Center 1967. Nagasaki Maru, Faculty of Fisheries, Nagasaki University, Japan, June 16-21, 1966, southern Sea of Japan. Prelim. Data Rept. CSK (KDC Ref. No. 49K032) 58: 8 pp. (multilithed)

Chemistry oxygen (data only)
Japan, Japanese Oceanographic Data Center, 1967. Kofu Maru, Hakodate Marine Observatory, Japan Meteorological Agency, Japan, June 30-July 10, 1966, east of Japan. Prelim. Data Rept., CSK (KDC Ref. 49K026) 52:14 pp.

chemistry, oxygen (data only)
Japan, Japanese Oceanographic Data Center 1967. Kaiyo, Hydrographic Division, M.S.A., Japan, August 10-30, 1966, S.E. of Japan. Prelim. Data Rept. CSK (KDC Ref. No. 49K025) 51: 15 pp. (multilithed)

Chemistry oxygen (data only)
Japan, Japanese Oceanographic Data Center, 1967. Seifu Maru, Maizuru Marine Observatory, Japan Meteorological Agency, Japan, August 21-September 19, 1966, Japan Sea. Prelim. Data Rept., CSK (KDC Ref. 49K029) 55:15pp.

Chemistry oxygen (data only)
Japan, Japanese Oceanographic Data Center, 1967. Takuyo, Hydrographic Division, Maritime Safety Agency, Japan, February 23-March 16, 1967, south and east of Japan. Prelim. Data Rept., CSK (KDC Ref. 49K039) 81: 17 pp.

Chemistry oxygen (data only)
Japan, Japanese Oceanographic Data Center, 1967. Tansei Maru, Ocean Research Institute, University of Tokyo, Japan, July 30-August 6, 1966, vicinity of Izu islands. Prelim. Data Rept., CSK (KDC Ref. No. 49K419): 79: 10 pp.

Chemistry oxygen (data only)
Japan, Japanese Oceanographic Data Center, 1967. Yang Ming, Chinese National Committee on Oceanic Research, Republic of China, September 10-October 14, 1966, Adjacent Sea of Taiwan. Prelim. Data Rept. CSK (KDC Ref. 21K003) 67:17pp.

chemistry oxygen (data only)
Japan, Japanese Oceanographic Data Center, 1967. Orlick, USSR, July 9-September 12, 1966, North-west of North Pacific Ocean. Prelim. Data Rept. CSK (KDC Ref. 90K009) 74:34 pp.

Chemistry oxygen (data only)
Japan, Japanese Oceanographic Data Center, 1966. Kerin Ho, Fisheries Research and Development Agency, Korea, September 12-October 13, 1965. East of Yellow Sea. Prelim. Data Rept., CSK, (KDC Ref. 24K005), 38: 11 pp. (multilithed).

Chemistry, oxygen (data only)
Japan, Japanese Oceanographic Data Center, 1966. Bukhansan Ho, Fisheries Research and Development Agency, Korea, December 2-12, 1965. Prelim. Data Rept., CSK, (KDC 24K004), 37:15 pp. (multilithed).

Chemistry, oxygen (data only)
Japan, Japanese Oceanographic Data Center, 1966. Seifu Maru, Maizuru Marine Observatory, Japan Meteorological Agency, Japan, February 12-February 28, 1966, Japan Sea. Prelim. Data Rept., CSK (KDC Ref. 48K021), 32:14pp. (Multilithed).

Chemistry, oxygen (data only)
Japan, Japanese Oceanographic Data Center, 1966. Chofu Maru, Nagasaki Marine Observatory, Japan Meteorological Agency, Japan, January 26-February 25, 1966, East China Sea. Prelim. Data Rept., CSK, (KDC Ref. 49K020), 31: 13 pp. (multilithed).

chemistry oxygen (data only)
Japan, Japanese Data Center, 1966. Preliminary data report of CSK, Shumpu Maru, Kobe Marine Observatory, Japan Meteorological Agency, February 19-24, 1966, Southern Sea of Japan, No. 30 (October 1966). KDC Ref. No. 49K019:6 pp. (multilithed)

chemistry oxygen (data only)
Japan, Japanese Oceanographic Data Center, 1966. Ryofu Maru, Japan Meteorological Agency, Japan, February 4-February 28, 1966, East of Japan. Prelim. Data Rept., CSK (KDC Ref. 49K017), 28: 17 pp. (multilithed).

chemistry oxygen (data only)
Japan, Japanese Oceanographic Data Center, 1966. Preliminary data report of CSK, Takuyo, Japanese Hydrographic Division, February 23-March 15, 1966, South eastern Sea of Japan, No. 27 (October 1966). KDC Ref. No. 49K106:17 pp. (multilithed).

chemistry oxygen (data only)
Japan, Japanese Oceanographic Data Center, 1966.
Shumpu Maru (KMO) May 11-12, 1965 49K301
Chofu Maru (NMO) Jul. 1-2, 1965 19K302
Chofu Maru (NMO) Aug. 2-13, 1965 19K303
Chofu Maru (NMO) Sep. 25-26, 1965 49K304
Chofu Maru (NMO) Nov. 1-2, 1965 49K305
Shumpu Maru (KMO) Nov. 11-12, 1965 49K306
Japan Meteorological Agency, Japan, South-East of Yakushima. Prelim. Data Rept., CSK (49K301 - 49K306), 26: 26 pp. (multilithed).

Chemistry, oxygen (data only)
Japan, Japanese Oceanographic Data Center, 1966. Uliana Gromova, USSR, July 20-September 3, 1965. E. & S. of Japan. Prelim. Data Rept., CSK, (KDC 90K003), 25:37 pp. (multilithed)

chemistry, oxygen, (data only)
Japan, Japanese Oceanographic Data Center, 1966. U.M. Shokalski, USSR, July 16-August 18, 1965, E. & S. of Japan. Prelim. Data Rept., CSK, (KDC Ref. 90K001), 23: 41 pp. (multilithed).

Chemistry oxygen (data only)
Japan, Japanese Oceanographic Data Center, 1966. Yang Ming, Chinese National Committee on Oceanic Research, Republic of China, August 10-October 13, 1965, Adjacent Sea of Taiwan. Prelim. Data Rept., CSK (KDC Ref. 21K001), 22: 18 pp. (multilithed).

chemistry oxygen (data only)
Japan, Japanese Oceanographic Data Center, 1966. Preliminary data report of CSK, Atlantis II, Woods Hole Oceanographic Institution, USA, August 4-September 23, 1965, Southern Sea of Japan, No. 20, (October 1966). KDC Ref. No. 31K001: 68 pp. (multilithed).

Chemistry oxygen (data only)
Japan, Japanese Oceanographic Data Center, 1966. Cape St. Mary, Fisheries Research Station, Hong Kong, October 2-October 10, 1965, South China Sea. Prelim. Data Rept., CSK, (KDC Ref. 74K001) 20:12pp. (multilithed).

chemistry oxygen (data only)
Japan, Japanese Oceanographic Data Center, 1966. Preliminary data report of CSK. Kofu Maru, Hakodate Marine Observatory, Japan Meteorological Agency, July 22-28, 1965, Eastern Sea of Japan, No. 14 (October 1966). KDC Ref. No. 49 K004:10 pp. (multilithed).

Chemistry oxygen (data only)
Japan, Japanese Oceanographic Data Center, 1966. Preliminary data report of CSK. Ryofu Maru, Marine Division, Japan Meteorological Agency, July 7-August 3, 1965, Eastern Sea of Japan. No. 10 (October 1966). KDC Ref. No. 49K003:31 pp. (multilithed).

Chemistry oxygen (data only)
Japan, Kagoshima University, Faculty of Fisheries, 1957. Oceanographical observation made during the International Cooperation Expedition, EQUAPAC, in July-August 1956, by M.S. Kagoshima-maru and M.S. Keitan-maru, 69 pp. (mimeographed).

Chemistry oxygen (data only)
Japan, Kobe Marine Observatory, 1961. Data of the oceanographic observations in the sea south of Honshu from February to March and in May, 1959. Bull. Kobe Mar. Obs., No. 167(27):99-108; 127-130; 149-152; 161-164; 205-218.

chemistry oxygen (data only)
Japan, Maizuru Marine Observatory, 1963. Data of the oceanographic observations (1960-1961) (35-36):115-272.

Chemistry, oxygen (data only)
Japan, Maritime Safety Agency, 1967. Results of oceanographic observations in 1965. Data Rept., Hydrogr. Obs., Ser. Oceanogr. (Publ. 792) Oct. 1967, 5:115 pp.

chemistry, oxygen (data only)
Japan, Maritime Safety Agency 1967. Results of oceanographic observations in 1964. Data Rept. Hydrogr. Obs. Ser. Oceanogr. Publ. 792 (4): 88 pp.

chemistry, oxygen (data only)
Japan, Maritime Safety Agency 1966.
Results of oceanographic observations in 1963
Data Rept. Hydrogr. Obs. Ser. Oceanogr.
Publ. 792 (3): 74pp.

chemistry, oxygen (data only)
Japan, Maritime Safety Agency 1965.
Results of oceanographic observations in 1962.
Data Rept. Hydrogr. Obs. Ser. Oceanogr.
Publ. 792 (1): 65 pp.

chemistry oxygen (data only)
Japan, Maritime Safety Agency, Tokyo, 1964.
Tables of results from oceanographic observations in 1961.
Hydrogr. Bull., Tokyo, (Publ. No. 981), No. 77: 82 pp.

chemistry oxygen (data only)
Japan, Maritime Safety Agency, Tokyo, 1964.
Tables of results from oceanographic observations in 1960.
Hydrogr. Bull., (Publ., No. 981), No. 75:86 pp.

chemistry, oxygen (data only)
Japan, Maritime Safety Board 1961.
Tables of results from oceanographic observation in 1959.
Hydrogr. Bull. Publ. 981 (68): 112 pp

chemistry, oxygen (data only)
Japan, Maritime Safety Board, 1961.
Tables of results from oceanographic observations in 1958.
Hydrogr. Bull. Tokyo, Publ. 981 (66): 153 pp.

chemistry, oxygen (data only)
Japan, Maritime Safety Board, 1960.
Tables of results from oceanographic observation in 1957.
Hydrogr. Bull. (Publ. No. 981), No. 64: 103 pp.

chemistry, oxygen (data only)
Japan, Maritime Safety Board, 1956.
Tables of results from oceanographic observations in 1952 and 1953. Hydrogr. Bull. (Publ. 981)(51):1-171.

chemistry, oxygen (data only)
Japan, Maritime Safety Board 1954.
Tables of results from oceanographic observations in 1951.
Hydrogr. Bull (Publ. 981) Spec. No. 15: 31-129.

chemistry oxygen (data)
Japan, Maritime Safety Board, 1954.
Tables of results from oceanographic observations in 1950. Publ. No. 981, Hydrogr. Bull., Maritime Safety Bd., Spec. Number (Ocean. Repts.) No. 14:26-164, 5 textfigs.

chemistry, oxygen (data only)
Japan, Nagasaki Marine Observatory
Tables 9, 10, 11, 22, 23, 24, 34, 35 and 44
Oceanogr. Met. Nagasaki mar. Obs. 15: 92-100, 176-180, 191-192, 213-

chemistry, oxygen (data only)
Japan, Nagasaki Marine Observatory, 1967
Report of the oceanographic observations in the sea west of Japan from July to September 1965. (In Japanese)
Oceanogr. Met. Nagasaki mar. Obs. 17 (241): 26 pp. (the dates are correct)

chemistry, oxygen (data only)
Japan, Nagasaki Marine Observatory, 1967.
Report of the Oceanographic observations in the sea west of Japan from January to February, 1969 (In Japanese)
Oceanogr. Met. Nagasaki Mar. Obs. 17 (242): 26 pp (The dates are correct!)

chemistry, oxygen (data only)
Japan, Nagasaki University Research Party for CSK, 1966.
The results of the CSK - NU65S.
Bull. Fac. Fish. Nagasaki Univ., No. 21:273-292.

chemistry oxygen (data only)
Japan, Shimonoseki University of Fisheries 1970.
Oceanographic surveys of the Kuroshio and its adjacent waters, 1967 and 1968.
Data. Oceanogr. Obs. Explor. Fish. 5: 117 pp.

chemistry, oxygen (data only)
Japan, Shimonoseki University of Fisheries, 1970.
Data of oceanographic observations and exploratory fishings. 5. Oceanographic surveys of the Kuroshio and its adjacent waters, 1967 and 1968: 117 pp.

chemistry oxygen (data only)
Japan, Shimonoseki University of Fisheries 1968
Oceanographic surveys of the Kuroshio and its adjacent waters 1965 and 1966.
Data Oceanogr. Obs. Explor. Fish. 4: 1-178.

chemistry oxygen (data only)
Japan, Shimonoseki University of Fisheries 1968.
Eastern Pacific Ocean cruise and central Atlantic Ocean cruise.
Data Oceanogr. Obs. Explor. Fish., 3: 1-145

chemistry oxygen (data only)
Japan, Shimonoseki University of Fisheries, 1965.
Data of oceanographic observations and exploratory fishings, International Indian Ocean Expedition 1962-63 and 1963-64. No. 1: 453 pp.

chemistry oxygen (data only)
Japan, Tokyo University of Fisheries, 1966.
J. Tokyo Univ. Fish. (Spec.ed) 8(3):1-44.

chemistry, oxygen (data only)
Jarrige, F., J. Merle et J. Noel, 1968.
Résultats des observations physico-chimiques de la croisière CYCLONE 4 du N.O. Coriolis.
Rapp. sci., O.R.S.T.O.M., Noumea, 25:22 pp.

K

chemistry, oxygen (data only)
Katsuura, Hiroshi, Hideo Akamatsu, Tsutomu Akiyama, 1964.
Preliminary report of the Japanese Expedition of Deep-Sea, the Eighth Cruise (JEDS-8).
Oceanogr. Mag., Tokyo, 16 (1/2):125-136.

chemistry, oxygen (data)
Kiilerich, A., 1957.
Galathea-Ekspeditionens arbejde i Phillipinergraven. Ymer, 1957(3):200-222.

chemistry oxygen (data)
King, J.E., T.S. Austin and M.S. Doty, 1957.
Preliminary report on Expedition EASTROPIC.
USFWS Spec. Sci. Rept., Fish., 201:155 pp.

chemistry oxygen (data)
Koczy, F., 1954.
Swedish observations.
Ann. Biol., Cons. Perm. Int. Expl. Mer, 10:134-136.

oxygen (data only)
Korea, Republic of, Fisheries Research and Development Agency, 1971
Annual report of oceanographic observations, 19:717 pp.

chemistry Oxygen (data only)
Korea, Fisheries Research and Development Agency, 1968.
Annual report of oceanographic observations, 16: ~~961~~ 691 pp.

chemistry, oxygen (data only)
Korea, Fisheries College, Pusan, 1968.
Baek-Kyung-Ho cruise to the central Pacific Ocean, 1967.
Data Rept. Oceanogr. Obs. Expl. Fish., 1:30 pp.

chemistry oxygen (data only)
Korea, Fisheries Research and Development Agency, 1967.
Annual report of oceanographic observations, 15 (1966):459 pp.

chemistry, oxygen (data only)
Korea (Republic of), Hydrographic Office, 1965.
Technical reports.
H.O. Pub., Korea, No. 1101:179 pp.

chemistry oxygen (data only)
Koroleff, Folke, 1959.
The Baltic cruise with R/V Aranda 1958, hydrographic data.
Merent. Julk. (Havforskningsinstitutets Skr.), No. 193:25 pp.

Chemistry, oxygen (data only)
Koroleff, Folke, and Aarno Voipio, 1963
The Finnish Baltic cruise 1960. Hydrographical data.
Havsforskningsinst. Skrift (Merent. Julk.), No. 204: 27 pp.

Chemistry, oxygen (data only)
Koroleff, Folke, and Aarno Voipio, 1961.
The Baltic cruise with R/V Aranda, 1959. Hydrographical data.
Merent. Julk. (Havforskningsinst. Skrift), No. 197:26 pp.

Chemistry - oxygen (data only)
Krauel, David P., 1969/
Bedford Basin data report, 1967.
Techn. Rept. Fish. Res. Bd., Can., 120:84 pp (multilithed).

L

chemistry, oxygen (data only)
Lacombe, Henri 1969.
Résultats des mesures d'hydrologie et de courants effectuées à bord de la Calypso. Projet "Mediterranean outflow" du Sous-Comité de Recherches Océanographiques du Conseil Scientifique de l'OTAN (septembre-octubre 1965).
Cah. océanogr. 21. (Suppl. 1):1-48.

Chemistry oxygen (data only)
Lemasson, L., et J.P. Rebert, 1968.
Observations de courants sur le plateau continental Ivoirien: mise en evidence d'un sous-courant. Doc. Sci. Provisoire, Min. Product. Animale Cote d'Ivoire, (022):66 pp. (mimeographed).

Chemistry, oxygen (data only)
Longhurst, A.R., 1961.
Cruise report of oceanographic cruise 6/61.
Federal Fisheries Service, Nigeria, 4 pp. (mimeographed).

Chemistry, oxygen (data only)
Love, C.M., 1966.
Physical, chemical, and biological data from the northeast Pacific Ocean: Columbia river effluent area, January-June 1963. 6. Brown Bear Cruise 326:13-23 June:CNAV Oshawa Cruise Oshawa-3:17-30 June.
Univ. Washington, Dept. Oceanogr., Tech. Rep., No. 134:230 pp.

Chemistry oxygen (data)
Liubimowa, E.M., 1959.
Vertical distribution of organic phosphorus in the Black Sea. Hydrometeorology. Hydrochemistry. (In Russian).
Trudy Morsk. Gidrofiz. Inst., 16:127-160.

M

chemistry oxygen (data only)
Magnier, Y., H. Rotschi, B. Voituriez et J. Merle 1968.
Résultats des observations physico-chimiques de la croisière CYCLONE 3 du N.O. Coriolis.
Rapp. sci., O.R.S.T.O.M., Nouméa, 21:23 pp. (mimeographed)

chemistry, oxygen (data only)
Magnier, Yves, Pierre Rual et Bruno Voituriez 1968.
Resultats des observations physico-chimiques de la croisière CYCLONE 6 du N.O. Coriolis.
Rapp. sci. ORSTOM Noumea 27: 23pp. (mimeographed)

Chemistry, oxygen (data only)
Marvin, K.T., 1955.
Oceanographic observations in west coast Florida waters, 1949-52. U.S. Fish & Wildlife Service, Spec. Sci. Rept., No. 149:6 pp., 2 figs., numerous pp tables.

Mc

Chemistry oxygen (data only)
McGary, J.W., 1954.
Mid-Pacific oceanography, Pt. 6. Hawaiian offshore waters, December 1949-November 1951.
Spec. Fish. Rept:Fish. No. 152:1-138, 33 figs.

Chemistry oxygen (data only)
McGary, James W., and Joseph J. Graham, 1960.
Biological and oceanographic observations in the central north Pacific, July-September 1958.
U.S.F.W.S. Spec. Sci. Rept., Fish., No. 358: 107 pp.

chemistry oxygen (data only)
McGary, J.W., E.C. Jones and T.S. Austin, 1955.
Mid-Pacific oceanography. IX. Operation NORPAC.
Spec. Sci. Repts.: Fish. 168:127 pp.

Chemistry oxygen (datá)
McGary, J.W., and E.D. Stroup, 1956.
Mid-Pacific oceanography. VIII. Middle latitude waters, January-March 1954.
USFWS Spec. Sci. Rept., Fish., 180:173 pp.

chemistry oxygen (data only)
Mexico, Secretaria de Marina, 1964.
Campaña oceanografica Atlantico "Operación Neptuno", plataforma continental del Golfo de México entre Punta Delgada y Punta Roca partida del 5 al 18 de mayo de 1963, 170 pp.

chemistry, oxygen (data)
Midttun, L., and J. Natvig, 1957.
Pacific Antarctic waters. Scientific results of the "Brategg" Expedition, 1947-48, No. 3.
Publikasjon, Komm. Chr. Christensens Hvalfangstmus., Sandefjord, No. 20:1-130, 39 figs.

Chemistry oxygen (data only)
Minas, H.J. 1971.
Résultats de la campagne "Mediprod I" du Jean Charcot.
Cah. océanogr. 23 (Suppl. 1):93-144.

Chemistry oxygen (data only)
Minas, H.J., 1963.
Contribution préliminaire à l'étude hydrologique et hydro-chimique du Golfe de Marseille.
Rec. Trav. Sta. Mar. Endoume, 28(43):3-16.

chemistry, oxygen (data only)
Mosby, Håkon 1969.
Norwegian Atlantic Current, March and December 1965.
Rept. Geophys Inst. Div. Phys. Oceanogr., Univ. Bergen 17: 53pp.

Chemistry oxygen (data only)
Mosby, Håkon, 1965.
Oceanographical tables from Weather Ship Station A (62° N, 33° W), 1954-1965.
Univ. i Bergen, Geofysisk Inst., 85 pp. (Mimeographed).

Chemistry oxygen (data only)
Mosby, Håkon, 1964.
Oceanographical tables from weather ship station M 1959-1963.
Geofysisk Institutt, Universitet i Bergen, 48 pp.

chemistry oxygen (data only)
Mosby, Håkon, 1963?
Oceanographical tables from Weather Ship Station M (66°N, 2°E).
Univ. i Bergen, Geofysisk Inst. 114 pp.

Also in: Collected Papers, Weather Station M, 66N, 2°E, Univ. Bergen, Geophys. Inst., 1963.

N

Chemistry, oxygen (data only)
*Nasu, Keiji, and Tsugio Shimano, 1966.
The physical results of oceanographic survey in the south east Indian Ocean in 1963/64.
J. Tokyo Univ. Fish. (Spec.ed.) 8(2):133-164.

Chemistry oxygen (data only)
NORPAC Committee, 1955.
Oceanic observation of the Pacific, 1955.
The NORPAC data, 532 pp.

oxygen (data only)
(NATO) Organisation du Traité de l'Atlantique Nord, Sous-Comité Océanographique, 1969
Résultats des mesures d'hydrologie et de courants faits à bord du navire océanographique français Calypso (1er septembre-1er octobre 1965). Résultats des mesures d'hydrologie faits à bord du navire océanographique français Origny. Projet Mediterranean Outflow (septembre-octobre 1965)
Rapp. Techn. OTAN H5: 192 pp. (multilithed)

chemistry, oxygen (data only)
Nishihira, Moritaka 1968.
Oceanographical conditions observed at definite station off Asamushi 1967.
Bull. mar. biol. Sta. Asamushi 13(2):151.

O

Chemistry oxygen (data only)
Oren, O.H., 1967.
The fifth cruise in the eastern Mediterranean, Cyprus-05, May 1967. Hydrographic data.
Bull.Sea Fish.Res.Stn Israel, 47:55-63.

Chemistry oxygen (data only)
Owen, R.W., Jr., 1967.
Northeast Pacific albacore- oceanography data, 1962-64.
U.S. Fish Wildl. Serv., Data Rep., 15:47 pp. (microfiche - 1 card).

P

Chemistry oxygen (data only)
Patten, Bernard C., David K. Young and Morris Roberts, 1963.
Suspended particulate material in the lower York River, Virginia, June 1961-July 1962.
Virginia Inst. Mar. Sci., Spec. Sci. Rept. (44): unnumbered pages (unpublished mss.)

chemistry, oxygen (data only)
Peluchon, Georges 1965.
Campagne Alboran I. Hydrologie en mer d'Alboran. Résultats des mesures faites à bord des navires Eupen (Belgique), Origny (France), Xauen et Segura (Espagne) en juillet-août 1962. Présentation des résultats.
Cah. océanogr. 17 (10) (Suppl. 2):219 pp.

chemistry, oxygen (data only)
Peluchon, Georges 1965.
Campagne Alboran I. Hydrologie en mer d'Alboran. I. Résultats des mesures faits à bord des navires Eupen (Belgique) et Origny (France) en juillet-août 1962. Présentation des résultats.
Cah. océanogr. 17 (Suppl. 1):1-88

chemistry, oxygen (data only)
Piton, B., et Y. Magnier 1971.
Observations physico-chimiques faites par le Vauban le long de la côte nord-ouest de Madagascar de janvier à septembre 1970.
Doc. scient. Cent. Nosy-Bé, Off. Rech. scient. techn. Outre-Mer 21: unnumbered pp. (mimeographed)

chemistry, oxygen (data only)
Platt, Trevor, and Brian Irwin 1971.
Phytoplankton production and nutrients in Bedford Basin, 1969-70.
Techn. Rept. Fish. Res. Bd Can. 247:172pp. (multilithed)

chemistry oxygen (data only)
Portugal, Instituto Hidrografico, 1965.
Resultados das observacões oceanograficas no Canal de Mocambique, Cruzeiro al 1/64:abril-maio 1964.
Servico de Oceanografia, Publ.1:73 pp.,46 figs.

chemistry oxygen (data only)
Portugal, Instituto Hidrográfico, Servico de Oceanografia, 1965.
Resultados das observações oceanograficas no Canal de Moçambique, Cruzeiro al 1/64: Abril-Maio 1964, 73 pp., 46 figs.

chemistry, oxygen (data only)
Postma, H., 1959.
Chemical results and a survey of water masses and currents. Tables: oxygen hydrogen ion, alkalinity and phosphate.
Snellius Exped., Eastern Part of the East Indian Archipelago, 1929-1930, Chem. Res., 4:35 pp., 1 fig.

Q

R

chemistry oxygen (data only)
Republica de Chile, Instituto Hidrográfico de la Armada,1966.
Operación Oceanográfica Marchile III: datos fisico-quimicos. Unnumbered pp.

chemistry oxygen (data only)
Republic of China, Chinese National Committee on Oceanic Research, Academia Sinica, 1968.
Oceanographic Data Report of CSK, 2: 126 pp.

chemistry, oxygen (data only)
Republic of China, Chinese National Committee on Oceanic Research, Academia Sinica 1966.
Oceanographic data report of CSK 1:129pp.

chemistry oxygen (data only)
Republica de Chile, Instituto Hidrográfico de la Armada,1965.
Operación Oceanográfica Marchile II: datos fisico-quimicos y bathimetria, realizada por el A.G.S. Yelcho entre el 5 de Julio y el 4 de Agosto de 1962, Unnumbered.

chemistry oxygen (data only)
Republic of Korea,Hydrographic Office,1967.
Hydrographic and oceanographic data.
Techn. Reps.1967,H.O.Pub.,Korea,1101:1-72.

chemistry, oxygen (data only)
Republic of Korea Fisheries Research and Development Agency 1966.
Annual Report of observations 1965, 14: 343pp.

chemistry, oxygen (data only)
Republic of Korea, Hydrographic Office, 1966.
The data for cooperative study of Kuroshio.
Tech. Rep. H.O. Pub. No. 1101: 41-73.

chemistry, oxygen (data only)
Republic of Korea, Hydrographic Office 1964.
Technical reports
H.O. Publ. 1101: 136pp.

chemistry oxygen (data only)
Rotschi, Henri, 1958
"Orsom III", Océanographie physique.
Rapp. Tech. de la Croisière 56-5, Inst. Français d'Océanie, Rapp. Sci., No. 5: 35 pp (mimeographed).

chemistry oxygen (data only)
Rotschi, Henri, 1957
"Orsom III", Oceanographie physique.
Rapp. Tech. de l'Expedition EQUAPAC (Croisiere 56-4), Inst. Francais d'Oceanie, Rapp. Sci., 3: 52 pp. (mimeographed).

chemistry oxygen (data only)
Rotschi, Henri, Michel Angot, Michel Legand and Roget Desrosieres, 1961
Orsom III, Resultats de la Croisiere "Dillon", 2eme Partie. Chimie et Biologie.
ORSTOM, Inst. Francais d'Oceanie, Centre d'Oceanogr., Rapp. Sci., No. 19: 105 pp. (mimeographed).

chemistry oxygen (data only)
Rotschi,H., Ph.Hisard,L.Lemasson,Y.Magnier, J. Noel et B.Piton,1967.
Resultats des observations physico-chimiques de la croisiere "Alize" du N.O. Coriolis.
Rapp.scient.,Off.Rech.scient.tech.Outre-Mer, Noumea, 2:56 pp. (mimeographed).

chemistry oxygen (data only)
Rotschi, H., Ph. Hisard, L. Lemasson, Y. Magnier, J. Noel et B. Piton, 1966.
Resultats des observations physico-chimiques de la croisiere "Alize".
Centre ORSTROM Noumea Rapp. Sci., No. 28:56 pp. (mimeographed).

chemistry oxygen (data only)
Rotschi, Henri, Michel Legand and Roger Desrosieres, 1961
Orsom III, Croisières diverses de 1960, physique chimie et biologie. ORSTOM, Inst. Francais d'Océanie, Centre d'Océanogr., Noumea, Rapp. Sci., No. 20: 59 pp. (mimeographed).

chemistry oxygen (data only)
Rotschi, Henri, Yves Magnier, Maryse Tirelli, et Jean Garbe, 1964.
Résultats des observations scientifiques de "La Dunkerquoise " (Croisière "Guadalcanal").
Océanographie, Cahiers O.R.S.T.R.O.M., 11(1):49-154.

chemistry oxygen (data only)
Rotschi,H.,G.Pickard,P.Hisard etJ.Cannevet,1968.
Résultats des observations physico-chimiques de la croisière CYCLONE 5 du N.O. Coriolis.
Rapp. sci., O.R.S.T.O.M., Noumea,26:23 pp.

S

chemistry oxygen(data only)
Saelen, O.H., 1962
The natural history of the Hardangerfjord.
3. The hydrographical observations 1955-1956
Sarsia, 6:1-25.

chemistry oxygen (data only)
Saloman, Carl H., Finucane, John H., and John A. Kelly, Jr., 1964.
Hydrographic ovservations of Tampa Bay, Florida and adjacent waters, August 1961 through December 1962.
U.S. Dept. Interion, Fish Wildl. Serv., Bur. Comm. Fish., Data Rept., 4:6 cards(microfiche).

chemistry, oxygen
Saloman, Carl H., and John L. Taylor 1968.
Hydrographic observations in Tampa Bay, Florida, and the adjacent Gulf of Mexico, 1965-1966.
Data Rept., U.S. Fish Wildl. Bur. Comm. Fish. 6 cards (microfiche)

chemistry oxygen(data only)
* Scaccini Cicatelli, Marta, 1967.
Distribuzione stagionale dei sali nutrivi in una zona dell'alto e medio Adriatico.
Boll. Pesca, Piscic. Idrobiol., n.s., 22(1): 49-82.

chemistry oxygen (data only)
Schmidt, J., 1912.
Hydrographical observations.
Rept. Danish Oceanogr. Exped., 1908-1910, 1:51-75.

chemistry, oxygen (data only)
Scripps Institution of Oceanography, 1949
Physical and chemical data, Cruise 2, March 28 to April 12, 1949. Physical and Chemical Data Report No.2:10 figs. (ozalid) tables of data (mimeographed).

chemistry, oxygen (data only)
Scripps Institution of Oceanography, 1949
Physical and chemical data, Cruise 1, February 28 to March 16, 1949. Marine Life Research Program. Physical and Chemical Data Report No.1:10 figs. (ozalid), tables of data (mimeographed).

chemistry oxygen (data only)
Scripps Institution of Oceanography, 1949
Marine Life Research Program. Progress Report to 30 April 1949, 25 pp. (mimeographed), numerous figs. (photo + ozalid). 1 May 1949.

chemistry oxygen (data chiefly)
Seckel, G.R., 1955.
MidPacific oceanography. VII. Hawaiian offshore waters, Sept. 1952-August 1953.
Spec. Sci. Repts., Fish., No. 164:250 pp., 38 figs

chemistry oxygen (data only)
Shipley, A.M., and P. Zoutendyk, 1964.
Hydrographic and plankton data collected in the south west Indian Ocean during the SCOR International Indian Ocean Expedition, 1962-1963.
Univ. Cape Town, Inst. Oceanogr., Data Rept., No. 2:210 pp.

chemistry, oxygen (data only)
South Africa, Division of Sea Fisheries, Department of Industry 1968.
Hydrological station list 1961-1962, 1: 342pp.

chemistry oxygen (data only)
South Africa, Department of Commerce and Industries, Division of Sea Fisheries, 1964.
Thirty-second Annual Report for the period 1st April, 1960 to 31st March 1961:267 pp.

chemistry oxygen (data only)
Spain, Instituto Español de Oceanografia, 1961.
Campañas biologicas del "Xauen" en las costas del Mediterraneo Marroqui, Mar de Alboran, Baleares y Noroeste y Cantabrico Españoles en los años 1952, 1953 y 1954.
Bol. Inst. Español Oceanogr., 103:1-130.

chemistry oxygen(data only)
Spain, Instituto Español de Oceanografia, 1955.
Campaña del "Xauen" en la costa noroeste de España en 1949 y 1950. Registro de operaciones.
Bol. Inst. Español Ocean., 71:72 pp.

chemistry, oxygen (data only)
Svansson, Artur, compiler 1970.
Hydrographical observations on Swedish lightships and fjord stations in 1968.
Rept. Fish. Bd. Swedn, Ser. Hydrogr. 24: 83pp.

Chemistry oxygen (data only)
Svansson, Artur, 1966.
Hydrographical observations on Swedish Lightships and fiord stations.
Fish. Bd., Sweden, Ser. Hydrogr., Rept. No. 18: 95 pp.

Chemistry oxygen (data only)
Sweden, Havsfiskelaboratoriet, Lysekil 1971.
Hydrographical data January-June 1970, R.V. Skagerak, R.V. Thetis and TV252, 1970.
Meddn. Hydrogr. avd. Göteborg, 104: unnumbered pp. (multilithed).

Chemistry oxygen (data only)
Sweden, Havsfiskelaboratoriet, Lysekil 1970.
Hydrographical data 1966, R.V. Skagerak, R.V. Thetis.
Meddn Hydrogr. avd. Göteborg 85: 255 pp.

Chemistry oxygen (data only)
Sweden, Havsfiskelaboratoriet, Lysekil 1970.
Hydrographical data, R.V. Skagerak, R.V. Thetis.
Hydrogr. avd. Göteborg
Meddn. 84: 296 pp. (multilithed).
pH (data only)

Chemistry oxygen (data only)
Sweden, Havsfiske laboratoriet, lysekil, 1969.
Hydrographical data, January-June 1968.
R.V. Skagerak, R.V. Thetis.
Meddn Havsfiskelab., Lysekil, hydrog. Avdeln., Goteborg 63: numerous unnumbered pp. (multilithed).

chemistry oxygen (data only)
Sweden, Havsfiskelaboratoriet, Lysekil, 1968.
Hydrographical data, Kuly-December 1967 R.V. Skagerak.
Meddn Havsfiskebal., Lysekil, 52: unnumbered pp. (mimeographed).

chemistry oxygen (data only)
Sweden, Havsfiskelaboratoriet, Lysekil, 1968.
Hydrographical data, Xuly-December 1967, R.V. Skagerak.
Meddn Havsfiskelab., Lysekil, Hydrogr. Avd., Göteborg, 52 : unnumbered pp. (multilithed).

chemistry oxygen (data only)
Sweden, Havsfiskelaboratoriet, Lysekil, Hydrografiska Avd., Göteborg, 1967.
Hydrographical data, January-June 1967, R.V. Thetis.
Meddn. Havsfiskelab., Lysekil, Hydrogr. Avd., Göteborg, 41: numerous pp. (mimeographed).

Chemistry oxygen (data only)
Sweden, Havsfiskelaboratoriet, Lysekil, Hydrografiska Avd., Göteborg, 1967.
Hydrographical data, January-June 1967, R.V. Skagerak.
Meddn. Havsfiskelab., Lysekil, Hydrogr. Avd., Göteborg, 38: numerous pp. (mimeographed).

Chemistry oxygen (data)
Szarejko-Lukaszewicz, D., 1957.
[Qualitative investigations of phytoplankton of Firth of Vistula in 1953.]
Prace Morsk. Inst. Rybacki, Gdyni, No. 9:439-451.

T

Chemistry oxygen (data only)
Tchernia, Paul, et Michèle Fieux 1971.
Résultats des observations hydrologiques executées à bord du N/O Jean Charcot pendant la campagne MEDOC 1969 (30 janvier - 28 février) (18 mars - 31 mars).
Cah. océanogr. 23 (Suppl. 1): 1-91.

Chemistry, oxygen (data only)
Trotti, L. 1967.
Crociera Golfo Palmas e Canale di Sardegna 1965: dati oceanografici.
Pubbl. Ist. Speriment. Talassogr. Trieste, 439: unnumbered pp. (quarto)

Chemistry, oxygen (data only)
Trotti, L. 1967.
Crociera Bocche di Bonifacio 1964.
Pubbl. Ist. Speriment. Talassogr. Trieste 438: unnumbered pp. (quarto).

Chemistry, oxygen (data only)
Trotti, L. 1967.
Dati oceanografici raccolti durante l'A.G.I. 1957-1958 dal Centro Talassografico Tirreno.
Pubbl. Ist. Speriment. Talassogr. Trieste 437: unnumbered pp. (quarto)

Chemistry oxygen (data only)
Trotti, L., 1954.
Risultati delle crociere talassografiche nel Mar Ligure e nell'alto Tirreno. Introduzione, osservazioni meteorologiche e idrografiche.
Centro Talassografico Tirreno, Pubbl. No. 14: 7 pp., photos, 12 fold-in tables.

Chemistry oxygen (data only)
Trotti, L., 1953.
Risultati delle crociere talassografiche nel Mar Ligure e nell'alto Tirreno. Introduzione, osservazione meteorologiche e idrografiche.
Centro Talassograf. Tirreno, Pubbl. No. 14:12 pp.

Chemistry oxygen (data only)
Tsubata, B., 1958.
Oceanographical conditions observed at definite station off Asamushi during 1956.
Bull. Mar. Biol. Sta., Asamushi, 9(1):43.

Chemistry oxygen (data only)
Tsubata, B., 1956.
Oceanographical observations off Asamushi during 1954. Bull. Mar. Biol. Sta., 8(1):43-48.

U

Chemistry, oxygen (data only)
Uda, Michitaka, Yoshima Morita and Makoto Ishino 1957.
Results from the oceanographic observations in the North Pacific (1955-56) with Umitaka Maru and Shinyo Maru.
Rec. Oceanogr. Wks, Japan (Spec. No.): 1-20.

Chemistry oxygen (data only)
Union of South Africa, Department of Commerce and Industry, 1965.
Station list.
Div. Sea Fish., Invest. Rept., No. 51:50-67.

Chemistry oxygen (data only)
U.S.A. Johns Hopkins University, Chesapeake Bay Institute, 1962.
Temperature and salinity data collected in the Chesapeake Bay and tributary estuaries during the period 1 February 1956 and 9 February 1956.
Data Report 49 (Ref. 62-23):22 pp. (unpublished mss.).

Chemistry oxygen (data only)
United States, Johns Hopkins University, Chesapeake Bay Institute, 1962.
Cruise 28, July 24-August 7, 1962, Data Rept., 48, Ref. 62-16:17 pp. (mimeographed).

Chemistry, oxygen (data only)
United States, National Oceanographic Data Center 1965.
Data report, EQUALANT III.
Publ. Nat. Oceanogr. Data Cent. Gen. Serv. G-7: 339 pp.

Chemistry oxygen (data only)
United States, National Oceanographic Data Center 1964.
Data report EQUALANT II.
NODC Gen. Ser., Publ. G-5: numerous pp. (not serially numbered; loose leaf - $5.00).

Chemistry oxygen (data only)
U.S. Navy Hydrographic Office, 1960.
Oceanographic observations, Arctic waters, Task Force Five and Six, summer-autumn 1956, USS Requisite (AGS-18), USS Eldorado (AGC-11), USS Atka (AGB-3) and USCGC Eastwind (WAGB-279).
U.S. Navy Hydrogr. Off., Techn. Rept., TR-58: 89 pp.

Chemistry oxygen (data only)
United States, U.S. Coast Guard, 1965.
Oceanographic cruise USCGC Northwind: Chukchi, East Siberian and Laptev seas, August-September, 1963.
U.S.C.G. Oceanogr. Rept., No. 6(CG373-6):69 pp.

Chemistry oxygen (data only)
U.S.A., U.S. Coast Guard, 1964.
Oceanographic cruise, USCGC NORTHWIND, Bering and Chukchi Sea, July-Sept. 1962.
U.S.C.G. Oceanogr. Rept., No. 1(CG373-1):104 pp.

Chemistry oxygen (data only)
University of California, Scripps Institution of Oceanography, 1960.
Oceanic observations of the Pacific, 1950, 508 pp.

V

Chemistry, oxygen (data only)
Valdez, Alberto J., Alberto Gomez, Aldo Orlando, y Andres J. Lusquiños 1969.
Datos y resultados de la campaña pesqueria V "Pesqueria V" (28 de agosto al 7 de setiembre de 1967).
Publ. Ser. Inf. técn. Mar del Plata 10(5): unnumbered pp.

Chemistry, oxygen (data only)
Valdez, Alberto J., Alberto Gomez, Aldo Orlando y Andres J. Lusquiños 1968.
Datos y resultados de la campaña pesqueria "Pesqueria IV" (7 de junio al 4 de julio de 1967).
Publ. Ser. Inf. técn. Mar del Plata, 10(4): 1-159.

Chemistry, oxygen (data only)
Vicariot, Jean, 1967.
Résultats des mesures faites à bord du navire Origny en Méditerranée occidentale sur le méridien 6° est (12 septembre 1962 - 7 mai 1963).
Cah. océanogr., 19(Suppl. 1):71-155.

chemistry, oxygen (data only)
Villanueva Sebastian F., Alberto Gomez, Aldo Orlando y Andres J. Lusquiños 1969.
Datos y resultados de la campaña pesquería "Pesquería VII" 16 de febrero al 1° de marzo de 1968.
Publ. Ser. Inf. Tecn. Proyecto Desarrollo Pesq. Mar del Plata (10/VII): unnumbered pp.

W

Chemistry, oxygen (data only)
Wilson, R.C., and M.O. Rinkel, 1957.
Marquesas area oceanographic and fishery data, January-March 1957.
USFWS Spec. Sci. Rept., Fish., No. 238:136 pp.

Chemistry oxygen (data)
Wojnicz, B., 1957.
[Deversement d'eaux salées dans la Baltique observé au mois de novembre 1951.] Przeglad Geofiz., Rocznik II(10)Zeszyt 1/2):53-58.

XYZ

chemistry, oxygen (data only)
Zoutendyk, P., and D. Sacks 1969.
Hydrographic and plankton data 1960-1965.
Data Rept. Inst. Oceanogr. Univ. Cape Town (5): 82 pp.

oxygen (data only)
Zvereva, A.A., Edit., 1959.
[Data, 2nd Marine Expedition, "Ob", 1956-1957.]
Arktich. i Antarkt. Nauchno-Issled. Inst. Mezhd. Geofiz. God, Sovetsk. Antarkt. Exped., 6: 1-387.

Oxygen, deep

Chemistry oxygen, deep
Brouardel, J., and L. Fage, 1955.
Variation, en mer, de la teneur en oxygène dissous au proche voisinage des sediments.
Pap. Mar. Biol. and Oceanogr., Deep-Sea Res., Suppl. to Vol. 3:40-44.

oxygen deficit

Chemistry oxygen deficit
Fedosov, M.V., 1955.
[The reasons for the origin of the oxygen deficit in the Sea of Azov.] Trudy VNIRO, 31:80-94.

chemistry oxygen deficit
Fonselius, Stig H., 1960.
Hydrography of the Baltic deep basins. III.
Rept., Fish. Bd., Sweden, Ser. Hydrogr., 23: 1-97.

Chemistry oxygen deficiency
Kändler, Rudolf 1971.
Untersuchungen über die Abwasserbelastung der Untertrave.
Kieler Meeresforsch. 27(1): 20-28.

Chemistry oxygen deficit
Kullenberg, Gunnar 1970.
On the oxygen deficit in the Baltic deep water.
Tellus 22(3): 357-

Chemistry Oxygen deficit
Richards, F.A., 1965.
Anoxic basins and fjords.
In: Chemical oceanography, J.P. Riley and G. Skirrow, editors, Academic Press, 1:611-645.

Chemistry oxygen deficiency
Tulkki, Paavo, 1965.
Disappearance of the benthic fauna from the basin of Bornholm (southern Baltic) due to oxygen deficiency.
Cahiers Biol. Mer. 6(4):455-463.

Chemistry oxygen deficit, effect of
Vinogradov, M.E., and N.M. Voronina, 1961.
[The influence of oxygen deficit upon the plankton distribution in the Arabian Sea.]
Okeanologiia, Akad. Nauk, SSSR, 1(4):670-678.

Oxygen demand

Chemistry oxygen demand
Grindley, J., and A.B. Wheatland, 1956.
Salinity and the biochemical oxygen demand of estuary water. Wat. Sanit. Eng., 6:10-14.

Abstr. in Anal. Abstr., 4(10):#3475.

Chemistry, oxygen demand
Wheatland, A.B., and J. Whaeatland, 1956.
Salinity and the biochemical oxygen demand of estuary water. Water & Sanit. Eng. 6:10-14.

Abstr. Publ. Health Eng. Abstr. 36(12):27.

Oxygen, density

Chemistry oxygen-density
Brazil, Diretoria de Hidrografia e Navegacao, 1957.
Ano Geofisico Internacional, Publicacao DG-06-VI: unnumered mimeographed pages.

Chemistry oxygen-density
Brasil, Diretoria de Hidrografia e Navegacao 1957.
Ano Geofisico Internacional. Publicacao DG-06-V: 3 pp., 18 figs. 3 pp. of tables

Chemistry, oxygen/density
Brazil, Diretoria de Hidrografia e Navegacão, 1957.
Ano Geofisico Internacional. Publicacão DG-06-II mimeographed.

Chemistry oxygen-density
Brasil, Diretoria de Hidrografia e Navegacao, 1957.
Ano Geofisico Internacional. Publ. DG-06-III (mimeographed)

chemistry, oxygen (density data)
Great Britain, Fisheries Laboratory, Lowestoft, 1954.
Research Vessel 'Ernest Holt', hydrographical observations:130 pp.

Chemistry oxygen-density
Ichiye, T., 1962.
Circulacion y distribucion de la masa de agua en el Golfo de Mexico. (In Spanish and English).
Geofisica Internacional, Rev. Union Geofis. Mexicana, Inst. Geofis., Univ. Nacional Autonoma de Mexico, 2(3):47-76, 22 figs.

Chemistry oxygen-density
Ichiye, T., 1956.
On the distributions of oxygen and their seasonal variations in the adjacent seas of Japan (IV).
Ocean. Mag., Tokyo, 8(1):1-27.

Chemistry oxygen-density correlation
Kawamoto, T., 1957.
On the oxygen-density correlation in the northwestern North Pacific Ocean. Ocean. Mag., Tokyo, 9(1):65-73.

Chemistry, oxygen-density
Matsudaira, Y., T. Kawamoto and H. Kitamura, 1956.
On the distributions of dissolved oxygen in the Kuroshio region. J. Fac. Fish. & Animal Husbandry, Hiroshima Univ., 1(2):297-301.

Chemistry oxygen-density
Richards, F.A., and A.C. Redfield, 1955.
Oxygen-density relationships in the western North Atlantic. Deep-Sea Res. 2(3):182-199, 14 textfigs.

Chemistry oxygen-density correlations
Stefansson, Unndteinn, and Francis A. Richards, 1964.
Distributions of dissolved oxygen, density and nutrients off the Washington and Oregon coasts.
Deep-Sea Res., 11(3):355-380.

Oxygen-density-temperature

Chemistry, oxygen/Sigma-t
Canadian Oceanographic Data Centre, 1965.
Data record, Seguenay and Gulf of St. Lawrence, August 19 to August 30, 1963.
1965 Data Record Series, No. 12:123 pp. (multilithed).

Oxygen depletion

Chemistry Oxygen depletion
De Decker, A.H.B., 1970.
Notes on an oxygen-depleted subsurface current off the west coast of South Africa.
Invest. Rept., Div. Sea Fish. S.Afr. 84: 1-24

Chemistry oxygen depletion
Okubo, A., 1956.
On an oxygen depletion-phosphate diagram (ΔO_2-P diagram) of the Kuroshio-Oyashio waters. 1. A hypothesis. Ocean. Mag., Tokyo, 8(1):79-91.

Chemistry, "oxygen depletion anomaly"
Sugiura, Yoshio 1966.
Reply to Dr. Fukai's comments.
J. Oceanogr. Soc. Japan 22(1): 19.

oxygen determination

oxygen determination
See chiefly: methods, chemistry (oxygen)

Chemistry, oxygen determination
*Solovyev, L.G., 1967.
Some examples and peculiarities of oxygen determinations in sea water with the aid of the oxmeter ICAN. (In Russian; English abstract)
Trudy Inst. Okeanol., 83(63-67.

Oxygen, effect of

Adrov, N.M., 1964. Chemistry oxygen, effect of
On the problem of the interrelationships of temperature, oxygen and fish concentrations in some commercial fishing areas of the Barents and Norwegian seas. (In Russian).
Trudy, Poliarn. Nauchno-Issled. i Proektn. Inst. Morsk. Ribn. Choz. i Okeanogr. im N.M. Knipovicha, 16:251-268.

Chemistry oxygen, effect of
Callame, B., 1954.
Periodes de fixation de quelques organismes marine sessiles, en rapport avec les conditions du milieu, dans le Port de la Pallace.
Trav. C.R.E.O. 1(7):8 pp., 12 textfigs.

Chemistry, oxygen effect of
Guinther, Eric B. 1971.
Ecologic observations on an estuarine environment at Fanning Atoll.
Pacific Science 25(2): 249-259.

Chemistry oxygen, effect of
Hallam, A., 1965.
Environmental causes of stunting in living and fossil marine invertebrates.
Paleontology, 8(1):132-155.

chemistry oxygen, effect of

McLeese, D.W., 1956.
Effects of temperature, salinity and oxygen on the survival of the American lobster.
J. Fish. Res. Bd., Canada, 13(2):247-272.

chemistry oxygen, effect of

Ramsey, W.L., 1962.
Bubble growth from dissolved oxygen near the sea surface.
Limnol. & Oceanogr., 7(1):1-7.

chemistry, oxygen, effect of

Theede, H. und A. Ponat, 1970.
Die Wirkung der Sauerstoffspannung auf die Druckresistenz einiger mariner Wirbelloser.
Marine Biol., 6(1): 66-73.

chemistry oxygen, effect of

Wise, J.P., 1958.
Cod and hydrography, a review.
USFWS Spec. Sci. Rept., Fish., 245:16 pp.

oxygen-errors of Winkler method

chemistry oxygen (errors of Winkler method)

Grasshoff, Klaus, 1962
Untersuchungen über die Sauerstoffbestimmung im Meerwasser.
Kieler Meeresf., 18(1):42-50.

oxygen estuarine

chemistry, oxygen (estuarine)

McHugh, J.L., 1967.
Estuarine nekton.
In: Estuaries, G.H. Lauff, editor, Publs Am. Ass. Advmt. Sci., 83:581-620.

oxygen exchange rates

chemistry oxygen exchange rates

Pytkowicz, Ricardo M., 1964.
Oxygen exchange rates off the Oregon coast.
Deep-Sea Res., 11(3):381-389.

oxygen fjords

chemistry oxygen (fjords)

Saelen, Odd H., 1967.
Some features of the hydrography of Norwegian fjords.
In: Estuaries, G.H. Lauff, editor, Publs Am. Ass. Advmt Sci., 83:63-70.

oxygen isotopes

chemistry, oxygen isotopes

Ault, Wayne U., 1959.
Oxygen isotope measurements on Arctic cores.
Sci. Studies, Fletcher's Ice Island, T-3.
Vol. 1, Air Force Cambridge Res. Center, Geophys. Res. Pap. No. 63:159-168.

chemistry, oxygen isotopes

Broecker, Wallace S. and Jan Van Donk 1970
Insolation changes, ice volumes, and the O^{18} record in deep-sea cores.
Rev. Geophys. Space Phys. 8(1): 169-198.

chemistry isotopic composition (oxygen)

Cortecci, G. and A. Longinelli, 1971.
$^{18}O/^{16}O$ ratios in sulfate from living marine organisms.
Earth planet. Sci. Letts 11(4): 273-276

chemistry oxygen-18

*Craig, H., 1969.
Geochemistry and origin of the Red Sea brines.
In: Hot brines and Recent heavy metal deposits in the Red Sea, E.T. Degens and D.A. Ross, editors, Springer-Verlag, New York, Inc., 208-242.

chemistry oxygen isotopes

Craig, H., 1961

Standard for reporting concentrations of deuterium and oxygen-18 in natural waters.
Science, 133(3467): 1833-1834.

chemistry, oxygen isotopes

Craig, H., and L.I. Gordon, 1965.
Isotopic oceanography: deuterium and oxygen 18 variations in the ocean and the marine atmosphere.
Narragansett Mar. Lab., Univ. Rhode Island, Occ. Publ., No. 3:277-374.

chemistry oxygen isotopes

Culkin, Frederick, 1965.
The major constituents of sea water.
In: Chemical oceanography, J.P. Riley and G. Skirrow, editors, Academic Press, 1: 121-161.

chemistry, oxygen isotopes

Degens, E.T., J.M. Hunt, J.H. Reuter and W.E. Reed 1964
Data in distribution of amino acids and oxygen isotopes in petroleum brine waters of various geologic ages.
Sedimentology, Elsevier Publ. Co. 3(3): 199-225

chemistry oxygen, heavy isotope

Dansgaard, W., 1960.

The content of heavy oxygen isotope in the water masses of the Philippine Trench. Deep-Sea Res., 6(4): 346-350.

chemistry, oxygen isotopes

Deuser, W.G., 1968.
Postdepositional changes in the oxygen isotope ratios of Pleistocene foraminifer tests in the Red Sea.
J. geophys. Res., 73(10):3311-3314.

chemistry, oxygen isotopes

Emiliani, C., 1954.
Depth habitats of some species of pelagic foraminifera as indicated by oxygen isotope ratios.
Am. J. Sci. 252:149-158, 4 textfigs.

chemistry oxygen isotopes

Epstein, Samuel, 1962.
The oxygen isotopic compositions of marine waters. (Abstract only).
J. Geophys. Res., 67(9):3555.

chemistry, oxygen isotopes

Epstein, S., and R.P. Sharp, 1959

Oxygen isotope studies. Trans. Amer. Geo. U., 40(1): 81-84. I.G.Y. Bull., No. 21: 81-84

chemistry, oxygen isotopes

Garlick, G.D. and J.R. Dymond 1970.
Oxygen isotope exchange between volcanic materials and ocean water.
Bull. geol. Soc. Am. 81(7): 2137-2142.

chemistry, oxygen isotopes

Longinelli, A., 1966.
Ratios of oxygen-18 : oxygen-16 in phosphate and carbonate from living and fossil marine organisms.
Nature, 211(5052):923-927.

chemistry oxygen-18

Longinelli, A., and H. Craig, 1967.
Oxygen-18 variations in sulfate ions in sea water and saline lakes.
Science, 156(3771):56-59.

Chemistry, oxygen isotope ratios

Longinelli, A. and S. Nuti 1968.
Oxygen isotope ratios in phosphate from fossil marine organisms.
Science 160 (3830): 879-882.

Chemistry, oxygen isotopes

Lloyd, R. Michael 1967.
Oxygen-18 composition of oceanic sulfate.
Science 156 (3779): 1228-1231.

Chemistry, oxygen isotopes

Lloyd, R.M. 1966.
Oxygen isotope enrichment of sea water by evaporation.
Geochim. Cosmochim. Acta 30(8): 801-814.

chemistry, oxygen isotope ratios

Muehlenbachs, K. and R.N. Clayton 1971.
Oxygen isotopes ratios of submarine diorites and their constituent minerals.
Can. J. Earth Sci. 8 (12): 1591-1595.

Chemistry, oxygen isotopes

Savin, Samuel M., and Samuel Epstein 1970.
The oxygen and hydrogen isotope geochemistry of ocean sediments and shales.
Geochim. cosmochim. Acta 34(1): 43-63

Chemistry, oxygen isotopes

Smith, P.B., and C. Emiliani 1968.
Oxygen-isotope analysis of recent tropical Pacific benthonic Foraminifera.
Science 160 (3834): 1335-1336.

Chemistry, oxygen isotopes

Weber, Jon N. 1965.
Concerning the O^{18}/O^{16} composition of ancient oceans. (In Russian).
Geokhimia (6): 674-680.

chemistry, oxygen-18

Weber, Jon N. and Peter M.J. Woodhead, 1972
Temperature dependence of oxygen-18 concentration in reef coral carbonates. J. geophys. Res. 77(3): 463-473.

chemistry, oxygen isotope fractionation

Mopper, Kenneth, and G.D. Garlick 1971.
Oxygen isotope fractionation between biogenic silica and ocean water.
Geochim. Cosmochim Acta 35(11): 1185-1187

Oxygen-maximum layer

Chemistry oxygen (maximum layer)

Le Pichon, Xavier, 1960

The deep water circulation in the southwest Indian Ocean. J. Geophys. Res., 65(12): 4061-4074.

oxygen, monthly values

Chemistry, oxygen(monthly means)
Bose, B.B., 1956.
Observations on the hydrology of the Hooghly Estuary. Indian J. Fish., 1(3):101-118.

Chemistry, oxygen(monthly mean) (data)
Gerard, R., 1964
Étude de l'eau de mer de surface dans une baie de Nosy-Bé.
Océanogr., Cahiers O.R.S.T.R.O.M., 11(2):5-26.

Chemistry, Oxygen, (monthly means)
Newell, B. S., 1966.
Seasonal changes in the hydrological and biological environments off Port Hacking, Sydney.
Australian J. Mar. freshw. Res., 17(1):77-91.

SUMMARY

Chemistry, oxygen, mean monthly values
Prasad, R. Raghu, 1956.
Further studies on the plankton of the inshore waters off Mandapam. Indian J. Fish., 3(1):1-42.

oxygen, monthly means
Slinn, D.J., and Gaenor Offlow (Mrs.M.Solly), 1968.
Chemical constituents of sea water off Port Erin during 1967.
Ann.Rept.mar.biol Stn.Port Erin, 80:37-42.

Chemistry, oxygen(mean monthly) (data)
Szarejko-Lukaszewicz, D., 1959
Hydrographic investigations on the Firth of Vistula in 1953-1954.
Prace Morsk. Inst. Ryback., Gdyni, 10A: 215-228.

Chemistry, oxygen minimum layer, effect of
Childress, James J., 1971.
Respiratory adaptations to the oxygen minimum layer in the bathyp elagic mysid Gnathophausia ingens. Biol. Bull. mar. biol. Lab. Woods Hole 141(1): 109-121.

oxygen-phosphate

Chemistry, O_2/PO_4^{-3}
Broenkow, William W., 1965.
The distribution of nutrients in the Costa Rica dome in the eastern tropical Pacific Ocean.
Limnology and Oceanography 10(1)40-52.

Chemistry, oxygen-phosphate
Inter-American Tropical Tuna Commission, 1961
Annual report for the year 1960: 183 pp.

Chemistry, oxygen-phosphate correlation
Pytkowicz, Ricardo M., 1964.
Oxygen exchange rates off the Oregon coast.
Deep-Sea Res., 11(3):381-389.

oxygen/phosphorus

Chemistry, oxygen/phosphorus
Forsbergh, Eric D., 1963
Some relationships of meteorological, hydrographic, and biological variables in the Gulf of Panama. (In English and Spanish).
Bull., Inter-American Tropical Tuna Comm., 7(1): 109 pp.

Chemistry, oxygen/phosphorus
Okuda, Taizo, 1960
Chemical oceanography in the South Atlantic Ocean, adjacent to north-eastern Brazil. (Portuguese and French resumés).
Trabalhos, Inst. Biol. Marit. e Oceanogr., Universidade do Recife, Brasil, 2(1):155-174.

Chemistry, oxygen reserves
Broecker, Wallace S., 1970.
Man's oxygen reserves. Science, 168(3939): 1537-1538.

Oxygen-salinity

Chemistry, oxygen/salinity
Rotschi, Henri, 1961
Sur certaines propriétés chimiques des eaux equatoriales et tropicales du Pacifique. Caractères generaux de la distribution de l'oxygène dissous. Cahiers Océanogr., C.C.O. E.C. 13(1): 14-31.

Chemistry, oxygen/salinity
Wüst, Georg, 1960.
Die Tiefenzirkulation des Mittelländischen Meeres in den Kernschichten des Zwischen- und des Tiefenwassers.
Deutsche Hydrogr. Zeits., 13(3):105-131.

oxygen saturation

Chemistry, oxygen saturation
Green, E.J. and D.E. Carritt 1967
New tables for oxygen saturation of seawater.
J. mar. Res. 25(2): 140-147

Chemistry, oxygen saturation
Kester, Dana R., and Ricardo M. Pytkowicz, 1968.
Oxygen saturation in the surface waters of the northeast Pacific ocean.
J. geophys. Res., 73(16):5421-5424.

oxygen solubility

Chemistry, oxygen (solubility of)
Gilbert, William E., Walter M. Rawley and Kilho Park 1967.
Carpenter's oxygen solubility table and nomograph for seawater as a function of temperature and salinity.
J. oceanogr. Soc. Japan 23(5):252-255.

chemistry, oxygen solubility
*Green, E.J., and D.E. Carritt, 1967.
Oxygen solubility in sea water: thermodynamic influences of sea salt.
Science, 157(3785):191-193.

Chemistry, oxygen solubility
Truesdale, G.A., and A.L.H. Gameson, 1957.
The solubility of oxygen in saline water.
J. du Cons., 22(2):163-166.

Chemistry, oxygen (supersaturation)
Codispoti, L.A. and F.A. Richards 1971.
Oxygen supersaturations in the Chukchi and East Siberian seas.
Deep-Sea Res. 18(3):341-351.

oxygen surface

Chemistry oxygen, surface
Murty, A.V.S., and P. Udaya Varma, 1964.
The hydrographical features of the waters of Palk Bay during March 1963.
J. Mar. Biol. Ass., India, 6(2):207-216.

Chemistry oxygen (surface, data)
Rotschi, H., 1961.
Chimie.
O.R.S.T.O.M., I.F.O., Rapp. Sci., 23:4-45.

Chemistry oxygen, surface, mean
Wiktor, K., and D. Zembrzuska, D., 1959.
Materials for hydrography of Firth of Szczecin.
Prace Morsk. Inst. Ryback., Gdyni, 10A:259-272

Oxygen, Surface

Chemistry oxygen, surface-bottom(data only)
Dietrich, G., Editor, 1960
Temperatur-, Salzgehalts-und Sauerstoff-Verteilung auf den Schnitten von F.F.S."Anton Dohrn" und V.F.S. "Gauss" im Internationalen Geophysikalischen Jahr 1957/1958.
Deutsche Hydrogr. Zeits., Ergänzungsheft, Reihe B, 4(3):1-103.

oxygen (monthly average, surface and bottom)
Jayaraman, R., 1954.
Seasonal variations in salinity dissolved oxygen and nutrient salts in the inshore waters of the Gulf of Mannar and Palk Bay near Mandapam (S. India). Indian J. Fish. 1(1/2):345-364, 11 textfigs.

Chemistry oxygen(surface bottom)
Krug, Joachim, 1963
Erneuerung des Wassers in der Kieler Bucht im Verlaufe eines Jahres am Beispiel 1960/61.
Kieler Meeresforsch., 19(2):158-174.

Oxygen-temperature

chemistry, oxygen-temperature
See principally: temperature-salinity (T/S)

Chemistry, oxygen-Temperature
Ichiye, T., 1960
On the deep water in the western North Pacific. Oceanogr. Mag., Tokyo, 11(2): 99-110.

Chemistry oxygen tension, effect of
LaRow, Edward J., 1970.
The effect of oxygen tension on the vertical migration of Chaoborus larvae. Limnol. Oceanogr 15(3): 357-362.

oxygen, utilization

Chemistry, oxygen utilization
Carey, Andrew G. Jr. 1967
Energetics of the benthos of Long Island Sound I. Oxygen utilization of sediment.
Bull. Bingham oceanogr. Coll. 19(2):136-144.

Oxygen, vertical distribution

Chemistry, oxygen, vertical distribution
Miyake, Y., and K. Saruhashi, 1956.
On the vertical distribution of the dissolved oxygen in the ocean. Deep-Sea Res., 3(4):242-247

oxygen, yearly range

Chemistry oxygen (yearly range)
Barnard, J.L., 1958.
Amphipod crustaceans as fouling organisms in Los Angeles-Long Beach harbors, with reference to the influence of seawater turbidity.
Calif. Fish & Game, 44(2):160-171.

ozone

Chemistry ozone

Aldaz, Luis 1969.
Flux measurements of atmospheric ozone over land and water. J. geophys. Res., 74(28): 6943-6946.

chemistry, ozone

Kuznetsov, G.I., 1966.
Atmospheric ozone over the tropical belt of the Atlantic. (In Russian).
Dokl. Akad. Nauk, SSSR, 171(3):587-589.

chemistry, ozone (atmosphere)

Kuznetsov, G.I., and A. Kh. Khrgian, 1966.
Atmospheric ozone and its variations correlated with the circulation over the Atlantic Ocean. (In Russian; English abstract).
Fisika Atmosferi i Okeana, Izv., Akad. Nauk, SSSR, 2(8):859-871.

palladium

*Harriss, Robert C., J.H. Crockett and M. Stainton, 1968.
Palladium, iridium and gold in deep-sea manganese nodules.
Geochim. cosmochim. Acta, 32(10):1049-1056.

paraffins

chemistry, paraffins

Clark, Robert C., Jr., and Max Blumer, 1967.
Distribution of n-paraffins in marine organisms and sediments.
Limnol. Oceanogr., 12(1):79-87.

paraffins

*Van der Weide, B., 1967.
Evolution des n-paraffines par traitement thermique de sédiments marins récents.
Bull. Centre Rech. PAU-SNPA, 1(1):161-164.

Chemistry, paraffins

Welte, Dietrich H., und Götz Ebhardt 1968.
Die Verteilung höherer, geradkettiger Paraffin und Fettsäuren in einem Sedimentprofil aus dem Persischen Golf.
Meteor Forschungsergebn. (C) 1:43-52.

n-paraffins

Welte, Dietrich H., and Götz Ebhardt, 1968.
Distribution of long chain n-paraffins and n-fatty acids from the Persian Gulf.
Geochim. cosmochim. Acta, 32(4):465-466

chemistry, pentanes

McIver, R.D., 1971.
Organic geochemical analyses of frozen samples from dsdp leg 8 cores. Initial Repts. Deep Sea Drilling Project, 8: 871-872

perchlorate

Chemistry perchlorate

Greenhalgh, R., and J.P. Riley, 1961.
Perchlorate in sea water.
J.M.B.A., U.K., 41(1):175-186.

Chemistry perchlorate

Greenhalgh, R., and J.P. Riley, 1960.
Alleged occurrence of the perchlorate ion in sea water.
Nature, 187(4743):1107-1108.

peridinin

Chemistry, peridinin

Jeffrey, S.W., and F.T. Haxo 1968.
Photosynthetic pigments of symbiotic dinoflagellates (zooxanthellae) from corals and clams.
Biol. Bull. mar. biol. Lab. Woods Hole 135(1): 149-165.

Chemistry peridinin

Parsons, T.E., and J.D.H. Strickland, 1963.
Discussion of spectrophotometric determination of marine-plant pigments with revised equations for ascertaining chlorophylls and carotenoids.
J. Mar. Res., 21(3):155-163.

Chemistry pesticides

Butler, Philip A., 1968.
Pesticide residues in estuarine mollusks. Proceedings of the National Symposium on Estuarine Pollution (August 1967), Stanford University, Stanford, Calif., p. 107-121, 1968. Also in: Collected Repr. Div. Biol. Bur. Comm. Fish., U.S. Fish Wildl. Serv., 1968, 1.

pesticides

Butler, Philip A., 1968.
Pesticides in the estuary. Proceedings of the Marsh and Estuary Management Symposium (July 1967), Baton Rouge, La.: 120-124. Also in: Collected Repr. Div. Biol. Res., Bur. Comm. Fish., U.S. Fish Wildl. Serv., 1968, 1.

pesticides

Butler, Philip A., 1966.
The problem of pesticides in estuaries.
Am. Fish. Soc. Spec. Publ., (3):110-115.

pesticides

Butler, Philip A., 1966.
Pesticides in the marine environment.
J. appl. Ecol., 3(Suppl.):253-259.

Chemistry, pesticides

Cox, James L. 1970.
Accumulation of DDT residues in Triphoturus mexicanus from the Gulf of California.
Nature, Lond. 227(5254): 192-193.

pesticides

Duffy, J.R., and D. O'Donnell, 1968.
DDT residues and metabolites in Canadian Atlantic coast fish.
J. Fish. Res. Bd., Can., 25(1):189-195.

chemistry, pesticides

Fougeras-Lavergnolle, Jean 1971.
Recherche des pesticides organochlorés dans les milieux littoraux.
Revue Trav. Inst. Pêches marit. 35(3): 367-371.

pesticides

*Holden, A.V., and K. Marsden, 1967.
Organoclorine pesticides in seals and porpoises.
Nature, Lond., 216(5122):1274-1276.

Chemistry, pesticides

Kobayashi, Kunio, Hiroshi Akitake and Tetuo Tomiyama 1970.
Studies on the metabolism of pentachlorophenate, a herbicide, in aquatic organisms. III. Isolation and identification of a conjugated PCP yielded by a shell-fish, Tapes philippinarum. (In Japanese; English abstract)
Bull. Jap. Soc. scient. Fish. 36(1): 103-108.

Chemistry pesticides

Kobayashi, Kunio, Hiroshi Akitake and Tetuo Tomiyama, 1970.
Studies on the metabolism of pentachlorophenate a herbicide in aquatic organisms. I. Biochemical change of PCP in sea water by detoxication mechanism of Tapes philippinarum.
Bull. Jap. Soc. scient. Fish. 36(1): 96-102 (In Japanese; English abstract)

chemistry, pesticides, effect of

Risebrough, R.W., 1971.
Chlorinated hydrocarbons. (10) In: Impingement of man on the oceans, D.W. Hood, editor, Wiley Interscience: 259-286.

pesticides

*Risebrough, R.W., R.J. Huggett, J.J. Griffin and E.D. Goldberg, 1968.
Pesticides: transatlantic movements in the northeast trades.
Science, 159(3820):1233-1235.

pesticides, effect of

Davis, Harry C., and Herbert Hidu, 1969.
Effects of pesticides on embryonic development of clams and oysters and on survival and growth of the larvae.
Fish. Bull. U.S. Dept. Comm. 67(2): 393-404

Chemistry - pesticides, effect of

Ernst, W. 1970.
Stoffwechsel von Pesticiden in marinen Organismen. III. Abbau und Speicherung von DDT-^{14}C in Plattfischen, Solea solea im Kurzzeitversuch.
Veröff. Inst. Meeresforsch. Bremerh. 12(3): 361-364.

chemistry, pesticides effect of

Ernst, W. 1970.
Stoffwechsel von Pesticiden in marinen Organismen. II. Biotransformation und Akkumulation von DDT-^{14}C in Plattfischen: Platichthys flesus.
Veröff. Inst. Meeresforsch. Bremerh. 12(3): 353-360.

Chemistry - pesticides, effect of

Jensen, S., A.G. Johnels, M. Olsson and G. Otterlind, 1969.
DDT and PCB in marine animals from Swedish waters.
Nature, Lond., 224(5216): 247-250.

Chemistry Pesticides, effect

Lane, Charles E., and Robert J. Livingston 1970.
Some acute and chronic effects of dieldrin on the sailfin molly, Poecilia latipinna.
Trans. Am. Fish. Soc. 99(3): 489-495

Chemistry Pesticides, effect of

Macek, Kenneth J., and Sidney Korn, 1970.
Significance of the food chain in DDT accumulation by fish.
J. Fish. Res. Bd., Can., 27(8): 1496-1498.

pesticides, effect of

Mason, James W., II, and Donald R. Rowe, 1969.
How pesticides are threatening our oysters.
Ocean Industry, 4(7):60-61.

pesticides, effect of

pH

A

pH

Ackefors, Hans 1965.
On the zooplankton fauna at Askö (The Baltic - Sweden).
Ophelia, 2(2): 269-280

Chemistry, pH
*Akiyama, Ysutomu, Takeshi Sagi, Takeshi Yura and Yoshisuke Maeda, 1968.
On the distribution of pH in situ and total alkalinity in the western North Pacific Ocean.
Oceanogrl Mag., 20(1):1-8.

Chemistry, pH
Akiyama, Tsutomu, Takeshi Sagi, Takeshi Yura and Ken Kimura, 1966
On the distribution of in situ pH in the adjacent seas of Japan.
Oceanogr. Mag., Jap. Met. Soc., 18(1/2):83-90.

Chemistry, pH
Alander, H., 1949
Hydrography. Ann. Biol., Int. Cons. 4: 139-141; text figs. 1-11.

Chemistry, pH (data)
Aleksandrovskaya, N.B., M.V. Kales and E.A. Yola, 1966.
Peculiarities of the hydrological regime of the Baltic Sea in 1964.
Annls. biol., Copenh., 21:43-45.

Chemistry, pH
Amit, O., and Y.K. Bentor, 1971.
pH-dilution curves of saline waters.
Chem. Geol. 7 (4): 307-313

Chemistry, pH
Aravio-Torre, J., 1946.
Estudio del pH en el agua. Notas y Res. Inst. Español, Ocean., 135L12, Fig. 6.

Chemistry, pH
Aravio-Torre Martínez de Murguía, J., 1946.
Estudio del electrodo de antimonio para la medida del pH en el agua del mar. Notas y Res., Inst. Español Oceanogr., 129:1-14.

Chemistry, pH
*Aubert, M., J. Aubert, S. Daniel, J.-P. Gambarotta, L.-A. Romey, J.*Ph. Angin, P.-F. Bulard, P. Irr et Y. Chevalier, 1967
Dynamique d'eaux résiduaires rejetées en bassin maritime.
Rev. intern. Océanogr. Med. 8:5-40.

B

Chemistry
pH
Băcescu, M., 1961
Cercetări fizico-chimice şi biologice româneşti la Marea Neagră, efectuate în perioda 1954-1959.
Hidrobiologia, Acad. Repub. Pop. Rom., (3): 17-46.

Chemistry, pH
Bakaev, V.G., editor, 1966.
Atlas Antarktiki, Sovetskaia Antarktichekaia Ekspeditsiia. 1.
Glabnoe Upravlenie Geodesii i Kartgrafii MG SSSR, Moskva-Leningrad, 225 charts.

Chemistry pH
Bal, D.V., L.B. Pradhan, and K.G. Gupte, 1946
A preliminary record of some of the chemical and physical conditions in waters of the Bombay harbour during 1944-45. Proc. Indian Acad. Sci. Sect. B24(2):60-73, 4 figs.

Chemistry pH (data)
Ballester, A., 1965.
Hidrografia y nutrientes de la Fosa de Cariaco.
Informe de Progresso del Estudio Hidrografico de la Fosa de Cariaco, Fundación La Salle de Ciencias Naturales Estación de Investigacion-es Marinas de Margarita, Caracas, Sept. 1965. (mimeographed):3- 12.

pH (data)
Bardin, I.P., 1958.
[Hydrological, hydrochemical, geological and biological studies, Research Ship "Ob", 1955-1957.] Trudy Kompleks. Antarkt. Exped., Akad. Nauk, SSSR, Mezh. Geofiz. God, Gidrometeorol. Izdatel., Leningrad, 217 pp.

Chemistry, pH
Barrett, E., and G. Brodin, 1955.
The acidity of Scandinavian precipitation.
Tellus 7(2):251-257, 6 textfigs.

Chemistry, pH
Berthois, L., 1968.
Contribution à l'étude sédimentologique du Kangerdlugssuaq, Côte ouest du Groenland.
Medde Grønland 187 (1): 185 pp.

Chemistry, pH
Berthois, Leopold, 1966.
Hydrologie et sedimentologie dans le Kangerd-lugssuaq (fjord a la cote ouest du Groenland).
C.r. hebd. seanc., Acad. Sci., Paris (D), 262(13): 1400-1402.

Chemistry pH (data)
Berthois, Léopold, 1965.
Remarques sur les propriétés physico-chimiques des eaux marines dans le Golfe de Gascogne, en Mai 1964.
Rev. Trav. Inst. Pêches Marit., 29(4):383-392.

Chemistry, pH
Berthois, L., and M. Barbier, 1953.
La sédimentation en Loire maritime en période d'étiage. C.R. Acad. Sci., Paris, 236(20):1984-1986, 1 fig.

pH (data)
Beyers, Robert J., James L. Larimer, Howard T. Odum, Richard B. Parker and Neal E. Armstrong, 1963
Directions for the determination of changes in carbon dioxide concentration from changes in pH.
Publ. Inst. Mar. Sci., Port Aransas, 9:454-489.

chemistry, pH
Bienati, Norberto L., y Rufino A. Comes 1970.
Variación estacional de la composicion fisico-quimica del agua de mar en Puerto Paraiso, Antartida occidental
Contrib. Inst. Antart. Argentino 130: 45pp.

Chemistry, pH
Black, W.A.P., and E.T. Dewar.
Correlation of some of the physical and chemical properties of the sea with the chemical composition of the algae. J.M.B.A. 28(3):673-699, 14 textfigs.

Chemistry, pH
Bogoyavlenskii, A.N., 1955.
[Chemical characteristics of the water in the region of the Kurile-Kamchatka Trench.]
Trudy Inst. Okeanol., 12:161-176.

Chemistry pH
Bogoiavlenskii, A.N., 1958.
[VI. Hydrochemical work.] Opisanie Exped. D/E "Ob", 1955-1956, MGG, Trudy Kompaeksnoi Antarkt. Exped., Akad. Nauk, USSR, 91-102.

Chemistry
pH
Bogoyavlensky, A.N., 1959
[Hydrochemical investigations.]
Arktich. i Antarktich. Nauchno-Issled. Inst., Mezhd. Geofiz. God, Sovetsk. Antarkt. Eksped., 5: 159-172.

Chemistry ph
Bogojavlensky, A.N. and O.V. Shishkina 1971.
On the hydrochemistry off Peru and Chile. (In Russian; English abstract)
Trudy Inst. Okeanol. P.P. Shirshova Akad. Nauk SSSR 89: 91-105

Chemistry pH
Bruevich, S.W., 1960
[Hydrochemical investigations on the White Sea.]
Trudy Inst. Okeanol., 42: 199-254.

pH(data)
Brujewicz, S.W., 1957.
On certain chemical features of water and sediments in north-west Pacific.
Proc. UNESCO Symp., Phys. Ocean., Tokyo, 1955: 277-292.

pH(data)
Bruevich, S.V., and I.A. Chikina, 1933
[Hydrochemical observations in the northern part of the Kola Fjord (Barents Sea) in summer 1931.]
Trudy Gosud. Okeanogr. Inst., 3(3): 120-124.

pH(data)
Central Meteorological Observatory, Japan, 1951.
The results of marine meteorological and oceanographical observations, July-Dec. 1948, No. 4: vi+414 pp., of 57 tables, 1 map, 1 photo.

Chemistry, pH (data)
Bruneau, L., N.G. Jerlov, F.F. Koczy, 1953.
Physical and chemical methods.
Repts. Swedish Deep-sea Exped., 1947-1948, Phys. Chem., 3(4):101-112, 7 textfigs., 2 tables. (1-iv)

Chemistry, pH
Brust, H.F., and C.F. Newcombe 1940.
Observations on the alkalinity of estuarine waters of the Chesapeake Bay.
J. mar. Res. 3(2): 105-111.

Chemistry, pH
Buch, K., 1949.
Über den biochemischen Stoffwechsel in der Ostsee
Kieler Meeresforschungen 6:31-44, 6 textfigs.

Chemistry, pH
Buch, K., 1939.
Beobachtungen über das Kohlensäure-gleichgeweit und über den Kohlensäure-austausch zurwischen Atmosphäre und Meer im Nordatlantischen Ozean. Acta Academiae Aboensis, Math. et Phys. XI.9:31 pp., 2 text figs.

Chemistry, pH
Buch, K., 1938.
Versuche über photoelektrische pH-Bestimmung im Meerwasser. C.R. Lab. Carlsberg, Ser. Chim. 22: 109-117, 2 textfigs.

Chemistry, pH
Buch, K., 1937.
Versuche über photoelektrische pH-Bestimmung im Meerwasser. Compt. Rendu., Lab. Carlsberg, ser. Chem., 22:109-117, 2 textfigs.

Chemistry, pH
Buch, K., 1934.
Beobachtungen über chemische Faktoren in der Nordsee, zwischen Nordsee und Island sowie auf dem Schelfgebiete nördlich von Island. Rapp. Proc. Verb. 89(3):13-31, 8 tables, 2 textfigs.

Chemistry, pH
Buch, K., 1930.
Die Kohlensäurefaktoren des Meereswassers.
Cons. Perm. Int. Expl. Mer, Rapp. Proc. Verb. 67:51-88. figs.

Chemistry pH
Buch, K., 1929.
Ueber die pH-Bestimmung des Wassers nach der Chinhydronmethode.
Finska Vetenskap. Comm. Phys. Math., 4(21):10 pp.
Transl. cited:
FWS Spec. Sci. Rept., Fish., 227.

Chemistry, pH
Buch, K., and S. Gripenberg, 1938.
Jahreszeitlicher Verlauf der chemischen und biologischen Faktoren im Meerwasser bei Hangö im Jahre 1935. Havforskningsinstitutets Skrift No. 118:26 pp., 5 textfigs.

chemistry, pH

Buch, K., and O. Nynäs, 1940.
Die pH-Bestimmung des Wassers mit dem photoelektrischen Kolorimeter und der Glaselektrode.
Acta Acad. Aboensis, Math. Phys., 13(4):19 pp.

chemistry pH

Buch, K., and O. Nynäs, 1939.
Studien über neuere pH-Methodik mit besonderer Berücksichtigung des Meerwassers. Acta Acad. Aboensis, Math. Phys., 12(3):41 pp., 6 textfigs.

chemistry, pH.

Burbanck, W.D., and George P. Burbanck 1967.
Parameters of interstitial water collected by a new sampler from the biotopes of Cyathura polita (Isopoda) in six southeastern states.
Chesapeake Sci. 8(1):14-27.

chemistry, pH

Burbanck, W.D., M.E. Pierce and G.C. Whiteley, Jr. 1956.
A study of the bottom fauna of Rand's Harbor, Massachusetts, an application of the ecotone concept. Ecol. Monogr., 26:213-243.

chemistry pH

Burkholder, Paul R., 1960.
Distribution of some chemical values in Lake Erie. Limnological survey of western and central Lake Erie, 1928-1929. USFWS Spec. Sci. Rept., Fish., No. 334:71-110.

C

chemistry pH (data)

California, Humboldt State College, 1964.
An oceanographic study between the points of Trinidad Head and Eel River.
State Water Quality Control Bd., Resources Agency, California, Sacremento, Publ., No. 25: 136 pp.

chemistry pH.

Calmels, Augusto Pablo y Hugo Alfredo Toffetani 1969.
Reconocimientos oceanographicos en la ria interior de la Bahia Blanca.
Contrib. Inst. Oceanogr. Univ. Nac. Sur, Bahia Blanca, Argentina 1969 (3):1-21.

chemistry pH

Carpenter, J.H., 1960
The Chesapeake Bay Institute study of the Baltimore Harbor.
Proc. 33rd Ann. Conf., Maryland-Delaware Water and Sewage Assoc., June 9-10, 1960: 62-78.
Also in:
Chesapeake Bay Inst., Collected Reprints, 5 (1963).

chemistry, pH

Caspers, H., 1951.
Quantitative Untersuchungen über die Bodentierwelt des Schwarzen Meeres im bulgarischen Küstenbereich. Arch. f. Hydrobiol. 45(1/2):192 pp., 66 textfigs.

chemistry, pH

Cavaliere, Antonio, 1963.
Studi sulla biologia e pesca di Xiphias gladius. II.
Boll. Pesca, Piscicolt. e Idrobiol., 8(2):143-170.

chemistry pH

Chacko, P.I., 1950.
Marine plankton from waters around Kusadai Island. Proc. Indian Acad. Sci., Sect. B, 31(3): 162-174, 1 textfig.

chemistry pH

Chalmers, G.V., and A.K. Sparks, 1959.
An ecological survey of the Houston ship channel and adjacent bays.
Publ. Inst. Mar. Sci., 6:213-250.

chemistry pH

Chau, Y.K. and R. Abesser, 1958
A preliminary study of the hydrology of Hong Kong territorial waters. Hong Kong Univ. Fish J. (2): 43-57.

chemistry pH

Chidambaram, K., A.D.I. Rajandran, and A.P. Valsan, 1951.
Certain observations on the hydrography and biology of the pearl bank, Tholayviam Paar off Tuticorin in the Gulf of Manaar in April 1949.
J. Madras Univ., Sect. B, 21(1):48-74, 2 figs.

chemistry, pH

Chippenfield, P.N.J., 1962.
Some thoughts on the discharge of industrial effluents to tidal waters.
Chem. Indust., 38:1660-1666.

chemistry pH (data)

Choe Sang 1969.
Phytoplankton studies in Korean waters III. Surface phytoplankton survey of the north-eastern Korea Strait in May of 1967.
(In Korean; English abstract).
J. oceanogr. Soc. Korea 4(1):1-8.

chemistry pH

*Choe, Sang, Tai Wha Chung and Hi-Sang Kwak, 1968.
Seasonal variations in primary productivity and pigments of downstream water of the Han River. (In Korean; English abstract).
J. oceanogr. Soc., Korea, 3(1):16-25.

pH

*Choe, Sang, Tai Wha Chung and Hi-Sang Kwak, 1968.
Seasonal variations in nutrients and principal ions contents of the Han River water and its water characteristics. (In Korea; English abstract
J. oceanogr. Soc., Korea, 3(1):26-38.

chemistry pH

Comité local d'Oceanographie et d'études des Côtes de l'Afrique Occidentale Francaise, 1949.
Travaux d'océanographie physique effectués en 1949 par la Section Technique des Pêches Maritimes de Dakar. A. Dakar et environs B. Côtes du Sénégal. C.L.O.E.C. de l'A.O.F., Annee. 1949:27-39, 4 fold-in charts.

chemistry pH

Cooper, L.H.N., 1948.
The nutrient balance in the sea. Research 1: 242-248, 2 textfigs.

chemistry, pH

Cooper, L.H.N., 1933.
Chemical constituents of biological importance in the English Channel, November 1930 to January 1932. II. Hydrogen ion concentration, excess base, carbon dioxide and oxygen. J.M.B.A., 18: 729-753.

chemistry pH

Copeland, B.J., 1967
Environmental characteristics of hypersaline lagoons.
Contrib. mar. Sci., Port Aransas, 12: 207-218.

chemistry pH(data)

Copenhagen, W.J., 1953.
The periodic mortality of fish in the Walvis Region. A phenomenon within the Benguela Current. Union of S. Africa, Div. Fish., Investig. Rept. No. 14:1-35, 9 pls.

chemistry, pH

*Culberson, C., and R.M. Pytkowicz, 1968.
Effect of pressure on carbonic acid, boric acid, and the pH in seawater.
Limnol. Oceanogr., 13(3):403-417.

D

chemistry, pH

Davidovitch, R.L., 1964.
Short chemical characteristics of the northwest Pacific Ocean. (In Russian).
Gosudarst. Kom. Sov. Ministr., SSSR, Ribn. Choz., Trudy WNIRO, 49, Izv. TINRO, 51:93-98.

chemistry, pH

Day, J.H., 1951.
The ecology of South African estuaries. 1. A review of estuarine conditions in general. Trans. R. Soc., S. African 33(1):53-91, 2 textfigs.

chemistry, pH

de Barros Machado, L., 1950.
Pequisas físicas e químicas do sistema hidrográfico da região lagunar de Cananéia. 1. Cursos de águas. Bol. Inst. Paulista Oceanogr. 1(1):45-67, 3 textfigs.

chemistry pH(data)

Denamur, J., 1955.
Étude physico-chimique des eaux du détroit de Bonifacio. Trav. C.R.E.O., 2(10-11):14 pp.

chemistry, pH

Distèche, A., and M. Dubuisson, 1960.
Mesures directes de pH aux grandes profondeurs sous-marines.
Bull. Inst. Océanogr., Monaco, 1174:8 pp.

chemistry pH (data)

Donnelly, P.V., J. Vuille, M.C. Jayaswal, R.A. Overstreet, J. Williams, M.A. Burklew and R.M. Ingle, 1966.
A study of contributory chemical parameters to red tide in Apalachee Bay.
Florida Bd., Conserv., St. Petersburg, Mar. Lab. Prof. Papers Ser. No. 8:43-83.

chemistry pH

*Düing, Walter, Klaus Grasshoff und Gunther Krause, 1967.
Hydrographische Beobachtungen auf einem Äquatorschnitt im Indischen Ozean.
"Meteor Forschungsergebnisse (A) (3):84-92.

chemistry pH

Dutta, N., J.C. Malhotra and B.B. Bose, 1954.
Hydrology and seasonal fluctuations of the plankton in the Hooghly estuary.
Symp. Mar. Fresh-water Plankton, Indo-Pacific, Bangkok, Jan. 25-26, 1954, FAO-UNESCO:35-47.

chemistry, pH

Dybern, Bernt I. 1967.
Topography and hydrography of Kviturdvikpollen and Vågsböpollen on the west coast of Norway.
Sarsia 30:1-27.

pH

*Dyrssen, David, et al., 1967.
Analysis of sea water in the Uddevalla Fjord system 1.
Rept. Chem. Sea Water, Univ. Göteborg, 4:1-8 pp. (mimeographed).

E

chemistry pH

Edmondson, W.T., and Y.H. Edmondson, 1947.
Measurements of production in fertilized salt-water. J. Mar. Res. 6(3):228-246, Figs. 55-62.

chemistry, pH

Eggvin, J., 1933.
A Norwegian fat-herring fjord. An oceanographical study of the Eidsfjord. Repts. Norwegian Fish. Mar. Invest. 4(6):22 pp., 11 textfigs.

chemistry pH(data)

El-Sayed, Sayed Z., and Enrique F. Mandelli, 1965
Primary production and standing crop of phytoplankton in the Weddell Sea and Drake Passage. In: Biology of Antarctic seas, II.
Antarctic Res. Ser., Amer. Geophys. Union, 5:87-106.

F

chemistry, pH
Faganelli, A., 1948/49.
Concentration en ions hydrogene dans le Porto Canale et la Lagune de Chiaggio. Atti Ist. Veneto, Cl. Sci. Mat. Nat. 107:51-55.

chemistry, pH
Fedosov, M.V., and L.A. Barsukova, 1959.
[Gas regime of the water masses of the northern Caspian.] Trudy VNIRO, 38:78-87.

chemistry pH
Fiske, John D., Clinton E. Watson and Philip G. Coates, 1966.
A study of the marine resources of the North River.
Comm. Mass., Div. Mar. Fish., Monogr. Ser. 3:53 pp.

chemistry pH
Fleming, J.A., C.C. Ennis, H.U. Sverdrup, S.L. Seaton, W.C. Hendrix, 1945
Observations and results in physical oceanography. Graphical and tabular summaries. Ocean. 1-B, Sci. Res. of Cruise VII of the Carnegie during 1928-1929 under command of Capt. J.P. Ault, Carnegie Inst., Washington, Publ. 545:315 pp., 5 tables, 254 figs.

chemistry pH
Fonselius, Stig H., 1960.
Hydrography of the Baltic deep basins. III.
Rept., Fish. Bd., Sweden, Ser. Hydrogr., 23: 1-97.

chemistry, pH.
Fonselius, S.H. 1968.
Observations at Swedish lightships and in the central Baltic 1966.
Annls biol. Copenh. 1966, 23: 74-78.

chemistry pH (data)
Fonselius, S.H., 1966.
Observations at Swedish lightships in the Central Baltic in 1964.
Annls biol., Copenh., 21:46-49.

chemistry pH
Fonselius, Stig H., 1962
Hydrography of the Baltic deep basins.
Fishery Board of Sweden, Ser. Hydrogr., Rept. 13: 40 pp.

chemistry pH
Brancis-Boeuf, C., 1942.
Mesures physico-chimiques des eaux de la Penzé Maritime (Finistère). Bull. Inst. Océan., Monaco, No. 829:1-16.

chemistry, pH
Francis-Boeuf, C., 1941.
Résultats des mesures physico-chimiques effectuées à bord du Chausseur 2 le long de la côte marocaine entre Port-Lyautey et Mazagan, au mois de janvier 1941. Bull. Inst. Océan., Monaco, 804:1-16.

chemistry, pH
Francis-Boeuf, C., 1941.
Observations sur les variations de quelques facteurs physicochimiques des eaux de la Penzé Maritime (Finistère). C. R. Acad. Sci., Paris, 212:805-080, 3 textfigs.

Chemistry - pH (data)
Franco, Paolo 1970.
Oceanography of northern Adriatic Sea. I. Hydrologic features: cruises July-August and October-November 1965.
Archo Oceanogr. Limnol. 16 (Suppl.): 1-93.

chemistry, pH (data)
Fuse, Shin,ichiro, and Eiji Harada, 1960
A study on the productivity of Tanabe Bay. (3) Result of the survey in the summer of 1958.
Rec. Oceanogr. Wks., Japan, Spec. No. 4: 13-28.
— Oceanographic conditions of Tanabe Bay.

G

Chemistry pH (data)
Gaarder, T., and R. Spärck, 1932.
Hydrographisch-biochemische Untersuchungen in norwegischen Austern-Pollen. Bergens Mus. Aarbok, Naturvidensk.-rekke, No. 1:5-144, 75 textfigs.

Chemistry, pH.
Galtsoff, P.S., ed., 1954.
Gulf of Mexico, its origin, waters and marine life. Fish. Bull., Fish and Wildlife Service, 55:1-604, 74 textfigs.

chemistry, pH.
Garkavaia, G.P., and L.E. Pozdniakova 1968.
Data on the hydrochemical regime in Chesha Bay, Barents Sea. (In Russian).
Trudy Murmansk. Morsk. Biol. Inst. 17 (21): 4-21.

Chemistry, pH.
Garland, C.F., 1954.
A study of water quality in Baltimore Harbor.
Chesapeake Biol. Lab. Publ. 96:1-132, figs.

Chemistry, pH.
Gheorghiu, V. G., and N. Calinicenco, 1946.
Daily variations of the pH of the water of the Black Sea. Rev. Stiintifica "V. Adamachi", 32: 248-249.

Abstract in Chem. Abst. 1947:64471

chemistry pH
Gheorgin, V., and N. Calinecenco, 1946.
[The pH value and salinity of the water of the Black Sea.] (In Roumanian). Rev. Stiintifica "V. Adamachi" 32:43-45, 127-129.

Abstr. in Chem. Abstr. 1947:5763e

chemistry pH
Gieskes, Joris M., 1969.
Effect of temperature on the pH of seawater.
Limnol. Oceanogr., 14(5): 679-685.

Chemistry, pH
Gillbricht, M., 1953.
Kolorimetrische pH-Messung im Seewasser unter Verwendung von Kresolrot. Kurze Mitt., Fischeribiol. Abt., Max-Planck-Inst. f. Meeresbiol., Wilhelmshaven, No. 2:1-19, 5 tables.

Chemistry, pH
Gorsline, Donn S., 1963.
Oceanography of Apalachicola Bay, Florida.
In: Essays in Marine Geology in honor of K.O. Emery, Thomas Clements, Editor, Univ. Southern California Press, 69-96.

Chemistry pH
Graham, H. W. and N. Bronikovsky, 1944
The genus Ceratium in the Pacific and North Atlantic Oceans. Sci. Res. Cruise VII of the Carnegie, 1928-1929 ----- Biol. V (565):209 pp., 54 charts, 27 figs., 54 tables.

chemistry, pH.
Graham, H.W., and E.G. Moberg 1944.
Chemical results of the last cruise of the Carnegie. Chemistry I. Scientific results of cruise VII of the Carnegie during 1928-1929 under command of Captain J.P. Ault.
Carnegie Inst. Washington 562: 58 pp., 7 maps.

Chemistry pH
Gripenberg, Stina, 1966.
Equilibria of the complexes formed by mannitol in sea water.
Commentat. physico-math., 32(1):3-38.

Chemistry, pH
Grossman, Stuart, 1967.
Ecology of Rhizopoda and Ostracoda of southern Pamlico Sound region, North Carolina.
I. Living and subfossil rhizopod and ostracode populations.
Paleontol. Contrib., Univ. Kansas, 44(1):7-82.

Chemistry, pH
Guarrera, S.A., 1950.
Estudios hidrobiologicos en el Rio de la Plata.
Rev. Inst. Nac., Invest. Cienc. Nat., Ciencias Botanicas 2(1):62 pp.

chemistry pH (data)
Gudkovich, Z.M., 1955.
[Results of a preliminary analysis of the deep-water hydrological observations.]
Material. Nabluid. Nauch.-Issledov. Dreifuius., 1950/51, Morskoi Transport 1:41-46;48-170.

David Knauss translator, AMS-Astia Doc. 117133

H

chemistry, pH (data)
Halim, Youssef 1960.
Etude quantitative et qualitative du cycle écologique des dinoflagellés dans les eaux de Villefranche-sur-Mer (1953-1955).
Ann. Inst. Océanogr. Monaco 38: 123-232

chemistry, pH.
Harvey, H.W., 1925.
Hydrography of the English Channel.
Cons. Perm. Int. Expl. Mer, Rapp. Proc. Verb. 37:59-89, Figs. 17-29, 1 pl.

chemistry, pH
Hutton, R.F., B. Eldred, K.D. Woodburn and R.M. Ingle, 1956.
The ecology of Boca Ciega Bay with special reference to dredging and filling operations.
Florida State Bd., Conserv., Mar. Lab., St. Petersburg, Tech. Ser., 17(1):87 pp.

Chemistry pH
Hung, Tsu-Chang 1970
Photo- and radiation effects on inorganic nitrogen compounds in pure water and sea water near Taiwan in the Kuroshio Current.
Bull. Inst. Chem. Acad. Sinica 18: 37-56

Chemistry pH
Hung, Tsu-Chang, and Ching-Wei Lee, 1967.
Research on chemical oceanography in the Kuroshio around Taiwan Island.
Bull. Inst. Chem., Acad. Sinica, Taiwan, 14:80-102.

I

Chemistry, pH
Ivanenkov, V.N., 1964.
Distribution of PCO", pH, alk/cl in the northern part of the Indian Ocean. Investigations in the Indian Ocean (33rd voyage of E/S "Vitiaz"). (In Russian).
Trudy Inst. Okeanol., Akad. Nauk, SSSR, 64:128-143.

chemistry pH
Ivanenkov, V.N., and F.A. Gebin, 1960.
Water masses and hydrochemistry in the western and southern parts of the Indian Ocean. Physics of the Sea. Hydrology. (In Russian).
Trudy Morsk. Gidrofiz. Inst., 22:33-115.

Translation: Scripta Technica, Inc., for Amer. Geophys. Union, 27-99.

chemistry, pH

Iyengar, M.O.P. and G.Venkataraman,1951.
The ecology and seasonal succession of the algae flora of the River Cooum at Madras with special reference to the Diatomaceae. J. Madras Univ. 21, Sect. B(1): 140-192, 1 pl of 4 figs., 11 text figs.

J

chemistry, pH

Jakleln, A., 1936
Oceanographic investigations in East Greenland waters in the summers of 1930-1932. Skr. om Svalbard og Ishavet No. 67:79 pp., Pl. 2, 28 text figs.

chemistry, pH

Jakobi, H., 1953.
The distribution of salinity and pH in the Bay of Guaratuba. Arq. Mus. Paranaense 10:4-35.

Chem. Abstr. (1955):14399a.

pH.

Japan, Central Meteorological Observatory 1949.
Antarctic Whaling Ground (1947-48).
Oceanogrl Mag. Japan 1(1): 49-58, 17 textfigs.

chemistry, pH

Japan, Kobe Marine Observatory 1967.
Report of the oceanographic observations in the sea south of Honshu from October to November 1963. (In Japanese)
Bull. Kobe Mar. Obs. No. 178: 41-49.

chemistry, pH

Japan Kobe Marine Observatory, 1967.
Report of the oceanographic observations in the sea south of Honshu and in the Seto-Naikai from May to June 1964. (In Japanese)
Bull. Kobe Mar. Obs. No. 178: 37-

chemistry, pH

Japan, Kobe Marine Observatory, 1967.
Report of the oceanographic observations in the sea south of Honshu from July to August 1963.
(In Japanese).
Bull. Kobe Mar. Obs. No. 178: 31-40

chemistry, pH

Japan Kobe Marine Observatory, 1967.
Report of the oceanographic observations in the sea south of Honshu from February to March 1964. (In Japanese)
Bull. Kobe Mar. Obs. No. 178: 27-

chemistry
pH

Japan, Kobe Marine Observatory, Oceanographical Section, 1964.
Report of the oceanographic observations in the Kuroshio and region east of Kyushu from October to November 1962. (In Japanese) Res. Mar. Meteorol. and Oceanogr., Japan. Meteorol. Agency, 32: 41-50.

Also in: Bull. Kobe Mar. Obs., 175. 1965.

chemistry
pH

Japan, Kobe Marine Observatory, Oceanographical Section, 1964.
Report of the oceanographic observations in the sea south of Honshu from February to March, 1963. Res. Mar. Meteorol. and Oceanogr., Japan Meteorol. Agency, 33: 27-32.

Also in: Bull. Kobe Mar. Obs., 175. 1965.

chemistry
pH

Japan, Kobe Marine Observatory, Oceanographical Section, 1964.
Report of the oceanographic observations in the sea south of Honshu from July to August, 1962. Res. Mar. Meteorol. and Oceanogr., Japan, Meteorol. Agency, 32: 32-40. (In Japanese).

Also in: Bull. Kobe Mar. Obs. 175.

chemistry
pH

Japan, Kobe Marine Observatory, Oceanographical Section, 1963
Report of the oceanographic observations in the sea south of Honshu from February to March, 1962. (In Japanese).
Res. Mar. Met. & Ocean., J.M.A., 31:37-44.

Also in:
Bull. Kobe Mar. Obs., No. 173(4):1964.

chemistry
pH

Japan, Kobe Marine Observatory, Oceanographical Section, 1963
Report of the oceanographic observations in the sea south of Honshu from May to June, 1962. (In Japanese).
Res. Mar. Met. & Ocean., J.M.A., 31:45-49.

Also in:
Bull. Kobe Mar. Obs., No. 173(4):1964.

chemistry
pH

Japan, Kobe Marine Observatory, Oceanographical Observatory, 1962
Report of the oceanographic observations in the cold water region off Enshu Nada in May, 1961. (In Japanese).
Res. Mar. Meteorol. and Oceanogr. Obs., Jan.-June, 1961, No. 29:28-35.

chemistry
pH

Japan, Kobe Marine Observatory, 1962
Report of the oceanographic observations in the cold water region south of Enshu Nada from October to November, 1960. (In Japanese).
Res. Mar. Meteorol. and Oceanogr., July-Dec., 1960, Japan Meteorol. Agency, No. 28: 43-51.

chemistry
pH

Japan, Kobe Marine Observatory, Oceanographical Section, 1962
Report of the oceanographical observations in the sea south of Honshu from July to August 1961. (In Japanese)
Res. Mar. Meteorol. & Oceanogr., 30:39-48.

Also in:
Bull. Kobe Mar. Obs., No. 173(3). 1964

chemistry
pH

Japan, Kobe Marine Observatory, Oceanographical Section, 1962
Report of the oceanographic observations in the sea south of Honshu from February to March, 1961. (In Japanese).
Res. Mar. Meteorol. and Oceanogr. Obs., Jan.-June, 1961, No. 29:22-27.

chemistry
pH

Japan, Maizuru Marine Observatory, 1963
Report of the oceanographic observations in the Japan Sea in June 1961. (In Japanese)
Bull. Maizuru Mar. Obs., No. 8:59-79.

chemistry
pH

Japan, Maizuru Marine Observatory and Hakodate Marine Observatory, Oceanographical Sections, 1962
Report of the oceanographic observations in the Japan Sea in June, 1961. (In Japanese).
Res. Mar. Meteorol. and Oceanogr. Obs., Jan.-June, 1961, No. 29:59-79.

chemistry pH

Japan, Nagasaki Marine Observatory 1971
Report of the oceanographic observations in the sea south of Kyushu in April, 1966 (In Japanese).
Oceanogr. Met. Nagasaki Mar. Obs. 18: 53-57.

chemistry, pH.

Japan, Nagasaki Marine Observatory 1971.
Report of the oceanographic observations in the sea south of Kyushu in March 1966. (In Japanese)
Oceanogr. Met. Nagasaki Mar. Obs. 18: 50-52.

chemistry, pH

Japan, National Committee for the International Geophysical Year 1960.
The results of the Japanese oceanographic project for the International Geophysical Year, 1957/58: 198 pp.

chemistry, pH

Jensen, A.J.C., 1944.
The hydrography of the Praestø Fjord.
Folia Geogr. Danica 3(2):47-55, 2 textfigs., 4 tables.

chemistry, pH

Jerlov, N.G., 1956.
The Equatorial currents in the Pacific Ocean.
Repts. Swedish Deep-Sea Exped. 3(Phys. Chem., 6) :129-154.

chemistry, pH

Jones, J.A., 1963.
Ecological studies of the southeastern Florida patch reefs. 1. Diurnal and seasonal changes in the environment.
Bull. Mar. Sci., Gulf and Caribbean, 13(2):282-307.

K

Chemistry, pH

Kabanov, V.V., 1962
On the application of the pH-meter LP-57M for the determination of the active reaction in sea water under vessel conditions aboard R/V 'Mikhail Lomonosov". (In Russian).
Okeanologiia, Akad. Nauk, SSSR, 2(6):1085-1092.
Abstr. in:
Soviet Bloc Res. Geophys. Astron. & Space, 52: 23.

Chemistry, pH (data)

Kaleis, M.V., and N.B. Aleksandrovskaya, 1965.
Hydrological régime of the Baltic Sea in 1963.
Ann. Biol., Cons. Perm. Int. Expl. Mer, 1963, 20:70-72.

chemistry, pH.

Kaleis, M.V., N.B. Alexandrovskaya and E.A. Yula 1968.
Some peculiarities in the oceanographical regime of the Baltic in 1966.
Annls. biol. Copenh. 1966, 23: 78-81.

Chemistry, pH

Kandler, R., 1953.
Hydrographische Untersuchungen zum Abwasserproblem in den Buchten und Förden der Ostseeküste Schleswig Holstein. Kieler Meeresf. 9(2):176-200, Pls. 9-14 (18 figs.).

Chemistry, pH

Kändler, R., 1930.
Beiträge zur Kenntnis über die Beziehungen zwischen Wasserstoffionen-Konzentration, freier Kohlensäure und Alkalinität im Meerwasser.
Cons. Perm. Int. Expl. Mer, Rapp. Proc. Verb. 67: 89-90.

pH (Data)

Kato, Kenji, 1966.
Studies on calcium content in sea water. III. Calcium in the waters of Cananéia lagoon and its adjacent regions, State of Sao Paulo, Brazil.
Bolm Inst. Oceanogr., S. Paulo, 15(1):41-45.

pH (data)
Kato, Kenji, 1966.
Geochemical studies on the mangrove region of Cananéia, Brazil. II. Physico-chemical observations on the reduction states.
Bolm Inst. Oceanogr., S. Paulo, 15(1):21-24.

pH (data)
Kato, Kenji, 1966.
Geochemical studies on the mangrove region of Cananéia, Brazil. 1. Tidal variations of water properties.
Bolm Inst. Oceanogr., S. Paulo, 15(1):13-20.

chemistry, pH
Kato, K. 1966
Chemical investigations on the hydrographical system of Cananéia lagoon.
Bolm Inst. Oceanogr. S. Paulo 15(1): 1-12.

chemistry
pH
Keeling, Charles D., and Bert Bolin, 1968
The simultaneous use of chemical tracers in oceanic studies. II. A three-reservoir model of the North and South Pacific oceans.
Tellus, 20(1): 17-54

chemistry, pH
Kijowski, S., 1937.
Nieco danych o skladzii chemicznym wod Zatoki Gdanskiej. (Quelques données sur la composition chimique des eaux du Golf de Dantzig). Bull. Sta. Mar. Hel 1(1):33-41, 1 fig.

chemistry, pH
Kitamura, Hiroyuki, and Takeshi Sagi, 1964.
On the chemical elements in the sea south of Honshu, Japan. (In Japanese; English abstract).
Bull. Kobe Mar. Obs., No. 172:6-54.

chemistry, pH (data)
Koczy, F.F., 1952.
Hydrography. Swedish observations. Cons. Perm. Int. Expl. Mer, Ann. Biol. 8:130-131.

chemistry
pH (data)
Kramer, J.R., 1961
Chemistry of Lake Erie.
Proc. Fourth Conf., Great Lakes Research, Great Lakes Res. Div., Inst. Sci. & Tech., Univ. Michigan, Publ., (7):27-56.

L

pH
Laktionov, A.F., V.A. Shamontyev and A. V. Yanes, 1960
Oceanographic characteristics of the North Greenland Sea.
Soviet Fish. Invest., North European Seas, VNIRO, PNIRO, Moscow, 1960:51-65.

chemistry - pH (data)
Lee, Min Jai, Jae Hyung Shim and Chong Kyuh Kim 1967.
Studies on the plankton of the neighboring seas of Korea. 1. On the marine conditions and phytoplankton of the Yellow Sea in summer.
Repts. Inst. mar. Biol. Seoul Nat. Univ. 1(6): 1-14.

chemistry, pH
Lefevre, S., E. Leloup and L. Van Meal, 1956.
Observations biologiques dans le port d'Ostende
K. Belg. Inst. Naturwet., Verhandl., [Inst. R. Sci. Nat. Belg., Mem.,] 133:157 pp.

chemistry, pH.
Leloup, E., editor, 1966.
Recherches sur l'ostreiculture dans le bassin d'Ostende en 1964.
Minist. Agricult. Comm. T.W.O.Z., Groupe de Travail "Ostreiculture": 58 pp.

chemistry, pH
Le Masson, L., et Y. Magnier, 1966.
Resultats des observations scientifiques de La Dunkerquoise sous le commandement du Capitaine de Corvette Brosset. Croisiere Hunter.
Cah. ORSTOM, Ser. Oceanogr., 4(1):3-78.

chemistry pH
Lisitzin, A.P., 1966.
Processes of Recent sedimentation in the Bering Sea. (In Russian).
Inst. Okeanol., Kom. Osad. Otdel Nauk o Zemle, Isdatel, Nauka, Moskva, 574 pp.

Chemistry
pH(data)
Littlepage, Jack L., 1965.
Oceanographic investigations in McMurdo Sound, Antarctica.
In: Biology of Antarctic seas, II.
Antarctic Res. Ser., Am. Geophys. Union, 5:1-37.

chemistry pH
Lyman, John, and Jacob Verduin, 1961.
Changes in pH and total CO2 in natural waters.
Limnol. & Oceanogr., 6(1):80-83.

M

chemistry pH
Machado, L. de B., 1952.
Pesquisas fisicas e quimicas do sistema hidrografico da regiao Lagunar de Cananeia. Bol. Inst. Ocean., Univ. Sao Paulo, 3(1/2):55-75, 1 chart

chemistry, pH
Machado, L. de B., 1951.
Resultados científicos do cruzeiro do "Baependi" e do "Vega" a Ilha de Trindade. Oceanografia física. Contribuição para o conhecimento das características físicas e químicas das águas.
Bol. Inst. Paulista Ocean. 2(2):95-110, 5 pls.

chemistry pH
Machado, L. de B., 1950.
Pesquisas fisicas e quimicas so sistema hidrográfico da região lagunar de Cananéia. 1. Cursos de água. Bol. Inst. Paulista Ocean. 1(1):45-68.

chemistry, pH(data)
Maeda, H., 1953.
Studies on Yoda-Naikai. 2. Considerations upon the range of the stagnation and the influences by the River Noda and the open sea.
J. Shimonoseki Coll. Fish. 3(2):133-140, 2 textfigs.

chemistry, pH(data)
Maeda, H., 1953.
Studies on Yoda-Naikai. 3. Analytical investigations on the influence of the River Noda and the benthonic communities. J. Shimonoseki Coll. Fish. 3(2):141-149, 3 textfigs.

Chemistry, pH (data)
Magazzu, Giuseppe, e Guglielmo Cavallaro 1969.
Rapporto sulle crociere di studio lungo le coste meridionali Calabresi e della Sicilia orientale (1967-1968). 1. Idrografia.
Programma Ricerca Risorse Mar. Fondo mar (B) 45: 5-70.

Chemistry-pH (data)
Magazzu, Giuseppe, Guglielmo Cavallaro e Letterio Guglielmo 1969.
Considerazioni preliminari sulle condizioni chimico-fisiche e sullo zooplancton delle acque costiere fra C. po Milazzo e C. po D'Orlando (Messina)
Programma Ricerca Risorse mar. Fondo mar. (B) 45: 71-90

chemistry, pH
Makkaveeva, N.S., 1965.
The hydrochemical investigations of the Korovinskaya Inlet. (In Russian).
Materiel. Ribochoz. Issled. Severn. Bassin. Poliarn. Nauchno-Issled. i Proekt. Inst. Morsk. Rib. Choz. i Okeanogr., N.M. Knipovich, 5:117-122

chemistry, pH
Martinez de Murguia, Jesus Arairo-Torre, 1946
Estudio del pH en al agua del mar, notas y resumenes, Ser. 2 No.135, 6 pp.

Chemistry pH (data)
Marukawa, H., and T. Kamiya, 1926.
Outline of the hydrographical features of the Japan Sea.
Annot. Oceanogr. Res., 1(1):1-7, tables, charts.

Chemistry pH (data)
Masuzawa, Jotaro, and Hideo Akamatsu, 1962.
1. Hydrographic observations during the JEDS-4.
Oceanogr. Mag., Japan Meteorol. Agency, 13(2): 122-130.

Chemistry
pH (data)
Matsudaira, Yasuo, 1964
Cooperative studies on primary productivity in the coastal waters of Japan, 1962-63. (In Japanese; English abstract).
Inform. Bull., Planktol., Japan, No. 11:24-73.

Chemistry pH
Matsue, Yoshiyuki, Yuzo Komaki and Masaaki Murano, 1957
On the distribution of minute nutrients in the North Japan Sea, during the close of August, 1955.
Bull. Japan Sea Reg. Fish. Res. Lab., 6:121-127
Abstr. in:
Rec. Res., Fac. Agric., Tokyo Univ., Mar. 1958, 7(80):57-58.

Chemistry
pH
Matsue, Yoshiyuki, Yuzo Komaki and Masaaki Murano, 1957
On the distribution of minute nutrients in the North Japan Sea; March-April, 1956.
Bull. Japan Sea Reg. Fish. Res. Lab., 6: 316-320.
Abstr. in:
Rec. Res., Fac. Agric., Tokyo Univ., Mar. 1958, 7(81): 58.

Chemistry pH
Maxwell, B.E., 1956.
Hydrobiological observations for Wellington Harbour. Trans. R. Soc., New Zealand, 83(3):493-503.

Mc

Chemistry, pH
Millard, N.A.H., and K.M.F. Scott, 1954.
The ecology of South African estuaries. VI. Milnerton estuary and the Diep River, Cape. Trans. R. Soc., S. Africa, 34(2):279-324, 8 textfigs.

Chemistry, pH
Milo di Villagrazia, Pietro, 1946
Il coefficiente di Sörensen (pH) dell'acqua di mare determinato a scopo orientativo col metodo dei fogli indicatori di Wulff.
Bollettino di Pesca, Piscicoltura e Idrobiologia, Anno XXII, Vol. 1 (ns), Fasc. 1, pp. 28-38, 2 textfigs.

Chemistry - pH
Mitchell, P.H., K. Buch and N.W. Rakestraw, 1936.
The effect of salinity and temperature upon the dissociation of cresol red and phenol red in sea water.
J. du Cons., 11(2):183-189.

chemistry - pH
Mitchell, P.H., and I.R. Taylor, 1935.
The dissociation constant of cresol red in sea water.
J. du Cons., 10(2):169-172.

Chemistry, pH
Moberg, E.G., and W.E. Allen, 1927.
Effect of tidal changes on physical, chemical, and biological conditions in the sea water of the San Diego region. 1. Observations on the effect of tidal changes on physical and chemical conditions of sea water in the San Diego region. 2. Half-hourly collections of marine microplankton taken at the Scripps Institution pier in 1923.
Bull. S.I.O., tech. ser., 1:1-17, 4 textfigs.

Chemistry, pH

Mohamed, A.F., 1940
Chemical and physical investigations. The distribution of hydrogen-ion concentration in the north-western Indian Ocean and adjacent waters. John Murrary Exped., 1933-34, Sci. Repts. 2(5):121-202, 34 textfigs.

Chemistry pH

Mokiyevskaya, V.V., 1961
[Some peculiarities of hydrochemistry in the northern Indian Ocean.] (In Russian; English abstract).
Okeanolog. Issled., Mezhd. Komitet Proved. Mezhd. Geofiz. Goda, Prezidiume, Akad. Nauk, SSSR, (4):50-61.

Chemistry pH(data)

Moreira da Silva, Paulo, 1957
Oceanografia do triangulo Cabo-Frio-Trinidade-Salvador.
Anais Hidrograficos, Brasil, 16:213-308.

Chemistry, pH

Mosby, H., 1938
Svalbard waters. Geofysiske Publ. 12(4): 85 pp., 34 textfigs.

Chemistry, pH

Motoda, S. 1940.
Comparison of the conditions of water in the bay, lagoon and open sea in Palao.
Palao Trop. Biol. Sta. Stud. (Tokyo) 2(1):41-48.

Chemistry, pH

Motoda, S., 1940.
The environment and the life of massivo reef coral, Goniastrea aspera Verrill, inhabiting the reef flat in Palao. Palao Trop. Biol. Sta. Studies 2(1):61-104, 9 textfigs.

Chemistry, pH

Motoda, S., 1940.
Comparison of the conditions of water in the bay lagoon and open sea in Palao.
Palao Trop. Biol. Sta. Studies 2(1):41-48, 2 textfigs.

Chemistry, pH

Motoda, S., 1940.
Organic matter in sea water of Palao, South Sea.
Trans. Sapporo Nat. Hist. Soc. 16(2):100-104.

chemistry, pH

Murray, J.W., 1966.
A study of the seasonal changes of the water mass of Christchurch Harbour, England.
Jour. mar. biol. Assoc., U.K., 46(3):561-578.

N

Chemistry pH(data)

Nagaya, Y., 1959
On the distribution of the nutrient salts in the Equatorial region of the Western Pacific, Jan.-Feb., 1958.
Publ. 981, Hydrogr. Bull., No. 59: 31-40.

Chemistry pH

Nasu, Keiju, 1963
Oceanography and whaling ground in the Sub-arctic region of the Pacific Ocean.
Sci. Repts., Whales Res. Inst., No. 14:105-155.

chemistry pH

Nasu, K., 1960
Oceanographic investigations in the Chukchi Sea during the summer of 1958.
Sci. Repts., Whales Res. Inst., 15: 143-157.

chemistry pH(data)

Navarro, F. de P., F. Lozano, J.M. Navaz, E. Otero, J. Sáinz Pardo and others, 1943.
La pesca de Arrastre en los fondos del Cabo Blanco y del Banco Arguin (Africa Sahariana).
Trab. Inst. Español Ocean., No. 18:225 pp., 38 pl.

Chemistry, pH

Nehring, Dietwart, und Karl-Heinz Rohde 1967.
Weitere Untersuchungen über anomale Ionenverhältnisse in der Ostsee.
Beitr. Meereskunde 20: 10-33.

Chemistry, pH

Neumann, A. Conrad, 1963.
Processes of recent carbonate sedimentation in Harrington Sound, Bermuda.
Mar. Sci. Center, Lehigh Univ., 130 pp. (Unpublished manuscript).

Chemistry, pH

Newell, B. S., 1966.
Seasonal changes in the hydrological and biological environments off Port Hacking, Sydney.
Australian J. Mar. freshw. Res., 17(1):77-91.

Chemistry pH

Newell, B.S., 1959
The hydrography of the British East African coastal waters. II. Colonial Off., Fish. Publ. London, No. 12: 18 pp.

chemistry, pH

Newell, N.D., J. Imbrie, E.G. Purdy and D.L. Thurber, 1959.
Organism communities and bottomfacies, Great Bahama Bank.
Bull. Amer. Mus. Nat. Hist., 117(4):183-228.

Chemistry, pH

Nichols, Maynard M., 1964.
Characteristics of sedimentary environments in Moriches Bay.
In: Papers in Marine Geology, R.L. Miller, Editor, Macmillan Co., N.Y., 363-383.

chemistry, pH (data)

Niemi, Åke, Heinrichs Skuja and Torbjörn Willén 1970
Phytoplankton from the Pojoviken-Tvärminne area, S. Coast of Finland.
Mem. Soc. Fauna Flora Fennica 46: 14-28.

Chemistry, pH

Nikiforov, E. G., E. V. Belysheva and N. I. Blinov, 1966.
On the structure of the water masses in the eastern part of the Arctic Basin. (In Russian).
Okeanologiia, Akad. Nauk, SSSR, 6(1):76-81

chemistry pH(data)

Nishida, Keizo, 1958
[On the oceanographic conditions of the lower cold water in the Japan Sea.]
J. Fac. Fish. and Animal Husbandry, Hiroshima Univ., 2(1): 1-65.

Chemistry, pH

Nishizawa, Tanzo and T. Muraki, 1940
Chemical studies in the sea adjacent to Palau. I. A survey crossing the sea from Palau to New Guinea. Kagaku Nanyo (Sci. of the South Sea) 2(3):1-7.

Chemistry pH

Noble, A., 1968.
Studies on sea-water off the North Kanara coast.
J. mar. biol. Ass., India, 10(2): 197-223.

Chemistry, pH

Norina, A.M., 1965.
Hydrochemical characteristics of the northern part of the Barents Sea. (In Russian).
Trudy, Gosudarst Okeanogr. Inst., No. 83:243-271.

O

Chemistry, pH

Odum, Howard T., Rene P. Cuzon du Rest, Robert J. Beyers and Clyde Allbaugh, 1963.
Diurnal metabolism, total phosphorus, Ohle anomaly and zooplankton diversity of abnormal marine ecosystems of Texas.
Publ. Inst. Mar. Sci., Port Aransas, 9:404-453.

chemistry pH

Okabe, Shiro, Yoshimasa Toyota and Takafumi Murakami, 1967.
Major aspects of chemical oceanography. (In Japanese; English abstract).
J.Coll.mar.Sci.Techn., Tokai Univ., (2):39-56.

Chemistry, pH

Okitsu, T., T. Tokui and B. Tsubata, 1954.
Oceanographical observations off Asamushi during 1952. Bull. Mar. Biol. Sta., Asamushi, 7(1): 21-25.

Chemistry pH

Okuda, Taizo, 1960
Metabolic circulation of phosphorus and nitrogen in Matsushima Bay (Japan) with special reference to exchange of these elements between sea water and sediments. (Portuguese, French resumés)
Trabalhos, Inst. Biol. Maritima e Oceanogr., Universidade do Recife, Brasil, 2(1):7-153.

Chemistry pH

Okuda, Taizo, Lourinaldo Cavalcanti and Manoel Pereira Borba, 1960.
Estudo da Barra das Jangadas. 2. Variacao do pH, Oxigenio dissolvido e consumo de permaganato durante o ano. (In Portuguese; English and French resumés).
Trabalhos, Inst. Biol. Marit. e Oceanogr., Universidade do Recife, Brasil, 2(1):193-205.

Chemistry, pH

Olson, F.C.W., 1953.
Tampa Bay studies.
Ocean. Inst., Florida State Univ., Rept. 1:27 pp. (mimeographed), 21 figs. (multilithed).
UNPUBLISHED.

Chemistry pH (data)

Oren, O.H., 1962
A note on the hydrography of the Gulf of Eylath. Contributions to the knowledge of the Red Sea No. 21.
Sea Fish. Res. Sta., Haifa, Israel, Bull. No. 30: 3-14.

Chemistry, pH

Orr, A.P., 1947
An experiment in marine fish cultivation: II. Some physical and chemical conditions in a fertilized sea-loch (Loch Craiglin, Argyll).
Proc. Roy. Soc., Edinburgh, Sect.B, 63, Pt.1 (2):3-20, 19 text figs.

Chemistry, pH

Orr, A.P. 1933.
Physical and chemical conditions in the sea in the neighborhood of the Great Barrier Reef.
Scient. Repts. Brit. Mus. (N.H.) Great Barrier Reef Exped. 1928-29, 2(3): 37-86.

chemistry pH
Orr, A.P., and F.W. Moorhouse, 1933.
(a) Variations in some physical and chemical conditions on and near Low Isles Reef.
(b) Temperature of the waters in the anchorage, Low Isles.
(c) Physical and chemical conditions in mangrove swamps.
Brit. Mus. (N.H.) Great Barrier Reef Exped., 1928-1929, Sci. Res., 2(4):87-110.

chemistry pH
Ottmann, Francois, and Taizo Okuda, 1961
Etude des conditions physico-chimiques des eaux de deux estuaires du Nord-est Bresilien.
Cahiers Oceanogr., C.C.O.E.C., 13(4): 234-242.

chemistry pH
Ottmann, François, et Jeanne-Marie Ottmann, 1959.
La marée de salinité dans le Capibaribe, Recife-Bresil.
Trabalhos Inst. Biol. Marit. e Oceanogr., Recife, 1(1):39-49.

P

chemistry, pH
Palitzsch, S., 1912.
Measurement of hydrogrn ion concentration in sea water. Rept. Danish Oceanogr. Exped., 1908-1910, 1:237-254, 2 pls.

chemistry pH
*Park, P. Kilho, 1968.
The processes contribution to the vertical distribution of apparent pH in the northeastern Pacific Ocean.
J. oceanogr. Soc., Korea, 3(1):1-7.

chemistry, pH.
Park Kilho, 1968.
Alkalinity and pH off the coast of Oregon.
Deep-Sea Res. 15(2):171-183.

chemistry, pH
Park, P. Kilho, 1968.
Seawater hydrogen-ion concentration: vertical distribution.
Science, 162(3851):357-358.

pH
Park, Kilho 1967
Chemical features of the subarctic boundary near 170°W.
J. Fish. Res. Bd. Can., 24 (5):899-908.

pH
Park, Kilho 1966.
Surface pH of the northeastern Pacific Ocean.
J. oceanol. Soc. Korea 1 (1/2):1-6.

chemistry, pH
Park, Kilho, 1966.
Deep-sea pH.
Science, 154(3756):1540-1542.

chemistry pH
Park, Kilho, 1964
Partial equivalent conductance of electrolytes in sea water.
Deep-Sea Res., 11(5):729-736.

pH
Patil, M.R. and C.P. Ramamirtham 1963
Hydrography of the Laccadives offshore waters - a study of the winter conditions.
J. mar. biol. Ass. India 5 (2):159-169.

chemistry, pH
Perez Sori, Jose A., 1961.
Salinidad, temperatura, presion atmospherica y pH en la playa Habana: II.
Centro Invest. Pesqueras, Cuba, Contrib. No. 13:19 pp.

chemistry pH
Phifer, L. D. and T. G. Thompson, 1937.
Seasonal variations in the surface waters of San Juan channel during the five year period, January 1931 to December 30, 1935. J. Mar. Res., 1(1):34-59, text figs. 12-18, 12 tables.

chemistry, pH
Picotti, Mario, 1965.
La crociera idrografico-talassografica di Capo Matapan.
Pubbl. Commissione Ital. Comitato int. Geofis., Ser. IGC, 42:63 pp.

chemistry pH(data)
Picotti, Mario, 1960
Crociera Talassografica Adriatica 1955. III. Tabelle delle osservazioni fisiche, chimiche, biologiche e psammografiche.
Archivio di Oceanograf. e Limnol., 11(3): 371-377, plus tables.

chemistry pH(data)
Picotti, M., and A. Vatova, 1942.
Osservazioni fisiche e chemiche periodiche nell' Alto Adriatico (1920-1938). Thalassia 5(1):1-157.

chemistry pH
Pirie, Robert Gordon, 1965.
Petrology and physical-chemical environment of bottom sediments of the Ribière Bonaventure-Chaleur Bay area, Quebec, Canada.
Rept. B.I.O. 65-10:182 pp. (multilithed).

chemistry pH
Piro, A. 1971.
Chemical environmental factors in marine radiocontamination.
CNEN Rept. RT/BIO (70)-11, M. Bernhard, editor: 11-22.

chemistry pH
Ponomarenko, L.S., 1966.
Hydrochemical conditions in the Norwegian and Greenland seas in summer (from the data of the June surveys 1954-1964). (In Russian).
Trudy, Poliarn. Nauchno-Issled. Procktn. Inst. Norsk. Ribn. Khoz. Okeanogr., N. M. Knipovich, PINRO, 18:125-145.

chemistry, pH
Pozdniakova, L.E., and V.N. Vinogradov, 1966.
Some peculiarities in the distribution of the hydrochemical elements in the southeaster Barents Sea in August-September 1959. (In Russian).
Trudy murmansk. biol. Inst., 11(15):140-156

chemistry, pH
Pratt, D.M., 1949.
Experiments in the fertilization of a salt water pond. J. Mar. Res. 8(1):36-59, 8 textfigs.

chemistry, pH
Pritchard, D. W., and R.E. Bunce, 1959.
Physical and chemical hydrography of the Magothy River.
Chesapeake Bay Inst., Tech Rept., (Ref. 59-2), 17:1-2. (Unpublished manuscript).

chemistry pH
Puri, Harbans S., Gioacchino Bonaduce and John Malloy, 1964.
Ecology of the Gulf of Naples.
Pubbl. Staz. Zool., Napoli, 33(Suppl.):87-199.

chemistry pH
Pytkowicz, Ricardo M., 1963
Calcium carbonate and the insitu pH.
Deep-Sea Res., 10(5):633-638.

chemistry, pH
Pytkowicz, R.M., and D.R. Kester and B.C. Burgener, 1966.
Reproducibility of pH measurements in seawater.
Limnol. Oceanogr., 11(3):417-419.

Q

R

chemistry pH
Rae, B.B., R. Johnston and J.A. Adams, 1965.
The incidence of dead and dying fish in the Moray Firth, September 1963.
Jour. Mar. Biol. Assoc., U.K., 45(1):29-47.

chemistry pH
Rakestraw, N. W. and D. E. Carritt, 1948
Some seasonal chemical changes in the open ocean. J. Mar. Res. 7(3):362-369, 6 text figs.

chemistry pH
Rakestraw, N.W., and H.P. Smith, 1937.
A contribution to the chemistry of the Caribbean and Cayman Seas. Bull. Bingham Ocean. Coll. 6(1): 1-41, 32 text-figs.

chemistry, pH (data)
Ramalho, A. de M., L. S. Dentinho, C. A. M. de Sousa, Fronteira, F. L. Mamede, and H. Vilela, 1935.
Observacoes oceanograficas. "Albacora" 1934. Trav. St. Biol. Mar., Lisbonne, No. 35: 35 pp., 1 fig.

chemistry, pH
Ramamurthy, S. 1963.
Studies on the hydrological factors in the north Kanara coastal waters.
Indian J. Fish. (A) 10 (1):75-93.

chemistry pH
Ramamurthy, S., 1954.
Hydrobiological studies in the Madras coastal waters. J. Madras Univ., B, 23(2):148-163, 3 textfigs.

chemistry pH
Ramamurthy, S., 1953.
Seasonal changes in the hydrogen-ion concentration and the dissolved oxygen content of the surface waters of the Madras coast.
J. Madras Univ., B, 23(1):52-60, 2 textfigs.

chemistry, pH
Rao, D.S., and N. Madhavan, 1964.
On some pH. measurements in the Arabian Sea along the west coast of India.
J. Mar. Biol. Assoc., India, 6(2):217-221.

chemistry, pH
Reid, G.K., jr., 1955.
A summer study of the biology and ecology of East Bay, Texas.
Texas J. Sci., 7(3):316-343, 6 textfigs.

chemistry pH (data)
Reish, D.J., and H.A. Winter, 1954.
The ecology of Alamitos Bay, California, with special reference to pollution.
Calif. Fish and Game 40(2):105-121, 1 textfig.

chemistry pH
Richards, Francis A., 1968.
Chemical and biological factors in the marine environment. Ch. 8 in: Ocean engineering: goals, environment, technology, J.F. Brahtz, editor, John Wiley & Sons, 259-303.

Chemistry pH

Rieke, H.H., III., and G.V. Chilingar, 1962. Note on pH of brines. Sedimentology, 1(1):75-79.

Chemistry, pH (water samples)

Rittenberg, S.C., K.O. Emery and W.L. Orr, 1955. Regeneration of nutrients in sediments of marine basins. Deep-Sea Res. 3(1):23-45.

Chemistry, pH

Roche, A., and A. Roubault, 1947. Observations sur les courants superficiels de la Mer de Monaco. Bull. Inst. Océan., Monaco, No. 909, 11 pp., 2 textfigs.

Chemistry, pH(data)

Rochford, D.J., 1953. Estuarine hydrological investigations in eastern and south-western Australia, 1951. Oceanographical station listof investigations made by the Division of Fisheries, Commonwealth Scientific and Industrial Research Organization, Australia, 12:111 pp.

Chemistry pH

Rochford, D.J., 1951. Studies in Australian estuarine hydrology. Australian J. Mar. and Freshwater Res. 2(1):1-116, 1 pl., 7 textfigs.

Chemistry, pH

*Roger, C.,1967. Contribution à la connaissance des euphausiacés du Pacificque équatorial. Cah. ORSTOM, Sér. Océanogr., 5(1):29-37.

Chemistry, pH

Rosfelder, A., 1955. Carte provisoire au 1/500,000 de la marge continental algérienne. Note de présentation. Publ. Service Carte Géol., Algérie, n.s., Bull. No. 5, Trav. des Coll., 1954:57-106, 6 figs., 1 pl., 1 map.

Chemistry pH

Rotschi, Henri, 1965. Le pH et l'alcalinité des eaux profondes de la Fosse des Hebrides et du Bassin des Fidji. Progress in Oceanography, 3:301-310.

Chemistry pH

Rotschi, Henri, 1961 Influence de la divergence des Salomon sur la répartition de certaines propriétés des eaux. C.R. Acad. Sci., Paris, 253:2559-2561.

Chemistry, pH(data)

Rotschi, H., 1961. Chimie. O.R.S.T.O.M., I.F.O., Rapp. Sci., 23:4-45.

Chemistry pH(data)

Rotschi, Henri, 1960 Orsom III, Resultats des Croisières diverses de 1959. Oceanographie physique. Centre d:Océanogr., Inst. Francais d'Océanie, Rapp. Sci., No. 17:59 pp.

Chemistry pH

Rotschi, Henri, 1960. Sur la distribution du pH en mer de Corail. C.R. Acad. Sci., Paris, 251:1223-1225.

Chemistry pH

Rotschi, Henri, 1959 Chimie, "OrsomIII", Resultats de la Croisière BOUSSOLE. O.R.S.T.O.M., Inst. Français d'Océanie, Rapp. Sci. 13: 3-60.

Chemistry pH (data)

Rotschi, Henri, Michel Angot and Roger Desrosieres, 1960. Orsom III, Resultats de la croisiere "Choiseul" 2ème partie. Chimie, productivité, phytoplancton qualitatif. Rapp. Sci., Noumeau, No. 16:91 pp. (mimeographed).

Chemistry pH (data)

Rotschi, Henri, Michel Angot, et Michel Legrand, 1959 Orsom III. Resultats de la croisiere "Astrolabe". 2. Chimie, productivité et zooplankton. Rapp. Sci., Inst. Francais d'Oceanie, No. 9: 97 pp. (mimeographed).

Chemistry pH(data)

Rotschi, Henri, Michel Angot, Michel LeGand, and H.R. Jitt, 1959 Chimie, productivité et zooplancton. "Orsom III Resultats de la Croisiere "Boussole". Resultats "production primaire" de la croisière 56-5. Rapp. Sci. Inst. Francais d'Oceanie, Centre d' Oceanogr., No. 13.

Chemistry, pH

Rouch, J., 1946 Traité d'Océanographie physique. L'eau de mer. Payot, Paris, 349 pp., 150 text figs.

Chemistry pH

Rouch, J., 1940. Observations océanographiques de surface dans l' océan Atlantique et dans la Mediterranée. Ann. Inst. Océan. 20(2):51-73, 11 textfigs.

Chemistry pH

Rouch, J., 1939. Observations océanographiques de surface dans l'océan Atlantique et dans l'océan Pacifique. Bull. Inst. Océan., Monaco, No. 781: 10 pp., 4 figs.

Chemistry pH

Rubinchik, E.E., 1959 [Data on the hydrochemistry of the southern Caspian.] Trudy V.N.I.R.O., 38:152-164.

Chemistry pH

Ruud, B., 1926. Quantitative investigations of plankton at Lofoten, March-April, 1922-1924. Preliminary report. Rept. Norwegian Fish. Mar. Invest. 3(7):30 pp., 3 charts, 5 diagrams.

S

Chemistry, pH

Saruhashi, K., 1953. On the total carbonaceous matter and hydrogen ion concentration in sea water - a study on the metabolism of natural water (1). Pap. Met. Geophys. 3(3):202-206, 2 textfigs.

Chemistry, pH

Saruhashi, K., 1952. [A study on pH and the total carbon dioxide in sea water.] Umi to Sora 30(1/2):26-29, 2 textfigs.

Chemistry, pH

Schachter, D., 1954. Contribution à l'étude hydrographique et hydrologique de l'étang de Berre (Bouches-du-Rhone). Bull. Inst. Océan., Monaco, No. 1048: 20 pp.

Chemistry pH

Schulz, B., 1934. Die Ergebnisse der Polarexpedition auf dem U-Boat "Nautilus". Ann. Hydrogr., usw., 62:147-152, Pl. 16, with 2 figs.

Chemistry pH(data)

Schulz, B., 1923. Hydrographische Untersuchungen besonders über den Durchlüftungszustand in der Ostsee in Jahre 1922. (Forschungsschiffe "Skagerak" und "Nautilus"). Arch. Deutschen Seewarte 41(1):1-64, 5 pls. 2 textfigs.

Chemistry, pH(data)

Schulz, B., 1922. Hydrographischen Beobachtungen inbesondere über die Kohlensäure in der Nord- und Ostsee im Sommer 1921. (Forschungsschiffe "Poseidon" und "Skagerak"). Arch. Deutschen Seewarte 40(2):1-44, 2 textfigs., 4 pls.

Chemistry, pH

Schulz, B., and A. Wulff, 1929 Hydrographie und Oberflächen plankton des westlichen Barentsmeeres im Sommer 1927. Ber. deutschen wissensch. Komm. F. Meeresforsch. n.s. 4(5):232-372, 13 tables, 25 text figs.

Chemistry, pH

Scott, K.M.F., A.D. Harrison and W. Macnae, 1952. The ecology of South African estuaries. II. The Klien River estuary, Harmanus, Cape. Trans. R. Soc., S. Africa, 33(3):283-332.

Chemistry, pH

Scruton, P.C., 1956. Oceanography of Mississippi delta sedimentary environments. Bull. Amer. Assoc. Petr. Geol., 40(12):2864-2952.

Chemistry pH(data)

Seibold, E., 1962. Untersuchungen zur Kalkfällung und Kalklösung am Westrand der Great Bahama Bank. Sedimentology, 1(1):50-74.

pH (data)

Shimomura, T., and K. Miyata, 1957. [The oceanographical conditions of the Japan Sea and its water systems, laying stress on the summer of 1955.] Bull. Japan Sea Reg. Fish. Res. Lab., No. 6 (General survey of the warm Tsushima Current 1): 23-120.

Chemistry, pH

Skirrow, Geoffrey, 1965. The dissolved gases - carbon dioxide. In: Chemical oceanography, J.P. Riley and G. Skirrow, editors, Academic Press, 1:227-322.

Chemistry pH

Smetanin, D.A., 1960 [Some features of the chemistry of water in the central Pacific.] Trudy Inst. Okeanol., 40: 58-71.

Chemistry pH

Smith, W.H., Jr., and Donald W. Hood, 1964. pH measurement in the ocean: a sea water secondary hiffer system. In: Recent researches in the fields of hydrosphere, atmosphere and nuclear geochemistry, Ken Sugawara festival volume, Y. Miyake and T. Koyama, editors, Maruzen Co., Ltd., Tokyo, 185-202.

Chemistry pH

Soika, A. Giordani, 1955 Ricerche sull'ecologia e sul popolamento della zona intercotidale delle Spiagge di Sabbia fina. Boll. Mus. Civico di Storia Nat., Venezia, 8: 7-151.

Chemistry, pH

Soloni Toural, F., 1954. Salinidad, temperatura y pH en la playa Habana (abril 1952-marzo 1953). Mem. Soc. Cub. Hist. Nat. 21:389-391.

Chemistry pH

Solov'eva, N.F., 1959. [Hydrochemical analyses of the Aral Sea in 1948.] Akad. Nauk, SSSR, Lab. Ozerovedeniya, Trudy, 8: 3-22.

5 Sept. 61 JPRS:10054
OTS-SLA $3.60 61-28414.

Chemistry, pH(data)
Soot-Ryen, T., 1956.
Report on the hydrographical conditions in West-Finnmark, March-April, 1935.
Acta Borealia, TromsøMus., A. Scientia, No. 10: 1-37.

Chemistry pH(data)
Soot-Ryen, T., 1947.
Hydrographical investigations in the Tromsø-district 1934-1938 (Tables). Tromsø Mus. Aarsheft., Naturhist. Avd. No. 33, 66(1943)(1): numerous pp. (unnumbered).

Chemistry pH (data)
Soot-Ryen, T., 1938.
Hydrographical investigations in the Tromsø district in 1931.
Tromsø Mus. Aarsheft., Naturhist. Avd. No. 10, 54(2):1-6 plus 41 pp. tables.

Chemistry, pH
Sourie, R., 1954.
Contribution à l'étude écologique des côtes rocheuses du Sénégal.
Mém. Inst. Francais d'Afrique Noire No. 38: 342 pp., 23 pls.

Chemistry, pH
*Smirnov, E.V., N.N. Mijailov y V.V. Rossov, 1967.
Algunas características hidroquinicas del Mediterráneo americano.
Estudios Inst. Oceanol., Acad. Ciencias, Cuba, 2(1): 51-74.

Chemistry, pH
Steemann-Nielsen, Einar, 1951
The marine vegetation of the Isefjord. A study on ecology and production. Medd. Komm. Danmarks Fiskeri-og Havundersøgelser. Ser. Plankton. 5(4); 114pp., 46 text figs.

Chemistry pH (data)
Sugawara, Ken, and Shiro Okabe, 1966.
Molybdenum and Vanadium determination of Umitaka-maru samples from her 1962-1963 and 1963-1964 cruises of the International Indian Ocean Expedition. (In Japanese; English abstract)
J. Tokyo Univ. Fish., (Spec.ed.)8(2):165-171.

chemistry, pH
Sugiura, Yoshio 1969.
Distribution of pH in the Kuroshio and Oyashio regions and its oceanographic significance.
Oceanogr. Mag 21(1): 39-46.

Chemistry pH
Svansson, Artur, 1964.
Observations at Swedish lightships and in the central Baltic.
Ann. Biol., Cons. Perm. Int. Expl. Mer, 1962, 19: 45-48.

Chemistry pH
Svansson, Artur, 1960(1958)
Observations at Swedish light vessels.
Ann. Biol., Cons. Perm. Int. Expl. Mer, 15: 43-45.

Chemistry, pH
Syazuki, K., 1953.
[Studies on the foul-water drained from factories 2. On the harmful components and the water pollution of the foul-water drained from the metal plants.] J. Shimonoseki Coll. Fish. 3(2):181-185, 2 textfigs.

T

Chemistry, pH
Tamura, T., and J. Suguira, 1950.
A report on oceanographical observations of the surface seawaters extending between Japan and Antarctic whaling grounds. Bull. Fac. Fish., Hokkaido Univ. 1(1):12-17, 2 figs. (In Japanese; English summary).

Chemistry pH
Tanita, S., K. Kato, and T. Okuda, 1951.
Studies on the environmental conditions of shell-fish-fields. In the case of Muroran Harbour (2).
Bull. Fac. Fish., Hokkaido Harbour, 2(3):220-230.

Chemistry, pH
Tanita, S., S. Kato, and T. Okuda, 1950.
Studies on the environmental conditions of shell-fish fields. 1. In the case of Hakodate Harbor.
Bull. Fac. Fish., Hokkaido Univ., 1(1):18-27, 7 textfigs. (In Japanese; English summary).

Chemistry, pH (data)
Teixeira, Clovis, and Miryam B. Kutner, 1962
Plankton studies in a mangrove environment. 1. First assessment of standing stock and principal ecological factors.
Bol. Inst. Oceanogr., Sao Paulo, Brasil, 12 (3):101-124.

Chemistry pH (data)
Teixeira, C., J. Tundisi and M.B. Kutner, 1965.
Plankton studies in a mangrove environment. II. The standing stock and some ecological factors.
Bolm Inst. Oceanogr., S Paulo, 14(1):13-41.

Chemistry pH
Theisen, E. 1946
Tanafjorden. Enfinmarksfjords oceanografi. Fiskeridirektoratets skrifter Ser. Havundersøkelser. (Repts. on Norwegian Fishery and Marine Investigations) VIII (8):1-77, 23 text figs., 8 pls.

Chemistry pH
Thompson, T.G., and D.H. Anderson 1940
The determination of the alkalinity of sea water
J. mar. Res. 3(3): 224-229.

Chemistry, pH
Toural, F.S., 1954.
Salinidad, temperatura y pH en la Playa Habana.
Mem. Soc. Cubana Hist. Nat. Felipe Poey, 22(4): 389.

Chemistry, pH (data)
Tsubata, B., 1956.
Oceanographical observations off Asamushi during 1954. Bull. Mar. Biol. Sta., 8(1):43-48.

Chemistry pH
Tsubata, B., and T. Numakunai, 1960.
Oceanographical conditions observed at definite Station off Asamushi during 1958 and 1959.
Bull. Mar. Biol. Sta., Asamushi, Tohoku Univ., 10(1):73

Chemistry, pH
Tully, J.P., and A.J. Dodimead, 1957.
Properties of the water in the Strait of Georgia, British Columbia, and influencing factors.
J. Fish. Res. Bd., Canada, 14(3):241-319.

U

Chemistry pH(data)
Uda, Michitaka, and Makoto Ishino, 1960
Researches on the currents of Kuroshio.
Rec. Oceanogr. Wks., Japan, Spec. No. 4: 59-72.

Chemistry, pH
Udaya Varma Thirupad, P., and C.V. Gangadhara Reddy, 1959.
Seasonal variations of the hydrological factors of the Madras coastal waters.
Indian J. Fish., 6(2):298-305.

Chemistry, pH (data)
University of Southern California, Allan Hancock Foundation, 1965.
An oceanographic and biological survey of the southern California mainland shelf.
State of California, Resources Agency, State Water Quality Control Board, Publ. No. 27*232 pp. Appendix, 445 pp.

Chemistry, pH (data)
*Usuki, Itaru, 1967.
A record of hydrographic conditions of inshore water AROUND the vicinity of the Sado Marine Biological Station during 1964 to 1966.
Sci. Rep. Niigata Univ. (D)4:87-107.

V

Chemistry, pH
Val Cordón, M. a Jesús, y E.O. Aenlle, 1941.
Resultados de una campaña oceanográfica verificada en la Ría de Vigo durante el mes de agosto de 1941, bajo le dirección del Jefe del Departmento de Química Aplicada, D. Ricardo Montequi. Investigaciones químicas y determinación de algunas constantes físicas. Notas y Resúmenes Inst. Español Oceanogr. 104:1-27, map.

Chemistry pH (data)
Van Goethem, C., 1951.
Étude physique et chimique du milieu marin. Rés. Sci., Exped. Océan. Belge dans les Eaux Côtières Africaines de l'Amerique Sud (1948-1949) 1:1-151, 1 pl.

Chemistry pH(data)
Varlet, F., 1960
Sur l'hydrologie du plateau continental africain du Cap des Palmes au Cap des Trois Pointes. Bull. Inst. Pêches Maritimes, Maroc, No. 5: 20 pp.

Chemistry pH
Varlet, F., 1958
Les traits essentials du régime côtise de l'Atlantique près d'Abidjan (Côte d'Ivoire).
Bull. d'I.F.A.N., (A), 20(4):1089-1102.

Chemistry, pH
Vatova, A., 1934.
L'anormale regime fisico-chimico dell' Alto Adriatico nel 1929 e le sue ripercussioni sulla fauna. Thalassia 1(8):49 pp., 3 fold-ins, 16 figs., tables.

Chemistry, pH (data)
Vatova, A., and P.M. di Villagrazia, 1950.
Sulle condizioni idrographiche di Canal di Leme in Istria. Nova Thalassia 1(8):24 pp., graphs.

Chemistry, pH
Vatova, A., and P. Milo di Villagrazia, 1948.
Sulle condizioni chimicofisiche del Canale di Lema presso Rovigno d'Istria.
Bol. Pesca, Piscicol., Idrobiol., n.s., 3(1): 5-27, 2 textfigs., 5 pls.

Chemistry, pH
Vercelli, F., and M. Picotti, 1926.
Pt. 2. Il regime fisico-chimico della acque nello Stretto di Messina. Crociere per lo studio dei fenomeni nello Stretto di Messina. (Campagne della R. Nave "Marsigli" negli anni 1922 e 1923).
Comm. Int. Medit., Delag. Ital., 161 pp., 40 figs.

Chemistry, pH
Villagrazia, P. Milo de, 1950.
Ulteriore contributo alla conOscenza delle condizioni idrografiche e biologiche del Canali di Lema (Istria). Bol. Pesca, Piscicolt. Idrobiol. 5(2):225-247, 3 textfigs.

Chemistry, pH
di Villagrazia, P.M., 1946.
Il coefficiente di Sörensen (pH) dell'acque di mare determinate a scopo orientatino col metodo dei fogli indicatori di Wulff. Boll. Pesc. Pisc. Idrobiol. 1 (n.s.):28-39.

Chemistry pH
Voit, S.S., D.A. Aksenov, M.M. Bogorodsky, Y.V. Sinukov and V.A. Vladimirzev, 1961
[Some peculiar features of water circulation in the Black Sea and the water regime in the Pre-Bosphorus area.]
Okeanologiia, Akad. Nauk, SSSR, 1(4):613-625.

W

Chemistry, pH

Waldichuk, Michael, 1965.
Water exchange in Port Moody, British Columbia, and its effect on waste disposal.
J. Fish. Res. Bd., Canada, 22(3):801-822.

Chemistry, pH

Waldichuk, M., 1956.
Basic productivity of Trevor Channel and Alberni Inlet from chemical measurements.
J. Fish. Res. Bd., Canada, 13(1):7-20.

Chemistry, pH

Waldichuk, M., 1954.
Effect of pulpmill waste in Alberni Harbour.
Fish. Res. Bd., Canada, Pacific Coast Stas., Prog. Repts. No. 101:23-26, 2 textfigs.

Chemistry, pH

Watts, J.C.D., 1958.
The hydrology of a tropical West African estuary.
Bull., I.F.A.N., 20(A):697-752.

Chemistry
pH

Wiktorowie, Jozef i Krystyna, 1962
Some hydrological properties of the Pomeranian Bay water.
Prace Morsk. Inst., Ryback., Gdyni, Oceanogr. Ichtiol., 11(A):113-136.

Chemistry, pH

Won, Chong Hun, 1964
Tidal variations of chemical constituents of the estuary water at the Lava bed in the Nack-Dong River from Nov. 1962-to Oct. 1963. (In Korean; English abstract).
Bull. Pusan Fish. Coll., 6(1):21-32.

Chemistry, pH

Won, Chong Hun, and Kil Soon Park, 1968.
Tidal variations of chlorinity and pH at the Yong-Ho Basin from Mar. 1 to Mar. 20, 1968.
(In Korean; English abstract).
Bull. Pusan Fish. Coll. (Nat. Sci.), 8(2):103-111.

XYZ

Chemistry pH(profiles)

Yamazi, I., 1953.
Plankton investigation in inlet waters along the coast of Japan. VII. The plankton collected during the cruises to the new Yamamoto Bank in the Sea of Japan. Publ. Seto Mar. Biol. Lab. 3(1): 75-108, 19 textfigs.

Chemistry, pH

Yamazi, I., 1951.
Plankton investigations in inlet waters along the coast of Japan. II. The plankton of Hakodate Harbour and Yoichi Inlet in Hokkaido.
Publ. Seto Mar. Biol. Sta., Kyoto Univ., 1(4): 185-194, 3 textfigs.

Chemistry, pH

Yamazi, I., 1952.
Plankton investigations in inlet waters along the coast of Japan. III. plankton of Imari Bay in Kyusyu. IV. The plankton of Nagasaki Bay and Harbour in Kyusyu. V. The plankton of Hiroshima Bay in the Seto-Nakai (Inland Sea)
Publ. Seto Mar. Biol. Lab., Kyoto Univ. 2(2): 289-304; 305-318; 319-330, 8 and 8 and 7 textfigs

Chemistry, pH

Yamazi, I., 1950.
Plankton investigations in inlet waters along the coast of Japan. 1. Publ. Seto Mar. Biol. Lab. 1(3):93-113, 14 textfigs.

pH, annual variation

Choe, Sang and Tai Wha Chung, 1971.
Oceanological characteristics of the Ko-ri sea area. 1. Annual cyclic changes in water temperature, salinity, pH and transparency.
J. oceanogr. Soc. Korea 6(1): 37-48.
(In Korean; English abstract)

pH anomalies

Chemistry pH anomalies

Adrov, M.M., 1966.
The problem of the practical use of the distribution of some oceanograhic parameters.
(In Russian).
Materiali, Sess. Uchen. Soveta PINRO Rez. Issled., 1964, Minist. Ribn. Khoz., SSSR, Murmansk, 148-162.

pH (data only)

Chemistry
P H (data only)

Akamatsu, Hideo, Tsutomu Akiyama and Tsutomu Sawara, 1965.
Preliminary report of the Japanese Expedition of Deep-Sea, the Tenth Cruise, 1965(JEDS-10).
Oceanogr. Mag., Tokyo, 17(1/2):49-68.

Chemistry pH (data only)

Anon., 1950.
Oceanographical observations taken in the North Pacific Ocean (east of Sanriku); physical and chemical data for stations occupied by R.M.S. "Yushio-maru" (continued). Res. Mar. Met., Ocean. Obs., Jan.-June 1947, No. 1:28-37, Table 4.

Chemistry, pH (data only)

Anon., 1950.
[Report of the oceanographical observations from Fisheries Experimental Stations (Sept. 1950)]
J. Ocean., Kobe Obs., 2nd ser., 1(2):5-11.

Chemistry, pH(data only)

Anon., 1950.
Table 14. Oceanographical observations taken in the Maizuru Bay; physical and chemical data for stations occupied by R.M.S. "Syunpu-maru". Res. Mar. Met., Ocean. Obs., Tokyo, July-Dec. 1947, No. 2:71-76.

Chemistry, pH(data only)

Anon., 1950.
Table 2. Oceanographical observations taken in the Uchiura Bay; physical and chemical data for stations occupied by R.M.S. "Kuroshio-maru". Res. Mar. Met., Ocean. Obs., Tokyo, July-Dec. 1947, No. 2:3-7.

Chemistry pH (data only)

Anon., 1950.
Table 1. Oceanographical observations taken in the Uchiura Bay; physical and chemical data for stations occupied by a fishing boat belonging to Mori Fishery Assoc. Res. Mar. Met., Ocean. Obs., Tokyo, July-Dec. 1947, No. 2:1-2.

Chemistry pH (data only)

Anonymous, 1947
Discovery investigations Station List 1937-1939. Discovery Report, XXIV: 197-422. pls IV-VI.

Chemistry,
pH (data only)

Aragno, Federico, Alberto Gomez, Aldo Orlando y Andres J. Lusguinos 1968.
Datos y resultados de las campañas pesquerias "Pesqueria II" (9 de noviembre al 12 diciembre de 1966).
Publ. (Ser. Informes técn) Mar del Plata, Argentina 10(2): 1-129.

Chemistry,
pH. (data only)

Aragno, Federico, Alberto Gomez, Aldo Orlando y Andres J. Lusguinos 1968.
Datos y resultados preliminares de las campañas pesquerias "Pesqueria I" (12 de agosto al 8 de setiembre de 1966).
Publ. (Ser. Informes técn) Mar del Plata, Argentina, 10(1): 1-159

chemistry
PH (data only)

Argentina, Secretaria de Marina, Servicio de Hidrografia Naval, 1961.
Trabajos oceanograficos realizados en la campana Antartica 1959/1960. Resultados preliminares
Publico, H. 623:unnumbered pp.

Chemistry pH (data only)

Argentina, Servicio de Hidrografia Naval, 1959.
Trabajos oceanograficos realizados en la campana Antartica, 1955-1956.
Servicio de Hidrografia Naval, Argentina, Publ., H. 620: numerous unnumbered pp.

Chemistry pH (data only)

Argentina, Servicio de Hidrografia Naval, 1959.
Operacion Oceanografica Meridiano. Resultados preliminares.
Servicio de Hidrografia Naval, Argentina, Publ., H. 617: numerous unnumbered pp.

Chemistry pH(data only)

Argentina, Servicio de Hidrografia Naval, 1959.
Trabajos oceanograficos realizados en la campana Antartica 1958/1959. Resultados preliminares.
Servicio de Hidrografia Naval, Argentina, Publ., H. 616:127 pp.

Chemistry pH (data only)

Argentina, Servicio de Hidrografia Naval, 1959.
Trabajos oceanograficos realizados en la campana Antartica, 1957-1958. Resultados preliminares.
Servicio de Hidrografia Naval, Argentina, Publ., H. 615: numerous unnumbered pp.

Chemistry pH (data only)

Argentina, Servicio de Hidrografia Naval, 1959.
Operacion Oceanografica Drake. 1. Resultados preliminares.
Servicio de Hidrografia Naval, Argentina, Publ., H. 613: numerous unnumbered pp.

Chemistry pH (data only)

Argentina, Servicio de Hidrografia Naval, 1959.
Operacion Oceanografica Atlantico Sur. Resultados preliminares.
Servicio de Hidrografia Naval, Argentina, Publ., H. 608: numerous unnumbered pp.

Chemistry
pH(data only)

Australia, Commonwealth Scientific and Industrial Research Organization, 1961
Coastal investigations at Port Hacking, New South Wales, 1959.
Div. Fish. & Oceanogr., Oceanogr. Sta. List, No. 47: 135 pp.

Chemistry
pH (data only)

Australia, Commonwealth Scientific and Industrial Research Organization, 1960
Coastal hydrological investigations in south-eastern Australia, 1958. Oceanogr. Sta. List. Div. Fish. and Oceanogr., 60 pp.

Chemistry pH(data only)

Australia, Commonwealth Scientific and Industrial Research Organization, 1960.
Coastal investigations at Port Hacking, New South Wales, 1958.
Oceanogr. Sta. List, Div. Fish. & Oceanogr., 42: 99 pp.

Chemistry pH (data only)

Australia, Commonwealth Scientific and Industrial Research Organization, 1957.
Estuarine hydrological investigations in eastern and south-western Australia.
Oceanogr. Sta. List., Div. Fish. and Oceanogr., 32:170 pp.

Chemistry
pH (data only)

Australia Commonwealth and Industrial Research Organization, 1957.
Onshore and oceanic hydrological investigations in eastern and southwestern Australia, 1956.
Oceanogr. Sta. List, Div. Fish. & Oceanogr. 30:79pp.

Chemistry
pH(data only)

Australia, Commonwealth Scientific and Industrial Research Organization, 1957.
Estuarine hydrological investigations in eastern and south-western Australia, 1955.
Ocean. Station List, 29:93 pp.

chemistry pH (data only)
Australia, Commonwealth Scientific and Industrial Research Organization, 1957.
Onshore and oceanic hydrological investigations in eastern and southwestern Australia, 1955.
Oceanogr. Sta. List, 27:145 pp.

chemistry pH (data only)
Australia, Commonwealth Scientific and Industrial Research Organization, 1956.
Onshore hydrological investigations in eastern and south-western Australia, 1954. Compiled by D.R. Rochford.
Ocean. Sta. List., Invest. Div. Fish., 24:119 pp.

chemistry pH(data only)
Australia, Commonwealth Scientific and Industrial Research Organization, 1954.
Onshore hydrological investigations ineastern and south-western Australia, 1953.
Ocean. Sta. List, Invest. Div. Fish. No. 18:64 pp

chemistry, pH (data only)
Australia, Commonwealth Scientific and Industrial Research Organization, 1953.
Onshore hydrological investigations in eastern and south-western Australia, 1951. Vol. 14:64 pp.

chemistry, pH (data only)
Australia, Commonwealth Scientific and Industrial Research Organization, Melbourne, 1952.
Oceanographical station list of investigations made by the Fisheries Division, C.S.I.R.O.
Vol. 5. Estuarine hydrological investigations in eastern Australia 1940-50:150 pp.
Vol. 6. Ibid.:137 pp.
Vol. 7. Ibid.:139 pp.
Vol. 8: Hydrological investigations in south-western Australia, 1944-50:152 pp.
Vol. 10. Records of twenty-four hourly hydrological observations at Shell Point, Georges River, New South Wales 1942-50: 134 pp.

chemistry, pH(data only)
Australia, Commonwealth Scientific and Industrial Research Organizations, 1951.
Oceanographic station list of investigations made by the Fisheries Division, C.S.I.R.O.
Vol. 4. Onshore hydrological investigations in eastern Australia, 1942-50:114 pp.

chemistry pH (data only)
Ballester, Antonio, 1965(1967).
Tablas hidrograficas.
Memoria Soc.Cienc.nat.La Salle, 25(70/71/72):41-137.

chemistry pH (data only)
~~Ballester Antonio Enrique Arias, Antonio~~ Cruzado, Dolores Blasco y José María Camps, 1967.
Estudio hidrográfico de la coste catelena de junio de 1965 a mayo de 1967.
Investigación, pesq. 31(3): 621-662

chemistry-pH (data only)
Brasil, Diretoria de Hidrografia e Navegação 1970.
IV, V e VI Comissões oceanográficas no Almirante Saldanha (29/5 a 4/6/1957) (5/10 a 17/10/1957) (26/11 a 4/12/1957).
DG 20-IV: 149 pp.

pH (data only)
Brasil, Diretoria de Hidrografia e Navegação, 1969
III Comissão oceanográfica no Almirante Saldanha (20/3 a 16/4/1957), DG 20-III: 73 pp.

chemistry, pH (data only)
Brasil, Diretoria de Hidrografia e Navegação, 1969.
II Comissão Oceanográfica no Almirante Saldanha (15 A 28/2/1957).
DG20-II:80 pp.

chemistry pH (data only)
Brasil, Diretoria de Hidrografia e Navegacao, 1968.
XXXV Comissao Oceanografica "Operacao Norte/Nordeste 1" noc Almirante Saldanha (14/9 a 16/12/1967).
DG 26-XI: 600 pp.

chemistry pH(data only)
Brasil, Diretoria de Hidrografia e Navegacao, 1961
Estudo das condições oceanográficas nas proximidades do Rio de Janeiro durante o mês de Dezembro.
DG-06-XIII:mimeographed sheets.

chemistry pH(data only)
Brasil, Diretoria de Hidrografia e Navegacao, Marinha do Brasil, 1961
Estudo das condições oceanográficas sobre a plataforma continental, entre Cabo-Frio e a Ponta do Boi, durante o mes de setembro (transição inverno-primervera).
DG-06-XII: unnumbered pp. (mimeographed).

Chemistry pH(data only)
Brasil, Marinha do Brasil, Diretoria de Hidrografia e Navegacao, 1960.
Estudo das condicoes oceanograficas na regiao profunda a Nornordeste de Natal, Estado do Rio Grande do Norte.
DG-06-XI (Sept. 1960):unnumbered pp. (mimeographed).

Chemistry pH (data only)
Brasil, Marinha do Brasil, Diretoria de Hidrografia e Navegação, 1960.
Estudo das condições oceanográficas sobre a plataforma continental, entre Cabo-Frio e Vitoria, durante o outono (abril-maio).
DG-06-X(junho): unnumbered pp. (mimeographed).

Chemistry, pH(data only)
Brasil, Marinha do Brasil, Diretoria de Hidrografia e Navegacão, 1959.
Levantamento oceanografico da costa nordeste.
DG-06-IX:unnumbered pp. (mimeographed).

chemistry, pH(data only)
Brazil, Diretoria de Hidrografia e Navegacao, 1958.
Ano Geofisico Internacional, Publ. DG-06-VII: mimeographed sheets.

chemistry, pH(data only)
Brazil, Diretoria de Hidrografia e Navegacao, 1957.
Ano Geofisico Internacional, Publicacao DG-06-VI: unnumbered mimeographed pages.

chemistry pH(data only)
Brasil, Diretoria de Hidrografia e Navegacao, 1957.
Ano Geofisico Internacional. Publicacao DG-06-V: 3 pp., 18 figs., 3 pp. of data.

chemistry, pH(data only)
Brazil, Diretoria de Hidrografia e Navegacao, 1957.
Ano Geofisico Internacional. Publ. DG-06-III and IV (mimeographed)

Chemistry, pH(data only)
Chesapeake Bay Institute, 1954.
Data Report 24. Choptank River winter cruise, 7 December to 10 December 1952. Ref. 54-11:1-37, 1 fig.

chemistry, pH(data only)
Chesapeake Bay Institute, 1954.
Data Report 23, Patuxent River Winter Cruise, 3 December - 7 December 1952. Ref. 54-10:1-44, 1 textfig.

Chemistry pH(data only)
Chesapeake Bay Institute, 1954.
Choptank River autumn cruise, 23 Sept-27 Sept 1952. Data Rept. 22, Ref. 54-9:37 pp., 1 fig.

Chemistry pH (data only)
Chesapeake Bay Institute, 1954.
Choptank River spring cruise, 28 April-1May 1952.
Ref. 54-1:37 pp.

chemistry pH (data only)
Chesapeake Bay Institute, 1951.
Data report, Cruise VIII, Jan. 10, 1951-Jan. 23, 1951. Rept. No. 6:29 pp. (duplicated).

Chemistry pH(data only)
Chesapeake Bay Institute, 1951.
Data report, Cruise VII, 14 Oct. 1950-2 Nov. 1950. Rept. No. 5:41 pp. (duplicated).

Chemistry pH (data only)
Chesapeake Bay Institute, 1951.
Data report. Cruise VII. October 14, 1950-November 2, 1950. Rept. No. 2:41 pp. (duplicated).

chemistry pH (data only)
Chesapeake Bay Institute, 1950.
Data Report, cruises V and VI, May 20, 1950-May 25, 1950, July 14, 1950-July 19, 1950. Rept. No. 4:51 pp. (mimeographed).

chemistry pH (data
Chesapeake Bay Institute, 1950.
Data Report. Cruise IV, March 25, 1950-April 25, 1950. Rept. No. 3:49 pp., (mimeographed).

chemistry, pH (data
Chesapeake Bay Institute 1949.
Data Report, Cruise III, October 10, 1949-October 25, 1949. Rept. 2:39 pp. (mimeographed).

Chemistry pH (data
Chesapeake Bay Institute, 1949.
Quarterly report, July 1-October 1, 1949.
Rept. No. 1: 121 pp. (multilith).

Chemistry, pH (data only)
Conseil International pour l'Exploration de la Mer, 1968.
Cooperative synoptic investigation of the Baltic 1964. Sweden.
ICES Oceanogr.Data Lists 4:181 pp.(mimeographed)

Chemistry, pH (data only)
Conseil International pour l'Exploration de la Mer, 1968.
Cooperative synoptic investigation of the Baltic 1964. 3. Poland.
ICES Oceanogr., Data Lists: 266 pp. (multilithed)

Chemistry, pH (data only)
Conseil International pour l'Exploration de la Mer, 1968.
Cooperative synoptic investigation of the Baltic 1964. 5. U.S.S.R.
ICES Oceanogr., Data Lists: 173 pp. (multilthed)

Chemistry, pH (data only)
Conseil Permanent International pour l'Exploration de la Mer, 1957.
Bulletin Hydrographique pour l'année 1953:167 pp.

Chemistry pH(data only)
Conseil Permanent International pour l'Exploration de la Mer, 1954.
Bulletin Hydrographique pour l'Année 1950: 114 pp., 5 charts.

Chemistry pH (data only)
Conseil Permanent International pour l'Exploration de la Mer, 1939
Bulletin Hydrographique pour l'annee 1937: 106 pp., 4 maps., 4 figs.

chemistry - pH (data only)
Conseil Permanent International pour l'Exploration de la Mer 1936.
Bulletin hydrographique pour l'année 1935: 105pp. 4 maps, 4 figs.

Chemistry pH(data only)
Conseil permanent international pour l'Exploration de la Mer, 1927.
Bulletin Hydrographique Trimestrial, No. 3, 1927 (Juillet-Septembre*:16 pp. (multilith), charts.

Chemistry pH (data only)
Conseil International pour l'Exploration de la Mer. Joint Skagerak Expedition 1966. 1. Oceanographic stations, temperature-salinity-oxygen content. 2. Oceanographic stations, chemical observations. ICES Oceanogr. Data lists: 250 pp; 209 pp.

chemistry pH (data only)
Côte d'Ivoire, Centre de Recherches Océanographiques, 1963.
Résultats hydrologiques effectués au large de la Côte d'Ivoire de 1956-1963. Première partie: observations, 1956-1963.
Trav. Centre Océanogr., Ministère de la Production Animale, Côte d'Ivoire, unnumbered pp. (multilithed).
de Recherches

Chemistry, pH (data only)
DeMaio, A., D. Bregant and E. Sanone, 1968.
Oceanographic data of the R.V. Bannock collected during the International NATO cruise in the Tyrrhenian Sea (16 September-24 October 1963).
Cah. océanogr., 20(Suppl.1):1-64.

Chemistry pH (data only)
Fonselius, S.H., 1967.
Hydrographical observations at Swedish lightships and in the central Baltic, 1965.
Annls. biol. Copenh.(1965)22:54-56.

Chemistry pH (data only)
France, Service Hydrographique de la Marine, 1962
Observations hydrologiques du bâtiment océanographe "Origny", Campagne Internationale à Gibraltar (15 mai - 15 juin 1961).
Cahiers Océanogr., C.C.O.E.C., 14(5):340-375

Chemistry pH(data only)
France, Service Hydrographique de la Marine, 1959.
Observations du "Pyrrhus" bâtiment de recherche de Travaux Publics d'Afrique Occidentale Française.
Cahiers Océanogr., C.C.O.E.C., 11(1):63-73.

Chemistry ph(data only)
France, Service Hydrographique de la Marine, 1959
Resultats d'observations hydrologiques. 1. Batiments de la Marine Nationale. 2. Observations du "Pyrrhus". 3. Observations du niveau marin.
Cahiers Oceanogr., 11(1):54-62.

chemistry - pH (data only)
France, Sous-Comité Océanographique de l'Organisation du Traité de l'Atlantique Nord 1969.
Projet Mer Tyrrhénienne (1963). Résultats des mesures faites à bord du navire océanographique italien Bannock (16 septembre - 24 octobre 1963) et à bord du navire océanographique français Origny (17 septembre - 15 octobre 1963).
Rapp. techn. OTAN, Serv. Hydrogr. de la Marine 44: 125 pp.

chemistry pH(data only)
Granqvist, G., 1955.
The summer cruise with M/S Aranda in the northern Baltic, 1954. Merent. Julk., No. 166:56 pp.

chemistry pH (data only)
Great Britain, Discovery Committee, 1947.
Station list, 1937-1939, Discovery Repts., 24: 198-422, Pls. 4-6.

chemistry pH (data only)
Great Britain, Discovery Committee, 1945.
Station list, 1935-1937.
Discovery Rept., 24:3-196.

Chemistry, pH (data only)
Great Britain, Discovery Committee, 1932.
Station list. Discovery Repts., 4:3-230, Pls. 1-5.

Chemistry pH (data only)
Great Britain, Discovery Committee, 1931.
Station list. Discovery Repts. 3:3-132, Pls. 1-10.

Chemistry pH (data only)
Great Britain, Discovery Committee, 1929.
Station list. Discovery Repts. 1:3-140.

Chemistry, pH (data only)
Guillén, Oscar, y Francisco Vásquez 1966.
Informe preliminar del Crucero 6602 (Cabo Blanco-Arica).
Inst. Mar, Peru, Informe, No. 12: 27pp.

Chemistry pH(data only)
Hapgood, William, 1959.
Hydrographic observations in the Bay of Naples (Golfo di Napoli), January 1957-January 1958 (station lists).
Pubbl. Sta. Zool., Napoli, 31(2):337-371.

chemistry pH(data only)
Hela, Ilmo, and F. Koroleff, 1959.
Hydrographical and chemical data collected in 1956 on board the R/V "Aranda" in the Baltic Sea.
Merent. Julk., No. 183:1-52.

Chemistry pH(data only)
Hela, Ilmo, and Folke Koroleff, 1958
Hydrographical and chemical data collected in 1957 on board the R/V Aranda in the Barents Sea.
Merent. Julk., No. 179: 67 pp.

Chemistry, pH (data only)
Hoshino, Zen-ichiro, 1970.
Oceanographical conditions observed at definite station off Asamushi during 1969.
Bull. mar. biol. Sta. Asamushi, 14(1): 63 (fold out)

Chemistry, pH (data only)
Hufford, Gary L. and James M. Seabrooke, 1970
Oceanography of the Weddell Sea in 1969 (IWSOE).
Oceanogr. Rept. U.S. Coast Guard 31 (CG 373-31) 32 pp.

Chemistry pH(data only)
India, Naval Hdqtrs., Office of Scientific Research and Development, 1957.
Indian Oceanographic Stations List No. 1/57: 5 pp. (mimeographed)

Chemistry pH (data)
*Ishino Makoto, Keiji Nasu, Yoshimi Morita and Makoto Hamada, 1968.
Oceanographic conditions in the west Pacific Southern Ocean in summer of 1964-1965. J. Tokyo Univ. Fish., 9(1): 115-208.

Chemistry, pH (data only)
Japan, Central Meteorological Observatory, 1952
The results of Marine Meteorological and oceanographical observations, July - Dec. 1951, No. 10:310 pp., 1 fig.

Chemistry, pH(data only)
Japan, Central Meteorological Observatory, 1952.
The results of Marine meteorological and oceanographical observations. July - Dec. 1950. No. 8: 299 pp.

Chemistry, pH(data only)
Japan, Central Meteorological Observatory, 1952.
The Results of Marine Meteorological and oceanographical observations. Jan. - June 1950. No. 7: 220 pp.

Chemistry, pH(data only)
Japan, Central Meteorological Observatory, 1951.
The results of marine meteorological and oceanographic observations. July - Dec. 1949. No. 6: 423 pp.

Chemistry, pH (data only)
Japan, Central Meteorological Observatory, 1951.
Table 14. Oceanographical observations taken in Miyazu Bay: physical and chemical data for stations occupied by R.M.S. "Asanagi-maru". Res. Mar. Met. Ocean. Obs., Jan.-June, 1949, No. 5: 77-84.

chemistry, pH (data only)
Japan, Central Meteorological Observatory, Japan, 1951.
The results of marine meteorological and oceanographical observations, Jan.-June 1948, No. 3: 256 pp.

Chemistry, pH (data only)
Japan, Central Meteorological Observatory, Japan, 1950.
The results of marine meteorological and oceanographical observations, Jan.-June 1947, No. 1: 113 pp., 19 tables.

Chemistry, pH (data only)
Japan, Fisheries Agency, Research Division, 1956.
Radiological survey of western area of the dangerous zone around Bikini-Eniwetok Atolls, investigated by the "Shunkotsu maru" in 1956, Part 1: 143 pp.

Chemistry, pH (data only)
Japan, Hokkaido University, Faculty of Fisheries 1970.
Data record of oceanographic observations and exploratory fishing 13: 406 pp.

Chemistry, pH (data only)
Japan, Hokkaido University, Faculty of Fisheries, 1968.
The Oshoro Maru cruise 23 to the east of Cape Erimo, Hokkaido, April 1967.
Data Record Oceanogr. Obs. Expl. Fish., 12: 115-169.

Chemistry, pH (data only)
Japan, Hokkaido University, Faculty of Fisheries, 1968.
The Oshoro Maru cruise 21 to the Southern Sea of Japan, January 1967.
Data Record Oceanogr. Obs. Explor. Fish. 12: 1-97; 113-119.

Chemistry
pH (data only)

Japan, Hokkaido University, Faculty of Fisheries, 1967.
The Oshoro Maru cruise 16 to the Great Australian Bight November 1965-February 1966.
Data Record Oceanogr. Obs. Explor. Fish., Fac. Fish., Hokkaido Univ. 11: 1-97; 113-119.

Chemistry, pH (data only)
Japan, Hokkaido University, Faculty of Fisheries, 1967.
Data record of oceanographic observations and exploratory fishing, 11:383 pp.

Chemistry, pH (data only)
Japan Faculty of Fisheries, Hokkaido University 1961.
Data record of oceanographic observations and exploratory fishing, No. 5: 391 pp.

Chemistry, pH (data only)

Japan, Hokkaido University, Faculty of Fisheries, 1959.
Data record of oceanographic observations and exploratory fishing, No. 3:296 pp.

Chemistry pH (data only)
Japan, Hokkaido University, Faculty of Fisheries, 1957.
Data record of oceanographic observations and exploratory fishing, No. 1:247 pp.

Chemistry, pH (data only)

Japan, Japan Meteorological Agency, 1970

The results of marine meteorological and oceanographical observations. (The results of the Japaneses Expedition of Deep Sea (JEDS-11); January-June 1967 41: 332 pp.

Chemistry, pH (data only)
Japan, Japan Meteorological Agency 1969.
The results of marine meteorological and oceanographical observations, Jan.-June 1966, 39: 349 pp. (multilithed).

Chemistry pH (data only)
Japan Meteorological Agency, 1968.
The results of marine meteorological and oceanographical observations, July-December, 1965, 38: 404 pp. (multilithed)

Chemistry, pH (data only)
Japan, Japanese Meteorological Agency, 1967.
The results of marine meteorological and oceanographical observations, July-December, 1964, 36: 367 pp.

Chemistry, pH (data only)
Japan, Japan Meteorological Agency, 1964.
The results of marine meteorological and oceanographical observations, January-June 1963, No. 33:289 pp.

Chemistry, pH(data only)
Japan, Japan Meteorological Agency, 1964.
The results of the Japanese Expedition of Deep Sea (JEDS-5).
Res. Mar. Meteorol. and Oceanogr. Obs., July-Dec. 1962, No. 32:328 pp.

Chemistry, pH (data only)
Japan, Japan Meteorological Agency, 1964.
Oceanographic observations.
Res. Mar. Met. & Oceanogr. Obs., No. 31:220 pp.

Chemistry
pH (data only)
Japan, Japan Meteorological Agency, 1962
The results of marine meteorological and oceanographical observations, July-December 1961, No. 30:326 pp.

Chemistry
pH (data only)
Japan, Meteorological Agency, 1962
The results of marine meteorological and oceanographical observations, July-December, 1960, No. 28: 304 pp.

Chemistry - pH (data only)

Japan, Japan Meteorological Agency, 1961
The results of marine meteorological and oceanographical observations, January-June 1960. The results of the Japanese Expedition of Deep-Sea (JEDS-2, JEDS-3), No. 27: 257 pp.

Chemistry - pH (data only)

Japan Meteorological Agency, 1960.
The results of marine meteorological and oceanographical observations. Supplement, 149 pp.

Chemistry, pH(data only)
Japan, Japan Meteorological Agency, 1959.
The results of marine meteorological and oceanographical observations, July-December, 1958, No. 24:289 pp.

Chemistry pH (data only
Japan, Maritime Safety Board, 1954.
[Tables of results from oceanographic observations in 1950.] Publ. No. 981, Hydrogr. Bull. Maritime Safety Bd., Spec. Number (Ocean. Repts.), No. 14:26-164, 5 textfigs.

Chemistry pH (data only)

Japan, Japanese Oceanographic Data Center, 1971.
Takuyo, Hydrographic Department, Maritime Safety Agency, Japan, 10-30 July 1970, south & east of Japan. Data Rept. CSK (KDC Ref. 49K126) 278: 19 pp. (multilithed).

Chemistry pH (data only)

Japan, Japanese Oceanographic Data Center, 1970
Ji Ri San, Fisheries Research and Development Agency, Republic of Korea, Feb. 18 - 26, 1970, West of the Japan Sea. Data Rept. CSK (KDC Ref. 24K038) 272: 16 pp. (multilithed).

Chemistry
pH (data only)

Japan, Japanese Oceanographic Data Center, 1970.
Han Ra San, Fisheries Research and Development Agency, Republic of Korea, Feb. 11 - 24, 1970, East of the Yellow Sea. Data Rept. CSK (KDC Ref. 24K037) 271: 14 pp. (multilithed).

Chemistry
pH (data only)

Japan, Japanese Oceanographic Data Center, 1970.
Baek Du San, Fisheries Research and Development Agency, Republic of Korea, Feb. 4 - 20, 1970, South of Korea & East China Sea. Data Rept. CSK (KDC Ref. 24K036) 270: 15 pp. (multilithed).

Chemistry
pH (data only)

Japan, Japanese Oceanographic Data Center, 1970.
Takuyo, Hydrographic Department, Maritime Safety Agency, Japan, Feb. 12 - Mar. 7, 1970. South and East of Japan. Data Rept. CSK (KDC Ref. 49K120) 264: 13 pp. (multilithed).

Chemistry
pH (data only)

Japan, Japanese Oceanographic Data Center, 1970.
Kaiyo, Hydrographic Department, Maritime Safety Agency Japan, Nov. 7 - 29, 1969, South of Japan. Data Rept. CSK (KDC Ref. 49K116) 257: 10 pp. (multilithed).

Chemistry
pH (data only)

Japan, Japanese Oceanographic Data Center, 1970.
Suro No. 3, Hydrographic Office, Republic of Korea, July 23 - August 19, 1969, South of Korea Data Rept. CSK (KDC Ref. 24K035) 240: 19 pp. (multilithed).

Chemistry
Ph (Data Only)

Japan, Japanese Oceanographic Data Center, 1970.
Ji Ri San, Fisheries Research and Development Agency, Republic of Korea, August 7-29, 1969, South of Korea and East China Sea, Data Rept. CSK (KDC REF. 24K034) 239: 15pp. (multilithed).

Chemistry
pH (DATA ONLY)

Japan, Japanese Oceanographic Data Center, 1970.
Takuyo, Hydrographic Department, Maritime Safety Agency, Japan, August 11 - September 2, 1969, South and East of Japan. Data Rept. CSK (KDC Ref. 49K112) 229: 21 pp. (multilithed).

Chemistry
pH (data only)

Japan, Japanese Oceanographic Data Center, 1970
Orlick, USSR, Jan. 27 - Mar. 6, 1969, Northwest of the North Pacific Ocean. Data Rept. CSK (KDC Ref. 90K023) 227: 30 pp. (multilithed).

Chemistry
pH (data only)

Japan, Japanese Oceanographic Data Center, 1970
G. Nevelskoy, USSR, Oct. 28, 1968 - Jan. 14 1969, Northwest of the North Pacific Ocean. Data Rept. CSK (KDC Ref. 90K024) 209:55 pp. (multilithed).

Chemistry
pH (data only)

Japan, Japanese Oceanographic Data Center, 1970.
SURO No. 3, Hydrographic Office, Republic of Korea, October 12 - November 8, 1968, South of Korea. Data Report CSK (KDC Ref. 24K027) 201: 15 pp. (multilithed).

Chemistry - pH (data only)
Japan, Japanese Oceanographic Data Center 1970.
Suro No. 3 Hydrographic Office, Republic of Korea, July 22 - August 7, 1968, South of Korea.
Data Rept. CSK (KDC Ref. 24K023) 197: 17 pp. (multilithed).

Chemistry
pH (data only)

Japan, Japanese Oceanographic Data Center, 1969

Cape St. Mary, Fisheries Research Station, UK (Hong Kong). May 16 - 23, 1968, South China Sea. Prelim. Data Rept. CSK (KDC Ref. No. 74K010) 203: 15 pp. (multilithed).

Chemistry - pH. (data only)
Japan, Japanese Oceanographic Data Center 1969
Orlick USSR July 21 - August 20, 1968, northwest North Pacific Ocean.
Data Rept. CSK (KDC Ref. 90K020) 191: 32 pp. (multilithed).

Chemistry - pH (data only)
Japan, Japanese Oceanographic Data Center 1969.
Iskatel USSR January 26 - March 11, 1968, northwest North Pacific Ocean.
Data Rept CSK (KDC Ref 90K019) 187: 32 pp. (multilithed).

Chemistry pH (data only)
Japan, Japanese Oceanographic Data Center, 1969
Takuyo, Hydrographic Division, Maritime Safety Agency, Japan. July 19-September 6, 1968, Kuroshio Extension Area. Prelim. Data Rept. CSK (KDC Ref. No. 49K082) 176: 27 pp. (multilithed).

Chemistry - pH. (data only)
Japan, Japanese Oceanographic Data Center 1969.
Hakuho Maru, Ocean Research Institute University of Tokyo, Japan, May 15 - June 8, 1968, southwest of Japan and the East China Sea.
Data Rept. CSK (KDC Ref. 49KD78) 172: 7 pp. (multilithed).

Chemistry - pH (data only)
Japan, Japanese Oceanographic Data Center, 1969. Takuyo, Hydrographic Division, Maritime Safety Agency, May 14-25, 1968, South of Japan, Data Rept. CSK (KDC Ref. 49K077) 171:16 pp.

Chemistry pH (data only
Japan, Japanese Oceanographic Data Center, 1969.
Cape St. Mary, Fisheries Research Station, UK (Hong Kong). January 6 - 14, 1968, South China Sea. Prelim. Data Rept. CSK (KDC Ref. No. 74K009) 165: 15 pp. (multilithed).

Chemistry pH (data only)
Japan, Japanese Oceanographic Data Center, 1969
U.M. Schokalsky, USSR, April 28 - June 3, 1967, Northwest of the North Pacific Ocean. Prelim. Data Rept. CSK (KDC Ref. 90K014) 121: 41 pp. (multilithed).

Chemistry, pH (data only)
Japan, Japanese Oceanographic Data Center, 1968. Tae Baek San, Fisheries Research and Development Agency, Republic of Korea, February 16-March 12, 1968, South of Korea. Data Rept. CSK (KDC Ref. 24K022) 163: 12 pp. (multilithed)

Chemistry, pH (data only)
Japan, Japanese Oceanographic Data Center, 1968. Han ra San, Fisheries Research and Development Agency Republic of Korea, February 15-March 12, 1968, East of the Yellow Sea. Data Rept. CSK (KDC Ref. 24K021) 162: 18 pp.

Chemistry, pH (data only)
Japan, Japanese Oceanographic Data Center., 1968. Baek du San, Fisheries Research and Development Agency, Republic of Korea, February 17-March 12, 1968, West of the Japan Sea. Data Rept. CSK (KDC Ref. 24K020) 161:23 pp. (multilithed)

Chemistry, pH (data only)
Japan, Japanese Oceanographic Data Center, 1968. Takuyo, Hydrographic Division, Maritime Safety Agency, Japan, February 19-March 10, 1968, south and east of Japan. Data Rept. CSK (KDC Ref. 49K068) 154: 14 pp.

Chemistry, pH (data only)
Japan, Japanese Oceanographic Data Center, 1968. Suro No. 1, Hydrographic Office, Korea, September 23-October 21, 1967, South of Korea. Data Rept. CSK (KDC Ref. 24K019) 136: 13 pp. (multilithed).

Chemistry, pH (data only)
Japan, Japanese Oceanographic Data Center, 1968. Kaiyo, Hydrographic Division, Maritime Safety Agency, Japan November 13-December 1, 1967, South of Japan. Data Rept. CSK (KCD Ref. 49K063) 128: 13 pp. (multilithed).

Chemistry, pH (data only)
Japan, Japanese Oceanographic Data Center, 1968
Orlick, USSR, July 29 - September 3, 1967, Northwest of the North Pacific Ocean. Prelim. Data Rept. CSK (KDC Ref. No. 90K015) 122: 31 pp. (multilithed).

Chemistry, pH (data only)
Japan, Japanese Oceanographic Data Center, 1968. Han Ra San, Fisheries Research and Development Agency, Korea, August 12 - September 4, 1967, North of East China Sea. Prelim. Data Rept. CSK (KDC Ref. 24K018) 117:12 pp. (multilithed).

Chemistry, pH (data only)
Japan, Japanese Oceanographic Data Center, 1968. Chun Ma San, Fisheries Research and Development Agency, Korea, August 13-21, 1967, East of Yellow Sea. Prelim. Data Rept. CSK (KDC Ref. 24K017) 116:16 pp. (multilithed).

Chemistry - pH (data only)
Japan, Japanese Oceanographic Data Center, 1968. Baek Du San, Fisheries Research and Development Agency, Korea, August 12-25, 1967, West of Japan Sea. Data Rept. CSK (KDC Ref. 24K016) 115 L 26 pp. (multilithed).

Chemistry - pH (data only)
Japan, Japanese Oceanographic Data Center, 1968. Hakuho Maru, Ocean Research Institute, University of Tokyo, Japan, September 6-20, 1967, South of Japan. Data Rept. CSK (KDC Ref. 49K059) 111: 9 pp. (multilithed).

Chemistry - pH (data only)
Japan, Japanese Oceanographic Data Center, 1968. Takuyo, Hydrographic Division, Maritime Safety Agency, July 12-August 30, 1967, Central Part of the North Pacific Ocean. Prelim. Data Rept. CSK (KDC Ref. 49K053) 105:25 pp. (multilithed).

Chemistry, pH (data only)
Japan, Japanese Oceanographic Data Center, 1968. Yang Ming, Chinese National Committee on Ocean Research, Republic of China, April 1; May 14, 1967, Surrounding waters of Taiwan. Data Rept. CSK (KDC Ref. 21K004) 102: 15 pp. (multilithed).

Chemistry pH (data only)
Japan, Japanese Oceanographic Data Center, 1968. Kaiyo, Hydrographic Division, Maritime Safety Agency, May 10-29, 1967, South of Japan. Prelim. Data Rept. CSK (KDC 49K047), 96:15 pp. (multilithed).

Chemistry, pH (data only)
Japan, Japanese Oceanographic Data Center, 1968. Orlick, USSR February 5-March 11, 1967, Northwest of North Pacific Ocean. Prelim. Data Rept. CSK (KDC Ref. 90K012) 94:29 pp.

Chemistry - pH (data only)
Japan, Japanese Oceanographic Data Center, 1968. Ryofu Maru, Marine Division, Japan Meteorological Agency, January 11-February 24, 1967, West of the North Pacific Ocean. Prelim. Data Rept. CSK (KDC 49K040) 82: 52 pp.

Chemistry, pH (data only)
Japan, Japanese Oceanographic Data Center, 1968. U.M. Schokalsky, USSR, July 20-August 23, 1966, North-west of North Pacific Ocean. Prelim. Data Rept. CSK (KDC Ref. 90K010) 75:41 pp. (multilithed).

Chemistry, pH (data only)
Japan, Japanese Oceanographic Data Center, 1968. G. Nevelskoy, USSR, July 13-September 17, 1966, Northwest of North Pacific Ocean. Prelim. Data Rept. CSK (KDC Ref. 90K008) 73: 32 pp. (multilithed).

Chemistry - pH (data only)
Japan, Japanese Oceanographic Data Center, 1967. Shumpu maru, Kobe Marine Observatory, Japan Meteorological Agency, Japan, May 13-14, 1967, South of Japan. Prelim Data Rept. CSK (KDC Ref. 49K050) 99: 5 pp.

Chemistry - pH (data only)
Japan, Japanese Oceanographic Data Center, 1967. Oshoro maru, the Faculty of Fisheries, Hokkaido University, Japan, January 15-February 1, 1967, South of Japan. Prelim. Data Rept. CSK (KDC Ref. 49K045) 87: 9 pp.

Chemistry pH (data only)
Japan, Japanese Oceanographic Data Center, 1967. Takuyo, Hydrographic Division, Maritime Safety Agency, Japan, February 23-March 16, 1967, south and east of Japan. Prelim. Data Rept., CSK (KDC Ref. 49K039) 81: 17 pp.

Chemistry - pH (data only)
Japan, Japanese Oceanographic Data Center, 1967. Orlick, USSR, July 9-September 12, 1966, North-West of North Pacific Ocean. Prelim. Data Rept. CSK (KDC Ref. 90K009) 74:34 pp.

pH (data only)
Japan, Japanese Oceanographic Data Center, 1967. Bukhansan, Fisheries Research and Development Agency, Korea, July 16- August 9, 1966, East of the Yellow Sea. Prelim. Data Rept., CSK (KDC Ref. 24K010) 68:17 pp.

Chemistry, pH (data only)
Japan, Japanese, Oceanographic Data Center, 1967 Baekdusan, Fisheries Research and Development Agency, Korea, July 14-28, 1966, West of the Japan Sea. Prelim. Data Rept. CSK (KDC Ref. 24K009) 65:22 pp.

Chemistry - pH (data only)
Japan, Japanese Oceanographic Data Center, 1967. Yang Ming, Chinese National Committee on Oceanic Research, Republic of China, September 10-October 14, 1966, Adjacent Sea of Taiwan. Prelim. Data Rept. CSK KDC Ref. 21K003) 67: 17pp.

Chemistry - pH (data only)
Japan, Japanese Oceanographic Data Center, 1966. Koyo Maru, Shimonoseki University of Fisheries, October 26-29, south of Japan. Prelim. Data Rept. CSK (KDC Ref. No. 49K038) 66: 7 pp. (multilithed).

Chemistry - pH (data only)
Japan, Japanese Oceanographic Data Center, 1967. Chofu Maru (NMO) Jul. 2-3, 1966; Chofu Maru (NMO), Aug. 27-29, 1966; Shumpu Maru (KMO), Oct. 27-31, 1966, Chofu Maru, Dec. 7-8, 1966, Japan Meteorological Agency of Japan, south-east of Yakushima. Prelim. Data Rept., CSK (KDC Ref. 49K311, 49K312, 49K313, 49K314) 61: 18 pp.

Chemistry - pH (data only)
Japan, Japanese Oceanographic Data Center 1967.
Kaiyo, Hydrographic Division, M.S.A. Japan, August 10-30, 1966, S.E. of Japan.
Prelim. Data Rept. CSK (KDC Ref. No. 49K025) 51:15 pp. (multilithed).

Chemistry, pH (data only)
Japan, Japanese Oceanographic Data Center, 1966. Kerin Ho, Fisheries Research and Development Agency, Korea, September 12-October 13, 1965, East of Yellow Sea. Prelim. Data Rept., CSK, (KDC Ref. 24K005), 38: 11 pp. (multilithed).

Chemistry, pH (data only)
Japan, Japanese Oceanographic Data Center, 1966. Bukhansan Ho, Fisheries Research and Development Agency, Korea, December 2-12, 1965. Prelim. Data Rept., CSK, (KDC 24K004), 37: 15 pp. (multilithed).

Chemistry pH (data only)
Japan, Japanese Oceanographic Data Center, 1966. Oshoro Maru, The Faculty of Fisheries, Hokkaido University, Japan, November 30, 1965-January 25, 1966, South of Japan. Prelim. Data Rept., CSK (KDC Ref. 49K022), 33: 11 pp. (multilithed).

Chemistry, pH (data only)
Japan, Japanese Oceanographic Data Center, 1966. Seifu Maru, Maizuru Marine Observatory, Japan Meteorological Agency, Japan, February 12-February 28, 1966, Japan Sea. Prelim. Data Rept., CSK (KDC Ref. 48Ko21), 32: 14pp. (multilithed).

Oceanographic pH (data only)
Japan, Japanese Data Center, 1966. Preliminary data report of CSK, Shumpu Maru, Kobe Marine Observatory, Japan Meteorological Agency, February 19-24, 1966, Southern Sea of Japan, No. 30 (October 1966). KDC Ref. No. 49K019: 6 pp. (multilithed).

Chemistry, pH (data only)
Japan, Japanese Oceanographic Data Center, 1966. Uliana Gromova, USSR, July 20-September 3, 1965. E. & S. of Japan. Prelim. Data Rept., CSK, (KDC 90K003), 25: 37pp. (multilithed).

Chemistry, pH (data only)
Japan, Japanese Oceanographic Data Center, 1966. Zhyemchug, USSR, July 28-September 19, 1965, E. & S. of Japan. Prelim. Data Rept., CSK, (KDC Ref. 90K002), 24: 32 pp. (multilithed).

Chemistry, pH, (data only)
Japan, Japanese Oceanographic Data Center, 1966. U.M. Shokalski, USSR, July 16-August 18, 1965, E. & S. of Japan. Prelim. Data Rept., CSK, (KDC Ref. 90K001), 23: 41 pp (multilithed).

Chemistry, pH (data only)
Japan, Japanese Oceanographic Data Center, 1966. Yang Ming, Chinese National Committee on Oceanic Research, Republic of China, August 10-October 13, 1965, Adjacent Sea of Taiwan. Prelim. Data Rept., CSK (KDC Ref. 21K001), 22: 18 pp. (multilithed).

Chemistry, pH (data only)
Japan Maritime Safety Agency 1967. Results of oceanographic observations in 1965. Data Rept. Hydrogr. Obs. Ser. Oceanogr. (Publ. 792) Oct. 1967, 5: 115 pp.

Chemistry - pH. (data only)
Japan, Maritime Safety Agency 1967. Results of oceanographic observations in 1964. Data Rept. Hydrogr. Obs. Ser. Oceanogr. Publ. 792 (4): 88 pp.

Chemistry, pH (data only)
Japan, Maritime Safety Agency 1966. Results of oceanographic observations in 1963. Data Rept. Hydrogr. Obs. Ser. Oceanogr. Publ. 792 (3): 74 pp.

Chemistry pH (data only)
Japan, Maritime Safety Board, 1956. Tables of results from oceanographic observations in 1952 and 1953. Hydrogr. Bull. (Publ. 981)(51): 1-171.

Chemistry pH (data only)
Japan, Maritime Safety Board, Tokyo, 1954. Tables of results from oceanographic observations in 1951. Publ. No. 981, Hydrogr. Bull., Spec. No. 15: 31-129.

Chemistry, pH (data only)
Japan, Nagasaki Marine Observatory Tables 9, 10, 11, 22, 23, 24, 34, 35 and 44
Oceanogr. Met. Nagasaki mar. Obs. 15: 92-120; 176-180; 191-192; 213-

Chemistry - pH (data only)
Japan, Shimonoseki University of Fisheries, 1970. Data of oceanographic observations and exploratory fishings. 5. Oceanographic surveys of the Kuroshio and its adjacent waters, 1967 and 1968: 117 pp.

Chemistry pH (data only)
Japan, Shimonoseki University of Fisheries 1968 Oceanographic surveys of the Kuroshio and its adjacent waters 1965 and 1966. Data Oceanogr. Obs. Explor. Fish. 4: 1-178.

Chemistry pH (data only)
Japan, Shimonoseki University of Fisheries, 1965. Data of oceanographic observations and exploratory fishings, International Indian Ocean Expedition 1962-63 and 1963-64. No. 1: 453 pp.

Chemistry pH (data only)
Japan, Tokyo University of Fisheries, 1966. J. Tokyo Univ. Fish. (Spec. ed.) 8(3): 1-44.

Chemistry, PH (data only)
Katsuura, Hiroshi, Hideo Akamatsu, Tsutomu Akiyama, 1964. Preliminary report of the Japanese Expedition of Deep-Sea, the Eighth Cruise (JEDS-8). Oceanogr. Mag., Tokyo, 16(1/2): 125-136.

Chemistry pH (data only)
Klepikova, V.V., Edit., 1961 Third Marine Expedition on the D/E "Ob", 1957-1958. Data. Trudy Sovetskoi Antarktich. Exped., Arktich. i Antarktich. Nauchno-Issled. Inst., Mezhd. Geofiz. God, 22: 1-234 pp.

Chemistry pH (data only)
Korea, Republic of, Fisheries Research and Development Agency, 1971 Annual report of oceanographic observations, 19: 717 pp.

Chemistry, pH (data only)
Koroleff, Folke, 1959. The Baltic cruise with R/V Aranda 1958, hydrographic data. Merent. Julk. (Havforskningsinstitutets Skr.), No. 193: 25 pp.

Chemistry pH (data only)
Koroleff, Folke, and Aarno Voipio, 1963 The Finnish Baltic cruise 1960. Hydrographical data. Havsforskningsinst. Skrift (Merent. Julk.), No. 204: 27 pp.

Chemistry, pH (data Only)
Koroleff, Folke, and Aarno Voipio, 1961. The Baltic cruise with R/V Aranda, 1959. Hydrographical data. Merent. Julk. (Havforskningsinst. Skrift), No. 197: 26 pp.

Chemistry, pH (data)
Marumo, Ryuzo, Editor, 1970. Preliminary report of the Hakuho Maru Cruise KH-69-4, August 12-November 13, 1969, The North and Equatorial Pacific Ocean. Ocean Res. Inst. Univ. Tokyo, 68 pp.

Chemistry - pH (data only)
Marvin, K.T., 1955. Oceanographic observations in west coast Florida waters, 1949-52. U.S. Fish. Wildlife Service, Spec. Sci. Rept., No. 149: 6 pp., 2 figs., numerous pp tables.

Chemistry - pH (data only)
Minas, H.J., 1963. Contribution préliminaire à l'étude hydrologique et hydro-chimique du Golfe de Marseille. Rec. Trav. Sta. Mar. Endoume, 28(43): 3-16.

Chemistry - pH (data only)
*Nasu, Keiji, and Tsugio Shimano, 1966. The physical results of oceanographic survey in the south east Indian Ocean in 1963/64. J. Tokyo Univ. Fish., (Spec. ed.) 8(2): 133-164.

Chemistry, pH. (data only)
Nishihira, Moritaka 1968. Oceanographical conditions observed at definite station off Asamushi during 1967. Bull. mar. biol. Sta. Asamushi 13 (2): 151 pp.

Chemistry - pH (data only)
NORPAC Committee, 1955. Oceanic observation of the Pacific, 1955. The NORPAC data, 532 pp.

Chemistry - pH (data only)
Oren, O.H., 1967. The fifth cruise in the eastern Mediterranean, Cyprus-05, May 1967. Hydrographic data. Bull. Sea Fish. Res. Stn Israel, 47: 55-63.

Chemistry - pH (data only)
Postma, H., 1959. Chemical results and a survey of water masses and currents. Tables: oxygen, hydrogen ion, alkalinity and phosphate. Snellius Exped., Eastern Part of the East Indian Archipelago, 1929-1930, Chem. Res., 4: 35 pp., 1 fig.

Chemistry, pH (data only)
Republic of China, Chinese National Committee on Oceanic Research, Academia Sinica, 1968. Oceanographic Data Report of CSK, 2: 126 pp.

Chemistry - pH (data only)
Republic of China, Chinese National Committee on Oceanic Research, Academia Sinica, 1966. Oceanographic data report of CSK 1: 123 pp.

Chemistry - pH (data only)
Republic of Korea, Fisheries Research and Development Agency 1966. Annual report of observations, 1965, 14: 343 pp.

Chemistry, pH (data only)
Republic of Korea, Hydrographic Office, 1967. Hydrographic and oceanographic data. Techn. Repts. 1967, H.O. Pub., Korea, 1-72.

chemistry - pH (data only)
Republic of Korea, Hydrographic Office, 1966.
The data for cooperative study of Kuroshio.
Tech. Rep. H.O. Pub. No. 1101:41-73.

chemistry pH (data only)
Korea (Republic of), Hydrographic Office, 1965.
Technical reports.
H.O.Pub., Korea, No.1101:179 pp.

chemistry - pH (data only)
Rotschi, Henri, Michel Angot, Michel Legand and Roget Desrosieres, 1961
Orsom III, Resultats de la Croisiere "Dillon", 2eme Partie. Chimie et Biologie.
ORSTOM, Inst. Francais d'Oceanie, Centre d'Oceanogr., Rapp. Sci., No. 19: 105 pp. (mimeographed).

chemistry - pH (data only)
Rotschi, H., Ph. Hisard, L. Lemasson, Y. Magnier, J. Noel et B. Piton, 1967.
Resultats des observations physico-chimiques de la croisiere "Alize" du N.O. Coriolis.
Rapp. scient., Off. Rech. scient, tech, Outre-Mer, Noumea, 2:56 pp. (mimeographed).

Chemistry, pH (data only)
Rotschi, H., Ph. Hisard, L. Lemasson, Y. Magnier, J. Noel, et B. Piton, 1966.
Resultats des observations physico-chimiques de la croisiere "Alize".
Centre ORSTROM Noumea Rapp. Sci., No. 28:56 pp. (mimeographed).

Chemistry pH (data only)
Rotschi, Henri, Michel Legand and Roger Desrosieres, 1961
Orsom III, Croisières diverses de 1960, physique chimie et biologie. ORSTOM, Inst. Francias d'Océanie, Centre d'Océanogr., Noumea, Rapp. Sci., No. 20: 59 pp. (mimeographed).

Chemistry pH (data only)
Rotschi, Henri, Yves Magnier, Maryse Tirelli, et Jean Garbé, 1964.
Résultats des observations scientifiques de "La Dunkerquoise" (Croisière "Guadalcanal").
Océanographie, Cahiers O.R.S.T.R.O.M., 11(1):49-154.

chemistry pH (data only)
Saloman, Carl H., and John L. Taylor 1968.
Hydrographic observations in Tampa Bay, Florida, and the adjacent Gulf of Mexico, 1965-1966.
Data Rept. U.S. Fish Wildl. Bur. Comm. Fish. 6 cards (microfiche).

chemistry - pH (data only)
Schmidt, J., 1912.
Hydrographical observations.
Rept. Danish Oceanogr. Exped., 1908-1910, 1:51-75.

chemistry - pH (data only)
Svansson, Artur 1970.
Hydrographical observations on Swedish lightships and fjord stations in 1968
Rept. Fish. Bd, Sweden, Ser. Hydrogr. 24: 83 pp.

Chemistry, pH (data only)
Svansson, Artur, 1966.
Hydrographical observations on Swedish Lightships and fiord stations.
Fish. Bd., Sweden, Ser. Hydrogr., Rept. No. 18: 95 pp.

Chemistry pH (data only)
Sweden, Havsfiskelaboratoriet, Lysekil 1970.
Hydrographical data 1966, R.V. Skagerak, R.V. Thetis.
Meddn Hydrogr. avd. Göteborg 85:255 pp

chemistry pH (data only)
Sweden, Havsfiskelaboratoriet, Lysekil 1970.
Hydrographical data, R.V. Skagerak, R.V. Thetis.
Hydrogr. avd. Göteborg
Meddn. 84: 296 pp. (multilithed).

chemistry pH (data only)
Sweden, Havsfiske laboratoriet, lysekil, 1969.
Hydrographica data, January-June 1968.
R.V. Skagerak, R.V. Thetis.
Meddn Havsfiskelab., Lysekil, hydrog. Avdeln., Goteborg 63: numerous unnumbered pp. (multilithed).

chemistry pH (data only)
Sweden, Havsfiskelaboratoriet, Lysekil, 1968.
Hydrographical data, July-December 1967, R.V. Skagerak.
Meddn Havsfiskebal., Lysekil, 52: unnumbered pp. (mimeographed).

Chemistry - pH (data only)
Sweden, Havsfiskelaboratoriet, Lysekil, Hydrografiska Avd., Göteborg, 1967.
Hydrographical data, January-June 1967, R.V. Skagerak.
Meddn, Havsfiskelab., Lysekil, Hydrogr. Avd., Göteborg, 38: numerous pp. (mimeographed)

chemistry - pH (data only)
Sweden, Havsfiskelaboratoriet, Lysekil, Hydrografiska Avd., Göteborg, 1967.
Hydrographical data, January-June 1967, R.V. Thetis.
Meddn. Havsfiskelab., Lysekil, Hydrogr. Avd., Göteborg, 41: numerous pp. (mimeographed).

chemistry - pH (data only)
Trotti L. 1967.
Crociera Golfo Palmas e Canale di Sardegna, 1965: dati oceanografici.
Pubbl. Ist. Speriment. Talassogr. Trieste, 439: unnumbered pp. (quarto).

chemistry - pH (data only)
Trotti, L. 1967.
Crociera Bocche di Bonifacio 1964.
Pubbl. Ist. Speriment. Talassogr. Trieste, 438: unnumbered pp. (quarto).

chemistry - pH (data only)
Tsubata, B., 1958.
Oceanographical observations observed at definite station off Asamushi during 1956.
Bull. Mar. Biol. Sta., Asamushi, 9(1):43.

chemistry - pH (data only)
United States, Johns Hopkins University, Chesapeake Bay Institute, 1962.
Cruise 28, July 24-August 7, 1962. Data Rept., 48, Ref. 62-16:17 pp. (mimeographed).

Chemistry - pH (data only)
U.S.A., Johns Hopkins University, Chesapeake Bay Institute, 1962.
Temperature and salinity data collected in the Chesapeake Bay and tributary estuaries during the period 1 February 1956 and 9 February 1956.
Data Rept. 49 (Ref. 62-23): 22 pp. (unpublished mss.).

chemistry - pH (data only)
United States, National Oceanographic Data Center, 1965.
Data report EQUALANT III.
Nat. Oceanogr. Data Cent., Gen. Ser., G-7:339pp $5.00
Publ.

Chemistry - pH (data only)
United States, National Oceanographic Data Center, 1964.
Data report EQUALANT II.
NODC Gen. ser., Publ. G-5: numerous pp. (not serially numbered; loose leaf - $5.00)

chemistry - pH (data only)
United States, U.S. Coast Guard, 1965.
Oceanographic cruise USCGC Northwind: Chukchi, East Siberian and Laptev seas, August-September, 1963.
U.S.C.G. Oceanogr. Rept., No. 6(CG-373-6):69 pp.

chemistry - pH (data only)
Valdez, Alberto J., Alberto Gomez, Aldo Orlando y Andres J. Lusquiños 1968.
Datos y resultados de las campañas pesquerias "Pesqueria IV" (7 de junio al 4 de julio de 1967).
Publ. (Ser. Informes técn.) Mar del Plata, Argentina 10(4):1-159.

chemistry - pH (data only)
Villanueva, Sebastian F., Alberto Gomez, Aldo Orlando y Andres J. Lusquiños 1969
Datos y resultados de las campañas pesquerias "Pesqueria VII", 16 de febrero al 1° de marzo de 1968.
Publ. Ser. Inf. técn. Proyecto Desarrollo Pesq., Mar del Plata (10/VII): unnumbered pp.

chemistry - pH (data only)
Zvereva, A.A., Edit., 1959.
[Data, 2nd Marine Expedition, "Ob", 1956-1957.]
Arktich. i Antarkt. Nauchno-Issled. Inst., Mezhd. Geofiz. God, Sovetsk. Antarkt. Exped., 6: 1-387.

pH-density

chemistry - pH-density
Brasil, Diretoria de Hidrografia e Navegacao, 1957.
Ano Geofisico Internacional. Publicacao DG-06-V: 3 pp., 18 figs., 3 pp. tables.

chemistry - pH-density
Brazil, Diretoria de Hidrografia e Navegacao, 1957.
Ano Geofisico Internacional, Publicacao DG-06-VI: unnumbered momeographed pages.

effects of pH

chemistry pH (effect of)
Costa, H.H. 1967.
Responses of Gammarus pulex (L.) to modified environment. II. Reactions to abnormal hydrogen ion concentrations.
Crustaceana 13(1):1-10.

chemistry pH, effect of
Dietrich, R., 1956.
Die Bedeutung der Wasserstoffionenkonzentration und des Ionenmilieus im Gebrauchswasser für die Qualität der Fischundustrieerzeugnisse.
Veröff. Inst. Meeresf., Bremerhaven, 4(2):117-126

chemistry PH, effect of
Duursma, E. K., and W. Sevenhuysen, 1966.
Note on chelation and solubility of certain metals in sea water at different PH values.
Netherlands J. Sea Res., 3(1):95-106.

Chemistry, pH, effect of
Kuwatani, Yukimasa and Tamotsu Nishii, 1969.
Effects of pH of culture water on the growth of the Japanese pearl oyster. (In Japanese; English abstract).
Bull. Jap. Soc. scient. Fish., 35(4): 342-350.

chemistry
effects of pH
Moore, H. B. and J. A. Kitching, 1939
The biology of Chthamalus stellatus (Poli).
JMBA 23:521-541.

chemistry- pH, effect of
Ogata, Eizi, and Toshio Matsui, 1965.
Photosynthesis in several marine plants of Japan as affected by salinity, drying and pH, with attention to their growth habitats.
Botanica Marina, 8(2/4):199-217.

chemistry- pH, effect of
Park, K., D.W. Hood, and H.T. Odum, 1958.
Diurnal variation in Texas bays and its application to primary production estimation.
Publ. Inst. Mar. Sci., 5:47-64.
Also:
Contr. in Oceanogr. and Meteorol., A. & M. College of Texas, Vol. 5, Contrib. No. 131.

chemistry- pH, effect of
Wise, J.P., 1958.
Cod and hydrography, a review.
U.S.F.W.S.SPec. Sci. Rept., Fish., No. 245:16 pp.

chemistry, pH/eH
*Hartmann, Martin, 1969.
Investigations of Atlantis II Deep samples taken by the FS Meteor.
In: Hot brines and Recent heavy metal deposits in the Red Sea, E.T. Degens and D.A. Ross, editors, Springer-Verlag, New York, Inc., 204-207.

pH, surface

chemistry - pH (surf. etc)
Park, Kilho, 1966.
Surface pH of the northeastern Pacific Ocean.
J. oceanogr. Soc., Korea, 1(1/2):1-6.

chemistry
pH, surface, mean
Wiktor, K., and D. Zembrzuska, D., 1959.
Materials for hydrography of Firth of Szczecin.
Prace Morsk. Inst. Ryback., Gdyni, 10A:259-272.

chemistry pH-temperature
Ben-Yaakov, S. and I.R. Kaplan, 1968.
pH-temperature profiles in ocean and lakes using an in situ probe.
Limnol. Oceanogr., 13(4):688-693.

pH, weekly average

chemistry
pH (weekly average)
Perez Sori, Jose A., 1962
Datos oceanograficos de la playa Habana: III.
Contribucion, Centro Invest. Pesqueras, Depart Pesca, Inst. Nacional de Reforma Agraria, Cuba No. 16: 28 pp.

phaeophytin

Chemistry
phaeophytin
Beers, John R. and Gene L. Stewart, 1971.
Micro-zooplankters in the plankton communities of the upper waters of the eastern tropical Pacific. Deep-Sea Res. 18(9): 861-883.

chemistry, phaeophytin
Kawarada Yutaka, and Akira Sano 1969.
Distribution of chlorophyll and phaeophytin in the western North Pacific.
Oceanogrl Mag. 21(2): 137-146.

Chemistry, phaeophytin
Lorenzen, Carl J., 1965.
A note on the chlorophyll and phaeophytin content of the chlorophyll maximum.
Limnol. Oceanogr., 10(3):482-483.

Chemistry - phaeophytin (data)
Platt, Trevor, and Brian Irwin, 1968.
Primary productivity measurements in St. Margaret's Bay, 1967.
Techn. Rept. Fish. Res. Bd., Can., 77: 123 pp.

chemistry - phaeophytin
Taylor, W. Rowland, and Conrad D. Gebelein, 1966.
Plant pigments and light penetration in intertidal sediments.
Helgoländer wiss Meeresunters. 13(3):229-237.

chemistry
phaeophytin
Yentsch, Charles S., and David W. Menzel, 1963
A method for the determination of phytoplankton chlorophyll and phaeophytin by fluorescence.
Deep-Sea Res., No. 1348.

chemistry, phaeophytin (data only)
Japan, Hokkaido University, Faculty of Fisheries 1970.
Data record of oceanographic observations and exploratory fishing 13:406 pp.

Chemistry
phaeophytin a (data only)
Japan, Hokkaido University, Faculty of Fisheries, 1968.
Data record of oceanographic observations and exploratory fishing.
No. 12:420 pp.

chemistry - phaeophytin (data only)
Japan, Hokkaido University, Faculty of Fisheries, 1968.
The Oshoro Maru cruise 23 to the east of Cape Erimo, Hokkaido, April 1967.
Data Record Oceanogr. Obs. Expl. Fish., 12: 115-169.

chemistry, phaeophytin (data only)
Platt, Trevor, and Brian Irwin 1970.
Primary productivity measurements in St. Margaret's Bay 1968-1970.
Techn. Rept. Fish. Res. Bd. Can. 203:67pp. (multilithed).

chemistry, phaeo-pigments
Tietjen, John H. 1968.
Chlorophyll and phaeo-pigments in estuarine sediments.
Limnol. Oceanogr. 13(1): 189-192.

chemistry, phaeophytin
Uyeno, Fukuzo, Kyoichi Kawaguchi, Nagao Terada and Tadashi Okada 1970.
Decomposition, effluent and deposition of phytoplankton in an estuarine pearl oyster area.
Rept. Fac. Fish. Prefect. Univ. Mie 7(1):7-41

pharmacological substances

chemistry, pharmacological substances
Barnes, W.J.P., and G.A. Horridge, 1965.
A neuropharmacologically active substance from jellyfish ganglia.
J. Exp. Biol. ,42(257-267).

phosphatase

Chemistry, phosphatases
Antia, N.J., and A. Watt, 1965.
Phosphatas activity in some species of marine phytoplankters.
J. Fish. Res. Bd., Canada, 22(3):793-799.

chemistry
phosphatase (acid)
Jennings, J.B. and Linda G. Halverson, 1971.
Variations in subcuticular acid phosphatase activity during the molting cycle of the euphausiid crustacean Thysanoessa raschii (M. Sars) Hansen. J. mar. Res. 29(2): 133-139.

chemistry
phosphatase
Wai, Nganshou, Tsu-Chang Hung and Yeh-Hunglu,
Alkaline phosphatase in Taiwan sea water.
Bull. Inst. Chem. Acad. Sinica, No. 3:1-10.

phosphate

Chemistry, phosphate

See also: Chemistry, phosphate (inorganic), etc.
Chemistry, phosphorus, etc.

A

chemistry
phosphate
Acara, A., 1961
Poor water masses in the North Pacific Ocean.
The description of the water mass. Pt. 1.
Hidrobiologi, Istanbul, (B);5(3/4):97-112.

Chemistry
phosphate
Acara, A., 1961.
Poor water masses in the North Pacific Ocean.
The distribution of the water mass. Pt. 2.
Hidrobiologi, Istanbul, (B), 5(3/4): 113-128.

chemistry
phosphate
Adrov, M.M., 1962.
Hydrologic investigations off West Greenland.
Sovetskie Riboch. Issledov. v. Severo-Zapadnói Atlant. Okeana, VINRO-PINRO, Moskva, 137-153.

Chemistry
phosphate
Akiyama Tsutomu 1968
The distribution of dissolved oxygen and phosphate phosphorus in the adjacent seas of Japan.
Oceanogrl Mag. 20(2): 147-172

chemistry, phosphate (data)
Aleksandrovskaya, N.B., M.V. Kales and E.A. Yola, 1966.
Peculiarities of the hydrological regime of the Baltic Sea in 1964.
Annls. biol., Copenh., 21:43-45.

Chemistry
phosphate
Aleksejev, A., B. Istoshin, W. Pachorukov, P. Myrland, H. Einarsson, F. Hermann and U. Stefansson, (1959) 1961.
International investigations on hydrographic conditions in the Norwegian Sea, June 1959.
Ann. Biol., Cons. Perm. Int. Expl. Mer, 16: 25.

chemistry
phosphate
Aleksejev, A., B. Istoshin, L. Ponomarenko (1959) 1961
Hydrographic conditions in the Norwegian Sea, November-December 1959.
Ann. Biol., Cons. Perm. Int. Expl. Mer, 16: 26-28.

chemistry phosphate
Alexander, J.E., J.H. Steels and E.F. Corcoran, 1962
The seasonal cycle of chlorophyll in the Florida Straits.
Proc. Gulf and Caribbean Fish. Inst., Inst. Mar. Sci., Univ. Miami, 14th Ann. Sess.: 63-67.

chemistry, phosphate
Andersen, K.P., 1957.
Hydrographic conditions in the southern North Sea, the Blødden Ground area in 1955. Ann. Biol., Cons. Perm. Int. Expl. Mer, 12:76, Figs. 24-34.

chemistry, phosphate
Andersen, K.P., 1954.
Hydrographic conditions in the southern part of the Norwegian Sea, 1953.
Ann. Biol., Cons. Perm. Int. Expl. Mer, 10:20-24, Textfigs. 7-20.

chemistry, phosphate
Antonov, A.E., and O.S. Roudneva, 1965.
Peculiarities of hydrological and hydrochemical conditions in the south Baltic during the IGY. (In Russian; English abstract).
Okeanolog. Issled. Rezult. Issled. Programme Mezhd. Geofiz. Goda, Mezhd. Geofiz. Komitet Presidiume, Akad. Nauk, SSSR, No. 13:90-95.

chemistry, phosphate (data)
Argentina, Servicio de Hidrografia Naval, 1960.
Operacion Oceanografica Drake II. Resultados preliminares.
Servicio de Hidrografia Naval, Argentina, Publ., H. 14: numerous unnumbered pp.

chemistry phosphate (data)
Argentina, Servicio de Hidrografia Naval, 1960.
Operacion Oceanografica Malvinas (Resultados preliminares).
Servicio de Hidrografia Naval, Argentina, Publ., H. 606: numerous unnumbered pp.

chemistry, phosphate (data)
Argentina, Servicio de Hidrografia Naval, 1959.
Trabajos oceanograficos realizados en la campana Antartico, 1955-1956.
Servicio de Hidrografia Naval, Argentina, Publ., H. 620: numerous unnumbered pp.

chemistry, phosphate(data)
Argentina, Servicio de Hidrografia Naval, 1959.
Operacion Oceanografica Meridiano. Resultados preliminares.
Servicio de Hidrografia Naval, Argentina, Publ., H. 617: numerous unnumbered pp.

chemistry, phosphate(data)
Argentina, Servicio de Hidrografia Naval, 1959.
Trabajos oceanograficos realizados en la campana Antartica, 1958/1959. Resultados preliminares.
Servicio de Hidrografia Naval, Argentina, Publ., H. 616: 127 pp.

chemistry, phosphate (data)
Argentina, Servicio de Hidrografia Naval, 1959.
Trabajos oceanograficos realizados en la campana Antartica, 1957-1958. Resultados preliminares.
Servicio de Hidrografia Naval, Argentina, Publ., H. 615: numerous unnumbered pp.

chemistry phosphate (data)
Argentina, Servicio de Hidrografia Naval, 1959.
Operacion Oceanografica Drake. 1. Resultados preliminares.
Servicio de Hidrografia Naval, Argentina, Publ., H. 613: numerous unnumbered pp.

chemistry phosphate (data)
Argentina, Servicio de Hidrografia Naval, 1959.
Operacion Oceanografica Atlantico Sur. Resultados preliminares.
Servicio de Hidrografia Naval, Argentina, Publ., H. 608: numerous unnumbered pp.

chemistry, phosphate(data)
Argentina, Servicio de Hidrografia Naval, 1959.
Operacion Oceanografica Cuenca. Resultados preliminares.
Servicio de Hidrografia Naval, Argentina, Publ. H. 607: numerous unnumbered pp.

chemistry, phosphate
Armstrong, F.A.J., and E.I. Butler, 1968.
Chemical changes in sea water off Plymouth during the years 1962 to 1965.
J. mar. biol. Ass., U.K., 48(1):153-160.

chemistry phosphate
Armstrong, F.A.J., and E.I. Butler, 1963
Chemical changes in sea water off Plymouth in 1963.
J. Mar. Biol. Assoc., U.K., 43(1):75-78.

chemistry, phosphate
Armstrong, F.A.J., and E.I. Butler, 1962.
Hydrographic surveys off Plymouth in 1959 and 1960.
J. Mar. Biol. Assoc., U.K., 42(2):445-463.

chemistry, phosphate
Armstrong, F.A.J., and E.I. Butler, 1962.
Chemical changes in sea water off Plymouth during 1960.
J. Mar. Biol. Assoc., U.K., 42(2):253-258.

chemistry phosphate
Armstrong, F.A.J., and E.I. Butler, 1960.
Chemical changes in the sea water off Plymouth during 1958. J.M.B.A., 39(2):299-302.

chemistry, phosphate
Arrhenius, O., 1948.
4. The phosphate content of some sediments from the Mediterranean. Medd. Ocean. Inst. Göteborg, 15 (Göteborgs Kungl. Vetenskaps- och Vitterhets-Samhälles Handlingar, Sjätte Följden, Ser. B, 5(13)):26-29, Fig. 4.

chemistry phosphate
Atkins, W.R.G., 1953.
Seasonal variations in the phosphate and silicate content of sea water. Pt. VI. 1948 compared with the 1923-25 period. J.M.B.A. 31(3):489-492, 4 textfigs.

chemistry phosphate
Atkins, W.R.G., 1930.
Seasonal variations in the phosphate and silicate content of sea water in relation to the phytoplankton crop. V. November 1927 to April 1929 compared with earlier years from 1923. J.M.B.A. 16:821-852.

chemistry phosphate
Atkins, W.R.G. 1928.
Seasonal variation in the phosphate and silicate content of sea water during 1926 and 1927 in relation to the phytoplankton crop.
J. mar. biol. Ass. U.K. 15:191-205.

chemistry phosphate
Atkins, W.R.G., 1925.
Seasonal changes in the phosphate content of sea water in relation to the growth of algal plankton during 1923 and 1924. J.M.B.A. 13:700-720.

phosphate
Atkins, W.R.G., 1923.
The phosphate content of fresh and salt waters in its relationship to the growth of the algal plankton. J.M.B.A. 13:119-150.

chemistry, phosphate
Austin, Thomas S., 1960
Oceanography of the east central equatorial Pacific as observed during expedition "Eastropic". U.S.F.&W.S., Fish. Bull. 168, Vol. 60: 257-282.

chemistry phosphate
Austin, Thomas S., 1958.
Variations with depth of oceanographic properties along the equator in the Pacific.
Trans. A.G.U., 39(6):1055-1063.

B

chemistry phosphate
Băcescu, M., M.T. Gomoiu, N. Bodeanu, A. Petran G.I. Müller si V. Chirila, 1965.
Dinamica populațiilor animale și vegetale din zona nisipurilor fine de la nord de Constanța în condițiile anilor 1962-1965.
In: Ecologie marină, M. Băcescu, redactor, Edit. Acad. Republ. Pop. Romane, București, 2: 7-167.

chemistry phosphate (data)
Băcescu, M., G. Müller, H. Skolka, A. Petran, V. Elian, M.T. Gomoiu, N. Bodeanu, și S. Stănescu, 1965.
Cercetări de ecologie marină în sectorul predeltaic în condițiile anilor 1960-1961.
In: Ecologie marină, M. Băcescu, redactor, Edit. Acad. Republ. Pop. Romane, București, 1:185-344.

chemistry, phosphate
Bakaev, V.G., editor, 1966.
Atlas Anterktiki, Sovetskaie Anterktich-eskaia Ekspeditsiie. 1.
Glabnoe Upravlenie Geodesii i Kartografii MG SSSR, Moskva-Leningrad, 225 charts.

chemistry phosphate
Bal, D.V., L.B. Pradhan, and K.G. Gupte, 1946
A preliminary record of some of the chemical and physical conditions in waters of the Bombay harbour during 1944-45. Proc. Indian Acad. Sci. Sect. B24(2):60-73, 4 figs.

chemistry phosphate (data)
Ballester, A., 1965.
Hidrografia y nutrientes de la Fosa de Cariaco.
Informe de Progresso del Estudio Hidrografico de la Fosa de Cariaco, Fundación La Salle de Ciencias Naturales Estación de Investigacion -es Marinas de Margarita, Caracas, Sept. 1965. (mimeographed):3-12.

chemistry, phosphate (data)
Ballester, Antonio, Enrique Arias, Antonio Cruzado, Dolores Blasco y José María Camps 1967.
Estudio hidrográfico de la costa catalana de junio de 1965 a mayo de 1967.
Investigación pesq. 31(3): 621-662.

chemistry, phosphate
Banse, Karl, 1960
Bemerkungen zu meereskundlichen Beobachtungen vor der Ostküste von Indien.
Kieler Meeresf., 16(1): 214-220.

chemistry, phosphate
Banse, K., 1957.
Ergebnisse eines hydrographisch-produktionsbiologischen Längsschnitts durch die Ostsee im Sommer 1956. II. Die Verteilung von Sauerstoff, Phosphat und suspendierten Substanz.
Kieler Meersf., 13(2):186-201.

chemistry, phosphate (data)
Bardin, I.P., 1958.
[Hydrological, hydrochemical, geological and biological studies, Research Ship "Ob", 1955-1957.] Trudy Kompleks. Antarkt. Exped., Akad. Nauk, SSSR, Mezh. Geofiz. God, Gidrometeorol. Izdatel., Leningrad, 217 pp.

chemistry phosphate
Bennett, Edward B., and Milner B. Schaefer, 1960
Studies of physical, chemical and biological oceanography in the vicinity of the Revilla Gigedo Islands during "Island Current Survey" of 1957.
Bull. Inter-American Tropical Tuna Comm., 4(5): 219-317. (Also in Spanish)

chemistry phosphate (data)
Bernhard, Michel, 1963.
1. Introduzione. 2. Chimica oceanografica.
Rapp. Attività Sci. e Tecn., Lab. Studio della Contaminazione Radioattiva del Mare, Fiascherino, La Spezia (maggio 1959-maggio 1962), Comit. Naz. Energia Nucleare, Roma, RT/BIO (63), 8:7-39.
(multilithed)

chemistry, phosphate (some data)
Bernard, F., 1956.
Contribution à la connaissance du détroit de Gibraltar. (Hydrographie et nannoplancton en juin 1954). Bull. Inst. Océan., Monaco, No. 1074:1-22.

chemistry, phosphates
Bernard, F., 1952.
Eaux atlantiques et mediterraneennes au large de l'Algerie. 1. Hydrographie, sals nutritifs, et phytoplancton en 1950. Ann. Hydrogr., Monaco, n.s., 27(1):48 pp., 15 textfigs.

chemistry, phosphates
Bernard, F., 1948
Recherches préliminaires sur la fertilité marine au large d'Alger. J. du Cons. 15(3): 260-267, 5 text figs.

chemistry phosphates
Bernard, M. F., 1938
Recherches récentes sur la densité du plancton méditerranéen. Rap. Proc. Verb des Réunion, Comm. Int. l'Expl. Sci. de la Mediterranée, n.s., XI:289-300.

chemistry phosphates
Bernard, F., 1937.
Résultats d'une annee de reacherches quantitative sur le phytoplancton de Monaco. Cons. Perm. Int. Expl. Mer, Rapp. Proc. Verb. 105:26-29, 8 text-figs.

chemistry phosphate
Bessonov, N.M., 1964
On some features of variations of hydrochemical characteristics in fishery areas off Dakar and Takoradi. (In Russian).
Okeanologiia, Akad. Nauk, SSSR, 4(5):813-824.

chemistry, phosphate
Bezrukov, P.L., 1959
[Oceanographic investigations in the North-West Pacific, August-October, 1954.] Trudy Inst. Okeanol., 16: 70-97.

chemistry, phosphates (data)
Bienati, Norberto L., y Rufino A. Comes 1970.
Variación estacional de la composicion fisico-quimica del agua de mar en Puerto Paraiso, Antartida occidental
Contrib. Inst. Antart. Argentino 130: 45pp.

chemistry, phosphate
Bigelow, H.B., and M. Leslie, 1930
Reconnaissance of the waters and plankton of Monterey Bay, July 1928. Bull. M.C.Z., 70(5):429-481, 43 text figs.

chemistry, phosphates
Bigelow, H.B., L.C. Lillick, and M. Sears, 1940.
Phytoplankton and planktonic protozoa of the offshore waters of the Gulf of Maine. Pt. 1. Numerical distribution. Trans. Am. Phil. Soc., n.s., 31(3):149-191, 10 textfigs.

chemistry, phosphate
Black, W.A.P., and E.T. Dewar, 1949.
Correlation of some of the physical and chemical properties of the sea with the chemical composition of the algae. J.M.B.A. 28(3):673-399, 14 textfigs.

chemistry, phosphates (data)
Blackburn, M., 1962.
An oceanographic study of the Gulf of Tehuantepec U.S.F.W.S. Spec. Sci. Rept., Fish., No. 404:28 pp

chemistry phosphate
Bogdanov, D.V., 1967.
Sobre la variabilidad de las condiciones oceanográficas en el Mar Caribe y el Golfo de Mexico. (In Russian; Spanish abstract).
Sovetsk.-Kubinsk. Ribokhoz. Issled. Vses. Nauchno-issled. Inst. Morsk. Ribn. Khoz. Okeanogr (VNIRO), Centr Ribokhoz. Issled. Natsion. Inst. Ribolov. Respibl. Kuba (TSRI), 2:21-38.

chemistry, phosphate
Bogdanov, D.V., 1965.
Algunos rasgos de la oceanografía del Golfo de México y del Mar Caribe. (In Russian; Spanish abstract).
Sovetsk. -Cub. Ribokhoz. Issled., VNIRO:Tsentr. Ribokhoz. Issled. Natsional. Inst. Ribolovsta. Republ. Cuba, 23-45.

chemistry, phosphate
Bogdanov, M.A., I.A. Ermachenko, S.I. Potaichuk, and M. S. Edelman, 1962.
On the hydrology of the Faroe-Iceland region. Trudy Vses. Nauchno-Issledov. Inst. Morsk. Ribn. Chos. i Okean., VNIRO, 46:61-64.

chemistry, phosphate
Bogoyavlensky, A.N., 1959
[Hydrochemical investigations.] Arktich. i Antarktich. Nauchno-Issled. Inst., Mezhd. Geofiz. God, Sovetsk. Antarkt. Eksped., 5: 159-172.

chemistry, phosphate
Bogoiavlenskii, A.N., 1958.
[VI Hydrochemical work.] Opisanie Exped. D/E "Ob", 1955-1956, MGG, Trudy Kompaeksion Antarkt. Exped., Akad. Nauk, USSR, 91-102.

chemistry, phosphates
Bogoyavlenskii, A.N., 1955.
[Chemical characteristics of the water in the region of the Kurile-Kamchatka Trench.] Trudy Inst. Okeanol., 12:161-176.

chemistry phosphates
Bogojavlensky, A.N. and O.V. Shishkina 1971.
On the hydrochemistry off Peru and Chile. (In Russian; English abstract)
Trudy Inst. Okeanol. P.P. Shirshova Akad. Nauk SSSR 89:96-105

chemistry, phosphate
Boto, R. G., 1945
Contribuicao para os estudios de oceanografia ao longo da costa de Portugal. Fosfatos e nitratos. Trav. St. Biol. Mar., Lisbonne, No. 49:102 pp., 57 figs. (Bumpus reprint).

chemistry, phosphate
Braarud, T., 1934
A note on the phytoplankton of the Gulf of Maine in the summer of 1933. Biol. Bull. 67(1):76-82. (Contribution No.46 of the Woods Hole Oceanographic Institution)

chemistry phosphate
Braarud, T., and Adam Bursa, 1939
On the phytoplankton of the Oslo Fjord, 1933-1934. Hvalrådets Skr. No.19:1-63; 9 text figs. Reviewed: J. du. Cons. 14(3): 418-420. A.C. Gardiner.

chemistry, phosphate (data)
Braarud, T., K.R. Gaarder and J. Grøntved, 1953.
The phytoplankton of the North Sea and adjacent waters in May 1948. Rapp. Proc. Verb., Cons. Perm Int. Expl. Mer, 133:1-87, 29 tables, Pls. A-B, 18 textfigs.

chemistry, phosphate
Braarud, T., and A. Klem, 1931.
Hydrographical and chemical investigations in the coastal waters off Møre and in the Romsdalfjord. Hvalrådets Skrifter, 1931(1):88 pp., 19 figs.

chemistry phosphates
Braarud, T. and J. T. Ruud, 1937
The Hydrographic conditions and aeration of the Oslo Fjord, 1933-1934. Hvalrådets Skr. No. 15:56 pp., 24 figs.

Reviewed: J. du Cons. XIV(3):406-408. J. N. Carruthers.

chemistry phosphate
Brandhorst, W., 1955.
Hydrographie des Nord- Ostsee Kanal. Kieler Meeresf. 11(2):174-187.

chemistry phosphate
Brandt, K., 1905.
On the production and the conditions of production in the sea. Cons. Perm. Int. Expl. Mer, Rapp. Proc. Verb., 3(1):12 pp.

chemistry phosphate (data)
Brasil, Diretoria de Hidrografia e Navegação, Marinha do Brasil, 1961
Estudo das condições oceanográficas sobre a plataforma continental, entre Cabo-Frio e a Ponta do Boi, durante o mes de setembro (transição inverno-primervera).
DG-06-XII: unnumbered pp. (mimeographed).

chemistry phosphate (data)
Bregant, Davide 1967 (1969).
Distribuzione dei sali nutritivi nell'area delle Bocche di Bonifacio e del Golfo dell'Asinara – Crociera Bannock 1964.
Boll. Pesca Piscic. Idrobiol., n.s. 22(2): 113-120

chemistry phosphate
Broenkow, William W., 1965.
The distribution of nutrients in the Costa Rica dome in the eastern tropical Pacific Ocean. Limnology and Oceanography, 10(1):40-52.

chemistry, phosphate
*Brosin, Hans-Jürgen, und Dietwart Nehring, 1968.
Der Aquatoriale Unterstrom im Atlantischen Ozean auf 29°30'W im September und Dezember 1966. Beiträge Meeresk., 22:5-17.

chemistry phosphate
Bruce, Herbert E., and Donald W. Hood, 1959.
Diurnal inorganic phosphate variations in Texas Bays.
Inst. Mar. Sci., Publ., 6:133-145.

chemistry phosphate
Bruevich, S.W., 1960
[Hydrochemical investigations on the White Sea.] Trudy Inst. Okeanol., 42: 199-254.

chemistry, phosphate
Brucks, John T., Merton C. Ingham and Thomas D. Leming 1968.
Oceanic conditions off Sierra Leone, 10 February to 2 March 1965 (part of Geronimo Cruise 5).
Data Rept, U.S. Dept. Interior, Bur. Comm. Fish. 28: 1 film (microfiche)

chemistry, phosphate
Brucks, John T., Merton C. Ingham and Thomas D. Leming 1968.
Oceanic conditions in the northwestern Gulf of Guinea, 14 to 30 March 1965 (part of Geronimo Cruise 5).
Data Rept., U.S. Dept. Interior, Bur. Comm. Fish. 27: 1 film (microfiche).

chemistry phosphate
Bruevich, S.W., A.N. Bogoyavelnsky, and V.V. Mokievskaya, 1960
[Hydrochemical features of the Okhotsk Sea.] Trudy Inst. Okeanol., 42: 125-198.

chemistry phosphate
Cooper, L.H.N. 1947.
Internal Waves and Upwelling of Oceanic Water from Mid-depths on to a Continental Shelf. Nature, 159, (4043): 579-580.

chemistry, phosphate
Cooper, L.H.N. 1938.
Phosphate in the English Channel 1933-38, with a comparison with earlier years, 1916 and 1923-32.
J. mar. biol. Ass. U.K. 23: 181-196.

chemistry phosphate
Cooper, L.H.N., 1938
Salt error determinations of phosphate in sea water. JMBA, 23:171-178

chemistry phosphate
Copenhagen, W.J., 1953.
The periodic mortality of fish in the Walvis Region. A phenomenon within the Benguela Current.
Union of S. Africa, Div. Fish., Investig. Rept. No. 14:1-35, 9 pls.

chemistry phosphate
Corcoran, E.F., and James E. Alexander, 1963.
Nutrient, chlorophyll and primary production studies in the Florida Current.
Bull. Mar. Sci., Gulf and Caribbean, 13(4):527-541.

chemistry, phosphate
Coste, Bernard 1969
Echange de sels nutritifs dissous entre la mer Méditerranée et l'ocean Atlantique.
Cah. océanogr. 21 (10): 943-963.

chemistry phosphate (data)
COSTE, B., et H.J. MINAS, 1968.
Production organique primaire et sels nutritifs au large des côtes occidentales Corso-Sardes en février 1966.
Rec. Trav. Sta. mar. Endoume, 44 (60): 49-61

chemistry, phosphate (data)
Coste, Bernard, et Hans-Joachim Minas 1967.
Premières observations sur la distribution des taux de productivité et des concentrations en sels nutritifs des eaux de surface du Golfe du Lion.
Cah. océanogr. 19(5): 417-429.

chemistry phosphate
Craig, R.E., 1956.
Hydrography, near northern seas and approaches.
Ann. Biol., Cons. Perm. Int. Expl. Mer, 11:33-36, Figs. 1-2.

chemistry phosphate
Craig, R.E., 1954.
A first study of the detailed hydrography of some Scottish West Highland sea lochs (Lochs Inchard, Kanaird and the Cairnbawn Group).
Ann. Biol., Cons. Perm. Int. Expl. Mer, 10:16-19, Textfigs. 4-5.

chemistry phosphate
Craig, R.E., 1952.
Hydrography of the Firth of Forth in the spring.
Cons. Perm. Int. Expl. Mer, Ann. Biol. 8:98-99, Text-fig. 1.

chemistry phosphate
Craig, R.E., and T. Lovegrove, 1953.
Hydrographic and plankton survey of the north-east coast of Scotland.
Ann. Biol., Cons. Perm. Int. Expl. Mer, 9:107-112, Figs. 11-13.

chemistry phosphate
Crehuet, Ramón Fernández y Maria Jesús del Val Cordón, 1960
Observaciones oceanograficas en la Bahia de Malaga (Marzo a marzo 1957). Bol. Inst. Esp. Ocean., 98: 1-29.

chemistry, phosphate (data)
Cromwell, T., 1954.
Mid-Pacific oceanography II. Transequatorial waters, June-August 1950, January-March 1951.
U.S.F.W.S. Spec. Sci. Rept. - Fish. No. 131:

chemistry, phosphate (data)
Cromwell, T., 1951.
Mid-Pacific oceanography, January through March 1950. Spec. Sci. Rept., Fish., No. 54:9 pp., 17 figs., station data.

D

chemistry, phosphates
Datsko, V.G., 1955.
[Dynamics of phosphates in the upper layers of the Black Sea.] Dokl. Akad. Nauk, SSSR, 100(6):1127.

phosphate
Davidovitch, R.L., 1964.
Short chemical characteristics of the northwest Pacific Ocean. (In Russian).
Gosudarst. Kom. Sov. Ministr., SSSR, Ribn. Choz., Trudy VNIRO, 49, Izv. TINRO, 51:93-98.

chemistry phosphate (data)
Davidovich, R.L., 1963
Hydrochemical features of the southern and southeastern parts of the Bering Sea. (In Russian).
Sovetsk. Ribokh. Issled. B Severo-Vostokh. Chasti Tikhogo Okeana, VNIRO 48, TINRO 50(1): 85-96.

chemistry, phosphate
Deacon, G. E. R., 1933
A general account of the hydrology of the South Atlantic Ocean. Discovery Repts. 7:173-238, pls.8-10.

chemistry phosphate
Defant, A., G. Böhnecke, H. Wattenberg, 1936.
I. Plan und Reiseberichte die Tiefenkarte das Beobachtungsmaterial. Die Ozeanographischen Arbeiten des Vermessungsschiffes "Meteor" in der Dänemarkstrasse und Irmingersee während der Fischereischutzfahrten 1929, 1930, 1933 und 1935. Veroffentlichungen des Instituts für Meereskunde, n.f., A. Geogr.-naturwiss. Reihe, 32:1-152 pp., 7 text figs., 1 plate.

chemistry phosphates (geochemical)
Degens, Egon T., 1965.
Geochemistry of sediments: a brief survey.
Prentice-Hall, Inc., 342 pp.

chemistry phosphate
Della Croce, Norberto, 1962
Zonazione zooplanctonica nel Golfo di Napoli.
Pubbl. Staz. Zool., Napoli, 32(Suppl.):368-379.

chemistry phosphate
Delsman, H.C., 1936.
Preliminary plankton investigations in the Java Sea. Treubia 17:139-181, 8 maps, 41 figs.

chemistry phosphates
Denigès, G., 1920.
Réaction de coloration extrêmement sensible des phosphates et des arseniates. C.R. Acad. Sci., Paris, 171:802-804.

chemistry, phosphate (data)
De Queiroz Santos, Neuzon 1967.
Principais nutrientes e alguns dados fisico-quimicos da região lagunar de Cananeia.
Bolm Inst. Biol. mar. Univ. Fed. Rio Grande Norte 4: 1-14.

chemistry phosphate (data)
Donnelly, P.V., J. Vuille, M.C. Jayaswal, R.A. Overstreet, J. Williams, M.A. Burklew and R.M. Ingle, 1966.
A study of contributory chemical parameters to red tide in Apalachee Bay.
Florida Bd., Conserv., St. Petersburg, Mar. Lab. Prof. Papers Ser. No. 8:43-83.

chemistry phosphate
Doty, Maxwell S., and Mikihiko Oguri, 1958
Primary production patterns in enriched areas.
Proc. Ninth Pacific Sci. Congr., Pacific Sci. Assoc., 1957, 16(Oceanogr.):94-97.

chemistry phosphate
Dragovich, Alexander, 1961
Relative abundance of plankton off Naples, Florida, and associated hydrographic data, 1956-57.
USFWS Spec. Sci. Rept., Fish., No. 372:41 pp.

chemistry phosphate
Dragovich, A., and B.Z. May, 1962
Hydrological characteristics of Tampa Bay tributaries.
U.S.F.W.S. Fish. Bull. (205) 62:163-176.

chemistry phosphate
*Düing, Walter, Klaus Grasshoff Und Gunther Krause, 1967.
Hydrographische Beobachtungen auf einem Äquatorschnitt in Indischen Ozean.
"Meteor Forschungsergebnisse (A)(3):84-92.

chemistry, phosphate
*Duxbury, Alyn C. and Noel B. McGary, 1968.
Local changes of salinity and nutrients off the mouth of the Columbia River.
Limnol. Oceanogr., 13(4):626-636.

E

chemistry, phosphate
Edmondson, W.T., and Y.H. Edmondson, 1947.
Measurements of production in fertilized salt-water. J. Mar. Res. 6(3):228-246, Figs. 55-62.

chemistry, phosphate
Eggvin, J., 1933.
A Norwegian fat-herring fjord. An oceanographical study of the Eidsfjord. Repts. Norwegian Fish. Mar. Invest. 4(6):22 pp., 11 textfigs.

chemistry phosphate
Egorova, V.A., 1955.
[Dynamics of phosphate distribution in the coastal areas of the Black Sea.] Dokl. Akad. Nauk, SSSR, 102(4):983.

chemistry, phosphate
Eisma, D. and A.J. van Bennekom, 1971.
Oceanographic observations on the eastern Surinam Shelf. Hydrogr. Newsletter, R. Netherlands Navy, Spec. Publ. 6: 25-29.

chemistry phosphate
Elizarov, A.A., 1968.
Preliminary results of oceanographic investigations of the west coast of India. (In Russian).
Trudy, Vses. Nauchno-Issled. Inst. Morsk. Ribn. Okeanogr (VNIRO) 64, Trudy Azovo-Chernomorsk. Nauchno-Issled. Inst. Morsk. Ribn. Khoz. Okeanogr. (AscherNIRO), 28: 94-101.

chemistry, phosphate (data)
El-Sayed, Sayed Z., and Enrique F. Mandelli, 1965.
Primary production and standing crop of phytoplankton in the Weddell Sea and Drake Passage.
In: Biology of Antarctic seas II.
Antarctic Res. Ser., Am. Geophys. Union, 5:87-106.

chemistry phosphate (data)
Emery, K.O., and Jobst Hulsemann, 1963
Submarine canyons of southern California.
Part 1. Topography, water, and sediments.
Allan Hancock Pacific Expeditions, 27(1): 1-80.

chemistry, Phosphate

Emery, K. O., and J. Hulsemann, 1961 (1962).
The relationships of sediments, life and water in a marine basin.
Deep-Sea Res., 8(3/4):165-180.

chemistry, phosphates

Ercegovic, A., 1934.
Température, salinité, oxygène et phosphates dans les eaux cotières de l'Adriatique oriental moyen. Acta Adriatica 1(5):1-51, 19 textfigs.

chemistry
phosphates (data)

Establier, R., and R. Margalef, 1964
Fitoplancton e hidrografía de las costas de Cádiz (Barbate), de junio de 1961 a agosto de 1962.
Invest. Pesquera, Barcelona, 25:5-31.

chemistry
phosphate

*Ewins, P.A., and C.P. Spencer, 1967.
The annual cycle of nutrients in the Menai Straits.
J. mar. biol. Ass., U.K., 47(3):533-542.

F

chemistry, Phosphate (data)

Faganelli, Armando, 1961.
Primi risultati relativi alla concentrazione dei sali nutritivi nelle acque del Mare Mediterraneo Centrale e mari adiacenti.
Archivio Oceanograf. e Limnol., 12(2):191-208.

Also:
Ist. Sperimentale Talassograf., Trieste, Pubbl. No. 377 (1961).

chemistry phosphate (data)

Faganelli, A., 1961.
Primi risultati relativi alla concentrazione dei sali nutritivi nelle acque del Mar Mediterraneo centrale et mari adiacenti.
Rapp. Proc. Verb., Réunions, Comm. Int. Expl. Sci., Mer Méditerranée, Monaco, 16(3):675-686.

chemistry phosphate

Fedosov, M. V., 1955
[Chemical basis for the foodstuffs in the Azov Sea and prognosis of its change in connection with water construction projects in rivers.]
Trudy VNIRO 31: 35-61.

chemistry phosphate

*Fedosov, M.V., and I. A. Ermachenko, 1967.
Hydrochemical characteristics of water masses and water exchange between Iceland and Faroes (hydrographical surveys in 1960).
Rapp. P.-V. Réun. Cons. perm. Int. Explor. Mer, 157:196-

chemistry, phosphate

Fedosov, M.V., and E.G. Vinogradova, 1955.
[Hydrological and hydrochemical regime, primary production in the Azov Sea and forecasting change. Fundamental basis of the hydrochemical regime in the Azov Sea.] Trudy VNIRO, 31(1):9-34.

chemistry, phosphate

Fedosov, M.V., and G. N. Zaitsev, 1960
[Water balance and the chemical regime in the Baltic Sea and its gulfs.]
Trudy Vses. Nauchno-Issled. Inst. Morsk. Ribnogo Chozia.i Oceanogr. (VNIRO) 42: 7-14.

chemistry phosphate

Finucane, John N., and Alexander Dragovich 1959.
Counts of red tide organisms Gymnodinium brevis and associated oceanographic data from Florida west coast, 1954-1957.
U.S. Fish. Wildl. Serv. Spec. scient. Rept. Fish. 289: 220 pp.

chemistry, phosphate

Fleming, J.A., C.C. Ennis, H.U. Sverdrup, S.L. Seaton, W.C. Hendrix, 1945
Observations and results in physical oceanography. Graphical and tabular summaries. Ocean. 1-B, Sci. Res. of Cruise VII of the Carnegie during 1928-1929 under command of Capt. J.P. Ault, Carnegie Inst., Washington, Publ. 545:315 pp., 5 tables, 254 figs.

chemistry, phosphate

Flores P., Luis Alberto 1967.
Informe preliminar del crucero 6611 de la primavera de 1966 (Cabo Blanco-Punta Coles).
Informe Inst. Mar, Peru 17: 16 pp.

chemistry, phosphate

Flores, Luis, Oscar Guillen y Rogelio Villanieva, 1966.
Informe preliminar del Crucero de invierno 1965 (Mancora-Morro Sama).
Inst. Mar, Peru, Informe, No. 11:1-34 (multilithed).

chemistry, phosphate

Flores, P., Luis Alberto, y Luis A. Poma Elias 1967.
Informe preliminar del crucero 6608-09 de invierno 1966 (Mancora-Ilo).
Informe Inst. Mar, Peru, 16: 24 pp.

chemistry, phosphate

Fonselius, Stig H. 1970.
On the stagnation and recent turnover of the water in the Baltic.
Tellus 22 (5): 533-544.

chemistry
phosphate

Fonselius, Stig H., 1969.
Hydrography of the Baltic deep basins. III.
Rept., Fish. Bd., Sweden, Ser. Hydrogr., 23: 1-97.

chemistry, phosphate

Fonselius, S.H. 1968.
Observations at Swedish lightships and in the central Baltic 1966
Annls biol. Copenh. 1966 23: 74-78

chemistry, phosphate

Fonselius, Stig H. 1967
Hydrography of the Baltic deep basins II
Rep. Fish. Bd. Sweden, Hydrogr. 20: 81 pp.

chemistry, phosphate (data)

Fonselius, S.H., 1966.
Observations at Swedish lightships in the Central Baltic in 1964.
Annls biol., Copenh., 21:46-49.

chemistry, phosphate

Fonselius, Stig H., 1965.
Observations at Swedish light vessels and in the Central Baltic.
Ann. Biol., Cons. Perm. Int. Expl. Mer, 1963, 20:66-68.

chemistry
Phosphate (data)

Forsbergh, Eric D., William W. Broenkow, 1965.

Observaciones oceanograficas del oceano Pacific oriental recolectadas por el barco Shoyo Maru, octubre 1963-marzo 1964.
Comision Interamericana del Atun Tropical, Bol. 10(2):85-237.

chemistry,
phosphate (data)

Forsbergh, Eric D., and James Joseph, 1964
Biological production in the eastern Pacific Ocean. (In English and Spanish).
Inter-American Tropical Tuna Commission, Bull., 8(9):479-527.

chemistry
phosphate

Frolander H.F. 1962
Quantitative estimations of temporal variations of zooplankton off the coast of Washington and British Columbia.
J. Fish. Res. Bd. Canada, 19(4):657-675.

chemistry, phosphate

Fukai, R., 1954.
Critical studies on the analytical methods for minor chemical constituents in sea water. Pt. 1. On the estimation of phosphate-phosphorus.
J. Ocean. Soc., Japan, 10(3):112-120, 5 textfigs.

chemistry, phosphate

Fukai, R., 1953.
On the distribution of nutrient salts in the Equatorial North Pacific.
Bull. Chem Soc., Japan, 26(9):485-489, 5 textfigs.,

Also reprinted in: Bull. Tokai Reg. Fish. Lab. 7(Contr. B) as Contr. 113.

chemistry, phosphate

Fukai, R., 1953.
Seasonal variations of minor chemical constituents in the waters of the Zunan-Kuroshio region. 1. On nitrogen-phosphorus content ratio with the behavior of nitrate and phosphate.
Bull. Chem. Soc., Japan, 25(5):323-325, 3 textfig

Also in:
Bull. Tokai Regional Fish. Res. Lab., Fish. Agency (Contrib. B) 6, 1953

chemistry,
phosphate

Fukuo, Yoshiaki, 1962
On the exchange of water and the productivity of a bay with special reference to Tanabe Bay (IV).
Rec. Oceanogr. Wks., Japan, Spec. No. 6:161-168.

chemistry phosphates (data)

Fuse, Shin-ichiro, and Eiji Harada, 1960

A study on the productivity of Tanabe Bay. (3) Result of the survey in the summer of 1958.
Rec. Oceanogr. Wks., Japan, Spec. No. 4: 13-28.
Oceanographic conditions of Tanabe Bay.

chemistry phosphate

Fuse, S., I. Yamazi and E. Harada, 1958.
A study on the productivity of the Tanabe Bay. (Part 1). 1. Oceanographic conditions of the Tanabe Bay. Results of the survey in the autumn of 1956.
Rec. Oceanogr. Wks., Japan, Spec. No. 2:3-9.

G

chemistry, phosphate

Gaarder, Karen Ringdal, 1938.
Phytoplankton studies from the Tromsø district, 1930-31. Tromsø Mus. Årshefter, Naturhist. Avd. 11, 55 (1):159 pp., 4 fold-in pls., 12 textfigs.

chemistry phosphate

Gaarder, T., and R. Spärck, 1932.
Hydrographisch-biochemische Untersuchungen in norwegische Austern-Pollen. Bergens Mus. Aarbok, Naturvidensk.-rekke, No. 1:5-144, 75 textfigs.

chemistry, phosphate (data)

Ganapati, P.N., E.C. La Fond, and P.V. Bhavanarayana, 1956.
On the vertical distribution of chemical constituents in the shelf waters off Waltair.
Proc. Indian Acad. Sci., 44:68-71.

Chemistry, phosphate
Ganapati, P.N., and D.V. Subba Rao, 1958.
Quantitative study of plankton off Lawson's Bay, Waltair.
Proc. Indian Acad. Sci., (B), 48(4):189-209.

Chemistry, phosphate
Ganapati, P.N., and D.V. Subba Rao, 1957.
On upwelling and productivity of the waters off Lawson's Bay, Waltair. Current Science, 26(11): 347-348.

Chemistry, phosphate
Ganapati, P.N., and D. Venkata Rama Sarma, 1958.
Hydrography in relation to the production of plankton off Waltair coast.
Andhra Univ. Mem. Oceanogr., 2:168-192.

Chemistry, phosphate (data)
Gennesseaux, Maurice, 1960
L'oxygène, les phosphates et les nitrates dissous dans les eaux du bassin sud-ouest de la Méditerranée occidentale (des Îles Baléares à Gibraltar). Trav. C.R.E.O. ns. 3(4): 5-12.
Also in: Trav. Lab. Geol. Sous-Marine 10 (1960)

phosphate
Gieskes, Joris M.T.M., and Klaus Grasshoff 1969.
A study of the variability in the hydrochemical factors in the Baltic Sea on the basis of two anchor stations, September 1967 and May 1968.
Kieler Meeresforsch. 25(1): 105-132

Chemistry, phosphate
Gillbricht, M., 1959.
Die Planktonverteilung in der Irminger See im Juni 1955.
Ber. Deutsch. Wiss. Komm. Meeresf., n.f., 15(3): 260-275.

Chemistry, phosphates
Gilmartin, M., 1962.
Annual cyclic changes in the physical oceanography of a British Columbia fjord.
J. Fish. Res. Bd., Canada, 19(5):921-974.

Chemistry, phosphate
Gilson, H. C., 1937
Chemical and Physical Investigations. The nitrogen cycle. John Murray Exped., 1933-34, Sci. Repts., 2(2):21-81, 16 text figs.

Chemistry, phosphate
Glowinska, A., 1967.
Polish observations in the southern Baltic 1965.
Annls. biol. Copenh. (1965)22:51-54.

Chemistry, phosphate (data)
Glowinska, A., 1964.
Polish observations in the southern Baltic, 1962.
Ann. Biol., Cons. Perm. Int. Expl. Mer, 1962, 19: 48-49.

Chemistry, phosphates
Glowinska, A., 1963.
Phosphates in the southern Baltic for the years 1947-1960. (In Polish; English and Russian summaries).
Prace Morsk. Inst. Ryback. w Gdyni 12A:7-21.

Chemistry, phosphate (data)
Glowinska, A., (1959) 1961
Polish observations in the southern Baltic in 1959.
Ann. Biol., Cons. Perm. Int. Expl. Mer, 16: 61-65.

Chemistry, phosphate (data)
Glowinska, A., 1954.
Hydrologic research in the southern Baltic in 1951. Repts. Sea Fish. Inst., Gdynia, No. 7:159-190, 27 textfigs.

Chemistry, phosphates
Glowinska, A., 1949
The Hydrography of the Gulf of Gdansk.
Ann. Biol. Int. Cons., 4:142-143, text fig.12.

Chemistry, phosphate
Goering, John J., and David Wallen, 1967.
The vertical distribution of phosphate and nitrite in the upper one-half meter in the southeast Pacific Ocean.
Deep-Sea Research, 14(1):29-33.

Chemistry, phosphate
Goldberg, E.D., T.J. Walker, and A. Whisenand, 1951.
Phosphate utilization by diatoms. Biol. Bull. 101(3):274-284, 5 textfigs.

Chemistry, phosphate
Gololobov, Ya. K., 1963.
Hydrochemical characteristics of the Aegean Sea during the autumn of 1959. (In Russian; English abstract).
Rez. Issled. Programme Mezhd. Geofiz. Goda, Okeanolog. Issled., Akad. Nauk, SSSR, No. 8: 90-96.

Chemistry, phosphate
Gomes, A.L., 1955.
Possibilidades da pesca maritima no Brasil.
Rev. Biol. Mar., Valparaiso, 6(1/3):6-20.

chemistry, phosphate
Goodman, J.R., J.H. Lincoln, T.G. Thompson and F.A. Zeusler, 1942.
Physical and chemical investigations: Bering Sea, Bering Strait, Chukchi Sea during the summers of 1937 and 1938.
Univ. Washington Publ. Oceanogr. 3(4): 105-169.

Chemistry, phosphates
*Gostan, Jacques, 1968.
Conditions hydrologiques observées pendant l'été entre la Riviera et la Corse.
Cah. océanogr., 20(1):37-66.

phosphates
Gostan, Jacques, et Paul Nival, 1967.
Relations entre la distribution des phosphates minéraux dissous et la répartition des pigments dans les eaux superficielles du Golfe de Gênes.
Cah. océanogr., 19(1):41-52.

Chemistry, phosphate (PO_4)
Graham, H. W. and N. Bronikovsky, 1944
The genus Ceratium in the Pacific and North Atlantic Oceans. Sci. Res. Cruise VII of the Carnegie, 1928-1929 ----- Biol. V (565):209 pp., 54 charts, 27 figs., 54 tables.

chemistry, phosphate
Graham, H.W. and E.G. Moberg 1944.
Chemical results of the last cruise of the Carnegie. Chemistry I. Scientific results of Cruise VII of the Carnegie during 1928-1929 under command of Captain J.P. Ault.
Carnegie Publ. Washington 562: 58pp.

chemistry, phosphate
Graham, M. 1938.
Phytoplankton and the herring. Pt. III Distribution of phosphate in 1934-36.
Min. Agric. Fish. Fish. Invest. (2) 16(3):

Chemistry, phosphates
Grall, Jean-René, et Guy Jacques, 1964.
Etude dynamique et variations saisonnières du plancton de la region de Roscoff. B. Phytoplancton.
Cahiers, Biol. Mar. Roscoff, 5:432-455.

Chemistry, phosphate
Gran, H.H., 1931.
On the conditions for the production of plankton in the sea. Rapp. Proc. Verb., Cons. Perm. Int. Expl. Mer, 75:37-46.

Chemistry, phosphates
Gran, H.H., and T. Braarud, 1935
A quantitative study of the phytoplankton in the Bay of Fundy and the Gulf of Maine (including observations on hydrography, chemistry, and turbidity).
J. Biol. Bd., Canada, 1(5):279-467, 69 text figs.

chemistry, phosphates
Great Britain, Marine Biological Association, 1949.
Phosphates
Annls Biol. Copenh. 5: 38.

Chemistry, phosphate (data)
Grøntved, J., 1952.
Investigations on the phytoplankton in the southern North Sea in May 1947. Medd. Komm. Danmarks Fisk.- og Havundersøgelser, Plankton ser., 5(5): 1-49, 1 pl., 21 tables, 24 textfigs.

Chemistry, phosphate
*Guillén, Oscar, 1967.
Anomalies in the waters off the Peruvian Coast during March and April 1965.
Stud. trop. Oceanogr., Miami, 5:452-465.

Chemistry, phosphate
Guillen G., Oscar, y Luis Alberto Flores P., 1968.
Informe preliminar del crucero 6702 del verano de 1967 (Cabo Blanco-Arica).
Informe Inst. Mar Peru 18: 17pp.

Chemistry, phosphate (data)
Guillén, Oscar, y Francisco Vásquez, 1966.
Informe preliminar del Crucero 6602 (Cabo Blanco - Arica).
Informe Inst. Mar Peru 12: 27pp.

Chemistry, phosphate
Gunther E.R. 1936.
A report on oceanographical investigations in the Peru Coastal Current.
Discovery Rept. 13: 107-276.

H

Chemistry, phosphate (data)
Halim, Youssef, 1960
Étude quantitative et qualitative du cycle écologique des dinoflagellés dans les eaux de Villefranche-sur-Mer. (1953-1955).
Ann. Inst. Océanogr., Monaco, 38:123-232.

Chemistry, phosphates
Halim, Y., 1960
Observations on the Nile bloom of phytoplankton in the Mediterranean. J. du Cons. 26(1): 57-67.

Chemistry, phosphate
Halim Youssef, and Selim A. Morcos 1965
Le rôle des particules en suspension dans l'eau du Nile en crue dans la répartition des sels nutritifs au large de ses embouchures.
Rapp. P.-v. Réun. Commn int. Explor. scient. Mer Médit. 18(3): 733-736.

Chemistry, phosphate

Hart, T. John, and Ronald I. Currie, 1960

The Benguela Current.
Discovery Repts., 31: 123-298.

Chemistry, phosphate

Harvey, H.W., 1948
The estimation of phosphate and of total phosphorus in sea waters. JMBA 27(2):337-359, 11 text figs.

Chemistry, phosphate

Harvey, H.W., 1947
Fertility of the Ocean. Proc. Linn. Soc., London 158 (2):82-85, 1 textfigs.

Chemistry, phosphate

Harvey, H.W., 1941.
On changes taking place in sea water during storage. J.M.B.A., n.s., 25:225-233, 1 textfig.

Chemistry, phosphate (data)

Hasegawa, Y., M. Yokoseki, E. Fukuhara, and K. Terai, 1952.
On the environmental conditions for the culture of laver in Usu-Bay, Iburi Prov., Hokkaido.
Bull. Hokkaido Regional Fish. Res. Lab. 6:1-24, 8 textfigs., 9 tables.

Chemistry phosphate

Hendey, N.I., 1951
Littoral diatoms of Chicester Harbour with special reference to fouling. J.Roy. Microscop. Soc. 71(1): 1-86, 18 pls.

Chemistry phosphate

Hermann, F., 1957.
Hydrographic conditions in the eastern part of the Labrador Sea and Davis Strait, 1955.
Ann. Biol., Cons. Perm. Int. Expl. Mer, 12:41-43, Figs., 62-70.

Chemistry, phosphate

Herman, F., 1956.
Hydrographic conditions in the e astern part of Labrador Sea and Davis Strait, 1954.
Ann. Biol., Cons. Perm. Int. Expl. Mer, 11:25-26, Figs. 29-38.

Chemistry phosphate

Hermann, F., 1954.
Hydrographic conditions in the e astern part of Labrador Sea and Davis Strait in 1953.
Ann. Biol., Cons. Perm. Int. Expl. Mer, 10:28-30 Textfigs. 30-38.

Chemistry phosphates

Hermann, F., 1951.
Section from the Faroe Islands to East Greenland.
Cons. Perm. Int. Expl. Mer, Ann. Biol. 7:18-19, Text-fig. 13.

Chemistry phosphate

Herrera, J., 1961.
Décroissance de la salinité et du phosphate dissous dans les eaux littoral de Castellon depuis 1956.
Rapp. Proc. Verb., Réunions, Comm. Int. Expl. Sci., Mer Méditerranée, Monaco, 16(3):669-672.

Chemistry phosphate

Herrara, Juan, and Ramon Margalef, 1961
Hidrografía y fitoplancton de las costas de Castellón de julio de 1958 a junio de 1959.
Inv. Pesq., Bacelona, 20:17-63.

Chemistry phosphate

*Hickel, Wolfgang, 1967.
Untersuchungen über die Phytoplanktonblüte in der westlichen Ostsee.
Helgoländer wiss. Meeresunters., 16(½):1-66.

Chemistry Phosphate

Hollan, Eckard, 1970.
Eine physikalische Analyse kleinräumiger Änderungen chemischer Parameter in den tiefen Wasserschichten der Gotlandsee.
Kieler Meeresforsch. 25(2): 255-267

Chemistry phosphate

Hulburt, Edward M., 1966.
The distribution of phytoplankton, and its relationship to hydrography, between southern New England and Venezuela.
J. Mar. Res., 24(1):67-81.

Chemistry phosphate

Hung, Tsu-Chang, and Ching-Wei Lee, 1967.
Research on chemical oceanography in the Kuroshio around Taiwan Island.
Bull. Inst. Chem., Acad. Sinica, Taiwan, 14:80-102.

Chemistry phosphate (data)

Husby, David M., and Gary L. Hufford 1971.
Oceanographic investigation of the northern Bering Sea and Bering Strait 8-21 June 1969.
Oceanogr. Rept. U.S. Cst Gd 42 (CG373-42): 55pp.

I

Chemistry, phosphates

Ichie, Takashi, 1951
[On the hydrography of the Kii-Suido (1951).]
Bull. Kobe Mar. Obs., No. 164: 253-278(top of page); 35-60(bottom of page).

Chemistry, phosphate

Ichie, T., 1949.
Report on the oceanographical observations on board R.M.S. "Syunpu Maru" in the Akashi Seto and the Yura Seto in March 1949. Papers and Rept. Ocean., Kobe Mar. Obs., Ocean. Dept., No. 4: 10 pp., numerous figs. and tables. (atlas-sized pages - mimeographed).

Chemistry, phosphate

Ichie, T., K. Tanioka, and T. Kawamoto, 1950.
Reports of the oceanographical observations on board the R.M.S. "Yushio Maru" off Shionomisaki (Aug. 1949). Papers and Repts., Ocean., Kobe Mar. Obs., Ocean. Dept., No. 5:15 pp., 33 figs. (Odd atlas-sized pages -mimeographed).

Chemistry, phosphate (data)

Ignatiades, Lydia, and Theano Becacos-Kontos, 1969.
Nutrient investigations in lower Saronicos Bay, Aegean Sea.
Vie Milieu (B) 20(1): 51-61

Chemistry, phosphates

Ilie, K. 1968.
Influence des eaux résiduaires sur la quantité des phosphates libres et du phosphore total dans le Golfe de Valdibora.
Revue int. Océanogr. Med. 12: 63-73.

Chemistry phosphate (data)

Ilyina, N.L., 1961
[Some characteristics of hydrochemistry of the Northern Part of the Norwegian and the Greenland seas (summer 1958)] (English abstract).
Mezhd. Kom. Mezhd. Geofiz. Goda, Presidiume Akad. Nauk, SSSR, Okeanol. Issled., (3):151-161.

Chemistry phosphate

Imai, Y., 1961
Oceanographical studies on the behaviour of chemical elements. V. On the influence of land water and sediment upon the quantity of inorganic phosphate contained in the sea water of Urado Bay. (In Japanese; English abstract).
J. Oceanogr. Soc., Japan, 17(3):157-160.

chemistry, phosphate

Ingham, Merton C., Julien R. Goulet, Jr. and John T. Brucks 1966.
Oceanic conditions in the northwestern Gulf of Guinea, Geronimo Cruise 4, 5 August to 13 October 1964.
Data Rept. U.S. Dept. Interior, Bur. Comm. Fish. 26: 1 film (microfiche).

Chemistry phosphate

Ivanenkov, V.N., and F.A. Gebin, 1960.
Water masses and hydrochemistry in the western and southern parts of the Indian Ocean. Physics of the Sea. Hydrology. (In Russian).
Trudy Morsk. Gidrofiz. Inst., 22:33-115.

Translation: Scripta Technica, Inc., for Amer. Geophys. Union, 27-99.

J

Chemistry phosphate (data)

Jacobs, Stanley S., Peter M. Bruchhausen and Edward B. Bauer, 1970
Cruises 32-36, 1968, hydrographic stations, bottom photographs, current measurements.
Eltanin Repts Lamont-Doherty Geol. Obs., Nat. Sci. Found., U.S. Antarctic Res. Program, 460 pp. (multilithed)

Chemistry phosphates

Jacques, Guy, 1964.
Etude dynamique et variations saisonnières du plancton de la région de Roscoff. A. Météorologie et hydrologie.
Cahiers, Biol. Mar., Roscoff, 5:423-431.

Chemistry phosphate

Japan, Central Meteorological Observatory, 1949
Report on sea and weather observation on Antarctic Whaling Ground (1947-48). Ocean. Mag., Japan, 1(1):49-88, 17 text figs.

Chemistry phosphate(data)

Japan, Fisheries Agency, Research Division, 1956.
Radiological survey of western area of the dangerous zone around the Bikini-Eniwetok Atolls, investigated by the "Shunkotsu maru" in 1956, Part 1:143 pp.

Japan, Hakodate Marine Observatory

chemistry, phosphate

Japan, Hakodate Marine Observatory 1970.
Report of the oceanographic observations in the sea east of Honshu and Hokkaido, and in the Tsugaru Straits from April to May 1966. (In Japanese)
Bull. Hakodate mar. Obs. 15: 11-16.

chemistry, phosphate

Japan, Hakodate Marine Observatory 1970.
Report of the oceanographic observations in the sea east of Honshu and Hokkaido from February to March 1966. (In Japanese).
Bull. Hakodate mar. Obs. 15: 3-10.

chemistry, phosphate

Japan, Hakodate Marine Observatory, 1969.
Report of the oceanographic observations in the sea south of Hokkaido and in the Sea of Okhotsk from October to November 1965 (In Japanese)
Bull. Hakodate mar. Obs. 14: 16-21

Chemistry, phosphate
Japan, Hakodate Marine Observatory, 1969.
Report of the oceanographic observations in the sea east of Hokkaido and Tohoku District from February to March 1965. (In Japanese)
Bull. Hakodate mar. Obs. 14:3-9

Chemistry phosphate
Japan, Hakodate Marine Observatory, 1967.
Report of the oceanographic observations in the sea east of Hokkaido and in the southern part of the Okhotsk Sea from May to June, 1964. (In Japanese).
Bull. Hakodate mar. Obs. 13:10-17.

phosphate
Japan, Hakodate Marine Observatory, 1964
Report of the oceanographic observations in the sea east of the Tohoku from February to March 1962. Report of the oceanographic observations in the Tsugaru Straits in April 1962.---in May 1962.---in June 1962. from August to September, 1962. Report of the oceanographic observations in the sea south of Hokkaido in June 1962. Report of the oceanographic observations in the sea west of Tsugaru Straits, the Tsugaru Straits and South of Hokkaido in May, 1962. (In Japanese).
Bull. Hakodate Mar. Obs.,
No. 11: misnumbered pp.

chemistry
phosphate
Japan, Hakodate Marine Observatory, Oceanographical Section, 1962
Report of the oceanographic observations in the sea east of Tohoku District from February to March, 1961. (In Japanese)
Res. Mar. Meteorol., and Oceanogr. Obs., Jan.-June, 1961, No. 29:3-8.

chemistry, phosphate
Japan, Hakodate Marine Observatory, 1961.
Report of the oceanographic observations in the sea east of Tohoku District from February to March 1961:3-8.
Report of the oceanographic observations in the sea east of Tohoku District in May 1961:9-12.
Report of the oceanographic observations in the Japan Sea in June 1961:59-79.
Report of the oceanographic observations in the western part of Wakasa Bay from January to May 1961:80.
Report of the oceanographic observations in the sea south of Hokkaido in July 1961:3-4.
In Japanese OVER

Chemistry phosphate
Japan, Hakodate Marine Observatory, 1961.
[Report of the oceanographic observations in the sea east of Tohoku District and in the Okhotsk Sea from August to September 1959.]
Bull. Hakodate Mar. Obs., (8):6-14.

chemistry
phosphate
Japan, Hakodate Marine Observatory, 1961
[Report of the oceanographic observations in the sea east of Tohoku District from February to March 1959.]
Bull. Hakodate Mar. Obs., (8):3-7.

chemistry, phosphate
Japan, Hakodate Marine Observatory, 1957.
[Report of the oceanographic observations in the sea east of Tohoku District from May to June 1956]
Bull. Hakodate Mar. Obs., No. 4:113-119.
9-15.

Chemistry, phosphate
Japan, Hakodate Marine Observatory, 1956.
[Report of the oceanograhical observations east off Tohoku District from August to September 1955.]
Bull. Hakodate Mar. Obs., (3):13-21.

Japan, Kobe Marine Observatory

Chemistry, phosphate
Japan, Kobe Marine Observatory 1967.
Report of the oceanographic observations in the sea south of Honshu from October to November 1963. (In Japanese)
Bull. Kobe Mar. Obs. 178:41-49.

Chemistry, phosphate
Japan, Kobe Marine Observatory, 1967.
Report of the oceanographic observations in the sea south of Honshu from July to August 1963. (In Japanese).
Bull. Kobe Mar. Obs. No. 178:31-40

chemistry phosphate
Japan, Kobe Marine Observatory 1967.
Report of the oceanographic observations in the sea south of Honshu and in the Seto-Naikai from May to June 1964. (In Japanese).
Bull. Kobe Mar. Obs. 178:37

chemistry, phosphate
Japan, Kobe Marine Observatory, 1967.
Report of the oceanographic observations in the sea south of Honshu from February to March 1964. (In Japanese)
Bull. Kobe Mar. Obs. No. 178:27

chemistry
phosphate
Japan, Kobe Marine Observatory, Oceanographical Section, 1964.
Report of the oceanographic observations in the Kuroshio and region east of Kyushu from October to November 1962. (In Japanese)
Res. Mar. Meteorol. and Oceanogr., Japan. Meteorol. Agency, 32: 41-50.

Also in: Bull. Kobe Mar. Obs., 175. 1965.

Chemistry, phosphate
Japan, Kobe Marine Observatory, Oceanographical Section, 1964.
Report of the oceanographic observations in the sea south of Honshu from February to March, 1963. Res. Mar. Meteorol. and Oceanogr., Japan Meteorol. Agency, 33: 27-32.

Also in: Bull. Kobe Mar. Obs., 175. 1965.

Chemistry
phosphate
Japan, Kobe Marine Observatory, Oceanographical Section, 1964.
Report of the oceanographic observations in the sea south of Honshu from July to August, 1962. Res. Mar. Meteorol. and Oceanogr., Japan. Meteorol. Agency, 32: 32-40. (In Japanese).

Also in: Bull. Kobe Mar. Obs. 175.

chemistry
phosphate
Japan, Kobe Marine Observatory, Oceanographical Section, 1962
Report of the oceanographic observations in the sea south of Honshu in May, 1960. (In Japanese).
Bull. Kobe Mar. Obs., No. 169(12):27-33.

Chemistry
phosphate
Japan, Kobe Marine Observatory, Oceanographical Section, 1962
Report on the oceanographic observations in the sea south of Honshu from July to August, 1959.
Bull. Kobe Mar. Obs., No. 169(11):37-43.
(In Japanese)

Chemistry phosphate
Japan, Kobe Marine Observatory, 1962
Report of the oceanographic observations in the cold water region south of Enshu Nada from October to November, 1960. (In Japanese).
Res. Mar. Meteorol. and Oceanogr., July-Dec., 1960, Japan Meteorol. Agency, No. 28: 43-51.

Chemistry
phosphate
Japan, Kobe Marine Observatory, 1961
[Report of the oceanographic observations in the sea south of Honshu in March 1958.]
Bull. Kobe Mar. Obs., No. 167(21-22):30-36.

--from May to June, 1958(21-22):37-42
--from July to September, 1958(23-24):34-40
--from October to December, 1958(23-24):41-47
--from February to March, 1959(25-26):33-47.

Chemistry, phosphate
Japan, Kobe Marine Observatory, 1956?
[Report on the oceanographical observations south off Honshu in 1954] J. Ocean., Kobe (2)7(2):46-69

Chemistry, phosphate
Japan, Kobe Marine Observatory, 1956.
[Report of the oceanographical observations in the sea south off Honshu in March 1955.]
J. Ocean., Kobe (2)7(2):17-24.

Chemistry, phosphate
Japan, Kobe Marine Observatory, 1956.
[Report of the oceanographical observations in the sea south off Honshu in May 1955.]
J. Ocean., Kobe, (2)7(2):17-25.

pagination of previous article repeated.

Chemistry, phosphate
Japan, Kobe Marine Observatory, 1956.
[Report of the oceanographic observations in the sea south of Honshu in August 1955.]
J. Ocean., Kobe (2)7(3):23-32.

Chemistry, phosphate
Japan, Kobe Marine Observatory, 1955.
[The outline of oceanographical observation on the Southern Sea of Japan on board R.M.S. "Syunpu-maru" (Jan. 1955).]
J. Ocean., Kobe Mar. Obs., (2)6(1):1-19, 10 figs.

Chemistry phosphate
Japan, Kobe Marine Observatory, 1954.
[The outline of the oceanographical observations in the southern area of Honshu on board the R.M.S. "Syunpu Maru".]
Aug. -Sept. 1954. J. Ocean. (2)5(9):1-44.
Oct. 1954. - J. Ocean. (2)5(10):1-44.

Chemistry phosphate
Japan, Kobe Marine Observatory, 1954.
[The outline of the oceanographical observations off Shionomisaki on board the R.M.S. "Syunpu-maru" (May 1954).] J. Ocean. (2)5(5):1-11, 14 figs.

Chemistry phosphate
Japan, Kobe Marine Observatory, 1954.
[The results of the regular monthly oceanographical observations on board the R.M.S. "Syunpu-maru" in the Kii Suido and Osaka Wan.] J. Ocean. (2)5(1):1-19, 16 figs.

Chemistry phosphate
Japan, Kobe Marine Observatory, 1953?
[The results of the regular monthly oceanographical observations on board the R.M.S. "Syunpu-maru" in the Kii Suido and Osaka Wan (Oct. 1953)]
J. Ocean., Kobe, Mar. Obs., 2nd ser., 4(11):1-21

Reports with similar titles in all other numbers of Vol. 4, this journal. These are not indexed.

phosphate
Japan, Kobe Marine Observatory, 1953.
[The results of the regular monthly oceanographical observations on board the R.M.S. "Syunpu-maru" in the Kii Suido and Osaka Wan.]
Jan. 1953 - J. Ocean. (2)4(1):1-10, 9 figs.
Feb. 1953 - J. Ocean. (2)4(2):1-9, 8 figs.
Mar. 1953 - J. Ocean. (2)4(3):1-10, 9 figs.
Apr. 1953 - J. Ocean. (2)4(5):1-9, 9 figs.
May 1953 - J. Ocean. (2)4(6):1-12, 9 figs.
June 15-22, 1953 - J. Ocean. (2)4(6):13-22, 8 figs.
July 1953 - J. Ocean. (2)4(8):1-25, 29 figs.
Aug. 1953 - J. Ocean. (2)4(9):1-30, 27 figs.
Sept. 1953 - J. Ocean. (2)4(10):1-13, 9 figs.
Oct. 1953 - J. Ocean. (2)4(11):1-21, 15 figs.
Nov. 11-18, 1953 - J. Ocean. (2)4(12):1-15, 12 figs.

Chemistry phosphate
Japan, Kobe Marine Observatory, 1953.
[On the chemical conditions in the northern part of Osaka Bay after heavy rain (July 1952).]
J. Ocean., (2)4(4):1-9, 5 figs.

Japan, Maizuru Marine Observatory

Chemistry, phosphate

Japan, Maizuru Marine Observatory, 1968.
Report of the oceanographical observations in the Japan Sea from August to September, 1964. (In Japanese).
Bull. Maizuru mar. Obs., 10:74-86.

Chemistry phosphate

Japan, Maizuru Marine Observatory, 1967.
Report of the oceanographic observations in the Japan Sea from May to June, 1964.
Bull. Maizuru mar. Obs., 10:65-76.
In Japanese)

Chemistry phosphate

Japan, Maizuru Marine Observatory, 1965.
Report of the oceanographic observations in the central part of the Japan Sea from February to March, 1962.---in the Japan Sea from June to July, 1962.---in the western part of Wakasa Bay from January to April, 1962.---in the central part of the Japan Sea from September to October, 1962.---in the western part of Wakasa Bay from May to November, 1962.---in the central part of the Japan Sea in March, 1963.---in the Japan Sea in June, 1963.---In Wakasa Bay in July, 1963.---in the central part of the Japan Sea in October, 1963. (In Japanese).
Bull. Maizuru Mar. Obs., No.9:67-73;74-88;89-95;71-80;81-87;59-65;66-77;80-84;85-91.

Chemistry phosphate

Japan, Maizuru Marine Observatory, 1963
Report of the oceanographic observations in the central part of the Japan Sea in October, 1961. (In Japanese).
Bull. Maizuru Mar. Obs., No. 8:78-88.

Chemistry phosphate

Japan, Maizuru Marine Observatory, 1963
Report of the oceanographic observations in the Japan Sea in June 1961. (In Japanese).
Bull. Maizuru Mar. Obs., No. 8:59-79.

Chemistry phosphate

Japan, Maizuru Marine Observatory, 1963
Report of the oceanographic observations in the central part of the Japan Sea in September, 1960. (In Japanese).
Bull. Maizuru Mar. Obs., No. 8:60-68.

Chemistry phosphate

Japan, Maizuru Marine Observatory, 1963
Report of the oceanographic observations in the Japan Sea from May to June, 1960. (In Japanese).
Bull. Maizuru Mar. Obs., No. 8:56-67.

Chemistry phosphate

Japan, Maizuru Marine Observatory, 1963
Report of the oceanographic observations in the Wakasa Bay from July to August, 1960. (In Japanese).
Bull. Maizuru Mar. Obs., No. 8:69-75.

Chemistry phosphate

Japan, Maizuru Marine Observatory, 1963
Report of the oceanographic observations in the western part of Wakasa Bay from January to May, 1961. (In Japanese).
Bull. Maizuru Mar. Obs., No. 8:80-90.

Chemistry phosphate

Japan, Maizuru Marine Observatory, 1963
Report of the oceanographic observations in the western part of Wakasa-Bay from January to June 1960. (In Japanese).
Bull. Maizuru Mar. Obs., No. 8:68-79, 59.

Chemistry phosphate

Japan, Maizuru Marine Observatory, 1963
Report of the oceanographic observations in the western part of Wakasa Bay from June to December, 1960. (In Japanese).
Bull. Maizuru Mar. Obs., No. 8:76-83.

Chemistry phosphate

Japan, Maizuru Marine Observatory and Hakodate Marine Observatory, Oceanographical Sections, 1962
Report of the oceanographic observations in the Japan Sea in June, 1961. (In Japanese).
Res. Mar. Meteorol. and Oceanogr. Obs., Jan.-June, 1961. No. 29:59-79.

Chemistry phosphate

Japan, Maizuru Marine Observatory, 1962
Report of the oceanographic observations in the central part of the Japan Sea in September, 1960. (In Japanese).
Res. Mar. Meteorol. and Oceanogr., July-Dec., 1960, Japan Meteorol. Agency, No. 28: 60-68.

Chemistry phosphate

Japan, Maizuru Marine Observatory, 1962
Report of the oceanographic observations in the Wasaka Bay from July to August, 1960. (In Japanese).
Res. Mar. Meteorol. and Oceanogr., July-Dec., 1960, Japan Meteorol. Agency, No. 28:69-75.

Chemistry phosphate

Japan, Maizuru Marine Observatory, Oceanographical Section, 1962
Report of the oceanographic observations in the western part of Wakasa Bay from June to May, 1961. (In Japanese).
Res. Mar. Meteorol. and Oceanogr. Obs., Jan.-June, 1961, No. 29:80-90.

Chemistry phosphate

Japan, Maizuru Marine Observatory, 1962
Report of the oceanographic observations in the western part of Wakasa Bay from June to December 1960. (In Japanese).
Res. Mar. Meteorol. and Oceanogr., July-Dec., 1960, Japan Meteorol. Agency, No. 28:76-83.

Chemistry phosphate

Japan, Maizuru Marine Observatory, Oceanographical Section, 1961
[Report of the oceanographic observations in the Japan Sea from June to July, 1959.]
Bull. Maizuru Mar. Obs., No. 7:57-64.

Chemistry phosphate

Japan, Maizuru Marine Observatory, Oceanographical Section, 1961
[Report of the oceanographic observations in the sea north of Kyoga-misaki from August to December 1958.]
Bull. Maizuru Mar. Obs., No. 7:60-63.

Chemistry phosphate

Japan, Maizuru Marine Observatory, Oceanographical Observatory, 1961
[Report of the Oceanographic observations in the western part of Wakasa Bay from August to December, 1959.]
Bull. Maizuru Mar. Obs., No. 7:68-74.

Chemistry phosphate

Japan, Maizuru Marine Observatory, Oceanographical Section, 1961
[Report of the oceanographic observations in the Wakasa Bay from July to August, 1959]
Bull. Maizuru Mar. Obs., No. 7:62-67.

Chemistry phosphate

Japan, Maizuru Marine Observatory, Oceanographical Section, 1961
[Report of the oceanographic observations off Kyoga-misaki and in the western part of Wakasa Bay from February to June, 1959.]
Bull. Maizuru Mar. Obs., No. 7:65-72, 61.

Chemistry phosphate

Japan, Maizuru Marine Observatory, Oceanographical Section, 1961
[Report of the oceanographic observations off Kyoga-misaki in the Japan Sea in April and May, 1958.]
Bull. Maizuru Mar. Obs., No. 7:55-60.

Chemistry phosphate

Japan, Maizuru Marine Observatory, 1958
[Report of the oceanographic observations in the sea north of Sanin and Hokuriju districts in August 1957.]
Bull. Maizuru Mar. Obs., No. 6: 245-253. [107-115.]

Chemistry phosphate

Japan, Maizuru Marine Observatory, 1958
[Report of the oceanographic observations north off Kyoga-misaki and off the east part of the Sanin district in April to June, 1957.]
Bull. Maizuru Mar. Obs., No. 6: 234-243. [95-105.]

Chemistry phosphate

Japan, Maizuru Marine Observatory, 1958
[Report of the oceanographic observations off Kyoga-misaki in the Japan Sea in January and March, 1957.]
Bull. Maizuru Mar. Obs., No. 6: 129-137. [45-53.]

Chemistry phosphate

Japan, Meteorological Agency, 1962
Report of the Oceanographic Observations in the sea south and west of Japan from October to November, 1960. (In Japanese).
Res. Mar. Meteorol. and Oceanogr., July-Dec., 1960, Japan Meteorol. Agency, No. 28:30-35.

Chemistry, phosphates

Japan, Japan Meteorological Agency 1959.
The results of marine meteorological and oceanographical observations, January-June 1958, No. 23: 240 pp.

Japan, Nagasaki Marine Observatory

Chemistry phosphate

Japan, Nagasaki Marine Observatory, Oceanographical Section, 1962
Report of the oceanographic observations in the sea west of Japan from April to May, 1961. (In Japanese).
Res. Mar. Meteorol. and Oceanogr. Obs., Jan.-June, 1961, No. 29: 45-53.

Chemistry phosphate

Japan, Nagasaki Marine Observatory, Oceanographical Section, 1962
Report of the oceanographic observations in the sea west of Japan from February to March 1961. (In Japanese).
Res. Mar. Meteorol. and Oceanogr. Obs., Jan.-June, 1961, No. 29:36-44.

Chemistry phosphate

Japan, Nagasaki Marine Observatory, 1962
Report of the oceanographic observations in the sea west of Japan from October to November, 1960. (In Japanese).
Res. Mar. Meteorol. and Oceanogr., July-Dec., 1960, Japan Meteorol. Agency, No. 28:52-59.

Chemistry, Phosphate

Japan, Nagasaki Marine Observatory, Oceanographical Section 1960.
Report of the oceanographic observation in the sea west of Japan from January to February, 1960. Report of the Oceanographic observation in the sea north-west of Kyushu from April to May, 1960.
Results Mar. Meteorol. & Oceanogr., J.M.A., 27:42-50; 51-67.

Also in:
Oceanogr. & Meteorol., Nagasaki Mar. Obs., (1961) 11 (202).

chemistry
phosphate

Japan, Nagasaki Marine Observatory, Oceanographical Section, 1960
[Report of the oceanographic observations in the sea west of Japan from June to July, 1959.]
Res. Mar. Meteorol. & Oceanogr., J.M.A., 26: 51-57.
Also in:
Oceanogr. & Meteorol., Nagasaki Mar. Obs., (1961), 11(200).

chemistry phosphate

Japan, Nagasaki Marine Observatory, 1959.
Report of the oceanographic observations in the sea west of Japan from June to July 1958 and in the sea north west of Kyushu in October 1958.
Results Mar. Meteorol. & Oceanogr., J.M.A., 24: 47-60.

Also in:
Oceanogr. & Meteorol., Nagasaki Mar. Obs., 10(195). 1960.

chemistry phosphate

Japan, Nagasaki Marine Observatory, 1971.
Report of the oceanographic observations in the sea southeast of Kyushu in September 1965. (In Japanese)
Oceanogr. Met. Nagasaki mar. Obs. 18: 75-77

chemistry phosphate

Japan, Nagasaki Marine Observatory 1971
Report of the oceanographic observations in the sea south of Kyushu in March 1966.
Oceanogr. Met. Nagasaki Mar. Obs. 18: 50-52
(In Japanese)

chemistry
phosphate

Japan, Nagasaki Marine Observatory 1971.
Report of the oceanographic observations in the sea west of Japan from January to February 1966. (In Japanese)
Oceanogr. Met. Nagasaki Mar. Obs. 18: 34-49.

chemistry, phosphate

Japan, Nagasaki Marine Observatory 1971.
Report of the oceanographic observations in the sea west of Japan from June to August, 1967. (In Japanese)
Oceanogr. Met. Nagasaki mar. Obs. 15: 1-12.

chemistry, phosphate

Japan, Nagasaki Marine Observatory 1971.
Report of the oceanographic observations in the sea west of Japan from January to February 1967. (In Japanese)
Oceanogr. Met. Nagasaki mar. Obs. 15: 1-10

chemistry, phosphate

Japan, National Committee for the International Geophysical Year 1960
The results of the Japanese oceanographic project for the International Geophysical Year 1957/58: 198 pp.

Japan, Science Council

chemistry
phosphate

Japan, Science Council, National Committee for IIOE, 1966.
General report of the participation of Japan in the International Indian Ocean Expedition.
Rec. Oceanogr. Wks., Japan, n.s. 8(2): 133 pp.

chemistry phosphate

Jayaraman, R., 1954.
Seasonal variations in salinity, dissolved oxygen and nutrient salts in the inshore waters of the Gulf of Manner and Palk Bay near Mandapam (S. India). Indian J. Fish. 1:345-364.

Chemistry phosphates

Jayaraman, R., 1951.
Observations on the chemistry of the waters of the Bay of Bengal off Madras City during 1948-1949. Proc. Indian Acad. Sci., B, 33(2):92-99, figs.

Chemistry
phosphate

Jayaraman, R., R. Viswanathan and S.S. Gogate, 1961.
Characteristics of sea water near the light house, Bombay.
J. Mar. Biol. Assoc., India, 3(1/2):1-5.

Chemistry
phosphate

Jeffries, Harry P., 1962
The atypical phosphate cycle of estuaries in relation to benthic metabolism.
The Environmental Chemistry of Marine Sediment Proc. Symp., Univ. R.I., Jan. 13, 1962, Occ. Papers, Narragansett Mar. Lab., No. 1:58-67.

Chemistry Phosphate

Jeffries, Harry P., 1962.
Environmental characteristics of Raritan Bay, a polluted estuary.
Limnol. and Oceanogr., 7(1):21-31.

Chemistry phosphate

Jerlov, N.G., 1956.
The Equatorial currents in the Pacific Ocean.
Repts. Swedish Deep-Sea Exped., 3(Phys. Chem. 6): 129-154.

Chemistry phosphate

Johnston, H.W., and R.B. Miller, 1959.
The solubilization of "insoluble" phosphates: Pt. 4. The reaction between organic acids and tricalcium.
New Zealand J. Sci., 2(1):109-120.

Chemistry phosphate

Johnston, R., 1956.
Chemical observations, northern North Sea.
Ann. Biol., Cons. Perm. Int. Expl. Mer, 11:41-43, Figs. 7-10.

Chemistry phosphate

Johnston, R., 1954.
Chemical observations.
Ann. Biol., Cons. Perm. Int. Expl. Mer, 10:84-86, Textfigs. 6-9.

chemistry phosphates

Johnston, R., 1953.
Northern North Sea and approaches. Chemical observations.
Ann. Biol., Cons. Perm. Int. Expl. Mer, 9:99-103, Figs. 3-6.

Chemistry phosphate

Johnston, R., and R.B. Burns, 1958(1960).
Chemical observations in the North Sea.
Cons. Perm. Int. Expl. Mer, Ann. Biol., 15:38-40.

Chemistry
phosphate

Johnston, R., and R.E. Craig, 1954.
Hydrographic conditions off the northwest of Scotland.
Ann. Biol., Cons. Perm. Int. Expl. Mer, 10:89-90, Fig. 12.

chemistry
phosphate

Jones, P.G.W., 1971.
The southern Benguela Current region in February 1966: Part I. Chemical observations with particular reference to upwelling. Deep-Sea Res., 18(2): 193-208.

Chemistry phosphate

Jones, P. G. W., 1966.
The nitrate and phosphate content of the eastern equatorial Atlantic Ocean.
Bull. I.F.A.N.(A), 28(2):444-449.

Chemistry, phosphate

Jones, P.G.W., and S.M. Haq, 1963.
The distribution of Phaeocystis in the Irish Sea.
J. du Conseil, 28(1):8-20.

K

Chemistry
phosphate

Kabanova, Yu. G., 1961
[Primary production and nutrients in the Indian Ocean.] (In Russian; English abstract).
Okeanolog. Issled., Mezhd. Komitet Proved. Mezhd. Geofiz. Goda, Prezidiume, Akad. Nauk, SSSR, (4):72-75.

chemistry, phosphate

Kaleis, M.V., N.B. Alexandrovskaya and E.A. Yula 1968.
Some peculiarities in the oceanographical regime of the Baltic in 1966.
Annls biol. Copenh. 1966, 23: 78-81

Chemistry phosphate (data)

Kaleis, M.V., and N.B. Aleksandrovskaye, 1965.
Hydrological régime of the Baltic Sea in 1963.
Ann. Biol., Cons. Perm. Int. Expl. Mer, 1963. 20:70-72.

Chemistry phosphate

Kalle, K., 1957.
Chemische Untersuchungen in der Irminger See im Juni 1955. Ber. Deutschen Wiss. Komm. f. Meeresf. n.f., 14(4):313-328.

Chemistry phosphate (data)

Kalle, K., 1956.
Chemisch-hydrographische Untersuchungen in der inneren Deutschen Bucht. Deutsche Hydrogr. Zeits. 9(2):55-65.

Chemistry phosphate

Kalle, K., 1943.
Die grosse Wasserumschichtung im Gotland-tief vom Jahre 1933-1934. Ann. Hydr., usw., 71:142-146, 6 textfigs.

Chemistry phosphate

Kalle, K., 1939
V. Die chemischen Arbeiten auf der "Meteor" Fahrt Januar bis Mai 1938. Bericht über die zweite Teilfahrt des Deutschen Nordatlantischen Expedition des Forschungs-und Vermessungs-schiffes "Meteor", Januar bis Juli, 1938. Ann. Hydro. u. Mar. Meteorol. 1939: 23-30, 6 text figs.

Chemistry
phosphate

Kandler, R., 1953.
Hydrographische Untersuchungen zum Abwasserproblem in den Buchten und Förden der Ostseeküste Schleswig-Holstein. Kieler Meeresf. 9(2):176-200, Pls. 9-14(18 figs.).

Chemistry, phosphate

Kato, Takeo, 1969.
An influence of the upper water of the Kuroshio upon the bottom water mass over the Continental Shelf in the East China Sea. (In Japanese; English abstract). Bull. Jap. Soc. fish. Oceanogr. Spec. No. (Prof Uda Commem. Pap.): 129-134.

Chemistry, phosphate

Kato, Kenji, 1966.
Geochemical studies on the mangrove region of Cananéia, Brazil. 1. Tidal variations of water properties.
Bolm Inst. Oceanogr., S. Paulo, 15(1):13-20.

chemistry, phosphate
Kato, K., 1966.
Chemical investigations on the hydrographic system of Cananéia lagoon.
Bolm Inst. Oceanogr., S. Paulo, 15(1):1-12.

chemistry Phosphate
Kato, Kenji, 1961.
Oceanographical studies in the Gulf of Cariaco.
1. Chemical and hydrographical observations in January, 1961.
Bol. Inst. Oceanograf., Univ. Oriente, Cumana, Venezuela, 1(1):49-72.

Chemistry phosphate(data)
Kato, Kenji, 1961
Some aspects on biochemical characteristics of sea water and sediments in Mochima Bay, Venezuela.
Bol. Inst. Oceanogr., Univ. de Oriente, Venezuela 1(2):343-358.

chemistry Phosphate
Kawai, Hideo, and Minoru Sasaki, 1961.
An example of the short-period fluctuation of the oceanographic condition in the vicinity of the Kuroshio front.
Bull. Tohoku Regional Fish. Res. Lab., No. 19:119-134.

chemistry, phosphate
Kay, H., 1954.
Untersuchungen zur Menge und Verteilung der organischen Substanz im Meerwasser.
Kieler Meeresf. 10(2):202-214, Figs. 18-25.

chemistry, phosphate
Ketchum, B.H., 1947
The biochemical Relations between Marine Organisms and their Environment. Ecol. Monogr., 17:309-315, 5 textfigs.

chemistry, phosphate
Khimitsa, V.A., 1971.
Phosphate distribution off the western coasts of Hindustan. (In Russian; English abstract).
Okeanologiia 11(4): 748-751.

chemistry, phosphate
Khimitsa, V.A., 1968.
Some characteristic features of phosphate distribution in the Gulf of Aden. (In Russian).
Trudy, Vses. Nauchno-Issled. Inst. Morks. Ribn. Okeanogr (VNIRO) 64, Trudy Azovo-Chernomorsk. Nauchno-Issled. Inst. Morsk. Ribn. Khoz. Okeanogr. (AscherNIRO), 28: 181-195.

chemistry, phosphate
Khovansky, Y.A., 1962.
Some features of the shelf-water dynamics off the South West Africa coast. (In Russian).
Baltisk. Nauchno-Issled. Inst. Morsk. Ribn. Khoz. i Okeanogr. (BALTIRO), Trudy, 9:57-69.

chemistry, phosphate
Khromov, N.S., 1965.
Some data on plankton in the Dakar-Freetown area. (Based on the material of the 10th and 11th cruise of the Mikhail Lomonosov, 1961-1962). (In Russian).
Trudy vses. nauchno-issled. Inst. morsk. ryb. Khoz. Okeanogr. (VNIRO), 57:393-404.

chemistry, phosphate
Khromov, N.S., 1965.
Some data on plankton in the Dakar-Freetown area. Investigations in line with the programme of the International Geophysical Year, 2. (In Russian).
Trudy, Vses. Nauchno-Issled. Inst. Morsk. Ribn. Choz. i Okeanogr. (VNIRO), 57:393-404.

chemistry, phosphate
Kijowski, S., 1937.
Nieco danych o skladzii chemicznym wod Zatoki Gdanskiej. (Quelques données sur la composition chimique des eaux du Golfe de Dantzig). Bull. Sta. Mar. Hel 1(1):33-41, 1 fig.

chemistry phosphate
King, Joseph E., 1958
Variation in abundance of zooplankton and forage organisms in the Central Pacific in respect to equatorial upwelling.
Proc. Ninth Pacific Sci. Congr., Pacific Sci. Assoc., 1957, 16(Oceanogr.):98-107.

Chemistry phosphate (data)
King, J.E., T.S. Austin and M.S. Doty, 1957.
Preliminary report on Expedition EASTROPIC.
USFWS Spec. Sci. Rept., Fish., 201:155 pp.

chemistry Phosphate
Kinzer, J., 1969.
Quantitative distribution of zooplankton in surface waters of the Gulf of Guinea during August and September 1963.
Actes Symp. Oceanogr. Ressources halieut. Atlant. trop., Abidjan, 20-28 Oct. 1966. UNESCO 231-240.

Chemistry Phosphate
Kitamura, H., 1962.
(Marine chemical studies on the phosphorus distributed in the North Pacific Ocean.)
Bull. Kobe Mar. Obs., No. 168:40-94.

chemistry phosphate
Kitamura, H., 1959.
Chemical oceanography in the North Pacific. VI. Phosphate in the equatorial Pacific Ocean.
J. Oceanogr. Soc., Japan, 15(3):131-135.

chemistry, phosphate
Kitamura, H., 1958.
On the distribution of phosphate in the western North Pacific. Mem. Kobe Mar. Obs., (2)12(2):15-20.

Chemistry phosphate
Kitamura, H., 1957.
On the phosphate distribution of the western North Pacific. Chemical oceanography of the western North Pacific (1). Umi to Sora, 33(4/5):65-69.

chemistry phosphate
Kitamura, Hiroyuki, and Takeshi Sagi, 1965.
On the chemical elements in the sea south of Honshu, Japan. II. (In Japanese; English abstract)
Bull. Kobe Mar. Obs., No. 174:39-55.

chemistry phosphate
Kitamura, Hiroyuki, and Takeshi Sagi, 1964.
On the chemical elements in the sea south of Honshu, Japan. (In Japanese; English abstract).
Bull. Kobe Mar. Obs., No. 172:6-54.

Chemistry phosphates
Klyashtorin, L.B., 1964
Primary productivity and phosphates in the Atlantic Ocean. (In Russian).
Okeanologiia, Akad. Nauk, SSSR, 4(2):311-312.

chemistry phosphate(data)
Koczy, F.F., 1952.
Hydrography. Swedish observations. Cons. Perm. Int. Expl. Mer, Ann. Biol. 8:130-131.

chemistry, phosphate
Koroleff, Folke, 1965.
The results of the phosphate intercalibration measurements.
UNESCO, Techn. Papers, Mar. Sci., No. 3:12-14. (mimeographed).

chemistry phosphate
Korringa, P., 1956.
Hydrographical, biological and ostreological observations in the Knysna Lagoon with notes on conditions in other South African waters.
S. Africa, Dept. Comm. & Fish., Div. Fish., Invest. Rept., No. 20:63 pp., 23 pls.

phosphate
Kostikova, A.N., 1965.
Hydrochemical characteristics of the Gulf of Alaska waters in the summer-autumn of 1962. (In Russian).
Sovetsk. Ribokhoz. Issled. Severo-Vostochn. Chasti Tikhogo Okeana, 4 (Vses. Nauchno-Issled. Inst., VNIRO Trudy 58: Tikhookean. Nauchno-Issled. Inst., TINRO, Trudy, 53):21-33.

Chemistry, phosphate
Krey, J., 1956.
Die Trophie küstennaher Meeresgebiete.
Kieler Meeresf., 12(1):46-64.

Chemistry, phosphates
Kruger, D., 1950.
Variations quantitatives des protistes marins au voisinage du Port d'Alger durant l'hiver 1949-1950. Bull. Inst. océanogr. Monaco, No. 978:20 pp., 5 textfigs.

chemistry phosphate
Kusunoki, Kou, 1962
Hydrography of the Arctic Ocean with special reference to the Beaufort Sea.
Contrib. Inst., Low Temp. Sci., Sec. A,(17): 1-74.

L

Laktionov, A.F., V.A. Shamontyev and A.V. Yanes, 1960 phosphates
Oceanographic characteristics of the North Greenland Sea.
Soviet Fish. Invest., North European Seas, VNIRO, PNIRO, Moscow, 1960:51-65.

Chemistry phosphates
Lecal, J., 1952.
Répartition en profondeur des coccolithophorides en quelques stations méditerranéenes occidentales. Bull. Inst. Océan., Monaco, No. 1018:13 pp., 4 textfigs.

chemistry phosphate
Legand, M., 1957.
Variations quantitatives du zooplancton récolté par l'Orsom III pendant la croisière EQUAPAC.
Off. Recherche Sci. et Tech. Outre-Mer, Inst. Français d'Océanie, Sect. Océanogr., Rapp. Sci., No. 2:31 pp. (mimeographed).

chemistry, phosphate
Leloup, E. editor, 1966.
Recherches sur l'ostreiculture dans le bassin d'Ostende en 1964.
Minist. Agricult. Comm. T.W.O.Z., Groupe de Travail "Ostreiculture": 58 pp.

Chemistry, phosphate
Leloup, E., Editor, with collaboratio of L. Van Meel, Ph. Polk, R. Halewyck and A. Gryson, undated.
Recherches sur l'ostreiculture dans le Bassin de Chasse d'Ostende en 1962.
Ministere de l'Agriculture, Commission T.W.O.Z., Groupe de Travail - "Ostreiculture", 58 pp.

chemistry, phosphate
Lillick, L.C. 1939.
Seasonal studies of the phytoplankton off Woods Hole, Massachusetts.
Biol. Bull. mar. biol. Lab. Woods Hole 73(3): 488-503.

Chemistry phosphate

Lindquist, Armin, 1959
Studien über das Zooplankton der Bottensee II. Zur Verbreitung und Zusammensetzung des Zooplanktons.
Inst. Mar. Res., Lysekil. Ser. Biol. Rept., No. 11:136 pp.

Chemistry phosphate

Longhurst, Alan R., 1962.
A review of the oceanography of the Gulf of Guinea.
Bull. Inst. Français Afrique Noire, 24(3):633-663.

Chemistry, phosphate

Lund, J.W.G., 1950.
Studies on Asterionella formosa Hass. II. Nutrient depletion and the spring maximum. Pt. II. Discussion. J. Ecol. 38(1):15-35.

Chemistry phosphate

Lund, J.W.G., 1950.
Studies on Asterionella formosa Hass. II. Nutrient depletion and the spring maximum. Pt. I. Observations on Windermere, Esthwaite water and Blelham Tarn. J. Ecol. 38(1):1-14, 7 textfigs.

Chemistry phosphate

Lüneberg, H., 1939.
Hydrochemische Untersuchungen in der Elbmündung mittels Elektrokolorimeter.
Arch. Deutschen Seewarte 59(1):1-27, 8 pls.

Chemistry phosphate

Lutsarev, S.V., and D.A. Smetanin, 1959
[Method of obtaining sea water free of silicon. and phosphate.] Trudy Inst. Okeanol., 35: 30-32

M

Chemistry, phosphate(data)

Maeda, H., 1953.
Studies on Yosa-Naikai. 3. Analytical investigations on the influence of the River Noda and the benthonic communities. J. Shimonoseki Coll. Fish. 3(2):141-149, 3 textfigs.

Chemistry, phosphate (data)

Maeda, H., 1953.
Studies on Yosa-Naikai. 2. Considerations upon the range of the stagnation and the influences by the River Noda and the open sea.
J. Shimonoseki Coll. Fish. 3(2):133-130, 2 textfigs.

Chemistry phosphate(data)

Mahadeva, N., 1962
Preliminary report on the experiments with Marutoku-B net and a 45 cm x 90 cm net of No. 0 bolting silk off Trincomalee, Ceylon, during October 1960.
Indo-Pacific Fish. Council, FAO, Proc., 9th Sess., (2/3):17-24.

Chemistry, phosphate

Makkaveeva, N.S., 1965.
The hydrochemical investigations of the Korovinskaya Inlet. (In Russian).
Material. Ribochoz. Issled. Severn. Bassin. Poliarn. Nauchno-Issled. i Proekt. Inst. Morsk. Rib. Choz. i Okeanogr., N.M. Knipovich, 5:117-122.

Chemistry, phosphate

Malone, Ph. G. and K. M. Towe 1970.
Microbial carbonate and phosphate precipitates from sea water cultures.
Marine Geol. 9(5): 301-309.

Chemistry, phosphate (data)

Marchand, J.M., 1957.
Twenty-seventh annual report for the period 1st April 1955 to 31st March 1956.
Comm. & Ind., S. Africa, July 1957:159 pp.

Chemistry, phosphate (chiefly data)

Marchand, J.M., 1956.
Twenty-sixth annual report for the period 1st April 1954 to 31st March 1955. Union of S. Africa, Dept. Comm. & Industr., Div. Fish.:183 p

Chemistry, Phosphate (data)

Marchand, J.M., 1955.
Twenty-fifth annual report for the period 1st April 1953 to 31st March 1954.
Commerce & Industry, S. Africa, July 1955:162 pp.

Chemistry, phosphates

Marchesoni, V., 1954.
Il trofismo della Laguna Veneta e la vivificazione marina. III. Ricerche sulle variazioni quantitative sul fitoplancton.
Arch. Oceanogr. e Limnol. 9(3):153-284.

Chemistry phosphate

Margalef, Ramón, 1965.
Distribución ecológica de las especies del fitoplancton marino en un área del Mediterráneo occidental.
Inv. Pesq., Barcelona, 28:117-131.

Chemistry phosphate(data)

Margalef, Ramón, and Juan Herrera, 1963
Hidrografía y fitoplancton de las costas de Castellón, de julio de 1959 a junio de 1960.
Inv. Pesq., Barcelona, 22: 49-109.

chemistry, phosphate (data)

Margalef, R., J. Herrera, M. Steyaert et J. Steyaert, 1966.
Distributions et caractéristiques des communautés phytoplanctoniques dans le bassin Tyrrhenien de la Méditerranée en fonction des facteurs ambiants et à la fin de la stratification estivale de l'année 1963.
Bull. Inst. r. Sci. nat. Belg., 42(5):1-56.

Chemistry phosphate

Marshall, S.M. and A.P. Orr, 1948
Further experiments on the fertilization of a sea loch (Loch Craiglin). The effect of different plant nutrients on the phytoplankton.
J.M.B.A. 27(2):360-379, 10 text figs.

Chemistry phosphate (data)

Marumo, Ryuzo, Editor, 1970.
Preliminary report of the Hakuho Maru Cruise KH-69-4, August 12-November 13, 1969, The North and Equatorial Pacific Ocean.
Ocean Res. Inst. Univ. Tokyo, 68 pp.

Chemistry phosphate

Matsue, Yoshiyuki, Yuzo Komaki and Masaaki Murano, 1957

On the distribution of minute nutrients in the North Japan Sea, during the close of August, 1955.
Bull. Japan Sea Reg. Fish. Res. Lab., 6:121-127
Abstr. in:
Rec. Res. Fac. Agric., Tokyo Univ., Mar. 1958, 7(80):57-58.

Chemistry phosphates

Matsue, Yoshiyuki, Yuzo Komaki and Masaaki Murano, 1957

On the distribution of minute nutrients in the North Japan Sea; March-April, 1956.
Bull. Japan Sea Reg. Fish. Res. Lab., 6:316-320.
Abstr. in:
Rec. Res., Fac. Agric., Tokyo Univ., Mar. 1958, 7(81): 58.

Mc

Chemistry phosphate (data)

McGary, James W., and Edward D. Stroup, 1958
Oceanographic observations in the central North Pacific, September 1954-August 1955.
U.S.F.W.S. Spec. Sci. Rept., Fish., No. 252: 250 pp.

Chemistry, phosphates

McGill, David A. 1965.
The relative supplies of phosphate nitrate and silicate in the Mediterranean Sea.
Rapp. P.-v. Réun. Commn int. Explor. scient. Mer Médit. 18(3): 737-744.

Chemistry, phosphate

McIntyre, A.D., and J.H. Steele, 1956.
Hydro-biological conditions in the Denmark Strait, May, 1954.
Ann. Biol., Cons. Perm. Int. Expl. Mer, 11:20-25, Figs., 25-28.

Chemistry phosphate

*McLain, Douglas R., 1968.
Oceanographic surveys of Traitors Cove, Revillagigedo Island, Alaska.
Spec. scient. Rep. U.S. Fish Wildl. Serv. Fish. 576: 15 pp. (multilithed).

Chemistry phosphate

Miller, S.M., H.B. Moore, and K.R. Kvammen, 1953.
Plankton of the Florida Current. 1. General conditions. Bull. Mar. Sci., Gulf and Caribbean 2(3): 465-485, 27 textfigs.

Chemistry phosphate (data)

*Minas, Hans Joachim, 1968.
A propos d'une remontée d'eaux "profondes" dans les parages du Golfe de Marseille (octobre 1964): conséquences biologiques.
Cah. océanogr., 20(8):647-674.

Chemistry phosphate

Minas, H. J., and M.J. Pizarro, 1962.
Quelques observations hydrologiques sur les eaux au large des côtes orientales de la Corse.
Bull. Inst. Océanogr., Monaco, 59(1232):1-23.

Chemistry phosphate

Miwa, Katsutoshi, 1962
Report on dissolved nutrients investigations in the North Pacific and the Bering Sea during June-August 1960. (In Japanese; English summary).
Bull. Hokkaido Reg. Fish. Res. Lab., (25): 58-65.

Chemistry phosphate

Miyake, Y., Y. Sugiura and K. Kameda, 1955.
On the distribution of radioactivity in the sea around Bikini Atoll in June 1954.
Pap. Met. Geophys., Tokyo, 5(3/4):253-262.

Chemistry phosphate

Miyake, Yasuo, Yoshio Sugiura, Yukio Sugimura, Tsutomu Akiyama and Hibrozo Yoshimura, 1962.
II. Chemistry in the Japanese Expedition to the Deep Sea (JEDS-4).
Oceanogr. Mag., Japan Meteorol. Agency, 13(2): 131-132.

JEDS Contrib. No. 30.

Chemistry phosphate

Mokiyevskaya, V.V., 1961
[Some peculiarities of hydrochemistry in the northern Indian Ocean.] (In Russian; English abstract).
Okeanolog. Issled., Mezhd. Komitet Proved. Mezhd. Geofiz. Goda, Prezidiume, Akad. Nauk, SSSR, (4):50-61.

Chemistry phosphate

Mokievskaja, V.V., 1959

[Nutrients salts in the upper layer of the Bering Sea.] Trudy Inst. Okeanol., 33: 87-113.

Chemistry Phosphate
Mommaerts, J.P., 1970.
On the unusual patterns of phosphate vertical distribution in the Tamar Estuary. J. mar. Biol. Ass., U.K., 50(3): 849-855.

Chemistry phosphate
Montgomery, R.B., and E.D. Stroup, 1962.
Equatorial waters and currents at 150° W in July -August 1952.
The Johns Hopkins Oceanogr. Studies, No. 1:68 pp.

Chemistry phosphate (data)
Morales, E., y E. Arias, 1965.
Ecología del puerto de Barcelona y desarrollo de adherencias orgánicas sobre placas sumergidas.
Inv. Pesq., Barcelona, 28:49-79.

Chemistry phosphate (data)
Moreira da Silva, Paulo, 1957
Oceanografia do triangulo Cabo-Frio-Trinidade-Salvador.
Anais Hidrograficos, Brasil, 16:213-308.

Chemistry, phosphates
Morii, Hideaki, Yusho Akishige, Ryoichi Kanagu and Tadanobu Fukuhara 1967.
Chemical studies of the Kuroshio and adjacent region. 1. On the nutrient salts and the dissolved oxygen. (In Japanese; English abstract).
Bull. Fac. Fish., Nagasaki Univ. 22:91-103.

Chemistry, phosphate (data)
Mosby, H., 1938
Svalbard waters. Geophys. Publ. 12(4): 85 pp., 34 text figs.

Chemistry phosphate
Murakami, A., 1954.
[Oceanography of Kasaoka Bay in Seto Inland Sea.]
Bull. Nakai Regional Fish. Res. Lab. 6:15-57, 42 textfigs.

Chemistry phosphate
Murphy, Garth I., and Richard S. Shomura, 1958
Variations in yellowfin abundance in the Central Equatorial Pacific.
Proc. Ninth Pacific Sci. Congr., Pacific Sci. Assoc., 1957, 16(Oceanogr.):108-113.

Chemistry phosphates
Murphy, J., & J.P. Riley, 1962
A modified single solution method for the determination of phosphate in natural waters.
Anal. Chim. Acta, 27(1): 31-36.

Chemistry phosphate
Murphy, J., and J.P. Riley, 1958.
A sinlge-solution method for the determination of seluble phosphate in sea water. J.M.B.A., 37(1):9-14.

N

Chemistry phosphate (data)
Nagaya, Y., 1959
On the distribution of the nutrient salts in the Equatorial region of the Western Pacific Jan.-Feb., 1958.
Publ. 981, Hydrogr. Bull., No. 59: 31-40.

Chemistry, phosphate (data)
Navarro, F. de P. 1947.
Exploración oceanografía del Africa occidental desde el Cabo Ghir al Cabo Judy:resultados de las campañas del "Malaspina" y del "Xauen" en Mayo 1946. Trab. Inst. Espagñol Ocean. No. 20:40 pp., 8 figs.

Chemistry, phosphate
Nehring, Dietwart, Sigund Schulz und Karl-Heinz Rohde 1969.
Untersuchungen über die Productivität der Ostsee. 1. Chemisch-biologisch Untersuchungen in der mittleren Ostsee und in der Bottensee im April/Mai 1967.
Beitr. Meeresk. 23: 5-36.

Chemistry, phosphate
Neumann, A. Conrad, and David A. McGill, 1961 (1962).
Circulation of the Red Sea in early summer.
Deep-Sea Res., 8(3/4):223-235.

chemistry phosphate
Newell, B.S., 1959
The hydrography of the British East African coastal waters. II. Colonial Off., Fish. Publ. London, No. 12: 18 pp.

Chemistry phosphate
Neyman, V.G., 1961.
[Factors conditioning the oxygen minimum in the subsurface waters of the Arabian Sea.]
Okeanol. Issledov., Mezhd. Komit., Mezhd. Geofiz. God, Presidiume, Akad. Nauk, SSSR,:62-65.

Chemistry, phosphate
Nichols, Maynard M., 1966.
A study of production and phosphate in a Sonoran lagoon.
Publs. Inst.mar.Sci.,Univ.Texas,Port Aransas, 11:159-167.

chemistry, phosphate (data)
Niemi, Åke, Heinrichs Skuja and Torbjörn Willén 1970
Phytoplankton from the Pojoviken-Tvärminne area, S. coast of Finland.
Mem. Soc. Fauna Flora Fennica 46: 14-28.

chemistry phosphate (data)
Nishida, Keizo, 1958
[On the oceanographic conditions of the lower cold water in the Japan Sea.]
J. Fac. Fish. and Animal Husbandry, Hiroshima Univ., 2(1): 1-65.

chemistry, phosphate
Nishizawa, Tanzo and T. Muraki, 1940
Chemical studies in the sea adjacent to Palau. I. A survey crossing the sea from Palau to New Guinea. Kagaku Nanyo (Sci. of the South Sea) 2(3):1-7.

chemistry phosphate (inorganic)
Noble, A., 1968.
Studies on sea-water off the North Kanara coast.
J. mar. biol. Ass., India, 10(2): 197-223.

Chemistry, phosphate
Norina, A.M., 1965.
Hydrochemical characteristics of the northern part of the Barents Sea. (In Russian).
Trudy, Gosudarst Okeanogr. Inst., No. 83:243-271.

chemistry, phosphate
NORPAC Committee, 1960
The NORPAC Atlas. Oceanic observations of the Pacific, 1955.

Chemistry phosphate
Nümann, W., 1941.
Der Nährstoffhaushalt in der nordöstlichen Adria. Thalassia 5(2):1-68, 12 textfigs.

O

Chemistry phosphate
Ohwada, Mamoru, and Hisanori Kon, 1963
A microplankton survey as a contribution to the hydrography of the North Pacific and adjacent seas. (II). Distribution of the microplankton and their relation to the character of water masses in the Bering Sea and northern North Pacific Ocean in the summer of 1960.
Oceanogr. Magazine, Japan Meteorol. Agency 14(2):87-99.

chemistry, phosphate
Ohwada, Mamoru, and Fumi Ogawa 1966.
Plankton in the Japan Sea.
Oceanogr. Mag. Jap. Met. Soc. 18(1/2):39-51.

chemistry, phosphate (data)
Okabe, Shiro, Yoshimasa Toyota and Takafumi Murakami, 1967.
Major aspects of chemical oceanography. (In Japanese; English abstract).
J.Coll.mar.Sci.Techn.,Tokai Univ., (2):39-56.

Chemistry, phosphate
Okitsu, T., 1952.
Seasonal change of the phosphate and silicate of the water of Mutsu Bay during 1950.
Bull. Mar. Biol. Sta., Asamushi, 5(1/4):61-64.

Chemistry phosphate
Okubo, A., 1956.
On an oxygen depletion-phosphate diagram (ΔO_2 -P diagram of the Kuroshio-Oyashio waters. 1. A hypothesis. Ocean. Mag., Tokyo, 8(1):79-91.

Chemistry phosphate
Okuda, Taizo, 1962
Physical and chemical oceanography over continental shelf between Cabo Frio and Vitoria (Central Brazil).
J. Oceanogr. Soc., Japan, 20th Ann. Vol.,514-540.

Chemistry, phosphate
Okuda, Taizo, 1960
Chemical oceanography in the South Atlantic Ocean, adjacent to north-eastern Brazil. (Portuguese and French resumés).
Trabalhos, Inst. Biol. Marit. e Oceanogr., Universidade do Recife, Brasil, 2(1):155-174.

Chemistry phosphate
Okuda, T., 1955.
[On the soluble nutrients in bay deposits. IV. An experiment of the behavior of phosphate-phosphorus between mud and sea water.]
Bull. Tohoku Reg. Fish. Res. Lab. (5):79-81.

Chemistry phosphate (data)
Okuda, Taizo, and Lourinaldo Barreto Cavalcanti, 1963.
Algumas condicoes oceanograficas na area nordeste de Natal. (Septembro 1960).
Trab. Inst. Oceanogr. Univ. do Recife, 3(1); 3-25.

chemistry phosphates
Okuda, Taizo, Angel José García, José Benetoz Alvarez, 1965.
Veriecion estacional de las elementos nutritions en el agua de la laguna y el rio Unare.
Bol. Inst., Oceanogr., Univ. Oriente, 4(1): 123-135.

chemistry phosphate
Oren, O.H., 1952.
Some hydrographical features observed off the coast of Israel. Bull. Inst. Océan., Monaco, No. 1017:9 pp., 4 textfigs.

chemistry, phosphate
Orr, A.P. 1933.
Physical and chemical conditions in the sea in the neighborhood of the Great Barrier Reef.
Sci. Repts. Brit. Mus. (N.H.) Great Barrier Reef Exped. 1928-29, 2(3): 37-86.

chemistry phosphate (data)
Orren, M.J., 1963.
Hydrological observations in the south west Indian Ocean.
Rept., S. Africa, Dept. Comm. Industr., Div. Sea Fish., Invest. Rept., No. 45:61 pp.
Reprinted from Commerce & Industry, May 1963.

P

chemistry, phosphate
Park, P. Kilho, Stephen W. Hager, Jacques E. Pirson and David S. Ball, 1968.
Surface phosphate and silicate distribution in the northeastern Pacific Ocean.
J.Fish.Res.Bd.,Can.,12(25):2739-2741.

chemistry, phosphate
Pavlychev, V.P. 1969.
Some features of the chemistry and thermal structure of the waters of the mixing region of the Kuroshio and Oyashio Currents. (In Russian).
Izv. Tichookean. nauchno-issled. Inst. ribn. Khoz. Okeanogr. (TINRO) 68: 33-44.

chemistry phosphates
Phifer, L. D. and T. G. Thompson, 1937.
Seasonal variations in the surface waters of San Juan channel during the five year period, January 1931 to December 30, 1935. J. Mar. Res., 1(1):34-59, text figs. 12-18, 12 tables.

chemistry phosphate
Piatek, Wanda, 1962
Initial elaboration of phosphate contents in the southern Baltic for the years 1948-1954.
Prace Morsk. Inst. Ryback., Gdyni, Oceanogr. Ichtiol., 11(A):65-79.
In Polish; Russian and English summary, p. 79.

chemistry phosphate
Pomeroy, Lawrence R., Henry M. Mathews and Hong Shik Min, 1963
Excretion of phosphate and soluble organic phosphorus compounds by zooplankton.
Limnology and Oceanography, 8(1):50-55.

chemistry, phosphates
Pomeroy, Lawrence, E.E. Smith, and Carol M. Grant, 1965.
The exchange of Phosphates between estuarine water and sediments.
Limnology and Oceanography, 10(2):167-172.

chemistry, phosphate
Ponomarenko, L.S., 1966.
Hydrochemical conditions in the Norwegian and Greenland seas in the summer of 1964. (In Russian).
Mater. Ribokhoz. Issled. severn. Basseina, Poliarn. Nauchno-Issled. Proektn. Inst. Morsk. Ribn. Khoz. Okeanogr. (PINRO), 7:155-167.

chemistry phosphates
Ponomarenko, L.S., and M.A. Istoshina, 1962
Hydrochemical studies in Newfoundland area. (In Russian; English summary).
Sovetskie Ribochoz. Issledov. Severo-Zapad. Atlant. Okeana, VNIRO-PINRO, 113-124.

chemistry, phosphate
Posner, G.S., 1957.
The Peru Current.
Bull. Bingham Oceanogr. Coll., 16(2):106-153.

chemistry, phosphate
Pozdniakova, L.E., and V.N. Vinogradov, 1966.
Some peculiarities in the distribution of the hydrochemical elements in the southeastern Barents Sea in August-September 1959. (In Russian).
Trudy murmansk. biol. Inst., 11(15):140-156

chemistry phosphates
Prasad, R. Raghu, 1956.
Further studies on the plankton of the inshore waters off Mandapam. Indian J. Fish., 3(1):1-42.

chemistry phosphate (average, monthly)
Prasad, R.R., 1954.
The characteristics of marine plankton at an inshore station in the Gulf of Mannar near Mandapam. Indian J. Fish. 1(1/2):1-36, 8 textfigs

chemistry phosphate
Pratt, David M., 1965.
The winter-spring diatom flowering in Narragansett Bay.
Limnology and Oceanography, 10(2):173-184.

chemistry phosphate
Pratt, D.M., 1949.
Experiments in the fertilizarion of a salt water pond. J. Mar. Res. 8(1):36-59, 8 textfigs.

chemistry, Phosphate
Proctor, C.M., and D.W. Hood, 1954.
Determination of inorganic phosphate in sea water by an iso-butanol extraction procedure.
J. Mar. Res. 13(1):122-132, 4 textfigs.

chemistry, phosphate
Pytkowicz, R.M. and D.R. Kester 1966.
Oxygen and phosphate as indicators for the deep intermediate waters in the northeast Pacific Ocean.
Deep-Sea Res. 13(3): 373-379.

Q

chemistry, phosphate
Qasim, S.Z., and C.V.G. Reddy, 1967.
The estimation of plant pigments of Cochin backwater during the monsoon months.
Bull. mar. Sci., Miami, 17(1):95-110.

R

chemistry phosphate
Rae, B.B., R. Johnston and J.A. Adams, 1965.
The incidence of dead and dying fish in the Moray Firth, September 1963.
Jour. Mar. Biol. Assoc., U.K., 45(1):29-47.

chemistry, phosphate
Rakestraw, N.W., 1933.
Studies on the biology and chemistry of the Gulf of Maine. 1. Chemistry of the waters of the Gulf of Maine in August, 1932. Biol. Bull. 64(2):149-158, 4 textfigs.

phosphate
Rakestraw, N. W., and H.P. Smith, 1937.
A contribution to the chemistry of the Caribbean and Cayman Seas. Bull. Bingham Ocean. Coll. 6(1): 1-41, 32 text-figs.

chemistry phosphate (data)
Ramalho, A. de M., R. G. Boto, B. C. Goncalves, H. Vilela, 1936.
Observacoes oceanograficas. "Albacora", 1935-1936. Trav. St. Biol. Mar. Lisbonne, No.39: 25 pp., 2 figs.

chemistry, phosphate (data)
Ramalho, A. de M., L. S. Dentinho, C. A. M. de Sousa, Fronteira, F. L. Mamede, and H. Vilela, 1935.
Observacoes oceanograficas. "Albacora" 1934. Trav. St. Biol. Mar., Lisbonne, No. 35: 35 pp., 1 fig.

chemistry, phosphate (data)
Ramalho, A. de M., B. C. Goncalves, R. G. Boto, Z. Vilela, 1938.
Observacoes oceanograficas. "Albacora", 1937. Trav. St. Biol. Mar., Lisbonne, No. 43: 30 pp., 1 fig.

chemistry phosphate
Ramamurthy, S., 1965.
Studies on the plankton of the North Canara coast in relation to the pelagic fishery.
J.mar.biol.Ass.India, 7(1):127-149.

phosphate
Ramamurthy S. 1963
Studies on the hydrological factors in the north Kanara coastal waters.
Indian J. Fish. (A) 10(1): 75-93

chemistry phosphate
Ramamurthy, S., 1954.
Hydrobiological studies in the Madras coastal waters. J. Madras Univ., B, 23(2):148-163, 3 textfigs.

chemistry phosphate
Raymont, J.E.G., and B.G.A. Carrie, 1964.
The production of zooplankton in Southampton Water.
Int. Rev. Ges. Hydrobiol., 49(2):185-232.

phosphate
Reddy, C.V. Gangadhara and V.N. Sankaranarayanan, 1969.
Distribution of phosphates and silicates in the Central Western north Indian Ocean, in relation to some hydrographical factors. Bull. natn. Inst. Sci., India, 38(1): 103-122.

phosphate
Redfield, A. C., 1942.
The processes determining the concentration of oxygen, phosphate and other organic derivatives within the depths of the Atlantic Ocean.
P.P.O.M. 9(4):1-22, 13 textfigs.

phosphate
Redfield, A.C., 1934.
On the proportions of organic derivatives in sea water and their relation to the composition of plankton. In: James Johnstone Memorial Volume:176-192, 5 textfigs.

phosphate
Redfield, A.C., and A.B. Keys, 1938.
The distribution of ammonia in the waters of the Gulf of Maine. Biol. Bull. 74(1):83-92, 6 textfigs.

chemistry phosphate
Reichelt, W., 1941.
Die ozeanographischen Verhältnisse bis zur warmen Zwischenschicht an der antarktischen Eisgrenze im Südsommer 1936/37. Nach Beobachtungen auf dem Walfangmutterschiff "Jan Wellem" in Weddell-Meer.
Archiv. Deutschen Seewarte 61(5):54 pp., 11 pls.

chemistry phosphate
Reid, Joseph L., Jr., 1965.
Intermediate waters of the Pacific Ocean.
The Johns Hopkins Oceanogr. Studies, No. 2: 1-85.

chemistry phosphate
Riley, Gordon A., 1959
Oceanography of Long Island Sound, 1954-1955
Bull. Bingham Oceanogr. Coll., 17(1): 9-30.

phosphate
Riley, G.A., 1952.
Phytoplankton of Block Island Sound, 1949.
Bull. Bingham Ocean. Coll. 13(3):40-64, 6 textfig

phosphate
Riley, G.A., 1951.
Oxygen, phosphate, and nitrate in the Atlantic Ocean. Bull. Bingham Ocean. Coll. 8(1):126 pp., 33 textfigs.

phosphates
Riley, G. A., 1941.
Plankton studies. IV. Georges Bank. Bull. Bingham Ocean. Coll. 7(4):73 pp.

phosphate
Riley, G. A., 1939.
Plankton studies. II. The western North Atlantic, May-June 1939. J. Mar. Res. 2(2):145-162, Textfigs. 49-51, 4 tables.

phosphate
Riley, G. A., 1937.
The significance of the Mississippi River drainage for biological conditions in the northern Gulf of Mexico. J. Mar. Res. 1(1):60-74, Textfigs. 19-25.

chemistry phosphate
Riley, G.A., and S.A.M. Conover, 1956.
Oceanography of Long Island Sound, 1952-1954. 3. Chemical Oceanography.
Bull. Bingham Oceanogr. Collection, 15:47-61.

Chemistry phosphate
Riley, G. A., H. Stommel, and D. F. Bumpus, 1949
Quantitative ecology of the plankton of the western North Atlantic. Bull. Bingham Ocean. Coll. 12(3):169 pp., 39 text figs.

phosphate
Riley, G. A., and R. von Arx, 1949.
Theoretical analysis of seasonal changes the phytoplankton of Husan Harbor, Korea. J. Mar. Res. 8(1):60-72, 5 textfigs.

chemistry phosphate
Riley, J.P., 1967.
The hot saline waters of the Red Sea bottom and their related sediments.
Oceanogr. Mar. Biol., Ann.Rev., H. Barnes, editor, George Allen and Unwin, Ltd., 5:141-157.

chemistry, phosphate (water samples)
total phosphate (bottom sedimen
Rittenberg, S.C., K.O. Emery and W.L. Orr, 1955.
Regeneration of nutrients in sediments of marine basins. Deep-Sea Res. 3(1):23-45.

Chemistry, phosphate
Robinson, R.J., 1941.
A method of freeing sea water of phosphate.
Science 93(2405):117-115.

chemistry, phosphates
Robinson, R.J., and T.G. Thompson, 1948
The determination of phosphates in sea water. J. Mar. Res. 7(1):33-41.

chemistry, phosphate
Robinson, R.J., and H.E. Wirth, 1934.
Photometric investigation of the cerulo-molybdate determination of phosphate in waters.
Industr. Engin. Chem. 7(3):147-150.

Rochford, D.J.

Chemistry, phosphate (inorganic)
Rochford, D.J. 1969.
Seasonal variation in the Indian Ocean along 110°E. 1. Hydrological structure of the upper 500 m.
Austr. J. mar. Freshwat. Res. 20(1): 1-50.

chemistry, phosphate
Rochford, D.J. 1967.
The phosphate levels of the major currents of the Indian Ocean.
Austr. J. mar. Freshwat. Res. 18(1): 1-22.

Chemistry, Phosphate
Rochford, D. J., 1966.
Some hydrological features of the eastern Arafura Sea and the Gulf of Carpenteria in August 1964.
Australian J. Mar. freshw. Res., 17(1):31-60.

Chemistry, phosphate
Rochford, D.J., 1966.
Hydrology. Port Philip survey 1957-1963.
Mem. nat. Mus. Vict., 27:107-118.

Chemistry phosphate
Rochford, D.J., 1964.
Hydrology of the Indian Ocean.
Australian J. mar. freshwater Res., 15(1):25-55.
Also in:
Collected Reprints, Int. Indian Ocean Exped., UNESCO, 3. 1966.

Chemistry, phosphate
Rochford, D.J., 1961.
Hydrology of the Indian Ocean. 1. The water masses in intermediate depths of the south-east Indian Ocean.
Australian J. Mar. Freshwater Res., 12(2):129-149.

Chemistry phosphate (data)
Rochford, D.J., 1953.
Analysis of bottom deposits in eastern and south-western Australia, 1951 and records of twenty-four hourly hydrological observations at selected stations in eastern Australian estuarine systems 1951. Oceanographical station list of investigations made by the Division of Fisheries, Commonwealth Scientific and Industrial Research Organization, Australia, 13:68 pp.

Chemistry, phosphate(data)
Rochford, D.J., 1953.
Estuarine hydrological investigations in eastern and south-western Australia, 1951. Oceanographical station list of investigations made by the Division of Fisheries, Commonwealth Scientific and Industrial Research Organization, Australia, 12:11 pp.

Chemistry phosphate
Rochford, D.J., 1952.
A comparison of the hydrological conditions of the eastern and western coasts of Australia.
Indo-Pacific Fish. Counc., Proc., 3rd meeting, 1-16 Feb. 1951, Sects. 2/3:61-68, 7 textfigs.

Chemistry phosphates
Rochford, D.J., 1951.
Summary to date of the hydrological work of the Fisheries Division, C.S.I.R.O. Proc. Indo-Pacific Fish. Cound., 17-28 Apr. 1950, Cronulla N.S.W., Australia, Sects. II-III:51-59.

Chemistry phosphate
Rochford, D.J., 1951.
Hydrology of the estuarine development. Proc. Indo-Pacific Fish. Counc., 17-28 Apr. 1950, Cronulla, N.S.W., Australia, Sects. II-III:157-168, 10 textfigs., Table 6.

chemistry phosphate
Rodhe, W., 1948
Environmental requirements of fresh-water plankton algae. Experimental studies in the ecology of phytoplankton. Symbolae Botanicae Upsalienses X(1):149 pp., 30 figs.

Chemistry phosphates
Roger, C., 1966.
Étude sur quelques espèces d'euphausiacés de l'est de l'Océan Indien (110°E.).
Cah. ORSTOM, Sér. Océanogr., 4(4):73-103.

Chemistry phosphate
Rotschi, Henri, 1962
Oxygène, phosphate et gaz carbonique total en Mer de Corail.
Deep-Sea Res., 8:181-195.

Chemistry phosphate (data)
Rotschi, H., 1961.
Chimie.
O.R.S.T.O.M., I.F.O., Rapp. Sci., 23:4-45.

chemistry phosphate
Rotschi, Henri, 1961
Contribution française en 1960 à la connaissance de la Mer de Coral: Océanographique physique.
Cahiers Océanogr., C.C.O.E.C., 13(7):434-455.

Chemistry phosphate
Rotschi, Henri, 1961(1962).
Oxygène, phosphate et gaz carbonique total en Mer de Corail.
Deep-Sea Res., 8(3/4):181-195.

chemistry phosphate (data)
Rotschi, Henri, 1960
Orsom III. Resultats des Croisières diverses de 1959. Oceanographie physique.
Centre d'Océanogr., Inst. Français d'Océanie, Rapp. Sci., No. 17: 59 pp.

Chemistry phosphate
Rotschi, Henri, 1960
Récents progrès des recherches océanographiques entreprises dans le Pacifique Sud-Ouest.
Cahiers Oceanogr., C.C.O.E.C., 12(4):248-267.

Chemistry phosphates
Rotschi, Henri, 1960
Sur la determination colorimetrique du phosphate.
Cahiers Oceanogr., 12(7):470-482.

Chemistry phosphate
Rotschi, Henri, 1959
Chimie, "Orsom III", Resultats de la Croisière BOUSSOLE.
O.R.S.T.O.M., Inst. Français d'Océanie, Rapp. Sci. 13: 3-60.

Chemistry phosphate (data)
Rotschi, Henri, Michel Angot and Roger Desrosieres, 1960.
Orsom III, Resultats de la croisiere "Choiseul" 2ème partie. Chimie, productivite, phytoplancton qualitatif.
Rapp. Sci., Noumeau, No. 16:91 pp. (mimeographed).

Chemistry phosphate(data)
Rotschi, Henri, Michel Angot, et Michel LeGand, 1959
Orsom III. Résultats de la croisière "Astrolabe". 2. Chimie, productivité et zooplankton.
Rapp. Sci., Inst. Français Océanie, No. 9: 97 pp. (mimeographed).

Chemistry phosphate(data)
Rotschi, Henri, Michel Angot, Michel LeGand, and H.R. Jitt, 1959
Chimie, productivité et zooplancton. "Orsom III". Resultats de la Croisiere "Boussole" Resultats "production primaire" de la croisière 56-5.
Rapp. Sci. Inst. Francais d'Oceanie, Centre d'Oceanogr., No. 13.

chemistry
phosphate

Rotschi, H., B. Wauthy and M. LeGand, 1961
Orsom III. Résultats de la croisière "Epi"
2eme partie, Chimie et biologie.
O.R.S.T.O.M., I.F.O., Rapp. Sci., (23): 4-111.

phosphate
Rounsefell, George A., 1964.
Preconstruction study of the fisheries of the estuarine areas traversed by the Mississippi River-Gulf outlet project.
U.S.F.W.S. Fish. Bull., 63(2):373-393.

Chemistry, phosphates (data)
Rozanov, A.G., 1964.
The distribution of phosphates and silicates in water of the northern part of the Indian Ocean. Investigations in the Indian Ocean (33rd voyage of E/S "Vitiaz"). (In Russian).
Trudy Inst. Okeanol., Akad. Nauk, SSSR, 64:102-104.

Chemistry
phosphate

Rubinchik, E.E., 1959
[Data on the hydrochemistry of the southern Caspian.]
Trudy V.N.I.R.O., 38: 152-164.

Chemistry Phosphate

Ruud, J. T., 1932.
On the biology of southern Euphausiidae.
Hvalrådets Skrifter No. 2:1-105, 37 text figs.

Chemistry phosphate

Ryther, John H., and John R. Hall, Allan K. Pease, Andrew Bakun and Mark M. Jones, 1966.
Primary organic production in relation to the chemistry and hydrography of the western Indian Ocean.
Limnol. Oceanogr., 11(3):371-380.

phosphate
Ryther, John H., D.W. Menzel and Nathaniel Corwin, 1967.
Influence of the Amazon River outflow on the ecology of the western tropical Atlantic. I. Hydrography and nutrient chemistry.
J.mar.Res., 25(1):69-83.

S

Chemistry phosphate
Saiki, A., 1952.
Some substances in sea water which disturb the colorimetric determination of phosphate.
J. Ocean. Soc., Japan, 8(2):73-78.

Chemistry phosphate (data)
San Feliu, J.M., y F. Muñoz, 1965.
Hidrografía y plancton del puerto de Castellón de junio de 1961 a enero de 1963.
Inv. Pesq., Barcelona, 28:3-48.

Chemistry phosphate
*Sapozhnikov, V.V., 1967.
The application of the computation parameter "Performed phosphate" to elucidating the basic regularities in the distribution of phosphate and oxygen along the meridional sections in the Pacific and Atlantic oceans. (In Russian; English abstract).
Trudy Inst. Okeanol., 83:5-19.

Chemistry
phosphate
Sapozhnikov, V.V., 1965.
Hydrochemical characteristics as applied to distinguishing the Cromwell Current. Methods of marine hydrochemical investigations. (In Russian)
Trudy, Inst. Okeanol., Akad. Nauk, SSSR, 79:87-96

Chemistry,
phosphate (data)
Satyanarayana Rao, T.S., and V. Chalapathi Rao, 1962.
Studies on diurnal variations in the hydrobiological conditions of the Waltair coast.
J. Mar. Biol. Assoc., India, 4(1):23-43.

Chemistry phosphate
Scaccini Cicatelli, Marta, 1968.
Un anno di osservazioni idrologiche in una stazione fissa nelle acque dell'Adriatico occidentale presso Fano.
Note Lab. Biol. mar. Pesca, Fano, Univ. Bologna. 2(9): 181-228.

Chemistry, phosphates
Schachter, D., 1954.
Contribution à l'étude hydrographique et hydrologique de l'étang de Berre (Bouches-du-Rhone).
Bull. Inst. Océan., Monaco, No. 1048:20 pp.

Chemistry, Phosphate
Schott, Friedrich, and Manfred Ehrhardt, 1970.
On fluctuations and mean relations of chemical parameters in the northwestern North Sea.
Kieler Meeresforsch. 25(2): 272-278

Chemistry, phosphate
Schulz, B., 1934.
Die Ergebnisse der Polarexpedition auf dem U-Boat "Nautilus". Ann. Hydrogr., usw., 62:147-152, Pl. 16 with 2 figs.

Chemistry, phosphates
Sebastian, Alfonso, Manual N. Llorca and Vitaliano B. Encina, 1965.
Oceanography of Lingayen Gulf.
Philipp. J. Fish., 7(1):13-33.

Chemistry phosphate
Seiwell, H. R., 1935.
Phosphate in the western basin of the North Atlantic. Nature 136:205, 1 textfig.

Chemistry, phosphates (data)
Serpoianu, Gheorghe, et Viorel Chirila, 1969.
Observations sur les particularités hydrologiques des eaux de la mer Noire dans la couche où la vie commence à disparaître.
Rapp. P.-v. Réun. Commn int. Explor. scient. Mer Mediterr., 19(4): 689-692.

chemistry
phosphate
Seryi, V.V., and V.A. Khimitza, 1963
On the hydrology and chemistry of the Gulf of Aden and the Arabian Sea. (In Russian).
Okeanologiia, Akad. Nauk, SSSR, 3(6):994-1003.

Chemistry, phosphate
Seshappa, G., 1953.
Phosphate content of mudbanks along the Malabar coast. Nature 171(4351):526-527.

Chemistry phosphate (surface)
Sette, O.E., et al., 1954.
Progress in Pacific Ocean Fishery Investigations 1950-53. Spec. Sci. Rept.: Fish. No. 116: 75 pp. 29 textfigs.

Chemistry
phosphate (data)
Sherman, Kenneth, and Robert P. Brown, 1961.
Oceanographic and biological data, Hawaiian waters, January-October, 1959.
U.S.F.W.S. Spec. Sci. Rept., Fish., No. 396:71 pp

Chemistry, phosphate (data)
Shimomura, T., and K. Miyaya, 1957.
[The oceanographical conditions of the Japan Sea and its water systems, laying stress on the summer of 1955.]
Bull. Japan Sea Reg. Fish. Res. Lab., No. 6 (General survey of the warm Tsushima Current 1): 23-120

Chemistry phosphate
Skola, H.V., et M.-T. Gomoiu, 1967.
Recherches océanologiques roumaines dans la mer Noire. (In Rumanian; French abstract).
Hydrobiologie, București 8:15-30.

Chemistry
phosphates
Skopintsev, B.A., 1968.
A study of some reduction and oxidation processes in the Black Sea. (In Russian; English abstract).
Okeanologiia, Akad. Nauk, SSSR, 8(3):412-426.

Chemistry phosphate
Sleggs, G.F., 1927.
Marine phytoplankton in the region of La Jolla, California during the summer of 1924. Bull. S.I.O., tech. ser., 1:93-117, 8 textfigs.

Chemistry phosphate
Slinn, D.J., 1966.
Chemical constituents in sea water off Port Erin during 1965.
Mar.biol.Sta., Univ.Liverpool.No.78:29-33.

Chemistry
phosphate
Slinn, D.J., 1962
Chemical constituents in sea water off Port Erin during 1961.
Mar. Biol. Sta., Univ. Liverpool, Port Erin, Ann. Rept., 74(1961):23-26.

Chemistry
phosphate
Slinn D.J. 1961
Chemical constituents in sea water off Port Erin in 1960.
Ann. Rept. Mar. Biol. Sta., Univ. Liverpool 1960, 73:23-28.

Chemistry
phosphate
Slinn, D.J., 1959
Chemical constituents in sea water off Port Erin during 1958.
Ann. Rept., Mar. Biol. Sta., Univ. Liverpool 1958 (No. 71):24-28.

Chemistry phosphate
Slinn, D.J., 1958.
Phosphate and oxygen in sea-water off Port Erin during 1957.
Ann. Rept., Mar. Biol. Sta., Univ. Liverpool, 70:27-30.

Chemistry phosphate
Slinn, D.J., 1956.
Phosphate and oxygen in sea-water off the Isle of Man, 1952-1954.
Ann. Rept. Mar. Biol., Sta., Isle of Man, 68:30-38.

Chemistry, phosphate
Slinn, D.J., and Winifred Chapman, 1965.
Chemical constituents in sea water off Port Erin in 1964.
Mar. Biol. Sta. Port Erin, Ann. Rept. (1964), No. 77:43-48.

Chemistry
phosphate
Slinn, D.J., and Winifred Chapman, 1964.
Chemical constituents of sea water off Port Erin during 1963.
Mar. Biol. Sta., Univ. Liverpool, Port Erin, Ann. Rept., No. 76(1963):26-31.

Chemistry phosphate
Slinn, D.J., and Gaenor Offlow (Mrs M. Solly), 1969.
Chemical constituents in sea water off Port Erin during 1968.
Ann. Rep. mar. biol. Sta. Port Erin, 81 (1968): 31-

Chemistry phosphate
Smayda, Theodore J., 1966.
A quantitative analysis of the phytoplankton of the Gulf of Panama. III General ecological conditions and the phytoplankton dynamics at 8o 45'N, 79o 23'W from November 1954 to May 1957.
Inter-Amer. Trop. Tuna Comm., Bull., 11(5): 355-612.

chemistry, phosphate
Smayda, T.J., 1957.
Phytoplankton studies in lower Narragansett Bay.
Limnol. & Oceanogr., 2(4):342-357.

chemistry phosphate
Smetanin, D.A., 1961.
[Some features of the chemistry of waters in the northeastern Pacific based on observations made in 1958-1959]
Trudy Inst. Okeanol., Akad. Nauk, SSSR, 45:130-141.

chemistry
phosphate
Smetanin, D.A., 1960
[Some features of the chemistry of water in the central Pacific.]
Trudy Inst. Okeanol., 40: 58-71.

chemistry phosphate
Smetanin, D.A., 1959
[Hydrochemistry of the Kuril-Kamchatka deep-sea trench.] Trudy Inst. Okeanol., 33: 43-86.

chemistry
phosphate
Smetanin, D.A., 1958
[Hydrochemistry in the region of the Kurile-Kamchatka Trench. 1. Certain questions of the hydrology and chemistry of the lower sub-arctic water in the region of the Kurile-Kamchatka Trench.]
Trudy Inst. Okeanol., 27: 22-54.

chemistry phosphate
*Smirnov,E.V., N.N.Mijailov y V.V. Rossov,1967.
Algunas características hidroquimicas del Mediterráneo americano.
Estudios Inst. Oceanol.,Acad.Ciencias,Cuba,2(1): 51-74.

chemistry, phosphate
Smith, F.G.W., R.H. Williams, and C.C. Davis, 1950.
An ecological survey of the subtropical inshore waters adjacent to Miami. Ecol. 31(1):119-146, 7 textfigs.

chemistry phosphates (data)
Soot-Ryen, T., 1947.
Hydrographical investigations in the Tromsø-district 1934-1938 (Tables.
Tromsø Mus. Aarsheft., Naturhist Avd. No. 33, 66(1943)(1): numperous pp. (unnumbered).

chemistry phosphate (data)
Soule, F.M., and J.E. Murray, 1956.
Physical oceanography of the Grand Banks and the Labrador Sea in 1955.
U.S.C.G. Bull., No. 41:59-114.

chemistry, phosphate
Spencer, R.S., 1956.
Studies in Australian estuarine hydrology.
Australian J. Mar. & Freshwater Res. 7(2):193-253.

chemistry phosphate
Steele, J.H., 1956.
Plant production on the Fladen Ground. J.M.B.A. 35(1):1-33.

chemistry Phosphate
Steele, J. H., and I. E. Baird, 1962.
Further relations between primary production, Chlorophyll and particulate carbon.
Limnol. & Oceanogr., 7(1):42-47.

chemistry, phosphate
Steemann-Nielsen, Einar, 1951
The marine vegetation of the Isefjord. A study on ecology and production. Medd. Komm. Danmarks Fiskeri-og Havundersøgelser. Ser. Plankton. 5(4); 114pp., 46 text figs.

chemistry, phosphate
Steemann Nielsen, E., 1940.
Die Produktionsbedingungen des Phytoplanktons im Übergangsgebiet zwischen der Nord- und Ostsee.
Medd. Komm. Havundersøgelser, Ser. Plankton, 3(4):55 pp., 17 textfigs., 22 tables.

chemistry phosphate
Steemann Nielsen, E., and E.A. Jensen, 1957.
Primary oceanic production. The autotrophic production of organic matter in the oceans.
Galathea Repts. No. 1:49-136.

chemistry phosphate
Stefansson,U.,1968.
Nitrate-phosphate relationships in the Irminger Sea.
J.Cons.perm.int.Explor.Mer,32(2):188-200.

chemistry, phosphate
Stefansson, Unstein, 1966.
Influence of the Surtsey eruption on the nutrient content of the surrounding seawater.
J. Mar. Res., 24(2):241-268.

chemistry phosphate
Stefansson, Unnstein, and Francis A. Richards, 1963.
Processes contributing to the nurtient distributions off the Columbia River and Strait of Juan de Fuca.
Limnology and Oceanography, 8(4):394-410.

chemistry
phosphate
Stephens, K., 1963
Determination of low phosphate concentrations in lake and marine waters.
Limnology and Oceanography, 8(3):361-362.

chemistry phosphate
Stephensen, W., 1951.
Preliminary observations upon the release of phosphate from estuarine mud. Proc. Indo-Pacific Fish Counc., 17-28 Apr. 1950, Cronulla N.S.W., Australia, Sects. II-III:184-189, 1 textfig.

chemistry phosphate
Stephensen, W., 1949.
Certain effects of agitation upon the release of phosphate from mud. J.M.B.A. 28(2):371-380.

chemistry, phosphate
Stroup, E.D., and J.H. Wood 1966.
Atlas of the distribution of turbidity, phosphate, and chlorophyll in Chesapeake Bay, 1949-1951.
Chesapeake Bay Inst, Johns Hopkins Univ., Ref. 66-1: 193pp. (Unpublished manuscript).

chemistry
phosphate
Subrahmanyan, R., 1959.
Studies on the phytoplankton of the west coast of India. 1. Quantitative and qualitative fluctuation of the total phytoplankton crop, the zooplankton crop and their interrelationship, with remarks on the magnitude of the standing crop and production of matter and their relationship to fish landings. 2. Physical and chemical factors influencing the production of phytoplankton, with remarks on the cycle of nutrients and on the relationship of the phosphate content to fish landings.
Proc. Indian Acad. Sci., 1:113-252.

chemistry Phosphate
Sugiura Yoshio, 1969.
Identification of the source region of the Kuroshio Water by the conservative phosphate concentration.
J. oceanogr. Soc. Japan, 25(5): 229-232.

Chemistry, phosphate
Sugiura, Yoshio 1969.
The relationship between phosphate and oxygen concentrations in the upwelling waters and its oceanographical significance.
Oceanogrl Mag. 21(1): 47-52.

chemistry
Phosphate
Sugiura Yoshio, 1965.
Some chemico-oceanographic properties of the Kuroshio and its adjacent regions.
Proc. Symp., Kuroshio, Tokyo, Oct. 29, 1963, Oceanogr. Soc., Japan, and UNESCO, 12-13.

Chemistry phosphate
Suguira, Yoshio, 1964.
Some chemico-oceanographical properties of the Kurishio and its adjacent regions.
In: Recent researches in the fields of hydrosphere, atmosphere and nuclear geochemistry, Ken Sugawara festival volume, Y. Miyake and T. Koyama, editors, Maruzen Co., Ltd., Tokyo, 49-63.

chemistry
phosphate
Sugiura, Yoshio, 1963
Some chemico-oceanographical properties of the water of Suruga Bay. (In Japanese; English abstract).
J. Oceanogr. Soc., Japan, 18(4):193-199.

chemistry, phosphate
Sugiura, Yoshio, Yukio Sugimura, Hirozo Yoshimura and Tsuguo Shimano, 1963
Preparation of a stable, dilute solution of phosphate standard for routine working aboard the ship.
Oceanogr. Mag., Tokyo, 15(1):1-6.

chemistry
phosphate
Sugiura, Yoshio, and Hirozo Yoshimura, 1964.
Distrubution and mutual relation of dissolved oxygen and phosphate in the Oyashio and the northern part of Kuroshio.
J. Oceanogr. Soc., Japan., 20(1):14-23.

chemistry, phosphate
Sugiyama, Teruyuki, Yoshio Miyake and Kuniyasu Fujisawa, 1971.
On the distribution of inorganic nutrients along the coastal waters of Okayama Prefecture (1970).
Bull. Fish. Exp. Sta. Okayama Pref. (1970): 22-28. (In Japanese)

chemistry
phosphate
Sutcliffe, William H., Jr., Edward R. Baylor and David W. Menzel, 1963
Sea surface chemistry and Langmuir circulation.
Deep-Sea Res., 10(3):233-243.

chemistry phosphate (data)
Svansson,A., 1966.
Hydrography of the Kattegat area,Swedish observations.
Annls biol., Copenh., 21:37-40.

Chemistry, phosphate
Svansson, Artur, 1964.
Observations at Swedish lightships and in the central Baltic.
Ann. Biol., Cons. Perm. Int. Expl. Mer, 1962, 19:46-48.

chemistry, phosphate (data).
Svansson, Artur, 1963.
Hydrography of the Kattegat area. Swedish data.
Ann. Biol., Cons. Perm. Int. Expl. Mer, 1961, 18: 44-46.

chemistry
phosphate
Svansson, Artur, 1960(1958)
Observations at Swedish light vessels.
Ann. Biol., Cons. Perm. Int. Expl. Mer, 15: 43-45.

chemistry phosphate
Sverdlov, A.I., 1965.
Distribution of hydrochemical elements in the area of Feroe-Shetland Strait in November-January 1963-1964. (In Russian).
Material. Ribochoz. Issled. Severn. Bassin. Poliarn. Nauchno-Issled. i Proekt. Inst. Morsk. Rib. Choz. i Okeanogr., N.M. Knipovich, 5:91-98.

Sverdrup, H. U., 1933
On vertical circulation in the ocean due to the action of the wind with application to conditions within the Antarctic Circumpolar Current. Discovery Repts. 7:139-169, 23 text figs.

Chemistry, phosphates
Sykes, J.B. and A.D Boney 1970.
Seasonal variations in inorganic phytoplankton nutrients in the inshore waters of Cardigan Bay.
J. mar. biol. Ass. U.K. 50 (3): 819-827.

Chemistry, phosphate
Szekielda, Karl-Heinz, 1968.
Ein chemisches Modell für den Auf-und Abbau organischen Materials und dessen Anwendung in der offenen See.
J. Cons.perm.int.Explor.Mer, 32(2):180-187.

T

phosphate(data)
Tanaka, Otohiko, Haruhiko Irie, Shozi Iizuka and Fumihiro Koga, 1961
The fundamnetal investigation on the biological productivity in the norty-west of Kyushu. 1. Rec . Oceanogr. Wks., Japan, Spec. No. 5: 1-58.

chemistry, phosphate
Taft, Bruce A., and John A. Knauss 1967.
The equatorial undercurrent of the Indian Ocean as observed by the Lusiad Expedition.
Bull. Scripps Inst. Oceanogr. 9:1-163.

Chemistry, phosphate
Tanita, S., S. Kato, and T. Okuda, 1950.
Studies on the environmental conditions of shellfish fields. 1. In the case of Hakodate Harbor. Bull. Fac. Fish., Hokkaido Univ., 1(1):18-27, 7 textfigs. (In Japanese; English summary).

Chemistry phosphate
Tanita, S., K. Kato, and T. Okuda, 1951.
Studies on the environmental conditions of shellfish-fields in the case of Muroran Harbour (2). Bull. Fac. Fish., Hokkaido Univ. 2(3):220-230.

Chemistry phosphate
Theisen, E. 1946
Tanafjorden. Enfinmarksfjords oceanografi. Fiskeridirektoratets skrifter Ser. Havundersøkelser. (Repts. on Norwegian Fishery and Marine Investigations) VIII (8):1-77, 23 text figs., 8 pls.

chemistry phosphate
Thomas, William H., 1966.
Surface nitrogenous nutrients and phytoplankton in the northeastern tropical Pacific Ocean.
Limnol. Oceanogr., 11(3):393-400.

Chemistry, phosphate
Thompson, E. F., 1939.
Chemical and physical investigations. The general hydrography of the Red Sea. John Murray Exped., 1933-34, Sci. Repts. 2(3):83-103, 12 textfigs.

Chemistry, phosphate
Thomsen, H., 1933.
The distribution of phosphate and nitrate in the North Sea in May 1932. Rapp. Proc. Verb., Cons. Perm. Int. Expl. Mer 85:42-46, 8 textfigs.

Chemistry phosphate
Thomsen, H., 1931.
Nitrate and phosphate content of Mediterranean water.
Rept. Danish Oceanogr. Exped., 1908-1910, 3(6): 14 pp., 2 figs., 1 pl.

chemistry, phosphate
Tomo, Aldo P. 1970.
Cadenas tróficas observadas en la bahía de puerto Paraiso (Peninsula Antarctica) en relación con las variaciones de la fertilidad de sus aguas
Contrib. Inst. antart. argentino 131: 14pp.

Chemistry
phosphate
Tranter, D.J., and B.S. Newell, 1963
Enrichment experiments in the Indian Ocean. Deep-Sea Res., 10(1/2): 1-9.

Chemistry
phosphate
Tratet, Gerard, 1964.
Variations du phytoplancton à Tanger.
Trav. Inst. Sci., Cherifien, Rabat, Ser. Botan., (29):204 pp.

Chemistry, phosphate
Tsuruta, Arao, Satoru Tawara and Tadashi Ueda 1969.
On the distribution of dissolved oxygen and nutrient salts in tuna fishing grounds of the central tropical Atlantic.
J. Shimonoseki Univ. Fish. 15(3): 295-305.

U

Chemistry
phosphate
Uda, Michitaka, 1962
Subarctic oceanography in relation to whaling and salmon fisheries.
Sci. Repts., Whales Res. Inst., No. 16:105-119.

Chemistry
phosphate
Udaya Varma Thirupad, P., and C.V. Gangadhara Reddy, 1959
Seasonal variations of the hydrological factors of the Madras coastal waters.
Indian J. Fish., 6(2):298-305.

Chemistry
phosphate (data)
University of Southern California, Allan Hancock Foundation, 1965.
An oceanographic and biological survey of the southern California mainland shelf.
State of California, Resources Agency, State Water Quality Control Board, Publ. No. 27:232 pp. Appendix, 445 pp.

Chemistry phosphate
Uppström, Leif, 1968.
Analysis of boron in sea water by a modified curcumin method.
Rept.Chem.Sea Water, Univ.Göteborg, 4:8-13. (mimeographed).

Chemistry
phosphate (data)
Uyeno, Fukuzo, 1966.
Nutrient and energy cycles in an estuarine oyster area.
J. Fish. Res. Bd., Can., 23(11):1635-1652.

Chemistry, phosphate
Uyeno, Fukuzo, 1964.
Relationships between production of foods and oceanographical condition of sea water in pearl farms. II. On the seasonal changes of sea water constituents and of bottom condition, and the effect of bottom cultivation. (In Japanese
J. Fac. Fish., Pref. Univ. Mie, 6(2):145-169.

V

Chemistry phosphate (data)
Vaccaro, Ralph F., 1963.
Available nitrogen and phosphorus and the biochemical cycle in the Atlantic off New England. J. Mar. Res., 21(3):284-301.

Chemistry phosphates
Vachon, A., R. Gaudry, and R. Bernard, 1938.
Contribution à l'étude des phosphates dans l'estuaire du Saint-Laurent. Publ. Sta. Biol., Saint-Laurent, Contr. No. 12:24 pp.

Chemistry, phosphate
Valdivia, Julio E., y Oscar Guillén, 1966.
Informe preliminar del Crucero de primavera 1965 (Cabo Blanco-Morro Sama).
Inst. Mar. Peru, Informe, No. 11:35-70.

Chemistry phosphates(data)
Varlet, F., 1960
Sur l'hydrologie du plateau continental africain du Cap des Palmes au Cap des Trois Pointes. Bull. Inst. Pêches Maritimes, Maroc, No. 5: 20 pp.

Chemistry, phosphate
Vasiliev, G.D., y Iu. A. Torin, 1965.
Carastica oceanografica pesquera y biologica del Golfo de México y del Mar Caribe.(In Russian;Spanish abstract).
Sovetsk.-Cub. Ribokhoz. Issled., VNIRO:Tsentr. Ribokhoz. Issled. Natsional. Inst. Ribolovsta Republ. Cuba, 241-266.

Chemistry, phosphate
Vatova, A. 1965.
Les conditions hydrographiques de la Mar Piccolo de Tarante pendant l'annee 1963.
Rapp. P.-v. Réun. Commn int. Explor. scient. Mer Méditerr. 18(3): 653-655.

Chemistry, phosphates (data)
Viorel, Chirilă, 1965.
Observații asupra condițiilor fizico- chimice ale mării la Mamaia, în anii 1959 și 1960.
In: Ecologie marină, M. Băcescu, redactor, Edit. Acad. Republ. Pop. Romane, București, 1:139-184.

Chemistry
phosphate(data)
Vives, Francisco, y Manuel López-Benito, 1957
El fitoplancton de la Ría de Vigo dèsde julio de 1955 à junio de 1956.
Inv. Pesq., Barcelona, 10: 45-146.

Chemistry
phosphate(data)
Voit, S.S., D.A. Aksenov, M.M. Bogorodsky, V.V. Sinukov and V.A. Vladimirzev, 1961
[Some peculiar features of water circulation in the Black Sea and the water regime in the Pre-Bosphorus area.]
Okeanologiia, Akad. Nauk, SSSR, 1(4):613-625.

Chemistry phosphates
*Volkovinsky, V.V., 1967.
Application of the two-layer system to the computations of the primary productivity of sea water. (In Russian;English abstract).
Oceanologiia, Akad.Nauk, SSSR, 7(6):1037-1052.

phosphate
Volokhonsky, A.G., 1966.
Hydrochemical conditions in the Faroe-Iceland Channel in November-December 1964. (In Russian).
Mater. Ribokhoz. Issled. severn. Basseina, Poliarn. Nauchno-Issled. Proektn. Inst. Morsk. Ribn. Khoz. Okeanogr. (PINRO), 7:146-154.

W

Chemistry, phosphate
Waldichuk, M., 1956.
Basic productivity of Trevor Channel and Alberni Inlet from chemical measurements.
J. Fish. Res. Bd., Canada, 13(1):7-20.

Chemistry, phosphate
Watanabe, Nobuo, Toshiyuki Hirano, Rinnosuke Fukai, Kozi Matsumoto, and Fumiko Shiokawa, 1957.
A preliminary report on the oceanographic survey in the "Kuroshio" area, south of Honshu, June-July 1966.
Rec. Oceanogr. Wks., Japan (Suppl.):197-

Chemistry, phosphate
Wattenberg, H., 1938.
Die Verteilung des Sauerstoffs und des Phosphats im Atlantischen Ozean. Wiss. Ergeb., Meteor, 9(1):132 pp.

Chemistry, phosphate
Wattenberg, H., 1937.
Methoden zur Bestimmung von Phosphat, Silicat, Nitrate und Ammoniak im See Wasser. Rapp. Proc. Verb. 103(1):1-26.

Translated by N.W. Rakestraw.

ms

Chemistry, phosphate
Wattenberg, H., 1937.
IV. Die chemischen Arbeiten auf der "Meteor"-Fahrt Februar bis Maj 1937. Ann. Hydr., usw., 17-22, 3 textfigs.

Chemistry phosphate
Wattenberg, H., 1929.
7. Die Phosphat- und Nitrat-Untersuchungen der Deutschen Atlantischen Expedition auf V.S. "Meteor".
Rapp. Proc. Verb., Cons. Perm. Int. Expl. Mer, 53:90-94.

Transl. cited:
USFWS Spec. Sci. Rept., Fish. 227.

Chemistry phosphate
Watts, J.C.D., 1958.
The hydrology of a tropical west African estuary.
Bull. Inst. Francais, Afrique Noire, 20(3):697-752.

Chemistry, phosphate
Weichart Günter 1970.
Registrierung des Phosphat-Gehalts im Oberflächenwasser der Nordsee und des Armelkanals.
Ber. dt. wiss. Komm. Meeresforsch. 21(1/4): 410-419

Chemistry phosphate
Wennekens, M.P., 1959
Water mass properties of the Straits of Florida and related waters.
Bull. Mar. Sci., Gulf and Caribbean, 9(1): 1-52.

Chemistry phosphate
Won, Chong Hun, 1963.
Distribution of chemical constituents of the estuary water in Gwang-Yang Inlet. (In Korean; English abstract).
Bull. Fish. Coll., Pusan Nat. Univ., 5(1):1-10.

Chemistry, phosphate
Wooster, W.S., 1951.
Distribution of oxygen and phosphate in the Arctic Sea. Proc. Alaskan Sci. Conf., Bull. Nat. Res. Coun. 122:81.

Chemistry, phosphate
Wooster, W.S., and T. Cromwell, 1958.
An oceanographic description of the eastern tropical Pacific. Bull. S.I.O., 7(3):169-282.

Chemistry phosphate
Wooster, W.S., and N.W. Rakestraw, 1951.
The estimation of dissolved phosphate in sea water. J. Mar. Res. 10(1):91-100, 5 textfigs.

Chemistry phosphate
Wooster, W.S., and J.L. Reid, Jr., 1963
11. Eastern boundary currents. In: The Sea, M.N. Hill, Edit., Vol. 2 (III) Currents, Interscience publishers, New York and London, 253-280.

Chemistry phosphates
Wynne, E.A., R.D. Burdick and L.H. Fine, 1961.
Determination of microgram amounts of phosphates. A new general technique.
Microchem. J., 5(2):185-191.

Chemistry phosphate
Wyrtki, Klaus, 1964.
Upwelling in the Costa Rica dome.
U.S.F.W.S. Fish. Bull., 63(2):355-372.

XYZ

Chemistry, phosphate (data)
Yamanaka, Hajime, Noborum Anraku and Jiro Morita 1965.
Seasonal and long-term variations in oceanographic conditions in the western North Pacific Ocean
Rept. Nankai Reg. Fish. Res. Lab., No. 22:35-70.

Chemistry, phosphates
Yamazi, I., 1955.
Plankton investigations in inlet waters along the coast of Japan. XVI. The plankton of Tokyo Bay in relation to water movement.
Publ. Seto Mar. Biol. Lab., 4(2/3):285-309, Pls. 19-20, 22 textfigs.

Chemistry, phosphate (profiles)
Yamazi, I., 1953.
Plankton investigation in inlet waters along the coast of Japan. VII. The plankton collected during the cruises to the new Yamamoto Bank in the Sea of Japan. Publ. Seto Mar. Biol. Lab. 3(1): 75-108, 19 textfigs.

Chemistry, phosphates
Yamazi, I., 1952.
Plankton investigations in inlet waters along the coast of Japan. III. The plankton of Imari Bay in Kyusyu. IV. The plankton of Nagasaki Bay and Harbour in Kyusyu. V. The plankton of Hiroshima Bay in the Seto-Nakai (Inland Sea).
Publ. Seto Mar. Biol. Lab., Kyoto Univ., 2(2): 289-304; 305-318; 319-330, 8 and 8 and 7 textfigs.

Chemistry, phosphate
Yamazi, I., 1951.
Plankton investigations in inlet waters along the coast of Japan. II. The plankton of Hakodate Harbour and Yoichi Inlet in Hokkaido.
Publ. Seto Mar. Biol. Sta., Kyoto Univ., 1(4): 185-194, 3 textfigs.

Chemistry, phosphate
Yamazi, I., 1950.
Plankton investigations in inlet waters along the coast of Japan. 1. Publ. Seto Mar. Biol. Lab. 1(3):93-113, 14 textfigs.

Chemistry, phosphate
Zlobin, V.S., M.F. Perlyuk, and N.G. Sapronetskaya 1968.
Some regularities in the distribution of phosphate and oxygen in the Davis Strait, summer and winter 1965.
Annls biol. Copenh. 1966 23: 13-20.

Chemistry, phosphate
Zlobin, V.S., N.G. Sapronetskaya, and A.P. Alekseeva, 1968.
Some regularities in the formation of the hydrochemical regime and primary productivity of the Norwegian Sea. (In Russian).
Mater. Ribokhoz. Issled. Severn. Basseina, 11: 133-144.

Chemistry, phosphate
Zlobin, V.S., and Yu. G. Zhilin 1970.
On the kinetic approach to the modelling of exchange processes in marine phytoplankton. (In Russian).
Poliarn. nauchno-issled. Proektn. Inst. Morsk. Ribn. Khoz. Okeanogr. N.M. Knipovich (PINRO) 14: 122-138.

Chemistry phosphate
Zuta, Salvador, y Oscar Guillen, 1970.
Oceanografia de las aguas costeras del Peru.
Bol. Inst. Mar., Peru, 2(5):157-324.

Phosphate adsorption of

Chemistry
phosphates, adsorption of
Baylor, E.R., W.H. Sutcliffe and D.S. Hirschfeld, 1962
Adsorption of phosphates onto bubbles.
Deep-Sea Res., 9(2):120-124.

Chemistry, phosphate absorption
McRoy, C. Peter, and Robert J. Barsdate 1970.
Phosphate absorption in eelgrass.
Limnol. Oceanogr. 15 (1): 6-13.

phosphate/AOU

Chemistry
phosphate/AOU
Fonselius, Stig H., 1967.
Hydrography of the Baltic deep basins II.
Rep. Fishery Bd., Swed., Hydrogr., 20:31 pp.

chemistry, phosphate cycle
Postma, H., 1971.
Distribution of nutrients in the sea and the oceanic nutrient cycle. (English and Portuguese abstracts). In: Fertility of the Sea, John D. Costlow, editor, Gordon Breach, 2: 337-349.

Phosphate, data only

A

Chemistry
phosphate (data only)
Akamatsu, Hideo, Tsutomu Akiyama and Tsutomu Sawara, 1965.
Preliminary report of the Japanese Expedition of Deep-Sea, the Tenth Cruise, 1965(JEDS-10).
Oceanogr. Mag., Tokyo, 17(1/2):49-68.

Chemistry
phosphate (data only)
Anderson, William W., and Jack W. Gehringer, 1959.
Physical oceanographic, biological and chemical, South Atlantic coast of the United States, M/V Theodore N. Gill cruise9.
USFWS Spec. Sci. Rept., Fish. No. 313:226 pp.

Chemistry phosphate (data only)
Anderson, W.W., and J.W. Gehringer, 1959.
Physical oceanographic, biological and chemical data South Atlantic coast of the United States, M/V Theodore N. Gill Cruise 8.
USFWS Spec. Sci. Rept., Fish., No. 303:227 pp.

chemistry, phosphate (data only)
Anderson, William W., and Jack W. Gehringer 1959.
Physical oceanography, biological and chemical data, South Atlantic coast of the United States M/V Theodore N. Gill Cruise 7.
U.S. Fish Wildl. Serv. Spec. Scient. Rept. Fish. 278: 277 pp.

Chemistry phosphate (data only)
Anderson, William W., and Jack W. Gehringer, 1958.
Physical oceanographic, biological and chemical data, south Atlantic coast of the United States, M/V Theodore N. Gill Cruise 6.
USFWS Spec. Sci. Rept., Fish. No. 265:99 pp.

Chemistry phosphate (data only)
Anderson, W.W., and J.W.Gehringer, 1958.
Physical oceanographic, biological and chemical data, South Atlantic coast of the United States, M/V Theodore N. Gill Cruise 5.
USFWS Spec. Sci. Rept., Fish., 248:220 pp.

Chemistry phosphate(data only)
Anderson, W.W., and J.W. Gehringer, 1957.
Physical oceanographic, biological and chemical data, South Atlantic coast of the United States, M/V Theodore N. Gill cruise 4.
USFWS Spec. Sci. Rept., Fish., 234:192 pp.

Chemistry, phosphate (data only)
Anderson, W.W., and J.W. Gehringer, 1957.
Physical oceanographic, biological and chemical data, South Atlantic coast of the United States, Theodore N. Gill Cruise.
USFWS Spec. Sci. Rept., Fish., No. 210:208 pp.

Chemistry phosphate (data only)
Anderson, W.W., J.W. Gehringer and E. Cohen, 1956
Physical oceanographic, biological and chemical data, South Atlantic coast of the United States, Theodore N. Gill, Cruise 2.
Spec. Sci. Rept., Fish., No. 198:270 pp.

Chemistry, phosphate (data only)
Anderson, W.W., J.W. Gehringer and E. Cohen, 1956.
Physical oceanographic, biological and chemical data, south Atlantic coast of the United States, M/V Theodore N. Gill, Cruise 1.
USFWS Spec. Fish. Rept., Fish.,No. 178:160 pp.

Chemistry phosphate (data only)
Anon., 1951.
Report of the oceanographical observations on board the R.M.S. "Syunpu Maru in the Kii-Suido (Oct. 1950) J. Ocean., Kobe Obs., 2nd ser., 1(6):1-15, 7 figs.

Chemistry, phosphate (data only)
Anon., 1951.
Bulletin of the Marine Biological Station of Asamushi 4(3/4): 15 pp.

Chemistry, phosphate (data only)
Anon., 1950.
Oceanographical observations taken off the Sanriku coast; physical and chemical data for stations occupied by R.M.S. "Oyashio-maru". Res. Mar. Met., Ocean. Obs., Jan.-June 1947, No. 1:1-11, Table I.

Chemistry phosphate (data only)
Anon., 1950.
Oceanographical observations taken in the North Pacific Ocean (east off Hideshima Is.); physical and chemical data for stations occupied by R.M.S. "Kuroshio-maru" (continued). Res. Mar. Met., Ocean. Obs., Jan.-June 1947, No. 1:12-23, Table 2.

Chemistry, phosphate (data only)
Anon., 1950.
Oceanographic observations taken in the Miyako Bay; physical and chemical data for stations occupied by R.M.S. "Kuroshio-maru". Res. Mar. Met., Ocean. Obs., Jan.-June 1947, No. 1:24-37, Table 3.

Chemistry, phosphate (data only)
Aragno, Federico J., Alberto Gomez, Aldo Orlando y Andres J. Lusquiños 1968.
Datos y resultados de las campañas pesqueras "Pesqueria III" (20 de febrero al 20 de marzo de 1967).
Publ. (Ser. Informes tecn.) Mar del Plata, Argentina 10(3): 1-162.

Chemistry, phosphate (data only)
Aragno, Federico, Alberto Gomez, Aldo Orlando y Andres J. Lusquiños 1968.
Datos y resultados de las campañas pesquerias "Pesqueria II" (9 de noviembre al 12 diciembre de 1966).
Publ. (Ser. Informes tecn.) Mar del Plata, Argentina 10(2): 1-129.

Chemistry phosphate (data only)
Argentina, Secretaria de Marina, Servicio de Hidrografia Naval, 1961.
Operacion oceanografica, Vema - Canepa 1. Resultados preliminares.
Publico, H. 628:30 pp.

Chemistry phosphate (data only)
Argentina, Secretaria de Marina, Servicio de Hidrografia Naval, 1961.
Trabajos oceanograficos realizados en la campana Antartica 1959/1960. Resultados preliminares
Publico, H., 623:unnumbered pp.

Chemistry phosphate (data only)
Argentina, Secretaria de Marina, Servicio de Hidrografia Naval, 1961.
Trabajos oceanograficos realizados en la campana Antartica 1960/1961. Resultados preliminares.
Publico, H. 629:unnumbered pp.

Chemistry phosphate(data only)
Austin, T.S., 1957.
Summary, oceanographic and fishery data, Marquesas Islands area, August-September 1956, (EQUAPAC). USFWS Spec. Sci. Rept., Fish., No. 217:186 pp.

Chemistry phosphate (data only)
Austin, T.S., 1954.
Mid-Pacific oceanography. V. Transequatorial waters, May-June 1952, August 1952.
Spec. Sci. Rept., Fish., No. 136:1-85, 31 figs.

Chemistry, phosphate (data only)
Austin, T.S., 1954.
Mid-Pacific oceanography III. Transequatorial waters, August-October 1951.
U.S.F.W.S. Spec. Sci. Rept. - Fish. No. 131: numerous pp. (unnumbered), 13 figs. (multilithed)

Chemistry phosphate (data only)
Australia, Commonwealth Scientific and Industrial Organization, 1968.
Coastal investigations off Port Hacking, New South Wales, in 1964.
Oceanigr. Sta. List, 84: 49 pp. (multilithed)

Chemistry phosphate (data only)
Australia, Commonwealth Scientific and Industrial Organization, 1968.
Coastal investigations off Port Hacking, New South Wales, in 1961.
Oceanogr. Sta. List, 81: 55 pp. (multilithed)

chemistry, phosphate (data only)
Australia, Commonwealth Scientific and Industrial Research Organization 1968.
Oceanographical observations in the Pacific Ocean in 1964, H.M.A.S. Gascoyne, Cruise G 6/64.
Oceanogr. Cruise Rept. 42: 53 pp.

Chemistry phosphate (data only)
Australia, Commonwealth Scientific and Industrial Research Organization, 1964.
Oceanographical observations in the Indian Ocean in 1961, H.M.A.S. Diamantina, Cruise Dm 3/61.
Div. Fish. and Oceanogr., Oceanogr. Cruise Rept., No. 11:215 pp.

Chemistry phosphate (data only)
Australia, Commonwealth Scientific and Industrial Research Organization, 1964.
Oceanographical observations in the Indian Ocean in 1962, H.M.A.S. Diamantina, Cruise Dm 2/62.
Oceanogr. Cruise Rept., Div. Fish. and Oceanogr. No. 15:117 pp.

Chemistry phosphate (data only)
Australia, Commonwealth Scientific and Industrial Research Organization, 1963
Oceanographical observations in the Indian Ocean in 1961, H.M.A.S. Diamantina Cruise Dm 2/61.
Oceanogr. Cruise Rept., Div. Fish. and Oceanogr., No. 9: 155 pp., 14 figs.

Chemistry phosphate (data only)
Australia, Commonwealth Scientific and Industrial Research Organization, 1963.
Oceanographical observations in the Pacific Ocean in 1961, H.M.A.S. Gascoyne, Cruise G 1/61.
Oceanogr. Cruise Repts., Div. Fish. and Oceanogr. No. 8:130 pp., 12 figs.

Chemistry phosphate (data only)
Australia, Commonwealth Scientific and Industrial Organization, 1962.
Oceanographical observations in the Indian Ocean in 1960, H.M.A.S.DIAMANTINA Cruise Dm 3/60.
Oceanogr. Cruise Rept., Div. Fish. and Oceanogr., No. 4:39 pp.

Chemistry phosphate (inorganic) (data only)
Australia, Commonwealth Scientific and Industrial Research Organization, Division of Fisheries and Oceanography, 1963.
Oceanographical observations in the Indian Ocean in 1960, H.M.A.S. Diamantina, Cruise Dm 2/60.
Oceanographical Cruise Report No. 3:347 pp.

Chemistry, phosphate (data only)
Australia, Commonwealth Scientific and Industrial Research Organization, 1960
Coastal hydrological investigations in eastern Australia, 1959.
Div. Fish. and Oceanogr., Oceanogr. Sta. List, 45: 24 pp.

Chemistry, phosphate (data only)
Australia, Commonwealth Scientific and Industrial Organization, 1960
Coastal hydrological investigations in south-eastern Australia, 1958. Oceanogr. Sta. List. Div. Fish. and Oceanogr., 60 pp.

Chemistry phosphate (data only)
Australia, Division of Fisheries and Oceanography, C.S.I.R.O., 1959.
Coastal hydrological investigations in the New South Wales tuna fishing area, 1958.
Oceanogr. Sta. Lists, 38:96 pp.

Chemistry phosphate (data only)

Australia, C.S.I.R.O., Division of Fisheries, 1959

Coastal hydrological investigations at Eden, New South Wales, 1957.
Oceanogr. Sta. List, 35: 36 pp.

Chemistry phosphate (data only)

Australia, C.S.I.R.O., Division of Fisheries, 1959

Coastal hydrological investigations at Port Hacking, New South Wales, 1957.
Oceanogr. Sta. List, 34:72 pp.

Chemistry phosphate (data only)
Australia, Commonwealth Scientific and Industrial Research Organization, 1957.
Estuarine hydrological investigations in eastern and south-western Australia, 1956.
Oceanogr. Sta. List, Div. Fish. and Oceanogr., 32:170 pp.

Chemistry, phosphate (data only)
Australia, Commonwealth Scientific and Industrial Research Organization, 1957.
Etsuarine hydrological investigations in eastern and south-western Australia, 1955.
Ocean. Sta. List, 29:93 pp.

Chemistry phosphate (data only)
Australia, Commonwealth Scientific and Industrial Research Organization, 1957.
Onshore and oceanic hydrological investigations in eastern and southwestern Australia, 1955.
Oceanogr. Sta. List, 27:145 pp.

Chemistry, phosphate (data only)
Australia, Commonwealth Scientific and Industrial Research Organization, 1956.
Onshore hydrological investigations in eastern and south-western Australia, 1954. Compiled by D.R. Rochford.
Ocean. Sta. List, Invest. Div. Fish., 24:119 pp.

Chemistry, phosphate (data only)
Australia, Commonwealth Scientific and Industrial Research Organization, 1954.
Onshore hydrological investigations in eastern and south-western Australia, 1953.
Ocean. Sta. List, Invest. Div. Fish. 18:64 pp.

Chemistry, phosphate(data)
Australia, Commonwealth Scientific and Industrial Research Organization, 1953.
Onshore hydrological investigations in eastern and south-western Australia, 1951. Vol. 14:64 pp.

Chemistry, phosphate (data only)
Australia, Commonwealth Scientific and Industrial Research Organization, 1952.
Records of twenty-four hourly hydrological observations at Shell Point, Georges River, New South Wales. (Compiled by D.R. Rochford).
Ocean. Sta. List, Invest. Fish. Div., No. 10:134 pp.

Chemistry, phosphate (data only)
Australia, Commonwealth Scientific and Industrial Research Organization, Melbourne, 1952.
Oceanographical station list of investigations made by the Fisheries Division, C.S.I.R.O.
Vol. 5. Estuarine hydrological investigations in eastern Australia, 1940-50:150 pp.
Vol. 6. Ibid.:137 pp.
Vol. 7. Ibid. 139 pp.
Vol. 8:Hydrological investigations in south-western Australia, 1944-50:152 pp.
Vol. 10. Records of twenty-four hourly hydrological observations at Shell Point, Georges River, New South Wales:1942-50:134 pp.

Chemistry, phosphate(data)
Australia, Commonwealth Scientific and Industrial Research Organization, 1951.
Oceanographic station list of investigations made by the Fisheries Division, C.S.I.R.O.
Vol. 4. Onshore hydrological investigations in eastern Australia, 1942-50:114 pp.

B

Chemistry phosphate (data only)
Ballester, Antonio, 1965(1967).
Tablas hidrograficas.
Memoria Soc.Cienc.nat.La Salle, 25(70/71/72):41-137.

Chemistry - phosphate (data only)
Ballester Antonio Enrique Arias Antonio Cruzado, Dolores Blasco y José María Camps, 1967.
Estudio hidrográfico de la costa catalana de junio de 1965 a mayo de 1967.
Investigación, pesq. 31(3): 621-662

Chemistry, phosphate (data only)
Brasil, Diretoria de Hidrografia e Navegação 1970.
IV, V e VI Comissões oceanográficas nó Almirante Saldanha (29/5 a 4/6/1957) (5/10 a 17/10/1957) (26/11 a 4/12/1957)
DG 20-IV: 149 pp.

phosphate (data only)
Brasil, Diretoria de Hidrografia e Navegação, 1969
III Comissão oceanográfica no Almirante Saldanha (20/3 a 16/4/1957), DG 20-III: 73 pp.

Chemistry, phosphate (data only)
Brasil, Diretoria de Hidrografia e Navegação, 1969.
II Comissão Oceanográfica no Almirante Saldanha (15 A 28/2/1957).
DG20-II:60 pp.

Chemistry
phosphate (data only)

Brasil, Diretoria de Hidrografia e Navegacao, 1968.
XXXV Comissao Oceanografica "Operacao Norte/Nordeste 1" noc Almirante Saldinha (14/9 a 16/12/1967).
DG 26-XI: 600 pp.

Chemistry phosphate (data only)
*Brazil, Diretoria de Hidrografia e Havegação, 1968.
XXXI Comissao oceanografica noc Almirante Saldahha (14/11 a 16/12/66, DG26-VIII:249 pp.

Chemistry phosphate (data only)
Brazil, Diretoria de Hidrografia e Navegação 1967.
XXXII Comissão Oceanográfica nó Almirante Saldanha (14/3 a 3/5/67) DG 26-X: 411 pp.

Chemistry
phosphate(data only)
Brasil, Marinha do Brasil, Diretoria de Hidrografia e Navegação, 1963
Operação, "TRIDENTE I", Estudo das condições oceanográficas entre o Rio de Janeiro e o Rio da Prata, durante o inverno (Agosto-Septembro) ano de 1962.
DG-06-XV:unnumbered pp. (Mimeographed).

Chemistry
phosphate(data only)
Brasil, Diretoria de Hidrografia e Navegação, Marinha do Brasil, 1962
Estudo das condições oceanográficas entre o Rio de Janeiro e o Rio da Prata, durante o outono (Maio de 1962).
DG-06-XIV:unnumbered pp., charts, (mimeographed

Chemistry
phosphate(data only)
Brasil, Diretoria de Hidrografia e Navegacao 1961
Estudo das condições oceanográficas nas proximidades do Rio de Janeiro durante o mês de Dezembro.
DG-06-XIII:mimeographed sheets.

Chemistry, phosphate (data only)
Brasil, Marinha do Brasil, Diretoria de Hidrografia e Navegacao, 1960.
Estudo das condicoes oceanograficas na regiao profunda a Nornordeste de Natal, Estado do Rio Grande do Norte.
DG-06-XI (Sept. 1960):unnumbered pp. (mimeographed).

Chemistry, phosphate (data only)
Brasil, Marinha do Brasil, Diretoria de Hidrografia e Navegação, 1959.
Levantamento oceanografico da costa nordeste.
DG-06-IX:unnumbered pp. (mimeographed).

Chemistry, phosphate (data only)
Brazil, Diretoria de Hidrografia e Navegacao, 1958.
Ano Geofisico Internacional, Publ. DG-06-VII: mimeographed sheets

Chemistry phosphate (data only)
Brazil, Diretoria de Hidrografia e Navegacao, 1957
Ano Geofisico Internacional, Publicacao DG-06-VI: unnumbered mimeographed pages.

Chemistry, phosphate(data)
Brazil, Diretoria de Hidrografia e Navegacao, 1957.
Ano Geofisico Internacional. Publicacao DG-06-V: 3 pp., 18 figs., 3 pp., of tables.

Chemistry, phosphate (data only)
Brazil, Diretoria de Hidrografia e Navegacao, 1957.
Ano Geofisico Internacional. Publ. DG-06-III and IV (mimeographed).

Chemistry phosphate (data only)
Brazil, Diretoria de Hidrografia e Navegacao, 1957.
Ano Geofisico Internacional. Publicacao DG-06-II mimeographed.

Chemistry, phosphate (data)
Brujewicz, S.W., 1957.
On certain chemical features of waters and sediments in north-west Pacific.
Proc. UNESCO Symp., Phys. Ocean., Tokyo, 1955: 277-292.

Chemistry Phosphate (data only)
Buljan, M., and M. Marinkovic, 1956.
Some data on hydrography of the Adriatic (1946-1951). Acta Adriatica 7(12):1-55.

Chemistry, phosphate (data only)
Buljan, Miljenko, and Mira Zore-Armanda 1966.
Hydrographic data on the Adriatic Sea collected in the period from 1952 through 1962.
Acta adriat. 12:438 pp.

C

Chemistry phosphate (data only)
Callaway, Richard J., and James W. McGary, 1959.
Northeastern Pacific albacore survey. 2. Oceanographic and meteorological observations.
USFWS Spec. Sci. Rept., Fish., No. 315:133 pp.

Chemistry, phosphate (data only)
Canada, Fisheries Research Board, 1959.
Data record, Ocean Weather Station "P" (Latitude 50 00'N, Longitude 145 00'W) January 22 - July 11, 1958.
MSS Rept. Ser. (Oceanogr. & Limnol.), No. 31: 112 pp.

Chemistry, phosphate (data only)
Canada, University of British Columbia 1969.
British Columbia inlet cruises 1968
Data Rept. 28: 59 pp. (multilithed).

Chemistry phosphate (data only)
Canada, University of British Columbia, Institute of Oceanography, 1956.
Queen Charlotte Strait, 1956. Data Rept.:20 pp. (mimeographed) No. 9

Chemistry, phosphate (data)
Carlsberg Foundation, 1937.
Hydrographical Observations made during the "Dana" Expedition with an introduction by Helge Thomsen. Dana Rept. No. 12: 46 pp.

Chemistry phosphate (data only)
*Charnell, Robert L., David W.K. Au and Gunter R. Seckel, 1967.
The Trade Wind Zone Oceanography Pilot Study 5. Townsend Cromwell cruises 14 and 15, March and April 1965.
Spec. scient. Rep. U.S. Fish Wildl. Serv., Fish. 556:54 pp.

Chemistry phosphate (data Only)
*Charnell, Robert L., David W.K. Au and Gunter R. Seckel, 1967.
The Trade Wind Zone Oceanography Pilot Study 6. Townsend Cromwell cruises 16, 17and 21.
Spec. Scient. Rep. U.S. Fish. Wildl. Serv., Fish., 557-59 pp.

Chemistry phosphate(data)
Chesapeake Bay Institute, 1954.
Choptank River spring cruise 28 April-1 May 1954 Ref. 54-1:37 pp.

Chemistry phosphate (data)
Chesapeake Bay Institute, 1952.
Data report, St. Mary's River cruise, June 19-July 18, 1952. Rept. 11, Ref. 52-19:115 pp.

Chemistry, phosphate (data only)
Chesapeake Bay Institute, 1951.
Data report, Cruise VIII, Jan. 10, 1951-Jan. 23, 1951. Rept. No. 6:29 pp. (duplicated).

Chemistry, phosphate (data only)
Chesapeake Bay Institute, 1951.
Data report, Cruise VII, 14 Oct. 1950-2 Nov. 1950. Rept. No. 5:41 pp. (duplicated).

Chemistry, phosphate (data only)
Chesapeake Bay Institute, 1951.
Data report. Cruise VII. October 14, 1950-November 2, 1950. Rept. No. 2:41 pp. (duplicated).

Chemistry, phosphate
Chesapeake Bay Institute, 1950.
Data Report, Cruises V and VI, May 20, 1950-May 25, 1950, July 14, 1950-July 19, 1950. Rept. No. 4:51 pp. (mimeographed).

Chemistry, phosphate (data)
Chesapeake Bay Institute, 1950.
Data Report. Cruise IV, March 25, 1950-April 25, 1950. Rept. No. 3:49 pp. (mimeographed).

Chemistry phosphate (data)
Chesapeake Bay Institute, 1949.
Data Report, Cruise III, October 10, 1949-October 25, 1949. Rept. No. 2:39 pp. (mimeographed).

Chemistry, phosphate (data only)
Chesapeake Bay Institute, 1949.
Quarterly report, Jult 1, 1949-October 1, 1949. Rept. No. 1: 121 pp. (multilith).

Chemistry phosphate (data only)
Conseil International pour l'Exploration de la Mer, 1968.
Cooperative synoptic investigation of the Baltic 1964. Sweden.
ICES Oceanogr. Data Lists 4:181 pp. (mimeographed)

Chemistry phosphate (data only)
Conseil International pour l'Exploration de la Mer, 1968.
Cooperative synoptic investigation of the Baltic 1964. 5. U.S.S.R.
ICES Oceanogr. Data Lists: 173 pp. (multilithed).

Chemistry, phosphate (data only)
Conseil International pour l'Exploration de la Mer, 1968.
Cooperative synoptic investigation of the Baltic 1964. 3. Poland.
ICES Oceanogr., Data Lists: 266 pp. (multilithed).

Chemistry phosphate (data only)
Conseil Permanent International pour l'Exploration de la Mer, 1967.
ICES Oceanographic data lists, 1960, No. 6: 295 pp. (multilithed).

Chemistry phosphate (data only)
Conseil Permanent International pour l'Exploration de la Mer, 1957.
Bulletin hydrographique pour l'année 1953:167 pp.

Chemistry, phosphate (data only)
Conseil Permanent International pour l'Exploration de la Mer, 1956.
Bulletin hydrographique pour l'année 1952:164 pp.

Chemistry, phosphate (data only)
Conseil Permanent International pour l'Exploration de la Mer, 1955.
Bulletin Hydrographique pour l'année 1951, 131 pp

Chemistry, phosphate -P (data)
Conseil Permanent International pour l'Exploration de la Mer, 1954.
Bulletin Hydrographique pour l'Année 1950: 114 pp., 5 charts.

Chemistry, phosphate (data)
Conseil Permanent International pour l'Exploration de la Mer, 1954.
Bulletin hydrographique pour l'Annee 1949, 85 pp., 4 charts.

Chemistry, phosphate(data)
Conseil Permanent International pour l'Exploration de la Mer, 1952.
Bulletin Hydrographique, 1948:87 pp.

Chemistry, phosphate(data only)
Conseil Permanent International pour l'Exploration de la Mer, 1950.
Bulletin hydrographiques pour les années 1940-1946, avec un appendix pour les années 1936-1939. 190 pp.

Chemistry, phosphate (data)
Conseil Permanent International pour l'Exploration de la Mer, 1939
Bulletin Hydrographique pour l'annee 1937: 106 pp., 4 maps., 4 figs.

Chemistry, phosphate (data only)
Conseil Permanent International pour l'Exploration de la Mer 1936.
Bulletin hydrographique pour l'année 1935: 105 pp.

Chemistry, phosphate (data only)
Côte d'Ivoire, Centre de Recherches Océanographiques, 1963.
Résultats hydrologiques effectuées au large de la Côte d'Ivoire de 1956-1963. Première partie: observations, 1956-1963.
Trav. Centre de Recherches Océanogr., Ministère de la Production Animale, Côte d'Ivoire, unnumbered pp. (multilithed)

D

Chemistry phosphate (data only)
Dragovich, Alexander, John H. Finucane and Billie Z. May, 1961.
Counts of red tide organisms, Gymnodinium breve, and associated oceanographic data from Florida west coast, 1957-1959.
USFWS Spec. Sci. Rept., Fish., No. 369:175 pp.

E

F

Chemistry, phosphate (data only)
Favorite, Felix, and Glenn M. Pedersen, 1959.
North Pacific and Bering Sea oceanography, 1957
USFWS Spec. Sci. Rept., Fish., No. 292:106 pp.

Chemistry phosphate(data only)
Favorite, Feli, John W. Schantz and Charles R. Hebard, 1961
Oceanographic observations in Bristol Bay and the Bering Sea 1939-41 (USCGT Redwing).
U.S.F.W.S. Spec. Sci. Rept., Fish. No. 381: 323 pp.

Chemistry phosphate (data only)
Fonselius, S.H., 1967.
Hydrographical observations at Swedish lightships and in the central Baltic, 1965.
Annls. biol. Copenh. (1965)22:54-56.

Chemistry phosphate (data only)
France, Centre de Recherches et d'Etudes Océanographiques, 1960.
Stations hydrologiques effectuées par le "Passeur du Printemps" dans le cadre des travaux de l'Année Géophysique Internationale.
Travaux, C.R.E.O., 3(4):17-22.

phosphate (data only)
France, Service Hydrographique de la Marine, 1959.
Observations du "Pyrrhus" bâtiment de recherche de Travaux Publics d'Afrique Occidentale Française.
Cahiers Océanogr., C.C.O.E.C., 11(1):63-73.

Chemistry phosphate(data only)
France, Service Hydrographique de la Marine, 1959
Resultats d'observations hydrologiques. 1. Batiments de la Marine Nationale. 2. Observations du "Pyrrhus". 3. Observations du niveau marin.
Cahiers Oceanogr., 11(1):54-62.

Chemistry phosphates (data)
Fuse, Shin-ichiro, 1959.
A study on the productivity of Tanabe Bay III.
Oceanographic conditions of Tanabe Bay (2) Stratification and fluctuation of hydrological conditions on two sectional survey lines.
Rec. Oceanogr. Wks., Japan, Spec. No. 3:31-45.

G

Chemistry phosphate(data only)
Grainger, E.H. 1971
Biological oceanographic observations in Frobisher Bay. I. Physical, nutrient and primary production data, 1967-1971.
Techn. Rept. Fish. Res. Bd Can. 265: 75 pp (multilithed)

Chemistry phosphates (data only)
Granqvist, G., 1955.
The summer cruise with M/S Aranda in the northern Baltic, 1954. Merent. Julk., No. 166:56 pp.

Chemistry phosphate (data only)
Great Britain, Discovery Committee, 1949.
Station list, R.R.S. "William Scoresby".
Discovery Rept., 25(144-280, Pls. 34-37.

Chemistry phosphate (data only)
Great Britain, Discovery Committee, 1949.
Station list, R.R.S. William Scoresby, 1931-1938
Discovery Repts., 25:143-280, Pls. 34-37.

Chemistry phosphate (data only)
Great Britain, Discovery Committee, 1947.
Station list, 1937-1939. Discovery Repts., 24:
198-422, Pls. 4-6.

Chemistry phosphate (data only)
Great Britain, Discovery Committee, 1945.
Station list, 1935-1937. Discovery Repts., 24:
3-196, Pls. 1-3.

Chemistry phosphate (data only)
Great Britain, Discovery Committee, 1942.
Station list, 1933-1935. Discovery Repts. 22:3-
196, Pls. 1-4.

Chemistry phosphate (data only)
Great Britain, Discovery Committee, 1932.
Station list. Discovery Repts., 4:3-230, Pls. 1-
5.

Chemistry phosphate (data only)
Great Britain, Discovery Committee, 1931.
Station list. Discovery Repts., 3:3-132, Pls. 1-
10.

Chemistry phosphate (data only)
Great Britain, Discovery Committee, 1929.
Station list. Discovery Repts. 1:3-140.

Chemistry phosphate (data only)
Great Britain, Discovery Committee, 1955.
Station list, R.R.S. "William Scoresby".
Discovery Repts., 26:212-258, Pls. 11-12.

Chemistry phosphate (data only)
Great Britain, Ministry of Agriculture, Fisheries and Food, 1953.
Fisheries Laboratory, Lowestoft, Research Vessel "Ernest Holt"; hydrographical observations.
Unnumbered pages.

Chemistry phosphate (data only)
Great Britain, Ministry of Agriculture, Fisheries and Food, 1949.
Fisheries Laboratory, Lowestoft, Research Vessel "Ernest Holt"; hydrographical observations.
Unnumbered pages.

Chemistry, phosphate (data only)
Guillén, Oscar, y Francisco Vásquez 1966.
Informe preliminar del Crucero 6602 (Cabo Blanco - Arica).
Informe, Inst. Mar Peru 12:27 pp.

H

Chemistry phosphate (data only)
Hela, Ilmo, and Folke Koroleff, 1958
Hydrographical and chemical data collected in 1957 on board the R/V Aranda in the Barents Sea.
Merent. Julk., No. 179: 67 pp.

Chemistry phosphate (data only)
Hela, Ilmo, and F. Koroleff, 1958.
Hydrographical and chemical data collected in 1956 on board the R/V"Aranda" in the Baltic Sea.
Merent. Julk., No. 183:1-52.

Chemistry, Phosphate (data only)
Hisard, P., F. Jarrige, et P. Rual, 1968.
Résultats des observations physico-chimiques de la croisière CYCLONE 2 du N.O. Coriolis.
Rapp. sci., O.R.S.T.O.M., Nouméa, 20:21 pp. (mimeographed).

chemistry phosphate (data only)
Holmes, Robert W., and Maruice Blackburn, 1960
Physical, chemical and biological observations in the eastern tropical Pacific Ocean, Scot Expedition, April-June 1958.
USFWS Spec. Sci. Rept. Fish., No. 345:106 pp.

Chemistry, phosphate (data only)
Hufford, Gary L., and James M. Seabrooke, 1970.
Oceanography of the Weddell Sea in 1969 (IWSOE).
Oceanogr. Rept. U.S. Coast Guard 31 (CG 373-31): 32 pp.

I

chemistry, phosphate (data only)
Ilahude, Abdul Gani 1971.
Oceanographic Station List 1963-1966.
Special Issue, Oceanogr. Cruise Rept. (1963-1966), Djakarta, 67 pp.

Chemistry phosphate (data only)
India, Naval Headquarters, New Delhi, 1958.
Indian oceanographic station list, Ser. No. 2: unnumbered pp. (mimeographed)

Chemistry phosphate (data only)
India, Naval Headquarters, Office of Scientific Research & Development, 1957.
Indian Oceanographic Stations List, No. 1/57: unnumbered pp. (mimeographed).

J

Chemistry phosphate (data only)
Japan, Central Meteorological Observatory, 1956.
The results of marine meteorological and oceanographic pbservations (NORPAC Expedition Spec. No.) 17:131 pp.

Chemistry phosphate (data only)
Japan, Central Meteorological Observatory, 1953
The results of Marine meteorological and oceanographical observations. Jan.-June 1952, No. 11:362, 1 fig.

Chemistry phosphate (data only)
Japan, Central Meteorological Observatory. 1952
The results of Marine Meteorological and oceanographical observations, July - Dec. 1951, No. 10:310 pp., 1 fig.

chemistry phosphate (data only)
Japan, Central Meteorological Observatory, 1952.
The results of Marine meteorological and oceanographical observations. July - Dec. 1950. No. 8: 299 pp.

Chemistry phosphate (data only)
Japan, Central Meteorological Observatory, 1951.
Table 20. Surface observations made on board the "Kuroshio-maru"; physical and chemical conditions of the surface water of the sea between Tokyo and Hachijo Is. Res. Mar. Met. Ocean. Obs., Jan.-June, 1949, No. 5:108-115.

Chemistry phosphate (data only)
Japan, Central Meteorological Observatory, 1951.
Table 19. Surface observations made on board the "Tokitsu-maru"; physical and chemical conditions of the surface water of the sea between Tokyo and Hakodate. Res. Mar. Met. Ocean. Obs., Jan.-June, 1949, No. 5:106-108.

Chemistry phosphate (data only)
Japan, Central Meteorological Observatory, 1951.
Table 14. Oceanographical observations taken in Miyazu Bay: physical and chemical data for stations occupied by R.M.S. "Asanagi-maru". Res. Mar. Met. Ocean. Obs., Jan.-June, 1949, No. 5:77-84.

Chemistry phosphate (data only)
Japan, Central Meteorological Observatory, 1951.
Table 10. Oceanographical observations taken in the East China Sea. (A) Physical and chemical data for stations occupied by the "Akebono-maru No. 9"; (B) ------ by the "Hatsutaka-maru". Res. Mar. Met. Ocean. Obs., Jan.-June 1949, No. 5:68-70.

chemistry phosphate (data only)
Japan, Central Meteorological Observatory, 1951.
The results of marine meteorological and oceanographic observations. July - Dec. 1949. No. 6: 423 pp.

chemistry phosphate (data only)
Japan, Central Meteorological Observatory, 1951.
Table 9. Oceanographical observations taken in the Goto-nada and the Tsushima Straits. (A). Physical and chemical data for stations occupied by the fishing boart "Daikoku-maru": (B) ------- R.S.S. "Umikaze-maru".
Res. Mar. Met. Ocean. Obs., Jan.-June, 1949, No. 5:61-67.

Chemistry phosphate (data only)
Japan, Central Meteorological Observatory, 1951.
Table 8. Oceanographical observations taken in the Ariake Sea; physical and chemical data for stations occupied by R.S.S. "Umikaze-maru".
Res. Mar. Met. Ocean. Obs., Jan.-June, 1949, No. 5:50-60.

Chemistry phosphate (data only)
Japan, Central Meteorological Observatory, 1951.
Table 6. Oceanographical observations taken in the Akashi-seto, the Yura-seto and the Kii Suido; physical and chemical data for stations occupied by R.M.S. "Shumpu-maru".
Res. Mar. Met. Ocean. Obs., Jan.-June 1949, No. 5:40-47.

Chemistry phosphate (data only)
Japan, Central Meteorological Observatory, 1951.
Table 3. Oceanographical observations taken in the North Pacific Ocean along "C" line; (A) Physical and chemical data for stations occupied by R.M.S. "Ukuru-maru"; (B) --- "Ikuna-maru"; (C) --- "Ryofu-maru"; (D) --- "Shinnan-maru"; (E) --- "Kung-maru"; (F) --- "Chikubu-maru"; (G) --- "Ryofu-maru"; (H) --- "Ikuna-maru"; (I) --- "Chikubu-maru"; (J) --- "Ukuru-maru"; (K) --- "Shinnan-maru"; (L) --- "Ukuru-maru"; (M) --- "Ryofu-maru". Res. Mar. Met. Ocean. Obs., Jan.-June, No. 5:13-30.

Chemistry, phosphate (data only)
Japan, Hokkaido University, Faculty of Fisheries 1970.
Data record of oceanographical observations and exploratory fishing, 13: 406 pp.

Chemistry phosphate (data only)
Japan, Hokkaido University, Faculty of Fisheries, 1968.
The Oshoro Maru cruise 23 to the east of Cape Erimo, Hokkaido, April 1967.
Data Record Oceanogr. Obs. Expl. Fish., 12: 115-169.

Chemistry phosphate (data only)
Japan, Hokkaido University, Faculty of Fisheries, 1967.
Data record of oceanographic observations and exploratory fishing, 11:383 pp.

Chemistry, phosphate (data only)
Japan, Hokkaido University, Faculty of Fisheries, 1967.
The Oshoro Maru cruise 18 to the east of Cape Erimo, Hokkaido, April 1966.
Data Record Oceanogr. Obs. Explor. Fish. 11: 121-164.

Chemistry phosphate (data only)
Japan, Hokkaido University, Faculty of Fisheries, 1967.
The Oshoro Maru cruise 16 to the Great Australian Bight November 1965-February 1966.
Data Record Oceanogr. Obs. Explor. Fish., Fac. Fish., Hokkaido Univ. 11: 1-97; 113-119.

Chemistry, phosphate (data only)
Japan Hokkaido University, Faculty of Fisheries 1961.
Data record of oceanographic observations and exploratory fishing No. 5:391 pp.

Chemistry, phosphate (data only)

Japan, Hokkaido University, Faculty of Fisheries, 1960.
Data record of oceanographic observations and exploratory fishing, No. 4:221 pp.

Chemistry phosphate (data only)

Japan, Hokkaido University, Faculty of Fisheries, 1959.
Data record of oceanographic observations and exploratory fishing, No. 3:296 pp.

chemistry phosphate, (data only)
Japan, Hokkaido University, Faculty of Fisheries, 1957.
Data record of oceanographic observations and exploratory fishing, No. 1:247 pp.

chemistry phosphate (data only)
Japan, Hokkaido University, Faculty of Fisheries, 1955.
Correction of data presented in "Hydrographic dat data obtained principally in the Bering Sea by Training Ship 'Oshoro Maru' in the summer of 1955 published in January 1956:8 pp.

chemistry phosphate (data only)
Japan, Japan Meteorological Agency, 1966.
The results of the Japanese Expedition of Deep Sea (JEDS-8).
Results mar.met.oceanogr.Obsns.Tokyo, 35:328 pp.

Japan, Japan Meteorological Agency

Chemistry phosphate (data only)
Japan, Japan Meteorological Agency 1971.
The results of marine meteorological and oceanographical observations, July - December 1969, 46:270 pp.

Chemistry phosphate (data only)
Japan, Japan Meteorological Agency, 1971.
The results of marine meteorological and oceanographical observations, 45: 338 pp.

Chemistry phosphate (data only)

Japan, Japan Meteorological Agency, 1970.
The results of marine meteorological and oceanographical observations, July-December, 1968, 44:311 pp.

Chemistry, Phosphate (data only)

Japan, Japan Meteorological Agency, 1970

The results of marine meteorological and oceanographical observations. (The results of the Japaneses Expedition of Deep Sea (JEDS-11); January-June 1967 41: 332 pp.

chemistry, phosphate (data only)
Japan, Japan Meteorological Agency 1970?
The results of marine meteorological and oceanographical observations, July-December 1966, 40: 336 pp.

chemistry, phosphate (data only)
Japan Meteorological Agency 1969.
The results of marine meteorological and oceanographical observations Jan.-June 1966, 39:349 pp. (multilithed)

chemistry Phosphate (data only)
Japan Meteorological Agency, 1968.
The results of marine meteorological and oceanographical observations, July-December, 1965, 38: 404 pp. (multilithed)

Chemistry phosphate (data only)
Japan, Japan Meteorological Agency, 1966.
The results of the Japanese Expedition of Deep Sea (JEDS-10).
Res.mar.met.oceanogr.Observ., Jan-June 1965:37: 385.

Chemistry phosphate (data only)
Japan, Japanese Meteorological Agency, 1967.
The results of marine meteorological and oceanographical observations, July-December, 1964, 36: 367 pp.

Chemistry phosphate (data only)
Japan, Japan Meteorological Agency, 1964.
The results of marine meteorological and oceanographical observations, January-June 1963, No. 33:289 pp.

chemistry phosphate (data only)
Japan, Japan Meteorological Agency, 1964.
The results of the Japanese Expedition of Deep-Sea (JEDS-5).
Res. Mar. Meteorol. and Oceanogr. Obs., July-Dec. 1962, No. 32:328 pp.

Chemistry phosphate (data only)
Japan, Japan Meteorological Agency, 1964.
Oceanographic observations.
Res. Mar. Met. & Oceanogr. Obs., No. 31:220 pp.

Chemistry Phosphate (data only)
Japan, Japan Meteorological Agency, 1962
The results of marine meteorological and oceanographical observations, July-December 1961, No. 30:326 pp.

Chemistry phosphate (data only)
Japan, Japan Meteorological Agency, 1962
The results of marine meteorological and oceanographical observations, January-June 1961, No. 29: 284 pp.

Chemistry phosphates (data only)
Japan, Meteorological Agency, 1962
The results of marine meteorological and oceanographical observations, July-December, 1960, No. 28: 304 pp.

Chemistry, phosphate (data only)
Japan, Japan Meteorological Agency, 1961
The results of marine meteorological and oceanographical observations, January-June 1960. The results of the Japanese Expedition of Deep-Sea (JEDS-2, JEDS-3), No. 27: 257 pp.

Chemistry phosphate (data only)

Japan, Meteorological Agency, 1960.
The results of marine meteorological and oceanographical observations. Supplement, 149 pp.

Chemistry phosphate (data only)
Japan, Japan Meteorological Agency, 1960.
The results of marine meteorological and oceanographical observations, July-December 1959, No. 26:256 pp.

Chemistry phosphate (data only)

Japan, Japan Meteorological Agency, 1960.
The results of marine meteorological and oceanographical observations, Jan-June 1959, No. 25: 258 pp.

Chemistry phosphate (data only)
Japan, Japan Meteorological Agency, 1959.
The results of marine meteorological and oceanographical observations, July-December, 1958, No. 24:289 pp.

Chemistry phosphate (data only)

Japan, Japan Meteorological Agency, 1958
The results of marine meteorological and oceanographical observations, July-December 1957, No. 22: 183 pp.

Chemistry phosphate (data only)
Japan, Japan Meteorological Agency, 1958.
The results of marine meteorological and oceanographical observations, January-June, 1957, No. 21:168 pp.

Chemistry phosphate (data only)

Japan, Japanese Meteorological Agency, 1957

The results of marine meteorological and oceanographical observations, Jan.-June, 1956: 184 pp.
July-December, No. 20: 191 pp.

Japan, Japanese Oceanographic Data Center

Chemistry phosphate (data only)
Japan, Japanese Oceanographic Data Center, 1968.
Yang Ming, Chinese National Committee on Ocean Research, Republic of China, April 1-May

Chemistry phosphate (data only)
Japan, Japanese Oceanographic Data Center, 1968.
Chofu Maru, Nagasaki Marine Observatory, Japan Meteorological Agency, Japan, February 2-4, 1968, East China Sea.
Data Rept.CSK (KDC Ref. 49K072) 158: 5 pp. (multilithed).

Chemistry phosphate (data only)
Japan, Japanese Oceanographic Data Center, 1968.
Shumpu Maru, Kobe Marine Observatory, Japan Meteorological Agency, Japan, February 17-28, 1968, South of Japan.
Data Rept.CSK (KDC Ref. 49K071) 157:6 pp. (multilithed).

Chemistry phosphate (data only)
Japan, Japanese Oceanographic Data Center, 1968.
Ryofu Maru, Marine Division, Japan Meteorological Agency, Japan, January 13-March 22, 1967, West of North Pacific Ocean.
Data Rept.CSK (KDC Ref. 49K069) 155: 70 pp. (multilithed).

Chemistry phosphate (data only)
Japan, Japanese Oceanographic Data Center, 1968.
Takuyo, Hydrographic Division, Maritime Safety Agency, Japan, February 19-March 10, 1968, south and east of Japan.
Data Rept. CSK (KDC Ref. 49K068) 154: 14 pp. (multilithed).

Chemistry phosphate (data only)
Japan, Japanese Oceanographic Data Center, 1968.
Jalanidhi, Naval Hydrographic Office, Indonesia, October 4-19, 1967, South China Sea.
Data Rept.CSK (KDC Ref. 42K002) 137: 15 pp. (multilithed).

Chemistry phosphate (data only)
Japan, Japanese Oceanographic Data Center, 1968.
Suro No. 1, Hydrographic Office, Korea, September 23-October 21, 1967, South of Korea.
Data Rept. CSK (KDC Ref. 24K019) 136: 13 pp. (multilithed).

Chemistry phosphates (data only)
Japan, Japanese Oceanographic Data Center, 1968.
Chofu Maru, Nagasaki Marine Observatory, Japan Meteorological Agency, Japan, July 3-4, 1967, Aug. 28-29, 1967, Oct. 6-7, 1967, Nov. 9-10, 1967.
Data Rept.CSK (KDC Refs. 49K319, 49K320, 49K321, 49K322). 134: 18 pp. (multilithed).

Chemistry phosphate (data only)
Japan, Japanese Oceanographic Data Center, 1968.
Nagasaki Maru, Faculty of Fisheries, Nagasaki University of Fisheries, Japan, November 7-29, 1967, East China Sea.
Data Rept. CSK (KDC Ref. 49K076) 133: 5 pp. (multilithed).

Chemistry phosphate (data only)
Japan, Japanese Oceanographic Data Center, 1968. Chofu Maru, Nagasaki Marine Observatory, Japan Meteorological Agency, Japan, October 5-6, 1967, East China Sea.
Data Rept. CSK (KCD Ref. 49K065) 130: 5 pp. (multilithed).

Chemistry phosphate (data only)
Japan, Japanese Oceanographic Data Center, 1968. Kaiyo, Hydrographic Division, Maritime Safety Agency, Japan, November 13-December 1, 1967, South of Japan.
Data Rept. CSK (KCD Ref. 49K063) 128: 13 pp. (multilithed).

Chemistry phosphate (data only)
Japan, Japanese Oceanographic Data Center, 1968. Baek Du San, Fisheries Research and Development Agency, Korea, August 12-25, 1967, West of Japan Sea. Data Rept. CSK (KDC Ref. 24K016) 115: 26 pp. (multilithed).

Chemistry phosphate (data only)
Japan, Japanese Oceanographic Data Center, 1968. Seifu maru, Maizuru Marine Observatory, Japan Meteorological Agency, Japan, August 2-24, 1967, Japan Sea.
Data Rept. CSK (KDC Ref. 49K058) 110: 9 pp. (multilithed).

Chemistry phosphorus (data only)
Japan, Japanese Oceanographic Data Center, 1968. Takuyo, Hydrographic Division, Maritime Safety Agency, July 12-August 30, 1967, Central Part of the North Pacific Ocean.
Prelim. Data Rept. CSK (KDC Ref. 49K053) 105: 25 pp. (multilithed)

Chemistry phosphate (data only)
Japan, Japanese Oceanographic Data Center, 1968. Yang Ming, Chinese National Committee on Ocean Research, Republic of China, April 1-May 14, 1967, Surrounding waters of Taiwan.
Data Rept. CSK (KDC Ref. 21K004) 102: 15 pp. (multilithed).

Chemistry phosphate (data only)
Japan, Japanese Oceanographic Data Center, 1968. Kofu Maru, Hakodate Marine Observatory, Japan Meteorological Agency, Japan, May 17-19, 1967, East of Japan.
Data Rept. CSK (KDC Ref. 49K049) 98: 6 pp. (multilithed).

Chemistry phosphorus (data only)
Japan, Japanese Oceanographic Data Center, 1968. Seifu Maru, Maizuru Marine Observatory, Japan Meteorological Agency, February 10-March 2, 1967, Japan Sea.
Prelim. Data Rept. CSK (KDC 49K044) 86: 14 pp. (multilithed)

Chemistry phosphate (data only)
Japan, Japanese, Oceanographic Data Center, 1967. Suro No. 1, Hydrographic Division, Korea, October 12-November 2, 1966, south of Korea.
Prelim. Data Rept., CSK (KDC Ref. 24K011) 70 12 pp.

Chemistry phosphate (data only)
Japan, Japanese Oceanographic Data Center, 1967. 69 Bukhansan, Fisheries Research and Development Agency, Korea, July 16- August 9, 1966, East of the Yellow Sea.
Prelim. Data Rept., CSK (KDC Ref. 24K010): 17 pp.

Chemistry phosphate (data only)
Japan, Japanese, Oceanographic Data Center, 1967. Baekdusan, Fisheries Research and Development Agency, Korea, July 14-28, 1966, West of the Japan Sea.
Prelim. Data Rept, CSK (KDC Ref. 24K009) 65: 22 pp.

chemistry, phosphate (data only)
Japan, Japanese Oceanographic Data Center 1967.
Ryofu Maru, Marine Division, Japan Meteorological Agency, Japan September 13-17, 1966, eastern Sea of Japan.
Prelim. Data Rept. CSK (KDC Ref. No. 49K033) 59: 11 pp. (multilithed)

chemistry, phosphate (data only)
Japan, Japanese Oceanographic Data Center 1967.
Nagasaki Maru, Faculty of Fisheries, Nagasaki University, Japan, June 16-21, 1966, southern Sea of Japan.
Prelim. Data Rept. CSK (KDC Ref. No. 49K032) 58: 8 pp. (multilithed)

chemistry, phosphate (data only)
Japan, Japanese Oceanographic Data Center 1967.
Kaiyo, Hydrographic Division, M.S.A. Japan August 10-30, 1966, S.E. of Japan.
Prelim. Data Rept. CSK (KDC Ref. No. 49K025) 51: 15 pp. (multilithed).

Chemistry phosphate (data only)
Japan, Japanese Oceanographic Data Center, 1966. Kerin Ho, Fisheries Research and Development Agency, Korea, September 12-October 13, 1965, East of Yellow Sea.
Prelim. Data Rept., CSK, (KDC Ref. 24K005), 38: 11 pp. (multilithed).

Chemistry phosphate (data only)
Japan, Japanese Oceanographic Data Center, 1966. Oshoro Maru, The Faculty of Fisheries, Hokkaido University, Japan, November 30, 1965-January 25, 1966, South of Japan.
Prelim. Data Rept., CSK (KDC Ref. 49K022), 33: 11 pp. (multilithed)

Chemistry phosphate (data only)
Japan, Japanese Oceanographic Data Center, 1966. Seifu Maru, Maizuru Marine Observatory, Japan Meteorological Agency, Japan, February 12-February 28, 1966, Japan Sea.
Prelim. Data Rept., CSK (KDC Ref. 48K021), 32: 14 pp. (multilithed).

Chemistry phosphate (data only)
Japan, Japanese Oceanographic Data Center, 1966. Chofu Maru, Nagasaki Marine Observatory, Japan Meteorological Agency, Japan, January 26-February 25, 1966, East China Sea.
Prelim. Data Rept., CSK, (KDC Ref. 49K020), 31: 13 pp. (multilithed)

Oceanographic phosphate (data only)
Japan, Japanese, Data Center, 1966.
Preliminary data report of CSK, Shumpu Maru, Kobe Marine Observatory, Japan Meteorological Agency, February 19-24, 1966, Southern Sea of Japan, No. 30 (October 1966).
KDC Ref. No. 49K019: 6 pp. (multilithed).

Chemistry Phosphate (data only)
Japan, Japanese Oceanographic Data Center, 1966. Ryofu Maru, Japan Meteorological Agency, Japan, February 4-February 28, 1966, East of Japan.
Prelim. Data Rept., CSK (KDC Ref. 49K017), 28: 17 pp. (multilithed).

Chemistry phosphate (data only)
Japan, Japanese Oceanographic Data Center, 1966. Preliminary data report of CSK, Takuyo, Japenese Hydrographic Division, February 23-March 15, 1966, South eastern Sea of Japan, No. 27 (October 1966).
KDC Ref. No. 49K016: 17 pp. (multilithed).

Chemistry phosphate, (data only)
Japan, Japanese Oceanographic Data Center, 1966. U.M. Shokalski, USSR, July 16-August 18, 1965, E. & S. of Japan.
Prelim. Data Rept., CSK, (KDC Ref. 90K001), 23: 41 pp. (multilithed).

Chemistry phosphate (data only)
Japan, Japanese Oceanographic Data Center, 1966. Yang Ming, Chinese National Committee on Oceanic Research, Republic of China, August 10-October 13, 1965, Adjacent Sea of Taiwan.
Prelim. Data Rept., CSK (KDC Ref. 21K001), 22: 18 pp. (multilithed).

Chemistry phosphate (data only)
Japan, Japanese Oceanographic Data Center, 1966. Preliminary data report of CSK. Kofu Maru, Hakodate Marine Observatory, Japan Meteorological Agency, July 22-28, 1965, Eastern Sea of Japan, No. 14 (October 1966.)
KDC Ref. No. 49K004: 10 pp. (multilithed).

Chemistry, phosphate (data only)
Japan, Japanese Oceanographic Data Center, 1966. Preliminary data report of CSK. Ryofu Maru, Marine Division, Japan Meteorological Agency, July 7-August 3, 1965, Eastern Sea of Japan. No. 10 (October 1966).
KDC Ref. No. 49K003: 31 pp. (multilithed).

Chemistry phosphate (data only)
Japan, Kobe Marine Observatory, 1961
Data of the oceanographic observations in the sea south of Honshu from February to March and in May, 1959.
Bull. Kobe Mar. Obs., No. 167(27): 99-108; 127-130; 149-152; 161-164; 205-218.

Chemistry-phosphate (data only)
Japan, Maritime Safety Agency 1967.
Results of oceanographic observations in 1965.
Data Rept. Hydrogr. Obs. Ser. Oceanogr. (Publ. 792) Oct. 1967, 5: 115 pp.

chemistry, phosphate (data only)
Japan Maritime Safety Agency 1967.
Results of oceanographic observations in 1964.
Data Rept. Hydrogr. Obs. Ser. Oceanogr. (Publ. 792) 4: 88 pp.

Chemistry, phosphate (data only)
Japan, Maritime Safety Agency 1966.
Results of oceanographic observations in 1963.
Data Rept. Hydrogr. Obs. Ser. Oceanogr. (Publ. 792) 3: 74 pp.

Chemistry phosphate (data only)
Japan, Maritime Safety Board, 1961
Tables of results from oceanographic observation in 1959.
Hydrogr. Bull. (Publ. No. 981), No. 68: 112 pp.

Chemistry, phosphate (data only)
Japan Maritime Safety Board 1961.
Tables of results from oceanographic observations in 1958.
Hydrogr. Bull. Tokyo, No. 66 (Publ No. 981): 153 pp.

Chemistry phosphate (data only)
Japan, Maritime Safety Board, 1956.
Tables of results from oceanographic observations in 1952 and 1953.
Hydrogr. Bull. (Publ. 981)(51): 1-171.

Chemistry phosphate (data only)
Japan, Maritime Safety Board, Tokyo, 1954.
Tables of results from oceanographic observations in 1951. Publ. No. 981, Hydrogr. Bull., Spec. No. 15: 31-129.

Chemistry, phosphate (data only)
Japan, Nagasaki Marin Observatory
Tables 9, 10, 11, 22, 23, 24, 34, 35 and 44
Oceanogr. Met. Nagasaki mar. Obs. 15: 92-120, 176-180, 191-192, 213-

Chemistry phosphate (data only)
Japan. Shimonoseki University of Fisheries. 1968
Oceanographic surveys of the Kuroshio and its adjacent waters 1965 and 1966.
Data Oceanogr. Obs. Explor. Fish. 4:1-178.

Chemistry phosphate (data only)
Japan, Shimonoseki University of Fisheries 1965.
Eastern Pacific Ocean Cruise and central Atlantic Ocean cruise.
Data Oceanogr. Obs. Explor. Fish., 3: 1-145

chemistry, phosphate (data only)
Japan Shimonoseki University of Fisheries 1965
International Indian Ocean Expedition 1962-63. and 1963-1964
Data Oceanogr. Obs. Explor. Fish. 1: 453 pp.

chemistry, phosphate (data only)
Japan, Tokai Regional Fisheries Research Laboratory 1959.
16X Physical and chemical data by the R.V. Soyo-maru 25 July - 14 September 1958: 17pp. (multilithed).

Chemistry phosphate (data only)
Japan, Tokyo University of Fisheries, 1966.
J. Tokyo Univ. Fish. (Spec.ed.) 8 (3):1-44.

Chemistry phosphate (data only)
Jarrige, F., J. Merle et J. Noel, 1968.
Résultats des observations physico-chimiques de la croisière CYCLONE 4 du N.O. Coriolis.
Rapp. sci., O.R.S.T.O.M., Noumea, 25: 22 pp.

K

Chemistry Phosphate (data only)
Katsuura, Hiroshi, Hideo Akamatsu, Tsutomu Akiyama, 1964.
Preliminary report of the Japanese Expedition of Deep-Sea, the Eighth Cruise (JEDS-8).
Oceanogr. Mag., Tokyo, 16(1/2):125-136.

Chemistry phosphate (data only)
Kollmeyer, R.C., J.W. McGary and R.M. Morse, 1964
Oceanography observations, Tropical Atlantic Ocean, EQUALANT II, August 1963.
U.S.C.G. Oceanogr. Rept., (CG373-4):96 pp.

phosphate (data only)
Korea, Republic of, Fisheries Research and Development Agency, 1971
Annual report of oceanographic observations, 19:717 pp.

Chemistry phosphate (data only)
Korea, Fisheries Research and Development Agency, 1968.
Annual report of oceanographic observations, 16: ~~961~~ 691 pp.

Chemistry phosphate (data only)
Korea, Fisheries Research and Development Agency, 1967.
Annual report of oceanographic observations, 15 (1966):459 pp.

chemistry, phosphate (data only)
Korea (Republic of), Fisheries Research and Development Agency 1966.
Annual Report of observations, 1965,

Chemistry phosphate (data only)
Republic of Korea, Hydrographic Office, 1967.
Hydrographic and oceanographic data.
Techn. Repts, 1967, H.O. Pub., Korea, 1:72.

Chemistry, phosphate (data only)
Republic of Korea, Hydrographic Office, 1966.
The data for cooperative study of Kuroshio.
Tech. Rep. H.O. Pub. No. 1101: 41-73.

Chemistry, phosphate (data only)
Korea (Republic of), Hydrographic Office, 1965.
Technical reports.
H.O. Pub., Korea, No. 1101:179 pp.

phosphate (data only)
Krauel, David P., 1969
Bedford Basin data report, 1967.
Techn. Rept. Fish. Res. Bd., Can., 120:84 pp (multilithed).

L

Chemistry, phosphate (data only)
Love, C.M., 1966.
Physical, chemical, and biological data from the northeast Pacific Ocean: Columbia river effluent area, January-June 1963. 6. Brown Bear Cruise 326:13-23 June: CNAV Oshawa Cruise Oshawa-3:17-30 June.
Univ. Washington, Dept. Oceanogr., Tech. Rep., No. 134:230 pp. (Unpublished manuscript).

M

Chemistry Phosphate (data only)
Magnier, Y., H. Rotschi, B. Voituriez et J. Merle 1968.
Résultats des observations physico-chimiques de la croisière CYCLONE 3 du N.O. Coriolis.
Rapp. sci., O.R.S.T.O.M., Nouméa, 21:23 pp. (mimeographed)

chemistry, phosphate (data only)
Magnier, Yves, Pierre Rual et Bruno Voituriez 1968.
Résultats des observations physico-chimiques de la croisière CYCLONE 6 du N.O. Coriolis.
Rapp. sci. ORSTOM, Noumea 27: 23 pp. (mimeographed).

Mc

Chemistry phosphate (data only)
McGary, J.W., 1955.
Mid-Pacific oceanography, Pt. 6. Hawaiian off-shore waters, December 1949-November 1951.
Spec. Fish. Rept.: Fish. No. 152:1-138, 33 figs.

Chemistry phosphate (data only)
McGary, James W., and Joseph J. Graham, 1960.
Biological and oceanographic observations in the central north Pacific, July-September 1958.
U.S.F.W.S. Spec. Sci. Rept., Fish., No. 358: 107 pp.

Chemistry phosphate (data only)
McGary, J.W., and E.D. Stroup, 1956.
Mid-Pacific oceanography. VIII. Middle latitude waters, January-March 1954.
USFWS Spec. Sci. Rept., Fish., 180:173 pp.

Chemistry phosphate (data only)
Minas, H.J. 1971.
Résultats de la campagne "Mediprod I" du Jean Charcot.
Cah. océanogr. 23 (Suppl. 1): 93-144.

K. Kato phosphate (data only)
Motoda, S., H. Koto and T. Fugii, 1956.
Hydrographic data obtained principally in the Bering Sea by Training Ship "Oshoro-maru" in the summer of 1955, 59 pp.

N

Chemistry, phosphate (data only)
*Nasu, Keiji, and Tsugio Shimano, 1966.
The physical results of oceanographic survey in the south east Indian Ocean in 1963/64.
J. Tokyo Univ. Fish., (Spec.ed.) 8(2):133-164.

Chemistry, phosphate (data only)
NORPAC Committee, 1955.
Oceanic observation of the Pacific, 1955.
The NORPAC data, 532 pp.

O

Chemistry phosphate (data only)
Oren, Oton Haim, 1967.
Croisière Chypre-04 en Méditerranée orientale, février-mars, 1965: résultats des observations hydrologiques.
Cah. océanogr., 19(9):783-798.

Chemistry phosphate (data only)
Oren, O.H., 1967.
Croisière Chypre-04 dans la Méditerranée orientale, février-mars 1965. Résultats des observations hydrographiques.
Bull. Sea Fish. Res. Stn. Israel, 47:37-54.

Chemistry phosphate (data only)
Oren, Oton Haim, 1966.
Croisière "Chypre-02" en Méditerranée orientale, juillet-août 1963. Résultats des observations hydrologiques.
Cah. océanogr., 18(Suppl.):1-17.

P

Chemistry phosphate (data only)
Postma, H., 1959.
Chemical results and a survey of water masses and currents. Tables: oxygen, hydrogen ion, alkalinity and phosphate.
Snellius Exped., Eastern Part of the East Indian Archipelago, 1929-1930, Chem. Res., 4:35pp., 1 fig.

Q

R

phosphate (data only)
Republic of China, Chinese National Committee on Oceanic Research, Academia Sinica, 1968.
Oceanographic Data Report of CSK, 2: 126 pp.

chemistry, phosphate (data only)
Republic of China, Chinese National Committee on Oceanic Research, Academia Sinica 1966.
Oceanographic data report of CSK 1: 123pp.

Chemistry phosphate (data only)
Rotschi, Henri, 1958
"Orsom III", Océanographie physique.
Rapp. Tech. de la Croisière 56-5, Inst. Francais d'Océanie, Rapp. Sci., No. 5: 35 pp. (miméographed).

Chemistry phosphate (data only)
Rotschi, Henri, 1957
"Orsom III", Oceanographie physique.
Rapp. Tech. de l'Expedition EQUAPAC (Croisiere 56-4), Inst. Francais d'Oceanie, Rapp. Sci., 3: 52 pp. (mimeographed).

Chemistry phosphate (data only)
Rotschi, Henri, Michel Angot, Michel Legand and Roget Desrosieres, 1961
Orsom III, Resultats de la Croisiere "Dillon" 2eme Partie. Chimie et Biologie.
ORSTOM, Inst. Francais d'Oceanie, Centre d'Oceanogr., Rapp. Sci., No. 19: 105 pp. (mimeographed).

Chemistry phosphate (data only)
Rotschi, H., Ph. Hisard, L. Lemasson, Y. Magnier, J. Noel, et B. Piton, 1966.
Resultats des observations physico-chimiques de la croisiere "Alize".
Centre ORSTROM Noumea Rapp. Sci., No. 28:56 pp. (mimeographed).

Chemistry phosphate (data only)
Rotschi, H., Ph. Hisard, L. Lemasson, Y. Magnier, J. Noel et B. Piton, 1967.
Resultats des observations physico-chimiques de la croisiere "Alize" du N.O. Coriolis.
Rapp. Scient., Off. Rech. scient. tech. Outre-Mer, Noumea, 2:56 pp. (mimeographed).

Chemistry phosphate (data only)
Rotschi, Henri, Michel Legand and Roger Desrosieres, 1961
Orsom III, Croisières diverses de 1960, physique chimie et biologie. ORSTOM, Inst. Francais d'Océanie, Centre d'Océanogr., Noumea, Rapp. Sci., No. 20: 59 pp. (mimeographed).

Chemistry, phosphate (data only)
Rotschi, H., G. Pickard, P. Hisard et J. Cannevet, 1968.
Résultats des observations physico-chimiques de la croisière CYCLONE 5 du N.O. Coriolis.
Rapp. sci., O.R.S.T.O.M., Noumea, 26:23 pp.

S

chemistry Phosphate (data only)
*Scaccini Cicatelli, Marta, 1967.
Distribuzione stagionale dei sali nutrivi in una zona dell'alto e medio Adriatico.
Boll. Pesca, Piscic. Idrobiol., n.s., 22(1): 49-62.

Chemistry phosphate (data only)
Scripps Institution of Oceanography, 1949
Physical and chemical data, Cruise 2, March 28 to April 12, 1949. Physical and Chemical Data Report No.2:10 figs. (ozalid) tables of data (mimeographed).

Chemistry phosphate (data only)
Scripps Institution of Oceanography, 1949
Physical and chemical data, Cruise 1, February 28 to March 16, 1949. Marine Life Research Program. Physical and Chemical Data Report No.1:10 figs. (ozalid), tables of data (mimeographed).

Chemistry phosphate (data only)
Scripps Institution of Oceanography, 1949
Marine Life Research Program. Progress Report to 30 April 1949, 25 pp. (mimeographed), numerous figs. (photo.+ozalid). 1 May 1949.

Chemistry phosphate (data chiefly)
Seckel, G.R., 1955.
Mid-Pacific oceanography. VII. Hawaiian offshore waters, Sept. 1952-August 1953.
Spec. Sci. Repts., Fish., No. 164:250 pp., 38 figs

Chemistry, phosphate (data only)
South Africa, Division of Sea Fisheries, Department of Industry 1968.
Hydrological station list, 1961-1962, 1:342 pp.

Chemistry phosphate (data only)
South Africa, Department of Commerce and Industries, Division of Sea Fisheries, 1964.
Thirty-second annual report for the period 1st April 1960, to 31st March 1961:267 pp.

Chemistry phosphate (data)
South Africa, Division of Fisheries, 1954.
Twenty-fourth annual report:1-199.

Chemistry phosphate (data only)
Spain, Instituto Español de Oceanografia, 1961.
Campañas biologicas del "Xauen" en las costas del Mediterraneo Marroqui, Mar de Alboran, Baleares y Noroeste y Cantabrico Españoles en los años 1953, 1953 y 1954.
Bol. Inst. Español, Oceanogr., 103:1-130.

Chemistry phosphate (data only)
Spain, Instituto Español de Oceanografia, 1955.
Campaña del "Xauen" en la costa noroeste de España en 1949 y 1950. Registro de operaciones.
Bol. Inst. Español Ocean., 71:72 pp.

Chemistry, phosphate (data only)
Stroup, E.D., 1954.
Mid-Pacific oceanography IV. Transequatorial waters, January-March 1952.
U.S.F.W.S. Spec. Sci. Rept. - Fish. No. 135: 1-52, 18 figs. (multilithed).

Chemistry, phosphate (data only)
Svansson, Artur 1970.
Hydrographical observations on Swedish lightships and fjord stations in 1968
Rept. Fish. Bd, Sweden, Ser. Hydrogr. 24: 83pp.

Chemistry phosphate (data only)
Sweden, Havsfiskelaboratoriet, Lysekil 1971.
Hydrographical data January-June 1970, R.V. Skagerak, R.V. Thetis and TV252, 1970.
Meddn. Hydrogr. avd. Göteborg, 104: unnumbered pp. (multilithed).

Chemistry phosphate (data only)
Sweden, Havsfiskelaboratoriet, Lysekil 1970.
Hydrographical data 1966, R.V. Skagerak, R.V. Thetis.
Meddn Hydrogr. avd. Göteborg 85:255pp

Chemistry phosphate (data only)
Sweden, Havsfiskelaboratoriet, Lysekil 1970.
Hydrographical data, R.V. Skagerak, R.V. Thetis.
Hydrogr. avd. Göteborg
Meddn. 84: 296 pp. (multilithed).
pH (data only)

Chemistry phosphate (data only)
Sweden, Havsfiske laboratoriet, lysekil, 1969.
Hydrographica data, January-June 1968.
R.V. Skagerak, R.V. Thetis.
Meddn Havsfiskelab., Lysekil, hydrog. Avdeln., Goteborg 63: numerous unnumbered pp. (multilithed).

Chemistry, phosphate (data only)
Sweden, Havsfiskelaboratoriet, Lysekil, 1968.
Hydrographical data, July-December 1967, R.V. Skagerak.
Meddn Havsfiskelab., Lysekil, Hydrogr. Avd., Göteborg, 52: unnumbered pp. (multilithed).

Chemistry phosphate (data only)
Sweden, Havsfiskelaboratoriet, Lysekil, 1968.
Hydrographical data, July-December 1967 R.V. Skagerak.
Meddn Havsfiskebal., Lysekil, 52: unnumbered pp. (mimeographed).

Chemistry phosphate (data only)
Sweden, Havsfiskelaboratoriet, Lysekil, Hydrografiska Avd., Göteborg, 1967.
Hydrographical data, January-June 1967, R.V. Skagerak.
Meddn. Havsfiskelab., Lysekil, Hydrogr. Avd., Göteborg, 38: numerous pp. (mimeographed).

Chemistry phosphate (data only)
Sweden, Havsfiskelaboratoriet, Lysekil, Hydrografiska Avd., Göteborg, 1967.
Hydrographical data, January-June 1967, R.V. Thetis.
Meddn. Havsfiskelab., Lysekil, Hydrogr. Avd., Göteborg, 41: numerous pp. (mimeographed).

T

U

Chemistry Phosphate (data only)
Uda, Michitaka, Yoshima Morita and Makoto Ishino, 1957.
Results from the oceanographic observations in the North Pacific (1955-56) with Umitaka Maru and Shinyo-Maru. Rec. Ocean. Wks., Japan (Spec. No.):1-20

Chemistry phosphate (data only)
United States, U.S. Coast Guard, 1965.
Oceanographic cruise USCGC Northwind: Chukchi, East Siberian and Laptev seas, August-September, 1965.
U.S.C.G. Oceanogr. Rept., No. 6(CG373-6):69 pp.

Chemistry, phosphate (data only)
United States, National Oceanographic Data Center, 1965.
Data report EQUALANT III.
Nat. Oceanogr. Data Cent., Gen. Ser., G-7:339pp $5.00
Publ.

Chemistry phosphate (data only)
United States, National Oceanographic Data Center, 1964.
Data report EQUALANT II.
NODC Gen. Ser., Publ. G-5: numerous pp. (not serially numbered; loose leaf - $5.00).

Chemistry phosphate (data only)
University of California, Scripps Institution of Oceanography, 1960
Oceanic observations of the Pacific, 1950, 508 pp.

V

Chemistry, phosphate (data only)
Valdez, Alberto J., Alberto Gomez, Aldo Orlando y Andres J. Lusquinos 1969.
Datos y resultados de las campañas pesqueras "Pesqueria V" (28 de agosto al 7 de setiembre de 1967.
Publ. (Ser. Informes técn.) Mar del Plata, Argentina 10(5): unnumbered pp.

Chemistry, phosphate (data only)
Valdez, Alberto J., Alberto Gomez, Aldo Orlando y Andres J. Lusquiños 1968.
Datos y resultados de las campañas pesqueras "Pesqueria IV" (9 de junio al 4 de julio de 1967.
Publ. (Ser. Informes técn.) Mar del Plata Argentina 10(4):1-159.

Chemistry, phosphate (data only)
Villanueva, Sebastian F., Alberto Gomez, Aldo Orlando y Andres J. Lusquiños 1969.
Datos y resultados de las campañas pesquerias "Pesqueria VII" (16 de febrero al 1° de marzo de 1968).
Publ. (Ser. Informes técn.) Proyecto Desarrollo Pesquero, Mar del Plata (10/VII): unnumbered pp.

chemistry, phosphate (data only)
Villanueva, Sebastian F., Alberto Gomez, Aldo Orlando y Andres J. Lusquiños 1969.
Datos y resultados de las campañas pesquerias "Pesqueria VI" (2 de noviembre al 6 de deciembre de 1967).
Publ. (Ser. Informes tecn.) Proyecto Dessarollo, Pesquero, Mar del Plata (10(x)): unnumbered pp.

W

chemistry phosphate (data only)
Wilson, R.C., and M.O. Rinkel, 1957.
Marquesas area oceanographic and fishery data, January-March 1957.
USFWS Spec. Sci. Rept., Fish. No. 238:136 pp.

XYZ

chemistry, phosphate (data only)
Zoutendyk, P., and D. Sacks 1969.
Hydrographic and plankton data, 1960-1965
Data Rept. Inst. Oceanogr. Univ. Cape Town (3): 82 pp.

chemistry phosphate (data only)
Zvereva, A.A., Edit., 1959.
Data, 2nd Marine Expedition, "Ob", 1956-1957.
Arktich. i Antarkt. Nauchno-Issled. Inst., Mezhd. Geofiz. God, Sovetsk. Antarkt. Exped., 6: 1-387.

phosphate, density

chemistry, phosphate-density
Brazil, Diretoria de Hidrografia e Navegação, 1957.
Ano Geofisico Internacional, Publicacao DG-06-VI: unnumbered mimeographed pages.

chemistry, phosphate-density
Brasil, Diretoria de Hidrografia e Navegação, 1957.
Ano Geofisico Internacional. Publicacao DG-06-V: 3 pp., 18 figs., 3 pp. tables.

chemistry phosphate-density
Brasil, Diretoria de Hidrografia e Navegacao, 1957.
Ano Geofisico Internacional. Publ. DG-06-III (mimeographed.

chemistry, phosphate-density
Brazil, Diretoria de Hidrografia e Navegação, 1957.
Ano Geofisical Internacional. Publicacão DG-06-II mimeographed.

chemistry phosphate-density
Ichiye, T., 1962.
Circulacion y distribucion de la masa de agua en el Golfo de Mexico. (In Spanish and English).
Geofisica Internacional, Rev. Union Geofis. Mexicana, Inst. Geofis., Univ. Nacional Autonoma, de Mexico, 2(3):47-76, 22 figs.

phosphate, effect of

chemistry phosphate, effect of
Peters, Nicolaus, 1934
Die Bevolkerung des Sudatlantischen Ozeans mit Ceratien. Biol. Sonderuntersuchungen 1.
Wiss Ergeb. Deutschen Atlantischen Exped.--- "Meteor" 1925-1927, 12(1): 1-69, 28 text figs.

chemistry, phosphate, effect of
*Thomas, William H., and Anne N. Dodson, 1968.
Effects of phosphate concentration on cell division rates and yield of a tropical oceanic diatom.
Biol. Bull. mar. biol. Lab., Woods Hole, 134(1):199-208.

phosphate, free

phosphate (free)
Burns, R.B., 1967.
Chemical observations in the North Sea in 1965.
Annls biol. Copenh. (1965)22:29-31.

chemistry phosphate, high
Reddy, C.V., P.S.N. Murty and V.N. Sankaranarayanan, 1968.
An incidence of very high phosphate concentrations in the waters around Andaman Islands.
Curr. Sci., 37(1): 17-19.
Also in: Coll. Repr., Int. Indian Ocean Exped., 6, Contrib. 472

phosphate, inorganic

chemistry phosphate (inorganic)
Bennett, Edward B., 1963.
An oceanographic atlas of the Eastern Tropical Pacific Ocean, based on data from EASTROPIC expedition. October-December 1955.
Bull. Inter-American Trop. Tuna Comm., 8(2):33-165.

chemistry inorganic phosphate (surface)
Clowes, A.J., 1954.
The temperature, salinity and inorganic phosphate of the surface layer near St. Helena Bay, 1950-52. Union of S. Africa, Dept. Commerce & Industries, Div. Fish., Invest. Rept. No. 16: 1-47, 17 pls.

Reprinted from "Commerce and Industry" Aug. 1954.

chemistry phosphate, inorganic (data)
Corwin, Nathaniel, and David A. McGill., 1963.
Nutrient distribution in the Labrador Sea and Baffin Bay.
U.S.C.G. Bull., No. 48:79-94; 95-153.

chemistry phosphate, inorganic
Cromwell, T., 1953.
Circulation in a meridional plane in the central equatorial Pacific. J. Mar. Res. 12(2):196-213, 9 textfigs.

chemistry phosphate (inorganic)
Darbyshire, J., 1964
A hydrological investigation of the Agulhas Current.
Deep-Sea Res., 11(5):781-815.

chemistry phosphate (inorganic)
deJager, B.v.D., 1957.
Variations in the phytoplankton of the St. Helena Bay area during 1954.
Union of S. Africa, Div. Fish., Invest. Rept., No. 25:78 pp.

phosphate, inorganic
Doe, L. A. E., 1965.
Physical conditions on the shelf near Karachi during the post-monsoonal calm, 1964. Ocean Sci. and Ocean Eng., Mar. Sci. Techn. Soc. - Amer. Soc. Limnol. Oceanogr., 1: 278-292.

chemistry, phosphate (inorganic) (data)
Dragovich, Alexander, 1961
Relative abundance of plankton off Naples, Florida, and associated hydrographic data, 1956-57.
USFWS Spec. Sci. Rept., Fish., No. 372:41 pp.

chemistry phosphate (inorganic) (data)
Dragovich, Alexander, John H. Finucane, John A. Kelly, Jr., and Billie Z. May, 1963.
Counts of red-tide organisms, Gymnodinium breve, and associated oceanographic data from Florida west coast, 1960-61.
U.S. Fish and Wildlife Service, Spec. Sci. Repts., Fish., No. 455:1-40.

chemistry phosphate (inorganic)
Fonselius, Stig H., 1962
Hydrography of the Baltic deep basins.
Fishery Board of Sweden, Ser. Hydrogr., Rept. 13: 40 pp.

chemistry, phosphate (inorganic)
Greenfield, L.J., and F.A. Kalber, 1954.
Inorganic phosphate measurements in sea water.
Bull. Mar. Sci., Gulf and Caribbean, 4(4):323-335, 3 textfigs.

chemistry, phosphate (inorganic)
Gilmartin, Malvern, 1967.
Changes in inorganic phosphate concentration occurring during seawater sample storage.
Limnol. Oceanogr., 12(2):325-326.

chemistry phosphate (inorganic) (data)
Herrera, Juan, y Ramón Margalef, 1963
Hidrografía y fitoplancton de la costa comprendida entre Castellón y la desembocadura del Ebro, de julio de 1960 a junio de 1961.
Inv. Pesq., Barcelona, 24:33-112.

chemistry phosphate (inorganic)
Holm-Hansen, Osmund, J.D.H. Strickland and P.M. Williams, 1966.
A detailed analysis of biologically important substances in a profile off southern California.
Limnol. Oceanogr., 11(4):548-561.

chemistry, phosphate, inorganic, & total
Hulburt, E.M., 1957.
Distribution of phosphorus in Grat Pond, Massach-usetts. J. Mar. Res., 195(3):181-192.

chemistry phosphate, inorganic
Japan, Hakodate Marine Observatory, 1967.
Report of the oceanographic observations in the Okhotsk Sea and east of the Kurile islands and Hokkaido from October to November. 1964. (In Japanese).
Bull. Hakodate mar. Obs., 13: 20-28.

chemistry phosphate (inorganic)
Japan, Hakodate Marine Observatory, 1967.
Report of the oceanographic observations in the Okhotsk Sea, east of the Kurile islands and Hokkaido and east of the Tohoku district from August to September, 1964. (In Japanese).
Bull. Hakodate mar. Obs., 13:7-19.

chemistry phosphate, inorganic
*Kirkwood, L.F., 1967.
Inorganic phosphate, organic phosphorus, and nitrate in Australian waters.
Tech. Pap. Div. Fish. Oceanogr., CSIRO, Australia, 25: 18 pp.

chemistry phosphorus, inorganic
Krey, J., 1963.
The components of phosphorus in Kiel Bay in 1961.
Ann. Biol., Cons. Perm. Int. Expl. Mer, 1961, 18: 54-56.

chemistry phosphate, inorganic (data)
Littlepage, Jack L., 1965.
Oceanographic investigations in McMurdo Sound, Antarctica.
In: Biology of Antarctic seas, II.
Antarctic Res. Ser., Am. Geophys. Union, 5:1-37.

chemistry phosphate (total & inorganic) (data only)
Marvin, K.T., 1955.
Oceanographic observations in west coast Florida waters, 1949-52. U.S. Fish & Wildlife Service, Spec. Sci. Rept., No. 149:6 pp., 2 figs., numerous pp tables.

chemistry phosphate, inorganic
Mostert, S.A., 1966.
Distribution of inorganic phosphate and dissolved oxygen in the south west Indian Ocean.
Repub. S. Africa, Dept. Comm. Indust., Div. Sea. Fish. Invest. Rept., No. 54:1-23.

Chemistry phosphate, inorganic
Murphy, J., and J.P. Riley, 1956.
The storage of sea water samples for the determination of dissolved inorganic phosphate.
Anal. Chim. Acta, 14(4):318-319.

Abstr. Analyt. Chem., 3(10):#3209.

Chemistry phosphate, inorganic
*Nellen, Walter, 1967.
Horizontale und vertikale Verteilung der Planktonproduktion im Golfe von Guinea und in angrenzenden Meeresgebieten während der Monate Februar bis Mai 1964.
Kieler Meeresforsch., 23(1):48-67/

Chemistry phosphate(inorganic) (data)
Nutt, D.C., and L.K. Coachman, 1956.
The oceanography of Hebron Fjord, Labrador.
J. Fish. Res. Bd., Canada, 13(5):709-758.

Chemistry, phosphate (inorganic)
Park, Kilho 1967.
Chemical features of the subarctic boundary near 170° W.
J. Fish. Res. Bd. Can. 24 (7): 899-908.

Chemistry phosphate, inorganic
Reid, Joseph L., Jr., 1965.
Intermeditate waters of The Pacific Ocean.
Johns Hopkins Oceanogr. Studies, 85 pp.

Chemistry phosphate(inorganic)
Reid, Joseph L., Jr., 1962
On circulation, phosphate-phosphorus content and zooplankton volumes in the upper part of the Pacific Ocean.
Limnol. and Oceanogr., 7(3):287-306.

Chemistry phosphate, inorganic
Richards, Francis A., Joel D. Cline, William W. Broenkow and Larry P. Atkinson, 1965.
Some consequences of the decomposition of organic matter in Lake Nitinat, an anoxic fjord.
Limnol. and Oceanogr., Redfield Vol., Suppl. to 10:R185-R201.

Chemistry phosphate, inorganic (data)
Rochford, D.J., 1962.
Hydrology of the Indian Ocean. II. The surface waters of the south-east Indian Ocean and Arafura Sea in the spring and summer.
Australian J. Mar. Freshwater Res., 13(2):226-251.

Chemistry, phosphate (inorganic)
Stefansson, U., 1957.
Hydrographic conditions north and north-east of Iceland, 1955.
Ann. Biol., Cons. Perm. Int. Expl. Mer, 12:28-33, Figs., 37-46.

Chemistry, phosphate, inorganic
Stefansson, U., 1956.
Hydrographic conditions north and northeast of Iceland, 1954.
Ann. Biol., Cons. Perm. Int. Expl. Mer, 11:15-20, Figs., 12-21.

Chemistry, inorganic phosphates
Stefansson, U., 1956.
Astand sjavar a sildveidisvaedinu nordanlands sumareid 1955. [Hydrographic conditions of the North Icelandic herring grounds during the summer 1955.] Fjölbrit Fiskideildar, 6:23 pp.

Chemistry, phosphate, inorganic
Tully, J.P., and A.J. Dodimead, 1957.
Properties of the water in the Strait of Georgia, British Columbia, and influencing factors.
J. Fish. Res. Bd., Canada, 14(3):241-319.

Chemistry, phosphate (inorganic)
Wyrtki, K., 1962.
The upwelling in the region between Java and Australia during the south-east monsoon.
Australian J. Mar. Freshwater Res., 13(3):217-225.

phosphate, inorganic data only

Chemistry, phosphate (inorganic) (data only)
Australia, Commonwealth Scientific and Industrial Research Organization, 1965.
Oceanographic observations in the Indian Ocean in 1963, H.M.A.S. DIAMANTINA Cruise DM 3/63.
Div. Fish. Oceanogr., Oceanogr. Cruise Rept. No. 25: 147 pp.

Chemistry - phosphate (inorganic) (data only)
Australia, Commonwealth Scientific and Industrial Organization 1965.
Oceanographical observations in the Indian Ocean in 1963, H.M.A.S. Diamantina, Cruise Dm 2/63.
Div. Fish. Oceanogr., Oceanogr. Cruise Rept., No. 24: 153 pp.

Chemistry, phosphate (inorganic) - (data only)
Australia, Commonwealth Scientific and Industrial Organization 1965.
Oceanographical observations in the Indian Ocean in 1963, H.M.A.S. Diamantina, Cruise Dm 1/63.
Div. Fish. Oceanogr., Oceanogr. Cruise Rept. No. 23: 175 pp.

Chemistry phosphate, inorganic (data only)
Australia, Commonwealth Scientific and Industrial Research Organization, 1965.
Oceanographic observations in the Indian Ocean in 1963, H.M.A.S. Gascoyne, Cruise G 1/63.
Div. Fish. and Oceanogr., Oceanogr. Cruise Rept., No. 21:135 pp.

Chemistry phosphate (data only) inorganic
Australia, Commonwealth Scientific and Industrial Research Organization, 1964.
Oceanographical observaions in the Indian Ocean in 1962, H.M.A.S. Diamantina, Cruise Dm 2/62.
Oceanogr. Cruise Rept., Div. Fish. and Oceanogr. No. 15:117 pp.

Chemistry phosphate, inorganic (data only)
Australia, Commonwealth Scientific and Industrial Research Organization, 1964.
Oceanographical observations in the Indian Ocean in 1961, H.M.A.S. Diamantina, Cruise Dm 3/61.
Div. Fish. and Oceanogr., Oceanogr. Cruise Rept. No. 11:215 pp.

Chemistry phosphate (inorganic)(data only)
Australia, Commonwealth Scientific and Industrial Research Organization, 1963.
Oceanographical observations in the Pacific Ocean in 1961, H.M.A.S. Gascoyne, Cruise G 1/61.
Oceanogr. Cruise Rept., Div. Fish. and Oceanogr., No. 8:130 pp., 12 figs.

Chemistry phosphate (inorganic) (data only)
Australia, Commonwealth Scientific and Industrial Research Organization, 1964.
Oceanographic observations in the Indian Ocean in 1962, H.M.A.S. Diamantina, Cruise, Dm 1/62.
Oceanogr. Cruise Rept., Div. Fish. Oceanogr., No. 10: 128 pp.

Chemistry phosphate, inorganic (data only)
Australia, Commonwealth Scientific and Industrial Organization, 1962.
Oceanographical observations in the Indian Ocean in 1960, H.M.A.S. DIAMANTINA Cruise Dm 3/60.
Oceanogr. Cruise Rept., Div. Fish. and Oceanogr. No. 4:39 pp.

Chemistry phosphate(inorganic data only)
Australia, Commonwealth Scientific and Industrial Research Organization, Division of Fisheries and Oceanography, 1962.
Oceanographic observations in the Pacific Ocean in 1960, H.M.A.S. Gascoyne, Cruises G 1/60 and G 2/60.
Oceanographical Cruise Report No. 5:255 pp.

Chemistry inorganic phosphate (data only)
Australia, Commonwealth Scientific and Industrial Research Organization, 1960.
Oceanic investigations in eastern Australia, H.M.A. Ships Queenborough, Quickmatch and Warrego, 1958.
Div. Fish. and Oceanogr., Oceanogr. Sta. List, 43:57 pp.

Chemistry phosphate (inorganic) (data only)
Collier, Albert, 1958.
Gulf of Mexico physical and chemical data from ALASKA cruises.
USFWS Spec. Sci. Rept., Fish., No. 249:417 pp.

Chemistry phosphate, inorganic (data only)
Dragovich, Alexander, John H. Finucane and Billie Z. May, 1961.
Counts of red tide organisms, Gymnodinium breve, and associated oceanographic data from Florida west coast, 1957-1959.
USFWS Spec. Sci. Rept., Fish., No. 369:175 pp.

Chemistry phosphate, inorganic (data only)
Portugal, Instituto Hidrografico, 1965.
Resultados das observacões oceanograficas no Canal de Mocambique, Cruzeiro al 1/64: abril-maio 1964.
Servico de Oceanografia, Publ. 1:73 pp., 46 figs

phosphate maximum layer

Chemistry phosphate maximum layer
*Sapozhnikov, V.V., and A.M. Chernyakova, 1967.
Oxygen minima and phosphate maxima in the Pacific Ocean. (In Russian; English abstract).
Trudy Inst. Okeanol., 83:166-176.

phosphate maximum-minimum

Chemistry - phosphate (maximum-minimum)
Guillén, Oscar, y Francisco Vasquez 1966.
Informe preliminar del crucero 6602 (Cabo Blanco - Arica)
Informe, Inst. Mar. Peru No. 12: 27 pp.

phosphate, monthly mean

Chemistry phosphate (monthly means)
Bose, B.B., 1956.
Observations on the hydrology of the Hooghly Etsuary. Indian J. Fish., 1(3):101-118.

Chemistry Phosphate (monthly means)
Newell, B. S., 1966.
Seasonal changes in the hydrological and biological environments off Port Hacking, Sydney.
Australian J. Mar. freshw. Res., 17(1):77-91.

Chemistry phosphate, monthly means
Slinn, D.J., and Gaenor Offlow (Mrs. M. Solly), 1968.
Chemical constituents of sea water off Port Erin during 1967.
Ann. Rept. mar. biol Stn. Port Erin, 80:37-42.

Chemistry phosphate(mean monthly)(data)
Szarejko-Lukaszewicz, D., 1959
Hydrographic investigations on the Firth of Vistula in 1953-1954.
Prace Morsk. Inst. Ryback., Gdyni, 10A: 215-228.

phosphate nitrate

Chemistry phosphate/nitrate
Kitamura, Hiroyuki, and Takeshi Sagi, 1965.
On the chemical elements in the sea south of Honshu, Japan. II. (IN Japanese; English abstract).
Bull. Kobe Mar. Obs., No. 174:39-55.

Chemistry phosphate/nitrate
Kitamura, Hiroyuki, and Takeshi Sagi, 1964.
On the chemical elements in the sea south of Honshu, Japan. (In Japanese; English abstract).
Bull. Kobe Mar. Obs., No. 172:6-54.

Chemistry, phosphate nodules

Pratt, Richard M., 1971.
Lithology of rocks dredged from the Blake Plateau.
SEast. Geol. 13(1):19-38.

phosphate, organic

Chemistry, organic phosphate (data only)
Australia, Commonwealth Scientific and Industrial Research Organization, 1960.
Oceanic investigations in eastern Australia, H.M.A. Ships Queenborough, Quickmatch and Warrego, 1958.
Div. Fish. and Oceanogr., Oceanogr. Sta. List, 43:57 pp.

Chemistry, PHosphate, organic (data)
Liubimova, E.M., 1959.
Vertical distribution of organic phosphorus in the Black Sea. Hydrometeorology. Hydrochemistry. (In Russian).
Trudy Morsk. Gidrofiz. Inst., 16:127-160.

phosphate/oxygen

Chemistry, P-O relations
Sugiura, Yoshio, and Hirozo Yoshimura, 1964.
Distribution and mutual relation of dissolved oxygen and phosphate in the Oyashio and the Northern part of Kuroshio.
J. Oceanogr. Soc., Japan., 20(1):14-23.

phosphate/pH

Chemistry, phosphate/pH
Fonselius, Stig H., 1967.
Hydrography of the Baltic deep basins II.
Rep. Fishery Bd., Swed., Hydrogr., 20:31 pp.

phosphate-phosphorus

phosphate, preformed

chemistry, phosphate, preformed
Alvarez-Borrego, Saul, and P. Kilho Park 1971.
AOU as indicator of water-flow direction in the central North Pacific.
J. Oceanogr. Soc. Japan 27 (4): 142-151

Chemistry, phosphate, preformed
Pytkowicz, Ricardo M., 1971.
On the apparent oxygen utilization and the preformed phosphate in the oceans. Limnol. Oceanogr. 16(1): 39-42.

phosphate (preformed)
*Sapozhnikov, V.V., 1967.
The application of the computation parameter "Performed phosphate" to elucidating the basic regularities in the distribution of phosphate and oxygen along the meridional sections in the Pacific and Atlantic ocean. (In Russian; English abstract).
Trudy Inst. Okeanol., 83:5-19.

Chemistry, phosphate (preformed)
Sugiura, Yoshio, 1965.
Distribution of reserved (preformed) phosphate in the subarctic Pacific region.
Papers, Meteorol. Geophys., Tokyo, 15(3/4):208-215.

Chemistry, phosphate regeneration
Motohashi, Keinosuke and Chikayoshi Matsudaira, 1959.
On the relation between the oxygen consumption and the phosphate regeneration from phytoplankton decomposing in stored sea water.
J. oceanogr. Soc., Japan, 25(5): 249-254

phosphate "reserved"

Chemistry, phosphate ("reserved")
Fukai, Rinnosuke 1966.
Comments on Dr. Sugiura's "reserved nutrients."
J. oceanogr. Soc. Japan 22(1): 17-19.

phosphate/silicate

Chemistry, phosphate/silicate
Kitamura, Hiroyuki, and Takeshi Sagi, 1965.
On the chemical elements in the sea south of Honshu, Japan. II. (In Japanese; English abstract)
Bull. Kobe Mar. Obs., No. 174:39-55.

phosphate, sorption

Chemistry, phosphate, sorption of
Hassenteufel, W., R. Jagitsch and F.F. Koczy, 1963
Impregnation of glass surface against sorption of phosphate traces.
Limnol. and Oceanogr., 8(2):152-156.

phosphate, surface + bottom

Chemistry, phosphate, surface (data only)
Australia, Commonwealth Scientific and Industrial Organization, 1962.
Surface sampling in the Coral and Tasman seas, 1960.
Oceanogr. Sta. List, Div. Fish. and Oceanogr., 50:172pp.

Chemistry, phosphate (surface)(data)
Callaway, R.J., 1957.
Oceanographic and meteorological observations in the northeast and central North Pacific, July-December 1956. USFWS Spec. Sci. Rept., Fish., No. 230:49 pp.

Chemistry, phosphate (surface) (data only)
Graham, Joseph J., and William L. Craig, 1961
Oceanographic observations made during a cooperative survey of albacore (Thunnus germo) off the North American west coast in 1959.
U.S.F.W.S. Spec. Sci. Rept., Fish., No. 386: 31 pp.

Chemistry, phosphates (monthly average surface and bottom)
Jayaraman, R., 1954.
Seasonal variations in salinity, dissolved oxygen and nutrient salts in the inshore waters of the Gulf of Mannar and Palk Bay near Mandapam. (S. India). Indian J. Fish. 1(1/2):345-364, 11 textfigs.

phosphate, toxicity

Chemistry, phosphate toxicity
Braginskii, L.P., 1956.
[Toxicity of phosphates for zooplankton.]
Priroda (3):116.

Transl. cited: USFWS Spec. Sci. Rept., Fish. 227.

phosphate utilization apparent

Chemistry, apparent phosphate utilization (APU)
Stefansson, Unsteinn, and Francis A. Richards, 1964.
Distributions of dissolved oxygen, density and nutrients off the Washington and Oregon coasts.
Deep-Sea Res., 11(3):355-380.

phosphoric acid

Chemistry, phosphoric acid
Kester, Dana R., and Ricardo M. Pytkowicz 1967.
Determination of the apparent dissociation constants of phosphoric acid in sea water.
Limnol. Oceanogr. 12(2): 243-252.

Chemistry, phosphoric acid
Matthews, D.J., 1916
On the amount of phosphoric acid in the sea water off Plymouth Sound. J.M.B.A., XI:122-130.

Chemistry, phosphoric acid
Raben, E., 1916.
[Quantitative determination of the phosphoric acid dissolved in sea water.]
Wiss. Meeresuntersuch., Kiel, n.s., 18(1):1-24.

Transl. cited: USFWS Spec. Sci. Rept., Fish., 227

Phosphorus

Chemistry, phosphorus

See also: Chemistry, phosphate

Chemistry, phosphorus
Abbott, W., 1957.
Unusual phosphorus source for plankton algae.
Ecology 38(1):152.

Chemistry, phosphorus
Ackman, R.G., R.F. Addison, P.J. Ke and J.C. Sipos, 1971.
Examination of bottom deposits, Long Harbour, Newfoundland, for elemental phosphorus and for fluorides.
Tech. Rept. Fish. Res. Bd, Canada, 233: 19pp.

phosphorus
Addison, R.F., R.G. Ackman, L.J. Cowley, D. Mascoluk and S. Pond, 1971.
Elemental phosphorus levels in water samples from Long Harbour, Nfld. (July 1969-June 1970).
Techn. Rept. Fish. Res. Bd. Can. 254: 41 pp. (multilithed)

Chemistry, phosphorus
*Akiyama, Tsutomu, 1968.
Phosphorus in the Antarctic Ocean and related seas.
J. oceanogr. Soc., Japan, 24(3):115-123.

Chemistry, phosphorus
Anon., 1950.
[The results of harmonic analysis of tidal currents, water temperature and chemical components in the Kii-Suido.] J. Ocean., Kobe Obs., 2nd ser. 2(1):28-31, 2 figs.

Chemistry, phosphorus
Anon., 1949-1950.
[The results of the regular monthly oceanographical observations on board the R.M.S. "Syunpu Maru" and "Takatori Maru" in the Osaka Wan.]
J. Ocean., Kobe Obs., 2nd ser., 2(1):1-28.

Chemistry, phosphorus
Anon., 1950.
[Report of the oceanographical observations from Fisheries Experimental Stations (Sept. 1950).]
J. Ocean., Kobe Obs., 2nd ser., 1(2):5-11.

Chemistry, phosphorus
Anon., 1950.
[The report of the oceanographical observations in Bungo Suido and off Ashizuri-Misaki in August 1950.] J. Ocean., Kobe Obs., 2nd ser., 1(1):15-44, 7 figs., tables.

Chemistry, phosphorus
Argentina, Servicio de Hidrografia Naval, 1959.
Quimica del agua del mar.
Servicio de Hidrografia Naval, Argentina, Publ., H. 604:140 pp.

chemistry phosphorus

Armstrong, F.A.J., 1965.
Phosphorus.
In: Chemical oceanography, J.P. Riley and G. Skirrow, editors, Academic Press, 1: 323-364.

chemistry phosphorus

Armstrong, F.A.J., 1957.
Phosphorus and silicon in sea water off Plymouth during 1955. J. Mar. Biol. Assoc., U.K., 36(2): 317-322.

chemistry phosphorus

Armstrong, F.A.J., 1955.
Phosphorus and silicon in sea water off Plymouth during 1954. J.M.B.A. 34(2):223-2228.

chemistry phosphorus

Armstrong, F.A.J., 1954.
Phosphorus and silicon in sea water off Plymouth during the years 1950 to 1953. J.M.B.A. 33(2): 381-392, 3 textfigs.

chemistry, phosphorus

Armstrong, F.A.J., 1949.
A source of error in the absorptiometric determination of inorganic and total phosphorus in sea water. J.M.B.A. 28(3):701-705.

Chemistry, phosphorus

Armstrong, F.A.J., and E.I. Butler, 1962.
Chemical changes in sea water off Plymouth during 1960.
J. Mar. Biol. Assoc., U.K., 42(2):253-258.

chemistry, phosphorus

Armstrong, F.A.J.' and E.I. Butler, 1960.
Chemical changes in the s ea water off Plymouth during 1958. J.M.B.A., 39(2):299-302.

chemistry, phosphorus

Armstrong, F.A.J., and H. W. Harvey, 1950.
The cycle of phosphorus in the waters of the English Channel. J.M.B.A. 29(1):145-162, 10 textfigs., 3 tables.

chemistry phosphorus

Atkins, W.R.G., and E.G. Wilson, 1927
The phosphorus and Arsenic compounds of sea-water. J.M.B.A. 14: 609-614.

chemistry phosphorus

Auninsh, E.A., 1966.
Biogenic elements in the Gulf of Riga. (In Russian).
Trudy, Gosudarst. Okeanogr. Inst., No. 83:172-206.

chemistry phosphorus

Austin, K. H., 1959.

Combined phosphorus in the surface waters of Departure Bay, B. C.
Mss. Rept. Ser. (Oceanogr. Limnol.), No. 32:13 pp.

chemistry, phosphorus

Bandel, W., 1940
Phytoplankton- und Nahrstoffgehalt der Ostsee im Gebeit der Darsser Schwelle. Internat. Rev. ges. Hydrobiol. u. Hydrogr., 40:249-304

chemistry phosphorus

Beers, John R., 1966.
Studies on the chemical composition of the major zooplankton groups in the Sargasso Sea.
Limnol. Oceanogr., 11(4):520-528.

Chemistry, phosphorus

Bein, S.J., 1957.
The relationship of total phosphorus concentration in sea water to red tide blooms.
Bull. Mar. Sci., Gulf and Caribbean, 7(4):316-329

Chemistry phosphorus

Berrit, G., and B. Dussart, 1950.
Dosage dans les eaux naturelles des composés minéraux du phosphore, (bibliographie). Circ. C.R.E.O., R.T.B., No. 4:11 pp.

chemistry phosphorus

Beyers, R. J., & H. T. Odum, 1959.

The use of carbon dioxide to construct pH curves for the measurement of productivity.
Limnol. & Ocean., 4(4):499-502.

chemistry, phosphorus

Bibik, V.A. 1964.
Dynamic characteristics of the water of the southeastern part of the Black Sea and the distribution of its oceanographical elements. (In Russian).
Trudy azov.-chernomorsk. nauchno-issled. Inst. morsk ryb. Khoz. okeanogr. 23:23-31.

chemistry phosphorus

Bogojavlensky, A. N., 1958.

Certain peculiarities of the oxygen, phosphorus and siliceous acid distribution in Antarctic waters.
Inform. Biull. Sovetsk. Antarkt. Exped., (3):19-20.

phosphorus

Bruevich, S.W., editor, 1966.
Chemistry of the Pacific Ocean.
Inst. Okeanol., Akad. Nauk, SSSR, Isdatel. Nauka, Moskva, 358 pp.

Chemistry phosphorus

Bruevich, S.V., and E.G. Vinogradova, 1946.
General features of sedimentation in the Caspian Sea according to the distribution of carbonates, Fe, Mn, and P. in sea deposits. Dokl. Akad. Nauk, SSSR, n.s., 52:789-792.

chemistry, phosphorus

Bsharah, L., 1957.
Plankton of the Florida Current 5. Environmental condition, standing crop, seasonal and diurnal changes at a station forty miles east of Miami.
Bull. Mar. Sci., Gulf & Caribbean, 7(3):201-251.

chemistry Phosphorus

California Academy of Sciences
California Division of Fish and Game
Scripps Institution of Oceanography
U. S. Fish and Wildlife Service
} 1950
California Cooperative Sardine Research Program.
Progress Rept. 1950:54 pp., 37 text figs.

phosphorus

Carpenter, Edward J., 1970.
Phosphorus requirements of two planktonic diatoms in a steady state culture.
J. Phycol. 6(1):28-30

chemistry phosphorus

Chu, S.P., 1947.
The utilization of organic phosphorus by phytoplankton. J.M.B.A. 26(3):285-295, 1 diagram

Chemistry phosphorus

Chu, S. P., 1946
The utilization of organic phosphorus by phytoplankton. JMBA, XXVI(3):285-295.

chemistry, phosphorus

Codispoti, Louis A. 1968.
Some results of an oceanographic survey in the northern Greenland Sea, summer 1964.
Techn. Rept. U.S.N. Oceanogr. Off TR202:49pp.

Chemistry phosphorus

Conover, R.J., 1961
The turnover of phosphorus by Calanus finmarchicus.
J. Mar. Biol. Assoc., U.K., 41:484-488.

Chemistry phosphorus

Cooper, L.H.N., 1952.
The physical and chemical oceanography of the waters bathing the continental slope of the Celtic Sea. J.M.B.A. 30: 465-510, 15 textfigs.

Chemistry phosphorus

Cooper, L.H.N., 1952.
Utilization of total phosphorus determinations in physical oceanography. Assoc. Océanogr. Phys. Proc. Verb. No. 5:199-200.

Chemistry Phosphorus

Cooper, L.H.N., 1948.
The nutrient balance in the sea. Research 1: 242-248, 2 textfigs.

Chemistry Phosphorus

Cooper, L.H.N., 1939.
Phosphorus, nitrogen, iron, and manganese in marine zooplankton. J.M.B.A. 23:387-390.

chemistry phosphorus

Cooper, L. H. N., 1934.
The determination of phosphorus and nitrogen in plankton. JMBA, XIX:755-759

Chemistry, phosphorus

Correll, David L., 1965.
Pelagic phosphorus metabolism in Antarctic waters.
Limnol. Oceanogr., 10(3):364-370.

chemistry, phosphorus

Corwin, Nathaniel, and David A. McGill, 1963.
Nutrient distribution in the Labrador Sea and Baffin Bay.
U.S.C.G. Bull., No. 48:79-94; 95-153.

chemistry phosphorus

Culliney, John L., 1970.
Measurements of reactive phosphorus associated with pelagic Sargassum in the northwest Sargasso Sea. Limnol. Oceanogr., 15(2): 304-306.

Chemistry phosphorus

Cushing, D.H., and H.F. Nicolson, 1963
Studies on a Calanus patch. IV. Nutrient salts off the north-east coast of England in the spring of 1954.
J. Mar. Biol. Assoc., U.K., 43(2):373-386.

Chemistry, phosphorus

Demolon, A., and P. Boischot, 1948.
Observations sur le cycle du phosphore dans le biosphère. C.R. Acad. Sci., Paris, 227(14):655-658.

chemistry phosphorus

Dmitrenko, O.I., and Pavlova, G.A., 1962.
About the chemistry of the phosphorus in the sea. (In Russian; English summary).
Trudy Inst. Okeanol., Akad. Nauk, SSSR, 54:100-114.

chemistry phosphorus

Dragovich, A., 1963. (data)
Hydrology and plankton of coastal waters at Naples, Florida.
Q.J. Florida Acad. Sci., 26(1):22-47.

chemistry, phosphorus

Duursma, E.K., 1961
Dissolved organic carbon, nitrogen and phosphorus in the sea.
Netherlands J. Sea Res., 1(1/2):1-147.

chemistry
phosphorus

Duursma, E.K., 1960

Dissolved organic carbon, nitrogen and phosphorus in the sea. Netherlands J. Mar. Res. 1(1): 1-148.

chemistry
phosphorus

Duursma, E.K., 1959
Über gelöste organische Substanz auf einem Meridionalschnitt durch den nördlichen Nordatlantischen Ozean im April 1958.
Deutsche Hydrogr. Zeits., Ergänzungsheft, Reihe B,4(3):69-72.

chemistry
phosphorus

El Wardani, Sayed A., 1960

Total and organic phosphorus in waters of the Bering Sea, Aleutian Trench and Gulf of Alaska.
Deep-Sea Res., 7(3): 201-207.

chemistry, phosphorus

Febres, Germán, Gilberto Rodríguez y Andrés Eloy Esteves, 1966.
Ch. 2. Química del agua.
Estudios Hidrobiologicos en el Estuario de Maracaibo, Inst. Venezolano de Invest. Cient., 21-65. (Unpublished manuscript).

chemistry, phosphorus

*Fedosov, M.V., and I.A. Ermachenko, 1967.
Hydrochemical characteristics of water masses and water exchange between Iceland and Faroes (hydrographical surveys in 1960).
Rapp. P.-V. Réun. Cons. perm. int. Explor. Mer, 157:196-

chemistry
phosphorus

Finucane, John H., and Alexander Dragovich, 1959

Counts of red tide organisms, Gymnodinium brevis and associated oceanographic data from Florida west coast, 1954-1957.
USFWS Spec. Sci. Rept., Fish. No. 289:220 pp.

chemistry, phosphorus

Fletcher, G.L., 1971.
Accumulation of yellow phosphorus by several marine invertebrates and seaweed.
J. Fish. Res. Bd. Can. 28(6): 793-796.

chemistry
phosphorus

Forsbergh, Eric D., 1963
Some relationships of meteorological, hydrographic, and biological variables in the Gulf of Panama. (In English and Spanish).
Bull., Inter-American Tropical Tuna Comm., 7(1): 109 pp.

chemistry, phosphorus

Fraga, F. y A. Ballester 1966.
Distribución vertical del nitrogeno organico y su relación con el fosforo total.
Mem. Soc. Cienc. nat. La Salle 26 (75): 274-282

chemistry phosphorus

Galtsoff, P.S., ed., 1954.
Gulf of Mexico, its origin, waters and marine life. Fish. Bull., Fish and Wildlife Service, 55:1-604, 74 textfigs.

chemistry, phosphorus

Garkavaia, G.P., and L.E. Pozdniakova, 1968.
Data on the hydrochemical regime in Chesha Bay, Barents Sea. (In Russian).
Trudy Murmansk. Morsk. Biol. Inst. 17(21): 4-21.

chemistry, phosphorus

Gololobov, Ia. K. 1964.
Analysis of the chemical compounds involved in biological production in the Black Sea and some peculiarities of their formation.
Trudy azov.-chernomorsk. nauchno-issled. Inst. morsk. ryb. Khoz. Okeanogr. 23: 33-47.

chemistry phosphorus

Graham, H.W., J.M. Amison and K.T. Marvin, 1954.
Phosphorus content of waters along the west coast of Florida. U.S.F.W.S. Spec. ~~Fish~~ Sci. Rept. 122: 1-43.

chemistry, phosphorus

Hansen, A.L., and R.J. Robinson, 1953.
The determination of organic phosphorus in sea water with perchloric acid oxidation.
J. Mar. Res. 12(1):31-42, 2 textfigs.

chemistry, phosphorus

Harvey, H.W., 1953.
Note on the absorption of organic phosphorus compounds by Nitzschia closterium in the dark.
J.M.B.A. 31(3#:475-476.

chemistry phosphorus

Hela, I., 1955.
Ecological observations on a locally limited red tide bloom. Bull. Mar. Sci., Gulf and Caribbean, 5(4):269-291, 16 textfigs.

chemistry phosphorus

Hoffman, C., 1956.
Untersuchungen über remineralisation des Phosphors im Plankton. Kieler Meeresf. 12(1):25-36.

chemistry phosphorus

Hoffmann, C., and M. Reinhardt, 1952.
Zur Frage der Remineralisation des phosphors bei Benthosalgen. Kieler Meeresf. 8:135-144.

chemistry, phosphorus

Holm-Hansen, Osmund, J.D.H. Strickland and P.M. Williams, 1966.
A detailed analysis of biologically important substances in a profile off southern California.
Limnol. Oceanogr., 11(4):548-561.

chemistry
phosphorus

Ichiye, Takashi, Sigeo Moriyasu and Hiroyuki Kitamura, 1951
On the hydrography near the estuaries.
Bull. Kobe Mar. Obs., No. 164: 349(top of page)-369; 53 (bottom of page)-75.

chemistry
phosphorus

Ishibashi, Masayoshi, and Toshio Yamamoto, 1960

Chemical studies on the ocean (Pts. 78-79).
Chemical studies on the seaweeds (VI&VII).
The content of calcium, magnesium and phosphorus in seaweeds. Iron content in seaweeds.
Rec. Oceanogr. Wks., Japan, Spec. No. 4: 73-78; 79-86.

chemistry
Phosphorus (data)

Ishino Makoto, Keiji Nasu, Yoshimi Morita and Makoto Hamada, 1968.
Oceanographic conditions in the west Pacific Southern Ocean in summer of 1964-1965. J. Tokyo Univ. Fish., 9(1): 115-208.

chemistry, phosphorus

Jacob, Otto Ernst, 1961.
Uber die kurzfristige Veranderlichkeit in der Vertikalverteilung ausgewahlter chemischer Faktoren in der ostlichen und mittleren Ostsee im Sommer 1960.
Kieler Meeresf., 17(2):154-158.

chemistry phosphorus

Japan, Kobe Marine Observatory, Oceanographical Section, 1964.?
Researches in chemical oceanography made at K.M.O. for the recent ten years (1954-1963). (In Japanese).
Weather Service Bull., Japan. Meteorol. Agency, 31(7):188-191.

Also in:
Bull. Kobe Mar. Obs., No. 175. 1965.

chemistry phosphorus

Japan, Kobe Marine Observatory, 1962
Report of the oceanographic observations in the sea south of Honshu from July to August, 1960. (In Japanese).
Res. Mar. Meteorol. and Oceanogr., July-Dec., 1960, Japan Meteorol. Agency, No. 28:36-42.

chemistry
phosphorus

Japan, Tokai Regional Fisheries Research Laboratory, 1959.

IGY Physical and chemical data by the R. V. Soyo-maru, 25 July-14 September 1958.
17 pp. (multilithed).

chemistry phosphorus

Jayaraman, R., and G. Seshappa, 1957.
Phosphorus cycle in the sea with particular reference to tropical inshore waters.
Proc. Indian Acad. Sci., B, 46(2):110-125.

chemistry phosphorus

Jayaraman, R., R. Viswanathan and S.S. Gogate, 1961.
Characteristics of sea water near the light house, Bombay.
J. Mar. Biol. Assoc., India, 3(1/2):1-5.

chemistry, phosphorus

Jespersen, P., 1954 (posthumous).
On the quantities of macroplankton in the North Atlantic. Medd. Danmarks Fisk. og Havundersøgel., n.s., 1(2):1-12.

chemistry, phosphorus

Johannes, R.E., 1964.
Uptake and release of dissolved organic phosphorus by representatives of a coastal ecosystem.
Limnology and Oceanography, 9(2):224-234.

chemistry
phosphorus

Jones, J.A., 1963.
Ecological studies of the southeastern Florida patch reefs. 1. Diurnal and seasonal changes in the environment.
Bull. Mar. Sci., Gulf and Caribbean, 13(2):282-307.

chemistry
phosphorus

Kabanova, Yu. G., 1961
[Primary production and nutrients in the Indian Ocean.] (In Russian; English abstract).
Okeanolog. Issled., Mezhd. Komitet Proved. Mezhd. Geofiz. Goda, Prezidiume, Akad. Nauk, SSSR, (4):72-75.

chemistry
phosphorus

Kändler, Rudolf, 1963
Hydrographische Untersuchungen über die Abwasserbelastung der Flensburger Förde.
Kieler Meeresforsch., 19(2):142-157.

chemistry phosphorus

Kawamoto, T., 1955.
The distribution of phosphorus in the waters of the Kuroshio Region (Preliminary report).
J. Ocean. Soc., Japan, 11(2):59-61.

chemistry phosphorus,

Keeling, Charles D., and Bert Bolin, 1968
The simultaneous use of chemical tracers in oceanic studies. II. A three-reservoir model of the North and South Pacific oceans.
Tellus, 20(1): 17-54

chemistry phosphorus

Ketchum, B.H., N. Corwin, and D.J. Keen, 1955.
The significance of organic phosphorus determination in ocean waters. Deep-Sea Res. 2(3):172-181, 3 textfigs.

chemistry, phosphorus

Ketchum, B.H., and J. Keen, 1948
Unusual phosphorus concentrations in the Florida "Red Tide" sea water. J. Mar. Res. 7(1):17-21, fig.2.

chemistry, phosphorus

Kirkwood, L.F., 1967.
Inorganic phosphate, organic phosphorus and nitrate in Australian waters.
Tech. Pap. Div. Fish. Oceanogr., CSIRO, Australia, 25: 18 pp.

Chemistry phosphorus

Koczy, F.F., 1950.
Zur Sedimentation und Geochemie im aequatorischen Atlantischen Ozean. Medd. Oceanografiska Inst. Göteborg 17 (Göteborgs Kungl. Vetenskaps- och Vitterhets Samhälles Handlingar, Sjätte Följden Ser. B, 6(1)):44 pp., 17 textfigs.

Chemistry phosphorus

Koshimizu, M., 1950.
[On the ratio of phosphorus and silicon of the waters masses around the banks in Japan Sea.]
Bull. Maizuru Mar. Obs., (1):1-3.

Chemistry, phosphorus (data)

Krey, J., 1963.
The components of phosphorus in Kiel Bay in 1961. Ann. Biol., Cons. Perm. Int. Expl. Mer, 1961, 18: 54-56.

chemistry, phosphorus

Krey, J., 1959.
Über den Gehalt an gelöstem anorganischem Phosphor in der Kieler Förde, 1952-1957.
Kieler Meersf., 15(1):17-28.

Chemistry phosphorus

Krey, J., 1957.
Ergebnisse eines hydrographisch-produktionsbiologischen Längsschnitts durch die Ostsee im Sommer 1956. Kieler Meeresf., 12(2):202-211.

III. Die Verteilung des Gesamtphosphors.

Chemistry phosphorus

Krey, J., 1956.
Die Trophie küstennaher Meeresgebiete.
Kieler Meeresf., 12(1):46-64.

chemistry, phosphorus

Krey, J., 1942.
NÄHrstoff und Chlorophylluntersuchungen in der Kieler Förd 1939. Kieler Meeresforsch. 4:1-17, textfigs.

Chemistry, phosphorus

Krey, Johannes, Peter H. Koske and Karl-Heinz Szekields, 1965.
Produktionsbiologische und hydrographische Untersuchengen in der Eckernförder Bucht.
Kieler Meeresforsch., 21(2):135-143.

chemistry, phosphorus

*Krishnamurthy, K., 1967.
Some aspects of chemical composition of plankton.
Kieler Meeresforsch., 23(2):99-104.

chemistry, phosphorus

*Krishnamurthy, K., 1967.
The cycle of nutrient salts in Porto Novo (India) water.
Int. Rev. ges. Hydrobiol., 52(3):427-436.

chemistry phosphorus

Krishnamurthy, K., 1963
Phosphorus in plankton. 1.
J. Zool. Soc., India, Calcutta, 14(2):161-164.

Also in:
Annamalai Univ. Mar. Biol. Sta. Porto Novo, S. India, Publ. 1961-1962.

Chemistry phosphorus

Kuenzler, Edward J., and Bostwick H. Ketchum, 1962
Rate of phosphorus uptake by Phaeodactylum tricornutum.
Biol. Bull., 123(1):134-145.

Chemistry phosphorus

Kumari, P. Santha, 1969.

Phosphorus fractions in Porto Novo waters (11°29'N -79°49'E) during 1965-66. Bull. natn. Inst. Sci., India, 38(1): 87-92.

Chemistry phosphorus

Levine, H., J.J. Rowe and F.S. Grimaldi, 1955.
Molybdenus blue reaction and determination of phosphorus in waters containing arsenic, silicom and germanium. Analyt. Chem. 27(2):258-262.

Chemistry phosphorus

Lisitzin, A.P., 1966.
Processes of Recent sedimentation in the Bering Sea. (In Russian).
Inst. Okeanol., Kom. Osad. Otdel Nauk o Zemle, Isdatel, Nauka, Moskva, 574 pp.

Chemistry phosphorus

Lisitsin, A.P., 1964.
Distribution and chemical composition of suspended matter in the waters of the Indian Ocean. (In Russian; English Abstract).
Rezult. Issled., Programme Mezhd. Geofiz. Goda, Mezhd. Geofiz. Kom., Presidume Akad. Nauk, SSSR, Okeanol., No. 10:136 pp.

Chemistry phosphorus

MacLeod, R.A., R.E.E. Jonas and E. Onofrey, 1960.
On the forms, balance and cycle of phosphorus observed in the coastal and oceanic waters of the northeastern Pacific.
J. Fish. Res. Bd., Canada, 17(3):337-345.

Chemistry phosphorus (data)

Marchand, J.M., 1953.
Pilchard-research programme. First Progress Report. Annexure "A", 23rd Ann. Rept. Div. Fish., Union S. Africa:17-181, 5 charts.

Chemistry phosphorus

Marino, Liana Bisi, 1962
La determinazione del fosforo nei sedimenti Adriatici.
Archiv. di Ocean. Limnol, XII(3):267-274
Istit. Sper. Talass., Trieste, Pubbl. No. 397

chemistry, phosphorus

Maslennikov, V.V., 1971.
Oceanographic surveys in the Andaman Sea and the northeastern Bay of Bengal. (In Russian)
Vses. nauchno-issled. Inst. morsk. Ribn. Khoz. Okeanogr. VNIRO, Trudy 72: 46-55

Chemistry phosphorus (data)

Masuzawa, Jotaro, and Hideo Akamatsu, 1962.
1. Hydrographic observations during the JEDS-4.
Oceanogr. Mag., Japan Meteorol. Agency, 13(2): 122-130.

chemistry, phosphorus

Masuzawa, Jotaro, Tsutomu Akiyama, Yutaka Kawarada and Tsutomu Sawara, 1970.
Preliminary report of the Ryofu Maru Cruise Ry7001 in January-March 1970. Oceanogrl Mag., 22(1): 1-25.

Chemistry phosphorus

Matida, Y., 1955.
On the cycle of phosphorus in Tokyo Bay.
Bull. Japan. Soc., Sci. Fish., 20(9):793-796, 2 textfigs.

Chemistry phosphorus

Matida, Y., 1953.
The cycles of phosphorus and nitrogen in Tokyo Bay. Bull. Japan. Soc. Sci. Fish. 19(4):429-434, 7 textfigs.

Chemistry phosphorus

Matida, Y., 1952.
[A critical study of the determination of total phosphorus in natural waters.]
Bull. Jap. Soc. Sci. Fish. 18(4):175-181.

Chemistry phosphorus

McGill, David A., 1964.
The distribution of phosphorus and oxygen in the Atlantic Ocean as observed during the I.G.Y., 1957-1958.
Progress in Oceanography, Pergamon Press, 2:127-211.

Chemistry, phosphorus (data)

McGill, David A., and Nathaniel Corwin, 1965.
The distribution of nutrients in the Labrador Sea, summer, 1964.
U.S.C.G. Rept., No. 10(CG 373-10):25-33; 277-285.
Oceanogr.

chemistry, phosphorus

McGill, David A., Nathaniel Corwin and Bostwick H. Ketchum, 1964
Organic phosphorus in the deep water of the western North Atlantic.
Limnology and Oceanography, 9(1):27-34.

Chemistry phosphorus

McGill, David A., Nathaniel Corwin and Bostwick H. Ketchum, 1959.

The temperature correction of oceanic inorganic phosphorus analyses.
Deep-Sea Res., 6(1):60-64.

Chemistry phosphorus

Megia, T.G., and R.G. Lao, 1955.
A report on the O:N:P ratio of Philippine and adjacent waters.
Philippine J. Fish., 3(1):55-61, 3 textfigs.

Chemistry phosphorus

Menzel, David W., and Nathaniel Corwin, 1965.
The measurement of total phosphorus in seawater based on the liberation of organically bound fractions by persulfate oxidation.
Limnology and Oceanography, 10(2):280-282.

Chemistry phosphorus

Meyer, P.F., and K. Kalle, 1950.
Die biologische Umstimmung der Ostsee in den letzten Jahrzehnten - eine Folge hydrographischer Wasserumschichtungen? Archiv f. Fischereiwiss. 2 Jahrg.:1-9, 6 textfigs.

Chemistry phosphorus

Miller, S.M., 1952.
Phosphorus exchange in a sub-tropical marine basin. Bull. Mar. Sci. Gulf & Carribean 1(4): 257-265, 5 textfigs.

Reviewed: J. du Cons. 19(1):94-95 by H.W. Harvey

Chemistry, phosphorus

Minkina, A. L., 1961.
[Organic nitrogen and phosphorus in the water masses of the Barents Sea.]
Mezhd. Kom. Mezhd. Geofiz. Goda, Presidium Akad. Nauk, SSSR, Okeanol. Issled., (3):162-171.

chemistry, phosphorus

Miyake, Yasuo, and Katsuko Saruhashi 1967.
The geochemical balance of nutrient matters in the oceans.
Pap. Met. Geophys. Tokyo 18(2): 89-94.

Chemistry, phosphorus

Muñoz, Felipe, y Jose M. San Feliu, 1965.
Hidrografia y fitoplancton de las costas de Castellon de agosto de 1962 a julio de 1963.
Inv. Pesq., Barcelona, 28:173-209.

Chemistry, phosphorus

Newcombe, C.L., and H. F. Brust, 1940.
Variations in the phosphorus content of estuarine waters of the Chesapeake Bay near Solomon's Island, Maryland. J. Mar. Res., 3(1): 76-88, 6 figs.

Chemistry phosphorus

Neumann, W., 1949.
Beiträge zur Hydrographie des Vrana-Sees (Insel Cherso), inbesondere Untersuchungen über organisches sowie anorganische Phosphor- und Stickstoffverbindungen. Nova Thalassia 1(6):17 pp.

Chemistry phosphorus

Nümann, W., 1949.
Kolorimetrische Methoden zur quantitativen Bestimmung von Silikat und organisch wie anorganisch gebundenem Phosphor und Stickstoff im Meerwasser unter Benutzung des Pulfrich-Photometers. Deut. Hydro. Zeit. 2(4):137-153, 3 textfigs.

chemistry Phosphorus

Nümann, W., 1941
Der Nährstoffhaushalt in der nordöstlichen Adria. Thalassia 5(2):1-68, 12 figs.

Chemistry, phosphorus

Nichols, Maynard M., 1964.
Characteristics of sedimentary environments in Moriches Bay.
In: Papers in Marine Geology, R.L. Miller, Editor, Macmillan Co., N.Y., 363-383.

Chemistry, phosphorus

Odum, Howard T., Rene P. Cuzon du Rest, Robert J. Beyers and Clyde Allbaugh, 1963.
Diurnal metabolism, total phosphorus, Ohle anomaly and zooplankton diversity of abnormal marine ecosystems of Texas.
Publ. Inst. Mar. Sci., Port Aransas, 9:404-453.

Chemistry phosphorus

Okuda, Taizo, 1960
Chemical oceanography in the South Atlantic Ocean, adjacent to north-eastern Brazil. (Portuguese and French resumés).
Trabalhos, Inst. Biol. Marit. e Oceanogr., Universidade do Recife, Brasil, 2(1):155-174.

Chemistry phosphorus

Okuda, Taizo, Lourinaldo Cavalcanti, and Manoel Pereira Borba, 1960
Estudo da Barra das Jangadas. 3. Variação de nitrogenio e fosfato durante o ano.
(In Portuguese; English and French abstracts)
Trabalhos, Inst. Biol. Marit. e Oceanogr., Universidade do Recife, Brasil, 2(1):207-218.

Chemistry, Phosphate

Orr, A.P., 1947
An experiment in marine fish cultivation: II. Some physical and chemical conditions in a fertilized sea-loch (Loch Craiglin, Argyll).
Proc. Roy. Soc., Edinburgh, Sect.B, 63, Pt.1 (2):3-20, 19 text figs.

chemistry phosphorus

Orr, A.P., and F.W. Moorhouse, 1933.
(a) Variations in some physical and chemical conditions on and near Low Isles Reef.
(b) Temperature of the waters in the anchorage, Low Isles.
(c) Physical and chemical conditions in mangrove swamps.
Brit. Mus. (N.H.) Great Barrier Reef Exped., 1928-1929, Sci. Res., 2(4):87-110.

chemistry phosphorus (data)

Parsons, T.R., 1960.
A data record and discussion of some observations made in 1958-1960 of significance to primary productivity.
Fish. Res. Bd., Canada, Manuscript Rept. Ser., (Oceanogr. & Limnol.), No. 81:19 pp. (multilithed).

chemistry phosphorus

Parsons, T.R., K. Stephens and J.D.H. Strickland, 1961
On the chemical composition of eleven species of marine phytoplankters.
J. Fish. Res. Bd., Canada, 18(6):1001-1016.

Chemistry phosphorus

Patten, B.C., R.A. Mulford and J.E. Warinner, 1963
An annual phytoplankton cycle in the lower Chesapeake Bay.
Chesapeake Science, 4(1):1-20.

Chemistry, phosphorus (data)

Platt, Trevor, and Brian Irwin, 1968.
Primary productivity measurements in St. Margaret's Bay, 1967.
Techn.Rept., Fish.Res.Bd., Can., 77: 123 pp.

Chemistry phosphorus (data)

Pomeroy, Lawrence R., 1960.
Residence time of dissolved phosphorus in natural waters.
Science, 131(3415):1731-1732.

Chemistry, phosphorus

*Pomeroy, L.R. and E.J. Kuenzler, 1969.
Phosphorus turnover by coral reef animals.
Symp. Radioecology, Proc. Second Natn. Symp., Ann Arbor, Mich., 1967 (Conf.670503):474-482.

Chemistry, phosphorus

Pomeroy, Lawrence R., Henry M. Mathews and Hong Shik Min, 1963
Excretion of phosphate and soluble organic phosphorus compounds by zooplankton.
Limnology and Oceanography, 8(1):50-55.

Chemistry phosphorus

Postma, H., 1954.
Hydrography of the Dutch Waddensea. A study of the relations between water movement, the transport of suspended materials and the production of organic matter. Arch. Néerl. Zool. 10(4):405-511, 55 textfigs.

Chemistry phosphorus

Pratt, D. M., 1950.
Experimental study of the phosphorus cycle in fertilized sea water. J. Mar. Res. 9(1):29-54, 10 textfigs.

Chemistry phosphorus

Rakestraw, N. W. and D. E. Carritt, 1948
Some seasonal chemical changes in the open ocean. J. Mar. Res. 7(3):362-369, 6 text figs.

Chemistry, phosphorus

Ramamurthy, V.D., K. Krishnamurthy and R. Seshadri, 1965.
Comparative hydrographical studies of the near shore and estuarine waters of Porto-Novo, S. India.
J. Annamalai Univ., 26:154-164.

Also in: Collected Reprints, Mar. Biol. Sta., Porto Novo, 1963/64.

chemistry, phosphorus

Ramamurthy, V.D., and R. Seshadri, 1966.
Phosphorus concentration during red-water phenomenon in the near shore waters of Port Novo (S. India).
Current Science, 35(4):100-101.

Chemistry phosphorus

Rao, S.V. Suryanarayana, 1957.
Preliminary observations on the total phosphorus content of the inshore waters of the Malabar coast of Calicut.
Proc. Indian Acad. Sci., (B), 45(2):77-85.

chemistry phosphorus

Rao, V. Chalapati and T.S. Satyanarayana Rao, 1969.

Distribution of total phosphorus in the Bay of Bengal. Bull. natn. Inst. Sci., India, 38(1): 93-102.

Chemistry, phosphorus

Redfield, A.C., 1960.
The distribution of phosphorus in the deep oceans of the world.
Assoc. Océanogr. Phys., Union Géodés. et Géophys. Int., Proc. Verb., No. 7(G16):189-193.

Chemistry phosphorus, total

Redfield, A.C., 1955.
The hydrography of the Gulf of Venezuela.
Pap. Mar. Biol. and Oceanogr., Deep-Sea Res., Suppl. to Vol. 3:115-133.

Chemistry phosphorus

Redfield, A.C., B.H. Ketchum and F.A. Richards, 1963
2. The influence of organisms on the composition of sea-water. In: The Sea, M.N. Hill, Edit., Vol. 2. The composition of sea water, Interscience Publishers, New York and London, 26-77.

Chemistry phosphorus

Redfield, A.C., H.P. Smith, and B.H. Ketchum, 1937.
The cycle of organic phosphorus in the Gulf of Maine. Biol. Bull. 73(3):421-443, 4 textfigs.

Reviewed: J. du Cons. XIV(3):405-406 by LHN Cooper.

Chemistry, phosphorus

Reimold, Robert J., and Franklin C. Daiber, 1967
Eutrophication of estuarine areas by rainwater.
Chesapeake Sci., 8(2):132-133.

Chemistry phosphorus

Renn, C.E., 1937.
Bacteria and the phosphorus cycle in the sea.
Biol. Bull. 72(2):190-195.

Chemistry, phosphorus

Rice, T.R., 1953.
Phosphorus exchange in marine phytoplankton.
Fish. Bull. 80, Vol. 54:77-89, 3 textfigs.

Chemistry phosphorus

Richards, F.A., 1957.
Some current aspects of chemical oceanography.
In: Physics and chemistry of the earth, Pergamon Press; 2:77-128.

Chemistry, phosphorus

Rochford, D.J., 1963.
Some features of organic phosphorus distribution in the south-east Indian and south-west Pacific oceans.
Australian J. Mar. Freshwater Res., 14(2):119-138

chemistry, phosphorus

Rochford, D., 1958.
Total phosphorus as a means of identifying East Australian water masses. Deep-Sea Res., 5:89-110.

Chemistry phosphorus

Rochford, D.J., 1951.
Studies in Australian estuarine hydrology.
Australian J. Mar. and Freshwater Res. 2(1):1-116, 1 pl., 7 textfigs.

Chemistry phosphorus

Rotschi, H., 1960.
Remarques sur la relation entre l'oxygène et le phosphore mineral dissous en Mer de Corail.
C. R. Acad. Sci., Paris, 250(13):2403-2405.

Resume in:
Cahiers du Pacifique, 1962(4):85.

chemistry, phosphorus

Rozanov, A.G., 1964.
The distribution of phosphates and silicates in water of the northern part of the Indian Ocean. Investigations in the Indian Ocean (33rd voyage of E/S"Vitiaz"). (In Russian).
Trudy Inst. Okeanol., Akad. Nauk, SSSR, 64:102-104.

chemistry phosphorus

Salnikov, N.E., 1965.
Investigaciones de economía pesquera en el Golfo de México y del Mar Caribe. (In Russian; Spanish abstract).
Sovetsk. -Cub. Ribokhoz. Issled., VNIRO:Tsentr. Ribokhoz. Issled. Natsional. Inst. Ribolovsta Republ. Cuba, 93-179.

chemistry phosphorus (data)

San Feliu, J.M., y F. Muñoz, 1965.
Hidrografía y plancton del puerto de Castellón de junio de 1961 a enero de 1963.
Inv. Pesq., Barcelona, 28:3-48.

chemistry, phosphorus

Sapozhnikov, V.V., 1971.
The role of advection and the influence of biochemical processes on vertical phosphorus distribution in the tropical ocean. (In Russian; English abstract). Okeanologiia 11(2): 223-230.

chemistry phosphorus(data)

San Feliu, J.M. y F. Muñoz 1967.
Hidrografía y fitoplancton de las costas de Castellón, de mayo de 1965 a julio de 1966.
Investigación pesq. 31(3):419-461.

chemistry phosphorus

Seiwell, H.R., 1935
The cycle of phosphorus in the Western Basin of the North Atlantic. I Phosphate phorphorus. P.P.O.M. 3(4):1-56, 27 textfigs.

chemistry phosphorus

Seiwell, H. R., 1935.
The annual organic production and nutrient phosphorus requirement in the tropical western North Atlantic. J. du Cons. 10(1):20-32, 1 text-fig.

chemistry phosphorus

Seiwell, H. R., and G. E. Seiwell, 1938.
The sinking of decomposing plankton in sea water and its relationship to oxygen consumption and phosphorus liberation. Proc. Am. Phil. Soc. 78(3):465-481, 5 textfigs.

chemistry phosphorus

Seiwell, H. R., and G. E. Seiwell, 1934.
Über den Gesamtphosphorgehalt des Seewassers im westlichen Nordatlantischen Ozean. Ann. Hydrogr. u. Maritimen Meteorol., 1934(1):7-13, 1 fold-in, 1 textfig.

chemistry phosphorus

Skopintsev, B.A., 1960.
Organic matter in waters of the sea. Marine chemistry, hydrology, oceanic geology. (In Russian).
Trudy Morsk. Gidrofiz. Inst., 19:3-20.

chemistry phosphorus

Skopintsev, B. A., 1948.
[The change in nitrogen and phosphorus content in terrigenous suspended particles in a water medium.] Izvest. Akad. Nauk SSSR, Ser. Geogr. Geofiz. 12:107-118.

chemistry phosphorus

Skopinzev, B.A., and E.S. Bruck, 1940
[Contribution to the study of regeneration of nitrogen and phosphorus compounds in the course of decomposition of dead plankton.] C.R. Acad. Sci., Leningrad, 26:807-810

chemistry, phosphorus

Skopintsev, B.A., S.N. Timofeeva, A.F. Danilenko and M.V. Sokolova, 1967.
Organic carbon, nitrogen, phosphorus and their mineralization products in the Black Sea water. (In Russian; English abstract).
Okeanologiia, Akad. Nauk, SSSR, 7(3):457-469.

chemistry phosphorus

Solov'eva, N.F., 1959.
[Hydrochemical analyses of the Aral Sea in 1948.] Akad. Nauk, SSSR, Lab. Ozerovedeniya, Trudy, 8: 3-22.

5 Sept. 61 JPRS:10054
OTS -SLA $3.60 61-28414.

chemistry, phosphorus

Soot-Ryen, T., 1956.
Report on the Hydrographical conditions in West-Finmark, March-April, 1935.
Acta Borealia, Tromsø Mus., A. Scientia, No. 10: 1-37.

chemistry phosphorus (data)

Soot-Ryen, T., 1938.
Hydrographical investigations in the Tromsø district in 1931.
Tromsø Mus. Aarsheft., Naturhist. Avd. No. 10, 54(2):1-6, plus 41 pp. of tables.

chemistry phosphorus (data)

Soule, F.M., A.J. Bush and J.E. Murray, 1955.
International ice observation and Ice Patrol Service in the North Atlantic Ocean. Season of 1953. U.S.C.G. Bull. No. 39:45-168, 43 textfigs.

chemistry, phosphorus

Strickland, J.D.H., and K.H. Austin, 1960
On the forms, balance and cycle of phosphorus observed in the coastal and oceanic waters of the northeastern Pacific.
J. Fish. Res. Bd., Canada, 17(3):337-345.

chemistry, phosphorus

Strickland, J.D.H., & K.H. Austin, 1959
The direct estimation of ammonia in sea water with notes on "reactive" iron, nitrate and inorganic phosphorus.
J. du Conseil, 24(3):446-451.

chemistry, phosphorus

Szekielda, Karl-Heinz 1971.
Organisch gelöster und partikulärer Kohlenstoff in einem Nebenmeer mit starken Salzgehaltsschwankungen (Ostsee).
Vie Milieu Suppl. 22 (2): 579-612

chemistry, phosphorus

Tagigi, T., and N. Hayashi, 1957.
Organic phosphorus in Pacific water off the eastern shore of northern Honshu (mainland of Japan). II. J. Chem. Soc., Japan, 78(4):491-494.

chemistry, phosphorus

Takeda, I., 1956.
Organic phosphorus in Pacific waters off the eastern shore of northern Honshu.
J. Chem. Soc., Japan, 77(8):1208-1212.

chemistry, phosphorus

Taylor, W. Rowland, H.H. Selinger, W.G. Fastie and W.D. McElroy, 1966.
Biological and physical observations on a phosphorescent bay in Falmouth Harbor, Jamaica, W.I.
J. Mar. Res., 24(1):28-43.

chemistry phosphorus

Timofeev, N.A., 1968.
Oxygen and phosphorus distribution in the Red Sea. (In Russian).
Trudy, Vses. Nauchno-Issled. Inst. Morsk. Ribn. Okeanogr (VNIRO) 64, Trudy Azovo-Chernomorsk. Nauchno-Issled. Inst. Morsk. Ribn. Khoz. Okeanogr. (AscherNIRO), 28: 152-161.

chemistry phosphorus

Toyota, Yoshimasa and Shiro Okabe 1967.
Vertical distribution of iron, aluminum, silicon and phosphorus in particulate matter collected in the western North Pacific Indian and Antarctic oceans. (In Japanese; English abstract)
J. oceanogr. Soc. Japan 23 (1): 1-9.

Also in: Coll. Repr. Coll. mar. Sci. Techn. Tokai Univ. 1967-68, 3: 335-343.

chemistry, phosphorus

Toyota, Yoshimasa and Shiro Okabe, 1967.
Surface distribution of iron, aluminum, silicon and phosphorus in the western North Pacific and eastern Indian Oceans. (In Japanese).
J. Coll. mar. Sci. Techn., Tokai Univ., (2):227-229.

chemistry phosphorus

*Tundisi, J., S. Krishnaswamy, 1967.
Total phosphorus content of Neomysis integer.
Bolm Inst. Oceanogr., S. Paulo, 16(1):99-100.

chemistry phosphorus

Vaccaro, Ralph F., 1963.
Available nitrogen and phosphorus and the biochemical cycle in the Atlantic off New England.
J. Mar. Res., 21(3):284-301.

chemistry phosphorus

Viswanathan, R. and A.K. Ganguly, 1969.
The distribution of phosphorus in northern India Ocean, 1962-63. Bull. natn. Inst. Sci., India, 38(1): 350-362.

chemistry phosphorus

Voronkov, P.P., 1948.
[Basic characteristics of the hydrochemical regime of the coastal zone of Barents Sea in the central Murman region.] Trudy Dal'niye Zelentsy, 1948(1):39-101, 25 diagrams, sketch map.

chemistry phosphorus

Vorontsova, R. V., 1964.
The hydrochemical conditions in the southern part of the Davis Strait in the spring and summer of 1963. (In Russian).
Material. Sess. Uchen. Sov., PINRO, Rez. Issled. 1962-1963., Murmansk, 130-135.

chemistry, phosphorus

Wellershaus, Stefan, 1964.
Die Schichtungsverhältnisse im Pelagial des Bornholmbeckens. Über den Jahresgang einiger biotischer Faktoren.
Kieler Meeresf., 20(2):148-156.

chemistry phosphorus

*Williams, P.M., 1967.
Sea surface chemistry: organic carbon and organic and inorganic nitrogen and phosphorus in surface films and subsurface waters.
Deep-Sea Res., 14(6):791-800.

chemistry phosphorus

Yamamoto, Katsumi, 1968.
The total and organic phosphorus in the Japan Sea.
Oceanogrl Mag., 20(1):39-50.

chemistry, phosphorus assimilation

Krupatkina (Akinina), D.K., 1971.
Phosphorus assimilation by planktonic algae in the dark and in the presence of faint light. (In Russian; English abstract). Okeanologiia 11(2): 270-275.

chemistry phosphorus, carbon bound

*Quinn, L.D., and F.A. Shelburne, 1969.
An examination of marine animals for the presence of carbon-bound phosphorus.
J. mar. Res., 27(1):73-84.

phosphorus-chlorinity

chemistry Phosphorus-chlorinity

Kitamura, H., 1962.
(Marine chemical studies on the phosphorus distributed in the North Pacific Ocean.)
Bull. Kobe Mar. Obs., No. 168:40-94.

Phosphorus cycle

phosphorus cycle

Corner, E.D.S. and Anthony G. Davies, 1971. Plankton as a factor in the nitrogen and phosphorus cycles in the sea. Adv. mar. Biol. 9: 101-204.

chemistry phosphorus cycle

Ketchum, Bostwick H., and Nathaniel Corwin, 1965. The cycle of phosphorus in a plankton bloom in the Gulf of Maine. Limnol. and Oceanogr., Redfield Vol., Suppl. to 10:R148-R161.

chemistry phosphorus cycle

Ketchum, B.H., R.F. Vaccaro and Nathaniel Corwin, 1958. The annual cycle of phosphorus and nitrogen in New England coastal waters. J. Mar. Res., 17:282-301

phosphorus cycle

Krishnamurthy, K., 1966. The cycle of phosphorus in the near-shore water of Porto Novo during 1961. Proc. Second All-Indian Congr. Zool., Varanasi, (1962) (2):313-316. Also in: Collected Reprints, Portonovo mar. biol. Sta. 1966-1967

chemistry, phosphorus cycle

Nellen, Walter 1970. 2.7. Phosphorkreislauf. Chemische, mikrobiologische und planktologische Untersuchungen in der Schlei im Hinblick auf deren Abwasserbelastung. Kieler Meeresforsch 26(2): 132-138.

chemistry, phosphorus cycle

Strickland, J.D.H., and K.H. Austin, 1960. On the forms, balance and cycle of phosphorus observed in the coastal and oceanic waters of the northeastern Pacific. J. Fish. Res. Bd., Canada, 17(3):337-345.

chemistry, phosphorus cycle

Watt, W.D., and F.R. Hayes, 1962. Tracer study of the phosphorus cycle in sea water. Limnol. and Oceanogr., 8(2):276-285.

Phosphorus, data only

chemistry phosphorus (data only)

Anderson, William W., and Jack W. Gehringer, 1959. Physical oceanographic, biological and chemical, South Atlantic coast of the United States, M/V Theodore N. Gill cruise 9. USFWS Spec. Sci. Rept., Fish. No. 313:226 pp.

chemistry, phosphorus (data only)

Anderson, William W. and Jack W. Gehringer 1959. Physical oceanography, biological and chemical data, South Atlantic coast of the United States, M/V Theodore N. Gill Cruise 7. Spec. scient. Rept. Fish, U.S. Fish Wildl. Serv. No. 278: 277 pp.

chemistry, phosphorus (data only)

Anderson, W.W., and J.W. Gehringer, 1957. Physical oceanographic, biological and chemical data, South Atlantic coast of the United States, M/V Theodore N. Gill cruise 4. USFWS Spec. Sci. Rept., Fish., 234:192 pp.

chemistry, phosphorus (data only)

Anon., 1950. Table 14. Oceanographicalm observations taken in the Maizuru Bay; physical and chemical data for stations occupied by R.M.S. "Syunpu-maru". Res. Mar. Met., Ocean. Obs., Tokyo, July-Dec. 1947, No. 2:71-76.

chemistry, phosphorus (data only)

Anon., 1950. Table 10. Oceanographical observations taken in the sea northwest of Kyushu; physical and chemical data for stations occupied by the "Fugen-maru". Res. Mar. Met., Ocean. Obs., Tokyo, July-Dec. 1947, No. 2:57-58.

chemistry, phosphorus (data only)

Anon., 1950. Table 4. Oceanographical observations taken in the North Pacific Ocean along the coast of Sanriku; physical and chemical data for stations occupied by R.M.S. "Kuroshio-maru". Res. Mar. Met., Ocean. Obs., Tokyom July-Dec. 1947, No. 2: 13-21.

chemistry, phosphorus (data only)

Anon., 1950. Table 2. Oceanographical observations taken in the Uchiura Bay; physical and chemical data for stations occupied by R.M.S. "Kuroshio-maru". Res. Mar. Met., Ocean. Obs., Tokyo, July-Dec. 1947, No. 2:3-7.

chemistry, phosphorus (data only)

Anon., 1950. Table 1. Oceanographical observations taken in the Uchiura Bay; physical and chemical data for stations occupied by a fishing boat belonging to Mori Fishery Assoc. Res. Mar. Met., Ocean. Obs., Tokyo, July-Dec. 1947, No. 2:1-2.

chemistry Phosphorus (data only)

Anonymous, 1947. Discovery investigations Station List 1937-1939. Discovery Report, XXIV: 197-422. pls IV-VI.

phosphorus (data only)

Australia Commonwealth Scientific and Industrial Research Organization, Division of Fisheries and Oceanography, 1970. Coastal investigations off Port Hacking, New South Wales in 1965. Oceanogr. Sta. List 85: 124 pp.

chemistry, phosphorus (data only)

Australia, Commonwealth Scientific and Industrial Research Organization, 1969. Oceanographical observations in the Indian Ocean in 1965 H.M.A.S. Diamantina Cruise Dm2/65. Div. Fish. Oceanogr. Oceanogr. Cruise Rept. 49: 57 pp. (multilithed)

chemistry, phosphorus (data only)

Australia, Commonwealth Scientific and Industrial Research Organization, 1969. Oceanographical observations in the Pacific Ocean in 1965 H.M.A.S. Gascoyne Cruise G3/65. Div. Fish. Oceanogr. Oceanogr. Cruise Rept. 44: 24 pp. (multilithed).

chemistry phosphorus (data only)

Australia, Commonwealth Scientific and Industrial Research Organization, 1969. Oceanographical observations in the Indian Ocean in 1964 H.M.A.S. Diamantina Cruise DM4/64 Div. Fish. Oceanogr. Oceangr. Cruise Rept. 38: 65 pp. (multilithed).

chemistry, phosphorus (data only)

Australia, Commonwealth Scientific and Industrial Research Organization 1968. Oceanographical observations in the Indian Ocean in 1965, H.M.A.S Gascoyne, Cruise G 2/65. Oceanogr. Cruise Rept. 43: 1-58.

chemistry phosphate (data only)

Australia Commonwealth Scientific and Industrial Research Organization 1968. Oceanographical observations in the Pacific Ocean in 1964, H.M.A.S. Gascoyne, Cruise G 6/64. Oceanogr. Cruise Rept. 42: 53 pp.

chemistry, phosphorus (data only)

Australia Commonwealth Scientific and Industrial Organization 1968. Oceanographical observations in the Indian Ocean in 1964, H.M.A.S. Gascoyne, Cruise G5/64. Oceanogr. Cruise Rept., Div. Fish. Oceanogr. 41: 52 pp.

chemistry phosphorus, (data only)

Australia, Commonwealth Scientific and Industrial Research Organization, 1968. Oceanographical observations in the Pacific Ocean in 1963, H.M.A.S. Gascoyne, Cruise G3/63. Oceanogr. Cruise Rept., Div. Fish. Oceanogr., 26: 134 pp.

Chemistry-phosphorus (data only)

Australia Commonwealth Scientific and Industrial Organization 1967. Oceanographical observations in the Indian Ocean in 1965, H.M.A.S. Gascoyne, Cruise G 5/65. Oceanogr. Cruise Rept. Div. Fish Oceanogr. 46: 62 pp.

chemistry, phosphorus (data only)

Australia, Commonwealth Scientific and Industrial Search Organization, 1967. Oceanographical observations in the Pacific Ocean in 1964, H.M.A.S. Gascoyne Cruise G4/64. Oceanogr. Cruise Rept., Div. Fish. Oceanogr. 39: 39 pp.

chemistry, phosphorus (data only)

Australia, Commonwealth Scientific and Industrial Organization, 1967. Oceanographical observations in the Indian Ocean in 1964, H.M.A.S. Diamantina, Cruise Dm 2/64. Oceanor. Cruise Rep. Div. Fish. Oceanogr. 36: 53 pp.

chemistry, phosphorus (data only)

Australia, Commonwealth Scientific and Industrial Organization, 1967. Oceanographical observations in the Indian and Pacific oceans in 1964, H.M.A.S. Gascoyne, Cruise G3/64. Oceanogr. Cruise Rep. Div. Fish. Oceanogr., 35: 40 pp.

chemistry, phosphorus (data only)

Australia, Commonwealth Scientific and Industrial Research Organization, 1967. Oceanographical observations in the Pacific Ocean in 1964, H.M.A.S. Gascoyne, Cruise G 1/64. Div. Fish. Oceanogr., Oceanogr. Cruise Rep. 32: 66 pp (multilithed).

chemistry, phosphorus (data only)

Australia, Commonwealth Scientific and Industrial Organization, 1967. Oceanographic observations in the Pacific Ocean in 1963, H.M.A.S. Gascoyne, Cruise G 5/63. Div. Fish. Oceanogr. Cruise Rep. 31: 57 pp.

chemistry, phosphorus (data only)

Australia, Commonwealth Scientific and Industrial Research Organisation, 1967. Oceanographical observations in the Pacific Ocean in 1963, H.M.A.S. Gascoyne, Cruise G 4/63. Div. Fish. Oceanogr., Oceanogr. Cruise Rep. 29: 64 pp. (multilithed).

chemistry, phosphorus (data only)
Australia, Commonwealth Scientific and Industrial Research Organization, 1967.
Oceanographic observations in the Indian Ocean in 1963, H.M.A.S. Gascoyne, Cruise G 2/63.
Div. Fish. Oceanogr., Oceanogr. Cruise Rep. 22: 51 pp.

chemistry, phosphorus (data only)
Australia, Commonwealth Scientific and Industrial Research Organization, 1967.
Oceanographical observations in the Indian Ocean in 1962, H.M.A.S. Diamantina Cruise Dm 4/62.
Div. Fish. Oceanogr., Oceanogr. Cruise Rep. 20: 138 pp.

Chemistry - phosphorus (data only)
Australia, Commonwealth Scientific and Industrial Organization 1967.
Oceanographical observations in the Pacific Ocean in 1962, H.M.A.S. Gascoyne, Cruise G 5/62.
Oceanogr. Cruise Rept. Div. Fish. Oceanogr. 19: 71 pp.

Chemistry - phosphorus (data only)
Australia, Commonwealth Scientific and Industrial Organization 1967.
Oceanographical observations in the Pacific and Indian oceans in 1962, H.M.A.S. Gascoyne cruises G 2/62 and G 3/62.
Div. Fish Oceanogr. Oceanogr. Cruise Rept 16: 90 pp.

chemistry, phosphorus (data only)
Australia, Commonwealth Scientific and Industrial Research Organization, 1967.
Oceanographical observations in the Pacific Ocean in 1962, H.M.A.S. Gascoyne, Cruise G 1/62.
Div. Fish. Oceanogr., Oceanogr. Cruise Rep. 13: 180 pp.

chemistry, phosphorus (data only)
Australia, Commonwealth Scientific and Industrial Research Organization, 1967.
Oceanographical observations in the Pacific Ocean in 1961, H.M.A.S. Gascoyne Cruise G 3/61.
Div. Fish. Oceanogr., Oceanogr. Cruise Rep., 12 126 pp.

chemistry, phosphorus (data only)
Australia, Commonwealth Scientific and Industrial Research Organization, 1966.
Oceanographical observations in the Indian Ocean in 1962 H.M.A.S. Gascoyne Cruise G 4/62.
Div. Fish. Oceanogr., Oceanogr. Cruise Rep. No. 17: 151 pp.

chemistry, phosphorus, (data only)
Australia, Commonwealth Scientific and Industrial Research Organization, 1965.

Oceanographic observations in the Indian Ocean in 1963, H.M.A.S. Gascoyne, Cruise G 1/63.

Div. Fish. and Oceanogr., Oceanogr. Cruise Rept., No. 21:135 pp.

Chemistry phosphorus (data only)
Australia, Commonwealth Scientific and Industrial Research Organization, 1964.
Oceanographic observations in the Indian Ocean in 1962, H.M.A.S. Diamantina, Cruise Dm 1/62.
Oceanogr. Cruise Rept., Div. Fish Oceanogr., No. 10: 128 pp.

chemistry, phosphorus (data only)
Australia, Commonwealth Scientific and Industrial Research Organization, 1963.
Coastal hydrological investigations in the New South Wales tuna fishing area, 1963.
Div. Fish. and Oceanogr., Oceanogr. Sta. List, 53:81 pp.

A.D. Crooks, compiler

chemistry, phosphorus (data only)
Australia, Commonwealth Scientific and Industrial Research Organization, 1963.
Coastal investigations at Port Hacking, New South Wales, 1960.
Div. Fish. and Oceanogr., Oceanogr. Sta. List, No. 52:135 pp.

A.D. Crooks, compiler

chemistry, phosphorus (data only)
Australia, Commonwealth Scientific and Industrial Research Organization, 1963
Coastal hydrological investigations in eastern Australia, 1960.
Div. Fish. and Oceanogr., Oceanogr. Sta. List 51: 46 pp.

chemistry, phosphorus (data only)
Australia, Commonwealth Scientific and Industrial Research Organization, 1963
Oceanographical observations in the Indian Ocean in 1961, H.M.A.S. Diamantina Cruise Dm 2/61.
Oceanogr. Cruise Rept., Div. Fish. and Oceanogr., No. 9:155 pp., 14 figs.

chemistry, phosphorus (data only)
Australia, Commonwealth Scientific and Industrial Organization, 1962
Oceanographical observations in the Indian Ocean in 1960, H.M.A.S. Diamantina, Cruise Dm 1/60.
Oceanogr. Cruise Rept., Div. Fish. and Oceanogr., No. 2:128 pp.

Chemistry, phosphorus (data only)
Australia, Commonwealth Scientific and Industrial Organization, 1962.
Oceanographical observations in the Indian Ocean in 1959, H.M.A.S. Diamantina, Cruises Dm 1/59 and Dm 2/59.
Oceanogr. Cruise Rept., Div. Fish. and Oceanogr., 1:134 pp.

chemistry, phosphorus
Australia, Commonwealth Scientific and Industrial Organization, 1962.
Oceanographic observations in the Indian Ocean in 1961. H.M.A.S. Diamantina.
Div. Fish. Oceanogr., Cruise DM 1/61:88 pp.

chemistry, phosphorus (data only)
Australia, Commonwealth Scientific and Industrial Organization, 1962.
Oceanographic observations in the Indian Ocean in 1961. H.M.A.S. Diamantina.
Div. Fish. Oceanogr., Cruise DM 1/61:88 pp.

Chemistry
phosphorus (data only)
Australia, Commonwealth Scientific and Industrial Research Organization, 1961
Coastal hydrological sampling at Rottnest Island, W.A., and Port Moresbu, Papua, during the I.G.Y. (1957-58), and surface sampling in the Tasman and Coral seas, 1959.
Div. Fish. & Oceanogr., Oceanogr. Sta. List, 48:239 pp.

chemistry
phosphorus
(data only)
Australia, Commonwealth Scientific and Industrial Research Organization, 1961
Oceanic investigations in Eastern Australian waters, F.R.V. "Derwent Hunter", 1959.
Div. Fish. and Oceanogr., Oceanogr. Sta. List, 48: 84 pp.

chemistry,
phosphorus (data only)
Australia, Commonwealth Scientific and Industrial Research Organization, 1961
Coastal investigations at Port Hacking, New South Wales, 1959.
Div. Fish. & Oceanogr., Oceanogr. Sta. List, No. 47: 135 pp.

chemistry, phosphorus (data only)
Australia, Commonwealth Scientific and Industrial Research Organization, 1961
Coastal hydrological investigations in the New South Wales tuna fishing area, 1959.
Div. Fish. & Oceanogr., Oceanogr. Sta. List, 46: 132 pp.

chemistry
phosphorus (data only)
Australia, Commonwealth Scientific and Industrial Research Organization, 1960

Oceanic observations in Antarctic waters, M.V. Magga Dan, 1959.
Div. Fish. and Oceanogr., Oceanogr. Sta. List, 44: 78 pp.

chemistry, phosphorus (data only)
Australia, Commonwealth Scientific and Industrial Research Organization, 1960.
Oceanic investigations in eastern Australia, H.M.A. Ships Queenborough, Quickmatch and Warrego, 1958.
Div. Fish. and Oceanogr., Oceanogr. Sta. List, 43:57 pp.

Chemistry, phosphorus (data only)
Australia, Commonwealth Scientific and Industrial Research Organization, 1960.
Coastal investigations at Port Hacking, New South Wales, 1958.
Oceanogr. Sta. List, Div. Fish. & Oceanogr., 42: 99 pp.

chemistry, phosphorus (data only)
Australia, Commonwealth Scientific and Industrial Organization, 1960

Oceanic investigations in eastern Australian waters, F.R.V. Derwent Hunter, 1958. Oceanogr. Sta. List., Div. Fish. and Oceanogr., No. 41: 232 pp.

chemistry, phosphorus (data only)
Australia, Commonwealth Scientific and Industrial Organization, 1960

Surface sampling in the Coral and Tasman Sea, 1958. Oceanogr. Sta. List, Div. Fish. and Oceanogr., No. 39: 276 pp.

chemistry phosphorus (data only)
Australia, Commonwealth Scientific and Industrial Research Organization 1959.
Hydrological investigations from F.R.V. Derwent Hunter 1957.
Oceanogr. Sta. List Div. Fish. Oceanogr. 37: 96 pp. (compiled by D. J. Rochford)

chemistry, phosphorus (data only)
Australia, C.S.I.R.O., Division of Fisheries, 1959.
Surface sampling in the Tasman and Coral Seas, 1957.
Oceanogr. Sta. List, 36:175 pp.

Chemistry
phosphorus (data only)

Australia, C.S.I.R.O., Division of Fisheries, 1959

Coastal hydrological investigations at Eden, New South Wales, 1957.
Oceanogr. Sta. List, 35: 36 pp.

Chemistry
phosphorus (data only)

Australia, C.S.I.R.O., Division of Fisheries, 1959

Coastal hydrological investigations at Port Hacking, New South Wales, 1957.
Oceanogr. Sta. List, 34:72 pp.

Chemistry, phosphorus (data only)

Australia, Commonwealth Scientific and Industrial Research Organization, Division of Fisheries and Oceanography, 1960.

F.R.V., "Derwent Hunter" Scientific Reports of Cruises 1-8, 1958: 57 pp. (Mimeographed)

Chemistry phosphorus (data only)
Australia, C.S.I.R.O., Division of Fisheries and Oceanography, 1959.
Scientific reports of a cruise on H.M.A. Ships "Queenborough" and "Quickmatch", March 24-April 26, 1959. Rept. No. 24: 24 pp. (mimeographed).

Chemistry, phosphorus (data only)
Australia, C.S.I.R.O., Division of Fisheries and Oceanography, Marine Biological Laboratory, Cronulla, 1959.
F.R.V. "Derwent Hunter", Scientific report of Cruise DH9/57, Aug. 19-25, 1957; Cruise DH10/57, Sept. 4-11, 1957; Cruise DH11/57, Sept. 18-21, 1957; Cruise DH12/57, Sept. 26-Oct. 11, 1957. CSIRO Div. Fish. & Ocean., Rept. No. 20:20pp. (mimeographed)

Chemistry, phosphorus (data only)
Australia, C.S.I.R.O., Division of Fisheries and Oceanography, 1958/1959.
FVR "Derwent Hunter". Rept. No. 19:16 pp.
No. 21:16 pp.
(mimeographed)

Chemistry phosphorus (data only)

Australia, Commonwealth Scientific and Industrial Research Organization, 1957.

Surface sampling in the Tasman and Coral Seas and the South-eastern Indian Ocean, 1956. Oceanogr. Stal List, Div. Fish. & Oceanogr. 31:150 pp.

Chemistry phosphorus (data only)

Australia Commonwealth and Industrial Research Organization, 1957.

Onshore and oceanic hydrological investigations in eastern and southwestern Australia, 1956. Oceanogr. Sta. List. Div. Fish. & Oceanogr. 30:79pp.

Chemistry phosphorus (data only)
Australia, Marine Biological Laboratory, Cronulla, Sydney, 1957.
F.R.V. "Derwent Hunter", scientific report of Cruise 3/56, September 19-October 5, 1956; ----- Cruise 4/56, October 9-November 6, 1956; ------- Cruise 5/56, November 8-December 3, 1956. C.S.I.R.O., Australia, Div. Fish. and Oceanogr., Rept., No. 5:16 pp. (mimeographed).

Chemistry, phosphorus (data only)
Australia, Marine Biological Laboratory, Cronulla, 1960.
F.R.V. "Derwent Hunter", scientific cruise of -- cruises10-20/58 -------.
C.S.I.R.O., Div. Fish. & Oceanogr., Rept., 30: 53 pp., numerous figs. (mimeographed).

For complete "title", see author card.

Chemistry - phosphorus (data only)
Blackburn, M., R.C. Griffiths, R.W. Holmes and W.H. Thomas, 1962.
Physical, chemical and biological observations in the eastern tropical Pacific Ocean: three cruises to the Gulf of Tehuantepec, 1958-1959.
U.S.F.W.S. Spec. Sci. Rept., Fish., No. 420:170pp

Chemistry, phosphorus (data only)
Buljan, Miljenko, and Mira Zore-Armanda 1966.
Hydrographic data on the Adriatic Sea collected in the period from 1952 through 1962.
Acta adriat. 12:438 pp.

Chemistry phosphorus (data only)
Brasil, Marinha do Brasil, Diretoria de Hidrografia e Navegação, 1960.
Estudo das condições oceanograficas sobre a plat aforma continental entre Cabo-Frio e Vitoria, durante o Outono (abril-maio).
DG-06-X(junho): unnumbered pp. (mimeographed).

Chemistry phosphorus (data only)
Brasil, Marinha do Brasil, Diretoria de Hidrografia e Navegação, 1959.
Levantamento oceanografico da costa nordeste.
DG-06-IX:unnumbered pp. (mimeographed).

Chemistry, phosphorus (data only)
Conseil Permanent International pour l'Exploration de la _er, 1957.
Bulletin Hydrographique pour l'Année 1953:167 pp.

Chemistry, phosphorus (data only)
Conseil Permanent International pour l'Exploration de la _er, 1954.
Bulletin Hydrographique pour l'Année 1950: 114 pp., 5 charts.

Chemistry - phosphorus (data only)
Dragovich, Alexander, and James E. Sykes 1967.
Oceanographic atlas for Tampa, Florida and adjacent waters of the Gulf of Mexico, 1958-61.
Circular Fish Wildl. Serv. Bur. Comm. Fish. 255: 466 pp. (quarto)

Chemistry phosphorus (data only)
Gostan, Jacques, 1967.
Résultats des observations hydrologiques effectuées entre les côtes de Provence et de Corse (6 août 1962- 30 juillet 1964).
Cah. océanogr. 19 (Suppl.1):1-69.

Chemistry phosphorus (data only)
Gostan, Jacques, 1967.
Résultats des observations hydrologiques effectuées entre les côtes de Provence et de Corse (6 août 1962-30 juillet 1964).
Cah. océanogr., 19(Suppl.1):1-69.

Chemistry phosphorus (data only)
Great Britain, Discovery Committee, 1957.
Station list, 1950-1951.
Discovery Rept. 28:300-398.

Chemistry, phosphorus (data only)
Great Britain, Discovery Committee, 1953.
Station list, R.R.S. "William Scoresby", 1950.
Discovery Repts. 26:211-258, Pls. 11-12.

Chemistry phosphorus (data only)
Great Britain, Discovery Committee, 1945.
Station list, 1935-1937.
Discovery Rept., 24:3-196.

Chemistry phosphorus (data only)
Hapgood, William, 1959.
Hydrographic observations in the Bay of Naples (Golfo di Napoli), January 1957-January 1958 (station lists).
Pubbl. Sta. Zool., Napoli, 31(2):337-371.

Chemistry, phosphorus (data only)
Ivanoff A., A. Morel, Mme Vesin-Couffinhal, C. Amiel, J. P. Bethoux, J. Boutler, C. Copin-Montegut, P. Courau, P. Geistdoerfer et F. Nyffeler 1969.
Résultats des observations effectuées en Mer Méditerranée orientale et en mer Tyrrhénienne à bord du navire Amalthée en mars 1967.
Cah. océanogr. 21 (Suppl.2): 245-263.

phosphorus (data only)
Japan, Central Meteorological Observatory, 1952.
The Results of Marine Meteorological and oceanographical observations. Jan. - June 1950. No. 7: 220 pp.

Chemistry, phosphorus (data only)
Japan, Central Meteorological Observatory, 1951.
Table 21. Surface observations made on board R.S.S. "Umikaze-maru"; physical and chemical conditions of the surface waters of the Tsushima Straits and Goto-nada. Res. Mar. Met., Ocean. Obs. Jan.-June 1949, No. 5:115-116.

Japan Chemistry, phosphorus (data only)
Central Meteorological Observatory, Japan, 1951.
The results of marine meteorological and oceanographical observations, July-Dec. 1948, No. 4: vi + 414 pp., of 57 tables, 1 map, 1 photo.

Japan Chemistry, phosphorus (data only)
Central Meteorological Observatory, Japan, 1951.
The results of marine meteorological and oceanographical observations, Jan.-June 1948, No. 3: 256 pp.

Japan Chemistry phosphorus (data only)
Central Meteorological Observatory, Japan, 1950.
The results of marine meteorological and oceanographical observations, Jan.-June 1947, 1:113 pp.

Chemistry phosphorus (data only)

Japan, Hokkaido University, Faculty of Fisherhies, 1968.
The Oshoro Maru cruise 21 to the Southern Sea of Japan, January, 1967.
Data Record Oceanogr. Obs. Explor. Fish. 12: 1-97; 113-119.

Chemistry phosphorus (data only)

Japan, Hokkaido University, Faculty of Fisheries, 1967.
The Oshoro Maru cruise 18 to the east of Cape Erimo, Hokkaido, April 1966.
Data Record Oceanogr. Obs. Explor. Fish. 11: 121-164.

Chemistry phosphorus (data only)

Japan, Hokkaido University, Faculty of Fisheries, 1967.
The Oshoro Maru cruise 16 to the Great Australian Bight November 1965-February 1966.
Data Record Oceanogr. Obs. Explor. Fish., Fac. Fish., Hokkaido Univ. 11: 1-97; 113-119.

Chemistry phosphorus (data only)
Japan, Hokkaido University, Faculty of Fisheries, 1967.
Data record of oceanographic observations and exploratory fishing, 11:383 pp.

Chemistry phosphorus (data only)
Japan, Japan Meteorological Agency, 1964.
The results of marine meteorological and oceanographical observations, January-June 1963, No. 33:289 pp.

Chemistry phosphorus (data only)
Japan, Japan Meteorological Agency, 1964.
The results of the Japanese Expedition of Deep Sea (JEDS-5).
Res. Mar. Meteorol. and Oceanogr. Obs., July-Dec. 1962, No. 32: 328 pp.

Chemistry, phosphorus (data only)
Japan, Japan, Meteorological Agency, 1964.
Oceanographic observations.
Res. Mar. Met. & Oceanogr. Obs., No. 31:220 pp.

Chemistry, phosphorus (data only)

Japan, Japan Meteorological Agency 1961. The results of marine meteorological and oceanoglaphical observations, January-June 1960. Results Japan Exped. Deep-Sea (JEDS-2, JEDS-3), No. 27: 257 pp.

Chemistry phosphorus (data only)

Japan, Japan Meteorological Agency, 1959. The results of marine meteorlogical and oceanographical observations, July-December, 1958, No. 24: 289 pp.

Chemistry, phosphorus (data only)

Japan, Japan Meteorological Agency 1959. The results of marine meteorological and oceanographical observations, January-June 1958, No. 23: 240 pp.

Chemistry phosphorus (data only)

Japan, Japanese Oceanographic Data Center, 1971. Shumpu Maru, Kobe Marine Observatory, Japan Meteorological Agency, 16 July - 8 August 1970, south of Japan. Data Rept. CSK (KDC Ref. 49K127) 279: 7 pp. (multilithed).

Chemistry Phosphorus (data only)

Japan, Japanese Oceanographic Data Center, 1971. Takuyo, Hydrographic Department, Maritime Safety Agency, Japan, 10-30 July 1970, south & east of Japan. Data Rept. CSK (KDC Ref. 49K126) 278: 19 pp. (multilithed).

Chemistry phosphorus (data only)

Japan, Japanese Oceanographic Data Center, 1971. Kofu Maru, Hadkodate Marine Observatory, Japan Meteorological Agency, 27 June - 9 August 1968, East of Japan & Okhotsk Sea. Data Rept. CSK (KDC Ref. 49K084) 178: 32 pp. (multilithed).

Chemistry Phosphorus (data only)

Japan, Japanese Oceanographic Data Center, 1970. Ji Ri San, Fisheries Research and Development Agency, Republic of Korea, Feb. 18 - 26, 1970, West of the Japan Sea. Data Rept. CSK (KDC Ref. 24K038) 272: 16 pp. (multilithed).

Chemistry phosphorus (data only)

Japan, Japanese Oceanographic Data Center, 1970. Han Ra San, Fisheries Research and Development Agency, Republic of Korea, Feb. 11 - 24, 1970, East of the Yellow Sea. Data Rept. CSK (KDC Ref. 24K037) 271: 14 pp. (multilithed).

Chemistry phosphorus (data only)

Japan, Japanese Oceanographic Data Center, 1970. Baek Du San, Fisheries Research and Development Agency, Republic of Korea, Feb. 4 - 20, 1970, South of Korea & East China Sea. Data Rept. CSK (KDC Ref. 24K036) 270: 15 pp. (multilithed).

Chemistry phosphorus (data only)

Japan, Japanese Oceanographic Data Center, 1970. Seifu Maru, Maizuru Marine Observatory, Japan Meteorological Agency, Feb. 6 - Mar. 16, 1970, Japan Sea. Data Rept. CSK (KDC Ref. 49K124) 268: 8 pp. (multilithed).

Chemistry Phosphorus (data only)

Japan, Japanese Oceanographic Data Center, 1970. Shumpu Maru, Kobe Marine Observatory, Japan Meteorological Agency, Feb. 21 - Mar. 11, 1970, South of Japan. Data Rept. CSK (KDC Ref. 49K122) 266: 4 pp. (multilithed).

Chemistry phosphorus (data only)

Japan, Japanese Oceanographic Data Center, 1970. Takuyo, Hydrographic Department, Maritime Safety Agency, Japan, Feb. 12 - Mar. 7, 1970. South and East of Japan. Data Rept. CSK (KDC Ref. 49K120) 264: 13 pp. (multilithed).

Chemistry phosphorus (data only)

Japan, Japanese Oceanographic Data Center, 1970. Kaiyo, Hydrographic Department, Maritime Safety Agency Japan, Nov. 7 - 29, 1969, South of Japan. Data Rept. CSK (KDC Ref. 49K116) 257: 10 pp. (multilithed).

Chemistry PHOSPHORUS (DATA ONLY)

Japan, Japanese Oceanographic Data Center, 1970. Suro No. 3, Hydrographic Office, Republic of Korea, July 23 - August 19, 1969, South of Korea Data Rept. CSK (KDC Ref. 24K035) 240: 19 pp. (multilithed).

Chemistry, phosphorus (data only)

Japan, Japanese Oceanographic Data Center 1970. Han Ra San, Fisheries Research and Development Agency, Republic of Korea, August 9-21, 1969, East of the Yellow Sea. Data Rept. CSK (KDC Ref. 24K033) 238: 17 pp. (multilithed).

Chemistry phosphorus (DATA ONLY)

Japan, Japanese Oceanographic Data Center, 1970. Ji Ri San, Fisheries Research and Development Agency, Republic of Korea, August 7 - 29, 1969, South of Korea and East China Sea, Data Rept. CSK (KDC Ref. 24K034) 239: 15 pp. (multilithed).

Chemistry PHOSPHORUS (DATA ONLY)

Japan, Japanese Oceanographic Data Center, 1970. Takuyo, Hydrographic Department, Maritime Safety Agency, Japan, August 11 - September 2, 1969, South and East of Japan. Data Rept. CSK (KDC Ref. 49K112) 239: 21 pp. (multilithed).

Chemistry, phosphorus (data only)

Japan, Japanese Oceanographic Data Center 1970. Shumpu Maru, Kobe Marine Observatory, Japan Meteorological Agency, July 18-August 14, 1969, South of Japan. Data Rept. CSK (KDC Ref. 49K111) 228: 7 pp. (multilithed).

Chemistry phosphorus (data only)

Japan, Japanese Oceanographic Data Center, 1970 Orlick, USSR, Jan. 27 - Mar. 6, 1969, Northwest of the North Pacific Ocean. Data Rept. CSK (KDC Ref. 90K023) 227: 30 pp. (multilithed).

Chemistry phosphorus (data only

Japan, Japanese Oceanographic Data Center, 1970. Oshoro Maru, Faculty of Fisheries, Hokkaido University, Japan, November 5, 1968 - January 20, 1969, Southwest of the North Pacific Ocean. Data Rept. CSK (KDC Ref. 49K104) 216: 13 pp. (multilithed).

Chemistry, phosphorus (data only)

Japan, Japanese Oceanographic Data Center 1969. Seifu Maru, Maizuru Marine Observatory, Japan Meteorological Agency, Japan, February 8-26, 1969, Japan Sea. Data Rept. CSK (KDC Ref. 49K103) 215: 7 pp. (multilithed).

Chemistry phosphorus (data only)

Japan, Japanese Oceanographic Data Center, 1969 Cape St. Mary, Fisheries Research Station, UK (Hong Kong). May 16 - 23, 1968, South China Sea. Prelim. Data Rept. CSK (KDC Ref. No. 74K010) 203: 15 pp. (multilithed).

Chemistry phosphorus (data only)

Japan, Japanese Oceanographic Data Center, 1969. Seifu Maru, Maizuru Marine Observatory, Japan Meteorological Agency. August 6 - 10, 1968, Japan Sea. Prelim. Data Rept. CSK (KDC Ref. No. 49K087) 181: 9 pp. (multilithed).

Chemistry - phosphorus (data only)

Japan, Japanese Oceanographic Data Center 1969. Orlick, USSR, July 21- August 20, 1968, northwest Pacific Ocean. Data Rept. CSK (KDC Ref. 90K020) 191: 32 pp. (multilithed).

Chemistry - phosphorus (data only)

Japan, Japanese Oceanographic Data Center 1969 Iskatel, USSR, January 26 - March 11, 1968, northwest, North Pacific Ocean. Data Rept. CSK (KDC Ref 90K019) 187: 32 pp. (multilithed).

Chemistry phosphorus (data only)

Japan, Japanese Oceanographic Data Center, 1969. Chofu Maru, Nagasaki Marine Observatory, Japan Meteorological Agency, August 4 - 5, 1968, East China Sea. Prelim. Data Rept. CSK (KDC Ref. No. 49K086) 180: 5 pp. (multilithed).

Chemistry, phosphorus (data only)

Japan, Japanese Oceanographic Data Center, 1969. Shumpu Maru, Kobe Marine Observatory, Japan Meteorological Agency, July 19-31, 1968, South of Japan. Data Rept. CSK (KDC Ref. 49K085) 179: 9 pp.

Chemistry phosphorus (data only)

Japan, Japanese Oceanographic Data Center 1969. Hakuho Maru, Ocean Research Institute, University of Tokyo, Japan, May 15-June 8, 1968, southwest of Japan and the East China Sea. Data Rept CSK (KDC Ref. 49K078) 172: 7 pp. (multilithed).

Chemistry - phosphorus (data only)

Japan, Japanese Oceanographic Data Center, 1969. Takuyo, Hydrographic Division, Maritime Safety Agency, May 14-25, 1968, South of Japan. Data Rept. CSK (KDC Ref. 49K077) 171: 16 pp.

Chemistry
Phosphorus (data only)

Japan, Japanese Oceanographic Data Center, 1969

Takuyo, Hydrographic Division, Maritime Safety Agency, Japan. July 19-September 6, 1968, Kuroshio Extension Area. Prelim. Data Rept. CSK (KDC Ref. No. 49K082) 176: 27 pp. (multilithed).

Chemistry
phosphorus (data only)

Japan, Japanese Oceanographic Data Center, 1969.

Cape St. Mary, Fisheries Research Station, UK (Hong Kong). January 6 - 14, 1968, South China Sea. Prelim. Data Rept. CSK (KDC Ref. No. 74K009) 165: 15 pp. (multilithed).

Chemistry phosphorus (data only)

Japan, Japanese Oceanographic Data Center, 1969. Kofu Maru, Hakodate Marine Observatory, Japan Meteorological Agency, February 7-26, 1968, east of Japan.
Data Rept. CSK (KDC Ref. 49K070) 156: 9 pp.

Chemistry, phosphorus (data only)

Japan, Japanese Oceanographic Data Center, 1969

Umitaka Maru, Tokyo University of Fisheries, Japan, November 1, 1967- February 23, 1968, West of the North Pacific Ocean. Prelim. Data Rept. CSK (KDC Ref. No. 49K066) 131: 13 pp. (multilithed).

Chemistry, phosphorus (data only)

Japan, Japanese Oceanographic Data Center, 1969.

Orlick, USSR, July 29 - September 3, 1967, Northwest of the North Pacific Ocean. Prelim. Data Rept. CSK (KDC Ref. No. 90K015) 122: 31 pp. (multilithed).

Chemistry, phosphorus (data only)

Japan, Japanese Oceanographic Data Center 1969. Orlick, USSR July 29- September 3, 1967, Northwest of the North Pacific Ocean. Preliminary Data Rept. CSK (KDC Ref. No. 90K015) 122: 31pp. (multilithed).

Chemistry
phosphorus (data only)

Japan, Japanese Oceanographic Data Center, 1969.

U.M. Schokalsky, USSR, April 28 - June 3, 1967, Northwest of the North Pacific Ocean. Prelim. Data Rept. CSK (KDC Ref. 90K014) 121: 41 pp. (multilithed).

Chemistry, phosphorus (data only)

Japan, Japanese Oceanographic Data Center, 1969.

U.M. Schokalsky, USSR, April 28 - June 3, 1967, Northwest of the North Pacific Ocean. Prelim. Data Rept. CSK (KDC Ref. 90K014) 121: 41 pp. (multilithed).

Chemistry - phosphorus (data only)

Japan, Japanese Oceanographic Data Center, 1968. Tae Baek San, Fisheries Research and Development Agency, Republic of Korea, February 16 - March, 12 1968, South of Korea.
Data Rept. CSK (KDC Ref. 24K022) 163: 12 pp. (Multilithed)

Chemistry, phosphorus (data only)

Japan, Japanese Oceanographic Data Center, 1968. Han ra San, Fisheries Research and Development Agency, Republic of Korea, Febrary 15- March 12, 1968, East of the Yellow Sea.
Data Rept. CSK (KDC Ref. 24K021) 162: 18 pp. (multilithed)

Chemistry, Phosphorus (data only)

Japan, Japanese Oceanographic Data Center, 1968. Baek du San, Fisheries Research and Development Agency, Republic of Korea, February 17-March 12, 1968, West of the Japan Sea.
Data Rept. CSK (KDC Ref. 24K020) 161:23 pp. (multilithed)

Chemistry phosphorus (data only)

Japan, Japanese Oceanographic Data Center, 1969. Seifu Maru, Maizuru Marine Observatory, Japan Meteorological Agency, February 17-24, 1968, Japan Sea.
Data Rept. CSK (KDC Ref. 49K073) 159:7 pp.

Chemistry, phosphorus (data only)

Japan, Japanese Oceanographic Data Center, 1968. Cape St. Mary, Fisheries Research Station, Hong Kong, September 19-October 11, 1967, South China Sea. Prelim. Data Rept. CSK (KDC Ref. 74K008) 138:16 pp. (multilithed).

Chemistry phosphorus (data only)

Japan, Japanese Oceanographic Data Center, 1968. Han Ra San, Fisheries Research and Development Agency, Korea, August 12-September 4, 1967, North of East China Sea.
Prelim. Data Rept. CSK (KDC Ref. 24K018) 117:12 pp. (multilithed).

Chemistry, phosphorus (data only)

Japan, Japanese Oceanographic Data Center, 1968. Chun Ma San, Fisheries Research and Development Agency, Korea, August 13-21, 1967, East of Yellow Sea. Prelim. Data Rept. CSK (KDC Ref. 24K017) 116-16 pp.

(multilithed).

Chemistry, phosphorus (data only)

Japan, Japanese Oceanographic Data Center, 1968. Orlick, USSR, February 5-March 11, 1967, Northwest of North Pacific Ocean.
Prelim. Data Rept. CSK (KDC Ref. 90K012) 94:29 pp.

Chemistry phosphorus (data only)

Japan, Japanese Oceanographic Data Center, 1967. Chofu Maru (NMO), Jan. 13-14, 1967; Mar. 19-20, 1967, Apr. 15-15, 1967, May 11-12, 1967, Japan Meteorological Agency, South-East of Yakushima. Prelim. Data Rept. CSK (KDC Ref. Nos. 49K315, 49K316, 49K317, 49K318) 104: 18 pp.

Chemistry, phosphorus (data only)

Japan, Japanese Oceanographic Data Center, 1967. Chofu maru, Nagasaki Marine Observatory, Japan Meteorological Agency, Japan, May 17-18, 1967, East China Sea.
Prelim. Data Rept. CSK (KDC Ref. 49K051) 100:5pp.

Chemistry phosphorus (data only)

Japan, Japanese Oceanographic Data Center, 1967. Shumpu maru, Kobe Marine Observatory, Japan Meteorological Agency, Japan, May 13-14, 1967, South of Japan.
Prelim. Data Rept. CSK (KDC Ref. 49K050) 99:5 pp.

Chemistry phosphorus (data only)

Japan, Japanese Oceanographic Data Center, 1967. Nagasaki maru, The Faculty of Fisheries, Nagasaki University, Japan, January 19-22, 1967, East China Sea.
Prelim. Data Rept. CSK (KDC Ref. 49K046) 88:5 pp.

Chemistry phosphorus (data only)

Japan, Japanese Oceanographic Data Center, 1967. Oshoro maru, the Faculty of Fisheries , Hokkaido University, Japan, January 15-February 1, 1967, South of Japan.
Prelim. Data Rept. CSK (KDC Ref. 49K045) 87: 9pp.

Chemistry phosphorus (data only)

Japan, Japanese Oceanographic Data Center, 1967. Chofu maru, Nagasaki Marine Observatory, Japan Meteorological Agency, Japan, January 20-February 22, 1967, East China Sea.
Prelim. Data Rept. CSK (KDC Ref. 49K043) 85:11 pp.

Chemistry phosphorus (data only)

Japan, Japanese Oceanogra hic Data Center, 1967. Shumpu maru, Kobe Marine Observatory, Japan Meteorological Agency, Japan, February 26-27, 1967, south of Japan.
Prelim. Data Rept. CSK (KDC Ref. 49K042) 84: 6 pp.

Chemistry. phosphorus (data only)

Japan, Japanese Oceanographic Data Center, 1967. Kofu Maru, Hakodate Marine Observatory, Japan Meteorological Agency, February 4-March 7, 1967, East of Japan.
Prelim. Data Rept. CSK (KDC Ref. No. 49K041) 83:15pp.

Chemistry
phosphorus (data only)

Japan, Japanese Oceanographic Data Center, 1967 Takuyo, Hydrographic Division, Maritime Safety Agency, Japan, February 23-March 16, 1967, south and east of Japan.
Prelim. Data Rept. CSK (KDC Ref. 49K039) 81:17 pp.

Chemistry, phosphorus (data only)

Japan, Japanese Oceanographic Data Center, 1967. Yang Ming, Chinese National Committee on Oceanic Research, Republic of China, September 10-October 14, 1966, Adjacent Sea of Taiwan. Prelim. Data Rept. CSK (KDC Ref. 21K003) 67: 17pp

Chemistry, phosphorus (data only)

Japan, Japanese Oceanographic Data Center, 1966. Koyo Maru, Shimonoseki University of Fisheries, October 26-29, south of Japan.
Prelim. Data Rept. CSK (KDC Ref. No. 49K038) 66: 7 pp. (multilithed).

Chemistry phosphorus (data only)

Japan, Japanese Oceanographic Data Center, 1967. Chofu Maru (NMO) Jul. 2-3, 1966; Chofu Maru (NMO), Aug. 27-29, 1966; Shumpu Maru (KMO), Oct. 27-31, 1966, Chofu Maru, Dec. 7-8, 1966, Japan Meteorological Agency of Japan, south-east of Yakushima.
Prelim. Data Rept., CSK (KDC Ref. 49K311, 49K312, 49K313, 49K314) 61: 18 pp.

Chemistry, phosphorus (data only)

Japan, Japanese Oceanographic Data Center, 1967. Kagoshima maru, the Faculty of Fisheries, Kagoshima University, Japan, August 5-14, 1966, East China Sea.
Prelim. Data Rept. CSK (KDC Ref. 49K031) 57:9pp.

Chemistry
phosphorus (data only)

Japan, Japanese Oceanographic Data Center, 1967. Seifu Maru, Maizuru Marine Observatory, Japan Meteorological Agency, Japan, August 21-September 19, 1966, Japan Sea.
Prelim. Data Rept., CSK (KDC Ref. 49K029) 55:15 pp.

phosphorus (data only)

Japan, Nagasaki Marine Observatory, 1967. Report of the Oceanographic observations in the sea west of Japan from January to February, 1969 (In Japanese) Oceanogr. Met. Nagasaki Mar. Obs. 17 (242): 26pp (dates correct as given)

phosphorus (data only)

Japan, Nagasaki Marine Observatory, 1967 Report of the oceanographic observations in the sea west of Japan from July to September 1965. (In Japanese) Oceanogr. Met. Nagasaki mar. Obs. 17 (241): 26 pp. (dates correct as given)

phosphorus (data only)

Japan, Nagasaki Marine Observatory, 1967. Report of the oceanographic observations in the sea southeast of Yakushima Island from April to May 1965. (In Japanese) Oceanogr. Met. Nagasaki mar. Obs., 17 (240): 26 pp. (dates correct as given)

Chemistry, phosphorus (data only)

Japan, Shimonoseki University of Fisheries 1965 International Indian Ocean Expedition 1962-63 and 1963-1964 Data Oceanogr. Obs. Explor. Fish. 1:453 pp.

Chemistry
phosphorus (data only)

Klepikova, V.V., Editor, 1961
Third Marine Expedition on the D/E "Ob", 1957-1958. Data.
Trudy Sovetskoi Antarktich. Exped., Arktich. i Antarktich. Nauchno-Issled. Inst., Mezhd. Geofiz. God, 22:1-234 pp.

Chemistry - phosphorus (data only)
Platt, Trevor and Brian Irwin 1971.
Phytoplankton production and nutrients in Bedford Basin 1969-70.
Techn. Rept. Fish Res. Bd. Can. 247:172 pp. (multilithed)

chemistry - phosphorus (data only)
Republic of China, Chinese National Committee on Oceanic Research, Academia Sinica 1966.
Oceanographic data report of CSK 1: 123 pp.

Chemistry, phosphorus (data only)
Saloman, Carl H., and John L. Taylor 1968.
Hydrographic observations in Tampa Bay, Florida and the adjacent Gulf of Mexico, 1965-1966.
Data Rept. U.S. Fish. Wildl. Bur. Comm. Fish. 6 cards, (microfiche).

Chemistry total phosphorus (data only)
Sweden, Havsfiskelaboratoriet, Lysekil 1971.
Hydrographical data, January-June 1970 R.V. Skagerak, R.V. Thetis and TV252, 1970.
Meddn. Hydrogr. avd. · Göteborg, 104: unnumbered pp. (multilithed).

chemistry phosphorus (data only)
Union of South Africa, Department of Commerce and Industry, 1965.
Station list.
Div. Sea Fish., Invest. Rept., No. 51:50-67.

chemistry phosphorus (data only)
South Africa, Division of Fisheries, 1950.
Station list - R.S. "Africana II", R.V. "Shipa" and P.B. "Palinurus". 22nd Ann. Rept.:21-169, 2 charts.

Chemistry, phosphorus (data only)
United States, National Oceanographic Data Center, 1965.
Data report EQUALANT III
Nat. Oceanogr. Data Cent., Gen. Ser., G-7:339pp. $5.00

phosphorus-density

phosphorus, density
Australia, Commonwealth Scientific and Industrial Research Organization, Division of Fisheries and Oceanography, 1960.
F.R.V., "Derwent Hunter" Scientific report of Cruises 1-8, 1958: 57 pp. (Mimeographed)

chemistry, phosphorus (effect of)
Yartseva, I.A., O.L. Solovieva, 1971.
Some data on phosphorus effect on photosynthesis of the Black Sea phyllophore. (In Russian).
Gidrobiol. Zh. 7(5): 75-78.

phosphorus excretion

Chemistry phosphorus (excretion)
Kuenzler, Edward J. 1970.
Dissolved organic phosphorus excretion by marine phytoplankton.
J. Phycol. 6(1):7-13.

chemistry phosphorus excretion
Johannes, R.E., 1964
Phosphorus excretion and body size in marine animals: microzooplankton and nutrient regeneration.
Science, 146(3646):923-924.

phosphorus, inorganic

phosphorus radioactive

Chemistry phosphorus, radioactive
Coffin, C.C., F.R. Hayes, L.H. Jodrey, and S.G. Whiteway, 1950.
Exchange of materials in a lake as studied by the addition of radioactive phosphorus. (Abstract).
Proc. N.S. Inst. Sci., 22(3):42.

Chemistry, phosphorus (radioactive)
Foster, R.F., 1959
Radioactive tracing of the movement of an essential element through an aquatic community with specific reference to radiophosphorus.
Pubb. Sta. Zool. Nap., 31(Suppl): 34-69.

chemistry, radiophosphorus
Harris, E., 1957.
Radiophosphorus metabolism in zooplankton and microorganisms. Canadian J. Zool., 35(6):769-782.

chemistry radiophosphorus
Hayes, F.R., 1955.
The effect of bacteria on the exchange of radiophosphorus at the mud-water interface.
Proc. Int. Assoc. Theor. Appl. Limnol., 12:111-

chemistry phosphorus, radioactive
Hayes, F.R., J.A. McCarter, M.L. Cameron, and D.A. Livingstone, 1950.
A further experiment on the addition of radioactive phosphorus to a lake. (Abstract). (Read at 6th Ord. Meeting.) Proc. N.S. Inst. Sci., 22(4): 61.

Chemistry, phosphorus, radioactive
Kobayashi, K., S. Oyama and T. Tomiyama, 1960.
Studies on absorption and metabolism of P32 in aquatic organisms. II. Incorporation of directly uptaken P32 into various acid-soluble phosphorus compounds of Rhizodrilus limosus.
Bull. Jap. Soc. Sci. Fish., 26(3):338-342.

Chemistry radioactive phosphorus
Marcolini, B.M., 1953.
Alcuni esperimenti per l'alimentazione di Cladoceri e Copepodi d'acque dolce con fosforo radioattivo (P32).
Bol. Pesca, Piscicolt., Idrebiol., n.s., 8(2): 181-200, 8 textfigs.

chemistry phosphorus, radioactive
McCarter, J.A., F.R. Hayes, L.H. Jodrey, and M.L. Cameron, 1949.
Exchange of materials in a lake studied by the addition of radioactive phosphorus below the thermocline (Abstract). (Read at N.S. Inst. Sci. 2ns Ordinary Meeting). Proc. N.S. Inst. Sci. 22(4):59.

Chemistry "photodieldrin"
Matsumura, F., K.C. Patil, and G.M. Boush, 1970
Formation of "photodieldrin" by microorganisms.
Science, 170(3963): 1206-1207.

Phycobilins

Chemistry, phycobilins
Eriksson, Caj E.A., and Per Halldal, 1965.
Purification of phycobilins from red algae and their fluorescence excitation spectra in visible and ultraviolet.
Physiol. Plantarum, 18:146-152.

chemistry, phycobilins
Lewin, Ralph A., Editor, 1962.
Physiology and biochemistry of algae.
Academic Press, New York and London, 929 pp.

chemistry, physical
Garrels R.M. and M.E. Thompson 1962
A chemical model for sea water at 25°C and one atmosphere total pressure.
Amer. J. Sci., 260(1):57-66.

phytadienes

Chemistry phytadienes
Blumer, Max, and David W. Thomas, 1965.
Phytadienes in zooplankton
Science, 147(3662): 1148-1149.

phytane

chemistry, phytane
Clark, Robert C., Jr., and Max Blumer, 1967.
Distribution of n-paraffins in marine organisms and sediments.
Limnol. Oceanogr., 12(1):79-87.

chemistry, phytanic acid
Hansen, R.P., 1969.
The occurrence of phytanic acid in Antarctic krill (Euphausia superba)
Aust. J. Sci., 32(4): 160-161.

Pigments

Angot, M., 1964. Chemistry - pigments (data)
Phytoplancton et production primaire de la région de Nosy-Bé, decembre 1963 a mars 1964.
Cahiers O.R.S.T.R.O.M., Océanogr., 11(4):99-125.

Angot, M., 1964. pigments, (data)
Production primaire de la région de Nosy-Be.
Août à novembre 1963.
Cahiers, O.R.S.T.R.O.M., Océanogr., 11(4):27-53.

Chemistry pigments
Angot, M., 1959.
Production primaire. Orsom III. Resultats de la croisière "Boussole".
ORSTROM III. Inst. Français d'Océanie, Rapp. Sc: 13:61-70.

Chemistry pigments (data only)
Australia, Commonwealth Scientific and Industrial Research Organization, 1964.
Oceanographical observations in the Indian Ocean in 1962, H.M.A.S. Diamantina, Cruise Dm 2/62.
Oceanogr. Cruise Rept., Div. Fish. and Oceanogr. No. 15:117 pp.

chemistry, pigments (tetrapyrrole)
Baker, Earl W., 1970.
Tetrapyrrole pigments. Initial Reports of the Deep Sea Drilling Project, Glomar Challenger, 4: 431-438.

Chemistry pigments
Blumer, Max, 1965.
Organic pigments: their long-term fate. Fossil pigments provide evidence of modification of chemical structure under geological conditions.
Science, 149(3685):722-726.

Chemistry pigments
Cassie, R. Morrison, 1963.
Relationship between plant pigments and gross primary production in Skeletonema costatum.
Limnology and Oceanography, 8(4):433-439.

Chemistry pigments
Castellvi, J., 1963
Variations dans la composition des pigments assimilateurs chez un Platymonas (Chlorophycee) marin.
Rapp. Proc. Verb., Reunions, Comm. Int. Expl. Sci. Mer Mediterranee, Monaco, 17(2):465-466.

chemistry, pigments
Ceccaldi, Hubert J., 1964.
Contribution à l'étude de dosages quantitatifs du plancton. 1. Introduction.
Rec. Trav. Sta. Mar., Endoume, 51(35):9-16.

pigments, carotenoids
Ceccaldi, H., 1962.
Contribution à l'étude des pigments carotenoides chez Ciona intestinalis (L.).
Bull. Inst. Océanogr., Monaco, 59(1238):1-36.

Chemistry, pigments
Ceccaldi, Hubert J., et Brigitte Berland, 1964.
Contribution à l'étude de dosages quantitatifs du plancton. 2. Lyophilisation ou filtration sur filtres "millipore" comparaison de la solubilité des pigments photosynthetiques de la diatomée Phaeodactylum tricornutum (Bohlin) par quelques solvants organiques à diverse concentrations.
Rec. Trav. Sta. Mar. Endoume, 51(35):17-42.

Chemistry pigment
Creitz, G.I., and F.A. Richards, 1955.
The estimation and characterization of plankton populations by pigment analysis. III. A note on the use of "millipore" membrane filters in the estimates of plankton pigments.
J. Mar. Res. 14(3):211-216.

Chemistry pigments
Currie, R.I., 1962.
Pigments in zooplankton faeces.
Nature, 193(4819):956-957.

Chemistry pigment
Duxbury, A.C., and C.S. Yentsch, 1956.
Plankton pigment nomographs. J. Mar. Res., 15(1): 92-101.

Chemistry pigments
Establier, R., and R. Margalef, 1964
Fitoplancton e hidrografía de las costas de Cádiz (Barbate), de junio de 1961 a agosto de 1962.
Invest. Pesquera, Barcelona, 25:5-31.

Chemistry pigments
*Finenko, Z.Z., and L.A. Lanskaya, 1968.
The amount of pigments in marine planktonic algae grown in the laboratory. (In Russian; English abstract).
Okeanologiia, Akad. Nauk, SSSR, 8(5):839-847.

chemistry pigments
Foxton, P., 1964
Seasonal variations in the plankton of Antarctic waters.
In: Biologie Antarctique, Proc. S.C.A.R. Symposium, Paris, 2-8 September 1962, Hermann, Paris, 311-318.

Also in:
Collected Reprints, Nat. Inst. Oceanogr., Wormley, 12. 1964

Chemistry pigments
Ganapati, P.N., and D.V. Subba Rao, 1958.
Quantitative study of plankton off Lawson's Bay, Waltair.
Proc. Indian Acad. Sci., (B), 48(4):189-209.

Chemistry - pigment extraction
Gardiner, A. C., 1943.
Measurement of phytoplankton population by the pigment extraction method. J.M.B.A., n.s., 25:739-744.

Chemistry cytochromes (pigments)
Ghiretti, F., A. Ghiretti-Magaldi, H. A. Rothschild, L. Tosi, 1958.
A study of the cytochromes of marine invertebrates.
Acta Physiol. Lat. Amer., 8(4):239-247.

chemistry pigments
Gostan, Jacques, et Paul Nival, 1967
Relations entre la distributions des phosphates minéraux dissous et la répartition des pigments dans les eaux superficielles du Golge de Gênes.
Cah. océanogr., 19(1):41-52.

Chemistry pigments
Grall, Jean-René, et Guy Jacques, 1964.
Etude dynamique et variations saisonnières du plancton de la region de Roscoff. B. Phytoplancton.
Cahiers, Biol. Mar., Roscoff, 5:432-455.

Chemistry pigments
Hart, T.J., 1962
Notes on the relation between transparency and plankton content of the surface waters of the Southern Ocean.
Deep-Sea Res., 9(2):109-114.

Chemistry pigments
Healy, F.P., J. Coombs and B.E. Volcani, 1967.
Changes in pigment content of the diatom, Navicula pelliculosa (Bréb.) Hilse in silicon starvation synchrony.
Archiv fur Mikrobiol., 59:131-142.

Chemistry pigments
Herring, P.J., 1967.
The pigments of plankton at the sea surface.
In: Aspects of Marine Zoology, N.B. Marshall, editor, Symp. Zool. Soc., Lond., 19: 215-235.

Chemistry - pigment, blue
Herring, P.J., 1965.
Blue pigment of a surface-living oceanic copepod.
Nature, 205(4966):103-104.

pigments
Koblents-Mishke, O.I., 1963.
A review of data on measurements of primary production in the northern part of the Pacific. (In Russian).
Mezhd. Geofiz. Kom. Prezidiume, Akad. Nauk, SSSR, Rezult. Issled. Programme Mezhd. Geofiz. Goda, Okeanol. Issled., (8):104-111.

Chemistry - pigment
Honjo, Tsuneo and Tsuku Hanaoka 1969.
Diurnal fluctuations of photosynthetic rate and pigment contents in marine phytoplankton. (In Japanese; English abstract)
J. Oceanogr. Soc. Japan, 25(4):182-190.

Chemistry pigments
Humphrey, G.F., and M. Wootton, 1966.
Comparison of the techniques used in the determination of phytoplankton pigments.
UNESCO, Monogr. Oceanogr. Methodol., 1:37-63.

Chemistry Pigments
Koutyurin, V. M., A. P. Lisitsin, 1961.
[Vegetable pigments in the suspended material and in bottom sediments of the Indian Ocean.]
Mezhd. Kom. Mezhd. Geofiz. Goda, Presidiume Akad. Nauk, SSSR, Okeanol. Issled., (3):90-116.

Chemistry pigments
Krasnovskii, A.A., Iu. E. Erokhin, Khun Iui-zun, 1962.
[Fluorescence of aggregated forms of bacterio-chlorophyll, bacterioviridin and Chlorophyll connected with the state of pigments in photosynthetic organisms.]
Doklady Akad. Nauk, SSSR, 143(2):456-459.

Krishnamurthy, K. 1971. chemistry, pigments
Phytoplankton pigments in Porto Novo waters (India).
Int. Revue ges. Hydrobiol. 56(2):273-282

Chemistry pigments
Kuroda, C., and M. Okajima, 1964.
Studies on the derivatives of naphthoquinones. XVII. The pigments of sea urchins. XII.
Proc. Japan. Acad., 40(10):836-839.

Chemistry pigments
Kutyurin, E.M., and A.P. Lisitsin, 1962
[Vegetative pigments present in suspensions and bottom sediments in the western part of the Indian Ocean. Communication. 2. Quantitative distributions and qualitative contents of pigments present in suspensions.]
Mezhd. Geofiz. Komitet, Prezidiume Akad. Nauk, SSSR, Rezult. Issled. Programme Mezhd. Geofiz. Goda, Okeanol. Issled., No. 5:112-129.

Laevastu, Taivo, 1958 pigments
The occurrence of pigments in marine sediments. J. Mar. Res., 17:325-334.

pigments
Liaci, Lidia, 1964.
Pigmenti e steroli negli invertebrati marini.
Archivio Zool. Ital., 49:281-300.

Chemistry pigments
Lorenzen, Carl J., 1967.
Vertical distribution of chlorophyll and phaeopigments: Baja California.
Deep-Sea Res., 14(6):735-745.

Chemistry pigments
*Mandelli, Enrique F., 1968.
Carotenoid pigments of the dinoflagellate Glenodinium foliaceum Stein.
J. Phycol., 4(4):347-348.

chemistry, pigments (data)
Margalef, R., J. Herrera, M. Steyaert et J. Steyaert, 1966.
Distributions et caractéristiques des communautes phytoplanctoniques dans la bassin Tyrrhenien de la Méditerranée en fonction des facteurs ambiants et à la fin de la stratification estivale de l'année 1963.
Bull. Inst. r. Sci. nat. Belg., 42(5):1- 56.

chemistry, pigments
Margalef, R., F. Saiz, J. Rodriguez-Rode, R. Toll, y J.M. Valles, 1952.
Plancton recogido por los laboratorios costeros. V. Fitoplancton de las costas de Castellon durante el ano 1951. Publ. Inst. Biol. Aplic. 10:133-143, 2 textfigs.

Chemistry pigment
Miller, S.M., H.B. Moore, and K.R. Kvammen, 1953.
Plankton of the Florida Current. 1. General conditions. Bull. Mar. Sci., Gulf and Caribbean 2(3): 465-485, 27 textfigs.

Chemistry - pigment, blue
Neuville, Dominique, et Philippe Daste, 1971.
Observations concernant la production de pigment bleu par la diatomée Navicula ostrearia (Gaillon) Bory maintenue en culture unialgale sur un milieu synthétique.
C.r. hebd. Séanc. Acad. Sci. Paris (D) 272 (16): 2232-2234

Chemistry pigments
Neuville, Dominique, et Philippe Daste, 1970.
Observations concernant la production de pigment bleu par la diatomée Navicula ostrearia (Gaillon) Bory maintenue en culture uni-algale.
C.r. hebd. Séanc., Acad. Sci. Paris 271(25): 2389-2391

Chemistry pigments
Pablo, I.S., & A.L. Tappel, 1961
Cytochromes of marine invertebrates.
J. Cell. & Comp. Physiol., 58(2): 185-194.

Chemistry pigments

Parsons, T.R., 1966.
The determination of photosynthetic pigments in sea-water. A survey of methods.
UNESCO, Monogr. Oceanogr. Methodol., 1:19-36.

Chemistry-pigments

Parsons, T.R., 1961.
On the pigment composition of eleven species of marine phytoplankton.
J. Fish. Res. Bd., Canada, 18(6):1017-1025.

pigments

Parsons, T.R., and D.J. Blackbourn, 1968
Pigments of the ciliate Mesodinium rubrum (Lohmann).
Netherlands J. Sea Res. 4(1): 27-31

Chemistry, pigments

Qasim, S.Z., and C.V.G. Reddy, 1967.
The estimation of plant pigments of Cochin backwater during the monsoon months.
Bull. mar. Sci., Miami, 17(1):95-110.

Chemistry pigments

Ramírez R., Boris, 1970.
Análisis de pigmentos vegetales marinos en un corte meridiano del Pacífico sur.
Inv. pesq. Barcelona, 34 (2): 309-318.

Chemistry pigments

Richards, F.A., 1952.
The estimation and characterization of plankton populations by pigment analyses. I. The absopption spectra of some pigments occurring in diatoms dinoflagellates and brown algae.
J. Mar. Res. 11(2):147-155.

Chemistry, pigments

Richards, F.A., and T.G. Thompson, 1952.
The estimation and characterization of plankton populations by pigment analyses. II. A spectrophotometric method for the estimation of plankton pigments. J. Mar. Res. 11(2):156-172, 3 textfigs.

Chemistry - pigment

Riley, G.A., 1952.
Phytoplankton of Block Island Sound, 1949.
Bull. Bingham Ocean. Coll. 13(3):40-64, 6 textfigs.

Chemistry pigments

Riley, G. A., 1942.
The relationship of vertical turbulence and spring diatom flowerings. J. Mar. Res., 5(1): 67-87.

Chemistry pigments

Riley, J.P. and D.A. Segar, 1969.
The pigments of some further marine phytoplankt species. J. mar. biol. Ass. U.K., 49(4): 1047-1056.

Chemistry, pigments

Riley, J.P., and T.R.S. Wilson, 1967.
The pigments of some marine phytoplankton species.
J. mar. biol. Ass., U.K., 4 (2):351-362.

Chemistry pigments

Rotschi, Henri, Michel Angot, Michel Legand and Roget Desrosieres, 1961

Orsom III, Resultats de la Croisiere "Dillon", 2eme Partie. Chimie et Biologie.
ORSTOM, Inst. Francais d'Oceanie, Centre d'Oceanogr., Rapp. Sci., No. 19: 105 pp. (mimeographed).

Chemistry pigments

Rotschi, H., B. Wauthy and M. LeGand, 1961
Orsom III. Résultats de la croisière "Epi" 2eme partie. Chimie et biologie.
O.R.S.T.O.M., I.F.O., Rapp. Sci., (23):4-111.

Chemistry pigments

San Feliu, J.M., y F. Muñoz, 1970.
Hidrografía y fitoplancton de las costas de Castellón, de julio de 1967 a junio de 1968.
Inv. pesq. Barcelona, 34 (2): 417-449.

Chemistry-pigments

SCOR-UNESCO Working Group, 17, 1966.
Determination of photosynthetic pigments.
UNESCO, Monogr. Oceanogr. Methodol., 1:9-18.

Chemistry - pigments, phytoplankton

Shah, M.N., 1968.
Certain features of diel variation of phytoplankton pigments and associated hydrographic conditions in the Laccadive Sea off Cochin.
A comparison of two seasons. Bull. Dept. mar. Biol. Oceanogr. Univ. Kerala, 4: 167-174.

pigments

Spencer, C.P., 1964.
The estimation of phytoplankton pigments.
J. du Conseil, 28(3):327-334.

Chemistry-pigments

Stephens, K., 1965.
Phytoplankton pigment nomographs.
J. Fish Res., Bd., Canada, 22(6):1575-1578.

Chemistry pigments

Strickland, J.D.H., 1965.
Production of organic matter in the primary stages of the marine food chain.
In: Chemical oceanography, J.P. Riley and G. Skirrow, editors, Academic Press, 1:477-610.

Chemistry pigments

Travers, A., 1962.
Recherches sur le phytoplancton du Golfe de Marseille. II. Étude quantitative des populations phytoplanctoniques du Golfe de Marseille.
Rec. Trav. Sta. Mar. Endoume, Bull., 28(41):70-140.

Chemistry - pigments

Travers, Marc, 1969.
Contribution à l'étude du phytoplancton et des tintinnides de la région de Tuléar (Madagascar). II. Les pigments planctoniques.
Rec. Trav. Sta. mar. Endoume, hors sér. Suppl. 9:49-57

Chemistry-pigment extraction

Tucker, A., 1949.
Pigment extraction as a method of quantitative s analysis of phytoplankton. Trans Am. Micros. Soc. 68(1):21-33, 6 textfigs.

Chemistry pigments

UNESCO, 1966.
Determination of photosynthetic pigments in sea water.
Monographs on Oceanographic Methodology, 1:69 pp.

Chemistry pigment concentrations

Ward, Ronald W., Valerie Vreeland, Charles H. Southwick and Anthony J. Reading, 1965.
Ecological studies related to plankton productivity in two Chesapeake Bay estuaries.
Chesapeake Science, 6(4):214-225.

pigments,

Wauthy, B., et J. Le Bourhis, 1966.
Considérations sur l'étude des pigments du phytoplancton marin en zone tropicale oligotrophe.
Cah. ORSTOM, Sér. Océanogr., 4(4):3-19.

Chemistry pigments, yellow

Yentsch, Charles S., and Carol A. Reichert, 1962.
The interrelationship between water-soluble yellow substances and chloroplastic pigments in marine algae.
Botanica Marina, 3(3/4):65-74.

Plankton, chemistry
See also: Chemistry holding under different biological groups.
See also: plankton pigments (under "P" in chemistry section)
See also: "biochemistry" under chemistry heading
See also: Chemistry, chemicals, concentration of chemistry, radioactivity, accumulation and concentration

Chemistry, plankton organisms

See also: Chemistry, concentration of chemicals within organisms
Cumulative Index of Marine Organisms under various animal groups, Etc. etc.

chemistry, plankton organisms

Bajkov, A., 1945
Utilization of marine plankton as food under emergency survival conditions. Memo Rept. AAF Technical Service Command, Personal Equipment Lab., Ser.No. TSEAL-5H-4-184, Expediture Order No.670-346, 39 pp. (mimeographed), 13 photos. 24 Sept. 1945.

Appendix I. Methods of catching plankton and data obtained at Woods Holo.
Appendix II. Plankton and its food value.
Appendix III. Program for determining nutritional value of plankton by Lt.Col. Irving 15 Aug. 1945.

chemistry-plankton organisms

Bajkov, A.D. and T.W. Robinson, 1947.
The nutritional value of plankton. Aero Medical Laboratory, Army Air Forces Air Material Command, Engineering Division Memo. Rept., Ser. No. TSEAA-691-3E, Expenditure Order 691-12, 11 pp. (ozalid), 4 photos. 29 Mar. 1947.

chemistry plankton

Baschei, Marie-Claude, Claude-Jean Bastard, Robert Raimondi et Christian Venet, 1968.
Contribution à l'étude biochimique du plancton. III. Composition qualitative et quantitative en acides aminés de quelques prélèvements du Golfe de Marseille.
Rapp. Proc.-verb. Réun. Comm. int. Explor. scient. Mer Méditerranée, 19 (5): 553-556.

Chemistry plankton

Battaglia, B., C. Mozzi, and A.M. Varagndo, 1961.
La distribuzione del plancton nell'Adriatico in rapporto con la concentrazione dei sali nutrivi.
Rapp. Proc. Verb., Réunions, Comm. Int. Expl. Sci. Mer Méditerranée, Monaco, 16(2):93-95.

Chemistry, plankton

Camps Mestre, José María, y Enrique Arias Serrano 1966.
Microdeterminación de proteínas y carbohidratos del plancton.
Inv. Pesq., Barcelona, 30:631-638.

Chemistry, plankton

Chari, S. T., and P. Anantha Pai, 1946.
Preservation of prawns and its effect on the nutritive value. Current Sci., 15(2):342-344.

Chemistry, plankton

Christomanos, An.An., A. Dimitriades and V. Gardiki, 1962
Contribution to plancton chemistry. (In English;Greek summary).
Chemia Chronika, Greece, 26A:23-26.

chemistry, plankton

Corner, E.D.S., and C.B. Cowey, 1964
Some nitrogenous constituents of the plankton
In: Oceanography and Marine Biology, Harold Barnes, Editor, George Allen & Unwin, Ltd., 2:147-167.

chemistry, plankton

del Riego, A.F., 1962.
El molibdeno en el planeton de la ría de Vigo. Algunas determinaciones de nitrógeno y fósforo: su relación y contenida en materia organica y humedad.
Bol. Inst. Espanol Oceanogr., Madrid, 108:24 pp.

chemistry of plankton

Derenbach, Jens B., 1970.
Zur Bestimmung der Ribonucleinsäure in planktischem Analysenmaterial.
Kieler Meeresforsch. 26(1): 79-84.

chemistry, plankton

Deuser, W.G., E.T. Degens and R.R.L. Guillard 1968.
Carbon isotope relationships between plankton and sea water.
Geochim. Cosmochim. Acta, 32(6): 657-660.

chemistry, plankton

Farkas, Tibor, and Sandor Herodek, 1964.
The effect of environmental temperature on the fatty acid composition of crustacean plankton.
J. Lipid Res., 5(3):369-373.

chemistry, plankton

Ferry, J. D., 1941.
Extraction of protein from various jellyfishes.
J. Biol. Chem., 140(1):xxxix-xl.

chemistry, plankton

Fujita, Tetsuo 1971.
Concentration of major chemical elements in marine plankton.
Geochem. J., Japan 4 (3): 143-156.

chemistry, plankton

Gastaud, J.M., 1961.
Contribution à la biochimie des éléments planctoniques. Étude de l'insaponifiable des lipides et de la fraction sterolique.
Rapp. Proc. Verb., Réunions, Comm. Int. Expl. Sci. Mer Méditerranée, Monaco, 16(2):251-254.

Chemistry plankton

Hagmeier, E., 1961
Plankton-Äquivalente (Auswertung von chemischen und mikroskopischen Analysen).
Kieler Meeresf., 17(1):32-47.

chemistry plankton

Harris, E., 1956.
Oceanography of Long Island Sound, 1952-1954. 8. Chemical composition of the plankton.
Bull. Bingham Oceanogr. Coll., 15:315-323.

chemistry, plankton

Kakimoto, D., and A. Kanazawa, 1957.
[Biochemical studies on the Sergestes phosphorus I. On general components. 2. On the sugars.]
Mem. Fac. Fish., Kagoshima Univ., 5:153-155; 156-159.

chemistry, plankton

*Hussein, M.Fawzy, R.Boulos and F.M. Hanna, 1967.
Studies on the chemical composition of plankton of Lake Qarun. I. Seasonal variations in the protein, lipids and carbohydrate content of plankton.
Bull.Fac.Sci.,Cairo Univ.Press, 40:121-131.

chemistry, plankton

Khalil, F., M.Fawzy Hussein, R. Boulos and F.M. Hanna, 1967.
Studies on the chemical composition of plankton of Lake Qarun. V. Seasonal variation in the calcium, magnesium, phosphorus and iron of plankton.
Bull.Fac.Sci.,Cairo Univ. Press, 40: 133-142.

chemistry, plankton

Krishnamurthy, K., 1967.
Nitrogen and phosphorus in plankton.
Hydrobiologia, 30(2):273-280.
Also in: Collected Reprints, Portonovo mar. biol. Sta., 1966-1967.

chemistry, plankton

Krishnamurthy, K., 1963
Phosphorus in plankton 1.
J. Zool. Soc., India, Calcutta, 14(2):161-164.

Also in:
Annamalai Univ. Mar. Biol. Sta. Porto Novo, S. India, Publ., 1961-1962.

chemistry, plankton

Lewis, R.W., 1969.

The fatty acid composition of Arctic marine phytoplankton and zooplankton with special reference to minor acids. Limnology and Oceanogr. 14(1): 35-40.

chemistry, plankton

Miquelis, E., 1961.
Étude chimique d'un échantillon de plancton lyophilise.
Les Cahiers du Centre d'Études et de Recherches de Biologie et d'Océanographie (C.E.R.B.O.M.) (2):14-18 (mimeographed).

chemistry, plankton

Nicholls, G.D., 1959

Spectrographic analyses of marine plankton.
Limnol. & Oceanogr., 4(4): 472-478.

chemistry, plankton

Parsons, T.R., K. Stephens and J.D.H. Strickland, 1961.
On the chemical composition of eleven species of marine phytoplankters.
J. Fish. Res. Bd., Canada, 18(6):1001-1016.

chemistry, plankton

Saiki, Masamichi, Sinchen Fang and Takajiro Mori, 1959

Studies on the lipid of plankton. 2. Fatty acid composition of lipids from Euphausiacea collected in the Antarctic and Northern Pacific Oceans. Bull. Jap. Soc. Sci. Fish., 24:837-839.

Sato, Tadao, Yoshishige Horiguchi, and Rokuro Adachi, 1965.
On the chemical composition of coarse (mainly plankton) and the fine (mainly detritus) suspended particles in the water of Matoyo Bay.
(In Japanese; English abstract).
Inf. Bull. Planktol., Japan, No. 12:66-71.

chemistry, plankton

Schreiber, B. 1965.
Études sur la radioactivité du plancton et des sédiments côtiers de la mer ligurienne.
Rapp. P.-v. Réun. Comm. int. Explor. scient. Mer Méditerr. 18(3): 883-892.

chemistry, plankton

Seshadri, B., 1958
Seasonal variations in the total biomass and total organic matter of the plankton in the marine zone of the Vellar Estuary.
J. Zool. Soc., India, Calcutta, 9(2):183-191.

chemistry, plankton

Skopintsev, B.A., 1961.
Investigation into mineralization of organic substance of dead plankton under anaerobic conditions. (Abstract).
Tenth Pacific Sci. Congr., Honolulu, 21 Aug.-6 Sept., 1961, Abstracts of Symposium Papers, 385-386.

chemistry, plankton

Sugawara, K., S. Okabe and M. Tanaka, 1961
Geochemistry of molybdenum in natural waters.
J. Earth Sci., Nagoya Univ., 9(1):114-128.

chemistry, plankton

Suschenia, L.M., 1961.
[The chlorophyll content of plankton in the Aegean, Ionic and Adriatic seas.]
Okeanologiia, Akad. Nauk, SSSR, 1(6):1039-1045.

chemistry, plankton

Tskhomelidze, O.I., 1958
[Biochemical composition of plankton in the eastern part of the Black Sea.]
Akad. Nauk Gruzinskoi SSR, Tiflis, Soobshcheniia, 21(2): 147-

chemistry, plankton

Verhaeghe, M., 1962.
Étude chimique de la biomasse planctonique.
Cahiers, Centre d'Études et de Recherches Biol., Océanogr. Med., (2):58-70. (mimeographed).

chemistry, plankton organisms

Vinogradov, A.P. 1938
Chemical composition of marine plankton.
Trans. Inst. Marine Fisheries and Oceanogr. of the USSR, 7:97-112. Moscow.

chemistry - plankton organisms

Vinogradov, A.P., 1935
Elementary Chemical Composition of Marine organisms. Laboratoires de Biochemie pres l'Acad. des Sci., U.S.S.R., Travaux, 3:63-278. Leningrad.

chemistry, plankton

Vinogradova, Z.A., and V.V. Koval'ski, 1962.
A study of the elementary composition of the Black Sea plankton. (In Russian).
Doklady, Akad. Nauk, SSSR, 147(6):1458-1460.

chemistry, plankton

Vinogradova, Z.A., 1957.
[The biochemical composition of the Black Sea plankton.] Doklady Akad. Nauk, SSSR, 116(4):688-690.

chemistry - plankton

Yudanova, O., 1940.
Chemical composition of Calanus finmarchicus in the Barents Sea. Compt. Rend (Doklady) Acad. Sci., U.S.S.R., 29:218-224.

chemistry, plutonium-239

Bowen, V.T., K.M. Wong and V.E. Noshkin, 1971.
Plutonium-239 in and over the Atlantic Ocean.
J. mar. Res., 29(1): 1-10.

chemistry, fallout, plutonium

Miyake, Y., Y. Katsuragi and Y. Sugimura, 1970.
A study on plutonium fallout. J. geophys. Res., 75(12): 2329-2330.

chemistry, plutonium

Miyake, Yasuo, and Yukio Sugimura 1969.
Plutonium content in the western North Pacific waters.
Pap. Met. Geophys. 19(3): 481-485.

chemistry, plutonium

Wong, Kai M. 1971.
Radiochemical determination of plutonium in sea water, sediments and marine organisms.
Analyt. Chim. Acta 56:355-364

polonium-210

chemistry, polonium-210
Shannon, L.V., and R.D. Cherry, 1967.
Polonium-210 in marine plankton.
Nature, Lond., 216(5113):352-353.

chemistry, polonium-210
Shannon, L.V., R.D. Cherry and M.J. Orren 1970
Polonium-210 and lead-210 in the marine environment.
Geochim. cosmochim. Acta 34(6): 701-711.

polycarbonylic acids

Cooper, L.H.N., 1965. polycarbonylic acids
Chemistry of the sea. 2. Organic.
Chemistry in Britain, 1965, 1:150-154.

chemistry, polymerization
Jackson, Togwell A., 1971.
Preferential polymerization and absorption of L-optical isomers of amino acids relative to D-optical isomers on kaolinite templates.
Chem. Geol. 7(4): 295-306

chemistry, polyphosphate
Solórzano, L., and J.D.H. Strickland, 1968.
Polyphosphate in seawater.
Limnol. Oceanogr., 13(3):515-518.

polysaccharides

polysaccharides
*Ernst, Wolfgang, 1967.
Dünnschichtchromatographische Trennung und quantitative Bestimmung von Galacturon- und Glucuronsäure in Polysaccharidfraktionen des Meeresbodens.
Veröff. Inst. Meeresforsch. Bremerh. 10(3):183-185.

chemistry, polysaccharides
Ernst, W., 1966.
Nachweis, Identifizierung und quantitative Bestimmung von Uronsäuren in Polysaccharidfraktionen mariner Sedimente der Deutschen Bucht.
Veröff. Inst. Meeresforsch, Bremerh., 10(2):81-92.

porphyrins

chemistry, porphyrins
Blumer, M., and W.D. Snyder 1967.
Porphyrins of high molecular weight in a Triassic oil shale: evidence by gel permeation chromatography.
Chem. Geol. 2(1): 35-45.

chemistry porphyrins
Liaci, Lidia, 1964.
Pigmenti e steroli negli invertebrati marini.
Archivio Zool. Ital., 49:281-300.

potassium

chemistry potassium
Atkins, W.R.G., 1948
Note on the Spectroscopic and biological detection of potassium in sea water and "potassium-free" artificial sea water. J. du Cons. 15(2):169-172.

chemistry potassium
Berthois, Léopold, 1966.
Hydrologie et sédimentologie dans le Kangerd-lugssuaq (fjord à la côte ouest du Groenland).
C.r. hebd. séanc., Acad. Sci., Paris, (D), 262(13): 1400-1402.

chemistry, potassium (data)
Berthois, L. et A. Crosnier 1966.
Étude dynamique de la sédimentation au large de l'estuaire de la Betsiboka.
Cah. ORSTOM, Sér. Océanogr. 4(2):49-130.

chemistry potassium
Brewer, Peter G., and Derek W. Spencer, 1969.
A note on the chemical composition of the Red Sea brines.
In: Hot brines and Recent heavy metal deposits in the Red Sea, E.T. Degens and D.A. Ross, editors, Springer-Verlag, New York, Inc., 174-179.

chemistry, potassium
Burovina, I.V., et al., 1963.
The content of lithium, sodium, potassium, rubidium and caesium in the muscles of marine animals of the Barents Sea and Black Sea. (In Russian).
Doklady, Akad. Nauk, SSSR, 149(2):413-415.

chemistry, potassium
Buyanov, N.I., 1966.
Content of potassium in sea water and marine organisms of the White Sea. (In Russian).
Mater. Ribokhoz. Issled. severn. Basseina, Poliarn. Nauchno Issled. Procktn. Inst. Morsk. Ribn. Khoz. Okeanogr. (PINRO), 7:183-191.

chemistry potassium
Chow, Tsaihwa J., 1964.
Flame photometric determination of potassium in sea water and marine organisms.
Analyt. Chimica Acta, 31(1):58-63.

chemistry, potassium
Chu, S. P., 1949
Experimental studies on the environmental factors influencing the growth of phytoplankton.
Sci. & Tech. in China 2(3):37-52.

chemistry potassium
Culkin, Frederick, 1965.
The major constituents of sea water.
In: Chemical oceanography, J.P. Riley and G. Skirrow, editors, Academic Press, 1: 121-161.

chemistry, potassium
Culkin, F., and R.A. Cox, 1966.
Sodium, potassium, magnesium, calcium and strontium in sea water.
Deep-Sea Res., 13(5):789-804.

chemistry potassium
Dyrssen, David, et al., 1967.
Analysis of sea water in the Uddevalla Fjord system 1.
Rept. Chem. Sea Water, Univ. Göteborg, 4:1-8 pp. (mimeographed).

chemistry, potassium
Fabricand, B.S. Imbimbo, M. E. Brey and J.A. Weston, 1966.
Atomic absorption for Li, Mg, K, Rb, and Sr in ocean waters.
J. geophys. Res., 71(6):3917-3921.

chemistry, potassium
Gololobov, Ia. K. 1964.
Analysis of the chemical compounds involved in biological production in the Black Sea and some peculiarities of their production. (In Russian)
Trudy azov.-chernomorsk. nauchno-issled. Inst. morsk. ryb. Khoz. okeanogr. 23:33-47.

chemistry potassium
Ishibashi, M., & T. Hara, 1959.
A systematic analysis of potassium, rubidium and cesium and its application to sea-muds.
Bull. Inst. Chem. Res. Kyoto Univ., 37(3):185-190.

chemistry potassium
Ishibashi, M, & T. Hara, 1959.
On the determination of potassium in dilute solution and its application in the analysis of sea-water.
Bull. Inst. Chem. Res. Kyoto Univ., 37(3):167-171.

chemistry potassium
Ishibashi, M., and F. Morii, 1957.
A study on the method of the extraction of potassium from bittern as perchlorate.
Rec. Ocean. Wks., Japan (Suppl.):46-49.

chemistry potassium
Jentoft, R.E., and R.J. Robinson, 1956.
The potassium -chlorinity ratio of ocean water.
J. Mar. Res., 15(2):170-180.

chemistry potassium (data)
Kramer, J.R., 1961
Chemistry of Lake Erie.
Proc. Fourth Conf., Great Lakes Research. Great Lakes Res. Div., Inst. Sci. & Tech., Univ. Michigan, Publ., (7):27-56.

chemistry potassium
Morse, Jean, et Jean Barret, 1963.
Dosage du potassium dans divers produits de pêcheries en vue de la mesure de la radioactivité
Rev. Trab. Inst. Pêche Marit., 27(2):235-240.

chemistry, potassium
Nishizawa, Tanzo and T. Muraki, 1940
Chemical studies in the sea adjacent to Palau. I. A survey crossing the sea from Palau to New Guinea. Kagaku Nanyo (Sci. of the South Sea) 2(3):1-7.

chemistry potassium
Rodhe, W., 1948
Environmental requirements of fresh-water plankton algae. Experimental studies in the ecology of phytoplankton. Symbolae Botanicae Upsalienses X(1):149 pp., 30 figs.

chemistry potassium
Sherstnev, A.I., and I.I. Baksheeva, 1962.
Maintenance of common potassium and the radio-active isotope potassium-40 in the organs of some different fish from the west coast of Africa. (In Russian).
Trudy, Baltiisk. Nauchno-Issled. Inst. Morsk. Ribn. Khoz. i Okeanogr. (Baltniro), 8:135-137.

chemistry potassium
*Sreekvmaran, C., K.C. Pillai and T.R. Folsom, 1968.
The concentrations of lithium, potassium, rubidium adn cesium in some western American rivers and marine sediments.
Geochim. cosmochim. Acta, 32(11):1229-1234.

chemistry potassium
Valyashko, M.G., 1951.
[Volume relations of the liquid and solid phases in the evaporation process of ocean water as a determining factor in the formation of layers of potassium salts.] Doklady Akad. Nauk, SSSR, 77(6):1055.

chemistry, potassium
Viswanathan R., S.M. Shah and C.K. Unni 1969.
Atomic absorption and fluorometric analyses of seawater in the Indian Ocean. Bull. natn. Inst. Sci., India 38(1): 284-288.

chemistry potassium
*Weaver, Charles E., 1967.
Potassium, illite and the ocean.
Geochim. cosmochim. Acta, 31(11):2181-2196.

chemistry potassium
Webb, D.A., 1939.
The sodium and potassium content of sea water.
J. Exp. Biol., 16:178-183.

potassium-chlorinity ratios

chemistry, potassium/chlorine ratio
Berthois, L. 1969.
Contribution à l'étude sédimentologique du Kangerdlugssuaq, côte ouest du Groenland. Meddr Grønland 187(1): 185 pp.

potassium/chlorinity ratio
Jentoft, R.E., and R.J. Robinson, 1956.
The potassium-chlorinity ratio of ocean water.
J. Mar. Res., 15(2):170-180.

Chemistry, potassium/chlorinity ratio
*Riley, J.P., and M. Tongudai, 1967.
The major cation/chlorinity ratios in sea water.
Chem. Geol., 2(3):263-269.

Chemistry, potassium/rubidium ratio
Jakeš P., and A.J.R. White, 1970.
K/Rb ratios of rocks from island arcs.
Geochim. cosmochim. Acta, 34(8): 849-856.

potassium-42

Chemistry potassium-42
Bryan, G.W., 1961.
The accumulation of radioactive caesium in crabs.
J. Mar. Biol. Assoc., U.K., 41(3):551-575.

precipitates

chemistry, precipitates
Jones, Galen E., 1967.
Precipitates from autoclaved seawater.
Limnol. Oceanogr., 12(1):165-167.

pristane

Chemistry pristane
Blumer, M., M.M. Mullin and D.W. Thomas, 1963.
Pristane in zooplankton.
Science, 140(3570):974.

chemistry, pristane
Clark, Robert C., Jr., and Max Blumer, 1967.
Distribution of n-paraffins in marine organisms and sediments.
Limnol. Oceanogr., 12(1):79-87.

Protactinium

Promethium

chemistry promethium
Bowen, Vaughan T., and Thomas T. Sugihara, 1965.
Oceanographic implications of radioactive fall-out distributions in the Atlantic Ocean: from 20°N to 25° S, from 1957-1961.
J. Mar. Res., 23(2):123-146.

Chemistry Protactinium
Heye, Dietrich 1970.
A system for detection of ionium thorium and protactinium to date deep-sea cores.
Geochim. cosmochim. Acta, 34(3): 389-397.

chemistry protactinum
Ku, Teh-Lung, and Wallace S. Broecker 1967.
Uranium, thorium and protactinium in a manganese nodule.
Earth Planet. Sci. Letters, 2(4): 317-320

chemistry, protactinium
Moore, Willard S., and William M. Sackett, 1964
Residence times of protactinium and thorium in sea water. (Abstract).
Trans. Amer. Geophys. Union, 45(1):119.

Chemistry proactinium-231
Sackett, W. M., 1960
Proactinium-231 content of ocean water and sediments. Science, 132(3441): 1761-1762.

Chemistry proactinium 231
Servant, Jean, 1966.
La radioactivité de l'eau de mer.
Cahier Océanogr., C.C.O.E.C., 18(4):277-318.

proteins

Chemistry protein (albumen equivalent) (data only)
Australia, Commonwealth Scientific and Industrial Research Organization, 1964.
Oceanographical observations in the Indian Ocean in 1961, H.M.A.S. Diamantina, Cruise Dm 3/61.
Div. Fish. and Oceanogr., Oceanogr. Cruise Rept., No. 11:215 pp.

chemistry, proteins
*Bogdanov, Yu.A., Yu.A. Grigorovich, and M.G. Shaposhnikova, 1968.
Protein determination in the water suspended matter. (In Russian; English abstract).
Okeanologiia, Akad. Nauk. SSSR, 8(6):1087-1090.

Chemistry proteins (data only)
Collier, Albert, 1958.
Gulf of Mexico physical and chemical data from ALASKA cruises.
USFWS Spec. Sci. Rept., Fish., No. 249:417 pp.

Chemistry Protein
Derenbach, Jens B., 1970.
Partikuläre Substanz und Plankton an Hand chemischer und biologischer Daten gemessen in den oberen Wasserschichten des Gotland-Tief im Mai 1968.
Kieler Meeresforsch. 25(2): 279-289

Chemistry protein (data)
Dragovich, Alexander, 1961
Relative abundance of plankton off Naples, Florida, and associated hydrographic data, 1956-'57.
USFWS Spec. Sci. Rept., Fish., No. 372:41 pp.

Chemistry proteins (tyrosine)
Finucane, John H., and Alexander Dragovich, 1959
Counts of red tide organisms, Gymnodinium brevis and associated oceanographic data from Florida west coast, 1954-1957.
USFWS Spec. Sci. Rept., Fish. No. 289:220 pp.

Chemistry protein content
Gundersen, K.R., 1953.
Zooplankton investigations in some fjords in western Norway, during 1950-1951.
Rept. Norwegian Fish. Mar. Invest. 10(6):1-54, 24 textfigs.

Chemistry proteins
*Hussein, M.Fawzy, R. Boulus and F.M. Hanna, 1967.
Studies on the chemical composition of plankton of Lake Qarun. 1. Seasonal variations in the protein, lipids and carbohydrate content of plankton.
Bull. Fac. Sci., Cairo Univ. Press, 40:121-131.

chemistry, proteins
Levine, Philip T., Melvin J. Glimcher, Jerome M. Seyer, James I. Huddleston and John W. Hein, 1966.
Noncollagenous nature of the proteins of shark enamel.
Science, 154(3753):1192-1193.

chemistry, protein (data only)
Krey, Johannes, Rolf Boje, Max Gillbricht und Jürgen Lenz 1971.
Planktologisch-chemische Daten der Meteor-Expedition in den Indischen Ozean 1964/65.
Meteor Forsch.-Ergebn. (D) 9:1-120.

Chemistry Protein
Matterne, Manfred, 1970.
Vergleich zwischen Primärproduktion und Syntheseraten organischer Zellbestandteile mariner Phytoplankter.
Kieler Meeresforsch. 25(2): 290-313.

Chemistry protein
Mayeda, Osamu and Tatsuichi Iwamura, 1968.
Biochemical approach to the ecology in aquatic environments: Measurement of protein and RNA in natural water. (In Japanese; English abstract). Bull. Plankton Soc., Japan, 15(1): 17-18.

Chemistry protein content
Nakai, Z., 1955.
The chemical composition, volume weight and size of the important marine plankton.
Tokai Reg. Fish. Res. Lab., Spec. Publ., 5:12-24.

Chemistry proteins
Raymont, J.E.G., J.Austin and Eileen Linford, 1967.
The biochemical composition of certain zooplankton decapods.
Deep-Sea Research, 14(1):113-115.

chemistry, protein
Raymont, J.E.G., R.T. Srinivasagam and J.K.B. Raymont, 1971.
Biochemical studies on marine zooplankton - VIII. Further investigations on Meganyctiphanes norvegica (M. Sars). Deep-Sea Res. 18(12): 1167-1178.

Chemistry protein
Raymont, J.E.G., R.T. Srinivasagam and J.K. B. Raymont, 1969.
Biochemical studies on marine zooplankton. IV. Investigation on Meganyctiphanes norvegica (M. Sars.) Deep-Sea Res., 16(2): 141-156.

chemistry, protein

Schmitt, Walter R. and John D. Isaacs, 1971.
Enhancement of marine protein production.
In: Fertility of the Sea, John D. Costlow, editor, Gordon Breach, 2: 455-462.

Chemistry proteins

Snyder, Donald G., 1966.
Marine protein concentrate.
Exploiting the Ocean, Trans. 2nd Mar. Techn. Soc. Conf., June 27-29, 1966, 530-534

chemistry proteins

*Wada, Koji, 1967.
Studies on the mineralization of the calcified tissue in mollusco. XIV. Modification of the amino acid pattern of proteins in the extra pallial fluid during the process of formation and mineralization of nacreous conchiolin in some bivalves.
Bull. Jap. Soc. scient. Fish., 33(7):613-617.

Protium

Chemistry protium
Greene, C.H., and R.J. Voskuyl, 1939.
The deuterium-protium ration. 1. The densities of natural waters from various sources.
J. Amer. Chem Soc., 61:1342-1349.

protoporphyrin

protoporphyrin
Townsley, P.M., R.A. Richy and P.C. Trussell, 1965.
The occurrence of protoporphyrin and myohlobin in the marine borer, Bankia setacea (Tryon).
Canadian J. Zool., 43(1):167-172.

pteridins

chemistry, pteridins
Liaci, Lidia, 1964.
Pigmenti e steroli negli invertebrati marini.
Archivio Zool. Ital., 49:281-300.

purines

Chemistry purines
Belser, W.L., 1963
9. Bioassay of trace substances. In: The Sea M.N. Hill, Edit., Vol. 2 (II) Fertility of the oceans, Interscience Publishers, New York and London, 220-231.

Chemistry purine
Rosenberg, E., 1964.
Purine and pyrimidines in sediments from the experimental Mohole.
Science, 146(3652):1680-1681.

pyrimidines

Chemistry pyrimidines
Belser, W.L., 1963
9. Bioassay of trace substances. In: The Sea, M.N. Hill, Edit., Vol. 2(II) Fertility of the oceans, Interscience Publishers, New York and London, 220-231.

Chemistry Pyrimidine
Rosenberg, E., 1964.
Purine and pyrimidines in sediments from the experimental Mohole.
Science, 146(3652):1680-1681.

Chemistry quinones

Droop, M.R. and J.F. Pennock, 1971.
Terpenoid quinones and steroids in the nutrition of Oxyrrhis marina. J. mar. biol. Ass. U.K. 51(2): 455-470.

radioactive contamination

Chemistry, radioactivity, contamination by

See also: Pollution (radioactive)

Chemistry, radioactive contamination
Belyaev, V.L., A.G. Kolesnikov and B.A. Nelepo, 1965.
Regularities of the radioactive contamination spreading in the ocean. (In Russian; English abstract).
Fisika Atmosfer. Okean. Izv. Akad. Nauk, SSSR, 3(10): 1092-1100.

radioactivity, contamination by
Kreps, E.M., 1959.
The problem of radioactive contamination of oceans and marine organisms. (In Russian).
Biull., Akad. Nauk, SSSR, Biol. (3):321-334.

radioactivity (contamination)
Nakai, Zinziro, Rinnosuke Fukai, Harumi Tozawa, Shigemasa Hattori, Jatsuo Okubo, and Takashi Kidachi, 1960
Radioactivity of marine organisms and sediments in the Tokyo Bay and its southern neighborhood.
Collected Reprints, Tokai Reg. Fish. Res. Lab., 1960(35): 18-38.

Chemistry radioactive contamination

Saiki, Masamichi, 1958.

Studies on the radioactive materials in the radiologically contaminated fishes. 5. Column chromatography of the radioelements in the liver of a big-eyed tuna Parathunnus mebachi, with an ion exchange resin.
Bull. Jap. Soc. Sci. Fish., 23:729-734.

Chemistry radioactive debris

Saha, Kshudiram, 1970.
Interhemispheric drift of radioactive debris and tropical circulation.
Tellus 22(6): 688-698

radioactive elements + isotopes

Chemistry radioactive elements
Agamirov, A. Sh., 1963.
Geochemical balance of radioactive elements in the Black Sea Basin. (In Russian).
Geokhimiya, (6):612-614.

Translation:
GEOCHEMISTRY (6):630-633.

Chemistry radioactive elements
Akamsin, A.D., 1961.
[Distribution of P32, S35, Sr90, Y91, Ce144 among marine waters and sediments under experimental conditions.]
Trudy Sevastopol Biol. Sta., (14):309-313.

Chemistry radioisotopes
Broecker, Wallace S., 1966.
Radioisotopes and the rate of mixing across the main thermoclines of the ocean.
J. geophys. Res., 71(24):5827-5836.

Chemistry radioactive isotopes

Gersht, E.P., 1957

[On the use of radioactive isotopes in meteorology and hydrology] Met. i Gidrol. (8): 62.

Chemistry radioactive elements
Hayes, F.R., 1955.
The effect of bacteria on the exchange of radiophosphorus at the mud-water interface.
Verh. Int. Ver. Limnol. 12:111-116, 4 textfigs.

Chemistry radioactive elements
Koczy, F.F., 1956.
Geochemistry of radioactive elements of the ocean. Deep-Sea Res. 3(2):93-103.

Chemistry radioactive isotopes
Miyake, Y., 1967.
Oceanographic studies by means of natural and artificaial radioactive isotopes and stable isotopes. (A lecture) (In Japanese).
J. oceanogr. Soc., Japan, 23(3):148-155.

Chemistry radioactive elements
Pettersson, H., 1939
Die radioaktiven Elemente im Meere.
Cons. Perm. Internat. p. l'Explor. de la Mer. Rapp. et Proc.-Verb. 109(3):66-67.

Chemistry radioactive elements
Polikarpov, G.G., 1961.
[Material on coefficient of accumulation of P32, S35, Sr90, Y91, Cs137, and Ce144 in marine organisms.]
Trudy Sevastopol Biol. Sta., (14):314-328.

radioactive fallout

Chemistry radioactive fallout
Beck, J.N., and P.K. Kuroda, 1966.
Radiostrontium fallout from the nuclear explosion of October 16, 1964.
J. geophys. Res., 71(10):2451-2456.

Chemistry radioactive fall-out.
Bowen, Vaughan T., and Thomas T. Sugihara, 1965.
Oceanographic implications of radioactive fall-out distributions in the Atlantic Ocean: from 20° N to 25° S, from 1957 to 1961.
J. Mar. Res., 23(2):123-146.

Chemistry radioactive fallout
Broecker, Wallace S., and Gregory G. Rocco, 1963
Direct comparison of radiocarbon and fall-out measurements in sea water. In: Nuclear Geophysics.
Nat. Acad. Sci.-Nat. Res. Council, Publ. No. 1075:150-151.

Chemistry, radioactive fallout
Broecker, Wallace S., Gregory G. Rocco and H.L. Volchok, 1966.
Strontium-90 fallout: comparison of rates over ocean and land.
Science, 152(3722):639-640.

Chemistry, radioactive fallout
Chesselet, Roger, et Claude Lalou, 1965.
Résultats d'une étude expérimentale de dispersion de débris radio-actifs atmosphériques dans l'eau de mer.
Bull. Inst. Océanogr., Monaco, 64(1328):14pp.

chemistry, radioactive fallout
Chipman, Walter, and Jean Thommeret 1970.
Manganese content and the occurrence of fallout ^{54}Mn in some marine benthos of the Mediterranean.
Bull. Inst. océanogr. Monaco, 69(1402):15pp

Chemistry, radioactive fallout
Crocker, G. R., 1966.

Radiochemical data correlations and coral-surface bursts.

Nature, 210(5040):1028-1031.

Chemistry, radioactive fallout

Folsom, T.R., and Katsuko Saruhashi, 1963
A comparison of analytical techniques used for determination of fallout cesium in sea water for oceanographic purpose.
J. Radiation Res., 4(1):39-53.

Chemistry radioactive fallout
Freiling, E.C., and N.E. Ballou, 1962.
Nature of nuclear debris in sea-water.
Nature, 195(4848):1283-1287.

Chemistry, radioactive fallout
Gross, M. Grant, 1967.
Sinking rates of radioactive fallout particles in the North East Pacific Ocean, 1961-1962.
Nature, Lond., 216(5116):670-672.

Chemistry, radioactive fallout
Jaworowski, Zbigniew, 1959.
Radioactive fallout measurements at Hornsund-fiorden in Spitzbergen.
Acta Geophysica Polonica, 7(2):130-133.

Chemistry
Radioactive fallout
Karol, I. L., Y. V. Krasnopevtsev, V. D. Vilensky and S. G. Malchov, 1966.
Comparative analysis of the conditions of world-wide fallout of the products from nuclear test explosions on the continents and into the oceans. (In Russian).
Trudy Inst. Okeanol., Akad. Nauk, SSSR, 82:56-71.

Chemistry fallout, radioactive
Kleinman, Michael T., and Herbert L. Volchok, 1969
Radionuclide concentrations in surface air: direct relationship to global fallout.
Science, 166 (3903):376-377

Chemistry
radioactive fallout
Kruger, Paul, and Albert Miller, 1964.
Radiochemical fallout study of a Pacific cyclonic storm.
J. Geophys. Res., 69(8):1469-1480.

radioactive fallout
Miyake, Yasuo and Teruko Kanazawa, 1967
Atmospheric ozone and radioactive fallout.
Pap. Met. Geophys. Tokyo, 18(4):311-326

Chemistry
radioactive fallout
Nagaya, Y., M. Shiozaki, and Y. Seto, 1964.
Radiological survey of sea water of adjacent Sea of Japan in 1963. (In Japanese; English abstract).
Hydrogr. Bull., Tokyo, (Publ. No. 981). No. 78:63-67.

chemistry, radioactive fallout
Pritchard, D.W., R.O. Reid, A. Okubo and H.H. Carter, 1971.
Physical processes of water movement and mixing
Radioactivity in the marine environment, U.S. Nat. Acad. Sci., 1971. 90-136.

Chemistry, radioactive fallout
Schroeder, B.W., and R.D. Cherry, 1962.
Caesium-137 in the seas off the Cape of Good Hope.
Nature, 194(4829):669.

chemistry, radioactive fallout
Shelby, Cader A., 1963.
Radioactive fallout in Sargassum drift on Texas Gulf coast beaches.
Publ. Inst. Mar. Sci., Port Aransas, 9:33-36.

Chemistry
radioactive fallout
Sotobayashi, Takeshi, and Seitaro Koyama 1966.
Strontium-90 fallout from surface and underground nuclear tests.
Science 152 (3725): 1059-1060.

Chemistry, radioactive fallout
*Volchok, H.L., M. Feiner, H.J. Simpson, W.S. Broecker, V.E. Noshkin, V.T. Bowen and E. Willis, 1970.
Ocean fallout - the Crater Lake experiment.
J. geophys. Res., 75(6): 1084-1091.

Chemistry, radioactive fallout
Wilgain, S., and E. Picciotto and W. De Breuck, 1965.
Strontium-90 fallout in Antarctica.
J. geophys. Res., 70(24):6023-6032.

chemistry, radioactive fallout
Wolfe, Douglas A., 1971.
Fallout cesium-137 in clams (Rangia cuneata) from the Neuse River Estuary, North Carolina.
Limnol. Oceanogr. 16(5): 797-805.

Chemistry ~~fallout, radioactive~~
Yoshimura, Hirozo, 1967.
The gross B-radioactivity of fallout collected at the southern weather station located in 29°N and 135°E.
Oceanogrl Mag., 19(2):169-172.

Chemistry, radioactive phosphorus
See under: chemistry, phosphorus (radioactive)
chemistry, phosphorus (isotopes), etc.

radioactive purification

Chemistry radioactive purification
Murano, Masaki, and Yoshiyuki Matsue, 1958.
The purification of radioactive sewage by activated sludge process. Industrial Water 4:22-24.
Resume in: Rec. Res., Fac. Agric., Univ. Tokyo, No. VIII (1957-1958):p. 60.

radioactive tracers, see also
CHEMISTRY { TRACERS
TRACERS, isotopes
TRACERS, radioactive

Chemistry radioactive tracers
Allen, F.H., and J. Grindley, 1957.
Radioactive tracers in the Thames Estuary, report on an experiment carried out in 1955.
Dock & Harbour Authority 37(435):302-306.

Chemistry radioactive tracers
Anon., 1965.
Radioactive tracer "captured" by huge water bag.
Naval Res. Rev., April, 22-24.

Chemistry radioactive tracers
Aubert, Maurice, 1965.
Le comportement des bacteries terrigenes en mer: relations avec la phytoplancton.
Cahiers, C.E.R.B.O.M.,19-20:285-pp.

Chemistry radioactive tracers
Chipman, Walter A., 1959
The use of radioisotopes in studies of the foods and feeding activities of marine animals. Pubb. Sta. Zool. Nap., 31(Suppl.): 154-175.

Chemistry, radioactive tracers
Cortelezzi, Cesar, horacio Cazenouve, Manual Levin y Felix Mouzo, 1965.
Estudio del movimiento de sedimentos en la zona del Puerto de Mar del Plata mediante el uso de radioisotopos.
Anais Acad. bras. Cienc., 37(Supl.):289-305.

Chemistry radioactive tracers
Craig, H., 1957.
Ch. 11:Isotopic tracer techniques for measurements of physical and chemical processes in the sea and the atmosphere. NAS-NRC Publ., 551:103-120.

chemistry, tracers (radioactive)
Duke, Thomas W., John P. Baptist and Donald E. Hoss, 1966
Bioaccumulation of radioactive gold used as a sedimenttracer in the estuarine environment.
Fishery Bull., Fish Wildl. Serv., U.S., 65(2):427-436.

Chemistry radiactive tracers,
Great Britain, Department of Scientific and Industrial Research, 1955.
Hydraulics research, 1954, 50 pp.

Chemistry, radioactive tracers
Foster, R.F., 1959
Radioactive tracing of the movement of an essential element through an aquatic community with specific reference to radiophosphorus.
Pubb. Sta. Zool. Nap., 31(Suppl): 34-69.

Chemistry radioactive tracers
Great Britain, Hydraulics Research Board, 1957.
Hydraulics research, 1956, 54 pp.

Chemistry
radioisotope tracers
Gutknecht, John, 1963
Zn 65 uptake by benthic marine algae.
Limnology and Oceanography, 8(1):31-38.

Chemistry
radioactive tracers
Kautsky, Hans, 1961
Über Herkunft, Verteilung und Nachweismöglichkeit kunstlich radioaktiver Stoffe im Meer.
Deutsche Hydrogr. Zeits., 14(4):121-135.

Chemistry Radioactive tracers
Kolesnikov, A.G., A.A. Pivovarov, E.S. Ivanova, 1961.
On calculation of rate of radioactivity spreading in depth of ocean. (Abstract).
Tenth Pacific Sci. Congr., Honolulu, 21 Aug.-6 Sept., 1961, Abstracts of Symposium Papers, 346-347.

Chemistry
radioisotopes tracers
Mishima, Jiro, and Eugene P. Odum, 1963
Excretion rate of Zn 65 by Littorina irrorata in relation to temperature and body size.
Limnology and Oceanography, 8(1):39-44.

Chemistry, radioactive tracers
Miyake, Y., 1963
3. Artificial radioactivity in the sea.
In: The Sea, M.N. Hill, Edit., Vol. 2. The composition of sea water, Interscience Publishers, New York and London, 78-87.

Chemistry radioactive tracers
Nelepo, B.A., 1960
[Determination of the coefficients of turbulent diffusion in the sea by use of radioactive tracers.] Vestnik Moskovskogo Universiteta (4): 64-70.

Chemistry,
radioactive tracers
Osterberg, Charles, Norman Cutshall and John Cronin 1965.
Chromium-51 as a radioactive tracer of Columbia River water at sea.
Science 150 (3703): 1585-1586.

Chemistry radioactive tracers
Reid, W.J., 1958.
Coast experiments with radioactive tracers.
Dock & Harbour Auth., 39(453):84-88.

Chemistry radioactive tracers
Rodina, A.G., 1957.
Application of the radioactive tracer method for the solution of the food selectivity problem in aquatic animals. Zool. Zhurnal, 36(3):337-343.

Chemistry
radioactive tracers
Schaefer, M.B. 1957.
Ch.12. Large scale biological experiments using radioactive tracers.
Nat. Acad. Sci.- Nat. Res. Counc. Publ. 551: 133-137

Chemistry radioactive tracers
Shirey, V.A., 1964.
Investigation of circulation in the ocean by the tracer cloud method. (In Russian).
Radioaktionaya Zagryaznennost' Morey i Okeanov, "Nauka", Moscow, 197-198.

Abstracted in:
Soviet Bloc Res., Geophys. Astron., Space, No. 95:51.

radioactive wastes + waste disposal

Chemistry, radioactive waste disposal

See: waste disposal
pollution, etc.

Chemistry
radioactive waste
Bogorov, V.G., and E.M. Kreps, 1959
Concerning the possibility of disposing of radioactive waste in ocean trenches.
Progress in Nuclear Energy, Ser. 12, 1 (Health Physics):583-590.

Chemistry radioactive waste.
Bogorov, V.G., and Krebs, E.M., 1958.
Is it possible to dispose of radioactive wastes on the bottom of the ocean? (In Russian).
Priroda, 47(9):45-

Chemistry
radioactive wastes
Bogorov, V.G., & B.A. Tareev, 1960
Ocean depths and the problem of burying radioactive wastes in them.
Izv. Akad. Nauk SSSR, Ser. Geofiz., (4):3-10.
Listed in Pub. Health Eng. Abs., 41(11): 26.

Chemistry radioactive waste
Brown, Robert M., and Donald W. Pritchard, 1961.
Radioactive wastes in the marine environment.
Amer. J. Pub. Health, 51(11):1647-1661.

Also in:
Collected Reprints, The Johns Hopkins Univ., Chesapeake Bay Inst. and Dept. Oceanogr., 6. 1965.

Chemistry radioactive wastes
Bryan, G.W., and Eileen Ward, 1962.
Potassium metabolism and the accumulation of 137 caesium by decapod Crustacea.
J. Mar. Biol. Assoc., U.K., 42(2):199-242.

Chemistry radioactive wastes
Butler, Alan M., Frederick G. Keyes, Albert Szent-Gyorgy, 1960
Sea disposal of atomic wastes. Bull. Atomic Scientists 16(4): 141

Chemistry radioactivity, waste disposal
Chipman, W.A., 1959.
Disposal of radioactive materials and its relation to fisheries.
Proc. Nat. Shellfish. Assoc., 1958, 49:5-12.

Chemistry radioactive wastes
Collins, J.C., Editor, 1960.
Radioactive wastes. Their treatment and disposal
E. & F.N. Spon Ltd., London, xxi 239 pp.

Reviewed by:
Dhaw, I.I., 1961. J. Mar. Biol. Assoc. U.K., 41(3):836.

Chemistry radioactive wastes
Committee on the Disposal and Dispersal of Radioactive Wastes, 1956.
Disposal and dispersal of radioactive wastes.
Science 124(3210):17-18.

Chemistry, radioactive waste
Faughn, J.L., T.R. Folsom, J.D. Isaacs, F.D. Jennings, DeC. Martin, Jr., L.E. Miller, and R.L. Wisner, 1958.
A preliminary radioactivity survey along the California coast through disposal areas.
Proc. Ninth Pacific Sci. Congr., Pacific Sci. Assoc., 1957, Oceanogr., 16:152-158.

Chemistry, radioactive waste
Folsom, Theodore, R., 1958.
Approximate dosages close to submerged radioactive layers of biological interest.
Proc. Ninth Pacific Sci. Congr., Pacific Sci. Assoc., 1957, Oceanogr., 16:170-175.

Chemistry radioactive waste
Ginell, W.S., J.J. Martin and L.P. Hatch, 1954.
Ocean disposal of radioactive waste.
Nucleonics 12(12) :54.

Chemistry radioactive wastes
Ginell, W.S., J.J. Martin and L.P. Hatch, 1954.
Ultimate disposal of radioactive wastes.
Nucleonics 12(12):14-18.

Chemistry radioactive wastes
India, Council of Scientific and Industrial Research, New Delhi, 1963.
High-level radioactive wasters: methods of treatment and storage - an international symposium.
J. Sci. Industr. Res., 22(2):73-76.

Chemistry
radioactive wastes
Jacob, J., 1962
Radioisotopes and radioactivity in marine biology.
Proc. First All-India Congress of Zool., 1959 Symposia, (3):32-36.

Also in:
Annamalai Univ. Mar. Biol. Sta., Porto Novo, S. India, Publ. 1961-1962.

Chemistry radioactive wastes, uptake of, by algae
Lewin, Ralph A., Editor, 1962.
Physiology and biochemistry of algae.
Academic Press, New York and London, 929 pp.

Chemistry
radioactivity, waste disposal
Martin, DeCourcey, Jr., 1958
The uptake of radioactive wastes by benthic organisms.
Proc. Ninth Pacific Sci. Congr., Pacific Sci. Assoc., 1957, 16 (Oceanogr):167-169.

Chemistry radioactive wastes
Renn, C.E., 1956.
Ultimate disposal of radioactive reactor wastes in the ocean. J. Amer. Water Wks. Assoc., 48: 535-537.

Abstr. in: Publ. Health Eng. Abstr., 37(3):28.

Chemistry radioactive waste
Kovringa, P., 1960
Problems arising from disposal of low-activity radioactive waste in the coastal waters of the Netherlands. Disposal of Radioactive Wastes. Internat. Atomic Energy Agency, Vienna: 51-56.

Chemistry radioactive waste
Mauchline, J., and W.L. Templeton, 1963.
Dispersion in the Irish Sea of the radioactive liquid effluent from Windscale works of the U.K. Atomic Energy Authority.
Nature, 198(4881):623-626.

Chemistry
radioactive waste disposal
Miyake, Y., 1963
3. Artificial radioactivity in the sea. In: The Sea, M.N. Hill, Edit., Vol. 2. The composition of sea water, Interscience Publishers, New York and London, 78-87.

Chemistry radioactive wastes
Newcombe, Curtis L., 1964.
The potential problem of radioactive wastes in the aquatic environment.
Turtox News, 42(4):98-101.

Chemistry radioactive waste
Picotti, M., 1961.
Expansion des déchets radioactifs et dynamiques des eaux de la Méditerranée.
Rapp. Proc. Verb., Réunions, Comm. Int. Expl. Sci., Mer Méditerranée, Monaco, 16(3):569-574.

Chemistry, radioactive waste
Pritchard, D.W., 1961
Disposal of radioactive wastes in the ocean.
Health Physics, 6:103-109.

Also in: Chesapeake Bay Inst., Collected Reprints, 5(1963)

Chemistry, radioactive waste
Pritchard, D.W., 1960
The application of existing oceanographic knowledge to the problem of radioactive waste disposal into the sea.
Disposal of Radioactive Wastes (Proc. IAEA Sci. Conf. on Disposal of Radioactive Wastes, Monaco, Nov., 16-21, 1959), Int. Atomic Energy 2:229-248.

Also in: Chesapeake Bay Inst., Collected Reprints, 5. (1963)

Chemistry radioactive wastes
Pritchard, D.W., 1959
Factors affecting the dispersal of fission products in estuarine and inshore environments.
Proc. 2nd Int. Conf., Peaceful Uses of Atomic Energy, Sept. 1-13, 1959(Session D-19):410-413.

Chemistry
radioactive wastes
Pritchard, D.W., 1959
Problems related to disposal of radioactive wastes in estuarine and coastal waters.
Trans. 2nd Seminar on Biol. Problems in Water Pollution, April 20-24, 1959: 11 pp.(unnumbered in COLLECTED REPRINTS, Chesapeake Bay Inst.).

Chemistry Waste Disposal, Radioactive
Pritchard, Donald W., and Arnold B. Joseph, 1963
Disposal of radioactive wasters: its history, status, and possible impact on the environment.
Radioecology, Proc. First Nat. Symposium on Radioecology, Colorado State Univ., Sept.11-15, 1961, Reinhold Publ. Corp. and A.I.B.S.,27-31.

Also in:
Collected Reprints, The Johns Hopkins Univ., Chesapeake Bay Inst. and Dept. Oceanogr., 6. 1965.

Renn, C.E., 1956. radioactive wastes
Ultimate disposal of radioactive reactor wastes in the ocean.
J. Amer. Water Works Assoc., 48:535-537.

Abstr. In:
Pub. Health Eng. Abstr., 37(3):28.

chemistry, wastes (radioactive)
Rice, Theodore R. and Douglas A. Wolfe, 1971.
Radioactivity - chemical and biological aspects.
In: Impingement of man on the oceans, D.W. Hood, editor, Wiley Interscience: 325-379.

chemistry, radioactive waste
Sasaki, Tadayoshi, Motoaki Kishino, Gohachiro Oshiba, Seiichi Watanabe and Moriyoshi Okazaki, 1965.
Studies on the containers for disposing radioactive wastes into the sea. IV. Experiments of throwing mortar containers into the sea (2). (In Japanese; English abstract).
J. oceanogr. Soc., Japan, 21(2):45-53.

chemistry, radioactive waste
Sasaki, Tadayoshi, Motoaki Kishino, Gohachiro Oshiba, Seiichi Watanabe and Moriyoshi Okazaki, 1964.
Studies on the container for the disposing radioactive wastes into the sea. III. Experiment of throwing mortar containers into the sea. (In Japanese; English abstract).
Jour. Oceanogr. Soc., Japan, 20(4):179-184.

chemistry radioactive, waste
Sasaki, Tadayoshi, Moriyoshi Okazaki, Seiichi Watanabe, Gohachiro Oshiba and Motoaki Kishno, 1964.
Studies on the container for disposing radioactive wastes into the sea. II. On the permeability and diffusion in mortar samples. (In Japanese; English abstract).
Jour. Oceanogr. Soc., Japan, 20(4):168-178.

Chemistry, radioactive waste
Sasaki, Tadayoshi Moriyoshi Okazaki, Seiichi Watanabe, Gohachiro Oshiba, Motoaki Kishino and Noburu Takematsu 1966.
Studies on the container for disposing radioactive wastes into the sea. VI. On the permeability and diffusion in water. (In Japanese; English abstract)
J. oceanogr. Soc. Japan 22(6):245-254

chemistry radioactive waste disposal
Sasaki, Tadayoshi, Seiichi Watanabe, Gohachiro Oshiba, Noboru Okami and Masahiro Kajihara, 1963.
Studies on the conditions for disposing radioactive wastes into the sea. 1. (In Japanese; English abstract).
J. Oceanogr. Soc., Japan, 19(1):16-26.

Chemistry radioactive waste
Schaefer, Milner B., 1960
New Research required in support of radioactive waste disposal.
Disposal of Radioactive Wastes, Int. Atomic Energy Agency, Vienna, pp. 265-282.

chemistry RADIOACTIVE WASTE DISPOSAL
Smith, J.M., Jr., 1959.
Disposal of radioactive liquids from nuclear-powered ships.
Sewage and Industr. Wastes, 31:1323-1326.
Abstr. in:
Publ. Health Eng. Abstr., 40(7):28.

chemistry, waste disposal (radioactive)
Templeton, W.L., 1965.
Ecological aspects of the disposal of radioactive wastes to the sea.
In: Ecology and the Industrial Society, Fifth Symposium of the British Ecological Society, Blackwell, Oxford, 65-97.

chemistry, radioactive waste
United Nations, 1958.
Waste treatment and environmental aspects of atomic energy.
Proc. 2nd United Nations Internat. Conf. on the Peaceful Uses of Atomic Energy, Geneva, 18: 624 pp.

Chemistry radioactive waste
Vodianitskii, V.A., 1958.
Is it admissible that atomic industry waste be dumped into the Black Sea. (In Russian).
Priroda, (2):119-

chemistry Radioactive Waste
Waldichuk, Michel, 1961.
The pollution problem arising from the disposal of radioactive materials.
Fish. Res. Bd., Canada, MSS Rept. Ser. (Oceanogr. Limnol.) No. 100:30 pp., figs., (multilithed).

chemistry radioactive wastes
Wallen, I. Eugene, 1964.
Atomic and other wastes in the sea.
Ann. Rept., Smithsonian Inst., 1963, Publ. 4530: 381-400.

Chemistry radioactive wastes
Weichart, Günter, 1967.
Berechnung der Vertikaldiffusion von natürlichen Stoffen und Abfallstoffen in der Iberischen Tiefsee-Ebene aus der vertikalen Konzentralionsverteilung der natürlichen Stoffe über dem Meeresboden.
Dt. hydrogr. Z., 19(6):266-284.

radioactivity

A

Chemistry radioactivity
Akiyama, Tsutomu, 1965.
On an instrument for in situ measurement of γ-ray activity in the deep water of the ocean.
Oceanogr. Mag., Tokyo, 17(1/2):69-75.

Chemistry radioactivity
Altman, E. N., 1959.
Experiment on the organizational radioactivity measurements on board ship of the Logger type.
Meteorol. i Gidiol. (6):42-44.

chemistry radioactivity
Argiero, L., G. Del Corso, S. Manfredini et G. Palmas, 1963.
Contrôle systématique de la radioactivité de la mer le long des côtes italiennes. Organiations des réseaux de prélèvement d'échantillons d'eau et de faune marine, techniques et données de mesure. Programme des études et des recherches.
Rapp. Proc. Verb., Réunions, Comm. Int., Expl. Sci., Mer Méditerranée, Monaco, 17(3):957-961.

chemistry radioactivity
Arrhenius, G., M.N. Bramlette and E. Picciotto, 1957.
Localization of radioactive and stable heavy nuclides in ocean sediments. Nature 180:85-86.

chemistry radioactivity
Arrhenius, G., and E.D. Goldberg, 1955.
Distribution of radioactivity in pelagic clays.
Tellus 7(2):226-231, 3 textfigs.

chemistry radioactivity
Anghanova, Z. A., A.F. Fedorov and V.P. Podymakhin 1966.
Application of the gamma spectrometric method for the analysis of radioactivity of samples of marine organisms. (In Russian)
Mater. Ribokhoz. Issled. severn. Basseina Poliarn. nauchno-issled. Proektn. Inst. morsk. ribn. Khoz. Okeanogr. (PINRO) 7: 177-182.

chemistry radioactivity
Aubert, M., R. Chesselet and D. Normann, 1962.
Mesures de radioactivité (α, β et γ) effectuées sur des échantillons de plancton et d'algues prelevés en Méditerranée dans la zone cotière de la ville de Nice.
Cahiers Centre d'Étude et de Recherches Biol. Océanogr. Med., (2):23-43. (mimeographed).

B

Chemistry radioactivity
Bagge, E., and S. Skorka, 1958.
Der Übergangseffekt der Ultrastrahlungsneutronen an der Grenzfläche Luft-Wasser. Zeits. f. Physik, 152(1):34-48.

chemistry, radioactivity
Baksheeva, I.P., A.D. Zemlyanoy, V.N. Markelov and B.A. Nelepo, 1971.
Water radioactivity in the northeastern Atlantic Ocean. (In Russian; English abstract)
Okeanologiia 11(6): 1041-1048.

Chemistry radioactivity
Baranov, V.I., and L.A. Khristianova, 1960.
Reply to the comments of S.M. Graschchenko, Yu. V Kuznetzov, V.K. Legin and D.S. Nikolaev, on "Radioactivity of Indian Ocean waters".
Geochemistry (7):788-789. (English Edition of Geokhimiia)

Chemistry radioactivity
Baranov, V.I., and L.A. Khristianova, 1959
Radioactivity of the waters of the Indian Ocean.
Geokhimiya (7):
English Transl. (7):765-769.

Chemistry, radioactivity
Bernhard, Michael 1965.
γ-spectra of marine organisms.
Rapp. P.-v. Réun. Comm. int. Explor. scient. Mer Méditerr. 18(3):899-905.

chemistry, radioactivity
Bernhard, Michael, 1965.
Remarks on the ecological problems in connection with marine radioactivity.
Rapp. P.-v. Réun. Comm. int. Explor. scient. Mer Méditerr. 18(3):893-897.

chemistry radioactivity
Bernhard, M., 1963.
7. Misure di radioattivita.
Rapp. Attivita Sci. e Tecn., Lab. Studio della Contaminazione Radioattiva del Mare, Fiascherino, La Spezia (maggio 1959-maggio 1961), Comit. Naz. Energia Nucleare, Roma, RT/BIO, (63), 8:159-161. (multilithed).

Chemistry, radioactivity
Blanc, Jean Joseph, 1962.
Remarques sur divers types de sédiments sous-marins de Méditerranée et leur radioactivité.
In: Océanographie Géologique et Géophysique de la Méditerranée Occidentale, Centre National de la Recherche Scientifique, Villefranche sur Mer, 4 au 8 avril 1960, 171-176.

Chemistry, radioactivity
Boroughs, H., S.J. Townsley and R.W. Hiatt, 1956.
Method for predicting amount of strontium-89 in marine fishes by external monitoring. Science 124(3230):1027-1028.

Chemistry radioactivity
Bowen, Vaughan T., and Thomas T. Sugihara, 1964.
Fission product concentration in the Chukchi Sea.
Arctic, 17(3):198-203.

Chemistry radioactivity
Bowen, Vaughan T., and Thomas T. Sugihara, 1962
Fallout fission products in the oceans. (Abstract).
J. Geophys. Res., 67(9):3543-3544.

Chemistry radioactivity
Broecker, W. S. & A. Walton, 1959
Radiocarbon from nuclear tests.
Science, 130(3371):309-314.

chemistry radioactivity
Buyanov, N.I., and V.P. Kilezhenko, 1964.
Potassium radioactivity of some marine organisms in the southern part of the Barents Sea. (In Russian).
Materiali, Ribochoz. Issled. Severn. Basseina, Poliarn. Nauchno-Issled. i Proektn. Inst. Morsk. Rib. Choz. i Okeanogr. im. N.M. Knipovicha, PINRO, Gosud. Proizvodst. Komm. Ribn. Choz. SSSR, 4:142-148.

Chemistry radioactivity
Byerly, John Robert, 1963.
The relationship between watershed geology and beach radioactivity.
Beach Erosion Bd., Techn. Memo., (135):32 pp.

C

chemistry radioactivity
Carritt, D.E., and H.H. Harley, 1957.
Ch. 6. Precipitation of fission product elements on the ocean bottom by physical, chemical and biological processes. NAS-NRC Publ. 551:60-68.

chemistry, radioactivity
Chaen, Masaaki, Tomio Hemmi and Tadao Takahashi, 1960.
Report of the Keiten-maru IGY cruise 1958.
Mem. Fac. Fish., Kagoshima Univ., 8: 87-88.

chemistry radioactivity
Chesselet, Roger, Tovy Grjebine, Gérard Lambert, et Daniel Nordemann, 1962.
Identification directe des nuclides radioactifs dans l'eau de mer par spectrographie.
In: Océanographie Géologique et Géophysique de la Méditerranée Occidentale, Centre National de la Recherche Scientifique, Villefranche sur Mer, 4 au 8 avril 1961, 183-210.

chemistry, radioactivity
Chesselet, Roger, et Claude Lalou 1965.
Étude de l'évolution dans le milieu marin des particules radioactives ayant pour origine les aérosols de la retombée atmosphérique.
Rapp. P.-v. Réun. Commn int. Explor. scient. Mer Méditerr. 18(3):851-856.

chemistry radioactivity
Chesselet, R., et Cl. Lalou, 1965.
Recherches récentes de la radioactivité du plancton et du détritus organique. 1. Étude de la radioactivité.
Cahiers du C.E.R.B.O.M., Nice, 2:67-85.

chemistry, radioactivity
Chesselet, Roger, Claude Lalou et Daniel Nordemann 1965.
Résultats de mesures récentes de spectrométrie gamma in situ en Méditerranée occidentale (Campagne Calypso - décembre 1963).
Rapp. P.-v. Réun. Comm. int. Explor. scient. Mer Méditerr. 18(3): 845-850.

chemistry radioactivity
Chesselet, R., C. Lalou et D. Nordemann, 1965.
Méthodes et résultats de mesure de radioactivité dans la mer.
Cahiers du C.E.R.B.O.M., Nice, 2:35-60.

chemistry radioactivity
Chesselet, Roger, et Daniel Nordemann, 1963.
Étude de la radioactivité des eaux de la Méditerranée occidentale par spectrometrie gamma in situ.
Rapp. Proc. Verb., Réunions, Comm. Int., Expl. Sci., Mer Méditerranée, Monaco, 17(3):939-946.

Chemistry, radioactivity of copepods and Copelata
Chiba, T., A. Tsuruta and H. Maeda, 1955.
Report on zooplankton samples hauled by larva-net during the cruise of Bikini-Expedition, with special reference to copepods.
J. Shimonoseki Coll., Fish., 5(3):189-213.

chemistry radioactivity
Coopey, R.W., 1953.
Radioactive plankton from the Columbia River.
Trans. Amer. Microsc. Soc. 72(4):315-327, 6 textfigs.

chemistry radioactivity
Craig, H., 1957.
Ch. 3. Disposal of radioactive wastes in the ocean: the fission product spectrum in the sea as a function of time and mixing characteristics
NAS-NRC Publ. 551:34-42.

D

chemistry radioactivity
Debyser, J., 1957.
La sédimentation dans le bassin d'Archachon.
Bull., C.E.R.S., Biarritz, 1(3):405-418.

chemistry radioactivity
Dera, J., B. Szczeblewski. & B. Lojicijevski, 1962
Radioactive contamination in sea water of the North European region.
Acta Geoph. Polonica, 10(2):173-182.

chemistry radioactivity
Devaputra, D., T.G. Thompson and C.L. Utterback, 1932.
The radioactivity of sea water. J. du Cons., 7: 358-366.

chemistry radioactivity
Drogaitsev, D.A., 1958.
Possible paths of the spread of atomic disintegration products from the Marshall Islands. (In Russian).
Priroda, 47(7):78-

E

chemistry radioactivity
Eisenbud, M., 1957.
Global distribution of radioactivity from nuclear detonations with special reference to strontium 90. J. Washington Acad. Sci., 47(6):180-188.

F

chemistry, radioactivity
Fabricand, E.S. Imbimbo, M.E. Brey and J.A. Weston 1966.
Atomic absorption for Li, Mg, K, Rb, and Sr in ocean waters.
J. geophys. Res. 71 (6): 3917-3921.

Federov, A.F.
Fyedorov, A.F.

chemistry, radioactivity
Federov, Anatol F., 1965.
Natural radioactivity in some ocean regions.
Bull. Inst. océanogr. Monaco 64 (1335): 59 pp.
Also: IAEA Radioactivity in the Sea, Publ. 15.

chemistry, radioactivity
Fyedorov, A.F., 1965.
On one of the criteria of radiation conditions in different sea areas. (In Russian).
Material. Ribochoz. Issled. Severn. Bassin. Poliern. Nauchno-Issled. i Proekt. Inst. Morsk. Rib. Choz. i Okeanogr., N.M. Knipovich, 5:142-144.

chemistry radioactivity
Federov, A.F., V.N. Podymakhin, V.P. Kilizhenko, N.I. Buyanov and E.M. Goloskova, 1964.
The radiation conditions in the fishery areas in the northern Atlantic. (In Russian).
Okeanologiia, Akad. Nauk, SSSR, 4(3):431-436.

Chemistry radioactivity
Federov, A.F., and G.V. Samokhin, 1964
The influence of natural radioactive background of sea water on the instrumental methods of control for radiological purity of fishery basins. (In Russian).
Materiali Vtoroi Konferentsii, Vzaimod. Atmosfer. i Gidrosfer. v Severn. Atlant. Okean., Mezhd. Geofiz. God, Leningrad. Gidrometeorol. Inst., 180-186.

Chemistry, radioactivity
Federov, A.F., and G.V. Samokhin, 1962.
Gamma-field intensity over sea surface.
Doklady Akad. Nauk, SSSR, 143(1):101-107.

Chemistry radioactivity
Fedorova, A.F., Kilezhenko, V.P., 1963
The radioactivity of bottom organisms in the Norwegian Sea. (In Russian).
Okeanologiia, Akad. Nauk, SSSR, 3(1):123-126.

chemistry radioactivity
Folsom, Theodore R., 1958.
Comparison of radioactive dosage from potassium with estimated dosage from uranium and radium in marine biospheres.
Proc. Ninth Pacific Sci. Congr., Pacific Sci. Assoc., 1957, Oceanogr., 16:176-178.

Chemistry radioactivity
Folsom, T.R., and H.H. Harley, 1957.
Ch. 2. Comparison of some natural radiations received by selected organisms. NAS-NRC Publ. 551:28-33.

chemistry radioactivity
Folsom, T.R., and A.C. Vine, 1957.
On the tagging of water masses for the study of physical processes in the oceans. In: The effects of atomic radiation on oceanography and fisheries Nat. Acad. Sci.-Nat. Res. Counc., Publ. 155:121-132.

chemistry radioactivity
Fonselius, S. and G. Östlund, 1959
Natural radiocarbon measurements on surface water from the North Atlantic and the Arctic Sea. Tellus, 11(1): 77-82.

chemistry radioactivity
Fontaine, M., 1958.
Oceanographie et radio elements.
Journees des 24 et 25 fevrier 1958: 91-107.
Publ. by Centre Belge d'Ocean. Recherches Sous-marines.

chemistry radioactivity
Fontaine, M., 1956.
Radiations atomiques et vie aquatique.
La Semaine des Hôpitaux No. 7:72-79.

Chemistry radioactivity
Ford, W.L., 1949
Radiological and Salinity relationships in the water at Bikini Atoll. Trans. Am. Geophys. Un., 30(1):46-54, 22 text figs.

Chemistry, radioactivity
Føyn, Ernst, B. Karlik, H. Pettersson and E. Rona 1939.
Radioactivity in sea water.
Göteborgs Kungl. Vetenskaps- och Vitterhets Samh. Handl., Femte Foljden, (B), 6(12):44 pp.

Chemistry radioactivity
French, W. R., Jr. and R. L. Chasson, 1959.
Atmospheric effects on the hard component of cosmic radiation near sea level.
J. Atm. & Terr. Phy., 14(1/2):1-18.

G

chemistry radioactivity
Gerard, Robert D., and Wallace S. Broecker, 1962
Natural radiocarbon in the subsurface waters of the Caribbean. (Abstract).
J. Geophys. Res., 67(9):3560.

Chemistry, radioactivity
Got, Henri, 1968.
La radioactivité naturelle des sédiments de la baie de Banyuls.
Cah. océanogr., 20(3):225-235.

Chemistry radioactivity
Graschenko, S.M., Yu. V. Kuznetsov, K.F. Lazarev, V.K. Legin, D.S. Nikolaev, 1960.
Discussion of the paper by V.I. Baranov and L.A. Khristianova:"Radioactivity of Indian Ocean waters".
Geochemistry, (7):785-787 (translation of Geochimiia).

Chemistry radioactivity
Gross, M. Grant, and C.A. Barnes, 1964.
Radioactivity of the Columbia River discharge in the northeast Pacific Ocean in August 1963. (Abstract).
Trans. Amer. Geophys. Union, 45(1):65.

Chemistry radioactivity
Gross, M.G., C.A. Barnes and G.K. Riel, 1965.
Radioactivity of the Columbia River effluent.
Science, 149(3688):1088-1090.

Chemistry radioactivity
*Gustafson, P.F., S.S. Brar, D.M. Nelson and S.E. Muniak, 1967.
Radioactive ^{54}Mn and ^{65}Zn in euryhaline fish.
Can. J. Zool., 45(5):729-735.

H

Chemistry, radioactivity
Harris, E., 1957.
Radiophosphorus metabolism in zooplankton and microorganisms. Canadian J. Zool., 35(6):769-782.

Chemistry radioactivity
Held, Edward E., 1960.
Land crabs and fission products at Eniwetok Atoll.
Pacific Science 14(1):18-27.

Chemistry radio activity
Herrmann, G., 1959.
Chemical monitoring of the distribution of radioactive substances.
Angew. Chemie, 71(18):561-571.

Chemistry radioactivity
Higano, Ryoji, Yutaka Nagaya, Masaru Shiozaki and Yoshio Seto, 1963.
On the artificial radioactivity in sea water. (In Japanese; English abstract).
J. Oceanogr. Soc., Japan, 18(4):200-207.

Chemistry radioactivity
Hishida, Kozo, and Katsumi Yamamoto, 1958.
On the radioactivity in the sea off Kyoga-misaki in the Sea of Japan.
Umi to Sora 33(12):85-86
34(4):

Chemistry radioactivity
Hiyama, Yoshio, 1960
An idea on the maximum permissible concentration of radio-active materials in sea-water. Rec. Oceanogr. Wks., Japan, n.s., 5(2): 98-104.

Chemistry radioactivity
Hiyama, Y., 1957.
Maximum permissible concentration of Sr 90 in food and its environment.
Rec. Oceanogr. Wks., Japan, n.s., 3(1):70-77.

Chemistry radioactivity
Hiyama, Yoshio, and Ryushi Ichikawa, 1958.
A measure on level of strontium-90 concentration in sea water around Japan at the end of 1956.
Proc. Ninth Pacific Sci. Congr., Pacific Sci. Assoc., 1957, Oceanogr., 16:141-145.

Chemistry radioactivity
Hiyama, Y., and R. Ichikawa, 1957.
A measure on level of strontium-90 concentration in sea-water around Japan, at the end of 1956.
Rec. Oceanogr. Wks., Japan, n.s., 4(1):49-54.

Chemistry radioactivity
Hiyama, Y., R. Ichikawa, T. Isogai, Y. Oshima, H. Koyama, S. Arasaki and K. Nozawa, 1956.
State and behavior of fission products in fresh-water and marine coastal systems in Japan, 1954-55. Research on effects and influences of the Nuclear Bomb Test Explosions. Jap. Soc. Prom. Sci., 1956:1027-1065.

chemistry, radioactivity
Hiyama, Y., R. Ichikawa and F. Yasuda 1956.
Studies on the processes of contamination of fish by the fission product. The research on the effects and influences of the nuclear bomb test explosions.
Jap. Soc. Prom. Sci. 1956: 1119-1113 (as given!)

Chemistry, radioactivity
Hiyama, Yoshio, Makoto Shimizu, Junko Matsubara, Tamiya Asari and Ryushi Ichikawa, 1960.
Sr 90 in marine organisms in Japan. Radioactive contamination of marine products in Japan.
Document for the Scientific Committee on the Effect of the Atomic Radiation of the United Nation 9-12.

chemistry radioactivity
Hood, Donald W., editor 1971
Impingement of man on the oceans.
Wiley-Interscience, 738 pp.

Chemistry radioactivity
Horrer, Paul L., 1962
Oceanographic studies during Operation "Wigwam". Physical oceanography of the test area.
Limnol. and Oceanogr., Suppl. to Vol. 7: vii-xxvi.

Chemistry radioactivity
Hours, R., W.D. Nesteroff and V. Romanovsky, 1955.
Utilisation d'un traceur radioactif dans l'étude de l'évolution d'une plage. C.R. Acad. Sci., (Paris) 240(18):1798-1800.

I

Chemistry radioactivity
Inoue, N. & M. Fukuda, 1956.
Report from the "Oshoro Maru" on oceanographic and biological investigations in the Bering Sea, and Northern Pacific in the summer of 1955.
II Radioactivity in sea water.
Bull of Fac. of Fish. Hokkaido Univ. Vol. 7, No. 3 p. 233-235.

Chemistry radioactivity
Israël, H., 1958.
Die naturlich Radioaktivität in Boden, Wasser und Luft. Beitr. z. Phys. d. Atmos., 30(2/3):177-188.

J

Chemistry, radioactivity
Jaffe, G., and J.M. Hughes, 1953.
The radioactivity of bottom sediments in Chesapeake Bay. Trans. Amer. Geophys. Union 34(4): 539-542, 2 textfigs.

Chemistry radioactivity
Japan, Committee on the Investigation on Artificial Radioactivity, Meteorological Society of Japan, 1957.
The thermo-nuclear experiment and its after effects on the atmosphere and the ocean.
Bull. Amer. Meteorol. Soc., 38(8):453-459.

Chemistry, radioactivity
Japan, Research Division, Fisheries Agency, 1956.
Radiological survey of western area of the danger-ous zone around the Bikini-Eniwetok Atolls, investigated by the "Shunkotsu maru" in 1956, Part 1:143 pp.

chemistry, radioactivity
Japan, Hokkaido University, Faculty of Fisheries 1959
Data record of oceanographic observations and exploratory fishing No. 1: 247 pp.

Chemistry radioactivity
Japan, Japan Meteorological Agency, Oceanographical Section, 1962
Report of the oceanographic observations in the sea east of Honshu from February to March, 1961. (In Japanese).
Res. Mar. Meteorol. and Oceanogr. Obs., Jan.-June, 1961, No. 29:13-21.

Chemistry radioactivity
Japan, Japan Meteorological Agency, 1958.
Present status of artificial radioactivity observations in Japan, 35 pp.

Chemistry radioactivity
Japan, Meteorological Society, Committee of the Investigation of Artificial Radioactivity, 1957.
The thermo-nuclear experiment and its after-effect on the atmosphere and the ocean.
Bull. Amer. Meteorol. Soc., 38(8):453-459.

Chemistry radioactivity
Japan, Prefectural University of Mie, 1956.
Studies on the radioactive marine organisms (especially Katsuwonus vagans) caused by nuclear detonation. J. Fac. Fish., Pref. Univ., Mie, 2(2):43-96.

Chemistry radioactivity
Japan, Maritime Safety Board, 1959.
The report of radio-active contamination at "Takuyo" "Satuma".
Hydrogr. Bull., Tokyo, (Publ. 981) (61):25-40.

Chemistry radioactivity
Jaworowski, Zbigniew, 1959.
Radioactive fallout measurements at Hornsund-fiorden in Spitzbergen.
Acta Geophysica Polonica, 7(2):130-133.

chemistry, radioactivity
Jelisavcic, Olga, Zvonimir Kolar, Petar Strohal, Stjepan Keckes and Stjepan Lulic 1969.
Gross beta activity in selected organisms from the North Adriatic.
Rapp. P.-v. Réun. Commn int. Explor. scient. Mer Méditerr. 19(5): 957-959.

Chemistry, radioactivity
Jurain, G., 1960
Moyens et résultats d'étude de la radioactivité due au radon dans les eaux naturelles. Geochim. et Cosmochim. Acta, 20(1): 51-82.

K

Chemistry radioactivity
Kanamori, M., T. Kuroki, T. Morita and T. Tanoue, 1957.
[V. On the treatments to decrease the radioactivity of fishing gear materials contaminated.]
Mem. Fac. Fish., Kagoshima Univ., 5:184-189.

Chemistry radioactivity
Kanamori, M., T. Morita, T. Tanoue and T. Kuroki, 1957.
[IV. On the inspection of radioactivity of sea-water in the several Southern Sea fronts and the Eastern China Sea.] Mem. Fac. Fish., Kagoshima Univ., 5:178-183.

Chemistry radioactivity
Kawabata, T., 1955.
Japan J. Med. Sci. & Biol., 8:347
Shunkostu-maru - Bikini 1954 - copepod radioactivity 1000x greater than sea water they inhabited.

Chemistry radioactivity
Ketchum, B.H., 1957.
The effects of the ecological system on the transport of elements in the sea. In: The effects of atomic radiation on oceanography and fisheries. Nat. Acad. Sci.-Nat. Res. Counc., Pub. 551:52-59.

Chemistry radioactivity
Khitrov, L.M., and K.A. Kotliarov, 1962
The abyssal gamma-radiometer and the measurement of radioactivity in the abyssal water layers of the Indian Ocean.
Okeanologiia, Akad. Nauk, SSSR, 2(2):334-345.

Chemistry radioactivity
Klocke, F.J., and H. Rahm, 1959.
No place to hide atomic wastes.
Fisheries Newsletter, 18(8):15.

Chemistry radioactivity
Koczy, F.F., 1963.
The natural radioactive series in organic material.
In: Radioecology, V. Schultz and A.W. Klement, Jr. editors, Reinhold Publishing Corp. and Amer. Inst. Biol. Sci., 611-613.

Chemistry Radioactivity
Koide, Minoru, and Edward D. Goldberg, 1965.
Uranium-234/uranium-238 ratios in sea water.
Progress in Oceanography, 3:173-177.

Chemistry radioactivity
Kopcewitz, Teodor, Roman Tomaszenko, and Jan Tomczak, 1959
La radioactivité des aerosols, des aerosols, des precipitations, de l'eau de la Vistuel et de la mer observée par les stations de l'Institut Hydrologique et Météorologique d'Etat en 1958.
Acta Geophysica Polonica, 7(2):103-129.

Chemistry radioactivity
Kospurov, G.I., and N.P. Ushakova, 1964.
Long-lived beta-activity in the atmosphere above the Atlantic Ocean in 1960. Hydrophysical investigations. (Results of the investigations of the seventh cruise of the Research Vessel "Mikhail Lomonosov"). (In Russian).
Trudy, Morsk. Gidrofiz. Inst., Akad. Nauk Ukrain. SSR, 29:13-21.

Chemistry radioactivity
Kotliarov, K.A., and L.M. Khitrov, 1964
The measurement of minute radioactivity values under field conditions. (In Russian).
Okeanologiia, Akad. Nauk, SSSR, 4(2):213-222.

Chemistry radioactivity
Kruger, Paul, Charles L. Hosler and Larry G. Davis 1962
Radiochemical study of two extratropical cyclones containing nuclear debris of different ages.
J. geophys. Res. 71(18):4257-4266.

Chemistry radioactivity
Kruger, Paul, and Albert Miller, 1966.
Transport of radioactivity in rain and air across the Trade Wind Inversion at Hawaii.
J. geophys. Res., 71(18):4243-4255.

Chemistry radioactivity
Krumholz, L.A., 1956.
Observations on the fish population of a lake contaminated by radioactive wastes.
Bull. Amer. Mus. Nat. Hist., 110(4):283-367.

Chemistry radioactivity
Krumholz, L.A., and R.F. Foster, 1957.
Ch. 9: Accumulation and retention of radioactivity from fission products and other radiomaterials by fresh-water organisms.
NAS -NRC Publ. 551: 88-95.

Chemistry radioactivity
Kurbatov, L.M., 1936.
Radioactivity of ferromanganese formations in seas and lakes of the USSR. Nature 136:871.

Chemistry radioactivity
Kuroki, T., and T. Tanoue, 1955.
II. On the radioactivity of fishes, planktons and sea-water in the sea off Kagoshima.
Mem. Fac. Fish., Kagoshima Univ., 4:143-150.

Chemistry radioactivity
Kusaka, T., 1956.
Comparison of the method of decontamination of fission products from the materials of fishing gears of the Fukuryu-maru No. 5. Research on the Effects and Influences of the Nuclear Bomb Test Explosion. Jap. Soc. Prom. Sci., 1956:1235-1243.

L

Chemistry radioactivity
Lambert, G., 1965.
Radioactivité de la mer due aux produits naturels et radioactivité artificielle amenée par les retombées.
Cahiers du C.E.R.B.O.M., Nice, 2:15-31.

Chemistry radioactivity
Lear, Donald W., Jr., and Carl H. Oppenheimer, Jr., 1962.
Oceanographic studies during Operation "Wigwam". Biological removal of radioisotopes Sr90 and Y90 from sea water by marine microorganisms.
Limnol. and Oceanogr., Suppl. to Vol. 7:xlivlxii.

Chemistry radioactivity
Libby, W.F., 1956.
Current research findings on radioactive fallout.
Proc. Nat. Acad. Sci., 42(12):945-961.

Chemistry radioactivity
Likens, Gene E., and Arthur D. Hasler, 1962.
Movements of radiosodium (Na 24) within an ice-covered lake.
Limnol. & Oceanogr., 7(1):48-56.

Chemistry radioactivity
Lockhart, L.B., Jr., 1958.
Concentrations of radioactive materials in the air during 1957.
Science 128(3332):1139.

Chemistry, radioactivity
Loveridge, B.A., 1959
Determination of radio-strontium in sea water.
A.E.R.E. Report C/M 380: 7 pp.
Abs. in Anal. Abs., 6(8): 3248.

Chemistry radioactivity
Loveridge, B.A., & A.M. Thomas, 1959
Determination of radio-ruthenium in effluent and sea water.
A.E.R.E. Report C/R 2828: 11 pp.
Abs. in Anal. Abs., 6(8): 3250.

M

Chemistry radioactivity
Machta, L., 1957.
Meteorological factors affecting spread of radioactivity.
J. Washington Acad. Sci., 47(6):169-179.

Chemistry radioactivity
Mahadevan, C., and V. Aswathanarayana, 1954.
Radioactivity of sea floor sediments off the east coast of India. Andhra Univ. Ocean. Mem. 1:36-50, 4 textfigs.

Chemistry, radioactivity,
Marcolini, B.M., 1953.
Alcuni esperimenti per l'alimentazione di cladoceri e copepodi d'acqua dolce con fosforo radioattivo (P32). Bol. Pesca, Piscicolt. e Idrobiol., n.sp., 8(2):181-200.

Chemistry radioactivity
Matsue, Y., 1956.
On the amount of the radioactive ash fallen on board the Fukuryu-maru No. 5. Research on the effect and influences of the nuclear bomb test explosions. Jap. Soc. Prom. Sci., 1956:993-996.

Chemistry radioactivity
Matsue, Y., and R. Hirano, 1956.
Accumulation of radioactivity in plankton cultured or kept in water infusions of the "Bikini" ash. Research on the Effects and Influences of the Nuclear Bomb Test Explosions.
Jap. Soc. Prom. Sci., 1956:1099-1104.

Chemistry radioactivity
Mauchline, John, Angela M. Taylor and Eric B. Ritson, 1964.
The radioecology of a beach.
Limnology and Oceanography, 9(2):187-194.

Mc

chemistry, radioactivity (artificial)
Miyake, Yasuo, 1971.
Radioactive models. In: Impingement of man on the oceans, D.W. Hood, editor, Wiley Interscience: 565-588.

Chemistry, radioactivity
Miyake Y. 1969.
Artificial radioactivity.
Annls int. geophys. Year, 46:187-190

Chemistry radioactivity
Miyake, Y., 1963
3. Artificial radioactivity in the sea. In: The Sea, M.N. Hill, Edit., Vol. 2. The composition of sea water, Interscience Publishers, New York and London, 78-87.

Chemistry radioactivity(artificial)
Miyake, Yasuo, 1963.
Artificial radioactivity in the Pacific Ocean. Radioactive tracers in oceanography. Symposium 10th Pacific Science Congress, 1961.
Union Géodes. Géophys. Int., Monogr., 20:21-30.

Chemistry radioactivity
Miyake, Yasuo, 1961.
Artificial radioactivity in oceanography. (Abstract).
Tenth Pacific Sci. Congr., Honolulu, 21 Aug.-6 Sept., 1961, Abstracts of Symposium Papers, 347.

Chemistry radioactivity
Miyake, Y., 1961
The distribution of man-made radioactivity in the North Pacific (through the summer of 1955). (Summary).
Ann. Int. Geophys. Year, 11:310-311.

Chemistry radioactivity
Miyake, Y., 1958.
The distribution of artificial radioactivity in the equatorial region of the Pacific in the summer of 1956. (Abstract).
Proc. Ninth Pacific Sci. Congr., Pacific Sci. Assoc., 1957, Oceanogr., 16:227.

Miyake, Yasuo, and Katsuko Saruhashi, 1958 radioactivity
Distribution of Man-made radioactivity in the North Pacific through summer 1955. J. Mar. Res., 17:383-389.

Chemistry radioactivity
Miyake, Y., K. Saruhashi and Y. Katsuragi, 1960
Strontium 90 in Western North Pacific surface waters.
Papers Meteorol. & Geophys., Tokyo, 11(1): 188-190.

Chemistry radioactivity
Miyake, Yasuo, Katsuko Saruhashi, Yukio Katsuragi and Teruko Kanazawa, 1962
Penetration of 90 Sr and 137 Cs in deep layers of the Pacific and vertical diffusion rate of deep water.
Repts. JEDS, Deep-Sea Res. Comm., Japan Soc. Promotion of Science, 3:141-147.
Originally published (1962):
J. Radiation Res., 3(3):141-147.
JEDS Contrib. No. 25

Chemistry radioactivity
Miyake, Y., and Y. Sugiura, 1958.
The method of measurement of radioactivity in sea water.
Pap. Meteorol. & Geophys., Tokyo, 9(1):48-50.

Chemistry radioactivity
Miyake, Y., and Y. Sugiura, 1955.
The radiochemical analysis of radio-nuclides in sea water collected near Bikini Atoll.
Pap. Meteorol. Geophys., Tokyo, 6(1):33-38.

Chemistry radioactivity
Miyake, Y., and Y. Sugiura, 1955.
The radiochemical analysis of radionuclides in sea water collected near Bikini Atoll.
Rec. Ocean. Wks., Japan, 2(2):108-112.

Chemistry radioactivity
Miyake, Y., Y. Sugiura and K. Kamada, 1955.
On the artificial radioactivity in the sea near Japan. Pap. Meteorol., Geophys., Tokyo, 6(1):90-92.

Chemistry radioactivity (data)
Miyake, Y., Y. Sugiura and K. Kameda, 1955.
On the distribution of radioactivity in the sea around Bikini Atoll in June 1954.
Pap. Met. Geophys., Tokyo, 5(3/4):253-262.

Chemistry radioactivity
Miyake, Y., Y. Sugiura and K. Kamada, 1955.
On the distribution of the radioactivity in the sea around Bikini Atoll in June 1954.
Rec. Ocean. Wks., Japan, 2(1):34-44, 8 textfigs.

Chemistry radioactivity
Miyoshi, H., S. Hori and S. Yoshida, 1955.
Drift and diffusion of radiologically contaminated water in the ocean. Rec. Ocean. Wks., Japan, 2(2):30-36.

Chemistry radioactivity
Mori, T., and M. Saiki, 1956.
Studies on the radioactive material in the radiologically contaminated fishes. Research on the Effects and Influence of the Nuclear Bomb Test Explosions. Jap. Soc. Prom. Sci., :889-894.

Chemistry radioactivity
Morita, T., K. Saito and T. Kenka, 1955.
[III. On the inspection of radioactivity in the Equatorial Sea and the Coral Sea.]
Mem. Fac. Fish., Kagoshima Univ., 4:151-156.

Chemistry Radioactivity
Morre, Jean, et Jean Barret, 1963.
Dosage du potassium dans divers produits de pêcheries en vue de la mésure de la radioactivité
Rev. Trav. Inst. Pêches Marit., 27(2):235-240.

Chemistry radioactivity
Munk, W.H., G.C. Ewing, and R.R. Revelle, 1949
Diffusion in Bikini Lagoon. Trans. Am. Geophys. Un., 30(1):59-66; 9 text figs.

N

Chemistry radioactivity
Nagahara, M., and E. Noguchi, 1959
Radiological survey of the Japan Sea in 1957 and 1958. Ann. Rept. Japan Sea Res. Fish. Res. Lab., (5): 177-183.

Chemistry radioacitvity
Nagaya, Y., M. Shiozaki, and Y. Seto, 1964.
Radiological survey of sea water of adjacent Sea of Japan in 1963. (In Japanese; English abstract).
Hydrogr. Bull. Tokyo, (Publ. No. 981), No. 78:63-67.

Chemistry radioactivity
Nagaya, Y., M. Shiozaki and Y. Seto, 1964.
On the radiological survey of harbours during 1960-1962. (In Japanese; English abstract).
Hydrogr. Bull., Tokyo, (Publ. No. 981), No. 78:53-62.

Chemistry radioactivity
Nagaya, Y., M. Siozaki, Y. Seto and S. Isizone, 1964.
On the unusual radioactive contamination of sea water collected in spring of 1963 in the Japan Sea. (In Japanese; English abstract).
Hydrogr. Bull., (Publ. No. 981), Tokyo, No. 76: 31-33.

Chemistry, radioactivity
National Academy of Science-National Research Council, 1957.
The effects of atomic radiation on oceanography and fisheries. NAS-NRC Publ. 551:137 pp.
Report of the Committee on Effects of Atomic radiation on Oceanography and Fisheries.

Chemistry radioactivity
Nelepo, B.A., 1962
The problems of radioactivity. The Xth Pacific Science Congress.
Okeanologiia, Akad. Nauk, SSSR, 2(3):457-463.

Chemistry radioactivity
Nelepo, B.A., 1961.
Gamma ray spectrometric measurements of artificial radioactivity in the upper layer of the ocean. (Abstract)
Tenth Pacific Sci. Congr., Honolulu, 21 Aug.-6 Sept., 1961, Abstracts of Symposium Papers, 347-348.

Chemistry radioactivity
Nelepo, B.A., 1960
[An investigation of the radioactivity of water in the Atlantic] Doklady Akad. Nauk, USSR, 134 (4): 810-811.

Chemistry radioactivity
Nelepo, B.A., 1960.
[Direct methods for the determination of radioactivity in oceanic waters in the Antarctic region of the Pacific Ocean.]
Trudy Okeanol., Komissii, Akad. Nauk, SSSR, 10(1):141-143.

Chemistry radioactivity
Nelepo, B.A., 1960
Gamma-spectrometric measurements of the radioactivity of the waters of the Atlantic Ocean and some results of the total radioactivity of oceanic waters in the Antarctic sector of the Pacific Ocean.
Moscow U. Vestnik, Ser. 3, Fiz. & Astron., USSR, 15(5): 36-46.
Abs. in Techn. Transl., 5(12): 675.

Chemistry radioactivity
Niino, Hiroshi, 1960
Study on the natural radioactivity of bottom deposits. II.
Rec. Oceanogr. Wks., Japan, Spec. No. 4: 165-170.

Chemistry radioactivity
Niino, Hiroshi, 1959
On the natural radioactivity of the marine sediments (1.).
Rec. Oceanogr. Wks., Japan, Spec. No. 3: 163-164.

O

Chemistry radioactivity
Osterberg, Charles, 1965.
Radioactivity from the Columbia River.
Ocean Sci. and Ocean Eng., Mar. Techn. Soc.,-Amer. Soc. Limnol. Oceanogr., 2:968-979.

chemistry, radioactivity
*Osterberg, C.L., N. Cutshall, V. Johnson, J. Cronin, D. Jennings, and L. Frederick, 1966.
Some non-biological aspects of Columbia River radioactivity.
Disposal of radioactive wastes into seas, oceans, and surface waters, Int. Atomic Energy Agency, 321-335.
Also in: Coll. Repr., Dep. Oceanogr., Oregon State Univ., 5.

Chemistry radioactivity
Osterberg, Charles, W.G. Pearcy and Herbert Curl, Jr., 1964
Radioactivity and its relationship to oceanic food chains.
J. Mar. Res., 22(1):2-12.

P

Chemistry, radioactivity
Park, Kelho, Marilyn J. George, Yasuo Miyake, Katsuko Saruhashi, Yukio Katsuragi and Teruko Kanazawa, 1964.
Strontium- 90 and caesium- 137 in Columbia River plume, July 1964.
Nature, 208(5015):1084-1085.

Chemistry radioactivity
Penzar, B., and I. Penzar, 1960
[The distribution of global radiation over Jugoslavia and the Adriatic Sea.]
Hidrografski Godisnjak, Split, 1959(30):151-171.

Chemistry, radioactivity
Pettersson, H., 1954.
The radioactivity of the ocean. Science 119:584.

Chemistry radioactivity
Picotti, M., 1963.
Ouverture de la séance speciale de travail sur la radioactivité.
Rapp. Proc. Verb., Réunions, Comm. Int. Expl. Sci., Mer Méditerranée, Monaco, 17(3):937-938.

Chemistry radioactivity
Piggot, C.S., and W.D. Urry, 1942.
Radioactivity of ocean sediments. IV. The radium content of sediments of the Cayman Trough.
Amer. J. Sci. 240:1-12.

chemistry radioactivity
Policarpov, G.C., 1961.
[On the stability of accumulation coefficients of Sr 90, Y 91 and Ce 144 in sea algae.]
Doklady Akad. Nauk, SSSR, 140(5):1192-1194

chemistry radioactivity
Polikarpov, G.G., and V.P. Parchevsky, 1961
[The radioactivity in algae of the Adriatic and Black Seas.]
Okeanologiia, Akad. Nauk, SSSR, (2):338-339.

chemistry radioactivity
Popov, N.I., 1964.
Natural radioactivity of ocean waters. (In Russian).
Okeanologiia, Akad. Nauk, SSSR, 4(2):223-231.

chemistry radioactivity
Proctor, Charles M., Emanuel Papadopulos, and Ralph H. Firminhac, 1962
Gamma scintillation probe for field use and measurements of radiation background in Puget Sound.
Limnol. and Oceanogr., 7(3):280-286.

Q

R

chemistry radioactivity
Rae, K.M., and R.G. Bader, 1959.
Clay-mineral sediments as a reservoir for radioactive materials in the sea.
Proc. Gulf and Caribbean Fish. Inst., 12th Sess., Nov. 1959:55-61.

chemistry radioactivity
Rafter, T.A., and G.J. Fergusson, 1957.
The atom bomb effect. Recent increase in the 14C content of the atmosphere, biosphere and the surface waters of the oceans.
New Zealand J. Sci. & Tech., B, 38(8):871-883.

chemistry radioactivity
Renn, C.E., 1957.
Ch. 1. Physical and chemical properties of wastes produced by atomic power industry. NAS-NRC Publ. 551:26-27.

chemistry radioactivity
Revelle, R., et al, 1959
Considerations on the disposal of radioactive wastes from nuclear-powered ships into the marine environment.
Nat. Acad. Sci.-Nat. Res. Coun. Pub. 658: 52 pp.

chemistry radioactivity
Revelle, Roger, and M.B. Schaefer, 1958
Oceanic research needed for safe disposal of radioactive wastes in the sea.
Proc. 2nd UN Geneva Conf. on the Peaceful Uses of Atomic Energy, Sept. 1-15, 1958, 18:364-370.

chemistry radioactivity
Revelle, R., and M.B. Schaefer, 1957.
General considerations concerning the ocean as a receptacle for artificially radioactive materials. The effects of atomic radiation on oceanography and fisheries. Nat. Acad. Sci.-Nat. Res. Counc., Publ., 551:1-25.

chemistry, radioactivity (artificial)
Rice, Theodore R. and Douglas A. Wolfe, 1971.
Radioactivity - chemical and biological aspects.
In: Impingement of man on the oceans, D.W. Hood, editor, Wiley Interscience: 325-379.

chemistry radioactivity
Riel, Gordon K., 1962
Radioactivity measurements in the ocean. (Abstract).
J. Geophys. Res., 67(9):3591.

chemistry, radioactivity
Rivière, André, et Solange Vernhet, 1965.
Études littorales. Contributions a l'étude des rivages du Golfe du Lion. Signification sédimentologique des radioactivités naturelles.
Cah. océanogr., 18(10):857-900.

chemistry radioactivity
Rohde, Karl-Heinz, 1963.
Vorläufige Ergebnisse von Radioaktivitätsmessungen in der westlichen Ostsee.
Beiträge z. Meereskunde, (10):27-28.

chemistry, radioactivity
Romankevich, E.A., P.L. Bezrukov, V.I. Baranov and L.A. Khristianova 1966.
Stratigraphy and absolute age of deep-sea sediments in the western Pacific. (In Russian; English abstract).
Reg. Issled. Meghd. Geofiz. Proekt. Okeanol. Meghd. Geofiz. Komitet, Presid. Akad. Nauk SSSR, No. 14: 165 pp.

chemistry radioactivity
Rona, E., 1944.
Radioactivity of sea water. Yearbook for 1943, Am. Phil. Soc.:136-138.

chemistry radioactivity
Rosenthal, H.L., 1957.
The metabolism of strontium-90 and calcium-45 by Lebistes. Biol. Bull., 113(3):442-450.

S

chemistry radioactivity
Saiki, M., K. Shirai, S. Ohno and T. Mori, 1957.
Studies on the radioelements in the radioactive contaminating fishs II. On skipjacks caught at the Pacific Ocean in 1956 (Part 1).
Bull. Jap. Soc. Sci. Fish., 22(10):645-650.

chemistry radioactivity
Saiki, Masamichi, Susumu Ohno, and Takajiro Mori, 1956.
Radioactivity and its decay of the plankton, seaweed, sea urchin, cuttle-fish and stomach contents of fish.
Repts. Res. Influence of the Nuclear Bomb Test Explosions, Ministry Health & Welfare, 1-4.
Abstr in:
Rec. Res. Fac. Agric., Tokyo Univ., Mar. 1958, 7(92):62.

chemistry radioactivity
Saiki, M., S. Yoshina, R. Ichikawa, Y. Hiyama and T. Mori, 1956.
Studies on the radioactivity of fishes caught from the Pacific Ocean in 1954. Research on the Effects and Influences of the Nuclear Bomb Test Explosions Jap. Soc. Prom. Sci., 1956:825-839.

chemistry radioactivity
Saito, Kaname, 1961
Radiological survey in the Indian Ocean in 1961.
Memoirs, Fac. Fish., Kagoshima Univ., 10: 9-14.

chemistry radioactivity
Saito, K., and M. Sameshima, 1955.
[1. Studies on the radiologically contaminated fishes caught at Kagoshima Sea Region.]
Mem. Fac. Fish., Kagoshima Univ., 4:124-142.

chemistry radioactivity
Saito, Kaname, Muneo Sameshima and Tomio Hemmi, 1960.
XI. Studies on the radiological contamination of sea water, planktons and fishes in the western region of the north equatorial current during the period from 1958 to 1959.
Mem. Fac. Fish., Kagoshima Univ., 8: 181-193.

chemistry radioactivity
Saito, K., M. Sameshima and T. Tanaka, 1958.
[Studies on the uptake of S35 by marine algae.]
Mem. Fac. Fish., Kagoshima Univ., 6:153-158.

chemistry radioactivity
Sameshima, Muneo, and Kaname Saito, 1961
Studies on the radiological contamination of fishes, sea water and plankton in the southern region of Kagoshima prefecture during the period from November 1959 to December 1960. (In Japanese).
Memoirs, Fac. Fish., Kagoshima, Univ., 10: 15-22.

chemistry radioactivity
Sastri, C.S., and Sivaramakrishnan, 1968.
Study of the radioactivity of separated minerals of beach sands of Manavalakurichi, Madras State.
Current Sci., 37(19):550.

chemistry radioactivity
Schaefer, M.B., 1957.
Large-scale biological experiments using radioactive tracers. The effects of atomic radiation on oceanography and fisheries. Nat. Acad. Sci.-Nat. Res. Counc., Publ. 551:133-137.

chemistry, radioactivity
Schofield, J.C. 1967.
Origin of radioactivity at Niue Island.
N.Z. Jl Geol. Geophys. 10(6):1362-1371

chemistry radioactivity
Schreiber, B., 1971
Radioecology research in Taranto Gulf (Taranto III). Appendix.
CNEN Rept. RT/BIO(70)-11, M. Bernhard, editor: 101-109

chemistry, radioactivity
Schreiber, B. 1965.
Études sur la radioactivité du plancton et des sédiments côtiers de la mer ligurienne.
Rapp. P.-v. Réun. Commn int. Explor. scient. Mer Méditerr. 18(3): 883-892.

radioactivity
Schreiber, B., and L. Tassi Pelati, 1965.
Beta and gamma radioactivity measurements, Sr90 determination in plankton and sea water samples collected during the oceanographic campaign of M/N "Bannock" - summer 1964.
Energia Nucleare, 12(6):330-332.

chemistry radioactivity
Schreiber, B., L. Pelati, E. Cerrai e C. Triulzi, 1964.
Gross beta radioactivity of litoral sediments of the Ligurian Sea.
Energia Nucleare, 11(11):616-624.

chemistry, radioactivity
Servant, Jean, 1966.
La radioactivité de l'eau de mer.
Cahier Océanogr., C.C.O.E.C., 18(4):277-318.

chemistry, radioactivity

Seymour, A.H., 1971.
Introduction. Radioactivity in the marine environment, U.S. Nat. Acad. Sci., 1971: 1-5.

Chemistry radioactivity

Seymour, A.H., et al., 1957.
Survey of radioactivity in the sea and in pelagic marine life west of the Marshall Islands Sept. 1-20, 1956. Appl. Fish. Lab., Univ. Washington, U.A. AEC Rept. UWFL-47:57 pp.

Chemistry radioactivity

Shannon, L.V., 1969
The alpha-activity of marine plankton.
Investl Rep. Div. Sea Fish. S Afr. 68:1-35

Chemistry radioactivity

Shannon, L.V., 1967.
Radioactivity in relation to oceanography.
Fish. Bull., Misc. Contr. Oceanogr. mar. Biol., S. Afr., 4:15-17.

Chemistry radioactivity

Sherstnev, A.I., I.I. Baksheeva, G.I. Dolbilina, and V.V. Dubinkina, 1962.
Radiological characteristics of marine organisms and their environment in the equatorial Atlantic (west coast of Africa). (In Russian).
Trudy, Baltiisk, Nauchno-Issled. Inst. Morsk. Ribn. Khoz. i Okeanogr., (BALTNIRO) 8:238-241.

Chemistry, radioactivity

Sherstnev, A.I. and G.I. Dolbilina 1963.
The examination of the radioactivity of sea and oceanic water. (In Russian).
Atlant. nauchno-issled. Inst. ribn. khoz. i Okeanogr. (ATLANTNIRO), Trudy 10: 217-220.

chemistry, radioactivity

Sherstnev. A.I., T.I. Gorshkova and N.A. Kuznetsova, 1963.
The radioactivity of the bottom sediments in the Baltic Sea, its bays, and the characteristics of some elements of their composition. (In Russian).
Atlantich. Nauchno-Issled. Inst. Ribn. Khoz. i Okeanogr. (ATLANTNIRO), Trudy, 10:209-212.

Chemistry radioactivity

Shiozaki, Masaru, Yoshiro Seto and Ryoji Higano 1964.
Oceanographic investigations of radioactive contamination of sea water at the Mid Pacific area.
J. Oceanogr. Soc. Japan, 20(2):81-88.

Chemistry radioactivity

Shirai, K., M. Saiki and S. Ohno, 1957.
Studies on the radioelements in the radioactive contaminating fishes. III. On skipjacks caught at the Pacific Ocean in 1956 (Part 2). On the presence of Cd 113M.
Bull. Jap. Soc. Sci. Fish., 22(10):651-655.

Chemistry radioactivity

Sievert, R.M., 1956.
Record of gamma radiation from the ground and beta radiation from radioactive debris in Sweden 1950-1955, Pt. 1. Tellus, 8(2):117-126.

Chemistry radioactivity

Simano, T., 1959.
Determination of gamma ray activity in sea water by a dipping scintillation counter.
Hydrogr. Bull., Tokyo, (Publ. 981), 61:41-44.

chemistry radioactivity

Sisoev, N.N., and Y.I. Kiriljuk, 1962
Some data on the radioactivity of the Pacific (In Russian).
Okeanologiia, Akad. Nauk SSSR, 2(4):743-744.

Chemistry radioactivity

Skorka, S., 1958.
Breiteneffekt der Nukleonen- und Mesonenkomponente der Ultrastrahlung in Meereshöhe im Indischen und Atlantischen Ozean.
Zeits. f. Physik, 151(5):630-645.

Chemistry radioactivity

Sorokin, J.I., 1960.
[Self-absorption of C14 radiation during counts on dried invertebrates.]
Trudy Inst. Biol. Bodo., Akad. Nauk, SSSR, 3(6): 103-105.

Chemistry radioactivity

Starik, I. Ye., D.S. Nikolayev, Yu. V. Kuznetsov, and V.K. Legin, 1961.
The radioactivity of Black Sea sediments. (In Russian)
Doklady, Akad. Nauk, SSSR, 139(6):1456-1459.

Translation:
Consultants Bureau for Amer. Geol. Inst. (Earth Sciences Section only), 139(1-6): 704-707.

Chemistry radioactivity

Starik, I.E., D.S. Nikolaev, Iu. V. Kuznetsov and V.K. Legin, 1961
[A comparative study of the radioactivity of Azov and Black Sea sediments.]
Doklady Akad. Nauk, SSSR, 139(2):456-460.

Chemistry, radioactivity

Strohal Petar, Stjepan Lulić, Zvonimir Kolar, Olga Jelisavcić and Stjepan Keckes 1969.
Gamma-spectrometric analysis of some North Adriatic organisms.
Rapp. P.-v. Réun. Commn int. Explor. scient. Mer Méditerr. 19(5): 953-955.

Chemistry radioactivity

Suyehiro, Y., and T. Hibiya, 1956.
Effects of radioactive materials upon blood-pictures of fish. Research on the effects and influences of the nuclear bomb test explosions.
Jap. Soc. Prom. Sci., 1231-1232.

T

radioactivity

Tage, N., 1956.
Preliminary studies on the removal of radioactive substances from polluted waters by biological methods. On the capacity of water bacteria for adsorption of fission products. Research on the effects and influences of nuclear bomb test explosions. Jap. Soc. Prom. Sci., 1956:1245-1250.

Chemistry radioactivity

Takenouti, Y., 1960

X. Oceanography. Japanese Contrib. I.G.Y., 1957/8, 2: 124-145.

Chemistry radioactivity

Takeuchi, Osamu, Hanato Tsuruga and Tadao Nitta, 1960.
[Qualitative radiochemical study of water in the Seto-Naikai.]
Bull. Nakai Reg. Fish. Res. Lab., Fish. Agency No. 13:39-41.

Chemistry radioactivity, sources of

Templeton, W.L., 1965
Ecological aspects of the disposal of radioactive wastes to the sea.
In: Ecology and the Industrial Society, Fifth Symposium of the British Ecological Society, Blackwell, Oxford, 65-97.

Chemistry radioactivity

Thuronyi, G., 1956.
Annotated bibliography on natural and artificial radioactivity in the atmosphere.
Met. Abstr. Bibl., A.M.S., 7(5):616-657.

Chemistry radioactivity (artificial)

Tkachenko, V.N. and S.A. Patin, 1971.
Artificial radioactivity of some hydrobionts in the Southern Ocean. (In Russian; English abstract). Okeanologiia 11(4): 705-710.

Chemistry radioactivity

Tomiyama, Tetuo, and Kunio Kobayashi, 1962
Direct uptake of radioisotopes by fish.
Proc. Ninth Pacific Sci. Congr., Pacific Sci. Assoc., 1957, 16(Oceanogr.):159-166.

Chemistry radioactivity

Tozawa, H. Y. Yokue and K. Amano, 1957.
[Studies on the radioactivity in certain pelagic fish caught in 1956. (1)]
Bull. Jap. Soc. Sci. Fish., 23(6):335-341.

U

Chemistry, radioactivity

Unni C.K. 1967.
Natural radioactivity of marine algae.
In: Proc. Seminar, Sea, Salt and Plants V. Krishnamurthy, editor, Bhavnagar, India, 264-272.

Chemistry, radioactivity

Urry, W. D., 1949.
Significance of radioactivity in geophysics - thermal history of the earth. Trans. Am. Geophys. Union 30(2):171-180, 6 textfigs.

Chemistry, radioactivity

Urry, W. D., 1949.
Radioactivity of ocean sediments. VI. Concentrations of the radioelements in marine sediments of the southern hemisphere. Amer. J. Sci., 247 (4):257-275, 4 textfigs.

Chemistry, radioactivity

Urry, W. D., 1948.
Radioactivity of ocean sediments. VII Rate of deposition of deep sea sediments. J. Mar. Res. 7(3):618-634, 6 textfigs.

Chemistry, radioactivity

Urry, W.D., and C.S. Piggot, 1942.
Radioactivity of ocean sediments. V. Concentration of the radioelements and their significance in red clay. Amer. J. Sci. 240:93-103.

Chemistry radioactivity

U.S., Nat. Acad. Sci., Nat. Res. Counc., Committee on Oceanography, 1959.
Oceanography, 1960-1970. 5. Artificial radioactivity in the marine environment.

Abstr. No. 2-1308 in GeoScience Abstr., 2(5):62.

V

Chemistry radioactivity

Vernhet, Solange, 1966.
Signification sédimentologique de la radioactivité naturelle des plages du Golfe du Lion.
Comptes Rendus, Acad. Sci. Paris, 262(4):440-443.

W

Chemistry, radioactivity

Waldichuk, M., 1961.
Fisheries off the coast of British Columbia in relation to any future sea disposal of radioactive wastes.
Fish. Res. Bd., Canada, Manuscript Rept. Ser., (Oceanogr. & Limnol), No. 86:31 pp.

Chemistry radioactivity

Watanabe, Nobuo, Toshiyuki Hirano, Rinnosuke Fukai, Kozi Matsumoto and Fumiko Shiokawa, 1957.
A preliminary report on the oceanographic survey in the "Kuroshio" area, south of Honshu, June-July 1955.
Rec. Oceanogr. Wks., Japan, (Suppl.):197-

Chemistry radioactivity
Wooster, W.S., and B.H. Ketchum, 1957.
Transport and dispersal of radioactive elements in the sea. In: The effects of atomic radiation on oceanography and fisheries.
Nat. Acad. Sci.-Nat. Res. Counc., Publ., 551: 43-51.

XYZ

Chemistry radioactivity
Zaitsev, Yu. P., G.G. Polikarpov, 1964
Problems of radioecology of hyponeuston. (In Russian).
Okeanologiia, Akad. Nauk, SSSR, 4(3):423-430.

Radioactivity accumulation + concentration

See also: Chemicals, Concentration of (by organisms)

Radioactivity in air

Chemistry radioactivity(air)
de Santis, L., 1961
Prime misure di radioattivita naturele dell' aria a Bari.
Boll. Geofis. Teorica ed Applicata, 3(9):1-16.

Chemistry radioactivity(air)
Fontan, J., D. Blanc et A. bouville, 1962.
Mesure de la radioactivité naturelle de l'air (radon et thoron) au-dessus de l'Océan Atlantique.
J. Mech. Phys. Atmos., (2) 4(15):107-113.

Chemistry radioactivity(air)
Garrigue, M., 1960
Sur la radioactivité de l'Atmosphère. La Météorologie, No. 58: 303-311.

Chemistry Radioactivity (air)
Parlier, Bruno, 1960.
La côte de Terre Adelie, Radioactivité de l'air et des precipitations.
Exped. Polaires Francaises, Publ., No. 218:41-44.

Chemistry radioactivity(air)
Popova, N.M., and G.V. Svinukhov, 1970.
Radioactivity of the atmosphere over the oceans in January-March of 1969 and its connection with aerosynoptic conditions. (In Russian; English abstract).
Met. Gidrol., 1970 (9): 62-67.

Radioactivity, control of

Chemistry Radioactivity, control of methods, radioactivity.
Kiefer, H., and R. Maushart, 1961.
Möglichkeiten zur Überwachung der Radioaktivität des Meereswassers.
Deutsche Hydrogr. Zeits., 14(1):11-16.

radioactivity (data)

Chemistry radioactivity (data only)
Finland, Finnish Meteorological Office, 1964.
Observations of radioactivity for the year 1963.
Radioaktiivisuushavaintoja (Observ. on Radioactivity), No. 3:47 pp.

Chemistry, radioactivity (data only)
Japan, Hokkaido University, Faculty of Fisheries 1961.
Data record of oceanographic observations and exploratory fishing, No. 5: 391 pp.

Chemistry radioactivity (data only)
Japan, Hokkaido University, Faculty of Fisheries, 1957.
Data record of oceanographic observations and exploratory fishing, No. 1:247 pp.

Radioactivity, effect of

chemistry, radioactivity, effect of
Chipman, Walter 1965.
Remarks on biological problems in relation to marine radioactivity.
Rapp. P.-v. Réun. Comm. int. Explor. scient. Mer Méditerr. 18(3): 857-859.

Chemistry radioactivity, effect of
de Loyola e Silva, Jayme, 1965.
Marine isopods from highly radioactive places (Guarapari, Espiritu Santo). (Abstract).
Anais Acad. bras. Cienc., 37 (Supl.):259.

Chemistry radioactivity, effect of
Donaldson, Lauren R., and R.E. Foster, 1957.
Ch. 10. Effects of radiation on aquatic organisms
NAS-NRC Publ. 551:96-102.

Chemistry radioactivity, effect of
Federov, Anatol F., 1965.
Radiation doses for some types of marine biota under present-day conditions.
Bull. Inst. Océanogr., Monaco, 64(1334):28 pp.

chemistry, radioactivity, effect of
Foster, R.F., I.L. Ophel and A. Preston, 1971.
Evaluation of human radiation exposure. Radioactivity in the marine environment, U.S. Nat. Acad. Sci., 1971: 240-260.

chemistry radioactivity (effect of)
Fukai, Rinnosuke 1965.
Remarks on the chemical problems in relation to the marine radioactivity.
Rapp. P.-v. Réun. Commn int. Explor. scient. Mer Méditerr. 18(3): 861-863.

Chemistry radioactivity, effect of
Grosch, Daniel S., 1962
The survival of Artemia populations in radioactive sea water.
Biol. Bull., 123(2):302-316.

Chemistry radioactivity, effect of
Hibiya, T., and T. Yagi, 1957.
Effects of fission materials upon the development of aquatic animals. Research on the Effects and Influences of the Nuclear Bomb Tests. Jap. Soc. Prom. Sci., 1956:1219-1224.

Chemistry radioactivity, effects of
Hiyama, Y., and R. Ichikawa, 1956.
Movement of fishing grounds where contaminated tuna were caught. Research on the Effects and Influences of the Nuclear Bomb Tests. Jap. Soc. Proc. Sci., 1956:1956-1079.

Chemistry Radioactivity, effect of
Rees, George H., 1962.
Effects of gamma radiation on two decapod crustaceans, Palaemonetes pugio and Uca pugnax.
Chesapeake Science, 3(1):29-34.

Chemistry, radioactivity, effect of
Saito, K., and M. Sameshima, 1957.
[Studies on the contamination of marine algae by fission products.]
Mem. Fac. Fish., Kagoshima Univ., 5:196-204.

Chemistry radioactivity, effect of
Takasa, K., and J. Nishimoto, 1957.
[II. The behavior of fission product for the fish meat.] Mem. Fac. Fish., Kagoshima Univ., 5:190-195

chemistry, radioactivity, effect of
Templeton, W.L., R.E. Nakatani, and E.E. Held, 1971.
Radiation effects. Radioactivity in the marine environment, U.S. Nat. Acad. Sci., 1971: 223-239.

Chemistry radioactivity, effect of
White, John C., and Joseph W. Angelovic, 1966.
Tolerances of several marine species to Co 60 irradiation.
Chesapeake Science, 7(1):36-39.

Chemistry radioactivity of organisms
Cmpion, J., 1968.
Relation entre les différents modes de conservation des organismes marins et leur radioactivité naturelle.
Revl Trav. Inst. (scient. tech.) Pêch. marit. 45(1): 71-78

radioactivity of plankton

Chemistry, radioactivity of plankton
Chesselet, R., et Cl. Lalou, 1965.
Recherches recentes de la radioactivité du plancton et du détritus organique. 1. Étude de la radioactivité.
Cahiers du C.E.R.B.O.M., Nice, 2:67-85.

Chemistry, radioactivity of plankton
Schreiber, B., and L. Tassi Pelati, 1965.
Beta and gamma radioactivity measurements, Sr90 determination in plankton and sea water samples collected during the oceanographic campaign of M/N"Bannock" - summer 1964.
Energia Nucleare, 12(6):330-332.

radioactivity in rain

Chemistry radioactivity in rain
Miyake, Y., 1955.
The artificial radioactivity in rain water observed in Japan, from Autumn 1954 to Spring 1955
Pap. Meteorol. Geophys., Tokyo, 6(1):26-32.

radioactivity, safety

Chemistry radioactivity, safety
Pawlikiewicz, J.M., 1961
[Safety aspect of marine nuclear propulsion.]
Nukleonika, 6(6): 423-432.
Abs. in Pub. Health Eng. Abs., 41(11): 25

chemistry, radioactivity (sources of)
Joseph, A.B., P.F. Gustafson, I.R. Russell, E.A. Schuert, H.L. Volchok and A. Tamplin, 1971.
Sources of radioactivity and their characteristics. Radioactivity in the marine environment, U.S. Nat. Acad. Sci., 1971: 6-41.

radiocarbon SEE ALSO: Carbon 14

Chemistry, radiocarbon
See also: chemistry, carbon-14, etc.

Chemistry radiocarbon
Bien, G.S., N.W. Rakestraw and H.E. Suess, 1966.
Radiocarbon in the Pacific and Indian Oceans and its relation to deep water movements.
Limnol. and Oceanogr., Redfield Vol., Suppl. to 10:25-R37.

Chemistry carbon (radio)
Bien, G.S., N.W. Rakestraw and H.E. Suess, 1963.
Natural radiocarbon in the Pacific and Indian Ocean.
Symposium on oceanic mixing. In: Nuclear Geophysics, Proc. Conf., Woods Hole, Massachusetts, June 709, 1962.
NAS-NRC, Publ., No. 1075:152-160.

Also: Nuclear Sci. Ser. Rept., No. 38.

Chemistry radiocarbon
Bien, G.S., N.W. Rakestraw and H. E. Suess, 1960.
Radiocarbon concentration in Pacific Ocean water
Tellus, 12(4):436-443.

Chemistry radiocarbon (data)
Bien, George S., Norris W. Rakestraw, and Hans E. Suess, 1963.
Radiocarbon dating of deep water of the Pacific and Indian oceans.
Bull. Inst. Oceanogr., Monaco, 61(1278):1-16.

Chemistry carbon (radio)
Broecker, W.S., 1963.
C14/C12 ratios in surface ocean water.
Symposium on oceanic mixing. In: Nuclear Geophysics, Proc. Conf., Woods Hole, Massachusetts, June 7-9, 1962.
NAS-NRC, Publ., No. 1075:138-149.

Also: Nuclear Sci. Ser. Rept., No. 38.

Broecker, Wallace S., Robert Gerard, Maurice Ewing, and Bruce C. Heezen, 1960 radiocarbon
Natural radiocarbon in the ocean. J. Geophys. Res., 65(9):2903-2932.

Chemistry carbon (radio)
Broecker, W.S., and G.G. Rocco, 1963.
Direct comparison of radiocarbon and fallout measurements in sea water.
Symposium on oceanic mixing. In Nuclear Geophysics Proc. Conf., Woods Hole, Massachusetts, June 7-9, 1962.
NAS-NRC, Publ., No. 1075:150-151.

Also, Nuclear Sci. Ser. Rept., No. 38.

Chemistry radiocarbon
Craig, H., 1969.
Abyssal carbon and radiocarbon in the Pacific.
J. geophys. Res., 74(23): 5491-5506.

Chemistry, radiocarbon
Craig, H., 1957.
The natural distribution of radiocarbon and the exchange time of carbon dioxide between atmosphere and sea. Tellus 9(1):1-17.

Chemistry, radiocarbon
Fairhall, A.W., R.W. Buddemeier, I.C. Yang and A.W. Young 1969.
Radiocarbon from nuclear testing and air-sea exchange of CO_2
Antarctic Jl U.S. 4(5):184-185.

Chemistry radiocarbon
Gerner, D.M., 1958.
A radiocarbon profile in the Tasman Sea.
Nature 182(4633):466-468.

chemistry, radiocarbon distribution
Holland, William R. 1971.
Ocean tracer distributions. 1. A preliminary numerical experiment.
Tellus 23 (4/5): 371-392

Chemistry radiocarbon
Hubbs, C.L., G.S. Bien, & H.E. Suess, 1962
La Jolla natural radiocarbon measurements II.
Amer. J. Sci., Suppl.; Radiocarbon, 4: 239-249

Chemistry, radiocarbon
Nydal, Reidar 1968.
Further investigation on the transfer of radiocarbon in nature.
J. geophys. Res. 73(12): 3617-3635.

Chemistry radiocarbon
Rusnak, Gene A., Albert L. Bowman and H. Göte Östlund, 1964
Miami natural radiocarbon measurements. III.
Radiocarbon, 6:208-214.

Chemistry radio-carbon
Skirrow, Geoffrey, 1965.
The dissolved gases - carbon dioxide.
In: Chemical oceanography, J.P. Riley and G. Skirrow, editors, Academic Press, 1:227-322.

chemistry, radiocarbon
Suess, H.E. and Edward Goldberg, 1971.
Comments on paper by H. Craig 'Abyssal carbon and radiocarbon in the Pacific'. J. geophys. Res. 76(21): 5131-5132.

Chemistry radiocarbon
*Szabo, B.J., F.F. Koczy and G. Östlund, 1967.
Radium and radiocarbon in Caribbean waters.
Earth Planet. Sci. Letters, 3(1):51-61.

chemistry, radiocarbon content
*Taylor, R.E., and Rainer Berger, 1967.
Radiocarbon content of marine shells from the Pacific coasts of Central and South America.
Science, 158(3805):1180-1182.

Chemistry radioactive carbon
Vinberg, G.G., 1954.
[Radio-active carbon and the photosynthesis of marine plankton.] Priroda (5):92.

Chemistry, radiocarbon
Williams, P.M, H. Oeschger, and P. Kinney 1969.
Natural radiocarbon activity of the dissolved organic carbon in the north-east Pacific Ocean.
Nature, Lond. 224 (5216): 256-258.

Chemistry radiocarbon
*Young, James A., and A.W. Fairhall, 1968.
Radiocarbon from nuclear weapons tests.
J. geophys. Res., 73(4):1185-1200.

Chemistry, radiocarbon dating
See also: age determination

radioelements

Chemistry, radioelements
Chesselet, Roger, et Claude Lalou, 1965.
Rôle du "détritus" dans la fixation de radio-éléments dans le milieu marin.
C.R., Acad. Sci., Paris, 260(4):1225-1227.

Chemistry radioelements
Naughton, John J., and I. Lynus Barnes, 1965.
Geochemical studies of Hawaiian rocks related to the study of the upper mantle.
Pacific Science, 19(3): 287-290.

Chemistry radio-elements
Riley, J.P., 1965.
Analytical chemistry of sea water.
In: Chemical oceanography, J.P. Riley and G. Skirrow, editors, Academic Press, 2:295-424.

radio-isotopes (use of) etc.

Chemistry Radioisotopes
Broecker, Wallace, 1962.
Ra226 to U238 ratios in marine shells of known radiocarbon age. (Abstract).
J. Geophys. Res., 67(4):1629-1630.

Chemistry, radioisotopes
Duursma, E.K., 1966.
Molecular diffusion of radioisotopes in interstitial water of sediments.
Disposal of Radioactive Wastes into Seas, Oceans and Surface Water, IAEA, 355-371. (SM-72/20)

Chemistry, Radioisotopes
Duursma, E.K. and C.J. Bosch, 1970.
Theoretical, experimental and field studies concerning diffusion of radioisotopes in sediments and suspended particles of the sea. B. Methods and experiments.
Neth. J. Sea Res. 4(4): 395-469.

Chemistry radioisotopes
Hela, Ilmo, 1964.
Alternative ways of expressing the concentration factors for radioactive substances in aquatic organisms.
Bull. Inst. Océanogr., Monaco, 61(1280):1-8.

Chemistry radioisotopes
Mauchline, J., and W.L. Templeton, 1964.
Artificial and natural radioisotopes in the marine environment.
In: Oceanography and Marine Biology, Harold Barnes, Editor, George Allen & Unwin, Ltd., 2: 229-279.

Chemistry, radio-isotopes
Noshkin, V.E., 1969.
Radio-isotopes.
Oceanus, 15(1). (popular)

Chemistry Radioisotopes
Patin, S. A., 1966.
On the question of the forms of existence of artificial radioisotopes and their migration within sea water. (In Russian).
Trudy Inst. Okeanol., Akad. Nauk, SSSR, 82:72-90

chemistry, radioisotopes
Yamagata, Noboru, and Kiyoshi Iwashima, 1963.
Monitoring of sea-water for important radioisotopes released by nuclear reactors.
Nature, 200(4901):52.

radionuclides

Chemistry, radio-isotopes
Anon., 1956.
The measurements of littoral drift by radioisotopes. Field experiments to aid design of new Japanese Port. Dock and Harbour Authority 36(423):284-288.

chemistry radionuclides
Barinov, H.V., 1966.
Kinetic regularities of accumulation and exchange of radionuclides by sea algae. (Abstract only).
Second Int. Oceanogr. Congr., 30 May-9 June 1966. Abstracts, Moscow:21-22.

chemistry, radionuclides
Baranova, D.D., 1966.
A comparative study of sorption and desorption of various radionuclides by marine shallow water sediments. (Abstract only).
Second Int. Oceanogr. Congr., 30 May-9 June 1966. Abstracts, Moscow:18-19.

Chemistry, radionuclides
Baranova, D.D., and G.G. Polikarpov, 1968.
Sorption and desorption of radionuclides by shallow-water sediments of the Black Sea. (In Russian; English abstract).
Okeanologiia, Akad. Nauk, SSSR, 8(3):427-430.

Chemistry radionuclides

Burton, J.D., 1965.
Radioactive nuclides in sea water, marine sediments and marine organisms.
In: Chemical oceanography. J.P. Riley and G. Skirrow, editors, Academic Press, 2:425-475.

chemistry, radionuclides

*Carey, A.G., W.G. Pearcy and C.L. Osterberg, 1966.
Artificial radionuclides in marine organisms in the northeast Pacific Ocean off Oregon.
Disposal of radioactive wastes into seas, oceans, and surface waters, Int. Atomic Energy Agency, 303-319.
Also in: Coll. Repr., Dep. Oceanogr., Oregon State Univ., 5.

Chemistry radio nuclides

Cerrai, E., B. Schreiber, C. Triulzi and L. Tassi-Pelati, 1964.
Contribution of Ce^{144} -Pr^{114} and Pm^{147} to the radioactivity of upper layers of coastal sediments of the Ligurian Sea.
Ist. Lombardo, Accad. Sci. e Lett., Rendiconti (B)

Chemistry, radionuclides

Chesselet, Roger, et Claude Lalou, 1964.
Concentrations en radionuclides émetteurs gamma presentées par des Holothuries prelevées dans la zone côtière d'Antibes au cap Ferrat en août 1962.
Bull. Inst. Océanogr., Monaco, 63(1305):16 pp.

Chemistry Radionuclides

de Santis, L., and F. Mosetti, 1961.
Alcuni ragguagli sulle ricerche idrologiche con traccianti.
Boll. Geofis. Teorica ed Applicata, Trieste, 3(10):1-16.

chemistry, radio isotopes

Duursma, E.K., and C. Hoede 1967.
Theoretical, experimental and field studies concerning molecular diffusion of radioisotopes in sediments and suspended solid particles of the sea. A. Theories and mathematical calculations.
Neth. J. Sea Res., 3(3):423-457

chemistry, radionuclides

Fukai, Rinnosuke 1969.
On the chelation of some radionuclides in a sea-water medium.
Rapp. P.-v. Réun. Commn int. Explor. scient. Mer Mediterr. 19(5):935-936.

Chemistry RADIONUCLIDES

Fukai, R., 1966.
Fate of some radionuclides fixed on ion-exchange resins in sea-water medium.
Disposal of Radioactive Wastes into Seas, Oceans and Surface Waters, IAEA (SM-72/29):483-496.

Chemistry radio-nuclides

Hiyama, Yoshio, 1960
An idea on the maximum permissible concentration of radio-active materials in sea-water. Rec. Oceanogr. Wks., Japan, n.s., 5(2): 98-104.

chemistry, radionuclides

Iorgulescu, Adriana, M. Oncescu, O. Serbanescu, and Florica Porumb, 1965.
Active fission γ (gamma) elements identified in marine plankton.
Rev. Roum. Biol. Zool., 10(5):373-378.

Chemistry radio nuclides

Jefferies, Douglas F., 1968.
Fission-product radionuclides in sediments from the north-east Irish Sea.
Helgoländer wiss. Meeresunters. 17: 280-290.

chemistry, radionuclides

Koczy, F.F., 1963.
Natural radio-nuclides in the ocean. Radioactive tracers in oceanography. Symposium 10th Pacific Science Contress.
Union. Geodes. Geophys. Int., Monogr., 20:1-9.

chemistry, radionuclides

Krumholz, L.A., E.D. Goldberg and H.A. Boroughs, 1957.
Ch. 7. Ecological factors involved in the uptake and loss of radionuclides by aquatic organisms
NAS-NRC Publ. 551:69-79.

chemistry, radionuclides

Lal, D., 1963.
Studies of oceanographic phenomena using radioactive nuclides produced by cosmic rays. Radioactive tracers in oceanography. Symposium of the 10th Pacific Science Congress, 1961.
Union Géodes. Géophys. Int. Monogr., 20:10-20.

Chemistry radionuclides

Lal, D., 1962
Cosmic ray produced radionuclides in the sea.
J. Oceanogr. Soc., Japan, 20th Ann. Vol., 600-614.

Chemistry radionuclides

Mauchline, John, and Angela M. Taylor, 1964.
The accumulation of radionuclides in the thornback ray, Raia clavata, L., in the Irish Sea.
Limnology and Oceanography, 9(3):303-309.

Chemistry radionuclides

Mauchline, J., and W.L. Templeton, 1964.
Artificial and natural radioisotopes in the marine environment.
In: Oceanography and Marine Biology, Harold Barnes, Editor, George Allen & Unwin, Ltd., 2: 229-279.

Chemistry radionuclides

Miyake, Yasuo, Katsuko Saruhashi, Yukio Katsuragi and Teruko Kanazawa, 1962
Penetration of 90 Sr and 137 Cs in deep layers of the Pacific and vertical diffusion rate of deep water.
Repts. JEDS, Deep-Sea Res. Comm., Japan Soc. Promotion of Science, 3:141-147.
Originally published (1962):
J. Radiation Res., 3(3):141-147.

JEDS Contrib. No. 25.

chemistry, radionuclides

Osterberg, C., 1962.
Fallout radionuclides in euphausids.
Science, 138(3539):529-530.

Chemistry radionuclides

Perkins, R. W., J. L. Nelson and W.L. Haushild, 1966.
Behavior and transport of radionuclides in the Columbia River between Hanford and Vancouver, Washington.
Limnol. Oceanogr., 11(2):235-248.

Chemistry, radionuclides

Riley, J.P. and G. Skirrow, editors, 1965.
Chemical oceanography.
Academic Press, Vol. 2:508 pp.

Chemistry radio nuclides

*Seki, Humitake, and Nobuo Taga, 1967.
Removal of radio nuclides at different multiplication phases of marine bacteria.
J. Oceanogr. Soc., Japan, 23(3):136-140.

Chemistry, radionuclides

Riley, J.P., 1965.
Analytical chemistry of sea water.
In: Chemical oceanography, J.P. Riley and G. Skirrow, editors, Academic Press, 2:295-424.

Chemistry radio nuclides

Servant, Jean, 1966.
La radioactivité de l'eau de mer.
Cahier Océanogr., C.C.O.E.C., 18(4):277-318.

Chemistry radionuclides

Slowey, J. Frank, David Hayes, Bryan Dixon and Donald W. Hood, 1965.
Distribution of gamma emitting radionuclides in the Gulf of Mexico.
Narragansett Mar. Lab., Univ. Rhode Island, Occ. Publ., No. 3:109-129.

chemistry, radionuclides

Strohal, Petar, Olga Jelisavčić, Zvonimir Kolar, Stjepan Lulić and Stjepan Kečkeš, 1969.
Hydrographic and biotical conditions in the north Adriatic - VIII: Radioecological analyses of selected biota. (Jugoslavian and Italian abstracts).
Thalassia Jugoslavica, 5: 377-382.

chemistry, radionuclides

Tennant, David A., and William O. Forster 1969.
Seasonal variation and distribution of ^{65}Zn, ^{54}Mn and ^{51}Cr in tissues of the crab Cancer magister Dana.
Health Physics 18: 649-657.
Also in: Coll. Repr. Dept. Oceanogr. Oregon State Univ. 9 (2) (1970).

Chemistry radionuclides

Tsuguga, Hanato, 1965.
Sequential analysis of radionuclides in marine organisms.
Bull. Jap. Soc. Sci. Fish., 31(9):651-658.

chemistry, radionuclides

Volchok, H.L., V.T. Bowen, T.R. Folsom, W.S. Broecker, E.A. Schuert and G.S. Bien, 1971.
Oceanic distributions of radionuclides from nuclear explosions. Radioactivity in the marine environment, U.S. Nat. Acad. Sci., 1971: 42-89.

Chemistry radio nuclides

Zesenko, A. Ya., A.A. Lyubimov, V.N. Ivanov, and L.N. Leshshenko, 1968.
Changes of the state of radionuclides in sea water and its influence on the accumulation by the sea organisms. (In Russian).
Okeanologiia, Akad. Nauk, SSSR, 8(6):1092.

Chemistry fallout

*Mamuro, T., T. Matsunami, A. Fujita, and K. Yoshikaw, 1969.
Radionuclide fractionation in fallout particles from a land surface burst.
J. geophys. Res., 74(6):1374-1387.

Radiosodium

Chemistry radiosodium

Likens, Gene E., and Arthur D. Hasler, 1962.
Movement of radiosodium (Na 24) within an ice-covered lake.
Limnol. & Oceanogr., 7(1):48-56.

radiostrontium

Chemistry, radiostrontium

See also: Chemistry, strontium isotopes etc.

Chemistry radiostrontium

Boroughs, Howard, 1958
The metabolism of radiostrontium by marine fishes.
Proc. Ninth Pacific Sic. Congr., Pacific Sci. Assoc., 1957, 16 (Oceanogr.):146-151.

chemistry, radiostrontium

Ichikawa, R., and Y. Hiyama, 1957.
The amount of up-take of strontium by marine fish from the environment by various concentration of calcium and strontium.
Rec. Oceanogr. Wks., Japan, n.s., 4(1):55-66.

Chemistry, radiostrontium

Ichikawa, R., and Y. Hiyama, 1956.
Deposition of radiostrontium in the various tissues of Jack mackerel, mackerel and tuna.
Research on the effects and influences of the nuclear bomb text explosions. Jap. Soc. Prom. Sci., 1956:1143-1150.

chemistry
radiostrontium

Martin, DeCourcey, and Edward D. Goldberg, 1962
Oceanographic studies during Operation "Wigwam". Uptake and assimilation of radiostrontium by Pacific salmon.
Limnol. and Oceanogr., Suppl., to Vol. 7: lxxvi-lxxxi.

chemistry, radiostrontium

Nelson, D.J., 1962
Clams as indicators of strontium-90.
Science, 137(3523):38-39.

radium

Chemistry
radium

Broecker, W., 1963
4. Radioisotopes and large oceanic mixing.
In: The Sea, M.N. Hill, Edit., Vol. 2. The composition of sea water, Interscience Publishers, New York and London, 88-108.

chemistry, radium

Broecker, W.S., and T.-H. Peng, 1971.
The vertical distribution of radon in the Bomex area.
Earth planet. Sci. Letts 11(2): 99-108.

chemistry
radium

Dostrovsky, I., S. Amiel and L. Winsberg, 1954.
Determination of radon and radium in water.
Bull. Res. Counc., Israel 4(1):94.

Chemistry radium

Evans, R.D., A. Kip and E. Moberg, 1938.
The radium and radon content of Pacific Ocean waters, life and sediments. Amer. J. Sci., (5), 36:241-259.

chemistry
radium

Folsom, Theodore R., 1958.
Comparison of radioactive dosage from potassium with estimated dosages from uranium and radium in marine biospheres.
Proc. Ninth Pacific Sci. Congr., Pacific Sci. Assoc., 1957, Oceanogr., 16:176-178.

chemistry radium

Graschenko, S.M., D.S. Nikolaev, L.B. Koliadin, Iu. V. Kuznetsov, K.F. Laz arev, 1960.
Radium concentration in the waters of the Black Sea. (In Russian).
Doklady, Akad. Nauk, SSSR, 132(5):1171-1172.

Chemistry radium

Heye, Dietrich, 1969.
Uranium, thorium, and radium in ocean water and deep-sea sediments
Earth planet. Sci. Letters 6(2): 112-116.

chemistry radium

Holland, H.D., and J.L. Kulp, 1954.
The mechanism of removal of ionium and radium from the oceans. Geochim. Cosmochim Acta 5:214-224.

Abstr: Chem. Abstr. 11855b, 1954.

chemistry radium

Holland, H.D., and J.L. Kulp, 1954.
The transport and deposition of uranium, ionium and radium in rivers, oceans and ocean sediments.
Geochim. Cosmochim. Acta 5:197-213.

Abstr. Chem. Abstr. 11855a, 1954.

chemistry
radium

Holland, H.D., and L.J. Kulp, 1954.
The mechanism of removal of ionium and radium from the oceans. Geochimica et Cosmochimica Acta 5:214-224, 6 textfigs.

Koczy, F.F., and Heinz Titze, 1958 radium

Radium content of carbonate shells. J. Mar. Res., 17 : 302-311.

Chemistry
radium distribution

Koczy, F., and B. Szabo, 1962
Renewal time of bottom water in the Pacific and Indian oceans.
J. Oceanogr. Soc., Japan, 20th Ann. Vol., 590-599.

Chemistry radium

Kroll, V.S., 1955.
The distribution of radium in deep-sea cores.
Rept. Swedish Deep-Sea Exped., 1947-1948, 10(1):1-32.

Chemistry radium

Kröll, V. St., 1955.
Radium in manganese crusts.
Göteborgs K. Vetenskaps- och Vitterhets-Samhälles Handl. (7) B 6(13):1-10.

Medd. Ocean. Inst., Göteborg 24.

Chemistry, radium

Ku, Teh-Lung, Yuan-Hui Li and Guy G. Mathieu 1969.
Radium and inorganic carbon in Antarctic waters.
Antarctic Jl U.S. 4(5): 186-187.

Chemistry, radium

Kuznetsov, Yu. A., Z.N. Simonyak, A.P. Lisitsyn and M.S. Frenklikh, 1968.
Uranium and radium in the surface layer of oceanic sediments. (In Russian)
Geokhimiya 1968(3): 323-333.
Translation: Geochem. int. 5(2): 306-311

Chemistry
radium

Lisitsin, A.P., 1964.
Distribution and chemical composition of suspended matter in the waters of the Indian Ocean. (In Russian: English Abstract).
Rezult. Issled., Programme Mezhd. Geofiz. Goda, Mezhd. Geofiz. Kom., Presidume Akad. Nauk, SSSR, Okeanol., No. 10:136 pp.

chemistry, radium

Miyaki, Y., K. Saruhashi, Y. Katsuraji, J. Kanazawa and Y. Sugimura, 1964.
Uranium, radium, thorium, ionium, strontium 90 and cesium 137 in coastal waters of Japan.
In: Recent researchers in the fields of hydrosphere, atmosphere and nuclear geochemistry, Ken Sugawara, festival volume, Y. Miyake and T. Koyama, editors, Maruzen Co., Ltd., Tokyo, 127-141.

chemistry radium

*Miyake, Yasuo, Katsuko Saruhashi and Yukio Sugimura, 1968.
Biogeochemical balance of natural radioactive elements in the Sea.
Rec. oceanogr. Wks, Japan, n.s., 9(2):180-187.

Chemistry
radium

Miyake, Y., and Y. Sugimura, 1964
Uranium and radium in the western North Pacific waters.
In: Studies on Oceanography dedicated to Professor Hidaka in commemoration of his sixtieth birthday, 274-278.

Chemistry, radium

Pettersson, H., 1955.
Manganese nodules and oceanic radium.
Pap. Mar. Biol. and Oceanogr., Deep-Sea Res., Suppl. to Vol. 3:335-345.

Chemistry radium

Pettersson, H., 1953.
Radium and deep sea. Amer. Sci. 41(2):245-255.

Chemistry radium

Pettersson, H., 1951.
Radium and deep-sea chronology. Nature 167:942.

Chemistry radium

Pettersson, H., 1949.
Exploring the bed of the ocean. Nature 164(4168) 468-470.

Chemistry radium

Pettersson, H., 1948
Three sediment cores from the Tyrrhenian Sea. Medd. Ocean. Inst., Göteborg 15 (Göteborgs Kungl. Vetenskaps-och Vitterhets-Samhälles Handlingar. Sjätte Följden, Ser.B 5(13)):94 pp.

Chemistry radium

Pettersson, H., 1939.
Die radioaktiven Elemente im Meer.
Rapp. Proc. Verb., Cons. Perm. Int. Expl. Mer, 109:66-67.

Chemistry radium

Piggot, C. S., 1944.
Scientific results of the Carnegie during 1928-1929 under command of Captain J.P. Ault. Oceanography. II. Radium content of ocean-bottom sediments. Carnegie Inst. Washington Publ. 556:183-196, Map, 1 pl.

Chemistry radium

Poole, J.H.J., 1938.
Appendix. Radium content of some sub-oceanic basalts from the floor of the Indian Ocean. John Murray Exped., 1933-34, Sci. Repts. 3(1): 28-30.

Chemistry
radium

Rao, S.R., S.M. Shah and R. Viswanathan, 1968.
Calcium, strontium and radium content of molluscan shells.
J. mar. biol. Ass., India, 10(1): 159-165.

Chemistry radium

Revelle, R.R., and C.S. Piggot, 1944.
1. Marine bottom samples collected in the Pacific Ocean by the "Carnegie" on its seventh cruise. II. Radium content of ocean bottom sediments.
Publ. Carnegie Inst., Washington, 556:5-196, 47 textfigs.

Chemistry radium

Rona, E., and W.O. Urry, 1952.
Radioactivity of ocean sediments. VIII. Radium and uranium content of ocean and river waters.
Am. J. Sci. 250:241-262.

chemistry radium

Sanderman, L.A., and C.L. Utterback, 1941.
Radium content of ocean bottom sediments from the Arctic Ocean, Bering Sea, Alaska peninsula and the coasts of southern Alaska and western Canada.
J. Mar. Res. 4(2):132-141, 2 figs.

chemistry
radium

St. Kroll, V., 1955.
Radium in manganese crusts. Göteborgs K. Vetenskaps- och Vitterhets-Samhalles Handl., Sjätte Följden, (B), 6(13):10 pp., 9 textfigs.

Also:
Medd. Oceanogr. Inst. Göteborg, No. 24:

Chemistry radium

St. Kröll., V., 1955.
The distribution of radium in deep-sea cores.
Repts. Swedish Deep-Sea Exped., 1947-48, Spec. Invest. 10(1):3-32.

Chemistry, radium

St. Kroll, V., 1953.
Vertical distribution of radium in deep-sea sediments. Nature 171(4356):742, 1 textfig.

Chemistry radium

Starik, I. Ye., A. P. Lisitsyn, and Yu. V. Kuznetsov, 1964.
Concerning the mechanism for removal of radium from sea water and its accumulation in sea and ocean sediments. (In Russian).
Antarktika, Moscow, 1962:70-133.

Abstracted in:
Soviet Bloc Res., Geophys., Astron and Space, 96:47.

chemistry radium
Sugimura, Y., and Hiroyuki Tsubota, 1963.
A new method for the chemical determination of radium in sea water.
J. Mar. Res., 21(2):74-80.

chemistry radium
Szabo, B.J., F.F.Koczy and G. Östlund, 1967.
Radium and radiocarbon in Caribbean waters.
Earth Planet-Sci. Letters, 3(1):51-61.

chemistry, radium
Ueda, S., 1957.
Chemical studies on the ocean. 66-67. Chemical studies of the shallow-water deposits. 19-20. Radium content of the shallow-water deposits. (1-2) J. Oceanogr. Soc., Japan, 13(2):61-66; 67-72.

chemistry, radium
Utterback, C.L., and L.A. Sanderamn, 1948.
Radium analyses of marine sediments in northern Pacific and adjacent waters. J. Mar. Res. 7(3): 635-643.

chemistry radium
Yoshimura, J., I. Hatayi, K. Tachibana & N. Yoza, 1961
Radium contents of the Tertiary rocks of Ariake Sea, Japan.
J. Chem. Soc. Japan, 82(9): 1159-1160.

radium-226

chemistry radium-226
*Broecker, Wallace S., Yuan Hui Li and John Cromwell, 1967.
Radium-226 and radon-222: concentration in Atlantic and Pacific oceans.
Science, 158(3806):1307-1310.

chemistry radium 226
Servant, Jean, 1966.
La radioactivité de l'eau de mer.
Cahier Océanogr., C.C.O.E.C., 18(4):277-318.

chemistry, radium-226
Szabo, Barney J., 1971.
Concentration of radium-226 in northeast Providence Channel and the Tongue of the Ocean, Bahamas. Bull. mar. Sci. Coral Gables, 21(3): 748-753.

chemistry radium - 226
Szabo, B.J., 1967.
Radium content in plankton and sea water in the Bahamas.
Geochim. Cosmochim. Acta, 31(8):1321-1331

chemistry, radium-228
Kaufman, Aaron 1969.
Ra^{228} in high latitude southern hemisphere waters.
Antarctic Jl U.S. 4(5): 187.

chemistry, radium-228
Moore, Willard S. 1969.
Oceanic concentrations of ^{228}radium.
Earth planet. Sci. Letts 6(6): 437-446.

radium/calcium

chemistry radium/calcium
Blanchard, Richard L., and Duke Oakes, 1965.
Relationships between uranium and radium in coastal marine shells and their environment.
J. Geophys. Res., 70(12):2911-2921.

radon

chemistry, radon
Björnsson, Sveinbjörn, 1968.
Radon and water in volcanic gas at Surtsey Island.
Geochim. Cosmochim. Acta 32(8): 815-820.

Chemistry radon
Broecker, Wallace S., 1965.
An application of natural radon to problems in ocean circulation.
Symposium, Diffusion in oceans and fresh waters, Lamont Geol. Obs., 31 Aug.-2 Sept. 1964., 116-145.

chemistry, radon
Broecker, W.S. and T.-H. Peng 1971.
The vertical distribution of radon in the BOMEX area.
Earth planet. Sci. Letts 11(2): 99-108.

radon
Desai, B.N. 1969.
Possible radon concentration over the Indian seas during the southwest monsoon season on the basis of climatic features of the area and utilization of the radon results for identifying air masses.
Indian J. Met. Geophys. 20(3): 253-256

chemistry radon
Desai, B.N. and Y.P. Rao, 1971.
Can natural radon data be used for delineating monsoon circulation? J. geophys. Res. 76(24): 5916-5919.

Chemistry radon
Dostrovsky, I., S. Amiel and L. Winsberg, 1954.
Determination of radon and radium in water.
Bull. Res. Counc., Israel 4(1):94.

Chemistry radon
Evans, R.D., A. Kip, and E. Moberg, 1938.
The radium and radon content of Pacific Ocean waters, life and sediments. Amer. J. Sci. (5), 36:241-259.

Chemistry radon
Rama, 1971.
Reply. J. geophys. Res. 76(24): 5920.

Chemistry radon
Rama, 1970.
Using natural radon for delineating monsoon circulation. J. geophys. Res., 75(12): 2227-2229.

chemistry radon
Rama and A.S. Chhabra, 1966.
Radon concentration in the monsoon air at Bombay.
Bull. Geochem. Soc., India, 1(3): 111-113

chemistry radon 222
Assaf, G., and J.R. Gat, 1970.
Sea surface as a sink of emanation products.
Earth Planet. Sci. Letts., 7(5): 385-388.

chemistry Radon-222
Mathieu, Guy G., 1969.
Rn^{222} in Antarctic near-surface waters.
Antarctic J., U.S. 4(5): 185-186

chemistry Radon-222
Prospero, Joseph M. and Toby N. Carlson, 1970.
Radon-222 in the North Atlantic trade winds: its relationship to dust transport from Africa.
Science, 167(3920): 974-977.

chemistry radon-222
Broecker, Wallace S., Yuan Hui Li and John Cromwell, 1967.
Radium-226 and radon-222: concentration in Atlantic and Pacific oceans.
Science, 158(3806):1307-1310.

rare earths

chemistry rare earths
Balashov, Yu. A., L.V. Dmitriev and A. Ya. Sharas'kin, 1970.
Distribution of the rare earths and yttrium in the bedrock of the ocean floor. (In Russian)
Geokhimiya (6): 647-660.
Translation: Geochem. Int. 7(3): 456-468.

Chemistry rare earths
Balashov, Yu. A., and L.M. Khitov, 1961.
Distribution of the rare earths in the waters of the Indian Ocean. (In Russian)
Geokhimiya, 9:796-806.
English Ed., 9:877-890.

Chemistry rare-earths
*Balashov, Ju.A., and A.P. Lisitsin, 1968.
Migration of rare-earth elements in the ocean. (In Russian; English abstract).
Okeanolog. Issled., Rez. Issled. Mezhd. Geofiz. Proekt. Mezhd. Geofiz. Komit., Pres. Akad. Nauk, SSSR, 18:213-282.

chemistry rare earths
Blokh, A.M., 1961.
Rare earths in the remains of Paleozoic fishes of the Russian platform.
Geokhimiia, (5):390-400.
Translation in: Geochemistry (5):404-415.

chemistry, rare earths
Fomina, L.S., 1966.
Accumulation and redistribution of rare-earth elements upon the formation of iron-mangenese concretions in the ocean.
Doklady, Akad. Nauk, SSSR, 170(5):1181-1184.

chemistry, rare earth
Fomina, L.S., and I.I. Volkov, 1969.
Rare-earth elements in ferromanganesian concretions of the Black Sea. (In Russian)
Dokl. Acad. Nauk, SSSR, 185(1): 188-191.

chemistry rare earths
Frey, Fred A., and Larry Haskin, 1964.
Rare earths in oceanic basalt.
J. Geophys. Res., 69(4):775-780.

chemistry rare earths
Goldberg, Edward D., Minoru Koide, R.A. Schmitt and R.H. Smith, 1963
Rare-earth distribution in the marine environment.
J. Geophys. Res., 68(14):4209-4217.

chemistry rare earths
Haskin, Larry, and Mary A. Gehl, 1962.
The rare-earth distribution in sediments.
J. Geophys. Res., 67(6):2537-2541.

chemistry
Rare earths

Jakeš, P., and J. Gill. 1970
Rare earth elements and the island arc tholeiitic series.
Earth Planet. Sci. Letts. 9(1):17-25.

chemistry
rare earths

Lisitsin, A.P., 1964.
Distribution and chemical composition of suspended matter in the waters of the Indian Ocean. (In Russian; English Abstract).
Rezult. Issled., Programme Mezhd. Geofiz. Goda, Mezhd. Geofiz. Kom., Presidume Akad. Nauk, SSSR, Okeanol., No. 10:136 pp.

Chemistry rare earths

Masuda, Akimasa, 1968.
Nature of the experimental Mohole basalt - redetermination of lanthanides.
J. geophys. Res., 73(16):5425-5428.

Chemistry rare earths

Masuda, A., 1957.
Simple regularity in the variation of relative abundances of rare earth elements.
J. Earth Sci., Nagoya Univ., 5(2):125-134.

Chemistry rare earths

Miyake, Yasuo, Hiroshi Hamaguchi, Masumi Osawa, Naoki Onuma, Ken Sugawara, Satoru Kanmori, Sadayuki Omori, Toshio Shimada, Kiyoshi Kawamura, Hirozo Yoshimura, and Takeshi Sagi, 1960.
Chemistry in the Japanese Expedition of Deep Sea. (JEDS-2)
Repts. of JEDS, 1:181-185.
JEDS Contrib. No. 4.
Also in Oceanogr. Mag., 11(2):181-185.

Chemistry rare earths

Ostroumov, E.A., A.A. Astmanuna and T.G. Shokhor, 1956.
[Method for the determination of rare earths in marine sediments.]
Trudy Inst. Okeanol., 19:297-303.

rare earths

Pucar, Zvonimir, and Ljerka Maregović, 1969.
Investigation of the physico-chemical behaviour of some microconstituents in sea water using electromigration techniques
J. Chromatog. 27:450-454
Croat. Chem. Acta. 35:183-191
Abstract in: Rapp. P.-v. Reun. Commn int. Explor. scient. Mer Méditerr., 19(4): 727.

Chemistry
rare earths

Schilling, J.-G., 1971.
Sea-floor evolution: rare-earth evidence.
Phil. Trans. R. Soc. Lond. (A)268(1192): 663-706

Chemistry
rare earths

Wildeman, Thomas R., and Larry Haskin, 1965.
Rare-earth elements in ocean sediments.
J. Geophys. Res., 70(12):2905-2910.

rare elements

chemistry
rare elements

Graf, Donald L., 1960
Geochemistry of carbonate sediments and sedimentary carbonate rocks. III Minor element distribution. Illinois State Geol. Survey, Circ. 301:71 pp.

Chemistry
rare elements

Naughton, John J., and I. Lynus Barnes, 1965.
Geochemical studies of Hawaiian rocks related to the study of the upper mantle.
Pacific Science, 19(3):287-290.

rare gases

Chemistry, rare gases

See also: chemistry, argon
helium
neon, etc.

chemistry, rare gases

Bieri, Rudolph H., Minoru Koide and Edward D. Goldberg, 1964.
Rare gas determinations in sea water. (Abstract).
Trans. Amer. Geophys. Union, 45(1):119.

Chemistry
rare gases

Mazor, E., G.J. Wasserburg, and H. Craig, 1964.
Rare gases in Pacific Ocean water.
Deep-Sea Res., 11(6):929-932.

Chemistry
rare gases (helium, argon)

Revelle, R., and H.E. Suess, 1962
Gases, Ch. 7, Sect. II, Interchange of properties between sea and air. In: The Sea, Interscience Publishers, Vol. 1, Physical Oceanography, 313-321.

rare metals

Chemistry rare metals

Carritt, D.E., 1953.
Separation and concentration of trace metals from natural waters. Anal. Chem., 25:1927.

Chemistry rare metals

Krauskopf, K.B., 1956.
Factors controlling the concentration of thirteen rare metals in sea water.
Geochim. Cosmochim. Acta, 9:1-32B.

chemistry, records

Saelen, Odd, 1969.
Data logging and storage. In: Chemical Oceanography, Rolf Lange, editor, Universitetsforlaget, Oslo: 104-110.

redox potential

redox potential (data)

Kato, Kenji, 1966.
Geochemical studies on the mangrove of Cananéia, Brazil. II. Physico-chemical observations on the reduction states.
Bolm Inst. Oceanogr., S. Paulo, 15(1):21-24.

Chemistry
redox processes

*Sevastyanov, V.F., and I.I. Volkov, 1967.
Redistribution of chemical elements in redox processes in the bottom sediments of the oxygen zone of the Black Sea. (In Russian; English abstract).
Trudy Inst. Okeanol., 83:115-134.

Chemistry
redox potentials

Spencer, Derek W. and Peter G. Brewer, 1971.
Vertical advection, diffusion and redox potentials as controls on the distribution of manganese and other trace metals dissolved in waters of the Black Sea. J. geophys. Res, 76(24): 5877-5892.

refractivity-chlorinity-temperature

refractivity-chlorinity-temperature

Utterback, C.L., T.G. Thompson and B.D. Thomas, Refractivity-chlorinity-temperature relationships of ocean water.
J. du Conseil, 9:35-38.

chemistry, retinochrome

Hara, Tomiyuki, and Reiko Hara 1967.
Rhodopsin and retinochrome in the squid retina.
Nature, Lond. 214(5088): 573-575.

Chemistry - retinochrome

Hara, Tomiyuki, Reiko Hara and Jitsuzo Takeuchi 1967.
Rhodopsin and retinochrome in the octopus retina.
Nature, Lond., 214(5088):572-573.

rH

Chemistry
rH

Rivière, A., and Solange Vernhet, 1959
États d'oxydo-reduction dans les milieux naturels. Technique de determination directe du rH. Quelques resultats en milieu lagunaire.
Cahiers Océan., C.C.O.E.C., 11(5):309-314.

rhamnosides

Chemistry, rhamnosides

Wangersky, P.J., 1952.
Isolation of ascorbic acid and rhamnosides from sea water. Science 115(2999):685.

rhenium

Chemistry
rhenium

Fukai, Rinnosuke, and W. Wayne Meinke, 1962
Activation analyses of vanadium, arsenic, molybdenum, tungsten, rhenium and gold in marine organisms.
Limnology and Oceanography, 7(2):186-200.

Chemistry, rhenium

Scadden E.M., 1969.
Rhenium: its concentration in Pacific Ocean surface waters.
Geochim. cosmochim. Acta, 33(5): 633-637.

Chemistry rhodamine-B

*Talbot, J.W., and J.L. Henry, 1968.
The adsorption of rhodamine-B onto materials carried in suspension by inshore waters.
J. Cons. perm. Int. Explor. Mer, 32(1):7-16.

chemistry, rhodopsin

Hara, Tomiyuki, Reiko Hara and Jitsuzo Takeuchi 1967.
Rhodopsin and retinochrome in the octopus retina.
Nature, Lond. 214(5088): 572-573.

chemistry rhodopsin

Hara, Tomiyuki, and Reiko Hara, 1967.
Rhodopsin and retinochrome in the squid retina.
Nature, Lond., 214(5088):573-575

Chemistry
Rhodopsin, deep sea

Lythgoe, J.N., and H.J.A. Dartnall, 1970
A "deep sea rhodopsin" in a mammal.
Nature, Lond., 227(5261): 955-956

ribonucleic acid

Chemistry, RNA

Derenbach, Jens B., 1970.
Zur Bestimmung der Ribonucleinsäure in planktischem Analysenmaterial.
Kieler Meeresforsch. 26(1): 79-84.

RNA

Matterne, Manfred, 1970
Vergleich zwischen Primärproduktion und Syntheseraten organischer Zellbestandteile mariner Phytoplankter.
Kieler Meeresforsch. 25(2):290-313.

Chemistry, RNA

Mayeda, Osamu, and Tatsuichi Iwamura, 1968.
Biochemical approach to the ecology in aquatic environments: Measurement of protein and RNA in natural water. (In Japanese; English abstract). Bull. Plankton Soc., Japan, 15(1): 17-18.

Chemistry ribonucleic acid

Sutcliffe, W.H., Jr., 1965.
Growth estimates from ribonucleic acid content in some small organisms.
Limnol. and Oceanogr., Redfield Vol., Suppl. to 10:R253-R258.

rubidium

Chemistry, rubidium

Biscaye, Pierre E. and E. Julius Basch, 1971.
The Rubidium, strontium, strontium-isotope system in deep-sea sediments: Argentine Basin.
J. geophys. Res, 76(21): 5087-5096.

Chemistry rubidium (data)

Bolter, Ernst, Kark K. Turekian and Donald F. Schutz, 1964.
The distribution of rubidium, cesium and barium in the oceans.
Geochimica et Cosmochimica Acta, 28(9):1459-1466.

Chemistry rubidium

Borovik-Romanova, T.F., 1944.
On the rubidium content of sea water.
Dokl. Akad. Nauk, SSSR, 42:216-218.

Chemistry rubidium

Burovina, I.V., et al., 1963.
The content of lithium, sodium, potassium, rubidium and caesium in the muscles of marine animals of the Barents Sea and Black Sea. (In Russian).
Doklady, Akad. Nauk, SSSR, 149(2):413-415.

chemistry, rubidium

Fabricand, E.S. Imbimbo, M. E. Brey and J.A. Weston, 1966.
Atomic absorption for Li, Mg, K, Rb, and Sr in ocean waters.
J. geophys. Res., 71(6):3917-3921.

Chemistry rubidium

Ishibashi, M., & T. Hara, 1959.
A systematic analysis of potassium, rubidium and cesium and its application to sea-muds.
Bull. Inst. Chem. Res. Kyoto Univ., 37(3):185-190.

Chemistry rubidium

Ishibashi, M., & T. Hara, 1959.
On concentrating rubidium and cesium from a large volume of aqueous solution.
Bull. Inst. Chem. Res. Kyoto Univ., 37(3):172-178.

Chemistry rubidium

Ishibashi, M., and T. Hara, 1955.
On the content of rubidium and casium in sea-water.
Rec. Ocean. Wks., Japan, 2(1):45-48.

Chemistry rubidium

Ishibashi, M., and T. Harada, 1942.
Chemical studies of the ocean. XI. Quantitative determination of rubidium in sea water and in bittern. J. Chem. Soc., Japan, 63(3):211-216.
Chem. Abstr., 1947, 41:2945.

Chemistry rubidium

Kovaleva, K.N., and E.S. Bukser, 1941.
Determination of rubidium in sea water.
Dopovidi Akad. Nauk, USSR, 1940(5):35-36.

chemistry, rubidium

Morozov, N.P., 1969.
Direct determinations of lithium and rubidium in the sea water and in the interstitial solutions with the flame-spectrophotometric technique. (In Russian; English abstract).
Okeanologiia 9(2): 353-358.

chemistry, rubidium

Morozov, N.P. 1968.
On the geochemistry of rare alkaline elements in the oceans and seas. (In Russian; English abstract)
Okeanologiia, 8(2): 216-224.

chemistry, rubidium

*Morozov, N.P., 1068.
Lithium and rubidium in the waters of inland seas. (In Russian; English abstract).
Okeanologiia, Akad. Nauk, SSSR, 8(4):612-615.

Chemistry rubidium

Parchevskiy, V.P., and K.M. Khailov, 1968.
Formation of Ce^{141}, Ru^{106}, Cs^{137}, and Zn^{65} complexes with hydrophyl high-molecular combinations dissolved in sea water. (In Russian)
Okeanologiia, Akad. Nauk, SSSR, 8(6):1092.

Chemistry rubidium

Seto, Yoshio, 1965
On the contribution of rubidium to the determination of radioactive cesium in deep-sea water.
J. Oceanogr. Soc., Japan, 21(5):202-205.

Chemistry rubidium

Smales, A.A., and L. Salmon, 1955.
Determination by radioactivation of small amounts of rubidium and caesium in sea-water and related materials of geological interest.
Analyst 80:37-50.
Abstr.:
Anal. Abstr. 2(5):1141.

Chemistry rubidium

Smales, A.A., and R.K. Webster, 1957.
The determination of rubidium in sea water by the stable-isotope dilution method.
Geochimica et Cosmochimica Acta, 11(1):139.

Chemistry rubidium

Smith, Raymond C., K.C. Pillai, Ysaihwa J. Chow and Theodore R. Folsom, 1965.
Determination of rubidium in sea water.
Limnology and oceanography, 10(2):226-232.

Chemistry rubidium

*Sreekvmaran, C., K.C. Pillai and T.R. Folsom, 1968
The concentrations of lithium, potassium, rubidium and cesium in some western American rivers and marine sediments.
Geochim. cosmochim. Acta, 32(11):1229-1234.

ruthenium

Chemistry ruthenium

Hiyama, Yoshio, and Junko Matsubara Khan, 1964
On the concentration factors of radioactive I, Co, Fe and Ru in marine organisms.
Rec. Oceanogr. Wks., Japan, 7(2):79-106.

Chemistry ruthenium (nitrosyl)

Jones, R.F., 1960.
The accumulation of nitrosyl ruthenium by fine particles and marine organisms.
Limnol. and Oceanogr., 5(3):312-325.

chemistry, Ruthenium

Keckeš, S., Z. Pučar and Lj. Marazović, 1966.
The influence of the physico-chemical form of ^{106}Ru on its uptake by mussels from sea water.
In: Radioecological concentration processes, Proc. Int. Symp., Stockholm, 25-29 Apr. 1966, Pergamon Press, 993-994.

Chemistry ruthenium

Loveridge, B.A., & A.M. Thomas, 1959
Determination of radio-ruthenium in effluent and sea water.
A.E.R.E. Report C/R 2828: 11 pp.
Abs. in Anal. Abs., 6(8): 3250.

chemistry, 106 ruthenium

Marazović, Lj., and Z. Pučar, 1967.
Two-dimensional electrochromatography of ^{106}ruthenium and some other radiomicroconstituents in sea water.
J. Chromatogr., 27:450-459.

Chemistry ruthenium

Parchevsky, V.P., 1965.
Radionuclides of cesium, ruthenium and zirconium in animal and plant organisms of the Black Sea. (In Russian).
Okeanologiia, Akad. Nauk, SSSR, 5(5):856-862.

Chemistry ruthenium uptake

Tsuruga, H., 1962.
The uptake of radioruthenium by several kinds of littoral seaweeds.
Bull. Jap. Soc. Sci. Fish., 28(12):1149-1154.

salinity

See also in main subject file for non-chemical aspects

Chemistry, salinity

Chiefly under: Salinity

chemistry, salinity

Grasshoff, K., 1965.
The results of the chlorinity and salinity intercalibration measurements during the informal meeting in Copenhagen.
UNESCO, Techn. Papers, Mar. Sci., 3:6-8. (mimeographed)

salt

chemistry, salts

Duedall, Iver W., and Peter K. Weyl, 1967.
The partial equivalent volumes of salts in seawater.
Limnol. Oceanogr., 12(1):52-59.

chemistry salt

Kirkland, D.W., and J.E. Gerhard, 1971.
Jurassic salt, Central Gulf of Mexico, and its temporal relation to circum-gulf evaporites.
Bull. Am. Ass. Petrol. Geol. 55(5): 680-686.

chemistry salt

Myers, D.M., and C.W. Bonython, 1958.
The theory of recovering salt from sea water by solar evaporation. J. Appl. Chem. 8(4):207-218.

chemistry salts

Sung, Yong Kil, and Mu Shik Jhon, 1970
On the hydration numbers and activity coefficients of some salts present in sea water.
J. oceanogr. Soc., Korea, 5(1):1-5.

chemistry salts in aqueous solutions
Zen, E-an, 1961.
Partial molar volumes of some salts in aqueous solutions.
Geochim. Cosmochim. Acta, 23(1/2):151.

Note: This is a correction of his paper in: Geochimica et Cosmochimica Acta, 1957, 12:103-122.

Salt bed

chemistry salt bed
*Morcos, Selim A., 1967.
The chemical composition of sea water from the Suez Canal region. I. The major anions.
Kieler Meeresforsch., 23(2):80-91.

Salts, connate

chemistry, salts, connate
Carlé, W., 1965.
Die Mineral- und Thermalwässer am Golf von Neapel
Geol. Rundschau, 54(2):1261-1313.

salts, effect of

chemistry salts, effect of
Lenoble, J., 1956.
Sur le role des principaux sels dans l'absorption ultraviolette de l'eau de mer.
C.R. Acad. Sci., Paris, 242(6):806-808.

SALTS IN AQUEOUS SOLUTIONS

samarium

chemistry samarium
Goldberg, Edward D., Minoru Koide, R.A. Schmitt and R.H. Smith, 1963
Rare-earth distribution in the marine environment.
J. Geophys. Res., 68(14):4209-4217.

chemistry, scavengers
Noshkin, V.E., 1969.
Radio-isotopes.
Oceanus, 15(1). (popular)

chemistry, sea water

See also: seawater

chemistry, sea water
Broecker W.S., 1971.
A kinetic model for the chemical composition of sea water.
Quatern. Res. 1(2): 188-207

chemistry, sea water
Sarma, T.P., G.R. Doshi, S.S. Gogate, T.M. Krishnamoorty, V.R. Neralla, M. Rama Rao, S.R. Rao, V.N. Sastry, S.M. Shah and C.K. Unni, 1969.
Geochemical investigations off Tarapur Coast.
Bull. natn. Inst. Sci., India, 38(1): 308-318.

selenium

chemistry selenium
Ishibashi, M., 1953.
Studies on minute elements in sea water.
Rec. Ocean. Wks., Japan, n.s. 1(1):88-92.

chemistry selenium
Ishibashi, M., T. Shigematsu and Y. Nakagawa, 1953.
Determination of selenium in sea water.
Rec. Ocean. Wks., Japan, n.s., 1(2):44-48, 1 textfig.

chemistry, selenium
Schutz, D.F., and K.K. Turekian, 1964.
The distribution of selenium, antimony, silver, cobalt and nickel in sea water. (Abstract).
Trans. Amer. Geophys. Union, 45(1):118.

Silica (silicon dioxide)

chemistry silica
Aleem, A.A., 1949.
A quantitative method for estimating the periodicity of diatoms. J.M.B.A. 28(3):713-717, 1 textfig., 1 pl.

chemistry, silica
Anon., 1951.
Bulletin of the Marine Biological Station of Asamushi 4(3/4): 15 pp.

chemistry silica
Armstrong, F.A.J., and E.I. Butler 1968.
Chemical changes in sea water off Plymouth during the years 1962 to 1965.
J. mar. biol. Ass. U.K. 48(1):153-160.

chemistry silica
Atkins, W.R.G., and M. Parke, 1951.
Seasonal changes in the phytoplankton as indicated by chlorophyll estimation. J.M.B.A. 29(3): 609-618.

chemistry silica
Auninsh, E.A., 1966.
Biogenic elements in the Gulf of Riga. (In Russian).
Trudy, Gosudarst. Okeanogr. Inst., No. 83:172-206.

chemistry silica
Ayers, John C., David C. Chandler, George H. Lauf, Charles F. Powers and E. Bennett Henson, 1958
Currents and water masses of Lake Michigan.
Great Lakes Res. Inst., Publ. No. 3:169 pp.

chemistry silica
Bal, D.V., L.B. Pradhan, and K.G. Gupte, 1946
A preliminary record of some of the chemical and physical conditions in waters of the Bombay harbour during 1944-45. Proc. Indian Acad. Sci. Sect. B24(2): 60-73, 4 figs.

chemistry, silica
Bien, G.S., 1958.
Salt effect correction in determining soluble silica in sea water by silicomplybdic acid method
Anal. Chem., 30(9):1525-1526.

chemistry, silica
Bogoiavlensky, A.N., 1966.
Distribution and migration of dissloved silica in the oceans. (In Russian).
In: Geochemistry of silica, N.M. Strakhov, editor, Isdatel. "Nauka", Moskva, 11-36.

chemistry silica
Bogoyavlenskii, A.N., 1955.
[Chemical characteristics of the water in the region of the Kurile-Kamchatka Trench.] Trudy Inst. Okeanol., 12:161-176.

chemistry silica, geochemisty of
Bruevich, S.V., 1953.
[On the geochemistry of silica in the sea.]
Izvest. Akad. Nauk, Ser. Geol., 4:67-79.

chemistry, silica
Bruevich, S.V., and L.K. Blinov, 1933.
[Determination of silica in seawater.]
Bull. Gos. Okeanogr. Inst. 14:

chemistry, silica
Calvert, S.E. 1971.
Nature of silica phases in deep sea cherts of the North Atlantic.
Nature, Lond. 234 (50): 133-134.

chemistry, silica
Calvert, S.E. 1968.
Silica balance in the ocean and diagenesis
Nature, London, 219 (5157): 919-920.

chemistry silica
Chu, S. P., and H. Y. Shin, 1949.
The variation of certain chemical constituents of biological importance and some other properties of sea water in Chiaochow Bay, August, 1948, to May, 1949. (Contrib. Fish. Res. Inst., Dept. Fish., Nat. Univ. Shantung No. 2) Sci. and Tech. in China 2(3):54-56.

chemistry silica
Chu, S.P., and K.C. Young, 1949.
The variation with depth of certain nutrient salts for phytoplankton growth and some other properties of water in the fishing ground east of Chusan Islands in the Chinese East Sea.
Science (China) 31(6):181-182, 1 fig.

chemistry, silica
Dienert, F., and Wandenbulcke, F., 1923.
Sur le dosage de la silica dans les eaux.
C.R. Acad. Sci., Paris, 176:1478-1480.

chemistry silica
Emelianov, E.M., 1966.
Distribution of authigenous silica in the suspension and in Recent sediments of the Mediterranean Sea. (In Russian).
In: Geochemistry of silica, N.M. Strakhov, editor, Isdatel. "Nauka", Moskva, 284-294.

chemistry, silica
Fanning, Kent A., and David R. Schink, 1969.
Interaction of marine sediments with dissolved silica. Limnol. Oceanogr., 14(1): 59-68.

chemistry silica(data)
Fedosov, M.V., and I.A. Ermachenko (1959) 1961.
Silica content of northern waters.
Ann. Biol., Cons. Perm. Int. Expl. Mer, 16:23-25

chemistry, silica
Fleming, J.A., C.C. Ennis, H.U. Sverdrup, S.L. Seaton, W.C. Hendrix, 1945
Observations and results in physical oceanography. Graphical and tabular summaries. Ocean. 1-B, Sci. Res. of Cruise VII of the Carnegie during 1928-1929 under command of Capt. J.P. Ault, Carnegie Inst., Washington, Publ. 545:315 pp., 5 tables, 254 figs.

chemistry silica
Farmer, D.M., and Wm. L. Ford, 1969.
Mid-Atlantic Ridge near 45° N. IV. Water properties in the median valley.
Can. J. Earth Sci., 6(6): 1359-1363

chemistry, silica
Garrels, Robert M., 1965.
Silica: role in the buffering of natural waters.
Science, 148(3666):69.

chemistry, silica
Gorsline, Donn S., 1963.
Oceanography of Apalachicila Bay, Florida.
In: Essays in Marine Geology in honor of K.O. Emery, Thomas Clements, Editor, Univ. Southern California Press, 69-96.

chemistry, silica
Harriss, Robert C., 1966.
Biological buffering of oceanic silica.
Nature, 212(5059):275-276.

Chemistry, silica

Iyengar, M.O.P. and G.Venkataraman,1951.
The ecology and seasonal succession of the algae flora of the River Cooum at Madras with special reference to the Diatomaceae. J. Madras Univ. 21, Sect. B(1): 140-192, 1 pl of 4 figs., 11 text figs.

Chemistry silica

Kamatani, A., 1971.
Physical and chemical characteristics of biogenous silica. Marine Biol., 8(2): 89-95.

Chemistry, silica

*Kato,Kikuo, and Yasushi Kitano,1968.
Solubility and dissolution rate of amorphous silica in distilled and sea water at 20°C.
J. Oceanogr.,Soc.,Japan,24(4):147-152.

chemistry, silica

Leloup,E. editor, 1966.
Recherches sur l'ostreiculture dans le bassin d'Ostende en 1964.
Minist. Agricult. Comm. T.W.O.Z, Groupe de Travail "Ostreiculture": 58 pp.

silica

Lisitzin, Alexander P., 1970.
Sedimentation and geochemical considerations.
In: Scientific exploration of the South Pacific, W.S. Wooster, editor Nat. Acad. Sci., 89-132.

Chemistry silica (amorphous)

Lisitsin, A.P., 1964.
Distribution and chemical composition of suspended matter in the waters of the Indian Ocean. (In Russian; English Abstract).
Rezult. Issled., Programme Mezhd. Geofiz. Goda, Mezhd. Geofiz. Kom., Presidume Akad. Nauk, SSSR, Okeanol., No. 1;136 pp.

Chemistry silica,amorphous

*Lisitsin,A.P., and Ju.A. Bogdanov,1968.
Suspended amorphous silica in the waters of the Pacific Ocean. (In Russian;English abstract).
Okeanolog.Issled.,Rez.Issled.Mezhd.Geofiz.Proekt. Mezhd.Geofiz.Komit.,Pres.Akad.Nauk,SSSR,18:5-41.

Chemistry silica

Lund, J.W.G., 1950.
Studies on Asterionella formosa Hass. II. Nutrient depletion and the spring maximum. Pt. II. Discussion. J. Ecol. 38(1):15-35.

Chemistry silica

Lund. J.W.G., 1950.
Studies on Asterionella formosa Hass. II. Nutrient depletion and the spring maximum. Pt. I. Observations on Windermere, Esthwaite water and Blelham Tarn. J. Ecol.38(1):1-14, 7 textfigs.

chemistry, silica

MacKenzie, Fred T.,and Robert M. Garrels, 1966.
Silica-bicarbonate balance in the ocean and early diagenesis.
J. Sedim. Petrol.,36(4):1075-1084.

chemistry, silica

MacKenzie,Fred T., Robert M. Garrels, Owen P. Bricker and Frances Brickley, 1967.
Silica in sea water: control by silica minerals.
Science, 155(3768):1404-1405.

chemistry, silica (biogenic)

Mopper, Kenneth, and G.D. Garlick 1971.
Oxygen isotope fractionation between biogenic silica and ocean water.
Geochim. Cosmochim Acta 35(11): 1185-1187

Chemistry, silica

Nishizawa, Tanzo and T. Muraki, 1940
Chemical studies in the sea adjacent to Palau. I. A survey crossing the sea from Palau to New Guinea. Kagaku Nanyo (Sci. of the South Sea) 2(3):1-7.

Chemistry, silica

Norina, A.M., 1965.
Hydrochemical characteristics of the northern part of the Barents Sea. (In Russian).
Trudy, Gosudarst Okeanogr. Inst., No. 83:243-271.

Chemistry, silica

Strakhov, N.M. editor, 1966.
Geochemistry of Silica. (In Russian).
Isdatel. "Nauka", Moskva, 421 pp.

Chemistry silica

Tanita, S., S. Kato, and T. Okuda, 1950.
Studies on the environmental conditions of shellfish fields. 1. In the case of Hakodate Harbor.
Bull. Fac. Fish., Hokkaido Univ., (16): 18-27, 7 textfigs. (In Japanese; English summary)

chemistry, silica

Turekian, Karl K., 1971
2. Rivers, tributaries and estuaries. In: Impingement of man on the oceans, D.W. Hood, editor, Wiley Interscience, 9-73.

Chemistry, silica

Wattenberg, H., 1937.
IV. Die chemischen Arbeiten auf der "Meteor"-Fahrt, Februar bis Maj 1937. Ann. Hydr., usw., 1937:17-22, 3 textfigs.

chemistry-silica

Wollast, R., and F. De Broeu, 1971.
Study of the behavior of dissolved silica in the estuary of the Scheldt.
Geochim. cosmochim. Acta 35(6): 613-620

Chemistry, silica

Wollast, Rowland, and Robert M. Garrels 1971.
Diffusion coefficient of silica in sea-water.
Nature, Lond., 229 (3): 94.

Chemistry silicate

Yamazi, I., 1950.
Plankton investigations in inlet waters along the coast of Japan. 1. Publ. Seto Mar. Biol. Lab. 1(3):93-113, 14 textfigs.

Silica budget

chemistry Silicabudget

Calvert, S. E., 1966.
Accumulation of diatomaceous silica in the sediments of the Gulf of California.
Geol. Soc., Am., Bull., 77(6):569-596.

chemistry, silica budget

Schink, David R., 1967.
Budget for dissolved silica in the Mediterranean Sea.
Geochim. cosmochim. Acta 31 (6): 987-999.

chemistry silica (data only)

Japan, Hokkaido University, Faculty of Fisheries, 1968.
The Oshoro Maru cruise 23 to the east of Cape Erimo, Hokkaido, April 1967.
Data Record Oceanogr. Obs. Expl. Fish., 12: 115-169.

Chemistry silica (data only)

Japan Meteorological Agency, 1968.
The results of marine meteorological and oceanographical observations, July-December, 1965, 38: 404 pp. (multilithed)

Silicate - (salt of silicic acid)

A

Chemistry silicate

Adrov, M.M., 1962.
Hydrologic investigations off West Greenland.
Sovetskie Riboch. Issledov. v Severo-Zapadnoi Atlant. Okeana, VINRO-PINRO, MOSKVA, 137-153.

Chemistry silicate

Akuyama, Tsutomu, and Shigeru Fukase, 1961
On the variation of silicate in sea water during its storage.
J. Oceanogr. Soc., Japan, 17(1):21-27.

Chemistry silicate

Aleksejev, A., B. Istoshin, L. Ponomarenko (1959) 1961
Hydrographic conditions in the Norwegian Sea, November-December 1959.
Ann. Biol., Cons. Perm. Int. Expl. Mer, 16: 26-28.

Chemistry silicate

Anon., 1951.
Report of the oceanographical observations on board the R.M.S. "Syunpu Maru" in the Kii-Suido (Oct. 1950). J. Ocean., Kobe Obs., 2nd ser., 1(6):1-15, 7 figs.

chemistry silicate

Antonov, A.E., and O.S. Roudneva, 1965.
Peculiarities of hydrological and hydrochemical conditions in the south Baltic during the IGY. (In Russian; English abstract).
Okeanolog. Issled. Rezult. Issled. Programme Mezhd. Geofiz. Goda, Mezhd. Geofiz. Komitet Presidiume, Akad. Nauk, SSSR, No. 13:90-95.

chemistry silicate (data)

Argentina, Servicio de Hidrografia Naval, 1960.
Operacion Oceanografica Drake II. Resultados preliminares.
Servicio de Hidrografia Naval, Argentina, Publ., H. 14: numerous unnumbered pp.

Chemistry silicate (data)

Argentina, Servicio de Hidrografia Naval, 1960.
Operacion Oceanografica Malvinas (Resultados preliminares).
Servicio de Hidrografia Naval, Argentina, Publ., H. 606: numerous unnumbered pp.

Chemistry silicate (data)

Argentina, Servicio de Hidrografia Naval, 1959.
Operacion Oceanografica Atlantico Sur. Resultados preliminares.
Servicio de Hidrografia Naval, Argentina, Publ., H. 608: numerous unnumbered pp.

Chemistry silicate (data)

Argentina, Servicio de Hidrografia Naval, 1959.
Operacion Oceanografica Cuenca. Resultados preliminares.
Servicio de Hidrografia Naval, Argentina, Publ., H. 607: numerous unnumbered pp.

Chemistry silicate (data)

Argentina, Servicio de Hidrografia Naval, 1959.
Operacion Oceanografica Drake, 1. Resultados preliminares.
Servicio de Hidrografia Naval, Argentina, Publ., H. 613: numerous unnumbered pp.

chemistry silicate(data)
Argentina, Servicio de Hidrografia Naval, 1959.
Operacion Oceanografiaa Meridiano. Resultados preliminares.
Servicio de Hidrografia Naval, Argentina, Publ., H. 617: numerous unnumbered pp.

chemistry silicate (data)
Argentina, Servicio de Hidrografia Naval, 1959.
Trabajos oceanograficos realizados en la campana Antartica, 1958/1959. Resultados preliminares.
Servicio de Hidrografia Naval, Argentina, Publ., H. 616:127 pp.

chemistry silicate (data)
Argentina, Servicio de Hidrografia Naval, 1959.
Trabajos oceanograficos realizados en la campana Antartica, 1957-1958. Resultados preliminares.
Servicio de Hidrografia Naval, Argentina, Publ., H. 615: numerous unnumbered pp.

chemistry silicate
Armstrong, F.A.J., 1965.
Silicon.
In: Chemical oceanography, J.P. Riley and G. Skirrow, editors, Academic Press, 1:409-432.

chemistry silicate
Armstrong, F.A.J., 1951.
The determination of silicate in sea water.
J.M.B.A. 30(1):149-160, 3 textfigs.

chemistry silicate
Armstrong, F.A.J., and E.I. Butler, 1963
Chemical changes in sea water off Plymouth in 1963.
J. Mar. Biol. Assoc., U.K., 43(1):75-78.

chemistry silicate
Armstrong, F.A.J., and E.I. Butler, 1962.
Hydrographic surveys off Plymouth in 1959 and 1960.
J. Mar. Biol. Assoc., U.K., 42(2):445-463.

chemistry silicate
Armstrong, F.A.J., and E.I. Butler, 1962.
Chemical changes in sea water off Plymouth during 1960.
J. Mar. Biol. Assoc., U.K., 42(2):253-258.

chemistry silicate
Armstrong, F.A.J., and E.I. Butler, 1960.
Chemical changes in sea water off Plymouth during 1958.
J.M.B.A., 39(2):299-302.

chemistry, silicate
Armstrong, F.A.J., and E.C. LaFond, 1966.
Chemical nutrient concentrations and their relationship to internal waves and turbidity off southern California.
Limnol. Oceanogr., 11(4):538-547.

chemistry silicate
Atkins, W.R.G., 1953.
Seasonal variations in the phosphate and silicate content of sea water. Pt. VI. 1948 compared with the 1923-25 period. J.M.B.A. 31(3):489-492, 4 textfigs.

chemistry silicate
Atkins, W.R.G., 1930.
Seasonal variations in the phosphate and silicate content of sea water in relation to the phytoplankton crop. V. November 1927 to April 1929 compared with earlier years from 1923. J.M.B.A. 16:821-852.

chemistry, silicate
Atkins, W.R.G., 1928
Seasonal variation in the phosphate and silicate content of sea water during 1926 and 1927 in relation to the phytoplankton crop. JMBA, 15: 191-205

B

chemistry, silicate (data)
Băcescu, M., G. Müller, H. Skolka, A. Petran, V. Elian, M.T. Gomoiu, N. Bodeanu, şi S. Stănescu, 1965.
Cercetări de ecologie marină în sectorul predeltaic în condiţiile anilor 1960-1961.
In: Ecologie marină, M. Băcescu, redactor, Edit. Acad. Republ. Pop. Romane, Bucuresti, 1: 185-344.

chemistry silicate
Băcescu, M., M.T. Gomoiu, N. Bodeanu, A. Petran, G.I. Müller si V. Chirila, 1965.
Dinamice populaţiilor animale şi vegetale din zone nisipurilor fine de la nord de Constanţa în condiţiile anilor 1962-1965.
In: Ecologie marină, M. Băcescu, redactor, Edit. Acad. Republ. Pop. Romane, Bucuresti, 2: 7-167.

chemistry, silicate (data)
Ballester, A. 1965.
Hidrografia y nutrientes de la Fosa de Cariaco.
Informe de Progresso del Estudio Hidrografico de la Fosa de Cariaco, Fundación La Salle de Ciencias Naturales Estación de Investigaciones Marinas de Margarita, Caracas, Sept. 1965: 3-12 (mimeographed)

chemistry silicate
*Banoub, M.W., and J.D. Burton, 1968.
The winter distribution of silicate in Southampton Water.
J. Cons. perm. int. Explor. Mer, 32(2):201-208.

chemistry silicate (data)
Bardin, I.P., 1958.
[Hydrological, hydrochemical, geological and biological studies, Research Vessel "Ob", 1955-1957.] Trudy Kompleks. Antarkt. Exped., Akad. Nauk, SSSR, Mezh. Geofiz. God, Gidrometeorol. Izdatel., Leningrad, 217 pp.

chemistry silicates
Bernard, M. F., 1938
Recherches récentes sur la densité du plancton méditerranéen. Rap. Proc. Verb des Réunion, Comm. Int. l'Expl. Sci. de la Méditerranée, n.s., XI:289-300.

chemistry, silicates
Bernard, F., 1937.
Résultats d'une annee de recherches quantitatives sur le phytoplancton de Monaco. Cons. Perm. Int. Expl. Mer, 105:28-39, 8 textfigs.
Rapp. Proc. Verb.

chemistry, silicate (data)
Bernhard, Michel, 1963.
1. Introduzione. 2. Chimica oceanografica.
Rapp. Attività Sci. e Tech., Lab. Studio della Contaminazione Radioattiva del Mare, Fiascherino, La Spezia (maggio 1959-maggio 1962), Comit. Naz. Energia Nucleare, Roma, RT/ BIO, (63), 8:7-39. (multilithed)

chemistry silicate
Bessonov, N.M., 1964
On some features of variations of hydrochemical characteristics in fishery areas off Dakar and Takoradi. (In Russian).
Okeanologiia, Akad. Nauk, SSSR, 4(5):813-824.

chemistry silicate
Bigelow, H.B., and M. Leslie, 1930
Reconnaissance of the waters and plankton of Monterey Bay, July 1928.
Bull. M.C.Z., 70(5):429-481, 43 text figs.

chemistry, silicate
Bogoyavlensky, A.N., 1959
[Hydrochemical investigations.] Arktich. i Antarktich. Nauchno-Issled. Inst., Mezhd. Geofiz. God, Sovetsk. Antarkt. Eksped., 5: 159-172.

chemistry silicate
Bogoiavlenskii, A.N., 1958.
[VI. Hydrochemical work.] Opisanie Exped. D/E "Ob", 1955-1956, MGG, Trudy Kompaeksnoi Antarkt. Exped., Akad. Nauk, USSR, 91-102.

chemistry silicate
Bogojavlensky, A.N. and O.V. Shishkina 1971.
On the hydrochemistry off Peru and Chile. (In Russian; English abstract)
Trudy Inst. Okeanol. P.P. Shirshova Akad. Nauk SSSR 89: 91-105

chemistry silicate
Broenkow, William W., 1965.
The distribution of nutrients in the Cosa Rica dome in the eastern tropical Pacific Ocean.
Limnology and Oceanography, 10(1):40-52.

chemistry silicate
Bruevich, S.W., 1960
[Hydrochemical investigations on the White Sea.] Trudy Inst. Okeanol., 42: 199-254.

chemistry, silicate(data)
Brujewicz, S.W., 1957.
On certain chemical features of waters and sediments in north-west Pacific.
Proc. UNESCO Symp., Phys. Ocean., Tokyo, 1955: 277-292.

chemistry silicate
Bruevich, S.W., A.N. Bogoyavlensky, and V.V. Mokievskaya, 1960
[Hydrochemical features of the Okhotsk Sea.] Trudy Inst. Okeanol., 42: 125-198.

chemistry silicate(data)
Bruevich, S.V., and I.A. Chikina, 1933
[Hydrochemical observations in the northern part of the Kola Fjord (Barents Sea) in summer 1931.] Trudy Gosud. Okeanogr. Inst., 3(3): 120-124.

chemistry silicate(data)
Bruneau, L., N.G. Jerlov, F.F. Koczy, 1953.
Physical and chemical methods.
Repts. Swedish Deep-sea Exped., 1947-1948, Phys. Chem., 3(4):101-112, i-iv, 7 textfigs., 2 tables.

chemistry silicate
Buch, K., 1934.
Beobachtungen über chemische Faktoren in der Nordsee, zwischen Nordsee und Island sowie auf dem Schelfgebiete nördlich von Island. Rapp. Proc. Verb. 89(3):13-31, 8 tables, 2 textfigs.

chemistry silicate
Burns, R.B., 1967.
Chemical observations in the North Sea in 1965.
Annls biol. Copenh. (1965)22:29-31.

chemistry, silicate
Burns, R.B., and R. Johnston, 1966.
Chemical observations from 1964 in the northern North Sea.
Annls. biol., Copenh., 21:29-36.

chemistry silicate
Burns, R.B., and R. Johnston, 1965.
Chemical observations on the waters of the Scottish west coast and the western approaches.
Ann. Biol., Cons. Perm. Int. Expl. Mer, 1963, 20:

chemistry silicate
Burns, R.B., and R. Johnston, 1965.
Chemical observations in the northern North Sea, 1963.
Ann. Biol., Cons. Perm. Int. Expl. Mer, 1963, 20:57-60.

Chemistry, silicate (data)
Burns, R.B., and R. Johnston, 1964.
Chemical observations.
Ann. Biol., Cons. Perm. Int. Expl. Mer, 1962, 19: 42-44.

Chemistry, silicate
Burns, R.B., and R. Johnston, 1964.
Chemical and hydrographic survey of the North Channel and Clyde Sea area in April, 1962.
Ann. Biol., Cons. Perm. Int. Expl. Mer, 1962, 19: 32-33.

silicate (data)
Burns, R.B., and R. Johnston, 1963.
Chemical observations.
Ann. Biol., Cons. Perm. Int. Expl. Mer, 1961, 18: 47-48.

C

Chemistry, silicate (data)
California, Humboldt State College, 1964.
An oceanographic study between the points of Trinidad Head and the Eel River.
State Water Quality Control Bd., Resources Agency of California, Sacramento, Publ., No. 25: 136 pp.

Chemistry silicate
Chidambaram, K., A.D.I. Rajandran, and A.P. Valsan, 1951.
Certain observations on the hydrography and biology of the pearl bank, Thoylayviam Paar off Tuticorin in the Gulf of Manaar in April 1949.
J. Madras Univ., Sect. B, 21(1):48-74, 2 figs.

Chemistry silicate
Choe, Sang, 1969.
Phytoplankton studies in Korean waters. III. Surface phytoplankton survey of the north-eastern Korea Strait in May of 1967. (In Korean; English abstract).
J. oceanogr. Soc. Korea, 4(1):1-8.

Chemistry silicate
*Choe, Sang, Tai Wha Chung and Hi-Sang Kwak, 1968.
Seasonal variations in nutrients and principal ions contents of the Han River water and its water characteristics. (In Korean; English abstract).
J. oceanogr. Soc., Korea, 3(1):26-38.

Chemistry, silicate
Chow, D.T.-W., and R.J. Robinson, 1953.
Forms of silicate available for colorimetric determination. Analyt. Chem. 25:646-648, 2 textfigs.

Chemistry chemistry, silicate
Chow, Tsaihwa J., and Arnold W. Mantyla, 1965.
Inorganic nutrient anions in deep ocean waters.
Nature, 206(4982):383-385.

Also in:
Collected Reprints, Int. Indian Ocean Exped., UNESCO, 3. 1966.

Chemistry, silicates
Clowes, A. J., 1938.
Phosphate and silicate in the Southern Ocean.
Discovery Repts. 19: 120 pp., 25 pls., 29 textfigs.

Chemistry, silicate
Codispoti, Louis A. 1968.
Some results of an oceanographic survey in the northern Greenland Sea, summer 1964.
Techn. Rept. U.S. Nav. Oceanogr. Off. TR 202: 49 pp.

chemistry silicate
Codispoti, L.A., and F.A. Richards 1968.
Micronutrient distributions in the East Siberian and Laptev seas during summer 1963.
Arctic 21(2): 67-83.

Chemistry, silicate (data)
Conseil Permanent International pour l'Exploration de la Mer, 1952.
Bulletin Hydrographique, 1948: 87 pp.

silicate
Cooper, L.H.N., 1952.
Factors affecting the distribution of silicate in the North Atlantic Ocean and the formation of North Atlantic Deep Water. J.M.B.A. 30:511-526, 2 textfigs.

Chemistry, SILICATES
Cooper, L.H.N., 1948.
The nutrient balance in the sea. Research 1: 242-248, 2 textfigs.

Chemistry silicate
Cooper, L.H.N., P.G.W. Jones and A.J. Lee, 1962
Hydrographic conditions along the 14°W meridian southwest of Ireland, July 1960.
Ann. Biol., Cons. Perm. Int. Expl. Mer, 17(1960): 73-76.

Chemistry, silicate
Correns, C.W., 1950.
Faktoren der Sedimentbildung erläutet an Kalk- und Kieselsedimenten. Deutsche Hydro. Zeits. 3(1/2):83-88, 2 textfigs.

Chemistry, silicate (data)
Corwin, Nathaniel, and David A. McGill, 1963.
Nutrient distribution in the Labrador Sea and Baffin Bay.
U.S.C.G., Bull., No. 48:79-94; 95-153.

Chemistry, silicate (data)
Coste, Bernard, et Hans-Joachim Minas 1967.
Premières observations sur la distribution des taux de productivité et des concentrations en sels nutritifs des eaux de surface du golfe du Lion.
Cah. océanogr. 19(5): 417-429.

Chemistry, silicate (data)
COSTE, B., et H.J. MINAS, 1968.
Production organique primaire et sels nutritifs au large des côtes occidentales Corso-Sardes en février 1966.
Rec. Trav. Sta. mar. Endoume, 44(60): 49-61

Chemistry silicate
Crehuet, Ramón Fernández y Maria Jesús del Val Cordón, 1960
Observaciones oceanograficas en la Bahia de Malaga (Marzo a marzo 1957). Bol. Inst. Esp. Ocean., 98: 1-29.

D

Chemistry silicate
Davidovitch, R.L., 1964.
Short chemical characteristics of the northwest Pacific Ocean. (In Russian).
Gosudarst. Kom. Sov. Ministr., SSSR, Ribn. Choz., Trudy VNIRO, 49, Izv. TINRO, 51:93-98.

Chemistry silicate (data)
Davidovich, R.L., 1963
Hydrochemical features of the southern and southeastern parts of the Bering Sea. (In Russian).
Sovetsk. Ribokh. Issled. B Severo-Vostokh. Chasti Tikhogo Okeana, VNIRO 48, TINRO 50(1): 85-96.

Chemistry silicates (geochemical)
Degens, Egon T., 1965.
Geochemistry of sediments: a brief survey.
Prentice-Hall, Inc., 342 pp.

Chemistry silica (data)
Donnelly, P.V., J. Vuille, M.S. Jayaswal, R.A. Overstreet, J. Williams, M.A. Burklew and R.M. Ingle, 1966.
A study of contributory chemical parameters to red tide in Apalachee Bay.
Florida Bd., Conserv., St. Petersburg, Mar. Lab. Prof. Papers Ser. No. 8:43-83.

Chemistry silicate
*Duxbury, Alyn C. and Noel B. McGary, 1968.
Local changes of salinity and nutrients off the mouth of the Columbia River.
Limnol. Oceanogr., 13(4):626-636.

E

Chemistry silicate (data)
El-Sayed, Sayed Z., and Enrique F. Mandelli, 1965.
Primary production and standing crop of phytoplankton in the Weddell Sea and Drake Passage.
In: Biology of Antarctic seas. II.
Antarctic Res. Ser., Am. Geophys. Union, 5:87-106.

Chemistry silicate
Emery, K. O., and J. Hulsemann, 1961 (1962).
The relationships of sediments, life and water in a marine basin.
Deep-Sea Res., 8(3/4):165-180.

Chemistry silicate
*Ewins, P.A., and C.P. Spencer, 1967.
The annual cycle of nutrients in the Menai Straits.
J. mar. biol. Ass., U.K., 47(3):533-542.

F

Chemistry silicate (data)
Faganelli, Armando, 1961.
Primi risultati relativi alla concentrazione dei sali nutritivi nelle acque del Mare Mediterraneo Centrale e mari adiacenti.
Archivio Oceanograf. e Limnol., 12(2):191-208.

Also:
Ist. Sperimentale Talassograf., Trieste, Pubbl. No. 377 (1961).

Chemistry silicate (data)
Faganelli, A., 1961.
Primi risultati relativi alla concentrazione dei sali nutritivi nelle acque del Mar Mediterraneo centrale et mari adiacenti.
Rapp. Proc. Verb., Réunions, Comm. Int. Expl. Sci., Mer Méditerranée, Monaco, 16(3):675-686.

Chemistry silicate
Fedosov, M.V., and E. G. Vinogradova, 1955.
Hydrological and hydrochemical regime, primary production in the Azov Sea and forecasting change. Fundamental basis of the hydrochemical regime in the Azov Sea. Trudy VNIRO, 31(1):9-34.

Chemistry silicate
Filatov, K.V., 1961.
Silicate water and its place in horizontal hydrochemical zoning. (In Russian).
Doklady, Akad. Nauk, SSSR, 138(3):663-666.

Translation: Consultants Bureau for Amer. Geol. Inst., (Earth Sciences Section only), 138(1-6):638-641.

Chemistry, silicate
Flores, P., Luis Alberto 1967.
Informe preliminar del crucero 6611 de la primavera de 1966 (Cabo Blanco - Punta Coles).
Informe Inst. Mar Peru 17: 16 pp.

Chemistry, silicate
Flores P., Luis Alberto, y Luis A. Pома Elias 1967.
Informe preliminar del crucero 6608-09 de invierno 1966 (Mancora-Ilo).
Informe. Inst. Mar, Peru 16: 24 pp.

Chemistry silicate
Fonselius, Stig H., 1960.
Hydrography of the Baltic deep basins. III.
Rept., Fish. Bd., Sweden, Ser. Hydrogr., 23: 1-97.

Chemistry
Silicate (data)

Forsbergh, Eric D., William W. Broenkow, 1965.

Observaciones oceanograficas del oceano Pacific oriental recolectadas por el barco Shoyo Maru, octubre 1963-marzo 1964. Comision Interamericana del Atun Tropical, Bol. 10(2):85-237.

Chemistry, silicates

Fukai, R., 1954.
Seasonal variation of minor chemical constituents in the waters of the Zunan-Kuroshio region. II. Silicates. Bull. Chem. Soc., Japan, 27:408-412.

Chemistry silicate

Fukai, R., 1954.
Critical studies on the analytical methods for minor chemical constituents in sea water. Pt. 2. Some notes on the determination of silicate. J. Ocean. Soc., Japan, 10(4):201-208, 6 textfigs.

Chemistry, silicates (data)

Fuse, Shin-ichiro, 1959
A study on the productivity of Tanabe Bay, III.
Oceanographic conditions of Tanabe Bay (2) Stratification and fluctuation of hydrological conditions on two sectional survey lines.
Rec. Oceanogr. Wks., Japan. Spec. No. 3:31-45.

Chemistry silicates (data)

Fuse, Shin-ichiro, and Eiji Harada, 1960

A study on the productivity of Tanabe Bay.(3) Result of the survey in the summer of 1958. Rec. Oceanogr. Wks., Japan, Spec. No. 4: 13-28.

Chemistry silicate

Fuse, S., I. Yamazi and E. Harada, 1958.
A study on the productivity of the Tanabe Bay. (Part 1). 1. Oceanographic conditions of the Tanabe Bay. Results of the survey in the autumn of 1956.
Rec. Oceanogr. Wks., Japan, Spec. No. 2:3-9.

G

Chemistry, silicate (data)

Ganapati, P.N., E.C. LaFond and P.V. Bhavanarayana, 1956.
On the vertical distribution of chemical constituents in the shelf waters off Waltair.
Proc. Indian Acad. Sci., 44:68-71.

Chemistry silicate

Ganapati, P.N., and D.V. Subba Rao, 1958.
Quantitative study of plankton off Lawson's Bay, Waltair.
Proc. Indian Acad. Sci., (B), 48(4):189-209.

Chemistry silicate

Ganapati, P.N., and D.V. Subba Rao, 1957.
On upwelling and productivity of the waters off Lawson's Bay, Waltair. Current Science, 26(11): 347-348.

Chemistry silicate

Ganapati, P.N., and D. Venkata Rama Sarma, 1958.
Hydrography in relation to the production of plankton off Waltair coast.
Andhra Univ. Mem. Oceanogr., 2:168-192.

Chemistry
silicate

Goodman, J.R., J.H. Lincoln, T.G. Thompson, and F.A. Zeusler, 1942.
Physical and chemical investigations: Bering Sea, Bering Strait, Chukchi Sea during the summers of 1937 and 1938. Univ. Washington Publ. Ocean., 3(4):105-169, 37 maps.

Chemistry, silicate

Graham, H. W., and E. G. Moberg, 1944
Chemical results of the last cruise of the Carnegie. Chemistry I. Scientific results of cruise VII of the Carnegie during 1928-1929 under command of Captain J. P. Ault. Carnegie Publ. Washington 562, 58 pp., 7 maps.

Chemistry
silica

Grasshoff, K., 1964
On the determination of silica in sea water.
Deep-Sea Res., 11(4):597-604.

---

chemistry, silicate

Grill, E.V., 1970.
A mathematical model for the marine dissolved silicate cycle. Deep-Sea Res., 17(2): 245-266.

Chemistry, silicate

Guillen G., Oscar, y Luis Alberto Flores P. 1968.
Informe preliminar del crucero 6702 del verano de 1967 (Cabo Blanco- Arica).
Informe Inst. Mar. Peru 18:17 pp.

---

chemistry, silicate (data)

Guillen, Oscar, y Francisco Vásquez 1966.
Informe preliminar del Crucero 6602 (Cabo Blanco - Arica).
Informe Inst. Mar Peru 12: 27 pp.

H

Chemistry silicates

Halim, Y., 1960

Observations on the Nile bloom of phytoplankton in the Mediterranean. J. du Cons. 26(1): 57-67.

Chemistry, silicate(data)

Hasegawa, Y., M. Yokoseki, E. Fukuhara, and K. Terai, 1952.
On the environmental conditions for the culture of laver in Usu-Bay, Iburi Prov., Hokkaido.
Bull. Hokkaido Regional Fish. Res. Lab. 6:1-24, 8 textfigs., 9 tables.

Chemistry, silicate

Hendey, N.I., 1951
Littoral diatoms of Chicester Harbour with special reference to fouling. J.Roy. Microscop. Soc. 71(1): 1-86, 18 pls.

Chemistry - Silicate

Hollan, Eckard, 1970.
Eine physikalische Analyse kleinräumiger Änderungen chemischer Parameter in den tiefen Wasserschichten der Gotlandsee.
Kieler Meeresforsch. 25(2): 255- 267

Chemistry, silicate

Holm-Hansen, Osmund, J.D.H. Strickland and P.M. Williams, 1966.
A detailed analysis of biologically important substances in a profile off southern California.
Limnol. Oceanogr., 11(4):548-561.

Chemistry, silicate

Hung, Tsu-Chang, and Ching-Wei Lee 1967.
Research on chemical oceanography in the Kuroshio around Taiwan Island.
Bull. Inst. Chem., Acad. Sinica, Taiwan 14: 80-102.

I

Chemistry
silicates

Ichie, Takashi, 1951
[On the hydrography of the Kii-Suido (1951)]
Bull. Kobe Mar. Obs., No. 164:253-278(top of page); 35-60(bottom of page).

Chemistry, silicate

Ichie, T., 1949.
Report on the oceanographical observations on board R.M.S. "Syunpu Maru" in the Akashi Seto and the Yura Seto in March 1949. Papers and Repts., Ocean., Kobe Mar. Obs., Ocean. Dept., No. 4:10 pp. numerous figs. and tables. (atlas-sized pages - mimeographed).

Chemistry silicate

Ichie, T., K. Tanioka, and T. Kawamoto, 1950.
Reports of the oceanographical observations on board the R.M.S. "Yushio Maru" off Shionomisaki (Aug. 1949). Papers and Repts., Ocean., Kobe Mar. Obs., Ocean. Dept., No. 5:15 pp., 33 figs. (Odd atlas-sized pages - mimeographed).

Chemistry, silicate

Ignatiades, Lydia, and Theano Becacos-Kontos 1969.
Nutrient investigations in lower Saronikos Bay, Aegean Sea.
Vie Milieu (B) 20(1):51-61.

J

Chemistry
silicate (data)

Jacobs, Stanley S., Peter M. Bruchausen and Edward B. Bauer, 1970
Cruises 32-36, 1968, hydrographic stations, bottom photographs, current measurements. Eltanin Repts, Lamont-Doherty Geol. Obs., Nat. Sci. Found., U.S. Antarctic Res. Program, 460 pp. (multilithed)

---

silicate

Japan,
Central Meteorological Observatory, 1949
Report on sea and weather observation on Antarctic Whaling Ground (1947-48). Ocean. Mag., Japan, 1(1):49-88, 17 text figs.

Japan Kobe Marine Observatory

Chemistry - Silicate

Japan Kobe Marine Observatory, 1967.
Report of the oceanographic observations in the sea south of Honshu from February to March 1964. (In Japanese)
Bull. Kobe Mar. Obs. No. 178: 27-

Chemistry, silicate

Japan, Kobe Marine Observatory, Oceanographical Section, 1964.?
Researches in chemical oceanography made at K.M.O. for the recent ten years (1954-1963). (In Japanese).
Weather Service Bull., Japan. Meteorol. Agency, 31(7):188-191.

Also in:
Bull. Kobe Mar. Obs., No. 175. 1965.

Chemistry
silicate

Japan, Kobe Marine Observatory, 1961
[Report of the oceanographic observations in the sea south of Honshu in March 1958.]
Bull. Kobe Mar. Obs., No. 167(21-22):30-36.
--from May to June, 1958(21-22):37-42
--fromJuly to September, 1958(23-24):34-40
--from October to December, 1958(23-24):41-47
--from February to March, 1959 (25-26):33-47.

Chemistry silicates

Japan, Kobe Marine Observatory, 1956.
[Report of oceanographic observations in the sea south of Honshu in August 1955.]
J. Ocean., Kobe (2)7(3):23-32.

Chemistry silicate

Japan, Kobe Marine Observatory, 1956.
[Report of the oceanographical observations in the sea south off Honshu in May 1955.]
J. Ocean., Kobe, (2)7(2):17-25.

Pagination of previous article repeated.

Chemistry silicate
Japan, Kobe Marine Observatory, 1956?
[Report on the oceanographical observations south off Honshu in 1954.] J. Ocean., Kobe (2)7(2):46-69.

Chemistry silicate
Japan, Kobe Marine Observatory, 1955.
[The outline of the oceanographical observation on the Southern Sea of Japan on board R.M.S. "Syunpu-maru" (Jan. 1955).]
J. Ocean., Kobe Mar. Obs., (2)6(1):1-19, 10 figs.

Chemistry silicate
Japan, Kobe, Marine Observatory, 1954.
[The outline of the oceanographical observations in the southern area of Honshu on board the R.M.S. "Syunpu maru".]
Aug.-Sept. 1954 - J. Ocean. (2)5(9):1-44
Oct. 1954 - J. Ocean. (2)5(10):1-44.

Chemistry silicate
Japan, Kobe Marine Observatory, 1954.
[The outline of the oceanographical observations off Shionomisaki on board the R.M.S. "Syunpu-maru" (May 1954).] J. Ocean. (2)5(5):1-11, 14 figs

Chemistry silicate
Japan, Kobe Marine Observatory, 1954.
[The results of the regular monthly oceanographical observations on board the R.M.S. "Syunpu-maru" in the Kii Suido and Osaka Wan.] J. Ocean. (2)5(1):1-19, 16 figs.

Chemistry silicate
Japan, Kobe Marine Observatory, 1953.
[On the chemical conditions in the northern part of Osaka Bay after heavy rain (July 1952).]
J. Ocean. (2)4(4):1-9, 5 figs.

Chemistry silicate
Japan, Kobe Marine Observatory, 1953.
[The results of the regular monthly oceanographical observations on board the R.M.S. "Syunpu-maru" in the Kii Suido and Osaka Wan.]
Jan. 1953 - J. Ocean. (2)4(1):1-10, 9 figs.
Feb. 1953 - J. Ocean. (2)4(2):1-9, 8 figs.
Mar. 1953 - J. Ocean. (2)4(3):1-10, 9 figs.
Apr. 1953 - J. Ocean. (2)4(5):1-9, 9 figs.
May 1953 - J. Ocean. (2)4(6):1-12, 9 figs.
June 15-22, 1953 - J. Ocean. (2)4(6):13-22, 8 figs.
July 1953 - J. Ocean. (2)4(8):1-25, 29 figs.
Aug. 1953 - J. Ocean. (2)4(9):1-30, 27 figs.
Sept. 1953 - J. Ocean. (2)4(10):1-13, 9 figs.
Oct. 1953 - J. Ocean. (2)4(11):1-21, 15 figs.
Nov. 11-18, 1953 - J. Ocean. (2)4(12):1-15, 12 figs.

Japan, Maizuru Marine Observatory

Chemistry silicate
Japan, Maizuru Marine Observatory, 1968.
Report of the oceanographical observations in the Japan Sea from August to September, 1964. (In Japanese).
Bull. Maizuru mar. Obs., 10:74-86.

Chemistry silicate
Japan, Maizuru Marine Observatory, 1967.
Report of the oceanographic observations in the Japan Sea from May to June, 1964.
Bull. Maizuru mar. Obs., 10:65-76.
(In Japanese)

Chemistry silicate
Japan, Maizuru Marine Observatory, 1965.
Report of the oceanographic observations in the central part of the Japan Sea from February to March, 1962.---in the Japan Sea from June to July 1962.---in the western part of Wakasa Bay from January to April, 1962.---in the central part of the Japan Sea from September to October, 1962.---in the western part of Wakasa Bay from May to November, 1962.---in the central part of the Japan Sea in March, 1963.---in the Japan Sea in June, 1963.---in Wakasa Bay in July, 1963.---in the central part of the Japan Sea in October, 1963. (In Japanese).
Bull. Maizuru Mar. Obs., No. 9:67-73; 74-88; 89-95; 71-80; 81-87; 59-65; 66-77; 80-84; 85-91.

Chemistry silicate
Japan, Maizuru Marine Observatory, 1963
Report of the oceanographic observations in the central part of the Japan Sea in October, 1961. (In Japanese).
Bull. Maizuru Mar. Obs., No. 8:78-88.

Chemistry silicate
Japan, Maizuru Marine Observatory, 1963
Report of the oceanographic observations in the central part of the Japan Sea in September, 1960. (In Japanese).
Bull. Maizuru Mar. Obs., No. 8:60-68.

Chemistry silicate
Japan, Maizuru Marine Observatory, 1963
Report of the oceanographic observations in the Japan Sea in June 1961. (In Japanese).
Bull. Maizuru Mar. Obs., No. 8:59-79.

Chemistry silicate
Japan, Maizuru Marine Observatory, 1963
Report of the oceanographic observations in the Japan Sea from May to June, 1960. (In Japanese).
Bull. Maizuru Mar. Obs., No. 8:56-67.

Chemistry silicate
Japan, Maizuru Marine Observatory, 1963
Report of the oceanographic observations in the Wakasa Bay in August, 1961. (In Japanese)
Bull. Maizuru Mar. Obs., No. 8:89-95.

Chemistry silicate
Japan, Maizuru Marine Observatory, 1963
Report of the oceanographic observations in the Wakasa Bay from July to August, 1960. (In Japanese).
Bull. Maizuru Mar. Obs., No. 8:69-75.

Chemistry silicate
Japan, Maizuru Marine Observatory, 1963
Report of the oceanographic observations in the western part of Wakasa Bay from January to May, 1961. (In Japanese).
Bull. Maizuru Mar. Obs., No. 8:80-90.

Chemistry silicate
Japan, Maizuru Marine Observatory, 1963
Report of the oceanographic observations in the western part of Wakasa-Bay from January to June 1960. (In Japanese)
Bull. Maizuru Mar. Obs., No. 8:68-79, 59.

Chemistry silicate
Japan, Maizuru Marine Observatory, 1963
Report of the oceanographic observations in the western part of Wakasa Bay from June to December, 1960. (In Japanese).
Bull. Maizuru Mar. Obs., No. 8:76-83.

Chemistry silicate
Japan, Maizuru Marine Observatory and Hakodate Marine Observatory, Oceanographical Sections, 1962.
Report of the oceanographic observations in the Japan Sea in June, 1961. (In Japanese).
Res. Mar. Meteorol. and Oceanogr. Obs., Jan.-June, 1961, No. 29:59-79.

Chemistry silicate
Japan, Maizuru Marine Observatory, Oceanographical Section, 1962
Report of the oceanographic observations in the western part of Wakasa Bay from June to May, 1961. (In Japanese).
Res. Mar. Meteorol. and Oceanogr. Obs., Jan.-June, 1961, No. 29:80-90.

Chemistry silicate
Japan, Maizuru Marine Observatory, 1962
Report of the oceanographic observations in the central part of the Japan Sea in September 1960. (In Japanese).
Res. Mar. Meteorol. and Oceanogr., July-Dec., 1960, Japan Meteorol. Agency, No. 28: 60-68.

Chemistry silicate
Japan, Meteorological Agency, 1962
Report of the Oceanographic observations in the sea south and west of Japan from October to November, 1960. (In Japanese).
Res. Mar. Meteorol. and Oceanogr., July-Dec., 1960, Japan Meteorol. Agency, No. 28:30-35.

Chemistry silicate
Japan, Maizuru Marine Observatory, 1962
Report of the oceanographic observations in the Wasaka Bay from July to August, 1960. (In Japanese).
Res. Mar. Meteorol. and Oceanogr., July-Dec., 1960, Japan Meteorol. Agency, No. 28:69-75.

Chemistry silicate
Japan, Maizuru Marine Observatory, 1962
Report of the oceanographic observations in the western part of Wakasa Bay from June to December 1960. (In Japanese).
Res. Mar. Meteorol. and Oceanogr., July-Dec., 1960, Japan Meteorol. Agency, No. 28:76-83.

Chemistry silicate
Japan, Maizuru Marine Observatory, Oceanographical Section, 1961
[Report of the oceanographic observations in the sea north of Kyoga-misaki from August to December 1958.]
Bull. Maizuru Mar. Obs., No. 7:60-63.

Chemistry silicate
Japan, Maizuru Marine Observatory, Oceanographical Observatory, 1961
[Report of the Oceanographic observations in the western part of Wakasa Bay from August to December, 1959.]
Bull. Maizuru Mar. Obs., No. 7:68-74.

Chemistry silicate
Japan, Maizuru Marine Observatory, Oceanographical Section, 1961
[Report of the oceanographic observations off Kyoga-misaki in the Japan Sea in April and May, 1958.]
Bull. Maizuru Mar. Obs., No. 7:55-60.

Chemistry silicate
Japan, Maizuru Marine Observatory, Oceanographical Section, 1961
[Report of the oceanographic observations in the Japan Sea from June to July, 1959.]
Bull. Maizuru Mar. Obs., No. 7:57-64.

Chemistry silicate
Japan, Maizuru Marine Observatory, Oceanographical Section, 1961
Report of the oceanographic observations in the Japan Sea from June to July, 1958.
Bull. Maizuru Mar. Obs., No. 7:60-67.

Chemistry silicate
Japan, Maizuru Marine Observatory, Oceanographical Section, 1961
[Report of the oceanographic observations off Kyoga-misaki and in the western part of Wakasa Bay from February to June 1959.]
Bull. Maizuru Mar. Obs., No. 7:65-72, 61.

Chemistry silicate
Japan, Maizuru Marine Observatory, Oceanographical Section, 1961.
[Report of the oceanographic observations in the sea north of San'in District in August. 1958.]
Bull. Maizuru Mar. Obs., No. 7: 64-68.

Chemistry silicate
Japan, Maizuru Marine Observatory, Oceanographical Section, 1961
[Report of the oceanographic observations in the Wakasa Bay from July to August, 1959.]
Bull. Maizuru Mar. Obs., No. 7:62-67.

Chemistry silicate
Japan, Maizuru Marine Observatory, Oceanographical Section, 1961
[Report of the oceanographic observations in the western part of Wakasa Bay from January to March, 1958.]
Bull. Maizuru Mar. Obs., No. 7:50-54.

Chemistry silicate
Japan, Maizuru Marine Observatory, Oceanographical Section, 1961
[Report of the oceanographic observations off Kyoga-misaki in Japan Sea from October to December, 1957.]
Bull. Maizuru Mar. Obs., No. 7:29-36.

Chemistry silicate
Japan, Maizuru Marine Observatory, 1958
[Report of the oceanographic observations off Kyoga-misaki in the Japan Sea in January and March, 1957.]
Bull. Maizuru Mar. Obs., No. 6: 129-137. [45-53.]

Chemistry silicate
Japan, Maizuru Marine Observatory, 1958
[Report of the oceanographic observations north off Kyoga-misaki and off the east part of the Sanin district in April to June, 1957.]
Bull. Maizuru Mar. Obs., No. 6: 234-243. [95-105.]

Chemistry silicate
Japan, Maizuru Marine Observatory, 1958
[Report of the oceanographic observations off Kyoga-misaki in Japan Sea in October and November, 1956.]
Bull. Maizuru Mar. Obs., No. 6: [37-44.]

Chemistry silicate
Japan, Maizuru Marine Observatory, 1958
[Report of the oceanographic observations in the sea north of Sanin and Hokuriku districts in August 1957.]
Bull. Maizuru Mar. Obs., No. 6: 245-253. [107-115.]

Chemistry silicate
Japan, Maizuru Marine Observatory, 1958
[Report of the oceanographic observations north of Sanin and Hokuriku districts in summer 1956]
Bull. Maizuru Mar. Obs., No. 6: 157-unnumbered. [53-85.]

Chemistry silicate
Japan, Maizuru Marine Observatory, 1956
[Report of the serial observations off Kyoga-misaki during the first half of 1955.]
Bull. Maizuru Mar. Obs., (5): 27-32.

Chemistry silicate
Japan, Maizuru Marine Observatory, 1956.
[Report of the oceanographical observations north off Sanin and Hokuriku districts in summer 1954.]
[Report of the oceanographical observations north off Sanin and Hokuriku districts in summer 1955.]
Bull. Maizuru Mar. Obs., (5):85-94; 49-56.

Chemistry silicate
Japan, Maizuru Marine Observatory, 1956
[Report of the oceanographic observations off Kyoga-misaki during the latter half of 1955.]
Bull. Maizuru Mar. Obs., (5): 31-37.

Chemistry silicates
Japan, Maizuru Marine Observatory, 1956.
[Report of the serial observations off Kyoyamisaki during the first half of 1955.]
Bull. Maizuru Mar. Obs., (5):27-32.

Chemistry silicate
Japan, Maizuru Marine Observatory, 1956.
[Report of the oceanographic observations off Kyoga-misaki during the latter half of 1955.]
Bull. Maizuru Mar. Obs., (5):31-37.

Chemistry silicate
Japan, Maizuru Marine Observatory, 1955.
[Report of the oceanographical observations taken] off San'in and Hokuriku (July-August, 1953).]
Bull. Maizuru Mar. Obs., (4):13-27. 217-231.

Japan Marine Safety Board

Chemistry silicate (data)
Japan, Maritime Safety Board, 1954.
[Tables of results from oceanographic observations in 1950.] Publ. 981, Hydrogr. Bull., Maritime Safety Bd., Spec. Number (Ocean. Repts) No. 14:26-164, 5 textfigs.

Chemistry silicate
Japan, Nagasaki Marine Observatory 1971
Report of the oceanographic observations in the sea south of Kyushu in April, 1966 (In Japanese).
Oceanogr. Met. Nagasaki mar. Obs. 18: 53-57.

Chemistry silicate
Japan National Committee for the International Geophysical Year 1960.
The results of the Japanese oceanographic project for the International Geophysical Year 1957/58: 198 pp.

Chemistry silicate
Japan, Science Council, National Committee for IIOE, 1966.
General report of the participation of Japan in the International Indian Ocean Expedition.
Rec. Oceanogr. Wks., Japan, n.s. 8(2): 133 pp.

Chemistry silicate
Japan, Tokai Regional Fisheries Research Laboratory, 1959.
IGY Physical and chemical data by the R. V. Soyo-maru, 25 July-14 September 1958.
17 pp. (multilithed).

Chemistry silicates
Jayaraman, R., 1951.
Observations on the chemistry of the waters of the Bay of Bengal off Madras City during 1948-1949.
Proc. Indian Acad. Sci., B, 33(2):92-99, figs.

Chemistry silicate
Jerlov, N.G., 1956.
The Equatorial currents in the Pacific Ocean.
Repts. Swedish Deep-Sea Exped., 3(Phys. Chem. 6):129-154.

Chemistry silicate
Jones, P.G.W., 1971.
The southern Benguela Current region in February 1966: Part I. Chemical observations with particular reference to upwelling. Deep-Sea Res., 18(2): 193-208.

Chemistry silicate
Jørgensen, E.G., 1953.
Silicate assimilation by diatoms.
Physiol. Plant., 6:301-315.

K

Chemistry silicate(data)
Kalle, K., 1956.
Chemisch-hydrographische Untersuchungen in der inneren Deutschen Bucht. Deutsche Hydrogr. Zeits. 9(2):55-65.

Chemistry silicate
Kalle, K., 1939
V. Die chemischen Arbeiten auf der "Meteor" Fahrt Januar bis Mai 1938. Bericht über die zweite Teilfahrt des Deutschen Nordatlantischen Expedition des Forschungs-und Vermessungsschiffes "Meteor", Januar bis Juli, 1938. Ann. Hydro. u. Mar. Meteorol. 1939: 23-30, 6 text figs.

Chemistry silicate
Kato, K., 1966.
Chemical investigations on the hydrographic system of Cananéia lagoon.
Bolm Inst. Oceanogr., S. Paulo, 15(1):1-12.

Chemistry silicate
Kato, Kenji, 1961.
Oceanographical studies in the Gulf of Cariaco.
1. Chemical and hydrographical observations in January, 1961.
Bol. Inst. Oceanograf., Univ. Oriente, Cumana, Venezuela, 1(1):49-72.

Chemistry silicate(data)
Kato, Kenji, 1961
Some aspects on biochemical characteristics of sea water and sediments in Mochima Bay, Venezuela.
Bol. Inst. Oceanogr., Univ. de Oriente, Venezuela 1(2):343-358.

Chemistry silicate
Kitamura, Hiroyuki, 1958.
[Seasonal variation of the silicate silicon in the Kuroshio. Chemical oceanography of the North Pacific. III.]
Umi to Sora, 33(8):51-57. 34(2)

Chemistry silicate
Kitamura, Hiroyuki, and Takeshi Sagi, 1965.
On the chemical elements in the sea south of Honshu, Japan, II. (In Japanese;English abstract).
Bull. Kobe Mar. Obs., No. 174:39-55.

Chemistry silicate
Kitamura, Hiroyuki, and Takeshi Saga, 1964.
On the chemical elements in the sea south of Honshu, Japan. (In Japanese; English abstract).
Bull. Kobe Mar. Obs., No. 172:6-54.

Chemistry silicate(data)
Klepikova, V.V., Edit., 1961
Third Marine Expedition on the D/E "Ob", 1957-1958. Data.
Trudy Sovetskoi Antarktich. Exped., Arktich. i Antarktich. Nauchno-Issled. Inst., Mezhd. Geofiz. God, 22:1-234 pp.

Chemistry silicate
*Krishnamurthy,K.,1967.
The cycle of nutrient salts in Porto Novo(India) water.
Int.Rev.ges.Hysrobiol., 52(3):427-436.

Chemistry silicate
Kusunoki, Kou, 1962
Hydrography of the Arctic Ocean with special reference to the Beaufort Sea.
Contrib. Inst., Low Temp. Sci., Sec. A,(17): 1-74.

L

silicates
Laktionov, A.F., V.A. Shamontyev and A.V. Yanes, 1960
Oceanographic characteristics of the North Greenland Sea.
Soviet Fish. Invest., North European Seas, VNIRO, PNIRO, Moscow, 1960: 51-65.

Chemistry silicate (data)
Lee, Min Jai, Jae Hyung Shim and Chong Kyun Kim, 1967
Studies on the plankton of the neighboring seas of Korea 1. On the marine conditions and phytoplankton of the Yellow Sea in summer.
Repts. Inst. mar. Biol., Seoul Nat. Univ. 1(6): 1-14.

chemistry
silicates
Leloup, E., Editor, with collaboration of L. Van Meel, Ph. Polk, R. Halewyck and A. Gryson, undated.
Recherches sur l'ostreiculture dans le Bassin de Chasse d'Ostende en 1962.
Ministere de l'Agriculture, Commission T.W.O.Z. Groupe de Travail - "Ostreiculture", 58 pp.

Chemistry silicates (data)
Littlepage, Jack L., 1965.
Oceanographic investigations in McMurdo Sound, Antarctica.
In: Biology of Antarctic seas, II.
Antarctic Res. Ser., Am. Geophys. Union, 5:1-37.

chemistry silicate
López-Benito, Manuel, 1970.
Silicatos en el agua de mar.
Inv. pesq. Barcelona, 34(2): 385-397.

silicate (data)
Lüneberg, H., 1939.
Hydrochemische Untersuchungen in der Elbemündе mittels Elektrokolorimeter.
Arch. Deutschen Seewarte 59(1):1-27, 8 pls.

M

Chemistry silicates
Mackenzie, Fred T., and Robert M. Garrels, 1965.
Silicates: reactivity with sea water.
Science, 150(3692):57-58.

chemistry silicate
Maeda, H., 1953.
The relation between chlorinity and silicate concentration of water observed in some estuaries.
J. Shimonoseki Coll. Fish. 3(2):167-180, 16 textfigs.

Chemistry silicate(data)
Maeda, H., 1953.
Studies on Yosa-Naikai. 3. Analytical investigations on the influence of the River Noda and the benthonic communities. J. Shimonoseki Coll. Fish. 3(2):141-149, 3 textfigs.

Chemistry, silicate(data)
Maeda, H., 1953.
Studies on Yosa-Naikai. 2. Considerations upon the range of the stagnation and the influences by the River Noda and the open sea.
J. Shimonoseki Coll. Fish. 3(2):133-140, 2 textfigs.

Chemistry silicate
Maeda, H., 1952.
The relation between chlorinity and silicate concentration of the water observed in some estuaries
Publ. Seto Mar. Biol. Lab., Kyoto Univ., 2(2): 249-255, 5 textfigs.

Chemistry silicate
Makimoto, H., H. Maeda and S. Era, 1955.
The relation between chlorinity and silicate concentration of water observed in some estuaries
Rec. Ocean. Wks., Japan, 2(1):106-112, 4 textfigs

Chemistry silicate(data)
Masuzawa, Jotaro and Hideo Akamatsu, 1962.
1. Hydrographic observations during the JEDS-4.
Oceanogr. Mag., Japan Meteorol. Agency, 13(2):

Chemistry, silicates
Matsue, Yoshiyuki, Yuzo Komaki and Masaaki Murano, 1957
On the distribution of minute nutrients in the North Japan Sea; March-April, 1956.
Bull. Japan Sea Reg. Fish. Res. Lab., 6: 316-320.
Abstr. in:
Rec. Res., Fac. Agric., Tokyo Univ., Mar. 1958 7(81): 58.

Chemistry
silicate
Matsue, Yoshiyuki, Yuzo Komaki and Masaaki Murano, 1957
On the distribution of minute nutrients in the North Japan Sea, during the close of August, 1955.
Bull. Japan Sea Reg. Fish. Res. Lab., 6:121-127
Abstr. in:
Rec. Res., Fac. Agric., Tokyo Univ., Mar. 1958 7(80): 57-58.

Mc

chemistry, silicate
McGill, David A. 1965.
The relative supplies of phosphate, nitrate and silicate in the Mediterranean Sea.
Rapp. P.-v. Réun. Commn int. Explor. scient. Mer Méditerr. 18(3): 737-744.

Chemistry silicate (data)
McGill, David A., and Nathaniel Corwin, 1965.
The distribution of nutrients in the Labrador Sea, summer, 1964.
U.S.C.G. Rept., No. 10(373-10):25-33;277-285.
[Oceanogr.

Chemistry silicate
*McLain, Douglas R., 1968.
Oceanographic surveys of Traitors Cove, Revillagigedo Island, Alaska.
Spec.scient.Rep.U.S.Fish Wildl.Serv.Fish, 576: 15 pp. (multilithed).

Chemistry silicate
*Metcalf, William G., 1969.
Dissolved silicate in the deep North Atlantic.
Deep-Sea Res., Suppl. 16: 139-145.

Chemistry silicate
Mijake, Y., Y. Sugiura, and K. Kamada, 1953.
A study of the property of the coastal waters around Hachijo Island. Rec. Ocean. Wks., Japan, n.s., 1(1):93-99, 10 textfigs.

Chemistry
silicate
Miwa, Katsutoshi, 1962
Report on dissolved nutrients investigations in the North Pacific and the Bering Sea during June-August 1960. (In Japanese; English summary).
Bull. Hokkaido Reg. Fish. Res. Lab., (25): 58-65.

Chemistry silicate
Miyake, Yasuo, Yoshio Sugiura, Yukio Sugimura, Tsutomu Akiyama and Hibrozo Yoshimura, 1962.
II. Chemistry in the Japanese Expedition to the Deep Sea (JEDS-4).
Oceanogr. Mag. Japan Meteorol. Agency, 13(2): 131-132.

JEDS Contrib. No. 30.

Chemistry - silicate
Mokiyevskaya, V.V., 1961.
[Some peculiarities of hydrochemistry in the northern Indian Ocean.]
Okeanol. Issledov., Mezhd. Komit., Mezhd. Geofiz. God, Presidiume, Akad. Nauk, SSSR, :50-61.

Chemistry
silicate
Mokievskaja, V.V., 1959
[Nutrients salts in the upper layer of the Bering Sea.] Trudy Inst. Okeanol., 33: 87-113.

Chemistry, silicate
Moreira de Silva, Paulo, y Federico José Aragno, 1966.
Significativa relacion entre los valores de salinidad y silicatos en la desembocadura del Rio de la Plata.
Bol Servicio Hidrogr. Naval, Publ., Argentina, H. 106:99-104.

Chemistry, silicate
Morii, Hideaki, Yusho Akishige, Ryoichi Kanazu and Tadanobu Fukuhara, 1967.
Chemical studies of the Kuroshio and adjacent region. 1. On the nutrient salts and the dissolved oxygen. (In Japanese; English abstract)
Bull. Fac. Fish., Nagasaki Univ. 22:91-103.

Chemistry, silicate
Mullin, J.B., and J.P. Riley, 1955.
The colorimatric determination of silicate with special reference to sea and natural waters.
Anal. Chim. Acta 12(2):162-176.

Abstr. Anal. Abs. 2(6):1686.

Chemistry silicate
Mullin, J.B., and J.P. Riley, 1955.
Storage of sea water samples for the determination of silicate. The Analyst 80(946):73-74.

Chemistry silicate
Mullin, J.B., and J.P. Riley, 1955.
The colorimetric determination of silicate with special reference to sea and natural waters.
Analytica Chimica Acta 12(2):162-176.

N

Chemistry silicates
Naguib, Monir, 1961.
Studies on the ecology of Lake Quarun.
Kieler Meeresf., 17(1):94-131.

Chemistry silicate
Neumann, W., 1949.
Beiträge zur Hydrographie des Vrana-Sees (Insel Cherso), inbesondere Untersuchungen über organisches sowie anorganische Phosphor- und Stickstoffverbindungen. Nova Thalassia 1(6):17 pp.

Chemistry
silicate(data)
Nishida, Keizo, 1958
[On the oceanographic conditions of the lower cold water in the Japan Sea.]
J. Fac. Fish. and Animal Husbandry, Hiroshima Univ., 2(1):1-65.

Chemistry silicate
Noble, A., 1968.
Studies on sea-water off the North Kanara coast.
J. mar. biol. Ass., India, 10(2): 197-223.

Chemistry silicate
Nümann, W., 1949.
Kolorimetrische Methoden zur quantitativen Bestimmung von Silikat und organisch wie anorganisch gebundenem Phosphor und Stickstoff im Meerwasser unter Benutzung des Pulfrich-Photometers. Deut. Hydro. Zeit. 2(4):137-153.

Chemistry, silicate
Nümann, W., 1941
Der Nährstoffhaushalt in der nordöstlichen Adria. Thalassia 5(2):1-68, 12 figs.

O

Chemistry
silicate
Ohwada, Mamoru, and Hisanori Kon, 1963
A microplankton survey as a contribution to the hydrography of the North Pacific and adjacent seas. (II). Distribution of the microplankton and their relation to the character of water masses in the Bering Sea and northern North Pacific Ocean in the summer of 1960.
Oceanogr. Magazine, Japan Meteorol. Agency, 14(2):87-99.

Chemistry, silicate (data)
Okabe, Shiro, Yoshimasa Toyota and Takafumi Murakami, 1967.
Major aspects of chemical oceanography. (In Japanese; English abstract).
J.Coll.mar. Sci.Techn., Tokai Univ., (2):39-56.

Chemistry silicate
Okitsu, T., 1952.
Seasonal change of the phosphate and silicate of the water of Mutsu Bay during 1950.
Bull. Mar. Biol. Sta., Asamushi, 5(1/4):61-64.

Chemistry silicates
Okubo, A., 1954.
A note on the decomposition of sinking remains of plankton organisms and its relationship to nutrient liberation. J. Ocean. Soc., Japan, 10(3):121-131, 4 textfigs.

Chemistry silicate
Okuda, Taizo, 1960
Metabolic circulation of phosphorus and nitrogen in Matsushima Bay (Japan) with special reference to exchange of these elements between sea water and sediments.
(Portuguese, French resumés)
Trabalhos, Inst. Biol. Maritima e Oceanogr., Universidade do Recife, Brasil, 2(1):7-153.
Resumé pp. 147-149.

Chemistry, silicates
Okuda, Taizo, Angel José García, José Benetoz Alvarez, 1965.
Variacion estacional de las elementos nutritions en el agua de la laguna y el rio Unare.
Bol; Inst. Oceanogr., Univ. Oriente, 4(1): 123-135.

Chemistry silicate
Onishi, T., 1964.
A study on the silicic acid in the natural waters (2). On fluctuation along time duration in the concentration of the silicic acid in waters at a definite point of the river. (In Japanese).
Sci. Repts., Kagoshima Univ., No. 13:29-34.

Chemistry, silicate
Orr, A.P. 1933.
Physical and chemical conditions in the sea in the neighborhood of the Great Barrier Reef.
Brit. Mus. (N.H.) Great Barrier Reef Exped 1928-29 Scient. Repts. 2(3):37-86.

P

Chemistry, silicate
Park, Kilho 1967.
Chemical features of the subarctic boundary near 170°W.
J. Fish. Res. Bd., Can. 24(5):899-908.

Chemistry silicate
Park, P. Kilho, Stephen W. Hager, Jacques E. Pirson and David S. Ball, 1968.
Surface phosphate and silicate distribution in the northeastern Pacific Ocean.
J.Fish.Res.Bd.,Can., 12(25):2739-2741.

Chemistry silicate (data)
Parsons, T.R., 1960.
A data record and discussion of some observations made in 1958-1960 of significance to primary productivity.
Fish. Res. Bd., Canada, Manuscript Rept. Ser., (Oceanogr. & Limnol.), No. 81:19 pp. (multilithed).

Chemistry silicate
Phelps, A. 1937
The variation in the silicate content of the water in Monterey Bay, California during 1932, 1933, and 1934. Trans. Am. Philos. Soc. 29(2):153-188.

Chemistry, silicates
Phifer, L. D. and T. G. Thompson, 1937.
Seasonal variations in the surface waters of San Juan channel during the five year period, January 1931 to December 30, 1935. J. Mar. Res., 1(1):34-59, text figs. 12-18, 12 tables.

Chemistry, silicate (data)
Platt, Trevor, and Brian Irwin, 1968.
Primary productivity measurements in St. Margaret's Bay, 1967.
Techn.Rept., Fish.Res.Bd., Can., 77: 123 pp.

Chemistry, silicate
Ponomarenko, L.S., 1966.
Hydrochemical conditions in the Norwegian and Greenland seas in the summer of 1964. (In Russian).
Mater. Ribokhoz. Issled. severn. Basseina, Poliarn. Nauchno-Issled. Proektn. Inst. Morsk. Ribn. Khoz. Okeanogr. (PINRO), 7:155-167.

Chemistry, silicate
Ponomarenko, L.S., 1966.
Hydrochemical conditions in the Norwegian and Greenland seas in summer (from the data of the June surveys 1954-1964). (In Russian).
Trudy, Poliarn. Nauchno-Issled. Proektn. Inst. Morsk. Ribn. Khoz. Okeanogr., N. M. Knipovich, PINRO, 18:125-145.

Chemistry silicate
Ponomarenko, L.S., and M.A. Istoshina, 1962.
[Hydrochemical studies in Newfoundland area.]
Sovetskie Ribokh. Issledov. v Severo-Zapadnoi Atlant. Okeana, VNIRO-PINRO, Moskva, 113-124.

Chemistry silicate
Pratt, David M., 1965.
The winter-spring diatom flowering in Narragansett Bay.
Limnology and Oceanography, 10(2):173-184.

Chemistry, silicates
Pucher-Petković, Tereza 1966.
Végétation des diatomées pélagiques de l'Adriatique moyenne.
Acta adriat. 13(1): 1-97.

Q

Chemistry silicate
Qasim, S.Z., and C.V.G. Reddy, 1967.
The estimation of plant pigments of Cochin backwater during the monsoon months.
Bull.mar.Sci., Miami, 17(1):95-110.

R

Chemistry silicate (data)
Bakestraw, Norris W., 1964.
Some observations on silicate and oxygen in the Indian Ocean.
In: Recent researches in the fields of hydrosphere, atmosphere and nuclear geochemistry, Ken Sugawara festival volume, Y. Miyake and T. Koyama, editors, Maruzen Co., Ltd., Tokyo, 243-255.

Chemistry silicate
Ramamurthy, S., 1965.
Studies on the plankton of the North Canara coast in relation to the pelagic fishery.
J. mar.biol. Ass.India, 7(1):127-149.

Chemistry silicate
Ramamurthy S. 1963.
Studies on the hydrological factors in the north Kanara coastal waters.
Indian J. Fish. (A) 10(1). 75-93

Chemistry silicate
Ramamurthy, S., 1954.
Hydrobiological studies in the Madras coastal waters. J. Madras Univ., B, 23(2):148-163, 3 textfigs.

Chemistry silicate (data)
Ramamurthy, V.D., K. Krishnamurthy and R. Seshadri, 1965.
Comparative hydrographical studies of the near shore and estuarine waters of Porto-Novo, S. India.
J. Annamalai Univ., 26:154-164.

Also in: Collected Reprints, Mar. Biol. Sta., Porto-Novo, 1963/64.

Chemistry silicate
Reddy, C.V. Gangadhara and V.N. Sankaranarayanan, 1969.
Distribution of phosphates and silicates in the Central Western north Indian Ocean, in relation to some hydrographical factors. Bull. natn. Inst. Sci., India, 38(1): 103-122.

Richards, F.A., 1958 silicate
Dissolved silicate and related properties of some western North Atlantic and Caribbean waters. J.Mar. Res., 17:449-465.

Chemistry silicate
Richards, Francis A., Joel D. Cline, William W. Broenkow and Larry P. Atkinson, 1965.
Some consequences of the decomposition of organic matter in Lake Nitinat, an anoxic fjord.
Limnol. and Oceanogr., Redfield Vol., Suppl. to 10:R185-R201.

Chemistry silicate
Riley, J.P., 1967.
The hot saline waters of the Red Sea bottom and their related sediments.
Oceanogr. Mar. Biol., Ann. Rev., H. Barnes, editor, George Allen and Unwin, Ltd., 5:141-157.

Chemistry (silicate)
Rittenberg, S.C., K.O. Emery and W.L. Orr, 1955.
Regeneration of nutrients in sediments of marine basins. Deep-Sea Res. 3(1):23-45.

Chemistry silicate
Robinson, R.J., and T.G. Thompson, 1948
The determination of silicate in sea water.
J. Mar. Res. 7(1):49-55.

Chemistry silicates (data)
Rozanov, A.G., 1964.
The distribution of phosphates and silicates in water of the norterhn part of the Indian Ocean.
Investigations in the Indian Ocean (33rd voyage of E/S "Vitiaz"). (In Russian).
Trudy Inst. Okeanol., Akad. Nauk, SSSR, 64:102-104.

Chemistry silicate
Rubinchik, E.E., 1959
[Data on the hydrochemistry of the southern Caspian.]
Trudy V.N.I.R.O., 38: 152-164.

Chemistry silicate
Ryther, John H., D.W. Menzel and Nathaniel Corwin, 1967.
Influence of the Amazon River outflow on the ecology of the western tropical Atlantic. I. Hydrography and nutrient chemistry.
J.mar.Res., 25(1):69-83.

S

Chemistry-silicate
Schott, Friedrich, and Manfred Ehrhardt 1970.
On fluctuations and mean relations of chemical parameters in the northwestern North Sea.
Kieler Meeresforsch. 25(2): 272-298

Chemistry silicate (data)
Shimomura, T., and K. Miyata, 1957.
[The oceanographical conditions of the Japan Sea and its water systems, laying stress on the summer of 1955.]
Bull. Japan Sea Reg. Fish. Res. Lab., No. 6 (General survey of the warm Tsushima Current 1): 23-120.

Chemistry silicate
Sholkovitz, Edward R. and Joris M. Gieskes, 1971.
A physical-chemical study of the flushing of the Santa Barbara Basin. Limnol. Oceanogr. 16(3): 479-489.

Chemistry silicates
Siever, Raymond, 1968.
Sedimentological consequences of a steady state-ocean-atmosphere.
Sedimentology, 11 (1/2): 5-29.

Chemistry, silicate
Skolka, H.V., et M.-T. Gomoiu 1967.
Recherches océanologiques roumaines dans la mer Noire. (In Rumanian, French abstract).
Hydrobiologie, Bucureşti 8:15-30

Chemistry silicate
Slinn, D.J., 1966.
Chemical constituents in sea water off Port Erin during 1965.
Mar. biol. Sta., Univ. Liverpool. No. 78. :29-33.

Chemistry silicate
Slinn, D.J., 1962
Chemical constituents in sea water off Port Erin during 1961.
Mar. Biol. Sta., Univ. Liverpool, Port Erin, Ann. Rept., 74(1961):23-26.

Chemistry silicate
Slinn D.J. 1961
Chemical constituents in sea water off Port Erin in 1960.
Ann. Rept. Mar. Biol. Sta. Univ. Liverpool 1960 73:23-28.

Chemistry silicate
Slinn, D.J., 1959
Chemical constituents in sea water off Port Erin during 1958.
Ann. Rept., Mar. Biol. Sta., Univ. Liverpool 1958 (No. 71):24-28.

Chemistry silicate
Slinn, D.J., and Winifred Chapman, 1965.
Chemical constituents in sea water off Port Erin in 1964.
Mar. Biol. Sta. Port Erin, Ann. Rept. (1964), No. 77:43-48.

Chemistry silicate
Slinn, D.J., and Winifred Chapman, 1964.
Chemical constituents of sea water off Port Erin during 1963.
Mar. Biol. Sta., Univ. Liverpool, Port Erin, Ann Rept., No. 76(1963):26-31.

Chemistry, silicate
Slinn, D.J., and Gaenor Offlow (Mrs. M. Solly) 1969.
Chemical constituents in sea water off Port Erin during 1968.
Ann. Rept. mar. biol. Sta. Port Erin 81(1968): 31-

Chemistry silicate
Smayda, Theodore J., 1966.
A quantitative analysis of the phytoplankton of the Gulf of Panama. III. General ecological conditions and the phytoplankton dynamics at 8o 45'N, 79o 23'W from November 1854 to May 1957.
Inter-Amer. Trop. Tuna Comm., Bull., 11(5): 355-612.

Chemistry silicate
Smetanin, D.A., 1960
[Some features of the chemistry of water in the central Pacific.]
Trudy Inst. Okeanol., 40: 58-71.

Chemistry silicate
Smetanin, D.A., 1959
[Hydrochemistry of the Kuril-Kamchatka deep-sea trench.] Trudy Inst. Okeanol., 33: 43-86.

Chemistry silicate
Stefansson, Unstein, 1966.
Influence of the Surtsey eruption on the nutrient content of the surrounding seawater.
J. Mar. Res., 24(2):241-268.

Chemistry silicate
Stefansson, Unnstein, and Francis A. Richards, 1963.
Processes contributing to the nutrient distributions off the Columbia River and Strait of Juan de Fuca.
Limnology and Oceanography, 8(4):394-410.

Chemistry silicate
Subrahmanyan, R., 1959.
Studies on the phytoplankton of the west coast of India. 1. Quantitative and qualitative fluctuation of the total phytoplankton crop, the zooplankton crop and their interrelationship, with remarks on the magnitude of the standing crop and production of matter and their relationship to fish landings. 2. Physical and chemical factors influencing the production of phytoplankton, with remarks on the cycle of nutrients and on the relationship of the phosphate content to fish landings.
Proc. Indian Acad. Sci., 1:113-252.

Chemistry silicate
Sugiura, Yoshio, and Suphachai Chaitiamvong, 1964
Relation of silicate concentration to dissolved oxygen amount in sea water collected in the northern frontal region of Kuroshio.
J. Oceanogr. Soc., Japan, 20(2):89-92.

chemistry, silicates
Sykes, J.B. and A.D. Boney, 1970.
Seasonal variations in inorganic phytoplankton nutrients in the inshore waters of Cardigan Bay. J. mar. biol. Ass., U.K., 50(3): 819-827.

T

Chemistry, silicate (data)
Tanaka, Otohiko, Haruhiko Irie, Shozi IIzuka and Fumihiro Koga, 1961
The fundamental investigation on the biological productivity in the north-west of Kyushu. 1. Rec. Oceanogr. Wks., Japan, Spec. No. 5: 1-58.

Chemistry, silicate
Tanita, S., K. Kato and T. Okuda, 1951.
Studies on the environmental conditions of shell-fish-fields. In the case of Muroran Harbour (2).
Bull. Fac. Fish., Hokkaido Univ., 2(3):220-230.

chemistry, silicate (data)
Tsuruta, Arao, Satoru Tawara and Tadashi Ueda 1967.
On the distribution of dissolved oxygen and nutrient salts in tuna fishing grounds of the central tropical Atlantic.
J. Shimonoseki Univ. Fish. 15(3): 295-305.

Chemistry, silicates
Tully, J.P., and A.J. Dodimead, 1957.
Properties of the water in the Strait of Georgia, British Columbia, and influencing factors.
J. Fish. Res. Bd., Canada, 14(3):241-319.

U

Chemistry silicate
Udaya Varma Thirupad, P., and C.V. Gangadhara Reddy, 1959
Seasonal variations of the hydrological factors of the Madras coastal waters.
Indian J. Fish., 6(2):298-305.

Chemistry, Silicate (data)
United States, Department of Commerce, Environmental Sciences Services Administration, 1965
International Indian Ocean Expedition, USC&GS Ship Pioneer - 1964.
Vol. 1. Cruise Narrative and scientific results 139 pp.
Vol. 2. Data report: oceanographic stations, BT observations, and bottom samples, 183 pp.

chemistry, silicate (data)
University of Southern California, Allan Hancock Foundation, 1965.
An oceanographic and biological survey of the southern California mainland shelf.
State of California, Resources Agency, State Water Quality Control Board, Publ. No. 27:232 pp. Appendix, 445 pp.

V

Chemistry - silicate
Vatova, A. 1965
Les conditions hydrographiques de la Mar Piccolo de Taranto pendant l'année 1963.
Rapp. P.-v. Réun. Commn int. Explor. scient. Mer Médit. 18(3): 653-655.

Chemistry silicates
Vatova, A., 1956.
Il dosaggio dei silicati nell'acqua di mare con l'elettrofotometro Elko II. Nova Thalassia 2(6): 3-20.

Chemistry silicates (data)
Viorel, Chirilă, 1965.
Observații asupra condițiilor fizico-chimice ale mării la Mamaia, în anii 1959 și 1960.
In: Ecologie marină, M. Băcescu, redactor, Edit. Acad. Republ. Pop. Române, Bucureşti, 1:139-184.

Chemistry silicate
Voipio, Aarno, 1961.
The silicate in the Baltic Sea.
Annales Acad. Sci. Fennicae, (A) II. Chemica, 106:15 pp.

Chemistry silicate
Volokhonsky, A.G., 1966.
Hydrochemical conditions in the Faroe-Iceland Channel in November-December 1964. (In Russian).
Mater. Ribokhoz. Issled. severn. Basseina, Poliarn. Nauchno-Issled. Proektn. Inst. Morsk. Ribn. Khoz. Okeanogr. (PINRO), 7:146-154.

Chemistry silicate
Vorontsovs, R. V., 1964.
The hydrochemical conditions in the southern part of the Davis Strait in the spring and summer of 1963. (In Russian).
Material. Sess. Uchen. Sov., PINRO, Rez. Issled. 1962-1963., Murmansk, 130-135.

W

Chemistry silicate
Wattenberg, H., 1937.
Methoden zur Bestimmung von Phosphat, Silicat, Nitrat, und Ammoniak im Seewasser. Rapp. Proc. Verb. 103(1):1-26.

chemistry, silicate

Won, Chung Hun 1964.
Tidal variations of chemical constituents of the estuary water at the lava bed in the Nack-Dong River from Nov. 1962 - Oct. 1963. (In Korean; English abstract).
Bull. Pusan Fish Coll., 6(1):21-32.

chemistry, silicate

Won, Chong Hun, 1963.
Distribution of chemical constituents of the estuary water in Gwang-Yang Inlet. (In Korean; English abstract).
Bull. Fish. Coll., Pusan Nat. Univ., 5(1):1-10.

chemistry, silicate

Wooster, W.S., and T. Cromwell, 1958.
An oceanographic description of the eastern tropical Pacific. Bull. S.I.O., 7(3):169-282.

XYZ

chemistry chemistry, silicon

Yelkovsky, Ralph, 1967.
Signs test applied to Caribbean Core A 176-6.
Science, 155(3768):1408-1409.

chemistry silicates

Yamazi, I., 1955.
Plankton investigations in inlet waters along the coast of Japan. XVI. The plankton of Tokyo Bay in relation to water movement.
Publ. Seto Mar. Biol. Lab., 4(2/3):285-309, Pls. 19-20, 22 textfigs.

chemistry, silicates

Yamazi, I., 1954.
Plankton investigations in inlet waters along the coast of Japan. XIV. The plankton of Turuga Bay on the Japan Sea coast.
Publ. Seto Mar. Biol. Lab. 4(1):115-126, 11 textfigs.

chemistry silicate (profiles)

Yamazi, I., 1953.
Plankton investigation in inlet waters along the coast of Japan. VII. The plankton collected during the cruises to the new Yamamoto Bank in the Sea of Japan. Publ. Seto Mar. Biol. Lab. 3(1):75-108, 19 textfigs.

chemistry silicate

Yamazi, I., 1951.
Plankton investigations in inlet waters along the coast of Japan. II. The plankton of Hakodate Harbour and Yoichi Inlet in Hokkaido.
Publ. Seto Mar. Biol. Sta., Kyoto Univ., 1(4): 185-194, 3 textfigs.

chemistry, silicate

Zuta, Salvador, y Oscar Guillen, 1970.

Oceanografia de las aguas costeras del Peru.
Bol. Inst. Mar., Peru, 2(5):157-324.

silicate - data only

chemistry
Silicate (data only)

Akamatsu, Hideo, Tsutomu Akiyama and Tsutomu Sawara, 1965.
Preliminary report of the Japanese Expedition of Deep-Sea, the Tenth Cruise, 1965(JEDS-10).
Oceanogr. Mag., Tokyo, 17(1/2):49-68.

Japanese Expedition of the Deep-Sea

chemistry, silicate (data only)

Aragno, Federico, Alberto Gomez, Aldo Orlando y Andrés J. Lusquiños 1968.
Datos y resultados de las campañas pesquerías "Pesquería III" (20 de febrero al 20 de marzo de 1967).
Publ. (Ser. Informes técn.) Mar del Plata, Argentina 10(3): 1-162.

chemistry, silicate (data only)

Aragno, Federico, Alberto Gomez, Aldo Orlando y Andres J. Lusquiños 1968.
Datos y resultados de las campañas pesquerías "Pesquería II" (19 de noviembre al 12 diciembre de 1966)
Publ. (Ser. Informes técn.) Mar del Plata, Argentina, 10(2): 1-129.

chemistry, silicate (data only)

Aragno, Federico, Alberto Gomez, Aldo Orlando y Andres J. Lusquiños 1968.
Datos y resultados preliminares de las campañas pesquerías "Pesquería I" (12 de agosto al 8 de setiembre de 1966).
Publ. (Ser. Informes técn.) Mar del Plata, Argentina 10(1): 1-159.

chemistry
silicate (data only)

Argentina, Secretaria de Marina, Servicio de Hidrografia Naval, 1961.
Trabajos oceanograficos realizados en la campana Antartica 1959/1960. Resultados preliminares
Publico, H. 623:unnumbered pp.

chemistry
silicate (data only)

Argentina, Secretaria de Marina, Servicio de Hidrografia Naval, 1961.
Operacion oceanografica, Vema - Canepa 1, Resultados preliminares.
Publico, H. 628:30 pp.

chemistry
silicate (data only)

Argentina, Secretaria de Marina, Servicio de Hidrografia Naval, 1961.
Trabajos oceanograficos realizados en la campana Antartica 1960/1961. Resultados preliminares.
Publico, H. 629:unnumbered pp.

chemistry, silicate (data only)

Ballester, Antonio, 1965(1967).
Tablas hidrograficas.
Memoria Soc.Cienc.nat.La Salle, 25(70/71/72):41-137.

chemistry
Silicate (data only)

Brasil, Diretoria de Hidrografia e Navegacao, 1968.
XXXV Comissao Oceanografica "Operacao Norte/Nordeste 1" noc Almirante Saldanha (14/9 a 16/12/1967).
DG 26-XI: 600 pp.

chemistry, silicate (data only)

Brasil, Diretoria de Hidrografia e Navegação 1969.
XXXII Comissão Oceanográfica noc Almirante Saldanha (14/3 a 3/5/67), DG 26-X: 411 pp.

chemistry, silicate (data only)

*Brazil, Diretoria de Hidrografia e Navegação, 1968.
XXXI Comissao oceanografica noc Almirante Saldanha (14/11 a 16/12/66, DG26-VIII:249 pp.

chemistry, silicate (data only)

Brasil, Marinha do Brasil, Diretoria de Hidrografia e Navagação, 1960.
Estudo das condições oceanográficas sobre a plataforma continental, entre Cabo-Frio e Vitoria, durante o outono (abril-maio).
DG-06-X(junho):unnumbered pp. (mimeographed).

chemistry silicate (data only)

Brasil, Marinha do Brasil, Diretoria de Hidrografia e Navegação, 1959.
Levantamento oceanografico da costa nordeste.
DG-06-IX:unnumbered pp. (mimeographed).

chemistry, silicate (data only)

Buljan, M., and M. Marinkovic, 1956.
Some data on hydrography of the Adriatic (1946-1951). Acta Adriatica 7(12):1-55.

chemistry, silicate (data only)

Canada, Canadian Oceanographic Data Center 1969.
East Greenland, Denmark Strait and Irminger Sea, January 16 to April 5, 1967.
1969 Data Rec. Ser. 4:158 pp. (multilithed)

chemistry, silicate (data only)

Canada, Canadian Oceanographic Data Centre, 1965.
Data Record, Grand Banks to the Azores, June 1 to July 15, 1964.
1965 Data Record Series, No. 13:221 pp. (mimeographed).

chemistry silicate (data only)

Canada, Fisheries Research Board, 1959.
Data record, Ocean Weather Station "P" (Latitude 50 00'N, Longitude 145 00'W) January 22 - July 11, 1958.
MSS Rept. Ser. (Oceanogr. & Limnol.), no. 31: 112 pp.

chemistry, silicate (data only)

Canada, Fisheries Research Board, 1959.
Physical and chemical data record, coastal seaways project, November 12 to December 5, 1958.
MSS Rept. Ser. (Oceanogr. & Limnol.), No. 36: 120 pp.

chemistry, silicate (data only)

Conseil Permanent International pour l'Exploration de la Mer, 1957.
Bulletin hydrographique pour l'année 1953:167 pp.

chemistry silicate (data only)

Conseil Permanent International pour l'Exploration de la Mer, 1956.
Bulletin hydrographique pour l'année 1952:164 pp

chemistry, silicate (data only)

Conseil Permanent International pour l'Exploration de la Mer, 1954.
Bulletin Hydrographique pour l'Année 1950: 114 pp., 5 charts.

chemistry silicate (data only)

Conseil Permanent International pour l'Exploration de la Mer, 1954.
Bulletin hydrographique pour l'Annee 1949, 85 pp., 4 charts.

chemistry
silicate (data only)

Favorite, Felix, John W. Schantz and Charles R. Hebard, 1961
Oceanographic observations in Bristol Bay and the Bering Sea 1939-41(USCGT Redwing).
U.S.F.W.S. Spec. Sci. Rept., Fish. No.381: 323 pp.

chemistry
silicate

Fukuo, Yoshiaki, 1962
On the exchange of water and the productivity of a bay with special reference to Tanabe Bay (IV).
Rec. Oceanogr. Wks., Japan, Spec. No. 6:161-168.

chemistry, silicates (data only)

Granqvist, G., 1955.
The summer cruise with M/S Aranda in the northern Baltic, 1954. Merent. Julk., No. 166:56 pp.

Chemistry silicate (data only)
Great Britain, Discovery Committee, 1957.
Station list, 1950-1951.
Discovery Rept. 28:300-398.

Chemistry silicate (data only)
Great Britain, Discovery Committee, 1947.
Station list, 1937-1939. Discovery Repts., 24:
198-422, Pls. 4-6.

Chemistry silicate (data only)
Great Britain, Discovery Committee, 1945.
Station list. 1935-1937.
Discovery Rept., 24:3-196.

Chemistry silicate (data only)
Great Britain, Discovery Committee, 1942.
Station list, 1933-1935. Discovery Repts., 22:3-
196, Pls. 1-4.

Chemistry silicate (data only)
Hapgood, William, 1959.
Hydrographic observations in the Bay of Naples
(Golfo di Napoli), January 1957-January 1958
(station lists).
Pubbl. Sta. Zool., Napoli, 31(2):337-371.

Chemistry silicate (data only)
Hela, Ilmo, and Folke Koroleff, 1958
Hydrographical and chemical data collected
in 1957 on board the R/V Aranda in the
Barents Sea.
Merent Julk., No. 179: 67 pp.

Chemistry silicate (data only)
Hela, Ilmo, and F. Koroleff, 1958.
Hydrographical and chemical data collected in
1956 on board the R/V"Aranda" in the Baltic Sea.
Merent. Julk., No. 183:1-52.

Chemistry, silicate (data only)
Hufford, Gary L. and James M. Seabrooke,
Oceanography of the Weddell Sea 1970
in 1969 (IWSOE).
Oceanogr. Rept. U.S. Coast Guard 31 (CG373-31)
32 pp.

Chemistry silicate (data only)
Japan, Central Meteorological Observatory, 1956.
The results of marine meteorological and oceanographic observations (NORPAC Expedition Spec.
No.) 17:131 pp.

Chemistry silicate (data only)
Japan, Central Meteorological Observatory,
1953
The results of Marine meteorological and
oceanographical observations. Jan.-June
1952, No. 11:362, 1 fig.

Chemistry silicate (data only)
Japan, Central Meteorological Observatory,
1952
The results of Marine Meteorological and
oceanographical observations, July - Dec.
1951, No. 10:310 pp., 1 fig.

Chemistry silicate (data only)
Japan, Central Meteorological Observatory,
1952.
The results of Marine meteorological and
oceanographical observations. July - Dec.
1950. No. 8: 299 pp.

Chemistry, silicate (data only)
Japan, Central Meteorological Observatory,
1952.
The Results of Marine Meteorological and
oceanographical observations. Jan. - June
1950. No. 7: 220 pp.

Chemistry, silicate (data only)
Japan, Central Meteorological Observatory,
1951.
The results of marine meteorological and
oceanographic observations. July - Dec.
1949. No. 6: 423 pp.

Chemistry, silicate (data only)
Japan, Central Meteorological Observatory, 1951.
Table 14. Oceanographical observations taken in
Miyazu Bay: physical and chemical data for
stations occupied by R.M.S. "Asanagi-maru".
Res. Mar. Met. Ocean. Obs., Jan.-June, 1949,
No. 5:77-84.

Chemistry, silicate (data only)
Japan, Central Meteorological Observatory, 1951.
Table 20, Surface observations made on board the
"Kuroshio-maru"; physical and chemical conditions
of the surface water of the sea between Tokyo and
Hachijo Is. Res. Mar. Met. Ocean. Obs., Jan.-
June, 1949, No. 5:108-115.

Chemistry, silicate (data only)
Japan, Central Meteorological Observatory, 1951.
Table 19. Surface observations made on board the
"Tokitsu-maru"; physical and chemical conditions
of the surface water of the sea between Tokyo and
Hakodate. Res. Mar. Met. Ocean. Obs., Jan.-June,
1949, No. 5:106-108.

Chemistry silicate (data only)
Japan, Central Meteorological Observatory, 1951.
Table 10. Oceanographical observations taken in
the East China Sea. (A) Physical and chemical
data for stations occupied by the "Akebono-maru
No. 9": (B) ------ by the "Hatsutaka-maru".
Res. Mar. Met. Ocean. Obs., Jan.-June, 1949,
No. 5:68-70.

Chemistry, silicate (data only)
Japan, Central Meteorological Observatory, 1951.
Table 9. Oceanographical observations taken in
the Goto-nada and the Tsushima Straits. (A)
Physical and chemical data for stations occupied
by the fishing boat "Daikoku-maru": (B) -------
R.S.S. "Umikaze-maru".
Res. Mar. Met. Ocean. Obs., Jan.-June, 1949,
No. 5:61-67.

Chemistry silicate (data only)
Japan, Central Meteorological Observatory, 1951.
Table 8. Oceanographical observations taken in
the Ariake Sea; physical and chemical data for
stations occupied by R.S.S. "Umikaze-maru".
Res. Mar. Met. Ocean. Obs., Jan.-June, 1949, No.5:
50-60.

Chemistry, silicate (data only)
Japan, Central Meteorological Observatory
1951.
Table 6. Oceanographical observations taken in
the Akashi-seto, the Yura-seto and the Kii Suido;
physical and chemical data for stations occupied
by R.M.S. "Shumpu-maru".
Res. Mar. Met. Ocean. Obs., Jan.-June 1949, No.
5:40-47.

Chemistry silicate (data only)
Japan, Central Meteorological Observatory, 1951.
Table 5. Oceanographical observations taken in
the sea area to the south of the Kii Peninsula
along "E"-line; physical and chemical data for
stations occupied by R.M.S. "Chikubu-maru".
Res. Mar. Met. Ocean. Obs., Jan.-June 1949, No.
5:37-40.

Chemistry silicate (data only)
Japan, Central Meteorological Observatory, 1951.
Table 4. Oceanographical observations taken in
Sagami Bay; physical and chemical data for
stations occupied by R.M.S. "Asashio-maru".
Res. Mar. Met. Ocean. Obs., Jan.-June, 1949, No.
5:30-37.

Chemistry silicate (data only)
Japan, Central Meteorological Observatory, 1951.
Table 3. Oceanographical observations taken in the
North Pacific Ocean along "C" line;(A) Physical
and chemical data for stations occupied by R.M.S.
"Ukuru-maru";(B) --- "Ikuna-maru";(C) --- "Ryofu-
maru";(D) --- "Shinnan-maru";(E) --- "Kung-maru";
(F) --- "Chikubu-maru";(G) --- "Ryofu-maru";(H) --
- "Ikuna-maru";(I) --- "Chikubu-maru";(J) ---
"Ukuru-maru";(L) --- "Ukuna-maru";(K) ---"Shinnan-
maru";(M) --- "Ryofu-maru". Res. Mar. Met. Ocean.
Obs., Jan.-June 1949, No. 5:13-30.

Chemistry, silicate (data only)
Japan, Fisheries Agency, Research Division, 1956.
Radiological survey of western area of the dangerous zone around the Bikini-Eniwetok Atolls,
investigated by the "Shunkotsu maru" in 1956,
Part 1:143 pp.

Chemistry
silicate (data only)
Japan, Hokkaido University, Faculty
of Fisheries, 1968.
Data record of oceanographic
observations and exploratory fishing.
No. 12:420 pp.

Chemistry, silicate
(data only)
Japan, Faculty of Fisheries, Hokkaido
University 1961.
Data record of oceanographic observations
and exploratory fishing, No. 5:391 pp.

Chemistry silicate (data only)
Japan, Hokkaido University, Faculty of Fisheries, 1959.
Data record of oceanographic observations and exploratory
fishing, No. 3:296 pp.

Chemistry silicate (data only)
Japan, Hokkaido University, Faculty of Fisheries,
1957.
Data record of oceanographic observations and
exploratory fishing, No. 1:247 pp.

Chemistry. silicate (data only)
Japan, Hokkaido University, Faculty of Fisheries,
1955.
Correction of data presented in "Hydrographic
data obtained principally in the Bering Sea by
Training Ship 'Oshoro Maru' in the summer of
1955" published in January 1956: 8 pp.

Chemistry, silicate
(data only)
Japan, Maritime Safety Agency 1967.
Results of oceanographic observations in
1965.
Data Rept. Hydrogr. Obs. Ser. Oceanogr.
(Publ. 792) Oct. 1967, 5:115 pp.

Chemistry, silicate
(data only)
Japan, Maritime Safety Agency 1967.
Results of oceanographic observations in
1964.
Data Rept. Hydrogr. Obs., Ser. Oceanogr.
(Publ. 792) No. 4:88 pp.

Chemistry, silicate
(data only)
Japan, Maritime Safety Agency 1966.
Results of oceanographic observations in
1963.
Data Rept. Hydrogr. Obs. Ser. Oceanogr. Publ.
792 (3): 74 pp.

Chemistry, silicate (data only)
Japan, Maritime Safety Board, 1961
Tables of results from oceanographic
observation in 1959.
Hydrogr. Bull. (Publ. No. 981), No. 68:
112 pp.

Chemistry, silicate
(data only)
Japan, Maritime Safety Board 1961.
Tables of results from oceanographic
observations in 1958.
Hydrogr. Bull., Tokyo, No. 66 (Publ. 981): 153 pp.

Chemistry silicate (data only)
Japan, Maritime Safety Board, 1956.
Tables of results from oceanographic observations in 1952 and 1953.
Hydrogr. Bull., (Publ. 981)(51):1-171.

Chemistry silicate (data only)
Japan, Maritime Safety Board, Tokyo, 1954.
Tables of results from oceanographic observations in 1951. Publ. No. 981, Hydrogr. Bull., Spec. No. 15:31-129.

Chemistry silicate (data only)
Japan, Japan Meteorological Agency 1971.
The results of marine meteorological and oceanographical observations, July - December 1969, 46:270 pp.

Chemistry Silicate (data only)
Japan, Japan Meteorological Agency, 1970
The results of marine meteorological and oceanographical observations. (The results of the Japaneses Expedition of Deep Sea (JEDS-11); January-June 1967 41: 332 pp.

Chemistry, silicate (data only)
Japan, Japan Meteorological Agency 1970: The results of marine meteorological and oceanographical observations, July-December, 1966, No. 40: 336 pp.

Chemistry silicate (data only)
Japan, Japanese Meteorological Agency, 1967.
The results of marine meteorological and oceanographical observations, July-December, 1964, 36: 367 pp.

Chemistry silicate (data only)
Japan, Japan Meteorological Agency, 1966.
The results of the Japanese Expedition of Deep Sea (JEDS-8).
Results mar. met. oceanogr. Obsns. Tokyo, 35:328 pp.

Chemistry silicate (data only)
Japan, Japan Meteorological Agency, 1964.
The results of marine meteorological and oceanographical observations, January-June 1963, No. 33:289 pp.

Chemistry silicate (data only)
Japan, Japan Meteorological Agency, 1964.
The results of the Japanese Expedition of Deep Sea (JEDS-5).
Res. Mar. Meteorol. and Oceanogr. Obs., July-Dec., 1962, No. 32:328 pp.

Chemistry silicate (data only
Japan, Japan Meteorological Agency, 1964.
Oceanographic observations.
Res. Mar. Meteorol. & Oceanogr. Obs., No. 31:220 pp.

Chemistry silicate (data only)
Japan, Japan Meteorological Agency, 1962
The results of marine meteorological and oceanographical observations, July-December 1961, No. 30:326 pp.

Chemistry silicate (data only)
Japan, Japan Meteorological Agency, 1962
The results of marine meteorological and oceanographical observations, January-June 1961, No. 29: 284 pp.

Chemistry silicate (data only)
Japan, Meteorological Agency, 1962
The results of marine meteorological and oceanographical observations, July-December, 1960, No. 28: 304 pp.

Chemistry silicate (data only)
Japan, Japan Meteorological Agency, 1961
The results of marine meteorological and oceanographical observations, January-June 1960. The results of the Japanese Expedition of Deep-Sea (JEDS-2, JEDS-3), No. 27: 257 pp.

Chemistry silicate (data only)
Japan, Japan Meteorological Agency, 1960.
The results of marine meteorological and oceanographical observations, July-December, 1959, No. 26:256 pp.

Chemistry, silicate (data only)
Japan, Japan Meteorological Agency, 1960.
The results of marine meteorological and oceanographical observations, Jan.-June. 1959, No. 25: 258 pp.

Chemistry, silicate (data only)
Japan, Meteorological Agency, 1960.
The results of marine meteorological and oceanographical observations. Supplement, 149 pp.

Chemistry, silicate (data only)
Japan, Japan Meteorological Agency 1959.
The results of marine meteorological and oceanographical observations, January-June 1958: 240 pp.

Chemistry silicate (data only)
Japan, Japan Meteorological Agency, 1959.
The results of marine meteorological and oceanographical observations, July-December, 1958, No. 24:289 pp.

Chemistry silicate (data only)
Japan, Japanese Meteorological Agency, 1957
The results of marine meteorological and oceanographical observations, Jan.-June, 1956: 184 pp.
July-December, No. 20: 191 pp.

Chemistry silicate (data only)
Japan, Japanese Oceanographic Data Center, 1970
Ji Ri San, Fisheries Research and Development Agency, Republic of Korea, Feb. 18 - 26, 1970, West of the Japan Sea. Data Rept. CSK (KDC Ref. 24K038) 272: 16 pp. (multilithed).

Chemistry silicate (data only)
Japan, Japanese Oceanographic Data Center, 1970.
Han Ra San, Fisheries Research and Development Agency, Republic of Korea, Feb. 11 - 24, 1970, East of the Yellow Sea. Data Rept. CSK (KDC Ref. 24K037) 271: 14 pp. (multilithed).

Chemistry silicate (data only)
Japan, Japanese Oceanographic Data Center, 1970.
Baek Du San, Fisheries Research and Development Agency, Republic of Korea, Feb. 4 - 20, 1970, South of Korea & East China Sea. Data Rept. CSK (KDC Ref. 24K036) 270: 15 pp. (multilithed).

Chemistry silicate (data only)
Japan, Japanese Oceanographic Data Center, 1970.
Shumpu Maru, Kobe Marine Observatory, Japan Meteorological Agency, Feb. 21 - Mar. 11, 1970, South of Japan. Data Rept. CSK (KDC Ref. 49K122) 266: 4 pp. (multilithed).

Chemistry silicate (data only)
Japan, Japanese Oceanographic Data Center, 1970.
Takuyo, Hydrographic Department, Maritime Safety Agency, Japan, Feb. 12 - Mar. 7, 1970. South and East of Japan. Data Rept. CSK (KDC Ref. 49K120) 264: 13 pp. (multilithed).

Chemistry silicate (data only)
Japan, Japanese Oceanographic Data Center, 1970.
Kaiyo, Hydrographic Department, Maritime Safety Agency Japan, Nov. 7 - 29, 1969, South of Japan. Data Rept. CSK (KDC Ref. 49K116) 257: 10 pp. (multilithed).

Chemistry silicate (data only)
Japan, Japanese Oceanographic Data Center, 1970
Orlick, USSR, Jan. 27 - Mar. 6, 1969, Northwest of the North Pacific Ocean. Data Rept. CSK (KDC Ref. 90K023) 227: 30 pp. (multilithed).

Chemistry silicate (data only)
Japan, Japanese Oceanographic Data Center, 1969
Cape St. Mary, Fisheries Research Station, UK (Hong Kong). May 16 - 23, 1968, South China Sea. Prelim. Data Rept. CSK (KDC Ref. No. 74K010) 203: 15 pp. (multilithed).

Chemistry silicate (data only)
Japan, Japanese Oceanographic Data Center, 1969.
Seifu Maru, Maizuru Marine Observatory, Japan Meteorological Agency. August 6 - 10, 1968, Japan Sea. Prelim. Data Rept. CSK (KDC Ref. No. 49K087) 181: 9 pp. (multilithed).

Chemistry silicate (data only)
Japan, Japanese Oceanographic Data Center, 1969.
Umitaka Maru, Tokyo University of Fisheries, Japan, November 1, 1967 - February 23, 1968, West of the North Pacific Ocean. Prelim. DAta Rept. CSK (KDC Ref. No. 49K066) 131: 13pp. (multilithed).

Chemistry silicate (data only)
Japan, Japanese Oceanographic Data Center, 1969.
Orlick, USSR, July 29 - September 3, 1967, Northwest of the North Pacific Ocean. Prelim. Data Rept. CSK (KDC Ref. No. 90K015) 122: 31 pp. (multilithed).

Chemistry, silicate (data only)
Japan, Japanese Oceanographic Data Center, 1969.
U.M. Schokalsky, USSR, April 28 - June 3, 1967, Northwest of the North Pacific Ocean. Prelim. Data Rept. CSK (KDC Ref. 90K014) 121: 41 pp. (multilithed).

Chemistry, silicate (data only)
Japan, Japanese Oceanographic Data Center, 1968.
Takuyo, Hydrographic Division, Maritime Safety Agency, July 12-August 30, 1967, Central Part of the North Pacific Ocean.
Prelim. Data Rept. CSK (KDC Ref. 49K053) 105:25 pp. multilithed)

Chemistry, silicate (data only)
Japan, Japanese Oceanographic Data Center, 1968.
Ryofu Maru, Marine Division, Japan Meteorological Agency, January 11-February 24, 1967, West of the North Pacific Ocean.
Prelim. Data Rept. CSK (KDC 49K040)82: 52 pp.

Chemistry, silicate (data only)
Japan, Japanese Oceanographic Data Center. 1967.
Ryofu Maru, Marine Division, Japan Meteorological Agency, Japan, September 13-17, 1966, eastern Sea of Japan.
Prelim. Data Rept. CSK (KDC Ref. No. 49K033) 59:11 pp. (multilithed)

Chemistry, silicate (data only)
Japan, Japanese Oceanographic Data Center. 1967.
Nagasaki Maru, Faculty of Fisheries, Nagasaki University, Japan, June 16-21, 1966, southern Sea of Japan.
Prelim. Data Rept. CSK (KDC Ref. No. 49K032) 58:8 pp. (multilithed)

Chemistry silicate (data only)
Japan, Japanese Oceanographic Data Center, 1967.
Kofu Maru, Hakodate Marine Observatory, Japan Meteorological Agency, Japan, June 30-July 10, 1966, east of Japan.
Prelim. Data Rept., CSK (KDC Ref. 49K026) 52:14 pp.

Chemistry silicate (data only)
Japan, Japanese Oceanographic Data Center, 1967.
Seifu Maru, Maizuru Marine Observatory, Japan Meteorological Agency, Japan, August 21-September 19, 1966, Japan Sea.
Prelim. Data Rept., CSK (KDC Ref. 49K029) 55:15 pp.

Chemistry, silicate (data only)
Japan, Japanese Oceanographic Data Center, 1967.
Chofu Maru (NMO) Jul. 2-3, 1966; Chofu Maru (NMO), Aug. 27-29, 1966; Shumpu Maru (KMO), Oct. 27-31, 1966, Chofu Maru, Dec. 7-8, 1966, Japan Meteorological Agency of Japan, south-east of Yakushima.
Prelim. Data Rept., CSK (KDC Ref. 49K311, 49K312 49K313, 49K314) 61: 18 pp.

Chemistry silicate (data only)
Japan, Japanese Oceanographic Data Center, 1967.
Kaiyo, Hydrographic Division, M.S.A., Japan, August 10-30, 1966, S.E. of Japan.
Prelim. Data Rept., CSK, (KDC Ref. No. 49K025) 51:15 pp. (multilithed).

silicata (data only)
Japan, Japanese Oceanographic Data Center, 1969.
Ryofu Maru, Marine Division, Japan Meteorological Agency, Japan. October 7 - November 9, 1968. East China Sea. Data Rept. CSK (KDC Ref. No. 49K449) 217: 39 pp. (multilithed).

Chemistry silicate (data only)
Japan, Japanese Oceanographic Data Center, 1969
Takuyo, Hydrographic Division, Maritime Safety Agency, Japan. July 19-September 6, 1968, Kuroshio Extension Area. Prelim. Data Rept. CSK (KDC Ref. No. 49K082) 176: 27 pp. (multilithed).

Chemistry, silicate (data only)
Japan, Japanese Oceanographic Data Center, 1969.
Takuyo, Hydrographic Division, Maritime Safety Agency, May 14-25, 1968, South of Japan.
Data Rept. CSK (KDC Ref. 49K077) 171: 16 pp.

Chemistry, silicate (data only)
Japan, Japanese Oceanographic Data Center, 1968.
Cape St. Mary, Fisheries Research Station, Hong Kong, September 19-October 11, 1967, South China Sea.
Prelim. Data Rept. CSK (KDC Ref. 74K008) 138:16 pp. (multilithed)

Chemistry silicate (data only)
Japan, Japanese Oceanographic Data Center, 1968.
Han Ra San, Fisheries Research and Development Agency, Korea, August 12-September 4, 1967, North of East China Sea.
Prelim. Data Rept. CSK (KDC Ref. 24K018) 117:12 pp. (multilithed).

Chemistry, silicate (data only)
Japan, Japanese Oceanographic Data Center, 1968.
Chun Ma San, Fisheries Research and Development Agency, Korea, August 13-21, 1967, East of Yellow Sea.
Prelim. Data Rept. CSK (KDC Ref. 24K017) 116:16 pp. (multilithed).

Chemistry silicate (data only)
Japan, Japanese Oceanographic Data Center, 1968.
Orlick, USSR, February 5 - March 11, 1967, Northwest of North Pacific Ocean.
Prelim. Sata Rept. CSK (KDC Ref. 90K012) 94:29 pp.

Chemistry silicate (data only)
Japan, Japanese Oceanographic Data Center, 1967.
Kofu Maru, Hakodate Marine Observatory, Japan Meteorological Agency, Febuary 4-March 7, 1967, East of Japan.
Prelim. Data Rept. CSK (KDC Ref. No. 49K041) 83:15 pp.

Chemistry silicate (data only)
Japan, Japanese Oceanographic Data Center, 1967.
Takuyo, Hydrographic Division, Maritime Safety Agency, Japan, February 23-March 16, 1967, south and east of Japan.
Prelim. Data Rept., CSK (KDC Ref. 49K039) 81:17 pp.

Chemistry silicate (data only)
Japan, Japanese Oceanographic Data Center, 1967.
Bukhansan, Fisheries Research and Development Agency, Korea, July 16- August 9, 1966, East of the Yellow Sea.
Prelim. Data Rept., CSK (KDC Ref. 24K010) 69:17 pp.

Chemistry silicate (data only)
Japan, Japanese, Oceanographic Data Center, 1967
Baekdusan, Fisheries Research and Development Agency, Korea, July 14-28, 1966, West of the Japan Sea.
Prelim. Data Rept. CSK (KDC Ref. 24K009) 65:22 pp.

Chemistry silicate (data only)
Japan, Japanese Oceanographic Data Center, 1967.
Yang Ming, Chinese National Committee on Oceanic Research, Republic of China, September 10-October 14, 1966, Adjacent Sea of Taiwan.
Prelim. Data Rept. CSK (KDC Ref. 21K003) 67:17 pp.

Chemistry silicate (data only)
Japan, Japanese Oceanographic Data Center, 1966.
Koyo Maru, Shimonoseki University of Fisheries, October 26-29, south of Japan.
Prelim. Data Rept. CSK (KDC Ref. No. 49K038) 66: 7 pp. (multilithed).

Chemistry silicate (data only)
Japan, Japanese Oceanographic Data Center, 1966.
Seifu Maru, Maizuru Marine Observatory, Japan Meteorological Agency, Japan, February 12-February 28, 1966, Japan Sea.
Prelim. Data Rept., CSK (KDC Ref. 48K021), 32:14 pp. (multilithed).

Chemistry silicate (data only)
Japan, Japanese Oceanographic Data Center, 1966.
Yang Ming, Chinese National Committee On Oceanic Research, Republic of China, August 10-October 13, 1965, Adjacent Sea of Taiwan.
Prelim. Data Rept., CSK (KDC Ref. 21K001), 22: 18 pp. (multilithed).

Chemistry, silicate (data only)
Japan, Japanese Oceanographic Data Center, 1966.
Preliminary data report of CSK. Ryofu Maru, Marine Division, Japan Meteorological Agency, July 7-August 3, 1965, Eastern Sea of Japan. No. 10 (October 1966).
KDC Ref. No. 49003:31 pp. (multilithed).

Chemistry silicate (data only)
Japan, Kobe Marine Observatory, 1961
Data of the oceanographic observations in the sea south of Honshu from February to March and in May, 1959.
Bull. Kobe Mar. Obs., No. 167(27):99-108; 127-130;149-152;161-164;205-218.

Chemistry silicate (data only)
Japan, Maizuru Marine Observatory, 1963
Data of the oceanographic observations (1960-1961) (35-36):115-272.

Chemistry, silicate (data only)
Japan, Nagasaki Marine Observatory
Tables 9, 10, 11, 22, 23, 24, 34, 35, and 44
Oceanogr. Met. Nagasaki mar. Obs. 15: 92-120; 176-180; 191-192; 213-

Chemistry, silicate (data only)
Japan, Shimonoseki University of Fisheries 1968
Oceanographic surveys of the Kuroshio and its adjacent waters 1965 and 1966.
Data Oceanogr. Obs. Explor. Fish. 4:1-178.

Chemistry silicate (data only)
Japan, Shimonoseki University of Fisheries 1968.
Eastern Pacific Ocean cruise and central Atlantic Ocean cruise.
Data Oceanogr. Obs. Explor. Fish., 3: 1-145

Chemistry silicate (data only)
Japan, Shimonoseki University of Fisheries, 1965.
Data of oceanographic observations and exploratory fishings, International Indian Ocean Expedition 1962-63 and 1963-64. No. 1: 453 pp.

Chemistry silicate (data only)
Katsuura, Hiroshi, Hideo Akamatsu, Tsutomu Akiyama, 1964.
Preliminary report of the Japanese Expedition of Deep-Sea, the Eighth Cruise (JEDS-8).
Oceanogr. Mag., Tokyo, 16(1/2):125-136.

Chemistry silicate (data only)
Korea, Republic of, Fisheries Research and Development Agency, 1971
Annual report of oceanographic observations, 19:717 pp.

Chemistry silicate (data only)
Korea, Fisheries Research and Development Agency, 1968.
Annual report of oceanographic observations, 16: 691 pp.

Chemistry silicate (data only)
Korea, Fisheries Research and Development Agency, 1967.
Annual report of oceanographic observations, 15 (1966):459 pp.

Chemistry, silicate (data only)
Krauel, David P., 1969
Bedford Basin data report, 1967.
Techn. Rept. Fish. Res. Bd., Can., 120:84 pp (multilithed).

Chemistry, silicate (data only)
Love, C.M., 1966.
Physical, chemical, and biological data from the northeast Pacific Ocean: Columbia river effluent area, January-June 1963. 6. Brown Bear Cruise 326:13-23 June: CNAV Oshawa Cruise Oshawa-3:17-30 June.
Univ. Washington, Dept. Oceanogr., Tech. Rep., No. 134:230 pp.

Chemistry silicate (data only)
Motoda, S., H. Koto, K. Kato and T. Fugii, 1956.
Hydrographic data obtained principally in the Bering Sea by Training Ship "Oshoro-maru" in the summer of 1955, 59 pp.

Chemistry silicate (data only)
*Nasu, Keiji, and Tsugio Shimano, 1966.
The physical results of oceanographic survey in the south east Indian Ocean in 1963/64.
J. Tokyo Univ. Fish., (Spec. ed.) 8(2):133-164.

Chemistry silicate (data only)
Oren, O.H., 1967.
Croisière Chypre-04 dans la Méditerranée orientale, février-mars 1965. Résultats des observations hydrographiques.
Bull. Sea Fish. Res. Stn Israel, 47:37-54.

Chemistry silicate (data only)
Oren, Oton Haïm, 1967.
Croisière Chypre-04 en Méditerranée orientale, février-mars, 1965: résultats des observations hydrologiques.
Cah. océanogr., 19(9):783-798.

Chemistry, silicate (data only)
Oren, Oton Haïm, 1966.
Croisière "Chypre-02" en Méditerranée orientale, juillet-août 1963. Résultats des observations hydrologiques.
Cah. océanogr., 18(Suppl.):1-17.

chemistry, silicate (data only)
Platt, Trevor, and Brian Irwin 1971.
Phytoplankton production and nutrients in Bedford Basin, 1969-70.
Techn. Rept. Fish. Res. Bd. Can. 247:172 pp. (multilithed).

Chemistry, silicate (data only)
Portugal, Instituto Hidrografico, 1965.
Resultados das observacões oceanográficas no Canal de Mocambique, Cruzeiro el 1/64: abril-maio 1964.
Servico de Oceanografia, Publ. 1:73 pp., 46 figs

Chemistry, silicate (data only)
Portugal, Instituto Hidrográfico, Servico de Oceanografio, 1965.
Resultados das observacões oceanográficas no Canal de Mocambique, Cruzeiro al 1/64: Abril-Maio 1964, 73 pp., 46 figs.

Chemistry, silicate (data only)
Republic of China, Chinese National Committee on Oceanic Research, Academia Sinica, 1968.
Oceanographic Data Report of CSK, 2: 126 pp.

Chemistry, silicate (data only)
Republic of China, Chinese National Committee on Oceanic Research, Academia Sinica 1966.
Oceanographic data report of CSK 1:123 pp.

Chemistry silicate (data only)
Republic of Korea, Hydrographic Office, 1967.
Hydrographic and oceanographic data.
Techn. Repts, 1967, H.O. Pub., Korea, 1-72.

Chemistry silicate (data only)
Sweden, Havsfiskelaboratoriet, Lysekil 1971.
Hydrographical data January-June 1970 R.V. Skagerak, R.V. Thetis and TV252, 1970.
Meddn. Hydrogr. avd. Göteborg, 104: unnumbered pp. (multilithed).

Chemistry silicate (data only)
Tabata, S., C.D. McAllister, R.L. Johnston, D.G. Robertson, J.H. Meikle, and H.J. Hollister, 1961
Data record. Ocean Weather station "P" (Latitude 50 00'N, Longitude 145 00'W, December 9, 1959 to January 19, 1961.
Fish. Res. Bd., Canada, MSS. Rept. Ser. (Ocean & Limnol.), No. 98:296 pp. (Multilithed).

Chemistry silicate (data only)
United States, National Oceanogrphic Data Center, 1965
Data report EQUALANT III
Nat. Oceanogr., Data Cent., Gen. Ser., G-7:339pp. $5.00

Chemistry silicate (data only)
United States, U.S. Coast Guard, 1965.
Oceanographic cruise, USCGC Northwind: Chukchi, East Siberian and Laptev seas, August-September, 1963.
U.S.C.G. Oceanogr. Rept., No. 6:CG373-6):69 pp.

chemistry silicate (data only)
United States, National Oceanographic Data Center, 1964.
Data report EQUALANT II.
NODC Gen. Ser., Publ. G-5: numerous pp. (not serially numbered; loose leaf - $5.00).

chemistry, silicate (data only)
Valdez, Alberto J., Alberto Gomez, Aldo Orlando y Andres J. Lusquiños 1969.
Datos y resultados de las campañas pesquerias "Pesqueria V" (28 de agosto al 7 de setiembre de 1967).
Publ. (Ser. Informes técn.) Mar del Plata, Argentina 10(5): unnumbered pp.

chemistry, silicate (data only)
Valdez, Alberto J., Alberto Gomez, Aldo Orlando y Andres J. Lusquiños 1968.
Datos y resultados de las campañas pesquerias "Pesqueria IV" (7 de junio al 4 de julio de 1967).
Publ. (Ser. Informes técn.) Mar del Plata, Argentina 10(4): 1-159.

chemistry, silicate (data only)
Villanueva, Sebastian F., Alberto Gomez, Aldo Orlando y Andres J. Lusquiños 1969.
Datos y resultados de las campañas pesquerias "Pesqueria VII", 16 de febrero al 1° de marzo de 1968.
Publ. (Ser. Informes técn.) Proyecto Desarollo Pesquero, Mar del Plata, Argentina (10|VII): unnumbered pp.

chemistry, silicate (data only)
Villanueva, Sebastian F., Alberto Gomez, Aldo Orlando y Andres J. Lusquiños 1969.
Datos y resultados de las campañas pesquerias "Pesqueria VI" (2 de noviembre al 6 de diciembre de 1967).
Publ. (Ser. Informes técn.) Proyecto Desarrollo Pesquero, Mar del Plata, Argentina (10|VI): unnumbered pp.

chemistry, silicate (data only)
Zoutendyk, P., and D. Sacks 1969.
Hydrographic and plankton data, 1960-1965.
Data Rept. Inst. Oceanogr. Univ. Cape Town, S Afr. (3): 82 pp.

Chemistry silicate (data only)
Zvereva, A.A., Edit., 1959.
Data, 2nd Marine Expedition, "Ob", 1956-1957.
Arktich. i Antarkt. Nauchno-Issled. Inst., Mezhd. Geofiz. God, Sovetsk. Antarkt. Exped., 6: 1-387.

Silicate - monthly avg. surface + bottom

chemistry, silicate (monthly means)
Bose, B.B., 1958.
Observations on the hydrology of the Hooghly Estuary. Indian J. Fish., 1(3):101-118.

silicates (monthly average, surface and bottom)
Jayaraman, R., 1954.
Seasonal variations in salinity, dissolved oxygen and nutrient salts in the inshore waters of the Gulf of Mannar and Palk Bay near Mandapam (S. India). Indian J. Fish. 1(1/2):345-364, 11 textfigs.

chemistry, silicates, mean monthly values
Prasad, R. Raghu, 1956.
Further studies on the plankton of the inshore waters off Mandapam. Indian J. Fish., 3(1):1-42.

Chemistry, silicate, monthly means
Slinn, D.J., and Gaenor Offlow (Mrs. M. Solly), 1968
Chemical constituents of sea water off Port Erin during 1967.
Ann. Rept. mar. biol Stn, Port Erin, 80:37-42.

silicate/oxygen A.O.U.

Chemistry, silicate/oxygen A.O.U.
Sugiura, Yoshio, and Suphachai Chaitiamvong, 1964.
Relation of silicate concentration to dissolved oxygen amount in sea water collected in the northern frontal region of Kuroshio.
J. Oceanogr. Soc., Japan, 20(2):89-92

Chemistry
silicate-sea water equilibria
Perry, Edward A., Jr., 1971.
Silicate-sea water equilibria in the ocean system: a discussion. Deep-Sea Res. 18(9): 921-924.

Silicic acid

Chemistry, Silicic acid
Bogojavlensky, A. N., 1958.
Certain peculiarities of the oxygen, phosphorus and siliceous acid distribution in Antarctic waters.
Inform. Biull. Sovetsk. Antarkt. Exped., (3):19-20.

Chemistry, silicic acid
Brandt, K., 1905.
On the production and the conditions of production in the sea. Cons. Perm. Int. Expl. Mer, Rapp. Proc. Verb. 3(1):12 pp.

chemistry, silicic acid
Eisma, D. and A.J. van Bennekom, 1971.
Oceanographic observations on the eastern Surinam Shelf. Hydrogr. Newsletter, R. Netherlands Navy, Spec. Publ. 6: 25-29.

Chemistry silicic acid
Ivanenkov, V.N., and F.A. Gebin, 1960.
Water masses and hydrochemistry in the western and southern parts of the Indian Ocean. Physics of the Sea. Hydrology. (In Russian).
Trudy Morsk. Gidrofiz. Inst., 22:33-115.

Translation: Scripta Technica, Inc., for Amer. Geophys. Union, 27-99.

Chemistry silicic acid
Mokiyevskaya, V.V., 1961
Some peculiarities of hydrochemistry in the northern Indian Ocean. (In Russian; English abstract).
Okeanolog. Issled., Mezhd. Komitet Proved. Mezhd. Geofiz. Goda, Prezidiume, Akad. Nauk, SSSR, (4):50-61.

Chemistry silicic acid content
Nakai, Z., 1955.
The chemical composition, volume weight and size of the important marine plankton.
Tokai, Reg. Fish. Res. Lab., Spec. Publ., 5:12-24

Chemistry silicic acid
Solov'eva, N.F., 1959.
Hydrochemical analyses of the Aral Sea in 1948.
Akad. Nauk, SSSR, Lab. Ozerovedeniya, Trudy, 8: 3-22.

silicon · Si

Silicon, inorganic

chemistry silicon

Armstrong, F.A.J., 1965.
Silicon.
In: Chemical oceanography, J.P. Riley and G. Skirrow, editors, Academic Press, 1:409-432.

chemistry silicon

Armstrong, F.A.J., 1957.
Phosphorus and silicon in sea water off Plymouth during 1955. J. Mar. Biol. Assoc., U.K., 36(2): 317-322.

chemistry silicon

Armstrong, F.A.J., 1955.
Phosphorus and silicon in sea water off Plymouth during 1954. J.M.B.A. 34(2):223-228.

chemistry silicon

Armstrong, F.A.J., 1954.
Phosphorus and silicon in sea water off Plymouth during the years 1950 to 1953. J.M.B.A. 33(2): 381-392, 3 textfigs.

chemistry, silicon

Bruevich, S.W., editor, 1966.
Chemistry of the Pacific Ocean.
Inst. Okeanol, Akad. Nauk, SSSR, Isdatel. Nauka, Moskva, 358 pp.

chemistry, silicon

Brujewicz, S. W. and E. G. Vinogradova, 1946
[Biogenic elements in the sediment solutions of the northern, middle and southern parts of the Caspian Sea.] Comptes Rendus (Doklady) Acad. Sci. URSS, LIV (5):419-422.

chemistry silicon

Burton, J.D. and T.M. Leatherland, 1970.
The reactivity of dissolved silicon in some natural waters. Limnol. Oceanogr., 15(3): 473-476.

chemistry, silicon

Burton, J.D., and P.S. Liss, 1968.
Ocean budget of dissolved silicon.
Nature, Lond., 220(5170):905-906.

chemistry silicon

Chester, R., 1965.
Elemental geochemistry of marine sediments.
In: Chemical oceanography, J.P. Riley and G. Skirrow, editors, Academic Press, 2:23-80.

chemistry, silicon

Chu, S. P., 1949
Experimental studies on the environmental factors influencing the growth of phytoplankton. Sci. & Tech. in China 2(3):37-52.

chemistry silicon

Corcoran, E.F., and James E. Alexander, 1963.
Nutrient, chlorophyll and primary production studies in the Florida Current.
Bull. Mar. Sci., Gulf and Caribbean, 13(4):527-541.

chemistry silicon

Cushing, D.H., and H.F. Nicolson, 1963
Studies on a Calanus patch. IV. Nutrient salts off the north-east coast of England in the spring of 1954.
J. Mar. Biol. Assoc., U.K., 43(2):373-386.

chemistry, silicon

Dragovich, Alexander, and James E. Sykes 1967
Oceanographic atlas for Tampa, Florida, and adjacent waters of the Gulf of Mexico, 1958-61.
Circular, Fish Wildl Serv. Bur. Comm. Fish (U.S.A.) 255: 466 pp. (quarto)

chemistry silicon

Fedosov, M.V., and N.A. Azova, 1965.
Hydrochemical constituents of the balance of biogenic matter in the Gulf of Alaska. (In Russian).
Sovetsk. Ribokhoz. Issled. Severo-Vostochn. Chasti Tikhogo Okeana, 4 (Vses. Nauchno-Issled. Inst. VNIRO Trudy 58: Tikhookean. Nauchno-Issled. Inst., TINRO Trudy 53):11-20.

chemistry silicon

*Fedosov, M.V., and I.A. Ermachenko, 1967.
Hydrochemical charecteristics of water masses and water exchange between Iceland and Faroes (hydrographical surveys in 1960).
Rapp. P.-V. Réun. Cons. perm. Int. Explor. Mer, 157:196.

chemistry, silicon

Gololobov, Ia. K. 1969.
Analysis of the chemical compounds involved in biological production in the Black Sea and some peculiarities of their formation. (In Russian)
Trudy azov.-chernomorsk. nauchno-issled. Inst. morsk. ryb. Khoz. okeanogr. 23: 33-47.

chemistry silicon

Ichiye, Takashi, Sigeo Moriyasu and Hiroyuki Kitamura, 1951
On the hydrography near the estuaries.
Bull. Kobe Mar. Obs., No. 164:349 (top of page) -369; 53 (bottom of page)-75.

chemistry silicon (data)

*Ishino Makoto, Keiji Nasu, Yoshimi Morita and Makoto Hamada, 1968.
Oceanographic conditions in the west Pacific Southern Ocean in summer of 1964-1965. J. Tokyo Univ. Fish., 9(1): 115-208.

Chemistry, silicon

Kitamura, H., 1962.
(Marine chemical studies on the phosphorus distributed in the North Pacific Ocean.)
Bull. Kobe Mar. Obs., No. 168:40-94.

chemistry silicon

Koshimizu, N., 1952.
[The variation of the chlorinity and silicon content of the surface water between Maizuru and Nakhodoka.] Bull. Maizuru Mar. Obs., (3):32-35.

chemistry, silicon

Koshimizu, N., 1950.
[On the ratio of phosphorus and silicon of the water masses around the banks in Japan Sea.]
Bull. Maizuru Mar. Obs., (1):1-3.

chemistry silicon

Kostikova, A.N., 1965.
Hydrochemical characteristics of the Gulf of Alaska waters in the summer-autumn of 1962. (In Russian).
Sovetsk. Ribokhoz. Issled. Severo-Vostochn. Chasti Tikhogo Okeana, 4 (Vses. Nauchno-Issled. Inst., VNIRO Trudy 58: Tikhookean. Nauchno-Issled. Inst., TINRO, Trudy, 53):21-33.

chemistry, silicon

Krey, J., 1942.
Nährstoff und Chlorophylluntersuchungen in der Kieler förd 1939. Kieler Meeresforsch. 4:1-17, textfigs.

chemistry silicon

Lisitzin, A.P., 1966.
Processes of Recent sedimentation in the Bering Sea. (In Russian).
Inst. Okeanol., Kom. Osad. Otdel Nauk o Zemle, Isdatel, Nauka, Moskva, 574 pp.

chemistry, silicon

Lisitzin, A.P., Yu. I. Beliaev, Yu. A. Bogdanov, and A.N. Bogoiavlensky, 1966.
Regularities in the distribution and forms of silicon suspended in the waters of the World Ocean. (In Russian).
In: Geochemistry of silica, N.M. Strakhov, editor, Isdatel. "Nauka", Moskva, 37-89.

chemistry silicon

Lutsarev, S.V., and D.A. Smetanin, 1959
[Method of obtaining sea water free of silicon and phosphate.] Trudy Inst. Okeanol., 35: 30-32.

chemistry, silicon

Miyake Yasuo and Katsuko Saruhashi 1967.
The geochemical balance of nutrient matter in the oceans.
Pap. Met. Geophys., Tokyo, 15(6): 79-94

chemistry, silicon

Nikiforov, E. G., E.V. Belysheva and N. I. Blinov, 1965.
On the structure of the water masses in the eastern part of the Arctic Basin. (In Russian).
Okeanologiia, Akad. Nauk, SSSR, 6(1):76-81.

chemistry silicon

Ostrowski, Stanislaw, 1962.
Silicon content of the Gdansk Bay waters.
(In Polish: English summary).
Acta Geophys. Polonica, 10(2):1665-171.

chemistry silicon

Parsons, T.R., K. Stephens and J.D.H. Strickland, 1961
On the chemical composition of eleven species of marine phytoplankters.
J. Fish. Res. Bd., Canada, 18(6):1001-1016.

chemistry silicon

Richards, F.A., 1957.
Some current aspects of chemical oceanography.
IN: Physics and chemistry of the earth, Pergamon Press, 2:77-128.

chemistry silicon

Saeki, A., 1950.
[Silicon in the water. 3. Variation of the values estimated by the colorimetric method in the fresh water and its mixed solution with sea water.] J. Ocean. Soc., Tokyo, 6(1):44-47, 1 textfig. (In Japanese, with English abstract).

chemistry silicon

Saeki, A., 1950.
[Silicon in the water. 2. Its colorimetric determination in sea water.] J. Ocean. Soc., Tokyo, 6(1):39-43. (In Japanese with English abstract).

chemistry silicon

Thompson, T.G., and H.G. Houlton, 1933.
Determination of silicon in sea water.
Industr. Engin. Chem. 5(6):417-418.

chemistry, silicon

Toyota Yoshimasa, and Shiro Okabe 1967
Vertical distribution of iron, aluminum, silicon and phosphorus in particulate matter collected in the western North Pacific, Indian and Antarctic oceans. (In Japanese; English abstract)
J. oceanogr. Soc. Japan 23(1): 1-9.
Also in: Coll. Repr. mar. Sci. Techn. Tokai Univ., 1967-68, 3: 335-343.

chemistry silicon

Toyota, Yoshimasa and Shiro Okabe, 1967.
Surface distribution of iron, aluminum, silicon and phosphorus in the western North Pacific and eastern Indian Oceans. (In Japanese).
J. Coll. mar. Sci. Techn., Tokai Univ., (2):227-229.

chemistry silicon

Wstepne, U., 1962
Silicon content of the Gdansk Bay waters.
Acta Geoph. Polonica 10(2):165-171.

Silicon (data only)

chemistry silicon (data only)
Anon., 1950.
[The results of harmonic analysis of tidal currents, water temperature and chemical components in the Kii-Suido.] J. Ocean., Kobe Obs., 2nd ser. 2(1):28-31, 2 figs.

chemistry silicon (data only)
Anon., 1950.
[Report of the oceanographical observations from Fisheries Experimental Stations (Sept. 1950).] J. Ocean., Kobe Obs., 2nd ser., 1(2):5-11.

chemistry silicon (data only)
Anon., 1950.
[The report of the oceanographical observations in the Bungo Suido and off Ashizuri-Misaki in August 1950.] J. Ocean., Kobe Obs., 2nd ser., 1(1):15-44, 7 figs., tables.

chemistry, silicon (data only)
Anon., 1950.
Table 14. Oceanographical observations taken in the Maizuru Bay; physical and chemical data for stations occupied by R.M.S. "Syunpu-maru". Res. Mar. Met., Ocean. Obs., Tokyo, July-Dec. 1947, No. 2:71-76.

chemistry, silicon (data only)
Anon., 1950.
Table 13. Oceanographical observations taken in the Nagasaki Harbour; physical and chemical data for stations occupied by a cutter. Res. Mar. Met. Ocean. Obs., Tokyo, July-Dec. 1947, No. 2:65-70.

Chemistry, silicon (data only)
Anon., 1950.
Table 10. Oceanographic observations taken in the sea northwest of Kyushu; physical and chemical data for stations occupied by the "Fugen-maru". Res. Mar. Met., Ocean. Obs., Tokyo, July-Dec. 1947, No. 2:57-58.

chemistry, silicon (data only)
Anon., 1950.
Table 4. Oceanographical observations taken in the North Pacific Ocean along the coast of Sanriku; physical and chemical data for stations occupied by R.M.S. "Kuroshio-maru". Res. Mar. Met., Ocean. Obs., Tokyo, July-Dec. 1947, No. 2: 13-21.

Chemistry, silicon (data only)
Anon., 1950.
Table 2. Oceanographical observations taken in the Uchiura Bay; physical and chemical data for stations occupied by R.M.S. "Kuroshio-maru". Res. Mar. Met., Ocean. Obs., Tokyo, July-Dec. 1947, No. 2:3-7.

chemistry silicon (data only)
Anon., 1950.
Oceanographic observations taken in the Miyako Bay; physical and chemical data for stations occupied by R.M.S. "Kuroshio-maru". Res. Mar. Met., Ocean. Obs., Jan.-June 1947, No. 1:24-27, Table 3.

chemistry, silicon (data only)
Anon., 1950.
Oceanographical observations taken in the North Pacific Ocean (east off Hideshima Is.); physical and chemical data for stations occupied by R.M.S. "Kuroshio-maru" (continued). Res. Mar. Met., Ocean. Obs., Jan.-June 1947, No. 1:12-23, Table 2

chemistry silicon (data only)
Anon., 1950.
Table 1. Oceanographical observations taken in the Uchiura Bay; physical and chemical data for stations occupied by a fishing boat belonging to Mori Fishery Assoc. Res. Mar. Met., Ocean. Obs., Tokyo, July-Dec. 1947, No. 2:1-2.

chemistry, silicon (data only)
Anon., 1950.
Oceanographical observations taken off the Sanriku coast; physical and chemical data for stations occupied by R.M.S. "Oyashio-maru". Res. Mar. Met. Ocean. Obs., Jan.-June 1947, No. 1:1-11, Table 1.

chemistry, silicon (data only)
Anonymous, 1947
Discovery investigations Station List 1937-1939. Discovery Report, XXIV: 197-422. pls IV-VI.

chemistry, silicon (data only)
Côte d'Ivoire, Centre de Recherches Océanographiques, 1963.
Résultats hydrologiques effectuées au large de la Côte d'Ivoire de 1956-1963. Première partie: observations 1956-1959.
Trav. Centre de Recherches Océanogr., Ministère de la Production Animale, Côte d'Ivoire, unnumbered pp. (multilithed).

Chemistry
Silicon (data only)
Great Britain, Discovery Committee, 1957.
Station List 1950-51. Discovery Repts. 28:299-398.

Chemistry, silicon (data only)
Ivanoff, A., A. Morel, Mme Vesin-Couffinhal, C. Amiel, J.P. Bethoux, J. Boutler, C. Copin-Montegut, P. Courau, P. Geistdoerfer et F. Nyffeler 1969.
Résultats des observations effectuées en Mer Méditerranée orientale et en mer Tyrrhénienne à bord du navire Amalthée en mars 1967. Cah. océanogr. 21 (Suppl. 2): 245-263.

Chemistry silicon (data only)
Japan, Central Meteorological Observatory, 1951.
Table 21. Surface observations made on board R.S.S. "Umikaze-maru"; physical and chemical conditions of the surface waters of the Tsushima Straits and Goto-nada. Res. Mar. Met., Ocean. Obs. Jan.-June 1949, No. 5:115-116.

Chemistry silicon (data only)
Japan, Central Meteorological Observatory, 1951.
Table 1. Oceanographical observations taken in the sea area south of Hokkaido; physical and chemical data for stations occupied by R.M.S. "Yushio-maru". Res. Mar. Met. Ocean. Obs., Jan.-June 1949, No. 5:2-3.

chemistry, silicon (data only)
Japan, Central Meteorological Observatory, Japan, 1951.
The results of marine meteorological and oceanographical observations, July-Dec. 1948, No. 4: vi+414 pp., of 57 tables, 1 map, 1 photo.

chemistry, silicon (data only)
Japan, Central Meteorological Observatory, Japan, 1951.
The results of marine meteorological and oceanographical observations, Jan. June 1948, No. 3: 256 pp.

chemistry silicon (data only)
Japan, Central Meteorological Observatory, Japan, 1950
The results of marine meteorological and oceanographical observations, Jan.-June 1948, 1:113 pp.

Chemistry silicon (data only)
Japan, Hokkaido University, Faculty of Fisheries, 1967.
Data record of oceanographic observations and exploratory fishing, 11:383 pp.

Chemistry
silicon (data only)
Japan, Japan Meteorological Agency, 1971.
The results of marine meteorological and oceanographical observations, 45: 338 pp.

Chemistry
silicon (data only)
Japan, Japanese Oceanographic Data Center, 1971.
Shumpu Maru, Kobe Marine Observatory, Japan Meteorological Agency, 16 July - 8 August 1970, south of Japan. Data Rept. CSK (KDC Ref. 49K127) 279: 7 pp. (multilithed).

Chemistry
silicon (data only)
Japan, Japanese Oceanographic Data Center, 1971.
Takuyo, Hydrographic Department, Maritime Safety Agency, Japan, 10-30 July 1970, south & east of Japan. Data Rept. CSK (KDC Ref. 49K126) 278: 19 pp. (multilithed).

Chemistry
SILICON
(DATA ONLY)
Japan, Japanese Oceanographic Data Center, 1970.
Ji Ri San, Fisheries Research and Development Agency, Republic of Korea, August 7 - 29, 1969, South of Korea and East China Sea, Data Rept. CSK (KDC Ref. 24K034) 239: 15 pp. (multilithed).

Chemistry, silicon (data only)
Japan, Japanese Oceanographic Data Center 1970.
Han Ra San, Fisheries Research and Development Agency, Republic of Korea, August 9-21, 1969, east of the Yellow Sea. Data Rept. CSK (KDC Ref. No. 24K033) 238: 17 pp. (multilithed).

Chemistry
SILICON (DATA ONLY)
Japan, Japanese Oceanographic Data Center, 1970.
Takuyo, Hydrographic Department, Maritime Safety Agency, Japan, August 11 - September 2, 1969, South and East of Japan. Data Rept. CSK (KDC Ref. 49K112) 229: 21 pp. (multilithed).

Chemistry silicon (data only)
Japan, Japanese Oceanographic Data Center, 1970
Shumpu Maru, Kobe Marine Observatory, Japan Meteorological Agency, July 18 - August 14, 196 South of Japan. Data Rept. CSK (KDC Ref. 49K11 7 pp. (multilithed).
228:

Chemistry
Silicon (data only)
Japan, Japanese Oceanographic Data Center, 1970.
Oshoro Maru, Faculty of Fisheries, Hokkaido University, Japan, November 5, 1968 - January 20, 1969, Southwest of the North Pacific Ocean. Data Rept. CSK (KDC Ref. 49K104) 216: 13 pp. (multilithed).

Chemistry, silicon (data only)
Japan, Japanese Oceanographic Data Center 1969.
Iskatel, USSR, January 26 - March 11, 1968, northwest North Pacific Ocean.
Data Rept. CSK (KDC Ref. 90K019): 187: 32 pp. (multilithed).

Chemistry, silicon (data only)
Japan, Japanese Oceanographic Data Center 1969.
Orlick, USSR, July 21 - August 20, 1968, northwest North Pacific Ocean.
Data Rept. CSK (KDC Ref 90K020) 191: 32 pp. (multilithed).

Chemistry silicon (data only)
Japan, Japanese Oceanographic Data Center, 1968.
Tae Bag San, Fisheries Research and Development Agency, Republic of Korea, February 16-March 12, 1968, South of Korea.
Data Rept. CSK (KDC Ref. 24K022) 163: 12 pp.

Chemistry silicon (data only)
Japan, Japanese Oceanographic Data Center, 1968. Han ra San. Fisheries Research and Development Agency, Republic of Korea, February 15- March 12, 1968, East of the Yellow Sea.
Data Rept. CSK (KDC Ref. 24K021) 162: 18 pp. (multilithed)

Chemistry silicon (data only)
Japan, Japanese Oceanographic Data Center, 1968. Baek du San, Fisheries Research and Development Agency, Republic of Korea, February 17-March 12, 1968, West of the Japan Sea.
Data Rept. CSK (KDC Ref. 24K020) 161:23 pp. (multilithed)

Chemistry silicon (data only)
Japan, Japanese Oceanographic Data Center,1968. Ryofu Maru,Marine Division,Japan Meteorological Agency,Japan,January 13-March 22,1968,west of North Pacific Ocean.
Data Rept.CSK (KDC Ref. 49K069) 155: 70 pp. (multilithed).

Chemistry silicon (data only)
Japan, Japanese Oceanographic Data Center,1968. Takuyo, Hydrographic Division, Maritime Safety Agency, Japan, February 19-March 10, 1968, south and east of Japan.
Data Rept. CSK (KDC Ref. 49K068) 154:14 pp. (multilithed).

Chemistry silicon (data only)
Japan, Japanese Oceanographic Data Center,1968. Suro No. 1, Hydrographic Office,Korea,September 23-October 21, 1967, South of Korea.
Data Rept.CSK (KDC Ref. 24K019) 136: 13 pp. (multilithed).

Chemistry silicon (data only)
Japan, Japanese Oceanographic Data Center,1968. Nagasaki Maru,Faculty of Fisheries,Nagasaki University of Fisheries,Japan,November 7-29, 1967,East China Sea,
Data Rept.CSK (KDC Ref.49K076) 133: 5 pp. (multilithed).

Chemistry silicon (data only)
Japan, Japanese Oceanographic Data Center,1968. Kaiyo, Hydrographic Division,Maritime Safety Agency, Japan, November 13-December 1, 1967, South of Japan.
Data Rept. CSK (KCD Ref. 49K063) 128: 13 pp. (multilithed).

Chemistry, silicon (data only)
Japan, Japanese Oceanographic Data Center,1968. Baek Du San,Fisheries Research and Development Agency,Korea,August 12-25,1967,West of Japan Sea.
Data Rept. CSK (KDC Ref.24K016) 115:26 pp. (multilithed).

Chemistry, silicon (data only)
Japan, Japanese Oceanographic Data Center,1968. Hakuho Maru, Ocean Research Institutem University of Tokyo, Japan, September 6-20,1967,South of Japan.
Data Rept.CSK (KDC Ref. 49K059) 111: 9 pp. (multilithed).

Chemistry silicon (data only)
Japan, Japanese Oceanographic Data Center,1968. Yang Ming,Chinese National Committee on Ocean Research, Republic of China,April 1-May 14,1967, Surrounding waters of Taiwan.
Data Rept.CSK (KDC Ref. 21K004) 102: 15 pp. (multilithed).

Chemistry silicon (data only)
Japan, Japanese Oceanographic Data Center,1968. Kofu Maru, Hakodate Marine Observatory, Japan Meteorological Agency, Japan, May 17-19,1967, East of Japan.
Data Rept. CSK (KDC Ref. 49K049) 98: 6 pp. (multilithed).

Chemistry, silicon (data only)
Japan, Shimonoseki University of Fisheries 1965.
International Indian Ocean Expedition 1962-63 and 1963-1964.
Data Oceanogr. Obs. Explor. Fish. 1: 453 pp.

chemistry, silicon (data only)
Japan, Tokyo University of Fisheries, 1966.
J. Tokyo Univ. Fish. (Spec. ed.) 8(3):1-44.

Chemistry silicon (data only)
Sweden, Havsfiske Laboratoriet, Lysekil, 1969.
Hydrographica data, January-June 1968.
R.V. Skagerak, R.V. Thetis.
Meddn Havsfiskelab., Lysekil, hydrog. Avdeln., Goteborg 63: numerous unnumbered pp. (multilithed).

chemistry, silicon, effect of
Darley, W.M. and B.E. Volcani 1969.
Role of silicon in diatom metabolism.
Exp. Cell Res. 58: 334-342.
Also in Coll. Repr. Scripps Inst. Oceanogr. NO: 870-878.

Chemistry, silicon metabolism
Lewin, Joyce, and Ching-hong Chen 1968
Silicon metabolism in diatoms. VI Silicic acid uptake by a colorless marine diatom, Nitzschia alba Lewin and Lewin.
J. Phycol. 4(2): 161-166

silicon, radioactive

Chemistry silicon, radioactive
Anon., 1965.
Radioactive tracer "captured" by huge water bag.
Naval Res. Rev., April, 22-24.

Chemistry silicon-32
Burton, J.D., 1965.
Radioactive nuclides in sea water, marine sediments and marine organisms.
In: Chemical oceanography, J.P. Riley and G. Skirrow, editors, Academic Press, 2:425-475.

Chemistry silicon-32
Lal, Devandra, E.D. Goldberg and Minoru Koide, 1960
Cosmic-ray produced silicon-32 in nature.
Silicon-32, discovered in marine sponges, shows promise as a means for dating oceanographic phenomena.
Science, 131(3397):332-337.

Chemistry silicon-32
Servant, Jean, 1966.
La radioactivite de l'eau de mer.
Cahier Oceanogr., C.C.O.E.C., 18(4):277-318.

Chemistry chemistry, silicon-32
Somayajulu, B.L.K., 1969
Cosmic ray produced silicon-32 in near-coastal waters.
Proc. Indian Acad. Sci. (A), 69 (6), 338-346.

Chemistry silicon 32
Schink, David, 1962
Dissolved silicon 32 in sea water. (Abstract).
J. Geophys. Res., 67(9):3596.

Chemistry silicon starvation
*Healy, F.P., J. Coombs and B.E. Volcani, 1967.
Changes in pigment content of the diatom, Navicula pelliculosa (Bréb.) Hilse in silicon starvation synchrony.
Archiv für Mikrobiol., 59:131-142.

silification

Chemistry silification
Lewin, Ralph A., Editor, 1962.
Physiology and biochemistry of algae.
Academic Press, New York and London, 929 pp.

silver

Chemistry, silver-110
Folsom, Theodore R., D.R. Young and L.E. Finnin 1965.
Sum coincidence gamma-ray spectrometry in tracing cobalt-60 and silver-110 in marine organisms.
In: Radioisotope sample measurement techniques in medicine and biology, Int. Atom. Energy Agency 1965: 57-69.

Chemistry silver
Redfield, A.C. and C.M. Weiss, 1948
The resistance of metallic silver to fouling.
Biol. Bull. 94(1):25-28.

silver
Schutz, Donald F., and Karl K. Turekian, 1965.
The distribution of cobalt, nickel, and silver in ocean water profiles around Pacific Antarctica.
J. geophys. Res., 70(22):5519-5528.

chemistry, silver
Schutz, D.F., and K.K. Turekian, 1964.
The distribution of selenium, antimony, silver, cobalt and nickel in sea water. (Abstract).
Trans. Amer. Geophys. Union, 45(1):118.

chemistry, silver
Turekian, Karl K., 1971
2. Rivers, tributaries and estuaries. In: Impingement of man on the oceans, D.W. Hood, editor, Wiley Interscience, 9-73.

chemistry, silver
Turekian, Karl K. 1968.
Deep-sea deposition of barium, cobalt and silver.
Geochim. cosmochim. Acta 32(6): 603-612.

chemistry, silver
Turekian, Karl K., Donald F. Schultz and David Johnson, 1966.
The distribution of Sr, Ba, Co, Ni, and Ag in ocean water profiles of the Pacific sector of the Antarctic seas.
Antarctic J., United States, 1(5):224.

Chemistry silver
Vech, H. Herbert, and Karl K. Turekian, 1968.
Cobalt, silver, and uranium concentrations of reef-building corals in the Pacific Ocean.
Limnol. Oceanogr., 13(2):304-308.

silver, effect of

Chemistry, silver, effect of
Soyer, J., 1963.
Contribution a l'etude des effets biologiques du mercure et de l'argent dans l'eau de mer.
Vie et Milieu, 14(1):1-36.

chemistry, siphonaxanthin
Jeffrey, S.W. 1968.
Pigment composition of Siphonales algae in the brain coral, Favia.
Biol. Bull. mar. biol. Lab. Woods Hole 135(1): 141-148.

Sodium

chemistry sodium

Ayers, John C., David C. Chandler, George H. Lauf, Charles F. Powers and E. Bennett Henson, 1958
Currents and water masses of Lake Michigan.
Great Lakes Res. Inst., Publ. No. 3:169 pp.

chemistry, sodium

Brewer, Peter G., and Derek W. Spencer, 1969.
A note on the chemical composition of the Red Sea brines.
In: Hot brines and Recent heavy metal deposits in the Red Sea, E.T. Degens and D.A. Ross, editors, Springer-Verlag, New York, Inc., 174-179.

chemistry sodium

Burovina, I.V., et al., 1963.
The content of lithium, sodium, potassium, rubidium and caesium in the muscles of marine animals of the Barents Sea and Black Sea. (In Russian).
Doklady, Akad. Nauk, SSSR, 149(2):413-415.

chemistry, sodium

Chu, S. P., 1949
Experimental studies on the environmental factors influencing the growth of phytoplankton.
Sci. & Tech. in China 2(3):37-52.

chemistry sodium

Culkin, Frederick, 1965.
The major constituents of sea water.
In: Chemical oceanography, J.P. Riley and G. Skirrow, editors, Academic Press, 1: 121-161.

chemistry, sodium

Culkin, F., and R.A. Cox, 1966.
Sodium, potassium, magnesium, calcium and strontium in sea water.
Deep-Sea Res., 13(5):789-804.

chemistry, sodium

*Dyrssen, David, et al., 1967.
Analysis of sea water in the Uddevalla Fjord system 1.
Rept. Chem. Sea Water, Univ. Göteborg, 4:1-8. pp. (mimeographed).

chemistry, sodium

Fukai, R., and F. Shiokawa, 1955.
On the main chemical components dissolved in the adjacent waters to the Aleutian Islands in the North Pacific. Bull. Chem. Soc., Japan, 28(9): 636-640.

Also: Bull. Tokai Reg. Fish. Res. Lab., 11

chemistry sodium(data)

Kramer J.R., 1961
Chemistry of Lake Erie.
Proc. Fourth Conf., Great Lakes Research, Great Lakes Res. Div., Inst. Sci. & Tech., Univ. Michigan, Publ., (7):27-56.

chemistry sodium (activity coefficient)

Platford, R.F., 1965.
Activity coefficient of the sodium ion in sea water.
J. Fish. Res. Bd., Canada, 22(4):885-889.

chemistry sodium

*Presley, B.J., R.R. Brooks and I.R. Kaplan, 1967.
Manganese and related elements in the interstitial water of marine sediments.
Science, 158(3803):906-910.

chemistry sodium

Webb, D.A., 1939.
The sodium and potassium content of sea water.
J. Exp. Biol., 16:178-183.

chemistry, sodium balance

Hagerman, Lars 1971
Osmoregulation and sodium balance in Crangon vulgaris (Fabricius) (Crustacea, Natantia) in varying salinities.
Ophelia 9(1):21-31

Sodium isotopes

chemistry sodium-24

Lobe, D.L., and D. Sam, 1962.
Radiochemical determination of sodium-24 and sulphur-35 in sea water.
Anal. Chem., 34(3):336-340.

Listed in Anal. Abstr., 9(9):#3936.

chemistry, sodium-22

Parker, Patrick L., 1966.
Movement of radioisotopes in a marine bay: cobalt-60, iron-59, manganese-54, zinc-65, soduim-22.
Publs., Inst. mar. Sci., Univ. Texas, Port Aransas, 11:102-107.

Sodium chloride

chemistry sodium chloride

Bieliayev, L.I., 1959.
Extraction of sodium chloride from sea water.
Trudy, Morsk Hydrophys. Inst., 15:106-116.

chemistry sodium chloride

Bieliayov, L.A., 1959.
Purification of sodium chloride extracted from sea water.
Trudy Morsk. Hydrophys. Inst., 15:117-135.

chemistry sodium chloride

Cunningham, G.L., and T.S. Oey, 1955.
The system sodium chloride-sodium chloride-water at various temperatures. J. Am. Chem. Soc. 77(3): 799-801.

chemistry sodium chloride

Platford, R.F., 1965.
The activity coefficient of sodium chloride in seawater.
J. Mar. Res., 23(2):55-62.

chemistry, sodium chloride

Schachter, D., 1954.
Contribution à l'étude hydrographique et hydrologique de l'étang de Berre (Bouches-du-Rhone).
Bull. Inst. Océan., Monaco, No. 1048:20 pp.

chemistry, sodium/chlorinity

*Riley, J.P., and M. Tongudai, 1967.
The major cation/chlorinity ratios in sea water.
Chem. Geol., 2(3):263-269.

Sodium hydroxide

chemistry sodium hydroxide (mfg.)

Kume, T., 1957.
Studies on the direct manufacture of sodium hydroxide from sea water.
Rec. Ocean. Wks., Japan (Suppl.):57.

chemistry sodium hydroxide

Kume, T., T. Kuno and T. Tanaka, 1957.
Studies on the direct manufacture of sodium hydroxide and other materials from sea water. 3. Effects of co-existing ions for electrolysis.
Rec. Ocean. Wks., Japan (Suppl.):50-52.

chemistry sodium hydroxide

Kume, T., and M. Okado, 1957.
Studies on the direct manufacture of sodium hydroxide and other materials from sea water. 2. Electrolysis by mercury process using a diaphragm. Rec. Ocean. Wks., Japan, (Suppl.):53.

Sodium ions

chemistry, sodium ions

Kikkawa, Kyozo, and Shiko Shiga, 1965.
The modified chemical constituent of sea water intruding into the coastal aquifer.
Spec. Contrib., Geophys. Inst., Kyoto Univ., No. 5:7-16.

chemistry, sodium regulation

Harris, R.R., 1972.
Aspects of sodium regulation in a brackish-water and a marine species of the isopod genus Sphaeroma. Mar. Biol. 12(1): 18-27.

Sodium sulphate

chemistry sodium sulfate

Kester, Dana R. and Ricardo M. Pytkowicz, 1969.
Sodium, magnesium and calcium sulfate ion-pairs in seawater at 25C. Limnol. Oceanogr., 14(5): 686-692.

chemistry sodium sulfate

Platford, R.F., and Thomas Defoe, 1965.
The activity coefficient of sodium sulfate.
J. Mar. Res., 23(2):63-68.

chemistry, solubility

Klotz, I.M., 1963.
Comment on the variation of solubility with depth.
Limnology and Oceanography, 8(4):486.

chemistry solubility

Klotz, Irving M., 1963
Variation of solubility with depth in the ocean: a thermodynamic analysis.
Limnol. and Oceanogr., 8(2):149-151.

chemistry solubility

Pytkowicz, Ricardo M., 1963.
Comment on the variation of solubility with depth.
Limnology and Oceanography, 8(4):486.

chemistry solutions

Gibbs, Ronald J., 1967.
Amazon river: environmental factors that control its dissolved and suspended load.
Science, 156(3783):1734-1737.

sorption processes

chemistry sorption processes

Carritt, D.E., and S. Goodgal, 1954.
Sorption reactions and some ecological implications. Deep-Sea Res. 1(4):224-243, 16 textfigs.

chemistry sorption

Erokhin, V.E., 1970.
On possibility of sorbtion accumulation of macromolecules solved in sea water by copepods Tigriopus brevicornis O.F. Müller and larvae of Balanus improvisus Darw. (In Russian).
Gidrobiol. Zh., 6(6): 94-98.

chemistry sorption

Siegel, Alvin, and Barbara Burke, 1965.
Sorption studies of cations on "bubble produced organic aggregates" in sea water.
Deep-Sea Res., 12(6):789-796.

Specific alkalinity/salinity

Standard sea water

chemistry standard water

Knudsen, M., 1903.
On the standard-water used in the hydrographical research until July 1903.
Publ. de Circ., Cons. Perm. Int. Expl. Mer, No. 2: 1-9.

standard sea water

Toll, R., and J.-Ma Vallès, 1952.
Sur l'emploi comme dissolution "Standard" pour la détermination de la chorinité des eaux de mer d'une dissolution de ClNa du même Cl/l que celui de l'eau normale internationale. Vie et Milieu, Suppl. 2, Océan. Médit., Jour. Études, Lab. Arago :292-296.

standard water

Toll, R., J.-M[a] Vallès, and F. Saiz, 1952. Sur l'utilisation d'une eau de mer quelconque comme eau auxiliaire pour la détermination de la chlorinité des eaux marins. Vie et Milieu, Suppl. 2, Océan. Medit., Jour. Etudes, Lab. Arago:282-291.

chemistry, starch

Bursa Adam S. 1968. Starch in the oceans. J. Fish. Res. Bd. Can. 25(6):1269-1284.

chemistry, statistics in

Palmork, Karsten H., 1969. Determination of reliability in marine chemistry by means of intercalibration and statistics. In: Chemical oceanography, Rolf Lange, editor, Universitetsforlaget, Oslo: 91-103.

Chemistry, steroids

Droop, M.R. and J.F. Pennock, 1971. Terpenoid quinones and steroids in the nutrition of Oxyrrhis marina. J. mar. biol. Ass. U.K. 51(2): 455-470.

Sterols

Chemistry Steroid glycosides

Mackie, A.M., 1970. Avoidance reactions of marine invertebrates to either steroid glycosides of starfish or synthetic surface-active agents. J. exp. mar. Biol. Ecol., 5(1): 63-69.

Chemistry chemistry- sterols

Fagerlund, U.H.M., 1962. Marine sterols. X. The sterol content of the plankton Euphausia pacifica. Canadian J. Biochem. Physiol., 40(2):1839-1840.

Chemistry sterols

Fox, D.L., and C.H. Oppenheimer, 1954. The riddle of sterol and carotenoid metabolism in muds of the ocean floor. Arch. Biochem. Biophys. 51(2):323-328.

chemistry, sterols

Kita, M., and Y. Toyama, 1960. On the sterols of twenty five species of marine invertebrates in Japanese waters. J. Chem. Soc., Japan, 81(3):485-489.

Chemistry sterols

Liaci, Lidia, 1964. Pigmenti e steroli negli invertebrati marini. Archivio Zool. Ital., 49:281-300.

chemistry sterol biosynthesis

*Whitney, Joanne O'Connell, 1970. Absence of sterol biosynthesis in the blue crab Callinectes sapidus Rathbun and in the barnacle Balanus nubilus Darwin. J. exp. mar. Biol. Ecol., 4(3): 229-237.

Strontium

chemistry strontium

Andersen, Neil R. 1967. Observed variations in the strontium concentration of sea water: a letter received. Chem. Geol. 2(1):77.

chemistry, strontium

Angino, E.E., G.K. Billings and N. Andersen. Observed variations in the Strontium concentration of sea water. Chem. Geol., 1(2):145-153.

chemistry, strontium

Armstrong, Richard Lee 1971. Glacial erosion and the variable isotopic composition of strontium in sea water. Nature, Lond. 230(14):132-133.

Chemistry strontium

Bernhard, M., 1964. Chemical composition and the radiocontamination of marine organisms. Proc. Symp. Nuclear Detonations and Marine Radioactivity, S.H. Small, editor, 137-150.

chemistry, strontium

Biscaye, Pierre E. and E. Julius Dasch, 1971. The Rubidium, strontium, strontium-isotope system in deep-sea sediments: Argentine Basin. J. geophys. Res, 76(21): 5087-5096.

chemistry, strontium

Bojanowski, Ryszard, and Stanislaw Ostrowski, 1968. On the strontium content of the southern Baltic waters. Acta Geophys. polon., 16(4): 351-356.

Chemistry strontium

Borovik-Romanova, J.F., M.V. Korolev and Y.I. Kutsenko, 1954. [Spectroscopic determination of strontium and lithium in natural waters.] Zhurn. Analit. Khim., 9(5):265.

Chemistry strontium

Bowen, H.J.M., 1956. Strontium and barium in sea water and marine organisms. J.M.B.A., 35(3):451-460.

Chemistry strontium

Bowen, H.J.M., 1956. The biogeochemistry of strontium. Adv. Sci., 12(49):585-588.

Chemistry strontium

Broecker, W., 1963 4. Radioisotopes and large scale oceanic mixing. In: The Sea, M.N. Hill, Edit. Vol.2. The composition of sea water, Interscience Publishers, New York and London, 88-108.

Chemistry strontium

Chester, R., 1965. Elemental geochemistry of marine sediments. In: Chemical oceanography, J.P. Riley and G. Skirrow, editors, Academic Press, 2:23-80.

chemistry strontium

Chow, T.J., and T.G. Thompson, 1955. Flame photometric determination of strontium in sea water. Analyt. Chem. 27(1):18-21.

chemistry strontium

Culkin, Frederick, 1965. The major constituents of sea water. In: Chemical oceanography, J.P. Riley and G. Skirrow, editors, Academic Press, 1: 121-161.

chemistry, strontium

Culkin, F., and R.A. Cox, 1966. Sodium, potassium, magnesium, calcium and strontium in sea water. Deep-Sea Res., 13(5):789-804.

chemistry, strontium

Dasch, E. Julius, and Pierre E. Biscaye 1971. Isotopic composition of strontium in Cretaceous-to-Recent pelagic Foraminifera. Earth planet. Sci. Letts 11(3): 201-204

Chemistry strontium

Dasch, E. Julius, F. Allan Hills and Karl K. Turekian, 1966. Strontium isotopes in deep-sea sediments. Science, 153(3733):295-297.

Chemistry strontium

Dezgrez, A., and J. Meunier, 1926. Recherche et dosage du strontium dans l'eau de mer. C.R. Acad. Sci., Paris, 183(17):689-691.

Transl. cited FWS Spec. Sci. Rept., Fish. 227

Chemistry strontium

Dodd, J. Robert, 1965. Environmental control of strontium and magnesium in Mytilus. Geochimica et Cosmochimica Acta, 29(5):385-398.

Chemistry strontium

*Dyrssen, David, et al., 1967. Analysis of sea water in the Uddevalla Fjord system 1. Rept. Chem. Sea Water, Univ. Göteborg, 4:1-8 pp. (mimeographed).

chemistry, strontium

Fabricand, E.S. Imbimbo, M. E. Brey and J.A. Weston, 1966. Atomic absorption for Li, Mg, K, Rb, and Sr in ocean waters. J. geophys. Res., 71(6):3917-3921.

chemistry, strontium

Faure, G., P.M. Hurley, and J. L. Powell, 1964. The isotopic composition of strontium in surface water from the North Atlantic Ocean. (Abstract). Trans. Amer. Geophys. Union, 45(1):113-114.

chemistry strontium

*Faure, Gunter, and Lois M. Jones, 1969. Anomalous strontium in the Red Sea brines. In: Hot brines and Recent heavy metal deposits in the Red Sea, E.T. Degens and D.A. Ross, editors, Springer-Verlag, New York, Inc., 243-250.

Chemistry, strontium

Hallam, A., and N.B. Price 1968 Environmental and biochemical control of strontium in shells of Cardium edule. Geochim. cosmochim. Acta 32(3):319-328.

chemistry, strontium

Hallam, A., and N.B. Price, 1966. Strontium contents of Recent and fossil aragonite cephalopod shells. Nature, Lond., 212(5057):25-27.

chemistry, strontium

Hamilton, E.I., 1966. The isotopic composition of strontium in Atlantic Ocean water. Earth Planet. Sci. Letters 1(6):435-436.

Chemistry strontium

Hamilton, E.I., 1965. Isotopic composition of strontium in a variety of rocks from Reunion Island. Nature, 207(5002):1188.

Chemistry Strontium

Harris, Reobert C., 1965. Trace element distribution in molluscan skeletal material. 1. Magnesium, iron, manganese and strontium. Bull. Mar. Sci., 15(2):265-273.

chemistry, strontium

Hildreth, Robert A., and William T. Henderson 1970. Comparison of Sr^{87}/Sr^{86} for sea-water strontium and the Eimer and Amend $SrCO_3$. Geochim. cosmochim. Acta, 35(2): 235-238

Chemistry strontium
Hiyama, Y., and R. Ichikawa, 1957.
A measure on level of strontium-90 concentration in sea-water around Japan, at the end of 1956.
Rec. Oceanogr. Wks., Japan, n.s., 4(1):49-54.

Chemistry strontium (uptake)
Hiyama, Y., and R. Ichikawa, 1957.
Up-take of strontium by marine fish from the environment. Rec. Oceanogr. Wks., Japan, n.s., 3(1):78-84.

Chemistry strontium
Hiyama, Yoshio, and Makoto Shimizu, 1964
On the concentration factors of radioactive Cs, Sr, Cd, Zn, and Ce in marine organisms.
Rec. Oceanogr. Wks., Japan, 7(2):43-77.

Chemistry strontium
Hiyama, Yoshio, Makoto Shimizu, Junko Matsubara, Tamiya Asari and Ryushi Ichikawa, 1960
Sr 90 in marine organisms in Japan. Radioactive contamination of marine products in Japan.
Document for the Scientific Committee on the Effect of the Atomic Radiation of the United Nations, 9-12.

Chemistry strontium
Hummel, R.W., and A.A. Smales, 1956.
Determination of strontium in sea water by using both radioactive and stable isotopes. Analyst 81(959):110-113.

Chemistry strontium
Ishibashi, Masayoshi, Shunzo Ueda and Yoshikazu Yamamoto, 1969.
The contents of strontium and strontium-90 in shallow water deposits (In Japanese; English abstract).
J. oceanogr. Soc. Japan 25(5): 233-238

Chemistry strontium
Ishibashi, Masayoshi, Taitiro Fujinaga, Tooru Kuwamota and Takahiro Kumamaru, 1960
Chemical studies on the ocean (83). Chemical studies on seaweeds. Quantitative determination of strontium in seaweeds. Rec. Oceanogr Wks., Japan, Spec. No. 4: 91-94.

Chemistry strontium
Ichikawa, R., and Y. Hiyama, 1957.
The amount of uptake of strontium by marine fish from the environment of various concentration of calcium and strontium.
Rec. Oceanogr. Wks., Japan, n.s., 4(1):55-66.

chemistry, strontium
Kautsky, Hans, 1970.
Auftreten und Veränderungen des Sr-90- und Cs-137-Gehaltes im Oberflächenwasser der Barentssee in den Jahren 1961 bis 1969
Dt. hydrogr. Z. 23(6): 264-268

Chemistry strontium
Kevern, Niles R., 1964.
Relative strontium and calcium uptake by green algae.
Science, 146(3650):1488.

Chemistry strontium
Khan, Khan Umuardaraz, and Yoshio Hiyama, 1964
Mutual effect of Sr-Ca upon their uptake by fish and freshwater plants.
Rec. Oceanogr. Wks., Japan, 7(2):107-122.

Chemistry strontium
*Kinsman, David J.J., and H.D. Holland, 1969.
The co-precipitation of cations with $CaCO_3$.
IV. The co-precipitation of Sr^{2+} with aragonite between 16o and 96oC.
Geochim. cosmochim. Acta, 33(1):1-17.

Chemistry strontium
Lerman, Abraham, 1965.
Strontium and magnesium in water and in Crassostrea calcite.
Science, 150(3697):745-751.

Chemistry strontium
Loveridge, B.A., 1959
Determination of radio-strontium in sea water.
A.E.R.E. Report C/M 380: 7 pp.
Abs. in Anal. Abs., 6(8): 3248.

Chemistry strontium
Lowenstam, Heinz A., 1964.
Coexisting calcites and aragonites from skeletal carbonates of marine organisms and their strontium and magnesium contents.
In: Recent researches in the fields of hydrosphere, atmosphere and nuclear geochemistry, Ken Sugawara, festival volume, Y. Miyake and T. Koyama, editors, Maruzen Co., Ltd., Tokyo, 373-404.

Chemistry strontium
MacKenzie, Fred T., 1964
Strontium content and variable strontium-chlorinity relationship of Sargasso Sea water.
Science, 146(3643):517-518.

Chemistry strontium
Mauchline, J., and W.L. Templeton, 1966.
Strontium, calcium and barium in marine organisms from the Irish Sea.
J. Cons. perm. int. Expl. Mer, 30(2):161-170.

Chemistry strontium
Miyake, Y., 1939.
Strontium in sea water. Geophys. Mag., Japan, 12:305-306.

Chemistry strontium
Müller, German, 1962.
Zur Geochemie des Strontium in ozeanen Evaporiten unter besonderer Berücksichtung der Sedimentären Cölestinlagerstätte von Hemmelte-West (Süd-Oldenburg).
Akad. Verlag, Berlin, 90 pp., 35 figs., 38 tables
R.v. by H. Heide, 1964. Chimie der Erde, 23 (3):222

Chemistry Strontium
*Murthy, V. Rama and E. Beiser, 1968.
Strontium isotopes in ocean water and marine sediments.
Geochim. cosmochim. Acta, 32(10):1121-1126.

chemistry, strontium
Nagaya, Yutaka, Kiyoshi Nakamura and Masamichi Saiki 1971.
Strontium concentrations and strontium-chlorinity ratios in sea water of the North Pacific and the adjacent seas of Japan.
J. oceanogr. Soc. Japan 27(1): 20-26.

Chemistry strontium
Nelson, Daniel J., 1965.
Strontium in calcite: new analyses.
Publ. Inst. Mar. Sci., Port Aransas, 10:76-79.

Chemistry strontium
Odum, H.T., 1957.
Strontium in natural waters.
Publ. Inst. Mar. Sci., Port Aransas, 4(2):22-37.

Chemistry strontium
Odum, H.T., 1957.
Biogeochemical deposition of strontium.
Publ. Inst. Mar. Sci., Port Aransas, 4(2):38-114

Chemistry strontium
Odum, H.T., 1951.
Notes on the strontium content of sea water, celestite radiolaria, and strontianite snail shells.
Science, 114:211-213.

Chemistry strontium
Paoletti, A., 1966
Captation du strontium radioactif par des Algues microscopiques.
Revue Int. Océanogr. Medicale, C.E.R.B.O.M., 1: 92-99

Chemistry Strontium
Parchevskii, V.P., G.G. Polycarpov and I.S. Zaburunova, 1965
Certain regularities in the accumulation of yttrium and strontium by marine organisms. (In Russian)
Doklady, Akad. Nauk, SSSR, 164(4):913-916

strontium
Peterman, Zell E., and Carl E. Hedge, 1971.
Related strontium isotopic and chemical variations in oceanic basalts.
Bull. geol. Soc. Am., 82(2): 493-500.

Chemistry strontium
Pilkey, Orrin H., and H.G. Goodell, 1963
Trace elements in Recent mollusk shells.
Limnol. and Oceanogr., 8(2):137-148.

chemistry, strontiums
*Price, N.B., and A. Hallam, 1967.
Variation of strontium content within shells of Recent Nautilus and Sepia.
Nature, 215(5107):1272-1274.

Chemistry, strontium
Pushkar, Paul, and M.N.A. Peterson, 1967.
Strontium isotopic analyses of the marine phillipsite from the Pacific Ocean.
Earth Planet. Sci. Letters, 2(4): 349-356.

Chemistry strontium
Rao, S.R., S.M. Shah and R. Viswanathan, 1968.
Calcium, strontium and radium content of molluscan shells.
J. mar. biol. Ass., India, 10(1): 159-165.

Chemistry strontium exchange
Rice, T.R., 1956.
The accumulation and exchange of strontium by marine planktonic algae. Limnol. & Oceanogr., 1(2):123-138.

Chemistry strontium
Schreiber, B., 1963.
Acantharia as "scavengers" for strontium and their role in the sedimentation of radioactive debris.
In: Nuclear detonations and marine radioactivity the report of a symposium held at the Norwegian Defence Research Establishment, 16-20 September, 1963. Forsvarets, Forskningsinstitutt, P.O. Box 25, Kjeller, Norge, 113-126. (multilithed)

Chemistry strontium
Schreiber, B., E. Bottazzi-Massera, A. Fano-Schreiber, F. Guerra, L. Pelati, 1962.
Ricerche sulla presenza dello Sr nel plancton marino in rapporto alla ecologia degli Acantari. Problemi ecologici delle zone litorale del Mediterraneo, 17-23 luglio 1961.
Pubbl. Staz. Zool., Napoli, 32(Suppl.):400-426.

chemistry-strontium
Servant, Jean, 1966.
La radioactivité de l'eau de mer.
Cahier Océanogr., C.C.O.E.C., 18(4):277-318.

Chemistry strontium
Smales, A.A., 1951.
Determination of strontium in sea water by a combination of flame photometry and radio-chemistry.
Analyst, 76:348-355.

Chemistry strontium
Sokolova, I.A., V.P. Parchevskiy, and N.V. Sokolova, 1968.
Methods and data of research of hydrobiosphere Sr90 pollutions. (In Russian).
Okeanologiia, Akad. Nauk, SSSR, 8(6):1092.

Chemistry strontium
Sugawara, K., 1957.
The distribution of some minor bio-elements in Western Pacific waters.
Proc. UNESCO Symp., Phys. Ocean., Tokyo, 1955: 169-174.

Chemistry strontium
Sugawara, K., and N. Kawasaki, 1958.
Strontium and calcium distribution in western Pacific, Indian and Antarctic Oceans.
Rec. Oceanogr. Wks., Japan, Spec. No. 2:227-242.

Chemistry strontium
Sugawara, Ken, Kikuo Terada, Satoru Kanamori, Nobuko Kanamori and Shiro Okabe, 1962
On different distribution of calcium, strontium, iodine, arsenic and Molybdenum in the northwestern Pacific, Indian and Antarctic Oceans.
J. Earth Sciences, Nagoya Univ., 10(1):34-50.

Chemistry strontium
Swan, E.F., 1956.
The meaning of strontium-calcium ratios.
Letter to the Editors. Deep-Sea Res., 4(1):71.

Chemistry strontium
Tamontiev, V.P., and S.W. Bruewicz, 1964.
Strontium in the waters of the Pacific and Indian oceans and of the Black Sea. Chemistry of the waters and sediments of seas and oceans. (In Russian; English abstract).
Trudy Inst. Okeanol., Akad. Nauk, SSSR, 67:41-55.

Chemistry strontium-90
*Taylor, C.B., 1968.
A comparison of tritium and strontium-90 fallout in the Southern Hemisphere.
Tellus, 20(4):559-576.

Chemistry-strontium
Thompson G., and H.D. Livingston 1970
Strontium and uranium concentrations in aragonite precipitated by some modern corals.
Earth Planet. Sci. Letts 8(6): 439-442.

chemistry, strontium
Turekian, Karl K., 1971
2. Rivers, tributaries and estuaries. In: Impingement of man on the oceans, D.W. Hood, editor, Wiley Interscience, 9-73.

Chemistry strontium
Turekian, Kark K., 1964.
The marine geochemistry of strontium.
Geochimica et Cosmochimica Acta, 28(9):1479-1496.

Chemistry strontium
Turekian, K.K., 1957.
The significance of variations in the strontium content of deep sea cores. Limnol. & Oceanogr., 2(4):309-314.

Chemistry strontium
Turekian, Karl K., and Richard L. Armstrong, 1960
Magnesium, strontium and barium concentrations and calcite-aragonite ratios of some recent molluscan shells. J. Mar. Res., 18(3): 133-151

Chemistry strontium
Turekian, Karl K., Donald F. Shutz, Peter Bower and David G. Johnson, 1967.
Alkalinity and strontium profiles in Antarctic waters.
Antarctic Jl, U.S.A., 2(5):186-188.

chemistry, strontium
Turekian, Karl K., Donald F. Schultz and David Johnson, 1966.
The distribution of Sr, Ba, Co, Ni, and Ag in ocean water profiles of the Pacific sector of the Antarctic seas.
Antarctic J., United States, 1(5):224.

Chemistry strontium
Uzumasa, Y., Y. Nasu, and T. Seo, 1960.
Flame photometric determination of strontium and calcium in natural waters.
J. Chem. Soc., Japan, 81(3):430-433.

Chemistry strontium
Viswanathan, R., S.M. Shah and C.K. Unni, 1969
Atomic absorption and fluorometric analyses of seawater in the Indian Ocean. Bull. natn. Inst. Sci., India, 38(1): 284-288.

Chemistry strontium-90
*Volchok, H.L., and M.T. Kleinman, 1969.
Strontium 90 yield of the 1967 Chinese thermonuclear explosion.
J. Geophys. Res., 74(6):1694-1696.

Chemistry strontium
Wangersky, Peter J., and Oiva Joensuu, 1964
Strontium, magnesium and manganese in fossil foraminiferal carbonates.
J. Geol., 72(4):477-483.

Chemistry strontium
Wattenberg, H., and E. Timmermann, 1938.
Die Löslichkeit von Magnesiumkarbonat und Strontiumkarbonat in Seewasser. Kieler Meeresf. 2:81-94.

Chemistry strontium
Williams, Louis G., 1964.
Relative strontium and calcium uptake by green algae.
Science, 146(3650):1488.

Chemistry strontium
Wiseman, J.D.H., M.W. Strong and H.J.M. Bowen, 1956.
Marine organisms and biogeochemistry. 1. The rates of accumulation of nitrogen and calcium carbonate on the equatorial Atlantic floor. 2. Marine iron bacteria as rock forming organisms. 3. The biogeochemistry of strontium.
Adv. Sci., 12(49):579-588.

chemistry, strontium
*Yamamoto, Toshio, Tatsuo Fujita, and Tsunenobu Shigematsu, 1969.
Chemical studies on the seaweeds (24): Strontium content in seaweeds.
Rec. oceanogr. Wks. Japan, 10(1):29-38.

strontium chlorinity ratio
Nagaya, Yutaka, Kiyoshi Nakamura and Masamichi Saiki 1971.
Strontium concentrations and strontium-chlorinity ratios in sea water of the North Pacific and the adjacent seas of Japan.
J. oceanogr. Soc. Japan 27(1): 20-26.

Strontium-calcium atom ratio

Chemistry, strontium-calcium atom ratio
Thompson, T.G., and T.J. Chow, 1955.
The strontium-calcium atom ratio in carbonate-secreting marine organisms.
Pap. Mar. Biol. and Oceanogr., Deep-Sea Res., Suppl. to Vol. 3:20-39.

Chemistry strontium/calcium
Turekian, Karl K., 1964.
The marine geochemistry of strontium.
Geochim. et Cosmochim. Acta, 28(9):1479-1496.

strontium-chlorinity

Chemistry Strontium-chlorinity ratio
Andersen, N.R., J.D. Gassaway and W.E. Maloney, 1970.
The relationship of the strontium: chlorinity ratio to water masses in the tropical Atlantic Ocean and Caribbean Sea. Limnol. Oceanogr., 15(3): 467-472.

chemistry strontium-chlorinity
MacKenzie, Fred T., 1964
Strontium content and variable strontium-chlorinity relationship of Sargasso Sea water.
Science, 146(3643):517-518.

chemistry, strontium/chlorinity
*Riley, J.P., and M. Tongudai, 1967.
The major cation/chlorinity ratios in sea water.
Chem. Geol., 2(3):263-269.

Chemistry strontium, effect of
Webb, D.A., 1938.
Strontium in sea water and its effect on calcium determinations.
Nature, 142(3599)751-752.

Chemistry, strontium fallout
Machta, L., K. Telegadas and D.L. Harris, 1970.
Strontium-90 fallout over Lake Michigan. J. geophys. Res., 75(6): 1092-1096.

chemistry, strontium fallout
Swindle, D.L. and P.K. Kuroda, 1969.
Variation of the Sr^{89} / Sr^{90} ratio in rain caused by the Chinese nuclear explosions of December 28, 1966, and June 17, 1967.
J. geophys. Res., 2136-2140.

Chemistry, strontium fallout
*Taylor, C.B., 1968.
A comparison of tritium and strontium-90 fallout in the Southern Hemisphere.
Tellus 20(4):559-576.

strontium isotopes

Chemistry strontium-90
Azhazha, E.G., and P.M. Chulkov, 1964.
Strontium-90 in the surface waters of the Atlantic Ocean in the first half of 1961. (In Russian).
Okeanologiia, Akad. Nauk, SSSR, 4(1):68-73.

Chemistry strontium-90
Baranova, D.D., and G.G. Polikarpov, 1965.
Sorption of strontium-90 and cesium-137 in the aleurite silts of the Black Sea. (In Russian).
Okeanologiia, Akad. Nauk, SSSR, 5(4):646-648.

chemistry, strontium-90
Belyaev, V.I., I.F. Doronin, A.I. Ermolenko, T.M. Ivanova, B.A. Nelepo and I.E. Timchenko, 1969.
Objective analysis of Sr-90 concentration field in the Atlantic Ocean. (In Russian; English abstract).
Fizika Atmosfer. Okean. Izv. Akad. Nauk SSSR 5(8): 860-867.

chemistry strontium-90
Belyaev, L.I., L. I. Gedeonov, and G.V. Yakovleva, 1966.
Estimation of Sr^{90} and Cs^{137} content in the Black Sea. (In Russian; English abstract).
Okeanologiia, Akad. Nauk, SSSR, 6(4): 641-644

chemistry, strontium isotopes

Biscaye, Pierre E. and E. Julius Basch, 1971. The Rubidium, strontium, strontium-isotope system in deep-sea sediments: Argentine Basin. J. geophys. Res, 76(21): 5087-5096.

Chemistry strontium-90
Bolter, Ernst, Karl K. Turekian and Donald E. Schutz, 1964. The distribution of rubidium, cesium and barium in the oceans. Geochimica et Cosmochimica Acta, 28(9):1459-1466.

Chemistry strontium 90
Boroughs, H., and D.F. Reid, 1958. The role of the blood in the transportation of strontium 90-yttrium 90 in teleost fish. Biol. Bull., 115(1):64-73.

Chemistry strontium-89
Boroughs, H., S.J. Townsley and R.W. Hiatt, 1956. Method for predicting amount of strontium-89 in marine fishes by external monitoring. Science 124(3230):1027-1028.

chemistry, strontium-90
Bowen, Vaughn T., and Victor E. Noshkin and Thomas T. Sugihara, 1966. Transport of strontium-90 towards the equator at mid-depths in the Atlantic Ocean. Nature, 212(5060):383-384.

chemistry, strontium-90
Bowen, Vaughan T., Victor E. Noshkin, Herbert L. Volchok and Thomas T. Sugihara, 1969 Strontium-90: concentrations in surface waters of the Atlantic Ocean. Science, 164(3881):825-827.

Chemistry strontium - 90
Bowen, Vaughan T., and Thomas T. Sugihara, 1965. Oceanographic implications of radioactive fall-out distributions in the Atlantic Ocean: from 20° N to 25°S, from 1957 to 1961. J. Mar. Res., 23(2):123-146.

Chemistry strontium 90
Bowen, V.T., and T.T. Sugihara, 1957. Strontium 90 in North Atlantic surface water. Proc. Nat. Acad. Sci., 43(7):576-580.

Chemistry strontium 90
Broecker, Wallace S., Erno R. Bonebakker and Gregory G. Rocco, 1966. The vertical distribution of cesium 137 and strontium 90 in oceans. 2. J. geophys. Res., 71(8):1999-2003.

Chemistry strontium-90
Broecker, Wallace S., Gregory G. Rocco and H.L. Volchok, 1966. Strontium-90 fallout: comparison of rates over ocean and land. Science, 152(3722):639-640.

Chemistry strontium-90
Corcoran, E.F., and J.F. Kimball, Jr., 1963. The uptake, accumulation and exchange of strontium-90 by open sea phytoplankton. Radioecology, V. Schultz and A.W. Klement, Jr., editors, Proc. First Nat. Symp., Radioecology, Sept. 1961, Reinhold Publ. Corp., and Amer. Inst. Biol. Sci., 187-191.

chemistry, strontium isotopes
Dasch, E. Julius, 1969. Strontium isotopes in weathering profiles, deep-sea sediments, and sedimentary rocks. Geochim. Cosmochim. Acta, 33(12): 1521-1552.

Chemistry, strontium 90
Eisenbud, M., 1957. Global distribution of radioactivity from nuclear detonations with special reference to strontium 90. J. Washington Acad. Sci., 47(6):180-188.

Chemistry, strontium 90
Fabian, P., W.F. Libby and C.E. Palmer, 1968. Stratospheric residence time and interhemispheric mixing of strontium 90 from fallout in rain. J. geophys. Res., 73(12):3611-3616.

Chemistry strontium isotopes
Faure, G., P.M. Hurley and J.L. Powell, 1965. The isotopic composition of strontium in surface waters from the North Atlantic Ocean. Geochimica et Cosmochimica Acta, 29(4):209-220.

Chemistry strontium - 90
Freudenthal, Peter C., 1970. Strontium 90 concentrations in surface air: North America versus Atlantic Ocean from 1966 to 1969. J. geophys. Res., 75(21): 4089-4096.

Chemistry strontium-90
Fukai, Rinnosuke, 1964. A note on the strontium-90 content of marine organisms as an index for variations of strontium-90 in sea water. Bull. Inst. Océanogr., Monaco, 63(1307):16 pp.

Chemistry strontium-90
Fukai, R., H. Suzuki and K. Watanabe, 1962. Strontium-90 in marine organisms during the period 1957-1961. Bull. Inst. Océanogr., Monaco, 59(1251):16 pp.

Chemistry strontium-90
Hiyama, Y., 1957. Maximum permissible concentration of Sr90 in food and its environment. Rec. Oceanogr. Wks., Japan, n.s., 3(1):70-77.

Chemistry strontium-90
Hiyama, Yoshio, and Ryushi Ichikawa, 1958. A measure on level of strontium-90 concentration in sea water around Japan at the end of 1956. Proc. Ninth Pacific Sci. Congr., Pacific Sci. Assoc., 1957, Oceanogr., 16:141-145.

chemistry, strontium 90
Koshlyakov, M.N., 1967. Application of the boxes method to the study of exchange processes and the dynamics of strontium-90 in the upper layer of the North Atlantic. (In Russian; English abstract). Fisika Atmosfer. Okean., Izv. Akad. Nauk, SSSR, 3(7):742-756.

chemistry, strontium - 91
Lerman, Abraham and Harry Taniguchi, 1972. Strontium-90 - diffusional transport in sediments of the Great Lakes. J. geophys. Res. 77(3): 474-481.

Chemistry strontium-90 fallout
Machta, L., K. Telegadas and D.L. Harris, 1970. Strontium-90 fallout over Lake Michigan. J. geophys. Res., 75(6): 1092-1096.

Chemistry Strontium 90
Malakhov, S.G. and I.B. Pudovkina, 1970. Strontium 90 fallout distribution at middle latitudes of the northern and southern hemispheres and its relation to precipitation. J. geophys. Res., 75(18): 3623-3628.

chemistry, strontium 85
Martin, J.-L.M., 1972. Etude de l'absorption, de la concentration et du métabolisme du strontium 85 chez le crustacé décapode Carcinus maenas. Mar. Biol. 12(2): 154-158.

Chemistry strontium-90
Menzel, R.G., 1960. Transport of strontium-90 in runoff. Science, 131(3399):499-500.

Chemistry strontium-90
Mironov, O. G., 1964. On the problem of migration of strontium-90 and cesium-137 into the organism of man in the course of consuming some species of marine animals. (In Russian). Okeanologiia, Akad. Nauk, SSSR. 4(6):1059-1061.

Chemistry strontium 90
Miyake, Y., and K. Saruhashi, 1957. The worl-wide strontium 90 deposition during the period from 1951 to thefall of 1955. Pap. Meteorol. & Geophys., Tokyo, 8(3):241-244.

Chemistry strontium 90
Miyake, Y., K. Saruhashi and Y. Katsuragi, 1960 Strontium 90 in Western North Pacific surface waters. Papers Meteorol. & Geophys., Tokyo, 11(1): 188-190.

Chemistry strontium-90
Miyake, Yasuo, Katsuko Saruhashi, Yukio Katsuragi and Teruko Kanazawa, 1962 Penetration of 90 Sr and 137 Cs in deep layers of the Pacific and vertical diffusion rate of deep water. Repts. JEDS, Deep-Sea Res. Comm., Japan Soc., Promotion of Science, 3:141-147. Originally published (1962): J. Radiation Res., 3(3):141-147.

JEDS Contrib. No. 25.

Chemistry strontium 90
Miyake, Y., K. Saruhashi, Y. Katsuragi and T. Kanazawa, 1961 Cesium 137 and strontium 90 in sea water. Papers in Meteorol. & Geophys., Tokyo, 12(1): 85-88.

strontium 90
Miyaki, Y., K. Saruhashi, Y. Katsuragi, J. Kanazawa and Y. Sugimura, 1964. Uranium, radium, thorium, ionium, strontium 90 and cesium 137 in coastal waters of Japan. In: Recent researches in the fields of hydrosphere, atmosphere and nuclear geochemistry, Ken Sugawara festival volume, Y. Miyake and T. Koyama, editors, Maruzen Co., Ltd., Tokyo, 127-141.

Chemistry strontium-90
Nelson, D.J., 1962 Clams as indicators of strontium-90. Science, 137(3523):38-39.

Chemistry strontium-90
Ozmidov, R.V., and N.I. Popov, 1966. The study of the vertical water exchange in the ocean using the strontium-90 distribution data. (In Russian; English abstract). Fizik. Atmosfer. i Okeana, 2(2):183-190.

Chemistry strontium - 90
Paakkola, Olli, and Aarno Voipio, 1965. Strontium - 90 in the Baltic Sea. Suomen Kemistilehti (B), 38:11-17.

Chemistry strontium-90
Paakkola, Olli, and Aarno Voipio, 1965. Strontium-90 in the waters of the Baltic Sea. Nature, 205(4968):274.

Chemistry strontium-90
Park, Kilho, Merilyn J. George, Yasuo Miyake, Katsuko Saruhashi, Yukio Katsuragi and Teruko Kanazawa, 1964. Strontium-90 and cesium-137 in Columbia River plume, July 1964. Nature, 208(5015):1084-1085.

Chemistry, strontium-90
Patin, S.A., 1965.
On the regional distribution of strontium-90 on the surface of the World Ocean. (In Russian).
Okeanologiia, Akad., Nauk, SSSR, 5(3):458-462.

Chemistry
Strontium-90
Patin, S. A., A. V. Alexandrov, and V. M. Orlov, 1966.
Strontium-90 on the surface of the Atlantic Ocean during the second part of 1961. (In Russian).
Trudy Inst. Okeanol., Akad. Nauk, SSSR, 82:32-34.

Chemistry, strontium isotopes
Peterman, Zell E., Carl E. Hedge and Harry A. Tourtelot, 1970.
Isotopic composition of strontium in sea water throughout Phanerozoic time.
Geochim. cosmochim. Acta 34(1): 105-120

Chemistry strontium-90
Polikarpov, G.G., 1961.
The role of detritus formation in the migration of strontium-90, cesium-137 and cerium-144. Experiments with the sea alga Cystoseira barbata.
Doklady Akad. Nauk, SSSR, 136(4):921-923.
English Edit., 1962, 136(1-6):11-13.

Chemistry
Strontium-90
Popov, N.I., E.G Ahzhazha, G.I. Kosourov, and A.A. Uzefovich, 1962
Strontium-90 in the surface waters of the Atlantic Ocean. (In Russian).
Okeanologiia, Akad. Nauk, SSSR, 2(5):845-849.

Chemistry
Strontium-90
Popov, N. I., V. M. Orlov and V. F. Dabija, 1966.
Strontium-90 in the waters of the Pacific Ocean. 4. The surface waters of the South China Sea. (In Russian)
Trudy Inst. Okeanol., Akad. Nauk, SSSR, 82:11-19.

Chemistry
Strontium-90
Popov, N. I., V. M. Orlov and S. A. Patin, 1966.
Strontium-90 in deep waters of the Indian Ocean (In Russian).
Trudy Inst. Okeanol., Akad. Nauk, SSSR, 82:24-31.

Chemistry strontium-90
Popov, N.I., V.M. Orlov, S.A. Patin and N.P. Ushakova, 1964.
Strontium-90 in the surface waters of the Indian Ocean in 1960-1961. (In Russian).
Okeanologiia, Akad. Nauk, SSSR, 4(3):418-422.

Chemistry
strontium-90
Popov, N.I., V.M. Orlov and V.A. Pchelin, 1963
Strontium-90 in the waters of the Pacific Ocean. (In Russian).
Okeanologiia, Akad. Nauk, SSSR, 3(4):666-668.

Chemistry
Strontium-90
Popov, N. I. and S. A. Patin, 1966.
The primary features of global distribution of strontium-90 on the surface of World Ocean (1960-1961). (In Russian).
Trudy Inst. Okeanol., Akad. Nauk, SSSR, 82:42-55.

Chemistry
Strontium-90
Popov, N. I., S. A. Patin, R. M. Polevoy and V. A. Konnov, 1966.
Strontium-90 in the waters of the Pacific Ocean 3. On vertical distribution in the central region. (In Russian).
Trudy, Inst. Okeanol, Acad. Nauk, SSSR, 82:5-15.

Chemistry
strontium-90
Popov, N. I., S. A. Patin, R. I. Polivoy, and V. A. Konnov, 1964.
Strontium-90 in the Pacific Ocean waters. II. Water surface in the central area for 1961.
Okeanologiia, Akad. Nauk, SSSR, 4(6):1026-1029.

Chemistry strontium isotopes
Pushkar, Paul, 1968.
Strontium isotope ratios in volcanic rocks of three island arc areas.
J. geophys. Res., 73(8):2701-2714.

Chemistry
Strontium 90
Rocco, Gregory G., and Wallace S. Broecker, 1963
The vertical distribution of Cesium 137 and Strontium 90 in the oceans.
J. Geophys. Res., 68(15):4501-4512.

chemistry, strontium 87
Rama Murthy, V., 1964.
The significance of Sr 87 in ocean water. (Abstract).
Trans. Amer. Geophys. Union, 45(1):113.

Chemistry
strontium-90
Sreekumaran, C., S.S. Gogate, G.R. Doshi, V.N. Sastry and R. Viswanathan, 1968.
Distribution of cesium-137 and strontium-90 in the Arabian Sea and Bay of Bengal.
Current Sci., 37(22): 629-631.

Chemistry strontium 90
Suyehiro, Y., S. Yoshino, Y. Tsukamoto, M. Akamatsu, K. Takahashi and T. Mori, 1956
Transmission and metabolism of strontium 90 in aquatic animals. Research on the effects and influences of the nuclear bomb test explosions.
Jap. Soc. Prom. Sci., 1956:1135-1142.

Chemistry, Strontium 90, transmission in metabolism of aquatic animals.
Suyshiro, Y., T. Hibiya, S. Hoshino, T. Yagi, Y. Tsukamoto, M. Akamatsu, K. Takahashi and T. Imamura, 1954.
Notes on the action of radioactive substances on aquatic animals. Science, Japan, 24:619-622.

chemistry, strontium-90
Volchok, H.L., V.T. Bowen, T.R. Folsom, W.S. Broecker, E.A. Schuert and G.S. Bien, 1971.
Oceanic distributions of radionuclides from nuclear explosions. Radioactivity in the marine environment, U.S. Nat. Acad. Sci., 1971: 42-89.

Chemistry, strontium-90
Wilgain, S., and E. Picciotto and W. De Breuck, 1965.
Strontium-90 fallout in Antarctica.
J. geophys. Res., 70(24):6023-6032.

Chemistry strontium-90
*Yakovleva, G.V., 1967.
The effect of trade-winds air transport on water pollution by strontium-90 and cesium-137 of the equatorial Atlantic. (In Russian; English abstract)
Okeanologiia, Akad. Nauk, SSSR, 7(4):617-622.

Chemistry STRONTIUM isotopes
*Zlobin, V.S., 1968.
Dynamics of radiostrontium accumulation by some brown algae and the influence of sea water salinity on the coefficients of accumulation. (In Russian; English abstract).
Okeanologiia, Akad. Nauk, SSSR, 8(1):78-85.

succinic dehydrogenase

Chemistry Succinic dehydrogenase
*Krishnaswamy, S., J.E.G. Raymont and J. Tundisi, 1967.
Succinic dehydrogenase activity in marine animals.
Int. Revue ges-Hydrobiol., 52(3):447-451.

chemistry succinic dehyrogenase
*Raymont, J.E.G., S. Krishnaswamy and J. Tundisi, 1967.
Biochemical studies on marine zooplankton. 4. Onvestigations on succinic dehydrogenase activity in zooplankton with special reference to Neomysis integer.
J. Cons. perm. int. Explor. Mer. 31(2):164-169.

Sugars

Chemistry sugars (amino)
Antia, N.J., and C.Y. Lee, 1964.
The determination of "free" amino sugars in sea-water.
Limnology and Oceanography, 9(2):261-262.

Chemistry Sugars
Derenbach, Jens B., 1970.
Partikulare Substanz und Plankton an Hand chemischer und biologischer Daten gemessen in den oberen Wasserschichten des Gotland-Tief im Mai 1968.
Kieler Meeresforsch. 25(2): 279-289

Chemistry
sugar transport
Krishnan, S., 1971.
Autoradiographic studies on sugar transport in the sea cucumber Holothuria scabra. Mar. Biol. 10(2): 189-191.

Chemistry
sugars
Krishnan, S. and S. Krishnaswamy, 1970.
Studies on the transport of sugars in the holothurian Holothuria scabra. Marine Biol., 5(4): 303-306.

Chemistry sugars
Vallentyne, J.R., and J.R. Whittaker, 1956.
On the presence of free sugars in filtered lake water. Science 124(3230):1026-1027.

Sulphate: Note: Cross ref. on cards may use either 'ph' or 'f' for sulphur + compounds.

Chemistry sulphate
Bather, J.M., and J.P. Riley, 1954.
The chemistry of the Irish Sea. Part 1. The sulphate-chlorinity ratio. J. du Cons. 20(2): 145-152.

Chemistry
sulfate
Beisova, M.P., and Krivkov, P.A., 1957
[Conductometric titration of sulfate in natural waters.] Gidrokhim. Materialy 26: 190.

sulfate ions
Berge, Hans und Lutz Brügmann, 1970
Die indirekte polarographische Bestimmung von Sulfationen im Meerwasser.
Beitr. Meeresk., 27: 5-13

chemistry, sulfate
Büchen, Mattias, und Hans Walter Georgii 1971.
Ein Beitrag zum atmosphärischen Schwefelhaushalt über dem Atlantik
Meteor-Forsch. Ergebnisse (B) 7: 71-77.

Chemistry sulphate
Caspers, H., 1951.
Quantitative Untersuchungen über die Bodentierwelt des Schwarzen Meeres im bulgarischen Küstenbereich. Arch. f. Hydrobiol. 45(1/2):192 pp., 66 textfigs.

Chemistry
sulphate
Culkin, Frederick, 1965.
The major constituents of sea water.
In: Chemical oceanography, J.P. Riley and G. Skirrow, editors, Academic Press, 1: 121-161.

chemistry sulfates (geochemical)
Degens, Egon T., 1965.
Geochemistry of sediments: a brief survey.
Prentice-Hall, Inc., 342 pp.

Chemistry sulfate
Fukai, R., and F. Shiokawa, 1955.
On the main chemical components dissolved in the adjacent waters to the Aleutian Islands in the North Pacific. Bull. Chem. Soc., Japan, 28(9): 636-640.

Also: Bull. Tokai Reg. Fish. Res. Lab. 11

Chemistry sulfate
*Hartmann, Martin, 1969.
Investigations of Atlantis II Deep samples taken by the FS Meteor.
In: Hot brines and Recent heavy metal deposits in the Red Sea, E.T. Degens and D.A. Ross, editors, Springer-Verlag, New York, Inc., 204-207.

chemistry sulfate (data)
Kramer J.R., 1961
Chemistry of Lake Erie.
Proc. Fourth Conf., Great Lakes Research, Great Lakes Res. Div., Inst. Sci. & Tech., Univ. Michigan, Publ., (7):27-56.

chemistry sulfate
Kwiecinski, Bogdan, 1965.
The sulfate content of Baltic water and its relation to the chlorinity.
Deep-Sea Res., 12(6):797-804.

chemistry, sulphate
Lloyd, R. Michael, 1967.
Oxygen-18 composition of oceanic sulfate.
Science, 156(3779):1228-1231.

chemistry, sulfate
Longinelli, A., and H. Craig, 1967.
Oxygen-18 variations in sulfate ions in sea water and saline lakes.
Science, 156(3771):56-59.

chemistry, sulphates
*Morcos, Selim A., 1967.
The chemical composition of sea water from the Suez Canal region. I. The major anions.
Kieler Meeresforsch., 23(2):80-91.

chemistry, sulphate
Nishizawa, Tanzo and T. Muraki, 1940
Chemical studies in the sea adjacent to Palau. I. A survey crossing the sea from Palau to New Guinea. Kagaku Nanyo (Sci. of the South Sea) 2(3):1-7.

chemistry sulfates
Robertson, A.J., 1907.
On hydrographical investigations in the Faroe-Shetland Channel, and the northern part of the North Sea during the years 1904-1905. Fish. Bd., Scotland, North Sea Fish. Invest. Comm., Northern Area, Rept. No. 2:1-140, 13 text-figs., 15 pls.

chemistry sulfate
Ruppin, E., 1911.
Bericht über das Verhältnis du Cl, SO3 und % Werte in einer Reihe von 14 Meerwasser Proben.
Zeits. Anorg. Allgem. Chem., 69:232-246.

Transl. cited: USFWS Spec. Sci. Rept., Fish., 227

chemistry sulphates
Schachter, D., 1954.
Contribution à l'étude hydrographique et hydrologique de l'étang de Berre (Bouches-du-Rhone).
Bull. Inst. Océan., Monaco, No. 1048:20 pp.

chemistry sulfate
Schmidt-Ries, H., 1948.
Kurzgefasste Hydrographie Griechenlands.
Arch. Hydrobiol. 43(1):95-141.

chemistry sulfate
Shishkina, O.V., 1954.
[Technique of determining sulfate in sea water.]
Trudy Inst. Okeanol., 8:253-268.

chemistry sulphates
Skopintsev, B.A., and A.V. Karpov, 1957.
[Conditions of conservation and also the determin-ation of sulphates in natural waters.]
Gidrokhim. Materialy, 26:230-

chemistry sulfate
Vardaro, Emilio, 1958.
Sul rapporto solfato-clorurі nelle acque del Mar Piccolo del Mar Grande e del Golfo di Taranto.
Thalassia Jonica, Ist. Speriment. Talassogr., Taranto, 1:61-64.

chemistry, sulfates
Villagrazia, P.M., di., 1942.
Il dosaggio volumterico dei solfati con la buretta Knudsen nella determinazione indiretta delle costanti dell'acque di mare. Boll. Pesca. Piscicolt. ed Idrobiol. 18(2/3):17-24.

chemistry sulfate
Watts, J.C.D., 1960
Sea-water as the primary source of sulfate in tidal swamp soils from Sierra Leone. Nature 186(721): 308-309.

chemistry sulfate
Watts, J.C.D., 1960
Sulfate absorption by muds of relatively high organic content from the Sierra Leone River Estuary. Bull. de l'I.F.A.N. (2) 22(4): 1153-1158.

chemistry, sulfate ion association
Millero, Frank J. 1971.
Effect of pressure on sulfate ion association in sea water.
Geochim. cosmochim. Acta 35(11): 1089-1098.

sulphate-chlorinity ratio

chemistry, sulphate-chlorinity ratio
Bather, J.M., and J.P. Riley, 1954.
The chemistry of the Irish Sea. 1. Sulphate-chlorinity ratio. J. du Cons. 20(2):145-152, 1 textfig.

chemistry sulphate/chlorinity ratio
Gupta, R. Sen, and S. Ananthanarayanan, 1963.
Volumetric estimation of sulphate in sea water using benzidine hydrochloride and determinations of sulphate-chlorinity ratio.
Indian J. Chem., 1(9):369-412.

chemistry sulfate/chlorinity (data)
Kato, Takeo 1969.
On the distribution of the water mass produced by water-mixing and the change of chemical composition. (Ratio $Ca^{2+}:Cl^-$, $SO_4^{2-}:Cl^-$) - Oceanographical studies on the East China Sea, 3. (In Japanese: English abstract) Umi to Sora, 44(2/3): 55-80

chemistry sulphate-chlorinity ratio
Lewis, G.J., Jr., and T.G. Thompson, 1950.
Effect of freezing on the sulphate chlorinity ratio in sea water.
J. Mar. Res., 9(3):211-217.

chemistry, sulphate/chlorinity
Morcos, Selim A., 1967.
The chemical composition of sea water from the Suez Canal region. I. The major anions.
Kieler Meeresforsch., 23(2):80-91.

chemistry, sulphate/chlorinity
Morris, A.W., and J.P. Riley, 1966.
The bromide/chlorinity and sulphate/chlorinity ratio in sea water.
Deep-Sea Res., 13(4):699-705.

chemistry, sulfate, effect of
*Lewin, Joyce, and William F. Busby, 1967.
The sulfate requirements of some unicellular marine algae.
Phycologia, 6(4):211-217.

sulphate reduction

sulfate reduction
Bansemir, Klaus, und Gerhard Rheinheimer 1970.
3.6. Bakterielle Sulfatreduktion und Schwefeloxydation. Chemische, mikrobiologische und planktologische Untersuchungen in der Schlei im Hinblick auf deren Abwasserbelastung. Kieler Meeresforsch 26(2): 170-173.

chemistry sulphate reduction
Gasanov, M.V., 1960
On the reduction of sulphates in mixtures of sea and layer water in the atmosphere of molecular hydrogen with the participation of sulphate reducing bacteria. Microbiologie, Akad. Nauk, SSSR 29(3): 419-421. (In Russian with English abstract)

sulfides

chemistry sulfides (geochemical)
Degens, Egon T., 1965.
Geochemistry of sediments: a brief survey.
Prentice-Hall, Inc., 342 pp.

chemistry sulfides
Hata, Yoshihiko, 1965.
Microbial production of sulfides in marine and estuarine sediments. (In Japanese; English abstract).
J. Shimonoseki Univ., Fish., 14(2):37-83.

chemistry sulfides
Kaplan, I.R., and S.C. Rittenberg, 1963.
23. Basin sedimentation and diagenesis.
In: The Sea, M.N. Hill, Editor, Interscience Publishers, 3:583-619.

chemistry sulphide
Kimata, M., H. Kadota, T. Hata and H. Miyoshi, 1958.
The formation of sulfide by sulfate-reducing bacteria in the estuarine zone of the river receiving a large quantity of organic drainage.
Rec. Oceanogr. Wks., Japan, Spec. No. 2:187-199.

chemistry sulfides
Kimata, M., H. Kadota, Y. Hana and H. Miyoshi, 1957.
Studies on the marine sulfate-reducing bacteria. 4. Production of sulfides in the estuarine region of the river receiving a large amount of organic drainage. (1). Bull. Jap. Soc. Sci. Fish., 22(11):701-707.

chemistry sulphides
Matudavis, T., 1942.
On the inorganic sulphides as a growth promoting ingredient for diatom.
Proc. Imp. Acad., Tokyo, 18(2):107-116.

chemistry sulphides
Richards, Francis A., and Edwin V. Grill, 1962
Some chemical observations in oxygen-free, sulfide-bearing marine environments. (Abstract)
J. Geophys. Res., 67(9):3590-3591.

Chemistry sulphide
Siyazuki, K., 1951.
[Studies on the foul-water drained from factories I. On the influence of foul-waters drained from factories by the coast on the water of Mitaziri Bay.] Contr. Simonoseki Coll. Fish. Pt. 1:155-158, 4 textfigs. (English summary).

Chemistry sulfide
Tomiyama, T., and Y. Kojima, 1953.
The errors and their removal in the determination of sulfide of bottom deposits.
Bull. Jap. Soc. Sci. Fish. 18(12):29-32.

Sulfur

chemistry, sulfur
Berner, Robert A., 1971.
Worldwide sulfur pollution of rivers. J. geophys. Res. 76(27): 6597-6600.

Chemistry sulfur
*Berner, Robert A., 1968.
Migration of iron and sulfur within anaerobic sediments during early diagenesis.
Am.J.Sci., 267(1):18-42.

Chemistry sulfur
Eriksson, Erik, 1960.
The yearly circulation of chloride and sulfur in nature; meteorological, geochemical and pedological implications. II.
Tellus, 12(1):63-109.

chemistry, sulphur
Kellogg, W.W., R.D. Cadle, E.R. Allen, A.L. Lazrus and E.A. Martell, 1972.
The sulfur cycle: man's contributions are compared to natural sources of sulfur compounds in the atmosphere and oceans. Science 175 (4022): 587-596.

Chemistry, sulfur
Koide, Minoru and Edward D. Goldberg, 1971.
Atmospheric sulfur and fossil fuel combustion.
J. geophys. Res. 76(27): 6589-6596.

Chemistry sulphur
Lewin, Ralph A., Editor, 1962.
Physiology and biochemistry of algae.
Academic Press, New York and London, 929 pp.

Chemistry sulphur
Nechiporenko, G.N., 1957.
[Determination of sulphates in natural waters by volumetric methods.]
Gidrokhem. Materialy, 26:207.

Chemistry sulphur
Nicol, H., 1951.
Sulphur from sea water. Nature 168:894-895.

Chemistry sulphur
Nishizawa, Tanzo and T. Muraki, 1940
Chemical studies in the sea adjacent to Palau. I. A survey crossing the sea from Palau to New Guinea. Kagaku Nanyo (Sci. of the South Sea) 2(3):1-7.

Chemistry sulphur
Ostroumov, E.A., 1953.
Method of determining the form of combined sulphur in the sediments of the Black Sea. (In Russian).
Trudy Inst. Okeanol., Akad. Nauk, SSSR, 7:56-69.

chemistry, sulfur
*Ostroumov, E.A., and I.I. Volkov, 1967.
Geochemical behaviour of sulfur in the bottom sediments of the Pacific. (In Russian; English abstract).
Trudy Inst. Okeanol., 83:68-82.

Chemistry Sulphur compounds
Skopintsev, B.A., A.V. Karpov and O.A. Vershinina, 1959.
Investigation of the dynamics of some sulphur compounds in the Black Sea under experimental conditions. (In Russian).
Trudy Morsk. Gidrofiz. Inst., Akad. Nauk, SSSR, 16:89-111.

Translation: Scripta Tecnica, Inc., for AGU, 55-73.

Chemistry sulfur
Sugawara, K., T. Koyama and A. Kozawa, 1954.
Distribution of various forms of sulfur in lake-, river-, and sea-muds. II.
J. Earth Sci., Nagoya Univ. 2:1-4.

Abstr. in Chem. Abstr. 11855e, 1954.

Chemistry sulfur
Szabo, A., A. Tudge, J. MacNamara and H.G. Thode, 1950.
Distribution of sulfur 34 in nature and the sulfur cycle.
Science, 111:464-465.

Chemistry sulphur
Thode, H.G., J. Monster, & H.B. Dunford, 1961
Sulphur isotope geochemistry.
Geochim. et Cosochim. Acta, 25(3):159-174.

sulphur cycle

Chemistry sulphur cycle
Baas Becking, L.G.M., and E. J. Ferguson Wood 1955.
Biological processes in the estuarine environment. 1. Ecology of the sulphur cycle. Proc. K. Nederl. Akad. Wetensk., Amsterdam, B., 58(3):161-181.

Chemistry sulphur cycle
Baas Becking, L.G.M., and E.J. Ferguson Wood, 1953.
Microbial ecology of the estuarine sulphuretum.
Atti VI Congreso Int. Microbiol., Roma. 6-12 settembre 1953, 7(22):379-388.

Chemistry sulphur cycle
Baas Becking, L.G.M., and I.R. Kaplan, 1956.
Biological processes in the estuarine environment. IV. Attempts at interpretation of observed Eh-pH relations of various members of the sulphur cycle. Proc. K. Nederl. Akad. Wetensk., Amsterdam, B, 59(2):97-108.

Chemistry sulphur cycle
Baas Becking, L.G.M., and I. R. Kaplan, 1956.
Biological processes in the estuarine environment. III. Electrochemical considerations regarding the sulphur cycle. K. Nederl. Akad. Wetens., Amsterdam, Proc., B, 59(2):85-96.

Chemistry sulfur cycle
Jensen, M.L., 1965.
Sulfur isotopes in the marine environment.
Narragansett Mar. Lab., Univ. Rhode Island, Occ. Publ., No. 3:131-148.

sulphur deposits

Chemistry sulphur deposits
Feely, H.W., and J.L. Kulp, 1957
Origin of Gulf coast salt-dome sulphur deposits.
Bull. Amer. Assoc. Petr. Geol., 41(8):1802-1853.

chemistry, sulfur dioxide
Büchen, Mattias, und Hans Walter Georgii 1971.
Ein Beitrag zum atmosphärischen Schwefelhaushalt über dem Atlantik.
Meteor-Forsch. Ergebnisse (B) 7: 71-77.

Chemistry sulfur dioxide
*Kühme, Heinrich, 1968.
Untersuchungen der Konzentration atmosphärischen Spurengase über dem Atlantik.
Meteor Forschungsergeb, (A)2:83-93.

Sulfur, effect of

Chemistry sulfur, effect of
Curl, Herbert, Jr., 1962
Effect of divalent sulfur and vitamin B12 in controlling the distribution of Skeletonema costatum.
Limnol. and Oceanogr., 7(3):422-424.

Sulphur isotopes

Chemistry sulphur isotopes
Culkin, Frederick, 1965.
The major constituents of sea water.
In: Chemical oceanography, J.P. Riley and G. Skirrow, editors, Academic Press, 1: 121-161.

chemistry, sulfur isotopes
Kaplan, I.R., R.E. Sweeney and Arie Nissenbaum, 1969.
Sulfur isotope studies on Red Sea geothermal brines and sediments.
In: Hot brines and Recent heavy metal deposits in the Red Sea, E.T. Degens and D.A. Ross, editors, Springer-Verlag, New York, Inc., 474-498.

Chemistry sulphur-35
Lobe, D.L., and D. Sam, 1962.
Radiochemical determinations of sodium-24 and sulphur-35 in sea water.
Anal. Chem., 34(3):336-340.

Listed in Anal. Abstr., 9(9):#3936.

Chemistry sulfur isotopes
Nakai, Nobuyuki, and Mead LeRoy Jensen, 1960
Biogeochemistry of sulfur isotopes.
J. Earth Sciences, Nagoya Univ., 8(2): 181-196.

sulphur isotopes
Rees, C.E., 1970.
The sulphur isotope balance of the ocean: an improved model.
Earth Planet. Sci. Letters 7(6): 366-370

sulphur-35
Sorokin, Yu.I., 1970.
Experimental studies of the rate and mechanism of oxidation of hydrogen sulphide in the Black Sea with the aid of sulphur-35. (In Russian; English abstract). Okeanologiia, 10(1): 51-62.

Chemistry, sulphur
Weeks, Lewis G., 1968.
The gas, oil and sulfur potentials of the sea.
Ocean Industry 3(6): 43-51.

sulphur, oxidation of
Bansemir, Klaus, und Gerhard Rheinheimer 1970.
3.6. Bakterielle Sulfatreduktion und Schwefeloxydation. Chemische, mikrobiologische und planktologische Untersuchungen in der Schlei im Hinblick auf deren Abwasserbelastung. Kieler Meeresforsch 26(2): 170-173.

Sulfuric acid

chemistry, surface
*Williams, P.M., 1967.
Sea surface chemistry: organic carbon and organic and inorganic nitrogen and phosphorus in surface films and subsurface waters.
Deep-Sea Res., 14(6):791-800.

surface active agents, effect of

Swedmark, M., B. Braaten, E. Emanuelsson and A. Granmo, 1971.
Biological effects of surface active agents on marine animals. Mar. Biol. 9(3): 183-201.

tantalum

chemistry, tantalum

Burton, J.D. and K.S. Massie, 1971.
The occurrence of tantalum in some marine organisms. J. mar. biol. Ass. U.K. 51(3): 679-683.

Chemistry sulfuric acid

Syazuki, K., 1953.
[Studies on the foul-water drained from factories. 2. On the harmful components and the water pollution of the foul-water drained from the metal plants.] J. Shimonoseki Coll. Fish. 3(2):181-188, 2 textfigs.

chemistry tantalum

Hamahuchi, Hiroshi, Rokuro Kuroda and Yoshichika Watanabe, 1963.
Tantalum contents of deep-sea sediments.
J. Chem. Soc., Japan, (Nippon Kagaku Zasshi), 84(9):725-726.

Chemistry tantalum

Kokubu, N., and T. Hidaka, 1965.
Tantalum and niobium in ascidians.
Nature, 205(4975):1028-1029.

chemistry, taurine

Allen, J.A. and M.R. Garrett, 1971.
Taurine in marine invertebrates. Adv. mar. Biol. 9: 205-253.

tellurium

Chemistry tellurium

Lakin, H.W., C.E. Thompson, and D.F. Davidson 1963
Tellurium content of marine manganese oxides and other manganese oxides.
Science 142(3599):1568-1569.

chemistry Thallium

Matthews, A.D., and J.P. Riley, 1970.
The occurrence of Thallium in sea water and marine sediments.
Chem. Geol. 6(2): 149-152.

thiamine

Chemistry thiamine

Carlucci, A.F., 1970
Vitamine B_{12}, thiamine, and biotin
Bull. Scripps. Inst. Oceanogr. 17: 23-31

chemistry, thiamin

Carlucci, A.F., S.B. Silbernagel and P.M. McNally 1969.
Influence of temperature and solar radiation on persistence of vitamin B_{12}, thiamin and biotin in sea water.
J. Phycol. 5(4): 302-305.

chemistry, thiamine

Gold, Kenneth 1968.
Some factors affecting the stability of thiamine.
Limnol. Oceanogr. 13(1): 185-188.

chemistry, thiamine

Gold, Kenneth, Oswald A. Roels and Harvey Bank, 1966.
Temperature dependent destruction of thiamine in seawater.
Limnol. Oceanogr., 11(3):410-413.

Chemistry, thiamine

Lewin, Joyce C., 1965.
The thiamine requirement of a marine diatom.
Phycologia, 4(3):141-144.

chemistry, thiamine

Natarajan, K.V., and R.C. Dugdale, 1966.
Bioassay and distribution of thiamine in the sea.
Limnol. Oceanogr., 11(4):621-629.

chemistry, thiamine

Tokuda, Hiroshi 1966.
Studies on the growth of a marine diatom Nitzschia closterium. 1. Its requirements for thiamine.
Bull. Jap. Soc. scient. Fish. 32(7): 565-567.

chemistry, thiamin

Vishniac, Helen S., 1961.
A biological assay for thiamine in sea water.
Limnol. & Oceanogr., 6(1):31-35.

Chemistry, thioether

Kadota, Hajime, and Yuzaburo Ishida 1968.
Evolution of volatile sulfur compounds from unicellular marine algae.
Bull. Misaki mar biol. Inst. Kyoto Univ. 12: 35-48.

thiosulphate

Chemistry thiosulphate

Volkov, I.I., 1957.
[Determination of thiosulphate in silt waters of the Black Sea deposits.]
Doklady Akad. Nauk, SSSR, 114(4):853-

thorium

chemistry, thorium

*Bernat, Michel, and Edward D. Goldberg, 1969.
Thorium isotopes in the marine environment.
Earth Planet Sci. Letters, 5(5):308-312.

chemistry, thorium

*Blanchard, R.L., M.H. Cheng and H.A. Potratz, 1967.
Uranium and thorium series disequilibria in Recent and fossil marine molluscan shells.
J. geophys. Res., 72(18):4745-4757.

chemistry, thorium

Cherry, R.D., I.H. Gericke, and L.V. Shannon, 1969.
Thorium-228 in marine plankton and sea-water.
Earth planet. Sci. Letts. 6(6): 451-456.

Chemistry thorium

Goldberg, E.D., and E. Picciotto, 1955.
Thorium determination in manganese nodules.
Science 121 (3147):613-614.

Chemistry thorium

Haskin, Larry A., and Fred A. Frey, 1966.
Discussion of paper by J.W. Morgan and J.F. Lovering, 'Uranium and thorium abundances in the basalt cored in the Mohole Project (Guadalupe Site)'.
J. geophys. Res., 71(2):688-689.

Chemistry thorium

Heye, Ditrich 1970.
A system for detection of ionium thorium and protactinium to date deep-sea cores.
Geochim. cosmochim. Acta, 34(3): 389-397.

Chemistry thorium

Heye, Dietrich 1969.
Uranium, thorium, and radium in ocean water and deep-sea sediments
Earth planet. Sci. Letters 6(2): 112-116.

Chemistry thorium

Higashi, S., 1959.
[Estimation of the microgram amount of thorium in sea water] J. Oceanogr. Soc., Japan, 15(2): 65-68.

Chemistry thorium

Higashi, S., 1959.
[On the thorium transfer from lithosphere to hydrosphere - in connection with the occurrence of thorium in sea water.] J. Oceanogr. Soc., Japan, 15(2): 69-75.

Chemistry thorium

Koczy, F.F., 1949.
Thorium in sea water and marine sediments. Geol. Fören. Förhandl. 71:238-242.

Chemistry thorium

Koczy, F.F., E. Picciotto, G. Poulaert and S. Wilgain, 1957.
Mesure des isotopes du thorium dans l'eau de mer
Geochimica et Cosmochimica Acta 11:103-129.

Chemistry thorium

Ku, Teh-Lung, and Wallace S. Broecker, 1967.
Uranium, thorium, and protactinium in a manganese nodule.
Earth Planet. Sci. Letters, 2(4): 317-320

thorium

Miyaki, Y., K. Saruhashi, Y. Katsuragi, J. Kanazawa and Y. Sugimura, 1964.
Uranium, radium, thorium, ionium, strontium 90 and cesium 137 in coastal waters of Japan.
In: Recent researches in the fields of hydrosphere, atmosphere and nuclear geochemistry, Ken Sugawara festival volume, Y. Miyake and T. Koyama, editors, Maruzen Co., Ltd., Tokyo, 127-141.

Chemistry thorium

*Miyake, Yasuo, Katsuko Saruhashi and Yukio Sugimura, 1968.
Biogeochemical balance of natural radioactive elements in the sea.
Rec. oceanogr. Wks, Japan, n.s., 9(2):180-187.

Chemistry thorium

Moore, Willard S., and William M. Sackett, 1964.
Uranium and thorium series inequilibrium in sea water.
J. Geophys. Res., 69(24):5401-5405.

chemistry, thorium

Moore, Willard S., and William M. Sackett, 1964
Residence times of protactinium and thorium in sea water. (Abstract).
Trans. Amer. Geophys. Union, 45(1): 119.

Chemistry thorium

Morgan, J. W., and J. F. Lovering, 1965.
Uranium and thorium abundances in the basalt cored in Mohole Project (Guadalupe Site).
J. geophys. Res., 70(18):4724-4725

Chemistry thorium

Nikolayev, D.S., K.F. Lazarev, and S.M. Graschenko, 1961.
The concentration of thorium isotopes in the water of the Sea of Azov. (In Russian).
Doklady, Akad. Nauk, SSSR, 138(3):674-676

Translation:
Consultant's Bureau for Amer. Geol. Inst., Earth Sci. Sect., 138(1-6):489-490 1962.

Chemistry thorium
Ostroumov, E.A., and I.I. Volkov, 1962.
Separation of titanium, zirconium and thorium from manganese, nickel, cobalt and zinc by means of cinnamic acid.
Trudy Inst. Okeanol., Akad. Nauk, SSSR, 54:170-181.

In Russian; English summary

Chemistry, thorium
Pettersson, H., 1939.
Die radioaktiven Elemente im Meer.
Rapp. Proc. Verb., Cons. Perm. Int. Expl. Mer, 109:66-67.

chemistry thorium
Picciotto, E., and S. Wilgain, 1954.
Thorium determination of deep-sea sediments.
Nature 173(4405):632-633.

Chemistry, thorium
Sackett, W.M., H.A. Potratz and E.D. Goldberg, 1958.
Thorium content of ocean water. Science 128(3317):204-205.

thorium isotopes

Chemistry thorium 232
Kaufman A., 1969.
Th 232 Th concentration of surface ocean water.
Geochim. cosmochim. Acta 33(6): 717-720

chemistry thorium isotopes
Koczy, F.F., E. Picciotto, G. Poulaert and S. Wilgain, 1957.
Measurement of Th isotopes in sea water.
Geochim. Cosmochim. Acta, 11:103-129.

Chemistry, thorium 232
Servant, Jean 1966.
La radioactivité de l'eau de mer.
Cah. océanogr. 18(4): 297-318.

Chemistry, thorium (isotopes)
Somayajulu, B.L.K., and E.D. Goldberg 1966.
Thorium and uranium isotopes in seawater and sediments.
Earth Planet Sci. Letts 1(3): 102-106.

Chemistry, thorium isotopes
Starik, I.E., K.F. Lazarev, D.S. Nikolaev, S.M. Grashchenko, L.B. Koladin, and Iu. V. Kuznetsov, 1959.
Concentration of thorium isotopes in the Black Sea water.
Doklady, Akad. Nauk, SSSR, 129(4):919-924.

thorium 230/protactinium 231

chemistry, thorium-230/protactinium-231
Sackett, William M., 1966.
Manganese nodules: thorium-230:protactinium-231 ratios.
Science, 154(3749):646-647.

234Th/238 U ratio
Bhat, S.G., S. Krishnaswamy, D. Lal, Rama and W.S. Moore, 1969.
234Th/238U ratios in the ocean.
Earth Planet. Sci. Letters 5(7): 483-491.

threonine

Chemistry threonine
Litchfield, Carol D., and Donald W. Hood, 1966.
Microbiological assay for organic compounds in seawater. II. Distribution of adenine, uracil, and threonine.
Appl. Microbiol., 14(2):145-151.

thynine

chemistry thynine
Rosenberg, E., 1964.
Purine and pyrimidines in sediments from the experimental Mohole.
Science, 146(3652):1680-1681.

tin

chemistry, tin
Smith, J. David 1970
Tin in organisms and water in the Gulf of Naples.
Nature, Lond. 225 (5227): 103-104.

chemistry – tin
Straczek, John A., 1968.
Problems in offshore exploration for tin in Thailand: a case history.
Occ. Publ. Narragansett Mar. Lab., Univ. R.I. 4:66-74.

titanium

Chemistry titanium
Collier, A., 1953.
Titanium and zirconium in bloom of Gymnodinium brevis Davis. Science 118(3064):329.

chemistry titanium
Griel, J.V., and R.J. Robinson, 1952.
Titanium in sea water. J. Mar. Res. 11(2):173-179.

Chemistry titanium
Lisitsin, A.P., 1964.
Distribution and chemical composition of suspended matter in the waters of the Indian Ocean. (In Russian: English Abstract).
Rezult. Issled., Programme Mezhd. Geofiz. Goda, Mezhd. Geofiz. Kom., Presidume Akad. Nauk, SSSR, Okeanol., No. 10:136 pp.

Chemistry titanium
Ostroumov, E.A., and I.I. Volkov, 1962.
Separation of titanium, zirconium and throium from magganese, nickel, cobalt and zinc by means of cinnamic acid.
Trudy Inst. Okeanol., Akad. Nauk, SSSR, 54:170-181.

In Russian; English summary

chemistry, titanium
Yamamoto, Toshio, Tetsuo Fujita and Masayoshi Ishibashi 1970.
Chemical studies on the seaweeds (25). Vanadium and titanium content in seaweeds.
Rec. oceanogr. Wks, Japan 10(2): 125-135.

titanomagnetic

chemistry, titanomagnetic
Vasilchikov, N.V., Yu. A. Pavlidis and N.P. Slovinsky-Sidak, 1966.
On the vanadium titanomagnetic sea-beach placers in the Far East. (In Russian; English abstract).
Okeanologiia, Akad. Nauk, SSSR, 6(5):823-829.

toxins
Abbott, B.C. and Paster, Z. 1970
Actions of toxins from Gymnodinium breve.
Toxicon 8:120.

Chemistry toxins
*Banner, Albert H., Judd C. Nevenzel and Webster R. Hudgins, 1969.
Marine toxins from the Pacific. 2. The contamination of the Wake Island Lagoon.
Atoll Res. Bull., 122: 1-12.

Chemistry - toxins
Blanquet, Richard, 1968.
Properties and composition of the nematocyst toxin of the sea anemone, Aiptasia pallida.
Comp. Biochem. Physiol., 25: 893-902.

chemistry toxins
Doty, Maxwell S. and Gertrudes Aguilar-Santos 1970.
Transfer of toxic algal substances in marine food chains.
Pacific Sci. 24(3): 351-355

Chemistry toxins
Doty, Maxwell S., and Gertrudes Aguilar-Santos, 1966.
Caulerpicin, a toxic constituent of Caulerpa.
Nature, 211 (5052):990.

Chemistry- toxins
Fogg, G.E., 1966.
The extracellular products of algae.
In: Oceanography and marine biology, H. Barnes, editor, George Allen & Unwin, Ltd., 4:195-212.

Chemistry toxins
Ghiretti, F., 1970.
Pharmacologically active compounds from marine organisms (selected topics).
Rev. int. Océanogr. méd. 18-19: 169-179.

Chemistry toxins
Ghiretti, F., and E. Rocca, 1963
Some experiments on ichthyotoxin.
In: Venomous and poisonous animals and noxious plants of the Pacific region. H.L. Keegan and M.V. MacFarlane, Editors. Pergamon Press, Inc. 211-216.

Chemistry Toxins
Hashimoto, Yoshiro, Takeshi Yasumoto, Hisao Kamiya, and Tamao Yoshida, 1969.
Occurrence of ciguatoxin and ciguaterin in ciguatoxic fishes in the Ryukyu and Amami islands.
Bull. Jap. Soc. scient. Fish. 35(3). 327-332.

Chemistry toxins
Hessel, Donald W., 1963
Marine biotoxins. III. The extraction and partial purification of Ciguatera toxin from Lutjanus bohar (Forskal); use of silicic acid chromatography.
In: Venomous and poisonous animals and noxious plants of the Pacific region, H.L. Keegan and M.V. MacFarlane, Editors, Pergamon Press Inc., 203-209.

Chemistry toxins
Hood, D.W., 1963
Chemical oceanography, In: Oceanography and Marine Biology, H. Barnes, Edit., George Allen & Unwin, 1:129-155.

Chemistry toxins
Ingham, H.R., James Mason and P.C. Wood, 1968.
Distribution of toxin in molluscan shellfish following the occurence of mussel toxicity in North-east England.
Nature, Lond., 220(5162):25-27.

Chemistry toxins
Lewin, Ralph A., 1970
Toxin secretion and tail autotomy by irritated Oxynoe panamensis (Opisthobranchia; Sacoglossa).
Pacific Sci., 24(3): 356-358.

Chemistry toxins
Martin, Dean F., and Ashvin B. Chatterjee, 1971.
Some chemical and physical properties of two toxins from the red-tide organism, Gymnodinium breve.
Fish. Bull. U.S. Nat. Ocean. Atmos. Adm. 68(3): 433-443

chemistry toxins, coelenterate
Middlebrook, Robert E., Lawrence W Wittle, Edward D. Scura and Charles E. Lane 1971.
Isolation and purification of a toxin from Millepora dichotoma.
Toxicon 9(4): 333-336

chemistry toxins
Nozawa, Koji, 1968
The effect of Peridinium toxins on other algae.
Bull. Misaki mar. biol. Inst., Kyoto Univ. 2:21-24

chemistry toxins
Ohshika, Hideyo 1971.
Marine toxins from the Pacific IX. Some effects of ciguatoxin on isolated mammalian atria.
Toxicon 9(4): 337-343

chemistry toxins
Paster, Z., 1968.
Prymnesin, the toxin of Prymnesium parvum Carter.
Rev. intern. Océanogr. Méd., 10:249-257

chemistry toxins
Rayner, Martin D., Morris H. Baslow and Thomas I. Kosaki, 1969.
Marine toxins from the Pacific - ciguatoxin not an in vivo anticholinesterase.
J. Fish. Res. Bd, Can., 26(8): 2208-2210

chemistry - Toxins
Rice, Nolan E. and W. Allan Powell, 1970.
Observations on three species of jellyfishes from Chesapeake Bay with special reference to their toxins. I. Chrysaora (Dactylometra) quinquecirrha. Biol. Bull. mar. biol. Lab., Woods Hole, 139(1): 180-187.

chemistry, toxins
Russell, Findlay E., 1965.
Marine toxins and venomous and poisonous marine animals.
In: Advances in Marine Biology, Sir Frederick S. Russell, editor, Academic Press, 3:255-384.

toxins
Trieff, N.M., J.J. Spikes, S.M. Ray and J.B. Nash 1970.
Isolation of Gymnodinium breve toxin
Toxicon 8:157-158

toxins
Welsh, J.H., 1955.
On the nature of coelenterate toxins.
Pap. Mar. Biol. and Oceanogr., Deep-Sea Res. Suppl. to Vol. 3:287-297.

chemistry toxins, coelenterates
Wittle, Lawrence W., Robert E. Middlebrook and Charles E. Lane 1971.
Isolation and partial purification of a toxin from Millepora alcicornis.
Toxicon 9(4): 327-331

chemistry toxins, echinoderm
Habermehl, Gerhard, and Gert Volkwein 1971.
Aglycones of the toxins from the Cuvierian organs of Holothuria forskåli and a new nomenclature for the aglycones from Holothurioideae.
Toxicon 9(4): 319-326.

chemistry toxins, effect of
Béress, László, und Rosemarie Béress 1971.
Reinigung zweier krabbenlähmender Toxine aus der Seeanemone Anemonia sulcata.
Kieler Meeresforsch. 27(1): 117-127

trace elements

chemistry trace elements
Bardet, T., A Tchakirian and R. LaGrange, 1938.
Recherche spectrographique des éléments existant à l'état de traces dans l'eau de mer.
C.R. Acad. Sci., Paris, 206:250-252.

chemistry trace elements
Beliaev, L.I., 1964.
On the greatest probable importance of the results of chemico-spectrum determination of microelements in sea water. Oceanographic investigations in the Atlantic Ocean and the Black Sea. (In Russian).
Trudy Morsk. Gidrofiz. Inst., Akad. Nauk Ukrain., SSR, 30: 137-147.

Chemistry trace substances
Belser, W.L., 1963
9. Bioassay of trace substances. In: The Sea, M.N. Hill, Edit., Vol. 2 (II) Fertility of the oceans, Interscience Publishers New York and London, 220-231.

trace elements
Bieri, Robert, and D. H. Krinsley, 1958.
Trace elements in the pelagic coelenterate, Velella lata.
J. Mar. Res., 16(3):246-254.

Chemistry trace elements
Black, W.A.P., and R.L. Marshall, 1952.
Trace elements in the common brown algae and in sea water. J.M.B.A. 30:575-584.

chemistry, trace elements
Branica, Marko, Milica Petek, Ante Barić and Ljubomir Jeftić 1969.
Polarographic characterisation of some trace elements in sea water.
Rapp. P.-v. Reun. Commn int. Explor. scient. Mer Mediterr. 19(5): 929-933

chemistry trace elements (data)
*Brooks, R.R., I.R. Kaplan, and M.N.A. Peterson, 1969.
Trace element composition of Red Sea geothermal brine and interstitial water.
In: Hot brines and Recent heavy metal deposits in the Red Sea, E.T. Degens and D.A. Ross, editors, Springer-Verlag, New York, Inc., 180-203.

chemistry trace elements
Brooks, Robert R., and Martin G. Rumsby, 1965.
The biogeochemistry of trace element uptake by some New Zealand bivalves.
Limnol. Oceanogr., 10(4):521-527.

Chemistry trace elements
Bruevich, S.W., editor, 1966.
Chemistry of the Pacific Ocean.
Inst. Okeanol., Akad. Nauk, SSSR, Isdatel. Nauka, Moskva, 358 pp.

chemistry, trace elements
Chester, R., and L.R. Johnson, 1971
Trace element geochemistry of North Atlantic aeolian dusts.
Nature, Lond., 231(5299): 176-178

trace elements
Cooper, L. H.N., 1958
A system for international exchange of samples for trace element analysis of ocean water. J. Mar. Res., 17: 128-132.

chemistry trace elements
Corless, James T., 1966.
A trace element study in Narragansett Bay
Maritimes, Univ. R.I., 10(1):4-6.

chemistry trace elements
Duke, Thomas W., James N. Willis and Thomas J. Price, 1966.
Cycling of trace elements in the estuarine environment. 1. Movement and distribution of zinc 65 and stable zinc in experimental ponds.
Chesapeake Science, 7(1):1-10.

Chemistry trace elements
Fitzgerald, W.F., 1969.
Trace elements.
Oceanus, 15(1):23-24. (popular)

chemistry, trace elements
Fonselius, Stig H., 1970.
Some trace metal analyses in the Mediterranean, the Red Sea and the Arabian Sea.
Bull. Inst. océanogr., Monaco, 69 (1407): 15 pp.
Also: IAEA Radioactivity in the Sea, Publ. 29.

Chemistry trace elements
Fukai, R., & W. W. Meinke, 1959.
Some activation analyses of six trace elements in marine biological ashes.
Nature, 184(4689): 815-816.

Chemistry trace elements
Fukai, R., & W. W. Meinke, 1959.
Trace analysis of marine organisms: A comparison of activation analysis and conventional methods
Limnol. & Ocean., 4(4):399-408.

Chemistry, trace elements
*Gast, Paul W., 1968.
Trace element fractionation of tholeiitic and alkaline magma types.
Geochim. cosmochim. Acta, 32(10):1057-1086.

chemistry trace elements
Graham, J. W., 1959.
Metabolically induced precipitation of trace elements from sea water.
Science 129(3360):1428-1429.

chemistry
Trade Elements
Harris, Robert C., 1965.
Trace element distribution in molluscan skeletal material. 1. Magnesium, iron, manganese and strontium. Bull. Mar. Sci., 15(2):265-273.

Chemistry trace elements
Hood, D.W., 1963
Chemical oceanography. In: Oceanography and Marine Biology, H. Barnes, Edit., George Allen & Unwin, 1:129-155.

chemistry trace elements
*Ishibashi, Masayoshi, Taitiro Fujinaga, Tooru Kuwamoto, Shinji Sugibayashi, Hiromichi Sawamoto, Yoshinori Ogino and Shigeo Murai, 1968.
Fundamental investigation on the dissolution and deposition of some transition elements in the sea.
Rec. oceanogr. Wks. Japan, n.s., 9(2):169-177.

Chemistry trace elements
Johnston, Robert, 1965.
The trace elements.
Ser. Atlas, Mar. Environment, Folio 8:4 pp., 2 pls.

Chemistry trace elements
Jones, James I., and Wayne D. Bock, 1964.
Trace-element distribution in some living and fossil Foraminifera from South Florida, Bahamian and Caribbean waters.
Geol. Soc., Amer., Special Paper, No. 76:88-89.

Chemistry, trace elements
Kilezhenko, V.P., and V.N. Podymakhin, 1964.
Accumulation of trace elements by some organisms in the southern part of the Barents Sea. (In Russian).
Materiali, Ribochoz. Issled. Severn. Basseina, Poliarn. Nauchono-Issled. i Proektn. Inst. Morsk. Rib. Choz. i Okeanogr. im N.M. Knipovicha, PINRO, Gosud. Proizvodst. Komm. Ribn. Choz. SSSR. 4:135-134.

chemistry, trace elements
Livingston, H.D. and Geoffrey Thompson, 1971.
Trace element concentrations in some modern corals. Limnol. Oceanogr. 16(5): 786-796.

chemistry, trace elements (basalts)
Melson, W.G., Geoffrey Thompson and Tjeerd H. Van Andel 1968.
Volcanism and metamorphism in the Mid-Atlantic Ridge, 22° N. Latitude
J. geophys. Res. 73 (18): 5925-5941

chemistry, trace elements
Picotti, M., 1956.
Les éléments oligodynamiques marins en relation avec la cuivre. Ist. Sperimentale Talassografico Trieste, Pubbl., 333:181-184.
Also:
Rapp. Proc. Verb., Comm. Int. Expl. Sci., Mer Medit., 13:181-184.

chemistry, trace elements
Phelps, D.C., R.J. Santiago, D. Luciano, and N. Irizarry, 1969.
Trace element composition of inshore and offshore benthic populations.
Symp. Radioecology, Proc. Second Natn. Symp., Ann Arbor, Mich., 1967 (Conf. 670503): 509-526.

Chemistry trace elements
Pilkey, Orrin H., and H.G. Goodell, 1963
Trace elements in Recent mollusk shells.
Limnol. and Oceanogr., 8(2):137-148.

Chemistry trace elements
Richards, F.A., 1956.
On the state of our knowledge of trace elements in the sea. Ltr. to the Editors, Geochimica et Cosmochimica Acta, 10(4):241-243.

Chemistry trace elements
Riley, J.P. and Igal Roth, 1971.
The distribution of trace elements in some species of phytoplankton grown in culture.
J. mar. biol. Ass. U.K., 51(1): 63-72.

Chemistry trace elements
Schutz, Donald F., and Karl K. Turekian, 1965.
The investigation of the geographical and vertical distribution of several trace elements in sea water neutron activation analysis.
Geochimica et Cosmochimica Acta, 29(4):259-314.

Chemistry trace elements
Segar, D.A., J.D. Collins and J.P. Riley, 1971.
The distribution of the major and some minor elements in marine animals. J. mar. biol. Ass. U.K., 51(1): 131-136.

Chemistry trace elements
Spencer, C.P., 1957.
Utilization of trace elements by marine unicellular algae. J. Gen. Microbiol., 16(1):282-285.

Chemistry trace elements
Stevenson, Robert A., 1965.
Differences in trace element composition in the sea urchins Tripneustes esculentus (Leske) and Echinometra lucunter (L.). (Abstract)
Ocean Sci. and Ocean Eng., Mar Techn. Soc., Amer. Soc. Limnol. Oceanogr., 1:300.

Chemistry trace elements
Szabo, B.J., 1968.
Trace element content of plankton population from the Bahamas.
Carib. J. Sci., 8(3/4):185-186.

chemistry trace elements
Thompson, Geoffrey, Donald C. Bankston and Susan M. Pesley, 1970.
Trace element data for reference carbonate rocks.
Chem. Geol. 6(2): 165-170.
WHOI Contrib. No. 2453

trace elements
*Ting, Robert Y. and V. Roman de Vega, 1969.
The nature of the distribution of trace elements in longnose anchovy (Anchoa lamprotaenia Hildebrand), Atlantic thread herring (Opisthonema oglinum LaSueur), and alga Udotea flabellum Lamouroux).
Symp. Radioecology, Proc. Second Natn. Symp., Ann Arbor, Mich., 1967, (Conf. 670503): 527-534.

chemistry, trace elements
Turekian, Karl K., 1971
2. Rivers, tributaries and estuaries. In: Impingement of man on the oceans, D.W. Hood, editor, Wiley Interscience, 9-73.

chemistry, trace elements
Turekian, Karl K., 1967.
Estimates of the average Pacific deep-sea clay accumulation rate from material balance calculations.
Progress in Oceanography, 4:227-244.

Chemistry - trace elements
Turekian, Karl K., 1965.
Some aspects of the geochemistry of marine sediments.
In: Chemical oceanography, J.P. Riley and G. Skirrow, editors, Academic Press, 2:81-126.

trace elements
Turekian, Karl K., Donald F. Schultz and David Johnson, 1966.
The distribution of Sr, Ba, Co, Ni, and Ag in ocean water profiles of the Pacific sector of the Antarctic seas.
Antarctic J., United States, 1(5):224.

Chemistry trace elements
Van Andel, Tj. H., 1971.
Tectonic evolution and trace element composition of basement rocks of the Mid-Atlantic Ridge: 8°S
Phil. Trans. R. Soc. Lond. (A) 268 (1192): 661. (abstract only).

Chemistry trace elements
Willis, J.P., & L.H. Ahrens, 1962
Some investigations on the composition of manganese nodules, with particular reference to certain trace elements.
Geochim. et Cosmochim. Acta, 26: 751-764.

trace metals

Chemistry trace metals
*Bender, Michael L. and Cynthia Schultz, 1969.
The distribution of trace metals in cores from a traverse across the Indian Ocean.
Geochim. cosmochim. Acta, 33(2):292-297.

Chemistry trace metals
Carritt, D.E., 1953.
Separation and concentration of trace metals from natural waters: partition chromatographic technique. Analyt. Chem. 25:1927.

chemistry, trace metals
Fonselius, Stig H. 1970.
Some trace metal analyses in the Mediterranean, the Red Sea and the Arabian Sea.
Bull. Inst. Océanogr. Monaco 69 (1407): 15pp.

Chemistry trace metals
Goldberg, E.D., 1957.
Biogeochemistry of trace metals. Ch. 12 in: Treatise on Marine Ecology and Paleoecology, Vol. 1. Mem. Geol. Soc., Amer., 67:345-358.

trace metals
Shuster, Carl N., Jr. and Benjamin H. Pringle, 1969.
Trace metal accumulation by the American eastern oyster Crassostrea virginica.
Proc. nat. Shellfish Ass., 59: 91-103

Chemistry trace metals
Spencer, Derek W. and Peter G. Brewer, 1971.
Vertical advection, diffusion and redox potentials as controls on the distribution of manganese and other trace metals dissolved in waters of the Black Sea. J. geophys. Res, 76(24): 5877-5892.

Chemistry trace metals
Windom, Herbert L., 1970.
Contribution of atmospherically transported trace metals to South Pacific sediments.
Geochim. cosmochim. Acta, 34 (4): 509-514

chemistry, trace metals
Windom, H.L., K.C. Beck, and R. Smith 1971
Transport of trace metals to the Atlantic Ocean by three southeastern rivers
SEast. Geol. 12(3): 169-181.

tracers, isotopes
Begemann, F., and W.F. Libby, 1957.
Continental water balance, ground water inventory and storage times, surface ocean mixing rates and world wide circulation patterns from cosmic ray and man-made tritium.
Int. Conf., Radioisotopes in Sci. Res., Paris, Sept., 1957, UNESCO/NS/RIC/221:20 pp.

tracers, radioactive
Burbanck, W.D., Robert Grabske and J.R. Comer, 1964.
The use of the radioisotope zinc 65, in a preliminary study of the population movements of the estuarine isopod, Cyathura polita (Stimpson, 1855).
Crustaceana, 7(1):17-20.

Chemistry tracers
Carritt, D.E., 1963
5. Chemical instrumentation. In: The Sea, M.N. Hill, Edit., Vol. 2. The composition of sea water, Interscience Publishers, New York and London, 109-123.

Chemistry tracers (C14)
Fedorova, G.V., 1963.
Absorption and elimination of C14 by fishes, as dependent on their age and sex, as well as on water temperature. (In Russian).
Doklady, Akad. Nauk, SSSR, 150(1):169-169.

Chemistry - tracers
Ford, W.L., 1949
Radiological and Salinity relationships in the water at Bikini Atoll. Trans. Am. Geophys. Un., 30(1):46-54, 22 text figs.

Chemistry - tracers (radioactive)
Foster, R.F., 1959
Radioactive tracing of the movement of an essential element through an aquatic community with specific reference to radiophosphorus.
Pubb. Sta. Zool. Nap., 31(Suppl): 34-69.

Chemistry tracers, radioactive
Marples, T.G., 1966.
A radionuclide tracer study of arthropod food chains in a Spartina salt marsh ecosystem.
Ecology, 47(2):270-277.

Chemistry
Tracers

Patin, S.A., 1963.
Possibilities of using tritium for the study of dynamic processes in the ocean and in the atmosphere. (In Russian).
Trudy, Morsk. Gidrofiz. Inst., Akad. Nauk Ukrain. SSR, 28:99

Translation:
Soviet Oceanography, Issue 2 - 1963 series.
Scripta Tecnica, Inc., for AGU, 69-70. 1965.

Chemistry tracers, radioactive
Babu Abraham, Y. Hiyama and F. Yasuda, 1962.
Quantitative relation between marine and fresh-water preys and predators determined by 32P
Bull. Jap. Soc. Sci. Fish., 28(11):1092-1098.

Chemistry
trimethylamine

Budd, J.A. 1969.
Catabolism of trimethylamine by a marine bacterium, Pseudomonas NCMB 1154.
Mar. Biol. 4(3): 257-266.

Chemistry ~~tripton~~
Small, Lawrence F., and Herbert Curl, Jr., 1968.
The relative contribution of particulate chlorophyll and river tripton to the extinction of light off the coast of Oregon.
Limnol. Oceanogr., 13(1):84-91.

tritium

Chemistry tritium
Bainbridge, A.E., 1963.
Tritium in surface waters of the North Pacific.
Symposium on oceanic mixing. In:Nuclear Geophysics, Proc. Conf., Woods Hole, Massachusetts, June 7-9, 1962.
NAS-NRC, Publ., 1075:129-137.

Also: Nuclear Sci. Ser. Rept., No. 38.

Chemistry tritium
Begemann, F., and W.F. Libby, 1957.
Continental water balance, ground water inventory and storage times, surface ocean mixing rates and world-wide water circulation patterns from cosmic ray and man-made tritium.
Int. Conf., Radioisotopes in Sci. Res., Paris, Sept., 1957, UNESCO/NS/RIC/221:20 pp.

Chemistry tritium
Burton, J.D., 1965.
Radioactive nuclides in sea water, marine sediments and marine organisms.
In:Chemical oceanography, J.P.Riley and G. Skirrow, editors, Academic Press, 2:425-475.

Chemistry tritium
Craig, Harmon and W.B. Clarke, 1970
Oceanic ^{3}He: contribution from cosmogenic tritium.
Earth Planet. Sci. Letts 9(1): 45-48.

Chemistry Tritium
de Santis, L., and F. Mosetti, 1961.
Alcuni ragguagli sulle ricerche idrologiche con traccianti.
Boll. Geofis. Teorica ed Applicata, Trieste, 3(10):1-16.

chemistry, tritium
*Dockins, K.O., A.E. Bainbridge, J.C. Houtermans and H.E. Suess.
Tritium in the mixed layer of the North Pacific Ocean.
In: Radioactive dating and methods of low-level counting.
Int.Atom.Energy Agency, Vienna, 1967:129-141.

Chemistry tritium
Eriksson, Erik, 1965.
An account of the major pulses of tritium and their affects in the atmosphere.
Tellus, 17(1):118-130.

Chemistry tritium
Giletti, Bruno J., Fernando Bazan and J. Lawrence Kulp, 1958
The geochemistry of tritium. Trans. Amer. Geophys. Union, 39(5): 807-818.

Chemistry
tritium
Giletti, B. J., and J. L. Kulp, 1959.
Tritium observations.
Sci. Studies, Fletcher's Ice Island, T-3, Vol. 1.
Air Force Cambridge Res. Center, Geophys. Res. Pap., No. 63:153-158.

chemistry, tritium
Kincaid, George P., Jr. and Edward R. Ibert, 1970.
Tritium production from nitrogen in fission reactors. Nature, 226(5241): 139-141. Also in: Contrib. Oceanogr. Texas A & M Univ., Coll. Geosciences, 443.

Chemistry
tritium
Libby, W.F., 1963
Moratorium tritium geophysics.
J. Geophys. Res., 68(15):4485-4494.

Chemistry
tritium
Libby, W.F., 1961
Tritium geophysics.
J. Geophys.Res., 66(11): 3767-3782.

Chemistry tritium
Libby, W.F., 1955.
Tritium in nature. J. Washington Acad. Sci., 45(10):301-314.

Chemistry tritium
Mirolli, Maurizio, 1965.
Tritium: distribution in Busycon canaliculatum (L.) injected with labeled reserpine.
Science, 149(3691):1503-1504.

Chemistry Tritium
Miyake, Yasuo, Hiroshi Hamaguchi, Masumi Osawa, Naoki Onuma, Ken Sugawara, Satoru Kanmori, Sadayuki Omori, Toshio Shimada, Kiyoshi Kawamura, Hirozo Yoshimura, and Takeshi Sagi, 1960.
Chemistry in the Japanese Expedition of Deep Sea. (JEDS-2)
Repts. of JEDS, 1: 181-185.
JEDS Contrib. No. 4
Also in Oceanogr. Mag., 11(2) : 181-185.

Chemistry, tritium
Östlund, H. Göte, 1970.
Hurricane tritium III: Evaporation of sea water in hurricane Faith 1966. J. geophys. Res., 75(12): 2303-23

Chemistry
tritium
*Östlund, H. Göte, 1968.
Hurricane tritium II: air-sea exchange of water in Betsy 1965.
Tellus, 20(4):577-594.

Chemistry, tritium
Östlund, H. Göte, 1967.
Hurricane tritium I: preliminary results on Hilda 1964 and Betsy 1965.
Geophys. Monogr., Am. Geophys. Un., 11:58-60.

chemistry, tritium
Östlund, H. Göte, Murice O. Rinkel and Claes Rooth, 1969.
Tritium in the equatorial Atlantic current system. J. geophys. Res., 74(18): 4535-4543.

Chemistry
Tritium
Patin, S.A., 1963.
Possibilities of using tritium for the study of dynamic processes in the ocean and in the atmosphere. (In Russian).
Trudy, Morsk. Gidrofiz. Inst., AAkad. Nauk Ukrain. SSR, 28:99-101

Translation:
Soviet Oceanography, Issue 2 - 1963 series.
Scrpta Tecnica, Inc., for AGU, 69-70. 1965.

Chemistry
tritium
Suess H.E., 1970.
The transfer of carbon 14 and tritium from the atmosphere to the ocean. J. geophys. Res., 75(12): 2363-2364.

Chemistry
tritium
*Taylor, C.B., 1968.
A comparison of tritium and strontium-90 fallout in the Southern Hemisphere.
Tellus, 20(4):559-576.

chemistry, tritium
Volchok, H.L., V.T. Bowen, T.R. Folsom, W.S. Broecker, E.A. Schuert and G.S. Bien, 1971.
Oceanic distributions of radionuclides from nuclear explosions. Radioactivity in the marine environment, U.S. Nat. Acad. Sci., 1971: 42-89.

Chemistry, tritium fallout
*Leventhal, J.S. and W.F. Libby, 1970.
Tritium fallout in the Pacific United States.
J. geophys. Res., 75(36): 7628-7633.

tungsten

Chemistry
tungsten
Fukai, Rinnousuke, and W. Wayne Meinke, 1962
Activation analyses of vanadium, arsenic, molybdenum, tungsten, rhenium and gold in marine organisms.
Limnology and Oceanography, 7(2):186-200.

Chemistry tungsten
Ishibashi, M., 1953.
Studies on minute elements in sea water.
Rec. Ocean. Wks., Japan, n.s., 1(1):88-92.

Chemistry
tungsten
Ishibashi, Masayoshi, Taitiro Fujinaga and Tooru Kuwamoto, 1962
Chemical studies on the ocean. Part 89.
Fundamental investigation on the dissolution and deposition of molybdenum, tungsten and vanadium in the sea.
Rec. Oceanogr. Wks., Japan, Spec. No. 6:215-218.

Chemistry tungsten
Pilipchuk, M.F., and I.I. Volkov, 1966.
Tungsten in modern sediments of the Black Sea. (In Russian).
Doklady, Akad. Nauk, SSSR, 167(2):430-433.

Tyrosine

Chemistry
tyrosine (data only)
Anderson, William W., and Jack W. Gehringer, 1959.
Physical oceanographic, biological and chemical, South Atlantic coast of the United States, M/V Theodore N. Gill cruise 9.
USFWS Spec. Sci. Rept., Fish. No. 313:226 pp.

Chemistry tyrosine (data only)
Anderson, W.W., and J.W. Gehringer, 1959.
Physical oceanographic, biological and chemical data South Atlantic Soast of the United States, M/V Theodore N. Gill Cruise 8.
USFWS Spec. Sci. Rept., Fish., No. 303:227 pp.

Chemistry tyrosine (data only)
Anderson, William W., and Jack W. Gehringer, 1959.
Physical oceanography, biological and chemical data, South Atlantic coast of the United States M/V Theodore N. Gill, Cruise 7.
USFWS Spec. Sci. Rept., Fish. No. 278:277 pp.

Chemistry tyrosine (data only)
Anderson, William W., and Jack W. Gehringer, 1958.
Physical oceanographic, biological and chemical data, south Atlantic coast of the United States, M/V Theodore N. Gill Cruise 6.
USFWS Spec. Sci. Rept., Fish., No. 265:99 pp.

Chemistry tyrosine (data only)
Anderson, W.W., and J.W. Gehringer, 1958.
Physical oceanographic, biological and chemical data, South Atlantic coast of the United States, M/V Theodore N. Gill cruise 5.
USFWS Spec. Sci. Rept., Fish., 248:220 pp.

Chemistry tyrosine (data only)
Anderson, W.W., and J.W. Gehringer, 1957.
Physical oceanographic, biological and chemical data, South Atlantic coast of the United States, M/V Theodore N. Gill cruise 4.
USFWS Spec. Sci. Rept., Fish., 234:192 pp.

Chemistry tyrosine (data only)
Anderson, W.W., J.W. Gehringer and E. Cohen, 1956
Physical oceanographic, biological and chemical data, South Atlantic coast of the United States, Theodore N. Gill, Cruise 2.
Spec. Sci. Rept., Fish., 198:270 pp.

Chemistry tyrosine (data only)
Anderson, W.W., J.W. Gehringer and E. Cohen, 1956.
Physical oceanographic, biological and chemical data, south Atlantic coast of the United States, M/V Theodore N. Gill, Cruise 1.
USFWS Spec. Sci. Rept., Fish., No. 178:160 pp.

Chemistry tyrosine
Dragovich, A., 1963.
Hydrology and plankton of coastal waters at Naples, Florida.
Q.J. Florida Acad. Sci., 26(1):22-47.

uracil

Chemistry, uracil
Litchfield, Carol D., and Donald W. Hood, 1966.
Microbiological assay for organic compounds in seawater. II. Distribution of adenine, uracil, and threonine.
Appl. Microbiol., 14(2):145-151.

chemistry uracil
Rosenberg, E., 1964.
Purine and pyrimidines in sedimentsfrom the experimental Mohole.
Science, 146(3652):1680-1681.

uranium

Chemistry uranium
Agamirov, S. Sh., 1963.
Precipitation of uranium on the bottom of the Black Sea. (In Russian).
Geochimiya, (2):92-93.

Translation:
Geochemistry, (1):104-106.

chemistry, uranium
Aumento, F., 1971.
Uranium content of mid-oceanic basalts.
Earth planet. Sci. Letts 11 (2): 90-94.

chemistry, uranium
Aumento, F., and R.D. Hyndman 1971.
Uranium content of the oceanic upper mantle.
Earth planet. Sci. Letts 12 (4): 373-384.

Chemistry uranium
Baturin, G.N., 1969.
Uranium in the surface sediment layer of the northwestern Indian Ocean. (In Russian; English abstract). Okeanologiia, 9(6): 1031-1037.

chemistry, uranium
Baturin, G.N. 1968.
Geochemistry of uranium in the Baltic.
Geokhimiya 1968(3): 377-381.
Translation: Geochem. int. 5(2): 344-348.

Chemistry uranium
Baturin, G.N., A.V. Kochenov and S.A. Kovaleve, 1966.
Certain peculiar features of uranium distribution in the waters of the Black Sea. (In Russian).
Doklady, Akad. Nauk, SSSR, 166(3):698-700.

chemistry, uranium
*Blanchard, R.L., M.H. Cheng and H.A. Potratz, 1967.
Uranium and thorium series disequilibria in Recent and fossil marine molluscan shells.
J. geophys. Res., 72(18):4745-4757.

Chemistry uranium
Chow, T.J., M. Tatsumoto and C.C. Patterson, 1962
Lead isotopes and uranium contents in experimental Mohole cores (Guadalupe site).
J. Sed. Petr., 32(4):866-869.

Chemistry uranium
Davankov, A.B., V.M. Laufer and A.V. Gordiyevskiy 1961.
Storehouses of the Atlantic Ocean.
Priroda, 1961(12):101-103.

Abstracted in:
Soviet-Bloc Res. Geophys., Astron. and Space, No. 34:22-23.

Chemistry uranium
Davies, R.V., J. Kennedy, R.W. McIlroy, R. Spence and K.M. Hill, 1964.
Extraction of uranium from sea water.
Nature, 203 (4950):1110-1115.

Chemistry uranium
Folsom, Theodore R., 1958.
Comparison of radioactive dosage from potassium with estimated dosages from uranium and radium in marine biospheres.
Proc. Ninth Pacific Sci. Congr., Pacific Sci. Assoc., 1957, Oceanogr., 16:176-178.

Chemistry uranium
Gloyna, E.E., and B.B. Ewing, 1957.
Uranium recovery from saline solutions of biological slimes. Nucleonics, 15(1):78-83.

Chemistry uranium
Goldzstein, M., et J. Ros, 1963.
Sur la teneur en uranium des sédiments du bassin occidental de la Méditerranée.
Rapp. Proc. Verb., Réunions, Comm. Int., Expl. Sci., Mer Méditerranée, Monaco, 17(3):1021-1028.

Chemistry uranium
Hahofer, E., and F. Hecht, 1954.
Determination of uranium in deep sea samples.
Microchimica Acta:417-434.

Chemistry uranium
Haskin, Larry A., and Fred A. Frey, 1966.
Discussion of paper by J.W. Morgan and J.F. Lovering, 'Uranium and thorium abundances in the basalt cored in the Mohole Project (Guadalupe Site)'.
J. geophys. Res., 71(2):688-689.

Chemistry uranium
Hernegger, F., and B. Karlik, 1935.
Uranium in sea water.
Göteborgs Kungl. Vetenskaps- och Vitterhets Samh. Handl. Femte Följden, 4(12):15 pp.

Chemistry uranium
Heye, Dietrich, 1969.
Uranium, thorium, and radium in ocean water and deep-sea sediments
Earth planet. Sci. Letters 6(2): 112-116.

Chemistry uranium
Holland, H.D., and J.L. Kulp, 1954.
The transport and deposition of uranium, ionium and radium in rivers, oceans and ocean sediments.
Geochim. Cosmochim. Acta 5:197-213.

Abstr. Chem. Abstr. 11855a, 1954.

Chemistry Uranium
Ishibashi, Masayoshi, Taitiro Fujinaga, Kosuke Izutsu, Tadashi Yamamoto and Hitoshi Tamura, 1961.
Determination of uranium in sea-water using a polarographic method.
Rec. Oceanogr. Wks., Japan, 6(1):106-111.

Chemistry uranium
Kaplan, I.R., and S.C. Rittenberg, 1963.
23. Basin sedimentation and diagenesis.
In: The Sea, M.N. Hill, Editor, Interscience Publishers, 3:583-619.

Chemistry uranium
Kitano, Yasushi, and Tamotsu Oomori 1971.
The coprecipitation of uranium with calcium carbonate.
J. oceanogr. Soc. Japan, 27(1): 34-42

uranium
*Kochenov, A.V., And G.N. Baturin, 1967.
Uranium distribution in the bottom sediments of the Aral Sea. (In Russian; English abstract).
Okeanologiia, Akad. Nauk, SSSR, 7(4):623-627.

Chemistry uranium
Koczy, F.F., E. Tomic & F. Hecht, 1957.
Geochemistry of uranium in the Baltic basin.
Geochim. Cosmochim. Acta. 11:86-102.

Chemistry uranium
Koczy, G., 1950.
Further uranium determinations in samples of sea water. Sitzb. Österr. Akad. Wiss., Math. Naturw. Kl., (IIa), 158:113-121.

Chemistry uranium
Koliadin, L.B., D.S. Nikolaev, S.M. Graschenko, Iu. V. Kuznetsov and K.F. Lazarev, 1960.
States of uranium detected in Black Sea waters.
Doklady, Akad. Nauk, SSSR, 132(4):915-917.

chemistry, uranium
Ku, Teh-Lung, and Wallace S. Broecker 1967.
Uranium, thorium and protactinium in a manganese nodule.
Earth Planet. Sci. Letts 2(4): 317-320.

Chemistry uranium
Lambet, M.S., and D.S. Nikolayev, 1962.
The forms of occurrence of uranium in the waters of the Sea of Azov and some estuaries and rivers of the Azov-Black Sea Basin.
Doklady, Akad. Nauk, SSSR, 142(3):681-682.

Translations:
Earth Sci. Sect. (Amer. Geol. Inst.), 142(1-6): 155-156.

Chemistry uranium

Milner, G.W.C., J.D. Wilson, G.A. Barnett, & A.A. Smales, 1961
Determination of uranium in sea water by pulse polarography.
J. Electroanal. Chem., 2(1): 25-38.
Abs. in Pub. Health Eng. Abs., 41(11): 25

uranium

Miyaki, Y., K. Sarwhashi Y. Katsuragi, J. Kanazawa and Y. Sugimura, 1964.
Uranium, radium, thorium, ionium strontium 90 and cesium 137 in coastal waters of Japan.
In: Recent researches in the fields of hydrosphere, atmosphere and nuclear geochemistry, Ken Sugawara festival volume, Y. Miyake and T. Koyama, editors, Maruzen Co., Ltd., Tokyo, 127-141.

Chemistry uranium

Miyake, Y., and Y. Sugimura, 1964.
Uranium and radium in the western North Pacific waters.
In: Studies on Oceanography dedicated to Professor Hidaka in commemoration of his sixtieth birthday, 274-278.

Chemistry uranium

Moore, Willard S., and William M. Sackett, 1964.
Uranium and thorium series inequilibrium in sea water.
J. Geophys. Res., 69(24):5401-5405.

Chemistry uranium

Morgan, J. W., and J. F. Lovering, 1965.
Uranium and thorium abundances in the basalt cored in Mohole Project (Guadalupe Site).
J. geophys. Res., 70(18):4724-4725

Chemistry uranium

Nakanishi, M., 1951.
Fluorometric microdetermination of uranium. V. The uranium content of sea water.
Bull. Chem. Soc., Japan, 24:36-38.

Chemistry uranium

Nikolaev, D.S., O.P. Koron, K.F. Lazarev, L.B. Koliadin, Iu. V. Kuznetsov, and S.M. Grashchenko, 1960.
The concentration of uranium in the waters of the Black Sea. (In Russian).
Doklady, Akad. Nauk, SSSR, 132(6):1411-1412.

Chemistry uranium

Nikolaev, D.S., K.F. Lazarev, O.P.Korn, M.I. Iakunin, V.M. Drozhzhin and A.G. Semartseva, 1965.
On the isotopic composition of the uranium contained in the waters and sediments of the Black and Azov seas. (In Russian).
Doklady, Akad. Nauk. SSSR, 165(1):187-189.

Chemistry uranium

Pettersson, H., 1949.
Exploring the bed of the ocean. Nature 164(4168) 468-470.

Chemistry uranium

Pettersson, H., 1939.
Die radioaktiven Elemente im Meer.
Rapp. Proc. Verb., Cons. Perm. Int. Expl. Mer, 109:66-67.

Chemistry uranium

Rona, E., L.O. Gilpatrick and L.M. Jeffrey, 1956.
Uranium determination in sea water.
Trans. Amer. Geophys. Union, 37(6):697-701.

Chemistry, uranium

Rona, E., and W.O. Urry, 1952.
Radioactivity of ocean sediments. VIII. Radium and uranium content of ocean and river waters.
Am. J. Sci. 250:241-262.

chemistry, uranium

Somayajulu, B.L.K., and E.D. Goldberg 1966.
Thorium and uranium isotopes in seawater and sediments.
Earth Planet. Sci. Letts 1(3): 102-106.

chemistry, uranium

Spalding, Roy F. and William M. Sackett, 1972.
Uranium in runoff from the Gulf of Mexico distributive province: anomalous concentrations
Science 175(4022): 629-631.

Chemistry uranium

Stewart, D.C., and W.C. Bentley, 1954.
Analysis of uranium in sea water. Science 120: 50-51.

Chemistry uranium

Sugimura, Y., 1964
Chemical behavior of uranium in marine environment. (In Japanese).
J. Oceanogr. Soc., Japan, 19(4):200-206.

Chemistry uranium

Sugimura, Y., T. Torii and S. Murata, 1964.
Uranium distribution in Drake Passage waters.
Nature, 204(4957):464-465.

Chemistry uranium

Tatsumoto, M., and E.D. Goldberg, 1959.
Some aspects of the marine geochemistry of uranium.
Geochim. et Cosmochim. Acta, 17(3/4):201-208.

Chemistry uranium

Thatcher, L.L., and F.B. Barker, 1957.
Determination of uranium in natural waters.
Anal. Chem., 29(11):1575-1578.

Chemistry uranium

Thompson, G., and H.D. Livingston 1970.
Strontium and uranium concentrations in aragonite precipitated by some modern corals.
Earth Planet. Sci. Letts 8(6): 439-442.

chemistry, uranium

Thurber, David L., 1964
The uranium content of sea water. (Abstract).
Trans. Amer. Geophys. Union, 45(1):119.

Chemistry uranium (data)

Torii, Tetsuya, and Sadao Murata, 1964.
Distribution of uranium in the Indian and the Southern Ocean waters.
In: Recent researches in the fields of hydrosphere, atmosphere and nuclear geochemistry, Ken Sugawara festival volume, Y. Miyake and T. Koyama editors, Maruzen Co., Ltd., Tokyo, 321-334.

chemistry, uranium

Viswanathan, R., S.M. Shah and C.K. Unni, 196

Atomic absorption and fluorometric analyses of seawater in the Indian Ocean. Bull. natn. Inst. Sci., India, 38(1): 284-288.

Chemistry, uranium

Rona, Elizabeth and Clara C. Dorta, 1969.
Geochemistry of uranium in the Cariaco Trench.
Trans. Gulf Coast Ass. geol. Socs, 19 : 264.

Chemistry, uranium

Sackett, W.M. and G. Cook, 1969.
Uranium geochemistry of the Gulf of Mexico.
Trans. Gulf Coast Ass. geol. Socs, 19 : 233-238

Chemistry, uranium

Veeh, H. Herbert, and Karl K. Turekian 1968.
Cobalt, silver and uranium concentrations of reef-building corals in the Pacific Ocean.
Limnol. Oceanogr. 13(2): 304-308.

uranium isotopes

Chemistry, uranium isotopes

Kolodny, Yehoshua, and I.R. Kaplan 1970.
Uranium isotopes in sea-floor phosphorites.
Geochim. cosmochim. Acta 34(1): 3-24.

Chemistry - uranium isotopes

Ku, Teh-Lung, 1969.
Uranium series isotopes in sediments from the Red Sea hot-brine area.
In: Hot brines and Recent heavy metal deposits in the Red Sea, E.T. Degen and D.A. Ross, editors Springer-Verlag, New York, Inc., 512-524.

chemistry, uranium isotopes (238, 235)

Servant, Jean 1966.
La radioactivité de l'eau de mer.
Cah. océanogr. 18(4): 279-318.

uranium ratios

Chemistry, uranium ratios

Blanchard, Richard L., 1965.
U 234/U238 ratios in coastal marine waters and calcium carbonates.
J. Geophys. Res., 70(16):4055-4061.

Chemistry, uranium/calcium

Blanchard, Richard L., and Duke Oakes, 1965.
Relationships between uranium and radium in coastal marine shells and their environment.
J. Geophys. Res., 70(12):2911-2921.

chemistry, uranium

Dorta, Clara C. and Elizabeth Rona, 1971.
Geochemistry of uranium in the Cariaco Trench.
Bull. mar. Sci. Coral Gables, 21(3): 754-765.

Chemistry Uranium ratios

Koide, Minoru, and Edward D. Goldberg, 1965.
Uranium-234/uranium-238 ratios in sea water.
Progress in Oceanography, 3:173-177.

Uranium, activity ratios

Krishnaswamy, S., D. Lal and B.L.K. Somayajulu, 1970.
U^{234}/U^{238} activity ratios in South Pacific Ocean waters.
Proc. Indian Acad Sci. (A) 71(5): 238-241

Uranium-234/ Uranium-238

Veeh, H. Herbert, and Kurt Boström, 1971.
Anomalous $^{234}U/^{238}U$ on the East Pacific Rise.
Earth Planet. Sci. Letts, 10(3): 372-374

Chemistry uranium ratios

Veeh, H. Herbert, 1968.
U^{234}/U^{238} in the East Pacific sector of the Antarctic Ocean and in the Red Sea.
Geochim. cosmochim. Acta, 32(1):117-119.

uranium ratios

Yaron, F., and R. Frenkel, 1969.
U^{234}/U^{238} ratio in the Red Sea.
Israel J. earth Sci., 18:149.

uranium decay

Chemistry uranium decay

Turekian, Karl K., 1965.
Some aspects of the geochemistry of marine sediments.
In: Chemical oceanography, J.P. Riley and G. Skirrow, editors, Academic Press, 2:81-126.

uranium deposition

Chemistry, uranium deposition

*Veeh, H. Herbert, 1967.
Deposition of uranium from the ocean.
Earth Planet. Sci. Letters, 3(2):145-150.

urea

chemistry, urea

Goldstein, Leon, 1967.
Urea biosynthesis in elasmobranchs. In: Sharks, skates, and rays, Perry W. Gilbert, Robert F. Mathewson and David P. Rall, editors, Johns Hopkins Univ. Press, 207-214.

Chemistry urea

Kimata, M., and Y. Hata, 1954.
Studies on urea-splitting bacteria in the sea.
Spec. Publ. Japan Sea Region. Fish. Res. Lab., 3rd Ann., 135-138.

chemistry, urea

Remsen, Charles C., 1971.
The distribution of urea in coastal and oceanic waters. Limnol. Oceanogr. 16(5): 732-740.

uric acid

Chemistry uric acid

Norton, D.R., M.A. Plunkett and F.A. Richards, 1954.
Spectrophotometric determination of uric acid and some redeterminations of its solubility.
Anal. Chem. 26(3):454-457, 5 textfigs.

uronic acid

chemistry, uronic acid

Ernst, W., 1966.
Nachweis, Identifizierung und quantitative Bestimmung von Uronsäuren in Polysaccharidfraktionen mariner Sedimente der Deutschen Bucht.
Veröff. Inst. Meeresforsch. Bremerh., 10(2):81-92.

Vanadium

Chemistry Vanadium

Bach, J.M., and R.A. Trelles, 1941.
Vanadium, its determination in water.
Bol. Obras. Sanit. Nacion, Buenos Aires, 5:127-128.

Chemistry vanadium

Baltscheffsky, H., and M. Baltscheffsky, 1953.
Vanadium content and distribution in Pallusia mamillata Cuvier. Pubbl. Staz. Zool., Napoli, 24(3):447-451.

Chemistry, vanadium

Bertrand, D., 1950.
The biogeochemistry of vanadium.
Bull. Amer. Mus. Nat. Hist. 94:407-455.

chemistry, Vanadium

Burton, J.D., 1966.
Some problems concerning the marine geochemistry of Vanadium.
Nature, Lond., 212(5066):976-978.

Vanadium

Carlisle, D.B., 1968.
Vanadium and other metals in ascidians.
Proc. R. Soc. B. 171(1022): 31-42.

Chemistry vanadium

Chester, R., 1965.
Elemental geochemistry of marine sediments.
In: Chemical oceanography, J.P. Riley and G. Skirrow, editors, Academic Press, 2:23-80.

Chemistry vanadium

Elroi, D., and B. Komoravosky, 1961
On the possible use of the fouling ascidian Ciona intestinalis as a source of vanadium cellulose and other products.
Proc. Gen. Fis. Counc., Medit., Proc. and Techn. Papers, 6:261-267.

Technical Paper No. 37.

Chemistry Vanadium

Ernst, T., and H. Hoermann, 1936.
Bestimmung von Vanadium, Nickel und Molybdän im Meerwassers. Nach. Ges. Wiss., Göttingen, Math. Phys. Kl., n.s., (4) 1(16):205-208.

Chemistry vanadium

Fukai, Rinnosuke, and W. Wayne Meinke, 1962
Activation analyses of vanadium, arsenic, molybdenum, tungsten, rhenium and gold in marine organisms.
Limnology and Oceanography, 7(2):186-200.

Chemistry vanadium

Goldberg, E.D., W. McBlair, and K.M. Taylor, 1951.
The uptake of vanadium by tunicates. Biol. Bull. 101(1):84-94, 5 textfigs.

Chemistry, vanadium

Ishibashi, M., 1953.
Studies on minute elements in sea water.
Rec. Ocean. Wks., Japan, n.s., 1(1):88-92.

Chemistry vanadium

Ishibashi, Masayoshi, Taitiro Fujinaga and Tooru Kuwamoto, 1962
Chemical studies on the ocean. Part 89.
Fundamental investigation on the dissolution and deposition of molybdenum, tungsten and vanadium in the sea.
Rec. Oceanogr. Wks., Japan, Spec. No. 6:215-218.

Chemistry, vanadium

Ishibashi, M., T. Shigematsu, Y. Nakagawa, and Y. Ishibashi, 1951.
The determination of vanadium in sea water.
Bull. Inst. Chem. Res., Kyoto Univ., 24:68-69.

vanadium

Kalk, Margaret, 1963.
Cytoplasmic transmission of vanadium compound in a tunicate oocyte visible with electron-microscopy.
Acta Embryologieae et Morphologiae Experimentalis 6:289-303.

Chemistry vanadium

Kalk, Margaret, 1963.
Absorption of vanadium by tunicates.
Nature, 198(4884):1010-1011.

Chemistry vanadium

Naito, H., and K. Sugawara, 1957.
A rapid spectrophotometric method for determination of a minute amount of vanadium in natural waters. Bull. Chem. Soc., Japan, 30(7):799-800.

chemistry, vanadium

Okabe, Shiro, and Toyoko Morinaga 1969.
Distribution of vanadium in the Kuroshio and adjacent regions. (In Japanese; English abstract).
J. Oceanogr. Soc. Japan 25(5): 223-228.

Vanadium

Okabe, Shiro, and Toyoko Morinaga, 1968.
Vanadium and molybdenum in the rivers and estuary water which pour into the Suruga Bay, Japan. (In Japanese; English abstract).
Nippon Kagaku Zasshi, 89(3).
Also in: Coll. Repr. Coll. mar. Sci. Techn., Tokai Univ. 1967-68, 3:329-332.

chemistry, vanadium

Okabe, Shiro, Yoshimasa Toyota and Toyoko Morinaga, 1970.
Pollution in coastwise waters of Japan. Kagaku-no-Ryoiki 24(1). Also in: Coll. Repr. Coll. mar. Sci. Techn. Tokai Univ. 4: 303-309.

Chemistry, vanadium

Sugawara, K., 1957.
The distribution of some minor bio-elements in Western Pacific waters.
Proc. UNESCO Symp., Phys. Ocean., Tokyo, 1955: 169-174.

Chemistry, vanadium

Sugawara, K., H. Naito and S. Yamada, 1956.
Geochemistry of vanadium in natural waters.
J. Earth Sci., Nagoya Univ., 4(1):44-61.

chemistry, Vanadium (data)

Sugawara, Ken and Shiro Okabe, 1966.
Molybdenum and vanadium determination of Umitaka-maru samples from her 1962-1963 and 1963-1964 cruises of the International Indian Ocean Expedition. (In Japanese; English abstract)
J. Tokyo Univ. Fish., (Spec. ed.) 8(2):165-171.

Chemistry, vanadium

Sugawara, Ken, and Shiro Okabe 1966.
Molybdenum and vanadium determination of Umitaka Maru samples from her 1962-1963 and 1963-1964 cruises of the International Indian Ocean Expedition (In Japanese; English abstract)
J. Tokyo Univ. Fish. (Spec. Ed.) 8(2): 25-31.
Reprinted in: Coll. Repr. Coll. mar. Sci. Technol. Tokai Univ. 2 (1966):

chemistry, vanadium

Webb, D.A., 1956.
The blood of tunicates and the biochemistry of vanadium. Pubbl. Staz. Zool., Napoli, 28:273-288.

Vanadium

Yamamoto, Toshio, Tetsuo Fujita and Masayoshi Ishibashi, 1970.
Chemical studies on the seaweeds (25).
Vanadium and titanium content in seaweeds.
Rec. oceanogr. Wks, Japan 10(2): 125-135

Vitamins

see also under names of vitamins.

Chemistry, vitamins

See also: chemistry, under the several names of vitamins

Chemistry vitamins

Aubert, M., 1963.
Possibilités d'application diététique de la biomasse planctonique.
Cahiers Centre d'Études Recherche Biol. Oceanogr. Medicale, Fasc. 2:19-34 (mimeographed).

Chemistry vitamins

Aubert, M., et C. Morgat, 1965.
Les facteurs probiotiques du milieu marin.
Cahiers du C.E.R.B.O.M., 17:43-48

Chemistry vitamins

Belser, W.L., 1963
9. Bioassay of trace substances. In: The Sea, M.N. Hill, Edit., Vol. 2(II) Fertility of the oceans, Interscience Publishers, New York and London, 220-231.

chemistry, vitamins

Burkholder, Paul R., and Seymour Lewis 1968.
Some patterns of B vitamin requirements among neritic marine bacteria.
Can. J. Microbiol. 14(5): 537-543

Chemistry, vitamins

*Droop, M.R., 1968.
Vitamin B_{12} and marine ecology. IV. The kinetics of uptake, growth and inhibition in Monochrysis lutheri.
J. mar. biol. Ass., 48(3):689-733.

Chemistry vitamins

Droop, M.R., 1955.
A pelagic marine diatom requiring cobalamin.
J.M.B.A. 34(2):229-231.

Chemistry vitamin B_{12}

Hirano, Toshiyuki, Takeaki Kikuchi, Kazuo Ino, Noboru Tanoue, Takeshi Taguchi and Ikunosulke Okada, 1964.
Contents of inorganic substance and vitamin B_{12} in Euphausia.
J. Tokyo Univ., Fish., 50(2):65-70.

Chemistry vitamins (B12)

Hirano, Tosniyuki, Takeaki Kikuchi, and Ikunosuke Okada, 1964.
Contents of inorganic substances and vitamin B12 in Euphausia. (In Japanese; English abstract).
Bull. Jap. Soc. Sci. Fish., 30(3):267-271.

Chemistry vitamin B_{12}

Holm-Hansen, Osmund, J.D.H. Strickland and P.M. Williams, 1966.
A detailed analysis of biologically important substances in a profile off southern California.
Limnol. Oceanogr., 11(4):548-561.

Chemistry vitamin B_{12}

Karlström, O., D. Callieri and K. Bäck, 1961.
Studies on vitamin B12 in algae.
Arkiv för Kemi, 16(3/4):299-307.

Chemistry vitamin B_{12}

Kashiwada, Ken-ichi, Daiichi Kakimoto and Akio Kanazawa, 1960
Studies of Vitamin B12 in natural water. Rec. Oceanogr. Wks., Japan, n.s., 5(2): 71-76.

Chemistry vitamin B_{12}

Kashiwada, K., D. Kakimoto and A. Kanazawa, 1957.
Studies on Vitamon B12 in sea water. 1. On the assay of Vitamin B12 in sea water.
Mem. Fac. Fish., Kagoshima Univ., 5:148-152.

Chemistry vitamin B_{12}

Kashiwada, K., D. Kakimoto and K. Kawagoe, 1957.
Studies on vitamin B12 in sea water. III. On the diurnal fluctuation of Vitamin B12 in the sea and its vertical distribution in the sea.
Bull. Jap. Soc., Sci. Fish., 23(7/8):450-453.

Chemistry vitamin B_{12}

Kashiwada, K., D. Kakimoto, T. Morita, A. Kanazawa and K. Kawagoe, 1957.
Studies on vitamin B12 in sea water. II. On the assay method and the distribution of this vitamin Bla in the ocean.
Bull. Jap. Soc. Sci. Fish., 22(10):637-640.

Chemistry Vitamin B_{12}

Kurata, Akira, 1970.
Vitamin B_{12} transudation from marine bottom muds. (In Japanese; English abstract).
J. oceanogr. Soc., Japan, 26(2): 81-86.

Chemistry, vitamin B_{12}

Lebedeva, M.N., E.M. Markianovich and L.G. Gutveib, 1971.
Spreading of heterotrophic bacteria auxotrophic in vitamin B_{12} in southern seas. (In Russian; English abstract). Gidrobiol. Zh., 7(2): 20-26.

Chemistry vitamin B12, effect of

Menzel, David W., and Jane P. Spaeth, 1962
Occurrence of Vitamin B12 in the Sargasso Sea.
Limnology and Oceanography, 7(2):151-154.

Chemistry vitamin B_{12}

Naterajan, K.V., and R.C. Dugdale, 1966.
Bioassay and distribution of thiamine in the sea.
Limnol. Oceanogr., 11(4):621-629.

Chemistry, vitamin B_{12}

Propp, L.N., 1970.
On the seasonal dynamics of B_{12} vitamin and phytoplankton variability in the Dalnezelenetskaya Cuba (inlet) of the Barents Sea. Oceanologiia, 10(5): 851-857.

Chemistry vitamin B_{12}

Provosoli, L., and I.J. Pintner, 1953.
Assay of vitamin B12 in sea water.
Proc. Soc. Protozool., 4:10.

Chemistry vitamin B12

Starr, T.J., 1956.
Relative amounts of vitamin B12 in detritus from oceanic and estuarine environments near Sapelo Island, Georgia. Ecology 37(4):658-664.

Chemistry vitamin B_{12}

Starr, T.J., M.E. Jones and D. Martinez, 1957.
The production of Vitamin B12 - active substance -s by marine bacteria. Limnol. & Oceanogr. 2(2):114-119.

Chemistry vitamin B_{12}

Stewart, Violet N., Harold Wahlquist and Richard Burket, 1967.
Occurrence of vitamin B_{12} along the gulf coast of Florida. Prof. Pap. Ser., Fla Bd Conserv. 9: 79-84.

chemistry, Vitamin B_{12}

Stewart, V.N., H. Wahlquist, R. Burquet and C. Wahlquist, 1966.
Observations of vitamin B_{12} distribution in Apalachee Bay, Florida.
Florida Bd., Conserv., St. Petersburg, Mar. Lab., Prof. Papers Ser., No. 8:34-38.

Chemistry vitamin B_{12}

Suprunov, A. T., and Z. A. Muravskaia, 1946.
The vitamin B12 content of the water of Sevastopol Bay and its possible ecological importance. (In Russian).
Trudy Sevastopol Biol. Sta., 7:342-345.

Chemistry vitamin B_{12}

Sweeney, B.M., 1954.
Gymnodinium splendens, a marine dinoflagellate requiring vitamin B12. Amer. J. Bot., 41:821-824.

Chemistry Vitamin B12

Tosneur, C., 1970.
Contribution à l'étude de la production d'antibiotiques et de vitamine B12 par des bactéries isolées des vases soumises à l'influence du milieu marin.
Cah. océanogr. 22(6): 613-614

Chemistry vitamin B_{12}

Van Baalen, C., 1961
Vitamin B12 requirement of a marine blue-green alga.
Science, 133(3468): 1922-1923.

vitamin D

Chemistry VITAMIN D

Johnson, N.G. and T. Leving, 1947
On the production of vitamin D in the sea.
Svenske Hydrografisk-Biologiske Kommissionens Skrifter, Tradje Serien: Hydrographi. 1(3):6 pp.

vitamin effect of

Chemistry vitamins, effect of

Curl, Herbert, Jr., 1962
Effect of divalent sulfur and vitamin B12 in controlling the distribution of Skeletonema costatum.
Limnol. and Oceanogr., 7(3):422-424.

chemistry, vitamin excretion

Provasoli, L., 1971.
Nutritional relationships in marine organisms.
In: Fertility of the Sea, John D. Costlow, editor, Gordon Breach, 2: 369-382.

Chemistry vitamin requirements

Guillard, R.R.L., and Vivienne Cassie, 1963
Minimum cyanocobalamin requirements of some marine centric diatoms.
Limnol. and Oceanogr., 8(2):161-165.

chemistry, xanthophyll

Jeffrey, S.W. 1968.
Pigment composition of Siphonales algae in the brain coral Favia.
Biol. Bull. mar. biol. Lab. Woods Hole 135(1): 141-148.

xenon

Chemistry Xenon

Bieri, Rudolph, Minoru Koide and Edward D. Goldberg, 1964
Noble gases in sea water.
Science, 146(3647):1035-1037.

Chemistry xenon

Mazor, E., G.J. Wasserburg, and H. Craig, 1964.
Rare gases in Pacific Ocean water.
Deep-Sea Res., 11(6):929-932.

yttrium 90

yttrium

Balashov, Yu. A., L.V. Dmitriev and A. Ya. Sharas'kin, 1970.
Distribution of the rare earths and yttrium in the bedrock of the ocean floor. (In Russian)
Geokhimiya (6): 647-660.
Translation: Geochem. Int. 7(3): 456-468.

Chemistry, yttrium 90

Boroughs, H., and D.F. Reid, 1958.
The role of the blood in the transportation of strontium 90- yttrium 90 in teleost fish.
Biol. Bull., 115(1):64-73.

chemistry, yttrium-91

Ivanov, V.N. and A.A. Lyubimov 1970.
Adsorption of microquantities of yttrium-91 from the sea water on fluoroplastic-4, polyethylene and paper. (In Russian; English abstract).
Okeanologiia 10(3): 546-551.

Chemistry yttrium-91

Lyubimov, A.A., A. Ya. Zesenko, and L.N. Leshchenko, 1970.
The effect of changing physical-chemical state of radiottrium on its accumulation by marine organisms. (In Russian; English abstract).
Okeanologiia, 10(6): 1001-1008.

Chemistry Yttrium

Parchevskii, V.P., G.G. Polycarpov and I.S. Zaburunova, 1965.
Certain regularities in the accumulation of yttrium and strontium by marine organisms. (In Russian)
Doklady, Akad. Nauk, SSSR, 164(4):913-916.

chemistry, yttrium- 91

Spitsyn, B.I., R.N. Bernovskaia, and N.I. Popov, 1969.
State of ultrasmall quantities of cerium- 144 and yttrium- 91 in sea water. (In Russian)
Dokl. Acad Nauk SSSR, 185(1): 111-114.

zinc

Chemistry zinc

Atkins, W.R.G., 1957.
The direct estimation of ammonia in sea water, with notes on nitrate, copper, zinc and sugars.
J. du Cons., 22(3):271-277.

Chemistry zinc
Atkins, W.R.G., 1936.
The estimation of zinc in sea water using sodium diethyldithiocarbamate. J.M.B.A., 20: 625.

Chemistry zinc
Belova, I.V., 1970.
Zinc in Recent Black Sea sediments (In Russian).
Dokl. Akad. Nauk SSSR 193(2): 433-436

Chemistry zinc
Bernhard, M., preparator, 1965.
Studies on the radioactive contamination of the sea, annual report 1964.
Com. Naz. Energ. Nucleare, La Spezia, Rept., No. RT/BIO (65) 18:35 pp.

Chemistry zinc
Bernhard, M., 1964.
Chemical composition and the radiocontamination of marine organisms.
Proc. Symp. Nuclear Detonations and Marine Radioactivity, S.H. Small, editor, 137-150.

Chemistry zinc
Bertrand, G., 1938.
Sur la quanitité de zinc contenue dans l'eau de mer. Ann. Inst. Pasteur, 62:571-575.

Also in:
Mem. Bull. Soc. Chim., 6:697-700.

Chemistry, zinc
Bertrand, G., 1938.
Sur la quantité de zinc contenue dans l'eau de mer. C.R. Acad. Sci., Paris, 207(24):1137-1141.

chemicals, zinc
Bhatt, Y.M., V.N. Sastry S.M.Shah and T.M. Krishnamoorthy, 1968.
Zinc, manganese and cobalt contents of some marine bivalves from Bombay.
Proc. nat. Inst. Sci., India(B), 34(6): 283-287.

Chemistry, zinc
Bougis, P., 1961.
Sur l'effet biologique du zinc en eau de mer.
C.R. Acad. Sci., Paris, 253(4):740-741.

Chemistry zinc
Brooks, R.R.,
Trace elements in New Zealand coastal waters.
Geochimica et Cosmochimica Acta, 29(12): 1369-1370.

Chemistry zinc
Bryan, G.W., 1964
Zinc regulation in the lobster Homarus vulgaris. 1. Tissue zinc and copper concentrations.
J. Mar. Biol. Assoc., U.K., 44(3):549-563.

Chemistry zinc
Chester, R., 1965.
Adsorption of zinc and cobalt on illite in sea-water.
Nature, 206(4987):884-886.

chemistry, zinc
Coombs, T.L., 1972.
The distribution of zinc in the oyster Ostrea edulis and its relation to enzymic activity and to other metals. Mar. Biol. 12(2): 170-178.

chemistry, zinc
Cross, Ford A., James N. Willis and John P. Baptist 1971.
Distribution of radioactive and stable zinc in an experimental marine ecosystem.
J. Fish. Res. Bd Can. 28 (11): 1783-1788

Chemistry zinc
Duursma, E. K., and W. Sevenhuysen, 1966.
Note on chelation and solubility of certain Metals in sea water at different Ph values.
Netherlands J. Sea Res., 3(1):95-106.

Chemistry zinc (data)
Fonselius, Stig H. 1970.
Some trace metal analyses in The Mediterranean, the Red Sea and the Arabian Sea.
Bull. Inst. Océanogr. Monaco 69(1407): 15pp.

chemistry, zinc
Fonselius, Stig H, 1970.
Some trace metal analyses in The Mediterranean, the Red Sea and the Arabian Sea.
Bull. Inst. océanogr., Monaco 69 (1407): 15pp.
Also: IAEA Radioactivity in the Sea, Publ. 29.

Chemistry zinc
Fonselius, Stig H., and Folke Koroleff, 1963.
Copper and zinc content of the water in the Ligurian Sea.
Bull. Inst. Oceanogr., Monaco, 61(1281):1-15.

Chemistry, zinc
Fukai, Rinnosuke, and Lang Huynh-ngoc 1968
Studies on the chemical behaviour of radionuclides in sea-water. I. General consideration and study of precipitation of trace amounts of chromium, manganese, iron, cobalt, zinc and cerium.
IAEA Radioactivity in the sea, Publ. 22:1-26.

Chemistry zinc
Gutknecht, John, 1962
Zn uptake by Ulva lactuca.
Limnology and Oceanography, 7(2):270-271.

Chemistry zinc
Hiyama, Yoshio, and Makoto Shimizu, 1964
On the concentration factors of radioactive Cs, Sr, Cd, Zn, and Ce in marine organisms.
Rec. Oceanogr. Wks., Japan, 7(2):43-77.

Chemistry zinc
Hood, D.W., 1963
Chemical oceanography. In: Oceanography and Marine Biology, H. Barnes, Edit., George Allen & Unwin, 1:129-155.

Chemistry, zinc
Ikuta, Kunio 1968.
Studies on accumulation of heavy metals in aquatic organisms. II. On accumulation of copper and zinc in oysters. III On accumulation of copper and zinc in the parts of oysters. (In Japanese; English abstract).
Bull. Jap. Soc. scient. Fish. 34(2): 112-116; 117-122

chemistry zinc
Ishibashi, Masayoshi, Taitiro Fujinaga, Fuji Morii, Yoshihiko Kanchiku, and Fumio Kamiyama, 1964.
Chemical studies on the ocean (Part 94). Chemical studies on the seaweeds (19). Determinatio of zinc, copper, lead, cadmium and nickel in seaweeds using dithizone extraction and polarographic method.
Rec. Oceanogr. Wks., Japan, 7(2):33-63.

Chemistry zinc
Ishibashi, Masayoshi, Sunzo Ueda and Yoshikazu Yamamoto, 1959.
Studies on the utilization of the shallow-water deposits (continued). On the zinc content of the shallow-water deposits.
Rec. Oceanogr. Wks., Japan, Spec. No. 3:123-133.

Chemistry zinc
Johnston, R., 1964
Sea water, the natural medium of phytoplankton. II. Trace metals and chelation, and general discussion.
J. Mar. Biol. Assoc., U.K., 44(1):87-109.

Chemistry zinc
Kamada, M., and T. Onishi, 1964.
The copper and zinc contents of river waters flowing into Ariake Sea and Yatusiro Bay. (In Japanese).
Sci. Repts., Kagoshima Univ., No. 13:17-27.

Chemistry Zinc
Macchi G., et P. Chemerd
Etude préliminaire sur la distribution du zinc ionique dans l'eau de mer.
Rapp. P.-v. Réun., Comm. int. Explor. scient. Mer Mediterr., 18(3):871-874.

Chemistry zinc
Mehran, A.R., and J.L. Tremblay, 1965.
Un aspect du metabolisme du zinc chez Littorina obtusata L., et Fucus edentata de la Pylaie.
Revue canadienne de Biologie, 24(3):157-161.

Chemistry zinc
Mitsunobu, T., 1957.
Chemical investigations of deep-sea deposits. 24-25. The contents of copper and zinc in sea sediments. J. Chem. Soc., Japan, 78(3):405-415.

Chemistry, zinc
Morita, Y., 1953.
Copper and zinc in Pacific waters.
Rec. Ocean. Wks., Japan, n.s., 1(2):49-51.

Chemistry zinc
Morita, Y., 1950.
Distribution of copper and zinc. IV. Copper and zinc contents of sea water. (In Japanese).
J. Chem. Soc., Japan, Pure Chem. Sect., 71:246-248.

Chem. Abstr., 1951, 45:4856.

Chemistry, zinc-65
Österberg, Charles, 1962
Zn 65 content of salps and euphausiids.
Limnol. and Oceanogr., 7(4):478-479.

Chemistry zinc
Ostroumov, E.A., and I.I. Volkov, 1962.
Separation of titanium, zirconium and thorium from manganese, nickel, cobalt and zinc by means of cinnamic acid.
Trudy Inst. Okeanol., Akad. Nauk, SSSR, 54:170-181.

In Russian; English summary

Chemistry, zinc
Parchevskiy, V.P., and K.M. Khailov, 1968.
Formation of Ce^{141}, Ru^{106}, Cs^{137}, and Zn^{65} complexes with hydrophyl high-molecular combinations dissolved in sea water. (In Russian).
Okeanologiia, Akad, Nauk, SSSR., 8(6):1092.

Chemistry zinc
*Pequegnat, John E., Scott W. Fowler and Lawrence F. Small, 1969.
Estimates of zinc requirements of marine organisms.
J. Fish. Res. Bd. Can., 26(1):145-150.

zinc

Petek, Milica, and Marko Branica, (1968) 1969.
Hydrographic and biotical conditions in North Adriatic. III. Distribution of zinc and iodate.
Thalassia jugosl. 5 (1968)
Abstract in: Rapp. P.-v. Reun. Commn int. Explor. scient. Mer Mediterr., 19(4): 765.

chemistry zinc

Petek, Milica, and Marko Branica, 1969.
Hydrographical and biotical conditions in north Adriatic - III. Distribution of zinc and iodate. (Jugoslavian and Italian abstracts).
Thalassia Jugoslavica 5: 257-263.

Chemistry zinc

Piro, A. 1971.
Chemical environmental factors in marine radiocontamination.
CNEN Rept. RT/BIO (70)-11, M. Bernhard, editor: 11-22.

chemistry, zinc

Piro, A., M. Verzi e C. Papucci 1969.
L'importanza dello stato fisico-chimico degli elementi per l'accumulo negli organismi marini. I. Lo stato chimico-fisico dello zinco in acqua di mare.
Pubbl. Staz. Zool. Napoli 37 (2 Suppl.): 298-310.

Chemistry zinc

Rona, Elizabeth, Donald W. Hood, Lowell Muse and Benjamin Buglio, 1962
Activation analysis of manganese and zinc in sea water.
Limnology and Oceanography, 7(2):201-206.

chemistry, zinc

Slowey, J. Frank, and Donald W. Hood, 1971.
Copper, manganese and zinc concentrations in Gulf of Mexico waters.
Geochim. Cosmochim. Acta 35 (2): 121-138.

chemistry, zinc

Spencer, Derek W., and Peter G. Brewer, 1969.
The distribution of copper, zinc and nickel in sea water of the Gulf of Maine and the Sargasso Sea.
Geochim. cosmochim. Acta 33(3): 325-339.

Chemistry zinc

Topping, Graham, 1969.
Concentrations of Mn, Co, Cu, Fe, and Zn in the northern Indian Ocean and Arabian Sea. J. mar. Res., 27(3): 318-326.

Chemistry, zinc (data)

Torii, Tetsuya, and Sadao Murata 1966.
The distribution of copper and zinc in the Indian and the Southern Ocean waters.
J. oceanogr. Soc. Japan 22(2): 56-60.

chemistry zinc

Verduin, Jacob, 1962
Zn uptake by Ulva lactuca.
Limnology and Oceanography, 7(2):270.

chemistry, zinc

Zattera, Antonio, e Michael Bernhard 1969.
L'importanza dello stato chimico-fisico degli elementi per l'accumulo negli organismi marini. II. Accumulo di zinco stabile e radioattivo in Phaeodactylum tricornutum.
Pubbl. Staz. Zool. Napoli, 37 (2 Suppl.): 386-399.

chemistry, zinc

Zirino, Alberto and Michael L. Healy, 1971.
Voltammetric measurement of zinc in the northeastern tropical Pacific Ocean. Limnol. Oceanogr. 16(5): 773-778.

chemistry, zinc complexes

Zirino, Alberto and Michael L. Healy, 1970.
Inorganic zinc complexes in seawater. Limnol. Oceanogr., 15(6): 956-958.

Zinc-65

Zinc 65

Bachmann, Rober W., and Eugene P. Odum, 1960
Uptake of Zn 65 and primary productivity in marine benthic algae. Limnol. & Oceanogr., 5(4):349-355.

Chemistry, zinc-65

Carey, Andrew G., Jr. 1969.
Zinc-65 in echinoderms and sediments in the marine environment off Oregon.
Symp. Radioecology, Proc. Second Natn. Symp., Ann Arbor, Mich., 1967 (Conf. 670503): 380-388.

chemistry, zinc-65

Duke, Thomas W., James N. Willis and Thomas J. Price, 1966.
Cycling of trace elements in the estuarine environment. 1. Movement and distribution of zinc 65 and stable zinc in experimental ponds. Chesapeake Science, 7(1):1-10.

chemistry, zinc-65

*Fowler, Scott W., and Lawrence F. Small, 1967.
Moulting of Euphausia pacifica as a possible mechanism for the vertical transport of zinc-65 in the sea.
Int. J. Oceanol. Limnol., 1(4):237-245.

zinc-65

Fowler, Scott W., Lawrence F. Small and John M. Dean 1970.
Distribution of ingested zinc-65 in the tissues of some marine crustaceans.
J. Fish. Res. Bd. Can. 27 (6): 1051-1055

Chemistry zinc 65

Gutknecht, John, 1965.
Uptake and retention of cesium 137 and zinc 65 by seaweeds.
Limnology and Oceanography, 10(1):58-66.

Chemistry zinc 65

Gutknecht, John, 1963
Zn 65 uptake by benthic marine algae.
Limnology and Oceanography, 8(1):31-38.

Chemistry zinc 65

Gutknecht, J., 1961.
Mechanisms of radioactive zinc uptake by Ulva lactuca.
Limnol. & Oceanogr., 6(4):426-431.

Chemistry zinc-65

Hoss, Donald E., 1964.
Accumulation of zinc-65 by flounder of the Genus Paralichthys.
Trans. Amer. Fish. Soc., 93(4):364-368.

Chemistry, zinc-65

Keckeš, Stjepan, Bartolo Ozretić and Mirjana Krajnović 1969.
Metabolism of Zn^{65} in mussels (Mytilus galloprovincialis Lam.) Uptake of Zn^{65}.
Rapp. P.-v. Reun. Commn int. Explor. scient. Mer Mediterr. 19(5): 949-952.

Chemistry, zinc-65

Lewis, Gary B., and Allyn H. Seymour, 1965.
Distribution of zinc-65 in plankton from offshore waters of Washington and Oregon, 1961-1963
Ocean Sci. and Ocean Eng., Mar. Techn. Soc.,-Amer. Soc. Limnol. Oceanogr., 2:956-967.

Chemistry zinc 65

Mishima, Jiro, and Eugene P. Odum, 1963
Excretion rate of Zn 65 by Littorina irrorata in relation to temperature and body size.
Limnology and Oceanography, 8(1):39-44.

chemistry, zinc-65

Nagaya, Yutaka, and Theodore R. Folsom, 1966.
Zinc-65 and other fallout nuclides in marine organisms of the California coast.
J. Radiation Res., 5(1):82-89.

Chemistry zinc-65

Osterberg, Charles, 1965.
Radioactivity from the Columbia River.
Ocean Sci. and Ocean Eng., Mar. Techn. Soc.,-Amer. Soc. Limnol. Oceanogr., 2:968-979.

Chemistry zinc 65

Osterberg, Charles, 1962
Zn 65 content of salps and euphausiids.
Limnol. and Oceanogr., 7(4):478-479.

Chemistry zinc-65

Osterberg, C., L.D. Kulm and J.V. Byrne, 1963.
Gamma emitters in marine sediments near the Columbia River.
Science, 139(3558):916-917.

Chemistry zinc 65

Osterberg, Charles, June Pattullo and William Pearcy, 1964
Zinc-65 in euphausiids as related to Columbia River water off the Oregon Coast.
Limnology and Oceanography, 9(2):249-257.

chemistry, zinc-65

Parker, Patrick L., 1966.
Movement of radioisotopes in a marine bay: cobalt-60, iron-59, manganese-54, zinc-65, soduim-22.
Publs. Inst. mar. Sci., Univ. Texas, Port Aransas, 11:102-107.

Chemistry, zinc-65

Pearcy, William G., and Charles L. Osterberg 1968.
Zinc-65 and manganese-54 in albacore Thunnus alalunga from the west coast of North America.
Limnol. Oceanogr. 13(3): 490-498.

Chemistry, zinc-65

Pearcy, William G., and Charles L. Osterberg, 1967.
Depth, diel, seasonal, and geographic variations in zinc-65 of midwater animals off Oregon.
Int. J. Oceanol. Limnol., 1(2):103-116.

Chemistry zinc-65

Pearcy, William G., and Charles L. Osterberg, 1964.
Vertical distribution of radionuclides as measured in oceanic animals.
Nature, 204(4957):440-441.

chemistry, zinc-65

Tennant, David A., and William O. Forster 1969.
Seasonal variation and distribution of ^{65}Zn, ^{54}Mn and ^{51}Cr in tissues of the crab Cancer magister Dana.
Health Physics 18: 649-657.
Also in: Coll. Repr. Dept. Oceanogr. Oregon State Univ. 9 (2) (1970).

chemistry zinc-65

Watson, D.G., J.J. Davis, and W.C. Hansen, 1961.
Zinc-65 in marine organisms along the Oregon and Washington coasts.
Science, 133(3467):1826-1828.

zirconium

chemistry zirconium

Chester, R., 1965.
Elemental geochemistry of marine sediments.
In: Chemical oceanography, J.P. Riley and G. Skirrow, editors, Academic Press, 2:23-80.

Chemistry zirconium

Collier, A., 1953.
Titanium and zirconium in bloom of Gymnodinium brevis Davis. Science 118(3064):329.

chemistry, zirconium

Glagoleva, M.A. 1970.
Zirconium in Recent sediments of the Black Sea. (In Russian).
Dokl. Akad. Nauk SSSR 193(4): 184-187.

chemistry zirconium

Hill, Patrick Arthur, and Andrew Parker 1970.
Tin and zirconium in the sediments around the British Isles: a preliminary reconnaissance.
Econ. Geol. 65(4): 409-416.

chemistry, zirconium

Holmes, Charles W., 1969.
In the Gulf of Mexico ---- geochemical exploration produces exciting results: sea floor samples lead to a surprising discovery of zirconium, titanium and heavy metals off the Texas coast.
Ocean Industry 4(6): 49-52 (popular)

chemistry zirconium

Ishibashi, Masayoshi, Shunzo Uedo, Yoshikazu Yamamoto, Fujii Morii, 1963.
Studies on the utilization of the shallow water deposits (Cont.). On the zirconium content of the shallow water deposits.
Rec. Oceanogr. Wks., Japan, n.s., 7(1):37-46.

chemistry, zirconium

Ostroumov, E.A., and I.I. Volkov, 1962.
Separation of titanium, zirconium and th ium from manganese, nickel, cobalt and zinc by means of cinnamic acid.
Trudy Inst. Okeanol., Akad. Nauk, SSSR, 54:170-181.

In Russian; English summary

Chemistry zirconium-95

Pearcy, William G., and Charles L. Osterberg, 1964.
Vertical distribution of radionuclides as measured in oceanic animals.
Nature, 204(4957):440-441.

chemistry zirconium

Parchevsky, V.P., 1965.
Radionuclides of cesium, ruthenium and zirconium in animal and plant organisms of the Black Sea. (In Russian).
Okeanologiia, Akad. Nauk, SSSR, 5(5):856-862.

chemistry, zirconium

Sastry, V.N., T.M. Krishnamoorthy and T.P. Sarma 1969.
Microdetermination of zirconium in the marine environment.
Current Sci. 38(12):279-281.

chemotaxis

Castilla, J.C., 1972.
Responses of Asterias rubens to bivalve prey in a Y-maze. Mar. Biol. 12(3): 222-228.

chemoreception

Gurin, S., and W.E. Carr 1971.
Chemoreception in Nassarius obsoletus: The role of specific stimulatory proteins.
Science 174(4006): 293-295.

chemotaxis

Pincemin, J.-M. 1971.
Télémédiateurs chimiques et équilibre biologique océanique. 3. Étude in vitro de relations entre populations phytoplanctoniques.
Rev. int. Océanogr. Méd. 22-23:165-196

Chert

chert

Calvert, S.E. 1971.
Nature of silica phases in deep sea cherts of the North Atlantic.
Nature, Lond. 234 (50): 133-134.

cherts

Heath, G.R. and Ralph Moberly, Jr., 1971.
Cherts from the western Pacific, Leg 7, Deep Sea Drilling Project. Initial Repts Deep Sea Drill. Proj. 7(2): 991-1007.

chert (freshwater)

Peterson, M. N. A., and C. C. Von Der Borch, 1965.
Chert: modern inorganic deposition in a carbonate-precipitating locality.
Science 149(3691):1501-1503.

chert

Von der Borch, C.C., J. Galehouse, and W.D. Nesteroff, 1971.
Silicified limestone-chert sequences cored during leg 8 of the deep sea drilling project.
Initial Repts Deep Sea Drilling Project 8: 819-827.

China clay, effect of

Howell, B.R. and R.G.J. Shelton, 1970.
The effect of China clay on the bottom fauna of St. Austell and Mevagissey bays. J. mar. biol. Ass., U.K., 50(3): 593-607.

China clay, effect of

Portmann, J.E., 1970.
The effect of China clay on the sediments of St. Austell and Mevagissey bays. J. mar. biol. Ass., U.K., 50(3): 577-591.

Chlorophyll

See: Chemistry, chlorophyll

Chromatophores

chromatophores

Aoto, Tomoji 1963.
The primary response of white chromatophores and the nauplius eye in the prawn, Palaemon paucidens.
J. Fac. Sci. Hokkaido Univ., 6 Zool. 15(2):177-189.

chromatophores

*Fernlund, P., and L. Josefson, 1968.
Chromactivating hormones of Pandalus borealis. On the bioassay of the distal retinal pigment hormone.
Marine Biol., 2(1):19-22.

chromatophores

*Fernlund, P., 1968.
Chromactivating hormones of Pandalus borealis. Bioassay of the red-pigment-concentrating hormone.
Marine Biol., 2(1):13-18.

chromatophores

*Ishikawa, Yusuke, Yoshio Miyake and Shigeki Yasuie, 1967.
Chromatophores and carotenoids in cultured karuma-prawn, Penaeus japonicus Bate, of different body colours. (In Japanese; English abstract).
Bull. Fish. Exp. Stn, Okayama Pref., 1966:18-24.

chromatophores

Kulakovskii, E.E., 1971.
Neurohormonal control for chromatophores in mysids. (In Russian).
Dokl. Akad. Nauk, SSSR 196(1):234-236.

chromatophores

*Nagabhushanam, R., 1967.
The endocrine control of white chromatophores of the crab, Uca annulipes (H. Milne Edwards).
Crustaceana, 13(3):292-298.

chromatophores

Nagabhushanam, R., 1964.
Physiology of the red chromatophores of the shrimp, Alpheus malabaricus.
Brotéria, Lisboa, 33(3/4):133-142.

chromatophores

Nagabhushanam, R., and R. Sarojini, 1968.
Chromatophore activating substances in the developing crab, Gelasimus annulipes.
Brotéria, 37 (3/4): 125-129

chromatophores

Nagabhushanam, R., and R. Sarojini, 1968.
Chromatophore physiology of the zoaea of the mud shrimp, Upogebia affinis.
Brotéria, 37 (3/4): 119-124.

chromatophores

Rao, K. Ranga, Milton Fingerman and Clelmer K. Bartell, 1967.
Physiology of the white chromatophores in the fiddler crab, Uca pugilator.
Biol. Bull., mar. biol. Lab., Woods Hole, 133(3):606-617.

Chromosomes

chromosomes

Colombera, D., 1971.
Number and morphology of chromosomes in three geographical populations of Ascidiella aspersa (Ascidiacea). Mar. Biol. 11(2): 149-151.

chromosomes

Niiyama Hidejiro, 1959.
A comparative study of the chromosomes in decapods, isopods and amphipods, with some remarks on cytotaxonomy and sex-determination in the Crustacea.
Mem. Fac. Fish., Hokkaido Univ., 7(1/2):1-71, 12 pls.

chromosomes

Nishikawa, S., 1960.
[Chromosomes of Balanus amphitrite albicostatus Pilsbry.]
Dobuts. Zasshi, (Zool. Mag.), 69(12):355-356.

CHRONOLOGY: See Age determination

Chronology,

See also: age determination (geological)
geological ages

Flemming, N.C., 1965 chronology
Derivation of Pleistocene marine chronology from morphometry of erosion profiles.
J. Geol. 76: 280-296
Also in: Coll. Repr. Nat. Inst. Oceanogr. Wormley, 16 (656)

Chronometry

chronometry

Kulp, J.L., 1955.
Geological chronometry by radioactive methods.
IN: Advances in Geophysics, 2:179-219.
Academic Press, N.Y., 286 pp.

chronostratigraphy

Berggren, W.A. 1971
Neogene chronostratigraphy, planktonic foraminiferal zonation and the radiometric time scale.
Földtani Közlöny [Bull. Hungarian Geol. Soc.] 101 (2/3): 162-169.